HANDBOOK OF PROCESS ALGEBRA

HANDBOOK OF PROCESS ALGEBRA

Edited by

J.A. Bergstra
University of Amsterdam and Utrecht University, The Netherlands

A. Ponse
University of Amsterdam and CWI, The Netherlands

S.A. Smolka
State University of New York at Stony Brook, NY, USA

2001

ELSEVIER
Amsterdam · London · New York · Oxford · Paris · Shannon · Tokyo

ELSEVIER SCIENCE B.V.
Sara Burgerhartstraat 25
P.O. Box 211, 1000 AE Amsterdam, The Netherlands

First edition 2001

Library of Congress Cataloging in Publication Data
A catalog record from the Library of Congress has been applied for.

ISBN: 0-444-82830-3

∞ The paper used in this publication meets the requirements of ANSI/NISO Z39.48-1992 (Permanence of Paper).

Printed in The Netherlands

Preface

1. Introduction

According to the Oxford English Dictionary (OED II CD-ROM), a *process* is a series of actions or events, and an *algebra* is a calculus of symbols combining according to certain defined laws. Completing the picture, a *calculus* is a system or method of calculation. Despite going back as far as the 13th Century, collectively, these definitions do a good job of accurately conveying the meaning of this Handbook's subject: *process algebra*.

A process algebra is a formal description technique for complex computer systems, especially those with communicating, concurrently executing components. A number of different process algebras have been developed – ACP [1], CCS [6], and TCSP [2] being perhaps the best-known – but all share the following key ingredients.

- **Compositional modeling.** Process algebras provide a small number of constructs for building larger systems up from smaller ones. CCS, for example, contains six operators in total, including ones for composing systems in parallel and others for choice and scoping.
- **Operational semantics.** Process algebras are typically equipped with a Plotkin-style [7] structural operational semantics (SOS) that describes the single-step execution capabilities of systems. Using SOS, systems represented as terms in the algebra can be "compiled" into labeled transition systems.
- **Behavioral reasoning via equivalences and preorders.** Process algebras also feature the use of behavioral relations as a means for relating different systems given in the algebra. These relations are usually equivalences, which capture a notion of "same behavior", or preorders, which capture notions of "refinement".

In a process-algebraic approach to system verification, one typically writes two specifications. One, call it SYS, captures the design of the actual system and the other, call it SPEC, describes the system's desired "high-level" behavior. One may then establish the correctness of SYS with respect to SPEC by showing that SYS behaves the "same as" SPEC (if using an equivalence) or by showing that it refines SPEC (if using a preorder).

Establishing the correctness of SYS with respect to SPEC can be done in a syntax-oriented manner or in a semantics-oriented manner. In the former case, an *axiomatization* of the behavioral relation of choice is used to show that one expression can be transformed into the other via syntactic manipulations. In the latter case, one can appeal directly to the definition of the behavioral relation, and to the operational semantics of the two expressions, to show that they are related. In certain cases, e.g., when SYS and SPEC are "finite-state", verification, be it syntax-based or semantics-based, can be carried out automatically.

The advantages to an algebraic approach are the following.

- **System designers need learn only one language** for specifications and designs.
- **Related processes may be substituted for one another** inside other processes. This makes process algebras particularly suitable for the *modular analysis* of complex systems, since a specification and a design adhering to this specification may be used interchangeably inside larger systems.
- **Processes may be minimized** with respect to the equivalence relation before being analyzed; this sometimes leads to orders of magnitude improvement in the performance of verification routines.

Process-algebraic system descriptions can also be verified using *model checking* [3], a technique for ascertaining if a labeled transition system satisfies a correctness property given as a temporal-logic formula. Model checking has enjoyed considerable success in application to hardware designs. Progress is now being seen in other application domains such as software and protocol verification.

2. Classical roots

Process algebra can be viewed as a generalization of the classical theory of formal languages and automata [4], focusing on system specification and behavior rather than language recognition and generation. Process algebra also embodies the principles of cellular automata [5] – cells receiving inputs from neighboring cells and then taking appropriate action – while adding a notion of programmability: nondeterminism, dynamic topologies, evolving cell behavior, etc.

Process algebra lays the groundwork for a rigorous system-design ideology, providing support for specification, verification, implementation, testing and other life-cycle-critical activities. Interest in process algebra, however, extends beyond the system-design arena, to areas such as programming language design and semantics, complexity theory, real-time programming, and performance modeling and analysis.

3. About this Handbook

This Handbook documents the fate of process algebra from its modern inception in the late 1970's to the present. It is intended to serve as a reference source for researchers, students, and system designers and engineers interested in either the theory of process algebra or in learning what process algebra brings to the table as a formal system description and verification technique.

The Handbook is divided into six parts, the first five of which cover various theoretical and foundational aspects of process algebra. Part 6, the final part, is devoted to tools for applying process algebra and to some of the applications themselves. Each part contains between two and four chapters. Chapters are self-contained and can be read independently of each other. In total, there are 19 chapters spanning roughly 1300 pages. Collectively, the Handbook chapters give a comprehensive, albeit necessarily incomplete, view of the field.

Part 1, consisting of four chapters, covers a broad swath of the **basic theory** of process algebra. In Chapter 1, *The Linear Time – Branching Time Spectrum I*, van Glabbeek gives a useful structure to, and an encyclopedic account of, the many behavioral relations that have been proposed in the process-algebra literature. Chapter 2, *Trace-Oriented Models of Concurrency* by Broy and Olderog, provides an in-depth presentation of trace-oriented models of process behavior, where a trace is a communication sequence that a process can perform with its environment. Aceto, Fokkink and Verhoef present a thorough account of *Structural Operational Semantics* in Chapter 3. Part 1 concludes with Chapter 4, *Modal Logics and Mu-Calculi: An Introduction* by Bradfield and Stirling. Modal logics, which extend classical logic with operators for possibility and necessity, play an important role in filling out the semantic picture of process algebra.

Part 2 is devoted to the sub-specialization of process algebra known as **finite-state processes**. This class of processes holds a strong practical appeal as finite-state systems can be verified in an automatic, push-button style. The two chapters in Part 2 address finite-state processes from an axiomatic perspective: Chapter 5, *Process Algebra with Recursive Operations* by Bergstra, Fokkink and Ponse; and from an algorithmic one: Chapter 6, *Equivalence and Preorder Checking for Finite-State Systems* by Cleaveland and Sokolsky.

Infinite-state processes, the subject of Part 3, capture process algebra at its most expressive. Chapter 7, the first of the three chapters in this part, *A Symbolic Approach to Value-Passing Processes* by Ingólfsdóttir and Lin, systematically examines the class of infinite-state processes arising from the ability to transmit data from an arbitrary domain of values. Symbolic techniques are proposed as a method for analyzing such systems. Chapter 8, by Parrow, is titled *An Introduction to the π-Calculus*. This chapter investigates the area of mobile processes, an enriched form of value-passing process that is capable of transmitting communication channels and even processes themselves from one process to another. Finally, Burkhart, Caucal, Moller and Steffen consider the equivalence-checking and model-checking problems for a large variety of infinite-state processes in Chapter 9, *Verification on Infinite Structures*.

The three chapters of Part 4 explore several **extensions to process algebra** that make it easier to model the kinds of systems that arise in practice. Chapter 10 focuses on real-time systems. *Process Algebra with Timing: Real Time and Discrete Time* by Middelburg and Baeten, presents a real-time extension of the process algebra ACP that extends ACP in a natural way. The final two chapters of Part 4 study the impact on process algebra of replacing the standard notion of "nondeterministically choose the next transition to execute" with one in which probability or priority information play pivotal roles. Chapter 11, *Probabilistic Extensions of Process Algebras* by Jonsson, Larsen and Yi, targets the probabilistic case, which is especially useful for modeling system failure, reliability, and performance. Chapter 12, *Priority in Process Algebra* by Cleaveland, Lüttgen and Natarajan, considers the case of priority, and shows how a process algebra with priority can be used to model interrupts, prioritized choice and real-time behavior.

Process algebra was originally conceived with the view that concurrency equals interleaving. That is, the concurrent execution of a collection of events can be modeled as their interleaved execution, in any order. More recent versions of process algebra known as **non-interleaving process algebras**, aim to model concurrency directly, for example, as embodied in Petri nets. The four chapters of Part 5 address this subject. Chapter 13, *Partial-Order Process Algebra* by Baeten and Basten, thoroughly considers the impact of a non-interleaving semantics on ACP. Chapter 14, *A Unified Model for Nets and Process*

Algebras by Best, Devillers and Koutny, examines a range of issues that arise when process algebra and Petri nets are combined together. Another kind of non-interleaving treatment of concurrency is put forth in Chapter 15, Castellani's *Process Algebras with Localities*. In this approach, "locations" are assigned to parallel components, resulting in what Castellani calls a "distributed semantics" for process algebra. Finally, in Chapter 16, Gorrieri and Rensink's *Action Refinement* gives a thorough treatment of process algebra with action refinement, the operation of replacing a high-level atomic action with a low-level process. The interplay between action refinement and non-interleaving semantics is carefully considered.

Part 6, the final part of the Handbook, contains three chapters dealing with **tools and applications** of process algebra. The first of these, Chapter 17, *Algebraic Process Verification* by Groote and Reniers, gives a close-up account of verification techniques for distributed algorithms and protocols, using process algebra extended with data (μCRL). Chapter 18, *Discrete Time Process Algebra and the Semantics of SDL* by Bergstra, Middelburg and Usenko, introduces a discrete-time process algebra that is used to provide a formal semantics for SDL, a widely used formal description technique for telecommunications protocols. Finally, Chapter 19, *A Process Algebra for Interworkings* by Mauw and Reniers, devises a process-algebra-based semantics for Interworkings, a graphical design language of Philips Kommunikations Industrie.

Acknowledgements

The editors gratefully acknowledge the constant support of Arjen Sevenster, our manager at Elsevier; without his efforts, this Handbook would not have seen the light of day. We are equally grateful to all the authors; their diligence, talent, and patience are greatly appreciated. We would also like to thank the referees, whose reports significantly enhanced the final contents of the Handbook. They are: Luca Aceto, Jos Baeten, Wan Fokkink, Rob Goldblatt, Hardi Hungar, Joost-Pieter Katoen, Alexander Letichevsky, Bas Luttik, Faron Moller, Uwe Nestmann, Nikolaj Nikitchenko, Benjamin Pierce, Piet Rodenburg, Mariëlle Stoelinga, P.S. Thiagarajan, and Yaroslav Usenko. Finally, we would like to thank Rance Cleaveland for his help in writing this preface.

Autumn 2000
Jan A. Bergstra (Amsterdam),
Alban Ponse (Amsterdam),
Scott A. Smolka (Stony Brook, New York)

References

[1] J.A. Bergstra and J.W. Klop, *Process algebra for synchronous communication*, Inform. and Control **60** (1/3) (1984), 109–137.

[2] S.D. Brookes, C.A.R. Hoare and A.W. Roscoe, *A theory of communicating sequential processes*, J. ACM **31** (3) (1984), 560–599.

[3] E.M. Clarke, E.A. Emerson and A.P. Sistla, *Automatic verification of finite-state concurrent systems using temporal logic specifications*, ACM TOPLAS **8** (2) (1986).
[4] J.E. Hopcroft and J.D. Ullman, *Introduction to Automata Theory, Languages, and Computation*, Addison-Wesley (1979).
[5] J. von Neumann, *Theory of self-reproducing automata*, A.W. Burks, ed., Urbana, University of Illinois Press (1966).
[6] R. Milner, *A Calculus of Communicating Systems*, Lecture Notes in Comput. Sci. 92, Springer-Verlag (1980).
[7] G.D. Plotkin, *A structural approach to operational semantics*, Report DAIMI FN-19, Computer Science Department, Aarhus University (1981).

Jan A. Bergstra[2,3], Alban Ponse[1,2], Scott A. Smolka[4]

[1] *CWI, Kruislaan 413, 1098 SJ Amsterdam, The Netherlands*
http://www.cwi.nl/

[2] *University of Amsterdam, Programming Research Group, Kruislaan 403, 1098 SJ Amsterdam, The Netherlands*
http://www.science.uva.nl/research/prog/

[3] *Utrecht University, Department of Philosophy, Heidelberglaan 8, 3584 CS Utrecht, The Netherlands*
http://www.phil.uu.nl/eng/home.htmlE-mail:

[4] *State University of New York at Stony Brook, Department of Computer Science*
Stony Brook, NY 11794-4400, USA
http://www.cs.sunysb.edu/

E-mails: janb@science.uva.nl, alban@science.uva.nl, sas@cs.sunysb.edu

List of Contributors

Aceto, L., *Aalborg University, Aalborg* (Ch. 3).
Baeten, J.C.M. *Eindhoven University of Technology, Eindhoven* (Chs. 10, 13).
Basten, T., *Eindhoven University of Technology, Eindhoven* (Ch. 13).
Bergstra, J.A., *University of Amsterdam, Amsterdam and Utrecht University, Utrecht* (Chs. 5, 18).
Best, E., *Carl von Ossietzky Universität, Oldenburg* (Ch. 14).
Bradfield, J.C., *University of Edinburgh, Edinburgh, UK* (Ch. 4).
Broy, M., *Technische Universität München, München* (Ch. 2).
Burkart, O., *Universität Dortmund, Dortmund* (Ch. 9).
Castellani, I., *INRIA, Sophia-Antipolis* (Ch. 15).
Caucal, D., *IRISA, Rennes* (Ch. 9).
Cleaveland, R., *SUNY at Stony Brook, Stony Brook, NY* (Chs. 6, 12).
Devillers, R., *Université Libre de Bruxelles, Bruxelles* (Ch. 14).
Fokkink, W.J., *CWI, Amsterdam* (Chs. 3, 5).
Glabbeek, R.J. van, *Stanford University, Stanford, CA* (Ch. 1).
Gorrieri, R., *Università di Bologna, Bologna* (Ch. 16).
Groote, J.F., *Eindhoven University of Technology, Eindhoven* (Ch. 17).
Ingólfsdóttir, A., *Aalborg University, Aalborg* (Ch. 7).
Jonsson, B., *Uppsala University, Uppsala* (Ch. 11).
Koutny, M., *University of Newcastle, Newcastle upon Tyne, UK* (Ch. 14).
Larsen, K.G., *Aalborg University, Aalborg* (Ch. 11).
Lin, H., *Institute of Software, Chinese Academy of Sciences, Republic of China* (Ch. 7).
Lüttgen, G., *NASA Langley Research Center, Hampton, VA* (Ch. 12).
Mauw, S., *Eindhoven University of Technology, Eindhoven* (Ch. 19).
Middelburg, C.A., *Eindhoven University of Technology, Eindhoven and Utrecht University, Utrecht* (Chs. 10, 18).
Moller, F., *University of Wales Swansea, Swansea, UK* (Ch. 9).
Natarajan, V., *IBM Corporation, Research Triangle Park, NC* (Ch. 12).
Olderog, E.-R., *Universität Oldenburg, Oldenburg* (Ch. 2).
Parrow, J., *Royal Institute of Technology, Stockholm* (Ch. 8).
Ponse, A., *University of Amsterdam and CWI, Amsterdam* (Ch. 5).
Reniers, M.A., *Eindhoven University of Technology, Eindhoven* (Chs. 17, 19).
Rensink, A., *University of Twente, Enschede* (Ch. 16).
Sokolsky, O., *University of Pennsylvania, Philadelphia, PA* (Ch. 6).
Steffen, B., *Universität Dortmund, Dortmund* (Ch. 9).
Stirling, C., *University of Edinburgh, Edinburgh, UK* (Ch. 4).

Usenko, Y.S., *CWI, Amsterdam* (Ch. 18).
Verhoef, C., *Free University of Amsterdam, Amsterdam* (Ch. 3).
Wang Yi, *Uppsala University, Uppsala* (Ch. 11).

Contents

Part 5: Non-Interleaving Process Algebra

Part 6: Tools and Applications

Part 1
Basic Theory

CHAPTER 1

The Linear Time – Branching Time Spectrum I.* The Semantics of Concrete, Sequential Processes

R.J. van Glabbeek

Computer Science Department, Stanford University, Stanford, CA 94305-9045, USA
E-mail: rvg@cs.stanford.edu

Contents

*This is an extension of [20]. The research reported in this paper has been initiated at CWI in Amsterdam, continued at the Technical University of Munich, and finalized at Stanford University. It has been supported by Sonderforschungsbereich 342 of the TU München and by ONR under grant number N00014-92-J-1974. Part of it was carried out in the preparation of a course Comparative Concurrency Semantics, given at the University of Amsterdam, Spring 1988. A coloured version of this paper is available at http://boole.stanford.edu/pub/spectrum1.ps.gz.

HANDBOOK OF PROCESS ALGEBRA
Edited by Jan A. Bergstra, Alban Ponse and Scott A. Smolka

Abstract

In this paper various semantics in the linear time – branching time spectrum are presented in a uniform, model-independent way. Restricted to the class of finitely branching, concrete, sequential processes, only fifteen of them turn out to be different, and most semantics found in the literature that can be defined uniformly in terms of action relations coincide with one of these fifteen. Several testing scenarios, motivating these semantics, are presented, phrased in terms of 'button pushing experiments' on generative and reactive machines. Finally twelve of these semantics are applied to a simple language for finite, concrete, sequential, nondeterministic processes, and for each of them a complete axiomatization is provided.

Introduction

Process theory. A *process* is the behaviour of a system. The system can be a machine, an elementary particle, a communication protocol, a network of falling dominoes, a chess player, or any other system. *Process theory* is the study of processes. Two main activities of process theory are *modelling* and *verification*. Modelling is the activity of representing processes, mostly by mathematical structures or by expressions in a system description language. Verification is the activity of proving statements about processes, for instance that the actual behaviour of a system is equal to its intended behaviour. Of course, this is only possible if a criterion has been defined, determining whether or not two processes are equal, i.e., two systems behave similarly. Such a criterion constitutes the *semantics* of a process theory. (To be precise, it constitutes the semantics of the equality concept employed in a process theory.) Which aspects of the behaviour of a system are of importance to a certain user depends on the environment in which the system will be running, and on the interests of the particular user. Therefore it is not a task of process theory to find the 'true' semantics of processes, but rather to determine which process semantics is suitable for which applications.

Comparative concurrency semantics. This paper aims at the classification of process semantics.[1] The set of possible process semantics can be partially ordered by the relation 'makes strictly more identifications on processes than', thereby becoming a complete lattice.[3] Now the classification of some useful process semantics can be facilitated by drawing parts of this lattice and locating the positions of some interesting process semantics, found in the literature. Furthermore the ideas involved in the construction of these semantics can be unravelled and combined in new compositions, thereby creating an abundance of new process semantics. These semantics will, by their intermediate positions in the semantic lattice, shed light on the differences and similarities of the established ones. Sometimes they also turn out to be interesting in their own right. Finally the semantic lattice serves as a map on which it can be indicated which semantics satisfy certain desirable properties, and are suited for a particular class of applications.

Most semantic notions encountered in contemporary process theory can be classified along four different lines, corresponding with four different kinds of identifications. First there is the dichotomy of linear time versus branching time: to what extent should one identify processes differing only in the branching structure of their execution paths? Secondly there is the dichotomy of interleaving semantics versus partial order semantics: to what extent should one identify processes differing only in the causal dependencies between their actions (while agreeing on the possible orders of execution)? Thirdly one encounters

---

[1] This field of research is called *comparative concurrency*[2] *semantics*, a terminology first used by Meyer in [36].

[2] Here *concurrency* is taken to be synonymous with process theory, although strictly speaking it is only the study of *parallel* (as opposed to *sequential*) processes. These are the behaviours of systems capable of performing different actions at the same time. In this paper the term concurrency is considered to include sequential process theory. This may be justified since much work on sequential processes is intended to facilitate later studies involving parallelism.

[3] The supremum of a set of process semantics is the semantics identifying two processes whenever they are identified by every semantics in this set.

different treatments of abstraction from internal actions in a process: to what extent should one identify processes differing only in their internal or silent actions? And fourthly there are different approaches to infinity: to what extent should one identify processes differing only in their infinite behaviour? These considerations give rise to a four-dimensional representation of the proposed semantic lattice.

However, at least three more dimensions can be distinguished. In this paper, stochastic and real-time aspects of processes are completely neglected. Furthermore it deals with *uniform concurrency*[4] only. This means that processes are studied, performing actions[5] $a, b, c, \ldots$ which are not subject to further investigations. So it remains unspecified if these actions are in fact assignments to variables or the falling of dominoes or other actions. If also the options are considered of modelling (to a certain degree) the stochastic and real-time aspects of processes and the operational behaviour of the elementary actions, three more parameters in the classification emerge.

Process domains. In order to be able to reason about processes in a mathematical way, it is common practice to represent processes as elements of a mathematical domain.[6] Such a domain is called a *process domain*. The relation between the domain and the world of real processes is mostly stated informally. The semantics of a process theory can be modelled as an equivalence on a process domain, called a *semantic equivalence*. In the literature one finds among others:

- *graph domains*, in which a process is represented as a *process graph*, or *state transition diagram*,
- *net domains*, in which a process is represented as a (labelled) *Petri net*,
- *event structure domains*, in which a process is represented as a (labelled) *event structure*,
- *explicit domains*, where a process is represented as a mathematically coded set of its properties,
- *projective limit domains*, which are obtained as projective limits of series of finite term domains,
- and *term domains*, in which a process is represented as a term in a system description language.

Action relations. Write $p \xrightarrow{a} q$ if the process p can evolve into the process q, while performing the action a. The binary predicates $\xrightarrow{a}$ are called *action relations*. The semantic equivalences which are treated in this paper will be defined entirely in terms of action relations. Hence these definitions apply to any process domain on which action relations are defined. Such a domain is called a *labelled transition system*. Furthermore they will be defined *uniformly* in terms of action relations, meaning that all actions are treated in the same way. For reasons of convenience, even the usual distinction between internal and external actions is dropped in this paper.

---

[4] The term uniform concurrency is employed by De Bakker et al. [8].

[5] Strictly speaking processes do not perform actions, but systems do. However, for reasons of convenience, this paper sometimes uses the word process, when actually referring to a system of which the process is the behaviour.

[6] I use the word *domain* in the sense of *universal algebra*; it can be any class of mathematical objects – typically the first component of an *algebra*; the other component being a collection of operators defined on this domain. Without further adjectives I do not refer to the more restrictive domains employed in *domain theory*.

Finitely branching, concrete, sequential processes. Being a first step, this paper limits itself to a very simple class of processes. First of all only *sequential* processes are investigated: processes capable of performing at most one action at a time. Furthermore, instead of dropping the usual distinction between internal and external actions, one can equivalently maintain to study *concrete* processes: processes in which no internal actions occur. For this simple class of processes the announced semantic lattice collapses in two out of four dimensions and covers only the *infinitary linear time – branching time spectrum*.

Moreover, the main interest is in *finitely branching* processes: processes having in each state only finitely many possible ways to proceed. The material pertaining to infinitely branching processes – coloured brown in the electronic version of this paper – can easily be omitted in first reading.

Literature. In the literature on uniform concurrency 12 semantics can be found which are uniformly definable in terms of action relations and different on the domain of finitely branching, sequential processes (see Figure 1). The coarsest one (i.e., the semantics making the most identifications) is *trace semantics*, as presented in Hoare [30]. In trace semantics only *partial traces* are employed. The finest one (making less identifications than any of the others) is *bisimulation semantics*, as presented in Milner [39]. Bisimulation semantics is the standard semantics for the system description language CCS (Milner [37]). The

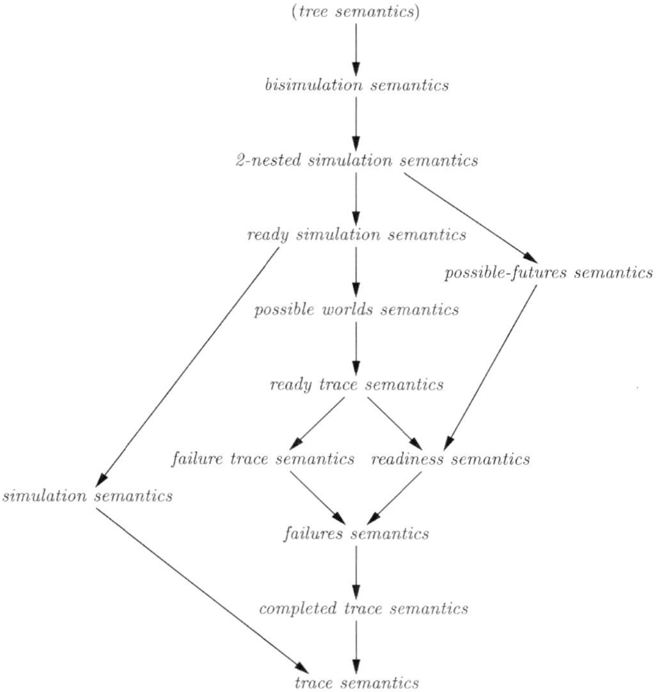

Fig. 1. The linear time – branching time spectrum.

notion of bisimulation was introduced in Park [41]. Bisimulation equivalence is a refinement of *observational equivalence*, as introduced by Hennessy and Milner in [27]. On the domain of finitely branching, concrete, sequential processes, both equivalences coincide. Also the semantics of De Bakker and Zucker, presented in [9], coincides with bisimulation semantics on this domain. Then there are ten semantics in between. First of all a variant of trace semantics can be obtained by using *complete traces* besides partial ones. In this paper it is called *completed trace semantics*. *Failures semantics* is introduced in Brookes, Hoare and Roscoe [13], and used in the construction of a model for the system description language CSP (Hoare [29,31]). It is finer than completed trace semantics. The semantics based on *testing equivalences*, as developed in De Nicola and Hennessy [17], coincides with failures semantics on the domain of finitely branching, concrete, sequential processes, as do the semantics of Kennaway [34] and Darondeau [15]. This has been established in De Nicola [16]. In Olderog and Hoare [40] *readiness semantics* is presented, which is slightly finer than failures semantics. Between readiness and bisimulation semantics one finds *ready trace semantics*, as introduced independently in Pnueli [43] (there called *barbed semantics*), Baeten, Bergstra and Klop [6] and Pomello [44] (under the name *exhibited behaviour semantics*). The natural completion of the square, suggested by failures, readiness and ready trace semantics yields *failure trace semantics*. For finitely branching processes this is the same as *refusal semantics*, introduced in Phillips [42]. *Simulation semantics*, based on the classical notion of *simulation* (see, e.g., Park [41]), is independent of the last five semantics. *Ready simulation semantics* was introduced in Bloom, Istrail and Meyer [12] under the name *GSOS trace congruence*. It is finer than ready trace as well as simulation semantics. In Larsen and Skou [35] a more operational characterization of this equivalence was given under the name $\frac{2}{3}$-*bisimulation equivalence*. The (denotational) notion of *possible worlds semantics* of Veglioni and De Nicola [49] fits between ready trace and ready simulation semantics. Finally 2-*nested simulation semantics*, introduced in Groote and Vaandrager [25], is located between ready simulation and bisimulation semantics, and *possible-futures semantics*, as proposed in Rounds and Brookes [46], can be positioned between 2-nested simulation and readiness semantics.

Tree semantics, employed in Winskel [50], is even finer than bisimulation semantics. However, a proper treatment requires more than mere action relations.

About the contents. The first section of this paper introduces labelled transition systems and process graphs. A labelled transition system is any process domain that is equipped with action relations. The domain of *process graphs* or *state transition diagrams* is one of the most popular labelled transition systems. In Sections 2–14 all semantic equivalences mentioned above are defined on arbitrary labelled transition systems. In particular these definitions apply to the domain of process graphs. Most of the equivalences can be motivated by the observable behaviour of processes, according to some testing scenario. (Two processes are equivalent if they allow the same set of possible observations, possibly in response to certain experiments.) I will try to capture these motivations in terms of *button pushing experiments* (cf. Milner [37], pp. 10–12). Furthermore the semantics will be partially ordered by the relation 'makes at least as many identifications as'. This yields the linear time – branching time spectrum. Counterexamples are provided, showing that on the graph domain this ordering cannot be further expanded. However, for deterministic processes the spectrum collapses, as was first observed by Park [41]. Section 6 describes various other classes of processes on which parts of the spectrum collapse. In Section 17, the semantics are applied to a simple language for finite, concrete, sequential, nondeterministic processes, and for twelve of them a complete axiomatization is provided. Section 18 applies a few criteria indicating which semantics are suitable for which applications. Finally, in Section 19 the work of this paper is extended to labelled transition systems that distinguish between deadlock and successful termination.

With each of the semantic equivalences treated in this paper (except for tree semantics) a preorder is associated that may serve as an implementation relation between processes. The results obtained for the equivalences are extended to the associated preorders as well.

1. Labelled transition systems and process graphs

1.1. *Labelled transition systems*

In this paper processes will be investigated that are capable of performing actions from a given set *Act*. By an *action* any activity is understood that is considered as a conceptual entity on a chosen level of abstraction. Actions may be instantaneous or durational and are not required to terminate, but in a finite time only finitely many actions can be carried out. Any activity of an investigated process should be part of some action $a \in Act$ performed by the process. Different activities that are indistinguishable on the chosen level of abstraction are interpreted as occurrences of the same action $a \in Act$.

A process is *sequential* if it can perform at most one action at the same time. In this paper only sequential processes will be considered. A class of sequential processes can often be conveniently represented as a labelled transition system. This is a domain $\mathbb{P}$ on which infix written binary predicates $\xrightarrow{a}$ are defined for each action $a \in Act$. The elements of $\mathbb{P}$ represent processes, and $p \xrightarrow{a} q$ means that p can start performing the action a and after completion of this action reach a state where q is its remaining behaviour. In a labelled transition system it may happen that $p \xrightarrow{a} q$ and $p \xrightarrow{b} r$ for different actions a and b or different processes q and r. This phenomenon is called *branching*. It need not be specified how the choice between the alternatives is made, or whether a probability distribution can be attached to it.

Certain actions may be synchronizations of a process with its environment, or the receipt of a signal sent by the environment. Naturally, these actions can only occur if the environment cooperates. In the labelled transition system representation of processes all these potential actions are included, so $p \xrightarrow{a} q$ merely means that there is an environment in which the action a can occur.

Notation. For any alphabet Σ, let Σ^* be the set of finite sequences and Σ^∞ the set of infinite sequences over Σ. $\Sigma^\omega := \Sigma^* \cup \Sigma^\infty$. Write ε for the empty sequence, $\sigma\rho$ for the concatenation of $\sigma \in \Sigma^*$ and $\rho \in \Sigma^\omega$, and a for the sequence consisting of the single symbol $a \in \Sigma$.

DEFINITION 1.1. A *labelled transition system* is a pair $(\mathbb{P}, \rightarrow)$ with $\mathbb{P}$ a class and $\rightarrow \subseteq \mathbb{P} \times Act \times \mathbb{P}$, such that for $p \in \mathbb{P}$ and $a \in Act$ the class $\{q \in \mathbb{P} \mid (p, a, q) \in \rightarrow\}$ is a set.

Most of this paper should be read in the context of a given labelled transition system $(\mathbb{P}, \rightarrow)$, ranged over by $p, q, r, \ldots$. Write $p \xrightarrow{a} q$ for $(p, a, q) \in \rightarrow$. The binary predicates $\xrightarrow{a}$ are called *action relations*.

DEFINITION 1.2 (*Remark that the following concepts are defined in terms of action relations only*).

- The *generalized action relations* $\xrightarrow{\sigma}$ for $\sigma \in Act^*$ are defined recursively by:
  (1) $p \xrightarrow{\varepsilon} p$, for any process p.
  (2) $(p, a, q) \in \rightarrow$ with $a \in Act$ implies $p \xrightarrow{a} q$ with $a \in Act^*$.
  (3) $p \xrightarrow{\sigma} q \xrightarrow{\rho} r$ implies $p \xrightarrow{\sigma\rho} r$.
  In words: the generalized action relations $\xrightarrow{\sigma}$ are the reflexive and transitive closure of the ordinary action relations $\xrightarrow{a}$. $p \xrightarrow{\sigma} q$ means that p can evolve into q, while performing the sequence σ of actions. Remark that the overloading of the notion $p \xrightarrow{a} q$ is quite harmless.
- A process $q \in \mathbb{P}$ is *reachable* from $p \in \mathbb{P}$ if $p \xrightarrow{\sigma} q$ for some $\sigma \in Act^*$.
- The set of *initial actions* of a process p is defined by: $I(p) = \{a \in Act \mid \exists q\colon\ p \xrightarrow{a} q\}$.
- A process $p \in \mathbb{P}$ is *finite* if the set $\{(\sigma, q) \in (Act^* \times \mathbb{P}) \mid p \xrightarrow{\sigma} q\}$ is finite.
- p is *image finite* if for each $\sigma \in Act^*$ the set $\{q \in \mathbb{P} \mid p \xrightarrow{\sigma} q\}$ is finite.
- p is *deterministic* if $p \xrightarrow{\sigma} q \wedge p \xrightarrow{\sigma} r \Rightarrow q = r$.
- p is *well-founded* if there is no infinite sequence $p \xrightarrow{a_1} p_1 \xrightarrow{a_2} p_2 \xrightarrow{a_3} \cdots$.
- p is *finitely branching* if for each q reachable from p, the set $\{(a, r) \in Act \times \mathbb{P} \mid q \xrightarrow{a} r\}$ is finite.

Note that a process $p \in \mathbb{P}$ is image finite iff for each $q \in \mathbb{P}$ reachable from p and each $a \in Act$, the set $\{r \in \mathbb{P} \mid q \xrightarrow{a} r\}$ is finite. Hence finitely branching processes are image finite. Moreover, by König's lemma a process is finite iff it is well-founded and finitely branching.

1.2. *Process graphs*

DEFINITION 1.3. A *process graph* over an alphabet Act is a rooted, directed graph whose edges are labelled by elements of Act. Formally, a process graph g is a triple $(\text{NODES}(g), \text{ROOT}(g), \text{EDGES}(g))$, where

- $\text{NODES}(g)$ is a set, of which the elements are called the *nodes* or *states* of g,
- $\text{ROOT}(g) \in \text{NODES}(g)$ is a special node: the *root* or *initial state* of g,
- and $\text{EDGES}(g) \subseteq \text{NODES}(g) \times Act \times \text{NODES}(g)$ is a set of triples (s, a, t) with $s, t \in \text{NODES}(g)$ and $a \in Act$: the *edges* or *transitions* of g.

If $e = (s, a, t) \in \text{EDGES}(g)$, one says that e *goes from* s *to* t. A (finite) *path* π in a process graph is an alternating sequence of nodes and edges, starting and ending with a node, such that each edge goes from the node before it to the node after it. If $\pi = s_0(s_0, a_1, s_1)s_1(s_1, a_2, s_2)\cdots(s_{n-1}, a_n, s_n)s_n$, also denoted as $\pi : s_0 \xrightarrow{a_1} s_1 \xrightarrow{a_2} \cdots \xrightarrow{a_n} s_n$, one says that π *goes from* s_0 *to* s_n; it *starts* in s_0 and *ends* in $end(\pi) = s_n$. Let $\text{PATHS}(g)$ be the set of paths in g starting from the root. If s and t are nodes in a process graph then

t *can be reached from* s if there is a path going from s to t. A process graph is said to be *connected* if all its nodes can be reached from the root; it is a *tree* if each node can be reached from the root by exactly one path. Let $\mathbb{G}$ be the domain of connected process graphs over a given alphabet Act.

DEFINITION 1.4. Let $g, h \in \mathbb{G}$. A *graph isomorphism* between g and h is a bijective function $f : \text{NODES}(g) \to \text{NODES}(h)$ satisfying
- $f(\text{ROOT}(g)) = \text{ROOT}(g)$, and
- $(s, a, t) \in \text{EDGES}(g) \Leftrightarrow (f(s), a, f(t)) \in \text{EDGES}(h)$.

Graphs g and h are *isomorphic*, notation $g \cong h$, if there exists a graph isomorphism between them.

In this case g and h differ only in the identity of their nodes. Remark that graph isomorphism is an equivalence relation on $\mathbb{G}$.

Connected process graphs can be pictured by using open dots ($\circ$) to denote nodes, and labelled arrows to denote edges, as can be seen further on. There is no need to mark the root of such a process graph if it can be recognized as the unique node without incoming edges, as is the case in all my examples. These pictures determine process graphs only up to graph isomorphism, but usually this suffices since it is virtually never needed to distinguish between isomorphic graphs.

DEFINITION 1.5. For $g \in \mathbb{G}$ and $s \in \text{NODES}(g)$, let g_s be the process graph defined by
- $\text{NODES}(g_s) = \{t \in \text{NODES}(g) \mid \text{there is a path going from } s \text{ to } t\}$,
- $\text{ROOT}(g_s) = s \in \text{NODES}(g_s)$, and
- $(t, a, u) \in \text{EDGES}(g_s)$ iff $t, u \in \text{NODES}(g_s)$ and $(t, a, u) \in \text{EDGES}(g)$.

Of course $g_s \in \mathbb{G}$. Note that $g_{\text{ROOT}(g)} = g$. Now on $\mathbb{G}$ action relations $\xrightarrow{a}$ for $a \in Act$ are defined by $g \xrightarrow{a} h$ iff $(\text{ROOT}(g), a, s) \in \text{EDGES}(g)$ and $h = g_s$. This makes $\mathbb{G}$ into a labelled transition system.

1.3. *Embedding labelled transition systems in* $\mathbb{G}$

Let $(\mathbb{P}, \to)$ be an arbitrary labelled transition system and let $p \in \mathbb{P}$. The *canonical graph* $G(p)$ of p is defined as follows:
- $\text{NODES}(G(p)) = \{q \in \mathbb{P} \mid \exists \sigma \in Act^*\colon p \xrightarrow{\sigma} q\}$,
- $\text{ROOT}(G(p)) = p \in \text{NODES}(G(p))$, and
- $(q, a, r) \in \text{EDGES}(G(p))$ iff $q, r \in \text{NODES}(G(p))$ and $q \xrightarrow{a} r$.

Of course $G(p) \in \mathbb{G}$. This means G is a function from $\mathbb{P}$ to $\mathbb{G}$.

PROPOSITION 1.1. *$G : \mathbb{P} \to \mathbb{G}$ is injective and satisfies, for $a \in Act$: $G(p) \xrightarrow{a} G(q) \Leftrightarrow p \xrightarrow{a} q$. Moreover, $G(p) \xrightarrow{a} h$ only if h has the form $G(q)$ for some $q \in \mathbb{P}$ (with $p \xrightarrow{a} q$).*

PROOF. Trivial. □

Proposition 1.1 says that G is an *embedding* of $\mathbb{P}$ in $\mathbb{G}$. It implies that any labelled transition system over *Act* can be represented as a subclass $G(\mathbb{P}) = \{G(p) \in \mathbb{G} \mid p \in \mathbb{P}\}$ of $\mathbb{G}$.

Since $\mathbb{G}$ is also a labelled transition system, G can be applied to $\mathbb{G}$ itself. The following proposition says that the function $G : \mathbb{G} \to \mathbb{G}$ leaves its arguments intact up to graph isomorphism.

PROPOSITION 1.2. *For* $g \in \mathbb{G}$, $G(g) \cong g$.

PROOF. Remark that $\text{NODES}(G(g)) = \{g_s \mid s \in \text{NODES}(g)\}$.

Now the function $f : \text{NODES}(G(g)) \to \text{NODES}(g)$ defined by $f(g_s) = s$ is a graph isomorphism. □

1.4. *Equivalences relations and preorders on labelled transition systems*

This paper studies semantics on labelled transition systems. Each of the semantics examined here (except for tree semantics) is defined or characterized in terms of a function $\mathcal{O}$ that associates with every process $p \in \mathbb{P}$ a set $\mathcal{O}(p)$. In most cases the elements of $\mathcal{O}(p)$ can be regarded as the possible observations one could make while interacting with the process p in the context of a particular testing scenario. The set $\mathcal{O}(p)$ then constitutes the observable behaviour of p. For every such $\mathcal{O}$, the equivalence relation $=_\mathcal{O} \in \mathbb{P} \times \mathbb{P}$ is given by $p =_\mathcal{O} q \Leftrightarrow \mathcal{O}(p) = \mathcal{O}(q)$, and the preorder $\sqsubseteq_\mathcal{O} \in \mathbb{P} \times \mathbb{P}$ by $p \sqsubseteq_\mathcal{O} q \Leftrightarrow \mathcal{O}(p) \subseteq \mathcal{O}(q)$. Obviously $p =_\mathcal{O} q \Leftrightarrow p \sqsubseteq_\mathcal{O} q \wedge q \sqsubseteq_\mathcal{O} p$. The semantic equivalence $=_\mathcal{O}$ partitions $\mathbb{P}$ into equivalence classes of processes that are indistinguishable by observation (using observations of type $\mathcal{O}$). The preorder $\sqsubseteq_\mathcal{O}$ moreover provides a partial order between these equivalence classes; one that could be taken to constitute an "implementation" relation. The associated *semantics*, also called $\mathcal{O}$, is the criterion that identifies two processes whenever they are $\mathcal{O}$-equivalent. Two semantics are considered the same if the associated equivalence relations are the same.

As the definitions of $\mathcal{O}$ are given entirely in terms of action relations, they apply to any labelled transition system $\mathbb{P}$. Moreover, the definitions of $\mathcal{O}(p)$ involve only action relations between processes reachable from p. Thus Proposition 1.1 implies that $\mathcal{O}(G(p)) = \mathcal{O}(p)$. This in turn yields

COROLLARY 1.1. $p \sqsubseteq_\mathcal{O} q$ *iff* $G(p) \sqsubseteq_\mathcal{O} G(q)$ *and* $p =_\mathcal{O} q$ *iff* $G(p) =_\mathcal{O} G(q)$.

Write $\mathcal{O} \preceq_\mathbb{P} \mathcal{N}$ if semantics $\mathcal{O}$ makes at least as much identifications as semantics $\mathcal{N}$. This is the case if the equivalence corresponding with $\mathcal{O}$ is equal to or coarser than the one corresponding with $\mathcal{N}$, i.e., if $p =_\mathcal{N} q \Rightarrow p =_\mathcal{O} q$ for all $p, q \in \mathbb{P}$. Let $\preceq$ abbreviate $\preceq_\mathbb{G}$. The following is then immediate by Corollary 1.1.

COROLLARY 1.2. $\mathcal{O} \preceq \mathcal{N}$ *iff* $\mathcal{O} \preceq_\mathbb{P} \mathcal{N}$ *for* each *labelled transition system* $\mathbb{P}$.

On the other hand, $\mathcal{O} \not\preceq \mathcal{N}$ *iff* $\mathcal{O} \not\preceq_\mathbb{P} \mathcal{N}$ *for* some *labelled transition system* $\mathbb{P}$.

Write $\mathcal{O} \preceq^*_{\mathbb{P}} \mathcal{N}$ if $p \sqsubseteq_{\mathcal{N}} q \Rightarrow p \sqsubseteq_{\mathcal{O}} q$ for all $p, q \in \mathbb{P}$, and let $\preceq^*$ abbreviate $\preceq^*_{\mathbb{G}}$. By definition $\mathcal{O} \preceq^* \mathcal{N} \Rightarrow \mathcal{O} \preceq \mathcal{N}$ for all semantics $\mathcal{O}$ and $\mathcal{N}$. The reverse does not hold by definition, but it will be shown to hold for all semantics discussed in this paper (cf. Section 15).

1.5. *Initial nondeterminism*

In a process graph it need not be determined in which state one ends after performing a nonempty sequence of actions. This phenomenon is called *nondeterminism*. However, process graphs as defined above are not capable of modelling *initial nondeterminism*, as there is only one initial state. This can be rectified by considering *process graphs with multiple roots*, in which $\text{ROOTS}(g)$ may be any nonempty subset of $\text{NODES}(g)$ – let $\mathbb{G}^{mr}$ be the class of such connected process graphs. A process graph with multiple roots can also be regarded as a nonempty set of process graphs with single roots. More generally, initial nondeterminism can be modelled in any labelled transition system $\mathbb{P}$ by regarding the nonempty subsets of $\mathbb{P}$ (rather than merely its elements) to be processes. The elements of a process $P \subseteq \mathbb{P}$ then represent the possible initial states of P.

Now any notion of observability $\mathcal{O}$ on $\mathbb{P}$ extends to processes with initial nondeterminism by defining $\mathcal{O}(P) = \bigcup_{p \in P} \mathcal{O}(p)$ for $P \subseteq \mathbb{P}$. Thus also the equivalences $=_{\mathcal{O}}$ and preorders $\sqsubseteq_{\mathcal{O}}$ are defined on such processes. Write $\mathcal{O} \preceq'_{\mathbb{P}} \mathcal{N}$ if $P =_{\mathcal{N}} Q \Rightarrow P =_{\mathcal{O}} Q$ for all nonempty $P, Q \subseteq \mathbb{P}$, and let $\preceq'$ abbreviate $\preceq'_{\mathbb{G}}$. Clearly, one has $\mathcal{O} \preceq' \mathcal{N} \Rightarrow \mathcal{O} \preceq \mathcal{N}$ for all semantics $\mathcal{O}$ and $\mathcal{N}$.

Let g be a process graph over Act with multiple roots. Let i be an action (*initialize*) which is not in Act. Define $\rho(g)$ as the process graph over $Act \cup \{i\}$ obtained from g by adding a new state $*$, which will be the root of $\rho(g)$, and adding a transition $(*, i, r)$ for every $r \in \text{ROOTS}(g)$. Now for every semantics $\mathcal{O}$ to be discussed in this paper it will be the case that $g \sqsubseteq_{\mathcal{O}} h \Leftrightarrow \rho(g) \sqsubseteq_{\mathcal{O}} \rho(h)$, as the reader may easily verify for each such $\mathcal{O}$. From this it follows that we have in fact $\mathcal{O} \preceq' \mathcal{N} \Leftrightarrow \mathcal{O} \preceq \mathcal{N}$ for all semantics $\mathcal{O}$ and $\mathcal{N}$ treated in this paper. This justifies focusing henceforth on process graphs with single roots and processes as mere elements of labelled transition systems.

2. Trace semantics

DEFINITION 2. $\sigma \in Act^*$ is a *trace* of a process p if there is a process q such that $p \xrightarrow{\sigma} q$. Let $T(p)$ denote the set of traces of p. Two processes p and q are *trace equivalent*, notation $p =_T q$, if $T(p) = T(q)$. In *trace semantics* (T) two processes are identified iff they are trace equivalent.

Testing scenario. Trace semantics is based on the idea that two processes are to be identified if they allow the same set of observations, where an observation simply consists of a sequence of actions performed by the process in succession.

Modal characterization

DEFINITION 2.1. The set $\mathcal{L}_T$ of *trace formulas* over *Act* is defined recursively by:
- $\top \in \mathcal{L}_T$.
- If $\varphi \in \mathcal{L}_T$ and $a \in Act$ then $a\varphi \in \mathcal{L}_T$.

The *satisfaction relation* $\models \subseteq \mathbb{P} \times \mathcal{L}_T$ is defined recursively by:
- $p \models \top$ for all $p \in \mathbb{P}$.
- $p \models a\varphi$ if for some $q \in \mathbb{P}$: $p \xrightarrow{a} q$ and $q \models \varphi$.

Note that a trace formula satisfied by a process p represents nothing more or less than a trace of p. Hence one has

PROPOSITION 2.1. $p =_T q \Leftrightarrow \forall \varphi \in \mathcal{L}_T (p \models \varphi \Leftrightarrow q \models \varphi)$.

Process graph characterization. Let $g \in \mathbb{G}^{mr}$ and $\pi : s_0 \xrightarrow{a_1} s_1 \xrightarrow{a_2} \cdots \xrightarrow{a_n} s_n \in$ PATHS(g). Then $T(\pi) := a_1 a_2 \cdots a_n \in Act^*$ is the *trace* of π. As $\mathbb{G}$ is a labelled transition system, $T(g)$ is defined above. Alternatively, it could be defined as the set of traces of paths of g. It is easy to see that these definitions are equivalent:

PROPOSITION 2.2. $T(g) = \{T(\pi) \mid \pi \in \text{PATHS}(g)\}$.

Explicit model. In trace semantics a process can be represented by a trace equivalence class of process graphs, or equivalently by the set of its traces. Such a *trace set* is always nonempty and prefix-closed. The next proposition shows that the domain $\mathbb{T}$ of trace sets is in bijective correspondence with the domain $\mathbb{G}/_{=_T}$ of process graphs modulo trace equivalence, as well as with the domain $\mathbb{G}^{mr}/_{=_T}$ of process graphs with multiple roots modulo trace equivalence. Models of concurrency like $\mathbb{T}$, in which a process is not represented as an equivalence class but rather as a mathematically coded set of its properties, are sometimes referred to as *explicit models*.

DEFINITION 2.2. The *trace domain* $\mathbb{T}$ is the set of subsets T of Act^* satisfying

T1 $\varepsilon \in \mathrm{T}$,
T2 $\sigma\rho \in \mathrm{T} \Rightarrow \sigma \in \mathrm{T}$.

PROPOSITION 2.3. $\mathrm{T} \in \mathbb{T} \Leftrightarrow \exists g \in \mathbb{G}$: $T(g) = \mathrm{T} \Leftrightarrow \exists g \in \mathbb{G}^{mr}$: $T(g) = \mathrm{T}$.

PROOF. Let $\mathrm{T} \in \mathbb{T}$. Define the *canonical graph* $G(\mathrm{T})$ of T by NODES$(G(\mathrm{T})) = \mathrm{T}$, ROOT$(G(\mathrm{T})) = \varepsilon$ and $(\sigma, a, \rho) \in$ EDGES$(G(\mathrm{T}))$ iff $\rho = \sigma a$. As T satisfies T2, $G(\mathrm{T})$ is connected, i.e., $G(\mathrm{T}) \in \mathbb{G}$. In fact, $G(\mathrm{T})$ is a tree. Moreover, for every path $\pi \in$ PATHS$(G(\mathrm{T}))$ one has $T(\pi) = end(\pi)$. Hence, using Proposition 2.2, $T(G(\mathrm{T})) = \mathrm{T}$.

For the remaining two implication, note that $\mathbb{G} \subseteq \mathbb{G}^{mr}$, and the trace set $T(g)$ of any graph $g \in \mathbb{G}^{mr}$ satisfies T1 and T2. □

$\mathbb{T}$ was used as a model of concurrency in Hoare [30].

Infinite processes. For infinite processes one distinguishes two variants of trace semantics: *(finitary) trace semantics* as defined above, and *infinitary trace semantics* (T^∞), obtained by taking infinite runs into account.

DEFINITION 2.3. $a_1 a_2 \cdots \in Act^\infty$ is an *infinite trace* of a process $p \in \mathbb{P}$ if there are processes $p_1, p_2, \ldots$ such that $p \xrightarrow{a_1} p_1 \xrightarrow{a_2} \cdots$. Let $T^\infty(p)$ denote the set of infinite traces of p. Two processes p and q are *infinitary trace equivalent*, notation $p =_T^\infty q$, if $T(p) = T(q)$ and $T^\infty(p) = T^\infty(q)$.

Clearly $p =_T^\infty q \Rightarrow p =_T q$. That on $\mathbb{G}$ the reverse does not hold follows from Counterexample 1: one has $T(\mathit{left}) = T(\mathit{right}) = \{a^n \mid n \in \mathbb{N}\}$, but $T^\infty(\mathit{left}) \neq T^\infty(\mathit{right})$, as only the graph at the right has an infinite trace.

However, with König's lemma one easily proves that for image finite processes finitary and infinitary trace equivalence coincide:

PROPOSITION 2.4. *Let p and q be image finite processes with $p =_T q$. Then $p =_T^\infty q$.*

PROOF. It is sufficient to show that $T^\infty(p)$ can be expressed in terms of $T(p)$ for any image finite process p. In fact, $T^\infty(p)$ consists of all those infinite traces for which all finite prefixes are in $T(p)$. One direction of this statement is trivial: if $\sigma \in T^\infty(p)$, all finite prefixes of σ must be in $T(p)$. For the other direction suppose that, for $i \in \mathbb{N}$, $a_i \in Act$ and $a_1 a_2 \cdots a_i \in T(p)$. With induction on $i \in \mathbb{N}$ one can show that there exists processes p_i such that $i = 0$ and $p_0 = p$, or $p_{i-1} \xrightarrow{a_i} p_i$, and for every $j \geqslant i$ one has $a_{i+1} a_{i+2} \cdots a_j \in T(p_i)$. The existence of these p_i's immediately entails that $a_1 a_2 a_3 \cdots \in T^\infty(p)$. The base case ($i = 0$) is trivial. Suppose the claim holds for certain i. For every $j \geqslant i + 1$ there must be a process q with $p_i \xrightarrow{a_{i+1}} q$ and $a_{i+2} a_{i+3} \cdots a_j \in T(q)$. As there are only finitely many processes q with $p_i \xrightarrow{a_{i+1}} q$, there must be one choice of q for which $a_{i+2} a_{i+3} \cdots a_j \in T(q)$ for infinitely many values of j. Take this q to be p_{i+1}. As $T(p_{i+1})$ is prefix-closed, one has $a_{i+2} a_{i+3} \cdots a_j \in T(p_{i+1})$ for *all* $j \geqslant i + 1$. □

An explicit representation of infinitary trace semantics is obtained by taking the subsets T of Act^ω satisfying T1 and T2.

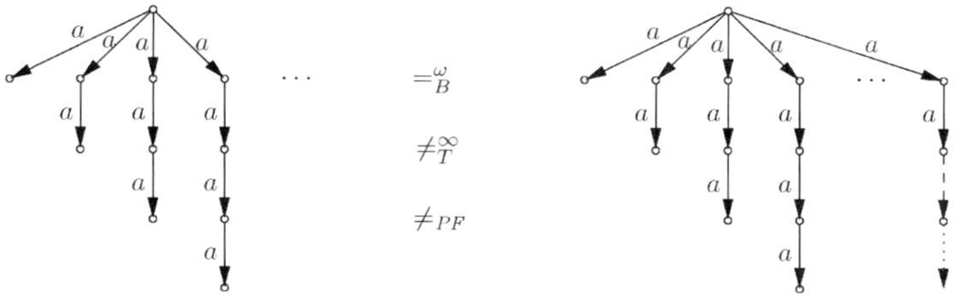

Counterexample 1. Finitary equivalent but not infinitary equivalent.

3. Completed trace semantics

DEFINITION 3. $\sigma \in Act^*$ is a *complete trace* of a process p, if there is a process q such that $p \xrightarrow{\sigma} q$ and $I(q) = \emptyset$. Let $CT(p)$ denote the set of complete traces of p. Two processes p and q are *completed trace equivalent*, notation $p =_{CT} q$, if $T(p) = T(q)$ and $CT(p) = CT(q)$. In *completed trace semantics* (CT) two processes are identified iff they are completed trace equivalent.

Testing scenario. Completed trace semantics can be explained with the following (rather trivial) *completed trace machine*. The process is modelled as a black box that contains as its interface to the outside world a display on which the name of the action is shown that is currently carried out by the process. The process autonomously chooses an execution path that is consistent with its position in the labelled transition system $(\mathbb{P}, \rightarrow)$. During this execution always an action name is visible on the display. As soon as no further action can be carried out, the process reaches a state of deadlock and the display becomes empty. Now the existence of an observer is assumed that watches the display and records the sequence of actions displayed during a run of the process, possibly followed by deadlock. It is assumed that an observation takes only a finite amount of time and may be terminated before the process stagnates. Hence the observer records either a sequence of actions performed in succession – a trace of the process – or such a sequence followed by deadlock – a completed trace. Two processes are identified if they allow the same set of observations in this sense.

The *trace machine* can be regarded as a simpler version of the completed trace machine, were the last action name remains visible in the display if deadlock occurs (unless deadlock occurs in the beginning already). On this machine traces can be recorded, but stagnation cannot be detected, since in case of deadlock the observer may think that the last action is still continuing.

Modal characterization

DEFINITION 3.1. The set $\mathcal{L}_{CT}$ of *completed trace formulas* over Act is defined recursively by:
- $\top \in \mathcal{L}_{CT}$.
- $0 \in \mathcal{L}_{CT}$.
- If $\varphi \in \mathcal{L}_{CT}$ and $a \in Act$ then $a\varphi \in \mathcal{L}_{CT}$.

The *satisfaction relation* $\models \subseteq \mathbb{P} \times \mathcal{L}_{CT}$ is defined recursively by:
- $p \models \top$ for all $p \in \mathbb{P}$.
- $p \models 0$ if $I(p) = \emptyset$.

Figure 2. The completed trace machine.

- $p \models a\varphi$ if for some $q \in \mathbb{P}$: $p \xrightarrow{a} q$ and $q \models \varphi$.

Note that a completed trace formula satisfied by a process p represents either a trace (if it has the form $a_1 a_2 \cdots a_n \top$) or a completed trace (if it has the form $a_1 a_2 \cdots a_n 0$). Hence one has

PROPOSITION 3.1. $p =_{CT} q \Leftrightarrow \forall \varphi \in \mathcal{L}_{CT}(p \models \varphi \Leftrightarrow q \models \varphi)$.

Also note the close link between the constructors of the modal formulas (corresponding to the three clauses in Definition 3.1) and the types of observations according to the testing scenario: $\top$ represents the act of the observer of terminating the observation, regardless of whether the observed process has terminated, 0 represents the observation of deadlock (the display becomes empty), and $a\varphi$ represents the observation of a being displayed, followed by the observation φ.

Process graph characterization. Let $g \in \mathbb{G}^{mr}$ and $s \in \text{NODES}(g)$. Then $I(s) := \{a \in Act \mid \exists t\colon (s, a, t) \in \text{EDGES}(g)\}$ is the *menu* of s. $CT(g)$ can now be characterized as follows.

PROPOSITION 3.2. $CT(g) = \{T(\pi) \mid \pi \in \text{PATHS}(g) \wedge I(end(\pi)) = \emptyset\}$.

Classification. Trivially $T \preceq CT$ (as in Figure 1). Counterexample 2 shows that the reverse does not hold: one has $T(left) = T(right) = \{\varepsilon, a, ab\}$, whereas $CT(left) \neq CT(right)$ (since $a \in CT(left) - CT(right)$). Hence the two process graphs are identified in trace semantics but distinguished in completed trace semantics. Thus $T \prec CT$: on $\mathbb{G}$ completed trace semantics makes strictly less identifications than trace semantics.

Explicit model. In completed trace semantics a process can be represented by a completed trace equivalence class of process graphs, or equivalently by the pair (T, CT) of its sets of traces and complete traces. The next proposition gives an explicit characterization of the domain $\mathbb{CT}$ of pairs of sets of traces and complete traces of process graphs with multiple roots.

DEFINITION 3.2. The *completed trace domain* $\mathbb{CT}$ is the set of pairs $(\mathrm{T}, \mathrm{CT}) \in \mathcal{P}(Act^*) \times \mathcal{P}(Act^*)$ satisfying

$$\mathrm{T} \in \mathbb{T} \quad \text{and} \quad \mathrm{CT} \subseteq \mathrm{T},$$
$$\sigma \in \mathrm{T} - \mathrm{CT} \Rightarrow \exists a \in Act\colon \sigma a \in \mathrm{T}.$$

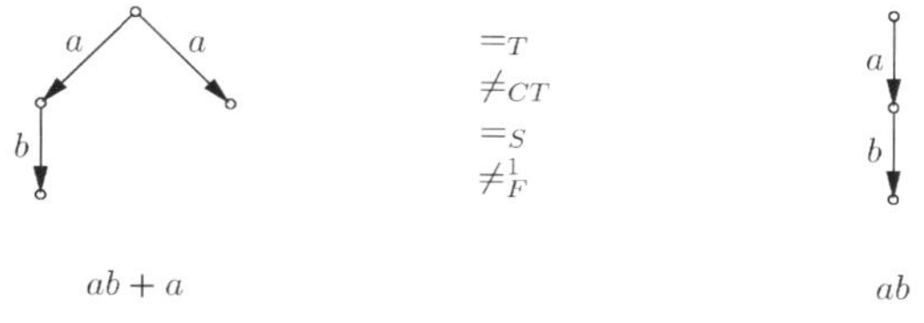

Counterexample 2. Trace and simulation equivalent, but not completed trace equivalent.

PROPOSITION 3.3. $(\mathrm{T},\mathrm{CT}) \in \mathbb{CT} \Leftrightarrow \exists g \in \mathbb{G}^{mr}\colon T(g) = \mathrm{T} \wedge CT(g) = \mathrm{T}$.

PROOF. Let $(\mathrm{T},\mathrm{CT}) \in \mathbb{CT}$. Define the *canonical graph* $G(\mathrm{T},\mathrm{CT})$ of (T,CT) by

- $\mathrm{NODES}(G(\mathrm{T},\mathrm{CT})) = \mathrm{T} \cup \{\sigma\delta \mid \sigma \in \mathrm{CT}\}$,
- $\mathrm{ROOTS}(G(\mathrm{T},\mathrm{CT})) = \{\varepsilon\} \cup \{\delta \mid \varepsilon \in \mathrm{CT}\}$, and
- $(\sigma, a, \rho) \in \mathrm{EDGES}(G(\mathrm{T}))$ iff $\rho = \sigma a \vee \rho = \sigma a \delta$.

As T satisfies T2, $G(\mathrm{T},\mathrm{CT})$ is connected, i.e., $G(\mathrm{T},\mathrm{CT}) \in \mathbb{G}^{mr}$. In fact, $G(\mathrm{T},\mathrm{CT})$ is a tree, except that it may have two roots. Using Propositions 2.2 and 3.2 it is easy to see that $T(G(\mathrm{T},\mathrm{CT})) = \mathrm{T}$ and $CT(G(\mathrm{T},\mathrm{CT})) = \mathrm{CT}$. □

The pairs obtained from process graphs with single roots are the ones moreover satisfying

$$\varepsilon \in \mathrm{CT} \Leftrightarrow \mathrm{T} = \{\varepsilon\}.$$

Infinite processes. Also for completed trace semantics one can distinguish a finitary and an infinitary variant. In terms of the testing scenario, the latter (CT^{∞}) postulates that observations may take an infinite amount of time.

DEFINITION 3.3. Two processes p and q are *infinitary completed trace equivalent*, notation $p =_{CT}^{\infty} q$, if $CT(p) = CT(q)$ and $T^{\infty}(p) = T^{\infty}(q)$. Note that in this case also $T(p) = T(q)$.

Proposition 2.4 implies that for image finite processes CT and CT^{∞} coincide, whereas Counterexample 1 shows that in general the two are different. In fact, $T \prec T^{\infty} \prec CT^{\infty}$ and $T \prec CT \prec CT^{\infty}$, and the two preceding counterexamples show that there are no further inclusions.

4. Failures semantics

Testing scenario. The *failures machine* contains as its interface to the outside world not only the display of the completed trace machine, but also a switch for each action $a \in Act$ (as in Figure 3). By means of these switches the observer may determine which actions are *free* and which are *blocked*. This situation may be changed any time during a run of the

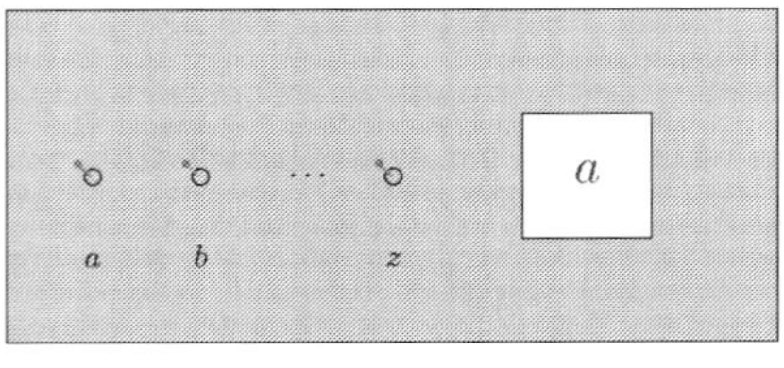

Figure 3. The failure trace machine.

process. As before, the process autonomously chooses an execution path that fits with its position in $(\mathbb{P}, \rightarrow)$, but this time the process may only start the execution of free actions. If the process reaches a state where all initial actions of its remaining behaviour are blocked, it can not proceed and the machine stagnates, which can be recognized from the empty display. In this case the observer may record that after a certain sequence of actions σ, the set X of free actions is refused by the process. X is therefore called a *refusal set* and $\langle \sigma, X \rangle$ a *failure pair*. The set of all failure pairs of a process is called its *failure set*, and constitutes its observable behaviour.

DEFINITION 4. $\langle \sigma, X \rangle \in Act^* \times \mathcal{P}(Act)$ is a *failure pair* of a process p if there is a process q such that $p \xrightarrow{\sigma} q$ and $I(q) \cap X = \emptyset$. Let $F(p)$ denote the set of failure pairs of p. Two processes p and q are *failures equivalent*, notation $p =_F q$, if $F(p) = F(q)$. In *failures semantics* (F) two processes are identified iff they are failures equivalent.

Note that $T(p)$ can be expressed in terms of $F(p)$: $T(p) = \{\sigma \in Act^* \mid \langle \sigma, \emptyset \rangle \in F(p)\}$; hence $p =_F q$ implies $T(p) = T(q)$.

DEFINITION 4.1. For $p \in \mathbb{P}$ and $\sigma \in T(p)$, let

$$Cont_p(\sigma) = \{a \in Act \mid \sigma a \in T(p)\},$$

the set of possible *continuations* of σ.

The following proposition says that the failure set $F(p)$ of a process p is completely determined by the set of failure pairs $\langle \sigma, X \rangle$ with $X \subseteq Cont_p(\sigma)$.

PROPOSITION 4.1. *Let $p \in \mathbb{P}$, $\sigma \in T(p)$ and $X \subseteq Act$. Then*

$$\langle \sigma, X \rangle \in F(p) \Leftrightarrow \langle \sigma, X \cap Cont_p(\sigma) \rangle \in F(p).$$

PROOF. If $p \xrightarrow{\sigma} q$ then $I(q) \subseteq Cont_p(\sigma)$. □

Modal characterization

DEFINITION 4.2. The set $\mathcal{L}_F$ of *failure formulas* over *Act* is defined recursively by:
- $\top \in \mathcal{L}_F$.
- $\widetilde{X} \in \mathcal{L}_F$ for $X \subseteq Act$.
- If $\varphi \in \mathcal{L}_F$ and $a \in Act$ then $a\varphi \in \mathcal{L}_F$.

The *satisfaction relation* $\models \subseteq \mathbb{P} \times \mathcal{L}_F$ is defined recursively by:
- $p \models \top$ for all $p \in \mathbb{P}$.
- $p \models \widetilde{X}$ if $I(p) \cap X = \emptyset$.
- $p \models a\varphi$ if for some $q \in \mathbb{P}$: $p \xrightarrow{a} q$ and $q \models \varphi$.

$\widetilde{X}$ represents the observation that the process refuses the set of actions X, i.e., that stagnation occurs in a situation where X is the set of actions allowed by the environment. Note

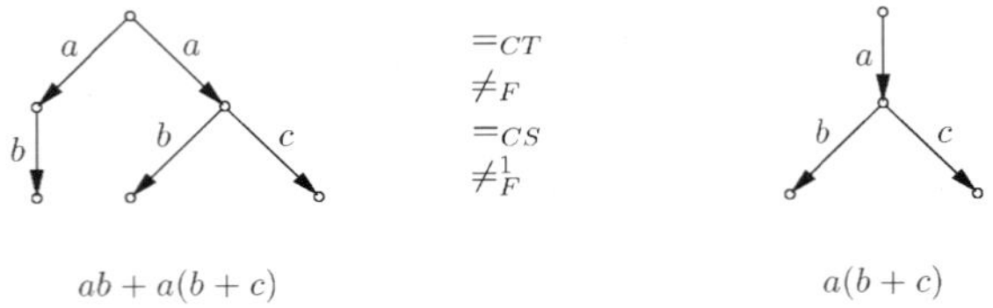

Counterexample 3. Completed trace and completed simulation equivalent, but not failures equivalent or even singleton-failures equivalent.

that a failure formula satisfied by a process p represents either a trace (if it has the form $a_1a_2\cdots a_n\top$) or a failure pair (if it has the form $a_1a_2\cdots a_n\widetilde{X}$). Hence one has

PROPOSITION 4.2. $p =_F q \Leftrightarrow \forall\varphi \in \mathcal{L}_F(p \models \varphi \Leftrightarrow q \models \varphi)$.

Process graph characterization. Let $g \in \mathbb{G}^{mr}$ and $\pi \in \text{PATHS}(g)$. Then

$$F(\pi) := \{\langle T(\pi), X\rangle \mid I(end(\pi)) \cap X = \emptyset\}$$

is the *failure set* of π. $F(g)$ can now be characterized as follows.

PROPOSITION 4.3. $F(g) = \bigcup_{\pi\in\text{PATHS}(g)} F(\pi)$.

Classification. $CT \prec F$.

PROOF. For "$CT \preceq F$" it suffices to show that also $CT(p)$ can be expressed in terms of $F(p)$:

$$CT(p) = \{\sigma \in Act^* \mid \langle\sigma, Act\rangle \in F(p)\}.$$

It also suffices to show that the modal language $\mathcal{L}_{CT}$ is a sublanguage of $\mathcal{L}_F$: $p \models 0 \Leftrightarrow p \models \widetilde{Act}$.

"$CT \not\succeq F$" follows from Counterexample 3: one has $CT(left) = CT(right) = \{ab, ac\}$, whereas $F(left) \neq F(right)$ (since $\langle a, \{c\}\rangle \in F(left) - F(right)$). □

Explicit model. In failures semantics a process can be represented by a failures equivalence class of process graphs, or equivalently by its failure set. The next proposition gives an explicit characterization of the domain $\mathbb{F}$ of failure sets of process graphs with multiple roots.

DEFINITION 4.3. The *failures domain* $\mathbb{F}$ is the set of subsets F of $Act^* \times \mathcal{P}(Act)$ satisfying

F1 $\langle\varepsilon, \emptyset\rangle \in \mathrm{F}$,
F2 $\langle\sigma\rho, \emptyset\rangle \in \mathrm{F} \Rightarrow \langle\sigma, \emptyset\rangle \in \mathrm{F}$,
F3 $\langle\sigma, Y\rangle \in \mathrm{F} \wedge X \subseteq Y \Rightarrow \langle\sigma, X\rangle \in \mathrm{F}$,
F4 $\langle\sigma, X\rangle \in \mathrm{F} \wedge \forall a \in Y(\langle\sigma a, \emptyset\rangle \notin \mathrm{F}) \Rightarrow \langle\sigma, X \cup Y\rangle \in \mathrm{F}$.

PROPOSITION 4.4. $\mathrm{F} \in \mathbb{F} \Leftrightarrow \exists g \in \mathbb{G}^{mr}$: $F(g) = \mathrm{F}$.

PROOF. "$\Leftarrow$": F1 and F2 follow from T1 and T2 in Section 2, as one has $\langle \sigma, \emptyset \rangle \in F(g) \Leftrightarrow \sigma \in T(g)$.

F3 follows immediately from the definitions, as $I(q) \cap Y = \emptyset \wedge X \subseteq Y \Rightarrow I(q) \cap X = \emptyset$.

F4 follows immediately from Proposition 4.1, as $\forall a \in Y(\langle \sigma a, \emptyset \rangle \notin F(g))$ iff $Y \cap Cont_g(\sigma) = \emptyset$.

For "$\Rightarrow$" let $\mathrm{F} \in \mathbb{F}$. For $\sigma \in Act^*$ write $Cont_{\mathrm{F}}(\sigma)$ for $\{a \in Act \mid \langle \sigma a, \emptyset \rangle \in \mathrm{F}\}$.

Define the *canonical graph* $G(\mathrm{F})$ of F by

- NODES$(G(\mathrm{F})) = \{\langle \sigma, X \rangle \in \mathrm{F} \mid X \subseteq Cont_{\mathrm{F}}(\sigma)\}$,
- ROOTS$(G(\mathrm{F})) = \{\langle \varepsilon, X \rangle \mid \langle \varepsilon, X \rangle \in \mathrm{F}\}$,
- EDGES$(G(\mathrm{F})) = \{(\langle \sigma, X \rangle, a, \langle \sigma a, Y \rangle) \mid \langle \sigma, X \rangle, \langle \sigma a, Y \rangle \in$ NODES$(G(\mathrm{F})) \wedge a \notin X\}$.

By F1, ROOTS$(G(\mathrm{F})) \neq \emptyset$. Using F3 and F2, any node $s = \langle a_1 \cdots a_n, X \rangle$ of $G(\mathrm{F})$ is reachable from a root by the path

$$\pi_s : \langle \varepsilon, \emptyset \rangle \xrightarrow{a_1} \langle a_1, \emptyset \rangle \xrightarrow{a_2} \cdots \xrightarrow{a_{n-1}} \langle a_1 \cdots a_{n-1}, \emptyset \rangle \xrightarrow{a_n} \langle a_1 \cdots a_n, X \rangle;$$

hence $G(F)$ is connected. So $G(\mathrm{F}) \in \mathbb{G}^{mr}$. I have to show that $F(G(\mathrm{F})) = \mathrm{F}$.

"$\supseteq$": Suppose $\langle \sigma, X \rangle \in \mathrm{F}$. Then, by F3, $s := \langle \sigma, X \cap Cont_{\mathrm{F}}(\sigma) \rangle \in$ NODES$(G(\mathrm{F}))$. By construction one has $T(\pi_s) = \sigma$ and $I(s) \cap X = \emptyset$. Hence $\langle \sigma, X \rangle \in F(\pi_s) \subseteq F(G(\mathrm{F}))$.

"$\subseteq$": With induction on the length of paths, it follows immediately from the definition of $G(\mathrm{F})$ that for $\pi \in$ PATHS$(G(\mathrm{F}))$, if $end(\pi) = \langle \rho, Y \rangle$ then $\rho = T(\pi)$ and

$$I(end(\pi)) = Cont_{\mathrm{F}}(\rho) - Y. \qquad (*)$$

Suppose $\langle \sigma, X \rangle \in F(G(\mathrm{F}))$. Then, by Proposition 4.3, there must be a path $\pi \in$ PATHS$(G(\mathrm{F}))$ with $\langle \sigma, X \rangle \in F(\pi)$. So $T(\pi) = \sigma$ and $I(end(\pi)) \cap X = \emptyset$. Let $end(\pi) := \langle \rho, Y \rangle \in \mathrm{F}$. By (*), $\rho = \sigma$ and $X \cap Cont_{\mathrm{F}}(\sigma) \subseteq Y$. By F3 it follows that $\langle \sigma, X \cap Cont_{\mathrm{F}}(\sigma) \rangle \in \mathrm{F}$, and F4 yields $\langle \sigma, X \rangle \in \mathrm{F}$. □

A variant of $\mathbb{F}$ was used as a model of concurrency in Hoare [31].[7]

If ROOTS(g) would be allowed to be empty, a characterization is obtained by dropping requirement F1. A characterization of the domain of failure sets of process graphs with single roots is given by adding to F1–F4 the requirement

$$\text{F5} \quad \langle \varepsilon, X \rangle \in \mathrm{F} \Rightarrow \forall a \in X\colon \langle a, \emptyset \rangle \notin \mathrm{F}.$$

That F5 holds follows from the observation that $I(\text{ROOT}(g)) = \{a \in Act \mid \langle a, \emptyset \rangle \in F(g)\}$ for $g \in \mathbb{G}$.

[7] There a process is given as a triple (A, F, D) with $A \subseteq Act$ a set of actions that may occur in the process, $F \in \mathbb{F}$ and D a set of so-called *divergencies*, traces that can lead along a state where an infinite sequence of internal actions is possible. As this paper considers only concrete, and hence divergence-free, processes, D is always empty here.

Alternative characterizations. In De Nicola [16] several equivalences, that were proposed in Kennaway [34], Darondeau [15] and De Nicola and Hennessy [17], are shown to coincide with failures semantics on the domain of finitely branching transition systems without internal moves. For this purpose he uses the following alternative characterization of failures equivalence.

DEFINITION 4.4. Write p *after* σ *MUST* X if for each $q \in \mathbb{P}$ with $p \xrightarrow{\sigma} q$ there is an $a \in I(q)$ with $a \in X$. Put $p \simeq q$ if for all $\sigma \in Act^*$ and $X \subseteq Act$: p *after* σ *MUST* $X \Leftrightarrow q$ after σ *MUST* X.

PROPOSITION 4.5. *Let* $p, q \in \mathbb{P}$. *Then* $p \simeq q \Leftrightarrow p =_F q$.

PROOF. p *after* σ *MUST* $X \Leftrightarrow \langle\sigma, X\rangle \notin F(p)$ [16]. □

Instead of the complement of the failure set of a process p, one can also take the complement $Cont_p(\sigma) - X$ of every refusal set X within a failure pair $\langle\sigma, X\rangle$ of p. In view of Proposition 4.1, the same information stored in $F(p)$ is given by the set of all pairs $\langle\sigma, X\rangle \in Act^* \times \mathcal{P}(Act)$ for which there is a process q such that $p \xrightarrow{\sigma} q$ and $I(q) \subseteq X \subseteq Cont_p(\sigma)$. In Hennessy [26], a model for nondeterministic behaviours is proposed in which a process is represented as an *acceptance tree*. An acceptance tree of a finitely branching process without internal moves is essentially the set of pairs described above, conveniently represented as a finitely branching, deterministic process tree, of which the nodes are labelled by collections of sets of actions. Thus acceptance trees constitute an explicit model of failures semantics.

Infinite processes. For infinite processes, three versions of failures semantics can be distinguished.

DEFINITION 4.5. Two processes p and q are *(finitary) failures equivalent* if $F(p) = F(q)$. p and q are *infinitary failures equivalent*, notation $p =_F^\infty q$, if $F(p) = F(q)$ and $T^\infty(p) = T^\infty(q)$. They are *finite-failures equivalent*, notation $p =_F^- q$, if $F^-(p) = F^-(q)$, where $F^-(p)$ denotes the set of failure pairs $\langle\sigma, X\rangle$ of p with X finite.

The original failures semantics of Brookes, Hoare and Roscoe [13] is F^-, i.e., what I call *finite-failures semantics*. They "adopt this view of distinguishability because [they] consider a *realistic* environment to be one that is at any time capable of performing only a finite number of events". In terms of the failures machine this means that at any time only finitely many switches can be set on free. Finitary failures semantics is the default version introduced at the beginning of this section. This can be regarded to be the semantics employed in Brookes and Roscoe [14] and Hoare [31]. Infinitary failures semantics was first discussed in Bergstra, Klop and Olderog [10]; it was proposed as a semantics for CSP in Roscoe [45]. The difference between the testing scenarios for F and F^∞ is that only the latter allows observations of infinite duration. Obviously, $F^- \preceq F \preceq F^\infty$. That the latter inclusion is strict follows from Counterexample 1; Counterexample 4 shows that also the former is strict: one has $F^-(\mathit{left}) = F^-(\mathit{right})$, whereas $F(\mathit{left}) \neq F(\mathit{right})$. In fact even

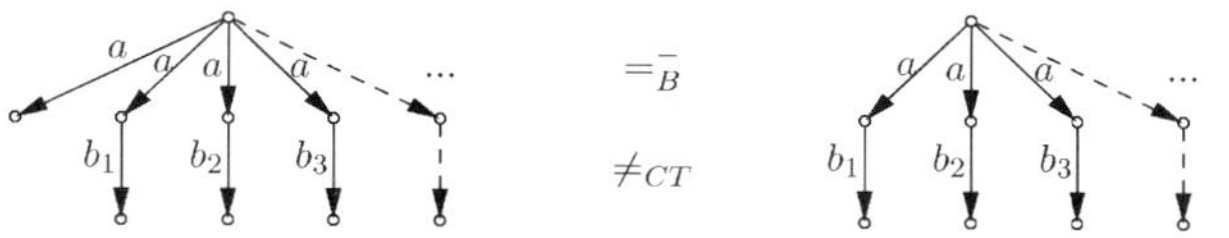

Counterexample 4. HML- and finite-failures equivalent, but not completed trace equivalent.

$CT(left) \neq CT(right)$, as $a \in CT(left) - CT(right)$. Thus, although $T \prec F^-$, $CT \prec F$ and $CT^\infty \prec F^\infty$, CT and F^- are independent, as are CT^∞ and F.

In addition to the three variants of Definition 4.5 one could also define a version of failures semantics based on infinite traces and finite refusal sets. Such a semantics would distinguish the two graphs of Counterexample 1, but identify the ones of Counterexample 4. As this semantics does not occur in the literature, and has no clear advantages over the other variants, I will not further consider it here.

PROPOSITION 4.6. *Let p and q be image finite processes. Then $p =_F^- q \Leftrightarrow p =_F q \Leftrightarrow p =_F^\infty q$.*

PROOF. "$\Leftarrow$" has been established for all processes, and the second "$\Rightarrow$" follows immediately from Proposition 2.4 (as $p =_F q \Rightarrow p =_T q \Rightarrow p =_T^\infty q$). So it remains to show that $p \neq_F q \Rightarrow p \neq_F^- q$. Suppose $F(p) \neq F(q)$, say there is a failure pair $\langle \sigma, X \rangle \in F(p) - F(q)$. By the image finiteness of q there are only finitely many processes r_i with $q \xrightarrow{\sigma} r_i$, and for each of them there is an action $a_i \in I(r_i) \cap X$ (as otherwise $\langle \sigma, X \rangle$ would be a failure pair of q). Let Y be the set of all those a_i's. Then Y is a finite subset of X, so $\langle \sigma, Y \rangle \in F^-(p)$. On the other hand, $a_i \in I(r_i) \cap Y$ for all r_i, so $\langle \sigma, Y \rangle \notin F^-(q)$. □

It is not hard to change the leftmost process in Counterexample 4 to an image finite one with the same failure pairs. Thus, in the first statement of Proposition 4.6 it is necessary that both processes are image finite. For the subclass of finitely branching processes a stronger result can be obtained.

PROPOSITION 4.7. *Let $p, q \in \mathbb{P}$ and p is finitely branching. Then $p =_F^- q \Leftrightarrow p =_F q$.*

PROOF. Suppose $p =_F^- q$. As p is finitely branching, $Cont_p(\sigma)$ is finite for all $\sigma \in T(p)$. And as $T(q) = T(p)$, $Cont_q(\sigma) = Cont_p(\sigma)$, which is finite, for all $\sigma \in T(q)$. Now for processes p with this property, $F(p)$ is completely determined by $F^-(p)$, as follows from Proposition 4.1. □

The second statement of Proposition 4.6 does not allow such a strengthening, as will follow from Counterexample 12.

5. Failure trace semantics

Testing scenario. The *failure trace machine* has the same layout as the failures machine, but is does not stagnate permanently if the process cannot proceed due to the circumstance that all actions it is prepared to continue with are blocked by the observer. Instead it idles – recognizable from the empty display – until the observer changes its mind and allows one of the actions the process is ready to perform. What can be observed are traces with idle periods in between, and for each such period the set of actions that are not blocked by the observer. Such observations can be coded as sequences of members and subsets of *Act*.

EXAMPLE. The sequence $\{a, b\}cdb\{b, c\}\{b, c, d\}a(Act)$ is the account of the following observation: At the beginning of the execution of the process p, only the actions a and b were allowed by the observer. Apparently, these actions were not on the menu of p, for p started with an idle period. Suddenly the observer canceled its veto on c, and this resulted in the execution of c, followed by d and b. Then again an idle period occurred, this time when b and c were the actions not being blocked by the observer. After a while the observer decided to allow d as well, but the process ignored this gesture and remained idle. Only when the observer gave the green light for the action a, it happened immediately. Finally, the process became idle once more, but this time not even one action was blocked. This made the observer realize that a state of eternal stagnation had been reached, and disappointed he terminated the observation.

A set $X \subseteq Act$, occurring in such a sequence, can be regarded as an offer from the environment, that is refused by the process. Therefore such a set is called a *refusal set*. The occurrence of a refusal set may be interpreted as a 'failure' of the environment to create a situation in which the process can proceed without being disturbed. Hence a sequence over $Act \cup \mathcal{P}(Act)$, resulting from an observation of a process p may be called a *failure trace* of p. The observable behaviour of a process, according to this testing scenario, is given by the set of its failure traces, its *failure trace set*. The semantics in which processes are identified iff their failure trace sets coincide, is called *failure trace semantics* (*FT*).

For image finite processes failure trace semantics is exactly the equivalence that originates from PHILLIPS notion of *refusal testing* [42]. (Image infinite processes are not considered in [42].) There it is called *refusal equivalence*.

DEFINITION 5.
- The *refusal relations* $\xrightarrow{X}$ for $X \subseteq Act$ are defined by: $p \xrightarrow{X} q$ iff $p = q$ and $I(p) \cap X = \emptyset$.
  $p \xrightarrow{X} q$ means that p can evolve into q, while being idle during a period in which X is the set of actions allowed by the environment.
- The *failure trace relations* $\xrightarrow{\sigma}$ for $\sigma \in (Act \cup \mathcal{P}(Act))^*$ are defined as the reflexive and transitive closure of both the action and the refusal relations. Again the overloading of notation is harmless.
- $\sigma \in (Act \cup \mathcal{P}(Act))^*$ is a *failure trace* of a process p if there is a process q such that $p \xrightarrow{\sigma} q$. Let $FT(p)$ denote the set of failure traces of p. Two processes p and q are *failure trace equivalent*, notation $p =_{FT} q$, if $FT(p) = FT(q)$.

Modal characterization

DEFINITION 5.1. The set $\mathcal{L}_{FT}$ of *failure trace formulas* over *Act* is defined recursively by:

- $\top \in \mathcal{L}_{FT}$.
- If $\varphi \in \mathcal{L}_{FT}$ and $X \subseteq Act$ then $\widetilde{X}\varphi \in \mathcal{L}_{FT}$.
- If $\varphi \in \mathcal{L}_{FT}$ and $a \in Act$ then $a\varphi \in \mathcal{L}_{FT}$.

The *satisfaction relation* $\models \subseteq \mathbb{P} \times \mathcal{L}_{FT}$ is defined recursively by:

- $p \models \top$ for all $p \in \mathbb{P}$.
- $p \models \widetilde{X}\varphi$ if $I(p) \cap X = \emptyset$ and $p \models \varphi$.
- $p \models a\varphi$ if for some $q \in \mathbb{P}$: $p \xrightarrow{a} q$ and $q \models \varphi$.

$\widetilde{X}\varphi$ represents the observation that the process refuses the set of actions X, followed by the observation φ. A modal failure trace formula satisfied by a process p represents exactly a failure trace as defined above. Hence one has

PROPOSITION 5.1. $p =_{FT} q \Leftrightarrow \forall \varphi \in \mathcal{L}_{FT}(p \models \varphi \Leftrightarrow q \models \varphi)$.

Process graph characterization. Let $g \in \mathbb{G}^{mr}$ and $\pi : s_0 \xrightarrow{a_1} s_1 \xrightarrow{a_2} \cdots \xrightarrow{a_n} s_n \in$ PATHS(g). Then the *failure trace set* of π, $FT(\pi)$, is the smallest subset of $(Act \cup \mathcal{P}(Act))^*$ satisfying

- $(Act - I(s_0))a_1(Act - I(s_1))a_2 \cdots a_n(Act - I(s_n)) \in FT(\pi)$,
- $\sigma X \rho \in FT(\pi) \Rightarrow \sigma\rho \in FT(\pi)$,
- $\sigma X \rho \in FT(\pi) \Rightarrow \sigma XX\rho \in FT(\pi)$,
- $\sigma X \rho \in FT(\pi) \wedge Y \subset X \Rightarrow \sigma Y \rho \in FT(\pi)$.

$FT(g)$ can now be characterized as follows.

PROPOSITION 5.2. $FT(g) = \bigcup_{\pi \in \mathrm{PATHS}(g)} FT(\pi)$.

Proposition 5.2 yields a technique for deciding that two process graphs are failure trace equivalent, without calculating their entire failure trace set.

Let $g, h \in \mathbb{G}^{mr}$, $\pi : s_0 \xrightarrow{a_1} s_1 \xrightarrow{a_2} \cdots \xrightarrow{a_n} s_n \in$ PATHS(g) and $\pi' : t_0 \xrightarrow{b_1} t_1 \xrightarrow{b_2} \cdots \xrightarrow{b_m} t_m \in$ PATHS(h). Path π' is a *failure trace augmentation* of π, notation $\pi \leqslant_{FT} \pi'$, if $FT(\pi) \subseteq FT(\pi')$. This is the case exactly when $n = m$, $a_i = b_i$ and $I(t_i) \subseteq I(s_i)$ for $i = 1, \ldots, n$. From this the following can be concluded.

COROLLARY 5.1. *Two process graphs $g, h \in \mathbb{G}^{mr}$ are failure trace equivalent iff*

- *for any path $\pi \in$ PATHS(g) in g there is a $\pi' \in$ PATHS(h) such that $\pi \leqslant_{FT} \pi'$*
- *and for any path $\pi \in$ PATHS(g) in h there is a $\pi' \in$ PATHS(g) such that $\pi \leqslant_{FT} \pi'$.*

If g and h are moreover without infinite paths, then it suffices to check the requirements above for maximal paths.

Classification. $F \prec FT$.

PROOF. For "$F \preceq FT$" it suffices to show that $F(p)$ can be expressed in terms of $FT(p)$:

$$\langle \sigma, X \rangle \in F(p) \Leftrightarrow \sigma X \in FT(p).$$

"$F \not\succeq FT$" follows from Counterexample 5; see Section 7 for details. □

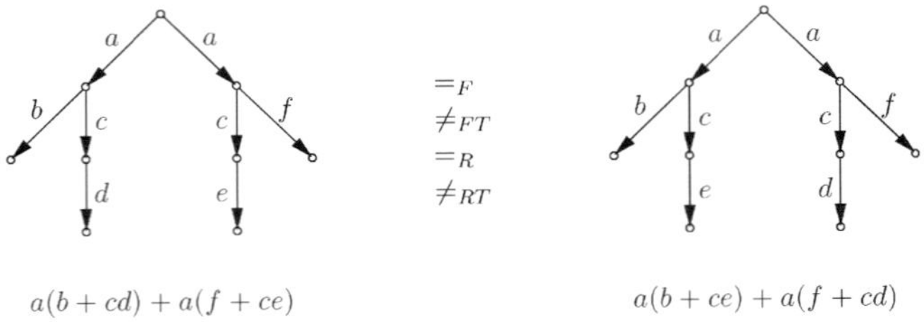

Counterexample 5. Failures and ready equivalent, but not failure trace or ready trace equivalent.

Infinite processes. As for failures semantics, three variants of failure trace semantics for infinite processes can be defined. Besides the default version (FT) there is an infinitary version (FT^∞), motivated by observations that may last forever, and a finite version (FT^-), motivated by an observer that may only set finitely many switches on free at any time.

DEFINITION 5.2. $\sigma_1\sigma_2\cdots \in (Act \cup \mathcal{P}(Act))^\infty$ is an *infinite failure trace* of a process $p \in \mathbb{P}$ if there are processes $p_1, p_2, \ldots$ such that $p \xrightarrow{\sigma_1} p_1 \xrightarrow{\sigma_2} \cdots$. Let $FT^\infty(p)$ denote the set of infinite failure traces of p. Two processes p and q are *infinitary failure trace equivalent*, notation $p =_{FT}^\infty q$, if $FT^\infty(p) = FT^\infty(q)$ and $FT(p) = FT(q)$. They are *finite-failure trace equivalent*, notation $p =_{FT}^- q$, if $FT^-(p) = FT^-(q)$, where $FT^-(p)$ denotes the set of failure traces of p in which all refusal sets are finite.

Clearly, $FT^- \prec FT \prec FT^\infty$; Counterexamples 1 and 4 show that the inclusions are strict. One also has $F^- \prec FT^-$, $F \prec FT$ and $F^\infty \prec FT^\infty$; here strictness follows from Counterexample 5.

PROPOSITION 5.3. *Let p and q be image finite processes. Then $p =_{FT}^- q \Leftrightarrow p =_{FT} q \Leftrightarrow p =_{FT}^\infty q$.*

PROOF. "$p =_{FT}^- q \Leftarrow p =_{FT} q \Leftarrow p =_{FT}^\infty q$" holds for all processes.

Note that the definition of $FT(p)$ is exactly like the definition of $T(p)$, except that the failure trace relations are used instead of the generalized action relations; the same relation exists between $FT^\infty(p)$ and $T^\infty(p)$. Moreover, a process $p \in \mathbb{P}$ is image finite in terms of the failure trace relations on $\mathbb{P}$ iff it is image finite in terms of terms of the (generalized) action relations on $\mathbb{P}$, as defined in Definition 1.2. Hence "$p =_{FT} q \Rightarrow p =_{FT}^\infty q$" follows immediately from Proposition 2.4.

"$p =_{FT}^- q \Rightarrow p =_{FT} q$": Suppose $FT(p) \neq FT(q)$, say $FT(p) - FT(q) \neq \emptyset$. Let σ be a failure trace in $FT(p) - FT(q)$ with at least one infinite refusal set. I will show that there must be a failure trace in $FT(p) - FT(q)$ with strictly fewer infinite refusal sets than σ. By applying this result a finite number of times, a failure trace $\rho \in FT(p) - FT(q)$ is found without infinite refusal sets, showing that $FT^-(p) \neq FT^-(q)$.

So let $\sigma = \sigma_1 X \sigma_2 \in FT(p) - FT(q)$ with X an infinite refusal set. Clearly $\sigma_1\sigma_2 \in FT(p)$. By the image finiteness of q there are only finitely many pairs of processes r_i, s_i with $q \xrightarrow{\sigma_1} r_i \xrightarrow{\sigma_2} s_i$, and for each of them there is an action $a_i \in I(r_i) \cap X$ (as otherwise

$\sigma_1 X \sigma_2$ would be a failure trace of q). Let Y be the set of all those a_i's. Then Y is finite. As Y is a subset of X, one has $\sigma_1 Y \sigma_2 \in FT(p)$. On the other hand, $a_i \in I(r_i) \cap Y$ for all r_i, so $\sigma_1 Y \sigma_2 \notin FT(q)$. □

Unlike the situation for failures semantics, in the first statement of Proposition 5.3 it is not necessary that *both* processes are image finite.

PROPOSITION 5.4. *Let $p, q \in \mathbb{P}$ and p is image finite. Then $p =_{FT}^{-} q \Leftrightarrow p =_{FT} q$.*

PROOF. More difficult, and omitted here. □

The second statement of Proposition 5.3 does not allow such a strengthening, as will follow from Counterexample 12.

6. Ready trace semantics

Testing scenario. The *ready trace machine* is a variant of the failure trace machine that is equipped with a lamp for each action $a \in Act$. Each time the process idles, the lamps of all actions the process is ready to engage in are lit. Of course all these actions are blocked by the observer, otherwise the process wouldn't idle. Now the observer can see which actions could be released in order to let the process proceed. During the execution of an action no lamps are lit. An observation now consists of a sequence of members and subsets of *Act*, the actions representing information obtained from the display, and the sets of actions representing information obtained from the lights. Such a sequence is called a *ready trace* of the process, and the subsets occurring in a ready trace are referred to as *menus*. The information about the free and blocked actions is now redundant. The set of all ready traces of a process is called its *ready trace set*, and constitutes its observable behaviour.

DEFINITION 6.

- The *ready trace relations* $\overset{\sigma}{*\!\!\longrightarrow}$ for $\sigma \in (Act \cup \mathcal{P}(Act))^*$ are defined recursively by:
  (1) $p \overset{\varepsilon}{*\!\!\longrightarrow} p$, for any process p.
  (2) $p \overset{a}{\longrightarrow} q$ implies $p \overset{a}{*\!\!\longrightarrow} q$.
  (3) $p \overset{X}{*\!\!\longrightarrow} q$ with $X \subseteq Act$ whenever $p = q$ and $I(p) = X$.

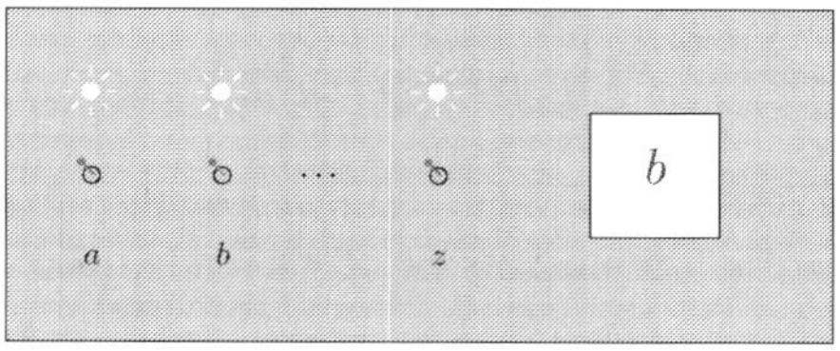

Figure 4. The ready trace machine.

(4) $p \overset{\sigma}{\ast\!\!\longrightarrow} q \overset{\rho}{\ast\!\!\longrightarrow} r$ implies $p \overset{\sigma\rho}{\ast\!\!\longrightarrow} r$.
The special arrow $\overset{\sigma}{\ast\!\!\longrightarrow}$ had to be used, since further overloading of $\xrightarrow{\sigma}$ would cause confusion with the failure trace relations.

- $\sigma \in (Act \cup \mathcal{P}(Act))^*$ is a *ready trace* of a process p if there is a process q such that $p \overset{\sigma}{\ast\!\!\longrightarrow} q$. Let $RT(p)$ denote the set of ready traces of p. Two processes p and q are *ready trace equivalent*, notation $p =_{RT} q$, if $RT(p) = RT(q)$. In *ready trace semantics* (RT) two processes are identified iff they are ready trace equivalent.

In Baeten, Bergstra and Klop [6], Pnueli [43] and Pomello [44] ready trace semantics was defined slightly differently. By Proposition 6.1 below, their definition yields the same equivalence as mine.

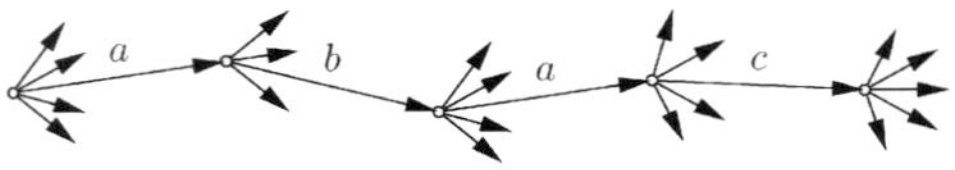

DEFINITION 6.1. $X_0 a_1 X_1 a_2 \cdots a_n X_n \in \mathcal{P}(Act) \times (Act \times \mathcal{P}(Act))^*$ is a *normal ready trace* of a process p if there are processes $p_1, \ldots, p_n$ such that $p \xrightarrow{a_1} p_1 \xrightarrow{a_2} \cdots \xrightarrow{a_n} p_n$ and $I(p_i) = X_i$ for $i = 1, \ldots, n$. Let $RT_N(p)$ denote the set of normal ready traces of p. Two processes p and q are ready trace equivalent in the sense of [6,43,44] if $RT_N(p) = RT_N(q)$.

PROPOSITION 6.1. *Let $p, q \in \mathbb{P}$. Then $RT_N(p) = RT_N(q) \Leftrightarrow RT(p) = RT(q)$.*

PROOF. The normal ready traces of a process are just the ready traces which are an alternating sequence of sets and actions, and vice versa the set of all ready traces can be constructed from the set of normal ready traces by means of doubling and leaving out menus. □

Modal characterization

DEFINITION 6.2. The set $\mathcal{L}_{RT}$ of *ready trace formulas* over *Act* is defined recursively by:
- $\top \in \mathcal{L}_{RT}$.
- If $\varphi \in \mathcal{L}_{RT}$ and $X \subseteq Act$ then $X\varphi \in \mathcal{L}_{RT}$.
- If $\varphi \in \mathcal{L}_{RT}$ and $a \in Act$ then $a\varphi \in \mathcal{L}_{RT}$.

The *satisfaction relation* $\models \subseteq \mathbb{P} \times \mathcal{L}_{RT}$ is defined recursively by:
- $p \models \top$ for all $p \in \mathbb{P}$.
- $p \models X\varphi$ if $I(p) = X$ and $p \models \varphi$.
- $p \models a\varphi$ if for some $q \in \mathbb{P}$: $p \xrightarrow{a} q$ and $q \models \varphi$.

$X\varphi$ represents the observation of a menu, followed by the observation φ. A ready trace formula satisfied by a process p represents exactly a ready trace as in Definition 6. Hence one has

PROPOSITION 6.2. $p =_{RT} q \Leftrightarrow \forall \varphi \in \mathcal{L}_{RT}(p \models \varphi \Leftrightarrow q \models \varphi)$.

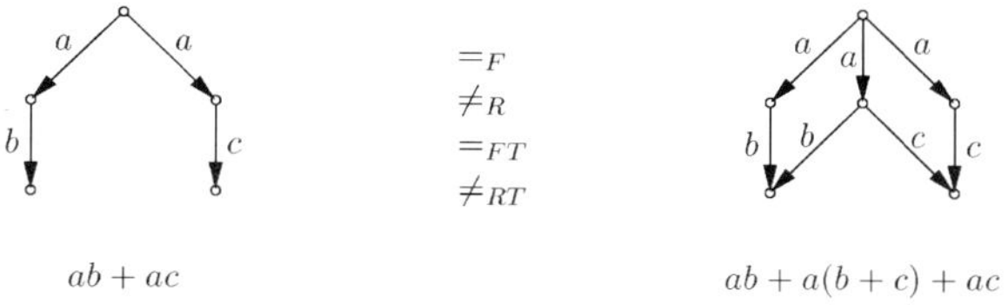

Counterexample 6. Failures and failure trace equivalent, but not ready or ready trace equivalent.

Process graph characterization. Let $g \in \mathbb{G}^{mr}$ and $\pi : s_0 \xrightarrow{a_1} s_1 \xrightarrow{a_2} \cdots \xrightarrow{a_n} s_n \in$ PATHS(g).

The *ready trace* of π is given by $RT_N(\pi) := I(s_0)a_1I(s_1)a_2 \cdots a_nI(s_n)$.

$RT_N(g)$ can now be characterized by:

PROPOSITION 6.3. $RT_N(g) = \{RT_N(\pi) \mid \pi \in \text{PATHS}(g)\}$.

Moreover, $RT(g)$ is the smallest subset of $(Act \cup \mathcal{P}(Act))^*$ containing $RT_N(g)$ and satisfying

$$\sigma X\rho \in RT(g) \Rightarrow \sigma\rho \in RT(g) \wedge \sigma XX\rho \in RT(g).$$

Classification. $FT \prec RT$.

PROOF. For "$FT \preceq RT$" it suffices to show that $FT(p)$ can be expressed in terms of $RT(p)$:

$$\begin{aligned}&\sigma = \sigma_1\sigma_2 \cdots \sigma_n \in FT(p)(\sigma_i \in Act \cup \mathcal{P}(Act)) \Leftrightarrow\\ &\exists\rho = \rho_1\rho_2 \cdots \rho_n \in RT(p)(\rho_i \in Act \cup \mathcal{P}(Act)) \text{ such that for } i = 1, \ldots, n \text{ either}\\ &\sigma_i = \rho_i \in Act \text{ or } \sigma_i, \rho_i \subseteq Act \text{ and } \sigma_i \cap \rho_i = \emptyset.\end{aligned}$$

"$FT \not\succeq RT$" follows from Counterexample 6; see Section 7 for details. □

Explicit model. In ready trace semantics a process can be represented by a ready trace equivalence class of process graphs, or equivalently by its ready trace set, possibly in the normal form of Definition 6.1. The next proposition gives an explicit characterization of the domain $\mathbb{RT}$ of ready trace sets in this form of process graphs with multiple roots.

DEFINITION 6.3. The *ready trace domain* $\mathbb{RT}$ is the set of subsets RT of $\mathcal{P}(Act) \times (Act \times \mathcal{P}(Act))^*$ satisfying

RT1 $\exists X(X \in \text{RT})$,
RT2 $\sigma X \in \text{RT} \wedge a \in X \Leftrightarrow \exists Y(\sigma XaY \in \text{RT})$.

PROPOSITION 6.4. $\text{RT} \in \mathbb{RT} \Leftrightarrow \exists g \in \mathbb{G}^{mr}$: $RT_N(g) = \text{RT}$.

PROOF. "$\Leftarrow$" is evident. For "$\Rightarrow$" let $\mathrm{RT} \in \mathbb{RT}$. Define the *canonical graph* $G(\mathrm{RT})$ of RT by

- $\mathrm{NODES}(G(\mathrm{RT})) = \mathrm{RT}$,
- $\mathrm{ROOTS}(G(\mathrm{RT})) = \{X \subseteq Act \mid X \in \mathrm{RT}\}$,
- $\mathrm{EDGES}(G(\mathrm{RT})) = \{(\sigma, a, \sigma a Y) \mid \sigma, \sigma a Y \in \mathrm{NODES}(G(\mathrm{RT}))\}$.

By RT1, $\mathrm{ROOTS}(G(\mathrm{RT})) \neq \emptyset$. Using R2, $G(\mathrm{RT})$ is connected. So $G(\mathrm{RT}) \in \mathbb{G}^{mr}$. Moreover, for every path $\pi \in \mathrm{PATHS}(G(\mathrm{RT}))$ one has $RT_N(\pi) = end(\pi)$. Hence $\mathrm{RT}_N(G(\mathrm{RT})) = \mathrm{RT}$. □

If $\mathrm{ROOTS}(g)$ would be allowed to be empty, a characterization is obtained by dropping requirement RT1. A characterization of the domain of ready trace sets of process graphs with single roots is given by strengthening RT1 to $\exists! X (X \in \mathrm{RT})$, where $\exists! X$ means "there is exactly one X such that".

Infinite processes. An infinitary version of ready trace semantics (RT^∞) is defined analogously to infinitary failure trace semantics. A finite version is not so straightforward; a definition will be proposed in the next section.

DEFINITION 6.4. $\sigma_1\sigma_2 \cdots \in (Act \cup \mathcal{P}(Act))^\infty$ is an *infinite ready trace* of a process $p \in \mathbb{P}$ if there are processes $p_1, p_2, \ldots$ such that $p \ast\!\xrightarrow{\sigma_1} p_1 \ast\!\xrightarrow{\sigma_2} \cdots$. Let $RT^\infty(p)$ denote the set of infinite ready traces of p. Two processes p and q are *infinitary ready trace equivalent*, notation $p =_{RT}^\infty q$, if $RT^\infty(p) = RT^\infty(q)$ and $RT(p) = RT(q)$.

Clearly, $RT \prec RT^\infty$; Counterexample 1 shows that the inclusion is strict. Moreover $FT^\infty \prec RT^\infty$.

PROPOSITION 6.5. *Let p and q be image finite processes. Then $p =_{RT} q \Leftrightarrow p =_{RT}^\infty q$.*

PROOF. Exactly as the corresponding part of Proposition 5.3. □

Counterexample 12 will show that in Proposition 6.5 *both* p and q need to be image finite.

7. Readiness semantics and possible-futures semantics

Testing scenario. The *readiness machine* has the same layout as the ready trace machine, but, like the failures machine, can not recover from an idle period. By means of the lights the menu of initial actions of the remaining behaviour of an idle process can be recorded, but this happens at most once during an observation of a process, namely at the end. An observation either results in a trace of the process, or in a pair of a trace and a menu of actions by which the observation could have been extended if the observer wouldn't have blocked them. Such a pair is called a *ready pair* of the process, and the set of all ready pairs of a process is its *ready set*.

DEFINITION 7. $\langle \sigma, X \rangle \in Act^* \times \mathcal{P}(Act)$ is a *ready pair* of a process p if there is a process q such that $p \xrightarrow{\sigma} q$ and $I(q) = X$. Let $R(p)$ denote the set of ready pairs of p. Two processes p and q are *ready equivalent*, notation $p =_R q$, if $R(p) = R(q)$. In *readiness semantics* (R) two processes are identified iff they are ready equivalent.

Modal characterization

DEFINITION 7.1. The set $\mathcal{L}_R$ of *readiness formulas* over Act is defined recursively by:
- $\top \in \mathcal{L}_R$.
- $X \in \mathcal{L}_R$ for $X \subseteq Act$.
- If $\varphi \in \mathcal{L}_R$ and $a \in Act$ then $a\varphi \in \mathcal{L}_R$.

The *satisfaction relation* $\models \subseteq \mathbb{P} \times \mathcal{L}_R$ is defined recursively by:
- $p \models \top$ for all $p \in \mathbb{P}$.
- $p \models X$ if $I(p) = X$.
- $p \models a\varphi$ if for some $q \in \mathbb{P}$: $p \xrightarrow{a} q$ and $q \models \varphi$.

X represents the observation of a menu. A readiness formula satisfied by a process p represents either a trace (if it has the form $a_1 a_2 \cdots a_n \top$) or a ready pair (if it has the form $a_1 a_2 \cdots a_n X$). Hence one has

PROPOSITION 7.1. $p =_R q \Leftrightarrow \forall \varphi \in \mathcal{L}_R (p \models \varphi \Leftrightarrow q \models \varphi)$.

Process graph characterization. Let $g \in \mathbb{G}^{mr}$ and $\pi \in \text{PATHS}(g)$. The *ready pair* of π is given by $R(\pi) := \langle T(\pi), I(end(\pi)) \rangle$. $R(g)$ can now be characterized by:

PROPOSITION 7.2. $R(g) = \{R(\pi) \mid \pi \in \text{PATHS}(g)\}$.

Classification. $F \prec R \prec RT$, but R and FT are independent.

PROOF. For "$F \preceq R$" it suffices to show that $F(p)$ can be expressed in terms of $R(p)$:

$$\langle \sigma, X \rangle \in F(p) \;\Leftrightarrow\; \exists Y \subseteq Act\colon \langle \sigma, Y \rangle \in R(p) \wedge X \cap Y = \emptyset.$$

For "$R \preceq RT$" it suffices to show that $R(p)$ can be expressed in terms of $RT(p)$:

$$\langle \sigma, X \rangle \in R(p) \;\Leftrightarrow\; \sigma X \in RT(p).$$

"$R \not\succeq FT$" (and hence "$R \not\succeq RT$" and "$F \not\succeq FT$") follows from Counterexample 5, in which $R(left) = R(right)$ but $FT(left) \neq FT(right)$. The first statement follows with Proposition 7.2. Both graphs have 9 paths starting from the root, and hence 9 ready pairs. These are easily seen to be the same at both sides; in the second graph only 4 ready pairs swapped places. The second statement follows since $a\{b\}ce \in FT(left) - FT(right)$.

"$R \not\preceq FT$" (and hence "$R \not\preceq F$" and "$RT \not\preceq FT$") follows from Counterexample 6, in which $FT(left) = FT(right)$ but $R(left) \neq R(right)$. The first statement follows from Corollary 5.1, since the new maximal paths at the right-hand side are both failure trace augmented by the two maximal paths both sides have in common. The second one follows since $\langle a, \{b, c\} \rangle \in R(right) - R(left)$. □

Explicit model. In readiness semantics a process can be represented by a ready equivalence class of process graphs, or equivalently by its ready set. The next proposition gives an explicit characterization of the domain $\mathbb{R}$ of ready sets of process graphs with multiple roots.

DEFINITION 7.2. The *readiness domain* $\mathbb{R}$ is the set of subsets R of $Act^* \times \mathcal{P}(Act)$ satisfying

$$\begin{array}{ll} \text{R1} & \exists X(\langle \varepsilon, X\rangle \in \mathrm{R}), \\ \text{R2} & \exists X(\langle \sigma, X \cup \{a\}\rangle \in \mathrm{R}) \Leftrightarrow \exists Y(\langle \sigma a, Y\rangle \in \mathrm{R}). \end{array}$$

PROPOSITION 7.3. $\mathrm{R} \in \mathbb{R} \Leftrightarrow \exists g \in \mathbb{G}^{mr}\colon\ R(g) = \mathrm{R}$.

PROOF. "$\Leftarrow$" is evident. For "$\Rightarrow$" let $\mathrm{R} \in \mathbb{R}$. Define the *canonical graph* $G(\mathrm{R})$ of R by
- $\text{NODES}(G(\mathrm{R})) = \mathrm{R}$,
- $\text{ROOTS}(G(\mathrm{R})) = \{\langle \varepsilon, X\rangle \mid \langle \varepsilon, X\rangle \in \mathrm{R}\}$,
- $\text{EDGES}(G(\mathrm{R})) = \{(\langle \sigma, X\rangle, a, \langle \sigma a, Y\rangle) \mid \langle \sigma, X\rangle, \langle \sigma a, Y\rangle \in \text{NODES}(G(\mathrm{R})) \wedge a \in X\}$.

By R1, $\text{ROOTS}(G(\mathrm{R})) \neq \emptyset$. Using R2, $G(\mathrm{R})$ is connected. Hence $G(\mathrm{R}) \in \mathbb{G}^{mr}$. Moreover, for every path $\pi \in \text{PATHS}(G(\mathrm{R}))$ one has $R(\pi) = end(\pi)$. From this it follows that $R(G(\mathrm{R})) = \mathrm{R}$. □

If $\text{ROOTS}(g)$ would be allowed to be empty, a characterization is obtained by dropping requirement R1. A characterization of the domain of ready sets of process graphs with single roots is given by strengthening R1 to $\exists! X(\langle \varepsilon, X\rangle \in \mathrm{R})$, where $\exists! X$ means "there is exactly one X such that".

Possible-futures and acceptance-refusal semantics. Readiness semantics was proposed by Olderog and Hoare [40]. Two preliminary versions stem from Rounds and Brookes [46]: in *possible-futures semantics* (PF) the menu consists of the entire trace set of the remaining behaviour of an idle process, instead of only the set of its initial actions; in *acceptance-refusal semantics* a menu may be any finite subset of initial actions, while also the finite refusal sets of Section 4 are observable.

DEFINITION 7.3. $\langle \sigma, X\rangle \in Act^* \times \mathcal{P}(Act^*)$ is a *possible future* of a process p if there is a process q such that $p \xrightarrow{\sigma} q$ and $T(q) = X$. Let $PF(p)$ denote the set of possible futures of p. Two processes p and q are *possible-futures equivalent*, notation $p =_{PF} q$, if $PF(p) = PF(q)$.

The modal and process graph characterizations of possible-future semantics are straightforward, but a plausible testing scenario has not been proposed. Trivially $R \preceq PF$. That the reverse does not hold, and even that $PF \npreceq RT$, will follow from Counterexample 10. Counterexample 7 shows that $FT \npreceq PF$. There $PF(left) = PF(right)$ but $FT(left) \neq FT(right)$. As for the first statement, both graphs have 18 paths starting from the root, and hence 18 possible futures. These are easily seen to be the same at both sides; in the second graph only 2 possible futures swapped places. The second statement follows since $a\{b\}a\{b\}cd \in$

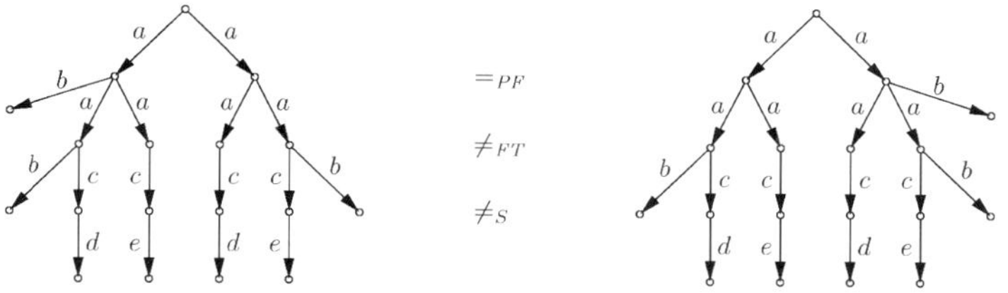

$a(b + a(b + cd) + ace) + a(acd + a(ce + b))$ $\quad$ $a(a(b + cd) + ace) + a(acd + a(ce + b) + b)$

Counterexample 7. Possible-futures equivalent, but not failure trace or simulation equivalent.

$FT(\mathit{left}) - FT(\mathit{right})$. Thus possible-future semantics is incomparable with failure trace and ready trace semantics.

DEFINITION 7.4. $\langle\sigma, X, Y\rangle \in \mathit{Act}^* \times \mathcal{P}(\mathit{Act}) \times \mathcal{P}(\mathit{Act})$ is an *acceptance-refusal triple* of a process p if X and Y are finite and there is a process q such that $p \xrightarrow{\sigma} q$, $X \subseteq I(q)$ and $Y \cap I(q) = \emptyset$. Let $AR(p)$ denote the set of acceptance-refusal triples of p. Two processes p and q are *acceptance-refusal equivalent*, notation $p =_{AR} q$, if $AR(p) = AR(q)$.

The modal and process graph characterizations are again straightforward. A motivating testing scenario would be the same as for readiness semantics, except that at any time only finitely many switches can be set on free, and only finitely many lamps can be investigated in a finite amount of time. Clearly $p =_R q \Rightarrow p =_{AR} q$, for

$$AR(p) = \big\{\langle\sigma, X, Y\rangle \mid \exists\langle\sigma, Z\rangle \in R(p) \mid X, Y \text{ finite } \wedge X \subseteq Z \wedge Y \cap Z = \emptyset\big\}.$$

That this implication is strict follows from Counterexample 4. It is not difficult to see that for finitely branching processes acceptance-refusal equivalence coincides with ready equivalence: $\langle\sigma, X\rangle$ is a ready pair of a process p iff p has an acceptance-refusal triple $\langle\sigma, X, Y\rangle$ with $X \cup Y = \mathit{Cont}_p(\sigma)$ (cf. Definition 4.1).

Infinite processes. Note that if in Definition 7.4 the sets X and Y are allowed to be infinite the resulting equivalence would be ready equivalence again. Namely $\langle\sigma, X\rangle$ is a ready pair of a process p iff p has such an acceptance-refusal triple $\langle\sigma, X, \mathit{Act} - Y\rangle$. Thus acceptance-refusal semantics can be regarded as the finite variant of readiness semantics, and will therefore be denoted R^-. The infinitary variant of readiness semantics (R^∞), motivated by observations that may last forever, is defined analogously to F^∞:

DEFINITION 7.5. p and q are *infinitary ready equivalent* if $R(p) = R(q)$ and $T^\infty(p) = T^\infty(q)$.

Clearly, $R \prec R^\infty$; by Counterexample 1 the inclusion is strict. Moreover, $F^\infty \prec R^\infty \prec RT^\infty$.

PROPOSITION 7.4. *Let p and q be image finite processes. Then* $p =_R q \Leftrightarrow p =_R^\infty q$.

PROOF. "$\Leftarrow$" has been established for all processes, and the second "$\Rightarrow$" follows immediately from Proposition 2.4 (as $p =_R q \Rightarrow p =_T q \Rightarrow p =_T^\infty q$). □

PROPOSITION 7.5. *Let $p, q \in \mathbb{P}$ and p is image finite. Then* $p =_{AR} q \Leftrightarrow p =_R q$.

PROOF. "$\Leftarrow$" holds for all process. I will prove "$\Rightarrow$" assuming that p has the property that for any $\sigma \in Act^*$ there are only finitely many ready pairs $\langle\sigma, X\rangle \in R(p)$. This property (call it *RIF*) is clearly implied by image finiteness. So suppose p has the RIF property and $AR(p) = AR(q)$. I will show that $R(p) = R(q)$.

Suppose $\langle\sigma, Y\rangle \notin R(p)$. By RIF there are only finitely many ready pairs $\langle\sigma, X_i\rangle \in R(p)$. For each of them choose an action $a_i \in Y - X_i$ or $b_i \in X_i - Y$. Let U be the set of all those a_i's, and V the set of the b_i's. Then $\langle\sigma, U, V\rangle \notin AR(p) = AR(q)$ and hence $\langle\sigma, Y\rangle \notin R(q)$.

It follows that $R(q) \subseteq R(p)$, and thus q has the property RIF as well. Now the same argument applies in the other direction, yielding $R(p) \subseteq R(q)$. □

Inspired by the definition of R^-, a finite version of ready trace semantics (RT^-) can be defined likewise. Here I will just give its modal characterization.

DEFINITION 7.6. The set $\mathcal{L}_{RT}^-$ of *finite ready trace formulas* over *Act* is given by:
- $\top \in \mathcal{L}_{RT}^-$.
- If $\varphi \in \mathcal{L}_{RT}^-$ and $X \subseteq_{fin} Act$ then $X\varphi \in \mathcal{L}_{RT}^-$ and $\widetilde{X}\varphi \in \mathcal{L}_{RT}^-$.
- If $\varphi \in \mathcal{L}_{RT}^-$ and $a \in Act$ then $a\varphi \in \mathcal{L}_{RT}^-$.

The *satisfaction relation* $\models \subseteq \mathbb{P} \times \mathcal{L}_{RT}^-$ is given by the usual clauses for $\top$ and $a\varphi$, and:
- $p \models X\varphi$ if $X \subseteq I(p)$ and $p \models \varphi$.
- $p \models \widetilde{X}\varphi$ if $I(p) \cap X = \emptyset$ and $p \models \varphi$.

Processes p and q are *finite-ready trace equivalent*, notation $p =_{RT}^- q$, if $\forall\varphi \in \mathcal{L}_{RT}^-(p \models \varphi \Leftrightarrow q \models \varphi)$.

As these formulas are expressible in terms of the ones of Definition 6.2, one has $RT^- \prec RT$; Counterexample 4 shows that the inclusion is strict. Also $FT^- \prec RT^-$ and $F^- \prec R^- \prec RT^-$.

PROPOSITION 7.6. *Let $p, q \in \mathbb{P}$ and p is image finite. Then* $p =_{RT}^- q \Leftrightarrow p =_{RT} q$.

PROOF. "$\Leftarrow$" holds for all process. "$\Rightarrow$" follows just as in Proposition 7.5, using the property that for any $a_1 a_2 \cdots a_n \in Act^\omega$ there are only finitely many normal ready traces $X_0 a_1 X_1 a_2 \cdots a_n X_n \in RT_N(p)$. □

Unlike the semantics T to RT, possible-futures semantics distinguishes between the two processes of Counterexample 1: $\langle a, a^*\rangle \in PF(right) - PF(left)$. Still, $T^\infty \not\prec PF$, as can be seen from the variant of Counterexample 1 in which the left-hand process is appended to the endnodes of both processes. The so obtained systems have the same possible futures, including $\{\langle a^n, a^*\rangle \mid n \in \mathbb{N}\}$, but only the right-hand side has an infinite trace.

For the sake of completeness I include a definition of infinitary possible-futures semantics (PF^∞), such that $PF \prec PF^\infty$ and $R^\infty \prec PF^\infty$. A finite variant of PF has not been explored.

DEFINITION 7.7. $\langle\sigma, X\rangle \in Act^* \times \mathcal{P}(Act^*)$ is an *infinitary possible future* of a process p if there is a process q such that $p \xrightarrow{\sigma} q$ and $T(q) \cup T^\infty(q) = X$. Let $PF^\infty(p)$ denote the set of infinitary possible futures of p. Two processes p and q are *infinitary possible-futures equivalent*, notation $p =^\infty_{PF} q$, if $PF^\infty(p) = PF^\infty(q)$.

8. Simulation semantics

The following concept of *simulation* occurs frequently in the literature (see, e.g., Park [41]).

DEFINITION 8. A *simulation* is a binary relation R on processes, satisfying, for $a \in Act$:
- if pRq and $p \xrightarrow{a} p'$, then $\exists q'$: $q \xrightarrow{a} q'$ and $p'Rq'$.

Process p *can be simulated by* q, notation $p \subsetneq\!\!\to q$, if there is a simulation R with pRq. p and q are *similar*, notation $p \rightleftarrows q$, if $p \subsetneq\!\!\to q$ and $q \subsetneq\!\!\to p$.

PROPOSITION 8.1. *Similarity is an equivalence relation on the domain of processes.*

PROOF. Symmetry is immediate, so it has to be checked that $p \subsetneq\!\!\to p$, and $p \subsetneq\!\!\to q \wedge q \subsetneq\!\!\to r \Rightarrow p \subsetneq\!\!\to r$.
- The identity relation is a simulation with pRp.
- If R is a simulation with pRq and S is a simulation with qSr, then the relation $R; S$, defined by $x(R; S)z$ iff $\exists y$: $xRy \wedge ySz$, is a simulation with $p(R; S)r$. □

Hence the relation will be called *simulation equivalence*. In *simulation semantics* (S) two processes are identified iff they are simulation equivalent.

Testing scenario and modal characterization. The testing scenario for simulation semantics resembles that for trace semantics, but in addition the observer is, at any time during a run of the investigated process, capable of making arbitrary many copies of the process in its present state and observe them independently. Thus an observation yields a tree rather than a sequence of actions. Such a tree can be coded as an expression in a simple modal language.

DEFINITION 8.1. The class $\mathcal{L}_S$ of *simulation formulas* over Act is defined recursively by:
- If I is a set and $\varphi_i \in \mathcal{L}_S$ for $i \in I$ then $\bigwedge_{i \in I} \varphi_i \in \mathcal{L}_S$.
- If $\varphi \in \mathcal{L}_S$ and $a \in Act$ then $a\varphi \in \mathcal{L}_S$.

The *satisfaction relation* $\models \subseteq \mathbb{P} \times \mathcal{L}_S$ is defined recursively by:
- $p \models \bigwedge_{i \in I} \varphi_i$ if $p \models \varphi_i$ for all $i \in I$.
- $p \models a\varphi$ if for some $q \in \mathbb{P}$: $p \xrightarrow{a} q$ and $q \models \varphi$.

Let $S(p)$ denote the class of simulation formulas satisfied by the process p: $S(p) = \{\varphi \in \mathcal{L}_S \mid p \models \varphi\}$. Write $p \sqsubseteq_S q$ if $S(p) \subseteq S(q)$ and $p =_S q$ if $S(p) = S(q)$.

Write $\top$ for $\bigwedge_{i \in \emptyset} \varphi_i$, and $\varphi_1 \wedge \varphi_2$ for $\bigwedge_{i \in \{1,2\}} \varphi_i$. It turns out that $\mathcal{L}_T$ is a sublanguage of $\mathcal{L}_S$.

PROPOSITION 8.2. *$p \subsetneq\!\!\rightarrow q \Leftrightarrow p \sqsubseteq_S q$. Hence $p \rightleftarrows q \Leftrightarrow p =_S q$.*

PROOF. For "$\Rightarrow$" I have to prove that for any simulation R and for all $\varphi \in \mathcal{L}_S$ one has

$$pRq \Rightarrow (p \models \varphi \Rightarrow q \models \varphi).$$

I will do so with structural induction on φ. Suppose pRq.
- Let $p \models a\varphi$. Then there is a $p' \in \mathbb{P}$ with $p \xrightarrow{a} p'$ and $p' \models \varphi$. As R is a simulation, there must be a $q' \in \mathbb{P}$ with $q \xrightarrow{a} q'$ and $p'Rq'$. So by induction $q' \models \varphi$, and hence $q \models a\varphi$.
- $p \models \bigwedge_{i \in I} \varphi_i \Leftrightarrow \forall i \in I(p \models \varphi_i) \overset{\text{ind.}}{\Rightarrow} \forall i \in I(q \models \varphi_i) \Leftrightarrow q \models \bigwedge_{i \in I} \varphi_i$.

For "$\Leftarrow$" it suffices to establish that $\sqsubseteq_S$ is a simulation.

Suppose $p \sqsubseteq_S q$ and $p \xrightarrow{a} p'$. I have to show that $\exists q' \in \mathbb{P}$ with $q \xrightarrow{a} q'$ and $p' \sqsubseteq_S q'$. Let Q' be

$$\{q' \in \mathbb{P} \mid q \xrightarrow{a} q' \wedge p' \not\sqsubseteq_S q'\}.$$

By Definition 1.1 Q' is a set. For every $q' \in Q'$ there is a formula $\varphi_{q'} \in S(p') - S(q')$. Now

$$a \bigwedge_{q' \in Q'} \varphi_{q'} \in S(p) \subseteq S(q),$$

so there must be a $q' \in \mathbb{P}$ with $q \xrightarrow{a} q'$ and $q' \notin Q'$, which had to be shown. □

Process graph characterization. Simulation equivalence can also be characterized by means of relations between the nodes of two process graphs, rather than between process graphs themselves.

DEFINITION 8.2. Let $g, h \in \mathbb{G}$. A *simulation* of g by h is a binary relation $R \subseteq$ NODES$(g) \times$ NODES(h), satisfying:
- ROOT$(g)R$ROOT(h).
- If sRt and $(s, a, s') \in$ EDGES(g), then there is an edge $(t, a, t') \in$ EDGES(h) such that $s'Rt'$.

This definition is illustrated in Figure 5. Solid lines indicates what is assumed, dashed lines what is required. It follows easily that $g \subsetneq\!\!\rightarrow h$ iff there exists a simulation of g by h. For process graphs with multiple roots, the first requirement of Definition 8.2 generalizes to
- $\forall s \in$ ROOTS$(g)\, \exists t \in$ ROOTS(h): sRt.

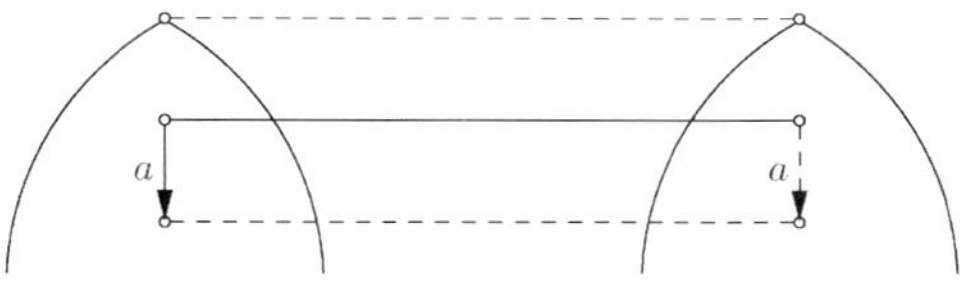

Figure 5. A simulation.

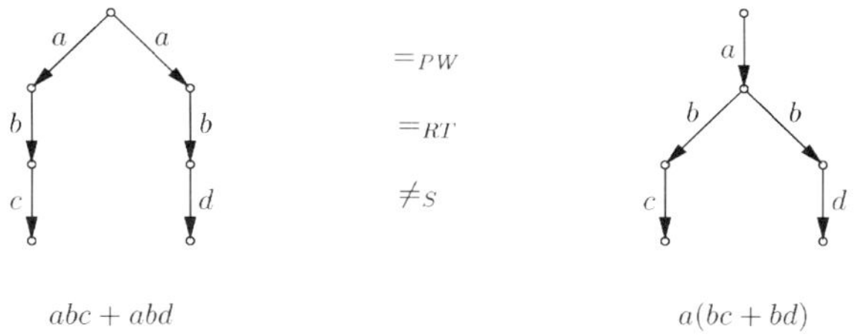

Counterexample 8. Possible worlds and ready trace equivalent, but not simulation equivalent.

Classification. Simulation semantics (S) is finer than trace semantics ($T \prec S$), but independent of CT, F, R, FT, RT and PF.

PROOF. "$T \preceq S$" follows since $\mathcal{L}_T$ is a sublanguage of $\mathcal{L}_S$.

"$S \not\preceq CT$" (and hence "$S \not\preceq RT$", "$S \not\preceq PF$" etc.) follows from Counterexample 2. There $left \neq_{CT} right$, although $left \rightleftarrows right$; the construction of the two simulations is left to the reader.

"$S \not\succeq RT$" (and hence "$S \not\succeq T$" etc.) follows from Counterexample 8. There $RT(left) = RT(right)$, but $S(left) \neq S(right)$. The first statement follows from Proposition 6.3 and the insight that it suffices to check the two ready traces contributed by the maximal paths; these are the same for both graphs. The second statement follows since $a(bc\top \wedge bd\top) \in S(right) - S(left)$.

"$S \not\succeq PF$" follows from Counterexample 7, where $PF(left) = PF(right)$ but $S(left) \neq S(right)$. The latter statement follows since $a(b\top \wedge a(b\top \wedge cd\top)) \in S(left) - S(right)$. □

Infinite processes. In order to make the testing scenario match its formalization in terms of the modal language $\mathcal{L}_S$ even for infinite processes, one has to assume that the amount of copies one can make at any time is infinite. Moreover, although no single copy can be tested forever, due to its infinite branching there may be no upperbound upon the duration of an observation.

One might consider an even more infinitary testing scenario by allowing observations to go on forever on some or all of the copies. However, this would not give rise to a more discriminating equivalence; ordinary simulation equivalences already preserves infinite traces.

PROPOSITION 8.3. *If $p \subsetneq\!\!\!\to q$ then $T^\infty(p) \subseteq T^\infty(q)$. Hence $T^\infty \prec S$.*

from the Sections 3–7 together with a *replicator*, an ingenious device by which one can replicate the machine whenever and as often as one wants. In order to represent observations, the modal languages from Sections 3–7 need to be combined with the one from Section 8.

DEFINITION 9. The language $\mathcal{L}_{CS}$ and the corresponding satisfaction relation is defined recursively by combining the clauses of Definition 3.1 (for $\mathcal{L}_{CT}$) with those of Definition 8.1 (for $\mathcal{L}_S$). Likewise, $\mathcal{L}_{FS}$ is obtained by combining $\mathcal{L}_F$ and $\mathcal{L}_S$; $\mathcal{L}_{FTS}$ by combining $\mathcal{L}_{FT}$ and $\mathcal{L}_S$; $\mathcal{L}_{RS}$ by combining $\mathcal{L}_R$ and $\mathcal{L}_S$; and $\mathcal{L}_{RTS}$ by combining $\mathcal{L}_{RT}$ and $\mathcal{L}_S$. For $p \in \mathbb{P}$ and $\mathcal{O} \in \{CS, FS, FTS, RS, RTS\}$ let $\mathcal{O}(p) = \{\varphi \in \mathcal{L}_{\mathcal{O}} \mid p \models \varphi\}$. Two processes $p, q \in \mathbb{P}$ are
- *completed simulation equivalent*, notation $p =_{CS} q$, if $CS(p) = CS(q)$;
- *failure simulation equivalent*, notation $p =_{FS} q$, if $FS(p) = FS(q)$;
- *failure trace simulation equivalent*, notation $p =_{FTS} q$, if $FTS(p) = FTS(q)$;
- *ready simulation equivalent*, notation $p =_{RS} q$, if $RS(p) = RS(q)$;
- *ready trace simulation equivalent*, notation $p =_{RTS} q$, if $RTS(p) = RTS(q)$.

It is obvious that failure trace simulation equivalence coincides with failure simulation equivalence and ready trace simulation equivalence with ready simulation equivalence ($p \models X\varphi \Leftrightarrow p \models X \wedge \varphi$). Also it is not difficult to see that failure simulation equivalence and ready simulation equivalence coincide ($p \models X \Leftrightarrow p \models \widetilde{Y} \wedge \bigwedge_{a \in X} a\top$, where $Y = Act - X$). So one has

PROPOSITION 9.1. $p =_{FS} q \Leftrightarrow p =_{FTS} q \Leftrightarrow p =_{RTS} q \Leftrightarrow p =_{RS} q$.

Relational characterizations. The two remaining equivalences can be characterized as follows:

DEFINITION 9.1. A *complete simulation* is a binary relation R on processes, satisfying, for $a \in Act$:
- if pRq and $p \xrightarrow{a} p'$, then $\exists q'$: $q \xrightarrow{a} q'$ and $p'Rq'$;
- if pRq then $I(p) = \emptyset \Leftrightarrow I(q) = \emptyset$.

PROPOSITION 9.2. *Two processes p and q are completed simulation equivalent if there exists a complete simulation R with pRq and a complete simulation S with qSp.*

PROOF. A trivial modification of the proof of Proposition 8.2. □

DEFINITION 9.2. A *ready simulation* is a binary relation R on processes, satisfying, for $a \in Act$:
- if pRq and $p \xrightarrow{a} p'$, then $\exists q'$: $q \xrightarrow{a} q'$ and $p'Rq'$;
- if pRq then $I(p) = I(q)$.

PROPOSITION 9.3. *Two processes p and q are ready simulation equivalent if there exists a ready simulation R with pRq and a ready simulation S with qSp.*

PROOF. A trivial modification of the proof of Proposition 8.2. □

A variant of ready simulation equivalence was originally proposed by Bloom, Istrail and Meyer [12] under the name *GSOS trace congruence*; they provided a modal characterization, to be discussed in Section 10. A relational characterization was first given by Larsen and Skou [35] under the name $\frac{2}{3}$*-bisimulation equivalence*. A $\frac{2}{3}$-bisimulation is defined just like a ready simulation, except that the second clause reads "if pRq and $\exists q' : q \xrightarrow{a} q'$ then $\exists p'$: $p \xrightarrow{a} p'$". This is clearly equivalent.

Classification. $RT \prec RS$, $CT \prec CS$ and $S \prec CS \prec RS$. CS is independent of F to RT.

PROOF. "$RT \preceq RS$" follows since $\mathcal{L}_{RT}$ is a sublanguage of $\mathcal{L}_{RTS}$, using Proposition 9.1.

"$CT \preceq CS$" and "$S \preceq CS \preceq RS$" follow since $\mathcal{L}_{CT}$ and $\mathcal{L}_S$ are sublanguages of $\mathcal{L}_{CS}$, which is a sublanguage of $\mathcal{L}_{FS}$.

"$RT \not\succeq RS$" follows from Counterexample 8, using "$RS \succeq S$"; similarly $RT \not\succeq CS$ and $CT \not\succeq CS$.

"$S \not\succeq CS$" follows from Counterexample 2, using "$CS \succeq CT$".

"$CS \not\succeq F$" (and hence "$CS \not\succeq RS$") follows from Counterexample 3, in which $F(\mathit{left}) \neq F(\mathit{right})$ but $\mathit{left} =_{CS} \mathit{right}$; the construction of the two complete simulations is left to the reader. □

PROPOSITION 9.4. *PF is incomparable with CS and RS.*

PROOF. "$CS \not\preceq PF$" (and hence "$RS \not\preceq PF$") follows from Counterexample 7, using "$CS \succeq S$".

"$RS \not\succeq PF$" (and hence "$CS \not\succeq PF$") follows from Counterexample 10, which shows two graphs that are ready simulation equivalent but not possible-futures equivalent. Concerning the first claim, note that there exists exactly one simulation of *right* by *left*, namely the one mapping *right* on the right-hand side of *left*. There also exists exactly one simulation of *left* by *right*, which relates the black node on the *left* to the black node on the *right*. Both simulations are ready simulations, as related nodes have the same menu of initial actions. The second claim follows since $\langle a, \{\varepsilon, b, bc\}\rangle \in PF(\mathit{left}) - PF(\mathit{right})$. □

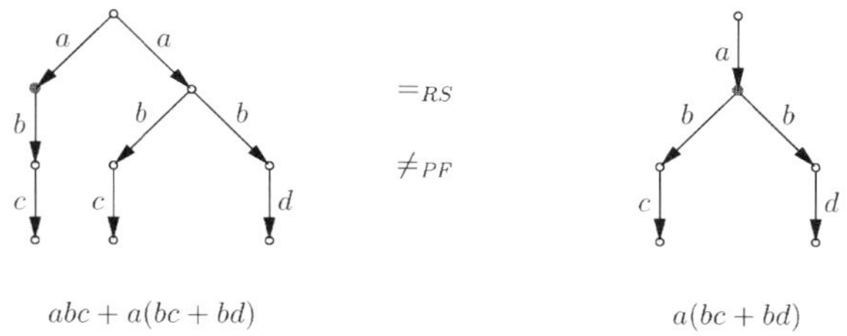

Counterexample 10. Ready simulation equivalent, but not possible-futures equivalent.

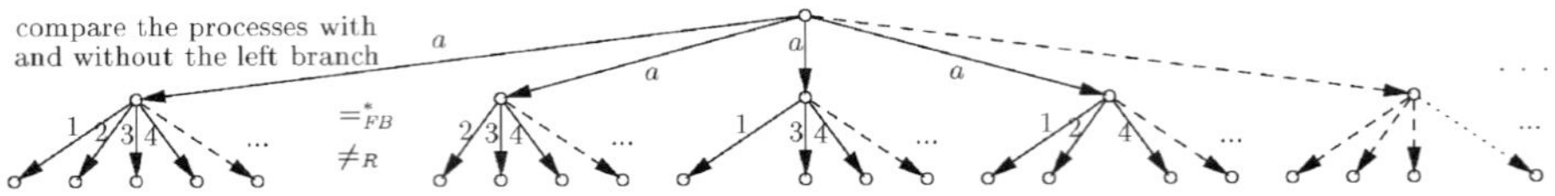

Counterexample 11. Finitary failure simulation equivalent, but not ready equivalent.

Infinite processes. For each of the semantics *CS*, *FS*, *FTS*, *RS* and *RTS* a finitary variant (superscripted with a *), motivated by allowing finite replication only, is defined by combining the modal languages $\mathcal{L}_{CT}$, $\mathcal{L}_F$, $\mathcal{L}_{FT}$, $\mathcal{L}_R$ and $\mathcal{L}_{RT}$, respectively, with $\mathcal{L}_S^*$. Likewise, an intermediate variant (superscripted with an ω), motivated by requiring any observation to be over within a finite amount of time, is defined by combining these languages with $\mathcal{L}_S^\omega$. Finally, a finite variant (superscripted with a $-$), motivated by observers that can only engage in finite replication, can only set finitely many switches on free, and can only inspect finitely many lamps in a finite time, is obtained by combining the (obvious) modal languages $\mathcal{L}_F^-$, $\mathcal{L}_{FT}^-$, $\mathcal{L}_R^-$ and $\mathcal{L}_{RT}^-$ with $\mathcal{L}_S^*$ (there is no CS^-). Exactly as in the case of Proposition 9.1 one finds:

PROPOSITION 9.5. *$FS^\omega = FTS^\omega = RTS^\omega = RS^\omega$ and $FS^- = FTS^- = RTS^- = RS^-$. Moreover, $FS^* = FTS^*$ and $RTS^* = RS^*$.*

However, as pointed out in Schnoebelen [47], FS^* and RS^* are different: in Counterexample 11 one has $FS^*(with) = FS^*(without)$, but $\langle a, \{1, 2, \ldots\}\rangle \in R(with) - R(without)$.

Clearly one has $CS^* \prec CS^\omega \prec CS$ and $RS^- \prec FS^* \prec RS^* \prec RS^\omega \prec RS$. The strictness of these inclusions is given by Counterexamples 4, 11, 9 and 1. In addition one has $RT^- \prec RS^-$, $S^* \prec RS^-$, $RT \prec RS^*$, $FT \prec FS^*$, $CT \prec CS^*$ and $S^* \prec CS^* \prec FS^*$; as well as $RT^\infty \prec RS$, $CT^\infty \prec CS$, $S^\omega \prec CS^\omega \prec RS^\omega$ and $S \prec CS \prec RS$. Counterexamples against further inclusions have already been provided.

PROPOSITION 9.6. *Let $p, q \in \mathbb{P}$ be image finite. Then $p =_{CS} q \Leftrightarrow p =_{CS}^* q$ and $p =_{RS} q \Leftrightarrow p =_{RS}^- q$.*

PROOF. Two trivial modifications of the proof of Proposition 8.4. In the second one, one uses that if $\forall \varphi \in \mathcal{L}_{RS}^-(p \models \varphi \Rightarrow q \models \varphi)$ then surely $I(p) = I(q)$. □

In fact, if it is merely known that only q is image finite it follows already that $p \sqsubseteq_{CS} q \Leftrightarrow p \sqsubseteq_{CS}^* q$ and $p \sqsubseteq_{RS} q \Leftrightarrow p \sqsubseteq_{RS}^- q$. However, the following variant of Counterexample 1 shows that in the statement of Proposition 9.6 it is essential that *both* p and q are image finite. In Counterexample 12 *right* is image finite – in fact, it is even finitely branching – but *left* is not. It turns out that *left* $=_{RS}^\omega$ *right* (and hence *left* $=_{RS}^-$ *right*, *left* $=_{CS}^*$ *right*, *left* $=_{RT}$ *right*, *left* $=_F$ *right*, etc.) but *left* $\neq_T^\infty$ *right* (and hence *left* $\neq_F^\infty$ *right*, *left* $\neq_{RT}^\infty$ *right*, *left* $\neq_{CS}$ *right*, *left* $\neq_{RS}$ *right*, etc.).

For general (non-image-finite) processes, no relational characterizations of the finite, finitary and intermediate equivalences are known.

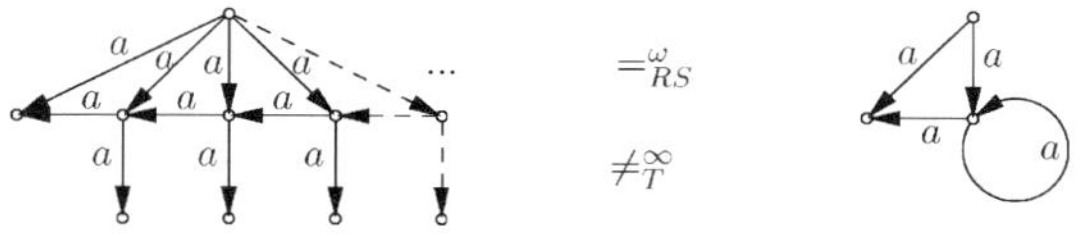

Counterexample 12. Finitary ready simulation equivalent but not infinitary equivalent.

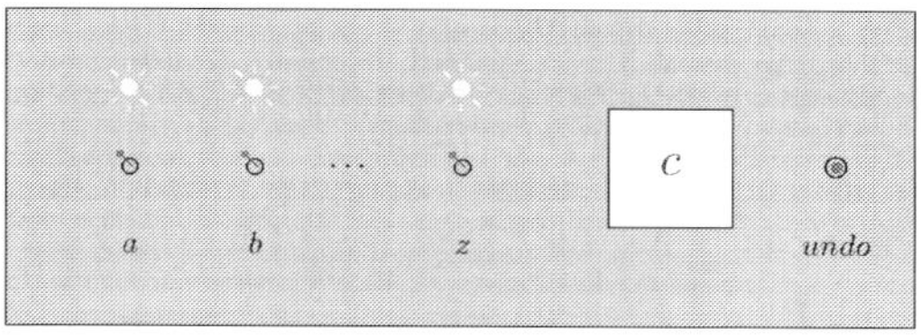

Figure 6. The ready simulation machine.

Testing scenario. An alternative and maybe more natural testing scenario for finitary ready simulation semantics (or simulation semantics) can be obtained by exchanging the replicator for an *undo*-button on the (ready) trace machine (Figure 6). It is assumed that all intermediate states that are past through during a run of a process are stored in a memory inside the black box. Now pressing the *undo*-button causes the machine to shift one state backwards. In the initial state pressing the button has no effect. An observation now consists of a (ready) trace, enriched with *undo*-actions. Such observations can easily be translated into finitary (ready) simulation formulas.

10. Reactive versus generative testing scenarios

In the testing scenarios presented so far, a process is considered to perform actions and make choices autonomously. The investigated behaviours can therefore be classified as *generative processes*. The observer merely restricts the spontaneous behaviour of the generative machine by cutting off some possible courses of action. An alternative view of the investigated processes can be obtained by considering them to react on stimuli from the environment and be passive otherwise. *Reactive machines* can be obtained out of the generative machines presented so far by replacing the switches by buttons and the display by a green light. Initially the process waits patiently until the observer tries to press one of the buttons. If the observer tries to press an a-button, the machine can react in two different ways: if the process cannot start with an a-action the button will not go down and the observer may try another one; if the process can start with an a-action it will do so and the button goes down. Furthermore the green light switches on. During the execution of a no buttons can be pressed. As soon as the execution of a is completed the light switches off, so that the observer knows that the process is ready for a new trial. Reactive machines as described above originate from Milner [37,38].

One family of testing scenarios with reactive machines can be obtained by allowing the observer to try to depress more than one button at a time. In order to influence a particular choice, the observer could already start exercising pressure on buttons during the execution of the preceding action (when no button can go down). When this preceding action is finished, at most one of the buttons will go down. These testing scenarios are equipotent with the generative ones: putting pressure on a button is equivalent to setting the corresponding switch on 'free'; moreover an action a appearing in the display is mimicked by the a-button going down, and the disappearance of a from the display by the green light going off.

Another family of testing scenarios is obtained by allowing the user to try only one button at a time. They are equipotent with those generative testing scenarios in which at any time only one switch can be set on 'free'. Next I will discuss the equivalences that originate from these scenarios.

First consider the reactive machine that resembles the failure trace machine, thus without menu-lights and *undo*-button. An observation on such a machine consists of a sequence of accepted and refused actions, indicating which buttons went down in a sequence of trials of the user. Such a sequence can be seen as a failure trace where all refusal sets are singletons. Call the resulting semantics FT^1. Clearly, the failure trace set of any process p satisfies

$$\sigma(X \cup Y)\rho \in FT(p) \Leftrightarrow \sigma XY\rho \in FT(p).$$

Thus, any failure trace $\sigma\{a_1, \ldots, a_n\}\rho$ can be rewritten as (contains the same information as) $\sigma\{a_1\}\{a_2\}\cdots\{a_n\}\rho$. It follows that the singleton-failure trace set $FT^1(p)$ of a process p contains as much information as its finite-failure trace set $FT^-(p)$, so the semantics FT^1 coincides with FT^-.

In order to arrive at a reactive counterpart to failures semantics, one could suppose that an observer continues an experiment only as long as all buttons he tries to depress actually go down; when a button refuses to go down, he will not try another one. This testing scenario gives rise to the variant F^1 of failures semantics in which all refusal sets are singletons.

DEFINITION 10. $\langle\sigma, a\rangle \in Act^* \times Act$ is a *singleton-failure pair* of a process p if there is a process q such that $p \xrightarrow{\sigma} q$ and $a \notin I(q)$. Let $F^1(p)$ denote the set of singleton-failure pairs of p. Two processes p and q are *singleton-failures equivalent*, $p =^1_F q$, if $T(p) = T(q)$ and $F^1(p) = F^1(q)$.

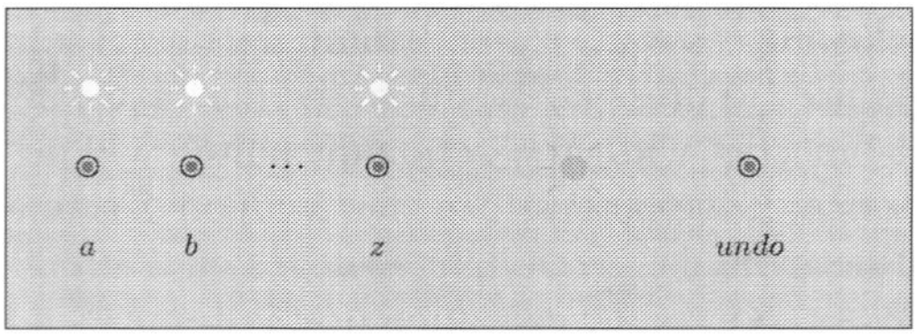

Figure 7. The reactive ready simulation machine.

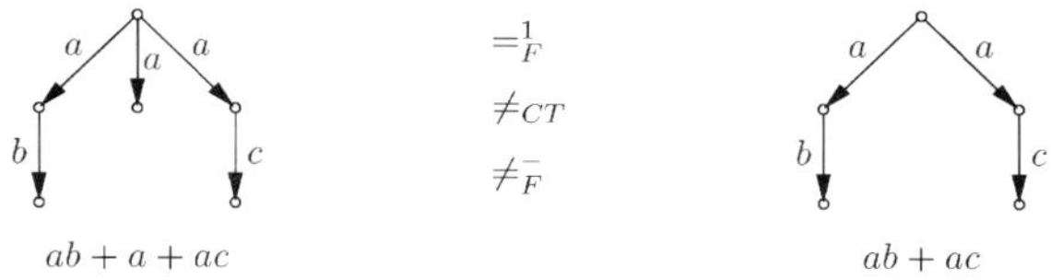

Counterexample 13. Singleton-failures equivalent, but not completed trace or failures equivalent.

Unlike for F and F^-, $F^1(p) = F^1(q)$ does not always imply that $T(p) = T(q)$, so one has to keep track of traces explicitly. These model observations ended by the observer before stagnation occurs.

Singleton-failures semantics (F^1) is situated strictly between trace (T) and finite-failures semantics (F^-). For Counterexample 2 shows two processes with $T(\mathit{left}) = T(\mathit{right})$ but $\langle a, b\rangle \in F^1(\mathit{left}) - F^1(\mathit{right})$, and Counterexample 13 shows two processes with $F^1(\mathit{left}) = F^1(\mathit{right})$ (both contain $\langle a, b\rangle$ and $\langle a, c\rangle$), but $\langle a, \{b, c\}\rangle \in F(\mathit{left}) - F(\mathit{right})$. Furthermore, F^1 is independent of CT, S and CS, for in Counterexample 13 one has $CT(\mathit{left}) \neq CT(\mathit{right})$, in Counterexample 8 one has $\mathit{left} =^1_F \mathit{right}$ but $\mathit{left} \neq_S \mathit{right}$, and in Counterexample 3 one has $\mathit{left} =_{CS} \mathit{right}$ but $\langle a, c\rangle \in F^1(\mathit{left}) - F^1(\mathit{right})$.

Adding the *undo*-button to the reactive failure trace machine gives a semantics FS^1 characterized by the modal language $\mathcal{L}^*_S$ to which has been added a modality "Can't(a)", with $p \models$ Can't(a) iff $a \notin I(p)$. This modality denotes a failed attempt to depress the a-button. If fact, Bloom, Istrail and Meyer studied the coarsest equivalence finer than trace equivalence that is a congruence for the class of so-called *GSOS-operators*, and characterized this *GSOS trace congruence* by the modal language above; its formulas were called *denial formulas*. As in the modal language $\mathcal{L}^-_{FS}$ one has $p \models \widetilde{X \cup Y} \Leftrightarrow p \models \widetilde{X} \wedge \widetilde{Y}$, and Can't($a$) is the same as $\widetilde{\{a\}}$, it follows that the language of denial formulas is equally expressive as $\mathcal{L}^-_{FS}$, and hence FS^1 coincides with FS^- and RS^-.

If the menu-lights are added to the reactive failure trace machine considered above one can observe ready trace sets, and the green light is redundant. Likewise, adding menu-lights to the reactive failure scenario would give readiness semantics, and adding them to the reactive failure simulation machine would yield ready simulation. If the green light (as well as the menu-lights) are removed from the reactive failure trace machine, one can only test trace equivalence, since any refusal may be caused by the last action not being ready yet. Likewise, removing the green light from the reactive failure simulation machine (with *undo*-button) yields (finitary) simulation semantics. Reactive machines on which only one button at a time is depressed appear to be unsuited for testing completed trace, completed simulation and failures equivalence.

11. 2-nested simulation semantics

2-*nested simulation equivalence* popped up naturally in Groote and Vaandrager [25] as the coarsest congruence with respect to a large and general class of operators that is finer than completed trace equivalence.

DEFINITION 11. A 2-*nested simulation* is a simulation contained in simulation equivalence ($\rightleftarrows$). Two processes p and q are 2-*nested simulation equivalent*, notation $p =_{2S} q$, if there exists a 2-nested simulation R with pRq and a 2-nested simulation S with qSp.

Modal characterization. A modal characterization of this notion is obtained by the fragment of the infinitary Hennessy–Milner logic (cf. Definition 12.1) without nested negations.

DEFINITION 11.1. The class $\mathcal{L}_{2S}$ of 2-*nested simulation formulas* over *Act* is defined recursively by:

- If I is a set and $\varphi_i \in \mathcal{L}_{2S}$ for $i \in I$ then $\bigwedge_{i\in I} \varphi_i \in \mathcal{L}_{2S}$.
- If $\varphi \in \mathcal{L}_{2S}$ and $a \in Act$ then $a\varphi \in \mathcal{L}_{2S}$.
- If $\varphi \in \mathcal{L}_S$ then $\neg\varphi \in \mathcal{L}_{2S}$.

Note that $\mathcal{L}_S \subseteq \mathcal{L}_{2S}$. The *satisfaction relation* $\models \subseteq \mathbb{P} \times \mathcal{L}_{2S}$ is defined recursively by:

- $p \models \bigwedge_{i\in I} \varphi_i$ if $p \models \varphi_i$ for all $i \in I$.
- $p \models a\varphi$ if for some $q \in \mathbb{P}$: $p \xrightarrow{a} q$ and $q \models \varphi$.
- $p \models \neg\varphi$ if $p \not\models \varphi$.

PROPOSITION 11.1. $p =_{2S} q \Leftrightarrow \forall\varphi \in \mathcal{L}_{2S}(p \models \varphi \Leftrightarrow q \models \varphi)$.

PROOF. A trivial modification of the proof of Proposition 8.2. □

Testing scenario. In order to obtain a testing scenario for this equivalence one has to introduce the rather unnatural notion of a *lookahead* [25]: The 2-*nested simulation machine* is a variant of the ready trace machine with replicator, where in an idle state the machine not only tells which actions are on the menu, but even which simulation formulas are (not) satisfied in the current state.

Classification. $RS \prec 2S$ and $PF \prec 2S$.

PROOF. For "$RS \preceq 2S$" it suffices to show that each 2-nested simulation is a ready simulation. This follows since $p \rightleftarrows q \Rightarrow I(p) = I(q)$. $PF \prec 2S$ is easily established using that $T \prec S$. That both inclusions are strict follows immediately from the fact that *RS* and *PF* are incomparable (Proposition 9.4). □

Infinite processes. Exactly as for ready simulations semantics, 5 versions of 2-nested simulation semantics can be defined that differ for infinite processes. $2S^-$ is the semantics whose modal characterization has the constructs $\top$, $\wedge$, $a\varphi$ and $\neg\varphi'$ with $\varphi' \in \mathcal{L}_S^*$. The constructs $\widetilde{X}$ and X for $X \subseteq_{fin} Act$ are expressible in this logic. $F2S^*$ additionally has the construct $\widetilde{X}$, and $R2S^*$ the construct X, for $X \subseteq Act$. Finally $2S^\omega$ is characterized by the class of 2-nested simulation formulas with a finite upperbound on the nesting of the $a\varphi$ construct. The constructs $\widetilde{X}$ and X for $X \subseteq Act$ are expressible in $\mathcal{L}_{2S}^\omega$, and hence also in $\mathcal{L}_{2S}$.

We have $2S^- \prec F2S^* \prec R2S^* \prec 2S^\omega \prec 2S$. The strictness of these inclusions is given by Counterexamples 4, 11, 9 and 1. In addition one has $RS^- \prec 2S^-$, $FS^* \prec F2S^*$, $RS^* \prec$

$R2S^*$, $RS^\omega \prec 2S^\omega$ and $RS \prec 2S$; as well as $PF^\infty \prec 2S$. Counterexample 1 shows that $PF \not\prec 2S^\omega$: $2S^\omega(left) = 2S^\omega(right)$ (cf. Proposition 12.10), but $\langle a, a^* \rangle \in PF(right) - PF(left)$.

PROPOSITION 11.2. *Let $p, q \in \mathbb{P}$ be image finite. Then $p =_{2S} q \Leftrightarrow p =_{\overline{2S}} q$.*

PROOF. An easy modification of the proof of Proposition 8.4, also using its result. □

12. Bisimulation semantics

The concept of *bisimulation equivalence* stems from Milner [37]. Its formulation below is due to Park [41].

DEFINITION 12. A *bisimulation* is a binary relation R on processes, satisfying, for $a \in Act$:
- if pRq and $p \xrightarrow{a} p'$, then $\exists q': q \xrightarrow{a} q'$ and $p'Rq'$;
- if pRq and $q \xrightarrow{a} q'$, then $\exists p': p \xrightarrow{a} p'$ and $p'Rq'$.

Two processes p and q are *bisimilar*, notation $p \underline{\leftrightarrow} q$, if there exists a bisimulation R with pRq.

The relation $\underline{\leftrightarrow}$ is again a bisimulation. As for similarity, one easily checks that bisimilarity is an equivalence relation on $\mathbb{P}$. Hence the relation will be called *bisimulation equivalence*. identified iff they are bisimulation equivalent. Note that the concept of bisimulation does not change if in the definition above the action relations $\xrightarrow{a}$ were replaced by generalized action relations $\xrightarrow{\sigma}$.

Modal characterization

DEFINITION 12.1. The class $\mathcal{L}_B$ of *infinitary Hennessy–Milner formulas* over Act is defined by:
- If I is a set and $\varphi_i \in \mathcal{L}_B$ for $i \in I$ then $\bigwedge_{i \in I} \varphi_i \in \mathcal{L}_B$.
- If $\varphi \in \mathcal{L}_B$ and $a \in Act$ then $a\varphi \in \mathcal{L}_B$.
- If $\varphi \in \mathcal{L}_B$ then $\neg \varphi \in \mathcal{L}_B$.

The *satisfaction relation* $\models \,\subseteq \mathbb{P} \times \mathcal{L}_B$ is defined recursively by:
- $p \models \bigwedge_{i \in I} \varphi_i$ if $p \models \varphi_i$ for all $i \in I$.
- $p \models a\varphi$ if for some $q \in \mathbb{P}$: $p \xrightarrow{a} q$ and $q \models \varphi$.
- $p \models \neg \varphi$ if $p \not\models \varphi$.

Let $B(p)$ denote the class of all infinitary Hennessy–Milner formulas satisfied by the process p: $B(p) = \{\varphi \in \mathcal{L}_B \mid p \models \varphi\}$. Write $p \sqsubseteq_B q$ if $B(p) \subseteq B(q)$ and $p =_B q$ if $B(p) = B(q)$.

PROPOSITION 12.1. $p \sqsubseteq_B q \Leftrightarrow p =_B q$.

PROOF. If $\varphi \in B(q) - B(p)$ then $\neg \varphi \in B(p) - B(q)$. □

PROPOSITION 12.2. $p \underline{\leftrightarrow} q \Leftrightarrow p =_B q$.

PROOF. For "$\Rightarrow$" I have to prove that for any bisimulation R and for all $\varphi \in \mathcal{L}_B$ one has

$$pRq \Rightarrow (p \models \varphi \Leftrightarrow q \models \varphi).$$

I will do so with structural induction on φ. Suppose pRq.
- Let $p \models a\varphi$. Then there is a $p' \in \mathbb{P}$ with $p \xrightarrow{a} p'$ and $p' \models \varphi$. As R is a bisimulation, there must be a $q' \in \mathbb{P}$ with $q \xrightarrow{a} q'$ and $p'Rq'$. So by induction $q' \models \varphi$, and hence $q \models a\varphi$.
  By symmetry one also obtains $q \models a\varphi \Rightarrow p \models a\varphi$.
- $p \models \bigwedge_{i \in I} \varphi_i \Leftrightarrow \forall i \in I (p \models \varphi_i) \stackrel{\text{ind.}}{\Leftrightarrow} \forall i \in I (q \models \varphi_i) \Leftrightarrow q \models \bigwedge_{i \in I} \varphi_i$.
- $p \models \neg\varphi \Leftrightarrow p \not\models \varphi \stackrel{\text{ind.}}{\Leftrightarrow} q \not\models \varphi \Leftrightarrow q \models \neg\varphi$.

For "$\Leftarrow$" it suffices to establish that $\sqsubseteq_B$ is a simulation (Proposition 12.1 then implies that $=_B = \sqsubseteq_B = \sqsubseteq_B^{-1}$ is a bisimulation). This goes exactly as in the proof of Proposition 8.2. □

Testing scenario. The testing scenario for bisimulation semantics, as presented in Milner [37], is the oldest and most powerful testing scenario, from which most others have been derived by omitting some of its features. It was based on a reactive failure trace machine with replicator, but additionally the observer is equipped with the capacity of *global testing*. Global testing is described in Abramsky [1] as: "the ability to enumerate all (of finitely many) possible 'operating environments' at each stage of the test, so as to guarantee that all nondeterministic branches will be pursued by various copies of the subject process". Milner [37] implemented global testing by assuming that

"(i) It is the *weather* at any moment which determines the choice of transition (in case of ambiguity [...]);
 (ii) The weather has only finitely many states – at least as far as choice-resolution is concerned;
 (iii) We can control the weather."

Now it can be ensured that all possible moves a process can perform in reaction on a given a-experiment will be investigated by simply performing the experiment in all possible weather conditions. Unfortunately, as remarked in Milner [38], the second assumption implies that the amount of different moves an investigated process can perform in response to any given experiment is bounded by the number of possible weather conditions (i.e., $\exists n \in \mathbb{N} \forall p \in \mathbb{P} \forall a \in Act$: $|\{q \in \mathbb{P} \mid p \xrightarrow{a} q\}| < n$). So for general application this condition has to be dropped, thereby losing the possibility of effective implementation of the testing scenario.

An observation in the global testing scenario can be represented as an infinitary Hennessy–Milner formula $\varphi \in \mathcal{L}_B$. This is essentially a simulation formula in which it is possible to indicate that certain branches are not present. A formula $\neg\varphi$ says that by making sufficiently many copies of the investigated process, and exposing them to all possible weather conditions, it can be observed that none of these copies permits the observation φ.

REMARK. Let $[a]\varphi$ denote $\neg a \neg\varphi$. Now the negation in $\mathcal{L}_B$ can be eliminated in favour of the modalities $[a]$ and infinitary disjunction $\bigvee_{i\in I}$. A formula $[a]\varphi$ says that in all possible weather conditions, after an a-move it is always possible to make the observation φ.

In order to justify the observations of $\mathcal{L}_B$ in a generative testing scenario no switches or menu-lights are needed; the architecture of the completed trace machine suffices. However, in order to warrant negative observations, one has to assume that actions take only a finite amount of time, and idling can be detected (either by observations that last forever, or by means of the display becoming empty). Adding switches and or menu-lights does not increase the discriminating power of the observers. It would give rise to observations that can be modelled as formulas in languages $\mathcal{L}_{FB}$, $\mathcal{L}_{RTB}$, etc., obtained by combining $\mathcal{L}_F$, $\mathcal{L}_{RT}$, etc. with $\mathcal{L}_B$. These observations can already be expressed in $\mathcal{L}_B$: $p \models \widetilde{X} \Leftrightarrow p \models \bigwedge_{a\in X} \neg a\top$ and $p \models X\varphi \Leftrightarrow p \models (\bigwedge_{a\notin X} \neg a\top) \wedge (\bigwedge_{a\in X} a\top) \wedge \varphi$.

A different implementation of global testing is given in Larsen and Skou [35]. They assumed that every transition in a transition system has a certain probability of being taken. Therefore an observer can with an arbitrary high degree of confidence assume that all transitions have been examined, simply by repeating an experiment many times.

As argued among others in Bloom, Istrail and Meyer [12], global testing in the above sense is a rather unrealistic testing ability. Once you assume that the observer is really as powerful as in the described scenarios, in fact more can be tested then only bisimulation equivalence: in the testing scenario of Milner also the correlation between weather conditions and transitions being taken by the investigated process can be recovered, and in that of Larsen and Skou one can determine the relative probabilities of the various transitions.

Process graph characterization. Also bisimulation equivalence can be characterized by means of relations between the nodes of two process graphs.

DEFINITION 12.2. Let $g, h \in \mathbb{G}$. A *bisimulation* between g and h is a binary relation $R \subseteq \mathrm{NODES}(g) \times \mathrm{NODES}(h)$, satisfying:

- $\mathrm{ROOT}(g) R \mathrm{ROOT}(h)$.
- If sRt and $(s, a, s') \in \mathrm{EDGES}(g)$, then there is an edge $(t, a, t') \in \mathrm{EDGES}(h)$ such that $s'Rt'$.
- If sRt and $(t, a, t') \in \mathrm{EDGES}(h)$, then there is an edge $(s, a, s') \in \mathrm{EDGES}(g)$ such that $s'Rt'$.

This definition is illustrated in Figure 8. Solid lines indicates what is assumed, dashed lines what is required. It follows easily that $g \underline{\leftrightarrow} h$ iff there exists a bisimulation between g and h.

For process graphs with multiple roots, the first requirement of Definition 12.2 generalizes to

- $\forall s \in \mathrm{ROOTS}(g)\, \exists t \in \mathrm{ROOTS}(h)$: sRt.
- $\forall t \in \mathrm{ROOTS}(h)\, \exists s \in \mathrm{ROOTS}(g)$: sRt.

Classification. $2S \prec B$.

PROOF. "$2S \preceq B$" follows since $\mathcal{L}_{2S}$ is a sublanguage of $\mathcal{L}_B$.

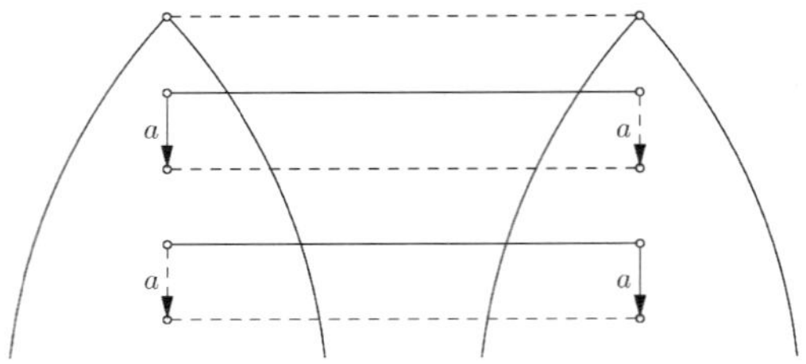

Figure 8. A bisimulation.

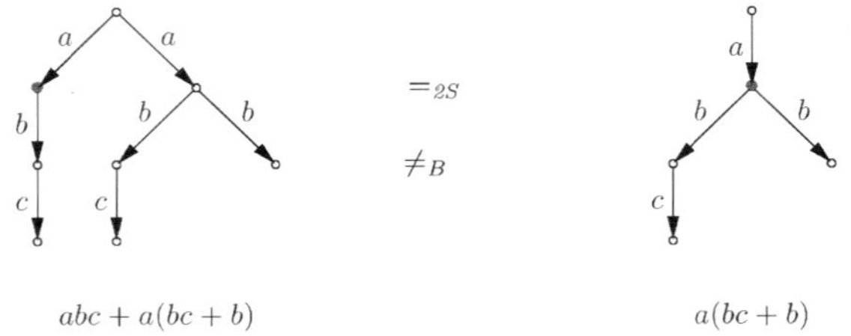

Counterexample 14. 2-nested simulation equivalent, but not bisimulation equivalent.

"$2S \not\preceq B$" follows from Counterexample 14, which shows two graphs that are 2-nested simulation equivalent, but not bisimulation equivalent. Concerning the first claim, as in Counterexample 10 there exists exactly one simulation of *left* by *right*, which relates the black node on the *left* to the black node on the *right*. Unlike in Counterexample 10, this simulation is 2-nested, for the two subgraphs originating from the two black nodes are simulation equivalent, as are the graphs *left* and *right* themselves. Likewise, the simulation mapping *right* on the right-hand side of *left* is also 2-nested. The second claim follows since $a\neg b\neg c\top \in B(\mathit{left}) - B(\mathit{right})$. □

Thus bisimulation equivalence is the finest semantic equivalence treated so far. The following shows however that on $\mathbb{G}$ graph isomorphism is even finer, i.e., isomorphic graphs are always bisimilar. In fact, a graph isomorphism can be seen as a bijective bisimulation. That not all bisimilar graphs are isomorphic will follow from Counterexample 15.

PROPOSITION 12.3. *For $g, h \in \mathbb{G}$, $g \cong h$ iff there exists a bisimulation R between g and h, satisfying*

- *If sRt and uRv then $s = u \Leftrightarrow t = v$.* (*)

PROOF. Suppose $g \cong h$. Let $f : \text{NODES}(g) \to \text{NODES}(h)$ be a graph isomorphism. Define $R \subseteq \text{NODES}(g) \times \text{NODES}(h)$ by sRt iff $f(s) = t$. Then it is routine to check that R satisfies all clauses of Definition 12.2 and (*). Now suppose R is a bisimulation between g and h satisfying (*). Define $f : \text{NODES}(g) \to \text{NODES}(h)$ by $f(s) = t$ iff sRt. Since g is connected it follows from the definition of a bisimulation that for each s such a t can be found. Furthermore direction "$\Rightarrow$" of (*) implies that $f(s)$ is uniquely determined. Hence f is well-defined. Now direction "$\Leftarrow$" of (*) implies that f is injective. From the connectedness of h if follows that f is also surjective, and hence a bijection. Finally, the clauses of Definition 12.2 imply that f is a graph isomorphism. $\square$

COROLLARY 12.1. *If $g \cong h$ then g and h are equivalent according to all semantic equivalences encountered so far.*

Non-well-founded sets. Another characterization of bisimulation semantics can be given by means of Aczel's universe $\mathcal{V}$ of non-well-founded sets [4]. This universe is an extension of the Von Neumann universe of well-founded sets, where the axiom of foundation (every chain $x_0 \ni x_1 \ni \cdots$ terminates) is replaced by an *anti-foundation axiom*.

DEFINITION 12.3. Let $\mathcal{B}$ denote the unique function $\mathcal{M} : \mathbb{P} \to \mathcal{V}$ satisfying

$$\mathcal{M}(p) = \{\langle a, \mathcal{M}(q)\rangle \mid p \xrightarrow{a} q\}$$

for all $p \in \mathbb{P}$. Two processes p and q are *branching equivalent* (my terminology) if $\mathcal{B}(p) = \mathcal{B}(q)$.

It follows from Aczel's anti-foundation axiom that such a function exists. In fact the axiom amounts to saying that systems of equations like the one above have unique solutions. In [4] there is also a section on communicating systems. There two processes are identified iff they are branching equivalent.

A similar idea underlies the semantics of De Bakker and Zucker [9], but there the domain of processes is a complete metric space and the definition of $\mathcal{B}$ above only works for finitely branching processes, and only if $=$ is interpreted as *isometry*, rather then equality, in order to stay in well-founded set theory. For finitely branching processes the semantics of De Bakker and Zucker coincides with the one of Aczel and also with bisimulation semantics. This is observed in Van Glabbeek and Rutten [22], where also a proof can be found of the next proposition, saying that bisimulation equivalence coincides with branching equivalence.

PROPOSITION 12.4. *Let $p, q \in \mathbb{P}$. Then $p \leftrightarroweq q \Leftrightarrow \mathcal{B}(p) = \mathcal{B}(q)$.*

PROOF. "$\Leftarrow$": Let B be the relation defined by pBq iff $\mathcal{B}(p) = \mathcal{B}(q)$; then it suffices to prove that B is a bisimulation. Suppose pBq and $p \xrightarrow{a} p'$. Then $\langle a, \mathcal{B}(p')\rangle \in \mathcal{B}(p) = \mathcal{B}(q)$. So by the definition of $\mathcal{B}(q)$ there must be a process q' with $\mathcal{B}(p') = \mathcal{B}(q')$ and $q \xrightarrow{a} q'$. Hence $p'Bq'$, which had to be proved. The second requirement for B being a bisimulation follows by symmetry.

"$\Rightarrow$": Let $\mathcal{B}^*$ denote the unique solution of $\mathcal{M}^*(p) = \{\langle a, \mathcal{M}^*(r')\rangle \mid \exists r\colon r \leftrightarroweq p \wedge r \xrightarrow{a} r'\}$. As for $\mathcal{B}$ it follows from the anti-foundation axiom that such a unique solution exists. From the symmetry and transitivity of $\leftrightarroweq$ it follows that

$$p \leftrightarroweq q \;\Rightarrow\; \mathcal{B}^*(p) = \mathcal{B}^*(q). \tag{1}$$

Hence it remains to be proven that $\mathcal{B}^* = \mathcal{B}$. This can be done by showing that $\mathcal{B}^*$ satisfies the equations $\mathcal{M}(p) = \{\langle a, \mathcal{M}(q)\rangle \mid p \xrightarrow{a} q\}$, which have $\mathcal{B}$ as unique solution. So it has to be established that $\mathcal{B}^*(p) = \{\langle a, \mathcal{B}^*(q)\rangle \mid p \xrightarrow{a} q\}$. The direction "$\supseteq$" follows directly from the reflexivity of $\leftrightarrows$. For "$\subseteq$", suppose $\langle a, X\rangle \in \mathcal{B}^*(p)$. Then $\exists r$: $r \leftrightarrows p$, $r \xrightarrow{a} r'$ and $X = \mathcal{B}^*(r')$. Since $\leftrightarrows$ is a bisimulation, $\exists p'$: $p \xrightarrow{a} p'$ and $r' \leftrightarrows p'$. From (1) it follows that $X = \mathcal{B}^*(r') = \mathcal{B}^*(p')$. Therefore $\langle a, X\rangle \in \{\langle a, \mathcal{B}^*(q)\rangle \mid p \xrightarrow{a} q\}$, which had to be established. □

Infinite processes. The following predecessor of bisimulation equivalence was proposed in Hennessy and Milner [27,28].

DEFINITION 12.4. Let $p, q \in \mathbb{P}$. Then:
- $p \sim_0 q$ is always true.
- $p \sim_{n+1} q$ if for all $a \in Act$:
  - $p \xrightarrow{a} p'$ implies $\exists q'$: $q \xrightarrow{a} q'$ and $p' \sim_n q'$;
  - $q \xrightarrow{a} q'$ implies $\exists p'$: $p \xrightarrow{a} p'$ and $p' \sim_n q'$.
- p and q are *observationally equivalent*, notation $p \sim q$, if $p \sim_n q$ for every $n \in \mathbb{N}$.

Hennessy and Milner provided the following modal characterization of observational equivalence on image finite processes.

DEFINITION 12.5. The set $\mathcal{L}_{HM}$ of *Hennessy–Milner formulas* over *Act* is defined recursively by:
- $\top \in \mathcal{L}_{HM}$.
- If $\varphi, \psi \in \mathcal{L}_{HM}$ then $\varphi \wedge \psi \in \mathcal{L}_{HM}$.
- If $\varphi \in \mathcal{L}_{HM}$ and $a \in Act$ then $a\varphi \in \mathcal{L}_{HM}$.
- If $\varphi \in \mathcal{L}_{HM}$ then $\neg\varphi \in \mathcal{L}_{HM}$.

The *satisfaction relation* $\models \subseteq \mathbb{P} \times \mathcal{L}_{HM}$ is defined recursively by:
- $p \models \top$ for all $p \in \mathbb{P}$.
- $p \models \varphi \wedge \psi$ if $p \models \varphi$ and $p \models \psi$.
- $p \models a\varphi$ if for some $q \in \mathbb{P}$: $p \xrightarrow{a} q$ and $q \models \varphi$.
- $p \models \neg\varphi$ if $p \not\models \varphi$.

The modal logic above is now known as the *Hennessy–Milner logic* (*HML*). Let $HM(p)$ denote the set of all Hennessy–Milner formulas that are satisfied by the process p: $HM(p) = \{\varphi \in \mathcal{L}_{HM} \mid p \models \varphi\}$. Two processes p and q are *HML-equivalent*, notation $p =_B^- q$, if $HM(p) = HM(q)$.

Theorem 2.2 in Hennessy and Milner [27,28] says that $\sim$ and $=_B^-$ coincide for image finite processes. This result will be strengthened by Proposition 12.6. Below I provide a modal characterization of $\sim$ that is valid for arbitrary processes.

DEFINITION 12.6. Let $\mathcal{L}_B^\omega = \bigcup_{n=0}^\infty \mathcal{L}_B^n$, where $\mathcal{L}_B^n$ is given by:
- If I is a set and $\varphi_i \in \mathcal{L}_B^n$ for $i \in I$ then $\bigwedge_{i \in I} \varphi_i \in \mathcal{L}_B^n$.
- If $\varphi \in \mathcal{L}_B^n$ and $a \in Act$ then $a\varphi \in \mathcal{L}_B^{n+1}$.
- If $\varphi \in \mathcal{L}_B^n$ then $\neg\varphi \in \mathcal{L}_B^n$.

Let $B^\omega(p) = \{\varphi \in \mathcal{L}_B^\omega \mid p \models \varphi\}$ and write $p =_B^\omega q$ if $B^\omega(p) = B^\omega(q)$.

PROPOSITION 12.5. *$p \sim_n q \Leftrightarrow \forall \varphi \in \mathcal{L}_B^n (p \models \varphi \Leftrightarrow q \models \varphi)$ for all $n \in \mathbb{N}$. Hence $p \sim q \Leftrightarrow p =_B^\omega q$.*

PROOF. *Induction base*: Formulas in $\mathcal{L}_B^0$ do not contain the construct $a\varphi$. Hence for such formulas ψ the statement $p \models \psi$ is independent of p. Thus $\forall p, q \in \mathbb{P} : \forall \varphi \in \mathcal{L}_B^0 (p \models \varphi \Leftrightarrow q \models \varphi)$.

Induction step: Suppose $p \sim_{n+1} q$. I now use structural induction on φ.

- Let $p \models a\varphi$ with $a\varphi \in \mathcal{L}_B^{n+1}$. Then there is a $p' \in \mathbb{P}$ with $p \xrightarrow{a} p'$ and $p' \models \varphi \in \mathcal{L}_B^n$. As $p \sim_{n+1} q$, there must be a $q' \in \mathbb{P}$ with $q \xrightarrow{a} q'$ and $p' \sim_n q'$. So by induction $q' \models \varphi$, and hence $q \models a\varphi$.
  By symmetry one also obtains $q \models a\varphi \Rightarrow p \models a\varphi$.
- $p \models \bigwedge_{i\in I} \varphi_i \Leftrightarrow \forall i \in I (p \models \varphi_i) \overset{\text{ind.}}{\Leftrightarrow} \forall i \in I (q \models \varphi_i) \Leftrightarrow q \models \bigwedge_{i \in I} \varphi_i$.
- $p \models \neg\varphi \Leftrightarrow p \not\models \varphi \overset{\text{ind.}}{\Leftrightarrow} q \not\models \varphi \Leftrightarrow q \models \neg\varphi$.

Now suppose $\forall \varphi \in \mathcal{L}_B^{n+1} (p \models \varphi \Leftrightarrow q \models \varphi)$ and $p \xrightarrow{a} p'$. Considering the symmetry in the definitions involved, all I have to show is that $\exists q' \in \mathbb{P}$ with $q \xrightarrow{a} q'$ and $p' \sim_n q'$. Let Q' be

$$\{q' \in \mathbb{P} \mid q \xrightarrow{a} q' \wedge p' \not\sim_n q'\}.$$

By Definition 1.1 Q' is a set. For every $q' \in Q'$ there must, by induction, be a formula $\varphi_{q'} \in \mathcal{L}_B^n$ with $p' \models \varphi_{q'}$ but $q' \not\models \varphi_{q'}$ (use negation if necessary). Now $p \models a \bigwedge_{q' \in Q'} \varphi_{q'} \in \mathcal{L}_B^{n+1}$ and therefore $q \models a \bigwedge_{q' \in Q'} \varphi_{q'}$. So there must be a $q' \in \mathbb{P}$ with $q \xrightarrow{a} q'$ and $q' \notin Q'$, which had to be shown. □

Comparing their modal characterizations ($=_B$ of $\leftrightarroweq$ and $=_B^\omega$ of $\sim$) one finds

$$p \leftrightarroweq q \;\Rightarrow\; p \sim q \;\Rightarrow\; p =_B^- q.$$

Theorem 2.1 in Hennessy and Milner [27,28] says, essentially, that for image finite processes the relation $\sim$ satisfies the defining properties of a bisimulation (cf. Definition 12). Inspired by this insight, Park [41] proposed the concise formulation of bisimulation equivalence employed in Definition 12. It follows immediately that if $p, q \in \mathbb{P}$ are image finite, then $p \leftrightarroweq q \Leftrightarrow p \sim q$. The following strengthening of this result is due to Hollenberg [32].

PROPOSITION 12.6. *Let $p, q \in \mathbb{P}$ and p is image finite. Then $p \leftrightarroweq q \Leftrightarrow p =_B^- q$.*

PROOF. Write pBq iff $p =_B^- q$ and p is image finite. It suffices to establish that B is a bisimulation.

- Suppose pBq and $q \xrightarrow{a} q'$. I have to show that $\exists r \in \mathbb{P}$ with $p \xrightarrow{a} r$ and $HM(r) = HM(q')$. Let R be

$$\{r \in \mathbb{P} \mid p \xrightarrow{a} r \wedge HM(r) \neq HM(q')\}.$$

As p is image finite, R is finite. For every $r \in R$ take a formula $\varphi_r \in HM(q') - HM(r)$ (note that if $\psi \in HM(r) - HM(q')$ then $\neg\psi \in HM(q') - HM(r)$). Now

$$a \bigwedge_{r \in R} \varphi_r \in HM(q) = HM(p),$$

so there must be a $r \in \mathbb{P}$ with $p \xrightarrow{a} r$ and $r \models \bigwedge_{r \in R} \varphi_r$. The latter implies $r \notin R$, i.e., $HM(r) = HM(q')$, which had to be shown.

- Suppose pBq and $p \xrightarrow{a} p'$. I have to show that $\exists q' \in \mathbb{P}$ with $q \xrightarrow{a} q'$ and $HM(p') = HM(q')$. Let S be

$$\{s \in \mathbb{P} \mid p \xrightarrow{a} s \wedge HM(s) \neq HM(p')\}.$$

As p is image finite, S is finite. For every $s \in S$ take a formula $\varphi_s \in HM(p') - HM(s)$. Now

$$a \bigwedge_{s \in S} \varphi_s \in HM(p) = HM(q),$$

so there must be a $q' \in \mathbb{P}$ with $q \xrightarrow{a} q'$ and $q' \models \bigwedge_{s \in S} \varphi_s$. By the previous item in this proof, $\exists r \in \mathbb{P}$ with $p \xrightarrow{a} r$ and $HM(r) = HM(q')$, hence $r \models \bigwedge_{s \in S} \varphi_s$. The latter implies $r \notin S$, so $HM(r) = HM(p')$. Thus $HM(p') = HM(q')$, which had to be shown. □

By Counterexample 12, a result like the one above does not hold for (ready) simulation semantics.

For the sake of completeness, two more variants of bisimulation equivalence can be considered. Let FB^* be characterized by the Hennessy–Milner logic augmented with formulas $\widetilde{X}$, and RB^* by the Hennessy–Milner logic augmented with formulas X, for $X \subseteq Act$.

Then $B^- \prec FB^* \prec RB^* \prec B^\omega \prec B$, and for image finite processes all five equivalences coincide. The strictness of these inclusions is given by Counterexamples 4, 11, 9 and 1:

PROPOSITION 12.7. *$CT \not\preceq B^-$, and hence $FB^* \not\preceq B^-$.*

PROOF. Counterexample 4 shows two processes with $CT(left) \neq CT(right)$. It remains to be shown that $HM(left) = HM(right)$, i.e., that for all $\varphi \in \mathcal{L}_{HM}$: $left \models \varphi \Leftrightarrow right \models \varphi$. Using Definition 12.5 it is sufficient to restrict attention to formulas φ which are of the form $a(\bigwedge_{i \in I} b_i \top \wedge \bigwedge_{j \in J} \neg b_j \top)$ with I and J finite sets of indices. It is not difficult to see that each such formula that is satisfied on one side is also satisfied on the other side. □

PROPOSITION 12.8. *$R \not\preceq FB^*$, and hence $RB^* \not\preceq FB^*$.*

PROOF. Counterexample 11, shows two processes with $R(with) \neq R(without)$. It remains to be shown that $FB^*(with) = FB^*(without)$. The argument is the same as in the previous proof, but this time focusing on formulas of the form $a(\widetilde{X} \wedge \bigwedge_{i \in I} i\top \wedge \bigwedge_{j \in J} \neg j\top)$ with I and J finite sets of numbers and X a possibly infinite set of numbers (= actions). □

PROPOSITION 12.9. *$S^\omega \not\preceq RB^*$, and hence $RB^\omega \not\preceq RB^*$.*

PROOF. Counterexample 9 shows two processes with $S^\omega(\mathit{with}) \neq S^\omega(\mathit{without})$. It remains to be shown that $RB^*(\mathit{with}) = RB^*(\mathit{without})$. The argument is the same as in the previous proofs – this time using formulas

$$a\left(\{b\} \wedge \bigwedge_{i \in I} bi\top \wedge \bigwedge_{j \in J} \neg bj\top\right)$$

with I and J finite sets of numbers. □

PROPOSITION 12.10. *$T^\infty \not\preceq B^\omega$, and hence $B \not\preceq B^\omega$. In addition, $PF \not\preceq B^\omega$.*

PROOF. Counterexample 1 shows two processes with $T^\infty(\mathit{left}) \neq T^\infty(\mathit{right})$. As remarked at the end of Section 7, also $PF(\mathit{left}) \neq PF(\mathit{right})$. It remains to be shown that $\mathit{left} =^\omega_B \mathit{right}$, i.e., that for all $n \in \mathbb{N}$: $\mathit{left} \sim_n \mathit{right}$. In order to establish $p \sim_n q$ for two trees p and q, the parts of p and q that are further than n edges away from the root play no rôle, and can just as well be omitted. As the cut versions of *left* and *right* are isomorphic, by Corollary 12.1 surely $\mathit{left} \sim_n \mathit{right}$. □

In addition one has $2S^- \prec B^-$, $F2S^* \prec FB^*$, $R2S^* \prec RB^*$, $2S^\omega \prec B^\omega$ and $2S \prec B$.

13. Tree semantics

DEFINITION 13. Let $g \in \mathbb{G}$. The *unfolding* of g is the graph $U(g) \in \mathbb{G}$ defined by
- NODES$(U(g))$ = PATHS(g),
- ROOT$(U(g))$ = ROOT(g), i.e., the empty path, starting and ending at the root of g,
- $(\pi, a, \pi') \in$ EDGES$(U(g))$ iff π' extends π by one edge, which is labelled a.

Two processes p and q are *tree equivalent*, notation $p =_U q$, if their unfoldings are isomorphic, i.e., if $U(G(p)) \cong U(G(p))$. In *tree semantics* (U) two processes are identified iff they are tree equivalent.

It is easy to see that the unfolding of any process graph is a tree, and the unfolding of a tree is isomorphic to itself. It follows that up to isomorphism every tree equivalence class of process graphs contains exactly one tree, which can be obtained from an arbitrary member of the class by means of unfolding.

PROPOSITION 13.1. *Let $g \in \mathbb{G}$. Then $U(g) \leftrightarroweq g$. Hence $g =_U h \Rightarrow g \leftrightarroweq h$.*

PROOF. As is easily verified, $\{(\pi, \mathit{end}(\pi)) \mid \pi \in \text{PATHS}(g)\}$ is a bisimulation between $U(g)$ and g. □

Tree semantics is employed in Winskel [50]. No plausible testing scenario or modal characterization is known for it. Proposition 13.1 shows that $B \preceq U$. That $B \not\succeq U$ follows from Counterexample 15.

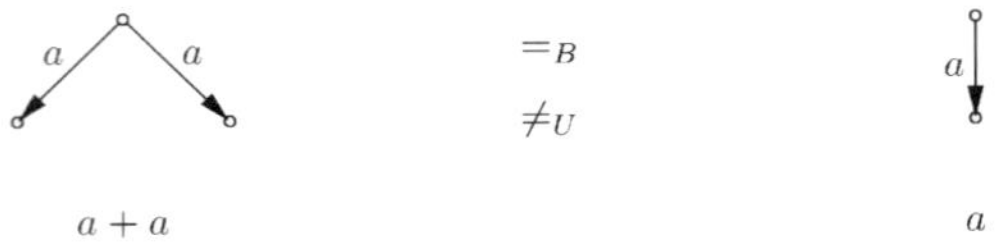

Counterexample 15. Bisimulation equivalent, but not tree equivalent.

Although above tree equivalence is defined entirely in terms of action relations, such a definition is in fact misleading, as action relations abstract from an aspect of system behaviour that tree semantics tries to capture. The problem can best be explained by considering the process 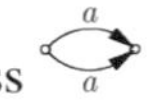 that can proceed from its initial to its final state by performing one of two different a-transitions. In tree semantics, such a process should be considered equivalent to the leftmost process of Counterexample 15, and hence different from the rightmost one. However, action relations only tell whether a process p can evolve into q by performing an a-action; they do not tell in how many ways this can happen. So in labelled transition systems as defined in this paper the mentioned process is represented as  and hence considered tree equivalent to the rightmost process of Counterexample 15. The mishap that ensues this way will be illustrated in Section 17.

Tree semantics on labelled transitions systems as in Section 1.1 is a sensible notion only if one knows that each transition in the system can be taken in only one way. In general, more satisfactory domains for defining tree equivalence are labelled transition systems in which the transitions (p, a, q) are equipped with a multiplicity, telling in how many different ways this transition can be taken, or process graphs $g = (\text{NODES}(g), \text{ROOT}(g), \text{EDGES}(g), \textit{begin}, \textit{end}, \textit{label})$ in which $\text{NODES}(g)$ and $\text{EDGES}(g)$ are sets, $\text{ROOT}(g) \in \text{NODES}(g)$, $\textit{begin}, \textit{end}: \text{EDGES}(g) \to \text{NODES}(g)$ and $\textit{label}: \text{EDGES}(g) \to \textit{Act}$. The functions *begin*, *end* and *label* associate with every edge a triple $(s, a, t) \in \text{NODES}(g) \times \textit{Act} \times \text{NODES}(g)$, but contrary to the situation in Definition 1.3 the identity of an edge is not completely determined by such a triple. On such process graphs, the notions PATHS, unfolding and tree equivalence are defined exactly as for the process graphs of Definition 1.3.

14. Possible worlds semantics

In Veglioni and De Nicola [49], a nondeterministic process is viewed as a set of deterministic ones: its *possible worlds*. Two processes are said to be *possible worlds equivalent* iff they have the same possible worlds. Two different approaches by which a nondeterministic process can be resolved into a set of deterministic ones need to be distinguished; I call them the *state-based* and the *path-based* approach. In the state-based approach a deterministic process h is obtained out of a nondeterministic process $g \in \mathbb{G}$ by choosing, for every state s of g and every action $a \in I(s)$ a single edge $s \xrightarrow{a} s'$. Now h is the reachable part of the subgraph of g consisting of the chosen edges. In the path-based approach on the other hand, one chooses for every path $\pi \in \text{PATHS}(g)$ and every action $a \in I(\textit{end}(\pi))$ a single edge $\textit{end}(\pi) \xrightarrow{a} s'$ to continue with. The chosen edges may now be different for different paths ending in the same state. The difference between the two approaches is illustrated in Counterexample 16. In the state-based approach, the process in the middle has two possible worlds, depending on which of the two b-edges is chosen. These worlds are essentially abc and ab^∞. In the path-based approach, the process in the middle has countably many possible worlds, namely ab^nc for $n \geqslant 1$ and ab^∞.

In [49], Veglioni and De Nicola take the state-based approach: "once we have resolved the underspecification present in a state s by saying, for example, $s \xrightarrow{a} s$, then, we cannot choose $s \xrightarrow{a} 0$ in the same possible world". However, they provide a denotational characterization of possible worlds semantics on finite processes, namely by inductively allocating sets of deterministic trees to BCCSP expressions (cf. Section 17), which can be regarded as path-based. In addition, they give an operational characterization of possible world semantics, essentially following the state-based approach outlined above. They claim that both characterizations agree. This, however, is not the case, as Counterexample 17 reveals a difference between the two approaches even on finite processes. In the path-based approach the process displayed has a possible world $acd + bce$ (i.e., a process with branches acd and bce), which it has not in the state-based approach. As it turns out, the complete axiomatization they provide w.r.t. BCCSP is correct for the path-based, denotational characterization, but is unsound for the state-based, operational characterization. To be precise: their operational semantics fails to be compositional w.r.t. BCCSP.

Counterexample 16 shows that a suitable formulation[8] of the state-based approach to possible worlds semantics is incomparable with any of the semantics encountered so far. The processes *left* and *middle* are state-based possible worlds equivalent, yet

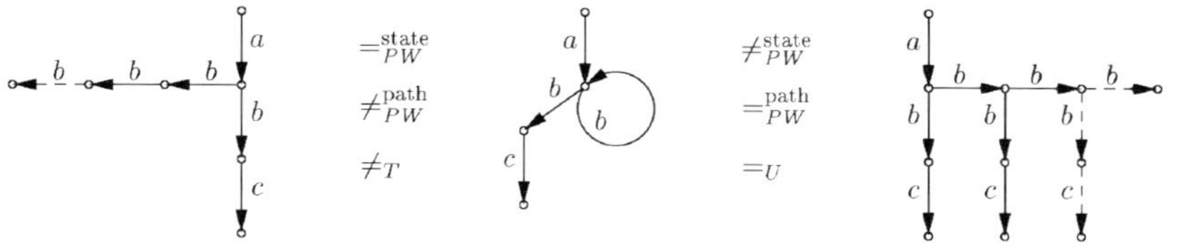

Counterexample 16. State-based versus path-based possible worlds equivalence.

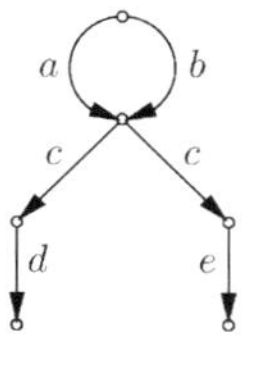

Counterexample 17. State-based versus path-based possible worlds equivalence for finite processes.

[8] Let two processes be *possible worlds equivalent* iff each possible world of the one is $\sim$-equivalent to a possible world of the other, where $\sim$ is any of the equivalences treated in this paper. Theorem 6 will imply that the choice of $\sim$ is immaterial.

$abbc \in T(middle) - T(left)$. Furthermore, the processes *right* and *middle* are tree equivalent, yet in the state-based approach one has $abbc \in PW(right) - PW(middle)$.

Below I propose a formalization of the path-based approach to possible worlds semantics that, on finite processes, agrees with the denotational characterization of [49].

DEFINITION 14. A process p is a *possible world* of a process q if p is deterministic and $p \sqsubseteq_{RS} q$. Let $PW(q)$ denote the class of possible worlds of q. Two processes q and r are *possible worlds equivalent*, notation $q =_{PW} r$, if $PW(q) = PW(r)$. In *possible worlds semantics* (*PW*) two processes are identified iff they are possible worlds equivalent. Write $q \sqsubseteq_{PW} r$ iff $PW(q) \subseteq PW(r)$.

It can be argued that the philosophy underlying possible worlds semantics is incompatible with the view on labelled transition systems taken in this paper. The informal explanation of the action relations in Section 1.1 implies for instance that the right-hand process graph of Counterexample 8 has a state in which a has happened already and both bc and bd are possible continuations. In the possible worlds philosophy on the other hand, this process graph is just a compact representation of the set of deterministic processes $\{abc, abd\}$. None of the two processes in this set has such a state.

This could be a reason not to treat possible worlds semantics on the same footing as the other semantics of this paper. However, one can give up on thinking of non-deterministic processes as sets of deterministic ones, and justify possible worlds semantics – at least the path-based version of Definition 14 – by an appropriate testing scenario. This makes it fit in the present paper.

Testing scenario. A testing scenario for possible worlds semantics can be obtained by making one change in the reactive testing scenario of failure simulation semantics. Namely in each state only as many copies of the process can be made as there are actions in *Act*, and, for $a \in Act$, the first test on copy p_a of p is pressing the a-button. If it goes down, one goes on testing that copy, but is has already changed its state; if it does not go down, the test on p_a ends.

Modal characterization. On well-founded processes, a modal characterization of possible worlds semantics can be obtained out of the modal characterization of ready simulation semantics by changing the modality $\bigwedge_{i \in I} \varphi_i$ into $\bigwedge_{a \in X} a\varphi_a$ with $X \subseteq Act$. Possible worlds of a well-founded process p can be simply encoded as modal formulas in the resulting language. Probably, this modal characterization applies to image finite processes as well. For processes that are neither well-founded nor image finite this characterization is not exact, as it fails to distinguish the two processes of Counterexample 1.

Classification. $RT \prec PW \prec RS$. *PW* is independent of *S*, *CS* and *PF*.

PROOF. "$PW \preceq RS$"[9] follows by the transitivity of $\sqsubseteq_{RS}$.

[9] The counterexample against "$PW \preceq RS$" given in [49] is incorrect. The two processes displayed there are not ready simulation equivalent.

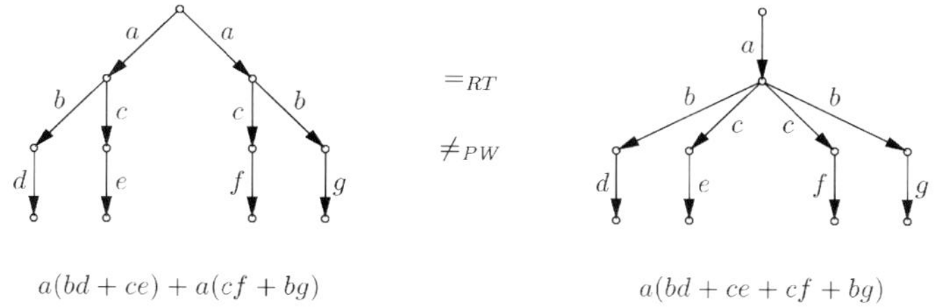

Counterexample 18. Ready trace equivalent, but not possible worlds equivalent.

"$RT \preceq PW$" holds as σ is a ready trace of $p \in \mathbb{P}$ iff it is a ready trace of a possible world of p.

"$S \not\preceq PW$" (and hence "$RS \not\preceq PW$") follows from Counterexample 8. There $S(\mathit{left}) \neq S(\mathit{right})$, but $PW(\mathit{left}) = PW(\mathit{right}) = \{abc, abd\}$.

"$PF \not\preceq PW$" follows since $PF \not\preceq RS$.

"$CS \not\succeq PW$" follows since $CS \not\succeq RT$, and "$PF \not\succeq PW$" since $PF \not\succeq RT$.

Finally, "$RT \not\succeq PW$" follows from Counterexample 18, taken from [49]. There the first process denotes two possible worlds, whereas the second one denotes four. □

Infinite processes. The version of possible worlds semantics defined above is the infinitary one. Note that $RT^\infty \prec PW$. Exactly as above one even establishes

$$p \sqsubseteq_{RS} q \Rightarrow p \sqsubseteq_{PW} q \Rightarrow p \sqsubseteq_{RT}^{\infty} q,$$

i.e., $RT^\infty \preceq^* PW \preceq^* RS$. Finitary versions could be defined by means of the modal characterization given above. I will not pursue this here.

15. Summary

In Sections 2–14 fifteen semantics were defined that are different for finitely branching processes. These are abbreviated by $T, CT, F^1, F, R, FT, RT, PF, S, CS, RS, PW, 2S, B$ and U. For each of these semantics $\mathcal{O}$, except U, a modal language $\mathcal{L}_\mathcal{O}$ (a set of modal formulas φ) has been defined:

$\mathcal{L}_T \quad \varphi ::= \top \mid a\varphi' (\varphi' \in \mathcal{L}_T)$
  the (*partial*) *trace formulas*

$\mathcal{L}_{CT} \quad \varphi ::= \top \mid a\varphi' (\varphi' \in \mathcal{L}_{CT}) \mid 0$
  the *completed trace formulas*

$\mathcal{L}_{F^1} \quad \varphi ::= \top \mid a\varphi' (\varphi' \in \mathcal{L}_{F^1}) \mid \widetilde{a} (a \in Act)$
  the *singleton-failure formulas*

$\mathcal{L}_F \quad \varphi ::= \top \mid a\varphi' (\varphi' \in \mathcal{L}_F) \mid \widetilde{X} (X \subseteq Act)$
  the *failure formulas*

obtained by allowing machines that are equipped with (only) those features corresponding with its modal characterization.

I write $\mathcal{O} \preceq \mathcal{N}$ if semantics $\mathcal{O}$ makes at least as much identifications as semantics $\mathcal{N}$, i.e., if $=_{\mathcal{O}} \supseteq =_{\mathcal{N}}$. Clearly, if $\mathcal{L}_{\mathcal{O}}$ is a sublanguage of $\mathcal{L}_{\mathcal{N}}$ it must be that $\mathcal{O} \preceq \mathcal{N}$. This immediately yields[11] the following theorem, whose proof has also appeared in the various subsections entitled "classification".

THEOREM 1. *$T \preceq CT \preceq F \preceq R \preceq RT$, $T \preceq F^1 \preceq F \preceq FT \preceq RT \preceq PW \preceq RS \preceq 2S \preceq B \preceq U$, $R \preceq PF \preceq 2S$, $T \preceq S \preceq CS \preceq RS$ and $CT \preceq CS$.*

Theorem 1 is illustrated in Figure 1. There, however, singleton-failures semantics and completed simulation semantics are missing, since they did not occur in the literature, and appear to be of minor interest. The theorem applies to any labelled transition system $(\mathbb{P}, \rightarrow)$. Whether the inclusions are strict depends on the choice of $(\mathbb{P}, \rightarrow)$. In the subsections "classification" a number of counterexamples have been presented, showing that on $\mathbb{G}$ all semantic notions mentioned in Theorem 1 are different and $\mathcal{O} \preceq \mathcal{N}$ holds only if this follows from that theorem. Moreover, all relevant examples use finite processes only.

Let $\mathbb{H}$ be the set of finite connected process graphs. Here *finite* is used in the sense of Definition 1.2; a process graph $g \in \mathbb{G}$ is finite iff PATHS(g) is finite, which is the case iff g is acyclic and has only finitely many nodes and edges. Now the next theorem follows.

THEOREM 2. *Let $\mathcal{O}, \mathcal{N} \in \{T, CT, F^1, F, R, FT, RT, PF, S, CS, RS, PW, 2S, B, U\}$. Then $\mathcal{O} \not\preceq \mathcal{N}$, and even $\mathcal{O} \not\preceq_{\mathbb{H}} \mathcal{N}$, unless $\mathcal{O} \preceq \mathcal{N}$ follows from Theorem* 1 (*and the fact that $\preceq$ is a partial order*).

The following theorem says that the inclusion hierarchy of the preorders T, CT, F^1, F, R, FT, RT, PF, S, CS, RS, PW, $2S$ and B is the same as the inclusion hierarchy of the corresponding equivalences (there is no preorder for U).

THEOREM 3. *Let $\mathcal{O}, \mathcal{N} \in \{T, CT, F^1, F, R, FT, RT, PF, S, CS, RS, PW, 2S, B\}$. Then $\mathcal{O} \preceq^* \mathcal{N}$ iff $\mathcal{O} \preceq \mathcal{N}$.*

PROOF. Clearly, if $\mathcal{L}_{\mathcal{O}}$ is a sublanguage of $\mathcal{L}_{\mathcal{N}}$ it must be that $p \sqsubseteq_{\mathcal{N}} q \Rightarrow p \sqsubseteq_{\mathcal{O}} q$, i.e., $\mathcal{O} \preceq^* \mathcal{N}$. This yields "if" (except for $RT \preceq^* PW \preceq^* RS$, which have been established in Section 14). "Only if" is immediate (cf. Section 1.4). □

When the restriction to finitely branching processes is dropped, there exists a finitary and an infinitary variant of each of these semantics, depending on whether or not infinite observations are taken into account (I do not consider the finitary version of PW or the infinitary version of F^1 though). These versions are notationally distinguished by means of superscripts "*" and "∞", respectively; the unsubscripted abbreviation will refer to the infinitary versions in case of simulation-like semantics (treated in Sections 8–12) and to the finitary versions for the *decorated trace semantics* (treated in Section 2–7). The modal

[11] The statements involving PW and U do not follow this way, but have been established in Sections 13 and 14.

characterizations summarized above apply to the default (= unsubscripted) versions. Modal characterizations of T^∞, CT^∞, F^∞, R^∞, respectively FT^∞ and RT^∞, are obtained by allowing traces, respectively failure traces or ready traces, of infinite length as modal formulas; a modal characterization of PF^∞ is obtained by replacing the reference to T by one to T^∞. Modal characterizations of S^*, CS^*, etc. are obtained by requiring the index sets I to be finite. For the simulation-like semantics also an intermediate variant is considered – superscripted with "ω" – based on the assumption that observers can investigate arbitrary many copies of a process in parallel, but have only a finite amount of time to do so. Modally, this corresponds to the restriction to modal formulas with a finite upperbound on the number of nestings of the $a\varphi$ construct. For the semantics that incorporate refusal sets, the finitary versions come in two variants, depending on whether the refusal sets are required to be finite (superscript "–") or not (the default assumption). A similar distinction is made for semantics where menus of actions can be observed: in R^-, RT^- and RS^- the modal formula X is replaced by $\bigwedge_{a\in Y} \neg a\top \wedge \bigwedge_{a\in Z} a\top$, where the sets of actions Y and Z are required to be finite. Finally, whereas failure simulation semantics, modally characterized by

$$\mathcal{L}_{FS} \quad \varphi ::= a\varphi'(\varphi' \in \mathcal{L}_{FS}) \mid \bigwedge_{i\in I} \varphi_i (\varphi_i \in \mathcal{L}_{FS}) \mid \widetilde{X}(X \subseteq Act)$$

the *failure simulation formulas*,

coincides with ready simulation semantics, its finitary version (FS^*) can be distinguished from RS^*. The intermediate notions FS^ω and RS^ω coincide again, as do FS^- and RS^-. By analogy, new semantics FB^* and RB^* can be defined by adding the modality $\widetilde{X}$ respectively X to $\mathcal{L}_B^-$. These modalities would be redundant on top of $\mathcal{L}_B^\omega$ or $\mathcal{L}_B$. A similar situation occurs for 2-nested simulation.

All semantics encountered are displayed in Figure 9, in which the $\preceq$-relation is represented by solid and dotted arrows.

THEOREM 4. *For all semantics $\mathcal{O}$ and $\mathcal{N}$ defined so far, the formula $\mathcal{O} \preceq \mathcal{N}$ holds iff there is a path $\mathcal{O} \to \cdots \to \mathcal{N}$ (consisting of solid and dotted arrows alike) in Figure 9. Furthermore, semantics connected by dotted arrows coincide for image finite processes.*

PROOF. That $T^\infty \preceq S$ has been established in Proposition 8.3; that $CT^\infty \preceq CS$, $RT^\infty \preceq RS$ and $PF^\infty \preceq 2S$ follows in the same way. $R^\infty \preceq PW \preceq RS$ has been established in Section 14. All other implications $\mathcal{O} \preceq \mathcal{N}$ follow from the observation that the modal language $\mathcal{L}_\mathcal{O}$ is included in $\mathcal{L}_\mathcal{N}$. The latter statement has been established in Propositions 2.4, 4.6, 5.3, 6.5, 7.4, 7.5, 7.6, 8.3, 9.6, 11.2 and 12.6 (except that the case of possible-futures semantics is left to the reader). In order to show that on $\mathbb{G}$ there are no inclusions that are not indicated in Figure 9, is suffices, in view of Theorem 2, the already established parts of Theorem 4, and the fact that $\preceq$ is a partial order, to show that $CT \not\preceq B^-$, $R \not\preceq FB^*$, $S^\omega \not\preceq RB^*$, $T^\infty \not\preceq B^\omega$, $PF \not\preceq B^\omega$ and $T^\infty \not\preceq PF$. This has been done in Propositions 12.7, 12.8, 12.9 and 12.10, and at the end of Section 7. □

Again, the inclusion hierarchy for the preorders is the same as for the equivalences.

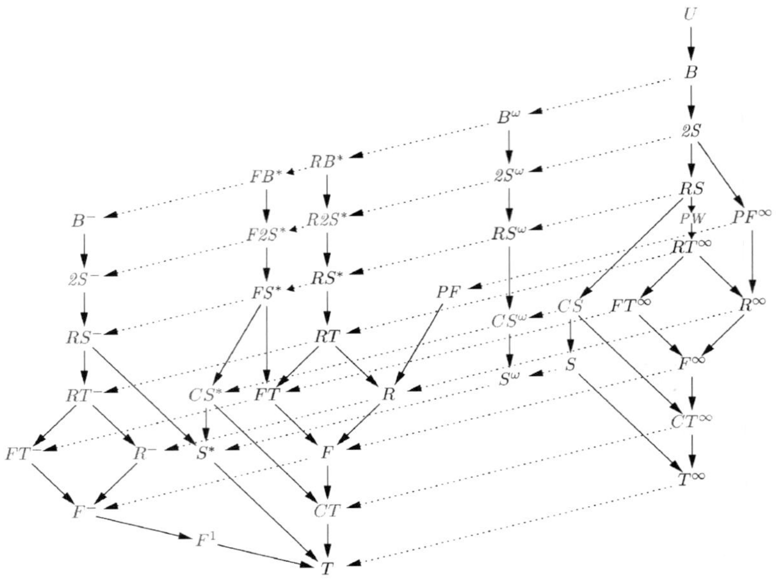

Figure 9. The infinitary linear time – branching time spectrum.

THEOREM 5. *For all semantics $\mathcal{O}$ and $\mathcal{N}$ defined so far, the formula $\mathcal{O} \preceq^* \mathcal{N}$ holds iff there is a path $\mathcal{O} \to \cdots \to \mathcal{N}$ (consisting of solid and dotted arrows alike) in Figure* 9.

PROOF. That $p \sqsubseteq_S q \Rightarrow p \sqsubseteq_T^{\infty} q$ has been established in Proposition 8.3; that $p \sqsubseteq_{CS} q \Rightarrow p \sqsubseteq_{CT}^{\infty} q$, $p \sqsubseteq_{RS} q \Rightarrow p \sqsubseteq_{RT}^{\infty} q$ and $p \sqsubseteq_{2S} q \Rightarrow p \sqsubseteq_{PF}^{\infty} q$ follows in the same way. In Section 14 it has been established that $p \sqsubseteq_{RS} q \Rightarrow p \sqsubseteq_{PW} q \Rightarrow p \sqsubseteq_{RT}^{\infty} q$. All other implications $p \sqsubseteq_{\mathcal{O}} q \Rightarrow p \sqsubseteq_{\mathcal{N}} q$ follow from the observation that the modal language $\mathcal{L}_{\mathcal{O}}$ is included in $\mathcal{L}_{\mathcal{N}}$. The "only if" part is an immediate consequence of Theorem 4. □

16. Deterministic and saturated processes

If the labelled transition system $\mathbb{P}$ on which the semantic equivalences of Section 15 are defined is large enough, then they are all different and $\mathcal{O} \preceq_{\mathbb{P}} \mathcal{N}$ holds only if this is indicated in Figure 9. However, for certain labelled transition systems much more identifications can be made. Is has been remarked already that for image finite processes all semantics that are connected by dotted arrows coincide. In this section various other classes of processes are examined on which parts of the linear time – branching time spectrum collapse. All results of this section, expect for Propositions 16.1 and 16.2, will be used in the completeness proofs in Section 17.

Recall that a process p is *deterministic* iff $p \xrightarrow{\sigma} q \wedge p \xrightarrow{\sigma} r \Rightarrow q = r$.

REMARK. If p is deterministic and $p \xrightarrow{\sigma} p'$ then also p' is deterministic. Hence any domain of processes on which action relations are defined, has a subdomain of deterministic processes with the inherited action relations. (A similar remark can be made for image finite processes.)

PROOF. Suppose $p' \xrightarrow{\rho} q$ and $p' \xrightarrow{\rho} r$. Then $p \xrightarrow{\sigma\rho} q$ and $p \xrightarrow{\sigma\rho} r$, so $q = r$. □

THEOREM 6 (Park [41]). *On a domain of deterministic processes all semantics in the infinitary linear time – branching time spectrum coincide.*

PROOF. Because of Theorem 4 it suffices to show that $g =_T h \Rightarrow g =_U h$ for any two deterministic process graphs $g, h \in \mathbb{G}$. Note that a process graph $g \in \mathbb{G}$ is deterministic iff for every trace $\sigma \in T(g)$ there is exactly one path $\pi \in \text{PATHS}(g)$ with $T(\pi) = \sigma$. Now let g and h be deterministic process graphs with $g =_T h$. Then the relation $i \subseteq \text{PATHS}(g) \times \text{PATHS}(h)$ that relates $\pi \in \text{PATHS}(g)$ with $\pi' \in \text{PATHS}(h)$ iff $T(\pi) = T(\pi')$ clearly is an isomorphism between $U(g)$ and $U(h)$. □

Thus, if two processes p and q are both deterministic, then $p =_T q \Leftrightarrow p =_F^1 q \Leftrightarrow p \leftrightarroweq q \Leftrightarrow p =_U q$. In case only one of them is deterministic, this cannot be concluded, for in Counterexamples 2 and 15 the right-hand processes are deterministic. However, in such cases one still has $p =_F^1 q \Leftrightarrow p \leftrightarroweq q$. In fact, a stronger statement holds: if q is deterministic, then $p \sqsubseteq_F^1 q \Leftrightarrow p \leftrightarroweq q$.

LEMMA 16.1. *If $p \sqsubseteq_F^1 q$ then $I(p) = I(q)$.*

PROOF. Let $p \sqsubseteq_F^1 q$, i.e., $T(p) \subseteq T(q)$ and $F^1(p) \subseteq F^1(q)$. Then $a \in I(p) \Leftrightarrow a \in T(p) \Rightarrow a \in T(q) \Leftrightarrow a \in I(q)$ and $a \notin I(p) \Leftrightarrow \langle \varepsilon, a \rangle \in F^1(p) \Rightarrow \langle \varepsilon, a \rangle \in F^1(q) \Leftrightarrow a \notin I(q)$. □

PROPOSITION 16.1. *If q is deterministic then $p \sqsubseteq_F^1 q \Leftrightarrow p \leftrightarroweq q$.*

PROOF. Let R be the binary relation on $\mathbb{P}$ defined by pRq iff q is deterministic and $p \sqsubseteq_F^1 q$, then it suffices to prove that R is a bisimulation. Suppose pRq and $p \xrightarrow{a} p'$. Then $a \in I(p) = I(q)$. So there is a process $q' \in \mathbb{P}$ with $q \xrightarrow{a} q'$. As q is deterministic, so is q'. Now let $\langle \sigma, b \rangle \in F^1(p')$. Then $\exists r$: $p' \xrightarrow{\sigma} r \wedge b \notin I(r)$. Hence $p \xrightarrow{a\sigma} r$ and $\langle a\sigma, b \rangle \in F^1(p) \subseteq F^1(q)$. So there must be a process s with $q \xrightarrow{a\sigma} s \wedge b \notin I(s)$. By the definition of the generalized action relations $\exists t$: $q \xrightarrow{a} t \xrightarrow{\sigma} s$, and since q is deterministic, $t = q'$. Thus $\langle \sigma, b \rangle \in F^1(q')$. From this it follows that $F^1(p') \subseteq F^1(q')$. Similarly one finds $T(p') \subseteq T(q')$, hence $p' \sqsubseteq_F^1 q'$.

Now suppose pRq and $q \xrightarrow{a} q'$. Then $a \in I(q) = I(p)$. So there is a process $p' \in \mathbb{P}$ with $p \xrightarrow{a} p'$. Exactly as above it follows that q' is deterministic and $p' \sqsubseteq_F^1 q'$. □

Call a process p *deterministic up to* $\equiv$, for $\equiv$ an equivalence relation or preorder, if there exists a deterministic process p' with $p \equiv p'$. Now the above proposition implies that determinism up to $\leftrightarrow$ coincides with determinism up to $=^1_F$, and even with determinism up to $\sqsubseteq^1_F$. In contrast, *any* process is deterministic up to $=_T$, as the canonical graphs constructed in the proof of Proposition 2.3 are deterministic. Furthermore, determinism up to $=_U$ is just determinism, for $g \in \mathbb{G}$ is deterministic iff $U(g)$ is, and determinism is preserved under isomorphism.

The following notion of *determinacy* was proposed in Engelfriet [18].

DEFINITION 16.1. Let $\equiv$ be an equivalence relation on $\mathbb{P}$. A process $p \in \mathbb{P}$ is $\equiv$-*determinate* if $p \xrightarrow{\sigma} q \wedge p \xrightarrow{\sigma} r \Rightarrow q \equiv r$.

Note that $=$-determinacy is determinism. Furthermore, if $\mathcal{O} \leqslant \mathcal{N}$ then $=_{\mathcal{O}}$-determinacy is implied by $=_{\mathcal{N}}$-determinacy. Besides $=_T$-determinacy, $=_F$-determinacy and $\leftrightarrow$-determinacy, Engelfriet also considers $=_I$-determinacy, where $=_I$ is given by $p =_I q$ iff $I(p) = I(q)$. Clearly $=_I$ is coarser than any of the equivalences of Section 15: $p =_T q \Rightarrow p =_I q$. Moreover, $=_I$ is even coarser than most of the preorders: $p \sqsubseteq^1_F q \Rightarrow p =_I q$, as established in Lemma 16.1.

Engelfriet established the following three results:

(1) $\leftrightarrow$-determinacy and $=_I$-determinacy are the same. Hence $\equiv$-determinacy is the same for all equivalences $\equiv$ of Section 15, except U. Therefore, he just calls this *determinacy*.

(2) For determinate processes, bisimulation equivalence and trace equivalence (and hence all equivalences in between) are the same.

(3) Determinacy is preserved under failures equivalence (and hence under $\leftrightarrow$). Even stronger, if q is determinate and $p \sqsubseteq_F q$, then p is determinate and $p \leftrightarrow q$. (In [18], $\sqsupseteq_F$ is written $\subseteq_{\mathrm{f}}$.)

Using Proposition 16.1 I show that both $=_I$-determinacy and $\leftrightarrow$-determinacy coincide with determinism up to $\leftrightarrow$, from which (1), (2) and (3) follow.

PROPOSITION 16.2. *Let $p \in \mathbb{P}$. The following are equivalent*:

(a) *p is $\leftrightarrow$-determinate*;

(b) *p is $=_I$-determinate*;

(c) *p is deterministic up to $=_R$*;

(d) *p is deterministic up to $\leftrightarrow$.*

PROOF. "(a) $\Rightarrow$ (b)" is immediate as $=_I$ is coarser than $\leftrightarrow$.

"(b) $\Rightarrow$ (c)": Suppose p is $=_I$-determinate. Let $G(T(p))$ be the canonical graph of the trace set of p as defined in the proof of Proposition 2.3. By construction, $G(T(p))$ is deterministic and $T(p) = T(G(T(p)))$. It remains to be shown that $p =_R G(T(p))$.

As p is $=_I$-determinate, one has $\langle \sigma, X\rangle, \langle \sigma, Y\rangle \in R(p) \Rightarrow X = Y$. Hence $\langle \sigma, X\rangle \in R(p)$ iff $\sigma \in T(p) \wedge X = \{a \in Act \mid \sigma a \in T(p)\}$, i.e., $R(p)$ is completely determined by $T(p)$. As also $G(T(p))$ is $=_I$-determinate (for it is even deterministic), also $R(G(T(p)))$ is completely determined by $T(G(T(p)))$: $\langle \sigma, X\rangle \in R(G(T(p)))$ iff $\sigma \in T(G(T(p))) \wedge X = \{a \in Act \mid \sigma a \in T(G(T(p)))\}$. It follows that $R(p) = R(G(T(p)))$.

"(c) $\Rightarrow$ (d)" has been established in Proposition 16.1.

"(d) $\Rightarrow$ (a)": Suppose $p \underline{\leftrightarrow} q$ and q is deterministic. Let $p \xrightarrow{\sigma} p'$ and $p \xrightarrow{\sigma} p''$. Then $\exists q'$: $q \xrightarrow{\sigma} q' \wedge p' \underline{\leftrightarrow} q'$ and $\exists q''$: $q \xrightarrow{\sigma} q'' \wedge p'' \underline{\leftrightarrow} q''$. As q is deterministic, $q' = q''$. Hence $p' \underline{\leftrightarrow} p''$. It follows that p is $\underline{\leftrightarrow}$-determinate. □

Now (1) is part of Proposition 16.2. (2) is a generalization of Theorem 6, that is now implied by it: Suppose p and q are determinate and $p =_T q$. By Proposition 16.2 there are deterministic processes p' with $p \underline{\leftrightarrow} p'$ and q' with $q \underline{\leftrightarrow} q'$. Hence $p' =_T q'$, so by Theorem 6 $p' \underline{\leftrightarrow} q'$. Thus $p \underline{\leftrightarrow} q$, yielding (2). (3) holds even for F^1 instead of F. For let q be determinate and $p \sqsubseteq_F^1 q$. Then there is a deterministic process q' with $q \underline{\leftrightarrow} q'$. Hence $p \sqsubseteq_F^1 q'$. By Proposition 16.1 $p \underline{\leftrightarrow} q'$, so p is determinate and $p \underline{\leftrightarrow} q$.

Note that a process p is deterministic iff for $\pi, \pi' \in \text{PATHS}(G(p))$ one has $T(\pi) = T(\pi') \Rightarrow \pi = \pi'$. For this reason, determinism could have been called *trace determinism*, and the notions of *ready trace determinism* and *completed trace determinism* can be defined analogously.

DEFINITION 16.2. A process p is *ready trace deterministic* if for $\pi, \pi' \in \text{PATHS}(G(p))$ one has $RT(\pi) = RT(\pi') \Rightarrow \pi = \pi'$. It is *completed trace deterministic* if for $\pi, \pi' \in \text{PATHS}(G(p))$ one has $T(\pi) = T(\pi') \wedge (I(end(\pi)) = \emptyset \Leftrightarrow I(end(\pi')) = \emptyset) \Rightarrow \pi = \pi'$.

A process $p \in \mathbb{P}$ is ready trace deterministic iff there is are no $p', q, r \in \mathbb{P}$ and $a \in Act$ such that p' is reachable from p, $p' \xrightarrow{a} q$, $p' \xrightarrow{a} r$, $I(q) = I(r)$ and $q \neq r$. For trace determinism the condition $I(q) = I(r)$ is dropped, and for completed trace determinism it is weakened to $I(q) = \emptyset \Leftrightarrow I(r) = \emptyset$. Note that if p is ready (or completed) trace deterministic and $p \xrightarrow{\sigma} p'$ then so is p'. Now the following variants of Theorem 6 can be established.

PROPOSITION 16.3. *If g and $h \in \mathbb{G}$ are ready trace deterministic then $g =_{RT} h \Leftrightarrow g =_U h$. Likewise, if g and $h \in \mathbb{G}$ are completed trace deterministic then $g =_{CT} h \Leftrightarrow g =_U h$.*

PROOF. Let g and h be ready trace deterministic process graphs with $g =_{RT} h$. Then the relation $i \subseteq \text{PATHS}(g) \times \text{PATHS}(h)$ that relates $\pi \in \text{PATHS}(g)$ with $\pi' \in \text{PATHS}(h)$ iff $RT(\pi) = RT(\pi')$ clearly is an isomorphism between $U(g)$ and $U(h)$. The proof of the second statement goes likewise. □

For completed trace deterministic processes, the equivalences $=_T$ and $=_{CT}$ are different, as can be seen from Counterexample 2. For ready trace deterministic processes, the equivalences $=_T$, $=_{CT}$, $=_F^1$, $=_F$, $=_{FT}$, $=_R$, $=_{RT}$, $=_S$ and $=_{CS}$ are all different, as can be seen from Counterexamples 2, 3, 5, 6 and 13. Theorem 6 and Proposition 16.3 do not generalize to the corresponding preorders, for in Counterexample 19 one finds two deterministic processes *middle* and *right* with *middle* $\sqsubseteq_{CT}$ *right* but *middle* $\not\sqsubseteq_B$ *right*, and in Counterexample 3 one finds two ready trace deterministic processes *right* and *left* with *right* $\sqsubseteq_{RT}$ *left* but *right* $\not\sqsubseteq_B$ *left*. However, the following variants of these results can be obtained.

PROPOSITION 16.4. *If q is ready trace deterministic then $p \sqsubseteq_{RT} q \Leftrightarrow p \sqsubseteq_{RS} q$.*

Likewise, if q is completed trace deterministic then $p \sqsubseteq_{CT} q \Leftrightarrow p \sqsubseteq_{CS} q$, and if q is (trace) deterministic then $p \sqsubseteq_T q \Leftrightarrow p \sqsubseteq_S q$.

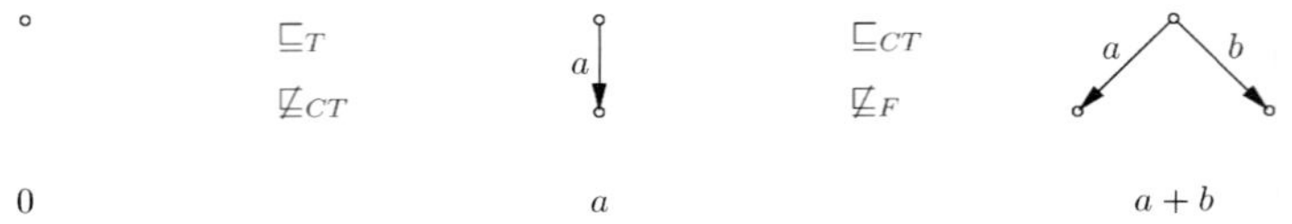

Counterexample 19. The trace, completed trace, and failures preorders are all different on deterministic processes.

PROOF. Let R be the binary relation on $\mathbb{P}$ given by pRq if q is ready trace deterministic and $p \sqsubseteq_{RT} q$. For the first statement it suffices to prove that R is a ready simulation. Clearly pRq implies $I(p) = I(q)$. Now suppose pRq and $p \xrightarrow{a} p'$. Let X be $I(p')$. Then $aX \in RT(p) \subseteq RT(q)$. So there is a process $q' \in \mathbb{P}$ with $q \xrightarrow{a} q'$ and $I(q') = X$. Now let $\sigma \in RT(p')$. Then $\exists r$: $p' \overset{\sigma}{\ast\!\!\longrightarrow} r$. Hence $p \overset{aX\sigma}{\ast\!\!\longrightarrow} r$ and $aX\sigma \in RT(p) \subseteq RT(q)$. So there must be a process s with $q \overset{aX\sigma}{\ast\!\!\longrightarrow} s$. By the definition of the ready trace relations $\exists t$: $q \xrightarrow{a} t \overset{\sigma}{\ast\!\!\longrightarrow} s \wedge I(t) = X$, and since q is ready trace deterministic, $t = q'$. Thus $\sigma \in RT(q')$. From this it follows that $RT(p') \subseteq RT(q')$, implying $p'Rq'$.

This finishes the proof of the first statement. The proofs of the other two statements go the same way (but involving a trivial case distinction for the completed trace deterministic one). □

Together, Propositions 16.1 and 16.4 imply that on a domain of deterministic processes only three of the preorders of Section 15 are different, namely $\sqsubseteq_T$, $\sqsubseteq_{CT}$ and $\sqsubseteq_F^1$, coinciding with $\sqsubseteq_S$, $\sqsubseteq_{CS}$ and $\sqsubseteq_B$, respectively. That these three are indeed different is shown in Counterexample 19.

DEFINITION 16.3. A process p is *cross saturated* if $p \xrightarrow{\sigma} q \xrightarrow{a} r \wedge p \xrightarrow{\sigma} s \xrightarrow{a} t \Rightarrow q \xrightarrow{a} t$.

Thus a process graph $g \in \mathbb{G}$ is cross saturated iff for any $\pi, \pi' \in \text{PATHS}(g)$ and $a \in Act$ such that $a \in I(end(\pi))$ and $T(\pi') = T(\pi)a$ one has $(end(\pi), a, end(\pi')) \in \text{EDGES}(g)$.

PROPOSITION 16.5. *If g and $h \in \mathbb{G}$ are cross saturated then $g =_R h \Leftrightarrow g \leftrightarroweq h$.*

PROOF. Let g and $h \in \mathbb{G}$ be cross saturated and suppose that $R(g) = R(h)$. Define the relation $R \subseteq \text{NODES}(g) \times \text{NODES}(h)$ by sRt iff there are $\pi \in \text{PATHS}(g)$ and $\rho \in \text{PATHS}(h)$ with $end(\pi) = s$, $end(\rho) = t$ and $R(\pi) = R(\rho)$. It suffices to show that R is a bisimulation between g and h.

As $I(\text{ROOT}(g)) = I(\text{ROOT}(h))$ one clearly has $\text{ROOT}(g) R \text{ROOT}(h)$.

Now suppose sRt and $(s, a, s') \in \text{EDGES}(g)$. Let π and ρ be such that $end(\pi) = s$, $end(\rho) = t$ and $R(\pi) = R(\rho)$, and let π' be the extension of π with (s, a, s'). Now $a \in I(end(\pi)) = I(end(\rho))$. Choose $\rho' \in \text{PATHS}(h)$ with $R(\rho') = R(\pi')$ (using that $R(g) = R(h)$). Then $T(\rho') = T(\pi') = T(\pi)a = T(\rho)a$, so $(t, a, end(\rho')) \in \text{EDGES}(h)$. Moreover, $s' R\, end(\rho')$.

The remaining requirement of Definition 12.2 follows by symmetry. □

PROPOSITION 16.6. *If $h \in \mathbb{G}$ is cross saturated then $g \sqsubseteq_R h \Leftrightarrow g \sqsubseteq_{RS} h$.*

PROOF. Exactly as above. □

DEFINITION 16.4. A process p is *saturated* if $\langle \sigma, X \rangle \in R(p) \wedge \langle \sigma, Y \cup Z \rangle \in R(p) \Rightarrow \langle \sigma, X \cup Y \rangle \in R(p)$.

PROPOSITION 16.7. *If p is finitely branching and q is saturated then $p \sqsubseteq_F q \Leftrightarrow p \sqsubseteq_R q$. Thus if both p and q are finitely branching and saturated then $p =_F q \Leftrightarrow p =_R q$.*

PROOF. Suppose p is finitely branching, q is saturated and $p \sqsubseteq_F q$. Let $\langle \sigma, Y \rangle \in R(p)$. Then Y is finite. In case $Y = \emptyset$ one has $\langle \sigma, Act \rangle \in F(p) \subseteq F(q)$, implying $\langle \sigma, \emptyset \rangle \in R(q)$, as desired. So assume $Y \neq \emptyset$. Then, for all $a \in Y$, $\langle \sigma a, \emptyset \rangle \in F(p) \subseteq F(q)$ so $\exists Z_a \subseteq Act$ with $\langle \sigma, \{a\} \cup Z_a \rangle \in R(q)$. Hence, using Definition 16.4 with $Z = \emptyset$, one obtains $\langle \sigma, Y \cup \bigcup_{a \in Y} Z_a \rangle \in R(q)$. As $\langle \sigma, Act - Y \rangle \in F(p) \subseteq F(q)$ it must be that $\langle \sigma, X \rangle \in R(q)$ for some $X \subseteq Y$. Now Definition 16.4 gives $\langle \sigma, Y \rangle \in R(q)$. □

DEFINITION 16.5. A process p is *RT-saturated* if

$$\sigma X \rho \in RT_N(p) \wedge \sigma Y \in RT_N(p) \Rightarrow \sigma (X \cup Y) \rho \in RT_N(p).$$

PROPOSITION 16.8. *If p is finitely branching and q is RT-saturated then $p \sqsubseteq_{FT} q \Leftrightarrow p \sqsubseteq_{RT} q$. Thus if both p and q are finitely branching and RT-saturated then $p =_{FT} q \Leftrightarrow p =_{RT} q$.*

PROOF. Suppose p is finitely branching, q is RT-saturated and $p \sqsubseteq_{FT} q$. With induction on $k \in \mathbb{N}$ I will show that whenever $X_0 a_1 X_1 a_2 \cdots a_n X_n \in RT(p)$ then there are $Y_i \subseteq X_i$ for $i = k+1, \ldots, n$ such that $X_0 a_1 X_1 a_2 \cdots a_k X_k a_{k+1} Y_{k+1} a_{k+2} \cdots a_n Y_n \in RT(q)$. The case $k = n$, together with Proposition 6.1, completes the proof of the proposition.

Induction base ($k = 0$): Let $X_0 a_1 X_1 a_2 \cdots a_n X_n \in RT(p)$. Write $\overline{X}$ for $Act - X$. Then $\overline{X_0} a_1 \overline{X_1} a_2 \cdots a_n \overline{X_n} \in FT(p) \subseteq FT(q)$. Hence there are $Y_i \subseteq X_i$ for $i = 0, \ldots, n$ such that $Y_0 a_1 Y_1 a_2 \cdots a_n Y_n \in RT(q)$. As $p \sqsubseteq_{FT} q \Rightarrow p \sqsubseteq_T q \Rightarrow I(p) \subseteq I(q)$, we have $Y_0 = X_0$.

Induction step: Take $k > 0$ and suppose the statement has been established for $k - 1$. Let $X_0 a_1 X_1 a_2 \cdots a_n X_n \in RT(p)$. Then, by induction, there are $Y_i \subseteq X_i$ for $i = k, \ldots, n$ such that

$$X_0 a_1 X_1 a_2 \cdots a_{k-1} X_{k-1} a_k Y_k a_{k+1} Y_{k+1} a_{k+2} \cdots a_n Y_n \in RT(q).$$

Moreover, for every $b \in X_k$, $X_0 a_1 X_1 a_2 \cdots a_k X_k b \in RT(p)$, so, again using the induction hypothesis, there must be a $Z_b \subseteq X_k$ such that $X_0 a_1 X_1 \cdots X_{k-1} a_k Z_b b \in RT(q)$, and hence $X_0 a_1 X_1 \cdots X_{k-1} a_k (Z_b \cup \{b\}) \in RT(q)$. As X_k is finite and $Y_k \cup \bigcup_{b \in X_k} (Z_b \cup \{b\}) = X_k$, the RT-saturation of q gives

$$X_0 a_1 \cdots a_k X_k a_{k+1} Y_{k+1} a_{k+2} \cdots a_n Y_n \in RT(q),$$

which had to be established. □

17. Complete axiomatizations

17.1. *A language for finite, concrete, sequential processes*

Consider the following basic CCS- and CSP-like language BCCSP for finite, concrete, sequential processes over a given alphabet *Act*:

inaction: 0 (called *nil* or *stop*) is a constant, representing a process that refuses to do any action.

action: a is a unary operator for any action $a \in Act$. The expression ap represents a process, starting with an a-action and proceeding with p.

choice: $+$ is a binary operator. $p + q$ represents a process, first being involved in a choice between its summands p and q, and then proceeding as the chosen process.

The set $\mathsf{T}(\text{BCCSP})$ of (closed) *process expressions* or *terms* over this language is defined as usual:

- $0 \in \mathsf{T}(\text{BCCSP})$,
- $ap \in \mathsf{T}(\text{BCCSP})$ for any $a \in Act$ and $p \in \mathsf{T}(\text{BCCSP})$,
- $p + q \in \mathsf{T}(\text{BCCSP})$ for any $p, q \in \mathsf{T}(\text{BCCSP})$.

Subterms $a0$ may be abbreviated by a. Brackets are used for disambiguation only, assuming associativity of $+$, and letting a bind stronger than $+$. If $P = \{p_1, \ldots, p_n\}$ is a finite nonempty multiset of BCCSP expressions, then $\sum P$ abbreviates $p_1 + \cdots + p_n$. This expression is determined only up to associativity and commutativity of $+$. Let $\sum \emptyset := 0$. An expression ap' is called a *summand* of p if, up to associativity and commutativity of $+$, p can be written as $\sum P$ with $ap' \in P$.

On $\mathsf{T}(\text{BCCSP})$ action relations $\xrightarrow{a}$ for $a \in Act$ are defined as the predicates on $\mathsf{T}(\text{BCCSP})$ generated by the *action rules* of Table 1. Here a ranges over Act and p and q over $\mathsf{T}(\text{BCCSP})$. A trivial structural induction shows that $p \xrightarrow{a} p'$ iff ap' is a summand of p. Now all semantic equivalences of Sections 2–14 are well-defined on $\mathsf{T}(\text{BCCSP})$, and for each of the semantics it is determined when two process expressions denote the same process.

The following theorem says that, apart from U, all these semantics are *compositional* with respect to BCCSP, i.e., all semantic equivalences are *congruences* for BCCSP.

THEOREM 7. *Let $p, q, r, s \in \mathsf{T}(\text{BCCSP})$ and let $\mathcal{O}$ be any of the semantics of Section* 15 *except U. Then*

$$p =_{\mathcal{O}} q \ \wedge\ r =_{\mathcal{O}} s \ \Rightarrow\ ap =_{\mathcal{O}} aq \ \wedge\ p + r =_{\mathcal{O}} q + s.$$

Table 1
Action rules for BCCSP

$ap \xrightarrow{a} p$	$\dfrac{p \xrightarrow{a} p'}{p+q \xrightarrow{a} p'}$	$\dfrac{q \xrightarrow{a} q'}{p+q \xrightarrow{a} q'}$

PROOF. Each of the semantics $\mathcal{O}$ has a modal characterization, given by $p =_{\mathcal{O}} q \Leftrightarrow \mathcal{O}(p) = \mathcal{O}(q)$, where $\mathcal{O}(p)$ is the set of modal formulas of the appropriate form satisfied by p. Let $\mathcal{O}^+(p) := \{a\varphi \mid a\varphi \in \mathcal{O}(p)\}$ be the set of such formulas which are of the form $a\varphi$. For each choice of $\mathcal{O}$ one easily verifies that $\mathcal{O}(p)$ is completely determined by $\mathcal{O}^+(p)$, i.e., $\mathcal{O}(p) = \mathcal{O}(q) \Leftrightarrow \mathcal{O}^+(p) = \mathcal{O}^+(q)$. One also verifies easily that $\mathcal{O}^+(0) = \emptyset$, $\mathcal{O}^+(ap) = \{a\varphi \mid \varphi \in \mathcal{O}(p)\}$ and $\mathcal{O}^+(p+q) = \mathcal{O}^+(p) \cup \mathcal{O}^+(q)$. From this the theorem follows immediately. □

For each such choice of $\mathcal{O}$ one easily verifies that moreover $\mathcal{O}(p) \subseteq \mathcal{O}(q) \Leftrightarrow \mathcal{O}^+(p) \subseteq \mathcal{O}^+(q)$. From this it follows that all the preorders of Section 15 are *precongruences* for BCCSP:

THEOREM 7b. *Let $p, q, r, s \in \mathsf{T}(\mathrm{BCCSP})$ and let $\mathcal{O}$ be any of the semantics of Section* 15 *but U. Then*

$$p \sqsubseteq_{\mathcal{O}} q \ \wedge \ r =_{\mathcal{O}} s \ \Rightarrow \ ap \sqsubseteq_{\mathcal{O}} aq \ \wedge \ p + r \sqsubseteq_{\mathcal{O}} q + s.$$

Tree semantics, when defined merely in terms of the action relations on $\mathsf{T}(\mathrm{BCCSP})$, fails to be compositional with respect to BCCSP. The expression $a0 + a0$ has only a single outgoing a-transition, namely to the expression 0. Thus, by Definition 13, $a0 + a0 =_U a0$. Likewise $b(a0 + a0) =_U ba0$. However, $b(a0 + a0) + b0 \neq_U ba0 + ba0$, as the first process has two outgoing b-transitions and the second process only one. If follows that tree equivalence as defined above is not compositional with respect to $+$.

$\mathsf{T}(\mathrm{BCCSP})$ can be turned into a labelled transition system with multiplicities by assuming a different transition $p \xrightarrow{a} q$ for every different proof of $p \xrightarrow{a} q$ from the action rules of Table 1. On such a transition system tree equivalence is compositional with respect to BCCSP.

A straightforward structural induction shows that any process $p \in \mathsf{T}(\mathrm{BCCSP})$ is finite in the sense of Definition 13. Hence the process graph $G(p)$ is finite as well. The next proposition establishes that moreover, up to bisimulation equivalence, any finite process graph can be represented by a BCCSP expression. In fact, all finite process graphs displayed in this paper have been annotated by their representing BCCSP expressions.

DEFINITION 17.1. Let $\langle\!\langle\cdot\rangle\!\rangle : \mathbb{H} \to \mathsf{T}(\mathrm{BCCSP})$ be a mapping satisfying $\langle\!\langle g\rangle\!\rangle = \Sigma\{a\langle\!\langle h\rangle\!\rangle \mid g \xrightarrow{a} h\}$.

A straightforward induction on the length of the longest path of finite process graphs teaches that such a mapping exists and is completely determined up to associativity and commutativity of $+$.

PROPOSITION 17.1. *Let $g \in \mathbb{H}$. Then there is a $p \in \mathsf{T}(\mathrm{BCCSP})$ with $G(p) \leftrightarroweq g$. In fact, $G(\langle\!\langle g\rangle\!\rangle) \leftrightarroweq g$.*

PROOF. It suffices to show that the relation $\{h, G(\langle\!\langle h\rangle\!\rangle) \mid h \in \mathbb{H}\}$ is a bisimulation. Suppose $h \xrightarrow{a} h'$. Then $a\langle\!\langle h'\rangle\!\rangle$ is a summand of $\langle\!\langle h\rangle\!\rangle$, so $\langle\!\langle h\rangle\!\rangle \xrightarrow{a} \langle\!\langle h'\rangle\!\rangle$, and by Proposition 1.1

$G(\langle\!\langle h \rangle\!\rangle) \xrightarrow{a} G(\langle\!\langle h' \rangle\!\rangle)$. Vice versa, let $G(\langle\!\langle h \rangle\!\rangle) \xrightarrow{a} h''$. Then, by Proposition 1.1, $h'' = G(p')$ for some $p' \in \mathsf{T}(\text{BCCSP})$ with $\langle\!\langle h \rangle\!\rangle \xrightarrow{a} p'$. Thus ap' must be a summand of $\langle\!\langle h \rangle\!\rangle$. By Definition 17.1 $p' = \langle\!\langle h' \rangle\!\rangle$ for some $h' \in \mathbb{H}$ with $h \xrightarrow{a} h'$. As h' is related to $h'' = G(\langle\!\langle h' \rangle\!\rangle)$, also this requirement is satisfied. □

COROLLARY 17.1. *Let* $p \in \mathsf{T}(\text{BCCSP})$. *Then* $p \leftrightarroweq \langle\!\langle G(p) \rangle\!\rangle$.

PROOF. By the above $G(\langle\!\langle G(p) \rangle\!\rangle) \leftrightarroweq G(p)$. Now apply Corollary 1.1. □

COROLLARY 17.2. *Let* $g, h \in \mathbb{H}$ *and let* $\mathcal{O}$ *be any of the semantics of Section* 15. *Then*

$$g \sqsubseteq_{\mathcal{O}} h \Leftrightarrow \langle\!\langle g \rangle\!\rangle \sqsubseteq_{\mathcal{O}} \langle\!\langle h \rangle\!\rangle \quad \textit{and} \quad g =_{\mathcal{O}} h \Leftrightarrow \langle\!\langle g \rangle\!\rangle =_{\mathcal{O}} \langle\!\langle h \rangle\!\rangle.$$

PROOF. Let $g \sqsubseteq_{\mathcal{O}} h$. By the above $G(\langle\!\langle g \rangle\!\rangle) \leftrightarroweq g \sqsubseteq_{\mathcal{O}} h \leftrightarroweq G(\langle\!\langle h \rangle\!\rangle)$. Now apply Corollary 1.1.

For "$\Leftarrow$" let $\langle\!\langle g \rangle\!\rangle \sqsubseteq_{\mathcal{O}} \langle\!\langle h \rangle\!\rangle$. By Corollary 1.1 and Proposition 17.1 $g \leftrightarroweq G(\langle\!\langle g \rangle\!\rangle) \sqsubseteq_{\mathcal{O}} G(\langle\!\langle h \rangle\!\rangle) \leftrightarroweq h$. □

17.2. *Axiomatizing the equivalences*

In Table 2, complete axiomatizations can be found for twelve of the fifteen semantic equivalences of this paper that differ on BCCSP. Axioms for singleton-failures, 2-nested simulation and possible-futures semantics are more cumbersome, and the corresponding testing

Table 2
Complete axiomatizations for the equivalences

	U	*B*	*RS*	*PW*	*RT*	*FT*	*R*	*F*	*CS*	*CT*	*S*	*T*
$(x+y)+z = x+(y+z)$	+	+	+	+	+	+	+	+	+	+	+	+
$x+y = y+x$	+	+	+	+	+	+	+	+	+	+	+	+
$x+0 = x$	+	+	+	+	+	+	+	+	+	+	+	+
$x+x = x$		+	+	+	+	+	+	+	+	+	+	+
$I(x) = I(y) \Rightarrow a(x+y) = a(x+y)+ay$			+	v	v	v	v	v	v	v	v	v
$a(bx+by+z) = a(bx+z)+a(by+z)$				+	v	v	v	v		v		v
$I(x) = I(y) \Rightarrow ax+ay = a(x+y)$					+	+	v	v		v		v
$ax+ay = ax+ay+a(x+y)$						+		v		v		v
$a(bx+u)+a(by+v) = a(bx+by+u)+a(by+v)$							+	+		v		v
$ax+a(y+z) = ax+a(x+y)+a(y+z)$								+		ω		v
$a(x+by+z) = a(x+by+z)+a(by+z)$									+	v	v	v
$a(bx+u)+a(cy+v) = a(bx+cy+u+v)$										+		v
$a(x+y) = a(x+y)+ay$											+	v
$ax+ay = a(x+y)$												+
$I(0) = 0$	+	+	+	+	+	+	+	+	+	+	+	+
$I(ax) = a0$	+	+	+	+	+	+	+	+	+	+	+	+
$I(x+y) = I(x)+I(y)$	+	+	+	+	+	+	+	+	+	+	+	+

notions are less plausible. Therefore they have been omitted. The axiomatization of tree semantics (U) requires action relations with multiplicities. Although rather trivial, I will not formally establish its soundness and completeness here. In order to formulate the axioms, variables have to be added to the language as usual. In the axioms they are supposed to be universally quantified. Most of the axioms are axiom schemes, in the sense that there is one axiom for each substitution of actions from Act for the parameters a, b, c. Some of the axioms are conditional equations, using an auxiliary operator I. Thus provability is defined according to the standards of either first-order logic with equality or conditional equational logic. I is a unary operator that calculates the set of initial actions of a process expression, coded as a process expression again.

THEOREM 8. *For each of the semantics $\mathcal{O} \in \{T, S, CT, CS, F, R, FT, RT, PW, RS, B\}$ two process expressions $p, q \in \mathsf{T}(\mathrm{BCCSP})$ are $\mathcal{O}$-equivalent iff they can be proved equal from the axioms marked with "+" in the column for $\mathcal{O}$ in Table 2. The axioms marked with "v" or "ω" are valid in $\mathcal{O}$-semantics but not needed for the proof.*

PROOF. "If" (*soundness*): In the light of Theorem 7 it suffices to show that the closed instances of the indicated axioms are valid in the corresponding semantics. This is straightforward.

"Only if" (*completeness*): Let $T_\mathcal{O}$ be the set of axioms marked with "+" in the column for $\mathcal{O}$. Write $T_\mathcal{O} \vdash p = q$ if the equation $p = q$ is provable from $T_\mathcal{O}$. I have to show that

$$p =_\mathcal{O} q \ \Rightarrow\ T_\mathcal{O} \vdash p = q \tag{2}$$

for any $p, q \in \mathsf{T}(\mathrm{BCCSP})$. For the cases $\mathcal{O} \in \{B, S, RS, CS\}$ I will show that

$$p \sqsubseteq_\mathcal{O} q \ \Rightarrow\ T_\mathcal{O} \vdash q = q + p \tag{3}$$

for any $p, q \in \mathsf{T}(\mathrm{BCCSP})$, from which (2) follows immediately. This will be done with structural induction on p and q. So assume $p \sqsubseteq_\mathcal{O} q$ and (3) has been proven for all pairs of smaller expressions $p', q' \in \mathsf{T}(\mathrm{BCCSP})$. Provided $T_\mathcal{O}$ contains at least the first four axioms of Table 2, one has $T_\mathcal{O} \vdash q = q + p$ iff $T_\mathcal{O} \vdash q = q + ap'$ for every summand ap' of p.

Take $\mathcal{O} = B$, so $p \sqsubseteq_B q$. Let ap' be a summand of p. Then $p \xrightarrow{a} p'$, so $\exists q'$: $q \xrightarrow{a} q'$ and $p' =_B q'$. By induction $T_B \vdash p' = p' + q' = q'$, using Proposition 12.1. Furthermore, aq' must be a summand of q, so $T_B \vdash q = q + aq' = q + ap'$ and therefore $T_B \vdash q = q + p$.

Take $\mathcal{O} = S$, so $p \sqsubseteq_S q$. Let ap' be a summand of p. Then $p \xrightarrow{a} p'$, so $\exists q'$: $q \xrightarrow{a} q'$ and $p' \sqsubseteq_S q'$. By induction $T_S \vdash q' = q' + p'$, so $T_S \vdash aq' = a(q' + p') = a(q' + p') + ap' = aq' + ap'$. Furthermore, aq' must be a summand of q, so $T_S \vdash q = q + aq' = q + ap'$ and thus $T_S \vdash q = q + p$.

Take $\mathcal{O} = RS$, so $p \sqsubseteq_{RS} q$. Let ap' be a summand of p. Then $p \xrightarrow{a} p'$, so $\exists q'$: $q \xrightarrow{a} q'$ and $p' \sqsubseteq_{RS} q'$. Now $I(p') = I(q')$ and hence $T_{RS} \vdash I(p') = I(q')$. By induction $T_{RS} \vdash q' = q' + p'$, so $T_{RS} \vdash aq' = a(q' + p') = a(q' + p') + ap' = aq' + ap'$. Furthermore, aq' must be a summand of q, so $T_{RS} \vdash q = q + aq' = q + ap'$ and thus $T_{RS} \vdash q = q + p$.

Take $\mathcal{O} = CS$, so $p \sqsubseteq_{CS} q$. Let ap' be a summand of p. Then $p \xrightarrow{a} p'$, so $\exists q' : q \xrightarrow{a} q'$ and $p' \sqsubseteq_{CS} q'$. In case $I(p') = \emptyset$ it must be that $I(q') = \emptyset$ as well, and hence $T_{CS} \vdash p' =$

$q' = 0$. Otherwise, $T_{CS} \vdash p' = bp'' + r$ and by induction $T_{CS} \vdash q' = q' + p'$, so $T_{CS} \vdash aq' = a(q' + p') = a(q' + bp'' + r) = a(q' + bp'' + r) + a(bp'' + r) = a(q' + p') + ap' = aq' + ap'$. Furthermore, aq' must be a summand of q, so in both cases $T_{CS} \vdash q = q + aq' = q + ap'$ and thus $T_{CS} \vdash q = q + p$.

Take $\mathcal{O} = PW$. Suppose $p =_{PW} q$. The axiom $a(bx + by + z) = a(bx + z) + a(by + z)$ allows to rewrite p and q to BCCSP expressions $p' = \sum_{i \in I} a_i p_i$ and $q' = \sum_{j \in J} a_j q_j$ with p_i and q_j deterministic. For expressions of this form it is easy to establish that $p' =_{PW} q' \Leftrightarrow p' \leftrightarroweq q'$. Using the soundness of the axiom employed, and the completeness of $T_B \subseteq T_{PW}$ for $\leftrightarroweq$, it follows that $T_{PW} \vdash p = p' = q' = q$.

For F and R (as well as B) a proof is given in Bergstra, Klop and Olderog [11] by means of *graph transformations*. A similar proof for *RT* can be found in Baeten, Bergstra and Klop [6]. This method, applied to semantics $\mathcal{O}$, requires the definition of a class $\mathbb{H}^*$ of finite process graphs that contains at least all finite process trees, and a binary relation $\overset{\mathcal{O}}{\rightsquigarrow} \subseteq \mathbb{H}^* \times \mathbb{H}^*$ – a system of *graph transformations* – such that the following can be established:

(1) $\overset{\mathcal{O}}{\rightsquigarrow}$, used as a rewriting system, is *terminating* on $\mathbb{H}^*$, i.e., any reduction sequence $g_0 \overset{\mathcal{O}}{\rightsquigarrow} g_1 \overset{\mathcal{O}}{\rightsquigarrow} \cdots$ leads (in finitely many steps) to a *normal form*, a graph that cannot be further transformed,

(2) if $g \overset{\mathcal{O}}{\rightsquigarrow} h$ then
   (a) $g =_{\mathcal{O}} h$, and
   (b) $T_{\mathcal{O}} \vdash \langle\!\langle g \rangle\!\rangle = \langle\!\langle h \rangle\!\rangle$, and

(3) two normal forms are bisimilar iff they are $\mathcal{O}$-equivalent.

Now the completeness proof goes as follows: Suppose $p =_{\mathcal{O}} q$. As PATHS$(G(p))$ and PATHS$(G(q))$ are finite, $U(G(p))$ and $U(G(q))$ belong to $\mathbb{H}^*$, and by requirement (1) they can be rewritten to normal forms g and h. Using Corollary 1.1, Proposition 13.1 and requirement 2(a) above

$$g =_{\mathcal{O}} U\big(G(p)\big) \leftrightarroweq G(p) =_{\mathcal{O}} G(q) \leftrightarroweq U\big(G(q)\big) =_{\mathcal{O}} h.$$

Thus, with requirement (3), $g \leftrightarroweq h$; and Corollaries 17.1 and 17.2 yield

$$p \leftrightarroweq \big\langle\!\big\langle G(p) \big\rangle\!\big\rangle \leftrightarroweq \big\langle\!\big\langle U(G(p)) \big\rangle\!\big\rangle, \quad \langle\!\langle g \rangle\!\rangle \leftrightarroweq \langle\!\langle h \rangle\!\rangle, \quad \big\langle\!\big\langle U(G(q)) \big\rangle\!\big\rangle \leftrightarroweq \big\langle\!\big\langle G(q) \big\rangle\!\big\rangle \leftrightarroweq q.$$

Requirement 2(b) and the completeness result for bisimulation semantics proved above finally give

$$T_{\mathcal{O}} \vdash p = \big\langle\!\big\langle U(G(p)) \big\rangle\!\big\rangle = \langle\!\langle g \rangle\!\rangle = \langle\!\langle h \rangle\!\rangle = \big\langle\!\big\langle U(G(q)) \big\rangle\!\big\rangle = q.$$

I will now apply this method to T, *RT*, *CT*, R, F and *FT*. In the cases of T, *RT* and *CT*, $\mathbb{H}^*$ is taken to be $\mathbb{H}^{tree}$, the class of finite process trees.

Take $\mathcal{O} = T$. Let $\overset{T}{\rightsquigarrow}$ be the graph transformation that converts g into h, notation $g \overset{T}{\rightsquigarrow} h$, iff g is a finite tree with edges (s, a, t) and (s, a, u) with $t \neq u$, and h is obtained by *identifying* t and u. Formally speaking, the nodes of h are those of g, except that t and u are omitted and a fresh node v has been added instead. Often v is taken to be the

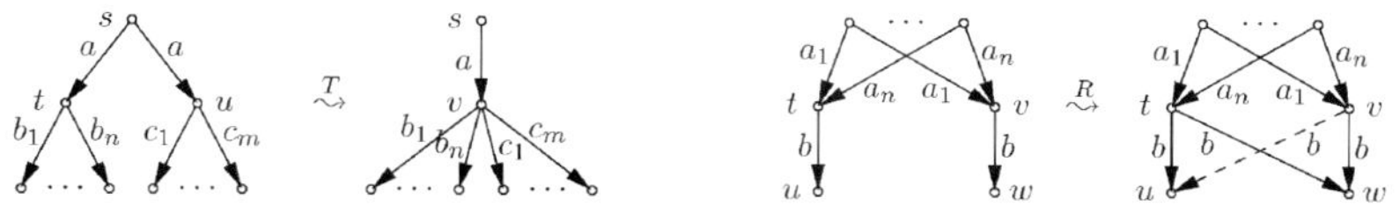

Figure 10. Graph transformations.

(equivalence) class $\{t, u\}$. Define the function $'$: NODES$(g) \to$ NODES(h) by $t' = v$, $u' = v$ and $w' = w$ for $w \neq t, u$. Now EDGES$(h) = \{(p', a, q') \mid (p, a, q) \in$ EDGES$(g)\}$ and ROOT$(h) =$ ROOT$(g)'$. This graph transformation is illustrated in Figure 10.

If g is a finite tree and $g \overset{T}{\rightsquigarrow} h$ then so is h. Moreover, h has fewer nodes than g. Hence $\overset{T}{\rightsquigarrow}$ is terminating on $\mathbb{H}^{tree}$. The normal forms are exactly the finite deterministic trees. Now requirement (3) has been established by Theorem 6. Requirement 2(a) is trivial, and for 2(b) observe that any application of $\overset{T}{\rightsquigarrow}$ corresponds to an application of the axiom $ax + ay = a(x + y)$.

Take $\mathcal{O} = RT$. Let $\overset{RT}{\rightsquigarrow}$ be the same graph transformation as $\overset{T}{\rightsquigarrow}$, except that it only applies if $I(t) = I(u)$. This time the normal forms are the ready trace deterministic trees, and requirement (3) has been established by Proposition 16.3. Again requirement 2(a) is easy to check, and for 2(b) it suffice to observe that any application of $\overset{RT}{\rightsquigarrow}$ corresponds to an application of the axiom $I(x) = I(y) \Rightarrow ax + ay \;=\; a(x + y)$.

Take $\mathcal{O} = CT$. Let $\overset{CT}{\rightsquigarrow}$ be the same graph transformation as $\overset{T}{\rightsquigarrow}$, except that it only applies if $I(t) = \emptyset \Leftrightarrow I(u) = \emptyset$. This time the normal forms are the completed trace deterministic trees, and again requirement (3) has been established by Proposition 16.3. Once more requirement 2(a) is easy to check, and for 2(b) observe that any application of $\overset{CT}{\rightsquigarrow}$ corresponds to an application of the law $(I(x) = 0 \Leftrightarrow I(y) = 0) \Rightarrow ax + ay = a(x + y)$. This law falls outside conditional equational logic, but it can be reformulated equationally by considering the two cases $I(x) = 0 = I(y)$ and $I(x) \neq 0 \neq I(y)$. In the first case it must be that $T_B \vdash x = 0 = y$ and hence the law follows from the third and fourth axiom of Table 2. In the second, observe that $I(p) \neq 0$ iff p has the form $bq + r$ with $b \in Act$. Hence the law can be reformulated as $a(bx + u) + a(cy + v) = a(bx + cy + u + v)$.

A process graph $g \in \mathbb{G}$ is called *history unambiguous* [11] if any two paths from the root to the same node give rise to the same trace, i.e., if for $\pi, \pi' \in$ PATHS(g) one has $end(\pi) = end(\pi') \Rightarrow T(\pi) = T(\pi')$. The *history* or *trace* $T(s)$ of a node s in such a graph g is defined as $T(\pi)$ for π an arbitrary path from the root of g to s. Observe that trees are history unambiguous. In the next two completeness proofs (the cases R and F) $\mathbb{H}^*$ is taken to be the class $\mathbb{H}^{hu}$ of finite, history unambiguous, connected process graphs. For $g \in \mathbb{H}^{hu}$ and $t, v \in$ NODES(g) let $t \sim v$ abbreviate

$$\forall s \in \text{NODES}(g),\ a \in Act\colon\ (s, a, t) \in \text{EDGES}(g) \Leftrightarrow (s, a, v) \in \text{EDGES}(g).$$

Take $\mathcal{O} = R$. Let $\overset{R}{\rightsquigarrow}$ be the graph transformation with $g \overset{R}{\rightsquigarrow} h$ iff g has edges (t, b, u) and (v, b, w) with $t \sim v$, and h is obtained by adding a new edge (t, b, w). This graph transformation is illustrated in Figure 10. Note that by applying $\overset{R}{\rightsquigarrow}$ twice, one can also add the edge (v, b, u) (indicated with a dashed arrow in Figure 10) if it isn't there already. If g is a finite, history unambiguous process graph and $g \overset{R}{\rightsquigarrow} h$ then so is h. Moreover, h has more edges than g. As there is an upperbound to the number of edges of graphs that can be obtained from a given graph $g \in \mathbb{H}^{hu}$ by applying $\overset{R}{\rightsquigarrow}$ (namely $n \times l \times n$, where n is the number of nodes in g, and l the number of different edge-labels occurring in g), $\overset{R}{\rightsquigarrow}$ is terminating on $\mathbb{H}^{hu}$ (requirement (1)). It is easy to see that $\overset{R}{\rightsquigarrow}$ does not add new ready pairs. This gives requirement 2(a). For 2(b) observe that an application of $\overset{R}{\rightsquigarrow}$ corresponds to a number of applications of $a(bx + u) + a(by + v) = a(bx + by + u) + a(by + v)$. Finally, requirement (3) follows from Proposition 16.5 and the following

CLAIM. *The normal forms with respect to* $\overset{R}{\rightsquigarrow}$ *are cross saturated.*

PROOF OF THE CLAIM. Let $g \in \mathbb{H}^{hu}$ be a normal form with respect to $\overset{R}{\rightsquigarrow}$. With induction to the length of $T(u)$ I will show that, for $u, w \in \text{NODES}(g)$,

$$\text{if } T(u) = T(w) \quad \text{then } u \sim w. \tag{4}$$

This implies that g is cross saturated, for if π, π' and a are as in the remark below Definition 16.5, there must be an edge $(end(\pi), a, u)$ in g. Now $T(u) = T(\pi)a = T(end(\pi'))$, so also $(end(\pi), a, end(\pi')) \in \text{EDGES}(g)$.

Induction base: If $length(T(u)) = 0$, one has $u = w = \text{ROOT}(g)$ and the statement is trivial.

Induction step: Let $T(u) = T(w) \neq \varepsilon$, and let $(t, b, u) \in \text{EDGES}(g)$. By symmetry, it suffices to show that $(t, b, w) \in \text{EDGES}(g)$. As g is connected and history unambiguous, there must be an edge (v, b, w) with $T(t) = T(v)$. By induction $t \sim v$. As g is in normal form it must have an edge (t, b, w).

Take $\mathcal{O} = F$. Let $\overset{fork}{\rightsquigarrow}$ be the graph transformation with $g \overset{fork}{\rightsquigarrow} h$ iff g has edges (s, a, t) and (s, a, u), $\exists Y \subseteq I(u)$ such that h is given by

- $\text{NODES}(h) = \text{NODES}(g) \dot{\cup} \{v\}$,
- $\text{ROOT}(h) = \text{ROOT}(g)$,
- $\text{EDGES}(h) = \text{EDGES}(g) \cup \{(s, a, v)\}$
  $\cup \{(v, b, w) \mid (t, b, w) \in \text{EDGES}(g)\}$
  $\cup \{(v, b, w) \mid (u, b, w) \in \text{EDGES}(g) \wedge b \in Y\}$

and $|R(h)| > |R(g)|$. This graph transformation is illustrated in Figure 11. Note that for any path $\pi \in \text{PATHS}(h)$ not ending in v, a path $\pi' \in \text{EDGES}(g)$ can be found with $T(\pi') = T(\pi)$ and $end(\pi') = end(\pi)$, namely by circumventing the possible portion through v along t or u. Thus, such paths do not give rise to new ready or failure pairs. For any path $\pi \in \text{PATHS}(h)$ ending in v there is a path $\pi' \in \text{EDGES}(g)$ with $T(\pi') = T(\pi)$ and $end(\pi') = t$. As $I(t) \subseteq I(v)$, also such paths do not give rise to new failure pairs. Hence one has $R(h) = R(g) \cup \{\langle T(t), I(t) \cup Y\rangle\}$ and $F(h) = F(g)$. Note that if $g \in \mathbb{H}^{hu}$ and $g \overset{fork}{\rightsquigarrow} h$, then also $h \in \mathbb{H}^{hu}$. Let $\overset{F}{\rightsquigarrow}$ be $\overset{R}{\rightsquigarrow} \cup \overset{fork}{\rightsquigarrow}$. As $g \overset{fork}{\rightsquigarrow} h \Rightarrow g =_F h$ and $g \overset{R}{\rightsquigarrow} h \Rightarrow g =_R$

Figure 11. Fork.

$h \Rightarrow g =_F h$, requirement 2(a) is satisfied. For 2(b) observe that an application of $\overset{fork}{\rightsquigarrow}$ corresponds to an application of the axiom $ax + a(y+z) = ax + a(x+y) + a(y+z)$.

The requirement $|R(h)| > |R(g)|$ says that the transformation may only take place if it actually increases the ready set of the transformed graph. Note that if $g \overset{F}{\rightsquigarrow} h$ then $T(g) = T(h)$. As there is an upperbound to the number of ready pairs of graphs g with a given trace set (namely $|T(g)| \times 2^l$, where l is the number of different edge-labels occurring in g), a reduction sequence $g_0 \overset{F}{\rightsquigarrow} g_1 \overset{F}{\rightsquigarrow} \cdots$ on $\mathbb{H}^{hu}$ can contain only finitely many occurrences of $\overset{fork}{\rightsquigarrow}$. After the last such occurrence it leads in finitely many steps to a normal form, because $\overset{R}{\rightsquigarrow}$ is terminating on $\mathbb{H}^{hu}$. Hence also $\overset{F}{\rightsquigarrow}$ is terminating on $\mathbb{H}^{hu}$ (requirement (1)).

Suppose g is a normal form with respect to $\overset{F}{\rightsquigarrow}$ and $\langle\sigma, X\rangle \in R(g) \wedge \langle\sigma, Y \cup Z\rangle \in R(g)$. Then g has nodes t and u with $T(t) = T(u) = \sigma$, $I(t) = X$ and $Y \subseteq I(u)$. As g must be a normal form with respect to $\overset{R}{\rightsquigarrow}$, it satisfies (4) and hence $t \sim u$. As g is connected, there are edges (s, a, t) and (s, a, u) in g. As g must also be a normal form with respect to $\overset{fork}{\rightsquigarrow}$, $\langle\sigma, X \cup Y\rangle \in R(g)$. Thus normal forms with respect to $\overset{F}{\rightsquigarrow}$ are saturated as well as cross saturated, and hence requirement (3) follows by Propositions 16.7 and 16.5.

Take $\mathcal{O} = FT$. Let $\overset{sf}{\rightsquigarrow}$ (*symmetric fork*) be the graph transformation consisting of those instances of $\overset{fork}{\rightsquigarrow}$ where $Y = I(u)$, but with the requirement $|R(h)| > |R(g)|$ relaxed to $|RT_N(h)| > |RT_N(g)|$. Let $\mathbb{H}^*$ be $\mathbb{H}^{tree}$, and define $\overset{sfu}{\rightsquigarrow}$ by $g \overset{sfu}{\rightsquigarrow} h$ if $g \overset{sf}{\rightsquigarrow} h'$ and $h = U(h')$. Thus $\overset{sfu}{\rightsquigarrow}$ is the variant of $\overset{sf}{\rightsquigarrow}$ in which the target is unfolded into a tree. Let $\overset{FT}{\rightsquigarrow}$ be $\overset{RT}{\rightsquigarrow} \cup \overset{sfu}{\rightsquigarrow}$. As there is an upperbound to the number of normal ready traces of graphs with a given finite trace set, $\overset{FT}{\rightsquigarrow}$ is terminating on $\mathbb{H}^*$ (requirement (1)). The normal forms are exactly the finite RT-saturated ready trace deterministic process trees, so requirement (3) follows from Propositions 16.3 and 16.8. It follows immediately from Corollary 5.1 that $g \overset{sf}{\rightsquigarrow} h \Rightarrow g =_{FT} h$. Hence Proposition 13.1 gives $g \overset{sfu}{\rightsquigarrow} h \Rightarrow g =_{FT} h$. Moreover, $g \overset{RT}{\rightsquigarrow} h \Rightarrow g =_{RT} h \Rightarrow g =_{FT} h$, which yields requirement 2(a). For 2(b) observe that an application of $\overset{sf}{\rightsquigarrow}$ corresponds to an application of the axiom $ax + ay = ax + ay + a(x+y)$, and as $h \leftrightarrows U(h)$ Corollary 17.2 gives $T_B \vdash \langle\langle h\rangle\rangle = \langle\langle U(h)\rangle\rangle$ for $h \in \mathbb{H}$. □

In Theorem 8 the fifth and seventh axioms of Table 2 may be replayed by

$$a\sum_{i=1}^{n}(b_i x_i + b_i y_i) = a\sum_{i=1}^{n}(b_i x_i + b_i y_i) + a\sum_{i=1}^{n} b_i y_i \quad \text{and}$$

$$a\sum_{i=1}^{n} b_i x_i + a\sum_{i=1}^{n} b_i y_i = a\sum_{i=1}^{n}(b_i x_i + b_i y_i).$$

These laws derive the same closed substitution instances. Thus none of the axiomatizations require the operator I, or conditional equations. However, the laws above are axiom schemes which have instances for any choice of $n \in \mathbb{N}$. Even if *Act* is finite, the axiomatizations involving these laws are infinite.

As observed by Stefan Blom, if *Act* is finite, ready simulation equivalence can be finitely axiomatized without using conditional equations or auxiliary operators. In fact, the first law above can be simplified to (is derivable from)

$$a(bx + by + u) = a(bx + by + u) + a(bx + u).$$

Moreover, the seventh axiom in Table 2 is derivable from the eighth and the fifth. Using this, it can be seen that if *Act* is finite also failure trace equivalence has a finite equational axiomatization. However, it is unknown whether the same holds for ready trace equivalence.

THEOREM 9. *Suppose Act is infinite. For each of the semantics $\mathcal{O} \in \{T, S, CT, F, R, FT, RT, RS, B, U\}$ two BCCSP expressions with variables are $\mathcal{O}$-equivalent iff they can be proved equal from the axioms marked with '+' or 'ω' in the column for $\mathcal{O}$ in Table 2. It follows that the axioms marked with 'v' are derivable.*

PROOF. For $\mathcal{O} \in \{T, CT, F, R, FT, RT, B\}$ this has been established in Groote [23]. His proof for F, R, FT and RT can be applied to S and RS as well. The proof for U is rather trivial, but omitted here. □

Groote also showed that if *Act* is finite, Theorem 9 does not hold for F, R, FT and RT. But for B and CT it suffices to assume that *Act* is nonempty, and for T it suffices to assume that *Act* has at least two elements. I do not know which cardinality restriction on *Act* is needed in the cases of S and RS. A complete axiomatization for open terms for completed simulation or possible worlds semantics has so far not been provided.

17.3. *Axiomatizing the preorders*

In Table 3, complete axiomatizations can be found for the eleven preorders corresponding to the equivalences axiomatized in Table 2 (there is no preorder for tree semantics (U)). This time provability is defined according to the standards of either first-order logic with inequality or conditional inequational logic, i.e., it may be used that $\sqsubseteq$ is reflexive and transitive and satisfies the precongruence properties of Theorem 7b. For any semantics $\mathcal{O}$ the $\mathcal{O}$-preorder and $\mathcal{O}$-equivalence are related by $p =_{\mathcal{O}} q \Leftrightarrow p \sqsubseteq_{\mathcal{O}} q \wedge q \sqsubseteq_{\mathcal{O}} p$. Thus either $p = q$ is taken to be an abbreviation of $p \sqsubseteq q \wedge q \sqsubseteq p$ or the conditional axioms $p = q \Rightarrow p \sqsubseteq q$ and $p \sqsubseteq q \wedge q \sqsubseteq p \Rightarrow p = q$ are considered part of the axiomatizations.

Table 3
Complete axiomatizations for the preorders

	B	RS	PW	RT	FT	R	F	CS	CT	S	T
$(x+y)+z=x+(y+z)$	+	+	+	+	+	+	+	+	+	+	+
$x+y=y+x$	+	+	+	+	+	+	+	+	+	+	+
$x+0=x$	+	+	+	+	+	+	+	+	+	+	+
$x+x=x$	+	+	+	+	+	+	+	+	+	+	+
$ax \sqsubseteq ax+ay$		+	+	+	+	+	+	v	v	v	v
$a(bx+by+z)=a(bx+z)+a(by+z)$			+	v	v	v	v		v		v
$I(x)=I(y) \Rightarrow ax+ay=a(x+y)$				+	v	v	v		v		v
$ax+ay \sqsupseteq a(x+y)$					+		v		v		v
$a(bx+u)+a(by+v) \sqsupseteq a(bx+by+u)$						+	v		v		v
$ax+a(y+z) \sqsupseteq a(x+y)$							+		v		v
$ax \sqsubseteq ax+y$								+	+	v	v
$a(bx+u)+a(cy+v)=a(bx+cy+u+v)$									+		v
$x \sqsubseteq x+y$										+	+
$ax+ay=a(x+y)$											+
$I(0)=0$	+	+	+	+	+	+	+	+	+	+	+
$I(ax)=a0$	+	+	+	+	+	+	+	+	+	+	+
$I(x+y)=I(x)+I(y)$	+	+	+	+	+	+	+	+	+	+	+

In the latter case, the axioms of Table 3 also constitute complete axiomatizations of the equivalences.

The three axioms in Table 3 in which the inequality is written "$\sqsubseteq$" represent strengthenings of the corresponding axioms in Table 2. The axioms in which the inequality is written "$\sqsupseteq$" are merely slick reformulations of the corresponding axioms in Table 2, and could be replaced by them. Unlike in Table 2, the characteristic axiom for the readiness preorder (the ninth) is now a substitution instance of the characteristic axiom for the failures preorder (the tenth).

Note that the characteristic axiom for the ready simulation preorder (the fifth) derives all closed instances of $I(x)=I(y) \Rightarrow ax \sqsubseteq a(x+y)$, which gives the fifth axiom of Table 2. Hence all closed instances of the characteristic axiom for the ready trace preorder (the seventh) are derivable from the fifth and eighth axioms. Thus conditional (in)equations, involving the operator I, or unbounded sums, are not needed in the axiomatizations of ready simulation and failure trace semantics.

THEOREM 10. *For each of the semantics* $\mathcal{O} \in \{T, S, CT, CS, F, R, FT, RT, PW, RS, B\}$ *one has* $p \sqsubseteq_{\mathcal{O}} q$ *for* $p, q \in \mathrm{T(BCCSP)}$ *iff* $p \sqsubseteq q$ *can be proved from the axioms marked with "+" in the column for* $\mathcal{O}$ *in Table* 3. *The axioms marked with "v" are valid in* $\mathcal{O}$-*semantics but not needed for the proof.*

PROOF. "If" (*soundness*): In the light of Theorem 7b it suffices to show that the closed instances of the indicated axioms are valid in the corresponding semantics. This is straightforward.

"Only if" (*completeness*): Let $T^*_{\mathcal{O}}$ be the set of axioms marked with "+" in the column for $\mathcal{O}$. Write $T^*_{\mathcal{O}} \vdash p \sqsubseteq q$ if the inequation $p \sqsubseteq q$ is provable from $T^*_{\mathcal{O}}$. I have to show that

$$p \sqsubseteq_{\mathcal{O}} q \Rightarrow T^*_{\mathcal{O}} \vdash p \sqsubseteq q \tag{5}$$

for any $p, q \in \mathsf{T}(\text{BCCSP})$. The case $\mathcal{O} = B$ follows from Proposition 12.1 and Theorem 8. For the cases $\mathcal{O} \in \{S, CS, RS\}$ (5) will be established with structural induction on p and q. So assume $p \sqsubseteq_{\mathcal{O}} q$ and (5) has been proven for all pairs of smaller expressions $p', q' \in \mathsf{T}(\text{BCCSP})$.

Take $\mathcal{O} = S$, so $p \sqsubseteq_S q$. Using the axiom $x \sqsubseteq x + y$ one finds that $T^*_S \vdash p \sqsubseteq q$ if for every summand ap' of p there is a summand aq' of q such that $T^*_S \vdash ap' \sqsubseteq aq'$. So let ap' be a summand of p. Then $p \xrightarrow{a} p'$, so $\exists q'$: $q \xrightarrow{a} q'$ and $p' \sqsubseteq_S q'$. Note that aq' is a summand of q. By induction $T^*_S \vdash p' \sqsubseteq q'$, so $T^*_S \vdash ap' \sqsubseteq aq'$.

Take $\mathcal{O} = CS$, so $p \sqsubseteq_{CS} q$. Using the axiom $ax \sqsubseteq ax + y$ one finds that $T^*_{CS} \vdash p \sqsubseteq q$ if $I(p) \neq \emptyset$ and for every summand ap' of p there is a summand aq' of q such that $T^*_{CS} \vdash ap' \sqsubseteq aq'$. In case $I(p) = \emptyset$ it must be that $I(q) = \emptyset$ as well, and hence $T^*_{CS} \vdash p = q = 0$. Otherwise, let ap' be a summand of p. Then $p \xrightarrow{a} p'$, so $\exists q'$: $q \xrightarrow{a} q'$ and $p' \sqsubseteq_{CS} q'$. Note that aq' is a summand of q. By induction $T^*_{CS} \vdash p' \sqsubseteq q'$, so $T^*_{CS} \vdash ap' \sqsubseteq aq'$.

Take $\mathcal{O} = RS$, so $p \sqsubseteq_{RS} q$. Using the first five axioms of Table 3 one finds that $T^*_{RS} \vdash p \sqsubseteq q$ if $I(p) = I(q)$ and for every summand ap' of p there is a summand aq' of q such that $T^*_{RS} \vdash ap' \sqsubseteq aq'$. As $p \sqsubseteq_{RS} q$ one has $I(p) = I(q)$. Let ap' be a summand of p. Then $p \xrightarrow{a} p'$, so $\exists q'$: $q \xrightarrow{a} q'$ and $p' \sqsubseteq_{RS} q'$. Note that aq' is a summand of q. By induction $T^*_{RS} \vdash p' \sqsubseteq q'$, so $T^*_{RS} \vdash ap' \sqsubseteq aq'$.

Take $\mathcal{O} = PW$. Suppose $p \sqsubseteq_{PW} q$. The axiom $a(bx + by + z) = a(bx + z) + a(by + z)$ allows to rewrite p and q to BCCSP expressions $p' = \sum_{i \in I} a_i p_i$ and $q' = \sum_{j \in J} a_j q_j$ with p_i and q_j deterministic. For expressions of this form it is easy to establish that $p' \sqsubseteq_{PW} q' \Leftrightarrow p' \sqsubseteq_{RS} q'$. Using the soundness of the axiom employed, and the completeness of $T_{RS} \subseteq T_{PW}$ for $\sqsubseteq_{RS}$, it follows that $T_{PW} \vdash p = p' \sqsubseteq q' = q$.

The remaining completeness proofs go by a variant of the method of graph transformations, where requirement (3) is replaced by

$$\text{if } g \text{ and } h \text{ are normal forms, then } g \sqsubseteq_{\mathcal{O}} h \Leftrightarrow g \sqsubseteq_{\mathcal{N}} h.$$

Here $\mathcal{N}$ should be a semantics finer than $\mathcal{O}$, for which the completeness theorem has already been established, and for which $T^*_{\mathcal{N}} \subseteq T^*_{\mathcal{O}}$. The reasoning now goes exactly as in the proof of Theorem 8: Suppose $p \sqsubseteq_{\mathcal{O}} q$. Rewrite $U(G(p))$ and $U(G(q))$ to normal forms g and h. Then

$$g =_{\mathcal{O}} U\big(G(p)\big) \leftrightarrows G(p) \sqsubseteq_{\mathcal{O}} G(q) \leftrightarrows U\big(G(q)\big) =_{\mathcal{O}} h.$$

Thus, with requirement (3), $g \sqsubseteq_{\mathcal{N}} h$. Corollary 17.2 yields $\langle\!\langle g \rangle\!\rangle \sqsubseteq_{\mathcal{N}} \langle\!\langle h \rangle\!\rangle$, and one obtains

$$T^*_{\mathcal{O}} \vdash p = \big\langle\!\!\big\langle U(G(p)) \big\rangle\!\!\big\rangle = \langle\!\langle g \rangle\!\rangle \sqsubseteq \langle\!\langle h \rangle\!\rangle = \big\langle\!\!\big\langle U(G(q)) \big\rangle\!\!\big\rangle = q.$$

For each of the six remaining completeness proofs, the class $\mathbb{H}^*$ and the graph transformations are the same as in the proof of Theorem 8. Thus requirements (1) and 2(a) are fulfilled. As (the closed instances of) the axioms for the respective equivalences from Table 2 are easily derivable from the ones for the corresponding preorders from Table 3, requirement 2(b) is fulfilled as well. Requirement (3), which used to follow from Theorem 6 and Propositions 16.3, 16.5, 16.7 and 16.8, now follows from Propositions 16.4, 16.6, 16.7 and 16.8. □

17.4. *A language for finite, concrete, sequential processes with internal choice*

Let BCSP be the language that extends BCCSP with a binary operator $\oplus$, modelling *internal choice*. Like $p+q$, the expression $p \oplus q$ represents a process, first being involved in a choice between its summands p and q, and then proceeding as the chosen process. However, whereas $+$ represents a choice that can be influenced by the environment of the process (an *external choice*), $\oplus$ represents one that is due to internal nondeterminism of the specified system. BCSP can be regarded as a basic fragment of the language CSP of Hoare [31].

The set $\mathsf{T}(\mathrm{BCSP})$ of (closed) terms over BCSP, or (closed) BCSP-*expressions*, and its subset $\mathsf{T}_1(\mathrm{BCSP})$ of *initially deterministic* BCSP-*expressions*, are defined by:

- $0 \in \mathsf{T}_1(\mathrm{BCSP}) \subseteq \mathsf{T}(\mathrm{BCSP})$,
- $aP \in \mathsf{T}_1(\mathrm{BCSP})$ for any $a \in Act$ and $P \in \mathsf{T}(\mathrm{BCSP})$,
- $p+q \in \mathsf{T}_1(\mathrm{BCSP})$ for any $p, q \in \mathsf{T}_1(\mathrm{BCSP})$,
- $P+Q \in \mathsf{T}(\mathrm{BCSP})$ for any $P, Q \in \mathsf{T}(\mathrm{BCSP})$,
- $P \oplus Q \in \mathsf{T}(\mathrm{BCSP})$ for any $P, Q \in \mathsf{T}(\mathrm{BCSP})$.

Again, subterms $a0$ may be abbreviated by a. Brackets are used for disambiguation only, assuming associativity of $+$ and $\oplus$, and letting a bind stronger than $+$ and $\oplus$. Semantically, BCSP-expressions represent nonempty, finite sets of initially deterministic BCSP expressions: for $P, Q \in \mathsf{T}(\mathrm{BCSP})$ let

$$[\![0]\!] := \{0\} \qquad [\![aP]\!] := \{aP\} \qquad [\![P+Q]\!] := \{p+q \mid p \in [\![P]\!],\ q \in [\![Q]\!]\}$$
$$[\![P \oplus Q]\!] := [\![P]\!] \cup [\![Q]\!].$$

On $\mathsf{T}_1(\mathrm{BCSP})$ action relations $\xrightarrow{a}$ for $a \in Act$ are defined as the predicates on $\mathsf{T}_1(\mathrm{BCSP})$ generated by the action rules of Table 4. Here a ranges over Act, P over $\mathsf{T}(\mathrm{BCSP})$ and p and q over $\mathsf{T}_1(\mathrm{BCSP})$. This makes $\mathsf{T}_1(\mathrm{BCSP})$ into a labelled transition system. Hence, in the light of Section 1.5 all semantic equivalences of Sections 2–12 and 14 are well-defined

Table 4
Action rules for BCSP

$$\frac{p \in [\![P]\!]}{aP \xrightarrow{a} p} \qquad \frac{p \xrightarrow{a} p'}{p+q \xrightarrow{a} p'} \qquad \frac{q \xrightarrow{a} q'}{p+q \xrightarrow{a} q'}$$

on $\mathsf{T}(\mathrm{BCSP})$, and for each of the semantics it is determined when two BCSP-expressions denote the same process.

The following theorem says that all these semantic equivalences are congruences for BCSP. Even stronger, all the preorders of this paper are precongruences for BCSP.

THEOREM 11. *Let $P, Q, R, S \in \mathsf{T}(\mathrm{BCSP})$ and let $\mathcal{O}$ be any of the semantics of Sections 2–12, 14. Then*

$$P =_{\mathcal{O}} Q \wedge R =_{\mathcal{O}} S \Rightarrow aP =_{\mathcal{O}} aQ \wedge P + R =_{\mathcal{O}} Q + S \wedge P \oplus R =_{\mathcal{O}} Q \oplus S,$$
$$P \sqsubseteq_{\mathcal{O}} Q \wedge R \sqsubseteq_{\mathcal{O}} S \Rightarrow aP \sqsubseteq_{\mathcal{O}} aQ \wedge P + R \sqsubseteq_{\mathcal{O}} Q + S \wedge P \oplus R \sqsubseteq_{\mathcal{O}} Q \oplus S.$$

PROOF. Each of the preorders $\mathcal{O}$ has a modal characterization, given by $P \sqsubseteq_{\mathcal{O}} Q \Leftrightarrow \mathcal{O}(P) \subseteq \mathcal{O}(Q)$, where $\mathcal{O}(P) = \bigcup_{p \in [\![P]\!]} \mathcal{O}(p)$ for $P \in \mathsf{T}(\mathrm{BCSP})$ and $\mathcal{O}(p) = \{\varphi \in \mathcal{L}_{\mathcal{O}} \mid p \models \varphi\}$ for $p \in \mathsf{T}_1(\mathrm{BCSP})$. Now $\mathcal{O}(P \oplus Q) = \mathcal{O}(P) \cup \mathcal{O}(Q)$. This immediately yields the compositionality of $\mathcal{O}$ with respect to $\oplus$: $P \sqsubseteq_{\mathcal{O}} Q \wedge R \sqsubseteq_{\mathcal{O}} S \Rightarrow P \oplus R \sqsubseteq_{\mathcal{O}} Q \oplus S$, and hence $P =_{\mathcal{O}} Q \wedge R =_{\mathcal{O}} S \Rightarrow P \oplus R =_{\mathcal{O}} Q \oplus S$.

Note that every formula in the infinitary Hennessy–Milner logic is logically equivalent to a disjunction of formulas of the form $\bigwedge_{i \in I} a_i\varphi_i \wedge \bigwedge_{j \in J} \neg a_j\varphi_j$. Let $\mathcal{O}'(P)$ be the class of formulas in $\mathcal{O}(P)$ of that form. It follows that $P \sqsubseteq_{\mathcal{O}} Q \Leftrightarrow \mathcal{O}'(P) \subseteq \mathcal{O}'(Q)$ for $P, Q \in \mathsf{T}(\mathrm{BCSP})$.

For $p, q \in \mathsf{T}_1(\mathrm{BCSP})$ one has $p + q \models \bigwedge_{i \in I} a_i\varphi_i \wedge \bigwedge_{j \in J} \neg a_j\varphi_j$ iff I can be written as $I_1 \cup I_2$ such that $p \models \bigwedge_{i \in I_1} a_i\varphi_i \wedge \bigwedge_{j \in J} \neg a_j\varphi_j$ and $q \models \bigwedge_{i \in I_2} a_i\varphi_i \wedge \bigwedge_{j \in J} \neg a_j\varphi_j$. Moreover, for each semantics $\mathcal{O}$ of this paper, if $\bigwedge_{i \in I} a_i\varphi_i \wedge \bigwedge_{j \in J} \neg a_j\varphi_j \in \mathcal{L}_{\mathcal{O}}$ and $I' \subset I$, then $\bigwedge_{i \in I'} a_i\varphi_i \wedge \bigwedge_{j \in J} \neg a_j\varphi_j \in \mathcal{L}_{\mathcal{O}}$.[12] Thus, for $P, Q \in \mathsf{T}(\mathrm{BCSP})$ and $\bigwedge_{i \in I} a_i\varphi_i \wedge \bigwedge_{j \in J} \neg a_j\varphi_j \in \mathcal{L}_{\mathcal{O}}$, one has $\bigwedge_{i \in I} a_i\varphi_i \wedge \bigwedge_{j \in J} \neg a_j\varphi_j \in \mathcal{O}'(P + Q)$ iff $I = I_1 \cup I_2$ such that $\bigwedge_{i \in I_1} a_i\varphi_i \wedge \bigwedge_{j \in J} \neg a_j\varphi_j \in \mathcal{O}'(P)$ and $\bigwedge_{i \in I_2} a_i\varphi_i \wedge \bigwedge_{j \in J} \neg a_j\varphi_j \in \mathcal{O}'(Q)$. This immediately yields the compositionality of $\mathcal{O}$ with respect to $+$.

The compositionality of $\mathcal{O}$ with respect to a is straightforward. □

If $P \in \mathsf{T}(\mathrm{BCSP})$, then $G([\![P]\!])$ is a finite process graph with multiple roots. Vice versa, any finite process graph with multiple roots $g \in \mathbb{G}^{mr}$ can be represented by a BCSP-expression $\langle\!\langle g \rangle\!\rangle \in \mathsf{T}(\mathrm{BCSP})$, such that $G(\langle\!\langle g \rangle\!\rangle) \leftrightarrows g$. Just extend Definition 17.1 by $\langle\!\langle g \rangle\!\rangle = \bigoplus_{r \in \mathsf{roots}(g)} \langle\!\langle g_r \rangle\!\rangle$.

Axioms. In Table 5, complete axiomatizations in terms of BCSP can be found for the same eleven semantics axiomatized in terms of BCCSP in Tables 2 and 3. The first two sections of the table apply to the equivalences and the first and last section to the preorders. These axioms are mild variations of the ones in Tables 2 and 3, and have been found by exploiting a close correspondence in semantic validity between BCSP and BCCSP expressions. First of all, using the definitions just given, the soundness of the axioms in the first section of Table 5 is easily established. Using these, any closed BCSP expression can be rewritten in the form $\bigoplus_{i=1}^{n} p_i$ with p_i closed BCCSP expressions. Now the following lemma reduces the validity of (in)equations over BCSP to that of (in)equations over BCCSP.

---

[12] At least when replacing the modality X of R, RT, PW and RS by $\bigwedge_{a \in Y} \neg a\top \wedge \bigwedge_{a \in Z} a\top$.

Table 5
Complete axiomatizations in terms of BCSP

	B	*RS*	*PW*	*RT*	*FT*	*R*	*F*	*CS*	*CT*	*S*	*T*
$(x \oplus y) \oplus z = x \oplus (y \oplus z)$	+	+	+	+	+	+	+	+	+	+	+
$x \oplus y = y \oplus x$	+	+	+	+	+	+	+	+	+	+	+
$x \oplus x = x$	+	+	+	+	+	+	+	+	+	+	+
$(x + y) + z = x + (y + z)$	+	+	+	+	+	+	+	+	+	+	+
$x + y = y + x$	+	+	+	+	+	+	+	+	+	+	+
$x + 0 = x$	+	+	+	+	+	+	+	+	+	+	+
$(x \oplus y) + z = (x + z) \oplus (y + z)$	+	+	+	+	+	+	+	+	+	+	+
$a(x \oplus y) = ax + ay$	+	+	+	+	+	+	+	+	+	+	+
$\sum_{i=1}^{n}(b_i x_i + b_i y_i) = \sum_{i=1}^{n}(b_i x_i + b_i y_i) \oplus \sum_{i=1}^{n} b_i y_i$		+	v	v	v	v	v	v	v	v	v
$bx + by + z = (bx + z) \oplus (by + z)$			+	v	v	v	v		v		v
$\sum_{i=1}^{n} b_i x_i \oplus \sum_{i=1}^{n} b_i y_i = \sum_{i=1}^{n}(b_i x_i + b_i y_i)$				+	+	v	v		v		v
$x + x = x$					+		v		v		v
$(bx + u) \oplus (by + v) = (bx + by + u) \oplus (by + v)$						+	+		v		v
$x \oplus (y + z) = x \oplus (x + y) \oplus (y + z)$							+		v		v
$x + by + z = (x + by + z) \oplus (by + z)$								+	v	v	v
$(bx + u) \oplus (cy + v) = bx + cy + u + v$									+		v
$x + y = (x + y) \oplus y$										+	v
$x \oplus y = x + y$											+
$x \sqsubseteq x \oplus y$		+	+	+	+	+	+	+	+	+	+
$bx + by + z = (bx + z) \oplus (by + z)$			+	v	v	v	v		v		v
$\sum_{i=1}^{n} b_i x_i \oplus \sum_{i=1}^{n} b_i y_i = \sum_{i=1}^{n}(b_i x_i + b_i y_i)$				+	v	v	v		v		v
$x + x = x$					+		v		v		v
$(bx + u) \oplus (by + v) \sqsupseteq bx + by + u$						+	v		v		v
$x \oplus (y + z) \sqsupseteq x + y$							+		v		v
$ax \sqsubseteq ax + y$								+	v	v	v
$(bx + u) \oplus (cy + v) = bx + cy + u + v$									+		v
$x \sqsubseteq x + y$										+	v
$x \oplus y = x + y$											+

LEMMA 17.1.

$$\bigoplus_{i=1}^{n} p_i \sqsubseteq_{\mathcal{O}} \bigoplus_{j=1}^{m} q_j \Leftrightarrow \sum_{i=1}^{n} ap_i \sqsubseteq_{\mathcal{O}} \sum_{j=1}^{m} aq_j$$

for $p_i, q_j \in \mathsf{T}(\text{BCCSP})$.

PROOF. $\varphi \in \mathcal{O}(\bigoplus_{i=1}^{n} p_i) \Leftrightarrow a\varphi \in \mathcal{O}(\sum_{i=1}^{n} ap_i)$. □

Most of the axioms in the last two sections of Table 5 can be recognized as restatements of the axioms of Tables 2 and 3, using the insight of Lemma 17.1. However, in BCSP it is not so clear how the set of initial actions of a process should be defined, and the obvious adaptations of the axioms involving the operator I would not be sound. Therefore the alternatives to those axioms discussed near the end of Section 17.2 are used. Moreover, in BCSP the axiom $x + x = x$ is not sound for readiness semantics. Substituting $a \oplus b$ for x, one derives $a \oplus (a + b) \oplus b = a \oplus b$, of which only the left-hand side has a ready pair $\langle \varepsilon, \{a, b\}\rangle$. However, in the setting of BCCSP all closed instances of $x + x = x$ are derivable from the law $ax + ax = ax$, which corresponds with the BCSP axiom $x \oplus x = x$. Following Lemma 17.1, the characteristic axiom for failure trace equivalence should be $x \oplus y = x \oplus y \oplus (x + y)$. This axiom is derivable from $x + x = x$, and all closed instances of $x + x = x$ are derivable from $x \oplus y = x \oplus y \oplus (x + y)$ and the axioms in the first section of Table 5.

Let $U_{\mathcal{O}}$ be the set of axioms marked with "+" in the column for $\mathcal{O}$ in the first two sections of Table 5, and $U^*_{\mathcal{O}}$ be the set of axioms marked with "+" in the column for $\mathcal{O}$ in the first and last section of Table 5. Write $S \vdash \Phi$ if the formula Φ is provable from the set of axioms S.

THEOREM 12. *For* $\mathcal{O} \in \{T, S, CT, CS, F, R, FT, RT, PW, RS, B\}$ *and* $P, Q \in \mathsf{T}(\text{BCSP})$ *one has* $P =_{\mathcal{O}} Q \Leftrightarrow U_{\mathcal{O}} \vdash P = Q$ *and* $P \sqsubseteq_{\mathcal{O}} Q \Leftrightarrow U^*_{\mathcal{O}} \vdash P \sqsubseteq Q$.

PROOF. "$\Leftarrow$" (*soundness*): In the light of Theorem 11 it suffices to show that the closed instances of the indicated axioms are valid in the corresponding semantics. In fact, one may restrict attention to the instances where expressions $\bigoplus_{i=1}^{n} p_i$ with p_i closed BCCSP expressions are substituted for the variables. It is not difficult to check, for each of these axioms, that such instances of it are derivable from the instances of it where simple closed BCCSP expressions are substituted for the variables (but taking $x \oplus y = x \oplus y \oplus (x + y)$ instead of $x + x = x$ to be the characteristic axiom for failure trace semantics). That the instances of the latter kind are valid in the corresponding semantics follows immediately from Lemma 17.1 and the soundness of the axioms for BCCSP.

"First $\Rightarrow$" (*completeness of the axioms for the equivalences*): Let $T'_{\mathcal{O}}$ be the set of axioms marked with "+" in the column for $\mathcal{O}$ in Table 2, but using $a\sum_{i=1}^{n} b_i x_i + a\sum_{i=1}^{n} b_i y_i = a\sum_{i=1}^{n}(b_i x_i + b_i y_i)$ and $a\sum_{i=1}^{n}(b_i x_i + b_i y_i) = a\sum_{i=1}^{n}(b_i x_i + b_i y_i) + a\sum_{i=1}^{n} b_i y_i$ instead of the axioms involving the operator I. As Theorem 8 establishes completeness for closed terms only, it holds for $T'_{\mathcal{O}}$ as well.

CLAIM. *If* $T'_{\mathcal{O}} \vdash p = \sum_{j=1}^{m} aq_j$ *for* $p, q_j \in \mathsf{T}(\text{BCCSP})$, *then, modulo applications of the first three axioms of Table* 2, p *has the form* $p = \sum_{i=1}^{n} ap_i$.

PROOF OF THE CLAIM. As all axioms in $T'_{\mathcal{O}}$ are equations, I may use induction on the proof of $p = \sum_{j=1}^{m} aq_j$ in equational logic. The case that $p = \sum_{j=1}^{m} aq_j$ is a closed instance of an axiom of $T'_{\mathcal{O}}$ proceeds by inspection of those axioms. The cases of placing an equation in a context, as well as reflexivity, symmetry and transitivity, are trivial. □

CLAIM.

$$T'_{\mathcal{O}} \vdash \sum_{i=1}^{n} ap_i = \sum_{j=1}^{m} aq_j \Rightarrow U_{\mathcal{O}} \vdash \bigoplus_{i=1}^{n} p_i = \bigoplus_{j=1}^{m} q_j \quad \textit{for any } p_i, q_j \in \mathsf{T}(\text{BCCSP}).$$

PROOF OF THE CLAIM. I use induction on the proof of $\sum_{i=1}^{n} ap_i = \sum_{j=1}^{m} aq_j$ from $T'_{\mathcal{O}}$ in equational logic. The case that $\sum_{i=1}^{n} ap_i = \sum_{j=1}^{m} aq_j$ is a closed instance of an axiom of $T'_{\mathcal{O}}$ proceeds by inspection of those axioms, taking into account the remark about $x \oplus y = x \oplus y \oplus (x + y)$ right before this theorem. The case of a closed instance of an axiom of $T'_{\mathcal{O}}$ in a context is straightforward, also using that all closed instances of axioms of $T'_{\mathcal{O}}$ are derivable from the ones of $U_{\mathcal{O}}$, taking into account the remark about $x + x = x$ right before this theorem. The cases of reflexivity and symmetry are trivial. Transitivity follows from the previous claim. □

COMPLETENESS PROOF. Suppose $P =_{\mathcal{O}} Q$ for certain $P, Q \in \mathsf{T}(\text{BCSP})$. Using the axioms in the first section of Table 5 one obtains $U_{\mathcal{O}} \vdash P = \bigoplus_{i=1}^{n} p_i$ and $U_{\mathcal{O}} \vdash Q = \bigoplus_{j=1}^{m} q_j$ with $p_i, q_j \in \mathsf{T}(\text{BCCSP})$. By the soundness of these axioms one has $\bigoplus_{i=1}^{n} p_i =_{\mathcal{O}} \bigoplus_{j=1}^{m} q_j$. Therefore $\sum_{i=1}^{n} ap_i =_{\mathcal{O}} a \bigoplus_{i=1}^{n} p_i =_{\mathcal{O}} a \bigoplus_{j=1}^{m} q_j =_{\mathcal{O}} \sum_{j=1}^{m} aq_j$ by the soundness of $a(x \oplus y) = ax + ay$ and Theorem 11, and hence $T'_{\mathcal{O}} \vdash \sum_{i=1}^{n} ap_i = \sum_{j=1}^{m} aq_j$ by the completeness of $T'_{\mathcal{O}}$. Now $U_{\mathcal{O}} \vdash P = Q$ follows by the claim above.

The second "⇒" (*completeness of the axioms for the preorders*) goes likewise, except that in the proof of the second claim, in order to handle the axioms $ax \sqsubseteq ax + y$ and $x \sqsubseteq x + y$, one uses the axiom $x \sqsubseteq x \oplus y$ of $U^*_{\mathcal{O}}$. Furthermore, $ax \sqsubseteq ax + y$ is derivable from U^*_{CT}, and $x \sqsubseteq x + y$ from U^*_T. □

18. Criteria for selecting a semantics for particular applications

Must testing. Assume the testing scenario of trace semantics: we are unable to influence the behaviour of an investigated system in any way and can observe the performed actions only. Not even deadlock is observable. In this case there appears to be no reason to distinguish the two processes of Counterexample 3, $ab + a(b + c)$ and $a(b + c)$. They have the same traces, and consequently allow the same observations. Likewise, one might see no reason to distinguish between the two processes of Counterexample 2, $ab + a$ and ab; also these have the same traces. However, when buying process ab, it may come with the guarantee that, in every run of the system, sooner or later it will perform the action b, at least if the action a is known to terminate. Such a guarantee cannot be given for $ab + a$. The distinction between ab and $ab + a$ alluded to here can be formalized with the concept of *must testing*, originally due to De Nicola and Hennessy [17]: ab *must* do a b, whereas $ab + a$ must not.

For finite processes, must testing could be formalized as follows. For $t \subseteq Act^*$ we say that a finite process $p \in \mathbb{P}$ *must* pass the test t if $CT(p) \subseteq t$. To test whether a process will sooner or later perform a b-action take t to be all sequences of actions containing a b. To test whether a process will always perform a b immediately after it does an a, take t to be all traces in which any a is immediately followed by a b. Now write $p \sqsubseteq_T^{\text{must}} q$ if for all tests $t \subseteq Act^*$ such that p must pass t, q must pass t as well. It is easy to see that, for finite processes p and q, $p \sqsubseteq_T^{\text{must}} q$ iff $q \sqsubseteq_{CT} p$.

All testing scenarios $\mathcal{O}$ sketched earlier in this paper can be regarded as forms of *may testing*: it is recorded whether an observation $\varphi \in \mathcal{L}_{\mathcal{O}}$ *may* be made for a process p, and one writes $p \sqsubseteq_{\mathcal{O}} q$ if any observation that may be made for p, may also be made for q.

In the context of a testing scenario $\mathcal{O}$ with $\mathcal{O} \succeq CT$, a plausible form of must testing can be defined as well, and for finite processes plausible formalizations yield that $p \sqsubseteq_{\mathcal{O}}^{\text{must}} q$ iff $q \sqsubseteq_{\mathcal{O}} p$.

For infinite processes there are several ways to formalize must testing, and analyzing the resulting preorders falls outside of the scope if this paper.

Deadlock behaviour. A process is said to reach a state of *deadlock* if it can do no further actions.[13] The process $ab+a$ for instance may deadlock right after performing an a-action, whereas the process ab may not. One could say that a semantics $\mathcal{O}$ *respects deadlock behaviour* iff $\mathcal{O} \succeq CT$. Counterexample 4 then shows that none of the semantics on the left in Figure 9 respects deadlock behaviour; only the left-hand process of Counterexample 4 can deadlock after an a-move. Respecting deadlock behaviour may be a requirement on semantics in applications where either deadlock is important in its own right, or where (implicitly) a form of must-testing is considered.

Full abstraction. Many testing scenarios mentioned in this paper employ the notion that an action can happen only if it is not blocked by the environment, that is, only if both the investigated process *and* the environment are ready to participate in it. Modelling both the investigated process and the responsible part of the environment as process graphs gives rise to the following binary *intersection operator* that allows an action to happen only if it can happen in both of its arguments.

DEFINITION 18.1. Let $\cap$ be the binary operator on process graphs defined by
- $\text{NODES}(g \cap h) = \text{NODES}(g) \times \text{NODES}(h)$,
- $\text{ROOTS}(g \cap h) = \text{ROOTS}(g) \times \text{ROOTS}(h)$,
- $((s,t), a, (s',t')) \in \text{EDGES}(g)$ iff $(s,a,s') \in \text{EDGES}(g) \wedge (t,a,t') \in \text{EDGES}(h)$.

In order to obtain a connected process graph, unreachable parts need to be removed.

This operator is also called *synchronous parallel composition* and is denoted $\|$ in Hoare [31]. It can be added to BCCSP or BCSP by employing the action rule

$$\frac{p \xrightarrow{a} p',\ q \xrightarrow{a} q'}{p \cap q \xrightarrow{a} p' \cap q'}.$$

Trace semantics turns out to compositional for the intersection operator, i.e., if $g =_T g'$ and $h =_T h'$ then $g \cap h =_T g' \cap h'$. For $T(g \cap h) = T(g) \cap T(h)$. So are failures and readiness semantics:

$$\langle \sigma, X \rangle \in F(g \cap h) \Leftrightarrow \exists \langle \sigma, Y \rangle \in F(g),\ \langle \sigma, Z \rangle \in F(h)\colon\ X = Y \cup Z$$
$$\langle \sigma, X \rangle \in R(g \cap h) \Leftrightarrow \exists \langle \sigma, Y \rangle \in R(g),\ \langle \sigma, Z \rangle \in R(h)\colon\ X = Y \cap Z.$$

In fact, it is not hard to see that all semantics of this paper are compositional for $\cap$, except for *CT* and *CS*, and their (in)finitary versions. The two processes of Counterexample 3,

[13] In settings were successful termination is modelled (cf. Section 19) a state of deadlock is only reached if moreover the process cannot terminate successfully.

$ab + a(b + c)$ and $a(b + c)$, are completed trace equivalent, even completed simulation equivalent, yet after intersecting them with ac only the first one has a completed trace a.

In applications where the intersection operator is used, one may require a suitable semantics to be compositional for it. This rules out *CT* and *CS*. If also deadlock behaviour is of importance, F appears to be the coarsest semantics to be considered, as least among the ones reviewed in this paper. As a matter of fact, it is the coarsest semantics even among the ones not reviewed here.

DEFINITION 18.2. An equivalence relation is called *fully abstract* with respect to a property if it is the coarsest equivalence with that property, i.e., if it has the property, and any other equivalence having that property is finer.

An equivalence is said to *fully abstract* with respect to another equivalence $\sim$ and some operators, if it is the coarsest equivalence finer than $\sim$ that is compositional with respect to those operators.

An equivalence $\approx$ on $\mathbb{G}$ is fully abstract with respect to an equivalence $\sim$ and a set L of operators on $\mathbb{G}$ iff

(1) it is compositional with respect to the operators in L, and

(2) for any two process graphs $g, h \in \mathbb{G}$ with $g \not\approx h$ there exists a context $C[\cdot]$ of operators from L such that $C[g] \not\sim C[h]$.

In fact, for every equivalence relation $\sim$ on $\mathbb{G}$ and every set L of operators on $\mathbb{G}$ there exists a unique equivalence relation $\approx$ that is fully abstract with respect to $\sim$ and the operators in L, namely the one defined by $g \approx h$ iff $C[g] \sim C[h]$ for every context $C[\cdot]$ of operators from L.

THEOREM 13. *Failures equivalence is fully abstract with respect to $=_{CT}$ and $\cap$, i.e., with respect to deadlock behaviour and intersection.*

PROOF. (1) has already been established. For (2), let $g \neq_F h$. Without loss of generality let $\langle\sigma, X\rangle \in F(g) - F(h)$. Let k be the process graph that is shaped like the failure pair $\langle\sigma, X\rangle$, i.e., the process that performs the actions of σ in succession, after which it offers a choice between the actions of X, and nothing else. Then $\sigma \in CT(g \cap k) - CT(h \cap k)$. □

Variants of Theorem 13 are abundant in the literature. See, e.g., [11].

Renaming. For every function $f : Act \to Act$ one can define a unary renaming operator on $\mathbb{G}$ that renames the labels of all transitions in its argument according to f. In case f is injective, all semantics of this paper are compositional for the associated renaming operator, as is trivial to check. Non-injective renaming operators are useful to express a degree of abstraction. Imagine a process that can do, among others, actions a_1 and a_2. At some level of abstraction, the difference between a_1 and a_2 may be considered irrelevant. This can be expressed by applying a renaming that relabels both a_1 and a_2 into the same action a. Naturally, if two processes are equivalent before applying such a renaming operator, one would expect them to still be equivalent afterwards, i.e., after abstracting from the difference between a_1 and a_2. It is for this reason that one might require semantics to be compositional

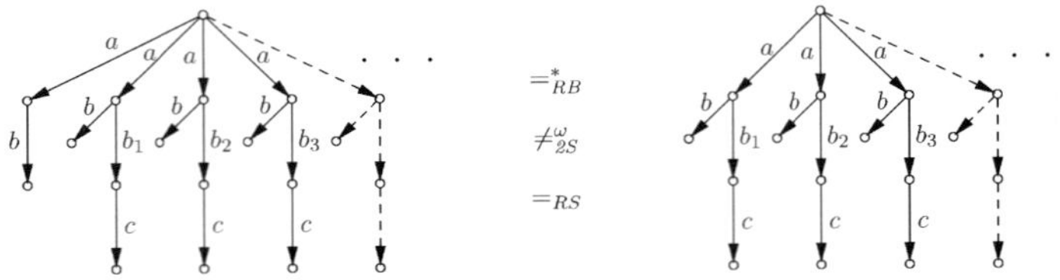

Counterexample 20. $F2S^*$, $R2S^*$, FB^* and RB^* are not compositional for renaming.

for (non-injective) renaming. As it happens, all semantics between F^1 and B^- fail this requirement. For the two processes of Counterexample 4 are HML-equivalent ($=^-_B$), but after renaming all actions b_i into b (for $i = 1, 2, \ldots$) the resulting processes are not even singleton-failures equivalent ($=^1_F$). For only the first one has a singleton-failure pair $\langle a, b\rangle$. This can be considered an argument against the semantics on the left of Figure 9.

Counterexample 20 shows that also $F2S^*$, $R2S^*$, FB^* and RB^* are not compositional for renaming. In this counterexample b is a shorthand for $\sum_{i=1}^{\infty} b_i$, in the sense that whenever a transition $p \xrightarrow{b} q$ is displayed, all the transitions $p \xrightarrow{b_i} q$ for $i \geqslant 1$ are meant to be present. With some effort one checks that both processes satisfy the same formulas in $\mathcal{L}^*_{RB}$. However, after renaming all actions b_i into b they are no longer $2S^-$-equivalent: only the first process satisfies $a\neg(bc\top)$. For all other semantics of Figure 9 it is rather easy to establish that they are compositional for renaming.

Other compositionality requirements. Many formal languages for the description of concurrent systems, including CCS [37], SCCS [39], CSP [31] and ACP [7], are *De Simone languages* (cf. [3]). This means that their operators (the *De Simone operators*) can be defined with action rules of a particular form (the *De Simone format*). Because De Simone languages are used heavily in algebraic system verification, semantic equivalences that are compositional for such languages are often desirable.

THEOREM 14. *The semantics T, T^∞, F, F^∞, R, R^∞, FT, FT^∞, RT, RT^∞, PF, PF^∞, S^*, S^ω, S, FS^*, RS^*, RS^ω, RS, $2S^\omega$, $2S$, B^ω and B are compositional with respect to all De Simone languages.*

PROOF. Omitted. □

For all the other semantics of Figure 9, which are displayed there in red (or shaded), there are counterexamples against such a result. Tree semantics fails to be compositional with respect to the $+$ of BCCSP, unless the action relations are upgraded with multiplicities, but that takes us outside of De Simone format. The semantics F^1, F^-, R^-, FT^-, RT^-, RS^-, $2S^-$, B^-, $F2S^*$, $R2S^*$, FB^* and RB^* fail to be compositional with respect to renaming, and CT, CT^∞, CS, CS^ω, CS fail to be compositional with respect to intersection. These are all De Simone operators. Finally, Counterexample 21 shows that PW is not compositional for the synchronization operator $\times$ of SCCS [39] – also a De Simone operator.

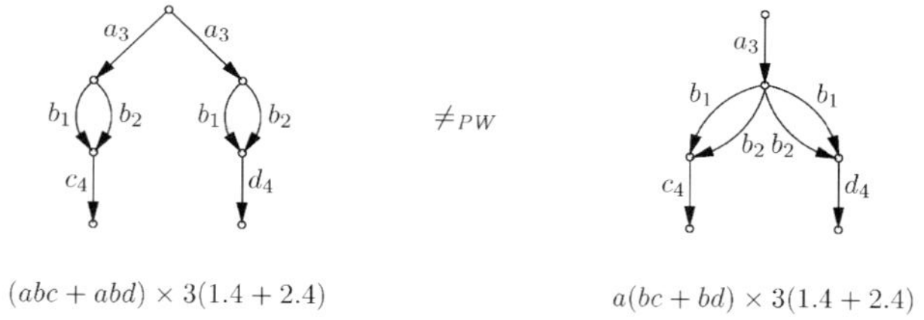

Counterexample 21. Possible worlds semantics is not compositional for synchronization.

This operator can be used to create a context, in which the two possible worlds equivalent processes of Counterexample 8 are converted into the two processes above. These are no longer possible worlds equivalent, for only the one on the right has a possible world $a_3(b_1c_4 + b_2d_4)$. The same counterexample can also be created with the inverse image operator of CSP [31].

In Baeten, Bergstra and Klop [6] a unary priority operator was defined on process graphs. This operator, which is not a De Simone operator, assumes a partial ordering $<$ on *Act*, i.e., there is one priority operator for each such ordering. The operator acts on graphs by removing all transitions (s, a, t) for which there is a transition (s, b, u) with $b > a$ (and unreachable parts are removed as well). Thus, in a choice between several actions, only the actions with maximal priority may be executed. It is known that *RT*, *RS*, *B* and *U* are compositional for the priority operators. I think that RT^∞, *PW*, RS^*, RS^ω, RB^* and B^ω are too. However, none of the other semantics of Figure 9 is. Thus, in applications where priority operators are used and algebraic reasoning makes compositionality essential, only semantics like *RT*, *RS* and *B* are recommendable.

Depending on the application, compositionality for other operators may be required as well, leading to various restrictions on the array of suitable semantics. More on which semantics are compositional for which operators can be found in Aceto, Fokkink and Verhoef [3] and the references therein.

The recursive specification principle. A *recursive specification* is an equation of the form $X = t$ with X a variable and t a term (in a language such as BCCSP) containing no other variables than X. (In the literature often recursive specifications are allowed to involve more variables and more such equations, but I do not need those here.) A recursive specification $X = t$ over BCCSP is *guarded* if every occurrence of X in t occurs in a subterm at' of t with $a \in Act$. Recursive specifications are meant to specify processes. A process p is said to be a *solution* of the recursive specification $X = t$, using the semantics $\mathcal{O}$, if the equation evaluates to a true statement when substituting p for X and interpreting $=$ as $=_\mathcal{O}$. The *recursive specification principle* (*RSP*) says that guarded recursive specifications have unique solutions. It has been established for bisimulation semantics by Milner [39] (using the language SCCS), and holds in fact for most semantics encountered in this paper. In process algebra, two processes are often proven semantically equivalent by showing that they are solutions of the same recursive specification (cf. [5]). For this purpose it is important to work with a semantics in which RSP holds. In the infinitary semantics between T^∞

and *PW* this is in fact not the case. For in those semantics the two different processes of Counterexample 1 are both solutions of the guarded recursive specification $X = aX + a$. For the finitary semantics this counterexample does not apply, because the two processes are identified, whereas in simulation semantics (of finer) these two processes fail to be solutions of the same recursive specification.

Other considerations. In general it depends on the kind of interactions that are permitted between a process and its environment (i.e., the testing scenario) which semantics is sufficiently discriminating for a particular application. When a range of appropriate semantics is found, also considering the criteria discussed earlier in the section, the question rises which of these semantics to actually use (e.g., in making a formal verification). A natural choice is the *coarsest* of the appropriate semantics, i.e., the one which is fully abstract with respect to the requirements it has to meet in order to be adequate in the context in which the investigated processes will be operating. In this semantics more equations are valid than in any other. If the goal is to prove that two processes are equivalent, this may succeed when using the fully abstract semantics, whereas it may not even be true in a finer one. Sometimes it is argued that the complexity of deciding equivalence between processes is too high for certain semantics; using them would give rise to too hard verifications. However, this cannot be an argument for rejecting a semantics in favour of a finer one. For doing the verification in the finer semantics is actually a *method* of establishing equivalence in the coarser semantics. In other words, when $\mathcal{O} \prec \mathcal{N}$, establishing $p =_{\mathcal{O}} q$ cannot be harder than establishing $p =_{\mathcal{N}} q$, as establishing $p =_{\mathcal{N}} q$ is one of the ways of establishing $p =_{\mathcal{O}} q$. If deciding $\mathcal{O}$-equivalence has a higher complexity than deciding $\mathcal{N}$-equivalence, the hard cases to decide must be the equations $p =_{\mathcal{O}} q$ for which $p =_{\mathcal{N}} q$ is not even true. It is especially for those applications that $\mathcal{O}$-semantics has a distinct advantage over $\mathcal{N}$-semantics. This argument has been made forcefully in Valmari [48].

In practice, it may not always be certain in what ways the environment can interact with investigated processes, and hence what constitutes their observable behaviour. Moreover, the processes under investigation may be transferred to more powerful environments long after their initial use. One of the ways this could happen is through the introduction of more operators for which the underlying semantics has to be compositional. A big disadvantage of semantics that are fully abstract with respect to non-stable notions of observability (or non-stable sets of operators) is that whenever a verification is carried out in a such a semantics, and one decides that the context in which the verified system will be working is such that actually a little bit more can be observed that what was originally accounted for, the verification has to be completely redone. Moreover, the correctness of the investigated systems keeps depending on the completeness of the underlying testing scenario. In such cases it is preferable to carry out verifications in the finest semantics for which this is convenient. This gives stronger equivalence results, which have a greater change of surviving in conditions where the environment gets more powerful than originally anticipated. Especially using bisimulation is safe bet, as it respects the internal structure of processes to such a degree that it is hard to imagine ever running into an environment that distinguishes bisimilar processes. In Bloom, Istrail and Meyer [12] it is argued that ready simulation semantics already respects the limits of observable behaviour, so this may be a good alternative. It should be pointed out, however, that most applications involve abstraction from internal actions (not treated in this paper), and hence require variants of the semantics treated here that accommodate such abstractions. In this setting, the question of which semantics represents the limit of observable behaviour is much harder.

19. Distinguishing deadlock and successful termination

Often researchers feel the need to distinguish two ways in which a process can end: successfully (by completing its mission) or unsuccessfully (for instance because its waits for an input from the environment that will never arrive). This distinction can be formally modelled in the context of labelled transition systems by considering triples $(\mathbb{P}, \rightarrow, \surd)$ in which $(\mathbb{P}, \rightarrow)$ is a labelled transition system as in Definition 1.1 and $\surd \subseteq \mathbb{P}$ is a predicate on processes expressing which ones can terminate successfully in their current state. It may or may not be required that the processes $p \in \mathbb{P}$ with $\surd(p)$ have no outgoing transitions. Likewise, in the setting of process graphs, one studies tuples $(\text{NODES}(g), \text{ROOT}(g), \text{EDGES}(g), \surd(g))$ with $\surd(g) \subseteq \text{NODES}(g)$. Now any labelled transition system over an alphabet *Act* equipped with such a successful termination predicate, can be encoded as an ordinary labelled transition system over an alphabet $Act \cup \{\surd\}$ with $\surd \notin Act$. Namely, instead of labelling the processes/states where successful termination occurs with $\surd$, one can view successful termination as a kind of action, and add $\surd$-labelled transitions from those processes/states to fresh endstates. Now any semantic equivalence defined on ordinary labelled transition systems extends to labelled transition systems with a successful termination predicate by declaring two processes equivalent iff they are equivalent in the encoded transition system. In fact, in the same way all equivalences and preorders of this paper extend to labelled transition systems equipped with arbitrary predicates $P \subseteq \mathbb{P}$. Below, three of the thusly defined equivalences are characterized explicitly in terms of $\surd$.

DEFINITION 19.1. Let $(\mathbb{P}, \rightarrow, \surd)$ be a labelled transition system with successful termination.

$\sigma \in Act^*$ is a *terminating trace* of a process p if there is a process q such that $p \xrightarrow{\sigma} q$ and $\surd(q)$. Let $L(p)$ denote the set of terminating traces of p (and let $T(p)$ and $CT(p)$ be defined as before).

Now two processes p and q are trace equivalent iff $T(p) = T(q)$ and $L(p) = L(q)$. They are completed trace equivalent iff $T(p) = T(q)$, $CT(p) = CT(q)$ and $L(p) = L(q)$. They are bisimulation equivalent iff there exists a binary relation R on $\mathbb{P}$ with pRq, satisfying, for $a \in Act$:

- if pRq and $p \xrightarrow{a} p'$, then $\exists q'$: $q \xrightarrow{a} q'$ and $p'Rq'$;
- if pRq and $q \xrightarrow{a} q'$, then $\exists p'$: $p \xrightarrow{a} p'$ and $p'Rq'$;
- if pRq, then $\surd(p) \Leftrightarrow \surd(q)$.

Language semantics. The nondeterministic *automata* studied in *automata theory* (cf. Hopcroft and Ullman [33]) can be regarded as process graphs with a termination predicate (except that in automata theory the focus is on *finite* automata). The states $s \in \text{NODES}(g)$

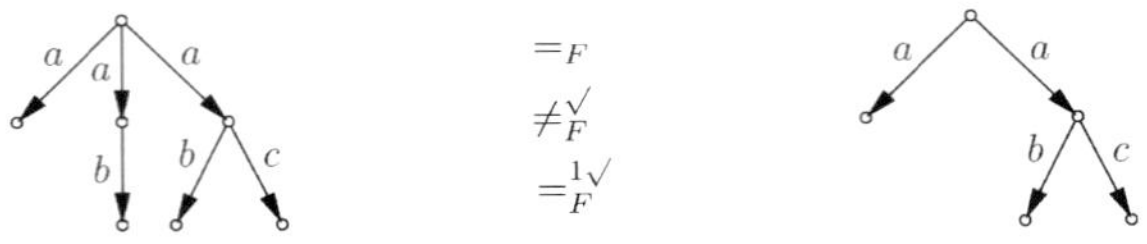

$a + ab + a(b + c)$ $a + a(b + c)$

Counterexample 22. Failures semantics is not compositional for sequencing.

Counterexample 22 shows that failures semantics is not compositional for sequencing. There *left* $=_F$ *right*, but *left*; $c \neq_F$ *right*; c. The same counterexample, with all endnodes successfully terminating, shows that singleton-failures semantics is not compositional for either sequencing or sequential composition. Likewise, Counterexample 14 shows that PF and $2S$ are not compositional for sequencing, and Counterexample 4 shows that none of the semantics between T and B^- are. All of the semantics studied in this paper, except for F^1, are compositional for sequential composition. As sequencing is the same as sequential composition on processes where all endstates, and only those, are considered to be successfully terminating, this implies that all the semantics $\mathcal{O}^{\surd}$, except for $F^{1\surd}$, are compositional for sequencing. If c is an action that does not occur in p or q, then $p; c =_{\mathcal{O}} q; c \Leftrightarrow p =_{\mathcal{O}}^{\surd} q$. (Think of c as $\surd$.) From this it follows that for all semantics $\mathcal{O}$, except F^1, $\mathcal{O}^{\surd}$ is fully abstract with respect to $\mathcal{O}$ and sequencing, at least for processes that can not execute every action in *Act*.

Concluding remarks

In this paper various semantic equivalences for concrete sequential processes are defined, motivated, compared and axiomatized. Of course many more equivalences can be given than the ones presented here. The reason for selecting just these, is that they can be motivated rather nicely and/or play a rôle in the literature on semantic equivalences. In Abramsky and Vickers [2] the observations which underly many of the semantics in this paper are placed in a uniform algebraic framework, and some general completeness criteria are stated and proved. They also introduce *acceptance semantics*, which can be obtained from acceptance-refusal semantics (Section 7) by dropping the refusals, and analogously *acceptance trace semantics*. I am not aware of any reasonable testing scenario for these notions.

In Section 9 I remarked that a testing scenario for simulation and ready simulation semantics can be obtained by adding an *undo*-button to the scenario's for trace and ready trace semantics. Likewise, Schnoebelen [47] investigates the addition of an *undo*-button to the testing scenarios for completed trace, readiness, failures and failure trace semantics, thereby obtaining 3 new equivalences $CT_\#$, $R_\#$ and $F_\#$. *Undo*-failure trace equivalence coincides with finitary failure simulation equivalence, just like *undo*-trace and *undo*-ready trace equivalence coincide with finitary simulation and finitary ready simulation equivalence. For image finite processes $R_\#$ coincides with $F_\#$. Furthermore $R \preceq R_\# \preceq RS^*$, $F \preceq F_\# \preceq FS^*$, $CT \preceq CT_\# \preceq CS^*$ and $S^* \preceq CT_\# \preceq F_\# \preceq R_\#$.

An interesting topic is the generalization of this work to a setting with silent moves and/or with parallelism. In both cases there turn out to be many interesting variations. The generalization to a setting with invisible actions will be tackled in [21]. Some work towards generalizing the spectrum to a setting with parallelism can be found for instance in [44] and [19].

Acknowledgement

My thanks to Tony Hoare for suggesting that the axioms of Table 2 could be simplified along the lines of Table 5.

References

[1] S. Abramsky, *Observation equivalence as a testing equivalence*, Theoret. Comput. Sci. **53** (1987), 225–241.

[2] S. Abramsky and S. Vickers, *Quantales, observational logic and process semantics*, Math. Structures Comput. Sci. **3** (1993), 161–227.

[3] L. Aceto, W.J. Fokkink and C. Verhoef, *Structural operational semantics*, Handbook of Process Algebra, J.A. Bergstra, A. Ponse and S.A. Smolka, eds, Elsevier, Amsterdam (2001), 197–292.

[4] P. Aczel, *Non-well-founded Sets*, CSLI Lecture Notes 14, Stanford University (1988).

[5] J.C.M. Baeten, ed., *Applications of Process Algebra*, Cambridge Tracts in Theoretical Comput. Sci. 17, Cambridge University Press (1990).

[6] J.C.M. Baeten, J.A. Bergstra and J.W. Klop, *Ready-trace semantics for concrete process algebra with the priority operator*, Comput. J. **30** (6) (1987), 498–506.

[7] J.C.M. Baeten and W.P. Weijland, *Process Algebra*, Cambridge Tracts in Theoret. Comput. Sci. 18, Cambridge University Press (1990).

[8] J.W. de Bakker, J.N. Kok, J.-J.Ch. Meyer, E.-R. Olderog and J.I. Zucker, *Contrasting themes in the semantics of imperative concurrency*, Current Trends in Concurrency, Lecture Notes in Comput. Sci. 224, J.W. de Bakker, W.P. de Roever and G. Rozenberg, eds, Springer (1986), 51–121.

[9] J.W. de Bakker and J.I. Zucker, *Processes and the denotational semantics of concurrency*, Inform. and Control **54** (1/2) (1982), 70–120.

[10] J.A. Bergstra, J.W. Klop and E.-R. Olderog, *Failure semantics with fair abstraction*, Report CS-R8609, CWI, Amsterdam (1986).

[11] J.A. Bergstra, J.W. Klop and E.-R. Olderog, *Readies and failures in the algebra of communicating processes*, SIAM J. Comput. **17** (6) (1988), 1134–1177.

[12] B. Bloom, S. Istrail and A.R. Meyer, *Bisimulation can't be traced*, J. ACM **42** (1) (1995), 232–268.

[13] S.D. Brookes, C.A.R. Hoare and A.W. Roscoe, *A theory of communicating sequential processes*, J. ACM **31** (3) (1984), 560–599.

[14] S.D. Brookes and A.W. Roscoe, *An improved failures model for communicating processes*, Seminar on Concurrency, Lecture Notes in Comput. Sci. 197, S.D. Brookes, A.W. Roscoe and G. Winskel, eds, Springer (1985), 281–305.

[15] P. Darondeau, *An enlarged definition and complete axiomatisation of observational congruence of finite processes*, Proceedings International Symposium on Programming: 5th Colloquium, Aarhus, Lecture Notes in Comput. Sci. 137, M. Dezani-Ciancaglini and U. Montanari, eds, Springer (1982), 47–62.

[16] R. De Nicola, *Extensional equivalences for transition systems*, Acta Inform. **24** (1987), 211–237.

[17] R. De Nicola and M. Hennessy, *Testing equivalences for processes*, Theoret. Comput. Sci. **34** (1984), 83–133.

[18] J. Engelfriet, *Determinacy* $\rightarrow$ *(observation equivalence* $=$ *trace equivalence)*, Theoret. Comput. Sci. **36** (1) (1985), 21–25.

[19] R.J. van Glabbeek, *The refinement theorem for ST-bisimulation semantics*, Proceedings IFIP TC2 Working Conference on Programming Concepts and Methods, Sea of Gallilee, Israel, M. Broy and C.B. Jones, eds, North-Holland (1990), 27–52.

[20] R.J. van Glabbeek, *The linear time – branching time spectrum*, Report CS-R9029, CWI, Amsterdam. Extended abstract in: Proceedings CONCUR '90, Theories of Concurrency: Unification and Extension, Amsterdam, August 1990, Lecture Notes in Comput. Sci. 458, J.C.M. Baeten and J.W. Klop, eds, Springer-Verlag (1990), 278–297.

[21] R.J. van Glabbeek, *The linear time – branching time spectrum II; The semantics of sequential systems with silent moves*, Manuscript (1993). Preliminary version available by ftp at ftp://boole.stanford.edu/pub/-spectrum.ps.gz. Extended abstract in: Proceedings CONCUR '93, 4th International Conference on Concurrency Theory, Hildesheim, Germany, August 1993, Lecture Notes in Comput. Sci. 715, E. Best, ed., Springer (1993), 66–81.

[22] R.J. van Glabbeek and J.J.M.M. Rutten, *The processes of De Bakker and Zucker represent bisimulation equivalence classes*, 25 Jaar Semantiek, Liber Amicorum, J.W. de Bakker, ed., CWI, Amsterdam (1989), 243–246.

[23] J.F. Groote, *A new strategy for proving ω-completeness with applications in process algebra*, Proceedings CONCUR 90, Amsterdam, Lecture Notes in Comput. Sci. 458, J.C.M. Baeten and J.W. Klop, eds, Springer (1990), 314–331.

[24] J.F. Groote and H. Hüttel, *Undecidable equivalences for basic process algebra*, Inform. and Control **115** (2) (1994), 354–371.

[25] J.F. Groote and F.W. Vaandrager, *Structured operational semantics and bisimulation as a congruence*, Inform. and Comput. **100** (2) (1992), 202–260.

[26] M. Hennessy, *Acceptance trees*, J. ACM **32** (4) (1985), 896–928.

[27] M. Hennessy and R. Milner, *On observing nondeterminism and concurrency*, Proceedings 7th ICALP, Noorwijkerhout, Lecture Notes in Comput. Sci. 85, J.W. de Bakker and J. van Leeuwen, eds, Springer (1980), 299–309.

[28] M. Hennessy and R. Milner, *Algebraic laws for nondeterminism and concurrency*, J. ACM **32** (1) (1985), 137–161.

[29] C.A.R. Hoare, *Communicating sequential processes*, Comm. ACM **21** (8) (1978), 666–677.

[30] C.A.R. Hoare, *Communicating sequential processes*, On the Construction of Programs – An Advanced Course, R.M. McKeag and A.M. Macnaghten, eds, Cambridge University Press (1980), 229–254.

[31] C.A.R. Hoare, *Communicating Sequential Processes*, Prentice Hall, Englewood Cliffs (1985).

[32] M.J. Hollenberg, *Hennessy–Milner classes and process algebra*, Modal Logic and Process Algebra: A Bisimulation Perspective, CSLI Lecture Notes 53, A. Ponse, M. de Rijke and Y. Venema, eds, CSLI Publications, Stanford, CA (1995), 187–216.

[33] J.E. Hopcroft and J.D. Ullman, *Introduction to Automata Theory, Languages and Computation*, Addison-Wesley (1979).

[34] J.K. Kennaway, *Formal semantics of nondetermism and parallelism*, Ph.D. Thesis, University of Oxford (1981).

[35] K.G. Larsen and A. Skou, *Bisimulation through probabilistic testing*, Inform. and Comput. **94** (1991).

[36] A.R. Meyer, *Report on the 5th international workshop on the semantics of programming languages in Bad Honnef*, Bull. European Assoc. Theoret. Comput. Sci. **27** (1985), 83–84.

[37] R. Milner, *A Calculus of Communicating Systems*, Lecture Notes in Comput. Sci. 92, Springer (1980).

[38] R. Milner, *Modal characterisation of observable machine behaviour*, Proceedings CAAP '81, Lecture Notes in Comput. Sci. 112, G. Astesiano and C. Bohm, eds, Springer (1981), 25–34.

[39] R. Milner, *Calculi for synchrony and asynchrony*, Theoret. Comput. Sci. **25** (1983), 267–310.

[40] E.-R. Olderog and C.A.R. Hoare, *Specification-oriented semantics for communicating processes*, Acta Inform. **23** (1986), 9–66.

[41] D.M.R. Park, *Concurrency and automata on infinite sequences*, Proceedings 5th GI Conference, Lecture Notes in Comput. Sci. 104, P. Deussen, ed., Springer (1981), 167–183.

[42] I.C.C. Phillips, *Refusal testing*, Theoret. Comput. Sci. **50** (1987), 241–284.

[43] A. Pnueli, *Linear and branching structures in the semantics and logics of reactive systems*, Proceedings 12th ICALP, Nafplion, Lecture Notes in Comput. Sci. 194, W. Brauer, ed., Springer (1985), 15–32.

[44] L. Pomello, *Some equivalence notions for concurrent systems – An overview*, Advances in Petri Nets 1985, Lecture Notes in Comput. Sci. 222, G. Rozenberg, ed., Springer (1986), 381–400.
[45] A.W. Roscoe, *Unbounded non-determinism in CSP*, J. Logic Comput. **3** (2) (1993), 131–172.
[46] W.C. Rounds and S.D. Brookes, *Possible futures, acceptances, refusals and communicating processes*, Proceedings 22th Annual Symposium on Foundations of Computer Science, Nashville, TN, IEEE, New York (1981), 140–149.
[47] Ph. Schnoebelen, *Experiments on processes with backtracking*, Proceedings CONCUR '91, Amsterdam, 527, J.C.M. Baeten and J.F. Groote, eds, Springer (1991), 80–94.
[48] A. Valmari, *Failure-based equivalences are faster than many believe*, Proceedings of the International Workshop on Structures in Concurrency Theory, Berlin, May 1995, Workshops in Computing, J. Desel, ed., Springer (1995), 326–340.
[49] S. Veglioni and R. De Nicola, *Possible worlds for process algebras*, Proceedings CONCUR '98, Nice, France, Lecture Notes in Comput. Sci. 1466, D. Sangiorgi and R. de Simone, eds, Springer (1998), 179–193.
[50] G. Winskel, *Synchronization trees*, Theoret. Comput. Sci. **34** (1/2) (1984), 33–82.

Subject index

CHAPTER 2

Trace-Oriented Models of Concurrency

Manfred Broy[1], Ernst-Rüdiger Olderog[2]

[1] *Institut für Informatik, Technische Universität München, D-80333 München, Germany*
E-mail: broy@informatik.tu-muenchen.de

[2] *Fachbereich Informatik, Universität Oldenburg, D-26111 Oldenburg, Germany*
E-mail: olderog@informatik.uni-oldenburg.de

Contents

1. Introduction . . . 103
2. Transitions and finite traces . . . 103
   2.1. Automata . . . 104
   2.2. Finite sequences or traces . . . 106
   2.3. Trace semantics . . . 108
3. Composition of processes . . . 110
   3.1. Process terms . . . 110
   3.2. Structured operational semantics . . . 113
   3.3. Synchrony versus asynchrony . . . 116
   3.4. Asynchronous communication . . . 117
4. Denotational approach . . . 120
   4.1. General setting . . . 120
   4.2. Denotational trace semantics . . . 124
   4.3. Asynchronous communication . . . 128
5. Algebraic approach . . . 130
   5.1. Algebraic laws for trace semantics . . . 130
   5.2. Algebraic verification . . . 134
6. Logical approach . . . 137
   6.1. Trace logic . . . 137
   6.2. Trace specifications . . . 141
   6.3. Process correctness . . . 144
7. Mixed term approach . . . 145
   7.1. Transformations on mixed terms . . . 146
   7.2. Milner's scheduling problem . . . 152
8. Extended trace models . . . 156
   8.1. Readiness model . . . 157
   8.2. Failures model . . . 158
9. Infinite traces – streams . . . 158
   9.1. The stream algebra . . . 158

HANDBOOK OF PROCESS ALGEBRA
Edited by Jan A. Bergstra, Alban Ponse and Scott A. Smolka

9.2. Predicates on streams . . . 160
9.3. Streams as a Scott domain . . . 162
10. Functions on streams . . . 162
10.1. Monotonicity and continuity . . . 162
10.2. Channels and their valuation . . . 164
10.3. Input and output actions as operations on stream processing functions . . . 165
11. Composition of stream processing functions . . . 166
11.1. Independent parallel composition . . . 166
11.2. Functional composition (pipelining) . . . 167
11.3. Feedback . . . 168
11.4. Hiding . . . 168
11.5. The algebra of stream processing functions . . . 169
12. Operational aspects of stream functions . . . 169
12.1. SOS for stream processing functions . . . 170
12.2. Stream processing functions and state machines . . . 170
12.3. Modelling discrete time . . . 172
13. Specification and nondeterminism . . . 173
13.1. Specifications . . . 173
13.2. Nondeterminism . . . 174
13.3. Operations on specifications . . . 175
14. Refinement . . . 175
14.1. Property refinement . . . 176
14.2. Interaction refinement . . . 177
14.3. Glass box refinement . . . 181
15. Stream processing and handshake communication . . . 181
15.1. Input enabledness and asynchrony . . . 182
15.2. Stream processing functions as trace sets . . . 183
15.3. From synchronous to asynchronous traces . . . 184
15.4. Composition of asynchronous and synchronous systems . . . 188
16. Conclusion . . . 189
References . . . 190
Subject index . . . 193

Abstract

This chapter provides an in-depth presentation of trace-oriented models of concurrent processes. We begin by introducing and investigating finite traces as a simple abstraction of the transition behaviour of automata. Using finite traces safety properties of processes can be modelled. Later infinite traces or *streams* together with stream processing functions are studied. Using infinite traces more advanced phenomena like fairness and liveness properties can be modelled. We discuss and relate operational, denotational, algebraic and logical approaches to trace-oriented models and explain methods for the specification and verification of process behaviour based on these models.

1. Introduction

In this chapter we consider a *process* as an object designed for a possibly persisting interaction with its user or environment which can be another process. Typical examples of processes are vending machines, communication protocols, operating systems, and controllers of software or physical systems. Such reactive systems are designed to run forever.

A process interaction can be an input or output of a value, but initially we just think of it abstractly as a *communication*. In between two subsequent communications the process usually engages in some *internal actions*. These proceed autonomously at a certain speed and are not visible to the user. However, as a result of such internal actions the process behaviour may appear *nondeterministic* to the user. Concurrency arises because there can be more than one user and inside the process there can be more than one active subprocess. The behaviour of a process is unsatisfactory for its user(s) if it does not communicate as desired. The reason can be that the process stops too early or that it engages in an infinite loop of internal actions. The first problem causes a *deadlock* with the user(s); the second one is known as *divergence*. Thus most processes are designed to communicate arbitrarily long without any danger of deadlock or divergence.

Since the behaviour of concurrent processes has so many facets, it is not surprising that many different semantic models have been proposed for their description. These models differ in the degree of how much they abstract from the operational details of the process behaviour. Such an abstraction is necessary to obtain tractable methods for analyzing and reasoning about the process behaviour. The most abstract models describe processes by the set of communication sequences or *traces* that it can perform with its environment.

In this chapter we provide an in-depth presentation of trace-oriented models of concurrent processes. In Sections 2–8 we introduce and investigate finite traces as a simple abstraction of the transition behaviour of automata. Using finite traces safety properties of processes can be modelled. In Sections 9–15 infinite traces or *streams* together with stream processing functions are studied. Using infinite traces more advanced phenomena like fairness and liveness properties can be modelled. We discuss and relate operational, denotational, algebraic and logical approaches to trace-oriented models and explain methods for the specification and verification of process behaviour based on these models.

2. Transitions and finite traces

To begin with we model processes as transition systems or automata. The advantage of this model is that communications and internal actions can be represented in great detail and other well-known models can be seen as abstractions of this model. We introduce two such abstractions, one based on isomorphisms and one based on traces, which are finite sequences of communications. Whereas isomorphisms abstract only from the names of states, traces abstract from all of the internal process behaviour.

2.1. *Automata*

We consider *labelled transition systems* [50] which are *automata* in the sense of classical automata theory [77] but with possibly infinitely many states and without designated sets of final states. In general, computations of such automata never stop.

We start from an infinite set *Comm* of *communications* and an element $\tau \notin \mathit{Comm}$ to build the set

$$\mathit{Act} = \mathit{Comm} \cup \{\tau\}$$

of *actions*. The intuition is that a transition labelled by a communication requires the participation of both the user and the process whereas a transition labelled by the action τ occurs spontaneously inside the process without participation of the user. Therefore the element τ is called *internal* action and the communications are called *external* actions. We let a, b, c range over *Comm* and u, v, w over *Act*.

In applications communications often possess a structure. For example, they may be pairs (ch, m) consisting of a channel name ch and a message m [43,61], or the set *Comm* may have an algebraic structure as in [15,62]. For simplicity we shall not consider such structures here.

By a *communication alphabet* or simply *alphabet* we mean a finite subset of *Comm*. We let A, B range over alphabets. Every process will have a communication alphabet associated with it. This alphabet describes the interface through which the process communicates with its user(s) or with other processes.

DEFINITION 2.1. A *transition system* or *automaton* is a structure

$$\mathcal{A} = (A, St, \to, q_0),$$

where

(1) A is a communication alphabet;
(2) St is a possibly infinite set of *states*;
(3) $\to \subseteq St \times (A \cup \{\tau\}) \times St$ is the *transition relation*;
(4) $q_0 \in St$ is the *initial state*.

Let *Aut* denote the class of all automata.

We let p, q, r range over St. An element $(p, u, q) \in \to$ is called a *transition* (*labelled with the action* u) and usually written as

$$p \xrightarrow{u} q.$$

In automata theory, the set A corresponds to the input alphabet and τ to a spontaneous ε-move of an automaton. For processes Milner's symbol τ (instead of ε) is common [61].

Every automaton $\mathcal{A} = (A, St, \to, q_0)$ has a graphical representation which is often easier to understand. We draw a rectangular box subdivided into an upper part displaying the alphabet A and a lower part displaying the *state transition diagram*. This is a directed, rooted graph with edges labelled by actions in $A \cup \{\tau\}$ which represents in an obvious way the remaining components St, $\rightarrow$, and q_0 of $\mathcal{A}$. We mark the root of a state diagram by an additional ingoing arc.

EXAMPLE 2.2. The following automaton describes a process with an alphabet consisting of the communications a, b, c and d, and with four states q_0, q_1, q_2, q_3 and three transitions as shown in the state transition diagram.

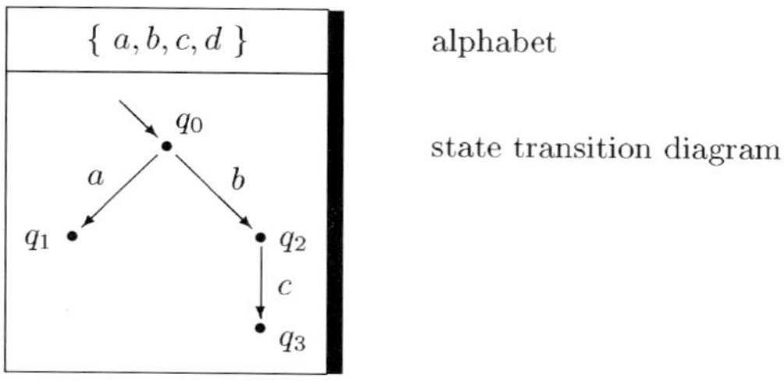

The state transition diagram tells us that initially the process is in state q_0 and thus ready to engage in the communications a or b but not in c or d. Which of the communications a or b occurs is the user's choice. If the user chooses a, the transition labelled with a occurs, after which no further communication is possible. If the user chooses b, the transition labelled with b occurs, after which the process is ready for the communication c.

Note that the communication d of the alphabet does not appear as a label of any transition. Consequently, the process described by this automaton can never engage in d.

Often we wish to abstract from some details of an automaton like unreachable states or even the names of states. To this end, we introduce the following definitions.

DEFINITION 2.3. Consider an automaton $\mathcal{A} = (A, St, \rightarrow, q_0)$. A *reachable state* of $\mathcal{A}$ is a state $q \in St$ for which there exist intermediate states $q_1, \ldots, q_n \in St$ and actions $u_1, \ldots, u_n \in A \cup \{\tau\}$ where $n \in \mathbb{N}$, the set of natural numbers including 0, such that

$$q_0 \xrightarrow{u_1} q_1 \xrightarrow{u_2} \cdots \xrightarrow{u_n} q_n = q$$

holds. By *reach*($\mathcal{A}$) we denote the restriction of $\mathcal{A}$ to its reachable states:

$$reach(\mathcal{A}) = (A, St_0, \rightarrow_0, q_0)$$

where St_0 is the set of all reachable states of $\mathcal{A}$, $\rightarrow_0$ is the restriction of $\rightarrow$ to St_0, i.e.,

$$\rightarrow_0 = \rightarrow \cap \big(St_0 \times (A \cup \{\tau\}) \times St_0\big).$$

DEFINITION 2.4. Two automata $\mathcal{A}_i = (A_i, St_i, \rightarrow_i, q_{0i})$, $i = 1, 2$, are *isomorphic*, abbreviated $\mathcal{A}_1 =_{isom} \mathcal{A}_2$, if $A_1 = A_2$ and there exists a bijection

$$\beta : St_1 \rightarrow St_2$$

on the set of states such that $\beta(q_{01}) = q_{02}$ holds and for all states $p, q \in St_1$ and all actions $u \in A_1 \cup \{\tau\}$ the following propositions are equivalent:

$$p \stackrel{u}{\longrightarrow}_1 q \quad \text{iff} \quad \beta(p) \stackrel{u}{\longrightarrow}_2 \beta(q).$$

The bijection β is called a *isomorphism between* $\mathcal{A}_1$ and $\mathcal{A}_2$.

Intuitively, isomorphic automata

$$\mathcal{A}_1 =_{isom} \mathcal{A}_2$$

have the same state transition diagrams but with different names of the states. Note that $=_{isom}$ is indeed an equivalence relation on automata.

2.2. *Finite sequences or traces*

Automata exhibit too many details about the process behaviour. In particular, all internal actions τ are still visible. We wish to describe the intended communication behaviour between user and process more abstractly guided by the following principle:

> The internal structure of a process is irrelevant as long
> as it exhibits the specified communication behaviour.

But what exactly *is* the communication behaviour? Many answers are possible and meaningful. We study here the simplest solution where the behaviour is modelled by a set of finite sequences of communications

$$tr \in Comm^*,$$

known as *histories* or *traces* [40]. Since traces are insensitive to intervening internal actions τ or concurrent process activities, this definition is independent of both internal activity and concurrency. Our viewpoint is here that internal activity and concurrency are only part of the process construction, not of the specified communication behaviour. Of course, other viewpoints are possible. For example, in the work of Mazurkiewicz [60] even the word "trace" is used for something more elaborate, viz. the equivalence class of finite communication sequences modulo an independence relation on communications expressing concurrency. To avoid confusion, we call these equivalence classes "Mazurkiewicz-traces" and reserve the word "trace" for finite sequences.

The domain of finite sequences. Mathematically, traces are just finite sequences. Therefore we first recall the basic notations and operations on the domain M^* of finite sequences over a set M of elements. Let m range over M and x, y, z over M^*. A sequence consisting of elements $m_1, \ldots, m_n \in M$ with $n \in \mathbb{N}$ in that order is denoted by $\langle m_1, \ldots, m_n \rangle$. For $n = 0$ we obtain the *empty sequence* denoted by $\langle\rangle$ and for $n = 1$ the *one element sequence* $\langle m_1 \rangle$.

On finite sequences the following operations are defined:

$\frown$	denotes the binary *concatenation* operation on sequences,
head	denotes a partially defined unary operation yielding the *first element* of a non-empty sequence,
tail	denotes a partially defined unary operation yielding the *rest* of a non-empty sequence obtained by eliminating its first element.

These operations satisfy the following algebraic axioms:

$$\langle\rangle^\frown x = x^\frown\langle\rangle = x$$
$$x^\frown z = x \Rightarrow z = \langle\rangle$$
$$x^\frown(y^\frown z) = (x^\frown y)^\frown z$$

$$head(\langle m\rangle^\frown x) = m$$
$$tail(\langle m\rangle^\frown x) = x.$$

Additionally, the following operations on sequences are helpful in specifications:

$\#x$	denotes the *length* of the sequence x,
$x(i)$	denotes the *ith element* in the sequence for $1 \leqslant i \leqslant \#x$,
$x \downarrow A$	denotes the operation of *filtering* out the subsequence of x of elements in the set A, also called *projection* of x onto A.

We may describe the meaning of these operations by algebraic axioms as follows:

$$\#\langle\rangle = 0$$
$$\#(\langle m\rangle^\frown x) = 1 + \#x$$
$$(\#x = \#y \wedge \forall i \leqslant \#x \bullet x(i) = y(i)) \Rightarrow x = y$$
$$(\langle m\rangle^\frown x)(1) = m$$
$$(\langle m\rangle^\frown x)(i+1) = x(i) \qquad \Leftarrow i > 0$$
$$\langle\rangle \downarrow A = \langle\rangle$$
$$(\langle m\rangle^\frown x) \downarrow A = x \downarrow A \qquad \Leftarrow m \notin A$$
$$(\langle m\rangle^\frown x) \downarrow A = \langle m\rangle^\frown(m \downarrow A) \qquad \Leftarrow m \in A.$$

The *prefix ordering* $\leqslant$ is a binary relation on sequences. Given two sequences $x, y \in M^*$ it is defined as follows:

$$x \leqslant y \Longleftrightarrow \exists z \bullet x^\frown z = y.$$

This definition of $\leqslant$ yields indeed a partial ordering on sequences with the least element $\langle\rangle$. This can easily be proved as follows:

The reflexivity of $x \leqslant y$ is immediately proved by choosing $z = \langle\rangle$ in the above definition. The antisymmetry is shown as follows: from $x \leqslant y \wedge y \leqslant x$ we may conclude that there exist sequences z and z' such that:

$$x^\frown z = y \wedge y^\frown z' = x.$$

Thus we obtain the equation

$$x^\frown z^\frown z' = x$$

and that $z^\frown z' = \langle\rangle$ and $z = z' = \langle\rangle$. Hence $x = x^\frown z$ and $x^\frown z = y$. This proves $x = y$.

Transitivity of the prefix relation is a direct consequence of the associativity of concatenation:

$$\begin{aligned}
& x \leqslant y \wedge y \leqslant z \\
& \Rightarrow \\
& \exists z', z'' \bullet x^\frown z' = y \wedge y^\frown z'' = z \\
& \Rightarrow \\
& \exists z', z'' \bullet (x^\frown z')^\frown z'' = z \\
& \Rightarrow \\
& \exists z', z'' \bullet x^\frown (z'^\frown z'') = z \\
& \Rightarrow \\
& x \leqslant z.
\end{aligned}$$

2.3. *Trace semantics*

We can now define a *trace semantics* for automata. This is a mapping

$$\mathcal{T}: Aut \rightarrow DOM_{\mathcal{T}}$$

which assigns to every $\mathcal{A} \in Aut$ an element $\mathcal{T}(\mathcal{A})$ in the so-called *trace domain* $DOM_{\mathcal{T}}$. It consists of pairs (A, Γ) where A is a communication alphabet and Γ is a *set of traces* over this alphabet:

$$DOM_{\mathcal{T}} = \{(A, \Gamma) \mid A \subseteq Comm \text{ is finite and } \Gamma \subseteq A^*\}.$$

DEFINITION 2.5. Consider an automaton $\mathcal{A} = (A, St, \rightarrow, q_0)$.

(1) For $p, q \in St$ and $tr \in Comm^*$ we write $p \stackrel{tr}{\Longrightarrow} q$ if there exist intermediate states $r_0, \ldots, r_n \in St$ and actions $u_1, \ldots, u_n \in A \cup \{\tau\}$ with $n \in \mathbb{N}$ and

$$p = r_0 \stackrel{u_1}{\longrightarrow} r_1 \stackrel{u_2}{\longrightarrow} \cdots \stackrel{u_n}{\longrightarrow} r_n = q$$

such that $tr = \langle u_1, \ldots, u_n \rangle \downarrow A$, i.e., tr results from the action sequence $\langle u_1, \ldots, u_n \rangle$ by *deleting* all internal action symbols τ.

(2) The *trace semantics* of $\mathcal{A}$ is given by

$$\mathcal{T}(\mathcal{A}) = \big(A, \{tr \in A^* \mid \exists q \in St \bullet q_0 \stackrel{tr}{\Longrightarrow} q\}\big).$$

Compared with classical automata theory the trace semantics $\mathcal{T}(\mathcal{A})$ corresponds to the language consisting of all words accepted by $\mathcal{A}$ provided we consider every state of $\mathcal{A}$ as an accepting state. Ignoring internal τ-actions in process theory corresponds to ignoring ε-moves in automata theory. The reason that we explicitly record the alphabet is that this facilitates dealing with parallel composition of processes in Section 4.

We investigate the basic properties of this semantics. For this purpose, we introduce some basic notations.

DEFINITION 2.6. A set T of traces is called *prefix closed* or *downward closed* if for each of its traces it contains all its prefixes, formally, for all traces tr, tr':

$$tr \in T \wedge tr' \leqslant tr \Rightarrow tr' \in T.$$

The *prefix closure* $PC(T)$ of a trace set T is the set of all traces that are prefixes of some trace in T:

$$PC(T) = \{tr' \mid \exists tr \in T \bullet tr' \leqslant tr\}.$$

Note that prefix closure is in fact a closure operation:

$$T \subseteq PC(T) \quad \text{and} \quad PC\big(PC(T)\big) = PC(T).$$

PROPOSITION 2.7. *For every automaton $\mathcal{A}$ the set of traces of its trace semantics $\mathcal{T}(\mathcal{A})$ is non-empty and prefix closed.*

DEFINITION 2.8. Two automata $\mathcal{A}_1$ and $\mathcal{A}_2$ are called *trace equivalent*, abbreviated $\mathcal{A}_1 =_{\mathcal{T}} \mathcal{A}_2$, if $\mathcal{T}(\mathcal{A}_1) = \mathcal{T}(\mathcal{A}_2)$.

Note that removing unreachable states or considering isomorphic automata is not observable under trace equivalence.

PROPOSITION 2.9. *For all automata $\mathcal{A}$, $\mathcal{A}_1$, $\mathcal{A}_2$ the following holds*:
- $\mathcal{A} =_{\mathcal{T}} reach(\mathcal{A})$.
- $\mathcal{A}_1 =_{isom} \mathcal{A}_2$ *implies* $\mathcal{A}_1 =_{\mathcal{T}} \mathcal{A}_2$.

Trace equivalence is the coarsest equivalence that one wishes to consider for processes. It corresponds to the classical language equivalence on automata. In [61] Milner observed that this equivalence is in fact too coarse when handshake communication between processes is considered. This led him to introduce a much stronger equivalence on processes called *bisimilarity* (cf. [34]). But there are also some simple variants of trace equivalence that avoid the shortcomings observed by Milner, viz. *readiness* and *failure equivalence*. These variants will be discussed in Section 8.

3. Composition of processes

Processes can be structured by suitable composition operators like nondeterministic choice, parallel composition and recursion. These composition operators permit to define a process language consisting of process terms. These terms are then linked to the basic automata model of processes via a *structured operational semantics*. Based on this semantics the concept of traces can be lifted to process terms. This poses the question whether the trace semantics of a process term can be determined directly from its syntax without first constructing the underlying automaton. This is the question for a *compositional* or more precisely *denotational trace semantics*. It can be shown that operational and denotational approach yield the same trace sets.

3.1. *Process terms*

The general syntactic scheme for process terms is very simple. Starting from a set of operator symbols op, each one with a certain arity $n \geqslant 0$, and a set of identifiers X, we first build so-called *recursive terms* by closing up with recursion in the form of Park's μ-operator of [10,73]. The context-free syntax of recursive terms P is given by the following production rules:

$$P ::= op(P_1, \ldots, P_n) \quad \text{where } op \text{ has arity } n \mid X \mid \mu X \bullet P.$$

The reason for separating recursion from the other operators op is that semantically, recursion will get an extra treatment.

Let us now explain how this scheme is applied to yield process terms. Recall that *Comm* denotes the infinite set of communications and $Act = Comm \cup \{\tau\}$ the set of actions. As before, letters a, b range over *Comm* and letters A, B over communication alphabets, which are finite subsets of *Comm*. The set of (*process*) *identifiers* is denoted by *Idf*; it is partitioned into sets Idf: $A \subseteq Idf$ of *identifiers with alphabet* A, one for each communication alphabet A. We let X, Y, Z range over *Idf*. By an *action morphism* we mean a mapping $\varphi : Act \to Act$ with $\varphi(\tau) = \tau$ and $\varphi(a) \neq a$ for only finitely many $a \in Comm$. Communications a with $\varphi(a) = \tau$ are said to be *hidden* via φ and communications a with $\varphi(a) = b$ for some $b \neq a$ are said to be *renamed* into b via φ.

DEFINITION 3.1. The set *Rec* of (*recursive*) *terms*, with typical elements P, Q, R, consists of all terms generated by the following context-free production rules:

$$\begin{array}{rlll} P ::= & & stop{:}A & \text{(deadlock)} \\ & \mid & div{:}A & \text{(divergence)} \\ & \mid & a.P & \text{(prefix)} \\ & \mid & P \; or \; Q & \text{(nondeterminism)} \\ & \mid & P + Q & \text{(choice)} \\ & \mid & P \parallel Q & \text{(parallelism)} \\ & \mid & P[\varphi] & \text{(morphism)} \\ & \mid & X & \text{(identifier)} \\ & \mid & \mu X \bullet P & \text{(recursion)} \end{array}$$

The intended interpretation is given as a commentary.

Note that for an easier distinction of the subterms we write in the above production rules $P\,or\,Q$, $P + Q$ and $P \parallel Q$ instead of $P\,or\,P$, $P + P$ and $P \parallel P$ as would be formally required by a context free grammar with non-terminal P.

More precisely, the intuitive meaning of process terms is as follows. The term $stop{:}A$ denotes *deadlock*, a process which neither engages in any communication nor in any internal action. The term $div{:}A$ denotes *divergence*, a process which pursues an infinite loop of internal actions. *Prefix* $a.P$ first engages in a communication a with its environment and then behaves like P. Thus prefix is a restricted form of sequential composition. *Nondeterminism* $P\,or\,Q$ behaves like P or Q, and the choice is not controllable by the environment. By contrast, the *choice* $P + Q$ is controllable; it also behaves like P or like Q but the choice depends on whether the first action is one of P or one of Q. If the first action belongs to both P and Q, this choice is again nondeterministic. *Parallel composition* combines concurrency and synchronisation: $P \parallel Q$ behaves like P and Q working independently or concurrently except that all communications which occur in the alphabet of both P and Q have to synchronize. *Morphism* $P[\varphi]$ behaves like P, but with all actions u changed into $\varphi(u)$. *Recursion* $\mu X.P$ behaves like P, but with every occurrence of X inside P denoting a recursive call to $\mu X.P$.

Formally, the signature of *Rec* consists of the nullary operator symbols $stop{:}A$ and $div{:}A$ for each communication alphabet A, a unary prefix symbol the symbol $a.\_$ for each communication a, a unary postfix symbol $\_\,[\varphi]$ for each morphism φ, and binary infix symbols or, $+$ and $\parallel$. Syntactic ambiguities are eliminated by using brackets and the following priorities among the operator symbols and recursion:

$$\begin{array}{lll} \text{priority } 0 & : & or,\ + \text{ and } \parallel, \\ \text{priority } 1 & : & a.\_ \text{ and } \mu X, \\ \text{priority } 2 & : & \_[\varphi]. \end{array}$$

Since $+$ and $\parallel$ turn out to be commutative and associative, we sometimes use the abbreviations

$$\sum_{i \in \{1,\dots,n\}} P_i = \sum_{i=1}^{n} P_i = P_1 + \cdots + P_n$$

and

$$\|_{i \in \{1,\dots,n\}} P_i = \|_{i=1}^{n} P_i = P_1 \parallel \cdots \parallel P_n.$$

For action morphisms which purely rename or hide communications we use the following notations:

- *Renaming*: For distinct $a_1, \dots, a_n$ and $n \geqslant 1$ let

$$P[b_1, \dots, b_n/a_1, \dots, a_n] =_{df} P[\varphi], \quad \text{where for } u \in Act$$

$$\varphi(u) = \begin{cases} b_i & \text{if } u = a_i \text{ for some } i \in 1, \ldots, n, \\ u & \text{otherwise.} \end{cases}$$

- *Hiding*: $P \setminus B =_{df} P[\varphi]$ where for $u \in Act$

$$\varphi(u) = \begin{cases} \tau & \text{if } u \in B, \\ u & \text{otherwise,} \end{cases}$$

and $P \setminus b =_{df} P \setminus \{b\}$.

We now introduce some auxiliary notions for terms. An occurrence of an identifier X in a term P is said to be *bound* if it occurs in P within a subterm of the form $\mu X \bullet Q$. Otherwise the occurrence is said to be *free*. Let *free*(P) denote the set of identifiers that occur freely in a term P. A term without free occurrences of identifiers is called *closed*. For terms P, Q and identifiers X let $P\{Q/X\}$ denote the result of *substituting* Q for every free occurrence of X in P. As usual, by a preparatory renaming of bound identifiers in P one has to ensure that no free identifier occurrence in Q gets bound in $P\{Q/X\}$. Thus $P\{Q/X\}$ is unique up to renaming of bound identifiers. This notation readily extends to simultaneous substitution of lists of terms for lists of distinct identifiers: $P\{Q_1, \ldots, Q_n/X_1, \ldots, X_n\}$. Note that we use brackets $\{\ldots\}$ for substitutions of process terms and brackets $[\ldots]$ for substitutions of actions like the renaming operators above.

A term P is called *action-guarded* if in every recursive subterm $\mu X \bullet Q$ of P every free occurrence of X in Q occurs within a subterm of the form $a.R$ of Q. For example,

$$\mu X \bullet a.X, \quad \mu X \bullet \mu Y \bullet a.Y, \quad (\mu X \bullet a.X) \setminus a$$

are all action-guarded, but $a.\mu X \bullet X$ is not. The example $(\mu X \bullet a.X) \setminus a$ shows that the guarding communication a may well be within the scope of a hiding operator $\setminus a$. That's why we talk of *action*-guardedness; later in Section 5 we shall also introduce *communication*-guardedness.

The term "guardedness" is due to [61]; it corresponds to the "Greibach condition" in formal language theory [66] if we identify terminal symbols with actions, and nonterminal symbols with process identifiers. The Greibach condition requires that the right-hand side of every production rule of a context-free grammar starts with a terminal symbol. For example, the production $X ::= a.X$ satisfies this condition whereas $X ::= Y$ does not.

To every term P we assign a communication alphabet $\alpha(P)$. It is defined inductively on the structure of P according to Definition 3.1:

$$\begin{aligned}
\alpha(stop{:}A) &= \alpha(div{:}A) = A, \\
\alpha(a.P) &= \{a\} \cup \alpha(P), \\
\alpha(P \, or \, Q) &= \alpha(P + Q) = \alpha(P \parallel Q) = \alpha(P) \cup \alpha(Q), \\
\alpha(P[\varphi]) &= \varphi\big(\alpha(P)\big) - \{\tau\}, \\
\alpha(X) &= A \quad \text{if } X \in Idf{:}A, \\
\alpha(\mu X \bullet P) &= \alpha(X) \cup \alpha(P).
\end{aligned}$$

DEFINITION 3.2. A *process term* is a term $P \in Rec$ which satisfies the following context-sensitive restrictions:

(1) P is action-guarded,

(2) every subterm $a.Q$ of P satisfies $a \in \alpha(Q)$,

(3) every subterm Q *or* R and $Q + R$ of P satisfies $\alpha(Q) = \alpha(R)$,

(4) every subterm $\mu X \bullet Q$ of P satisfies $\alpha(X) = \alpha(Q)$.

Let *Proc* denote the set of all process terms and *CProc* the set of all closed process terms.

The context-sensitive restrictions for process terms simplify their subsequent semantic treatment. Here we notice that the alphabet of a process term P depends only on the alphabet of its subterms, not on their syntactic structure.

REMARK 3.3. For every n-ary operator symbol *op* of *Proc* there exists a set-theoretic operator op_α such that

$$\alpha\big(op(P_1, \dots, P_n)\big) = op_\alpha\big(\alpha(P_1), \dots, \alpha(P_n)\big).$$

Consequently, by context restriction (4),

$$\alpha(\mu X \bullet P) = \alpha\big(P\{\mu X \bullet P/X\}\big)$$

for all recursive process terms $\mu X \bullet P$.

3.2. *Structured operational semantics*

Here we make the informal semantics of process terms precise by introducing a *structured operational semantics* (SOS for short) in the sense of Plotkin [75,76] for process terms. Operational means that for every process term the states and transitions of an abstract machine, here an automaton, are described. Plotkin's idea is that states are denoted by terms and that transitions are defined by structural induction using a deductive system. Parallelism is modelled by interleaving the transitions of the individual components.

Altogether this semantics assigns to every closed process term P an automaton $\mathcal{A}[\![P]\!]$ and is based on a transition relation

$$\rightarrow \subseteq CProc \times Act \times CProc.$$

DEFINITION 3.4. In the following let P, P', Q, Q' range over *CProc*, X over *Idf*, a over *Comm*, u over *Act*, and A over alphabets. Then the *transition relation* $\rightarrow$ consists of all transitions that are deducible by the following transition axioms and rules:

(1) *Prefix.*

$$a.P \xrightarrow{a} P.$$

(2) *Divergence.*

$$div{:}A \xrightarrow{\tau} div{:}A.$$

(3) *Nondeterminism.*

$$P \; or \; Q \xrightarrow{\tau} P, \qquad P \; or \; Q \xrightarrow{\tau} Q.$$

(4) *Choice.*

$$\frac{P \xrightarrow{u} P'}{P + Q \xrightarrow{u} P',\ Q + P \xrightarrow{u} P'}.$$

(5) *Parallel Composition.*
Asynchrony:

$$\frac{P \xrightarrow{u} P'}{P \| Q \xrightarrow{u} P' \| Q,\ Q \| P \xrightarrow{u} Q \| P'} \quad \text{where } u \notin \alpha(Q).$$

Synchrony:

$$\frac{P \xrightarrow{a} P',\ Q \xrightarrow{a} Q'}{P \| Q \xrightarrow{a} P' \| Q'} \quad \text{where } a \in \alpha(P) \cap \alpha(Q).$$

(6) *Morphism.*

$$\frac{P \xrightarrow{u} Q}{P[\varphi] \xrightarrow{\varphi(u)} Q[\varphi]}.$$

(7) *Recursion.*

$$\frac{P\{\mu X \bullet P/X\} \xrightarrow{u} Q}{\mu X \bullet P \xrightarrow{u} Q}.$$

The above transition rules are all standard [61,63,70]. The transition rule for parallel composition can be found in [63]. However, there and also in [62] a difficulty arises in knowing whether $\alpha(P) \cap \alpha(Q)$ always determines the right synchronisation set. In our set-up the synchronisation set $\alpha(P) \cap \alpha(Q)$ of $P \| Q$ is invariant under transitions of P and Q. This is a consequence of the following lemma, which resolves the difficulties mentioned in [62,63].

ALPHABET LEMMA 3.5. *For all $P, Q \in CProc$ and $u \in Act$*

$$P \xrightarrow{u} Q \quad \textit{implies} \quad \alpha(P) = \alpha(Q).$$

PROOF. We use induction on the structure of transitions, thereby using Definition 3.2 and Remark 3.3. □

The operational semantics of process terms associates with every closed process term P an automaton $\mathcal{A}[\![P]\!]$ where the transition relation is obtained by restricting the general transition relation $\rightarrow$ of Definition 3.4 to the alphabet of P and the states that are reachable from P.

DEFINITION 3.6. The *operational semantics* for process terms is a mapping

$$\mathcal{A}[\![\_]\!] : CProc \rightarrow Aut,$$

which assigns to every $P \in CProc$ the automaton

$$\mathcal{A}[\![P]\!] = reach\big(\alpha(P), CProc, \rightarrow\downarrow\alpha(P), P\big).$$

Here $\rightarrow$ is the transition relation of 3.4 and

$$\rightarrow\downarrow\alpha(P) = \rightarrow \bigcap\big(CProc \times \big(\alpha(P) \cup \{\tau\}\big) \times CProc\big)$$

its restriction to the communications in $\alpha(P)$ and τ.

To construct $\mathcal{A}[\![P]\!]$, we start from P and explore all transitions that are successively applicable. In this way, we will automatically restrict ourselves to the communications in $\alpha(P)$ and the reachable states.

EXAMPLE 3.7. Let us construct the automaton $\mathcal{A}[\![P]\!]$ of the process term $P = (a.stop\colon \{a, c\} \parallel b.stop\colon\{b, c\}) + c.stop\colon\{a, b, c\}$:

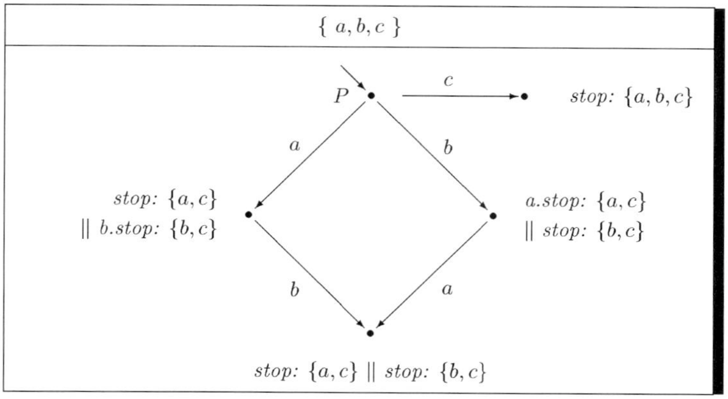

To explain the construction, the process terms are retained as names of the states.

By the following lemma each morphism operator applied to a process term can be decomposed into a sequence of simple renaming and hiding operations.

MORPHISM LEMMA 3.8. *For each process term P and each morphism φ there exists a sequence $\varphi_1, \ldots, \varphi_n$ with $n \in \mathbb{N}$ such that each φ_i is of the form of a renaming operator $\_[b_i/a_i]$ or a hiding operator $\_\setminus b_i$ and*

$$\mathcal{A}[\![P[\varphi]]\!] =_{isom} \mathcal{A}[\![P[\varphi_1]\ldots[\varphi_n]]\!].$$

PROOF. The proof relies on the fact that P has a finite alphabet $\alpha(P)$ and that we may use auxiliary communications names outside this alphabet. For example,

$$\mathcal{A}[\![P[a,b/b,a]]\!] =_{isom} \mathcal{A}[\![P[x/a][a/b][b/x]]\!],$$

where $x \notin \alpha(P)$. □

We shall use this lemma to state some definitions concerning morphisms only for the case of simple renaming and hiding operations. The extension to the general case is then straightforward by the above lemma.

Operational trace semantics. In Section 2 we defined the trace semantics for automata. We lift this definition to closed process terms by introducing a mapping

$$\mathcal{T}[\![\_]\!] : CProc \rightarrow DOM_{\mathcal{T}}.$$

This mapping is defined by *first* applying the operational semantics for process terms yielding an automaton and *then* retrieving the traces from this automaton. Therefore we talk of an *operational* trace semantics for process terms.

DEFINITION 3.9. The (*operational*) *trace semantics* of a closed process term P is given by $\mathcal{T}[\![P]\!] = \mathcal{T}(\mathcal{A}[\![P]\!])$.

If $\mathcal{T}[\![P]\!] = (A, \Gamma)$, we also write $tr \in \mathcal{T}[\![P]\!]$ instead of $tr \in \Gamma$. We say that P *may engage* in tr if $tr \in \mathcal{T}[\![P]\!]$ holds.

3.3. *Synchrony versus asynchrony*

The issue of synchrony versus asynchrony arises for different process concepts. First of all, the *parallel composition* of processes may be interpreted synchronously or asynchronously. In the synchronous case the processes would proceed in lock-step as if triggered by one global clock. This interpretation would be suitable for modelling synchronous circuits. It is embodied in Milner's process algebra SCCS [62]. In the asynchronous case the processes would proceed independently of each other: a step of one process would have no effect on the other processes in the parallel composition.

We have chosen the parallel composition $P \parallel Q$ of CSP [42] that first appeared in Lauer's formalism COSY [56]. Here $P \parallel Q$ combines synchronous and asynchronous parallelism. Internal actions and communications that appear only in one of the processes P or Q may proceed asynchronously, communications that appear in both P and Q must proceed synchronously. This is most clearly expressed by the operational semantics given in Definition 3.4.

Secondly, the *communication* between processes may be interpreted synchronously or asynchronously. Using synchronous communication the sender of a message can deliver it only when the receiver is ready to accept it at the same moment. This form of communication is realised in the telephone system. Using asynchronous communication a sender can always deliver its message. It will be buffered until the receiver collects it. This form of communication is realised in the mail system.

The parallel composition $P \parallel Q$ requires that communications that appear in both P and Q can occur only simultaneously. Without further additions this models synchronous communication. However, asynchronous communication can be modelled by synchronous communication if a buffer that always accepts input is introduced as an explicit component (see, e.g., [37]). This stipulates a distinction between input and output. In the following we will show how this can be achieved by extending the previous setting of process terms.

3.4. *Asynchronous communication*

So far we did not distinguish between input and output communications. Input and output are helpful notions for component models. In traces, communications may be pure input or pure output but not both. The distinction into input and output indicates clearly by whom the selection of the message communicated is done. Input is determined by the environment and output by the system.

This distinction will be used to model *asynchronous communication* between processes where processes are always able to accept input. This is achieved by adding an explicit input buffer to each process. To this end, we extend the syntax of recursive terms given in Definition 3.1 by one operator:

$$P ::= \ldots \mid P \lhd a \quad \text{(buffering)}.$$

The term $P \lhd a$ denotes a process that has the communication a as a buffered input. Call the resulting set Rec_{asyn}. In the following we consider only terms $P \in Rec_{asyn}$ satisfying the context sensitive restrictions of Definition 3.2.

The calculation of the sets of input and output communications of a process term requires some care. For each alphabet A we assume a decomposition into a set of input communications A_I and a set of output communications A_O with

$$A = A_I \cup A_O \quad \text{and} \quad A_I \cap A_O = \emptyset.$$

We express this fact as $A = A_I : A_O$. Thus instead of *stop*:A and *div*:A and $X \in \mathit{Idf} : A$ we write

$$\mathit{stop}{:}(A_I : A_O) \quad \text{and} \quad \mathit{div}{:}(A_I : A_O) \quad \text{and} \quad X \in \mathit{Idf} : (A_I : A_O).$$

Correspondingly, the alphabet $\alpha(P)$ of each process P is decomposed into a set $\alpha_I(P)$ of input communications and a set $\alpha_O(P)$ of output communications with

$$\alpha(P) = \alpha_I(P) \cup \alpha_O(P) \quad \text{and} \quad \alpha_I(P) \cap \alpha_O(P) = \emptyset.$$

The input alphabet is defined inductively as follows:

$$\begin{aligned}
\alpha_I\big(\mathit{stop}{:}(A_I : A_O)\big) &= \alpha_I\big(\mathit{div}{:}(A_I : A_O)\big) = A_I, \\
\alpha_I(P \lhd a) &= \{a\} \cup \alpha_I(P), \\
\alpha_I(a.P) &= \{a\} \cup \alpha_I(P) \quad \text{if } a \in \alpha_I(P), \\
\alpha_I(a.P) &= \alpha_I(P) \qquad\qquad \text{if } a \notin \alpha_I(P), \\
\alpha_I(P \mathbin{\mathit{or}} Q) &= \alpha_I(P + Q) = \alpha_I(P) \cup \alpha_I(Q), \\
\alpha_I(P \| Q) &= \big(\alpha_I(P) \backslash \alpha_O(Q)\big) \cup \big(\alpha_I(Q) \backslash \alpha_O(P)\big), \\
\alpha_I\big(P[\varphi]\big) &= \varphi\big(\alpha_I(P)\big) - \{\tau\}, \\
\alpha_I(X) &= A_I \qquad\qquad\qquad \text{if } X \in \mathit{Idf} : (A_I : A_O), \\
\alpha_I(\mu X \bullet P) &= \alpha_I(X) \cup \alpha_I(P).
\end{aligned}$$

The output alphabet is defined similarly:

$$\begin{aligned}
\alpha_O\big(\mathit{stop}{:}(A_I : A_O)\big) &= \alpha_O\big(\mathit{div}{:}(A_I : A_O)\big) = A_O, \\
\alpha_O(P \lhd a) &= \alpha_O(P), \\
\alpha_O(a.P) &= \{a\} \cup \alpha_O(P) \quad \text{if } a \in \alpha_O(P), \\
\alpha_O(a.P) &= \alpha_O(P) \qquad\qquad \text{if } a \notin \alpha_O(P), \\
\alpha_O(P \mathbin{\mathit{or}} Q) &= \alpha_O(P + Q) = \alpha_O(P) \cup \alpha_O(Q), \\
\alpha_O(P \| Q) &= \big(\alpha_O(Q) \backslash \alpha_I(P)\big) \cup \big(\alpha_O(P) \backslash \alpha_I(Q)\big), \\
\alpha_O\big(P[\varphi]\big) &= \varphi\big(\alpha_O(P)\big) - \{\tau\}, \\
\alpha_O(X) &= A_O \qquad\qquad\qquad \text{if } X \in \mathit{Idf} : (A_I : A_O), \\
\alpha_O(\mu X \bullet P) &= \alpha_O(X) \cup \alpha_O(P).
\end{aligned}$$

DEFINITION 3.10. An *asynchronous process term* is a term $P \in \mathit{Rec}_{asyn}$ which satisfies the context sensitive restrictions of Definition 3.2 and the following additional ones:

(1) every subterm $Q \lhd a$ of P satisfies $a \in \alpha_I(Q)$,
(2) every subterm $Q \mathbin{\mathit{or}} R$ and $Q + R$ of P satisfies

$$\alpha_I(Q) = \alpha_I(R) \quad \text{and} \quad \alpha_O(Q) = \alpha_O(R),$$

(3) every subterm $Q \| R$ of P satisfies

$$\alpha_I(P) \cap \alpha_I(Q) = \alpha_O(P) \cap \alpha_O(Q) = \emptyset,$$

(4) every subterm $\mu X \bullet Q$ of P satisfies

$$\alpha_I(X) = \alpha_I(Q) \quad \text{and} \quad \alpha_O(X) = \alpha_O(Q).$$

Let $Proc_{asyn}$ denote the set of all asynchronous process terms and $CProc_{asyn}$ the set of all closed asynchronous process terms.

The intended behaviour of asynchronous process terms can be made precise by a structured operational semantics. It is defined in the same way as before but refers to the following transition rules. For convenience, we separate the morphism operators into renaming and hiding. By Lemma 3.8, this can be done without loss of generality.

DEFINITION 3.11. For asynchronous process terms we reuse the transition rules (2)–(6) of Definition 3.4 and apply the following additional transition rules:

(A1) Buffering of input.

$$P \xrightarrow{a} P \lhd a \quad \text{where } a \in \alpha_I(P).$$

(A2) Input from buffer.

$$(a.P) \lhd a \xrightarrow{\tau} P.$$

(A3) Output prefix.

$$b.P \xrightarrow{b} P \quad \text{where } b \in \alpha_O(P).$$

(A3) Output over buffers.

$$\frac{P \xrightarrow{b} Q}{P \lhd a \xrightarrow{b} Q \lhd a} \quad \text{where } a \in \alpha_I(P) \text{ and } b \in \alpha_O(P) \cup \{\tau\}.$$

(A4) Local buffering.

$$\begin{aligned}
(P+Q) \lhd a &\xrightarrow{\tau} (Q \lhd a) + (P \lhd a) && \text{where } a \in \alpha_I P \cap \alpha_I(Q),\\
(P \| Q) \lhd a &\xrightarrow{\tau} (P \lhd a) \| Q && \text{where } a \in \alpha_I(P),\\
(Q \| P) \lhd a &\xrightarrow{\tau} (Q \| P \lhd a) && \text{where } a \in \alpha_I(P),\\
\big(P[b/a]\big) \lhd b &\xrightarrow{\tau} (P \lhd a)[b/a], &&\\
\big(P[b/a]\big) \lhd c &\xrightarrow{\tau} (P \lhd c)[b/a] && \text{where } b \neq c \in \alpha_I\big(P[b/a]\big),\\
(P \setminus b) \lhd a &\xrightarrow{\tau} (P \lhd a) \setminus b. &&
\end{aligned}$$

(A5) Recursion.

$$\mu X \bullet P \xrightarrow{\tau} P\{\mu X \bullet P/X\}.$$

These rules make sure that an asynchronous process is always ready to accept input. We will come back to this issue in Section 15 under the concept of *input enabledness*. Note that an input prefix is processed using two transitions:

$$a.P \xrightarrow{a} (a.P) \lhd a \xrightarrow{\tau} P.$$

The first transition adds a to the input buffer, the second one consumes a from this buffer as an internal action.

4. Denotational approach

In this section we provide an alternative, denotational definition $\mathcal{T}^*$ of the trace semantics. Following the approach of Scott and Strachey [84,88,10,35], a semantics of a programming language is called *denotational* if each phrase of the language is given a *denotation*; that is, an abstract mathematical object that represents the contribution of the phrase to the meaning of any complete program in which it occurs. It is required that the denotation of each phrase is determined in a compositional fashion from the denotations of its subphrases. In general, the discipline of denotational semantics is concerned with giving mathematical models for programming languages, and with developing canonical techniques for constructing such models.

In our case the programming language is the set of process terms. The "phrases" of a process term are its subterms. As the denotations of process terms we wish to take elements of the trace domain $DOM_{\mathcal{T}}$. A denotational definition of the semantic mapping from process terms into this domain should be compositional and use fixed point techniques when dealing with recursion. These techniques stipulate a certain structure of the composition operators and the semantic domain. We will use here the standard set-up and work with monotonic or continuous operators over (chain-) complete partial orders (cpo's). Let us briefly recall the basic definitions and properties. Further details can be found in the literature, for instance in [5,57].

4.1. *General setting*

Let DOM be a set of values, usually called a *semantic domain*, and $\sqsubseteq$ be a binary relation on DOM. Then $\sqsubseteq$ is called a *partial order* on DOM if $\sqsubseteq$ is reflexive, antisymmetric and transitive. If $x \sqsubseteq y$ for some $x, y \in DOM$, we say that x is *less than* y and that y is *greater* or *better than* x.

Suppose now that $\sqsubseteq$ is a partial order on DOM. Let $x \in DOM$ and $D \subseteq DOM$. Then x is called the *least* element of D if $x \in D$ and $x \sqsubseteq d$ for all $d \in D$. The element x is called an *upper bound* of D if $d \sqsubseteq x$ for all $d \in D$. Note that $x \in D$ is not required here. Let U be the set of all upper bounds of D and x be the least element of U. Then we write $x = \bigsqcup D$.

A mapping $\Phi : DOM \to DOM$ is called $\sqsubseteq$*-monotonic* if it preserves the partial order $\sqsubseteq$, i.e., if for all $x, y \in DOM$

$$x \sqsubseteq y \quad \text{implies} \quad \Phi(x) \sqsubseteq \Phi(y).$$

More generally, an n-ary mapping

$$\Phi : \underbrace{DOM \times \ldots \times DOM}_{n \text{ times}} \to DOM$$

is $\sqsubseteq$-monotonic if it is $\sqsubseteq$-monotonic in every argument.

The partial order $\sqsubseteq$ on DOM is called *chain-complete* or simply *complete* if DOM contains a least element, often denoted by $\perp$, and if every ascending chain

$$x_0 \sqsubseteq x_1 \sqsubseteq x_2 \cdots \tag{1}$$

of elements $x_n \in DOM$, $n \in \mathbb{N}$, has a *limit*

$$\bigsqcup_{n \in \mathbb{N}} x_n$$

which is defined as the least upper bound of the set of elements in (1), i.e.,

$$\bigsqcup_{n \in \mathbb{N}} x_n = \bigsqcup \{x_0, x_1, x_2, \ldots\}.$$

If $\sqsubseteq$ is a complete partial order on DOM, we also say that DOM is a complete order *under* $\sqsubseteq$.

Suppose now that DOM is a complete partial order under $\sqsubseteq$. A mapping $\Phi : DOM \to DOM$ is $\sqsubseteq$*-continuous* if it is $\sqsubseteq$-monotonic and if it preserves limits such that for every ascending chain (1) the equation

$$\Phi\Big(\bigsqcup_{n \in \mathbb{N}} x_n\Big) = \bigsqcup_{n \in \mathbb{N}} \Phi(x_n) \tag{2}$$

holds. Note that the limit on the right-hand side of (2) exists because the elements $\Phi(x_n)$ form an ascending chain

$$\Phi(x_0) \sqsubseteq \Phi(x_1) \sqsubseteq \Phi(x_2) \sqsubseteq \cdots$$

thanks to the $\sqsubseteq$-monotonicity of Φ. More generally, an n-ary mapping

$$\Phi : \underbrace{DOM \times \cdots \times DOM}_{n \text{ times}} \to DOM$$

is $\sqsubseteq$-continuous if it is $\sqsubseteq$-continuous in every argument.

Consider a mapping $\Phi : DOM \to DOM$. An element $x \in DOM$ is called a *fixed point* of Φ if $\Phi(x) = x$. Let *FIX* be the set of all fixed points of Φ. An element $x \in DOM$ is called the *least fixed point* of Φ if it is the least element of *FIX*, then $x \in FIX$ and $x \sqsubseteq d$ for all $d \in FIX$. Then we write $x = \mathit{fix}\,\Phi$.

We wish to define the denotational trace semantics of a recursive process term as the least fixed point of a certain mapping. To this end, we need sufficient conditions for the existence and representation of such fixed points. We use the following two theorems.

KNASTER–TARSKI'S FIXED POINT THEOREM 4.1 ([89]). *Let DOM be a complete partial order under $\sqsubseteq$ and $\Phi : DOM \to DOM$ be a $\sqsubseteq$-monotonic mapping. Then Φ has a least fixed point fix Φ.*

This fixpoint is uniquely characterized by the following two axioms:

$$\Phi(\mathit{fix}\,\Phi) = \mathit{fix}\,\Phi \qquad \text{(fixpoint property)}$$
$$x = \Phi(x) \Rightarrow \mathit{fix}\,\Phi \sqsubseteq x \qquad \text{(least fixpoint property).}$$

If Φ is not only monotonic but continuous, the least fixed point of Φ can be represented as the limit of its *n-fold iterations* Φ^n where $n \in \mathbb{N}$. These are defined inductively as follows:

$$\Phi^0(x) = x \quad \text{and} \quad \Phi^{n+1}(x) = \Phi\big(\Phi^n(x)\big)$$

for every $x \in DOM$.

KLEENE'S FIXED POINT THEOREM 4.2 ([51]). *Let DOM be a complete partial order under $\sqsubseteq$ with least element $\bot$, and let $\Phi : DOM \to DOM$ be a $\sqsubseteq$-continuous mapping. Then Φ has a least fixed point fix Φ, and moreover,*

$$\mathit{fix}\,\Phi = \bigsqcup_{n \in \mathbb{N}} \Phi^n(\bot).$$

To apply this general fixed point theory to the specific format of our programming language, viz. process terms with alphabets, we need a few extra details which we explain now.

DEFINITION 4.3. A *denotational semantics $\mathcal{M}$ for process terms* is a mapping

$$\mathcal{M} : \mathit{Proc} \to (\mathit{Env}_{\mathcal{M}} \to DOM_{\mathcal{M}}),$$

which satisfies the following conditions:

(1) The semantic domain $DOM_{\mathcal{M}}$ is equipped with a partial order $\sqsubseteq$. We stipulate that the elements of $DOM_{\mathcal{M}}$ are of the form (A, Γ) where A is some alphabet and Γ is some information about the process behaviour, e.g., a set of traces. For each

alphabet A we consider the subdomain $DOM_{\mathcal{M}} : A \subseteq DOM_{\mathcal{M}}$ consisting of all pairs $(A, \Gamma) \in DOM_{\mathcal{M}}$. Thus

$$DOM_{\mathcal{M}} = \bigcup_{A \text{ alphabet}} DOM_{\mathcal{M}} : A.$$

We assume that $\sqsubseteq$ is complete on the subdomain $DOM_{\mathcal{M}} : A$ but not necessarily on the full domain $DOM_{\mathcal{M}}$.

(2) The set $Env_{\mathcal{M}}$ of *environments* consists of mappings

$$\rho : Idf \to DOM_{\mathcal{M}},$$

which assign to every process identifier an element of the semantic domain and which respect alphabets, i.e., where $\alpha(X) = A$ implies $\rho(X) \in DOM_{\mathcal{M}} : A$.

(3) For every n-ary operator symbol *op* of *Proc* there exists a corresponding semantic operator

$$op_{\mathcal{M}} : \underbrace{DOM_{\mathcal{M}} \times \cdots \times DOM_{\mathcal{M}}}_{n \text{ times}} \to DOM_{\mathcal{M}}$$

which is $\sqsubseteq$-monotonic and satisfies

$$op_{\mathcal{M}}\big((A_1, \Gamma_1), \ldots, (A_n, \Gamma_n)\big) \in DOM_{\mathcal{M}} : op_\alpha(A_1, \ldots, A_n).$$

Here op_α is the set-theoretic operator on alphabets that corresponds to *op* (cf. Remark 3.3).

(4) The definition of $\mathcal{M}$ proceeds by induction on the structure of process terms and obeys the following principles:

- *Environment technique*:

$$\mathcal{M}[\![X]\!](\rho) = \rho(X).$$

- *Compositionality*:

$$\mathcal{M}\big[\!\big[op(P_1, \ldots, P_n)\big]\!\big](\rho) = op_{\mathcal{M}}\big(\mathcal{M}[\![P_1]\!](\rho), \ldots, \mathcal{M}[\![P_n]\!](\rho)\big).$$

- *Fixed point technique*:

$$\mathcal{M}[\![\mu X \bullet P]\!](\rho) = \mathit{fix}\, \Phi_{P,\rho}.$$

Here $\mathit{fix}\, \Phi_{P,\rho}$ denotes the least fixed point of the mapping

$$\Phi_{P,\rho} : DOM_{\mathcal{M}} : B \to DOM_{\mathcal{M}} : B$$

defined by

$$\Phi_{P,\rho}(B, \Gamma) = \mathcal{M}[\![P]\!]\big(\rho\big[(B, \Gamma)/X\big]\big),$$

where $B = \alpha(X)$ and $\rho[(B, \Gamma)/X]$ is the environment that agrees with ρ except for the identifier X where its value is (B, Γ).

By the conditions (1)–(3) above and Knaster–Tarski's fixed point Theorem 4.1, the inductive definition of $\mathcal{M}$ in (4) is well-defined and yields a value

$$\mathcal{M}[\![P]\!](\rho) \in DOM_{\mathcal{M}} : \alpha(P)$$

for every process term P.

Environments ρ assign values to the free identifiers in process terms, analogously to Tarski's semantic definition for predicate logic. Thus for closed process terms P the environment parameter ρ of $\mathcal{M}$ is not needed; in that case we shall write $\mathcal{M}[\![P]\!]$ instead of $\mathcal{M}[\![P]\!](\rho)$.

REMARK 4.4. Note that the semantics $\mathcal{M}$ is uniquely determined by the domain $DOM_{\mathcal{M}}$, its partial order $\sqsubseteq$, and the semantic operators $op_{\mathcal{M}}$.

4.2. *Denotational trace semantics*

We specialize now the general definition of a denotational semantics $\mathcal{M}$ for the trace model of process terms thus obtaining a denotational definition of the trace semantics. To this end, we need only to introduce an appropriate partial order $\sqsubseteq$ and monotonic semantic operators $op_{\mathcal{T}}$ on the trace domain $DOM_{\mathcal{T}}$ of Section 2.

As partial ordering we take the following relation on pairs $(A, \Gamma), (B, \Delta) \in DOM_{\mathcal{T}}$:

$$(A, \Gamma) \sqsubseteq (B, \Delta) \quad \text{if } A = B \text{ and } \Gamma \subseteq \Delta.$$

Thus for every ascending chain $(A_0, \Gamma_0) \sqsubseteq (A_1, \Gamma_1) \sqsubseteq (A_2, \Gamma_2) \sqsubseteq \cdots$ the alphabets are identical and the least upper bound in $DOM_{\mathcal{T}}$ exists, viz.

$$\bigsqcup_{n\in\mathbb{N}}(A_n, \Gamma_n)(A_n, \Gamma_n) = \left(A_0, \bigcup_{n\in\mathbb{N}} \Gamma_n\right).$$

There is no least element in $DOM_{\mathcal{T}}$, but one in every subdomain $DOM_{\mathcal{T}}: A$, viz. $(A, \{\langle\rangle\})$. Hence we can state:

PROPOSITION 4.5. *Every subdomain $DOM_{\mathcal{T}}: A$ is a complete partial order under the relation $\sqsubseteq$.*

We now introduce the semantic operators $op_{\mathcal{T}}$ on the trace domain $DOM_{\mathcal{T}}$. For convenience, we separate the morphism operators into renaming and hiding. By Lemma 3.8, this can be done without loss of generality.

DEFINITION 4.6. Consider $(A, \Gamma), (B, \Delta) \in DOM_{\mathcal{T}}$. Then the operators $op_{\mathcal{T}}$ are defined as follows:

(1) *Deadlock.*

$$stop{:}A_{\mathcal{T}} = \big(A, \{\langle\rangle\}\big).$$

This process never engages in any communication and hence only the empty trace can be observed.

(2) *Divergence.*

$$div{:}A_{\mathcal{T}} = \big(A, \{\langle\rangle\}\big).$$

This process engages only in internal actions. Thus its observable trace behaviour is the same as for the *stop* process.

(3) *Prefix.*

$$a._{\mathcal{T}}(A, \Gamma) = \big(A \cup \{a\},\ \{\langle\rangle\} \cup \big\{\langle a\rangle^\frown tr \mid tr \in \Gamma\big\}\big).$$

Initially, this process can engage in the communication a and after a it behaves as described in Γ.

(4) *Nondeterminism.*

$$(A, \Gamma)\ or_{\mathcal{T}}\ (B, \Delta) = (A \cup B,\ \Gamma \cup \Delta).$$

The process can do whatever Γ or Δ can.

(5) *Choice.*

$$(A, \Gamma) +_{\mathcal{T}} (B, \Delta) = (A \cup B,\ \Gamma \cup \Delta).$$

This process yields the same traces as the nondeterministic one.

(6) *Parallelism.*

$$(A, \Gamma) \parallel_{\mathcal{T}} (B, \Delta) = \big(A \cup B,\ \{tr \in (A \cup B)^* \mid tr \downarrow A \in \Gamma \wedge tr \downarrow B \in \Delta\}\big),$$

where $tr \downarrow A$ and $tr \downarrow B$ are the projections of the trace tr onto the alphabets A and B, respectively. A parallel composition may engage in a trace $tr \in (A \cup B)^*$ if its components Γ and Δ may engage in the projections $tr \downarrow A$ and $tr \downarrow B$.

(7) *Renaming.*

$$(A, \Gamma)[b/a]_{\mathcal{T}} = \big(A\{b/a\},\ \{tr\{b/a\} \mid tr \in \Gamma\}\big),$$

where $A\{b/a\}$ and $tr\{b/a\}$ result from A and tr by replacing every occurrence of a by b (cf. Section 4.1) .

(8) *Hiding.*

$$(A, \Gamma) \setminus b_{\mathcal{T}} = \big(A - \{b\},\ \{tr \downarrow (A - \{b\}) \mid tr \in \Gamma\}\big).$$

PROPOSITION 4.7. *The above operators $op_{\mathcal{T}}$ are all monotonic and continuous with respect to $\sqsubseteq$.*

PROOF. Monotonicity is obvious from the form of the operator definitions: for larger trace sets as input the operators generate larger trace sets as output. Also continuity is easy to check. □

By instantiating the general Definition 4.3 with 4.5–4.7, we obtain a denotational trace semantics

$$\mathcal{T}^*[\![\_]\!] : Proc \to (Env_{\mathcal{T}} \to DOM_{\mathcal{T}}).$$

To become familiar with this new semantics, let us compute it for an example of a recursive process.

EXAMPLE 4.8. Consider $\mu X \bullet a.X$ where $\alpha(X) = \{a\}$. By Definition 4.3,

$$\mathcal{T}^*[\![\mu X \bullet a.X]\!] = \mathcal{T}^*[\![\mu X \bullet a.X]\!](\rho) = \mathit{fix}\,\Phi_{a.X,\rho},$$

where ρ is an arbitrary environment and where the mapping

$$\Phi_{a.X,\rho} : DOM_{\mathcal{T}}{:}\{a\} \to DOM_{\mathcal{T}}{:}\{a\}$$

can be calculated follows: for any pair $(\{a\}, \Gamma) \in DOM_{\mathcal{T}}{:}\{a\}$

$$\begin{aligned}\Phi_{a.X,\rho}(\{a\}, \Gamma) &= \mathcal{T}^*[\![a.X]\!]\big(\rho\big[(\{a\}, \Gamma)/X\big]\big)\\ &= \big(\{a\}, \{\langle\rangle\} \cup \{\langle a\rangle^\frown tr \mid tr \in \Gamma\}\big).\end{aligned}$$

The least fixed point of $\Phi_{a.X,\rho}$ is $(\{a\},\ \{\langle a\rangle^n \mid n \in \mathbb{N}\})$. Thus

$$\mathcal{T}^*[\![\mu X \bullet a.X]\!] = \big(\{a\},\ \{\langle a\rangle^n \mid n \in \mathbb{N}\}\big).$$

By construction the denotational trace semantics yields trace sets that are *prefix closed* (cf. Definition 2.6). Thus for every process term P with

$$(A, \Gamma) = \mathcal{T}^*[\![P]\!]$$

we have for all traces tr, tr':

$$tr \in \Gamma \wedge tr' \leqslant tr \Rightarrow tr' \in \Gamma.$$

This shows that this version of a denotational semantics only captures what is called safety properties of a process. A more detailed explanation of safety and its counterpart liveness is given later.

Note also that the trace semantics does not distinguish the operationally different operators *or* (nondetermism) and $+$ (choice). This simplifies the treatment of processes but – as first observed by Milner [61] – has disadvantages when these operators are used in the context of parallel composition. We come back to this issue in Section 8.

Now that we have given two different approaches to trace semantics, an operational one in Section 3 and a denotational one above, the question arises what the relationship is between these two semantics. Similarly as in [69] one can prove the following result.

EQUIVALENCE THEOREM 4.9. *For every closed process term P the operational and denotational trace semantics coincide*:

$$\mathcal{T}[\![P]\!] = \mathcal{T}^*[\![P]\!].$$

PROOF (*Sketch*). One first proves that the operational trace semantics $\mathcal{T}$ behaves *compositonally* over all process operators, i.e., for all n-ary operator symbols op of $CProc$ and all process terms $P_1, \ldots, P_n \in CProc$ the equation

$$\mathcal{T}\big[\!\big[op(P_1, \ldots, P_n)\big]\!\big] = op_{\mathcal{T}}\big(\mathcal{T}[\![P_1]\!], \ldots, \mathcal{T}[\![P_n]\!]\big)$$

holds. An immediate consequence of this result is that operational and denotational trace semantics coincide on all *non-recursive* process terms $P \in CProc$. This follows by induction on the structure of P.

To extend this result to *all* closed process terms, we represent the semantics of each recursive term as the limit of its non-recursive approximations. More precisely, given a recursive term $\mu X.P \in CProc$ and another term $R \in CProc$ with $\alpha(X) = \alpha(R)$, we define for $n \in \mathbb{N}$ the term $P^n(R)$ inductively as follows:

$$P^0(R) = R \quad \text{and} \quad P^{n+1}(R) = P\big\{P^n(R)/X\big\}.$$

With this notation, we can state the desired approximation property: for every recursive term $\mu X.P \in CProc$ the equation

$$\mathcal{T}^*[\![\mu X.P]\!] = \bigsqcup_{n \in \mathbb{N}} \mathcal{T}^*\big[\!\big[P^n\big(div{:}\alpha(X)\big)\big]\!\big]$$

holds, i.e., the denotational trace semantics of $\mu X.P$ can be represented as the semantic limit of its syntactic approximations $P^n(div{:}\alpha(X))$. This is a consequence of the continuity of the process operators and Kleene's fixed point Theorem 4.2.

One can show that this approximation result holds also for the operational trace semantics $\mathcal{T}$ instead of $\mathcal{T}^*$. This allows us to prove the desired result: For every closed process term P the operational and denotational readiness semantics coincide:

$$\mathcal{T}[\![P]\!] = \mathcal{T}^*[\![P]\!].$$

We proceed by induction on the number m of occurrences of μ in P.

Induction basis $m = 0$: This is the non-recursive case.

Induction step $m \to m+1$: Suppose P contains $m+1$ occurrences of μ. Then P can be represented as

$$P = P_0\{\mu Y.P_1, P_2, \dots, P_k / X_1, \dots, X_k\},$$

where $P_0 \in Proc$ is non-recursive and $P_1, \dots, P_k$ together contain m occurrences of μ. Then

$$\begin{aligned}
&\mathcal{T}[\![P]\!]\\
&\quad = \{\mathcal{T} \text{ and } \mathcal{T}^* \text{ agree on non-recursive processes}\}\\
&\quad\ \ \mathcal{T}^*[\![P_0]\!]\big(\rho\big[\mathcal{T}[\![\mu Y.P_1]\!], \mathcal{T}[\![P_2]\!], \dots, \mathcal{T}[\![P_k]\!] / X_1, \dots, X_k\big]\big)\\
&\quad = \{\text{approximation of } \mathcal{T}\}\\
&\quad\ \ \mathcal{T}^*[\![P_0]\!]\Big(\rho\Big[\bigsqcup_{n\in\mathbb{N}} \mathcal{T}\big[\![P_1^n(div{:}\alpha(Y))\big]\!], \mathcal{T}[\![P_2]\!], \dots, \mathcal{T}[\![P_k]\!] / X_1, \dots, X_k\Big]\Big)\\
&\quad = \{\text{induction hypothesis}\}\\
&\quad\ \ \mathcal{T}^*[\![P_0]\!]\Big(\rho\Big[\bigsqcup_{n\in\mathbb{N}} \mathcal{T}^*\big[\![P_1^n(div{:}\alpha(Y))\big]\!], \mathcal{T}^*[\![P_2]\!], \dots, \mathcal{T}^*[\![P_k]\!] / X_1, \dots, X_k\Big]\Big)\\
&\quad = \{\text{approximation of } \mathcal{T}^*\}\\
&\quad\ \ \mathcal{T}^*[\![P_0]\!]\big(\rho\big[\mathcal{T}^*[\![\mu Y.P_1]\!], \mathcal{T}^*[\![P_2]\!], \dots, \mathcal{T}^*[\![P_k]\!] / X_1, \dots, X_k\big]\big)\\
&\quad = \{\mathcal{T}^* \text{ denotational}\}\\
&\quad\ \ \mathcal{T}^*[\![P]\!]
\end{aligned}$$

as required. □

As a consequence of the Equivalence Theorem we shall henceforth write $\mathcal{T}$ for both the operational and the denotational version of trace semantics.

4.3. *Asynchronous communication*

Also for the case of asynchronous communication, where processes are always able to accept input, we can define a denotational trace semantics. It follows the general pattern of Section 4.1 but with the process operators *op* interpreted differently.

DEFINITION 4.10. Consider $(A, \Gamma), (B, \Delta) \in DOM_{\mathcal{T}}$ with $A = (A_I : A_O)$ and $B = (B_I : B_O)$. Then the asynchronous operators op_{asyn} are defined as follows:

(1) *Deadlock.*

$$stop{:}A_{asyn} = \big(A, A_I^*\big).$$

This process accepts any input of its input alphabet A_I but never engages in any output nor any internal action.

(2) *Divergence.*

$$div{:}A_{asyn} = \big(A,\ A_I^*\big).$$

This process engages only in inputs and internal actions. Its observable trace behaviour is the same as for the *stop* process.

(3) *Prefix.* For $a \in A$ the prefix operator yields

$$a \cdot_{asyn} (A, \Gamma) = \big(A,\ A_I^* \cup \{\langle a\rangle^\frown tr \mid tr \in \Gamma\}\big).$$

This process can accept any input but if it first accepts a then it can subsequently behave as described by Γ. This is true for both a being input or output.

(4) *Buffering.* For $a \in A_I$ the buffer operator yields

$$(A, \Gamma) \lhd_{asyn} a = \big(A,\ A_I^* \cup \{tr \mid \langle a\rangle^\frown tr \in \Gamma\}\big).$$

This process can accept any input and otherwise behaves as Γ *after* the input of a.

(5) *Nondeterminism.*

$$(A, \Gamma)\ or_{asyn}\ (B, \Delta) = (A \cup B,\ \Gamma \cup \Delta).$$

The process can do whatever Γ or Δ can.

(6) *Choice.*

$$(A, \Gamma) +_{asyn} (B, \Delta) = (A \cup B,\ \Gamma \cup \Delta).$$

This operator yields the same traces as the nondeterministic one.

(7) *Parallelism.* Let $Int = (A_I \cup B_I) \cap (A_O \cup B_O)$ describe the internal communications and $C = (A \cup B) \setminus Int$ the visible communications of the asynchronous parallel composition. Then

$$\begin{aligned}(A, \Gamma) \parallel_{asyn} (B, \Delta) \\ = \big(C,\ \{tr' \in C^* \mid \exists tr \in (A \cup B)^* \bullet tr \downarrow A \in \Gamma \wedge tr \downarrow B \in \Delta \wedge tr' = tr \downarrow C\}\big),\end{aligned}$$

where $tr \downarrow A$ and $tr \downarrow B$ are the projections of the trace tr onto the alphabets A and B, respectively. As in the synchronous case, a parallel composition may engage in a trace $tr \in (A \cup B)^*$ if its components Γ and Δ may engage in the projections $tr \downarrow A$ and $tr \downarrow B$. The difference is that here the common communications that are input of one component and output of the other are considered as internal communications and thus filtered out to yield the trace tr'.

(8) *Renaming.*

$$(A, \Gamma)[b/a]_{asyn} = \big(A[b/a],\ \{tr[b/a] \mid tr \in \Gamma\}\big),$$

where $A\,[b/a]$ and $tr[b/a]$ result from A and tr by replacing every occurrence of a by b (cf. Section 4.1).

(9) *Hiding.*

$$(A, \Gamma) \setminus b_{asyn} = \big(A - \{b\},\ \{tr \downarrow (A - \{b\}) \mid tr \in \Gamma\}\big).$$

5. Algebraic approach

The heart of process algebra is the use of algebraic techniques for the specification and verification of communicating processes. Specifications and implementions of communicating processes are described in the same language, by suitable process terms. Verification is done by proving equations between specifications and implementations using algebraic laws describing properties of the process operators.

5.1. *Algebraic laws for trace semantics*

We consider here algebraic laws as equational theorems about the trace semantics of process terms. They can be expressed as deduction rules of the form

$$(\mathcal{D}) \qquad \frac{P_1 = Q_1, \ldots, P_n = Q_n}{P_{n+1} = Q_{n+1}} \quad \text{where } Cond$$

where $n \geqslant 0$ and *Cond* is a condition on the syntax of process terms $P_1, Q_1, \ldots, P_n, Q_n, P_{n+1}, Q_{n+1}$. The law $\mathcal{D}$ states that if the condition *Cond* is satisfied and the equations $P_1 = Q_1, \ldots, P_n = Q_n$ are true, then the equation $P_{n+1} = Q_{n+1}$ also holds. If $n = 0$, the notation simplifies to

$$P_1 = Q_1 \quad \text{where } Cond.$$

The syntactic condition *Cond* will always be a simple decidable property, e.g., concerning alphabets.

We say that a law $\mathcal{D}$ is *sound for the trace semantics* if the theorem denoted by $\mathcal{D}$ is true when each equation $P_i = Q_i$ with $1 \leqslant i \leqslant n+1$ in $\mathcal{D}$ is interpreted as $\mathcal{T}[\![P_i]\!] = \mathcal{T}[\![Q_i]\!]$.

DEFINITION 5.1. In the following algebraic laws, letters P, Q, R range – unless stated otherwise – over closed process terms in *CProc*, X, Y over process identifiers in *Idf*, a, b over communications in *Comm*, and A, B over alphabets.

(1) *Equation Laws.*
(Reflexivity)

$$P = P.$$

(Symmetry)

$$\frac{P = Q}{Q = P}.$$

(Transitivity)

$$\frac{P = Q,\ \ Q = R}{P = R}.$$

(Congruence in Context)

$$\frac{Q = R}{P\{Q/X\} = P\{R/X\}} \quad \text{where } P \in \mathit{Proc} \text{ and } P\{Q/X\},\ P\{R/X\} \in \mathit{CProc}.$$

(2) *Deadlock and Divergence.*

$$\mathit{stop}{:}A = \mathit{div}{:}A.$$

(3) *Laws for Nondeterminism and Choice.*
(Nondeterminism)

$$P \ \mathit{or}\ Q = P + Q.$$

(Commutativity)

$$P + Q = Q + P.$$

(Associativity)

$$P + (Q + R) = (P + Q) + R.$$

(Idempotence)

$$P + P = P.$$

(Deadlock)

$$P + \mathit{stop}{:}\alpha(P) = P.$$

(Distributivity over Prefix)

$$a.(P + Q) = a.P + a.Q.$$

(4) *Laws for Parallel Composition.*
(Commutativity)

$$P \| Q = Q \| P.$$

(Associativity)

$$P \| (Q \| R) = (P \| Q) \| R.$$

(Distributivity over Choice)

$$P \| (Q + R) = (P \| Q) + (P \| R).$$

(Synchronous Communication)

$$a.P \| a.Q = a.(P \| Q) \quad \text{where } a \in \alpha(P) \cap \alpha(Q).$$
$$a.P \| b.Q = \mathit{stop}{:}\alpha(P) \cup \alpha(Q) \quad \text{where } a, b \in \alpha(P) \cap \alpha(Q) \text{ and } a \neq b.$$
$$a.P \| \mathit{stop}{:}A = \mathit{stop}{:}A \cup \alpha(P) \quad \text{where } a \in A.$$

(Asynchronous Communication)

$$a.P \| b.Q = a.(P \| b.Q) + b.(a.P \| Q) \quad \text{where } a, b \notin \alpha(P) \cap \alpha(Q).$$
$$a.P \| \mathit{stop}{:}A = a.(P \| \mathit{stop}{:}A) \quad \text{where } a \notin A.$$
$$\mathit{stop}{:}A \| \mathit{stop}{:}B = \mathit{stop}{:}(A \cup B).$$

(5) *Laws for Morphism.*
(Renaming)

$$(a.P)[\varphi] = b.P[\varphi] \quad \text{where } \varphi(a) = b \neq \tau.$$

(Hiding)

$$(a.P)[\varphi] = P[\varphi] \quad \text{where } \varphi(a) = \tau.$$

(Distributivity over Choice)

$$(P + Q)[\varphi] = P[\varphi] + Q[\varphi].$$

(Partial Distributivity over Parallel Composition)

$$(P \| Q)[\varphi] = P[\varphi] \| Q[\varphi]$$

where for all $a \in \mathit{Comm}$ the following holds:
$\varphi(a) \neq a$ implies $\varphi(a) \notin \alpha(P) \cap \alpha(Q)$.

(Composition)

$$(P[\varphi_1])[\varphi_2] = P[\varphi_1 \circ \varphi_2].$$

(Stop)

$$(stop{:}A)[\varphi] = stop{:}\big(\varphi(A) - \{\tau\}\big).$$

(6) *Laws for Recursion.*
(Renaming)

$$\mu X \bullet P = \mu Y \bullet P\{Y/X\}$$
where $P \in Proc$ and $\mu X \bullet P \in CProc$ and $\alpha(X) = \alpha(Y)$.

(Fixed Point)

$$\mu X \bullet P = P\{\mu X \bullet P/X\}$$
where $P \in Proc$ and $\mu X \bullet P \in CProc$.

(Unique Fixed Points)

$$\frac{Q = P\{Q/X\},\ \ R = P\{R/X\}}{Q = R}$$
where $P \in Proc$ and $\mu X \bullet P \in CProc$ is communication-guarded and $\alpha(Q) = \alpha(X) = \alpha(R)$ holds.

(Koomen's Fair Abstraction Rule: KFAR)

$$\frac{P = b.P + Q}{P \setminus b = Q \setminus b}.$$

(7) *Laws for Buffering.*

$$(a.P) \lhd a = P \qquad \text{where } a \in \alpha_I(P).$$
$$(P \; or \; Q) \lhd a = (P \lhd a) \text{ or } (P \lhd a) \qquad \text{where } a \in \alpha_I(P) \cap \alpha_I(Q).$$
$$(P + Q) \lhd a = (P \lhd a) + (Q \lhd a) \qquad \text{where } a \in \alpha_I(P) \cap \alpha_I(Q).$$
$$(P \| Q) \lhd a = (P \lhd a) \| Q \qquad \text{where } a \in \alpha_I(P).$$

The law for unique fixed points depends on the concept of *communication-guardedness* which informally says that before each resursive call in a process at least one communication has occurred. The following definition gives a sufficient syntactic condition for this dynamic concept.

DEFINITION 5.2. Let $A \subseteq Comm$. A process term $P \in Proc$ is *communication-guarded by* A if for every recursive subterm $\mu X \bullet Q$ of P, every free occurrence of X in Q occurs within a subterm of Q of the form $a.R$ with $a \in A$, but not within a subterm of Q of the form $R[\varphi]$ with $\varphi(a) = \tau$ or $\varphi(a) = b$ where $a \in A$ and $b \notin A$. P is called *communication-guarded* if there exists a set A of communications such that P is communication-guarded by A.

The process term $\mu X \bullet a.X$ studied in Example 4.8 is communication-guarded. More elaborate examples of communication-guarded process terms with $\alpha(X) = \alpha(Y) = \{a, b\}$ are the following ones:

$$P_1 = \mu X \bullet a.\ \mu Y \bullet (b.\ X + a.\ b.\ Y),$$
$$P_2 = \big((\mu X \bullet a.\ b.\ X)[c/b] \parallel (\mu Y \bullet a.\ b.\ Y)[c/a]\big)\backslash c.$$

By contrast,

$$Q_1 = \mu X \bullet a.\ \big((X[c/a])\backslash c \parallel stop{:}\{a, b\}\big)$$

is not communication-guarded.

Koomen's Fair Abstraction Rule was first used in a formula manipulation system for Milner's CCS [52]. The formal analysis of the rule is due to [9]. Fair abstraction means here that the process $P \setminus b$ where we hide or abstract from b will eventually exit the hidden b-cycle to behave like $Q \setminus b$.

Altogether the following theorem can be proved.

SOUNDNESS THEOREM 5.3. *The algebraic laws of Definition* 5.1 *are sound for the trace semantics.*

A set of algebraic laws is called complete if all equations in a given semantic model can be proved. The question of *completeness* is out of the scope of the present handbook chapter. We refer the reader to [14].

5.2. *Algebraic verification*

By the algebraic laws we can prove the equivalence of process terms. We speak of algebraic verification. Algebraic verification of processes proceeds in three steps:

(1) Specify the desired process behaviour by some process term *SPEC*.
(2) Represent the implemented process behaviour by another, possibly more elaborate process term *IMPL* with $\alpha(SPEC) \subseteq \alpha(IMPL)$.
(3) Show that the implementation *IMPL* satisfies the specification *SPEC* by proving the equation

$$IMPL \setminus B = SPEC$$

with the algebraic laws where $B = \alpha(IMPL) - \alpha(SPEC)$.

As an example of an algebraic verification we consider a simple communication protocol due to J. Parrow [74].

EXAMPLE 5.4. As the specification of the communication protocol we take

$$SPEC = \mu W \bullet send.rec.W$$

where $\alpha(W) = \{send, rec\}$. It requires that every message sent is also received. The protocol *PROT* itself is the implementation of the service and consists of the parallel composition

$$PROT = S \| M \| R$$

of three components, a sender

$$S = \mu X \bullet send.\mu X1 \bullet in.(ack.X + error.X1),$$

a lossy transmission medium

$$M = \mu Y.in.(out.Y \; or \; error.Y),$$

and a receiver

$$R = \mu Z \bullet out.rec.ack.Z,$$

where

$$\alpha(X) = \alpha(X1) = \{send, in, ack, error\},$$
$$\alpha(Y) = \{t, in, out, error\},$$
$$\alpha(Z) = \{out, rec, ack\}.$$

The above equations model the following scenario. The sender S takes a message sent by the environment, inputs it to the transmission medium and then waits for an acknowledgement or an error message from the medium. In case of an error message the message is retransmitted, i.e., input once more to the transmission medium. In case of an acknowledgement the sender waits for a new message sent by the environment. The transmission medium M inputs a message from the sender and nondeterministically outputs it correctly to the receiver or produces an error. The receiver R waits for a message output by the transmission medium, receives this message properly and acknowledges the receipt to the medium.

The aim is now to verify that this protocol satisfies the specification in the trace semantics by deriving the equation

$$PROT \setminus B = SPEC \tag{3}$$

where $B = \{in, out, ack, error\}$. To this end, we show that $PROT \setminus B$ and $SPEC$ satisfy the same communication-guarded equation, viz.

$$\begin{aligned} SPEC &= send.rec.SPEC, \\ PROT \setminus B &= send.rec.(PROT \setminus B). \end{aligned}$$

Then by the law of Unique Fixed Points the desired equation (3) follows.

The equation for $SPEC$ follows by applying the Fixed Point law to its definition. The equation for $PROT \setminus B$ requires a detailed calculation involving the parallel composition operator. For this purpose, we need the alphabets of S, M, R:

$$\begin{aligned} &\alpha(S) = \{send, in, ack, error\}, \\ &\alpha(M) = \{in, out, error\}, \\ &\alpha(R) = \{out, rec, ack\}. \end{aligned}$$

Note that the communications *send* and *rec* occur only in one of the protocol components. By the definition of parallel composition, they can occur asynchronously. For all other communications two of the three protocol components have to synchronize.

Let us introduce the abbreviation

$$PROT1 = S1 \| M \| R$$

where $S1 = \mu X1 \bullet in.(ack.S + error.X1)$ holds. Then

$$PROT \setminus B = send.PROT1 \setminus B \tag{4}$$

holds as the following calculation shows:

$$\begin{aligned} PROT \setminus B &= \{\text{Definition of } PROT\} \\ &\quad (S \| M \| R) \setminus B \\ &= \{\text{Fixed Point: definition of } S\} \\ &\quad \big(send.\mu X1 \bullet in.(ack.S + error.X1) \| M \| R\big) \setminus B \\ &= \{\text{Morphism}\} \\ &\quad send.\big(\mu X1 \bullet in.(ack.S + error.X1) \| M \| R\big) \setminus B \\ &= \{\text{Definition of } PROT1\} \\ &= send.PROT1 \setminus B. \end{aligned}$$

It remains to show the following equation:

$$PROT1 \setminus B = rec.PROT \setminus B. \tag{5}$$

Putting $B1 = \{in, out, ack\}$ one can prove

$$PROT1 \setminus B1 = error.PROT1 \setminus B1 + rec.PROT \setminus B1. \tag{6}$$

From (6) we derive (5) using Koomen's Fair Abstraction Rule KFAR as follows:

$$\begin{aligned} \mathit{PROT}1 \setminus B &= \{\text{Definition of } B \text{ and } B1\} \\ &\quad (\mathit{PROT}1 \setminus B1) \setminus \mathit{error} \\ &= \{\text{Using (6)}\} \\ &\quad (\mathit{error}.\mathit{PROT}1 \setminus B1 + \mathit{rec}.\mathit{PROT} \setminus B1) \setminus \mathit{error} \\ &= \{\text{KFAR}\} \\ &\quad (\mathit{rec}.\mathit{PROT} \setminus B1) \setminus \mathit{error} \\ &= \{\text{Morphism}\} \\ &\quad \mathit{rec}.(\mathit{PROT} \setminus B1) \setminus \mathit{error} \\ &= \{\text{Definition of } B \text{ and } B1\} \\ &\quad \mathit{rec}.\mathit{PROT} \setminus B. \end{aligned}$$

Using (4) and (5) we obtain

$$\begin{aligned} \mathit{PROT} \setminus B &= \{\text{Using (4)}\} \\ &\quad \mathit{send}.\mathit{PROT}1 \setminus B \\ &= \{\text{Using (5)}\} \\ &\quad \mathit{send}.\mathit{rec}.\mathit{PROT} \setminus B. \end{aligned}$$

This proves that $\mathit{PROT} \setminus B$ satisfies the same recursive equation as *SEND* and thus completes the verification of (3) by the law of unique fixpoints.

6. Logical approach

The logical approach describes the desired properties of processes, but not how they are implemented. It is thus more abstract than the previously discussed approaches. As specification language for trace sets we will use a *many-sorted first-order predicate logic*. Since its main sort is "trace", it is called *trace logic* and its formulas are called *trace formulas*. Informal use of trace logic appeared in a number of papers (e.g., [94,65,71,86,79,91]). Precise syntax and semantics were first given by Zwiers [96,95]. We adopt Zwiers' proposal, but we need only a simplified version of it because we deal here only with atomic communications instead of messages sent along channels (cf. Section 2).

6.1. *Trace logic*

We adopt syntax and semantics of trace logic from Zwiers [95]. However, since we do not consider channels, messages and termination signals, simplified definitions suffice so that in our case trace logic is a many-sorted predicate logic with the following sorts:

trace	(finite sequences of communications)
nat	(natural numbers)

comm (communications)
log (logical values).

Trace logic then consists of sorted expressions built up from sorted constants, variables and operator symbols. For notational convenience, trace formulas count here as expressions of sort *log*.

We use the following notation. All communications a, b appear as constants of sort *trace* and of sort *comm*, and all natural numbers $k > 0$ appear as constants of sort *nat*. The set *Var* of *variables* is partitioned into a set *Var:trace* of variables t of sort *trace* and a set *Var:nat* of variables n of sort *nat*. For all communication alphabets A and all communications a, b there are unary operator symbols $\_\downarrow A$ and $\_[b/a]$ of sort

$$\begin{array}{lll} \_\downarrow A & : & \mathit{trace} \to \mathit{trace} \quad \text{(projection)} \\ \_[b/a] & : & \mathit{trace} \to \mathit{trace} \quad \text{(renaming).} \end{array}$$

Further on, there are binary operator symbols $\_\frown\_$ and $\_(\_)$ of sort

$$\begin{array}{lll} \_\frown\_ & : & \mathit{trace} \times \mathit{trace} \to \mathit{trace} \quad \text{(concatenation)} \\ \_(\_) & : & \mathit{trace} \times \mathit{nat} \to \mathit{comm} \quad \text{(selection)} \end{array}$$

and a unary operator symbol $\#\_$ of sort

$$\#\_ : \mathit{trace} \to \mathit{nat} \quad \text{(length).}$$

The intended interpretation is given as a commentary. The remaining symbols used in trace logic are all standard.

DEFINITION 6.1. The syntax of trace logic is given by a set

$$\mathit{Exp} = \mathit{Exp} : \mathit{trace} \cup \mathit{Exp} : \mathit{nat} \cup \mathit{Exp} : \mathit{comm} \cup \mathit{Exp} : \mathit{log}$$

of *expressions* ranged over by xe. The constituents of *Exp* are defined as follows.

(1) The set *Exp* : *trace* of *trace expressions* consists of all expressions te that are generated by the production rules

$$te ::= \langle\rangle \mid a \mid t \mid te_1 \frown te_2 \mid te \downarrow A \mid te[b/a]$$

and that satisfy the following context-sensitive restriction: every trace variable t in a trace expression te occurs within a subexpression of the form $te_0 \downarrow A$ of te.

(2) The set *Exp* : *nat* of *natural number expressions* consists of the following expressions ne:

$$ne ::= k \mid n \mid ne_1 + ne_2 \mid ne_1 * ne_2 \mid \# te.$$

(3) The set $Exp : comm$ of *communication expressions* consists of the following expressions ce:

$$ce ::= a \mid te(ne).$$

(4) The set $Exp : log$ of *trace formulas* or *logical expressions* consists of the following expressions le:

$$\begin{aligned} le \ ::= \ & true \mid te_1 \leqslant te_2 \mid ne_1 \leqslant ne_2 \mid ce_1 = ce_2 \mid \\ & \neg le \mid le_1 \wedge le_2 \mid \exists t \bullet le \mid \exists n \bullet le. \end{aligned}$$

Examples of trace expressions are $t \downarrow \{up\}$ and $((\langle up \rangle ^\frown t) \downarrow \{up\})^\frown dn$; counterexamples are t and $\langle up \rangle ^\frown t ^\frown \langle dn \rangle$ because the context-sensitive restriction on trace expressions is not satisfied. This restriction is needed to prove properties like the Projection Lemma 6.7 and the Renaming Lemma 6.9.

Free and *bound* (occurrences of) variables in an expression xe are defined as usual. Let $free(xe)$ denote the set of free variables in xe. The notation $xe\{te/t\}$ denotes the result of *substituting* the trace expression te for every free occurrence of the trace variable t in xe. As for process terms, substitution requires a preparatory renaming of the bound variables in xe to avoid clashes with the free variables in te. Simultaneous substitution of a list $te_1, \ldots, te_n$ of trace expressions for a list of distinct trace variables $t_1, \ldots, t_n$ is denoted by $xe\{te_1, \ldots, te_n/t_1, \ldots, t_n\}$. Furthermore, let $xe\{b/a\}$ denote the result of *renaming* every occurrence of the communication a in xe into b. Simultaneous renaming of distinct communications $a_1, \ldots, a_n$ in xe into $b_1, \ldots, b_n$ is denoted by $xe\{b_1, \ldots, b_n/a_1, \ldots, a_n\}$.

Note that by Definition 6.1, we have the following inclusions:

$$Comm \subseteq Comm^* \subseteq Exp : trace \quad \text{and} \quad \mathbb{N} \subseteq Exp : nat.$$

Thus every trace is also a trace expression. Hence we may also write substitutions of the form $xe\{tr/t\}$ or $xe\{tr_1, \ldots, tr_n/t_1, \ldots, t_n\}$. Note that we carefully distinguish traces, ranged over by tr, from trace variables, ranged over by t.

Further on, we make use of several abbreviations.

(1) Natural number expressions *counting* communications in a trace:

$$\begin{aligned} A \# te \ &=_{df} \ \#(te \downarrow A), \\ a \# te \ &=_{df} \ \{a\} \# te. \end{aligned}$$

(2) Communication expressions *selecting specific* elements of a trace:

$$\begin{aligned} head\, te \ &=_{df} \ te(1), \\ last\, te \ &=_{df} \ te(\# te). \end{aligned}$$

(3) Extended syntax for logical expressions: As usual disjunction ($\vee$), implication ($\Rightarrow$), equivalence ($\Longleftrightarrow$) and universal quantification ($\forall$) can be introduced as abbreviations.

Syntactic ambiguities are resolved by using brackets and some priorities among the operator symbols. Within trace expressions the unary symbols $\_\downarrow A$ and $\_[b/a]$ have a higher priority than the binary concatenation $\_\frown\_$; within natural number expressions $*$ has a higher priority than $+$; within logical expressions we stipulate the following priorities:

$$\begin{aligned}&\text{priority } 0 : \Rightarrow \text{ and } \Longleftrightarrow\\&\text{priority } 1 : \wedge \text{ and } \vee\\&\text{priority } 2 : \neg \text{ and } \exists \text{ and } \forall.\end{aligned}$$

Additionally, we save brackets by relying on the associativity of $\wedge$ and $\vee$.

Semantics. The *standard semantics* or *interpretation* $\Im$ *of trace logic* is introduced along the lines of Tarski's semantic definition for predicate logic. It assumes for each sort a set of values, called the *semantic domain* of that sort.

- The semantic domain of sort *trace* is $Comm^*$, the set of finite sequences of communications. We let *tr* range over $Comm^*$.
- The semantic domain of sort *nat* is $\mathbb{N}$, the set of natural numbers.
- The semantic domain of sort *comm* is $Comm \cup \{\bot\}$, the set consisting of all communications and the special value $\bot$, called *bottom*. This value is needed to interpret communication expressions that do not yield a proper communication value. For example,

  $$\langle up, up, dn\rangle(5)$$

  denotes the fifth value of a trace $\langle up, up, dn\rangle$, which is $\bot$.
- The semantic domain of sort *log* is {true, false}, the set of Boolean values.

The semantic domain $DOM_\Im$ of $\Im$ is defined as the union of the semantic domains of sort *trace, nat, comm* and *log*:

$$DOM_\Im = Comm^* \cup \mathbb{N} \cup Comm \cup \{\bot\} \cup \{\text{true}, \text{false}\}.$$

The standard semantics of logic is a mapping

$$\Im : Exp \to (Env_\Im \to DOM_\Im)$$

that assigns to every expression in *Exp* a value in the semantics domain $DOM_\Im$ with the help of so-called *environments*. These are mappings

$$\rho : Var \to DOM_\Im$$

that assign values to the free variables in expressions. The set $Env_\Im$ consists of all environments ρ that respect sorts, i.e., where trace variables t get values $\rho(t) \in Comm^*$ and natural number variables n get values $\rho(n) \in \mathbb{N}$. We omit the formal definition of $\Im$ (see [69] for details).

DEFINITION 6.2. A trace formula *le* is called *valid*, abbreviated

$$\models le,$$

if $\Im[\![le]\!](\rho) = \text{true}$ for all environments ρ.

Reasoning about expressions is simplified by introducing a normal form where trace projections $- \downarrow A$ are adjacent to the trace variables [68,95].

DEFINITION 6.3. A trace expression *te* is called *normal* if it can be generated by the following syntax rules:

$$te ::= \varepsilon \mid a \mid t \downarrow A \mid te_1 {}^\frown te_2 \mid te[b/a].$$

An arbitrary expression *xe* is normal if every maximal trace expression *te* in *xe* is normal. Maximal means that *te* is not contained in a larger trace expression in *xe*.

Thanks to the context-sensitive restriction imposed on trace expressions by Definition 6.1, every expression *xe* can be converted into a unique normal expression, called its *normal form* and denoted by xe_{norm}.

6.2. *Trace specifications*

How to use trace logic for the specification of trace sets? We restrict ourselves to a certain subset of trace formulas. To this end, we introduce a distinguished trace variable called h standing for *history*.

DEFINITION 6.4. The set *Spec* of *trace specifications*, ranged over by letters S, T, U, consists of all trace formulas where at most the distinguished variable h of sort *trace* is free.

The logical value $\Im[\![S]\!](\rho)$ of a trace specification S depends only on the trace value $\rho(h)$. We therefore say that a trace $tr \in Comm^*$ *satisfies* S and write

$$tr \models S$$

if $\Im[\![S]\!](\rho) = \text{true}$ for $\rho(h) = tr$. Note the following relationship between satisfaction and validity:

$$tr \models S \quad \text{iff} \quad \models S\{tr/h\}.$$

Recall that $S\{tr/h\}$ is the result of substituting the trace tr for every free occurrence of the trace variable h in S. Thus a trace specification S specifies the set of all traces satisfying S. In fact, whether or not a trace satisfies a trace specification S depends only on the trace value within the so-called projection alphabet $\alpha(S)$.

DEFINITION 6.5. For normal trace expressions *te* the *projection alphabet* or simply *alphabet* $\alpha(te)$ is defined inductively as follows:

$$\begin{aligned}
\alpha(\langle\rangle) &= \emptyset, \\
\alpha(\langle a\rangle) &= \emptyset, \\
\alpha(h \downarrow A) &= A, \\
\alpha(t \downarrow A) &= \emptyset \quad \text{if } t \neq h, \\
\alpha(te_1 {}^\frown te_2) &= \alpha(te_1) \cup \alpha(te_2), \\
\alpha(te[b/a]) &= \alpha(te).
\end{aligned}$$

For arbitrary trace expressions *te* the alphabet is given by

$$\alpha(te) = \alpha(te_{norm}).$$

For trace specifications S the alphabet is

$$\alpha(S) = \bigcup \alpha(te),$$

where the union is taken over all maximal trace expressions *te* in S that contain an occurrence of h that is free in S. If such a trace expression does not exist, the alphabet $\alpha(S)$ is empty.

EXAMPLE 6.6. Consider the trace specification

$$S =_{df} \#(c^\frown h) \downarrow \{b\} \leqslant \#(c^\frown h) \downarrow \{a\}.$$

Then, for example, $\langle\rangle$, $\langle a\rangle$, $\langle a, b\rangle$, $\langle a, a, b\rangle \models S$, but $\langle b\rangle \models S$ does not hold. To calculate the projection alphabet $\alpha(S)$, we determine the maximal trace expressions of S. These are

$$te1 =_{df} (c^\frown h) \downarrow \{b\} \quad \text{and} \quad te2 =_{df} (c^\frown h) \downarrow \{a\}.$$

Using the projection laws, they are transformed into normal form:

$$te1_{norm} = h \downarrow \{b\} \quad \text{and} \quad te2_{norm} = h \downarrow \{a\}.$$

Thus we obtain $\alpha(S) = \{a, b\}$.

The following lemma states that each trace specification determines the shape of traces only inside its projection alphabet.

PROJECTION LEMMA 6.7. *Let S be a trace specification. Then*

$$tr \models S \quad \textit{iff} \quad tr \downarrow \alpha(S) \models S$$

for all traces $tr \in Comm^*$.

This lemma is needed in the proof of Soundness Theorem 7.5. There we also need the subsequent Renaming Lemma based on the following definition.

DEFINITION 6.8. The *extended alphabet* $\alpha\alpha(S)$ of a trace specification S is the set of all communications appearing somewhere in S. By convention, whenever $te \downarrow A$ is a subexpression in S then $A \subseteq \alpha\alpha(S)$.

RENAMING LEMMA 6.9. *Consider a trace specification S and communications a, b with $b \notin \alpha\alpha(S)$. Then*

$$tr \models S \quad \textit{iff} \quad tr\{b/a\} \models S\{b/a\}$$

for all traces $tr \in \alpha(S)$.

EXAMPLE 6.10. Consider once more the trace specification S of Example 6.6. We calculated there the projection alphabet as $\alpha(S) = \{a, b\}$. Thus by the Projection Lemma, we can deduce, e.g.,

$$\langle a, happy, a, birth, b, day\rangle \models S$$

because $\langle a, happy, a, birth, b, day\rangle \downarrow \alpha(S) = \langle a, a, b\rangle$ and $\langle a, a, b\rangle \models S$ as stated in Example 6.6.

The extended alphabet of S is $\alpha\alpha(S) = \{c, b, a\}$. Consider the renaming of the communication b into e. Then

$$S\{e/b\} = \#(c.h) \downarrow \{e\} \leqslant \#(c.h) \downarrow \{a\}$$

and $\langle a, a, b\rangle\{e/b\} = \langle a, a, e\rangle$. Since $e \notin \alpha\alpha(S)$ and $a.a.b \models S$, the Renaming Lemma yields

$$\langle a, a, e\rangle \models S\{e/b\}.$$

To see that the condition of the Renaming Lemma is needed, consider the renaming of b into c. Since $c \in \alpha\alpha(S)$, the Renaming Lemma is not applicable. Indeed, we have $a.b \models S$ but not $\langle a, b\rangle\{c/b\} \models S\{c/b\}$.

Since trace logic includes the standard interpretation of Peano arithmetic, viz. the model $(\mathbb{N}, 0, 1, +, *, =)$, trace specifications are very expressive [95].

EXPRESSIVENESS THEOREM 6.11. *Let $\mathcal{L} \subseteq A^*$ be a recursively enumerable set of traces over the alphabet A. Then there exists a trace specification $TRACE(\mathcal{L})$ with projection alphabet $\alpha(TRACE(\mathcal{L})) = A$ such that*

$$tr \in \mathcal{L} \quad \textit{iff} \quad tr \models TRACE(\mathcal{L})$$

for all traces $tr \in A^*$. The same is true for sets $\mathcal{L} \subseteq A^*$ whose complement in A^* is recursively enumerable.

PROOF (*Idea*). Using coding functions and standard techniques from recursion theory, every recursively enumerable set $\mathcal{L} \subseteq A^*$ can be represented by a first order formula *le* in Peano arithmetic [85]. From *le* the trace specification $TRACE(\mathcal{L})$ can be constructed easily. Obviously, negation deals with the complements of recursively enumerable sets. □

6.3. *Process correctness*

In this section we relate process terms to trace specification by introducing a notion of process correctness based on trace inclusion.

DEFINITION 6.12. Consider a closed process term P and a trace specification S. We say that P is *correct* with respect to S, or P *satisfies* S, abbreviated

$$P \; sat \; S,$$

if $\alpha(P) = \alpha(S)$ and for every trace $tr \in \alpha(P)^*$ whenever P may engage in tr then $tr \models S$.

Process correctness can also be expressed in terms of the partial order $\sqsubseteq$ on the trace domain introduced in Section 4. For a trace specification S let

$$\mathcal{T}[\![S]\!] = \big(\alpha(S), \{tr \in \alpha(S)^* \mid pref \; tr \models S\}\big).$$

The notation $pref\; tr \models S$ means that tr and all its prefixes satisfy S.

PROPOSITION 6.13. *$P \; sat \; S$ iff $\mathcal{T}[\![P]\!] \sqsubseteq \mathcal{T}[\![S]\!]$.*

EXAMPLE 6.14. A *counter* is a process that stores a natural number which can be incremented and decremented provided the storage bound is not exceeded. Incrementing is done by a communication *up*, decrementing is done by a communication *dn*. To specify the allowed traces of *up*'s and *dn*'s of a counter of capacity k, consider the following trace specification:

$$S_k =_{df} 0 \leqslant up\#h - dn\#h \leqslant k.$$

Let us examine how a process P satisfying S_k should behave. Since $P \; sat \; S_k$ implies $\alpha(P) = \alpha(S_k) = \{up, dn\}$, P should engage only in the communications *up* and *dn*.

For every communication trace tr that P may engage in, the difference between the number of *up*'s and *dn*'s is between 0 and k. Thus S_k specifies that P should behave like a bounded counter of capacity k which can internally store a natural number n with $0 \leqslant n \leqslant k$. After a communication trace tr, the number stored is

$$n = up\#tr - dn\#tr.$$

Initially, when tr is empty, n is zero. Communicating *up* increments n and communicating *dn* decrements n. Of course, these communications are possible only if the resulting changes of n do not exceed the counter bound k.

Consider the case $k = 2$. A simple process satisfying S_2 is

$$P =_{df} \mu X \bullet up.\mu Y \bullet (dn.X + up.dn.Y).$$

Another process involving parallelism satisfying S_2 is

$$Q =_{df} \big((\mu X \bullet up.dn.X)[lk/dn] \parallel (\mu X \bullet up.dn.X)[lk/up]\big)\backslash lk.$$

Here, after each *up*-transition, the process has to engage in an internal action τ before it is ready for the corresponding *dn*-transition. Since τ-actions occur autonomously, readiness for the next *dn* is guaranteed.

7. Mixed term approach

We present now an approach to the top-down construction of process terms from trace specifications. The trace specification describes what the desired communication behaviour is and the process term describes how this communication behaviour is realised. During the construction we wish to proceed by *stepwise refinement* as advocated by Dijkstra and Wirth [32,93]. Thus at each step of the construction we wish to replace a part of the specification by a bit of process syntax. To formalize this idea, we use so-called *mixed terms* (cf. [67, 68]). These are syntactic constructs mixing process terms with trace specifications. An example of a mixed term is

$$a.(S \parallel T).$$

It describes a process where some parts are already constructed (prefix and parallel composition) whereas other parts are only specified (S and T) and remain to be constructed in future refinement steps.

Formally, mixed terms are introduced by extending the syntax of process terms given in Definitions 3.1 and 3.2 by the clause

$$P ::= S$$

where S is a trace specification. Thus from now on letters P, Q, R will range over mixed terms whereas letters S, T, U continue to range over the set *Spec* of trace specifications.

DEFINITION 7.1. The set $Proc + Spec$ of *mixed terms* consist of all terms P that are generated by the context-free production rules

$P ::=$		S	(specification)
	|	$stop{:}A$	(deadlock)
	|	$div{:}A$	(divergence)
	|	$a.P$	(prefix)
	|	$P \text{ or } Q$	(nondeterminism)
	|	$P + Q$	(choice)
	|	$P \parallel Q$	(parallelism)

$$\begin{array}{lll} | & P[\varphi] & \text{(morphism)} \\ | & X & \text{(identifier)} \\ | & \mu X \bullet P & \text{(recursion)} \end{array}$$

and that satisfy the context-sensitive restrictions stated in Definition 3.2. The alphabet $\alpha(P)$ of a mixed term P is defined as in Section 3.1 where in case of a trace specification S the alphabet $\alpha(S)$ is as in Definition 6.5. Mixed terms without free process identifiers are called *closed*. The set of all closed mixed terms is denoted by $CProc + Spec$.

Semantically, mixed terms denote elements of the trace domain $DOM_{\mathcal{T}}$. In fact, since trace specifications S have a trace semantics $\mathcal{T}[\![S]\!]$ (cf. Section 6.3) and since for every operator symbol op of $Proc$ there exists a corresponding semantic operator $op_{\mathcal{T}}$ on $DOM_{\mathcal{T}}$ (cf. Definition 4.6), we immediately obtain a denotational trace semantics

$$\mathcal{T}[\![\cdot]\!] : Proc + Spec \to (Env_{\mathcal{T}} \to DOM_{\mathcal{T}})$$

for mixed terms which is defined according to the principles of Definition 4.3. As before, environments $\rho \in Env_{\mathcal{T}}$ assign values to the free identifiers in mixed terms. Thus for closed mixed terms the environment parameter ρ of $\mathcal{T}[\![\cdot]\!]$ is not needed. We therefore write $\mathcal{T}[\![P]\!]$ instead of $\mathcal{T}[\![P]\!](\rho)$.

Under the trace semantics $\mathcal{T}$, process terms, trace specifications and mixed terms are treated on an equal footing so that they are easy to compare. To do so, we introduce the following notation.

DEFINITION 7.2. For $P, Q \in CProc + Spec$ we write:

$$\begin{array}{ll} P \equiv> Q & \text{if } \mathcal{T}[\![P]\!] \sqsubseteq \mathcal{T}[\![Q]\!], \\ P \equiv Q & \text{if } \mathcal{T}[\![P]\!] = \mathcal{T}[\![Q]\!]. \end{array}$$

We call $P \equiv> Q$ a *semantic implication* and $P \equiv Q$ a *semantic equation* in the trace semantics. If $P \equiv> Q$ holds we also say that P *refines* Q.

7.1. *Transformations on mixed terms*

A top-down construction of a process term P from a trace specification S will be presented as a sequence

$$\begin{array}{l} S \equiv Q_1 \\ \quad\Uparrow \\ \quad\vdots \\ \quad\Uparrow \\ \quad Q_n \equiv P \end{array}$$

of semantics implications (and in special cases: equations) between mixed terms $Q_1, \ldots,$ Q_n where Q_1 is the given trace specification S and Q_n is the constructed process term P. The transitivity of $\equiv>$ ensures $P \equiv> S$, the desired result.

The sequence of equations is generated by applying the principles of *transformational programming* as, e.g., advocated in the Munich Project CIP [12,13]. Thus each implication

$$Q_{i+1} \equiv> Q_i$$

in the sequence is obtained by applying a *transformation rule* to a subterm of Q_i, usually a trace specification, say S_i. Most of the transformation rules state that under certain conditions a semantic implication of the form

$$op(S_{i_1}, \ldots, S_{i_n}) \equiv> S_i$$

holds, expressing that the trace specification S_i can be refined into a mixed term consisting of an n-ary process operator op applied to trace specifications $S_{i_1}, \ldots, S_{i_n}$ that are logically related to S_i. Then Q_{i+1} results from Q_i by replacing its subterm S_i by $op(S_{i_1}, \ldots, S_{i_n})$.

Transformation rules are theorems about the trace semantics of mixed terms that can be expressed as deduction rules of the form

$$(\mathcal{D}) \quad \frac{\models S_1, \ldots, \models S_m \qquad P_1 \equiv> Q_1, \ldots, P_n \equiv> Q_n}{P \equiv> Q} \quad \text{where } Cond$$

where $m, n \geqslant 0$ and *Cond* is a condition on the syntax of the trace specifications $S_1, \ldots, S_m$ and the mixed terms $P_1, Q_1, \ldots, P_n, Q_n, P, Q$. The rule $\mathcal{D}$ states that if the condition *Cond* is satisfied, the trace specifications $S_1, \ldots, S_m$ are valid and the semantic implications $P_1 \equiv> Q_1, \ldots, P_n \equiv> Q_n$ are true, then the semantic implication $P \equiv> Q$ also holds. If $m = n = 0$, the notation simplifies to

$$P \equiv> Q \quad \text{where } Cond.$$

We say that a deduction rule $\mathcal{D}$ is *sound* if the theorem denoted by $\mathcal{D}$ is true.

The syntactic condition *Cond* will always be a simple decidable property, for instance, concerning alphabets. By contrast, the validity of trace specifications is in general not decidable due to the Expressiveness Theorem 6.11 Thus in general the transformation rules $\mathcal{D}$ are effectively applicable only *relative* to the logical theory of trace specifications, i.e., the set $Th(Spec) = \{S \in Spec \mid \models S\}$.

To enhance the readability of transformation rules, we will sometimes deviate from the form $\mathcal{D}$ above and use equivalent ones. For example, we will often require that certain traces tr satisfy a trace specification S, but this is of course equivalent to a validity condition:

$$tr \models S \quad \text{iff} \quad \models S\{tr/h\}.$$

Some rules will be classified as *derived rules* because they can be deduced from other rules. Their purpose is to organize the presentation of process constructions more clearly.

To state the semantic and syntactic conditions of our transformation rules, we need a few additional notions. The *prefix kernel* of a trace specification S is given by the formula

$$kern(S) =_{df} \forall t \bullet t \leqslant h \downarrow \alpha(S) \Rightarrow S\{t/h\}.$$

Thus $kern(S)$ is a trace specification which denotes the largest prefix closed set of traces satisfying S (cf. [95]). We summarize its basic properties:

PROPOSITION 7.3. *For all $S \in Spec$ and $tr \in Comm^*$*

(1) $\alpha(kern(S)) = \alpha(S)$.
(2) $\models kern(S) \Rightarrow S$.
(3) $tr \models kern(S)$ *iff* $pref\ tr \models kern(S)$.
(4) $tr \models kern(S)$ *iff* $pref\ tr \models S$.
(5) $pref\ tr \models kern(S)$ *iff* $pref\ tr \models S$.

By contrast, the *prefix closure* of a trace specification S denotes the smallest prefix closed set of traces *containing* those which satisfy S. It can be expressed by the formula

$$PC(S) =_{df} \exists t \bullet S\{h^\frown t/h\}$$

(cf. Definition 2.6). The set of *initial communications* of a trace specification S is given by

$$init(S) = \{a \in \alpha(S) \mid pref\ a \models S\}.$$

Recall that $pref\ a \models S$ is equivalent to requiring $\langle\rangle \models S$ and $a \models S$.

We can now introduce the announced transformation rules for mixed terms.

DEFINITION 7.4. In the following transformation rules, letters P, Q, R range over $CProc + Spec$ unless stated otherwise. As usual, letters S, T; X; a, b; A range over $Spec$, Idf, $Comm$, and alphabets, respectively.

(1) *Implication Rules.*

(Reflexivity)

$$P \equiv> P.$$

(Transitivity)

$$\frac{P \equiv> Q,\ Q \equiv> R}{P \equiv> R}.$$

(Monotonicity under Context)

$$\frac{Q \equiv> R}{P\{Q/X\} \equiv> P\{R/X\}}$$

where $P \in Proc + Spec$ and $P\{Q/X\}$, $P\{R/X\} \in CProc + Spec$.

(2) *Specification Rules.*
(Kernel)

$$S \equiv kern(S).$$

(Logic)

$$\frac{\models S \Rightarrow T}{S \equiv> T} \quad \text{where } \alpha(S) = \alpha(T).$$

(3) *Construction Rules.*
(Deadlock)

$$stop{:}A \equiv> S \quad \text{where } \alpha(S) = A.$$

(Divergence)

$$div{:}A \equiv> S \quad \text{where } \alpha(S) = A.$$

(Prefix)

$$\frac{a \in init(S)}{a.S\{a^\frown h/h\} \equiv> S}.$$

(Choice)

$$S + T \equiv> S \vee T \quad \text{where } \alpha(S) = \alpha(T).$$

(Parallelism)

$$S \parallel T \equiv> S \wedge T.$$

(Renaming)

$$\frac{\forall tr \in \alpha(S)^* \bullet tr \models S \text{ implies } tr\{b/a\} \models T}{S[b/a] \equiv> T} \quad \text{where } \alpha(T) = \alpha(S)\{b/a\}.$$

(Hiding)

$$\frac{\models S \Rightarrow T}{S \backslash B \equiv> T} \quad \text{where } \alpha(T) = \alpha(S) - B.$$

(Recursion)

$$\frac{P\{S/X\} \equiv> S}{\mu X \bullet P \equiv> S} \quad \text{where } \mu X \bullet P \in CProc + Spec \text{ and } \alpha(S) = \alpha(X).$$

(Expansion: derived)

$$\frac{a_1, \ldots, a_n \in init(S)}{\sum_{i=1}^{n} a_i . S\{a_i . h / h\} \equiv> S}$$

where $a_1, \ldots, a_n$ with $n \geqslant 2$ are distinct communications.

(Disjoint renaming: derived)

$$S[b/a] \equiv S\{b/a\} \quad \text{where } b \notin \alpha\alpha(S).$$

The simple conjunction rule for parallelism was one of our aims when selecting the operators for process terms. That is why we have chosen COSY's version of parallel composition [56] that is also present in CSP [42]. By contrast, the parallel composition of CCS [64] would be extremely difficult to deal with in our trace logic approach because it combines in a single operator the effect of COSY's parallelism, renaming and hiding. Therefore, we follow CSP and treat all these issues separately.

For the renaming operator $S[b/a]$ we have presented one very simple rule for the case of disjoint renaming: if $b \notin \alpha\alpha(S)$, the extended alphabet of S, then the semantics of $S[b/a]$ is captured by a simple syntactic substitution on S, viz. $S\{b/a\}$. In general, however, $S[b/a]$ and $S\{b/a\}$ differ. Consider for example

$$S \equiv h \downarrow \{a\} \leqslant a \wedge h \downarrow \{b\} \leqslant b.b \quad \text{with } \alpha\alpha(S) = \{a, b\}.$$

Then

$$S[b/a] \equiv h \downarrow \{b\} \leqslant b.b.b \quad \text{whereas } S\{b/a\} \equiv h \downarrow \{b\} \leqslant b.$$

This difficult case of renaming is an instance of *aliasing* where two originally different communications a and b get the same name, viz. b. It is covered by the more complex general renaming rule.

Note that the recursion rule is just a formalisation of Park's fixed point induction (cf. for example [92]). Altogether we have the following theorem.

SOUNDNESS THEOREM 7.5. *The mixed term transformation rules of Definition* 7.4 *are sound.*

PROOF. See essentially [69]. □

As a first illustration of this approach we shall construct a *counter*, which is a process that can store a natural number which can be incremented and decremented with the help of communications *up* and *dn*.

EXAMPLE 7.6. Consider the trace specification

$$S_1 =_{df} 0 \leqslant up\#h - dn\#h \leqslant 1$$

of a counter with bound 1. We wish to construct a process term P_1 satisfying S_1 such that with $P_1 \equiv> S_1$. The idea of the construction is to use the prefix rule in order to discover a recursive semantic equation which can then be dealt with by the recursion rule.

We begin by determining the initial communications of S_1. Since $init(S_1) = \{up\}$, the prefix rule yields as the first construction step:

$$up.S_1\{up^\frown h/h\} \equiv> S_1. \tag{7}$$

We calculate the result of the substitution in S_1 using the logic rule:

$$\begin{array}{c} S_1\{up^\frown h/h\} \\ \Uparrow\!\!\! \vert \quad \{\text{definition}\} \\ 0 \leqslant up\#(up^\frown h) - dn\#(up^\frown h) \leqslant 1 \\ \Uparrow\!\!\! \vert \quad \{\text{calculus}\} \\ 0 \leqslant 1 + up\#h - dn\#h \leqslant 1. \end{array} \tag{8}$$

Call the last trace specification $S_{1,up}$. Combining (7) and (8) by the transitivity and the context rule yields

$$up.S_{1,up} \equiv> S_1. \tag{9}$$

We now examine the initial communications of $S_{1,up}$. Since $init(S_{1,up}) = \{dn\}$, another application of the prefix rule yields the second construction step:

$$dn.S_{1,up}\{dn^\frown h/h\} \equiv> S_{1,up}. \tag{10}$$

We calculate the result of the substitution in $S_{1,up}$ and find

$$S_1 \equiv> S_{1,up}\{dn^\frown h/h\}. \tag{11}$$

Combining (9)–(11) by the transitivity and the context rule, we can summarize the two construction steps as follows:

$$\begin{array}{c} S_1 \\ \Uparrow\!\!\! \vert \\ up.S_{1,up} \\ \Uparrow\!\!\! \vert \\ up.dn.S_1. \end{array}$$

Thus by two applications of the prefix rule, we have discovered the recursive semantic equation

$$up.dn.S_1 \equiv> S_1.$$

Let X be a process identifier with $\alpha(X) = \{up, dn\}$. Since $\mu X \bullet up.dn.X$ is communication-guarded, the recursion rule is applicable and yields

$$\mu X.up.dn.X \equiv> S_1.$$

Thus we can choose $P_1 = \mu X \bullet up.dn.X$ to achieve $P_1 \equiv S_1$.

In the previous example a process term satisfying a given trace specification was constructed in a rather mechanical way which we shall call the *expansion strategy*. The basic idea of this strategy is as follows. Given a trace specification S use the rules for deadlock, prefix and expansion to explore the initial communications of S and generate an equation of the form

$$stop{:}A \equiv> S \quad \text{or} \quad \sum_{i=1}^{n} a_i.S_i \equiv> S.$$

Repeat this for all newly generated trace specifications S_i until each of these new S_i is semantically equivalent to a previously encountered specification. This repetition terminates provided the trace set of the original specification is *regular* in the sense of formal language theory [44]. Using the equation and specification rules, we substitute the above equations into each other to obtain finitely many equations

$$P\{S/X\} \equiv> S$$

which are solved one by one using the recursion rule. This yields a process term satisfying the original trace specification.

7.2. *Milner's scheduling problem*

In his books on CCS, Milner introduced a small scheduling problem [61,64]. Informally speaking, this problem deals with $k \geqslant 1$ machines $M_1, \ldots, M_k$ processing some items. Each machine M_i with $1 \leqslant i \leqslant k$ signals by a communication a_i when it starts processing an item and by a communication b_i when it finishes processing this item.

The task is to schedule these machines in such a way that

(1) individually, each machine M_i is prepared to work indefinitely, i.e., is ready to communicate a_i, b_i cyclicly, and

(2) together, the k machines $M_1, \ldots, M_k$ start sequentially; thus a_2 can occur only after a_1 has happened, a_3 only after a_2, etc., and finally a new turn of a_1 is possible only after a_k has occurred.

In [61,64] the formal specification and implementation of the scheduler are described by CCS process terms; the proof that the implementation meets the specification is purely algebraic, i.e., by application of the algebraic rules of CCS.

Here we perform a process construction starting from the following trace specification of Milner's scheduling problem:

$$SPEC_k =_{df} \bigwedge_{i=1}^{k} h \downarrow \{a_i, b_i\} \in pref(a_i.b_i)^* \\ \wedge\, h \downarrow \{a_1, \ldots, a_k\} \in pref(a_1 \ldots a_k)^*$$

where $k \geqslant 1$. We use regular expressions to specify a process that can engage in the communications $a_1, \ldots, a_k$ and $b_1, \ldots, b_k$. Besides the usual operators like Kleene star we also use the prefix operator: for a regular expression *re* the notation *pref*(*re*) describes the set of all prefixes of the traces denoted by *re*. Thus $SPEC_k$ requires that when observed through the window $\{a_i, b_i\}$ the process engages in a small cycle $a_i b_i$ and when observed through the window $\{a_1, \ldots, a_k\}$ it engages in a large cycle $a_1 \ldots a_k$.

Consider for distinct communications $c_1, \ldots, c_n$ the trace specification

$$S =_{df} tr \downarrow \{c_1, \ldots, c_n\} \in pref(c_1, \ldots, c_n)^*$$

and

$$T =_{df} c_n \# h \leqslant \cdots \leqslant c_1 \# h \leqslant 1 + c_n \# h.$$

Since $\alpha(S) = \alpha(T)$ and $\models S \Leftrightarrow kern(T)$, the kernel and logic rule yield

$$S \equiv T. \tag{12}$$

Thus we obtain

$$SPEC_k \equiv \bigwedge_{i=1}^{k} b_i \# h \leqslant a_i \# h \leqslant 1 + b_i \# h \\ \wedge\, a_k \# h \leqslant \cdots \leqslant a_1 \# h \leqslant 1 + a_k \# h.$$

We find regular expressions easier to read, but inequalities between the number of occurrences of certain communications easier to calculate with. We now present three different constructions of process terms from $SPEC_k$.

First construction. Since $SPEC_k$ describes a regular set of traces, the expansion strategy is successful. For $k = 1$ this strategy yields a copy of a 1-counter:

$$\begin{array}{c} SPEC_1 \\ \Lleftarrow \\ a_1.SPEC_1\{a_1.h/h\} \\ \Lleftarrow \\ a_1.b_1.SPEC_1 \\ \Lleftarrow \\ \mu X_1 \bullet a_1.b_1.X_1. \end{array}$$

However, already for $k = 2$ things get more complicated. In general, the expansion strategy leads to process terms whose structure is difficult to grasp. To obtain easier to understand process terms, we consider a construction involving parallelism.

Second construction. Since $SPEC_k$ is given as a conjunction of several conditions, the application of the parallelism is straightforward. Define

$$PROC_i =_{df} h \downarrow \{a_i, b_i\} \in pref(a_i.b_i)^*$$

for $i = 1, \ldots, k$ and

$$SCH_k =_{df} h \downarrow \{a_1, \ldots, a_k\} \in pref(a_1 \ldots a_k)^*$$

standing for "process i" and "scheduler of size k". Then by successive applications of the parallelism rule, we obtain:

$$\left(\|_{i=1}^{k} PROC_i\right) \parallel SCH_k \equiv> SPEC_k. \tag{13}$$

Process terms satisfying the specifications $PROC_i$ and SCH_k are constructed by the expansion strategy. We record only the results of these constructions:

$$\begin{array}{ccc} PROC_i & & SCH_k \\ \Uparrow\!\!| & \text{and} & \Uparrow\!\!| \\ \mu XX_i \bullet a_i.b_i.XX_i & & \mu YY_k \bullet a_1 \ldots a_k.YY_k \end{array} \tag{14}$$

where $\alpha(XX_i) = \{a_i, b_i\}$ and $\alpha(YY_i) = \{a_1, \ldots, a_k\}$. Combining (13) and (14) completes the second construction:

$$\begin{array}{c} SPEC_k \\ \Uparrow\!\!| \\ \left(\|_{i=1}^{k} PROC_i\right) \parallel SCH_k \\ \Uparrow\!\!| \\ \left(\|_{i=1}^{k} \mu XX_i \bullet a_i.b_i.XX_i\right) \parallel \mu YY_k \bullet a_1 \ldots a_k.YY_k. \end{array}$$

Third construction. Milner's aim was a *modular* scheduler which for arbitrary problem size could be built as a "ring of elementary identical components" [61]. This idea can be realised very neatly using transformations on mixed terms. The point is to break down the large cycle $a_1 \ldots a_k$ of SCH_k into $k + 1$ cycles of size 2, viz. $a_1a_2, a_2a_3, \ldots, a_{k-1}a_k$ and a_1a_k.

Formally, this idea is justified by simple calculations with inequalities based on the representation (12) of cyclic regular expressions. We represent the large cycle $a_1 \ldots a_k$ by

$$CYC_k =_{df} a_k \# h \leqslant \cdots \leqslant a_1 \# h \leqslant 1 + a_k \# h$$

and for $1 \leqslant i < j \leqslant k$ we represent a cycle $a_i a_j$ by

$$SCH_{i,j} =_{df} a_j \# h \leqslant a_i \# h \leqslant 1 + a_j \# h.$$

Then the following logical equivalence holds:

$$\models CYC_k \Longleftrightarrow \left(\bigwedge_{i=1}^{k-1} SCH_{i,i+1} \right) \wedge SCH_{1,k}. \tag{15}$$

By (12), we have $SCH_k \equiv CYC_k$ and hence by (15), the logic rule yields

$$SCH_k \equiv \left(\bigwedge_{i=1}^{k-1} SCH_{i,i+1} \right) \wedge SCH_{1,k}.$$

Thus successive applications of the parallelism rule lead to the following decomposition of the large cycle SCH_k into small cycles $SCH_{i,j}$:

$$\left(\|_{i=1}^{k} SCH_{i,i+1} \right) \| SCH_{1,k} \equiv> SPEC_k. \tag{16}$$

Combining (16) with (13) of the previous construction we obtain:

$$\begin{array}{c} SPEC_k \\ \Uparrow \\ \left(\|_{i=1}^{k} PROC_i \right) \| SCH_k \\ \Uparrow \\ \left(\|_{i=1}^{k} PROC_i \right) \| \left(\|_{i=1}^{k-1} SCH_{i,i+1} \right) \| SCH_{1,k}. \end{array} \tag{17}$$

Note that each of the specifications $SCH_{i,j}$ is equivalent to

$$0 \leqslant a_i \# h - a_j \# h \leqslant 1.$$

Moreover, by (12) of the specifications $PROC_i$ is equivalent to

$$0 \leqslant a_i \# h - b_i \# h \leqslant 1.$$

Thus $SCH_{i,j}$ and $PROC_i$ are all renamed copies of the 1-counter specified by the trace specification S_1 of Example 7.6. Formally,

$$\begin{array}{c} PROC_i \\ \Uparrow \\ S_1[a_i, b_i/up, dn] \end{array} \quad \text{and} \quad \begin{array}{c} SCH_{i,j} \\ \Uparrow \\ S_1[a_i, a_j/up, dn] \end{array} \tag{18}$$

by the logic and disjoint renaming rule. Recall from Example 7.6 that

$$\mu X \bullet up.dn.X \equiv> S_1. \tag{19}$$

Thus combining (17)–(19) we obtain that the process term

$$\begin{aligned}&\big(\|_{i=1}^{k}(\mu X \bullet up.dn.X)[a_i, b_i/up, dn]\big)\\&\quad\| \big(\|_{i=1}^{k-1}(\mu X \bullet up.dn.X)[a_i, a_{i+1}/up, dn]\big)\\&\quad\| (\mu X \bullet up.dn.X)[a_1, a_k/up, dn]\end{aligned}$$

consisting of the parallel composition of $2k$ copies of the 1-counter satisfies $SPEC_k$.

8. Extended trace models

The semantic model based on finite traces abstracts from a number of aspects of the operational process behaviour.

(1) Traces are insensitive against *nondeterministic choices*. Milner observed in [61] that this is unsatisfactory in the context of parallel composition with handshake communication. His famous example compares the processes

$$P = a.(b.stop{:}A + c.stop{:}A) \quad \text{and} \quad Q = a.b.stop{:}A + a.c.stop{:}A$$

in the context of the process $R = a.b.stop{:}A$ where $A = \{a, b, c\}$.
Although both P and Q have the same trace semantics, the operational behaviour of the parallel composition with R differs. In fact, according to the operational semantics of Definition 3.4 $P \| R$ can always engage in the trace $a.b$ whereas $Q \| R$ can also deadlock after the communication a.

(2) Traces are insensitive against *divergence*. Most strikingly, the processes $stop{:}A$ and $div{:}A$ are identified in the trace model.

(3) The transformational approach presented in Section 7 contains miracles. A *miracle* is a process term satisfying *every* specification. In our approach both $stop{:}A$ and $div{:}A$ represent miracles. Indeed the transformation rules in Definition 7.4 show that

$$stop{:}A \equiv> S \quad \text{and} \quad div{:}A \equiv> S$$

hold for all specifications S with $\alpha(S) = A$. In our examples we avoided these trivial solutions by applying the prefix or expansion rule to the *maximal* set of initial communications. But in principle nothing prevents us from just delivering the deadlocking or divergent process as "implementations".

(4) Sets of finite traces cannot express essential limit properties like *liveness* and *fairness*. For example, a process that can engage in communications a and b but where after every sequence of communications a eventually a communication b occurs cannot be specified in a model based on finite traces.

(5) Parallelism is reduced to *interleaving* of traces. For example, we do not distinguish between

$$a.stop{:}\{a\} \parallel b.stop{:}\{b\} \quad \text{and} \quad \big(a.b.stop{:}\{a,b\}\big) + \big(b.a.stop{:}\{a,b\}\big).$$

To deal with points (1)–(3), the finite trace model has been extended by recording after each trace some information about communications that are possible or not possible afterwards. To deal with (4), *infinite traces* or *streams* will be considered in Sections 9–15 of this handbook chapter. To deal with (5), stream functions are considered that process tuples of streams.

8.1. *Readiness model*

In the *readiness model* [41,70,69] instead of simple traces tr pairs

$$(tr, Rdy)$$

are recorded where $Rdy \subseteq Comm$ is a set of communications that are *ready* after the trace tr has occurred. Therefore Rdy is called a *ready set* and (tr, Rdy) a *ready pair*. The readiness model of a process records all ready pairs that are retrievable from its operational semantics.

For example, the processes P and Q mentioned in item (1) above can be distinguished in the readiness model. For P this model records the ready pairs

$$(\langle\rangle, \{a\}), \quad (\langle a\rangle, \{a,b\}), \quad (\langle a,b\rangle, \emptyset), \quad (\langle a,c\rangle, \emptyset).$$

For Q instead of $(\langle a\rangle, \{a,b\})$ the ready pairs

$$(\langle a\rangle, \{a\}) \quad \text{and} \quad (\langle a\rangle, \{b\})$$

appear. They describe that after the communication a the process Q can nondetermistically decide to be ready only for a or only for b. By contrast process P is after a ready for both a and b as the ready set $\{a,b\}$ indicates. A ready pair $(tr, \emptyset)$ describes that after the trace tr no further communication is possible, i.e., a deadlock has occurred. Thus both P and Q deadlock after the traces $\langle a,b\rangle$ and $\langle a,c\rangle$.

Another aspect that is recorded in the readiness model is *divergence*. If a process can diverge, i.e., engage in an infinite sequence of internal τ-actions after a trace tr, this is recorded by a so-called *divergence point*

$$(tr, \uparrow).$$

In this way processes like $stop{:}A$ and $div{:}A$, that are identified in the trace model, can be distinguished in the readiness model. For $stop{:}A$ the model records only the ready pair $(\langle\rangle, \emptyset)$ but for $div{:}A$ also the divergence point $(\langle\rangle, \uparrow)$.

For further details of this model we refer to [41,70,69]. A mixed term transformational approach based on the readiness model has been described in [69]. In this approach there are no miracles.

8.2. *Failures model*

In the *failures model* [20] ready pairs are replaced by so-called *failures* of the form

$$(tr, Ref)$$

where $Ref \subseteq Comm$ is a set of communications that can be *refused* after the trace tr has occurred. Therefore Ref is called a *refusal set* . Refusal sets are not just the complements of ready sets but the *downward closures* of the complements. Thus if a process can refuse a set Y of communications after a trace tr, it can also refuse every set X with $X \subseteq Y$ after tr. As a consequence the failures model identifies more processes than the readiness model.

The failures model has been extended to the so-called *failures/divergences model* where also divergence points are recorded. This model is taken as the standard semantics for CSP [42,81]. It is interesting to note that the failure/divergence model induces the same equivalence on processes than the so-called *acceptance model* developed by De Nicola and Hennessy [31,38] as a result of their approach to testing of processes.

9. Infinite traces – streams

Processes typically represent systems or system components that are designed to run forever. So far we have modelled this behaviour by infinite sets of *finite* traces with unbounded length. Another option are *infinite* traces. As discussed above, a number of characteristic phenomena of interactive systems can only be studied when looking at infinite traces. Typical examples are specific liveness properties such as fairness. Therefore, we work in the following with streams, which are finite or infinite sequences of messages or actions.

9.1. *The stream algebra*

Given a set M of messages or actions we can form the set M^ω of *streams*, i.e., of finite and infinite sequences over M:

$$M^\omega = M^* \cup M^\infty.$$

In the following let m range over the set M and x, y, z over M^ω. An infinite stream $x \in M^\infty$ is represented by a mapping $x : \mathbb{N}^+ \to M$ where $\mathbb{N}^+ = \mathbb{N}\backslash\{0\}$. For every index $i \in \mathbb{N}^+$ we denote by $x(i)$ and also $x.i$ the *ith element* in the stream x. Of course, for finite streams of length n, the term $\alpha.i$ is only defined if $1 \leqslant i \leqslant n$.

On streams we work, similar to the operations on traces in Section 2, with a number of characteristic symbols and operations introduced in the following:

$\langle\rangle$	denotes the *empty stream*,
$\langle m\rangle$	denotes the *one element stream*,
$\frown$	denotes the binary *concatenation* operation on streams,
head	denotes the unary operation yielding the *first element* of a non-empty stream,
tail	denotes the unary operation yielding the *rest* of a non-empty stream obtained by eliminating its first element.

These operations are easily specified by the following algebraic axioms:

$$\langle\rangle^\frown x = x^\frown\langle\rangle = x$$
$$x^\frown(y^\frown z) = (x^\frown y)^\frown z$$
$$head(\langle m\rangle^\frown x) = m$$
$$tail(\langle m\rangle^\frown x) = x.$$

Note that head and tail are partial functions which are not defined for the empty stream. Additional operations on streams that are helpful in specifications are listened in the following:

$\#x$	denotes the *length* of the stream x yielding an element of $\mathbb{N}\cup\{\infty\}$,
$x\downarrow A$	denotes the operation of *filtering* out the substream of x of elements in the set $A\subseteq M$.

Again we may describe the meaning of these operations by algebraic axioms easily as follows:

$$\#\langle\rangle = 0$$
$$\#(\langle m\rangle^\frown x) = 1 + \#x$$
$$\#x = \infty \Rightarrow x^\frown y = x$$
$$(\forall i \in \mathbb{N} : x.i = y.i) \Rightarrow x = y$$
$$(\langle m\rangle^\frown x).1 = m$$
$$(\langle m\rangle^\frown x).(i+1) = x.i \qquad \Leftarrow i > 0$$
$$\langle\rangle \downarrow A = \langle\rangle$$
$$(\langle m\rangle^\frown x) \downarrow A = x \downarrow A \qquad \Leftarrow m \notin A$$
$$(\langle m\rangle^\frown x) \downarrow A = \langle m\rangle^\frown(m \downarrow A) \qquad \Leftarrow m \in A.$$

The *prefix ordering* $\leqslant$ is a binary relation on streams. Given two streams $x, y \in M^\omega$ it is defined as follows:

$$x \leqslant y \Longleftrightarrow \exists z \bullet x^\frown z = y.$$

This relation is, in fact, a partial ordering on streams with the least element $\langle\rangle$. The proof is similar to the case of finite sequences in Section 2.

Often it is useful to work with tuples of streams. We use the following notation for n-tuples of streams. The set of n-tuples of streams with $n \in \mathbb{N}$ is denoted by

$$(M^{\omega})^n.$$

All operations introduced so far can be extended elementwise to stream tuples. Given a tuple $x \in (M^{\omega})^n$ of streams where $x = (x_1, \ldots, x_n)$ we denote for $i \in \mathbb{N}$ with $1 \leqslant i \leqslant n$ by the expression

$$(i : a)^\frown x$$

the stream tuple in $(M^{\omega})^n$ specified by the equation

$$(i : a)^\frown x = \bigl(x_1, \ldots, x_{i-1}, \langle a\rangle^\frown x_i, x_{i+1}, \ldots, x_n\bigr).$$

Given tuples $x \in (M^{\omega})^n$ and $y \in (M^{\omega})^m$ we denote by

$$x \| y$$

the $(n+m)$-tuple $x \| y \in (M^{w})^{n+m}$ defined by the axiom

$$(x_1, \ldots, x_n) \| (y_1, \ldots, y_m) = (x_1, \ldots, x_n, y_1, \ldots, y_m).$$

Thus $x \| y$ is the tuple in $(M^{\omega})^{n+m}$ that is the result of concatenating the tuples x and y.

9.2. *Predicates on streams*

In this section we study predicates on streams. A predicate on streams is a Boolean function

$$P : M^{\omega} \to \mathbb{B}.$$

Such predicates describe properties of streams and therefore can be used to specify systems that are modelled by streams of actions. Accordingly, a system specification is a logical formula defining a predicate on streams. It characterizes the behaviour of a system by its set of action streams.

We refrain from introducing a detailed syntax for predicates on streams as we did for traces, but in examples we reuses as much as possible the syntax of trace logic as defined in Section 6.

EXAMPLE 9.1 (*Queue*). Let *Data* be the set of data elements. We choose the actions $put(d)$ with $d \in Data$ and *get* as the actions of a queue. On that basis, we may characterize some properties of the streams of actions of the queue over the set of actions

$$M = PUT \cup \{get\} \quad \text{where } PUT = \{put(d) \mid d \in Data\}$$

by the predicate $Q : M^{\omega} \to \mathbb{B}$ defined as follows:

$$Q(x) \iff \big(\forall z \in M^{*} \bullet z \leqslant x \Rightarrow get\#z \leqslant PUT\#z\big) \wedge get\#x = PUT\#x.$$

The predicate essentially says that for every get operation a corresponding number of put operations has previously occurred and for every element that is put into the queue a corresponding get operation will eventually occur. Note that it is not yet specified which values the get operation delivers. This aspect of the queue behaviour will be specified in Example 13.1.

Predicates can be used to specify the properties of systems described by streams. Often it is helpful to classify such predicates into *safety* and *liveness* properties. Safety properties express that certain unwanted patterns of behaviors that can be observed by looking at finite traces only do not occur. Liveness properties make sure that certain patterns of behaviours eventually take place.

Since this informal characterization of safety and liveness is vague and imprecise, we give a mathematical definition. A predicate P on streams is called a *safety property* if the following formula is valid:

$$P(x) \iff \big(\forall z \in M^{*} \bullet z \leqslant x \Rightarrow P(z)\big).$$

For the above example of a queue the predicate Q' defined by

$$Q'(x) \iff \forall z \in M^{*} \bullet z \leqslant x \Rightarrow get\#z \leqslant PUT\#z$$

is a safety property. A predicate P is called a *liveness property* if

$$\forall z \in M^{*} \bullet \exists x \in M^{\omega} \bullet z \leqslant x \wedge P(x).$$

For our example above the predicate Q'' with

$$Q''(x) \iff (get\#x = PUT\#x)$$

is a liveness property.

It is well-known that every predicate can be decomposed into a safety and a liveness property [4]. For our example we get

$$Q(x) \iff Q'(x) \wedge Q''(x).$$

The safety property P' contained in an arbitrary predicate P is easily extracted by the following logical equivalence

$$P'(x) \iff \forall z \in M^{*} \bullet z \leqslant x \Rightarrow \exists z' \in M^{\omega} \bullet z \leqslant z' \wedge P(z').$$

This definition is equivalent to the set of finite traces in the prefix closure of the set of traces. The liveness part P'' contained in a predicate P is easily isolated by the logical equivalence

$$P''(x) \Longleftrightarrow \big(P'(x) \Rightarrow \exists z \in M^{\omega} \bullet x \leqslant z \wedge P(z)\big),$$

where $P'(x)$ is the above safety predicate. Often it is helpful in the specification or verification of system requirements to decompose system properties into liveness and safety parts.

9.3. *Streams as a Scott domain*

As shown in the previous section, the set of streams is partially ordered by the prefix order $\leqslant$. The empty stream $\langle\rangle$ is its least element. The infinite streams are the maximal elements in the set of streams under this order.

A chain of streams is a set $X = \{x_i \in M^{\omega} \mid i \in \mathbb{N}\}$ with $x_i \leqslant x_{i+1}$ for all $i \in \mathbb{N}$. The partially ordered set $(M^{\omega}, \leqslant)$ is *chain complete* under $\leqslant$. According to Section 4 this means that M^{ω} contains a least element and that every chain X has a least upper bound $\bigsqcup X$. Note that every infinite stream can be described by the least upper bound of a chain of finite streams.

Thus $(M^{\omega}, \leqslant)$ is a complete partial order (cpo) and in fact a *domain* in the sense of Scott (see [92]). As described in Section 4 this is the basis of the definition of least fixed points for monotonic functions; here it will be used in Section 11 to give a meaning to the feedback operator.

10. Functions on streams

In this section we consider functions between streams. They can be used to model the behaviour of components with input and output ports.

As introduced in Section 3, we distinguish input and output of messages. This leads to a functional view of systems in terms of stream processing functions which are continuous functions on streams. For them simple but powerful composition forms are available such as parallel composition, sequential composition and feedback.

In principle, we could also study functions between finite sequences instead of stream processing functions. However, including infinite sequences seem to be more appropriate, since for them the classical fixed point construction is available to define the semantics of feedback loops.

10.1. *Monotonicity and continuity*

A function on tuples of streams has the functionality

$$f : \left(M^{\omega}\right)^{n} \rightarrow \left(M^{\omega}\right)^{m}.$$

It provides a *history based model* of systems where a separate "history" represented by a stream is used for each input and output channel. This is in contrast to the *trace model* of processes as studied in Section 2 where a single common history (called a trace) is kept for all the channels that is obtained by interleaving all messages or actions on the channels.

Not all functions on streams are appropriate to model the behaviour of components. We require certain properties. A function f on streams is called *monotonic* if we have

$$x \leqslant z \Rightarrow f(x) \leqslant f(z).$$

A function f is called *continuous* if f is monotonic and for every chain X we have:

$$f\Big(\bigsqcup X\Big) = \bigsqcup\big\{f(x) \mid x \in X\big\}.$$

A function

$$f : (M^{\omega})^{n} \to (M^{\omega})^{m}$$

is called a *stream processing function* of arity (n, m) if it is monotonic and continuous. Such functions can be used to model the behaviour of a deterministic *data flow* component that communicates in an asynchronous way with its environment. In particular, a stream processing function models the behaviour of a data flow component. The monotonicity reflects the fact that prefixes of the output are produced on prefixes of the input and that this partial output sequence cannot be changed but at most increased by increasing the input.

Continuity makes sure that the function is already determined by its behaviour on finite streams. A monotonic function

$$f : (M^{*})^{n} \to (M^{*})^{m}$$

on finite sequences can be extended canonically into a function

$$\hat{f} : (M^{\omega})^{n} \to (M^{\omega})^{m}$$

by the equation

$$f(x) = \bigsqcup\big\{y \mid \exists x' \in (M^{*})^{n} \bullet x' \leqslant x \wedge y = f(x')\big\}.$$

Monotonicity has a crucial consequence. By the Knaster–Tarski Fixed Point Theorem 4.1 we obtain that every stream processing function

$$f : (M^{\omega})^{n} \to (M^{\omega})^{n}$$

has a least fixed point denoted by *fix* f. By Kleene's Fixed Point Theorem 4.2, due to the continuity of the function f, the fixed point *fix* f is identical to the least upper bound of the chain of streams obtained by iterating the function f starting with the n-tuple of empty streams. We get therefore

$$\mathit{fix}\, f = \bigsqcup \{ f^k(\langle\rangle, \ldots, \langle\rangle) \mid k \in \mathbb{N} \},$$

where

$$f^0(x) = x \quad \text{and} \quad f^{k+1}(x) = f\big(f^k(x)\big).$$

This iteration yields a procedure to calculate better and better approximations for the fixed point.

10.2. *Channels and their valuation*

Often it is more convenient to work with families of named streams instead of tuples of streams. Hence in the following we use *channel identifiers* or *channels* for short to name streams. Let C be a set of channels. A *channel valuation* is a mapping

$$x : C \to M^\omega.$$

For every channel $c \in C$ we denote by x_c the stream $x(c)$. Actually a stream n-tuple can be also seen as a set of streams named by numbers from $\{1, \ldots, n\}$. A channel is nothing but an identifier for a stream. Stream concatenation easily extends to channel valuations by pointwise concatenation. Let the channel valuations

$$x, y : C \to M^\omega$$

be given; then for every channel $c \in C$ we define the concatenation $x ^\frown y$ of channel valuations x and y as follows:

$$(x ^\frown y)_c = x_c ^\frown y_c.$$

With every channel $c \in C$ we associate an individual type $M_c \subseteq M$ of messages. Given a set C of typed channels we introduce what we call a *typed channel valuation*. The set $\vec{C}$ of typed channel valuations for the channels in C is defined by the equation:

$$\vec{C} = \{ x : C \to M^\omega \mid \forall c \in C \bullet x_c \in M_c^\omega \}.$$

All notions we have introduced for streams so far carry over schematically to channel valuations.

Given two sets of channels, the set I of input channels and the set O of output channels, we denote by

$$\vec{I} \to \vec{O}$$

the set of functions between the sets of streams named by the channels in I and O. By the pair (I, O) we denote the arity of the function f.

For every channel $c \in C$, every message $a \in M_c$ and every stream valuation $x \in \vec{C}$ we denote by

$$(c : a)^\frown x$$

the channel valuation that is characterized by the following equations. For $e \in C$ we have:

$$\begin{aligned} \big((c : a)^\frown x\big)_c &= \langle a\rangle^\frown (x_c) \\ \big((c : a)^\frown x\big)_e &= x_e \qquad \Leftarrow e \neq c. \end{aligned}$$

For all channel valuations $x \in \vec{C}$, all channels $c, e \in C$ and all messages $a \in M_c$, $b \in M_e$ we obtain the following equation:

$$c : a^\frown e : b^\frown x = e : b^\frown c : a^\frown x \Leftarrow e \neq c.$$

This is a basic rule of *asynchrony*. Messages on different channels are independent.

The idea of channels combined with this notation allows us to speak about atomic actions of input and output. Each pair $c : a$ of a channel c and a message a represents an input or an output action.

10.3. *Input and output actions as operations on stream processing functions*

In this section we show how to deal with atomic steps of input and output in the connection with stream processing functions. For a stream processing function f of arity (I, O) we write for a channel $c \in I$ and a message $d \in M_c$:

$$f \lhd c : d$$

for the stream processing function that is specified by the equation (assuming $c \in I$).

$$(f \lhd c : d)(x) = f(c : d^\frown x).$$

This equation describes an input transition. If $c \notin I$ we specify

$$(f \lhd c : d)(x) = f(x).$$

In a similar way we can handle output. For a channel $e \in O$ and a message $d \in M_j$ we write

$$e : d \lhd f$$

for the stream processing function specified by the equation

$$(e:d \lhd f)(x) = e:d \frown f(x).$$

This equation is called the output transition. By these operations and equations we get an algebra of stream processing functions. We combine these operations with specific operations that correspond to ways composing systems in the following section. In fact, we may speak again of a *process algebra* here.

11. Composition of stream processing functions

We may compose stream processing functions to construct the behaviour of composed systems. We study four forms of composition, namely *parallel composition*, *functional composition (pipelining)*, *feedback*, and *hiding*. These operations lead to composed systems and can be graphically illustrated by dataflow networks.

Given two disjoint sets of channels C_1 and C_2 we define a join operation $\|$ for the valuations $x \in \vec{C}_1$, $y \in \vec{C}_2$ by the following equations:

$$(x \| y)_c = x_c \quad \text{if } c \in \vec{C}_1 \quad \text{and}$$
$$(x \| y)_c = y_c \quad \text{if } c \in \vec{C}_2.$$

Let, in the following, for $k = 1, 2$, f_k be a stream processing function of arity (I_k, O_k).

11.1. *Independent parallel composition*

Let O_1 and O_2 be disjoint sets of channels; by

$$f_1 \| f_2$$

we denote the stream processing function of arity $(I_1 \cup I_2, O_1 \cup O_2)$ obtained by the parallel composition of f_1 and f_2 as visualized in Figure 1. We use the operation symbol $\|$ both for the join operation and the parallel composition. Its interpretation should always be clear from the context. Parallel composition is specified by the equation (here we assume for simplicity $I_1 \cap I_2 = \emptyset$).

$$(f_1 \| f_2)(x_1 \| x_2) = f_1(x_1) \| f_2(x_2).$$

We may describe parallel composition also by the following axioms. Parallel composition is commutative and associative:

$$(f_1 \| f_2) \| f_3 = f_1 \| (f_2 \| f_3),$$
$$f_2 \| f_1 = f_1 \| f_2.$$

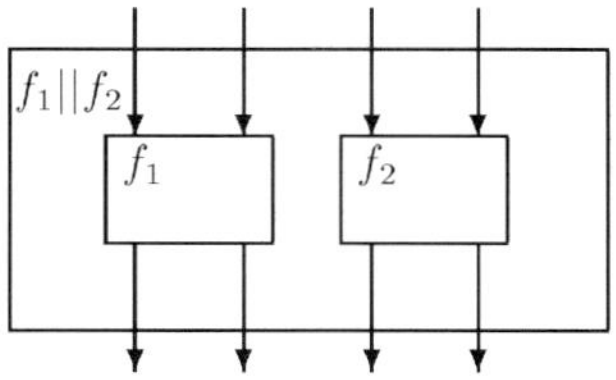

Fig. 1. Parallel composition.

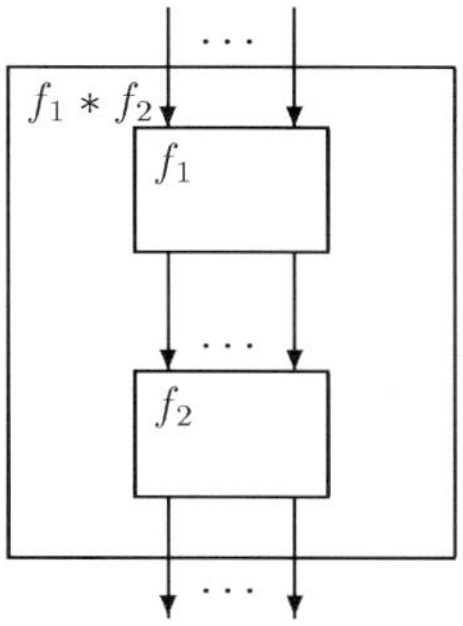

Fig. 2. Functional composition.

Input and output actions of systems composed in parallel are shared according to the following equations:

$$(f_1 \| f_2) \lhd i : m = (f_1 \lhd i : m) \| (f_2 \lhd i : m),$$
$$(j : m \lhd f_1) \| f_2 = j : m \lhd (f_1 \| f_2).$$

Note again that the sets of output channels are assumed to be disjoint. Therefore, we do not run into logical inconsistencies by the second equation due to the axiom of asynchrony.

11.2. *Functional composition (pipelining)*

If $O_1 = I_2$ then by

$$f_2 * f_1$$

we denote the stream processing function of arity (I_1, O_2) as visualized in Figure 2. It is specified by the equation

$$(f_2 * f_1)(x) = f_2\big(f_1(x)\big).$$

We obtain the following axioms for input and output transitions for functional composition:

$$(f_2 * f_1) \lhd i : m = f_2 * (f_1 \lhd i : m),$$
$$f_2 * (i : m \lhd f_1) = (f_2 \lhd i : m) * f_1,$$
$$(i : m \lhd f_2) * f_1 = i : m \lhd (f_2 * f_1).$$

Functional composition is usually also called function composition. Operationally it models the pipelining of messages through the two components.

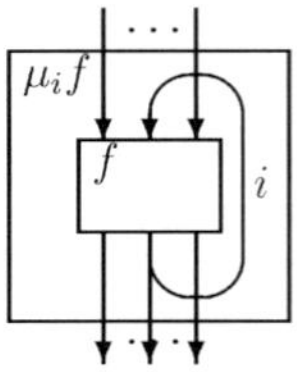

Fig. 3. Feedback.

11.3. *Feedback*

Let $f : \vec{I} \to \vec{O}$ and channel $i \in I \cap O$ be given; by

$$\mu_i(f) : \vec{I'} \to \vec{O} \quad \text{where } I' = I \backslash \{i\}$$

we denote a stream processing function of arity $(I\backslash\{i\}, O)$. It is specified by the equation

$$\mu_i(f)(x) = f(x \| i : s),$$

where

$$x \in \overrightarrow{I\backslash\{i\}} \quad \text{and} \quad s \in M_i^{\omega}$$

is the least fixed point of the function $\lambda s\ .\ f(x\|i : s)_i$. For a stream s we denote by $i : s$ the channel valuation $z \in \{\vec{i}\}$ with the property $z_i = s$.

Feedback is specified by the following axioms for input and output transitions:

$$\begin{aligned} &\mu_i(f) \lhd k : m = \mu_i(f \lhd k : m) && \Leftarrow k \neq i,\\ &\mu_i(k : m \lhd f) = k : m \lhd \mu_i(f) && \Leftarrow k \neq i,\\ &\mu_i(i : m \lhd f) = i : m \lhd \mu_i(f \lhd i : m). \end{aligned}$$

In addition, we get the classical fixed point rules for feedback that we mentioned already for continuous functions. For $x \in \vec{I'}$, $y \in \vec{O}$ these are

$$\begin{aligned} &y = (\mu_i f).x \Rightarrow y = f(x\|i : y_i) && \{\text{fixed point property}\},\\ &y \geqslant f(x\|i : y_i) \Rightarrow y \geqslant \mu_i(f)(x) && \{\text{least fixed point property}\}. \end{aligned}$$

These rules allow us to prove properties about the behaviour of components constructed by feedback loops.

11.4. *Hiding*

Given a function $f : \vec{I} \to \vec{O}$ by

$$f\backslash i : \vec{I'} \to \vec{O'} \quad \text{where } I' = I\backslash\{i\},\ O' = O\backslash\{i\}$$

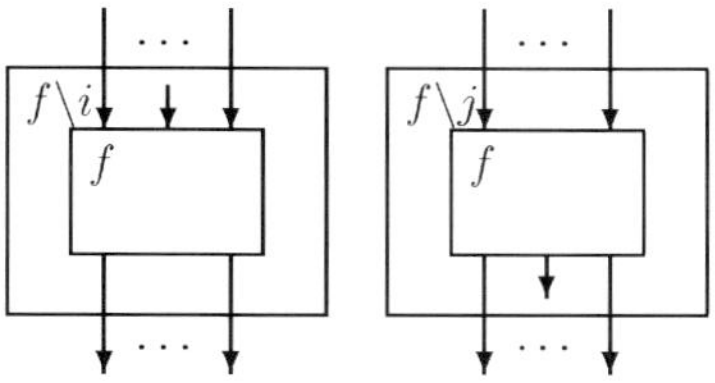

Fig. 4. Hiding of input and output channels.

we denote a function where the channel i occurring as an input channel in I or as an output channel in O is hidden. Its meaning is easily defined by the axioms for input and output transitions ($k \neq i$)

$$
\begin{aligned}
&(f\backslash i) \lhd i : a = f\backslash i,\\
&(i : a \lhd f)\backslash i = f\backslash i,\\
&(f\backslash i) \lhd k : a = (f \lhd k : a)\backslash i \Leftarrow k \neq i,\\
&(k : a \lhd f)\backslash i) = k : a \lhd (f\backslash i) \Leftarrow k \neq i.
\end{aligned}
$$

For the function $f\backslash i$ the channel i is *hidden* and thus no longer visible.

With the operations introduced so far we can form all kinds of data flow nets from given components. For a careful investigation of the completeness of algebraic equations for data flow sets, see [28].

11.5. *The algebra of stream processing functions*

As we have seen, we have a rich set of rules of the compositional forms for stream processing functions. Typical examples are the following equations stating the associativity of parallel and sequential composition

$$
\begin{aligned}
&f_1 * (f_2 * f_3) = (f_1 * f_2) * f_3,\\
&f_1 \| (f_2 \| f_3) = (f_1 \| f_2) \| f_3.
\end{aligned}
$$

Accordingly, stream processing functions form an algebra, in fact a *process algebra*. We do not go deeper into the discussion of the algebraic structure here. For more details, see [28].

12. Operational aspects of stream functions

In this section, we define a labelled transition system to describe the operational behaviour of stream processing functions. It is based on input and output transitions and hence is a special case of a labelled transition relation.

12.1. *SOS for stream processing functions*

For describing the operational behaviour of stream functions by labelled rewrite rules we use the following two basic axioms of input and output transitions (let $f : \vec{I} \to \vec{O}$ be a stream processing function, $i \in I$, $j \in O$ be channels, and m be a message):

$$f \xrightarrow{i?m} f \lhd i : m$$

$$j : m \lhd f \xrightarrow{j!m} f.$$

These rules describe steps of asynchronous communication where input and output are independent actions. The first rule indicates that stream processing functions are always input enabled.

If we add these two rules to the algebraic calculus for input and output transitions, as introduced by the equations given above, we do not need to give more rules to evaluate terms describing stream processing functions by labelled rewriting. In particular, we get further rewrite rules like, for instance, the following one for feedback by combining these rules with the algebraic equations:

$$f \xrightarrow{j!m} f' \Rightarrow (\mu_j f) \xrightarrow{j!m} \mu_j(f' \lhd j : m).$$

This rule describes the communication of output to input in feedback loops.

EXAMPLE 12.1 (*Store*). We define a store f with the input channel in and the output channel out by the equations

$$f \lhd in : put(d) \lhd in : get = out : d \lhd f \lhd in : put(d),$$

$$f \lhd in : put(d) \lhd in : put(e) = f \lhd in : put(e).$$

With these two equations we get the following rewriting sequence

$$f \xrightarrow{in?put(d)}$$

$$f \lhd in : put(d) \xrightarrow{in?put(e)} f \lhd in : put(d) \lhd in : put(e) =$$

$$f \lhd in : put(e) \xrightarrow{in?get} f \lhd in : put(e) \lhd in : get =$$

$$out : e \lhd f \lhd in : put(e) \xrightarrow{out!e} f \lhd in : put(e).$$

This shows how equations and the rewriting rules work together.

Of course, we can also give an axiomatisation of the operational behaviour of the store exclusively by rewrite rules without referring to the defining equations.

12.2. *Stream processing functions and state machines*

Stream processing functions relate closely to state machines with input and output. Given a state space Σ we define a stream processing function parameterized by a state as follows

$$h:\Sigma \to (\vec{I} \to \vec{O}).$$

Let us assume we have a state machine given by the state transition function

$$\Delta:\Sigma \times (I \times M) \to \Sigma \times (O \times M^{\omega})$$

which fulfills the following property of asynchrony for its input channels $i_1, i_2 \in I$:

$$\begin{array}{ll} & i_1 \neq i_2 \\ \wedge & \Delta(\sigma, (i_1, m_1)) = (\sigma_1, o_1) \wedge \Delta(\sigma_1, (i_2, m_2)) = (\sigma_3, o_3) \\ \wedge & \Delta(\sigma, (i_2, m_2)) = (\sigma_2, o_2) \wedge \Delta(\sigma_2, (i_1, m_1)) = (\sigma_4, o_4) \\ \Rightarrow & \sigma_3 = \sigma_4 \wedge o_1 {}^\frown o_3 = o_2 {}^\frown o_4. \end{array}$$

This property expresses that transitions with input on different channels are independent and therefore commute. Under this condition, we can associate a monotonic stream processing function

$$h(\sigma):\vec{I} \to \vec{O}$$

with every state $\sigma \in \Sigma$ by the transition equation

$$h(\sigma)(i : m {}^\frown x) = z {}^\frown y,$$

where

$$(\sigma', z) = \Delta(\sigma, (i, m)) \quad \text{and} \quad y = h(\sigma')(x).$$

This way we get a close relationship between state machines and stream processing functions. Every state machine describes for every state in its state space a stream processing function.

Vice versa, every stream processing function $f:\vec{I} \to \vec{O}$ defines a state machine. We use as its state space

$$\Sigma = [\vec{I} \to \vec{O}]$$

the set of stream processing functions. The transition function Δ is specified by the equation

$$\Delta(f, (i, m)) = (f', y)$$

provided $f(i : m {}^\frown x) = y {}^\frown f'(x)$ and $f'(z)(o) = \langle\rangle$ for all $o \in O$ where $z \in \vec{I}$ with $z(i) = \langle\rangle$ for all $i \in I$. Note that y and f' are uniquely determined by these equations.

EXAMPLE 12.2 (*Store as state machine*). In the example of the store above, we can represent the state by the set D of data elements. We get

$$\Delta : D \times (\{in\} \times \{put(d) \mid d \in D\} \cup \{get\}) \to D \times (\{out\} \times D^{\omega})$$

and obtain the following equations:

$$\Delta\big(d, (in, put(e))\big) = \big(e, (out, \langle\rangle)\big),$$
$$\Delta\big(d, (in, get)\big) = \big(d, (out, \langle d\rangle)\big).$$

The first equation describes the writing and the second the reading of the store.

The approach can be extended to nondeterministic state machines as shown in [25].

12.3. *Modelling discrete time*

It is easy to extend the algebra of stream processing functions to a system model with discrete time. To do this, we introduce a time tick signal $\surd$ as a special message that we add to every message set. We define that each complete timed stream contains an infinite number of time ticks. According to this definition, every complete timed stream is infinite but possibly carries only a finite number of messages from M. Between two consecutive time ticks we may observe an arbitrary but finite number of messages.

We denote the set of *timed streams* over a set M of messages by the set

$$M^{\underline{\omega}} = \big\{x \in (M \cup \{\surd\})^{\omega} \mid \surd\#x = \infty\big\}.$$

In principle, timed streams are just a special case of the streams introduced so far. Therefore, all our definitions easily carry over.

EXAMPLE 12.3 (*Timer*). For every number $k \in \mathbb{N}$ a timer $T(k)$ is a function

$$T(k) : \big(\{set(n) \mid n \in \mathbb{N}\} \cup \{®\}\big)^{\underline{\omega}} \to \{timeout\}^{\underline{\omega}}$$

that receives the messages $set(n)$ with $n \in \mathbb{N}$ to set the timer and the message ® to reset the timer. $T(0)$ is the unset timer. If the time that was set has elapsed, the timer sends a *timeout* message. This behaviour can be specified by the following equations:

$$\begin{aligned}
T(0) \lhd \surd &= \surd \lhd T(0), \\
T(n) \lhd set(k) &= T(k), \\
T(1) \lhd \surd &= \surd \lhd timeout \lhd T(0), \\
T(k+1) \lhd \surd &= \surd \lhd T(k) && \Leftarrow k > 0, \\
T(k) \lhd ® &= T(0).
\end{aligned}$$

Note that we use here the notation of Section 10.3 for the special case where there is just one stream to consider for input and output and, therefore, channel identifiers are superficial. In such a case, for a stream processing function $f : M^{\underline{\omega}} \to M^{\underline{\omega}}$ we assume

$$(f \lhd m)(x) = f(\langle m\rangle ^\frown x),$$
$$(m \lhd f)(x) = \langle m\rangle ^\frown f(x)$$

for all $m \in M \cup \{\surd\}$ and all $x \in M^{\underline{\omega}}$.

Given a timed stream $x \in M^{\omega}$ and $k \in \mathbb{N}$ we write $x :: k$ to denote the largest prefix that contains at most k time ticks:

$$x :: k = \bigsqcup \{ z \mid z \leqslant x \wedge \surd \# z \leqslant k \}.$$

This condition ensures that output produced till time k does not depend on input received only later.

To guarantee a proper time flow we require the following condition (in addition to prefix monotonicity):

$$x :: k = z :: k \Rightarrow f(x) :: k = f(z) :: k$$

for timed stream functions. This equation expresses that the output at time k cannot depend on input arriving after time k.

To deal with time in an adequate way we have to redefine our rule of parallel composition for the time signals. Our time is global. Of course, parallel components have to share their time on all channels. This means that time ticks occur simultaneously on all channels. We therefore do not write $c : \surd$ with an individual channel c for the time tick action but rather only write $\surd$ to express that a time tick occurs on all channels. We get the rules

$$(\surd \lhd f_1) \| (\surd \lhd f_2) = \surd \lhd (f_1 \| f_2),$$
$$(f_1 \| f_2) \lhd \surd = (f_1 \lhd \surd) \| (f_2 \lhd \surd).$$

These are the rules of synchronous time for message asynchronous communication between the components. For an extended discussion on time models see [26].

13. Specification and nondeterminism

So far we have only considered stream processing functions that model deterministic systems. Now we study nondeterministic systems and their specifications. We formalize specifications as well as nondeterministic components by sets of stream processing functions described by predicates.

13.1. *Specifications*

A specification of a system is a predicate Q that describes a set of stream processing functions. All operations on stream processing functions carry over to sets of functions by a pointwise application.

EXAMPLE 13.1 (*Queue revisited*). We extend Example 9.1 by distinguishing input and output. Then the queue can be described by a set of stream processing functions $f : (M^{\omega} \to Data^{\omega})$ characterized by a predicate

$$R : (M^{\omega} \to Data^{\omega}) \to \mathbb{B}$$

14.1. *Property refinement*

There are many different proposals for formalizing the notion of refinement (see, e.g., [2, 3,5,8,24,39,80]). We choose here the most basic logical notion of refinement of specifications, namely logical implication: a behaviour specification Q is called a *behaviour refinement* of the behaviour specification P if both P and Q have the same syntactic interface and, in addition, we have

$$Q(f) \Rightarrow P(f)$$

for all functions f; we then write $Q \Rightarrow P$. Accordingly a behaviour refinement never introduces new observable interactions, but just restricts the behaviour by adding properties. An inconsistent specification is a refinement for every specification with the same syntactic interface. It is, however, not a very useful refinement, since it cannot be refined into an implementation.

We understand all other classes of refinements considered in the following as special forms of behaviour refinements where Q and P in addition are in a more specific syntactic or semantic relationship. Concepts of refinement for data structures and their characteristic operations are well-known and well-understood in the framework of algebraic specification (see, e.g., [27]). In the modelling of distributed interactive systems data structures are used to represent

- the messages passed between the components,
- the histories of interactions between components (streams of messages),
- the states of the system.

In all three cases we may use the very general notion of data structure refinement. As it will be demonstrated in the sequel, several concepts of system refinement can be obtained by variations of data structure refinement.

We consider two versions of refinement of the black box view: refinement of the syntactic interface (by changing the number and the names as well as the sorts of the channels) of a system and refinement of the behaviour of a system. If the syntactic interface is refined then a concept is needed for relating the behaviours of the original and the refined system. This can be done by appropriate mappings (for another approach to refinement, see [6] and [7]).

A behaviour refinement is obtained by sharpening the requirements formalized in the specification. If we have the specifying predicate

$$P : [\vec{I} \to \vec{O}] \to \mathbb{B}$$

a behaviour refinement is any specifying predicate

$$\hat{P} : [\vec{I} \to \vec{O}] \to \mathbb{B},$$

where

$$\hat{P} \Rightarrow P$$

(or more precisely $\forall f \bullet \hat{P}(f) \Rightarrow P(f)$). Of course, a refinement is only practically helpful if the refined specification $\hat{P}$ is consistent, more formally, if we have

$$\exists f \bullet \hat{P}(f).$$

From a methodological point of view there are many different reasons and motivations for performing a behaviour refinement. Examples are development steps adding properties to specification in the course of requirements engineering or carrying out design decisions.

14.2. *Interaction refinement*

In this section we treat refinements that change the syntactic interface of a component. The syntactic interface is determined by the number of input and output channels as well as their message sorts. These are often refinements between different levels of abstractions. We study refinement steps from an abstract level to a (more) concrete level. Then the behaviours of the refined component have to be related to the behaviours of the original component. This can be achieved by translating communication histories for the channels on the abstract level to communication histories for the channels on the concrete level.

In our setting a syntactic interface is given by a set C of (input or output) channels with their corresponding sorts M of messages. Communication histories for these channels are given by the elements of the set $\vec{C}$. Given another set of channels C', a refinement is a mapping

$$\vec{C} \rightarrow \vec{C}'.$$

We speak of a *communication history refinement*. In such a refinement, a channel valuation representing a communication history is replaced by another channel valuation. For doing that we specify the translation. In our framework such a translation is defined by specifications *Rep* ("*representation specification*") and *Abs* ("*abstraction specification*") defining translations between the communication histories of the abstract level and the concrete level. This translation leads to commuting diagrams of the structure illustrated in Figure 5.

A pair of predicates specifying interactive system components

$$Rep: [\vec{C} \rightarrow \vec{C}'] \rightarrow \mathbb{B} \quad \textit{representation specification}$$
$$Abs: [\vec{C}' \rightarrow \vec{C}] \rightarrow \mathbb{B} \quad \textit{abstraction specification}$$

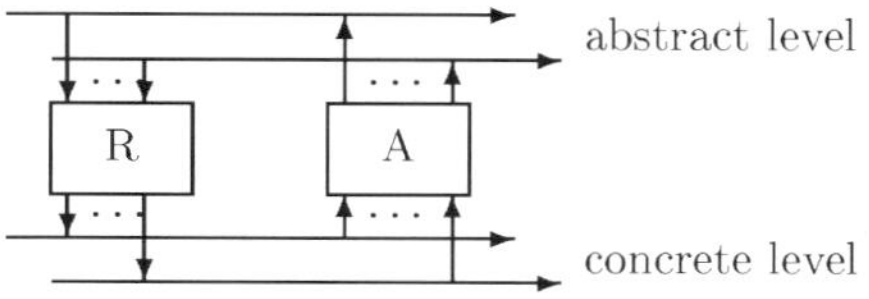

Fig. 5. Communication history refinement.

is called *refinement pair* for the communication histories from $\vec{C}$ with representation elements in $\vec{C}'$ if the abstraction specification *Abs* and representation specification *Rep* are consistent, that is the formula

$$\exists \alpha \bullet Abs(\alpha) \wedge \exists \rho \bullet Rep(\rho)$$

holds and for all functions α and ρ we have that

$$Abs(\alpha) \wedge Rep(\rho) \Rightarrow \alpha * \rho = Id$$

or, for short,

$$Abs * Rep = Id,$$

where I denotes the identity function (or more precisely the predicate that characterizes the identity function). Then *Abs* is called *abstraction specification* and *Rep* is called *representation specification*.

By the functions ρ with $Rep(\rho)$ a communication history $x \in \vec{C}$ can be rewritten into a communication history $\rho(x) \in \vec{C}'$. From $\rho(x)$ we can obtain x again by applying any function α with $Abs(\alpha)$ since $x = \alpha(\rho(x))$. Therefore $\rho(x)$ can be viewed as a (possibly "less abstract") representation of the communication history x.

In many approaches a relation is used instead of the representation specification *Rep*. The specification *Rep* is, however, very similar to a relation. Every function ρ with $Rep(\rho)$ represents one choice of representations for the abstract communication histories. In the sequel we use the following notion for *restricting* a function to a particular subdomain. Given a function

$$f : D \rightarrow E$$

for $B \subseteq D$ we denote by $f|B$ the function f restricted to arguments from B. Formally, the restriction of the function f to the set B yields the function

$$f|B : B \rightarrow E$$

where

$$(f|B)(x) = f(x) \quad \text{for } x \in B.$$

PROPOSITION 14.1. *The following properties hold for communication history refinement pairs Abs and Rep*:

(1) *all abstraction functions α with $Abs(\alpha)$ are surjective*;

(2) *A is deterministic on the image of Rep, that is, all functions α with $Abs(\alpha)$ behave identical on the image of Rep*; *formally expressed defining the image of Rep by a set $J \subseteq (M^{\omega})^{n'}$, where*

$$J = \{\rho(x) \in \vec{C}' \mid x \in \vec{C} \wedge Rep(\rho)\},$$

we have for all functions α *and* α'

$$Abs(\alpha) \wedge Abs(\alpha') \Rightarrow \alpha | J = \alpha' | J;$$

(3) *all functions* ρ *with* $Rep(\rho)$ *are injective. More generally, we have for all* x *and* x' *and all representation functions* ρ *and* ρ':

$$Rep(\rho) \wedge Rep(\rho') \wedge \rho(x) = \rho'(x') \Rightarrow x = x'.$$

PROOF. Property (1) immediately follows by the surjectivity of the identity *Id*.

Property (2) can easily be proved as follows: assume $Abs(\alpha)$ and $Abs(\alpha')$; then for $y = \rho(x)$ with $x \in \vec{I}$ we have

$$\alpha(y) = \alpha(\rho(x)) = x = \alpha'(\rho(x)) = \alpha'(y).$$

Property (3) immediately follows by the injectivity of *Id*. □

Based on the concept of refinement pairs, we introduce a more general notion of refinement. We consider a specification of an interactive component

$$P : [\vec{I} \rightarrow \vec{O}] \rightarrow \mathbb{B}$$

and predicates

$$Rep : [\vec{I} \rightarrow \vec{I'}] \rightarrow \mathbb{B},$$
$$\hat{P} : [\vec{I'} \rightarrow \vec{O'}] \rightarrow \mathbb{B},$$
$$\widetilde{Abs} : [\vec{O'} \rightarrow \vec{O}] \rightarrow \mathbb{B},$$

where *Rep* is a representation specification (with a corresponding abstraction specification *Abs*) and $\widetilde{Abs}$ is an abstraction specification (for a representation specification $\widetilde{Rep}$); the triple

$$(Rep, \hat{P}, \widetilde{Abs})$$

of predicates is called *interaction refinement* for the component specification P with representation specification *Rep* and abstraction specification $\widetilde{Abs}$, if the following condition is fulfilled (also called U-*simulation*):

$$Rep * \hat{P} * \widetilde{Abs} \Rightarrow P \quad \text{U-}\textit{simulation}.$$

This condition is graphically expressed by the commuting diagram given in Figure 6.

This concept of refinement between levels of abstraction is formalized as follows. Given representation specifications and corresponding abstraction specifications as defined above then $\hat{P}$ is called a *refinement of* P *under the representation specifications* R and $\widetilde{R}$ if

$$Rep * \hat{P} \Rightarrow P * \widetilde{Rep}.$$

This formula is visualized by the commuting diagram in Figure 7.

As demonstrated in Sections 2–7, synchronous communication is usually modelled by interleaving leading to sets of finite traces. This leads to a compositional model for safety. To achieve compositionality also for liveness conditions, one adds ready sets or refusal sets, as shown in Section 8. Asynchronous communication is modelled by sets of stream processing functions. They include liveness properties and explicit concurrency since messages are exchanged concurrently on the channels.

For handshake communication, working with traces is essential. Each action, be it a send or a receive action, may influence the future behaviour of the process with respect to all its channels. In particular, a process is not necessarily input enabled in all states. Actions may change the set of enabled input or output channels. With each process communicating by handshake (with actions separated into input and output actions) a set of traces is associated to model its safety properties.

In fact, also with each stream processing function and each specification of such functions a set of traces can be associated. This way stream processing functions model the behaviour by the same mathematical structure as it is used for handshake communication.

In the following, we study the relationship between handshake and buffered communication relating their models of (finite) traces. For safety properties we need, in both cases, only prefix closed sets of finite traces. In the case of message asynchrony these sets are input enabled. We define these notions formally in the following section.

15.1. *Input enabledness and asynchrony*

For simplicity we study only systems with disjoint sets A_I and A_O of input and output actions. We define the set of actions A as follows:

$$A = A_I \cup A_O.$$

Let $T \subseteq A^*$ be a set of finite traces over the action set A. A prefix closed trace set T (cf. Definition 2.6) is called *input enabled* if in its traces input actions may always occur ("are always accepted"). Mathematically expressed, the following property holds:

$$t \in T \wedge a \in A_I \Rightarrow t^\frown \langle a \rangle \in T.$$

A not necessarily prefix closed trace set is also called input enabled if its prefix closure is input enabled.

Given the partitioning of the actions into input and output actions, we may speak about the *asynchrony of traces*. Asynchrony means that the order of input with respect to output actions in the trace matters only in one case, namely when input is causal for output. Therefore, restricting ourselves to safety, we can shift in a valid trace input actions to the left and output actions to the right, as long as we respect the ordering of the actions for the each of the individual channels and still keep valid traces. This models the effect of implicit buffers.

Given traces s and t, we say that t is *more responsive* than s if $s \sqsubseteq_B t$, where the relation $\sqsubseteq_B$ is specified as follows:

$$\begin{aligned} s \sqsubseteq_B t \iff & \left(\forall s' \leqslant s\ \exists t' \leqslant t \bullet t' \downarrow A_I \leqslant s' \downarrow A_I \wedge s' \downarrow A_O \leqslant t' \downarrow A_O\right) \\ & \wedge s \downarrow A_I = t \downarrow A_I \\ & \wedge s \downarrow A_O = t \downarrow A_O. \end{aligned}$$

Informally, $s \sqsubseteq_B t$ means that in the trace t output may be generated a bit earlier than in the trace s and input received a bit later. In other words, t is *more responsive* than s if in s some output may be delayed compared to t and some input in s may be received earlier. Therefore, the trace s is called *less responsive* than t. Hence the trace s can be generated from t by shifting some of the output actions to the right and some of the input actions to the left.

The relation $\sqsubseteq_B$ is a partial order. A prefix closed trace set T is called *asynchronous* or *output buffered* if it is downward closed with respect to $\sqsubseteq_B$. Mathematically expressed, we have

$$t \in T \wedge t' \sqsubseteq_B t \Rightarrow t' \in T.$$

We define the asynchronous closure $T^{@}$ by

$$T^{@} = \left\{t' \mid \exists t \in T \bullet t' \sqsubseteq_B t\right\}.$$

Again, $T^{@}$ is a closure operation for trace sets T. We have

$$\left(T^{@}\right)^{@} = T^{@} \quad \text{and} \quad T \subseteq T^{@}.$$

$T^{@}$ is the least set that is asynchronous and includes T. If T is input enabled, so is its closure $T^{@}$.

15.2. *Stream processing functions as trace sets*

Let a stream processing function

$$f : A_I^{\omega} \to A_O^{\omega}$$

be given, which produces finite streams on finite input streams. We associate a trace set $T_f \subseteq A^*$ with the stream processing function f by the following definition:

$$\begin{aligned} T_f = \big\{t \in A^* \mid\ & t \downarrow A_O = f(t \downarrow A_I) \wedge \\ & \forall t' \in A^* \bullet t' \leqslant t \Rightarrow t' \downarrow A_O \leqslant f(t' \downarrow A_I)\big\}. \end{aligned}$$

The first part of the formula covers the liveness condition and the second part the safety condition induced by f.

THEOREM 15.1. *For every stream processing function f, the trace set T_f is input enabled and output buffered.*

PROOF. Input enabledness is straightforward by the monotonicity of f. Output bufferedness is a direct consequence of the definition of T_f. □

We easily define a trace set T_Q for any specification Q that specifies a set of stream processing functions by the equation

$$T_Q = \{t \in T_f \mid Q(f)\}.$$

We get the corollary:

THEOREM 15.2. *For every specification Q the trace set T_Q is input enabled and output buffered.*

PROOF. The union of input enabled and output buffered sets leads to sets with the same properties. □

This shows that asynchronous systems can be represented by a certain subclass of traces.

15.3. *From synchronous to asynchronous traces*

In this section we study how to relate synchronous and asynchronous systems. The traces we associate with systems communicating by handshake are neither input enabled nor asynchronous, in general. Traces of systems working by buffered communication are always input enabled and asynchronous. To go from an asynchronous system to a synchronous one therefore means to go from an input enabled asynchronous trace set to one that is no longer asynchronous nor necessarily input enabled.

Vice versa, we can go from a synchronous to an asynchronous one in a canonical way by adding traces such that the set gets input enabled and asynchronous. In the following, we show how to construct from an arbitrary prefix closed set an input enabled set and finally an asynchronous trace set in a canonical way. This is done by closures. To start with, we restrict ourselves to safety aspects and therefore to prefix closed trace sets.

15.3.1. *Safety.* Consider a synchronous process that is represented by a failure set

$$F \subseteq A^* \times \mathcal{P}(A)$$

in the sense of Section 8.2. Then the safety part of F is the trace set $F_S \subseteq A^*$ defined by projection as follows:

$$F_S = \{t \in A^* \mid \exists Ref \subseteq A \bullet (t, Ref) \in F\}.$$

The safety part of an asynchronous system specified by the predicate Q and represented by the trace set $T_Q \subseteq A^\omega$ is defined by its prefix closure

$$T_S = PC(T_Q)$$

(cf. Definition 2.6). Of course, in both steps we loose information, namely the liveness part. Note that the trace set F_S is neither input enabled nor asynchronous, in general. In fact, F_S may exhibit some information about the refusal structure. For instance, if none of the traces starts with a certain input action $a \in A_I$, then this action will certainly be refused in the initial state of the system.

Given a prefix closed trace set T, we construct the least set $T^{©}$ that is input enabled and includes T. It is defined by the equation

$$T^{©} = T^\frown A_I^* = \{t^\frown s \mid t \in T \wedge s \in A_I^*\}.$$

We observe that $T^{©}$ is prefix closed since T is prefix closed and whenever $t \in T^{©}$ we have $t^\frown\langle a\rangle \in T^{©}$ if $a \in A_I$. This shows that a prefix closed trace set T is input enabled if and only if we have

$$T = T^\frown A_I^*.$$

Given a prefix closed set of traces T to construct the *asynchronous closure* $T^{@}$ is simple. Note that $T^{@}$ is input enabled if T is input enabled and prefix closed if T is prefix closed.

15.3.2. *Liveness.* So far we did not consider liveness properties when relating synchrony to asynchrony. For synchronous systems, liveness is captured by the refusal sets, for asynchronous systems by the output that is guaranteed. A simple way to relate asynchronous and synchronous systems with respect to liveness is a one-to-one translation. This way we get a synchronous system that is input enabled and guarantees output whenever it is guaranteed by all traces of the asynchronous system.

In this section we study the relationship between asynchronous systems represented by trace sets that are input enabled and asynchronous, and synchronous systems represented by refusals.

Going from a synchronous to an asynchronous trace set can be understood as introducing unbounded buffers for all input and output channels [37]. This way, input is always enabled and may come earlier, before the system is actually ready to receive it (and then it is buffered), and output can be produced later until the environment is actually ready to receive it (and then it is buffered till then).

Given a synchronous process represented as a set of failures

$$F \subseteq A^* \times \mathcal{P}(A)$$

in the sense of Section 8.2, we obtain an asynchronous process by adding buffers for the input and output actions. Such a modification results in a failure set F_B defined as follows.

In asynchronous processes we add buffers for their input actions. More technically, we add a buffer process that is always enabled to receive input and – as long as it is not empty – always offers buffered input following the FIFO-principle to the process. Formally, this means that for the process constructed that way, input actions cannot be refused. This leads to the following definition of the failure set F_B:

$$\begin{aligned}(t, \mathit{Ref}) \in F_B \iff & \exists t' \in A^*,\ n \in A_I^*,\ s \in A^*,\ r \in A_O^*,\ \mathit{Ref}' \subseteq A \bullet t \sqsubseteq_B s^\frown n \\ & \wedge\quad s^\frown r \sqsubseteq_B t'\end{aligned}$$

$$\begin{aligned} &\wedge \quad (t', Ref') \in R \\ &\wedge \quad \big((r = \langle\rangle \wedge Ref = Ref' \backslash A_I) \\ &\qquad \vee Ref = Ref' \backslash \big(A_I \cup \{head(r) \mid r \neq \langle\rangle\}\big)\big). \end{aligned}$$

This expresses that a failure (t, Ref) is in F_B if and only if there is buffered input action that is refused by the original process and the trace of actions accepted by the original process has a refusal set that does not accept the buffered input. *Ref* is obtained by the refusal set of the trace eliminating all input actions. The trace n in the formula above represents the buffered input. The trace r denotes the buffered output. This is output generated by the synchronous process but buffered by the asynchronous process and therefore not forwarded to the environment. The trace s denotes those actions that are executed by the synchronous process and no longer buffered.

The readiness of output actions is not weakened by introducing buffers for output channels. If the original process is ready to do some output, the output is buffered and the process with the buffers stays output enabled for that channel until the output action is executed. The process becomes output stable.

EXAMPLE 15.3 (*Even/odd number generator*). Let $A_I = \{even, odd\}$ and $A_O = \mathbb{N}$. We describe a synchronous process that accepts the signals *even* or *odd* and generates an even or odd natural number resp. in response. It can be defined by the failure set

$$F = \big\{(t, Ref) \in A^* \times \mathcal{P}(A) \mid ok(t) \wedge Ref \subseteq \big\{a \in A^* \mid \neg ok(t^\frown \langle a\rangle)\big\}\big\},$$

where $A = A_I \cup A_O$ and where for $a, b \in A$ and $x \in A^*$

$$ok: A^* \to \mathbb{B}$$

is specified by the equations

$$\begin{aligned} ok(\langle\rangle) = ok(\langle a\rangle) &= true \quad \text{for } a \in A_I \\ ok(\langle a\rangle^\frown\langle b\rangle^\frown x) &= \big((a = even \wedge b \in \mathbb{N} \wedge b \,\mathbf{mod}\, 2 = 0) \vee \\ &\qquad (a = odd \wedge b \in \mathbb{N} \wedge b \,\mathbf{mod}\, 2 = 1)\big) \wedge \\ &\qquad\qquad ok(x). \end{aligned}$$

We obtain for instance

$$\big(\langle even\rangle^\frown\langle 4\rangle^\frown\langle odd\rangle^\frown\langle 11\rangle^\frown\langle odd\rangle^\frown\langle 7\rangle^\frown\langle even\rangle^\frown\langle 8\rangle, \mathbb{N}\big) \in F$$

and thus for F_B as defined above

$$\big(\langle even\rangle^\frown\langle odd\rangle^\frown\langle odd\rangle^\frown\langle 4\rangle^\frown\langle 11\rangle, \mathbb{N}\backslash\{7\}\big) \in F_B.$$

So far we have shown how to go from a synchronous process given by a failure set

$$F \subseteq A^* \times \mathcal{P}(A)$$

to an asynchronous, input enabled trace set. We finally arrive at the trace set

$$T = \left[\left(\{t \mid \exists Ref \bullet (t, Ref) \in F\}\right)^{©}\right]^{@}.$$

This construction leads to a prefix closed, input enabled, and output buffered trace T that does not reflect anymore the refusal structure of the original system described by the failure set F.

EXAMPLE 15.4 (*Partial identity*). As a second simple example we study the so-called *partial identity*. It is specified by the predicate $Pid(f)$ on stream functions $f : M^{\omega} \to M^{\omega}$ as follows:

$$Pid(f) \iff \forall x \in M^{\omega} \bullet f(x) \leqslant x.$$

Thus a stream processing function f is a partial identity if it issues a prefix of its input stream as output. We associate with Q the trace set

$$T = \{t \in A^* \mid t_o \leqslant t_i\}$$

where $A = A_I \cup A_O$ with $A_I = \{m? \mid m \in M\}$ and $A_O = \{m! \mid m \in M\}$. By $t_o \in M^*$ we abbreviate $t \downarrow A_O$ and by $t_i \in M^*$ we abbreviate $t \downarrow A_I$ after eliminating the tags "!" and "?". The trace set T is input enabled and asynchronous.

Now we add refusals to T. For instance, we get a one element buffer by

$$\begin{aligned} B_1 = \{(t, Ref) \mid \exists m \in M \bullet\ & (t_o = t_i \vee t_o \frown \langle m\rangle = t_i) \wedge \\ & (t_o = t_i \Rightarrow Ref \subseteq A_O) \wedge \\ & (t_o \frown \langle m\rangle = t_i \Rightarrow Ref \subseteq A_I \cup (A_O \backslash \{m!\}))\}. \end{aligned}$$

An unbounded buffer is characterized by B_∞ specified as follows:

$$\begin{aligned} B_\infty = \{(t, Ref) \mid\ & t_o \leqslant t_i \wedge \\ & (t_o = t_i \Rightarrow Ref \subseteq A_o) \wedge \\ & (\forall m \in M \bullet t_o \frown \langle m\rangle \leqslant t_i \Rightarrow Ref \subseteq A_O \backslash \{m!\})\}. \end{aligned}$$

Of course we may select many more refusal sets for the trace set T to characterize other versions of buffers.

From both failure sets B_1 and B_∞ we get back the original trace set T by adding an input buffer. We get the same trace set by extracting the trace sets from B_1 or B_∞ respectively and applying the closure with respect to input enabledness and asynchrony.

The example shows how to derive a trace set from an asynchronous specification and how to enrich this trace set by refusals to get synchronous process descriptions. This leads to the following relationship between message asynchrony and message synchrony. Message asynchrony defines input enabled, asynchronous traces but does not define nor need a refusal structure. This information can be given in addition, to obtain message synchronous systems. This demonstrates that asynchronous systems can be considered as abstractions of synchronous systems.

From a methodological point of view, this relationship between asynchrony and synchrony can be exploited by carrying out a system development in two steps: in a first step we define an input enabled asynchronous behaviour, and in a second step we refine it by selecting a subset of this trace set and by adding refusals to it.

Adding buffers for input channels is enough to guarantee input enabledness, but this does not guarantee stable asynchrony nor output stability. If output actions are ready, then by taking them we may change the readiness set. To guarantee asynchrony, we add buffers for the output channels, too. This makes sure that an enabled output channel stays enabled independent of input and output on other channels until the output has been forwarded.

The introduction of buffers for all input channels of a *CSP* process has serious consequences. This way we obtain a process

- that is input enabled,
- for which we cannot conclude by the fact that it accepts input that it is ready to process this input.

This shows that we lose this way one of the fundamental ideas of handshake communication, namely to control the choices of behaviour of processes guided by the fact whether both communication partners are ready to communicate.

The notions introduced above indicate that asynchronous systems can be seen as a special case of synchronous systems, namely those that are input enabled, output stable, and asynchronous. For this subclass of synchronous systems we get the mathematically simpler model of asynchrony formed by stream processing functions. Vice versa, we can also encode synchronous systems by asynchronous systems by introducing messages that indicate acceptance and rejection.

15.4. *Composition of asynchronous and synchronous systems*

In this section we study ways to compose message asynchronous and message synchronous systems. Synchrony with its handshake communication suggests a specific form of parallel composition as introduced in the previous sections. We can only specify the behaviour of a system composed from a set of processes communicating via handshake precisely if we are given their traces with refusal sets. Traces alone do not suffice since they do not provide enough information about possible deadlocks. If the system communicates via buffers the refusal information is no longer needed. A deadlock may only occur if all processes are no longer capable of producing output, since otherwise at least one process is ready to produce output, and since this output can always be buffered, it may do so.

Finally, we show how to introduce buffers for input explicitly via processes. Given a set A of actions decomposed into a set A_I of input actions and A_O of output actions, we define a set A'_I of actions which contains for every action $a \in A_I$ a copy $a' \in A'_I$, such that A_I and A'_I are disjoint. Formally, we define a mapping

$$\varphi : A_I \to A'_I \quad \text{with } \varphi(a) = a'.$$

Thus φ allows us to rename the actions in A_I; this mapping can easily be extended to traces by pointwise application. A buffer process B for the action set A_I is defined over the alphabet

$$\alpha(B) = A_I \cup A'_I.$$

The buffer handles the actions $a \in A_I$ as input, buffers them, and offers the actions in its buffer as a'. Its behavior is captured by the following failure set:

$$B = \{(t, \mathit{Ref}) \mid \exists a \in A_I \bullet (t \downarrow A'_I)^\frown\langle a'\rangle \leqslant \varphi(t \downarrow A_I) \wedge \mathit{Ref} \subseteq A'_I \backslash \{a'\}) \\ \vee \big(t \downarrow A'_I = \varphi(t \downarrow A_I) \wedge \mathit{Ref} \subseteq A'_I\big)\}.$$

The first line in the definition of B treats the case where input is buffered, while the second line treats the case where no input is buffered (the buffer is empty). Except for a different alphabet, B specifies the same behaviour as the unbounded buffer B_∞ in Example 15.4.

We get a buffered version of a synchronous process P for the actions A_I by

$$\big(P[\varphi] \parallel B\big)[\varphi'],$$

where the morphism φ' hides the actions in A'_I:

$$\varphi' : A \cup A'_I \to A \cup \{\tau\}$$

with

$$\varphi'(a) = a \quad \text{for } a \in A,$$
$$\varphi'(a') = \tau \quad \text{for } a' \in A'_I.$$

Given a process term P that does not contain a parallel composition and where the action set A is decomposed into A_I and A_O, we may turn P into an asynchronous process term P' by replacing

$$\mathit{stop}{:}A \quad \text{by} \quad \mathit{stop}{:}(A_I : A_O)$$
$$\mathit{div}{:}A \quad \text{by} \quad \mathit{div}{:}(A_I : A_O).$$

Then

$$P' \quad \text{and} \quad (P[\varphi] \parallel B)[\varphi']$$

are equivalent. In other words: asynchronous communication is identical with synchronous communication after adding buffers for the input actions.

16. Conclusion

Traces provide history based system models. A trace is the history of steps taken by a system. Trace-based system description shows a rich structure reflecting notions as states, composition, and refinement. Various process algebras can be based on the concept of traces.

We have introduced the basic models of message passing systems, namely asynchronous message passing and synchronous message passing. Both models are based on sets of traces. Sets of traces are sufficient for expressing safety properties in both cases. For liveness properties additional information is required, considering infinite sequences or failures and readiness sets.

References

[1] M. Abadi and L. Lamport, *Composing specifications*, ACM TOPLAS **15** (1) (1993), 73–132.

[2] L. Aceto, *Action Refinement in Process Algebras*, Cambridge University Press (1992).

[3] L. Aceto and M. Hennessy, *Adding action refinement to a finite process algebra*, Proc. ICALP'91, Lecture Notes in Comput. Sci. 510, Springer-Verlag (1991), 506–519.

[4] B. Alpern and F.B. Schneider, *Defining liveness*, Inform. Proc. Lett. **21** (1985), 181–185.

[5] R.J.R. Back, *Correctness preserving refinements: Proof theory and applications*, Technical Report, Mathematical Centre Tracts 131, Mathematical Centre, Amsterdam (1980).

[6] R.J.R. Back, *Refinement calculus, Part I: Sequential nondeterministic programs*, REX Workshop, Stepwise Refinement of Distributed Systems, Lecture Notes in Comput. Sci. 430, J.W. de Bakker, W.-P. de Roever and G. Rozenberg, eds, Springer-Verlag (1990), 42–66.

[7] R.J.R. Back, *Refinement calculus, Part II: Parallel and reactive programs*, REX Workshop, Stepwise Refinement of Distributed Systems, Lecture Notes in Comput. Sci. 430, J.W. de Bakker, W.-P. de Roever and G. Rozenberg, eds, Springer-Verlag (1990), 67–93.

[8] R.J.R. Back, *Refinement of parallel and reactive systems*, Program Design Calculi, Springer NATO ASI Series, Series F: Computer and System Sciences 118, M. Broy, ed. (1993).

[9] J.C.M. Baeten, J.A. Bergstra and J.W. Klop, *On the consistency of Koomen's fair abstraction rule*, Theoret. Comput. Sci. **51** (1987), 129–176.

[10] J.W. de Bakker, *Mathematical Theory of Program Correctness*, Prentice-Hall, London (1980).

[11] J.W. de Bakker, W.-P. de Roever and G. Rozenberg, eds, *Stepwise Refinement of Distributed Systems*, Lecture Notes in Comput. Sci. 430, Springer-Verlag (1990).

[12] F.L. Bauer et al., *The Munich Project CIP, Vol. I: The Wide Spectrum Language CIP-L*, Lecture Notes in Comput. Sci. 183, Springer-Verlag (1985).

[13] F.L. Bauer et al., *The Munich Project CIP, Vol. II: The Program Transformation System CIP-S*, Lecture Notes in Comput. Sci. 292, Springer-Verlag (1987).

[14] J.A. Bergstra, W. Fokkink and A. Ponse, *Process algebra with recursive operations*, Handbook of Process Algebra, J.A. Bergstra, A. Ponse and S.A. Smolka, eds, Elsevier, Amsterdam (2001), 333–389.

[15] J.A. Bergstra and J.W. Klop, *Algebra of communicating processes*, Proc. CWI Symposium on Mathematics and Computer Science, CWI Monograph I, J.W. de Bakker, M. Hazewinkel and J.K. Lenstra, eds, North-Holland, Amsterdam (1986), 89–138.

[16] J.A. Bergstra, J.W. Klop and E.-R. Olderog, *Failures without chaos: A new process semantics for fair abstraction*, Formal Description of Programming Concepts III, M. Wirsing, ed., North-Holland, Amsterdam (1987), 77–101.

[17] J.A. Bergstra, J.W. Klop and E.-R. Olderog, *Readies and failures in the algebra of communicating processes*, SIAM J. Comput. **17** (1988), 1134–1177.

[18] F.S. de Boer, J.N. Kok, C. Palamidessi and J.J. Rutten, *The failure of failures in a paradigm for asynchronous communication*, Proc. CONCUR'91, Lecture Notes in Comput. Sci. 527, J. Baeten and J. Groote, eds, Springer-Verlag (1991), 111–126.

[19] J.D. Brock and W.B. Ackermann, *Scenarios: A model of nondeterminate computation,* Lecture Notes in Comput. Sci. 107, J. Diaz and I. Ramos, eds, Springer-Verlag (1981), 225–259.

[20] S.D. Brookes, C.A.R. Hoare and A.W. Roscoe, *A theory of communicating sequential processes*, J. ACM **31** (1984), 560–599.

[21] M. Broy, *Algebraic methods for program construction: The project CIP*, Program Transformation and Programming Environments, NATO ASI Series, Series F: 8, P. Pepper, ed., Springer-Verlag (1984), 199–222.

[22] M. Broy, *Semantics of finite or infinite networks of communicating agents*, Distrib. Comput. **2** (1987), 13–31.

[23] M. Broy, *Functional specification of time sensitive communicating systems*, REX Workshop, J.W. de Bakker, W.-P. de Roever and G. Rozenberg, eds, Stepwise Refinement of Distributed Systems, Lecture Notes in Comput. Sci. 430, Springer-Verlag (1990), 153–179.

[24] M. Broy, *Compositional refinement of interactive systems*, DIGITAL Systems Research Center, SRC 89 (1992). Also in: J. ACM **44** (6) (1997), 850–891.

[25] M. Broy, *Functional specification of time sensitive communicating systems*, ACM Trans. Software Engrg. Meth. **2** (1) (1993), 1–46.

[26] M. Broy, *Refinement of time*, Transformation-Based Reactive System Development, Proc. ARTS'97, Mallorca 1997, Lecture Notes in Comput. Sci. 1231, M. Bertran and Th. Rus, eds, Springer-Verlag (1997), 44–63. To appear in TCS.

[27] M. Broy, B. Möller, P. Pepper and M. Wirsing, *Algebraic implementations preserve program correctness*, Sci. Comput. Programming **8** (1986), 1–19.

[28] M. Broy and G. Stefanescu, *The algebra of stream processing functions*, Technische Universität München, Institut für Informatik, TUM-I9620 (Mai 1996). To appear in TCS.

[29] K.M. Chandy and J. Misra, *Parallel Program Design: A Foundation*, Addison Wesley (1988).

[30] J. Coenen, W.P. de Roever and J. Zwiers, *Assertional data reification proofs: Survey and perspective*, Christian-Albrechts-Universität Kiel, Institut für Informatik und Praktische Mathematik, Bericht Nr. 9106 (1991).

[31] R. DeNicola and M. Hennessy, *Testing equivalences for processes*, Theoret. Comput. Sci. **34** (1984), 83–134.

[32] E.W. Dijkstra, *A Discipline of Programming*, Prentice-Hall, Englewood Cliffs, NJ (1976).

[33] D.L. Dill, *Trace Theory for Automatic Hierarchical Verification of Speed-Independent Circuits*, MIT Press, Cambridge, MA (1989).

[34] R.J. van Glabbeek, *The linear time – branching time spectrum I. The semantics of concrete, sequential processes*, Handbook of Process Algebra.

[35] C.A. Gunter, P.D. Mosses and D.S. Scott, *Semantic domains and denotational semantics*, Technical Report DAIMI PB-276, Comput. Sci. Dept., Aarhus University (1989).

[36] D. Harel and A. Pnueli, *On the development of reactive systems*, Logic and Models of Concurrent Systems, K.R. Apt, ed., Springer-Verlag (1985), 477–498.

[37] J. He, M.B. Josephs and C.A.R. Hoare, *A theory of synchrony and asynchrony*, Programming Concepts and Methods, M. Broy and C.B. Jones, eds, North-Holland-Elsevier, Amsterdam (1990), 459–478.

[38] M. Hennessy, *Algebraic Theory of Processes*, MIT Press, Cambridge, MA (1988).

[39] C.A.R. Hoare, *Proofs of correctness of data representations*, Acta Inform. **1** (1972), 271–281.

[40] C.A.R. Hoare, *A model for communicating sequential processes*, On the Construction of Programs, R.M. McKeag and A.M. McNaghton, eds, Cambridge University Press (1980), 229–243.

[41] C.A.R. Hoare, *A calculus of total correctness for communicating processes*, Sci. Comput. Progr. **1** (1981), 44–72.

[42] C.A.R. Hoare, *Communicating Sequential Processes*, Prentice-Hall, London (1985).

[43] C.A.R. Hoare, *Programs are predicates*, Mathematical Logic and Programming Languages, C.A.R. Hoare and J.C. Shepherdson, eds, Prentice-Hall, London (1985), 141–155.

[44] J.E. Hopcroft and J.D. Ullman, *Formal Languages and Their Relation to Automata Theory*, Addison-Wesley, Reading, MA (1969).

[45] W. Janssen, M. Poel and J. Zwiers, *Action systems and action refinement in the development of parallel systems – an algebraic approach*, CONCUR'91, Lecture Notes in Comput. Sci. 527, J.C.M. Baeten and J.F. Groote, eds, Springer-Verlag (1991), 298–345.

[46] C.B. Jones, *Systematic Program Development Using VDM*, Prentice Hall, London (1986).

[47] B. Jonsson, *A model and proof system for asynchronous networks*, Proc. 4th ACM Symposium on Principles of Distributed Computing (1985), 49–58.

[48] B. Jonsson, *Compositional verification of distributed systems*, Ph.D. Thesis, Department of Computer Systems, Uppsala University, Uppsala, Sweden (1987).

[49] G. Kahn, *The semantics of a simple language for parallel programming*, Information Processing **74** (1974), 471–475.

divergence, 110, 111, 114, 125, 129, 149

environment, 123, 140
– technique, 123
equivalence
– theorem, 127
expansion, 150
– strategy, 152
expression, 138
– communication, 139
– logical, 139
– natural number, 138
– trace, 138
expressiveness, 143

failures model, 158
fair abstraction, 133
feedback, 168
fixed point, 122
– least, 122, 163
– technique, 123
– theorem, 122
– unique, 133
functional composition, 167

Greibach condition, 112

hiding operator, 112, 126, 130, 149, 168
history
– of communications, 106
– variable, 141

identifier, 110
– bound, 112
– free, 112
input enabled, 182
isomorphic, 105
isomorphism, 106

length
– of a stream, 159
– of a trace, 138
limit
– of a chain, 121
liveness property, 161, 185

miracle, 156
mixed term, 145
modularity, 154
monotonicity, 121
more responsive, 183
morphism, 110, 111, 114
mu-operator, 110

nondeterminism, 110, 111, 125, 129, 174

output buffered, 183

parallel composition, 110, 111, 114, 125, 129, 149, 166
– independent, 166
partial identity, 187
partial order, 120
– complete, 121
– for streams, 159
– for trace semantics, 124
Peano arithmetic, 143
pipelining, 167
prefix
– closed, 109
– closure, 109, 148
– kernel, 148
– operator, 110, 111, 113, 125, 129, 149
– ordering, 107, 159
priority, 111, 140
process
– asynchronous, 119
– correctness, 144
– synchronous, 184
– term, 113
projection
– alphabet, 142
– operator, 138

queue, 160, 173

readiness model, 157
ready set, 157
recursion, 110, 111, 114, 123, 126, 149
recursive terms, 110
recursively enumerable, 143
refinement, 175
– glas box, 181
– of interaction, 177
– of properties, 176
– stepwise, 145
refusal set, 158
renaming, 139
– disjoint, 150
renaming operator, 111, 125, 129, 138, 149
rule
– algebraic, 130
– transformation, 147
– transition, 113

safety property, 161, 184
satisfaction relation
– for process terms, 144

– for traces, 141
scheduling problem, 152
Scott domain, 162
semantic
– equation, 146
– implication, 146
– operator, 123
semantic domain, 120
– for streams, 162
– for trace logic, 140
– for trace semantics, 108
semantics
– denotational, 120, 122
– operational, 115, 116, 170
– SOS, 113, 170
– trace, 109
sequence
– finite, 106
– infinite, 158
simulation, 179
– downward, 180
– upward, 180
soundness, 130, 147
– theorem, 134, 150
specification, 141, 149, 173
state, 104
– initial, 104
– reachable, 105
state machine, 170
state transition diagram, 104
store, 170
stream, 158
– processing function, 163
– timed, 172
substitution, 112, 139
synchrony, 114

term, 110
– action-guarded, 112
– closed, 112
– communication-guarded, 134
– mixed, 145
– process, 113
– recursive, 110
time
– discrete, 172
timeout, 172
trace
– domain, 108
– equivalence, 109
– expression, 138
– finite, 106
– formula, 139
– infinite, 158
– logic, 137
– Mazurkiewicz-, 106
– specification, 141
trace semantics
– algebraic laws, 130
– denotational, 126, 146
– of automata, 109
– of mixed terms, 146
– of process terms, 116
– operational, 116
transformation rule, 147
transformational programming, 147
transition, 104
– labelled, 104
– relation, 104
– system, 104

upper bound, 120

validity
– of trace formulas, 141
variable, 138
– bound, 139
– free, 139

CHAPTER 3

Structural Operational Semantics

Luca Aceto[1,*], Wan Fokkink[2,†], Chris Verhoef[3]

[1] *BRICS (Basic Research in Computer Science), Centre of the Danish National Research Foundation, Department of Computer Science, Aalborg University, Fredrik Bajers Vej 7-E, DK-9220 Aalborg Ø, Denmark*
E-mail: luca@cs.auc.dk

[2] *CWI, Department of Software Engineering, Kruislaan 413, 1098 SJ Amsterdam, The Netherlands*
E-mail: wan@cwi.nl

[3] *University of Amsterdam, Department of Computer Science, Programming Research Group, Kruislaan 403, 1098 SJ Amsterdam, The Netherlands*
E-mail: x@wins.uva.nl

Contents

1. Introduction 199
2. Preliminaries 201
   2.1. Labelled transition systems 201
   2.2. Behavioural equivalences and preorders 202
   2.3. Hennessy–Milner logic 205
   2.4. Term algebras 205
   2.5. Transition system specifications 206
   2.6. Examples of TSSs 207
3. The meaning of TSSs 209
   3.1. Model-theoretic answers 209
   3.2. Proof-theoretic answers 212
   3.3. Answers based on stratification 214
   3.4. Evaluation of the answers 215
   3.5. Applications 216
4. Conservative extension 217
   4.1. Operational conservative extension 218
   4.2. Implications for three-valued stable models 220
   4.3. Applications to axiomatizations 221
   4.4. Applications to rewriting 225
5. Congruence formats 227
   5.1. Panth format 229
   5.2. Ntree format 230

*Partially supported by a grant from the Italian CNR, Gruppo Nazionale per l'Informatica Matematica (GNIM).
†Partially supported by a grant from the Nuffield Foundation.

HANDBOOK OF PROCESS ALGEBRA
Edited by Jan A. Bergstra, Alban Ponse and Scott A. Smolka

5.3. De Simone format . . . 231
5.4. GSOS format . . . 237
5.5. RBB safe format . . . 254
5.6. Precongruence formats for behavioural preorders . . . 258
5.7. Trace congruences . . . 262
6. Many-sorted higher-order languages . . . 264
6.1. The actual world . . . 265
6.2. The formal world . . . 266
6.3. Actual and formal transition rules . . . 267
6.4. Operational conservative extension . . . 268
7. Denotational semantics . . . 270
7.1. Preliminaries . . . 271
7.2. From recursive GSOS to denotational semantics . . . 276
References . . . 280
Subject index . . . 289

Abstract

Structural Operational Semantics (SOS) provides a framework to give an operational semantics to programming and specification languages, which, because of its intuitive appeal and flexibility, has found considerable application in the theory of concurrent processes. Even though SOS is widely used in programming language semantics at large, some of its most interesting theoretical developments have taken place within concurrency theory. In particular, SOS has been successfully applied as a formal tool to establish results that hold for whole classes of process description languages. The concept of rule format has played a major role in the development of this general theory of process description languages, and several such formats have been proposed in the research literature. This chapter presents an exposition of existing rule formats, and of the rich body of results that are guaranteed to hold for any process description language whose SOS is within one of these formats. As far as possible, the theory is developed for SOS with features like predicates and negative premises.

1. Introduction

The importance of giving precise semantics to programming and specification languages was recognized since the sixties with the development of the first high-level programming languages (cf., e.g., [29,205] for some early accounts). The use of operational semantics – i.e., of a semantics that explicitly describes how programs compute in stepwise fashion, and the possible state-transformations they perform – was already advocated by McCarthy in [146], and elaborated upon in references like [140,141]. Examples of full-blown languages that have been endowed with an operational semantics are Algol 60 [138], PL/I [171], and CSP [176].

Structural operational semantics (SOS) [175] provides a framework to give an operational semantics to programming and specification languages. In particular, because of its intuitive appeal and flexibility, SOS has found considerable application in the study of the semantics of concurrent processes, where, despite successful work by, among others, de Bakker, Zucker, Hennessy, and Abramsky (see, e.g., [1,30,116,119,121,124,149]), the methods of denotational semantics appear to be difficult to apply in general. SOS generates a labelled transition system, whose states are the closed terms over an algebraic signature, and whose transitions between states are obtained inductively from a collection of so-called transition rules of the form $\frac{\text{premises}}{\text{conclusion}}$. A typical example of a transition rule is

$$\frac{x \xrightarrow{a} x'}{x \| y \xrightarrow{a} x' \| y}$$

stipulating that if $t \xrightarrow{a} t'$ holds for certain closed terms t and t', then so does $t \| u \xrightarrow{a} t' \| u$ for each closed term u. In general, validity of the premises of a transition rule, under a certain substitution, implies validity of the conclusion of this rule under the same substitution.

Recently, SOS has been successfully applied as a formal tool to establish results that hold for classes of process description languages. This has allowed for the generalization of well-known results in the field of process algebra, and for the development of a meta-theory for process calculi based on the realization that many of the extant results in this field only depend upon general semantic properties of language constructs. The concept of a rule format has played a major role in the development of the meta-theory of process description languages, and several such formats have been proposed in the research literature. A principal aim of this chapter is to give an exposition on existing rule formats. Each of the formats surveyed here comes equipped with a rich body of results that are guaranteed to hold for any process calculus whose SOS is within that format.

Predicates in SOS semantics can be coded as binary relations [111]. Moreover, negative premises can often be expressed positively using predicates [26]. However, in the literature we see more and more that SOS definitions are decorated with predicates and/or negative premises. For example, predicates are used to express matters like (un)successful termination, convergence, divergence [10], enabledness [41], maximal delay, and side conditions [163]. Negative premises are used to describe, e.g., deadlock detection [135], sequencing [55], priorities [24,64], probabilistic behaviour [137], urgency [58], and various real [134] and discrete time [23,126,222] settings. Since predicates and negative premises are so pervasive, and often lead to cleaner semantic descriptions for many features and constructs of interest, we present the theory of SOS in a setting that deals explicitly with these notions as much as possible. We hope that this makes this chapter a useful reference guide to the literature on the use of SOS in process algebra.

The organization of this chapter is as follows. Section 2 presents the preliminaries of SOS theory, and contains some standard SOS definitions that serve as running examples. Section 3 gives an overview of the different ways to give meaning to SOS definitions. Section 4 presents syntactic constraints under which an extension of an SOS definition does not influence some properties of the original SOS definition. Section 5 studies a wide range of syntactic formats for SOS definitions that guarantee that the semantics of a term is determined by the semantics of its arguments, and focuses on the connection between SOS semantics and complete proof systems. Section 6 describes a formalism to deal with variable binders explicitly. Finally, Section 7 pays attention to the automatic generation of fully abstract denotational models of process calculi from their SOS semantics.

On terminology: Structural versus structured operational semantics. As mentioned above, in this chapter we shall use the acronym SOS to stand for *Structural Operational Semantics*. The adjective *structural* was used by Plotkin in the title of his seminal set of lecture notes [175] as this approach to giving formal semantics for programming and specification languages places great emphasis on defining the effect of running a program in terms of its structure. Moreover, the term Structural Operational Semantics is the most commonly used in the literature on semantics of programming languages and in various textbooks on this topic (see, e.g., [113,118,166]). The form of semantics we describe in this chapter is sometimes also called "Plotkin-style" operational semantics because of the aforementioned influential DAIMI report of Plotkin [175] and several papers in which he used this kind of specification. Some authors (see, e.g., [113]) prefer to use the term *transition semantics* to emphasize that transitions between program states are the main objects of study in this form of semantics. This terminology, albeit more descriptive in this context than "structural" or "Plotkin-style", has the drawback of being applicable to a range of operational semantics – such as those for automata and Petri nets [182] – that are rather different in nature from those that we deal with in this chapter. In [110,111], Groote and Vaandrager used the acronym SOS to stand for *Structured Operational Semantics*. Their aim was to emphasize that a transition system specification that leads to a transition system for which bisimulation equivalence [169] is not a congruence should not be called *structured*, even though it is possibly compositional on the level of concrete transition systems. We have shunned from adopting their terminology as it is only used in the process algebra literature, and may be construed as suggesting that other forms of operational semantics are unstructured.

Disclaimer. In this chapter, we focus on the results on the theory of SOS that, we feel, have the most interest from the point of view of process algebra. It is, however, a sign of the maturity of this field that SOS has found applications in many other settings. The original motivation for the development of SOS was to give semantics to programming languages, and the success of this endeavour is witnessed by the growing number of real-life programming languages that have been given *usable* semantic descriptions by means of SOS (see, e.g., [43,161,171,176,184]). As other applications of SOS, we limit ourselves to mentioning here that:

- the operational approach to type soundness, pioneered in [234], is now the preferred choice over methods based upon denotational semantics;
- the correctness of hardware implementations of real-life programming languages, and of compilation techniques, has been established using SOS [43,213,232];
- the fit between reasonable operational extensions for the language PCF [173] and Scott's original lattice model for it has been studied in [45] within the framework of SOS;
- it has been observed that SOS is an appropriate style for static program analysis [139];
- the derivation of proof rules for functional languages from their operational specifications has been investigated in [193], building upon the work in [8] (cf. Section 5.4.5).

These are only a few of the many interesting examples of applications of SOS that are not covered in this chapter. We hope that the reader will be tempted to explore them, and possibly to contribute to this fascinating research area.

Acknowledgements. Our thoughts on the theory of SOS have been shaped by the inspiring work of, and collaborations with, many researchers. We cannot thank them all explicitly here. However, it will be evident to the readers of this chapter that the theory we survey, and the presentation we give of it, would not have been possible without the work of our colleagues. In particular, the ideas and work of Bard Bloom, Rob van Glabbeek (on whose work Section 3 is heavily based), Jan Friso Groote, Robert de Simone and Frits Vaandrager have been most influential. We hope that the list of references will prove useful in guiding the interested readers to the original sources for our subject matter. Finally, we thank Davide Marchignoli, Simone Tini and an anonymous referee for their thorough reading of a draft of this chapter.

2. Preliminaries

In this section we present the basic notions from process theory that are needed in the remainder of this chapter. The presentation is necessarily brief, and the interested reader is warmly encouraged to consult the references for much more information and motivation on the background material to our subject matter. We hope, however, that the basic definitions and results mentioned in this section will help the reader go through the material presented in this chapter with some ease.

2.1. *Labelled transition systems*

We begin by reviewing the model of *labelled transition systems* [132,175], which are used to express the operational semantics of many process calculi. They consist of binary relations between states, carrying an action label, and predicates on states. Intuitively, $s \xrightarrow{a} s'$ expresses that state s can evolve into state s' by the execution of action a, while sP expresses that predicate P holds in state s. For convenience of terminology, we refer to both binary relations and predicates on states as *transitions*.

DEFINITION 2.1 (*Labelled transition system*). A *labelled transition system* (LTS) is a quadruple $(\mathsf{Proc}, \mathsf{Act}, \{\xrightarrow{a} \mid a \in \mathsf{Act}\}, \mathsf{Pred})$, where:

- Proc is a set of *states*, ranged over by s;
- Act is a set of *actions*, ranged over by a, b;
- $\xrightarrow{a} \subseteq \mathsf{Proc} \times \mathsf{Proc}$ for every $a \in \mathsf{Act}$. As usual, we use the more suggestive notation $s \xrightarrow{a} s'$ in lieu of $(s, s') \in \xrightarrow{a}$, and write $s \stackrel{a}{\nrightarrow}$ if $s \xrightarrow{a} s'$ for no state s';
- $P \subseteq \mathsf{Proc}$ for every $P \in \mathsf{Pred}$. We write sP (respectively $s\neg P$) if state s satisfies (respectively does not satisfy) predicate P.

Binary relations $s \xrightarrow{a} s'$ and unary predicates sP in an LTS are called *transitions*.

In what follows, we shall sometimes identify an LTS with the set of its transitions. We trust that the meaning will always be clear from the context.

DEFINITION 2.2 (*Finiteness constraints on LTSs*). An LTS is:
- *finitely branching* if for every state s there are only finitely many outgoing transitions $s \xrightarrow{a} s'$;
- *regular* if it is finitely branching and each state can reach only finitely many other states;
- *finite* if it is finitely branching and there is no infinite sequence of transitions $s_0 \xrightarrow{a_0} s_1 \xrightarrow{a_1} \cdots$.

REMARK. The conditions of regularity and finiteness defined above are usually used at the level of *process graphs*, i.e., transition systems with a distinguished initial state from which all other states are reachable in zero or more transitions. In particular, the above definition ensures that an LTS is finite or regular if so are all the process graphs obtained by choosing an arbitrary state as the initial one, removing all the states that are unreachable from it, and restricting the transition relations to the set of reachable states. Note that the notion of regularity defined above is a purely "syntactic" one. For instance, the LTS defined by

$$\{n \xrightarrow{a} n+1 \mid n \in \mathbb{N}\}$$

is not regular according to the above definition, even though it is the unfolding of the regular LTS $0 \xrightarrow{a} 0$. To define more semantic notions of regularity one has to work modulo some notion of behavioural equivalence. (See the following section and [102] in this Handbook for information on behavioural equivalences over states of LTSs.)

2.2. *Behavioural equivalences and preorders*

LTSs describe the operational behaviour of processes in great detail. In order to abstract away from irrelevant information on the way that processes compute, a wealth of notions of behavioural equivalence (i.e., a relation that is reflexive, transitive, and symmetric) and preorder (i.e., a relation that is reflexive and transitive) over the states of an LTS have been studied in the literature on process theory. A systematic investigation of these notions is presented in [94,97] (see also [93, Chapter 1], and [102] in this Handbook), where van Glabbeek presents the linear time/branching time spectrum. This lattice contains all the known behavioural equivalences and preorders over LTSs, ordered by inclusion. We investigate only a fragment of this spectrum, which we now proceed to present for the sake of completeness.

DEFINITION 2.3 (*Simulation, ready simulation, and bisimulation*). Assume an LTS.

- A binary relation $\mathcal{R}$ on states is a *simulation* if whenever $s_1 \mathcal{R} s_2$:
  - if $s_1 \xrightarrow{a} s_1'$, then there is a transition $s_2 \xrightarrow{a} s_2'$ such that $s_1' \mathcal{R} s_2'$;
  - if $s_1 P$, then $s_2 P$.
- A binary relation $\mathcal{R}$ on states is a *ready simulation* if it is a simulation with the property that, whenever $s_1 \mathcal{R} s_2$:
  - if $s_1 \stackrel{a}{\not\longrightarrow}$, then $s_2 \stackrel{a}{\not\longrightarrow}$;
  - if $s_1 \neg P$, then $s_2 \neg P$.
- A *bisimulation* is a symmetric simulation.

We write $s_1 \sqsubseteq_S s_2$ (respectively $s_1 \sqsubseteq_{RS} s_2$) if there is a simulation (respectively a ready simulation) $\mathcal{R}$ with $s_1 \mathcal{R} s_2$. Two states s_1, s_2 are *bisimilar*, written $s_1 \underline{\leftrightarrow} s_2$, if there is a bisimulation relation that relates them. Henceforth the relation $\underline{\leftrightarrow}$ is referred to as *bisimulation equivalence*.

Bisimulation equivalence [155,169] relates two states in an LTS precisely when they have the same branching structure. Simulation (see, e.g., [169]) and ready simulation (also known as $\frac{2}{3}$ bisimulation) [55,136] relax this requirement to different degrees.

We present seven more preorders, which are induced by yet further ways of abstracting away from the full branching structure of LTSs. They are based on (decorated) versions of traces.

DEFINITION 2.4 (*Trace semantics*). Given an LTS, a sequence

$$\varsigma = a_1 \cdots a_n \in \mathsf{Act}^*,$$

for $n \in \mathbb{N}$, is a *trace* of state s_0 if there exist states $s_1, \ldots, s_n$ such that $s_0 \xrightarrow{a_1} s_1 \xrightarrow{a_2} \cdots \xrightarrow{a_n} s_n$ (abbreviated by $s_0 \xrightarrow{\varsigma} s_n$). Moreover, ςP with $\varsigma \in \mathsf{Act}^*$ and $P \in \mathsf{Pred}$ is a *trace* of state s if there exists a state s' such that $s \xrightarrow{\varsigma} s' P$. We write $s \sqsubseteq_T s'$ if the set of traces of s is included in that of s'.

For a state s we define (here, and in what follows, we use the symbol $\triangleq$ to stand for "equals by definition"):

$$\mathsf{initials}(s) \triangleq \{a \in \mathsf{Act} \mid \exists s' \in \mathsf{Proc}\,(s \xrightarrow{a} s')\} \cup \{P \in \mathsf{Pred} \mid s P\}.$$

DEFINITION 2.5 (*Decorated trace semantics*). Assume an LTS, with $\surd$ as one of its predicates.

- *Ready traces*. A sequence $X_0 a_1 X_1 \cdots a_n X_n$, where $X_i \subseteq \mathsf{Act} \cup \mathsf{Pred}$ and $a_i \in \mathsf{Act}$ for $i = 0, \ldots, n$, is a *ready trace* of state s_0 if $s_0 \xrightarrow{a_1} s_1 \xrightarrow{a_2} \cdots \xrightarrow{a_n} s_n$ and $\mathsf{initials}(s_i) = X_i$ for $i = 0, \ldots, n$. We write $s \sqsubseteq_{RT} s'$ if the set of ready traces of s is included in that of s'.
- *Failure traces*. A sequence $X_0 a_1 X_1 \cdots a_n X_n$ or $X_0 a_1 X_1 \cdots a_n X_n P$, where $X_i \subseteq \mathsf{Act} \cup \mathsf{Pred}$, $a_i \in \mathsf{Act}$ for $i = 0, \ldots, n$ and $P \in \mathsf{Pred}$, is a *failure trace* of state s_0 if $s_0 \xrightarrow{a_1}$

$s_1 \xrightarrow{a_2} \cdots \xrightarrow{a_n} s_n$ or $s_0 \xrightarrow{a_1} s_1 \xrightarrow{a_2} \cdots \xrightarrow{a_n} s_n P$, respectively, and $\mathsf{initials}(s_i) \cap X_i = \emptyset$ for $i = 0, \ldots, n$. We write $s \underset{\sim}{\sqsubset}_{FT} s'$ if the set of failure traces of s is included in that of s'.
- *Readies.* A pair (ς, X) with $\varsigma \in \mathsf{Act}^*$ and $X \subseteq \mathsf{Act} \cup \mathsf{Pred}$ is a *ready* of state s_0 if $s_0 \xrightarrow{\varsigma} s_1$ for some state s_1 with $\mathsf{initials}(s_1) = X$. We write $s \underset{\sim}{\sqsubset}_R s'$ if the set of readies of s is included in that of s'.
- *Failures.* A pair (ς, X) with $\varsigma \in \mathsf{Act}^*$ and $X \subseteq \mathsf{Act} \cup \mathsf{Pred}$ is a *failure* of state s_0 if $s_0 \xrightarrow{\varsigma} s_1$ for some state s_1 with $\mathsf{initials}(s_1) \cap X = \emptyset$. We write $s \underset{\sim}{\sqsubset}_F s'$ if the sets of failures and of traces of s are included in that of s'.
- *Completed traces.* $\varsigma \in \mathsf{Act}^*$ is a *completed trace* of state s_0 if $s_0 \xrightarrow{\varsigma} s_1$ for some state s_1 with $\mathsf{initials}(s_1) = \emptyset$. Moreover, ςP with $\varsigma \in \mathsf{Act}^*$ and $P \in \mathsf{Pred}$ is a *completed trace* of state s_0 if $s_0 \xrightarrow{\varsigma} s_1 P$ for some state s_1. We write $s \underset{\sim}{\sqsubset}_{CT} s'$ if the set of completed traces of s is included in that of s'.
- *Accepting traces.* $\varsigma \in \mathsf{Act}^*$ is an *accepting trace* of state s_0 if $s_0 \xrightarrow{\varsigma} s_1 \surd$ for some state s_1. We write $s \underset{\sim}{\sqsubset}_{AT} s'$ if the set of accepting traces of s is included in that of s'.

The decorated trace semantics defined above take predicates into account. However, most of the uses of these semantics in the literature on process theory occur in settings without predicates. Our definition of completed trace preorder agrees with the definition of this notion in, e.g., [143,111], but differs from the one in [94], where it is required that not only the completed traces but also the traces of s are included in that of s'. The notion of an accepting trace is standard in formal language theory (see, e.g., [192]), but has not received widespread treatment in the literature on process theory.

For $\Theta \in \{S, RS, CT, RT, FT, F, R, L, T\}$, the relation $\underset{\sim}{\sqsubset}_\Theta$ is a preorder over states in arbitrary LTSs. Its kernel is denoted by $\simeq_\Theta$; i.e., $s \simeq_\Theta s'$ iff $s \underset{\sim}{\sqsubset}_\Theta s'$ and $s' \underset{\sim}{\sqsubset}_\Theta s$. The following result is a standard one in process theory (cf., e.g., [94]).

PROPOSITION 2.6. *In any LTS (with $\surd$ as one of its predicates),*

$$
\begin{array}{ccccccccccc}
 & & & & \underset{\sim}{\sqsubset}_S & & \underset{\sim}{\sqsubset}_{FT} & & & & \\
 & & & \nearrow & & \nearrow & & \searrow & & & \\
\underline{\leftrightarrow} & \rightarrow & \underset{\sim}{\sqsubset}_{RS} & \rightarrow & \underset{\sim}{\sqsubset}_{RT} & & & & \underset{\sim}{\sqsubset}_F & \rightarrow & \underset{\sim}{\sqsubset}_{CT} \rightarrow \underset{\sim}{\sqsubset}_{AT} \\
 & & & & & \searrow & & \nearrow & & & \\
 & & & & & & \underset{\sim}{\sqsubset}_R & & & &
\end{array}
$$

where a directed edge from one relation to another means that the source of the edge is included in the target. Moreover,
- *$\underset{\sim}{\sqsubset}_S$ is included in $\underset{\sim}{\sqsubset}_{AT}$ and $\underset{\sim}{\sqsubset}_T$, and*
- *$\underset{\sim}{\sqsubset}_F$ is included in $\underset{\sim}{\sqsubset}_T$.*

The same inclusions hold for the kernels of the preorders.

All the inclusions presented in the previous proposition are proper if the LTS under consideration includes, modulo bisimulation equivalence, the finite synchronization trees [152] (see also Section 7.1.3) used in the examples presented in [94].

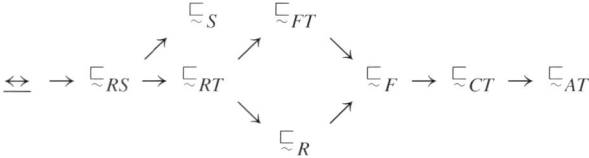

2.3. *Hennessy–Milner logic*

Modal and temporal logics of reactive programs have found considerable use in the theory and practice of concurrency (see, e.g., [76,177,209]). One of the earliest and most influential connections between logics of reactive programs and behavioural relations was given by Hennessy and Milner [123], who introduced a multi-modal logic and showed that it characterized bisimulation equivalence. We limit ourselves to briefly recalling the basic definitions and results on Hennessy–Milner logic. The interested reader is referred to, e.g., [123,208] for more details and motivation. The following definition is standard, apart from the use of atomic propositions to cater for the presence of predicates in LTSs.

DEFINITION 2.7 (*Hennessy–Milner logic*). The set of *HML formulae* is given by the BNF grammar [18]

$$\varphi ::= \mathsf{true} \mid P \mid \neg\varphi \mid \varphi_1 \wedge \varphi_2 \mid \langle a \rangle \varphi$$

where a and P range over Act and Pred, respectively.

Given an LTS, the states s that satisfy HML formula ϕ, written $s \models \phi$, are defined inductively by:

$$\begin{aligned} s &\models \mathsf{true} \\ s \models P &\iff sP \\ s \models \neg\varphi &\iff \text{not } s \models \varphi \\ s \models \varphi_1 \wedge \varphi_2 &\iff s \models \varphi_1 \text{ and } s \models \varphi_2 \\ s \models \langle a \rangle \varphi &\iff s' \models \varphi \text{ for some } s' \text{ such that } s \xrightarrow{a} s'. \end{aligned}$$

Using negation and conjunction in HML, one can define the other standard Boolean connectives. Two states s, s' are considered equivalent with respect to HML, written $s \sim_{\mathrm{HML}} s'$, iff for all HML formulae ϕ: $s \models \phi \iff s' \models \phi$. The following seminal result is due to Hennessy and Milner [123].

THEOREM 2.8. *The equivalence relations* $\underline{\leftrightarrow}$ *and* $\sim_{\mathrm{HML}}$ *coincide over finitely branching LTSs.*

The restriction to finitely branching LTSs in Theorem 2.8 can be dropped if infinitary conjunctions are allowed in the syntax of HML.

2.4. *Term algebras*

This section reviews the basic notions of term algebras that will be needed in this chapter. We start from a countably infinite set Var of variables, ranged over by x, y, z.

DEFINITION 2.9 (*Signature*). A *signature* Σ is a set of *function symbols*, disjoint from Var, together with an *arity* mapping that assigns a natural number $ar(f)$ to each function symbol f. A function symbol of arity zero is called a *constant*, while function symbols of arity one and two are called *unary* and *binary*, respectively.

The arity of a function symbol represents its number of arguments.

DEFINITION 2.10 (*Term*). The set $\mathbb{T}(\Sigma)$ of (*open*) *terms* over a signature Σ, ranged over by t, u, is the least set such that:
- each $x \in \mathsf{Var}$ is a term;
- $f(t_1, \ldots, t_{ar(f)})$ is a term, if f is a function symbol and $t_1, \ldots, t_{ar(f)}$ are terms.

$\mathrm{T}(\Sigma)$ denotes the set of *closed terms* over Σ, i.e., terms that do not contain variables.

For a constant a, the term $a()$ is abbreviated to a. By convention, whenever we write a term-like phrase (e.g., $f(t, u)$), we intend it to be a term (i.e., f is binary).

A *substitution* is a mapping $\sigma : \mathsf{Var} \to \mathbb{T}(\Sigma)$. A substitution is *closed* if it maps each variable to a closed term in $\mathrm{T}(\Sigma)$. A substitution extends to a mapping from terms to terms as usual; the term $\sigma(t)$ is obtained by replacing occurrences of variables x in t by $\sigma(x)$. A *context* $C[x_1, \ldots, x_n]$ denotes an open term in which at most the distinct variables $x_1, \ldots, x_n$ appear. The term $C[t_1, \ldots, t_n]$ is obtained by replacing all occurrences of variables x_i in $C[x_1, \ldots, x_n]$ by t_i, for $i = 1, \ldots, n$.

DEFINITION 2.11 (*Congruence*). Assume a signature Σ. An equivalence relation (respectively preorder) $\mathcal{R}$ over $\mathrm{T}(\Sigma)$ is a *congruence* (respectively *precongruence*) if, for all $f \in \Sigma$,

$$t_i \,\mathcal{R}\, u_i \text{ for } i = 1, \ldots, ar(f) \quad \text{implies} \quad f(t_1, \ldots, t_{ar(f)}) \,\mathcal{R}\, f(u_1, \ldots, u_{ar(f)}).$$

2.5. *Transition system specifications*

In the remainder of this chapter, the set Proc of states will in general consist of the closed terms over some signature. We proceed to introduce the main object of study in the field of SOS, viz. a *transition system specification*, being a collection of inductive proof rules to derive the transitions over the set of closed terms.

DEFINITION 2.12 (*Transition system specification*). Let Σ be a signature, and let t and t' range over $\mathbb{T}(\Sigma)$. A *transition rule* ρ is of the form H/α, with H a set of *positive premises* $t \xrightarrow{a} t'$ and tP, and of *negative premises* $t \stackrel{a}{\not\longrightarrow}$ and $t\neg P$. Moreover, the *conclusion* α is of the form $t \xrightarrow{a} t'$ or tP. The left-hand side of the conclusion is the *source* of ρ, and if the conclusion is of the form $t \xrightarrow{a} t'$, then its right-hand side is the *target* of ρ. A transition rule is closed if it does not contain variables.

A *transition system specification* (TSS) is a set of transition rules. A TSS is *positive* if its transition rules do not contain negative premises.

For the sake of clarity, transition rules will often be displayed in the form $\frac{H}{\alpha}$, and the premises of a transition rule will not always be presented using proper set notation. The first systematic study of TSSs may be found in [198], while the first study of TSSs with negative premises appeared in [54].

We proceed to define when a transition is provable from a TSS. The following notion of a proof from [100] generalizes the standard definition (see, e.g., [111]) by allowing the derivation of closed transition rules. The derivation of a transition α corresponds to the derivation of the closed transition rule H/α with $H = \emptyset$. The case $H \neq \emptyset$ corresponds to the derivation of α under the assumptions in H.

DEFINITION 2.13 (*Literal*). *Positive literals* are transitions $t \xrightarrow{a} t'$ and tP, while *negative literals* are expressions $t \stackrel{a}{\nrightarrow}$ and $t\neg P$, where t and t' range over the collection of closed terms. A *literal* is a positive or negative literal.

DEFINITION 2.14 (*Proof*). Let T be a TSS. A *proof* of a closed transition rule H/α from T is an upwardly branching tree without infinite branches, whose nodes are labelled by literals, where the root is labelled by α, and if K is the set of labels of the nodes directly above a node with label β, then

(1) either $K = \emptyset$ and $\beta \in H$,

(2) or K/β is a closed substitution instance of a transition rule in T.

If a proof of H/α from T exists, then H/α is *provable* from T, notation $T \vdash H/\alpha$.

2.6. *Examples of TSSs*

In this section we present some TSSs from the literature, which will serve as running examples in sections to come. Abundant examples of the systematic use of SOS can be found, e.g., in [27,222] and elsewhere in this handbook. Hartel [114] recently developed a tool environment LATOS for the animation of such TSSs, based on functional programming languages.

2.6.1. *Basic process algebra with empty process.* The signature of Basic Process Algebra with empty process [228], denoted by BPA_ε, consists of the following operators:

- a set Act of constants, representing indivisible behaviour;
- a special constant ε, called *empty process*, representing successful termination;
- a binary operator $+$, called *alternative composition*, where a term $t_1 + t_2$ represents the process that executes either t_1 or t_2;
- a binary operator $\cdot$, called *sequential composition*, where a term $t_1 \cdot t_2$ represents the process that executes first t_1 and then t_2.

So the BNF grammar for BPA_ε is (with $a \in \mathsf{Act}$):

$$t ::= a \mid \varepsilon \mid t_1 + t_2 \mid t_1 \cdot t_2.$$

The intuition above for the operators in BPA_ε is formalized by the transition rules in Table 1 from [28], which constitute the TSS for BPA_ε. This TSS defines transitions $t \xrightarrow{a} t'$ to

Table 1
Transition rules for BPA$_\varepsilon$

$$\frac{}{a \xrightarrow{a} \varepsilon} \qquad \frac{}{\varepsilon\surd}$$

$$\frac{x\surd}{x+y\surd} \qquad \frac{x \xrightarrow{a} x'}{x+y \xrightarrow{a} x'} \qquad \frac{y\surd}{x+y\surd} \qquad \frac{y \xrightarrow{a} y'}{x+y \xrightarrow{a} y'}$$

$$\frac{x\surd \quad y\surd}{x\cdot y\surd} \qquad \frac{x\surd \quad y \xrightarrow{a} y'}{x\cdot y \xrightarrow{a} y'} \qquad \frac{x \xrightarrow{a} x'}{x\cdot y \xrightarrow{a} x'\cdot y}$$

Table 2
Transition rules for the priority operator

$$\frac{x\surd}{\theta(x)\surd} \qquad \frac{x \xrightarrow{a} x' \quad x \not\xrightarrow{b} \text{ for } a<b}{\theta(x) \xrightarrow{a} \theta(x')}$$

Table 3
Transition rules for discrete time

$$\frac{}{\sigma_d(x) \xrightarrow{\sigma} x} \qquad \frac{x\surd \quad y \xrightarrow{\sigma} y'}{x\cdot y \xrightarrow{\sigma} y'} \qquad \frac{x \xrightarrow{\sigma} x'}{x\cdot y \xrightarrow{\sigma} x'\cdot y}$$

$$\frac{x \xrightarrow{\sigma} x' \quad y \xrightarrow{\sigma} y'}{x+y \xrightarrow{\sigma} x'+y'} \qquad \frac{x \xrightarrow{\sigma} x' \quad y \not\xrightarrow{\sigma}}{x+y \xrightarrow{\sigma} x'} \qquad \frac{y \xrightarrow{\sigma} y' \quad x \not\xrightarrow{\sigma}}{x+y \xrightarrow{\sigma} y'}$$

express that term t can evolve into term t' by the execution of action $a \in \mathsf{Act}$, and transitions $t\surd$ to express that term t can terminate successfully. The variables x, x', y, and y' in the transition rules range over the collection of closed terms, while the a ranges over Act.

2.6.2. *Priorities.* The language BPA$_{\varepsilon\theta}$ is obtained by adding the priority operator θ from [24] to BPA$_\varepsilon$. This function symbol assumes a partial order $<$ on Act. Intuitively, the process $\theta(t)$ is obtained by eliminating all transitions $s \xrightarrow{a} s'$ from the process t for which there is a transition $s \xrightarrow{b} s''$ with $a<b$. For example, if $a<b$ then $\theta(a+b)$ can execute the action b but not the action a. The semantics of the priority operator is captured by the transition rules in Table 2. The TSS for BPA$_{\varepsilon\theta}$ consists of the transition rules in Tables 1 and 2.

2.6.3. *Discrete time.* Our final example is the TSS for an extension of BPA$_\varepsilon$ with relative discrete time, denoted by BPA$_\varepsilon^{\mathrm{dt}}$ [23]. Time progresses in distinct time steps, where a transition $t \xrightarrow{\sigma} t'$ denotes passing to the next time slice. The syntax of BPA$_\varepsilon^{\mathrm{dt}}$ consists of the operators from BPA$_\varepsilon$ together with a unary operator σ_d to represent a delay of one time unit. That is, a term $\sigma_d(t)$ can execute all transitions of t delayed by one time step. A term $t+t'$ can evolve into the next time slice if t or t' can evolve into the next time slice. The transition rules dealing with time steps are presented in Table 3. The TSS for BPA$_\varepsilon^{\mathrm{dt}}$ consists of the transition rules in Tables 1 and 3.

3. The meaning of TSSs

A positive TSS specifies an LTS in a straightforward way as the set of all provable transitions (cf. Definition 2.14). However, as Groote [107,108] pointed out, it is much less trivial to associate an LTS with a TSS containing negative premises. Several solutions were investigated in [56,57,107,108], mostly originating from logic programming. This section presents an overview of how to associate one or more LTSs with a TSS. Our presentation here is heavily based upon the excellent systematic analysis of the meaning of TSSs by van Glabbeek [98,100], and we heartily refer the reader to *op. cit.* for more details.

To see that it is sometimes unclear what the meaning of a TSS with negative premises is, consider the TSS T_1 consisting of the constant a and transition rules

$$T_1 \qquad \boxed{\frac{a \neg P_2}{a\, P_1} \quad \frac{a \neg P_1}{a\, P_2}}$$

T_1 can be regarded as an example of a TSS that does not specify a well-defined LTS. The above example suggests that some TSSs may indeed be meaningless. Hence there are two questions to answer:

Which TSSs are meaningful, (1)

and which LTSs can be associated with them? (2)

The papers [98,100] present several possible answers to these questions, each consisting of a class of TSSs and a mapping from this class to LTSs. Two such answers are consistent if they agree upon which LTS to associate with a TSS in the intersection of their domains. Answer S_2 extends answer S_1 if the class of meaningful TSSs according to S_2 extends that of S_1, and the two are consistent.

The collection of answers proposed by van Glabbeek in *op. cit.* can be grouped into those with a model-theoretic and those with a proof-theoretic flavour. These we now proceed to present.

3.1. *Model-theoretic answers*

ANSWER 1. A first answer to questions (1) and (2) is to take the class of positive TSSs as the meaningful ones, and associate with each positive TSS the LTS consisting of the provable transitions.

Since negative premises make it possible to give a clean description of important constructs found in programming and specification languages, the above answer is not really satisfactory. More general answers to questions (1) and (2) have been proposed in the literature. Before reviewing them, we recall two criteria from [55,56] that can be imposed on reasonable answers.

DEFINITION 3.1 (*Entailment*). For an LTS L and a set of literals H, we write $L \models H$ if:
- $\alpha \in L$ for all positive literals α in H;
- $t \xrightarrow{a} t' \notin L$ for all negative literals $t \stackrel{a}{\nrightarrow}$ in H and all closed terms t';
- $tP \notin L$ for all negative literals $t\neg P$ in H.

DEFINITION 3.2 (*Supported model*). Let T be a *TSS* and L an LTS.
- L is a *model* of T if $\alpha \in L$ whenever there is a closed substitution instance H/α of a transition rule in T with $L \models H$.
- L is *supported* by T if whenever $\alpha \in L$ there is a closed substitution instance H/α of a transition rule in T with $L \models H$.

The first requirement, of being a model, says that L contains all transitions for which T offers a justification. The second requirement, of being supported, says that L only contains transitions for which T offers a justification. Note that the LTS containing all possible transitions is a model of any TSS, while the LTS containing no transitions is supported by any TSS.

The following result is standard, and has its roots in the classic theory of inductive definitions.

PROPOSITION 3.3. *Let T be a positive TSS and L the set of transitions provable from T. Then L is a supported model of T. Moreover, L is the least model of T.*

Starting from Proposition 3.3, there are two ways to generalize Answer 1 to TSSs with negative premises.

ANSWER 2. A TSS is meaningful iff it has a least model.

ANSWER 3. A TSS is meaningful iff it has a least supported model.

Note that, in general, no unique least (supported) model may exist. A counter-example is given by the TSS T_1, which has two least models, namely $\{aP_1\}$ and $\{aP_2\}$, both of which are supported. Answers 2 and 3 are incomparable. For example, the TSS T_2 below has $\{aP_1\}$ as its least model, but no supported models. On the other hand, the TSS T_3 has two least models, namely $\{aP_1\}$ and $\{aP_2\}$, of which only the first one is supported, and this is its least supported model.

$$T_2 \quad \boxed{\frac{a\neg P_1}{aP_1}} \qquad\qquad T_3 \quad \boxed{\frac{a\neg P_2}{aP_1}}$$

Answers 2 and 3 both extend Answer 1, but they are inconsistent with each other. For example, the TSS T_4 below has a least model $\{aP_1\}$ and a least supported model $\{aP_1, aP_2\}$.

$$T_4 \quad \boxed{\frac{a\neg P_1}{aP_1} \quad \frac{aP_2}{aP_1} \quad \frac{aP_2}{aP_2}}$$

In [54,55] the following answer was proposed.

ANSWER 4. A TSS is meaningful iff it has a unique supported model.

The positive TSS T_5 below has two supported models, viz. $\emptyset$ and $\{a P_1\}$, so Answer 4 does not extend Answer 1.

$$T_5 \quad \boxed{\frac{a P_1}{a P_1}}$$

For the *GSOS languages* considered in *op. cit.* (cf. Section 5.4), Answer 4 coincides with all acceptable answers mentioned in this section. Note, however, that the least supported model of T_4 is also its unique supported model. This seems to entail that Answer 4 is not satisfactory for TSSs in general.

Fages [78] proposed a strengthening of the notion of support, in the setting of logic programming. Being supported means that a transition may only be present if there is a non-empty proof of its presence, starting from transitions that are also present. These premises in the proof may include the transition under derivation, thereby allowing loops, as in the case of T_4. The notion of a *well-supported* model is based on the idea that the absence of a transition may be assumed a priori, provided that this assumption is consistent, but the presence of a transition needs to be proven, in the sense of Definition 2.14, building upon a set of assumptions that only contains negative literals.

DEFINITION 3.4 (*Well-supported model*). An LTS L is a *well-supported model* of a TSS T if it is a model of T and for each transition α in L, T proves a closed transition rule N/α where N only contains negative literals and $L \models N$.

A *stable* model, developed by Gelfond and Lifschitz [90] in the area of logic programming, and adapted to TSSs in [56,57], only allows transitions that are well-supported.

DEFINITION 3.5 (*Stable model*). An LTS L is a *stable model* of a TSS T if a transition α is in L iff T proves a closed transition rule N/α where N contains only negative literals and $L \models N$.

An LTS is a stable model of a TSS T iff it is a well-supported model of T [98,100].

ANSWER 5. A TSS is meaningful iff it has a unique stable model.

Answer 5 extends Answer 1, and it improves upon Answers 3 and 4 by rejecting the TSS T_4 as meaningless. It also improves upon Answer 2 by rejecting the TSS T_2 (whose least model was not supported). Furthermore, Answer 5 gives meaning to perfectly acceptable TSSs that could not be handled by Answers 1–4. As an example consider the TSS T_6 below.

$$T_6 \quad \boxed{\frac{a P_1}{a P_1} \quad \frac{a \neg P_1}{a P_2}}$$

Since there is no compelling way to obtain aP_1, we would expect that aP_1 does not hold, and consequently that aP_2 holds. Indeed, $\{aP_2\}$ is the unique stable model of this TSS. However, T_6 has two least models, both of which are supported, namely $\{aP_1\}$ and $\{aP_2\}$.

A *three-valued* stable model, introduced by Przymusinski [181] in logic programming, partitions the set of transitions into three disjoint subsets: the set C of transitions that are *certainly* true, the set U of transitions for which it is *unknown* whether or not they are true, and the set F of remaining transitions that are *false*.

DEFINITION 3.6 (*Three-valued stable model*). A disjoint pair of sets of transitions $\langle C, U\rangle$ constitutes a *three-valued stable model* for a TSS T if:
- a transition α is in C iff T proves a closed transition rule N/α where N contains only negative literals and $C \cup U \models N$;
- a transition α is in $C \cup U$ iff T proves a closed transition rule N/α where N contains only negative literals and $C \models N$.

Each TSS has one or more three-valued stable models. For example, the TSS T_1 has $\langle\{aP_1\}, \emptyset\rangle$, $\langle\{aP_2\}, \emptyset\rangle$, and $\langle\emptyset, \{aP_1, aP_2\}\rangle$ as its three-valued stable models. Each TSS T affords an *(information-)least* three-valued stable model $\langle C, U\rangle$, in the sense that the set U is maximal. Przymusinski [181] showed that this least three-valued stable model coincides with the so-called well-founded model that was introduced by van Gelder, Ross, and Schlipf [88,89] in logic programming.

ANSWER 6. A TSS is meaningful iff its least three-valued stable model does not contain unknown transitions. The associated LTS consists of the true transitions in this three-valued stable model.

Answer 6 extends Answer 1 and is extended by Answer 5, but it is inconsistent with Answers 2–4. In particular, the TSSs T_1, T_2, and T_4 are outside its domain, while it associates $\{aP_1\}$ to T_3, $\emptyset$ to T_5, and $\{aP_2\}$ to T_6. In Section 5, Answer 6 will stand us in good stead, as it is used in the formulation of congruence results in the presence of negative premises (cf. Theorems 5.3, 5.49, and 5.53), where Answers 1–5 would all be unsatisfactory.

3.2. *Proof-theoretic answers*

Note for the Reader. In this section only we extend the notion of negative literals to expressions of the form $t \stackrel{a}{\nrightarrow} t'$. Intuitively, this expression denotes that term t cannot evolve to term t' by the execution of action a.

This section reviews possible answers to questions (1) and (2) based on a generalization of the concept of proof. In [100], van Glabbeek proposed two generalizations of the concept of a proof in Definition 2.14, to enable the derivation of negative literals. These two generalizations are based on the notions of supported model (Definition 3.2) and well-supported model (Definition 3.4), respectively.

DEFINITION 3.7 (*Denying literal*). The following pairs of literals *deny* each other:

- $t \xrightarrow{a} t'$ and $t \stackrel{a}{\nrightarrow} t'$;
- $t \xrightarrow{a} t'$ and $t \stackrel{a}{\nrightarrow}$;
- tP and $t\neg P$.

DEFINITION 3.8 (*Supported proof*). A *supported proof* of a literal α from a TSS T is like a proof (see Definition 2.14), but with one extra clause:

(3) β is negative, and for each closed substitution instance H'/γ of a transition rule in T such that γ denies β, a literal in H' denies one in K.

We write $T \vdash_s \alpha$ if a supported proof of α from T exists.

DEFINITION 3.9 (*Well-supported proof*). A *well-supported proof* of a literal α from a TSS T is like a proof (Definition 2.14), but with one extra clause:

(3) β is negative, and for each set N of negative literals such that $T \vdash N/\gamma$ for γ a literal denying β, a literal in N denies one in K.

We write $T \vdash_{ws} \alpha$ if a well-supported proof of α from T exists.

Clause (3) in Definition 3.9 allows one to infer $t \stackrel{a}{\nrightarrow} t'$ or $t\neg P$ whenever it is manifestly impossible to infer $t \xrightarrow{a} t'$ or tP, respectively. Clause (3) in Definition 3.8 allows such inferences only if the impossibility to derive $t \xrightarrow{a} t'$ or tP can be detected by examining all possible proofs that consist of one step only. As a consequence, for each TSS, $\vdash_s$ is included in $\vdash_{ws}$. The following results stem from [100].

PROPOSITION 3.10. *For each TSS, the induced relation $\vdash_{ws}$ does not contain denying literals.*

PROPOSITION 3.11. *For any TSS T and literal α:*

(1) *$T \vdash_s \alpha$ implies $L \models \alpha$ for each supported model L of T;*
(2) *$T \vdash_{ws} \alpha$ implies $L \models \alpha$ for each well-supported model L of T.*

Following [100], we now introduce the concept of a *complete* TSS, in which every transition is either provable or refutable.

DEFINITION 3.12 (*Completeness*). A TSS T is x-*complete* ($x \in \{s, ws\}$) if for any transition $t \xrightarrow{a} t'$ (respectively tP) either $T \vdash_x t \xrightarrow{a} t'$ (respectively $T \vdash_x tP$) or $T \vdash_x t \stackrel{a}{\nrightarrow} t'$ (respectively $T \vdash_x t\neg P$).

ANSWER 7. A TSS is meaningful iff it is s-complete. The associated LTS consists of the s-provable transitions.

ANSWER 8. A TSS is meaningful iff it is ws-complete. The associated LTS consists of the ws-provable transitions.

From now on, by 'complete' we shall mean 'ws-complete'.

A TSS is complete iff its least three-valued stable model does not contain any unknown transitions (see [98]), so Answer 6 agrees with Answer 8. Moreover, Answer 8 extends

Answer 7. In [56], an example in the area of process theory was given (viz. the modelling of a priority operator in basic process algebra with silent step) that can be handled by Answer 8 but not by Answer 7, showing that the full generality of Answer 8 can be useful in applications.

We proceed to show how to associate an LTS with any TSS, using the concept of a well-supported proof. As illustrated by the TSSs T_1 and T_2, such an LTS cannot always be a supported model. Since being a model is a basic requirement, van Glabbeek [100] proposed a universal answer that gives up the requirement of supportedness. Let us examine T_1. Since the associated LTS should be a model, it must contain either aP_1 or aP_2. By symmetry, the associated LTS should include both transitions. As there is no reason to include any more transitions, the LTS associated with T_1 should be $\{aP_1, aP_2\}$. These considerations lead to the following proposal.

ANSWER 9. Any TSS is meaningful. The associated LTS consists of the transitions for which none of the denying negative literals are *ws*-provable.

Answer 9 is inspired by the observation in [100] that for each TSS, the set of transitions for which none of the denying negative literals are *ws*-provable constitutes a model. Answer 9 extends Answer 8, but it is inconsistent with Answers 2–5. Answer 9 associates the LTS $\{aP_1, aP_2\}$ with T_1, and the LTS $\{aP_1\}$ with T_2 and T_4.

3.3. *Answers based on stratification*

Finally, we review two methods to assign meaning to TSSs based on the technique of *(local) stratification*, as proposed in the setting of logic programming by Przymusinski [180]. This technique was adapted to TSSs in [107,108].

DEFINITION 3.13 (*Stratification*). A mapping S from transitions to ordinal numbers is a *stratification* of a TSS T if for every transition rule H/α in T and every closed substitution σ:

- for positive premises β in H, $S(\sigma(\beta)) \leqslant S(\sigma(\alpha))$; and
- for negative premises $t \overset{a}{\nrightarrow}$ and $t\neg P$ in H, $S(\sigma(t) \overset{a}{\longrightarrow} t') < S(\sigma(\alpha))$ for all closed terms t' and $S(\sigma(t)P) < S(\sigma(\alpha))$, respectively.

A stratification is *strict* if $S(\sigma(\beta)) < S(\sigma(\alpha))$ also holds for all the positive premises β in H. A TSS with a (strict) stratification is (*strictly*) *stratifiable*.

In a stratifiable TSS no transition depends negatively on itself. An LTS associated with such a TSS may be built one stratum of transitions with the same S-value at a time. A transition with S-value zero is present only if it is provable in the sense of Definition 2.14, and as soon as the validity of all transitions with S-value no greater than κ is known for some ordinal number κ, the validity of closed instantiations of negative premises that could occur in a proof of a transition with S-value $\kappa + 1$ is known, which determines the validity of those transitions.

Let T be a TSS with a stratification S. The stratum L_κ of transitions, for an ordinal number κ, is defined thus (using ordinal induction):

$$\alpha \in L_\kappa \quad \text{iff} \quad S(\alpha) = \kappa \text{ and } T \text{ proves a closed transition rule } H/\alpha \text{ with } \bigcup_{\mu<\kappa} L_\mu \models H.$$

Similarly, for a TSS T with a strict stratification S, the stratum M_κ of transitions, for an ordinal number κ, is defined thus (using ordinal induction):

$$\alpha \in M_\kappa \quad \text{iff} \quad S(\alpha) = \kappa \text{ and there is a closed substitution instance } H/\alpha \text{ of a transition rule in } T \text{ with } \bigcup_{\mu<\kappa} M_\mu \models H.$$

Groote [107,108] proved that the sets $\bigcup_\kappa L_\kappa$ and $\bigcup_\kappa M_\kappa$ are independent of the chosen (strict) stratification. This justifies the following answers to questions (1) and (2).

ANSWER 10. A TSS is meaningful iff it is stratifiable. The associated LTS is $\bigcup_\kappa L_\kappa$.

ANSWER 11. A TSS is meaningful iff it is strictly stratifiable. The associated LTS is $\bigcup_\kappa M_\kappa$.

Answer 10 extends Answer 1 and is extended by Answer 8. Answer 11 is extended by Answers 7 and 10.

3.4. *Evaluation of the answers*

We have presented several possible answers to the questions of which TSSs are meaningful and which LTSs are associated with them.

Answer 1 (positive) is the classical interpretation of TSSs without negative premises, and Answers 2 (least model) and 3 (least supported model) are two straightforward generalizations. Answer 4 (unique supported model) stems from [54], where it was used to ascertain that TSSs in GSOS format (cf. Section 5.4) are meaningful. The TSS T_4, however, shows that Answer 4 yields counter-intuitive results in general. Fortunately, TSSs in GSOS format are even strictly stratifiable, which is one of the most restrictive criteria for meaningful TSSs considered. For GSOS languages with recursion, however, it is no longer straightforward to find an associated LTS (see, e.g., [55]). A solution for this, involving a special divergence predicate, will be discussed in Section 5.4.6.

Answer 5 (unique stable) is generally considered to be the most general acceptable answer available. Answer 8 (complete) is the most general answer without undesirable properties. Answer 8 is based on a concept of provability incorporating the notion of *negation as failure* of Clark [63]. Answer 6 (no unknown transitions) agrees with Answer 8; i.e., a TSS is complete iff its least three-valued stable model does not contain unknown transitions. Answer 7 (complete with support) only yields unique supported models. Moreover, it is based on a notion of provability that is somewhat simpler to apply, and only incorporates the notion of *negation as finite failure* [63].

Answer 9 (irrefutable), which gives a meaning to each TSS, has the disadvantage that it sometimes yields unstable models, and even unsupported models. A good example from process algebra of a TSS without supported models is BPA with the priority operator, unguarded recursion, and renaming, as defined in [107,108]. Although Answer 9 gives a meaning to this TSS, it appears rather arbitrary and not very useful. In particular, recursively defined processes do not satisfy their defining equations – a highly undesirable feature by all accounts.

Answer 10 (stratification) is perhaps the best known answer in logic programming. A variant that only allows TSSs with a unique supported model is Answer 11 (strict stratification). Answers 10 and 11 are of practical importance, because they are extended by Answer 8. Thus, giving a (strict) stratification is a useful tool for showing that a TSS is complete, and this technique is applied in several examples in the remainder of this chapter.

3.5. *Applications*

We show that the three TSSs from Section 2.6 are complete, using a stratification. We use the fact that Answer 8 extends Answers 1 and 10.

BPA with empty process. The TSS for BPA_ε is positive.

Priorities. The TSS for $\mathrm{BPA}_{\varepsilon\theta}$ is complete, which can be seen by giving a suitable stratification S, counting the number of occurrences of the priority operator in the left-hand side of a transition. That is, if the closed term t contains n occurrences of θ, then $S(t \xrightarrow{a} t') = n$ and $S(t\surd) = n$. Consider for instance the second transition rule in Table 2:

$$\frac{x \xrightarrow{a} x' \quad x \stackrel{b}{\not\longrightarrow} \text{ for } a < b}{\theta(x) \xrightarrow{a} \theta(x')}.$$

Clearly $S(t \xrightarrow{a} t') < S(\theta(t) \xrightarrow{a} \theta(t'))$ and $S(t \xrightarrow{b} u) < S(\theta(t) \xrightarrow{a} \theta(t'))$ for all closed terms t, t', and u, because $\theta(t)$ contains one more occurrence of the priority operator than t. In a similar fashion it can be verified for the other transition rules of $\mathrm{BPA}_{\varepsilon\theta}$ that S is a stratification. Hence, the TSS for $\mathrm{BPA}_{\varepsilon\theta}$ is complete.

Discrete time. The TSS for $\mathrm{BPA}_\varepsilon^{\mathrm{dt}}$ is complete, which can be seen by giving a suitable stratification S, counting the occurrences of alternative composition on the left-hand side of a timed transition. That is, if the closed term t contains n occurrences of $+$, then $S(t \xrightarrow{\sigma} t') = n$. Moreover, $S(t \xrightarrow{a} t') = 0$ for $a \in \mathsf{Act}$. Consider for instance the last transition rule in Table 3:

$$\frac{y \xrightarrow{\sigma} y' \quad x \stackrel{\sigma}{\not\longrightarrow}}{x + y \xrightarrow{\sigma} y'}.$$

Clearly $S(u \xrightarrow{\sigma} u') < S(t + u \xrightarrow{\sigma} u')$ and $S(t \xrightarrow{\sigma} t') < S(t + u \xrightarrow{\sigma} u')$ for all closed terms t, t', u, and u', because $t + u$ contains more occurrences of the alternative composition than t and u. In a similar fashion it can be verified for the other transition rules of $\mathrm{BPA}_\varepsilon^{\mathrm{dt}}$ that S is a stratification. Hence, the TSS for $\mathrm{BPA}_\varepsilon^{\mathrm{dt}}$ is complete.

4. Conservative extension

Over and over again, process calculi such as CCS [157], CSP [186], and ACP [28] have been extended with new features, and the original TSSs, which provide the semantics for these process algebras, were extended with transition rules to describe these features; see, e.g., [27] for a systematic approach. A question that arises naturally is whether or not the LTSs associated with the original and with the extended TSS contain the same transitions $t \xrightarrow{a} t'$ and tP for closed terms t in the original domain. Usually it is desirable that an extension is operationally conservative, meaning that the provable transitions for an original term are the same both in the original and in the extended TSS.

Groote and Vaandrager [111, Theorem 7.6] proposed syntactic restrictions on a TSS, which automatically yield that an extension of this TSS with transition rules that contain fresh function symbols in their sources is operationally conservative (cf. the notion of a disjoint extension from [8] in Definition 5.31). Bol and Groote [57,108] supplied this conservative extension format with negative premises. Verhoef [226] showed that, under certain conditions, a transition rule in the extension can be allowed to have an original term as its source. D'Argenio and Verhoef [68,69] formulated a generalization in the context of inequational specifications. Fokkink and Verhoef [85] relaxed the syntactic restrictions on the original TSS, and lifted the operational conservative extension result to higher-order languages (see Section 6.4).

Operational conservative extension seems such a natural notion that in the literature this property is often a hidden assumption: its formulation and proof are omitted without justification. For example, this happens in the design of process algebras, and in applications of a strategy to prove ω-completeness mentioned in Section 4.3.3. Paying attention to operational conservative extension not only leads to more accurate contemplations on concurrency theory, but is also beneficial in other respects. Namely, operational conservative extension can be applied to obtain results in process algebra that are much harder to obtain using more classical term rewriting approaches or customized techniques.

The organization of this section is as follows. Section 4.1 presents syntactic constraints to ensure that an extension of a TSS is operationally conservative. Section 4.2 studies the relation between the three-valued stable models of a TSS and of its operational conservative extension. Sections 4.3 and 4.4 show how operational conservative extension can be applied to derive useful properties concerning axiomatizations and term rewriting systems.

As related work we mention that Mosses [164] introduced the concept of Modular SOS, in which transition labels are the arrows of a category, and adjacent labels in computations are required to be composable. Intuitively, transition labels are then used to represent information processing steps. Mosses argues that Modular SOS ensures a high degree of modularity: when one extends or changes the described language, its Modular SOS can be extended or changed accordingly, without reformulation. Degano and Priami [73,179] introduced the concept of Enhanced Operational Semantics, in which transitions are labelled by encodings of their proofs. Enhanced Operational Semantics supports parametricity, as it enables to express different views on the same system that are consistent with one another, and to retrieve these views from a single concrete specification.

4.1. *Operational conservative extension*

Often one wants to add new operators and rules to a given TSS. Therefore, a natural operation on TSSs is to take their componentwise union. The following definition stems from [111].

DEFINITION 4.1 (*Sum of TSSs*). Let T_0 and T_1 be TSSs whose signatures Σ_0 and Σ_1 agree on the arity of the function symbols in their intersection. We write $\Sigma_0 \oplus \Sigma_1$ for the union of Σ_0 and Σ_1.

The *sum* of T_0 and T_1, notation $T_0 \oplus T_1$, is the TSS over signature $\Sigma_0 \oplus \Sigma_1$ containing the rules in T_0 and T_1.

An *operational conservative extension* requires that an original TSS and its extension prove exactly the same closed transition rules that have only negative premises and an original closed term as their source. This notion of an operational conservative extension is related to an equivalence notion for TSSs in [83,98] (see also Theorem 5.6): two TSSs are equivalent if they prove exactly the same closed transition rules that have only negative premises. Such a definition is inspired by the notion of a well-supported proof in Definition 3.9.

DEFINITION 4.2 (*Operational conservative extension*). A TSS $T_0 \oplus T_1$ is an *operational conservative extension* of TSS T_0 if for each closed transition rule N/α such that:
- N contains only negative literals;
- the left-hand side of α is in $\mathrm{T}(\Sigma_0)$;
- $T_0 \oplus T_1 \vdash N/\alpha$;

we have that $T_0 \vdash N/\alpha$.

We proceed to define the notion of a *source-dependent* variable [85,96], which will be an important ingredient of a rule format to ensure that an extension of a TSS is operationally conservative (see Theorem 4.4). In order to conclude that an extended TSS is operationally conservative over the original TSS, we need to know that the variables in the original transition rules are source-dependent. In the literature this criterion is sometimes neglected. For example, in [165] an extended TSS is considered in which each transition rule in the extension contains a fresh operator in its source, and from this fact alone it is concluded that the extension is operationally conservative. In general, however, this characteristic is not sufficient, as is shown in the next example.

EXAMPLE. Let a and b be constants. Consider the TSS over signature $\{a\}$ that consists of the transition rule xP/aP. Extend this TSS with the TSS over signature $\{b\}$ that consists of the transition rule $\emptyset/bP$, which contains the fresh constant b in its source. The transition aP can be proven in the extended TSS, but not in the original one, so this extension is not operationally conservative.

DEFINITION 4.3 (*Source-dependency*). The *source-dependent* variables in a transition rule ρ are defined inductively as follows:
- all variables in the source of ρ are source-dependent;
- if $t \xrightarrow{a} t'$ is a premise of ρ and all variables in t are source-dependent, then all variables in t' are source-dependent.

A transition rule is *source-dependent* if all its variables are.

Note that the transition rule xP/aP from the example above is not source-dependent, because its variable x is not.

Theorem 4.4 below, which stems from [85], formulates sufficient criteria for a TSS $T_0 \oplus T_1$ to be an operational conservative extension of TSS T_0. We say that a term in $\mathbb{T}(\Sigma_0 \oplus \Sigma_1)$ is *fresh* if it contains a function symbol from $\Sigma_1 \backslash \Sigma_0$. Similarly, an action or predicate symbol in T_1 is *fresh* if it does not occur in T_0.

THEOREM 4.4. *Let T_0 and T_1 be TSSs over signatures Σ_0 and Σ_1, respectively. Under the following conditions, $T_0 \oplus T_1$ is an operational conservative extension of T_0.*

(1) *Each $\rho \in T_0$ is source-dependent.*

(2) *For each $\rho \in T_1$,*
- *either the source of ρ is fresh,*
- *or ρ has a premise of the form $t \xrightarrow{a} t'$ or tP, where:*
  - $t \in \mathbb{T}(\Sigma_0)$;
  - *all variables in t occur in the source of ρ;*
  - *t', a, or P is fresh.*

We apply Theorem 4.4 to our running examples from Section 2.6.

BPA with empty process. The transition rules for BPA_ε are all source-dependent. For example, consider the third transition rule for sequential composition in Table 1:

$$\frac{x \xrightarrow{a} x'}{x \cdot y \xrightarrow{a} x' \cdot y}.$$

The variables x and y are source-dependent, because they occur in the source. Moreover, since x is source-dependent, the premise $x \xrightarrow{a} x'$ ensures that x' is source-dependent. Since the three variables x, x', and y in this transition rule are source-dependent, the transition rule is source-dependent.

BPA with empty process and silent step. The process algebra $\text{BPA}_{\varepsilon\tau}$ is obtained by extending the syntax of BPA_ε with a fresh constant τ called the silent step (see Section 5.5 for details on the intuition behind this constant). The TSS for $\mathrm{BPA}_{\varepsilon\tau}$ is the TSS for $\mathrm{BPA}_{\varepsilon}$ in Table 1, with the proviso that a ranges over $\mathsf{Act} \cup \{\tau\}$. We make the following observations concerning the extra transition rules in the TSS for $\mathrm{BPA}_{\varepsilon\tau}$:

- the source of the transition rule $\emptyset/\tau \xrightarrow{\tau} \varepsilon$ for the silent step contains the fresh constant τ;
- each transition rule for alternative or sequential composition with τ-transitions, such as

$$\frac{x \xrightarrow{\tau} x'}{x + y \xrightarrow{\tau} x'}$$

  contains a premise with the fresh relation symbol $\xrightarrow{\tau}$ and with as left-hand side a variable from the source.

Hence, since the transition rules for $\mathrm{BPA}_{\varepsilon}$ are source-dependent, Theorem 4.4 implies that $\mathrm{BPA}_{\varepsilon\tau}$ is an operational conservative extension of $\mathrm{BPA}_{\varepsilon}$.

Priorities. The two transition rules for the priority operator in Table 2 contain the fresh function symbol θ in their sources. Hence, since the transition rules for $\mathrm{BPA}_{\varepsilon}$ are source-dependent, Theorem 4.4 implies that $\mathrm{BPA}_{\varepsilon\theta}$ is an operational conservative extension of $\mathrm{BPA}_{\varepsilon}$.

Discrete time. As for the TSS for $\mathrm{BPA}_{\varepsilon}^{\mathrm{dt}}$ in Table 3, we make the following observations. First, the transition rule for the delay operator contains the fresh operator σ_d in its source. Second, the transition rules for sequential composition and the three transition rules for alternative composition (which do not have a fresh operator in their sources) all contain the premise $x \xrightarrow{\sigma} x'$ or $y \xrightarrow{\sigma} y'$, where the relation symbol $\xrightarrow{\sigma}$ is fresh and the variable on the left-hand side occurs in the source. Hence, since the transition rules for $\mathrm{BPA}_{\varepsilon}$ are source-dependent, Theorem 4.4 implies that $\mathrm{BPA}_{\varepsilon}^{\mathrm{dt}}$ is an operational conservative extension of $\mathrm{BPA}_{\varepsilon}$.

4.2. *Implications for three-valued stable models*

In [85] it was noted that the operational conservative extension notion as formulated in Definition 4.2 implies a conservativity property for three-valued stable models (cf. Definition 3.6). If an extended TSS is operationally conservative over the original TSS, in the sense of Definition 4.2, and if a three-valued stable model of the extended TSS is restricted to those transitions that have an original term as left-hand side, then the result is a three-valued stable model of the original TSS.

PROPOSITION 4.5. *Let $T_0 \oplus T_1$ be an operational conservative extension of T_0. If $\langle C, U\rangle$ is a three-valued stable model of $T_0 \oplus T_1$, then*

$$C' \triangleq \{\alpha \in C \mid \textit{the left-hand side of } \alpha \textit{ is in } \mathrm{T}(\Sigma_0)\},$$
$$U' \triangleq \{\alpha \in U \mid \textit{the left-hand side of } \alpha \textit{ is in } \mathrm{T}(\Sigma_0)\}$$

is a three-valued stable model of T_0.

The converse of Proposition 4.5 also holds, in the following sense. If an extended TSS is operationally conservative over the original TSS, then each three-valued stable model of the original TSS can be obtained by restricting some three-valued stable model of the extended TSS to those transitions that have an original term as left-hand side.

PROPOSITION 4.6. *Let $T_0 \oplus T_1$ be an operational conservative extension of T_0. If $\langle C, U\rangle$ is a three-valued stable model of T_0, then there exists a three-valued stable model $\langle C', U'\rangle$ of $T_0 \oplus T_1$ such that*

$$C \triangleq \{\alpha \in C' \mid \textit{the left-hand side of } \alpha \textit{ is in } \mathrm{T}(\Sigma_0)\},$$
$$U \triangleq \{\alpha \in U' \mid \textit{the left-hand side of } \alpha \textit{ is in } \mathrm{T}(\Sigma_0)\}.$$

COROLLARY 4.7. *Let $T_0 \oplus T_1$ be an operational conservative extension of T_0. If $\langle C, U\rangle$ is the least three-valued stable model of $T_0 \oplus T_1$, then*

$$C' \triangleq \{\alpha \in C \mid \textit{the left-hand side of } \alpha \textit{ is in } \mathrm{T}(\Sigma_0)\},$$
$$U' \triangleq \{\alpha \in U \mid \textit{the left-hand side of } \alpha \textit{ is in } \mathrm{T}(\Sigma_0)\}$$

is the least three-valued stable model of T_0.

It is easy to see that Proposition 4.5 also holds for stable models (cf. Definition 3.5). The following example, however, shows that Proposition 4.6 does not hold for stable models.

EXAMPLE. Let T_0 be the empty TSS. Obviously, the empty LTS is a stable model of T_0. Let a be a constant, and let T_1 consist of the single transition rule $a\neg P/aP$. According to Theorem 4.4, $T_0 \oplus T_1$ is an operational conservative extension of T_0. However, $T_0 \oplus T_1$ does not have a stable model (but only the three-valued stable model $\langle\emptyset, \{aP\}\rangle$).

4.3. *Applications to axiomatizations*

This section discusses how operational conservative extension can be used to derive that an extension of an axiomatization is so-called axiomatically conservative, or that an axiomatization is complete or ω-complete with respect to some behavioural equivalence.

4.3.1. *Axiomatic conservative extension*

DEFINITION 4.8 (*Axiomatization*). A (*conditional*) *axiomatization* over a signature Σ consists of a set of (conditional) equations, called *axioms*, of the form $t_0 = u_0 \Leftarrow t_1 = u_1, \ldots, t_n = u_n$ with $t_i, u_i \in \mathbb{T}(\Sigma)$ for $i = 0, \ldots, n$.

An axiomatization gives rise to a binary equality relation = on $\mathbb{T}(\Sigma)$ thus:

- if $t_0 = u_0 \Leftarrow t_1 = u_1, \ldots, t_n = u_n$ is an axiom, and σ a substitution such that $\sigma(t_i) = \sigma(u_i)$ for $i = 1, \ldots, n$, then $\sigma(t_0) = \sigma(u_0)$;

- the relation $=$ is closed under reflexivity, symmetry, and transitivity;
- if f is a function symbol and $u = u'$, then

$$f(t_1, \ldots, t_{i-1}, u, t_{i+1}, \ldots, t_{ar(f)}) = f(t_1, \ldots, t_{i-1}, u', t_{i+1}, \ldots, t_{ar(f)}).$$

DEFINITION 4.9 (*Soundness and completeness*). Assume an axiomatization $\mathcal{E}$, together with an equivalence relation $\sim$ on $T(\Sigma)$.
(1) $\mathcal{E}$ is *sound* modulo $\sim$ iff $t = u$ implies $t \sim u$ for all $t, u \in T(\Sigma)$.
(2) $\mathcal{E}$ is *complete* modulo $\sim$ iff $t \sim u$ implies $t = u$ for all $t, u \in T(\Sigma)$.

Note that the above definitions of soundness and completeness, albeit standard in the literature on process algebras, are weaker than the classic ones in logic and universal algebra, where they are required to apply to arbitrary *open* expressions.

DEFINITION 4.10 (*Axiomatic conservative extension*). Let $\mathcal{E}_0$ and $\mathcal{E}_1$ be axiomatizations over signatures Σ_0 and $\Sigma_0 \oplus \Sigma_1$, respectively. Their union $\mathcal{E}_0 \cup \mathcal{E}_1$ is an *axiomatic conservative extension* of $\mathcal{E}_0$ if every equality $t = u$ with $t, u \in T(\Sigma_0)$ that can be derived from $\mathcal{E}_0 \cup \mathcal{E}_1$ can also be derived from $\mathcal{E}_0$.

The next theorem from [226] can be used to derive that an extension of an axiomatization is axiomatically conservative.

THEOREM 4.11. *Let $\sim$ be an equivalence relation on* $T(\Sigma_0 \oplus \Sigma_1)$. *Assume axiomatizations $\mathcal{E}_0$ and $\mathcal{E}_1$ over Σ_0 and $\Sigma_0 \oplus \Sigma_1$, respectively, such that*:
(1) *$\mathcal{E}_0 \cup \mathcal{E}_1$ is sound over* $T(\Sigma_0 \oplus \Sigma_1)$ *modulo* $\sim$;
(2) *$\mathcal{E}_0$ is complete over* $T(\Sigma_0)$ *modulo* $\sim$.
Then $\mathcal{E}_0 \cup \mathcal{E}_1$ is an axiomatic conservative extension of $\mathcal{E}_0$.

The idea behind Theorem 4.11 is as follows. Suppose $t = u$ can be derived from $\mathcal{E}_0 \cup \mathcal{E}_1$ for $t, u \in T(\Sigma_0)$. Soundness of $\mathcal{E}_0 \cup \mathcal{E}_1$ (requirement (1)) yields $t \sim u$. Hence, completeness of $\mathcal{E}_0$ (requirement (2)) yields that $t = u$ can be derived from $\mathcal{E}_0$.

Theorem 4.11 is particularly helpful in the case of an operational conservative extension of a TSS. Namely, assume TSSs T_0 and T_1 over signatures Σ_0 and $\Sigma_0 \oplus \Sigma_1$, respectively, where $T_0 \oplus T_1$ is an operational conservative extension of T_0. Moreover, let $\sim$ be an equivalence relation on states in LTSs. Since the states in the LTSs associated with T_0 and $T_0 \oplus T_1$ are closed terms, the equivalence relation $\sim$ carries over to $T(\Sigma_0)$ and $T(\Sigma_0 \oplus \Sigma_1)$, respectively. Owing to operational conservativity, the equivalence relation $\sim$ on $T(\Sigma_0)$ as induced by T_0 agrees with this equivalence relation on $T(\Sigma_0)$ as induced by $T_0 \oplus T_1$. Applications of Theorem 4.11 in process algebra, in the presence of an operational conservative extension of a TSS, are abundant in the literature; we give a typical example.

EXAMPLE. Using Theorem 4.4 it is easily seen that the process algebra ACP_θ [24] is an operational conservative extension of ACP. Baeten, Bergstra, and Klop presented in *op. cit.* an axiomatization $\mathcal{E}_0$ that is complete over ACP modulo bisimulation equivalence, and an axiomatization $\mathcal{E}_0 \cup \mathcal{E}_1$ that is sound over ACP$_\theta$ modulo bisimulation equivalence. Hence, Theorem 4.11 says that $\mathcal{E}_0 \cup \mathcal{E}_1$ is an axiomatic conservative extension of $\mathcal{E}_0$. (In [24], fifteen pages were needed to prove this fact for the more general case of open terms, by means of a term rewriting analysis.)

4.3.2. *Completeness of axiomatizations* The next theorem from [226] can be used to derive that an axiomatization is complete.

THEOREM 4.12. *Let $\sim$ be an equivalence relation on* $\mathrm{T}(\Sigma_0 \oplus \Sigma_1)$. *Assume axiomatizations $\mathcal{E}_0$ and $\mathcal{E}_1$ over Σ_0 and $\Sigma_0 \oplus \Sigma_1$, respectively, such that:*

(1) *$\mathcal{E}_0 \cup \mathcal{E}_1$ is sound over* $\mathrm{T}(\Sigma_0 \oplus \Sigma_1)$ *modulo* $\sim$;

(2) *$\mathcal{E}_0$ is complete over* $\mathrm{T}(\Sigma_0)$ *modulo* $\sim$;

(3) *for each $t \in \mathrm{T}(\Sigma_0 \oplus \Sigma_1)$ there is a $t' \in \mathrm{T}(\Sigma_0)$ such that $t = t'$ can be derived from $\mathcal{E}_0 \cup \mathcal{E}_1$.*

Then $\mathcal{E}_0 \cup \mathcal{E}_1$ is complete over $\mathrm{T}(\Sigma_0 \oplus \Sigma_1)$ *modulo* $\sim$.

The idea behind Theorem 4.12 is as follows. Let $t, u \in \mathrm{T}(\Sigma_0 \oplus \Sigma_1)$ with $t \sim u$. There exist terms $t', u' \in \mathrm{T}(\Sigma_0)$ such that $\mathcal{E}_0 \cup \mathcal{E}_1$ proves $t = t'$ and $u = u'$ (requirement (3)). Soundness of $\mathcal{E}_0 \cup \mathcal{E}_1$ (requirement (1)) yields $t \sim t'$ and $u \sim u'$, which together with $t \sim u$ implies $t' \sim u'$. Finally, owing to completeness of $\mathcal{E}_0$ over $\mathrm{T}(\Sigma_0)$ (requirement (2)), we may derive $t' = u'$, and thus $t = t' = u' = u$.

Similar to Theorem 4.11, Theorem 4.12 is particularly helpful in the case of an operational conservative extension of a TSS. In order to clarify the link between Theorem 4.12 and operational conservative extensions, we reiterate the following observation from Section 4.3.1. Assume TSSs T_0 and T_1 over signatures Σ_0 and $\Sigma_0 \oplus \Sigma_1$, respectively, where $T_0 \oplus T_1$ is an operational conservative extension of T_0. Moreover, let $\sim$ be an equivalence relation on states in LTSs. Since the states in the LTSs associated with T_0 and $T_0 \oplus T_1$ are closed terms, the equivalence relation $\sim$ carries over to $\mathrm{T}(\Sigma_0)$ and $\mathrm{T}(\Sigma_0 \oplus \Sigma_1)$, respectively. Owing to operational conservativity, the equivalence relation $\sim$ on $\mathrm{T}(\Sigma_0)$ as induced by T_0 agrees with this equivalence relation on $\mathrm{T}(\Sigma_0)$ as induced by $T_0 \oplus T_1$. Applications of Theorem 4.12 in process algebra, in the presence of an operational conservative extension of a TSS, are abundant in the literature; we give a typical example.

EXAMPLE. Using Theorem 4.4 it is easily seen that the process algebra ACP [37] is an operational conservative extension of BPA$_\delta$. Bergstra and Klop presented in *op. cit.* an axiomatization $\mathcal{E}_0$ that is complete over BPA$_\delta$ modulo bisimulation equivalence, and an axiomatization $\mathcal{E}_0 \cup \mathcal{E}_1$ that is sound over ACP modulo bisimulation equivalence, and that satisfies requirement (3) above. Hence, Theorem 4.12 says that $\mathcal{E}_0 \cup \mathcal{E}_1$ is complete over ACP modulo bisimulation equivalence.

For the precise proofs of Theorems 4.11 and 4.12, and for more detailed information such as generalizations of these results to axiomatizations based on inequalities, the reader is referred to [68,69,226].

4.3.3. *ω-Completeness of axiomatizations*

DEFINITION 4.13 (*ω-completeness*). An axiomatization $\mathcal{E}$ over a signature Σ is *ω-complete* if an equation $t = u$ with $t, u \in \mathbb{T}(\Sigma)$ can be derived from $\mathcal{E}$ whenever $\sigma(t) = \sigma(u)$ can be derived from $\mathcal{E}$ for all closed substitutions σ.

Milner [158] introduced a technique to derive ω-completeness of an axiomatization using SOS. The idea is to give a semantics to open (as opposed to closed) terms; in particular, variables need to be incorporated in the transition rules. See, e.g., [9,95] for further applications of this technique in the realm of process algebra.

The next theorem can be used to derive that an axiomatization is ω-complete.

THEOREM 4.14. *Let $\sim$ be an equivalence relation on $\mathbb{T}(\Sigma)$. Suppose that for all $t, u \in \mathbb{T}(\Sigma)$, $t \sim u$ whenever $\sigma(t) \sim \sigma(u)$ for all closed substitutions σ. If $\mathcal{E}$ is an axiomatization over Σ such that*

(1) $\mathcal{E}$ is sound over $\mathrm{T}(\Sigma)$ *modulo $\sim$, and*

(2) $\mathcal{E}$ is complete over $\mathbb{T}(\Sigma)$ modulo $\sim$,

then $\mathcal{E}$ is ω-complete.

The idea behind Theorem 4.14 is as follows. Let $t, u \in \mathbb{T}(\Sigma)$ and suppose $\sigma(t) = \sigma(u)$ can be derived from $\mathcal{E}$ for all closed substitutions σ. Soundness of $\mathcal{E}$ over $\mathrm{T}(\Sigma)$ modulo $\sim$ (requirement (1)) yields $\sigma(t) \sim \sigma(u)$ for all closed substitutions σ, so $t \sim u$. Then completeness of $\mathcal{E}$ over $\mathbb{T}(\Sigma)$ modulo $\sim$ (requirement (2)) yields that $t = u$ can be derived from $\mathcal{E}$.

Theorem 4.14 is particularly helpful in the case of an operational conservative extension of a TSS. Namely, assume a TSS T_0 over a signature Σ, and let T_0 be extended with a TSS T_1 that provides semantics to variables; thus, $T_0 \oplus T_1$ gives semantics to open terms in $\mathbb{T}(\Sigma)$. Suppose $T_0 \oplus T_1$ is an operational conservative extension of T_0. Moreover, let $\sim$ be an equivalence relation on states in LTSs. Since the states in the LTSs associated with T_0 and $T_0 \oplus T_1$ are closed and open terms, respectively, the equivalence relation $\sim$ carries over to $\mathrm{T}(\Sigma)$ and $\mathbb{T}(\Sigma)$. Owing to operational conservativity, the equivalence relation $\sim$ on $\mathrm{T}(\Sigma)$ as induced by T_0 agrees with this equivalence relation on $\mathrm{T}(\Sigma)$ as induced by $T_0 \oplus T_1$. Applications of Theorem 4.14 in process algebra are abundant in the literature; we give a typical example.

EXAMPLE. Extend the TSS for BPA_ε in Table 1 by letting the symbol a range not only over the set Act of actions, but also over the set Var of variables. In a sense this means that variables are considered to be constants. This extension is operationally conservative, which follows from Theorem 4.4 by the following facts:

- the transition rules for BPA_ε are source-dependent;
- the sources of transition rules $z \xrightarrow{z} \varepsilon$ for variables z are fresh;
- each transition rule for alternative or sequential composition with z-transitions, such as

$$\frac{x \xrightarrow{z} x'}{x + y \xrightarrow{z} x'}$$

contains a premise with the fresh relation symbol $\xrightarrow{z}$ and as left-hand side a variable from the source.

Furthermore, the following properties can be derived for the axiomatization $\mathcal{E}$ of BPA_ε in [228]:

(1) $\mathcal{E}$ is sound over BPA_ε modulo bisimulation equivalence;

(2) open terms t and u in BPA_ε are bisimilar whenever $\sigma(t)$ and $\sigma(u)$ are bisimilar for all closed substitutions σ;

(3) $\mathcal{E}$ is complete over the open terms in BPA_ε modulo bisimulation equivalence.

So Theorem 4.14 implies that $\mathcal{E}$ is ω-complete over BPA_ε modulo bisimulation equivalence.

4.4. *Applications to rewriting*

This section discusses how operational conservative extension can be used to derive that an extension of a conditional term rewriting system is so-called rewrite conservative, or that a conditional term rewriting system is ground confluent.

4.4.1. *Rewrite conservative extension*

DEFINITION 4.15 (*Conditional term rewriting system*). Assume a signature Σ. A *conditional term rewriting system* (CTRS) [17,38] over Σ consists of a set of *rewrite rules*

$$t_0 \rightarrow u_0 \Leftarrow t_1 \rightarrow^* u_1, \ldots, t_n \rightarrow^* u_n$$

with $t_i, u_i \in \mathbb{T}(\Sigma)$ for $i = 0, \ldots, n$.

Intuitively, a rewrite rule is a directed axiom that can only be applied from left to right. A CTRS induces a binary rewrite relation $\rightarrow^*$ on terms, similar to the way that an axiomatization induces an equality relation on terms (the only difference is that the rewrite relation is not closed under symmetry), thus:

- if $t_0 \rightarrow u_0 \Leftarrow t_1 \rightarrow^* u_1, \ldots, t_n \rightarrow^* u_n$ is a rewrite rule, and σ a substitution such that $\sigma(t_i) \rightarrow^* \sigma(u_i)$ for $i = 1, \ldots, n$, then $\sigma(t_0) \rightarrow^* \sigma(u_0)$;
- the relation $\rightarrow^*$ is closed under reflexivity and transitivity;
- if f is a function symbol and $u \rightarrow^* u'$, then

$$f(t_1, \ldots, t_{i-1}, u, t_{i+1}, \ldots, t_{ar(f)}) \rightarrow^* f(t_1, \ldots, t_{i-1}, u', t_{i+1}, \ldots, t_{ar(f)}).$$

DEFINITION 4.16 (*Rewrite conservative extension*). Let R_0 and R_1 be CTRSs over signatures Σ_0 and $\Sigma_0 \oplus \Sigma_1$, respectively. Their union $R_0 \oplus R_1$ is a *rewrite conservative extension* of R_0 if every rewrite relation $t \rightarrow^* u$ with $t \in \mathrm{T}(\Sigma_0)$ that can be derived from $R_0 \oplus R_1$ can also be derived from R_0.

The conservative extension theorem for TSSs, Theorem 4.4, applies to CTRSs just as well; see [86] for more details, and for applications of this result in the realm of software renovation (see, e.g., [60]). Note that the definition of source-dependent variables

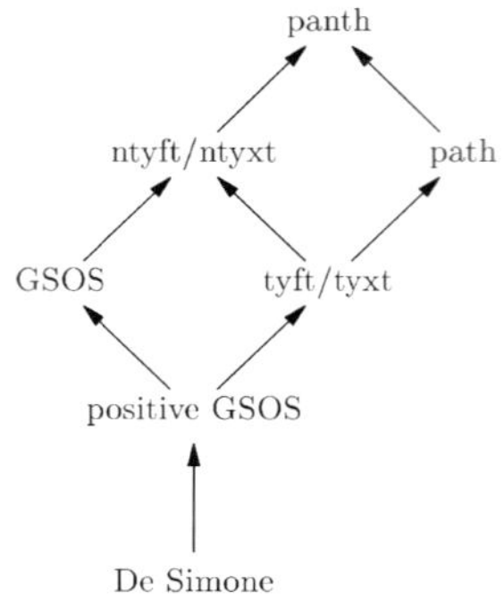

Fig. 1. Lattice of congruence formats.

premises. Finally, the *path* format [26] generalizes the tyft/tyxt format with predicates, while the *panth* format [225,227] extends the ntyft/ntyxt format with predicates. Figure 1 presents the lattice of congruence formats for bisimulation equivalence. An arrow from one rule format to another indicates that all transition rules in the first format are also in the second format. If there are no arrows connecting two rule formats, then they are (syntactically) incomparable.

If a TSS is both panth and (ws-)complete (in the sense of Definition 3.12), then bisimulation equivalence is a congruence with respect to all the function symbols in the signature. The panth format is the most general known syntactic format to guarantee that bisimulation equivalence is a congruence. However, more restrictive rule formats such as De Simone and GSOS guarantee other nice properties. Therefore, these rule formats are treated separately in this section.

For each TSS in panth format there exists an equivalent TSS in *ntree* format [83]. This result facilitates reasoning about the panth format, because it is often much easier to prove a theorem for TSSs in ntree format than for TSSs in panth format. For example, this is the case for the congruence theorem for bisimulation equivalence itself. Furthermore, the reduction of panth to ntree made it possible to remove the well-foundedness criterion on premises from an earlier version of the congruence theorem, owing to the fact that TSSs in ntree format satisfy this well-foundedness criterion by default; see [83].

For the sake of presentation, the lattice in Figure 1 focuses on the rule formats that are of practical importance; that is, we left out the ntree format and its derived (unnamed) formats that disallow predicates and/or negative premises. Other rule formats that are not mentioned in Figure 1 are those that deal with silent actions explicitly. Of particular interest here are the rule formats presented in, e.g., [50,82,215,218].

The organization of this section is as follows. Section 5.1 presents the panth format, while Section 5.2 deals with the equally expressive ntree format. Sections 5.3 and 5.4 study De Simone languages and GSOS languages, respectively. Section 5.5 introduces a congruence format in the presence of silent actions, while Section 5.6 presents congruence formats for a wide range of behavioural preorders. Finally, Section 5.7 studies the completed trace congruence induced by a number of congruence formats.

5.1. *Panth format*

This section presents the panth format, and states a congruence theorem (cf. Definition 2.11) from [57,111,227] with respect to bisimulation equivalence (cf. Definition 2.3).

DEFINITION 5.1 (*Panth format*). A transition rule ρ is in *panth format* if it satisfies the following three restrictions:

(1) for each positive premise $t \xrightarrow{a} t'$ of ρ, the right-hand side t' is a single variable;
(2) the source of ρ contains no more than one function symbol;
(3) the variables that occur as right-hand sides of positive premises or in the source of ρ are all distinct.

A TSS is in *panth format* if it consists of panth rules only.

The three subformats path, ntyft/ntyxt, and tyft/tyxt of the panth format do not allow negative premises and/or predicates.

DEFINITION 5.2 (*Path, ntyft/ntyxt, and tyft/tyxt format*). A TSS is in *path format* if it is in panth format and positive.

A TSS is in *ntyft/ntyxt format* if it is in panth format and its transition rules do not contain predicates.

A TSS is in *tyft/tyxt format* if it is both path and ntyft/ntyxt.

A TSS is in *tyft format* if it is tyft/tyxt and the source of each rule contains exactly one function symbol.

THEOREM 5.3. *If a TSS is complete and panth, then bisimulation equivalence is a congruence with respect to the LTS associated with it.*

The interested reader is referred to [83,111,227] for a proof of Theorem 5.3. Groote and Vaandrager [111] presented a string of examples of TSSs which show that all syntactic requirements of the panth format are essential for the congruence result in Theorem 5.3. We give an example to show that the restriction in Theorem 5.3 to complete TSSs is essential. In particular, it cannot be relaxed to TSSs that have exactly one (not necessarily least) three-valued stable model that does not contain unknown transitions. The example is derived from [57, Example 8.12].

EXAMPLE. Let the signature consist of constants a, b and of a unary function symbol f. Moreover, let P, Q_1, and Q_2 be predicates. Consider the following TSS in panth format:

$$\frac{}{aP} \qquad \frac{}{bP} \qquad \frac{xP \quad f(x)\neg Q_1 \quad f(a)\neg Q_2}{f(x)Q_2} \qquad \frac{xP \quad f(x)\neg Q_2 \quad f(b)\neg Q_1}{f(x)Q_1}.$$

Its least three-valued stable model contains the following unknown transitions: $f(a)Q_1$, $f(a)Q_2$, $f(b)Q_1$, and $f(b)Q_2$. So the TSS is not complete.

However, the TSS does have a three-valued stable model in which the set of unknown transitions is empty, and aP, bP, $f(a)Q_1$, and $f(b)Q_2$ are the true transitions. $a \underline{\leftrightarrow} b$ but $f(a) \not\underline{\leftrightarrow} f(b)$ with respect to this three-valued stable model.

In [167], van Oostrom and de Vink generalized the tyft/tyxt format (so in the absence of predicates and negative premises) to the *stalk format*, by somewhat relaxing the constraint that sources of transition rules can contain no more than one function symbol. They proved that the congruence result in Theorem 5.3 holds for the stalk format.

We apply the congruence theorem for the panth format to the running examples from Section 2.6.

Basic process algebra. The TSS for BPA_ε in Table 1 is panth. For example, the transition rule

$$\frac{x \xrightarrow{a} x'}{x \cdot y \xrightarrow{a} x' \cdot y}$$

for sequential composition is panth, because the right-hand side of its premise is a single variable x', its source contains only one function symbol (sequential composition), and the variables in the right-hand side of its premise (x') and in its source (x and y) are distinct. It is left to the reader to verify that the remaining transition rules in Table 1 are panth.

The TSS for BPA_ε is positive, so it is complete. Since this TSS is both panth and complete, Theorem 5.3 says that bisimulation equivalence is a congruence with respect to BPA_ε.

Priorities. It is not hard to check that the TSS for $\mathrm{BPA}_{\varepsilon\theta}$ in Table 2 is panth. Furthermore, it was noted in Section 3.5 that the TSS for $\mathrm{BPA}_{\varepsilon\theta}$ is complete. Hence, by Theorem 5.3, bisimulation equivalence is a congruence with respect to $\mathrm{BPA}_{\varepsilon\theta}$.

Discrete time. It is not hard to check that the TSS for $\mathrm{BPA}_\varepsilon^{\mathrm{dt}}$ in Table 3 is panth. Furthermore, it was noted in Section 3.5 that the TSS for $\mathrm{BPA}_\varepsilon^{\mathrm{dt}}$ is complete. Hence, by Theorem 5.3, bisimulation equivalence is a congruence with respect to $\mathrm{BPA}_\varepsilon^{\mathrm{dt}}$.

5.2. *Ntree format*

In this section we present a result from [80,83] to the effect that for each TSS in panth format there exists an equivalent TSS in the more restrictive ntree format. The following terminology originates from [55,111].

DEFINITION 5.4 (*Variable dependency graph*). The *variable dependency graph* of a set S of premises is a directed graph, with the set of variables as vertices, and with as edges the set

$$\bigl\{\langle x, y\rangle \mid \text{there is a } t \xrightarrow{a} t' \text{ in } S \text{ such that } x \text{ occurs in } t \text{ and } y \text{ in } t'\bigr\}.$$

S is *well-founded* if any backward chain of edges in its variable dependency graph is finite.

A transition rule is *pure* if its set of premises is well-founded and moreover each variable in the rule occurs in the source or as the right-hand side of a positive premise.

Typical examples of sets of premises that are not well-founded are $\{y \xrightarrow{a} y\}$, $\{y_1 \xrightarrow{a} y_2,\ y_2 \xrightarrow{b} y_1\}$, and $\{y_{i+1} \xrightarrow{a} y_i \mid i \in \mathbb{N}\}$.

DEFINITION 5.5 (*Ntree format*). A transition rule ρ is in *ntree format* if it satisfies the following three criteria:

(1) ρ is panth;
(2) ρ is pure;
(3) the left-hand sides of positive premises in ρ are single variables.

A TSS is in *ntree format* if it consists of ntree rules only.

For example, the TSSs for BPA_ε, $\mathrm{BPA}_{\varepsilon\theta}$, and $\mathrm{BPA}_\varepsilon^{\mathrm{dt}}$ from Section 2.6 are in ntree format. The following theorem originates from [83].

THEOREM 5.6. *For each TSS T in panth format there exists a TSS T' in ntree format such that for any closed transition rule N/α where N contains only negative literals, $T \vdash N/\alpha \Longleftrightarrow T' \vdash N/\alpha$.*

5.3. *De Simone format*

A *De Simone language* [198,200] consists of a signature together with a TSS whose transition rules are in *De Simone format*, extended with transition rules for recursion. Most process description languages encountered in the literature, including CCS [157], SCCS [155], CSP [186], ACP [28], and MEIJE [16], are De Simone languages.

5.3.1. *De Simone languages.* For consistency with subsequent developments, amongst the various definitions of De Simone languages presented in the literature we adopt the one given by Vaandrager [221].

DEFINITION 5.7 (*De Simone format*). Let Σ be a signature. A transition rule ρ is in *De Simone format* if it has the form

$$\frac{\{x_i \xrightarrow{a_i} y_i \mid i \in I\}}{f(x_1, \ldots, x_{ar(f)}) \xrightarrow{a} t},$$

where $I \subseteq \{1, \ldots, ar(f)\}$ and the variables x_i and y_i are all distinct and the only variables that occur in ρ. Moreover, the target $t \in \mathbb{T}(\Sigma)$ does not contain variables x_i for $i \in I$ and has no multiple occurrences of variables. We say that f is the *type* and a the *action* of ρ.

In conjunction with the signature Σ, we assume a countably infinite set of *recursion variables*, ranged over by X, Y. The recursive terms over Σ are given by the BNF grammar

$$t ::= X \mid f(t_1, \ldots, t_{ar(f)}) \mid \mathrm{fix}(X = t),$$

where X is any recursion variable, f any function symbol in Σ, and fix a binding construct. The latter construct gives rise to the usual notions of free and bound recursion variables in recursive terms. We use $t[u/X]$ to denote the recursive term t in which each occurrence of the recursion variable X has been replaced by u (after possibly renaming bound recursion variables in t). For every recursive term $\text{fix}(X = t)$ and action a, we introduce a transition rule

$$\frac{t[\text{fix}(X = t)/X] \xrightarrow{a} y}{\text{fix}(X = t) \xrightarrow{a} y}.$$

The reader is referred to Section 6 for a formal treatment of such transition rules that incorporate binding constructs. A *De Simone language* is a set of De Simone rules, extended with the transition rules above for recursion.

REMARK. In De Simone's original definition for his rule format (cf. [200, Definition 1.9]), transition rules carried side conditions $Pr(a_1, \ldots, a_n)$ where Pr is a predicate on actions (not to be confused with the predicates on states allowed in LTSs). No particular syntax for such predicates was considered in De Simone's work, but natural computability restrictions were imposed on the allowed sets of tuples. Most subsequent literature on rule formats for transition rules has abstracted from such predicates.

5.3.2. *Expressiveness of De Simone languages.* The main original motivation for the development of the De Simone format was to gain insight into the expressive power of the process calculi SCCS and MEIJE, in the semantic realm of LTSs. In particular, what De Simone was aiming at in his seminal papers [198,200] was an expressive completeness result for the aforementioned process calculi that would fully justify the choice of basic operators made by their developers. After all, the motivation for the study of foundational calculi for concurrency stems from Milner's idea that for a proper understanding of the basic issues in the behaviour of concurrent systems it would be helpful to look for a simple language "with as few operators or combinators as possible, each of which embodies some distinct and intuitive idea, and which together give completely general expressive power" [155, p. 264]. Before embarking on such an investigation, however, one has to choose an appropriate measure of expressiveness for a calculus. As argued by Vaandrager [221], there are at least three different ways in which a language can have completely general expressive power:

(1) each Turing machine can be simulated in lock step;
(2) each recursively enumerable LTS can be specified up to some notion of behavioural equivalence;
(3) each operation in a "natural" class of operations is realizable by means of the primitive operations in the language up to some notion of behavioural equivalence.

Since most languages that have been proposed in the literature are completely expressive in the first sense, this criterion does not offer a useful means to classify the expressiveness of languages. (This is what Meyer [148] calls the "Turing tarpit".) The remaining two criteria have been investigated by De Simone in *op. cit.*, and, since then, by several other authors. Indeed, this kind of expressiveness results has, to the best of our knowledge, only been developed for De Simone languages, and similar investigations are lacking for more general rule formats. For this reason, the rest of this section is devoted to a brief review of such results. The other results that have been developed for this format are instances of theorems for the more general formats we survey.

The question of whether there exists a process description language which is completely expressive with respect to the collection of operations definable by means of De Simone rules has been addressed by several authors since De Simone's original work. In *op. cit.*, De Simone showed the following result.

THEOREM 5.8. *Let* Act *be a finite set of actions. Let f be specified by a De Simone language containing only finitely many rules. Then f can be expressed, up to bisimulation equivalence, in the calculi SCCS and* MEIJE.

As a corollary of this expressiveness result pertaining to the class of operators specifiable in SCCS and MEIJE, De Simone was able to prove that the calculi SCCS and MEIJE, and, *a fortiori*, reasonably expressive De Simone languages, can denote every recursively enumerable LTS up to graph isomorphism.

DEFINITION 5.9 (*Properties of LTSs*). An LTS is:

- *countably branching* if for every state s there are at most countably many outgoing transitions $s \xrightarrow{a} s'$;
- *recursively enumerable* if there exists an algorithm enumerating all transitions $s \xrightarrow{a} s'$;
- *decidable* if there exists an algorithm that determines for every transition whether it is in the LTS;
- *computable* [25] if there exists an algorithm that computes for every state s its complete finite list of outgoing transitions $s \xrightarrow{a} s'$;
- *primitive recursive* [178] if there is such an algorithm that is primitive recursive.

Note that "computable" is a stronger requirement than "decidable and finitely branching" (cf. Definition 2.2). The following result is from [200, Theorem 3.2].

THEOREM 5.10. *Every recursively enumerable LTS can be realized by a recursive term in SCCS-*MEIJE *up to graph isomorphism.*

Several variations on Theorem 5.10 have been presented in the literature. Before stating some of these results, we need to introduce some preliminary definitions from [221].

DEFINITION 5.11 (*Testing arguments*). Assume a De Simone language. A function symbol f *tests* its ith argument if one of the De Simone rules has a source $f(x_1, \ldots, x_{ar(f)})$ and a premise $x_i \xrightarrow{a_i} y_i$.

DEFINITION 5.12 (*Guardedness*). Assume a De Simone language. For recursive terms t, the sets $U(t)$ of *unguarded* recursion variables are defined thus:

$$U(X) \triangleq \{X\},$$

PROPOSITION 5.18. *Any De Simone language satisfying certain properties at the left side is expressible in aprACP$_R$ with the corresponding features at the right.*

finitary	*with guarded recursion*
image-finite	*with image-finite renaming*
functional	*with functional renaming*
–	*recursively enumerable*
primitive width-effective	*primitive effective*

In what follows we use two superscripts. The superscript r.e. denotes the recursively enumerable version of a language, i.e., its version that only includes a partial recursive communication function (see [99, Section 3.4] for details). The superscript p.e. denotes the primitive effective version of a language (see [99, Definition 15] for details). Proposition 5.18 establishes several expressiveness results. Since virtually all De Simone languages encountered in practice are finitary, the most significant of these results are the following.

(1) Any finitary De Simone language is expressible in aprACP$_R$ with guarded recursion.
(2) Any finitary image-finite De Simone language is expressible in aprACP$_R$ with guarded recursion and image-finite renamings.
(3) Any finitary functional De Simone language is expressible in aprACP$_F$ with guarded recursion.
(4) Any finitary recursively enumerable De Simone language is expressible in aprACP$_R^{\text{r.e.}}$ with guarded recursion.
(5) Any finitary recursively enumerable image-finite De Simone language is expressible in aprACP$_R^{\text{r.e.}}$ with guarded recursion and image-finite renamings.
(6) Any finitary recursively enumerable functional De Simone language is expressible in aprACP$_F^{\text{r.e.}}$ with guarded recursion.
(7) Any finitary primitive effective De Simone language is expressible in aprACP$_R^{\text{p.e.}}$ with guarded recursion.
(8) Any finitary primitive effective functional De Simone language is expressible in aprACP$_F^{\text{p.e.}}$ with guarded recursion.

Result 4 in the above list generalizes the original theorem by De Simone, saying that any finitary recursively enumerable De Simone language with recursion is expressible in the recursively enumerable version of MEIJE with recursion. The generalization is that, under the assumption that the source languages have only guarded recursion, the target language (now aprACP$_R$) can be required to use only guarded recursion as well.

Using the constant U yields an even stronger result for recursively enumerable De Simone languages, viz. without requiring finitariness. This result from [99] has no effective counterpart.

THEOREM 5.19. *Assume a recursively enumerable De Simone language T. Every closed recursive term in the LTS associated with T is bisimilar to a closed guarded recursive term in the LTS associated with aprACP$_U$.*

CSP (and SCCS and Meije) can specify infinitely branching processes. To do so in CSP, one uses the inverse image operation, which is similar to the relational renaming operator. The following result, to the effect that the process algebra CSP [61] is completely expressive with respect to operations definable using De Simone rules, stems from [130].

THEOREM 5.20. *Assume a De Simone language T. Every closed recursive term in the LTS associated with T is bisimilar to a closed CSP term in the LTS associated with CSP.*

The interested reader will find further expressiveness results for variations on De Simone languages and several notions of "expressiveness" in, e.g., [20,70,75,99,170,221].

5.4. *GSOS format*

This section introduces one of the most thoroughly studied rule formats, viz. the GSOS format of Bloom, Istrail, and Meyer [55]. We present some of the many results that have been developed for this rule format, focusing on its sanity properties, and its connections with axiomatizations modulo bisimulation equivalence.

5.4.1. *GSOS languages*

DEFINITION 5.21 (*GSOS format*). A transition rule ρ is in *GSOS format* if it has the form

$$\frac{\{x_i \xrightarrow{a_{ij}} y_{ij} \mid 1 \leqslant i \leqslant ar(f), 1 \leqslant j \leqslant m_i\} \cup \{x_i \stackrel{b_{ik}}{\nrightarrow} \mid 1 \leqslant i \leqslant ar(f), 1 \leqslant k \leqslant n_i\}}{f(x_1, \ldots, x_{ar(f)}) \xrightarrow{c} t},$$

where $m_i, n_i \in \mathbb{N}$, and the variables x_i and y_{ij} are all distinct and the only variables that occur in ρ.

A (*finitary*) *GSOS language* is a finite set of GSOS rules over a finite signature, and a finite set Act of actions.

Every De Simone rule is also a GSOS rule. Unlike De Simone rules, however, GSOS rules allow negative premises, as well as multiple occurrences of variables in left-hand sides of premises and in the target.

An example of a GSOS rule with negative premises is the second transition rule for the priority operator θ; see Table 2 in Section 2.6. For actions a that are not a supremum with respect to the ordering on Act, the second transition rule for θ contains negative premises. The priority operator cannot be expressed, up to bisimulation equivalence, using De Simone rules. Namely, θ does not preserve trace equivalence (see Definition 2.4) unlike any operator expressible using De Simone rules (cf. Theorem 5.65). For example, $a\cdot(b+c)$ and $a\cdot b + a\cdot c$ are trace equivalent, but $\theta(a\cdot(b+c))$ and $\theta(a\cdot b + a\cdot c)$ are not, if $c > b$.

A notable example of a GSOS rule that uses the same variable in the left-hand side of a premise and in the target is a transition rule for the *binary Kleene star* [133]:

$$\frac{x \xrightarrow{a} x'}{x^*y \xrightarrow{a} x' \cdot (x^*y)}.$$

This operator has been studied in the realm of process algebra in, e.g., [33,87,156] (see also [34], Chapter 5 in this Handbook).

Each GSOS language allows a stratification (cf. Definition 3.13), and is therefore complete (cf. Definition 3.12). LTSs associated with GSOS languages are computable (cf. Definition 5.9) and finitely branching (cf. Definition 2.2). By contrast, there exist TSSs in tyft/tyxt format (cf. Definition 5.2), consisting of only finitely many transition rules with only finitely many premises, such that the associated LTSs are neither computable (see [44, 70]) nor finitely branching (see [111, p. 258]). It is not straightforward to associate LTSs to GSOS languages with recursion (see, e.g., [55]). A solution for this problem, involving a special divergence predicate $\uparrow$, is discussed in Section 5.4.6.

5.4.2. *Junk rules.* The definition of a GSOS language does not exclude *junk* rules, i.e., transition rules that support no transitions in the associated LTS. For example, the transition rule

$$\frac{x \xrightarrow{a} y \qquad x \stackrel{a}{\nrightarrow}}{f(x) \xrightarrow{a} f(y)}$$

has contradictory premises, so under no closed substitution do these premises both hold. Furthermore, the seemingly innocuous transition rule

$$\frac{x \xrightarrow{a} y}{f(x) \xrightarrow{b} f(y)}$$

does not support any transition if the associated LTS does not contain a-transitions. We present an unpublished result by Aceto, Bloom, and Vaandrager [6] to the effect that it is decidable whether a transition rule in a GSOS language is junk. This decision procedure for "rule junkness" allows a language designer to check whether or not any of the transition rules describing some language features are used. Since this result has not appeared in the literature before, we present its proof here.

THEOREM 5.22. *It is decidable whether a transition rule ρ in a GSOS language T is junk.*

PROOF. Let Σ denote the signature. It is not hard to determine which GSOS rules in T are junk once we have computed the set

$$\mathsf{initials}(\mathrm{T}(\Sigma)) = \{\mathsf{initials}(t) \mid t \in \mathrm{T}(\Sigma)\},$$

where we recall from Section 2.1 that $\mathsf{initials}(t)$ denotes $\{a \in \mathsf{Act} \mid \exists t' \in \mathrm{T}(\Sigma)\,(t \xrightarrow{a} t')\}$. The premises of a GSOS rule as in Definition 5.21 are satisfiable iff there exist closed terms $t_1, \ldots, t_{ar(f)} \in \mathrm{T}(\Sigma)$ such that $\{a_{ij} \mid j = 1, \ldots, m_i\} \subseteq \mathsf{initials}(t_i)$ and $\{b_{ik} \mid k = 1, \ldots, n_i\} \cap \mathsf{initials}(t_i) = \emptyset$ for $i = 1, \ldots, ar(f)$.

So we are left to give an effective way of computing the set $\mathsf{initials}(\mathrm{T}(\Sigma))$ for any GSOS language. Note that each function symbol f determines a computable function

$$\hat{f}:\underbrace{2^{\mathsf{Act}}\times\cdots\times 2^{\mathsf{Act}}}_{\mathrm{ar}(f)\text{ times}}\to 2^{\mathsf{Act}}$$

by $\hat{f}(S_1,\ldots,S_{ar(f)})\triangleq S$, where for all $c\in\mathsf{Act}$, $c\in S$ iff there exists a GSOS rule as in Definition 5.21 (with type f and action c) such that, for $i=1,\ldots,ar(f)$, $\{a_{ij}\mid j=1,\ldots,m_i\}\subseteq S_i$ and $\{b_{ik}\mid 1\leqslant k\leqslant n_i\}\cap S_i=\emptyset$. Now, for each $\mathcal{S}\subseteq 2^{\mathsf{Act}}$, let $\mathcal{G}(\mathcal{S})$ be given by

$$\mathcal{G}(\mathcal{S})\triangleq\left\{\hat{f}(S_1,\ldots,S_{ar(f)})\mid f\in\Sigma,\ S_1,\ldots,S_{ar(f)}\in\mathcal{S}\right\}.$$

For each $\mathcal{S}\subseteq 2^{\mathsf{Act}}$ the set $\mathcal{G}(\mathcal{S})$ can be effectively computed, and $\mathcal{S}\subseteq\mathcal{S}'$ implies $\mathcal{G}(\mathcal{S})\subseteq\mathcal{G}(\mathcal{S}')$.

The set $\mathsf{initials}(\mathrm{T}(\Sigma))$ can be computed by dividing $\mathrm{T}(\Sigma)$ into sets U_i of closed terms that contain no more than i function symbols, and computing the non-decreasing sequence

$$\mathsf{initials}(U_1)\subseteq\mathsf{initials}(U_2)\subseteq\cdots$$

until it stabilizes. Obviously, this sequence stabilizes in a finite number of steps, as Act is finite.

The set of terms U_0 is empty, so $\mathsf{initials}(U_0)=\emptyset$. Now suppose we want to compute $\mathsf{initials}(U_{i+1})$, given that we already have $\mathsf{initials}(U_i)$. We claim that $\mathsf{initials}(U_{i+1})=\mathcal{G}(\mathsf{initials}(U_i))$. In fact, each term in U_{i+1} is of the form $f(t_1,\ldots,t_{ar(f)})$, where the t_i's are all in U_i. Thus we know $\mathsf{initials}(t_i)$ for all $i=1,\ldots,\mathrm{ar}(f)$, and that is exactly what is needed to determine for each transition rule of type f under which closed instantiations its premises hold. Hence we can compute $\mathsf{initials}(U_1)$, and each $\mathsf{initials}(U_{i+1})$ can be computed from $\mathsf{initials}(U_i)$ using the monotonic and effective operation $\mathcal{G}$. □

Clearly, junk rules can be removed from a GSOS language T without altering the associated LTS. Note, moreover, that it is legitimate to eliminate all the junk rules from T at once. This is because whenever ρ_1 and ρ_2 are junk rules in T, then ρ_2 is still junk in the GSOS language obtained from T by removing ρ_1, as the two GSOS languages are associated with the same LTS.

5.4.3. *Coding a universal* 2*-counter machine.* Despite the finiteness restrictions imposed on GSOS languages, they are a Turing powerful model of computation. We exhibit a GSOS language with, for each n, a term $\mathrm{U2CM}_n$ that behaves as a universal 2-counter machine on input n. Then $\mathrm{U2CM}_n \leftrightarrow a^\omega$, where $a^\omega\xrightarrow{a}a^\omega$, iff the 2-counter machine diverges on input n, a prototypical undecidable problem.

Suppose the 2-counter machine has code of the form:

l_1: if $I=0$ `goto` l_5
l_2: `inc I`
l_3: `dec J`

l_4: `goto` l_7

$\vdots$

l_k: `halt`

We assume a toy process algebra containing the inactive constant $\mathbf{0}$ and the unary prefix multiplication operators $zero \cdot \_$ and $succ \cdot \_$, while $\mathsf{Act} \triangleq \{a, succ, zero\}$. Since $\mathbf{0}$ does not exhibit any behaviour, it does not have any transition rules. The transition rules for prefix multiplication are

$$\frac{}{succ \cdot x \xrightarrow{succ} x} \qquad \frac{}{zero \cdot x \xrightarrow{zero} x}.$$

Intuitively, $succ$ and $zero$ represent the successor function and the zero for the counters: a natural number n is encoded by the term $succ^n \cdot zero \cdot \mathbf{0}$ (where $succ^n$ denotes n nestings of the prefix multiplication function $succ$). Thus, if a closed term t codes n and $t \xrightarrow{succ} t'$, then $n > 0$ and t' codes $n - 1$. Also, if t codes n, then $t \xrightarrow{zero} \mathbf{0}$ iff $n = 0$. The action a is the pulse emitted by a process as it performs a computation step.

Finally, the syntax also contains binary function symbols $l_1, \ldots, l_k$ to code the states of the 2-counter machine: $l_i(succ^m \cdot zero \cdot \mathbf{0}, succ^n \cdot zero \cdot \mathbf{0})$ codes the machine at label l_i, with the two counters $\mathtt{I} = m$ and $\mathtt{J} = n$. The transition rules for these function symbols are as follows.

- If the ith instruction is of the form `if I=0 goto` l_j, then

$$\frac{x \xrightarrow{zero} x'}{l_i(x, y) \xrightarrow{a} l_j(zero \cdot x', y)} \qquad \frac{x \xrightarrow{succ} x'}{l_i(x, y) \xrightarrow{a} l_{i+1}(succ \cdot x', y)}.$$

- If the ith instruction is `goto` l_j, then

$$\frac{}{l_i(x, y) \xrightarrow{a} l_j(x, y)}.$$

- If the ith instruction is `inc I`, then

$$\frac{}{l_i(x, y) \xrightarrow{a} l_{i+1}(succ \cdot x, y)}.$$

- If the ith instruction is `dec I`, then

$$\frac{x \xrightarrow{zero} x'}{l_i(x, y) \xrightarrow{a} l_{i+1}(zero \cdot x', y)} \qquad \frac{x \xrightarrow{succ} x'}{l_i(x, y) \xrightarrow{a} l_{i+1}(x', y)}.$$

Commands that deal with the other counter `J` are similar. There are no transition rules for labels of `halt` commands, as these cause the automaton to halt. We define $\mathrm{U2CM}_n \triangleq l_1(succ^n \cdot zero \cdot \mathbf{0}, zero \cdot \mathbf{0})$. The reader will not find it hard to see that $\mathrm{U2CM}_n \leftrightarrow a^\omega$ iff the universal machine diverges on input n.

5.4.4. *Infinitary GSOS languages inducing regular LTSs.* Regular LTSs (cf. Definition 2.2) may be used to describe many interesting concurrent systems – e.g., several communication protocols and mutual exclusion algorithms [229] – and form the basis of semantic-based automated verification tools like those presented in, e.g., [65,79,201]. As (subsets of) programming languages that can be given semantics by means of regular LTSs are, at least in principle, amenable to automated verification techniques, it is interesting to develop techniques to check whether languages give rise to regular LTSs. Moreover, since such a property is in general undecidable, it is useful to single out sufficient syntactic restrictions on the transition rules in a TSS, to ensure regularity of the associated LTS.

We saw in Section 5.4.3 that GSOS languages can specify a universal 2-counter machine, and are therefore Turing powerful. In this section, we study an infinitary version of GSOS languages, in which the finiteness restrictions on the signature, set of actions, and transition rules are (temporarily) relaxed to countability restrictions. We present a restricted version of infinitary GSOS languages, which are guaranteed to give rise to regular LTSs.

DEFINITION 5.23 (*Infinitary GSOS*). An infinitary GSOS language is a countable set of GSOS rules over a countable signature and a countable set of actions.

In order to ensure that the associated LTSs are regular, it is necessary to impose restrictions on the class of infinitary GSOS languages, ensuring that the LTS is finitely branching and that the set of closed terms reachable from any closed term is finite. We recall that the LTS associated with a finitary GSOS language is finitely branching. However, an infinitary GSOS language such as $\emptyset/a_0 \xrightarrow{a_i} a_i$ for $i \in \mathbb{N}$ gives rise to an LTS that is infinitely branching.

DEFINITION 5.24 (*Positive trigger*). The *positive trigger* of a GSOS rule as in Definition 5.21 is a tuple $\langle e_1, \ldots, e_{ar(f)} \rangle$ of subsets of Act, where

$$e_i = \{a_{ij} \mid 1 \leqslant j \leqslant m_i\} \quad (\text{for } i = 1, \ldots, ar(f)).$$

DEFINITION 5.25 (*Boundedness*). Assume a function symbol f in the signature of an infinitary GSOS language. We say that:

- f is *bounded* if for each positive trigger, the corresponding set of GSOS rules of type f is finite;
- f is *uniformly bounded* if there exists a finite upper bound on the number of GSOS rules of type f having the same positive trigger.

As far as we know, all standard operations in process algebras that occur in the literature are uniformly bounded. The notion of a bounded function symbol was originally developed by Vaandrager [221] for De Simone languages (see Definition 5.14), and was extended to infinitary GSOS languages in [5]. The notion of a uniformly bounded function symbol stems from [4], and will reoccur in the definition of a regular GSOS language (see Definition 5.38). The following result is from [5].

PROPOSITION 5.26. *If each function symbol in the signature of an infinitary GSOS language is bounded, then the associated LTS is finitely branching.*

We introduce a further restriction on infinitary GSOS languages from [5], to ensure that in the associated LTSs each state can reach only finitely many other states.

DEFINITION 5.27 (*Simple GSOS*). A GSOS rule is *simple* if its target contains at most one function symbol. A GSOS language is *simple* if each of its transition rules is.

Rule formats similar to the simple GSOS rules have emerged in work by several researchers, e.g., [20], [70, p. 230], and [170, Definition 13]. Most of the standard operations in process algebras are given operational semantics by means of simple GSOS rules. An exception is the binary Kleene star, which was discussed in Section 5.4.1. Two further exceptions are the "desynchronizing" Δ operation present in the early versions of Milner's SCCS [155] studied in [115,154], and the parallel composition operation in the π-calculus [160]. The Δ operation has GSOS rules

$$\frac{x \xrightarrow{a} x'}{\Delta x \xrightarrow{a} \delta \Delta x'}$$

for $a \in \mathsf{Act}$, where δ is the *delay* operation of SCCS. The GSOS rules for the parallel composition operation of the π-calculus dealing with so-called scope extrusion (see [160, part II]) take the form

$$\frac{x \xrightarrow{\overline{v}(w)} x' \quad y \xrightarrow{v(w)} y'}{x \mid y \xrightarrow{\tau} (w)(x' \mid y')}$$

where (w) denotes the restriction operation of the π-calculus and τ a silent step (cf. Section 5.5).

The following result can be shown by structural induction on closed terms, following the lines of [5, Theorem 5.5].

THEOREM 5.28. *Assume a simple infinitary GSOS language. If each function symbol in its signature is bounded and depends on only finitely many function symbols* (*cf. Definition* 5.16), *then the associated LTS is regular.*

The above result would not hold if we allowed GSOS rules with more than one function symbol in their targets, as the following example shows.

EXAMPLE. Consider a GSOS language with action a, constants b and c, a unary function symbol f, and transition rules

$$\frac{}{c \xrightarrow{a} b} \qquad \frac{}{c \xrightarrow{a} f(c)} \qquad \frac{}{f(x) \xrightarrow{a} x} \qquad \frac{x \xrightarrow{a} y}{f(x) \xrightarrow{a} f(y)}.$$

Note that the second transition rule of type c is not simple, as its target carries two function symbols. It is not hard to see that c can reach infinitely many states $f^n(c)$ and $f^n(b)$ for

$n \in \mathbb{N}$, because $f^n(c) \xrightarrow{a} f^{n+1}(c)$ and $f^n(c) \xrightarrow{a} f^n(b)$. Moreover, these states are all non-bisimilar.

Madelaine and Vergamini [142] studied syntactic conditions on De Simone rules [198, 199] to ensure that the associated LTS is regular. They identify two classes of well-behaved function symbols, which they call *non-growing operations* and *sieves*. Intuitively, non-growing operations are function symbols which, when fed with (terms denoting) regular LTSs, build regular LTSs. Sieves are a special class of unary non-growing operations whose transition rules have the form

$$\frac{x \xrightarrow{a} x'}{f(x) \xrightarrow{b} f(x')}.$$

For example, standard process algebra operations like CCS restriction and renaming [157] and CSP hiding [127] are sieves. Note that transition rules for sieves are simple. In view of Theorem 5.28, all function symbols in an infinitary GSOS language given by means of simple transition rules are non-growing in the sense of Madelaine and Vergamini.

The syntactic condition used by Madelaine and Vergamini to establish that some operations are non-growing is based on term rewriting techniques, to find a so-called *simplification ordering* over terms (see [145, Definition 4]). This is similar in spirit to showing that linear GSOS languages (see Definition 5.34) are syntactically well-founded (see Definition 5.35); the interested reader is referred to Section 5.4.5 and [8, Section 6] for more information. Unfortunately, the existence of a simplification ordering compatible with a set of rewrite rules is not decidable even for finitary GSOS languages.

Specialized techniques which can be used to show that certain closed terms can reach only finitely many other closed terms have been proposed for CCS and related languages. The interested reader is invited to consult [71] and the references therein. Not surprisingly, these specialized methods tend to be more powerful than general syntactic ones, as they rely on language-dependent semantic information. For instance, a method to check the regularity of a large set of CCS terms based on abstract interpretation techniques (see, e.g., [2]) has been proposed in [71].

5.4.5. *Turning GSOS rules into equations.* There are several methods for specifying and verifying processes behaviour, e.g., modal formulae [208] and variants of Hoare logic [168, 207]. A fairly successful verification technique is to approximate the specification by a (not necessarily implementable) term in some process algebra. In this setting, a set of axioms can be applied to try and show that the term is behaviourally equivalent to, or in some other sense a suitable approximation of, the required specification. Indeed, one of the major schools of theoretical concurrency and its applications, that of ACP [21,28], takes the notion of behavioural equivalence as primary, and defines operational semantics to fit its axioms.

A logic of programs is complete (relative to a programming language) if all true formulas of the language are provable in the logic. As properties of interest are generally non-recursive, we are often obliged to have infinitary or other non-recursive rules in our logics to achieve completeness.

This section presents results from [7,8], which offer an algorithmic solution to the problem of computing a sound and complete axiomatization (possibly including one infinitary conditional axiom) for any GSOS language, modulo bisimulation equivalence. That is, two closed terms can be equated by the axiom system iff they are bisimilar in the associated LTS (cf. Definition 4.9). The procedure introduces fresh function symbols as needed. Completeness results for axiomatizations have become rather standard in many cases. The generalization of extant completeness results given in [8] shows that, at least in principle, this burden can be completely removed if one gives GSOS rules for a process algebra. Of course, this does not mean that there is nothing to do on specific process algebras. For instance, sometimes it may be possible to eliminate some of the auxiliary function symbols (we will see an example of this later on in this section), or the infinitary conditional axiom. (The interested reader will find many results on complete axiomatizations of behavioural equivalences over several process algebras elsewhere in this handbook.)

We first define the GSOS language T_{FINTREE}, which is a fragment of CCS suitable for expressing finite LTSs (cf. Definition 2.2). Its signature $\Sigma_{\mathsf{FINTREE}}$ consists of:
- the constant $\mathbf{0}$, denoting the inactive process;
- binary alternative composition $x + y$, which chooses non-deterministically between x and y;
- unary prefix multiplication $a{\cdot}x$ for $a \in \mathsf{Act}$, which executes action a and thereafter behaves as x.

The constant $\mathbf{0}$ does not exhibit any behaviour and consequently does not have any transition rules. The transition rules of alternative composition and prefix multiplication have been formulated earlier in this chapter (see Table 1 in Section 2.6, and Section 5.4.3, respectively). Most process algebras contain the function symbols above, either directly or as derived operations. The following completeness result (cf. Definition 4.9) is well-known [122,157].

PROPOSITION 5.29. *Let $\mathcal{E}_{\mathsf{FINTREE}}$ denote the axiomatization*

$$x + y = y + x, \tag{3}$$
$$(x + y) + z = x + (y + z), \tag{4}$$
$$x + x = x, \tag{5}$$
$$x + \mathbf{0} = x. \tag{6}$$

$\mathcal{E}_{\mathsf{FINTREE}}$ is sound and complete modulo bisimulation equivalence as induced by T_{FINTREE}.

Following [8], we show how to find for any GSOS language T extending T_{FINTREE}, an axiomatization $\mathcal{E}$, extending $\mathcal{E}_{\mathsf{FINTREE}}$, that is sound and complete modulo bisimulation equivalence. That is, two closed terms are bisimilar as states in the LTS associated with T iff they can be equated by $\mathcal{E}$.

Moller [162] has shown that bisimulation equivalence over a subset of CCS with the interleaving operation $\|$, which can be defined in GSOS, cannot be completely characterized by any finite unconditional axiomatization over that language. Thus, the algorithm to produce $\mathcal{E}$ may require the addition of auxiliary function symbols to the signature, and of GSOS rules for these auxiliary function symbols to T.

We start with a typical example of the way in which the completeness result for T_{FINTREE} in Proposition 5.29 is used.

EXAMPLE. Consider the GSOS language T_{∂^1} that is obtained by extending the signature of T_{FINTREE} with unary function symbols ∂^1_H [22], where H is a finite set of actions, and adding the transition rules

$$\frac{x \xrightarrow{a} y}{\partial^1_H(x) \xrightarrow{a} y} \quad a \notin H.$$

In other words, the process $\partial^1_H(t)$ behaves like t, except that it cannot do any actions from H in its first move. Note that ∂^1_H is different from the CCS restriction operation, as CCS restriction is persistent while ∂^1_H disappears after one transition.

The following result, whose proof may be found in [8], is a corollary of Theorem 4.12 on completeness of axiomatizations over extended signatures, and a blueprint for developments to follow. The idea is that, using the axioms below, the completeness problem for a super-language of T_{FINTREE} can be reduced to the completeness problem for T_{FINTREE}, which has been solved in Proposition 5.29.

PROPOSITION 5.30. *Let $\mathcal{E}_{\partial^1}$ be the axiomatization that extends $\mathcal{E}_{\mathsf{FINTREE}}$ with the axioms*

$$\begin{aligned} \partial^1_H(x+y) &= \partial^1_H(x) + \partial^1_H(y), \\ \partial^1_H(a \cdot x) &= a \cdot x \quad \textit{if } a \notin H, \\ \partial^1_H(a \cdot x) &= \mathbf{0} \quad \textit{if } a \in H, \\ \partial^1_H(\mathbf{0}) &= \mathbf{0}. \end{aligned}$$

$\mathcal{E}_{\partial^1}$ is sound and complete modulo bisimulation equivalence as induced by T_{∂^1}.

Proposition 5.30 follows from Theorem 4.12, owing to the observations that T_{∂^1} is an operational conservative extension (cf. Definition 4.2) of T_{FINTREE} (this follows from Theorem 4.4), $\mathcal{E}_{\partial^1}$ is sound over the extended signature, $\mathcal{E}_{\mathsf{FINTREE}}$ is complete over $\Sigma_{\mathsf{FINTREE}}$ (Proposition 5.29), and each closed term over the extended signature can be equated to a closed term over $\Sigma_{\mathsf{FINTREE}}$ by means of the axiomatization $\mathcal{E}_{\partial^1}$.

The approach that yielded a sound and complete axiomatization for T_{∂^1} modulo bisimulation equivalence can be generalized to arbitrary GSOS super-languages of T_{FINTREE}. That is, we want to find axioms, on top of $\mathcal{E}_{\mathsf{FINTREE}}$, that allow us to eliminate all extra function symbols from closed terms. This requires a variety of methods, in which Proposition 5.30 plays an important role, because the ∂^1_H operator can be used to encode negative premises.

The following notion from [8] is needed in presenting the main result on the automatic generation of axiomatizations modulo bisimulation equivalence for GSOS languages.

DEFINITION 5.31 (*Disjoint extension*). A GSOS language T_1 is a *disjoint extension* of a GSOS language T_0 if the signature and transition rules of T_1 include those of T_0, and the types of transition rules in T_1, which were not present in T_0, are not in the signature of T_0.

Note that disjoint extension is a partial order. If T_1 disjointly extends T_0, then $T_0 \oplus T_1$ is an operational conservative extension (see Definition 4.2) of T_0. This follows immediately from Theorem 4.4, owing to the fact that GSOS rules are by default source-dependent (cf. Definitions 4.3 and 5.21), and the sources of transition rules in T_1 are fresh.

Before presenting the main results of [8], we need to discuss a subtlety. We want to know that the axioms in $\mathcal{E}_{\mathsf{FINTREE}}$ are sound modulo bisimulation equivalence as induced by any disjoint extension T of T_{FINTREE}. In general it is not the case that validity of axioms is preserved by taking disjoint extensions. For instance, consider the trivial GSOS language T_{NIL} consisting of the constant $\mathbf{0}$ and no transition rules. The axiom $x = y$ is sound modulo bisimulation equivalence as induced by T_{NIL}, but clearly this law is not sound modulo bisimulation equivalence as induced by T_{FINTREE}, even though T_{FINTREE} is a disjoint extension of T_{NIL}. Fortunately, soundness of the axioms in $\mathcal{E}_{\mathsf{FINTREE}}$, and also all the other axioms that are needed in the developments of [8], is preserved by taking disjoint extensions of T_{FINTREE}.

Semantically well-founded GSOS languages. We start with the generation of sound and complete axiomatizations modulo bisimulation equivalence for the limited class of *semantically well-founded* GSOS languages, generating finite LTSs (cf. Definition 2.2). For such languages we can do without infinitary conditional axioms.

DEFINITION 5.32 (*Semantic well-foundedness*). A GSOS language is *semantically well-founded* if its associated LTS is finite.

The class of semantically well-founded GSOS languages contains the recursion-free finite-alphabet sublanguages of most of the standard process algebras.

In [8] an algorithm is presented that, given a disjoint extension T of T_{FINTREE}, constructs a disjoint extension T' of T and T_{∂^1}, and a finite (unconditional) axiomatization $\mathcal{E}$ that is sound modulo bisimulation equivalence as induced by T', such that each closed term over the signature of T' can be equated by $\mathcal{E}$ to a closed term of the form $a_1 \cdot t_1 + \cdots + a_n \cdot t_n$ (where the empty sum represents $\mathbf{0}$). For semantically well-founded GSOS languages it is possible to iterate this reduction a finite number of times, thereby eliminating all function symbols that are not in $\Sigma_{\mathsf{FINTREE}}$. As in the completeness proof for T_{∂^1}, this reduces completeness of $\mathcal{E}$ with respect to T' to completeness of $\mathcal{E}_{\mathsf{FINTREE}}$ with respect to T_{FINTREE}, which has been solved in Proposition 5.29. Thus we obtain the following result.

THEOREM 5.33. *There is an algorithm that, given a GSOS language T, which is a semantically well-founded disjoint extension of T_{FINTREE}, constructs a disjoint extension T' of T and T_{∂^1}, and a finite unconditional axiomatization $\mathcal{E}$, such that $\mathcal{E}$ is sound and complete modulo bisimulation equivalence as induced by T'.*

The requirement that T is a disjoint extension of T_{FINTREE} is necessary in Theorem 5.33, because otherwise the quoted algorithm might not preserve semantic well-foundedness. Namely, the combination of a semantically well-founded GSOS language with T_{FINTREE} is in general not semantically well-founded; see [8].

EXAMPLE. An operation found in many process algebras is the parallel composition $\|$ without communication, which is defined by the transition rules

$$\frac{x \xrightarrow{a} x'}{x \| y \xrightarrow{a} x' \| y} \qquad \frac{y \xrightarrow{a} y'}{x \| y \xrightarrow{a} x \| y'}$$

for $a \in \mathsf{Act}$. This is an intuitively reasonable definition of parallel composition, and the transition rules are easy to explain. It is somewhat harder to see how to describe it equationally. Some axioms are clear enough – the operation $\|$ is commutative and associative, and the stopped process is its identity – but the first finite equational description did not appear until [35]. This equational characterization required an additional function symbol "left merge" $\mathbin{\|\mkern-6mu\raisebox{-1pt}{\_}}$. Intuitively, $t \mathbin{\|\mkern-6mu\raisebox{-1pt}{\_}} u$ behaves as $t \| u$ except that its first move must be taken by t. For each $a \in \mathsf{Act}$, left merge has a transition rule

$$\frac{x \xrightarrow{a} x'}{x \mathbin{\|\mkern-6mu\raisebox{-1pt}{\_}} y \xrightarrow{a} x' \| y}.$$

The axioms for $\|$ and $\mathbin{\|\mkern-6mu\raisebox{-1pt}{\_}}$ are:

$$\begin{aligned} x \| y &= (x \mathbin{\|\mkern-6mu\raisebox{-1pt}{\_}} y) + (y \mathbin{\|\mkern-6mu\raisebox{-1pt}{\_}} x), \\ (x + y) \mathbin{\|\mkern-6mu\raisebox{-1pt}{\_}} z &= (x \mathbin{\|\mkern-6mu\raisebox{-1pt}{\_}} z) + (y \mathbin{\|\mkern-6mu\raisebox{-1pt}{\_}} z), \\ (a \cdot x) \mathbin{\|\mkern-6mu\raisebox{-1pt}{\_}} y &= a \cdot (x \| y), \\ \mathbf{0} \mathbin{\|\mkern-6mu\raisebox{-1pt}{\_}} x &= \mathbf{0}. \end{aligned}$$

These axioms for $\|$ and $\mathbin{\|\mkern-6mu\raisebox{-1pt}{\_}}$, together with the axioms in $\mathcal{E}_{\partial^1}$ for $+$, $a \cdot \_$, $\mathbf{0}$, and ∂^1_H, form a sound and complete axiomatization for the closed terms over this signature modulo bisimulation equivalence.

T_{FINTREE} with parallel composition is a semantically well-founded GSOS language and a disjoint extension of T_{FINTREE}. The auxiliary operator $\mathbin{\|\mkern-6mu\raisebox{-1pt}{\_}}$, and the axioms above for parallel composition and the left merge, are also produced by the algorithm from [8] that was mentioned in Theorem 5.33. In fact, due to the symmetric character of the parallel composition operator, the algorithm from [8] actually produces two auxiliary operators $\mathbin{\|\mkern-6mu\raisebox{-1pt}{\_}}$ and $\mathbin{\raisebox{-1pt}{\_}\mkern-6mu\|}$, where $t \mathbin{\raisebox{-1pt}{\_}\mkern-6mu\|} u$ behaves as $t \| u$ except that its first move must be taken by u. Parallel composition is then axiomatized by

$$x \| y = (x \mathbin{\|\mkern-6mu\raisebox{-1pt}{\_}} y) + (x \mathbin{\raisebox{-1pt}{\_}\mkern-6mu\|} y).$$

However, since $t \mathbin{\raisebox{-1pt}{\_}\mkern-6mu\|} u \leftrightarrow u \mathbin{\|\mkern-6mu\raisebox{-1pt}{\_}} t$, the right merge $\mathbin{\raisebox{-1pt}{\_}\mkern-6mu\|}$ can be expressed by means of the left merge $\mathbin{\|\mkern-6mu\raisebox{-1pt}{\_}}$.

Syntactic well-foundedness. Since GSOS languages are Turing powerful, it is undecidable whether a GSOS language is semantically well-founded. However, for an interesting subclass of GSOS languages there exist effective syntactic constraints on GSOS rules that ensure semantic well-foundedness.

DEFINITION 5.34 (*Linear GSOS*). A GSOS rule as in Definition 5.21 is *linear* if each variable occurs at most once in the target t and, for each argument i that is tested positively (cf. Definition 5.11), x_i does not occur in the target and at most one of the y_{ij}'s does.
A GSOS language is *linear* if all its transition rules are.

As far as we know, all transition rules for standard process algebras are linear.

DEFINITION 5.35 (*Syntactic well-foundedness*). A GSOS language is *syntactically well-founded* if there exists a weight mapping w from function symbols to natural numbers such that, for each GSOS rule ρ with type f and target t:
- if ρ has no positive premise then $W(t) < w(f)$, and
- $W(t) \leqslant w(f)$ otherwise,

where $W(t)$ adds the weights of all function symbols in t:

$$\begin{aligned} W(x) &\triangleq 0, \\ W\big(g(t_1, \ldots, t_{\mathrm{ar}(g)})\big) &\triangleq w(g) + W(t_1) + \cdots + W(t_{ar(g)}). \end{aligned}$$

PROPOSITION 5.36. *If a GSOS language is linear and syntactically well-founded, then it is semantically well-founded. Moreover, it is decidable whether a GSOS language is syntactically well-founded.*

General GSOS languages. It follows from some recursion theoretic considerations, discussed in [8] and based upon the programming exercise in Section 5.4.3, that the extension of the completeness result given in Theorem 5.33 to general GSOS languages requires some proof rules beyond purely equational logic. However, it is possible to extend the completeness result to the whole class of GSOS languages in a rather standard way. Bisimulation equivalence over finitely branching LTSs supports a powerful induction principle, known as the *Approximation Induction Principle* (AIP) [39,92]. Since the LTS associated with a GSOS language is finitely branching, AIP applies.

We introduce a family of unary function symbols $\pi_n(\_)$ for $n \in \mathbb{N}$, with transition rules

$$\frac{x \xrightarrow{a} y}{\pi_{n+1}(x) \xrightarrow{a} \pi_n(y)}$$

for $a \in \mathsf{Act}$. These function symbols are known as *projection* operators in the literature on ACP [28]. Intuitively, $\pi_n(t)$ allows t to perform n moves freely, and then stops it. AIP is the infinitary conditional axiom

$$\text{AIP} \qquad \forall n \in \mathbb{N}\big(\pi_n(x) = \pi_n(y)\big) \Rightarrow x = y.$$

Intuitively, this proof rule states a "continuity" property of bisimulation equivalence over finitely branching LTSs: if two states are bisimilar at any finite depth, then they are bisimilar.

The projection operators are somewhat heavy-handed, as there are infinitely many of them, and GSOS languages are defined to be finite. It is, however, possible to mimic the projection operators by means of a binary function symbol $\_/\_$. Intuitively, a closed term t/u executes t (where u also silently takes a step) until the "hourglass" process u runs out and halts the execution. That is, for all actions $a, b \in \mathsf{Act}$ we have the following transition rule for $\_/\_$:

$$\frac{x \xrightarrow{a} x' \quad y \xrightarrow{b} y'}{x/y \xrightarrow{a} x'/y'}. \tag{7}$$

For each $n \in \mathbb{N}$, $\pi_n(t) \underline{\leftrightarrow} t/c^n$ with c an arbitrarily chosen action. In this formulation, we may rephrase AIP as follows:

$$\text{AIP}' \qquad \forall n \in \mathbb{N}(x/c^n = y/c^n) \Rightarrow x = y.$$

Now we are ready to formulate the analogue of Theorem 5.33 for GSOS languages that need not be semantically well-founded.

THEOREM 5.37. *There is an algorithm that, given a GSOS language T, constructs a disjoint extension T' of T, T_{∂^1}, and the operation $\_/\_$ defined by* (7)*, and a finite unconditional axiomatization $\mathcal{E}$, such that $\mathcal{E}$ and AIP′ together are sound and complete modulo bisimulation equivalence as induced by T'.*

Term rewriting properties of the axiomatizations generated by (variations on) the methods we have just surveyed have been studied by Bosscher [59]. Ulidowski [216] proposed a modification of the approach in [8] that produces complete axiomatizations for a subclass of the De Simone languages, modulo the refusal simulation preorder from [215] that takes into account the silent step τ (cf. Section 5.5).

Regular GSOS languages. In [4] it was shown that for a subclass of infinitary recursive GSOS languages generating regular LTSs it is possible to obtain sound and complete axiomatizations modulo bisimulation equivalence that do not rely on infinitary proof rules like AIP. The following definition introduces the class of *regular* GSOS languages that is considered in *op. cit.*, which is a subclass of the simple infinitary GSOS languages (cf. Definition 5.27).

DEFINITION 5.38 (*Regular GSOS*). An infinitary GSOS language is *regular* if it is simple and for each function symbol f in its signature:
(1) f is uniformly bounded (cf. Definition 5.25);
(2) f depends on only finitely many function symbols (cf. Definition 5.16);
(3) there is a finite upper bound on the number of positive premises in transition rules with type f.

Most of the TSSs for standard process algebras in the literature are regular. It follows immediately from the theory outlined in Section 5.4.4 that every regular GSOS language induces a regular LTS. In [4] it was shown how to reduce the completeness problem for regular GSOS languages to that for regular LTSs, which was solved earlier in [40,156].

We recall from Section 5.3.1 on De Simone languages that the recursive terms over a signature Σ are given by the BNF grammar

$$t ::= X \mid f(t_1, \ldots, t_{ar(f)}) \mid \mathrm{fix}(X = t),$$

where X ranges over a countably infinite set of recursion variables, f ranges over the signature Σ, and fix is a binding construct. The latter construct gives rise to the usual notions of free and bound recursion variables in recursive terms. For every recursive term $\mathrm{fix}(X = t)$ there is a transition rule

$$\frac{t[\mathrm{fix}(X = t)/X] \xrightarrow{a} y}{\mathrm{fix}(X = t) \xrightarrow{a} y}.$$

We consider the subset of guarded recursive terms (cf. Definition 5.12) without free recursion variables.

Bisimulation equivalence over guarded recursive terms has been completely axiomatized by Milner [156] and Bergstra and Klop [40]. The following proof rules, called the *Recursive Definition Principle* (RDP) and the *Recursive Specification Principle* (RSP), play a key role in these completeness proofs. Let t denote a guarded recursive term with no other free recursion variables than X, and let u denote a guarded recursive term without free recursion variables:

$$\begin{array}{ll} \text{RDP} & \mathrm{fix}(X = t) = t[\mathrm{fix}(X = t)/X] \\ \text{RSP} & u = t[u/X] \Rightarrow u = \mathrm{fix}(X = t). \end{array}$$

THEOREM 5.39. *There is an algorithm that, given a regular GSOS language T, constructs a disjoint extension T' of T over guarded recursive terms, and an unconditional axiomatization $\mathcal{E}$, such that $\mathcal{E}$, RDP, and RSP together are sound and complete modulo bisimulation equivalence as induced by T'.*

The algorithm used in the proof of Theorem 5.39 does not work for GSOS languages that are not regular; see [4] for details.

Apart from the work we have just reviewed, the automatic generation of complete axiomatizations from TSSs has received a good deal of attention in the literature. Further results on this line of research may be found in, e.g., [12,50,217].

5.4.6. *From recursive GSOS to LTSs with divergence.* This section considers GSOS languages for recursive terms (cf. Section 5.3.1 and the end of the previous section) over a signature Σ. Let $\mathsf{CREC}(\Sigma)$ denote the set of recursive terms that do not contain free recursion variables.

We recall that, in the absence of recursion, GSOS languages are strictly stratified (cf. Definition 3.13), yielding one of the most restrictive criteria for meaningful TSSs considered in Section 3. For recursive GSOS languages, however, it is no longer straightforward to find an associated LTS. In this section it is shown how this problem can be overcome by the use of a special divergence predicate in the definition of LTSs. The interested reader is referred to, e.g., [115,125,153,230] for motivation and more information on (variations on) this semantic model for reactive systems.

DEFINITION 5.40 (*LTS with divergence*). An *LTS with divergence* is an LTS, in the sense of Definition 2.1, whose only predicates are the *divergence* predicate $\uparrow$ and the *convergence* predicate $\downarrow$.

In [11,12] it is shown how a recursive GSOS language specifies an LTS with divergence over $\mathsf{CREC}(\Sigma)$. The recursive GSOS languages considered in *op. cit.* contain a special constant Ω that is akin to the constant $\mathbf{0}$ from CCS; i.e., there are no transition rules of type Ω. (The difference between $\mathbf{0}$ and Ω is that whereas the former denotes the convergent process without transitions, the latter stands for the divergent stopped process.) The transition rules in a recursive GSOS language are used to define a divergence (or under-specification) predicate on $\mathsf{CREC}(\Sigma)$. In fact, as is common practice in the literature on process algebras, we first define the notion of convergence, and use it to define the divergence predicate.

DEFINITION 5.41 (*Convergence*). Assume a recursive GSOS language over a signature Σ. The convergence predicate $\downarrow$ is defined to be the least predicate over the set of closed recursive terms $\mathsf{CREC}(\Sigma)$ that satisfies the following clauses:

(1) $f(t_1, \ldots, t_{ar(f)}) \downarrow$ if
  (a) $f \neq \Omega$, and
  (b) for every argument i of f, if f tests i (cf. Definition 5.11) then $t_i \downarrow$;

(2) $\mathrm{fix}(X = t) \downarrow$ if $t[\mathrm{fix}(X = t)/X] \downarrow$.

We write $t \uparrow$ if it is not the case that $t \downarrow$.

The motivation for the above definition is the following: a term t is divergent if its initial transitions are not fully specified. This occurs either when the initial behaviour of term t depends on under-specified arguments like Ω, or in the presence of unguarded recursive definitions (cf. Definition 5.12). For example, the terms $\mathrm{fix}(X = X)$ and $\mathrm{fix}(X = a.\mathbf{0} + X)$ are not convergent, as the initial behaviours of these terms depend on themselves. We remark here that, when applied to SCCS [155] and the version of CCS considered in [230], the above definition delivers exactly the convergence predicates given by Hennessy [115] and Walker [230], respectively.

We hint at how an LTS with divergence can be associated with a recursive GSOS language – the interested reader is referred to [11,12] for full technical details. We want the LTS with divergence to be at least a supported model in the sense of Definition 3.2. Moreover, reminiscent of positive TSSs, we want to associate the least such LTS with the recursive GSOS language in question. As described in, e.g., [55], this is not possible in the absence of divergence. However, the extra structure given by the convergence predicate can be put to good use in giving a simple way of constructing the desired LTS with divergence over $\mathsf{CREC}(\Sigma)$, in two steps. First, the transitions emanating from convergent terms are derived by induction on the convergence predicate. This is done according to the standard approach for GSOS languages outlined in Section 5.4, and using the transition rule for recursion to derive the transitions of recursive terms. Next, the information about the transitions that are possible for convergent terms is used to determine the outgoing transitions for all terms in $\mathsf{CREC}(\Sigma)$. The key point in the construction of the associated LTS is that negative premises can be satisfied by convergent terms only. Intuitively, to know that a closed term cannot initially perform a given action, we need to find out precisely all the initial actions that it can perform. If a closed term is divergent, then its set of initial actions is not fully specified; thus we cannot be sure whether such a term satisfies a negative premise or not.

The reader familiar with the literature on Hennessy–Milner logics (cf. Definition 2.7) for prebisimulation-like relations (cf. Definition 7.2 to follow) may have noticed that the notion of satisfaction for negative premises discussed above is akin to that for formulae of the form $[a]\varphi$ given in, e.g., [1,10,153,206,208]. In those references, the new interpretation is necessary to obtain monotonicity of the satisfaction relation with respect to the appropriate notion of prebisimulation. The intuitionistic interpretation of negative premises given in [11,12] is crucial to obtain operations that are monotonic with respect to the notion of prebisimulation $\lesssim$ presented in Definition 7.2. Basically, it ensures that, for a closed term t, the transition formula $t \stackrel{a}{\not\rightarrow}$ holds iff $u \stackrel{a}{\not\rightarrow}$ holds for every closed term u with $t \lesssim u$.

The following result is from [11,12], where the details of the construction of the desired LTS with divergence may be found.

PROPOSITION 5.42. *For every recursive GSOS language with the inert constant Ω, there exists a least sound and supported LTS with divergence over* $\mathsf{CREC}(\Sigma)$.

EXAMPLE. Consider the term $\mathrm{fix}(X = odd(X))$, where the unary function symbol odd is defined by the transition rule

$$\frac{x \stackrel{a}{\not\rightarrow}}{odd(x) \stackrel{a}{\rightarrow} \mathbf{0}}.$$

This operation is a standard example used in the literature to show that negative premises and unguarded recursive definitions can lead to inconsistent specifications (see, e.g., [44]). The reason for this phenomenon is that, if we follow the standard GSOS approach, the recursive equation

$$X = odd(X)$$

does not have any solution modulo bisimulation equivalence. In fact, with the standard operational interpretation of GSOS languages and general TSSs with negative premises, a term t solving the above recursive equation modulo bisimulation equivalence (i.e., $t \leftrightarrow odd(t)$) would have to exhibit an initial a-transition iff it does not have one. In the approach of [11,12], instead, the above recursive equation has a unique solution modulo prebisimulation equivalence (see Definition 7.2). Namely, $\mathrm{fix}(X = odd(X))$ is a divergent term. Since negative premises are interpreted over convergent terms only, the above transition rule cannot be applied to derive a transition for $\mathrm{fix}(X = odd(X))$, so this term is prebisimilar to the inert constant Ω. Thus Ω is the unique solution of $X = odd(X)$ modulo prebisimulation equivalence.

5.4.7. *Other results for GSOS languages.* The theory of GSOS languages is rather rich in results, and we have only presented a sample of the body of work documented in the literature on this rule format. We end this overview of the work on GSOS languages with pointers to other interesting results.

Ruloids. Bloom, Istrail, and Meyer [55] observed that the behaviour of each open term $C[x_1, \ldots, x_N]$ is completely determined by a finite set of derived transition rules of the form

$$\frac{\{x_i \xrightarrow{a_{ij}} y_{ij},\ x_i \stackrel{b_{ik}}{\not\longrightarrow} \mid 1 \leqslant i \leqslant N, 1 \leqslant j \leqslant m_i, 1 \leqslant k \leqslant n_i\}}{C[x_1, \ldots, x_N] \xrightarrow{c} t}.$$

These derived transition rules are referred to as *ruloids*, to distinguish them from the GSOS rules defining the operational semantics of the language under consideration. Ruloids facilitate the development of theory for SOS. For instance, the use of ruloids simplifies the proof of the congruence result for TSSs, Theorem 5.3, in the restricted case of GSOS languages. The following result may be found in [55, Theorem 7.6].

THEOREM 5.43. *Let T be a GSOS language. For every open term $C[x_1, \ldots, x_N]$ there exists a finite set R of ruloids such that*:
- *every ruloid in R has $C[x_1, \ldots, x_N]$ as its source*;
- *the LTS associated with T is a model of R (cf. Definition* 3.2);
- *if $C[t_1, \ldots, t_N] \xrightarrow{c} u$, then there exists a ruloid in R supporting that transition in the sense of Definition* 3.2.

EXAMPLE. Consider the process algebra BPA with the priority operator θ from Section 2.6. The set of ruloids determining the operational semantics of the term $\theta(a.x + y)$ consists of

$$\frac{y \stackrel{b}{\not\longrightarrow} \text{ for } b > a}{\theta(a.x + y) \xrightarrow{a} \theta(x)} \qquad \frac{y \xrightarrow{c} y' \quad y \stackrel{b}{\not\longrightarrow} \text{ for } b > c}{\theta(a.x + y) \xrightarrow{c} \theta(y')}$$

where the second ruloid is only present for actions c that are not smaller than a.

Protean specification languages. Case studies in the literature on process algebras often use mechanisms to define new operations on terms. Vaandrager [219] formulated the "fresh atom principle" to formalize a standard practice in process algebra proofs, namely, the introduction of fresh constants. Verhoef [223,224] introduced the "operator definition principle" (similar to RDP as discussed at the end of Section 5.4.5) to facilitate the specification of new unary function symbols in process algebra.

Bloom, Cheng, and Dsouza [51,52] advocated the use of "Protean" languages to enhance the expressiveness, and ease of use, of specification languages. Intuitively, when writing specifications by means of a process algebra, one is often faced with the choice between the use of a few basic operations with a clear semantics, or the introduction of ad hoc operations that simplify the writing of specifications and enhance their readability, but which complicate reasoning about the resulting high-level description of the behaviour. The aforementioned paper argues that the use of SOS, combined with the theory presented in this overview work, enables the systematic extension of process algebras in a way that is guaranteed to preserve the semantic properties of the original language.

Compositional proof systems for HML. Proof systems for modal logics enable to give formal proofs that (states in) LTSs satisfy certain requirements. A desirable feature of such proof systems is that they should allow a *compositional* style of proof development. Informally, a proof system is compositional if it builds a proof for a property of an LTS out of proofs for properties of certain sub-LTSs.

The work presented in [202] is based upon the realization that, in the context of pure first-order logic, the issue of compositionality was addressed by Gentzen [91] in his work on the sequent calculus. There, compositionality is obtained via cut-elimination. In [202], Simpson developed a sequent calculus for showing that closed terms in a process algebra with its operational semantics specified in the GSOS format satisfy assertions of the modal logic HML (see Section 2.3). Such process algebras provide interesting examples, because of the well-known difficulties in giving proof rules for flavours of parallel composition [206,233]. As usual, the benefit of working with an arbitrary GSOS language is that one obtains a generic proof system that is applicable to a wide class of process algebras.

Binary decision diagrams from GSOS languages. Binary decision diagrams [62] are widely used to represent LTSs symbolically in the second generation of verification tools for concurrent processes – see, e.g., McMillan's textbook on the model checker SMV [147]. A binary decision diagram for such an application is often generated from an LTS over the closed terms in some process calculus, which in turn is generated using the transition rules defining the operational semantics of this calculus. Such a two-step approach can be avoided by using a direct construction of the binary decision diagram from the transition rules. This is the approach followed in [74] for GSOS languages. The results in *op. cit.* suggest that this general procedure yields binary decision diagrams that are of comparable quality to those generated for specific process calculi by ad hoc methods (cf., e.g., [77]).

5.5. *RBB safe format*

In order to abstract away from internal actions, Milner [152] introduced a special action τ, called the *silent step*. The relation symbol $\xrightarrow{\tau}$ intuitively represents an internal computation. A number of equivalence notions have been developed to identify states in LTSs that incorporate silent steps, such as weak bisimulation [123] and branching bisimulation [103, 104]. In [50,215,218], rule formats have been introduced to ensure that weak and branching bisimulation equivalence are a congruence (cf. Definition 2.11). However, in general such equivalences are not a congruence with respect to most process algebras, because of the pre-emptive power of silent steps (see, e.g., [157, Section 2.3] for an intuitive discussion of this phenomenon). For this reason it has become standard practice to impose a rootedness condition on equivalences for the silent step [157].

Bloom [50] presented a rule format to ensure that rooted branching bisimulation is a congruence, imposing additional requirements on the GSOS format. The rule format recognizes so-called *patience* rules, via which a closed term can inherit the τ-transitions of its arguments. The *RBB safe format* [82] relaxes some syntactic restrictions of Bloom's rule format, imposing additional requirements on the panth format. Notably, certain arguments of function symbols are labelled 'liquid', and this labelling is used to restrict occurrences of variables in targets and in left-hand sides of premises. If a TSS is complete (cf. Definition 3.12) and satisfies the syntactic restrictions of the RBB safe format, then rooted branching bisimulation with respect to the associated LTS is a congruence.

Rooted branching bisimulation. We assume that Act is extended with a special action τ, representing the silent step. The reflexive and transitive closure of the relation $\xrightarrow{\tau}$ is denoted by $\xrightarrow{\varepsilon}$. First, we define the notion of branching bisimulation [103,104].

DEFINITION 5.44 (*Branching bisimulation*). Assume an LTS. A binary relation $\mathcal{B}$ on states is a *branching bisimulation* if it is symmetric and, whenever $s_1 \,\mathcal{B}\, s_2$:
- if $s_1 \xrightarrow{a} s_1'$, then either
  (1) $a = \tau$ and $s_1' \,\mathcal{B}\, s_2$, or
  (2) there are transitions $s_2 \xrightarrow{\varepsilon} s_2' \xrightarrow{a} s_2''$ such that $s_1 \,\mathcal{B}\, s_2'$ and $s_1' \,\mathcal{B}\, s_2''$;
- if $s_1 P$, then there are transitions $s_2 \xrightarrow{\varepsilon} s_2' P$ such that $s_1 \,\mathcal{B}\, s_2'$.

Two states s_1, s_2 are *branching bisimilar*, written $s_1 \mathrel{\underline{\leftrightarrow}_b} s_2$, if there exists a branching bisimulation relation that relates them.

Branching bisimulation is an equivalence relation; see [32]. However, branching bisimulation equivalence is not a congruence with respect to some standard operations in process algebras. For example, in $\mathrm{BPA}_{\varepsilon\tau}$ (see Section 2.6) with constants a and c, $a \mathrel{\underline{\leftrightarrow}_b} a$ and $c \mathrel{\underline{\leftrightarrow}_b} \tau c$, but $a + c$ and $a + \tau c$ are not branching bisimilar. Therefore, we introduce a rootedness condition.

DEFINITION 5.45 (*Rooted branching bisimulation*). Assume an LTS. A binary relation $\mathcal{R}$ on states is a *rooted branching bisimulation* if it is symmetric and, whenever $s_1 \,\mathcal{R}\, s_2$,
- if $s_1 \xrightarrow{a} s_1'$, then there is a transition $s_2 \xrightarrow{a} s_2'$ such that $s_1' \mathrel{\underline{\leftrightarrow}_b} s_2'$;
- if $s_1 P$, then $s_2 P$.

Two states s_1, s_2 are *rooted branching bisimilar*, written $s_1 \mathrel{\underline{\leftrightarrow}_{rb}} s_2$, if there exists a rooted branching bisimulation relation that relates them.

Since branching bisimulation is an equivalence relation, it is easy to see that rooted branching bisimulation is also an equivalence relation.

RBB safe. We proceed to present a congruence format for rooted branching bisimulation equivalence from [82]. Let $C[\,]$ denote a context, viz. a term with one occurrence of the context symbol $[\,]$ (cf. Section 2.4). We assume that every argument of each function symbol is labelled either *frozen* or *liquid*. A context is *liquid* if the context symbol occurs inside a nested string of liquid arguments.

DEFINITION 5.46 (*Liquid context*). The set of *liquid* contexts over a signature Σ is defined inductively by:
(1) $[\,]$ is liquid;
(2) if $C[\,]$ is liquid, and argument i of function symbol $f \in \Sigma$ is liquid, then $f(t_1, \ldots, t_{i-1}, C[\,], t_{i+1}, \ldots, t_{ar(f)})$ is liquid.

A patience rule for an argument i of a function symbol f implies that a closed term $f(t_1, \ldots, t_{ar(f)})$ inherits the τ-transitions of its argument t_i.

DEFINITION 5.47 (*Patience rule*). A *patience* rule for the ith argument of a function symbol f is a GSOS rule of the form

$$\frac{x_i \xrightarrow{\tau} y}{f(x_1, \ldots, x_{ar(f)}) \xrightarrow{\tau} f(x_1, \ldots, x_{i-1}, y, x_{i+1}, \ldots, x_{ar(f)})}.$$

Now we are in a position to present the RBB safe format, which imposes additional restrictions on the panth format (cf. Definition 5.1).

DEFINITION 5.48 (*RBB safe*). A TSS T in panth format is *RBB safe* if there exists a frozen/liquid labelling of arguments of function symbols such that each of its transition rules ρ is
(1) either a patience rule for a liquid argument of a function symbol,
(2) or a rule with source $f(x_1, \ldots, x_{ar(f)})$ for which the following requirements are fulfilled:
- right-hand sides of positive premises do not occur in left-hand sides of premises of ρ;
- if argument i of f is liquid and does not have a patience rule in T, then x_i does not occur in left-hand sides of premises of ρ;
- if argument i of f is liquid and has a patience rule in T, then x_i occurs no more than once in the left-hand side of a premise of ρ, where this premise
  - is positive,
  - does not contain the relation symbol $\xrightarrow{\tau}$, and
  - has left-hand side x_i;
- right-hand sides of positive premises and variables x_i for i a liquid argument of f only occur at liquid positions in the target of ρ.

THEOREM 5.49. *If a complete TSS is RBB safe, then the rooted branching bisimulation equivalence that it induces is a congruence.*

See [82] for a string of examples of complete TSSs to show that all syntactic requirements of the RBB safe format are essential for the congruence result in Theorem 5.49.

Computation of frozen/liquid labels. The crux in determining whether a TSS T is RBB safe is to find a suitable frozen/liquid labelling of arguments of function symbols. Assuming that the signature Σ is finite, there exists an efficient procedure that computes a frozen/liquid labelling Λ witnessing that T is RBB safe if and only if one such a labelling exists.

Procedure "Compute Liquid Labels for Σ and T":
The red/green directed graph G consists of vertices $\langle f, i\rangle$ for $f \in \Sigma$ and $1 \leqslant i \leqslant ar(f)$. There is an edge from $\langle f, i\rangle$ to $\langle g, j\rangle$ in G iff there is a transition rule in T with its conclusion of the form

$$f(x_1, \ldots, x_{ar(f)}) \xrightarrow{a} C\big[g(t_1, \ldots, t_{j-1}, D[x_i], t_{j+1}, \ldots, t_{ar(g)})\big].$$

A vertex $\langle g, j\rangle$ is red iff there is a transition rule in T with its target of the form

$$C\big[g(t_1, \ldots, t_{j-1}, D[y], t_{j+1}, \ldots, t_{ar(g)})\big],$$

where y is the right-hand side of a positive premise of this rule. All other vertices in G are coloured green.
The procedure turns green vertices in G red as follows. If a vertex $\langle f, i\rangle$ is red, and there exists an edge in G from $\langle f, i\rangle$ to a green vertex $\langle g, j\rangle$, then $\langle g, j\rangle$ is coloured red.
The procedure terminates if none of the green vertices can be coloured red anymore, at which point it outputs the red/green directed graph.

Λ labels argument i of function symbol f 'liquid' iff the vertex $\langle f, i\rangle$ in the output graph of the procedure above is red.

We proceed to apply Theorem 5.49 to two of the TSSs from Section 2.6, extended with the silent step.

BPA with empty process and silent step. The process algebra $\mathrm{BPA}_{\varepsilon\tau}$ is obtained from $\mathrm{BPA}_{\varepsilon}$ by extending Act with the silent step τ. The TSS for $\mathrm{BPA}_{\varepsilon\tau}$ is the TSS for $\mathrm{BPA}_{\varepsilon}$ in Table 1, with the proviso that a ranges over $\mathsf{Act} \cup \{\tau\}$.

It was noted in Section 5.1 that the TSS in Table 1 is panth. The procedure to calculate a frozen/liquid labelling for TSS produces the following result: the first argument of sequential composition is liquid (because of the target $x' \cdot y$ in the third transition rule for sequential composition), while both arguments of alternative composition and the second argument of sequential composition are frozen. The TSS in Table 1, with a ranging over $\mathsf{Act} \cup \{\tau\}$, is RBB safe with respect to this frozen/liquid labelling. As an example, for sequential composition we have that:

- its third transition rule with $a = \tau$ constitutes a patience rule for the first argument of sequential composition;
- in its first two transition rules, and in its third transition rule with $a \neq \tau$, the variable x in the liquid argument of the source occurs as the left-hand side of one positive premise, which does not contain the relation symbol $\xrightarrow{\tau}$;
- in its third transition rule, the variable x' in the right-hand side of the premise occurs in a liquid position of the target.

It is left to the reader to verify that the remaining transition rules in Table 1 are RBB safe. It was proven in Section 3.5 that the TSS in Table 1 is complete. Hence, according to Theorem 5.49, rooted branching bisimulation equivalence is a congruence with respect to $\mathrm{BPA}_{\varepsilon\tau}$.

Priorities with silent step. In general, rooted branching bisimulation equivalence is not a congruence with respect to the priority operator. For example, suppose $b < c$; then $a\cdot(\tau\cdot(b+c)+b)$ and $a\cdot(b+c)$ are rooted branching bisimilar, but $\theta(a\cdot(\tau\cdot(b+c)+b))$ and $\theta(a\cdot(b+c))$ are not rooted branching bisimilar. Consequently, in view of Theorem 5.49, the TSS for $\mathrm{BPA}_{\varepsilon\tau\theta}$ in Table 2 (with a and b ranging over $\mathsf{Act}\cup\{\tau\}$) cannot be in the RBB safe format.

Since the second transition rule in Table 2 has target $\theta(x')$, the procedure in Section 5.5 labels the argument of θ liquid. So, assuming there are one or more actions b greater than action a with respect to the ordering on Act, the liquid argument x in the source of this transition rule occurs as the left-hand side of the negative premises $x \overset{b}{\nrightarrow}$. This violates the RBB safe format.

5.6. *Precongruence formats for behavioural preorders*

The literature on SOS is particularly rich in results on (pre)congruence formats for behavioural equivalences and preorders, mostly in the absence of predicates. This section presents precongruence formats for simulation and ready simulation from [96], for decorated trace preorders from [53], for accepting trace preorder from [81], and for trace preorder from [53]. Vaandrager [220] moreover showed that the De Simone format constitutes a precongruence format for the external trace, external failure, and must [72,117] preorders. We note that if a preorder is a precongruence, then its kernel is a congruence.

5.6.1. *Simulation.* Path (cf. Definition 5.2) is a precongruence format for simulation preorder (cf. Definition 2.3). (According to van Glabbeek [96], path is actually a congruence format for all n-nested simulation equivalences [111].)

THEOREM 5.50. *If a TSS is in path format, then the simulation preorder that it induces is a precongruence.*

For example, since the TSS for $\mathrm{BPA}_{\varepsilon}$ from Section 2.6 is in path format, Theorem 5.50 implies that simulation preorder is a precongruence with respect to $\mathrm{BPA}_{\varepsilon}$.

5.6.2. *Ready simulation.* A precongruence format for ready simulation preorder (cf. Definition 2.3) is obtained by disallowing *look-ahead* in panth rules.

DEFINITION 5.51 (*No look-ahead*). A transition rule has *no look-ahead* if the variables occurring in the right-hand sides of its positive premises do not occur in the left-hand sides of its premises.

DEFINITION 5.52 (*Ready simulation format*). A panth rule is in *ready simulation* if it has no look-ahead. A TSS is in *ready simulation format* if all its transition rules are.

THEOREM 5.53. *If a complete TSS is in ready simulation format, then the ready simulation preorder that it induces is a precongruence.*

For example, the TSS for BPA_ε from Section 2.6 is positive and so complete. Furthermore, it is in path format, and none of its transition rules contains look-ahead. Hence, Theorem 5.53 implies that ready simulation preorder is a precongruence with respect to BPA_ε.

The following counter-example shows that the omission of look-ahead from the ready simulation format is essential.

EXAMPLE. Let $\mathsf{Act} = \{a, b, c\}$ and let f be a unary function symbol. Extend the TSS for BPA_ε with

$$\frac{x \xrightarrow{a} y}{f(x) \xrightarrow{a} f(y)} \qquad \frac{x \xrightarrow{b} y \quad y \xrightarrow{c} z}{f(x)\surd}.$$

Note that the premises of the second transition rule contain look-ahead.

Clearly, $a\cdot(b + b\cdot c) + a\cdot b$ and $a\cdot(b + b\cdot c)$ are ready simulation equivalent. However, $f(a\cdot(b + b\cdot c) + a\cdot b)$ and $f(a\cdot(b + b\cdot c))$ are not ready simulation equivalent. Namely, the transition $f(a\cdot(b + b\cdot c) + a\cdot b) \xrightarrow{a} f(b)$ can only be simulated by $f(a\cdot(b + b\cdot c)) \xrightarrow{a} f(b + b\cdot c)$, but $f(b)$ has no initial transitions while $f(b + b\cdot c)\surd$.

5.6.3. *Decorated traces.* This section presents precongruence formats for ready trace, failure trace, readies and failures preorder (cf. Definition 2.5) from [53]. These formats, which re-use the notion of a liquid context (cf. Definition 5.46), were obtained by a careful study of the modal characterizations of the preorders in question. The ready trace, readies, and failures formats generalize earlier formats by van Glabbeek [96], Bloom [47], and De Simone [220], respectively. (In [96], van Glabbeek sketched a congruence format for failures equivalence; that format, however, is flawed [101].)

DEFINITION 5.54 (*Ntytt*). An *ntytt rule* is a transition rule in which the right-hand sides of positive premises are variables that are all distinct, and that do not occur in the source.

DEFINITION 5.55 (*Propagation and polling*). An occurrence of a variable in an ntytt rule is *propagated* if the occurrence is either in the target, or in the left-hand side of a positive premise of which the right-hand side occurs in the target. An occurrence of a variable in an ntytt rule is *polled* if the occurrence is in the left-hand side of a premise that does not have a right-hand side occurring in the target.

Intuitively, the precongruence formats for the four decorated trace preorders operate by keeping track of which variables represent running processes, and which do not. For example, it is semantically reasonable to copy a process before it starts. However, copying a running process would give information about the branching structure of the process, which is incompatible with any form of decorated trace semantics. A *floating* variable may represent a running process. This notion assumes a predicate Λ on arguments of function symbols; if $\Lambda(f, i)$ then we say that argument i of f is *liquid*, and otherwise it is *frozen* (cf. Section 5.5).

DEFINITION 5.56 (*Floating variable*). A variable in an ntytt rule is *floating* if either it occurs as the right-hand side of a positive premise, or it occurs exactly once in the source, at a liquid position (cf. Definition 5.46).

DEFINITION 5.57 (*Decorated trace safe*). Let Λ be a predicate on arguments of function symbols. An ntytt rule is *Λ-ready trace safe* if
- it has no look-ahead (cf. Definition 5.51), and
- each floating variable is propagated at most once, and at a liquid position.

The rule is *Λ-readies safe* if
- it is Λ-ready trace safe, and
- each floating variable is not both propagated and polled.

The rule is *Λ-failure trace safe* if
- it is Λ-readies safe, and
- each floating variable is polled at most once, at a liquid position in a positive premise.

The second restriction on "Λ-ready trace safe" guarantees that a running process is never copied, and continued to be marked as running after it has executed. The "Λ-readies safe" restriction ensures that only at the end of its execution a running process is tested multiple times. The "Λ-failure trace safe" restriction further limits to a positive test on a single action or predicate.

DEFINITION 5.58 (*Decorated trace formats*). A TSS is in *ready trace format* if it is in ntyft/ntyxt format and its rules are Λ-ready trace safe with respect to some Λ. A TSS is in *readies format* if it is in ntyft/ntyxt format and its rules are Λ-readies safe with respect to some Λ. A TSS is in *failure trace format* if it is in ntyft/ntyxt format and its rules are Λ-failure trace safe with respect to some Λ.

THEOREM 5.59. *If a TSS is in ready trace format, then the ready trace preorder that it induces is a precongruence.*

THEOREM 5.60. *If a TSS is in readies format, then the readies preorder that it induces is a precongruence.*

THEOREM 5.61. *If a TSS is in failure trace format, then the failure trace and failure preorders that it induces are precongruences.*

5.6.4. *Accepting traces.* Similar to the RBB safe format, a precongruence format for accepting trace preorder (cf. Definition 2.5) from [81] is based on a frozen/liquid labelling of arguments of function symbols.

DEFINITION 5.62 (*L cool*). A TSS in path format is *L cool* if there exists a frozen/liquid labelling of arguments of function symbols such that each of its transition rules ρ satisfies the following syntactic restrictions:

- each variable in ρ that occurs in a liquid argument of the source, or as the right-hand side of a premise, occurs exactly once either as the left-hand side of a premise or at a liquid position (see Definition 5.46) in the target of ρ;
- the variable dependency graph (see Definition 5.4) of the set of premises of ρ does not contain an infinite forward chain of edges.

THEOREM 5.63. *If a TSS is in L cool format, then the accepting trace preorder that it induces is a precongruence.*

The following counter-example shows that the L cool format cannot allow an infinite forward chain of edges in the variable dependency graph of the premises of a transition rule.

EXAMPLE. Let $\mathsf{Act} = \{a\}$, f a unary function symbol, and b and c constants. Consider the TSS

$$\frac{}{b \xrightarrow{a} b} \qquad \frac{\{x_i \xrightarrow{a} x_{i+1} \mid i \in \mathbb{N}\}}{f(x_0)\surd}.$$

Note that the edges $\langle x_i, x_{i+1}\rangle$ for $i \in \mathbb{N}$ in the variable dependency graph of the premises of the second transition rule form an infinite forward chain.

The terms b and c are accepting trace equivalent, because they both have no accepting traces. However, $f(b)$ and $f(c)$ are not accepting trace equivalent, as $f(b)$ has the accepting empty trace ε while $f(c)$ has no accepting traces.

See [81] for further examples of TSSs showing that all the syntactic requirements of the L cool format are essential for the precongruence result in Theorem 5.63. Similar to the procedure for the RBB safe format in Section 5.5, there exists an efficient procedure to compute a frozen/liquid labelling Λ witnessing that T is L cool if and only if such a labelling exists; see [81].

EXAMPLE. The TSS for BPA_ε from Section 2.6 is in path format, and for each transition rule the variable dependency graph of its premises does not contain an infinite forward chain of edges. Take the first argument of sequential composition to be liquid, and the two arguments of alternative composition and the second argument of sequential composition to be frozen. It is not hard to see, for the TSS of BPA_ε, that if a variable occurs in a liquid argument of the source or as the right-hand side of a premise of a transition rule, then it occurs exactly once either as the left-hand side of a premise or at a liquid position in the target of this transition rule. Hence, Theorem 5.63 implies that accepting trace preorder is a precongruence with respect to BPA_ε.

THEOREM 5.70. *Assume a stratifiable TSS* (*cf. Definition* 3.13) *in pure ntyft/ntyxt format containing at least one constant in its signature. Then, for every pair of closed terms* t, u,

$$t \simeq_{CT}^{\text{pure ntyft/ntyxt}} u \iff t \sim_{\text{HML}} u.$$

In view of Theorem 2.8 this means that the completed trace congruence induced by the pure ntyft/ntyxt format coincides with bisimulation equivalence if the LTS associated with the TSS in question is finitely branching.

The use of negative premises appears to be necessary in order to test for bisimulation equivalence. Indeed, Groote and Vaandrager [111] characterized the completed trace congruence induced by the tyft/tyxt format thus.

THEOREM 5.71. *Assume a TSS in pure tyft/tyxt format whose associated LTS is finitely branching. Then, for every pair of closed terms* t, u,

$$t \simeq_{CT}^{\text{pure tyft/tyxt}} u \iff t \text{ and } u \text{ are 2-nested simulation equivalent.}$$

We refer the interested reader to *op. cit.* (and [102] in this Handbook) for the definition of 2-nested simulation equivalence, and for more details on completed trace congruences.

6. Many-sorted higher-order languages

This section presents a formal framework to describe TSSs in the style of Plotkin, allowing one to express many-sortedness, general binding mechanisms, and substitutions. Such variable binding mechanisms are widely used in SOS semantics for, e.g., concurrent and functional programming languages [31,161,176,184,196], the π-calculus [160], value-passing process algebras [109,120,121], process algebras with recursion [127,157], and timed process algebra [84]. See [106] for a collection of articles about recent developments in operational semantics for higher-order programming languages.

Several concepts in the setting of operational semantics with variable binding, which seem to be intuitively clear at first sight, turn out to be ambiguous when studied more carefully. In order to obtain a formal framework in which transition rules with a variable binding mechanism can be expressed rigorously, we distinguish between actual and formal variables, following conventions from programming languages, and formalize the binding construct $t[u/x]$ in transition rules. In many programming languages there are actual parameters and formal parameters. The formal parameters are used to define procedures or functions; the actual parameters are the "real" variables to be used in the main program. In the main program the formal parameters are bound to the actual parameters. When discussing procedures on a conceptual level, it is often useful to introduce a notational distinction between formal and actual parameters; see for instance [231]. A transition rule can be thought of as a procedure to establish a transition relation by means of substituting (actual) terms for the (formal) variables. The following example illustrates that it is useful to make a notational distinction between actual and formal variables.

EXAMPLE. Consider the transition rule

$$\frac{y[z/x]P_1}{y P_2}$$

where x, y, z are variables and $y[z/x]$ is a standard notation that binds the x in y and replaces it by z. Application of a substitution σ to this transition rule yields

$$\frac{\sigma(y)[\sigma(z)/x]P_1}{\sigma(y) P_2}.$$

The example above highlights two matters.

(1) The expression $y[z/x]$ is not a substitution (for then it would equal y), but a syntactic construct with a suggestive form, called a *substitution harness*. Only after application of a substitution σ, the result $\sigma(y)[\sigma(z)/x]$ can be evaluated to a term.

(2) Substitutions only apply to part of the variables that occur in a transition rule. In order to distinguish such *formal* variables in a transition rule, they are marked with an asterisk.

Hence, the transition rule above takes the form

$$\frac{y^*[z^*/x]P_1}{y^* P_2}.$$

The use of formal variables in SOS with variable binding was proposed in, e.g., [85,128, 194].

The organization of this section is as follows. Section 6.1 introduces actual terms, while Section 6.2 introduces formal terms. Section 6.3 describes the framework for many-sorted higher-order SOS definitions. Finally, Section 6.4 explains how the operational conservative extension format from Theorem 4.4 carries over to this higher-order setting.

Binding mechanisms exist in many and diverse forms. Here, these mechanisms are described using a notational approach, based on [15], for the Nuprl proof development system [66]. An alternative formalism would of course be the λ-calculus [31].

6.1. *The actual world*

We assume a set of sorts together with a countably infinite set Var of sorted actual variables. The actual world contains actual terms, actual substitutions, and so forth. Let $\vec{O}$ denote a sequence $O_1 \cdots O_k$, and $\vec{O}_i$ a sequence $O_{i1} \cdots O_{ik}$, with $k \in \mathbb{N}$.

DEFINITION 6.1 (*Many-sorted higher-order signature*). A *many-sorted higher-order signature* Σ is a set of function symbols

$$f : \vec{S}_1.S_1 \times \cdots \times \vec{S}_{ar(f)}.S_{ar(f)} \to S,$$

where the S_{ij}, the S_i, and S are sorts.

The intuitive idea embodied by the above definition is that a function symbol f denotes an operation that takes functions of type $\vec{S_i} \to S_i$ as arguments, and delivers a result of sort S.

DEFINITION 6.2 (*Actual term*). Let Σ be a many-sorted higher-order signature. The collection $\mathbb{A}(\Sigma)$ of *actual terms* over Σ is the least set satisfying:

- each actual variable from Var is in $\mathbb{A}(\Sigma)$;
- for each function symbol $f : \vec{S_1}.S_1 \times \cdots \times \vec{S}_{ar(f)}.S_{ar(f)} \to S$, the expression $f(\vec{x}_1.t_1, \ldots, \vec{x}_{ar(f)}.t_{ar(f)})$ is an actual term of sort S, if
  - the t_i are actual terms of sort S_i, and
  - each sequence $\vec{x}_i$ consists of distinct actual variables in Var of sorts $\vec{S_i}$.

Free occurrences of actual variables in actual terms are defined inductively as expected:

- x occurs free in x for each $x \in \mathsf{Var}$;
- if x occurs free in t_i, and x does not occur in the sequence $\vec{x}_i$, then x occurs free in $f(\vec{x}_1.t_1, \ldots, \vec{x}_{ar(f)}.t_{ar(f)})$.

An actual term is *closed* if it does not contain any free occurrences of actual variables.

An *actual* substitution is a sort preserving mapping $\sigma : \mathsf{Var} \to \mathbb{A}(\Sigma)$, where sort preserving means that x and $\sigma(x)$ are always of the same sort. An actual substitution extends to a mapping from actual terms to actual terms; the actual term $\sigma(t)$ is obtained by replacing free occurrences of actual variables x in t by $\sigma(x)$. As usual, $\_[t/x]$ is the postfix notation for the actual substitution that maps x to t and is inert otherwise. Such postfix denoted actual substitutions are called *explicit* (as opposed to *implicit* actual substitutions σ).

In the definition of actual substitutions on actual terms there is a well-known complication. Namely, consider an actual term $\sigma(t)$, and let x occur free in t. After x in t has been replaced by $\sigma(x)$, actual variables y that occur in $\sigma(x)$ are bound in actual subterms such as $f(y.u)$ of t. A solution for this problem, which originates from the λ-calculus, is to allow unrestricted substitution by applying α-conversion; that is, by renaming bound actual variables. From now on, actual terms are considered modulo α-conversion, and when an actual substitution is applied, bound actual variables are renamed. Stoughton [212] presented a clean treatment of this technique.

6.2. *The formal world*

We argued that it is a good idea to distinguish between formal and actual variables when discussing transition rules with variable bindings and substitutions on an abstract level. A *formal* term t^* is an actual term with possible occurrences of formal variables and substitution harnesses.

Assume a many-sorted higher-order signature Σ. The set Var^* of *formal* variables is defined as $\{x^* \mid x \in \mathsf{Var}\}$, where x^* and x are of the same sort.

DEFINITION 6.3 (*Formal term*). The collection $\mathbb{F}(\Sigma)$ of *formal terms* over a many-sorted higher-order signature Σ is the least set satisfying:

- each actual variable from Var is in $\mathbb{F}(\Sigma)$;

- each formal variable from Var^* is in $\mathbb{F}(\Sigma)$;
- for each function symbol $f : \vec{S}_1.S_1 \times \cdots \times \vec{S}_{ar(f)}.S_{ar(f)} \to S$, the expression $f(\vec{x}_1.t_1^*, \ldots, \vec{x}_{ar(f)}.t_{ar(f)}^*)$ is a formal term of sort S, if
  - the t_i^* are formal terms of sort S_i, and
  - each $\vec{x}_i$ consists of distinct actual variables in Var of sorts $\vec{S}_i$,
- if t^* and u^* are formal terms of sorts S_0 and S_1 respectively, and $x \in \mathsf{Var}$ is of sort S_1, then $t^*[u^*/x]$ is a formal term of sort S_0.

A *formal* substitution is a sort preserving mapping $\sigma^* : \mathsf{Var}^* \to \mathbb{A}(\Sigma)$. It extends to a mapping $\sigma^* : \mathbb{F}(\Sigma) \to \mathbb{A}(\Sigma)$, where the actual term $\sigma^*(t^*)$ is obtained from the formal term t^* as follows. First replace each formal variable x^* in t^* by $\sigma^*(x^*)$. Then the substitution harnesses in t^* become explicit actual substitutions, so that the result evaluates to an actual term.

EXAMPLE. An example of a formal term is $y^*[z^*/x]$, which evaluates to the actual constant a after application of a formal substitution σ^* with $\sigma^*(z^*) = a$ and $\sigma^*(y^*) = x$. Namely, the implicit formal substitution σ^* turns the substitution harness $y^*[z^*/x]$ into the actual term $x[a/x]$, where $\_[a/x]$ is an explicit actual substitution, which evaluates to a.

We summarize the various notions of substitutions, and briefly discuss their differences. There are four notions in two worlds: implicit and explicit actual substitutions (which are semantically the same), formal substitutions, and substitution harnesses.

- Implicit actual substitutions σ and explicit actual substitutions $\_[t/x]$ both denote mappings from actual variables to actual terms.
- Formal substitutions σ^* are mappings from formal variables to actual terms.
- A substitution harness $t^*[u^*/x]$ is *not* a substitution, but a piece of syntax with a suggestive form. If a formal substitution σ^* is applied to it, then the result is an expression $\sigma^*(t^*)[\sigma^*(u^*)/x]$, containing an explicit actual substitution, so that it can be evaluated to an actual term.

Substitution harnesses are used to formulate in a precise way how a formal substitution is to act on a transition rule. The formal and actual substitutions are used to move from transition rules to a proof tree (cf. Definition 2.14).

6.3. *Actual and formal transition rules*

Before presenting the basic definitions of SOS for higher-order languages, we first consider as an example the recursive μ-construct, which combines formal variables, a binding mechanism, and a substitution harness. The μ-operator is similar to the construct $\mathrm{fix}(X = t)$ that was incorporated in De Simone languages (cf. Section 5.3.1).

EXAMPLE. Intuitively, a closed actual term $\mu x.t$ executes t until it encounters an expression x, in which case it starts executing $\mu x.t$ again. This intuition is expressed in the following transition rule, which we call the μ-rule:

$$\frac{y^*[\mu x.y^*/x] \xrightarrow{a} z^*}{\mu x.y^* \xrightarrow{a} z^*}.$$

The transition $\mu x.a{\cdot}x \xrightarrow{a} \mu x.a{\cdot}x$ with $a{\cdot}\_$ the prefix multiplication operator from CCS can be derived from the μ-rule together with the standard transition rule for prefix multiplication: $\emptyset/a{\cdot}w^* \xrightarrow{a} w^*$ (cf. Section 5.4.3). Namely, after application of the formal substitution σ^* to the μ-rule with $\sigma^*(y^*) = a{\cdot}x$ and $\sigma^*(z^*) = \mu x.a{\cdot}x$, the premise takes the form $a{\cdot}x[\mu x.a{\cdot}x/x] \xrightarrow{a} \mu x.a{\cdot}x$, which evaluates to $a{\cdot}\mu x.a{\cdot}x \xrightarrow{a} \mu x.a{\cdot}x$. Since this is an instance of the transition rule for prefix multiplication, with $\mu x.a{\cdot}x$ for w^*, we conclude that the σ^*-instantiation of the conclusion of the μ-rule is valid: $\mu x.a{\cdot}x \xrightarrow{a} \mu x.a{\cdot}x$. The proof of $\mu x.a{\cdot}x \xrightarrow{a} \mu x.a{\cdot}x$ is depicted below.

instance of prefixing rule
$$\left\{ \begin{array}{c} \emptyset \\ \downarrow \\ a \cdot \mu x.a \cdot x \xrightarrow{a} \mu x.a \cdot x \end{array} \right.$$

$$\left. \begin{array}{c} a \cdot \mu x.a \cdot x \xrightarrow{a} \mu x.a \cdot x \\ \downarrow \\ \mu x.a \cdot x \xrightarrow{a} \mu x.a \cdot x \end{array} \right\}$$
instance of μ-rule

DEFINITION 6.4 (*Actual transition rule*). An *actual transition rule* is an expression of the form H/α, where H is a set of literals (cf. Definition 2.13), and α is a positive literal.

Actual transition rules are deduced by means of *formal* transition rules. The formal transition rules are the ones that are presented in the literature; they are the recipes that enable one to deduce an LTS. For instance, in the example above, the actual transition rule

$$\frac{a{\cdot}\mu x.a{\cdot}x \xrightarrow{a} \mu x.a{\cdot}x}{\mu x.a{\cdot}x \xrightarrow{a} \mu x.a{\cdot}x}$$

was deduced from the μ-rule, which is a formal transition rule.

DEFINITION 6.5 (*Formal transition rule*). A *formal transition rule* is an expression H^*/α^*, where:
- H^* is a set of premises of the form $t^* \xrightarrow{a} u^*$, t^*P, $t^* \stackrel{a}{\nrightarrow}$, and $t^*\neg P$;
- α^* is the conclusion of the form $t^* \xrightarrow{a} u^*$ or t^*P;

where $t^*, u^* \in \mathbb{F}(\Sigma)$, $a \in \mathsf{Act}$, and P denotes any predicate. A *higher-order TSS* is a set of formal transition rules.

6.4. *Operational conservative extension*

Only few rule formats for higher-order languages have appeared in the literature. Sands [193] introduced the GDSOS format for SOS specifications of functional languages and established some proof principles that are sound for all languages expressible in this format.

Howe [128] presented a congruence format for higher-order TSSs with respect to bisimulation equivalence, which shows a strong resemblance with the tyft/tyxt format from Groote and Vaandrager (cf. Definition 5.2). Rensink [183] obtained congruence results for three extensions of bisimilation equivalence to open terms, in the presence of recursion. Interestingly, Bernstein [42] showed that in many cases the semantics of a higher-order language can be captured by a first-order TSS with terms as transition labels. It appears that existing rule formats can be extended with terms as transition labels in a straightforward manner; see [42,85]. Thus, it might be the case that current first-order rule formats are sufficient to deal with higher-order languages. Another reference of interest in this area is [48].

We proceed to present a generalization from [85] of the operational conservative extension format (see Section 4) to a higher-order setting. This generalization is based on an adaptation of the notion of source-dependency (cf. Definition 4.3), requiring a distinction between occurrences of formal variables in- and outside the substitution harnesses of a formal term. $FV(t^*)$ denotes the set of formal variables that occur in the formal term t^*.

DEFINITION 6.6 (*The set $FV(t^*)$*). The sets $FV(t^*)$ are defined inductively by:

$$\begin{aligned} FV(x^*) &\triangleq x^*, \\ FV\big(f(\vec{x}_1.t_1^*, \ldots, \vec{x}_{ar(f)}.t_{ar(f)}^*)\big) &\triangleq FV(t_1^*) \cup \cdots \cup FV(t_{ar(f)}^*), \\ FV\big(t^*[s^*/x]\big) &\triangleq FV(t^*) \cup FV(s^*). \end{aligned}$$

For example, $FV(f(v.x^*[y^*/w])) = \{x^*, y^*\}$. By contrast, $EV(t^*)$ denotes a more restricted set of formal variables in the formal term t^*, which does not take into account formal variables that occur inside a substitution harness.

DEFINITION 6.7 (*The set $EV(t^*)$*). The sets $EV(t^*)$ are defined inductively by:

$$\begin{aligned} EV(x^*) &\triangleq x^*, \\ EV\big(f(\vec{x}_1.t_1^*, \ldots, \vec{x}_{ar(f)}.t_{ar(f)}^*)\big) &\triangleq EV(t_1^*) \cup \cdots \cup EV(t_{ar(f)}^*), \\ EV\big(t^*[s^*/x]\big) &\triangleq EV(t^*). \end{aligned}$$

For example, $EV(f(v.x^*[y^*/w])) = \{x^*\}$. The sets $FV(t^*)$ and $EV(t^*)$ are used in the definition of source-dependent variables in a formal transition rule.

DEFINITION 6.8 (*Source dependency*). For a formal transition rule ρ^*, the formal variables in ρ^* that are *source-dependent* are defined inductively by:

(1) if t^* is the source of ρ^*, then all formal variables in $EV(t^*)$ are source-dependent in ρ^*;

(2) if $t^* \xrightarrow{a} u^*$ is a premise of ρ^*, and all formal variables in $FV(t^*)$ are source-dependent in ρ^*, then all formal variables in $EV(u^*)$ are source-dependent in ρ^*.

A formal transition rule ρ^* is *source-dependent* if so are all the variables in $FV(\rho^*)$.

Theorem 6.9 formulates sufficient criteria for a higher-order TSS $T_0 \oplus T_1$ to be an operational conservative extension of T_0 (see Definition 4.2); it extends Theorem 4.4 to higher-order languages. We say that a formal term in $\mathbb{F}(\Sigma_1)$ is *fresh* if it contains a function symbol

dered (respectively preordered) set and that the operators be monotonic. The notion of Σ-homomorphism extends to the ordered Σ-structures in the obvious way by requiring that such maps preserve the underlying order-theoretic structure as well as the Σ-structure.

The interpretation $\mathcal{A}[\![\cdot]\!]$ of $\mathrm{T}(\Sigma, \mathsf{RVar})$ in a Σ-algebra $\mathcal{A}$ associates each term in $\mathrm{T}(\Sigma, \mathsf{RVar})$ with a mapping from substitutions (going from recursion variables to $\mathcal{A}$) to $\mathcal{A}$. This interpretation is defined by induction as follows, where σ is any mapping from recursion variables to $\mathcal{A}$:

$$\begin{aligned} \mathcal{A}[\![X]\!]\sigma &\triangleq \sigma(X), \\ \mathcal{A}[\![f(t_1, \ldots, t_n)]\!]\sigma &\triangleq f_{\mathcal{A}}\big(\mathcal{A}[\![t_1]\!]\sigma, \ldots, \mathcal{A}[\![t_n]\!]\sigma\big), \end{aligned}$$

where $n = ar(f)$. We recall that the recursive terms over Σ are given by the BNF grammar

$$t ::= X \mid f(t_1, \ldots, t_{ar(f)}) \mid \mathrm{fix}(X = t),$$

where X is any recursion variable, f any function symbol in Σ, and fix a binding construct. The latter construct gives rise to the usual notions of free and bound recursion variables in recursive terms. $\mathsf{CREC}(\Sigma)$ denotes the set of recursive terms that do not contain free recursion variables. Furthermore, $t[u/X]$ denotes the recursive term t in which each occurrence of the recursion variable X has been replaced by u. If $\mathcal{A}$ is a Σ-domain, then the interpretation $\mathcal{A}[\![\cdot]\!]$ extends to the set of recursive terms over Σ by

$$\mathcal{A}[\![\mathrm{fix}(X = t)]\!]\sigma \triangleq \mathsf{Y}\lambda d.\,\mathcal{A}[\![t]\!]\sigma',$$

where Y denotes the least fixed-point operator, d is a metavariable ranging over $\mathcal{A}$, $\sigma'(X) \triangleq d$, and $\sigma'(Y) \triangleq \sigma(Y)$ for $Y \neq X$. Note that for each $t \in \mathsf{CREC}(\Sigma)$, $\mathcal{A}[\![t]\!]\sigma$ does not depend on σ.

In what follows, we make use of some general results about the semantic mappings defined above, which may be found in [67,112,117]. The first result states that for any recursive term t (possibly containing free recursion variables) there is a sequence of *finite approximations* $t_n \in \mathrm{T}(\Sigma, \mathsf{RVar})$ for $n \in \mathbb{N}$ such that, for any Σ-domain $\mathcal{A}$,

$$\mathcal{A}[\![t]\!] = \bigsqcup_{n \in \mathbb{N}} \mathcal{A}[\![t_n]\!].$$

(That is, the interpretation of the term t in $\mathcal{A}$ is the least upper bound of the interpretations of its finite approximations.) The second result states that if $\leqslant_\Omega$ is the least precongruence (cf. Definition 2.11) satisfying

$$\begin{aligned} \mathrm{fix}(X = t) &\leqslant_\Omega t\big[\mathrm{fix}(X = t)/X\big], \\ t\big[\mathrm{fix}(X = t)/X\big] &\leqslant_\Omega \mathrm{fix}(X = t), \\ \Omega &\leqslant_\Omega X, \end{aligned}$$

then $t_n \leqslant_\Omega t$ for every $n \in \mathbb{N}$.

For any binary relation $\mathcal{R}$ over $\mathsf{CREC}(\Sigma)$, the algebraic part of $\mathcal{R}$, denoted by $\mathcal{R}^A$, is defined as follows [117]:

$$t\,\mathcal{R}^A\,u \iff \forall n \in \mathbb{N}\ \exists m \in \mathbb{N}\ (t_n\,\mathcal{R}\,u_m).$$

We say that $\mathcal{R}$ is *algebraic* if $\mathcal{R}$ is equal to $\mathcal{R}^A$. Intuitively, a relation is algebraic if it is completely determined by how it behaves on recursion-free terms. Every denotational interpretation $\mathcal{A}[\![\cdot]\!]$ induces a preorder $\sqsubseteq_{\mathcal{A}}$ over $\mathsf{CREC}(\Sigma)$ by:

$$t \sqsubseteq_{\mathcal{A}} u \iff \mathcal{A}[\![t]\!] \sqsubseteq_{\mathcal{A}} \mathcal{A}[\![u]\!].$$

The following result characterizes a class of denotational interpretations which induce relations over terms that are algebraic.

LEMMA 7.1. *Let $\mathcal{A}$ be a Σ-domain. If $\mathcal{A}[\![t]\!]$ is a compact element in $\mathcal{A}$ for every $t \in \mathrm{T}(\Sigma)$ (see, e.g., [117, p. 130]), then $\sqsubseteq_{\mathcal{A}}$ is algebraic.*

In view of the above general lemma, the relations over the recursive terms in $\mathsf{CREC}(\Sigma)$ induced by a denotational semantics are always algebraic, provided that the denotations of the recursion-free terms in $\mathrm{T}(\Sigma)$ are compact elements in the algebraic cpo $\mathcal{A}$.

7.1.2. *Prebisimulation.* We consider LTSs with divergence from Definition 5.40, which include a special divergence predicate $\uparrow$. The convergence predicate $\downarrow$ holds in a state iff the divergence predicate does not hold in this same state (cf. Definition 5.41). The behavioural relation over LTSs with divergence that we study in this section is that of *prebisimulation* [115,125,153,230] (also known as *partial bisimulation* [1]). We recall (from Definition 2.1) that Proc and Act denote the sets of states and actions, respectively, of the LTS with divergence in question. Let $\mathsf{Rel}(\mathsf{Proc})$ denote the set of binary relations over Proc.

DEFINITION 7.2 (*Prebisimulation*). Assume an LTS with divergence. The functional $G : \mathsf{Rel}(\mathsf{Proc}) \to \mathsf{Rel}(\mathsf{Proc})$ is defined as follows. Given a relation $\mathcal{R} \in \mathsf{Rel}(\mathsf{Proc})$, we have $s_1\,G(\mathcal{R})\,s_2$ whenever:
- if $s_1 \xrightarrow{a} s_1'$, then there is a transition $s_2 \xrightarrow{a} s_2'$ such that $s_1'\,\mathcal{R}\,s_2'$;
- if $s_1 \downarrow$, then $s_2 \downarrow$;
- if $s_1 \downarrow$ and $s_2 \xrightarrow{a} s_2'$, then there is a transition $s_1 \xrightarrow{a} s_1'$ such that $s_1'\,\mathcal{R}\,s_2'$.

A relation $\mathcal{R}$ is a *prebisimulation* iff $\mathcal{R} \subseteq G(\mathcal{R})$. We write $s_1 \lesssim s_2$ if there exists a prebisimulation $\mathcal{R}$ such that $s_1\,\mathcal{R}\,s_2$.

The relation $\lesssim$ is a preorder over Proc; its kernel is denoted by $\simeq$. Prebisimulation is similar in spirit to the notion of bisimulation (cf. Definition 2.3). Intuitively, $s_1 \lesssim s_2$ if the behaviour of s_2 is at least as specified as that of s_1, and s_1 and s_2 can simulate each other when restricted to the part of their behaviour that is fully specified. A divergent state s with no outgoing transition intuitively corresponds to a process whose behaviour is totally unspecified – essentially an operational version of the bottom element $\perp$ in Scott's theory of domains [174,197,210].

The following precongruence result for $\lesssim$ with respect to recursive GSOS languages including the inert constant Ω originates from [11,12].

PROPOSITION 7.3. *$\lesssim$ is a precongruence with respect to the LTS with divergence over* $\mathsf{CREC}(\Sigma)$ *associated with a recursive GSOS language including the inert constant Ω (cf. Proposition* 5.42).

7.1.3. *Finite synchronization trees.* A useful source of examples for LTSs with divergence is the set of *finite synchronization trees* [152].

DEFINITION 7.4 (*Finite synchronization tree*). The set of *finite synchronization trees* over a set of actions Act, denoted by ST(Act), is defined inductively by:

(1) $\emptyset \in \mathsf{ST}(\mathsf{Act})$;
(2) if $S \in \mathsf{ST}(\mathsf{Act})$, then $S \cup \{\bot\} \in \mathsf{ST}(\mathsf{Act})$;
(3) if $a_1, \ldots, a_n \in \mathsf{Act}$ and $S_1, \ldots, S_n \in \mathsf{ST}(\mathsf{Act})$, then

$$\{\langle a_1, S_1\rangle, \ldots, \langle a_n, S_n\rangle\} \in \mathsf{ST}(\mathsf{Act}).$$

The symbol $\bot$ is used to represent that a finite synchronization tree is divergent. The set of finite synchronization trees ST(Act) can be turned into an LTS with divergence by stipulating that, for $S \in \mathsf{ST}(\mathsf{Act})$:

- $S \uparrow$ iff $\bot \in S$;
- $S \xrightarrow{a} S'$ iff $\langle a, S'\rangle \in S$.

When relating behavioural semantics based upon bisimulation-like relations with denotational semantics (usually based upon algebraic domains), one is faced with the following mismatch:

(1) on the one hand, in an algebraic domain $d \sqsubseteq e$ iff every compact element smaller than or equal to d is also dominated by e;
(2) on the other hand, there are closed terms that have the same finite approximations, but that are not bisimilar.

This implies that bisimulation is not algebraic, and thus cannot be captured in a standard domain theoretic framework. One way to address this problem is to study its finitary part. To this end, it is generally agreed upon in the literature that finite synchronization trees are a natural operational counterpart of compact elements.

In what follows, we are interested in relating the notion of prebisimulation to a preorder on finite synchronization trees induced by a denotational semantics given by means of an algebraic domain. As such preorders are completely determined by how they act on finite processes, we are interested in comparing them with the "finitely observable", or finitary, part of the bisimulation in the sense of, e.g., [112,115]. The following definition from [1] is inspired by property (1) above for algebraic domains.

DEFINITION 7.5 (*Finitary preorder*). The *finitary preorder* $\lesssim^F$ is defined on any LTS by

$$s_1 \lesssim^F s_2 \Longleftrightarrow \forall S \in \mathsf{ST}(\mathsf{Act})\ (S \lesssim s_1 \Rightarrow S \lesssim s_2).$$

Since it is, in general, technically difficult to work with $\lesssim^F$, it is common practice to try and obtain an alternative characterization of the finitary preorder (see, e.g., [13]). An alternative method for using the functional $G : \mathsf{Rel}(\mathsf{Proc}) \to \mathsf{Rel}(\mathsf{Proc})$ from Definition 7.2 to obtain a preorder is to apply it inductively as follows:

- $\lesssim_0 \triangleq \mathsf{Proc} \times \mathsf{Proc}$,
- $\lesssim_{n+1} \triangleq G(\lesssim_n)$,

and finally $\lesssim_\omega \triangleq \bigcap_{n\in\mathbb{N}} \lesssim_n$. Intuitively, the preorder $\lesssim_\omega$ is obtained by restricting the prebisimulation relation to observations of finite depth. The preorders $\lesssim$, $\lesssim_\omega$, and $\lesssim^F$ are related thus:

$$\lesssim \subseteq \lesssim_\omega \subseteq \lesssim^F.$$

Moreover the inclusions are, in general, strict. The interested reader is referred to [1] for a wealth of examples distinguishing these preorders, and for a thorough analysis of their general relationships and properties.

The following comparison of preorders with respect to recursive GSOS languages including the inert constant Ω originates from [11,12].

PROPOSITION 7.6. *The preorders $\lesssim^F$ and $\lesssim_\omega$ coincide over the LTS with divergence associated with a recursive GSOS language including the inert constant Ω (cf. Proposition* 5.42).

7.1.4. *A domain of synchronization trees.* The canonical domain used in [11,12] to give a denotational semantics to a class of recursive GSOS languages is the domain of *synchronization trees* over a countably infinite set Act of actions, as considered by Abramsky [1]. This is defined to be the initial solution $\mathcal{D}$ (in the category **SFP**, cf. [172]) of the domain equation

$$D = (\mathbf{1})_\perp \oplus P\left[\sum_{a\in\mathsf{Act}} D\right]$$

where $(\_)_\perp$ denotes lifting, $\mathbf{1}$ a one-point domain used to model $\mathbf{0}$, $\oplus$ the coalesced sum, $\sum$ a separated sum, and $P[\_]$ the Plotkin powerdomain construction (see [172,174] for details on these domain theoretic operations). Intuitively one constructs the least fixed-point $\mathcal{D}$ of the domain equation above by starting with the one-point domain $\mathbf{1}$, and at the nth iterative step building the finite synchronization trees of height n.

To streamline the presentation we abstract away from the domain theoretic description of $\mathcal{D}$ given by the domain equation above. Our description of the domain of synchronization trees $\mathcal{D}$ follows the one given in [129], and we rely on results presented in that reference showing how to construct $\mathcal{D}$ starting from a suitable preorder on the set of finite synchronization trees $\mathsf{ST}(\mathsf{Act})$. The reconstruction of $\mathcal{D}$ is given in three steps.

(1) First, we define a preorder $\sqsubseteq$ on the set of finite synchronization trees $\mathsf{ST}(\mathsf{Act})$. (This preorder is a reformulation of the Egli-Milner preorder over $\mathsf{ST}(\mathsf{Act})$ presented in [129]; see Proposition 7.8.)

(2) Next, we relate the poset of compact elements of $\mathcal{D}$ to the poset of equivalence classes induced by $(\mathsf{ST}(\mathsf{Act}), \sqsubseteq)$.
(3) Finally, we use the fact that $\mathcal{D}$ is the ideal completion of its poset of compact elements to relate it to $(\mathsf{ST}(\mathsf{Act}), \sqsubseteq)$.

This approach allows us to factor the definition of the continuous algebra structure [105, 112,117] on $\mathcal{D}$ in three similar steps.

DEFINITION 7.7 (*ST-preorder*). $\sqsubseteq$ is the least binary relation over $\mathsf{ST}(\mathsf{Act})$ such that the following conditions are satisfied if $S_1 \sqsubseteq S_2$:
(1) if $\langle a, S_1' \rangle \in S_1$, then $S_1' \sqsubseteq S_2'$ for some $\langle a, S_2' \rangle \in S_2$;
(2) if $\bot \in S_2$, then $\bot \in S_1$;
(3) if $\langle a, S_2' \rangle \in S_2$, then either $\bot \in S_1$ or $S_1' \sqsubseteq S_2'$ for some $\langle a, S_1' \rangle \in S_1$.

The relation $\sqsubseteq$ so defined is easily seen to be a preorder over $\mathsf{ST}(\mathsf{Act})$. Moreover, it coincides with $\lesssim$ over $\mathsf{ST}(\mathsf{Act})$. We proceed to relate the preorder $(\mathsf{ST}(\mathsf{Act}), \sqsubseteq)$ with the poset of compact elements of $\mathcal{D}$ in a way that allows us to define, in a canonical way, continuous operations on $\mathcal{D}$ from monotonic ones on $(\mathsf{ST}(\mathsf{Act}), \sqsubseteq)$.

First of all, we recall from [1] that $\mathcal{D}$ is, up to isomorphism, the algebraic cpo whose poset of compact elements $(\mathcal{K}(\mathcal{D}), \sqsubseteq_{\mathcal{K}(\mathcal{D})})$ is given as follows.

- $\mathcal{K}(\mathcal{D})$ is defined inductively by:
  - $\emptyset \in \mathcal{K}(\mathcal{D})$;
  - $\{\bot\} \in \mathcal{K}(\mathcal{D})$;
  - $a \in \mathsf{Act} \wedge d \in \mathcal{K}(\mathcal{D}) \Rightarrow \{\langle a, d \rangle\} \in \mathcal{K}(\mathcal{D})$;
  - $d, e \in \mathcal{K}(\mathcal{D}) \Rightarrow \mathsf{Con}(d \cup e) \in \mathcal{K}(\mathcal{D})$, where Con denotes the convex closure operation (see, e.g., [1, p. 170]).
- $\sqsubseteq_{\mathcal{K}(\mathcal{D})}$ is defined by:

$$d \sqsubseteq_{\mathcal{K}(\mathcal{D})} e \Longleftrightarrow d = \{\bot\} \vee d \sqsubseteq_{EM} e$$

  where $\sqsubseteq_{EM}$ denotes the Egli–Milner preorder (see, e.g., [1, Definition 3.3]).

From the above definitions it follows that $\mathcal{K}(\mathcal{D})$ is a subset of $\mathsf{ST}(\mathsf{Act})$. Hence it makes sense to compare the relations $\sqsubseteq$ and $\sqsubseteq_{\mathcal{K}(\mathcal{D})}$ over it. The following result from [11,12] lends credence to our previous claims.

PROPOSITION 7.8. *For all $d, e \in \mathcal{K}(\mathcal{D})$, $d \sqsubseteq e$ iff $d \sqsubseteq_{\mathcal{K}(\mathcal{D})} e$.*

As a consequence of this result, to ease the presentation of the technical results to follow, from now on we use $\sqsubseteq$ as our notion of preorder on $\mathcal{K}(\mathcal{D})$.

7.2. *From recursive GSOS to denotational semantics*

We now characterize a class of GSOS languages, incorporating the inert constant Ω for which there are no transition rules, that map finite LTSs to finite LTSs (cf. Definition 2.2). The semantic counterparts of these function symbols have the property of being compact in the sense of [129], i.e., of mapping compact elements in the Plotkin powerdomain of synchronization trees to compact elements. In view of Lemma 7.1, denotational interpretations for the resulting languages induce preorders over terms that are algebraic.

DEFINITION 7.9 (*Compact GSOS*). A GSOS language including the inert constant Ω is *compact* if it is linear (cf. Definition 5.34) and syntactically well-founded (cf. Definition 5.35).

For each function symbol f we introduce a mapping $\mathbf{f}_{\mathsf{ST}}$, mapping $ar(f)$ finite synchronization trees to a finite synchronization tree.

DEFINITION 7.10 (*The operation* $\mathbf{f}_{\mathsf{ST}}$). Assume a compact GSOS language, and consider a function symbol f. The operation $\mathbf{f}_{\mathsf{ST}} : \mathsf{ST}(\mathsf{Act})^{ar(f)} \to \mathsf{ST}(\mathsf{Act})$ is defined inductively by stipulating that, for every $S_1, \ldots, S_{ar(f)} \in \mathsf{ST}(\mathsf{Act})$:

- $\perp \in \mathbf{f}_{\mathsf{ST}}(S_1, \ldots, S_{ar(f)})$ iff $f = \Omega$ or there is an argument i of f such that f tests its ith argument (see Definition 5.11) and $\perp \in S_i$;
- $\langle c, S\rangle \in \mathbf{f}_{\mathsf{ST}}(S_1, \ldots, S_{ar(f)})$ iff there are a GSOS rule

$$\frac{\{x_i \xrightarrow{a_{ij}} y_{ij},\ x_i \stackrel{b_{ik}}{\not\rightarrow} \mid 1 \leqslant i \leqslant ar(f), 1 \leqslant j \leqslant m_i, 1 \leqslant k \leqslant n_i\}}{f(x_1, \ldots, x_{ar(f)}) \xrightarrow{c} C[x_1, \ldots, x_{ar(f)}, y_{11}, \ldots, y_{ar(f)m_{ar(f)}}]}$$

and finite synchronization trees $S'_{i1}, \ldots, S'_{im_i}$ for $i = 1, \ldots, ar(f)$ such that:
(1) $\langle a_{ij}, S'_{ij}\rangle \in S_i$ for $i = 1, \ldots, ar(f)$ and $j = 1, \ldots, m_i$;
(2) if $n_i > 0$, then $\perp \notin S_i$ and $\langle b_{ik}, S'\rangle \notin S_i$ for $S' \in \mathsf{ST}(\mathsf{Act})$ and $k = 1, \ldots, n_i$;
(3) $\mathbf{C}_{\mathsf{ST}}[S_1, \ldots, S_{ar(f)}, S'_{11}, \ldots, S'_{ar(f)m_{ar(f)}}] = S$, where $\mathbf{C}_{\mathsf{ST}}$ denotes the derived semantic operation associated with the target of the GSOS rule.

The above definition, which is discussed at length in [11,12], endows the preorder of finite synchronization trees $\mathsf{ST}(\mathsf{Act})$ with a Σ-preorder structure (where Σ is the signature under consideration), in the sense of [117]. Since the poset of compact elements of the domain $\mathcal{D}$ is a substructure of the preorder $(\mathsf{ST}(\mathsf{Act}), \sqsubseteq)$, this is enough to give a denotational interpretation for the recursion-free terms in a compact GSOS language in terms of compact elements in $\mathcal{D}$.

THEOREM 7.11. *Assume a compact GSOS language. For all* $t, u \in \mathrm{T}(\Sigma)$, $t \lesssim_\omega u$ *iff* $\mathcal{K}(\mathcal{D})[\![t]\!] \sqsubseteq \mathcal{K}(\mathcal{D})[\![u]\!]$.

The above full abstraction result can be extended to the whole of the language $\mathsf{CREC}(\Sigma)$, for any compact recursive GSOS language. In order to define an interpretation of programs in $\mathsf{CREC}(\Sigma)$ as elements of $\mathcal{D}$, we need to define a continuous Σ-algebra structure on $\mathcal{D}$.

From the theory of powerdomains [129,172,204], we know that the domain of synchronization trees $\mathcal{D}$ is, up to isomorphism, the ideal completion of the poset of compact elements $\mathcal{K}(\mathcal{D})$. (The construction of the ideal completion of a poset and a discussion of its basic properties can be found in [117, p. 139–145].) Let $\sqsubseteq_{\mathcal{D}}$ be standard set inclusion on $\mathcal{D}$. Since $\mathcal{D}$ is the ideal completion of $\mathcal{K}(\mathcal{D})$, the monotonic function $\mathbf{f}_{\mathsf{ST}} : (\mathcal{K}(\mathcal{D}), \sqsubseteq)^{ar(f)} \to (\mathcal{K}(\mathcal{D}), \sqsubseteq)$ for any function symbol f can be extended to a continuous function $\mathbf{f}_{\mathcal{D}} : (\mathcal{D}, \sqsubseteq_{\mathcal{D}})^{ar(f)} \to (\mathcal{D}, \sqsubseteq_{\mathcal{D}})$ by:

$$\mathbf{f}_{\mathcal{D}}(e_1, \ldots, e_{ar(f)}) \triangleq \bigcup \bigl\{\mathbf{f}_{\mathsf{ST}}(d_1, \ldots, d_{ar(f)}) \mid d_1 \in e_1, \ldots, d_{ar(f)} \in e_{ar(f)}\bigr\},$$

where $e_1, \ldots, e_{ar(f)}$ are ideals in $(\mathcal{K}(\mathcal{D}), \sqsubseteq_{\mathcal{K}(\mathcal{D})})$, and we identify an element of $(\mathcal{K}(\mathcal{D}), \sqsubseteq_{\mathcal{K}(\mathcal{D})})$ with the principal ideal it generates (see [117, p. 139]). The interested reader is invited to consult, e.g., [117, Section 3.3] for a discussion of the properties afforded by this canonical extension. By the general theory of algebraic semantics we then have that, for all closed terms t, u,

$$\mathcal{D}[\![t]\!] \sqsubseteq_{\mathcal{D}} \mathcal{D}[\![u]\!] \iff \mathcal{K}(\mathcal{D})[\![t]\!] \sqsubseteq \mathcal{K}(\mathcal{D})[\![u]\!].$$

In view of Theorem 7.11, the desired full abstraction result follows if we prove that the preorder $\lesssim_\omega$ is algebraic. Namely, owing to our constructions each closed term t is interpreted as a compact element of $\mathcal{D}$, so Lemma 7.1 implies that the relation $\sqsubseteq_{\mathcal{D}}$ is algebraic. And two algebraic relations that coincide over the collection of closed terms $\mathrm{T}(\Sigma)$ do, in fact, coincide over the whole of $\mathsf{CREC}(\Sigma)$.

The key to the proof of algebraicity of $\lesssim_\omega$ is the following general theorem from [11, 12], providing a partial completeness result for $\lesssim$ in the sense of Hennessy [10,115] for arbitrary compact GSOS languages. This axiomatizability result uses the fact that $\lesssim$ is a precongruence with respect to the LTS with divergence associated with a recursive GSOS language including the inert constant Ω; see Proposition 7.3.

PROPOSITION 7.12. *Let G be a compact recursive GSOS language. Then there exists a compact recursive GSOS language H over a signature Σ', being a disjoint extension* (*cf. Definition* 5.31) *of G and T_{FINTREE}* (*cf. Section* 5.4.5), *together with a set $\mathcal{I}$ of inequalities between recursive terms over Σ', such that for all $t \in \mathrm{T}(\Sigma')$ and $u \in \mathsf{CREC}(\Sigma')$:*

$$H \text{ induces } t \lesssim u \quad \text{iff} \quad \mathcal{I} \vdash t \leqslant u.$$

Apart from its intrinsic interest, the main consequence of Proposition 7.12 is the following key result from [11,12], essentially stating that, for any compact GSOS language, finite synchronization trees are compact elements with respect to the preorder $\lesssim$.

PROPOSITION 7.13. *Assume a compact recursive GSOS language. If $S \in \mathsf{ST}(\mathsf{Act})$ and $t \in \mathsf{CREC}(\Sigma)$, then $S \lesssim^F t$ iff there exists a finite approximation t_n of t such that $S \lesssim^F t_n$.*

The above result, in conjunction with Proposition 7.6, yields that $\lesssim_\omega$ is indeed algebraic.

PROPOSITION 7.14. *$\lesssim_\omega$ is algebraic over the LTS with divergence associated with a compact recursive GSOS language.*

In light of Theorem 7.11 and Proposition 7.14, for any compact recursive GSOS language the induced denotational semantics over $\mathsf{CREC}(\Sigma)$ is fully abstract with respect to $\lesssim_\omega$.

THEOREM 7.15. *Assume a compact recursive GSOS language. For all* $t, u \in \mathsf{CREC}(\Sigma)$, $t \lesssim_\omega u$ *iff* $\mathcal{D}[\![t]\!] \sqsubseteq_{\mathcal{D}} \mathcal{D}[\![u]\!]$.

When applied to the version of SCCS considered by Abramsky [1], the techniques we have presented deliver a denotational semantics that is exactly the one given in *op. cit.*

Related work. The work reported in this section is by no means the only attempt to systematically derive denotational models from SOS language specifications. The main precursors to this work in the field of the meta-theory of process description languages may be found in the work by Bloom [46] and Rutten [188–191]. In his unpublished paper [46], Bloom gives operational, logical, relational, and three denotational semantics for GSOS languages without negative premises and unguarded recursion, and shows that they coincide. Bloom's work is based on the behavioural notion of simulation [123], and two of his denotational semantics are given by means of Scott domains based on finite synchronization trees. On the other hand, the work by Rutten [188–190] gives methods for deriving a denotational semantics based on complete metric spaces and Aczel's non-well-founded sets [14] for languages specified by means of subformats of the tyft/tyxt format (cf. Definition 5.2). In particular, the reference [190] gives a detailed and clear introduction to a technique, called "processes as terms", for the definition of operations on semantic models from transition rules. Rutten's general "processes as terms" approach could have been applied to yield an equivalent formulation of the semantic operations on finite synchronization trees given above. The work presented in the aforementioned papers has been generalized by Rutten and Turi [191]. *Ibidem* it is shown how TSSs in tyft/tyxt format induce a denotational semantics, and the essential properties of semantic domains that make their definitions possible are investigated in a categorical perspective.

Abramsky and Vickers [3] consider various notions of process observations in a uniform algebraic framework provided by the theory of *quantales* (see, e.g., [187]). The methods developed in [3] yield, in a uniform fashion, observational logics and denotational models for each notion of process observation they consider. Their work is, however, semantic in nature, and ignores the algebraic structure of process expressions.

In the area of the semantics of functional programs, developments that are somewhat similar in spirit to those given above are presented in [144,203]. Those papers study natural notions of preorder over programs written in a simple functional programming language, and show how any ordering on programs with certain basic properties can be extended to a term model that is fully abstract with respect to it.

The issue of defining abstract mathematical models *for*, rather than from, operational semantics has also received some attention. We refer the interested reader to, e.g., [19, 214], and the references therein, for details on this line of investigation.

References

[1] S. Abramsky, *A domain equation for bisimulation*, Inform. and Comput. **92** (1991), 161–218.

[2] S. Abramsky and C. Hankin, *Abstract Interpretation of Declarative Languages*, Ellis Horwood (1987).

[3] S. Abramsky and S. Vickers, *Quantales, observational logic and process semantics*, Math. Structures Comput. Sci. **3** (1993), 161–227.

[4] L. Aceto, *Deriving complete inference systems for a class of GSOS languages generating regular behaviours*, Proceedings 5th Conference on Concurrency Theory, Uppsala, Sweden, Lecture Notes in Comput. Sci. 836, B. Jonsson and J. Parrow, eds, Springer-Verlag (1994), 449–464.

[5] L. Aceto, *GSOS and finite labelled transition systems*, Theoret. Comput. Sci. **131** (1994), 181–195.

[6] L. Aceto, B. Bloom and F. Vaandrager, *Checking equations in GSOS systems* (1992). Unpublished working paper.

[7] L. Aceto, B. Bloom and F. Vaandrager, *Turning SOS rules into equations*, Proceedings 7th Symposium on Logic in Computer Science, Santa Cruz, CA, IEEE Computer Society Press (1992), 113–124. Preliminary version of [8].

[8] L. Aceto, B. Bloom and F. Vaandrager, *Turning SOS rules into equations*, Inform. and Comput. **111** (1994), 1–52.

[9] L. Aceto, W. Fokkink, R.v. Glabbeek and A. Ingólfsdóttir, *Axiomatizing prefix iteration with silent steps*, Inform. and Comput. **127** (1996), 26–40.

[10] L. Aceto and M. Hennessy, *Termination, deadlock and divergence*, J. Assoc. Comput. Mach. **39** (1992), 147–187.

[11] L. Aceto and A. Ingólfsdóttir, *CPO models for a class of GSOS languages*, Proceedings 6th Conference on Theory and Practice of Software Development, Århus, Denmark, Lecture Notes in Comput. Sci. 915, P. Mosses, M. Nielsen and M. Schwartzbach, eds, Springer-Verlag (1995), 439–453. Preliminary version of [12].

[12] L. Aceto and A. Ingólfsdóttir, *CPO models for compact GSOS languages*, Inform. and Comput. **129** (1996), 107–141.

[13] L. Aceto and A. Ingólfsdóttir, *A characterization of finitary bisimulation*, Inform. Process. Lett. **64** (1997), 127–134.

[14] P. Aczel, *Non-well-founded Sets*, CSLI Lecture Notes 14, Stanford University (1988).

[15] S. Allen, R. Constable, D. Howe and W. Aitken, *The semantics of reflected proof*, Proceedings 5th Symposium on Logic in Computer Science, Philadelphia, PA, IEEE Computer Society Press (1990), 95–105.

[16] D. Austry and G. Boudol, *Algèbre de processus et synchronisations*, Theoret. Comput. Sci. **30** (1984), 91–131.

[17] F. Baader and T. Nipkow, *Term Rewriting and All That*, Cambridge University Press (1998).

[18] J. Backus, *The syntax and semantics of the proposed international algebraic language of the Zurich ACM-GAMM conference*, Proceedings ICIP, UNESCO (1960), 125–131.

[19] E. Badouel, *Conditional rewrite rules as an algebraic semantics of processes*, Research Report 1226, INRIA Rennes (1990).

[20] E. Badouel and P. Darondeau, *Structural operational specifications and trace automata*, Proceedings 3rd Conference on Concurrency Theory, Stony Brook, NY, Lecture Notes in Comput. Sci. 630, R. Cleaveland, ed., Springer-Verlag (1992), 302–316.

[21] J. Baeten, ed., *Applications of Process Algebra*, Cambridge Tracts in Theoret. Comput. Sci. 17, Cambridge University Press (1990).

[22] J. Baeten and J. Bergstra, *A survey of axiom systems for process algebras*, Report P9111, University of Amsterdam, Amsterdam (1991).

[23] J. Baeten and J. Bergstra, *Discrete time process algebra*, Proceedings 3rd Conference on Concurrency Theory, Stony Brook, NY, Lecture Notes in Comput. Sci. 630, R. Cleaveland, ed., Springer-Verlag (1992), 401–420.

[24] J. Baeten, J. Bergstra and J.W. Klop, *Syntax and defining equations for an interrupt mechanism in process algebra*, Fundamenta Informaticae **IX** (1986), 127–168.

[25] J. Baeten, J. Bergstra and J.W. Klop, *On the consistency of Koomen's fair abstraction rule*, Theoret. Comput. Sci. **51** (1987), 129–176.

[26] J. Baeten and C. Verhoef, *A congruence theorem for structured operational semantics with predicates*, Proceedings 4th Conference on Concurrency Theory, Hildesheim, Germany, Lecture Notes in Comput. Sci. 715, E. Best, ed., Springer-Verlag (1993), 477–492.

[27] J. Baeten and C. Verhoef, *Concrete process algebra*, Handbook of Logic in Computer Science, Vol. IV, S. Abramsky, D. Gabbay and T. Maibaum, eds, Oxford University Press (1995), 149–268.

[28] J. Baeten and P. Weijland, *Process Algebra*, Cambridge Tracts in Theoret. Comput. Sci. 18, Cambridge University Press (1990).

[29] J.d. Bakker, *Semantics of programming languages*, Adv. Inform. Systems Sci. **2** (1969), 173–227.

[30] J.d. Bakker and J. Zucker, *Processes and the denotational semantics of concurrency*, Inform. and Control **54** (1982), 70–120.

[31] H. Barendregt, *The Lambda Calculus, Its Syntax and Semantics*, Studies in Logic and the Foundation of Mathematics 103, North-Holland, Amsterdam (1984).

[32] T. Basten, *Branching bisimilarity is an equivalence indeed!*, Inform. Process. Lett. **58** (1996), 141–147.

[33] J. Bergstra, I. Bethke and A. Ponse, *Process algebra with iteration and nesting*, Computer J. **37** (1994), 243–258.

[34] J. Bergstra, W. Fokkink and A. Ponse, *Process algebra with recursive operations*, Handbook of Process Algebra, J.A. Bergstra, A. Ponse and S.A. Smolka, eds, Elsevier, Amsterdam (2001), 333–389.

[35] J. Bergstra and J.W. Klop, *Fixed point semantics in process algebras*, Report IW 206, Mathematisch Centrum, Amsterdam (1982).

[36] J. Bergstra and J.W. Klop, *The algebra of recursively defined processes and the algebra of regular processes*, Proceedings 11th Colloquium on Automata, Languages and Programming, Antwerp, Belgium, Lecture Notes in Comput. Sci. 172, J. Paredaens, ed., Springer-Verlag (1984), 82–95.

[37] J. Bergstra and J.W. Klop, *Process algebra for synchronous communication*, Inform. and Control **60** (1984), 109–137.

[38] J. Bergstra and J.W. Klop, *Conditional rewrite rules: Confluence and termination*, J. Comput. System Sci. **32** (1986), 323–362.

[39] J. Bergstra and J.W. Klop, *Verification of an alternating bit protocol by means of process algebra*, Spring School on Mathematical Methods of Specification and Synthesis of Software Systems, Wendisch-Rietz, Germany, Lecture Notes in Comput. Sci. 215, W. Bibel and K. Jantke, eds, Springer-Verlag (1986), 9–23.

[40] J. Bergstra and J.W. Klop, *A complete inference system for regular processes with silent moves*, Proceedings Logic Colloquium 1986, F. Drake and J. Truss, eds, North-Holland, Hull (1988), 21–81.

[41] J. Bergstra, A. Ponse and J.v. Wamel, *Process algebra with backtracking*, Proceedings REX School/Symposium on A Decade of Concurrency: Reflections and Perspectives, Noordwijkerhout, The Netherlands, Lecture Notes in Comput. Sci. 803, J.d. Bakker, W.d. Roever and G. Rozenberg, eds, Springer-Verlag (1994), 46–91.

[42] K. Bernstein, *A congruence theorem for structured operational semantics of higher-order languages*, Proceedings 13th Symposium on Logic in Computer Science, Indianapolis, Indiana, IEEE Computer Society Press (1998), 153–164.

[43] G. Berry, *A hardware implementation of Pure Esterel*, Rapport de Recherche 06/91, Ecole des Mines, CMA, Sophia-Antipolis, France (1991).

[44] B. Bloom, *Ready simulation, bisimulation and the semantics of CCS-like languages*, Ph.D. Thesis, Department of Electrical Engineering and Computer Science, Massachusetts Institute of Technology (1989).

[45] B. Bloom, *Can LCF be topped? Flat lattice models of typed λ-calculus*, Inform. and Comput. **87** (1990), 263–300.

[46] B. Bloom, *Many meanings of monosimulation: Denotational, operational and logical characterizations of a notion of simulation of concurrent processes* (1991). Unpublished manuscript.

[47] B. Bloom, *Ready, set, go: Structural operational semantics for linear-time process algebras*, Report TR 93-1372, Cornell University, Ithaca, New York (1993).

[48] B. Bloom, *CHOCOLATE: Calculi of Higher Order COmmunication and LAmbda TErms*, Conference Record 21st ACM Symposium on Principles of Programming Languages, Portland, OR (1994), 339–347.

[49] B. Bloom, *When is partial trace equivalence adequate?*, Formal Aspects of Computing **6** (1994), 317–338.

[50] B. Bloom, *Structural operational semantics for weak bisimulations*, Theoret. Comput. Sci. **146** (1995), 25–68.

[51] B. Bloom, *Structured operational semantics as a specification language*, Conference Record 22nd ACM Symposium on Principles of Programming Languages, San Francisco, CA (1995), 107–117.

[52] B. Bloom, A. Cheng and A. Dsouza, *Using a protean language to enhance expressiveness in specification*, IEEE Trans. Software Engrg. **23** (1997), 224–234.

[53] B. Bloom, W. Fokkink and R.v. Glabbeek, *Precongruence formats for decorated trace preorders*, Proceedings 15th Symposium on Logic in Computer Science, Santa Barbara, CA, IEEE Computer Society Press (2000), 107–118.

[54] B. Bloom, S. Istrail and A. Meyer, *Bisimulation can't be traced: preliminary report*, Conference Record 15th ACM Symposium on Principles of Programming Languages, San Diego, CA (1988), 229–239. Preliminary version of [55].

[55] B. Bloom, S. Istrail and A. Meyer, *Bisimulation can't be traced*, J. Assoc. Comput. Mach. **42** (1995), 232–268.

[56] R. Bol and J.F. Groote, *The meaning of negative premises in transition system specifications (extended abstract)*, Proceedings 18th Colloquium on Automata, Languages and Programming, Madrid, Spain, Lecture Notes in Comput. Sci. 510, J. Leach Albert, B. Monien and M. Rodríguez, eds, Springer-Verlag (1991), 481–494. Preliminary version of [57].

[57] R. Bol and J.F. Groote, *The meaning of negative premises in transition system specifications*, J. Assoc. Comput. Mach. **43** (1996), 863–914.

[58] T. Bolognesi and F. Lucidi, *Timed process algebras with urgent interactions and a unique powerful binary operator*, Proceedings REX Workshop on Real-Time: Theory in Practice, Mook, The Netherlands, June 1991, Lecture Notes in Comput. Sci. 600, J.d. Bakker, C. Huizing, W.d. Roever and G. Rozenberg, eds, Springer-Verlag (1992), 124–148.

[59] D. Bosscher, *Term rewriting properties of SOS axiomatisations*, Proceedings 2nd Symposium on Theoretical Aspects of Computer Software, Sendai, Japan, Lecture Notes in Comput. Sci. 789, M. Hagiya and J. Mitchell, eds, Springer-Verlag (1994), 425–439.

[60] M. v.d. Brand, P. Klint and C. Verhoef, *Reverse engineering and system renovation – annotated bibliography*, Software Engineering Notes **22** (1997), 56–67.

[61] S. Brookes, C. Hoare and A. Roscoe, *A theory of communicating sequential processes*, J. Assoc. Comput. Mach. **31** (1984), 560–599.

[62] R. Bryant, *Graph-based algorithms for boolean function manipulation*, IEEE Trans. Comput. **C-35** (1986), 677–691.

[63] K. Clark, *Negation as failure*, Logic and Databases, H. Gallaire and J. Minker, eds, Plenum Press, New York (1978), 293–322.

[64] R. Cleaveland and M. Hennessy, *Priorities in process algebras*, Inform. and Comput. **87** (1990), 58–77.

[65] R. Cleaveland, J. Parrow and B. Steffen, *The concurrency workbench: A semantics-based verification tool for finite state systems*, ACM Trans. Prog. Lang. Syst. **15** (1993), 36–72.

[66] R. Constable, S. Allen, H. Bromley, R. Cleaveland, J. Cremer, R. Harper, D. Howe, T. Knoblock, N. Mendler, P. Panangaden, J. Sasaki and S. Smith, *Implementing Mathematics with the Nuprl Proof Development System*, Prentice Hall (1986).

[67] B. Courcelle and M. Nivat, *Algebraic families of interpretations*, Proceedings 17th Symposium on Foundations of Computer Science, Houston, TX, IEEE (1976), 137–146.

[68] P. D'Argenio, *A general conservative extension theorem in process algebras with inequalities*, Proceedings 2nd Workshop on the Algebra of Communicating Processes, Eindhoven, The Netherlands, Report CS-95-14, A. Ponse, C. Verhoef and B.v. Vlijmen, eds, Eindhoven University of Technology (1995), 67–79.

[69] P. D'Argenio and C. Verhoef, *A general conservative extension theorem in process algebras with inequalities*, Theoret. Comput. Sci. **177** (1997), 351–380.

[70] P. Darondeau, *Concurrency and computability*, Semantics of Systems of Concurrent Processes, Proceedings LITP Spring School on Theoret. Computer Science, La Roche Posay, France, Lecture Notes in Comput. Sci. 469, I. Guessarian, ed., Springer-Verlag (1990), 223–238.

[71] N. De Francesco and P. Inverardi, *Proving finiteness of CCS processes by non-standard semantics*, Acta Inform. **31** (1994), 55–80.

[72] R. De Nicola and M. Hennessy, *Testing equivalences for processes*, Theoret. Comput. Sci. **34** (1984), 83–133.

[73] P. Degano and C. Priami, *Enhanced operational semantics*, ACM Comput. Surveys **28** (1996), 352–354.

[74] A. Dsouza and B. Bloom, *Generating BDD models for process algebra terms*, Proceedings 7th Conference on Computer Aided Verification, Liege, Belgium, Lecture Notes in Comput. Sci. 939, P. Wolper, ed., Springer-Verlag (1995), 16–30.

[75] A. Dsouza and B. Bloom, *On the expressive power of CCS*, Proceedings 15th Conference on Foundations of Software Technology and Theoret. Comput. Sci., Bangalore, India, Lecture Notes in Comput. Sci. 1026, P. Thiagarajan, ed., Springer-Verlag (1995), 309–323.

[76] A. Emerson, *Automated temporal reasoning about reactive systems*, Logics for Concurrency: Structure versus Automata, Lecture Notes in Comput. Sci. 1043, F. Moller and G. Birtwistle, eds, Springer-Verlag (1996), 41–101.

[77] R. Enders, T. Filkorn and D. Taubner, *Generating BDDs for symbolic model checking in CCS*, Distrib. Comput. **6** (1993), 155–164.

[78] F. Fages, *A new fixpoint semantics for general logic programs compared with the well-founded and the stable model semantics*, New Generation Computing **9** (1991), 425–443.

[79] J.-C. Fernandez, *Aldébaran: A tool for verification of communicating processes*, Technical Report SPECTRE c14, LGI–IMAG, Grenoble (1989).

[80] W. Fokkink, *The tyft/tyxt format reduces to tree rules*, Proceedings 2nd Symposium on Theoret. Aspects of Computer Software, Sendai, Japan, Lecture Notes in Comput. Sci. 789, M. Hagiya and J. Mitchell, eds, Springer-Verlag (1994), 440–453. Preliminary version of [83].

[81] W. Fokkink, *Language preorder as a precongruence*, Theoret. Comput. Sci. **243** (2000), 391–408.

[82] W. Fokkink, *Rooted branching bisimulation as a congruence*, J. Comput. System Sci. **60** (2000), 13–37.

[83] W. Fokkink and R.v. Glabbeek, *Ntyft/ntyxt rules reduce to ntree rules*, Inform. and Comput. **126** (1996), 1–10.

[84] W. Fokkink and S. Klusener, *An effective axiomatization for real time ACP*, Inform. and Comput. **122** (1995), 286–299.

[85] W. Fokkink and C. Verhoef, *A conservative look at operational semantics with variable binding*, Inform. and Comput. **146** (1998), 24–54.

[86] W. Fokkink and C. Verhoef, *Conservative extension in positive/negative conditional term rewriting with applications to software renovation factories*, Proceedings 2nd Conference on Fundamental Approaches to Software Engineering, Amsterdam, The Netherlands, Lecture Notes in Comput. Sci. 1577, J.-P. Finance, ed., Springer-Verlag (1999), 98–113.

[87] W. Fokkink and H. Zantema, *Basic process algebra with iteration: Completeness of its equational axioms*, Computer J. **37** (1994), 259–267.

[88] A.v. Gelder, K. Ross and J. Schlipf, *Unfounded sets and well-founded semantics for general logic programs*, Proceedings 7th ACM Symposium on Principles of Database Systems, Austin, TX, ACM (1988), 221–230. Preliminary version of [89].

[89] A.v. Gelder, K. Ross and J. Schlipf, *The well-founded semantics for general logic programs*, J. Assoc. Comput. Mach. **38** (1991), 620–650.

[90] M. Gelfond and V. Lifschitz, *The stable model semantics for logic programming*, Proceedings 5th Conference on Logic Programming, Seattle, WA, R. Kowalski and K. Bowen, eds, MIT Press (1988), 1070–1080.

[91] G. Gentzen, *Investigations into logical deduction*, The Collected Papers of Gerhard Gentzen, M. Szabo, ed., North-Holland (1969), 68–128.

[92] R.v. Glabbeek, *Bounded nondeterminism and the approximation induction principle in process algebra*, Proceedings 4th Symposium on Theoret. Aspects of Computer Science, Passau, Germany, Lecture Notes in Comput. Sci. 247, F. Brandenburg, G. Vidal-Naquet and M. Wirsing, eds, Springer-Verlag (1987), 336–347.

[93] R.v. Glabbeek, *Comparative concurrency semantics and refinement of actions*, Ph.D. Thesis, Free University, Amsterdam (1990).

[94] R.v. Glabbeek, *The linear time – branching time spectrum*, Proceedings 1st Conference on Concurrency Theory, Amsterdam, The Netherlands, Lecture Notes in Comput. Sci. 458, J. Baeten and J.W. Klop, eds, Springer-Verlag (1990), 278–297.

[95] R.v. Glabbeek, *A complete axiomatization for branching bisimulation congruence of finite-state behaviours*, Proceedings 18th Symposium on Mathematical Foundations of Computer Science 1993, Gdansk, Poland, Lecture Notes in Comput. Sci. 711, A. Borzyszkowski and S. Sokołowski, eds, Springer-Verlag (1993), 473–484.

[146] J. McCarthy, *Towards a mathematical science of computation*, Information Processing 1962, C. Popplewell, ed. (1963), 21–28.

[147] K. McMillan, *Symbolic Model Checking*, Kluwer Academic Publishers (1993).

[148] A. Meyer, *Semantical paradigms: Notes for an invited lecture*, Proceedings 3rd Symposium on Logic in Computer Science, Edinburgh, IEEE Computer Society Press (1988), 236–242.

[149] G. Milne and R. Milner, *Concurrent processes and their syntax*, J. Assoc. Comput. Mach. **26** (1979), 302–321.

[150] R. Milner, *Processes: A mathematical model of computing agents*, Proceedings Logic Colloquium 1973, Bristol, UK, H. Rose and J. Shepherdson, eds, North-Holland (1973), 158–173.

[151] R. Milner, *Fully abstract models of typed λ-calculi*, Theoret. Comput. Sci. **4** (1977), 1–22.

[152] R. Milner, *A Calculus of Communicating Systems*, Lecture Notes in Comput. Sci. 92, Springer-Verlag (1980).

[153] R. Milner, *A modal characterisation of observable machine behaviour*, Proceedings 6th Colloquium on Trees in Algebra and Programming, Genoa, Italy, Lecture Notes in Comput. Sci. 112, E. Astesiano and C. Böhm, eds, Springer-Verlag (1981), 25–34.

[154] R. Milner, *On relating synchrony and asynchrony*, Technical Report CSR-75-80, Department of Computer Science, University of Edinburgh (1981).

[155] R. Milner, *Calculi for synchrony and asynchrony*, Theoret. Comput. Sci. **25** (1983), 267–310.

[156] R. Milner, *A complete inference system for a class of regular behaviours*, J. Comput. System Sci. **28** (1984), 439–466.

[157] R. Milner, *Communication and Concurrency*, Prentice-Hall International, Englewood Cliffs (1989).

[158] R. Milner, *A complete axiomatisation for observational congruence of finite-state behaviors*, Inform. and Comput. **81** (1989), 227–247.

[159] R. Milner, *Elements of interaction (Turing Award Lecture)*, Comm. ACM **36** (1993), 78–89.

[160] R. Milner, J. Parrow and D. Walker, *A calculus of mobile processes, part I + II*, Inform. and Comput. **100** (1992), 1–77.

[161] R. Milner, M. Tofte, R. Harper and D. MacQueen, *The Definition of Standard ML (Revised)*, MIT Press (1997).

[162] F. Moller, *The importance of the left merge operator in process algebras*, Proceedings 17th Colloquium on Automata, Languages and Programming, Warwick, UK, Lecture Notes in Comput. Sci. 443, M. Paterson, ed., Springer-Verlag (1990), 752–764.

[163] F. Moller and C. Tofts, *A temporal calculus of communicating systems*, Proceedings 1st Conference on Concurrency Theory, Amsterdam, The Netherlands, Lecture Notes in Comput. Sci. 458, J. Baeten and J.W. Klop, eds, Springer-Verlag (1990), 401–415.

[164] P. Mosses, *Foundations of modular SOS (extended abstract)*, Proceedings 24th Symposium on Mathematical Foundations of Computer Science, Szklarska Poreba, Poland, Lecture Notes in Comput. Sci. 1672, M. Kutyłowski, L. Pacholski and T. Wierzbicki, eds, Springer-Verlag (1999), 70–80.

[165] X. Nicollin and J. Sifakis, *The algebra of timed processes, ATP: Theory and application*, Inform. and Comput. **114** (1994), 131–178.

[166] H. Nielson and F. Nielson, *Semantics with Applications: A Formal Introduction*, Wiley Professional Computing, John Wiley & Sons, Chichester, England (1992).

[167] V.v. Oostrom and E.d. Vink, *Transition system specifications in stalk format with bisimulation as a congruence*, Proceedings 11th Symposium on Theoret. Aspects of Computer Science, Caen, France, Lecture Notes in Comput. Sci. 775, P. Enjalbert, E. Mayr and K. Wagner, eds, Springer-Verlag (1994), 569–580.

[168] S. Owicki and D. Gries, *An axiomatic proof technique for parallel programs*, Acta Inform. **6** (1976), 319–340.

[169] D. Park, *Concurrency and automata on infinite sequences*, Proceedings 5th GI Conference, Karlsruhe, Germany, Lecture Notes in Comput. Sci. 104, P. Deussen, ed., Springer-Verlag (1981), 167–183.

[170] J. Parrow, *The expressive power of parallelism*, Future Generation Computer Systems **6** (1990), 271–285.

[171] PL/I Definition Group, *Formal definition of PL/I version 1*, Report TR25.071, American Nat. Standards Institute (1986).

[172] G. Plotkin, *A powerdomain construction*, SIAM J. Comput. **5** (1976), 452–487.

[173] G. Plotkin, *LCF considered as a programming language*, Theoret. Comput. Sci. **5** (1977), 223–256.

[174] G. Plotkin, *Lecture notes in domain theory*, University of Edinburgh (1981).

[175] G. Plotkin, *A structural approach to operational semantics*, Report DAIMI FN-19, Computer Science Department, Aarhus University (1981).

[176] G. Plotkin, *An operational semantics for CSP*, Proceedings IFIP TC2 Working Conference on Formal Description of Programming Concepts – II, Garmisch-Partenkirchen, Germany, D. Bjørner, ed., North-Holland (1983), 199–225.

[177] A. Pnueli, *The temporal logic of programs*, Proceedings 18th Symposium on Foundations of Computer Science, Providence, RI, IEEE (1977), 46–57.

[178] A. Ponse, *Computable processes and bisimulation equivalence*, Formal Aspects of Computing **8** (1996), 648–678.

[179] C. Priami, *Enhanced operational semantics for concurrency*, Ph.D. Thesis, Department of Computer Science, University of Pisa (1996).

[180] T. Przymusinski, *On the declarative semantics of deductive databases and logic programs*, Foundations of Deductive Databases and Logic Programming, J. Minker, ed., Morgan Kaufmann, Los Altos, CA (1988), 193–216.

[181] T. Przymusinski, *The well-founded semantics coincides with the three-valued stable semantics*, Fundamenta Informaticae **13** (1990), 445–463.

[182] W. Reisig, *Petri Nets – An Introduction*, EATCS Monographs on Theoret. Computer Science, Vol. 4, Springer-Verlag (1985).

[183] A. Rensink, *Bisimilarity of open terms*, Proceedings 4th Workshop on Expressiveness in Concurrency, Santa Margherita Ligure, Italy, Electronic Notes in Theoret. Computer Science 7, C. Palamidessi and J. Parrow, eds, Elsevier (1997).

[184] J. Reppy, *CML: A higher-order concurrent language*, Programming Language Design and Implementation, SIGPLAN Notices **26** (1991), 293–305.

[185] H. Rogers, *Theory of Recursive Functions and Effective Computability*, McGraw-Hill Book Co. (1967).

[186] A. Roscoe, *The Theory and Practice of Concurrency*, Prentice-Hall International (1998).

[187] K. Rosenthal, *Quantales and their Applications*, Research Notes in Mathematics, Pitman, London (1990).

[188] J. Rutten, *Deriving denotational models for bisimulation from structured operational semantics*, Ten Years of Concurrency Semantics: Selected Papers of the Amsterdam Concurrency Group, J.d. Bakker and J. Rutten, eds, World Scientific (1992), 425–441.

[189] J. Rutten, *Nonwellfounded sets and programming language semantics*, Proceedings 7th Conference on Mathematical Foundations of Programming Semantics, Pittsburgh, PA, Lecture Notes in Comput. Sci. 598, S. Brookes, M. Main, A. Melton, M. Mislove and D. Schmidt, eds, Springer-Verlag (1992), 193–206.

[190] J. Rutten, *Processes as terms: Non-well-founded models for bisimulation*, Math. Structures Comput. Sci. **2** (1992), 257–275.

[191] J. Rutten and D. Turi, *Initial algebra and final coalgebra semantics for concurrency*, Proceedings REX School/Symposium on A Decade of Concurrency: Reflections and Perspectives, Noordwijkerhout, The Netherlands, Lecture Notes in Comput. Sci. 803, J.d. Bakker, W.d. Roever and G. Rozenberg, eds, Springer-Verlag (1994), 530–581.

[192] A. Salomaa, *Theory of Automata*, International Series of Monographs in Pure and Applied Mathematics 100, Pergamon Press Oxford (1969).

[193] D. Sands, *From SOS rules to proof principles: An operational metatheory for functional languages*, Conference Record 24th ACM Symposium on Principles of Programming Languages, Paris, France (1997), 428–441.

[194] D. Sangiorgi, *The lazy lambda calculus in a concurrency scenario*, Inform. and Comput. **111** (1994), 120–153.

[195] D. Sangiorgi, *πI: A symmetric calculus based on internal mobility*, Proceedings 6th Conference on Theory and Practice of Software Development, Århus, Denmark, Lecture Notes in Comput. Sci. 915, P. Mosses, M. Nielsen and M. Schwartzbach, eds, Springer-Verlag (1995), 172–186.

[196] S. Schneider, *An operational semantics for timed CSP*, Inform. and Comput. **116** (1995), 193–213.

[197] D. Scott and C. Strachey, *Towards a mathematical semantics for computer languages*, Proceedings Symposium on Computers and Automata, Microwave Research Institute Symposia Series 21 (1971).

[198] R.d. Simone, *Calculabilité et expressivité dans l'algèbre de processus parallèles Meije*, Thèse de 3e cycle, Univ. Paris 7 (1984).

[199] R.d. Simone, *On Meije and SCCS: Infinite sum operators vs. non-guarded definitions*, Theoret. Comput. Sci. **30** (1984), 133–138.

[200] R.d. Simone, *Higher-level synchronising devices in Meije–SCCS*, Theoret. Comput. Sci. **37** (1985), 245–267.

[201] R.d. Simone and D. Vergamini, *Aboard AUTO*, Technical Report 111, INRIA, Centre Sophia-Antipolis, Valbonne Cedex (1989).

[202] A. Simpson, *Compositionality via cut-elimination: Hennessy–Milner logic for an arbitrary GSOS*, Proceedings 10th Symposium on Logic in Computer Science, San Diego, CA, IEEE Computer Society Press (1995), 420–430.

[203] S. Smith, *From operational to denotational semantics*, Proceedings 7th Conference on Mathematical Foundations of Programming Semantics, Pittsburgh, PA, Lecture Notes in Comput. Sci. 598, S. Brookes, M. Main, A. Melton, M. Mislove and D. Schmidt, eds, Springer-Verlag (1992), 54–76.

[204] M. Smyth, *Powerdomains*, J. Comput. System Sci. **16** (1978), 23–36.

[205] T. Steel, ed., *Formal Language Description Languages for Computer Programming*, Proceedings IFIP Working Conference on Formal Language Description Languages, North-Holland (1966).

[206] C. Stirling, *Modal logics for communicating systems*, Theoret. Comput. Sci. **49** (1987), 311–347.

[207] C. Stirling, *A generalization of Owicki–Gries's Hoare logic for a concurrent while-language*, Theoret. Comput. Sci. **58** (1988), 347–359.

[208] C. Stirling, *Modal and temporal logics*, Handbook of Logic in Computer Science, Vol. 2, S. Abramsky, D. Gabbay and T. Maibaum, eds, Oxford University Press (1992), 477–563.

[209] C. Stirling, *Modal and temporal logics for processes*, Logics for Concurrency: Structure versus Automata, Lecture Notes in Comput. Sci. 1043, F. Moller and G. Birtwistle, eds, Springer-Verlag (1996), 149–237.

[210] V. Stoltenberg-Hansen, I. Lindström and E. Griffor, *Mathematical Theory of Domains*, Cambridge Tracts in Theoret. Comput. Sci. 22, Cambridge University Press (1994).

[211] A. Stoughton, *Fully abstract models of programming languages*, Research Notes in Theoret. Comput. Sci., Pitman, London (1988).

[212] A. Stoughton, *Substitution revisited*, Theoret. Comput. Sci. **59** (1988), 317–325.

[213] S. Tini, *Structural operational semantics for synchronous languages*, Ph.D. Thesis, Department of Computer Science, University of Pisa (2000).

[214] D. Turi and G. Plotkin, *Towards a mathematical operational semantics*, Proceedings 12th Symposium on Logic in Computer Science, Warsaw, Poland, IEEE Computer Society Press (1997), 280–291.

[215] I. Ulidowski, *Equivalences on observable processes*, Proceedings 7th Symposium on Logic in Computer Science, Santa Cruz, CA, IEEE Computer Society Press (1992), 148–159.

[216] I. Ulidowski, *Axiomatisations of weak equivalences for De Simone languages*, Proceedings 6th Conference on Concurrency Theory, Philadelphia, PA, Lecture Notes in Comput. Sci. 962, I. Lee and S. Smolka, eds, Springer-Verlag (1995), 219–233.

[217] I. Ulidowski, *Finite axiom systems for testing preorder and De Simone process languages*, Proceedings 5th Conference on Algebraic Methodology and Software Technology, Munich, Germany, Lecture Notes in Comput. Sci. 1101, M. Wirsing and M. Nivat, eds, Springer-Verlag (1996), 210–224.

[218] I. Ulidowski and I. Phillips, *Formats of ordered SOS rules with silent actions*, Proceedings 7th Conference on Theory and Practice of Software Development, Lille, France, Lecture Notes in Comput. Sci. 1214, M. Bidoit and M. Dauchet, eds, Springer-Verlag (1997), 297–308.

[219] F. Vaandrager, *Process algebra semantics of POOL*, Applications of Process Algebra, Cambridge Tracts in Theoret. Comput. Sci. 17, J. Baeten, ed., Cambridge University Press (1990), 173–236.

[220] F. Vaandrager, *On the relationship between process algebra and input/output automata (extended abstract)*, Proceedings 6th Symposium on Logic in Computer Science, Amsterdam, The Netherlands, IEEE Computer Society Press (1991), 387–398.

[221] F. Vaandrager, *Expressiveness results for process algebras*, Proceedings REX Workshop on Semantics: Foundations and Applications, Beekbergen, The Netherlands, June 1992, Lecture Notes in Comput. Sci. 666, J.d. Bakker, W.d. Roever and G. Rozenberg, eds, Springer-Verlag (1993), 609–638.

[222] J.J. Vereijken, *Discrete-time process algebra*, Ph.D. Thesis, Eindhoven University of Technology (1997).

[223] C. Verhoef, *An operator definition principle (for process algebras)*, Report P9105, Programming Research Group, University of Amsterdam (1991).

[224] C. Verhoef, *Linear unary operators in process algebra*, Ph.D. Thesis, University of Amsterdam (1992).

[225] C. Verhoef, *A congruence theorem for structured operational semantics with predicates and negative premises*, Proceedings 5th Conference on Concurrency Theory, Uppsala, Sweden, Lecture Notes in Comput. Sci. 836, B. Jonsson and J. Parrow, eds, Springer-Verlag (1994), 433–448. Preliminary version of [227].

[226] C. Verhoef, *A general conservative extension theorem in process algebra*, Proceedings 3rd IFIP Working Conference on Programming Concepts, Methods and Calculi, San Miniato, Italy, IFIP Transactions A-56, E.-R. Olderog, ed., Elsevier (1994), 149–168.

[227] C. Verhoef, *A congruence theorem for structured operational semantics with predicates and negative premises*, Nordic J. Comput. **2** (1995), 274–302.

[228] J. Vrancken, *The algebra of communicating processes with empty process*, Theoret. Comput. Sci. **177** (1997), 287–328.

[229] D. Walker, *Automated analysis of mutual exclusion algorithms using CCS*, Formal Aspects of Computing **1** (1989), 273–292.

[230] D. Walker, *Bisimulation and divergence*, Inform. and Comput. **85** (1990), 202–241.

[231] D. Watt, *Programming Concepts and Paradigms*, Prentice Hall (1990).

[232] S. Weber, B. Bloom and G. Brown, *Compiling Joy into silicon: An exercise in applied structural operational semantics*, Proceedings REX Workshop on Semantics: Foundations and Applications, Beekbergen, The Netherlands, June 1992, Lecture Notes in Comput. Sci. 666, J.d. Bakker, W.d. Roever and G. Rozenberg, eds, Springer-Verlag (1993), 639–659.

[233] G. Winskel, *A complete proof system for SCCS with modal assertions*, Fundamenta Informaticae **IX** (1986), 401–420.

[234] A. Wright and M. Felleisen, *A syntactic approach to type soundness*, Inform. and Comput. **115** (1994), 38–94.

Subject index

$\triangleq$, 203
$\xrightarrow{a}$, 201
$\sqsubseteq_S$, 203
$\sqsubseteq_{RS}$, 203
$\underline{\leftrightarrow}$, 203
$\xrightarrow{\varsigma}$, 203
$\sqsubseteq_T$, 203
$\surd$, 203
$\sqsubseteq_{RT}$, 203
$\sqsubseteq_{FT}$, 204
$\sqsubseteq_R$, 204
$\sqsubseteq_F$, 204
$\sqsubseteq_{CT}$, 204
$\sqsubseteq_{AT}$, 204
$\models$, 205, 210
$\sim_{\mathrm{HML}}$, 205
Σ, 206
$ar(f)$, 206
$\mathbb{T}(\Sigma)$, 206
$\mathrm{T}(\Sigma)$, 206
$\stackrel{a}{\nrightarrow}$, 206
$\neg P$, 206
$\vdash$, 207
ε, 207
θ, 208
ς_d, 208
$\vdash_s$, 213
$\vdash_{ws}$, 213
$\oplus$, 218
$=$, 221
$\rightarrow^*$, 225
$U(t)$, 233
ρ_R, 235
$\mathbf{0}$, 235
T_{FINTREE}, 244
$\Sigma_{\mathsf{FINTREE}}$, 244
$\mathcal{E}_{\mathsf{FINTREE}}$, 244
T_{∂^1}, 245
∂^1_H, 245
$\mathcal{E}_{\partial^1}$, 245
$\|$, 247
$\lfloor\!\lfloor$, 247
$\rfloor\!\rfloor$, 247
π_n, 248
$\uparrow$, 251
$\downarrow$, 251
Ω, 251
τ, 254
$\xrightarrow{\varepsilon}$, 255
$\underline{\leftrightarrow}_b$, 255

$\underline{\leftrightarrow}_{rb}$, 255
$\simeq^{\mathcal{F}}_{CT}$, 263
$\sim_{\mathcal{D}}$, 263
$\mathbb{A}(\Sigma)$, 266
$\mathbb{F}(\Sigma)$, 266
$FV(t^*)$, 269
$EV(t^*)$, 269
$f_{\mathcal{A}}$, 271
$\mathsf{T}(\Sigma, \mathsf{RVar})$, 271
$\sqsubseteq_{\mathcal{A}}$, 271
$\mathcal{A}[\![\cdot]\!]$, 272
$\lesssim$, 273
$\simeq$, 273
$\mathsf{ST}(\mathsf{Act})$, 274
$\lesssim^F$, 274
$\lesssim_{\omega}$, 275
$\sqsubseteq$, 276
$\mathcal{K}(\mathcal{D})$, 276
$\sqsubseteq_{\mathcal{K}(\mathcal{D})}$, 276
$\sqsubseteq_{EM}$, 276
$\mathbf{f}_{\mathsf{ST}}$, 277
$\mathbf{f}_{\mathcal{D}}$, 278

accepting trace, 204
Act, 201
action, 202, 231
– fresh, 219
AIP, 248
algebraic relation, 273
alternative composition, 207
Approximation Induction Principle, *see* AIP
aprACP$_F$, 235
aprACP$_R$, 235
aprACP$_U$, 235
argument, 206
– frozen, 256
– liquid, 256
– test of an, 233
arity, 206
axiom, 221
axiomatization, 221
– complete, 222
– ω-complete, 224
– sound, 222

Basic Process Algebra, 207
behavioural equivalence, 202
behavioural preorder, 202
binary decision diagram, 254
binary Kleene star, 237
bisimulation, 203
BPA$_\varepsilon$, 207
BPA$_\varepsilon^{\mathrm{dt}}$, 208
BPA$_\theta$, 208

branching bisimulation, 255
– rooted, 255

complete partial order, 271
completed trace, 204
conclusion, 206
conditional term rewriting system, *see* CRTS
congruence, 206
– completed trace, 263
conservative extension
– axiomatic, 222
– operational, 218
– rewrite, 225
constant, 206
context, 206
– liquid, 256
$\mathsf{CREC}(\Sigma)$, 250
CTRS, 225

De Simone format, 231
De Simone language, 232
– bounded, 234
– coeffective, 234
– – primitive, 234
– effective, 234
– – primitive, 234
– finitary, 235
– functional, 235
– image-finite, 235
– recursively enumerable, 234
– width-effective, 235
– – primitive, 235
– width-finitary, 235
denial formula, 263
disjoint extension, 245

Egli–Milner preorder, 276
empty process, 207
encapsulation, 262
equality relation, 221

failure, 204
failure trace, 203
failure trace format, 260
finitary preorder, 274
fix$(X = t)$, 231
function symbol, 206
– binary, 206
– bounded, 241
– – uniformly, 241
– unary, 206

ground confluence, 226
GSOS format, 237

GSOS language, 237
– compact, 277
– infinitary, 241
– linear, 248
– regular, 249
– semantically well-founded, 246
– simple, 242
– syntactically well-founded, 248

Hennessy–Milner logic, 205
HML formula, 205

initials(), 203
initials( ), 203

L cool format, 261
labelled transition system, *see* LTS
left merge, 247
literal, 207
– denying, 212
– negative, 207
– positive, 207
look-ahead, 258
LTS, 201
– computable, 233
– countably branching, 233
– decidable, 233
– finite, 202
– finitely branching, 202
– primitive recursive, 233
– recursively enumerable, 233
– regular, 202
– with divergence, 251

model, 210
– stable, 211
– – three-valued, 212
– supported, 210
– well-supported, 211

ntree format, 231
ntyft/ntyxt format, 229
ntytt rule, 259
– Λ-failure trace safe, 260
– Λ-readies safe, 260
– Λ-ready trace safe, 260

operator dependency, 235

panth format, 229
parallel composition, 247
path format, 229
polling, 259
prebisimulation, 273
precongruence, 206
Pred, 201
predicate
– convergence, 251
– divergence, 251
– fresh, 219
prefix multiplication, 235
premise, 206
– negative, 206
– positive, 206
priority, 208
Proc, 201
projection, 248
proof, 207
– supported, 213
– well-supported, 213
propagation, 259
Protean language, 253

quantales, 279

RBB safe format, 256
RDP, 250
readies format, 260
ready, 204
ready simulation, 203
– format, 259
ready trace, 203
ready trace format, 260
Recursive Definition Principle, *see* RDP
Recursive Specification Principle, *see* RSP
relational renaming, 235
rewrite relation, 225
rewrite rule, 225
RSP, 250
rule format, 227
ruloid, 253

sequential composition, 207
Σ-algebra, 271
Σ-domain, 271
Σ-homomorphism, 271
Σ-poset, 271
Σ-preorder, 271
signature, 206
– many-sorted higher-order, 265
silent step, 254
simplification ordering, 243
simulation, 203
source, 206
stalk format, 230
state, 202
stratification, 214
– strict, 214

substitution, 206
– actual, 266
– closed, 206
– formal, 267
synchronization tree, 275
– finite, 274

target, 206
term, 206
– actual, 266
– closed, 206
– formal, 266
– fresh, 219
– open, 206
– recursive, 231
– – guarded, 234
trace, 203
trace format, 262
transition, 202
transition rule, 206
– actual, 268
– junk, 238
– patience, 256
– pure, 230
– source-dependent, 219
transition system specification, *see TSS*
trigger, 234
– positive, 241
TSS, 206
– complete, 213
– positive, 206
– *s*-complete, 213
– stratifiable, 214
– – strictly, 214
– *ws*-complete, 213
tyft format, 229
tyft/tyxt format, 229
type, 231

Var, 205
Var*, 266
variable, 205
– actual, 265
– floating, 260
– formal, 266
– recursion, 231
– – unguarded, 233
– source-dependent, 219
variable dependency graph, 230
– well-founded, 230

CHAPTER 4

Modal Logics and mu-Calculi: An Introduction

Julian Bradfield, Colin Stirling

Laboratory for Foundations of Computer Science, University of Edinburgh, King's Buildings, Edinburgh EH9 3JZ, United Kingdom

E-mail: jcb@dcs.ed.ac.uk, cps@dcs.ed.ac.uk

Contents

1. Introduction . . . 295
   1.1. Background . . . 295
   1.2. The general framework . . . 297
2. Early logics . . . 298
   2.1. The small model property . . . 299
   2.2. CTL model-checking . . . 300
3. Introduction to mu-calculi . . . 301
   3.1. Fixpoints as recursion . . . 301
   3.2. Approximating fixpoints and μ as 'finitely' . . . 303
   3.3. Syntax of modal mu-calculus . . . 304
   3.4. Semantics of modal mu-calculus . . . 305
   3.5. Examples . . . 306
   3.6. Modal mu-calculus as a process logic . . . 309
   3.7. Linear time mu-calculus . . . 310
4. Theory of modal mu-calculus . . . 311
   4.1. Fixpoint regeneration and the 'fundamental semantic theorem' . . . 311
   4.2. The finite model property and decidability . . . 313
   4.3. Axiomatization . . . 314
   4.4. Model-checking . . . 315
   4.5. Alternation . . . 317
5. More on model-checking . . . 319
   5.1. Local model-checking 319
   5.2. ... for infinite systems . . . 321
   5.3. Decidable model-checking for infinite processes . . . 322
   5.4. Equation systems and efficient local model-checking . . . 322
6. Automata, games and logic . . . 323
   6.1. Automata . . . 323
   6.2. Monadic second-order logic . . . 325

HANDBOOK OF PROCESS ALGEBRA
Edited by Jan A. Bergstra, Alban Ponse and Scott A. Smolka

6.3. Games 326
References 327
Subject index 329

Abstract

We briefly survey the background and history of modal and temporal logics. We then concentrate on the modal mu-calculus, a modal logic which subsumes most other commonly used logics. We provide an informal introduction, followed by a summary of the main theoretical issues. We then look at model-checking, and finally at the relationship of modal logics to other formalisms.

1. Introduction

In the chapter, we shall first briefly survey the background, history and some of the issues arising from the use of modal logics for processes. We briefly introduce earlier modal logics such as Hennessy–Milner logic, PDL and CTL. We then devote the major part of the chapter to the modal mu-calculus. It is a logic which, although much studied and also widely used, has a reputation for being hard to understand compared to, say, CTL. We therefore devote a section to introducing the logic and developing an intuition for it. In Section 4 we look at the theoretical basis, and major results, in a little more detail. However, much of the detailed technical work in this area has already been covered in other handbook articles [62,21], or in easily accessible original papers; we have therefore chosen to give few results in fine detail, but rather to provide outline proofs which, we hope, cover the essentials, and serve as a helpful map to the detailed proofs elsewhere, but yet remain readable for those who have no wish to go into detail.

We follow this with a survey of results specifically about model-checking, including the notions of local model-checking, and the work on model-checking classes of infinite processes.

Section 6 looks, quite briefly, at the relationship of modal mu-calculus to other formalisms, in particular automata and games. Much of this is of primarily mathematical interest; but it is also of practical interest, as automata have been used for model-checking, and recently games have also been found to have uses in tools.

1.1. *Background*

The application of modal and temporal logics to process calculi is part of a line of program verification going back to the 1960s and program schemes and Floyd–Hoare logic. Originally the emphasis was on *proof*: Floyd–Hoare logic allows one to make assertions about programs, and there is a proof system to verify these assertions. This line of work has, of course, continued and flourished, and today there are highly sophisticated theories for proving properties of programs, and equally sophisticated machine support for these theories. However, the use of proof systems has some disadvantages, and one hankers after a more purely algorithmic approach to simple problems. One technique was pioneered by Manna and Pnueli [45], who turned program properties into questions of satisfiability or validity in first order logic, which can then be attacked by means that are not just proof-theoretic; this idea was later applied by them to linear temporal logics.

During the 1970s, the theory of program correctness was extended by investigating more powerful logics, and studying them in a manner more similar to the traditions of mathematical logic. A family of logics which received much attention was that of dynamic logics, which can be seen as extending the ideas of Hoare logic [56]. Dynamic logics are modal logics, where the different modalities correspond to the execution of different programs – the formula $\langle\alpha\rangle\phi$ is read as 'it is possible for α to execute and result in a state satisfying ϕ'. The programs may be of any type of interest; the variety of dynamic logic most often referred to is a propositional language in which the programs are built from atomic programs by regular expression constructors; henceforth, Propositional Dynamic Logic, PDL, refers to this logic. PDL is interpreted with respect to a model on a Kripke structure, formalizing the notion of the global state in which programs execute and which they change – each point in the structure corresponds to a possible state, and programs determine a relation between states giving the changes effected by the programs.

Once one has the idea of a modal logic defined on a Kripke structure, it becomes quite natural to think of the finite case and write programs which just check whether a formula is satisfied. This idea was developed in the early 80s by Clarke, Emerson, Sistla and others. They worked with a logic that has much simpler modalities than PDL – in fact, it has just a single 'next state' modality – but which has built-in temporal connectives such as 'until'. This logic is CTL, and it and its extensions remain some of the most popular logics for expressing properties of systems.

Meanwhile, the theory of process calculi was being developed in the late 70s, most notably by Milner [47]. An essential component was the use of labelled Kripke structures ('labelled transition systems') as a raw model of concurrent behaviour. An important difference between the use of Kripke structures here and their use in program correctness was that the states are the behaviour expressions, which model concurrent systems, themselves and the labels on the accessibility relation (the transitions) are simple actions (and not programs). The criterion for behavioural equivalence of process expressions was defined in terms of observational equivalence (and later in terms of bisimulation relations). Hennessy and Milner introduced a primitive modal logic in which the modalities refer to actions: $\langle a \rangle \phi$ 'it is possible to do an a action and then have ϕ be true', and its dual $[a]\phi$ 'ϕ holds after every a action'. Together with the usual Boolean connectives, this gives Hennessy–Milner logic [30], HML, which was introduced as an alternative exposition of observational equivalence. However, as a logic HML is obviously inadequate to express many properties, as it has no means of saying 'always in the future' or other temporal connectives – except by allowing infinitary conjunction. Using an infinitary logic is undesirable both for the obvious reason that infinite formulae are not amenable to automatic processing, and because infinitary logic gives much more expressive power than is needed to express temporal properties.

In 1983, Dexter Kozen published a study of a logic that combined simple modalities, as in HML, with fixpoint operators to provide a form of recursion. This logic, the modal mu-calculus, has become probably the most studied of all temporal logics of programs. It has a simple syntax, an easily given semantics, and yet the fixpoint operators provide immense power. Most other temporal logics, such as the CTL family, can be seen as fragments of the modal mu-calculus. Moreover, this logic lends itself to transparent model-checking algorithms.

Another 'root' to understanding modal logics is the work in the 60s on automata over infinite words and trees by Büchi [14] and Rabin [59]. The motivation was decision questions of monadic second-order logics. Büchi introduced automata as a normal form for such formulas. This work founded new connections with logic and automata theory. Later it was realised that modal logics are merely sublogics of appropriate monadic second-order logic, and that the automata normal forms provide a very powerful framework within which to study properties of modal logic. A further development was the use of games by Gurevich and Harrington [28] as an alternative to automata.

There is also an older game-theoretic tradition due to Ehrenfeucht and Fraïssé, for understanding the expressive power of logics. These techniques are also applicable within process calculi. For instance bisimulation equivalence can be naturally rendered as such a game, and expressivity of modal logics can be understood using game-theoretic techniques.

1.2. *The general framework*

There is an enormous variety of process calculi, as illustrated in other chapters of this Handbook. For the most part they involve the following: a syntactic description of a process as an expression built from a few combinators, a behavioural meaning, a labelled transition system, determined by structural operational rules, and an equivalence (or better a congruence) to stipulate in what circumstances two expressions count as describing the same system. Using the operational rules a process expression generates a transition graph which may be infinite-state. This graph reflects the potential behaviour that an expression may exhibit.

One systematic approach to capturing 'correctness' in process calculi is through the use of an equivalence; a complex description of a system in terms of all its components may be shown to be equivalent to a small description. However, for many complex systems either their descriptions are too large to formally manipulate or there is not a small intended equivalent description. In these cases one may be more interested in ascertaining whether crucial properties hold, such as a variety of liveness and safety properties.

The primary application of modal logics to processes is therefore: given a process expression E, does it have a particular modal property ϕ, that is does $E \models \phi$? More traditional questions about a modal logic may also be interesting. For instance the question 'is ϕ satisfiable?', that is, 'is there a process E such that $E \models \phi$?', could be useful if ϕ captures the requirements of a system. Another traditional question is whether a formula ϕ is valid, that is, does every process E have property ϕ ?

There are many issues involved in the verification question. What kind of expression is E and what is its structure? Which 'control structures' does it employ and what extras does it contain such as time, probability, and locality? What sort of property is ϕ, and which property language does it belong to? Also there are different approaches to the basic problem, does $E \models \phi$?

(1) Direct approach. Develop techniques which can verify directly whether or not $E \models \phi$. Note that there are two kinds of structure, the logical structure of ϕ which may contain Boolean operators, modal and temporal operators and fixed points, and the process structure which may contain concurrent composition, synchronization and abstraction mechanisms. Methods for verification may or may not employ 'structured reasoning'.

(2) Indirect approach I. Reduce the verification problem to a validity problem. Code the process E as a formula ψ_E (such as a 'characteristic formula') in the logic. The verification problem thereby reduces to the validity problem, is $\psi_E \rightarrow \phi$ valid?

(3) Indirect approach II. This is similar to (2) but dispenses with logic altogether! A process E is very close to an automaton (or to a game). Consider the characteristic automaton (game) associated with E, call it A_E and consider the automaton (game) for ϕ. This gives a containment problem: is A_E contained in A_ϕ? (Alternatively take the complement automaton of ϕ, A^c_ϕ, and find out if the intersection of A_E and A^c_ϕ is nonempty.)

It is straightforward to define an inductive procedure for this, which is polynomial in the size of the formula and of the system. For example, to determine the truth of $\langle \alpha^* \rangle \phi$, one computes the *-closure of the α relation, and then checks for an α^*-successor satisfying ϕ. These results give an NTIME(c^n) upper bound for the satisfiability problem. By a reduction to alternating Turing machines, [26] also gave a lower bound of DTIME($c^{n/\lg n}$). A closer to optimal technique for satisfiability due to Pratt uses tableaux [57].

Although CTL, CTL* and modal mu-calculus all have the finite model property, the filtration technique does not apply. If one filters **T** through a finite set Γ containing $\forall \mathbf{F} Q$ unintended loops may be added. For instance if **T** is $E_i \xrightarrow{a} E_{i+1}$ for $1 \leqslant i < n$ and Q is only true at state E_n then $E_i \models \forall \mathbf{F} Q$ for each i. But when n is large enough the filtered model will have at least one transition $[E_j] \xrightarrow{a} [E_i]$ when $i \leqslant j < n$, with the consequence that $[E_i] \not\models \forall \mathbf{F} Q$. The initial approach to showing the finite model property utilized semantic tableaux where one explicitly builds a model for a satisfiable formula which has small size. But such a technique is very particular, and more sophisticated methods based on automata are used for optimal results, as we shall mention later.

2.2. *CTL model-checking*

CTL has some obvious differences from PDL. Firstly, although it is a state-based logic, in the sense that the formulae one writes are defined on states, it uses path operators internally: evaluating the formula $\forall[\phi \mathbf{U} \psi]$ at a state involves considering all paths from that state. Thus, at first sight, one might expect to lose the obvious exponential upper bound on model-checking. However, this turns out not to happen, and in fact CTL is very easy to model-check. This was shown by a direct construction in [18]; it also follows from the fact that, as we shall see later, CTL is a particularly simple fragment of the modal mu-calculus.

The model-checking procedure of [18] is an example of an apparently global technique. It proceeds by model-checking subformulae from the bottom up, doing a full pass over the state space for subformulae before considering the superformula. In the original paper, the algorithm is presented as imperative pseudo-code, and it takes a little thought to see why it works. Here is an English outline, which extracts the idea of the code:

- to check tt, $\neg\phi$, $\phi_1 \wedge \phi_2$, check the subformulae and perform the Boolean operation;
- to check $\langle a \rangle \phi$, $[a]\phi$, check ϕ, and then apply the semantic definitions;
- to check $\exists[\phi \mathbf{U} \psi]$, check the subformulae, then find the states at which ψ holds, and trace backwards along paths on which ϕ holds; these states satisfy $\exists[\phi \mathbf{U} \psi]$;
- to check $\forall[\phi \mathbf{U} \psi]$, check the subformulae; then make a depth-first traversal of the system, applying the following: if a state satisfies ψ, mark it as satisfying $\forall[\phi \mathbf{U} \psi]$; otherwise, if it fails ϕ, mark it as failing $\forall[\phi \mathbf{U} \psi]$; otherwise, after processing the successors, mark it as satisfying $\forall[\phi \mathbf{U} \psi]$ iff all its successors do.

In following sections, we shall see CTL translated into modal mu-calculus, and we shall see global, backward-looking model-checking algorithms, and local, forward-looking algorithms. It is interesting, from today's perspective, to see that this early CTL algorithm has elements of both: the code for $\exists[\phi \mathbf{U} \psi]$ is doing exactly the computation by approximants (see Section 4.4) of the mu-calculus translation; but the code for $\forall[\phi \mathbf{U} \psi]$ is doing tableau model-checking (see Section 5.1).

3. Introduction to mu-calculi

The defining feature of mu-calculi is the use of fixpoint operators.[1] The use of fixpoint operators in program logics goes back at least to De Bakker, Park and Scott [52]. However, their use in modal logics of programs dates from work of Pratt, Emerson and Clarke and Kozen. Pratt's version [58] used a fixpoint operator like the minimization operator of recursion theory, and this has not been further studied. Emerson and Clarke added fixed points to a temporal logic to capture fairness and other correctness properties [22]. Kozen's [37] paper introduced the modal mu-calculus as we use it today, and established a number of basic results.[2]

Fixpoint logics are notorious for being incomprehensible. Indeed, it has been known for several reasonably expert people to spend an hour or two trying to work out whether a one line formula means what it was intended to mean. Furthermore, a full understanding of the mathematical intricacies of the modal mu-calculus requires familiarity with mathematics that, although not difficult, is not covered in most undergraduate mathematics or computer science programmes; this can be a source of irritation for the practitioner new to the area. Fortunately, one can get a very good intuitive understanding, which is even provably equivalent to the formal semantics, without knowing the full details. Moreover, this intuition is all but essential for understanding the deep theory as well.

In this section, therefore, we shall give an introduction to the modal mu-calculus that avoids the harder and more technical details, and concentrates rather on developing the intuition. In the first two subsections, we shall be so informal as to use notation and terminology well in advance of their definitions. In the following section, the semantics of the logic is examined in a little more detail, and the intuitions given a precise mathematical meaning.

3.1. *Fixpoints as recursion*

Suppose that $\mathcal{S}$ is the state space of some system. For example $\mathcal{S}$ could be the set of all processes reachable from by arbitrary length sequences of transitions from some initial process E. One way to provide semantics of a state-based modal logic is to map formulae ϕ to sets of states, that is to elements of $\wp\mathcal{S}$. For any formula ϕ this mapping is given by $\|\phi\|$.[3] The idea is that this mapping tells us at which states each formula holds. If we allow our logic to contain variables with interpretations ranging over $\wp\mathcal{S}$, then we can view the semantics of a formula with a free variable, $\phi(Z)$, as a function $f : \wp\mathcal{S} \to \wp\mathcal{S}$. If we take the usual lattice structure on $\wp\mathcal{S}$, given by set inclusion, and if f is a monotonic function, then by the Knaster–Tarski theorem we know that f has fixed points, and indeed has a

---

[1] Whether one should say 'fixpoint', 'fix-point', or 'fixed point' is one of the most controversial questions in the theory of mu-calculi. One of us (Bradfield) takes the view that one has 'fixpoint operators', which generate 'fixed points' of functions. Fortunately, few people enunciate with sufficient clarity to distinguish the two in conversation.

[2] Kozen called the logic 'propositional mu-calculus' but this would be more appropriate for propositional logic + fixpoint operators, a logic we shall comment on later.

[3] The mapping can be either given directly (inductively) or indirectly as the set $\{s \in \mathcal{S}\colon\ s \models \Phi\}$.

unique maximal and a unique minimal fixed point. So we could extend our basic logic with a minimal fixpoint operator μ, so that $\mu Z.\phi(Z)$ is a formula whose semantics is the least fixed point of f; and similarly a maximal fixpoint operator ν, so that $\nu Z.\phi(Z)$ is a formula whose semantics is the greatest fixed point of f.

The reason we might want to do this is that, as in domain theory for example, it provides a semantics for *recursion*. Writing recursive modal logic formulae may not be an immediately obvious thing to do, but it provides a neat way of expressing all the usual operators of temporal logics. For example, consider the CTL formula $\forall \mathbf{G}\phi$, 'always ϕ'. Another way of expressing this is to say that it is a property X such that if X is true, then ϕ is true, and wherever we go next, X remains true; so X satisfies the modal equation

$$X = \phi \wedge [-]X$$

where $=$ is logical (truth-table) equivalence,[4] and $[-]X$ means that X is true at every immediate successor (see Section 3.3). So we could write this formula as $?X.\phi \wedge [-]X$. But what is '?'? Which fixed point is the right solution of the equation – least, greatest, or something in between? We may argue thus: if a state satisfies *any* solution X' of the equation, then surely it satisfies $\forall \mathbf{G}\phi$. Hence the meaning of the formula is the largest solution, $\nu X.\phi \wedge [-]Z$.

On the other hand, consider the CTL property $\exists \mathbf{F}\phi$, 'there exists a path on which ϕ eventually holds'. We could write this recursively as 'Y holds if either ϕ holds now, or there's some successor on which Y is true':

$$Y = \phi \vee \langle - \rangle Y.$$

This time, it is perhaps less obvious which solution is correct. However, we can argue that if a state satisfies $\exists \mathbf{F}\phi$, then it surely satisfies any solution Y' of the equation; and so we want the least such solution.

The astute reader will by now have noticed that the equations do not capture the meaning of the English recursive definitions – we should have written

$$X \Rightarrow \phi \wedge [-]X$$

and

$$Y \Leftarrow \phi \vee \langle - \rangle Y,$$

or in terms of the semantic function, $\|X\| \subseteq \|\phi \wedge [-]X\|$ and $\|Y\| \supseteq \|\phi \vee \langle - \rangle Y\|$. So we are not really interested in fixed points, but rather in respectively *post-fixed points* and *pre-fixed points* of the functions defined by the formulae on the right of the (in)equations; and since there is only one canonical post-fixed point and one canonical pre-fixed point, it becomes obvious which is required. It is an easy fact that the greatest post-fixed point and the least pre-fixed point of a monotonic function are also the greatest and least fixed points,

[4] We use double arrows $\Rightarrow$, $\Leftarrow$ for logical implication; single arrows $\rightarrow$ are used for transition relations, as well as with their normal mathematical use.

and so we may continue to use the notation μ and ν, and work with fixpoints. However, it is usual to emphasize this point when writing the semantics of fixpoint logics by writing, for example,

$$\|\mu Z.\phi(Z)\| = \bigcap\{S \mid S \supseteq \|\phi(S)\|\}$$

with inclusion rather than equality.

At this point, the reader who does not already know the answer might like to apply a similar argument to produce the fixpoint translation of the CTL formula $\forall[\phi\ \mathbf{U}\ \psi]$.

3.2. *Approximating fixpoints and μ as 'finitely'*

The other key idea is that of approximants and unfolding. The standard theory tells us that if f is a monotonic function on a lattice, we can construct the least fixed point of f by iterating f on the bottom element $\perp$ of the lattice to form an increasing chain whose limit is the fixed point. The length of the iteration is in general transfinite, but is bounded at worst by the cardinal after cardinality of the lattice, and in the special case of a powerset lattice $\wp\mathcal{S}$, by the cardinal after the cardinality of $\mathcal{S}$. So if f is monotonic on $\wp\mathcal{S}$, we have

$$\mu f = \bigcup_{\alpha<\kappa} f^{\alpha}(\emptyset)$$

and similarly

$$\nu f = \bigcap_{\alpha<\kappa} f^{\alpha}(\mathcal{S})$$

– or in terms of a infinitary logic, $\mu Z.\phi(Z) = \bigvee_{\alpha<\kappa} \phi^{\alpha}(\mathrm{ff})$ – where κ is at worst $|\mathcal{S}|+1$ for finite $\mathcal{S}$, or $\aleph_1$ for countable $\mathcal{S}$. So for a minimal fixpoint $\mu Z.\phi(Z)$, if a state s satisfies the fixpoint, it satisfies some approximant, say for convenience the $(\beta+1)$th so that $s \models \phi^{\beta+1}(\mathrm{ff})$. Now if we *unfold* this formula once, we get $s \models \phi(\phi^{\beta}(\mathrm{ff}))$. That is, the fact that s satisfies the fixpoint depends, via ϕ, on the fact that other states satisfy the fixpoint *at smaller approximants than s does*. These states in turn depend on states that satisfy the fixpoint at still smaller approximants, bringing to mind the well-known doggerel (in the version of [20])

Great fleas have little fleas
  upon their backs to bite 'em,
and little fleas have smaller fleas,
  and so *ad infinitum*.

However, in this case the doggerel is wrong: the process cannot continue *ad infinitum*, because the ordinals are well-founded and there is a smallest flea, namely $\phi(\mathrm{ff})$.[5] So if

[5] The seldom quoted following lines give a better picture: 'And the great fleas themselves, in turn, have greater fleas to go on; While these again have greater still, and greater still, and so on'.

one follows a chain of dependencies, the chain terminates. This is the strict meaning behind the slogan "μ means 'finite looping' ", which, with a little refinement, is sufficient to understand the mu-calculus.

On the other hand, for a maximal fixpoint $\nu Z.\phi(Z)$, there is no such decreasing chain: $s \models \nu Z.\phi(Z)$ iff $s \models \phi(\nu Z.\phi(Z))$, and we may loop for ever, as in the CCS process $P \stackrel{\text{def}}{=} a.P$, which satisfies $\nu Z.\langle a \rangle Z$. (However, if a state *fails* a maximal fixpoint, then there is a descending chain of failures.) Instead, we have the somewhat mystical principle of fixpoint induction: if by assuming that a set $S \models Z$, we can show that $S \models \phi(Z)$, then we have shown that $S \models \nu Z.\phi$ (compare the recursive formulation of $\forall \mathbf{G}\phi$ in the previous section).

So in summary, one may understand fixpoints by the slogan 'ν means looping, and μ means finite looping'. This slogan provides an alternative means of explaining why a minimal fixpoint is required in the translation of $\exists \mathbf{F}\phi$. This formula means that there is a path on which ϕ eventually holds: that is, on the chosen path, ϕ holds within finite time. Hence the 'equation' $Y = \phi \vee \langle - \rangle Y$ must only be applied a finite number of times, and so by the slogan we should use a minimal fixpoint.

In the case of formulae with alternating fixpoints (which we shall examine a little later), the slogan remains valid, but requires a little more care in application. It is essential to almost all proofs about modal mu-calculus: the notion of 'well-founded premodel' with which Streett and Emerson [68] proved the finite model property, is an example of the slogan; so are the tableau model-checking approaches of Stirling and Walker [66], and Bradfield and Stirling [13].

3.3. *Syntax of modal mu-calculus*

Let Var be an (infinite) set of *variable names*, typically indicated by $Z, Y, \ldots$; let Prop be a set of *atomic propositions*, typically indicated by $P, Q, \ldots$; and let $\mathcal{L}$ be a set of *labels*, typically indicated by $a, b, \ldots$. The set of modal mu-calculus formulae (with respect to Var, Prop, $\mathcal{L}$) is defined in parsimonious form as follows:

- P is a formula.
- Z is a formula.
- If ϕ_1 and ϕ_2 are formulae, so is $\phi_1 \wedge \phi_2$.
- If ϕ is a formula, so is $[a]\phi$.
- If ϕ is a formula, so is $\neg\phi$.
- If ϕ is a formula, then $\nu Z.\phi$ is a formula, provided that every free occurrence of Z in ϕ occurs positively, i.e., within the scope of an even number of negations. (The notions of free and bound variables are as usual, where ν is the only binding operator.)

If a formula is written as $\phi(Z)$, it is to be understood that the subsequent writing of $\phi(\psi)$ means ϕ with ψ substituted for all free occurrences of Z. There is no suggestion that Z is the only free variable of ϕ; and variable capture is not avoided.

The positivity requirement on the fixpoint operator is a syntactic means of ensuring that $\phi(Z)$ denotes a functional monotonic in Z, and so has unique minimal and maximal fixpoint. This requires irritating book-keeping, and it is usually more convenient to introduce derived operators defined by de Morgan duality, and work in positive form:

- $\phi_1 \vee \phi_2$ means $\neg(\neg\phi_1 \wedge \neg\phi_2)$.
- $\langle a \rangle \phi$ means $\neg[a]\neg\phi$.
- $\mu Z.\phi(Z)$ means $\neg\nu Z.\neg\phi(\neg Z)$.

Note the triple use of negation in μ, which is required to maintain the positivity. A formula is said to be in *positive form* if it is written with the derived operators so that $\neg$ only occurs applied to atomic propositions. It is in *positive normal form* if in addition all bound variables are distinct. Any formula can be put into positive normal form by use of de Morgan laws and α-conversion. So we shall often assume positive normal form, and when doing structural induction on formulae will often take the derived operators as primitives.

For the concrete syntax, we shall assume that modal operators have higher precedence than Boolean, and that fixpoint operators have lowest precedence, so that the scope of a fixpoint extends as far to the right as possible.

There are a few extensions to the syntax which are convenient in presenting examples, and in practice. The most useful is to allow modalities to refer not just to single actions, but to sets of actions. The most useful set is 'all actions except a'. So:

- $s \models [K]\phi$ iff $\forall a \in K.s \models [a]\phi$, and $[a, b, \ldots]\phi$ is short for $[\{a, b, \ldots\}]\phi$.
- $[-K]\phi$ means $[\mathcal{L} - K]\phi$, and set braces may be omitted.

Thus $[-]\phi$ means just $[\mathcal{L}]\phi$.[6]

3.4. *Semantics of modal mu-calculus*

A modal mu-calculus *structure* $\mathcal{T}$ (over Prop, $\mathcal{L}$) is a labelled transition system, namely a set $\mathcal{S}$ of states and a transition relation $\rightarrow \subseteq \mathcal{S} \times \mathcal{L} \times \mathcal{S}$ (as usual we write $s \xrightarrow{a} t$), together with an interpretation $\mathcal{V}_{\text{Prop}} : \text{Prop} \rightarrow \wp\mathcal{S}$ for the atomic propositions.

Given a structure $\mathcal{T}$ and an interpretation $\mathcal{V} : \text{Var} \rightarrow \wp\mathcal{S}$ of the variables, the set $\|\phi\|^{\mathcal{T}}_{\mathcal{V}}$ of states satisfying a formula ϕ is defined as follows:

$$\begin{aligned}
\|P\|^{\mathcal{T}}_{\mathcal{V}} &= \mathcal{V}_{\text{Prop}}(P), \\
\|Z\|^{\mathcal{T}}_{\mathcal{V}} &= \mathcal{V}(Z), \\
\|\neg\phi\|^{\mathcal{T}}_{\mathcal{V}} &= \mathcal{S} - \|\phi\|^{\mathcal{T}}_{\mathcal{V}}, \\
\|\phi_1 \wedge \phi_2\|^{\mathcal{T}}_{\mathcal{V}} &= \|\phi_1\|^{\mathcal{T}}_{\mathcal{V}} \cap \|\phi_2\|^{\mathcal{T}}_{\mathcal{V}}, \\
\|[a]\phi\|^{\mathcal{T}}_{\mathcal{V}} &= \{s \mid \forall t.s \xrightarrow{a} t \Rightarrow t \in \|\phi\|^{\mathcal{T}}_{\mathcal{V}}\}, \\
\|\nu Z.\phi\|^{\mathcal{T}}_{\mathcal{V}} &= \bigcup \{S \subseteq \mathcal{S} \mid S \subseteq \|\phi\|^{\mathcal{T}}_{\mathcal{V}[Z:=S]}\},
\end{aligned}$$

where $\mathcal{V}[Z := S]$ is the valuation which maps Z to S and otherwise agrees with $\mathcal{V}$. If we are working in positive normal form, we may add definitions for the derived operators by duality:

[6] Beware that many authors use '$[]\phi$' to mean '$[\mathcal{L}]\phi$', rather than the (vacuous) '$[\emptyset]\phi$' that it means in our notation.

$$\begin{aligned}
\|\phi_1 \vee \phi_2\|_{\mathcal{V}}^{\mathcal{T}} &= \|\phi_1\|_{\mathcal{V}}^{\mathcal{T}} \cup \|\phi_2\|_{\mathcal{V}}^{\mathcal{T}},\\
\|\langle a\rangle\phi\|_{\mathcal{V}}^{\mathcal{T}} &= \{s \mid \exists t. s \xrightarrow{a} t \wedge t \in \|\phi\|_{\mathcal{V}}^{\mathcal{T}}\},\\
\|\mu Z.\phi\|_{\mathcal{V}}^{\mathcal{T}} &= \bigcap \{S \subseteq \mathcal{S} \mid S \supseteq \|\phi\|_{\mathcal{V}[Z:=S]}^{\mathcal{T}}\}.
\end{aligned}$$

We drop $\mathcal{T}$ and $\mathcal{V}$ whenever possible; and write $s \models \phi$ for $s \in \|\phi\|$.

We have discussed informally the importance of approximants; let us now define them. If $\mu Z.\phi(Z)$ is a formula, then for $\alpha \in \mathrm{On}$, let $\mu Z^{\alpha}.\phi$ and $\mu Z^{<\alpha}.\phi$ be formulae, with semantics given, with simultaneous induction on α, by:

$$\begin{aligned}
\|\mu Z^{<\alpha}.\phi\|_{\mathcal{V}}^{\mathcal{T}} &= \bigcup_{\beta<\alpha} \|\mu Z^{\beta}.\phi\|_{\mathcal{V}}^{\mathcal{T}},\\
\|\mu Z^{\alpha}.\phi\|_{\mathcal{V}}^{\mathcal{T}} &= \|\phi\|_{\mathcal{V}[Z:=\|\mu Z^{<\alpha}.\phi\|_{\mathcal{V}}^{\mathcal{T}}]}^{\mathcal{T}}.
\end{aligned}$$

The approximants of a maximal fixpoint are defined dually:

$$\begin{aligned}
\|\nu Z^{<\alpha}.\phi\|_{\mathcal{V}}^{\mathcal{T}} &= \bigcap_{\beta<\alpha} \|\nu Z^{\beta}.\phi\|_{\mathcal{V}}^{\mathcal{T}},\\
\|\nu Z^{\alpha}.\phi\|_{\mathcal{V}}^{\mathcal{T}} &= \|\phi\|_{\mathcal{V}[Z:=\|\nu Z^{<\alpha}.\phi\|_{\mathcal{V}}^{\mathcal{T}}]}^{\mathcal{T}}.
\end{aligned}$$

Note that $\mu Z^{<0}.\phi \iff \mathrm{ff}$ and $\nu Z^{<0}.\phi \iff \mathrm{tt}$. By abuse of notation, we write Z^{α} or ϕ^{α} to mean $\substack{\mu\\\nu} Z^{\alpha}.\phi$; of course this only makes sense when one knows which fixpoint and variable is meant.

We should remark here that most literature on the modal mu-calculus uses a slightly different definition, putting $\mu Z^0.\phi = \mathrm{ff}$, $\mu Z^{\alpha+1}.\phi = \phi(\mu Z^{\alpha}.\phi)$, and $\mu Z^{\lambda}.\phi = \bigcup_{\beta<\lambda} \mu Z^{\beta}.\phi$ for limit λ – which in effect is writing α for our $<\alpha$. That notation is inherited from set theory; its advantage is that a limit approximant is the limit of approximants. Our notation is inherited from more recent set theory; its advantages are that it sometimes reduces the number of trivial case distinctions in inductive proofs. However, the difference is not significant.

3.5. *Examples*

We have seen, both informally and in the formal semantics, the meaning of the fixpoint operators, and we have seen some simple examples of the modal mu-calculus translating CTL. We now consider some examples of modal mu-calculus formulae in their own right, which express properties one might meet in practice.

There is a well-known 'classification' [40] of basic properties into safety and liveness. In mu-calculus terms, it is not unreasonable to say that μ is liveness and ν is safety. Consider first simple ν formulae. For example:

$$\nu Z.P \wedge [a]Z$$

is a relativized 'always' formula: 'P is true along every a-path'. Slightly more complex is the relativized 'while' formula

$$\nu Z.Q \vee (P \wedge [a]Z)$$

'on every a-path, P holds while Q fails'. Both formulae can be understood directly via the fixpoint construction, or via the idea of 'ν as looping': for example the second formula is true if either Q holds, or if P holds and wherever we go next (via a), the formula is true, and ..., and because the fixpoint is maximal, we can repeat forever. So in particular, if P is always true, and Q never holds, the formula is true.

μ formulae, in contrast, require something to happen, and thus are liveness properties. For example

$$\mu Z.P \vee [a]Z$$

is 'on all infinite length a-paths, P eventually holds'; and

$$\mu Z.Q \vee (P \wedge \langle a \rangle Z)$$

is 'on some a-path, P holds until Q holds (and Q *does* eventually hold)'. Again, these can be understood by 'μ as finite looping': in the second case, we are no longer allowed to repeat the unfolding forever, so we must eventually 'bottom out' in the Q disjunct.

This level of complexity suffices to translate CTL, since we have $\mu Z.Q \vee (P \wedge \langle - \rangle Z)$ as a translation of $\exists[P \mathbf{U} Q]$, and $\mu Z.Q \vee (P \wedge [-]Z \wedge \langle - \rangle \mathrm{tt})$ as a translation of $\forall[P \mathbf{U} Q]$; and obviously we can nest formulae inside one another, such as

$$\nu Z.(\mu Y.P \vee \langle - \rangle Y) \wedge [-]Z$$

'it is always possible that P will hold', or $\forall \mathbf{G}(\exists \mathbf{F} P)$. Equally obviously, we can write formulae with no CTL translation, such as

$$\mu Z.[a]\mathrm{ff} \vee \langle a \rangle \langle a \rangle Z$$

which asserts the existence of a maximal a-path of even length; a formula which is, incidentally, expressible in PDL. This is, however, a fairly simple extension; much more interesting is the power one gets from mixing fixpoints that depend on one another. Consider the formula

$$\mu Y.\nu Z.(P \wedge [a]Y) \vee (\neg P \wedge [a]Z).$$

This formula is likely to stymie the average newcomer to the modal mu-calculus, but it has a simple meaning, which can be seen by using the slogans. $\mu Y. \ldots$ is true if $\nu Z. \ldots$ is true if $(P \wedge [a]Y) \vee (\neg P \wedge [a]Z)$, which is true if either P holds and at the next (a)-states we loop back to $\mu Y. \ldots$, or P fails, and at the next states we loop back to $\nu Z. \ldots$. By the slogan 'μ means finitely', we can only loop through $\mu Y. \ldots$ finitely many times on any path, and hence P is true only finitely often on any path.

A variant on this theme illustrates the ease with which the modal mu-calculus can talk about actions, which from a traditional process calculus viewpoint are all that one can observe. By similar reasoning to the above, one can see that

$$\mu Y.\nu Z.[a]Y \wedge [-a]Z$$

asserts that a process does only finitely many a actions, whatever else it does: it may even diverge. For example, consider the divergent value-passing CCS process

$$A(n) \stackrel{\text{def}}{=} \tau.A(n+1) + \tau.B(n) \qquad B(n) = \textbf{if } n > 0 \textbf{ then } a.B(n-1).$$

$A(0)$ satisfies the formula. To see that this really is true, as the slogans suggest it is, we can examine the semantics. As this is an example of a nested fixpoint, let us consider the approximants. The first approximant of the outer minimal fixpoint is

$$\mu Y^0\ldots = \nu Z.[a]\big(\mu Y^{<0}\ldots\big) \wedge [-a]Z = \nu Z.[a]\mathrm{ff} \wedge [-a]Z.$$

We may also calculate this maximal fixpoint by approximants. Its first approximant is $\nu Z^0\ldots = [a]\mathrm{ff} \wedge [-a]\mathrm{tt}$, which is, by inspection, the processes $A(i)$ $(i \geqslant 0)$ and $B(0)$. The next approximant is $\nu Z^1\ldots = [a]\mathrm{ff} \wedge [-a](\nu Z^{<1}\ldots)$, which again by inspection is just $B(0)$; and here we stabilize. So we are now in a position to calculate the next outer approximant:

$$\mu Y^1\ldots = \nu Z.[a]\big(\mu Y^{<1}\ldots\big) \wedge [-a]Z = \nu Z.[a]\{B(0)\} \wedge [-a]Z.$$

The reader may compute this by approximants to obtain $B(0)$, $B(1)$. Continuing, we obtain $\mu Y^i\ldots = \{B(j) \mid j \leqslant i\}$. The next approximant is where everything happens:

$$\mu Y^\omega\ldots = \nu Z.[a]\big(\mu Y^{<\omega}\ldots\big) \wedge [-a]Z = \nu Z.[a]\{B(i)\} \wedge [-a]Z.$$

A further calculation shows that here $\nu Z^0\ldots$ is all processes, and so we stabilize; now the outer approximant chain has also converged at all processes.

It is also worth briefly considering the negative of this, namely 'infinitely often', as it illustrates the finite unfolding of μ within an outer ν. Consider the formula

$$\nu Y.\mu Z.\langle a\rangle Y \vee \langle \tau\rangle Z.$$

Applying the slogans, we see that this is true if either we can do an a and loop to Y, or do a τ, and loop to Z. We can only loop to Z finitely often, so we can only do finitely many consecutive τs; but if we do an a and loop back out to Y, we then re-enter the inner fixpoint, free to do another independent finitely many τs. The *really* enthusiastic reader might like to evaluate this formula by approximants on the modified set of processes given by putting $B(n) = \tau.A(2n) + \textbf{if } n > 0 \textbf{ then } a.B(n-1)$.

We shall see in a later section that this so-called alternation of fixpoint operators does indeed give ever more expressive power as the number of alternations increases. It also appears to increase the complexity of model-checking: all known algorithms are exponential in the alternation, but whether this is necessarily the case is the main remaining open problem about the modal mu-calculus. (See Section 4.4 for more on the complexity of model-checking.) However, the practitioner who is alarmed by the thought of even more complex formulae, can take comfort in the widely asserted proposition that one never actually needs more than two alternating fixpoints, and even two is a bit unusual. In fact, many would go further and say that in real life we are only interested in safety properties. There are two main reasons why alternating formulae might appear. The first is if one has a front end working in CTL* (or another complex logic), and is translating into modal mu-calculus for the main engine. This is theoretically possible, by work of Mads Dam [19], but the translation is sufficiently complex that it would be surprising to find a tool doing this. The second is if one needs to express *fairness*. Fairness properties, whether weak (an event continuously enabled must happen) or strong (an event enabled infinitely often must happen), can be written in the modal mu-calculus, and require mixing fixpoints. For example, to say that an action a is fairly treated means that there are no paths on which a is enabled infinitely often, but occurs only finitely often; this can be written as

$$\nu X.\mu Y.\nu Z.[a]X \wedge (\langle a\rangle \mathrm{tt} \Rightarrow [-a]Y) \wedge [-a]Z.$$

Of course, we may be in a situation where the system is inherently unfair – e.g., it is a normal interleaving of concurrent components – and in that case we can use a similar idea to express properties such as 'P eventually holds on all paths, provided that a is fairly treated', which is left as an exercise for the reader. However, fairness can also be handled by other means, such as defining the set of fair paths 'outside' the logic, and adapting algorithms to apply only to fair paths.

3.6. *Modal mu-calculus as a process logic*

So far, we have described the modal mu-calculus as a logic on labelled transition systems; but this is a handbook of process algebra. Of course, process algebras, or at least simple traditional process algebras, are equipped with an operational semantics that defines a labelled transition system, in which the states are usually pieces of syntax. The easy way to interpret any modal logic on a process algebra is just to use this generated transition system. However, in theory one's notion of process is not just a piece of syntax, but some *behaviour*, and one wishes to treat identically 'processes' that are behaviourally equivalent. In the CCS world view, we believe that processes that are strongly bisimilar are equivalent for all practical purposes – and most non-practical purposes – and in many situations we might treat observational congruence or weak bisimulation similarly. In more exotic situations we might view any of the hundreds of known equivalences as process equality. This raises the obvious question of how our logic behaves with respect to such equivalences, and what should we do if it behaves badly. The first answer is that the logic behaves well:

If two processes (i.e., states in transition systems) are strongly bisimilar, then they satisfy exactly the same modal mu-calculus formulae.

This is proved by an easy induction. In fact, for finite-branching systems (more precisely, image-finite systems) the converse is also true provided the atomic propositions respect bisimulation, so that modal mu-calculus (or indeed its small subset HML) characterizes strong bisimulation; this fails for arbitrary systems. An inductive proof of this result is presented in [64].

However, if we prefer, as we usually do, to work with a weaker equivalence, this of course is no longer true, since the modal mu-calculus as we have defined it treats τ just as any other action. There are several ways to deal with this. One is just to accept it, and when writing properties in the mu-calculus to be very careful to write explicitly a formula that does the desired thing with silent actions; this has obvious drawbacks. A more natural option, provided that whatever tool one is using supports it, is to work with the weak transition system, namely the transition system defined by[7] $s \stackrel{a}{\Longrightarrow} t$ if $s \xrightarrow{\tau^*}\xrightarrow{a}\xrightarrow{\tau^*} t$. A compromise is to stick with the strong transition system, but to define derived modalities which take account of τ. For example, one can define $\langle\langle a \rangle\rangle\phi$ to mean $\mu Z.\langle\tau\rangle Z \vee \langle a\rangle \mu Z.\phi \vee \langle\tau\rangle Z$, which captures in the strong logic the diamond modality in the weak transition system.

Another issue that affects practice is the atomic propositions. We have presented the logic with arbitrary valuations for atomic propositions, but this is neither possible nor desirable in practice. Typically there will be some small language of atomic propositions; for example, in value-passing CCS one wants to have basic expressions over the parameters of a process. One might imagine that by restricting this language one could obtain better results on decidability etc.; although this is, loosely speaking, true in practice, in the sense that 'sensible' properties of 'sensible' systems can be expressed and often proved entirely within a 'sensible' language, it is not generally true in a mathematical sense, since the basic mu-calculus operators are sufficient to expose any undecidability that may be lurking in a system – for example, the halting property is just $\mu Z.[-]Z$.

3.7. *Linear time mu-calculus*

The modal mu-calculus is a pure branching time logic, although as we have seen in the examples it can express properties that are usually found in mixed branching/linear logics such as CTL*. One may also add fixpoint operators to linear time logic, to get a pure linear time mu-calculus. The modality in this case is a 'next' operator @, and the logic is interpreted not over states of a system, but over *runs* or maximal paths. The formula $@\phi$ is satisfied by the runs $\sigma = s_0 \xrightarrow{a_1} s_1 \xrightarrow{a_2} \cdots$ such that $a_1 = a$ and $s_1 \xrightarrow{a_2} \cdots$ satisfies ϕ. Adding negation, conjunction and maximal fixpoint to this operator gives the linear time mu-calculus.

Just as in the modal case, the linear mu-calculus encompasses most other linear time logics. The linear 'until' operator $\phi\ \mathbf{U}\ \psi$ is just $\mu Z.\psi \vee (\phi \wedge \bigcirc Z)$, and linear fairness operators can be expressed in a similar manner to their branching counterparts.

At first sight, the linear mu-calculus is much harder to handle than the modal logic: we are interpreting over runs, and even a finite-state system generates continuum-many runs. So we are taking fixpoints in a lattice of size 2^{2^ω}, and our approximants may, at first sight, not close until $(2^\omega)^+$! Of course, this doesn't actually happen, and in fact for a finite-state system fixpoints close at a finite approximant. However, this approximant may

7 This double arrow is *not* logical implication!

be large, and model-checking is exponential [71]. It is possible to give a direct procedure, and even a direct local procedure, to model-check the linear time mu-calculus [66,12], but they are complicated, and for all but the smallest examples, automata-theoretic techniques are superior.

On the other hand, the linear time mu-calculus is in some ways simpler than the modal mu-calculus. In particular, increasing alternation does not give increasing expressive power: any formula is equivalent to one of alternation depth two. Indeed, any formula is equivalent to an ω-regular expression. An elegant direct proof of this result is contained in [53] (which doesn't mention temporal logic and uses fixpoints of linear systems of equations[8]). An alternative proof uses automata. Büchi automata and linear time mu-calculus are equi-expressive, and a Büchi automaton is easily transformed into a formula whose alternation depth is at most two, see for example [51].

Generally speaking, the linear time mu-calculus has not offered the same advantages as the modal mu-calculus, and those who use a linear time logic use the traditional logic of 'next' and 'until', and its fragments.

4. Theory of modal mu-calculus

4.1. *Fixpoint regeneration and the 'fundamental semantic theorem'*

In the informal description of the meaning of fixpoints, we used the idea of the dependency of s at ϕ on t at ψ. We now make this precise. Assume a structure $\mathcal{T}$, and a formula ϕ. Suppose that we annotate the states with sets of subformulae, such that the sets are locally consistent: that is, s is annotated with a conjunction iff it is annotated with both conjuncts; s is annotated with a disjunction iff it is annotated with at least one disjunct; if s is annotated with $[a]\psi$ (respectively, $\langle a\rangle\psi$), then all (respectively at least one) a-successor is annotated with ψ; if s is annotated with a fixpoint or fixpoint variable, it is annotated with the body of the fixpoint. We call such an annotated structure a *quasi-model*.

A *choice function* f is a function which for every disjunctive subformula $\psi_1 \vee \psi_2$ and every state annotated with $\psi_1 \vee \psi_2$ chooses one disjunct $f(s, \psi_1 \vee \psi_2)$; and for every subformula $\langle a\rangle\psi$ and every state s annotated with $\langle a\rangle\psi$ chooses one a-successor $t = f(s, \langle a\rangle\psi)$ annotated with ψ.

A *pre-model* is a quasi-model equipped with a choice function.

Given a pre-model with choice function f, the *dependencies* of a state s that satisfies a subformula ψ are defined thus: $s@\psi_1 \wedge \psi_2 \succ s@\psi_i$ for $i = 1, 2$; $s@[a]\psi \succ t@\psi$ for every t such that $s \xrightarrow{a} t$; $s@\psi_1 \vee \psi_2 \succ s@f(s, \psi_1 \vee \psi_2)$; $s@\langle a\rangle\psi \succ f(s, \langle a\rangle\psi)@\psi$; $s@\genfrac{}{}{0pt}{}{\mu}{\nu}Z.\psi \succ s@\psi$; $s@Z \succ s@\psi$ where Z is bound by $\genfrac{}{}{0pt}{}{\mu}{\nu}Z.\psi$. A *trail* is a maximal chain of dependencies.

If every trail has the property that the highest (i.e., with the outermost binding fixpoint) variable occurring infinitely often is a ν-variable, the pre-model is *well-founded*. (Equivalently: in any trail, a μ-variable can only occur finitely often unless a higher variable is encountered.)

The fundamental theorem on the semantics of the modal mu-calculus can now be stated:

[8] This is the same paper where bisimulation first appears.

THEOREM 1. *A well-founded pre-model is a model: in a well-founded pre-model, if s is annotated with ψ, then indeed $s \models \psi$.*

The theorem in this form is due to Streett and Emerson in [68], from which the term 'well-founded pre-model' is taken. Stirling and Walker [65] presented a tableau system for model-checking on finite structures, and the soundness theorem for this is essentially a finite version of the fundamental theorem with a more relaxed notion of choice; the later infinite-state version of [13,8] is the fundamental theorem, again with a slight relaxation on choice.

A converse is also true:

THEOREM 2. *If in some structure $s \models \phi$, then there is a locally consistent annotation of the structure and a choice function which make the structure a well-founded pre-model.*

The fundamental theorem, in its various guises, is the precise statement of the slogan 'μ means finite looping'. To explain why it is true, we need to make a finer analysis of approximants.

Assume a structure $\mathcal{T}$, valuation $\mathcal{V}$, and formula ϕ in positive normal form. Let $Y_1, \ldots, Y_n$ be the μ-variables of ϕ, in an order compatible with formula inclusion: that is, if $\mu Y_j.\psi_j$ is a subformula of $\mu Y_i.\psi_i$, then $i \leqslant j$. If Y_i is some inner fixpoint, then its denotation depends on the meaning of the fixpoints enclosing it: for example, in the formula $\mu Y_1.\langle a\rangle \mu Y_2.(P \vee Y_1) \vee \langle b\rangle Y_2$, to calculate the inner fixpoint μY_2 we need to know the denotation of Y_1. We may ask: what is the least approximant of Y_1 that could be plugged in to make the formula true? Having fixed that, we can then ask what approximant of Y_2 is required. This idea is the notion of *signature*. A signature is a sequence $\sigma = \alpha_1, \ldots, \alpha_n$ of ordinals, such that the i least fixpoint will be interpreted by its α_ith approximant (calculated relative to the outer approximants).

The definition and use of signatures inevitably involves some slightly irritating book-keeping, and they appear in several forms in the literature. In [68], the Fischer–Ladner closure of ϕ was used, rather than the set of subformulae. The closure is defined as for PDL (see Section 2.1), with the addition of the rule that if $\genfrac{}{}{0pt}{}{\mu}{\nu} Z.\,\psi(Z) \in \mathrm{cl}(\phi)$, then $\psi(\genfrac{}{}{0pt}{}{\mu}{\nu} Z.\,\psi) \in \mathrm{cl}(\phi)$. The signatures were defined by syntactically unfolding fixpoints, rather than by semantic approximants. In [65] and following work, a notion of *constant* was used, which allows some of the book-keeping to be moved into the logic. Although all the notions and proofs using them are interconvertible, the 'constant' variant is perhaps easier to follow, and has the advantage that it adapts easily to the modal equation system presentation of modal mu-calculus, which we shall see below. Indeed, it arises more naturally from that system.

Add to the language a countable set of *constants* $U, V, \ldots$. Constants will be defined to stand for maximal fixpoints or approximants of minimal fixpoints. Specifically, given a formula ϕ, let $Y_1, \ldots, Y_n$ be the μ-variables as above, let $Z_1, \ldots, Z_m$ be the ν-variables, let $\sigma = \alpha_1, \ldots, \alpha_n$ be a signature, and let $U_1, \ldots, U_n, V_1, \ldots, V_m$ be constants, which will be associated with the corresponding variables. They are given semantics thus: if Y_i is bound by $\mu Y_i.\psi_i$, then $\|U_i\|_\sigma$ is $\|\mu Y_i^{\alpha_i}.\psi_i'\|_\sigma$, where ψ_i' is obtained from ψ_i by substituting the corresponding constants for the free fixpoint variables of $\mu Y_i.\psi_i$. If Z_i is bound by

$\nu Z_i.\psi_i$, its semantics is $\|\nu Z_i.\psi_i'\|_\sigma$. Given an arbitrary subformula ψ of ϕ, we say a state s satisfies ψ with signature σ, written $s \models_\sigma \psi$, if $s \in \|\psi'\|_\sigma$, where ψ' is ψ with its free fixpoint variables substituted by the corresponding constants.

Order signatures lexicographically. Now, given a pre-model for ϕ, extend the annotations so that each subformula at s is accompanied by a signature – write $s@\psi[\sigma]$. Such an extended annotation is said to be locally consistent if the signature is unchanged or decreases by passing through Boolean, modal, or ν-variable dependencies, and when passing through $s@Y_i$ it strictly decreases in the ith component and is unchanged in the $1, \ldots, (i-1)$th components. It can now be shown, by a slightly delicate but not too difficult induction, that if $s@\psi[\sigma]$, then $s \models_\sigma \psi$. Furthermore, given a well-founded pre-model, one can construct a locally consistent signature annotation – essentially, the Y_i component of σ in $s@\psi[\sigma]$ is the maximum 'number' (in the transfinite sense) of Y_i occurrences without meeting a higher variable in trails from $s@\psi$, and so on; the well-foundedness of the pre-model guarantees that this is well-defined. This gives the fundamental theorem. (A little care is required to get the details of this argument correct, as will be seen from an inspection of the proofs in [68,65,8].)

The converse is quite easy: given a model, annotate the states by the subformulae they satisfy; for $s@\psi$ assign the least σ such that $s \models_\sigma \psi$; and choose a choice function that always chooses the successor with least signature. It is easy to show that this is a well-founded pre-model and signature assignment.

4.2. *The finite model property and decidability*

The notion of well-founded pre-model was introduced by Streett and Emerson in order to establish the

THEOREM 3. *The modal mu-calculus is decidable; that is, it is decidable whether a formula has some model.*

As a corollary of this proof, they obtained the small model property

THEOREM 4. *If a modal mu-calculus formula has a model, it has a finite model, of size exponential in the size of the formula.*

These results were obtained by appealing to automata theory. First of all, one can easily show that if a formula has a model, it has a model that is a tree (just unravel the original model), and has bounded branching degree (just cut off all the branches that are not actually required by some diamond subformula; this leaves at most (number of diamond subformulae) branches at each node). We can now construct an automaton that accepts such bounded-branching tree models, by combining a finite-state automaton to check the local consistency (i.e., to check that the putative model is a pre-model), and a Rabin automaton (see Section 6) to check that the pre-model is well-founded. Thus the formula is satisfiable if this product automaton accepts some tree. Now established automata theory tells us that (a) this question is decidable (b) if such an automaton accepts some tree, it accepts a regular tree, that is, one that is the unravelling of a finite system; this gives the results.

One can, however, obtain a small model property, and thence decidability, by a more direct argument on models. Take a pruned tree model as above, and equip it with an annotation and choice function to make it a well-founded pre-model. Consider an outermost maximal fixpoint $\nu Z.\psi$, say. Look at the first occurrence of a state s labelled by Z; let F be the set of formula annotating s. Now, if there is a state t below s that is also annotated by F, prune the tree at t, and identify t with s. Repeat for all Z labelled states with repetitions below them. On the remaining infinite branches of the tree, Z does not occur infinitely often; and by the well-foundedness of the pre-model, no immediately inferior least fix-point occurs infinitely often either; so we still have a well-founded pre-model. Now repeat the procedure with the next maximal fixpoint in, and so on. When all fixpoints have been processed, the result is a finite graph that is still a well-founded pre-model. The repetition detection is reminiscent of the filtration technique we saw for PDL. Unfortunately, filtration is not sound for minimal fixpoints, so we have this less constructive procedure. It generates, naively, a model of triply exponential size; but it is intuitively easy, and of course the existence of even a triply exponential model is enough to give decidability. One can improve the complexity somewhat by being less simple-minded, but to obtain the exponential upper bound of [68], which is in fact also the lower bound, one needs the more sophisticated techniques of automata or games.

A different method for obtaining the small model property proceeds via a normal form result for modal-mu calculus. A formula is in "automaton normal form", anf, if it belongs to the following sublogic.

- P, $\neg P$ and Z are anfs.
- If ϕ_1 and ϕ_2 are anfs, so is $\phi_1 \vee \phi_2$.
- If ϕ is an anf, then so are $\nu Z.\phi$, $\mu Z.\phi$.
- If each Γ_i is a finite set of anfs and $a_i \neq a_j$ when $i \neq j$ and Σ is a finite set of atomic propositions and their negations, then $(a_1)\Gamma_1 \wedge \cdots \wedge (a_n)\Gamma_n \wedge \Sigma$ is an anf.

We need to explain the auxiliary subformula $(a)\Gamma$, which abbreviates

$$(a)\Gamma = \bigwedge_{\phi\in\Gamma} \langle a\rangle\phi \wedge [a] \bigvee_{\phi\in\Gamma} \phi.$$

Walukiewicz in [74] (and see also [33]) proved that every (closed) modal mu-calculus formula is equivalent to an anf formula. Walukiewicz calls anfs "disjunctive formulae". However they are very close to characteristic automata (and in fact games). The proof of the normal form result uses automata. What is interesting is that an anf formula is satisfiable iff all its fixed point subformulae $\mu Z.\phi(Z)$ are replaced with $\phi(\mathrm{ff})$ and all subformulae $\nu Z.\phi(Z)$ are replaced with $\phi(\mathrm{tt})$. The result of these substitutions is a modal mu-calculus formula without fixed points, for which proof of the finite model property is easy.

4.3. *Axiomatization*

A related problem to decidability is the question of providing an axiomatization of the theory of the modal mu-calculus. In his original paper, Kozen presented the following axiomatization of the equational theory, where $\phi \leqslant \psi$ means $\phi \vee \psi = \psi$, taking $\langle\rangle$ and μ as primitives and defining $[]$ and ν by duality:

(1) axioms for Boolean algebras;
(2) $\langle a\rangle\phi \vee \langle a\rangle\psi = \langle a\rangle(\phi \vee \psi)$;
(3) $\langle a\rangle\phi \wedge [a]\psi \leqslant \langle a\rangle(\phi \wedge \psi)$;
(4) $\langle a\rangle \mathrm{ff} = \mathrm{ff}$;
(5) $\phi(\mu X.\phi(X)) \leqslant \mu X.\phi(X)$;
(6) $\dfrac{\phi(\psi) \leqslant \psi}{\mu X.\phi(X) \leqslant \psi}$.

Axiom (5) is the axiom of fixed point induction, in dual form; rule (6) says that μ is indeed the least pre-fixed point. (An equivalent presentation of the axiom system is as an extension of minimal multi-modal logic K, with the additional axiom $\phi(\mu X.\phi(X)) \Rightarrow \mu X.\phi(X)$ and the additional inference rule if $\phi(\psi) \Rightarrow \psi$ then $\mu X.\phi(X) \Rightarrow \psi$.)

However, despite the naturalness of this axiomatization, Kozen was unable to show that it was complete. He was, however, able to show completeness for a restricted language, the language of *aconjunctive* formulae, in which (roughly) fixpoint variables are not allowed to occur in both branches of a conjunction.

Completeness for the full language remained open for more than a decade, until it was finally solved by Igor Walukiewicz in [74], who established that Kozen's axiomatization is indeed complete. The proof is very involved and utilises automata normal form described above (which generalises the aconjunctive fragment). It is reasonably straightforward to show using tableaux that if an anf formula ϕ is consistent[9] then it has a model. Much harder to prove is that every (closed) formula is provably equivalent within the axiom system to an anf formula. Walukiewicz utilises automata and games to show this. More information can also be found in the notes [75].

4.4. *Model-checking*

Decidability and axiomatization are standard questions for logicians; but for the practioner, the important question is model-checking: given a system $\mathcal{T}$ with some initial state s and a formula ϕ, is it the case that $s \models \phi$? For the modal mu-calculus, the decidability of this question is obvious, for finite systems; the problem is, what is its complexity? There have been great technological advances in the model-checking of finite-state systems by the use of clever data structures such as BDDs. There have also been useful advances for fragments of the modal mu-calculus. However, there has been only one notable improvement in our knowledge of the complexity of the general problem, and a full answer to the question remains something that haunts our dreams – literally, for at least one of the authors!

To discuss the complexity of model-checking, we need to introduce the notion of *alternation depth* of a formula. We shall define this more precisely in the following subsection, which examines alternation. For the moment, it is sufficient to define it loosely as the maximum number of μ/ν alternations in a chain of nested fixpoints. In fact we usually add the

---

[9] A formula ϕ is consistent with respect to an axiom system if $\phi \Rightarrow \mathrm{ff}$ is not derivable within the axiom system. Completeness of an axiom system is equivalent to the statement that every consistent formula has a model (is satisfiable).

proviso that the nesting is genuine interdependence of the 'infinitely often' kind rather than mere inclusion of closed subformulae of the 'always eventually' kind (see Section 3.5); but the proviso may be ignored without changing the essentials of the arguments.

Suppose that we simply 'model-check' by brute force calculation of the approximants. Let the size (number of states plus number of transitions) of the system be n, the length of the formula be m. Every fixpoint closes at, at worst, the nth approximant. Calculating the booleans takes time $O(n)$; calculating modalities takes time $O(n)$. For an innermost fixpoint, given a valuation of the free variables, we may need to evaluate the body n times to close. So the innermost fixpoint takes time $O(n \cdot (mn))$. Now the next fixpoint out may also need to be evaluated n times, each of which involves evaluating the inner fixpoint (and some non-fixpoint operators); and so that takes $O(n \cdot n \cdot (mn))$. And so on: so we end up with $O(m \cdot n^{d+1})$, where d is the depth of fixpoint operators.

It turns out that by general results on fixpoints in lattices [7], adjacent fixpoints of the same type, such as $\mu Y.\mu Z.\phi(Y, Z)$ are not much worse than one, for the following reason: since $\|\mu Z.\phi(Y^{\alpha+1}, Z)\| \supseteq \|\mu Z.\phi(Y^{\alpha}, Z)\|$, and the formulae are monotone, when one evaluates the ith outer approximant, one need not start evaluating the inner fixpoint from 0, but from wherever one had reached on finishing the $(i-1)$th outer approximant. Alternatively, one can turn $\mu Y.\mu Z.\phi$ into a single fixpoint over a vector: $\mu(X, Y).\phi$. Thus we obtain the theorem of Emerson and Lei [25]:

THEOREM 5. *The model-checking problem for a formula of size m and alternation depth d on a system of size n is* $O(m \cdot n^{d+1})$.

This theorem holds for the simple-minded notion of alternation, for the somewhat more refined version that Emerson and Lei used, and for the stronger notion of Niwiński [50].

So model-checking appears to be exponential in alternation depth. If, in the usual way, one defines the size of the problem to be the size of the system plus the size of the formula, this means that the problem seems to be exponential. Of course, it is a very well-behaved exponential, both theoretically, as by stratifying the formulae according to alternation depth one gets a chain of polynomial problems of increasing degree; and practically, because systems are typically large, formulae are typically small, and alternation depths are typically 1 or 2. However, it does not appear that the model-checking problem is polynomial.

On the other hand, it is readily apparent that the problem is in NP. This was first remarked by Emerson, Jutla and Sistla in [24], who used an automata-theoretic argument. However, it can be seen directly [63]: take a model, and guess the right choice function and annotation to make it a well-founded pre-model. Now to check that it really is a well-founded pre-model, we need to check local consistency of the annotation, which is polynomial (and in fact quadratic); and we need to check well-foundedness. But to check well-foundedness on a finite pre-model, it suffices to show that there are no bad loops in the dependencies: if a minimal fixpoint occurs infinitely often without a higher fixpoint appearing, then there must be some state that occurs twice (and indeed infinitely often) labelled with that fixpoint. To check this, we can do strongly connected component analysis: starting from the bottom up, for each minimal fixpoint, cut the dependency graph at all higher fixpoints, and then look for a strongly connected component involving the said minimal fixpoint. This analysis is polynomial, of low degree; and we do it for each fixpoint; so the whole check is polynomial.

Since model-checking is trivially closed under complementation, the problem is in NP $\cap$ co-NP. This places it in the very small class of such problems that are not known to be polynomial. Opinion is divided on the likely true complexity; some believe it to be polynomial, but although the P faction includes some most accomplished complexity theorists, no polynomial algorithm has yet been found, either for the mu-calculus itself, or for the stochastic pay-off games to which it can be reduced and which the complexity theorists are really interested in. Others, including exactly one of the authors, believe it to be non-polynomial; although this would make the world a more interesting place, this belief has the obvious disadvantage of being impossible to confirm definitely without a resolution of $P = NP$. Of course, even a proof of non-polynomiality subject to $P \neq NP$ would be highly interesting.

We mentioned that there has been one advance in the complexity of global model-checking. This is due to Long et al. [43]. Essentially, they observed that the monotonicity argument that we applied above to adjacent least fixpoints, can also be applied to a formula of shape $\mu Y.\nu Z.\phi(Y, Z)$; by a careful and quite subtle analysis, it can be shown that actually this requires quadratic, not cubic, time to compute. Unfortunately, this does not allow all alternation to be collapsed, but they showed that it generalizes to the extent of proving

THEOREM 6. *Model-checking is* $O(m \cdot n^{\lceil d/2 \rceil + 1})$.

4.5. *Alternation*

We now look at the notion of alternation in more detail. As we have said, the idea is to count alternations of minimal and maximal fixpoint operators, but to do so in a way that only counts real dependency. The paradigm is 'always eventually' versus 'infinitely often': the 'always eventually' formula

$$\nu Y.(\mu Z.P \vee \langle a \rangle Z) \wedge \langle a \rangle Y$$

is, using the brute-force model-checker above, really no worse to compute than two disjoint fixpoints, since the inner fixpoint can be computed once and for all, rather than separately on each outer approximant; on the other hand, the 'infinitely often' formula

$$\nu Y.\mu Z.(P \vee \langle a \rangle Z) \wedge \langle a \rangle Y$$

really does need the full double induction on approximants.

The definition of Emerson and Lei takes care of this by observing that the 'eventually' subformula is a closed subformula, and giving a definition that ignores closed subformulae when counting alternations. The stronger notion of Niwiński, which also has the advantage of being robust under translation to modal equation systems, also observes that, for example, $\mu X.\nu Y.[-]Y \wedge \mu Z.[-](X \vee Z)$ although it looks like a $\mu/\nu/\mu$ formula, is morally a μ/ν formula, since the inner fixpoint does not refer to the middle fixpoint.

The alternation depth referred to in the complexity of model-checking is a measure of alternation that is symmetric in μ and ν. It is possible to give algorithms that compute the alternation depth of a formula [25,1,36], and this is how the notion was presented by Emerson and Lei. However, for our purposes it is easier to start from a definition of classes for formula, formalizing the idea of 'a $\mu/\nu/\mu$ formula' etc.; such a definition is analogous to the usual definition of quantifier alternation for predicate logic, an analogy which will be exploited later. This was how Niwiński [50] presented the notion of alternation, and we follow his presentation.

Assuming positive form, a formula ϕ is said to be in the classes $\Sigma_0^{N\mu}$ and $\Pi_0^{N\mu}$ iff it contains no fixpoint operators. To form the class $\Sigma_{n+1}^{N\mu}$ (respectively $\Pi_{n+1}^{N\mu}$, take $\Sigma_n^{N\mu} \cup \Pi_n^{N\mu}$, and close under the following rules:

(1) if $\phi_1, \phi_2 \in \Sigma_{n+1}^{N\mu}$ (respectively $\Pi_{n+1}^{N\mu}$), then $\phi_1 \vee \phi_2, \phi_1 \wedge \phi_2, \langle a\rangle\phi_1, [a]\phi_1 \in \Sigma_{n+1}^{N\mu}$ (respectively $\Pi_{n+1}^{N\mu}$);
(2) if $\phi \in \Sigma_{n+1}^{N\mu}$ (respectively $\Pi_{n+1}^{N\mu}$), then $\mu Z.\phi \in \Sigma_{n+1}^{N\mu}$ (respectively $\nu Z.\phi \in \Pi_{n+1}^{N\mu}$);
(3) if $\phi(Z), \psi \in \Sigma_{n+1}^{N\mu}$ (respectively $\Pi_{n+1}^{N\mu}$), then $\phi(\psi) \in \Sigma_{n+1}^{N\mu}$ (respectively $\Pi_{n+1}^{N\mu}$), *provided* that no free variable of ψ is captured by a fixpoint operator in ϕ.

If we omit the last clause, we get the definition of 'simple-minded' alternation $\Sigma_n^{S\mu}$, that just counts syntactic alternation; if we modify the last clause to read '... *provided* that ψ is a closed formula', we obtain the Emerson–Lei notion $\Sigma_n^{EL\mu}$. (We write just Σ_n^{μ} when the distinctions are not important, or when we are making a statement that applies to all versions.)

To get the symmetrical notion of *alternation depth* of ϕ, we can define it be the least n such that $\phi \in \Sigma_{n+1}^{\mu} \cap \Pi_{n+1}^{\mu}$. To make these definitions clear, consider the following examples:

- The 'always eventually' formula is $\Pi_2^{S\mu}$, but not $\Sigma_2^{S\mu}$, and so its simple alternation depth is 2. However, in the Emerson–Lei notion, it is also $\Sigma_2^{EL\mu}$, since $\nu Y.W \wedge \langle a\rangle Y$ is $\Pi_1^{EL\mu}$ and so $\Sigma_2^{EL\mu}$, and by substituting the closed $\Sigma_2^{EL\mu}$ (and in fact $\Sigma_1^{EL\mu}$) formula $\mu Z.P \vee \langle a\rangle Z$ for W we get 'always eventually' in $\Sigma_2^{EL\mu}$; hence its Emerson–Lei (and Niwiński) alternation depth is 1.
- The 'infinitely often' formula is Σ_2^{μ} but not Π_2^{μ}, in all three definitions, and so has alternation depth 2.
- The formula $\mu X.\nu Y.[-]Y \wedge \mu Z.[-](X \vee Z)$ is $\Sigma_3^{S\mu}$, but not $\Pi_3^{S\mu}$; it is also $\Sigma_3^{EL\mu}$ but not $\Pi_3^{EL\mu}$, since there are no closed subformulae to bring the substitution clause into play. However, in the Niwiński definition, it is actually $\Sigma_2^{N\mu}$: $\nu Y.[-]Y \wedge W$ is $\Pi_1^{N\mu}$ and so $\Sigma_2^{N\mu}$; we can substitute the $\Sigma_1^{N\mu}$ formula $\mu Z.[-](X \vee Z)$ for W without variable capture, and so $\nu Y.[-]Y \wedge \mu Z.[-](X \vee Z)$ is $\Sigma_2^{N\mu}$; and now we can add the outer fixpoint, still remaining in $\Sigma_2^{N\mu}$.

We have already discussed the role alternation depth plays in the complexity of model-checking. A natural question is whether the hierarchy of properties definable by Σ_n^{μ} formulae is actually a strict hierarchy, or whether the hierarchy collapses at some point so that no further alternation is needed. This problem remained open for a while; by 1990, it was known that $\Sigma_2^{N\mu} \neq \Pi_2^{N\mu}$ [3]. No further advance was made until 1996, when the strictness of the hierarchy was established by Bradfield [10].

THEOREM 7. *For every n, there is a formula $\phi \in \Sigma_n^\mu$ which is not equivalent to any Π_n^μ formula.*

Bradfield established this for $\Sigma_n^{N\mu}$, which implies the result for the other two notions. At the same time, Giacomo Lenzi [41] independently established a slightly weaker hierarchy theorem for $\Sigma_n^{EL\mu}$.

Lenzi's proof is technically complex, and the underlying stratagem is not easy. Bradfield's proof appears technically complex, but most of the complexity is really just routine recursion-theoretic coding; the underlying stratagem is quite simple, and in some ways surprising. If one takes first-order arithmetic, one can add fixpoint operators to it, and one can then define a fixpoint alternation hierarchy in arithmetic. A standard coding and diagonalization argument shows that this hierarchy is strict [11]. The trick now is to transfer this hierarchy to the modal mu-calculus. Simply by writing down the semantics, it is clear (give or take some work to deal with the more complex notions of alternation) that if one takes a recursively presented transition system and codes it into the integers, then for a modal formula $\phi \in \Sigma_n^\mu$, its denotation $\|\phi\|$ is describable by an arithmetic Σ_n^μ formula. However, it is also possible, given any arithmetic fixpoint formula χ, to build a transition system and a modal formula ϕ, of the same alternation depth as χ, such that $\|\phi\|$ is characterized by χ. If we take χ to be a strict Σ_n^μ arithmetic formula, then no Π_n^μ arithmetic formula is equivalent to it, and therefore no Π_n^μ modal formula can be equivalent to ϕ. The transition system that is constructed is infinite, but by the finite model property, the hierarchy transfers down to the class of finite models.

Both proof techniques construct explicit examples of hard formulae. Bradfield's examples are:

THEOREM 8. *The Σ_n^μ formula*

$$\mu X_n.\nu X_{n-1}.\dots.\mu X_1.[c]X_1 \vee \langle a_1\rangle X_1 \vee \dots \vee \langle a_n\rangle X_n$$

is not equivalent to any Π_n^μ formula.

5. More on model-checking

In the last section, we saw the approximant approach to model-checking finite systems. In this section, we shall look at some other techniques that have been proposed, and at some of the extensions to deal with classes of infinite processes.

5.1. *Local model-checking ...*

The approximant calculation of model-checking is inherently global; calculating a fixpoint approximant involves applying a function over the state space. A variant approach is to use *local* techniques [65], that is, techniques where to determine the truth of $s \models \phi$, one looks at the immediate neighbourhood of s as required by the formula; if there are fixpoints, one will of course move further and further away from s, but one can stop as soon as the truth or falsity of ϕ at s is established, rather than working with the whole state space. This is attractive when a state is a process expression E belonging to a process calculus, and one doesn't want to build its transition graph explicitly. The formulation adopted was inspired by tableau methods in logic. In outline, suppose we want to check $E \models \phi$. We start with a *sequent* $E \vdash \phi$, and then apply rules according to the structure of ϕ. For example, if $\phi = \phi_1 \wedge \phi_2$, there is an $\wedge$-rule which is:

$$\frac{E \vdash \phi_1 \wedge \phi_2}{E \vdash \phi_1 \quad E \vdash \phi_2}$$

and if we get to a sequent $s \vdash \langle a \rangle \psi$, we apply the $\langle\rangle$-rule

$$\frac{E \vdash \langle a \rangle \psi}{F \vdash \psi} \qquad E \stackrel{a}{\longrightarrow} F.$$

The transition $E \stackrel{a}{\longrightarrow} F$ is derived from the transition rules for the particular process calculus that E belongs to. If we reach a fixpoint formula, or a variable Z which is bound by $\substack{\mu \\ \nu} Z.\, \psi$, we apply the appropriate fixpoint rule[10]

$$\frac{E \vdash \substack{\mu \\ \nu} Z.\, \psi}{E \vdash Z} \qquad \frac{E \vdash Z}{E \vdash \psi}.$$

Thus we build up (or rather down, since the rules grow downwards) a goal-directed tree of sequents to be established. The question is, when do we stop? and how do we know we've succeeded? Obviously we stop on an a-modality if the process has no a-transitions, in which case $E \vdash [a]\psi$ succeeds and $E \vdash \langle a \rangle \psi$ fails; and we stop at $E \vdash P$, when we look E up in the proposition valuation to determine success. The key is the treatment of fixpoints. If we get to $E \vdash \phi$, and we have already seen the same sequent higher up in the tree, then we stop and examine the highest fixed point variable Z between these repeat sequents; the leaf sequent is successful if Z is bound by ν and unsuccessful if it is bound by μ. Once we have a finite tree, we propagate success and failure back up to the root via the rules.

This proof technique is sound and complete for finite systems. The proof of this fact was another development of the fundamental semantic theorem, for finite models. Indeed, it is probably obvious to the reader (at least the reader who has read Section 4.1) that the construction of a successful tableau gives a well-founded pre-model, and that a well-founded pre-model is a tableau. This statement needs to be interpreted carefully. In the basic tableau construction, sequents $E \vdash \langle a \rangle \psi$ and $E \vdash \phi \vee \psi$ may occur many times on different branches, with different successors; however this could easily be outlawed, and indeed should be outlawed if one wishes to develop an algorithm from the proof method.

A variant perspective on tableau is provided by games [63]. Consider two players Verifier and Refuter. A sequent is now a game position. At $E \vdash \phi \wedge \psi$ Refuter chooses $E \vdash \phi$

[10] This is a variant presentation. In the original [65] system, fixpoints were first replaced by constants, and then unfolded. This allowed the termination condition to be expressed simply, but at the expense of efficiency.

or $E \vdash \psi$ and similarly at $E \vdash [a]\phi$ Refuter chooses a possible next position $F \vdash \phi$ when $E \stackrel{a}{\longrightarrow} F$. Verifier chooses when the main connective of the sequent is $\vee$ or $\langle a \rangle$. A play of a game is just a sequence of sequents. The winning conditions are as for tableaux. For instance Refuter wins $E \vdash \langle a \rangle \phi$ if E has no a-transitions. A play halts if a position is repeated, and the winner is determined by the highest fixpoint variable. From the fundamental theorem it follows that $E \models \phi$ ($E \not\models \phi$) iff Verifier (Refuter) has a history-free winning strategy for the game which starts at $E \vdash \phi$.[11] Model checking games are instances of parity games (which are briefly mentioned later). A history-free winning strategy, unlike a tableau, is linear in the size of the model checking problem. Games provide a basis for efficient model checking when one asks, is it the case $s \models \phi$? and why? If the answer is returned with a winning strategy, a user can be the opponent and play against the machine which always wins! This can be helpful in understanding why the property holds – or more importantly, perhaps, in understanding why a desired property fails.

The proof rules for tableau mentioned above are guided by the structure of the formulae. However when E is a process expression it also has structure, built from the combinators of the particular process calculus. It is possible, for instance [2], to have rules which are guided by process structure. A simple example is to inductively define the effect of the CCS restriction operator $\backslash a$ on modal mu-calculus formulas. For instance, $(\phi \vee \psi)\backslash a$ is $\phi\backslash a \vee \psi\backslash a$ and $(\nu Z.\phi)\backslash a$ is $\nu Z.\phi\backslash a$. More important is the effect on modalities $([K]\phi)\backslash a$ is $[K - \{a, \overline{a}\}]\phi\backslash a$ (where $\overline{a}$ is a's partner). The idea is that $\backslash a$ on formulas is an *inverse* of its application to processes. Parallel composition is more problematic. However one can define a "slicing" operator $/F$, where F is a process expression operator, on formulae which is an inverse of parallel composition [2]. Tableau rules as follows are thereby justified

$$\frac{E\backslash a \vdash \phi}{E \vdash \phi\backslash a} \qquad \frac{E \mid F \vdash \phi}{E \vdash \phi/F}.$$

5.2. ... *for infinite systems*

Apart from its locality, another feature of the goal-directed tableau approach is that it can be generalized to use symbolic methods, so allowing the proof of properties of infinite systems (or just large systems). The idea is to use sets of states in sequents, rather than single states; then in the $\vee$ and $\langle\rangle$-rules, an explicit choice function is given; termination happens when we construct a sequent $S \vdash Z$ such that we have already seen a sequent $S' \vdash Z$ for $S' \supseteq S$, with no higher fixpoint intervening; and success involves giving a proof that all the dependency trails are well-founded. This latter is typically done by defining some well-founded measure on processes, and showing that it decreases as one passes through minimal fixpoints, and doesn't increase elsewhere. The canonical such measure is the signature, as described in Section 4.1, but simpler measures may suffice. Thus an infinite state tableau constructs a finite presentation of a well-founded quasi- or pre-model. Details may be found in [8,13], and an example of a Petri net implementation in [9].

[11] A winning strategy is a set of rules which tell a player how to move. A player uses a strategy in a play if all her moves obey the rules in it. A history-free strategy is one which is independent of previous positions in a play. A strategy is winning for a player if she wins every play in which she uses it.

5.3. *Decidable model-checking for infinite processes*

The infinite-state tableau system just described makes no claims for decidability: it allows arbitrary mathematical reasoning to be used in establishing the termination and success conditions. An obvious question is to ask what can be done for infinite processes while retaining decidability; the answer is quite limited.

Clearly model-checking is undecidable for any Turing-powerful class of processes, since $\mu Z.[-]Z$ expresses the halting property. It turns out to be easy to code halting in the modal mu-calculus, or even in small fragments of linear temporal logic, for many classes that are not Turing-powerful by themselves. For example, if we try to model a register machine as a Petri net, we need to test for zero, which nets cannot do; but if d is a transition that can fire when a particular place is non-empty, then $[d]\mathrm{ff}$ expresses the emptiness of that place, and so one can write a mu-calculus formula which says 'the net halts, if we look at those "honest" computations where a branch-on-zero is only taken when the corresponding counter really is zero'. Similarly, it is unnecessary for the registers actually to synchronize with the finite-state control, as we can encode the synchronization into the formula. The upshot of this is that model-checking, even with just one fixpoint, is undecidable for BPP, Basic Parallel Processes – a result which might be surprising, given the decidability of bisimulation for BPP.

On the other hand, for BPA, Basic Process Algebra, and for its generalization pushdown processes, model-checking is decidable. This follows from a classical result of Muller and Schupp [48] on the monadic second-order theory of pushdown graphs; but it has been shown by a direct and elegant argument by Burkart and Steffen [16]. They view pushdown processes as 'higher-order' transition graphs. For example, the BPA process system given by $A = aBA + c$ and $B = bBA + c$ is viewed as a system where the A process does an a transition followed by a B transition, understood as a recursive call to B, and returns to its initial state (if B terminates), or does a c to terminate. Now, if we are interested in a particular formula ϕ, each transition, including the 'higher-order' transitions, is viewed as a predicate transformer on the subformulae of ϕ: assuming that some subformula hold at the terminal state of the process, which formulae hold at the initial state? Obviously there are only finitely many possibilities, and the result can be calculated by an approximant-style procedure. If one now plugs into the A transformer the formulae that are true at a real terminal state, one obtains the formulae true of A.

Further detail on all these results can be found in the chapter by Burkart et al. later in this Handbook [15].

5.4. *Equation systems and efficient local model-checking*

Recall that in Section 3.1 we presented fixpoints as modal equations. We have also seen mention of vectorial fixpoints, allowing the collapse of adjacent fixpoints of the same sign into one. Combining these ideas, we say a *modal equation system* is a set $\{X_{ij} = \phi_{ij} \mid 1 \leqslant i \leqslant m, 1 \leqslant j \leqslant n_i\}$ of equations, where each ϕ_{ij} is a modal (no fixpoints) formula in the variables X_{ij}; for each i, the set $\{X_{ij} = \phi_{ij} \mid 1 \leqslant j \leqslant n_i\}$ is a *block*, and each block is either maximal or minimal; say, wlog, that odd blocks are minimal and even blocks are maximal.

There is no semantic ordering within a block, but the set of blocks is partially ordered; the ordering $1, \ldots, m$ of block indices is a linearization of this partial order. The semantics of such a system is given by treating each block as a vectorial fixpoint of the appropriate sign, with the nesting of fixpoints given by the partial ordering on blocks. By the obvious replacement of higher-numbered right-hand side variables by their defining equations, a modal equation system can be turned into a modal mu-calculus formula defining X_{11}.

Modal equation systems are a concise way of representing properties, as the direct use of variables referring to equations instead of explicit fixpoint formulae allows sharing between equations. For example, if $X_{11} = \phi_{11}$ and $X_{12} = \phi_{12}$ both refer to $X_{21} = \phi_{21}$, then to write the system in linear form we have to write the text of ϕ_{21} twice, once in each ϕ_{1i}; and so one can obtain an exponential (in the number of blocks) blowup by going from equation systems to normal modal mu-calculus. However, results about the modal mu-calculus in its linear form normally transfer without cost (save in intellectual effort) to equation systems.

A *Boolean equation system* (*BES*) is similar, except that the variables are just booleans, not states, and the equations contain only Boolean operators. Thus a BES stands to a formula of Boolean logic with fixpoints[12] as a modal equation system stands to a modal mu-calculus formula. A modal equation system together with a modal structure can be turned into a Boolean equation system, by treating a (state, modal variable) pair as a Boolean variable. Perhaps surprisingly, this is sometimes worth doing: BES's have the advantage that they strip away the modal operators and the artificial linearization imposed by text, laying bare the problem of computing alternating fixpoints. BES's have been studied by, among others, Andersen [1], Vergauwen and Lewi [73] and Mader [44]. Lui, Ramakrishnan and Smolka [42] presented a BES-based local model-checking algorithm that is quite short, relatively simple, and of similar efficiency to global model-checking.

6. Automata, games and logic

There are other formalisms that are intimately related to fixpoint logics, and in this section we shall give an overview of automata and games, which are the main examples. The relationship benefits both parties: the modal mu-calculus provides simple proofs of some automata-theoretic results, and automata theory is used in the proof of modal mu-calculus results.

6.1. *Automata*

The history of automata-theoretic approaches to verification and associated fields has involved ever more invention of new types of automata, as it becomes apparent that some particular restriction is technically inconvenient and can be lifted without harm. In order to explain the main results, we shall need to introduce several of these varieties; let us start at the beginning.

A *deterministic automaton on finite words* is the usual deterministic finite state machine, accepting finite words over its input alphabet; a *run* of such an automaton is a path

[12] A suitable candidate for the title propositional mu-calculus.

through the machine matching the input, and it accepts if the final state is accepting. A *non-deterministic* automaton on finite words is the usual notion, namely that the transition function may specify more than one successor state for a given state and input letter; a *run* of such an automaton requires choosing one successor at each non-deterministic choice, and the automaton accepts its input if some run accepts that input.

There are now several orthogonal generalizations of this idea. Firstly, we may allow infinite words as well as finite words. In this case, we must specify *acceptance conditions* which determine when an infinite run is successful. There are many of these; the most important are as follows. A *Büchi* acceptance condition specifies a set G of states, such that we infinitely often meet a G-state on the run (mnemonic: a green light must flash infinitely often). A *Rabin* acceptance condition specifies k sets R_i and k sets G_i, and requires that for some i, we see G_i infinitely often, and don't see R_i infinitely often (mnemonic: k sets of lights; there must be one set where we see green often and don't see red often). In a *Mostowski* or *parity* acceptance condition, each state of the automaton is assigned an integer *rank*, and the highest rank occurring infinitely often must be even (there is no general agreement on highest/lowest or even/odd in this definition).

In the next dimension, we may feed the automata trees rather than words. To start with, assume that the branching degree is fixed at n. The automaton transition function now specifies n successors, and notionally the automaton splits into n copies as it passes to the successors, so a run is a tree matching the input tree. The run is successful if *all* paths through the run satisfy the acceptance condition, which is specified as above. In the case of a non-deterministic tree automaton, the transition is chosen non-deterministically, but each transition specifies n successors. We can generalize this to automata on trees whose branching degree varies according to the label, or to trees whose branching degree is freely variable (so-called amorphous automata).

Finally, we can complicate the success conditions as follows: instead of saying just that every path through the run accepts, we can say that the subrun rooted at a node n is accepted if some Boolean combination, depending on the state, $f(\{T_i\})$ holds, where T_i is the statement that the subrun rooted at n's ith child is accepted, and that the run is accepted if the subrun starting at the root is accepted, provided the global acceptance condition is met. The global acceptance condition is most easily formulated game-theoretically, as we shall see later, but one can think of it in mu-calculus terms as being: there must exist a choice function which for each node selects a satisfying assignment for $f(\{T_i\})$, and thus selects some children; then every path through the selected nodes must meet the Rabin etc. acceptance condition. These are *alternating automata*. Warning: this alternation is that between $\exists$ and $\forall$; nothing to do with fixpoint alternation!

If the Boolean combination is just conjunction, $f(\{T_i\}) = \bigwedge_i T_i$, we get a normal tree automaton; if it is disjunction, we get a tree automaton that accepts if *some* path accepts, rather than if all paths accept, giving a form of non-determinism. It is probably not much of a surprise that:

THEOREM 9. *Modal mu-calculus formulae are equivalent to alternating amorphous parity automata (on trees).*

The equivalence is simple: a diamond modality corresponds to a node whose Boolean success combination is disjunction, a box modality to one whose success combination is conjunction; and if one assigns ranks to the fixpoint operators that are consistent with subformula inclusion, with even ranks for maximal fixpoints and odd for minimal, the parity condition asserts that the obviously induced tree pre-model is well-founded.

A rather more surprising fact is

THEOREM 10. *Ordinary Rabin automata are equivalent to alternating parity automata (on trees).*

It is obvious that a Rabin automaton can be turned into an alternating parity automaton. However, the reverse construction is not so easy. It is quite easy to turn a parity condition into a Rabin condition, but it is harder to remove the alternation, and a large size blowup is required. In terms of modal mu-calculus formulas this is precisely the transformation into automaton normal form, anf, mentioned earlier, which has no $\forall/\exists$ alternation. As a corollary, we have

THEOREM 11. *Rabin automata are equivalent to modal mu-calculus formulae (on trees).*

The theory of automata on infinite objects is highly developed. Wolfgang Thomas [70] provides a survey of the entire area, and Niwiński's [51] is a fundamental study of automata and fixpoint logics, including much useful background material. Here we shall just mention a few more of the immediate connections with modal mu-calculus.

- In Rabin's original paper, one of the hardest lemmas was proving that Rabin automata are closed under complementation. Given the above, it is now 'obvious', since modal mu-calculus is closed under complementation by definition.
- Rabin automata have the nice property that if they accept some tree, they accept some regular tree: that is, a tree that is the unravelling of a finite system. This gives the finite model property of the modal mu-calculus, as we have seen.
- The emptiness problem for Rabin automata is decidable. Hence one can model-check by forming the product of the system with the Rabin automaton, and checking for emptiness, as we have also seen.

We have mentioned mostly Rabin automata. Büchi automata on trees are strictly less expressive than Rabin automata. However, on words, parity, Rabin and Büchi automata are equivalent, and one has

THEOREM 12. *Rabin/Büchi/parity automata on infinite words are equivalent to linear time mu-calculus formulae.*

The fact that Büchi and parity automata are equivalent on words is the automata-theoretic analogue of the fact that the fixpoint alternation hierarchy collapses in linear-time mu-calculus. On trees, alternating Büchi automata correspond to Π_2^μ modal mu-calculus formulae.

6.2. *Monadic second-order logic*

To continue the story, we bring in monadic second-order logic, studied by Rabin in his original paper. SnS is the monadic second-order logic of the n-ary tree, so that elements are nodes of the tree, the n successor relations are in the logic, first-order quantification over nodes is allowed, and quantification over sets, but not relations, of nodes is allowed. Rabin showed that SnS is equivalent to Rabin automata. It follows that on trees, SnS is equivalent to modal mu-calculus. This is a very curious result, since at first sight, SnS offers much more power. (It is obvious that mu-calculus is a sublogic of SnS since the definitions of greatest and least fixed point are monadic second-order definitions.) For one thing, we can talk about ancestors and cousins of nodes; for another, arbitrary second-order quantification looks more powerful than just fixpoints. But in fact neither of these gives more power.

If we consider transition systems rather than trees, David Janin and Igor Walukiewicz [33,34] have recently developed appropriate generalizations of this theory. We already know that modal mu-calculus respects bisimulation; it now turns out that

THEOREM 13. *A formula in the monadic second-order language of transition systems is expressible in modal mu-calculus exactly if it respects bisimulation.*

6.3. *Games*

Games have played an important role in many areas of mathematics, but especially in logic. The so-called Ehrenfeucht–Fraïssé games characterize first-order logic, and extensions and restrictions of them characterize various extensions and restrictions of first-order logic. In a similar manner, we can define games for the modal mu-calculus. Technically, such games are fairly trivial reformulations of automata or tableaux; but they have good explanatory power, and, more surprisingly, provide useful approaches to model-checking. Owing to the extensive use of games in finite model theory, game formulations of mu-calculus problems assist in exploring the close relationships with finite model theory.

A parity game is played by two players, Abelard and Eloise.[13] An arena for the game is a graph, in which each node is labelled as either an Abelard or Eloise node, and also labelled with a rank, as in parity automata. A play of the game consists of a sequence of moves in which the appropriate player chooses a successor of the current node; Eloise wins if the play ends in Abelard getting stuck, or if the play is infinite and the highest rank occurring infinitely often is even. It may be convenient to designate certain nodes as immediate Abelard/Eloise wins, for example to deal with atomic propositions in the mu-calculus.

A strategy for one player is a function which given a partial play from which the player is due to move, gives the next move. A winning strategy is one which if followed guarantees a win. A strategy is history-free (or memoryless) if it depends only on the current position in the game.

[13] Game presentations are plagued by the variety of names for the players. In Ehrenfeucht–Fraïssé games, they are traditionally called Player I and Player II. In finite-model theory they are called Duplicator and Spoiler, but the meaning of these names does not transfer happily to all situations. Other possibilities are Player and Opponent, or Player 0 and Player 1, or for model checking Verifier and Refuter. Our choice, introduced by Wilfrid Hodges, has the advantage of being formally meaningless, but with a mnemonic for remembering which player is which in applications: ∀belard and ∃loise.

It should be obvious that, for example, modal mu-calculus model-checking can be viewed as a parity game from the discussion in the previous section: the game graph comprises pairs of states and formulae, with transitions given by the dependencies (that is, it is a quasi-model, in the terminology of Section 4.1). Abelard owns the conjunction, box, and maximal fixpoint nodes; Eloise the others; the ranks are assigned as for parity automata. The statement 'Eloise has a winning strategy' is equivalent to the statement that the corresponding alternating parity automaton has an accepting run is equivalent to the statement that the structure satisfies the formula. Equipping the structure with a choice function corresponds to giving Eloise a history-free strategy; and so by the fundamental theorem and duality, a player has a winning strategy iff they have a history-free winning strategy.

References

[1] H.R. Andersen, *Verification of temporal properties of concurrent systems*, DAIMI PB-445, Computer Science Dept., Aarhus University (1993).

[2] H.R. Andersen, C. Stirling and G. Winskel, *A compositional proof system for the modal μ-calculus*, Proc. 9th IEEE LICS (1994), 144–153.

[3] A. Arnold and D. Niwiński, *Fixed point characterization of Büchi automata on infinite trees*, J. Inform. Process. Cybernet. **EIK 26** (1990), 451–459.

[4] J.W. de Bakker, *Mathematical Theory of Program Correctness*, Prentice-Hall, Englewood Cliffs, NJ (1980).

[5] B. Banieqbal and H. Barringer, *Temporal logic with fixed points*, Temporal Logic in Specification, Lecture Notes in Comput. Sci. 398 (1989), 62–74.

[6] H. Barringer, R. Kuiper and A. Pnueli, *Now you may compose temporal logic specifications*, Proc. 16th ACM STOC (1984), 51–63.

[7] H. Bekič, *Definable operations in general algebras, and the theory of automata and flow charts*, Programming Languages and Their Definition, Lecture Notes in Comput. Sci. 177 (1984).

[8] J.C. Bradfield, *Verifying Temporal Properties of Systems*, Birkhäuser, Boston (1991).

[9] J.C. Bradfield, *A proof assistant for symbolic model-checking*, Proc. CAV '92, Lecture Notes in Comput. Sci. 663 (1993), 316–329.

[10] J.C. Bradfield, *The modal mu-calculus alternation hierarchy is strict*, Theoret. Comput. Sci. **195** (1997), 133–153.

[11] J.C. Bradfield, *Simplifying the modal mu-calculus alternation hierarchy*, Proc. STACS '98, Lecture Notes in Comput. Sci. 1373 (1998), 39–49.

[12] J.C. Bradfield, J. Esparza and A. Mader, *An effective tableau system for the linear time mu-calculus*, Proc. ICALP '96, Lecture Notes in Comput. Sci. 1099 (1996), 98–109.

[13] J.C. Bradfield and C.P. Stirling, *Local model checking for infinite state spaces*, Theoret. Comput. Sci. 96 (1992), 157–174.

[14] J.R. Büchi, *On a decision method in restricted second order arithmetic*, Logic, Methodology and Philosophy of Science, Proc. 1960 Congress, Stanford Univ. Press, Stanford, CA (1962), 1–11.

[15] O. Burkart, D. Caucal, F. Moller and B. Steffen, *Verification on infinite structures*, Handbook of Process Algebra, J.A. Bergsta, A. Ponse and S.A. Smolka, eds, Elsevier, Amsterdam (2001), 545–623.

[16] O. Burkart and B. Steffen, *Model checking for context-free processes*, Proc. CONCUR '92, Lecture Notes in Comput. Sci. 630 (1992), 123–137.

[17] E.M. Clarke and E.A. Emerson, *Design and synthesis of synchronization skeletons using branching time temporal logic*, Lecture Notes in Comput. Sci. 131 (1981), 52–71.

[18] E.M. Clarke, E.A. Emerson and A.P. Sistla, *Automatic verification of finite-state concurrent systems using temporal logic specifications*, ACM Trans. Programming Languages and Systems **8** (1986), 244–263.

[19] M. Dam, *CTL* and ECTL* as fragments of the modal mu-calculus*, Theoret. Comput. Sci. **126** (1994), 77–96.

[20] A. de Morgan, *A Budget of Paradoxes*, Longmans, Green, and Co., London (1872).

[21] E.A. Emerson, *Temporal and modal logic*, Handbook of Theoretical Computer Science, Vol. B., J. van Leeuwen, ed., Elsevier (1990), 995–1072.

[22] E.A. Emerson and E.M. Clarke, *Characterizing correctness properties of parallel programs using fixpoints*, Procs ICALP '80, Lecture Notes in Comput. Sci. 85 (1980), 169–181.

[23] E.A. Emerson and C.S. Jutla, *Tree automata, mu-calculus and determinacy*, Proc. 32nd IEEE FOCS (1990), 368–377.

[24] E.A. Emerson, C.S. Jutla and A.P. Sistla, *On model checking for fragments of μ-calculus*, Proc. CAV '93, Lecture Notes in Comput. Sci. 697 (1993), 385–396.

[25] E.A. Emerson and C.-L. Lei, *Efficient model checking in fragments of the propositional mu-calculus*, Proc. 1st IEEE LICS (1986), 267–278.

[26] M.J. Fischer and R.E. Ladner, *Propositional dynamic logic of regular programs*, J. Comput. System Sci. **18** (1979), 194–211.

[27] R. Floyd, *Assigning meanings to programs*, Mathematical Aspects of Computer Science, J.T. Schwartz, ed., Amer. Math. Soc., Providence, RI (1967), 19–32.

[28] Y. Gurevich and L. Harrington, *Trees, automata and games*, Proc. 14th ACM STOC (1982), 60–65.

[29] D. Harel, A. Pnueli and J. Stavi, *Propositional dynamic logic of context-free programs*, Proc. 22nd IEEE FOCS (1981), 310–321.

[30] M. Hennessy and R. Milner, *On observing nondeterminism and concurrency*, Proc. ICALP '80, Lecture Notes in Comput. Sci. 85 (1980), 295–309.

[31] C.A.R. Hoare, *An axiomatic basis for computer programming*, Comm. ACM **12** (1969), 576–580.

[32] H. Hüttel, *SnS can be modally characterized*, Theoret. Comput. Sci. **74** (1990), 239–248.

[33] D. Janin and I. Walukiewicz, *Automata for the μ-calculus and related results*, Proc. MFCS '95, Lecture Notes in Comput. Sci. 969 (1995), 552–562.

[34] D. Janin and I. Walukiewicz, *On the expressive completeness of the propositional mu-calculus with respect to monadic second order logic*, Proc. CONCUR '96, Lecture Notes in Comput. Sci. 1119 (1996), 263–277.

[35] R. Kaivola, *On modal mu-calculus and Büchi tree automata*, Inform. Process. Lett. **54** (1995), 17–22.

[36] R. Kaivola, *Using automata to characterise fixpoint temporal logics*, Ph.D. Thesis, University of Edinburgh (1997).

[37] D. Kozen, *Results on the propositional mu-calculus*, Theoret. Comput. Sci. **27** (1983), 333–354.

[38] D. Kozen, *A finite model theorem for the propositional μ-calculus*, Studia Logica **47** (1988), 233–241.

[39] D. Kozen and R. Parikh, *An elementary proof of the completeness of PDL*, Theoret. Comput. Sci. **14** (1981), 113–118.

[40] L. Lamport, *Proving the correctness of multiprocess programs*, IEEE Trans. Software Engrg. **2** (1977), 125–143.

[41] G. Lenzi, *A hierarchy theorem for the mu-calculus*, Proc. ICALP '96, Lecture Notes in Comput. Sci. 1099 (1996), 87–109.

[42] X. Liu, C.R. Ramakrishnan and S.A. Smolka, *Fully local and efficient evaluation of alternating fixed points*, Proc. TACAS '98, Lecture Notes in Comput. Sci. 1384 (1998), 5–19.

[43] D. Long, A. Browne, E. Clarke, S. Jha and W. Marrero, *An improved algorithm for the evaluation of fixpoint expressions*, Proc. CAV '94, Lecture Notes in Comput. Sci. 818 (1994), 338–350.

[44] A. Mader, *Verification of modal properties using Boolean equation systems*, Ph.D. Thesis, Technical University of Munich (1997).

[45] Z. Manna and A. Pnueli, *Formalization of properties of recursively defined functions*, Proc. ACM STOC (1969), 201–210.

[46] Z. Manna and A. Pnueli, *How to cook a temporal proof system for your pet language*, Proc. 10th ACM POPL (1983), 141–154.

[47] R. Milner, *A Calculus of Communicating Systems*, Lecture Notes in Comput. Sci. 92 (1980).

[48] D. Muller and P. Schupp, *The theory of ends, pushdown automata and second order logic*, Theoret. Comput. Sci. **37** (1985), 51–75.

[49] P. Niebert, *A ν-calculus with local views for systems of sequential agents*, Proc. MFCS '95, Lecture Notes in Comput. Sci. 969 (1995), 563–573.

[50] D. Niwiński, *On fixed point clones*, Proc. ICALP '86, Lecture Notes in Comput. Sci. 226 (1986), 464–473.

[51] D. Niwiński, *Fixed point characterization of infinite behavior of finite state systems*, Theoret. Comput. Sci. **189** (1997), 1–69.

[52] D. Park, *Fixpoint induction and proofs of program properties*, Machine Intelligence **5** (1969), 59–78.
[53] D. Park, *Concurrency and automata on infinite sequences*, Lecture Notes in Comput. Sci. 104 (1981), 167–183.
[54] W. Penczek, *Temporal logics for trace systems*, Internat. J. Found. Comput. Sci. **4** (1) (1993), 31–67.
[55] A. Pnueli, *Temporal semantics of concurrent programs*, Theoret. Comput. Sci. **27** (1983), 333–354.
[56] V. Pratt, *Semantical considerations of Floyd–Hoare logic*, Proc. 16th IEEE FOCS (1976), 109–121.
[57] V. Pratt, *A near-optimal method for reasoning about action*, J. Comput. System Sci. **20** (1980), 231–254.
[58] V. Pratt, *A decidable mu-calculus*, Proc. 22nd IEEE FOCS (1982), 421–427.
[59] M.O. Rabin, *Decidability of second-order theories and automata on infinite trees*, Trans. Amer. Math. Soc. **141** (1969), 1–35.
[60] M.O. Rabin, *Weakly definable relations and special automata*, Mathematical Logic and Foundations of Set Theory, Y. Bar-Hillel, ed., North-Holland, Amsterdam (1970), 1–23.
[61] A.P. Sistla and E.M. Clarke, *Complexity of propositional temporal logics*, J. ACM **32** (1986), 733–749.
[62] C. Stirling, *Modal and temporal logics*, Handbook of Logic in Computer Science, Vol. 2, S. Abramsky, D. Gabbay and T. Maibaum, eds, Oxford University Press (1991), 477–563.
[63] C. Stirling, *Local model checking games*, Proc. Concur '95, Lecture Notes in Comput. Sci. 962 (1995), 1–11.
[64] C. Stirling, *Modal and temporal logics for processes*, Logics for Concurrency, Lecture Notes in Comput. Sci. 1043 (1996), 149–237.
[65] C. Stirling and D. Walker, *Local model checking in the modal mu-calculus*, Theor. Comput. Sci. **89** (1991), 161–177.
[66] C. Stirling and D. Walker, *CCS, liveness, and local model checking in the linear time mu-calculus*, Proc. First International Workshop on Automatic Verification Methods for Finite State Systems, Lecture Notes in Comput. Sci. 407 (1990), 166–178.
[67] R. Streett, *Propositional dynamic logic of looping and converse*, Proc. 13th ACM STOC (1981), 375–383.
[68] R.S. Streett and E.A. Emerson, *An automata theoretic decision procedure for the propositional mu-calculus*, Inform. and Comput. **81** (1989), 249–264.
[69] P.S. Thiagarajan, *A trace based extension of linear time temporal logic*, Proc. 9th IEEE LICS (1994), 438–447.
[70] W. Thomas, *Languages, automata, and logic*, Handbook of Formal Language Theory, Vol. III, G. Rozenberg and A. Salomaa, eds, Springer-Verlag, New York (1998), 389–455.
[71] M.Y. Vardi, *A temporal fixpoint calculus*, Proc. 15th POPL (1988), 250–259.
[72] M.Y. Vardi and P. Wolper, *Yet another process logic*, Logics of Programs, Lecture Notes in Comput. Sci. 164 (1983), 501–512.
[73] B. Vergauwen and J. Lewi, *Efficient local correctness checking for single and alternating Boolean equation systems*, Proc. ICALP '94, Lecture Notes in Comput. Sci. 820 (1994), 302–315.
[74] I. Walukiewicz, *Completeness of Kozen's axiomatisation of the propositional μ-calculus*, Proc. 10th IEEE LICS (1995), 14–24.
[75] I. Walukiewicz, *Notes on the propositional μ-calculus: Completeness and related results*, BRICS Notes Series BRICS-NS-95-1, http://www.brics.dk/BRICS/NS/95/1/BRICS-NS-95-1/index.html.

Subject index

alternation (in automata), 324
alternation depth, 318
alternation, fixpoint, 318
anf, 314
annotation, 311
approximant, 303
automaton normal form, 314

choice function, 311

CTL, 298

disjunctive formula, 314

equation system, 322

fairness, 309
filtration, 299

CHAPTER 5

Process Algebra with Recursive Operations

Jan A. Bergstra[2,3], Wan Fokkink[1], Alban Ponse[1,3]

[1] *CWI, Kruislaan 413, 1098 SJ Amsterdam, The Netherlands*
http://www.cwi.nl

[2] *Utrecht University, Department of Philosophy, Heidelberglaan 8, 3584 CS Utrecht, The Netherlands*
http://www.phil.uu.nl/eng/home.html

[3] *University of Amsterdam, Programming Research Group, Kruislaan 403, 1098 SJ Amsterdam, The Netherlands*
http://www.science.uva.nl/research/prog

E-mails: janb@science.uva.nl, wan@cwi.nl, alban@science.uva.nl

Contents

1. Introduction . . . 335
2. Preliminaries: Axioms and operational semantics . . . 338
 2.1. ACP-based systems . . . 338
 2.2. Transition rules and operational semantics . . . 341
 2.3. Recursive specifications . . . 344
3. Axiomatisation of the binary Kleene star . . . 345
 3.1. Preliminaries . . . 346
 3.2. Completeness . . . 347
 3.3. Irredundancy of the axioms . . . 354
 3.4. Negative results . . . 355
 3.5. Extensions of $BPA^*(A)$ . . . 357
4. Axiomatisations of other iterative operations . . . 358
 4.1. Axiomatisation of no-exit iteration . . . 358
 4.2. Axiomatisation of multi-exit iteration . . . 359
 4.3. Axiomatisation of prefix iteration . . . 361
 4.4. Axiomatisation of string iteration . . . 363
 4.5. Axiomatisation of flat iteration . . . 363
5. Expressivity results . . . 365
 5.1. Expressivity of the binary Kleene star . . . 365
 5.2. Expressivity of multi-exit iteration . . . 370
 5.3. Expressivity of string iteration . . . 371
 5.4. Expressivity of flat iteration . . . 372
6. Non-regular recursive operations . . . 374
 6.1. Process algebra with a push-down operation . . . 375

HANDBOOK OF PROCESS ALGEBRA
Edited by Jan A. Bergstra, Alban Ponse and Scott A. Smolka

6.2. Expressing a stack 378
6.3. Undecidability results 381
7. Special constants 382
7.1. Silent step and fairness 383
7.2. Empty process 384
Acknowledgements 385
References 385
Subject index 388

Abstract

This chapter provides an overview of the addition of various forms of *iteration*, i.e., recursive operations, to process algebra. Of these operations, (the original, binary version of) the Kleene star is considered most basic, and an equational axiomatisation of its combination with basic process algebra is explained in detail.

The focus on iteration in process algebra raised interest in a number of variations of the Kleene star operation, of which an overview, including various completeness and expressivity results, is presented. Though most of these variations concern regular (iterative) operations, also the combination of process algebra and some non-regular operations is discussed, leading to undecidability and stronger expressivity results. Finally, some attention is paid to the interplay between iteration and the special process algebra constants representing the silent step and the empty process.

1. Introduction

In process algebra, a (potentially) infinite process is usually represented as the solution of a system of guarded recursive equations, and proof theory and verification tend to focus on reasoning about such recursive systems. Although specification and verification of concurrent processes defined in this way serve their purposes well, recursive *operations* give a more direct representation and are easier to comprehend. The *Kleene star* $*$ can be considered as the most fundamental recursive operation. In process algebra, the defining equation for the binary Kleene star reads

$$x^*y = x \cdot (x^*y) + y$$

where $\cdot$ models sequential composition and $+$ models non-deterministic choice ($\cdot$ binds stronger than $+$). In terms of operational semantics, the process x^*y chooses between x and y, and upon termination of x has this choice again. For example, the expression a^*b for atomic actions a and b can be depicted by

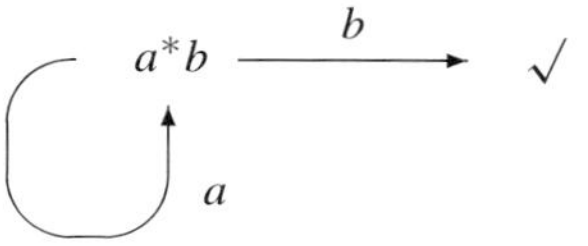

where $\surd$ expresses successful termination. This chapter discusses the Kleene star in the setting of process algebra, and considers some derived recursive operations, with a focus on axiomatisations and expressiveness hierarchies.

In the summer of 1951, S.C. Kleene was supported by the RAND Corporation, leading to Research Memorandum "Representation of Events in Nerve Nets and Finite Automata" [52]. The material in that paper, based on the fundamental paper [59],[1] was republished under the same title five years later [53]. In this seminal work, Kleene introduced the binary operation $*$ for describing 'regular events'. He defined *regular expressions*, which correspond to finite automata, and gave algebraic transformation rules for these, notably

$$E^*F = E(E^*F) \vee F$$

(E^*F being the *iterate* of E on F). Kleene noted the correspondence with conventions of algebra, treating $E \vee F$ as the analogue of $E + F$, and EF as the product of E and F.

In 1958, Copi, Elgot, and Wright [30] showed interest in the results from [53]. However, they judged Kleene's theorems on analysis[2] and synthesis[3] obscured both by the complexity of his basic concepts and by the nature of the elements used in his nets. They introduced

[1] Kleene judged the theory for nerve nets with circles in [59] (McCulloch and Pitts, 1943) to be obscure, and proceeded independently. He came up with "regular events" and the major correspondence results in automata theory.

[2] Theorem 5, stating that finite automata model regular events.

[3] Theorem 3, stating that each regular event can be described by a finite automaton ("nerve net").

simpler and stronger nets (in a sense weakening Kleene's synthesis result, but stating that this "brings the essential nature of the result into sharper focus"), and simpler operations. In particular they introduced a unary $*$ operation

> "[...] because the operation Kleene uses seems "essentially" singulary and because the singulary operation simplifies the algebra of regular events. It should be noted that the singulary and binary star operations are interdefinable." [30, p. 195]

This contradicts Kleene's original argument in [52, p. 50] that the length of an event is at least one, and that for this reason he did not define E^* as a unary operation.

Four years later, Redko [69] proved that there does not exist a sound and complete finite *equational* axiomatisation for regular expressions. (This proof was simplified and corrected by Pilling; see [31, Chapter 11].) In 1966, Salomaa [70] presented a sound and complete finite axiomatisation for regular expressions, with as basic ingredient an *implicational* axiom dating back to Arden [9], namely (in process algebra notation):

$$x = (y \cdot x) + z \Rightarrow x = y^* z$$

if y does not have the so-called *empty word property*. According to Kozen [57], this last property is not algebraic, in the sense that it is not preserved under action refinement; he proposed two alternative implicational axioms that do not have this drawback. Krob [58] settled two conjectures by Conway [31], to obtain an *infinite* sound and complete equational axiomatisation for regular expressions. Bloom and Ésik [25] developed an alternative infinite equational axiomatisation for regular expressions, within the framework of iteration theories.

In 1984, Milner [62] was the first to consider the unary Kleene star in process algebra, modulo *strong bisimulation equivalence* [66]. In contrast with regular expressions, this setting is not sufficiently expressive to describe all *regular* (i.e., finite-state) processes. Moreover, the *merge* operator $x \parallel y$ [60] that executes its two arguments in parallel, and which is fundamental to process algebra, cannot always be eliminated from process terms in the presence of the Kleene star. Milner presented an axiomatisation for the unary Kleene star in strong bisimulation semantics, being a subset of Salomaa's axiomatisation, and asked whether his axiomatisation is complete. Fokkink [39] solved this question for *no-exit iteration* $x^*\delta$, where the special constant δ, called *deadlock*, does not exhibit any behaviour; since δ blocks the exit, $x^*\delta$ executes x infinitely often. Bloom, Ésik, and Taubner [26] presented a complete axiomatisation for regular synchronisation trees modulo strong bisimulation equivalence, within the framework of iteration theories. Milner also asked for a characterisation of those recursive specifications that can be described modulo strong bisimulation semantics in process algebra with the unary Kleene star. Bosscher [28] solved this question in the absence of the deadlock. The two questions by Milner in their full generality remain unsolved.

The unary Kleene star naturally gives rise to the *empty process* [74], which does not combine well with the merge operator. Therefore, Bergstra, Bethke, and Ponse [16,17] returned in 1993 to the binary version x^*y of the Kleene star in process algebra, naming the operator after its discoverer. Troeger [73] introduced a process specification language with iteration, in which he introduced a striking axiom for the binary Kleene star, here presented in process algebra notation:

$$x^*\big(y \cdot \big((x+y)^*z\big) + z\big) = (x+y)^*z.$$

Fokkink and Zantema [38,44] gave an affirmative answer to an open question from [17], namely, that the defining axiom and Troeger's axiom for the binary Kleene star, together with

$$x^*(y \cdot z) = (x^*y) \cdot z$$

and the five standard axioms on + and · for *basic process algebra* [19], provide an equational characterisation of strong bisimulation equivalence.

Corradini, De Nicola, and Labella studied the unary Kleene star in the presence of deadlock modulo *resource bisimulation equivalence*. In [32,33] they presented a complete axiomatisation based on Kozen's conditional axioms. In [34] they came up with a complete equational axiomatisation including Troeger's axiom.

Aceto, Fokkink, and Ingólfsdóttir [4] proved that a whole range of process semantics coarser than strong bisimulation do not allow a finite equational characterisation of the binary Kleene star. Furthermore, Sewell [72] showed that there does not exist a finite equational characterisation of the binary Kleene star modulo strong bisimulation equivalence in the presence of the deadlock δ, due to the fact that $(a^k)^*\delta$ is strongly bisimilar to $a^*\delta$ for positive integers k.

Several variations of the binary Kleene star were introduced, to obtain particular desirable properties.

- In order to increase the expressive power of the binary Kleene star in strong bisimulation semantics, Bergstra, Bethke, and Ponse [16] proposed *multi-exit iteration* $(x_1, \ldots, x_k)^*(y_1, \ldots, y_k)$ for positive integers k, with as defining equation

  $$(x_1, \ldots, x_k)^*(y_1, \ldots, y_k) = x_1 \cdot \big((x_2, \ldots, x_k, x_1)^*(y_2, \ldots, y_k, y_1)\big) + y_1.$$

  Aceto and Fokkink [1] presented an axiomatic characterisation of multi-exit iteration in basic process algebra, modulo strong bisimulation equivalence.
- *Prefix iteration* (similar to the *delay* operation from Hennessy [51]) is obtained by restricting the left-hand side of the binary Kleene star to atomic actions. Aceto, Fokkink, and Ingólfsdóttir [5,36] presented finite equational characterisations of prefix iteration in *basic CCS* [60], modulo a whole range of process semantics.
- *String iteration* [7] is obtained by restricting the left-hand side of the binary Kleene star to non-empty finite strings of atomic actions. Aceto and Groote [7] presented an equational characterisation of string iteration in basic CCS, modulo strong bisimulation equivalence.
- Bergstra, Bethke, and Ponse [16] introduced *flat iteration*, which is obtained by restricting the left-hand side of the binary Kleene star to sums of atomic actions. Unlike the binary Kleene star, the merge operator can be eliminated from process terms in the presence of flat iteration. In [48], van Glabbeek presented a complete finite equational

characterisation of flat iteration in basic CCS extended with the *silent step* τ [60], modulo four rooted weak bisimulation semantics that take into account the silent nature of the special constant τ (identifying $x\tau$ and x).

This chapter also concerns expressivity of (subsystems of) the *algebra of communicating processes* (ACP) [19] extended with the constant τ and *abstraction* operators [20], and enriched with the binary Kleene star or variants thereof. The τ, which represents a silent step, in combination with abstraction operators, which rename actions into τ, enables one to abstract from internal behaviour. In weaker semantics that identify $x\tau$ and x, each *regular* process can be specified in ACP_τ (i.e., ACP with abstraction) and the binary Kleene star, using only *handshaking* (i.e., two-party communication) [22] and some auxiliary actions. Furthermore, in such a semantics, each *computable* process can be specified in ACP_τ with abstraction and a single recursive, non-regular operation, using only handshaking and some auxiliary actions.

This chapter is set up as follows. Section 2 introduces the preliminaries. Section 3 contains an exposition on axiomatisations for the binary Kleene star, while Section 4 discusses axiomatisations for related iterative operations. Section 5 compares the expressivity of these iterative operations, and Section 6 studies the expressivity of some non-regular recursive operations. Finally, Section 7 touches upon topics such as fairness and the empty process.

2. Preliminaries: Axioms and operational semantics

In this section we recall various process algebra axiom systems, structural operational semantics, behavioural equivalences, and recursive specifications. For a detailed introduction to these matters see, e.g., Baeten and Weijland [15].

2.1. *ACP-based systems*

Let A be a finite set of *atomic actions* $a, b, c, \ldots$, and let δ be a constant not in A. We write A_δ for $A \cup \{\delta\}$. Let furthermore $\_ \mid \_ : A_\delta \times A_\delta \to A_\delta$ be a *communication function* that is commutative,

$$a \mid b = b \mid a \quad \text{for all } a, b \in A_\delta,$$

associative,

$$a \mid (b \mid c) = (a \mid b) \mid c \quad \text{for all } a, b, c \in A_\delta,$$

and that satisfies $\delta \mid a = \delta$ for all $a \in A_\delta$. The communication function $\mid$ will be used to define *communication actions*: in the case that $a \mid b = c \in A$, the simultaneous execution of actions a and b results in communication action c. The actions in A and the communication function $\mid$ can be regarded as parameters of the process algebra axiom systems that are presented below.

The process algebraic framework ACP$(A, |, \tau)$ stands for a particular signature over fixed A and communication function $|$, and a set of axioms over this signature. Let $\mathcal{P}$ denote the set of *process terms* over this signature:

sorts:	A	(the given, finite set of atomic actions),
	$\mathcal{P}$	(the set of process terms; $A \subseteq \mathcal{P}$),
operations:	$+ : \mathcal{P} \times \mathcal{P} \to \mathcal{P}$	(non-deterministic choice or sum),
	$\cdot : \mathcal{P} \times \mathcal{P} \to \mathcal{P}$	(sequential composition),
	$\| : \mathcal{P} \times \mathcal{P} \to \mathcal{P}$	(merge, parallel composition),
	$\mathbin{⌊\!⌊} : \mathcal{P} \times \mathcal{P} \to \mathcal{P}$	(left merge),
	$\mid : \mathcal{P} \times \mathcal{P} \to \mathcal{P}$	(communication merge, extending the given communication function),
	$\partial_H : \mathcal{P} \to \mathcal{P}$	(encapsulation, $H \subseteq A$),
	$\tau_I : \mathcal{P} \to \mathcal{P}$	(abstraction, $I \subseteq A$),
constants:	$\delta \in \mathcal{P}$	(deadlock),
	$\tau \in \mathcal{P}$	(silent step).

Intuitively, an action a represents indivisible behaviour, δ represents inaction, and τ represents invisible internal behaviour. Moreover, $P + Q$ executes either P or Q, $P \cdot Q$ first executes P and at its successful termination proceeds to execute Q, and $P \| Q$ executes P and Q in parallel allowing communication of actions from P and Q. The operators $P \mathbin{⌊\!⌊} Q$ and $P \mid Q$ both capture part of the behaviour of $P \| Q$: $P \mathbin{⌊\!⌊} Q$ takes its first transition from P, while the first transition of $P \mid Q$ is a communication of actions from P and Q. Finally, in $\partial_H(P)$ all actions from H in P are blocked, while in $\tau_I(P)$ all actions from I in P are renamed into τ.

We take $\cdot$ to be the operation that binds strongest, and $+$ the one that binds weakest. As usual in algebra, we tend to write xy instead of $x \cdot y$. For $\Box \in \{+, \cdot, \|, |\}$ we will assume that expressions $P_0 \Box \cdots \Box P_n$ associate to the right. Furthermore, for $k \geqslant 1$ we define x^{k+1} as $x \cdot x^k$, and x^1 as x.

In Table 1 the axioms of the system ACP$(A, |, \tau)$ are collected. Note that $+$ and $\cdot$ are associative, and that + is moreover commutative and idempotent. In the case that

$$a \mid (b \mid c) = \delta \quad \text{for all } a, b, c \in A,$$

while $|$ itself defines a communication, we speak of *handshaking*. We will study the following subsystems of ACP$(A, |, \tau)$:

- BPA(A). The signature of BPA(A) contains the elements of A, non-deterministic choice, and sequential composition. The axioms of BPA(A) are (A1)–(A5).
- BPA$_\delta(A)$. The signature of BPA$_\delta(A)$ is the signature of BPA(A) extended with the deadlock. The axioms of BPA$_\delta(A)$ are (A1)–(A7).

Table 1
The axioms for ACP$(A, |, \tau)$, where $a, b \in A_{\delta\tau}$ and $H, I \subseteq A$

(A1)	$x + y$	$=$	$y + x$
(A2)	$x + (y + z)$	$=$	$(x + y) + z$
(A3)	$x + x$	$=$	x
(A4)	$(x + y)z$	$=$	$xz + yz$
(A5)	$(xy)z$	$=$	$x(yz)$
(A6)	$x + \delta$	$=$	x
(A7)	δx	$=$	δ
(C1)	$a \mid b$	$=$	$b \mid a$
(C2)	$(a \mid b) \mid c$	$=$	$a \mid (b \mid c)$
(C3)	$\delta \mid a$	$=$	δ
(CM1)	$x \parallel y$	$=$	$(x \mathbin{⌊\!⌊} y + y \mathbin{⌊\!⌊} x) + x \mid y$
(CM2)	$a \mathbin{⌊\!⌊} x$	$=$	ax
(CM3)	$ax \mathbin{⌊\!⌊} y$	$=$	$a(x \parallel y)$
(CM4)	$(x + y) \mathbin{⌊\!⌊} z$	$=$	$x \mathbin{⌊\!⌊} z + y \mathbin{⌊\!⌊} z$
(CM5)	$ax \mid b$	$=$	$(a \mid b)x$
(CM6)	$a \mid bx$	$=$	$(a \mid b)x$
(CM7)	$ax \mid by$	$=$	$(a \mid b)(x \parallel y)$
(CM8)	$(x + y) \mid z$	$=$	$x \mid z + y \mid z$
(CM9)	$x \mid (y + z)$	$=$	$x \mid y + x \mid z$
(D1)	$\partial_H(a)$	$=$	a if $a \notin H$
(D2)	$\partial_H(a)$	$=$	δ if $a \in H$
(D3)	$\partial_H(x + y)$	$=$	$\partial_H(x) + \partial_H(y)$
(D4)	$\partial_H(xy)$	$=$	$\partial_H(x)\partial_H(y)$
(B1)	$x\tau$	$=$	x
(TI1)	$\tau_I(a)$	$=$	a if $a \notin I$
(TI2)	$\tau_I(a)$	$=$	τ if $a \in I$
(TI3)	$\tau_I(x + y)$	$=$	$\tau_I(x) + \tau_I(y)$
(TI4)	$\tau_I(xy)$	$=$	$\tau_I(x)\tau_I(y)$

- PA(A). The signature of PA(A) is the signature of BPA(A) extended with the merge and left merge. The axioms of PA(A) are (A1)–(A5), (CM2)–(CM4) (with a ranging over A), and

$$\text{(M1)} \quad x \parallel y = x \mathbin{⌊\!⌊} y + y \mathbin{⌊\!⌊} x.$$

- PA$_\delta(A)$. The signature of PA$_\delta(A)$ is the signature of PA(A) extended with the deadlock. The axioms of PA(A) are (A1)–(A7), (M1), and (CM2)–(CM4) (with a ranging over A_δ).
- ACP$(A, |)$. The signature of ACP$(A, |)$ is the signature of ACP$(A, |, \tau)$ *without* the silent step and abstraction operators. The axioms of ACP$(A, |)$ are (A1)–(A7), (CF1)–(CF2), (CM1)–(CM9), and (D1)–(D4) (with a, b ranging over A_δ).

We note that in PA(A) and PA$_\delta(A)$, commutativity of the merge ($x \parallel y = y \parallel x$) can be derived from axioms (A1) and (M1).

The binary equality relation $=$ on process terms induced by an axiom system is obtained by taking all closed instantiations of axioms, and closing it under equivalence (i.e., under reflexivity, symmetry, and transitivity) and under contexts.

2.2. *Transition rules and operational semantics*

We define a structural operational semantics in the style of Plotkin [68], to relate each process term to a labelled transition system. Then we define strong bisimulation as an equivalence between labelled transition systems, which carries over to process terms. The operational semantics and strong bisimulation are used in the proofs on axiomatisations in Sections 3 and 4, and on classification results in Section 5.

A *labelled transition system* (LTS) is a tuple $\langle S, \{\xrightarrow{a}, \xrightarrow{a} \surd \mid a \in A\}, s\rangle$, where

S is a set of *states*,
$\xrightarrow{a}$ for $a \in A$ is a binary relation between states,
$\xrightarrow{a} \surd$ for $a \in A$ is a unary predicate on states,
$s \in S$ is the *initial state*.

Expressions $s \xrightarrow{a} s'$ and $s \xrightarrow{a} \surd$ are called *transitions*.

Intuitively, $s \xrightarrow{a} s'$ denotes that from state s one can evolve to state s' by the execution of action a, while $s \xrightarrow{a} \surd$ denotes that from state s one can terminate successfully by the execution of action a ($\surd$ is pronounced "tick").

Consider one of the process algebra axiom systems BPA(A) – ACP(A, |, τ), and let $\mathcal{P}$ represent all process terms given by its signature. We want to relate each process term in $\mathcal{P}$ to a labelled transition system. We take the process terms in $\mathcal{P}$ as the set of states, and the atomic actions in A as the set of labels. (Note that atomic actions can denote both states and labels.) Exploiting the syntactic structure of process terms, the transition relations $\xrightarrow{a}$ and $\xrightarrow{a} \surd$ for $a \in A$ are defined by means of inductive proof rules called *transition rules* $\frac{S}{c}$. Validity of the *premises* in S, under a certain substitution, implies validity of the *conclusion* c under the same substitution.

The transition rules in Table 2 define the labelled transition system associated to a process term in ACP(A, |, τ). The signature and parameters of $\mathcal{P}$ (possibly including a communication function |) determine which transition rules are appropriate. For example, the last two transition rules for $\|$ (i.e., with $x \parallel y$ in the left-hand sides of their conclusions) are not relevant for $\mathrm{PA}_\delta(A)$. Note that the deadlock δ has no outgoing transitions. The labelled transition system related to process term P has P itself as initial state. Often we will write simply P for the labelled transition system related to P.

Intuitively, *strong bisimulation* relates two states if the LTSs rooted at these states have the same branching structure. This semantics does not take into account the silent nature of τ.

DEFINITION 2.2.1 (*Strong bisimulation*). A *strong bisimulation* is a binary, symmetric relation $\mathcal{R}$ over the set of states that satisfies

$$P \,\mathcal{R}\, Q \wedge P \xrightarrow{a} P' \Rightarrow \exists Q'(Q \xrightarrow{a} Q' \wedge P' \,\mathcal{R}\, Q'),$$
$$P \,\mathcal{R}\, Q \wedge P \xrightarrow{a} \surd \Rightarrow Q \xrightarrow{a} \surd.$$

Two states P and Q are *strongly bisimilar*, notation $P \leftrightarrows Q$, if there exists a strong bisimulation relation $\mathcal{R}$ with $P \,\mathcal{R}\, Q$.

Table 2
Transition rules for $\mathrm{ACP}(A, |, \tau)$, where $a, b \in A_\tau,\ H, I \subseteq A$

$$a \xrightarrow{a} \surd, \quad a \in A_\tau$$

$$\frac{x \xrightarrow{a} x'}{x + y \xrightarrow{a} x'} \qquad \frac{y \xrightarrow{a} y'}{x + y \xrightarrow{a} y'} \qquad \frac{x \xrightarrow{a} \surd}{x + y \xrightarrow{a} \surd} \qquad \frac{y \xrightarrow{a} \surd}{x + y \xrightarrow{a} \surd}$$

$$\frac{x \xrightarrow{a} x'}{xy \xrightarrow{a} x'y} \qquad \frac{x \xrightarrow{a} \surd}{xy \xrightarrow{a} y}$$

$$\frac{x \xrightarrow{a} x'}{x \parallel y \xrightarrow{a} x' \parallel y} \qquad \frac{y \xrightarrow{a} y'}{x \parallel y \xrightarrow{a} x \parallel y'} \qquad \frac{x \xrightarrow{a} \surd}{x \parallel y \xrightarrow{a} y} \qquad \frac{y \xrightarrow{a} \surd}{x \parallel y \xrightarrow{a} x}$$

$$\frac{x \xrightarrow{a} x' \quad y \xrightarrow{b} y'}{x \parallel y \xrightarrow{a|b} x' \parallel y'} \ \text{if } a \mid b \in A \qquad \frac{x \xrightarrow{a} \surd \quad \xrightarrow{b} y\surd}{x \parallel y \xrightarrow{a|b} \surd} \ \text{if } a \mid b \in A$$

$$\frac{x \xrightarrow{a} \surd \quad y \xrightarrow{b} y'}{x \parallel y \xrightarrow{a|b} y'} \ \text{if } a \mid b \in A \qquad \frac{x \xrightarrow{a} x' \quad y \xrightarrow{b} \surd}{x \parallel y \xrightarrow{a|b} x'} \ \text{if } a \mid b \in A$$

$$\frac{x \xrightarrow{a} x'}{x \mathbin{\lfloor\!\lfloor} y \xrightarrow{a} x' \parallel y} \qquad \frac{x \xrightarrow{a} \surd}{x \mathbin{\lfloor\!\lfloor} y \xrightarrow{a} y}$$

$$\frac{x \xrightarrow{a} x' \quad y \xrightarrow{b} y'}{x \mid y \xrightarrow{a|b} x' \parallel y'} \qquad \frac{x \xrightarrow{a} \surd \quad y \xrightarrow{b} y'}{x \mid y \xrightarrow{a|b} y'} \qquad \frac{x \xrightarrow{a} x' \quad y \xrightarrow{b} \surd}{x \mid y \xrightarrow{a|b} x'} \qquad \frac{x \xrightarrow{a} \surd \quad y \xrightarrow{b} \surd}{x \mid y \xrightarrow{a|b} \surd}$$

$$\frac{x \xrightarrow{a} x'}{\partial_H(x) \xrightarrow{a} \partial_H(x')} \ \text{if } a \notin H \qquad \frac{x \xrightarrow{a} \surd}{\partial_H(x) \xrightarrow{a} \surd} \ \text{if } a \notin H$$

$$\frac{x \xrightarrow{a} x'}{\tau_I(x) \xrightarrow{\tau} \tau_I(x')} \ \text{if } a \in I \qquad \frac{x \xrightarrow{a} \surd}{\tau_I(x) \xrightarrow{\tau} \surd} \ \text{if } a \in I$$

$$\frac{x \xrightarrow{a} x'}{\tau_I(x) \xrightarrow{a} \tau_I(x')} \ \text{if } a \notin I \qquad \frac{x \xrightarrow{a} \surd}{\tau_I(x) \xrightarrow{a} \surd} \ \text{if } a \notin I$$

THEOREM 2.2.2.

(1) (Equivalence) *It is not hard to see that strong bisimulation is an* equivalence *relation over* $\mathrm{ACP}(A, |)$.

(2) (Congruence) *Strong bisimulation equivalence is a* congruence *relation up to* $\mathrm{ACP}(A, |)$, *meaning that* $P_0 \leftrightarrow Q_0$ *and* $P_1 \leftrightarrow Q_1$ *implies* $P_0 + P_1 \leftrightarrow Q_0 + Q_1$, $P_0P_1 \leftrightarrow Q_0Q_1$, $P_0 \parallel P_1 \leftrightarrow Q_0 \parallel Q_1$, $P_0 \mathbin{\lfloor\!\lfloor} P_1 \leftrightarrow Q_0 \mathbin{\lfloor\!\lfloor} Q_1$, $P_0 \mid P_1 \leftrightarrow Q_0 \mid Q_1$, *and* $\partial_H(P_0) \leftrightarrow \partial_H(Q_0)$. *This follows from the fact that the transition rules in Table* 2 *are in* path *format; see* [14,43,50] *or Chapter* 3 [6] *in this handbook.*

(3) (Soundness) *Up to and including* $\mathrm{ACP}(A, |)$, *all axiom systems are* sound *with respect to strong bisimilation equivalence, meaning that* $P = Q$ *implies* $P \leftrightarrow Q$. *Since strong bisimulation is a congruence, soundness follows from the fact that all closed instantiations of axioms in* $\mathrm{ACP}(A, |)$ *are valid in strong bisimulation semantics.*

(4) (Completeness) *Up to and including* $\mathrm{ACP}(A, |)$, *all axiom systems are complete with respect to strong bisimilation equivalence, meaning that* $P \leftrightarrow Q$ *implies* $P = Q$; *see, e.g.,* [15,42].

We proceed to define some more semantic equivalence relations on states in a labelled transition system that do not take into account the silent nature of τ. These definitions use the following notions, assuming an underlying LTS.

Let $P \xrightarrow{\sigma}\!\!\!\!\rightarrow Q$ for $\sigma \in A^*$ denote that state P can evolve to state Q by the execution of the sequence of actions σ. This binary relation on states is defined as follows (with ε denoting the empty string):

- $P \xrightarrow{\varepsilon}\!\!\!\!\rightarrow P$;
- $\dfrac{P \xrightarrow{\sigma}\!\!\!\!\rightarrow Q \quad Q \xrightarrow{a} R}{P \xrightarrow{\sigma a}\!\!\!\!\rightarrow R}$;
- $\dfrac{P \xrightarrow{\sigma}\!\!\!\!\rightarrow Q \quad Q \xrightarrow{a} \surd}{P \xrightarrow{\sigma a}\!\!\!\!\rightarrow \surd}$.

A string $\sigma \in A^*$ is a *trace* of P if $P \xrightarrow{\sigma}\!\!\!\!\rightarrow Q$ for some state Q or $P \xrightarrow{\sigma}\!\!\!\!\rightarrow \surd$.

DEFINITION 2.2.3 (*Substate*). Q is a *substate* of P if $P \xrightarrow{\sigma}\!\!\!\!\rightarrow Q$ for some $\sigma \in A^*$. Q is a *proper substate* of P if $P \xrightarrow{\sigma}\!\!\!\!\rightarrow Q$ for some $\sigma \in A^* \setminus \{\varepsilon\}$.

DEFINITION 2.2.4 (*Ready simulation*). A *simulation* is a binary relation $\mathcal{R}$ over the set of states that satisfies:

$$P\,\mathcal{R}\,Q \wedge P \xrightarrow{a} P' \Rightarrow \exists Q'(Q \xrightarrow{a} Q' \wedge P'\,\mathcal{R}\,Q'),$$
$$P\,\mathcal{R}\,Q \wedge P \xrightarrow{a} \surd \Rightarrow Q \xrightarrow{a} \surd.$$

A simulation $\mathcal{R}$ is a *ready simulation* if whenever $P\,\mathcal{R}\,Q$ and a is a trace of Q, then a is a trace of P.

Two states P and Q are *simulation equivalent*, notation $P \simeq_S Q$, if $P\,\mathcal{R}_1\,Q$ and $Q\,\mathcal{R}_2\,P$ for simulations $\mathcal{R}_1$ and $\mathcal{R}_2$.

Two states P and Q are *ready simulation equivalent*, notation $P \simeq_{RS} Q$, if $P\,\mathcal{R}_1\,Q$ and $Q\,\mathcal{R}_2\,P$ for ready simulations $\mathcal{R}_1$ and $\mathcal{R}_2$.

DEFINITION 2.2.5 (*Language equivalence*). Two states P and Q are *language equivalent*, notation $P \simeq_L Q$, if for each trace $P \xrightarrow{\sigma}\!\!\!\!\rightarrow \surd$ there is a trace $Q \xrightarrow{\sigma}\!\!\!\!\rightarrow \surd$, and vice versa.

DEFINITION 2.2.6 (*Trace equivalence*). Two states P and Q are *trace equivalent*, notation $P \simeq_T Q$, if they give rise to the same set of traces.

In [45], van Glabbeek gave a comparison of a wide range of behavioural equivalences and *preorders* (i.e., relations that are in general not symmetric) that do not take into account the silent nature of τ. Apart from the equivalence relation discussed above, he studied *completed trace* preorder and a variety of *decorated trace* preorders, which are based on decorated versions of traces. These preorders are coarser than strong bisimulation and ready simulation, but more refined than language and trace equivalence. In Section 4.3, prefix iteration is axiomatized with respect to some of the decorated trace preorders. The reader is referred to [45] for the definitions of these preorders.

3.1. *Preliminaries*

$BPA^*(A)$ is obtained by extending $BPA(A)$ with BKS. Bergstra, Bethke, and Ponse [17] introduced an axiomatisation for $BPA^*(A)$ modulo strong bisimulation equivalence, which consists of axioms (A1)–(A5) for $BPA(A)$, extended with the axioms (BKS1)–(BKS) for BKS in Table 4. The axiom (BKS3) stems from Troeger [73].

In order to provide process terms over $BPA^*(A)$ with an operational semantics, we introduce transition rules for BKS. The transition rules for BKS in Table 5 express that x^*y repeatedly executes x until it executes y. Together with the transition rules for $BPA(A)$ in Table 2 they provide labelled transition systems to process terms over $BPA^*(A)$. Note that by the first two transition rules in Table 5, a state P can have itself as a proper substate. For example, $a^*b \xrightarrow{a} a^*b$.

The transition rules for BKS are in *path* format [14,50]. Hence, strong bisimulation equivalence is a congruence with respect to $BPA^*(A)$; see [43] or Chapter 3 [6] in this handbook. Furthermore, its axiomatisation is sound for $BPA^*(A)$ modulo strong bisimulation equivalence. Since strong bisimulation equivalence is a congruence, this can be verified by checking soundness for each axiom separately. It can be easily shown that the BKS axioms are valid in strong bisimulation.

Fokkink and Zantema [44] proved that the axiomatisation for $BPA^*(A)$ is complete modulo strong bisimulation equivalence. Their proof is based on a *term rewriting* analysis (see, e.g., [10]), in a quest to reduce bisimilar process terms to the same *ground normal form*, which does not reduce any further. Since this aim cannot be fulfilled for BKS, this operator is replaced by an operator representing $x(x^*y)$, and the BKS axioms are adopted to fit this new operator. Those axioms are turned into conditional rewrite rules, which are applied *modulo associativity and commutativity* of the + (see, e.g., [67]). *Knuth–Bendix completion* [54] is applied to make the conditional rewrite system *weakly confluent*. *Termination* of the resulting conditional term rewriting system is obtained by means of the technique of *semantic labelling* from Zantema [75]. Hence, each process term is provably equal to a normal form. Finally, a careful case analysis learns that if two normal forms are strongly bisimilar, then they are syntactically equal modulo associativity and commutativity of the +. This observation yields the desired completeness result.

Table 4
Axioms for BKS

(BKS1)	$x(x^*y) + y = x^*y$
(BKS2)	$x^*(yz) = (x^*y)z$
(BKS3)	$x^*(y((x+y)^*z) + z) = (x+y)^*z$

Table 5
Transition rules for BKS

$$\frac{x \xrightarrow{a} x'}{x^*y \xrightarrow{a} x'(x^*y)} \qquad \frac{x \xrightarrow{a} \surd}{x^*y \xrightarrow{a} x^*y} \qquad \frac{y \xrightarrow{a} y'}{x^*y \xrightarrow{a} y'} \qquad \frac{y \xrightarrow{a} \surd}{x^*y \xrightarrow{a} \surd}$$

An alternative completeness proof was proposed in [38], based on induction on the structure of process terms. That proof method is more general, and was later on applied to obtain completeness results for axiomatisations of iteration operations in [1,34,39]. In the light of the generic applicability of this proof method and the significance of the completeness result in the realm of this chapter, we present the proof from [38] in some detail.

Following Milner [64] (see also [46]), the latter proof strategy can also be used to derive *ω-completeness* of the axiomatisation for $\mathrm{BPA}^*(A)$. That is, if P and Q are *open* terms over $\mathrm{BPA}^*(A)$, which may contain variables, and if $\sigma(P)=\sigma(Q)$ holds for all closed instantiations σ, then $P=Q$ can be derived from the axioms. Often, ω-completeness can be proved by providing variables with an operational semantics, such that $P \leftrightarrow Q$ holds with respect to this new operational semantics if and only if $\sigma(P) \leftrightarrow \sigma(Q)$ holds for all closed instantiations σ with respect to the original operational semantics. In [38], completeness of the axiomatisation for $\mathrm{BPA}^*(A)$ modulo bisimulation equivalence is derived for open terms, which immediately implies ω-completeness of the axiomatisation. Here, we present the proof from [38] for closed (instead of open) terms. The reader is referred to [38] for a proof of ω-completeness. (The motivation to refrain from this generalisation here is clarity of presentation; we prefer to work in an unambiguous semantic framework throughout this chapter.)

3.2. *Completeness*

We note that each process term over $\mathrm{BPA}^*(A)$ has only finitely many substates. In the sequel, process terms are considered modulo associativity and commutativity of the $+$, and we write $P =_{\mathrm{AC}} Q$ if P and Q can be equated by axioms (A1) and (A2). As usual, $\sum_{i=1}^{n} P_i$ represents $P_1+\cdots+P_n$. We take care to avoid empty sums (where $\sum_{i=1}^{0} P_i + Q$ is not considered empty and equals Q).

For each process term P, its collection of possible transitions is non-empty and finite, say $\{P \xrightarrow{a_i} P_i \mid i=1,\ldots,m\} \cup \{P \xrightarrow{b_j} \surd \mid j=1,\ldots,n\}$. We call

$$\sum_{i=1}^{m} a_i P_i + \sum_{j=1}^{n} b_j$$

the *HNF-expansion* of P (Head Normal Form expansion, cf. [15]). The process terms $a_i P_i$ and b_j are the *summands* of P.

LEMMA 3.2.1. *Each process term is provably equal to its HNF-expansion.*

PROOF. Straightforward, by structural induction, using (A4), (A5), and (BKS1). □

Process terms in $\mathrm{BPA}^*(A)$ are *normed*, which means that they are able to terminate in finitely many transitions. The *norm* [12] of a process yields the length of the shortest termination trace of this process. Norm can be defined inductively by

$$|a| = 1,$$

$$
\begin{aligned}
|P+Q| &= \min\{|P|, |Q|\}, \\
|PQ| &= |P| + |Q|, \\
|P^* Q| &= |Q|.
\end{aligned}
$$

We note that strongly bisimilar processes have the same norm. The following lemma, due to Caucal [29], is typical for normed processes.

LEMMA 3.2.2. *Let $PQ \leftrightarrow RS$. By symmetry we may assume $|Q| \leqslant |S|$. We can distinguish two cases*:

- *either $P \leftrightarrow R$ and $Q \leftrightarrow S$;*
- *or there is a proper substate P' of P such that $P \leftrightarrow RP'$ and $P'Q \leftrightarrow S$.*

PROOF. This lemma follows from the following Facts A and B.

FACT A. *If $PQ \leftrightarrow RS$ and $|Q| \leqslant |S|$, then either $Q \leftrightarrow S$, or there is a proper substate P' of P such that $P'Q \leftrightarrow S$.*

PROOF. We prove Fact A by induction on $|P|$. First, let $|P| = 1$. Then $P \xrightarrow{a} \surd$ for some a, so $PQ \xrightarrow{a} Q$. Since $PQ \leftrightarrow RS$, and $|R'S| > |S| \geqslant |Q|$ for all substates R' of R, it follows that $R \xrightarrow{a} \surd$ and $Q \leftrightarrow S$. Then we are done.

Next, suppose we have proved the case for $|P| \leqslant n$, and let $|P| = n+1$. Then there is a P' with $|P'| = n$ and $P \xrightarrow{a} P'$, which implies $PQ \xrightarrow{a} P'Q$. Since $PQ \leftrightarrow RS$, we have two options:

(1) $R \xrightarrow{a} \surd$ and $P'Q \leftrightarrow S$. Then we are done.

(2) $R \xrightarrow{a} R'$ and $P'Q \leftrightarrow R'S$. Since $|P'| = n$, induction yields either $Q \leftrightarrow S$ or $P''Q \leftrightarrow S$ for a proper substate P'' of P'. Again, we are done.

This concludes the proof of Fact A. □

FACT B. *If $PQ \leftrightarrow RQ$, then $P \leftrightarrow R$.*

PROOF. Define a binary relation $\mathcal{B}$ on process terms by $T \mathcal{B} U$ iff $TQ \leftrightarrow UQ$. We show that $\mathcal{B}$ constitutes a strong bisimulation relation between P and R.

- Since $\leftrightarrow$ is symmetric, so is $\mathcal{B}$.
- $PQ \leftrightarrow RQ$, so $P \mathcal{B} R$.
- Suppose $T \mathcal{B} U$ and $T \xrightarrow{a} \surd$. Then $TQ \xrightarrow{a} Q$. Since $TQ \leftrightarrow UQ$, and $|U'Q| > Q$ for all substates U' of U, it follows that $UQ \xrightarrow{a} Q$. In other words, $U \xrightarrow{a} \surd$.
- Suppose $T \mathcal{B} U$ and $T \xrightarrow{a} T'$. Then $TQ \xrightarrow{a} T'Q$. Since $TQ \leftrightarrow UQ$, and $|Q| < |T'Q|$, it follows that there is a transition $U \xrightarrow{a} U'$ with $T'Q \leftrightarrow U'Q$. Hence, $T' \mathcal{B} U'$.

This concludes the proof of Fact B. □

Finally, we show that Facts A and B together prove the lemma. Let $PQ \leftrightarrow RS$ with $|Q| \leqslant |S|$. According to Fact A we can distinguish two cases.

- $Q \leftrightarrow S$. Then $PQ \leftrightarrow RS \leftrightarrow RQ$, so Fact B yields $P \leftrightarrow R$.

- $P'Q \leftrightarrow S$ for some proper substate P' of P. Then $PQ \leftrightarrow RS \leftrightarrow RP'Q$, so Fact B yields $P \leftrightarrow RP'$. □

We construct a set $\mathbb{B}$ of *basic terms*, such that each process term is provably equal to a basic term. The completeness theorem is proved by showing that strongly bisimilar basic terms are provably equal.

$$\begin{aligned}(x+y)z &\to xz+yz,\\ (xy)z &\to x(yz),\\ (x^*y)z &\to x^*(yz).\end{aligned}$$

The term rewriting system above consists of directions of the axioms (A4), (A5), and (BKS2), pointing from left to right. Its rewrite rules are to be interpreted modulo associativity and commutativity of the $+$. The term rewriting system is terminating, meaning that there are no infinite reductions. This follows from the following weight function w in the natural numbers:

$$\begin{aligned}w(a) &= 2,\\ w(P+Q) &= w(P)+w(Q),\\ w(PQ) &= w(P)^2 w(Q),\\ w(P^*Q) &= w(P)+w(Q).\end{aligned}$$

It is not hard to see that if P reduces to Q in one or more rewrite steps, then $w(P) > w(Q)$. Since the ordering on the natural numbers is well-founded, we conclude that the term rewriting system is terminating. Let $\mathbb{G}$ denote the collection of *ground normal forms*, i.e., the collection of process terms that cannot be reduced by any of the three rewrite rules. Since the term rewriting system is terminating, and since its rewrite rules are directions of axioms, it follows that each process term is provably equal to a process term in $\mathbb{G}$. The elements in $\mathbb{G}$ are defined by:

$$P ::= a \mid P+P \mid aP \mid P^*P.$$

$\mathbb{G}$ is not yet our desired set of basic terms, due to the fact that there exist process terms in $\mathbb{G}$ which have a substate outside $\mathbb{G}$. We give an example.

EXAMPLE 3.2.3. Let $A = \{a, b, c\}$. Clearly, $(a^*b)^*c \in \mathbb{G}$, and

$$(a^*b)^*c \xrightarrow{a} (a^*b)\big((a^*b)^*c\big).$$

The substate $(a^*b)((a^*b)^*c)$ is not in $\mathbb{G}$, because the third rewrite rule in $\mathcal{R}$ reduces this process term to $a^*(b((a^*b)^*c))$.

In order to overcome this complication, we introduce the following collection of process terms:

$$\mathbb{H} = \{P^*Q,\ P'(P^*Q) \mid P^*Q \in \mathbb{G} \text{ and } P' \text{ is a proper substate of } P\}.$$

We define an equivalence relation $\cong$ on $\mathbb{H}$ by putting $P'(P^*Q) \cong P^*Q$ for proper substates P' of P, and taking the reflexive, symmetric, transitive closure of $\cong$.

The set $\mathbb{B}$ of *basic terms* is the union of $\mathbb{G}$ and $\mathbb{H}$.

LEMMA 3.2.4. *If $P \in \mathbb{B}$ and $P \xrightarrow{a} P'$, then $P' \in \mathbb{B}$.*

PROOF. By induction on the structure of P.

If $P \in \mathbb{H} \setminus \mathbb{G}$, then it is of the form $Q'(Q^*R)$ for some $Q^*R \in \mathbb{G}$. So P' is of the form either Q^*R or $Q''(Q^*R)$ for some proper substate Q'' of Q'. In both cases, $P' \in \mathbb{B}$.

If $P \in \mathbb{G}$, then it is of the form $\sum_i a_i Q_i + \sum_j R_j^* S_j + \sum_k b_k$, where the Q_i, R_j, and S_j are in $\mathbb{G}$. So P' is of the form Q_i, $R_j^* S_j$, $R'_j(R_j^* S_j)$, or S'_j, which are all basic terms (in the last case, this follows by structural induction). □

The *L-value* [44] of a process term is defined by

$$L(P) = \max\{|P'| \mid P' \text{ proper substate of } P\}.$$

$L(P) < L(PQ)$ and $L(P) < L(P^*Q)$, because for each proper substate P' of P, $P'Q$ is a proper substate of PQ and $P'(P^*Q)$ is a proper substate of P^*Q. Since norm is preserved under strong bisimulation, it follows that the same holds for L-value; i.e., if $P \leftrightarrow Q$ then $L(P) = L(Q)$.

We define an ordering $\prec$ on $\mathbb{B}$ as follows:

- $P \prec Q$ if $L(P) < L(Q)$;
- $P \prec Q$ if P is a substate of Q but Q is not a substate of P;
- if $P \prec Q$ and $Q \prec R$, then $P \prec R$.

Note that if $P, Q \in \mathbb{H}$ with $P \cong Q$, then P and Q have the same proper substates, and so $L(P) = L(Q)$. These observations imply that the ordering $\prec$ on $\mathbb{B}$ respects the equivalence $\cong$ on $\mathbb{H}$, that is, if $P \cong Q \prec R \cong S$, then $P \prec S$.

LEMMA 3.2.5. *$\prec$ is a well-founded ordering on $\mathbb{B}$.*

PROOF. If P is a substate of Q, then all proper substates of P are proper substates of Q, so $L(P) \leqslant L(Q)$. Hence, if $P \prec Q$ then $L(P) \leqslant L(Q)$.

Assume, toward a contradiction, that there exists an infinite backward chain $\cdots \prec P_2 \prec P_1 \prec P_0$. Since $L(P_{n+1}) \leqslant L(P_n)$ for all $n \in \mathbb{N}$, there is an N such that $L(P_n) = L(P_N)$ for all $n > N$. Since $P_n \prec P_N$ for $n > N$, it follows that P_n is a substate of P_N for $n > N$. Each process term has only finitely many substates, so there are $m, n > N$ with $n > m$ and $P_n =_{AC} P_m$. Then $P_n \not\prec P_m$, so we have found a contradiction. Hence, $\prec$ is well-founded. □

In the proofs of the next two lemmas, we need a weight function g in the natural numbers, which is defined inductively by

$$\begin{aligned} g(a) &= 0, \\ g(P+Q) &= \max\{g(P), g(Q)\}, \\ g(PQ) &= \max\{g(P), g(Q)\}, \\ g(P^*Q) &= \max\{g(P), g(Q)+1\}. \end{aligned}$$

It is not hard to see, by structural induction, that if $P \xrightarrow{a} P'$, then $g(P) \geqslant g(P')$.

LEMMA 3.2.6. *Let $P^*Q \in \mathbb{B}$. If Q' is a proper substate of Q, then $Q' \prec P^*Q$.*

PROOF. Q' is a substate of Q, so $g(Q') \leqslant g(Q) < g(P^*Q)$, which implies that P^*Q cannot be a substate of Q'. On the other hand, Q' is a substate of P^*Q, so $Q' \prec P^*Q$. □

LEMMA 3.2.7. *If $P \in \mathbb{B}$ and $P \xrightarrow{a} P'$, then either $P' \prec P$, or $P, P' \in \mathbb{H}$ and $P \cong P'$.*

PROOF. This lemma follows from the following Facts A and B.

FACT A. *If $P \in \mathbb{B}$ and $P \xrightarrow{a} P'$, then either $P' \in \mathbb{H}$ or P' has smaller size than P.*

PROOF. We prove Fact A by induction on the structure of P. Let

$$P =_{\mathrm{AC}} \sum_i a_i Q_i + \sum_j R_j^* S_j + \sum_k b_k.$$

Since $P \xrightarrow{a} P'$, P' is of one of the following forms.

- $P' =_{\mathrm{AC}} Q_i$ for some i. Then P' has smaller size than P.
- $P' =_{\mathrm{AC}} R_j'(R_j^* S_j)$ or $P' =_{\mathrm{AC}} R_j^* S_j$ for some j. Then $P' \in \mathbb{H}$.
- $S_j \xrightarrow{a} P'$ for some j. Then induction yields that either $P' \in \mathbb{H}$, or P' has smaller size than S_j, which in turn has smaller size than P.

This concludes the proof of Fact A. □

FACT B. *If $P \in \mathbb{H}$ and $P \xrightarrow{a} P'$, then either $g(P) > g(P')$, or $P' \in \mathbb{H}$ and $P \cong P'$.*

PROOF. Since $P \in \mathbb{H}$, either $P =_{\mathrm{AC}} Q'(Q^*R)$ or $P =_{\mathrm{AC}} Q^*R$ for some Q and R. Hence, $P' =_{\mathrm{AC}} Q''(Q^*R)$, $P' =_{\mathrm{AC}} Q^*R$, or $P' =_{\mathrm{AC}} R'$ for a proper substate R' of R. In the first two cases $P' \in \mathbb{H}$ and $P \cong P'$, and in the last case $g(P') = g(R') \leqslant g(R) < g(Q^*R) = g(P)$. This concludes the proof of Fact B. □

Finally, we show that Facts A and B together prove the lemma. Let $P \xrightarrow{a} P'$ with $P' \not\prec P$; we prove that $P, P' \in \mathbb{H}$ and $P \cong P'$.

Since P' is a substate of P and $P' \not\prec P$, P is a substate of P'. So there exists a sequence of transitions

$$P_0 \xrightarrow{a_1} P_1 \xrightarrow{a_2} \cdots \xrightarrow{a_n} P_n \quad (n \geqslant 1)$$

where $P_0 =_{AC} P$, $P_1 =_{AC} P'$, and $P_n =_{AC} P$.

Suppose $P_k \notin \mathbb{H}$ for all k. Then according to Fact A, P_{k+1} has smaller size than P_k for $k = 0, \ldots, n-1$, so P_n has smaller size than P_0. This contradicts $P_0 =_{AC} P =_{AC} P_n$. Hence, $P_l \in \mathbb{H}$ for some l.

Since each P_k is a substate of each $P_{k'}$, $g(P_k)$ must be the same for all k. Then it follows from Fact B, together with $P_l \in \mathbb{H}$, that $P_k \in \mathbb{H}$ for all k and $P_0 \cong P_1 \cong \cdots \cong P_n$. □

Now we are ready to prove the desired completeness result for $\mathrm{BPA}^*(A)$.

THEOREM 3.2.8. (A1)–(A5), (BKS1)–(BKS3) *is complete for* $\mathrm{BPA}^*(A)$ *modulo strong bisimulation equivalence.*

PROOF. Each process term is provably equal to a basic term, so it is sufficient to show that strongly bisimilar basic terms are provably equal. Assume $P, Q \in \mathbb{B}$ with $P \leftrightarrow Q$; we show that $P = Q$, by induction on the ordering $\prec$. To be precise, we assume that we have already dealt with strongly bisimilar pairs $R, S \in \mathbb{B}$ with $R \prec P$ and $S \prec Q$, or $R \prec P$ and $S \cong Q$, or $R \cong P$ and $S \prec Q$.

First, assume that P or Q is not in $\mathbb{H}$, say $P \notin \mathbb{H}$. By the induction hypothesis, together with Lemma 3.2.7, for all transitions $P \xrightarrow{a} P'$ and $Q \xrightarrow{a} Q'$ with $P' \leftrightarrow Q'$ we have $P = Q$. Since $P \leftrightarrow Q$, axiom (A3) can be used to adapt the HNF-expansions of P and Q to the form

$$P = \sum_{i=1}^{m} a_i P_i + \sum_{j=1}^{n} b_j, \qquad Q = \sum_{i=1}^{m} a_i Q_i + \sum_{j=1}^{n} b_j,$$

where $P_i = Q_i$ for $i = 1, \ldots, m$. Hence, $P = Q$.

Next, assume $P, Q \in \mathbb{H}$. We distinguish three cases.

(1) Let $P =_{AC} R^*S$ and $Q =_{AC} T^*U$. We prove $R^*S = T^*U$.
We spell out the HNF-expansions of R and T:

$$R = \sum_{i \in I} R_i, \qquad T = \sum_{j \in J} T_j,$$

where the R_i and the T_j are of the form either aV or a.
Since $R^*S \leftrightarrow T^*U$, each $R_i(R^*S)$ for $i \in I$ is strongly bisimilar either to $T_j(T^*U)$ for a $j \in J$ or to a summand of U. We distinguish these two cases.

(a) $R_i(R^*S) \leftrightarrow T_j(T^*U)$ for a $j \in J$. Then $R_i(R^*S) \leftrightarrow T_j(R^*S)$, so by Lemma 3.2.2 $R_i \leftrightarrow T_j$.

(b) $R_i(R^*S) \leftrightarrow aU'$ for a transition $U \xrightarrow{a} U'$.

Thus, I can be divided into the following, not necessarily disjoint, subsets:

$$I_0 = \{i \in I \mid \exists j \in J(R_i \leftrightarrows T_j)\},$$
$$I_1 = \{i \in I \mid \exists U \xrightarrow{a} U' \ (R_i(R^*S) \leftrightarrows aU')\}.$$

Similarly, J can be divided:

$$J_0 = \{j \in J \mid \exists i \in I \ (T_j \leftrightarrows R_i)\},$$
$$J_1 = \{j \in J \mid \exists S \xrightarrow{a} S'(T_j(T^*U) \leftrightarrows aS')\}.$$

If both I_1 and J_1 are non-empty, then $U'' \leftrightarrows R^*S$ for a proper substate U'' of U, and $S'' \leftrightarrows T^*U$ for a proper substate S'' of S, and so $U'' \leftrightarrows S''$. By Lemma 3.2.6, $S'' \prec R^*S$ and $U'' \prec T^*U$, so induction yields $R^*S = U'' = S'' = T^*U$, and we are done. Hence, we may assume that either I_1 or J_1 is empty, say $J_1 = \emptyset$. We proceed to derive

$$\sum_{i \in I_1} R_i(R^*S) + S = U. \tag{1}$$

We show that each summand at the left-hand side of the equality sign is provably equal to a summand of U, and vice versa.

- By definition of I_1, for each $R_i(R^*S)$ with $i \in I_1$ there is a summand aU' of U such that $R_i(R^*S) \leftrightarrows aU'$. By Lemma 3.2.6 $U' \prec T^*U$, so induction yields $R_i(R^*S) = aU'$.
- Consider a summand aS' of S. Since $R^*S \leftrightarrows T^*U$ and $J_1 = \emptyset$, it follows that aS' is strongly bisimilar to a summand aU' of U, so induction yields $aS' = aU'$.
- Finally, summands a of S correspond with summands a of U.
- By the converse arguments it follows that each summand of U is provably equal to a summand at the left-hand side of the equality sign.

This concludes the derivation of (1).
Since $J_1 = \emptyset$, it follows that $J_0 \neq \emptyset$, so $I_0 \neq \emptyset$. By the definitions of I_0 and $J_0 = J$, each R_i with $i \in I_0$ is strongly bisimilar to a T_j with $j \in J$, and vice versa. Since $L(R_i) \leqslant L(R) < L(R^*S)$, induction yields $R_i = T_j$. Hence,

$$\sum_{i \in I_0} R_i = T. \tag{2}$$

Finally, we derive

$$\begin{aligned} R^*S &\overset{(A3)}{=} \Bigl(\sum_{i \in I_0} R_i + \sum_{i \in I_1} R_i\Bigr)^* S \\ &\overset{(BKS3),(A3)}{=} \Bigl(\sum_{i \in I_0} R_i\Bigr)^* \Bigl(\sum_{i \in I_1} R_i(R^*S) + S\Bigr) \\ &\overset{(1),(2)}{=} T^*U. \end{aligned}$$

time spectrum [45] that are finer than language equivalence and coarser than ready simulation, and which constitute congruence relations over $\mathrm{BPA}^*(A)$, are failure semantics, ready semantics, failure trace semantics, and ready trace semantics.

The result above follows from the existence of an infinite set of equations that cannot all be proved by means of any finite set of equations that is sound modulo language equivalence. This family of equations consists of

$$\mathrm{E}.n \quad a^*(a^n) + (a^n)^*(a + \cdots + a^n) = (a^n)^*(a + \cdots + a^n)$$

for $n \geqslant 1$, where a is some action. Ready simulation is the finest semantics in the linear/branching time spectrum in which the E.n are sound. Note that for $n > 1$, none of the equations E.n is sound in strong bisimulation equivalence.

Given a finite set of equations that is sound with respect to language equivalence, Aceto, Fokkink, and Ingólfsdóttir construct a model $\mathcal{A}_p$ for these equations in which equation E.p fails, for some prime number p. The model that is used for this purpose is based on an adaptation of a construction due to Conway [31], who used it to obtain a new proof of a theorem, originally due to Redko [69], saying that $\mathrm{BPA}^*(A)$ is not finitely based modulo language equivalence.

Let a be an action. For p a prime number, the carrier A_p of the algebra $\mathcal{A}_p$ consists of non-empty formal sums of $a^0, a^1, \ldots, a^{p-1}$, together with the formal symbol a^*, that is,

$$\left\{\sum_{i\in I} a^i \mid \emptyset \subset I \subseteq \{0, \ldots, p-1\}\right\} \cup \{a^*\}.$$

The syntax of $\mathcal{A}_p$ contains three more operators, which are semantic counterparts of the binary function symbols in $\mathrm{BPA}^*(A)$. In order to avoid confusion, circled symbols denote the operators in the algebra $\mathcal{A}_p$: $\oplus$, $\odot$, and $\circledast$ represent the semantic counterparts of $+$, $\cdot$, and $*$, respectively. Table 6 presents an axiomatisation for $\mathcal{A}_p$.

Process terms over $\mathrm{BPA}^*(A)$ are mapped to $\mathcal{A}_p$ as expected: every action in A is mapped to the symbol a^1, while $+$, $\cdot$, and $*$ are mapped to $\oplus$, $\odot$, and $\circledast$, respectively. For a process

Table 6
Axiomatisation for the algebra $\mathcal{A}_p$

$$\begin{aligned}
a^* \oplus x &= a^* \\
x \oplus a^* &= a^* \\
\sum_{i\in I} a^i \oplus \sum_{j\in J} a^j &= \sum_{h\in I\cup J} a^h \\
\\
a^* \odot x &= a^* \\
x \odot a^* &= a^* \\
\sum_{i\in I} a^i \odot \sum_{j\in J} a^j &= \sum_{h\in\{(i+j) \bmod p \mid (i,j)\in I\times J\}} a^h \\
\\
x^{\circledast} y &= \begin{cases} y & \text{if } x = a^0 \\ a^* & \text{otherwise} \end{cases}
\end{aligned}$$

term P, the denotation of P in the algebra $\mathcal{A}_p$ is represented by $\mathcal{A}_p[\![P]\!]$. We note that equation E.p fails in $\mathcal{A}_p$. Namely,

$$\mathcal{A}_p[\![a^*(a^p) + (a^p)^*(a + \cdots + a^p)]\!] = a^* \neq \sum_{i=0}^{p-1} a^i = \mathcal{A}_p[\![(a^p)^*(a + \cdots + a^p)]\!].$$

The following theorem is the key to the nonaxiomatisability result from [4].

THEOREM 3.4.1. *For every finite set $\mathcal{E}$ of equations that are sound with respect to $\simeq_L$, there exists a prime number p such that all equations in $\mathcal{E}$ are valid in $\mathcal{A}_p$.*

COROLLARY 3.4.2. *No congruence relation over* BPA$^*(A)$ *that is included in $\simeq_L$ and satisfies E.n for all $n \geqslant 1$ has a complete finite equational axiomatisation.*

A process semantics that is coarser than language equivalence is *trace equivalence* $\simeq_T$, where two process terms are considered equivalent if they give rise to the same (not necessarily terminating) traces (see Definition 2.2.6). If $|A| > 1$, then trace equivalence is not a congruence relation over BPA(A); e.g., $a + aa \simeq_T aa$, but $(a + aa)b \not\simeq_T aab$. However, if the set A of actions is a singleton $\{a\}$, then trace equivalence constitutes a congruence relation over BPA$^*(A)$. In contrast with their negative results on the finite axiomatisability of BPA$^*(A)$ modulo process semantics between ready simulation and language equivalence, Aceto, Fokkink, and Ingólfsdóttir showed that BPA$^*(\{a\})$ modulo trace equivalence is axiomatized completely by the five axioms for BPA$(\{a\})$ together with the three axioms in Table 7.

THEOREM 3.4.3. (A1)–(A5), (T1)–(T3) *is complete for* BPA$^*(\{a\})$ *modulo trace equivalence.*

3.5. *Extensions of* BPA$^*(A)$

The signature of BPA$^*_\delta(A)$ is obtained by extending BPA$_\delta(A)$ with BKS. Its axioms are those of BPA$^*(A)$ and of BPA$_\delta(A)$, i.e., (A1)–(A7) in Table 1 for BPA$_\delta(A)$, and (BKS1)–(BKS3) for BKS. Sewell [72] showed that there does not exist a complete equational ax-

Table 7
Axioms for trace equivalence ($A = \{a\}$)

(T1)	$x + (y^*z)$	$=$	a^*a
(T2)	$x + xy$	$=$	xy
(T3)	xy	$=$	yx

Table 8
Axioms for BKS with encapsulation and abstraction

(BKS4)	$\partial_H(x^*y) = \partial_H(x)^*\partial_H(y)$
(BKS5)	$\tau_I(x^*y) = \tau_I(x)^*\tau_I(y)$

iomatisation for $BPA_\delta^*(A)$ modulo strong bisimulation equivalence. This motivates the introduction of the implicational axiom

$$(\text{RSP}^*) \quad \frac{x = yx + z}{x = y^*z}.$$

It remains an open question, dating back to Milner [62], whether (A1)–(A7), (BKS1), (RSP*) is complete modulo strong bisimulation equivalence. We note that (BKS2) and (BKS3) can be derived from this axiomatisation.

The signature of $PA^*(A)$ is obtained by extending $PA(A)$ with BKS. The axioms of $PA^*(A)$ are those of $PA(A)$ and (BKS1)–(BKS3). The system $PA_\delta(A)$ can be extended in a similar way to $PA_\delta^*(A)$. The system $ACP^*(A, |)$ is defined by inclusion of (BKS1)–(BKS4); see Table 8. Note that (BKS4) can be derived using (RSP*). Finally, $ACP^*(A, |, \tau)$ is obtained by inclusion into $ACP(A, |, \tau)$ of (BKS1)–(BKS5); see Table 8. Note that (RSP*) is not sound for $ACP^*(A, |, \tau)$; e.g.,

$$\tau = \tau\tau + \delta,$$

but $\tau = \tau^*\delta$ is not a desirable identity in any process semantics.

4. Axiomatisations of other iterative operations

This section considers four restricted versions and one generalised version of BKS. We discuss the different advantages of each of these operators, and formulate various axiomatisations and completeness results.

4.1. *Axiomatisation of no-exit iteration*

No-exit iteration (NEI) x^ω is bisimilar to $x^*\delta$. No-exit iteration can be used to formally describe programs that repeat a certain procedure without end. Many communication protocols can be expressed, and shown correct, using no-exit iteration. An explanation is that (concurrent) components of such protocols often perform repetitive behaviour in the following style (receive/process/send-repetition):

$$\Big(\sum_{d\in D} r_i(d)\, P\, s_j(d)\Big)^\omega \quad \text{or} \quad \Big(\sum_{d\in D} r_i(d)\, P\, s_j(d) + Q\Big)^\omega,$$

Table 9
Axioms for NEI

(NEI1)	$x^{\omega} = x(x^{\omega})$
(RSP$^{\omega}$)	$\dfrac{x = yx}{x = y^{\omega}}$

Table 10
Transition rules for NEI

$\dfrac{x \xrightarrow{a} x'}{x^{\omega} \xrightarrow{a} x'(x^{\omega})}$	$\dfrac{x \xrightarrow{a} \surd}{x^{\omega} \xrightarrow{a} x^{\omega}}$

where Q handles an exceptional situation. A standard example in process algebra is the *alternating bit protocol* (see, e.g., [21]), specified as the concurrent execution of four components, each of which can be specified in the style above. Further examples of this specification and verification style can be found in [18,76].

Table 9 presents two axioms for NEI. (NEI1) is its defining axiom, while (RSP$^{\omega}$) is an adaptation of (RSP*). The axiomatisations for BPA$^{\omega}(A)$ and BPA$^{\omega}_{\delta}(A)$ are obtained by extending BPA(A) and BPA$_{\delta}(A)$ with (NEI1) and (RSP$^{\omega}$).

In order to provide process terms over BPA$^{\omega}_{\delta}(A)$ with an operational semantics, we introduce transition rules for NEI. Together with the transition rules for BPA(A) in Table 2 they provide labelled transition systems to process terms over BPA$^{\omega}_{\delta}(A)$.

The transition rules for NEI are in path format. Hence, strong bisimulation equivalence is a congruence with respect to BPA$^{\omega}_{\delta}(A)$. Furthermore, its axiomatisation is sound for BPA$^{\omega}_{\delta}(A)$ modulo strong bisimulation equivalence. Since strong bisimulation equivalence is a congruence, this can be verified by checking soundness for each axiom separately. It is easily verified that (NEI1) and (RSP$^{\omega}$) are indeed sound modulo strong bisimulation equivalence.

The following two completeness results for no-exit iteration originate from [39]. Their proofs, which are omitted here, are based on the proof strategy from [38].

THEOREM 4.1.1. (A1)–(A5), (NEI1), (RSP$^{\omega}$) *is complete for* BPA$^{\omega}(A)$ *modulo strong bisimulation equivalence.*

THEOREM 4.1.2. (A1)–(A7), (NEI1), (RSP$^{\omega}$) *is complete for* BPA$^{\omega}_{\delta}(A)$ *modulo strong bisimulation equivalence.*

The observation by Sewell [72] that there does not exist a complete finite equational axiomatisation for BPA$^*_{\delta}(A)$ modulo strong bisimulation equivalence, is based on the fact that a^{ω} is strongly bisimilar to $(a^k)^{\omega}$ for $k \geqslant 1$. This argument can be copied to conclude that there do not exist complete finite equational axiomatisations for BPA$^{\omega}(A)$ and BPA$^{\omega}_{\delta}(A)$. Hence, the implicational axiom (RSP$^{\omega}$) is irredundant. It is not difficult to see that axiom (NEI1) is irredundant as well.

4.2. *Axiomatisation of multi-exit iteration*

Milner [62] noted that not every regular process can be described in BPA$^*(A)$, up to strong bisimulation equivalence. The limited expressive power of BKS was highlighted in [16],

Table 11
Axioms for MEI

(MEI1)	$x_1((x_2,\ldots,x_k,x_1)^*(y_2,\ldots,y_k,y_1))+y_1=(x_1,\ldots,x_k)^*(y_1,\ldots,y_k)$
(MEI2)	$((x_1,\ldots,x_k)^*(y_1,\ldots,y_k))z=(x_1,\ldots,x_k)^*(y_1z,\ldots,y_kz)$
(MEI3)	$(z_0,x_2,\ldots,x_k)^*(y_1+z_1((x_2,\ldots,x_k,z_0+z_1)^*(y_2,\ldots,y_k,y_1)),y_2,\ldots,y_k)$ $=(z_0+z_1,x_2,\ldots,x_k)^*(y_1,\ldots,y_k)$
(MEI4)	$(z_0,z_1,x_2,\ldots,x_k)^*(y_1,z_2((x_2,\ldots,x_k,z_0(z_1+z_2))^*(y_2,\ldots,y_k,y_1)),y_2,\ldots,y_k)$ $=(z_0(z_1+z_2),x_2,\ldots,x_k)^*(y_1,\ldots,y_k)$
(MEI5)	$((x_1,\ldots,x_k)^\ell)^*((y_1,\ldots,y_k)^\ell)=(x_1,\ldots,x_k)^*(y_1,\ldots,y_k)$

where it was shown that the process described by the recursive specification

$$\begin{aligned} X_1 &= aX_2+a, \\ X_2 &= aX_1+b, \end{aligned}$$

cannot be expressed in $\mathrm{BPA}^*(A)$ modulo strong bisimulation equivalence. See Section 5 for more information on the expressive power of iterative operators.

Bergstra, Bethke, and Ponse [16] introduced *multi-exit iteration* (MEI) as a more expressive variant of iteration. For every $k \geqslant 1$, and process terms P_i and Q_i $(1 \leqslant i \leqslant k)$, the process term $(P_1,\ldots,P_k)^*(Q_1,\ldots,Q_k)$ denotes a solution to the recursion variable X_1 in the recursive specification

$$\begin{aligned} X_1 &= P_1X_2+Q_1, \\ &\vdots \\ X_{k-1} &= P_{k-1}X_k+Q_{k-1}, \\ X_k &= P_kX_1+Q_k. \end{aligned}$$

Aceto and Fokkink [1] introduced the axiom system $\mathrm{BPA}^{me*}(A)$, which is obtained by adding the MEI axioms (MEI1)–(MEI5) in Table 11 to $\mathrm{BPA}(A)$. The first three MEI axioms are adaptations of the three BKS axioms. The last two MEI axioms relate process terms of distinct exit degrees. (MEI4) is the multiplicative counterpart of (MEI3), while (MEI5) enables to reduce repetitive patterns at the left- and right-hand side of MEI.

In order to provide process terms over $\mathrm{BPA}^{me*}(A)$ with an operational semantics, we introduce transition rules for MEI. Together with the transition rules for $\mathrm{BPA}(A)$ in Table 2 they provide labelled transition systems to process terms over $\mathrm{BPA}^{me*}(A)$.

The transition rules for MEI are in path format. Hence, strong bisimulation equivalence is a congruence with respect to $\mathrm{BPA}^{me*}(A)$. Furthermore, its axiomatisation is sound for $\mathrm{BPA}^{me*}(A)$ modulo strong bisimulation equivalence. Since strong bisimulation equivalence is a congruence, this can be verified by checking soundness for each axiom separately. It is easily verified that (MEI1), (MEI2), and (MEI5) are sound modulo strong

Table 12
Transition rules for MEI

$$\frac{x_1 \xrightarrow{a} x_1'}{(x_1,\ldots,x_k)^*(y_1,\ldots,y_k) \xrightarrow{a} x_1'((x_2,\ldots,x_k,x_1)^*(y_2,\ldots,y_k,y_1))}$$

$$\frac{x_1 \xrightarrow{a} \surd}{(x_1,\ldots,x_k)^*(y_1,\ldots,y_k) \xrightarrow{a} (x_2,\ldots,x_k,x_1)^*(y_2,\ldots,y_k,y_1)}$$

$$\frac{y_1 \xrightarrow{a} y_1'}{(x_1,\ldots,x_k)^*(y_1,\ldots,y_k) \xrightarrow{a} y_1'} \qquad \frac{y_1 \xrightarrow{a} \surd}{(x_1,\ldots,x_k)^*(y_1,\ldots,y_k) \xrightarrow{a} \surd}$$

bisimulation equivalence. See [1] for a detailed proof that (MEI3) and (MEI4) are sound modulo strong bisimulation equivalence.

In [1] it is proved that the axiomatisation $\mathrm{BPA}^{me*}(A)$ is complete for $\mathrm{BPA}^{me*}(A)$ modulo strong bisimulation equivalence. The completeness proof, which is omitted here, is based on the proof strategy from [38].

THEOREM 4.2.1. (A1)–(A5), (MEI1)–(MEI5) *is complete for* $\mathrm{BPA}^{me*}(A)$ *modulo strong bisimulation equivalence.*

4.3. *Axiomatisation of prefix iteration*

Prefix iteration (PI) [36] is a variation of BKS, obtained by restricting its first argument to single atomic actions. The advantage of PI over BKS is twofold:

(1) PI can be axiomatized in a setting with *prefix multiplication* of CCS [63], which is obtained from sequential composition by restricting its first argument to single atomic actions;

(2) PI allows a complete equational axiomatisation modulo strong bisimulation equivalence in the presence of the deadlock δ.

We note that, in general, sequential composition can be restricted to prefix multiplication without loss of expressivity.

$\mathrm{BPA}_\delta^{p*}(A)$ consists of $\mathrm{BPA}(A)$, with sequential composition xy restricted to prefix multiplication ax from CCS, extended with PI. Table 13 presents a collection of axioms for PI. First of all, (PI1)–(PI2) from [36] axiomatize PI with respect to strong bisimulation.

THEOREM 4.3.1. (A1)–(A3), (A6), (PI1)–(PI2) *is complete for* $\mathrm{BPA}_\delta^{p*}(A)$ *modulo strong bisimulation equivalence.*

The remaining equations and inequalities in Table 13 originate from Aceto, Fokkink, and Ingólfsdóttir [5], who proved completeness results for PI in a variety of behavioural equivalences and preorders in the linear/branching time spectrum [45]. These axiomatisations for $\mathrm{BPA}_\delta^{p*}(A)$ all incorporate axioms (A1)–(A3), (A6) for $\mathrm{BPA}_\delta(A)$ with prefix

Table 15
Axioms for FI

(FA4)	$(\alpha+\beta)x = \alpha x+\beta x$
(FB1)	$\alpha\tau = \alpha$
(FB2)	$\alpha(x+\tau y) = \alpha(x+\tau y)+\alpha y$
(FI1)	$\alpha^*(\beta((\alpha+\beta)^*x)+x) = (\alpha+\beta)^*x$
(FI2)	$\delta^*x = x$

restricting the left-hand side of BKS to sums of atomic actions. Similarly, *flat multiplication* is obtained by restricting the left-hand side of sequential composition to sums of atomic actions. The transition rules for these operators are simply the transition rules for BKS and sequential composition with the left-hand sides restricted to sums of atomic actions. $\mathrm{BPA}_\delta^{f*}(A)$ is obtained by adding FI to $\mathrm{BPA}_\delta(A)$ and restricting sequential composition to flat multiplication. The merge can be eliminated from process terms that contain FI. For example, typically, $(a^*b) \parallel (c^*d)$ is strongly bisimilar to

$$(a+c+a\mid c)^*\big((d+a\mid d)(a^*b)+(b+b\mid c)(c^*d)+b\mid d\big).$$

For a detailed discussion on this expressivity claim the reader is referred to Section 5.2.

In [48], van Glabbeek presented complete axiomatisations for $\mathrm{BPA}_\delta^{f*}(A)$ extended with the silent step τ, modulo four rooted bisimulation semantics that take into account the silent nature of τ: rooted *branching* bisimulation, rooted *delay* bisimulation, rooted η-bisimulation and rooted *weak* bisimulation. Of these four equivalences, rooted branching bisimulation constitutes the finest relation, while rooted weak bisimulation constitutes the coarsest relation; rooted delay bisimulation and rooted η-bisimulation are incomparable. All four equivalences constitute congruence relations over $\mathrm{BPA}_{\delta\tau}^{f*}(A)$ (in the case of rooted branching bisimulation equivalence this follows from the fact that the transition rules for $\mathrm{BPA}_{\delta\tau}^{f*}(A)$ are in *RBB safe* format [41]). The axiomatisations for FI are adaptations of axiomatisations introduced by Aceto, Fokkink, van Glabbeek, and Ingólfsdóttir [8,37,2] for PI.

Table 15 presents adaptations to prefix multiplication of axiom (A4) and of two standard axioms (B1)–(B2) for τ. Furthermore, (FI1) is an adaptation of (BKS3) to FI, while (FI2) expresses the interplay of FI with the deadlock δ. In the axioms, α and β range over sums of atomic actions (the empty sum representing δ). Note that the defining equation of FI,

$$\beta^*x = \beta(\beta^*x)+x$$

can be derived from (FI1) by taking α to be δ.

Table 16 presents axioms for the interplay of FI with the silent step τ, modulo the four aforementioned equivalence relations. (FT1) is an instantiation of *Koomen's fair abstraction rule* (KFAR) [55,11] for rooted branching and η-bisimulation, while (FT4) serves this same purpose for rooted delay and weak bisimulation. In [48], van Glabbeek proved completeness for $\mathrm{BPA}_{\delta\tau}^{f*}(A)$ with respect to rooted branching bisimulation; see Theorem 4.5.1.

Table 16
Axioms for FI with the silent step

(FT1)	$(\alpha+\tau)^*x$	$=$	$\alpha^*x+\tau(\alpha^*x)$
(FT2)	$\alpha(\beta^*(\tau(\beta^*(x+y))+x))$	$=$	$\alpha(\beta^*(x+y))$
(FT3)	$\alpha^*(x+\tau y)$	$=$	$\alpha^*(x+\tau y+\alpha y)$
(FT4)	$(\alpha+\tau)^*x$	$=$	$\tau(\alpha^*x)$

The complete axiomatisations for $\mathrm{BPA}^{f*}_{\delta\tau}(A)$ modulo rooted delay, η-, and weak bisimulation equivalence can then be obtained from the complete axiomatisation modulo rooted branching bisimulation equivalence, using a reduction technique from van Glabbeek and Weijland [49]; see Theorems 4.5.2, 4.5.3, and 4.5.4.

THEOREM 4.5.1. (A1)–(A3), (FA4), (A6)–(A7), (FB1), (FI1)–(FI2), (FT1)–(FT2) *is complete for* $\mathrm{BPA}^{f*}_{\delta\tau}(A)$ *modulo rooted branching bisimulation equivalence.*

THEOREM 4.5.2. (A1)–(A3), (FA4), (A6)–(A7), (FB1), (FI1)–(FI2), (FT4) *is complete for* $\mathrm{BPA}^{f*}_{\delta\tau}(A)$ *modulo rooted delay bisimulation equivalence.*

THEOREM 4.5.3. (A1)–(A3), (FA4), (A6)–(A7), (FB1)–(FB2), (FI1)–(FI2), (FT1)–(FT3) *is complete for* $\mathrm{BPA}^{f*}_{\delta\tau}(A)$ *modulo rooted* η*-bisimulation equivalence.*

THEOREM 4.5.4. (A1)–(A3), (FA4), (A6)–(A7), (FB1)–(FB2), (FI1)–(FI2), (FT3)–(FT4) *is complete for* $\mathrm{BPA}^{f*}_{\delta\tau}(A)$ *modulo rooted weak bisimulation equivalence.*

5. Expressivity results

This section concerns expressivity of process algebra with recursive operations, to categorize what can be specified with the various recursive operations. Of course, answers to these questions depend on the particular process semantics one adopts.

5.1. *Expressivity of the binary Kleene star*

In [17], Bergstra, Bethke, and Ponse showed that the expressivity of systems with BKS can be analyzed by establishing properties of cycles in labelled transition systems. These results were strengthened by Boselie [27]. We recall these results, and first introduce some further terminology. A state Q is a *successor* of state P if $P \xrightarrow{a} Q$. A *cycle* is a sequence of distinct states $(P_0, \ldots, P_n)$ such that P_{i+1} is a successor of P_i for $i = 0, \ldots, n-1$ and P_0 is a successor of P_n. An action a is an *exit action* of state P if $P \xrightarrow{a} \surd$. We use $\equiv$ to denote that two terms are syntactically the same.

LEMMA 5.1.1. *Let C be a cycle in a labelled transition system associated to a process term over* $\mathrm{ACP}^*(A, |, \tau)$. *Then C has one of the following forms, for* $n \in \mathbb{N}$:

(i) $C = (P_0 Q, P_1 Q, \ldots, P_n Q)$;
(ii) $C = (P^* Q, P_1(P^* Q), \ldots, P_n(P^* Q))$, *or any cyclic permutation thereof*;
(iii) $C = (P_0 \parallel Q_0, P_1 \parallel Q_1, \ldots, P_n \parallel Q_n)$;
(iv) $C = (\partial_H(P_0), \partial_H(P_1), \ldots, \partial_H(P_n))$.

PROOF. Let $C = (C_0, \ldots, C_n)$. We apply case distinction on C_0. Clearly C_0 is not a single atomic action, and as $+, \mathbin{\|\!\_}, |$ do not occur as the first operation in right-hand sides of conclusions of transition rules, it follows that $C_0 \not\equiv P \diamond Q$ for $\diamond \in \{+, \mathbin{\|\!\_}, |\}$.

Suppose $C_0 \equiv RS$. If S is not a state in C, then $C = (RS, R_1 S, \ldots, R_n S)$, which corresponds to case (i). If S is a state in C, then $S \xrightarrow{\sigma}\!\!\!\!\rightarrow RS$ for some $\sigma \in A^*$. It is not hard to see that only the first transition rule for BKS can give rise to a transition $T \xrightarrow{a} T'$ where T is a proper subterm of T'. Hence, S is of the form $P^* Q$, and the first transition in the sequence $S \xrightarrow{\sigma}\!\!\!\!\rightarrow RS$ is invoked by the first transition rule for BKS. This yields form (ii).

Suppose $C_0 \equiv R^* S$. Analogous to the case $C_0 \equiv RS$, we see that C is of form (ii).

Suppose $C_0 \equiv R \parallel S$. As $R \parallel S$ is not a substate of R or S, it follows from the transition rules of the merge that C must be of form (iii).

Suppose $C_0 \equiv \partial_H(R)$. Since only the first transition rule for ∂_H can have been used, it follows that C is of form (iv). □

Lemma 5.1.1 can be used to derive further properties of cycles.

LEMMA 5.1.2. *Let C be a cycle in a labelled transition system associated to a process term over* $\mathrm{BPA}^*_\delta(A)$. *Then there is at most one state P in C that has a successor Q such that P is not a proper substate of Q.*

PROOF. As C belongs to a process term over $\mathrm{BPA}^*_\delta(A)$, it must be of the form (i) or (ii) in Lemma 5.1.1. We apply induction with respect to the size of C.

Suppose $C = (P_0 Q, \ldots, P_n Q)$. By induction, the cycle $(P_0, \ldots, P_n)$ contains at most one state P_i that has a successor R such that P_i is not a proper substate of R. This implies that $P_i Q$ is the only state in C that may have a successor S such that $P_i Q$ is not a proper substate of S.

Suppose $C = (P^* Q, P_1(P^* Q), \ldots, P_n(P^* Q))$, or any cyclic permutation thereof. Then $P^* Q$ is the only state in C that may have a successor R such that $P^* Q$ is not a proper substate of R. □

LEMMA 5.1.3. *Let C be a cycle in a labelled transition system associated to a process term over* $\mathrm{PA}^*_\delta(A)$. *If there is a state in C with an exit action, then every other state in C has only successors in C.*

PROOF. As C belongs to a process term over $\mathrm{PA}^*_\delta(A)$, this cycle must be of the form (i), (ii), or (iii) in Lemma 5.1.1.

Suppose $C = (P_0 Q, \ldots, P_n Q)$. Then none of the states in C has an exit action.

Suppose $C = (P^* Q, P_1(P^* Q), \ldots, P_n(P^* Q))$, or any cyclic permutation thereof. Then $P^* Q$ is the only state in C that may have an exit action, and the other states in C have only successors in C.

Suppose $C = (P_0 \parallel Q_0, \ldots, P_n \parallel Q_n)$. Since the communication merge is excluded from $\mathrm{PA}^*_\delta(A)$, none of the states in C has an exit action. □

LEMMA 5.1.4. *Let C be a cycle in a labelled transition system associated to a process term over* $\mathrm{ACP}^*(A, |)$. *Then there is at most one state in C with an exit action.*

PROOF. As C belongs to a process term over $\mathrm{ACP}^*(A, |)$, this cycle must be of the form (i), (ii), (iii), or (iv) in Lemma 5.1.1. We apply induction with respect to the size of C.

Suppose $C = (P_0 Q, \ldots, P_n Q)$. Then none of the states in C has an exit action.

Suppose $C = (P^*Q, P_1(P^*Q), \ldots, P_n(P^*Q))$, or any cyclic permutation thereof. Then P^*Q is the only state in C that may have an exit action.

Suppose $C = (P_0 \parallel Q_0, \ldots, P_n \parallel Q_n)$. Assume $P_i \parallel Q_i$ and $P_j \parallel Q_j$ both have an exit action. Then by induction $P_i \parallel Q_i$ represent the same state, so $i = j$.

Suppose $C = (\partial_H(P_0), \ldots, \partial_H(P_n))$. By induction, the cycle $(P_0, \ldots, P_n)$ contains at most one state P_i that has an exit action. So $\partial_H(P_i)$ is the only state in C that may have an exit action. □

We have the following expressivity hierarchy for process algebra with BKS.

THEOREM 5.1.5.

$$\mathrm{BPA}^*_\delta(A) \overset{1}{\prec} \mathrm{PA}^*_\delta(A) \overset{2}{\prec} \mathrm{ACP}^*(A, |) \overset{4}{\prec} \mathrm{ACP}^*(A, |, \tau)$$

where $\overset{k}{\prec}$ means "less expressive than, provided A contains at least k actions" modulo strong bisimulation equivalence, except for the last inequality, which requires the presence of τ and soundness of (B1) *(i.e., $x\tau = x$). If one does not restrict to handshaking,*

$$\mathrm{PA}^*_\delta(A) \overset{1}{\prec} \mathrm{ACP}^*(A, |).$$

The same inclusions hold in the absence of δ.

PROOF. $\mathrm{BPA}^*(A) \overset{1}{\prec} \mathrm{PA}^*(A)$ and $\mathrm{BPA}^*_\delta(A) \overset{1}{\prec} \mathrm{PA}^*_\delta(A)$. Consider the $\mathrm{PA}^*(A)$ process term $P = (aa)^*a \parallel a$, which can be depicted as follows:

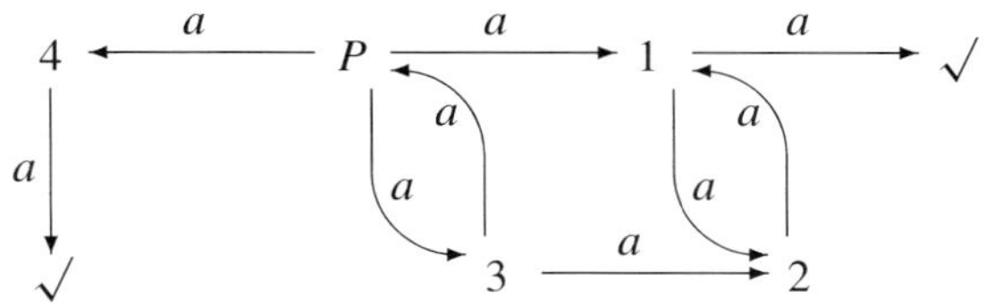

where

$$\begin{array}{lll} 1 & \text{abbreviates} & (aa)^*a, \\ 2 & \text{abbreviates} & a((aa)^*a), \\ 3 & \text{abbreviates} & a((aa)^*a) \parallel a, \text{ and} \\ 4 & \text{abbreviates} & a. \end{array}$$

According to Lemma 5.1.2, P cannot be specified in $\mathrm{BPA}_\delta^*(A)$. Namely, the states P and 3 are not strongly bisimilar and form a cycle, while both states have a successor from which one cannot return to this (nor to any strongly bisimilar) cycle.

$\mathrm{PA}_\delta^*(A) \overset{i}{\prec} \mathrm{ACP}^*(A, |)$, where $i = 1$ in a setting without handshaking and $i = 2$ otherwise. Take P as in the previous case, and let $a \mid a$ be defined (either as a, thus no handshaking, or as $b \neq a$). Then on top of the picture above, the labelled transition system associated to P contains the following transitions: $P \xrightarrow{a|a} \surd$, $P \xrightarrow{a|a} 2$, and $3 \xrightarrow{a|a} 1$. According to Lemma 5.1.3, P cannot be specified in $\mathrm{PA}_\delta^*(A)$. Namely, the states P and 3 are not strongly bisimilar and form a cycle, while P has an exit action and 3 has a successor from which one cannot return to this (nor to any strongly bisimilar) cycle.

$\mathrm{ACP}^*(A, |) \overset{4}{\prec} \mathrm{ACP}^*(A, |, \tau)$. Take the recursive specification

$$X_1 = aX_2 + a,$$
$$X_2 = aaX_1 + a.$$

Assume auxiliary actions b, c, and d, with $c \mid c \overset{\text{def}}{=} b$ and $d \mid d \overset{\text{def}}{=} b$ the only communications defined. Let the process term P be defined by:

$$\begin{aligned} P &= \tau_{\{b\}} \circ \partial_{\{c,d\}}\big(\big(a(ad + ac) + ac\big)Q \parallel R\big), \\ Q &= \big(c\big(a\big(a(ad + ac) + ac\big)\big)\big)^* d, \\ R &= (dc)^* cd. \end{aligned}$$

It can be derived from the axioms of $\mathrm{ACP}^*(A, |)$ together with (RSP) and (B1) that $P = X_1$. Hence, X_1 is expressible in $\mathrm{ACP}^*(A, |, \tau)$ modulo process semantics that respect (B1). According to Lemma 5.1.4, X_1 cannot be specified in $\mathrm{ACP}^*(A, |, \tau)$ modulo strong bisimulation equivalence, even if τ is allowed to occur in A as a non-silent action. Namely, X_1 and X_2 are not strongly bisimilar and form a cycle, while both X_1 and X_2 have an exit action. □

It is an open question whether $\mathrm{ACP}^*(A, |) \overset{3}{\prec} \mathrm{ACP}^*(A, |, \tau)$ holds.

The following theorem emphasizes the expressive power of $\mathrm{ACP}^*(A, |, \tau)$.

THEOREM 5.1.6. *For each regular process P there is a finite extension A^{ext} of A such that P can be expressed in $\mathrm{ACP}^*(A^{ext}, |, \tau)$, even if one restricts to handshaking and the actions in A are not subject to communication.*

PROOF. P is a solution for the recursion variable X_1 in a recursive specification $X_i = \sum_{j=1}^{n} \alpha_{i,j} X_j + \beta_i$ for $i = 1, \ldots, n$, where $\alpha_{i,j}$ and β_i are finite sums of actions or δ.

Define A^{ext} as the extension of A with the following $2n + 3$ fresh atomic actions:

$$in, \qquad r_j,\ s_j \quad (j = 0, \ldots, n).$$

Let $r_j \mid s_j \stackrel{\text{def}}{=} in$ (for $j = 0, \ldots, n$) be the only communications defined (so we have handshaking, and the actions in A are not subject to communication). Furthermore, let $H \stackrel{\text{def}}{=} \{r_j, s_j \mid i = 0, \ldots, n\}$, and let

$$\begin{array}{lll} G_i & \text{abbreviate} & \sum_{j=1}^{n} \alpha_{i,j} s_j + \beta_i s_0 \quad \text{for } i = 1, \ldots, n, \\ Q & \text{abbreviate} & \left(\sum_{j=1}^{n} r_j G_j\right)^* r_0, \\ M & \text{abbreviate} & \left(\sum_{j=1}^{n} r_j s_j\right)^* (r_0 s_0). \end{array}$$

We derive

$$\begin{aligned} \partial_H(G_i Q \parallel M) &= \partial_H\left(\left(\sum_{j=1}^{n} \alpha_{i,j} s_j Q + \beta_i s_0 Q\right) \parallel M\right) \\ &= \sum_{j=1}^{n} \alpha_{i,j} \partial_H(s_j Q \parallel M) + \beta_i \partial_H(s_0 Q \parallel M) \\ &= \sum_{j=1}^{n} \alpha_{i,j}\, in\, \partial_H(Q \parallel s_j M) + \beta_i\, in\, \partial_H(Q \parallel s_0) \\ &= \sum_{j=1}^{n} \alpha_{i,j}\, in\, in\, \partial_H(G_j Q \parallel M) + \beta_i\, in\, in. \end{aligned}$$

Hence,

$$\tau_{\{in\}} \circ \partial_H(G_i Q \parallel M) = \sum_{j=1}^{n} \alpha_{i,j} \tau_{\{in\}} \circ \partial_H(G_j Q \parallel M) + \beta_i.$$

Consequently, $\tau_{\{in\}} \circ \partial_H(G_i Q \parallel M)$ satisfies the recursive equation for X_i (for $i = 1, \ldots, n$). By (RSP) it follows that $X_1 = \tau_{\{in\}} \circ \partial_H(G_1 Q \parallel M)$. □

5.2. *Expressivity of multi-exit iteration*

We first note that in the extension of $\mathrm{BPA}_\delta(A)$ with multi-exit iteration one cannot describe all regular processes modulo strong bisimulation equivalence. For example, the process described by

$$\begin{aligned} X &= aY + aZ, \\ Y &= aZ + a, \\ Z &= aX + aa, \end{aligned}$$

cannot be expressed, as from the state X the two non-bisimilar exits a and aa can be reached in a single step.

In [1] it was shown that for every $k \geqslant 1$ there is a process over a single action that can be specified using $(k+1)$-exit iteration, but not using h-exit iteration with $h \leqslant k$. We proceed to sketch their argumentation. For $k \geqslant 1$, $\mathrm{BPA}^{me*(\leqslant k)}(A)$ denotes the set of process terms over $\mathrm{BPA}^{me*}(A)$ that only use h-exit iteration with $h \leqslant k$.

The set of *termination options* of a process term P over $\mathrm{BPA}^{me*}(A)$, is the smallest collection of process terms satisfying:

- if $P \xrightarrow{a} \surd$, then a is a termination option of P;
- if $P \xrightarrow{a} Q$ and Q does not contain occurrences of MEI, then aQ is a termination option of P.

LEMMA 5.2.1. *Let C be a cycle in a labelled transition system associated to a process term over* $\mathrm{BPA}^{me} * (\leqslant k)(A)$. *Then C contains at most k states with distinct, non-empty sets of termination options.*

PROOF. Let $C = (C_0, \ldots, C_n)$. We apply structural induction on C_0. Clearly C_0 is not a single atomic action, and as $+$ does not occur as the first operation in right-hand sides of conclusions of transition rules, it follows that C_0 is not of the form $P + Q$.

(1) $C_0 \equiv P_0 Q_0$. There are two possibilities.

(a) Q_0 is not a state in C. Then there is a cycle $(P_0, \ldots, P_n)$ such that $C_i \equiv P_i Q_0$ for $i = 0, \ldots, n$.

If Q_0 contains occurrences of MEI, then all states in C have an empty set of termination options.

If Q_0 does not contain occurrences of MEI, then the set of termination options of C_i (for $i = 0, \ldots, n$) is

$$\{R Q_0 \mid R \text{ is a termination option of } P_i\}.$$

The inductive hypothesis yields that there are at most k process terms P_i with distinct, non-empty sets of termination options. Hence, there are at most k states in C with distinct, non-empty sets of termination options.

(b) Q_0 is a state C_l in C. By induction there are at most k states in the cycle $(C_l, \ldots, C_n, C_0, \ldots, C_{l-1})$ with distinct, non-empty sets of termination options. So the same holds for C.

(2) $C_0 \equiv (P_1, \ldots, P_h)^*(Q_1, \ldots, Q_h)$, where $h \leqslant k$.

Clearly, substates of the Q_i cannot be in C. Thus the only states in C with possibly non-empty sets of termination options are

$$(P_i, \ldots, P_h, P_1, \ldots, P_{i-1})^*(Q_i, \ldots, Q_h, Q_1, \ldots, Q_{i-1})$$

for $i = 1, \ldots, h$. □

THEOREM 5.2.2. $\mathrm{BPA}^{me*(\leqslant k)}(A) \overset{1}{\prec} \mathrm{BPA}^{me*(\leqslant k+1)}(A)$ *for* $k \geqslant 1$.

PROOF. Let $a \in A$. It suffices to show that the $(k+1)$-exit iteration term

$$(a, a, \ldots, a)^*\big(a, a^2, \ldots, a^{k+1}\big)$$

cannot be specified in $\mathrm{BPA}^{me*(\leqslant k)}(A)$ modulo strong bisimulation equivalence. This follows from Lemma 5.2.1, because this process term induces a cycle that traverses the process term

$$(a, \ldots, a)^*\big(a^i, \ldots, a^{k+1}, a, \ldots, a^{i-1}\big),$$

which has $\{a^i\}$ as set of termination options, for $i = 1, \ldots, k+1$. Clearly, $a^i \not\leftrightarrow a^j$ if $i \neq j$. □

5.3. *Expressivity of string iteration*

In this section it is shown that for every $k \geqslant 1$ there is a process over a single action that can be specified by SI using a string of length $k+1$, but not by SI using strings of length at most k. For $k \geqslant 1$, $\mathrm{BPA}^{s*(\leqslant k)}(A)$ denotes the set of process terms over $\mathrm{BPA}^{s*}(A)$ that only use strings of length at most k.

LEMMA 5.3.1. *Let C be a cycle in a labelled transition system associated to a process term over* $\mathrm{BPA}^{s*(\leqslant k)}(A)$. *Then C contains at most k distinct states.*

PROOF. Let $C = (C_0, \ldots, C_n)$. We apply structural induction on C_0. Clearly C_0 is not a single atomic action, and as $+$ does not occur as the first operation in right-hand sides of conclusions of transition rules, it follows that C_0 is not of the form $P + Q$.

(1) $C_0 \equiv wP$.
Clearly, P is a state C_l in C. By induction there are at most k distinct states in the cycle $(C_l, \ldots, C_n, C_0, \ldots, C_{l-1})$. So the same holds for C.

(2) $C_0 \equiv (a_1 \cdots a_h)^* P$, where $h \leqslant k$.
Clearly, substates of P cannot be in C. Thus the only distinct states in C are

$$(a_{i+1} \cdots a_h)\big((a_1 \cdots a_h)^* P\big) \quad (\text{for } i = 1, \ldots, h).$$

□

THEOREM 5.3.2. $\mathrm{BPA}^{s*(\leqslant k)}(A) \overset{1}{\prec} \mathrm{BPA}^{s*(\leqslant k+1)}(A)$ *for* $k \geqslant 1$.

PROOF. Let $a \in A$. It suffices to show that the $(k+1)$-string iteration term

$$(a^{k+1})^* a$$

cannot be specified by a process term over $\mathrm{BPA}^{s*(\leqslant k)}(A)$ modulo strong bisimulation equivalence. This follows from Lemma 5.3.1, because the process term above induces a cycle that traverses the $k+1$ non-bisimilar process terms $a^i((a^{k+1})^* a)$ for $i = 0, \ldots, k$. □

5.4. *Expressivity of flat iteration*

This section presents some expressivity results on FI from [16]. $\mathrm{BPA}^{f*}(A)$, $\mathrm{BPA}_\delta^{f*}(A)$, $\mathrm{PA}^{f*}(A)$, and $\mathrm{ACP}^{f*}(A, |)$ are obtained by adding FI to $\mathrm{BPA}(A)$, $\mathrm{BPA}_\delta(A)$, $\mathrm{PA}(A)$, and $\mathrm{ACP}(A, |)$, respectively, and restricting sequential composition to flat multiplication.

As stated below, restricting sequential composition to prefix multiplication gives no loss of expressivity. *Flat iterative basic terms* over $\mathrm{BPA}^{f*}(A)$ are defined by the BNF grammar

$$P ::= a \mid P + P \mid aP \mid \alpha^* P$$

where $a \in A$ and α is an atomic sum. Flat iterative basic terms over $\mathrm{BPA}_\delta^{f*}(A)$ are defined by adding δ to the BNF grammar.

LEMMA 5.4.1. *Each process term over* $\mathrm{BPA}^*(A)$ [$\mathrm{BPA}_\delta^*(A)$] *with BKS restricted to FI is bisimilar to a flat iterative basic term over* $\mathrm{BPA}^{f*}(A)$ [$\mathrm{BPA}_\delta^{f*}(A)$].

PROOF. By structural induction, using the axioms of $\mathrm{BPA}_\delta(A)$ and those in Table 15. □

With respect to expressivity of systems with FI in strong bisimulation semantics we have the following results.

THEOREM 5.4.2.

(1) $\mathrm{BPA}^{f*}(A) \overset{1}{\prec} \mathrm{BPA}^*(A)$ *and* $\mathrm{BPA}_\delta^{f*}(A) \overset{1}{\prec} \mathrm{BPA}_\delta^*(A)$,

(2) $\mathrm{BPA}^{f*}(A)$ *is as expressive as* $\mathrm{PA}^{f*}(A)$, *and*

(3) $\mathrm{BPA}_\delta^{f*}(A)$ *is as expressive as* $\mathrm{ACP}^{f*}(A, |)$.

PROOF. Fact (1) is trivially true, as FI does not give rise to cycles of length greater than one. For example, the process term $(aa)^* a$ over $\mathrm{BPA}^*(A)$, which has a cycle of length two, cannot be expressed in $\mathrm{BPA}_\delta^{f*}(A)$ modulo strong bisimulation equivalence.

We proceed to present the proof of Fact (2). Fact (3) can be proved in a similar fashion.

From Lemma 5.4.1 it follows that $\mathrm{BPA}^{f*}(A)$ is as expressive as $\mathrm{PA}^{f*}(A)$ if all process terms $P \parallel Q$ and $P \mathbin{⫗} Q$ with P and Q flat iterative basic terms are expressible in $\mathrm{BPA}^{f*}(A)$. Expressibility of $P \parallel Q$ and $P \mathbin{⫗} Q$ in $\mathrm{BPA}^{f*}(A)$ can be proved in parallel, using induction on the size of such terms. We focus on the case $P \parallel Q$; the case $P \mathbin{\|\!\_} Q$ can be dealt with in a similar fashion. We consider three cases, depending on whether P and Q are of the form $\alpha^* R$.

(1) Let $P \equiv \alpha^* R$ and $Q \equiv \beta^* S$. Then we derive (using commutativity of $\parallel$ in $\mathrm{PA}^*(A)$)

$$P \parallel Q = (\alpha + \beta)(P \parallel Q) + R \mathbin{\|\!\_} Q + S \mathbin{\|\!\_} P.$$

The process terms $R \mathbin{\|\!\_} Q$ and $S \mathbin{\|\!\_} P$ have sizes smaller than $P \parallel Q$, so by induction they can be expressed in $\mathrm{BPA}^{f*}(A)$, say by U and V, respectively. By (RSP*),

$$P \parallel Q = (\alpha + \beta)^*(U + V)$$

so $P \parallel Q$ is expressible in $\mathrm{BPA}^{f*}(A)$.

(2) Let $P =_{\mathrm{AC}} \sum_i \alpha_i^* R_i + \sum_j a_j S_j + \sum_k b_k$ (with P not of the form $\alpha^* R$) and $Q \equiv \beta^* T$. Then we derive

$$\begin{aligned} P \parallel Q \;=\; & \beta(P \parallel Q) + T \mathbin{\|\!\_} P + \sum_i \alpha_i\big((\alpha_i^* R_i) \parallel Q\big) + \sum_j a_j(S_j \parallel Q) \\ & + \sum_k b_k Q. \end{aligned}$$

The process terms $T \mathbin{\|\!\_} P$, $(\alpha_i^* R_i) \parallel Q$, and $S_j \parallel Q$ have sizes smaller than $P \parallel Q$, so by induction they can be expressed in $\mathrm{BPA}^{f*}(A)$, say by U, V_i, and W_j, respectively. By (RSP*),

$$P \parallel Q \;=\; \beta^*\Big(U + \sum_i \alpha_i V_i + \sum_j a_j W_j + \sum_k b_k Q\Big)$$

so $P \parallel Q$ is expressible in $\mathrm{BPA}^{f*}(A)$.

(3) Let $P =_{\mathrm{AC}} \sum_i \alpha_i^* R_i + \sum_j a_j S_j + \sum_k b_k$ (with P not of the form $\alpha^* R$) and $Q =_{\mathrm{AC}} \sum_\ell \beta_\ell^* T_\ell + \sum_m c_m U_m + \sum_n d_n$ (with Q not of the form $\beta^* T$). Then we derive

$$\begin{aligned} P \parallel Q \;=\; & \sum_i \alpha_i\big((\alpha_i^* R_i) \parallel Q\big) + \sum_j a_j(S_j \parallel Q) + \sum_k b_k Q \\ & + \sum_\ell \beta_\ell\big((\beta_\ell^* T_\ell) \parallel P\big) + \sum_m c_m(U_m \parallel P) + \sum_n d_n P. \end{aligned}$$

The process terms $(\alpha_i^* R_i) \parallel Q$, $S_j \parallel Q$, $(\beta_\ell^* T_\ell) \parallel P$, and $U_m \parallel P$ have sizes smaller than $P \parallel Q$, so by induction they can be expressed in $\mathrm{BPA}^{f*}(A)$. Hence, $P \parallel Q$ is expressible in $\mathrm{BPA}^{f*}(A)$.

Owing to commutativity of the merge, the three cases above cover all possible forms of $P \parallel Q$. So we conclude that $P \parallel Q$ is expressible in $\mathrm{BPA}^{f*}(A)$. □

6.2. *Expressing a stack*

We provide recursive specifications of a stack over a finite data type in $\mathrm{ACP}^{\$}(A^{ext}, |, \tau)$, with the help of a regular control process and two counters. Let $D = \{\mathtt{d}_1, \ldots, \mathtt{d}_N\}$ for some $N \geqslant 1$ be a finite set of data elements, ranged over by $\mathtt{d}$. Let furthermore D^* be the set of finite strings over D, ranged over by σ, and let ε denote the empty string. The stack $S(\varepsilon)$ over D with empty-testing and termination option is defined by the infinite recursive specification

$$S(\varepsilon) = \left(\sum_{j=1}^{N} r(\mathtt{d}_j)S(\mathtt{d}_j)\right) + s(empty)S(\varepsilon) + r(sto),$$

$$S(\mathtt{d}\sigma) = \left(\sum_{j=1}^{N} r(\mathtt{d}_j)S(\mathtt{d}_j\mathtt{d}\sigma)\right) + s(\mathtt{d})S(\sigma).$$

Here the contents of the stack is represented by the argument of S: $S(\mathtt{d}\sigma)$ is the stack that contains $\mathtt{d}\sigma$ with $\mathtt{d}$ on top. Action $r(\mathtt{d}_i)$ (receive $\mathtt{d}_i$) models the push of $\mathtt{d}_i$ onto the stack, and action $s(\mathtt{d}_i)$ (send $\mathtt{d}_i$) represents deletion of $\mathtt{d}_i$ from the stack. Action $s(empty)$ models empty-testing of the (empty) stack, and action $r(stop)$ models termination of the (empty) stack. A non-terminating or non-empty-testing stack over D can be obtained by leaving out the concerning summand. In case $N = 1$ ($D = \{\mathtt{d}_1\}$), the recursive equations above specify a counter: the stack contents then models the counter value.

The following theorem is the second (and last) cornerstone of the universal expressivity result for $\mathrm{ACP}^{\$}(A, |, \tau)$.

THEOREM 6.2.1. *Each stack over a finite data type D with actions from A can be expressed in $\mathrm{ACP}^{\$}(A^{ext}, |, \tau)$ with A^{ext} a finite extension of A, even if one restricts to handshaking and the actions in A are not subject to communication.*

PROOF. Let a stack $S(\varepsilon)$ be given as described above. Without loss of generality, assume $D = \{\mathtt{d}_1, \ldots, \mathtt{d}_N\}$ for some $N > 1$ (if $N = 1$, then a counter does the job). Our approach is to encode the contents of the stack, i.e., elements from D^*, by natural numbers according to the following Gödel numbering $\ulcorner\cdot\urcorner : D^* \to \mathbb{N}$:

$$\ulcorner\varepsilon\urcorner \stackrel{\text{def}}{=} 0,$$

$$\ulcorner\mathtt{d}_j\sigma\urcorner \stackrel{\text{def}}{=} j + N \cdot \ulcorner\sigma\urcorner.$$

This encoding is a bijection with inverse $decode : \mathbb{N} \to D^*$ (let $\star$ denote concatenation of strings):

$$decode(n) \stackrel{\text{def}}{=} \begin{cases} \varepsilon & \text{if } n = 0, \\ \mathtt{d}_N \star decode(\frac{n-N}{N}) & \text{if } n \neq 0,\ n \ mod\ N = 0, \\ \mathtt{d}_{(n \ mod\ N)} \star decode(\frac{n-(n \ mod\ N)}{N}) & \text{otherwise.} \end{cases}$$

For example, if $N = 3$, then $\ulcorner \mathtt{d}_3\mathtt{d}_1\mathtt{d}_2 \urcorner = 24$ and $decode(32) = \mathtt{d}_2\mathtt{d}_1\mathtt{d}_3 \in \{\mathtt{d}_1, \mathtt{d}_2, \mathtt{d}_3\}^*$.

Next, we define two counters to specify $S(\varepsilon)$ in $\mathrm{ACP}^{\$}(A^{ext}, |, \tau)$:

$$C_j = \left(\overline{a}_j(\overline{a}_j^{\$}\overline{b}_j) + \overline{c}_j\right)^* \overline{d}_j \quad (j = 1, 2),$$

with add-action $\overline{a}_j$, subtract-action $\overline{b}_j$, zero-testing $\overline{c}_j$, and stop-action $\overline{d}_j$, all in $A^{ext} \setminus A$. We shall use the following abbreviations (for $n \in \mathbb{N}$):

$$\begin{aligned} C_j(0) &\stackrel{\text{def}}{=} C_j, \\ C_j(n+1) &\stackrel{\text{def}}{=} (\overline{a}_j{}^{\$}\overline{b}_j)C_j(n). \end{aligned}$$

We further define a regular control process X_ε with actions $a_j, b_j, c_j, d_j \in A^{ext} \setminus A$ and those of the stack. In combination with the C_j, the process X_ε is used to define $S(\varepsilon)$. Note that the coding discussed above does not occur explicitly in this recursive specification.

$$X_\varepsilon = \left(\sum_{j=1}^{N} r(\mathtt{d}_j) a_1^j X_j\right) + s(empty) X_\varepsilon + r(stop) d_1 d_2,$$

and for $k = 1, \ldots, N$:

$$\begin{aligned}
X_k &= \left(\sum_{j=1}^{N} r(\mathtt{d}_j)\,\mathrm{Push}_j\right) + s(\mathtt{d}_k)\,\mathrm{Pop}_k, \\
\mathrm{Push}_k &= (\text{Shift 1 to 2}) a_1^k (N \text{ Shift 2 to 1}) X_k, \\
\mathrm{Pop}_k &= b_1^k \left(\frac{1}{N} \text{ Shift 1 to 2}\right) \mathrm{Test}_\varepsilon, \\
\text{Shift 1 to 2} &= (b_1 a_2)^* c_1 \quad \text{(shift the contents of } C_1 \text{ to } C_2\text{)}, \\
N \text{ Shift 2 to 1} &= (b_2 a_1^N)^* c_2 \quad \text{(shift the } N\text{-fold of } C_2 \text{ to } C_1\text{)}, \\
\frac{1}{N} \text{ Shift 1 to 2} &= \left(b_1^N a_2\right)^* c_1 \quad \text{(shift the number of } N\text{-folds of } C_1 \text{ to } C_2\text{)}, \\
\mathrm{Test}_\varepsilon &= b_2 a_1 \mathrm{Test}_1 + c_2 X_\varepsilon \quad \text{(determine whether the stack is empty,} \\
\mathrm{Test}_1 &= b_2 a_1 \mathrm{Test}_2 + c_2 X_1 \quad \text{or which } D\text{-element is on top),} \\
\mathrm{Test}_2 &= b_2 a_1 \mathrm{Test}_3 + c_2 X_2, \\
&\ \vdots \\
\mathrm{Test}_N &= b_2 a_1 \mathrm{Test}_1 + c_2 X_N.
\end{aligned}$$

Let $|$ for $j = 1, 2$ be defined on $(A^{ext} \setminus A)^2$ by $a_j \mid \overline{a}_j = b_j \mid \overline{b}_j = c_j \mid \overline{c}_j = d_j \mid \overline{d}_j = in \in A^{ext} \setminus A$, and let $H = \{a_j, \overline{a}_j, b_j, \overline{b}_j, c_j, \overline{c}_j, d_j, \overline{d}_j \mid j = 1, 2\}$. We show that

$$\tau_{\{in\}} \circ \partial_H\left(X_\varepsilon \parallel C_1(0) \parallel C_2(0)\right)$$

behaves as $S(\varepsilon)$, the empty stack:

$$\begin{aligned}
&\tau_{\{in\}} \circ \partial_H\big(X_\varepsilon \parallel C_1(0) \parallel C_2(0)\big) \\
&\quad = \left(\sum_{j=1}^{N} r(\mathtt{d}_j)\tau_{\{in\}} \circ \partial_H\big(a_1^j X_j \parallel C_1(0) \parallel C_2(0)\big)\right) \\
&\qquad + s(empty)\tau_{\{in\}} \circ \partial_H\big(X_\varepsilon \parallel C_1(0) \parallel C_2(0)\big) \\
&\qquad + r(stop)\tau_{\{in\}} \circ \partial_H\big(d_1 d_2 \parallel C_1(0) \parallel C_2(0)\big) \\
&\quad = \left(\sum_{j=1}^{N} r(\mathtt{d}_j)\tau^{j+1}\tau_{\{in\}} \circ \partial_H\big(X_j \parallel C_1(j) \parallel C_2(0)\big)\right) \\
&\qquad + s(empty)\tau_{\{in\}} \circ \partial_H\big(X_\varepsilon \parallel C_1(0) \parallel C_2(0)\big) \\
&\qquad + r(stop)\tau\tau \\
&\quad = \left(\sum_{j=1}^{N} r(\mathtt{d}_j)\tau_{\{in\}} \circ \partial_H\big(X_j \parallel C_1(\ulcorner \mathtt{d}_j \urcorner) \parallel C_2(0)\big)\right) \\
&\qquad + s(empty)\tau_{\{in\}} \circ \partial_H\big(X_\varepsilon \parallel C_1(0) \parallel C_2(0)\big) + r(stop).
\end{aligned} \tag{3}$$

We are done if $\tau_{\{in\}} \circ \partial_H(X_j \parallel C_1(\ulcorner \mathtt{d}_j\sigma \urcorner) \parallel C_2(0))$ behaves as $S(\mathtt{d}_j\sigma)$ for some $\sigma \in D^*$. We prove this by first omitting the $\tau_{\{in\}}$-operation, and analyzing the behaviour of $\partial_H(X_j \parallel C_1(\ulcorner \mathtt{d}_j\sigma \urcorner) \parallel C_2(0))$. This analysis is arranged in a graphical style in Figure 1, where $P \xrightarrow{a} Q$ represents the statement $P = aQ$ for some $a \in A$, $P \overset{\sigma}{\twoheadrightarrow} Q$ represents $P = \sigma Q$, and branching represents an application of $+$. So the uppermost expression in Figure 1 with its arrows and resulting expressions represents the obviously derivable equation

$$\begin{aligned}
\partial_H\big(X_j \parallel C_1(\ulcorner \mathtt{d}_j\sigma \urcorner) \parallel C_2(0)\big) = &\left(\sum_{k=1}^{N} r(\mathtt{d}_k)\partial_H\big(\mathrm{Push}_k \parallel C_1(\ulcorner \mathtt{d}_j\sigma \urcorner) \parallel C_2(0)\big)\right) \\
&+ s(\mathtt{d}_j)\partial_H\big(\mathrm{Pop}_j \parallel C_1(\ulcorner \mathtt{d}_j\sigma \urcorner) \parallel C_2(0)\big).
\end{aligned}$$

By the axiom (B1), identity (3) above, and the derivation displayed in Figure 1 it follows that

$$\tau_{\{in\}} \circ \partial_H\big(X_\varepsilon \parallel C_1(0) \parallel C_2(0)\big) \quad \text{and} \quad \tau_{\{in\}} \circ \partial_H\big(X_j \parallel C_1(\ulcorner \mathtt{d}_j\sigma \urcorner) \parallel C_2(0)\big)$$

satisfy the recursive equations for $S(\varepsilon)$ and $S(\mathtt{d}_j\sigma)$, respectively ($j = 1, \ldots, N$ and $\sigma \in D^*$). By (RSP) it follows that

$$S(\varepsilon) = \tau_{\{in\}} \circ \partial_H(X_\varepsilon \parallel C_1 \parallel C_2).$$

By Theorem 6.1.1 and Lemma 6.1.2 it follows that once D is fixed, X_ε and hence the empty stack $S(\varepsilon)$ can be expressed in $\mathrm{ACP}^{\$}(A^{ext}, |, \tau)$ with handshaking for some $A^{ext} \supseteq A$. □

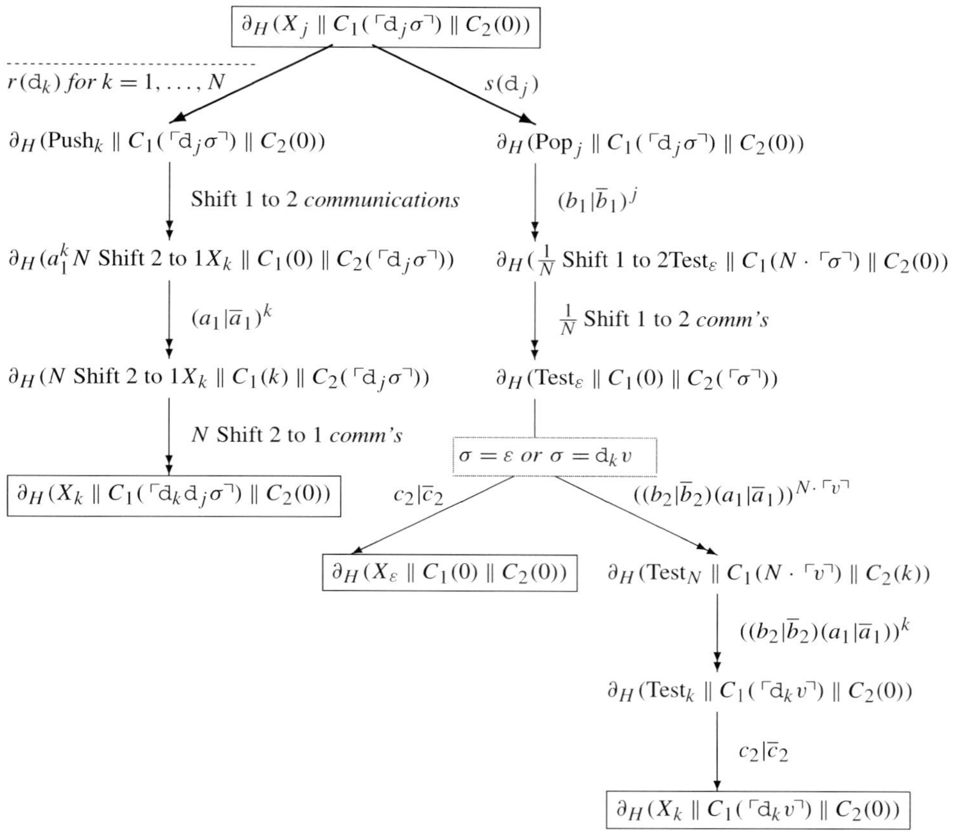

Fig. 1. Calculations with $\partial_H(X_j \parallel C_1(\ulcorner \mathtt{d}_j\sigma \urcorner) \parallel C_2(0))$.

Baeten, Bergstra, and Klop [11] showed that Turing machines can be specified in process algebra by means of two stacks and a regular control process. In view of Theorem 6.2.1, this yields that $\mathrm{ACP}^{\$}(A, |, \tau)$ is universally expressive; see [24] for details.

6.3. *Undecidability results*

We now sketch the undecidability result mentioned above for $\mathrm{ACP}^{*\$}(A, |)$. The idea is that in this signature one can 'implement' register machine computability in the following way.

(1) Registers (counters) have a straightforward definition in $\mathrm{ACP}^{*\$}(A, |)$, namely $(a(a^{\$}b) + c)^{*}d$ (cf. Section 6.1).

(2) Starting from a universal programming language for two-register machines (cf. Minsky in [65]), one can define a process algebraic representation of each program in $\mathrm{BPA}^{*}(A)$ (using a third register for I/O, and a fourth one as "program-line counter").

(3) Defining encapsulation in an appropriate way, this yields for any computable function $f:\mathbb{N}\to\mathbb{N}$ a process term P and a computable function $g:\mathbb{N}\to\mathbb{N}\setminus\{0\}$ such that the equation

$$\partial_H\big(Px \parallel C_0(n) \parallel C_1 \parallel C_2 \parallel C_3\big) = in^{g(n)}\partial_H\big(x \parallel C_0\big(f(n)\big) \parallel C_1 \parallel C_2 \parallel C_3\big)$$

can be derived from the axioms for $\mathrm{ACP}^{*\$}(A,|)$ if and only if $f(n)$ is defined, and the left-hand side equals an infinite *in*-trace otherwise. Here *in* is the result of a communication between the program (process term) P and the registers.

Now let W_{e_1}, W_{e_2} be recursively inseparable sets, and let $f:\mathbb{N}\to\mathbb{N}$ be the partial recursive function defined by

$$f(n) = \begin{cases} 0 & \text{if } n \in W_{e_1},\\ 1 & \text{if } n \in W_{e_2},\\ \text{undefined} & \text{otherwise.}\end{cases}$$

Choose P as described in item (3) above, and let P_1', P_2' and $P_1(n)$, $P_2(n)$ be defined by

$$\begin{aligned} P_1' &= P(s_3^*d_3)(s_2^*d_2)(s_1^*d_1)(s_0^*d_0),\\ P_2' &= P(s_3^*d_3)(s_2^*d_2)(s_1^*d_1)\big((s_0(s_0^*c_0))^*d_0\big),\\ P_i(n) &= \partial_H\big(P_i' \parallel C_0(n) \parallel C_1 \parallel C_2 \parallel C_3\big) \quad (i=1,2).\end{aligned}$$

Then we find

$$\begin{aligned} n\in W_{e_1} &\Rightarrow f(n)=0 \Rightarrow P_1(n)=P_2(n) \quad \big(=in^{g(n)+4}\big),\\ n\in W_{e_2} &\Rightarrow f(n)=1 \Rightarrow P_1(n)\neq P_2(n) \quad \big(in^{g(n)+5}\neq in^{g(n)+6}\big).\end{aligned}$$

As to the latter implication: assume otherwise, i.e., $in^k = in^{k+1}$ for some $k \geqslant 1$. Then by Lemma 2.2.2, $in^k \leftrightarrows in^{k+1}$, which clearly is a contradiction.

Thus, decidability of $P_1(n) = P_2(n)$ provides a recursive separation of W_{e_1} and W_{e_2}, which is contradictory. All details and a more precise explanation can be found in [24]. A similar proof strategy can be applied for $\mathrm{ACP}^{*\sharp}(A,|)$ and $\mathrm{ACP}^{*\leftrightarrows}(A,|)$, where counter-like processes are used instead.

7. Special constants

This section provides some last comments on two particular constants. First we shortly consider the silent step τ in relation to *fairness*, dealing with infinite τ-traces (cf. Definition 2.3.3 and Theorems 4.5.1–4.5.4). Finally, we briefly discuss the *empty process* in the context of iteration.

7.1. *Silent step and fairness*

Due to the character of τ, one would want to be able to abstract from infinite sequences of τ steps. Depending on the kind of process semantics adopted, different solutions have been found. In the case of rooted branching bisimulation, with next to (B1) the extra axiom

$$\text{(B2)} \quad x\big(\tau(y+z)+y\big)=x(y+z)$$

a general solution is provided by *Koomen's fair abstraction rule* [55,11]. For each n and each set of equations, there is a version KFAR_n^b that is valid in rooted branching bisimulation. For example, the axiom KFAR_1^b reads as follows:

$$\frac{x=ix+y \quad (i\in I)}{\tau\tau_I(x)=\tau\tau_I(y)}$$

(so the infinite τ sequence induced by ix is reduced to a single τ step). By definition of BKS we now have an immediate representation of the process in the premise of KFAR_1^b, namely i^*y. Henceforth we can represent KFAR_1^b by the law

$$\tau\tau_I(i^*y)=\tau\tau_I(y) \quad (i\in I).$$

Given the distribution law

$$\text{(BKS5)} \quad \tau_I(x^*y)=\tau_I(x)^*\tau_I(y)$$

(see Table 8), we can even represent KFAR_1^b simply by

$$(\mathrm{FIR}_1^b) \quad \tau(\tau^*x)=\tau x.$$

(taking x for $\tau_I(y)$), where FIR abbreviates Fair Iteration Rule.

EXAMPLE 7.1.1. A particular consequence of FIR_1^b is the case where x as above is replaced by τx:

$$\tau^*(\tau x)=\tau x, \tag{4}$$

the proof of which is trivial: $\tau^*(\tau x)=\tau(\tau^*(\tau x))+\tau x \overset{\mathit{FIR}_1^b}{=} \tau\tau x+\tau x=\tau x$.

As a small example of the use of FIR_1^b consider a statistic experiment which models the tossing of a coin until head comes up (cf. [15]). This process can be described by:

$$(\mathtt{throw\,tail})^*\mathtt{throw\,head}.$$

for actions `throw`, `tail`, and `head`. We assume that the probability of tossing heads is larger than 0. Thus we exclude the infinite trace that alternately executes `throw` and `tail`. Abstracting from just the two atomic actions in $I\overset{\text{def}}{=}\{\mathtt{throw},\mathtt{tail}\}$, FIR_1^b yields

$$\tau_I\big((\mathtt{throw\,tail})^*\mathtt{throw\,head}\big)=\tau\,\mathtt{head}.$$

First, observe $\tau_I(\texttt{throw tail}) = \tau$. Then, using (4), it easily follows that

$$\tau_I\big((\texttt{throw tail})^*\texttt{throw head}\big) = \tau\,\texttt{head}.$$

This expresses that head eventually comes up, and thus excludes the infinite sequence of τ-steps present in $\tau_I((\texttt{throw tail})^*\texttt{throw head})$.

7.2. *Empty process*

Let the symbol ε denote the *empty process*, introduced as a unit for sequential composition by Koymans and Vrancken in [56] (see also [15,74]). Obvious as ε may be (being a unit for $\cdot$), its introduction is nontrivial because at the same time it must be a unit for $\|$ as well. In the design of BPA, PA, ACP and related axiom systems, it has proved useful to study versions of the theory, both with and without ε. Just for this reason the star operation with its (original) defining equation as given by Kleene in [53] was introduced in process algebra.

Taking $y = \varepsilon$ in x^*y, one obtains $x^*\varepsilon$ which satisfies

$$x^*\varepsilon = x(x^*\varepsilon) + \varepsilon. \tag{5}$$

The unary operation $\_^*\varepsilon$ is a plausible candidate for the unary version of Kleene's star operation in process algebra. Moreover, taking $x = \delta$ in (5) implies that $\delta^*\varepsilon = \varepsilon$ (by the identities $\delta x = \delta$ and $\delta + x = x$), and hence that $\_^*\varepsilon$ cannot be used in a setting without having ε available as a separate process (once δ is accepted as one). So with ε, the interdefinability of the unary and the binary star, noted in [30], is preserved.

Milner [62] formulated an axiomatisation for the unary Kleene star in BPA with deadlock and empty process, modulo strong bisimulation equivalence. It remains an open question whether this axiomatisation is complete. Fokkink showed that Milner's axiomatisation adapted to no-exit iteration (NEI, see Section 4.1) is complete modulo strong bisimulation equivalence, in the presence of empty process.

A particular consequence of Milner's axiomatisation is (in our notation, using binary Kleene star)

$$\varepsilon^*x = x,$$

which seems a natural identity. Turning to the non-regular operations (see Section 6), the identity $\varepsilon^\sharp x = x$ seems as natural. The other two non-regular operations, i.e., the push-down \$ and the back and forth operation $\leftrightarrows$, have a more surprising effect when combined with ε. Using recursive specifications we find that $\varepsilon^\$ a$ is a solution of the recursive equation

$$X = X^2 + a,$$

and $\varepsilon^{\leftrightarrows} a$ is a solution of

$$X = Xa + a.$$

Both these recursive specifications are easily associated with infinitely branching processes. The (unguarded) specification $X = Xa + a$ occurs in [15] as an example specification that has two distinct solutions: $\sum_{i=1}^{\omega} a^i$ and $a^\omega + \sum_{i=1}^{\omega} a^i$. The transition rules for recursive specifications (see Table 3) as well as those for $\leftrightarrows$ yield the first solution. The interplay of recursive operations with empty process is apparently nontrivial and deserves further study.

Acknowledgements

We thank Faron Moller for providing useful comments.

References

[1] L. Aceto and W.J. Fokkink, *An equational axiomatization for multi-exit iteration*, Inform. and Comput. **137** (2) (1997), 121–158.

[2] L. Aceto, W.J. Fokkink, R.J. van Glabbeek and A. Ingólfsdóttir, *Axiomatizing prefix iteration with silent steps*, Inform. and Comput. **127** (1) (1996), 26–40.

[3] L. Aceto, W.J. Fokkink and A. Ingólfsdóttir, *On a question of A. Salomaa: The equational theory of regular expressions over a singleton alphabet is not finitely based*, Theoret. Comput. Sci. **209** (1/2) (1998), 163–178.

[4] L. Aceto, W.J. Fokkink and A. Ingólfsdóttir, *A menagerie of non-finitely based process semantics over BPA*: From ready simulation to completed traces*, Math. Struct. Comput. Sci. **8** (3) (1998), 193–230.

[5] L. Aceto, W.J. Fokkink and A. Ingólfsdóttir, *A Cook's tour of equational axiomatizations for prefix iteration*, Proceedings 1st Conference on Foundations of Software Science and Computation Structures (FoSSaCS'98), Lisbon, Lecture Notes in Comput. Sci. 1378, M. Nivat, ed., Springer-Verlag (1998), 20–34.

[6] L. Aceto, W.J. Fokkink and C. Verhoef, *Structural operational semantics*, Handbook of Process Algebra, J.A. Bergstra, A. Ponse and S.A. Smolka, eds, Elsevier, Amsterdam (2001), 197–292.

[7] L. Aceto and J.F. Groote, *A complete equational axiomatization for MPA with string iteration*, Theoret. Comput. Sci. **211** (1/2) (1999), 339–374.

[8] L. Aceto and A. Ingólfsdóttir, *An equational axiomatization of observation congruence for prefix iteration*, Proceedings 5th Conference on Algebraic Methodology and Software Technology (AMAST'96), Munich, Lecture Notes in Comput. Sci. 1101, M. Wirsing and M. Nivat, eds, Springer-Verlag (1996), 195–209.

[9] D.N. Arden, *Delayed logic and finite state machines*, Theory of Computing Machine Design, University of Michigan Press (1960), 1–35.

[10] F. Baader and T. Nipkow, *Term Rewriting and All That*, Cambridge University Press (1998).

[11] J.C.M. Baeten, J.A. Bergstra and J.W. Klop, *On the consistency of Koomen's fair abstraction rule*, Theoret. Comput. Sci. **51** (1/2) (1987), 129–176.

[12] J.C.M. Baeten, J.A. Bergstra and J.W. Klop, *Decidability of bisimulation equivalence for processes generating context-free languages*, J. ACM **40** (3) (1993), 653–682.

[13] J.C.M. Baeten and R.J. van Glabbeek, *Another look at abstraction in process algebra*, Proceedings 14th Colloquium on Automata, Languages and Programming (ICALP'87), Karlsruhe, Lecture Notes in Comput. Sci. 267, T. Ottmann, ed., Springer-Verlag (1987), 84–94.

[14] J.C.M. Baeten and C. Verhoef, *A congruence theorem for structured operational semantics with predicates*, Proceedings 4th Conference on Concurrency Theory (CONCUR'93), Hildesheim, Lecture Notes in Comput. Sci. 715, E. Best, ed., Springer-Verlag (1993), 477–492.

[15] J.C.M. Baeten and W.P. Weijland, *Process Algebra*, Cambridge Tracts in Theoretical Computer Science 18, Cambridge University Press (1990).

[16] J.A. Bergstra, I. Bethke and A. Ponse, *Process algebra with iteration*, Report P9314, Programming Research Group, University of Amsterdam (1993).

[17] J.A. Bergstra, I. Bethke and A. Ponse, *Process algebra with iteration and nesting*, Comput. J. **37** (4) (1994), 243–258.

[18] J.A. Bergstra, J.A. Hillebrand and A. Ponse, *Grid protocols based on synchronous communication*, Sci. Comput. Programming **29** (1/2) (1997), 199–233.

[19] J.A. Bergstra and J.W. Klop, *Process algebra for synchronous communication*, Inform. and Comput. **60** (1/3) (1984), 109–137.

[20] J.A. Bergstra and J.W. Klop, *Algebra of communicating processes with abstraction*, Theoret. Comput. Sci. **37** (1) (1985), 77–121.

[21] J.A. Bergstra and J.W. Klop, *Verification of an alternating bit protocol by means of process algebra*, Proceedings Spring School on Mathematical Methods of Specification and Synthesis of Software Systems 85, Wendisch-Rietz, Lecture Notes in Comput. Sci. 215, W. Bibel and K.P. Jantke, eds, Springer-Verlag (1986), 9–23.

[22] J.A. Bergstra, J.W. Klop and J.V. Tucker, *Algebraic tools for system construction*, Proceedings 4th Workshop on Logics of Programs, Pittsburgh, Lecture Notes in Comput. Sci. 164, E. Clarke and D. Kozen, eds, Springer-Verlag (1984), 34–44.

[23] J.A. Bergstra and A. Ponse, *Two recursive generalizations of iteration in process algebra*, Report P9808, Programming Research Group, University of Amsterdam (1998). Extended version to appear in: Theoret. Comput. Sci.

[24] J.A. Bergstra and A. Ponse, *Register machine based processes*, Programming Research Group, University of Amsterdam, June 13 (1999).

[25] S.L. Bloom and Z. Ésik, *Equational axioms for regular sets*, Math. Struct. Comput. Sci. **3** (1) (1993), 1–24.

[26] S.L. Bloom, Z. Ésik and D.A. Taubner, *Iteration theories of synchronization trees*, Inform. and Comput. **102** (1) (1993), 1–55.

[27] J. Boselie, *Expressiveness results for process algebra with iteration*, Master's Thesis, University of Amsterdam (1995).

[28] D.J.B. Bosscher, *Grammars modulo bisimulation*, Ph.D. Thesis, University of Amsterdam (1997).

[29] D. Caucal, *Graphes canoniques de graphes algébriques*, Theoret. Inform. Appl. **24** (4) (1990), 339–352.

[30] I.M. Copi, C.C. Elgot and J.B. Wright, *Realization of events by logical nets*, J. ACM **5** (1958), 181–196.

[31] J.H. Conway, *Regular Algebra and Finite Machines*, Chapman and Hall (1971).

[32] F. Corradini, R. De Nicola and A. Labella, *Fully abstract models for nondeterministic regular expressions*, Proceedings 6th Conference on Concurrency Theory (CONCUR'95), Philadelphia, Lecture Notes in Comput. Sci. 962, I. Lee and S.A. Smolka, eds, Springer-Verlag (1995), 130–144.

[33] F. Corradini, R. De Nicola and A. Labella, *Models of nondeterministic regular expressions*, J. Comput. System Sci. **59** (3) (1999), 412–449.

[34] F. Corradini, R. De Nicola and A. Labella, *A finite axiomatization of nondeterministic regular expressions*, Theoret. Inform. Appl. **33** (4) (1999), 447–466.

[35] S. Crvenković, I. Dolinka and Z. Ésik, *A note on equations for commutative regular languages*, Inform. Process. Lett. **70** (6) (1999), 265–267.

[36] W.J. Fokkink, *A complete equational axiomatization for prefix iteration*, Inform. Process. Lett. **52** (6) (1994), 333–337.

[37] W.J. Fokkink, *A complete axiomatization for prefix iteration in branching bisimulation*, Fundamenta Informaticae **26** (2) (1996), 103–113.

[38] W.J. Fokkink, *On the completeness of the equations for the Kleene star in bisimulation*, Proceedings 5th Conference on Algebraic Methodology and Software Technology (AMAST'96), Munich, Lecture Notes in Comput. Sci. 1101, M. Wirsing and M. Nivat, eds, Springer-Verlag (1996), 180–194.

[39] W.J. Fokkink, *Axiomatizations for the perpetual loop in process algebra*, Proceedings 24th Colloquium on Automata, Languages and Programming (ICALP'97), Bologna, Lecture Notes in Comput. Sci. 1256, P. Degano, R. Gorrieri and A. Marchetti-Spaccamela, eds, Springer-Verlag (1997), 571–581.

[40] W.J. Fokkink, *Language preorder as a precongruence*, Theoret. Comput. Sci. **243** (1/2) (2000), 391–408.

[41] W.J. Fokkink, *Rooted branching bisimulation as a congruence*, J. Comput. System Sci. **60** (1) (2000), 13–37.

[42] W.J. Fokkink, *Introduction to Process Algebra*, Springer-Verlag (2000).

[43] W.J. Fokkink and R.J. van Glabbeek, *Ntyft/ntyxt rules reduce to ntree rules*, Inform. and Comput. **126** (1) (1996), 1–10.

[44] W.J. Fokkink and H. Zantema, *Basic process algebra with iteration: Completeness of its equational axioms*, Comput. J. **37** (4) (1994), 259–267.

[45] R.J. van Glabbeek, *The linear time – branching time spectrum*, Proceedings 1st Conference on Concurrency Theory (CONCUR'90), Amsterdam, Lecture Notes in Comput. Sci. 458, J.C.M. Baeten and J.W. Klop, eds, Springer-Verlag (1990), 278–297. See also: Handbook of Process Algebra, J.A. Bergstra, A. Ponse and S.A. Smolka, eds, Elsevier, Amsterdam (2001), 3–99.

[46] R.J. van Glabbeek, *A complete axiomatization for branching bisimulation congruence of finite-state behaviours*, Proceedings 18th Symposium on Mathematical Foundations of Computer Science (MFCS'93), Gdansk, Lecture Notes in Comput. Sci. 711, A. Borzyszkowski and S. Sokołowski, eds, Springer-Verlag (1993), 473–484.

[47] R.J. van Glabbeek, *The linear time – branching time spectrum II: The semantics of sequential systems with silent moves*, Proceedings 4th Conference on Concurrency Theory (CONCUR'93), Hildesheim, Lecture Notes in Comput. Sci. 715, E. Best, ed., Springer-Verlag (1993), 66–81.

[48] R.J. van Glabbeek, *Axiomatizing flat iteration*, Proceedings 8th Conference on Concurrency Theory (CONCUR'98), Warsaw, Lecture Notes in Comput. Sci. 1243, A. Mazurkiewicz and J. Winkowski, eds, Springer-Verlag (1997), 228–242.

[49] R.J. van Glabbeek and W.P. Weijland, *Branching time and abstraction in bisimulation semantics*, J. ACM **43** (3) (1996), 555–600.

[50] J.F. Groote and F.W. Vaandrager, *Structured operational semantics and bisimulation as a congruence*, Inform. and Comput. **100** (2) (1992), 202–260.

[51] M. Hennessy, *A term model for synchronous processes*, Inform. and Control **51** (1) (1981), 58–75.

[52] S.C. Kleene, *Representation of events in nerve nets and finite automata*, Research Memorandum RM-704, U.S. Air Force Project RAND, The RAND Cooperation (15 December 1951).

[53] S.C. Kleene, *Representation of events in nerve nets and finite automata*, Automata Studies, C. Shannon and J. McCarthy, eds, Princeton University Press (1956), 3–41.

[54] D.E. Knuth and P.B. Bendix, *Simple word problems in universal algebras*, Computational Problems in Abstract Algebra, J. Leech, ed., Pergamon Press (1970), 263–297.

[55] C.J. Koomen, *A structure theory for communication network control*, Ph.D. Thesis, Delft Technical University (1982).

[56] C.P.J. Koymans and J.L.M. Vrancken, *Extending process algebra with the empty process ε*, Logic Group Preprint Series Nr. 1, CIF, Utrecht University (1985).

[57] D. Kozen, *A completeness theorem for Kleene algebras and the algebra of regular events*, Inform. and Comput. **110** (2) (1994), 366–390.

[58] D. Krob, *Complete systems of B-rational identities*, Theoret. Comput. Sci. **89** (2) (1991), 207–343.

[59] W.S. McCulloch and W. Pitts, *A logical calculus of ideas immanent in nervous activity*, Bull. Math. Biophys. **5** (1943), 115–133.

[60] R. Milner, *A Calculus of Communicating Systems*, Lecture Notes in Comput. Sci. 92, Springer-Verlag (1980).

[61] R. Milner, *A modal characterisation of observable machine-behaviour*, Proceedings 6th Colloquium on Trees in Algebra and Programming (CAAP'81), Genoa, Lecture Notes in Comput. Sci. 112, E. Astesiano and C. Böhm, eds, Springer-Verlag (1981), 25–34.

[62] R. Milner, *A complete inference system for a class of regular behaviours*, J. Comput. System Sci. **28** (3) (1984), 439–466.

[63] R. Milner, *Communication and Concurrency*, Prentice Hall (1989).

[64] R. Milner, *A complete axiomatisation for observational congruence of finite-state behaviours*, Inform. and Comput. **81** (2) (1989), 227–247.

[65] M.L. Minsky, *Computation: Finite and Infinite Machines*, Prentice Hall (1967).

[66] D.M.R. Park, *Concurrency and automata on infinite sequences*, Proceedings 5th GI (Gesellschaft für Informatik) Conference, Karlsruhe, Lecture Notes in Comput. Sci. 104, P. Deussen, ed., Springer-Verlag (1981), 167–183.

[67] D.A. Plaisted, *Equational reasoning and term rewriting systems*, Handbook of Logic in Artificial Intelligence and Logic Programming, Vol. 1, D. Gabbay and J. Siekmann, eds, Oxford University Press (1993), 273–364.

[68] G.D. Plotkin, *A structural approach to operational semantics*, Report DAIMI FN-19, Computer Science Department, Aarhus University (1981).

[69] V.N. Redko, *On defining relations for the algebra of regular events*, Ukrain. Mat. Zh. **16** (1964), 120–126, in Russian.

[70] A. Salomaa, *Two complete axiom systems for the algebra of regular events*, J. ACM **13** (1) (1966), 158–169.

[71] A. Salomaa, *Theory of Automata*, International Series of Monographs in Pure and Applied Mathematics 100, Pergamon Press (1969).

[72] P.M. Sewell, *Nonaxiomatisability of equivalences over finite state processes*, Ann. Pure Appl. Logic **90** (1/3) (1997), 163–191.

[73] D.R. Troeger, *Step bisimulation is pomset equivalence on a parallel language without explicit internal choice*, Math. Struct. Comput. Sci. **3** (1) (1993), 25–62.

[74] J.L.M. Vrancken, *The algebra of communicating processes with empty process*, Theoret. Comput. Sci. **177** (2) (1997), 287–328.

[75] H. Zantema, *Termination of term rewriting by semantic labelling*, Fundamenta Informaticae **24** (1/2) (1995), 89–105.

[76] M.B. van der Zwaag, *Some verifications in process algebra with iota*, Proceedings 3rd Workshop on Formal Methods for Industrial Critical Systems (FMICS'98), Amsterdam, J.F. Groote, S.P. Luttik and J.J. van Wamel, eds, Stichting Mathematisch Centrum (1998), 347–368.

Subject index

$+$, 339
$, 374
$*$, 335, 345
$\leftrightarrows$, 374
$\underline{\leftrightarrow}$, 341
$\cdot$, 339
δ, 339
∂_H, 339
ω, 358
ω-completeness, 347
$\|$, 339
$\llfloor$, 339
$|$, 339
$\sharp$, 374
τ, 339
τ-convergence, 345
τ-guarded, 344
τ_I, 339
ε, 384

abstraction, 339
ACP, 338, 339
action, 338
– communication, 338
– exit, 365
algebra of communicating processes, 338, 339

back and forth operation, 374
binary Kleene star, 345
bisimulation equivalence
– (rooted) branching, 344
– (rooted) weak, 344
– strong, 341
bisimulation relation, 341
BKS, 345
BPA, 339

communication
– action, 338
– handshaking, 339
– merge, 339
completeness, 342, 346, 347, 358, 361
congruence relation, 342
counter, 377
cycle, 365

deadlock, 339

empty process, 384
encapsulation, 339
equivalence
– language, 343
– ready simulation, 343
– simulation, 343
– strong bisimulation, 341
– trace, 343
exit action, 365

fairness, 383
FIR, 383
flat iteration, 363

handshaking, 339
HNF-expansion, 347

iteration
– flat, 363
– multi-exit, 360
– no-exit, 358
– prefix, 361
– string, 363

KFAR, 383
Kleene star
– binary, 345
– unary, 384

labelled transition system, 341
language equivalence, 343
left merge, 339
loop, *see* cycle
LTS, 341

merge, 339
– communication merge, 339
– left merge, 339
multi-exit iteration, 360

nesting operation, 374
no-exit iteration, 358
non-deterministic choice, 339
normed process term, 347

PA, 340
parallel composition, 339
prefix iteration, 361
process term, 339
– normed, 347
push-down operation, 374

ready simulation equivalence, 343
ready simulation relation, 343
recursive specification, 344
– solution of a, 344
recursive specification principle, *see* RSP
regular process, 345
RSP, 345
– $RSP^{\leftrightarrows}$, 374
– $RSP^{\$}$, 375
– RSP^{*}, 358
– RSP^{ω}, 359
– $RSP^{\sharp}$, 374

sequential composition, 339
silent step, 339
simulation equivalence, 343
simulation relation, 343
state, 341
– initial, 341
– proper substate, 343
– substate, 343
– successor of a, 365
string iteration, 363
strong bisimulation equivalence, 341
substate, 343
sum, 339

trace, 343
trace equivalence, 343
transition, 341
– labelled transition system, 341
– rule, 341

Abstract

This chapter surveys algorithms for computing semantic equivalences and refinement relations, or *preorders*, over states in finite-state labeled transitions systems. Methods for calculating a general equivalence, namely bisimulation equivalence, and a general preorder are described and shown to be useful as a basis for calculating other semantic relations as well. Two general classes of algorithms are considered: global ones, which require the *a priori* construction of the state space but are generally more efficient in the asymptotic case, and local, or on-the-fly ones, which avoid the construction of unnecessary states while incurring some additional computational overhead.

1. Introduction

Research in process algebra has focused on the use of behavioral relations such as equivalences and refinement orderings as a basis for establishing system correctness (see [16] in this Handbook and [3,2,20,22,23]). In the process-algebraic framework specifications and implementations are both given as terms in the same algebra; the intuition is that a specification describes the desired high-level behavior the system should exhibit, while the implementation details the proposed means for achieving this behavior. One then uses an appropriate equivalence or preorder to establish that the implementation conforms to the specification. In the case of equivalence-based reasoning, conformance means "has the same behavior as"; in this case an implementation is correct if its behavior is indistinguishable from that of the specification. Refinement (or preorder) relations, on the other hand, typically embody a notion of "better than": an implementation conforms to (or refines) a specification if the behavior of the former is "at least as good as" that stipulated by the specification. The benefits of such process-algebraic approaches include the following.

- Users need only learn one language in order to write specifications and implementations.
- The algebra provides explicit support for *compositional* specifications and implementations, allowing the specification (implementation) of a system to be built up from the specifications (implementations) of its components.
- Specifications include information about what is disallowed as well as what is allowed.

Consequently, a number of different process algebras have been studied, and a variety of different equivalences and refinement relations capturing different aspects of behavior have been developed.

A hallmark of process-oriented behavioral relations is that they are usually defined with respect to *labeled transition systems*, which form a semantic model of systems, rather than with respect to a particular syntax of process descriptions. This style of definition permits notions of equivalence and refinement to be applied to any algebra with a semantics given in terms of labeled transition systems. It also means techniques for establishing these relations may be given in terms of labeled transition systems. If, in addition, these labeled transition systems are finitary, then the relations may be calculated in a purely mechanical manner: tools may then be developed for automatically checking that (finitary) implementations conform to (finitary) specifications.

This paper surveys algorithms for calculating behavioral relations for a particular class of finitary labeled transition systems, namely, those consisting of a finite number of states and transition labels. We focus on relations that are sensitive only to the degree of nondeterminism systems may exhibit; we do not consider relations sensitive to other aspects of system behavior such as timing, probability, priority, or parallelism. See Chapters 10–13 of this volume, respectively, for a treatment of these features. This decision is due to the fact that models of nondeterminism have a longer history and hence are more settled, and to the fact that techniques for computing these relations find direct application in the computation of the others.

The remainder of this chapter has the following structure. The next section introduces the basic definitions and notations used throughout the remainder of the survey. The discussion then breaks into two major parts. Section 3 presents algorithms for calculating behavioral equivalences using *partition refinement* and *symbolic* techniques. Section 4 considers an algorithm for behavioral preorders. Section 5 introduces *local*, or *on-the-fly* techniques. Local algorithms aim to reduce the amount of work required by avoiding the *a priori* construction of the entire state space. Section 6 presents some of the tools that implement the algorithms described in this chapter, while Section 7 concludes.

2. Basic definitions

This section contains definitions of concepts and notations used in the remainder of the chapter.

2.1. *Labeled transition systems*

Semantically, systems are modeled as labeled transition systems, which may be defined as follows.

DEFINITION 2.1. A *labeled transition system* (LTS) is a triple $\langle \mathcal{S}, \mathcal{A}, \rightarrow \rangle$, where $\mathcal{S}$ is a set of states, $\mathcal{A}$ is a set of actions, and $\rightarrow \subseteq \mathcal{S} \times \mathcal{A} \times \mathcal{S}$ is the transition relation.

Intuitively, an LTS $\langle \mathcal{S}, \mathcal{A}, \rightarrow \rangle$ defines a computational framework, with S representing the set of states that systems may enter, A the actions systems may engage in, and $\rightarrow$ the execution steps system undergo as they perform actions. In what follows we generally write $s \xrightarrow{a} s'$ in lieu of $\langle s, a, s' \rangle \in \rightarrow$, and we say that s' is an a-derivative of s. We use $\xrightarrow{a}{}^{*}$ to denote the transitive closure of $\xrightarrow{a}$. We define a *process* to be a quadruple $\langle \mathcal{S}, \mathcal{A}, \rightarrow, s_I \rangle$ where $\langle \mathcal{S}, \mathcal{A}, \rightarrow \rangle$ is an LTS and $s_I \in S$ is the start state.

Let $\langle \mathcal{S}, \mathcal{A}, \rightarrow \rangle$ be an LTS, and let $s \in \mathcal{S}$ be a state and $a \in \mathcal{A}$ an action. We use the following terminology and notations in what follows.

- $\langle \mathcal{S}, \mathcal{A}, \rightarrow \rangle$ is finite-state if S and A are both finite sets.
- $s \xrightarrow{a}$ holds if $s \xrightarrow{a} s'$ for some $s' \in \mathcal{S}$.
- $\{\bullet \xrightarrow{a} s\} \subseteq \mathcal{S}$, the *preset* of s with respect to a, is the set $\{r \in \mathcal{S} \mid r \xrightarrow{a} s\}$.
- $\{s \xrightarrow{a} \bullet\} \subseteq \mathcal{S}$, the *postset* of s with respect to a, is the set $\{t \in \mathcal{S} \mid s \xrightarrow{a} t\}$.
- $\{s \xrightarrow{\bullet}\} \subseteq \mathcal{A}$, the *initial actions* of s, is the set $\{a \in \mathcal{A} \mid s \xrightarrow{a}\}$.
- $\{s \rightarrow^{*} \bullet\} \subseteq \mathcal{S}$, the *reachable* set of states from s, is the smallest set satisfying the following:
  - $s \in \{s \rightarrow^{*} \bullet\}$.
  - If $t \in \{s \rightarrow^{*} \bullet\}$ and $t \xrightarrow{a} t'$ for some $a \in \mathcal{A}$ then $t' \in \{s \rightarrow^{*} \bullet\}$.

These notions may be lifted to sets of states by taking unions in the obvious manner. Thus if $S \subseteq \mathcal{S}$ then we have the following:

$$\{\bullet \xrightarrow{a} S\} = \bigcup_{s \in S} \{\bullet \xrightarrow{a} s\},$$

$$\{S \xrightarrow{a} \bullet\} = \bigcup_{s \in S} \{s \xrightarrow{a} \bullet\},$$

$$\{S \xrightarrow{\bullet}\} = \bigcup_{s \in S} \{s \xrightarrow{\bullet}\},$$

$$\{S \rightarrow^* \bullet\} = \bigcup_{s \in S} \{s \rightarrow^* \bullet\}.$$

The traditional approach to defining the semantics of process algebras involves constructing an LTS in the following manner. Firstly, the syntax of the algebra includes a set $\mathcal{A}$ of actions and a set $\mathcal{P}$ of process terms. Then a transition relation $\rightarrow \subseteq \mathcal{P} \times \mathcal{A} \times \mathcal{P}$ is defined inductively in the SOS style using proof rules [25] (but also see [1] in this Handbook). The structure $\langle \mathcal{P}, \mathcal{A}, \rightarrow \rangle$ constitutes an LTS that in essence encodes all possible behavior of all processes. Of course, this LTS is not usually finite-state, so one may wonder how algorithms for finite-state systems could be used for determining if two process terms in a given algebra are semantically related. The answer lies in the fact that in general, one does not need to consider the entire LTS of the algebra; it typically suffices to consider only the terms reachable from the ones in question. If this reachable set is finite (and typically one may give syntactic characterizations of terms satisfying this property) then one may apply the algorithms presented in this chapter to the LTS induced by the finite set of reachable states. We return to this point later.

2.2. *Bisimulation equivalence*

Bisimulation equivalence is interesting in its own right as a basis for relating processes; it also may be seen as a basis for defining other relations as well. Bisimulation and other behavioral equivalences are treated in more detail in [16] in this Handbook.

DEFINITION 2.2 (*Bisimulation equivalence*). Let $\langle \mathcal{S}, \mathcal{A}, \rightarrow \rangle$ be an LTS.

- A relation $R \subseteq \mathcal{S} \times \mathcal{S}$ is a *bisimulation* if whenever $\langle s_1, s_2 \rangle \in R$ then the following hold for all $a \in \mathcal{A}$:
  1. If $s_1 \xrightarrow{a} s_1'$ then there is an s_2' such that $s_2 \xrightarrow{a} s_2'$ and $\langle s_1', s_2' \rangle \in R$.
  2. If $s_2 \xrightarrow{a} s_2'$ then there is an s_1' such that $s_1 \xrightarrow{a} s_1'$ and $\langle s_1', s_2' \rangle \in R$.
- Two states $s_1, s_2 \in \mathcal{S}$ are *bisimulation equivalent*, written $s_1 \sim s_2$, if there exists a bisimulation R such that $\langle s_1, s_2 \rangle \in R$.

Intuitively, two states in an LTS are bisimulation equivalent if they can "simulate" each other's transitions. Under this interpretation a bisimulation indicates how transitions from related states may be matched in order to ensure that the "bi-simulation" property holds.

Bisimulation equivalence enjoys a number of mathematical properties. Firstly, it is indeed an equivalence relation in that it is reflexive, symmetric and transitive. Secondly, it is itself a bisimulation, and in fact is the largest bisimulation with respect to set containment. Finally, it may be seen as the greatest fixpoint of the following function mapping relations to relations.

DEFINITION 2.3. Let $\mathcal{L} = \langle \mathcal{S}, \mathcal{A}, \rightarrow \rangle$ be an LTS. Then $\mathcal{F}_\mathcal{L} : 2^{\mathcal{S}\times\mathcal{S}} \rightarrow 2^{\mathcal{S}\times\mathcal{S}}$ is given by:

$$\mathcal{F}_\mathcal{L}(R) = \Big\{ \langle s_1, s_2 \rangle \mid \forall a \in \mathcal{A}. \forall s' \in \mathcal{S}. \, \big(s_1 \xrightarrow{a} s' \Rightarrow \exists t' \in \mathcal{S}. s_2 \xrightarrow{a} t' \wedge \langle s', t' \rangle \in R\big) \wedge \big(s_2 \xrightarrow{a} s' \Rightarrow \exists r' \in \mathcal{S}. s_1 \xrightarrow{a} r' \wedge \langle r', s' \rangle \in R\big) \Big\}.$$

THEOREM 2.4. *Let $\mathcal{L} = \langle \mathcal{S}, \mathcal{A}, \rightarrow \rangle$ be an LTS. Then $\sim = \mathcal{F}_\mathcal{L}(\sim)$, and for any R such that $R = \mathcal{F}_\mathcal{L}(R)$, $R \subseteq \sim$.*

The proof of this theorem relies on the Tarski–Knaster fixpoint theorem, which gives characterizations of the fixpoints of monotonic functions over complete lattices. In this case, the complete lattice in question is the set $2^{\mathcal{S}\times\mathcal{S}}$ of binary relations ordered by set inclusion; in this lattice it is easy to see that $\mathcal{F}_\mathcal{L}$ is monotonic (the more pairs there are in R, the more pairs there are in $F_\mathcal{L}(R)$), and hence the Tarski–Knaster theorem is applicable.

The characterization of bisimulation equivalence in Theorem 2.4 suggests that in order to compute the relation over a given LTS $\mathcal{L}$, it suffices to calculate the greatest fixpoint of $\mathcal{F}_\mathcal{L}$. The next result suggests how this might be done for finite-state LTSs.

DEFINITION 2.5. An LTS $\langle \mathcal{S}, \mathcal{A}, \rightarrow \rangle$ is *image-finite* if for every $s \in \mathcal{S}$ and $a \in \mathcal{A}$, the set $\{s \xrightarrow{a} \bullet\}$ is finite.

In other words, an LTS is image-finite if every state has a finite number of outgoing transitions for any given action. Certainly, any finite-state LTS is also image-finite.

Now, when an LTS $\mathcal{L}$ is image-finite, the function $\mathcal{F}_\mathcal{L}$ turns out to be continuous, and its greatest fixpoint (i.e., $\sim$) has the following "iterative" characterization.

THEOREM 2.6. *Let $\mathcal{L} = \langle \mathcal{S}, \mathcal{A}, \rightarrow \rangle$ be an image-finite LTS. Then $\sim = \bigcap_{i=0}^{\infty} \sim_i$, where the $\sim_i$ are defined as follows:*

$$\begin{aligned} \sim_0 &= \mathcal{S} \times \mathcal{S}, \\ \sim_{i+1} &= \mathcal{F}_\mathcal{L}(\sim_i) \quad \textit{for } i \geqslant 0. \end{aligned}$$

This characterization provides the basis for the algorithms discussed later in the chapter.

Parameterized bisimulation equivalence. We will show that other semantic equivalences may be computed by combining appropriate transformations on labeled transition systems with a bisimulation algorithm. For this purpose, it turns out that a slight modification of the definition of bisimulation equivalence is useful.

DEFINITION 2.7. Let $\langle \mathcal{S}, \mathcal{A}, \rightarrow \rangle$ be an LTS, and let $E \subseteq \mathcal{S} \times \mathcal{S}$ be an equivalence relation.

1. A relation $R \subseteq \mathcal{S} \times \mathcal{S}$ is an *E-bisimulation* if R is a bisimulation and $R \subseteq E$.
2. Two states $s_1, s_2 \in S$ are *E-bisimulation equivalent*, written $s_1 \sim^E s_2$, if there is an E-bisimulation R with $\langle s_1, s_2 \rangle \in R$.

The relation $\sim^E$ differs from $\sim$ in that it requires equivalent states to be related by E in addition to satisfying the transition conditions imposed by $\sim$. It is easy to show that $\sim$

coincides with $\sim^U$, where U is the universal relation relating every state to every other state. Theorems 2.4 and 2.6 have obvious analogs for $\sim^E$.

2.3. *Simulation-based refinement orderings*

Refinement orderings relate states in an LTS on the basis of the relative "quality" of their behavior. The (forward) simulation ordering represents one such notion that plays an important algorithmic role. Its definition is as follows.

DEFINITION 2.8. Let $\langle \mathcal{S}, \mathcal{A}, \rightarrow \rangle$ be an LTS.

- A relation $R \subseteq \mathcal{S} \times \mathcal{S}$ is a *forward simulation* if whenever $\langle s_1, s_2 \rangle \in R$ then the following holds for all $a \in \mathcal{A}$.

$$\text{If } s_1 \xrightarrow{a} s_2 \text{ then there is an } s_2' \text{ such that } s_2 \xrightarrow{a} s_2' \text{ and } \langle s_1', s_2' \rangle \in R.$$

- $s_1 \sqsubseteq s_2$ if there is a forward simulation R such that $\langle s_1, s_2 \rangle \in R$.

Note that $s_1 \sqsubsim s_2$ holds when s_2 is able to track, or "simulate", the transitions that s_1 is capable of. The relation $\sqsubsim$ turns out to be a preorder (i.e., reflexive and transitive), and like bisimulation it turns out to be the greatest fixpoint of an appropriately defined function over relations.

DEFINITION 2.9. Let $\mathcal{L} = \langle \mathcal{S}, \mathcal{A}, \rightarrow \rangle$ be an LTS. Then $\mathcal{G}_\mathcal{L} : 2^{\mathcal{S}\times\mathcal{S}} \rightarrow 2^{\mathcal{S}\times\mathcal{S}}$ is defined as follows

$$\mathcal{G}_\mathcal{L}(R) = \left\{ \langle s_1, s_2 \rangle \mid \forall a \in \mathcal{A}. \forall s' \in \mathcal{S}. \left(s_1 \xrightarrow{a} s' \Rightarrow \exists t'. s_2 \xrightarrow{a} t' \wedge \langle s', t' \rangle \in R \right) \right\}.$$

That $\sqsubsim$ is the greatest fixpoint of $\mathcal{G}_\mathcal{L}$ follows from the same line of reasoning offered for Theorem 2.4. In addition, one may give the following analog of Theorem 2.6, which provides a basis for algorithms presented later in the chapter.

THEOREM 2.10. *Let $\mathcal{L} = \langle \mathcal{S}, \mathcal{A}, \rightarrow \rangle$ be an image-finite LTS. Then $\sqsubsim = \bigcap_{i=0} \sqsubsim_i$, where the $\sqsubsim_i$ are defined as follows.*

$$\begin{aligned} \sqsubsim_0 &= \mathcal{S} \times \mathcal{S}, \\ \sqsubsim_{i+1} &= \mathcal{G}_\mathcal{L}(\sqsubsim_i) \quad \textit{for } i \geqslant 0. \end{aligned}$$

The backward simulation ordering $\sqsupsim$ is the inverse of $\sqsubsim$: $s_1 \sqsupsim s_2$ if and only if $s_2 \sqsubsim s_1$.

2.4. *A parameterized semantic relation*

We close this section with the definition of a parameterized semantic relation that proves useful as a basis for calculating other simulation and bisimulation relations.

$P := \{\mathcal{S}\}$
changed := *true*
while *changed* do
begin
  changed := *false*
  for each $B \in P$ do
  begin
    for each $a \in \mathcal{A}$ do
    begin
      SortTransitions(a, B)
      if *split*$(B, a, P) \neq \{B\}$
      then begin
        $P := P - \{B\} \cup split(B, a, P)$
        changed := *true*
        break
      end
    end
  end
end

Fig. 3. Algorithm *KS_PARTITIONING*.

In order to compare the postsets of the states of B efficiently, we need to order the transitions of s. For this purpose, we impose an ordering on the blocks of P. The transitions of s are lexicographically ordered by their labels. Further, for each label a, the transitions are ordered by the containing block of the target state of the transition. When a block is split, the ordering of transitions in states that have transitions into that block can be violated. Therefore, one needs to sort the a-transitions of all states of a block immediately before attempting to split the block. Procedure *SortTransitions*(a, B) uses lexicographic sorting to reorder the a-transitions of block B.

Finally, we present the main loop of *KS_PARTITIONING* in Figure 3. The algorithm iteratively attempts splitting of every block in P with respect to every $a \in A$ until no more blocks can be split.

Correctness of *KS_PARTITIONING* relies on the fact that when *changed* is *false*, there is no splitter for any of the blocks in P. Therefore, $P = \mathcal{F}_{\mathcal{L}}(P)$ and, by Theorem 2.4, $R \subseteq \sim$. Moreover, if we denote by P_i the partition after ith iteration of the main loop of *KS_PARTITIONING*, we have $\sim \subseteq \sim_i \subseteq P_i$. Thus we have that at termination of the algorithm, $P = \sim$.

The complexity of *KS_PARTITIONING* is given by the following theorem.

THEOREM 3.1. *Given a finite-state LTS* $\langle \mathcal{S}, \mathcal{A}, \rightarrow \rangle$ *with* $|\mathcal{S}| = n$ *and* $|\rightarrow| = m$, *algorithm KS_PARTITIONING takes* $O(n \cdot m)$ *time.*

PROOF. The main loop of the algorithm is repeated at most n times. Within one iteration of the main loop, procedure *split* is called for each block at most once for each action a.

In turn, *split* considers each transition of every state in the block at most once. Therefore, the calls to *split* within one iteration of the main loop take $O(m)$ time. The calls to *SortTransitions* collectively take $O(|\mathcal{A}| + m)$ time, or $O(m)$ when the set of labels is bounded by a constant. □

3.2. *The Paige–Tarjan algorithm for bisimulation equivalence*

Performance of the basic partition refinement algorithm can be significantly improved through the use of more complex data structures. Paige and Tarjan [24] proposed an algorithm that utilizes information about previous splits to make future splits more efficient. To simplify the presentation of the algorithm, we describe the case where the alphabet of the LTS is a singleton. An extension of the algorithm to handle multiple actions is straightforward. First, we introduce the notion of *stability* of blocks and partitions.

DEFINITION 3.2.

- A block B is stable with respect to a block S if either $B \subseteq \{\bullet \to S\}$ or $B \cap \{\bullet \to S\} = \emptyset$.
- A partition P is stable with respect to a block S if every $B \in P$ is stable with respect to S.
- A partition P is stable with respect to partition Q if P is stable with respect to every $S \in Q$.
- A partition is stable if it is stable with respect to itself.

Clearly, $\sim$ is the coarsest stable partition.

The Paige–Tarjan algorithm is based on the following observation. Let B be stable with respect to S, and let S be partitioned into S_1 and S_2. Then, if $B \cap S = \emptyset$, B is stable with respect to both S_1 and S_2. Otherwise, B can be split into *three* blocks:

$$\begin{aligned} B_1 &= B - \{\bullet \to S_2\}, \\ B_{12} &= B \cap \{\bullet \to S_1\} \cap \{\bullet \to S_2\}, \\ B_2 &= B - \{\bullet \to S_1\}. \end{aligned}$$

This three-way splitting is illustrated in Figure 4.

The improvement in complexity that the Paige–Tarjan algorithm provides over the basic partition refinement algorithm of the previous section stems from the fact that three-way splitting can be performed in time proportional to the size of the smaller of the two blocks S_1, S_2. To do so, with every state $s \in S$ and for every block S, we keep a variable $count(s, S) = |\{s' \in S \mid s \xrightarrow{s'}\}$. Now, we can decide in constant time to which of the sub-blocks of B a state $s \in B$ belongs. We have three cases: (1) if $count(s, S_1) = count(s, S)$, then $s \in B_1$; (2) if $0 < count(s, S_1) < count(s, S)$, then $s \in B_{12}$; (3) if $count(s, S_1) = 0$, then $s \in B_2$.

In addition to maintaining the *count* variables, we have to store the information about the history of prior splits. If a block has been used as a splitter, it is stable with respect to a partition. If such a block has been split itself, the smaller of its sub-blocks can be used in three-way splitting. We use an additional data structure to store split history: X is

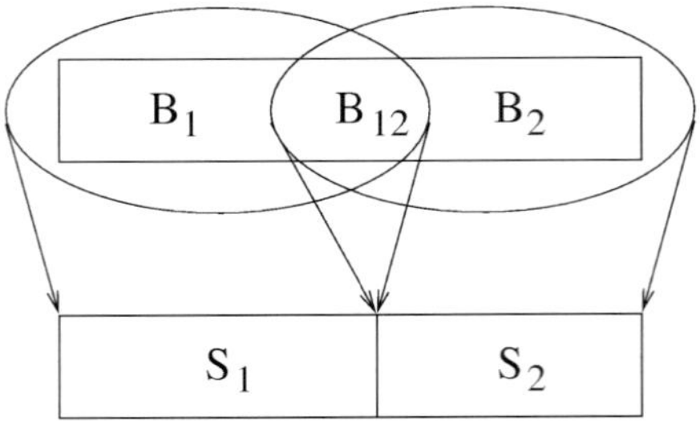

Fig. 4. Efficient splitting of a block.

```
P := {S}
X := P
while X ≠ ∅ do
begin
   choose S ∈ X
   X := X − S
   {* choose splitter *}
   if S is compound then
   begin
      Sp := the smaller of the children of S
      {* compound children of S stay in X *}
      X := X ∪ compound(S)
   end
   else
      Sp := S
   for each B ∈ {• → Sp} do
   begin
      P' = split(B, Sp)
      update X(B, P')
   end end
```

Fig. 5. Algorithm *PT_PARTITIONING*.

a forest of binary trees. Each node in a tree is a block satisfying the following conditions:

- For each leaf block B, $B \in P$.
- Each non-leaf block is the union of its children.
- For each root block B, P is stable with respect to B.

Roots of the trees are used as splitters for the current partition until X is empty. A splitter is called *compound* if it is the root of a non-trivial tree; otherwise it is called *simple*. For a compound splitter S, *compound*(S) is the set of the non-trivial subtrees rooted at the children of S.

The algorithm, called *PT_PARTITIONING*, is shown in Figure 5. It repeatedly chooses a splitter *Sp* from X. A compound splitter is chosen if one exists. Then, every block $B \in P$

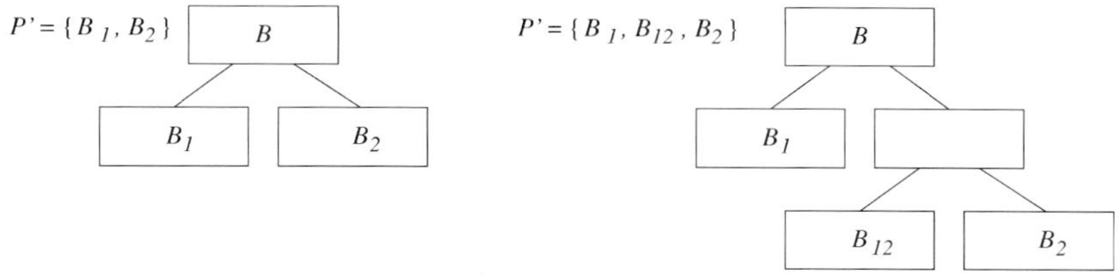

Fig. 6. Splitting a block.

that has transitions into *Sp* is split with respect to the chosen splitter. We use the notation $B \in \{\bullet \to Sp\}$ to mean $\{B \in P \mid \exists s \in B.s \to s' \wedge s' \in Sp\}$. The operation is denoted $split(B, Sp)$. The result is a set of blocks P' containing one, two, or three blocks, depending on how B was split. By $S_{P'}$ we denote P' represented as a tree. The shape of $S_{P'}$ is given in Figure 6.

Operation $split(B, Sp)$ also updates the variables $count(s, S)$ for all $s \in B$. To do this, *split* scans the set $\{\bullet \to S\}$ twice. On the first pass, counts are computed, and splitting itself is done on the second pass.

When P' is a singleton, no splitting has occurred and no further work is necessary. Otherwise, X needs to be updated to reflect the splitting of B. Three cases are possible:

1. B is a simple splitter in X. In this case, B is removed from X and all elements of P' are added to X as simple splitters.
2. B is a leaf in one of the trees in X. The tree is extended at B with $S_{P'}$.
3. B does not appear in any tree in X. $S_{P'}$ is added to X as a new tree.

The three cases are illustrated in Figure 7. We denote by $updateX(B, P')$ the operation of updating X as just described.

Correctness of *PT_PARTITIONING* is established by means of the following invariant. For $B \in P$, if B is not a simple splitter in X and B is not a leaf of a compound splitter in X, then P is stable with respect to B. Indeed, every newly split block is placed in one of the trees in X, and it remains there until it is used in splitting. At every iteration of the outer loop of the algorithm, at most two blocks are removed from X: the one used as a splitter in that iteration, and the sibling of the splitter, if one exists. After the iteration is complete, P is stable with respect to both blocks. Subsequent refinements do not alter stability with respect to a block: if P is stable with respect to a block B and P' is a refinement of P, then P' is stable with respect to B. Therefore, when X is empty, P is stable with respect to itself.

The worst-case running time of *PT_PARTITIONING* is $O(m \cdot \log_2 n)$. For each state in a splitter *Sp*, the algorithm visits every incoming edge twice, performing $O(1)$ work for each edge. Because the smaller of the two sub-nodes of a compound splitter is used, each state can appear in at most $\log_2 n + 1$ splitters.

3.3. *OBDD-based equivalence checking*

A major drawback of the algorithms presented in the preceding sections is that they require the LTS to be fully constructed in advance. When a process term is constructed from

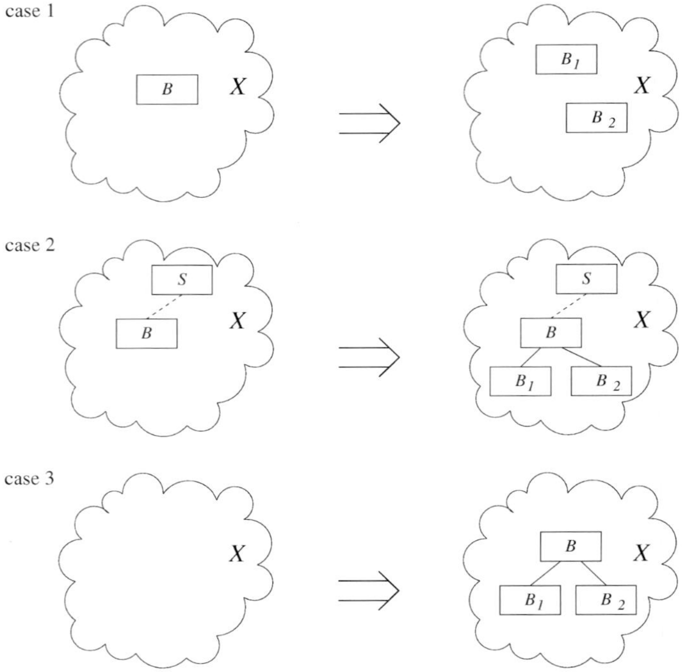

Fig. 7. Updating the splitters.

n subprocesses by parallel composition, the size of the resulting LTS may grow exponentially large in n, rendering equivalence checking infeasible. Several approaches have been proposed to alleviate this problem. A group of algorithms called "local", which construct only as much of the LTS as needed to determine equivalence (or inequivalence), will be introduced later in this chapter. This section is concerned with a different approach, one that uses *ordered binary decision diagrams* (OBDDs) to succinctly represent the LTS. OBDDs [6] are widely used in symbolic analysis algorithms such as model checking. We present an OBDD-based bisimulation checking algorithm based on the one by Bouali and de Simone [4].

An OBDD is a representation of a Boolean function by a rooted acyclic directed graph with respect to a fixed variable ordering. Terminal nodes of the graph are labeled with Boolean constants, non-terminal nodes are labeled with input variables. Each non-terminal node has two outgoing edges for the two possible values of the variable labeling the node, indicating which node is evaluated next. Every path from the root of the OBDD to a terminal node has to respect the ordering of variables. In addition, an OBDD does not contain duplicate terminals or non-terminals, nor redundant tests (i.e., nodes with both outgoing edges leading to the same node). Figure 8 shows the OBDD for the function $(x_1 \vee x_2) \wedge x_3$ with the ordering $x_1 < x_2 < x_3$.

For a fixed variable ordering, OBDDs offer a canonical representation of Boolean functions; that is, two OBDDs representing the same function are isomorphic. Commonly used

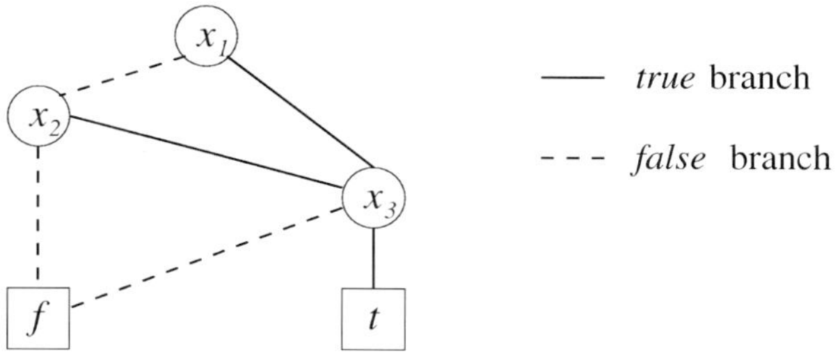

Fig. 8. An example OBDD.

operations on Boolean functions, such as Boolean operations, functional composition, and variable quantification can be computed with OBDDs in time polynomial in the size of the OBDD representations for the component functions. The size of the OBDD for a Boolean function depends on the chosen variable ordering. In the worst case, the size is exponential in the number of variables. Still, many functions have much more compact representations. For a detailed discussion of OBDD properties we refer the reader to [7].

The OBDD representation of an LTS. An LTS $\langle \mathcal{S}, \mathcal{A}, \rightarrow \rangle$ is represented by OBDDs in the following way. The states of the LTS are represented by an encoding of their enumeration, that is, by means of a function $\sigma : \mathcal{S} \rightarrow \{0, 1\}^k$ that associates with each state a distinct Boolean vector. Thus, representation of n states in an LTS requires $k = \log_2 n$ Boolean variables. We let $\vec{s}$ represent the encoding of a state s. The alphabet of the LTS is encoded in the same way, and $\vec{a}$ is the encoding of a label a. *States*(x) is the characteristic function of the set of states in the LTS. The transition relation of an LTS is represented by a characteristic function $\delta(\vec{a}, \vec{s}, \vec{s'})$, which returns *true* when $s \xrightarrow{a} s'$. Since all Boolean functions considered in this section operate on encodings of state, labels, etc., and not on the objects themselves, we will use x for $(\vec{x})$ where no confusion can arise.

Symbolic computation of the bisimulation equivalence. Like the algorithms presented previously, the algorithm for symbolic bisimulation checking is based on the characterization of Theorem 2.6. In order to compute the iterative fixed point symbolically, we need to express the function $\mathcal{F}_{\mathcal{L}}$ in terms of efficient OBDD operations.

First, we introduce several auxiliary operations on Boolean functions. A *restriction* of function f with respect to variable x is $f|_{x \leftarrow k}(x_1, \ldots, x, \ldots, x_n) = f(x_1, \ldots, k, \ldots, x_n)$. The *smoothing* operator is defined by $S_x(f) = f|_{x \leftarrow 0} \vee f|_{x \leftarrow 1}$. We have $S_x(f) = \exists x\, f$. The smoothing operator is extended to sets of variables in the obvious way. *Substitution* of a vector $X = \{x_1, \ldots, x_n\}$ by a vector $Y = \{y_1, \ldots, y_n\}$ within the OBDD for the function f is the simultaneous replacement of variables from X by respective variables from Y. Substitution is defined as

$$[Y \leftarrow X]f = S_X\Bigl(\bigvee_{i \in \{1,\ldots,n\}} (y_i \leftrightarrow x_i) \wedge f\Bigr).$$

$R(x, y) := States(x) \wedge States(y)$
$New(x, y) := R(x, y)$
`while` $New(x, y) \neq false$ `do`
`begin`
  $R^+(x, y) := \mathcal{F}_{\mathcal{L}}(R)$
  $New(x, y) := R^+(x, y) \wedge \overline{R(x, y)}$
  $R(x, y) := R^+(x, y)$
`end`

Fig. 9. Symbolic computation of bisimulation.

With these definitions, we can give the definition of $\mathcal{F}_{\mathcal{L}}(R)$ suitable for symbolic computation. We represent R as the characteristic function of the Cartesian product of the equivalence classes of R, denoted $R(x, y)$. $R(x, y)$ returns *true* if x and y belong to the same equivalence class.

An auxiliary function

$$E_a(x, z) = S_y\big(R(x, y) \wedge \delta(a, z, y)\big)$$

represents the relationship "z has an a-transition into the equivalence class of x". With this function, we can express the relationship that two states cannot be in the same equivalence class:

$$Bad(x, y) = [x \leftarrow z]\Big(\bigvee_{a \in \mathcal{A}} S_x\big([y \leftarrow z]E_a(x, z) \leftrightarrow \overline{E_a(x, z)}\big)\Big).$$

Finally, we have

$$\mathcal{F}_{\mathcal{L}}(R) = R(x, y) \wedge \overline{Bad(x, y)}.$$

Note that the use of substitutions in the definition of $Bad(x, y)$ allows us to construct E_a once per the application of $\mathcal{F}_{\mathcal{L}}$. Construction of E_a, which computes the inverse of the transition relation of the LTS, is a much more expensive operation than substitution.

The algorithm to compute the bisimulation equivalence symbolically, shown in Figure 9, is now a straightforward iterative fixed point computation that applies $\mathcal{F}_{\mathcal{L}}$ to the current partition in each iteration. Initially, $R = \mathcal{S} \times \mathcal{S}$ or, in a functional representation, $R(x, y) = States(x) \wedge States(y)$.

Performance issues. The time and space efficiency of the OBDD-based algorithm depend on the OBDD variable ordering and certain details of the LTS encoding. Significant performance improvements can be achieved in the case when the LTS under consideration is obtained as a product of several communicating processes. Then, in addition to representing and manipulating the global LTS, we need to represent the LTSs of the components, which we refer to as *local* LTSs, and compute the global LTS from the local LTSs symbolically.

States in the global LTS are represented as tuples of local states. There are two natural ways to order the variables used in the representation of local states. One ordering groups together all variables representing a state in a local LTS and uses the order of local states in the tuple to separate the states from different local LTSs. The other ordering groups local states together bit by bit. That is, the ordering places the first bit of the encodings of all local states before the second bit, etc. Experimental results presented in [4] suggest that the former ordering yields more compact representation during the initial stages of the algorithm, when the equivalence relation is coarse. When there are many equivalence classes, the latter ordering is a better choice. Dynamic reordering of variables may allow one to take advantage of both orderings.

In addition, the order of local states within global state tuples has its effect on the size of OBDDs. As a rule of thumb, it is advantageous to group together the processes that actively communicate with each other. For example, let the global process be composed of three local processes, p_1, p_2, p_3. Assume that p_1 communicates with both p_2 and p_3, but there is little or no communication between p_2 and p_3. Then, the optimal order of local processes is to place p_1 in the middle, for example $\langle p_2, p_1, p_3 \rangle$.

3.4. *Computing other equivalences via process transformations*

The algorithms presented above can be used to decide several other behavioral equivalences in addition to bisimulation equivalence. Examples of equivalences that can be decided by partition refinement include weak bisimulation [23] and testing equivalence [19]. In order to obtain a partition refinement algorithm for an equivalence relation R, we need to define a suitable *process transformation* $\mathcal{T}_R$ and an equivalence relation E_R that will be used by the algorithm.

Given an LTS $\mathcal{L} = \langle \mathcal{S}, \mathcal{A}, \rightarrow \rangle$, a process transformation is a function

$$\mathcal{T}_R(\mathcal{L}) = \langle \mathcal{S}', \mathcal{A}', \rightarrow' \rangle$$

from the set of LTSs to itself. Abusing notation, we will also use $\mathcal{T}_R$ to denote a mapping from $\mathcal{S}$ to $\mathcal{S}'$ that relates the states of P and $\mathcal{T}_R(\mathcal{L})$. A process transformation can be used to decide an equivalence relation R if there is an equivalence relation E_R such that the following condition holds:

$$s_1 \, R \, s_2 \quad \text{iff} \quad \mathcal{T}_R(s_1) \sim^{E_R} \mathcal{T}_R(s_2).$$

In this case, we can construct $\mathcal{T}_R(\mathcal{L})$ and then apply an equivalence-checking algorithm to compute $\sim^{E_R}$.

Weak bisimulation equivalence. As an example of process transformation, we present $\mathcal{T}_\approx$ that is used to decide $\approx$, *weak bisimulation* equivalence or *observational* equivalence. Weak bisimulation equivalence abstracts away from the special unobservable action $\tau \in \mathcal{A}$ by allowing transitions to be matched by sequences of transitions having the same observable (non-τ) content.

DEFINITION 3.3. Let $\langle \mathcal{S}, \mathcal{A}, \rightarrow \rangle$ be an LTS.
- A symmetric relation $R \subseteq \mathcal{S} \times \mathcal{S}$ is a *weak bisimulation* if whenever $\langle s_1, s_2 \rangle \in R$ and $s_1 \xrightarrow{a} s_1'$ then one of the following holds:
  - $a = \tau$ and there is an s_2' such that $s_2 \xrightarrow{\tau} {}^* s_2'$ and $\langle s_1', s_2' \rangle \in R$, or
  - there is an s_2' such that $s_2 \xrightarrow{\tau} {}^* \xrightarrow{a} \xrightarrow{\tau} {}^* s_2'$ and $\langle s_1', s_2' \rangle \in R$.
- Two states $s_1, s_2 \in \mathcal{S}$ are *weak bisimulation equivalent*, written $s_1 \approx s_2$, if there exists a weak bisimulation R such that $\langle s_1, s_2 \rangle \in R$.

The transformation used to decide weak bisimulation is $\mathcal{T}_{\approx}(\langle \mathcal{S}, \mathcal{A}, \rightarrow \rangle) = \langle \mathcal{S}, \mathcal{A} - \{\tau\} \cup \varepsilon, \Rightarrow \rangle$. The *weak* transition relation $\Rightarrow$ is defined as follows:
- $s \overset{\varepsilon}{\Longrightarrow} s'$ when $s \xrightarrow{\tau} {}^* s'$;
- for $a \neq \tau$, $s \overset{a}{\Longrightarrow} s'$ when $s \xrightarrow{\tau} {}^* \xrightarrow{a} \xrightarrow{\tau} {}^* s'$.

We compute $\Rightarrow$ in two steps. First, $\overset{\varepsilon}{\Longrightarrow}$ is constructed by applying a transitive-closure algorithm to the τ-transitions of the original LTS. Then, the composition of relations $\overset{\varepsilon}{\Longrightarrow}$ and $\xrightarrow{a}$, $a \neq \varepsilon$, produces all observable weak transitions.

Weak bisimulation equivalence can now be computed as U-bisimulation over $\mathcal{T}_{\approx}(\langle \mathcal{S}, \mathcal{A}, \rightarrow \rangle)$. The correctness of this approach follows directly from Definition 3.3 and the construction of $\Rightarrow$.

Other equivalences. Several other process transformations for equivalence checking have been presented in the literature. In particular, *observational congruence* [23] is computed as U-bisimulation over *congruence transition systems* (see [11] for details).

Testing equivalence [19] can also be computed by process transformation. It turns out to be an $\mathcal{A}$-bisimulation over *acceptance graphs*. Both the process transformation and the acceptance set equivalence $\mathcal{A}$ are described in [10].

3.5. *Computing branching bisimulation equivalence*

Branching bisimulation equivalence, introduced by [17], is similar to weak bisimulation equivalence in the sense that it abstracts away from the unobservable action τ. However, branching bisimulation equivalence preserves the branching structure of processes; that is, the non-deterministic choices are resolved by the equivalent processes in the same order. Consequently, branching bisimulation equivalence is finer than weak bisimulation equivalence and possesses an appealing set of algebraic properties.

Despite its close relation to bisimulation equivalence, branching bisimulation equivalence requires a modification of the partition refinement algorithm. The extended algorithm *BB_PARTITIONING* has been presented by Groote and Vaandrager [18].

DEFINITION 3.4. Let $\langle \mathcal{S}, \mathcal{A}, \rightarrow \rangle$ be an LTS.
- A symmetric relation $R \subseteq \mathcal{S} \times \mathcal{S}$ is a *branching bisimulation* if whenever $\langle s_1, s_2 \rangle \in R$ and $s_1 \xrightarrow{a} s_1'$ then one of the following holds:
  - $a = \tau$ and $\langle s_1', s_2 \rangle \in R$, or
  - there exist s_2', s_2'' such that $s_2 \xrightarrow{\tau} {}^* s_2' \xrightarrow{a} s_2''$ and $\langle s_1', s_2'' \rangle \in R$.

- Two states $s_1, s_2 \in \mathcal{S}$ are *branching bisimilar*, written $s_1 \underline{\leftrightarrow} s_2$, if there exists a branching bisimulation R such that $\langle s_1, s_2 \rangle \in R$.

Before presenting the algorithm, we introduce some additional terminology. Given a partition P, a transition $s \xrightarrow{\tau} s'$ is called P-*inert* if s and s' belong to the same block in P. A *bottom* state s in a block B is such that for all outgoing transitions of s, $s \xrightarrow{\tau} s' \Rightarrow s' \notin B$.

We will assume that the LTS does not contain cycles of unobservable events. This assumption is not a restriction. Indeed, if $s \xrightarrow{\tau} {}^{*}s'$ and $s' \xrightarrow{\tau} s$, then $s \underline{\leftrightarrow} s'$. Then, states strongly connected by unobservable transitions will always be within the same equivalence class, and we can collapse them into the same state before the algorithm is run.

A new notion of splitter is taken directly from Definition 3.4. For $B, B' \in P$ and $a \in \mathcal{A}$, let

$$\begin{aligned} reach_a(B, B') = \big\{ s \in B \mid \exists n \geqslant 0 \exists s_0, \ldots, s_n \in B \exists s' \in B'.\\ s_0 = s \wedge (\forall 0 < i \leqslant n. s_{i-1} \xrightarrow{\tau} s_i)\\ \wedge \big(s_n \xrightarrow{a} s' \vee (a = \tau \wedge s_n = s')\big)\big\}. \end{aligned}$$

A block B' is an a-splitter for B if $\emptyset \neq reach_a(B, B') \neq B$.

A partition refinement algorithm that uses this definition of splitter directly would be impractical. It would have to maintain information about sequences of inert τ-transitions in each block and update this information as blocks are split. Instead, a different characterization of splitter is used, as stipulated by the following theorem.

THEOREM 3.5. *Let P be a partition of $\mathcal{S}$, $B, B' \in P$, and $a \in \mathcal{A}$. Then, B' is an a-splitter of B if*

(1) $a \neq \tau$ *or* $B \neq B'$.
(2) *There exists* $s \in B$ *such that for some* $s' \in B'$, $s \xrightarrow{a} s'$.
(3) *There exists a* bottom *state* $s \in B$ *such that for no* $s' \in B'$, $s \xrightarrow{a} s'$.

PROOF. Suppose B' is an a-splitter of B. If $a = \tau$, then clearly $B \neq B'$, since $reach_\tau(B, B')$ is trivially equal to B, satisfying the first condition. By definition of a splitter, $reach_\tau(B, B') \neq \emptyset$, which means that condition (2) holds. Finally, assume that for every bottom state $s \in B$ there is an a-transition into B'. Choose an arbitrary state $t \in B$. Since there are no τ-cycles in the LTS, then there is a (possibly trivial) τ-path from t to a bottom state s. This means that $t \in reach_a(B, B')$. Since t was chosen arbitrarily, $reach_a(B, B') = B$, which is a contradiction, and condition (3) holds. For the reverse direction, $reach_a(B, B') \neq \emptyset$ because of condition (2), and $reach_a(B, B') \neq B$ because of conditions (1) and (3). □

This characterization of a splitter allows us to find splitters in P efficiently. To do this, we need the following data structures. For each block B, we keep two lists of states. One list holds bottom states of the block, the other holds non-bottom ones. Moreover, we assume that non-bottom states are topologically sorted according to τ-transitions. That is,

```
select S ∈ working
pre = ∅
for each a ∈ 𝒜 do
begin
   for each a-transition t ∈ in(S) do
      setMarks(t)
   for each B ∈ pre do
      if S is a splitter of B then
      begin
         remove B from its list
         changed := split(B, B1, B2)
         add B1, B2 to working
         if changed then
            append stable to working
      end
end
working := working − {S}
stable := stable ∪ {S}
```

Fig. 10. Refinement step of *BB_PARTITIONING*.

if $s \xrightarrow{\tau} s'$, then s' precedes s in the list of non-bottom states. This ordering is always well-defined since there are no τ-cycles. Each non-bottom state s contains a list of inert transitions $inert(s)$ originating in this state. Each block B, in addition to the lists of states, contains a list $in(B)$ of non-inert transitions that end in B. Non-inert transitions are lexicographically sorted by their label. Every state $s \in \mathcal{S}$ has a Boolean variable $mark(s)$, initialized to *false*.

Blocks of P are stored in two disjoint lists, *stable* and *working*. P is stable with respect to blocks in the first list; blocks in the second list can be splitters. When *working* becomes empty, the partition is stable and the algorithm terminates.

The pseudo-code of the refinement step is shown in Figure 10, where *pre* is the set of blocks that have a transition into B. Procedure *setMarks* accomplishes two things: (1) if s is the source state of transition t, set $mark(s)$ to *true*; (2) add the block to which s belongs to *pre*. For each $B \in pre$, S is *not* a splitter if all bottom states of B are marked.

If there are unmarked bottom states in B, procedure $split(B, B_1, B_2)$ partitions B into two blocks B_1, B_2 in the following way. First, marked bottom states of B become bottom states of B_1 and unmarked states become bottom states of B_2. Next, non-bottom states of B are scanned. If a state s is unmarked and does not have an outgoing transition that leads to a state in B_1, then s becomes a non-bottom state of B_2. All transitions in $inert(s)$ must lead to a state in B_2 and thus no adjustment of $inert(s)$ is needed. Here, the ordering of the non-bottom states is important (note that the ordering is preserved by splitting). Otherwise, s is placed in B_1. Its inert transitions are scanned and, if a transition leads to a state in B_2, it is removed from $inert(s)$ and placed into $in(B_2)$. If $inert(s)$ becomes empty, s is placed into the list of bottom states of B_1. Otherwise, it remains a non-bottom states of B_1. As each state is moved from B to its new block, its mark is cleared. Finally, $in(B)$ is distributed to $in(B_1)$ and $in(B_2)$ according to the target state of each transition.

Procedure *split* returns a Boolean value, which is *true* if the set of bottom states changed. It can be easily seen that, unless the set of bottom states in a partition has changed, stability of a block is preserved after refinement. Otherwise, every block in *stable* has to be considered as a splitter again. Thus, if *split* returns *true*, the contents of *stable* are moved to *working*.

THEOREM 3.6. *Given a finite-state LTS* $\langle \mathcal{S}, \mathcal{A}, \rightarrow \rangle$ *with* $|\mathcal{S}| = n$ *and* $|\rightarrow| = m$, *algorithm* BB_PARTITIONING *computes the stable partition in* $O(|\mathcal{A}| + m \cdot n)$ *time.*

PROOF. Initialization of the data structures in the algorithm is accomplished in $O(|\mathcal{A}|+m)$ time. This includes the computation of strongly connected components of $\mathcal{S}$ and topological sorting of inert transitions in each block of the initial partition. Both can be accomplished in $O(m)$ time. Lexicographic sorting of non-inert transitions takes $O(|A|+m)$ time. There can be at most $n - 1$ refinement steps, each of which takes $O(m)$ time. □

4. Global preorder algorithms

In this section we present an efficient algorithm to compute the parameterized semantic relations $\sqsubseteq^{\Pi}_{\Phi_1,\Phi_2}$ introduced in Section 2.4. The algorithm is due to Celikkan and Cleaveland [9]. We first introduce an auxiliary function on relations similar to the one given in Definition 2.3.

DEFINITION 4.1. Let $\mathcal{L} = \langle \mathcal{S}, \mathcal{A}, \rightarrow \rangle$ be an LTS and $R \subseteq \mathcal{S} \times \mathcal{S}$ be a relation between states. Then $\mathcal{F}^{\sqsubseteq}_{\mathcal{L}} : 2^{\mathcal{S}\times\mathcal{S}} \rightarrow 2^{\mathcal{S}\times\mathcal{S}}$ is given by:

$$\begin{aligned}\mathcal{F}^{\sqsubseteq}_{\mathcal{L}}(R) = \big\{\langle p,q\rangle \mid p\Pi q \wedge \forall a \in \mathcal{A}.\ &\langle p,q,a\rangle \in \Phi_1 \Rightarrow \big[p \xrightarrow{a} p' \Rightarrow \exists q'.\ q \xrightarrow{a} q' \wedge \langle p',q'\rangle \in R\big] \wedge \\ &\langle p,q,a\rangle \in \Phi_2 \Rightarrow \big[q \xrightarrow{a} q' \Rightarrow \exists p'.\ p \xrightarrow{a} p' \wedge \langle p',q'\rangle \in R\big]\big\}.\end{aligned}$$

Note that relation R is a $\langle \Pi, \Phi_1, \Phi_2\rangle$-bisimulation *iff* $R \subseteq \mathcal{F}^{\sqsubseteq}_{\mathcal{L}}(R)$. Correspondingly, if $R - \mathcal{F}^{\sqsubseteq}_{\mathcal{L}}(R) \neq \emptyset$ then R is not a $\langle \Pi, \Phi_1, \Phi_2\rangle$-bisimulation. It is easy to see that the algorithm *EFF_PREORDER* in Figure 11 computes $\sqsubseteq^{\Pi}_{\Phi_1,\Phi_2}$ over $\mathcal{S} \times \mathcal{S}$.

In order to efficiently implement this algorithm, one must maintain the value of *ToDelete* $= R - \mathcal{F}^{\sqsubseteq}_{\mathcal{L}}(R)$, as removing an element from R may change the value of $\mathcal{F}^{\sqsubseteq}_{\mathcal{L}}(R)$. Now from the definition of $\mathcal{F}^{\sqsubseteq}_{\mathcal{L}}$ it follows that if $\langle p,q\rangle \in R - \mathcal{F}^{\sqsubseteq}_{\mathcal{L}}(R)$ then one of the three conditions below must hold.

1. $\langle p,q\rangle \notin \Pi$;
2. $\langle p,q,a\rangle \in \Phi_1 \wedge \exists p'.\ p \xrightarrow{a} p' \wedge \forall q'.(q \xrightarrow{a} q' \Rightarrow \langle p',q'\rangle \notin R)$;
3. $\langle p,q,a\rangle \in \Phi_2 \wedge \exists q'.\ q \xrightarrow{a} q' \wedge \forall p'(p \xrightarrow{a} p' \Rightarrow \langle p',q'\rangle \notin R)$.

```
while ToDelete ≠ ∅ do begin
  Choose⟨p', q'⟩ ∈ ToDelete;
  R := R − {⟨p', q'⟩};
/* First condition */
  foreach a ∈ {—•→ q'} do begin
    foreach q ∈ {• —a→ q'} such that ⟨p, q, a⟩ ∈ Φ1 do begin
      HighCount(a, p', q) := HighCount(a, p', q) − 1;
        if HighCount(a, p', q) = 0 then begin
        foreach p ∈ {• —a→ p'} do
          if ⟨p, q⟩ ∈ R and ⟨p, q⟩ ∉ ToDelete then
            ToDelete := ToDelete ∪ {⟨p, q⟩};
      end;
    end;
  end;
/* Second condition */
  foreach a ∈ {—•→ p'} do begin
    foreach p ∈ {• —a→ p'} such that ⟨p, q, a⟩ ∈ Φ2 do begin
      LowCount(a, p, q') := LowCount(a, p, q') − 1;
      if LowCount(a, p, q') = 0 then begin
        foreach q ∈ {• —a→ q'} do
          if ⟨p, q⟩ ∈ R and ⟨p, q⟩ ∉ ToDelete then
            ToDelete := ToDelete ∪ {⟨p, q⟩};
      end;
    end;
  end;
  ToDelete := ToDelete − {⟨p', q'⟩};
end
```

Fig. 15. Main loop of algorithm *EFF_PREORDER*.

no a-derivative of q related to p'; thus all pairs $\langle p, q\rangle$ such that $p \xrightarrow{a} p'$ should be inserted into the set *ToDelete*. Note that the insertion of a pair $\langle p, q\rangle$ into the set *ToDelete* is guarded by the predicates Φ_1 and Φ_2. A $HighCount(a, p', q)$ value of 0 causes a pair $\langle p, q\rangle$ to be inserted into *ToDelete* only if $\langle p, q, a\rangle$ is in Φ_1. *LowCount* is employed analogously. Figure 15 contains the code for the iteration phase of the algorithm.

We now remark on the time complexity of the preorder-checking algorithm given in Figures 12 and 15. If $\mathcal{L} = \langle \mathcal{S}, \mathcal{A}, \rightarrow\rangle$ is an LTS, let $|\mathcal{L}| = |\mathcal{S}| + |{\rightarrow}|$. Also let t_{init} be the time spent on computing Π, Φ_1 and Φ_2. We also assume that sets are implemented using hash tables and thus that insertion, deletion and set membership may be computed in O(1) time.

THEOREM 4.2. *The code given in Figure* 12 *takes* $O(t_{\text{init}} + |\mathcal{S}| \times |\mathcal{L}|)$ *time.*

PROOF. Initialization of *ToDelete* and R takes $O(t_{\text{init}} + |\mathcal{S}|^2)$ time units. The nested "**foreach**" loops require $O(|\mathcal{S}| \times |\mathcal{L}|)$ set membership operations. This follows from the fact that

$$\sum_{p\in\mathcal{S}}\sum_{q\in\mathcal{S}}\sum_{a\in(\{p\stackrel{\bullet}{\longrightarrow}\}\cup\{q\stackrel{\bullet}{\longrightarrow}\})} O(1) \leqslant \sum_{p\in\mathcal{S}}\sum_{q\in\mathcal{S}}\big|\{p\stackrel{\bullet}{\longrightarrow}\}\big| + \big|\{q\stackrel{\bullet}{\longrightarrow}\}\big|$$
$$\leqslant \sum_{p\in\mathcal{S}}\sum_{q\in\mathcal{S}}\big|\{q\stackrel{\bullet}{\longrightarrow}\}\big| + \sum_{p\in\mathcal{S}}\sum_{q\in\mathcal{S}}\big|\{p\stackrel{\bullet}{\longrightarrow}\}\big|$$
$$\leqslant \sum_{p\in\mathcal{S}}|\mathcal{L}| + \sum_{q\in\mathcal{S}}|\mathcal{L}|$$
$$\leqslant O\big(|\mathcal{S}|\times|\mathcal{L}| + |\mathcal{S}|\times|L|\big)$$

and checking predicates Φ_1 and Φ_2 may be done in constant time after initializing them properly. Then the total amount of time spent is $O(t_{\text{init}} + |\mathcal{S}|^2 + |\mathcal{S}|\times|\mathcal{L}|)$. This reduces to $O(t_{\text{init}} + |\mathcal{S}|\times|\mathcal{L}|)$. □

THEOREM 4.3. *The code given in Figure* 15 *takes* $O(|\mathcal{S}|^2 + |\mathcal{S}|\times|\mathcal{L}|)$ *time.*

PROOF. We first note that in the worst case, the outer loop executes at most $O(|\mathcal{S}|^2)$ times. Let T_1 represent the total amount of time spent executing the first "`foreach`" loop over all iterations of the outermost "`while`" loop and T_2 represent the time spent in the second "`foreach`" loop. T_1 may be further decomposed into:

$$T_1 = T_i + T_r,$$

where T_i represents the total amount of time spend in the innermost "`foreach` $p\in\{\bullet\stackrel{a}{\longrightarrow} p'\}$ `do begin`" loop and T_r represents the time spent in the rest of the loop (e.g., in decrementing *HighCount*(a, p, q') and performing the test in the `if-then` statement). From the structure of the loops, we have the following

$$T_r = \sum_{\langle p',q'\rangle\in\mathcal{S}\times\mathcal{S}}\ \sum_{a\in\{\stackrel{\bullet}{\longrightarrow}q'\}}\ \sum_{q\in\{\bullet\stackrel{a}{\longrightarrow}q'\}} O(1) = \sum_{p'\in\mathcal{S}}\sum_{q'\in\mathcal{S}}\ \sum_{a\in\{\stackrel{\bullet}{\longrightarrow}q'\}}\ \sum_{q\in\{\bullet\stackrel{a}{\longrightarrow}q'\}} O(1)$$
$$= \sum_{p'\in\mathcal{S}} O\big(|\mathcal{L}|\big) = O\big(|\mathcal{S}|\times|\mathcal{L}|\big).$$

Regarding T_i, first note that for any a, p' and q, *HighCount*(a, p', q) has its value changed to zero at most once. Thus the total amount of time spent in the innermost loop is:

$$T_i = \sum_{\langle a,p',q\rangle\in Act\times\mathcal{S}\times\mathcal{S}}\ \sum_{p\in\{\bullet\stackrel{a}{\longrightarrow}p'\}} O(1) = \sum_{q\in\mathcal{S}}\ \sum_{\langle a,p'\rangle\in Act\times\mathcal{S}}\ \sum_{p\in\{\bullet\stackrel{a}{\longrightarrow}p'\}} O(1)$$
$$= \sum_{q\in\mathcal{S}} O\big(|\mathcal{L}|\big) = O\big(|\mathcal{S}|\times|\mathcal{L}|\big).$$

Thus $T_1 = O(|\mathcal{S}|\times|\mathcal{L}|)$. Using a similar argument, we may also infer that $T_2 = O(|\mathcal{S}|\times|\mathcal{L}|)$, and the theorem then follows. □

Note that the time spent in the initialization phase of this algorithm asymptotically dominates the time devoted to the second phase. We thus have the following.

THEOREM 4.4. *Given a finite-state LTS* $\langle \mathcal{S}, \mathcal{A}, \rightarrow \rangle$ *with* $|\mathcal{S}| = n$ *and* $|\rightarrow| = m$, *algorithm EFF_PREORDER takes* $O(t_{\text{init}} + |\mathcal{S}| \times |\mathcal{L}|)$ *time.*

Speeding up the algorithm. In general, when computing $\sqsubseteq^{\Pi}_{\Phi_1,\Phi_2}$ we are in fact really interested in whether the two given states, p and q, are related. If they are found not to be, then it is possible to terminate the algorithm immediately. Therefore if $p \not\sqsubseteq^{\Pi}_{\Phi_1,\Phi_2} q$, then we would like to determine this as quickly as possible. This immediately suggests an optimization: if $\langle p, q \rangle$ is ever inserted into *ToDelete* then the algorithm can terminate.

Another heuristic for early termination involves processing elements in *ToDelete* in a particular order. Note that when $\langle p', q' \rangle \in$ *ToDelete* is processed, it may induce the removal of $\langle p, q \rangle$ from R, where $p \xrightarrow{a} p'$ and $q \xrightarrow{a} q'$. Therefore while processing the pairs in *ToDelete*, the pair that is "closest" to the pair $\langle p, q \rangle$ that we are interested in will be the most promising candidate to process. Formally, define $l(p')$ to be the length of the shortest path from p to p', and $l(q')$ to be the length of the shortest path from q to q'. Also let $h(p', q') = l(p') + l(q')$. Then among the pairs in *ToDelete* the pair that minimizes $h(p', q')$ should be chosen for processing.

Another way to speed up the algorithm is to avoid entirely the corresponding loops when either Φ_1 or Φ_2 is empty. If one of these relations is empty there is no need check the condition involving this relation, since that condition is trivially satisfied.

5. Local algorithms

Algorithms for computing semantic equivalences and refinement replations usually consist of two steps [11]. In the first, the state space is generated and stored (possibly symbolically), while the second then manipulates the state space to determine whether an appropriate relation exists. The algorithms presented in the previous sections are examples of such "global" algorithms. Global algorithms often perform poorly in practice because of the requirement that the state space be generated in advance. In particular, in many cases, one may be able to determine that one process fails to be related to another by examining only a fraction of the state space. In a design setting where one is repeatedly changing a design and checking it against a specification, one would like a verification algorithm exploiting this fact.

On-the-fly verification algorithms combine the checking of a system's correctness with the generation of the system's state space [13]. In this section we present the efficient on-the-fly preorder-checking algorithm proposed in [8]. To determine whether two states p_0, q_0 are related, the algorithm attempts to construct a relation relating the states of the LTS incrementally starting with the pair $\langle p_0, q_0 \rangle$. In order to achieve the desired complexity we represent the relation being built as an edge-labeled graph whose vertices are pairs of related states. Formally let $\mathcal{L} = \langle \mathcal{S}, \mathcal{A}, \rightarrow \rangle$ be an LTS and $p_0, q_0 \in \mathcal{S}$. We have the following.

THEOREM 5.1. $p_0 \sqsubseteq^{\Pi}_{\Phi_1,\Phi_2} q_0$ *iff there exists a graph* $\mathcal{G} = \langle V, E \rangle$ *with* $V \subseteq \mathcal{S} \times \mathcal{S}$ *and* $E \subseteq V \times \mathcal{A} \times V$ *such that*

- $(p_0, q_0) \in V$;
- *whenever* $(p, q) \in V$ *then* $\langle p, q\rangle \in \Pi$;
- *whenever* $(p, q) \in V$ *and* $\langle p, q, a\rangle \in \Phi_1$ *and* $p \xrightarrow{a} p'$ *then there exists a* q' *such that* $q \xrightarrow{a} q'$ *with* $(p', q') \in V$ *and* $((p', q'), a_1, (p, q)) \in E$;
- *whenever* $(p, q) \in V$ *and* $\langle p, q, a\rangle \in \Phi_2$ *and* $q \xrightarrow{a} q'$ *then there exists a* p' *such that* $p \xrightarrow{a} p'$ *with* $(p', q') \in V$ *and* $((p', q'), a_2, (p, q)) \in E$.

If such a graph exists then V represents a $\langle\Pi, \Phi_1, \Phi_2\rangle$-bisimulation relating p_0 and q_0. If $\langle p, q\rangle \in V(\mathcal{G})$, then the edges leading into $\langle p, q\rangle$ may be thought of as the "justification" for including $\langle p, q\rangle$ in the preorder. The index i of the action a indicates which of the conditions of Definition 4.1 the justification is based on. If $i = 1$ then the basis of justification is condition 2; if $i = 2$ then it is condition 3.

Given LTS $\mathcal{L} = \langle\mathcal{S}, \mathcal{A}, \rightarrow\rangle$ and two states $p_0, q_0 \in \mathcal{S}$, the algorithm works by attempting to build a graph as described in Theorem 5.1. This construction proceeds incrementally using a *depth-first search* of $\mathcal{S} \times \mathcal{S}$, starting with $\langle p_0, q_0\rangle$. When the algorithm is invoked on p and q, a vertex for $\langle p, q\rangle$ is added to the graph (if one does not exist already) provided that $\langle p, q\rangle \in \Pi$. The algorithm then attempts to add nodes relating each a-derivative of p to some a-derivative of q whenever $\langle p, q, a\rangle \in \Phi_1$, with edges being added from these nodes to $\langle p, q\rangle$ to indicate that the inclusion of $\langle p, q\rangle \in \mathcal{G}$ currently depends on the presence of these nodes. If no matching a-derivative of q can be found for some a-derivative p' of p, then p and q cannot be related. In this case $\langle p, q\rangle$ must be removed from the graph, and the nodes depending on $\langle p, q\rangle$ must be re-examined to determine whether they should also be removed. Note that once p and q are found not to be related, they need not be processed again; no graph can be constructed that includes $\langle p, q\rangle$ as a node. A similar action is taken while matching the each a-derivative of q to some a-derivative of p. Note that if $\langle p, q\rangle \in \Pi$ and $\langle p, q, a\rangle \notin \Phi_1$ and $\langle p, q, a\rangle \notin \Phi_2$ then $\langle p, q\rangle$ becomes a leaf vertex of $\mathcal{G}$.

The following three data structures are used:

- The graph $\mathcal{G}$.
- A set $\overline{R}$ that stores all the state pairs that have been determined not to be related.
- A set of pointers of the form $high_{p',q,a}$ and $low_{q',p,a}$. Intuitively, $high_{p',q,a}$ is the index of the state in $\{q \xrightarrow{a} \bullet\}$ that is currently matched to p' and $low_{q',p,a}$ is the index of the state in $\{p \xrightarrow{a} \bullet\}$ that is currently matched to q'. By using these pointers we ensure that p' will be compared to each a-transition of q at most once and similarly q' will be compared to each a-transition of p at most once.
- A set A which records the pairs that need to be re-examined. It contains tuples of the sort $\langle r \xrightarrow{a} r', s, type\rangle$. r, r', and s are states, and *type* indicates which condition needs to be rechecked. If *type* is equal to 1 then r and s are candidates for being related by $r \sqsubseteq^{\Pi}_{\Phi_1,\Phi_2} s$ and r' needs to be matched to some a-derivative of s using the *high* pointers, namely $high_{r',s,a}$. Dually, if *type* is equal to 2 then r and s are supposed to be related by $s \sqsubseteq^{\Pi}_{\Phi_1,\Phi_2} r$ and *low* pointers will be used to find a match for r'.

The algorithm consists of three procedures. Given a transition $p \xrightarrow{a} p'$ and state q, procedure *SEARCH_HIGH* tries to match p' to some a-derivative of q, under the assump-

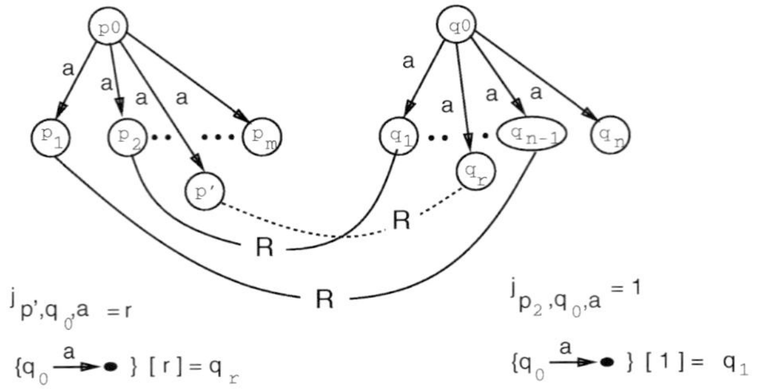

Fig. 16. Illustration of the pointers $high_{p',q,a}$.

tions that $high_{p',q,a}$ is the index of the first potential candidate and $\langle p, q, a\rangle \in \Phi_1$. If it finds a match then it adds the edge $\langle\langle p', q'\rangle, a_1, \langle p, q\rangle\rangle$ into $\mathcal{G}$ to indicate that the status of $\langle p, q\rangle$ depends on the status of $\langle p', q'\rangle$ — any status change of $\langle p', q'\rangle$ requires the re-analysis of state $\langle p, q\rangle$. Procedure *SEARCH_LOW* performs the dual task of procedure *SEARCH_HIGH*. Given states p and q, main procedure *PREORDER* tries to match each a-derivative of p with some a-derivative of q by calling *SEARCH_HIGH* and each a-derivative of q with some a-derivative of p by calling *SEARCH_LOW*. *PREORDER* does not invoke *SEARCH_HIGH* or *SEARCH_LOW* if $\langle p, q\rangle \notin \Pi$. When *SEARCH_HIGH* or *SEARCH_LOW* returns a *not_related* status (*SEARCH_HIGH* returns a *not_related* status if some a-derivative of p cannot be matched to any a-derivative of q) then vertex $\langle p, q\rangle$ is removed from $\mathcal{G}$ and inserted into $\overline{R}$. Then all the vertices having an incoming edge from $\langle p, q\rangle$ are re-analyzed. These vertices are stored in the set A. Pseudo-code for these procedures may be found in Figures 17, 18 and 19.

THEOREM 5.2. *Let p, q be states in a finite-state LTS, and assume* $\overline{R} = \emptyset$ *and* $\mathcal{G} = \langle\emptyset, \emptyset\rangle$. *Then PREORDER*$(p, q)$ *terminates, and PREORDER*$(p, q) =$ *related iff* $p \sqsubseteq^{\Pi}_{\Phi_1,\Phi_2} q$.

PROOF. Each possible vertex and edge is added and deleted at most once from $\mathcal{G}$. Since the number of potential edges and vertices is bounded, *PREORDER* eventually terminates. As for correctness, we first note that when *PREORDER*(p, q) terminates $\mathcal{G}$ satisfies the conditions laid out in Theorem 5.1. Also, *PREORDER*(p, q) returns a *related* result if and only if $\langle p, q\rangle$ is a vertex in $\mathcal{G}$. Therefore, if *PREORDER*$(p, q) =$ *related* then $p \sqsubseteq^{\Pi}_{\Phi_1,\Phi_2} q$. Now note that *PREORDER*$(p, q) =$ *not_related* if and only if $\langle p, q\rangle \in \overline{R}$ at termination. Also, $\langle p, q\rangle$ is added to $\overline{R}$ only if some a-transition of p can not be matched to any a-transition of q in such a way that the resulting $\mathcal{G}$ satisfies Theorem 5.1. Therefore, if *PREORDER*$(p, q) =$ *not_related* then $p \not\sqsubseteq q$. □

The time complexity of *PREORDER* may now be characterized as follows. We assume that the calculation of the transitions from a state takes unit time (that is, the set of transitions of a state is stored in the state rather than being computed by the algorithm).

```
R̄ := ∅;
𝒢 := ⟨∅, ∅⟩;
PREORDER(p, q) → (result: {related, not_related})
{* p and q are not related. *}
   if ⟨p, q⟩ ∈ R̄ then return (not_related);
   if ⟨p, q⟩ ∉ Π then
      R̄ := R̄ ∪ {⟨p, q⟩};
      return (not_related);
   end;
{* Pair ⟨p, q⟩ has already been inserted into V(𝒢). Its status is assumed related. *}
   if ⟨p, q⟩ ∈ V(𝒢) then return (related);
{* Pair ⟨p, q⟩ is going to be analyzed for the first time. *}
   𝒢 := ⟨V(𝒢) ∪ {⟨p, q⟩}, E(𝒢)⟩;
   status := related;
{* Match each a-derivative of p with some a-derivative of q. *}
   foreach a ∈ {p —•→} while status = related do
      foreach p′ ∈ {p —a→ •} while status = related do
         status := not_related;
         if high_{p′,q,a} does not exist then
            create high_{p′,q,a}
            high_{p′,q,a} := 1;
         end;
{* Match p′ with some a-derivative of q. *}
         status := SEARCH_HIGH(p —a→ p′, q);
         if status = not_related then
{* A contains vertices that are going to be affected as a result of
   the removal of ⟨p, q⟩ from 𝒢. *}
               A := {⟨r —a→ p, s, i⟩ | ⟨⟨p, q⟩, a_i, ⟨r, s⟩⟩ ∈ E(𝒢)};
      end;
   end;
{* Match each a-derivative of q with some a-derivative of p. *}
   foreach a ∈ {q —•→} while status = related do
      foreach p′ ∈ {q —a→ •} while status = related do
         status := not_related;
         if low_{q′,p,a} does not exist then
            create low_{q′,p,a}
            low_{q′,p,a} := 1;
         end;
{* Match q′ with some a-derivative of p. *}
         status := SEARCH_LOW(q —a→ q′, p);
         if status = not_related then A := {⟨s —a→ q, r, i⟩ | ⟨⟨p, q⟩, a_i, ⟨r, s⟩⟩ ∈ E(𝒢)};
      end;
   end;
```

Fig. 17. "On-the-fly" *PREORDER*.

```
{* Re-analyze all the states that are affected from the removal of vertex ⟨p, q⟩ from 𝒢.
   The set A contains all such pairs. *}
   if status = not_related then
{* Remove incoming and outgoing edges from vertex ⟨p, q⟩. *}
      𝒢 := ⟨V(𝒢) − {⟨p, q⟩}, E(𝒢) − {⟨π1, a_i, π2⟩ | (π1 = ⟨p, q⟩ ∨ π2 = ⟨p, q⟩) ∧ i ∈ {1, 2}}⟩;
{* Add ⟨p, q⟩ to the set of unrelated state pairs. *}
      R̄ := R̄ ∪ {⟨p, q⟩};
      while A ≠ ∅ do
         Choose ⟨r -a-> r', s, type⟩ ∈ A;
         A := A − {⟨r -a-> r', s, type⟩};
         if type = 1 then begin
            high_{r',s,a} := high_{r',s,a} + 1;
            {* Match r' to the next a-derivative of s. *}
            status := SEARCH_HIGH(r -a-> r', s);
            if status = not_related then
               A := A ∪ {⟨rr -b-> r, ss, i⟩ | ⟨⟨r, s⟩, b_i, ⟨rr, ss⟩⟩ ∈ E(𝒢)};
               𝒢 := ⟨V(𝒢) − {⟨r, s⟩}, E(𝒢) − {⟨π1, b_i, π2⟩ | π1 = ⟨r, s⟩ ∨ π2 = ⟨r, s⟩}⟩;
               R̄ := R̄ ∪ {⟨r, s⟩};
            end;
         end;
         if type = 2 then begin
            low_{r',s,a} := low_{r',s,a} + 1;
            status := SEARCH_LOW(r -a-> r', s);
            if status = not_related then
               A := A ∪ {⟨rr -b-> r, ss, i⟩ | ⟨⟨s, r⟩, b_i, ⟨ss, rr⟩⟩ ∈ E(𝒢)};
               𝒢 := ⟨V(𝒢) − {⟨s, r⟩}, E(𝒢) − {⟨π1, b_i, π2⟩ | π1 = ⟨s, r⟩ ∨ π2 = ⟨s, r⟩}⟩;
               R̄ := R̄ ∪ {⟨s, r⟩};
            end;
         end;
      end;
   endif;
   return(status);
end PREORDER;
```

Fig. 18. "On-the-fly" *PREORDER* (continued).

THEOREM 5.3. *Let* $\mathcal{L} = \langle \mathcal{S}, \mathcal{A}, \rightarrow \rangle$ *with* $|\mathcal{L}| = |\mathcal{S}| + |{\rightarrow}|$. *PREORDER takes time proportional to that required by* $\mathrm{O}(t_{\mathrm{init}} + m)$ *set membership operations, where* $m \leqslant |\mathcal{L}|^2$ *and* t_{init} *is the time required to compute* Π, Φ_1, *and* Φ_2.

PROOF. Ignoring the time consumed by recursive calls, the time spent to match each a-derivative of a state p to some a-derivative of q and to match each a-derivative of a state q to some a-derivative of p can be characterized as

$$\sum_{a\in\mathcal{A}} \bigl|\{p \xrightarrow{a} \bullet\}\bigr| * t(\mathit{SEARCH\_HIGH}) + \sum_{a\in\mathcal{A}} \bigl|\{q \xrightarrow{a} \bullet\}\bigr| * t(\mathit{SEARCH\_LOW}),$$

```
SEARCH_HIGH(p --a--> p', q) → (result: {related, not_related})
status := not_related;
if ⟨p, q, a⟩ ∉ Φ1 then status := related;
while status = not_related and high_{p',q,a} ⩽ |{q --a--> •}| do
{* Start searching beginning from the first potential candidate referenced by high_{p',q,a}. *}
   status := PREORDER(p', {q --a--> •}[high_{p',q,a}]);
   if status = related then
      G := (V(G), E(G) ∪ {⟨⟨p', {q --a--> •}[high_{p',q,a}]⟩, a1, ⟨p, q⟩⟩});
   else
{* A match is not found. Process the next candidate. *}
      high_{p',q,a} := high_{p',q,a} + 1;
end;
return (status);
end SEARCH_HIGH;

SEARCH_LOW(q --a--> q', p) → (result: {related, not_related})
status := not_related;
if ⟨p, q, a⟩ ∉ Φ2 then status := related;
while status = not_related and low_{q',p,a} ⩽ |{p --a--> •}| do
{* Start searching beginning from the first potential candidate referenced by low_{q',p,a}. *}
   status := PREORDER({p --a--> •}[low_{q',p,a}], q');
   if status = related then
      G := (V(G), E(G) ∪ {⟨⟨{p --a--> •}[low_{q',p,a}], q'⟩, a2, ⟨p, q⟩⟩});
   else
{* A match is not found. Process the next candidate. *}
      low_{q',p,a} := low_{q',p,a} + 1;
end;
return (status);
end SEARCH_LOW;
```

Fig. 19. Procedures *SEARCH_HIGH* and *SEARCH_LOW*.

where $t(\mathit{SEARCH\_HIGH}) = \mathrm{O}(|\{q \xrightarrow{a} \bullet\}|)$ and $t(\mathit{SEARCH\_LOW}) = \mathrm{O}(|\{p \xrightarrow{a} \bullet\}|)$. Then the total time spent to execute the nested `foreach` loops in Figure 17 over all the recursive calls is bounded by

$$\sum_{(p,q)\in\mathcal{S}\times\mathcal{S}} \sum_{a\in\mathcal{A}} \left(\left|\{p \xrightarrow{a} \bullet\}\right| * \left|\{q \xrightarrow{a} \bullet\}\right| + \left|\{q \xrightarrow{a} \bullet\}\right| * \left|\{p \xrightarrow{a} \bullet\}\right|\right) \leqslant \mathrm{O}\left(|\mathcal{L}|^2\right).$$

When a state pair (p, q) is found to be unrelated, all the state pairs that depend on this state pair need to be re-analyzed. The time spent to re-analyze these state pairs ignoring the time consumed by recursive calls is

$$\mathrm{O}\left(\sum_{a\in\mathcal{A}} \left|\{\bullet \xrightarrow{a} p\}\right| * \left|\{\bullet \xrightarrow{a} q\}\right|\right).$$

In the worst case each state pair needs to be re-analyzed, and total time required to re-analyze the state pairs is bounded by

$$\sum_{(p,q)\in\mathcal{S}\times\mathcal{S}} \sum_{a\in\mathcal{A}} \left|\{\bullet \xrightarrow{a} p\}\right| * \left|\{\bullet \xrightarrow{a} q\}\right| \leqslant \mathrm{O}\left(|\mathcal{L}|^2\right).$$

Then it follows that the time it takes to execute *PREORDER* is proportional to that required by $\mathrm{O}(t_{\mathrm{init}} + m)$ set membership operations. □

As noted in Section 2.4, $\sqsubseteq^{\Pi}_{\Phi_1,\Phi_2}$ can represent various behavioral equivalences and preorders by choosing Φ_1, Φ_2, and Π appropriately. Therefore, *PREORDER* yields an efficient local checking algorithm for all these relations.

6. Tools

Over the last decade, a number of tools that use process-algebraic formalisms for specification and analysis of systems have been developed. Several of these tools implement one or more of the algorithms presented in this chapter.

The first successful tool in this category was the Concurrency Workbench (CWB) [11], developed as a joint project between the University of Edinburgh, the University of Sussex, and Uppsala University. The CWB implements global algorithms for several behavioral equivalences and preorders, including bisimulation, observational equivalence, and branching bisimulation.

A continuation of the Concurrency Workbench effort resulted in CWB-NC, the Concurrency Workbench of the New Century [12]. In addition to improved efficiency and an extended set of algorithms, CWB-NC features the ability to "plug in" user-defined system description languages, and uses language-independent analysis algorithms on a uniform internal representation of processes.

The FC2 toolset [5] gives the user the choice between explicit representation of the state space with partition refinement analysis techniques and symbolic OBDD-based representation. A graphical user interface provided by the Autograph visual editor gives a more intuitive specification language compared with text-based process algebra terms. The Fc2 *interchange format makes* it possible to share specifications with other tools that support this format, such as Aldébaran, described below.

CADP (CÆSAR/ALDÉBARAN Development Package) [14] also supports both partition refinement techniques based on the Paige–Tarjan algorithm and symbolic BDD-based algorithms. The toolset is oriented towards analysis of systems expressed in the high-level process-algebraic languages LOTOS and E-LOTOS [15]. A number of interchange formats are supported, allowing the user to work with other high-level languages such as SDL as well as low-level representations.

The commercially available tool FDR2 [26] is based on the CSP [20] process algebra. FDR2's main analysis technique is based on establishing refinements (behavioral preorders) between processes. Analysis of real-time systems, including schedulability and resource requirements, is supported by the PARAGON toolset [27]. PARAGON calculates a resource-sensitive version of bisimulation equivalence using a partition-refinement algorithm.

7. Conclusions

This chapter presented several algorithms for the analysis of finite-state process-algebraic specifications. The algorithms calculate behavioral equivalences and preorders; equivalences can be used to compare two processes for indistinguishable behavior and to minimize the state space of a system in a behavior-preserving manner. Preorders can be used to establish refinements, that is, specification/implementation relationships between processes.

Finite-state systems represent an important class of system specifications, amenable to fully automatic formal analysis. Finite-state systems are used in the specification and verification of hardware and software designs, communication and security protocols, real-time systems, etc. Even when a specification has an infinite state space, some of its components may be finite-state. In this case, the inherent modularity of process-algebraic specifications allows one to apply automatic equivalence and preorder checking techniques to these components.

References

[1] L. Aceto, W. Fokkink and C. Verhoef, *Structural operational semantics*, Handbook of Process Algebra, J.A. Bergstra, A. Ponse and S.A. Smolka, eds, Elsevier, Amsterdam (2001), 197–292.

[2] J.C.M. Baeten, J.A. Bergstra and J.W. Klop, *Syntax and defining equations for an interrupt in mechanism process algebra*, Fundamenta Informatica **9** (1986), 127–168.

[3] T. Bolognesi and E. Brinksma, *Introduction to the ISO specification language LOTOS*, Computer Networks and ISDN Systems **14** (1987), 25–59.

[4] A. Bouali and R. de Simone, *Symbolic bisimulation minimization*, Proc. CAV '91, Lecture Notes in Comput. Sci. 663 (1992), 96–108.

[5] A. Bouali, A. Ressouche, V. Roy and R. de Simone, *The Fc2Tools set*, Proc. of Computer-Aided Verification (CAV '96), Lecture Notes in Comput. Sci. 1102 (1996), 441–445.

[6] R.E. Bryant, *Graph-based algorithms for Boolean function manipulation*, IEEE Trans. Comput. **C-35** (6) (1986), 677–691.

[7] R.E. Bryant, *Symbolic Boolean manipulation with ordered binary-decision diagrams*, ACM Computing Surveys **24** (3) (1992), 293–318.

[8] U. Celikkan, *Semantic preorders in the automated verification of concurrent systems*, Ph.D. thesis, North Carolina State University, Raleigh (1995).

[9] U. Celikkan and R. Cleaveland, *Generating diagnostic information for behavioral preorders*, Distrib. Comput. **9** (1995), 61–75.

[10] R. Cleaveland and M. Hennessy, *Testing equivalence as a bisimulation equivalence*, Formal Aspects of Computing **5** (1) (1993), 1–20.

[11] R. Cleaveland, J. Parrow and B. Steffen, *The Concurrency Workbench: A semantics-based tool for the verification of concurrent systems*, ACM TOPLAS **15** (1) (1993).

[12] R. Cleaveland and S. Sims, *The NCSU Concurrency Workbench*, Proc. of Computer-Aided Verification (CAV '96), Lecture Notes in Comput. Sci. 1102 (1996), 394–397.

[13] C. Courcoubetis, M. Vardi, P. Wolper and M. Yannakakis, *Memory efficient algorithms for the verification of temporal properties*, Proc. of Computer-Aided Verification (CAV '90) (1990).

[14] J.-C. Fernandez, H. Garavel, A. Kerbrat, R. Mateescu, L. Mounier and M. Sighireanu, *CADP: A protocol validation and verification toolbox*, Proc. of Computer-Aided Verification (CAV '96), Lecture Notes in Comput. Sci. 1102 (1996), 437–440.

[15] H. Garavel and M. Sighireanu, *Towards a second generation of formal description techniques – rationale for the design of e-lotos*, Proc. FMICS'98 (1998), 187–230.

[16] R.J. van Glabbeek, *The linear time – branching time spectrum I. The semantics of concrete, sequential process*, Handbook of Process Algebra, J.A. Bergstra, A. Ponse and S.A. Smolka, eds, Elsevier, Amsterdam (2001), 3–99.

[17] R.J. van Glabbeek and W.P. Weijland, *Branching time and abstraction in bisimulation semantics,* J. ACM **43** (3) (1996), 555–600.

[18] J.F. Groote and F.W. Vaandrager, *An efficient algorithm for branching bisimulation and stuttering equivalence*, Proc. 17th International Colloquium on Automata, Languages and Programming, Lecture Notes in Comput. Sci. 443, Springer-Verlag (1990), 626–638.

[19] M. Hennessy, *An Algebraic Theory of Processes*, MIT Press (1988).

[20] C.A.R. Hoare, *Communicating Sequential Processes*, Prentice-Hall (1985).

[21] P.C. Kanellakis and S.A. Smolka, *CCS expressions, finite state processes and three problems of equivalence*, Inform. and Comput. **86** (1) (1990), 43–68.

[22] R. Milner, *A Calculus of Communicating Systems*, Lecture Notes in Comput. Sci. 92 (1980).

[23] R. Milner, *Communication and Concurrency*, Prentice-Hall (1989).

[24] R. Paige and R.E. Tarjan, *Three partition refinement algorithms*, SIAM J. Comput. **16** (6) (1987), 973–989.

[25] G. Plotkin, *A structural approach to operational semantics*, Aarhus University, Computer Science Department (1981).

[26] A.W. Roscoe, *Model-checking CSP*, A Classical Mind: Essays in Honour of C.A.R. Hoare, Prentice-Hall (1994).

[27] O. Sokolsky, I. Lee and H. Ben-Abdallah, *Specification and analysis of real-time systems with PARAGON*, Ann. Software Engineering **7** (1999), 211–234.

Subject index

$\sim$, 395, 396, 398
$\sim^E$, 396, 398, 407
$\underline{\leftrightarrow}$, 408, 409
$\approx$, 407
$\precsim^{\Pi}_{\Phi_1,\Phi_2}$, 398, 411, 412, 416–418, 422
$\precsim$, 397, 398

algorithm
– *BB_PARTITIONING*, 408–411
– *EFF_PREORDER*, 411–416
– Kanellakis–Smolka, *see KS_PARTITIONING*
– *KS_PARTITIONING*, 398–401
– local, 416–422
– on-the-fly, *see* local
– Paige–Tarjan, *see PT_PARTITIONING*
– *PREORDER*, 417, 422
– *PT_PARTITIONING*, 401–403
– symbolic, 405–407

equivalence
– bisimulation, 395, 398
– – fixpoint characterization, 396
– – iterative characterization, 396
– – parameterized, 396, 398
– branching bisimulation, 408
– observational, 407
– observational congruence, 408
– testing, 408
– weak bisimulation, 407

labeled transition system, 394
– finite-state, 394
– image-finite, 396
LTS, *see* labeled transition system

OBDD, 404–407

parameterized semantic relation, 397, 411, 416
partition refinement, 398
preorder
– forward simulation, 397
process transformation, 407

refinement ordering, *see* preorder

splitter, 398, 401, 404, 409, 411
– compound, 402

Part 3
Infinite-State Processes

CHAPTER 7

A Symbolic Approach to Value-Passing Processes

Anna Ingólfsdóttir[1], Huimin Lin[2]

[1] *Institute of Computer Science, Aalborg University, Denmark*
E-mail: annai@cs.auc.dk

[2] *Laboratory for Computer Science, Institute of Software, Chinese Academy of Sciences, Republic of China*
E-mail: lhm@ox.ios.ac.cn

Contents

1. Introduction . . . 429
   1.1. Some examples . . . 429
   1.2. Late and early semantics . . . 431
   1.3. Infinite branching . . . 434
   1.4. Axiomatization . . . 436
   1.5. Outline of the chapter . . . 437
2. The value-passing language . . . 437
   2.1. The syntax . . . 437
   2.2. Operational semantics . . . 439
3. Bisimulation equivalences . . . 440
   3.1. Early bisimulation . . . 440
   3.2. Late bisimulation . . . 442
   3.3. Translating into pure processes . . . 446
4. Symbolic bisimulations . . . 448
   4.1. Symbolic operational semantics . . . 449
   4.2. Early symbolic bisimulation . . . 451
   4.3. Late symbolic bisimulation . . . 454
   4.4. Weak symbolic equivalences . . . 456
5. Proof systems . . . 459
   5.1. The Core system . . . 459
   5.2. The axioms for early bisimulation . . . 464
   5.3. The τ laws . . . 465
   5.4. Adding other operators . . . 467
   5.5. Reasoning about recursion . . . 468
6. Summary and related work . . . 475
Acknowledgements . . . 475
References . . . 475
Subject index . . . 477

HANDBOOK OF PROCESS ALGEBRA
Edited by Jan A. Bergstra, Alban Ponse and Scott A. Smolka

Abstract

In this chapter we give a survey of the semantic theory for value-passing processes, focusing on bisimulation equivalences. The emphasis is on the symbolic method. Both operational and proof theoretic approaches are described and their expressiveness compared.

1. Introduction

In the pioneering work of Milner [29] on *CCS* and Hoare [22] on *CSP*, exchange of data appeared as a natural feature of these languages. Subsequently, although some work was done on their semantics, see, e.g., [6,21,28], most of the research effort was devoted to the pure versions of these languages such as *pure CCS* [29,32] and *theoretical CSP*, *TCSP* [2]. The focus was on the behavioural properties of processes where no data are involved and the only form of communication considered was pure synchronization. In particular, structural operational semantics by means of inference rules [40] and more abstract behavioral equivalences based on these were defined and their properties studied in detail.

About a decade after the introduction of *CCS* and *CSP*, the theory of pure processes was considered as being well established and research on semantic theories for value-passing processes was back in focus [13–17]. Not so surprisingly, it turned out that allowing transmission of elements from a possibly infinite value domain introduced some technical complications, with the consequences that some of the already established theories for the pure calculi could not be applied directly.

In this chapter we will try to give the reader some insight into the theoretical complications that arise when such value-passing processes are described semantically, as well as solutions to those. Our main focus will be on techniques for checking semantic equivalences between processes. In particular, we are interested in investigating equivalences based on the idea of *bisimulation* due to Milner and Park [30,37]. For this purpose *symbolic operational semantics* and the corresponding notion of *symbolic bisimulation* are introduced. These are abstractions of the standard semantics which allow us to perform case analysis on the input values and thus, in many cases, to resolve the infinity arising from an infinite value space.

We start the chapter by providing the readers with some motivation, mainly inspired by examples. Based on the acquired intuition we give an informal introduction to our example language, which is basically the original *CCS* as it was introduced by Milner in [29].

1.1. *Some examples*

In standard sequential programming, manipulating and exchanging values with the environment occurs naturally, as illustrated by the following example.

EXAMPLE 1.1. The program Prog below is a simple sequential program that reads two numbers, calculates their sum and multiple, divides the sum by the multiple and outputs the result:

```
Prog:
begin
      input(x, y);
      t := x + y; u := x · y;
      output(t/u)
end.
```

This program interacts or communicates with its environment, i.e., the user, by receiving the input values x and y from him and giving him back the result of the computation.

In concurrent computations, value-passing occurs just as naturally and processes cooperate with each other by exchanging messages, typically through communication channels or ports. A synchronous communication event, which is the type of communication we consider in this chapter, usually consists of two complementary actions: input and output. A process *rec*, which is ready to receive or input a message from a channel c, may be given by $rec = c?x.rec'$ where typically x occurs free in the sequel rec'; when inputting a value v from channel c this process evolves into $rec'[v/x]$, i.e., the process rec' with all free occurrences of the variable x replaced by the value v. Similarly a process *send* that prepares to send or output a value v on the same channel and by sending evolves into its sequel $send'$, may be expressed as $send = c!v.send'$. When these two processes are running in parallel, denoted by $rec \mid send$, a communication event can take place, with the consequence that the composite system evolves into a new state $rec'[v/x] \mid send'$ where the "passing" of the value v from *send* to *rec* is modeled by substitution, as explained before. Furthermore, once these processes have agreed to communicate, the communication is private and no other processes can take part in it; this kind of communication is called *handshaking*. More formally the event described above may be given by a transition relation induced by

$$rec \xrightarrow{c?v} rec'[v/x] \text{ and } send \xrightarrow{c!v} send' \text{ imply } rec \mid send \xrightarrow{\tau} rec'[v/x] \mid send'.$$

Here τ is a special invisible action indicating that synchronization and exchange of values between processes is not observable and therefore it can not take part in a further synchronization with the environment.

EXAMPLE 1.2. In the above example, noting that the commands $t := x + y$ and $u := x \cdot y$ may be performed independently, the program `Prog` may be rewritten in the process description language *CCS* as a concurrent process by

$$p = \big(in_1?x.in_2?y.\big(c!(x+y) \mid d!(x \cdot y) \mid d?u.c?t.out!(t/u)\big)\big)\backslash\{c, d\},$$

where the restriction operator $\_\backslash\{c, d\}$ indicates that p can use c and d only for internal communication, but that it is not allowed to communicate with its environment via these channels. We note that the process calculus *CCS* is applicative in the sense that it does not allow for assignments. Thus the results of the calculations of $x + y$ and $x \cdot y$ are output directly without assigning them to a variable first.

The next example shows how processes with infinite behaviour may be given using recursive definitions parameterized over data.

EXAMPLE 1.3. A *counter* is a process that outputs the natural numbers, one after the other in increasing order on a given channel, say c. To express this we let

$$C(n) \Leftarrow c!n.C(n+1)$$

meaning that the parameterized recursive process $C(n)$ behaves like the right hand side of the defining equality, i.e., it outputs the number n on channel c and afterwards behaves like $C(n+1)$. It is not difficult to see that $C(0)$ has the expected behaviour of a counter as described above. To obtain two distinguishable copies of the counter, instead of giving two copies of similar recursive definitions, we apply a so-called *renaming operator* on $C(n)$. Thus $C(n)[c \mapsto c']$, the process $C(n)$ with c renamed to c', behaves like $C(n)$ with all occurrences of c replaced by c', i.e., like the counter that makes use of the channel c' instead of c.

The following example shows how parameterized recursive processes over an inductively defined data domain may be defined inductively.

EXAMPLE 1.4. In what follows let $\mathit{Stack}(b_1 b_2 \ldots b_n)$ denote a stack over an alphabet L. We represent the content of the stack as a string $b_1 b_2 \ldots b_n \in L^*$ where b_1 is the top element of the stack. Let ε denote the empty string, $b \in L$ and $s \in L^*$. Furthermore, let $\mathit{push}?x$ denote the action of inputting an element from channel *push* and assigning it to x, and $\mathit{pop}!b$ the action of outputting the element b on channel *pop*. Then we may model $\mathit{Stack}(s)$, $s \in L^*$, inductively as follows:

$$\mathit{Stack}(\varepsilon) \Leftarrow \mathit{push}?x.\mathit{Stack}(x),$$
$$\mathit{Stack}(bs) \Leftarrow \mathit{pop}!b.\mathit{Stack}(s) + \mathit{push}?x.\mathit{Stack}(xbs).$$

Here the operator $+$ (*choice*) indicates that the the process $\mathit{Stack}(bs)$ nondeterministically chooses between behaving like one of its components, i.e., either like $\mathit{pop}!b.\mathit{Stack}(s)$ or like $\mathit{push}?x.\mathit{Stack}(xbs)$. The intuitive behaviour of the process *Stack* is that if it is empty ($\mathit{Stack}(\varepsilon)$) its only possible action is to push onto it an element received from the environment via the channel *push*. If the stack is not empty, i.e., it contains a string of the form bs ($\mathit{Stack}(bs)$) with b as its top element, then either the element b can be popped off the stack and returned to the environment via *pop*, or the environment can push a new element x onto the stack via *push* whereby it evolves to the stack $\mathit{Stack}(xbs)$.

1.2. *Late and early semantics*

A standard way of describing each computational step of a concurrent or reactive system is by assigning to it a so-called *Structural Operational Semantics* as described in [7] and [1] in this Handbook; the semantics we have introduced informally in this section is based on this approach. However it is a widespread opinion amongst researchers in the area that this kind of semantic description is in general too concrete to reason about the behaviour of processes on a more abstract level, for instance to decide when two processes are behaviourally or semantically equivalent. In [30] Milner overcomes this inadequacy by describing a general idea to model observational equivalence of processes referred to as *black box testing* (see [7] in this Handbook for further motivation); a process is considered to be a black box provided with an interface containing buttons, one for each (observable) action it can perform, plus possibly some other control devices such as lights or buttons. An external observer can interact with, and thus perform experiments on, the process by pressing these buttons and observing its reactions. Two processes are said to be behaviourally equivalent if the observer is not able to distinguish between their reactions to his experiments. A variety of preorders and equivalences are defined in the literature based on the black box model, but with different notions of observers, observable actions and experiments. A standard class of equivalences, and probably the most studied one, consists of Milner's equivalences based on the idea of *bisimulation*, which may briefly be described as follows (see also [7] in this Handbook for a more detailed explanation):

> Two processes p and q are bisimilar if whenever one of them, say p, reacts when the button labelled by a is pressed by evolving to some state p' (i.e., $p \xrightarrow{a} p'$) then q also reacts when some button, labelled by a, is pressed and by doing so it evolves to some state q' where p' and q' are again bisimilar.

Originally the idea of bisimulation only concerned pure processes. Extending the idea to value-passing processes raises some questions about what we mean by an (observable) action and consequently how processes are supposed to synchronize. We will try to explain this by the following example.

EXAMPLE 1.5. Let c, c_1 and c_2 be three different channel names and the processes p, p_1, p_2 and q, q_1, q_2 be defined by

$$\begin{aligned}
p_1 &= c?x.\mathtt{if}\ x \leqslant 5\ \mathtt{then}\ c_1!x\ \mathtt{else}\ c_2!x,\\
p_2 &= c?x.\mathtt{if}\ x \leqslant 5\ \mathtt{then}\ c_2!x\ \mathtt{else}\ c_1!x,\\
q_1 &= c?x.c_1!x,\\
q_2 &= c?x.c_2!x,\\
p &= p_1 + p_2,\\
q &= q_1 + q_2.
\end{aligned}$$

Obviously p and q are bisimilar according to Milner's original semantics as they have isomorphic transition graphs. However we also note that different moves of p due to the same component, say p_1, have to be matched by q by moves coming from different components. Thus

$$p \xrightarrow{c?v} \mathtt{if}\ v \leqslant 5\ \mathtt{then}\ c_1!v\ \mathtt{else}\ c_2!v \text{ for } v = 1 \text{ and } v = 6$$

because

$$p_1 \xrightarrow{c?v} \mathtt{if}\ v \leqslant 5\ \mathtt{then}\ c_1!v\ \mathtt{else}\ c_2!v \text{ for } v = 1 \text{ and } v = 6.$$

On the other hand, to match these moves by q we need one move from each of its components q_1 and q_2, i.e.,

$$q \xrightarrow{c?1} c_1!1 \quad \text{because} \quad q_1 \xrightarrow{c?1} c_1!1$$

and

$$q \xrightarrow{c?6} c_2!6 \quad \text{because} \quad q_2 \xrightarrow{c?6} c_2!6.$$

In terms of black box testing the semantic approach described in the examples above corresponds to the situation in which the buttons for the input actions are marked with $c?v$, i.e., with labels consisting of both $c?$, to indicate that the value should be received along the channel c, and also the *concrete value* v that is to be received along that channel. This semantic approach has been referred to as *early semantics* due to the early instantiation of the value variable that is involved in the communication.

Another approach would be to consider input actions of the form $c?$ for the action of inputting *some value* on the channel c. In the black box interpretation this means that we only mark the corresponding buttons for input actions with labels of the form $c?$ and leave out the concrete value v. This entails that the result of pressing such a button or performing an action of the form $c?$ is an incomplete process as the information about the concrete value the process is about to receive from the environment is still missing. In this introduction we consider this incomplete state as a function that inputs a value and delivers a completed process. (In the main body of this chapter these incomplete states are given as process terms with possibly free value variables.) Such functions are referred to as *abstractions* in [34]. Syntactically abstractions are denoted by $(x)p$ where x is a value variable and p is a process term. They can be interpreted as a shorthand for the λ-abstraction $\lambda v.p[v/x]$. Two such abstractions match each other semantically if they do so pointwise, i.e., if whenever they are both applied to the same value, the results are indistinguishable. This semantic approach has been termed *late* semantics due to the delayed instantiation of the value variables.

EXAMPLE 1.6. To show that p and q from the previous example are not strongly bisimilar according to the late approach, we proceed as follows. The process p has the late transition

$$p \xrightarrow{c?} \lambda v.\mathtt{if}\ v \leqslant 5\ \mathtt{then}\ c_1!v\ \mathtt{else}\ c_2!v = f.$$

The possible $c?$ transitions for q are

$$q \xrightarrow{c?} \lambda v.c_1!v = f_1 \quad \text{and} \quad q \xrightarrow{c?} \lambda v.c_2!v = f_2.$$

For p and q to be bisimilar in the late sense it is necessary that either f and f_1 match completely pointwise or otherwise f and f_2 must do. So either we have that for all v, $f(v)$ and $f_1(v)$ are bisimilar or that for all v, $f(v)$ and $f_2(v)$ are. To see that this is not the case we observe that

$$f(6) = \mathtt{if}\ 6 \leqslant 5\ \mathtt{then}\ c_1!6\ \mathtt{else}\ c_2!6 \xrightarrow{c_2!6} \quad \text{while} \quad f_1(6) \stackrel{c_2!6}{\not\longrightarrow}$$

and that

$$f(1) \xrightarrow{c_1!1} \quad \text{while} \quad f_2(1) \stackrel{c_1!}{\not\longrightarrow},$$

where $p \xrightarrow{a}$ means that $p \xrightarrow{a} p'$ for some p' and $p \stackrel{a}{\not\longrightarrow}$ means that no such p' exists.

The two semantic approaches described above lead to two different notions of bisimulation: the *early* bisimulation and the *late* one, which is strictly finer. However we want to point out here that if we look at the rules for communication, it turns out that the early and the late semantics yield exactly the same result.

EXAMPLE 1.7. In the example above, if $r = c!v$ then the late semantics yields

$$q \mid r \xrightarrow{\tau} f_1(v) \mid \mathbf{0} = c_1!v \mid \mathbf{0} \quad \text{and} \quad q \mid r \xrightarrow{\tau} f_2(v) \mid \mathbf{0} = c_2!v \mid \mathbf{0}.$$

It is not difficult to see that the early semantics gives exactly the same transitions.

In more general terms only the intermediate states which the processes reach immediately after an input action can be distinguished by the late and early semantics but after a completed communication they are syntactically the same for these two approaches. This implies, as shown in [24], that although behavioural equivalences that involve investigation of these intermediate states are different for the late and early semantics, the equivalences that do not involve observations of states that occur as a result of input actions cannot be told apart. An example of a semantics, where the late and early versions coincide is the testing based semantics due to De Nicola and Hennessy.

1.3. *Infinite branching*

As pointed out at the beginning of this section, the main technical difficulties associated with introducing value-passing into the semantic theory of pure processes is due to the infinite branching of the input transitions. For example, the process $c?x.p$ has the transitions $c?x.p \xrightarrow{c?v} p[v/x]$ where v ranges over the value domain which in general may be assumed to be infinite. However, in *CCS* the only control structure that allows branching on input values is the construction `if_then_else_` and obviously each process can only contain finitely many of those. Therefore, by applying a case analysis on the input values, in many cases this infinite branching can be resolved and reduced to finite branching. To explain this point let us first have a look at the following example from sequential programming.

EXAMPLE 1.8. Let us consider the following two sequential programs.

```
Program1;
input x;
if x > 1 then x := x − 1
          else x := x + 1;
output x.

Program2;
input x;
if x ≥ 3 then x := x − 1
          else (if x = 1 then x := 2
                          else x := 1);
output x.
```

The actions of these two programs are given by the data flow diagrams below.

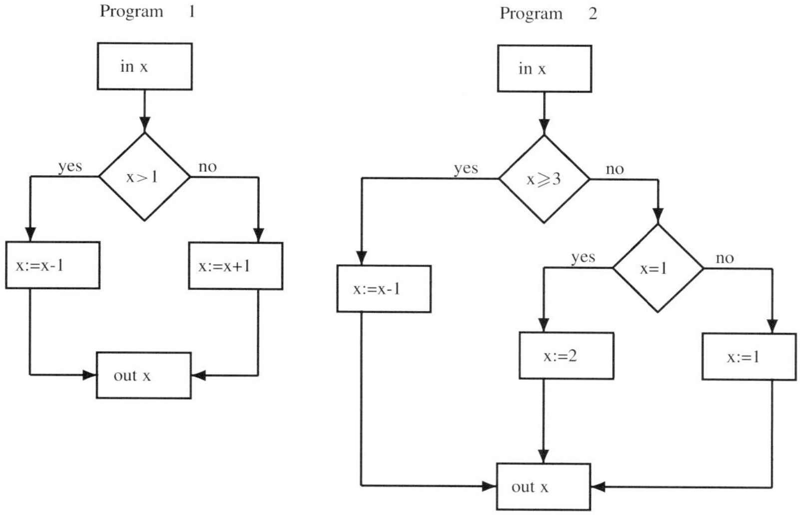

By case analysis on the input value for the variable x, considering the cases $0 \leqslant x < 1$, $x = 1$, $1 < x < 3$ and $x \geqslant 3$ the reader should have no difficulties in convincing himself that these two programs are equivalent in the sense that for a given input x they will both return exactly the same output.

As pointed out before, *CCS* does not allow for assignments. However the action of these two programs may be simulated by the following *CCS* processes.

$$p_1 = in?x.\texttt{if } x > 1 \texttt{ then } out!(x-1) \texttt{ else } out!(x+1),$$
$$p_2 = in?x.\texttt{if } x \geqslant 3 \texttt{ then } out!(x-1)$$
$$\texttt{else (if } x = 1 \texttt{ then } out!2 \texttt{ else } out!1).$$

Based on the same idea as the data flow diagrams, an abstract or *symbolic* transition graph for these processes may be given as below. We note that these graphs have finite branching.

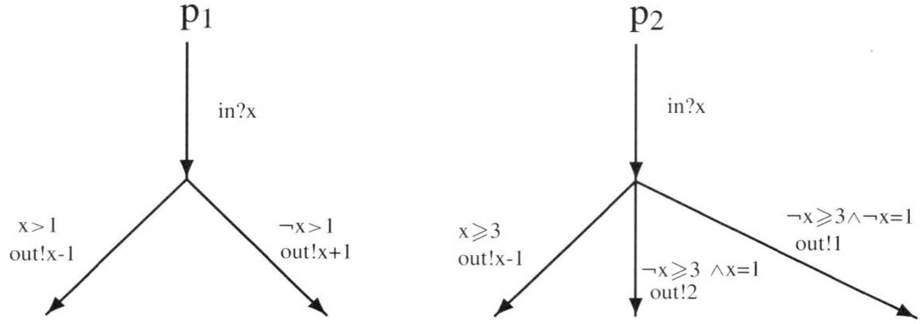

Again, using the same kind of case analysis, based on the same cases as before, the reader should be able to check that these two processes are bisimilar both in the early and the late sense.

One of the main purposes of this chapter is to describe formally a *symbolic semantics* that realizes the ideas described above and thus provides a theoretical framework that allows us to formally reason about equivalences between value-passing processes by only investigating finitely branching transition graphs.

1.4. *Axiomatization*

Providing sound and complete axiomatizations for various equivalence relations has been one of the major research topics in the development of process theories. A complete axiomatization not only allows us to reason about process behaviours by syntactic manipulation, but also helps to understand the properties of the operators used to build complex processes from simpler components.

Traditionally, axiomatizations in process calculi are based on pure equational reasoning, namely replacing equals by equals. For value-passing processes, however, this approach is in general not adequate, because, for instance, input prefixing introduces the notion of bound variables which is beyond the realm of pure equational reasoning. To be able to deal with such "non-algebraic" constructs we generalize equational reasoning by introducing inference rules while preserving the equational laws that are at the core of the axiomatizations in pure calculi.[1]

As data play an important role in value-passing processes, it seems that any axiomatization for such processes would have to involve axiomatizations for the data domain. However, it turns out that we can factor out data reasoning from process reasoning, by employing *conditional equations* of the form

$$b \rhd t = u,$$

where t and u are process terms and b is a Boolean expression representing the condition on the data domain under which t and u are equal. An example of a proof rule of our inference system is

$$\text{COND} \quad \frac{b' \wedge b \rhd t = u \quad b' \wedge \neg b \rhd \mathbf{0} = u}{b' \rhd \mathtt{if}\ b\ \mathtt{then}\ t = u}.$$

It captures the intuitive meaning of the (one-armed) conditional construct: $\mathtt{if}\ b\ \mathtt{then}\ t$ behaves like t when b is true, and like $\mathbf{0}$ (the "empty" process) otherwise. In fact this rule is all we need to know about the $\mathtt{if}\ \_\ \mathtt{then}\ \_$ construct when manipulating syntactical terms. Looked upon from a "goal-directed" point of view, it moves the parts involving data (b) from the process term ($\mathtt{if}\ b\ \mathtt{then}\ t$) to the conditional part of the judgment. Such conditions can be used to discharge constructs involving data when some other inference rules are applied. For instance, the following rule for output prefixing:

$$\text{OUTPUT} \quad \frac{b \models e = e', \quad b \rhd t = u}{b \rhd c!e.t = c!e'.u}$$

[1] For a treatment of these matters that stays closer to equational logic see [8].

allows us to conclude that $c!e.t$ and $c!e'.u$ are equal under b, provided that t and u are equal under b and that $e = e'$ is implied by b. As the last condition only relies on reasoning over data, it is factored out from the main proof system. This is manifested by using the notation $b \models e = e'$.

Using inference rules in this style, together with the usual equational axioms, a spectrum of proof systems for value-passing processes can be obtained which are sound and complete (relative to data domain reasoning) with respect to the various bisimulation-based equivalences.

1.5. *Outline of the chapter*

The rest of this chapter is organized as follows. In Section 2 we formally introduce the syntax and operational semantics for our example language, full *CCS*, whereas Section 3 is devoted to defining the various bisimulation based equivalences we investigate. Symbolic operational semantics and symbolic bisimulations are studied in Section 4. In Section 5 we formulate proof systems to reason about value-passing processes, one for each of the congruences mentioned above, and we give some soundness and completeness results. The main content of this chapter is summarized in Section 6 where pointers to relevant works are also given.

We will omit detailed proofs as much as possible in this chapter by referring to the literature, and will only supply them when they are not available in published papers.

2. The value-passing language

In this section we introduce the syntax for the language we consider in this chapter together with its operational semantics. The language is essentially the same as the one originally proposed by Milner in [29, Section 2.8], often referred to as *full CCS* or *value-passing CCS*.

2.1. *The syntax*

The language we consider is rather abstract as we want to focus on aspects which do not depend on the properties of the data domain. Therefore we simply assume a predefined syntactic category *DExp* of data expressions, ranged over by e, e', etc. This should include, at least, an infinite set *Val* of values, ranged over by v, v', etc. and a countably infinite set *Var* of data variables, ranged over by x, y, etc., which is disjoint from *Val*. We also assume a similar syntactic category *BExp* of Boolean expressions, ranged over by b, b', etc. with the usual set of operators *true*, *false*, $\neg$, $\vee$, $\wedge$ and $\rightarrow$. Furthermore, for every pair of data expressions e, e' we assume that $e = e'$ is a Boolean expression. We also assume that each data expression e and each Boolean expression b has associated to it a set of free data variables, $fv(e)$ and $fv(b)$ respectively, and that only data variables can occur free in these expressions.

$$t ::= \mathbf{0} \mid \alpha.t \mid t + t \mid \texttt{if } b \texttt{ then } t \mid t \mid t \mid t \backslash L \mid t[f] \mid P(\overline{e})$$
$$\alpha ::= \tau \mid c?x \mid c!e$$

Fig. 1. Syntax.

In general we do not worry about the expressive power of these languages although the results on symbolic bisimulations require that the language for Boolean expressions is sufficiently expressive to characterize any collection of evaluations over which two open terms are bisimilar.

The definition of the language also assumes a countably infinite set of *channel names* *Chan*, ranged over by c, c', etc., which is disjoint from *Val* and *Var*. To define recursive processes we use a set of *process identifiers* $\mathcal{K}$, ranged over by P, Q, etc. Each identifier is assigned an *arity*, a non-negative integer, indicating the number of parameters it should take to form a process term.

Our language is given by the BNF-definition in Figure 1 and contains the operators listed below. $\mathbf{0}$ is the inactive process capable of doing nothing; $\alpha.t$ is action prefixing where we have three kinds of actions: the invisible action τ, representing internal communication, and the input and output actions $c?x$ and $c!e$ respectively; $\texttt{if } b \texttt{ then } t$ is the (one armed) conditional construction; $t + u$ is nondeterministic choice (or summation); $t \mid u$ is parallel composition; $t \setminus L$, where L is a finite subset of *Chan*, is channel restriction; $t[f]$, where f is a partial function from *Chan* to *Chan* with finite domain, is channel renaming. $P(\overline{e})$, where $P \in \mathcal{K}$ and the length of $\overline{e}$ equals the arity of P, is a process which is defined recursively by associating to each such identifier a defining equation of the form

$$P(x_1, \ldots, x_n) \Leftarrow t.$$

This way of giving the syntax of recursive processes turns out to be quite intuitive and therefore easy to apply. However, for theoretical purposes, it is sometimes more convenient to have recursion directly in the syntax by introducing an operator of the form $\mathbf{fix}\, X._{-}$ where X is a process variable. This will be discussed in Section 5.5 and for the time being we will use the version based on identifiers and defining equations. In the sequel we will often abbreviate $\alpha.\mathbf{0}$ as α.

In our studies we need the notion of an *evaluation* ρ which is a total mapping from *Var* to *Val*. We use the standard notation $\rho\{v/x\}$ to denote the evaluation which differs from ρ only in that it maps x to v. We assume that an application of ρ to a data expression e, denoted $[\![e]\!]\rho$, always yields a value from *Val* and similarly for Boolean expressions, $[\![b]\!]\rho$ is either *true* or *false*. Applications of evaluations also generalize to process terms, denoted $t\rho$, in the obvious way. We assume that these evaluations satisfy some standard properties such as that whenever ρ and ρ' agree on $fv(e)$ then $[\![e]\!]\rho = [\![e]\!]\rho'$. If an expression e has no free variables, namely it is *closed*, then $[\![e]\!]\rho$ is independent of ρ. Consequently we simply write $[\![e]\!]$ to denote its value and use a similar convention for Boolean expressions. We will use the suggestive notation $\rho \models b$ to indicate that $[\![b]\!]\rho = true$, and $b \models b'$ to indicate that for every evaluation ρ, $\rho \models b$ implies $\rho \models b'$. Of course we could equally well say that $b \rightarrow b'$ is a logical theorem but our notation emphasizes the fact that we wish to describe our theory modulo the semantics of expressions. We will write $b = b'$ for $b \models b'$ and $b' \models b$.

We will also refer to substitutions of data expressions for data variables, and assume that they satisfy the expected properties. We use $e[e'/x]$ to denote the result of substituting e' for all free occurrences of x in e. More generally a substitution σ is a partial mapping from *Var* to *DExp* and we use $e\sigma$ to denote the result of an application of σ to the expression e.

An input prefix is a binding operator in the sense that $c?x.t$ binds x in t. This gives rise to the usual notions of bound and free variables. We will write $bv(t)$ and $fv(t)$ for the sets of bound and free variables in t, respectively. Bound variables lead to the standard notion of α-equivalence $\equiv_\alpha$ over terms and to that of substitution, $t[e/x]$ denoting the result of substituting all free occurrences of x in t by e. The definition of substitution in process terms obviously relies on the definition of substitution in data expressions and may involve renaming of bound variables when necessary to avoid variable capture. A term is *closed* if it contains no free variables, and is *open* otherwise. In a defining equation $P(x_1, \ldots, x_n) \Leftarrow t$ it is required that $fv(t) \subseteq \{x_1, \ldots, x_n\}$.

We let $\mathcal{T}$ denote the set of all process terms, ranged over by t, u etc., and let $\mathcal{P}$ denote the set of all closed terms, ranged over by p, q etc., which we refer to as *processes*.

We assign the following precedence to the operators (in decreasing order):

restriction and renaming, prefix, conditional, parallel composition, choice.

Although we only have the one-armed conditional $\texttt{if}\ b\ \texttt{then}\ \_$ in the language, with the help of $+$ the conventional two-armed $\texttt{if}\ b\ \texttt{then}\ \_\ \texttt{else}\ \_$ expression can be defined by

$$\texttt{if}\ b\ \texttt{then}\ t\ \texttt{else}\ u = \texttt{if}\ b\ \texttt{then}\ t + \texttt{if}\ \neg b\ \texttt{then}\ u.$$

In what follows we will use the `if_then_else_` construction freely without further comments.

2.2. *Operational semantics*

Our language is given an operational semantics in the usual way by assigning to it a *labelled transition system*, i.e., a triple $\langle \mathtt{S}, \mathtt{L}, \rightarrow \rangle$ where $\mathtt{S}$ is a set of *states*, $\mathtt{L}$ is a set of *labels* and $\rightarrow \subseteq \mathtt{S} \times \mathtt{L} \times \mathtt{S}$ is a *transition relation*. Usually we use the more intuitive notation $s_1 \xrightarrow{l} s_2$ in place of $(s_1, l, s_2) \in \rightarrow$.

The standard operational semantics for our language is obtained by taking the set $\mathtt{S}$ to be $\mathcal{P}$, the set $\mathtt{L}$ to be the set of actions Act_τ, where $Act = \{c!v, c?v \mid c \in Chan,\ v \in Val\}$ and $Act_\tau = Act \cup \{\tau\}$, ranged over by a, and the transition relation to be the relation $\rightarrow_{\mathtt{e}} \subseteq \mathcal{P} \times Act_\tau \times \mathcal{P}$ defined by the rules in Figure 2. The function *chan* used in the definition is defined by $chan(c?v) = chan(c!v) = \{c\}$ and $chan(\tau) = \emptyset$. As explained in the introduction this semantic approach is usually referred to as the *early* operational semantics (hence the

$$\text{TAU} \quad \frac{}{\tau.t \xrightarrow{\tau}_{\mathsf{e}} t} \qquad \text{OUTPUT} \quad \frac{}{c!e.t \xrightarrow{c!v}_{\mathsf{e}} t} \text{ where } v = [\![e]\!]$$

$$\text{INPUT} \quad \frac{}{c?x.t \xrightarrow{c?v}_{\mathsf{e}} t} \, v \in \mathit{Val} \qquad \text{COND} \quad \frac{t \xrightarrow{a}_{\mathsf{e}} t'}{\texttt{if } b \texttt{ then } t \xrightarrow{a}_{\mathsf{e}} t'} \, [\![b]\!] = \mathit{true}$$

$$\text{SUM1} \quad \frac{t \xrightarrow{a}_{\mathsf{e}} t'}{t + u \xrightarrow{a}_{\mathsf{e}} t'} \qquad \text{SUM2} \quad \frac{u \xrightarrow{a}_{\mathsf{e}} u'}{t + u \xrightarrow{a}_{\mathsf{e}} u'}$$

$$\text{PAR1} \quad \frac{t \xrightarrow{a}_{\mathsf{e}} t'}{t \mid u \xrightarrow{a}_{\mathsf{e}} t' \mid u} \qquad \text{PAR2} \quad \frac{u \xrightarrow{a}_{\mathsf{e}} u'}{t \mid u \xrightarrow{a}_{\mathsf{e}} t \mid u'}$$

$$\text{COM} \quad \frac{t \xrightarrow{c?v}_{\mathsf{e}} t' \quad u \xrightarrow{c!v}_{\mathsf{e}} u'}{t \mid u \xrightarrow{\tau}_{\mathsf{e}} t' \mid u'} \qquad \text{RES} \quad \frac{t \xrightarrow{a}_{\mathsf{e}} t'}{t \backslash L \xrightarrow{a}_{\mathsf{e}} t' \backslash L} \, \mathit{chan}(a) \cap L = \emptyset$$

$$\text{REN} \quad \frac{t \xrightarrow{a}_{\mathsf{e}} t'}{t[f] \xrightarrow{f(a)}_{\mathsf{e}} t'[f]} \qquad \text{REC} \quad \frac{t[[\![\overline{e}]\!]/\overline{x}] \xrightarrow{a}_{\mathsf{e}} t'}{P(\overline{e}) \xrightarrow{a}_{\mathsf{e}} t'} \, P(\overline{x}) \Leftarrow t$$

Fig. 2. (Early) operational semantics.

subscript e), because, when an input term such as $c?x.u$ performs the input action, the value received is immediately substituted for the variable x in u. In Section 3.2 we will consider its *late* counterpart where this substitution is delayed.

By transition induction it is easy to show that α-equivalent processes have the same transitions (up to α-equivalence) in the following sense.

LEMMA 2.1. *If* $p \equiv_\alpha q$ *and* $p \xrightarrow{a}_{\mathsf{e}} p'$ *then* $q \xrightarrow{a}_{\mathsf{e}} q'$ *for some* q' *such that* $p' \equiv_\alpha q'$.

3. Bisimulation equivalences

In this section we introduce the bisimulation based equivalences investigated in this chapter. We shall consider both the late and the early semantics and in each case we also distinguish between the strong version of bisimulation, where τ actions are considered observable, and the weak one where they are not. In what follows we refer to the equivalences introduced in this section as the *concrete equivalences* as opposed to their more abstract versions, the *symbolic equivalences*, that will be introduced in the following section. We start with the early case.

3.1. *Early bisimulation*

Using the operational semantics introduced in the previous section we can define the notion of bisimulation in the standard way.

DEFINITION 3.1. A symmetric relation R over $\mathcal{P}$ is an *early strong bisimulation* if it satisfies: $(p, q) \in R$ implies that

if $p \xrightarrow{a}_{\mathrm{e}} p'$ then $q \xrightarrow{a}_{\mathrm{e}} q'$ for some q' such that $(p', q') \in R$.

We write $p \sim_{\mathrm{e}} q$ if $(p, q) \in R$ for some early strong bisimulation R.

Like in the pure case, it is easy to show that the union of any collection of early strong bisimulations is again a bisimulation, and that $\sim_{\mathrm{e}}$ is the union of all early strong bisimulations and therefore the largest such bisimulation. The relation $\sim_{\mathrm{e}}$ turns out to be an equivalence relation and is preserved by all the operators of the language. The proof of these properties is exactly the same as in the pure case and may be found in [32, Section 4.4].

From Lemma 2.1 we get directly that α-equivalent processes are bisimilar in the early strong sense.

PROPOSITION 3.2. *If* $p \equiv_\alpha q$ *then* $p \sim_{\mathrm{e}} q$.

A more abstract and more realistic way of relating processes behaviourally is by means of *weak* bisimulation where τ-transitions representing internal communications are abstracted away. The standard way of defining this relation is to replace the original transitions with the corresponding weak transition relations in the definition of bisimulation. For this purpose let $\Rightarrow_{\mathrm{e}}$ be $\xrightarrow{\tau}{}^{*}_{\mathrm{e}}$ and $\overset{a}{\Longrightarrow}_{\mathrm{e}}$ be $\Rightarrow_{\mathrm{e}}\xrightarrow{a}_{\mathrm{e}}\Rightarrow_{\mathrm{e}}$ for $a \in Act_\tau$. Let also $\overset{\hat{\tau}}{\Longrightarrow}_{\mathrm{e}}$ be $\Rightarrow_{\mathrm{e}}$ and $\overset{\hat{a}}{\Longrightarrow}_{\mathrm{e}}$ be $\overset{a}{\Longrightarrow}_{\mathrm{e}}$ for $a \in Act$. The early weak bisimulation is then defined as usual.

DEFINITION 3.3. A symmetric relation R over $\mathcal{P}$ is an *early weak bisimulation* if it satisfies: $(p, q) \in R$ implies that

if $p \xrightarrow{a}_{\mathrm{e}} p'$ then $q \overset{\hat{a}}{\Longrightarrow}_{\mathrm{e}} q'$ for some q' such that $(p', q') \in R$.

We write $p \approx_{\mathrm{e}} q$ if $(p, q) \in R$ for some early weak bisimulation R.

Again, like in the strong case, $\approx_{\mathrm{e}}$ is the largest early weak bisimulation and is obtained as the union of all early weak bisimulations.

PROPOSITION 3.4. $\approx_{\mathrm{e}}$ *is an equivalence relation.*

PROOF. See [32, Section 5.2, Proposition 2]. □

As in the case of the pure calculus, $\approx_{\mathrm{e}}$ is not a congruence since it is not preserved by $+$. Therefore we introduce the following modified relation:

DEFINITION 3.5. For closed terms p and q, $p \simeq_{\mathrm{e}} q$ if and only if
- whenever $p \xrightarrow{a}_{\mathrm{e}} p'$ then $q \overset{a}{\Longrightarrow}_{\mathrm{e}} q'$ for some q' such that $p' \approx_{\mathrm{e}} q'$;
- whenever $q \xrightarrow{a}_{\mathrm{e}} q'$ then $p \overset{a}{\Longrightarrow}_{\mathrm{e}} p'$ for some p' such that $p' \approx_{\mathrm{e}} q'$.

Note that the only difference between the definition of the relations $\approx_e$ and $\simeq_e$ is that in the latter $q \stackrel{\hat{a}}{\Longrightarrow}_e q'$ is replaced by $q \stackrel{a}{\Longrightarrow}_e q'$ on the "top level" of the definition, i.e., the matching move for an initial τ-move has to be a real τ-move.

PROPOSITION 3.6. *$\simeq_e$ is preserved by all the operators in the language and is the largest congruence relation contained in $\approx_e$.*

PROOF. See [32, Section 7.2]. □

We refer to $\simeq_e$ as *early observational congruence*.

3.2. *Late bisimulation*

In the definition of early bisimulation in the previous section, input variables are instantiated immediately when input actions take place and thus *before* processes are compared for equivalences. As a consequence, input transitions of the form $c?v$ for different values v, arising from a single input prefix $c?x$ from one process, can be matched by transitions inferred from more than one input prefix from the other process. Let us examine this in more detail with an example.

EXAMPLE 3.7. Consider the following two processes where x ranges over integers.

$$\begin{aligned} P &\Leftarrow (c?x.\mathtt{if}\ x = 0\ \mathtt{then}\ a_1!x\ \mathtt{else}\ a_2!x) + c?x.a_3!x, \\ Q &\Leftarrow (c?x.\mathtt{if}\ x = 0\ \mathtt{then}\ a_3!x\ \mathtt{else}\ a_2!x) \\ &\quad + (c?x.\mathtt{if}\ x = 0\ \mathtt{then}\ a_1!x\ \mathtt{else}\ a_3!x), \end{aligned}$$

where a_1, a_2 and a_3 are different channel names.

Obviously P and Q are bisimilar in the early strong sense as they have isomorphic transition graphs. For instance, a move $P \xrightarrow{c?v} a_3!v$ can be matched by $Q \xrightarrow{c?v} a_3!v$ using Q's first summand when v is 0, and using its second summand when v is different from 0.

In the late semantics we take a more restrictive view of input transitions by requiring that an input prefix gives rise to a *single* transition. Thus we replace the rule INPUT in Figure 2 with

$$\text{LATE-INPUT}\ \frac{}{c?x.t \xrightarrow{c?x} t}.$$

We note that the resulting term t is an open term in which x may occur free. Thus the states in the resulting transition system are elements from $\mathcal{T}$, the actions are from the set of *late actions*

$$Act_\tau^l = \{c!v \mid c \in Chan,\ v \in Val\} \cup \{c?x \mid c \in Chan,\ x \in Var\} \cup \{\tau\}$$

also ranged over by a, and the late transition relation $\rightarrow_1$, defined by the modified rules, has the functionality

$$\rightarrow_1 \subseteq \mathcal{P} \times Act_\tau^l \times \mathcal{T}.$$

The free variables in the residuals of input transitions are instantiated when two terms are compared for equivalence, i.e., *after* the transitions have taken place (hence the adjective *late*).

DEFINITION 3.8. A symmetric relation R between closed terms is a *late strong bisimulation* if it satisfies: $(p, q) \in R$ implies that

- whenever $p \xrightarrow{c?x}_1 t$ then $q \xrightarrow{c?y}_1 u$ for some u such that $(t[v/x], u[v/y]) \in R$ for all $v \in Val$;
- whenever $p \xrightarrow{a}_1 p'$, $a \in \{\tau,\ c!v \mid c \in Chan,\ v \in Val\}$, then $q \xrightarrow{a}_1 q'$ for some q' such that $(p', q') \in R$.

We write $p \sim_1 q$ if $(p, q) \in R$ for some late strong bisimulation R.

Note that, although $\rightarrow_1$ is defined between open terms, the relation $\sim_1$ is between closed terms only, because input variables get instantiated when input transitions are matched for bisimulation.

According to this definition the two processes P and Q in the above example are *not* bisimilar, because, for instance, Q can not match $P \xrightarrow{c?x} a_3!x$ using either of its two input transitions.

It is not difficult to see that the union of any collection of late strong bisimulations is again a late strong bisimulation. Furthermore $\sim_1$ is the union of all late strong bisimulations and thus the largest such bisimulation. It is also easy to see that $\equiv_\alpha$ is included in $\sim_1$, and that $\sim_1$ is included in $\sim_e$, i.e., that late bisimulation is a finer relation than its early counterpart. This inclusion is strict as the above example demonstrates.

It can also be shown that $\sim_1$ is an equivalence relation which is preserved by all the operators of the language. The proofs of these two facts are similar (and simpler) than those of Propositions 3.12, and 3.15, and are omitted.

If we exchange the existential and universal quantifiers in the first clause in the above definition, that is, if we change the clause to

- whenever $p \xrightarrow{c?x}_1 t$ then for all $v \in Val$, $q \xrightarrow{c?y}_1 u$ for some u and y such that $(t[v/x], u[v/y]) \in R$,

then it can be shown that the resulting relation, which we refer to as $\forall\exists$-bisimulation here, coincides with the early bisimulation of Definition 3.1.

PROPOSITION 3.9. *A relation R over $\mathcal{P}$ is a $\forall\exists$-bisimulation if and only if it is an early bisimulation.*

PROOF. First, by a simple transition induction, we can prove that the following relationship holds between the early and the late transitions:

1. If $p \xrightarrow{c?x}_{1} t$ then $p \xrightarrow{c?v}_{\mathrm{e}} t[v/x]$ for any $v \in Val$.
   If $p \xrightarrow{c?v}_{\mathrm{e}} p'$ then $p \xrightarrow{c?x}_{1} t$ for some x and t with $p' \equiv t[v/x]$.
2. For $a \in \{\tau,\ c!v \mid c \in Chan,\ v \in Val\}$, $p \xrightarrow{a}_{1} q$ if and only if $p \xrightarrow{a}_{\mathrm{e}} q$.

The statement of the proposition then follows easily from these properties. □

Now we turn towards defining the weak version of the late bisimulation. As usual we shall base our definition on a weak version of the transition relations. However here care must be taken in the case of input actions, because now the residual may be an open term which has no transitions. Thus, in the definition of the late weak transitions, an input transition *can not* absorb τ moves that occur after it. These τ moves will be absorbed when processes are compared for equivalence.

Let $\Rightarrow_1$ be $\xrightarrow{\tau}{}_{1}^{*}$, $\xRightarrow{c?x}_{1}$ be $\Rightarrow_1 \xrightarrow{c?x}_{1}$, and $\xRightarrow{a}_{1}$ be $\Rightarrow_1 \xrightarrow{a}_{1} \Rightarrow_1$ for $a \in \{\tau, c!v \mid v \in Val\}$. These weak transition relations can now be used to define the late version of the weak bisimulation relation.

DEFINITION 3.10. A symmetric relation R over $\mathcal{P}$ is a *late weak bisimulation* if it satisfies: $(p, q) \in R$ implies that

- whenever $p \xrightarrow{c?x}_{1} t$ then $q \xRightarrow{c?y}_{1} u$ for some u such that for all $v \in Val$ there exists q' such that $u[v/y] \Rightarrow q'$ and $(t[v/x], q') \in R$;
- whenever $p \xrightarrow{a}_{1} p'$, where $a \in \{\tau,\ c!v \mid c \in Chan,\ v \in Val\}$, then $q \xRightarrow{\hat{a}}_{1} q'$ for some q' such that $(p', q') \in R$.

We write $p \approx_1 q$ if $(p, q) \in R$ for some late weak bisimulation R.

Note in this definition how τ moves following an input transition are absorbed *after* the input variable gets instantiated.

Again the union of any collection of late weak bisimulations is also a late weak bisimulation. Hence, as $\approx_1$ is the the union of all late weak bisimulations, it is the largest such bisimulation.

To show that $\approx_1$ is an equivalence relation we need a lemma.

LEMMA 3.11. *Suppose that R is a late weak bisimulation and that $(p, q) \in R$. Then the following holds*:

1. *If $p \xRightarrow{c?x}_{1} t$ then $q \xRightarrow{c?y}_{1} u$ for some u such that for all $v \in Val$ there exists q' such that $u[v/y] \Rightarrow_1 q'$ and $(t[v/x], q') \in R$.*
2. *If $p \xRightarrow{\hat{a}}_{1} p'$, where $a \in \{\tau,\ c!v \mid c \in Chan,\ v \in Val\}$, then $q \xRightarrow{\hat{a}}_{1} q'$ for some q' such that $(p', q') \in R$.*

PROOF. Follows by induction on the definition of the late weak transition relation. □

Now we have the following.

PROPOSITION 3.12. $\approx_1$ *is an equivalence relation.*

PROOF. The only non-trivial part is to prove that $\approx_1$ is transitive. To this end we show that if both R and S are late weak bisimulations then so is their composition RS. Assume $(p,q) \in RS$. Then there is an r such that $(p,r) \in R$ and $(r,q) \in S$. We need to show that every transition from p can be matched properly by q. We only consider the case when the transition is an input transition.

Assume that $p \xrightarrow{c?x}_1 t$. Since $(p,r) \in R$, $r \overset{c?z}{\Longrightarrow}_1 w$ for some w and z such that for all $v \in \mathit{Val}$ there is an r' with $w[v/z] \Rightarrow_1 r'$ and $(t[v/x], r') \in R$. Since $(r,q) \in S$, by Lemma 3.11 we have that $q \overset{c?y}{\Longrightarrow}_1 u$ for some u and y such that for all $v \in \mathit{Val}$ there is a q'' with $u[v/x] \Rightarrow_1 q''$ and $(w[v/x], q'') \in S$. Since $w[v/x] \Rightarrow_1 r'$, by Lemma 3.11, there is a q' such that $q'' \Rightarrow_1 q'$ and $(r', q') \in S$. Furthermore we have that $u[v/x] \Rightarrow_1 q'$ because $u[v/x] \Rightarrow_1 q''$ and $q'' \Rightarrow_1 q'$. Finally $(t[v/x], q') \in RS$ follows from the fact that $(t[v/x], r') \in R$ and $(r', q') \in S$. □

PROPOSITION 3.13. *$\approx_1$ is preserved by all operators in the language except $+$.*

PROOF. We only prove that $\approx_1$ is preserved by the parallel operator as the remaining cases are similar. Assume that R is a late weak bisimulation. We show that the relation $R' = \{(p \mid r, q \mid r) \mid (p,q) \in R,\ r \in \mathcal{P}\}$ is also a late weak bisimulation. To this end assume $p \mid r \xrightarrow{a}_1 p' \mid r'$. The only interesting case is when a has the form $c?x$. There are two possibilities:

- $p \xrightarrow{c?x}_1 p'$ and $r \equiv r'$. Since $(p,q) \in R$, there exist y and q' such that $q \overset{c?y}{\Longrightarrow}_1 q'$ and for all $v \in \mathit{Val}$ there is some q'' such that $q'[v/y] \Rightarrow_1 q''$ and $(p'[v/x], q'') \in R$. This implies $q \mid r \overset{c?y}{\Longrightarrow}_1 q' \mid r$ and, since r is a closed term, $(q' \mid r)[v/y] \equiv q'[v/y] \mid r \Rightarrow_1 q'' \mid r$. Furthermore $((p' \mid r)[v/x], q'' \mid r) \in R'$ because $(p' \mid r)[v/x] \equiv p'[v/x] \mid r$ and $(p'[v/x], q'') \in R$.
- $p \equiv p'$ and $r \xrightarrow{c?x}_1 r'$. Straightforward.

The relation $\approx_1$ is not a congruence; it is not preserved by $+$ as shown by the following standard example:

$$\tau.c?x \approx_1 c?x \quad \text{but} \quad \tau.c?x + d?x \not\approx_1 c?x + d?x.$$

□

To obtain a relation which is preserved by all the operators we follow the same approach as in the early case, introducing a modified version of $\approx_1$:

DEFINITION 3.14. For $p, q \in \mathcal{P}$, $p \simeq_1 q$ if and only if

- whenever $p \xrightarrow{c?x}_1 t$ then $q \overset{c?y}{\Longrightarrow}_1 u$ for some y and u such that for all $v \in \mathit{Val}$ there exists some q' such that $u[v/y] \Rightarrow_1 q'$ and $t[v/x] \approx_1 q'$;
- whenever $a \in \{\tau,\ c!v \mid c \in \mathit{Chan},\ v \in \mathit{Val}\}$ and $p \xrightarrow{a}_1 p'$ then $q \overset{a}{\Longrightarrow}_1 q'$ for some q' such that $p' \approx_1 q'$

and similarly for q.

The following proposition gives an alternative characterization of $\simeq_1$ in terms of $\approx_1$.

PROPOSITION 3.15. *$p \simeq_1 q$ if and only if $p + r \approx_1 q + r$ for all $r \in \mathcal{P}$.*

PROOF. The "only if" direction can be established by showing that $\{(p + r, q + r) \mid p, q, r \in \mathcal{P}$ and $p \simeq_1 q\}$ is a late weak bisimulation, which is straightforward.

For the "if" direction, we prove the contrapositive. Assume that $p \not\simeq_1 q$; we will prove that $p + r \not\approx_1 q + r$ for some r. By the assumption there exist, for example, a and t such that $p \xrightarrow{a}_1 t$ and the following holds:

- If a is τ or an output action, then $q \overset{a}{\Longrightarrow}_1 u$ implies $t \not\approx_1 u$;
- If $a \equiv c?x$ and $q \overset{c?y}{\Longrightarrow}_1 u$, then there exists some $v \in \mathit{Val}$ such that $u[v/y] \Rightarrow_1 q'$ implies $t[v/x] \not\approx_1 q'$.

Now choose r to be $d?x.\mathbf{0}$ where d does not occur in p or q or in the codomain of any renaming functions used in these terms. We show $p + r \not\approx_1 q + r$. From the assumption we have $p + r \xrightarrow{a}_1 t$ and we will prove that this move can not be matched properly by $q + r$. There are two cases to consider.

- If $a \equiv \tau$ and $u \equiv q + r$, then $t \not\approx_1 u$ because the transition $u \xrightarrow{d?x}_1 \mathbf{0}$ cannot be matched by t.
- Otherwise,
  - If a is not an input action then for all u such that $q + r \overset{a}{\Longrightarrow}_1 u$ we also have $q \overset{a}{\Longrightarrow}_1 u$ and therefore, by the assumption, $t \not\approx_1 u$;
  - If a is an input action of the form $c?x$ then $c \neq d$. Therefore for any u such that $q + r \overset{c?y}{\Longrightarrow}_1 u$ it is also the case that $q \overset{c?y}{\Longrightarrow}_1 u$. Hence there is some $v \in \mathit{Val}$ such that whenever $u[v/y] \Rightarrow_1 q'$ then $t[v/x] \not\approx_1 q'$.

This proves that $p + r \not\approx_1 q + r$ and thus the proposition follows by contraposition. □

It follows from Proposition 3.15 that $\simeq_1$ is a congruence relation and that it is the largest such relation contained in $\approx_1$. We refer to $\simeq_1$ as *late observational congruence*.

There is an alternative way of presenting the late semantics. Instead of using the input transition rule $c?x.t \xrightarrow{c?x}_1 t$ where the input variable is carried in the transition and the residual is an open term, one can define $c?x.t \xrightarrow{c?}_1 (x)t$ where the input variable is abstracted away from the transition and $(x)t$ is a new syntactical category called *abstraction*. Semantically an abstraction is a function from values to processes. Then the input clause in Definition 3.2 is changed accordingly:

If $p \xrightarrow{c?}_1 f$ then, for some g, $q \xrightarrow{c?}_1 g$ and $(fv, gv) \in R$ for all $v \in \mathit{Val}$.

Abstractions have been used extensively in the study of the π-calculus [34]. We have chosen to use the open terms presentation because it is in harmony with the symbolic approach, the main subject of this chapter.

The bisimulation-based equivalence relations we have studied so far are summarized in Figure 3 where arrows stand for inclusion.

3.3. *Translating into pure processes*

In this section we explain briefly how the language for value-passing processes may be translated into the "pure" language, as proposed in [32, Section 2.8]. We assume the stan-

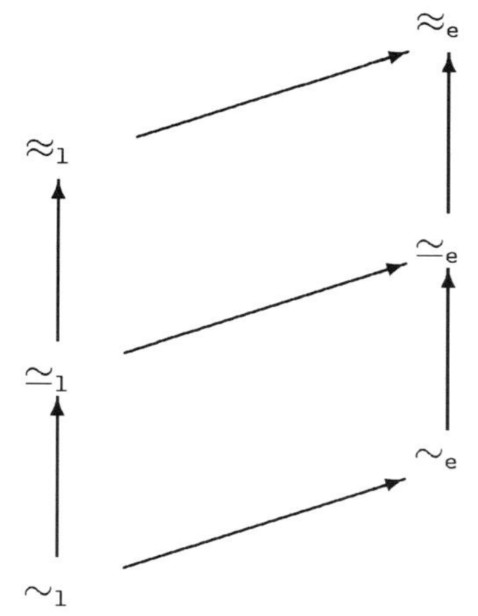

Fig. 3. The relationship between equivalences.

dard notations for pure CCS as in [32]. The key step in the translation is to compile an input prefix into a summation over all possible values in the data domain. So, for each channel name c we introduce a set of action labels $\{c?v,\ c!v \mid c \in \mathit{Chan},\ v \in \mathit{Val}\}$ in the pure calculus. The complement of $c?v$ is $c!v$, and vice versa. Also for each process identifier P of arity n in the value-passing language we introduce a set of indexed process identifiers $\{P_{\overline{v}} \mid \overline{v} \in \mathit{Val}^n\}$ in the pure language. The translation is defined by induction on the structure of process expressions and is given in Figure 4 where p is the original value-passing process and $\hat{p}$ its translated form.

Furthermore, each parameterized definition $P(\overline{x}) \Leftarrow t$ is translated into a set of non-parameterized defining equations

$$\left\{P_{\overline{v}} \Leftarrow \widehat{t[\overline{v}/\overline{x}]} \mid \overline{v} \in \mathit{Val}^n\right\}.$$

Note that this translation is only defined for closed terms. This is sufficient because the only construction which may introduce free variables is input prefixing, and in the translation each input variable is instantiated with a value. Also note that, when *Val* is infinite, an input prefixing is translated into an infinite summation, and a parameterized definition into an infinite set of non-parameterized equations.

Now the resulting terms of the translation are terms in the underlying pure process calculus (pure CCS in this case) which is equipped with a well-defined operational semantics and the standard notion of bisimulation. This provides a simple way of giving semantics to value-passing processes via translation (at the cost of infinity). It is easy to check that the semantics so obtained coincides with the early semantics. To be precise, let $\rightarrow$ be the transition relation corresponding to the standard operational semantics of pure CCS (as defined in [32, Section 2.5]). We have the following exact correspondence between the early operational semantics defined in Figure 2 and $\rightarrow$:

p	$\hat{p}$
$c?x.t$	$\sum_{v \in Val} c?v.t[\hat{v}/x]$
$c!e.t$	$c![\![e]\!].\hat{t}$
$\tau.t$	$\tau.\hat{t}$
$t+u$	$\hat{t}+\hat{u}$
$t \mid u$	$\hat{t} \mid \hat{u}$
$t \backslash L$	$\hat{t}\backslash\{c?v, c!v \mid c \in L\}$
$t[f]$	$\hat{t}[\hat{f}]$, where $\hat{f}(c?v) = f(c)?v$, $\hat{f}(c!v) = f(c)!v$
$\texttt{if}\ b\ \texttt{then}\ t$	$\begin{cases} \hat{t} & \text{if } [\![b]\!] = \mathit{true} \\ \mathbf{0} & \text{otherwise} \end{cases}$
$P(\overline{e})$	$P_{\overline{[\![e]\!]}}$

Fig. 4. The translation rules.

LEMMA 3.16. *For $p, q \in \mathcal{P}$ and $a \in Act_\tau$, $p \xrightarrow{a}_{\mathrm{e}} q$ if and only if $\widehat{p} \xrightarrow{a} \widehat{q}$.*

PROOF. Follows by transition induction, using the definition of $\widehat{p}$. □

Let $\sim$ be the bisimulation relation as defined in [32, Section 4.2]. The following proposition describes the relationship between $\sim_{\mathrm{e}}$ and $\sim$:

PROPOSITION 3.17. *For $p, q \in \mathcal{P}$, $p \sim_{\mathrm{e}} q$ if and only if $\widehat{p} \sim \widehat{q}$.*

PROOF. Suppose R is an early bisimulation with $(p, q) \in R$. Define $\widehat{R} = \{(\widehat{p}, \widehat{q}) \mid (p, q) \in R\}$. By Lemma 3.16 it is immediate that $\widehat{R}$ is a pure-CCS bisimulation.

For the other direction, let $\widehat{R}$ be a pure-CCS bisimulation. Define $R = \{(p, q) \mid (\widehat{p}, \widehat{q}) \in \widehat{R}\}$. Again it is immediate that R is an early bisimulation. □

The translation approach has been implemented in a tool named VP [3] which acts as a front-end for the verification tool Concurrency Workbench. It takes an input file written in full CCS (with finite data domain) and translates it into pure CCS in a format that can be accepted by the Concurrency Workbench. The translation is carried out exactly according to the rules in Figure 4.

4. Symbolic bisimulations

As explained in the introduction, the main problem in connection with the traditional semantics for value-passing processes is that in general it leads to infinitely branching transition graphs. Therefore, to check for equivalence, infinitely many transitions have to be investigated. The aim of this section is to give a more abstract version of the concrete semantics we have considered so far which allows us to check for equivalence by only investigating finitely branching graphs.

4.1. *Symbolic operational semantics*

To motivate the rules for the symbolic operational semantics let us look at a simple example. Let P be defined as

$$P \Leftarrow c?x.\texttt{if}\ x > 0\ \texttt{then}\ d!x\ \texttt{else}\ d!(-x),$$

where x ranges over integers. According to the operational semantics in Figure 2, P has a transition $\xrightarrow{c?n}$ for *each* integer n, hence its transition graph has infinitely many branches. This seems counter-intuitive since all P can do is first to input a number from channel c and then output on channel d either the number itself or its negation, depending on whether it is positive or not. This is a *finitary* description and conceptually we could take $P \xrightarrow{c?x} \texttt{if}\ x > 0\ \texttt{then}\ d!x\ \texttt{else}\ d!(-x)$ as a *single* abstract transition, without instantiating x. This kind of transition has been used in defining late bisimulation in Section 3.2. However, there the instantiation is only delayed and in the input clause of the definition of late bisimulation, infinitely many comparisons for the residuals are still required, one for each possible value of the data domain. Here we want to go further, to avoid such infinite comparisons as well. As the residual of such an abstract input transition is in general an open term, we will neither be able to evaluate Boolean conditions nor data expressions in output prefixes. Instead we shall allow these expressions to be carried around unevaluated, i.e., *symbolically*, in the transitions. So in the above example we shall have that $\texttt{if}\ x > 0\ \texttt{then}\ d!x\ \texttt{else}\ d!(-x) \xrightarrow{x>0,d!x} \mathbf{0}$ and $\texttt{if}\ x > 0\ \texttt{then}\ d!x\ \texttt{else}\ d!(-x) \xrightarrow{\neg(x>0),d!(-x)} \mathbf{0}$. This leads to the notion of *symbolic transition relations* which have the form $\xrightarrow{b,\alpha}$ where b is a Boolean expression and α a symbolic action, namely

$$\alpha \in Act_\tau^{symb} = \{\tau,\ c!e,\ c?x \mid c \in Chan,\ e \in DExp,\ x \in Var\}.$$

Intuitively $t \xrightarrow{b,\alpha} u$ means that, under circumstances in which b holds, t can perform α and thereby evolve into u. We will abbreviate $\xrightarrow{true,\alpha}$ as $\xrightarrow{\alpha}$. The set of bound variables of a symbolic action is defined thus: $bv(c?x) = \{x\}$ and $bv(\tau) = bv(c!e) = \emptyset$. Formally the symbolic operational semantics for the language is obtained by taking $\mathcal{T}$ as the set of states, $BExp \times Act_\tau^{symb}$ as the set of labels and $\rightarrow \subseteq \mathcal{T} \times (BExp \times Act_\tau^{symb}) \times \mathcal{T}$ as defined in Figure 5. These relations are defined up-to α-equivalence. The rule PAR1 has a side condition which ensures that, in the case when α is an input action of the form $c?x$, the variable x which is local in t will not be confused with any free variables in u. This comment applies to PAR2 as well. The rule COND also has a side condition to make sure that whenever $t \xrightarrow{b,c?x} t'$ the Boolean condition b does not mention x; the occurrence of $c?x$ in the label of the transition means that typically we are about to introduce x as a

and for each sub-case the move can be properly matched by Q. Such a set of sub-cases can be given by a finite set of Booleans B such that $b = \bigvee B$. For example if we are trying to check if $P_1 \sim^{true} Q_1$ then we can consider the cases given by $B = \{x = 0,\ x \neq 0\}$; in the case $x = 0$ the move is matched by $Q_1 \xrightarrow{\tau} Q_{11}$ while in the case $x \neq 0$ it is matched by $Q_1 \xrightarrow{\tau} Q_{12}$.

A finite set of Boolean expressions B is called a *b-partition* if $\bigvee B = b$. Semantically B can be regarded as a partition of the value space represented by b, or more precisely for any evaluation ρ, $\rho \models b$ if and only if $\rho \models b'$ for some $b' \in B$.

In what follows we write $\alpha =^b \alpha'$ to mean that if $\alpha \equiv c!e$ then α' has the form $c!e'$ with $b \models e = e'$, and $\alpha' \equiv \alpha$ otherwise.

DEFINITION 4.1. A *BExp*-indexed family of symmetric relations $\mathbf{S} = \{S^b \mid b \in BExp\}$ is an *early strong symbolic bisimulation* if it satisfies: $(t, u) \in S^b$ implies that

if $t \xrightarrow{b_1, \alpha} t'$ with $bv(\alpha) \cap fv(b, t, u) = \emptyset$, then there is a $b \wedge b_1$-partition B such that for each $b' \in B$ there exist b_2, α' and u' with $b' \models b_2$, $\alpha =^{b'} \alpha'$, $u \xrightarrow{b_2, \alpha'} u'$ and $(t', u') \in S^{b'}$.

We write $t \sim^b_{\mathrm{E}} u$ if $(t, u) \in S^b \in \mathbf{S}$ for some early symbolic bisimulation $\mathbf{S}$.

As an illustration let us consider again the example introduced at the beginning of this section. Let

$$A_1 = \{(P, Q), (P_1, Q_1)\},$$
$$A_2 = \{(P_{11}, Q_{11}), (P_{12}, Q_{12})\},$$
$$A_3 = \{P_{11}, Q_{12}), (P_{12}, Q_{11})\}.$$

Set

$$S^{true} = A_1 \cup A_1^{-1} \cup \mathbf{id}, \quad S^{x=0} = A_2 \cup A_2^{-1} \cup \mathbf{id}, \quad S^{x \neq 0} = A_3 \cup A_3^{-1} \cup \mathbf{id},$$

where **id** is the identity relation. Then $\mathbf{S} = \{S^{true},\ S^{x=0},\ S^{x \neq 0}\}$ is a symbolic early bisimulation. To check this let us investigate $(P_1, Q_1) \in S^{true}$. We know that $P_1 \xrightarrow{true, \tau} P_{11}$ and we must show that Q_1 can match this move properly. For this purpose we use $B = \{x = 0, x \neq 0\}$ as the partition for *true*. Then we have two cases to investigate: $x = 0$ and $x \neq 0$. As an example we consider the first one, $x = 0$. If we take $b_2 = true$ in the definition then we see that $x = 0 \models true$, $\tau =^{x=0} \tau$, $Q_1 \xrightarrow{true, \tau} Q_{11}$ and $(P_{11}, Q_{11}) \in S^{x=0}$, as required by the definition. The reader is encouraged to investigating the other cases keeping in mind that $B = \emptyset$ is a partition for *false* or any Boolean equivalent to *false*.

In the above example, the need for a partition arises from the Boolean guards carried around in the symbolic transitions. However, this is not the only reason why partitions are needed. Consider, for instance, the following definitions (where x ranges over integers):

$$P_1(x) \Leftarrow c!x + c!|x| + c!-|x|,$$
$$P_2(x) \Leftarrow c!|x| + c!-|x|.$$

To establish $P_1(x) \sim_{\mathrm{E}}^{true} P_2(x)$ we need to use a partition $\{x \geqslant 0,\ x \leqslant 0\}$ (or, equivalently, $\{x = |x|,\ x = -|x|\}$) for *true*, although all symbolic transitions involved carry only the trivial Boolean guard *true*.

It is worth pointing out that when α is an input action of the form $c?x$, i.e., when $t \xrightarrow{b_1, c?x} t'$, the input variable x may occur *free* in the partition B. This allows one to perform a case analysis on the value space associated with x *before* the input transition takes place, thus permitting to use different matching transitions from u for different cases. This reflects the "early instantiation" strategy. As an illustration let us consider again Example 3.7 introduced at the beginning of Section 3.2:

$$\begin{aligned} P &\Leftarrow (c?x.\mathtt{if}\ x = 0\ \mathtt{then}\ a_1!x\ \mathtt{else}\ a_2!x) + c?x.a_3!x, \\ Q &\Leftarrow (c?x.\mathtt{if}\ x = 0\ \mathtt{then}\ a_3!x\ \mathtt{else}\ a_2!x) \\ &\quad + (c?x.\mathtt{if}\ x = 0\ \mathtt{then}\ a_1!x\ \mathtt{else}\ a_3!x). \end{aligned}$$

As mentioned before P and Q are early bisimilar. To show that $P \sim_{\mathrm{E}}^{true} Q$, we need to match the symbolic move $P \xrightarrow{true, c?x} a_3!x$ by moves from Q. We can use the partition $B = \{x = 0,\ x \neq 0\}$ for *true*. In the case of $x = 0$, this move is matched by

$$Q \xrightarrow{true, c?x} \mathtt{if}\ x = 0\ \mathtt{then}\ a_3!x\ \mathtt{else}\ a_2!x$$

while in the case of $x \neq 0$ it is matched by

$$Q \xrightarrow{true, c?x} \mathtt{if}\ x = 0\ \mathtt{then}\ a_1!x\ \mathtt{else}\ a_3!x.$$

Here it is essential to use a partition in which x occurs free. Should this be disallowed it would not be possible to establish $P \sim_{\mathrm{E}}^{true} Q$.

Next we will determine the relationship between the symbolic and the concrete versions of the bisimulations. We start by showing the connection between the symbolic transitions and the concrete ones. Since symbolic transitions are defined for open terms, we need an evaluation mapping which assigns values to data variables.

LEMMA 4.2. *For all t, u, p and ρ,*

(1) *$t\rho \xrightarrow{\tau}_{\mathrm{e}} p$ if and only if there exist b, t' such that $\rho \models b$, $p \equiv_\alpha t'\rho$ and $t \xrightarrow{b,\tau} t'$,*

(2) *$t\rho \xrightarrow{c!v}_{\mathrm{e}} p$ if and only if there exist b, e, t' such that $\rho \models b$, $[\![e]\!]\rho = v$, $p \equiv_\alpha t'\rho$ and $t \xrightarrow{b,c!e} t'$,*

(3) *$t\rho \xrightarrow{c?v}_{\mathrm{e}} p$ if and only if there exist b, t' and $x \notin fv(t)$ such that $\rho \models b$, $p \equiv_\alpha t'\rho\{v/x\}$ and $t \xrightarrow{b,c?x} t'$.*

PROOF. Follows by transition induction. The details of the proof can be found in [17, p. 389]. □

The early strong symbolic bisimulation corresponds to its concrete counterpart in the following sense:

THEOREM 4.3. *$t \sim_{\mathrm{E}}^{b} u$ if and only if $t\rho \sim_{\mathrm{e}} u\rho$ for every evaluation ρ such that $\rho \models b$.*

PROOF (*Outline*). To prove the "only if" direction assume that $\mathbf{S} = \{S^b \mid b \in BExp\}$ is an early strong symbolic bisimulation and let

$$R_{\mathbf{S}} = \{(t\rho, u\rho) \mid \exists b.\ \rho \models b \text{ and } (t, u) \in S^b\}.$$

The result follows immediately if we can prove that $R_{\mathbf{S}}$ is a bisimulation.

For the "if" direction, assume that R is an early strong bisimulation and let

$$S_R^b = \{(t, u) \mid \forall \rho.\ \rho \models b \text{ implies } (t\rho, u\rho) \in R\}$$

for any Boolean expression b. Again the result follows if we can show that $\mathbf{S}_R = \{S_R^b \mid b \in BExp\}$ is a symbolic bisimulation. Details can be found in [16, Theorem 6.5]. □

It follows from this theorem that for closed terms p and q, $p \sim_{\mathrm{e}} q$ if and only if $p \sim_{\mathrm{E}}^{true} q$.

As a further corollary to this theorem, if $t \sim_{\mathrm{E}}^{b} u$ then $t[y/x] \sim_{\mathrm{E}}^{b[y/x]} u[y/x]$ for any $y \notin fv(b, t, u)$. Thus *symbolic bisimulations are invariant under renaming of free variables*.

Definition 4.1 requires that, for t and u to be symbolically bisimilar, if $t \xrightarrow{b_1, c?x} t'$ then u must match this transition with an input transition $u \xrightarrow{b_2, c?x} u'$, on the same channel, using the same input variable. Since the rule PRE in Figure 5 asserts $\alpha.t \xrightarrow{true, \alpha} t$, the reader may wonder how to establish that, for instance, $c?x.d!x$ and $c?y.(d!y + d!y)$ are symbolically bisimilar according to this definition. Since symbolic transitional semantics is defined up to α-equivalence, we may choose, for instance, $c?x.d!x \equiv_\alpha c?y.d!y \xrightarrow{true, c?y} d!y$ which can be matched by $c?y.(d!y + d!y) \xrightarrow{true, c?y} d!y + d!y$. Of course we may use *any* data variable z in the input action, provided it is not free in the terms under investigation. The important fact is that, when comparing a pair of input transitions for symbolic bisimulation, *only one* such variable is needed. It is immaterial which one is chosen. In general, if $t \xrightarrow{b_1, c?x} t'$, $u \xrightarrow{b_2, c?x} u'$ and $t' \sim_{\mathrm{E}}^{b'} u'$, then for any $z \notin fv(t, u)$, $t \xrightarrow{b_1, c?z} t'[z/x]$, $u \xrightarrow{b_2, c?z} u'[z/x]$, and we also have $t'[z/x] \sim_{\mathrm{E}}^{b'[z/x]} u'[z/x]$, because symbolic bisimulations are invariant under renaming of free variables. This remark applies to all symbolic equivalence relations discussed in this chapter.

4.3. *Late symbolic bisimulation*

Recall that in Section 3.2, when defining late bisimulations, we have to modify the transition semantics for input prefixing to allow abstract input transitions of the form $\xrightarrow{c?x}$, so that the instantiation of values for input variables can be deferred after input actions are matched for bisimulation. Since the symbolic operational semantics of Figure 5 already uses abstract actions, the same rules can be adopted to define the late version of symbolic bisimulation.

DEFINITION 4.4. A *BExp*-indexed family of symmetric relations $\mathbf{S} = \{S^b \mid b \in BExp\}$ is a *late strong symbolic bisimulation* if it satisfies: $(t, u) \in S^b$ implies that

> if $t \xrightarrow{b_1,\alpha}_{\mathrm{e}} t'$ with $bv(\alpha) \cap fv(b, t, u) = \emptyset$, then there is a $b \wedge b_1$-partition B with $fv(B) \subseteq fv(b, t, u)$ such that for each $b' \in B$ there exist b_2, α' and u' with $b' \models b_2$, $\alpha =^{b'} \alpha'$, $u \xrightarrow{b_2,\alpha'}_{\mathrm{e}} u'$ and $(t', u') \in S^{b'}$.

We write $t \sim_{\mathrm{L}}^b u$ if $(t, u) \in S^b \in \mathbf{S}$ for some late symbolic bisimulation $\mathbf{S}$.

It is important to note that we now require $fv(B) \subseteq fv(b, t, u)$ (and this is actually the only difference between the definitions of the late and the early symbolic bisimulation relations). This implies that, when $\alpha \equiv c?x$, it is guaranteed that $x \notin fv(B)$. Therefore case analysis on the value space associated with x is *not* allowed before input actions are compared for bisimulation. As a consequence, an input transition from one process has to be matched by a *single* such transition from the other. This reflects the "late instantiation" strategy, and makes all the differences between late and early bisimulations!

As an illustration let us consider Example 3.7 again:

$$\begin{aligned} P &\Leftarrow (c?x.\mathtt{if}\ x = 0\ \mathtt{then}\ a_1!x\ \mathtt{else}\ a_2!x) + c?x.a_3!x, \\ Q &\Leftarrow (c?x.\mathtt{if}\ x = 0\ \mathtt{then}\ a_3!x\ \mathtt{else}\ a_2!x) \\ &\quad + (c?x.\mathtt{if}\ x = 0\ \mathtt{then}\ a_1!x\ \mathtt{else}\ a_3!x). \end{aligned}$$

Now it is not allowed for Q to match the move $P \xrightarrow{true,c?x} a_3!x$, as any partition that permits this move to be matched by Q inevitably involves x which is not allowed according to the new definition.

As in the early case, late strong symbolic bisimulation captures late strong bisimulation in the following sense:

THEOREM 4.5. *$t \sim_{\mathrm{L}}^b u$ if and only if $t\rho \sim_1 u\rho$ for every evaluation ρ such that $\rho \models b$.*

PROOF. See [16, Theorem 4.5]. □

In Definition 4.4 we note that, whenever $t \sim_{\mathrm{L}}^b u$ holds, each symbolic transition from t may induce a different partition of b. Similarly, different transitions from u may induce different partitions of b. Sometimes it is more convenient to combine these partitions into a fine grain one which works uniformly for all symbolic transitions from both terms (as in the proof of Proposition 5.19). This is possible because there are only finitely many symbolic transitions from each term, as shown in the following lemma.

LEMMA 4.6. *Assume that $t \equiv \sum_I \alpha_i.t_i$ and $u \equiv \sum_J \beta_j.u_j$ where $bv(\alpha_i) \cap bv(\beta_j) \cap fv(b, t, u) = \emptyset$. Then $t \sim_{\mathrm{L}}^b u$ if and only if there exists a disjoint b-partition B with $fv(B) \subseteq fv(b, t, u)$ such that for each $b' \in B$ the following holds:*
- *For each $i \in I$ there is a $j \in J$ such that $\alpha_i =^{b'} \beta_j$ and $t_i \sim_{\mathrm{L}}^{b'} u_j$;*
- *For each $j \in J$ there is an $i \in I$ such that $\alpha_i =^{b'} \beta_j$ and $t_i \sim_{\mathrm{L}}^{b'} u_j$.*

PROOF. The "if" part of the proof is trivial and we only consider the "only if" direction here. Since $t \sim^b_{\mathrm{L}} u$, for each $i \in I$ there exists a b-partition B_i with $fv(B_i) \subseteq fv(t, u)$ such that for each $b_i \in B_i$ there exists $j \in J$ with $\alpha_i =^{b_i} \beta_j$ and $t_i \sim^{b_i}_{\mathrm{L}} u_j$; similarly for each $j \in J$ there exists a b-partition B'_j with $fv(B'_j) \subseteq fv(b, t, u)$ such that for each $b'_j \in B'_j$, there exists $i \in I$ with $\alpha_i =^{b'_j} \beta_j$ and $t_i \sim^{b'_j}_{\mathrm{L}} u_j$.

Let D_I denote the set of Booleans $\{\wedge_{i \in I} b_i \mid b_i \in B_i\}$, D_J the set $\{\wedge_{j \in J} b'_j \mid b'_j \in B_j\}$ and set $B = \{b_1 \wedge b_2 \mid b_1 \in D_I, b_2 \in D_J\}$. Then $\bigvee B = b$, $fv(B) \subseteq fv(b, t, u)$ and each $b' \in B$ has the form $(\wedge_i b_i) \wedge (\wedge_j b'_j)$ with $b_i \in B_i$, $b'_j \in B'_j$. For each $i \in I$, $b' \models b_i$ for some $b_i \in B_i$ and therefore there is a $j \in J$ such that $\alpha_i =^{b'} \beta_j$ and $t_i \sim^{b'}_{\mathrm{L}} u_j$. For each $j \in J$ $b' \models b'_j$ for some $b'_j \in B'_j$ and thus there is an $i \in I$ such that $\alpha_i =^{b'} \beta_j$ and $t_i \sim^{b'}_{\mathrm{L}} u_j$. □

A similar lemma holds for early symbolic bisimulation as well.

4.4. *Weak symbolic equivalences*

Our next task is to define the symbolic counterparts of the weak equivalences we have defined in Section 3 and study their relationships. Unlike in the strong case (but like in the concrete semantics) we need slightly different versions of the weak transition relations to define each of the early and late versions of weak symbolic bisimulation. Therefore we index the early weak transition relations with E and its late counterpart with L.

The early weak symbolic transition relations are defined as the least relations over $\mathcal{T}$ which satisfy:

- $t \overset{true,\varepsilon}{\Longrightarrow}_{\mathrm{E}} t$;
- $t \overset{b,\alpha}{\longrightarrow} u$ implies $t \overset{b,\alpha}{\Longrightarrow}_{\mathrm{E}} u$;
- $t \overset{b,\tau}{\longrightarrow} \overset{b',\alpha}{\Longrightarrow}_{\mathrm{E}} u$, where $bv(\alpha) \cap fv(b) = \emptyset$, implies $t \overset{b \wedge b',\alpha}{\Longrightarrow}_{\mathrm{E}} u$;
- $t \overset{b,\alpha}{\Longrightarrow}_{\mathrm{E}} \overset{b',\tau}{\longrightarrow} u$ implies $t \overset{b \wedge b',\alpha}{\Longrightarrow}_{\mathrm{E}} u$.

Concerning bound variables we now have:

LEMMA 4.7. *If* $t \overset{b,c?x}{\Longrightarrow}_{\mathrm{E}} t'$ *then* $fv(b) \subseteq fv(t) \cup \{x\}$ *and* $x \notin fv(t)$.

We have the following relationship between the concrete and the symbolic early weak transitions, similar to the strong case.

LEMMA 4.8. *For all* t, u, b *and* ρ,

(1) $t\rho \overset{\varepsilon}{\Longrightarrow}_{\mathrm{e}} p$ *if and only if there exist* b, t' *such that* $\rho \models b$, $p \equiv_\alpha t'\rho$ *and* $t \overset{b,\varepsilon}{\Longrightarrow}_{\mathrm{E}} t'$.

(2) $t\rho \overset{\tau}{\Longrightarrow}_{\mathrm{e}} p$ *if and only if there exist* b, t' *such that* $\rho \models b$, $p \equiv_\alpha t'\rho$ *and* $t \overset{b,\tau}{\Longrightarrow}_{\mathrm{E}} t'$.

(3) $t\rho \overset{c!v}{\Longrightarrow}_{\mathrm{e}} p$ *if and only if there exist* b, e, t' *such that* $\rho \models b$, $[\![e]\!]\rho = v$, $p \equiv_\alpha t'\rho$ *and* $t \overset{b,c!e}{\Longrightarrow}_{\mathrm{E}} t'$.

(4) $t\rho \overset{c?v}{\Longrightarrow}_{\mathrm{e}} p$ *if and only if there exist* b, t' *and* $x \notin fv(t)$ *such that* $fv(b) \subseteq fv(t) \cup \{x\}$, $\rho\{v/x\} \models b$, $p \equiv_\alpha t'\rho\{v/x\}$ *and* $t \overset{b,c?x}{\Longrightarrow}_{\mathrm{E}} t'$.

PROOF. Follows by induction on the definitions of the weak transitions. See [17, Lemmas 4.3–4.6] for the details. □

DEFINITION 4.9. A *BExp*-indexed family of symmetric relations, $\mathbf{S} = \{S^b \mid b \in BExp\}$, over $\mathcal{T}$ is an *early weak symbolic bisimulation* if it satisfies: $(t, u) \in S^b$ implies that

> if $t \xrightarrow{b_1, \alpha} t'$ with $bv(\alpha) \cap fv(b, t, u) = \emptyset$, then there is a $b \wedge b_1$-partition B such that for each $b' \in B$ there exist b_2, α' and u' with $b' \models b_2$, $\alpha =^{b'} \alpha'$, $u \xRightarrow{b_2, \hat{\alpha}'}_{\mathrm{E}} u'$ and $(t', u') \in S^{b'}$.

We write $t \approx^b_{\mathrm{E}} u$ if $(t, u) \in S^b \in \mathbf{S}$ for some early weak symbolic bisimulation $\mathbf{S}$.

Again we have to modify $\approx_{\mathrm{E}}$ so that it is preserved by $+$.

DEFINITION 4.10. For $t, u \in \mathcal{T}$, $t \simeq^b_{\mathrm{E}} u$, if and only if

> whenever $t \xrightarrow{b_1, \alpha} t'$ with $bv(\alpha) \cap fv(b, t, u) = \emptyset$, then there is a $b \wedge b_1$-partition B such that for each $b' \in B$ there exist b_2, α' and u' with $b' \models b_2$, $\alpha =^{b'} \alpha'$, $u \xRightarrow{b_2, \alpha'}_{\mathrm{E}} u'$ and $t' \approx^{b'}_{\mathrm{E}} u'$

and symmetrically for u.

We refer to $\simeq^b_{\mathrm{E}}$ as *early symbolic observational congruence with respect to b*.

Note that in the definition above it is still essential to use partitions when matching moves. For example

$$\tau.p \simeq^{true}_{\mathrm{E}} \texttt{if}\ b\ \texttt{then}\ \tau.p\ \texttt{else}\ \tau.\tau.p$$

but the symbolic move $\tau.p \xrightarrow{true, \tau} p$ of the left hand side can not be matched properly by a single symbolic move from the right hand side.

The two versions of early weak bisimulation equivalence/congruence can be related as in the case of strong bisimulation.

THEOREM 4.11.
(1) $t \approx^b_{\mathrm{E}} u$ *if and only if* $t\rho \approx_{\mathrm{e}} u\rho$ *for every* ρ *such that* $\rho \models b$.
(2) $t \simeq^b_{\mathrm{E}} u$ *if and only if* $t\rho \simeq_{\mathrm{e}} u\rho$ *for every* ρ *such that* $\rho \models b$.

PROOF. See [17, Theorem 4.2]. □

In the rest of this section we introduce the late version of weak symbolic bisimulation. Again, care must be taken when defining the corresponding weak symbolic transition relations and, like in the definition of $\Rightarrow_1$, we should not allow input moves to absorb τ moves that occur after them.

The late weak symbolic transition relations are defined as the least binary relations over $\mathcal{T}$ satisfying the following rules.

EQUIV	$\dfrac{}{t=t} \quad \dfrac{t=u}{u=t} \quad \dfrac{t=u \quad u=v}{t=v}$
α-CONV	$\dfrac{}{c?x.t=c?y.t[y/x]} \quad y \notin fv(t)$
AXIOM	$\dfrac{}{true \rhd t=u} \quad t=u$ is an axiom instance
CHOICE	$\dfrac{b \rhd t=t'}{b \rhd t+u=t'+u}$
INPUT	$\dfrac{b \rhd t=u}{b \rhd c?x.t=c?x.u} \quad x \notin fv(b)$
OUTPUT	$\dfrac{b \models e=e', \quad b \rhd t=u}{b \rhd c!e.t=c!e'.u}$
TAU	$\dfrac{b \rhd t=u}{b \rhd \tau.t=\tau.u}$
COND	$\dfrac{b \wedge b' \rhd t=u \quad b \wedge \neg b' \rhd \mathbf{0}=u}{b \rhd \texttt{if } b' \texttt{ then } t=u}$
PARTITION	$\dfrac{b \models b_1 \vee b_2, \quad b_1 \rhd t=u \quad b_2 \rhd t=u}{b \rhd t=u}$
ABSURD	$\dfrac{}{false \rhd t=u}$

Fig. 8. The inference rules.

expressions, instead of pure equations we employ *conditional equations* of the form $b \rhd t=u$, where the condition b is a Boolean expression, as judgments of our proof system. The intended meaning of $b \rhd t=u$ is that "$t \sim^b u$", or "t and u are bisimilar when the variables that occur free in these terms are instantiated with values satisfying b". In this setting ordinary equations are regarded as special conditional equations with the condition being *true*, i.e., $t=u$ is taken to be an abbreviation for $true \rhd t=u$.

Reasoning about value-passing processes will inevitably involve reasoning about data. However, instead of inventing proof rules for all possible data domains, we would like to factor out reasoning about data from reasoning about processes as much as possible. Therefore our proof system will be parameterized over data reasoning of the form $b \models b'$, with the intuitive meaning that whenever b is true then so is b'. As a consequence, all the completeness results in the following sections are relative to data domain reasoning. The proposed inference rules are listed in Figure 8. Before starting to investigate properties of this proof system, some informal discussion about these rules are in order.

The set of inference rules we put forward can be seen as a natural generalization of pure equational reasoning. For each construct in our language there is a corresponding introduction rule. The OUTPUT rule refers to reasoning about data domains by including $b \models e = e'$ in the premise. It says that if we can deduce a statement $b \rhd t = u$ from this proof system, and we know that b implies $e = e'$ from the knowledge about data, then we can conclude that $b \rhd c!e.t = c!e'.u$. The INPUT rule requires a side condition which guarantees the input variable to be "fresh". This is crucial to the soundness of the rule. Without this proviso we could, for instance, derive

$$x = 1 \rhd c?x.c!1 = c?x.c!x$$

from

$$x = 1 \rhd c!1 = c!x.$$

The premise is true but the conclusion is false. With the side condition this derivation is not possible because x is free in the condition $x = 1$. The introduction rule for the conditional construct is COND which performs a case analysis; an identity of the form $\texttt{if}\ b\ \texttt{then}\ t = u$ can be established by considering two cases, one when b is true and the other when it is false. Finally there are two "structural rules", PARTITION and ABSURD, which do not deal with any specific constructs in the language, but are useful to "glue" pieces of derivations together. The PARTITION rule allows partitioning of the current value space, represented by b, into two subspaces, represented by b_1 and b_2, respectively. Like OUTPUT, this rule also requires the establishment of facts about the data domain. Taking $b_1 \equiv b_2 \equiv b'$ it specializes to a useful rule

$$\text{CONSEQUENCE}\ \frac{b \models b',\ b' \rhd t = u}{b \rhd t = u}.$$

We write $\vdash b \rhd t = u$ to mean that $b \rhd t = u$ can be derived from axioms S1–S4 using the rules in Figure 8. With a slight abuse of notation we will also use $\vdash$ to denote this set of rules and axioms.

The soundness of $\vdash$ is stated below:

THEOREM 5.1. *If* $\vdash b \rhd t = u$ *then* $t\rho \sim_l u\rho$ *for any* ρ *such that* $\rho \models b$.

PROOF. The proof follows by induction on the derivation of $b \rhd t = u$ and a case analysis on the last rule used. □

Before embarking on the task of proving the completeness result, we first establish some useful facts about the proof system.

PROPOSITION 5.2.

(1) *If* $b \models \neg b'$ *then* $\vdash b \rhd \texttt{if}\ b'\ \texttt{then}\ t = \mathbf{0}$;

(2) *if* $b \models b'$ *then* $\vdash b \rhd t = \texttt{if}\ b'\ \texttt{then}\ t$;

(3) $\vdash \texttt{if}\ b\ \texttt{then}\ (\texttt{if}\ b'\ \texttt{then}\ t) = \texttt{if}\ b \wedge b'\ \texttt{then}\ t$;
(4) $\vdash t = t + \texttt{if}\ b\ \texttt{then}\ t$;
(5) $\vdash b \wedge b' \rhd t = u$ *implies* $\vdash b \rhd \texttt{if}\ b'\ \texttt{then}\ t = \texttt{if}\ b'\ \texttt{then}\ u$;
(6) $\vdash \texttt{if}\ b\ \texttt{then}\ (t + u) = \texttt{if}\ b\ \texttt{then}\ t + \texttt{if}\ b\ \texttt{then}\ u$;
(7) $\vdash \texttt{if}\ b\ \texttt{then}\ u + \texttt{if}\ b'\ \texttt{then}\ u = \texttt{if}\ b \vee b'\ \texttt{then}\ u$;
(8) *if* $fv(b) \cap bv(\alpha) = \emptyset$ *then* $\vdash \texttt{if}\ b\ \texttt{then}\ \alpha.t = \texttt{if}\ b\ \texttt{then}\ \alpha.(\texttt{if}\ b\ \texttt{then}\ t)$.

PROOF. We only give the details of the proof for three of these statements.

(1) An application of COND reduces the goal to

$$b \wedge b' \rhd t = \mathbf{0} \quad \text{and} \quad b \wedge \neg b' \rhd \mathbf{0} = \mathbf{0}.$$

The first subgoal can be settled by CONSEQUENCE and ABSURD, while the second follows by EQUIV.

(4) Because $true \models b \vee \neg b$, by the PARTITION rule it is sufficient to prove the two statements

$$\vdash b \rhd t = t + \texttt{if}\ b\ \texttt{then}\ t \quad \text{and} \quad \vdash \neg b \rhd t = t + \texttt{if}\ b\ \texttt{then}\ t. \qquad (1)$$

By S2 and CHOICE the first part of (1) can be reduced to $\vdash b \rhd t = \texttt{if}\ b\ \texttt{then}\ t$. By COND this can be further reduced to

$$\vdash b \wedge b \rhd t = t \quad \text{and} \quad b \wedge \neg b \rhd t = \mathbf{0}$$

which are simple consequences of EQUIV and ABSURD, respectively.

The second part of (1) can be derived from S1 and Proposition 5.2.1.

(8) Two applications of COND and ABSURD reduce this to

$$\vdash b \rhd \alpha.t = \alpha.(\texttt{if}\ b\ \texttt{then}\ t).$$

The proof then proceeds by a case analysis on the form of α. For example if it is $c?x$, by assumption $x \notin fv(b)$ and therefore, by INPUT, it can be reduced to $\vdash b \rhd t = \texttt{if}\ b\ \texttt{then}\ t$, which follows from Proposition 5.2.2. If α is τ or $c!e$ the reasoning is even more straightforward. □

This proof system is complete for late strong bisimulation over the recursion-free subset of our value-passing language, as stated in the following theorem.

THEOREM 5.3 (Completeness of $\vdash$). *$t \sim_{\mathrm{L}}^{b} u$ implies $\vdash b \rhd t = u$.*

As in the case of pure CCS, the proof of this theorem proceeds by induction on the heights of terms and relies on the notion of a normal form.

DEFINITION 5.4. The height of t, denoted $|t|$, is defined inductively as follows:

(1) $|\mathbf{0}| = 0$;

(2) $|t+u| = \max\{|t|, |u|\}$;
(3) $|\mathtt{if}\ b\ \mathtt{then}\ t| = |t|$;
(4) $|\alpha.t| = 1 + |t|$.

DEFINITION 5.5. A process term t is a *normal form*, or is *in normal form*, if it has the form $\sum_{i \in I} \mathtt{if}\ b_i\ \mathtt{then}\ \alpha_i.t_i$ where each t_i is in normal form (by convention the sum above is taken to be $\mathbf{0}$ when the index set I is empty).

LEMMA 5.6. *For every term t there exists a normal form t' such that $fv(t) = fv(t')$, $|t| = |t'|$ and $\vdash t = t'$.*

PROOF. By structural induction using the elementary facts about the proof system given in Proposition 5.2. □

PROOF OF THEOREM 5.3. By Lemma 5.6 we may assume that t and u are in normal form:

$$t \equiv \sum_{i \in I} \mathtt{if}\ c_i\ \mathtt{then}\ \alpha_i.t_i, \qquad u \equiv \sum_{j \in J} \mathtt{if}\ d_j\ \mathtt{then}\ \beta_j.u_j.$$

The proof proceeds by induction on $|t| + |u|$ where the base case follows immediately. For the inductive step we first show that

$$\vdash b \wedge c_i \rhd u + \mathtt{if}\ c_i\ \mathtt{then}\ \alpha_i.t_i = u \tag{2}$$

for each $i \in I$. For the proof of (2) we only examine the case $\alpha_i \equiv c?x$ (the remaining cases are similar). By α-conversion we may assume $x \notin fv(b, t, u)$. Since $t \sim^b_{\mathrm{L}} u$ and $t \xrightarrow{c_i, c?x} t_i$, there exists a $b \wedge c_i$-partition B such that $fv(B) \subseteq fv(b, t, u)$ and for each $b' \in B$ there is a $j \in J$ such that $b' \models d_j$, $u \xrightarrow{d_j, c?x} u_j$ and $t_i \sim^{b'}_{\mathrm{L}} u_j$. By induction

$$\vdash b' \rhd t_i = u_j.$$

Since $x \notin fv(b')$, applying INPUT we obtain

$$\vdash b' \rhd c?x.t_i = c?x.u_j.$$

Since $b' \models c_i$ and $b' \models d_j$, by Proposition 5.2.2 and transitivity

$$\vdash b' \rhd \mathtt{if}\ c_i\ \mathtt{then}\ c?x.t_i = \mathtt{if}\ d_j\ \mathtt{then}\ c?x.u_j.$$

Using S2 we derive that

$$\vdash b' \rhd u + \mathtt{if}\ c_i\ \mathtt{then}\ c?x.t_i = u.$$

Since $\bigvee B = b \wedge c_i$, by a repeated use of PARTITION

$$\vdash b \wedge c_i \rhd u + \texttt{if } c_i \texttt{ then } c?x.t_i = u.$$

This is the required (2) above.

Now we proceed with the completeness proof as follows. By Proposition 5.2.1

$$\vdash b \wedge \neg c_i \rhd \texttt{if } c_i \texttt{ then } \alpha_i.t_i = \mathbf{0}$$

and thus, by CHOICE and S1

$$\vdash b \wedge \neg c_i \rhd u + \texttt{if } c_i \texttt{ then } \alpha_i.t_i = u. \tag{3}$$

By (2), (3) and PARTITION

$$\vdash b \rhd u + \texttt{if } c_i \texttt{ then } \alpha_i.t_i = u.$$

This is true for each $i \in I$. By repeated use of this property we get

$$\vdash b \rhd u + t = u.$$

Symmetrically we can derive

$$\vdash b \rhd u + t = t$$

and the result follows by transitivity. □

5.2. *The axioms for early bisimulation*

As explained in Section 4.2, early bisimulation differs from late bisimulation only in that, when matching an input transition from one process, it is possible to use input transitions from different components of the other for different input values. Reflected at the syntactical level, to obtain a proof system for early bisimulation from the late one, what is needed is a rule which allows us to "break" the scopes of input prefixes at the same level. This can be achieved in two ways: either by adding a rule schema or an axiom schema to the proof system for late bisimulation. The axiom schema for early bisimulation was first proposed in [39] for an axiomatization of the π-calculus, and is adapted for value-passing processes in [17]. The rule schema was used in [17]. Here we first include into our system the axiom schema which has the following form:

$$\text{EA} \quad c?x.t + c?x.u = c?x.t + c?x.u + c?x.(\texttt{if } b \texttt{ then } t + \texttt{if } \neg b \texttt{ then } u).$$

Note that here b is an arbitrary Boolean expression. It is not difficult to see that the proof system is sound with respect to $\sim^b_{\mathrm{E}}$. Let us write $\vdash_{\mathrm{E}} b \rhd t = u$ to mean that $b \rhd t = u$ can be deduced from the proof system of the previous section extended with EA. EA can be generalized to allow arbitrary summations as follows.

PROPOSITION 5.7. *For any finite set of Booleans* $\{b_i \mid 1 \leqslant i \leqslant n\}$ *such that* $\bigvee_{1\leqslant i\leqslant n} b_i = b$ *and* $b_i \wedge b_j = \mathit{false}$ *for* $i \neq j$,

$$\vdash_{\mathrm{E}} b \rhd \sum_{1\leqslant i\leqslant n} c?x.t_i = \sum_{1\leqslant i\leqslant n} c?x.t_i + c?x.\sum_{1\leqslant i\leqslant n} \mathtt{if}\ b_i\ \mathtt{then}\ t_i.$$

PROOF. See [17, Proposition 3.1]. □

Now we have the following counterpart of Theorem 5.3.

THEOREM 5.8 (Completeness of $\vdash_{\mathrm{E}}$). $t \sim^b_{\mathrm{E}} u$ *implies* $\vdash_{\mathrm{E}} b \rhd t = u$.

PROOF. The proof follows the same pattern as the proof of Theorem 5.3, using Proposition 5.7. See [17, Theorem 3.2] for the details. □

Another way of dealing with the early case is to replace EA by the following rule schema

$$\text{E-INPUT}\quad \frac{b \rhd \sum_{i\in I} \tau.t_i = \sum_{j\in J} \tau.u_j}{b \rhd \sum_{i\in I} c?x.t_i = \sum_{e} j \in J c?x.u_j},\qquad x \notin fv(b).$$

This also is quite general but at least its application only depends on the structure of the terms in the proof being elaborated.

Note that the use of τ in this rule is essential. If τ is omitted then the rule becomes

$$\frac{b \rhd \sum_{i\in I} t_i = \sum_{j\in J} u_j}{b \rhd \sum_{i\in I} c?x.t_i = \sum_{j\in J} c?x.u_j},\qquad x \notin fv(b)$$

which is not sound with respect to early strong bisimulation. For example since $(p+q)+r = p+(q+r)$ we could use this rule to derive

$$c?x.(p+q) + c?x.r = c?x.p + c?x.(q+r)$$

for which there are obvious counterexamples.

With the help of COND and PARTITION, EA can be easily derived from E-INPUT. Hence the proof system obtained by replacing the axiom schema EA with E-INPUT is also complete.

5.3. *The τ laws*

One of the classical results in the theory of bisimulation is that, adding the three τ laws, presented in Figure 9, to the complete axiomatization of strong bisimulation results in a complete axiomatization of observational congruence. The aim of this section is to show that this result carries over to value-passing processes. We start with the late equivalence.

T1	$\alpha.\tau.X = \alpha.X$
T2	$X + \tau.X = \tau.X$
T3	$\alpha.(X + \tau.Y) + \alpha.Y = \alpha.(X + \tau.Y)$

Fig. 9. The τ laws.

Let us write $\vdash_{\mathrm{wL}} b \rhd t = u$ to mean that $b \rhd t = u$ can be deduced from the core system extended with the τ laws.

The structure of the completeness proof below follows closely a corresponding proof in pure CCS [32]. As in that reference, the proof relies on a notion of *full normal forms* which are τ-saturated. But again here we need to be careful with the τ transitions occurring after an input action; they should not be absorbed. These τ actions will be dealt with in the proof of the main theorem.

DEFINITION 5.9. A normal form $t \equiv \sum_{i \in I} \texttt{if}\ b_i\ \texttt{then}\ \alpha_i.t_i$ is a *late full normal form* if

$t \overset{b,\alpha}{\Longrightarrow}_{\mathrm{L}} t'$ implies $t \xrightarrow{b,\alpha} t'$ and each t_i is in late full normal form.

To show that every term can be transformed into a late full normal form we need a lemma permitting the absorption of τ actions by syntactic manipulation. This lemma requires a generalization of T3:

LEMMA 5.10. *If* $fv(b) \cap bv(\alpha) = \emptyset$ *then* $\vdash_{\mathrm{wL}} \alpha.(t + \texttt{if}\ b\ \texttt{then}\ \tau.u) = \alpha.(t + \texttt{if}\ b\ \texttt{then}\ \tau.u) + \texttt{if}\ b\ \texttt{then}\ \alpha.u$.

PROOF. See [17, Lemma 4.7]. □

LEMMA 5.11 (Absorption). *If* $t \overset{b,\alpha}{\Longrightarrow}_{\mathrm{L}} t'$, *with* $fv(b) \cap bv(\alpha) = \emptyset$, *then* $\vdash_{\mathrm{wL}} t = t + \texttt{if}\ b\ \texttt{then}\ \alpha.t'$.

PROOF. See [17, Lemma 4.8]. □

With the absorption lemma it can be easily shown that every term is equivalent to a late full normal form:

LEMMA 5.12. *For any normal form* t *there is a full normal form* t' *such that* $fv(t) = fv(t')$, $|t| = |t'|$ *and* $\vdash_{\mathrm{wL}} t = t'$.

By this lemma and Lemma 5.6, every term can be transformed into a full normal form of equal height.

We need one more ingredient for the completeness proof.

PROPOSITION 5.13. $t \approx^b_{\mathrm{E}} u$ *if and only if there is a* b*-partition* B *such that for all* $b' \in B$, $t \simeq^{b'}_{\mathrm{E}} u$ *or* $t \simeq^{b'}_{\mathrm{E}} \tau.u$ *or* $\tau.t \simeq^{b'}_{\mathrm{E}} u$.

PROOF. See [17, Proposition 4.2]. □

THEOREM 5.14 (Completeness of $\vdash_{\mathrm{wL}}$). *$t \simeq_{\mathrm{L}}^{b} u$ implies $\vdash_{\mathrm{wL}} b \rhd t = u$.*

PROOF. See [17, Theorem 5.3]. □

As in the case of strong equivalence, to obtain a complete proof system for early observation congruence, all we need is to add either the axiom schema EA or rule schema E-INPUT to the proof system. For the sake of variation here we use E-INPUT. So let us write $\vdash_{\mathrm{wE}} b \rhd t = u$ to mean that $b \rhd t = u$ can be deduced from the proof system for $\vdash_{\mathrm{wL}}$ plus E-INPUT. As can be expected, the proof of the completeness result for early congruence differs from that for late only in the case of input prefixing. The E-INPUT rule has a generalized form:

PROPOSITION 5.15. *Suppose $x \notin fv(b, c_i, d_j)$, $i \in I$, $j \in J$. Then from*

$$\vdash_{\mathrm{wE}} b \rhd \sum_{i \in I} \mathtt{if}\ c_i\ \mathtt{then}\ \tau.t_i = \sum_{j \in J} \mathtt{if}\ d_j\ \mathtt{then}\ \tau.u_j$$

we can infer

$$\vdash_{\mathrm{wE}} b \rhd \sum_{i \in I} \mathtt{if}\ c_i\ \mathtt{then}\ c?x.t_i = \sum_{j \in J} \mathtt{if}\ d_j\ \mathtt{then}\ c?x.u_j.$$

PROOF. See [17, Proposition 4.3]. □

THEOREM 5.16 (Completeness of $\vdash_{\mathrm{wE}}$). *$t \simeq_{\mathrm{E}}^{b} u$ implies $\vdash_{\mathrm{wE}} b \rhd t = u$.*

PROOF. See [17, Theorem 4.3]. □

5.4. *Adding other operators*

So far we have concentrated on the core language which does not include the restriction, renaming and parallel composition operators. These operators are however easy to axiomatize. The equations characterizing the restriction and renaming operator are the same as for pure processes and are listed in Figure 10. These laws are all fairly standard and need no further explanation. The expansion law for the parallel operator, shown in Figure 11, is an adaptation of the one proposed in [39] for the π-calculus.

We know from the previous sections that $\sim_{\mathtt{e}}$, $\simeq_{\mathtt{e}}$, $\sim_{\mathtt{l}}$ and $\simeq_{\mathtt{l}}$ are all preserved by the new operators. It is also easy to check that the new equations are sound for these congruence relations. Moreover, by adding these new equations to the core proof system, each term in the extended language can be transformed to a term in the core language. This in turn ensures the completeness of this extended proof system over the extended language with respect to each of the equivalences mentioned above.

$$\mathbf{0}\backslash L = \mathbf{0} \qquad\qquad \mathbf{0}[f] = \mathbf{0}$$

$$(X+Y)\backslash L = X\backslash L + Y\backslash L \qquad\qquad (X+Y)[f] = X[f] + Y[f]$$

$$(\alpha.X)\backslash L = \begin{cases} \alpha.(X\backslash L) & \text{if } Chan(\alpha) \notin L \\ \mathbf{0} & \text{if } Chan(\alpha) \in L \end{cases} \qquad\qquad (\alpha.X)[f] = f(\alpha).X[f]$$

Fig. 10. Axioms for restriction and renaming.

Let $t = \sum_i \alpha_i.t_i$ and $u = \sum_j \beta_j.u_j$, with $fv(t) \cap bv(u) = fv(u) \cap bv(t) = \emptyset$. Then

$$t \mid u = sync\_move(t,u) + async\_move(t,u),$$

where

$$\begin{aligned} sync\_move(t,u) &= \sum \{\tau.(t_i[e/x] \mid u_j) \mid \alpha_i \equiv c?x,\ \beta_j \equiv c!e\} \\ &\quad + \sum \{\tau.(t_i \mid u_j[e/x]) \mid \alpha_i \equiv c!e,\ \beta_j \equiv c?x\}, \\ async\_move(t,u) &= \sum_i \alpha_i.(t_i \mid u) + \sum_j \beta_j.(t \mid u_j). \end{aligned}$$

Fig. 11. The expansion law.

5.5. *Reasoning about recursion*

The proof systems considered so far are complete with respect to the corresponding equivalence relations over finite processes only. Processes exhibiting infinite behaviours are defined recursively. To be able to reason about infinite processes we need inference rules to handle recursion. For this purpose it is more convenient to have recursion terms in the language, instead of using process identifiers and defining equations. In pure CCS recursive terms can be constructed using process variables and the **fix** operator: if X is a process variable and t is a process term then $\mathbf{fix}\, Xt$ is a recursive term. However, in the value-passing calculus, a process variable with non-zero arity needs to take data parameters to form a process term. With the notation $\mathbf{fix}\, Xt$ it is not clear how free data variables in t are bound, and what are the free variables and parameters of this term. To avoid these confusions we will use *process abstractions* instead of process terms to form recursions.

Let *PVar* be a countably infinite set of *process abstraction variables* (or just *process variables* for short), ranged over by X, Y, Z etc. Each process abstraction variable is associated with a non-negative *arity* indicating the number of data expressions it should take to form a process term. We add a new kind of process term and a new clause for abstractions to the language introduced in Section 2. Also we shall omit parallel composition, restriction and renaming operators. The resulting language is known as *regular value-passing*

CCS and has the following BNF grammar:

$$
\begin{aligned}
t &::= \mathbf{0} \,|\, \alpha.t \,|\, t + t \,|\, \texttt{if}\ b\ \texttt{then}\ t \,|\, F\overline{e} \\
F &::= X \,|\, (\overline{x})t \,|\, \mathbf{fix}\, X F \\
\alpha &::= \tau \,|\, c?x \,|\, c!e
\end{aligned}
$$

An abstraction is either a process variable X, a process abstraction $(\overline{x})t$ (where $\overline{x}$ is a vector of distinct names), or a recursion abstraction $\mathbf{fix}\, XF$. The *degree* of an abstraction is defined as follows: The degree of X is the arity of X; the degree of $(\overline{x})t$ is the length of $\overline{x}$; the degree of $\mathbf{fix}\, XF$ is that of X which is required to be equal to the degree of F. Abstractions can be applied to data expressions to form process terms, and in $F\overline{e}$ we always require the length of $\overline{e}$ to be equal to the degree of F.

In $(\overline{x})t$ each data variable in $\overline{x}$ is bound in t. In a recursively defined term, $\mathbf{fix}\, XF$, the process variable X is bound in F. As usual a variable which is not bound in t is called *free* in t. We will write $BV(t)$ and $FV(t)$ for the sets of bound and free process variables in t. A term is *data closed* if it contains no free data variables. A term is *process closed* if it contains no free process variables. A term is *closed* if it is both data and process closed. A term is *data open* if it is not data closed and it is *process open* if it is not process closed; it is *open* if it is either data open or process open. We will only use data closed abstractions, so in $(\overline{x})t$ it is always required that $fv(t) \subseteq \{\overline{x}\}$. A process variable X is *(weakly) guarded* in a process term or an abstraction if each free occurrence of X is within the scope of some prefix operator $\alpha.\_$.

Using data closed abstractions to form recursion terms has two main advantages: In a recursion term $(\mathbf{fix}\, X(\overline{x})t)\overline{e}$ the free data variables are exactly those of $\overline{e}$, and the passing of the actual data parameters $\overline{e}$ to free occurrences of the recursion variable X in t is taken care of by β-conversion. As a consequence the inference rules for recursion become very simple, as simple as those for pure calculi. For instance, the rule for recursion unfolding is simply $\mathbf{fix}\, XF = F[\mathbf{fix}\, XF/X]$. Below is an example demonstrating how this rule acts (where F stands for $(x, y)(c!(x + y).d?z.X(z, y))$):

$$
\begin{aligned}
&(\mathbf{fix}\, XF)(u + 1, v - 1) \\
&\quad \overset{\text{REC}}{=} \big((x, y)\big(c!(x + y).d?z.(\mathbf{fix}\, XF)(z, y)\big)\big)(u + 1, v - 1) \\
&\quad \overset{\text{BETA}}{=} \big(c!(x + y).d?z.(\mathbf{fix}\, XF)(z, y)\big)[u + 1, v - 1/x, y] \\
&\quad = c!(u + 1 + v - 1).d?z.(\mathbf{fix}\, XF)(z, v - 1) \\
&\quad = c!(u + v).d?z.(\mathbf{fix}\, XF)(z, v - 1)
\end{aligned}
$$

The concrete and symbolic transitional semantics for recursion are given by the rules FIX and FIX, respectively, in Figure 12. The definitions of all the versions of bisimulations are the same as before, but since now we have a new syntactical category, namely abstractions, we need to extend the notions of bisimulation to it. We may wish to extend bisimulation to open process terms as well. This can be easily done by requiring that two open terms are bisimilar if and only if all their closed instantiations are, and two abstractions are bisimilar if and only if all their applications to data parameters are. To be precise, let $\sim$ be the

$$\text{FIX}\ \frac{t[\mathbf{fix}\,X(\overline{x})t/X][\overline{v}/\overline{x}] \xrightarrow{a} t'}{(\mathbf{fix}\,X(\overline{x})t)\overline{e} \xrightarrow{a} t'}\ \text{where } \overline{v} = [\![\overline{e}]\!]$$

$$\text{FIX}\ \frac{t[\mathbf{fix}\,X(\overline{x})t/X][\overline{e}/\overline{x}] \xrightarrow{b,\alpha} t'}{(\mathbf{fix}\,X(\overline{x})t)\overline{e} \xrightarrow{b,\alpha} t'}$$

Fig. 12. Operational semantics for recursion.

$$\text{REC}\ \frac{}{\mathbf{fix}\,XF = F[\mathbf{fix}\,XF/X]}$$

$$\text{CONGR-}\mathbf{fix}\ \frac{F = G}{\mathbf{fix}\,XF = \mathbf{fix}\,XG}$$

$$\text{UFI}\ \frac{G = F[G/X]}{G = \mathbf{fix}\,XF}\ X \text{ guarded in } F$$

$$\text{ABS}\ \frac{t = u}{(\overline{x})t = (\overline{x})u}$$

$$\text{APP}\ \frac{F = G}{F\overline{e} = G\overline{e}}$$

$$\text{BETA}\ \frac{}{((\overline{x})t)\overline{e} = t[\overline{e}/\overline{x}]}$$

Fig. 13. Inference rules for recursion.

generic symbol for different bisimulations. Then $t \sim u$ if and only if $t[\overline{p}/\overline{X}] \sim u[\overline{p}/\overline{X}]$, where $\overline{X} = FV(t, u)$, for any closed terms $\overline{p}$. $F \sim G$ if and only if $F\overline{e} \sim G\overline{e}$ for all vectors of values (in the case of concrete bisimulations) or data expressions (in the case of symbolic bisimulations) $\overline{e}$.

The proof rules for recursion are presented in Figure 13. REC is the standard rule for recursion unfolding and CONGR-**fix** expresses that all bisimulation equivalences we consider are congruent with respect to **fix**. Rules ABS and APP allow us to switch between abstractions and process terms. The most interesting rule is UFI, for Unique Fixpoint Induction. It asserts that every guarded abstraction has exactly one fixpoint. Syntactically it appears the same as its counterpart in the pure calculus [31]. However, here it works at the level of abstractions, while in the pure case it works at the level of processes. The latter can be viewed as a special case of the former when the arity of the process abstraction involved is 0. Let us write $\vdash_r b \rhd t = u$ (where t and u may also be abstractions) to mean that $b \rhd t = u$ can be derived using the core rules and axioms in Section 5.1 together with the rules in Figure 13. We state the soundness of $\vdash_r$:

THEOREM 5.17. *If* $\vdash_r b \rhd t = u$ *then* $t\rho \sim_1 u\rho$ *for any* $\rho \models b$.

For the completeness result we shall confine ourselves to guarded terms. As we have seen in the previous sections, to obtain a completeness result, a key instrument is the notion of a normal form whose syntax directly reflects its symbolic transition semantics. The completeness proof can then be performed by induction on the "height" (the maximum nesting depth of action prefixes) of the normal forms. This works fine for finite terms. But since the language we are considering now includes recursion we can not rely on such simple notion of a normal form. Instead we use *standard equation sets*.

Let $\overline{X} = \{X_1, X_2, \ldots, X_m\}$ and $\overline{W} = \{W_1, W_2, \ldots, W_n\}$ be disjoint sets of process variables. Then

$$E : X_i(\overline{x}_i) = H_i, \quad 1 \leqslant i \leqslant m,$$

where $fv(H_i) \subseteq \{\overline{x}_i\}$ and $FV(H_i) \subseteq \overline{X} \cup \overline{W}$, is an equation set with formal process variables in $\overline{X}$ and free process variables in $\overline{W}$. E is *process closed* if $\overline{W} = \emptyset$. E is *standard* if each H_i has the form

$$\sum_{k \in K_i} \texttt{if}\ b_{ik}\ \texttt{then} \left(\sum_{p \in P_{ik}} \alpha_{ikp}.X_{f(i,k,p)}(\overline{e}_{ikp}) + \sum_{p' \in P'_{ik}} W_{f'(i,k,p')}(\overline{w}_{ikp'}) \right).$$

In $X_{f(i,k,p)}$, the notation $f(i, k, p)$ is used to indicate that the subscript depends on i, k and p. An abstraction F provably satisfies equation set E if there are $\overline{F} = \{F_1, \ldots, F_n\}$ with $\vdash_{\mathrm{r}} F = F_1$ such that

$$\vdash_{\mathrm{r}} b_i \rhd F_i = (\overline{x}_i) H_i\big[\overline{F}/\overline{X}\big].$$

A process term t provably b-satisfies equation set E if there exist $\overline{b}$ and $\overline{t}$ with $fv(b_i) \cup fv(t_i) \subseteq \{\overline{x}_i\}$, $1 \leqslant i \leqslant m$, $b_1 = b$ and $\vdash_{\mathrm{r}} b_1 \rhd t_1 = t$, such that

$$\vdash_{\mathrm{r}} b_i \rhd t_i = H_i\big[(\overline{x}_k)(\texttt{if}\ b_k\ \texttt{then}\ t_k)/X_k \mid 1 \leqslant k \leqslant m\big], \quad 1 \leqslant i \leqslant m.$$

We will simply say that "t provably satisfies E" when $b \equiv true$.

Following [31], the proof of the completeness theorem consists of three major steps: First we show that each guarded term provably satisfies a standard equation set; then we show that if two terms are symbolically bisimilar over b then a single guarded equation set can be constructed which is provably b-satisfied by both terms; finally we show that if two terms t and u both provably satisfy a guarded equation set over b then $\vdash_{\mathrm{r}} b \rhd t = u$.

PROPOSITION 5.18. *Any guarded process term t provably satisfies a standard equation set E. Moreover, if t is process closed then so is E.*

PROOF. By mutual induction on the structures of process terms and abstractions. □

PROPOSITION 5.19. *Assume that t and u are guarded process terms. If $t \sim^b_{\mathrm{L}} u$ then there exists a standard equation set E such that both t and u provably b-satisfy E.*

$$
\begin{aligned}
&\stackrel{5.2}{=} \sum_{(p,q)\in I^{b'}} \alpha_{ikp}.\mathtt{if}\ b'\ \mathtt{then}\ \mathtt{if}\ b_{f(i,k,p)g(j,l,q)}\big[\overline{e}_{ikp}, \overline{e}'_{jlq}/\overline{x}_{f(i,k,p)}, \overline{y}_{g(j,l,q)}\big] \\
&\qquad\qquad \mathtt{then}\ t_{f(i,k,p)}[\overline{e}_{ikp}/\overline{x}_{f(i,k,p)}] \\
&\stackrel{(6),5.2}{=} \sum_{(p,q)\in I^{b'}} \alpha_{ikp}.\mathtt{if}\ b'\ \mathtt{then}\ t_{f(i,k,p)}[\overline{e}_{ikp}/\overline{x}_{f(i,k,p)}] \\
&\stackrel{5.2}{=} \sum_{(p,q)\in I^{b'}} \alpha_{ikp}.t_{f(i,k,p)}[\overline{e}_{ikp}/\overline{x}_{f(i,k,p)}] \\
&\stackrel{S2-S4}{=} \sum_{p\in P_{ik}} \alpha_{ikp}.t_{f(i,k,p)}[\overline{e}_{ikp}/\overline{x}_{f(i,k,p)}] \\
&\equiv t_{ik}.
\end{aligned}
$$

The step last but one relies on $I^{b'}$ being total.

Symmetrically, using the fact $I^{b'}$ is surjective, we can show E is provably b-satisfied by u. □

PROPOSITION 5.20. *Let $E : X_i(\overline{x}_i) = H_i$, $1 \leqslant i \leqslant m$, be a standard equation set. If t and u both provably b-satisfy E then $\vdash_{\mathrm{r}} b \rhd t = u$.*

PROOF. Similar to the proof of Theorem 5.7 in [31]. □

Summarizing the above three propositions gives the completeness theorem:

THEOREM 5.21. *For guarded process terms t and u, $t \sim^b_{\mathrm{L}} u$ implies $\vdash_{\mathrm{r}} b \rhd t = u$.*

It can also be shown that adding EA or E-INPUT to $\vdash_{\mathrm{r}}$ gives rise to a proof system for early bisimulation, complete over guarded processes. Furthermore, with the addition of the three τ-laws these proof systems can be extended to observation congruences. The details are rather involved and are omitted here.

So far we have only considered guarded recursions. For pure CCS there are elegant laws which are sufficient to remove all unguarded recursions [33] and thus extending the axiomatization to the whole regular language. However, in value-passing calculi the combination of unguarded recursion and the conditional construct may result in non-trivial situations. For example let us consider processes P and Q defined as

$$
\begin{aligned}
P &\Leftarrow c?x.d!0.P, \\
Q &\Leftarrow c?x.Q'(x), \\
Q'(x) &\Leftarrow \mathtt{if}\ x=0\ \mathtt{then}\ d!x.Q\ \mathtt{else}\ Q'(x-1),
\end{aligned}
$$

where x ranges over natural numbers. Clearly P and Q are equivalent, but in order to derive this we need to rely on the inductive nature of natural numbers, and close interaction between reasoning about processes and reasoning about data becomes inevitable.

6. Summary and related work

We have discussed the complications arising from generalizing the standard theory of pure process algebra to the value-passing setting, and presented an approach to tackle these problems. The main focus is on the theory of bisimulation. We have demonstrated how the data aspect can be separated from processes so that a theory of bisimulation for value-passing processes can be developed without depending on any particular theory of data. This "symbolic" approach manipulates open terms with Boolean expressions constraining their free data variables, in contrast to the traditional process algebra theory which only deal with closed terms. As a result the problem of infinite branching, due to instantiating input variables with values, is avoided, while the standard bisimulation equivalences are captured.

We have also developed finitary proof systems for the various bisimulation-based equivalences in our value-passing language. Again, reasoning about data is separated from reasoning about processes. These proof systems preserve the standard equational axioms for bisimulation-based equivalences while generalizing equational reasoning by introducing inference rules.

There are many aspects of value-passing calculi that are not covered in this chapter. Denotational semantics for value-passing processes is discussed at length in [14,15]. Various modal logics have been proposed for such processes [19,43]. Algorithmic aspects are addressed in [26,36,27]. The "symbolic" style proof systems presented in Section 5 has been implemented in the interactive proof assistant VPAM [25].

In our language, channel names have been separated from values and hence are not allowed to be transmitted between processes. Lifting this restriction results in a very expressive language known as π-calculus which is covered in [38], Chapter 8 of this Handbook.

For other approaches to value-passing processes, we want to mention μCRL which extends ACP with data introduced by algebraic specifications [9,10]. The theory of μCRL is described in detail in [11], Chapter 17 of this Handbook.

In [45] *parameterized graphs*, which are similar to symbolic transition graphs, were proposed as a model for value-passing processes.

Acknowledgements

The work presented in this chapter is the result of a long term collaboration with Matthew Hennessy, and we are indebted to him for many original ideas and insights. Thanks also go to Joachim Parrow for his constructive comments on the very first draft of this chapter, to Julian Rathke and Luca Aceto for correcting many English mistakes and suggesting several improvements. Comments from the anonymous referee were also helpful.

References

[1] L. Aceto, W.J. Fokkink and C. Verhoef, *Structural operational semantics*, Handbook of Process Algebra, J.A. Bergsta, A. Ponse and S.A. Smolka, eds, Elsevier, Amsterdam (2001), 197–292.

[2] S.D. Brookes, C.A.R. Hoare and A.W. Roscoe, *A theory of communicating sequential processes*, J. ACM **31** (3) (1984), 560–599.
[3] G. Bruns, *Distributed Systems Analysis with CCS*, Prentice-Hall (1996).
[4] R. Cleaveland, J. Parrow and B. Steffen, *The concurrency workbench: A semantics based verification tool for finite state systems*, ACM Transactions on Programming Systems **15** (1989), 36–72.
[5] R. DeNicola and M. Hennessy, *Testing equivalences for processes*, Theoret. Comput. Sci. **24** (1984), 83–113.
[6] N. Francez, D. Lehman and A. Pnueli, *A linear history of semantics for languages with distributed processing*, Theoret. Comput. Sci. **32** (1984), 25–46.
[7] R.J. van Glabbeek, *The linear time – branching time spectrum I. The semantics of concrete, sequential processes*, Handbook of Process Algebra, J.A. Bergstra, A. Ponse and S.A. Smolka, eds, Elsevier, Amsterdam (2001), 3–99.
[8] J.F. Groote and S.P. Luttik, *Undecidability and completeness results for process algebras with alternative quantification over data*, Technical Report SEN-R9806, CWI, Amsterdam (1998).
[9] J.F. Groote and A. Ponse, *The syntax and semantics of μCRL*, Technical Report CS-R9076, CWI, Amsterdam (1990).
[10] J.F. Groote and A. Ponse, *Proof theory for μCRL*, Technical Report CS-R9138, CWI, Amsterdam (1991).
[11] J.F. Groote and M.A. Reniers, *Algebraic process verification*, Handbook of Process Algebra, J.A. Bergstra, A. Ponse and S.A. Smolka, eds, Elsevier, Amsterdam (2001), 1151–1208.
[12] M. Hennessy, *An Algebraic Theory of Processes*, MIT Press (1988).
[13] M. Hennessy, *A proof system for communicating processes with value-passing*, Formal Aspects of Computer Science **3** (1991), 346–366.
[14] M. Hennessy and A. Ingólfsdóttir, *Communicating processes with value-passing and assignment*, J. Formal Aspects of Computing **5** (1993), 432–466.
[15] M. Hennessy and A. Ingólfsdóttir, *A theory of communicating processes with value-passing*, Inform. and Comput. **107** (2) (1993), 202–236.
[16] M. Hennessy and H. Lin, *Symbolic bisimulations*, Theoret. Comput. Sci. **138** (1995), 353–389.
[17] M. Hennessy and H. Lin, *Proof Systems for message passing processes*, Formal Aspects of Computing **8** (1996), 379–407.
[18] M. Hennessy and H. Lin, *Unique fixpoint induction for message-passing process calculi*, CATS'97, Proc. Computing: the Australasian Theory Symposium (1997), 122–131.
[19] M. Hennessy and X. Liu, *A modal logic for message passing processes*, Proc. 5th International Conference on Computer Aided Verification, Lecture Notes in Comput. Sci. 697 (1993), 359–370.
[20] M. Hennessy and R. Milner, *Algebraic laws for nondeterminism and concurrency*, J. ACM **32** (1) (1985), 137–161.
[21] M. Hennessy and G. Plotkin, *A Term Model for CCS*, Lecture Notes in Comput. Sci. 88, Springer-Verlag (1980).
[22] C.A.R. Hoare, *Communicating sequential processes*, Comm. ACM **21** (8) (1978), 666–677.
[23] C.A.R. Hoare and A.W. Roscoe, *The laws of occam*, Technical Report PRG Monograph 53, Oxford University Computing Laboratory (1986.)
[24] A. Ingólfsdóttir, *Late and early semantics coincide for testing*, Theoret. Comput. Sci. **146** (1–2) (1995), 341–349.
[25] H. Lin, *A verification tool for value-passing processes*, Proc. 13th International Symposium on Protocol Specification, Testing and Verification, IFIP Transactions, North-Holland (1993).
[26] H. Lin, *Symbolic transition graphs with assignment*, Proc. CONCUR 96, Pisa, Lecture Notes in Comput. Sci. 1119, U. Montanari and V. Sassone, eds, Springer-Verlag (1996), 50–65.
[27] H. Lin, *"On-the-fly" instantiation of value-passing processes*, Proc. FORTE/PSTV'98, Paris (1998).
[28] G. Milne and R. Milner, *Concurrent processes and their syntax*, J. ACM **26** (2) (1979), 302–321.
[29] R. Milner, *A Calculus of Communicating Systems*, Lecture Notes in Comput. Sci. 92, Springer-Verlag, Berlin (1980).
[30] R. Milner, *Calculi for synchrony and asynchrony*, Theoret. Comput. Sci. **25** (1983), 267–310.
[31] R. Milner, *A complete inference system for a class of regular behaviours*, J. Comput. System Sci. **28** (3) (1984), 439–466.

[32] R. Milner, *Communication and Concurrency*, Prentice-Hall (1989).

[33] R. Milner, *A complete axiomatisation for observational congruence of finite-state behaviours*, Inform. and Comput. **81** (1989), 227–247.

[34] R. Milner, *The polyadic pi-calculus*, Proc. CONCUR 92, Stony Brook, New York, Lecture Notes in Comput. Sci. 630, W.R. Cleaveland, ed., Springer-Verlag (1992), 1.

[35] R. Milner, J. Parrow and D. Walker, *A calculus of mobile processes, Part I and II*, Inform. and Comput. **100** (1) (1992), 1–40.

[36] P. Paczkowski, *Characterizing bisimilarity of value-passing parameterised processes*, The Infinity Workshop, Pisa, Italy (1996), 47–55.

[37] D. Park, *Concurrency and Automata on Infinite Sequences*, Theoretical Computer Science VII, Lecture Notes in Comput. Sci. 104, Springer-Verlag (1981). DAIMI FN-19, Computer Science Department, Aarhus University (1981).

[38] J. Parrow, *An introduction to the π-calculus*, Handbook of Process Algebra, J.A. Bergstra, A. Ponse and S.A. Smolka, eds, Elsevier, Amsterdam (2001), 479–543.

[39] J. Parrow and D. Sangiorgi, *Algebraic theories for name-passing calculi*, Inform. and Comput. **120** (2) (1995), 174–197.

[40] G. Plotkin, *A structural approach to operational semantics*, Report DAIMI FN-19, Computer Science Department, Aarhus University (1981).

[41] J. Rathke, *Unique fixpoint induction for value-passing processes*, Proc. 12th IEEE Symposium on Logic in Computer Science, IEEE Press (1997).

[42] J. Rathke, *Symbolic techniques for value-passing calculi*, Ph.D. Thesis, University of Sussex (1997).

[43] J. Rathke and M. Hennessy, *Local model checking for value-passing processes*, Proc. TACS'97, International Symposium on Theoretical Aspects of Computer Software, Sendai, Springer-Verlag (1997).

[44] D. Sangiorgi, *A theory of bisimulation for π-calculus*, Acta Inform. **33** (1996).

[45] M.Z. Schreiber, *Value-passing process calculi as a formal method*, Ph.D. thesis, Imperial College, London (1994).

Subject index

$\equiv_\alpha$, 439
$\rightarrow_{\mathtt{e}}$, 439
$\rightarrow_{\mathtt{l}}$, 443
$\stackrel{a}{\Longrightarrow}_{\mathtt{e}}$, 441
$\stackrel{a}{\Longrightarrow}_{\mathtt{l}}$, 444
$\approx_{\mathtt{e}}$, 441
0, 438
Act, 439
Act_τ, 439
Act_τ^l, 442
BExp, 437
$b \models b'$, 438
Chan, 438
chan, 439
$c!e$, 438
$c?x$, 438
DExp, 437
$e[e'/x]$, 439
$[\![e]\!]\rho$, 438
$e\sigma$, 439
fix$X._$, 438
fv(b), 437
fv(e), 437
`if` b `then` t, 438
$\mathcal{K}$, 438
$P(\overline{e})$, 438
$P(x_1, \ldots, x_n) \Leftarrow t$, 438
$p \simeq_{\mathrm{e}} q$, 441
$p \simeq_{\mathtt{l}} q$, 445
$p \sim_{\mathrm{e}} q$, 441
$p \approx_{\mathtt{l}} q$, 444
$\mathcal{P}$, 439
$t + u$, 438
$t \xrightarrow{b,\alpha} u$, 456
$t \simeq_{\mathrm{E}}^b u$, 457
$t \simeq_{\mathrm{L}}^b u$, 458
$t \stackrel{b,\alpha}{\Longrightarrow}_{\mathrm{L}} u$, 458
$t \mid u$, 438
$t \sim_{\mathrm{E}}^b u$, 452
$t[f]$, 438
$t \xrightarrow{b,\alpha} u$, 449
$t \backslash L$, 438
$t\rho$, 438
$t \sim_{\mathrm{L}}^b u$, 455

$t \approx^b_{\mathrm{E}} u$, 457, 458
$\mathcal{T}$, 439
Val, 437
Var, 437
$(x)t$, 446
α-equivalence, 439
$\alpha.t$, 438
ρ, 438
$\rho \models b$, 438
$\rho\{v/x\}$, 438
σ, 439
τ, 438
τ laws, 465
$\forall\exists$-bisimulation, 443

abstraction, 446, 469
action prefixing, 438
arity, 438

binding operator, 439
bisimulation equivalences, 440
bound, 439

channel names, 438
channel renaming, 438
channel restriction, 438
closed, 438
concrete equivalences, 440
conditional equation, 460

defining equation, 438

early strong bisimulation, 441
early strong symbolic bisimulation, 452
early weak bisimulation, 441
early weak symbolic bisimulation, 457
evaluation, 438

free variables, 439
full normal form, 466

guarded, 469

inference rule, 460
invisible action, 438

labelled transition system, 439
late actions, 442
late observational congruence, 446
late strong bisimulation, 443
late strong symbolic bisimulation, 455
late weak bisimulation, 444
late weak symbolic bisimulation, 458

nondeterministic choice, 438
normal form, 463

operational semantics, 439

parallel composition, 438
process identifiers, 438
proof system, 459

standard equation set, 471
symbolic bisimulations, 448
symbolic equivalences, 440
symbolic operational semantics, 449

transition relation, 439

CHAPTER 8

An Introduction to the π-Calculus

Joachim Parrow

Department Teleinformatics, Royal Institute of Technology, Stockholm, Sweden
E-mail: joachim@it.kth.se

Contents

HANDBOOK OF PROCESS ALGEBRA
Edited by Jan A. Bergstra, Alban Ponse and Scott A. Smolka

printer "moves" to the client, since after the interaction nothing else can access it. For this reason the π-calculus has been called a calculus of "mobile" processes. But the calculus is much more general than that. The printer may have many links that make it do different things, and the server can send these links to different clients to establish different access capabilities to a shared resource.

At first sight it appears as if the π-calculus is just a specialized form of a value-passing process algebra where the values are links. In such a comparison the calculus may be thought rather poor since there are no data types and no functions defined on the names; the transferable entities are simple atomic things without any internal structure. The reason that the π-calculus nevertheless is considered more expressive is that it admits migrating local scopes. This important point deserves an explanation here.

Most process algebras have a way to declare a communication link local to a set of processes. For example in CCS the fact that P and Q share a private port a is symbolized by $(P \mid Q)\backslash a$, where the operator $\backslash a$ is called *restriction* on a. The significance is that no other process can use the local link a, as if it were a name distinct from all other names in all processes.

In the π-calculus this restriction is written $(\boldsymbol{\nu} a)(P \mid Q)$. It is similar in that no other process can use a immediately as a link to P or Q. The difference is that the name a is also a transferable object and as such can be sent, by P or Q, to another process which then can use the restricted link. Returning to the example above suppose that a is a local link between the server and the printer. Represent the printer by R, then this is captured by $(\boldsymbol{\nu} a)(\overline{b}a \,.\, S \mid R)$. The server is still free to send a along b to the client. The result would be a private link shared between all three processes, but still distinct from any other name in any other process, and the transition is consequently written

$$(\boldsymbol{\nu} a)(\overline{b}a \,.\, S \mid R) \mid b(c) \,.\, \overline{c}d \,.\, P \xrightarrow{\tau} (\boldsymbol{\nu} a)(S \mid R \mid \overline{a}d \,.\, P).$$

So, although the transferable objects are simple atomic things they can also be declared local with a defined scope, and in this way the calculus transcends the ordinary value-passing process algebras. This is also the main source of difficulty in the development of the theory because the scope of an object, as represented by the operands of its restriction, must migrate with the object as it is transferred between processes.

The π-calculus is far from a single well defined body of work. The central idea, a process algebraic definition of link-passing, has been developed in several directions to accommodate specific applications or to determine the effects of various semantics. Proliferation is certainly a healthy sign for any scientific area although it poses problems for those who wish to get a quick overview. Presumably some readers new to the π-calculus will be satisfied with a compact presentation of a single version, while other may be interested in the spectrum of variations.

This paper aims to serve both these needs. In the following, the even-numbered sections develop a single strand of the calculus. Section 2 presents the syntax and give some small examples of how it is used. In Section 4 we proceed to the semantics in its most common form as a labelled transition system. In Section 6 we consider one of the main definitions of bisimulation and the congruence it induces, and in Section 8 we look at their axiomatizations through syntactic equalities of agents. These sections do not depend on

the odd-numbered sections and can be considered as a basic course of the calculus. There will be full definitions and formulations of the central results, and sketches that explain the ideas and structure of the proofs.

Each odd-numbered section presents variations on the material in the preceding one. Thus, in Section 3 we explore different versions of the calculus, such as the effect of varying the operators, and the asynchronous, polyadic, and higher-order calculus. Section 5 treats alternative ways to define the semantics, with different versions of labelled and unlabelled transitions. Section 7 defines a few other common bisimulation equivalences (the π-calculus, like any process algebra, boasts a wide variety of equivalences but in this paper we concentrate on the aspects particular to π), and their axiomatizations are treated in Section 9. In these sections we do not always get a full formal account, but hopefully enough explanations that the reader will gain an understanding of the basic ideas. Finally, Section 10 contains references to other work. We give a brief account of how the calculus evolved and mention other overviews and introductory papers. We also indicate sources for the material treated in this paper.

It must be emphasized that there are some aspects of the π-calculus we do not treat at all, such as modal logics, analysis algorithms, implementations, and ways to use the calculus to model concurrent systems and languages. Also, the different variants can be combined in many ways, giving rise to a large variety of calculi. I hope that after this introduction a reader can explore the field with some confidence.

2. The π-calculus

We begin with a sequence of definitions and conventions. The reader who makes it to Section 2.3 will be rewarded with small but informative examples.

2.1. *Basic definitions*

We assume a potentially infinite set of *names* $\mathcal{N}$, ranged over by $a, b, \ldots, z$, which will function as all of communication ports, variables and data values, and a set of (*agent*) *identifiers* ranged over by A, each with a fixed nonnegative arity. The *agents*, ranged over by $P, Q, \ldots$ are defined Table 1. From that table we see that the agents can be of the following forms:

1. The empty agent $\mathbf{0}$, which cannot perform any actions.
2. An *Output Prefix* $\overline{a}x \,.\, P$. The intuition is that the name x is sent along the name a and thereafter the agent continues as P. So $\overline{a}$ can be thought of as an output port and x as a datum sent out from that port.
3. An *Input Prefix* $a(x) \,.\, P$, meaning that a name is received along a name a, and x is a placeholder for the received name. After the input the agent will continue as P but with the newly received name replacing x. So a can be thought of as an input port and x as a variable which will get its value from the input along a.
4. A *Silent Prefix* $\tau \,.\, P$, which represents an agent that can evolve to P without interaction with the environment. We use α, β to range over $a(x)$, $\overline{a}x$ and τ and call

Table 1
The syntax of the π-calculus

Prefixes	$\alpha ::= \overline{a}x$	Output
	$a(x)$	Input
	τ	Silent
Agents	$P ::= \mathbf{0}$	Nil
	$\alpha . P$	Prefix
	$P + P$	Sum
	$P \mid P$	Parallel
	$\texttt{if}\ x = y\ \texttt{then}\ P$	Match
	$\texttt{if}\ x \neq y\ \texttt{then}\ P$	Mismatch
	$(\boldsymbol{\nu} x)P$	Restriction
	$A(y_1, \ldots, y_n)$	Identifier
Definitions	$A(x_1, \ldots, x_n) \stackrel{\text{def}}{=} P$	(where $i \neq j \Rightarrow x_i \neq x_j$)

them *Prefixes*, and we say that $\alpha . P$ is a *Prefix form*, or sometimes just *Prefix* when this cannot cause confusion.

5. A *Sum* $P + Q$ representing an agent that can enact either P or Q.
6. A *Parallel Composition* $P \mid Q$, which represents the combined behaviour of P and Q executing in parallel. The components P and Q can act independently, and may also communicate if one performs an output and the other an input along the same port.
7. A *Match* $\texttt{if}\ x = y\ \texttt{then}\ P$. As expected this agent will behave as P if x and y are the same name, otherwise it does nothing.
8. A *Mismatch* $\texttt{if}\ x \neq y\ \texttt{then}\ P$. This agent will behave as P if x and y are *not* the same name, otherwise it does nothing.
9. A *Restriction* $(\boldsymbol{\nu} x)P$. This agent behaves as P but the name x is local, meaning it cannot immediately be used as a port for communication between P and its environment. However, it can be used for communication between components within P.
10. An *Identifier* $A(y_1, \ldots, y_n)$ where n is the arity of A. Every Identifier has a *Definition* $A(x_1, \ldots, x_n) \stackrel{\text{def}}{=} P$ where the x_i must be pairwise distinct, and the intuition is that $A(y_1, \ldots, y_n)$ behaves as P with y_i replacing x_i for each i. So a Definition can be thought of as a process declaration, $x_1, \ldots, x_n$ as formal parameters, and the Identifier $A(y_1, \ldots, y_n)$ as an invocation with actual parameters $y_1, \ldots, y_n$.

The operators are familiar from other process algebras so we shall in the following concentrate on some important aspects particular to the π-calculus, trusting the reader to be confident with the more general principles.

The forms Nil, Sum and Parallel have exactly the same meaning and use as in other process algebras, and the Prefix forms are as in the algebras that admit value-passing. The `if` constructs Match and Mismatch may appear limited in comparison with value-passing algebras which usually admit arbitrary Boolean expressions (evaluating to either true or false). But on closer consideration it is apparent that combinations of Match and Mismatch are the only possible tests that can be performed in the π-calculus: the objects transmitted are just names and these have no structure and no operators are defined on them, so the only thing we can do is compare names for equality. We can combine such tests conjunctively by nesting them, for example

$$\texttt{if}\ x = y\ \texttt{then}\ \texttt{if}\ u \neq v\ \texttt{then}\ P$$

behaves as P if both $x = y$ and $u \neq v$ hold. We can combine them disjunctively by using Sum, for example

$$\texttt{if}\ x = y\ \texttt{then}\ P + \texttt{if}\ u \neq v\ \texttt{then}\ P$$

behaves as P if at least one of $x = y$ and $u \neq v$ hold. Sometimes we shall use a binary conditional

$$\texttt{if}\ x = y\ \texttt{then}\ P\ \texttt{else}\ Q$$

as an abbreviation for $\texttt{if}\ x = y\ \texttt{then}\ P + \texttt{if}\ x \neq y\ \texttt{then}\ Q$.

As in other algebras we say that P is *guarded* in Q if P is a proper subterm of a Prefix form in Q. Also, the input Prefix $a(x)\,.\,P$ is said to *bind* x in P, and occurrences of x in P are then called *bound*. In contrast the output Prefix $\overline{a}x\,.\,P$ does not bind x. These Prefixes are said to have *subject* a and *object* x, where the object is called *free* in the output Prefix and *bound* in the input Prefix. The silent Prefix τ has neither subject nor object.

The Restriction operator $(\boldsymbol{\nu}x)P$ also binds x in P. Its effect is as in other algebras (where it is written $\backslash x$ in CCS and δ_x in ACP) with one significant difference. In ordinary process algebras the things that are restricted are port names and these cannot be transmitted between agents. Therefore the restriction is static in the sense that the scope of a restricted name does not need to change when an agent executes. In the π-calculus there is no difference between "port names" and "values", and a name that represents a port can indeed be transmitted between agents. If that name is restricted the scope of the restriction must change, as we shall see, and indeed almost all of the increased complexity and expressiveness of the π-calculus over value-passing algebras come from the fact that restricted things move around. The reader may also think of $(\boldsymbol{\nu}x)P$ as "new x in P", by analogy with the object-oriented use of the word "new", since this construct can be thought of as declaring a new and hitherto unused name, represented by x for the benefit of P.

In summary, both input Prefix and Restriction bind names, and we can define the *bound names* $\texttt{bn}(P)$ as those with a bound occurrence in P and the *free names* $\texttt{fn}(P)$ as those with a not bound occurrence, and similarly $\texttt{bn}(\alpha)$ and $\texttt{fn}(\alpha)$ for a Prefix α. We sometimes write $\texttt{fn}(P, Q)$ to mean $\texttt{fn}(P) \cup \texttt{fn}(Q)$, and just α for $\texttt{fn}(\alpha) \cup \texttt{bn}(\alpha)$ when it is apparent that it represents a set of names, such as in "$x \in \alpha$". In a Definition $A(x_1, \ldots, x_n) \stackrel{\text{def}}{=} P$ we assume that $\texttt{fn}(P) \subseteq \{x_1, \ldots, x_n\}$. In some examples we shall elide the parameters of Identifiers and Definitions when they are unimportant or can be inferred from context.

A *substitution* is a function from names to names. We write $\{x/y\}$ for the substitution that maps y to x and is identity for all other names, and in general $\{x_1 \ldots x_n / y_1 \ldots y_n\}$, where the y_i are pairwise distinct, for a function that maps each y_i to x_i. We use σ to range over substitutions, and sometimes write $\tilde{x}$ for a sequence of names when the length is unimportant or can be inferred from context. The agent $P\sigma$ is P where all free names x are replaced by $\sigma(x)$, with alpha-conversion wherever needed to avoid captures. This means that bound names are renamed such that whenever x is replaced by $\sigma(x)$ then the so obtained occurrence of $\sigma(x)$ is free. For example,

$$\big(a(x)\,.\,(\boldsymbol{\nu}b)\overline{x}b\,.\,\overline{c}y\,.\,\mathbf{0}\big)\{xb/yc\} \text{ is } a(z)\,.\,(\boldsymbol{\nu}d)\overline{z}d\,.\,\overline{b}x\,.\,\mathbf{0}.$$

A process algebra fan may have noticed that one common operator is not present in the π-calculus: that of relabelling (in CCS written $[a/b]$). The primary use of relabelling is to define instances of agents from other agents, for example, if B is a buffer with ports i and o then $B[i'/i, o'/o]$ is a buffer with ports i' and o'. In the π-calculus we will instead define instances through the parameters of the Identifiers, so for example a buffer with ports i and o is $B(i, o)$, and with ports i' and o' it is $B(i', o')$. For injective relabellings this is just another style of specification which allows us to economize on one operator. (A reader familiar with the CCS relabelling should be warned that it has the same effect as port substitution only if injective. In general they are different.)

Finally some notational conventions: A sum of several agents $P_1 + \cdots + P_n$ is written $\sum_{i=1}^{n} P_i$, or just $\sum_j P_j$ when n is unimportant or obvious, and we here allow the case $n = 0$ when the sum means $\mathbf{0}$. A sequence of distinct Restrictions $(\boldsymbol{\nu}x_1)\cdots(\boldsymbol{\nu}x_n)P$ is often abbreviated to $(\boldsymbol{\nu}x_1 \cdots x_n)P$. In a Prefix we sometimes elide the object if it is not important, so $a\,.\,P$ means $a(x)\,.\,P$ where x is a name that is never used, and similarly for output. And we sometimes elide a trailing $\mathbf{0}$, writing α for the agent $\alpha\,.\,\mathbf{0}$, where this cannot cause confusion. We give the unary operators precedence over the binary and $|$ precedence over $+$, so for example $(\boldsymbol{\nu}x)P \mid Q + R$ means $(((\boldsymbol{\nu}x)P) \mid Q) + R$.

2.2. *Structural congruence*

The syntax of agents is in one sense too concrete. For example, the agents $a(x)\,.\,\overline{b}x$ and $a(y)\,.\,\overline{b}y$ are syntactically different, although they only differ in the choice of bound name and therefore intuitively represent the same behaviour: an agent that inputs something along a and then sends that along $\overline{b}$. As another example the agents $P \mid Q$ and $Q \mid P$ represent the same thing: a parallel composition of the agents P and Q. Our intuition about parallel composition is that it is inherently unordered, and we are forced to syntactically distinguish between $P \mid Q$ and $Q \mid P$ only because our language is linear.

We therefore introduce a *structural congruence* to identify the agents which intuitively represent the same thing. It should be emphasized that this has nothing to do with the traditional behavioural equivalences in process algebra which are defined in terms of the behaviour exhibited by an agent under some operational semantics. We have yet to define a semantics, and the structural congruence identifies only agents where it is immediately obvious from their *structure* that they are the same.

The reader will here correctly object that "represent the same thing" and "immediately obvious" are not formally defined concepts, and indeed several different versions of the structural congruence can be found in the literature; there is no canonical definition and each has different merits. In Section 5.1 we will meet some of them and explore their

Table 2
The definition of structural congruence

The structural congruence $\equiv$ is defined as the smallest congruence satisfying the following laws:

1. If P and Q are variants of alpha-conversion then $P \equiv Q$.
2. The Abelian monoid laws for Parallel: commutativity $P \mid Q \equiv Q \mid P$, associativity $(P \mid Q) \mid R \equiv P \mid (Q \mid R)$, and $\mathbf{0}$ as unit $P \mid \mathbf{0} \equiv P$; and the same laws for Sum.
3. The unfolding law $A(\tilde{y}) \equiv P\{\tilde{y}/\tilde{x}\}$ if $A(\tilde{x}) \overset{\text{def}}{=} P$.
4. The scope extension laws

$(\boldsymbol{\nu}x)\mathbf{0}$	$\equiv \mathbf{0}$	
$(\boldsymbol{\nu}x)(P \mid Q)$	$\equiv P \mid (\boldsymbol{\nu}x)Q$	if $x \notin \mathtt{fn}(P)$
$(\boldsymbol{\nu}x)(P + Q)$	$\equiv P + (\boldsymbol{\nu}x)Q$	if $x \notin \mathtt{fn}(P)$
$(\boldsymbol{\nu}x)\mathtt{if}\ u = v\ \mathtt{then}\ P$	$\equiv \mathtt{if}\ u = v\ \mathtt{then}\ (\boldsymbol{\nu}x)P$	if $x \neq u$ and $x \neq v$
$(\boldsymbol{\nu}x)\mathtt{if}\ u \neq v\ \mathtt{then}\ P$	$\equiv \mathtt{if}\ u \neq v\ \mathtt{then}\ (\boldsymbol{\nu}x)P$	if $x \neq u$ and $x \neq v$
$(\boldsymbol{\nu}x)(\boldsymbol{\nu}y)P$	$\equiv (\boldsymbol{\nu}y)(\boldsymbol{\nu}x)P$	

consequences. Until then we adopt a particular structural congruence. The definition is given in Table 2. We briefly comment on the clauses in the definition.

1. Alpha-conversion, i.e., choice of bound names, identifies agents like $a(x).\bar{b}x$ and $a(y).\bar{b}y$.
2. The Abelian monoid laws mean that Parallel and Sum are unordered. For example, when we think of a composition of three agents P, Q, R it does not matter if we write it as $(P \mid Q) \mid R$ or $(R \mid Q) \mid P$. The same holds for Sum. The fact that $\mathbf{0}$ is a unit means that $P \mid \mathbf{0} \equiv P$ and $P + \mathbf{0} \equiv P$, something which follows from the intuition that $\mathbf{0}$ is empty and therefore contributes nothing to a Parallel composition or Sum.
3. The unfolding just says that an Identifier is the same as its Definition, with the appropriate parameter instantiation.
4. The scope extension laws come from our intuition that $(\boldsymbol{\nu}x)P$ just says that x is a new unique name in P; it can be thought of as marking the occurrences of x in P with a special colour saying that this is a local name. It then does not really matter where the symbols "$(\boldsymbol{\nu}x)$" are placed as long as they mark the same occurrences. For example, in $\mathbf{0}$ there are no occurrences so the Restriction can be removed at will. In Parallel composition, if all occurrences are in one of the components then it does not matter if the Restriction covers only that component or the whole composition.

Note that we do *not* have that $(\boldsymbol{\nu}x)(P \mid Q) \equiv (\boldsymbol{\nu}x)P \mid (\boldsymbol{\nu}x)Q$. The same occurrences are restricted in both agents, but in $(\boldsymbol{\nu}x)(P \mid Q)$ they are restricted by the *same* binder (or if you will, coloured by the same colour), meaning that P and Q can interact using x, in contrast to the situation in $(\boldsymbol{\nu}x)P \mid (\boldsymbol{\nu}x)Q$.

Through a combination of these laws we get that $(\boldsymbol{\nu}x)P \equiv P$ if $x \notin \mathtt{fn}(P)$:

$$P \equiv P \mid \mathbf{0} \equiv P \mid (\boldsymbol{\nu}x)\mathbf{0} \equiv (\boldsymbol{\nu}x)(P \mid \mathbf{0}) \equiv (\boldsymbol{\nu}x)P.$$

So as a special case we get $(\boldsymbol{\nu}x)(\boldsymbol{\nu}x)P \equiv (\boldsymbol{\nu}x)P$ for all P.

Another key fact is that all unguarded Restrictions can be pulled out to the top level of an agent:

PROPOSITION 1. *Let P be an agent where $(\boldsymbol{\nu}x)Q$ is an unguarded subterm. Then P is structurally congruent to an agent $(\boldsymbol{\nu}x')P'$ where P' is obtained from P by replacing $(\boldsymbol{\nu}x)Q$ with $Q\{x'/x\}$, for some name x' not occurring in P.*

The proof is by alpha-converting all bound names so that they become syntactically distinct, and then applying scope extension (from right to left) to move the Restriction to the outermost level. This corresponds to the intuition that instead of declaring something as local it can be given a syntactically distinct name: the effect is the same in that nothing else can access the name.

Our scope extension laws are in fact chosen precisely such that Proposition 1 holds. For example, we have not given any scope extension law for Prefixes and can therefore only pull out unguarded Restrictions. The reader may have expected a law like $(\boldsymbol{\nu}x)\alpha\,.\,P \equiv \alpha\,.\,(\boldsymbol{\nu}x)P$ for $x \notin \alpha$. Indeed such a law would be sound, in the sense that it conforms to intuition and does not disrupt any of the results in this paper, and it will hold for the behavioural equivalences explored later in Sections 6 and 7. But it will not be necessary at this point, in particular it is not necessary to prove Proposition 1.

Structural congruence is much stronger, i.e., identifies fewer agents, than any of the behavioural equivalences. The structural congruence is used in the definition of the operational semantics, which in turn is used to define the behavioural equivalences. The main technical reasons for taking this route are that many of the following definitions and explanations become simpler and that we get a uniform treatment for those variants of the calculus that actually require a structural congruence. In Section 5.1 we comment on the possibility to define the calculus without a structural congruence.

2.3. *Simple examples*

Although we shall not present the operational semantics just yet (a reader who wishes to look at it now will find it in Section 4) it might be illuminating to see some examples of the scope migration mentioned in Section 1, that Restrictions move with their objects. Formally, scope migration is a consequence of three straightforward postulates. The first is the usual law for inferring interactions between parallel components. This is present in most process algebras and implies that

$$a(x)\,.\,\overline{c}x \mid \overline{a}b \xrightarrow{\tau} \overline{c}b \mid \mathbf{0}$$

or in general

$$a(x)\,.\,P \mid \overline{a}b\,.\,Q \xrightarrow{\tau} P\{b/x\} \mid Q.$$

The second postulate is that Restrictions do not affect silent transitions. $P \xrightarrow{\tau} Q$ represents an interaction between the components of P, and a Restriction $(\boldsymbol{\nu}x)P$ only restricts interactions between P and its environment. Therefore $P \xrightarrow{\tau} Q$ implies $(\boldsymbol{\nu}x)P \xrightarrow{\tau} (\boldsymbol{\nu}x)Q$. The third postulate is that structurally congruent agents should never be distinguished and thus any semantics must assign them the same behaviour. Now what are the implications for restricted objects? Suppose that b is a restricted name, i.e., that we are considering a composition

$$a(x) . \overline{c}x \mid (\boldsymbol{\nu} b)\overline{a}b.$$

Will there be an interaction between the components and if so what should it be? Structural congruence gives the answer, because b is not free in the left hand component so the agent is by scope extension structurally congruent to

$$(\boldsymbol{\nu} b)\big(a(x) . \overline{c}x \mid \overline{a}b\big)$$

and this agent has a transition between the components: because of

$$a(x) . \overline{c}x \mid \overline{a}b \xrightarrow{\tau} \overline{c}b \mid \mathbf{0}$$

we get that

$$(\boldsymbol{\nu} b)\big(a(x) . \overline{c}x \mid \overline{a}b\big) \xrightarrow{\tau} (\boldsymbol{\nu} b)(\overline{c}b \mid \mathbf{0})$$

and the rightmost $\mathbf{0}$ can be omitted by the monoid laws. So by identifying structurally congruent agents we obtain that

$$a(x) . \overline{c}x \mid (\boldsymbol{\nu} b)\overline{a}b \xrightarrow{\tau} (\boldsymbol{\nu} b)\overline{c}b$$

or in general that, provided $b \notin \mathtt{fn}(P)$,

$$a(x) . P \mid (\boldsymbol{\nu} b)\overline{a}b . Q \xrightarrow{\tau} (\boldsymbol{\nu} b)\big(P\{b/x\} \mid Q\big).$$

In other words, the scope of $(\boldsymbol{\nu} b)$ "moves" with b from the right hand component to the left. This phenomenon is sometimes called scope extrusion. If $b \in \mathtt{fn}(P)$ a similar interaction is possible by first alpha-converting the bound b to some name $b' \notin \mathtt{fn}(P)$, and we would get

$$a(x) . P \mid (\boldsymbol{\nu} b)\overline{a}b . Q \xrightarrow{\tau} \big(\boldsymbol{\nu} b'\big)\big(P\{b'/x\} \mid Q\{b'/b\}\big).$$

So $P\{b'/x\}$ still contains b free and it is not the same as the received restricted name b'.

For another example consider:

$$\big((\boldsymbol{\nu} b)a(x) . P\big) \mid \overline{a}b . Q.$$

Here the right hand component has a free b which should not be the same as the bound b to the left. Is there an interaction between the components? We cannot immediately extend the scope to the right hand component since it has b free. But we can first alpha-convert the bound b to some new name b' and then extend the scope to obtain

$$\big(\boldsymbol{\nu} b'\big)\big(a(x) . P\{b'/b\} \mid \overline{a}b . Q\big)$$

and it is clear that we have a transition

$$(\boldsymbol{\nu} b')\big(a(x) . P\{b'/b\} \mid \overline{a}b . Q\big) \xrightarrow{\tau} (\boldsymbol{\nu} b')P\{b'/b\}\{b/x\} \mid Q.$$

So the restricted name, now b', will still be local to the left hand component; the attempt to intrude the scope is thwarted by an alpha-conversion. In summary, through alpha-conversion and scope extension we can send restricted names as objects, and Restrictions will always move with the objects and never include free occurrences of that name.

This ability to send scopes along with restricted names is what makes the calculus convenient for modelling exchange of private resources. For example, suppose we have an agent R representing a resource, say a printer, and that it is controlled by a server S which distributes access rights to R. In the simplest case the access right is just to execute R. This can be modelled by introducing a new name e as a trigger, and guarding R by that name, as in

$$(\boldsymbol{\nu} e)(S \mid e . R).$$

Here R cannot execute until it receives a signal on e. The server can invoke it by performing an action $\overline{e}$, but moreover, the server can send e to a client wishing to use R. For example, suppose that a client Q needs the printer. It asks S along some predetermined channel c for the access key, here e, to R, and only upon receipt of this key can R be executed. We have

$$c(x) . \overline{x} . Q \mid (\boldsymbol{\nu} e)(\overline{c}e . S \mid e . R) \xrightarrow{\tau} (\boldsymbol{\nu} e)(\overline{e} . Q \mid S \mid e . R) \xrightarrow{\tau} (\boldsymbol{\nu} e)(Q \mid S \mid R).$$

The first transition means that Q receives an access to R and the second that this access is used. We can informally think of this as if the agent R is transmitted (represented by its key e) from S to Q, so in a sense this gives us the power of a higher-order communication where the objects are agents and not only names. But our calculus is more general since a server can send e to many clients, meaning that these will share R (rather than receiving separate copies of R). And R can have several keys that make it do different things, for example R can be $e_1 . R_1 \mid e_2 . R_2 \cdots$, and the server can send only some of the keys to clients and retain some for itself, or send different keys to different clients representing different access privileges.

A related matter is if S wishes to send two names d and e to a client, and insure that the same client receives both names. If there are several clients then the simple solution of transmitting d and e along predetermined channels may mean that one client receives d and another e. A better solution is to first establish a private channel with a client and then send d and e along that channel. The private channel is simply a restricted name:

$$(\boldsymbol{\nu} p)\overline{c}p . \overline{p}d . \overline{p}e . S.$$

A client interacting with C must be prepared to receive a name, and then along that name receive d and e:

$$c(p) . p(x) . p(y) . Q.$$

Now, even if we have a composition with several clients and a server, the only possibility is that d and e end up with the same client. This feature is so common that we introduce an abbreviation for it:

$$\begin{array}{lll} \overline{c}\langle e_1 \cdots e_n \rangle . P & \text{means} & (\boldsymbol{\nu} p)\overline{c}p . \overline{p}e_1 . \cdots . \overline{p}e_n . P, \\ c(x_1 \cdots x_n) . Q & \text{means} & c(p) . p(x_1) . \cdots . p(x_n) . Q, \end{array}$$

where we choose $p \notin \mathtt{fn}(P, Q)$ and all x_i are pairwise distinct. We will then have

$$\overline{c}\langle e_1 \cdots e_n \rangle . P \mid c(x_1 \cdots x_n) . Q \xrightarrow{\tau} \cdots \xrightarrow{\tau} P \mid Q\{e_1 \ldots e_n / x_1 \ldots x_n\}.$$

The idea to establish private links in this way has many other uses. Suppose for example that Q wishes to execute P by transmitting on its trigger e, and then also wait until P has completed execution. One way to represent this is to send to P a private name for signalling completion, as in

$$(\boldsymbol{\nu} r)\overline{e}r . r . Q \mid e(x) . P \xrightarrow{\tau} (\boldsymbol{\nu} r)\big(r . Q \mid P\{r/x\}\big).$$

Here Q must wait until someone signals on r before continuing. This someone can only be P since no other is in the scope of r. This scheme is quite general, for example P can delegate to another agent the task to restart Q, by sending r to it as an object in an interaction.

The π-calculus has been used to succinctly describe many aspects of concurrent and functional programming, and also of high-level system description where mobility plays an important role. We shall not attempt an overview of all applications here. In the rest of this paper we concentrate on some central aspects of the theory of the calculus.

3. Variants of the calculus

The calculus can be varied in many ways. There are many useful subcalculi which imply a somewhat simpler theory, and we shall also briefly consider two popular extensions: the polyadic and the higher-order calculi.

3.1. *Match and Mismatch*

The Match and Mismatch operators are absent in many presentations of the calculus. This means a certain loss of expressiveness. But in a sense, equality test of names can be implemented without Match. Consider the following typical use of a test: an agent P receives a name and continues as Q if the name is y and as R if the name is z:

$$P = a(x) . (\mathtt{if}\ x = y\ \mathtt{then}\ Q + \mathtt{if}\ x = z\ \mathtt{then}\ R).$$

Without the Match a similar effect can be achieved exploiting the Parallel operator, where a communication is possible only if the subjects of the input and output actions are the same:

$$P = a(x) . \big(\overline{x} \mid (y . Q + z . R)\big).$$

If the received name is y then, after reception, a communication between the first and second parallel component is possible, after which Q can execute. Similarly, if it is z then a communication between the first and third component enables R. This will work provided no other agent can interfere by interacting on the names y and z. So it is not a general encoding of Match, although many specific instances can be emulated in this way. Consider for example an implementation of Boolean values: There are agents $True_a$ and $False_a$, emitting Boolean values along a, and an agent $Case_a(Q, R)$, receiving a Boolean along a and enacting Q or R depending on the value of the Boolean. A straightforward encoding would use two special names t and f:

$$\begin{aligned} &True_a &&= \overline{a}t, \\ &False_a &&= \overline{a}f, \\ &Case_a(Q, R) &&= a(x) . (\texttt{if } x = t \texttt{ then } Q + \texttt{if } x = f \texttt{ then } R). \end{aligned}$$

So, e.g., $Case_a(Q, R) \mid True_a \xrightarrow{\tau} \texttt{if } t = t \texttt{ then } Q + \texttt{if } t = f \texttt{ then } R$, and this agent will behave as Q. The same effect can be achieved without Match, at the expense of an extra communication. The idea is that *Case* emits two new names and *True* and *False* respond by signalling on one of them:

$$\begin{aligned} &True_a &&= a(xy) . \overline{x}, \\ &False_a &&= a(xy) . \overline{y}, \\ &Case_a(Q, R) &&= (\boldsymbol{\nu} xy)\overline{a}\langle xy\rangle . (x . Q + y . R), \end{aligned}$$

where x and y do not occur in Q or R. It will now hold that $Case_a(Q, R) \mid True_a$ evolves to

$$(\boldsymbol{\nu} xy)\big(\overline{x} \mid (x . Q + y . R)\big)$$

and the *only* possible continuation, in any environment, is a τ to Q, since x and y are local. In this way many instances of the Match construct, including all that aim to capture tests over ordinary data types, can be encoded.

The Mismatch $\texttt{if } x \neq y \texttt{ then}$ cannot be implemented in a similar way. Omitting Mismatch may be thought to drastically reduce the expressive power but it turns out the effect is not very severe. For example, if the possible values of a certain data type are u, v and w, the Mismatch $\texttt{if } x \neq u \texttt{ then } P$ can be replaced by $\texttt{if } x = v \texttt{ then } P + \texttt{if } x = w \texttt{ then } P$. In practice, the data types where equality test is admitted almost always have a predetermined finite range, or can be modelled through a finite set of constructors. Mismatch also makes some portions of the theory a little more complicated. For these reasons many of the versions of the π-calculus do not use it. However, it appears to be necessary for many axiomatizations (cf. Sections 8 and 9).

3.2. *Sum*

The Sum operator $+$ is sometimes regarded as unrealistic from an implementation perspective. The problem is that it represents a form of synchronous global choice. Consider:

$$(a\,.\,P_1 + \overline{b}\,.\,P_2) \mid (\overline{a}\,.\,Q_1 + b\,.\,Q_2).$$

Clearly this agent can evolve to either $P_1 \mid Q_1$ (through a communication along a) or to $P_2 \mid Q_2$ (along b). The choice of which path should be taken involves both parallel components, and is resolved synchronously. If parallel components are distant then this needs a non-trivial protocol, and it can therefore be argued that the general form of Sum is not a realistic primitive operator.

There are two main reasons for including the Sum operator. One is that it admits representations of automata in a straightforward way. Automata have proved useful for high-level descriptions of communicating systems and are present in many modelling languages. The basic idea is that the behaviour of a component is thought of as a directed graph where the nodes represent the reachable states and the edges, labelled by actions, represent the possibilities to move between states. This can be represented in the π-calculus as follows: For each state choose a unique Identifier, and introduce the Definition

$$A \stackrel{\text{def}}{=} \sum_{i=1}^{n} \alpha_i\,.\,A_i$$

for each state A where the outgoing edges are labelled $\alpha_1, \ldots, \alpha_n$ leading to states $A_1, \ldots, A_n$.

The other reason is related to axiomatizations of the behavioural equivalences (see Section 8). All axiomatizations so far use a version of the expansion law, which replaces Parallel by Sum, as in

$$a \mid b = a\,.\,b + b\,.\,a.$$

No complete axiomatization is known for a calculus with Parallel and without Sum.

In situations where high-level modelling is not called for and where complete axiomatization is not an issue, variants of the π-calculus without Sum have been used, for example to describe the semantics of programming languages. Apart from omitting the computationally questionable primitive of synchronous global choice, which would make an implementation of the calculus difficult, it simplifies the theory to only have one binary operator.

Many presentations of the calculus use *guarded Sums* in place of Sum, meaning that the operands of the Sum must be Prefix forms. The general format of a guarded sum is $\sum_{i=1}^{n} \alpha_i\,.\,P_i$ (i.e., it is an n-ary operator). So for example $a\,.\,P + \overline{b}\,.\,Q$ is a guarded Sum, but $a\,.\,P + (Q \mid R)$ is not. With guarded Sum some parts of the theory become simpler, notably the weak bisimulation equivalence is a congruence (see Section 7.4) for the same reason as in ordinary process algebra. The loss of expressiveness is not dramatic:

It should be read as follows: In order to establish that $a(x_1 \dots x_n) . P$ conforms to Δ, find the object sorts $S_1 \dots S_n$ of a according to Δ, and verify that P conforms to Δ where also each x_i is assigned sort S_i. The rule for output is:

$$\frac{ob(\Delta(a)) = \Delta(y_1) \dots \Delta(y_n),\ \ \Delta \vdash P}{\Delta \vdash \overline{a}\tilde{y} . P}.$$

In order to establish that $\overline{a}\langle y_1 \dots y_n \rangle . P$ conforms to Δ it is enough to show that it assigns $y_1 \dots y_n$ the object sorts of a, and that P conforms to Δ. In this way agents such as $\overline{a}\, a . P$, where a name is sent along itself, can also be given a sorting: if S is the sort of a then $ob(S) = \langle S \rangle$.

There is a large variety of type systems which include more information, such as if names can be used for input or output or both, and there are notions of subtypes and polymorphism. Binding occurrences in Restrictions are either explicitly sorted as in $(\boldsymbol{\nu} x : S)P$, or a sort inference system is used to compute a suitable sort. Whether the latter is possible depends on the details of the sort system.

In conclusion, although polyadic interactions do not really increase the expressiveness it adds convenience and clarity when using the calculus, and efficiently implementable sort systems can ascertain that no mismatching arities will ever occur when agents evolve.

3.4. *Recursion and replication*

In the π-calculus as in most other process algebras the mechanism for describing iterative or arbitrarily long behaviour is recursion: a recursive Definition

$$A(\tilde{x}) \stackrel{\text{def}}{=} P$$

where A occurs in P can be thought of as the definition of a recursive procedure A with formal parameters $\tilde{x}$, and the agent $A(\tilde{y})$ is then an invocation with actual parameters $\tilde{y}$. Sometimes this is notated through an explicit fixpoint constructor: if P is an agent then $\text{fix}\, X . P$ is an agent. Here the *agent variable* X may occur in P, and $\text{fix}\, X . P$ means the same as the agent Identifier A with the Definition $A \stackrel{\text{def}}{=} P\{A/X\}$. Fixpoints and Definitions are thus only notational variants. In large specifications the Definitions tend to be more readable, but the fixpoints sometimes allow a more optimal formulation for theory development.

For some purposes the special case of *Replication* is convenient. If P is an agent then $!P$, the Replication of P, is given by the definition

$$!P \stackrel{\text{def}}{=} P \mid !P,$$

or using fixpoints: $!P$ is the agent

$$\text{fix}\, X . (P \mid X).$$

In other words, $!P$ represents an unbounded number of copies of P – the recursion can obviously be unfolded an arbitrary number of times:

$$!P \equiv P \mid !P \equiv P \mid P \mid !P \equiv P \mid P \mid P \mid !P \quad \text{etc.}$$

For example, an agent which can receive inputs along i and forward them on o is:

$$M = !i(x) . \overline{o}x .$$

Suppose this agent receives first u and then v along i. Then, by unfolding the Replication, $M \equiv i(x) . \overline{o}x \mid i(x) . \overline{o}x \mid M$, the agent will evolve to $\overline{o}u \mid \overline{o}v \mid M$, ready to receive more messages along i but also to emit u and v in arbitrary order.

Any agent using a finite family of Definitions can be encoded by Replication as follows. Consider

$$\begin{aligned} A_1(\tilde{x}_1) &\stackrel{\text{def}}{=} P_1, \\ &\vdots \\ A_n(\tilde{x}_n) &\stackrel{\text{def}}{=} P_n, \end{aligned}$$

used in an agent Q. Then there is a corresponding agent Q' which behaves as Q and only uses Replication. Q' is constructed as follows. Introduce names $a_1, \ldots, a_n$, one for each Identifier, and let $\widehat{P}$ be the agent obtained by replacing any invocation $A_i(\tilde{y})$ with an output $\overline{a}_i\langle\tilde{y}\rangle$ in P. Let

$$D = !a_1(\tilde{x}_1) . \widehat{P_1} \mid \cdots \mid !a_n(\tilde{x}_n) . \widehat{P_n}.$$

Thus D is an agent which emulates all the Definitions in the sense that it can interact with any such output. Finally

$$Q' = (\boldsymbol{\nu} a_1 \cdots a_n)\big(D \mid \widehat{Q}\big)$$

will then behave in the same way as Q, with the only difference that an unfolding of a Definition will be emulated by an interaction between $\widehat{Q}$ and D. For example, let $N \stackrel{\text{def}}{=} i(x) . \overline{o}x . N$, which differs from M above in that messages are delivered in order and only one message is stored at a time. The agent with only Replication which behaves as N is

$$(\boldsymbol{\nu} a)\big(!a . i(x) . \overline{o}x . \overline{a} \mid \overline{a}\big).$$

3.5. *The asynchronous calculus*

The π-calculus, as most other process algebras, is based on the paradigm of synchronous communication; an interaction means that one component emits a name at the same time as another component receives it. In contrast, in asynchronous communication there is an unpredictable delay between output and input, during which the message is in transit. This can be modelled by inserting an agent representing an asynchronous communication medium between sender and receiver. The properties of the medium (whether it has a bound on the capacity, whether it preserves the order of messages etc.) is then determined by its definition. For example, an unbounded medium not preserving the order of messages is:

$$M \stackrel{\text{def}}{=} i(x) \,.\, (\overline{o}x \mid M).$$

(The same behaviour is expressed with Replication in Section 3.4.) When M receives u along i it evolves to $\overline{o}u \mid M$, and can at any time deliver u along o, and also continue to accept more messages.

Interestingly, this particular form of asynchrony is also captured by a subcalculus of π in which there is no need to explicitly represent media. The subcalculus consists of the agents satisfying the following requirements:

1. Only **0** can follow an output Prefix.
2. An output Prefix may not occur as an unguarded operand of $+$.

The first requirement disallows agents such as $\overline{a}x \,.\, \overline{b}y$, where an agent other than **0** follows $\overline{a}x$. The second requirement disallows $\overline{a}x + b(y)$, but allows $\tau \,.\, \overline{a}x + b(y)$.

This subcalculus is known as the *asynchronous* π-calculus, and the rationale behind it is as follows. An unguarded output Prefix $\overline{a}x$ occurring in a term represents a message that has been sent but not yet received. The action of sending the message is placing it in an unguarded position, as in the following

$$\tau \,.\, (\overline{a}x \mid P) \xrightarrow{\tau} \overline{a}x \mid P.$$

After this transition $\overline{a}x$ can interact with a receiver and the sender proceeds concurrently as P. Because of requirement 1 the fact that a message has been received cannot be detected by P unless the receiver explicitly sends an acknowledgement. Because of requirement 2 a message cannot disappear unless it is received. Therefore, $\tau \,.\, (\overline{a}x \mid P)$ can be paraphrased as "send $\overline{a}x$ asynchronously and continue as P".

Of course, with a scheme for sending explicit acknowledgements synchronous communication can be emulated, so the loss of expressiveness from the full π-calculus is largely pragmatic. In agents like

$$\tau \,.\, (\boldsymbol{\nu} b)\big(\overline{a}b \mid b(x) \,.\, P\big)$$

the scope of b is used to protect this kind of acknowledgement. Here the agent can do an asynchronous output of $\overline{a}b$, but it is blocked from continuing until it receives an acknowledgement along b:

$$(\boldsymbol{\nu} b)\big(\overline{a}b \mid b(x) \,.\, P\big) \mid a(x) \,.\, Q \xrightarrow{\tau} (\boldsymbol{\nu} b)\big(b(x) \,.\, P \mid Q\{b/x\}\big).$$

The acknowledgement can only arrive from the recipient (Q) of the message, since there is no other agent that can send along the restricted name b.

The asynchronous calculus has been used successfully to model programming languages where the underlying communication discipline is asynchronous. The theory is slightly different since only "asynchronous" observers are deemed relevant. For example, the agent $a(x).\overline{a}x$ is semantically the same as τ since the actions of receiving a datum and then emitting it on the same channel cannot be detected by any other asynchronous process or observer. An interesting consequence is that a general form of Sum, the input-guarded Sum $y_1(x).P_1 + \cdots + y_n(x).P_n$ can be encoded as:

$$\begin{array}{l}(\boldsymbol{\nu}a)\big(\overline{a}t \mid \big(y_1(x).a(z).(\overline{a}f \mid \texttt{if } z=t \texttt{ then } P_1 \texttt{ else } \overline{y}_1 x) \\ \qquad\quad \mid \\ \qquad\quad \vdots \\ \qquad\quad \mid \\ \qquad\quad y_n(x).a(z).(\overline{a}f \mid \texttt{if } z=t \texttt{ then } P_n \texttt{ else } \overline{y}_n x)\big)\big).\end{array}$$

The idea is that a acts as a lock. Initially t is emitted on a; the first component to interact with a will continue to emit f on a, all other components will resend the datum along y_i, thereby undoing the input along y_i. This only works in the asynchronous calculus, where receiving and then emitting the same message is, from an observer's point of view, the same as doing nothing. (An alert reader will complain that `if...then...else` is defined using Sum. This is no great matter, it can be taken as primitive or defined as the Parallel composition of the branches.)

3.6. *The higher-order calculus*

Finally we shall look at an extension of the π-calculus called the *higher-order* π-calculus. Here the objects transmitted in interactions can also be agents. A higher-order output Prefix form is of the kind

$$\overline{a}\langle P\rangle.Q$$

meaning "send the agent P along a and then continue as Q". The higher-order input Prefix form is of the kind

$$a(X).Q$$

meaning "receive an agent for X and continue as Q". Of course Q may here contain X, and the received agent will then replace X in Q. For example, an interaction is

$$\overline{a}\langle\overline{b}u.\mathbf{0}\rangle.b(x) \mid a(X).(X \mid \overline{c}v) \xrightarrow{\tau} b(x) \mid \overline{b}u \mid \overline{c}v.$$

The first component sends $\overline{b}u.\mathbf{0}$ along a, the second component
it in parallel with $\overline{c}v$. The transmitted agent $\overline{b}u$ is then free
of the first component $b(x)$, sending u to it along b
be objects in an interaction, a sorting discipline

sort for “agent”, can be employed to weed out unwanted combinations such as $\overline{a}\langle P\rangle . Q \mid a(x) . x(u)$.

Formally, we introduce a new category of *agent variables*, ranged over by X, and extend the definition of agents to include the agent variables and the higher-order Prefix forms. The notion of replacing an agent variable by an agent, $P\{Q/X\}$, is as expected (and the same as in the fixpoint construct of Section 3.4), and the higher-order interaction rule gives that

$$a(X) . P \mid \overline{a}\langle Q\rangle . R \xrightarrow{\tau} P\{Q/X\} \mid R.$$

The substitution $P\{Q/X\}$ is defined with alpha-conversion such that free names in Q do not become bound in $P\{Q/X\}$. To see the need for this consider

$$a(X) . b(c) . X \mid \overline{a}\big\langle c(z) . \mathbf{0}\big\rangle . R.$$

Clearly the name c in the right hand component is not the same as the bound c to the left. The agent is alpha-equivalent to

$$a(X) . b(d) . X \mid \overline{a}\big\langle c(z) . \mathbf{0}\big\rangle . R,$$

where these names are dissociated. Therefore a transition to $b(c) . c(z) \mid R$ is not possible. In this sense the binding corresponds to *static* binding: the scope of a name is determined by where it occurs in the agent. An alternative scheme of *dynamic* binding, where the scope is determined only when the name is actually used, is possible but much more complicated semantically. For example, such bindings cannot use alpha-conversion.

Note that an a ariable may occur more than once in an agent. For example,

$$\overline{a}\langle Q\rangle . R \xrightarrow{\tau} Q \mid Q \mid R,$$

wh has been “duplicated” by the interaction. This lends the
e expressive power. Recursion and Replication are now
a-calculus where the fixpoint combinator **Y** is express-

agent and also retransmit it along a. Now

ber of P:

$$\mid R_P \xrightarrow{\tau} \cdots .$$

ler constructs (such as
between sites) the

higher-order calculus is suitable. However, its theory is considerably more complicated. In some situations it will then help to encode it into the usual calculus with the device from Section 2.3. Instead of transmitting an agent P we transmit a new name which can be used to trigger P. Since the receiver of P might invoke P several times (because the corresponding agent variable occurs at several places) P must be replicated. The main idea of the encoding $[\![\bullet]\!]$ from higher-order to the ordinary calculus is as follows, where we assume a previously unused name x for each agent variable X:

$$\begin{aligned} [\![(\overline{a}\langle P\rangle \,.\, Q)]\!] &= (\boldsymbol{\nu} p)\overline{a}p \,.\, ([\![Q]\!] \mid !p \,.\, [\![P]\!]), \quad \text{where } p \notin \mathtt{fn}(P, Q), \\ [\![a(X) \,.\, P]\!] &= a(x) \,.\, [\![P]\!], \\ [\![X]\!] &= \overline{x}. \end{aligned}$$

Consider the example above, $a(X) \,.\, (X \mid X) \mid \overline{a}\langle Q\rangle \,.\, R \xrightarrow{\tau} Q \mid Q \mid R$, where Q and R in turn contain no higher-order Prefixes. Using the encoding, and assuming q is not free in Q or R, we get a similar behaviour:

$$\begin{aligned} & a(x) \,.\, (\overline{x} \mid \overline{x}) \mid (\boldsymbol{\nu} q)\overline{a}q \,.\, (R \mid !q \,.\, Q) \\ \xrightarrow{\tau}\ & (\boldsymbol{\nu} q)(\overline{q} \mid \overline{q} \mid R \mid !q \,.\, Q) \\ \xrightarrow{\tau}\xrightarrow{\tau}\ & (\boldsymbol{\nu} q)(\mathbf{0} \mid \mathbf{0} \mid R \mid Q \mid Q \mid !q \,.\, Q) \\ \equiv\ & R \mid Q \mid Q \mid (\boldsymbol{\nu} q)!q \,.\, Q, \end{aligned}$$

where the rightmost component $(\boldsymbol{\nu} q)!q \,.\, Q$ will never be able to execute because it is guarded by a private name, so the whole term will behave as $Q \mid Q \mid R$ as expected.

4. Operational semantics

The standard way to give an operational semantics to a process algebra is through a labelled transition system, where transitions are of kind $P \xrightarrow{\alpha} Q$ for some set of actions ranged over by α. The π-calculus follows this norm and most of the rules of transitions are similar to other algebras. As expected, for an agent $\alpha \,.\, P$ there will be a transition labelled α leading to P. Also as expected, the Restriction operator will not permit an action with the restricted name as subject, so $(\boldsymbol{\nu} a)\overline{a}u \,.\, P$ has no transitions and is therefore semantically equivalent to $\mathbf{0}$. But what action should

$$(\boldsymbol{\nu} u)\overline{a}u \,.\, P$$

have? Clearly it must have *some* action; inserted in a context $a(x) \,.\, Q \mid (\boldsymbol{\nu} u)\overline{a}u \,.\, P$ it will enable an interaction since, assuming $u \notin \mathtt{fn}(Q)$, this term is structurally congruent to $(\boldsymbol{\nu} u)(a(x) \,.\, Q \mid \overline{a}u \,.\, P)$ and there is an interaction between the components. So $(\boldsymbol{\nu} u)\overline{a}u \,.\, P$ is not, intuitively, something that behaves as $\mathbf{0}$. On the other hand it is clearly distinct from $\overline{a}u \,.\, P$ and it can therefore not be given the action $\overline{a}u$. For example,

$$\big(a(x) \,.\, \mathtt{if}\ x = u\ \mathtt{then}\ Q\big) \mid \overline{a}u \xrightarrow{\tau} \mathtt{if}\ u = u\ \mathtt{then}\ Q$$

which can continue as Q, while

$$\begin{aligned}&\big(a(x)\,.\,\mathtt{if}\ x=u\ \mathtt{then}\ Q\big)\mid(\boldsymbol{\nu}u)\overline{a}u\equiv\\&(\boldsymbol{\nu}v)\big(\big(a(x)\,.\,\mathtt{if}\ x=u\ \mathtt{then}\ Q\big)\mid\overline{a}v\big)\xrightarrow{\tau}\\&(\boldsymbol{\nu}v)\mathtt{if}\ v=u\ \mathtt{then}\ Q\end{aligned}$$

and there are no further actions (remember that scope extension requires that u is alpha-converted before the scope is extended).

The solution is to give $(\boldsymbol{\nu}u)\overline{a}u$ a new kind of action called *bound output* written $\overline{a}\boldsymbol{\nu}u$. The intuition is that a local name represented by u is transmitted along a, extending the scope of u to the recipient. In summary, the *actions* ranged over by α consists of four classes:

1. The internal action τ.
2. The (free) output actions of kind $\overline{a}x$.
3. The input actions of kind $a(x)$.
4. The bound output actions of kind $\overline{a}\boldsymbol{\nu}x$.

The three first kinds correspond precisely to the Prefixes in the calculus. For the sake of symmetry we introduce a fourth kind of Prefix $\overline{a}\boldsymbol{\nu}x$, for $a \neq x$, corresponding to the bound output action. The bound output Prefix is merely a combination of Restriction and output as defined by $\overline{a}\boldsymbol{\nu}x\,.\,P = (\boldsymbol{\nu}x)\overline{a}x\,.\,P$. In this way we can continue to let α range over both actions and Prefixes, where $\mathtt{fn}(\overline{a}\boldsymbol{\nu}x) = \{a\}$ and $\mathtt{bn}(\overline{a}\boldsymbol{\nu}x) = \{x\}$.

An input transition $P \xrightarrow{a(x)} Q$ means that P can receive some name u along a, and then evolve to $Q\{u/x\}$. In that action x does not represent the value received, rather it is a reference to the places in Q where the received name will appear. When examining further transitions from Q, all possible instantiations for this x must be considered. Those familiar with functional programming and the lambda-calculus might think of the transition as $P \xrightarrow{a} \lambda x\, Q$, making it clear that the derivative, after the arrow, has a functional parameter x. (The reader may at this point wonder about an alternative way to treat input by including the received name in the action, as in $a(x)\,.\,P \xrightarrow{au} P\{u/x\}$. This is a viable alternative and will be discussed in Section 5.3.)

Similarly, the bound output transition $P \xrightarrow{\overline{a}\nu x} Q$ signifies an output of a local name and x indicates where in Q this name occurs. Here x is not a functional parameter, it just represents something that is distinct from all names in the environment.

The labelled transition semantics is given in Table 3. The rule STRUCT makes explicit our intuition that structurally congruent agents count as the same for the purposes of the semantics. This simplifies the system of transition rules. For example, SUM is sufficient as it stands in Table 3. The dual rule

$$\mathrm{SUM}_2\ \frac{Q \xrightarrow{\alpha} Q'}{P + Q \xrightarrow{\alpha} Q'}$$

is redundant since it can be inferred from SUM and STRUCT. Similar arguments hold for the rules PAR and COM.

In the rule PAR, note the extra condition that Q does not contain a name bound in α. This conforms to the intuition that bound names are just references to occurrences; in the

Table 3
The operational semantics

STRUCT	$\dfrac{P' \equiv P,\ P \xrightarrow{\alpha} Q,\ Q \equiv Q'}{P' \xrightarrow{\alpha} Q'}$
PREFIX	$\dfrac{}{\alpha\,.\,P \xrightarrow{\alpha} P}$
SUM	$\dfrac{P \xrightarrow{\alpha} P'}{P + Q \xrightarrow{\alpha} P'}$
MATCH	$\dfrac{P \xrightarrow{\alpha} P'}{\mathtt{if}\ x = x\ \mathtt{then}\ P \xrightarrow{\alpha} P'}$
MISMATCH	$\dfrac{P \xrightarrow{\alpha} P',\ x \neq y}{\mathtt{if}\ x \neq y\ \mathtt{then}\ P \xrightarrow{\alpha} P'}$
PAR	$\dfrac{P \xrightarrow{\alpha} P',\ \mathtt{bn}(\alpha) \cap \mathtt{fn}(Q) = \emptyset}{P \mid Q \xrightarrow{\alpha} P' \mid Q}$
COM	$\dfrac{P \xrightarrow{a(x)} P',\ Q \xrightarrow{\overline{a}u} Q'}{P \mid Q \xrightarrow{\tau} P'\{u/x\} \mid Q'}$
RES	$\dfrac{P \xrightarrow{\alpha} P',\ x \notin \alpha}{(\boldsymbol{\nu}x)P \xrightarrow{\alpha} (\boldsymbol{\nu}x)P'}$
OPEN	$\dfrac{P \xrightarrow{\overline{a}x} P',\ a \neq x}{(\boldsymbol{\nu}x)P \xrightarrow{\overline{a}\boldsymbol{\nu}x} P'}$

conclusion $P \mid Q \xrightarrow{\alpha} P' \mid Q$ the action should not refer to any occurrence in Q. To see the need for the condition consider the inference

$$\text{PAR}\ \frac{a(x)\,.\,P \xrightarrow{a(x)} P}{(a(x)\,.\,P) \mid Q \xrightarrow{a(x)} P \mid Q}.$$

Combined with an output $\overline{a}u\,.\,R \xrightarrow{\overline{a}u} R$ we get

$$\text{COM}\ \frac{(a(x)\,.\,P) \mid Q \xrightarrow{a(x)} P \mid Q,\ \overline{a}u\,.\,R \xrightarrow{\overline{a}u} R}{((a(x)\,.\,P) \mid Q) \mid \overline{a}u\,.\,R \xrightarrow{\tau} (P \mid Q)\{u/x\} \mid R}.$$

This is clearly only correct if $x \notin \mathtt{fn}(Q)$, otherwise the free x in Q would be affected by the substitution. If $x \in \mathtt{fn}(Q)$ then a similar derivation is possible after first alpha-converting x in $a(x)\,.\,P$ to some name not free in Q.

The rule OPEN is the rule generating bound outputs. It is interesting to note that bound output actions do not appear in the COM rule and therefore cannot interact directly with inputs. Such interactions are instead inferred by using structural congruence to pull the Restriction outside both interacting agents (possibly after an alpha-conversion), as in the following where we assume that $u \notin \mathtt{fn}(P)$:

$$\text{STRUCT}\,\frac{\text{RES}\,\dfrac{a(x)\,.\,P \mid \overline{a}u\,.\,Q \xrightarrow{\tau} P\{u/x\} \mid Q}{(\boldsymbol{\nu}u)(a(x)\,.\,P \mid \overline{a}u\,.\,Q) \xrightarrow{\tau} (\boldsymbol{\nu}u)(P\{u/x\} \mid Q)}}{a(x)\,.\,P \mid (\boldsymbol{\nu}u)\overline{a}u\,.\,Q \xrightarrow{\tau} (\boldsymbol{\nu}u)(P\{u/x\} \mid Q)}.$$

In view of this it might be argued that bound output transitions and the rule OPEN can be omitted altogether, since they have no impact on inferring interactions. Although technically this argument is valid there are other reasons for including the bound output. One is of philosophical nature: we think of the agent $(\boldsymbol{\nu}u)\overline{a}u$ as, intuitively, being able to do something, namely exporting a local name, and if we do not dignify that with a transition we introduce an incompleteness in the semantics in that not all behaviour is manifested by transitions. Another reason is of technical convenience: when it comes to developing behavioural equivalences (see Section 6) the bound output transitions will turn out to be indispensable.

5. Variants of the semantics

We will here consider alternative operational semantics. Some are mere presentational variants. But different semantics have different strengths, for example one may be suitable for an automatic tool and another may boost intuition, so there is a point to the diversity.

5.1. *The role of structural congruence*

Through the rule STRUCT we can regard an inference of $P \equiv Q$ as a step in an inference of a transition. If the *only* purpose of the congruence is to facilitate inference of labelled transitions then there is a possible trade-off between what to include in the structural congruence and what to include in the transition rules. As we saw the dual SUM_2 of SUM is unnecessary, but an alternative solution is to omit commutativity from the structural congruence and introduce SUM_2. A similar choice occurs for the Parallel combinator: structural commutativity can be omitted at the price of introducing the duals of PAR and COM.

For some operators there is even a choice between defining it entirely through the congruence or through the transition rules. For example, the Match operator can either be defined with the rule MATCH or with an additional structural rule

$$\mathtt{if}\ x = x\ \mathtt{then}\ P \equiv P.$$

In Table 3 we chose to give it with a transition rule because of the symmetry with MISMATCH. Note that Mismatch cannot be defined through the perhaps expected structural rule

$$\mathtt{if}\ x \neq y\ \mathtt{then}\ P \equiv P \quad \text{if } x \neq y$$

since that rule would be obviously unsound for a Mismatch under an input Prefix:

$$a(x)\,.\,\mathtt{if}\ x \neq y\ \mathtt{then}\ P \quad \text{and} \quad a(x)\,.\,P$$

represent two different behaviours!

Another such choice occurs for Identifiers. They are here defined through a structural congruence rule $A(\tilde{y}) \equiv P\{\tilde{y}/\tilde{x}\}$ if $A(\tilde{x}) \stackrel{\text{def}}{=} P$. Alternatively we could have given the transition rule:

$$\text{IDE}\ \frac{P\{\tilde{y}/\tilde{x}\} \xrightarrow{\alpha} P',\ A(\tilde{x}) \stackrel{\text{def}}{=} P}{A(\tilde{y}) \xrightarrow{\alpha} P'}.$$

Perhaps the most dramatic part of the structural congruence is the scope extension which, together with COM, make possible transitions like

$$a(x)\,.\,P \mid (\nu u)\overline{a}u\,.\,Q \xrightarrow{\tau} (\nu u)\big(P\{u/x\} \mid Q\big)$$

(provided $u \notin \mathtt{fn}(P)$). Interestingly, there is a way to achieve the same effect without the scope extension. It involves a new transition rule CLOSE which represents an interaction between a bound output and an input:

$$\text{CLOSE}\ \frac{P \xrightarrow{a(u)} P',\ Q \xrightarrow{\overline{a}\nu u} Q'}{P \mid Q \xrightarrow{\tau} (\nu u)(P' \mid Q')}.$$

Let us see how the transition from $a(x)\,.\,P \mid (\nu u)\overline{a}u\,.\,Q$ mentioned above is derived. It involves an alpha-conversion of x to u in the first operand:

$$\text{STRUCT}\ \frac{a(x)\,.\,P \equiv a(u)\,.\,P\{u/x\},\ a(u)\,.\,P\{u/x\} \xrightarrow{a(u)} P\{u/x\}}{a(x)\,.\,P \xrightarrow{a(u)} P\{u/x\}}.$$

OPEN in the second component gives:

$$\text{OPEN}\ \frac{\overline{a}u\,.\,Q \xrightarrow{\overline{a}u} Q}{(\nu u)\overline{a}u\,.\,Q \xrightarrow{\overline{a}\nu u} Q}$$

and combining the two conclusions we get:

$$\text{CLOSE}\ \frac{a(x)\,.\,P \xrightarrow{a(u)} P\{u/x\},\ (\nu u)\overline{a}u\,.\,Q \xrightarrow{\overline{a}\nu u} Q}{a(x)\,.\,P \mid (\nu u)\overline{a}u\,.\,Q \xrightarrow{\tau} (\nu u)(P\{u/x\} \mid Q)}.$$

Note how the scoping represented by ν moves from the second operand onto the transition arrow and reappears in the term in the conclusion!

In this way it is possible to remove all of the structural congruence except alpha-conversion, at the expense of introducing more transition rules. In fact, it is even possible to remove the alpha-conversion by building it into a special rule, or into the rules that generate bound actions. For example, the rule for input would become

$$\frac{}{a(x)\,.\,P \xrightarrow{a(u)} P\{u/x\}}$$

for any $u \notin \mathtt{fn}(P)$, and similarly the rule OPEN would need a modification.

Therefore, the choice whether to use a structural congruence at all and, if so, how many rules it should contain is largely one of convenience. Historically, the first presentations of the calculus did not use structural congruence and an advantage is then that some proofs by induction over inference of transition become clearer. Today most presentations use some form of structural congruence, at least including alpha-conversion, since the definitions become more compact and transparent. For example, the scope extension in structural congruence is easier to understand than the effect of the CLOSE and OPEN rules.

5.2. *Symbolic transitions*

In the transitional semantics an input action $P \xrightarrow{a(x)} Q$ represents that anything can be received along a. Consequently, if we want to explore further transitions from Q we must take into account all possible substitutions for x. Since there are infinitely many names it appears we have to explore the behaviour of $Q\{u/x\}$ for infinitely many names u. In general, when a number of such inputs have been performed, we may be interested in the behaviour not merely of an agent P but of $P\sigma$ for substitutions σ involving several names. This kind of infinite branching is awkward for proof systems and prohibits efficient tool support for automated analysis.

Therefore, an alternative way to present the semantics is to let a transition from P contain information about the behaviour not only of P itself but also of $P\sigma$ for any σ. The key observation is that even though there are infinitely many names a syntactically finite agent can only subject them to a bounded number of tests, and the conditions which enable a transition can be recorded. For example, consider

$$P = \bar{a}u\,.\,Q \mid b(x)\,.\,R.$$

We can schematically express the transitions of $P\sigma$ succinctly as follows. For any σ the agent $P\sigma$ will have an input and an output transition (with subject $\sigma(a)$ and $\sigma(b)$ respectively). In addition, if $\sigma(a) = \sigma(b)$ it will have a τ-transition arising from a communication between the components. This is the idea behind the so called *symbolic* semantics, where these transitions come out as:

$$P \xrightarrow{\mathbf{true},\bar{a}u} Q \mid b(x)\,.\,R,$$

$$P \xrightarrow{\mathbf{true},b(x)} \bar{a}u\,.\,Q \mid R,$$

$$P \xrightarrow{a=b,\tau} Q \mid R\{u/x\}.$$

The label on a transition here has two components. The first is a condition which must hold for the transition to be enabled, and the second is as usual the action of the transition.

In the π-calculus the conditions have a particularly simple form. The only tests an agent can perform on names are for equality (arising from the Parallel combinator, as above, and from Match) and inequality (arising from Mismatch). Of course several of these tests may affect a transition, so a condition will in general be a conjunction of equalities and inequalities on names. For example, we will have that

$$\texttt{if}\ x = y\ \texttt{then}\ \overline{a}u\,.\,Q \mid \texttt{if}\ x \neq b\ \texttt{then}\ b(y)\,.\,R \xrightarrow{x=y\wedge x\neq b\wedge a=b,\tau} Q \mid R\{u/y\}.$$

Formally we let M, N range over conjunctions of name equalities and inequalities, including the empty conjunction which is written **true**. Symbolic transitions are of the form $P \xrightarrow{M,\alpha} Q$, meaning "if M holds then P has an action α leading to Q". We let μ range over pairs (M, α), so a symbolic transition is written $P \xrightarrow{\mu} Q$. For $\mu = (M, \alpha)$ we let $\mathrm{bn}(\mu)$ mean $\mathrm{bn}(\alpha)$ and we let $\mu \wedge N$ mean $(M \wedge N, \alpha)$. The rules for symbolic transitions are given in Table 4 and are similar in structure to the rules in Section 4 for ordinary transitions. The only subtle point is the effect of Restriction on a condition in the rules S-RES and S-OPEN. Suppose P has a symbolic transition

$$P \xrightarrow{M,\alpha} Q$$

with a condition M that mentions x in a conjunct $x \neq y$. This means that the transition from P only holds for substitutions assigning x and y different names, perhaps because P is of the form $\texttt{if}\ x \neq y\ \texttt{then}\ \ldots$. What is then the condition M' of the inferred symbolic transition

$$(\boldsymbol{\nu}x)P \xrightarrow{M',\alpha'} Q'.$$

To see what it should be we must ask how substitutions involving x can enable that transition. The requirement from M is that x should not be assigned the same name as y. But since substitutions only affect free names, no substitution will assign y the same name as the bound x in $(\boldsymbol{\nu}x)P$. Therefore, the inferred transition from $(\boldsymbol{\nu}x)P$ will be possible no matter how x is treated by the substitution. Hence the conjunct $x \neq y$ should not be present in M'.

Suppose instead that M says that x is the same as some other name y, i.e., M contains a conjunct $x = y$, meaning "this transition only holds for substitutions assigning x and y the same name". Then the inferred transition from $(\boldsymbol{\nu}x)P$ will hold for *no* substitution, for the same reason. In that case M' can be chosen to be any unsatisfiable condition, for example $z \neq z$. The only other types of conjuncts in M mentioning x are the trivial $x = x$, which is always true and can be removed at will, and the contradictory $x \neq x$, which implies that the condition is unsatisfiable.

In conclusion, the condition M' can be defined as M where conjuncts $x = x$ and $x \neq y$ are removed, and conjuncts $x = y$ are replaced with an unsatisfiable conjunct $z \neq z$, for all y different from x. We denote this by "$M - x$" in Table 4. For example $(x \neq y \wedge z = u) - x$ is $(z = u)$, and $(x = y \wedge z = u) - x$ is $z \neq z$ (it is not satisfiable).

Table 4
The symbolic transition semantics

S-STRUCT	$\dfrac{P' \equiv P,\ P \xrightarrow{\mu} Q,\ Q \equiv Q'}{P' \xrightarrow{\mu} Q'}$
S-PREFIX	$\dfrac{}{\alpha\,.\,P \xrightarrow{\mathbf{true},\alpha} P}$
S-SUM	$\dfrac{P \xrightarrow{\mu} P'}{P + Q \xrightarrow{\mu} P'}$
S-MATCH	$\dfrac{P \xrightarrow{\mu} P',\ x, y \notin \mathrm{bn}(\mu)}{\mathtt{if}\ x = y\ \mathtt{then}\ P \xrightarrow{\mu \wedge x=y} P'}$
S-MISMATCH	$\dfrac{P \xrightarrow{\mu} P',\ x, y \notin \mathrm{bn}(\mu)}{\mathtt{if}\ x \neq y\ \mathtt{then}\ P \xrightarrow{\mu \wedge x\neq y} P'}$
S-PAR	$\dfrac{P \xrightarrow{\mu} P', \mathrm{bn}(\mu) \cap \mathtt{fn}(Q) = \emptyset}{P \mid Q \xrightarrow{\mu} P' \mid Q}$
S-COM	$\dfrac{P \xrightarrow{M,a(x)} P',\ Q \xrightarrow{N,\bar{b}u} Q'}{P \mid Q \xrightarrow{M \wedge N \wedge a=b,\tau} P'\{u/x\} \mid Q'}$
S-RES	$\dfrac{P \xrightarrow{M,\alpha} P',\ x \notin \alpha}{(\boldsymbol{\nu} x)P \xrightarrow{M-x,\alpha} (\boldsymbol{\nu} x)P'}$
S-OPEN	$\dfrac{P \xrightarrow{M,\bar{a}x} P',\ a \neq x}{(\boldsymbol{\nu} x)P \xrightarrow{M-x,\bar{a}\boldsymbol{\nu} x} P'}$

As an example inference consider the symbolic transition from

$$(\boldsymbol{\nu} x)\mathtt{if}\ x \neq y\ \mathtt{then}\ \alpha\,.\,P.$$

First by S-PREFIX

$$\alpha\,.\,P \xrightarrow{\mathbf{true},\alpha} P$$

and then by S-MISMATCH

$$\mathtt{if}\ x \neq y\ \mathtt{then}\ \alpha\,.\,P \xrightarrow{x\neq y,\alpha} P$$

and finally by S-RES, since $(x \neq y) - x = \mathbf{true}$ and assuming $x \notin \alpha$:

$$(\boldsymbol{\nu} x)\mathtt{if}\ x \neq y\ \mathtt{then}\ \alpha\,.\,P \xrightarrow{\mathbf{true},\alpha} (\boldsymbol{\nu} x)P.$$

In a similar way we get that

$$(\nu x)\mathtt{if}\ x = y\ \mathtt{then}\ \alpha\,.\,P \xrightarrow{z\neq z,\alpha} (\nu x)P.$$

In other words, the first transition holds for all substitutions and the second for no substitution, i.e., is never possible. This is as it should since no substitution can make x and y equal in an agent where x is bound.

We can now state how the symbolic semantics corresponds to the transitions in Section 4. Write $\sigma \models M$ to denote that for any conjunct $a = b$ in M it holds that $\sigma(a) = \sigma(b)$ and for any conjunct $a \neq b$ it holds that $\sigma(a) \neq \sigma(b)$. The correspondence between the semantics is that $P \xrightarrow{M,\alpha} P'$ implies that for all σ such that $\sigma \models M$ there is a transition $P\sigma \xrightarrow{\alpha\sigma} P'\sigma$. Conversely, if $P\sigma \xrightarrow{\beta} P''$ then for some M, α, P' it holds $P \xrightarrow{M,\alpha} P'$ where $\sigma \models M$ and $P'' \equiv P'\sigma$ and $\beta = \alpha\sigma$. The proof is through induction over the inference systems for transitions. Therefore, a tool or proof system is justified in using the symbolic semantics.

5.3. *The early semantics*

As mentioned in Section 4 there is an alternative way to treat the semantics for input, the so called *early* semantics, by giving the input Prefix the rule for all u:

$$\text{E-INPUT}\ \frac{}{a(x)\,.\,P \xrightarrow{au} P\{u/x\}}.$$

Here there is a fifth kind of action, the *free input* action of kind au, meaning "input the name u along a". Note the difference between

$P \xrightarrow{a(x)} Q$ meaning P inputs something to replace x in Q, and
$P \xrightarrow{ax} Q$ meaning P receives the name x and continues as Q.

In the early semantics the COM rule needs a modification, since the input action in the premise refers to the transmitted name:

$$\text{E-COM}\ \frac{P \xrightarrow{au} P',\ Q \xrightarrow{\bar{a}u} Q'}{P \mid Q \xrightarrow{\tau} P' \mid Q'}.$$

There is no substitution in the conclusion since that has already been performed at an earlier point in the inference of this transition, namely when the input action was inferred. Hence the name "early" semantics. Consequently the semantics in Section 4 is called the *late* semantics: the substitution emanating from an interaction is inferred as late as possible, namely when inferring the interaction in the COM-rule. In the early semantics there is no need for the bound input actions, so there will still only be four kinds of action. However, if scope extension in the structural congruence is omitted in favour of the CLOSE rule as described in Section 5.1 then the early semantics also needs a side condition $u \notin \mathtt{fn}(P)$ in that rule.

To see the similarity and difference between late and early, consider the inference of an interaction between $a(x)\,.\,P$ and $\overline{a}u\,.\,Q$. Both semantics yield the same τ-transition. First the late semantics:

$$\text{COM}\ \frac{\text{PREFIX}\ \dfrac{}{a(x)\,.\,P \xrightarrow{a(x)} P} \qquad \text{PREFIX}\ \dfrac{}{\overline{a}u\,.\,Q \xrightarrow{\overline{a}u} Q}}{a(x)\,.\,P \mid \overline{a}u\,.\,Q \xrightarrow{\tau} P\{u/x\} \mid Q}.$$

Then the early semantics:

$$\text{E-COM}\ \frac{\text{E-INPUT}\ \dfrac{}{a(x)\,.\,P \xrightarrow{au} P\{u/x\}} \qquad \text{PREFIX}\ \dfrac{}{\overline{a}u\,.\,Q \xrightarrow{\overline{a}u} Q}}{a(x)\,.\,P \mid \overline{a}u\,.\,Q \xrightarrow{\tau} P\{u/x\} \mid Q}.$$

The τ-transitions that can be inferred with the early semantics are exactly those that can be inferred with the late semantics. The proof of this is by induction over the depth of inference of the transition. The induction hypothesis needs to state that not only τ actions but also input and output actions correspond in the two semantics, where the correspondence between input and free input is that $P \xrightarrow{au} P'$ iff $\exists P'', w : P \xrightarrow{x(w)} P'' \wedge P' \equiv P''\{u/w\}$.

In view of this it is a matter of taste which semantics to adopt. It could be argued that the early semantics more closely follows an operational intuition since, after all, an agent performs an input action only when it actually receives a particular value. On the other hand, experimental evidence indicates that proof systems and decision procedures using the late semantics are slightly more efficient.

5.4. *Reductions*

Another idea for the operational semantics is to represent interactions using an unlabelled transition system. The idea is that $P \rightarrow Q$, pronounced "P reduces to Q", is the same as $P \xrightarrow{\tau} Q$. The difference is that reductions are inferred directly from the syntax of the agent, as opposed to τ-transitions which are inferred from input and output transitions. This is accomplished by a counterpart to the COM rule which explicitly mentions the Prefixes that give rise to the reduction:

$$\text{R-COM}\ \frac{}{(\cdots + a(x)\,.\,P) \mid (\cdots + \overline{a}u\,.\,Q) \rightarrow P\{u/x\} \mid Q}.$$

In addition, the rules PAR and RES need counterparts for reductions; they are quite simple since there is no label on the transition:

$$\text{R-PAR}\ \frac{P \rightarrow P'}{P \mid Q \rightarrow P' \mid Q} \qquad \text{R-RES}\ \frac{P \rightarrow P'}{(\boldsymbol{\nu}x)P \rightarrow (\boldsymbol{\nu}x)P'}$$

and of course the structural rule is as before:

$$\text{R-STRUCT}\ \frac{P' \equiv P,\ P \rightarrow Q,\ Q \equiv Q'}{P' \rightarrow Q'}.$$

That's it! Those four rules suffice for a subcalculus, formed by the agents satisfying:

1. All Sums are guarded Sums, i.e., of kind $\alpha_1 . P_1 + \cdots + \alpha_n . P_n$.
2. There are no unguarded `if`-operators.
3. There are no τ-Prefixes.

These restrictions are not very severe. In practice Sums are almost always guarded (this excludes agents like $(P \mid Q) + R$ which are seldom used). The second condition is easy to satisfy since any `if`-operator not under an input Prefix can be evaluated; the agent `if` $x = y$ `then` P can be replaced by P if $x = y$ and by $\mathbf{0}$ otherwise, and conversely for `if` $x \neq y$ `then` P. The importance of the two first conditions is that structural congruence can always be used to pull the two active Prefixes into a position where R-COM is possible. The third condition is also not dramatic. The silent Prefix is derivable in the sense that $(\boldsymbol{\nu} a)(a . P \mid \overline{a})$ behaves exactly like $\tau . P$.

The fact that $\rightarrow$ corresponds to $\xrightarrow{\tau}$ in the ordinary semantics is then easily proved by induction on inference of $\rightarrow$. Each such inference only uses R-COM once and R-STRUCT a number of times.

Clearly, for ease of presentation the reduction semantics is superior to the labelled transition semantics, and is therefore often taken as a point of departure when describing the π-calculus. It is then important to remember that the price for the simplicity is loss of information: the reduction semantics records only the completed interactions within an agent, and ignores the potential that agent shows for interacting with an environment. For example $\overline{a}u . \mathbf{0}$ and $\overline{b}u . \mathbf{0}$ and $(\boldsymbol{\nu} u)\overline{a}u . \mathbf{0}$ all have no reductions and are therefore in some sense semantically identical, even though they have different "potential" interactions.

For this reason proof systems and decision procedures based directly on the reduction semantics are not optimal. However, one great advantage is that it works equally well for the higher-order calculus (Section 3.6), where a labelled semantics is complicated with agents decorating the arrows.

5.5. *Abstractions and concretions*

The input transition $P \xrightarrow{a(x)} Q$ can intuitively be thought of as $P \xrightarrow{a} \lambda x\, Q$ where $\lambda x\, Q$ is a function from names to agents. This idea can be elaborated to give an alternative presentation of the transition system. It requires a few more definitions but the rules become simpler.

Define *agent abstractions*, ranged over by F, to be of kind $\lambda x\, P$, and dually *agent concretions*, ranged over by C, to be of kind $[x]P$. The latter is just a notation for the pair (x, P) and allows the output transition $P \xrightarrow{\overline{a}x} Q$ to be written $P \xrightarrow{\overline{a}} [x]Q$. We write an input Prefix $a(x) . P$ as $a . F$, where $F = \lambda x\, P$, and similarly an output Prefix $\overline{a}x . P$ as $\overline{a} . C$, with $C = [x]P$. Let agents, abstractions and concretions collectively be called the *extended agents*, ranged over by E, and let c range over names x, overlined names $\overline{x}$, and τ. A transition will then be written $P \xrightarrow{c} E$.

We further allow Restrictions $(\boldsymbol{\nu} x)$ to operate on abstractions and concretions with the additional scope extension laws

$$(\boldsymbol{\nu} y)\lambda x\, P \equiv \lambda x(\boldsymbol{\nu} y)P \quad \text{and} \quad (\boldsymbol{\nu} y)[x]P \equiv [x](\boldsymbol{\nu} y)P$$

Table 5
Transitions for extended agents

X-STRUCT	$\dfrac{P' \equiv P,\ P \xrightarrow{c} E,\ E \equiv E'}{P' \xrightarrow{c} Q'}$
X-PREFIX	$\dfrac{}{c\,.\,E \xrightarrow{c} E}$
X-SUM	$\dfrac{P \xrightarrow{c} E}{P + Q \xrightarrow{c} E}$
X-MATCH	$\dfrac{P \xrightarrow{c} E}{\mathtt{if}\ x = x\ \mathtt{then}\ P \xrightarrow{c} E}$
X-MISMATCH	$\dfrac{P \xrightarrow{c} E,\ x \neq y}{\mathtt{if}\ x \neq y\ \mathtt{then}\ P \xrightarrow{c} E}$
X-PAR	$\dfrac{P \xrightarrow{c} E}{P \mid Q \xrightarrow{c} E \mid Q}$
X-COM	$\dfrac{P \xrightarrow{a} \lambda x P',\ Q \xrightarrow{\overline{a}} [u]Q'}{P \mid Q \xrightarrow{\tau} P'\{u/x\} \mid Q'}$
X-RES	$\dfrac{P \xrightarrow{c} E,\ x \neq c,\ \overline{x} \neq c}{(\boldsymbol{\nu}x)P \xrightarrow{c} (\boldsymbol{\nu}x)E}$

provided $x \neq y$; we also allow alpha-conversion in the usual sense, where λx (but not $[x]$) counts as a binding occurrence of x. And we define Parallel composition for extended agents by

$$\begin{aligned} P \mid \lambda x Q &\equiv \lambda x(P \mid Q), \\ P \mid [u]Q &\equiv [u](P \mid Q), \\ P \mid (\boldsymbol{\nu}x)[x]Q &\equiv (\boldsymbol{\nu}x)[x](P \mid Q), \end{aligned}$$

where $x \notin \mathtt{fn}(P)$. With these conventions, abstractions are structurally congruent to forms of kind $\lambda x P$, and concretions are congruent to one of the two forms $[x]P$ or $(\boldsymbol{\nu}x)[x]P$, where P is an agent.

The transition rules for extended agents are given in Table 5. It is interesting to compare with the standard semantics in Table 3. The new rules are simpler mainly in two ways. First, since transitions carry no objects, X-PAR does not need to deal with a side condition on bound objects. Second, the rule X-RES fulfills the function of both RES and OPEN. The price for this simplification is the added complexity of the structural congruence which here has extra rules dealing with abstractions and concretions. To see this connection consider a derivation of

$$(\nu u)\big(\overline{a}u\,.\,P \mid u(b)\,.\,Q\big) \xrightarrow{\overline{a}\nu u} P \mid u(b)\,.\,Q$$

which uses PREFIX, PAR and OPEN in the standard semantics. With extended agents the inference goes by X-PREFIX and X-PAR to derive

$$\overline{a}\,.\,[u]P \mid u\,.\,\lambda bQ \xrightarrow{\overline{a}} [u]P \mid u\,.\,\lambda bQ$$

and then through X-STRUCT, the new part about Parallel composition of extended agents,

$$\overline{a}\,.\,[u]P \mid u\,.\,\lambda bQ \xrightarrow{\overline{a}} [u](P \mid u\,.\,\lambda bQ)$$

and finally X-RES

$$(\nu u)\big(\overline{a}\,.\,[u]P \mid u\,.\,\lambda bQ\big) \xrightarrow{\overline{a}} (\nu u)[u](P \mid u\,.\,\lambda bQ).$$

In conclusion, the use of extended agents is conceptually attractive since some of the "bookkeeping" activity in an inference, like keeping track of bound names and their scope, is factored out from the transition rules and included in the structural congruence. Also, the representation of transitions in these rules has proved suitable for automatic tools, in particular if used in conjunction with a symbolic semantics as in Section 5.2, where transitions are of the kind $P \xrightarrow{M,c} E$. The drawback is that the format is more unfamiliar to those accustomed to other process algebras.

6. Bisimilarity and congruence

We shall here look at one of the most fundamental behavioural equivalences, namely strong bisimilarity. It turns out not to be a congruence – it is not preserved by input prefix – but fortunately the largest contained congruence has a simple characterization.

6.1. *Bisimilarity*

In most process algebras a family of equivalence relations on agents is based on bisimulations, and the π-calculus is no exception. The generic definition of a bisimulation is that it is a symmetric binary relation $\mathcal{R}$ on agents satisfying

$$P\mathcal{R}Q \text{ and } P \xrightarrow{\alpha} P' \quad \text{implies} \quad \exists Q' : Q \xrightarrow{\alpha} Q' \wedge P'\mathcal{R}Q'.$$

The intuition is that if P can do an action then Q can do the same action and the derivatives lie in the same relation. Two agents are said to be bisimilar if they are related by some bisimulation; this means that they can indefinitely mimic the transitions of each other.

For the π-calculus extra care has to be taken for actions with bound objects. Consider

$$P = a(u), \qquad Q = a(x).(\boldsymbol{\nu} v)\overline{v}u.$$

Intuitively these represent the same behaviour: they can do an input along a and then nothing more. However, Q has the name u free where P has not. Therefore, x in Q cannot be alpha-converted to u, and the transition $P \xrightarrow{a(u)} \mathbf{0}$ cannot be simulated by Q. Such a difference between P and Q is not important since, if P has an action $a(u)$, then by alpha-conversion it also has a similarly derived action $a(w)$ for infinitely many w. Clearly it is sufficient for Q to simulate only the bound actions where the bound object is not free in Q. This argument applies to both input and bound output actions.

Also, input (but not bound output) actions mean that the bound object is a placeholder for something to be received. Therefore, if $P \xrightarrow{a(x)} P'$ then the behaviour of P' must be considered under all substitutions $\{u/x\}$, and we must require that for each such substitution Q' is related to P', or in other words, that they are related for each value received.

In the following we will use the phrase "$\mathtt{bn}(\alpha)$ is fresh" in a definition to mean that the name in $\mathtt{bn}(\alpha)$, if any, is different from any free name occurring in any of the agents in the definition.

DEFINITION 1. A *(strong) bisimulation* is a symmetric binary relation $\mathcal{R}$ on agents satisfying the following: $P\mathcal{R}Q$ and $P \xrightarrow{\alpha} P'$ where $\mathtt{bn}(\alpha)$ is fresh implies that

(i) If $\alpha = a(x)$ then $\exists Q' : Q \xrightarrow{a(x)} Q' \wedge \forall u : P'\{u/x\}\mathcal{R}Q'\{u/x\}$.

(ii) If α is not an input then $\exists Q' : Q \xrightarrow{\alpha} Q' \wedge P'\mathcal{R}Q'$.

P and Q are *(strongly) bisimilar*, written $P \dot{\sim} Q$, if they are related by a bisimulation.

It follows that $\dot{\sim}$, which is the union of all bisimulations, is a bisimulation. We also immediately have that $P \equiv Q$ implies $P \dot{\sim} Q$, by virtue of the rule STRUCT.

At this point it might be helpful to consider a few examples. Requirement (i) is quite strong as demonstrated by

$$P_1 = a(x).P + a(x).\mathbf{0}, \qquad P_2 = a(x).P + a(x).\mathtt{if}\ x = u\ \mathtt{then}\ P.$$

Assume that $P \dot{\not\sim} \mathbf{0}$. Then $P_1 \dot{\not\sim} P_2$ since the transition $P_1 \xrightarrow{a(x)} \mathbf{0}$ cannot be simulated by P_2. For example, $P_2 \xrightarrow{a(x)} \mathtt{if}\ x = u\ \mathtt{then}\ P$ does not suffice since, for the substitution $\{u/x\}$, the derivatives are not bisimilar. Similarly,

$$P_1 = a(x).P + a(x).\mathbf{0}, \qquad P_2 = a(x).P + a(x).\mathbf{0} + a(x).\mathtt{if}\ x = u\ \mathtt{then}\ P$$

are not bisimilar because of the transition $P_2 \xrightarrow{a(x)} \mathtt{if}\ x = u\ \mathtt{then}\ P$. Neither $\mathbf{0}$ nor P is bisimilar to $\mathtt{if}\ x = u\ \mathtt{then}\ P$ for *all* substitutions of x, since $\mathbf{0}$ fails for $\{u/x\}$ and P fails for the identity substitution. Using the intuition that an input $P \xrightarrow{a(x)} Q$ can be thought of as $P \xrightarrow{a} \lambda x\, Q$, clause (i) says that the derivatives (which are functions from names to agents) must be pointwise bisimilar.

As expected we have

$$a \mid \overline{b} \dot{\sim} a.\overline{b} + \overline{b}.a$$

and

$$a \mid \overline{a} \dot{\sim} a.\overline{a} + \overline{a}.a + \tau.$$

This demonstrates that $\dot{\sim}$ is not in general closed under substitutions, i.e., from $P \dot{\sim} Q$ we cannot conclude that $P\sigma \dot{\sim} Q\sigma$. For this reason we have

$$c(a).(a \mid \overline{b}) \dot{\not\sim} c(a).(a.\overline{b} + \overline{b}.a)$$

demonstrating that $\dot{\sim}$ is not preserved by input Prefix, i.e., that from $P \dot{\sim} Q$ we cannot conclude that $a(x).P \dot{\sim} a(x).Q$.

However, we have the following results:

PROPOSITION 2. *If $P \dot{\sim} Q$ and σ is injective then $P\sigma \dot{\sim} Q\sigma$.*

PROPOSITION 3. *$\dot{\sim}$ is an equivalence.*

PROPOSITION 4. *$\dot{\sim}$ is preserved by all operators except input Prefix.*

Proposition 2 is proved by establishing that transitions are preserved by injective substitutions, i.e., that for injective σ,

$$P \xrightarrow{\alpha} P' \quad \text{implies} \quad P\sigma \xrightarrow{\alpha\sigma} P'\sigma,$$

where the substitution σ applied to an action α is defined not to affect $\mathtt{bn}(\alpha)$. This result is established by induction of the inference of $P \xrightarrow{\alpha} P'$ and each rule in Table 3 gives rise to one induction step. It is then straightforward to show that $\{(P\sigma, Q\sigma) : P \dot{\sim} Q, \sigma \text{injective}\}$ is a bisimulation.

Proposition 3 is not difficult. Reflexivity and symmetry are immediate and for transitivity it suffices to show that $\dot{\sim}\dot{\sim}$ is a strong bisimulation.

The proof of Proposition 4 goes by examining each operator in turn. It is a bit complicated for the cases of Restriction and Parallel. For example, we would like to show that

$$\{((\boldsymbol{\nu}x)P, (\boldsymbol{\nu}x)Q) : P \dot{\sim} Q\} \cup \dot{\sim}$$

is a bisimulation, by examining the possible transitions from $(\boldsymbol{\nu}x)P$, where the rules RES and OPEN are applicable. Unfortunately this simple idea does not quite work since also the STRUCT rule is applicable, so we have to include all structurally congruent pairs in the bisimulation. A proof idea is the following. Say that an agent is *normal* if all bound names are distinct and all unguarded Restrictions are at the top level, i.e., of the form $(\boldsymbol{\nu}\tilde{x})P$ where P has no unguarded Restrictions. By Proposition 1 any P is structurally congruent to a normal agent. Now extend the transition rules with the symmetric counterparts of PAR and COM; in view of the commutativity in the structural congruence this does not affect the transitions. That makes it possible to prove the lemma

> If $P \xrightarrow{\alpha} P'$ and $P \equiv N$ where N is normal, then by an inference of no greater depth, $N \xrightarrow{\alpha} N'$ and $P' \equiv N'$.

The proof is through induction on the inference of $P \equiv N$ and involves a tedious examination of all rules for the structural congruence. Now we can show that the relation on normal agents

$$\{((\nu\tilde{x})(P \mid Q),\ (\nu\tilde{x})(P' \mid Q')) : P \dot{\sim} P',\ Q \dot{\sim} Q'\}$$

is a bisimulation up to $\dot{\sim}$ (as explained below), by showing that an action from one side is simulated by an action from the other side, through induction on the inference of the action. There are eight inductive steps through combinations of the PAR, COM, RES and OPEN laws to consider. All cases are routine. Because of the lemma, actions derived through the STRUCT rule can be ignored in the induction since they also have a shorter inference. Finally we can take $P = P' = \mathbf{0}$ to show that Restriction preserves bisimilarity, and $\tilde{x}$ the empty sequence to show the same for Parallel.

The notion of "bisimulation up to" is a standard proof trick in these circumstances. It involves modifying Definition 1 by replacing $\mathcal{R}$ in the consequents of (i) and (ii) by $\dot{\sim}\mathcal{R}\dot{\sim}$, i.e., the relational composition of $\dot{\sim}$, $\mathcal{R}$, and $\dot{\sim}$. The requirement is thus weakened: the derivatives must be bisimilar to a pair in $\mathcal{R}$, in contrast to Definition 1 which requires that the derivatives themselves lie in $\mathcal{R}$. It then holds that if two agents are related by a bisimulation up to $\dot{\sim}$ then they are bisimilar. The proof of this is very similar to the corresponding result for other process algebras.

6.2. *Congruence*

The fact that bisimilarity is not preserved by input Prefix makes us seek the largest congruence included in bisimilarity. The definition is fortunately simple:

DEFINITION 2. Two agents P and Q are (*strongly*) *congruent*, written $P \sim Q$, if $P\sigma \dot{\sim} Q\sigma$ for all substitutions σ.

For an example, although $a \mid \overline{b} \dot{\sim} a\,.\overline{b} + \overline{b}\,.a$ these agents are not strongly congruent since the substitution $\{a/b\}$ makes them non-bisimilar. However, we do have

$$a \mid \overline{b} \sim a\,.\overline{b} + \overline{b}\,.a + \texttt{if}\ a = b\ \texttt{then}\ \tau$$

and the reader may by this example realize the crucial role played by the Match operator in an expansion law which reduces a Parallel composition to a Sum.

PROPOSITION 5. *Strong congruence is the largest congruence in bisimilarity.*

We first have to show that strong congruence is a congruence, i.e., that it is preserved by all operators. For all operators except input Prefix this is immediate from the definition and Proposition 4. For input Prefix we show that

$$\{(a(x)\,.\,P, a(x)\,.\,P') : P \sim P'\} \cup \sim$$

is a strong bisimulation and closed under substitutions. The latter is trivial since $\sim$ is closed under substitutions. For the bisimulation part, the transition to consider is $a(x)\,.\,P \xrightarrow{a(x)} P$ which obviously is simulated by $a(x)\,.\,P' \xrightarrow{a(x)} P'$; we have to show that for all u, $P\{u/x\}$ is related to $P'\{u/x\}$, and this follows from $P \sim P'$.

We next show that it is the largest such congruence. Assume that P and Q are related by some congruence $\sim'$ in $\dot{\sim}$. We shall show that this implies that $P\sigma \dot{\sim} Q\sigma$ for any σ, in other words that $\sim' \subseteq \sim$. Without loss of generality we can assume that $\sigma(x) \neq x$ for only finitely many names x since the effect of σ on names not free in P or Q is immaterial. So let $\sigma = \{y_1 \ldots y_n / x_1 \ldots x_n\}$. Choose $a \notin \mathtt{fn}(P, Q)$. By $P \sim' Q$ and the fact that $\sim'$ is a congruence we get

$$(\boldsymbol{\nu} a)\big(a(x_1 \ldots x_n)\,.\,P \mid \overline{a}\langle y_1 \ldots y_n\rangle\big) \sim' (\boldsymbol{\nu} a)\big(a(x_1 \ldots x_n)\,.\,Q \mid \overline{a}\langle y_1 \ldots y_n\rangle\big)$$

and by the fact that $\sim'$ is in $\dot{\sim}$ we get, following the τ-transition from both sides, that $P\sigma \dot{\sim} Q\sigma$.

7. Variants of bisimilarity

There are several ways in which the definition of equivalence can be varied. We here consider some variants that are particular to the π-calculus, in the sense that they differ in how the name substitutions arising from input are treated. We also briefly consider the weak bisimulation equivalences.

7.1. *Early bisimulation*

Bisimulations can also be defined from the early semantics in Section 5.3. This leads to an equivalence which is slightly larger than $\dot{\sim}$, i.e., it equates more agents, although that difference is hardly ever of practical significance.

For clarity let the transitions in the early semantics be indexed by E, so for example we have

$$a(x)\,.\,P \xrightarrow{au}_{\mathrm{E}} P\{u/x\}.$$

As mentioned in that section the τ-transitions are the same with the late and early semantics, in other words $\xrightarrow{\tau}$ is the same as $\xrightarrow{\tau}_{\mathrm{E}}$, and similarly for output transitions. So the distinction is only relevant for input transitions.

It is then easy to prove that a relation is an early bisimulation precisely when it is an early bisimulation with late semantics. So it is a matter of convenience which of the two definitions to use. The converse idea (to capture late bisimilarity using early semantics) is also possible.

Early bisimilarity satisfies the same propositions as late bisimilarity in Section 6: it is preserved by injective substitutions and all operators except input Prefix; and *early congruence* $\sim_E$, defined by $P \sim_E Q$ if for all σ, $P\sigma \dot{\sim}_E Q\sigma$, is the largest congruence in $\dot{\sim}_E$. The proofs are very similar. Early congruence is larger than late congruence, i.e., $\sim \subseteq \sim_E$, and the example in Figure 1 suffices to prove that the inclusion is strict.

It is hard to imagine any practical situation where the difference between late and early congruence is important. Since the theoretical properties are very similar the choice between them is largely a matter of taste. It could be argued that early congruence to a higher degree conforms to our operational intuition, and that the definition is less complex. On the other hand, late congruence seems to lead to more efficient verifications in automated tools.

7.2. *Barbed congruence*

In some sense the notion of bisimilarity, be it late or early, assumes the observer is unrealistically powerful: since there is a difference between the actions $\overline{a}u$ and $\overline{a}\boldsymbol{\nu}u$ it is possible to directly detect if an emitted name is local. It is therefore natural to ask what is the effect of reducing this discriminatory power. This leads to the idea of *barbed bisimulation* where, intuitively, the observer is limited to sensing if an action is enabled on a given channel. Although barbed bisimilarity is weaker (i.e., equates more agents) than early bisimilarity it turns out that the congruence obtained from barbed bisimulation is the same as early congruence.

DEFINITION 5. A name a is *observable* at P, written $P \downarrow a$, if a is the subject of some action from P. A *barbed bisimulation* is a symmetric binary relation $\mathcal{R}$ on agents satisfying the following: $P\mathcal{R}Q$ implies that

(i) If $P \xrightarrow{\tau} P'$ then $\exists Q' : Q \xrightarrow{\tau} Q' \wedge P'\mathcal{R}Q'$.

(ii) If $P \downarrow a$ then $Q \downarrow a$.

P and Q are *barbed bisimilar*, written $P \dot{\sim}_B Q$, if they are related by a barbed bisimulation.

The idea behind a barbed bisimulation is that the only unquestionable execution steps are the internal ones, i.e., the τ-actions, since these do not require participation of the environment. Therefore labelled transitions are not needed to define barbed bisimulation. It is enough with the reduction relation of Section 5.4 and the observability predicate, which like reductions can be defined directly from the syntax as $P \downarrow a$ if P contains an unguarded Prefix with subject a not under the scope of a Restriction $\boldsymbol{\nu}a$.

Barbed bisimilarity is uninteresting by itself. For example,

$$\overline{a}u \dot{\sim}_B \overline{a}v \dot{\sim}_B (\boldsymbol{\nu}u)\overline{a}u$$

identifying three clearly different agents. Of more interest is the congruence it generates.

DEFINITION 6. Two agents P and Q are *barbed congruent*, written $P \sim_{\mathrm{B}} Q$, if for all contexts $\mathcal{C}$ it holds that $\mathcal{C}(P) \dot{\sim}_{\mathrm{B}} \mathcal{C}(Q)$.

In other words, barbed congruence is the largest congruence in barbed bisimilarity. Clearly the three agents above are not barbed congruent. Take the context

$$\mathcal{C}(P) = P \mid a(x) . \overline{x}.$$

Then $\mathcal{C}(\overline{a}u) \xrightarrow{\tau} \overline{u} \downarrow u$ and this cannot be simulated by $\mathcal{C}(\overline{a}v)$ and not by $\mathcal{C}((\boldsymbol{\nu} u)\overline{a}u)$. In fact barbed congruence coincides with early congruence:

PROPOSITION 6. $\sim_{\mathrm{B}} = \sim_{\mathrm{E}}$.

This lends considerable support to early congruence as the natural equivalence relation. The main idea is to prove that $\sim_{\mathrm{B}}$ is an early bisimulation. Suppose $P \sim_{\mathrm{B}} Q$ and $P \xrightarrow{\alpha} P'$, then a cushioning context $\mathcal{C}$ can be found such that $\mathcal{C}(P) \xrightarrow{\tau} \mathcal{C}'(P')$, so $\mathcal{C}(Q)$ must simulate this transition. But $\mathcal{C}$ is cleverly constructed so that the only possible way that $\mathcal{C}(Q)$ can do this is because of a transition $Q \xrightarrow{\alpha} Q'$. We shall not go further into the proof here which is quite involved.

Variants of barbed congruence use variants of the observability predicate, leading to a spectrum of possibilities. For example, a weaker variant is to use $P \downarrow$ to mean $\exists a : P \downarrow a$, and amend clause (ii) in the definition of barbed bisimilarity to $P \downarrow$ iff $Q \downarrow$. With this definition Proposition 6 still holds, though its counterpart for weak bisimulation (cf. Section 7.4) is less clear. Omitting clause (ii) entirely Proposition 6 no longer holds. For example, then $!\tau \mid Q$ and $!\tau \mid R$ would be barbed congruent for all Q and R.

Also, in subcalculi of π the barbed congruence may be different (because there are fewer contexts). As one example, in the asynchronous subcalculus (Section 3.5) and where observations $P \downarrow a$ mean that P can do an output action along a (in the asynchronous calculus it makes sense to regard inputs as unobservable) it holds that

$$a(x) . \overline{a}x + \tau \sim_{\mathrm{B}} \tau.$$

7.3. *Open bisimulation*

As has been demonstrated neither late nor early bisimilarity is a congruence; to gain a congruence we must additionally require bisimilarity under all substitutions of names. Interestingly, there is a way to define a congruence directly through bisimulations. This is the idea behind *open* bisimulation, which incorporates the quantification over all substitutions in the clause for bisimulation. This leads to a congruence finer than late congruence (i.e., it equates fewer agents) although that difference is not very significant in practice.

To begin we consider a subcalculus without Restriction and Mismatch. The definition will subsequently be extended to accommodate Restriction (no simple extension for Mismatch is known).

DEFINITION 7. A (*strong*) *open bisimulation* is a symmetric binary relation $\mathcal{R}$ on agents satisfying the following for all substitutions σ: $P\mathcal{R}Q$ and $P\sigma \xrightarrow{\alpha} P'$ where $\mathtt{bn}(\alpha)$ is fresh implies that

$$\exists Q' : Q\sigma \xrightarrow{\alpha} Q' \text{ and } P'\mathcal{R}Q'.$$

P and Q are (*strongly*) *open bisimilar*, written $P \dot{\sim}_{\mathrm{O}} Q$, if they are related by an open bisimulation.

A technical remark: the condition on $\mathtt{bn}(\alpha)$ is unnecessary in view of the fact that we deal with a subcalculus without Restriction, but it is harmless and is included for the sake of symmetry with other similar definitions.

In open bisimulation there is no need for a special treatment of input actions since the quantification over substitutions recurs anyway when further transitions are examined: If $P \xrightarrow{a(x)} P'$ is simulated by $Q \xrightarrow{a(x)} Q'$, then the requirement implies that all transitions from $P'\sigma$ must be simulated by $Q'\sigma$ for all σ, including those substitutions that instantiate x to another name. This also demonstrates the pertinent difference between open and late: where late requires the agents to continue to bisimulate under all substitutions for the bound input object, open requires them to bisimulate under *all* substitutions – not only those affecting the bound input object.

It is straightforward to show that an open bisimulation is also a late bisimulation, and hence $P \dot{\sim}_{\mathrm{O}} Q$ implies $P \dot{\sim} Q$. Furthermore, directly from the definition we get that open bisimilarity is closed under substitution, in the sense that $P \dot{\sim}_{\mathrm{O}} Q$ implies $P\sigma \dot{\sim}_{\mathrm{O}} Q\sigma$; together these facts give us that $P \dot{\sim}_{\mathrm{O}} Q$ implies $P \sim Q$. But the converse is not true as demonstrated by the following example:

$$Q \equiv \tau + \tau\,.\,\tau,$$
$$R \equiv \tau + \tau\,.\,\tau + \tau\,.\,\mathtt{if}\; x = y\; \mathtt{then}\; \tau.$$

Q and R are late congruent. To see that $Q\sigma$ is late bisimilar to $R\sigma$ for all σ, consider two possibilities for σ. Either $\sigma(x) = \sigma(y)$, in which case $R\sigma \xrightarrow{\tau} \mathtt{if}\; \sigma(x) = \sigma(y)\; \mathtt{then}\; \tau$ is simulated by $Q\sigma \xrightarrow{\tau} \tau$; or $\sigma(x) \neq \sigma(y)$, in which case it is simulated by $Q\sigma \xrightarrow{\tau} \mathbf{0}$. But Q and R are not open bisimilar. To see this, choose σ as the identity substitution and consider $R \xrightarrow{\tau} \mathtt{if}\; x = y\; \mathtt{then}\; \tau$. There are two possibilities for Q to make a τ transition. One is $Q \xrightarrow{\tau} \mathbf{0}$. But then open bisimulation requires that the two derivatives $\mathtt{if}\; x = y\; \mathtt{then}\; \tau$ and $\mathbf{0}$ bisimulate again under *all* substitutions σ', and this does not hold for $\sigma' = \{x/y\}$ since $(\mathtt{if}\; x = y\; \mathtt{then}\; \tau)\{x/y\} \xrightarrow{\tau} \mathbf{0}$ whereas $\mathbf{0}\{x/y\}$ has no transition. Similarly, the other transition $Q \xrightarrow{\tau} \tau$ fails for the identity substitution, since $\mathtt{if}\; x = y\; \mathtt{then}\; \tau$ has no transition, in contrast to τ.

This example may seem artificial and, as for the difference between late and early, the difference between open and late is probably of little practical significance. Open bisimilarity is an equivalence and a congruence, so by analogy with the other equivalences, writing $\sim_{\mathrm{O}}$ for the largest congruence in $\dot{\sim}_{\mathrm{O}}$, we get $\dot{\sim}_{\mathrm{O}} = \sim_{\mathrm{O}}$ (the proofs are technically similar to the proof for late congruence). It can also be characterized in a few other informative ways. One is through substitution closed ground bisimulations. A ground bisimulation is a relation $\mathcal{R}$ satisfying

$$P\mathcal{R}Q \text{ and } P \xrightarrow{\alpha} P' \quad \text{implies} \quad \exists Q' : Q \xrightarrow{\alpha} Q' \text{ and } P'\mathcal{R}Q'.$$

In other words, it does not introduce substitutions at all. Now it holds that $\dot{\sim}_{\mathrm{O}}$ is the largest substitution closed (i.e., $P\mathcal{R}Q$ implies $P\sigma\mathcal{R}Q\sigma$) ground bisimulation. Another characterization is with *dynamic* bisimulations: these are relations satisfying

$$P\mathcal{R}Q \text{ and } \mathcal{C}(P) \xrightarrow{\alpha} P' \quad \text{implies} \quad \exists Q' : \mathcal{C}(Q) \xrightarrow{\alpha} Q' \text{ and } P'\mathcal{R}Q'$$

for all contexts $\mathcal{C}$. It can be shown that dynamic bisimilarity is the same as open bisimilarity.

At first it may seem that the Restriction operator behaves well with open bisimulation since the substitutions only affect free names. Consider

$$P = (\boldsymbol{\nu}x)\mathtt{if}\ x = y\ \mathtt{then}\ \tau \quad \text{and} \quad Q = \mathbf{0}.$$

Neither $P\sigma$ nor $Q\sigma$ have any transitions, for any substitution σ, since the substitution cannot affect the bound name x. Therefore $P \dot{\sim}_{\mathrm{O}} Q$. Similarly with

$$P = (\boldsymbol{\nu}x)\tau\,.\,\mathtt{if}\ x = y\ \mathtt{then}\ \tau \quad \text{and} \quad Q = \tau$$

we have $P \dot{\sim}_{\mathrm{O}} Q$, since $P \xrightarrow{\tau} (\boldsymbol{\nu}x)\mathtt{if}\ x = y\ \mathtt{then}\ \tau$ and this derivative is similarly immune to substitutions. The problem arises because of bound output actions. Consider

$$P = (\boldsymbol{\nu}x)\overline{a}x\,.\,\mathtt{if}\ x = y\ \mathtt{then}\ \tau \quad \text{and} \quad Q = (\boldsymbol{\nu}x)\overline{a}x.$$

Here $P \xrightarrow{\overline{a}\boldsymbol{\nu}x} \mathtt{if}\ x = y\ \mathtt{then}\ \tau$. Clearly Q cannot simulate by $Q \xrightarrow{\overline{a}\boldsymbol{\nu}x} \mathbf{0}$ because the two derivatives, $\mathtt{if}\ x = y\ \mathtt{then}\ \tau$ and $\mathbf{0}$, are not bisimilar for the substitution $\{x/y\}$. Here the bound output action has lifted the Restriction so x becomes vulnerable to a substitution.

But we would certainly want to identify P and Q. The name x in P is a local name and no matter what an environment does on receiving it, that name cannot become the same as y, so the `if` test will never be positive. Also P and Q are late congruent. Therefore in the presence of Restriction, Definition 7 is too discriminating: it considers too many substitutions.

One way to amend it is to use the auxiliary concept of *distinctions*. The intuition behind a distinction is to record those pairs of names that will always be distinct, no matter what actions take place. Definition 7 is modified so that it only considers substitutions that respect distinctions, in the sense that two names required to be distinct are not substituted by the same name. Following a bound output transition the derivatives are related under a larger distinction, representing the fact that the bound object will be kept different from any other name.

Formally a distinction is a symmetric and irreflexive binary relation on names. In the following we let D range over finite distinctions (infinite distinctions are not necessary and may cause problems with alpha-conversion if they include all names). We say that a substitution σ *respects* a distinction D if aDb implies $\sigma(a) \neq \sigma(b)$. We further let $D\sigma$ be the relation $\{(\sigma(a), \sigma(b)) : aDb\}$. When N is a set of names we let $D + (x, N)$ represent $D \cup (\{x\} \times N) \cup (N \times \{x\})$, in other words it is D extended with the fact that x is distinct from all names in N.

DEFINITION 8. A *distinction-indexed* family of binary relations on agents is a set of binary agent relations $\mathcal{R}_D$, one for each distinction D. A (*strong*) *open bisimulation* is such a family satisfying the following for all substitutions σ which respect D: $P\mathcal{R}_D Q$ and $P\sigma \xrightarrow{\alpha} P'$ where $\mathrm{bn}(\alpha)$ is fresh implies that

(i) If $\alpha = \overline{a}\nu x$ then $\exists Q' : Q\sigma \xrightarrow{\overline{a}\nu x} Q' \wedge P'\mathcal{R}_{D'}Q'$, where $D' = D\sigma + (x, \mathtt{fn}(P\sigma, Q\sigma))$.

(ii) If α is not a bound output then $\exists Q' : Q\sigma \xrightarrow{\alpha} Q' \wedge P'\mathcal{R}_{D\sigma}Q'$.

P and Q are *open D-bisimilar*, written $P \dot{\sim}^D_{\mathrm{O}} Q$, if they are related by R_D in an open bisimulation, and they *open bisimilar*, written $P \dot{\sim}_{\mathrm{O}} Q$ if they are *open* $\emptyset$*-bisimilar*.

Here the proviso that $\mathrm{bn}(\alpha)$ is fresh means it is not in $\mathtt{fn}(P\sigma, Q\sigma)$. The overloading of "$\dot{\sim}_{\mathrm{O}}$" is motivated by the fact that for the subcalculus without Restriction Definitions 7 and 8 coincide. In essence, the distinction keeps track of what substitutions are admissible, excluding those that do not respect the distinction. After any action except a bound output, the distinction recurs in that the substituted names must still be distinct. After a bound output, additionally the bound object must be distinct from any free name in the involved agents.

As an example we can now demonstrate that $P \dot{\sim}_{\mathrm{O}} Q$ where

$$P = (\boldsymbol{\nu}x)\overline{a}x\,.\,\mathtt{if}\ x = y\ \mathtt{then}\ \tau \quad \text{and} \quad Q = (\boldsymbol{\nu}x)\overline{a}x.$$

Consider the transition $P \xrightarrow{\overline{a}\nu x} \mathtt{if}\ x = y\ \mathtt{then}\ \tau$ and the simulating transition $Q \xrightarrow{\overline{a}\nu x} \mathbf{0}$. The definition then requires, by clause (i), that $\mathtt{if}\ x = y\ \mathtt{then}\ \tau \dot{\sim}^D_{\mathrm{O}} \mathbf{0}$ for $D = \{(x, y), (y, x)\}$ and this is indeed the case: for a substitution σ respecting D, the agent $(\mathtt{if}\ x = y\ \mathtt{then}\ \tau)\sigma$ has no transitions.

Distinctions will however not admit Mismatch in a sensible way. For example, we would want to equate the agents

$$\mathtt{if}\ x \neq y\ \mathtt{then}\ \tau\,.\,\mathtt{if}\ x \neq y\ \mathtt{then}\ \tau \quad \text{and} \quad \mathtt{if}\ x \neq y\ \mathtt{then}\ \tau\,.\,\tau$$

but the requirement that derivatives bisimulate under all substitutions will make them inequivalent. In order to admit Mismatch a more fundamental redefinition of the semantics is necessary, where the Mismatch is explicitly noted in the transition, for example by using a symbolic semantics as in Section 5.2.

In conclusion, the definition of $\dot{\sim}_{\mathrm{O}}$ loses some of its simplicity in the presence of Restriction and it is still not applicable with Mismatch. It gains interest in that it appears more efficient for automatic verification and in that distinctions sometimes have an independent motivation to represent "constant" names such as values in some global data type or ports which are necessarily distinct. For example, if Booleans are represented by two globally available names t and f then clearly it is uninteresting to consider substitutions making

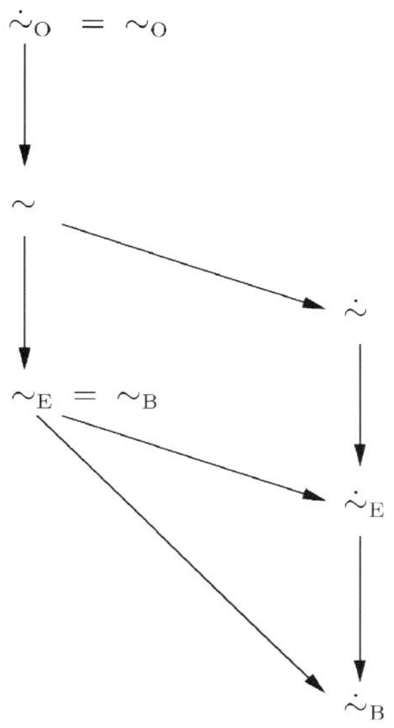

Fig. 2. Barbed, early, late and open bisimilarity and the corresponding congruences. An arrow means strict inclusion, and when it goes from the left column to the right it also means "largest congruence in".

them equal. Similarly, the question whether a buffer $B(i, o)$ with input port i and output o behaves as another buffer $B_2(i, o)$ might be relevant only for the case that $i \neq o$.

The various definitions of strong bisimilarity and congruence are related in Figure 2. In summary, barbed bisimilarity is much too weak to be interesting, while the difference between early and late is very small and has to do with the precise interpretation of input transitions. Barbed congruence is the same as early congruence. The early and late congruences are obtained as bisimilarity for all substitutions, and again the difference between them is small. Open bisimilarity (defined only for the subcalculus without Mismatch) is a congruence and is more discriminating than late congruence since it requires bisimilarity for all substitutions after every transition.

7.4. *Weak bisimulation*

The main idea of weak bisimulation is that τ transitions are regarded as unobservable. We therefore define $\Longrightarrow$ to mean $(\xrightarrow{\tau})^*$, i.e., zero or more τ transitions, $\overset{\alpha}{\Longrightarrow}$ to mean $\Longrightarrow\xrightarrow{\alpha}\Longrightarrow$, and $\overset{\widehat{\alpha}}{\Longrightarrow}$ to mean $\overset{\alpha}{\Longrightarrow}$ if $\alpha \neq \tau$ and $\Longrightarrow$ if $\alpha = \tau$. The intention is to modify strong bisimilarity in the standard way, replacing $\xrightarrow{\alpha}$ by $\overset{\widehat{\alpha}}{\Longrightarrow}$. The only complication is in the treatment of input actions. These call for substitutions of the bound object, and when simulating $a(x)$ by $\xrightarrow{\tau} \cdots \xrightarrow{\tau}\xrightarrow{a(x)}\xrightarrow{\tau} \cdots \xrightarrow{\tau}$ it turns out to be important that the substitution is applied immediately after the input action $\xrightarrow{a(x)}$, before further τ transitions.

DEFINITION 9. A *weak* (*late*) *bisimulation* is a symmetric binary relation $\mathcal{R}$ on agents satisfying the following: $P\mathcal{R}Q$ and $P \xrightarrow{\alpha} P'$ where $\mathrm{bn}(\alpha)$ is fresh implies that

(i) If $\alpha = a(x)$ then $\exists Q'' : Q \Longrightarrow\xrightarrow{a(x)} Q'' \wedge \forall u \exists Q' : Q''\{u/x\} \Longrightarrow Q' \wedge P'\{u/x\}\mathcal{R}Q'$.

(ii) If α is not an input then $\exists Q' : Q \stackrel{\widehat{\alpha}}{\Longrightarrow} Q' \wedge P'\mathcal{R}Q'$.

P and Q are *weakly* (*late*) *bisimilar*, written $P \dot{\approx} Q$, if they are related by a weak bisimulation.

Clause (i) says how an input transition should be simulated: first do any number of τ, then the input reaching Q'', and then, for each name u received, continue to some Q'. The effect is that the choice of Q'' is independent of u, and the choice of Q' may depend on u. Let us see an example of the significance of this:

$$\begin{aligned} P_1 &= a(x)\,.\,\mathtt{if}\ x = v\ \mathtt{then}\ P + a(x)\,.\,(\tau + \tau\,.\,P + \tau\,.\,\mathtt{if}\ x = v\ \mathtt{then}\ P),\\ P_2 &= a(x)\,.\,(\tau + \tau\,.\,P + \tau\,.\,\mathtt{if}\ x = v\ \mathtt{then}\ P),\\ P_3 &= a(x)\,.\,(\tau + \tau\,.\,P). \end{aligned}$$

First, observe that P_2 is even *strongly* bisimilar to P_3, since

$$\tau + \tau\,.\,P \sim \tau + \tau\,.\,P + \tau\,.\,\mathtt{if}\ x = v\ \mathtt{then}\ P$$

(any substitution makes the third summand behave as one of the first two). Now consider a transition

$$P_1 \xrightarrow{a(x)} \mathtt{if}\ x = v\ \mathtt{then}\ P.$$

Can P_2 simulate this? Yes, by the transition

$$P_2 \xrightarrow{a(x)} \tau + \tau\,.\,P + \tau\,.\,\mathtt{if}\ x = v\ \mathtt{then}\ P$$

and now applying $\{u/x\}$ by one τ-transition to $(\mathtt{if}\ x = v\ \mathtt{then}\ P)\{u/x\}$. Since $P_2 \sim P_3$ we also expect P_3 to simulate it, and this requires the full glory of clause (i), with different Q' for different u. The simulating transition is

$$P_3 \xrightarrow{a(x)} \tau + \tau\,.\,P$$

and now applying $\{u/x\}$, by one τ transition to either P or $\mathbf{0}$, depending on if $u = v$ or not. So here we have $P_1 \dot{\approx} P_2 \dot{\approx} P_3$.

This example also highlights why a simplification of clause (i) to

$$\exists Q' : Q \stackrel{a(x)}{\Longrightarrow} Q' \wedge \forall u : P'\{u/x\}\mathcal{R}Q'\{u/x\}$$

fails. With such a clause we would have that $P_1 \dot{\approx} P_2$ and $P_2 \dot{\approx} P_3$, yet $P_1 \not\dot{\approx} P_3$. In other words, weak bisimilarity would not be transitive. The reason is that P_3 no longer can simulate $P_1 \xrightarrow{a(x)} \texttt{if}\ x = v\ \texttt{then}\ P$, since neither $\mathbf{0}$, P, nor $\tau + \tau\,.\,P$ is bisimilar to $\texttt{if}\ x = v\ \texttt{then}\ P$ for all substitutions of x.

Another attempt at simplification is to replace $\xrightarrow{\alpha}$ by $\stackrel{\widehat{\alpha}}{\Longrightarrow}$ everywhere, also in the antecedent:

$P\mathcal{R}Q$ and $P \stackrel{\widehat{a(x)}}{\Longrightarrow} P'$ where x is fresh implies that $\exists Q' : Q \stackrel{\widehat{a(x)}}{\Longrightarrow} Q'$ and $\forall u : P'\{u/x\}\mathcal{R}Q'\{u/x\}$.

Then the so defined weak bisimilarity would be transitive. But it is inadequate for another reason. Consider

$$\begin{aligned} P_1 &= a(x)\,.\,\texttt{if}\ x \neq v\ \texttt{then}\ (P + \tau\,.\,\texttt{if}\ x \neq v\ \texttt{then}\ R),\\ P_2 &= a(x)\,.\,\texttt{if}\ x \neq v\ \texttt{then}\ (P + \tau\,.\,R). \end{aligned}$$

Here P_1 and P_2 are even *strong* late bisimilar. But with the attempted simplification they would not be weak late bisimilar since $P_1 \stackrel{a(x)}{\Longrightarrow} \texttt{if}\ x \neq v\ \texttt{then}\ R$ cannot be simulated by P_2.

So it seems that clause (i) in Definition 9 is unavoidable. Then $\dot{\approx}$ is an equivalence and every strong bisimulation is also a weak bisimulation, as expected. Of course $\dot{\approx}$ is not a congruence, for two independent reasons. The first is that it is not preserved by input Prefix, similarly to the situation for $\dot{\sim}$. The second is that it is not preserved by $+$, similarly to the situation for observation equivalence in CCS. These two concerns can be addressed independently in the standard ways: as in Section 6 closure under all substitutions is required to preserve input, and as in CCS simulating an initial τ transition by $\stackrel{\tau}{\Longrightarrow}$ (rather than $\Longrightarrow$) gives a relation preserved by +. In conclusion *weak late congruence* is defined as follows.

DEFINITION 10. P and Q are *weak* (*late*) *congruent* if, for all substitutions σ, $P\sigma \xrightarrow{\alpha} P'$ where $\mathrm{bn}(\alpha)$ is fresh implies that

(i) If $\alpha = a(x)$ then $\exists Q'' : Q\sigma \Longrightarrow\xrightarrow{a(x)} Q'' \wedge$
$\qquad \forall u \exists Q' : Q''\{u/x\} \Longrightarrow Q' \wedge P'\{u/x\} \dot{\approx} Q'$.

(ii) If α is not an input then $\exists Q' : Q\sigma \stackrel{\alpha}{\Longrightarrow} Q' \wedge P' \dot{\approx} Q'$

and conversely $Q\sigma \xrightarrow{\alpha} Q'$ implies similar transitions from $P\sigma$.

Other strong bisimulation equivalences also have weak counterparts in a similar way. For example weak early bisimilarity is obtained by replacing $Q \xrightarrow{\alpha} Q'$ in Definition 3 by $Q \stackrel{\widehat{\alpha}}{\Longrightarrow} Q'$; since there are no bound inputs, the complication with clause (i) in Definition 9 above does not arise. Weak barbed congruence is obtained by replacing $Q \xrightarrow{\tau} Q'$ by $Q \Longrightarrow Q'$ and $Q \downarrow a$ by $Q \Longrightarrow\downarrow a$ in Definition 5. The weak barbed congruence coincides with weak early congruence for image-finite agents (this is a mild technical condition which roughly means that the agent can only reach a finite number of different derivatives after $\stackrel{\alpha}{\Longrightarrow}$). Also weak open bisimulation is obtained by replacing $Q\sigma \xrightarrow{\alpha} Q'$ by $Q\sigma \stackrel{\widehat{\alpha}}{\Longrightarrow} Q'$ in Definition 7.

8. Algebraic theory

In this section we consider algebraic axiomatizations of late strong bisimilarity and congruence. This means that we identify a set of axioms for equality between agents, and that together with equational reasoning these imply all true equalities. We get one such set of axioms for bisimilarity and another set for congruence. These results only hold for the finite subcalculus, i.e., we do not include Identifiers or Replication. (In the whole calculus such a result cannot be obtained since bisimilarity, and also congruence, is not recursively enumerable and hence has no decidable axiomatization.)

8.1. *Bisimilarity*

Initially we restrict attention to the finite subcalculus without Parallel composition. The axioms for strong late bisimilarity are given in Table 6. We also implicitly use the laws of equational reasoning, i.e., that equality between agents is reflexive, symmetric and transitive. Note that substitutivity (that an agent can replace an equal agent in any expression) is *not* implied, since bisimilarity is not a congruence.

The axioms deserve little comment. STR says that all laws for structural congruence can be used. CONGR1 says that all operators except input Prefix preserve bisimilarity, and CONGR2 says that to infer bisimilarity of input Prefixes it is sufficient to establish bisimilarity under substitutions for the bound object. Note that CONGR2 mentions only a finite number of such substitutions: names free in the agents under consideration plus one more name represented by x. In view of Proposition 2 this is sufficient since all other names can be obtained through an injective substitution on x. Laws M1–MM2 serve to evaluate `if` constructs, and therefore a clause for Match and Mismatch in CONGR1 is unnecessary. The reader may remember from Section 5.1 that MM2 is not admissible in the structural congruence and may therefore be surprised to see this apparently unsuitable law in Table 6.

Table 6
Axioms for strong late bisimilarity

STR	If $P \equiv Q$ then $P = Q$	
CONGR1	If $P = Q$ then $\overline{a}u\,.\,P = \overline{a}u\,.\,Q$	
	$\tau\,.\,P = \tau\,.\,Q$	
	$P + R = Q + R$	
	$(\nu x)P = (\nu x)Q$	
CONGR2	If $P\{y/x\} = Q\{y/x\}$ for all $y \in \mathtt{fn}(P, Q, x)$ then $a(x)\,.\,P = a(x)\,.\,Q$	
S	$P + P = P$	
M1	$\mathtt{if}\ x = x\ \mathtt{then}\ P = P$	
M2	$\mathtt{if}\ x = y\ \mathtt{then}\ P = \mathbf{0}$	if $x \neq y$
MM1	$\mathtt{if}\ x \neq x\ \mathtt{then}\ P = \mathbf{0}$	
MM2	$\mathtt{if}\ x \neq y\ \mathtt{then}\ P = P$	if $x \neq y$
R1	$(\nu x)\alpha\,.\,P = \alpha\,.\,(\nu x)P$	if $x \notin \alpha$
R2	$(\nu x)\alpha\,.\,P = \mathbf{0}$	if x is the subject of α
R3	$(\nu x)(P + Q) = (\nu x)P + (\nu x)Q$	

The explanation is that MM2 is unsuitable for a congruence, since prefiguring both sides of it by an input prefix may invalidate it, and that bisimilarity is not a congruence; prefiguring with an input prefix requires the law CONGR2.

R1–R2 mean that a Restriction can be pushed through a Prefix or disappear; the only exception is when x is the free object of α in which case neither R1 nor R2 applies. Observe that R1 can be regarded as scope extension over Prefix; as was mentioned in Section 2.2 an option is to have this law as part of the structural congruence. Finally R3 means that Restriction distributes over Sum. This law is more powerful than scope extension over sum, since it splits one binder into two.

It is easily seen that all laws are sound, so if $P = Q$ is provable then it must hold that $P \dot{\sim} Q$. We shall now also prove the converse. In the following it is important to remember that α ranges also over bound output Prefixes of kind $\overline{a}\nu x$. Let the *depth* of an agent be the maximal nesting of its Prefixes.

PROPOSITION 7. *Using the axioms in Table* 6 *every agent P is provably equal to a* head normal form *(hnf) of kind $\sum_i \alpha_i \,.\, P_i$ of no greater depth.*

The proof is by induction over the structure of the agent and all cases are easy. If P is a Match or Mismatch then M1–MM2 applies; if it is a Restriction then R3 is used to distribute it onto the summands and R1–R2 to push it through the Prefixes or form part of a bound output Prefix.

PROPOSITION 8. *If $P \dot{\sim} Q$ then $P = Q$ is provable from the axioms in Table* 6.

The proof is by induction on the depths of P and Q. By Proposition 7 we can assume P and Q are head normal forms. The base case $P = Q = \mathbf{0}$ is trivial. For the inductive step we prove that for each summand in P there is a provably equal summand in Q and vice versa. For example, take a summand $a(x) \,.\, P'$ in P. Assume by alpha-conversion that all top-level input actions have the same bound object x. Then from $P \dot{\sim} Q$ and $P \xrightarrow{a(x)} P'$ we get $Q \xrightarrow{a(x)} Q'$ such that $P'\{u/x\} \dot{\sim} Q'\{u/x\}$ for all u. By induction they are also provably equal for all u. So $a(x) \,.\, Q'$ is a summand of Q and from CONGR2 we get that it is provably equal to $a(x) \,.\, P'$. The other cases are similar and simpler. So, each summand of P is provably equivalent to a summand of Q and therefore, by the law S (and STR), P is provably equivalent to Q.

We now turn to the Parallel operator. The idea behind the axiomatization is to introduce an expansion law through which the composition of two head normal forms is provably equal to a head normal form. With this law Proposition 7 will continue to hold for the calculus with Parallel, and therefore the proof of Proposition 8 needs not change. The expansion law is given in Table 7.

The proof of Proposition 7 is now extended with an inductive step for $P = Q \mid R$ where Q and R are hnfs. First use scope extension (Proposition 1) to pull any unguarded Restrictions in $Q \mid R$ to top level, obtaining an agent of kind $(\boldsymbol{\nu}\tilde{x})(Q' \mid R')$ where Q' and R' are hnfs with no bound output Prefixes. Then apply EXP to $Q' \mid R'$ gaining an agent of type $(\boldsymbol{\nu}\tilde{x})P'$ where P' is a hnf, and finally use scope extension to push back the Restrictions inwards, gaining a hnf.

Table 7
Expansion law for strong bisimilarity

EXP Let $P = \sum_i \alpha_i \,.\, P_i$ and $Q = \sum_j \beta_j \,.\, Q_j$ where $\mathrm{bn}(\alpha_i) \cap \mathrm{fn}(Q) = \emptyset$ and $\mathrm{bn}(\beta_j) \cap \mathrm{fn}(P) = \emptyset$ for all i, j, and none of α_i or β_j is a bound output Prefix. Then

$$P \mid Q = \sum_i \alpha_i \,.\, (P_i \mid Q) + \sum_j \beta_j \,.\, (P \mid Q_j) + \sum_{\alpha_i \, comp \, \beta_j} \tau \,.\, R_{ij}$$

where the relation $\alpha_i \; comp \; \beta_j$ and R_{ij} are defined as follows: either $\alpha_i = a(x)$ and $\beta_j = \overline{a}u$ in which case $R_{ij} = P_i\{u/x\} \mid Q_j$, or conversely $\alpha_i = \overline{a}u$ and $\beta_j = a(x)$ in which case $R_{ij} = P_i \mid Q_j\{u/x\}$.

8.2. *Congruence*

Following Definition 2 we immediately obtain an axiomatization of $\sim$ by adding that definition as a law. Interestingly, there is an alternative axiomatization that does not involve quantification over substitutions and does not refer back to bisimilarity. Again we begin with the subcalculus without Parallel.

It is informative to begin by examining the axioms in Table 6 to see what needs to change. The rule STR is still valid and useful, and the rules CONGR1–CONGR2 can be replaced by a simpler rule saying that $\sim$ is a congruence. The problematic laws are M2 and MM2 which are unsound for congruence. For example, even though $\mathtt{if}\ x = y\ \mathtt{then}\ P$ and $\mathbf{0}$ are bisimilar when $x \neq y$ they are not necessarily congruent since a substitution $\{x/y\}$ makes them non-bisimilar. In an axiom system for $\sim$ we cannot rely on axioms which eliminate unguarded Match and Mismatch operators.

A concise presentation of the axioms depends on the notion of a *Generalized* match operator written $\mathtt{if}\ M\ \mathtt{then}\ P$, where M is a conjunction of conditions of type $x = y$ and $x \neq y$. The Generalized match is simply defined to be a nested sequence of Matches and Mismatches, one for each condition:

$$\mathtt{if}\ m_1 \wedge \cdots \wedge m_k\ \mathtt{then}\ P = \mathtt{if}\ m_1\ \mathtt{then}\ (\cdots(\mathtt{if}\ m_k\ \mathtt{then}\ P)\cdots),$$

where each m_i is of type $x = y$ or $x \neq y$. We say that M logically implies N if all substitutions that make the conditions in M true also make the conditions in N true, and that they are logically equivalent, written $M \Leftrightarrow N$, if they imply each other. This allows us to use the compact law GM1:

$$\mathtt{if}\ M\ \mathtt{then}\ P = \mathtt{if}\ N\ \mathtt{then}\ P \quad \text{if } M \Leftrightarrow N$$

in place of a set of laws for nestings of Matches and Mismatches. The axioms for strong late congruence are given in Table 8.

We comment briefly on the new axioms. GM2 is a form of case analysis, allowing us to split an agent to a Sum of two mutually exclusive conditions. Writing out the definition of `if ...then ...else` this law is

$$\mathtt{if}\ x = y\ \mathtt{then}\ P + \mathtt{if}\ x \neq y\ \mathtt{then}\ P = P.$$

Table 8
Axioms for strong late congruence

STR	If $P \equiv Q$ then $P = Q$	
CONGR	"=" is preserved by all operators	
S	$P + P = P$	
MM1	$\mathtt{if}\ x \neq x\ \mathtt{then}\ P = \mathbf{0}$	
GM1	$\mathtt{if}\ M\ \mathtt{then}\ P = \mathtt{if}\ N\ \mathtt{then}\ P$	if $M \Leftrightarrow N$
GM2	$\mathtt{if}\ x = y\ \mathtt{then}\ P\ \mathtt{else}\ P = P$	
GM3	$\mathtt{if}\ M\ \mathtt{then}\ (P_1 + P_2) = \mathtt{if}\ M\ \mathtt{then}\ P_1 + \mathtt{if}\ M\ \mathtt{then}\ P_2$	
GM4	$\mathtt{if}\ M\ \mathtt{then}\ \alpha\,.\,P = \mathtt{if}\ M\ \mathtt{then}\ (\alpha\,.\,\mathtt{if}\ M\ \mathtt{then}\ P)$	if $\mathrm{bn}(\alpha) \notin M$
GM5	$\mathtt{if}\ x = y\ \mathtt{then}\ \alpha\,.\,P = \mathtt{if}\ x = y\ \mathtt{then}\ (\alpha\{x/y\})\,.\,P$	
GM6	$(\boldsymbol{\nu} x)\mathtt{if}\ x = y\ \mathtt{then}\ P = \mathbf{0}$	if $x \neq y$
R1	$(\boldsymbol{\nu} x)\alpha\,.\,P = \alpha\,.\,(\boldsymbol{\nu} x)P$	if $x \notin \alpha$
R2	$(\boldsymbol{\nu} x)\alpha\,.\,P = \mathbf{0}$	if x is the subject of α
R3	$(\boldsymbol{\nu} x)(P + Q) = (\boldsymbol{\nu} x)P + (\boldsymbol{\nu} x)Q$	

GM3 is a kind of distributive law for Generalized match over Sum. GM4 says that a test, once passed, can be done again after an action α, since that action cannot invalidate the outcome of the test. Here the side condition on $\mathrm{bn}(\alpha)$ is important. GM5 embodies the essence of a match: if x and y have been deemed equal then one can substitute the other (the substitution is defined to not affect $\mathrm{bn}(\alpha)$). Since $x = y \Leftrightarrow y = x$ we get from GM1 that it does not matter whether the substitution is $\{y/x\}$ or $\{x/y\}$. And combined with GM3 and GM4 we get all instances of a stronger law $\mathtt{if}\ x = y\ \mathtt{then}\ \alpha\,.\,P = \mathtt{if}\ x = y\ \mathtt{then}\ (\alpha\,.\,P)\{x/y\}$ where the substitution affects the whole Prefix form. Finally GM6 is the essence of Restriction: a restricted name can never be made equal to another name so the Match will always come out as false.

As can be seen the law MM1 is inherited from the axioms for bisimilarity. Of course also M1 is valid but it is not necessary as an axiom since it can be derived from GM2 and MM1. Similarly, a counterpart for Mismatch to GM6, namely

$$\text{GM6*} \quad (\boldsymbol{\nu} x)\mathtt{if}\ x \neq y\ \mathtt{then}\ P = (\boldsymbol{\nu} x)P \quad \text{if } x \neq y$$

is derivable from GM2 and GM6. Thus a Restriction can always be pushed through a generalized match. Also, note that $\mathtt{if}\ x = y\ \mathtt{then}\ \mathbf{0} = \mathbf{0}$ is derivable from MM1 and GM2.

For an example of a derivation consider $(\boldsymbol{\nu} x)\overline{a}x\,.\,\mathtt{if}\ x = y\ \mathtt{then}\ P$ which, as argued in Section 7.3, is congruent to $(\boldsymbol{\nu} x)\overline{a}x\,.\,\mathbf{0}$. The proof from axioms is:

$$
\begin{array}{ll}
(\boldsymbol{\nu} x)\overline{a}x\,.\,\mathtt{if}\ x = y\ \mathtt{then}\ P = & [\text{GM6*}] \\
(\boldsymbol{\nu} x)\mathtt{if}\ x \neq y\ \mathtt{then}\ \overline{a}x\,.\,\mathtt{if}\ x = y\ \mathtt{then}\ P = & [\text{GM4}] \\
(\boldsymbol{\nu} x)\mathtt{if}\ x \neq y\ \mathtt{then}\ \overline{a}x\,.\,\mathtt{if}\ x \neq y\ \mathtt{then}\ \mathtt{if}\ x = y\ \mathtt{then}\ P = & [\text{GM6*}] \\
(\boldsymbol{\nu} x)\overline{a}x\,.\,\mathtt{if}\ x \neq y\ \mathtt{then}\ \mathtt{if}\ x = y\ \mathtt{then}\ P = & \\
(\boldsymbol{\nu} x)\overline{a}x\,.\,\mathtt{if}\ x \neq y \wedge x = y\ \mathtt{then}\ P = & [\text{GM1}] \\
(\boldsymbol{\nu} x)\overline{a}x\,.\,\mathtt{if}\ x \neq x\ \mathtt{then}\ P = & [\text{MM1}] \\
(\boldsymbol{\nu} x)\overline{a}x\,.\,\mathbf{0}. &
\end{array}
$$

It is easy to establish that all laws are sound for congruence. The proof of completeness uses another kind of head normal form and is more involved than the corresponding proof for bisimilarity, and we shall here only sketch it. It relies on the notion of *complete* conjunctions. Formally, a conjunction M is complete on a set of names V if M implies either $x = y$ or $x \neq y$ for all names x, y in V. In other words, M expresses unambiguously which names are the same and which are not. We say that M *agrees* with a substitution σ if M implies $x = y$ if and only if $\sigma(x) = \sigma(y)$. The main use of complete conjunctions is that if M is complete on $\mathtt{fn}(P, Q)$ and agrees with σ, then $(\mathtt{if}\ M\ \mathtt{then}\ P) \sim (\mathtt{if}\ M\ \mathtt{then}\ Q)$ holds if and only if $P\sigma \dot{\sim} Q\sigma$. In this way the substitutions arising from the transitional semantics can be internalized and represented by Generalized matches.

DEFINITION 11. P is in *head normal form on a finite set of names* V (*V-hnf*) if

$$P = \sum_i \mathtt{if}\ M_i\ \mathtt{then}\ \alpha_i \,.\, P_i,$$

where for all i, $\mathtt{bn}(\alpha_i) \notin V$, and each M_i is complete on V.

So a V-hnf is a Sum of Generalized matches where each conjunction is complete on V.

PROPOSITION 9. *For any V, any agent P is provably equivalent to a V-hnf of no greater depth.*

The proof is by induction over the structure of P. If $P = \alpha \,.\, P'$ then use GM2 repeatedly to generate a Sum $\sum_i \mathtt{if}\ M_i\ \mathtt{then}\ \alpha \,.\, P'$ where the M_i are complete. If $P = (\boldsymbol{\nu} x)P'$ then by induction P' is a V-hnf $\sum_i \mathtt{if}\ M_i\ \mathtt{then}\ \alpha_i \,.\, P_i$. Use R3 to distribute $(\boldsymbol{\nu} x)$ onto all summands, GM6 and GM6* and STR to push the Restriction through all M_i and finally R1–R2 to push it through all α_i (unless x is the free object of α_i, in which case $(\boldsymbol{\nu} x)$ forms part of a bound output Prefix). The other cases are simple from the distributive laws.

PROPOSITION 10. *If $P \sim Q$ then $P = Q$ is provable from the axioms in Table* 8.

The proof is by induction on the depths of P and Q and by the previous proposition we can assume that P and Q are $\mathtt{fn}(P, Q)$-hnfs. The idea of the inductive step is as before to show that for each summand in P there is a provably equal summand in Q. We indicate the main lines for free output summands – the case for bound actions is a little bit more involved though the idea is the same. So let $\mathtt{if}\ M\ \mathtt{then}\ \alpha \,.\, P'$ be a summand in P with α a free output action. Choose a substitution σ that agrees with M. Then $P\sigma \xrightarrow{\alpha\sigma} P'\sigma$. From $P \sim Q$ we get that $Q\sigma$ must simulate this transition. Let $\mathtt{if}\ N\ \mathtt{then}\ \beta \,.\, Q'$ be the summand of Q that generates the simulating transition, which implies that $P'\sigma \dot{\sim} Q'\sigma$. Then N also agrees with σ, and since both M and N are complete and agree with σ we get $M \Leftrightarrow N$. Therefore by GM1 we can replace N by M in the summand of Q. And by GM5 we can replace β by α, because they can only differ in names identified by σ, meaning that M implies the equality of those names. There thus remains to prove $\mathtt{if}\ M\ \mathtt{then}\ \alpha \,.\, P' = \mathtt{if}\ M\ \mathtt{then}\ \alpha \,.\, Q'$. We know that $P'\sigma \dot{\sim} Q'\sigma$. Since M is complete it holds that this implies $\mathtt{if}\ M\ \mathtt{then}\ P' \sim \mathtt{if}\ M\ \mathtt{then}\ Q'$, whence by induction

Table 9
Expansion law for strong congruence

EXP2 Let $P = \sum_i \mathtt{if}\ M_i\ \mathtt{then}\ \alpha_i \,.\, P_i$ and $Q = \sum_j \mathtt{if}\ N_j\ \mathtt{then}\ \beta_j \,.\, Q_j$ where $\mathtt{bn}(\alpha_i) \cap \mathtt{fn}(Q) = \emptyset$ and $\mathtt{bn}(\beta_j) \cap \mathtt{fn}(P) = \emptyset$ for all i, j, and none of α_i or β_j is a bound output Prefix. Then

$$\begin{aligned} P \mid Q = & \sum_i \mathtt{if}\ M_i\ \mathtt{then}\ \alpha_i \,.\, (P_i \mid Q) + \sum_j \mathtt{if}\ N_j\ \mathtt{then}\ \beta_j \,.\, (P \mid Q_j) \\ & + \sum_{\alpha_i\ opp\ \beta_j} \mathtt{if}\ M_i \wedge N_j \wedge a_i = b_j\ \mathtt{then}\ \tau \,.\, R_{ij}, \end{aligned}$$

where the relation $\alpha_i\ opp\ \beta_j$, a_i, b_j and R_{ij} are defined as follows: either $\alpha_i = a_i(x)$ and $\beta_j = \overline{b}_j u$ in which case $R_{ij} = P_i\{u/x\} \mid Q_j$, or conversely $\alpha_i = \overline{a}_i u$ and $\beta_j = b_j(x)$ in which case $R_{ij} = P_i \mid Q_j\{u/x\}$.

these are provably equivalent. We now apply GM4 (for the first time in the proof!) to write $\mathtt{if}\ M\ \mathtt{then}\, \alpha \,.\, P'$ as $\mathtt{if}\ M\ \mathtt{then}\, \alpha \,.\, \mathtt{if}\ M\ \mathtt{then}\ P'$. That means it is provably equivalent to $\mathtt{if}\ M\ \mathtt{then}\, \alpha \,.\, \mathtt{if}\ M\ \mathtt{then}\ Q'$, and again by GM4 to $\mathtt{if}\ M\ \mathtt{then}\, \alpha \,.\, Q'$.

Finally we consider also the Parallel operator. By analogy with the case for bisimilarity it is enough to include an axiom which implies that the Parallel composition of two V-hnfs is a V-hnf. Unfortunately the axiom EXP in Table 7 is not sound for congruence, since a substitution may identify two names and thereby enable a communication. The amended expansion law for congruence is given in Table 9. Note the essential use of Matches to determine if a communication is possible. With this law Propositions 9 and 10 hold. Incidentally, the law is sound and complete also for bisimilarity, i.e., it can be used in place of EXP for the purpose of Proposition 8.

9. Variants of the theory

We here briefly mention the axiomatizations of the equivalences treated in Section 7.

9.1. *Early bisimilarity and congruence*

Early bisimilarity equates more agents than late and therefore we seek more axioms. It turns out that we need to add only one axiom to Table 6 to obtain a complete system. This characteristic axiom for early is:

$$\begin{aligned} \text{EARLY}\quad & a(x) \,.\, P + a(x) \,.\, Q = \\ & a(x) \,.\, P + a(x) \,.\, Q + a(x) \,.\, \mathtt{if}\ x = y\ \mathtt{then}\ P\ \mathtt{else}\ Q. \end{aligned}$$

The example from Section 7.1:

$$P_1 = a(x) \,.\, P + a(x) \,.\, \mathbf{0}, \qquad P_2 = a(x) \,.\, P + a(x) \,.\, \mathbf{0} + a(x) \,.\, \mathtt{if}\ x = u\ \mathtt{then}\ P$$

is an instance of EARLY with $Q=\mathbf{0}$. So clearly this axiom is unsound for late bisimilarity. The soundness for early bisimilarity is straightforward from the definition.

If EARLY is adjoined to Table 6 we get a complete axiomatization of early bisimilarity. The proof of this is different from the corresponding proof of Proposition 8 since it is no longer the case that each summand of P has a bisimilar summand of Q. Therefore, the hnfs corresponding to P and Q must first be saturated by applying EARLY, from left to right, as much as possible to generate new summands. Although the axiom can be applied ad infinitum it turns out that there are only a finite number of distinct new summands that need be generated this way. If instead EARLY is adjoined to Table 8 then we similarly get a complete axiomatization of early congruence. The proof is a combination of the above idea and the proof of Proposition 10.

Interestingly, there is an alternative to EARLY as a characteristic law for early congruence, namely

$$\text{EARLY2} \quad \text{if } \sum_i \tau\,.\,P_i=\sum_j \tau\,.\,Q_j \text{ then } \sum_i a(x)\,.\,P_i=\sum_j a(x)\,.\,Q_j.$$

We can see that EARLY and EARLY2 are equipotent as follows. Consider

$$\tau\,.\,P+\tau\,.\,Q=\tau\,.\,P+\tau\,.\,Q+\tau\,.\,\mathtt{if}\;x=y\;\mathtt{then}\;P\;\mathtt{else}\;Q.$$

This equation is certainly true for $\sim$, since any substitution will make the third summand behave as either $\tau\,.\,P$ or $\tau\,.\,Q$. Hence all instances are derivable from the laws in Table 8. Applying EARLY2 to the equation we get EARLY. So any equation that can be derived with EARLY can also be derived with EARLY2. The converse follows by the fact that EARLY yields a complete axiomatization for $\sim_{\mathrm{E}}$ and EARLY2 is sound.

Similarly, a version of EARLY2 for early bisimilarity is

$$\text{EARLY2}' \quad \text{if } \forall u: \sum_i \tau\,.\,P_i\{u/x\}=\sum_j \tau\,.\,Q_j\{u/x\} \text{ then } \sum_i a(x)\,.\,P_i =\sum_j a(x)\,.\,Q_j.$$

This may appear more complicated than EARLY but it also replaces CONGR2 and an interesting point is that it does not introduce a Mismatch operator. In the subcalculus without Mismatch we can therefore axiomatize late and early bisimilarity (just drop MM1 and MM2 from Table 6). In contrast, for late and early congruence the Mismatch operator plays a significant role in forming complete Generalized matches, so axiomatizations without Mismatch must take a different route.

9.2. *Open bisimilarity*

The remaining interesting equivalence in Figure 2, open bisimilarity (which also is open congruence), is special in that it is only defined for the subcalculus without Mismatch. The way to axiomatize it goes through the relations $\dot{\sim}_{\mathrm{O}}^{D}$ (open bisimilarity under distinction D) so the axiomatization schematically gives rules for all D. The axioms in Table 8 need some modifications. All axioms mentioning Mismatch are dropped; this means that MM1 and GM2 are omitted and in GM1,3,4 M ranges over Generalized matches with no negative conjuncts. A new axiom

$$\text{GM7} \quad P + \texttt{if } x = y \texttt{ then } P = P$$

is added (in the original Table 8 this is derivable through S and GM2). Finally for the interplay between different distinctions we add the following laws, where $=_D$ means provable equality under distinction D,

$$\begin{array}{ll} \text{D1 } \texttt{if } x = y \texttt{ then } P =_D \mathbf{0} & \text{if } x D y, \\ \text{D2 } P =_D Q \text{ implies } P =_{D'} Q & \text{if } D \subseteq D', \\ \text{D3 } P =_D Q \text{ implies } (\boldsymbol{\nu} x) P =_{D-x} (\boldsymbol{\nu} x) Q. & \end{array}$$

Here $D - x$ means the distinction where all pairs containing x are removed from D. Because of these laws GM6 is derivable and needs not be taken as an axiom.

With these changes we obtain a complete axiomatization of $\dot{\sim}_{\mathrm{O}}^{D}$ for all D. The proof is substantially more complicated than for Proposition 10. It uses a notion of head normal form with summands of type $\texttt{if } M \texttt{ then } \alpha . P$, and with the extra requirement that there are no unnecessary summands (a summand P is unnecessary if there is another summand Q and $Q \dot{\sim}_{\mathrm{O}} Q + P$). The proof that all agents are provably equal to such head normal forms uses, among other things, the new law GM7 in order to remove unnecessary summands. The idea of the completeness proof is that given two bisimilar head normal forms, each summand in one of them is provably equal to a summand in the other.

9.3. *Weak congruence*

Finally, for the weak late and early congruences (axiomatizations of weak open congruence have not yet been investigated) it is enough to add the three standard so called τ-laws:

$$\begin{array}{ll} \text{T1} & \alpha . \tau . P = \alpha . P, \\ \text{T2} & P + \tau . P = \tau . P, \\ \text{T3} & \alpha . (P + \tau . Q) = \alpha . (P + \tau . Q) + \alpha . Q. \end{array}$$

The proof follows the same lines as the corresponding proof in CCS: the head normal forms must be saturated in the sense that if $P \stackrel{\alpha}{\Longrightarrow} P'$ then P has a summand $\alpha . P'$. To see an example derivation which demonstrates the subtlety of the interplay between τ-actions and the laws for Match and Mismatch, and highlights a technique used in the completeness proof, suppose we want to prove

$$a(x).(P + \texttt{if } x = y \texttt{ then } \tau . P) = a(x) . P.$$

Clearly this holds, since $(P + \texttt{if } x = y \texttt{ then } \tau . P)\{u/x\}$ is weakly bisimilar to $P\{u/x\}$ for all u. The complication is that $(P + \texttt{if } x = y \texttt{ then } \tau . P)$ is not weakly congruent to P,

Table 10
Axiomatizations of the bisimilarities and congruences.

	strong bisimilarity	strong congruence	weak congruence
late	Table 6	Table 8	Table 8 *and* T1, T2, T3
early	Table 6 *and* EARLY *or* EARLY2′	Table 8 *and* EARLY *or* EARLY2	Table 8 *and* EARLY *or* EARLY2 *and* T1, T2, T3
open	Table 8 *except laws for Mismatch* *and* GM7, D1, D2, D3		*not investigated*

since the former has an initial τ-transition for the substitution $\{y/x\}$. The proof technique is therefore to first infer

$$\tau\,.\,(P + \mathtt{if}\ x = y\ \mathtt{then}\ \tau\,.\,P) = \tau\,.\,P$$

and then use CONGR, prefixing both sides by $a(x)$, and T1 to remove the τ. The inference of $\tau\,.\,(P + \mathtt{if}\ x = y\ \mathtt{then}\ \tau\,.\,P) = \tau\,.\,P$ is as follows:

$$\begin{aligned}
&\tau\,.\,(P + \mathtt{if}\ x = y\ \mathtt{then}\ \tau\,.\,P) = && (\text{GM2})\\
&\mathtt{if}\ x = y\ \mathtt{then}\ \tau\,.\,(P + \mathtt{if}\ x = y\ \mathtt{then}\ \tau\,.\,P)\\
&\qquad + \mathtt{if}\ x \neq y\ \mathtt{then}\ \tau\,.\,(P + \mathtt{if}\ x = y\ \mathtt{then}\ \tau\,.\,P) = && (\text{GM1,3,4})\\
&\mathtt{if}\ x = y\ \mathtt{then}\ \tau\,.\,(P + \tau\,.\,P) + \mathtt{if}\ x \neq y\ \mathtt{then}\ \tau\,.\,P = && (\text{T2})\\
&\mathtt{if}\ x = y\ \mathtt{then}\ \tau\,.\,\tau\,.\,P + \mathtt{if}\ x \neq y\ \mathtt{then}\ \tau\,.\,P = && (\text{T1})\\
&\mathtt{if}\ x = y\ \mathtt{then}\ \tau\,.\,P + \mathtt{if}\ x \neq y\ \mathtt{then}\ \tau\,.\,P = && (\text{GM2})\\
&\tau\,.\,P.
\end{aligned}$$

The different axiomatizations are summarized in Table 10.

10. Sources

Early developments. Computational models where mobility plays a predominant role have been present at least since the mid 1970's in the so called actor systems by Carl Hewitt and Gul Agha [18,2]. Technically the π-calculus has its roots in process algebras like Robin Milner's CCS, originally developed in the late 1970's [22,23]. The first efforts to extend CCS with the ability to pass communication channels were by Egidio Astesiano and Elena Zucca (1984) and by Uffe Engberg and Mogens Nielsen (1986) [4,12]. These calculi turned out to be quite complex. The π-calculus in its present form was developed in the late 1980's by Milner, David Walker and myself; it was first presented at seminars in 1987, the first comprehensive technical reports appeared in 1989 and the journal publication in 1992 [30]. It builds on the article by Engberg and Nielsen, and simplifies it by using a single syntactical class of names to represent values, variables and channels. There it is shown how to encode data types and a version of the lambda calculus; it also explores the operational semantics (late semantics without use of structural congruence), defines late and early bisimilarity and congruence and gives the axiomatization of late bisimilarity. The calculus is very similar to the one presented here, though there is no Mismatch operator and some differences in notation. For example Restriction $(\boldsymbol{\nu} x)$ is written (x), and $\texttt{if}\ x = y\ \texttt{then}\ P$ is written more compactly $[x = y]P$, a notation that many papers on the π-calculus follow.

Introductions and overviews. Today the π-calculus and related theories make up a large and diverse field with hundreds of published papers. Perhaps the main sign of vitality is not the theoretical developments, but rather the numerous calculi, computational models and languages that borrow semantics or central concepts from π and are aimed at more focussed areas of application, such as PICT, Facile, Join, Ambients, Spi, POOL, ... the list can be made very long [45,46,8,15,11,1,57]. A brief overview from 1998 by Uwe Nestmann and Björn Victor indicate the main issues and is accompanied by a well maintained and exhaustive searchable on-line bibliography founded in 1995 [35]. Nestmann also currently maintains the web page *Calculi for Mobile processes* at `http://www.move.to/mobility` with links to introductory papers, active researchers, departments, projects and other resources. In view of these efforts it seems excessive to here attempt a substantial overview or list of references. If a single introductory article for the layman should be mentioned it must be Milner's 1991 Turing Award Lecture [26]. A newcomer to the field may also appreciate my article from 1993 using graphs instead of formulas, with an emphasis on how various basic computational structures are represented through name-passing [39]. Otherwise a standard text is Milner's tutorial on the polyadic π-calculus (1991), and also his recent book (1999) is aimed at a non-specialist audience [27,29]. A short introduction to a simple variant of the π-calculus in comparison to the lambda-calculus is given by Benjamin Pierce (1996) [43].

A reader seeking more detailed information concerning some specific topic is well advised to first consult Kohei Honda's annotated on-line bibliography (reachable from the URL above); it is not exhaustive and not updated as frequently as the one by Nestmann and Victor, but focuses on a few central issues and explains the impact of each paper.

Below I briefly mention the main sources for the aspects of the calculus elaborated in this introduction. The chronology refers to the first written account of a piece of work known to me; the corresponding conference paper or journal article is often published a couple of years later.

Variants of the calculus. The Mismatch operator first appeared in 1990 and was used by Davide Sangiorgi and me in 1993 [38,41]. The role of the Sum operator has been studied by

Catuscia Palamidessi (1997) where she shows that it is necessary for the representation of certain distributed algorithms, and by Nestmann (1997) and Nestmann and Pierce (1996) where encodings of Sum for special cases are analyzed (the encoding in Section 3.5 comes from the latter article) [37,33,34]. The polyadic calculus with the sort system given in Section 3.3 was introduced by Milner in his 1991 tutorial. The ideas for encoding the polyadic calculus into the monadic have been known since the first papers on the π-calculus but were not studied in detail until 1996 by Nobuko Yoshida and 1998 by Paola Quaglia and Walker [58,47]. Type inference algorithms were first presented independently by Simon Gay and by Vasco Vasconcelos and Honda in 1993 [16,55]. More elaborate systems for types and sorts have been presented by many others. Prominent issues are more refined types, for example distinguishing different uses of a channel for input and output, subtyping, polymorphism and higher-order types. A survey of such systems remains to be written although Honda's bibliography mentioned above explains many of the contributions. The slides of a good introductory tutorial by Pierce (1998) are available on-line from the author's home page (reachable from the URL for the mobility home page above) [44].

Replication in place of recursion was first suggested by Milner in a paper on encodings of the lambda-calculus 1990, and although the relationship between lambda and π has subsequently been treated by many authors that paper remains the principal reference [25]. The issue is to some extent connected to higher-order process calculi. Bent Thomsen, Gérard Boudol and Flemming Nielson were the first to study such calculi in detail (independently 1989, these calculi differ significantly from the π-calculus) [53,9,36,54]. The connection with the lambda-calculus was later clarified by Sangiorgi, who also studied the encoding of the higher-order π-calculus into the standard calculus [49].

The significance of the asynchronous calculus was discovered independently by Kohei Honda and Mario Tokoro 1991 and by Boudol 1992 [19,10]. A particularly interesting variant is the subcalculus without free output actions, which has a simpler theory and retains a surprising expressive power, as demonstrated by Sangiorgi and Michele Boreale (1996) [50,6]. Sangiorgi has recently written a tutorial on the subject [52].

Variants of the semantics. The first presentation of a structural congruence and reduction semantics is by Milner (1990) [25], inspired by the chemical abstract machines of Gérard Berry and Boudol from 1989 [5] and Milner also gave the semantics for abstractions and concretions (with a slightly different notation; Milner prefers $P \succ c\,.\,E$ to $P \stackrel{c}{\longrightarrow} E$) in his 1991 tutorial. The early semantics was first presented by Milner, Walker and myself in 1991 [31].

Symbolic transitions and their use in decision procedures for bisimulation equivalences were introduced by Matthew Hennessy and Huimin Lin in 1992 for a process algebra where values (but not channel names) are transmitted between agents [17]. Another chapter in this handbook by Anna Ingólfsdóttir and Lin treats this issue in full. The semantics was subsequently adapted by Lin (1994) to the π-calculus [20,21]. Similar ideas were developed in parallel by Boreale and Rocco de Nicola [7]. Symbolic transitions are also used by Faron Moller and Björn Victor in an automatic verification tool (1994) [56].

Variants of bisimilarity. Early bisimilarity was mentioned in the first paper on the calculus and was studied in more depth by Milner, Walker and myself (1991) along with modal logics which clarify the relationship between late and early bisimulation [31]. The results on barbed congruence are primarily due to Milner and Sangiorgi (1992) and are developed in Sangiorgi's Ph.D. thesis [32,48]. A version of barbed congruence for the asynchronous subcalculus is defined and axiomatized by Roberto Amadio, Ilaria Castellani and Sangiorgi (1996) [3]; the example at the very end of Section 7.2 is from that paper.

Open bisimulation (both strong and weak) and its axiomatization (only strong) is due to Sangiorgi 1993 who also uses a symbolic semantics [51]. The weak late and early bisimulation equivalences and congruences were first formulated by Milner (1990) [24].

Algebraic theory. Strong late bisimilarity was axiomatized in the first paper on the calculus. Sangiorgi and I axiomatized strong late congruence and early bisimilarity and congruence (through the law EARLY) in 1993; the axiomatizations given here mainly follow that work [41]. Axiomatizations of the weak congruences (late and early) have been treated by Lin in a style slightly different from what has been presented here [21]. The law EARLY2 also comes from this line of work. Inferences are there of kind $C \rhd P = Q$ meaning "under condition C it holds that $P = Q$", which we would express as $\texttt{if}\ C\ \texttt{then}\ P = \texttt{if}\ C\ \texttt{then}\ Q$, making the condition part of the agents. The fact that this yields an isomorphic proof system was only recently established [40]. An alternative presentation of weak late congruence and its axiomatization is by Gianluigi Ferrari, Ugo Montanari and Paola Quaglia (1995) [14].

Unifying efforts. In this growing and diversifying field it is also appropriate to mention a few efforts at unification. The action calculi by Milner (emanating from the work on action structures 1992) separate the concerns of parametric dependency and execution control [28]. The tiles structures by Ferrari and Montanari (emanating from work with Fabio Gaducci 1995) generalize the concepts of context and context composition [13]. The Fusion Calculus by Victor and myself (begun in 1997) identifies a single binding operator which can be used to derive both input and Restriction [42]. Although each has made some progress it is clear that much work remains to be done.

Acknowledgements

I thank Huimin Lin, Davide Sangiorgi, Thomas Noll, Gunnar Övergaard, and the anonymous referees for many helpful comments.

References

[1] M. Abadi and A.D. Gordon, *A calculus for cryptographic protocols: The Spi calculus*, J. Inform. Comput. **143** (1999), 1–70. An extended abstract appeared in the Proc. Fourth ACM Conference on Computer and Communications Security, Zürich (1997). An extended version of this paper appears as Research Report 149, Digital Equipment Corporation Systems Research Center (1998), and, in preliminary form, as Technical Report 414, University of Cambridge Computer Laboratory (1997).

[2] G. Agha, *Actors: A Model of Concurrent Computation in Distributed Systems*, MIT Press (1986).

[3] R.M. Amadio, I. Castellani and D. Sangiorgi, *On bisimulations for the asynchronous π-calculus*, Theoret. Comput. Sci. **195** (2) (1998), 291–324. An extended abstract appeared in Proc. CONCUR '96, Lecture Notes in Comput. Sci. 1119, 147–162.

[4] E. Astesiano and E. Zucca, *Parametric channels via label expressions in CCS*, Theoret. Comput. Sci. **33** (1984), 45–64.

[5] G. Berry and G. Boudol, *The chemical abstract machine*, Theoret. Comput. Sci. **96** (1992), 217–248.

[6] M. Boreale, *On the expressiveness of internal mobility in name-passing calculi*, Theoret. Comput. Sci. **195** (2) (1998), 205–226. An extended abstract appeared in Proc. CONCUR '96, Lecture Notes in Comput. Sci. 1119, 163–178.

[7] M. Boreale and R. De Nicola, *A symbolic semantics for the π-calculus*, J. Inform. Comput. **126** (1) (1996), 34–52. Available as Report SI 94 RR 04, Università "La Sapienza" di Roma; an extended abstract appeared in Proc. CONCUR '94, Lecture Notes in Comput. Sci. 836, 299–314.

[8] R. Borgia, P. Degano, C. Priami, L. Leth and B. Thomsen, *Understanding mobile agents via a non interleaving semantics for Facile*, Proc. SAS '96, Lecture Notes in Comput. Sci. 1145, R. Cousot and D.A. Schmidt, eds, Springer (1996), 98–112. Extended version as Technical Report ECRC-96-4, 1996.

[9] G. Boudol, *Towards a lambda-calculus for concurrent and communicating systems*, Proc. TAPSOFT '89, Vol. 1, Lecture Notes in Comput. Sci. 351, J. Díaz and F. Orejas, eds, Springer (1989), 149–161.

[10] G. Boudol, *Asynchrony and the π-calculus (note)*, Rapport de Recherche 1702, INRIA Sophia-Antipolis (1992).

[11] L. Cardelli and A.D. Gordon, *Mobile ambients*, Proc. FoSSaCS '98, Lecture Notes in Comput. Sci. 1378, M. Nivat, ed., Springer (1998), 140–155.

[12] U. Engberg and M. Nielsen, *A calculus of communicating systems with label-passing*, Technical Report DAIMI PB-208, Comp. Sc. Department, Univ. of Aarhus, Denmark (1986).

[13] G. Ferrari and U. Montanari, *A tile-based coordination view of asynchronous π-calculus*, Proc. MFCS '97, Lecture Notes in Comput. Sci. 1295, I. Prívara and P. Ružička, eds, Springer (1994).

[14] G. Ferrari, U. Montanari and P. Quaglia, *The weak late π-calculus semantics as observation equivalence*, Proc. CONCUR '95, Lecture Notes in Comput. Sci. 962, I. Lee and S.A. Smolka, eds, Springer (1995), 57–71.

[15] C. Fournet and G. Gonthier, *The reflexive chemical abstract machine and the join-calculus*, Proc. POPL '96, J.G. Steele, ed., ACM (1996), 372–385.

[16] S.J. Gay, *A sort inference algorithm for the polyadic π-calculus*, Proc. POPL '93, ACM (1993).

[17] M. Hennessy and H. Lin, *Symbolic bisimulations*, Theoret. Comput. Sci. **138** (2) (1995), 353–389. Earlier version as Technical Report 1/92, School of Cognitive and Computing Sciences, University of Sussex, UK.

[18] C. Hewitt, *Viewing control structures as patterns of passing messages*, J. Artif. Intell. **8** (1977), 323–364.

[19] K. Honda and M. Tokoro, *An object calculus for asynchronous communication*, Proc. ECOOP '91, Lecture Notes in Comput. Sci. 512, P. America, ed., Springer (1991), 133–147.

[20] H. Lin, *Symbolic bisimulation and proof systems for the π-calculus*, Technical Report 7/94, School of Cognitive and Computing Sciences, University of Sussex, UK (1994).

[21] H. Lin, *Complete inference systems for weak bisimulation equivalences in the π-calculus*, Proc. TAPSOFT '95, Lecture Notes in Comput. Sci. 915, P.D. Mosses, M. Nielsen and M.I. Schwarzbach, eds, Springer (1995), 187–201. Available as Technical Report ISCAS-LCS-94-11, Institute of Software, Chinese Academy of Sciences (1994).

[22] R. Milner, *A Calculus of Communicating Systems*, Lecture Notes in Comput. Sci. 92, Springer (1980).

[23] R. Milner, *Communication and Concurrency*, Prentice-Hall (1989).

[24] R. Milner, *Weak bisimilarity: Congruences and equivalences*, π-calculus note RM10, Manuscript (1990).

[25] R. Milner, *Functions as processes*, J. Math. Structures in Comput. Sci. **2** (2) (1992), 119–141. Previous version as Rapport de Recherche 1154, INRIA Sophia-Antipolis, 1990 and in Proc. ICALP '91, Lecture Notes in Comput. Sci. 443.

[26] R. Milner, *Elements of interaction*, Comm. ACM **36** (1) (1993), 78–89. Turing Award Lecture.

[27] R. Milner, *The Polyadic π-Calculus: A Tutorial*, Logic and Algebra of Specification, Vol. 94 Series F, NATO ASI, F.L. Bauer, W. Brauer and H. Schwichtenberg, eds, Springer (1993). Available as Technical Report ECS-LFCS-91-180, University of Edinburgh (1991).

[28] R. Milner, *Calculi for interaction*, Acta Inform. **3** (8) (1996), 707–737.

[29] R. Milner, *Communicating and Mobile Systems: the π-Calculus*, Cambridge University Press (1999).

[30] R. Milner, J. Parrow and D. Walker, *A calculus of mobile processes, Part* I/II, J. Inform. Comput. **100** (1992), 1–77.
[31] R. Milner, J. Parrow and D. Walker, *Modal logics for mobile processes*, Theoret. Comput. Sci. **114** (1993), 149–171.
[32] R. Milner and D. Sangiorgi, *Barbed bisimulation*, Proc. ICALP '92, Lecture Notes in Comput. Sci. 623, W. Kuich, ed., Springer (1992), 685–695.
[33] U. Nestmann, *What is a 'good' encoding of guarded choice?* Proc. EXPRESS '97, ENTCS 7, C. Palamidessi and J. Parrow, eds, Elsevier Science Publishers (1997). Full version as report BRICS-RS-97-45, Universities of Aalborg and Århus, Denmark, 1997. Revised version accepted for J. Inform. Comput. (1998).
[34] U. Nestmann and B.C. Pierce, *Decoding choice encodings*, Proc. CONCUR '96, Lecture Notes in Comput. Sci. 1119, U. Montanari and V. Sassone, eds, Springer (1996), 179–194. Revised full version as report ERCIM-10/97-R051, European Research Consortium for Informatics and Mathematics (1997).
[35] U. Nestmann and B. Victor, *Calculi for mobile processes: Bibliography and web pages*, Bulletin of the EATCS **64** (1998), 139–144.
[36] F. Nielson, *The typed λ-calculus with first-class processes*, Proc. PARLE'89, Lecture Notes in Comput. Sci. 366, Springer-Verlag (1989), 357–373.
[37] C. Palamidessi, *Comparing the expressive power of the synchronous and the asynchronous π-calculus*, Proc. POPL '97, ACM (1997), 256–265.
[38] J. Parrow, *Mismatching and early equivalence*, π-calculus note JP13, Manuscript (1990).
[39] J. Parrow, *Interaction diagrams*, Nordic J. Computing **2** (1995), 407–443. A previous version appeared in Proc. A Decade in Concurrency, Lecture Notes in Comput. Sci. 803 (1993), 477–508.
[40] J. Parrow, *On the relationship between two proof systems for the pi-calculus*, π-calculus note JP15, Manuscript (1999). Available from the author.
[41] J. Parrow and D. Sangiorgi, *Algebraic theories for name-passing calculi*, J. Inform. Comput. **120** (2) (1995), 174–197. A previous version appeared in Proc. of A Decade in Concurrency, Lecture Notes in Comput. Sci. 803 (1993), 477–508.
[42] J. Parrow and B. Victor, *The fusion calculus: Expressiveness and symmetry in mobile processes*, Proc. LICS '98, IEEE, Computer Society Press (1998), 176–185.
[43] B.C. Pierce, *Foundational calculi for programming languages*, Handbook of Computer Science and Engineering, A.B. Tucker, ed., CRC Press (1996), Chapter 139.
[44] B.C. Pierce, *Type systems for concurrent calculi*, Invited tutorial at CONCUR, Nice, France (1998).
[45] B.C. Pierce and D.N. Turner, *Concurrent objects in a process calculus*, Proc. TPPP '94, Lecture Notes in Comput. Sci. 907, T. Ito and A. Yonezawa, eds, Springer (1995), 187–215.
[46] B.C. Pierce and D.N. Turner, *Pict: A programming language based on the pi-calculus*, Proof, Language and Interaction: Essays in Honour of Robin Milner, G. Plotkin, C. Stirling and M. Tofte, eds (1999). To appear.
[47] P. Quaglia and D. Walker, *On encoding $p\pi$ in $m\pi$*, 18th Conference on Foundations of Software Technology and Theoretical Computer Science (Chennai, India, December 17–19, 1998), Lecture Notes in Comput. Sci., V. Arvind and R. Ramanujam, eds (1998).
[48] D. Sangiorgi, *Expressing Mobility in Process Algebras: First-Order and Higher-Order Paradigms*, Ph.D. thesis, LFCS, University of Edinburgh (1993). CST-99-93 (also published as ECS-LFCS-93-266).
[49] D. Sangiorgi, *Bisimulation in higher-order process calculi*, J. Inform. Comput. **131** (1996), 141–178. Available as Rapport de Recherche RR-2508, INRIA Sophia-Antipolis, 1995. An early version appeared in Proc. PROCOMET'94, IFIP, North-Holland, 207–224.
[50] D. Sangiorgi, *π-calculus, internal mobility and agent-passing calculi*, Theoret. Comput. Sci. **167** (1,2) (1996), 235–274. Also as Rapport de Recherche RR-2539, INRIA Sophia-Antipolis, 1995. Extracts of parts of the material contained in this paper can be found in Proc. TAPSOFT '95 and ICALP '95.
[51] D. Sangiorgi, *A theory of bisimulation for the π-calculus*, Acta Inform. **33** (1996), 69–97. Earlier version published as Report ECS-LFCS-93-270, University of Edinburgh. An extended abstract appeared in the Proc. CONCUR '93, Lecture Notes in Comput. Sci. 715.
[52] D. Sangiorgi, *Asynchronous process calculi: The first-order and higher-order paradigms*, Theoret. Comput. Sci. (1999), to appear.
[53] B. Thomsen, *A calculus of higher order communicating systems*, Proc. POPL '89, ACM (1989), 143–154.
[54] B. Thomsen, *Plain CHOCS. A second generation calculus for higher order processes*, Acta Inform. **30** (1) (1993), 1–59.

[55] V.T. Vasconcelos and K. Honda, *Principal typing schemes in a polyadic π-calculus*, Proc. CONCUR '93, Lecture Notes in Comput. Sci. 715, E. Best, ed., Springer (1993), 524–538.
[56] B. Victor and F. Moller, *The Mobility Workbench – a tool for the π-calculus*, Proc. CAV '94, Lecture Notes in Comput. Sci. 818, D. Dill, ed., Springer (1994), 428–440.
[57] D. Walker, *Objects in the π-calculus*, J. Inform. Comput. **116** (2) (1995), 253–271.
[58] N. Yoshida, *Graph types for monadic mobile processes*, Proc. FSTTCS '96, Lecture Notes in Comput. Sci. 1180, V. Chandru and V. Vinay, eds, Springer (1996), 371–386. Full version as Technical Report ECS-LFCS-96-350, University of Edinburgh.

Subject index

τ-laws, 535

actions, 502
agent abstractions, 511
agent concretions, 511
agent identifier, 483
agent variable, 496
agents, 483
agrees with, 532
alpha-conversion, 487
asynchronous calculus, 498
automata, 493

barbed bisimulation, 520
barbed congruence, 521
bisimulation, 513
bisimulation up to, 516
Boolean values, 492
bound names, 485
bound output, 502

CLOSE, 505
COM, 503
complete conjunction, 532
CONGR, 531
CONGR1–2, 528

Definition, 484
depth, 529
distinction, 523
dynamic bisimulation, 523

E-COM, 509
E-INPUT, 509
EARLY, 533
early bisimulation, 518
early congruence, 520, 534
early semantics, 509
EARLY2, 534
expansion law, 530, 533
extended agents, 511

fixpoint, 496
free names, 485

Generalized match, 530
GM1–6, 531
GM6*, 531
ground bisimulation, 523
guarded, 485
guarded Sum, 493

head normal form, 529, 532
higher-order calculus, 499

Identifier, 484
Input Prefix, 483

late bisimulation, 518

M1–2, 528
Match, 484
MATCH, 503
Mismatch, 484
MISMATCH, 503
MM1–2, 528
monoid laws, 487

names, 483

object, 485
object sort, 495
observable, 520
OPEN, 503
open bisimilation, 534
open bisimulation, 522, 524
operational semantics, 503
Output Prefix, 483

PAR, 503
Parallel Composition, 484
polyadic calculus, 494
PREFIX, 503

R1–3, 528
reductions, 510
Replication, 496
RES, 503
Restriction, 484

S, 528
STR, 528
scope extension, 487
Silent Prefix, 483
sort, 495
sort context, 495
strong bisimulation, 514
strong congruence, 516
STRUCT, 503
Structural congruence, 486
subject, 485
substitution, 485
Sum, 484
SUM, 503
symbolic semantics, 506

unfolding, 487

weak bisimulation, 526
weak congruence, 527, 535

CHAPTER 9

Verification on Infinite Structures

Olaf Burkart[1], Didier Caucal[2], Faron Moller[3], Bernhard Steffen[4]

[1] *Lehrstuhl Informatik V, Universität Dortmund, Baroper Straße 301, D-442 21 Dortmund, Germany*
E-mail:burkart@cs.uni-dortmund.de

[2] *IRISA, Campus de Beaulieu, F-350 42 Rennes, France*
E-mail: caucal@irisa.fr

[3] *Department of Computer Science, University of Wales Swansea, Singleton Park, Swansea, SA2 8PP, United Kingdom*
E-mail: F.G.Moller@swansea.ac.uk

[4] *Lehrstuhl Informatik V, Universität Dortmund, D-442 21 Dortmund, Germany*
E-mail: steffen@cs.uni-dortmund.de

Contents

Introduction . . . 547
Equivalence checking . . . 548
Model checking . . . 549
1. A taxonomy of infinite state processes . . . 550
1.1. Rewrite transition systems . . . 550
1.2. Languages and bisimilarity . . . 556
1.3. Expressivity results . . . 561
1.4. Further classes of processes . . . 564
2. The equivalence checking problem . . . 565
2.1. Decidability results for BPA and BPP . . . 565
2.2. Polynomial time algorithms for BPA and BPP . . . 572
2.3. Computing a finite bisimulation base for BPA . . . 578
2.4. Undecidability results for MSA . . . 586
2.5. Summary of results . . . 589
3. The model checking problem . . . 591
3.1. Temporal logics . . . 592
3.2. Algorithms for branching-time logics . . . 602
3.3. Algorithms for linear-time logics . . . 611
3.4. Summary . . . 615
Acknowledgements . . . 617
References . . . 617
Subject index . . . 623

HANDBOOK OF PROCESS ALGEBRA
Edited by Jan A. Bergstra, Alban Ponse and Scott A. Smolka

Abstract

In this chapter, we present a hierarchy of infinite-state systems based on the primitive operations of sequential and parallel composition; the hierarchy includes a variety of commonly-studied classes of systems such as context-free and pushdown automata, and Petri net processes. We then examine the equivalence and regularity checking problems for these classes, with special emphasis on bisimulation equivalence, stressing the structural techniques which have been devised for solving these problems. Finally, we explore the model checking problem over these classes with respect to various linear- and branching-time temporal logics.

Introduction

The study of automated (sequential) program verification has an inherent theoretical barrier in the guise of the halting problem and formal undecidability. The simplest programs which manipulate the simplest infinite data types such as integer variables immediately fall foul of these theoretical limitations. During execution, such a program may evolve into any of an infinitude of states, and knowing if the execution of the program will lead to any particular state, such as the halting state, will in general be impossible to determine. However, this has not prevented a very successful attack on the problem of proving program correctness, and there are now elegant and accepted techniques for the semantic analysis of software.

The history behind the modelling of concurrent systems, in particular hardware systems, has followed a different course. Here, systems have been modelled strictly as finite-state systems, and formal analysis tools have been developed for completely and explicitly exploring the reachable states of any given system, for instance with the goal of detecting whether or not a halting, i.e., deadlocked, state is accessible. This abstraction has been warranted up to a point. Real hardware components are indeed finite entities, and protocols typically behave in a regular fashion irrespective of the diversity of messages which they may be designed to deliver.

In specifying concurrent systems, it is not typical to explicitly present the state spaces of the various components, for example by listing out the states and transition function, but rather to specify them using some higher-level modelling language. Such formalisms for describing concurrent systems are not usually so restrictive in their expressive power. For example, the typical process algebra can encode integer registers, and with them compute arbitrary computable functions; and Petri nets constitute a graphical language for finitely presenting typically infinite-state systems. However, tools which employ such formalisms generally rely on techniques for first assuring that the state space of the system being specified is semantically, if not syntactically, finite. For example, a given process algebra tool might syntactically check that no static operators such as parallel composition appear within the scope of a recursive definition; and a given Petri net tool might check that a net is safe, that is, that no place may acquire more than one token. Having verified the finiteness of the system at hand, the search algorithm can, at least in principle, proceed.

The problem with the blind search approach, which has thwarted attempts to provide practical verification tools, is of course that of state space explosion. The number of reachable states of a system will typically be on the order of exponential in the number of components which make up the system. Hence a great deal of research effort has been expended on taming this state space, typically by developing intelligent search strategies. Various promising techniques have been developed which make for the automated analysis of extremely large state spaces feasible; one popular approach to this problem is through the use of BDD (binary decision diagram) encodings of automata [20]. However, such approaches are inherently bound to the analysis of finite-state systems.

Recently, interest in addressing the problem of analyzing infinite-state systems has blossomed within the concurrency theory community. The practical motivation for this has been both to provide for the study of parallel program verification, where infinite data types are manipulated, as well as to allow for more faithful representations of concurrent systems. For example, real-time and probabilistic models have come into vogue during the last decade to reflect for instance the temporal and nondeterministic behaviour of asynchronous hardware components responding to continuously-changing analogue signals; and models have been developed which allow for the dynamic reconfiguration of a system's structure. Such enhancements to the expressive power of a modelling language immediately give rise to infinite-state models, and new paradigms not based on state space search need to be introduced to successfully analyze systems expressed in such formalisms.

In this survey, we explore the two major streams in system verification: *equivalence checking*, which aims at establishing some semantic equivalence between two systems, one of which is typically considered to represent the implementation of the specification given by the other; and *model checking*, which aims at determining whether or not a given system satisfies some property which is typically presented in some modal or temporal logic. In the survey we concentrate exclusively on discrete systems, and neglect infinite-state structures related to, e.g., timed and hybrid systems. The variety of process classes which we consider is catalogued in Section 1, where we detail the relative expressive powers of these classes. The two approaches to system verification are summarized here, where we also sketch the contents of the two relevant sections of the survey.

Equivalence checking

Equivalence checking, that is, determining when two (infinite-state) systems are in some semantic sense equal, is clearly a particularly relevant problem in system verification. Indeed, such questions have a long tradition in the field of (theoretical) computer science. Since the proof by Moore [121] in 1956 of the decidability of language equivalence for finite-state automata, formal language theorists have been studying the equivalence problem over classes of automata which express languages which are more expressive than the class of regular languages generated by finite-state automata. Bar-Hillel, Perles and Shamir [6] were the first to demonstrate, in 1961, that the class of languages defined by context-free grammars was too wide to admit a decidable theory for language equivalence. Shortly after this, Korenjak and Hopcroft [95] demonstrated that language equivalence between simple (deterministic) grammars is decidable. Only recently has the long-open problem of language equivalence between deterministic push-down automata (DPDA) been settled (positively) by Sénizergues [132,133].

Decidability questions for Petri nets were addressed already two decades ago, with the thesis of Hack [63]. However, it has only been in the much more recent past that a more concerted effort has been focussed on such questions, with the interest driven in part by analogies drawn between classes of concurrent system models and classes of generators for families of formal languages. In [114] Milner exploits the relationship between regular (finite-state) automata as discussed by Salomaa in [130] and regular behaviours to present the decidability and a complete axiomatization of bisimulation equivalence for finite-state behaviours, whilst in his textbook [115] he demonstrates that the halting problem for Turing machines can be encoded as a bisimulation question for the full CCS calculus thus demonstrating undecidability in general. This final feat is carried out elegantly using finite representations of counters in the thesis of Taubner [141]. These results are as expected; however, real interest was generated with the discovery by Baeten, Bergstra and Klop [4,5] that bisimulation equivalence is decidable for a family of infinite-state automata generated by a general class of context-free grammars.

In Section 2, we present an overview of various results obtained regarding decidability and complexity, with particular emphasis on bisimulation equivalence, focussing on the various techniques exploited in each case. Our interest in bisimulation equivalence stems predominantly from its mathematical tractability. Apart from being the fundamental notion of equivalence for several process algebraic formalisms, bisimulation equivalence possesses several pleasing mathematical properties, not least of which being – as we shall discover – that it is decidable over process classes for which all other common equivalences remain undecidable, in particular over the class of processes defined by context-free grammars. Furthermore in a particularly interesting class of processes – namely the normed deterministic processes – all of the standard equivalences coincide, so it is sensible to concentrate on the most mathematically tractable equivalence when analyzing properties of another equivalence. In particular, by studying bisimulation equivalence we can rediscover standard theorems about the decidability of language equivalence, as well as provide more efficient algorithms for these decidability results than have previously been presented. We expect that the structural techniques which can be exploited in the study of bisimulation equivalence will prove to be useful in tackling various other language theoretic problems. Indeed, while Sénizergues' proof of the decidability of DPDA is extremely long and complex, developed over 70 pages of a 166 page journal submission, Stirling [138] has since presented a far simpler proof of this result using variations on some of the structural analysis techniques explored in Section 2.

Apart from the basic equivalence checking problem, we also survey related questions regarding *regularity checking*, that is, determining if a given system is semantically finite-state. Within a typical application domain such as the modelling of protocols, we might analyze infinite-state systems which we intend to semantically represent finite-state behaviours: our specification of the system is finite-state, but the state space of the implementation is (syntactically) infinite. A positive answer for the regularity checking problem would allow the equivalence checking algorithm to exploit well-developed techniques for analyzing finite-state systems (assuming that the equivalent finite-state system could be extracted from from this answer).

Model checking

Apart from equivalence checking, model checking provides perhaps the most promising approach to the formal verification of distributed, reactive systems. In this approach, one uses formulae of a temporal logic to specify the desired properties of a system, and a (semi-)decision procedure then checks whether the given system is a model of the formula at hand. In particular, in the case of finite-state systems model checking is, at least theoretically, always applicable, since an exhaustive traversal through the reachable state space of the system under consideration can effectively provide enough information to solve the verification problem. In fact, using BDDs, finite-state model checking has had some prominent industrial success stories, despite the omni-present state space explosion threat, making it an indispensable tool for hardware design. The practical success of model checking can be witnessed by the immediate popularity of the newly-published textbook by Clarke, Grumberg and Peled [42].

Again, as algorithms for finite-state model checking typically involve the exhaustive traversal of the state space, they are inherently incapable of verifying infinite-state systems, and must be replaced by algorithms based on different techniques. The new techniques which have been developed for model checking have mainly been influenced by formal language theory, where there exists a long tradition of reasoning about finitely-presented infinite objects, in this case formal languages described by automata or grammars. Consequently, notions from formal language theory form a recurring theme in the verification of infinite-state systems, e.g., as a means for the description of such systems, or in characterizing the expressive power of certain temporal logics.

One general problem is, however, the trade-off between expressiveness and decidability. Expressive models of computation, in conjunction with powerful specification formalisms, inevitably lead to an undecidable model checking problem. However, taking weaker models and/or formalisms in order to recover decidability often greatly diminishes its practical value. The goal of this survey is to provide a systematic presentation of the results obtained so far concerning decidability and complexity issues in model checking infinite-state systems, thereby identifying feasible combinations of process classes and temporal logics.

In Section 3.1 we provide a brief introduction to the temporal logics we consider, together with their branching time and linear time classifications. We then proceed by presenting decidability and complexity results about model checking for various classes of infinite-state systems, first for branching time logics in Section 3.2, and then for linear time logics in Section 3.3. Due to the wealth of techniques which have been devised for use in conjunction with the wide variety of logics which one may consider, we only sketch the main ideas of the relevant results in this survey. This gives a conceptual and easy to read overview, which is complemented by tight links to the original literature, thus enabling the interested readers to access all the technical details.

1. A taxonomy of infinite state processes

1.1. *Rewrite transition systems*

Concurrent systems are modelled semantically in a variety of ways. They may be defined for example by the infinite traces or executions which they may perform, or by the entirety of the properties which they satisfy in some particular process logic, or as a particular algebraic model of some equational specification. In any case, a fundamental unifying view is to interpret such systems as edge-labelled directed graphs, whose nodes represent the states in which a system may exist, and whose transitions represent the possible behaviour of the system originating in the state represented by the node from which the transition emanates; the label on a transition represents an event corresponding to the execution of that transition, which will typically represent an interaction with the environment. The starting point for our study will thus be such graphs, which will for us represent processes.

DEFINITION 1. A *labelled transition system* is a tuple $\langle \mathcal{S}, \Sigma, \rightarrow, \alpha_0, F \rangle$ where

- $\mathcal{S}$ is a set of *states*.
- Σ is a finite set of *labels*.
- $\rightarrow \subseteq \mathcal{S} \times \Sigma \times \mathcal{S}$ is a *transition relation*, written $\alpha \xrightarrow{a} \beta$ for $\langle \alpha, a, \beta \rangle \in \rightarrow$.
- $\alpha_0 \in \mathcal{S}$ is a distinguished *start state*.
- $F \subseteq \mathcal{S}$ is a finite set of *final states* which are *terminal*, meaning that for each $\alpha \in F$ there is no $a \in \Sigma$ and $\beta \in \mathcal{S}$ such that $\alpha \xrightarrow{a} \beta$.

This notion of a labelled transition system differs from the standard definition of a finite-state automaton (as for example given in [74]) in that the set of states need not be finite, and final states must not have any outgoing transitions. This last restriction is mild and justified in that a final state refers to the successful termination of a concurrent system. This contrasts with unsuccessful termination (i.e., deadlock) which is represented by all non-final terminal states. We could remove this restriction, but only at the expense of Theorem 4 below which characterizes a wide class of labelled transition systems as push-down automata which accept on empty stack. (An alternative approach could be taken to recover Theorem 4 based on PDA which accept by final state, but we do not pursue this alternative here.)

We follow the example set by Caucal [32] and consider the families of labelled transition systems defined by various rewrite systems. Such an approach provides us with a clear link between well-studied classes of formal languages and transition system generators, a link which is of particular interest when it comes to exploiting process-theoretic techniques in solving problems in classical formal language theory.

DEFINITION 2. A *sequential labelled rewrite transition system* is a tuple $\langle V, \Sigma, P, \alpha_0, F \rangle$ where

- V is a finite set of *variables*; the elements of V^* are referred to as *states*.
- Σ is a finite set of *labels*.
- $P \subseteq V^* \times \Sigma \times V^*$ is a finite set of *rewrite rules*, written $\alpha \xrightarrow{a} \beta$ for $\langle \alpha, a, \beta \rangle \in P$, which are extended by the *prefix rewriting rule*: if $\alpha \xrightarrow{a} \beta$ then $\alpha\gamma \xrightarrow{a} \beta\gamma$.
- $\alpha_0 \in V^*$ is a distinguished *start state*.
- $F \subseteq V^*$ is a finite set of *final states* which are terminal.

A *parallel labelled rewrite transition system* is defined precisely as above, except that the elements of V^* are read modulo commutativity of concatenation, which is thus interpreted as parallel, rather than sequential, composition. We can thus consider states as monomials $X_1^{k_1} X_2^{k_2} \cdots X_n^{k_n}$ over the variables $V = \{X_1, X_2, \ldots, X_n\}$. With this in mind, we shall be able to exploit the following result due to Dickson [48] which is easily proved by induction on n.

LEMMA 3 (Dickson's lemma). *Given an infinite sequence of vectors of natural numbers* $\vec{x}_1, \vec{x}_2, \vec{x}_3, \ldots \in \mathbb{N}^n$ *we can always find indices* i *and* j *with* $i < j$ *such that* $\vec{x}_i \leqslant \vec{x}_j$ *(where* $\leqslant$ *is considered pointwise).*

The state $X_1^{k_1} X_2^{k_2} \cdots X_n^{k_n}$ can be viewed as the vector $(k_1, k_2, \ldots, k_n) \in \mathbb{N}^n$. Hence, Dickson's lemma says that, given any infinite sequence $\alpha_1, \alpha_2, \alpha_3, \ldots$ of such states, we can always find two of these, α_i and α_j with $i < j$, such that the number of occurrences of each variable X in α_j is at least as great as in α_i.

We shall freely extend the transition relation $\rightarrow$ homomorphically to finite sequences of actions $w \in \Sigma^*$ so as to write $\alpha \stackrel{\varepsilon}{\longrightarrow} \alpha$ and $\alpha \stackrel{aw}{\longrightarrow} \beta$ whenever $\alpha \stackrel{a}{\longrightarrow} \cdot \stackrel{w}{\longrightarrow} \beta$. Also, we refer to the set of states α into which the initial state can be rewritten, that is, such that $\alpha_0 \stackrel{w}{\longrightarrow} \alpha$ for some $w \in \Sigma^*$, as the *reachable* states. Although we do not insist that all states be reachable, we shall assume that all variables in V are accessible from the initial state, that is, that for all $X \in V$ there is some $w \in \Sigma^*$ and $\alpha, \beta \in V^*$ such that $\alpha_0 \stackrel{w}{\longrightarrow} \alpha X \beta$.

This definition is slightly more general than that given in [32], which does not take into account final states nor the possibility of parallel rewriting as an alternative to sequential rewriting. By doing this, we expand the study of the classes of transition systems which are defined, and extend some of the results given by Caucal, notably in the characterization of arbitrary sequential rewrite systems as push-down automata.

The families of transition systems which can be defined by restricted rewrite systems can be classified using a form of Chomsky hierarchy. This hierarchy provides an elegant classification of several important classes of transition systems which have been defined and studied independent of their appearance as particular rewrite systems. This classification is presented in Figure 1. (Type 1 rewrite systems, corresponding to context-sensitive grammars, do not feature in this hierarchy since the rewrite rules by definition are only applied to the prefix of a composition.) In the remainder of this section, we explain the classes of transition systems which are represented in this table, working upwards starting with the most restrictive class.

FSA represents the class of finite-state automata. Clearly if the rules are restricted to be of the form $A \stackrel{a}{\longrightarrow} B$ or $A \stackrel{a}{\longrightarrow} \varepsilon$ with $A, B \in V$, then the reachable states of both the sequential and parallel transition systems will be elements of the finite set $\{\alpha \in V^*\colon |\alpha| \leqslant |\alpha_0|\}$.

EXAMPLE 0. In the following we present two type 3 (regular) rewrite systems along with the FSA transition systems which the initial states X and A, respectively, denote.

$X \stackrel{a}{\longrightarrow} Y$

$Y \stackrel{b}{\longrightarrow} \varepsilon$

$Y \stackrel{c}{\longrightarrow} \varepsilon$

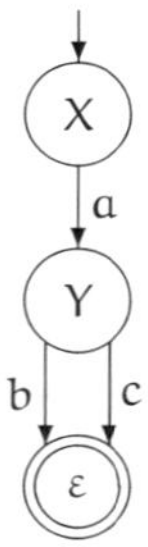

$A \stackrel{a}{\longrightarrow} B$

$A \stackrel{a}{\longrightarrow} C$

$B \stackrel{b}{\longrightarrow} \varepsilon$

$C \stackrel{c}{\longrightarrow} \varepsilon$

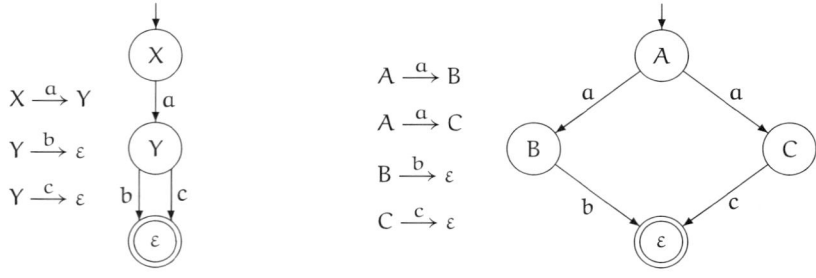

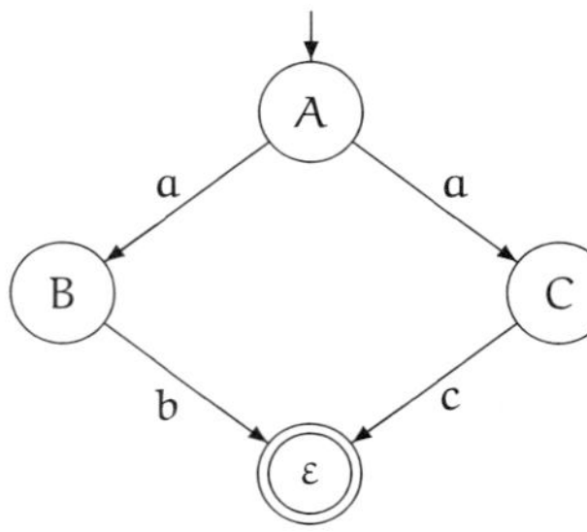

These two automata both recognize the same (regular) language $\{ab, ac\}$. However, they are substantially different automata.

	Restriction on the rules $\alpha \xrightarrow{a} \beta$ of P	Restriction on F	Sequential composition	Parallel composition
Type 0	*none*	*none*	PDA	PN
Type $1\frac{1}{2}$	$\alpha \in Q\Gamma$ and $\beta \in Q\Gamma^*$ where $V = Q \uplus \Gamma$	$F = Q$	PDA	MSA
Type 2	$\alpha \in V$	$F = \{\varepsilon\}$	BPA	BPP
Type 3	$\alpha \in V$ and $\beta \in V \cup \{\varepsilon\}$	$F = \{\varepsilon\}$	FSA	FSA

Fig. 1. A hierarchy of transition systems.

BPA represents the class of Basic Process Algebra processes of Bergstra and Klop [9], which are the transition systems associated with Greibach normal form (GNF) context-free grammars in which only left-most derivations are permitted.

EXAMPLE 1. In the following we present a type 2 (GNF context-free grammar) rewrite system along with the BPA transition system which the initial state X denotes.

$$X \xrightarrow{a} XB \qquad X \xrightarrow{c} \varepsilon \qquad B \xrightarrow{b} \varepsilon$$

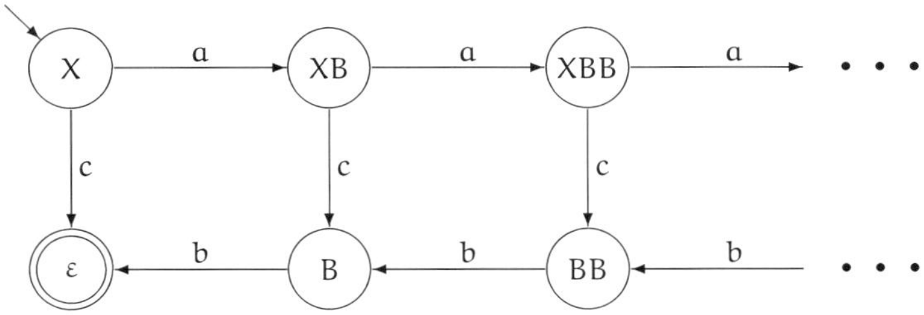

This automaton recognizes the context-free language $\{a^n cb^n \colon n \geqslant 0\}$.

BPP represents the class of Basic Parallel Processes introduced by Christensen [35] as a parallel analogy to BPA, and are defined by the transition systems associated with GNF context-free grammars in which arbitrary grammar derivations are permitted.

EXAMPLE 2. The type 2 rewrite system from Example 1 gives rise to the following BPP transition system with initial state X.

$$X \xrightarrow{a} XB$$
$$X \xrightarrow{c} \varepsilon$$
$$B \xrightarrow{b} \varepsilon$$

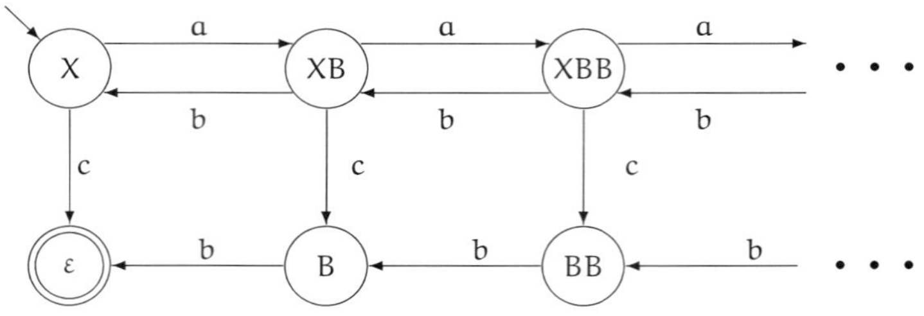

This automaton recognizes the language consisting of all strings of the form $(a+b)^*cb^*$ which contain an equal number of a's and b's in which no prefix contains more b's than a's.

We shall assume that for all type 2 rewrite systems, ε is the only terminal state. Combined with our previous assumption that each variable is reachable from the initial state, this implies that each variables is on the left hand side of at least one rule.

PDA represents the class of push-down automata which accept on empty stack. To present such PDA as a restricted form of rewrite system, we first assume that the variable set V is partitioned into disjoint sets Q (finite control states) and Γ (stack symbols). The rewrite rules are then of the form $pA \xrightarrow{a} q\beta$ with $p, q \in Q$, $A \in \Gamma$ and $\beta \in \Gamma^*$, which represents the usual PDA transition which says that while in control state p with the symbol A at the top of the stack, you may read the input symbol a, move into control state q, and replace the stack element A with the sequence β. Finally, the set of final states is given by Q, which represent the PDA configurations in which the stack is empty.

Caucal [32] demonstrates that, disregarding final states, any unrestricted (type 0) sequential rewrite system can be presented as a PDA, in the sense that the transition systems are isomorphic up to the labelling of states. The stronger result, in which final states are taken into consideration, actually holds as well. The idea behind the encoding is as follows. Given an arbitrary rewrite system $\langle V, \Sigma, P, \alpha_0, F\rangle$, take n satisfying

$$n \geqslant \max\{|\alpha|: \alpha \xrightarrow{a} \beta \in P\}; \quad \text{and}$$

$$n > \max\{|\alpha|: \alpha \xrightarrow{a} \beta \text{ with } \beta \in F\}.$$

Then let

$$Q = \{p_\alpha: \alpha \in V^* \text{ and } |\alpha| < n\}; \quad \text{and}$$

$$\Gamma = \{Z_\alpha: \alpha \in V^* \text{ and } |\alpha| = n\} \cup \{\Lambda\}.$$

Every final transition state $\alpha \in F$ is represented by the PDA state p_α, that is, by the PDA being in control state p_α with an empty stack denoting acceptance; and every non-final transition system state $\gamma\alpha_1\alpha_2\cdots\alpha_k \notin F$ with $|\gamma| < n$ and $|\alpha_i| = n$ for $1 \leqslant i \leqslant k$, is represented in the PDA by $p_\gamma Z_{\alpha_1} Z_{\alpha_2} \cdots Z_{\alpha_k}\Lambda$, that is, by the PDA being in control state p_γ

with with the sequence $Z_{\alpha_1} Z_{\alpha_2} \cdots Z_{\alpha_k} \Lambda$ on its stack. Then every rewrite rule introduces appropriate PDA rules which mimic it and respect this representation. Thus we arrive at the following result.

THEOREM 4. *Every sequential labelled rewrite transition system can be represented (up to the labelling of states) by a PDA transition system.*

EXAMPLE 3. The BPP transition system of Example 2 is given by the following sequential rewrite system.

$$X \xrightarrow{a} XB \qquad X \xrightarrow{c} \varepsilon \qquad B \xrightarrow{b} \varepsilon \qquad XB \xrightarrow{b} X$$

By the above construction, this gives rise to the following PDA with initial state $p_X\Lambda$. (We omit rules corresponding to the unreachable states.)

$$\begin{array}{llll}
\underline{X \xrightarrow{a} XB}: & p_X\Lambda \xrightarrow{a} p_\varepsilon Z_{XB}\Lambda & p_\varepsilon Z_{XB} \xrightarrow{a} p_X Z_{BB} & p_X Z_{BB} \xrightarrow{a} p_\varepsilon Z_{XB} Z_{BB} \\
\underline{X \xrightarrow{c} \varepsilon}: & p_X\Lambda \xrightarrow{c} p_\varepsilon & p_\varepsilon Z_{XB} \xrightarrow{c} p_B & p_X Z_{BB} \xrightarrow{c} p_\varepsilon Z_{BB} \\
\underline{B \xrightarrow{b} \varepsilon}: & p_B\Lambda \xrightarrow{b} p_\varepsilon & p_\varepsilon Z_{BB} \xrightarrow{b} p_B & p_B Z_{BB} \xrightarrow{b} p_\varepsilon Z_{BB} \\
\underline{XB \xrightarrow{b} X}: & p_\varepsilon Z_{XB} \xrightarrow{b} p_X & p_X Z_{BB} \xrightarrow{b} p_\varepsilon Z_{XB} &
\end{array}$$

This is expressed more simply by the following PDA with initial state $p\Lambda$.

$$\begin{array}{lll}
p\Lambda \xrightarrow{a} pB\Lambda & pB \xrightarrow{a} pBB & q\Lambda \xrightarrow{b} q \\
p\Lambda \xrightarrow{c} q & pB \xrightarrow{b} p & qB \xrightarrow{b} q \\
 & pB \xrightarrow{c} q &
\end{array}$$

Note that BPA coincides with the class of single-state PDA. However, we shall see in Section 1.3 that any PDA presentation of the transition system of Example 2 must have at least 2 control states: this transition system is not represented by any BPA.

MSA represents the class of multiset automata, which can be viewed as "parallel" or "random-access" push-down automata; they are defined as above except that they have random access capability to the stack.

EXAMPLE 4. The BPA transition system of Example 1 is isomorphic to that given by the following MSA with initial state pX.

$$pX \xrightarrow{a} pBX \qquad pX \xrightarrow{c} q \qquad qB \xrightarrow{b} q$$

Note that when the stack alphabet has only one element, PDA and MSA trivially coincide. Also note that BPP coincides with the class of single-state MSA. We shall see in Section 1.3

that any MSA presentation of the transition system of Example 1 must have at least 2 control states: this transition system is not represented by any BPP.

PN represents the class of (finite, labelled, weighted place/transition) Petri nets, as is evident by the following interpretation of unrestricted parallel rewrite systems. The variable set V represents the set of places of the Petri net, and each rewrite rule $\alpha \xrightarrow{a} \beta$ represents a Petri net transition labelled a with the input and output places represented by α and β respectively, with the weights on the input and output arcs given by the relevant multiplicities in α and β. Note that a BPP is a communication-free Petri net, one in which each transition has a unique input place.

EXAMPLE 5. The following unrestricted parallel rewrite system with initial state X and final state Y

$$X \xrightarrow{a} XA \qquad XAB \xrightarrow{c} X \qquad YA \xrightarrow{a} Y$$

$$X \xrightarrow{b} XB \qquad X \xrightarrow{d} Y \qquad YB \xrightarrow{b} Y$$

describes the Petri net which in its usual graphical representation net would be rendered as follows. (The weights on all of the arcs is 1.)

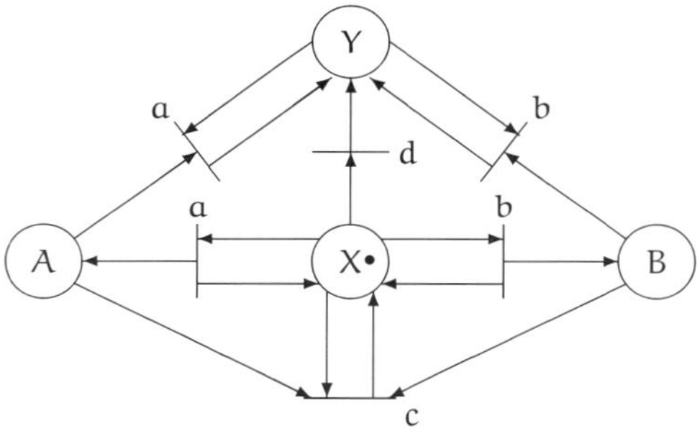

The automaton represented by this Petri net recognizes the language consisting of all strings from $(a+b+c)^*d(a+b)^*$ in which the number of c's in any prefix is bounded above by both the number of a's and the number of b's; and in which the number of a's (respectively b's) before the occurrence of the d minus the number of c's equals the number of a's (respectively b's) after the occurrence of the d.

Although in the sequential case, PDA constitutes a normal form for unrestricted rewrite transition systems, we shall see in Section 1.3 that this result fails to hold in the parallel case. We shall in fact demonstrate that the automaton associated with the above PN is not the automaton of any MSA.

1.2. *Languages and bisimilarity*

Given a labelled transition system $T = \langle S, \Sigma, \rightarrow, \alpha_0, F \rangle$, we can define its *language* $L(T)$ to be the language generated by its initial state α_0, where the language generated by a state is defined in the usual fashion as the sequences of actions which label rewrite transitions leading from the given state to a final state.

DEFINITION 5. $L(\alpha) = \{w \in \Sigma^*\colon \alpha \stackrel{w}{\longrightarrow} \beta$ for some $\beta \in F\}$, and $L(T) = L(\alpha_0)$. α and β are *language equivalent*, written $\alpha \sim_L \beta$, iff they generate the same language: $L(\alpha) = L(\beta)$.

With respect to the languages generated by rewrite systems, if a rewrite system is in the process of generating a word, then the partial word should be extendible to a complete word. That is, from any reachable state of the transition system, a final state should be reachable. If the transition system satisfies this property, it is said to be *normed*; otherwise it is *unnormed*.

DEFINITION 6. We define the *norm* of any state α of a labelled transition system, written $n(\alpha)$, to be the length of a shortest rewrite transition sequence which takes α to a final state, that is, the length of a shortest word in $L(\alpha)$. By convention, we define $n(\alpha) = \infty$ if there is no sequence of transitions from α to a final state, that is, $L(\alpha) = \emptyset$. The transition system is *normed* iff every reachable state α has a finite norm; otherwise it is *unnormed*.

Note that, due to the assumption following Definition 2 on the accessibility of all the variables, if a type 2 rewrite transition system is normed, then all of its variables must have finite norm. The following then is a basic fact about the norms of type 2 (BPA and BPP) states.

LEMMA 7. *Given any state $\alpha\beta$ of a type 2 rewrite transition systems (BPA or BPP), $n(\alpha\beta) = n(\alpha) + n(\beta)$.*

PROOF. For the sequential case, $L(\alpha\beta) = L(\alpha) \cdot L(\beta) = \{uv\colon u \in L(\alpha)$ and $v \in L(\beta)\}$. For the parallel case, $L(\alpha\beta) = L(\alpha) \parallel L(\beta) = \{u_1v_1u_2v_2 \cdots u_nv_n\colon u_1u_2 \cdots u_n \in L(\alpha)$ and $v_1v_2 \cdots v_n \in L(\beta)\}$. The result follows easily from this. □

A further common property of transition systems is that of *determinacy*.

DEFINITION 8. T is *deterministic* iff for every reachable state α and every label a there is at most one state β such that $\alpha \stackrel{a}{\longrightarrow} \beta$.

For example, the two finite state automata presented in Example 0 are both normed transition systems, while only the first is deterministic. All other examples which we have presented have been both normed and deterministic.

In the realm of concurrency theory, language equivalence is generally taken to be too coarse an equivalence. For example, it equates the two transition systems of Example 0 which generate the same language $\{ab, ac\}$ yet demonstrate different deadlocking capabilities due to the nondeterministic behaviour exhibited by the second transition system. Many finer equivalences have been proposed, with *bisimulation equivalence* being perhaps the finest behavioural equivalence studied. (Note that we do not consider here any so-called 'true concurrency' equivalences such as those based on partial orders.) Bisimulation equivalence was defined by Park [125] and used to great effect by Milner [113,115]. Its definition, in the presence of final states, is as follows.

DEFINITION 9. A binary relation $\mathcal{R}$ on states of a transition system is a *bisimulation* iff whenever $(\alpha, \beta) \in \mathcal{R}$ we have that
- if $\alpha \xrightarrow{a} \alpha'$ then $\beta \xrightarrow{a} \beta'$ for some β' with $(\alpha', \beta') \in \mathcal{R}$;
- if $\beta \xrightarrow{a} \beta'$ then $\alpha \xrightarrow{a} \alpha'$ for some α' with $(\alpha', \beta') \in \mathcal{R}$;
- $\alpha \in \mathsf{F}$ iff $\beta \in \mathsf{F}$.

α and β are *bisimulation equivalent* or *bisimilar*, written $\alpha \sim \beta$, iff $(\alpha, \beta) \in \mathcal{R}$ for some bisimulation $\mathcal{R}$. That is, $\sim = \bigcup\{\mathcal{R}\colon\ \mathcal{R}$ is a bisimulation relation$\}$.

LEMMA 10. *$\sim$ is the largest bisimulation relation.*

PROOF. An arbitrary union of bisimulations is itself a bisimulation. □

LEMMA 11. *$\sim$ is an equivalence relation.*

PROOF. Reflexivity holds since the identity relation is a bisimulation; symmetry holds since the inverse of a bisimulation is a bisimulation; and transitivity holds since the composition of two bisimulations is a bisimulation. □

Bisimulation equivalence has an elegant characterization in terms of certain two-player games [90,136]. Starting with a pair of states $\langle\alpha, \beta\rangle$, the two players alternate moves according to the following rules.

1. If exactly one of the pair of states is a final state, then player I is deemed to be the winner. Otherwise, player I chooses one of the states and makes some transition from that state (either $\alpha \xrightarrow{a} \alpha'$ or $\beta \xrightarrow{a} \beta'$). If this proves impossible, due to both states being terminal, then player II is deemed to be the winner.
2. Player II must respond to the move made by player I by making an identically-labelled transition from the other state (either $\beta \xrightarrow{a} \beta'$ or $\alpha \xrightarrow{a} \alpha'$). If this proves impossible, then player I is deemed to be the winner.
3. The play then repeats itself from the new pair $\langle\alpha', \beta'\rangle$. If the game continues forever, then player II is deemed to be the winner.

The following result is then immediately evident.

FACT 12. *$\alpha \sim \beta$ iff Player II has a winning strategy in the bisimulation game starting with the pair $\langle\alpha, \beta\rangle$.*

Conversely, $\alpha \not\sim \beta$ iff Player I has a winning strategy in the bisimulation game starting with the pair $\langle\alpha, \beta\rangle$.

PROOF. Any bisimulation relation defines a winning strategy for player II for the bisimulation game starting from a pair in the relation: the second player merely has to respond to moves by the first in such a way that the resulting pair is contained in the bisimulation.

Conversely, a winning strategy for player II for the bisimulation game starting from a particular pair of states defines a bisimulation relation containing that pair, namely the collection of all pairs which appear after every exchange of moves during any and all games in which player II uses this strategy. □

EXAMPLE 6. The states X and A from Example 0 are language equivalent, as they define the same language $\{ab, ac\}$. However, they are not bisimulation equivalent: there is no bisimulation relation which relates them. To see this, we can demonstrate an obvious winning strategy for player I in the bisimulation game starting with the pair of states $\langle X, A\rangle$. After the first exchange of moves, the new pair of states must be either $\langle Y, B\rangle$ or $\langle Y, C\rangle$; in the former case, player I may make the transition $Y \xrightarrow{c} \varepsilon$ to which player I cannot respond from B, and in the latter case player I may make the transition $Y \xrightarrow{b} \varepsilon$ to which player I cannot respond from C.

A transition system is *finite-branching* if there are only a finite number of transitions from any given reachable state; and it is *image-finite* if there are only a finite number of transitions with a given label from any given reachable state. For image-finite transition systems, we have the following stratified characterization of bisimulation equivalence [115].

DEFINITION 13. The *stratified bisimulation relations* $\sim_n$ are defined as follows.

- $\alpha \sim_0 \beta$ for all states.
- $\alpha \sim_{k+1} \beta$ iff
  - if $\alpha \xrightarrow{a} \alpha'$ then $\beta \xrightarrow{a} \beta'$ for some β' with $\alpha' \sim_k \beta'$;
  - if $\beta \xrightarrow{a} \beta'$ then $\alpha \xrightarrow{a} \alpha'$ for some α' with $\alpha' \sim_k \beta'$;
  - $\alpha \in F$ iff $\beta \in F$.

In terms of the game characterization of bisimilarity, $\alpha \sim_n \beta$ iff Player I cannot force a win within the first n exchanges of moves.

LEMMA 14. *If α and β are image-finite, then $\alpha \sim \beta$ iff $\alpha \sim_n \beta$ for all $n \geqslant 0$.*

PROOF. If $\alpha \sim \beta$ then an induction proof gives that $\alpha \sim_n \beta$ for each $n \geqslant 0$.

Conversely, $\{(\alpha, \beta)$: α and β are image-finite and $\alpha \sim_n \beta$ for all $n \geqslant 0\}$ is a bisimulation. □

It is clear that rewrite transition systems are image-finite, and that this lemma applies. It is equally clear that each of the relations $\sim_n$ is decidable, and that therefore non-bisimilarity is semi-decidable over rewrite transition systems. Hence the decidability results would follow from demonstrating the semi-decidability of bisimilarity.

EXAMPLE 7. Consider the following infinite-branching transition system.

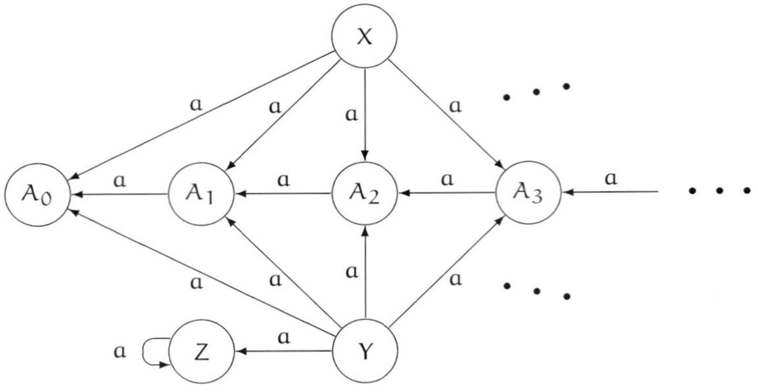

The states X and Y are clearly not bisimilar, as the state Z cannot be bisimilar to A_n for any $n \geqslant 0$. However, $X \sim_n Y$ for every $n \geqslant 0$ as $Z \sim_n A_n$ for each $n \geqslant 0$.

An immediately evident yet important property is the following lemma with its accompanying corollary relating bisimulation equivalence to language equivalence.

LEMMA 15. *If $\alpha \sim \beta$ and $\alpha \xrightarrow{w} \alpha'$ with $w \in \Sigma^*$, then $\beta \xrightarrow{w} \beta'$ for some β' such that $\alpha' \sim \beta'$.*

PROOF. Given a bisimulation $\mathcal{R}$ containing (α, β), an induction on the length of $w \in \Sigma^*$ demonstrates that if $\alpha \xrightarrow{w} \alpha'$ then $\beta \xrightarrow{w} \beta'$ with $(\alpha', \beta') \in \mathcal{R}$. □

COROLLARY 16. *If $\alpha \sim \beta$ then $\alpha \sim_L \beta$.*

PROOF. If $\alpha \sim \beta$ then $w \in L(\alpha)$ iff $\alpha \xrightarrow{w} \alpha'$ where α' is a final state, which – by the previous lemma – holds iff $\beta \xrightarrow{w} \beta'$ where β' is a final state, which in turn holds iff $w \in L(\beta)$. □

Apart from being the fundamental notion of equivalence for several process algebraic formalisms, bisimilarity has several pleasing mathematical properties not shared by other equivalences such as language equivalence. Furthermore as given by the following lemma, language equivalence and bisimilarity coincide over the class of normed deterministic processes.

LEMMA 17. *For states α and β of a normed deterministic transition system, if $\alpha \sim_L \beta$ then $\alpha \sim \beta$. Thus, taken along with Corollary 16, $\sim_L$ and $\sim$ coincide.*

PROOF. It suffices to demonstrate that the relation $\{(\alpha, \beta)\colon\ \alpha \sim_L \beta$ and α, β are states of a normed deterministic transition system$\}$ is a bisimulation relation. □

Hence it is sensible to concentrate on the more mathematically tractable bisimulation equivalence when investigating decidability results for language equivalence for deterministic language generators. We shall see examples of this in Section 2.

1.3. *Expressivity results*

The hierarchy from above gives us the following classification of processes.

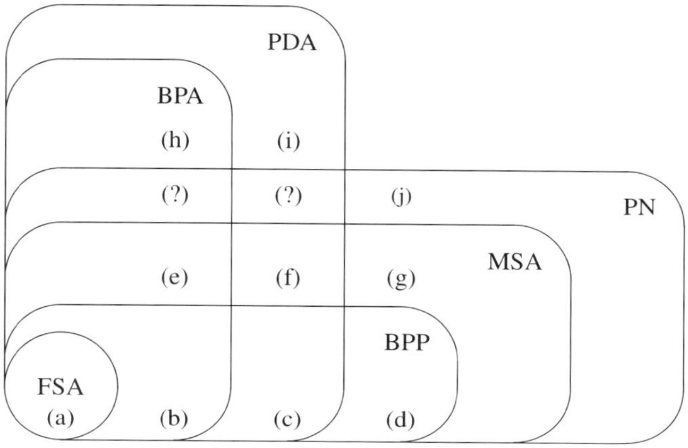

In this section we demonstrate the strictness of this hierarchy by providing example transition systems which lie precisely in the gaps (a)–(j) indicated in the classification. (We leave open the question regarding the final two gaps.) We in fact do more than this by giving examples of normed deterministic transition systems which separate all of these classes up to bisimulation (and hence also language) equivalence. These results complement those presented for the taxonomy described by Burkart, Caucal and Steffen [24].

(a) The first automaton in Example 0 provides a normed deterministic FSA.

(b) The type 2 rewrite system with the two rules $A \xrightarrow{a} AA$ and $A \xrightarrow{b} \varepsilon$ gives rise to the same transition system regardless of whether the system is sequential or parallel; this is an immediate consequence of the fact that it involves only a single variable A. This transition system is depicted as follows.

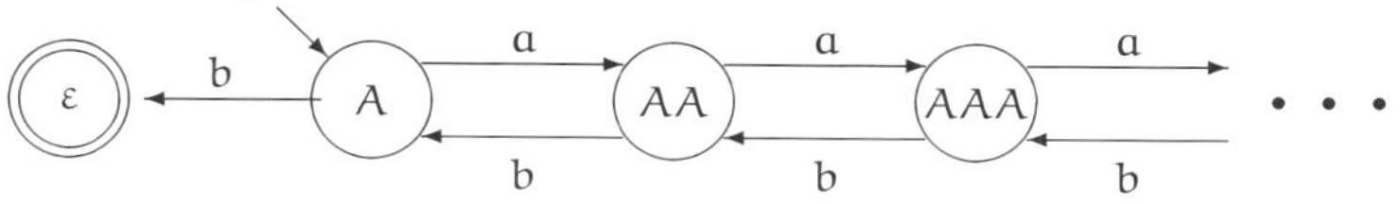

This is an example of a normed deterministic transition system which is both a BPA and a BPP but not an FSA.

(c) Examples 2 and 3 provide a transition system which can be described by both a BPP (Example 2) and a PDA (Example 3). However, it cannot be described up to bisimilarity by any BPA. To see this, suppose that we have a BPA which represents this transition system up to bisimilarity, and let m be at least as large as the norm of any of its variables. Then the BPA state corresponding to XB^m in Example 2 must be of the form $A\alpha$ where $A \in V$ and $\alpha \in V^+$. But then *any* sequence of $n(A)$ norm-reducing transitions must lead to the BPA state α, while the transition system in Example 2 has two such non-bisimilar derived states, namely XB^{k-1} and B^k where $k = n(\alpha)$.

(d) The following BPP with initial state X

$$X \xrightarrow{a} XB \qquad X \xrightarrow{c} XD \qquad X \xrightarrow{e} \varepsilon \qquad B \xrightarrow{b} \varepsilon \qquad D \xrightarrow{d} \varepsilon$$

is not language equivalent to any PDA, as its language is easily confirmed not to be context free. (The words in this language of the form $a^*c^*b^*d^*e$ are exactly those of the form $a^kc^nb^kd^ne$, which is clearly not a context-free language.)

(e) Examples 1 and 4 provide a transition system which can be described by both a BPA (Example 1) and a MSA (Example 4). However, the context-free language which it generates, $\{a^ncb^n\colon n \geqslant 0\}$, cannot be generated by any BPP, so this transition system is not even language equivalent to any BPP. To see this, suppose that $L(X) = \{a^ncb^n\colon n \geqslant 0\}$ for some BPP state X. (As the process has norm 1, the state must consist of a single variable X.) Let k be at least as large as the norm of any of the finite-normed variables of this BPP, and consider a transition sequence accepting the word a^kcb^k:

$$X \xrightarrow{a^k} Y\alpha \xrightarrow{c} \beta\alpha \xrightarrow{b^k} \varepsilon,$$

where the c-transition is generated by the transition rule $Y \xrightarrow{c} \beta$. We must have $n(Y\alpha) = k+1 > n(Y)$, so $\alpha \neq \varepsilon$; hence $\alpha \xrightarrow{b^i} \varepsilon$ and $\beta \xrightarrow{b^{k-i}} \varepsilon$ for some $i > 0$. Thus we have

$$X \xrightarrow{a^k} Y\alpha \xrightarrow{b^i} Y \xrightarrow{c} \beta \xrightarrow{b^{k-i}} \varepsilon$$

from which we get the desired contradiction: $a^kb^icb^{k-i} \in L(X)$ for some $i > 0$.

(f) The following PDA with initial state pX

$$pX \xrightarrow{a} pXX \qquad pX \xrightarrow{b} q \qquad pX \xrightarrow{b} r \qquad qX \xrightarrow{c} q \qquad rX \xrightarrow{d} r$$

coincides with the MSA which it defines, since there is only one stack symbol. This transition system is depicted as follows.

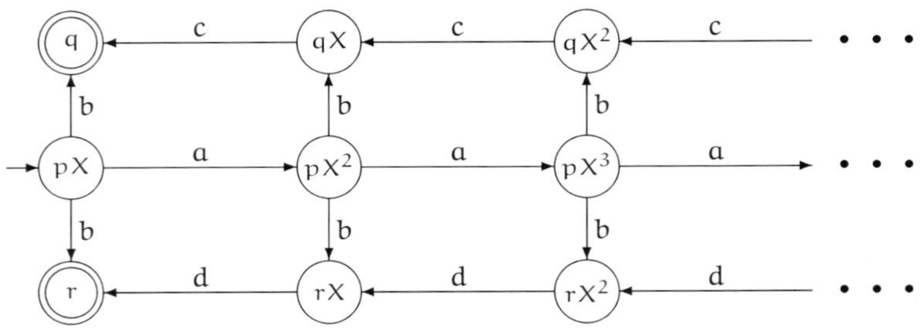

However, this transition system cannot be bisimilar to any BPA, due to a similar argument as for (c), nor language equivalent to any BPP, due to a similar argument as for (e).

(g) The following MSA with initial state pX

$$pX \xrightarrow{a} pA \qquad pA \xrightarrow{a} pAA \qquad qA \xrightarrow{b} qB \qquad rA \xrightarrow{c} r$$
$$pA \xrightarrow{b} qB \qquad qB \xrightarrow{c} r \qquad rB \xrightarrow{c} r$$

generates the language $\{a^n b^k c^n\colon 0 < k \leqslant n\}$, and hence cannot be language equivalent to any PDA, as it is not a context-free language, nor to any BPP, due to a similar argument as for (e).

(h) The following BPA with initial state X

$$X \xrightarrow{a} XA \qquad X \xrightarrow{b} XBX \xrightarrow{c} \varepsilon \qquad A \xrightarrow{a} \varepsilon \qquad B \xrightarrow{b} \varepsilon$$

generates the language $\{wcw^R\colon w \in \{a, b\}^*\}$ and hence is not language equivalent to any PN [126].

(i) The following PDA with initial state pX

$$\begin{array}{lllll}
pX \xrightarrow{a} pAX & pA \xrightarrow{a} pAA & pB \xrightarrow{a} pAB & qA \xrightarrow{a} q & rA \xrightarrow{a} r \\
pX \xrightarrow{b} pBX & pA \xrightarrow{b} pBA & pB \xrightarrow{b} pBB & qB \xrightarrow{b} q & rB \xrightarrow{b} r \\
pX \xrightarrow{c} qX & pA \xrightarrow{c} qA & pB \xrightarrow{c} qB & qX \xrightarrow{a} q & rX \xrightarrow{b} r \\
pX \xrightarrow{d} rX & pA \xrightarrow{d} rA & pB \xrightarrow{d} rB & &
\end{array}$$

is constructed by combining the ideas from (f) and (h). It can be schematically pictured as follows.

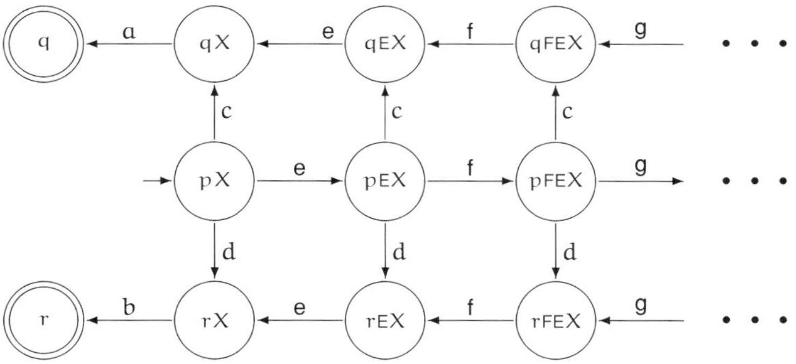

In this picture, e, f, g, ... $\in \{a, b\}$ and E, F, G, ... $\in \{A, B\}$ correspond in the obvious way. The language this PDA generates is $\{wcw^R a, wcw^R b\colon w \in \{a, b\}^*\}$ and hence as in (h) above it is not language equivalent to any PN; and as in (c) above it is not bisimilar to any BPA.

(j) The Petri net from Example 5 cannot be language equivalent to any PDA, as its language is easily confirmed not to be context-free. (The words in this language of the form $a^*b^*c^*d$ are exactly those of the form $a^n b^n c^n d$, which is clearly not a context-free language.)

More importantly, this Petri net cannot be bisimilar to any MSA [119]. To see this, suppose that the net is bisimilar to some MSA state pA. (As the process has norm 1, the stack must consist of a single symbol A.) Consider performing an indefinite sequence of a-transitions from pA. By Dickson's lemma (Lemma 3), we must eventually pass through two states $q\alpha$ and $q\alpha\beta$ in which the control states are equal and the stack of the first is contained in the stack of the second. This implies is that we can perform the following execution sequence.

$$pA \xrightarrow{a^k} q\alpha \xrightarrow{a^k} q\alpha\beta \xrightarrow{a^k} q\alpha\beta^2 \xrightarrow{a^k} \cdots$$

(We can assume that the period of the cycle is of the same length as the initial segment. If this isn't already given by the lemma, then we can merely extend the initial segment to the next multiple of the length of the cycle given by the Lemma, and use this multiple as the cycle length.) Considering now an indefinite sequence of b-transitions from $q\alpha$, a second application of Dickson's lemma gives us the following execution sequence.

$$q\alpha \xrightarrow{b^k} r\gamma \xrightarrow{b^k} r\gamma\delta \xrightarrow{b^k} r\gamma\delta^2 \xrightarrow{b^k} \cdots$$

(We can assume again by the same reasoning as above that the period of the cycle is of the same length as the initial sequence. Furthermore, we can assume that this is the same as the cycle length of the earlier a-sequence, by redefining the cycle lengths to be a common multiple of the two cycle lengths provided by the lemma.) Now there must be a state $s\sigma$ such that

$$pA \xrightarrow{a^k} q\alpha \xrightarrow{b^k} r\gamma \xrightarrow{c^k} s\sigma \stackrel{c}{\not\longrightarrow} .$$

Consider then the following sequence of transitions.

$$pA \xrightarrow{a^{2k}} q\alpha\beta \xrightarrow{b^{2k}} r\gamma\delta\beta \xrightarrow{c^k} s\sigma\delta\beta \xrightarrow{c}.$$

There must be a rule for $sX \xrightarrow{c}$ for some X which appears in either δ or β. But considering the following sequence of transitions

$$pA \xrightarrow{a^k} q\alpha \xrightarrow{b^{2k}} r\gamma\delta \xrightarrow{c^k} s\sigma\delta \stackrel{c}{\not\longrightarrow}$$

we must deduce that this X cannot appear in δ. Equally, considering the following sequence of transitions

$$pA \xrightarrow{a^{2k}} q\alpha\beta \xrightarrow{b^k} r\gamma\beta \xrightarrow{c^k} s\sigma\beta \stackrel{c}{\not\longrightarrow}$$

we must deduce that this X cannot appear in β. We thus have the desired contradiction.

We here summarize again these separation results in the following theorem.

THEOREM 18. *There exist* (*normed and deterministic*) *labelled transition systems lying precisely in the gaps* (a)–(j) *in the figure above.*

1.4. *Further classes of processes*

For concreteness, in this chapter we concentrate mainly on the classes of processes which fall within the described hierarchy. However, we shall also summarize various decidability and algorithmic results pertaining to other classes of processes which have been intensively studied. These are either extensions or restrictions of classes within our hierarchy, and are summarized below.

PA. Historically, the class BPA was first defined as a process algebra [9] with prefixing, nondeterministic choice, sequential composition, and recursion. Similarly, the class BPP was introduced by Christensen [35] as the parallel analogue to BPA, where sequential composition is replaced by parallel composition without communication. A natural extension of both classes then is their least common generalization, obtained by allowing both kinds of composition simultaneously. This class of processes is known simply as *Process Algebra* (PA).

ONE-COUNTER AUTOMATA. If we restrict the stack alphabet of a PDA so that it contains only one symbol (apart from a special bottom-of-stack marker), then we arrive at the class of *one-counter automata*. This is the class defined by finite-state automata augmented with a single counter variable; transitions may increment, decrement, or leave unchanged the value of the counter, and such a transition may be determined by a test-for-zero of the counter value. Note that with two counters we describe the full power of Turing machines, and hence we cannot hope for decidability for any semantic notions. Restricting to a single counter aims to delimit the decidability boundary.

ONE-COUNTER NETS. The class of Petri nets with (at most) one unbounded place gives rise to the class of *one-counter nets*. These differ from one-counter automata in that they cannot test-for-zero, but rather only test-for-nonzero. That is, if a transition may occur if the counter is zero, then it can equally occur if the counter is non-zero. As we shall see, various problems will be undecidable for Petri nets with only two unbounded places; this being the case, considering Petri nets with only a single unbounded place makes for a natural restriction.

2. The equivalence checking problem

2.1. *Decidability results for BPA and BPP*

In this section we describe several positive results which have been established regarding the decidability of bisimilarity between type 2 rewrite transition systems. In particular, we shall briefly describe the techniques behind the following results:

1. Bisimilarity is decidable for BPA [38] and BPP [36,37].
2. Bisimilarity is decidable in polynomial time for normed BPA [69,70] and normed BPP [71].

These results contrast with those regarding the undecidability of language equivalence for both BPA and BPP. The negative result for BPA [6] follows from the fact that BPA effectively defines the class of context-free languages; and that for BPP [66] follows from a modification by Hirshfeld of a technique of Jančar which is described in Section 2.4. Both arguments can be shown to hold for the class of normed systems. Also, for both BPA and

BPP, this undecidability extends to all equivalences which lie in Glabbeek's spectrum [59] between bisimilarity and language equivalence [61,79,82].

Baeten, Bergstra and Klop [4,5] presented the first such decidability result, that bisimilarity between normed BPA is decidable. Their lengthy proof exploits the periodicity which exists in normed BPA transition systems, and several simpler proofs exploiting structural properties were soon recorded, notably by Caucal [31], Hüttel and Stirling [80], and Groote [60]. Huynh and Tian [81] demonstrated that this problem has a complexity of Σ_2^P by providing a nondeterministic algorithm which relies on an NP oracle; Hirshfeld, Jerrum and Moller [69,70] refined this result by providing a polynomial algorithm, thus showing the problem to be in P. A generally more efficient, though worst-case exponential, algorithm is presented by Hirshfeld and Moller [72]. Finally, Christensen, Hüttel and Stirling [38,39] demonstrated that bisimilarity between arbitrary BPA systems is decidable, whilst Burkart, Caucal and Steffen [23] demonstrated an elementary decision procedure. For the parallel case, Christensen, Hirshfeld and Moller [37,36] demonstrated the general decidability result for BPP, whilst Hirshfeld, Jerrum and Moller [71] presented a polynomial algorithm for the normed case.

As we noted in Section 1.2, one direction for the decidability results, determining non-bisimilarity, is automatic using the stratified bisimulation relations of Definition 13. It therefore behaves us merely to prove that bisimilarity is semi-decidable.

2.1.1. *Composition and decomposition.* An important property of bisimulation equivalence which we shall exploit is the following congruence result, which is valid for both BPA and BPP.

LEMMA 19. *$\sim$ is a congruence (with respect to concatenation) over both BPA and BPP; that is, if $\alpha \sim \beta$ and $\alpha' \sim \beta'$ then $\alpha\alpha' \sim \beta\beta'$.*

PROOF. $\{(\alpha\alpha', \beta\beta'): \alpha \sim \beta \text{ and } \alpha' \sim \beta'\}$ is a bisimulation. □

Given this congruence property, we have an obvious potential technique for the analysis of (a state of) a type 2 rewrite transition system: decompose the state into simpler components. In general, this idea is not suitably provided for by this congruence property; states may indeed decompose, but not necessarily into simpler components, nor necessarily in any unique fashion. However, for the class of normed transition systems we have a unique factorization result for both BPA and BPP, of the form first explored by Milner and Moller [116]. Given a type 2 rewrite transition system, we say that a variable $X \in V$ is *prime* (with respect to bisimilarity $\sim$) iff $\alpha = \varepsilon$ whenever $X \sim Y\alpha$; and both normed BPA states and normed BPP states can be decomposed in a unique fashion into such prime components. We shall demonstrate these two results here separately. For the BPA case, we start with the following cancellation lemma.

LEMMA 20 (Cancellation lemma for normed BPA). *If α, β and γ are states of a BPA and if $n(\gamma) < \infty$, then $\alpha\gamma \sim \beta\gamma$ implies $\alpha \sim \beta$.*

PROOF. $\{(\alpha, \beta): \alpha\gamma \sim \beta\gamma \text{ for some } \gamma \text{ with } n(\gamma) < \infty\}$ is a bisimulation. □

The assumption in this lemma that γ is normed cannot be dropped, as can be readily seen by considering the following counterexample. Given the two rewrite rules $X \xrightarrow{a} \varepsilon$ and $Y \xrightarrow{a} Y$, we can immediately deduce that $XY \sim Y$, but that $X \not\sim \varepsilon$.

THEOREM 21 (Unique factorization theorem for normed BPA). *Every normed state α of a BPA decomposes uniquely (up to bisimilarity) into prime components.*

PROOF. The existence of a prime decomposition may be established by induction on the norm.

For uniqueness, suppose that $\alpha = X_1 \cdots X_p \sim Y_1 \cdots Y_q$ are prime decompositions, and that we have established the uniqueness of prime decompositions for all states β with $n(\beta) < n(\alpha)$. If $p = 1$ or $q = 1$ then uniqueness is immediate. Otherwise suppose that $X_1 \xrightarrow{a} \gamma$ is a norm-reducing transition that is matched by $Y_1 \xrightarrow{a} \delta$, so that $\gamma X_2 \cdots X_p \sim \delta Y_2 \cdots Y_q$. By the inductive hypothesis, the prime decompositions of these two states are equal (up to $\sim$), entailing $X_p \sim Y_q$. Hence, by Lemmas 19 and 20, $X_1 \cdots X_{p-1} \sim Y_1 \cdots Y_{q-1}$, and uniqueness then follows from a second application of the inductive hypothesis. $\square$

Notice that this theorem fails for unnormed BPA states, a fact which is immediately apparent from the observation that $\alpha \sim \alpha\beta$ whenever $n(\alpha) = \infty$.

THEOREM 22 (Unique factorization theorem for normed BPP). *Every normed state of a BPP decomposes uniquely (up to bisimilarity) into prime components.*

PROOF. Again, the existence of a prime decomposition may be established by induction on the norm.

For uniqueness, suppose that $\alpha = P_1^{k_1} P_2^{k_2} \cdots P_m^{k_m} \sim P_1^{l_1} P_2^{l_2} \cdots P_m^{l_m} = \beta$ represents a counterexample of smallest norm; that is, all states γ with $n(\gamma) < n(\alpha)$ have unique prime decompositions, the P_is are distinct (with respect to $\sim$) primes, but there is some i such that $k_i \neq l_i$. We may assume that the P_is are ordered by nondecreasing norms, and then we may choose this i so that $k_j = l_j$ whenever $j > i$. We shall furthermore assume without loss of generality that $k_i > l_i$. We distinguish three cases, and in each case show that the state α may perform a norm-reducing transition $\alpha \xrightarrow{a} \alpha'$ that cannot be matched by any transition $\beta \xrightarrow{a} \beta'$ with $\alpha' \sim \beta'$ (or vice versa with the roles of α and β reversed), which will supply the desired contradiction. Observe that by minimality of the counterexample, if α' and β' are to be bisimilar then their prime decompositions must be identical.

Case I. If $k_j > 0$ for some $j < i$, then we may let α perform some norm-reducing transition via the prime component P_j. β cannot match this transition, as it cannot increase the exponent l_i without decreasing the exponent of some prime with norm greater than that of P_i.

Case II. If $k_j > 0$ for some $j > i$, then we may let α perform a norm-reducing transition via the prime component P_j that maximizes (after reduction into primes) the increase in the exponent k_i. Again β is unable to match this transition.

Case III. If the state $\alpha = P_i^{k_i}$ is a prime power, then note that $l_j = 0$ for all $j > i$ by choice of i, and that $k_i \geqslant 2$ by the definition of "prime". If $l_i > 0$, then we may let β perform a norm-reducing transition via P_i; this transition cannot be matched by α, since it would require the exponent k_i to decrease by at least two. If $l_i = 0$ on the other hand, then we may let α perform a norm-reducing transition via P_i; this transition cannot be matched by β, since β is unable to increase the exponent l_i.

These cases are inclusive, so the theorem is proved. □

The following then is an immediate corollary.

LEMMA 23 (Cancellation lemma for normed BPP). *If α, β and γ are normed states of a BPP, then $\alpha\gamma \sim \beta\gamma$ implies $\alpha \sim \beta$.*

Notice that for unnormed BPP, unique decomposition and cancellation again each fail, for similar reasons to the sequential case.

Given a binary relation $\mathcal{R}$ on states of a type 2 rewrite transition system, let $\stackrel{\mathcal{R}}{\equiv}$ denote the least congruence containing $\mathcal{R}$; that is, $\stackrel{\mathcal{R}}{\equiv}$ is the least equivalence relation which contains $\mathcal{R}$ and the pair $(\alpha\alpha', \beta\beta')$ whenever it contains each of (α, β) and (α', β').

DEFINITION 24. A binary relation $\mathcal{R}$ on states of a type 2 rewrite transition system is a *bisimulation base* iff whenever $(\alpha, \beta) \in \mathcal{R}$ we have that

- if $\alpha \xrightarrow{a} \alpha'$ then $\beta \xrightarrow{a} \beta'$ for some β' with $\alpha' \stackrel{\mathcal{R}}{\equiv} \beta'$;
- if $\beta \xrightarrow{a} \beta'$ then $\alpha \xrightarrow{a} \alpha'$ for some α' with $\alpha' \stackrel{\mathcal{R}}{\equiv} \beta'$.

Hence the definition of a bisimulation base differs from that of a bisimulation only in how the derivative states α' and β' are related; in defining $\mathcal{R}$ to be a bisimulation, we would need these derivative states to be related by $\mathcal{R}$ itself and not just by the (typically much larger) congruence $\stackrel{\mathcal{R}}{\equiv}$. A bisimulation base then is in some sense a basis for a bisimulation. The importance of this idea is encompassed in the following theorem due originally (in the normed BPA case) to Caucal.

LEMMA 25. *If $\mathcal{R}$ is a bisimulation base for a BPA or BPP transition system, then $\stackrel{\mathcal{R}}{\equiv}$ is a bisimulation, and hence $\stackrel{\mathcal{R}}{\equiv} \subseteq \sim$.*

PROOF. If $\alpha \stackrel{\mathcal{R}}{\equiv} \beta$ then the two clauses given by the definition of $\stackrel{\mathcal{R}}{\equiv}$ being a bisimulation hold true. This can be demonstrated by a straightforward induction on the depth of inference of $\alpha \stackrel{\mathcal{R}}{\equiv} \beta$. □

COROLLARY 26. *$\alpha \sim \beta$ iff $(\alpha, \beta) \in \mathcal{R}$ for some bisimulation base $\mathcal{R}$.*

PROOF. Immediate from Lemma 25. □

It now becomes apparent that in order to demonstrate bisimilarity between terms, we needn't produce a complete (infinite) bisimulation relation which contains the pair; rather it suffices simply to produce a bisimulation base which contains the pair. What we shall demonstrate is that this corollary can be strengthened to a finite characterization of bisimulation, in that we shall prove the existence for both BPA and BPP of a *finite* relation $\mathcal{R}$ satisfying $\overset{\mathcal{R}}{\equiv} = \sim$. These relations $\mathcal{R}$ will clearly be bisimulation bases, and the semi-decidability results (and hence the decidability results) will be established, taking into account the following.

LEMMA 27. *It is semi-decidable whether a given finite binary relation $\mathcal{R}$ over V^* is a bisimulation base.*

PROOF. We need simply check that each pair (α, β) of the finite relation $\mathcal{R}$ satisfies the two clauses of the definition of a bisimulation base, which requires testing (in parallel) if each transition for one of α and β has a matching transition from the other. This matching test – that is, checking if the derivative states are related by $\overset{\mathcal{R}}{\equiv}$ – is itself semi-decidable, as the relation $\overset{\mathcal{R}}{\equiv}$ is semi-decidable. □

The semi-decision procedure for checking $\alpha \sim \beta$ would then consists of enumerating all finite binary relations over V^* containing the pair (α, β) and checking (in parallel) if any one of them is a bisimulation base. We thus concentrate on defining the finite relation $\mathcal{R}$ satisfying $\overset{\mathcal{R}}{\equiv} = \sim$.

2.1.2. *A finite bisimulation base for BPA.* The finite bisimulation base for BPA which we present here is taken from Christensen, Hüttel and Stirling [38,39]. We first present some technical results, starting with the following unique solutions lemma.

LEMMA 28 (Unique solutions lemma). *If $\alpha \sim \gamma\alpha$ and $\beta \sim \gamma\beta$ for some $\gamma \neq \varepsilon$ then $\alpha \sim \beta$.*

PROOF. Let $\mathcal{R} = \{(\delta\alpha, \delta\beta)\colon\ \alpha \sim \gamma\alpha \text{ and } \beta \sim \gamma\beta \text{ for some } \gamma \neq \varepsilon\}$. We may demonstrate straightforwardly that $\sim\mathcal{R}\sim$ is a bisimulation, from which we may deduce the desired result. □

An important finiteness result on which the construction of the finite bisimulation base hangs is given by the following.

LEMMA 29. *If $\alpha\gamma \sim \beta\gamma$ for infinitely many non-bisimilar γ, then $\alpha \sim \beta$.*

PROOF. We can show that $\mathcal{R} = \{(\alpha, \beta)\colon\ \alpha\gamma \sim \beta\gamma \text{ for infinitely many non-bisimilar } \gamma\}$ is a bisimulation, from which the result follows. □

We may split the set of variables into two disjoint sets $V = V_N \cup V_U$ with the variables in V_N having finite norm and those in V_U having infinite norm. The motive in this is based on the following lemma.

LEMMA 30. *If* X *has infinite norm then* $X\alpha \sim X$ *for all* α.

PROOF. We can immediately verify that $\{(X, X\alpha)\colon X \text{ has infinite norm}\}$ is a bisimulation. □

Hence we need only ever consider states $\alpha \in V_N{}^* \cup V_N{}^*V_U$, the others being immediately transformed into such a bisimilar state by erasing all symbols following the first infinite-norm variable.

The construction relies on recognizing when a term may be broken down into a composition of somehow simpler terms. To formalize this concept we start with the following definition.

DEFINITION 31. A pair $(X\alpha, Y\beta)$ satisfying $X\alpha \sim Y\beta$ is *decomposable* if X and Y have finite norm, and for some γ,
- $X \sim Y\gamma$ and $\gamma\alpha \sim \beta$; or
- $Y \sim X\gamma$ and $\gamma\beta \sim \alpha$.

The situation would be clear if all bisimilar pairs were decomposable; indeed we shall exploit this very property of normed transition systems – which follows there from the unique decomposability result – in Section 2.2.1. However we can demonstrate that there is in some sense only a finite number of ways that decomposability can fail. This crucial point in the argument is formalized in the following lemma. We consider two pairs $(X\alpha, Y\beta)$ and $(X\alpha', Y\beta')$ to be *distinct* if $\alpha \not\sim \alpha'$ or $\beta \not\sim \beta'$.

LEMMA 32. *For any* $X, Y \in V$, *any set* $\mathcal{R}$ *of the form*

$$\Big\{(X\alpha, Y\beta)\colon X\alpha,\ Y\beta \in V_N{}^* \cup V_N{}^*V_U,\ X\alpha \sim Y\beta, \\ \textit{and } (X\alpha, Y\beta) \textit{ not decomposable}\Big\}$$

which contains only distinct pairs must be finite.

PROOF. If $X, Y \in V_U$ then clearly $\mathcal{R}$ can contain at most the single pair (X, Y).

If $X \in V_U$ and $Y \stackrel{w}{\longrightarrow} \varepsilon$ and $\mathcal{R} = \{(X, Y\beta_i)\colon i \in I\}$ then for each $i \in I$ we must have that $X \stackrel{w}{\longrightarrow} \alpha_i$ such that $\alpha_i \sim \beta_i$. But then by image-finiteness there can be only a finite number of non-bisimilar such β_i.

Suppose then that $X, Y \in V_N$ and that $\mathcal{R} = \{(X\alpha_i, Y\beta_i)\colon i \in I\}$ is infinite. Without loss of generality, assume that $n(Y) \leqslant n(X)$, and that $Y \stackrel{w}{\longrightarrow} \varepsilon$ with $\mathrm{length}(w) = n(Y)$. Then for each $i \in I$ we must have that $X \stackrel{w}{\longrightarrow} \gamma_i$ such that $\gamma_i\alpha_i \sim \beta_i$. By image-finiteness, we can have only finitely many such γ_i, so we must have that $X \stackrel{w}{\longrightarrow} \gamma$ for some γ such that $\gamma\alpha_i \sim \beta_i$ holds for infinitely many $i \in I$; by distinctness these α_is must all be non-bisimilar. For these $i \in I$ we must then have that $X\alpha_i \sim Y\gamma\alpha_i$, But then by Lemma 29 we must have that $X \sim Y\gamma$, contradicting non-decomposability. □

We are now ready to demonstrate the main result, that there is a finite relation $\mathcal{R}$ satisfying $\stackrel{\mathcal{R}}{\equiv} = \sim$. This will be done by induction using the following well-founded ordering $\preceq$.

DEFINITION 33. Define the *finite prefix norm* $\mathrm{n_f}(\alpha)$ of $\alpha \in V^*$ as follows:

$$\mathrm{n_f}(\alpha) = \max\bigl\{\mathrm{n}(\beta)\colon \mathrm{n}(\beta) < \infty \text{ and } \alpha = \beta\gamma \text{ for some } \gamma\bigr\}.$$

We then define the following preorder on pairs: $(\alpha_1, \alpha_2) \preceq (\beta_1, \beta_2)$ iff $\max(\mathrm{n_f}(\alpha_1), \mathrm{n_f}(\alpha_2)) \leqslant \max(\mathrm{n_f}(\beta_1), \mathrm{n_f}(\beta_2))$.

LEMMA 34. *Let* $\mathcal{R}_0 = \{(X, \alpha)\colon X \in V_N \text{ and } X \sim \alpha\}$, *and let* $\mathcal{R}_1$ *be the largest set of the form*

$$\bigl\{(X\alpha, Y\beta)\colon X\alpha,\ Y\beta \in {V_N}^* V_U \cup {V_N}^*,\ X\alpha \sim Y\beta, \\ \text{and } (X\alpha, Y\beta) \text{ is not decomposable}\bigr\}$$

which contains only distinct pairs, and containing minimal elements with respect to $\preceq$. *Then* $\mathcal{R} = \mathcal{R}_0 \cup \mathcal{R}_1$ *is a finite relation satisfying* $\overset{\mathcal{R}}{\equiv} = \sim$.

PROOF. Firstly, $\mathcal{R}_0$ and $\mathcal{R}_1$ must both be finite. Also we must have $\overset{\mathcal{R}}{\equiv} \subseteq \sim$. We will demonstrate by induction on $\preceq$ that $X\alpha \sim Y\beta$ implies $X\alpha \overset{\mathcal{R}}{\equiv} Y\beta$.

If $(X\alpha, Y\beta)$ is decomposable, then $X, Y \in V_N$ and (without loss of generality) assume that $X \sim Y\gamma$ and $\gamma\alpha \sim \beta$. Then $\mathrm{n_f}(\gamma\alpha) < \mathrm{n_f}(Y\gamma\alpha) = \mathrm{n_f}(X\alpha)$ and $\mathrm{n_f}(\beta) < \mathrm{n_f}(Y\beta)$, so $(\gamma\alpha, \beta) \preceq (X\alpha, Y\beta)$. Hence by induction $\gamma\alpha \overset{\mathcal{R}}{\equiv} \beta$. Then from $(X, Y\gamma) \in \mathcal{R}_0$ we get $X\alpha \overset{\mathcal{R}}{\equiv} Y\gamma\alpha \overset{\mathcal{R}}{\equiv} Y\beta$.

Suppose then that $(X\alpha, Y\beta)$ is not decomposable. Then $(X\alpha', Y\beta') \in \mathcal{R}_1$ for some $\alpha' \sim \alpha$ and $\beta' \sim \beta$ with $(\alpha', \beta') \preceq (\alpha, \beta)$.

- If $X, Y \in V_N$, then $(\alpha, \beta), (\alpha', \beta') \prec (X\alpha, Y\beta)$, so $(\alpha, \alpha'), (\beta, \beta') \prec (X\alpha, Y\beta)$. Thus by induction $\alpha \overset{\mathcal{R}}{\equiv} \alpha'$ and $\beta \overset{\mathcal{R}}{\equiv} \beta'$, so $X\alpha \overset{\mathcal{R}}{\equiv} X\alpha' \mathcal{R} Y\beta' \overset{\mathcal{R}}{\equiv} Y\beta$.
- If $X \in V_N$ and $Y \in V_U$, then $\beta = \beta' = \varepsilon$ and $X\alpha \sim Y$. Also $\mathrm{n_f}(\alpha') \leqslant \mathrm{n_f}(\alpha) < \mathrm{n_f}(X\alpha)$, so $(\alpha, \alpha') \prec (X\alpha, Y)$. Thus by induction $\alpha \overset{\mathcal{R}}{\equiv} \alpha'$, so $X\alpha \overset{\mathcal{R}}{\equiv} X\alpha' \overset{\mathcal{R}}{\equiv} Y$. A symmetric argument applies for the case when $X \in V_U$ and $Y \in V_N$.
- If $X, Y \in V_U$, then $\alpha = \alpha' = \beta = \beta' = \varepsilon$ and $(X, Y) \in \mathcal{R}_1$, so $X\alpha \overset{\mathcal{R}}{\equiv} Y\beta$. □

With this, we have demonstrated the desired result.

THEOREM 35. *Bisimulation equivalence is decidable for BPA.*

2.1.3. *A finite bisimulation base for BPP.* The finite bisimulation base for BPP which we describe here is taken from Hirshfeld [67], which is itself a revised presentation using Dickson's lemma (Lemma 3) of the result of Redie [129] that every congruence on a finitely generated commutative semigroup (such as bisimilarity over BPP) is finitely generated.

Firstly, assuming that the set of variables is $V = \{X_1, \ldots, X_n\}$, we note that a state α is a monomial $X_1^{k_1} \cdots X_n^{k_n}$ over V. We then define the following two orderings on states.

DEFINITION 36.
- $X_1^{k_1} \cdots X_n^{k_n} < X_1^{l_1} \cdots X_n^{l_n}$ iff for some i: $k_i < l_i$ and $k_j = l_j$ for all $j < i$.
- $X_1^{k_1} \cdots X_n^{k_n} \sqsubset X_1^{l_1} \cdots X_n^{l_n}$ iff for some i: $k_i < l_i$ and $k_j \leqslant l_j$ for all j.

Note that $\alpha \sqsubset \beta$ implies that $\alpha < \beta$, and that these orderings are both well-founded: we cannot have an infinite decreasing sequence of states with respect to either ordering. Furthermore, $<$ is a total ordering: for any $\alpha \neq \beta$, either $\alpha < \beta$ or $\beta < \alpha$. Finally, both orderings are precongruences: if $\alpha < \beta$ and $\alpha' < \beta'$ ($\alpha \sqsubset \beta$ and $\alpha' \sqsubset \beta'$) then $\alpha\alpha' < \beta\beta'$ ($\alpha\alpha' \sqsubset \beta\beta'$).

DEFINITION 37. A pair (α, β) of states is *minimal* iff $\alpha \neq \beta$, and for any $(\alpha', \beta') \sqsubset (\alpha, \beta)$ with $\alpha' \sim \beta'$ we have $\alpha' = \beta'$. (By $(\alpha', \beta') \sqsubset (\alpha, \beta)$ we mean $(\alpha', \beta') \neq (\alpha, \beta)$ but $\alpha' \sqsubseteq \alpha$ and $\beta' \sqsubseteq \beta$; note that this implies that $\alpha'\beta' \sqsubset \alpha\beta$.)

Let $\mathcal{R}$ be the collection of all minimal bisimilar pairs.

LEMMA 38. *$\mathcal{R}$ is finite.*

PROOF. Suppose we have an infinite sequence of distinct pairs $(\alpha_i, \beta_i) \in \mathcal{R}$. We can take an infinite subsequence in which the exponents of the variable X_1 in the states α_i form a nondecreasing sequence, and then we can take a further infinite subsequence in which the exponents of the variable X_2 in the states α_i form a nondecreasing sequence, and continue in this fashion so that we are left with an infinite subsequence $(\alpha_i, \beta_i) \in \mathcal{R}$ with $(\alpha_1, \beta_1) \sqsubset (\alpha_2, \beta_2) \sqsubset \cdots$. But then (α_i, β_i) (for $i > 1$) cannot be minimal, contradicting its inclusion in $\mathcal{R}$. □

LEMMA 39. $\stackrel{\mathcal{R}}{\equiv} = \sim$.

PROOF. If $\alpha \stackrel{\mathcal{R}}{\equiv} \beta$ then clearly $\alpha \sim \beta$ since $\sim$ is a congruence and $\mathcal{R}$ by definition only contains bisimilar pairs.

Suppose then that $\alpha \sim \beta$ but that $\alpha \stackrel{\mathcal{R}}{\not\equiv} \beta$ (in particular, $\alpha \neq \beta$), and that this is a minimal counterexample in that if $\alpha' \sim \beta'$ and $\alpha'\beta' < \alpha\beta$ then $\alpha' \stackrel{\mathcal{R}}{\equiv} \beta'$.

Since $(\alpha, \beta) \notin \mathcal{R}$ there must be $(\alpha', \beta') \sqsubset (\alpha, \beta)$ with $\alpha' \sim \beta'$ and $\alpha' \neq \beta'$. Then $\alpha'\beta' \sqsubset \alpha\beta$, so $\alpha'\beta' < \alpha\beta$, and hence by the minimality assumption we have that $\alpha' \stackrel{\mathcal{R}}{\equiv} \beta'$.

Without loss of generality, assume $\alpha' < \beta'$. Let β'' be such that $\beta = \beta'\beta''$. Then $\alpha \sim \beta = \beta'\beta'' \sim \alpha'\beta''$. But $\alpha\alpha'\beta'' < \alpha\beta'\beta'' = \alpha\beta$, so $\alpha \stackrel{\mathcal{R}}{\equiv} \alpha'\beta'' \stackrel{\mathcal{R}}{\equiv} \beta'\beta'' = \beta$, contradicting the assumption that $\alpha \stackrel{\mathcal{R}}{\not\equiv} \beta$. □

Hence we have again demonstrated the desired result.

THEOREM 40. *Bisimulation equivalence is decidable for BPP.*

2.2. *Polynomial time algorithms for BPA and BPP*

In Section 2.1 we demonstrated the decidability of bisimulation equivalence over both BPA and BPP. However, the computational complexity of the algorithms which we presented shows them to be of little practical value. To overcome this deficiency we concentrate in this section on developing efficient algorithms for deciding bisimilarity within these classes of transition systems. What we demonstrate in fact are polynomial algorithms for the problem of deciding equivalences over the subclasses of normed transition systems. These algorithms will both be based on an exploitation of the decomposition properties enjoyed by normed transition systems; however, despite the apparent similarity of the two problems, different methods appear to be required.

For the algorithms, we fix a normed type 2 rewrite transition system with variable set V. The problem then is to determine efficiently – that is, in time which is polynomial in the the number n of symbols in the rewrite rules – whether or not $\alpha \sim \beta$ for $\alpha, \beta \in V^*$, where we interpret the transition system first as a BPA and then as a BPP. The first algorithm we describe for BPA is taken from Hirshfeld, Jerrum and Moller [69,70]. The second one for BPP is taken from Hirshfeld, Jerrum and Moller [71].

2.2.1. *A polynomial time algorithm for normed BPA.* The basic idea behind the algorithm we present here for deciding bisimilarity between normed BPA states is to exploit the unique prime decomposition theorem by decomposing states sufficiently far to be able to establish or refute the equivalence we are considering. Further, we try to construct these decompositions by a refinement process which starts with an overly generous collection of candidate decompositions. As the algorithm progresses, invalid decompositions will gradually be weeded out.

Assume that the variables V are ordered by non-decreasing norm, so that $X < Y$ implies $n(X) \leqslant n(Y)$. A *base* is a set $\mathcal{B}$ of pairs $(Y, X\alpha)$, where $X, Y \in V$, $\alpha \in V^*$, $X \leqslant Y$ and $n(Y) = n(X\alpha)$. We insist that $\mathcal{B}$ contains at most one pair of the form $(Y, X\alpha)$ for each choice of variables X and Y, so that the cardinality of $\mathcal{B}$ is at most $O(n^2)$. A base $\mathcal{B}$ is *full* iff whenever $Y \sim X\beta$ with $Y \geqslant X$ there exists a pair $(Y, X\alpha) \in \mathcal{B}$ such that $\alpha \sim \beta$. In particular, $(X, X) \in \mathcal{B}$ for all $X \in V$. The key observation is that bisimilarity is the congruence closure of a finite base.

At the heart of the algorithm is the definition of a polynomial-time (in the size of $\mathcal{B}$) relation $\equiv_{\mathcal{B}} \subseteq \overset{\mathcal{B}}{\equiv}$ which satisfies $\sim \subseteq \equiv_{\mathcal{B}}$ whenever $\mathcal{B}$ is full. Central to the definition of the relation $\equiv_{\mathcal{B}}$ is the idea of a decomposing function. A function $g : V \to V^*$ is a *decomposing function* if either $g(X) = X$ or $g(X) = X_1 X_2 \cdots X_p$ with $X_i < X$ for each $1 \leqslant i \leqslant p$. Such a function g can be extended to the domain V^* in the obvious fashion by defining $g(\varepsilon) = \varepsilon$ and $g(X\alpha) = g(X)g(\alpha)$. We then define $g^*(\alpha)$ for $\alpha \in V^*$ to be the limit of $g^t(\alpha)$ as $t \to \infty$; owing to the restricted form of g we know that it must be *eventually idempotent*, that is, that this limit must exist. The notation $g[X \mapsto \alpha]$ will be used to denote the function that agrees with g at all points in V except X, where its value is α.

The definition of the relation $\equiv_{\mathcal{B}}$ may now be given. For base $\mathcal{B}$ and decomposing function g, the relation $\alpha \equiv^g_{\mathcal{B}} g\beta$ is defined by the following decision procedure:

- if $g^*(\alpha) = g^*(\beta)$ then return true;
- otherwise let X and Y (with $X < Y$) be the leftmost mismatching pair of symbols in the words $g^*(\alpha)$ and $g^*(\beta)$;
  - if $(Y, X\gamma) \in \mathcal{B}$ then return $\alpha \equiv^g_{\mathcal{B}} g'\beta$, where $g' = g[Y \mapsto X\gamma]$;
  - otherwise return false.

Finally, $\equiv_{\mathcal{B}}$ is defined to be $\equiv_{\mathcal{B}}^{g}$ *Id* where *Id* is the identity function. (Note that this procedure terminates, as it calls itself at most $|V|$ times, each time updating the value of the parameter function g to decompose another variable.)

LEMMA 41. *$\equiv_{\mathcal{B}} \subseteq \overset{\mathcal{B}}{\equiv}$ and $\sim \subseteq \equiv_{\mathcal{B}}$ whenever $\mathcal{B}$ is full.*

PROOF. The first inclusion is easily confirmed, since for any g constructed by the algorithm for computing $\equiv_{\mathcal{B}}$, it is the case that $X \overset{\mathcal{B}}{\equiv} g(X)$ for each $X \in V$.

For the second inclusion, suppose that $\alpha \sim \beta$ and at some point in the procedure for deciding $\alpha \equiv_{\mathcal{B}} \beta$ we have that $g^*(\alpha) \neq g^*(\beta)$, and that we have only ever updated g with mappings $X \mapsto \gamma$ satisfying $X \sim \gamma$. Let X and Y (with $X < Y$) be the leftmost mismatching pair. Then $Y \sim X\gamma$ must hold for some γ, and so, by fullness, $(Y, X\gamma) \in \mathcal{B}$ for some γ with $Y \sim X\gamma$. So the procedure does not terminate with a false result, but instead updates g with this new semantically sound mapping and continues. □

To demonstrate that $\equiv_{\mathcal{B}}$ can be decided efficiently, we are left with the problem of deciding $g^*(\alpha) = g^*(\beta)$ in time polynomial in the size of $\mathcal{B}$ (and finding the leftmost mismatch in case of inequality); all other elements in the definition of $\equiv_{\mathcal{B}}$ are algorithmically undemanding. Note that the words $g^*(\alpha)$ and $g^*(\beta)$ will in general be of exponential length, so we cannot afford to compute them explicitly. However, the relation can indeed be computed in polynomial time with a dynamic programming algorithm which exploits standard string alignments tricks based on ideas from Knuth, Morris and Pratt [94]. The details of this algorithm can be found in [70].

The main task now is to discover a small full base that contains only semantically sound decomposition pairs. To do this, we start with a small full base, and then proceed to refine the base iteratively whilst maintaining fullness. Informally, we are proposing that at any instant the current base should consist of pairs (X, α) representing candidate decompositions, that is, pairs such that the relationship $X \sim \alpha$ is consistent with information gained so far. The refinement step is as follows.

Given a base $\mathcal{B}$, define the sub-base $\widehat{\mathcal{B}} \subseteq \mathcal{B}$ to be the set of pairs $(X, \alpha) \in \mathcal{B}$ such that

- if $X \xrightarrow{a} \beta$ then $\alpha \xrightarrow{a} \gamma$ with $\beta \equiv_{\mathcal{B}} \gamma$, and
- if $\alpha \xrightarrow{a} \gamma$ then $X \xrightarrow{a} \beta$ with $\beta \equiv_{\mathcal{B}} \gamma$.

LEMMA 42. *If $\mathcal{B}$ is full then $\widehat{\mathcal{B}}$ is full.*

PROOF. Suppose $Y \sim X\beta$ with $Y \geqslant X$. By fullness of $\mathcal{B}$, there exists a pair $(Y, X\alpha) \in \mathcal{B}$ such that $\alpha \sim \beta$. We show that the pair $(Y, X\alpha)$ survives the refinement step, to be included in $\widehat{\mathcal{B}}$. Note that, since $\sim$ is a congruence, $Y \sim X\alpha$. Thus, if $Y \xrightarrow{a} \gamma$ then $X\alpha \xrightarrow{a} \delta$ for some δ satisfying $\delta \sim \gamma$. By fullness of $\mathcal{B}$ and Lemma 41, $\delta \equiv_{\mathcal{B}} \gamma$. Similarly, if $X\alpha \xrightarrow{a} \delta$ then $Y \xrightarrow{a} \gamma$ with $\gamma \sim \delta$, and hence $\gamma \equiv_{\mathcal{B}} \delta$. The pair $(Y, X\alpha)$ therefore satisfies the conditions for inclusion in $\widehat{\mathcal{B}}$. □

In general, the refinement step makes progress, i.e., the new base $\widehat{\mathcal{B}}$ is strictly contained in the base $\mathcal{B}$ from which it was derived. If, however, no progress occurs, an important deduction may be made.

LEMMA 43. *If* $\widehat{\mathcal{B}} = \mathcal{B}$ *then* $\equiv_{\mathcal{B}} \subseteq \sim$.

PROOF. The relation $\equiv_{\mathcal{B}}$ is contained in $\overset{\mathcal{B}}{\equiv}$, the congruence closure of $\mathcal{B}$; and if $\widehat{\mathcal{B}} = \mathcal{B}$ then $\mathcal{B}$ must be a bisimulation base, in which case by Theorem 25, $\equiv_{\mathcal{B}} \subseteq \overset{\mathcal{B}}{\equiv} \subseteq \sim$. □

Note that by iteratively applying the refinement step $\mathcal{B} := \widehat{\mathcal{B}}$ to a full initial base, we are guaranteed by the preceding three lemmas to stabilize at some full base $\mathcal{B}$ for which $\equiv_{\mathcal{B}} = \sim$.

Finally, we are left with the task of constructing the initial base $\mathcal{B}_0$. This is achieved as follows. For each $X \in V$ and each $0 \leqslant \nu \leqslant n(X)$, let $[X]_\nu$ be some state that can be reached from X via a sequence of ν norm-reducing transitions. (Note that some norm-reducing transition is available to every state.)

LEMMA 44. *The base* $\mathcal{B}_0 = \{(Y, X[Y]_{n(X)})\colon X, Y \in V$ *and* $X \leqslant Y\}$ *is full.*

PROOF. Suppose $Y \sim X\beta$ with $X \leqslant Y$, and let $\nu = n(X)$; then $(Y, X[Y]_\nu) \in \mathcal{B}_0$ for some $[Y]_\nu$ such that $Y \xrightarrow{s} [Y]_\nu$ in ν norm-reducing steps, where $s \in A^\nu$. But the norm-reducing sequence $Y \xrightarrow{s} [Y]_\nu$ can only be matched by $X\beta \xrightarrow{s} \beta$. Hence $[Y]_\nu \sim \beta$, and $\mathcal{B}_0$ must be full. □

The basic structure of the procedure for deciding bisimilarity between normed states α and β is now clear: simply iterate the refinement procedure $\mathcal{B} := \widehat{\mathcal{B}}$ from the initial base $\mathcal{B} = \mathcal{B}_0$ until it stabilizes at the desired base $\mathcal{B}$, and then test $\alpha \equiv_{\mathcal{B}} \beta$. By the preceding four lemmas, this test is equivalent to $\alpha \sim \beta$.

We have not been specific about which state $[X]_\nu$ is to be selected among those reachable from X via a sequence of ν norm-reducing transitions. A suitable choice is provided by the following recursive definition. For each variable $X \in V$, let $\alpha_X \in V^*$ be some state reachable from X by a single norm-reducing transition $X \xrightarrow{a} \alpha_X$. Then,

$$[\alpha]_0 = \alpha,$$
$$[X\beta]_p = \begin{cases} [\beta]_{p-n(X)}, & \text{if } p \geqslant n(X), \\ [\alpha_X]_{p-1}\beta, & \text{if } p < n(X). \end{cases}$$

LEMMA 45. *With this definition for* $[\cdot]_\nu$, *the base* $\mathcal{B}_0$ *introduced in Lemma* 44 *may be explicitly constructed in polynomial time*; *in particular, every pair in* $\mathcal{B}_0$ *has a compact representation as an element of* $V \times V^*$.

PROOF. It is easily checked that the natural recursive algorithm based on the definition is polynomial-time bounded. □

We have already observed that $\mathcal{B}_0$ contains $O(n^2)$ pairs, so the refinement procedure is iterated at most $O(n^2)$ times. Hence we have succeeded in demonstrating our result.

THEOREM 46. *There is a polynomial-time* (*in the lengths of the states* α *and* β, *and the size of the rewrite rules*) *procedure for deciding bisimilarity between two states* α *and* β *of a normed type* 2 *sequential rewrite transition system.*

Simple context-free grammars. Recall that a simple grammar is a context-free grammar in Greibach normal form such that for every variable X and terminal symbol a there is at most one production of the form $X \to a\alpha$; these correspond to deterministic type 2 sequential rewrite transition systems. The decision procedure given by Korenjak and Hopcroft [95] for deciding language equivalence between simple grammars is doubly exponential; this time complexity was improved by Caucal [30] to be singly exponential. The following result demonstrates that this problem has a polynomial-time solution.

THEOREM 47. *There is a polynomial-time algorithm for deciding equivalence of simple grammars.*

PROOF. To obtain a polynomial-time decision procedure for deciding language equivalence of simple context-free grammars (deterministic type 2 sequential rewrite transition systems), we merely recall from Lemma 17 that in the case of normed deterministic transition systems, language equivalence and bisimulation equivalence coincide. We can restrict attention to normed grammars, as any unnormed grammar can be transformed into a language-equivalent normed grammar by removing productions containing unnormed nonterminals. (Note that this transformation does not preserve bisimulation equivalence, which makes it inapplicable for reducing the unnormed case to the normed case in checking bisimilarity.) Thus language equivalence of simple grammars may be checked in polynomial time by the procedure presented in the previous two sections. □

2.2.2. *A polynomial time algorithm for normed BPP.* To demonstrate that we can decide bisimilarity between normed BPP in polynomial time, we require a vastly different technique than that used for the BPA case; nonetheless the technique still relies completely on the unique factorization property.

To start off, we assume without loss of generality that the variables are given in order of non-decreasing norm, so that $n(X_1) \leqslant n(X_2) \leqslant \cdots \leqslant n(X_n)$. Define the *size* of a monomial $\alpha \in V^*$ to be the sum of the lengths of the binary encodings of the various exponents appearing in the monomial; the size of a production $X \xrightarrow{a} \beta$ to be the length of the triple (X, a, β), encoded in binary; and the size of a rewrite transition system to be the sum of the sizes of all the rewrite rules contained within it.

To prepare for the description of the algorithm, and the proof that it runs in polynomial time, we require some definitions and a few preparatory lemmas. We omit the (generally straightforward) proofs of these lemmas; they can be found in [71].

Suppose $\mathcal{R}$ is any relation on V^*. We say that a pair $(\alpha, \beta) \in V^* \times V^*$ *satisfies (norm-reducing) expansion in* $\mathcal{R}$ if

- if $\alpha \xrightarrow{a} \alpha'$ is a (norm-reducing) transition then $\beta \xrightarrow{a} \beta'$ for some β' with $\alpha' \mathcal{R} \beta'$; and
- if $\beta \xrightarrow{a} \beta'$ is a (norm-reducing) transition then $\alpha \xrightarrow{a} \alpha'$ for some α' with $\alpha' \mathcal{R} \beta'$.

Observe that a relation $\mathcal{R}$ is a bisimulation if every pair $(\alpha, \beta) \in \mathcal{R}$ satisfies expansion in $\sim$. Observe also that if $\mathcal{R}$ is an equivalence relation (respectively, congruence) then the relations "satisfies expansion in $\mathcal{R}$" and "satisfies norm-reducing expansion in $\mathcal{R}$" are both equivalence relations (respectively, congruences).

Define a *unique decomposition base*, $\mathcal{D}$, to be a pair (Π, Γ), where $\Pi = \Pi(\mathcal{D}) = \{P_1, \ldots, P_r\} \subseteq V$ is a set of *primes*, and $\Gamma = \Gamma(\mathcal{D})$ is a set of pairs $(X, P_1^{x_1} \ldots P_r^{x_r})$, one for each non-prime variable $X \in V - \Pi$. The set Γ may be viewed as specifying, for each non-prime variable X, a decomposition of X into primes. (Note that these "primes" are not in general the primes with respect to the maximal bisimulation, which were the subject of Theorem 22.) A unique decomposition base defines an equivalence relation $\equiv_{\mathcal{D}}$ on V^*: the relation $\alpha \equiv_{\mathcal{D}} \beta$ holds between $\alpha, \beta \in V^*$ if the prime decompositions of α and β are equal (as monomials).

LEMMA 48. *Let $\mathcal{D}$ be a unique decomposition base. Then:*

(i) *the equivalence relation $\equiv_{\mathcal{D}}$ is a congruence with cancellation (that is, $\alpha \equiv_{\mathcal{D}} \beta$ whenever $\alpha\gamma \equiv_{\mathcal{D}} \beta\gamma$) which coincides with $\stackrel{\mathcal{D}}{\equiv}$, the smallest congruence containing $\Gamma(\mathcal{D})$;*

(ii) *there is a polynomial-time (in the size of α and β) algorithm to decide $\alpha \equiv_{\mathcal{D}} \beta$ for arbitrary $\alpha, \beta \in V^*$;*

(iii) *the relation $\equiv_{\mathcal{D}}$ is a bisimulation provided every pair in $\Gamma(\mathcal{D})$ satisfies expansion within $\equiv_{\mathcal{D}}$; this condition may be checked by a polynomial-time algorithm;*

(iv) *the maximal bisimulation $\sim$ coincides with the congruence $\equiv_{\mathcal{D}}$, where $\mathcal{D}$ represents the unique decomposition in $\sim$.*

The next lemma allows us to shrink a congruence, defined by a unique decomposition base, whenever it is strictly larger than the maximal bisimulation.

LEMMA 49. *Let $\mathcal{D}$ be a unique decomposition base such that the congruence $\equiv_{\mathcal{D}}$ is norm-preserving and strictly contains the maximal bisimulation $\sim$. Define the relation $\equiv$ as follows: for all $\alpha, \beta \in V^*$, the relationship $\alpha \equiv \beta$ holds iff $\alpha \equiv_{\mathcal{D}} \beta$ and the pair (α, β) satisfies expansion in $\equiv_{\mathcal{D}}$. Then it is possible, in polynomial time, to find (a representation of) a relation $\equiv$ on V^* such that:*

(i) *the relation $\alpha \equiv \beta$ is decidable in polynomial time (in the sum of the sizes of α and β);*

(ii) *the relation $\equiv$ is a congruence;*

(iii) *there is a variable $X \in V$ that is decomposable in $\equiv_{\mathcal{D}}$ but not in $\equiv$;*

(iv) *the inclusions $\sim \subseteq \equiv \subset \equiv_{\mathcal{D}}$ hold.*

The final lemma allows us to "smooth out" an unmanageable congruence into a congruence defined by a unique decomposition base.

LEMMA 50. *Let $\equiv$ be a norm-preserving, polynomial-time computable congruence satisfying $\sim \subseteq \equiv$, where $\sim$ denotes maximal bisimulation. Then there is a decomposition base $\mathcal{D}$, computable in polynomial time, such that $\sim \subseteq \equiv_{\mathcal{D}} \subseteq \equiv$.*

With the three preceding lemmas in place, the procedure for deciding bisimulation equivalence is clear.

(1) Let the congruence $\equiv$ be defined by $\alpha \equiv \beta$ iff $n(\alpha) = n(\beta)$.

(2) Compute a decomposition base $\mathcal{D}$ with $\sim \subseteq \equiv_{\mathcal{D}} \subseteq \equiv$, using Lemma 50.

(3) If $\equiv_{\mathcal{D}}$ is a bisimulation – a condition that can be checked in polynomial time using Lemma 48 – then halt and return the relation $\equiv_{\mathcal{D}}$.

(4) Compute a congruence $\equiv$ satisfying $\sim \subseteq \equiv \subset \equiv_{\mathcal{D}}$, using Lemma 49. Go to step 2.
The main result is then virtually immediate.

THEOREM 51. *Given a normed BPP over variable set* V, *there is a polynomial-time (in the size of* α, β, *and the transition rules) algorithm to decide* $\alpha \sim \beta$ *for arbitrary* $\alpha, \beta \in V^*$.

PROOF. On each iteration of the loop formed by lines (2)–(4), the number of primes increases by at least one. Thus the number of iterations is bounded by n, and each iteration requires only polynomial time by the three preceding lemmas. □

2.3. *Computing a finite bisimulation base for BPA*

In Section 2.1.2, we demonstrated the existence of a finite bisimulation base for BPA; this permitted us to infer the decidability of bisimilarity between BPA states using two semi-decision procedures. In this section, we describe an algorithm from [23] for constructing a general bisimulation base, a finite relation $\mathcal{R}$ in which the least congruence $\overset{\mathcal{R}}{\equiv}$ containing $\mathcal{R}$ is equal to $\sim$. From this we deduce a stronger result than the decidability of bisimulation: the largest bisimulation $\sim$ is an effective rational relation [24]. Furthermore, the algorithm that we describe will demonstrate that the equivalence problem with respect to bisimilarity of BPA states can be solved in double exponential time.

Given any binary relation $\mathcal{R}$ on V^*, we define $E(\mathcal{R})$ to be the set of pairs in $\mathcal{R}$ which satisfy expansion in $\overset{\mathcal{R}}{\equiv}$; that is, $(\alpha, \beta) \in E(\mathcal{R})$ iff $(\alpha, \beta) \in \mathcal{R}$ and

- if $\alpha \xrightarrow{a} \alpha'$ then $\beta \xrightarrow{a} \beta'$ for some β' with $\alpha' \overset{\mathcal{R}}{\equiv} \beta'$; and
- if $\beta \xrightarrow{a} \beta'$ then $\alpha \xrightarrow{a} \alpha'$ for some α' with $\alpha' \overset{\mathcal{R}}{\equiv} \beta'$.

In particular, $\mathcal{R}$ is a bisimulation base (cf. Definition 24) if $\mathcal{R} = E(\mathcal{R})$, in which case $\overset{\mathcal{R}}{\equiv}$ is a bisimulation (cf. Lemma 25). Our intention is to construct a finite base $\mathcal{R}$ generating the largest bisimulation: $\sim = \overset{\mathcal{R}}{\equiv}$. To do this, we mimic the construction from Section 2.2.1:

(1) we find a *finite* relation $\mathcal{R}_0$ which is *complete*, meaning that $\sim = \overset{\mathcal{R}_0 \cap \sim}{\equiv}$ (this is a refinement of the notion of *fullness* defined in Section 2.2.1);
(2) we then repeatedly apply E to compute $\mathcal{R}_0 \supseteq E(\mathcal{R}_0) \supseteq E^2(\mathcal{R}_0) \supseteq \cdots$ until we find that $E^n(\mathcal{R}_0) = E^{n+1}(\mathcal{R}_0)$; we then take $\mathcal{R} = E^n(\mathcal{R}_0)$. (Note that n can be no greater than the size of $\mathcal{R}_0$.)

That this procedure provides the desired finite base $\mathcal{R}$ satisfying $\sim = \overset{\mathcal{R}}{\equiv}$ follows from the fact that completeness is preserved by the application of E, as demonstrated in the next result (cf. Lemma 42).

LEMMA 52. *If* $\mathcal{B}$ *is complete, then* $\mathcal{B} \cap \sim = E(\mathcal{B}) \cap \sim$, *so* $E(\mathcal{B})$ *is complete.*

PROOF. Assume that $\mathcal{B}$ is complete, so that $\sim = \stackrel{\mathcal{B}\cap\sim}{\equiv} \subseteq \stackrel{\mathcal{B}}{\equiv}$. Then since E is monotonic, we get $\sim = \mathsf{E}(\sim) \subseteq \mathsf{E}(\stackrel{\mathcal{B}}{\equiv})$, and hence

$$\begin{aligned}
\mathcal{B}\cap\sim &= \mathcal{B}\cap \mathsf{E}(\stackrel{\mathcal{B}}{\equiv})\cap\sim && \big(\text{since } \sim\subseteq \mathsf{E}(\stackrel{\mathcal{B}}{\equiv})\big)\\
&= \mathcal{B}\cap \mathsf{E}(\mathcal{B})\cap\sim && \big(\text{since } \mathcal{B}\cap \mathsf{E}(\stackrel{\mathcal{B}}{\equiv}) = \mathcal{B}\cap \mathsf{E}(\mathcal{B})\big)\\
&= \mathsf{E}(\mathcal{B})\cap\sim && \big(\text{since } \mathsf{E}(\mathcal{B})\subseteq\mathcal{B}\big).
\end{aligned}$$

□

We shall apply this method to an arbitrary BPA system. As before we split the set of variables into two disjoint sets $\mathsf{V} = \mathsf{V}_\mathsf{N}\cup\mathsf{V}_\mathsf{U}$ with the variables in V_N having finite norm and those in V_U having infinite norm, and make use of the *finite prefix norm* from Definition 33:

$$\mathfrak{n}_\mathsf{f}(\alpha) = \max\big\{\mathfrak{n}(\beta)\colon\ \mathfrak{n}(\beta) < \infty \text{ and } \alpha = \beta\gamma \text{ for some } \gamma\big\}.$$

Given the following three relations

$$\begin{aligned}
\mathcal{R}_0 &= \big\{(\mathsf{A}, \mathsf{AB})\colon\ \mathsf{A}\in\mathsf{V}_\mathsf{U} \text{ and } \mathsf{B}\in\mathsf{V}\big\},\\
\mathcal{R}_\mathsf{f} &= \big\{(\mathsf{A},\alpha)\in\mathsf{V}_\mathsf{N}\times{\mathsf{V}_\mathsf{N}}^*\colon\ \mathfrak{n}(\mathsf{A}) = \mathfrak{n}(\alpha)\big\},\\
\mathcal{R}_\infty &= {\mathsf{V}_\mathsf{N}}^*\mathsf{V}_\mathsf{U}\times{\mathsf{V}_\mathsf{N}}^*\mathsf{V}_\mathsf{U},
\end{aligned}$$

we shall construct a finite complete relation $\mathcal{R} = \mathcal{R}_0\cup\mathcal{R}_\mathsf{f}\cup\mathcal{R}'$ with $\mathcal{R}'\subseteq\mathcal{R}_\infty$. Then by Lemmas 52 and 53 below, we shall be able to construct $\mathcal{R}\cap\sim$.

LEMMA 53. *For any finite $\mathcal{S}\subseteq\mathcal{R}_\mathsf{f}\cup\mathcal{R}_\infty$ with $\mathcal{R}_0\subseteq\mathcal{S}$, the relation $\stackrel{\mathcal{S}}{\equiv}$ is decidable; hence we can compute* $\mathsf{E}(\mathcal{S})$.

PROOF. Let

$$\begin{aligned}
\mathcal{S}_\mathsf{f} &= \mathcal{S}\cap\mathcal{R}_\mathsf{f}, \qquad \text{and}\\
\mathcal{S}_\infty &= \big\{(\alpha',\beta')\colon\ \alpha\stackrel{\mathcal{S}_\mathsf{f}}{\equiv}\alpha' \text{ and } \beta\stackrel{\mathcal{S}_\mathsf{f}}{\equiv}\beta' \text{ for some } (\alpha,\beta)\in\mathcal{S}\cap\mathcal{R}_\infty\big\}.
\end{aligned}$$

As $\mathcal{S}_\mathsf{f}$ is finite and norm-preserving, $\mathcal{S}_\infty$ must be finite as well. Now,

$$\begin{aligned}
\stackrel{\mathcal{S}}{\equiv} &= \big\{(\alpha,\beta)\colon\ \alpha,\beta\in{\mathsf{V}_\mathsf{N}}^* \text{ and } \alpha\stackrel{\mathcal{S}_\mathsf{f}}{\equiv}\beta\big\}\\
&\quad\cup\big\{(\alpha\gamma,\beta\delta)\colon\ \alpha,\beta\in{\mathsf{V}_\mathsf{N}}^*\mathsf{V}_\mathsf{U} \text{ and } \alpha\stackrel{\mathcal{S}_\infty}{\equiv}\beta\big\}.
\end{aligned}$$

Note that $\stackrel{\mathcal{S}_\mathsf{f}}{\equiv}$ restricted to ${\mathsf{V}_\mathsf{N}}^*$ is decidable: $\mathcal{S}_\mathsf{f}$ is norm-preserving, and hence any equivalence class is an effective finite set. Furthermore $\stackrel{\mathcal{S}_\infty}{\equiv}$ restricted to ${\mathsf{V}_\mathsf{N}}^*\mathsf{V}_\mathsf{U}$ is decidable: it is the suffix derivation of $\mathcal{S}_\infty\cup\mathcal{S}_\infty^{-1}$, and hence any equivalence class is an effective rational set [19]. □

For any $n \geqslant 1$, we define the relation $\mathcal{R}_n$ as follows:

$$\mathcal{R}_n = \mathcal{R}_0 \cup \mathcal{R}_f \cup \{(\alpha, \beta) \in \mathcal{R}_\infty: n_f(\alpha), n_f(\beta) < n\}.$$

We shall compute a constant e such that $\mathcal{R}_e$ is complete, after which we can compute $\mathcal{R}_e \cap \sim$ and hence infer that $\sim$ is decidable. In fact, we get a stronger result: we can construct a transducer (a finite automaton labelled by pairs of states) recognizing the bisimulation.

THEOREM 54. *Bisimilarity is an effective rational relation over any BPA system.*

PROOF. We assume that we have our promised integer e such that $\mathcal{R}_e$ is complete. By Lemmas 52 and 53, we compute $\mathcal{R}_e \cap \sim = \mathcal{R}_0 \cup (\mathcal{R}_f \cap \sim) \cup \mathcal{S}_e$, where

$$\mathcal{S}_e = \{(\alpha, \beta): \alpha, \beta \in V_N{}^* V_U, \alpha \sim \beta \text{ and } n_f(\alpha), n_f(\beta) < e\}.$$

First we show that the set of bisimilar pairs of infinite norm is an effective rational relation. We have

$$\begin{aligned} \sim \cap V^* V_U V^* \times V^* V_U V^* &= \overset{\mathcal{R}_e \cap \sim}{\equiv} \cap V^* V_U V^* \times V^* V_U V^* \\ &= (\overset{\mathcal{S}_e}{\equiv} \cap V_N{}^* V_U \times V_N{}^* V_U) \cdot (V^* \times V^*). \end{aligned}$$

But $(\overset{\mathcal{S}_e}{\equiv} \cap V_N{}^* V_U \times V_N{}^* V_U)$ is the suffix derivation according to $\mathcal{S}_e$ restricted to the recognizable relation $V_N{}^* V_U \times V_N{}^* V_U$, and hence is an effective rational relation [32].

It then remains to show that the set of bisimilar pairs of finite norm is an effective rational relation. For this, we can exploit the Unique factorization theorem 21. Let $V_P \subseteq V_N$ be the set of all primes without repetition up to $\sim$; that is, for every $\alpha \in V_N{}^*$ there is a unique $\beta \in V_P{}^*$ such that $\alpha \sim \beta$. Now take

$$\mathcal{S} = \{(X, \alpha): X \in V_N, \ \alpha \in V_P{}^*, \ \text{and } X \sim \alpha\}.$$

The set of bisimilar pairs of finite norm can then be expressed as the effective rational set $\mathcal{S}^* \circ (\mathcal{S}^{-1})^*$. □

It now remains to explain the computation of the constant e such that $\mathcal{R}_e$ is complete. A useful tool is the following *divergence* function:

$$\begin{aligned} \mathrm{Div}(\alpha, \beta) &= \min(\{n : \alpha \not\sim_n \beta\} \cup \{\infty\}) \\ &= 1 + \max\{n : \alpha \sim_n \beta\}. \end{aligned}$$

An important property is that $1/\mathrm{Div}$ is an ultrametric distance.

LEMMA 55.

(a) $1 \leqslant \mathrm{Div}(\alpha, \beta) = \mathrm{Div}(\beta, \alpha)$;

(b) $\min(\mathit{Div}(\alpha, \beta), \mathit{Div}(\beta, \gamma)) \leqslant \mathit{Div}(\alpha, \gamma)$;
(c) *If* $\mathit{Div}(\alpha, \beta) > \mathit{Div}(\alpha, \gamma)$ *then* $\mathit{Div}(\beta, \gamma) = \mathit{Div}(\alpha, \gamma)$;
(d) *If* $n(\alpha) < n(\beta)$ *then* $\mathit{Div}(\alpha, \beta) \leqslant 1 + n(\alpha)$;
(e) *If* $\alpha \xrightarrow{u} \alpha'$ *and* $|u| < \mathit{Div}(\alpha, \beta)$ *then* $\beta \xrightarrow{u} \beta'$ *such that* $\mathit{Div}(\alpha, \beta) \leqslant |u| + \mathit{Div}(\alpha', \beta')$.

We extend Div to any binary relation R over states as follows.

$$\mathit{minDiv}(R) = \min\{\mathit{Div}(\alpha, \beta) : \alpha R \beta\}.$$

With this, we may now generalize Lemma 55.

LEMMA 56. *If* $\alpha R^* \alpha'$, $\beta R^* \beta'$, *and* $\mathit{Div}(\alpha, \beta) < \mathit{minDiv}(R)$ *then*

$$\mathit{Div}(\alpha, \beta) = \mathit{Div}(\alpha', \beta').$$

The above properties of the divergence function are true for arbitrary transition systems. Henceforth we consider only BPA systems, in which case the divergence function satisfies the following additional properties.

LEMMA 57. *For every* $\alpha, \beta, \gamma, \delta \in V^*$, *we have the following.*
(a) *If* $\gamma \sim \delta$ *then* $\mathit{Div}(\alpha, \beta) \leqslant \mathit{Div}(\alpha\gamma, \beta\delta) \leqslant \mathit{Div}(\alpha, \beta) + n(\gamma) \leqslant \mathit{Div}(\gamma\alpha, \delta\beta)$.
(b) *If* $\alpha\gamma \sim \beta\delta$ *and* $1 < \mathit{Div}(\alpha, \beta) < \infty$ *then* $\alpha \xrightarrow{a} \alpha'$ *and* $\beta \xrightarrow{a} \beta'$ *such that* $\alpha'\gamma \sim \beta'\delta$ *and* $\mathit{Div}(\alpha', \beta') < \mathit{Div}(\alpha, \beta)$.
(c) *If* $\alpha \overset{\mathcal{R}}{\equiv} \gamma$, $\beta \overset{\mathcal{R}}{\equiv} \delta$, *and* $\mathit{Div}(\alpha, \beta) < \mathit{minDiv}(R)$ *then* $\mathit{Div}(\alpha, \beta) = \mathit{Div}(\gamma, \delta)$.

If we can compute an integer $a_{\alpha,\beta}$ for any α, β such that $a_{\alpha,\beta} > \mathit{Div}(\alpha, \beta)$ when $\alpha \not\sim \beta$, then we would have the basis of an algorithm for deciding $\sim$. In fact, we are not able to find such a bound for an arbitrary BPA system; but for any BPA system, we shall show that there exists an equivalent BPA system in a form allowing to compute such a bound.

Consider any BPA system with rewrite rules P. For each $\alpha \in \mathit{Im}(P)$ with $n(\alpha) = \infty$ (that is, for each unnormed sequence α which appears as the right hand side of some rule of P) we associate a new variable X_α in such a way that $X_\alpha = X_\beta$ whenever $\alpha \sim \beta$. (This association is not effective, but we are interested only in the divergence function.) For each such variable X_α we also introduce a new label a_α. We then define a new BPA system with the following set $\widehat{P}$ of rewrite rules:

$$\begin{aligned}\widehat{P} &= \{A \xrightarrow{a} \alpha : A \xrightarrow{a} \alpha \in P \text{ and } n(\alpha) < \infty\} \\ &\cup \{A \xrightarrow{a} X_\alpha,\ X_\alpha \xrightarrow{a_\alpha} X_\alpha : A \xrightarrow{a} \alpha \in P \text{ and } n(\alpha) = \infty\}.\end{aligned}$$

The following then attests that these two BPA systems are equivalent.

LEMMA 58. *For* $\alpha, \beta \in {V_N}^*$, $\alpha \sim_P \beta$ *iff* $\alpha \sim_{\widehat{P}} \beta$.

We have that $\mathrm{Div}_{\widehat{P}}(\alpha, \beta) \leqslant \mathrm{Div}_P(\alpha, \beta)$, but equality is not true in general. The new system $\widehat{P}$ has the following restriction: each right hand side is either a normed sequence or an unnormed variable. Furthermore, $\mathrm{Div}(A, B) = 1$ for every $A, B \in V_U$ with $A \not\sim_{\widehat{P}} B$. These restrictions allow us to find an upper bound to the divergence for non-bisimilar pairs. Before proceedings, we need first to introduce some further notations. For any finite set $E \subseteq V^*$, we let $|E|$ denote its cardinality; $\text{max-n}(E) = \max\{n(\alpha)\colon \alpha \in E\}$; $\text{min-n}(E) = \min\{n(\alpha)\colon \alpha \in E\}$; $\text{max-n}_f(E) = \max\{n_f(\alpha)\colon \alpha \in E\}$; and $\text{min-n}_f(E) = \min\{n_f(\alpha)\colon \alpha \in E\}$. We shall also freely apply these max and min functions to lists as well as sets. We let $P_f = \{A \xrightarrow{a} \alpha \in P\colon n(\alpha) < \infty\}$ be the restriction of a system P to its rules of finite norm. Then for any $\alpha, \beta \in V^*$, we define

$$a_{\alpha,\beta} = \text{min-n}(\alpha, \beta) + \big(|V_N| - 1\big) \cdot \text{max-n}\big(\mathrm{Im}(P_f)\big) + \text{max-n}(V_N).$$

LEMMA 59. *For any BPA system* P *and any* $\alpha, \beta \in {V_N}^*$, *if* $\alpha \not\sim \beta$ *then* $\mathrm{Div}_{\widehat{P}}(\alpha, \beta) \leqslant a_{\alpha,\beta}$.

PROOF. Using Lemmas 55, 56 and 57, we generalize directly the construction given in [30] for simple grammars to obtain the same bound $a_{\alpha,\beta}$ for any BPA system in which each right hand side is either a normed sequence or an unnormed variable; and two unnormed non-bisimilar variables are of divergence one (this last condition is not strictly necessary, but without it the bound must be changed). □

For a normed system P, $\widehat{P} = P$, and $a_{\alpha,\beta}$ is optimal when $P = V \times \Sigma \times \{\varepsilon\}$, in which case $\text{max-n}(\mathrm{Im}(P)) = 0$ and $\text{max-n}(V) = 1$, so $a_{\alpha,\beta} = \min(|\alpha|, |\beta|) + 1 = \mathrm{Div}(\alpha, \beta)$ whenever $\alpha \not\sim \beta$. This bound $a_{\alpha,\beta}$ on $\mathrm{Div}_{\widehat{P}}(\alpha, \beta)$ also provides a bound on the length of a derivation of P from (α, β) to a pair with distinct norms, and which preserves a given unification, as formulated in the following.

LEMMA 60. *For* $\alpha, \beta \in {V_N}^*$, *if* $\alpha\gamma \sim \beta\delta$ *and* $\alpha \not\sim \beta$ *then for some* α', β' *and* u *with* $|u| < a_{\alpha,\beta}$ *we have that* $\alpha \xrightarrow{u} \alpha'$, $\beta \xrightarrow{u} \beta'$, $n(\alpha') \neq n(\beta')$, *and* $\alpha'\gamma \sim \beta'\delta$.

PROOF. By Lemma 58, $\alpha \not\sim_{\widehat{P}} \beta$, and by Lemma 59, $\mathrm{Div}_{\widehat{P}}(\alpha, \beta) \leqslant a_{\alpha,\beta}$. By Lemma 57, using induction on $\mathrm{Div}_{\widehat{P}}(\alpha, \beta)$, there exist two derivations

$$\begin{aligned} \alpha &= \alpha_1 \xrightarrow{a_1}_{\widehat{P}} \alpha_2 \xrightarrow{a_2}_{\widehat{P}} \cdots \xrightarrow{a_{n-1}}_{\widehat{P}} \alpha_n \quad \text{and} \\ \beta &= \beta_1 \xrightarrow{a_1}_{\widehat{P}} \beta_2 \xrightarrow{a_2}_{\widehat{P}} \cdots \xrightarrow{a_{n-1}}_{\widehat{P}} \beta_n \end{aligned}$$

such that $\alpha_i\gamma \sim \beta_i\delta$ for every $1 \leqslant i \leqslant n$, and

$$\mathrm{Div}_{\widehat{P}}(\alpha_1, \beta_1) > \mathrm{Div}_{\widehat{P}}(\alpha_2, \beta_2) > \cdots > \mathrm{Div}_{\widehat{P}}(\alpha_n, \beta_n) = 1.$$

In particular, $n \leqslant \mathrm{Div}_{\widehat{P}}(\alpha, \beta) \leqslant a_{\alpha,\beta}$. Since $\mathrm{Div}_{\widehat{P}}(\alpha_n, \beta_n) = 1$ and $\alpha_n\gamma \sim \beta_n\delta$, either $\alpha_n = \varepsilon \neq \beta_n$ or $\alpha_n \neq \varepsilon = \beta_n$, so $n(\alpha_n) \neq n(\beta_n)$. Thus $m = \min\{i\colon n(\alpha_i) \neq n(\beta_i)\}$ exists, and we take $u = a_1 \ldots a_{m-1}$, so $|u| = m - 1 < n \leqslant a_{\alpha,\beta}$. By symmetry, we may assume that $n(\alpha_m) < n(\beta_m)$.

- If $n(\beta_m) < \infty$, then it suffices to take $\alpha' = \alpha_m$ and $\beta' = \beta_m$.
- If $n(\beta_m) = \infty$, then $m > 1$ and $\beta_{m-1} \in V_N{}^+$. Let $B\mu = \beta_{m-1}$; then $B \xrightarrow{a_{m-1}}_P \rho$ with $X_\rho \mu = \beta_m$. Thus it suffices to take $\alpha' = \alpha_m$ and $\beta' = \rho\mu$; in particular,

$$\alpha'\gamma = \alpha_m\gamma \sim \beta_m\delta = X_\rho\mu\delta \sim \rho \sim \rho\mu\delta = \beta'\delta. \qquad \square$$

If $\gamma = \delta$ in the above lemma, then δ is *repetitive*: $\delta \sim \rho\delta$ for some $\rho \neq \varepsilon$, and we can find such a ρ with a finite norm bounded by $b_{\text{max-n}(\alpha,\beta)}$, where

$$b_n = (n + |V_N|) \cdot \text{max-n}\big(Im(P_f)\big) \cdot \big(\text{max-n}_f\big(Im(P)\big)\big)^2.$$

LEMMA 61. *For* $\alpha, \beta \in V_N{}^*$, *if* $\alpha\delta \sim \beta\delta$ *and* $\alpha \not\sim \beta$ *then* $\delta \sim \gamma\delta$ *for some* $\gamma \neq \varepsilon$ *with* $n_f(\gamma) < b_{\text{max-n}(\alpha,\beta)}$.

PROOF. By Lemma 60, there are derivations $\alpha \xrightarrow{u} \alpha'$ and $\beta \xrightarrow{u} \beta'$ such that $|u| < a_{\alpha,\beta}$, $n(\alpha') \neq n(\beta')$ and $\alpha'\delta \sim \beta'\delta$. Thus there are derivations

$$\big(\alpha' \xrightarrow{v} \varepsilon \text{ and } \beta' \xrightarrow{v} \gamma\big) \quad \text{or} \quad \big(\alpha' \xrightarrow{v} \gamma \text{ and } \beta' \xrightarrow{v} \varepsilon\big)$$

such that $|v| = \text{min-n}(\alpha', \beta')$ and $\delta \sim \gamma\delta$. As $\text{max-n}(V_N) \leqslant \text{max-n}(Im(P_f)) + 1$, we have that

$$a_{\alpha,\beta} - 1 \leqslant \text{min-n}(\alpha, \beta) + |V_N| \cdot \text{max-n}\big(Im(P_f)\big).$$

Furthermore,

$$\begin{aligned}
n_f(\alpha') &\leqslant n(\alpha) + |u| \cdot \big(\text{max-n}_f\big(Im(P)\big) - 1\big),\\
n_f(\beta') &\leqslant n(\beta) + |u| \cdot \big(\text{max-n}_f\big(Im(P)\big) - 1\big),\\
n_f(\gamma) &\leqslant \text{max-n}_f(\alpha', \beta') + |v| \cdot \big(\text{max-n}_f\big(Im(P)\big) - 1\big).
\end{aligned}$$

So

$$\begin{aligned}
n_f(\gamma) \leqslant\ & \text{max-n}(\alpha, \beta) + (a_{\alpha,\beta} - 1) \cdot \big(\text{max-n}\big(Im(P_f)\big) - 1\big)\\
& + \big(\text{min-n}(\alpha, \beta) + (a_{\alpha,\beta} - 1) \cdot \big(\text{max-n}\big(Im(P_f)\big) - 1\big)\big)\\
& \times \big(\text{max-n}_f\big(Im(P)\big) - 1\big)\\
\leqslant\ & \big(\text{max-n}(\alpha, \beta) + (a_{\alpha,\beta} - 1) \cdot \big(\text{max-n}\big(Im(P_f)\big) - 1\big)\big) \cdot \text{max-n}_f\big(Im(P)\big)\\
\leqslant\ & \big(\text{max-n}(\alpha, \beta) \cdot \text{max-n}\big(Im(P_f)\big)\\
& + |V_N| \cdot \text{max-n}\big(Im(P_f)\big) \cdot \big(\text{max-n}\big(Im(P_f)\big) - 1\big)\big) \cdot \text{max-n}_f\big(Im(P)\big)\\
<\ & \big(\text{max-n}(\alpha, \beta) + |V_N|\big) \cdot \text{max-n}\big(Im(P_f)\big) \cdot \big(\text{max-n}_f\big(Im(P)\big)\big)^2\\
=\ & b_{\text{max-n}(\alpha,\beta)}. \qquad \square
\end{aligned}$$

We then define the integer $c = b_{\text{max-n}(V_N)\cdot(1+\text{max-n}_f(Im(P)))}$ to obtain a sequence of decreasing repetitive states.

LEMMA 62. *Let* $\delta \sim \gamma\delta$ *with* $0 < n(\gamma) < c$ *and* $\text{max-n}(V_N) < n_f(\delta)$. *Then there exist* $\widehat{\gamma}$, $\widehat{\delta}$ *and* ρ *such that* $\widehat{\delta} \sim \widehat{\gamma}\widehat{\delta}$ *with* $0 < n(\widehat{\gamma}) < c$ *and* $\delta \sim \rho\widehat{\delta}$ *with* $n_f(\delta) = n_f(\rho\widehat{\delta})$ *and* $0 < n(\rho) < 2\,\text{max-n}(V_N)$.

PROOF. As $0 < n(\gamma) < c$, we have that $\gamma = A\gamma''$ with $A \in V_N$. Let $A \xrightarrow{u} \varepsilon$ with $|u| = n(A)$. Then $\delta = \delta'\delta'' \xrightarrow{u} \mu\delta''$ such that $\mu\delta'' \sim \gamma''\delta$ where $|\delta'|$ is minimal. As $n_f(\delta) > \text{max-n}(V_N) \geqslant |u|$, we have that $\delta'' \neq \varepsilon$ and $\delta' \in {V_N}^*$. By the minimality of $|\delta'|$, we have that $\delta' \xrightarrow{u'} B \xrightarrow{u''} \mu$ with $B = \delta'(|\delta'|)$, $u'u'' = u$ and $u'' \neq \varepsilon$. Hence $n(\delta') \leqslant |u'| + n(B) < 2\,\text{max-n}(V_N)$. Furthermore, $n_f(\mu) \leqslant n(B) + |u''| \cdot (\text{max-n}_f(Im(P) + 1) \leqslant \text{max-n}(V_N) \cdot \text{max-n}_f(Im(P))$. We have that $\delta'\delta'' = \delta \sim A\gamma''\delta \sim A\mu\delta''$.

Suppose that $\delta' \sim A\mu$. In particular, $\mu \in {V_N}^*$, and it suffices to take $\widehat{\gamma} = \gamma''A$, $\widehat{\delta} = \mu\delta''$ and $\rho = A$:

- $\widehat{\delta} = \mu\delta'' \sim \gamma''\delta = \gamma''\delta'\delta'' \sim \gamma''A\mu\delta'' = \widehat{\gamma}\widehat{\delta}$.
- $0 < n(\widehat{\gamma}) = n(\gamma''A) = n(A\gamma'') = n(\gamma) < c$.
- $\delta = \delta'\delta'' \sim A\mu\delta'' = \rho\widehat{\delta}$.
- $n_f(\delta) = n_f(\delta'\delta'') = n(\delta') + n_f(\delta'') = n(A) + n(\mu) + n_f(\delta'') = n_f(A\mu\delta'') = n_f(\rho\widehat{\delta})$.
- $0 < n(\rho) = n(A) \leqslant \text{max-n}(V_N) < 2\,\text{max-n}(V_N)$.

Assume then that $\delta' \not\sim A\mu$. Then we have that $\delta'\delta'' \sim A\mu\delta''$ with $n(\delta') < 2\,\text{max-n}(V_N)$ and $n(A\mu) \leqslant \text{max-n}(V_N) \cdot (1 + \text{max-n}_f(Im(P)))$.

- If $\mu \in {V_N}^*$, then by Lemma 61, there is $\widehat{\gamma}$ such that $\delta'' \sim \widehat{\gamma}\delta''$ and

$$n_f(\widehat{\gamma}) < b_{\max(2\,\text{max-n}(V_N),\,\text{max-n}(V_N)\cdot(1+\text{max-n}_f(Im(P))))} = c.$$

- If $n(\mu) = \infty$, then $\delta'\delta'' \sim A\mu$. Let $\delta' \xrightarrow{v} \varepsilon$ with $|v| = n(\delta')$. There is $A\mu \xrightarrow{v} \widehat{\gamma}$ with $\delta'' \sim \widehat{\gamma} \sim \widehat{\gamma}\delta''$. Then $n_f(\widehat{\gamma}) \leqslant n_f(A\mu) + |v| \cdot (\text{max-n}_f(Im(P)) - 1)$. Thus $n_f(\widehat{\gamma}) < \text{max-n}(V_N) \cdot (1 + \text{max-n}_f(Im(P))) + 2\,\text{max-n}(V_N) \cdot (\text{max-n}_f(Im(P)) - 1) < c$. So it suffices to take $\widehat{\delta} = \gamma''$ and $\rho = \gamma'$. □

We bound the finite norm of repetitive states with the following integer:

$$d = 2\,\text{max-n}(V_N) \cdot |V_N|^c.$$

LEMMA 63. *Let* $\delta \sim \gamma\delta$ *with* $n_f(\gamma) < c$. *Then there is* $\widehat{\delta} \sim \delta$ *with* $n_f(\widehat{\delta}) < d$.

PROOF. If $n(\gamma) = \infty$ then it suffices to take $\widehat{\delta} = \delta$ (since $c < d$). Let $\delta \sim \gamma\delta$ with $n(\gamma) < c$. We may assume that $n_f(\delta)$ is minimal. By repeated applications of Lemma 62, there is a finite sequence $(\gamma_i, \delta_i, \rho_i)$ with $1 \leqslant i \leqslant n$ such that

- $\gamma_1 = \gamma$ and $\delta_1 = \delta$;
- $n_f(\delta_n) \leqslant \text{max-n}(V_N)$;
- for each $1 \leqslant i \leqslant n$: $\delta_i \sim \gamma_i\delta_i$ and $0 < n(\rho_i) < 2\,\text{max-n}(V_N)$; and
- for each $1 \leqslant i < n$: $\delta_i \sim \rho_{i+1}\delta_{i+1}$ and $n_f(\delta_i) = n_f(\rho_{i+1}\delta_{i+1})$.

Assume that $\gamma_i = \gamma_j$ for $i < j$. Then $\delta_i \sim \delta_j$ with $n_f(\delta_j) < n_f(\delta_i)$. If $\widehat{\delta} = \rho_2 \ldots \rho_i\delta_j$ then $\widehat{\delta} \sim \rho_2 \ldots \rho_i\delta_i \sim \delta_1 = \delta$, but $n_f(\delta) = n_f(\rho_2 \ldots \rho_i\delta_i) > n_f(\widehat{\delta})$. This contradiction to the minimality of δ implies that the γ_i are pairwise distinct. Hence

$$\begin{aligned} n &\leqslant \left|\{\gamma \in V_N{}^*\colon n(\gamma) < c\}\right| \leqslant 1 + |V_N| + \cdots + |V_N|^{c-1} \\ &= \left(|V_N|^c - 1\right)/\left(|V_N| - 1\right) < |V_N|^c. \end{aligned}$$

Finally

$$\begin{aligned} n_f(\delta) &= n_f(\rho_2 \ldots \rho_n \delta_n) < 2(n-1) \cdot \text{max-n}(V_N) + \text{max-n}(V_N) \\ &< 2n\, \text{max-n}(V_N) < 2\, \text{max-n}(V_N) \cdot |V_N|^c = d. \end{aligned}$$ □

The integer $e = d + 1 + \text{max-n}(V_N) \cdot \text{max-n}(Im(P_f))$ is then a suitable bound.

THEOREM 64. *The relation* R_e *is complete:* $\overset{R_e \cap \sim}{\equiv} = \sim$.

PROOF. We have $\overset{R_e \cap \sim}{\equiv} \subseteq \overset{\sim}{\equiv} = \sim$. To prove the reverse inclusion and using R_0 it suffices to show that for every $\alpha, \beta \in V_N{}^* \cup V_N{}^* V_U$, if $\alpha \sim \beta$ then $\alpha \overset{R_e \cap \sim}{\equiv} \beta$. We may also assume that $Im(P) \subseteq V_N{}^* \cup V_N{}^* V_U$. In the sequel, we let $\equiv = \overset{R_e \cap \sim}{\equiv}$.

The proof is carried out by induction on $\preceq$ as defined in Definition 33: $(\alpha, \beta) \preceq (\gamma, \delta)$ iff $\text{max-n}_f(\alpha, \beta) \leqslant \max n_f(\gamma, \delta)$.

Suppose first that $\text{max-n}_f(\alpha, \beta) = 0$, that is, $\alpha, \beta \in V_U \cup \{\varepsilon\}$. As $\alpha \sim \beta$, either $\alpha = \beta = \varepsilon$, in which case $\alpha \equiv \beta$; or $\alpha, \beta \in V_U$, in which case $\alpha (R_e \cap \sim) \beta$.

Assume then that $\text{max-n}_f(\alpha, \beta) > 0$. First, let us verify the following property of normed sequences $\lambda, \mu \in V_N{}^*$:

($\star$) If $\mu \sim \lambda$ and $n_f(\mu) < e$ and $(\lambda, \lambda) \prec (\alpha, \beta)$ then $\mu \equiv \lambda$.

Either $n(\lambda) < \infty$ in which case $(\mu, \lambda) \prec (\alpha, \beta)$ and by the induction hypothesis $\mu \equiv \lambda$; or $n(\lambda) = \infty$ in which case either $n_f(\lambda) < e$ so $\mu (R_e \cap \sim) \lambda$, or $n_f(\lambda) \geqslant e$ in which case $(\mu, \lambda) \prec (\lambda, \lambda) \prec (\alpha, \beta)$, and by the induction hypothesis $\mu \equiv \lambda$.

Now let $A\alpha' = \alpha$ and $B\beta' = \beta$. By symmetry, we may assume that $n(B) \leqslant n(A)$, so $B \in V_N$. Let $B \xrightarrow{u} \varepsilon$ with $|u| = n(B)$. Then $A \xrightarrow{u} \gamma$ such that $\gamma\alpha' \sim \beta'$. In particular,

$$\begin{aligned} n_f(\gamma) &\leqslant |u| \cdot \left(\text{max-n}_f\left(Im(P)\right) - 1\right) + 1 \\ &\leqslant \text{max-n}(V_N) \cdot \left(\text{max-n}_f\left(Im(P)\right) - 1\right) + 1 < e. \end{aligned}$$

Furthermore, $A\alpha' \sim B\gamma\alpha'$.

Suppose $n(\gamma) = \infty$. Then $\gamma \sim \beta'$, and by Property ($\star$), $\gamma \equiv \beta'$. Furthermore $A\alpha' \sim B\gamma$ and $n_f(B\gamma) \leqslant \text{max-n}(V_N) \cdot \text{max-n}_f(Im(P)) + 1 < e$.

- If $A \in V_U$ then $\alpha' = \varepsilon$ and hence $\alpha = A(R_e \cap \sim) B\gamma \equiv B\beta' = \beta$.
- If $A \in V_N$ then $A \xrightarrow{v} \varepsilon$ such that $|v| = n(A)$. Hence $B\gamma \xrightarrow{v} \widehat{\alpha}$ such that $\alpha' \sim \widehat{\alpha}$. In particular, $n_f(\widehat{\alpha}) \leqslant \text{max-n}(V_N) \cdot (\text{max-n}_f(Im(P)) - 1) + n_f(B\gamma) \leqslant 2 \cdot \text{max-n}(V_N) \cdot \text{max-n}_f(Im(P)) < e$ and by Property ($\star$), $\widehat{\alpha} \equiv \alpha'$. Furthermore, $A\widehat{\alpha} \sim B\gamma$ and $n_f(A\widehat{\alpha}) \leqslant \text{max-n}(V_N) \cdot (2 \cdot \text{max-n}_f(Im(P)) + 1) < e$. Hence $A\widehat{\alpha}(R_e \cap \sim) B\gamma$. Finally, $\alpha = A\alpha' \equiv A\widehat{\alpha} \equiv B\gamma \equiv B\beta' = \beta$.

Assume then that $n(\gamma) < \infty$. Then $A \in V_N$, and since $n(\gamma\alpha') = n(\beta')$, by induction, $\gamma\alpha' \equiv \beta'$. Furthermore, $n(\gamma) \leqslant |u| \cdot (\text{max-n}(Im(P_f)) - 1) + 1 \leqslant \text{max-n}(V_N) \cdot (\text{max-n}(Im(P_f)) - 1) + 1$.

- If $A \sim B\gamma$ then $A(R_e \cap \sim)B\gamma$. Hence $\alpha = A\alpha' \equiv B\gamma\alpha' \equiv B\beta' = \beta$.
- If $A \not\sim B\gamma$ then $A\alpha' \sim B\gamma\alpha'$. By Lemma 61, $\alpha' \sim \delta\alpha'$ for some δ with $n_f(\delta) < b_{\text{max-n}(A,B\gamma)} < c$. By Lemma 63, $\alpha' \sim \widehat{\delta}$ for some $\widehat{\delta}$ with $n_f(\widehat{\delta}) < d$. By Property ($\star$), $\alpha' \equiv \widehat{\delta}$. Furthermore, $A\widehat{\delta}(R_e \cap \sim)B\gamma\widehat{\delta}$, since $A\widehat{\delta} \sim B\gamma\widehat{\delta}$ with $n_f(A\widehat{\delta}) < e$ and

$$\begin{aligned} n_f\big(B\gamma\widehat{\delta}\big) &< \text{max-n}(V_N) + n(\gamma) + d \\ &\leqslant \text{max-n}(V_N) \cdot \text{max-n}\big(Im(P_f)\big) + 1 + d = e. \end{aligned}$$

Finally, $\alpha = A\alpha' \equiv A\widehat{\delta} \equiv B\gamma\widehat{\delta} \equiv B\gamma\alpha' \equiv B\beta' = \beta$. □

Although we have a direct decision procedure to decide bisimilarity for any BPA system, the complexity is not polynomial: constant e is exponential in the size of the system. A close analysis would demonstrates that the complexity of the underlying algorithm is in fact doubly exponential in the size of the BPA system.

2.4. *Undecidability results for MSA*

In this section we outline Jančar's technique [83] for demonstrating undecidability results for bisimilarity through mimicking Minsky machines in a weak fashion but faithfully enough to capture the essence of the halting problem. Jančar first employed his ideas to demonstrate the undecidability of bisimulation equivalence over the class of Petri nets. Hirshfeld [66] modified the argument to demonstrate the undecidability of trace equivalence over BPP. Jančar and Moller [88] then used the technique to demonstrate the undecidability of regularity checking of Petri nets with respect to both simulation and trace equivalence.

In the following we generalize Jančar's result by demonstrating the undecidability of (normed) MSA. Jančar's technique applies ideally in this case, as MSA offer precisely the ingredients present in Petri nets which are needed to mimic Minsky machines.

Minsky machines [117] are simple straight-line programs which make use of only two counters. Formally, a *Minsky machine* is a sequence of labelled instructions

$$\begin{aligned} &X_0 && : \mathtt{comm}_0, \\ &X_1 && : \mathtt{comm}_1, \\ & && \dots \\ &X_{n-1} && : \mathtt{comm}_{n-1}, \\ &X_n && : \mathtt{halt}, \end{aligned}$$

where each of the first n instructions is either of the form

$$X_\ell : c_0 := c_0 + 1;\ \mathtt{goto}\ X_j \quad \text{or} \quad X_\ell : c_1 := c_1 + 1;\ \mathtt{goto}\ X_j$$

or of the form

$$\mathtt{X}_\ell : \mathtt{if}\ \mathtt{c}_0 = 0\ \mathtt{then}\ \mathtt{goto}\ \mathtt{X}_j \quad \text{or} \quad \mathtt{X}_\ell : \mathtt{if}\ \mathtt{c}_1 = 0\ \mathtt{then}\ \mathtt{goto}\ \mathtt{X}_j$$
$$\mathtt{else}\ \mathtt{c}_0 := \mathtt{c}_0 - 1;\ \mathtt{goto}\ \mathtt{X}_k \qquad \mathtt{else}\ \mathtt{c}_1 := \mathtt{c}_1 - 1;\ \mathtt{goto}\ \mathtt{X}_k$$

A Minsky machine M starts executing with the value 0 in the counters c_0 and c_1 and the control at the label X_0. When the control is at label X_ℓ $(0 \leqslant \ell < n)$, the machine executes instruction $\mathtt{comm}_\ell$, modifying the contents of the counters and transferring the control to the appropriate label as directed by the instruction. The machine halts if and when the control reaches the $\mathtt{halt}$ instruction at label X_n. We recall now the fact that the halting problem for Minsky machines is undecidable: there is no algorithm which decides whether or not a given Minsky machine halts.

A Minsky machine as presented above gives rise to the following MSA.

- The input alphabet is $\Sigma = \{i, d, z, \omega\}$.
- The control states are $Q = \{p_0, p_1, \ldots, p_{n-1}, p_n, q_0, q_1, \ldots, q_{n-1}, q_n\}$.
- The stack alphabet is $\Gamma = \{Z, 0, 1\}$.
- For each machine instruction

  $$\mathtt{X}_\ell : \mathtt{c}_b := \mathtt{c}_b + 1;\ \mathtt{goto}\ \mathtt{X}_j$$

  we have the MSA rules

  $$p_\ell Z \xrightarrow{i} p_j bZ \quad \text{and} \quad q_\ell Z \xrightarrow{i} q_j bZ.$$

- For each machine instruction

  $$\mathtt{X}_\ell : \mathtt{if}\ \mathtt{c}_b = 0\ \mathtt{then}\ \mathtt{goto}\ \mathtt{X}_j$$
  $$\mathtt{else}\ \mathtt{c}_b := \mathtt{c}_b - 1;\ \mathtt{goto}\ \mathtt{X}_k$$

  we have the MSA rules

  $$\begin{array}{lll} p_\ell b \xrightarrow{d} p_k & p_\ell Z \xrightarrow{z} p_j Z & p_\ell b \xrightarrow{z} q_j b \\ q_\ell b \xrightarrow{d} q_k & q_\ell Z \xrightarrow{z} q_j Z & q_\ell b \xrightarrow{z} p_j b \end{array}$$

- We have the one final MSA rule

  $$p_n Z \xrightarrow{\omega} p_n$$

The two states $p_0 Z$ and $q_0 Z$ of this MSA each mimic the machine M in the following sense.

- When M is at the command labelled X_ℓ with the values x and y in its counters, this is reflected by the MSA being in state $p_\ell 0^x 1^y Z$ (or $q_\ell 0^x 1^y Z$).
- If this command is an increment, then the MSA has only one transition available from the control state p_ℓ (q_ℓ), which is labelled by i (for 'increment'), resulting in the MSA state reflecting the state of the machine upon executing the increment command.

- If this command is a successful test for zero (that is, the relevant counter has the value 0), then the MSA has only one transition available from the control state p_ℓ (q_ℓ), which is labelled z (for 'zero'), again resulting in the MSA state reflecting the state of the machine upon executing the test for zero command.
- If this command is a decrement (that is, a failed test for zero), then the MSA in control state p_ℓ (q_ℓ) has three possible transitions, exactly one of which is labelled d (for 'decrement') which would once again result in the MSA state reflecting the state of the machine upon executing the decrement command.
- In this last instance, the MSA has the option to disregard the existence of a relevant counter symbol in the stack and behave as if the program counter was zero. This reflects the weakness of Petri nets (and hence MSA) in their inability to test for zero (a weakness which works in their favour with respect to several important positive decidability results such as the reachability problem [106]). In this case, the MSA in control state p_ℓ (q_ℓ) may make a z transition in either of two ways: either by "honestly" cheating using the rule $p_\ell Z \xrightarrow{z} p_j Z$ ($q_\ell Z \xrightarrow{z} q_j Z$), or by "knowingly" cheating using the rule $p_\ell b \xrightarrow{z} q_j b$ ($q_\ell b \xrightarrow{z} p_j b$) thus moving the control state over into the domain of the other MSA mimicking M.

FACT 65. $p_0 Z \sim q_0 Z$ *iff the Minsky machine* M *does not halt.*

PROOF. If M halts, then a winning strategy for player I in the bisimulation game would be to mimic the behaviour of M in either of the two MSA states. Player II's only option in response would be to do the same with the other MSA state. Upon termination, the game states will be $p_n 0^x 1^y Z$ and $q_n 0^x 1^y Z$ for some values x and y. Player I may then make the transition $p_n 0^x 1^y Z \xrightarrow{\omega} p_n 0^x 1^y$ which cannot be answered by player II from the state $q_n 0^x 1^y Z$. Hence $p_0 Z$ and $q_0 Z$ cannot be bisimilar.

If M fails to halt, then a winning strategy for player II would be to mimic player I's moves for as long as player I mimics M, and to cheat knowingly or honestly, respectively, in the instance that player I cheats honestly or knowingly, respectively, so as to arrive at the situation where the two states are identical; from here player II can copy every move of player I verbatim. Hence $p_0 Z$ and $q_0 Z$ must be bisimilar. □

We thus have undecidability of bisimulation equivalence over a very restricted class of Petri nets: those with only two unbounded places and a minimal degree of nondeterminism. Note that this nondeterminism is essential: Jančar [83] shows that bisimulation equivalence is decidable between two Petri nets when one of them is deterministic up to bisimilarity.

The above MSA can be made into a normed rewrite transition system by adding a new input symbol n along with the following MSA rules (one for each $\ell = 0 \ldots n$ and each $X = Z, 0, 1$).

$$p_\ell X \xrightarrow{n} p_\ell \qquad q_\ell X \xrightarrow{n} q_\ell$$
$$p_\ell X \xrightarrow{n} q_\ell \qquad q_\ell X \xrightarrow{n} p_\ell$$

These moves allow the MSA to exhaust its stack at any point during its execution, and continues to allow player II to produce a pair of identical states if player I elects to take one of these non-M-mimicking transitions. The same argument can then be made to show that p_0Z and q_0Z are bisimilar exactly when the Minsky machine M does not halt.

THEOREM 66. *Bisimilarity is undecidable over the class of normed MSA.*

This result contrasts with that of Stirling [137] regarding the decidability of bisimilarity over the class of normed PDA.

2.5. *Summary of results*

The tables in Figures 2 and 3 summarize a variety of decidability and algorithmic results for bisimulation problems over several classes of infinite-state systems, many of which being as described above, as well as give pointers to where to find these results in the literature. The first column of the first table lists results regarding bisimilarity as discussed in this chapter; the second column lists results regarding the problem of comparing a system to a finite-state system; and the third column lists results regarding the *regularity problem*, determining if the system is equivalent to some unspecified finite-state system. The second table presents the same collection of results, but this time in relation to *weak bisimulation equivalence*. To define weak bisimilarity, we assume that there is some label τ which represents an unobservable transition label, and we define a new transition relation $\Rightarrow$ by:

	$\sim$	$\sim$ FSA	$\sim$-regularity
BPA	2EXPTIME [23] PTIME in the normed case [70]	PTIME [99]	2EXPTIME [24]
PDA	decidable [134] PSPACE-hard [112]	EXPTIME [86] PSPACE-hard [112]	PSPACE-hard [112]
BPP	decidable [36] co-NP-hard [111] PTIME in the normed case [71]	PSPACE [86]	decidable [55] co-NP-hard [111]
PA	co-NP-hard [111] decidable in the normed case [68]	decidable [86]	co-NP-hard [111]
PN	undecidable [84]	decidable [55] EXPSPACE-hard [112]	decidable [55] EXPSPACE-hard [112]

Fig. 2. Results for bisimilarity.

	$\approx$	$\approx$ FSA	$\approx$-regularity
BPA	PSPACE-hard [139]	PTIME [99]	?
PDA	PSPACE-hard [139]	EXPTIME [86] PSPACE -hard [112]	PSPACE -hard [112]
BPP	NP-hard [139] Π_2^P-hard [111]	PSPACE [86]	Π_2^P-hard [111]
PA	PSPACE-hard [139]	decidable [86]	Π_2^P-hard [111]
PN	undecidable [84]	undecidable [55]	undecidable [55]

Fig. 3. Results for weak bisimilarity.

$\stackrel{\tau}{\Longrightarrow} = \stackrel{\tau}{\longrightarrow}^*$, and for $a \neq \tau$, $\stackrel{a}{\Longrightarrow} = \stackrel{\tau}{\longrightarrow}^* \cdot \stackrel{a}{\longrightarrow} \cdot \stackrel{\tau}{\longrightarrow}^*$. Then the definition of weak bisimilarity is the same as for bisimilarity except using this new transition relation $\Longrightarrow$ in place of the original $\rightarrow$ relation.

One-counter automata and one-counter nets. The discussion of equivalence checking in this chapter only touches the surface of this vast topic, and many related studies have been carried out which consider other classes of systems and other equivalences. In particular, there has recently been extensive work done on one-counter automata and one-counter nets, for both bisimulation equivalence and *simulation preorder/equivalence*.

A *simulation relation* is defined as a "one-sided" bisimulation relation, in the sense that we omit the second clause in Definition 9, and in the third clause we only demand that β be a final state whenever α is, but not vice versa. We then say that α is *simulated* by β if the pair (α, β) is contained in some simulation relation; that is, the simulation preorder is defined to be the union of all simulation relations (and hence the largest simulation relation). Finally, two states are *simulation equivalent* if they simulate each other.

The first relevant result is that of Jančar [85], demonstrating that bisimulation equivalence, as well as the regularity problem with respect to bisimilarity, are decidable for one-counter automata. Although this is subsumed by the later result of Sénizergues [132], the technique used by Jančar in his proof of the special case is elegant, employing novel ideas involving representing bisimilarity as a colouring of the plane and demonstrating the existence of a periodicity which can be detected.

Abdulla and Čerāns [1] recently outlined an extensive and technically-challenging proof of the decidability of the simulation preorder for one-counter nets. This result was subsequently given a concise and intuitive proof by Jančar and Moller [89], again using colourings of the plane.

For the slightly wider class of one-counter automata, Jančar, Moller and Sawa [91] demonstrate the undecidability of simulation equivalence, and hence also simulation preorder; for the equivalence problem, we can even assume that one of the automata is deterministic, and for the preorder, we can assume that both are deterministic. Contrasting with these results are, firstly, the decidability of the equivalence problem for deterministic one-counter automata, a result which can be extracted from the early results of Valiant and Paterson [143]; and secondly, the recent result of Jančar, Kučera and Moller [87] that equivalence is decidable between a deterministic one-counter automaton (in fact, a deterministic PDA even) and a one-counter net.

Finally, it is also demonstrated in [87] that the regularity problem for one-counter nets with respect to simulation equivalence is decidable. (The corresponding problem is undecidable for both Petri nets [88] and PA [98].) This proof is carried out by presenting an elegant general reduction from simulation problems over one-counter nets to bisimulation problems over one-counter automata, and then exploiting the original result of Jančar [85].

3. The model checking problem

In this section we survey the state-of-the-art in model-checking infinite-state systems. Again, we concentrate exclusively on discrete infinite-state systems and neglect, e.g., dense infinite-state structures related to timed or hybrid systems, which are covered, for instance, in [3] and [2]. Due to the broad scope of existing model checking algorithms, we have chosen to sketch only the main ideas behind the relevant results complemented by appropriate references: for decidable model checking problems we briefly present the developed algorithms together with their complexity; for undecidable problems, we outline the undecidability proof. This style of presentation is intended to provide readers interested in more technical details with access to the original papers, and ensures at the same time that the survey remains readable for those who wish only to glance over the research done in this area.

The remainder of this section is organized as follows. We briefly introduce in Section 3.1 the temporal logics we consider together with their branching and linear time classifications; this introduction can be deepened by consulting Chapter 4 in this Handbook [17]. We then proceed by presenting decidability and complexity results about model checking for various classes of infinite-state systems, first for branching time logics in Section 3.2, and subsequently for linear time logics in Section 3.3. The presentation is organized according to the two central parameters of the model checking problem:

- the size of the representation of the transition system; and
- the size of the temporal formula.

This allows us to distinguish three different complexities:

- the general case;
- the case where the formula is fixed; and
- the case where the system is fixed.

As in practice the size of the formula is usually very small compared to the size of the system representation, we focus on the first two cases here and neglect the last one in our complexity considerations.

3.1. *Temporal logics*

A successful approach to the verification of programs relies on a suitable formalism for specifying central aspects of the intended system behaviour. In the case of *sequential* programs where the input/output behaviour together with termination plays a predominant role such formalisms are traditionally based on state transformer semantics. In fact, sequential programs are typically verified by considering their (partial) correctness, specified in terms of pre- and postconditions [58].

Reactive systems, on the other hand, are typically *nonterminating*, as they maintain an ongoing interaction with the environment. Hence, verification methods which rely intrinsically on the existence of a final state are, usually, not applicable and must be replaced by radically different approaches (see [100] for a survey). Pnueli [127] was the first to recognize the need for formalisms which support reasoning about nonterminating behaviour. He proposed the use of *temporal logic* as a language for the specification of concurrent program properties. Temporal logics are tailored to expressing many important correctness properties characteristic for reactive systems, in particular, liveness properties which assert that something will eventually happen.

Since Pnueli's landmark paper a plethora of systems of temporal logic have been investigated concerning their expressiveness and suitability for verification purposes. The majority and better known of these logics belong to the class of point-based, future-tense, and propositional logics. In this survey we will also focus on this class, while elaborating on the differences between branching-time logics and linear-time logics. The reader interested in technical details beyond this exposition is advised to consult [120] and Chapter 4 in this Handbook [17].

Within their respective spectra, branching and linear-time logics can be classified according to their expressive power. Figure 4 shows the respective hierarchies of temporal logics we will consider in the following. In this figure an arrow $L_1 \to L_2$ indicates that the logic L_2 is strictly more expressive than the logic L_1.

Due to their difference in expressive power the various logics have also different importance for verification purposes. The weakest logics, Hennessy–Milner logic and weak linear-time logic, can only express properties about a finite prefix of a system, and play consequently only a minor role in themselves. They constitute, however, the basis for the more expressive logics and possess some interesting theoretical properties. Most importantly, for finite-branching processes, i.e., for processes where each state admits only finitely many transitions, it is known that two processes are bisimulation equivalent iff they satisfy the same set of Hennessy–Milner logic formulae [65]. To increase the expressiveness of these basic logics, which require infinite sets of formulae for characterizing infinite behaviour, the addition of the "until" operator has been proposed which can express that a property ϕ should hold *until* a second property ψ holds. On the branching-time side this leads to Computation Tree Logic (CTL) [40], while on the linear-time side this yields Linear Temporal Logic (LTL) [127]. Since both logics arise quite naturally and can express practically relevant properties they are widely used in tools for automatic verification. But even the logics EF and EG, complementary fragments of CTL, have stirred up some interest. They both have been successfully used to clarify the borderline between decidability and undecidability, as well as to establish lower complexity bounds for model checking various classes of

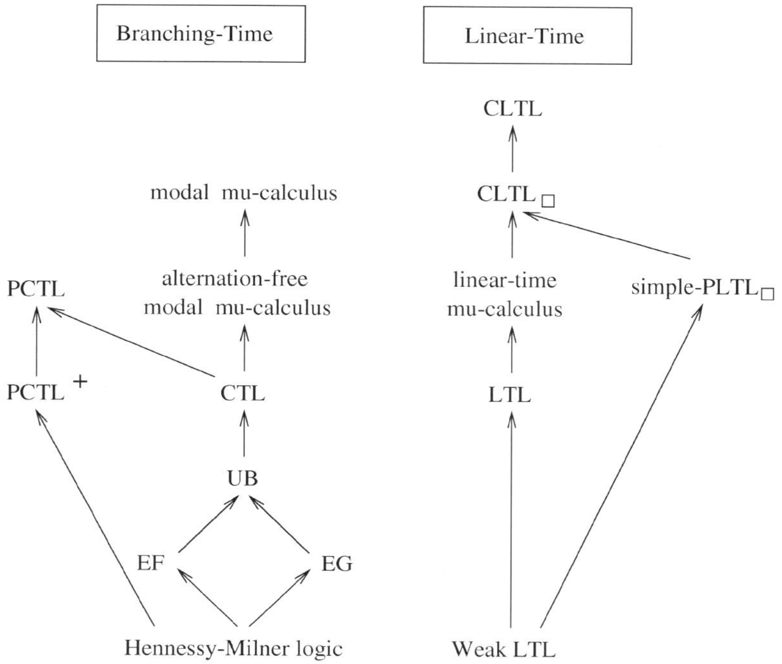

Fig. 4. Branching and linear time logics.

infinite-state systems. The logic UB is simply the smallest common generalization of EF and EG.

Even more expressive temporal logics are obtained by adding least and greatest fixpoint operators to the basic logics. This extension results in the branching-time modal μ-calculus and the linear-time μ-calculus. Both fixpoint logics play a central role in the theory of model checking due to their syntactic minimality and their pleasing mathematical properties. Accordingly, we will present both logics together with their respective model checking algorithms in greater detail.

All the logics mentioned so far have in common that they may express only *regular properties*. This means that in the branching-time setting the set of states, respectively in the linear-time setting the set of traces, satisfying a given property form regular sets. Whereas the meaning of regularity is clear for traces, regular sets of states need to be explained: a state in a transition system characterizes a tree, and for trees regularity is well-defined [128]. There have, however, been extensions proposed which incorporate non-regular features like counting. The most prominent examples are Presburger Computation Tree Logic (PCTL) [11] which combines CTL with Presburger arithmetic, and Constrained Linear Temporal Logic (CLTL) [10], which enhances LTL by Presburger arithmetic and constraints expressed by finite-state automata. Despite the fact that both logics in their general form are undecidable already for finite-state systems, decidability can be regained for nontrivial fragments.

For the following formal introduction of characteristic temporal logics let $\mathcal{T} = (\mathcal{S}, \Sigma, \rightarrow)$ be a labelled transition system without start and final states, Σ^*, resp. Σ^ω, denote the set of finite, resp. infinite words over Σ, and $\Sigma^\infty =_{\text{df}} \Sigma^* \cup \Sigma^\omega$. Moreover, given a word $\sigma = a_1 a_2 \ldots \in \Sigma^\infty$, $\sigma(2)$ will denote the first action of σ, i.e., a_1, and σ^1 the word $a_2 a_3 \ldots$.

A *path* $\pi = s_1 \xrightarrow{a_1} s_2 \xrightarrow{a_2} \cdots$ in $\mathcal{T}$ is a finite or infinite transition sequence such that $s_i \xrightarrow{a_i} s_{i+1}$, for all i, is in the transition relation. By $\pi(i)$ we denote its i-th state s_i, whereas π^i denotes the part of π starting at s_i. Moreover, we denote by $\pi(\vdash)$ its first state, and, if π is finite, by $\pi(\dashv)$ its last state, respectively. A *run* is a maximal path, i.e., a path that is either infinite or terminates in a state without successors. Furthermore, we denote by $\texttt{paths}(s)$ the set of paths starting in s, while $\texttt{runs}(s)$ will denote the set of runs starting in s. The set of finite paths from s is written $\texttt{paths}^*(s)$. Finally, we write $\texttt{prefs}(\pi)$ for the set of finite prefixes of the path π.

3.1.1. *Branching-time logics.* The main characteristic of branching-time logics is that they model the potential of possible futures in a tree-like structure: each moment in time has at least one, but may have even infinitely many possible successor moments. Together with the temporal order on moments in time, this structure therefore defines infinite trees, which, in a natural way, correspond to computation trees of concurrent processes. Consequently, formulae of a branching-time temporal logic can be conveniently used to loosely specify the behaviour of concurrent systems according to their structural operational semantics. Moreover, as many process algebras assume a set Σ of observable actions, modal operators of branching-time logics often quantify over action-labelled transitions allowing to specify which property should hold after a specific action has been observed.

Hennessy–Milner Logic (HML). The weakest branching-time logic is *Hennessy–Milner logic* [65], built out of "true", negation, conjunction, and the relativized existential next operator $\langle a \rangle$. Formally, formulae have the following syntax:

$$\phi ::= \texttt{tt} \mid \neg\phi \mid \phi_1 \wedge \phi_2 \mid \langle a \rangle \phi,$$

where a is an action of Σ. As is characteristic for a branching-time logic, the semantics $[\![\phi]\!]^{\mathcal{T}}$ of a formula ϕ is defined with respect to a labelled transition system $\mathcal{T}$, and denotes the subset of the set of states $\mathcal{S}$ for which the formula holds. In particular, for formulae of Hennessy–Milner logic the semantics is defined inductively as follows:

$$\begin{aligned}
[\![\texttt{tt}]\!]^{\mathcal{T}} &=_{\text{df}} \mathcal{S}, \\
[\![\neg\phi]\!]^{\mathcal{T}} &=_{\text{df}} \mathcal{S} - [\![\phi]\!]^{\mathcal{T}}, \\
[\![\phi_1 \wedge \phi_2]\!]^{\mathcal{T}} &=_{\text{df}} [\![\phi_1]\!]^{\mathcal{T}} \cap [\![\phi_2]\!]^{\mathcal{T}}, \\
[\![\langle a \rangle \phi]\!]^{\mathcal{T}} &=_{\text{df}} \left\{ s \in \mathcal{S} \mid \exists s' \in \mathcal{S}. s \xrightarrow{a} s' \text{ and } s' \in [\![\phi]\!]^{\mathcal{T}} \right\}.
\end{aligned}$$

Instead of $s \in [\![\phi]\!]^{\mathcal{T}}$ we also write $s \models \phi$. To ease the notation further, the dual notions *false*, disjunction and universal next operator are defined using the standard abbreviations

$$\texttt{ff} =_{\text{df}} \neg\texttt{tt},$$

$$\phi_1 \vee \phi_2 =_{\mathrm{df}} \neg(\neg\phi_1 \wedge \neg\phi_2),$$
$$[a]\phi =_{\mathrm{df}} \neg\langle a\rangle\neg\phi.$$

Under the customary assumption that atomic propositions are closed under negation, these abbreviations allow us to transform every formula of Hennessy–Milner logic into *positive normal form*, i.e., an equivalent formula without negation, by driving negation inwards.

From a conceptual point of view Hennessy–Milner logic is easy to deal with as validity of a formula at a state s may readily be checked for any process whose computation tree is finite-branching, i.e., has only a finite number of alternatives for each observable action. It simply suffices to investigate the computation tree of the labelled transition graph under consideration up to depth $|\phi|$, where $|\phi|$ denotes the size of ϕ defined in the usual way by counting the number of operators.

The logics EF, EG and UB. The logic EF [50] is obtained from Hennessy–Milner logic by adding the operator $\mathsf{EF}\phi$, meaning "there exists a path such that eventually ϕ holds". This addition increases the expressive power of the logic significantly, as it becomes possible to reason about paths of arbitrary length. In fact, it already suffices to express the practically relevant *reachability* properties and the practically even more important dual *invariance* or *safety* properties: $\neg\mathsf{EF}\phi$ expresses that a state satisfying a certain undesirable property ϕ cannot be reached.

The addition of the operator $\mathsf{EG}\phi$ to HML, meaning "there exists a path such that ϕ always holds" yields, symmetrically, the logic EG, which allows to express *inevitability properties*, e.g., $\neg\mathsf{EG}\phi$ means that the considered system is guaranteed to eventually violate property ϕ. The combination of the logics EF and EG yields the obviously even more expressive *Unified System of Branching-Time Logic* (UB) [7].

Formally the semantics of the operators EF and EG are given as follows:

$$[\![\mathsf{EF}\phi]\!]^{\mathcal{T}} =_{\mathrm{df}} \{s \in \mathcal{S} \mid \exists\pi \in \mathtt{paths}^*(s).\ \pi(\dashv) \in [\![\phi]\!]^{\mathcal{T}}\},$$
$$[\![\mathsf{EG}\phi]\!]^{\mathcal{T}} =_{\mathrm{df}} \{s \in \mathcal{S} \mid \exists\pi \in \mathtt{runs}(s).\ \forall\pi' \in \mathtt{prefs}(\pi).\ \pi'(\dashv) \in [\![\phi]\!]^{\mathcal{T}}\}.$$

Both operators have also a dual operator which in the case of EF is defined as $\mathsf{AG}\phi =_{\mathrm{df}} \neg\mathsf{EF}\neg\phi$, and in the case of EG as $\mathsf{AF}\phi =_{\mathrm{df}} \neg\mathsf{EG}\neg\phi$. Intuitively, they express properties of the form "always in the future ϕ holds" and "eventually in the future ϕ holds".

Finally, a useful complexity measure for formulae in EF, respectively EG, is their *nesting depth* which measures the nesting of EF, respectively EG operators.

Computation Tree Logic (CTL). *Computation Tree Logic* (CTL) [40] is one of the earliest proposed branching-time logics. It can be considered as the branching-time counterpart of Linear Temporal Logic (LTL) [127], which was proposed earlier and which will be introduced in the next section. CTL adds to Hennessy–Milner logic an *until* operator which is either existentially quantified $\mathsf{E}[\phi\,\mathsf{U}\,\psi]$ or universally quantified $\mathsf{A}[\phi\,\mathsf{U}\,\psi]$. The formal semantics of the new operators are as follows.

$$[\![\mathsf{E}[\phi\,\mathsf{U}\,\psi]]\!]^{\mathcal{T}} =_{\mathrm{df}} \{s \in \mathcal{S} \mid \exists\pi \in \mathtt{paths}^*(s).\ \forall\,\pi' \in \mathtt{prefs}(\pi),$$
$$(\pi' \neq \pi \Rightarrow \pi'(\dashv) \in [\![\phi]\!]^{\mathcal{T}}) \wedge \pi(\dashv) \in [\![\psi]\!]^{\mathcal{T}}\},$$

$$[\![\mathsf{A}[\phi\,\mathsf{U}\,\psi]]\!]^{\mathcal{T}} =_{\mathrm{df}} \{s \in \mathcal{S}\ \forall\, \pi \in \mathtt{paths}^*(s).\ \forall \pi' \in \mathtt{prefs}(\pi).$$
$$(\pi' \neq \pi \Rightarrow \pi'(\dashv) \in [\![\phi]\!]^{\mathcal{T}}) \wedge \pi(\dashv) \in [\![\psi]\!]^{\mathcal{T}}\}.$$

Intuitively, $\mathsf{E}[\phi\,\mathsf{U}\,\psi]$ means that there exists a path on which ϕ holds until, eventually, ψ holds, while $\mathsf{A}[\phi\,\mathsf{U}\,\psi]$ expresses that on all paths ϕ holds until, eventually, ψ holds. The until operators are expressive enough to encode both EF and EG by means of the following formulae.

$$\mathsf{EF}\phi \equiv \mathsf{E}[\mathtt{tt}\,\mathsf{U}\,\phi], \qquad \mathsf{EG}\phi \equiv \neg\mathsf{A}[\mathtt{tt}\,\mathsf{U}\,\neg\phi].$$

As a consequence, CTL subsumes the logic UB.

An example that demonstrates the practical applicability of CTL is the specification of the alternation bit protocol, a well-known communication protocol for exchanging messages between a sender process and a receiver process. The alternation bit protocol could be specified by the CTL formulae

$$\begin{aligned}
&\mathsf{AG}(\mathrm{RcvMsg} \Rightarrow \mathsf{A}[\mathrm{RcvMsg}\,\mathsf{U}(\neg\mathrm{RcvMsg} \wedge \mathsf{A}[\neg\mathrm{RcvMsg}\,\mathsf{U}\,\mathrm{SndMsg}])]),\\
&\mathsf{AG}(\mathrm{SndMsg} \wedge \mathrm{Snd0} \Rightarrow\\
&\qquad \mathsf{A}[\mathrm{SndMsg}\,\mathsf{U}(\neg\mathrm{SndMsg} \wedge \mathsf{A}[\neg\mathrm{SndMsg}\,\mathsf{U}(\mathrm{RcvMsg} \wedge \mathrm{Rcv0})])]),\\
&\mathsf{AG}(\mathrm{SndMsg} \wedge \mathrm{Snd1} \Rightarrow\\
&\qquad \mathsf{A}[\mathrm{SndMsg}\,\mathsf{U}(\neg\mathrm{SndMsg} \wedge \mathsf{A}[\neg\mathrm{SndMsg}\,\mathsf{U}(\mathrm{RcvMsg} \wedge \mathrm{Rcv1})])])
\end{aligned}$$

taken from [41]. The formulae express the fact that sending a message (SndMsg) strictly alternates with receiving a message (RcvMsg), and that if a message with bit 0 (Snd0), resp. with bit 1 (Snd1), is sent, then a message with bit 0 (Rcv0), resp. with bit 1 (Rcv1), is received.

The modal μ*-calculus.* The modal μ-calculus as introduced by Kozen [97] is one of the most powerful branching-time logics found in the literature. Based on mathematical fixpoint theory, it combines standard modal logic with least and greatest fixpoint operators. Similar to the λ-calculus for functional languages, and due to the extreme power of the fixpoint operators, on the one hand it allows us to express very complex (temporal) properties within a sparse syntactic formalism, while on the other hand, due to its syntactic minimalism, it also provides a good basis for the conceptual study of numerous theoretical intricacies which have attracted a lot of research. However, there is a price to be paid for these properties: even for experts, it is often hard to understand the meaning of a one-line μ-calculus formula. This is the reason for practitioners to prefer syntactically richer and less expressive derived logics. The μ-calculus is nevertheless also practically relevant, as it provides a convenient "assembly language" for temporal logics.

The main characteristic which distinguishes the modal μ-calculus from most other branching-time logics is the possibility to specify *recursive* properties, which adds tremendous expressive power. Keeping in mind that the semantics of branching-time formulae are subsets of the state set $\mathcal{S}$, such recursive properties can be viewed as equation systems which have to be solved over $2^{\mathcal{S}}$. Although, in general, several solutions may exist there are two prominent solutions with respect to the subset ordering on state sets, namely the smallest and the largest ones. Due to a theorem of Tarski [140], their existence can be guaranteed whenever the equation system satisfies an easily verifiable monotonicity condition. The desired solution can then be expressed by means of a least or greatest fixpoint operator, where $\mu X.\phi(X)$, resp. $\nu X.\phi(X)$, denotes the smallest, resp. greatest, solution of $X = \phi(X)$.

More formally, the modal μ-calculus is an extension of Hennessy–Milner logic which introduces a (countable) set of variables *Var*, and a least fixpoint operator $\mu X.\phi$ binding X in ϕ. Its semantics is defined with respect to a valuation $\mathcal{V}$ mapping variables to subsets of $\mathcal{S}$. To ensure the above-mentioned monotonicity condition, it is customary to impose the syntactic restriction for expressions $\mu X.\phi$ that any occurrence of X in ϕ must occur within the scope of an even number of negations. The semantics for variables and the least fixpoint operator are then given as follows, where $\mathcal{V}[X \mapsto \mathcal{E}]$ is the valuation obtained from $\mathcal{V}$ by updating the binding of X to $\mathcal{E}$.

$$\begin{aligned} [\![X]\!]^{\mathcal{T}}_{\mathcal{V}} &=_{\mathrm{df}} \mathcal{V}(X), \\ [\![\mu X.\phi]\!]^{\mathcal{T}}_{\mathcal{V}} &=_{\mathrm{df}} \bigcap\{\mathcal{E} \subseteq \mathcal{S} \mid [\![\phi]\!]^{\mathcal{T}}_{\mathcal{V}[X\mapsto\mathcal{E}]} \subseteq \mathcal{E}\}. \end{aligned}$$

The notions of bound and free occurrences of variables are defined in the usual way, and a formula is said to be closed if it does not contain any free variable occurrence. Using negation, it is possible to introduce also the greatest fixpoint operator $\nu X.\phi$ by

$$\nu X.\phi =_{\mathrm{df}} \neg\mu X.\neg\phi[\neg X/X],$$

where $\phi[\neg X/X]$ denotes the simultaneous replacement of all free occurrences of X by $\neg X$. The semantics of $\nu X.\phi$ can also be defined directly by

$$[\![\nu X.\phi]\!]^{\mathcal{T}}_{\mathcal{V}} =_{\mathrm{df}} \bigcup\{\mathcal{E} \subseteq \mathcal{S} \mid \mathcal{E} \subseteq [\![\phi]\!]^{\mathcal{T}}_{\mathcal{V}[X\mapsto\mathcal{E}]}\}.$$

The clauses for the fixpoints are reformulations of the characterization in the Tarski–Knaster theorem [140] which states that the least fixpoint is the intersection of all pre-fixpoints and the greatest fixpoint is the union of all post-fixpoints. The fixpoint property also ensures that states satisfy a fixpoint formula iff they satisfy the *unfolding* of the formula, i.e.,

$$s \in [\![\sigma X.\phi]\!]^{\mathcal{T}}_{\mathcal{V}} \text{ iff } s \in \big[\!\big[\phi[\sigma X.\phi/X]\big]\!\big]^{\mathcal{T}}_{\mathcal{V}}, \quad \text{where } \sigma \in \{\mu, \nu\}.$$

The following encoding of the until operators

$$\begin{aligned} E[\phi \, U \, \psi] &\equiv \mu X.\psi \vee \big(\phi \wedge \big(\textstyle\bigvee_{a\in\Sigma}\langle a\rangle X\big)\big), \\ A[\phi \, U \, \psi] &\equiv \mu X.\psi \vee \big(\phi \wedge \big(\textstyle\bigwedge_{a\in\Sigma}[a]X\big)\big) \end{aligned}$$

shows that the modal μ-calculus is strictly more expressive than CTL. Formulae resulting from such a translation are rather special as they possess only one sort of fixpoint, namely minimal fixpoints. The modal μ-calculus, however, permits users to write formulae containing alternating intertwined fixpoint operators, i.e., formulae like the following, which intuitively simply requires the existence of an infinite a-path on which the atomic proposition P is guaranteed to hold infinitely often.

$$\Phi_0 = \nu X.\big(\mu Y.\big(\langle a\rangle Y \vee \big(P \wedge \langle a\rangle X\big)\big)\big).$$

Here it is characteristic that a fixpoint expression for a variable X contains another fixpoint expression of the other kind (e.g., greatest instead of least), which itself contains X free. Formulae with such alternations are not only hard to understand, but also hard to check algorithmically: already for finite-state systems, the best known algorithms are exponential in the so-called *alternation depth* which is defined as follows [123].

DEFINITION 67 (*Alternation depth*). A formula Φ is said to be in the classes Σ_0 and Π_0 iff it contains no fixpoint operators. To form the class Σ_{n+1}, take $\Sigma_n \cup \Pi_n$, and close under (i) Boolean and modal combinators, (ii) $\mu X.\Phi$, for $\Phi \in \Sigma_{n+1}$, and (iii) substitution of $\Phi' \in \Sigma_{n+1}$ for a free variable of $\Phi \in \Sigma_{n+1}$ provided that no free variable of Φ' is captured by Φ; and dually for Π_{n+1}. The (Niwinski) *alternation depth* of a formula Φ, denoted by $\mathrm{ad}(\Phi)$, is then the least n such that $\Phi \in \Sigma_{n+1} \cap \Pi_{n+1}$, or equivalently $\Phi \in \mathrm{CBM}(\Sigma_n \cup \Pi_n)$ where CBM denotes the closure under Boolean and modal combinators.

Niwinski's inductive definition is stronger than the more popular and easier to understand definition of Emerson and Lei [49]. In order to illustrate Niwinski's quite technical definition, we give a few examples. The following subformulae of Φ_0 given above have, for instance, alternation depth 0, 1, and 2.

$$\langle a\rangle Y \vee \big(P \wedge \langle a\rangle X\big) \in \Sigma_0 \cap \Pi_0,$$
$$\mu Y.\big(\langle a\rangle Y \vee \big(P \wedge \langle a\rangle X\big)\big) \in \Sigma_1,$$
$$\nu X.\big(\mu Y.\big(\langle a\rangle Y \vee \big(P \wedge \langle a\rangle X\big)\big)\big) \in \Pi_2.$$

The fragment of formulae with alternation depth one, called the *alternation-free modal* μ*-calculus*, has stirred up some greater interest. It captures a good deal of the desired system properties (in particular, the whole of CTL), while permitting efficient model checking: e.g., for finite-state systems model checking is linear in both the system and the formula size [45] (cf. Section 3.2). On the other hand, Bradfield [16] (and, independently, Lenzi [101]) showed that alternation depth induces an infinite sequence of sublogics of strictly increasing expressive power. This, and other interesting details about the modal μ-calculus and its associated theory can be found in Chapter 4 in this Handbook [17].

Presburger Computation Tree Logic (*PCTL*). It is well known that many relevant properties of reactive systems involve constraints on the numbers of occurrences of events. However, traditional temporal logics, like the ones we have considered so far, can only express regular properties of processes. This means that the set of models satisfying a given property form regular sets, and in particular in the branching-time setting are a regular set of trees. To overcome this limitation, Bouajjani, Echahed, and Robbana [11] introduced

Presburger Computation Tree Logic (PCTL) which extends a propositional version of CTL with counting constraints on actions expressed by Presburger arithmetic formulae.

More formally, CTL is extended with state formulae π, i.e., Boolean combinations of atomic propositions, Presburger arithmetic formulae f, existential quantification over integers written as $\exists x.\varphi$, and a binding operator $[x : \pi].\varphi$. The binding operator associates the variable x with the state formula π, and starting from the current state the variable x then counts the number of states satisfying π.

These new constructs permit to specify, for example, typical communication protocol properties like

> "*between the beginning and the end of every session, there are exactly the same numbers of requests and acknowledgements*" [11]

by the PCTL formula

$$AG\big(\mathrm{BEGIN} \Rightarrow [x : \mathrm{REQ}].[y : \mathrm{ACK}].AG\big(\mathrm{END} \Rightarrow (x = y)\big)\big).$$

It has turned out, however, that PCTL is too expressive for automatic verification purposes, as it is undecidable already for finite-state systems. The authors therefore also introduce the fragment PCTL$^+$ which differs from PCTL in that the until operators must be of the restricted form $E[\pi \, U \, \phi]$ and $A[\phi \, U \, \pi]$ where π is required to be a state formula. Despite this syntactic restriction, PCTL$^+$ is still quite expressive and contains, e.g., the formula given above. Moreover, it is decidable for BPA.

3.1.2. *Linear-time logics.* In contrast to branching-time logics, linear-time logics assume that at each moment in time there is only one determined possible future. This point of view concerning the semantics of time leads immediately to models where the semantic entities are sequences of events along a single time line, called *runs*. Depending on whether the underlying program structures may or may not contain deadlocks, i.e., states in which no further action is possible, these semantic models then either take into consideration both the finite, as well as the infinite runs, or restrict themselves to the infinite runs only.

In either case, formulae of a linear-time logic are then interpreted over the set of all (considered) runs, and the semantics $[\![\phi]\!]$ of a formula ϕ is the set of all runs π for which the formula holds, i.e., $[\![\phi]\!] = \{\pi \mid \pi \models \phi\}$.

Also in the case of linear-time logics, for system verification one is typically interested whether a specific state satisfies a certain property. This led to the following convention: a state s of a transition system satisfies a formula if *every* run starting at s satisfies it.

In the following we introduce the linear-time logics most relevant for the study of infinite-state systems. For a better comparison with branching-time logics we will focus on *action-based* linear-time logics which have *relativized* next operators, one for each possible action.

Weak Linear Temporal Logic (*WL*). The weakest linear-time logic, called WL, is built in analogy to Hennessy–Milner logic out of *true*, negation, conjunction, and a relativized next operator $\langle a \rangle \phi$, one for each action $a \in \Sigma$. Formally, WL formulae have the following syntax.

$$\phi ::= \mathtt{tt} \mid \neg\phi \mid \phi_1 \wedge \phi_2 \mid \langle a \rangle \phi.$$

The inductive definition below stipulates when a run π has the property ϕ, written as $\pi \models \phi$.

$$\begin{array}{ll} \pi \models \mathtt{tt}, & \\ \pi \models \neg\phi & \text{if } \pi \models \phi \text{ does not hold,} \\ \pi \models \phi_1 \wedge \phi_2 & \text{if } \pi \models \phi_1 \text{ and } \pi \models \phi_2, \\ \pi \models \langle a \rangle \phi & \text{if } \pi(1) \xrightarrow{a} \pi(2) \wedge \pi^2 \models \phi. \end{array}$$

Intuitively, a run satisfies $\langle a \rangle \phi$ if its first action is a, and the suffix run obtained after chopping off the first state and the first action satisfies ϕ.

Like for HML, we would like to remark that WL is mainly of theoretical interest due to its limited expressive power which only captures finite behaviour. In particular this means that the validity of a formula may readily be checked independently of the type of the considered (finite-branching) system by investigating all runs up to length $|\phi|$.

Linear Temporal Logic (LTL). *Linear Temporal Logic* (LTL) [127] can be seen as the "standard" linear-time logic. It is widely used in tools for automatic verification of reactive systems [105]. Similar to CTL, LTL is obtained by adding to WL an *until* operator $\phi \,\mathrm{U}\, \psi$. This extension increases the expressive power significantly, as it admits to reason about runs of unbounded length. The formal semantics of the until operator is given as follows.

$$\pi \models \phi \,\mathrm{U}\, \psi \quad \text{if } \exists i: \pi^i \models \psi \text{ and } \forall j < i\text{: } \pi^j \models \phi.$$

Intuitively, a run $s_0 \xrightarrow{a_0} s_1 \xrightarrow{a_1} \cdots$ satisfies $\phi \,\mathrm{U}\, \psi$ if ϕ holds until eventually ψ holds, i.e., if there exists some suffix $s_i \xrightarrow{a_i} s_{i+1} \xrightarrow{a_{i+1}} \cdots$ satisfying ψ, and all preceding suffixes $s_j \xrightarrow{a_j} s_{j+1} \xrightarrow{a_{j+1}} \cdots$ with $j < i$ satisfy ϕ.

It is convenient to use the derived operators $\mathrm{F}\phi = \mathtt{tt} \,\mathrm{U}\, \phi$ ("eventually ϕ") and its dual $\mathrm{G}\phi = \neg\mathrm{F}\neg\phi$ ("always ϕ"). They directly and concisely express simple liveness properties like "ϕ eventually happens" as $\mathrm{F}\phi$, and simple safety properties like "ϕ never happens" as $\mathrm{G}\neg\phi$.

Despite its expressive power, there are, however, comparatively basic properties like "at every *even* moment ϕ holds" that cannot be expressed in LTL [146]. The reason for this inability becomes clear when switching to a formal language point of view. Obviously, the set of runs which satisfy a given LTL formula can also be interpreted as a formal ω-language, i.e., as a set of infinite words. Thomas [142] has shown that the LTL definable ω-languages correspond to the class of star-free ω-regular languages, an interesting subclass of the ω-regular languages which results from restricting the monadic second order logic over infinite words (S1S) to first-order logic.

Because of this lack of expressiveness of LTL, various more expressive logics have been studied, most notably the linear-time μ-calculus which we consider next.

The Linear-Time μ-Calculus (LTμ). The *linear-time μ-calculus* [144] is the linear-time analogue to the modal μ-calculus: it is a powerful fixpoint logic in which all the usual linear-time operators like "always", "eventually", and "until" can be described. In particular, the logic is equivalent in its expressive power to Büchi automata [144,47], and hence also to monadic second order logic over infinite words (S1S) [18]. Consequently, the linear-time μ-calculus characterizes the full class of ω-regular languages and is thus strictly more expressive than LTL which may only describe star-free ω-regular languages.

This expressiveness is gained from adding variables to WL, together with a least fixpoint operator $\mu X.\phi$, which binds the variable X. In order to ensure vital monotonicity properties, X is only allowed to occur in ϕ within the scope of an even number of negations.

The formal development proceeds structurally similarly as in the branching-time case. The semantics of the linear-time μ-calculus is defined with respect to a valuation $\mathcal{V}$ mapping variables to sets of runs. The denotations for variables and the fixpoint operator are inductively defined by the following rules, where $\mathcal{R}$ denotes the set of all runs of a given labelled transition system.

$$\begin{aligned}
[\![X]\!]_{\mathcal{V}} &= \mathcal{V}(X), \\
[\![\mu X.\phi]\!]_{\mathcal{V}} &= \bigcap\bigl\{R \subseteq \mathcal{R} \mid R \subseteq [\![\phi]\!]_{\mathcal{V}[R\mapsto X]}\bigr\}.
\end{aligned}$$

Monotonicity arguments guarantee that $[\![\mu X.\phi]\!]_{\mathcal{V}}$ is the least fixpoint of the function which assigns to a set R of runs the set $[\![\phi]\!]_{\mathcal{V}[R\mapsto X]}$. Dual operators can be defined just as in the branching-time case, allowing to transform every formula into positive normal form, which is very convenient for verification purposes.

Finally, the alternation depth of LTμ formulae is defined as for the modal μ-calculus. From a semantic point of view, there is, however, an important difference, as alternation depth does not yield a hierarchy of strictly more expressive sublogics. This follows from the fact that LTμ is equally expressive as Büchi automata, and a Büchi automaton can be encoded by an LTμ formulae of alternation depth at most two [124].

Constrained Linear Temporal Logic (CLTL). The most expressive linear-time logic considered here is *Constrained Linear Temporal Logic* (CLTL), which was introduced in [10]. In analogy to PCTL (see Section 3.1.1) it allows to define nonregular properties, i.e., properties which characterize nonregular sets of runs, and that are therefore not definable by finite-state ω-automata. This expressiveness is achieved by extending LTL with two kinds of constraints: *counting constraints*, which use Presburger arithmetic formulae to express constraints on the number of occurrences of events, and *pattern constraints*, which use finite-state Rabin–Scott automata to impose structural constraints on runs.

More formally, LTL is first extended with state formulae π, i.e., Boolean combinations of atomic propositions.

Counting constraints are then introduced, as in the case of PCTL, by Presburger arithmetic formulae f, existential quantification over integers written as $\exists x.\varphi$, and a binding operator $[x:\pi].\varphi$. Through the binding operator the variable x is associated with the state formula π, and from the current state on the variable x then counts the number of states satisfying π. For example, the formula

$$[x_1:\pi_1].[x_2:\pi_2].G\bigl(P \Rightarrow (x_1 \leqslant x_2)\bigr)$$

specifies the property that in every computation starting from the current state we have the invariant "whenever P holds, the number of states satisfying π_2 is greater or equal than the number of states satisfying π_1".

Finally, pattern constraints are introduced using formulae of the form $\downarrow u.\varphi$, and A^u where A is a deterministic finite-state automaton. The intended meaning is that a formula $\downarrow u.\varphi$ associates the current state with the position variable u, and if A^u is a subformula of φ, then A^u holds whenever the trace of the computation from the state u to the current state is accepted by the automaton A. For instance, the formula

$$\downarrow u.[x_1 : \pi_1].[x_2 : \pi_2].G(A^u \Rightarrow (x_1 \leqslant x_2))$$

means that in every computation starting from the current state which is accepted by A, the number of states satisfying π_2 is greater than or equal to the number of states satisfying π_1.

As in the branching-time case, already for finite-state systems this combination of counting and pattern constraints yields undecidability. On the other hand, two important fragments of CLTL which still allow us to specify a wide range of nonregular properties while retaining decidability have been identified: $CLTL_\Box$, where counting constraints cannot be introduced in eventuality formulae, and the less expressive simple-$PLTL_\Box$, which extends LTL with counting constraints, where only propositional formulae may be used in eventuality formulae.

$CLTL_\Box$, and therefore also CLTL, are strictly more expressive than LTL, since, e.g., the typical property of runs "every finite prefix contains at least as many a actions as b actions" is expressible in $CLTL_\Box$ but not even in the linear-time μ-calculus. On the other hand, simple-$PLTL_\Box$ is incomparable to LTL, but allows us to express the complement of simple ω-regular languages, which are ω-languages definable by a finite-state Büchi automaton where every loop in its transition graph is a self-loop.

3.2. *Algorithms for branching-time logics*

In the branching-time setting we have to distinguish two rather different situations.

First, for sequential type infinite-state systems like BPA, PDA and some extensions thereof monadic second order logic (MSOL) is known to be decidable by a complicated reduction to the emptiness problem for regular languages on trees definable in monadic second order logic with n successors (SnS) [122]. As a consequence, the modal μ-calculus which can be strictly embedded in MSOL is also decidable, and may thus be used effectively for the specification of properties for these systems. A drawback is, however, that all decision procedures based on this approach are nonelementary. To palliate this problem in the last years some simpler and more direct algorithms have been developed which will be presented in Section 3.2.1 and 3.2.2.

Second, for infinite-state models of concurrent computation like BPP, MSA, or Petri nets already the weak logic EG is undecidable [56]. This is mainly due to the possibility of these models to describe grid-like structures, an easy cause of undecidability. The only positive result for this class of infinite-state models concerns the logic EF and the process class BPP [52] which turn out to impose a PSPACE-complete model checking problem [107,109].

3.2.1. *BPA*

Hennessy–Milner logic. Since BPA processes are finite-branching, model checking HML is trivially decidable. As for all finite-branching processes, the given BPA process has only to be unfolded up to depth $|\phi|$, for a HML formula ϕ at hand, as validity of ϕ can then readily be checked. The encoding of the unfolded tree as a BPA process is, however, rather concise. More concretely, Mayr [110] has shown that the general model checking problem for HML and BPA is PSPACE-complete by a reduction from the problem of quantified Boolean formulae (QBF).

	General	Fixed formula
HML	PSPACE-complete	PTIME

The modal μ-calculus. The combination of expressive power and syntactic minimality makes the modal μ-calculus an ideal formalism for conceptual studies. It is therefore not surprising that it is indeed the most studied logic for BPA processes and its extensions. In particular, this line of research has produced several different model checking algorithms which can be classified into being either *global* or *local* in nature, and into algorithms dealing with the full logic or only with a fragment thereof.

The first algorithm given for BPA processes was a global one handling the alternation-free fragment [27]. Although the original algorithm works on formulations of BPA processes and alternation-free formulae which are rather different from those used in this survey, an adaptation to our setting is straightforward. The model checking algorithm proceeds by iteratively computing a formula specific *property transformer* for each nonterminal of the given BPA system. To be more precise, let ϕ be a fixed μ-calculus formula and D_ϕ be the power set of subformulae of ϕ. A property transformer T_A for a nonterminal A is then a mapping $D_\phi \to D_\phi$ where $T_A(\Delta) = \Delta'$ means that Δ' is the set of subformulae of ϕ valid at A under the assumption that all subformulae in Δ are valid after termination of A, i.e., at the final state ε from which no action can occur. Once the property transformers have been computed by means of a fixpoint iteration, the model checking problem is solved by taking the property transformer of the nonterminal corresponding to the initial state of the BPA process, and applying it to the set of subformulae of ϕ satisfied by ε (which can be easily *locally* computed as ε has no successor states).

Let us illustrate this approach further by a simple example.

EXAMPLE 8. We take as the system to be model checked the BPA system from Example 1 in Section 1.1 which is defined by the following three rewrite rules

$$X \xrightarrow{a} XB,\ X \xrightarrow{c} \varepsilon,\ B \xrightarrow{b} \varepsilon.$$

For this system we want to automatically verify that from the root X there exists an infinite a-path on which the action c is always enabled. This property can formally be expressed by the alternation-free modal μ-calculus formula

$$\phi =_{df} \nu Z.\langle a\rangle Z \wedge \langle c\rangle \mathtt{tt}.$$

The first step of the second-order model checking algorithm consists now of transforming the formula into an equivalent equational form [43] which explicitly names all subformulae, yielding in our case

$$\nu\{Z_1 = Z_2 \wedge Z_3,\ Z_2 = \langle a\rangle Z_1,\ Z_3 = \langle c\rangle Z_4,\ Z_4 = \mathtt{tt}\}.$$

Defining $\mathcal{Z} =_{\mathrm{df}} \{Z_1, Z_2, Z_3, Z_4\}$, the property transformers to be computed for the non-terminals X and B are thus mappings $D_\phi \to D_\phi$ where $D_\phi = 2^{\mathcal{Z}}$. For notational convenience, we split these property transformers further into component property transformers $T_N^{Z_i} : D_\phi \to 2^{\{Z_i\}}$, for $i \in \{1, \ldots, 4\}$ and $N \in \{X, B\}$. These component transformers are just the projections on the Z_i, i.e., $T_N^{Z_i}(M) =_{\mathrm{df}} t^{Z_i}(T_N(M))$ where $t^{Z_i}(M) =_{\mathrm{df}} \{Z_i\}$ if $Z_i \in M$, and $t^{Z_i}(M) =_{\mathrm{df}} \emptyset$ otherwise. Hence we have $T_N(M) = \bigcup_{i=1}^{4} T_N^{Z_i}(M)$. The next step combines now the BPA system with the formula in equational form yielding the following equations for the component property transformers

$$\nu \left\{ \begin{array}{ll} T_X^{Z_1} = T_X^{Z_2} \sqcap T_X^{Z_3} & T_B^{Z_1} = T_B^{Z_2} \sqcap T_B^{Z_3} \\ T_X^{Z_2} = T_X^{Z_1} \circ T_B & T_B^{Z_2} = t^{\perp} \\ T_X^{Z_3} = t^{Z_4} & T_B^{Z_3} = t^{\perp} \\ T_X^{Z_4} = t^{\top} & T_B^{Z_4} = t^{\perp} \end{array} \right\},$$

where $\sqcap$ means argumentwise intersection, and $\circ$ functional composition. Furthermore, we use $t^{\top}$, respectively $t^{\perp}$, to denote the function which maps every subset of D_ϕ to $\mathcal{Z}$, respectively $\emptyset$.

As we are looking for the largest solution, we have to initialize every component transformer with $t^{\top}$, and get the following fixpoint iteration.

	0	1	2	3
$T_X^{Z_1} = T_X^{Z_2} \sqcap T_X^{Z_3}$	$t^{\top}$	$t^{\top}$	t^{Z_4}	t^{Z_4}
$T_X^{Z_2} = T_X^{Z_1} \circ T_B$	$t^{\top}$	$t^{\top}$	$t^{\top}$	$t^{\top}$
$T_X^{Z_3} = t^{Z_4}$	$t^{\top}$	t^{Z_4}	t^{Z_4}	t^{Z_4}
$T_X^{Z_4} = t^{\top}$	$t^{\top}$	$t^{\top}$	$t^{\top}$	$t^{\top}$
$T_B^{Z_1} = T_B^{Z_2} \sqcap T_B^{Z_3}$	$t^{\top}$	$t^{\top}$	$t^{\perp}$	$t^{\perp}$
$T_B^{Z_2} = t^{\perp}$	$t^{\top}$	$t^{\perp}$	$t^{\perp}$	$t^{\perp}$
$T_B^{Z_3} = t^{\perp}$	$t^{\top}$	$t^{\perp}$	$t^{\perp}$	$t^{\perp}$
$T_B^{Z_4} = t^{\perp}$	$t^{\top}$	$t^{\top}$	$t^{\top}$	$t^{\top}$

The final step consists of two parts. First, computing the set of all subformulae of ϕ valid for ε, which in this example gives $\{Z_4\}$. Second, applying the property transformer of X to $\{Z_4\}$, which results in

$$T_X(\{Z_4\}) = \mathcal{Z}$$

tells us that, in particular, Z_1 and therefore ϕ holds at X. Thus, as one might have expected, the considered system does indeed have an infinite path from X on which the action c is always enabled. For more details about second-order model checking see [21].

This second order algorithm has been generalized in [29] to handle the full modal μ-calculus. Although the actual model checker results directly from a combination of the alternation-free variant and some kind of backtracking, as known from finite-state model checking algorithms (cf. [44]), the corresponding correctness proof requires a stronger framework, which uses *dynamic environments* [29]. They are needed to explicitly model valuations of *free* variables during the iteration process.

Both model checking algorithms have in common that they are *global*, as they provide complete information about which of the subformulae of ϕ are satisfied by the states of the BPA process under consideration. It is, in fact, even possible to explicitly represent the set of all states S_ϕ of a given transition system satisfying a given formula ϕ by means of a finite automaton constructed from the obtained property transformers. In particular, this shows that S_ϕ is always a regular set over the set of nonterminals [21].

The two algorithms have a time complexity which is *polynomial* in the size of the system, while *exponential* only in the size of the formula. From a practical point of view this means that the verification of BPA processes should be still feasible, since realistic system properties can usually be expressed by small formulae.

Mayr [110] has shown that this complexity is essentially optimal by giving a matching lower bound for it. He obtained this lower bound by reducing the acceptance problem for linearly space bounded alternating Turing-machines to model checking BPA and the alternation-free μ-calculus.

Summarizing, it follows that model checking the modal μ-calculus, or its alternation-free fragment, for BPA is EXPTIME-complete.

	General	Fixed formula
Alternation-free modal μ-calculus	EXPTIME-complete	PTIME
Full modal μ-calculus	EXPTIME-complete	PTIME

Hungar and Steffen [78] present a local model checking alternative to [27] in form of a remarkably simple tableau system. The sequents of the tableaux are again inspired by the notion of property transformer. They are of the form $A \vdash \langle \phi, \Delta \rangle$, with the intended meaning that ϕ holds at state A under the assumption that all formulae of Δ hold at the final state. The heart of the tableau system is the following composition rule, which is inspired by the sequential composition rule of Hoare logic:

$$\frac{\alpha\beta \vdash \langle \phi, \Delta \rangle}{\alpha \vdash \langle \phi, \Gamma \rangle \quad \beta \vdash \langle \Gamma, \Delta \rangle}.$$

The rule has the following intended meaning. In order to show that ϕ holds at $\alpha\beta$ under the assumption that Δ holds at the final state, we guess an *intermediate assertion* Γ, and prove the following:

- if Δ holds at the final state, then Γ holds before the execution of β;
- if Γ holds after execution of α, then ϕ holds immediately before its execution.

Hungar [75] extended this tableau system for parallel compositions of BPA processes and the alternation-free modal μ-calculus. In contrast to BPP where no synchronization is possible between processes, he considers a model where BPA processes communicate on common actions. The tableau system presented is sound, but, in general, not complete, as the parallel composition of two BPA processes (with communication) can simulate a Turing machine. It is, however, shown to be decidable for the special case in which at most one of the communicating processes is infinite-state. This latter result coincides with the observation of Burkart and Steffen [28] that the synchronous parallel composition of a BPA process with a finite-state process is essentially a pushdown automaton for which the model checking problem is known to be decidable.

The logic PCTL. Finally, we mention that the verification of nonregular properties for BPA processes has also been considered. Bouajjani, Echahed and Robbana introduce in [11] the logic PCTL, an extension of CTL with Presburger arithmetic (see Section 3.1.1). First of all they show that the logic PCTL is undecidable even for finite-state systems. This negative result is proved by a reduction to the halting problem for Minsky machines and underpins the expressive power of PCTL. Weakening, subsequently, PCTL to its fragment $PCTL^+$ they regain on the other hand decidability of the model checking problem by reduction to the validity problem of Presburger arithmetic.

3.2.2. *Pushdown processes and extensions.* Verifying Hennessy–Milner Logic is essentially independent of the considered process class, as long as the transitions systems which have to be considered are finite-branching. Moreover, the considered logics expressing nonregular properties are too powerful to be decidable for the class of pushdown processes and even more expressive models. Our discussion of the process classes beyond BPA will therefore focus on the alternation-free and full μ-calculus.

Pushdown processes. In formal language theory, automata are used as a "device" to accept sets of words. This goal leaves the free choice between two equivalent acceptance criteria: acceptance with final control states and acceptance with empty stack. For process theory, however, where automata are considered as a general computational model, acceptance with empty stack provides a much better conceptual match: Reaching a PDA configuration where the stack is empty then simply corresponds to termination of the process associated with the considered pushdown automaton.

After the observation that the process class PDA, the class of transition graphs defined by pushdown automata accepting with empty stack, is strictly more expressive than the class BPA [34] there has been a spurt of activity in the development of verification algorithms for such pushdown automata and their extensions.

Roughly, these algorithms can be classified as being

- global, iterative, or
- based on games, or
- based on reachability analysis.

The first algorithm developed for PDA [28] can handle the alternation-free μ-calculus and is a generalization of the global, iterative one for BPA [27]. It takes into account the finite control $Q = \{q_1, \ldots, q_n\}$ of the given pushdown automaton, by observing that starting from a configuration qA the PDA can terminate in any of the configurations $q_1\varepsilon, \ldots, q_n\varepsilon$ representing the empty stack. As a consequence, the property transformer for the nonterminal A must now be parameterized by $q \in Q$, and instead of a single assertion the property transformer must take now $|Q|$ assertions into account, one for each of the terminating configurations $q_i\varepsilon$. Besides this more general domain for the property transformers the algorithm is similar to [27], and can be extended along the lines of [29] to handle also the full modal μ-calculus.

A second, quite different approach is to consider games on the transition graphs of PDAs. Walukiewicz has shown that model checking a modal μ-calculus formula ϕ for a PDA P can be reduced to a parity game on a pushdown tree obtained from the combination of P and the syntax tree for ϕ [149]. Furthermore, he proved that the winning strategy for a player in such a game can be realized by a pushdown strategy automaton. Using this strategy automaton, it is then possible to reduce the existence of a winning strategy in the pushdown game to the existence of a winning strategy in a finite parity game associated with the strategy automaton. The last problem can, finally, be decided by any model checking algorithm for finite-state systems. It has to be pointed out, however, that the size of the finite game constructed is exponential in the size of the pushdown automaton. More specifically, Walukiewicz has shown that model checking the alternation-free μ-calculus for PDAs is EXPTIME-complete, and that this result even holds for a fixed formula.

The third technique for model checking pushdown automata is to analyze reachable sets of configurations as introduced by Bouajjani, Esparza, and Maler [12]. Their approach is based on *alternating pushdown automata*, a generalization of the classical notion of PDA which admits AND/OR transition steps. The main idea of the algorithm is, given a PDA P and an alternation-free μ-calculus formula ϕ, to construct the product of P and ϕ by means of an alternating pushdown automaton $AP_{P\times\phi}$. This representation allows us to exploit that for alternating pushdown automata the set of all predecessors $pre^*(C)$ for a regular set of configurations C is always regular and can effectively be computed in terms of alternating finite-state automata. In the case of $AP_{P\times\phi}$ this means that starting with the regular set C_{tt} of all pairs of configurations c and atomic formulae ψ such that $c \models \psi$, it is possible to effectively compute the set of all configurations which satisfy ϕ, and hence to solve the model checking problem.

The known complexity results for model checking PDA are summarized in the following table.

	General	Fixed formula
Alternation-free modal μ-calculus	EXPTIME-complete	EXPTIME-complete
Full modal μ-calculus	EXPTIME-complete	EXPTIME-complete

Regular graphs. Although we have emphasized in this survey the rewriting-based approach for the description of infinite transition systems, other natural representations have also been considered. Especially in graph theory (cf. [8]), a variety of graph generation mechanisms have been studied. A particularly attractive candidate in this field is the framework of deterministic graph grammars. They consist of an initial finite hypergraph together with a set of hyperedge replacement rules where hyperedges play the roles of nonterminals,

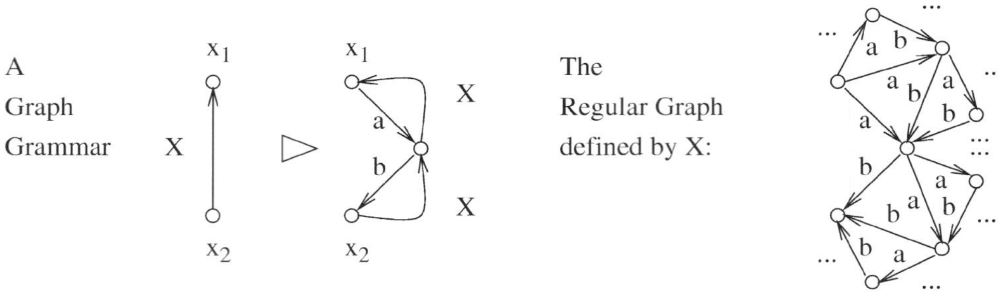

Fig. 5. An example graph grammar.

and finite hypergraphs the roles of right-hand side expressions. A particular deterministic graph grammar represents then a single infinite graph which is obtained as the limit of the finite expansions of the initial hypergraph.

Figure 5 shows an example graph grammar with a single rule where the initial hypergraph consists of the single edge X.

Graph grammars characterize the class of *regular graphs* which is equally expressive as the class of pushdown automata when unguarded rewrite rules are allowed [32]. Nevertheless, Courcelle has shown that regular graphs still possess a decidable monadic second order theory [46]. As already for BPA, this decidability result yields, however, only a nonelementary complexity upper bound.

Subsequently, Burkart and Quemener [26] have developed a direct model checking algorithm for the alternation-free μ-calculus and regular graphs which generalizes the global, iterative model checking algorithm of [27,28]. They extend the method of computing property transformers for nonterminals to whole finite hypergraphs where the glue vertices play the roles of start and end states. Surprisingly, this generalization has the same worst-case complexity as model checking pushdown automata. Along the lines of [29] the algorithm can even be adapted to cope with the full modal μ-calculus.

REC_{Rat} graphs. A still more expressive class of transition systems was introduced by Caucal in [33]. He considers the class of infinite transition graphs REC_{Rat} defined by rewrite systems where in rewrite rules the left-hand side, as well as the right-hand side may be regular languages. A rewrite rule $L_1 \xrightarrow{a} L_2$ is then interpreted as the infinite collection of "ordinary" rules $\alpha \xrightarrow{a} \beta$ where α is a word of the regular language L_1 and β is a word of the regular language L_2. Since these rules may only be applied with respect to prefix rewriting, REC_{Rat} belongs to an extension of sequential rewrite transition systems in which the set of rules may be infinite. Caucal's main results are that this class properly extends the class of regular graphs, and that REC_{Rat} has a decidable monadic second order theory.

By exploiting an alternative representation of REC_{Rat} graphs in terms of inverse regular morphisms and regular restrictions also given in [33] Burkart [22] has developed a more direct model checking algorithm for the full modal μ-calculus. The idea is to combine the given formula ϕ, the inverse regular morphism h^{-1}, and the regular restriction L into a new

μ-formula $\Phi(\phi, h^{-1}, L)$ which has then to be checked for validity on an infinite, complete tree with backward edges. The latter problem can readily be encoded as a BPA model checking problem for the full modal μ-calculus, thereby resulting in a single exponential model checking algorithm for REC_{Rat}.

Macro processes. Another natural extension of BPA are *macro processes* as considered by Hungar in [76]. This framework extends the interpretation of BPA processes as procedures, first given in [27], by allowing transitions with "higher-order procedure calls". Macro processes add to the simple idea that nonterminals correspond to procedure identifiers and right-hand sides to procedure bodies the concept of typed formal procedure parameters: left-hand sides in the grammar have typed formal parameters, which can be instantiated by nonterminals of appropriate type.

The typing discipline establishes a strict "is allowed to be passed as a parameter to" hierarchy among the nonterminals, which, in particular, excludes the possibility of self application: it is not possible to pass a procedure to itself as a parameter. This is essential to maintain the decidability of the model checking problem.

The concept of higher-order recursion is quite powerful. In fact, the class of macro processes belonging to the first level in the hierarchy corresponds to the class of potentially infinite-branching pushdown processes, and thus covers the class of regular graphs, while macro processes of higher levels are strictly more expressive [77].

For these macro processes, Hungar developed a local as well as global iterative model checking algorithm for the alternation-free μ-calculus. Both algorithms have complexity $O(tower(k))$ for a type hierarchy of depth k where $tower(0) =_{\mathrm{df}} 0$ and $tower(n+1) =_{\mathrm{df}} 2^{tower(n)}$. In particular, this shows that his algorithm for the general problem, where k remains unspecified, has nonelementary time complexity.

3.2.3. *BPP and Petri nets.* Going from models for sequential computation to models for concurrent computation the picture concerning decidability of model checking changes dramatically in the branching-time setting.

The logic EG. In [56], Kiehn and Esparza show that the model checking problem for the logic EG is undecidable for BPP, and consequently, also for all the other considered branching-time logics except EF. This result is obtained by a reduction from the halting problem for Minsky two counter machines which is known to be undecidable [117]. Given a Minsky two counter machine M, it is possible to construct a BPP P_M which models M in the weak sense that P_M may simulate the computations of M, but may also "cheat" at some points during the simulation. Nevertheless, it is still possible to characterize the "honest runs" of the system, as well as reachability of the halting state, in the logic EG, although the formula becomes quite complicated. A closer inspection of the proof reveals that the result even holds for the restricted class of deterministic BPPs, and that the formula expressing that the halting state can be reached along an honest computation is independent of the argument process.

The logic EF. The only branching-time logic not covered by the negative result for the logic EG is the logic EF. For this logic Esparza has shown in [52] that the model checking problem for BPPs is decidable. The result is based on the fact that the reachability relation for BPPs is effectively semilinear, i.e., can be characterized by a formula in Presburger arithmetic. Combining then this Presburger formula with the given EF formula ϕ yields again a Presburger formula characterizing the set of reachable states satisfying ϕ. Based on a decision procedure for the validity of Presburger arithmetic formulae one obtains a double exponential time model checking algorithm for EF and BPPs.

On the other hand, Esparza also proved a lower bound for the problem by a reduction of the validity problem for quantified Boolean formula (QBF), which is known to be PSPACE-complete. This PSPACE-hardness result, which even holds for BPPs that describe finite-state systems, was complemented by Mayr [107,109] who showed that the model checking problem for full BPP only requires polynomial space, thereby establishing that it is PSPACE-complete. Thus adding recursion to finite-state BPPs does not increase the complexity of model checking. A further consequence of Mayr's result is that for a fixed formula of nesting depth k the problem lies in Σ_k^P, the k-th level of the polynomial hierarchy. Overall, model checking BPP's with respect to EF has the following complexities.

	General	Fixed formula
EF	PSPACE-complete	Σ_k^P

Going up in the process hierarchy to the more expressive class of Petri Nets costs decidability for EF as shown by Esparza in [51]. Undecidability is shown by a reduction from the undecidable containment problem for Petri nets, which is defined as follows: determine for every two Petri nets N_1 and N_2 with the same number of places and each bijection f between their places, whether for each reachable marking M of N_1 it is guaranteed that $f(M)$ is reachable in N_2.

3.2.4. *PA.* As PA properly contains BPP, the logic EG is immediately undecidable for PA. Hence, the only branching-time logic not covered by this result is the logic EF for which Mayr [108] has proved decidability. He presents a sound and complete tableau system which is based on process decomposition. This algorithm has complexity $O(tower(k))$ for formulae of nesting-depth k. Moreover, when the formula is fixed, the complexity lies in Σ_k^P. Later, Lugiez and Schnoebelen [104] proved decidability of the logic EF for PA by a completely different method using tree-automata to represent sets of configurations. The complexity of their algorithm is, however, the same as that of Mayr's algorithm. Since the best known lower bound for the problem is PSPACE-hardness, the exact complexity of this model-checking problem is still unknown.

	General	Fixed formula
EF	PSPACE-hard	Σ_k^P

3.2.5. *General processes.* We close this section by mentioning that Bradfield has proposed in [15] a sound and complete tableau system for the modal μ-calculus and arbitrary infinite transition systems. This tableau system is of course highly undecidable, but provides a general guideline for interactive proofs.

3.3. *Algorithms for linear-time logics*

From a theoretical point of view linear-time logics are quite attractive as they have a natural connection to automata on infinite strings. More specifically, the set of runs denoted by formulae of classical linear-time logics like, e.g., LTL or the linear-time μ-calculus are always ω-regular sets. Consequently, most of the model checking algorithms for linear-time logics follow the automata-based approach where linear-time formulae are modelled as Büchi automata [146,145]. Intuitively this approach works as follows. Given a process and a linear-time formula ϕ a Büchi automaton $\mathcal{B}_{\neg\phi}$ is constructed, which accepts exactly all runs not satisfying ϕ. Building the product of this Büchi automaton with the Büchi automaton for the process yields a Büchi automaton, whose language is empty, if and only if it satisfies Φ. This reduces the model checking problem to the emptiness problem for Büchi automata.

Two different methods for the construction of a “property automaton” $\mathcal{B}_{\phi}$ have been proposed.

The first method could be termed *global*, as it constructs two automata $\mathcal{B}_{\mathrm{L}}$ and $\overline{\mathcal{B}_{\mathrm{R}}}$ such that $\mathcal{L}(\mathcal{B}_{\phi}) = \mathcal{L}(\mathcal{B}_{\mathrm{L}}) \cap \mathcal{L}(\overline{\mathcal{B}_{\mathrm{R}}})$. Here the automaton $\mathcal{B}_{\mathrm{L}}$ checks some local conditions on the model, while $\overline{\mathcal{B}_{\mathrm{R}}}$ is the complement automaton of $\mathcal{B}_{\mathrm{R}}$ which checks for non-well-foundedness of a regeneration sequence related to minimal fixpoints [144].

The second method takes the *compositional* approach where with each connective in the logic a corresponding construction on automata is associated. The final automaton $\mathcal{B}_{\phi}$ is then obtained by starting with the basic automata for the variables in ϕ, and applying the corresponding automata constructions inductively according to the structure of ϕ. The main difficulty of this method, which has been developed by Dam for the linear-time μ-calculus [47], is concerned with the automata constructions related to the fixpoint operators.

All algorithms considered in the sequel follow the automata-based approach and may use either the global or the compositional construction for the property Büchi automaton. Since Büchi automata are equally expressive as the linear-time μ-calculus all algorithms directly deal with the more expressive linear-time μ-calculus, and subsume model checking LTL as a special case.

Before we proceed with the specifics of the infinite-state case, we mention that LTL, respectively LTμ, model checking for finite-state systems has intensively been studied. For instance, it is well-known that model checking LTL, respectively LTμ, is PSPACE-complete for finite-state systems [135,144]. Both results follow from a polynomial reduction to satisfiability testing which is shown to be PSPACE-complete itself. Fortunately, the complexity is exponential only in the size of the formula which is in practice often small, but linear in the system size [102]. Despite these at first sight negative results, the feasibility of linear-time model checking for finite-state systems has been demonstrated by practical verification tools like SPIN [73], PROD [147] or PEP [148].

3.3.1. *BPA and pushdown processes*

LTL and LTμ. The decidability of the model checking problem for pushdown processes and the linear-time μ-calculus follows immediately from the decidability of the corresponding problem for the modal μ-calculus, since the linear-time μ-calculus can be interpreted in its branching-time equivalent, however at the price of an exponential blowup. This blowup in the translation can, however, be avoided by a more direct approach which applies automata-theoretic techniques.

Given a PDA $\mathcal{P}$ and a linear-time formula ϕ, one constructs as usual a Büchi automaton $\mathcal{B}_{\neg\phi}$ accepting exactly all runs which do not satisfy ϕ. Combining then this Büchi automaton with the PDA $\mathcal{P}$ yields a Büchi PDA $\mathcal{P} \times \mathcal{B}_{\neg\phi}$ representing the synchronized product of $\mathcal{P}$ and $\mathcal{B}_{\neg\phi}$ where a run is accepted if it visits a certain set of control locations infinitely often. Accordingly, all runs starting in a configuration c of $\mathcal{P}$ satisfy ϕ iff c has no accepting run in $\mathcal{P} \times \mathcal{B}_{\neg\phi}$. Bouajjani, Esparza, and Maler have shown in [12] that the set $C_{\neg\phi}$ of all configurations of this Büchi PDA possessing an accepting run can effectively be computed by means of a certain kind of finite automaton, called *multi-automaton*. The model checking problem then simply reduces to checking whether the initial configuration of the PDA $\mathcal{P}$ is a member of $C_{\neg\phi}$.

This algorithm shows that model checking pushdown processes with LTL, as well as the linear-time μ-calculus, is in EXPTIME. A reduction of the acceptance problem for linearly bounded alternating Turing machines, which is known to be EXPTIME-complete [93], shows, furthermore, that the problem is EXPTIME-hard. Overall, these results establish that the model checking problem for pushdown automata wrt. LTL, as well as the linear-time μ-calculus, is EXPTIME-complete. Fixing a formula ϕ, on the other hand, the model checking problem has only polynomial complexity in the size of the system representation. Thus the problem is only slightly harder than in the finite-state case which is PSPACE-complete, and it is also polynomial in the system size for a fixed formula.

Shortly afterwards Mayr [109] has also settled the exact complexity of model checking LTL for BPA processes which is not covered by the previous results. He has shown, by generalizing the proof of Bouajjani, Esparza, and Maler, that model checking the weaker class of context-free processes with LTL is EXPTIME-complete as well.

Summarizing, we have the following complexity results for model checking LTL, respectively LTμ, on sequential processes.

	General	Fixed formula
FSA	PSPACE-complete	PTIME
BPA	EXPTIME-complete	PTIME
PDA	EXPTIME-complete	PTIME

In addition, Esparza et al. [54] have developed efficient algorithms for model checking pushdown processes when the negation of the specification is already given in form of a Büchi automaton. This circumvents the exponential conversion from LTμ to Büchi automata.

The logic CLTL. Finally, decidability of nonregular linear-time properties has been investigated by Bouajjani and Habermehl in [13]. They consider CLTL, a linear-time logic which extends LTL with pattern constraints expressed by finite-state automata and counting constraints using Presburger arithmetic formulae. As full CLTL is undecidable even for finite-state systems [10], its fragment $\text{CLTL}_\square$ which essentially does not allow eventuality formulae with counting constraints is introduced. It turns out that $CLTL_{\Box}$ is decidable for pushdown automata which is shown in four steps:

- First, since $CLTL_{\Box}$ is not closed under negation, the syntactic complement of $CLTL_{\Box}$, called $CLTL_{\Diamond}$, is introduced. Both fragments are related by the property that for each $\phi \in CLTL_{\Box}$ there exists $\phi' \in CLTL_{\Diamond}$ such that $\neg\phi$ and ϕ' have the same semantics, and vice versa.
- Second, it is shown that any formula of $CLTL_{\Diamond}$ can be transformed into a normal form which is essentially a conjunction of an ω-regular property expressed in terms of a linear-time μ-calculus formula and a pure counting constraint eventuality formula.
- Third, it is proved that satisfiability of a formula $\phi \in CLTL_{\Diamond}$ relatively to a set of sequences S can be reduced to a constrained emptiness problem. If $\widetilde{S}$, a technical set construction related to S, has the additional property that its intersection with an ω-regular language always yields a semilinear set, then the satisfiability problem can further be reduced to the nonemptiness problem of semilinear sets which itself is known to be decidable.
- Fourth and last, the fact that pushdown processes generate ω-context-free languages, which are semilinear and closed under intersection with ω-regular languages shows that the satisfiability problem for $CLTL_{\Diamond}$ is decidable, and consequently so is the model-checking problem for $CLTL_{\Box}$.

3.3.2. *BPP and Petri nets*

LTL and LTμ. Here, only the model checking problem for the more expressive class of Petri nets and the stronger logic LTμ has been considered. As it turns out, the decidability of the model checking problem for LTμ and Petri nets is sensitive to whether atomic propositions beyond just *true* are allowed in the logic or not. As shown by Esparza, the case with general atomic propositions is undecidable [53], while in case that *true* is the only allowed atomic proposition this problem becomes decidable [51].

The given decision procedure is based on an automata-theoretic characterization of the logic which was introduced by Vardi and Wolper [146]. Refining a technique of [47] Esparza has shown that, given a closed formula ϕ, there exist two finite automata $A_{\neg\phi}$ and $B_{\neg\phi}$ accepting the finite, respectively infinite, words that do not satisfy ϕ. Consequently, the Petri net N satisfies ϕ iff $\mathcal{L}(N) \cap \mathcal{L}(A_{\neg\phi}) = \emptyset$ and $\mathcal{L}(N) \cap \mathcal{L}(B_{\neg\phi}) = \emptyset$. In order to decide these two properties the finite automata are transformed into certain Petri nets and combined with N in a way similar to the classical automata product yielding $N \times N_{A_{\neg\phi}}$ and $N \times N_{B_{\neg\phi}}$ such that $\mathcal{L}(N \times N_{A_{\neg\phi}}) = \mathcal{L}(N) \cap \mathcal{L}(N_{A_{\neg\phi}})$ and $\mathcal{L}(N \times N_{B_{\neg\phi}}) = \mathcal{L}(N) \cap \mathcal{L}(N_{B_{\neg\phi}})$. Esparza then proves that

(1) $\mathcal{L}(N) \cap \mathcal{L}(A_{\neg\phi}) \neq \emptyset$ iff the Petri net $N \times N_{A_{\neg\phi}}$ has a reachable dead marking which marks some place of a certain set S.

(2) $\mathcal{L}(N) \cap \mathcal{L}(B_{\neg\phi}) \neq \emptyset$ iff the Petri net $N \times N_{B_{\neg\phi}}$ has a run which contains infinitely many occurrences of transitions of a certain set T.

Now the existence of a reachable dead marking in (1) can be decided solving an exponential number of instances of the reachability problem for Petri nets. The reachability problem is known to be decidable since the early eighties when Kosaraju [96], as well as Mayr [106], presented a non-primitive recursive algorithm. Recently, Bouziane [14] claimed a doubly-EXPSPACE algorithm, which approaches the EXPSPACE lower bound of Lipton [103]; however, the correctness of this algorithm is unclear.

The existence of a run in (2) containing infinitely many occurrences of transitions of a given set T was shown to be decidable by Jantzen and Valk [92]. Later Yen has shown that this can be decided also within exponential space [150].

Overall, this yields an EXPSPACE-algorithm. Since on the other hand the reachability problem for Petri nets can be encoded in LTL, the model checking problem is also EXPSPACE-hard, proving the problem to be EXPSPACE-complete.

For reactive systems, which often are assumed to provide a certain service forever without ever terminating, another, non-standard semantics is of interest, which only takes infinite runs into account. When restricted to these ω-semantics, the space complexity of Esparza's decision procedure is exponential in the size of the Petri net and double exponential in the size of the formula. Recently, Habermehl [62] has improved on this complexity by showing that the space complexity can be reduced to be polynomial in the size of the formula, which is known to be a lower bound already for finite-state systems. However, also in the ω-setting the problem remains EXPSPACE-hard in the size of the system.

Even more surprisingly, the same complexity is obtained already for the weaker model of BPP and the weaker logic LTL [64]. In this case EXPSPACE-hardness can be shown in five steps:

- First, the EXPSPACE-complete problem whether an exponentially space-bounded Turing-machine accepts a given input, is reduced to the problem whether the exponentially space-bounded universal Turing-machine accepts a given input.
- Second, the universal Turing-machine is encoded in an exponentially space-bounded universal Minsky n-counter machine, for which, consequently, the acceptance problem is EXPSPACE-hard.
- Third, this universal Minsky n-counter machine can be encoded as a parallel composition of a BPP and a finite automaton yielding a PPDA similar to the construction in [103] or [53]. The size of this finite automaton is a function in the size of the finite control of the universal counter machine and thus fixed and finite.
- Fourth, the problem if the resulting PPDA has an infinite run can be encoded in a model checking problem for BPP and an LTL-formula. The LTL-formula depends only on the fixed finite automaton and is thus also fixed. This encoding of the automaton in LTL is possible, since, by choosing the atomic actions accordingly, the automaton can be made star-free.

Summarizing, we have the following complexities when model checking LTL, respectively LTμ.

LTL and LTμ with standard semantics		
	General	Fixed formula
BPP/PN	EXPSPACE-complete	EXPSPACE-complete

LTL and LTμ with ω-semantics		
	Fixed system	Fixed formula
BPP/PN	PSPACE-complete	EXPSPACE-complete

The logic CLTL. Pursuing their research on CLTL, Bouajjani and Habermehl have shown that $CLTL_{\Box}$ is also decidable for Petri nets [13]. Since Petri nets are in contrast to pushdown processes not semilinear, the proof for the decidability of $CLTL_{\Box}$ on pushdown processes does not carry over. This time satisfiability of $CLTL_{\Diamond}$ is reduced in several steps to the reachability problem in Petri nets.

- First, the regular part of the $CLTL_{\Diamond}$ formula ϕ is transformed into an equivalent Büchi automaton $\mathcal{B}$ which is then combined with the net N such that $N \times \mathcal{B}$ represents their synchronized product.
- Second, one can show that the set of markings M of $N \times \mathcal{B}$ from which the Büchi automaton accepts every run is semilinear.
- Third, since the counting constraint part of ϕ also describes a semilinear set of markings M', and since semilinear sets are closed under intersection, the satisfiability problem reduces to the reachability problem for the semilinear set of markings $M \cap M'$.
- Fourth and last, the previous reachability problem for semilinear sets can be reduced to the classical reachability problem for single markings in Petri nets.

3.3.3. *PA*

LTL and LTμ. Although LTL is decidable for BPA, as well as for BPP, this result does not longer hold in their common generalization PA. As shown in [10], a Minsky machine can be simulated by three BPA processes in parallel, such that their "honest" synchronization, as well as the reachability of the halting state, is expressible by a LTL formula. More technically, the halting of the Minsky machine is encoded as the nonemptiness of the intersection of an ω-star-free language with a PA ω-language. Undecidability of the model checking problem for LTL and PA follows then immediately from the undecidability of the halting problem for Minsky machines [117].

The logic CLTL. As a corollary, the undecidability of model checking LTL also proves that PA ω-languages are not closed under intersection with ω-star-free languages. On the other hand, it is shown in [13] that PA ω-languages are closed under intersection with the smaller class of simple ω-regular languages, i.e., ω-languages definable by a finite-state Büchi automaton where every loop in its transition graph is a self-loop. Since the fragment simple-$PLTL_{\Box}$ of CLTL, which is incomparable to LTL, expresses exactly the simple ω-regular languages, it can, finally, be shown that simple-$PLTL_{\Box}$ is indeed decidable for PA.

3.4. *Summary*

From a bird's perspective we may conclude that sequential-type systems, like BPA and PDA, are inherently easier to analyze than parallel-type systems. This conclusion is supported by both the (un)decidability as well as the complexity results:

For BPA and PDA, most branching and linear-time logics are decidable, while for parallel-type systems, like BPP and Petri nets, except for the very restrictive branching-time logics HML and EF only linear-time logics are decidable. In particular the modal

Class	Logic	Complexity
BPA	HML	G: PSPACE-complete [110] F: PTIME
BPA	modal μ-calculus	G: EXPTIME-complete [149,110] F: PTIME [29]
PDA	modal μ-calculus	G: EXPTIME-complete [149] F: EXPTIME-complete [149]
BPP	EF	G: PSPACE-complete [52,107,109] F: Σ_k^P [52,107,109]
PA	EF	G: PSPACE-hard [108,109] F: Σ_k^P [109]

Class	Logic	Complexity
FSA	LTL / LTμ	G: PSPACE-complete [135,144] F: PTIME
BPA	LTL / LTμ	G: EXPTIME-complete [109] F: PTIME [12]
PDA	LTL / LTμ	G: EXPTIME-complete [12] F: PTIME [12]
BPP / PN	LTL / LTμ	G: EXPSPACE-complete [51,109] F: EXPSPACE-complete [51,109]

Fig. 6. Complexities of model checking problems.

μ-calculus as well as LTL (and even $\mathrm{CLTL}_{\Box}$) are decidable for PDA, while on the other hand all branching-time logics except HML are undecidable for Petri nets, and merely the linear-time μ-calculus with true as only atomic proposition is decidable for Petri nets.

Considering the complexity results for model checking the picture is similar. For example on the sequential side model checking the modal μ-calculus and the linear-time μ-calculus are both EXPTIME-complete for PDA, while on the parallel side the model checking problem for Petri nets with respect to the linear-time μ-calculus is already EXPSPACE-complete. The following corresponding tables summarize these results and the pointers to the relevant literature. To abbreviate the notation, G: indicates the complexity of the general problem, while F: specifies the complexity of the problem when the formula is fixed.

The central results presented in this chapter concern worst-case estimations within the classical complexity class structure. This classification, although constituting a very good guideline also for practical purposes, must often be refined in order to provide a real basis for implementation decisions. We therefore propose to investigate

- finer complexity class structures, e.g., in many cases people only accept linear or at most quadratic algorithms, and even constant factors may play a significant role;
- average or 'in practice' behaviours: systems developed by people are often much better behaved as suggested by worst-case analyzes; and
- special restrictive but practical system/property combinations, which can be treated efficiently.

Whereas the first two points do not require any further explanation, the third should become clearer by considering program analysis as an illustrating example. Here the so-called *bitvector problems* constitute a highly relevant but tractable class. In fact, along the lines of [108,57,131] it is straightforward to deduce that interprocedural bitvector analyzes, even for programs with fork/join parallelism and shared variables, admit model checking in linear time. Special cases like this, which depend on rather specific patterns of formulae and/or models, will always have to be considered case by case. However, we are convinced that the overview of the general complexities provides beneficial guidelines even for such refined complexity considerations, as one cannot construct a scenario avoiding 'bottlenecks' without knowing exactly where they are located.

Acknowledgements

Sections 1 and 2 are based on the survey *Infinite Results* [118] presented at CONCUR'96, and the origins of Section 3 are found in the survey *More Infinite Results* [25] presented at INFINITY'96. However, there has been a flood of new results in the intervening years, and these earlier surveys are vastly extended here. We wish to thank Richard Mayr and Markus Mueller-Olm for detailed comments on earlier drafts of this survey.

References

[1] P.A. Abdulla and K. Čerāns, *Simulation is decidable for one-counter nets*, CONCUR'98, Nice, Lecture Notes in Comput. Sci. 1466, D. Sangiorgi and R. de Simone, eds, Springer (1998), 253–268.

[2] R. Alur, C. Courcoubetis, T.A. Henzinger, N. Halbwachs, P.H. Ho, X. Nicollin, A. Olivero, J. Sifakis and S. Yovine, *The algorithmic analysis of hybrid systems*, Theoret. Comput. Sci. **138** (1995), 3–34.

[3] R. Alur and T.A. Henzinger, *Real-time systems = discrete systems + clock variables*, Software Tools for Technology Transfer **1** (1997), 86–109.

[4] J.C.M. Baeten, J.A. Bergstra and J.W. Klop, *Decidability of bisimulation equivalence for processes generating context-free languages*, PARLE'87, Eindhoven, Lecture Notes in Comput. Sci. 259, J.W. de Bakker, A.J. Nijman and P.C. Treleaven, eds, Springer (1987), 94–113.

[5] J.C.M. Baeten, J.A. Bergstra and J.W. Klop, *Decidability of bisimulation equivalence for processes generating context-free languages*, J. ACM **40** (3) (1993), 653–682.

[6] Y. Bar-Hillel, M. Perles and E. Shamir, *On formal properties of simple phrase structure grammars*, Zeitschrift für Phonetik, Sprachwissenschaft, und Kommunikationsforschung **14** (1961), 143–177.

[7] M. Ben-Ari, Z. Manna and A. Pnueli, *The temporal logic of branching time*, Acta Inform. **20** (3) (1983), 207–226.

[8] C. Berge, *Graphs and Hypergraphs*, North-Holland, Amsterdam (1973).

[9] J.A. Bergstra and J.W. Klop, *Algebra of communicating processes with abstraction*, Theoret. Comput. Sci. **37** (1985), 77–121.

[10] R. Bouajjani, R. Echahed and P. Habermehl, *On the verification of nonregular properties for nonregular processes*, LICS'95, San Diego, IEEE Comput. Soc. Press (1995), 123–133.

[11] A. Bouajjani, R. Echahed and R. Robbana, *Verification of nonregular temporal properties for context-free processes*, CONCUR'94, Uppsala, Lecture Notes in Comput. Sci. 836, B. Jonsson and J. Parrow, eds, Springer (1994), 81–97.

[12] A. Bouajjani, J. Esparza and O. Maler, *Reachability analysis of pushdown automata: Application to model-checking*, CONCUR'97, Warsaw, Lecture Notes in Comput. Sci. 1243, A. Mazurkiewicz and J. Winkowski, eds, Springer (1997), 135–150.

[13] A. Bouajjani and P. Habermehl, *Constrained properties, semilinear systems and Petri nets*, CONCUR'96, Pisa, Lecture Notes in Comput. Sci. 1119, U. Montanari and V. Sassone, eds, Springer (1996), 481–497.

[14] Z. Bouziane, *A primitive recursive algorithm for the general Petri net reachability problem*, FOCS'98, Palo Alto, IEEE Comput. Soc. Press (1998), 130–136.

[15] J.C. Bradfield, *Verifying Temporal Properties of Systems*, Birkhäuser (1992).

[16] J.C. Bradfield, *The modal mu-calculus alternation hierarchy is strict*, CONCUR'96, Pisa, Lecture Notes in Comput. Sci. 1119, U. Montanari and V. Sassone, eds, Springer (1996), 233–246.

[17] J.C. Bradfield and C.P. Stirling, *Modal logics and mu-calculi: An introduction*, Handbook of Process Algebra, J.A. Bergstra, A. Ponse and S.A. Smolka, eds, Elsevier, Amsterdam (2001), 293–330.

[18] R. Büchi, *On a decision method in restricted second order arithmetic*, Proc. of the 1960 International Congress for Logic, Methodology and Philosophy of Science, Berkeley, Stanford University Press (1962), 1–11.

[19] R. Büchi, *Regular canonical systems*, Archive für Mathematische Logik und Grundlagenforschung **6** (1964), 91–111.

[20] J.R. Burch, E.M. Clarke, K.L. McMillan, D.L. Dill and L.J. Hwang, *Symbolic model checking* 10^{20} *states and beyond*, LICS'90, Philadelphia, IEEE Comput. Soc. Press (1990), 428–439.

[21] O. Burkart, *Automatic Verification of Sequential Infinite-State Processes*, Lecture Notes in Comput. Sci. 1354, Springer (1997).

[22] O. Burkart, *Model-checking rationally restricted right closures of recognizable graphs*, Electron. Notes in Theoret. Comput. Sci. **9** (1997).

[23] O. Burkart, D. Caucal and B. Steffen, *An elementary bisimulation decision procedure for arbitrary context-free processes*, MFCS'95, Prague, Lecture Notes in Comput. Sci. 969, J. Wiedermann and P. Hájek, eds, Springer (1995), 423–433.

[24] O. Burkart, D. Caucal and B. Steffen, Bisimulation collapse and the process taxonomy, CONCUR'96, Pisa, Lecture Notes in Comput. Sci. 1119, U. Montanari and V. Sassone, eds, Springer (1996), 247–262.

[25] O. Burkart and J. Esparza, *More infinite results*, Electron. Notes in Theoret. Comput. Sci. **5** (1997).

[26] O. Burkart and Y.-M. Quemener, *Model-checking of infinite graphs defined by graph grammars*, Electron. Notes in Theoret. Comput. Sci. **5** (1997).

[27] O. Burkart and B. Steffen, *Model checking for context-free processes*, CONCUR'92, Stony Brook, Lecture Notes in Comput. Sci. 630, W.R. Cleaveland, ed., Springer (1992), 123–137.

[28] O. Burkart and B. Steffen, *Composition, decomposition and model-checking of pushdown processes*, Nordic J. Computing **2** (1995), 89–125.

[29] O. Burkart and B. Steffen, *Model-checking the full-modal mu-calculus for infinite sequential processes*, ICALP'97, Bologna, Lecture Notes in Comput. Sci. 1256, P. Degano, R. Gorrieri and A. Marchetti-Spaccamela, eds, Springer (1997), 419–429.

[30] D. Caucal, *A fast algorithm to decide on the equivalence of stateless* dpda, RAIRO **27** (1) (1990), 23–48.

[31] D. Caucal, *Graphes canoniques de graphes algébriques*, RAIRO **24** (4) (1990), 339–352.

[32] D. Caucal, *On the regular structure of prefix rewriting*, Theoret. Comput. Sci. **10** (1992), 61–86.

[33] D. Caucal, *On infinite transition graphs having a decidable monadic theory*, ICALP'96, Paderborn, Lecture Notes in Comput. Sci. 1099, F. Meyer auf der Heide and B. Monien, eds, Springer (1996), 194–205.

[34] D. Caucal and R. Monfort, *On the transition graphs of automata and grammars*, Graph-Theoretic Concepts in Computer Science, Berlin, Lecture Notes in Comput. Sci. 484, R.H. Möhring, ed., Springer (1990), 311–337.

[35] S. Christensen, *Decidability and Decomposition in Process Algebras*, Ph.D. thesis, The University of Edinburgh, Department of Computer Science, ECS-LFCS-93-278 (1993).

[36] S. Christensen, Y. Hirshfeld and F. Moller, *Bisimulation equivalence is decidable for basic parallel processes*, CONCUR'93, Hildesheim, Lecture Notes in Comput. Sci. 715, E. Best, ed., Springer (1993), 143–157.

[37] S. Christensen, Y. Hirshfeld and F. Moller, *Decomposability, decidability and axiomatisability for bisimulation equivalence on basic parallel processes*, LICS'93, Montreal, IEEE Comput. Soc. Press (1993), 386–396.

[38] S. Christensen, H. Hüttel and C. Stirling, *Bisimulation equivalence is decidable for all context-free processes*, CONCUR'92, Stony Brook, Lecture Notes in Comput. Sci. 630, W.R. Cleaveland, ed., Springer (1992), 138–147.

[39] S. Christensen, H. Hüttel and C. Stirling, *Bisimulation equivalence is decidable for all context-free processes*, Inform. and Comput. **12** (2) (1995), 143–148.

[40] E.M. Clarke and E.A. Emerson, *Design and synthesis of synchronization skeletons using branching time temporal logic*, Proc. of the Workshop on Logics of Programs, Lecture Notes in Comput. Sci. 131, Yorktown Heights, Springer (1981), 52–71.

[41] E.M. Clarke, E.A. Emerson and A.P. Sistla, *Automatic verification of finite state concurrent systems using temporal logic specifications*, ACM Trans. Programming Languages and Systems **8** (2) (1986), 244–263.

[42] E.M. Clarke, O. Grumberg and D.A. Peled, *Model Checking*, MIT Press (1999).

[43] R. Cleaveland, M. Klein and B. Steffen, *Faster model checking for the modal mu-calculus*, CAV'92, Lecture Notes in Comput. Sci. 663, Montreal, G. Bochmann and D.K. Probst, eds, Springer (1992), 410–422.

[44] R. Cleaveland and O. Sokolsky, *Equivalence and preorder checking for finite-state systems*, Handbook of Process Algebra, J.A. Bergstra, A. Ponse and S.A. Smolka, eds, Elsevier, Amsterdam (2001), 391–424.

[45] R. Cleaveland and B. Steffen, *Computing behavioural relations, logically*, ICALP'91, Lecture Notes in Comput. Sci. 510, Madrid, J.L. Albert, B. Monien and Rodríguez, eds, Springer (1991), 127–138.

[46] B. Courcelle, *Graph rewriting: An algebraic and logic approach*, Handbook of Theoretical Computer Science, J. van Leeuwen, ed., Elsevier Science Publisher B.V. (1990), 193–242.

[47] M. Dam, *Fixpoints of Büchi automata*, FSTTCS'92, New Delhi, Lecture Notes in Comput. Sci. 652, R. Shyamasundar, ed., Springer (1992), 39–50.

[48] L.E. Dickson, *Finiteness of the odd perfect and primitive abundant numbers with distinct factors*, Amer. J. Math. **35** (1985), 413–422.

[49] E.A. Emerson and C.-L. Lei, *Efficient model checking in fragments of the propositional mu-calculus*, LICS'86, Cambridge, Massachusetts, IEEE Comput. Soc. Press (1986), 267–278.

[50] E.A. Emerson and Srinivasan, *Branching time temporal logic*, Linear Time, Branching Time and Partial Order in Logics and Models for Concurrency, Lecture Notes in Comput. Sci. 354, J.W. de Bakker, W.-P. de Roever and G. Rozenberg, eds, Springer (1989), 123–172.

[51] J. Esparza, *On the decidability of model checking for several μ-calculi and Petri nets* CAAP'94, Edinburgh, Lecture Notes in Comput. Sci. 787, S. Tison, ed., Springer (1994), 115–129.

[52] J. Esparza, *Decidability of model-checking for concurrent infinite-state systems*, Acta Inform. **34** (1997), 85–107.

[53] J. Esparza, *Decidability and complexity of Petri net problems: An introduction,* Lectures on Petri Nets I: Basic Models, Lecture Notes in Comput. Sci. 1491, W. Reisig and G. Rozenberg, eds, Springer (1998), 374–428.

[54] J. Esparza, D. Hansel, P. Rossmanith and S. Schwoon, *Efficient algorithms for model checking pushdown systems*, Submitted for publication.

[55] J. Esparza, P. Jančar and F. Moller, *Petri nets and regular behaviours*, J. Comput. and System Sci. **59** (3) (1999), 476–503.

[56] J. Esparza and A. Kiehn, *On the model checking problem for branching time logics and basic parallel processes*, CAV'95, Liege, Lecture Notes in Comput. Sci. 939, P. Wolper, ed., Springer (1995), 353–366.

[57] J. Esparza and A. Podelski, *Efficient algorithms for pre** and *post** *in interprocedural parallel flow graphs*, POPL'00, Boston, ACM Press (2000), 1–12.

[58] R.W. Floyd, *Assigning meanings to programs,* Symposium on Applied Mathematics, Vol. 19, J.T. Schwartz, ed., Amer. Math. Soc. (1967), 19–32.

[59] R.J. van Glabbeek, *The linear time – branching time spectrum*, CONCUR'90, Amsterdam, Lecture Notes in Comput. Sci. 458, J.C.M Baeten and J.W. Klop, eds, Springer (1990), 278–297.

[60] J.F. Groote, *A short proof of the decidability of bisimulation for normed BPA-processes*, Inform. Process. Lett. **42** (1991), 167–171.

[61] J.F. Groote and H. Hüttel, *Undecidable equivalences for basic process algebra*, Inform. and Comput. **11** (2) (1994), 354–371.

[62] P. Habermehl, *On the complexity of the linear time mu-calculus for Petri nets*, 18th International Conference on Application and Theory of Petri Nets, Toulouse, Lecture Notes in Comput. Sci. 1248, P. Azéma and G. Balbo, eds, Springer (1997), 102–116.

[63] M. Hack, *Decidability Questions for Petri Nets*, Ph.D. thesis, Massachusetts Institute of Technology, Laboratory for Computer Science (1976).

[64] P. Hebermehl and R. Mayr, *Personal communications* (January 2000).

[65] M. Hennessy and R. Milner, *Algebraic laws for nondeterminism and concurrency*, J. ACM **32** (1) (1985) 137–161.

[66] Y. Hirshfeld, *Petri nets and the equivalence problem*, CSL'93, Swanseea, Lecture Notes in Comput. Sci. 832, E. Börger, Y. Gurevich and K. Meinke, eds, Springer (1993), 165–174.

[67] Y. Hirshfeld, *Congruences in commutative semigroups*, Technical Report ECS-LFCS-94-291, Department of Computer Science, University of Edinburgh, UK (1994).

[68] Y. Hirshfeld and M. Jerrum, *Bisimulation equivalence is decidable for normed Process Algebra*, ICALP'99, Prague, Lecture Notes in Comput. Sci. 1644, J. Wiedermann, P. van Emde Boas and M. Nielsen, eds, Springer (1999), 412–421.

[69] Y. Hirshfeld, M. Jerrum and F. Moller, *A polynomial algorithm for deciding bisimilarity of normed context-free processes*, FOCS'94, Santa Fe, IEEE Comput. Soc. Press (1994), 623–631.

[70] Y. Hirshfeld, M. Jerrum and F. Moller, *A polynomial algorithm for deciding bisimilarity of normed context-free processes*, Theoret. Comput. Sci. **15** (1996), 143–159.

[71] Y. Hirshfeld, M. Jerrum and F. Moller, *A polynomial algorithm for deciding bisimulation equivalence of normed basic parallel processes*, Math. Structures Comput. Sci. **6** (1996), 251–259.

[72] Y. Hirshfeld and F. Moller, *A fast algorithm for deciding bisimilarity of normed context-free processes*, CONCUR'94, Uppsala, Lecture Notes in Comput. Sci. 836, B. Jonsson and J. Parrow, eds, Springer (1994), 48–63.

[73] G.J. Holzmann, *Design and Validation of Computer Programs*, Prentice-Hall, Englewood Cliffs, NJ (1991).

[74] J.E. Hopcroft and J.D. Ullman, *Introduction to Automata Theory, Languages and Computation*, Addison-Wesley (1979).

[75] H. Hungar, *Local model checking for parallel compositions of context-free processes*, CONCUR'94, Uppsala, Lecture Notes in Comput. Sci. 836, B. Jonsson and J. Parrow, eds, Springer (1994), 114–128.

[76] H. Hungar, *Model-checking of macro processes*, CAV'94, Stanford, Lecture Notes in Comput. Sci. 818, D.L. Dill, ed., Springer (1994), 169–181.

[77] H. Hungar, *Model-checking and higher-order recursion*, MFCS'99, Szklarska Poreba, Lecture Notes in Comput. Sci. 1672, L. Pacholski, ed., Springer (1999), 149–159.

[78] H. Hungar and B. Steffen, *Local model-checking for context-free processes*, Nordic J. Comput. **1** (3) (1994), 364–385.

[79] H. Hüttel, *Undecidable equivalences for basic parallel processes*, TACS'94, Sendai, Lecture Notes in Comput. Sci. 789, M. Hagiya and J.C. Mitchell, eds, Springer (1994), 454–464.

[80] H. Hüttel and C. Stirling, *Actions speak louder than words: Proving bisimilarity for context-free processes*, LICS'91, Amsterdam, IEEE Comput. Soc. Press (1991), 376–386.

[81] D.T. Huynh and L. Tian, *Deciding bisimilarity of normed context-free processes is in* Σ_2^P, Theoret. Comput. Sci. **12** (1994), 183–197.

[82] D.T. Huynh and L. Tian, *On deciding readiness and failure equivalences for processes*, Inform. and Comput. **11** (1995), 193–205.

[83] P. Jančar, *Decidability questions for bisimilarity of Petri nets and some related problems*, STACS'94, Caen, Lecture Notes in Comput. Sci. 775, P. Enjalbert, E.W. Mayr and K.W. Wagner, eds, Springer (1994), 581–592.

[84] P. Jančar, *Undecidability of bisimilarity for Petri nets and some related problems*, Theoret. Comput. Sci. **148** (1995), 281–301.

[85] P. Jančar, *Bisimulation equivalence is decidable for one-counter processes*, ICALP'97, Bologna, Lecture Notes in Comput. Sci. 1256, P. Degano, R. Gorrieri and A. Marchetti-Spaccamela, eds, Springer (1997), 549–559.

[86] P. Jančar, A. Kučera and R. Mayr, *Deciding bisimulation-like equivalences with finite-state processes*, ICALP'98, Aalborg, Lecture Notes in Comput. Sci. 1443, K.G. Larsen, S. Skyum and G. Winskel, eds, Springer (1998), 200–211.

[87] P. Jančar, A. Kučera and F. Moller, *Simulation and bisimulation over one-counter processes*, STACS'2000, Lecture Notes in Comput. Sci. 1770, H. Reichel and S. Tison, eds, Springer (2000).

[88] P. Jančar and F. Moller, *Checking regular properties of Petri nets*, CONCUR'95, Philadelphia, Lecture Notes in Comput. Sci. 962, I. Lee and S.A. Smolka, eds, Springer (1995), 348–362.

[89] P. Jančar and F. Moller, *Simulation of one-counter nets via colouring* (*abstract*), Workshop Journées Systèmes Infinis, A. Finkel, ed., Cachan, (1998), 1–6. Report CNRS URA 2236, Ecole Normale Supérieure

de Cachan. Full 10-page paper available as Uppsala University CSD Report No. 159 at http://www.csd.uu.se/papers/report.html.
[90] P. Jančar and F. Moller, *Techniques for decidability and undecidability for bisimilarity*, CONCUR'99, Eindhoven, Lecture Notes in Comput. Sci. 1664, J.C.M. Baeten and S. Mauw, eds, Springer (1999), 30–45.
[91] P. Jančar, F. Moller and Z. Sawa, *Simulation problems for one-counter machines*, SOFSEM'99, Milovy, Lecture Notes in Comput. Sci. 1725, J. Pavelka, G. Tel and M. Bartosek, eds, Springer (1999), 398–407.
[92] M. Jantzen and R. Valk, *The residue of vector sets with applications to decidability problems in Petri nets*, Acta Inform. **21** (1985), 643–674.
[93] D. Johnson, *A catalog of complexity classes*, Handbook of Theoretical Computer Science, J. van Leeuwen, ed., Vol. A, Elsevier (1990), 67–161.
[94] D.E. Knuth, J.H. Morris and V.R. Pratt, *Fast pattern matching in strings*, SIAM J. Comput. **6** (1977), 323–350.
[95] A.J. Korenjak and J.E. Hopcroft, *Simple deterministic languages*, 7th Annual IEEE Symposium on Switching and Automata Theory, Berkeley, IEEE (1966), 36–46.
[96] S.R. Kosaraju, *Decidability of reachability in vector addition systems*, STOC'82, San Fransisco (1982), 267–281.
[97] D. Kozen, *Results on the propositional* μ*-calculus*, Theoret. Comput. Sci. **27** (1983), 333–354.
[98] A. Kučera and R. Mayr, *Simulation preorder on simple process algebras*, ICALP'99, Prague, Lecture Notes in Comput. Sci. 1644, J. Wiedermann, P. van Emde Boas and M. Nielsen, eds, Springer (1999), 503–512.
[99] A. Kučera and R. Mayr, *Weak bisimilarity with infinite state systems can be decided in polynomial time*, CONCUR'99, Eindhoven, Lecture Notes in Comput. Sci. 1664, J.C.M. Baeten and S. Mauw, eds, Springer (1999), 368–382.
[100] L. Lamport, *Verification and specification of concurrent programs*, A Decade of Concurrency, Noordwijkerhout, Lecture Notes in Comput. Sci. 803, J.W. de Bakker, W.-P. de Roever and G. Rozenberg, eds, Springer (1994), 347–374.
[101] G. Lenzi, *A hierarchy theorem for the mu-calculus*, ICALP'96, Paderborn, Lecture Notes in Comput. Sci. 1099, F. Meyer auf der Heide and B. Monien, eds, Springer (1996), 87–109.
[102] O. Lichtenstein and A. Pnueli, *Checking that finite state concurrent programs satisfy their linear specification*, POPL'85, New Orleans, ACM Press (1985), 97–107.
[103] R. Lipton, *The reachability problem requires exponential space*, Technical Report 62, Yale University (1976).
[104] D. Lugiez and Ph. Schnoebelen, *The regular viewpoint on PA-processes*, CONCUR'98, Nice, Lecture Notes in Comput. Sci. 1466, D. Sangiorgi and R. de Simone, eds, Springer (1998), 50–66.
[105] Z. Manna and A. Pnueli, *Temporal Logic of Reactive and Concurrent Systems: Specification*, Springer (1992).
[106] E.W. Mayr, *An algorithm for the general Petri net reachability problem*, SIAM J. Comput. **13** (3) (1984), 441–459.
[107] R. Mayr, *Weak bisimulation and model checking for basic parallel processes*, FSTTCS'96, Hyderabad, Lecture Notes in Comput. Sci. 1180, V. Chandru and V. Vinay, eds, Springer (1996), 88–99.
[108] R. Mayr, *Model checking PA-processes*, CONCUR'97, Warsaw, Lecture Notes in Comput. Sci. 1243, A. Mazurkiewicz and J. Winkowski, eds, Springer (1997), 332–346.
[109] R. Mayr, *Decidability and Complexity of Model Checking Problems for Infinite-State Systems*, Ph.D. thesis, Technische Universität München (1998).
[110] R. Mayr, *Strict lower bounds for model checking BPA*, Electron. Notes in Theoret. Comput. Sci. **18** (1998).
[111] R. Mayr, *On the complexity of bisimulation problems for basic parallel processes*, Submitted for publication.
[112] R. Mayr, *On the complexity of bisimulation problems for pushdown automata*, Submitted for publication.
[113] R. Milner, *A Calculus of Communicating Systems*, Lecture Notes in Comput. Sci. 92, Springer (1980).
[114] R. Milner, *A complete inference system for a class of regular behaviours*, J. Comput. and System Sci. **28** (1984), 439–466.
[115] R. Milner, *Communication and Concurrency*, Prentice-Hall (1989).
[116] R. Milner and F. Moller, *Unique decomposition of processes*, Theoret. Comput. Sci. **107** (1993), 357–363.

[117] M. Minsky, *Computation: Finite and Infinite Machines*, Prentice-Hall (1967).
[118] F. Moller, *Infinite results*, CONCUR'96, Pisa, Lecture Notes in Comput. Sci. 1119, U. Montanari and V. Sassone, eds, Springer (1996), 195–216.
[119] F. Moller, *A taxonomy of infinite state processes*, Electron. Notes in Theoret. Comput. Sci. **18** (1998).
[120] F. Moller and G. Birtwistle, eds, *Logics for Concurrency*, Lecture Notes in Comput. Sci. 1043, Springer (1996).
[121] E.F. Moore, *Gedanken experiments on sequential machines*, Automata Studies (1956), 129–153.
[122] D.E. Muller and P.E. Schupp, *The theory of ends, pushdown automata and second-order logic*, Theoret. Comput. Sci. **37** (1985), 51–75.
[123] D. Niwiński, *On fixed-point clones,* ICALP'86, Rennes, Lecture Notes in Comput. Sci. 226, L. Kott, ed., Springer (1986), 464–473.
[124] D. Niwiński, *Fixed point characterization of infinite behaviour of finite state systems*, Theoret. Comput. Sci. **189** (1997), 1–69.
[125] D. Park, *Concurrency and automata on infinite sequences*, 5th GI Conference on Theoret. Comput. Sci., Karlsruhe, Lecture Notes in Comput. Sci. 104, P. Deussen, ed., Springer (1981), 167–183.
[126] J.L. Peterson, *Petri Net Theory and the Modelling of Systems*, Prentice-Hall (1981).
[127] A. Pnueli, *The temporal logic of programs*, FOCS'77, Providence, IEEE Comput. Soc. Press (1977), 46–57.
[128] R.O. Rabin, *Decidability of second-order theories and automata on infinite trees*, Trans. Amer. Math. Soc. **141** (1969), 1–35.
[129] L. Redei, *The Theory of Finitely Generated Commutative Semigroups*, Oxford University Press (1965).
[130] A. Salomaa, *Two complete axiom systems for the algebra of regular events*, J. ACM **13** (1966), 158–169.
[131] H. Seidel and B. Steffen, *Constraint-based inter-procedural analysis of parallel programs*, ESOP'00, Berlin, Lecture Notes in Comput. Sci. 1782, G. Smolka, ed., Springer (2000), 135–150.
[132] G. Sénizergues, *The equivalence problem for deterministic pushdown automata is decidable*, ICALP'97, Bologna, Lecture Notes in Comput. Sci. 1256, P. Degano, R. Gorrieri and A. Marchetti-Spaccamela, eds, Springer (1997), 671–681.
[133] G. Sénizergues, L(A) = L(B)? Technical Report 1161-97, LaBRI, Bordeaux, France (1997).
[134] G. Sénizergues, *Decidability of bisimulation equivalence for equational graphs of finite out-degree*, FOCS'98, Palo Alto, IEEE Comput. Soc. Press (1998), 120–129.
[135] A.P. Sistla and E.M. Clarke, *The complexity of propositional linear temporal logic*, J. ACM **32** (3) (1985), 733–749.
[136] C. Stirling, *Local model checking games*, CONCUR'95, Philadelphia, Lecture Notes in Comput. Sci. 962, I. Lee and S.A. Smolka, eds, Springer (1995), 1–11.
[137] C. Stirling, *Decidability of bisimulation equivalence for normed pushdown processes*, CONCUR'96, Pisa, Lecture Notes in Comput. Sci. 1119, U. Montanari and V. Sassone, eds, Springer (1996), 217–232.
[138] C. Stirling, *Decidability of dpda equivalence*, Technical Report ECS-LFCS-99-411, Department of Computer Science, University of Edinburgh, UK (1999).
[139] J. Stříbrná, *Hardness results for weak bisimilarity of simple process algebras*, Electron. Notes in Theoret. Comput. Sci. **18** (1998).
[140] A. Tarski, *A lattice-theoretical fixpoint theorem and its applications*, Pacific J. Math. **5** (1955), 285–309.
[141] D. Taubner, *Finite Representations of CCS and TCSP Programs by Automata and Petri Nets*, Lecture Notes in Comput. Sci. 369, Springer (1989).
[142] W. Thomas, *A combinatorial approach to the theory of ω-automata*, Inform. and Control **48** (1981), 261–283.
[143] L.G. Valiant and M.S. Paterson, *Deterministic one-counter automata,* J. Comput. and System Sci. **10** (1975), 340–350.
[144] M.Y. Vardi, *A temporal fixpoint calculus*, POPL'88, San Diego, ACM Press (1988), 250–259.
[145] M.Y. Vardi, *Alternating automata and program verification*, Computer Science Today: Recent Trends and Developments, Lecture Notes in Comput. Sci. 1000, J. van Leeuwen, ed., Springer (1995), 471–485.
[146] M.Y. Vardi and P. Wolper, *Automata-theoretic techniques for modal logics of programs*, J. Comput. System Sci. **32** (1986), 183–221.

[147] K. Varpaaniemi, *Prod* 3.3.02: *An advanced tool for efficient reachability analysis*, Technical report, Department of Computer Science and Engineering, Helsinki University of Technology, Finland, 1998. Available at http://www.tcs.hut.fi/pub/prod.

[148] F. Wallner, *Model checking LTL using net unfoldings*, CAV'98, Vancouver, Lecture Notes in Comput. Sci. 1427, A.J. Hu and M.Y. Vardi, eds, Springer (1998), 207–218.

[149] I. Walukiewicz, *Pushdown processes: Games and model-checking*, CAV'96, New Brunswick, Lecture Notes in Comput. Sci. 1102, R. Alur and T.A. Henzinger, eds, Springer (1996), 62–74.

[150] H. Yen, *A unified approach for deciding the existence of certain Petri net paths*, Inform. and Comput. **96** (1) (1992), 119–137.

Subject index

alternation depth, 598

base, 573
Basic Parallel Processes (BPP), 553
Basic Process Algebra (BPA), 553
bisimulation base, 568
bisimulation equivalence, 558
bisimulation game, 558
bisimulation relation, 558

Cancellation Lemma
– for normed BPA, 566
– for normed BPP, 568
Chomsky hierarchy of process classes, 552, 561
complete base, 578
Computation tree logic (CTL), 595
Constrained Linear Temporal Logic (CLTL), 601

decomposing function, 573
determinacy, 557
Dickson's lemma, 551

EF, 595
EG, 595

finite-branching, 559
finite-state automata (FSA), 552
full base, 573

graph grammar, 608

halting problem, 587
Hennessy–Milner Logic (HML), 594

image-finite, 559

labelled transition system (LTS), 551
language, 556
language equivalence, 557
Linear Temporal Logic (LTL), 600
Linear-Time μ-Calculus (LTμ), 601

macro processes, 609
Minskey machines, 586
Modal μ-calculus, 596
multiset automata (MSA), 555

nesting depth, 595
norm, 557

one-counter automata, 565
one-counter nets, 565

parallel labelled rewrite transition, 551
path, 594
Petri nets (PN), 556
positive normal form, 595
Presburger Computation Tree Logic (PCTL), 598
prime process, 566
Process Algebra (PA), 565
push-down automata (PDA), 554

REC_{Rat} graphs, 608
regular graphs, 607
run, 594

sequential labelled rewrite transition system, 551
simple grammars, 576
simulation equivalence, 590
simulation preorder, 590
simulation relation, 590
stratified bisimulation relations, 559

UB, 595
unique decomposition base, 576
Unique Factorization Theorem
– for normed BPA, 567
– for normed BPP, 567
Unique solutions lemma, 569

weak bisimulation equivalence, 589
Weak Linear Temporal Logic (WL), 599

Part 4
Extensions

CHAPTER 10

Process Algebra with Timing: Real Time and Discrete Time

J.C.M. Baeten[1], C.A. Middelburg[1,2]

[1] *Computing Science Department, Eindhoven University of Technology, P.O. Box 513, 5600 MB Eindhoven, The Netherlands*

[2] *Department of Philosophy, Utrecht University, P.O. Box 80126, 3508 TC Utrecht, The Netherlands*

E-mails: josb@win.tue.nl, keesm@win.tue.nl

Contents

1. Introduction 629
2. Real time process algebra: absolute timing 631
   2.1. Basic process algebra 632
   2.2. Algebra of communicating processes 640
   2.3. Time-stamped actions 643
3. Extension of ACP^{sat} 645
   3.1. Integration 646
   3.2. Initial abstraction 648
   3.3. Standard initialization axioms 652
4. Real time process algebra: relative timing 652
   4.1. Basic process algebra 652
   4.2. Algebra of communicating processes 656
   4.3. Embedding ACP^{srt} in $ACP^{sat}\surd$ 657
5. Discrete time process algebra 660
   5.1. Discrete time process algebra: absolute timing 661
   5.2. Discrete time process algebra: relative timing 666
   5.3. Embedding $ACP^{dat}\surd$ in $ACP^{sat}I\surd$ 669
6. Concluding remarks 673
A. Proofs of theorems 674
   A1. Theorem 6 674
   A2. Theorem 12 678
References 681
Subject index 683

HANDBOOK OF PROCESS ALGEBRA
Edited by Jan A. Bergstra, Alban Ponse and Scott A. Smolka

Abstract

We present real time and discrete time versions of ACP with absolute timing and relative timing. The starting-point is a new real time version with absolute timing, called ACP^{sat}, featuring urgent actions and a delay operator. The discrete time versions are conservative extensions of the discrete time versions of ACP being known as ACP_{dat} and ACP_{drt}. The principal version is an extension of ACP^{sat} with integration and initial abstraction to allow for choices over an interval of time and relative timing to be expressed. Its main virtue is that it generalizes ACP without timing and most other versions of ACP with timing in a smooth and natural way. This is shown for the real time version with relative timing and the discrete time version with absolute timing.

1. Introduction

Algebraic concurrency theories such as ACP [13,11,10], CCS [33,34] and CSP [18, 28] have been extended to deal with time-dependent behaviour in various ways. First of all, timing is either absolute or relative and the time scale on which time is measured is either continuous or discrete. Besides, execution of actions and passage of time are either separated or combined. Separation corresponds to the two-phase scheme of modeling time-dependent behaviour and combination corresponds to the time-stamping scheme.

Absolute timing and relative timing have been studied in the framework of ACP for both a continuous time scale and a discrete time scale. See, e.g., [2] and [7]. The versions of ACP with timing where time is measured on a continuous time scale are usually called real time versions. In the remainder of this chapter, we adhere to this terminology. In the principal real time versions of ACP, viz. ACPρ and ACPrρ, which were both introduced in [2], and ACP$\rho\sigma$, which was introduced in [4], execution of actions and passage of time are combined. On the contrary, they are separated in the principal discrete time versions of ACP, viz. ACP$_{\mathrm{dat}}$ and ACP$_{\mathrm{drt}}$, which were both introduced in [7]. A real time version where execution of actions and passage of time are separated is ACPst, which was introduced in [6], and [8] focusses on discrete time versions where they are combined.

Measuring time on a discrete time scale does not mean that the execution of actions is restricted to discrete points in time. In the discrete time versions of ACP, time is divided into time slices and timing of actions is done with respect to the time slices in which they are performed – within a time slice there is only the order in which actions are performed. Thus, the discrete time versions permit to consider systems at a more abstract level than the real time case, a level where time is measured with finite precision. This also occurs in practice: software components of a system are executed on processors where the measure of time is provided by a discrete clock and, in case a physical system is controlled, the state of the physical system is sampled and adjusted at discrete points in time. In any case, the abstraction made in the discrete time versions makes the time-dependent behaviour of programs better amenable to analysis.

ACP can simply be embedded in the discrete time versions ACP$_{\mathrm{dat}}$ and ACP$_{\mathrm{drt}}$ [7] by projecting the untimed process a (for each action a) onto the delayable process a – a delayable process a is capable of performing the action a in any time slice. Similarly, ACP can be embedded in the real time versions ACPρ and ACPrρ [2]. In other words, these discrete time and real time theories generalize the time free theory smoothly. Furthermore, in the discrete time case as well as the real time case, the relative time version can simply be embedded in the absolute time version extended with an initial abstraction operator to deal with relative timing.

However, the real time versions do not generalize the discrete time versions as smoothly as they generalize the time free theory. It turns out, as shown in [3], that the discrete time processes correspond to the real time processes for which the following holds: (1) if an action can be performed at some time $p \in \mathbb{R}$ such that $n < p < n + 1$ ($n \in \mathbb{N}$), it can also be performed at any other time $p' \in \mathbb{R}$ such that $n < p' < n + 1$; (2) no actions can be performed at times $p \in \mathbb{N}$. Clearly, such an embedding seriously lacks naturalness. The real time versions ACPρ and ACPrρ as well as the discrete time versions ACP$_{\text{dat}}$ and ACP$_{\text{drt}}$ are generalizations of ACP by intention. Since the real time versions were developed in advance of the discrete time versions, the former versions were not intentionally developed as generalizations of the latter versions. This explains at least partially the contrived embedding.

In this chapter, we present a new real time version of ACP with absolute timing which originates from ACPsρ, a real time version introduced in [6]. In this version, which features urgent actions and a delay operator, execution of actions and passage of time are separated. We explain how execution of actions and passage of time can be combined in this version. We further add an integration operator, with which a choice over an interval of time can be expressed, and an initial abstraction operator, with which relative timing can be expressed, to this version. We show how a real time version of ACP with relative timing, which originates from ACPst [6], can be embedded in the extended real time version with absolute timing. We also present discrete time versions of ACP with absolute timing and relative timing which are conservative extensions of ACP$_{\text{dat}}$ and ACP$_{\text{drt}}$ [7]. We add an initial abstraction operator to the discrete time version with absolute timing as well. Showing how the discrete time version with relative timing can be embedded in the extended discrete time version with absolute timing, can be done similarly to the real time case. We show that the extended real time version generalizes the extended discrete time version smoothly. In this case, the following holds for those real time processes that correspond to the discrete time processes: if an action can be performed at some time $p \in \mathbb{R}$ such that $n \leqslant p < n + 1$ ($n \in \mathbb{N}$), it can also be performed at any other time $p' \in \mathbb{R}$ such that $n \leqslant p' < n + 1$.

The main virtue of the extended real time version of ACP presented here is that it generalizes time free ACP as well as most other versions of ACP with timing in a smooth and natural way. The lack of a real time version of ACP with these characteristics was our main motivation to develop it. Different from the real time versions of [2] and [4], this version does not exclude the possibility of two or more actions to be performed consecutively at the same point in time. That is, it includes urgent actions, similar to ATP [37] and the different versions of CCS with timing [19,35,44]. This is useful in practice when describing and analyzing systems in which actions occur that are entirely independent. This is, for example, the case for actions that happen at different locations in a distributed system. In [2] and [4], the main idea was that it is difficult to imagine that actions are performed consecutively at the same point in time. But yet, this way of representing things is perfectly in line with modeling parallelism by interleaving. In point of fact it allows for independent actions to be handled faithfully.

In [2] and [4], ways to deal with independent actions are proposed where such actions take place at the same point in time by treating it as a special case of communication. This is, however, a real burden in the description and the analysis of the systems concerned. Of course, this does not limit the practical usefulness of ACPρ and ACPrρ for systems in which no independent actions occur. The real time versions ACPsρ and ACPst of [6] simply do not exclude the possibility of two or more actions to be performed consecutively at the same point in time. Embedding in ACPρ and ACPrρ, respectively, is obtained by extending the time domain to a domain that includes non-standard real numbers. We conjecture that the real time version presented in this chapter, which originates from ACPsρ, can be embedded in ACPρ as well.

We do not intend to give in this chapter a comprehensive overview of existing algebraic concurrency theories that deal with time-dependent behaviour. As suggested by the above, our aim is instead to present a coherent collection of algebraic concurrency theories that deal with time-dependent behaviour in different ways.

The structure of this chapter is as follows. First of all, in Section 2, we present the new real time version of ACP with absolute timing. We also explain how execution of actions and passage of time can be combined in this version. Then, in Section 3, we add integration and initial abstraction to this real time version of ACP. Next, in Section 4, we first present a real time version of ACP with relative timing and then show that it can be embedded in the real time version of ACP with absolute timing presented in Sections 2 and 3. After that, in Section 5, we first present conservative extensions of the discrete time versions ACP$_{\mathrm{dat}}$ and ACP$_{\mathrm{drt}}$ of [7] and then show that the presented discrete time version with absolute timing can be embedded in the real time version with absolute timing presented in Sections 2 and 3. Finally, in Section 6, we make some concluding remarks.

2. Real time process algebra: absolute timing

In this section, we give the signature, axioms and term model of ACP^{sat}, a standard real time process algebra with absolute timing. In this theory, the non-negative standard real numbers ($\mathbb{R}_{\geqslant 0}$) are used as the time domain. ACP^{sat} originates from the theory ACPsρ, presented in [6]. Unlike ACPsρ, it separates execution of actions and passage of time.

In case of ACP^{sat}, it is assumed that a theory of the non-negative real numbers has been given. Its signature has to include the constant $0 : \to \mathbb{R}_{\geqslant 0}$, the operator $+ : \mathbb{R}_{\geqslant 0} \times \mathbb{R}_{\geqslant 0} \to \mathbb{R}_{\geqslant 0}$, and the predicates $\leqslant : \mathbb{R}_{\geqslant 0} \times \mathbb{R}_{\geqslant 0}$ and $= : \mathbb{R}_{\geqslant 0} \times \mathbb{R}_{\geqslant 0}$. In addition, this theory has to include axioms that characterize $+$ as a commutative and associative operation with 0 as a neutral element and $\leqslant$ as a total ordering that has 0 as its least element and that is preserved by $+$.

In ACP^{sat}, as in the other versions of ACP with timing presented in this chapter, it is assumed that a fixed but arbitrary set A of *actions* has been given. It is also assumed that a fixed but arbitrary *communication function*, i.e. a partial commutative and associative function $\gamma : \mathsf{A} \times \mathsf{A} \to \mathsf{A}$, has been given. The function γ is regarded to give the result of the synchronous execution of any two actions for which this is possible, and to be undefined otherwise. The weak restrictions on γ allow many kinds of communication between parallel processes to be modeled.

First, in Section 2.1, we treat BPA^{sat}, basic standard real time process algebra with absolute timing, in which parallelism and communication are not considered. After that, in Section 2.2, BPA^{sat} is extended to ACP^{sat} to deal with parallelism and communication as well. Finally, we demonstrate in Section 2.3 how one can combine execution of actions and passage of time in ACP^{sat}.

2.1. *Basic process algebra*

In $\mathrm{BPA}^{\mathrm{sat}}$, we have the sort P of (absolute time) processes, the constants $\tilde{a}$ (one for each $a \in \mathsf{A}$), $\tilde{\delta}$ and $\dot{\delta}$, and the operators σ_{abs} (absolute delay), $\cdot$ (sequential composition) and $+$ (alternative composition). The constants $\tilde{a}$ stand for a at time 0. Similarly, the constant $\tilde{\delta}$ stands for a deadlock at time 0. The constant $\dot{\delta}$ stands for an immediate deadlock, a process that exhibits inconsistent timing at time 0. This means that $\dot{\delta}$, different from $\tilde{\delta}$, is not existing at time 0. The process $\sigma^{p}_{\mathsf{abs}}(x)$ is the process x shifted in time by p. Thus, the process $\sigma^{p}_{\mathsf{abs}}(\tilde{a})$ is capable of first idling from time 0 to time p and then upon reaching time p performing action a, immediately followed by successful termination. The process $\sigma^{p}_{\mathsf{abs}}(\tilde{\delta})$ is only capable of idling from time 0 to time p. Time p can be reached by $\sigma^{p}_{\mathsf{abs}}(\tilde{\delta})$. This is the difference with the process $\sigma^{p}_{\mathsf{abs}}(\dot{\delta})$, which can only idle upto, but not including, time p. So $\sigma^{p}_{\mathsf{abs}}(\dot{\delta})$ can not reach time p. The process $x \cdot y$ is the process x followed upon successful termination by the process y. The process $x + y$ is the process that proceeds with either the process x or the process y, but not both. As in the untimed case, the choice is resolved upon execution of the first action, and not before. We also have the auxiliary operators υ_{abs} (absolute time-out) and $\overline{\upsilon}_{\mathsf{abs}}$ (absolute initialization). The process $\upsilon^{p}_{\mathsf{abs}}(x)$ is the part of x that starts to perform actions before time p. The process $\overline{\upsilon}^{p}_{\mathsf{abs}}(x)$ is the part of x that starts to perform actions at time p or later.

A real time version of ACP with absolute timing where the notation $\tilde{a}$ was used earlier for urgent actions is ACPsρ [6], but there it always carries a time-stamp. The binary operator σ_{abs} generalizes the unary operator σ_{abs} of $\mathrm{ACP}_{\mathrm{dat}}$ [7] in a real time setting: for a real time process x that corresponds to a discrete time process x', $\sigma^{1}_{\mathsf{abs}}(x)$ corresponds to $\sigma_{\mathsf{abs}}(x')$. In earlier papers, including [2–4,6], the notations $x \gg p$ and $p \gg x$ were used instead of $\upsilon^{p}_{\mathsf{abs}}(x)$ and $\overline{\upsilon}^{p}_{\mathsf{abs}}(x)$, respectively. Besides, the time-out operator and the initialization operator were sometimes called the bounded initialization operator and the time shift operator, respectively.

It can be proved, using the axioms of $\mathrm{BPA}^{\mathrm{sat}}$, that each process expressed using the auxiliary operators υ_{abs} and $\overline{\upsilon}_{\mathsf{abs}}$ is equal to a process expressed without them. In other words, in $\mathrm{BPA}^{\mathrm{sat}}$, all processes can be constructed from the constants using absolute delay, alternative composition and sequential composition only.

SIGNATURE OF $\mathrm{BPA}^{\mathrm{sat}}$. The signature of $\mathrm{BPA}^{\mathrm{sat}}$ consists of the *urgent action* constants $\tilde{a} : \to \mathsf{P}$ (for each $a \in \mathsf{A}$), the *urgent deadlock* constant $\tilde{\delta} : \to \mathsf{P}$, the *immediate deadlock* constant $\dot{\delta} : \to \mathsf{P}$, the *alternative composition* operator $+ : \mathsf{P} \times \mathsf{P} \to \mathsf{P}$, the *sequential composition* operator $\cdot : \mathsf{P} \times \mathsf{P} \to \mathsf{P}$, the *absolute delay* operator $\sigma_{\mathsf{abs}} : \mathbb{R}_{\geqslant 0} \times \mathsf{P} \to \mathsf{P}$, the *absolute time-out* operator $\upsilon_{\mathsf{abs}} : \mathbb{R}_{\geqslant 0} \times \mathsf{P} \to \mathsf{P}$, and the *absolute initialization* operator $\overline{\upsilon}_{\mathsf{abs}} : \mathbb{R}_{\geqslant 0} \times \mathsf{P} \to \mathsf{P}$.

We assume that an infinite set of variables (of sort P) has been given. Given the signature of $\mathrm{BPA}^{\mathrm{sat}}$, terms of $\mathrm{BPA}^{\mathrm{sat}}$, also referred to as process expressions, are constructed in the usual way. We will in general use infix notation for binary operators. The need to use parentheses is further reduced by ranking the precedence of the binary operators. Throughout this chapter we adhere to the following precedence rules: (i) the operator $\cdot$

has the highest precedence amongst the binary operators, (ii) the operator + has the lowest precedence amongst the binary operators, and (iii) all other binary operators have the same precedence. We will also use the following abbreviation. Let $(t_i)_{i \in \mathcal{I}}$ be an indexed set of terms of BPA^{sat} where $\mathcal{I} = \{i_1, \ldots, i_n\}$. Then we write $\sum_{i \in \mathcal{I}} t_i$ for $t_{i_1} + \cdots + t_{i_n}$. We further use the convention that $\sum_{i \in \mathcal{I}} t_i$ stands for $\dot{\delta}$ if $\mathcal{I} = \emptyset$.

We denote variables by $x, x', y, y', \ldots$. An important convention is that we use $a, a', b, b', \ldots$ to denote elements of $\mathsf{A} \cup \{\delta\}$ in the context of an equation, and elements of A in the context of an operational semantics rule. Furthermore, we use H to denote a subset of A. We denote elements of $\mathbb{R}_{\geqslant 0}$ by p, p', q, q' and elements of $\mathbb{R}_{>0}$ by r, r'. We write A_δ for $\mathsf{A} \cup \{\delta\}$.

AXIOMS OF BPA^{sat}. The axiom system of BPA^{sat} consists of the equations given in Tables 1 and 2.

Axioms A1–A5 are common to ACP and all real and discrete time versions of ACP. Axioms A6ID and A7ID are simple reformulations of the axioms A6 and A7 of ACP: the constant δ has been replaced by the constant $\dot{\delta}$ – which is introduced because the intended interpretation of δ in ACP^{sat} differs from $\dot{\delta}$. These axioms or similar reformulations of A6 and A7 are found in all real and discrete time versions of ACP. Axiom SAT1, and a few axioms treated later, become easier to understand by realizing that in BPA^{sat}, as well as in ACP^{sat}, the equation $\overline{\upsilon}^{0}_{\mathsf{abs}}(t) = t$ is derivable for all closed terms t. This equation expresses that initialization at time 0 has no effect on processes with absolute timing. To accommodate for the extension with initial abstraction in Section 3.2, we have used $\overline{\upsilon}^{0}_{\mathsf{abs}}(x)$ instead of x where the former is needed in the extension. Axioms SAT1 and SAT2 point out that a time shift by 0 has no effect in case of absolute timing and that consecutive time shifts add up. Axioms SAT3–SAT5 represent the interaction of absolute delay with alternative composition and sequential composition. Axiom SAT3, called the time factorization axiom, shows that passage of time by itself can not determine a choice. Axioms SAT4 and SAT5 express that if a process terminates successfully at some point in time, it can only be followed by the part of another process that starts to perform actions at the same time or later. Axiom SAT6 is a generalization of axiom A7ID. Using axioms A6SAa and A6SAb, the equation $t + \tilde{\delta} = t$ can be derived for all closed terms t unless $t = \dot{\delta}$ –

Table 1
Axioms of BPA with immediate deadlock

$x + y = y + x$	A1
$(x + y) + z = x + (y + z)$	A2
$x + x = x$	A3
$(x + y) \cdot z = x \cdot z + y \cdot z$	A4
$(x \cdot y) \cdot z = x \cdot (y \cdot z)$	A5
$x + \dot{\delta} = x$	A6ID
$\dot{\delta} \cdot x = \dot{\delta}$	A7ID

Table 2
Additional axioms for $\mathrm{BPA}^{\mathrm{sat}}$ ($a \in \mathsf{A}_\delta,\ p, q \geqslant 0,\ r > 0$)

$\sigma^0_{\mathsf{abs}}(x) = \overline{\upsilon}^0_{\mathsf{abs}}(x)$	SAT1
$\sigma^p_{\mathsf{abs}}(\sigma^q_{\mathsf{abs}}(x)) = \sigma^{p+q}_{\mathsf{abs}}(x)$	SAT2
$\sigma^p_{\mathsf{abs}}(x) + \sigma^p_{\mathsf{abs}}(y) = \sigma^p_{\mathsf{abs}}(x + y)$	SAT3
$\sigma^p_{\mathsf{abs}}(x) \cdot \upsilon^p_{\mathsf{abs}}(y) = \sigma^p_{\mathsf{abs}}(x \cdot \dot{\delta})$	SAT4
$\sigma^p_{\mathsf{abs}}(x) \cdot (\upsilon^p_{\mathsf{abs}}(y) + \sigma^p_{\mathsf{abs}}(z)) = \sigma^p_{\mathsf{abs}}(x \cdot \overline{\upsilon}^0_{\mathsf{abs}}(z))$	SAT5
$\sigma^p_{\mathsf{abs}}(\dot{\delta}) \cdot x = \sigma^p_{\mathsf{abs}}(\dot{\delta})$	SAT6
$\tilde{a} + \tilde{\delta} = \tilde{a}$	A6SAa
$\sigma^r_{\mathsf{abs}}(x) + \tilde{\delta} = \sigma^r_{\mathsf{abs}}(x)$	A6SAb
$\tilde{\delta} \cdot x = \tilde{\delta}$	A7SA
$\upsilon^p_{\mathsf{abs}}(\dot{\delta}) = \dot{\delta}$	SATO0
$\upsilon^0_{\mathsf{abs}}(x) = \dot{\delta}$	SATO1
$\upsilon^r_{\mathsf{abs}}(\tilde{a}) = \tilde{a}$	SATO2
$\upsilon^{p+q}_{\mathsf{abs}}(\sigma^p_{\mathsf{abs}}(x)) = \sigma^p_{\mathsf{abs}}(\upsilon^q_{\mathsf{abs}}(x))$	SATO3
$\upsilon^p_{\mathsf{abs}}(x + y) = \upsilon^p_{\mathsf{abs}}(x) + \upsilon^p_{\mathsf{abs}}(y)$	SATO4
$\upsilon^p_{\mathsf{abs}}(x \cdot y) = \upsilon^p_{\mathsf{abs}}(x) \cdot y$	SATO5
$\overline{\upsilon}^0_{\mathsf{abs}}(\dot{\delta}) = \dot{\delta}$	SAI0a
$\overline{\upsilon}^r_{\mathsf{abs}}(\dot{\delta}) = \sigma^r_{\mathsf{abs}}(\dot{\delta})$	SAI0b
$\overline{\upsilon}^0_{\mathsf{abs}}(\tilde{a}) = \tilde{a}$	SAI1
$\overline{\upsilon}^r_{\mathsf{abs}}(\tilde{a}) = \sigma^r_{\mathsf{abs}}(\dot{\delta})$	SAI2
$\overline{\upsilon}^{p+q}_{\mathsf{abs}}(\sigma^p_{\mathsf{abs}}(x)) = \sigma^p_{\mathsf{abs}}(\overline{\upsilon}^q_{\mathsf{abs}}(\overline{\upsilon}^0_{\mathsf{abs}}(x)))$	SAI3
$\overline{\upsilon}^p_{\mathsf{abs}}(x + y) = \overline{\upsilon}^p_{\mathsf{abs}}(x) + \overline{\upsilon}^p_{\mathsf{abs}}(y)$	SAI4
$\overline{\upsilon}^p_{\mathsf{abs}}(x \cdot y) = \overline{\upsilon}^p_{\mathsf{abs}}(x) \cdot y$	SAI5

obviously $\dot{\delta} + \tilde{\delta} = \tilde{\delta}$. Axiom A7SA is another simple reformulation of axiom A7 of ACP. Axioms SATO0–SATO5 and SAI0–SAI5 reflect the intended meaning of the time-out and initialization operators clearly. Axioms SATO1 and SAI2 make precise what happens if a part that starts to perform actions before the time-out time and a part that starts to perform actions at the initialization time or later, respectively, do not exist. Equations SATO3′ and SAI3′ given in Table 3 are derivable from the axioms of $\mathrm{BPA}^{\mathrm{sat}}$. In $\mathrm{BPA}^{\mathrm{sat}}$ and $\mathrm{ACP}^{\mathrm{sat}}$, and also in the further extension with initial abstraction, axiom SATO1 can be replaced by equation SATO3′ just as well. In $\mathrm{BPA}^{\mathrm{sat}}$ and $\mathrm{ACP}^{\mathrm{sat}}$, but not in the further extension with initial abstraction, axioms SAI0a, SAI1 and SAI3 together can be replaced by the equations SAI1″ and SAI3″ given in Table 3. The absolute initialization operator could have been added later with the addition of the initial abstraction operator. However, having it available in $\mathrm{BPA}^{\mathrm{sat}}$ and $\mathrm{ACP}^{\mathrm{sat}}$ makes it possible to express interesting properties of real time processes with absolute timing such as the properties presented in Lemmas 1 and 3 below.

We can prove that the auxiliary operators υ_{abs} and $\overline{\upsilon}_{\mathsf{abs}}$ can be eliminated in closed terms of $\mathrm{BPA}^{\mathrm{sat}}$. We can also prove that sequential compositions in which the form of the

Table 3
Some derivable equations and alternative axioms ($p, q \geqslant 0$)

$\upsilon^{p}_{\mathsf{abs}}(\sigma^{p+q}_{\mathsf{abs}}(x)) = \sigma^{p}_{\mathsf{abs}}(\dot{\delta})$	SATO3′
$\overline{\upsilon}^{p}_{\mathsf{abs}}(\sigma^{p+q}_{\mathsf{abs}}(x)) = \sigma^{p+q}_{\mathsf{abs}}(x)$	SAI3′
$\overline{\upsilon}^{0}_{\mathsf{abs}}(x) = x$	SAI1″
$\overline{\upsilon}^{p+q}_{\mathsf{abs}}(\sigma^{p}_{\mathsf{abs}}(x)) = \sigma^{p}_{\mathsf{abs}}(\overline{\upsilon}^{q}_{\mathsf{abs}}(x))$	SAI3″

first operand is not $\tilde{a}$ ($a \in \mathsf{A}$) and alternative compositions in which the form of the first operand is $\sigma^{p}_{\mathsf{abs}}(t)$ can be eliminated in closed terms of BPA^{sat}. The terms that remain after exhaustive elimination are called the *basic terms* over BPA^{sat}. Because of this elimination result, we are permitted to use induction on the structure of basic terms over BPA^{sat} to prove statements for all closed terms of BPA^{sat}.

EXAMPLES. We give some examples of a closed term of BPA^{sat} and the corresponding basic term:

$$\sigma^{5}_{\mathsf{abs}}(\tilde{a}) \cdot \sigma^{4.9}_{\mathsf{abs}}(\tilde{b}) = \sigma^{5}_{\mathsf{abs}}(\tilde{a} \cdot \dot{\delta})$$
$$\sigma^{5}_{\mathsf{abs}}(\tilde{a}) \cdot (\sigma^{4.9}_{\mathsf{abs}}(\tilde{b}) + \sigma^{5.1}_{\mathsf{abs}}(\tilde{c})) = \sigma^{5}_{\mathsf{abs}}(\tilde{a} \cdot \sigma^{0.1}_{\mathsf{abs}}(\tilde{c}))$$
$$\upsilon^{5}_{\mathsf{abs}}(\sigma^{4.9}_{\mathsf{abs}}(\tilde{a}) + \sigma^{5.1}_{\mathsf{abs}}(\tilde{b})) = \sigma^{4.9}_{\mathsf{abs}}(\tilde{a} + \sigma^{0.1}_{\mathsf{abs}}(\dot{\delta}))$$
$$\overline{\upsilon}^{5}_{\mathsf{abs}}(\sigma^{4.9}_{\mathsf{abs}}(\tilde{a}) + \sigma^{5.1}_{\mathsf{abs}}(\tilde{b})) = \sigma^{5.1}_{\mathsf{abs}}(\tilde{b})$$

The following lemmas are also useful in proofs. They are, for example, used in the proof of Theorem 12 (embedding of ACP$^{\text{dat}}\surd$ in ACP^{sat}I$\surd$). These lemmas, as most other lemmas in this chapter, call for proofs by induction on the structure of basic terms. The proofs are generally straightforward, but long and tedious. For that reason, we will present for each such proof only one of the cases to be treated. The selected case is usually typical of the proof and relatively hard. We write $\stackrel{\text{IH}}{=}$ to indicate that the induction hypothesis of the proof is used.

LEMMA 1. *In* BPA^{sat} *and* ACP^{sat}, *as well as in the further extensions with restricted integration and initial abstraction introduced in Section* 3:
1. *the equation* $t = \upsilon^{p}_{\mathsf{abs}}(t) + \overline{\upsilon}^{p}_{\mathsf{abs}}(t)$ *is derivable for all closed terms* t *such that* $t = \overline{\upsilon}^{0}_{\mathsf{abs}}(t)$ *and* $t = t + \sigma^{p}_{\mathsf{abs}}(\dot{\delta})$;
2. *the equations* $t = \upsilon^{p}_{\mathsf{abs}}(t)$ *and* $\overline{\upsilon}^{p}_{\mathsf{abs}}(t) = \sigma^{p}_{\mathsf{abs}}(\dot{\delta})$ *are derivable for all closed terms* t *such that* $t = \overline{\upsilon}^{0}_{\mathsf{abs}}(t)$ *and* $t \neq t + \sigma^{p}_{\mathsf{abs}}(\dot{\delta})$.

PROOF. It is straightforward to prove both 1 and 2 by induction on the structure of t.

1. We present only the case that t is of the form $\sigma^{q}_{\mathsf{abs}}(t')$. The other cases are similar, but simpler, and do not require case distinction.

Case $p \leqslant q$: $\sigma_{\mathrm{abs}}^{q}(t') + \sigma_{\mathrm{abs}}^{p}(\dot{\delta}) \stackrel{\mathrm{A1}}{=} \sigma_{\mathrm{abs}}^{p}(\dot{\delta}) + \sigma_{\mathrm{abs}}^{q}(t') \stackrel{\mathrm{SATO3}',\mathrm{SAI3}'}{=} \upsilon_{\mathrm{abs}}^{p}(\sigma_{\mathrm{abs}}^{q}(t')) + \overline{\upsilon}_{\mathrm{abs}}^{p}(\sigma_{\mathrm{abs}}^{q}(t'))$

Case $p > q$: $\sigma_{\mathrm{abs}}^{q}(t') + \sigma_{\mathrm{abs}}^{p}(\dot{\delta}) \stackrel{\mathrm{SAT2}}{=} \sigma_{\mathrm{abs}}^{q}(t') + \sigma_{\mathrm{abs}}^{q}(\sigma_{\mathrm{abs}}^{p-q}(\dot{\delta})) \stackrel{\mathrm{SAT3}}{=} \sigma_{\mathrm{abs}}^{q}(t' + \sigma_{\mathrm{abs}}^{p-q}(\dot{\delta})) \stackrel{\mathrm{IH}}{=}$
$\sigma_{\mathrm{abs}}^{q}(\upsilon_{\mathrm{abs}}^{p-q}(t') + \overline{\upsilon}_{\mathrm{abs}}^{p-q}(t')) \stackrel{\mathrm{SAT3}}{=} \sigma_{\mathrm{abs}}^{q}(\upsilon_{\mathrm{abs}}^{p-q}(t')) + \sigma_{\mathrm{abs}}^{q}(\overline{\upsilon}_{\mathrm{abs}}^{p-q}(t')) \stackrel{\mathrm{SATO3},\mathrm{SAI3}}{=}$
$\upsilon_{\mathrm{abs}}^{p}(\sigma_{\mathrm{abs}}^{q}(t')) + \overline{\upsilon}_{\mathrm{abs}}^{p}(\sigma_{\mathrm{abs}}^{q}(t'))$

In applying SAI3 we assume that $t' = \overline{\upsilon}_{\mathrm{abs}}^{0}(t')$. In case of BPA^{sat}, ACP^{sat} and ACP^{sat} with integration, this equation is derivable for all closed terms t'. The assumption is also justified in case of extension with initial abstraction. In that case, we are permitted, because of elimination results presented in Section 3.2, to consider here only closed terms of the form $\sigma_{\mathrm{abs}}^{q}(t')$ where no initial abstraction occurs in t'.

2. Observe that $\overline{\upsilon}_{\mathrm{abs}}^{p}(t) = \sigma_{\mathrm{abs}}^{p}(\dot{\delta})$ follows immediately from $t = \upsilon_{\mathrm{abs}}^{p}(t)$ by axiom SI3. So it suffices to prove only $t = \upsilon_{\mathrm{abs}}^{p}(t)$. Again, we present only the case that t is of the form $\sigma_{\mathrm{abs}}^{q}(t')$.

Case $p \leqslant q$: $\sigma_{\mathrm{abs}}^{q}(t') \stackrel{\mathrm{SAT2}}{=} \sigma_{\mathrm{abs}}^{p}(\sigma_{\mathrm{abs}}^{q-p}(t')) \stackrel{\mathrm{A6ID}}{=} \sigma_{\mathrm{abs}}^{p}(\sigma_{\mathrm{abs}}^{q-p}(t') + \dot{\delta}) \stackrel{\mathrm{SAT3}}{=}$
$\sigma_{\mathrm{abs}}^{p}(\sigma_{\mathrm{abs}}^{q-p}(t')) + \sigma_{\mathrm{abs}}^{p}(\dot{\delta}) \stackrel{\mathrm{SAT2}}{=} \sigma_{\mathrm{abs}}^{q}(t') + \sigma_{\mathrm{abs}}^{p}(\dot{\delta})$
So $\sigma_{\mathrm{abs}}^{q}(t') \neq \sigma_{\mathrm{abs}}^{q}(t') + \sigma_{\mathrm{abs}}^{p}(\dot{\delta})$ does not hold in case $p \leqslant q$

Case $p > q$: $\sigma_{\mathrm{abs}}^{q}(t') \neq \sigma_{\mathrm{abs}}^{q}(t') + \sigma_{\mathrm{abs}}^{p}(\dot{\delta}) \stackrel{\mathrm{SAT2},\mathrm{SAT3}}{\Rightarrow} t' \neq t' + \sigma_{\mathrm{abs}}^{p-q}(\dot{\delta})$
By the induction hypothesis, $\sigma_{\mathrm{abs}}^{q}(t') = \sigma_{\mathrm{abs}}^{q}(\upsilon_{\mathrm{abs}}^{p-q}(t')) \stackrel{\mathrm{SATO3}}{=} \upsilon_{\mathrm{abs}}^{p}(\sigma_{\mathrm{abs}}^{q}(t'))$ □

From Lemma 1 we readily conclude the following.

COROLLARY 2. *In* BPA^{sat} *and* ACP^{sat}, *as well as in the further extensions with restricted integration and initial abstraction introduced in Section* 3, *the equation* $\sigma_{\mathrm{abs}}^{p}(t) \cdot t' = \sigma_{\mathrm{abs}}^{p}(t) \cdot \overline{\upsilon}_{\mathrm{abs}}^{p}(t')$ *is derivable for all closed terms* t *and* t' *such that* $t' = \overline{\upsilon}_{\mathrm{abs}}^{0}(t')$.

LEMMA 3. *In* BPA^{sat} *and* ACP^{sat}, *as well as in the further extensions with restricted integration and initial abstraction introduced in Section* 3, *for each* $p \in \mathbb{R}_{\geqslant 0}$ *and each closed term* t, *there exists a closed term* t' *such that* $\overline{\upsilon}_{\mathrm{abs}}^{p}(t) = \sigma_{\mathrm{abs}}^{p}(t')$ *and* $t' = \overline{\upsilon}_{\mathrm{abs}}^{0}(t')$. *In subsequent proofs, we write* $t_{[p]}$ *for a fixed but arbitrary closed term* t' *that fulfills these conditions.*

PROOF. It is straightforward to prove this by induction on the structure of t. We present only the case that t is of the form $\sigma_{\mathrm{abs}}^{q}(t'')$. Again, the other cases are similar, but simpler, and do not require case distinction.

Case $p \leqslant q$: $\overline{\upsilon}_{\mathrm{abs}}^{p}(\sigma_{\mathrm{abs}}^{q}(t'')) \stackrel{\mathrm{SAI3}'}{=} \sigma_{\mathrm{abs}}^{q}(t'') \stackrel{\mathrm{SAT2}}{=} \sigma_{\mathrm{abs}}^{p}(\sigma_{\mathrm{abs}}^{q-p}(t''))$ and
$\sigma_{\mathrm{abs}}^{q-p}(t'') \stackrel{\mathrm{SAT2}}{=} \sigma_{\mathrm{abs}}^{0}(\sigma_{\mathrm{abs}}^{q-p}(t'')) \stackrel{\mathrm{SAT1}}{=} \overline{\upsilon}_{\mathrm{abs}}^{0}(\sigma_{\mathrm{abs}}^{q-p}(t''))$

Case $p > q$: $\overline{\upsilon}_{\mathrm{abs}}^{p}(\sigma_{\mathrm{abs}}^{q}(t'')) \stackrel{\mathrm{SAI3}}{=} \sigma_{\mathrm{abs}}^{q}(\overline{\upsilon}_{\mathrm{abs}}^{p-q}(t'')) \stackrel{\mathrm{IH}}{=} \sigma_{\mathrm{abs}}^{q}(\sigma_{\mathrm{abs}}^{p-q}(t''_{[p-q]})) \stackrel{\mathrm{SAT2}}{=} \sigma_{\mathrm{abs}}^{p}(t''_{[p-q]})$ and
$t''_{[p-q]} = \overline{\upsilon}_{\mathrm{abs}}^{0}(t''_{[p-q]})$

In applying SAI3 we assume that $t'' = \overline{\upsilon}_{\mathrm{abs}}^{0}(t'')$. As described in the previous proof, this assumption is justified in all cases. □

Lemma 1 indicates that a process that is able to reach time p can be regarded as being the alternative composition of the part that starts to perform actions before p and the part that starts to perform actions at p or later. Lemma 3 shows that the part of a process that starts to perform actions at time p or later can always be regarded as a process shifted in time by p.

SEMANTICS OF BPA^{sat}. A *real time transition system* over A consists of a set of *states* S, a *root state* $\rho \in \mathsf{S}$ and four kinds of relations on states:

a binary relation	$\langle\_, p\rangle \xrightarrow{a} \langle\_, p\rangle$	for each $a \in \mathsf{A}$, $p \in \mathbb{R}_{\geqslant 0}$,
a unary relation	$\langle\_, p\rangle \xrightarrow{a} \langle\surd, p\rangle$	for each $a \in \mathsf{A}$, $p \in \mathbb{R}_{\geqslant 0}$,
a binary relation	$\langle\_, p\rangle \overset{r}{\mapsto} \langle\_, q\rangle$	for each $r \in \mathbb{R}_{>0}$, $p, q \in \mathbb{R}_{\geqslant 0}$ where $q = p + r$,
a unary relation	$\mathsf{ID}(\_, p)$	for each $p \in \mathbb{R}_{\geqslant 0}$;

satisfying

(1) if $\langle s, p\rangle \overset{r+r'}{\longmapsto} \langle s', q\rangle$, $r, r' > 0$, then there is a s'' such that $\langle s, p\rangle \overset{r}{\mapsto} \langle s'', p+r\rangle$ and $\langle s'', p+r\rangle \overset{r'}{\longmapsto} \langle s', q\rangle$;

(2) if $\langle s, p\rangle \overset{r}{\mapsto} \langle s'', p+r\rangle$ and $\langle s'', p+r\rangle \overset{r'}{\longmapsto} \langle s', q\rangle$, then $\langle s, p\rangle \overset{r+r'}{\longmapsto} \langle s', q\rangle$.

The four kinds of relations are called *action step*, *action termination*, *time step* and *immediate deadlock* relations, respectively. We write $\mathsf{RTTS}(\mathsf{A})$ for the set of all real time transition systems over A.

We shall associate a transition system $\mathsf{TS}(t)$ in $\mathsf{RTTS}(\mathsf{A})$ with a closed term t of BPA^{sat} by taking the set of closed terms of BPA^{sat} as set of states, the closed term t as root state, and the action step, action termination, time step and immediate deadlock relations defined below using rules in the style of Plotkin [38]. A semantics given in this way is called a structural operational semantics. On the basis of these rules, the operators of BPA^{sat} can also be directly defined on the set of real time transition systems in a straightforward way. Note that, by taking closed terms as states, the relations can be explained as follows:

$\langle t, p\rangle \xrightarrow{a} \langle t', p\rangle$:	process t is capable of first performing action a at time p and then proceeding as process t';
$\langle t, p\rangle \xrightarrow{a} \langle\surd, p\rangle$:	process t is capable of first performing action a at time p and then terminating successfully;
$\langle t, p\rangle \overset{r}{\mapsto} \langle t', q\rangle$:	process t is capable of first idling from time p to time q and then proceeding as process t';
$\mathsf{ID}(t, p)$:	process t is not capable of reaching time p.

The rules for the operational semantics have the form $\dfrac{h_1, \ldots, h_m, s}{c_1, \ldots, c_n}$, where s is optional. They are to be read as "if h_1 and ... and h_m then c_1 and ... and c_n, provided s". As customary, $h_1, \ldots, h_m$ and $c_1, \ldots, c_n$ are called the premises and the conclusions, respectively. The conclusions of a rule are positive formulas of the form $\langle t, p\rangle \xrightarrow{a} \langle t', p\rangle$, $\langle t, p\rangle \xrightarrow{a} \langle\surd, p\rangle$, $\langle t, p\rangle \overset{r}{\mapsto} \langle t', q\rangle$ or $\mathsf{ID}(t, p)$, where t and t' are open terms of BPA^{sat}. The premises of a rule are positive formulas of the above forms or negative formulas of the form $\neg\mathsf{ID}(t, p)$. The rules are actually rule schemas. The optional s is a side-condition restricting the actions over which a, b and c range and the non-negative real numbers over which p, q and r range. Within the framework of term deduction systems introduced in [9], the instances of the rule schemas that satisfy the stated side-conditions should be taken as the rules under consideration. For the rest, we continue to use the word rule in the broader sense.

The signature of $\mathrm{BPA}^{\mathrm{sat}}$ together with the rules that will be given constitute according to the definitions of [43] a *strictly stratifiable* term deduction system. For a term deduction system having rules with negative premises, it is not immediately clear whether the term deduction system is meaningful. That is, it is not clear whether there exist relations for which exactly those formulas hold that can be derived using the rules of the term deduction system (see, e.g., [25] and [15]). However, if the term deduction system is stratifiable then there exist such relations. If it is strictly stratifiable then there exist unique such relations. The restriction to stratifiable term deduction systems is essential for the congruence theorem mentioned below.

Term deduction systems support only unary and binary relations on closed terms over some one-sorted signature. Generalization to the many-sorted case is harmless if it is confined to unary and binary relations on closed terms of one of the sorts. Therefore, we chose to have, for instance, many binary action step relations for each action in A, viz. one for each element of $\mathbb{R}_{\geqslant 0}$, instead of one ternary or quaternary relation.

For the operational semantics of $\mathrm{BPA}^{\mathrm{sat}}$, as well as $\mathrm{ACP}^{\mathrm{sat}}$ and the further extensions described in Section 3, we will only define time step relations for which $\langle t, p\rangle \overset{r}{\mapsto} \langle t', q\rangle$ holds only if $t \equiv t'$. The given rules define relations for which exactly those formulas hold that can be derived using the rules. Consequently, there are no rules with conclusions of the form $\langle x, p\rangle \overset{r}{\mapsto} \langle x', q\rangle$ where $x \not\equiv x'$. Hence, it makes no difference if in a rule a premise of the form $\langle x, p\rangle \overset{r}{\mapsto} \langle x, q\rangle$ is replaced by $\langle x, p\rangle \overset{r}{\mapsto} \langle x', q\rangle$ if x' is a variable different from the variables occurring in the rule. We prefer the premises of the form $\langle x, p\rangle \overset{r}{\mapsto} \langle x, q\rangle$ because they are better suited to a natural explanation of the rules. However, the replacements are usually needed whenever general results about term deduction systems, e.g., results of [43], are used. In the remainder of this chapter, we will refrain from making mention of the need for the replacements because they are trivial and make no difference with respect to the relations defined.

The structural operational semantics of $\mathrm{BPA}^{\mathrm{sat}}$ is described by the rules given in Table 4.

These rules are easy to understand. We will only explain the rules for the absolute delay operator (σ_{abs}). The first pair of rules expresses that the action related capabilities of a process $\sigma^{0}_{\mathsf{abs}}(x)$ at time p include those of process x at time p. The second pair of rules expresses that the action related capabilities of a process $\sigma^{r}_{\mathsf{abs}}(x)$ at time $p+r$ include those of process x at time p shifted in time by r ($p \geqslant 0$, $r > 0$). The third pair of rules expresses that the time related capabilities of a process $\sigma^{q}_{\mathsf{abs}}(x)$ at time $p+q$ include those of process x at time p shifted in time by q ($q \geqslant 0$). The fourth pair of rules expresses that a process $\sigma^{r}_{\mathsf{abs}}(x)$ can idle from any time $p \geqslant 0$ to any time $q < r$ and that it can also idle to time r provided that process x can reach time 0.

By identifying bisimilar processes we obtain our preferred model of $\mathrm{BPA}^{\mathrm{sat}}$. One process is (strongly) bisimilar to another process means that if one of the processes is capable of doing a certain step, i.e. performing a certain action at a certain time or idling from a certain time to another, and next going on as a certain subsequent process then the other process is

Table 4
Rules for operational semantics of $\mathrm{BPA}^{\mathrm{sat}}$ $(a \in \mathsf{A},\ r > 0,\ p, q \geqslant 0)$

$$\frac{}{\mathsf{ID}(\dot{\delta}, p)} \quad \frac{}{\mathsf{ID}(\tilde{\delta}, r)} \qquad \frac{}{\langle \tilde{a}, 0\rangle \xrightarrow{a} \langle \surd, 0\rangle} \quad \frac{}{\mathsf{ID}(\tilde{a}, r)}$$

$$\frac{\langle x, p\rangle \xrightarrow{a} \langle x', p\rangle}{\langle \sigma^{0}_{\mathsf{abs}}(x), p\rangle \xrightarrow{a} \langle x', p\rangle} \qquad \frac{\langle x, p\rangle \xrightarrow{a} \langle \surd, p\rangle}{\langle \sigma^{0}_{\mathsf{abs}}(x), p\rangle \xrightarrow{a} \langle \surd, p\rangle}$$

$$\frac{\langle x, p\rangle \xrightarrow{a} \langle x', p\rangle}{\langle \sigma^{r}_{\mathsf{abs}}(x), p+r\rangle \xrightarrow{a} \langle \sigma^{r}_{\mathsf{abs}}(x'), p+r\rangle} \qquad \frac{\langle x, p\rangle \xrightarrow{a} \langle \surd, p\rangle}{\langle \sigma^{r}_{\mathsf{abs}}(x), p+r\rangle \xrightarrow{a} \langle \surd, p+r\rangle}$$

$$\frac{\langle x, p\rangle \stackrel{r}{\mapsto} \langle x, p+r\rangle}{\langle \sigma^{q}_{\mathsf{abs}}(x), p+q\rangle \stackrel{r}{\mapsto} \langle \sigma^{q}_{\mathsf{abs}}(x), p+q+r\rangle} \qquad \frac{\mathsf{ID}(x, p)}{\mathsf{ID}(\sigma^{q}_{\mathsf{abs}}(x), p+q)}$$

$$\frac{q > p}{\langle \sigma^{q+r}_{\mathsf{abs}}(x), p\rangle \stackrel{r}{\mapsto} \langle \sigma^{q+r}_{\mathsf{abs}}(x), p+r\rangle} \qquad \frac{\neg\mathsf{ID}(x, 0)}{\langle \sigma^{q+r}_{\mathsf{abs}}(x), q\rangle \stackrel{r}{\mapsto} \langle \sigma^{q+r}_{\mathsf{abs}}(x), q+r\rangle}$$

$$\frac{\langle x, p\rangle \xrightarrow{a} \langle x', p\rangle}{\langle x+y, p\rangle \xrightarrow{a} \langle x', p\rangle,\ \langle y+x, p\rangle \xrightarrow{a} \langle x', p\rangle} \qquad \frac{\langle x, p\rangle \xrightarrow{a} \langle \surd, p\rangle}{\langle x+y, p\rangle \xrightarrow{a} \langle \surd, p\rangle,\ \langle y+x, p\rangle \xrightarrow{a} \langle \surd, p\rangle}$$

$$\frac{\langle x, p\rangle \stackrel{r}{\mapsto} \langle x, p+r\rangle}{\langle x+y, p\rangle \stackrel{r}{\mapsto} \langle x+y, p+r\rangle,\ \langle y+x, p\rangle \stackrel{r}{\mapsto} \langle y+x, p+r\rangle} \qquad \frac{\mathsf{ID}(x, p),\ \mathsf{ID}(y, p)}{\mathsf{ID}(x+y, p)}$$

$$\frac{\langle x, p\rangle \xrightarrow{a} \langle x', p\rangle}{\langle x \cdot y, p\rangle \xrightarrow{a} \langle x' \cdot y, p\rangle} \qquad \frac{\langle x, p\rangle \xrightarrow{a} \langle \surd, p\rangle}{\langle x \cdot y, p\rangle \xrightarrow{a} \langle y, p\rangle}$$

$$\frac{\langle x, p\rangle \stackrel{r}{\mapsto} \langle x, p+r\rangle}{\langle x \cdot y, p\rangle \stackrel{r}{\mapsto} \langle x \cdot y, p+r\rangle} \qquad \frac{\mathsf{ID}(x, p)}{\mathsf{ID}(x \cdot y, p)}$$

$$\frac{\langle x, p\rangle \xrightarrow{a} \langle x', p\rangle,\ q > p}{\langle \upsilon^{q}_{\mathsf{abs}}(x), p\rangle \xrightarrow{a} \langle x', p\rangle} \qquad \frac{\langle x, p\rangle \xrightarrow{a} \langle \surd, p\rangle,\ q > p}{\langle \upsilon^{q}_{\mathsf{abs}}(x), p\rangle \xrightarrow{a} \langle \surd, p\rangle}$$

$$\frac{\langle x, p\rangle \stackrel{r}{\mapsto} \langle x, p+r\rangle,\ q > p+r}{\langle \upsilon^{q}_{\mathsf{abs}}(x), p\rangle \stackrel{r}{\mapsto} \langle \upsilon^{q}_{\mathsf{abs}}(x), p+r\rangle} \qquad \frac{\mathsf{ID}(x, p),\ q > p}{\mathsf{ID}(\upsilon^{q}_{\mathsf{abs}}(x), p)}$$

$$\frac{q \leqslant p}{\mathsf{ID}(\upsilon^{q}_{\mathsf{abs}}(x), p)}$$

$$\frac{\langle x, p\rangle \xrightarrow{a} \langle x', p\rangle,\ q \leqslant p}{\langle \overline{\upsilon}^{q}_{\mathsf{abs}}(x), p\rangle \xrightarrow{a} \langle x', p\rangle} \qquad \frac{\langle x, p\rangle \xrightarrow{a} \langle \surd, p\rangle,\ q \leqslant p}{\langle \overline{\upsilon}^{q}_{\mathsf{abs}}(x), p\rangle \xrightarrow{a} \langle \surd, p\rangle}$$

$$\frac{\langle x, p\rangle \stackrel{r}{\mapsto} \langle x, p+r\rangle,\ q \leqslant p+r}{\langle \overline{\upsilon}^{q}_{\mathsf{abs}}(x), p\rangle \stackrel{r}{\mapsto} \langle \overline{\upsilon}^{q}_{\mathsf{abs}}(x), p+r\rangle} \qquad \frac{\mathsf{ID}(x, p),\ q \leqslant p}{\mathsf{ID}(\overline{\upsilon}^{q}_{\mathsf{abs}}(x), p)}$$

$$\frac{q > p}{\langle \overline{\upsilon}^{q+r}_{\mathsf{abs}}(x), p\rangle \stackrel{r}{\mapsto} \langle \overline{\upsilon}^{q+r}_{\mathsf{abs}}(x), p+r\rangle} \qquad \frac{\neg\mathsf{ID}(x, q+r)}{\langle \overline{\upsilon}^{q+r}_{\mathsf{abs}}(x), q\rangle \stackrel{r}{\mapsto} \langle \overline{\upsilon}^{q+r}_{\mathsf{abs}}(x), q+r\rangle}$$

capable of doing the same step and next going on as a process bisimilar to the subsequent process. More precisely, a *bisimulation* on RTTS(A) is a symmetric binary relation R on the set of states S such that:

(1) if $R(s,t)$ and $\langle s, p\rangle \xrightarrow{a} \langle s', p\rangle$, then there is a t' such that $\langle t, p\rangle \xrightarrow{a} \langle t', p\rangle$ and $R(s', t')$;

(2) if $R(s,t)$, then $\langle s, p\rangle \xrightarrow{a} \langle \surd, p\rangle$ iff $\langle t, p\rangle \xrightarrow{a} \langle \surd, p\rangle$;

(3) if $R(s,t)$ and $\langle s, p\rangle \overset{r}{\mapsto} \langle s', q\rangle$, then there is a t' such that $\langle t, p\rangle \overset{r}{\mapsto} \langle t', q\rangle$ and $R(s', t')$;

(4) if $R(s,t)$, then $\mathsf{ID}(s, p)$ iff $\mathsf{ID}(t, p)$.

We say that two closed terms s and t are *bisimilar*, written $s \leftrightarrow t$, if there exists a bisimulation R such that $R(s,t)$.

It is known from [43] that if a stratifiable term deduction system is in panth format, bisimulation equivalence as defined here is a congruence for the operators in the signature concerned. This collection of constraints on the form of the rules of a term deduction system is defined in [43] for the one-sorted case, but in case of the real time and discrete time versions of ACP presented in this chapter we have in addition to the sort of processes also the sort of non-negative real numbers or the sort of natural numbers. In order to conform to the panth format as defined in [43], the number of constants and operators in the first argument of a conclusion must be either zero or one. In case of most term deduction systems given in this chapter, this condition is only fulfilled if we disregard constants and operators that do not yield processes. Careful checking of the proof of the congruence theorem for the panth format given in [43], the only result about the panth format that we will use, shows that the result goes through for the many-sorted case if the above-mentioned condition is relaxed in such a way that only the number of constants and operators that yield processes is restricted to zero or one. Therefore, we will refer to this panth-like format in the remainder of this chapter as the panth format. We note here that checking of the proof of the congruence theorem includes checking of the proofs of many related lemmas and theorems given in [15] and [43]. For a comprehensive introduction to rule formats guaranteeing that bisimulation equivalence is a congruence, the reader is referred to [1].

The signature of BPA^{sat} together with the rules for the operational semantics of BPA^{sat} constitute a stratifiable term deduction system in panth format. Consequently, bisimulation equivalence is a congruence for the operators of BPA^{sat}. For this reason, the operators of BPA^{sat} can be defined on the set of bisimulation equivalence classes. We can prove that this results in a model for BPA^{sat}, i.e. all equations derivable in BPA^{sat} hold. In other words, the axioms of BPA^{sat} form a sound axiomatization for the model based on bisimulation equivalence classes. As in the case of the other axiomatizations presented in this chapter, we leave it as an open problem whether the axioms of BPA^{sat} form a complete axiomatization for this model.

2.2. *Algebra of communicating processes*

In ACP^{sat}, we have, in addition to sequential and alternative composition, parallel composition of processes. The process $x \parallel y$ is the process that proceeds with the processes x and y in parallel. Furthermore, we have the encapsulation operators ∂_H (one for each $H \subseteq \mathsf{A}$)

which turns all urgent actions $\tilde{a}$, where $a \in H$, into $\tilde{\delta}$. As in ACP, we also have the auxiliary operators $\mathbin{|\!\!\lfloor}$ (left merge) and $|$ (communication merge) to get a finite axiomatization of the parallel composition operator. The processes $x \mathbin{|\!\!\lfloor} y$ and $x \parallel y$ are the same except that $x \mathbin{|\!\!\lfloor} y$ must start to perform actions by performing an action of x. The processes $x \mid y$ and $x \parallel y$ are the same except that $x \mid y$ must start to perform actions by performing an action of x and an action of y synchronously. In case of $\mathrm{ACP}^{\mathrm{sat}}$, an additional auxiliary operator ν_{abs} (absolute urgent initialization) is needed. The process $\nu_{\mathsf{abs}}(x)$ is the part of process x that starts to perform actions at time 0.

The operator ν_{abs} of $\mathrm{ACP}^{\mathrm{sat}}$ is simply the operator ν_{abs} of $\mathrm{ACP}^{\tau}_{\mathrm{dat}}$ [5] lifted to the real time setting.

SIGNATURE OF $\mathrm{ACP}^{\mathrm{sat}}$. The signature of $\mathrm{ACP}^{\mathrm{sat}}$ is the signature of $\mathrm{BPA}^{\mathrm{sat}}$ extended with the *parallel composition* operator $\parallel : \mathsf{P} \times \mathsf{P} \to \mathsf{P}$, the *left merge* operator $\mathbin{|\!\!\lfloor} : \mathsf{P} \times \mathsf{P} \to \mathsf{P}$, the *communication merge* operator $| : \mathsf{P} \times \mathsf{P} \to \mathsf{P}$, the *encapsulation* operators $\partial_H : \mathsf{P} \to \mathsf{P}$ (for each $H \subseteq \mathsf{A}$), and the *absolute urgent initialization* operator $\nu_{\mathsf{abs}} : \mathsf{P} \to \mathsf{P}$.

AXIOMS OF $\mathrm{ACP}^{\mathrm{sat}}$. The axiom system of $\mathrm{ACP}^{\mathrm{sat}}$ consists of the axioms of $\mathrm{BPA}^{\mathrm{sat}}$ and the equations given in Table 5.

Axioms CM1, CM4, CM8, CM9, D3 and D4 are common to ACP and all real and discrete time versions of ACP. Axioms CF1SA, CF2SA, CM2SA, CM3SA, CM5SA–CM7SA, D1SA and D2SA are simple reformulations of the axioms CF1, CF2, CM2, CM3, CM5–CM7, D1 and D2 of ACP: constants a ($a \in \mathsf{A}_\delta$) have been replaced by constants $\tilde{a}$, and in addition to that certain variables x have been replaced by $x + \tilde{\delta}$ in CM2SA and CM3SA. Recall that $x + \tilde{\delta} = x$ if $x \neq \dot{\delta}$, and $\dot{\delta} + \tilde{\delta} = \tilde{\delta}$. This means that $x + \tilde{\delta}$ never stands for $\dot{\delta}$. Axioms SACM1 and SACM2 represent the interaction of absolute delay with left merge. Axiom SACM2 shows that if two parallel processes start to perform actions by performing an action of one of them and that process starts to perform actions at a certain time, only the part of the other process proceeds that starts to perform actions at the same time or later. What happens if such a part does not exist, is reflected more clearly by a generalization of axiom SACM1 than by that axiom itself. This generalization, which is derivable from the axioms of $\mathrm{ACP}^{\mathrm{sat}}$, is equation SACM1$'$ given in Table 6. Note that a term of the form $\upsilon^{p}_{\mathsf{abs}}(y)$ stands for an arbitrary process that starts to perform actions before time p; and that a term of the form $\sigma^{p}_{\mathsf{abs}}(\nu_{\mathsf{abs}}(z) + \tilde{\delta})$ stands for an arbitrary process that starts to perform actions at time p or deadlocks at time p. So equation SACM1$'$ expresses that if the process that would perform the first action can only do so after the ultimate time to start performing actions or to deadlock for the other process, the result will be a deadlock at this ultimate starting time. Note that in case of sequential processes, the process that would first perform actions can always do so, irrespective of the ultimate starting time for the other process. This difference is apparent from equation SAT4$'$, given in Table 6, which is derivable from the axioms of $\mathrm{ACP}^{\mathrm{sat}}$ and the standard initialization axioms SI13 and SI16 (Table 14, page 651). Axioms SACM3–SACM5 represent the interaction of absolute delay with communication merge. Axioms SACM4 and SACM5 are similar to axioms SACM1 and SACM2. Axiom SACM3 is needed as well because communication merge requires that both processes concerned start performing actions at the same time.

Table 5
Additional axioms for ACP^{sat}
$(a, b \in \mathsf{A}_\delta,\ c \in \mathsf{A},\ p \geqslant 0,\ r > 0)$

$\tilde{a} \mid \tilde{b} = \tilde{c}$ if $\gamma(a, b) = c$	CF1SA
$\tilde{a} \mid \tilde{b} = \tilde{\delta}$ if $\gamma(a, b)$ undefined	CF2SA
$x \parallel y = (x \mathbin{\underline{\parallel}} y + y \mathbin{\underline{\parallel}} x) + x \mid y$	CM1
$\dot{\delta} \mathbin{\underline{\parallel}} x = \dot{\delta}$	CMID1
$x \mathbin{\underline{\parallel}} \dot{\delta} = \dot{\delta}$	CMID2
$\tilde{a} \mathbin{\underline{\parallel}} (x + \tilde{\delta}) = \tilde{a} \cdot (x + \tilde{\delta})$	CM2SA
$\tilde{a} \cdot x \mathbin{\underline{\parallel}} (y + \tilde{\delta}) = \tilde{a} \cdot (x \parallel (y + \tilde{\delta}))$	CM3SA
$\sigma^r_{\mathsf{abs}}(x) \mathbin{\underline{\parallel}} (\nu_{\mathsf{abs}}(y) + \tilde{\delta}) = \tilde{\delta}$	SACM1
$\sigma^p_{\mathsf{abs}}(x) \mathbin{\underline{\parallel}} (\upsilon^p_{\mathsf{abs}}(y) + \sigma^p_{\mathsf{abs}}(z)) = \sigma^p_{\mathsf{abs}}(x \mathbin{\underline{\parallel}} z)$	SACM2
$(x + y) \mathbin{\underline{\parallel}} z = x \mathbin{\underline{\parallel}} z + y \mathbin{\underline{\parallel}} z$	CM4
$\dot{\delta} \mid x = \dot{\delta}$	CMID3
$x \mid \dot{\delta} = \dot{\delta}$	CMID4
$\tilde{a} \cdot x \mid \tilde{b} = (\tilde{a} \mid \tilde{b}) \cdot x$	CM5SA
$\tilde{a} \mid \tilde{b} \cdot x = (\tilde{a} \mid \tilde{b}) \cdot x$	CM6SA
$\tilde{a} \cdot x \mid \tilde{b} \cdot y = (\tilde{a} \mid \tilde{b}) \cdot (x \parallel y)$	CM7SA
$(\nu_{\mathsf{abs}}(x) + \tilde{\delta}) \mid \sigma^r_{\mathsf{abs}}(y) = \tilde{\delta}$	SACM3
$\sigma^r_{\mathsf{abs}}(x) \mid (\nu_{\mathsf{abs}}(y) + \tilde{\delta}) = \tilde{\delta}$	SACM4
$\sigma^p_{\mathsf{abs}}(x) \mid \sigma^p_{\mathsf{abs}}(y) = \sigma^p_{\mathsf{abs}}(x \mid y)$	SACM5
$(x + y) \mid z = x \mid z + y \mid z$	CM8
$x \mid (y + z) = x \mid y + x \mid z$	CM9
$\partial_H(\dot{\delta}) = \dot{\delta}$	D0
$\partial_H(\tilde{a}) = \tilde{a}$ if $a \notin H$	D1SA
$\partial_H(\tilde{a}) = \tilde{\delta}$ if $a \in H$	D2SA
$\partial_H(\sigma^p_{\mathsf{abs}}(x)) = \sigma^p_{\mathsf{abs}}(\partial_H(x))$	SAD
$\partial_H(x + y) = \partial_H(x) + \partial_H(y)$	D3
$\partial_H(x \cdot y) = \partial_H(x) \cdot \partial_H(y)$	D4
$\nu_{\mathsf{abs}}(\dot{\delta}) = \dot{\delta}$	SAU0
$\nu_{\mathsf{abs}}(\tilde{a}) = \tilde{a}$	SAU1
$\nu_{\mathsf{abs}}(\sigma^r_{\mathsf{abs}}(x)) = \tilde{\delta}$	SAU2
$\nu_{\mathsf{abs}}(x + y) = \nu_{\mathsf{abs}}(x) + \nu_{\mathsf{abs}}(y)$	SAU3
$\nu_{\mathsf{abs}}(x \cdot y) = \nu_{\mathsf{abs}}(x) \cdot y$	SAU4

Axiom SACM5 is simpler than axiom SACM2 just because of the left distributivity of the communication merge (axiom CM9). Equations SACM3′ and SACM4′ given in Table 6 generalize axioms SACM3 and SACM4 like equation SACM1′ generalizes axiom SACM1. Axiom SAD represents the (lack of) interaction of absolute delay with encapsulation. Axioms SAU0–SAU4 reflect the intended meaning of the urgent initialization operator clearly.

Table 6
Some derivable equations ($p \geqslant 0,\ r > 0$)

$\sigma_{\mathrm{abs}}^{p+r}(x) \cdot (\upsilon_{\mathrm{abs}}^{p}(y) + \sigma_{\mathrm{abs}}^{p}(\nu_{\mathrm{abs}}(z) + \tilde{\delta})) = \sigma_{\mathrm{abs}}^{p+r}(x \cdot \dot{\delta})$	SAT4′
$\sigma_{\mathrm{abs}}^{p+r}(x) \mathbin{\lfloor\!\lfloor} (\upsilon_{\mathrm{abs}}^{p}(y) + \sigma_{\mathrm{abs}}^{p}(\nu_{\mathrm{abs}}(z) + \tilde{\delta})) = \sigma_{\mathrm{abs}}^{p}(\tilde{\delta})$	SACM1′
$\sigma_{\mathrm{abs}}^{p}(\nu_{\mathrm{abs}}(x) + \tilde{\delta}) \mid \sigma_{\mathrm{abs}}^{p+r}(y) = \sigma_{\mathrm{abs}}^{p}(\tilde{\delta})$	SACM3′
$\sigma_{\mathrm{abs}}^{p+r}(x) \mid \sigma_{\mathrm{abs}}^{p}(\nu_{\mathrm{abs}}(y) + \tilde{\delta}) = \sigma_{\mathrm{abs}}^{p}(\tilde{\delta})$	SACM4′

We can prove that the operators $\|$, $\lfloor\!\lfloor$, $|$, ∂_H and ν_{abs} can be eliminated in closed terms of $\mathrm{ACP}^{\mathrm{sat}}$. Because of the elimination result for $\mathrm{BPA}^{\mathrm{sat}}$, we are permitted to use induction on the structure of basic terms over $\mathrm{BPA}^{\mathrm{sat}}$ to prove statements for all closed terms of $\mathrm{ACP}^{\mathrm{sat}}$.

EXAMPLES. We give some examples of a closed term of $\mathrm{ACP}^{\mathrm{sat}}$ and the corresponding basic term (in case $\gamma(a, b)$ and $\gamma(a, c)$ are undefined):

$$\sigma_{\mathrm{abs}}^{5.1}(\tilde{a}) \| \sigma_{\mathrm{abs}}^{5.1}(\tilde{b}) \cdot \sigma_{\mathrm{abs}}^{4.9}(\tilde{c}) = \sigma_{\mathrm{abs}}^{5.1}(\tilde{a} \cdot \tilde{b} \cdot \dot{\delta} + \tilde{b} \cdot \dot{\delta})$$
$$\sigma_{\mathrm{abs}}^{5.1}(\tilde{a}) \| \sigma_{\mathrm{abs}}^{4.9}(\tilde{b}) \cdot \sigma_{\mathrm{abs}}^{5.1}(\tilde{c}) = \sigma_{\mathrm{abs}}^{4.9}(\tilde{b} \cdot \sigma_{\mathrm{abs}}^{0.2}(\tilde{a} \cdot \tilde{c} + \tilde{c} \cdot \tilde{a}))$$
$$\sigma_{\mathrm{abs}}^{5}(\tilde{a}) \| \sigma_{\mathrm{abs}}^{4.9}(\tilde{b}) \cdot \sigma_{\mathrm{abs}}^{5.1}(\tilde{c}) = \sigma_{\mathrm{abs}}^{4.9}(\tilde{b} \cdot \sigma_{\mathrm{abs}}^{0.1}(\tilde{a} \cdot \sigma_{\mathrm{abs}}^{0.1}(\tilde{c})))$$
$$\nu_{\mathrm{abs}}(\sigma_{\mathrm{abs}}^{5.1}(\tilde{a}) \| \sigma_{\mathrm{abs}}^{4.9}(\tilde{b}) \cdot \sigma_{\mathrm{abs}}^{5.1}(\tilde{c})) = \tilde{\delta}$$

SEMANTICS OF $\mathrm{ACP}^{\mathrm{sat}}$. The structural operational semantics of $\mathrm{ACP}^{\mathrm{sat}}$ is described by the rules for $\mathrm{BPA}^{\mathrm{sat}}$ and the rules given in Table 7.

These rules are easy to understand. We will only mention that the first two rules for the parallel composition operator ($\|$) express that a process x loses all its action related capabilities at time p if it is put in parallel with a process y that can not reach time p, and that the last rule for this operator expresses that in such cases the parallel composition can not reach time p either. As in the case of $\mathrm{BPA}^{\mathrm{sat}}$, we obtain a term deduction system in panth format that is stratifiable, so bisimulation equivalence is also a congruence for the additional operators of $\mathrm{ACP}^{\mathrm{sat}}$. Therefore, these operators can be defined on the set of bisimulation equivalence classes as well. As in the case of $\mathrm{BPA}^{\mathrm{sat}}$, we can prove that this results in a model for $\mathrm{ACP}^{\mathrm{sat}}$.

2.3. *Time-stamped actions*

The real time versions ACPρ, ACPrρ and ACP$\rho\sigma$ [2,4] feature time-stamped actions – and thus combine execution of actions and passage of time. The time-stamped actions defined below are more closely related to the ones in ACPsρ [6], the real time version of ACP from which $\mathrm{ACP}^{\mathrm{sat}}$ originates. This is because ACPsρ, like $\mathrm{ACP}^{\mathrm{sat}}$ and unlike ACPρ, ACPrρ and ACP$\rho\sigma$, does not exclude the possibility of two or more actions to be performed consecutively at the same point in time.

Time-stamped actions are defined in terms of urgent actions and the delay operator in Table 8. We also define a time-stamped version of immediate deadlock. In ACPρ and ACPrρ,

Table 7
Additional rules for $\mathrm{ACP}^{\mathrm{sat}}$ $(a, b, c \in \mathsf{A},\ r > 0,\ p \geqslant 0)$

$$\frac{\langle x, p\rangle \xrightarrow{a} \langle x', p\rangle,\ \neg\mathsf{ID}(y, p)}{\langle x \parallel y, p\rangle \xrightarrow{a} \langle x' \parallel y, p\rangle,\ \langle y \parallel x, p\rangle \xrightarrow{a} \langle y \parallel x', p\rangle,\ \langle x \mathbin{\lfloor\!\lfloor} y, p\rangle \xrightarrow{a} \langle x' \parallel y, p\rangle}$$

$$\frac{\langle x, p\rangle \xrightarrow{a} \langle \surd, p\rangle,\ \neg\mathsf{ID}(y, p)}{\langle x \parallel y, p\rangle \xrightarrow{a} \langle y, p\rangle,\ \langle y \parallel x, p\rangle \xrightarrow{a} \langle y, p\rangle,\ \langle x \mathbin{\lfloor\!\lfloor} y, p\rangle \xrightarrow{a} \langle y, p\rangle}$$

$$\frac{\langle x, p\rangle \xrightarrow{a} \langle x', p\rangle,\ \langle y, p\rangle \xrightarrow{b} \langle y', p\rangle,\ \gamma(a, b) = c}{\langle x \parallel y, p\rangle \xrightarrow{c} \langle x' \parallel y', p\rangle,\ \langle x \mid y, p\rangle \xrightarrow{c} \langle x' \parallel y', p\rangle}$$

$$\frac{\langle x, p\rangle \xrightarrow{a} \langle x', p\rangle,\ \langle y, p\rangle \xrightarrow{b} \langle \surd, p\rangle,\ \gamma(a, b) = c}{\begin{array}{c}\langle x \parallel y, p\rangle \xrightarrow{c} \langle x', p\rangle,\ \langle y \parallel x, p\rangle \xrightarrow{c} \langle x', p\rangle,\\ \langle x \mid y, p\rangle \xrightarrow{c} \langle x', p\rangle,\ \langle y \mid x, p\rangle \xrightarrow{c} \langle x', p\rangle\end{array}}$$

$$\frac{\langle x, p\rangle \xrightarrow{a} \langle \surd, p\rangle,\ \langle y, p\rangle \xrightarrow{b} \langle \surd, p\rangle,\ \gamma(a, b) = c}{\langle x \parallel y, p\rangle \xrightarrow{c} \langle \surd, p\rangle,\ \langle x \mid y, p\rangle \xrightarrow{c} \langle \surd, p\rangle}$$

$$\frac{\langle x, p\rangle \overset{r}{\mapsto} \langle x, p + r\rangle,\ \langle y, p\rangle \overset{r}{\mapsto} \langle y, p + r\rangle}{\begin{array}{c}\langle x \parallel y, p\rangle \overset{r}{\mapsto} \langle x \parallel y, p + r\rangle,\ \langle x \mathbin{\lfloor\!\lfloor} y, p\rangle \overset{r}{\mapsto} \langle x \mathbin{\lfloor\!\lfloor} y, p + r\rangle,\\ \langle x \mid y, p\rangle \overset{r}{\mapsto} \langle x \mid y, p + r\rangle\end{array}}$$

$$\frac{\mathsf{ID}(x, p)}{\begin{array}{c}\mathsf{ID}(x \parallel y, p),\ \mathsf{ID}(y \parallel x, p),\ \mathsf{ID}(x \mathbin{\lfloor\!\lfloor} y, p),\\ \mathsf{ID}(y \mathbin{\lfloor\!\lfloor} x, p),\ \mathsf{ID}(x \mid y, p),\ \mathsf{ID}(y \mid x, p)\end{array}}$$

$$\frac{\langle x, p\rangle \xrightarrow{a} \langle x', p\rangle,\ a \notin H}{\langle \partial_H(x), p\rangle \xrightarrow{a} \langle \partial_H(x'), p\rangle} \qquad \frac{\langle x, p\rangle \xrightarrow{a} \langle \surd, p\rangle,\ a \notin H}{\langle \partial_H(x), p\rangle \xrightarrow{a} \langle \surd, p\rangle}$$

$$\frac{\langle x, p\rangle \overset{r}{\mapsto} \langle x, p + r\rangle}{\langle \partial_H(x), p\rangle \overset{r}{\mapsto} \langle \partial_H(x), p + r\rangle} \qquad \frac{\mathsf{ID}(x, p)}{\mathsf{ID}(\partial_H(x), p)}$$

$$\frac{\langle x, 0\rangle \xrightarrow{a} \langle x', 0\rangle}{\langle \nu_{\mathsf{abs}}(x), 0\rangle \xrightarrow{a} \langle x', 0\rangle} \qquad \frac{\langle x, 0\rangle \xrightarrow{a} \langle \surd, 0\rangle}{\langle \nu_{\mathsf{abs}}(x), 0\rangle \xrightarrow{a} \langle \surd, 0\rangle}$$

$$\frac{\mathsf{ID}(x, 0)}{\mathsf{ID}(\nu_{\mathsf{abs}}(x), 0)} \qquad \frac{}{\mathsf{ID}(\nu_{\mathsf{abs}}(x), r)}$$

Table 8
Definitions of time-stamped actions and immediate deadlock $(a \in \mathsf{A}_\delta,\ p \geqslant 0)$

$$\tilde{a}(p) = \sigma^{p}_{\mathsf{abs}}(\tilde{a})$$
$$\dot{\delta}(p) = \sigma^{p}_{\mathsf{abs}}(\dot{\delta})$$

Table 9
Additional axioms for time-stamped actions
($a \in A_\delta$, $p, q \geqslant 0$, $r > 0$)

$x + \dot{\delta}(0) = x$	A6TSIDa
$\tilde{\delta}(p) + \dot{\delta}(p) = \tilde{\delta}(p)$	A6TSIDb
$\dot{\delta}(p) \cdot x = \dot{\delta}(p)$	A7TSID
$\tilde{a}(p) \cdot x = \tilde{a}(p) \cdot \overline{\upsilon}^{p}_{\mathsf{abs}}(x)$	SATTS
$\tilde{a}(p) + \tilde{\delta}(p) = \tilde{a}(p)$	A6TSa
$\tilde{\delta}(p+r) + \tilde{\delta}(p) = \tilde{\delta}(p+r)$	A6TSb
$\tilde{\delta}(p) \cdot x = \tilde{\delta}(p)$	A7TS
$\overline{\upsilon}^{p}_{\mathsf{abs}}(\dot{\delta}(p+q)) = \dot{\delta}(p+q)$	SATSI1
$\overline{\upsilon}^{p+r}_{\mathsf{abs}}(\dot{\delta}(p)) = \dot{\delta}(p+r)$	SATSI2
$\overline{\upsilon}^{p}_{\mathsf{abs}}(\tilde{a}(p+q)) = \tilde{a}(p+q)$	SATSI3
$\overline{\upsilon}^{p+r}_{\mathsf{abs}}(\tilde{a}(p)) = \dot{\delta}(p+r)$	SATSI4
$\overline{\upsilon}^{p}_{\mathsf{abs}}(x+y) = \overline{\upsilon}^{p}_{\mathsf{abs}}(x) + \overline{\upsilon}^{p}_{\mathsf{abs}}(y)$	SATSI5
$\overline{\upsilon}^{p}_{\mathsf{abs}}(x \cdot y) = \overline{\upsilon}^{p}_{\mathsf{abs}}(x) \cdot y$	SATSI6

which exclude the possibility of two or more actions to be performed consecutively at the same point in time, there is no reason to distinguish, for instance, between the processes $a(p+r) \cdot \delta(p+r)$ and $a(p+r) \cdot b(p)$ $(r > 0)$. In ACP^{sat}, unlike in ACPsρ, distinction is made in comparable cases by introducing immediate deadlock to deal with timing inconsistencies. Therefore, we have to introduce time-stamped immediate deadlock here.

We now consider the signature of BPA^{sat} with time-stamped actions, i.e. the signature of BPA^{sat}, but with the urgent action constants $\tilde{a}$, the urgent deadlock constant $\tilde{\delta}$, the immediate deadlock constant $\dot{\delta}$ and the delay operator σ_{abs} replaced by the time-stamped action constants $\tilde{a}(p)$ and the time-stamped deadlock constants $\tilde{\delta}(p)$ and $\dot{\delta}(p)$. From the axioms of BPA^{sat} and the definitions of time-stamped actions and immediate deadlock, we can easily derive the equations given in Table 9 for closed terms. Axioms A1–A5 from Table 1 and the equations from Table 9 together can be considered to form the axioms of BPA^{sat} with time-stamped actions. The differences with the axioms of BPAs$\rho\delta$ in [6] are all due to the different treatment of timing inconsistencies. Extension of this version to ACP is left to the reader.

3. Extension of ACP^{sat}

In this section, we describe the extension of ACP^{sat} with integration and initial abstraction. The extension with integration is needed to be able to embed discrete time process algebras, as exemplified in Section 5.3. The extension with initial abstraction is needed to be able to embed process algebras with relative timing, as illustrated in Section 4.3.

Integration and initial abstraction are both variable binding operators. Following, e.g., [24], we will introduce *variable binding operators* by a declaration of the form

$f : S_{11}, \ldots, S_{1k_1} \,.\, S_1 \times \cdots \times S_{n1}, \ldots, S_{nk_n} \,.\, S_n \to S$. Hereby is indicated that f combines an operator $f^* : ((S_{11} \times \cdots \times S_{1k_1}) \to S_1) \times \cdots \times ((S_{n1} \times \cdots \times S_{nk_n}) \to S_n) \to S$ with λ-calculus-like functional abstraction, binding k_i variables ranging over $S_{i1}, \ldots, S_{ik_i}$ in the ith argument ($0 \leqslant i \leqslant n$). Applications of f have the following form: $f(x_{11}, \ldots, x_{1k_1} \,.\, t_1, \ldots, x_{n1}, \ldots, x_{nk_n} \,.\, t_n)$, where each x_{ij} is a variable of sort S_{ij} and each t_i is a term of sort S_i.

Integration requires a more extensive theory of the non-negative real numbers than the minimal theory sketched at the beginning of Section 2 (page 631). In the first place, it has to include a theory of sets of non-negative real numbers that makes it possible to deal with set membership and set equality. Besides, the theory should cover suprema of sets of non-negative real numbers.

First, in Section 3.1, ACP^{sat} is extended with integration. After that, in Section 3.2, initial abstraction is added. Finally, some useful additional axioms, derivable for closed terms, are given in Section 3.3.

3.1. *Integration*

We add the integration operator $\int$ to ACP^{sat}. It provides for alternative composition over a continuum of alternatives. That is, $\int_{v \in V} P$, where v is a variable ranging over $\mathbb{R}_{\geqslant 0}$, $V \subseteq \mathbb{R}_{\geqslant 0}$ and P is a term that may contain free variables, proceeds as one of the alternatives $P[p/v]$ for $p \in V$. The resulting theory is called ACP^{sat}I. Obviously, we could first have added integration to BPA^{sat}, resulting in BPA^{sat}I, and then have extended BPA^{sat}I to deal with parallelism and communication.

SIGNATURE OF ACP^{sat}I. The signature of ACP^{sat}I is the signature of ACP^{sat} extended with the *integration* (variable-binding) operator $\int : \mathcal{P}(\mathbb{R}_{\geqslant 0}) \times \mathbb{R}_{\geqslant 0} \,.\, \mathsf{P} \to \mathsf{P}$.

We assume that an infinite set of *time variables* ranging over $\mathbb{R}_{\geqslant 0}$ has been given, and denote them by $v, w, \ldots$. Furthermore, we use $V, W, \ldots$ to denote subsets of $\mathbb{R}_{\geqslant 0}$. We denote terms of ACP^{sat}I by $P, Q, \ldots$. We will use the following notational convention. We write $\int_{v \in V} P$ for $\int(V, v \,.\, P)$.

AXIOM SYSTEM OF ACP^{sat}I. The axiom system of ACP^{sat}I consists of the axioms of ACP^{sat} and the equations given in Table 10.

Axiom INT1 is similar to the α-conversion rule of λ-calculus. Axioms INT2–INT6 are the crucial axioms of integration. They reflect the informal explanation that $\int_{v \in V} P$ proceeds as one of the alternatives $P[p/v]$ for $p \in V$. The remaining axioms are all easily understood by realizing that $\int$ stands for an infinite alternative composition.

We can prove that the auxiliary operators υ_{abs} and $\overline{\upsilon}_{\mathsf{abs}}$, as well as sequential compositions in which the form of the first operand is not $\tilde{a}$ ($a \in \mathsf{A}$) and alternative compositions in which the form of the first operand is $\sigma^{p}_{\mathsf{abs}}(t)$, can be eliminated in closed terms of BPA^{sat}I with a restricted form of integration. Basically, this restriction means that in terms of the form $\int_{v \in V} P$, V is an interval of which the bounds are given by linear expressions over

Table 10
Axioms for integration ($p \geqslant 0$, v not free in R)

$\int_{w \in V} R = \int_{v \in V} R[v/w]$	INT1
$\int_{v \in \emptyset} P = \dot{\delta}$	INT2
$\int_{v \in \{p\}} P = P[p/v]$	INT3
$\int_{v \in V \cup W} P = \int_{v \in V} P + \int_{v \in W} P$	INT4
$V \neq \emptyset \Rightarrow \int_{v \in V} R = R$	INT5
$(\forall p \in V \bullet P[p/v] = Q[p/v]) \Rightarrow \int_{v \in V} P = \int_{v \in V} Q$	INT6
$V \neq \emptyset \Rightarrow \int_{v \in V} \sigma^{v}_{\mathsf{abs}}(\dot{\delta}) = \sigma^{\sup V}_{\mathsf{abs}}(\dot{\delta})$	INT7
$V \neq \emptyset,\ \sup V \notin V \Rightarrow \int_{v \in V} \sigma^{v}_{\mathsf{abs}}(\tilde{\delta}) = \sigma^{\sup V}_{\mathsf{abs}}(\dot{\delta})$	INT8
$\sup V \in V \Rightarrow \int_{v \in V} \sigma^{v}_{\mathsf{abs}}(\tilde{\delta}) = \sigma^{\sup V}_{\mathsf{abs}}(\tilde{\delta})$	INT9
$\int_{v \in V} \sigma^{p}_{\mathsf{abs}}(P) = \sigma^{p}_{\mathsf{abs}}(\int_{v \in V} P)$ if $p \neq v$	INT10
$\int_{v \in V} (P + Q) = \int_{v \in V} P + \int_{v \in V} Q$	INT11
$\int_{v \in V} (P \cdot R) = (\int_{v \in V} P) \cdot R$	INT12
$\int_{v \in V} (P \mathbin{⌊\!⌊} R) = (\int_{v \in V} P) \mathbin{⌊\!⌊} R$	INT13
$\int_{v \in V} (P \mid R) = (\int_{v \in V} P) \mid R$	INT14
$\int_{v \in V} (R \mid P) = R \mid (\int_{v \in V} P)$	INT15
$\int_{v \in V} \partial_H(P) = \partial_H(\int_{v \in V} P)$	INT16
$\upsilon^{p}_{\mathsf{abs}}(\int_{v \in V} P) = \int_{v \in V} \upsilon^{p}_{\mathsf{abs}}(P)$ if $p \neq v$	SATO6
$\overline{\upsilon}^{p}_{\mathsf{abs}}(\int_{v \in V} P) = \int_{v \in V} \overline{\upsilon}^{p}_{\mathsf{abs}}(P)$ if $p \neq v$	SAI6
$\nu_{\mathsf{abs}}(\int_{v \in V} P) = \int_{v \in V} \nu_{\mathsf{abs}}(P)$	SAU5

time variables and P is of the form $\sigma^{v}_{\mathsf{abs}}(\tilde{a})$ or $\sigma^{v}_{\mathsf{abs}}(\tilde{a}) \cdot t$ ($a \in \mathsf{A}_{\delta}$). This restricted form of integration is essentially the same as prefix integration from [29] (see also [22,23]). The terms that remain after exhaustive elimination are called the basic terms over BPA^{sat} with restricted integration. We can also prove that the operators $\|$, $⌊\!⌊$, $|$, ∂_H and ν_{abs} can be eliminated in closed terms of ACP^{sat} with restricted integration. Because of these elimination results, we are permitted to use induction on the structure of basic terms over BPA^{sat} with restricted integration to prove statements for all closed terms of ACP^{sat} with restricted integration.

EXAMPLES. We give some examples of a closed term of ACP^{sat} with restricted integration and the corresponding basic term:

$$\int_{v \in [4.9,5.1)} \sigma^{v}_{\mathsf{abs}}(\nu_{\mathsf{abs}}(\sigma^{0.9}_{\mathsf{abs}}(\tilde{a}) \parallel \sigma^{1.8}_{\mathsf{abs}}(\tilde{b}) \cdot \sigma^{2.7}_{\mathsf{abs}}(\tilde{c}))) = \sigma^{5.1}_{\mathsf{abs}}(\dot{\delta})$$
$$\int_{v \in [4.9,5.1)} \sigma^{v}_{\mathsf{abs}}(\tilde{a}) + \int_{v \in [4.9,5.1)} \sigma^{v}_{\mathsf{abs}}(\tilde{b}) = \int_{v \in [4.9,5.1)} \sigma^{v}_{\mathsf{abs}}(\tilde{a} + \tilde{b})$$
$$(\int_{v \in [4.9,5.1)} \sigma^{v}_{\mathsf{abs}}(\tilde{a})) \mid (\int_{v \in [4.9,5.1)} \sigma^{v}_{\mathsf{abs}}(\tilde{b})) = \int_{v \in [4.9,5.1)} \sigma^{v}_{\mathsf{abs}}(\tilde{c}) \quad \text{if } \gamma(a,b) = c$$
$$(\int_{v \in [4.9,5.1)} \sigma^{v}_{\mathsf{abs}}(\tilde{a})) \mid (\int_{v \in [4.9,5.1)} \sigma^{v}_{\mathsf{abs}}(\tilde{b})) = \int_{v \in [4.9,5.1)} \sigma^{v}_{\mathsf{abs}}(\tilde{\delta}) \quad \text{if } \gamma(a,b) \text{ undefined}$$

SEMANTICS OF ACP^{sat}I. The structural operational semantics of ACP^{sat}I is described by the rules for ACP^{sat} and the rules given in Table 11.

Table 11
Rules for integration ($a \in \mathsf{A},\ r > 0,\ p, q \geqslant 0$)

$$\frac{\langle P[q/v], p\rangle \xrightarrow{a} \langle P', p\rangle,\ q \in V}{\langle \int_{v \in V} P, p\rangle \xrightarrow{a} \langle P', p\rangle} \qquad \frac{\langle P[q/v], p\rangle \xrightarrow{a} \langle \surd, p\rangle,\ q \in V}{\langle \int_{v \in V} P, p\rangle \xrightarrow{a} \langle \surd, p\rangle}$$

$$\frac{\langle P[q/v], p\rangle \overset{r}{\longmapsto} \langle P[q/v], p+r\rangle,\ q \in V}{\langle \int_{v \in V} P, p\rangle \overset{r}{\longmapsto} \langle \int_{v \in V} P, p+r\rangle} \qquad \frac{\mathsf{ID}(P[q/v], p) \text{ for all } q \in V}{\mathsf{ID}(\int_{v \in V} P, p)}$$

The rules for integration are simple generalizations of the rules for alternative composition to the infinite case.

The panth format does not cover variable binding operators such as integration. The integration operator combines an ordinary operator $\int^* : \mathcal{P}(\mathbb{R}_{\geqslant 0}) \times (\mathbb{R}_{\geqslant 0} \to \mathsf{P}) \to \mathsf{P}$ with λ-calculus-like functional abstraction, binding one variable ranging over $\mathbb{R}_{\geqslant 0}$ in the second argument. This implies that we have additional sorts here, including the sort of functions from non-negative real numbers to processes. These functions, denoted by closed terms of the form $v\,.\,P$, where v may occur free in P, cannot be dealt with in the same way as the non-negative real numbers: (1) they are denoted using open terms of the sort of processes and (2) their application is syntactically represented by means of substitution. As for (1), we can define bisimulation equivalence on closed terms $v\,.\,P$ and $w\,.\,Q$ as well as on open terms P and Q in such a way that it corresponds to the pointwise extension of the original bisimulation equivalence: $v\,.\,P \leftrightarroweq w\,.\,Q$ iff $P[p/v] \leftrightarroweq Q[p/w]$ for all $p \in \mathbb{R}_{\geqslant 0}$; $P \leftrightarroweq Q$ iff $\sigma(P) \leftrightarroweq \sigma(Q)$ for all substitutions σ of non-negative real numbers for time variables. As for (2), we have to make a clear distinction between variables that range over a semantic domain, such as v in the rules for integration, and meta-variables that range over a syntactic domain, such as P and P' in the rules for integration. In [24], the former variables are called actual variables and the latter ones formal variables. Careful checking of the proof of the congruence theorem for the panth format given in [43] shows that the result goes through for the case with variable binding operators if the condition about the number of constants and operators in the first argument of a conclusion is as in the many-sorted case without variable binding operators: the number of constants and operators that yield processes is restricted to zero or one.

The signature of ACP^{sat}I together with the rules for the operational semantics of ACP^{sat}I constitute a stratifiable term deduction system in this generalized panth format, so bisimulation equivalence is also a congruence for the integration operator. Hence, this operator can be defined on the set of bisimulation equivalence classes as well. As in the case of BPA^{sat} and ACP^{sat}, we can prove that this results in a model for ACP^{sat}I. We will call this model $\mathsf{M_A}$.

3.2. *Initial abstraction*

We add the initial abstraction operator $\surd_{\mathsf{s}}$ to ACP^{sat}I. It provides for (simple) parametric timing: $\surd_{\mathsf{s}}\, v\,.\,F$, where v is a variable ranging over $\mathbb{R}_{\geqslant 0}$ and F is a term that may contain free variables, proceeds as $F[p/v]$ if initialized at time $p \in \mathbb{R}_{\geqslant 0}$. This means that $\surd_{\mathsf{s}}\, v \,.\, F$ denotes a function $f : \mathbb{R}_{\geqslant 0} \to \mathsf{P}$ that satisfies $f(p) = \overline{\upsilon}^{p}_{\mathsf{abs}}(f(p))$ for all $p \in \mathbb{R}_{\geqslant 0}$. In the resulting theory, called $\mathrm{ACP^{sat}I}\surd$, the sort P of (absolute time) processes is replaced by the sort P^* of parametric time processes. Of course, it is also possible to add the initial abstraction operator to $\mathrm{ACP^{sat}}$, resulting in a theory $\mathrm{ACP^{sat}}\surd$.

SIGNATURE OF $\mathrm{ACP^{sat}I}\surd$. The signature of $\mathrm{ACP^{sat}I}\surd$ is the signature of $\mathrm{ACP^{sat}I}$ extended with the *initial abstraction* (variable-binding) operator $\surd_{\mathsf{s}} : \mathbb{R}_{\geqslant 0} \,.\, \mathsf{P}^* \to \mathsf{P}^*$.

We now use $x, y, \ldots$ to denote variables of sort P^*. Terms of $\mathrm{ACP^{sat}I}\surd$ are denoted by $F, G, \ldots$. We will use the following notational convention. We write $\surd_{\mathsf{s}}\, v \,.\, F$ for $\surd_{\mathsf{s}}(v \,.\, F)$.

AXIOM SYSTEM OF $\mathrm{ACP^{sat}I}\surd$. The axiom system of $\mathrm{ACP^{sat}I}\surd$ consists of the axioms of $\mathrm{ACP^{sat}I}$ and the equations given in Table 12.

Axioms SIA1 and SIA2 are similar to the α- and β-conversion rules of λ-calculus. Axiom SIA3 points out that multiple initial abstractions can simply be replaced by one. Axiom SIA4 shows that processes with absolute timing can be treated as special cases of processes with parametric timing: they do not vary with different initialization times. Axiom SIA5 is an extensionality axiom. Axiom SIA6 expresses that in case a process performs an action and then proceeds as another process, the initialization time of the latter process is the time

Table 12
Axioms for standard initial abstraction ($p \geqslant 0$, v not free in G)

$\surd_{\mathsf{s}}\, w \,.\, G = \surd_{\mathsf{s}}\, v \,.\, G[v/w]$	SIA1
$\overline{\upsilon}^{p}_{\mathsf{abs}}(\surd_{\mathsf{s}}\, v \,.\, F) = \overline{\upsilon}^{p}_{\mathsf{abs}}(F[p/v])$	SIA2
$\surd_{\mathsf{s}}\, v \,.\, (\surd_{\mathsf{s}}\, w \,.\, F) = \surd_{\mathsf{s}}\, v \,.\, F[v/w]$	SIA3
$G = \surd_{\mathsf{s}}\, v \,.\, G$	SIA4
$(\forall p \in \mathbb{R}_{\geqslant 0} \bullet \overline{\upsilon}^{p}_{\mathsf{abs}}(x) = \overline{\upsilon}^{p}_{\mathsf{abs}}(y)) \Rightarrow x = y$	SIA5
$\sigma^{p}_{\mathsf{abs}}(\tilde{a}) \cdot x = \sigma^{p}_{\mathsf{abs}}(\tilde{a}) \cdot \overline{\upsilon}^{p}_{\mathsf{abs}}(x)$	SIA6
$\sigma^{p}_{\mathsf{abs}}(\surd_{\mathsf{s}}\, v \,.\, F) = \sigma^{p}_{\mathsf{abs}}(F[0/v])$	SIA7
$(\surd_{\mathsf{s}}\, v \,.\, F) + G = \surd_{\mathsf{s}}\, v \,.\, (F + \overline{\upsilon}^{v}_{\mathsf{abs}}(G))$	SIA8
$(\surd_{\mathsf{s}}\, v \,.\, F) \cdot G = \surd_{\mathsf{s}}\, v \,.\, (F \cdot G)$	SIA9
$\upsilon^{p}_{\mathsf{abs}}(\surd_{\mathsf{s}}\, v \,.\, F) = \surd_{\mathsf{s}}\, v \,.\, \upsilon^{p}_{\mathsf{abs}}(F)$ if $p \neq v$	SIA10
$(\surd_{\mathsf{s}}\, v \,.\, F) \mathbin{\lfloor\!\lfloor} G = \surd_{\mathsf{s}}\, v \,.\, (F \mathbin{\lfloor\!\lfloor} \overline{\upsilon}^{v}_{\mathsf{abs}}(G))$	SIA11
$G \mathbin{\lfloor\!\lfloor} (\surd_{\mathsf{s}}\, v \,.\, F) = \surd_{\mathsf{s}}\, v \,.\, (\overline{\upsilon}^{v}_{\mathsf{abs}}(G) \mathbin{\lfloor\!\lfloor} F)$	SIA12
$(\surd_{\mathsf{s}}\, v \,.\, F) \mid G = \surd_{\mathsf{s}}\, v \,.\, (F \mid \overline{\upsilon}^{v}_{\mathsf{abs}}(G))$	SIA13
$G \mid (\surd_{\mathsf{s}}\, v \,.\, F) = \surd_{\mathsf{s}}\, v \,.\, (\overline{\upsilon}^{v}_{\mathsf{abs}}(G) \mid F)$	SIA14
$\partial_H(\surd_{\mathsf{s}}\, v \,.\, F) = \surd_{\mathsf{s}}\, v \,.\, \partial_H(F)$	SIA15
$\nu_{\mathsf{abs}}(\surd_{\mathsf{s}}\, v \,.\, F) = \surd_{\mathsf{s}}\, v \,.\, \nu_{\mathsf{abs}}(F)$	SIA16
$\textstyle\int_{v \in V}(\surd_{\mathsf{s}}\, w \,.\, F) = \surd_{\mathsf{s}}\, w \,.\, (\int_{v \in V} F)$ if $v \neq w$	SIA17

at which the action is performed. Notice that the equation $\tilde{a} \cdot x = \tilde{a} \cdot \overline{\upsilon}^{0}_{\mathsf{abs}}(x)$ is a special case of axiom SIA6. The related equation $\sigma^{p}_{\mathsf{abs}}(x) = \sigma^{p}_{\mathsf{abs}}(\overline{\upsilon}^{0}_{\mathsf{abs}}(x))$ follows immediately from axioms SAT1 and SAT2 (Table 2, page 634). Axioms SIA7–SIA17 become easier to understand by realizing that $\surd_{\mathsf{s}}\, v \,.\, F$ denotes a function $f : \mathbb{R}_{\geqslant 0} \to \mathsf{P}$ such that $f(p) = \overline{\upsilon}^{p}_{\mathsf{abs}}(f(p))$ for all $p \in \mathbb{R}_{\geqslant 0}$. This is reflected by the equation

$$\surd_{\mathsf{s}}\, v \,.\, F = \surd_{\mathsf{s}}\, v \,.\, \overline{\upsilon}^{v}_{\mathsf{abs}}(F) \quad \text{SIAI}$$

which can be derived using axioms SIA2 and SIA5 and a useful special case of standard initialization axiom SI2 presented in Section 3.3, viz. $\overline{\upsilon}^{p}_{\mathsf{abs}}(\overline{\upsilon}^{p}_{\mathsf{abs}}(x)) = \overline{\upsilon}^{p}_{\mathsf{abs}}(x)$.

The elimination results for $\mathrm{ACP}^{\mathrm{sat}}\mathrm{I}\surd$ with the restricted form of integration mentioned in Section 3.1 are essentially the same as the ones for $\mathrm{ACP}^{\mathrm{sat}}\mathrm{I}$ with the restricted form of integration. Besides, all closed terms of $\mathrm{ACP}^{\mathrm{sat}}\mathrm{I}\surd$ with this restricted form of integration can be written in the form $\surd_{\mathsf{s}}\, v \,.\, F$ where F is a basic term over $\mathrm{BPA}^{\mathrm{sat}}$ with restricted integration.

EXAMPLES. We give some examples of a closed term of $\mathrm{ACP}^{\mathrm{sat}}\mathrm{I}\surd$ with restricted integration, the corresponding term of the form $\surd_{\mathsf{s}}\, v \,.\, F$ where F is a basic term and, if possible, the corresponding basic term without initial abstraction:

$$\surd_{\mathsf{s}}\, v \,.\, \upsilon^{v+2.3}_{\mathsf{abs}}(\surd_{\mathsf{s}}\, w \,.\, \sigma^{w}_{\mathsf{abs}}(\tilde{a})) = \surd_{\mathsf{s}}\, v \,.\, \sigma^{v}_{\mathsf{abs}}(\tilde{a})$$
$$\surd_{\mathsf{s}}\, v \,.\, \overline{\upsilon}^{v+2.3}_{\mathsf{abs}}(\surd_{\mathsf{s}}\, w \,.\, \sigma^{w}_{\mathsf{abs}}(\tilde{a})) = \surd_{\mathsf{s}}\, v \,.\, \sigma^{v+2.3}_{\mathsf{abs}}(\tilde{a})$$
$$\overline{\upsilon}^{3.9}_{\mathsf{abs}}(\surd_{\mathsf{s}}\, v \,.\, \overline{\upsilon}^{v+2.3}_{\mathsf{abs}}(\textstyle\int_{w \in [6,6.1)} \sigma^{w}_{\mathsf{abs}}(\tilde{a}))) = \surd_{\mathsf{s}}\, v \,.\, \sigma^{6.2}_{\mathsf{abs}}(\dot{\delta}) = \sigma^{6.2}_{\mathsf{abs}}(\dot{\delta})$$
$$\overline{\upsilon}^{3.6}_{\mathsf{abs}}(\surd_{\mathsf{s}}\, v \,.\, \overline{\upsilon}^{v+2.3}_{\mathsf{abs}}(\textstyle\int_{w \in [6,6.1)} \sigma^{w}_{\mathsf{abs}}(\tilde{a}))) = \surd_{\mathsf{s}}\, v \,.\, \textstyle\int_{w \in [6,6.1)} \sigma^{w}_{\mathsf{abs}}(\tilde{a}) = \textstyle\int_{w \in [6,6.1)} \sigma^{w}_{\mathsf{abs}}(\tilde{a})$$

Table 13
Definition of operators on RTTS^* ($\varphi : \mathbb{R}_{\geqslant 0} \to \mathsf{RTTS}^*(\mathsf{A})$, $a \in \mathsf{A}_{\delta}$, $p \in \mathbb{R}_{\geqslant 0}$)

$$\dot{\delta} = \lambda t \,.\, \dot{\delta}$$
$$\tilde{a} = \lambda t \,.\, \overline{\upsilon}^{t}_{\mathsf{abs}}(\tilde{a})$$
$$\sigma^{p}_{\mathsf{abs}}(f) = \lambda t \,.\, \overline{\upsilon}^{t}_{\mathsf{abs}}(\sigma^{p}_{\mathsf{abs}}(f(0)))$$
$$f + g = \lambda t \,.\, (f(t) + g(t))$$
$$f \cdot g = \lambda t \,.\, (f(t) * g)$$
$$\upsilon^{p}_{\mathsf{abs}}(f) = \lambda t \,.\, \overline{\upsilon}^{t}_{\mathsf{abs}}(\upsilon^{p}_{\mathsf{abs}}(f(t)))$$
$$\overline{\upsilon}^{p}_{\mathsf{abs}}(f) = f(p)$$

$$f \parallel g = \lambda t \,.\, (f(t) \parallel g(t))$$
$$f \mathbin{\underline{\parallel}} g = \lambda t \,.\, (f(t) \mathbin{\underline{\parallel}} g(t))$$
$$f \mid g = \lambda t \,.\, (f(t) \mid g(t))$$
$$\partial_H(f) = \lambda t \,.\, \partial_H(f(t))$$
$$\nu_{\mathsf{abs}}(f) = \lambda t \,.\, \overline{\upsilon}^{t}_{\mathsf{abs}}(\nu_{\mathsf{abs}}(f(t)))$$

$$\textstyle\int_{v \in V}(f) = \lambda t \,.\, \int_{v \in V}(f(t))$$
$$\surd^{*}_{\mathsf{s}}\, \varphi = \lambda t \,.\, \overline{\upsilon}^{t}_{\mathsf{abs}}(\varphi(t))$$

On the basis of the rules for its operational semantics, the operators of $\mathrm{ACP^{sat}I}$ can also be directly defined on real time transition systems in a straightforward way. In the following, we will describe a model of $\mathrm{ACP^{sat}I}\surd$ in terms of these operators.

SEMANTICS OF $\mathrm{ACP^{sat}I}\surd$. We have to extend RTTS(A) to the function space

$$\mathsf{RTTS}^*(\mathsf{A}) = \{f : \mathbb{R}_{\geqslant 0} \to \mathsf{RTTS}(\mathsf{A}) \mid \forall p \in \mathbb{R}_{\geqslant 0} \bullet f(p) = \overline{\upsilon}^{p}_{\mathsf{abs}}(f(p))\}$$

of *real time transition systems with parametric timing*. We use $f, g, \ldots$ to denote elements of $\mathsf{RTTS}^*(\mathsf{A})$. In Table 13, the constants and operators of $\mathrm{ACP^{sat}I}\surd$ are defined on $\mathsf{RTTS}^*(\mathsf{A})$. We use λ-notation for functions – here t is a variable ranging over $\mathbb{R}_{\geqslant 0}$. We write $f(t) * g$ for the real time transition system obtained from $f(t)$ by replacing $\langle s, p\rangle \xrightarrow{a} \langle \surd, p\rangle$ by $\langle s, p\rangle \xrightarrow{a} \langle s', p\rangle$, where s' is the root state of $g(p)$, whenever s is reachable from the root state of $f(t)$.

We say that $f, g \in \mathsf{RTTS}^*(\mathsf{A})$ are bisimilar if for all $p \in \mathbb{R}_{\geqslant 0}$, there exists a bisimulation R such that $R(f(p), g(p))$. It is easy to see that bisimulation equivalence as defined here is a congruence for the operators of $\mathrm{ACP^{sat}I}\surd$. We obtain a model of $\mathrm{ACP^{sat}I}\surd$ by defining all operators on the set of bisimulation equivalence classes. We will call this model $\mathsf{M}^*_{\mathsf{A}}$. Notice that $f \in \mathsf{RTTS}^*(\mathsf{A})$ corresponds to a process that can be written with the constants and operators of $\mathrm{ACP^{sat}I}$ only iff $\overline{\upsilon}^{0}_{\mathsf{abs}}(f) = f$. In fact, M_{A} is isomorphic to a subalgebra of $\mathsf{M}^*_{\mathsf{A}}$.

Table 14
Standard initialization axioms ($p, q, q' \geqslant 0$, $r > 0$)

$\overline{\upsilon}^{p}_{\mathsf{abs}}(\upsilon^{p+r}_{\mathsf{abs}}(x)) = \upsilon^{p+r}_{\mathsf{abs}}(\overline{\upsilon}^{p}_{\mathsf{abs}}(x))$	SI1
$\overline{\upsilon}^{p}_{\mathsf{abs}}(\overline{\upsilon}^{p+q}_{\mathsf{abs}}(x)) = \overline{\upsilon}^{p+q}_{\mathsf{abs}}(x)$	SI2
$\overline{\upsilon}^{p+q}_{\mathsf{abs}}(\upsilon^{p}_{\mathsf{abs}}(x)) = \sigma^{p+q}_{\mathsf{abs}}(\dot{\delta})$	SI3
$\upsilon^{p}_{\mathsf{abs}}(\overline{\upsilon}^{p+q}_{\mathsf{abs}}(x)) = \sigma^{p}_{\mathsf{abs}}(\dot{\delta})$	SI4
$\sigma^{p}_{\mathsf{abs}}(\dot{\delta}) + \overline{\upsilon}^{p}_{\mathsf{abs}}(x) = \overline{\upsilon}^{p}_{\mathsf{abs}}(x)$	SI5
$\sigma^{p}_{\mathsf{abs}}(\tilde{\delta}) + \overline{\upsilon}^{p}_{\mathsf{abs}}(x + \tilde{\delta}) = \overline{\upsilon}^{p}_{\mathsf{abs}}(x + \tilde{\delta})$	SI6
$\overline{\upsilon}^{r}_{\mathsf{abs}}(x) + \tilde{\delta} = \overline{\upsilon}^{r}_{\mathsf{abs}}(x)$	SI7
$\upsilon^{p}_{\mathsf{abs}}(\upsilon^{q}_{\mathsf{abs}}(x)) = \upsilon^{\min(p,q)}_{\mathsf{abs}}(x)$	SI8
$\overline{\upsilon}^{p}_{\mathsf{abs}}(\overline{\upsilon}^{q}_{\mathsf{abs}}(\overline{\upsilon}^{q'}_{\mathsf{abs}}(x))) = \overline{\upsilon}^{\max(p,q)}_{\mathsf{abs}}(\overline{\upsilon}^{q'}_{\mathsf{abs}}(x))$	SI9
$\overline{\upsilon}^{p}_{\mathsf{abs}}(x \mathbin{⌊⌊} y) = \overline{\upsilon}^{p}_{\mathsf{abs}}(x) \mathbin{⌊⌊} \overline{\upsilon}^{p}_{\mathsf{abs}}(y)$	SI10
$\overline{\upsilon}^{p}_{\mathsf{abs}}(x \mid y) = \overline{\upsilon}^{p}_{\mathsf{abs}}(x) \mid \overline{\upsilon}^{p}_{\mathsf{abs}}(y)$	SI11
$\overline{\upsilon}^{p}_{\mathsf{abs}}(\partial_H(x)) = \partial_H(\overline{\upsilon}^{p}_{\mathsf{abs}}(x))$	SI12
$\overline{\upsilon}^{0}_{\mathsf{abs}}(\nu_{\mathsf{abs}}(x)) = \nu_{\mathsf{abs}}(\overline{\upsilon}^{0}_{\mathsf{abs}}(x))$	SI13
$\overline{\upsilon}^{r}_{\mathsf{abs}}(\nu_{\mathsf{abs}}(x)) = \sigma^{r}_{\mathsf{abs}}(\dot{\delta})$	SI14
$\nu_{\mathsf{abs}}(\overline{\upsilon}^{r}_{\mathsf{abs}}(x)) = \tilde{\delta}$	SI15
$\upsilon^{r}_{\mathsf{abs}}(\nu_{\mathsf{abs}}(x)) = \nu_{\mathsf{abs}}(x)$	SI16
$\nu_{\mathsf{abs}}(\upsilon^{r}_{\mathsf{abs}}(x)) = \nu_{\mathsf{abs}}(x)$	SI17

3.3. *Standard initialization axioms*

In Table 14, some equations concerning initialization and time-out are given that hold in the model $\mathsf{M}_{\mathsf{A}}^{*}$, and that are derivable for closed terms of $\mathrm{ACP}^{\mathrm{sat}}\mathrm{I}\surd$.

We will use these axioms in proofs in subsequent sections. Notice that the very useful equation $\overline{\upsilon}^{p}_{\mathsf{abs}}(\overline{\upsilon}^{p}_{\mathsf{abs}}(x)) = \overline{\upsilon}^{p}_{\mathsf{abs}}(x)$ is a special case of axiom SI2. We can easily prove by means of the standard initialization axioms, using axioms SIA2 and SIA5 (Table 12, page 649), that initial abstraction distributes over $+$, $\|$, $⌊\!⌊$ and $|$:

$$(\surd_{\mathsf{s}}\, v\,.\,F)\,\Box\,(\surd_{\mathsf{s}}\, v\,.\,F') = \surd_{\mathsf{s}}\, v\,.\,(F\,\Box\,F') \quad \text{DISTR}\Box$$

for $\Box = +, \|, ⌊\!⌊, |$. Using this fact shortens many of the calculations needed in the proof of Theorem 6 (embedding of $\mathrm{ACP}^{\mathrm{srt}}$ in $\mathrm{ACP}^{\mathrm{sat}}\surd$).

4. Real time process algebra: relative timing

In this section, we give the signature, axioms and term model of $\mathrm{ACP}^{\mathrm{srt}}$, a standard real time process algebra with relative timing. $\mathrm{ACP}^{\mathrm{srt}}$ originates from the theory ACPst, presented in [6]. Like ACPst, it separates execution of actions and passage of time.

First, in Section 4.1, we treat $\mathrm{BPA}^{\mathrm{srt}}$, basic standard real time process algebra with relative timing, in which parallelism and communication are not considered. After that, in Section 4.2, $\mathrm{BPA}^{\mathrm{srt}}$ is extended to $\mathrm{ACP}^{\mathrm{srt}}$ to deal with parallelism and communication as well. Finally, we show in Section 4.3 how $\mathrm{ACP}^{\mathrm{srt}}$ can be embedded in $\mathrm{ACP}^{\mathrm{sat}}\surd$.

4.1. *Basic process algebra*

In $\mathrm{BPA}^{\mathrm{srt}}$, we have the constants $\tilde{\tilde{a}}$ and $\tilde{\tilde{\delta}}$ instead of $\tilde{a}$ and $\tilde{\delta}$, and the operator σ_{rel} (relative delay) instead of σ_{abs} (absolute delay). The constants $\tilde{\tilde{a}}$ and $\tilde{\tilde{\delta}}$ stand for a without any delay and a deadlock without any delay, respectively. The process $\sigma^{p}_{\mathsf{rel}}(x)$ is the process x delayed for a period of time p. We also have relative counterparts of the absolute time-out and initialization operators: υ_{rel} (relative time-out) and $\overline{\upsilon}_{\mathsf{rel}}$ (relative initialization). The process $\upsilon^{p}_{\mathsf{rel}}(x)$ is the part of x that starts to perform actions after a period of time shorter than p. The process $\overline{\upsilon}^{p}_{\mathsf{rel}}(x)$ is the part of x that starts to perform actions after a period of time longer than or equal to p. In $\mathrm{BPA}^{\mathrm{srt}}$, the sort P of (absolute time) processes is replaced by the sort P^{r} of relative time processes.

The notation $\tilde{\tilde{a}}$ for urgent actions in case of relative timing was also used in ACPst [6], the theory from which $\mathrm{ACP}^{\mathrm{srt}}$ originates.

SIGNATURE OF $\mathrm{BPA}^{\mathrm{srt}}$. The signature of $\mathrm{BPA}^{\mathrm{srt}}$ consists of the *urgent action* constants $\tilde{\tilde{a}} : \to \mathsf{P}^{\mathsf{r}}$ (for each $a \in \mathsf{A}$), the *urgent deadlock* constant $\tilde{\tilde{\delta}} : \to \mathsf{P}^{\mathsf{r}}$, the *immediate deadlock* constant $\dot{\delta} : \to \mathsf{P}^{\mathsf{r}}$, the *alternative composition* operator $+ : \mathsf{P}^{\mathsf{r}} \times \mathsf{P}^{\mathsf{r}} \to \mathsf{P}^{\mathsf{r}}$, the *sequential composition* operator $\cdot : \mathsf{P}^{\mathsf{r}} \times \mathsf{P}^{\mathsf{r}} \to \mathsf{P}^{\mathsf{r}}$, the *relative delay* operator $\sigma_{\mathsf{rel}} : \mathbb{R}_{\geqslant 0} \times \mathsf{P}^{\mathsf{r}} \to \mathsf{P}^{\mathsf{r}}$, the *relative time-out* operator $\upsilon_{\mathsf{rel}} : \mathbb{R}_{\geqslant 0} \times \mathsf{P}^{\mathsf{r}} \to \mathsf{P}^{\mathsf{r}}$, and the *relative initialization* operator $\overline{\upsilon}_{\mathsf{rel}} : \mathbb{R}_{\geqslant 0} \times \mathsf{P}^{\mathsf{r}} \to \mathsf{P}^{\mathsf{r}}$.

Table 15
Additional axioms for BPA^{srt}
$(a \in A_\delta,\ p, q \geqslant 0,\ r > 0)$

$\sigma^0_{rel}(x) = x$	SRT1
$\sigma^p_{rel}(\sigma^q_{rel}(x)) = \sigma^{p+q}_{rel}(x)$	SRT2
$\sigma^p_{rel}(x) + \sigma^p_{rel}(y) = \sigma^p_{rel}(x + y)$	SRT3
$\sigma^p_{rel}(x) \cdot y = \sigma^p_{rel}(x \cdot y)$	SRT4
$\tilde{\tilde{a}} + \tilde{\tilde{\delta}} = \tilde{\tilde{a}}$	A6SRa
$\sigma^r_{rel}(x) + \tilde{\tilde{\delta}} = \sigma^r_{rel}(x)$	A6SRb
$\tilde{\tilde{\delta}} \cdot x = \tilde{\tilde{\delta}}$	A7SR
$\upsilon^p_{rel}(\dot{\delta}) = \dot{\delta}$	SRTO0
$\upsilon^0_{rel}(x) = \dot{\delta}$	SRTO1
$\upsilon^r_{rel}(\tilde{\tilde{a}}) = \tilde{\tilde{a}}$	SRTO2
$\upsilon^{p+q}_{rel}(\sigma^p_{rel}(x)) = \sigma^p_{rel}(\upsilon^q_{rel}(x))$	SRTO3
$\upsilon^p_{rel}(x + y) = \upsilon^p_{rel}(x) + \upsilon^p_{rel}(y)$	SRTO4
$\upsilon^p_{rel}(x \cdot y) = \upsilon^p_{rel}(x) \cdot y$	SRTO5
$\overline{\upsilon}^p_{rel}(\dot{\delta}) = \sigma^p_{rel}(\dot{\delta})$	SRI0
$\overline{\upsilon}^0_{rel}(x) = x$	SRI1
$\overline{\upsilon}^r_{rel}(\tilde{\tilde{a}}) = \sigma^r_{rel}(\dot{\delta})$	SRI2
$\overline{\upsilon}^{p+q}_{rel}(\sigma^p_{rel}(x)) = \sigma^p_{rel}(\overline{\upsilon}^q_{rel}(x))$	SRI3
$\overline{\upsilon}^p_{rel}(x + y) = \overline{\upsilon}^p_{rel}(x) + \overline{\upsilon}^p_{rel}(y)$	SRI4
$\overline{\upsilon}^p_{rel}(x \cdot y) = \overline{\upsilon}^p_{rel}(x) \cdot y$	SRI5

AXIOMS OF BPA^{srt}. The axiom system of BPA^{srt} consists of the equations given in Tables 1 and 15.

The axioms of BPA^{srt} are to a large extent simple reformulations of the axioms of BPA^{sat}. That is, constants $\tilde{a}$ ($a \in A_\delta$) have been replaced by constants $\tilde{\tilde{a}}$, and the operators σ_{abs}, υ_{abs} and $\overline{\upsilon}_{abs}$ have been replaced by σ_{rel}, υ_{rel} and $\overline{\upsilon}_{rel}$, respectively. Striking is the replacement of the axioms SAT4, SAT5 and SAT6 by the simple axiom SRT4. This axiom reflects that timing is relative to the most recent execution of an action. Axioms SRI0–SRI5 are reformulations, in the above-mentioned way, of alternative axioms for axioms SAI0–SAI5 – which, unlike axioms SAI0–SAI5, do not accommodate the addition of initial abstraction (see also Section 2.1).

Similar to the case of BPA^{sat}, we can prove that the auxiliary operators υ_{rel} and $\overline{\upsilon}_{rel}$, as well as sequential compositions in which the form of the first operand is not $\tilde{\tilde{a}}$ ($a \in A$) and alternative compositions in which the form of the first operand is $\sigma^p_{rel}(t)$, can be eliminated in closed terms of BPA^{srt}. The terms that remain after exhaustive elimination are called the basic terms over BPA^{srt}. Because of this elimination result, we are permitted to use induction on the structure of basic terms over BPA^{srt} to prove statements for all closed terms of BPA^{srt}.

EXAMPLES. We give some examples of a closed term of BPA^{srt} and the corresponding basic term:

$$\sigma^{5}_{\mathsf{rel}}(\tilde{\tilde{a}}) \cdot \sigma^{4.9}_{\mathsf{rel}}(\tilde{\tilde{b}}) = \sigma^{5}_{\mathsf{rel}}(\tilde{\tilde{a}} \cdot \sigma^{4.9}_{\mathsf{rel}}(\tilde{\tilde{b}}))$$
$$\sigma^{5}_{\mathsf{rel}}(\tilde{\tilde{a}}) \cdot (\sigma^{4.9}_{\mathsf{rel}}(\tilde{\tilde{b}}) + \sigma^{5.1}_{\mathsf{rel}}(\tilde{\tilde{c}})) = \sigma^{5}_{\mathsf{rel}}(\tilde{\tilde{a}} \cdot \sigma^{4.9}_{\mathsf{rel}}(\tilde{\tilde{b}} + \sigma^{0.2}_{\mathsf{rel}}(\tilde{\tilde{c}})))$$
$$\upsilon^{5}_{\mathsf{rel}}(\sigma^{4.9}_{\mathsf{rel}}(\tilde{\tilde{a}}) + \sigma^{5.1}_{\mathsf{rel}}(\tilde{\tilde{b}})) = \sigma^{4.9}_{\mathsf{rel}}(\tilde{\tilde{a}} + \sigma^{0.1}_{\mathsf{rel}}(\dot{\delta}))$$
$$\overline{\upsilon}^{5}_{\mathsf{rel}}(\sigma^{4.9}_{\mathsf{rel}}(\tilde{\tilde{a}}) + \sigma^{5.1}_{\mathsf{rel}}(\tilde{\tilde{b}})) = \sigma^{5.1}_{\mathsf{rel}}(\tilde{\tilde{b}})$$

SEMANTICS OF BPA^{srt}. In case of relative timing, we can use a simple kind of real time transition system. A *real time transition system with relative timing* over A consists of a set of *states* S, a *root state* $\rho \in \mathsf{S}$ and four kinds of relations on states:

a binary relation $\_ \xrightarrow{a} \_$ for each $a \in \mathsf{A}$,
a unary relation $\_ \xrightarrow{a} \surd$ for each $a \in \mathsf{A}$,
a binary relation $\_ \overset{r}{\mapsto} \_$ for each $r \in \mathbb{R}_{>0}$,
a unary relation $\mathsf{ID}(\_)$;

satisfying

(1) if $s \overset{r+r'}{\longmapsto} s'$, $r, r' > 0$, then there is a s'' such that $s \overset{r}{\mapsto} s''$ and $s'' \overset{r'}{\mapsto} s'$;
(2) if $s \overset{r}{\mapsto} s''$ and $s'' \overset{r'}{\mapsto} s'$, then $s \overset{r+r'}{\longmapsto} s'$.

We write $\mathsf{RTTS}^{\mathsf{r}}(\mathsf{A})$ for the set of all real time transition systems with relative timing over A.

We shall associate a transition system $\mathsf{TS}^{\mathsf{r}}(t)$ in $\mathsf{RTTS}^{\mathsf{r}}(\mathsf{A})$ with a closed term t of BPA^{srt} like before in the case of absolute timing. In case of relative timing, the action step, action termination, time step and immediate deadlock relations can be explained as follows:

$t \xrightarrow{a} t'$: process t is capable of first performing action a without the least delay and then proceeding as process t';
$t \xrightarrow{a} \surd$: process t is capable of first performing action a without the least delay and then terminating successfully;
$t \overset{r}{\mapsto} t'$: process t is capable of first idling for a time period r and then proceeding as process t';
$\mathsf{ID}(t)$: process t is not capable of reaching the present time.

The structural operational semantics of BPA^{srt} is described by the rules given in Table 16. In one of the rules for the alternative composition operator, a negative formula of the form $t \overset{r}{\not\mapsto}$ is used as a premise. A negative formula $t \overset{r}{\not\mapsto}$ means that for all closed terms t' of BPA^{srt} not $t \overset{r}{\mapsto} t'$. Hence, $t \overset{r}{\not\mapsto}$ is to be read as "process t is not capable of idling for a time period r".

Clearly, changing from absolute timing to relative timing leads to a significant simplification of the operational semantics. However, note that there are two rules now for the alternative composition operator concerning time related capabilities of a process $x + y$. These rules have complementary premises. Together they enforce that the choice between two idling processes is postponed till at least one of the processes cannot idle any longer.

Table 16
Rules for operational semantics of $\mathrm{BPA}^{\mathrm{srt}}$
$(a \in \mathsf{A},\ r > 0,\ p \geqslant 0)$

$$\frac{}{\mathsf{ID}(\dot{\delta})} \qquad \frac{}{\tilde{\tilde{a}} \xrightarrow{a} \surd}$$

$$\frac{x \xrightarrow{a} x'}{\sigma^{0}_{\mathsf{rel}}(x) \xrightarrow{a} x'} \qquad \frac{x \xrightarrow{a} \surd}{\sigma^{0}_{\mathsf{rel}}(x) \xrightarrow{a} \surd}$$

$$\frac{\mathsf{ID}(x)}{\mathsf{ID}(\sigma^{0}_{\mathsf{rel}}(x))} \qquad \frac{x \stackrel{r}{\longmapsto} x'}{\sigma^{p}_{\mathsf{rel}}(x) \stackrel{p+r}{\longmapsto} x'}$$

$$\frac{p > 0}{\sigma^{p+r}_{\mathsf{rel}}(x) \stackrel{r}{\longmapsto} \sigma^{p}_{\mathsf{rel}}(x)} \qquad \frac{\neg\mathsf{ID}(x)}{\sigma^{r}_{\mathsf{rel}}(x) \stackrel{r}{\longmapsto} x}$$

$$\frac{x \xrightarrow{a} x'}{x + y \xrightarrow{a} x',\ y + x \xrightarrow{a} x'} \qquad \frac{x \xrightarrow{a} \surd}{x + y \xrightarrow{a} \surd,\ y + x \xrightarrow{a} \surd}$$

$$\frac{x \stackrel{r}{\longmapsto} x',\ y \stackrel{r}{\not\longmapsto}}{x + y \stackrel{r}{\longmapsto} x',\ y + x \stackrel{r}{\longmapsto} x'} \qquad \frac{x \stackrel{r}{\longmapsto} x',\ y \stackrel{r}{\longmapsto} y'}{x + y \stackrel{r}{\longmapsto} x' + y'}$$

$$\frac{\mathsf{ID}(x),\ \mathsf{ID}(y)}{\mathsf{ID}(x + y)}$$

$$\frac{x \xrightarrow{a} x'}{x \cdot y \xrightarrow{a} x' \cdot y} \qquad \frac{x \xrightarrow{a} \surd}{x \cdot y \xrightarrow{a} y}$$

$$\frac{x \stackrel{r}{\longmapsto} x'}{x \cdot y \stackrel{r}{\longmapsto} x' \cdot y} \qquad \frac{\mathsf{ID}(x)}{\mathsf{ID}(x \cdot y)}$$

$$\frac{x \xrightarrow{a} x'}{\upsilon^{r}_{\mathsf{rel}}(x) \xrightarrow{a} x'} \qquad \frac{x \xrightarrow{a} \surd}{\upsilon^{r}_{\mathsf{rel}}(x) \xrightarrow{a} \surd}$$

$$\frac{x \stackrel{r}{\longmapsto} x',\ p > 0}{\upsilon^{p+r}_{\mathsf{rel}}(x) \stackrel{r}{\longmapsto} \upsilon^{p}_{\mathsf{rel}}(x')} \qquad \frac{\mathsf{ID}(x)}{\mathsf{ID}(\upsilon^{r}_{\mathsf{rel}}(x))}$$

$$\frac{}{\mathsf{ID}(\upsilon^{0}_{\mathsf{rel}}(x))}$$

$$\frac{x \xrightarrow{a} x'}{\overline{\upsilon}^{0}_{\mathsf{rel}}(x) \xrightarrow{a} x'} \qquad \frac{x \xrightarrow{a} \surd}{\overline{\upsilon}^{0}_{\mathsf{rel}}(x) \xrightarrow{a} \surd}$$

$$\frac{x \stackrel{r}{\longmapsto} x',\ p \leqslant r}{\overline{\upsilon}^{p}_{\mathsf{rel}}(x) \stackrel{r}{\longmapsto} x'} \qquad \frac{x \stackrel{r}{\longmapsto} x',\ p > 0}{\overline{\upsilon}^{p+r}_{\mathsf{rel}}(x) \stackrel{r}{\longmapsto} \overline{\upsilon}^{p}_{\mathsf{rel}}(x')}$$

$$\frac{x \stackrel{r}{\not\longmapsto},\ p > 0}{\overline{\upsilon}^{p+r}_{\mathsf{rel}}(x) \stackrel{r}{\longmapsto} \overline{\upsilon}^{p}_{\mathsf{rel}}(\dot{\delta})} \qquad \frac{\mathsf{ID}(x)}{\mathsf{ID}(\overline{\upsilon}^{0}_{\mathsf{rel}}(x))}$$

Also the notion of bisimulation becomes simpler in case of relative timing. A *bisimulation* on $\mathsf{RTTS}^{\mathsf{r}}(\mathsf{A})$ is a symmetric binary relation R on the set of states S such that:

(1) if $R(s,t)$ and $s \xrightarrow{a} s'$, then there is a t' such that $t \xrightarrow{a} t'$ and $R(s',t')$;
(2) if $R(s,t)$, then $s \xrightarrow{a} \surd$ iff $t \xrightarrow{a} \surd$;
(3) if $R(s,t)$ and $s \overset{r}{\mapsto} s'$, then there is a t' such that $t \overset{r}{\mapsto} t'$ and $R(s',t')$;
(4) if $R(s,t)$, then $\mathsf{ID}(s)$ iff $\mathsf{ID}(t)$.

As in the case of absolute timing, we obtain a model for BPA^{srt} by identifying bisimilar processes.

4.2. *Algebra of communicating processes*

In ACP^{srt}, we have a relative counterpart of the absolute urgent initialization operator: ν_{rel} (relative urgent initialization). The process $\nu_{\mathsf{rel}}(x)$ is the part of process x that starts to perform actions without any delay. Like before in the case of absolute timing, we use the relative urgent initialization operator to axiomatize the parallel composition operator.

SIGNATURE OF ACP^{srt}. The signature of ACP^{srt} is the signature of BPA^{srt} extended with the *parallel composition* operator $\|: \mathsf{P}^{\mathsf{r}} \times \mathsf{P}^{\mathsf{r}} \to \mathsf{P}^{\mathsf{r}}$, the *left merge* operator $\mathbin{⌊\!⌊}: \mathsf{P}^{\mathsf{r}} \times \mathsf{P}^{\mathsf{r}} \to \mathsf{P}^{\mathsf{r}}$, the *communication merge* operator $|: \mathsf{P}^{\mathsf{r}} \times \mathsf{P}^{\mathsf{r}} \to \mathsf{P}^{\mathsf{r}}$, the *encapsulation* operators $\partial_H : \mathsf{P}^{\mathsf{r}} \to \mathsf{P}^{\mathsf{r}}$ (for each $H \subseteq \mathsf{A}$), and the *relative urgent initialization* operator $\nu_{\mathsf{rel}} : \mathsf{P}^{\mathsf{r}} \to \mathsf{P}^{\mathsf{r}}$.

AXIOMS OF ACP^{srt}. The axiom system of ACP^{srt} consists of the axioms of BPA^{srt} and the equations given in Table 17.

The additional axioms of ACP^{srt} are just simple reformulations of the additional axioms of ACP^{sat}. That is, constants $\tilde{a}$ ($a \in \mathsf{A}_\delta$) have been replaced by constants $\tilde{\tilde{a}}$, and the operators σ_{abs}, υ_{abs} and ν_{abs} have been replaced by σ_{rel}, υ_{rel} and ν_{rel}, respectively.

Similar to the case of ACP^{sat}, we can prove that the operators $\|$, $\mathbin{⌊\!⌊}$, $|$, ∂_H and ν_{rel} can be eliminated in closed terms of ACP^{srt}. Because of the elimination result for BPA^{srt}, we are permitted to use induction on the structure of basic terms over BPA^{srt} to prove statements for all closed terms of ACP^{srt}.

EXAMPLES. We give some examples of a closed term of ACP^{srt} and the corresponding basic term (in case $\gamma(a,c)$ is undefined):

$$\sigma_{\mathsf{rel}}^{5}(\tilde{\tilde{a}}) \parallel \sigma_{\mathsf{rel}}^{5.1}(\tilde{\tilde{b}}) \cdot \sigma_{\mathsf{rel}}^{0.3}(\tilde{\tilde{c}}) = \sigma_{\mathsf{rel}}^{5}(\tilde{\tilde{a}} \cdot \sigma_{\mathsf{rel}}^{0.1}(\tilde{\tilde{b}} \cdot \sigma_{\mathsf{rel}}^{0.3}(\tilde{\tilde{c}})))$$
$$\sigma_{\mathsf{rel}}^{5.1}(\tilde{\tilde{a}}) \parallel \sigma_{\mathsf{rel}}^{5}(\tilde{\tilde{b}}) \cdot \sigma_{\mathsf{rel}}^{0.3}(\tilde{\tilde{c}}) = \sigma_{\mathsf{rel}}^{5}(\tilde{\tilde{b}} \cdot \sigma_{\mathsf{rel}}^{0.1}(\tilde{\tilde{a}} \cdot \sigma_{\mathsf{rel}}^{0.2}(\tilde{\tilde{c}})))$$
$$\sigma_{\mathsf{rel}}^{5.1}(\tilde{\tilde{a}}) \parallel \sigma_{\mathsf{rel}}^{4.8}(\tilde{\tilde{b}}) \cdot \sigma_{\mathsf{rel}}^{0.3}(\tilde{\tilde{c}}) = \sigma_{\mathsf{rel}}^{4.8}(\tilde{\tilde{b}} \cdot \sigma_{\mathsf{rel}}^{0.3}(\tilde{\tilde{a}} \cdot \tilde{\tilde{c}} + \tilde{\tilde{c}} \cdot \tilde{\tilde{a}}))$$

SEMANTICS OF ACP^{srt}. The structural operational semantics of ACP^{srt} is described by the rules for BPA^{srt} and the rules given in Table 18.

Changing from absolute timing to relative timing also leads to a simplification of the additional rules for parallel composition, left merge, etc. As in the previous cases, we obtain a model for ACP^{srt} by identifying bisimilar processes.

Table 17
Additional axioms for $\mathrm{ACP}^{\mathrm{srt}}$
$(a, b \in \mathsf{A}_\delta,\ c \in \mathsf{A},\ p \geqslant 0,\ r > 0)$

$\tilde{\tilde{a}} \mid \tilde{\tilde{b}} = \tilde{\tilde{c}}$ if $\gamma(a, b) = c$	CF1SR
$\tilde{\tilde{a}} \mid \tilde{\tilde{b}} = \tilde{\tilde{\delta}}$ if $\gamma(a, b)$ undefined	CF2SR
$x \parallel y = (x \mathbin{\|\!\lfloor} y + y \mathbin{\|\!\lfloor} x) + x \mid y$	CM1
$\dot{\delta} \mathbin{\|\!\lfloor} x = \dot{\delta}$	CMID1
$x \mathbin{\|\!\lfloor} \dot{\delta} = \dot{\delta}$	CMID2
$\tilde{\tilde{a}} \mathbin{\|\!\lfloor} (x + \tilde{\tilde{\delta}}) = \tilde{\tilde{a}} \cdot (x + \tilde{\tilde{\delta}})$	CM2SRID
$\tilde{\tilde{a}} \cdot x \mathbin{\|\!\lfloor} (y + \tilde{\tilde{\delta}}) = \tilde{\tilde{a}} \cdot (x \parallel (y + \tilde{\tilde{\delta}}))$	CM3SRID
$\sigma^r_{\mathsf{rel}}(x) \mathbin{\|\!\lfloor} (\nu_{\mathsf{rel}}(y) + \tilde{\tilde{\delta}}) = \tilde{\tilde{\delta}}$	SRCM1ID
$\sigma^p_{\mathsf{rel}}(x) \mathbin{\|\!\lfloor} (\upsilon^p_{\mathsf{rel}}(y) + \sigma^p_{\mathsf{rel}}(z)) = \sigma^p_{\mathsf{rel}}(x \mathbin{\|\!\lfloor} z)$	SRCM2ID
$(x + y) \mathbin{\|\!\lfloor} z = x \mathbin{\|\!\lfloor} z + y \mathbin{\|\!\lfloor} z$	CM4
$\dot{\delta} \mid x = \dot{\delta}$	CMID3
$x \mid \dot{\delta} = \dot{\delta}$	CMID4
$\tilde{\tilde{a}} \cdot x \mid \tilde{\tilde{b}} = (\tilde{\tilde{a}} \mid \tilde{\tilde{b}}) \cdot x$	CM5SR
$\tilde{\tilde{a}} \mid \tilde{\tilde{b}} \cdot x = (\tilde{\tilde{a}} \mid \tilde{\tilde{b}}) \cdot x$	CM6SR
$\tilde{\tilde{a}} \cdot x \mid \tilde{\tilde{b}} \cdot y = (\tilde{\tilde{a}} \mid \tilde{\tilde{b}}) \cdot (x \parallel y)$	CM7SR
$(\nu_{\mathsf{rel}}(x) + \tilde{\tilde{\delta}}) \mid \sigma^r_{\mathsf{rel}}(y) = \tilde{\tilde{\delta}}$	SRCM3ID
$\sigma^r_{\mathsf{rel}}(x) \mid (\nu_{\mathsf{rel}}(y) + \tilde{\tilde{\delta}}) = \tilde{\tilde{\delta}}$	SRCM4ID
$\sigma^p_{\mathsf{rel}}(x) \mid \sigma^p_{\mathsf{rel}}(y) = \sigma^p_{\mathsf{rel}}(x \mid y)$	SRCM5
$(x + y) \mid z = x \mid z + y \mid z$	CM8
$x \mid (y + z) = x \mid y + x \mid z$	CM9
$\partial_H(\dot{\delta}) = \dot{\delta}$	D0
$\partial_H(\tilde{\tilde{a}}) = \tilde{\tilde{a}}$ if $a \notin H$	D1SR
$\partial_H(\tilde{\tilde{a}}) = \tilde{\tilde{\delta}}$ if $a \in H$	D2SR
$\partial_H(\sigma^p_{\mathsf{rel}}(x)) = \sigma^p_{\mathsf{rel}}(\partial_H(x))$	SRD
$\partial_H(x + y) = \partial_H(x) + \partial_H(y)$	D3
$\partial_H(x \cdot y) = \partial_H(x) \cdot \partial_H(y)$	D4
$\nu_{\mathsf{rel}}(\dot{\delta}) = \dot{\delta}$	SRU0
$\nu_{\mathsf{rel}}(\tilde{\tilde{a}}) = \tilde{\tilde{a}}$	SRU1
$\nu_{\mathsf{rel}}(\sigma^r_{\mathsf{rel}}(x)) = \tilde{\tilde{\delta}}$	SRU2
$\nu_{\mathsf{rel}}(x + y) = \nu_{\mathsf{rel}}(x) + \nu_{\mathsf{rel}}(y)$	SRU3
$\nu_{\mathsf{rel}}(x \cdot y) = \nu_{\mathsf{rel}}(x) \cdot y$	SRU4

4.3. *Embedding* $\mathrm{ACP}^{\mathrm{srt}}$ *in* $\mathrm{ACP}^{\mathrm{sat}}\surd$

Consider two theories T and T'. An *embedding* of T in T' is a term structure preserving injective mapping ϵ from the terms of T to the terms of T' such that for all closed terms s, t of T, $s = t$ is derivable in T implies $\epsilon(s) = \epsilon(t)$ is derivable in T'. If there exists an embedding of T in T', we say that T can be embedded in T'. It roughly means that what

Table 18
Additional rules for ACP^{srt} $(a, b, c \in \mathsf{A},\ r > 0)$

$$\frac{x \xrightarrow{a} x',\ \neg\mathsf{ID}(y)}{x \parallel y \xrightarrow{a} x' \parallel y,\ y \parallel x \xrightarrow{a} y \parallel x',\ x \mathbin{⌊\!⌊} y \xrightarrow{a} x' \parallel y}$$

$$\frac{x \xrightarrow{a} \surd,\ \neg\mathsf{ID}(y)}{x \parallel y \xrightarrow{a} y,\ y \parallel x \xrightarrow{a} y,\ x \mathbin{⌊\!⌊} y \xrightarrow{a} y}$$

$$\frac{x \xrightarrow{a} x',\ y \xrightarrow{b} y',\ \gamma(a,b) = c}{x \parallel y \xrightarrow{c} x' \parallel y',\ x \mid y \xrightarrow{c} x' \parallel y'} \qquad \frac{x \xrightarrow{a} \surd,\ y \xrightarrow{b} \surd,\ \gamma(a,b) = c}{x \parallel y \xrightarrow{c} \surd,\ x \mid y \xrightarrow{c} \surd}$$

$$\frac{x \xrightarrow{a} x',\ y \xrightarrow{b} \surd,\ \gamma(a,b) = c}{x \parallel y \xrightarrow{c} x',\ y \parallel x \xrightarrow{c} x',\ x \mid y \xrightarrow{c} x',\ y \mid x \xrightarrow{c} x'}$$

$$\frac{x \overset{r}{\mapsto} x',\ y \overset{r}{\mapsto} y'}{x \parallel y \overset{r}{\mapsto} x' \parallel y',\ x \mathbin{⌊\!⌊} y \overset{r}{\mapsto} x' \mathbin{⌊\!⌊} y',\ x \mid y \overset{r}{\mapsto} x' \mid y'}$$

$$\frac{\mathsf{ID}(x)}{\mathsf{ID}(x \parallel y),\ \mathsf{ID}(y \parallel x),\ \mathsf{ID}(x \mathbin{⌊\!⌊} y),\ \mathsf{ID}(y \mathbin{⌊\!⌊} x),\ \mathsf{ID}(x \mid y),\ \mathsf{ID}(y \mid x)}$$

$$\frac{x \xrightarrow{a} x',\ a \notin H}{\partial_H(x) \xrightarrow{a} \partial_H(x')} \qquad \frac{x \xrightarrow{a} \surd,\ a \notin H}{\partial_H(x) \xrightarrow{a} \surd}$$

$$\frac{x \overset{r}{\mapsto} x'}{\partial_H(x) \overset{r}{\mapsto} \partial_H(x')} \qquad \frac{\mathsf{ID}(x)}{\mathsf{ID}(\partial_H(x))}$$

$$\frac{x \xrightarrow{a} x'}{\nu_{\mathsf{rel}}(x) \xrightarrow{a} x'} \qquad \frac{x \xrightarrow{a} \surd}{\nu_{\mathsf{rel}}(x) \xrightarrow{a} \surd} \qquad \frac{\mathsf{ID}(x)}{\mathsf{ID}(\nu_{\mathsf{rel}}(x))}$$

is expressible in T remains expressible in T' and what is derivable in T remains derivable in T'. The requirement that ϵ is term structure preserving means that, for all terms t of T with free variables among $x_1, \ldots, x_n$ and all closed terms $t_1, \ldots, t_n$ of T of appropriate sorts, $\epsilon(t[t_1, \ldots, t_n/x_1, \ldots, x_n]) = \epsilon(t)[\epsilon(t_1), \ldots, \epsilon(t_n)/x_1, \ldots, x_n]$.

Let f be an operator that is not in the signature of theory T. An *explicit definition* of f in T is an equation $f(x_1, \ldots, x_n) = t$ where t is a term of T that does not contain other free variables than $x_1, \ldots, x_n$. An extension of theory T with constants and operators defined by explicit definitions in T is called a definitional extension of T.

Consider again two theories T and T'. Suppose that the constants and operators in the signature of T that are not in the signature of T' can be defined in T' by explicit definitions. Let T'' be the resulting definitional extension of T'. Suppose further that the axioms of T are derivable for closed terms in T''. Then T can be embedded in T'. The explicit definitions induce the following embedding:

$$
\begin{aligned}
&\epsilon(x) = x \\
&\epsilon(f(t_1,\ldots,t_n)) = f(\epsilon(t_1),\ldots,\epsilon(t_n)) && \text{if } f \text{ in the signature of } T'; \\
&\epsilon(f(t_1,\ldots,t_n)) = t[\epsilon(t_1),\ldots,\epsilon(t_n)/x_1,\ldots,x_n] && \text{if the explicit definition of} \\
& && f \text{ is } f(x_1,\ldots,x_n) = t.
\end{aligned}
$$

In this chapter, we will show the existence of embeddings in the way outlined above.

The explicit definitions needed to show that $\mathrm{ACP^{srt}}$ can be embedded in $\mathrm{ACP^{sat}}\surd$ are given in Table 19. The following lemma presents an interesting property of processes with relative timing.

LEMMA 4. *For each closed term t of $\mathrm{ACP^{sat}}\surd$ generated by the embedded constants and operators of $\mathrm{ACP^{srt}}$, $\overline{\upsilon}^p_{\mathsf{abs}}(t) = \sigma^p_{\mathsf{abs}}(t)$.*

PROOF. It is straightforward to prove this by induction on the structure of t. We present only the case that t is of the form $\tilde{\tilde{a}} \cdot t'$. The other cases are similar, but simpler.

$$
\begin{aligned}
&\overline{\upsilon}^p_{\mathsf{abs}}(\tilde{\tilde{a}} \cdot t') = \overline{\upsilon}^p_{\mathsf{abs}}((\surd_{\mathsf{s}}\, v \,.\, \sigma^v_{\mathsf{abs}}(\tilde{a})) \cdot t') \overset{\text{SAI5}}{=} \overline{\upsilon}^p_{\mathsf{abs}}((\surd_{\mathsf{s}}\, v \,.\, \sigma^v_{\mathsf{abs}}(\tilde{a}))) \cdot t' \overset{\text{SIA2}}{=} \\
&\overline{\upsilon}^p_{\mathsf{abs}}(\sigma^p_{\mathsf{abs}}(\tilde{a})) \cdot t' \overset{\text{SAI3}'}{=} \sigma^p_{\mathsf{abs}}(\tilde{a}) \cdot t' \overset{\text{SIA6}}{=} \sigma^p_{\mathsf{abs}}(\tilde{a}) \cdot \overline{\upsilon}^p_{\mathsf{abs}}(t') \overset{\text{IH}}{=} \sigma^p_{\mathsf{abs}}(\tilde{a}) \cdot \sigma^p_{\mathsf{abs}}(t') \overset{\text{SAT5}}{=} \\
&\sigma^p_{\mathsf{abs}}(\tilde{a} \cdot \overline{\upsilon}^0_{\mathsf{abs}}(t')) \overset{\text{SAT1,2}}{=} \sigma^p_{\mathsf{abs}}(\overline{\upsilon}^0_{\mathsf{abs}}(\sigma^0_{\mathsf{abs}}(\tilde{a} \cdot \overline{\upsilon}^0_{\mathsf{abs}}(t')))) \overset{\text{SIA2}}{=} \\
&\sigma^p_{\mathsf{abs}}(\overline{\upsilon}^0_{\mathsf{abs}}(\surd_{\mathsf{s}}\, v \,.\, \sigma^v_{\mathsf{abs}}(\tilde{a} \cdot \overline{\upsilon}^0_{\mathsf{abs}}(t')))) \overset{\text{SAT1,2}}{=} \sigma^p_{\mathsf{abs}}(\surd_{\mathsf{s}}\, v \,.\, \sigma^v_{\mathsf{abs}}(\tilde{a} \cdot \overline{\upsilon}^0_{\mathsf{abs}}(t'))) \overset{\text{SAT5,IH}}{=} \\
&\sigma^p_{\mathsf{abs}}(\surd_{\mathsf{s}}\, v \,.\, (\sigma^v_{\mathsf{abs}}(\tilde{a}) \cdot \overline{\upsilon}^v_{\mathsf{abs}}(t'))) \overset{\text{SIA6,9}}{=} \sigma^p_{\mathsf{abs}}((\surd_{\mathsf{s}}\, v \,.\, \sigma^v_{\mathsf{abs}}(\tilde{a})) \cdot t') = \sigma^p_{\mathsf{abs}}(\tilde{\tilde{a}} \cdot t') \qquad \square
\end{aligned}
$$

Lemma 4 expresses that, for a process with relative timing, absolute initialization of the process at time p is the same as shifting the process in time from time 0 to time p – which implies preceding absolute initialization at time 0. It follows from Lemma 4 and axiom SAI5 that for each pair of closed terms t, t' of $\mathrm{ACP^{sat}}\surd$ generated by the embedded constants and operators of $\mathrm{ACP^{srt}}$, $\sigma^p_{\mathsf{abs}}(t \cdot t') = \sigma^p_{\mathsf{abs}}(t) \cdot t'$. The condition that t is a term generated by the embedded constants and operators of $\mathrm{ACP^{srt}}$ can not be dropped here. However, Lemma 5 points out that this condition is not a necessary one, since $\nu_{\mathsf{abs}}(t)$ is not equal to a term generated by the embedded constants and operators of $\mathrm{ACP^{srt}}$ – unless $t = \dot{\delta}$. Lemma 5 is needed in the proof of Theorem 6.

Table 19
Definitions of relative time operators
$(a \in \mathsf{A}_\delta)$

$$
\begin{aligned}
&\tilde{\tilde{a}} = \surd_{\mathsf{s}}\, v \,.\, \sigma^v_{\mathsf{abs}}(\tilde{a}) \\
&\sigma^p_{\mathsf{rel}}(x) = \surd_{\mathsf{s}}\, v \,.\, \overline{\upsilon}^{v+p}_{\mathsf{abs}}(x) \\
&\upsilon^p_{\mathsf{rel}}(x) = \surd_{\mathsf{s}}\, v \,.\, \upsilon^{v+p}_{\mathsf{abs}}(\overline{\upsilon}^v_{\mathsf{abs}}(x)) \\
&\overline{\upsilon}^p_{\mathsf{rel}}(x) = \surd_{\mathsf{s}}\, v \,.\, \overline{\upsilon}^{v+p}_{\mathsf{abs}}(\overline{\upsilon}^v_{\mathsf{abs}}(x)) \\
&\nu_{\mathsf{rel}}(x) = \surd_{\mathsf{s}}\, v \,.\, \sigma^v_{\mathsf{abs}}(\nu_{\mathsf{abs}}(x))
\end{aligned}
$$

LEMMA 5. *For each pair of closed terms t, t' of* $\mathrm{ACP}^{\mathrm{sat}}\surd$ *generated by the embedded constants and operators of* $\mathrm{ACP}^{\mathrm{srt}}$, $\sigma^p_{\mathsf{abs}}(\nu_{\mathsf{abs}}(t) \cdot t') = \sigma^p_{\mathsf{abs}}(\nu_{\mathsf{abs}}(t)) \cdot t'$.

PROOF. It is straightforward to prove this by induction on the structure of t. The proof is extremely long. We present only the case that t is of the form $\sigma^q_{\mathsf{rel}}(t'')$. The other cases are similar, but simpler, and do not require case distinction.

Case $q = 0$: $\sigma^p_{\mathsf{abs}}(\nu_{\mathsf{abs}}(\sigma^q_{\mathsf{rel}}(t'')) \cdot t') \overset{\text{see I}}{=} \sigma^p_{\mathsf{abs}}(\nu_{\mathsf{abs}}(\sigma^q_{\mathsf{abs}}(t'')) \cdot t') \overset{\text{SAT1}}{=}$
$\sigma^p_{\mathsf{abs}}(\nu_{\mathsf{abs}}(\overline{\upsilon}^q_{\mathsf{abs}}(t'')) \cdot t') \overset{\text{SI13}}{=} \sigma^p_{\mathsf{abs}}(\overline{\upsilon}^q_{\mathsf{abs}}(\nu_{\mathsf{abs}}(t'')) \cdot t') \overset{\text{SAI5}}{=}$
$\sigma^p_{\mathsf{abs}}(\overline{\upsilon}^q_{\mathsf{abs}}(\nu_{\mathsf{abs}}(t'') \cdot t')) \overset{\text{SAT1,2}}{=} \sigma^p_{\mathsf{abs}}(\nu_{\mathsf{abs}}(t'') \cdot t') \overset{\text{IH}}{=} \sigma^p_{\mathsf{abs}}(\nu_{\mathsf{abs}}(t'')) \cdot t' \overset{\text{SAT1,2}}{=}$
$\sigma^p_{\mathsf{abs}}(\overline{\upsilon}^q_{\mathsf{abs}}(\nu_{\mathsf{abs}}(t''))) \cdot t' \overset{\text{SI13}}{=} \sigma^p_{\mathsf{abs}}(\nu_{\mathsf{abs}}(\overline{\upsilon}^q_{\mathsf{abs}}(t''))) \cdot t' \overset{\text{SAT1}}{=}$
$\sigma^p_{\mathsf{abs}}(\nu_{\mathsf{abs}}(\sigma^q_{\mathsf{abs}}(t''))) \cdot t' \overset{\text{see II}}{=} \sigma^p_{\mathsf{abs}}(\nu_{\mathsf{abs}}(\sigma^q_{\mathsf{rel}}(t''))) \cdot t'$

Case $q > 0$: $\sigma^p_{\mathsf{abs}}(\nu_{\mathsf{abs}}(\sigma^q_{\mathsf{rel}}(t'')) \cdot t') \overset{\text{see I}}{=} \sigma^p_{\mathsf{abs}}(\nu_{\mathsf{abs}}(\sigma^q_{\mathsf{abs}}(t'')) \cdot t') \overset{\text{SAU2}}{=} \sigma^p_{\mathsf{abs}}(\tilde{\delta} \cdot t') \overset{\text{SAU1}}{=}$
$\sigma^p_{\mathsf{abs}}(\nu_{\mathsf{abs}}(\tilde{\delta}) \cdot t') \overset{\text{see III}}{=} \sigma^p_{\mathsf{abs}}(\nu_{\mathsf{abs}}(\tilde{\tilde{\delta}}) \cdot t') \overset{\text{IH}}{=} \sigma^p_{\mathsf{abs}}(\nu_{\mathsf{abs}}(\tilde{\tilde{\delta}})) \cdot t' \overset{\text{see IV}}{=}$
$\sigma^p_{\mathsf{abs}}(\nu_{\mathsf{abs}}(\tilde{\delta})) \cdot t' \overset{\text{SAU1}}{=} \sigma^p_{\mathsf{abs}}(\tilde{\delta}) \cdot t' \overset{\text{SAU2}}{=} \sigma^p_{\mathsf{abs}}(\nu_{\mathsf{abs}}(\sigma^q_{\mathsf{abs}}(t''))) \cdot t' \overset{\text{see II}}{=}$
$\sigma^p_{\mathsf{abs}}(\nu_{\mathsf{abs}}(\sigma^q_{\mathsf{rel}}(t''))) \cdot t'$

I. $\sigma^p_{\mathsf{abs}}(\nu_{\mathsf{abs}}(\sigma^q_{\mathsf{rel}}(t'')) \cdot t') = \sigma^p_{\mathsf{abs}}(\nu_{\mathsf{abs}}(\surd_{\mathsf{s}}\, v \,.\, \overline{\upsilon}^{v+q}_{\mathsf{abs}}(t'')) \cdot t') \overset{\text{SIA16}}{=}$
$\sigma^p_{\mathsf{abs}}((\surd_{\mathsf{s}}\, v \,.\, \nu_{\mathsf{abs}}(\overline{\upsilon}^{v+q}_{\mathsf{abs}}(t''))) \cdot t') \overset{\text{SIA9}}{=} \sigma^p_{\mathsf{abs}}(\surd_{\mathsf{s}}\, v \,.\, (\nu_{\mathsf{abs}}(\overline{\upsilon}^{v+q}_{\mathsf{abs}}(t'')) \cdot t')) \overset{\text{SIA7}}{=}$
$\sigma^p_{\mathsf{abs}}(\nu_{\mathsf{abs}}(\overline{\upsilon}^q_{\mathsf{abs}}(t'')) \cdot t') \overset{\text{Lemma 4}}{=} \sigma^p_{\mathsf{abs}}(\nu_{\mathsf{abs}}(\sigma^q_{\mathsf{abs}}(t'')) \cdot t')$

II, III and IV: The proofs are similar to the proof of I – axioms SAT1 and SAI1 are used in addition in III and IV. □

The existence of an embedding of $\mathrm{ACP}^{\mathrm{srt}}$ in $\mathrm{ACP}^{\mathrm{sat}}\surd$ is established by proving the following theorem.

THEOREM 6 (Embedding $\mathrm{ACP}^{\mathrm{srt}}$ in $\mathrm{ACP}^{\mathrm{sat}}\surd$). *For closed terms, the axioms of* $\mathrm{ACP}^{\mathrm{srt}}$ *are derivable from the axioms of* $\mathrm{ACP}^{\mathrm{sat}}\surd$ *and the explicit definitions of the constants and operators* $\tilde{\tilde{a}}$, σ_{rel}, υ_{rel}, $\overline{\upsilon}_{\mathsf{rel}}$, *and* ν_{rel} *in Table* 19.

PROOF. The proof of this theorem is given in Appendix A1. The proof is a matter of straightforward calculations. Equations SIAI (page 650) and DISTR□ (page 652), the standard initialization axioms (Table 14, page 651), and Lemmas 4 and 5 (pages 659, 660) are very useful in the proof. □

5. Discrete time process algebra

In this section, we present $\mathrm{ACP}^{\mathrm{dat}}$ and $\mathrm{ACP}^{\mathrm{drt}}$, discrete time process algebras with absolute timing and relative timing, respectively. $\mathrm{ACP}^{\mathrm{dat}}$ and $\mathrm{ACP}^{\mathrm{drt}}$ are conservative extensions of $\mathrm{ACP}_{\mathrm{dat}}$ and $\mathrm{ACP}_{\mathrm{drt}}$ [7], respectively. First, in Section 5.1, we present $\mathrm{ACP}^{\mathrm{dat}}$ and $\mathrm{ACP}^{\mathrm{dat}}\surd$, the extension of $\mathrm{ACP}^{\mathrm{dat}}$ with initial abstraction. After that, in Section 5.2, we present $\mathrm{ACP}^{\mathrm{drt}}$. Finally, we show in Section 5.3 how $\mathrm{ACP}^{\mathrm{dat}}\surd$ can be embedded in $\mathrm{ACP}^{\mathrm{sat}}\mathrm{I}\surd$.

5.1. *Discrete time process algebra: absolute timing*

In this subsection, we give the signature, axioms and term model of $\mathrm{ACP}^{\mathrm{dat}}$, a discrete time process algebra with absolute timing. $\mathrm{ACP}^{\mathrm{dat}}$ is a conservative extension of the theory $\mathrm{ACP}_{\mathrm{dat}}$, presented in [7]. Like $\mathrm{ACP}^{\mathrm{sat}}$, it separates execution of actions and passage of time. In $\mathrm{ACP}^{\mathrm{dat}}$, time is measured on a discrete time scale. The discrete time points divide time into time slices and timing of actions is done with respect to the time slices in which they are performed – "in time slice $n+1$" means "at some time point p such that $n \leqslant p < n+1$".

First, we treat $\mathrm{BPA}^{\mathrm{dat}}$, basic discrete time process algebra with absolute timing, in which parallelism and communication are not considered. After that, $\mathrm{BPA}^{\mathrm{dat}}$ is extended to $\mathrm{ACP}^{\mathrm{dat}}$ to deal with parallelism and communication as well. Finally, initial abstraction is added.

Basic process algebra. In $\mathrm{BPA}^{\mathrm{dat}}$, we have the constants $\underline{a}$ and $\underline{\delta}$ instead of $\tilde{a}$ and $\tilde{\delta}$. The constants $\underline{a}$ and $\underline{\delta}$ stand for a in time slice 1 and a deadlock in time slice 1, respectively. The operators σ_{abs}, υ_{abs} and $\overline{\upsilon}_{\mathsf{abs}}$ have a natural number instead of a non-negative real number as their first argument. The process $\sigma^{n}_{\mathsf{abs}}(x)$ is the process x shifted in time by n on the discrete time scale. The process $\upsilon^{n}_{\mathsf{abs}}(x)$ is the part of x that starts to perform actions before time slice $n+1$. The process $\overline{\upsilon}^{n}_{\mathsf{abs}}(x)$ is the part of x that starts to perform actions in time slice $n+1$ or a later time slice. Recall that time point n is the starting-point of time slice $n+1$.

In $\mathrm{ACP}_{\mathrm{dat}}$ [7], the notation $\mathsf{fts}(a)$ was used for actions in the first time slice. A discrete time version of ACP with absolute timing where the notation $\underline{a}$ was used earlier for actions in a time slice is ACPdρ [3], but there it always carries a time-stamp.

SIGNATURE OF $\mathrm{BPA}^{\mathrm{dat}}$. The signature of $\mathrm{BPA}^{\mathrm{dat}}$ consists of the *undelayable action* constants $\underline{a} : \to \mathsf{P}$ (for each $a \in \mathsf{A}$), the *undelayable deadlock* constant $\underline{\delta} : \to \mathsf{P}$, the *immediate deadlock* constant $\dot{\delta} : \to \mathsf{P}$, the *alternative composition* operator $+ : \mathsf{P} \times \mathsf{P} \to \mathsf{P}$, the *sequential composition* operator $\cdot : \mathsf{P} \times \mathsf{P} \to \mathsf{P}$, the *absolute delay* operator $\sigma_{\mathsf{abs}} : \mathbb{N} \times \mathsf{P} \to \mathsf{P}$, the *absolute time-out* operator $\upsilon_{\mathsf{abs}} : \mathbb{N} \times \mathsf{P} \to \mathsf{P}$, and the *absolute initialization* operator $\overline{\upsilon}_{\mathsf{abs}} : \mathbb{N} \times \mathsf{P} \to \mathsf{P}$.

We denote elements of $\mathbb{N}$ by m, m', n, n'.

AXIOMS OF $\mathrm{BPA}^{\mathrm{dat}}$. The axiom system of $\mathrm{BPA}^{\mathrm{dat}}$ consists of the equations given in Tables 1 and 20.

The axioms of $\mathrm{BPA}^{\mathrm{dat}}$ are to a large extent simple reformulations of the axioms of $\mathrm{BPA}^{\mathrm{sat}}$. That is, constants $\tilde{a}$ ($a \in \mathsf{A}_{\delta}$) have been replaced by constants $\underline{a}$, and the first argument of the operators σ_{abs}, υ_{abs} and $\overline{\upsilon}_{\mathsf{abs}}$ has been restricted to elements of $\mathbb{N}$. Striking is the new axiom DAT7. This axiom makes the reformulations of axioms A6SAb and A7SA, i.e. $\sigma^{n+1}_{\mathsf{abs}}(x) + \underline{\delta} = \sigma^{n+1}_{\mathsf{abs}}(x)$ and $\underline{\delta} \cdot x = \underline{\delta}$, derivable. Axiom DAT7 expresses that an immediate deadlock shifted in time by 1 is identified with an undelayable deadlock in the first time slice.

Table 21
Additional axioms for $\mathrm{ACP}^{\mathrm{dat}}$ $(a, b \in \mathsf{A}_\delta,\ c \in \mathsf{A})$

$\underline{a} \mid \underline{b} = \underline{c}$ if $\gamma(a,b) = c$	CF1DA
$\underline{a} \mid \underline{b} = \underline{\delta}$ if $\gamma(a,b)$ undefined	CF2DA
$x \parallel y = (x \mathbin{\lfloor\!\lfloor} y + y \mathbin{\lfloor\!\lfloor} x) + x \mid y$	CM1
$\dot{\delta} \mathbin{\lfloor\!\lfloor} x = \dot{\delta}$	CMID1
$x \mathbin{\lfloor\!\lfloor} \dot{\delta} = \dot{\delta}$	CMID2
$\underline{a} \mathbin{\lfloor\!\lfloor} (x + \underline{\delta}) = \underline{a} \cdot (x + \underline{\delta})$	CM2DA
$\underline{a} \cdot x \mathbin{\lfloor\!\lfloor} (y + \underline{\delta}) = \underline{a} \cdot (x \parallel (y + \underline{\delta}))$	CM3DA
$\sigma^n_{\mathsf{abs}}(x) \mathbin{\lfloor\!\lfloor} (\upsilon^n_{\mathsf{abs}}(y) + \sigma^n_{\mathsf{abs}}(z)) = \sigma^n_{\mathsf{abs}}(x \mathbin{\lfloor\!\lfloor} z)$	DACM2
$(x + y) \mathbin{\lfloor\!\lfloor} z = x \mathbin{\lfloor\!\lfloor} z + y \mathbin{\lfloor\!\lfloor} z$	CM4
$\dot{\delta} \mid x = \dot{\delta}$	CMID3
$x \mid \dot{\delta} = \dot{\delta}$	CMID4
$\underline{a} \cdot x \mid \underline{b} = (\underline{a} \mid \underline{b}) \cdot x$	CM5DA
$\underline{a} \mid \underline{b} \cdot x = (\underline{a} \mid \underline{b}) \cdot x$	CM6DA
$\underline{a} \cdot x \mid \underline{b} \cdot y = (\underline{a} \mid \underline{b}) \cdot (x \parallel y)$	CM7DA
$(\upsilon^1_{\mathsf{abs}}(x) + \underline{\delta}) \mid \sigma^{n+1}_{\mathsf{abs}}(y) = \underline{\delta}$	DACM3
$\sigma^{n+1}_{\mathsf{abs}}(x) \mid (\upsilon^1_{\mathsf{abs}}(y) + \underline{\delta}) = \underline{\delta}$	DACM4
$\sigma^n_{\mathsf{abs}}(x) \mid \sigma^n_{\mathsf{abs}}(y) = \sigma^n_{\mathsf{abs}}(x \mid y)$	DACM5
$(x + y) \mid z = x \mid z + y \mid z$	CM8
$x \mid (y + z) = x \mid y + x \mid z$	CM9
$\partial_H(\dot{\delta}) = \dot{\delta}$	D0
$\partial_H(\underline{a}) = \underline{a}$ if $a \notin H$	D1DA
$\partial_H(\underline{a}) = \underline{\delta}$ if $a \in H$	D2DA
$\partial_H(\sigma^n_{\mathsf{abs}}(x)) = \sigma^n_{\mathsf{abs}}(\partial_H(x))$	DAD
$\partial_H(x + y) = \partial_H(x) + \partial_H(y)$	D3
$\partial_H(x \cdot y) = \partial_H(x) \cdot \partial_H(y)$	D4

SEMANTICS OF $\mathrm{ACP}^{\mathrm{dat}}$. Like for the rules for the operational semantics of $\mathrm{BPA}^{\mathrm{dat}}$, the additional rules for $\mathrm{ACP}^{\mathrm{dat}}$ differ from the corresponding rules for the real time case only in that all numbers involved are restricted to $\mathbb{N}$. Therefore, we refrain again from giving the rules.

Again, we obtain a model for $\mathrm{ACP}^{\mathrm{dat}}$ by identifying bisimilar processes.

Initial abstraction. We add the initial abstraction operator $\surd_{\mathsf{d}}$ to $\mathrm{ACP}^{\mathrm{dat}}$. This operator is the discrete counterpart of $\surd_{\mathsf{s}}$. This means that $\surd_{\mathsf{d}}\, i\, .\, F$, where i is a variable ranging over $\mathbb{N}$ and F is a term that may contain free variables, denotes a function $f : \mathbb{N} \to \mathsf{P}$ that satisfies $f(n) = \overline{\upsilon}^n_{\mathsf{abs}}(f(n))$ for all $n \in \mathbb{N}$. In the resulting theory, called $\mathrm{ACP}^{\mathrm{dat}}\surd$, the sort P of (absolute time) processes is replaced by the sort P^* of parametric time processes.

SIGNATURE OF $\mathrm{ACP}^{\mathrm{dat}}\surd$. The signature of $\mathrm{ACP}^{\mathrm{dat}}\surd$ is the signature of $\mathrm{ACP}^{\mathrm{dat}}$ extended with the *initial abstraction* (variable-binding) operator $\surd_{\mathsf{d}} : \mathbb{N}\, .\, \mathsf{P}^* \to \mathsf{P}^*$.

Table 22
Axioms for discrete initial abstraction (i not free in G)

$\surd_{\mathsf{d}}\, j\,.\,G = \surd_{\mathsf{d}}\, i\,.\,G[i/j]$	DIA1
$\overline{\upsilon}^n_{\mathsf{abs}}(\surd_{\mathsf{d}}\, i\,.\,F) = \overline{\upsilon}^n_{\mathsf{abs}}(F[n/i])$	DIA2
$\surd_{\mathsf{d}}\, i\,.\,(\surd_{\mathsf{d}}\, j\,.\,F) = \surd_{\mathsf{d}}\, i\,.\,F[i/j]$	DIA3
$G = \surd_{\mathsf{d}}\, i\,.\,G$	DIA4
$(\forall n \in \mathbb{N} \bullet \overline{\upsilon}^n_{\mathsf{abs}}(x) = \overline{\upsilon}^n_{\mathsf{abs}}(y)) \;\Rightarrow\; x = y$	DIA5
$\sigma^n_{\mathsf{abs}}(\underline{a}) \cdot x = \sigma^n_{\mathsf{abs}}(\underline{a}) \cdot \overline{\upsilon}^n_{\mathsf{abs}}(x)$	DIA6
$\sigma^n_{\mathsf{abs}}(\surd_{\mathsf{d}}\, i\,.\,F) = \sigma^n_{\mathsf{abs}}(F[0/i])$	DIA7
$(\surd_{\mathsf{d}}\, i\,.\,F) + G = \surd_{\mathsf{d}}\, i\,.\,(F + \overline{\upsilon}^i_{\mathsf{abs}}(G))$	DIA8
$(\surd_{\mathsf{d}}\, i\,.\,F) \cdot G = \surd_{\mathsf{d}}\, i\,.\,(F \cdot G)$	DIA9
$\upsilon^n_{\mathsf{abs}}(\surd_{\mathsf{d}}\, i\,.\,F) = \surd_{\mathsf{d}}\, i\,.\,\upsilon^n_{\mathsf{abs}}(F)$ if $n \neq i$	DIA10
$(\surd_{\mathsf{d}}\, i\,.\,F) \mathbin{⌊\!⌊} G = \surd_{\mathsf{d}}\, i\,.\,(F \mathbin{⌊\!⌊} \overline{\upsilon}^i_{\mathsf{abs}}(G))$	DIA11
$G \mathbin{⌊\!⌊} (\surd_{\mathsf{d}}\, i\,.\,F) = \surd_{\mathsf{d}}\, i\,.\,(\overline{\upsilon}^i_{\mathsf{abs}}(G) \mathbin{⌊\!⌊} F)$	DIA12
$(\surd_{\mathsf{d}}\, i\,.\,F) \mid G = \surd_{\mathsf{d}}\, i\,.\,(F \mid \overline{\upsilon}^i_{\mathsf{abs}}(G))$	DIA13
$G \mid (\surd_{\mathsf{d}}\, i\,.\,F) = \surd_{\mathsf{d}}\, i\,.\,(\overline{\upsilon}^i_{\mathsf{abs}}(G) \mid F)$	DIA14
$\partial_H(\surd_{\mathsf{d}}\, i\,.\,F) = \surd_{\mathsf{d}}\, i\,.\,\partial_H(F)$	DIA15

We assume that an infinite set of variables ranging over $\mathbb{N}$ has been given, and denote them by $i, j, \ldots$. We denote terms of $\mathrm{ACP}^{\mathrm{dat}}\surd$ by $F, G, \ldots$.

AXIOM SYSTEM OF $\mathrm{ACP}^{\mathrm{dat}}\surd$. The axiom system of $\mathrm{ACP}^{\mathrm{dat}}\surd$ consists of the axioms of $\mathrm{ACP}^{\mathrm{dat}}$ and the equations given in Table 22.

The axioms for discrete initial abstraction are simple reformulations of the axioms for standard initial abstraction. That is, the operator $\surd_{\mathsf{s}}$ has been replaced by $\surd_{\mathsf{d}}$, and the variables ranging over $\mathbb{R}_{\geqslant 0}$ have been replaced by variables ranging over $\mathbb{N}$.

As in the case of $\mathrm{ACP}^{\mathrm{sat}}\surd$, all closed terms of $\mathrm{ACP}^{\mathrm{dat}}\surd$ can be written in the form $\surd_{\mathsf{d}}\, i\,.\,F$ where F is a basic term over $\mathrm{BPA}^{\mathrm{dat}}$.

EXAMPLES. We give some examples of a closed term of $\mathrm{ACP}^{\mathrm{dat}}\surd$, the corresponding term of the form $\surd_{\mathsf{d}}\, i\,.\,F$ where F is a basic term and, if possible, the corresponding basic term without initial abstraction:

$$\sigma^2_{\mathsf{abs}}(\surd_{\mathsf{d}}\, i\,.\,\sigma^{i+3}_{\mathsf{abs}}(\underline{a})) = \surd_{\mathsf{d}}\, i\,.\,\sigma^5_{\mathsf{abs}}(\underline{a}) = \sigma^5_{\mathsf{abs}}(\underline{a})$$
$$\upsilon^2_{\mathsf{abs}}(\surd_{\mathsf{d}}\, i\,.\,\sigma^{i+3}_{\mathsf{abs}}(\underline{a})) = \surd_{\mathsf{d}}\, i\,.\,\sigma^2_{\mathsf{abs}}(\dot{\delta}) = \sigma^2_{\mathsf{abs}}(\dot{\delta})$$
$$\surd_{\mathsf{d}}\, i\,.\,(\surd_{\mathsf{d}}\, j\,.\,\sigma^{i+j+3}_{\mathsf{abs}}(\underline{a})) = \surd_{\mathsf{d}}\, i\,.\,\sigma^{2i+3}_{\mathsf{abs}}(\underline{a})$$

SEMANTICS OF $\mathrm{ACP}^{\mathrm{dat}}\surd$. We have to extend $\mathsf{DTTS(A)}$ to the function space

$$\mathsf{DTTS}^*(\mathsf{A}) = \{f : \mathbb{N} \to \mathsf{DTTS(A)} \mid \forall n \in \mathbb{N} \bullet f(n) = \overline{\upsilon}^n_{\mathsf{abs}}(f(n))\}$$

The constants and operators of $\mathrm{ACP}^{\mathrm{dat}}\surd$ can be defined on $\mathsf{DTTS}^*(\mathsf{A})$ in the same way as for the real time case.

We say that $f, g \in \mathsf{DTTS}^*(\mathsf{A})$ are bisimilar if for all $n \in \mathbb{N}$, there exists a bisimulation R such that $R(f(n), g(n))$. Like before, we obtain a model of $\mathrm{ACP}^{\mathrm{dat}}\surd$ by defining all operators on the set of bisimulation equivalence classes.

5.2. *Discrete time process algebra: relative timing*

In this subsection, we give the signature, axioms and term model of $\mathrm{ACP}^{\mathrm{drt}}$, a discrete time process algebra with relative timing. $\mathrm{ACP}^{\mathrm{drt}}$ is a conservative extension of the theory $\mathrm{ACP}_{\mathrm{drt}}$, presented in [7]. Like $\mathrm{ACP}_{\mathrm{drt}}$, it separates execution of actions and passage of time.

First, we treat $\mathrm{BPA}^{\mathrm{drt}}$, basic discrete time process algebra with relative timing, in which parallelism and communication are not considered. After that, $\mathrm{BPA}^{\mathrm{drt}}$ is extended to $\mathrm{ACP}^{\mathrm{drt}}$ to deal with parallelism and communication as well.

Basic process algebra. In $\mathrm{BPA}^{\mathrm{drt}}$, we have the constants $\underline{\underline{a}}$ and $\underline{\underline{\delta}}$ instead of $\underline{a}$ and $\underline{\delta}$, and the operator σ_{rel} instead of σ_{abs}. The constants $\underline{\underline{a}}$ and $\underline{\underline{\delta}}$ stand for a in the current time slice and a deadlock in the current time slice, respectively. The process $\sigma^n_{\mathsf{rel}}(x)$ is the process x delayed for a period of time n on the discrete time scale, i.e. till the n-th next time slice. We have relative counterparts of the absolute time-out and initialization operators as well: υ_{rel} and $\overline{\upsilon}_{\mathsf{rel}}$. The process $\upsilon^n_{\mathsf{rel}}(x)$ is the part of x that starts to perform actions before the n-th next time slice. The process $\overline{\upsilon}^n_{\mathsf{rel}}(x)$ is the part of x that starts to perform actions in the n-th next time slice or a later time slice. As in Section 4, we use P^{r} for the sort of relative time processes.

In some presentations of $\mathrm{ACP}_{\mathrm{drt}}$, including [7], the notation $\mathsf{cts}(a)$ was used instead of $\underline{\underline{a}}$. The notation $\underline{\underline{a}}$ for actions in the current time slice was first used in ACPdt [3].

SIGNATURE OF $\mathrm{BPA}^{\mathrm{drt}}$. The signature of $\mathrm{BPA}^{\mathrm{drt}}$ consists of the *undelayable action* constants $\underline{\underline{a}} : \to \mathsf{P}^{\mathsf{r}}$ (for each $a \in \mathsf{A}$), the *undelayable deadlock* constant $\underline{\underline{\delta}} : \to \mathsf{P}^{\mathsf{r}}$, the *immediate deadlock* constant $\dot{\delta} : \to \mathsf{P}^{\mathsf{r}}$, the *alternative composition* operator $+ : \mathsf{P}^{\mathsf{r}} \times \mathsf{P}^{\mathsf{r}} \to \mathsf{P}^{\mathsf{r}}$, the *sequential composition* operator $\cdot : \mathsf{P}^{\mathsf{r}} \times \mathsf{P}^{\mathsf{r}} \to \mathsf{P}^{\mathsf{r}}$, the *relative delay* operator $\sigma_{\mathsf{rel}} : \mathbb{N} \times \mathsf{P}^{\mathsf{r}} \to \mathsf{P}^{\mathsf{r}}$, the *relative time-out* operator $\upsilon_{\mathsf{rel}} : \mathbb{N} \times \mathsf{P}^{\mathsf{r}} \to \mathsf{P}^{\mathsf{r}}$, and the *relative initialization* operator $\overline{\upsilon}_{\mathsf{rel}} : \mathbb{N} \times \mathsf{P}^{\mathsf{r}} \to \mathsf{P}^{\mathsf{r}}$.

AXIOMS OF $\mathrm{BPA}^{\mathrm{drt}}$. The axiom system of $\mathrm{BPA}^{\mathrm{drt}}$ consists of the equations given in Tables 1 and 23.

The axioms of $\mathrm{BPA}^{\mathrm{drt}}$ are to a large extent simple reformulations of the axioms of $\mathrm{BPA}^{\mathrm{dat}}$. That is, constants $\underline{a}$ ($a \in \mathsf{A}_\delta$) have been replaced by constants $\underline{\underline{a}}$, and the operators σ_{abs}, υ_{abs} and $\overline{\upsilon}_{\mathsf{abs}}$ have been replaced by σ_{rel}, υ_{rel} and $\overline{\upsilon}_{\mathsf{rel}}$, respectively. The replacement of the axioms DAT4, DAT5 and DAT6 by the simple axiom DRT4 as well as the replacement of the axioms DAI0–DAI5 by the axioms DRI0–DRI5 are strongly reminiscent of the real time case.

Table 23
Additional axioms for $\mathrm{BPA}^{\mathrm{drt}}$ ($a \in \mathsf{A}_\delta$)

$\sigma^0_{\mathsf{rel}}(x) = x$	DRT1
$\sigma^m_{\mathsf{rel}}(\sigma^n_{\mathsf{rel}}(x)) = \sigma^{m+n}_{\mathsf{rel}}(x)$	DRT2
$\sigma^n_{\mathsf{rel}}(x) + \sigma^n_{\mathsf{rel}}(y) = \sigma^n_{\mathsf{rel}}(x+y)$	DRT3
$\sigma^n_{\mathsf{rel}}(x) \cdot y = \sigma^n_{\mathsf{rel}}(x \cdot y)$	DRT4
$\sigma^1_{\mathsf{rel}}(\dot{\delta}) = \underline{\underline{\delta}}$	DRT7
$\underline{\underline{a}} + \underline{\underline{\delta}} = \underline{\underline{a}}$	A6DRa
$\upsilon^n_{\mathsf{rel}}(\dot{\delta}) = \dot{\delta}$	DRTO0
$\upsilon^0_{\mathsf{rel}}(x) = \dot{\delta}$	DRTO1
$\upsilon^{n+1}_{\mathsf{rel}}(\underline{\underline{a}}) = \underline{\underline{a}}$	DRTO2
$\upsilon^{m+n}_{\mathsf{rel}}(\sigma^n_{\mathsf{rel}}(x)) = \sigma^n_{\mathsf{rel}}(\upsilon^m_{\mathsf{rel}}(x))$	DRTO3
$\upsilon^n_{\mathsf{rel}}(x+y) = \upsilon^n_{\mathsf{rel}}(x) + \upsilon^n_{\mathsf{rel}}(y)$	DRTO4
$\upsilon^n_{\mathsf{rel}}(x \cdot y) = \upsilon^n_{\mathsf{rel}}(x) \cdot y$	DRTO5
$\overline{\upsilon}^n_{\mathsf{rel}}(\dot{\delta}) = \sigma^n_{\mathsf{rel}}(\dot{\delta})$	DRI0
$\overline{\upsilon}^0_{\mathsf{rel}}(x) = x$	DRI1
$\overline{\upsilon}^{n+1}_{\mathsf{rel}}(\underline{\underline{a}}) = \sigma^n_{\mathsf{rel}}(\underline{\underline{\delta}})$	DRI2
$\overline{\upsilon}^{m+n}_{\mathsf{rel}}(\sigma^n_{\mathsf{rel}}(x)) = \sigma^n_{\mathsf{rel}}(\overline{\upsilon}^m_{\mathsf{rel}}(x))$	DRI3
$\overline{\upsilon}^n_{\mathsf{rel}}(x+y) = \overline{\upsilon}^n_{\mathsf{rel}}(x) + \overline{\upsilon}^n_{\mathsf{rel}}(y)$	DRI4
$\overline{\upsilon}^n_{\mathsf{rel}}(x \cdot y) = \overline{\upsilon}^n_{\mathsf{rel}}(x) \cdot y$	DRI5

Similar to the case of $\mathrm{BPA}^{\mathrm{dat}}$, we can prove that the auxiliary operators υ_{rel} and $\overline{\upsilon}_{\mathsf{rel}}$, as well as sequential compositions in which the form of the first operand is not $\underline{\underline{a}}$ ($a \in \mathsf{A}$) and alternative compositions in which the form of the first operand is $\sigma^n_{\mathsf{rel}}(t)$, can be eliminated in closed terms of $\mathrm{BPA}^{\mathrm{drt}}$. The terms that remain after exhaustive elimination are called the basic terms over $\mathrm{BPA}^{\mathrm{drt}}$.

SEMANTICS OF $\mathrm{BPA}^{\mathrm{drt}}$. In case of relative timing, we can use a simple kind of discrete time transition system. A *discrete time transition system with relative timing* over A consists of a set of *states* S, a *root state* $\rho \in \mathsf{S}$ and four kinds of relations on states:

a binary relation $\_ \xrightarrow{a} \_$ for each $a \in \mathsf{A}$,
a unary relation $\_ \xrightarrow{a} \surd$ for each $a \in \mathsf{A}$,
a binary relation $\_ \overset{n}{\longmapsto} \_$ for each $n \in \mathbb{N}_{>0}$,
a unary relation $\mathsf{ID}(\_)$;

satisfying

(1) if $s \overset{n+n'}{\longmapsto} s'$, $n, n' > 0$, then there is a s'' such that $s \overset{n}{\longmapsto} s''$ and $s'' \overset{n'}{\longmapsto} s'$;
(2) if $s \overset{n}{\longmapsto} s''$ and $s'' \overset{n'}{\longmapsto} s'$, then $s \overset{n+n'}{\longmapsto} s'$.

We write $\mathsf{DTTS}^{\mathsf{r}}(\mathsf{A})$ for the set of all discrete time transition systems with relative timing over A. Associating a transition system in $\mathsf{DTTS}^{\mathsf{r}}(\mathsf{A})$ with a closed term t of $\mathrm{BPA}^{\mathrm{drt}}$ proceeds in essentially the same way as associating a transition system in $\mathsf{RTTS}^{\mathsf{r}}(\mathsf{A})$ with a closed term t of $\mathrm{BPA}^{\mathrm{srt}}$. The only difference is that in the rules for the operational semantics of $\mathrm{BPA}^{\mathrm{drt}}$ all numbers involved are restricted to $\mathbb{N}$. Therefore, we refrain from giving the rules.

Bisimulation on $\mathsf{DTTS}^{\mathsf{r}}(\mathsf{A})$ is defined as on $\mathsf{RTTS}^{\mathsf{r}}(\mathsf{A})$. As in the real time cases, we obtain a model for $\mathrm{BPA}^{\mathrm{drt}}$ by identifying bisimilar processes.

Algebra of communicating processes. Like in $\mathrm{ACP}^{\mathrm{dat}}$, we do not have a discrete time counterpart of ν_{rel} in $\mathrm{ACP}^{\mathrm{drt}}$. We can use $\upsilon^{1}_{\mathsf{rel}}$ instead.

SIGNATURE OF $\mathrm{ACP}^{\mathrm{drt}}$. The signature of $\mathrm{ACP}^{\mathrm{drt}}$ is the signature of $\mathrm{BPA}^{\mathrm{drt}}$ extended with the *parallel composition* operator $\|: \mathsf{P}^{\mathsf{r}} \times \mathsf{P}^{\mathsf{r}} \to \mathsf{P}^{\mathsf{r}}$, the *left merge* operator $\mathbin{|\!\!\lfloor}: \mathsf{P}^{\mathsf{r}} \times \mathsf{P}^{\mathsf{r}} \to \mathsf{P}^{\mathsf{r}}$, the *communication merge* operator $|: \mathsf{P}^{\mathsf{r}} \times \mathsf{P}^{\mathsf{r}} \to \mathsf{P}^{\mathsf{r}}$, and the *encapsulation* operators $\partial_H : \mathsf{P}^{\mathsf{r}} \to \mathsf{P}^{\mathsf{r}}$ (for each $H \subseteq \mathsf{A}$).

Table 24
Additional axioms for $\mathrm{ACP}^{\mathrm{drt}}$ $(a, b \in \mathsf{A}_{\delta},\ c \in \mathsf{A})$

$\underline{\underline{a}} \mid \underline{\underline{b}} = \underline{\underline{c}}$ if $\gamma(a, b) = c$	CF1DR
$\underline{\underline{a}} \mid \underline{\underline{b}} = \underline{\underline{\delta}}$ if $\gamma(a, b)$ undefined	CF2DR
$x \parallel y = (x \mathbin{|\!\!\lfloor} y + y \mathbin{|\!\!\lfloor} x) + x \mid y$	CM1
$\dot{\delta} \mathbin{|\!\!\lfloor} x = \dot{\delta}$	CMID1
$x \mathbin{|\!\!\lfloor} \dot{\delta} = \dot{\delta}$	CMID2
$\underline{\underline{a}} \mathbin{|\!\!\lfloor} (x + \underline{\underline{\delta}}) = \underline{\underline{a}} \cdot (x + \underline{\underline{\delta}})$	CM2DRID
$\underline{\underline{a}} \cdot x \mathbin{|\!\!\lfloor} (y + \underline{\underline{\delta}}) = \underline{\underline{a}} \cdot (x \parallel (y + \underline{\underline{\delta}}))$	CM3DRID
$\sigma^{n+1}_{\mathsf{rel}}(x) \mathbin{|\!\!\lfloor} (\upsilon^{n+1}_{\mathsf{rel}}(y) + \sigma^{n+1}_{\mathsf{rel}}(z)) = \sigma^{n+1}_{\mathsf{rel}}(x \mathbin{|\!\!\lfloor} z)$	DRCM2
$(x + y) \mathbin{|\!\!\lfloor} z = x \mathbin{|\!\!\lfloor} z + y \mathbin{|\!\!\lfloor} z$	CM4
$\dot{\delta} \mid x = \dot{\delta}$	CMID3
$x \mid \dot{\delta} = \dot{\delta}$	CMID4
$\underline{\underline{a}} \cdot x \mid \underline{\underline{b}} = (\underline{\underline{a}} \mid \underline{\underline{b}}) \cdot x$	CM5DR
$\underline{\underline{a}} \mid \underline{\underline{b}} \cdot x = (\underline{\underline{a}} \mid \underline{\underline{b}}) \cdot x$	CM6DR
$\underline{\underline{a}} \cdot x \mid \underline{\underline{b}} \cdot y = (\underline{\underline{a}} \mid \underline{\underline{b}}) \cdot (x \parallel y)$	CM7DR
$(\upsilon^{1}_{\mathsf{rel}}(x) + \underline{\underline{\delta}}) \mid \sigma^{n+1}_{\mathsf{rel}}(y) = \underline{\underline{\delta}}$	DRCM3ID
$\sigma^{n+1}_{\mathsf{rel}}(x) \mid (\upsilon^{1}_{\mathsf{rel}}(y) + \underline{\underline{\delta}}) = \underline{\underline{\delta}}$	DRCM4ID
$\sigma^{n+1}_{\mathsf{rel}}(x) \mid \sigma^{n+1}_{\mathsf{rel}}(y) = \sigma^{n+1}_{\mathsf{rel}}(x \mid y)$	DRCM5
$(x + y) \mid z = x \mid z + y \mid z$	CM8
$x \mid (y + z) = x \mid y + x \mid z$	CM9
$\partial_H(\dot{\delta}) = \dot{\delta}$	D0
$\partial_H(\underline{\underline{a}}) = \underline{\underline{a}}$ if $a \notin H$	D1DR
$\partial_H(\underline{\underline{a}}) = \underline{\underline{\delta}}$ if $a \in H$	D2DR
$\partial_H(\sigma^{n}_{\mathsf{rel}}(x)) = \sigma^{n}_{\mathsf{rel}}(\partial_H(x))$	DRD
$\partial_H(x + y) = \partial_H(x) + \partial_H(y)$	D3
$\partial_H(x \cdot y) = \partial_H(x) \cdot \partial_H(y)$	D4

AXIOMS OF $\mathrm{ACP}^{\mathrm{drt}}$. The axiom system of $\mathrm{ACP}^{\mathrm{drt}}$ consists of the axioms of $\mathrm{BPA}^{\mathrm{drt}}$ and the equations given in Table 24.

The additional axioms of $\mathrm{ACP}^{\mathrm{drt}}$ are just simple reformulations of the additional axioms of $\mathrm{ACP}^{\mathrm{dat}}$. That is, constants $\underline{a}$ ($a \in \mathsf{A}_\delta$) have been replaced by constants $\underline{\underline{a}}$, and the operators σ_{abs} and υ_{abs} have been replaced by σ_{rel} and υ_{rel}, respectively.

As in the case of $\mathrm{ACP}^{\mathrm{dat}}$, we can prove that the operators $\|$, $\lfloor\!\lfloor$, $|$ and ∂_H can be eliminated in closed terms of $\mathrm{ACP}^{\mathrm{drt}}$.

SEMANTICS OF $\mathrm{ACP}^{\mathrm{drt}}$. Like for the rules for the operational semantics of $\mathrm{BPA}^{\mathrm{drt}}$, the additional rules for $\mathrm{ACP}^{\mathrm{drt}}$ differ from the corresponding rules for the real time case only in that all numbers involved are restricted to $\mathbb{N}$. Therefore, we refrain again from giving the rules.

Again, we obtain a model for $\mathrm{ACP}^{\mathrm{drt}}$ by identifying bisimilar processes.

5.3. *Embedding* $\mathrm{ACP}^{\mathrm{dat}}\surd$ *in* $\mathrm{ACP}^{\mathrm{sat}}\mathrm{I}\surd$

In this subsection, we will show that $\mathrm{ACP}^{\mathrm{dat}}\surd$ can be embedded in $\mathrm{ACP}^{\mathrm{sat}}\mathrm{I}\surd$. We will do so in the way outlined in Section 4.3. The explicit definitions needed are given in Table 25. Notice that the operators σ_{abs}, υ_{abs} and $\overline{\upsilon}_{\mathsf{abs}}$ of $\mathrm{ACP}^{\mathrm{dat}}\surd$ are simply defined as the operators σ_{abs}, υ_{abs} and $\overline{\upsilon}_{\mathsf{abs}}$ of $\mathrm{ACP}^{\mathrm{sat}}\mathrm{I}\surd$ restricted in their first argument to $\mathbb{N}$. We will establish the existence of an embedding by proving that for closed terms the axioms of $\mathrm{ACP}^{\mathrm{dat}}\surd$ are derivable from the axioms of $\mathrm{ACP}^{\mathrm{sat}}\mathrm{I}\surd$ and the explicit definitions given in Table 25. However, we first take another look at the connection between $\mathrm{ACP}^{\mathrm{sat}}\mathrm{I}\surd$ and $\mathrm{ACP}^{\mathrm{dat}}\surd$ by introducing the notions of a discretized real time process and a discretely initialized real time process.

In Section 3.2, we have introduced the model $\mathsf{M}^*_{\mathsf{A}}$ of $\mathrm{ACP}^{\mathrm{sat}}\mathrm{I}\surd$. The model of $\mathrm{ACP}^{\mathrm{dat}}\surd$ outlined in Section 5.1 is isomorphic to the subalgebra of $\mathsf{M}^*_{\mathsf{A}}$ generated by the embedded constants and operators of $\mathrm{ACP}^{\mathrm{dat}}\surd$. The domain of this subalgebra consists of those real time processes, i.e. elements of the domain of $\mathsf{M}^*_{\mathsf{A}}$, that are discretized. We define the notion of a discretized real time process in terms of the auxiliary *discretization* operator $\mathcal{D} : \mathsf{P}^* \to \mathsf{P}^*$ of which the defining axioms are given in Table 26. A real time process x is a *discretized*

Table 25
Definitions of discrete time operators ($a \in \mathsf{A}_\delta$)

$$\underline{a} = \textstyle\int_{v \in [0,1)} \sigma^{v}_{\mathsf{abs}}(\tilde{a})$$
$$\sigma^{n}_{\mathsf{abs}}(x) = \sigma^{n}_{\mathsf{abs}}(x)$$
$$\upsilon^{n}_{\mathsf{abs}}(x) = \upsilon^{n}_{\mathsf{abs}}(x)$$
$$\overline{\upsilon}^{n}_{\mathsf{abs}}(x) = \overline{\upsilon}^{n}_{\mathsf{abs}}(x)$$
$$\surd_{\mathsf{d}}\, i \,.\, F = \surd_{\mathsf{s}}\, v \,.\, F[\lfloor v \rfloor / i]$$

Table 26
Definition of discretization ($a \in \mathsf{A}_\delta$)

$$\mathcal{D}(\dot{\delta}) = \dot{\delta}$$
$$\mathcal{D}(\tilde{a}) = \underline{a}$$
$$\mathcal{D}(\sigma^{p}_{\mathsf{abs}}(x)) = \sigma^{\lfloor p \rfloor}_{\mathsf{abs}}(\mathcal{D}(x))$$
$$\mathcal{D}(x + y) = \mathcal{D}(x) + \mathcal{D}(y)$$
$$\mathcal{D}(x \cdot y) = \mathcal{D}(x) \cdot \mathcal{D}(y)$$
$$\mathcal{D}(\textstyle\int_{v \in V} F) = \int_{v \in V} \mathcal{D}(F)$$
$$\mathcal{D}(\surd_{\mathsf{s}}\, v \,.\, F) = \surd_{\mathsf{s}}\, v \,.\, \mathcal{D}(F)$$

Table 27
Properties of discretized processes ($a \in \mathsf{A}_\delta$)

$$\dot{\delta},\ \underline{a} \in \mathsf{DIS}$$
$$x \in \mathsf{DIS} \Rightarrow \sigma^{n}_{\mathsf{abs}}(x),\ \upsilon^{n}_{\mathsf{abs}}(x),\ \overline{\upsilon}^{n}_{\mathsf{abs}}(x),\ \partial_H(x) \in \mathsf{DIS}$$
$$x, y \in \mathsf{DIS} \Rightarrow x + y,\ x \cdot y,\ x \parallel y,\ x ⌊⌊ y,\ x \mid y \in \mathsf{DIS}$$
$$(\forall n \in \mathbb{N} \bullet F[n/i] \in \mathsf{DIS}) \Rightarrow \surd_{\mathsf{d}}\, i \,.\, F \in \mathsf{DIS}$$
$$(\forall p \in V \bullet F[p/v] \in \mathsf{DIS}) \Rightarrow \textstyle\int_{v \in V} F \in \mathsf{DIS}$$
$$x \in \mathsf{DIS} \Rightarrow \mathcal{D}(x) \in \mathsf{DIS}$$

real time process, written $x \in \mathsf{DIS}$, if $x = \mathcal{D}(x)$. The properties given in Table 27 express that the set of all discretized real time processes is closed under the operators of $\mathrm{ACP}^{\mathrm{dat}}\surd$, integration and discretization.

For elements f of $\mathsf{RTTS}^*(\mathsf{A})$, the discretization of f, $\mathcal{D}(f)$, is obtained as follows ($t \in \mathbb{R}_{\geqslant 0}$, $q = p + r$ and $q' = p + r'$):

(1) for each t, if $\langle s, p\rangle \xrightarrow{a} \langle s', p\rangle$ in $f(t)$, then $\langle s, p'\rangle \xrightarrow{a} \langle s', p'\rangle$ in $\mathcal{D}(f)(t)$ for each $p' \in [\lfloor p \rfloor, \lfloor p + 1 \rfloor)$;
(2) for each t, if $\langle s, p\rangle \xrightarrow{a} \langle \surd, p\rangle$ in $f(t)$, then $\langle s, p'\rangle \xrightarrow{a} \langle \surd, p'\rangle$ in $\mathcal{D}(f)(t)$ for each $p' \in [\lfloor p \rfloor, \lfloor p + 1 \rfloor)$;
(3) for each t, if $\langle s, p\rangle \overset{r}{\longmapsto} \langle s, q\rangle$ in $f(t)$, then $\langle s, p\rangle \overset{r'}{\longmapsto} \langle s, q'\rangle$ in $\mathcal{D}(f)(t)$ for each $q' \in [q, \lfloor q + 1 \rfloor)$;
(4) for each t, if $\mathsf{ID}(s, p)$ in $f(t)$, then $\mathsf{ID}(s, p)$ in $\mathcal{D}(f)(t)$;
(5) for each t, if neither $\mathsf{ID}(s, p)$ in $f(t)$ nor either $\langle s, p\rangle \xrightarrow{a} \langle s', p\rangle$ in $f(t)$ for some a, s' or $\langle s, p\rangle \overset{r}{\longmapsto} \langle s, q\rangle$ in $f(t)$ for some r, then $\langle s, p\rangle \overset{r'}{\longmapsto} \langle s, q'\rangle$ in $\mathcal{D}(f)(t)$ for each $q' \in (p, \lfloor p + 1 \rfloor)$.

Hence, for real time processes corresponding to discrete time processes, the following holds: if an action can be performed at some time p such that $n \leqslant p < n + 1$, it can also be performed at any other time p' such that $n \leqslant p' < n + 1$.

A real time process x is a *discretely initialized* real time process, written $x \in \mathsf{DIP}$, if $x = \surd_{\mathsf{d}}\, i \,.\, \overline{\upsilon}^{i}_{\mathsf{abs}}(x)$. It follows immediately that $x \in \mathsf{DIP} \Leftrightarrow x = \surd_{\mathsf{s}}\, v \,.\, \overline{\upsilon}^{\lfloor v \rfloor}_{\mathsf{abs}}(x)$. It is easy to show by induction on the term structure that all discretized processes are discretely initialized, i.e. $x \in \mathsf{DIS} \Rightarrow x \in \mathsf{DIP}$. Not all discretely initialized processes are discretized, e.g., $\surd_{\mathsf{s}}\, v \,.\, \sigma^{\lfloor v+1 \rfloor}_{\mathsf{abs}}(\tilde{a}) \in \mathsf{DIP}$ and $\surd_{\mathsf{s}}\, v \,.\, \sigma^{\lfloor v+1 \rfloor}_{\mathsf{abs}}(\tilde{a}) \notin \mathsf{DIS}$. This means that for real time processes corresponding to discrete time processes, the initialization time can always be taken to be a discrete point in time; and that there are real time processes not corresponding to discrete time processes for which the initialization time can always be taken to be a discrete point in time.

LEMMA 7. *For each closed term t of* $\mathrm{ACP^{sat}I}\surd$ *generated by the embedded constants and operators of* $\mathrm{ACP^{dat}}\surd$, $t = \surd_{\mathsf{s}}\, v \,.\, \overline{\upsilon}^{\lfloor v \rfloor}_{\mathsf{abs}}(t)$.

PROOF. From the properties given in Table 27, we know that each process x generated by the embedded constants and operators of $\mathrm{ACP^{dat}}\surd$ is discretized, i.e. $x \in \mathsf{DIS}$. Because $x \in \mathsf{DIS} \Rightarrow x \in \mathsf{DIP}$ and $x \in \mathsf{DIP} \Leftrightarrow x = \surd_{\mathsf{s}}\, v \,.\, \overline{\upsilon}^{\lfloor v \rfloor}_{\mathsf{abs}}(x)$, the result immediately follows. □

The following lemmas present other useful properties of discrete time processes.

LEMMA 8. *For each closed term t of* $\mathrm{ACP^{sat}I}\surd$ *generated by the embedded constants and operators of* $\mathrm{ACP^{dat}}\surd$, *there exists a term t' containing no other free variable than v such that for each $p \in \mathbb{R}_{\geqslant 0}$: $\overline{\upsilon}^{p}_{\mathsf{abs}}(t) = \sigma^{p}_{\mathsf{abs}}(t'[p/v])$, $t'[p/v] = \overline{\upsilon}^{0}_{\mathsf{abs}}(t'[p/v])$, and if $p \in [0, 1)$ and $t \neq \mathring{\delta}$, $t'[p/v] = t'[p/v] + \sigma^{1-p}_{\mathsf{abs}}(\mathring{\delta})$ and $\overline{\upsilon}^{p}_{\mathsf{abs}}(t + \underline{\delta}) = \sigma^{p}_{\mathsf{abs}}(t'[p/v] + \tilde{\delta})$. In subsequent proofs, we write $t_{[v]}$ for a fixed but arbitrary term t' that fulfills these conditions.*

PROOF. Observe that, if $p \in [0, 1)$ and $t \neq \mathring{\delta}$, $\overline{\upsilon}^{p}_{\mathsf{abs}}(t + \underline{\delta}) = \sigma^{p}_{\mathsf{abs}}(t'[p/v] + \tilde{\delta})$ follows directly from $\overline{\upsilon}^{p}_{\mathsf{abs}}(t) = \sigma^{p}_{\mathsf{abs}}(t'[p/v])$ and $t'[p/v] = t'[p/v] + \sigma^{1-p}_{\mathsf{abs}}(\mathring{\delta})$. Observe further that, if t' is a term that fulfills all above-mentioned conditions but $t'[p/v] = \overline{\upsilon}^{0}_{\mathsf{abs}}(t'[p/v])$, $\overline{\upsilon}^{0}_{\mathsf{abs}}(t')$ is a term that fulfills all conditions. Consequently, it suffices to prove that there exists a t' such that $\overline{\upsilon}^{p}_{\mathsf{abs}}(t) = \sigma^{p}_{\mathsf{abs}}(t'[p/v])$ and, if $p \in [0, 1)$ and $t \neq \mathring{\delta}$, $t'[p/v] = t'[p/v] + \sigma^{1-p}_{\mathsf{abs}}(\mathring{\delta})$. It is straightforward to prove this by induction on the structure of t. We present only the case that t is of the form $\underline{a}$. The other cases are simpler or similar to corresponding cases in the proof of Lemma 3.

$$\overline{\upsilon}^{p}_{\mathsf{abs}}(\underline{a}) = \overline{\upsilon}^{p}_{\mathsf{abs}}(\textstyle\int_{w\in[0,1)} \sigma^{w}_{\mathsf{abs}}(\tilde{a})) \overset{\mathrm{SAI6}}{=} \int_{w\in[0,1)} \overline{\upsilon}^{p}_{\mathsf{abs}}(\sigma^{w}_{\mathsf{abs}}(\tilde{a})) \overset{\mathrm{INT4}}{=}$$
$$\textstyle\int_{w\in[0,p)} \overline{\upsilon}^{p}_{\mathsf{abs}}(\sigma^{w}_{\mathsf{abs}}(\tilde{a})) + \int_{w\in[p,1)} \overline{\upsilon}^{p}_{\mathsf{abs}}(\sigma^{w}_{\mathsf{abs}}(\tilde{a})) \overset{\mathrm{SAI1,2,3,SAI3'}}{=}$$
$$\textstyle\int_{w\in[0,p)} \sigma^{p}_{\mathsf{abs}}(\mathring{\delta}) + \int_{w\in[p,1)} \sigma^{w}_{\mathsf{abs}}(\tilde{a}) \overset{\mathrm{INT5}}{=} \sigma^{p}_{\mathsf{abs}}(\mathring{\delta}) + \int_{w\in[p,1)} \sigma^{w}_{\mathsf{abs}}(\tilde{a}) \overset{\mathrm{INT5}}{=}$$
$$\textstyle\int_{w\in[p,1)} \sigma^{p}_{\mathsf{abs}}(\mathring{\delta}) + \int_{w\in[p,1)} \sigma^{w}_{\mathsf{abs}}(\tilde{a}) \overset{\mathrm{INT11}}{=} \int_{w\in[p,1)} (\sigma^{p}_{\mathsf{abs}}(\mathring{\delta}) + \sigma^{w}_{\mathsf{abs}}(\tilde{a})) \overset{\mathrm{SAT2}}{=}$$
$$\textstyle\int_{w\in[p,1)} (\sigma^{p}_{\mathsf{abs}}(\mathring{\delta}) + \sigma^{p}_{\mathsf{abs}}(\sigma^{w-p}_{\mathsf{abs}}(\tilde{a}))) \overset{\mathrm{SAT3}}{=} \int_{w\in[p,1)} \sigma^{p}_{\mathsf{abs}}(\mathring{\delta} + \sigma^{w-p}_{\mathsf{abs}}(\tilde{a})) \overset{\mathrm{A6ID}}{=}$$
$$\textstyle\int_{w\in[p,1)} \sigma^{p}_{\mathsf{abs}}(\sigma^{w-p}_{\mathsf{abs}}(\tilde{a})) \overset{\mathrm{INT10}}{=} \sigma^{p}_{\mathsf{abs}}(\int_{w\in[p,1)} \sigma^{w-p}_{\mathsf{abs}}(\tilde{a})) = \sigma^{p}_{\mathsf{abs}}(\int_{w\in[0,1-p)} \sigma^{w}_{\mathsf{abs}}(\tilde{a})) =$$
$$\textstyle\sigma^{p}_{\mathsf{abs}}((\int_{w\in[0,1-v)} \sigma^{w}_{\mathsf{abs}}(\tilde{a}))[p/v]) \text{ and}$$

$\int_{w\in[0,1-p)}\sigma^{w}_{\mathsf{abs}}(\tilde{a}) \overset{\text{A6SAa}}{=} \int_{w\in[0,1-p)}\sigma^{w}_{\mathsf{abs}}(\tilde{a}+\tilde{\delta}) \overset{\text{SAT3}}{=} \int_{w\in[0,1-p)}(\sigma^{w}_{\mathsf{abs}}(\tilde{a})+\sigma^{w}_{\mathsf{abs}}(\tilde{\delta})) \overset{\text{INT11}}{=}$
$\int_{w\in[0,1-p)}\sigma^{w}_{\mathsf{abs}}(\tilde{a}) + \int_{w\in[0,1-p)}\sigma^{w}_{\mathsf{abs}}(\tilde{\delta}) \overset{\text{INT8}}{=} \int_{w\in[0,1-p)}\sigma^{w}_{\mathsf{abs}}(\tilde{a}) + \sigma^{1-p}_{\mathsf{abs}}(\dot{\delta})$ □

LEMMA 9. *For each $p\in\mathbb{R}_{\geqslant 0}$ and closed term t of $\mathrm{ACP^{sat}I}\surd$ generated by the embedded constants and operators of $\mathrm{ACP^{dat}}\surd$, there exists a closed term t' such that $\overline{\upsilon}^{p}_{\mathsf{abs}}(t)=\sigma^{p}_{\mathsf{abs}}(t')$, $t'=\overline{\upsilon}^{0}_{\mathsf{abs}}(t')$, and if $p\in[0,1)$ and $t\neq\dot{\delta}$, $t'=t'+\sigma^{1-p}_{\mathsf{abs}}(\dot{\delta})$ and $\overline{\upsilon}^{p}_{\mathsf{abs}}(t+\underline{\delta})=\sigma^{p}_{\mathsf{abs}}(t'+\tilde{\delta})$. In subsequent proofs, we write $t_{[p]}$ for a fixed but arbitrary closed term t' that fulfills these conditions – like in case of applications of Lemma* 3.

PROOF. This follows immediately from Lemma 8. □

LEMMA 10. *For each closed term t of $\mathrm{ACP^{sat}I}\surd$ generated by the embedded constants and operators of $\mathrm{ACP^{dat}}\surd$, there exists a closed term t' such that $\upsilon^{1}_{\mathsf{abs}}(t+\underline{\delta})=\int_{v\in[0,1)}\sigma^{v}_{\mathsf{abs}}(\nu_{\mathsf{abs}}(t')+\tilde{\delta})$. In subsequent proofs we write t° for a fixed but arbitrary closed term t' that fulfills this condition.*

PROOF. It is straightforward to prove this by induction on the structure of t. We present only the case that t is of the form $\underline{a}\cdot t''$. The other cases are simpler.

$\upsilon^{1}_{\mathsf{abs}}(\underline{a}\cdot t''+\underline{\delta}) = \upsilon^{1}_{\mathsf{abs}}((\int_{v\in[0,1)}\sigma^{v}_{\mathsf{abs}}(\tilde{a}))\cdot t''+\int_{v\in[0,1)}\sigma^{v}_{\mathsf{abs}}(\tilde{\delta})) \overset{\text{SATO4,5}}{=}$
$\upsilon^{1}_{\mathsf{abs}}(\int_{v\in[0,1)}\sigma^{v}_{\mathsf{abs}}(\tilde{a}))\cdot t''+\upsilon^{1}_{\mathsf{abs}}(\int_{v\in[0,1)}\sigma^{v}_{\mathsf{abs}}(\tilde{\delta})) \overset{\text{SATO2,3,6}}{=}$
$(\int_{v\in[0,1)}\sigma^{v}_{\mathsf{abs}}(\tilde{a}))\cdot t''+\int_{v\in[0,1)}\sigma^{v}_{\mathsf{abs}}(\tilde{\delta}) \overset{\text{INT12}}{=} \int_{v\in[0,1)}(\sigma^{v}_{\mathsf{abs}}(\tilde{a})\cdot t'')+\int_{v\in[0,1)}\sigma^{v}_{\mathsf{abs}}(\tilde{\delta}) \overset{\text{SIA6}}{=}$
$\int_{v\in[0,1)}(\sigma^{v}_{\mathsf{abs}}(\tilde{a})\cdot\overline{\upsilon}^{v}_{\mathsf{abs}}(t''))+\int_{v\in[0,1)}\sigma^{v}_{\mathsf{abs}}(\tilde{\delta}) \overset{\text{INT6,Lemma 8}}{=}$
$\int_{v\in[0,1)}(\sigma^{v}_{\mathsf{abs}}(\tilde{a})\cdot\sigma^{v}_{\mathsf{abs}}(t''_{[v]}))+\int_{v\in[0,1)}\sigma^{v}_{\mathsf{abs}}(\tilde{\delta}) \overset{\text{SAT5}}{=}$
$\int_{v\in[0,1)}\sigma^{v}_{\mathsf{abs}}(\tilde{a}\cdot\overline{\upsilon}^{0}_{\mathsf{abs}}(t''_{[v]}))+\int_{v\in[0,1)}\sigma^{v}_{\mathsf{abs}}(\tilde{\delta}) \overset{\text{INT11}}{=}$
$\int_{v\in[0,1)}(\sigma^{v}_{\mathsf{abs}}(\tilde{a}\cdot\overline{\upsilon}^{0}_{\mathsf{abs}}(t''_{[v]}))+\sigma^{v}_{\mathsf{abs}}(\tilde{\delta})) \overset{\text{SAT3}}{=} \int_{v\in[0,1)}\sigma^{v}_{\mathsf{abs}}(\tilde{a}\cdot\overline{\upsilon}^{0}_{\mathsf{abs}}(t''_{[v]})+\tilde{\delta}) \overset{\text{SAU1}}{=}$
$\int_{v\in[0,1)}\sigma^{v}_{\mathsf{abs}}(\nu_{\mathsf{abs}}(\tilde{a})\cdot\overline{\upsilon}^{0}_{\mathsf{abs}}(t''_{[v]})+\tilde{\delta}) \overset{\text{SAU4}}{=} \int_{v\in[0,1)}\sigma^{v}_{\mathsf{abs}}(\nu_{\mathsf{abs}}(\tilde{a}\cdot\overline{\upsilon}^{0}_{\mathsf{abs}}(t''_{[v]}))+\tilde{\delta})$ □

Lemmas 7–10 are used to shorten the calculations in the proof of Theorem 12. The following lemma is also used in the proof of that theorem.

LEMMA 11. *For $p\in[0,1)$, the equation $\overline{\upsilon}^{p}_{\mathsf{abs}}(\underline{\delta})=\sigma^{1}_{\mathsf{abs}}(\dot{\delta})$ is derivable from the axioms of $\mathrm{ACP^{sat}I}\surd$ and the explicit definition of the constants and operators in Table* 25.

PROOF.
$\overline{\upsilon}^{p}_{\mathsf{abs}}(\underline{\delta})=\overline{\upsilon}^{p}_{\mathsf{abs}}(\int_{v\in[0,1)}\sigma^{v}_{\mathsf{abs}}(\tilde{\delta})) \overset{\text{SAI6}}{=} \int_{v\in[0,1)}\overline{\upsilon}^{p}_{\mathsf{abs}}(\sigma^{v}_{\mathsf{abs}}(\tilde{\delta})) \overset{\text{INT4}}{=}$
$\int_{v\in[0,p)}\overline{\upsilon}^{p}_{\mathsf{abs}}(\sigma^{v}_{\mathsf{abs}}(\tilde{\delta}))+\int_{v\in[p,1)}\overline{\upsilon}^{p}_{\mathsf{abs}}(\sigma^{v}_{\mathsf{abs}}(\tilde{\delta})) \overset{\text{SAI3,SAI3}'}{=}$
$\int_{v\in[0,p)}\sigma^{v}_{\mathsf{abs}}(\overline{\upsilon}^{p-v}_{\mathsf{abs}}(\overline{\upsilon}^{0}_{\mathsf{abs}}(\tilde{\delta})))+\int_{v\in[p,1)}\sigma^{v}_{\mathsf{abs}}(\tilde{\delta}) \overset{\text{SAI1,SAI2}}{=}$
$\int_{v\in[0,p)}\sigma^{v}_{\mathsf{abs}}(\sigma^{p-v}_{\mathsf{abs}}(\dot{\delta}))+\int_{v\in[p,1)}\sigma^{v}_{\mathsf{abs}}(\tilde{\delta}) \overset{\text{SAT2}}{=} \int_{v\in[0,p)}\sigma^{p}_{\mathsf{abs}}(\dot{\delta})+\int_{v\in[p,1)}\sigma^{v}_{\mathsf{abs}}(\tilde{\delta}) \overset{\text{INT5,8}}{=}$
$\sigma^{p}_{\mathsf{abs}}(\dot{\delta})+\sigma^{1}_{\mathsf{abs}}(\dot{\delta}) \overset{\text{SAT2,3,A6ID}}{=} \sigma^{1}_{\mathsf{abs}}(\dot{\delta})$ □

The existence of an embedding of $\mathrm{ACP}^{\mathrm{dat}}\surd$ in $\mathrm{ACP}^{\mathrm{sat}}\mathrm{I}\surd$ is now established by proving the following theorem.

THEOREM 12 (Embedding $\mathrm{ACP}^{\mathrm{dat}}\surd$ in $\mathrm{ACP}^{\mathrm{sat}}\mathrm{I}\surd$). *For closed terms, the axioms of* $\mathrm{ACP}^{\mathrm{dat}}\surd$ *are derivable from the axioms of* $\mathrm{ACP}^{\mathrm{sat}}\mathrm{I}\surd$ *and the explicit definitions of the constants and operators* $\underline{a}$, σ_{abs}, υ_{abs}, $\overline{\upsilon}_{\mathsf{abs}}$ *and* $\surd_{\mathsf{d}}$ *in Table* 25.

PROOF. The proof of this theorem is given in Appendix A2. The proof is a matter of straightforward calculations. Lemmas 1 and 3 (pages 635–636) and Lemmas 7–11 (pages 671–672) are very useful in the proof. □

6. Concluding remarks

We presented real time and discrete time versions of ACP with both absolute timing and relative timing, starting with a new real time version of ACP with absolute timing called $\mathrm{ACP}^{\mathrm{sat}}$. We demonstrated that $\mathrm{ACP}^{\mathrm{sat}}$ extended with integration and initial abstraction generalizes the presented real time version with relative timing and the presented discrete time version with absolute timing. We focussed on versions of ACP with timing where execution of actions and passage of time are separated, but explained how they can be combined in these versions. The material resulted from a systematic study of some of the most important issues relevant to dealing with time-dependent behaviour of processes – viz. absolute vs relative timing, continuous vs discrete time scale, and separation vs combination of execution of actions and passage of time – in the setting of ACP.

All real time and discrete time versions of ACP presented in this chapter include the immediate deadlock constant $\dot{\delta}$. This constant enables us to distinguish timing inconsistencies from incapabilities of performing actions as well as idling. This is certainly relevant to versions with absolute timing because timing inconsistencies readily arise. The usefulness of the immediate deadlock constant in practice is not yet clear for versions with relative timing. Minor adaptations of the versions of ACP with relative timing presented in this chapter are needed to obtain versions without the immediate deadlock constant.

The discrete time versions of ACP presented in this chapter are conservative extensions of the discrete time versions of [7]. The real time versions presented in this chapter, unlike the real time versions of [2] and [4], do not exclude the possibility of two or more actions to be performed consecutively at the same point in time. This feature seems to be essential to obtain simple and natural embeddings of discrete time versions as well as useful in practice when describing and analyzing distributed systems where entirely independent actions happen at different locations.

We did not extend the different versions of ACP with timing presented in this chapter with recursion, abstraction, and other features that are important to make these versions suitable for being applied. This has been done for the earlier versions of ACP with timing referred to in this chapter. Some of those versions have been successfully used for describing and analyzing systems and protocols of various kinds, see, e.g., [16,27,31,32,40,41,

SRI3 : $\overline{\upsilon}^{p+q}_{\mathsf{rel}}(\sigma^{p}_{\mathsf{rel}}(t)) = \surd_{\mathsf{s}}\, v\,.\,\overline{\upsilon}^{v+p+q}_{\mathsf{abs}}(\overline{\upsilon}^{v}_{\mathsf{abs}}(\surd_{\mathsf{s}}\, w\,.\,\overline{\upsilon}^{w+p}_{\mathsf{abs}}(t))) \overset{\text{SIA2}}{=}$
$\surd_{\mathsf{s}}\, v\,.\,\overline{\upsilon}^{v+p+q}_{\mathsf{abs}}(\overline{\upsilon}^{v}_{\mathsf{abs}}(\overline{\upsilon}^{v+p}_{\mathsf{abs}}(t))) \overset{\text{SI2}}{=} \surd_{\mathsf{s}}\, v\,.\,\overline{\upsilon}^{v+p}_{\mathsf{abs}}(\overline{\upsilon}^{v+p+q}_{\mathsf{abs}}(\overline{\upsilon}^{v+p}_{\mathsf{abs}}(t))) \overset{\text{SIA2}}{=}$
$\surd_{\mathsf{s}}\, v\,.\,\overline{\upsilon}^{v+p}_{\mathsf{abs}}(\surd_{\mathsf{s}}\, w\,.\,\overline{\upsilon}^{w+q}_{\mathsf{abs}}(\overline{\upsilon}^{w}_{\mathsf{abs}}(t))) = \sigma^{p}_{\mathsf{rel}}(\overline{\upsilon}^{q}_{\mathsf{rel}}(t))$

SRI4 : $\overline{\upsilon}^{p}_{\mathsf{rel}}(t + t') = \surd_{\mathsf{s}}\, v\,.\,\overline{\upsilon}^{v+p}_{\mathsf{abs}}(\overline{\upsilon}^{v}_{\mathsf{abs}}(t + t')) \overset{\text{SAI4}}{=}$
$\surd_{\mathsf{s}}\, v\,.\,(\overline{\upsilon}^{v+p}_{\mathsf{abs}}(\overline{\upsilon}^{v}_{\mathsf{abs}}(t)) + \overline{\upsilon}^{v+p}_{\mathsf{abs}}(\overline{\upsilon}^{v}_{\mathsf{abs}}(t'))) \overset{\text{SI2}}{=}$
$\surd_{\mathsf{s}}\, v\,.\,(\overline{\upsilon}^{v+p}_{\mathsf{abs}}(\overline{\upsilon}^{v}_{\mathsf{abs}}(t)) + \overline{\upsilon}^{v}_{\mathsf{abs}}(\overline{\upsilon}^{v+p}_{\mathsf{abs}}(\overline{\upsilon}^{v}_{\mathsf{abs}}(t')))) \overset{\text{SIA2}}{=}$
$\surd_{\mathsf{s}}\, v\,.\,(\overline{\upsilon}^{v+p}_{\mathsf{abs}}(\overline{\upsilon}^{v}_{\mathsf{abs}}(t)) + \overline{\upsilon}^{v}_{\mathsf{abs}}(\surd_{\mathsf{s}}\, w\,.\,\overline{\upsilon}^{w+p}_{\mathsf{abs}}(\overline{\upsilon}^{w}_{\mathsf{abs}}(t')))) \overset{\text{SIA8}}{=}$
$(\surd_{\mathsf{s}}\, v\,.\,\overline{\upsilon}^{v+p}_{\mathsf{abs}}(\overline{\upsilon}^{v}_{\mathsf{abs}}(t))) + (\surd_{\mathsf{s}}\, w\,.\,\overline{\upsilon}^{w+p}_{\mathsf{abs}}(\overline{\upsilon}^{w}_{\mathsf{abs}}(t'))) = \overline{\upsilon}^{p}_{\mathsf{rel}}(t) + \overline{\upsilon}^{p}_{\mathsf{rel}}(t')$

SRI5 : $\overline{\upsilon}^{p}_{\mathsf{rel}}(t \cdot t') = \surd_{\mathsf{s}}\, v\,.\,\overline{\upsilon}^{v+p}_{\mathsf{abs}}(\overline{\upsilon}^{v}_{\mathsf{abs}}(t \cdot t')) \overset{\text{SAI5}}{=} \surd_{\mathsf{s}}\, v\,.\,(\overline{\upsilon}^{v+p}_{\mathsf{abs}}(\overline{\upsilon}^{v}_{\mathsf{abs}}(t)) \cdot t') \overset{\text{SIA9}}{=}$
$(\surd_{\mathsf{s}}\, v\,.\,\overline{\upsilon}^{v+p}_{\mathsf{abs}}(\overline{\upsilon}^{v}_{\mathsf{abs}}(t))) \cdot t' = \overline{\upsilon}^{p}_{\mathsf{rel}}(t) \cdot t'$

Next, we show that the additional axioms for $\mathrm{ACP}^{\mathrm{srt}}$ are derivable for closed terms.

CF1SR : $\tilde{\tilde{a}} \mid \tilde{\tilde{b}} = (\surd_{\mathsf{s}}\, v\,.\,\sigma^{v}_{\mathsf{abs}}(\tilde{a})) \mid (\surd_{\mathsf{s}}\, v\,.\,\sigma^{v}_{\mathsf{abs}}(\tilde{b})) \overset{\text{DISTR}\mid,\text{SACM5}}{=} \surd_{\mathsf{s}}\, v\,.\,\sigma^{v}_{\mathsf{abs}}(\tilde{a} \mid \tilde{b}) \overset{\text{CF1SA}}{=}$
$\surd_{\mathsf{s}}\, v\,.\,\sigma^{v}_{\mathsf{abs}}(\tilde{c}) = \tilde{\tilde{c}}$ if $\gamma(a, b) = c$

CF2SR : The proof is similar to the proof of axiom CF1SR – axiom CF2SA is used instead of axiom CF1SA.

CM2SRID : $\tilde{\tilde{a}} \mathbin{\|\!\_} (t + \tilde{\tilde{\delta}}) = (\surd_{\mathsf{s}}\, v\,.\,\sigma^{v}_{\mathsf{abs}}(\tilde{a})) \mathbin{\|\!\_} (t + (\surd_{\mathsf{s}}\, v\,.\,\sigma^{v}_{\mathsf{abs}}(\tilde{\delta}))) \overset{\text{SIA8},\|\!\_}{=}$
$\surd_{\mathsf{s}}\, v\,.\,(\sigma^{v}_{\mathsf{abs}}(\tilde{a}) \mathbin{\|\!\_} (\overline{\upsilon}^{v}_{\mathsf{abs}}(t) + \sigma^{v}_{\mathsf{abs}}(\tilde{\delta}))) \overset{\text{Lemma 4}}{=}$
$\surd_{\mathsf{s}}\, v\,.\,(\sigma^{v}_{\mathsf{abs}}(\tilde{a}) \mathbin{\|\!\_} (\sigma^{v}_{\mathsf{abs}}(t) + \sigma^{v}_{\mathsf{abs}}(\tilde{\delta}))) \overset{\text{SAT3,SACM2}}{=}$
$\surd_{\mathsf{s}}\, v\,.\,\sigma^{v}_{\mathsf{abs}}(\tilde{a} \mathbin{\|\!\_} (t + \tilde{\delta})) \overset{\text{CM2SA}}{=} \surd_{\mathsf{s}}\, v\,.\,\sigma^{v}_{\mathsf{abs}}(\tilde{a} \cdot (t + \tilde{\delta})) \overset{\text{SAI1,SAT1,SIA6}}{=}$
$\surd_{\mathsf{s}}\, v\,.\,\sigma^{v}_{\mathsf{abs}}(\tilde{a} \cdot \overline{\upsilon}^{0}_{\mathsf{abs}}(t + \tilde{\delta})) \overset{\text{SAT5}}{=} \surd_{\mathsf{s}}\, v\,.\,(\sigma^{v}_{\mathsf{abs}}(\tilde{a}) \cdot \sigma^{v}_{\mathsf{abs}}(t + \tilde{\delta})) \overset{\text{SAT3,Lemma 4}}{=}$
$\surd_{\mathsf{s}}\, v\,.\,(\sigma^{v}_{\mathsf{abs}}(\tilde{a}) \cdot (\overline{\upsilon}^{v}_{\mathsf{abs}}(t) + \sigma^{v}_{\mathsf{abs}}(\tilde{\delta}))) \overset{\text{SAI3}'}{=}$
$\surd_{\mathsf{s}}\, v\,.\,(\sigma^{v}_{\mathsf{abs}}(\tilde{a}) \cdot (\overline{\upsilon}^{v}_{\mathsf{abs}}(t) + \overline{\upsilon}^{v}_{\mathsf{abs}}(\sigma^{v}_{\mathsf{abs}}(\tilde{\delta})))) \overset{\text{SIA2}}{=}$
$\surd_{\mathsf{s}}\, v\,.\,(\sigma^{v}_{\mathsf{abs}}(\tilde{a}) \cdot (\overline{\upsilon}^{v}_{\mathsf{abs}}(t) + \overline{\upsilon}^{v}_{\mathsf{abs}}(\surd_{\mathsf{s}}\, w\,.\,\sigma^{w}_{\mathsf{abs}}(\tilde{\delta})))) \overset{\text{SAI4}}{=}$
$\surd_{\mathsf{s}}\, v\,.\,(\sigma^{v}_{\mathsf{abs}}(\tilde{a}) \cdot \overline{\upsilon}^{v}_{\mathsf{abs}}(t + \surd_{\mathsf{s}}\, w\,.\,\sigma^{w}_{\mathsf{abs}}(\tilde{\delta}))) \overset{\text{SIA6}}{=}$
$\surd_{\mathsf{s}}\, v\,.\,(\sigma^{v}_{\mathsf{abs}}(\tilde{a}) \cdot (t + \surd_{\mathsf{s}}\, w\,.\,\sigma^{w}_{\mathsf{abs}}(\tilde{\delta}))) \overset{\text{SIA9}}{=}$
$(\surd_{\mathsf{s}}\, v\,.\,\sigma^{v}_{\mathsf{abs}}(\tilde{a})) \cdot (t + \surd_{\mathsf{s}}\, w\,.\,\sigma^{w}_{\mathsf{abs}}(\tilde{\delta})) = \tilde{\tilde{a}} \cdot (t + \tilde{\tilde{\delta}})$

CM3SRID : The proof is similar to the proof of axiom CM2SRID – axiom CM3SA is used instead of axiom CM2SA.

SRCM1ID : $\sigma^{r}_{\mathsf{rel}}(t) \mathbin{\|\!\_} (\nu_{\mathsf{rel}}(t') + \tilde{\tilde{\delta}}) =$
$(\surd_{\mathsf{s}}\, v\,.\,\overline{\upsilon}^{v+r}_{\mathsf{abs}}(t)) \mathbin{\|\!\_} ((\surd_{\mathsf{s}}\, v\,.\,\sigma^{v}_{\mathsf{abs}}(\nu_{\mathsf{abs}}(t'))) + (\surd_{\mathsf{s}}\, v\,.\,\sigma^{v}_{\mathsf{abs}}(\tilde{\delta}))) \overset{\text{DISTR}}{=}$
$\surd_{\mathsf{s}}\, v\,.\,(\overline{\upsilon}^{v+r}_{\mathsf{abs}}(t) \mathbin{\|\!\_} (\sigma^{v}_{\mathsf{abs}}(\nu_{\mathsf{abs}}(t')) + \sigma^{v}_{\mathsf{abs}}(\tilde{\delta}))) \overset{\text{Lemma 4,SAT3}}{=}$
$\surd_{\mathsf{s}}\, v\,.\,(\sigma^{v+r}_{\mathsf{abs}}(t) \mathbin{\|\!\_} \sigma^{v}_{\mathsf{abs}}(\nu_{\mathsf{abs}}(t') + \tilde{\delta})) \overset{\text{SACM1}'}{=} \surd_{\mathsf{s}}\, v\,.\,\sigma^{v}_{\mathsf{abs}}(\tilde{\delta}) = \tilde{\tilde{\delta}}$

SRCM2ID : $\sigma^{p}_{\mathsf{rel}}(t) \mathbin{\lfloor\!\lfloor} (\upsilon^{p}_{\mathsf{rel}}(t') + \sigma^{p}_{\mathsf{rel}}(t'')) =$
$(\surd_{\mathsf{s}}\, v \,.\, \overline{\upsilon}^{v+p}_{\mathsf{abs}}(t)) \mathbin{\lfloor\!\lfloor} ((\surd_{\mathsf{s}}\, v \,.\, \upsilon^{v+p}_{\mathsf{abs}}(\overline{\upsilon}^{v}_{\mathsf{abs}}(t'))) + (\surd_{\mathsf{s}}\, v \,.\, \overline{\upsilon}^{v+p}_{\mathsf{abs}}(t''))) \overset{\text{DISTR}}{=}$
$\surd_{\mathsf{s}}\, v \,.\, (\overline{\upsilon}^{v+p}_{\mathsf{abs}}(t) \mathbin{\lfloor\!\lfloor} (\upsilon^{v+p}_{\mathsf{abs}}(\overline{\upsilon}^{v}_{\mathsf{abs}}(t')) + \overline{\upsilon}^{v+p}_{\mathsf{abs}}(t''))) \overset{\text{Lemma 4}}{=}$
$\surd_{\mathsf{s}}\, v \,.\, (\sigma^{v+p}_{\mathsf{abs}}(t) \mathbin{\lfloor\!\lfloor} (\upsilon^{v+p}_{\mathsf{abs}}(\overline{\upsilon}^{v}_{\mathsf{abs}}(t')) + \sigma^{v+p}_{\mathsf{abs}}(t''))) \overset{\text{SACM2}}{=}$
$\surd_{\mathsf{s}}\, v \,.\, \sigma^{v+p}_{\mathsf{abs}}(t \mathbin{\lfloor\!\lfloor} t'') \overset{\text{Lemma 4}}{=} \surd_{\mathsf{s}}\, v \,.\, \overline{\upsilon}^{v+p}_{\mathsf{abs}}(t \mathbin{\lfloor\!\lfloor} t'') = \sigma^{p}_{\mathsf{rel}}(t \mathbin{\lfloor\!\lfloor} t'')$

CM5SR : $\tilde{\tilde{a}} \cdot t \mid \tilde{\tilde{b}} = ((\surd_{\mathsf{s}}\, v \,.\, \sigma^{v}_{\mathsf{abs}}(\tilde{a})) \cdot t) \mid (\surd_{\mathsf{s}}\, v \,.\, \sigma^{v}_{\mathsf{abs}}(\tilde{b})) \overset{\text{SIA9,DISTR}\mid}{=}$
$\surd_{\mathsf{s}}\, v \,.\, ((\sigma^{v}_{\mathsf{abs}}(\tilde{a}) \cdot t) \mid \sigma^{v}_{\mathsf{abs}}(\tilde{b})) \overset{\text{SIA6,Lemma 4}}{=}$
$\surd_{\mathsf{s}}\, v \,.\, ((\sigma^{v}_{\mathsf{abs}}(\tilde{a}) \cdot \sigma^{v}_{\mathsf{abs}}(t)) \mid \sigma^{v}_{\mathsf{abs}}(\tilde{b})) \overset{\text{SAT5}}{=}$
$\surd_{\mathsf{s}}\, v \,.\, (\sigma^{v}_{\mathsf{abs}}(\tilde{a} \cdot \overline{\upsilon}^{0}_{\mathsf{abs}}(t)) \mid \sigma^{v}_{\mathsf{abs}}(\tilde{b})) \overset{\text{SACM5}}{=} \surd_{\mathsf{s}}\, v \,.\, \sigma^{v}_{\mathsf{abs}}(\tilde{a} \cdot \overline{\upsilon}^{0}_{\mathsf{abs}}(t) \mid \tilde{b}) \overset{\text{CM5SA}}{=}$
$\surd_{\mathsf{s}}\, v \,.\, \sigma^{v}_{\mathsf{abs}}((\tilde{a} \mid \tilde{b}) \cdot \overline{\upsilon}^{0}_{\mathsf{abs}}(t)) \overset{\text{SAT5}}{=} \surd_{\mathsf{s}}\, v \,.\, (\sigma^{v}_{\mathsf{abs}}(\tilde{a} \mid \tilde{b}) \cdot \sigma^{v}_{\mathsf{abs}}(t)) \overset{\text{Lemma 4,SIA6}}{=}$
$\surd_{\mathsf{s}}\, v \,.\, (\sigma^{v}_{\mathsf{abs}}(\tilde{a} \mid \tilde{b}) \cdot t) \overset{\text{SIA9}}{=} (\surd_{\mathsf{s}}\, v \,.\, \sigma^{v}_{\mathsf{abs}}(\tilde{a} \mid \tilde{b})) \cdot t \overset{\text{SACM5,DISTR}\mid}{=}$
$((\surd_{\mathsf{s}}\, v \,.\, \sigma^{v}_{\mathsf{abs}}(\tilde{a})) \mid (\surd_{\mathsf{s}}\, v \,.\, \sigma^{v}_{\mathsf{abs}}(\tilde{b}))) \cdot t = (\tilde{\tilde{a}} \mid \tilde{\tilde{b}}) \cdot t$

CM6SR and CM7SR :
The proofs are similar to the proof of axiom CM5SR – axioms CM6SA and CM7SA are used instead of axiom CM5SA.

SRCM3ID : $(\nu_{\mathsf{rel}}(t) + \tilde{\tilde{\delta}}) \mid \sigma^{r}_{\mathsf{rel}}(t') =$
$((\surd_{\mathsf{s}}\, v \,.\, \sigma^{v}_{\mathsf{abs}}(\nu_{\mathsf{abs}}(t))) + (\surd_{\mathsf{s}}\, v \,.\, \sigma^{v}_{\mathsf{abs}}(\tilde{\delta}))) \mid (\surd_{\mathsf{s}}\, v \,.\, \overline{\upsilon}^{v+r}_{\mathsf{abs}}(t)) \overset{\text{DISTR}}{=}$
$\surd_{\mathsf{s}}\, v \,.\, ((\sigma^{v}_{\mathsf{abs}}(\nu_{\mathsf{abs}}(t)) + \sigma^{v}_{\mathsf{abs}}(\tilde{\delta})) \mid \overline{\upsilon}^{v+r}_{\mathsf{abs}}(t)) \overset{\text{SAT3,Lemma 4}}{=}$
$\surd_{\mathsf{s}}\, v \,.\, (\sigma^{v}_{\mathsf{abs}}(\nu_{\mathsf{abs}}(t) + \tilde{\delta}) \mid \sigma^{v+r}_{\mathsf{abs}}(t)) \overset{\text{SACM3}'}{=} \surd_{\mathsf{s}}\, v \,.\, \sigma^{v}_{\mathsf{abs}}(\tilde{\delta}) = \tilde{\tilde{\delta}}$

SRCM4ID : The proof is similar to the proof of axiom SRCM3ID – axiom SACM4′ is used instead of axiom SACM3′.

SRCM5 : $\sigma^{p}_{\mathsf{rel}}(t) \mid \sigma^{p}_{\mathsf{rel}}(t') = (\surd_{\mathsf{s}}\, v \,.\, \overline{\upsilon}^{v+p}_{\mathsf{abs}}(t)) \mid (\surd_{\mathsf{s}}\, v \,.\, \overline{\upsilon}^{v+p}_{\mathsf{abs}}(t')) \overset{\text{DISTR}\mid\text{,SI11}}{=}$
$\surd_{\mathsf{s}}\, v \,.\, \overline{\upsilon}^{v+p}_{\mathsf{abs}}(t \mid t') = \sigma^{p}_{\mathsf{rel}}(t \mid t')$

D1SR : $\partial_H(\tilde{\tilde{a}}) = \partial_H(\surd_{\mathsf{s}}\, v \,.\, \sigma^{v}_{\mathsf{abs}}(\tilde{a})) \overset{\text{SIA15,SAD}}{=} \surd_{\mathsf{s}}\, v \,.\, \sigma^{v}_{\mathsf{abs}}(\partial_H(\tilde{a})) \overset{\text{D1SA}}{=}$
$\surd_{\mathsf{s}}\, v \,.\, \sigma^{v}_{\mathsf{abs}}(\tilde{a}) = \tilde{\tilde{a}}$ if $a \notin H$

D2SR : The proof is similar to the proof of axiom D1SR – axiom D2SA is used instead of axioms D1SA.

SRD : $\partial_H(\sigma^{p}_{\mathsf{rel}}(t)) = \partial_H(\surd_{\mathsf{s}}\, v \,.\, \overline{\upsilon}^{v+p}_{\mathsf{abs}}(t)) \overset{\text{SIA15,Lemma 4,SAD}}{=} \surd_{\mathsf{s}}\, v \,.\, \overline{\upsilon}^{v+p}_{\mathsf{abs}}(\partial_H(t)) =$
$\sigma^{p}_{\mathsf{rel}}(\partial_H(t))$

SRU0 : $\nu_{\mathsf{rel}}(\dot{\delta}) = \surd_{\mathsf{s}}\, v \,.\, \sigma^{v}_{\mathsf{abs}}(\nu_{\mathsf{abs}}(\dot{\delta})) \overset{\text{SAU0,SAI0}}{=} \surd_{\mathsf{s}}\, v \,.\, \overline{\upsilon}^{v}_{\mathsf{abs}}(\dot{\delta}) \overset{\text{SIA1,SIA4}}{=} \dot{\delta}$

SRU1 : $\nu_{\mathsf{rel}}(\tilde{\tilde{a}}) = \surd_{\mathsf{s}}\, v \,.\, \sigma^{v}_{\mathsf{abs}}(\nu_{\mathsf{abs}}(\surd_{\mathsf{s}}\, w \,.\, \sigma^{w}_{\mathsf{abs}}(\tilde{a}))) \overset{\text{SIA16}}{=}$
$\surd_{\mathsf{s}}\, v \,.\, \sigma^{v}_{\mathsf{abs}}(\surd_{\mathsf{s}}\, w \,.\, \nu_{\mathsf{abs}}(\sigma^{w}_{\mathsf{abs}}(\tilde{a}))) \overset{\text{SIA7}}{=} \surd_{\mathsf{s}}\, v \,.\, \sigma^{v}_{\mathsf{abs}}(\nu_{\mathsf{abs}}(\sigma^{0}_{\mathsf{abs}}(\tilde{a}))) \overset{\text{SAT1,SAI1}}{=}$
$\surd_{\mathsf{s}}\, v \,.\, \sigma^{v}_{\mathsf{abs}}(\nu_{\mathsf{abs}}(\tilde{a})) \overset{\text{SAU1}}{=} \surd_{\mathsf{s}}\, v \,.\, \sigma^{v}_{\mathsf{abs}}(\tilde{a}) = \tilde{\tilde{a}}$

II. Suppose $p \in [0, q]$, $q \in [0, 1)$, $t = \dot{\delta}$.

Then $\overline{\upsilon}^{p}_{\mathsf{abs}}(\sigma^{q}_{\mathsf{abs}}(\tilde{a})) \mathbin{⌊\!⌊} \overline{\upsilon}^{p}_{\mathsf{abs}}(t + \underline{\delta}) \overset{\text{SAI3}',\text{A6ID}}{=} \sigma^{q}_{\mathsf{abs}}(\tilde{a}) \mathbin{⌊\!⌊} \overline{\upsilon}^{p}_{\mathsf{abs}}(\underline{\delta}) \overset{\text{Lemma 11}}{=}$

$\sigma^{q}_{\mathsf{abs}}(\tilde{a}) \mathbin{⌊\!⌊} \sigma^{1}_{\mathsf{abs}}(\dot{\delta}) \overset{\text{SAT2,SACM2}}{=} \sigma^{q}_{\mathsf{abs}}(\tilde{a} \mathbin{⌊\!⌊} \sigma^{1-q}_{\mathsf{abs}}(\dot{\delta})) \overset{\text{CM2SA}}{=} \sigma^{q}_{\mathsf{abs}}(\tilde{a} \cdot \sigma^{1-q}_{\mathsf{abs}}(\dot{\delta})) \overset{\text{SAI3}',\text{SAT2.5}}{=}$

$\sigma^{q}_{\mathsf{abs}}(\tilde{a}) \cdot \sigma^{1}_{\mathsf{abs}}(\dot{\delta}) \overset{\text{INT8,A6ID}}{=} \sigma^{q}_{\mathsf{abs}}(\tilde{a}) \cdot (t + \underline{\delta}) \overset{\text{SAI3}'}{=} \overline{\upsilon}^{p}_{\mathsf{abs}}(\sigma^{q}_{\mathsf{abs}}(\tilde{a})) \cdot (t + \underline{\delta})$

III. Suppose $p \in (q, \infty)$, $q \in [0, 1)$.

Then $\overline{\upsilon}^{p}_{\mathsf{abs}}(\sigma^{q}_{\mathsf{abs}}(\tilde{a})) \mathbin{⌊\!⌊} \overline{\upsilon}^{p}_{\mathsf{abs}}(t + \underline{\delta}) \overset{\text{Lemma 9}}{=} \overline{\upsilon}^{p}_{\mathsf{abs}}(\sigma^{q}_{\mathsf{abs}}(\tilde{a})) \mathbin{⌊\!⌊} \sigma^{p}_{\mathsf{abs}}(t_{[p]} + \tilde{\delta}) \overset{\text{SAI1,2,3}}{=}$

$\sigma^{q}_{\mathsf{abs}}(\sigma^{p-q}_{\mathsf{abs}}(\dot{\delta})) \mathbin{⌊\!⌊} \sigma^{p}_{\mathsf{abs}}(t_{[p]} + \tilde{\delta}) \overset{\text{SAT2,SACM2,CMID1}}{=} \sigma^{p}_{\mathsf{abs}}(\dot{\delta}) \overset{\text{SAT6}}{=}$

$\sigma^{p}_{\mathsf{abs}}(\dot{\delta}) \cdot (t + \underline{\delta}) \overset{\text{SAT2}}{=} \sigma^{q}_{\mathsf{abs}}(\sigma^{p-q}_{\mathsf{abs}}(\dot{\delta})) \cdot (t + \underline{\delta}) \overset{\text{SAI1,2,3}}{=} \overline{\upsilon}^{p}_{\mathsf{abs}}(\sigma^{q}_{\mathsf{abs}}(\tilde{a})) \cdot (t + \underline{\delta})$

CM3DA : The proof is similar to the proof of axiom CM2DA – axiom CM3SA is used instead of axiom CM2SA.

CM5DA, CM6DA and CM7DA :
The proofs are similar to the proof of axiom CF1DA – axiom SIA6 is used in addition and axioms CM5SA, CM6SA and CM7SA are used instead of axiom CF1SA.

DACM3 : $(\upsilon^{1}_{\mathsf{abs}}(t) + \underline{\delta}) \mid \sigma^{n+1}_{\mathsf{abs}}(t') \overset{\text{SATO2,3,6}}{=} (\upsilon^{1}_{\mathsf{abs}}(t) + \upsilon^{1}_{\mathsf{abs}}(\underline{\delta})) \mid \sigma^{n+1}_{\mathsf{abs}}(t') \overset{\text{SATO4}}{=}$

$\upsilon^{1}_{\mathsf{abs}}(t + \underline{\delta}) \mid \sigma^{n+1}_{\mathsf{abs}}(t') \overset{\text{Lemma 10}}{=} (\textstyle\int_{v\in[0,1)} \sigma^{v}_{\mathsf{abs}}(\nu_{\mathsf{abs}}(t^{\circ}) + \tilde{\delta})) \mid \sigma^{n+1}_{\mathsf{abs}}(t') \overset{\text{INT5,11,14}}{=}$

$\textstyle\int_{v\in[0,1)} (\sigma^{v}_{\mathsf{abs}}(\nu_{\mathsf{abs}}(t^{\circ}) + \tilde{\delta}) \mid \sigma^{n+1}_{\mathsf{abs}}(t')) \overset{\text{SACM3}'}{=} \textstyle\int_{v\in[0,1)} \sigma^{v}_{\mathsf{abs}}(\tilde{\delta}) = \underline{\delta}$

DACM4 : The proof is similar to the proof of axiom DACM3 – axioms INT15 and SACM4′ are used instead of axioms INT14 and SACM3′.

D1DA : $\partial_H(\underline{a}) = \partial_H(\textstyle\int_{v\in[0,1)} \sigma^{v}_{\mathsf{abs}}(\tilde{a})) \overset{\text{INT16,SAD}}{=} \textstyle\int_{v\in[0,1)} \sigma^{v}_{\mathsf{abs}}(\partial_H(\tilde{a})) \overset{\text{D1SA}}{=}$

$\textstyle\int_{v\in[0,1)} \sigma^{v}_{\mathsf{abs}}(\tilde{a}) = \underline{a}$ if $a \notin H$

D2DA : The proof is similar to the proof of axiom D1DA – axiom D2SA is used instead of axioms D1SA.

Finally, we show that the additional axioms for discrete initial abstraction are derivable for closed terms.

DIA1 : $\surd_{\mathsf{d}} j \,.\, G = \surd_{\mathsf{s}} w \,.\, G[\lfloor w \rfloor / j] \overset{\text{SIA1}}{=} \surd_{\mathsf{s}} v \,.\, G[\lfloor w \rfloor / j][v/w] = \surd_{\mathsf{s}} v \,.\, G[\lfloor v \rfloor / j] =$

$\surd_{\mathsf{s}} v \,.\, G[i/j][\lfloor v \rfloor / i] = \surd_{\mathsf{d}} i \,.\, G[i/j]$

DIA2 : $\overline{\upsilon}^{n}_{\mathsf{abs}}(\surd_{\mathsf{d}} i \,.\, F) = \overline{\upsilon}^{n}_{\mathsf{abs}}(\surd_{\mathsf{s}} v \,.\, F[\lfloor v \rfloor / i]) \overset{\text{SIA2}}{=} \overline{\upsilon}^{n}_{\mathsf{abs}}(F[\lfloor v \rfloor / i][n/v]) = \overline{\upsilon}^{n}_{\mathsf{abs}}(F[n/i])$

DIA3 : $\surd_{\mathsf{d}} i \,.\, (\surd_{\mathsf{d}} j \,.\, F) = \surd_{\mathsf{s}} v \,.\, (\surd_{\mathsf{s}} w \,.\, F[\lfloor w \rfloor / j])[\lfloor v \rfloor / i] =$

$\surd_{\mathsf{s}} v \,.\, (\surd_{\mathsf{s}} w \,.\, F[\lfloor w \rfloor / j][\lfloor v \rfloor / i]) \overset{\text{SIA3}}{=} \surd_{\mathsf{s}} v \,.\, F[\lfloor w \rfloor / j][\lfloor v \rfloor / i][v/w] =$

$\surd_{\mathsf{s}} v \,.\, F[\lfloor v \rfloor / j][\lfloor v \rfloor / i] = \surd_{\mathsf{s}} v \,.\, F[i/j][\lfloor v \rfloor / i] = \surd_{\mathsf{d}} i \,.\, F[i/j]$

DIA4 : $G \overset{\text{SIA4}}{=} \surd_{\mathsf{s}} v \,.\, G = \surd_{\mathsf{s}} v \,.\, G[\lfloor v \rfloor / i] = \surd_{\mathsf{d}} i \,.\, G$

DIA5 : Suppose $p \in \mathbb{R}_{\geq 0}$ and $\forall n \in \mathbb{N} \bullet \overline{\upsilon}^{n}_{\mathsf{abs}}(F) = \overline{\upsilon}^{n}_{\mathsf{abs}}(F')$.

Then $\overline{\upsilon}^{p}_{\mathsf{abs}}(F) \overset{\text{Lemma 7}}{=} \overline{\upsilon}^{p}_{\mathsf{abs}}(\surd_{\mathsf{s}} v \,.\, \overline{\upsilon}^{\lfloor v \rfloor}_{\mathsf{abs}}(F)) \overset{\text{SIA2}}{=} \overline{\upsilon}^{p}_{\mathsf{abs}}(\overline{\upsilon}^{\lfloor p \rfloor}_{\mathsf{abs}}(F)) = \overline{\upsilon}^{p}_{\mathsf{abs}}(\overline{\upsilon}^{\lfloor p \rfloor}_{\mathsf{abs}}(F')) \overset{\text{SIA2}}{=}$

$\overline{\upsilon}^{p}_{\mathsf{abs}}(\surd_{\mathsf{s}} v \,.\, \overline{\upsilon}^{\lfloor v \rfloor}_{\mathsf{abs}}(F')) \overset{\text{Lemma 7}}{=} \overline{\upsilon}^{p}_{\mathsf{abs}}(F')$.

By SIA5, $F = F'$.

DIA6 : $\sigma^n_{\mathrm{abs}}(\underline{a}) \cdot F = \sigma^n_{\mathrm{abs}}(\int_{v\in[0,1)} \sigma^v_{\mathrm{abs}}(\tilde{a})) \cdot F \overset{\mathrm{INT10,12,SAT2}}{=} \int_{v\in[0,1)} (\sigma^{n+v}_{\mathrm{abs}}(\tilde{a}) \cdot F) \overset{\mathrm{Lemma\,7,SIA6}}{=}$
$\int_{v\in[0,1)} (\sigma^{n+v}_{\mathrm{abs}}(\tilde{a}) \cdot \overline{\upsilon}^{n+v}_{\mathrm{abs}}(\surd_{\mathsf{s}}\, w\,.\, \overline{\upsilon}^{\lfloor w \rfloor}_{\mathrm{abs}}(F))) \overset{\mathrm{SIA2,6}}{=}$
$\int_{v\in[0,1)} (\sigma^{n+v}_{\mathrm{abs}}(\tilde{a}) \cdot \overline{\upsilon}^n_{\mathrm{abs}}(F)) \overset{\mathrm{SAT2,INT10,12}}{=} \sigma^n_{\mathrm{abs}}(\int_{v\in[0,1)} \sigma^v_{\mathrm{abs}}(\tilde{a})) \cdot \overline{\upsilon}^n_{\mathrm{abs}}(F) =$
$\sigma^n_{\mathrm{abs}}(\underline{a}) \cdot \overline{\upsilon}^n_{\mathrm{abs}}(F)$

DIA7 : $\sigma^n_{\mathrm{abs}}(\surd_{\mathsf{d}}\, i\,.\, F) = \sigma^n_{\mathrm{abs}}(\surd_{\mathsf{s}}\, v\,.\, F[\lfloor v \rfloor / i]) \overset{\mathrm{SIA7}}{=} \sigma^n_{\mathrm{abs}}(F[\lfloor v \rfloor / i][0/v]) = \sigma^n_{\mathrm{abs}}(F[0/i])$

DIA8 : $(\surd_{\mathsf{d}}\, i\,.\, F) + G = (\surd_{\mathsf{s}}\, v\,.\, F[\lfloor v \rfloor / i]) + G \overset{\mathrm{Lemma\,7}}{=}$
$(\surd_{\mathsf{s}}\, v\,.\, F[\lfloor v \rfloor / i]) + (\surd_{\mathsf{s}}\, v\,.\, \overline{\upsilon}^{\lfloor v \rfloor}_{\mathrm{abs}}(G)) \overset{\mathrm{DISTR+}}{=} \surd_{\mathsf{s}}\, v\,.\, (F[\lfloor v \rfloor / i] + \overline{\upsilon}^{\lfloor v \rfloor}_{\mathrm{abs}}(G)) =$
$\surd_{\mathsf{s}}\, v\,.\, (F + \overline{\upsilon}^i_{\mathrm{abs}}(G))[\lfloor v \rfloor / i] = \surd_{\mathsf{d}}\, i\,.\, (F + \overline{\upsilon}^i_{\mathrm{abs}}(G))$

DIA9 : $(\surd_{\mathsf{d}}\, i\,.\, F) \cdot G = (\surd_{\mathsf{s}}\, v\,.\, F[\lfloor v \rfloor / i]) \cdot G \overset{\mathrm{SIA9}}{=} \surd_{\mathsf{s}}\, v\,.\, (F[\lfloor v \rfloor / i] \cdot G) =$
$\surd_{\mathsf{s}}\, v\,.\, (F \cdot G)[\lfloor v \rfloor / i] = \surd_{\mathsf{d}}\, i\,.\, (F \cdot G)$

DIA10 : $\upsilon^n_{\mathrm{abs}}(\surd_{\mathsf{d}}\, i\,.\, F) = \upsilon^n_{\mathrm{abs}}(\surd_{\mathsf{s}}\, v\,.\, F[\lfloor v \rfloor / i]) \overset{\mathrm{SIA10}}{=} \surd_{\mathsf{s}}\, v\,.\, \upsilon^n_{\mathrm{abs}}(F[\lfloor v \rfloor / i]) =$
$\surd_{\mathsf{s}}\, v\,.\, \upsilon^n_{\mathrm{abs}}(F)[\lfloor v \rfloor / i] = \surd_{\mathsf{d}}\, i\,.\, \upsilon^n_{\mathrm{abs}}(F)$

DIA11, DIA12, DIA13 and DIA14 :
The proofs are similar to the proof of axiom DIA8 – axioms SIA11, SIA12, SIA13 and SIA14 are used instead of axioms SIA8.

DIA15 : The proof is similar to the proof of axiom DIA10 – axiom SIA15 is used instead of axioms SIA10.

□

References

[1] L. Aceto, W.J. Fokkink and C. Verhoef, *Structural operational semantics*, Handbook of Process Algebra, J.A. Bergstra, A. Ponse and S.A. Smolka, eds, Elsevier, Amsterdam (2001), 197–292.

[2] J.C.M. Baeten and J.A. Bergstra, *Real time process algebra*, Formal Aspects of Computing **3** (2) (1991), 142–188.

[3] J.C.M. Baeten and J.A. Bergstra, *Discrete time process algebra (extended abstract)*, CONCUR'92, Lecture Notes in Comput Sci. 630, W.R. Cleaveland, ed., Springer-Verlag (1992), 401–420. Full version: Report P9208b, Programming Research Group, University of Amsterdam.

[4] J.C.M. Baeten and J.A. Bergstra, *Real space process algebra*, Formal Aspects of Computing **5** (6) (1993), 481–529.

[5] J.C.M. Baeten and J.A. Bergstra, *Discrete time process algebra with abstraction*, Fundamentals of Computation Theory, Lecture Notes in Comput Sci. 965, H. Reichel, ed., Springer-Verlag (1995), 1–15.

[6] J.C.M. Baeten and J.A. Bergstra, *Real time process algebra with infinitesimals*, Algebra of Communicating Processes 1994, A. Ponse, C. Verhoef and S.F.M. van Vlijmen, eds, Workshop in Computing Series, Springer-Verlag (1995), 148–187.

[7] J.C.M. Baeten and J.A. Bergstra, *Discrete time process algebra*, Formal Aspects of Computing **8** (2) (1996), 188–208.

[8] J.C.M. Baeten and J.A. Bergstra, *Discrete time process algebra: Absolute time, relative time and parametric time*, Fund. Inform. **29** (1/2) (1997), 51–76.

[9] J.C.M. Baeten and C. Verhoef, *A congruence theorem for structured operational semantics with predicates*, CONCUR'93, Lecture Notes in Comput Sci. 715, E. Best, ed., Springer-Verlag (1993), 477–492.

[10] J.C.M. Baeten and C. Verhoef, *Concrete process algebra*, Handbook of Logic in Computer Science, Vol. IV, S. Abramsky, D. Gabbay and T.S.E. Maibaum, eds, Oxford University Press (1995), 149–268.

[11] J.C.M. Baeten and W.P. Weijland, *Process Algebra*, Cambridge Tracts in Theoretical Computer Science 18, Cambridge University Press (1990).

[12] J.A. Bergstra and P. Klint, *The discrete time* TOOLBUS – *A software coordination architecture*, Sci. Comput. Programming **31** (1998), 205–229.

[13] J.A. Bergstra and J.W. Klop, *The algebra of recursively defined processes and the algebra of regular processes*, Proc. 11th ICALP, Lecture Notes in Comput Sci. 172, Springer-Verlag (1984), 82–95.

[14] J.A. Bergstra, C.A. Middelburg and Y.S. Usenko, *Discrete time process algebra and the semantics of SDL*, Handbook of Process Algebra, J.A. Bergstra, A. Ponse and S.A. Smolka, eds, Elsevier, Amsterdam (2001), 1209–1268.

[15] R.N. Bol and J.F. Groote, *The meaning of negative premises in transition system specifications*, J. ACM **43** (1996), 863–914.

[16] S.H.J. Bos and M.A. Reniers, *The I^2C-bus in discrete-time process algebra*, Sci. Comput. Programming **29** (1997), 235–258.

[17] J. van den Brink and W.O.D. Griffioen, *Formal semantics of discrete absolute timed interworkings*, Algebra of Communicating Processes 1994, A. Ponse, C. Verhoef and S.F.M. van Vlijmen, eds, Workshop in Computing Series, Springer-Verlag (1995), 106–123.

[18] S.D. Brookes, C.A.R. Hoare and A.W. Roscoe, *A theory of communicating sequential processes*, J. ACM **31** (1984), 560–599.

[19] L. Chen, *An interleaving model for real-time systems*, Symposium on Logical Foundations of Computer Science, Lecture Notes in Comput Sci. 620, A. Nerode and M. Taitslin, eds, Springer-Verlag (1992), 81–92.

[20] F. Corradini, D. D'Ortenzio and P. Inverardi, *On the relationships among four timed process algebras*, Fund. Inform. **38** (4) (1999), 377–395.

[21] J. Davies et al., *Timed CSP: Theory and practice*, Real Time: Theory and Practice, Lecture Notes in Comput Sci. 600, J.W. de Bakker, C. Huizing, W.P. de Roever, G. Rozenberg, eds, Springer-Verlag (1992), 640–675.

[22] W.J. Fokkink, *An elimination theorem for regular behaviours with integration*, CONCUR'93, Lecture Notes in Comput Sci. 715, E. Best, ed., Springer-Verlag (1993), 432–446.

[23] W.J. Fokkink and A.S. Klusener, *An effective axiomatization for real time ACP*, Inform. Comput. **122** (1995), 286–299.

[24] W.J. Fokkink and C. Verhoef, *A conservative look at operational semantics with variable binding*, Inform. Comput. **146** (1998), 24–54.

[25] J.F. Groote, *Transition system specifications with negative premises*, Theoret. Comput. Sci. **118** (1993), 263–299.

[26] M. Hennessy and T. Regan, *A process algebra for timed systems*, Inform. Comput. **117** (1995), 221–239.

[27] J.A. Hillebrand, *The ABP and CABP — a comparison of performances in real time process algebra*, Algebra of Communicating Processes 1994, A. Ponse, C. Verhoef and S.F.M. van Vlijmen, eds, Workshop in Computing Series, Springer-Verlag (1995), 124–147.

[28] C.A.R. Hoare, *Communicating Sequential Processes*, Prentice-Hall (1985).

[29] A.S. Klusener, *Completeness in real-time process algebra*, CONCUR'91, Lecture Notes in Comput Sci. 527, J.C.M. Baeten and J.F. Groote, eds, Springer-Verlag (1991), 376–392.

[30] A.S. Klusener, *Models and axioms for a fragment of real time process algebra*, Ph.D. thesis, Eindhoven University of Technology, Department of Computing Science (1993).

[31] M.J. Koens and L.H. Oei, *A real time μCRL specification of a system for traffic regulation at signalized intersections*, Algebra of Communicating Processes 1994, A. Ponse, C. Verhoef and S.F.M. van Vlijmen, eds, Workshop in Computing Series, Springer-Verlag (1995), 252–279.

[32] J.M.S. van den Meerendonk, *Specification and verification of a circuit in ACP_{drt}-ID*, M.Sc. thesis, Eindhoven University of Technology, Department of Mathematics and Computing Science (1996).

[33] R. Milner, *A Calculus of Communicating Systems*, Lecture Notes in Comput Sci. 92, Springer-Verlag (1980).

[34] R. Milner, *Communication and Concurrency*, Prentice-Hall (1989).

[35] F. Moller and C. Tofts, *A temporal calculus of communicating systems*, CONCUR'90, Lecture Notes in Comput Sci. 458, J.C.M. Baeten and J.W. Klop, eds, Springer-Verlag (1990), 401–415.

[36] F. Moller and C. Tofts, *Relating processes with respect to speed*, CONCUR'91, Lecture Notes in Comput Sci. 527, J.C.M. Baeten and J.F. Groote, eds, Springer-Verlag (1991), 424–438.

[37] X. Nicollin and J. Sifakis, *The algebra of timed processes ATP: Theory and application*, Inform. Comput. **114** (1994), 131–178.

[38] G.D. Plotkin, *A structural approach to operational semantics*, Technical Report DAIMI FN-19, University of Aarhus, Department of Computer Science (1981).

[39] J. Quemada, D. de Frutos and A. Azcorra, *TIC: A timed calculus*, Formal Aspects of Computing **5** (3) (1993), 224–252.

[40] A. Stins and A. Schoneveld, *Specification of a bank account with process algebra*, Report P9307, University of Amsterdam, Programming Research Group (1993).

[41] J.J. Vereijken, *Fischer's protocol in timed process algebra*, Algebra of Communicating Processes 1995, A. Ponse, C. Verhoef and S.F.M. van Vlijmen, eds, Report 95-14, Eindhoven University of Technology, Department of Computing Science (1995), 245–284.

[42] J.J. Vereijken, *Discrete time process algebra*, Ph.D. thesis, Eindhoven University of Technology, Department of Computing Science (1997).

[43] C. Verhoef, *A congruence theorem for structured operational semantics with predicates and negative premises*, Nordic J. Comput. **2** (1995), 274–302.

[44] Wang Yi, *Real-time behaviour of asynchronous agents*, CONCUR'90, Lecture Notes in Comput Sci. 458, J.C.M. Baeten and J.W. Klop, eds, Springer-Verlag (1990), 502–520.

[45] A. van Waveren, *Specification of remote sensing mechanisms in real space process algebra*, Report P9220, University of Amsterdam, Programming Research Group (1992).

[46] LOTOS — *a formal description technique based on the temporal ordering of observational behaviour*, International Standard ISO 8807 (1989).

Subject index

absolute timing, 631, 661
$\mathrm{ACP^{dat}}$, 663
$\mathrm{ACP^{dat}}\surd$, 664
$\mathrm{ACP^{drt}}$, 668
$\mathrm{ACP^{sat}}$, 640
$\mathrm{ACP^{sat}I}$, 646
$\mathrm{ACP^{sat}I}\surd$, 649
$\mathrm{ACP^{srt}}$, 656
action, 631
– undelayable, 661, 666
– urgent, 632, 652
action step relation, 637, 654
action termination relation, 637, 654
alternative composition, 632, 652, 661, 666
axiom system
– of $\mathrm{ACP^{dat}}$, 663
– of $\mathrm{ACP^{drt}}$, 669
– of $\mathrm{ACP^{sat}}$, 641
– of $\mathrm{ACP^{srt}}$, 656
– of $\mathrm{BPA^{dat}}$, 661
– of $\mathrm{BPA^{drt}}$, 666
– of $\mathrm{BPA^{sat}}$, 633
– of $\mathrm{BPA^{srt}}$, 653

basic term, 635, 647, 653, 662, 667
bisimilar, 640, 651, 666
bisimulation
– on DTTS(A), 663
– on DTTS*(A), 666
– on $\mathsf{DTTS^r}$(A), 668
– on RTTS(A), 640
– on RTTS*(A), 651
– on $\mathsf{RTTS^r}$(A), 656
$\mathrm{BPA^{dat}}$, 661
$\mathrm{BPA^{drt}}$, 666
$\mathrm{BPA^{sat}}$, 632
$\mathrm{BPA^{srt}}$, 652

communication function, 631
communication merge, 641, 656, 663, 668
congruence theorem, 640

deadlock
– immediate, 632, 652, 661, 666
– undelayable, 661, 666
– urgent, 632, 652
delay
– absolute, 632, 661
– relative, 652, 666
discrete time process algebra
– with absolute timing, 661
– – extensions of, 664
– with relative timing, 666
discrete time transition system, 662
– with parametric timing, 666
– with relative timing, 667
discretely initialized real time process, 671
discretization, 669

discretized real time process, 670
DTTS(A), 663
DTTS*(A), 666
$\mathsf{DTTS^r}$(A), 667

elimination result, 635, 643, 647, 650, 653, 656, 662, 663
embedding, 657
– of ACP^{dat} in $ACP^{sat}I\surd$, 669–673
– of ACP^{srt} in $ACP^{sat}\surd$, 657–660
encapsulation, 641, 656, 663, 668
explicit definition, 658

immediate deadlock, 632, 652, 661, 666, 673
immediate deadlock relation, 637, 654
initial abstraction
– discrete, 664–666
– standard, 648–651
initialization
– absolute, 632, 661
– relative, 652, 666
– urgent
– – absolute, 641
– – relative, 656
integration, 646–648
– prefix, 647

left merge, 641, 656, 663, 668

panth format, 640
parallel composition, 641, 656, 663, 668
parametric timing, 651, 666
process
– absolute time, 632
– parametric time, 649
– relative time, 652
process algebra
– discrete time
– – with absolute timing, 661–664
– – with relative timing, 666–669
– real time
– – with absolute timing, 631–645
– – with relative timing, 652–656

real time process algebra
– with absolute timing, 631
– – extensions of, 645
– with relative timing, 652
real time transition system, 637
– with parametric timing, 651
– with relative timing, 654
relative timing, 652, 666
root state, 637, 654, 662, 667
RTTS(A), 637
RTTS*(A), 651
$\mathsf{RTTS^r}$(A), 654

semantics
– of ACP^{dat}, 664
– of ACP^{drt}, 669
– of ACP^{sat}, 643
– of ACP^{srt}, 656
– of BPA^{dat}, 662
– of BPA^{drt}, 667
– of BPA^{sat}, 637
– of BPA^{srt}, 654
sequential composition, 632, 652, 661, 666
signature
– of ACP^{dat}, 663
– of ACP^{drt}, 668
– of ACP^{sat}, 641
– of ACP^{srt}, 656
– of BPA^{dat}, 661
– of BPA^{drt}, 666
– of BPA^{sat}, 632
– of BPA^{srt}, 652
standard initialization axioms, 652
state, 637, 654, 662, 667
stratifiable, 638
– strictly, 638
structural operational semantics, 637

term deduction system, 638
time factorization, 633
time slice, 661
time step relation, 637, 654
time variable, 646
time-out
– absolute, 632, 661
– relative, 652, 666
time-stamped actions, 643
time-stamping scheme, 629
transition system
– discrete time, 662
– – with parametric timing, 666
– – with relative timing, 667
– real time, 637
– – with parametric timing, 651
– – with relative timing, 654
two-phase scheme, 629

undelayable action, 661, 666
undelayable deadlock, 661, 666
urgent action, 632, 652
urgent deadlock, 632, 652
urgent initialization
– absolute, 641
– relative, 656

variable binding operator, 645

CHAPTER 11

Probabilistic Extensions of Process Algebras*

Bengt Jonsson[1], Wang Yi[1], Kim G. Larsen[2]

[1] *Department of Computer Systems, Uppsala University, Sweden,*
E-mails: bengt@docs.uu.se, yi@docs.uu.se

[2] *Department of Computer Science, Aalborg University, Denmark,*
E-mail: kgl@cs.auc.dk

Contents

1. Introduction . . . 687
2. Preliminaries . . . 688
3. Probabilistic models . . . 690
   3.1. Probabilistic transition systems . . . 690
   3.2. Variants of probabilistic transition systems . . . 691
4. Operators of probabilistic process algebras . . . 692
5. Probabilistic bisimulation and simulation . . . 695
   5.1. Probabilistic bisimulation . . . 695
   5.2. Probabilistic simulation . . . 697
   5.3. Congruence properties . . . 698
6. Testing preorders . . . 699
   6.1. Related work . . . 700
   6.2. Tests, testing systems and preorders . . . 700
   6.3. Characterization of testing preorders . . . 702
7. Probabilistic logics . . . 704
   7.1. Characterizing preorders . . . 704
   7.2. Model checking probabilistic temporal logics . . . 706
8. Conclusion and trends . . . 706
References . . . 707
Subject index . . . 710

*This chapter is dedicated to the fond memory of Linda Christoff.

HANDBOOK OF PROCESS ALGEBRA
Edited by Jan A. Bergstra, Alban Ponse and Scott A. Smolka

Abstract

In this chapter, we adopt Probabilistic Transition Systems as a basic model for probabilistic processes, in which probabilistic and nondeterministic choices are independent concepts. The model is essentially a nondeterministic version of Markov decision processes or probabilistic automata of Rabin. We develop a general framework to define probabilistic process languages to describe probabilistic transition systems. In particular, we show how operators for nonprobabilistic process algebras can be lifted to probabilistic process algebras in a uniform way similar to de Simone format. To establish a notion of refinement, we present a family of preorders including probabilistic bisimulation and simulation, and probabilistic testing preorders as well as their logical or denotational characterization. These preorders are shown to be precongruences with respect to the algebraic operators that can be defined in our general framework. Finally, we give a short account of the important work on extending the successful field of model checking to probabilistic settings and a brief discussion on current research in the area.

1. Introduction

Classic process, algebras such as CCS, CSP and ACP, are well-established techniques for modelling and reasoning about functional aspects of concurrent processes. The motivation for studying probabilistic extensions of process algebras is to develop techniques dealing with non-functional aspects of process behavior, such as performance and reliability. We may want to investigate, e.g., the average response time of a system, or the probability that a certain failure occurs. An analysis of these and similar properties requires that some form of information about the stochastic distribution over the occurrence of relevant events is put into the model. For instance, performance evaluation is often based on modeling a system as a continuous-time Markov process, in which distributions over delays between actions and over the choice between different actions are specified. Similarly, reliability can be analyzed quantitatively only if we know some probability of the occurrence of events related to a failure. Performance evaluation and reliability analysis are well-established topics, and it is not the aim to contribute in these areas. Rather, we should try to see what the process algebraic approach can offer to these fields. Process algebras has contributed to our understanding of

- how to describe (model) communicating systems compositionally,
- how to formulate correctness properties of systems,
- how properties of a system relate to properties of its components, and
- what it means for a description of a system component to be a correct implementation of another component description.

A solution to these problems would be very useful, e.g., in a stepwise development process. An abstract model can be analyzed by proving properties in some logic (for the nonprobabilistic case, see e.g., [11], Chapter 4 in this Handbook). The abstract model can then be refined in a sequence of steps, where correctness is preserved in each step by establishing a preorder relation between the refined system and the refining one (techniques for the nonprobabilistic case are described in [18], Chapter 6 of this Handbook).

In this chapter, we will study the above issues in the context of a simple yet general model of probabilistic processes. In the nonprobabilistic setting, labeled transition systems are well-established as a basic semantic model for concurrent and distributed systems (e.g., [47,50]). In the literature, the model of transition systems has been extended to the probabilistic case by adding a mechanism for representing probabilistic choice (e.g., [33,32, 35,45,48,51,52,54]). We will adopt a model of probabilistic transition systems, in which probabilistic and nondeterministic choice are independent concepts. Nondeterminism can be used to represent underspecification, which can then be partly removed in refinement steps. For example, nondeterminism can be used to specify the allowed probabilities of failure of a medium, and a refinement can decrease the set of allowed failure rates [41]. Nondeterminism can also represent incomplete information on the parameters of system behavior, such as Milner's "weather conditions" [47].

Our model is essentially a nondeterministic version of Markov decision processes [27] or the probabilistic automata of Rabin [54]. In the area of process algebra, the model has been put forward by Vardi under the term concurrent Markov chain [58], by Wang and Larsen [60] and by Hansson and Jonsson as the alternating model [35], and by Segala and Lynch [56,55]. A deterministic version has been proposed as the reactive model by Larsen and Skou [45]. There are several other models of probabilistic transition systems proposed in the literature. A short summary will be provided in Section 3.2.

The rest of the chapter will be organized as follows: Section 2 introduces notation and basic concepts from probability theory. Section 3 presents Probabilistic Transition Systems (PTS) as a basic model for probabilistic processes and summarizes variants of PTS proposed in the literature. Section 4 will consider how operators can be defined on probabilistic processes. We will, in particular, see how operators for nonprobabilistic process algebras can be lifted to the probabilistic case in a uniform way similar based on de Simone format. Probabilities are added only by means of a probabilistic choice construct. We will thereafter, in the following sections, consider different preorders between probabilistic processes, and how these preorders interact with the operators for constructing processes, i.e., (pre)congruence results. Section 5 is devoted to the development of probabilistic versions of bisimulation and simulation. These are preorders based on relating states or distributions to each other, i.e., relations on branching (or tree) structures. Section 6 presents testing preorders, another family of preorders that are defined in terms of an operational notion of testing, and characterize them in terms of simulations. Section 7 presents a basic probabilistic modal logic and show how it relates to and characterizes the various behavioural preorders. Also, we give a short account of the important work on extending the successful field of model checking to probabilistic settings. Section 8 concludes the chapter with a brief discussion on current research in the area.

2. Preliminaries

In this section, we introduce some notation and definitions from probability theory.

A *probability distribution* on a countable set S is a function $\pi : S \to [0, 1]$ such that $\sum_{s \in S} \pi(s) = 1$. More generally, a *weighting* on a set S is a function $\pi : S \to \mathcal{R}_{\geqslant 0}$ from S to nonnegative real numbers. Note that a *probability distribution* on a finite set S is a weighting π on S such that $\pi(S) = 1$. The *support* of a distribution or weighting π on S, denoted support(π) is the set of elements s such that $\pi(s) > 0$. For a subset $S' \subseteq S$, we define $\pi(S') = \sum_{s \in S'} \pi(s)$. Let $Dist(S)$ denote the set of probability distributions on S. If π is a probability distribution on S and ρ is a probability distribution on T, then their *product* $\sigma = \pi \times \rho$ is a probability distribution on $S \times T$, defined by $\sigma(\langle s, t \rangle) = \pi(s) * \rho(t)$. For simplicity, we shall write $\sigma(s, T)$ for $\sigma(\{s\} \times T)$ and $\sigma(S, t)$ for $\sigma(S \times \{t\})$.

We will next define a general way, proposed by Jonsson and Larsen [41], to lift a relation between two countable sets to a relation between distributions on these sets.

DEFINITION 1 (*Lifting*). Let $\approx \subseteq S \times T$ be a relation between the sets S and T, π be a probability distribution on S and ρ be a probability distribution on T. We define $\pi \approx^* \rho$ iff there is a distribution $\alpha \in Dist(S \times T)$ on $S \times T$ such that

- $\alpha(s, T) = \pi(s)$ for each $s \in S$,
- $\alpha(S, t) = \rho(t)$ for each $t \in T$, and
- $\alpha(s, t) = 0$ if $s \not\approx t$.

We shall write $\pi \approx \rho$ whenever $\pi \approx^* \rho$ and it is understood from the context.

Intuitively, $\pi \approx^* \rho$ means that there is a distribution on $S \times T$ whose projection onto S is π, whose projection onto T is ρ, and whose support is in $\approx$. The relation $\pi \approx^* \rho$ thus holds if for each $s \in S$, it is possible to distribute the probability $\pi(s)$ over elements of T that are $\approx$-related to s, in such a way that the sum of these distributed probabilities, weighted by π, is the distribution ρ.

There is a simpler way of lifting equivalence relations on countable sets to the distributions on these sets.

THEOREM 1. *Let $\approx$ be an equivalence relation over the set S. Then $\pi \approx^* \rho$ iff $\pi([s]) = \rho([s])$ for all equivalence classes $[s] \subseteq S$ of relation $\approx$.*

PROOF. *If*: Assume that $\pi([s]) = \rho([s])$ for all $s \in S$. Let $\alpha(s,t) = \pi(s) * \rho(t)/\pi([s])$ if $s \approx t$ and $\pi(s) > 0$, and 0 otherwise. Then $\alpha(s, S) = \pi(s) * \rho([s])/\pi([s]) = \pi(s)$ as $\pi([s]) = \rho([s])$. Similarly, $\alpha(S,t) = \rho(t)$. The third condition holds immediately from the definition of α.

Only If: Assume that $\pi \approx^* \rho$ and further assume that α is as in Definition 1. Then $\pi([s]) = \sum_{s \in [s]} \alpha(s, S) = \alpha([s], S) = \alpha([s],[s])$ since $\alpha(s,t) = 0$ for $t \notin [s]$. Symmetrically, $\rho([s]) = \alpha([s],[s])$. Thus $\pi([s]) = \rho([s])$. □

The lifting operation on relations preserves the characteristic properties of preorders and equivalences.

THEOREM 2. *Let $\approx$ be a relation on the set S. Then $\approx$ is a preorder implies that $\approx^*$ is a preorder on Dist(S).*

PROOF. We only show the transitivity of $\approx^*$. Assume that $\pi \approx \rho$ and $\rho \approx \varrho$. Then there are distributions α and β on $S \times S$ satisfying the three conditions given in Definition 1. Now let $\gamma(s,t) = \sum_{s' \in S} \alpha(s,s')/\, \rho(s') * \beta(s',t)$. We check the three conditions in Definition 1. First

$$\begin{aligned}\gamma(s,S) &= \sum_{s' \in S} \alpha(s,s')/\rho(s') * \beta(s',S) = \sum_{s' \in S} \bigl(\alpha(s,s')/\rho(s') * \rho(s')\bigr)\\ &= \sum_{s' \in S} \alpha(s,s') = \pi(s).\end{aligned}$$

Second,

$$\begin{aligned}\gamma(S,t) &= \sum_{s' \in S} \bigl(\alpha(S,s')/\rho(s') * \beta(s',t)\bigr) = \sum_{s' \in S} \bigl(\rho(s')/\rho(s') * \beta(s',t)\bigr)\\ &= \sum_{s' \in S} \beta(s',t) = \varrho(t).\end{aligned}$$

For the third condition, note that if $\gamma(s,t) > 0$ then there must be s' such that $\alpha(s,s') * \beta(s',t) > 0$. This implies that $s \approx s'$ and $s' \approx t$. By the transitivity of $\approx$, $s \approx t$. □

The above result can be extended to equivalence relations.

THEOREM 3. *Let $\approx$ be a relation on the set S. Then $\approx$ is an equivalence implies that $\approx^*$ is an equivalence on $Dist(S)$.*

PROOF. Immediate from Theorem 1. $\square$

Equivalences and preorders will be of particular interest in the rest of this chapter. We extend a relation $\approx$ on the set S to the Cartesian product $S \times S$ in the usual way. That is, $\langle s, t\rangle \approx \langle s', t'\rangle$ whenever $s \approx s'$ and $t \approx t'$. Then $\approx^*$ is preserved by the product operation on probability distributions.

THEOREM 4. *Let $\approx$ be a preorder (or an equivalence relation) and π, ρ, π' and ρ' be probabilistic distributions on S. Then $\pi \approx^* \rho$ and $\pi' \approx^* \rho'$ imply that $\pi \times \pi' \approx^* \rho \times \rho'$.*

PROOF. By the transitivity of $\approx^*$, we only need to establish that $\pi \approx \rho$ implies $\pi \times \varrho \approx^* \rho \times \varrho$.

Following Definition 1, we construct a distribution β over $(S \times S) \times (S \times S)$. First note that $\pi \approx^* \rho$. Thus, there exists a distribution α on $S \times S$ such that $\alpha(s, S) = \pi(s)$, $\alpha(S, s') = \varrho(t)$ and $\alpha(s, s') = 0$ for $s \not\approx s'$.

Now let $\beta(\langle s, t\rangle, \langle s', t'\rangle) = \alpha(s, s') * \varrho(t)$ whenever $t = t'$ and 0 otherwise. Then

$$\begin{aligned}\beta\big(\langle s, t\rangle, S \times S\big) &= \sum_{s', t' \in S} \big(\alpha(s, s') * \varrho(t)\big) = \rho(t) * \sum_{s', t' \in S} \alpha(s, s') \\ &= \rho(t) * \sum_{s' \in S} \alpha(s, s') = \rho(t) * \pi(s).\end{aligned}$$

Thus β holds for the first condition in Definition 1. Similarly, we can establish the second condition for β. That is $\beta(S \times S, \langle s', t'\rangle) = \rho(s') * \varrho(t')$. The third condition is obvious. If $\langle\langle s, t\rangle, \langle s', t'\rangle\rangle \not\approx$ then $s \not\approx s'$. Thus $\alpha(s, t) = 0$ and therefore $\beta((s, t), (s', t')) = \alpha(s, s') * \varrho(lt) = 0$. $\square$

3. Probabilistic models

We consider a model of probabilistic transition systems, containing probabilistic and non-deterministic choices as independent concepts.

3.1. *Probabilistic transition systems*

We assume a set *Act* of *actions*, ranged over by a and b.

DEFINITION 2. A *Probabilistic Transition System* (*PTS*) is a tuple $\langle S, \rightarrow, \pi_0\rangle$, where
- S is a non-empty finite set of *states*,
- $\rightarrow \subseteq S \times Act \times Dist(S)$ is a finite *transition relation*, and
- $\pi_0 \in Dist(S)$ is an *initial distribution* on S.

We shall use $s \xrightarrow{a} \pi$ to denote that $\langle s, a, \pi \rangle \in \rightarrow$. We use $s \xrightarrow{s}$ to denote that there is a π such that $s \xrightarrow{a} \pi$, and $s \not\xrightarrow{a}$ to denote that there is no π such that $s \xrightarrow{a} \pi$. We say that a state s is *terminal* (written $s \not\rightarrow$) if $s \not\xrightarrow{a}$ for all $a \in \mathcal{A}ct$.

This definition occurs with minor variations in [58,60,35,42,56,55,45]. In each state, a probabilistic transition system can perform a number of possible actions. Each action leads to a distribution over successor states. In many cases when it is understood from the context, we will identify a state s with the distribution that assigns probability 1 to the state s.

An *initial state* of a process $\mathcal{P} = \langle S, \rightarrow, \pi_0 \rangle$ is a state $s \in S$ such that $\pi_0(s) > 0$. A state s is *reachable* in $\mathcal{P}$ if there is a sequence $s_0 s_1 \ldots s_n$ where s_0 is initial, $s = s_n$, and for each $0 \leqslant i < n$ there is a distribution π_{i+1} such that $s_i \xrightarrow{a_i} \pi_{i+1}$ and $\pi_{i+1}(s_{i+1}) > 0$. A distribution $\pi \in \mathit{Dist}(S)$ is *reachable* in $\mathcal{P}$ if it is either the initial distribution or if $s \xrightarrow{a} \pi$ for some a and state s which is reachable in $\mathcal{P}$.

We use $s \xrightarrow{a}\rightsquigarrow s'$ to denote that there is a π such that $s \xrightarrow{a} \pi$ and $\pi(s') > 0$, and $s \rightarrow \rightsquigarrow s'$ to denote that there is an a such that $s \xrightarrow{a}\rightsquigarrow s'$. A *finite process* is a process $\langle S, \rightarrow, \pi_0 \rangle$ with a finite number of states, in which the relation $\rightarrow\rightsquigarrow$ is acyclic.

A scheduler will decide which action should be taken in each state and then makes a probabilistic choice. Thus each scheduler corresponds to a *probabilistic execution* of a process. In the following sections, we shall study how to compare processes in terms of such executions.

3.2. *Variants of probabilistic transition systems*

The reactive model. In [44,45,33] a simple class of probabilistic transition systems is identified as the *reactive models*. It is the class of probabilistic transition systems where all states s are *deterministic* in the sense that for each action a, whenever $s \xrightarrow{a} \pi_1$ and $s \xrightarrow{a} \pi_2$, then $\pi_1 = \pi_2$. Systems in the reactive models have the same structure as Markov Decision Processes [27].

The generative model. A definition of a generative probabilistic transition system differs from the one given above in that each transition is an element in $S \times \mathit{Dist}(\mathcal{A}ct \times S)$, i.e., the probability distribution also includes a distribution over the possible actions. If all actions in a distribution are identical, we get the above definition of a probabilistic transition system.[1]

Several researchers (e.g., [34,20] have noticed that it is not trivial to define a symmetric parallel composition operator in the generative model. One source of difficulty is that in a generative model, a probabilistic transition system defines in each of its states a probability distribution over a set of enabled actions. This view makes sense if the set of enabled actions is offered by, e.g., the environment. If two probabilistic transition systems are composed in parallel, then each of them defines a separate probability distribution over a set of

[1] Note that the above notion of generative model still allows some nondeterminism in the sense that a state may have transitions to several distributions. In [33], states are allowed at most one transition.

enabled actions. It is not clear how the set of "enabled actions" is to be defined, nor how the two probability distributions should be composed. Approaches to this problem can be found in, e.g., [20,17].

Another interpretation of a distribution over different actions in the generative model is that the choice is under control of the process itself. This could be the case if the actions are "output" actions, which communicate with corresponding "input actions" in a process that does not constrain the choice of the "outputting" process. A natural resulting model is then a probabilistic version of I/O automata, defined as *Probabilistic I/O automata* [59]. Probabilistic I/O automata use continuous time and rates to define transition probabilities. An model in a discrete-time framework that captures an analogous distinction between input and could, but not the time-dependent behavior modeled by the rates of Probabilistic I/O-automata, can be obtained from the model in Definition 2, as follows. For each input action, there is an enabled transition from each state. In each state, at most one output action may be enabled. The choice between different output actions must have been made in the previous transitions.

4. Operators of probabilistic process algebras

In this section, we consider how operators can be defined in probabilistic process algebras. We will present a general framework that allows essentially any nonprobabilistic process algebra to be extended to a probabilistic process algebra by introduction of a probabilistic (internal) choice operator. The operators of the given nonprobabilistic process algebra may be lifted to operators in the probabilistic process algebra in a completely uniform way, under certain assumptions. Later we will instantiate the framework to particular process algebras.

Terms in the resulting probabilistic algebra will denote probabilistic processes. A set of terms can be used to form a probabilistic transition systems as in Definition 2. Some of the terms will correspond to states, and the others will correspond to distributions over states.

Terms of the nonprobabilistic process algebra are assumed to be formed in the usual way by the constant NIL, and a number of operators, each with a certain arity. NIL denotes a state which has no outgoing transitions.

Also, the meaning of an n-ary operator *op* of the nonprobabilistic process algebra, is assumed to be defined by a finite set of rules in the so-called de Simone format [26] (see also Aceto, Fokkink and Verhoef, [2], Chapter 3 in this Handbook):

$$\frac{p_{i_1} \rightarrow^{a_1} q_{i_1} \qquad \cdots \qquad p_{i_k} \rightarrow^{a_k} q_{i_k}}{op(p_1, \ldots, p_n) \rightarrow^{a} t} \tag{1}$$

where $p_1, \ldots, p_n$ and $q_{i_1}, \ldots, q_{i_k}$ are all distinct process expression variables, a and $a_1, \ldots, a_k$ are actions, $\{p_{i_1}, \ldots, p_{i_k}\}$ is a subset of $\{p_1, \ldots, p_n\}$, and t is a linear term[2] over the process expression variables $\{p_1, \ldots, p_n\} \setminus \{p_{i_1}, \ldots, p_{i_k}\} \cup \{q_{i_1}, \ldots, q_{i_k}\}$.

[2] In a linear term each variable occurs at most once.

In the above rule (1), we say that the ith argument is *initially active* if i is among $i_1, \ldots, i_k$. For an n-ary operator *op*, the ith argument is said to be *initially active* if it is initially active in any of the defining rules for *op*.

The probabilistic extension of the process algebra is obtained by an extension of the syntax allowing for distributions to be expressed. We introduce a single additional binary operator, *internal probabilistic choice* $\oplus_p$, parameterized by a real number $0 \leqslant p \leqslant 1$. The term $E \oplus_p F$ denotes a distribution which assigns the probability $p \cdot E(s) + (1 - p) \cdot F(s)$ to each state s (note that we regard E and F as distributions and $E(s)$ and $F(s)$ are probabilities assigned by E and F on state s respectively). The choice is *internal* in the sense that the term denotes a probability distribution over states.

Now, terms of the extended process algebra are build using the original operators *op* together with the new additional operator $\oplus_p$. Our guiding principle for separating these terms into those denoting states and those denoting distributions over states is, that all probabilistic choices should be resolved before any nondeterministic transition is taken. Thus, terms denoting distributions may be inductively defined as either terms of the form $E \oplus_p F$, or terms of the form $op(t_1, \ldots, t_n)$, where for some initially active position i the t_i denotes a distribution. Formally this becomes:

DEFINITION 3. The set of process expressions that denote states is defined as the smallest set that satisfies:

– $op(t_1, \cdots, t_n)$ denotes a state if all terms $t_{i_1}, \ldots, t_{i_k}$ in initially active positions denote states.

An arbitrary process expression $op(t_1, \cdots, t_n)$, in which $i_1, \ldots, i_k$ are the initially active positions, and where each t_{i_j} denotes a distribution $[\![t_{i_j}]\!]$ over states, denotes a distribution, which assigns the probability $[\![t_{i_1}]\!](s_{i_1}) * \cdots * [\![t_{i_k}]\!](s_{i_k})$ to the state $op(t_1, \ldots, s_{i_1}, \ldots, s_{i_k}, \ldots, t_n)$, where $op(t_1, \ldots, s_{i_1}, \ldots, s_{i_k}, \ldots, t_n)$ is obtained from $op(t_1, \cdots, t_n)$ by replacing the terms $t_{i_1}, \ldots, t_{i_k}$ in initially active positions by states $s_{i_1}, \ldots, s_{i_k}$.

To understand our general framework it may be instructive to consider the following examples:

– NIL is a process expression, which denotes a state without outgoing transitions.

– The prefixing operator is defined by the rule

$$\frac{}{a.p \rightarrow^a p}.$$

Thus the term $a.E$ has no initially active positions, and hence it denotes a state.

– The nondeterministic choice operator is defined by the rules

$$\frac{p \rightarrow^a p'}{p + q \rightarrow^a p'} \qquad \frac{q \rightarrow^a q'}{p + q \rightarrow^a p'}.$$

Thus, both positions are initially active. Figure 1 illustrates the behaviour of the term $(a \oplus_{0.3} b) + (c \oplus_{0.4} d)$. Note that this term (due to the presence of $\oplus$ in active positions) denotes a distribution. In the figure, we use dotted lines to indicate probabilistic choices

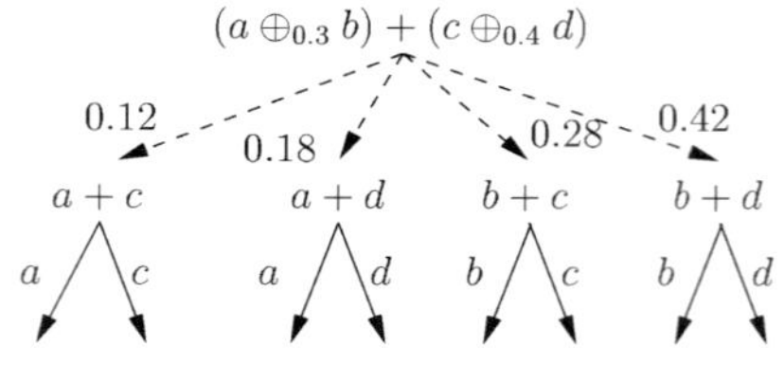

Fig. 1. Behaviour of $(a \oplus_{0.3} b) + (c \oplus_{0.4} d)$.

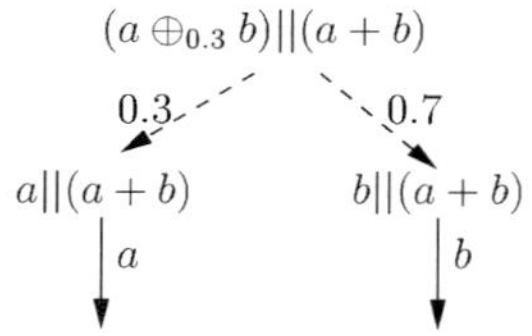

Fig. 2. Behaviour of $(a \oplus_{0.3} b) \| (a + b)$.

from distributions, and filled lines to indicate action transitions from states. Also, we are omitting trailing occurrences of NIL, writing a for a.NIL.

- The operator for synchronous parallel composition a'la CSP may be defined by the following rule:[3]

$$\frac{p \to^a p' \quad q \to^a q'}{p \| q \to^a p' \| p'}.$$

Clearly, both positions are initially active. Figure 2 illustrates the behaviour of the parallel term (denoting a distribution) $(a \oplus_{0.3} b) \| (a + b)$.

As usual, process constants P may be introduced by recursive definitions, $P \stackrel{\text{def}}{=} E$, with transitions of P being inferred from its definition E.

Along similar lines, we can present a probabilistic version of SCCS, which is similar to the calculus PCCS presented by Giacalone et al. [32]. As in SCCS, let $(Act, \times, 1,)$ be the Abelian group of *atomic actions*. Intuitively, actions of the form $\alpha \times \beta$ represent the simultaneous, atomic occurrence of the actions α and β. We will often write $\alpha\beta$ instead of $\alpha \times \beta$. The action 1 is the "idle action", and $\overline{\alpha}$ is the dual action of α. The action $\alpha \times \overline{\alpha}$ represents a synchronized communication between complementary actions, and $\alpha \times \overline{\alpha} = 1$.

The syntax of PCCS is given by

$$E ::= \mathsf{nil} \mid \mathsf{X} \mid \alpha.\mathsf{E} | \mathsf{E} \oplus_{\mathsf{p}} \mathsf{F} \mid \mathsf{E} \times \mathsf{F} \mid \mathsf{E} \backslash \mathsf{A} \mid \mathsf{E}[\mathsf{f}] \mid \mathbf{rec}\mathsf{X}.\mathsf{E}$$

[3] We have chosen to illustrate a synchronous parallel operator, as we will need it later for introducing testing preorders. However, asynchronous parallel composition as in CCS may be extended in a similar manner.

Table 1
Inference rules for PCCS

$$\frac{}{\alpha.P \xrightarrow{\alpha} P}$$

$$\frac{P \xrightarrow{\alpha} P' \qquad Q \xrightarrow{\beta} Q'}{P \times Q \xrightarrow{\alpha\beta} P' \times Q'}$$

$$\frac{P \xrightarrow{\alpha} P'}{P\backslash A \xrightarrow{\alpha} P'\backslash A}\ (\alpha, \bar{\alpha} \notin A) \qquad \frac{P \xrightarrow{\alpha} P'}{P[f] \xrightarrow{f(\alpha)} P'[f]}$$

where A is a subset of Act such that $1 \in A$ and $f : Act \mapsto Act$ is a group morphism. Note the absence of nondeterministic choice in the above definition. The operational semantics of PCCS for action transitions is given in Table 1.

In [32], the presentation is slightly different in appearance. In the operational semantics, transitions are of the form $P \xrightarrow{\alpha[p]} P'$ meaning that "P can perform the action α with probability p and become process P'". In this representation, P is a distribution over states, which assigns probability p to the set of states that can perform α and become P'.

5. Probabilistic bisimulation and simulation

In this section, we will present a series of bisimulation-preorders between probabilistic processes. These are intended to be generalizations of corresponding preorders on non-probabilistic processes (see also Cleaveland and Sokolsky, [18], Chapter 6 in this Handbook).

5.1. *Probabilistic bisimulation*

A bisimulation can be understood as an equivalence relation, where two processes are equivalent if each process can mimick whatever the other process can do. We will first consider probabilistic bisimulation defined as an equivalence over states, not over distributions.

DEFINITION 4 (*Bisimulation*). An equivalence relation R over a set of states S is a *bisimulation* if sRt implies that whenever $s \xrightarrow{a} \pi$ for some action a and distribution π, then there is a distribution ρ such that $t \xrightarrow{a} \rho$ and $\pi R \rho$.

In Figure 3 is an example of two bisimilar processes.

In the process to the left, the action *in* leads to a distribution over two states. From the left state, the process can perform the action *out*, meaning that the action *in* has been performed normally. From the right state, only the action *err* can be performed, meaning

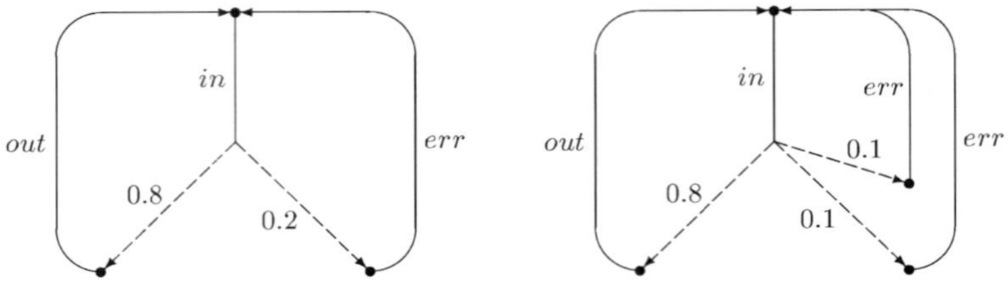

Fig. 3. Two bisimilar processes.

that the previous action *in* was not successful. In the left process, the probability of an unsuccessful *in* action is 0.2. In the right process, the intention is to model two ways, represented by the states to the right, for the action *in* to be unsuccessful. Each of the states is reached with probability 0.1. In this process, both of these states are bisimilar, since they can perform equivalent actions, leading to equivalent states. Therefore, also the two start states (from which the action *in* can be performed) are equivalent, since they both lead to the successful equivalence class with probability 0.8 and to the unsuccessful equivalence class with probability 0.2).

The above definition of bisimulation was introduced by Larsen and Skou [45] for reactive systems and it is the most commonly used generalization of bisimulation to the probabilistic setting.

For models that embody both probabilistic and nondeterministic choice, in the way defined in this chapter, it can be argued that the above definition of probabilistic bisimulation is too restrictive. Namely, recall that we could let nondeterministic choices be resolved by schedulers, and that schedulers can be probabilistic. But the above definition of bisimulation considers only deterministic choices by schedulers. Consider for example the two processes of Figure 4.

Note that ignoring the middle *in*-transition of the right process, these two processes are identical and the middle *in*-transition is a convex combination of the other two (see Definition 5). Assuming that schedulers can be probabilistic, the extra transition can be generated as a probabilistic choice over the two other transitions. It therefore does not add anything to the process, meaning that the two processes ought to be equivalent. The definition of probabilistic bisimulation can thus be weakened by including combined transitions as follows:

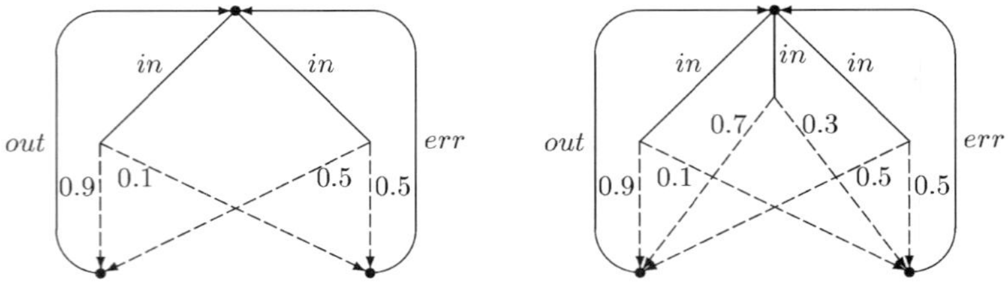

Fig. 4. Two probabilistic bisimilar processes.

DEFINITION 5 (*Combined transition*). Let t be a state of a probabilistic transition system and let a be an action. We say that $t \xrightarrow{a} \rho$ is a *combined* transition of t if ρ is a convex combination of the set $\Pi = \{\rho_i \mid t \xrightarrow{a} \rho_i\}$ of distributions, that is, for each ρ_i, there is a nonnegative real number λ_i such that $\sum_i \lambda_i = 1$ and $\rho = \sum_{\rho_i \in \Pi} \lambda_i * \rho_i$.

DEFINITION 6 (*Probabilistic bisimulation*). An equivalence relation R over S is a *probabilistic bisimulation* if $s\,R\,t$ implies that whenever $s \xrightarrow{a} \pi$ for some action a and distribution π, then there is a distribution ρ such that $t \xrightarrow{a} \rho$ is a combined transition with $\pi\,R\,\rho$.

We use $\simeq$ to denote the largest probabilistic bisimulation i.e., the union of all probabilistic bisimulations over S and $Dist(S)$.

Now we have a weaker equivalence relation than bisimulation. For example, consider the two processes shown in Figure 4. Note that the two processes are not bisimilar according to Definition 4. But they are bisimilar according to Definition 6 because the middle *in*-transition is a combined transition of the other two.

This definition was introduced by Segala and Lynch [56]. Following their terminology, we use the term *bisimulation* for the more restrictive definition, and the term *probabilistic bisimulation* for the definition which considers also combined transitions. With the definition of probabilistic bisimulation, we can use a combination of two or more transitions to denote a range of allowed probabilities for the outcome of an action, namely those that are in the convex closure of the outcomes of the actions actually described. In this way, it is similar to the use of intervals by Jonsson and Larsen [41].

5.2. *Probabilistic simulation*

A simulation can be understood as a relation, which captures the idea that one of the processes can mimick whatever the other process can perform. In contrast to bisimulation, a simulation need not be symmetric, since the mimicking capability is required only of one of the processes.

DEFINITION 7 (*Simulation*). A preorder R over S is a simulation if $s\,R\,t$ implies that whenever $s \xrightarrow{a} \pi$ for some action a and distribution π, then there is a distribution ρ such that $t \xrightarrow{a} \rho$ and $\pi\,R^*\,\rho$.

As an example, consider again the two processes in Figure 4. The right process simulates the left one since it has more transitions. However, the left process does not simulate the right one . But as in the case of bisimulation, we can allow the matching transition $t \xrightarrow{a} \rho$ to be a combined transition. We can then use simulation to let a range of possible outcomes of an action be simulated by an even wider range of possible outcomes. Figure 5 shows a simple example of this. According to previous definition, the left process is not simulated by the right process. For example, its first *in*-transitions is not simulated by any of the *in* transitions of the right process. However, the *in*-transition is simulated by the convex

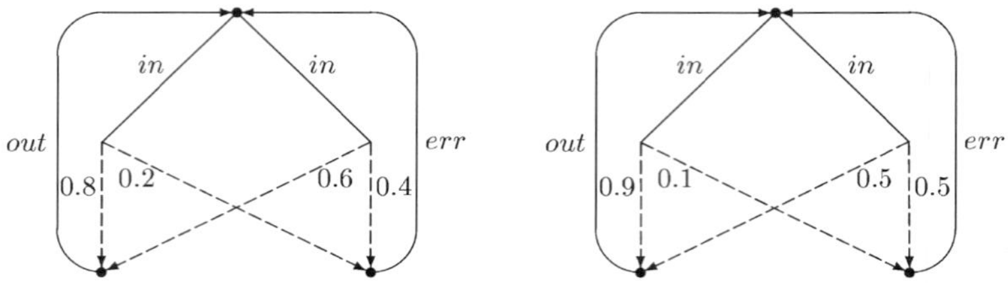

Fig. 5. Probabilistic simulation.

combination of the two *in*-transitions of the right process, resulted by multiplying the two distributions by $3/4$ and $1/4$ respectively.

It can thus be argued that a more appropriate definition of probabilistic simulation is the following:

DEFINITION 8 (*Probabilistic simulation*). A preorder R over S is a probabilistic simulation if $s R t$ implies that whenever $s \xrightarrow{a} \pi$ for some action a and distribution π, then there is a distribution ρ such that $t \xrightarrow{a} \rho$ is a combined transition with $\pi R^* \rho$.

We use $\sim$ to denote the largest simulation, i.e., the union of all probabilistic simulations over S and $Dist(S)$.

For the nonprobabilistic case, we know that a bisimulation is a symmetric simulation and vice versa. This is also true in the probabilistic setting.

THEOREM 5. *A probabilistic simulation is symmetric iff it is a probabilistic bisimulation.*

PROOF. The result follows from Theorem 1. □

5.3. *Congruence properties*

The bisimulation and simulation relations that have been introduced in this section, can all be checked on finite-state processes by a standard iterative procedure that repeatedly refines the universal relation until it satisfies the appropriate definition of (bi-)simulation. At each step, the condition for (bi-)simulation may in general involve checking for inclusion between convex polytopes, spanned by different possible outcomes of a nondeterministic choice. Decision procedures for checking probabilistic (bi-)simulation can be found in, e.g., [4,5].

These relations may be used to reason about probabilistic processes by means of algebraic operators in a compositional manner.

A process expression may be generated by process constants, process variables and algebraic operators described in Section 4. We define a process context to be a process expression containing free process variables. We use $\mathcal{C}$ to denote the set of process contexts. Let C be a process context that contains free process variables $t_1 \dots t_m$. Let $p_1 \dots p_m$ be process expressions that denote states or distributions. We write $C[p_1/t_1 \ldots p_m/t_m]$ for the process expression C in which all t_i are substituted with p_i. As desired, the largest (bi-)simulation relations are all preserved by the algebraic operators described in Section 4.

THEOREM 6. *Let C be a process context containing free process variables $t_1 \ldots t_m$. Let $p_1 \ldots p_m$ and $q_1 \ldots q_m$ be process expressions that may denote states or distributions.*
– *$p_i \simeq q_i$ for all i implies that $C[p_1/t_i \cdots p_m/t_m] \simeq C[q_1/t_i \cdots q_m/t_m]$;*
– *$p_i \sim q_i$ for all i implies that $C[p_1/t_i \cdots p_m/t_m] \sim C[q_1/t_i \cdots q_m/t_m]$.*

PROOF. By the transitivity of $\sim$ and $\simeq$, the case when $m > 1$ can be transformed to the case of $m = 1$. For example, if $p[p_1/t_1, p_2/t_2] \sim p[q_1/t_1, p_2/t_2]$ and $p[q_1/t_1, p_2/t_2] \sim p[q_1/t_1, q_2/t_2]$, we have $p[p_1/t_1, p_2/t_2] \sim p[q_1/t_1, q_2/t_2]$. Thus, we consider only process contexts C with a single variable t. Let $R = \{\langle C[p/t], C[q/t]\rangle \mid p \simeq q\}$. We will show that R is a probabilistic bisimulation.

First we observe that, whenever $\pi \simeq \rho$, then $C[\pi/t] R C[\rho/t]$. Now C will be of the form $op(\pi_1, \ldots, \pi_n, t, p_1, \ldots, p_k)$ for some derived operator op, with $\pi_1, \ldots, \pi_n$, and possibly t, denoting the initially active arguments. Assuming t is initially active in C, $C[\pi/t]$ denotes the distribution, which assigns the probability $\pi(q) * \pi_1(q_1) * \cdots * \pi_n(q_n)$ to the state $op(q_1, \ldots, q_n, q, p_1, \ldots, p_k)$. Similarly, $C[\rho/t]$ assigns the probability $\rho(q) * \pi_1(q_1) * \cdots * \pi_n(q_n)$ to the same state. As $\pi \simeq \rho$ it follows that $C[\pi/t]$ and $C[\rho/t]$ assign the same probabilities to equivalence classes of R, which are sets of the form $\{D[p/t] \mid p \simeq p_0\}$ for some state p_0. Thus, as claimed, $C[\pi/t]\ R\ C[\rho/t]$.

Now let $\langle C[p/t], C[q/t]\rangle \in R$ and assume that $C[p/t] \xrightarrow{b} \pi$. We must find a matching, combined transition for $C[q/t]$. The transition of $C[p/t]$ will be of one of the two following forms:

(i) $\pi = C'[p/t]$;
(ii) $\pi = C'[\pi'/t]$ where t is active in C and $p \xrightarrow{a} \pi'$ for some a.

In case (i), the "de Simone"-format allows us to infer that $C[q/t] \xrightarrow{b} C'[q/t]$. Since $p \simeq q$ (both as states and singleton distributions), it follows from our first observation that $C'[p/t] R C'[q/t]$.

In case (ii), it follows that $\pi' \simeq \rho'$ for some distribution ρ', where $q \xrightarrow{a} \rho'$ is a combined transition (as $p \simeq q$). Now, due to the "de Simone"-format, it may be argued that also $C[q/t] \xrightarrow{b} C'[\rho'/t]$ is a combined transition. From our first observation, it follows that $C'[\pi'/t]\ R\ C'[\rho'/t]$.

As R is symmetric, this suffices to demonstrate that R is a probabilistic bisimulation. The case of simulation can be established by similar argument using Theorem 4. □

6. Testing preorders

In the previous section, we have defined preorders which are based on relating states or distributions to each other. For nonprobabilistic systems, there is another rich family of preorders, which are based on the comparison between execution sequences. These preorders can be based on traces, refusals, ready-sets, behaviors, etc. In this section, we are going to give a partial answer to the question how these preorders generalize to a setting with probabilistic processes. However, we are not going to define first some analogue of, e.g., traces. The reason is that such an analogue would not lead to a compositional preorder, as shown, e.g., by Segala [55]. Instead, we will use another method for arriving at a weaker compositional preorder. We will use the framework of testing, as defined by de Nicola and Hennessy [25]. The idea of this framework is that processes are compared by their ability to pass a specified set of "tests".

We will generalize the framework of may-testing to probabilistic processes, and then characterize the resulting preorder. It will turn out that the resulting preorder will be rather different from the nonprobabilistic trace inclusion. In fact, it will be closer to the nonprobabilistic simulation preorder (in fact, it reduces to simulation in the nonprobabilistic case), for the reason that in the framework we will define, the probabilistic choices will induce an effect which is similar to "copying" of a process state. In this section, we give a simplified presentation by limiting tests to be finite processes.

6.1. *Related work*

Testing-based preorders of probabilistic processes have also been studied by Christoff [12] and by Cleaveland, Smolka, and Zwarico [17] and by Yuen et al. [53,53,59]. These works consider a pure probabilistic model [33], and therefore their preorders do not capture the notion of refinement in the sense of being "less nondeterministic". On the other hand, they can be efficiently checked in the finite-state case, as demonstrated by Christoff and Christoff [13], using the polynomial-time algorithm for checking equivalence between probabilistic automata by Tzeng [57]. The work which is closest to the current one is by Segala [55], who define essentially the same testing preorders as in this work. However, Segala does not develop an explicit characterization of the testing preorder.

6.2. *Tests, testing systems and preorders*

Following Wang and Larsen [60], we define tests as finite trees with a certain subset of the terminal states being "accepting states".

DEFINITION 9. A (*probabilistic*) *test* is a tuple $\langle T, \rightarrow, \rho_0, \mathcal{F}\rangle$, where $\langle\langle T, \rightarrow\rangle, \rho_0\rangle$ is a finite tree, and $\mathcal{F} \subseteq T$ is a set of *success-states*, each of which is terminal.

A test $\mathcal{T}$ is applied to a process $\mathcal{P}$ by putting the process $\mathcal{P}$ in parallel with the test $\mathcal{T}$ and observing whether the test reaches a success state.

We define a testing system as the parallel composition of a process and a test.

DEFINITION 10. Let $\mathcal{P} = \langle S, \rightarrow, \pi_0\rangle$ be a process and $\mathcal{T} = \langle T, \rightarrow, \rho_0, \mathcal{F}\rangle$ be a test. The composition of $\mathcal{P}$ and $\mathcal{T}$, denoted $\mathcal{P}\|\mathcal{T}$ is a so-called *testing system*, defined as the process $\langle S, \rightarrow, \pi_0\rangle \| \langle\langle T, \rightarrow\rangle, \rho_0\rangle$ with success states $S \times \mathcal{F}$.

Our intention is that a testing system defines a probability of reaching a success-state. However, since from each state there may be several outgoing transitions, such a probability is not uniquely defined. We will be interested in the maximal probabilities of success. These can be defined inductively on the structure of the testing system.

DEFINITION 11. Let $\mathcal{P}\|\mathcal{T}$ be a testing system, composed of the process $\mathcal{P} = \langle S, \rightarrow, \pi_0\rangle$ and the test $\mathcal{T} = \langle T, \rightarrow, \rho_0, \mathcal{F}\rangle$. For each state $s\|t$ of $\mathcal{P}\|\mathcal{T}$ we define its *maximal probability of success*, denoted $t\lceil s\rceil$ and its *minimal probability of success*, denoted $t\lfloor s\rfloor$, inductively by

- If $s\|t$ is terminal, then $t\lceil s\rceil = t\lfloor s\rfloor = 1$ if t is a success-state, else $t\lceil s\rceil = t\lfloor s\rfloor = 0$.
- If $s\|t$ is not terminal, then

$$t\lceil s\rceil = \max_{s\|t \xrightarrow{a} \pi\times\rho} \left(\sum_{s'\|t'} (\pi \times \rho)(s'\|t') * t'\lceil s'\rceil\right)$$

and

$$t\lfloor s\rfloor = \min_{s\|t \xrightarrow{a} \pi\times\rho} \left(\sum_{s'\|t'} (\pi \times \rho)(s'\|t') * t'\lfloor s'\rfloor\right).$$

For a distribution π on S and a distribution ρ on T, we define

$$\rho\lceil\pi\rceil = \sum_{s\|t} (\pi \times \rho)(s\|t) * t\lceil s\rceil$$

and

$$\rho\lfloor\pi\rfloor = \sum_{s\|t} (\pi \times \rho)(s\|t) * t\lfloor s\rfloor.$$

We define $\mathcal{T}\lceil\mathcal{P}\rceil = \pi_0\lceil\rho_0\rceil$. and $\mathcal{T}\lfloor\mathcal{P}\rfloor = \pi_0\lfloor\rho_0\rfloor$.

We note that, using the definition of $\rho\lceil\pi\rceil$, we can make a simpler definition of $t\lceil s\rceil$ as

$$t\lceil s\rceil = \max_{s\|t \xrightarrow{a} \pi\times\rho} \rho\lceil\pi\rceil.$$

We now define preorders of testing, which abstract from the set of possible expected outcomes when testing a process $\mathcal{P}$ by a test $\mathcal{T}$: *may* testing considers only maximal possible expected outcome of $\mathcal{P}\|\mathcal{T}$ and *must* testing considers only minimal possible outcome.

DEFINITION 12. Given two processes $\mathcal{P}$ and $\mathcal{Q}$, we define

(1) $\mathcal{P} \sqsubseteq_{may} \mathcal{Q}$ if $\forall\mathcal{T}$: $\mathcal{T}\lceil\mathcal{P}\rceil \leqslant \mathcal{T}\lceil\mathcal{Q}\rceil$;

(2) $\mathcal{P} \sqsubseteq_{must} \mathcal{Q}$ if $\forall\mathcal{T}$: $\mathcal{T}\lfloor\mathcal{P}\rfloor \leqslant \mathcal{T}\lfloor\mathcal{Q}\rfloor$.

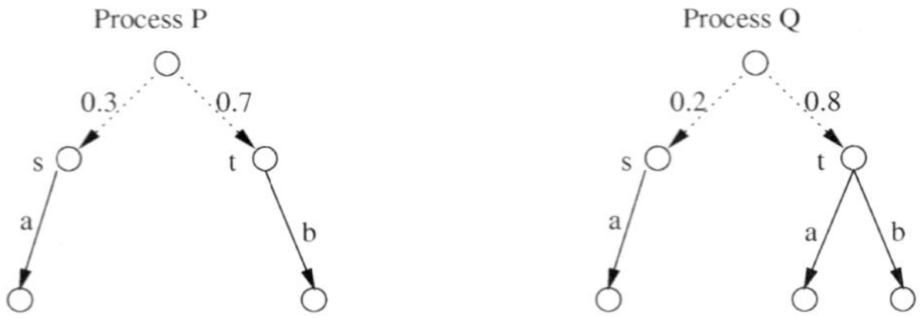

Fig. 6. $P \sqsubseteq_{may} Q$.

The intention, for example, behind the definition of $\sqsubseteq_{may}$ is that intuitively, $\mathcal{P} \sqsubseteq_{may} \mathcal{Q}$ should means that $\mathcal{P}$ refines $\mathcal{Q}$ with respect to "safety properties". We can regard the success-states of a test as states defining when the tester has observed some "bad" or "unacceptable" behavior. A process then refines another one if it has a smaller potential for "bad behavior" with respect to any test. In the definition of $\mathcal{P} \sqsubseteq_{may} \mathcal{Q}$, this means that the maximal probability of observing bad behavior of $\mathcal{P}$ should not exceed the maximal probability of observing bad behavior of $\mathcal{Q}$.

For example, consider process P and Q in Figure 6. The probability that P may pass a test is always less or equal to the probability Q may pass the same test; therefore $P \sqsubseteq_{may} Q$.

A useful property of $\sqsubseteq_{may}$ and $\sqsubseteq_{must}$ is that it is compositional in the sense that they are precongruences with respect to our parallel composition operator.

PROPOSITION 1. *For arbitrary processes P, Q, R,*
(1) $P \sqsubseteq_{may} Q$ *implies* $P \| R \sqsubseteq_{may} Q \| R$;
(2) $P \sqsubseteq_{must} Q$ *implies* $P \| R \sqsubseteq_{must} Q \| R$.

6.3. *Characterization of testing preorders*

In the following, we show that testing preorders defined in Definition 12 can be characterized by variants of probabilistic simulation. When restricted to nonprobabilistic processes, probabilistic may-testing preorder coincides with ordinary simulation; whereas probabilistic must-testing preorder with refusal simulation [40]. We shall only consider may-testing. It may seem a little surprising that a preorder defined in terms of testing, which is a "linear-time" activity, is characterized by a simulation relation, which is a "branching-time" relation. The explanation is that the probabilistic choices of tests have the effect of "copying" the process under test into a number of copies, and that the testing of each copy is performed independently [1].

Recall that a probabilistic process is essentially a distribution over states. Such a distribution gives rise to a number of possibilities for choosing the next action and next distribution. We will capture these possibilities in a notion of *step* corresponding to a transition in the nonprobabilistic setting.

DEFINITION 13. Let $\langle S, \rightarrow \rangle$ be a probabilistic transition system. A *step* is a weighting on $\mathcal{A}ct \times Dist(S)$. We say that a step ϕ is a step of distribution π on S if there is a function $h : (\mathcal{A}ct \times Dist(S)) \mapsto S$ such that
– $s \xrightarrow{S}$ whenever s is in the support of $h(\langle a, \pi' \rangle)$, and
– $(\sum_{h(\langle a, \pi' \rangle) = s} \phi(\langle a, \pi' \rangle)) \leqslant \pi(s)$ for each $s \in S$.
We say that a step ϕ is an *a-step* if $a' = a$ for all $\langle a', \pi' \rangle$ in the support of ϕ.

Intuitively, a step represents a combination of next transitions that can be made by a process. A step from a distribution π is a weighting over possible outgoing transitions, which is consistent with π in the sense that it can be obtained by choosing for each state in the support of π a subdistribution over outgoing transitions. Note that for a given distribution, there may be infinitely many steps possible. A step is *normal* if the function h in Definition 13 can be chosen such that for each nonterminal s in the support of π, there is a $\langle a, \pi' \rangle \in \phi$ such that $h(\langle a, \pi' \rangle) = s$ and $\pi(h(\langle a, \pi' \rangle)) = \phi(\langle a, \pi' \rangle)$.

That is, a normal step is obtained by choosing a unique transition from each state, satisfying the above condition. Since each state in a distribution in general has several outgoing transitions, there are many (but finitely many) normal steps from each distribution.

We define *post* on steps by

$$post(\phi) = \sum_{\langle a, \pi' \rangle} \phi(\langle a, \pi' \rangle) * \pi',$$

i.e., *post*(ϕ) is the weighting obtained by projecting a step onto the "next" distribution in its transitions. The notion of post weighting is analogous to the notions of next state in the nonprobabilistic setting. We can now define the notion of step-simulation between weightings.

DEFINITION 14 (*Probabilistic step-simulation*). Let $\langle S, \rightarrow \rangle$ and $\langle T, \rightarrow \rangle$ be two probabilistic transition systems. A relation $\lhd \subseteq (Weight(S) \times Weight(T))$ between weightings on S and weightings on T is a *probabilistic step-simulation* if $\pi \lhd \rho$ implies that
– $\pi(S) \leqslant \rho(T)$, and
– for each normal step ϕ from π there is a step ψ from ρ and a function $h : \text{support}(\phi) \mapsto Weight(\mathcal{A}ct \times Dist(R))$ from pairs $\langle a, \pi' \rangle$ in the support of ϕ to steps from ρ such that
  - h maps each $\langle a, \pi' \rangle$ to an a-step from ψ,
  - $h(\phi) \leqslant \psi$, i.e., the image of ϕ under h is "covered" by ψ, and
  - for each pair $\langle a, \pi' \rangle$ in the support of ϕ we have

$$\pi' \lhd post\big(h(\langle a, \pi' \rangle)\big).$$

For two probabilistic processes $\mathcal{P} = \langle S, \rightarrow, \pi_0 \rangle$ and $\mathcal{Q} = \langle T, \rightarrow, \rho_0 \rangle$, we say that $\mathcal{P}$ *is simulated by* $\mathcal{Q}$ if there is a probabilistic step-simulation $\lhd$ between $\langle S, \rightarrow \rangle$ and $\langle T, \rightarrow \rangle$ such that $\pi_0 \lhd \rho_0$.

Intuitively, a weighting π is simulated by a weighting ρ if the total "mass" of π is at most that of ρ (first condition), and if each step ϕ from π can be simulated by a step ψ from

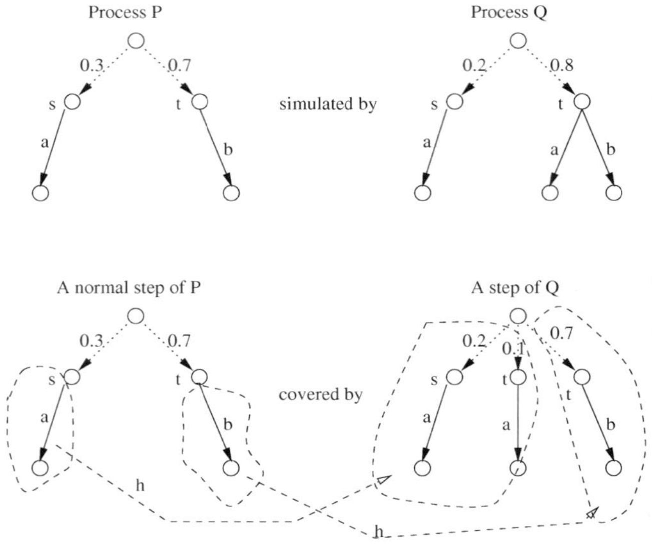

Fig. 7. A proof for $P \lhd Q$.

ρ in the sense that each "next transition" $\langle a, \pi' \rangle$ in the support of ϕ can be covered by an a-step from ρ, such that the weighted sum (weighted with respect to ϕ) of all the weightings $h(\langle a, \pi' \rangle)$ is covered by ψ, and such that π' is simulated by the next-state distribution obtained from $h(\langle a, \pi' \rangle)$. In Figure 7, we illustrate why process P is simulated by process Q. Note that $P \sqsubseteq_{may} Q$ as shown in Figure 6.

The following characterization theorem was stated and proven by Jonsson and Wang in [43].

THEOREM 7. *$\pi \sqsubseteq_{may} \rho$ if and only if there is a probabilistic step-simulation $\lhd$ such that $\pi \lhd \rho$.*

7. Probabilistic logics

7.1. *Characterizing preorders*

For non-probabilistic transition systems, several behavioural preorders have been characterized by simple modal logics. That is, the particular preorder has been shown to be completely captured by inclusion between the sets of logical properties satisfied by states (see also Bradfield and Stirling, [11], Chapter 4 in this Handbook).

Bisimilarity between nonprobabilistic transition systems is characterized by the so-called Hennessy–Milner Logic [37]. In [44–46], this characterization has been extended to probabilistic bisimulation for reactive probabilistic systems by identification of a suitable probabilistic extension of Hennessy–Milner Logic. In [10,28,29], a further generalization to non-discrete probabilistic systems (Labelled Markov Processes) is given, and in particular it is shown that probabilistic bisimulation can be characterized by a very weak, negation-free modal logic. In addition, and in contrast to [44–46], the characterization offered by [10,28,29] requires no finite branching assumptions.

Here, we adapt the probabilistic modal logic, PML, of [44–46] to the probabilistic model studied in this paper. As for states, the formulas of the logic come in two flavours: nondeterministic formulas (ranged over by $F, F_i, \ldots$) and probabilistic formulas (ranged over by $\varphi, \varphi_i, \ldots$), given by the following abstract syntax (where $p \in [0, 1]$):

$$F ::= tt \mid F_1 \wedge F_2 \mid \neg F \mid \langle a \rangle \varphi,$$
$$\varphi ::= tt \mid \varphi_1 \wedge \varphi_2 \mid \neg \varphi \mid \Diamond_p F.$$

The interpretation is relative to a probabilistic transition system, $\langle S, \rightarrow, \pi_0 \rangle$. More precisely, the semantics of a formula F (respectively φ) is a set of nondeterministic states $[\![F]\!]$ (respectively of probabilistic states $[\![\varphi]\!]$), defined inductively as follows:

$$\begin{array}{llll} \text{(i)} & [\![tt]\!] = S & \text{(ii)} & [\![F_1 \wedge F_2]\!] = [\![F_1]\!] \cap [\![F_2]\!] \\ \text{(iii)} & [\![\neg F]\!] = S \backslash [\![F]\!] & \text{(iv)} & [\![\langle a \rangle \varphi]\!] = \{s \mid \exists a, \pi \in [\![\varphi]\!].\, s \xrightarrow{a} \pi\} \end{array}$$

$$\begin{array}{llll} \text{(i}'\text{)} & [\![tt]\!] = Dist(S) & \text{(ii}'\text{)} & [\![\varphi_1 \wedge \varphi_2]\!] = [\![\varphi_1]\!] \cap [\![\varphi_2]\!] \\ \text{(iii}'\text{)} & [\![\neg \varphi]\!] = Dist(S) \backslash [\![\varphi]\!] & \text{(iv}'\text{)} & [\![\Diamond_p F]\!] = \{\pi \mid \pi([\![F]\!]) \geqslant p\}. \end{array}$$

For F a nondeterministic property, we write $[\![F]\!]_C$ to denote the set of nondeterministic states satisfying F, when we use the combined transition relation in (iv). For a state s we denote by $Sat(s)$ and $Sat_C(s)$ the set of properties satisfied by s with respect to $[\![\,]\!]$ and $[\![\,]\!]_C$. Bisimulation and probabilistic bisimulation as presented in Section 5 may now be characterized as follows:

THEOREM 8. *Let $\langle S, \rightarrow, \pi_0 \rangle$ be an image-finite*[4] *probabilistic transition system. Then two states s, t are bisimilar respectively probabilistic bisimilar if and only if $Sat(s) = Sat(t)$ respectively $Sat_C(s) = Sat_C(t)$.*

Let $NSat(s)$ and $NSat_C(s)$ denote the set of negation-free properties satisfied by the state s with respect to $[\![\,]\!]$ and $[\![\,]\!]_C$. Then simulation and probabilistic simulation as presented in Section 5 are characterized as follows:

THEOREM 9. *Let $\langle S, \rightarrow, \pi_0 \rangle$ be an image-finite probabilistic transition system. Then two states s, t are in the simulation preorder respectively probabilistic simulation preorder if and only if $NSat(s) \subseteq NSat(t)$ respectively $NSat_C(s) \subseteq NSat_C(t)$.*

---

[4] A probabilistic transition system $\langle S, \rightarrow, \pi_0 \rangle$ is said to be image-finite if for all reachable nondeterministic states s, the set $\{\pi \mid \exists a, s.s \xrightarrow{a} \pi\}$ is finite, and for all reachable probabilistic states π, the set $\{s \mid \pi(s) > 0\}$ is finite.

As an example reconsider Figure 4. Here the two processes are probabilistic bisimilar but not bisimilar. The lack of bisimilarity may, according to Theorem 8, be 'explained' by a distinguishing property, e.g.:

$$\langle in \rangle \big(\Diamond_{0.7} \langle out \rangle\, tt \wedge \Diamond_{0.3} \langle err \rangle\, tt\big)$$

which is satisfied by the right process but not by the left processes (with respect to $[\![]\!]$). The above property is clearly negation-free, thus according to Theorem 9, this also demonstrates that the right process is not even simulated by the left one.

Reconsidering Figure 5, the property

$$\langle in \rangle\, \Diamond_{0.9}\, \langle out \rangle\, tt$$

is satisfied by the right process but not by the left processes (with respect to $[\![]\!]_C$). Thus the right process is not probabilistically simulated by the left one.

7.2. *Model checking probabilistic temporal logics*

Several probabilistic extensions of temporal logics such as CTL and CTL* [15,16] have been suggested for the formal specification of probabilistic properties of systems. In addition, associated model checking algorithms have in many cases been offered.

The first probabilistic extension of branching-time logics for expressing properties of probabilistic system was proposed by Hansson and Jonsson [35,36]. Formulas of the resulting logic PCTL are obtained by adding subscripts and superscripts to CTL formulas, as in $\Diamond^{\leqslant 15}_{\geqslant 0.6}\, \varphi$, which expresses that the property φ will hold within 15 transition-steps with probability at least 0.6. The presented model-checking algorithms rely on results on Markov chains and dynamic programming. Later, the logic PCTL was extended to systems including nondeterminism by Hansson in [34] and Segala and Lynch in [56]. Christoff and Christoff [14] adapt a restricted form of the modal mu-calculus. A new probabilistic semantics for the mu-calculus has been developed by Narasimha et al. [49].

Aziz et al. [3] introduces pCTL* a probabilistic extension of CTL*. Here the model checking algorithm is based on early results due to Courcoubetis and Yannakakis [19]. The logic was later extended to systems with nondeterminism by Bianco and de Alfaro [9, 24]. Symbolic model-checking algorithms of these logics was presented by Baier et al. in [7].

For more detailed information on the interesting topic of model checking probabilistic systems, we refer the reader to the excellent works by Alfaro [23] and Baier [5].

8. Conclusion and trends

In this chapter, we have dealt with a number of classical process algebraic issues in a rather general setting allowing both discrete probabilistic choice as well as nondeterminism. In particular, we have

– shown how nonprobabilistic process algebraic operators, in a uniform manner, may be extended to this probabilistic setting;
– offered a range of probabilistic extensions of well-known behavioural preorders such as simulation, bisimulation and testing;
– established congruence properties for these preorders;
– provided alternative characterizations of the probabilistic preorders, either in terms of "trace"- or "tree"-based denotational models, or in terms of probabilistic modal logics.

Current research considers further extensions of the process algebraic framework to settings with continuous-time [39,38]. One goal is to combine the contributions of process algebra, viz. compositionality and the use of logics to specify and analyze properties, with the work on efficient algorithms for analyzing performance of stochastic processes [6,21]. Also, the basic notions of probabilistic bisimulation and probabilistic modal logics have been recast and analysed in settings with continuous-space probability distributions [10, 28,22,30,31]. From this work it may be inferred that the negation-free version of the logic in Section 7 suffices in order to characterize probabilistic bisimulation. In this chapter we have not considered the difficult problem of extending probabilistic bisimulation to allow for abstraction from internal computation (resulting in probabilistic versions of weak bisimulation). For the most promising suggestions in this direction we refer the reader to [8].

References

[1] S. Abramsky, *Observation equivalence as a testing equivalence*, Theoret. Comput. Sci. **53** (2,3) (1987), 225–241.

[2] L. Aceto, W.J. Fokkink and C. Verhoef, *Structural operational semantics*, Handbook of Process Algebra, J.A. Bergstra, A. Ponse and S.A. Smolka, eds, Elsevier, Amsterdam (2001), 197–292.

[3] A. Aziz, V. Singhal, F. Balarin, R.K. Brayton and A.L. Sangiovanni-Vincentelli, *It usually works: The temporal logic of stochastic systems*, Proc. CAV'95, Lecture Notes in Comput. Sci. 939, Springer-Verlag (1995).

[4] C. Baier, *Polynomial time algorithms for testing probabilistic bisimulation and simulation*, Proc. 8th Internat. Conference on Computer Aided Verification, New Brunswick, USA, 1996, Lecture Notes in Comput. Sci. 1102, R. Alur and T. Henzinger, eds, Springer-Verlag (1996), 38–49.

[5] C. Baier, *On algorithmic verification methods for probabilistic systems*, Habilitation, Thesis, University of Mannheim (1999).

[6] C. Baier, J.-P. Katoen and H. Hermanns, *Approximate symbolic model checking of continuous-time Markov chains*, Proc. CONCUR'99, 9th Internat. Conference on Concurrency Theory, Lecture Notes in Comput. Sci. 1664, Springer-Verlag (1999), 146–162.

[7] C. Baier, M. Kwiatkowska, M. Ryan, E. Clarke and V. Hartonas Garmhausen, *Symbolic model checking for probabilistic processes*, Proc. ICALP'97, Lecture Notes in Comput. Sci. 1256, Springer-Verlag (1997).

[8] C. Baier and H. Hermanns, *Weak bisimulation for fully probabilistic processes*, Lecture Notes in Comput. Sci. 1254, Springer-Verlag (1997).

[9] A. Bianco and L. De Alfaro, *Model checking of probabilistic and nondeterministic systems*, Proc. FSTTCS, Lecture Notes of Comput. Sci. 1026, Springer-Verlag (1995).

[10] R. Blute, J. Desharnais, A. Edalat and P. Panangaden, *Bisimulation for labelled Markov processes*, Proc. 12th IEEE Symposium on Logic in Computer Science (1997).

[11] J.C. Bradfield and C. Stirling, *Modal logics and mu-calculi: An introduction*, Handbook of Process Algebra, J.A. Bergstra, A. Ponse and S.A. Smolka, eds, Elsevier, Amsterdam (2001), 293–330.

[12] I. Christoff, *Testing equivalences and fully abstract models for probabilistic processes*, Proc. CONCUR, Amsterdam, Lecture Notes in Comput. Sci. 458, Baeten, ed., Springer-Verlag (1990), 126–140.

[13] L. Christoff and I. Christoff, *Efficient algorithms for verification of equivalences for probabilistic processes*, Proc. Workshop on Computer Aided Verification, Lecture Notes in Comput. Sci. 575, Larsen and Skou, eds, Springer-Verlag (1991).

[14] L. Christoff and I. Christoff, *Reasoning about safety and liveness properties for probabilistic properties*, Foundations of Software Technology and Theoretical Computer Science, Lecture Notes in Comput. Sci. 652, R.K. Shyamasundar, ed., Springer-Verlag (1992), 342–355.

[15] E.M. Clarke, E.A. Emerson and A.P. Sistla, *Automatic verification of finite-state concurrent systems using temporal logics specification: A practical approach*, Proc. 10th ACM Symp. on Principles of Programming Languages (1983), 117–126.

[16] E.M. Clarke and E.A. Emerson, *Design and synthesis of synchronization skeletons using branching time temporal logic*, Proc. IBM Workshop on Logics of Programs, Lecture Notes in Comput. Sci. 131, D. Kozen, ed. (1982). Also as Aiken Computation Lab TR-12-81, Harvard University (1981).

[17] R. Cleaveland, S. Smolka and A. Zwarico, *Testing preorders for probabilistic processes*, Proc. ICALP'92 (1992).

[18] R. Cleaveland and O. Sokolsky, *Equivalence and preorder checking for finite-state systems*, Handbook of Process Algebra, J.A. Bergstra, A. Ponse and S.A. Smolka, eds, Elsevier, Amsterdam (2001), 391–424.

[19] C. Courcoubetis and M. Yannakakis, *The complexity of probabilistic verification*, Proc. 29th Annual Symposium Foundations of Computer Science (1988), 338–345.

[20] P.R. D'Argenio, H. Hermanns and J.-P. Katoen, *On generative parallel composition*, Proc. Workshop on Probabilistic Methods in Verification (1998).

[21] P.R. D'Argenio, J.-P. Katoen and E. Brinksma, *General purpose discrete-event simulation using SPADES*, Proc. 6th Internat. Workshop on Process Algebra and Performance Modelling (1998).

[22] P.R. D'Argenio, J. Katoen and E. Brinksma, *An algebraic approach to the specification of stochastic systems*, PROCOMET'98, Chapmann and Hall (1998).

[23] L. de Alfaro, *Formal verification of probabilistic systems*, Ph.D. Thesis, Dept. of Computer Sciences, Stanford University (1997).

[24] L. de Alfaro, *Temporal logics for the specification of performance and reliability*, Proc. STACS'97, Lecture Notes in Comput. Sci. 1200 (1997).

[25] R. de Nicola and M. Hennessy, *Testing equivalences for processes*, Theoret. Comput. Sci. **34** (1984), 83–133.

[26] R. de Simone, *Higher-level synchronising devices in MEIJE-SCCS*, Theoret. Comput. Sci. **37** (3) (1985), 245–267.

[27] C. Derman, *Finite State Markovian Decision Processes*, Academic Press (1970).

[28] J. Desharnais, A. Edalat and P. Panangaden, *A logical characterization of bisimulation for labeled Markov processes*, Proc. 13th IEEE Symposium on Logic in Computer Science (1998).

[29] J. Desharnais, A. Edalat and P. Panangaden, *Bisimulation for labelled Markov processes*, Proc. 12th IEEE Symposium on Logic in Computer Science (1997). To appear in: Inform. and Comput. (2000).

[30] J. Desharnais, A. Edalat and P. Panangaden, *A logical characterization of bisimulation for labeled Markov processes*, Proc. 13th IEEE Symposium on Logic in Computer Science (1998).

[31] J. Desharnais, V. Gupta, R. Jagadeesan and P. Panangaden, *Metrics for labeled Markov systems*, Lecture Notes in Comput. Sci. 1664 (1999).

[32] A. Giacalone, C. Jou and S.A. Smolka, *Algebraic reasoning for probabilistic concurrent systems*, Proc. IFIP TC2 Working Conference on Programming Concepts and Methods, Sea of Galilee (1990).

[33] R. van Glabbeek, S.A. Smolka, B. Steffen and C. Tofts, *Reactive, generative, and stratified models of probabilistic processes*, Proc. 5th IEEE Internat. Symposium on Logic in Computer Science (1990), 130–141.

[34] H. Hansson, *Time and probabilities in formal design of distributed systems*, Real-Time Safety Critical Systems, Elsevier (1994).

[35] H. Hansson and B. Jonsson, *A calculus for communicating systems with time and probabilities*, Proc. 11th IEEE Real-Time Systems Symposium, Orlando, FL (1990).

[36] H. Hansson and B. Jonsson, *A logic for reasoning about time and reliability*, Formal Aspects of Computing **6** (1994), 512–535.

[37] M. Hennessy and R. Milner, *Algebraic laws for nondeterminism and concurrency*, J. ACM **32** (1) (1985), 137–161.

[38] H. Hermanns, *Interactive Markov chains*, Ph.D. Thesis, University of Erlangen–Nürnberg (1998).

[39] J. Hillston, *A compositional approach to performance modelling*, Ph.D. Thesis, University of Edinburgh (1994). Published in Cambridge University Press (1996).

[40] B. Jonsson, *Simulations between specifications of distributed systems*, Proc. CONCUR'91, Theories of Concurrency: Unification and Extension, Amsterdam, Holland, Lecture Notes in Comput. Sci. 527, Springer-Verlag (1991).

[41] B. Jonsson and K.G. Larsen, *Specification and refinement of probabilistic processes*, Proc. 6th IEEE Internat. Symposium on Logic in Computer Science, Amsterdam, Holland (1991).

[42] B. Jonsson and W. Yi, *Compositional testing preorders for probabilistic processes*, Proc. 10th IEEE Internat. Symposium on Logic in Computer Science (1995), 431–441.

[43] B. Jonsson and W. Yi, *Testing preorders for probabilistic processes can be characterized by simulations*, Technical report, DoCS, Uppsala University (2000). Submitted.

[44] K.G. Larsen and A. Skou, *Bisimulation through probabilistic testing*, Proc. 16th ACM Symposium on Principles of Programming Languages (1989).

[45] K.G. Larsen and A. Skou, *Bisimulation through probabilistic testing*, Inform. and Control **94** (1) (1991), 1–28.

[46] K.G. Larsen and A. Skou, *Compositional verification of probabilistic processes*, Proc. CONCUR'92, Theories of Concurrency: Unification and Extension, Lecture Notes in Comput. Sci. 630, Cleaveland, ed., Springer-Verlag (1992).

[47] R. Milner, *Communication and Concurrency*, Prentice-Hall (1989).

[48] M.K. Molloy, *Performance analysis using stochastic Petri nets*, IEEE Trans. Comput. **C-31** (9) (1982), 913–917.

[49] M. Narasimha, R. Cleaveland and P. Iyer, *Probabilistic temporal logics via the modal mu-calculus*, Foundations of Software Science and Computation Structures, Lecture Notes in Comput. Sci. 1578, W. Thomas, ed., Springer-Verlag (1999).

[50] G. Plotkin, *A structural approach to operational semantics*, Technical Report DAIMI FN-19, Computer Science Department, Aarhus University, Denmark (1981).

[51] A. Pnueli and L. Zuck, *Verification of multiprocess probabilistic protocols*, Distrib. Comput. **1** (1) (1986), 53–72.

[52] S. Purushothaman and P.A. Subrahmanyam, *Reasoning about probabilistic behavior in concurrent systems*, IEEE Trans. Software Engineering **SE-13** (6) (1989), 740–745.

[53] S.A. Smolka, R. Cleaveland, Z. Dayar and S. Yuen, *Testing preorders for probabilistic processes*, Inform. and Comput. **2** (152) (1999), 93–148.

[54] M.O. Rabin, *Probabilistic automata*, Inform. and Control **6** (1963), 230–245.

[55] R. Segala, *A compositional trace-based semantics for probabilistic automata*, Proc. CONCUR'95, 6th Internat. Conference on Concurrency Theory, Lecture Notes in Comput. Sci. 962, Springer-Verlag (1995), 234–248.

[56] R. Segala and N.A. Lynch, *Probabilistic simulations for probabilistic processes*, Nordic J. Comput. **2** (2) (1995), 250–273.

[57] W.-G. Tzeng, *A polynomial-time algorithm for the equivalence of probabilistic automata*, SIAM J. Comput. **21** (2) (1992), 216–227.

[58] M.Y. Vardi, *Automatic verification of probabilistic concurrent finite-state programs*, Proc. 26th Annual Symposium Foundations of Computer Science (1985), 327–338.

[59] S.-H. Wu, S.A. Smolka and E.W. Stark, *Composition and behaviors of probabilistic I/O-automata*, Theoret. Comput. Sci. **176** (1–2) (1997), 1–37.

[60] W. Yi and K.G. Larsen, *Testing probabilistic and nondeterministic processes*, Protocol Specification, Testing, and Verification XII (1992).

5.1. Real-time semantics . . . 749
5.2. Dynamic priority semantics . . . 751
5.3. Relating dynamic priority and real-time semantics . . . 752
5.4. Concluding remarks and related work . . . 753
6. Priority in other process-algebraic frameworks . . . 754
7. Conclusions and directions for future work . . . 756
8. Sources and acknowledgments . . . 758
9. Glossary . . . 758
References . . . 762
Subject index . . . 765

Abstract

This chapter surveys the semantic ramifications of extending traditional process algebras with notions of *priority* that allow some transitions to be given precedence over others. The need for these enriched formalisms arises when one wishes to model system features such as *interrupts*, *prioritized choice*, or *real-time behavior*.

Approaches to priority in process algebras can be classified according to whether the induced notion of *pre-emption* on transitions is *global* or *local* and whether priorities are *static* or *dynamic*. Early work in the area concentrated on global pre-emption and static priorities and led to formalisms for modeling interrupts and aspects of real-time, such as *maximal progress*, in *centralized* computing environments. More recent research has investigated localized notions of pre-emption in which the *distribution* of systems is taken into account, as well as dynamic priority approaches. The latter allows priority values to change as systems evolve and enables an efficient encoding of real-time semantics.

Technically, this chapter studies the different models of priorities by presenting extensions of Milner's *Calculus of Communicating Systems* (CCS) with static and dynamic priority as well as with notions of global and local pre-emption. For each extension, the operational semantics of CCS is modified appropriately, behavioral theories based on *strong* and *weak bisimulation* are given, and related approaches for other process-algebraic settings are discussed.

1. Introduction

Traditional *process algebras* [7,41,44,58] provide a framework for reasoning about the *communication potential* of *concurrent* and *distributed systems*. Such theories typically consist of a simple *calculus* with a well-defined *operational semantics* [1,69] given as *labeled transition systems*; a *behavioral equivalence* is then used to relate implementations and specifications, which are both formalized as terms in the calculus. In order to facilitate *compositional reasoning*, in which systems are verified component by component, researchers have concentrated on developing *behavioral congruences*, which allow the substitution of "equals for equals" inside larger systems.

Over the past decade, process-algebraic approaches to system modeling and verification have been applied to a number of case studies (see, e.g., [3]). Nevertheless, many systems in practice cannot be modeled accurately within this framework. One reason is that traditional process algebras focus exclusively on expressing the potential *nondeterminism* that the interplay of concurrent processes may exhibit; they do not permit the encoding of differing levels of *urgency* among the transitions that might be enabled from a given system state. In practice, however, some system transitions are intended to take precedence of others; examples include:

- **interrupts**, where non-urgent transitions at a state are *pre-empted* whenever an interrupt is raised;
- **programming language constructs**, such as the `PRIALT` construct in `occam` [45], that impose a priority order on transition choices;
- **real-time behavior** that is founded on the well-known *synchrony hypothesis* [13] or *maximal progress assumption* [79]; and
- **scheduling algorithms** which also rely on the concept of pre-emption.

In each of these cases urgency provides a means for *restricting nondeterminism*. This mechanism is simply ignored in traditional process algebras. Hence, the resulting system models are often not faithful since they contain spurious paths that cannot be traversed by the real-world systems themselves [16,29].

As an illustration of the need for integrating notions of urgency into process algebra consider the interrupt-based system depicted in Figure 1. The system consists of two processes: A, which flips back and forth between two states, and B, which prompts A to report its status. Whenever B receives a check message from the environment it instructs A of this via *interrupt port* $\underline{i}$; A in turn issues its response on its ok channel. In the absence of an indication that a communication on $\underline{i}$ is more urgent than one on back and forth, process A is not required to respond immediately to a request from B. Indeed, the system's behavior is not even *fair* [34,40] since A may ignore such requests indefinitely. It should be noted that fairness is a weaker notion than priority [8], as it only requires *eventual* rather than *immediate* responses.

1.1. *Classification of approaches to priority*

A number of approaches have been proposed for introducing priority to process algebras [5, 12,16,22–24,26,28,29,33,36,38,47,48,54,55,64,65,71,73,74]. One may classify these approaches according to the following two criteria.

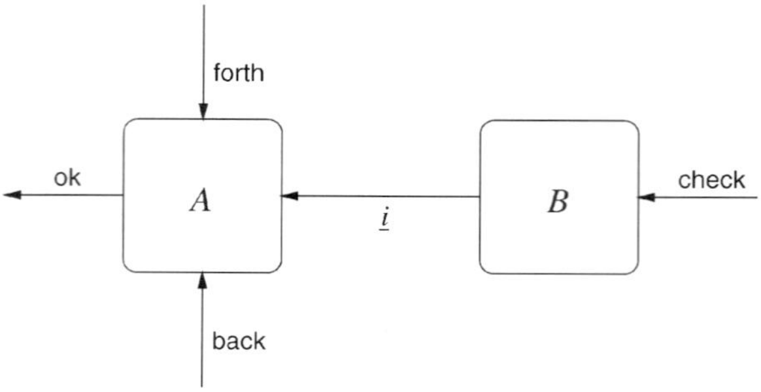

Fig. 1. Back-and-forth example.

Static priority versus dynamic priority:
In *static* approaches transition priorities may be inferred from the syntax of the system description before the system commences execution; they do not change during computation. These schemes find application in the modeling of interrupts or prioritized choice constructs. In the former case, interrupts have a fixed urgency level associated with them; in the latter priorities of transitions are fixed by the static program syntax. Almost all priority approaches to process algebra published so far deal with static priorities. The exceptions are [16,22], which present models that allow priority values of transitions to change as systems evolve. Such *dynamic* behavior is useful in modeling scheduling algorithms and real-time semantics.

Global pre-emption versus local pre-emption:
This criterion refers to the *scope* of priority values. In the case of centralized systems, priorities generally have a *global* scope in the sense that transitions in one process may pre-empt transitions in another. We refer to this kind of pre-emption, which has been advocated by Baeten, Bergstra, and Klop [5] and by Cleaveland and Hennessy [26] in the late Eighties, as *global pre-emption*. In contrast, in a distributed system containing multiple sites, transitions should only be allowed to pre-empt those at the same site or *location*. This kind of pre-emption, which was first studied by Camilleri and Winskel [24] in the early Nineties, is called *local pre-emption*.

Based on the above classification scheme, the body of this chapter investigates the following different semantics for a prototypical process-algebraic language: static/global, static/local, and dynamic/global. Only the combination of dynamic priority and local pre-emption is omitted, since research has not yet been carried out on this topic.

Some caveats about terminology are in order here. Other process algebra researchers have used the term "pre-emption" in a setting without priorities [17]; in their usage pre-emption occurs when the *execution* of one transition removes the possibility of others. In our priority-oriented framework, we say that pre-emption occurs when the *availability* of one transition disables others. Berry [12] refers to this latter notion as *must pre-emption* and to the former as *may pre-emption*. In this chapter, whenever we speak of "pre-emption", we mean "must pre-emption". It should also be noted that our concept of global pre-emption and dynamic priority differs from the notion of *globally dynamic* priority found in [73]; as the distinction is somewhat technical we defer further discussion on this point to later in the chapter.

1.2. *Summary of this chapter*

This chapter surveys published work on priority in process algebras. To illustrate the semantic issues involved, we introduce a single process calculus that we then equip with different notions of pre-emption. This calculus extends Milner's *Calculus of Communicating Systems* (CCS) [58] by attaching priority values to actions. For this language three different semantics are given: one reflecting static priorities and global pre-emption, one for static priorities and local pre-emption, and one capturing dynamic priorities. The common language allows for a detailed comparison of the semantic concepts; in addition, the classification scheme presented above helps categorize most existing approaches to priority. These have been proposed for a variety of well-known process algebras, such as the already mentioned CCS, the *Algebra of Communicating Processes* (ACP) [9], *Communicating Sequential Processes* (CSP) [44], the *Calculus of Broadcasting Systems* (CBS) [70], *Synchronous CCS* (SCCS) [58], and *Asynchronous Communicating Shared Resources* (ACSR) [23].

For the static-priority settings in this chapter we develop semantic theories based on the notion of bisimulation [58,67]. Our aim is to carry over the standard algebraic results from CCS [58], including full-abstractness theorems as well as axiomatic, logical, and algorithmic characterizations. In particular, we investigate both *strong* and *weak bisimulations* that are based on naive adaptations of the standard definitions as given by Milner; we focus on characterizing the *largest congruences* contained in these relations. Such full-abstractness results indicate that the behavioral relations are compositional and thus useful for formally reasoning about concurrent and distributed systems. Then we present sound and complete *axiomatizations* for the obtained strong bisimulations with respect to finite processes, i.e., those which do not contain recursion. These axiomatizations testify to the mathematical tractability of the semantic theories presented here. We also characterize the attendant notions of prioritized strong and weak bisimulations as standard bisimulations via enriched transition labels, so that well-known *partition-refinement algorithms* [30,50, 66] for their computation become applicable. Such accounts also provide support for logical characterizations of the behavioral relations, as they give direct insights into how the *Hennessy–Milner logic* [19,58] should be adapted. In case of the dynamic priority semantics, we prove a one-to-one correspondence with traditional *real-time semantics* [10,61,79] in terms of strong bisimulation. Because of this close relationship, semantic theories developed for real-time process algebras can be carried over to the dynamic priority setting.

1.3. *Organization of this chapter*

The remainder of this chapter is organized as follows. The next section introduces our extension of CCS, defines some formal notation used throughout the chapter, and discusses basic design decisions we have taken. Section 3 presents a semantics of the language based on static priority and global pre-emption; Section 4 then develops a semantics based on static priority and local pre-emption. A dynamic priority approach is illustrated in Section 5. Related work is referred to in each of the last three sections, while Section 6 surveys several priority approaches adopted for process-algebraic frameworks other than CCS. Section 7 contains our conclusions and suggestions for future work. The final section points to the sources of the research compiled in this chapter. A glossary of notation and symbols used may be found at the end.

2. Basic language and notation

As mentioned above, the language considered here is an extension of Milner's CCS [58], a process algebra characterized by *handshake communication* and *interleaving semantics* for parallel composition. Syntactically, CCS includes notations for *visible actions*, which are either sends or receives on ports, and a distinguished *invisible*, or *internal* action. The semantics of CCS is then given via a transition relation that labels execution steps with actions. When a sender and receiver synchronize, the resulting action is internal. Consequently, transitions labeled by visible actions can be seen as representing only "potential" computation steps, since in order for them to occur they require a contribution from the environment. Transitions labeled by internal actions describe complete synchronizations and, therefore, should be viewed as "actual" computation steps.

In order to capture priorities, the syntax of our language differs from CCS in that the *port* set exhibits a priority scheme, i.e., priorities are attached to ports. Our notion of pre-emption then stipulates that a system cannot engage in transitions labeled by actions with a given priority if it is able to perform a transition labeled by an *internal* action of higher priority. In this case we say that the lower-priority transition is pre-empted by the higher-priority internal transition. In accordance with the above discussion visible actions *never* have pre-emptive power over actions of lower priority, because visible actions only indicate the potential for execution. An algebraic justification of this design decision can be found in Section 3.5.

Priority values are taken from some finite domain equipped with a total order. For the sake of simplicity we use finite initial intervals $\mathcal{N}$ of the natural numbers $\mathbb{N}$ in what follows. We adopt the convention that smaller numbers mean higher priorities; so 0 is the highest priority value.

Let $\{\Lambda_k \mid k \in \mathcal{N}\}$ denote an $\mathcal{N}$-indexed family of countably infinite, pairwise disjoint sets of ports. Intuitively, Λ_k contains the ports with priority k over which processes may synchronize. Then the set of *actions* $\mathcal{A}_k$ with priority k may be defined by $\mathcal{A}_k =_{\mathrm{df}} \Lambda_k \cup \overline{\Lambda}_k \cup \{\tau_k\}$, where $\overline{\Lambda}_k =_{\mathrm{df}} \{\overline{\lambda}_k \mid \lambda_k \in \Lambda_k\}$ and $\tau_k \notin \Lambda_k$. An action $\lambda_k \in \Lambda_k$ may be thought of as representing the receipt of an input on port λ that has priority k, while $\overline{\lambda}_k \in \overline{\Lambda}_k$ constitutes the deposit of an output on λ. The invisible action τ_k represent internal computation steps with priority k. For better readability we write $\lambda : k$, if $\lambda_k \in \Lambda_k$, $\overline{\lambda} : k$, if $\overline{\lambda}_k \in \Lambda_k$, and $\tau : k$ for τ_k. The sets of all ports Λ, all complementary ports $\overline{\Lambda}$, and all actions $\mathcal{A}$ are defined by $\bigcup\{\Lambda_k \mid k \in \mathcal{N}\}$, $\bigcup\{\overline{\Lambda}_k \mid k \in \mathcal{N}\}$, and $\bigcup\{\mathcal{A}_k \mid k \in \mathcal{N}\}$, respectively. In what follows, we use $\alpha : k, \beta : k, \ldots$ to range over $\mathcal{A}$ and $a : k, b : k, \ldots$ to range over $\Lambda \cup \overline{\Lambda}$. We also extend complementation to all visible actions $a:k$ by $\overline{\overline{a}}:k =_{\mathrm{df}} a:k$. Finally, if $L \subseteq \Lambda \cup \overline{\Lambda}$ then $\overline{L} = \{\overline{a}:k \mid a:k \in L\}$. The syntax of our language is defined by the following BNF.

$$P ::= \mathbf{0} \mid x \mid \alpha:k.P \mid P + P \mid P \mid P \mid P[f] \mid P \setminus L \mid \mu x.P$$

Here, f is a *finite relabeling*, i.e., a mapping on $\mathcal{A}$ with $f(\tau:k) = \tau:k$, for all $k \in \mathcal{N}$, $\overline{f(\overline{a}:k)} = f(a:k)$, for all $a:k \in \Lambda \cup \overline{\Lambda}$, and $|\{\alpha:k \mid f(\alpha:k) \neq \alpha:k\}| < \infty$. A relabeling is also required to preserve priority values, i.e., for all $a:k \in \Lambda \cup \overline{\Lambda}$, we have $f(a:k) = b:k$, for some $b:k \in \Lambda_k \cup \overline{\Lambda}_k$. The *restriction set* L is a subset of $\Lambda \cup \overline{\Lambda}$, and x is a *variable* taken from a set $\mathcal{V}$. Sometimes it is convenient to write $C \overset{\text{def}}{=} P$ for $\mu C.P$, where identifier C is interpreted as a variable. We adopt the standard definitions for *sort* of a process, *free* and *bound variables*, *open* and *closed terms*, *guarded recursion*, and *contexts* [58]. We refer to closed and guarded terms as *processes* and use $P, Q, R, \ldots$ to range over the set $\mathcal{P}$ of processes. Finally, we denote syntactic equality by $\equiv$.

Although our framework allows for multi-level *priority schemes* we generally restrict ourselves to a two-level priority framework, i.e., we choose $\mathcal{N} = \{0, 1\}$. The reason is that even in this simple setting most semantic and technical issues regarding the introduction of priority to process algebra can be illustrated. However, we also discuss how the obtained results can be carried over to multi-level priority schemes. In order to improve readability within the two-level priority scheme we often write $\underline{\alpha}$ for the prioritized action $\alpha:0$, α for the unprioritized action $\alpha:1$, $\underline{A}$ for $\mathcal{A}_0$, and A for $\mathcal{A}_1$. We let δ and γ represent elements taken from $\underline{A} \cup A$. We emphasize that $\underline{\alpha}$ and α are considered to be different ports; i.e., the priority value is part of a port name. Thus, in a CCS-based framework only complementary actions having the *same* priority value can engage in a communication. We discuss the consequences of lifting this restriction in Section 3.7 for frameworks involving global pre-emption and in Section 4.6 for those involving local pre-emption. It should be noted that the dynamic priority approach presented in Section 5 also differs in its interpretation of ports, actions, and priority values. Finally, our language does not provide any operator for changing priority values of actions. However, we will discuss in Section 3.5 the effect of introducing such operators, called *prioritization* and *deprioritization*, which increase and decrease the priority values of actions, respectively.

3. Static priority and global pre-emption

In this section we introduce a semantics of our language that features static priorities, global pre-emption and a two-level priority scheme on actions. We refer to our language with this semantics as $\mathsf{CCS}^{\mathsf{sg}}$ (CCS with static priority and global pre-emption) and develop its semantic theory along the lines mentioned in Section 1.2.

The organization of this section is as follows. Section 3.1 formally introduces the operational semantics for $\mathsf{CCS}^{\mathsf{sg}}$. Sections 3.2 and 3.3 show how to adapt the notions of strong bisimulation and observational congruence to $\mathsf{CCS}^{\mathsf{sg}}$, respectively. Section 3.4 applies the semantic theory to the introductory back-and-forth example illustrated in Figure 1. The consequences of adding prioritization and deprioritization operators to $\mathsf{CCS}^{\mathsf{sg}}$

are discussed in Section 3.5. Finally, Section 3.6 comments on the extension of $\mathsf{CCS}^{\mathsf{sg}}$ to multi-level priority schemes, and Section 3.7 discusses the design decisions in our theory and presents related work.

3.1. *Operational semantics*

The *semantics* of a process $P \in \mathcal{P}$ is given by a labeled transition system $\langle \mathcal{P}, \mathcal{A}, \rightarrow, P \rangle$, where $\mathcal{P}$ is the set of *states*, $\mathcal{A}$ is the *alphabet*, $\rightarrow \subseteq \mathcal{P} \times \mathcal{A} \times \mathcal{P}$ is the *transition relation*, and P is the *start state*. We write $P \xrightarrow{\gamma} P'$ instead of $\langle P, \gamma, P' \rangle \in \rightarrow$ and say that P *may engage in action* γ *and thereafter behave like process* P'. Moreover, we let $P \xrightarrow{\gamma}$ stand for $\exists P' \in \mathcal{P}.\, P \xrightarrow{\gamma} P'$. The presentation of the operational rules requires *prioritized initial action sets* $\underline{\mathcal{I}}(P)$ which are defined as the least sets satisfying the equations in Table 1. Intuitively, $\underline{\mathcal{I}}(P)$ denotes the set of all prioritized actions in which P can initially engage. For convenience, we also write $\underline{\mathcal{I}\!\mathcal{I}}(P)$ for the set $\underline{\mathcal{I}}(P) \setminus \{\underline{\tau}\}$ of prioritized external actions.

The rules in Plotkin-style notation [69] in Table 2 formally define the transition relation and capture the following operational intuitions. Process $\gamma.P$ may engage in action γ and then behave like P. The *summation operator* $+$ denotes *nondeterministic choice*. Process $P + Q$ may behave like process P (Q) if Q (P) does not pre-empt an unprioritized transition by performing a prioritized internal transition. $P[f]$ behaves exactly as process P with the actions renamed with respect to *relabeling* f. Process $P \mid Q$ represents the *parallel composition* of P and Q, which are executed according to an *interleaving semantics* with *synchronized communication* on complementary actions on the same priority level that results in the internal action $\underline{\tau}$ or τ. However, if Q (P) is capable of engaging in a prioritized internal transition, then unprioritized transitions of P (Q) are pre-empted. The *restriction operator* $\setminus L$ prohibits in $P \setminus L$ the execution of transitions labeled by actions in $L \cup \overline{L}$ and, thus, permits the *scoping* of actions. Finally, $\mu x.P$ denotes a *recursively defined* process that is a distinguished solution to the equation $x = P$.

3.2. *Semantic theory based on strong bisimulation*

The semantic theory for $\mathsf{CCS}^{\mathsf{sg}}$ is based on the notion of *bisimulation* [58,67]. First, *strong bisimulation* [58] is adapted from CCS to our setting as follows; we refer to this relation as *prioritized strong bisimulation*.

Table 1
Prioritized initial action sets for $\mathsf{CCS}^{\mathsf{sg}}$

$\underline{\mathcal{I}}(\underline{\alpha}.P) = \{\underline{\alpha}\}$	$\underline{\mathcal{I}}(\mu x.P) = \underline{\mathcal{I}}(P[\mu x.P/x])$
$\underline{\mathcal{I}}(P + Q) = \underline{\mathcal{I}}(P) \cup \underline{\mathcal{I}}(Q)$	$\underline{\mathcal{I}}(P \mid Q) = \underline{\mathcal{I}}(P) \cup \underline{\mathcal{I}}(Q) \cup \{\underline{\tau} \mid \underline{\mathcal{I}}(P) \cap \overline{\underline{\mathcal{I}}(Q)} \neq \emptyset\}$
$\underline{\mathcal{I}}(P[f]) = \{f(\underline{\alpha}) \mid \underline{\alpha} \in \underline{\mathcal{I}}(P)\}$	$\underline{\mathcal{I}}(P \setminus L) = \underline{\mathcal{I}}(P) \setminus (L \cup \overline{L})$

Table 2
Operational semantics for CCS^{sg}

<u>Act</u>	$\dfrac{-}{\underline{\alpha}.P \xrightarrow{\underline{\alpha}} P}$	Act	$\dfrac{-}{\alpha.P \xrightarrow{\alpha} P}$
<u>Sum1</u>	$\dfrac{P \xrightarrow{\underline{\alpha}} P'}{P + Q \xrightarrow{\underline{\alpha}} P'}$	Sum1	$\dfrac{P \xrightarrow{\alpha} P'}{P + Q \xrightarrow{\alpha} P'}\ \underline{\tau} \notin \underline{\mathcal{I}}(Q)$
<u>Sum2</u>	$\dfrac{Q \xrightarrow{\underline{\alpha}} Q'}{P + Q \xrightarrow{\underline{\alpha}} Q'}$	Sum2	$\dfrac{Q \xrightarrow{\alpha} Q'}{P + Q \xrightarrow{\alpha} Q'}\ \underline{\tau} \notin \underline{\mathcal{I}}(P)$
<u>Com1</u>	$\dfrac{P \xrightarrow{\underline{\alpha}} P'}{P \mid Q \xrightarrow{\underline{\alpha}} P' \mid Q}$	Com1	$\dfrac{P \xrightarrow{\alpha} P'}{P \mid Q \xrightarrow{\alpha} P' \mid Q}\ \underline{\tau} \notin \underline{\mathcal{I}}(P \mid Q)$
<u>Com2</u>	$\dfrac{Q \xrightarrow{\underline{\alpha}} Q'}{P \mid Q \xrightarrow{\underline{\alpha}} P \mid Q'}$	Com2	$\dfrac{Q \xrightarrow{\alpha} Q'}{P \mid Q \xrightarrow{\alpha} P \mid Q'}\ \underline{\tau} \notin \underline{\mathcal{I}}(P \mid Q)$
<u>Com3</u>	$\dfrac{P \xrightarrow{\underline{a}} P' \quad Q \xrightarrow{\underline{\overline{a}}} Q'}{P \mid Q \xrightarrow{\underline{\tau}} P' \mid Q'}$	Com3	$\dfrac{P \xrightarrow{a} P' \quad Q \xrightarrow{\overline{a}} Q'}{P \mid Q \xrightarrow{\tau} P' \mid Q'}\ \underline{\tau} \notin \underline{\mathcal{I}}(P \mid Q)$
<u>Rel</u>	$\dfrac{P \xrightarrow{\underline{\alpha}} P'}{P[f] \xrightarrow{f(\underline{\alpha})} P'[f]}$	Rel	$\dfrac{P \xrightarrow{\alpha} P'}{P[f] \xrightarrow{f(\alpha)} P'[f]}$
<u>Res</u>	$\dfrac{P \xrightarrow{\underline{\alpha}} P'}{P \setminus L \xrightarrow{\underline{\alpha}} P' \setminus L}\ \underline{\alpha} \notin L \cup \overline{L}$	Res	$\dfrac{P \xrightarrow{\alpha} P'}{P \setminus L \xrightarrow{\alpha} P' \setminus L}\ \alpha \notin L \cup \overline{L}$
<u>Rec</u>	$\dfrac{P[\mu x.P/x] \xrightarrow{\underline{\alpha}} P'}{\mu x.P \xrightarrow{\underline{\alpha}} P'}$	Rec	$\dfrac{P[\mu x.P/x] \xrightarrow{\alpha} P'}{\mu x.P \xrightarrow{\alpha} P'}$

DEFINITION 3.1 (*Prioritized strong bisimulation*). A symmetric relation $\mathcal{R} \subseteq \mathcal{P} \times \mathcal{P}$ is called a *prioritized strong bisimulation* if for every $\langle P, Q\rangle \in \mathcal{R}$ and $\gamma \in \mathcal{A}$ the following condition holds.

$$P \xrightarrow{\gamma} P' \quad \text{implies} \quad \exists Q'.\, Q \xrightarrow{\gamma} Q' \text{ and } \langle P', Q'\rangle \in \mathcal{R}.$$

We write $P \simeq Q$ if $\langle P, Q\rangle \in \mathcal{R}$ for some prioritized strong bisimulation $\mathcal{R}$.

It is easy to see that $\simeq$ is an equivalence and that it is the *largest* prioritized strong bisimulation. The following result, which enables compositional reasoning, can be proved using standard techniques [1,26,77].

THEOREM 3.2. *$\simeq$ is a congruence.*

Table 3
Axiomatization of $\simeq$

(A1)	$t+u=u+t$		(A2)	$t+(u+v)=(t+u)+v$
(A3)	$t+t=t$		(A4)	$t+\mathbf{0}=t$
(E)	Let $t=\sum_i \gamma_i.t_i$ and $u=\sum_j \delta_j.u_j$. Then $t \mid u = \sum_i \gamma_i.(t_i \mid u) + \sum_j \delta_j.(t \mid u_j)$ $+ \sum_{\gamma_i \equiv \overline{\delta_j}} \{\underline{\tau}.(t_i \mid u_j) \mid \gamma_i \in \underline{A}\}) + \sum_{\gamma_i \equiv \overline{\delta_j}} \{\tau.(t_i \mid u_j) \mid \gamma_i \in A\})$			
(Res1)	$\mathbf{0} \setminus L = \mathbf{0}$		(Rel1)	$\mathbf{0}[f]=\mathbf{0}$
(Res2)	$(\gamma.t) \setminus L = \mathbf{0}$	$(\gamma \in L \cup \overline{L})$	(Rel2)	$(\gamma.t)[f]=f(\gamma).(t[f])$
(Res3)	$(\gamma.t) \setminus L = \gamma.(t \setminus L)$	$(\gamma \notin L \cup \overline{L})$	(Rel3)	$(t+u)[f]=t[f]+u[f]$
(Res4)	$(t+u) \setminus L = (t \setminus L) + (u \setminus L)$		(P)	$\underline{\tau}.t+\alpha.u=\underline{\tau}.t$

An *axiomatization* of $\simeq$ for *finite* processes, i.e., guarded and closed CCS^{sg} terms not containing recursion, can be developed closely along the lines of [26]. We write $\vdash t = u$ if process term t can be rewritten to u using the axioms in Table 3. These axioms correspond to the ones presented in [58], except that Axiom (P) dealing with global pre-emption is added. In the *Expansion Axiom* (E) the symbol $\sum$ stands for the indexed version of $+$, where the empty sum denotes the inaction process $\mathbf{0}$. The next theorem states that our equations characterize prioritized strong bisimulation for finite CCS^{sg} processes. Its proof can be found in [26]; it uses techniques described in [58].

THEOREM 3.3. *Let t, u be finite processes. Then $t \simeq u$ if and only if $\vdash t = u$.*

3.3. *Semantic theory based on weak bisimulation*

The behavioral congruence developed in the previous section is too strong for verifying systems in practice, as it requires that two equivalent terms match each other's transitions exactly, even those labeled by internal actions. The process-algebraic remedy for this problem involves the development of a semantic congruence that abstracts from internal transitions. In pursuit of such a relation we start off with the definition of a *naive prioritized weak bisimulation*, which is an adaptation of Milner's *observational equivalence* [58].

DEFINITION 3.4 (*Naive prioritized weak transitions*).

(1) $\hat{\underline{\tau}} =_{\mathrm{df}} \hat{\tau} =_{\mathrm{df}} \varepsilon$, $\hat{\underline{a}} =_{\mathrm{df}} \underline{a}$, and $\hat{a} =_{\mathrm{df}} a$.

(2) $\stackrel{\varepsilon}{\Longrightarrow}_{\times} =_{\mathrm{df}} (\stackrel{\underline{\tau}}{\longrightarrow} \cup \stackrel{\tau}{\longrightarrow})^*$.

(3) $\stackrel{\gamma}{\Longrightarrow}_{\times} =_{\mathrm{df}} \stackrel{\varepsilon}{\Longrightarrow}_{\times} \circ \stackrel{\gamma}{\longrightarrow} \circ \stackrel{\varepsilon}{\Longrightarrow}_{\times}$.

Observe that this transition relation ignores priority levels for $\stackrel{\varepsilon}{\Longrightarrow}_{\times}$. This is in accordance with the fact that a priority value is part of an action name and, thus, unobservable for internal actions.

DEFINITION 3.5 (*Naive prioritized weak bisimulation*). A symmetric relation $\mathcal{R} \subseteq \mathcal{P} \times \mathcal{P}$ is a *naive prioritized weak bisimulation* if for every $\langle P, Q\rangle \in \mathcal{R}$ and $\gamma \in \mathcal{A}$ the following condition holds.

$$P \xrightarrow{\gamma} P' \quad \text{implies} \quad \exists Q'.\, Q \overset{\hat{\gamma}}{\Longrightarrow}_{\times} Q' \text{ and } \langle P', Q'\rangle \in \mathcal{R}.$$

We write $P \approx_{\times} Q$ if $\langle P, Q\rangle \in \mathcal{R}$ for a naive prioritized weak bisimulation $\mathcal{R}$.

Naive prioritized weak bisimulation can be shown to be an equivalence. Unfortunately, $\approx_{\times}$ is not a congruence for $\mathsf{CCS}^{\mathsf{sg}}$ with respect to parallel composition and summation. Whereas the compositionality defect for summation is similar to the one for observational equivalence in CCS [58], the problem with parallel composition is due to pre-emption. As an example consider the processes $P \overset{\text{def}}{=} a.\mathbf{0}+\underline{b}.\mathbf{0}$ and $Q \overset{\text{def}}{=} a.\mathbf{0}+\tau.(a.\mathbf{0}+\underline{b}.\mathbf{0})$. It is easy to see that $P \approx_{\times} Q$. However, when composing these processes in parallel with process $\overline{\underline{b}}.\mathbf{0}$, $Q \mid \overline{\underline{b}}.\mathbf{0} \xrightarrow{a} \mathbf{0} \mid \overline{\underline{b}}.\mathbf{0}$ whereas $P \mid \overline{\underline{b}}.\mathbf{0} \overset{a}{\not\Longrightarrow}_{\times}$, i.e., $P \mid \overline{\underline{b}}.\mathbf{0} \not\approx_{\times} Q \mid \overline{\underline{b}}.\mathbf{0}$. In order to obtain a congruence, more care must be taken when defining the prioritized weak transition relation. Transitions labeled by visible actions may turn into internal transitions when composed with an environment and, thus, may gain pre-emptive power. An adequate notion of weak transitions must take the processes' potential of engaging in visible prioritized transitions into account.

3.3.1. *Prioritized weak bisimulation.* Despite its lack of compositionality, the definition of $\approx_{\times}$ reflects an intuitive approach to abstracting from internal computation, and consequently we wish to develop a congruence that is as "close" to $\approx_{\times}$ as possible. That such a relation exists is due to the following fact from universal algebra.

PROPOSITION 3.6. *For an equivalence $\mathcal{R}$ in some algebra $\mathfrak{R}$ the largest congruence $\mathcal{R}^{+} \subseteq \mathcal{R}$ exists and $\mathcal{R}^{+} = \{\langle P, Q\rangle \mid \forall \mathfrak{R}\text{-contexts } C[X].\, \langle C[P], C[Q]\rangle \in \mathcal{R}\}$. Here, a $\mathfrak{R}$-context $C[X]$ is a term in $\mathfrak{R}$ with one free occurrence of variable X.*

We therefore know that $\approx_{\times}$ contains a largest congruence $\approx_{\times}^{+}$ for $\mathsf{CCS}^{\mathsf{sg}}$ and devote the rest of this section to giving an operational characterization of this congruence. We begin by defining a new weak transition relation that takes pre-emptability into account.

DEFINITION 3.7 (*Prioritized transitions*). Let $L \subseteq \underline{A} \setminus \{\underline{\tau}\}$.

(i) $\hat{\underline{\tau}} =_{\text{df}} \underline{\varepsilon}$, $\hat{\tau} =_{\text{df}} \varepsilon$, $\hat{\underline{a}} =_{\text{df}} \underline{a}$, and $\hat{a} =_{\text{df}} a$.

(ii) $P \xrightarrow[L]{\alpha} P'$ if $P \xrightarrow{\alpha} P'$ and $\underline{\mathcal{I}}(P) \subseteq L$.

(iii) $\overset{\underline{\varepsilon}}{\Longrightarrow} =_{\text{df}} (\xrightarrow{\underline{\tau}})^{*}$.

(iv) $\overset{\varepsilon}{\underset{L}{\Longrightarrow}} =_{\text{df}} (\xrightarrow{\underline{\tau}} \cup \xrightarrow[L]{\tau})^{*}$.

(v) $\overset{\underline{\alpha}}{\Longrightarrow} =_{\text{df}} \overset{\underline{\varepsilon}}{\Longrightarrow} \circ \xrightarrow{\underline{\alpha}} \circ \overset{\underline{\varepsilon}}{\Longrightarrow}$.

(vi) $\overset{\alpha}{\underset{L}{\Longrightarrow}} =_{\text{df}} \overset{\varepsilon}{\underset{L}{\Longrightarrow}} \circ \xrightarrow[L]{\alpha} \circ \overset{\underline{\varepsilon}}{\Longrightarrow}$.

Intuitively, we have made the transition relation sensitive to pre-emption by introducing conditions involving prioritized initial action sets and by preserving priority levels of internal actions. In particular, $P \xrightarrow[L]{\alpha} P'$ holds when P can evolve to P' by performing the unprioritized action α, provided the environment does not offer any prioritized communication involving an action in L, which contains the prioritized actions P may initially engage in. In the remainder, we show that such prioritized initial action sets are an adequate means for measuring pre-emption potentials.

DEFINITION 3.8 (*Prioritized weak bisimulation*). A symmetric relation $\mathcal{R} \subseteq \mathcal{P} \times \mathcal{P}$ is a *prioritized weak bisimulation* if for every $\langle P, Q \rangle \in \mathcal{R}$, $\underline{\alpha} \in \underline{A}$, and $\alpha \in A$ the following conditions hold.

(1) $\underline{\tau} \notin \underline{\mathcal{I}}(P)$ implies $\exists Q'.\, Q \overset{\varepsilon}{\underset{L}{\Longrightarrow}} Q'$, $\underline{\mathcal{II}}(Q') \subseteq L$ where $L = \underline{\mathcal{II}}(P)$,

$$\underline{\tau} \notin \underline{\mathcal{I}}(Q'), \text{ and } \langle P, Q' \rangle \in \mathcal{R}.$$

(2) $P \xrightarrow{\underline{\alpha}} P'$ implies $\exists Q'.\, Q \overset{\hat{\underline{\alpha}}}{\Longrightarrow} Q'$, and $\langle P, Q' \rangle \in \mathcal{R}$.

(3) $P \xrightarrow{\alpha} P'$ implies $\exists Q'.\, Q \overset{\hat{\alpha}}{\underset{L}{\Longrightarrow}} Q'$, where $L = \underline{\mathcal{II}}(P)$, and $\langle P', Q' \rangle \in \mathcal{R}$.

We write $P \approxeq Q$ if $\langle P, Q \rangle \in \mathcal{R}$ for some prioritized weak bisimulation $\mathcal{R}$.

This new version of weak bisimulation is algebraically more robust than the naive one, as the next result indicates. It should be noted that Condition (1) of Definition 3.8 is necessary for achieving compositionality with respect to parallel composition [55].

PROPOSITION 3.9. *The equivalence $\approxeq$ is a congruence with respect to prefixing, parallel composition, relabeling, and restriction. Moreover, $\approxeq$ is the largest congruence contained in $\approx_\times$ with respect to the sub-algebra of* $\mathsf{CCS}^{\mathsf{sg}}$ *induced by these operators and recursion.*

Although $\approxeq$ is itself not a congruence for $\mathsf{CCS}^{\mathsf{sg}}$, this relation provides the basis for obtaining a congruence, as is made precise in the next section.

3.3.2. *Prioritized observational congruence.* The compositionality defect of $\approxeq$ with respect to summation is handled by the following notion of *prioritized observational congruence*. Unfortunately, the summation fix presented in [58], which requires an initial internal transition to be matched by a nontrivial internal weak transition, is not sufficient in order to achieve a congruence based on prioritized weak bisimulation. To see why, let $D \stackrel{\text{def}}{=} \underline{\tau}.E$ and $E \stackrel{\text{def}}{=} \tau.D$. Now define $P \stackrel{\text{def}}{=} \tau.D$ and $Q \stackrel{\text{def}}{=} \underline{\tau}.E$. By Definition 3.8 we may observe $P \approxeq Q$, but $P + a.\mathbf{0} \not\approxeq Q + a.\mathbf{0}$ since the former can perform an a-transition whereas the latter cannot. It turns out that we must require that observationally congruent processes must possess the same prioritized initial action sets; a requirement which is stronger than Condition (1) of Definition 3.8.

DEFINITION 3.10. Define $P \approxeq^{\mathrm{l}} Q$ if for all $\underline{\alpha} \in \underline{A}$ and $\alpha \in A$ the following conditions and their symmetric counterparts hold.

(1) $\underline{\mathcal{I}}(P) \supseteq \underline{\mathcal{I}}(Q)$.
(2) $P \xrightarrow{\alpha} P'$ implies $\exists Q'.\, Q \stackrel{\alpha}{\Longrightarrow} Q'$ and $P' \approxeq Q'$.
(3) $P \xrightarrow{\alpha} P'$ implies $\exists Q'.\, Q \underset{L}{\stackrel{\alpha}{\Longrightarrow}} Q'$, where $L = \underline{\underline{\mathcal{II}}}(P)$, and $P' \approxeq Q'$.

The following theorem states the desired algebraic result for $\approxeq^{1}$.

THEOREM 3.11. *$\approxeq^{1}$ is the largest congruence contained in $\approx_{\times}$, i.e., $\approxeq^{1} = \approx_{\times}^{+}$.*

Whereas the proof of the congruence property of $\approxeq^{1}$ is standard (cf. [58]), the "largest" part is proved by using the following fact from universal algebra.

PROPOSITION 3.12. *Let $\mathcal{R}_1$ and $\mathcal{R}_2$ be equivalences in some arbitrary algebra $\mathfrak{R}$ such that $\mathcal{R}_1^{+} \subseteq \mathcal{R}_2 \subseteq \mathcal{R}_1$. Then $\mathcal{R}_1^{+} = \mathcal{R}_2^{+}$.*

For the purposes of this section, we choose $\mathcal{R}_1 = \approx_{\times}$ and $\mathcal{R}_2 = \approxeq$. The following theorem, which establishes $\mathcal{R}_2^{+} = \approxeq^{1}$, can be proved along the lines of a corresponding theorem in [58]; for details see [55].

THEOREM 3.13. *$\approxeq^{1}$ is the largest congruence contained in $\approxeq$.*

In order to apply Proposition 3.12, the inclusions $\approx_{\times}^{+} \subseteq \approxeq \subseteq \approx_{\times}$ needs to be established. The inclusion $\approxeq \subseteq \approx_{\times}$ immediately follows from the definition of the naive prioritized weak and the prioritized weak transition relations. One is left with $\approx_{\times}^{+} \subseteq \approxeq$. This inclusion turns out to be difficult to show directly. Instead, the auxiliary relation $\approxeq_a =_{\mathrm{df}} \{\langle P, Q\rangle \mid C_{PQ}[P] \approx_{\times} C_{PQ}[Q]\}$ satisfying the inclusions $\approx_{\times}^{+} \subseteq \approxeq_a \subseteq \approxeq$ is defined. Using the abbreviation $\underline{S}$ for the union of the finite prioritized sorts of P and Q, we define $C_{PQ}[X] \stackrel{\text{def}}{=} X \mid H_{PQ}$, where

$$H_{PQ} \stackrel{\text{def}}{=} \underline{c}.\mathbf{0} + \sum_{\substack{L \subseteq \overline{\underline{S}},\\ \underline{b} \in \underline{S}}} \underline{\tau}.\left(\begin{array}{c} \underline{d}_{L,\underline{b}}.H_{PQ} + \\ D_L + \underline{e}.H_{PQ} + \\ \overline{\underline{b}}.H_{PQ} \end{array}\right)$$

and D_L is defined as $\sum_{\underline{\alpha} \in L} \underline{\alpha}.\mathbf{0}$. Actions $\underline{c}, \underline{d}_{L,\underline{b}}, \underline{e}$, for all $L \subseteq \overline{\underline{S}}$ and $\underline{b} \in \underline{S}$, and their complements, are assumed to be "fresh" actions, i.e., not in $\underline{S} \cup \overline{\underline{S}}$. By Proposition 3.6 we may conclude $\approx_{\times}^{+} \subseteq \approxeq_a$. The other necessary inclusion $\approxeq_a \subseteq \approxeq$ is proved by showing that $\approxeq_a$ is a prioritized weak bisimulation. Summarizing, Theorem 3.11 is a consequence of Proposition 3.12 when assembling our results as illustrated in Figure 2, where an arrow from relation R_1 to relation R_2 indicates that $R_1 \subseteq R_2$.

3.3.3. *Operational characterization.* The aim of this section is to show how prioritized weak bisimulation can be computed by adapting standard *partition-refinement algorithms* [30,50,66] developed for strong bisimulation [58]. To this end, we provide an operational characterization of prioritized weak bisimulation as strong bisimulation by introducing an *alternative prioritized weak transition relation*.

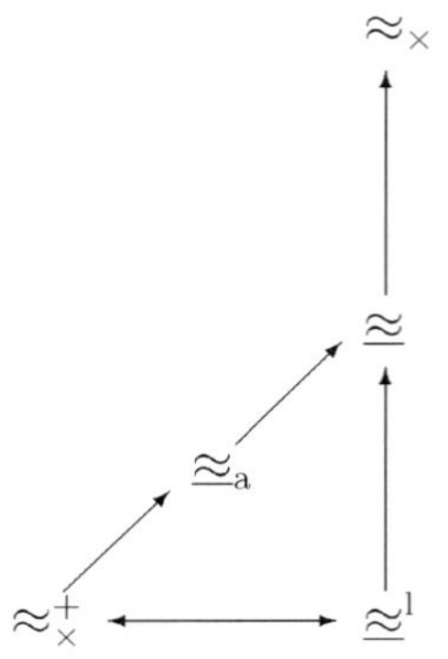

Fig. 2. Proof sketch of Theorem 3.11.

DEFINITION 3.14. Let $P, P' \in \mathcal{P}$, $\alpha \in A$, and $\underline{\alpha} \in \underline{A}$.

(1) $\overset{\hat{\alpha}}{\Longrightarrow}_* =_{\mathrm{df}} \overset{\hat{\alpha}}{\Longrightarrow}$, and

(2) $P \overset{\hat{\underline{\alpha}}}{\Longrightarrow}_* P'$ if $\exists P''.\ \underline{\tau} \notin \underline{\mathcal{I}}(P'')$ and $P \overset{\varepsilon}{\Longrightarrow} P'' \overset{\hat{\underline{\alpha}}}{\underset{L}{\Longrightarrow}} P'$, for $L = \underline{\underline{\mathcal{II}}}(P'')$.

Observe that the alternative prioritized weak transition relation is not parameterized by prioritized initial action sets. It can be computed efficiently by using *dynamic programming* techniques.

DEFINITION 3.15. A symmetric relation $\mathcal{R} \subseteq \mathcal{P} \times \mathcal{P}$ is called an *alternative prioritized weak bisimulation* if for all $\langle P, Q \rangle \in \mathcal{R}$ and $\gamma \in \mathcal{A}$ the following condition holds.

$$P \overset{\hat{\gamma}}{\Longrightarrow}_* P' \quad \text{implies} \quad \exists Q'.\ Q \overset{\hat{\gamma}}{\Longrightarrow}_* Q' \text{ and } \langle P', Q' \rangle \in \mathcal{R}.$$

We write $P \underline{\approx}_* Q$ if $\langle P, Q \rangle \in R$ for some alternative prioritized weak bisimulation $\mathcal{R}$.

THEOREM 3.16 (Operational characterization). $\underline{\approx} = \underline{\approx}_*$.

The proof of this characterization result is straightforward [55]. It should be mentioned that the presented characterization may also serve as a basis for defining a Hennessy–Milner logic for prioritized weak bisimulation along the lines of [58].

3.4. *Example*

As a simple example, we return to the back-and-forth system introduced in Section 1, which can be formalized in CCS^{sg} as follows: $\mathsf{Sys} \overset{\mathrm{def}}{=} (A \mid B) \setminus \{\underline{i}\}$ where $A \overset{\mathrm{def}}{=} \mathsf{back}.A' + \underline{i}.\underline{\tau}.\overline{\mathsf{ok}}.\bar{\underline{i}}.A$, $A' \overset{\mathrm{def}}{=} \mathsf{forth}.A + \underline{i}.\underline{\tau}.\overline{\mathsf{ok}}.\bar{\underline{i}}.A'$, and $B \overset{\mathrm{def}}{=} \mathsf{check}.\bar{\underline{i}}.\underline{i}.B$. Intuitively, $\underline{i}$ is an internal *interrupt*, and thus prioritized and restricted (via $\setminus\{\underline{i}\}$), which is invoked whenever check is

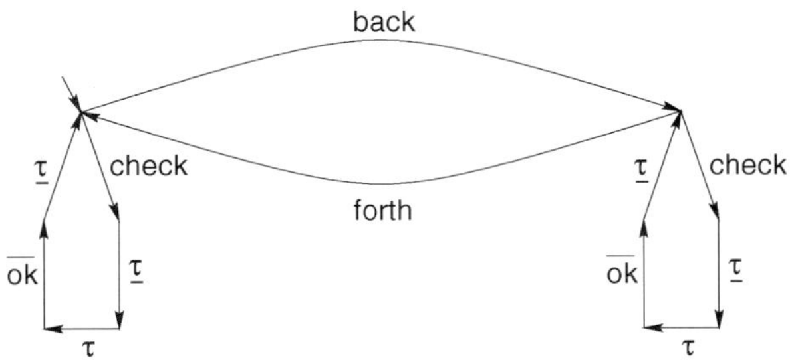

Fig. 3. Semantics of Sys.

Table 4
A relation whose symmetric closure is a prioritized weak bisimulation

$\{\, \langle\, \mathsf{Sys}$	$,\ \mathsf{Spec}\ \rangle,$	$\langle\, (A' \mid B) \setminus \{\underline{i}\}$	$,\ \mathsf{Spec}'\ \rangle,$
$\langle\, (A \mid \bar{\underline{i}}.\underline{i}.B) \setminus \{\underline{i}\}$	$,\ \overline{\mathsf{ok}}.\mathsf{Spec}\ \rangle,$	$\langle\, (\tau.\overline{ok}.\bar{\underline{i}}.A \mid \underline{i}.B) \setminus \{\underline{i}\}$	$,\ \overline{\mathsf{ok}}.\mathsf{Spec}\,\rangle,$
$\langle\, (\overline{ok}.\bar{\underline{i}}.A \mid \underline{i}.B) \setminus \{\underline{i}\}$	$,\ \overline{\mathsf{ok}}.\mathsf{Spec}\ \rangle,$	$\langle\, (\bar{\underline{i}}.A \mid \underline{i}.B) \setminus \{\underline{i}\}$	$,\ \mathsf{Spec}\ \rangle,$
$\langle\, (A' \mid \bar{\underline{i}}.\underline{i}.B) \setminus \{\underline{i}\}$	$,\ \overline{\mathsf{ok}}.\mathsf{Spec}'\,\rangle,$	$\langle\, (\tau.\overline{ok}.\bar{\underline{i}}.A' \mid \underline{i}.B) \setminus \{\underline{i}\}$	$,\ \overline{\mathsf{ok}}.\mathsf{Spec}\,\rangle,$
$\langle\, (\overline{ok}.\bar{\underline{i}}.A' \mid \underline{i}.B) \setminus \{\underline{i}\}$	$,\ \overline{\mathsf{ok}}.\mathsf{Spec}'\,\rangle,$	$\langle\, (\bar{\underline{i}}.A' \mid \underline{i}.B) \setminus \{\underline{i}\}$	$,\ \mathsf{Spec}'\ \rangle\ \}$

executed. Hence, processes A and A' cannot engage in a transition labeled by back and forth, respectively, according to our pre-emptive operational semantics, but must accept the communication on prioritized port $\underline{i}$. One can think of the τ-action in the definition of process A as representing some internal activities determining the current status of the system. The $\mathsf{CCS}^{\mathsf{sg}}$ semantics of Sys is shown in Figure 3.

In the sequel, we prove that Sys meets its specification Spec, which is given by

$$\mathsf{Spec} \stackrel{\text{def}}{=} \mathsf{back}.\mathsf{Spec}' + \mathsf{check}.\overline{\mathsf{ok}}.\mathsf{Spec},$$
$$\mathsf{Spec}' \stackrel{\text{def}}{=} \mathsf{forth}.\mathsf{Spec} + \mathsf{check}.\overline{\mathsf{ok}}.\mathsf{Spec}'.$$

First, the validity of $\mathsf{Sys} \approxeq \mathsf{Spec}$ is demonstrated by the relation presented in Table 4, whose symmetric closure is a prioritized weak bisimulation that contains $\langle \mathsf{Sys}, \mathsf{Spec} \rangle$.

In addition, both processes only possess visible initial actions, and their prioritized initial action sets are identical. Hence, we may conclude $\mathsf{Sys} \approxeq^{1} \mathsf{Spec}$.

3.5. *Prioritization and deprioritization operators*

There are several other language constructs of interest in the setting of priorities. Of particular interest are the *prioritization* and *deprioritization* operators introduced in [26]. The former operator may be used to raise the priority of a visible unprioritized action and is written $\lceil a$, for $a \not\equiv \tau$; the latter lowers the priority of a visible prioritized action and is written $\lfloor \underline{a}$, for $\underline{a} \not\equiv \underline{\tau}$. The operational semantics of these operators is formally defined in Table 5. For them to be well-defined there must be a one-to-one correspondence between prioritized ports $\underline{a}$ and unprioritized ports a; in addition every relabeling f must satisfy $f(\underline{a}) = \underline{f(a)}$. Intuitively, $P\lceil a$ prioritizes all a-transitions that P can perform, while $P\lfloor \underline{a}$ deprioritizes all $\underline{a}$-transitions in which P can engage. Observe that the notion of priority is still static and not dynamic since the prioritization and deprioritization operators are static operators and therefore define a scope.

Including prioritization and deprioritization operators with CCS^{sg} does not have any consequences for prioritized strong bisimulation since it is compositional with respect to these operators [26]. The axiomatization of prioritized strong bisimulation for finite processes can also be extended to cover the new operators. The necessary additional axioms are stated in Table 6.

The presence of the prioritization and the deprioritization operators suggests a formal justification for the design decision that only prioritized internal actions have pre-emptive power over unprioritized actions. For this purpose assume that

Table 5
Semantics for the prioritization and the deprioritization operator

$$\text{Prio1}\ \frac{P \stackrel{a}{\longrightarrow} P'}{P\lceil a \stackrel{\underline{a}}{\longrightarrow} P'\lceil a} \qquad \text{Deprio1}\ \frac{P \stackrel{\underline{a}}{\longrightarrow} P'}{P\lfloor \underline{a} \stackrel{a}{\longrightarrow} P'\lfloor \underline{a}}\ \underline{\tau} \notin \underline{\mathcal{I}}(P) \qquad \text{Deprio3}\ \frac{P \stackrel{\gamma}{\longrightarrow} P'}{P\lfloor \underline{a} \stackrel{\gamma}{\longrightarrow} P'\lfloor \underline{a}}\ \gamma \not\equiv \underline{a}$$

$$\text{Prio2}\ \frac{P \stackrel{\gamma}{\longrightarrow} P'}{P\lceil a \stackrel{\gamma}{\longrightarrow} P'\lceil a}\ \gamma \not\equiv a \qquad \text{Deprio2}\ \frac{P \stackrel{\underline{a}}{\longrightarrow} P'}{P\lfloor \underline{a} \stackrel{a}{\longrightarrow} P'\lfloor \underline{a}}\ \underline{\tau} \in \underline{\mathcal{I}}(P)$$

Table 6
Axioms for the prioritization and the deprioritization operator

(Prio1)	$\mathbf{0}\lceil a$	$=$	$\mathbf{0}$	
(Prio2)	$(a.t)\lceil a$	$=$	$\underline{a}.(t\lceil a)$	
(Prio3)	$(\gamma.t)\lceil a$	$=$	$\gamma.(t\lceil a)$	$\gamma \not\equiv a$
(Prio4)	$(t + \underline{\tau}.u + \underline{b}.v)\lceil a$	$=$	$(t + \underline{\tau}.u)\lceil a + \underline{b}.(v\lceil a)$	
(Prio5)	$(t + \delta.u + \gamma.v)\lceil a$	$=$	$(t + \delta.u)\lceil a + (t + \gamma.v)\lceil a$	$\delta, \gamma \in \mathcal{A} \setminus \{\underline{\tau}\}$
(Deprio1)	$\mathbf{0}\lfloor \underline{a}$	$=$	$\mathbf{0}$	
(Deprio2)	$(\underline{a}.t)\lfloor \underline{a}$	$=$	$a.(t\lfloor \underline{a})$	
(Deprio3)	$(\gamma.t)\lfloor \underline{a}$	$=$	$\gamma.(t\lceil \underline{a})$	$\gamma \not\equiv \underline{a}$
(Deprio4)	$(t + \underline{\tau}.u + \underline{b}.v)\lfloor \underline{a}$	$=$	$(t + \underline{\tau}.u)\lfloor \underline{a} + \underline{b}.(v\lfloor \underline{a})$	
(Deprio5)	$(t + \delta.u + \gamma.v)\lfloor \underline{a}$	$=$	$(t + \delta.u)\lfloor \underline{a} + (t + \gamma.v)\lfloor \underline{a}$	$\delta, \gamma \in \mathcal{A} \setminus \{\underline{\tau}\}$

(i) pre-emption is not encoded in the side conditions of the operational rules but, equivalently, in the notion of bisimulation [26] and that
(ii) the naive view of pre-emption gives all prioritized actions pre-emptive power.

Thus, a naive bisimulation $\sim_{\mathrm{n}}$ demands the following condition for equivalent processes $P \sim_{\mathrm{n}} Q$ and unprioritized actions $\alpha \in A$:

$$\big(P \stackrel{\alpha}{\longrightarrow} P' \text{ and } \nexists \underline{\beta}.\, P \stackrel{\underline{\beta}}{\longrightarrow}\big) \quad \text{implies} \quad \big(\exists Q'.\, Q \stackrel{\alpha}{\longrightarrow} Q',\ \nexists \underline{\beta}.\, Q \stackrel{\underline{\beta}}{\longrightarrow}, \text{ and } P' \sim_{\mathrm{n}} Q'\big)$$

and vice versa. The condition for prioritized actions can be adopted from standard strong bisimulation. It turns out that $\sim_{\mathrm{n}}$ is not a congruence; for example, $a.\mathbf{0} + \underline{b}.\mathbf{0} \sim_{\mathrm{n}} \underline{b}.\mathbf{0}$ but $(a.\mathbf{0}+\underline{b}.\mathbf{0}) \setminus \{\underline{b}\} \not\sim_{\mathrm{n}} (\underline{b}.\mathbf{0}) \setminus \{\underline{b}\}$ since the former process can engage in an a-transition while the latter is deadlocked. Then the question arises what the largest congruence contained in $\sim_{\mathrm{n}}$ is; it turns out that it is prioritized strong congruence as defined above (see [26] for details). This shows that in a pre-emptive semantics only prioritized internal actions may pre-empt unprioritized actions. However, the above algebraic result is only correct if we include the deprioritization operator in our language. An operational characterization of the largest congruence contained in $\sim_{\mathrm{n}}$ with respect to $\mathsf{CCS}^{\mathsf{sg}}$ is an open problem.

For the language extended by the prioritization and the deprioritization operator, an observational congruence together with an axiomatic characterization with respect to finite processes has been developed in [64,65], which is briefly reviewed here. For this purpose, we need to refine the prioritized weak transition relation. First, we re-define $\stackrel{a}{\Longrightarrow}$ to $\stackrel{\varepsilon}{\Longrightarrow} \circ \stackrel{a}{\longrightarrow} \circ \stackrel{\varepsilon}{\Longrightarrow}$, i.e., a weak unprioritized a-transition consists of an a-transition that is preceded and trailed by *prioritized* internal transitions only. Moreover, we replace $\underline{\mathcal{II}}(P)$ by $\underline{\mathcal{II}}(P) \cup \mathcal{II}(P)$ in the definition of $\stackrel{\varepsilon}{\Longrightarrow}$ since one has to take into account the fact that unprioritized actions may turn into prioritized ones when applying the prioritization operator. Finally, we write $P \stackrel{\tau}{\underset{L}{\Longrightarrow}} P'$ whenever $P \stackrel{\varepsilon}{\underset{L}{\Longrightarrow}} P'$ and $P \not\equiv P'$. Consequently, visible weak unprioritized transitions only abstract from prioritized internal actions. The reason for this restriction is that, otherwise, prioritized weak bisimulation would not be compositional with respect to the prioritization and the deprioritization operator. In contrast, the original prioritized weak transition relation allows an α-transition to be preceded by any sequence of $\underline{\tau}$- and τ-transitions (satisfying a condition on prioritized initial action sets) and only to be trailed by $\underline{\tau}$-transitions.

The notions of prioritized weak bisimulation and prioritized observational congruence are defined in [64,65] as follows, where $P\underline{\Downarrow}$ stands for $\exists P'.\, P \stackrel{\varepsilon}{\Longrightarrow} P'$ and $P' \stackrel{\underline{\tau}}{\not\longrightarrow}$.

DEFINITION 3.17. A symmetric relation $\mathcal{R} \subseteq \mathcal{P} \times \mathcal{P}$ is an *extended prioritized weak bisimulation* if for every $\langle P, Q\rangle \in \mathcal{R}$ and $\gamma \in \mathcal{A} \setminus \{\tau\}$ the following conditions hold.

(1) $P\underline{\Downarrow}$ implies $Q\underline{\Downarrow}$.

(2) $P \stackrel{\gamma}{\longrightarrow} P'$ implies $\exists Q'.\, Q \stackrel{\hat{\gamma}}{\Longrightarrow} Q'$ and $\langle P, Q'\rangle \in \mathcal{R}$.

(3) $P \stackrel{\tau}{\longrightarrow} P'$ implies $\exists Q'.\, Q \stackrel{\varepsilon}{\underset{L}{\Longrightarrow}} Q'$, $L = \underline{\mathcal{II}}(P) \cup \mathcal{II}(P)$, and $\langle P', Q'\rangle \in \mathcal{R}$.

We write $P \approxeq_{\mathrm{pd}} Q$ if there exists an extended prioritized weak bisimulation $\mathcal{R}$ such that $\langle P, Q\rangle \in \mathcal{R}$.

Table 7
Axioms for the τ-laws

($\underline{\tau}1$)	$\gamma.(1.t + t) = \gamma.t$	$1 \in \{\tau, \underline{\tau}\}$
($\underline{\tau}2$)	$\underline{\tau}.t = \underline{\tau}.t + t$	
($\underline{\tau}3$)	$\gamma.(t + \underline{\tau}.u) = \gamma.(t + \underline{\tau}.u) + \gamma.u$	
($\tau 1$)	$t + \tau.(u + \tau.v) = t + \tau.(u + \tau.v) + \tau.v$	$\vdash_I t \sqsubseteq_{\mathrm{i}} v$

Table 8
Axiomatization of $\sqsubseteq_{\mathrm{i}}$ (Axioms I)

(iC1) $\underline{\alpha}.t \sqsubseteq_{\mathrm{i}} \underline{\alpha}.u$	(iC2) $\mathbf{0} \sqsubseteq_{\mathrm{i}} \gamma.t \quad \gamma \in \mathcal{A} \setminus \{\underline{\tau}\}$	(iC3) $\alpha.t \sqsubseteq_{\mathrm{i}} \mathbf{0}$

DEFINITION 3.18. We define $P \approxeq^1_{\mathrm{pd}} Q$ if for all $\gamma \in \mathcal{A} \setminus \{\tau\}$ the following conditions and their symmetric counterparts hold.

(1) $P \xrightarrow{\gamma} P'$ implies $\exists Q'\, Q \xRightarrow{\gamma} Q'$ and $P' \approxeq_{\mathrm{pd}} Q'$.

(2) $P \xrightarrow{\tau} P'$ implies $\exists Q'.\, Q \overset{\tau}{\underset{L}{\Longrightarrow}} Q'$, where $L = \underline{\mathcal{II}}(P) \cup \mathcal{II}(P)$, and $P' \approxeq_{\mathrm{pd}} Q'$.

The observational congruence $\approxeq^1_{\mathrm{pd}}$ possesses nice algebraic properties for our language extended by the prioritization and the deprioritization operators, including a largest congruence result similar to Theorem 3.11 and a sound and complete axiomatization for finite processes. For the latter, the axiomatization for prioritized strong bisimulation is augmented with suitable τ-laws as shown in Table 7 (cf. [58]). The relation $\sqsubseteq_{\mathrm{i}}$, occurring in the side condition of Axiom ($\tau 1$), is the pre-congruence on finite processes generated by the axioms presented in Table 8 using the laws of inequational reasoning. We write $\vdash_I t \sqsubseteq_{\mathrm{i}} u$ if t can be related to u by Axioms (iC1), (iC2), and (iC3). Intuitively, $\vdash_I t \sqsubseteq_{\mathrm{i}} u$ holds, whenever

(i) $\underline{\tau} \in \underline{\mathcal{I}}(t)$ if and only if $\underline{\tau} \in \underline{\mathcal{I}}(u)$, and

(ii) $\underline{\mathcal{II}}(t) \subseteq \underline{\mathcal{II}}(u)$.

Finally, it should be noted that applications underline the importance of the additional abstraction from internal transitions gained by leaving out the prioritization and deprioritization operators. Indeed, the observational congruence $\approxeq^1_{\mathrm{pd}}$ does not relate processes Sys and Spec of our back-and-forth example [55]. This is due to the presence of action τ in process Sys.

3.6. *Extension to multi-level priority schemes*

We now remark on the extension of $\mathsf{CCS}^{\mathsf{sg}}$ to a multi-level priority scheme. We first alter the definition of prioritized initial action sets to capture the priority-level of actions by introducing sets $\mathrm{I}^k(P)$ for process P with respect to priority value k as shown in Table 9.

Table 9
Potential initial action sets for $\mathsf{CCS}^{\mathsf{sg}}$

$\mathrm{I}^k(\alpha:l.P) = \{\alpha:l \mid l = k\}$	$\mathrm{I}^k(P[f]) = \{f(\alpha:l) \mid \alpha:l \in \mathrm{I}^k(P)\}$
$\mathrm{I}^k(\mu x.P) = \mathrm{I}^k(P[\mu x.P/x])$	$\mathrm{I}^k(P+Q) = \mathrm{I}^k(P) \cup \mathrm{I}^k(Q)$
$\mathrm{I}^k(P \setminus L) = \mathrm{I}^k(P) \setminus (L \cup \overline{L})$	$\mathrm{I}^k(P \mid Q) = \mathrm{I}^k(P) \cup \mathrm{I}^k(Q) \cup \{\tau:k \mid \mathrm{I}^k(P) \cap \overline{\mathrm{I}^k(Q)} \neq \emptyset\}$

Table 10
Prioritized weak transition relation

$\stackrel{\varepsilon:0}{\Longrightarrow} =_{\mathrm{df}} (\stackrel{\tau:0}{\longrightarrow})^*$	$P \xrightarrow[L]{\alpha:k} P'$ if $P \stackrel{\alpha:k}{\longrightarrow} P'$ and $\mathcal{II}^l(P) \subseteq L$ for all $l < k$
$\stackrel{\varepsilon:k}{\underset{L}{\Longrightarrow}} =_{\mathrm{df}} (\{\xrightarrow[L]{\tau:l} \mid l \leqslant k\})^*$	$\stackrel{\alpha:k}{\underset{L}{\Longrightarrow}} =_{\mathrm{df}} \stackrel{\varepsilon:k}{\underset{L}{\Longrightarrow}} \circ \xrightarrow[L]{\alpha:k} \circ \stackrel{\varepsilon:0}{\Longrightarrow}$

Using this definition and the notation "$\tau \notin \mathrm{I}^{<k}(P)$ if $\nexists l < k.\ \tau:l \in \mathrm{I}^l(P)$", the operational semantics depicted in Table 2 can be re-stated in the manner illustrated by the following modification to Rule (Com3).

$$\mathsf{Com3}\ \frac{P \stackrel{a:k}{\longrightarrow} P' \quad Q \stackrel{\overline{a}:k}{\longrightarrow} Q'}{P \mid Q \stackrel{\tau:k}{\longrightarrow} P' \mid Q'}\ \tau \notin \mathrm{I}^{<k}(P \mid Q).$$

Observe that the sets $\mathrm{I}^k(P)$ may contain actions in which P cannot initially engage, since their definition does not consider pre-emption. In fact, the set $\mathcal{I}^k(P)$ of actions with priority value k in which P can indeed initially engage is given by $\{\alpha:k \in \mathrm{I}^k(P) \mid \tau:l \notin \mathrm{I}^l(P)$, for all $l < k\}$. However, it is easy to show that $\tau \notin \mathrm{I}^{<k}(P)$ if and only if $\tau \notin \mathcal{I}^{<k}(P)$ [55]. Thus, the side condition of Rule (Com3) captures our intuition that $P \mid Q$ cannot engage in a more urgent internal transition.

The re-development of the bisimulation-based semantic theory proceeds along the lines of the above sections and does not raise any new semantic issues. For example, the notion of prioritized observational congruence is defined as follows, where

(i) the prioritized weak transition relation is given by the rules in Table 10,
(ii) $\mathcal{II}^k(P) =_{\mathrm{df}} \mathcal{I}^k(P) \setminus \{\tau:k\}$,
(iii) the relation $\approxeq_{\mathsf{ml}}$ is the adaption of prioritized weak bisimulation to the multi-level priority scheme,
(iv) $\mathcal{I}(P) =_{\mathrm{df}} \bigcup\{\mathcal{I}^k(P) \mid k \in \mathcal{N}\}$, and
(v) $\mathcal{II}^{<k}(P) =_{\mathrm{df}} \mathcal{I}^{<k}(P) \setminus \{\tau:l \mid l < k\}$.

DEFINITION 3.19. Processes P and Q are *prioritized observational congruent* if for all actions $\alpha:k$ the following conditions and their symmetric counterparts hold.
(1) $\mathcal{I}(P) \supseteq \mathcal{I}(Q)$,
(2) $P \stackrel{\alpha:k}{\longrightarrow} P'$ implies $\exists Q'.\ Q \stackrel{\alpha:k}{\underset{L}{\Longrightarrow}} Q'$, where $L = \mathcal{II}^{<k}(P)$, and $P' \approxeq_{\mathsf{ml}} Q'$.

Details of this extension of $\mathsf{CCS}^{\mathsf{sg}}$ can be found in [55].

3.7. *Concluding remarks and related work*

We conclude by first commenting on the design decision that priority values are considered to be part of port names, which implies that only complementary actions having the same priority can synchronize. Lifting this design decision by allowing $a:k$ and $\overline{a}:l$, where $k \neq l$, to synchronize leads to the question of which priority value to assign to the resulting τ. One can imagine several obvious choices for this function, e.g., maximum or minimum. On the other hand, [36,38] recommend using the sum of the priority values of the actions involved. Unfortunately, while a specific function may be suitable for certain examples, it often appears *ad hoc* in general. In the next section we will see that such a function is unnecessary when dealing with *local pre-emption*.

Regarding related work, Gerber and Lee developed a real-time process algebra, the *Calculus of Communicating Shared Resources* (CCSR) [35], that explicitly takes the availability of system resources into account. Synchronizations between processes are modeled in an interleaving fashion using instantaneous transitions, whereas the access of resources is truly concurrent and consumes time. In CCSR a priority structure may be defined over resources in order to indicate their importance, e.g., for ensuring that deadlines are met. The underlying concept of priority is similar to that of $\mathsf{CCS}^{\mathsf{sg}}$ in that priorities are static and pre-emption is global. In [36] a resource-based prioritized (strong) bisimulation for CCSR together with axiomatizations for several classes of processes [21] is presented.

Prasad extended his *Calculus of Broadcasting Systems* (CBS) [70] for dealing with a notion of static priority [71]. He refers to the priority calculus as PCBS. For PCBS nice semantic theories based on Milner's strong and weak bisimulation [58] have been developed. Remarkably, these theories do not suffer from the technical subtleties which have been encountered for $\mathsf{CCS}^{\mathsf{sg}}$, although the concept of pre-emption is basically the same. The reason is that PCBS uses a much simpler model for communication which is based on the principle of *broadcasting*. In this setting, priority values are only attached to output actions, which cannot be restricted or hidden as in traditional process algebras. Finally, it should be mentioned that PCBS contains an operator, called *translate*, which enables the prioritization and the deprioritization of actions.

4. Static priority and local pre-emption

This section provides a new semantics, $\mathsf{CCS}^{\mathsf{sl}}$ (CCS with static priority and local pre-emption), for the language introduced in Section 2. The new semantics is distinguished from the one developed in the previous section in that it only allows actions to pre-empt others at the same "location" and, therefore, captures a notion of *localized precedence*. This constraint reflects an essential intuition about distributed systems, namely, that the execution of a process on one processor should not affect the behavior of a process on another processor unless the designer explicitly builds an interaction, e.g., a synchronization, between them.

The following example illustrates why locations should be taken into account when reasoning about priority within distributed systems. The system in question consists of an application that manipulates data from two blocks of memory (see Figure 4). In order to

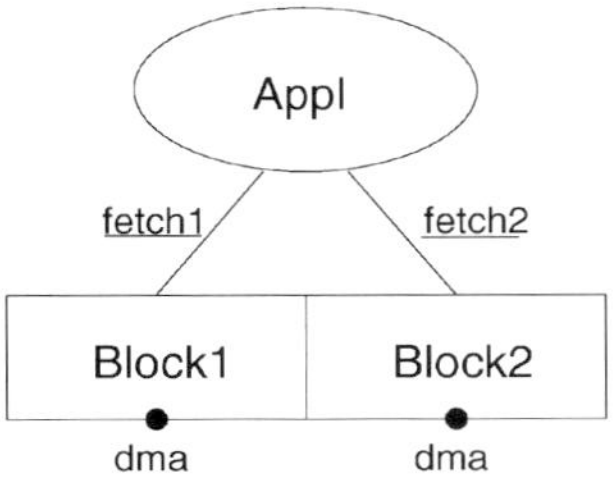

Fig. 4. Example system.

improve efficiency, each memory block is connected to a direct-memory-access (DMA) controller. Assume further, that application Appl fetches pieces of data alternately from each memory block, i.e., $\mathsf{Appl} \stackrel{\text{def}}{=} \overline{\mathsf{fetch1}}.\overline{\mathsf{fetch2}}.\mathsf{Appl}$. Both Block1 and Block2, are continuously able to serve the application or to allow the external DMA controller to access the memory via channel dma, with the former having priority over the latter; whence, $\mathsf{Block1} \stackrel{\text{def}}{=} \underline{\mathsf{fetch1}}.\mathsf{Block1} + \mathsf{dma}.\mathsf{Block1}$ and $\mathsf{Block2} \stackrel{\text{def}}{=} \underline{\mathsf{fetch2}}.\mathsf{Block2} + \mathsf{dma}.\mathsf{Block2}$. The overall system Sys is given by:

$$\mathsf{Sys} \stackrel{\text{def}}{=} (\mathsf{Appl} \mid \mathsf{Block1} \mid \mathsf{Block2}) \setminus \{\underline{\mathsf{fetch1}}, \underline{\mathsf{fetch2}}\}.$$

Since the application uses the memory blocks alternately, the DMA is expected to be allowed to access the one memory block which is currently not serving the application. However, when using the approach to priority involving global pre-emption presented in Section 3, all dma-transitions in the labeled transition system of Sys are pre-empted, since the application can indefinitely engage in a prioritized communication, i.e., direct-memory-access is never granted.

Generally speaking, one would expect that priorities at different *sites* of a distributed system do not influence the behavior of each other, i.e., priorities at different sites are supposed to be incomparable. The semantics given in Section 3 does not permit this distinction to be made. Hence, it precludes too many transitions that one would expect to find in a distributed setting. In the following, we show how to remedy the shortcoming of the semantics of Section 3 for distributed systems. Our elaboration is based on the concept of *local pre-emption* [24,28].

The remainder of this section is organized as follows. Section 4.1 introduces a notion of locations that is used in Section 4.2 in the definition of the operational semantics of $\mathsf{CCS}^{\mathsf{sl}}$ for a two-level priority scheme. Sections 4.3 and 4.4 develop the semantic theories based on strong and weak bisimulation, respectively, while Section 4.5 re-considers the direct-memory-access example. The consequences of lifting some design decisions in $\mathsf{CCS}^{\mathsf{sl}}$ are discussed in Section 4.6. After extending $\mathsf{CCS}^{\mathsf{sl}}$ to a multi-level priority scheme in Section 4.7 and presenting another calculus to priority taken from [24] in Section 4.8, a formal comparison of the two calculi is given in Section 4.9. Finally, Section 4.10 concludes with some additional remarks and with comments on related work.

4.1. *Locations*

A location represents the "address" of a subterm inside a larger term; when a system performs an action, CCS^{sl} semantics will also note the locations of the subterm that "generate" this action. Because of the potential for synchronization more than one subterm may be involved in an action. Our account of locations closely follows that of [28,62].

Let $\mathcal{A}_{\text{addr}} =_{\text{df}} \{L, R, l, r\}$ be the *address alphabet*, and let $\bullet$ be a special symbol not in $\mathcal{A}_{\text{addr}}$. Then $\mathit{Addr} =_{\text{df}} \{\bullet s \mid s \in \mathcal{A}^*_{\text{addr}}\}$ represents the set of (process) *addresses* ranged over by v and w. Intuitively, an element of *Addr* represents the address of a subterm, with $\bullet$ denoting the current term, l (r) representing the left (right) subterm of $+$, and L (R) the left (right) subterm of $|$. For example, in process $(a.\mathbf{0} \mid b.\mathbf{0}) + c.\mathbf{0}$ the address of $a.\mathbf{0}$ is $\bullet Ll$, of $b.\mathbf{0}$ is $\bullet Rl$, and of $c.\mathbf{0}$ is $\bullet r$. If $\bullet s_1$ and $\bullet s_2$ are addresses, then we write $\bullet s_1 \cdot \bullet s_2 = \bullet s_1 s_2$ to represent address concatenation (where $s_1 s_2$ represents the usual concatenation of elements in $\mathcal{A}^*_{\text{addr}}$). Further, if $V \subseteq \mathit{Addr}$ and $\varsigma \in \mathcal{A}_{\text{addr}}$, then we write $V \cdot \varsigma$ for $\{v \cdot \varsigma \mid v \in V\}$. Occasionally, we omit $\bullet$ from addresses.

As mentioned previously, we adopt the view that processes at different sides of the parallel composition operator are logically – not necessarily physically – executed on different processors. Thus, priorities on different sides of the parallel composition operator are distributed and, therefore, should be incomparable. However, priorities on different sides of the summation operator should be comparable since argument processes of summation are logically scheduled on the same processor. This intuition is formalized in the following *location comparability relation* on addresses which is adapted from [38].

DEFINITION 4.1 (*Location comparability relation*). The location comparability relation $\bowtie$ on addresses is the smallest reflexive and symmetric subset of $\mathit{Addr} \times \mathit{Addr}$ such that for all $v, w \in \mathit{Addr}$:

(1) $\langle v \cdot l, w \cdot r \rangle \in \bowtie$, and
(2) $\langle v, w \rangle \in \bowtie$ implies $\langle v \cdot \varsigma, w \cdot \varsigma \rangle \in \bowtie$, for $\varsigma \in \mathcal{A}_{\text{addr}}$.

In the sequel, we write $v \bowtie w$ instead of $\langle v, w \rangle \in \bowtie$. If $v \in \mathit{Addr}$ then we use $[v]$ to denote the set $\{w \in \mathit{Addr} \mid v \bowtie w\}$. Note that the location comparability relation is not transitive, e.g., we have $Ll \bowtie r$ and $r \bowtie Rl$, but $Ll \not\bowtie Rl$, since $L \not\bowtie R$.

We now define the set $\mathcal{L}oc$ of (transition) *locations* as $\mathit{Addr} \cup (\mathit{Addr} \times \mathit{Addr})$. Intuitively, a transition location records the addresses of the components in a term that par-

Table 11
Distributed prioritized initial action sets for CCS^{sl}

$\underline{\mathcal{I}}_m(\mu x.P) = \underline{\mathcal{I}}_m(P[\mu x.P/x])$	$\underline{\mathcal{I}}_\bullet(\underline{\alpha}.P) = \{\underline{\alpha}\}$
$\underline{\mathcal{I}}_{m\cdot l}(P+Q) = \underline{\mathcal{I}}_m(P)$	$\underline{\mathcal{I}}_{n\cdot r}(P+Q) = \underline{\mathcal{I}}_n(Q)$
$\underline{\mathcal{I}}_m(P[f]) = \{f(\underline{\alpha}) \mid \underline{\alpha} \in \underline{\mathcal{I}}_m(P)\}$	$\underline{\mathcal{I}}_m(P \setminus L) = \underline{\mathcal{I}}_m(P) \setminus (L \cup \overline{L})$
$\underline{\mathcal{I}}_{m\cdot L}(P\mid Q) = \underline{\mathcal{I}}_m(P)$	$\underline{\mathcal{I}}_{n\cdot R}(P\mid Q) = \underline{\mathcal{I}}_n(Q)$
$\underline{\mathcal{I}}_{\langle m\cdot L, n\cdot R\rangle}(P\mid Q) = \{\underline{\tau} \mid \underline{\mathcal{I}}_m(P) \cap \overline{\underline{\mathcal{I}}_n(Q)} \neq \emptyset\}$	

ticipate in the execution of a given action. In our language, transitions are performed by single processes or pairs of processes (in the case of a synchronization). We define $\langle v, w\rangle \cdot \zeta =_{\mathrm{df}} \langle v \cdot \zeta, w \cdot \zeta\rangle$ and $[\langle v, w\rangle] =_{\mathrm{df}} [v] \cup [w]$, where $v, w \in \mathit{Addr}$ and $\zeta \in \mathcal{A}_{\mathrm{addr}}$. We use $m, n, o, \ldots$ to range over $\mathcal{L}oc$ in what follows.

4.2. *Operational semantics*

The operational semantics of a $\mathsf{CCS}^{\mathsf{sl}}$ process P is given by a labeled transition system $\langle \mathcal{P}, \underline{A} \cup (\mathcal{L}oc \times A), \rightarrow, P\rangle$. The transition relation $\rightarrow\, \subseteq \mathcal{P} \times (\mathcal{L}oc \times A)) \times \mathcal{P}$ with respect to unprioritized actions is defined in Table 12 using Plotkin-style operational rules [69] whereas for prioritized actions the same rules as for $\mathsf{CCS}^{\mathsf{sg}}$ apply (cf. Table 2). We write $P \xrightarrow{m,\alpha} P'$ if $\langle P, \langle m, \alpha\rangle, P'\rangle \in\, \rightarrow$, and say that P *may engage in action* α *offered from location* m *and thereafter behave like process* P'.

The presentation of the operational rules requires *distributed prioritized initial action sets* which are defined as the least sets satisfying the rules in Table 11. Intuitively, $\underline{\mathcal{I}}_m(P)$ denotes the set of all prioritized initial actions of P from location m. Note that these sets are either empty or contain exactly one initial action. $\underline{\mathcal{I}}_m(P) = \emptyset$ means that either m is not a location of P or P is incapable of performing a prioritized action at location m. Additionally, let us denote the set $\bigcup\{\underline{\mathcal{I}}_m(P) \mid m \in M\}$ of all distributed prioritized initial actions of P from locations $M \subseteq \mathcal{L}oc$ by $\underline{\mathcal{I}}_M(P)$ and the set $\underline{\mathcal{I}}_{\mathcal{L}oc}(P)$ of all distributed prioritized initial actions of P by $\underline{\mathcal{I}}(P)$. We also define analogous sets restricted to visible actions: $\underline{\mathcal{II}}_M(P) =_{\mathrm{df}} \underline{\mathcal{I}}_M(P) \setminus \{\underline{\tau}\}$ and $\underline{\mathcal{II}}(P) =_{\mathrm{df}} \underline{\mathcal{I}}(P) \setminus \{\underline{\tau}\}$, respectively.

The side conditions of the operational rules guarantee that a process does not perform an unprioritized action if it can engage in a prioritized synchronization or in an internal computation, i.e., a $\underline{\tau}$-transition, from a comparable location. In contrast to the global

Table 12
Operational semantics for $\mathsf{CCS}^{\mathsf{sl}}$

$$\textsf{Act}\ \frac{-}{\alpha.P \xrightarrow{\bullet,\alpha} P} \qquad \textsf{Sum1}\ \frac{P \xrightarrow{m,\alpha} P'}{P + Q \xrightarrow{m\cdot l,\alpha} P'}\ \underline{\tau} \notin \underline{\mathcal{I}}(Q)$$

$$\textsf{Rel}\ \frac{P \xrightarrow{m,\alpha} P'}{P[f] \xrightarrow{m,f(\alpha)} P'[f]} \qquad \textsf{Sum2}\ \frac{Q \xrightarrow{n,\alpha} Q'}{P + Q \xrightarrow{n\cdot r,\alpha} Q'}\ \underline{\tau} \notin \underline{\mathcal{I}}(P)$$

$$\textsf{Res}\ \frac{P \xrightarrow{m,\alpha} P'}{P \setminus L \xrightarrow{m,\alpha} P' \setminus L}\ \alpha \notin L \cup \overline{L} \qquad \textsf{Com1}\ \frac{P \xrightarrow{m,\alpha} P'}{P|Q \xrightarrow{m\cdot L,\alpha} P'|Q}\ \underline{\mathcal{II}}_{[m]}(P) \cap \overline{\underline{\mathcal{II}}(Q)} = \emptyset$$

$$\textsf{Rec}\ \frac{P[\mu x.P/x] \xrightarrow{m,\alpha} P'}{\mu x.P \xrightarrow{m,\alpha} P'} \qquad \textsf{Com2}\ \frac{Q \xrightarrow{n,\alpha} Q'}{P|Q \xrightarrow{n\cdot R,\alpha} P|Q'}\ \underline{\mathcal{II}}_{[n]}(Q) \cap \overline{\underline{\mathcal{II}}(P)} = \emptyset$$

$$\textsf{Com3}\ \frac{P \xrightarrow{m,a} P' \quad Q \xrightarrow{n,\overline{a}} Q'}{P|Q \xrightarrow{\langle m\cdot L, n\cdot R\rangle,\tau} P'|Q'}\ \begin{array}{l}\underline{\mathcal{II}}_{[m]}(P) \cap \overline{\underline{\mathcal{II}}(Q)} = \emptyset\ \wedge \\ \underline{\mathcal{II}}_{[n]}(Q) \cap \overline{\underline{\mathcal{II}}(P)} = \emptyset\end{array}$$

notion of pre-emption defined in Section 3, the local notion here is much weaker since $\underline{\mathcal{I}\!\mathcal{I}}_{[m]}(P) \subseteq \underline{\mathcal{I}\!\mathcal{I}}(P)$, for all $m \in \mathcal{L}oc$ and $P \in \mathcal{P}$. In other words, local pre-emption does not pre-empt as many transitions as global pre-emption does. The difference between $\mathsf{CCS}^{\mathsf{sl}}$ and $\mathsf{CCS}^{\mathsf{sg}}$ semantics manifests itself in the side conditions of the rules for parallel composition with respect to unprioritized transitions. Since locations on different sides of parallel composition $P|Q$ are incomparable, $\underline{\tau}$'s arising from a location of P cannot pre-empt the execution of a transition, even an unprioritized one, of Q. Only if P engages in a prioritized synchronization with Q can unprioritized actions from a comparable location of P be pre-empted.

4.3. *Semantic theory based on strong bisimulation*

As in Section 3, we present an equivalence relation for $\mathsf{CCS}^{\mathsf{sl}}$ processes based on bisimulation [67]. Our aim is to characterize the largest congruence contained in the "naive" adaptation of strong bisimulation [58], which we refer to as *naive distributed prioritized strong bisimulation*, to our framework obtained by ignoring location information. The definition of naive distributed prioritized strong bisimulation, $\simeq$, is identical to the one presented in Definition 3.1, except that our transition relation $\xrightarrow{\gamma}$ is defined as mentioned above.

Although $\simeq$ is an equivalence, it is unfortunately – in contrast to Section 3.2 – not a congruence. The lack of compositionality is demonstrated by the following example, which embodies the traditional view that "*parallelism* = *nondeterminism*". We have $a.\underline{b}.\mathbf{0} + \underline{b}.a.\mathbf{0} \simeq a.\mathbf{0} \mid \underline{b}.\mathbf{0}$ but $(a.\underline{b}.\mathbf{0} + \underline{b}.a.\mathbf{0}) \mid \overline{\underline{b}}.\mathbf{0} \not\simeq (a.\mathbf{0} \mid \underline{b}.\mathbf{0}) \mid \overline{\underline{b}}.\mathbf{0}$; the latter process can perform an a-transition, while the corresponding a-transition of the former is pre-empted because the right process in the summation can engage in a prioritized communication. The above observation is not surprising since the distribution of processes influences the pre-emption of transitions and, consequently, the bisimulation. However, we know by Proposition 3.6 that $\simeq$ includes a largest congruence $\simeq^{+}$ for $\mathsf{CCS}^{\mathsf{sl}}$.

4.3.1. *Distributed prioritized strong bisimulation.* To develop a characterization of $\simeq^{+}$ we need to take *local* pre-emption into account.

DEFINITION 4.2. A symmetric relation $\mathcal{R} \subseteq \mathcal{P} \times \mathcal{P}$ is a *distributed prioritized strong bisimulation* if for every $\langle P, Q \rangle \in \mathcal{R}$, $\underline{\alpha} \in \underline{A}$, $\alpha \in A$, and $m \in \mathcal{L}oc$ the following conditions hold.

(1) $P \xrightarrow{\underline{\alpha}} P'$ implies $\exists Q'.\, Q \xrightarrow{\underline{\alpha}} Q'$ and $\langle P', Q' \rangle \in \mathcal{R}$.

(2) $P \xrightarrow{m,\alpha} P'$ implies $\exists Q', n.\, Q \xrightarrow{n,\alpha} Q'$, $\underline{\mathcal{I}\!\mathcal{I}}_{[n]}(Q) \subseteq \underline{\mathcal{I}\!\mathcal{I}}_{[m]}(P)$, and $\langle P', Q' \rangle \in \mathcal{R}$.

We write $P \simeq^{l} Q$ if $\langle P, Q \rangle \in \mathcal{R}$ for some distributed prioritized strong bisimulation $\mathcal{R}$.

Intuitively, the distributed prioritized initial action set of a process with respect to some location is a measure of the pre-emptive power of the process relative to that location. Thus, the second condition of Definition 4.2 states that an unprioritized action α from some location m of process P must be matched by the same action from some location n of Q and that the pre-emptive power of Q relative to n is at most as strong as the pre-emptive power of P relative to m. The next theorem is the main result of this section.

THEOREM 4.3. $\simeq^{\mathrm{l}}$ *is the* largest congruence *contained in* $\simeq$.

We refer to [55] for the proof; the context needed in the largest congruence proof is similar to the one used in Section 3.3.

4.3.2. *Axiomatic characterization.* In this section we present an axiomatization of $\simeq^{\mathrm{l}}$ for *finite* processes, for which we introduce a new binary summation operator $\oplus$ to the process algebra $\mathsf{CCS}^{\mathsf{sl}}$. This operator is called *distributed summation operator* and is needed for giving an *Expansion Axiom* (cf. Axiom (E) in Table 13). Its operational semantics is defined below and differs form the nondeterministic choice operator $+$ in that a location in its left argument is never comparable to one in its right argument.

$$\mathsf{dSum1}\ \frac{t \xrightarrow{\underline{\alpha}} t'}{t \oplus u \xrightarrow{\underline{\alpha}} t'} \qquad \mathsf{dSum1}\ \frac{t \xrightarrow{m,\alpha} t'}{t \oplus u \xrightarrow{m\cdot L,\alpha} t'}$$

$$\mathsf{dSum2}\ \frac{u \xrightarrow{\underline{\alpha}} u'}{t \oplus u \xrightarrow{\underline{\alpha}} u'} \qquad \mathsf{dSum2}\ \frac{u \xrightarrow{n,\alpha} u'}{t \oplus u \xrightarrow{n\cdot R,\alpha} u'}$$

It can easily be checked that $\simeq^{\mathrm{l}}$ is compositional with respect to $\oplus$.

Now, we turn to the axiom system for distributed prioritized strong bisimulation. We write $\vdash_E t = u$ if term t can be rewritten to u using the axioms in Tables 13 and 14 as well as Axioms (A1)–(A4), Axioms (Res1)–(Res4), Axioms (Rel1)–(Rel3), and Axiom (P) from Table 3. Axioms (Ic1), (D1), (S2), and (S3) involve side conditions. Regarding Axiom (Ic1), we introduce the unary predicate $\natural$ over processes (of the form $\sum_{j\in J} \gamma_j.t_j$, for some nonempty index set J) together with the following proof rules:

(i) $\natural\underline{\alpha}.t$, and

(ii) $\natural t$ and $\natural u$ implies $\natural(t+u)$.

Intuitively, $\natural(\sum_{j\in J} \gamma_j.t_j)$ if and only if $\gamma_j \in \underline{A}$ for all $j \in J$. The relation $\sqsubseteq_{\mathrm{i}}$ is defined as in Section 3.5 (cf. Table 8). The axioms in Table 13 are those given in Table 3 and augmented

Table 13
Axiomatization of $\simeq^{\mathrm{l}}$ (Axioms E)

(iA1)	$t \oplus u = u \oplus t$	(iA2)	$t \oplus (u \oplus v) = (t \oplus u) \oplus v$
(iA3)	$t \oplus t = t$	(iA4)	$t \oplus \mathbf{0} = t$

(E) $t \equiv \bigoplus_i \sum_j \gamma_{ij}.t_{ij}$ and $u \equiv \bigoplus_k \sum_l \delta_{kl}.u_{kl}$ implies

$$\begin{aligned} t \mid u = \textstyle\bigoplus_i \sum_j (\gamma_{ij}.(t_{ij} \mid u) &+ \textstyle\sum_k \sum_l \{\tau.(t_{ij} \mid u_{kl}) \mid \gamma_{ij} \equiv \overline{\delta}_{kl},\ \gamma_{ij}, \delta_{kl} \in A\} \\ &+ \textstyle\sum_k \sum_l \{\underline{\tau}.(t_{ij} \mid u_{kl}) \mid \gamma_{ij} \equiv \overline{\delta}_{kl},\ \gamma_{ij}, \delta_{kl} \in \underline{A}\}) \oplus \\ \textstyle\bigoplus_k \sum_l (\delta_{kl}.(t \mid u_{kl}) &+ \textstyle\sum_i \sum_j \{\tau.(t_{ij} \mid u_{kl}) \mid \gamma_{ij} \equiv \overline{\delta}_{kl},\ \gamma_{ij}, \delta_{kl} \in A\} \\ &+ \textstyle\sum_i \sum_j \{\underline{\tau}.(t_{ij} \mid u_{kl}) \mid \gamma_{ij} \equiv \overline{\delta}_{kl},\ \gamma_{ij}, \delta_{kl} \in \underline{A}\}) \end{aligned}$$

(iRes4)	$(t \oplus u) \setminus L = (t \setminus L) \oplus (u \setminus L)$	(iRel3)	$(t \oplus u)[f] = t[f] \oplus u[f]$

Table 14
Axioms E (continued)

(D1)	$(t \oplus t') + (u \oplus u') = ((t \oplus t') + u') \oplus ((u \oplus u') + t')$	
	$(\vdash_I t \sqsubseteq_{\mathrm{i}} t',\ \vdash_I u \sqsubseteq_{\mathrm{i}} u')$	
(D2)	$(t \oplus u) + \underline{\alpha}.v = (t + \underline{\alpha}.v) \oplus (u + \underline{\alpha}.v)$	
(Ic1)	$t \oplus \underline{\alpha}.u = t + \underline{\alpha}.u$	$(\natural t)$
(Ic2)	$(\underline{\alpha}.t + u) = (\underline{\alpha}.t + u) \oplus \underline{\alpha}.t$	
(S1)	$(t + \underline{\alpha}.u) \oplus (t' + \underline{\alpha}.u') = (t + \underline{\alpha}.u + \underline{\alpha}.u') \oplus (t' + \underline{\alpha}.u')$	
(S2)	$(t + \alpha.v) \oplus (u + \alpha.v) = (t + \alpha.v) \oplus u$	$(\vdash_I t \sqsubseteq_{\mathrm{i}} u)$
(S3)	$t \oplus u = t + u$	$(\vdash_I t =_{\mathrm{i}} u)$

with the corresponding axioms for the distributed summation operator. Moreover, the Expansion Axiom has been adapted for our algebra (cf. Axiom (E), where $\sum$ is the indexed version of $+$, and $\bigoplus$ is the indexed version of $\oplus$). Note that parallelism in $\mathsf{CCS}^{\mathsf{sl}}$ cannot be resolved in nondeterminism by using operator $+$ only, since priorities on different sides of $|$ are incomparable, but priorities on different sides of $+$ are comparable. This is the motivation for introducing operator $\oplus$. The axioms in Table 14 are new and show how we may "restructure" locations. They deal with the *distributivity* of the summation operators (Axioms (D1) and (D2)), the *interchangeability* of the summation operators (Axioms (Ic1) and (Ic2)), and the *saturation* of locations (Axioms (S1), (S2), and (S3)), respectively.

THEOREM 4.4. *Let t, u be finite processes. Then $\vdash_E t = u$ if and only if $t \simeq^{\mathrm{l}} u$.*

The proof of this theorem can be found in [28,55].

4.3.3. *Operational characterization.* The following definition introduces the equivalence $\simeq_*$ which characterizes $\simeq^{\mathrm{l}}$ as a standard strong bisimulation [55]. It uses the notation $P \xrightarrow[L]{\alpha} P'$, for $P, P' \in \mathcal{P}$, $\alpha \in A$, and $L \subseteq \underline{A} \setminus \{\underline{\tau}\}$, whenever $\exists m \in \mathcal{L}oc.\ P \xrightarrow{m,\alpha} P'$ and $\underline{\mathcal{II}}_{[m]}(P) \subseteq L$. Note that these enriched transitions take local pre-emption potentials into account, thereby avoiding the explicit annotation of transitions with locations.

DEFINITION 4.5. A symmetric relation $\mathcal{R} \subseteq \mathcal{P} \times \mathcal{P}$ is an *alternative distributed prioritized strong bisimulation* if for every $\langle P, Q\rangle \in \mathcal{R}$, $\alpha \in A$, $\underline{\alpha} \in \underline{A}$, and $L \subseteq \underline{A} \setminus \{\underline{\tau}\}$ the following conditions hold.

(1) $P \xrightarrow{\underline{\alpha}} P'$ implies $\exists Q'.\ Q \xrightarrow{\underline{\alpha}} Q'$ and $\langle P', Q'\rangle \in \mathcal{R}$.

(2) $P \xrightarrow[L]{\alpha} P'$ implies $\exists Q'.\ Q \xrightarrow[L]{\alpha} Q'$ and $\langle P', Q'\rangle \in \mathcal{R}$.

We write $P \simeq_* Q$ if $\langle P, Q\rangle \in \mathcal{R}$ for some alternative distributed prioritized strong bisimulation $\mathcal{R}$.

Similar to Section 3.3.3 we obtain an operational characterization result.

THEOREM 4.6 (Operational characterization). $\simeq^{\mathrm{l}} = \simeq_*$.

4.4. *Semantic theory based on weak bisimulation*

As for $\mathsf{CCS}^{\mathsf{sg}}$, we develop a coarser bisimulation-based congruence by abstracting from internal actions. We start off with the definition of a *naive distributed prioritized weak bisimulation* which is an adaptation of observational equivalence [58].

DEFINITION 4.7 (*Naive distributed prioritized weak transitions*).
(1) $\hat{\gamma} =_{\mathrm{df}} \varepsilon$, if $\gamma \in \{\underline{\tau}, \tau\}$, and $\hat{\gamma} =_{\mathrm{df}} \gamma$, otherwise.
(2) $\stackrel{\varepsilon}{\Longrightarrow}_{\times} =_{\mathrm{df}} (\stackrel{\underline{\tau}}{\longrightarrow} \cup \bigcup \{\stackrel{m,\tau}{\longrightarrow} \mid m \in \mathcal{L}oc\})^*$.
(3) $\stackrel{\alpha}{\Longrightarrow}_{\times} =_{\mathrm{df}} \stackrel{\varepsilon}{\Longrightarrow}_{\times} \circ \stackrel{\alpha}{\longrightarrow} \circ \stackrel{\varepsilon}{\Longrightarrow}_{\times}$.
(4) $\stackrel{m,\alpha}{\Longrightarrow}_{\times} =_{\mathrm{df}} \stackrel{\varepsilon}{\Longrightarrow}_{\times} \circ \stackrel{m,\alpha}{\longrightarrow} \circ \stackrel{\varepsilon}{\Longrightarrow}_{\times}$.

In the following we write $P \stackrel{\alpha}{\Longrightarrow}_{\times} P'$ for $\exists m \in \mathcal{L}oc.\ P \stackrel{m,\alpha}{\Longrightarrow}_{\times} P'$.

DEFINITION 4.8 (*Naive distributed prioritized weak bisimulation*). A symmetric relation $\mathcal{R} \subseteq \mathcal{P} \times \mathcal{P}$ is a *naive distributed prioritized weak bisimulation* if for every $\langle P, Q\rangle \in \mathcal{R}$ and $\gamma \in \mathcal{A}$ the following condition holds.

$$P \stackrel{\gamma}{\longrightarrow} P' \quad \text{implies} \quad \exists Q'.\ Q \stackrel{\hat{\gamma}}{\Longrightarrow}_{\times} Q' \text{ and } \langle P', Q'\rangle \in \mathcal{R}.$$

We write $P \approx_{\times} Q$ if $\langle P, Q\rangle \in \mathcal{R}$ for some naive distributed prioritized weak bisimulation $\mathcal{R}$.

It is fairly easy to see that $\approx_{\times}$ is not a congruence for $\mathsf{CCS}^{\mathsf{sl}}$. One compositionality defect arises with respect to parallel composition and is similar to the one mentioned for naive distributed prioritized strong bisimulation. Another defect, which is carried over from CCS, is concerned with the summation operators.

4.4.1. *Distributed prioritized weak bisimulation.* We devote the rest of Section 4.4 to operationally characterizing the largest congruence contained in naive distributed prioritized weak bisimulation. To do so, we first re-define the weak transition relation.

DEFINITION 4.9 (*Distributed prioritized weak transitions*). For $L, M \subseteq \underline{\mathcal{A}} \setminus \{\underline{\tau}\}$ we define the following notations.
(1) $\hat{\underline{\tau}} =_{\mathrm{df}} \varepsilon$, $\hat{\underline{a}} =_{\mathrm{df}} \underline{a}$, $\hat{\tau} =_{\mathrm{df}} \varepsilon$, and $\hat{a} =_{\mathrm{df}} a$.
(2) $P \xrightarrow[L]{m,\alpha} P'$ if $P \stackrel{m,\alpha}{\longrightarrow} P'$ and $\underline{\mathcal{I}\!\mathcal{I}}_{[m]}(P) \subseteq L$.
(3) $\stackrel{\varepsilon}{\Longrightarrow} =_{\mathrm{df}} (\stackrel{\underline{\tau}}{\longrightarrow} \cup \bigcup \{\xrightarrow[\emptyset]{m,\tau} \mid m \in \mathcal{L}oc\})^*$.
(4) $\stackrel{\alpha}{\Longrightarrow} =_{\mathrm{df}} \stackrel{\varepsilon}{\Longrightarrow} \circ \stackrel{\alpha}{\longrightarrow} \circ \stackrel{\varepsilon}{\Longrightarrow}$.
(5) $\xRightarrow[L]{\varepsilon} =_{\mathrm{df}} (\stackrel{\underline{\tau}}{\longrightarrow} \cup \bigcup \{\xrightarrow[L]{m,\tau} \mid m \in \mathcal{L}oc\})^*$.
(6) $P \xRightarrow[L,M]{m,\alpha} P'$ if $\exists P'', P'''.\ P \xRightarrow[L]{\varepsilon} P'' \xrightarrow[L]{m,\alpha} P''' \stackrel{\varepsilon}{\Longrightarrow} P'$ and $\underline{\mathcal{I}\!\mathcal{I}}(P'') \subseteq M$.

$P \xrightarrow[L]{m,\alpha} P'$ means that P can engage in action α at location m to P' *provided that* the environment does not offer a prioritized communication involving actions in L. If the environment were to offer such a communication, the result would be a $\underline{\tau}$ at a location comparable to m in P, which would pre-empt the α. Similarly, $P \stackrel{\varepsilon}{\Longrightarrow} P'$ holds if P can evolve to P' via a nonpre-emptable sequence of internal transitions, regardless of the environment's behavior. These internal transitions should therefore involve either $\underline{\tau}$, which can never be pre-empted, or τ, in which case no prioritized actions should be enabled at the same location. Likewise, $P \xRightarrow[L]{\varepsilon} P'$ means that, so long as the environment does not offer to synchronize with P on a prioritized action in L, process P may engage in a sequence of internal computation steps and become P'. Finally, the M-parameter in $\xRightarrow[L,M]{m,\alpha}$ provides a measure of the pre-emptive impact that a process has on its environment. $P \xRightarrow[L,M]{m,\alpha} P'$ is true if P can engage in some internal computation followed by α, so long as the environment refrains from synchronizations in L, and then engage in some nonpre-emptable internal computation to arrive at P'. In addition, the state at which α is enabled should only offer prioritized communications in M.

Note that the definition of $P \xRightarrow[L]{\varepsilon} P'$ is in accordance with our intuition that internal actions, and therefore their locations, are unobservable. Moreover, the environment of P is not influenced by internal actions performed by P, since priorities arising from different sides of the parallel composition operator are incomparable. Therefore, the parameter M is unnecessary in the definition of the relation $\xRightarrow[L]{\varepsilon}$. Finally, for notational convenience, $\xRightarrow[L,M]{m,\varepsilon}$ is interpreted as $\xRightarrow[L]{\varepsilon}$.

DEFINITION 4.10 (*Distributed prioritized weak bisimulation*). A symmetric relation $\mathcal{R} \subseteq \mathcal{P} \times \mathcal{P}$ is a *distributed prioritized weak bisimulation* if for every $\langle P, Q\rangle \in \mathcal{R}$, $\alpha \in A$, $\underline{\alpha} \in \underline{A}$, and $m \in \mathcal{L}oc$ the following conditions hold.

(1) $\exists Q', Q''.\, Q \stackrel{\varepsilon}{\Longrightarrow} Q'' \stackrel{\varepsilon}{\Longrightarrow} Q'$, $\underline{\mathit{II}}(Q'') \subseteq \underline{\mathit{II}}(P)$, and $\langle P, Q'\rangle \in \mathcal{R}$.

(2) $P \stackrel{\underline{\alpha}}{\longrightarrow} P'$ implies $\exists Q'.\, Q \stackrel{\hat{\underline{\alpha}}}{\Longrightarrow} Q'$, and $\langle P', Q'\rangle \in \mathcal{R}$.

(3) $P \stackrel{m,\alpha}{\longrightarrow} P'$ implies $\exists Q', n.\, Q \xRightarrow[L,M]{n,\hat{\alpha}} Q'$, $L = \underline{\mathit{II}}_{[m]}(P)$, $M = \underline{\mathit{II}}(P)$, and $\langle P', Q'\rangle \in \mathcal{R}$.

We write $P \underline{\approx} Q$ if $\langle P, Q\rangle \in \mathcal{R}$ for some distributed prioritized weak bisimulation $\mathcal{R}$.

Condition (4.10) of Definition 4.10 guarantees that distributed prioritized weak bisimulation is compositional with respect to parallel composition. Its necessity is best illustrated by the following example. The processes $P \stackrel{\text{def}}{=} \underline{\tau}.\underline{a}.\mathbf{0}$ and $Q \stackrel{\text{def}}{=} \underline{a}.\mathbf{0}$ would be considered equivalent if Condition (4.10) were absent. However, the context $C[X] \stackrel{\text{def}}{=} X \mid (\overline{\underline{a}}.\mathbf{0} + b.\mathbf{0})$ distinguishes them. The following proposition is the CCS^{sl} equivalent of Proposition 3.9.

PROPOSITION 4.11. *The equivalence relation* $\underline{\approx}$ *is a congruence with respect to prefixing, parallel composition, relabeling, and restriction. Moreover,* $\underline{\approx}$ *is characterized as the largest congruence contained in* $\approx_\times$, *in the sub-algebra of* $\mathsf{CCS}^{\mathsf{sl}}$ *induced by these operators and recursion.*

4.4.2. *Distributed prioritized observational congruence.* As in Section 3, the summation fix presented in [58] is not sufficient in order to achieve a congruence relation.

DEFINITION 4.12. We define $P \underline{\approx}^{\mathsf{l}} Q$ if for all $\alpha \in A$, $\underline{\alpha} \in \underline{A}$, and $m \in \mathcal{L}oc$ the following conditions and their symmetric counterparts hold.

(1) $\underline{\mathcal{I}}(P) \supseteq \underline{\mathcal{I}}(Q)$

(2) $P \xrightarrow{\alpha} P'$ implies $\exists Q'.\, Q \stackrel{\alpha}{\Longrightarrow} Q'$ and $P' \underline{\approx} Q'$.

(3) $P \xrightarrow{m,\underline{\alpha}} P'$ implies $\exists Q', n.\, Q \underset{L,M}{\overset{n,\underline{\alpha}}{\Longrightarrow}} Q'$, $L = \underline{\mathcal{II}}_{[m]}(P)$, $M = \underline{\mathcal{II}}(P)$, and $P' \underline{\approx} Q'$.

Now, we can state the main result of this section, which can be proved by using the technique already presented in Section 3.3.2; see [55] for details.

THEOREM 4.13. $\underline{\approx}^{\mathsf{l}}$ *is the largest congruence contained in* $\approx_\times$.

4.4.3. *Operational characterization.* We now characterize distributed prioritized weak bisimulation as standard bisimulation over an appropriately defined transition relation. To begin with, we introduce a family of relations $\underset{M}{\Longrightarrow}$ on processes, where $M \subseteq \underline{A} \setminus \{\underline{\tau}\}$, by defining $P \underset{M}{\Longrightarrow} P'$ if $\exists P''.\, P \stackrel{\varepsilon}{\Longrightarrow} P'' \stackrel{\varepsilon}{\Longrightarrow} P'$ and $\underline{\mathcal{II}}(P'') \subseteq M$. Moreover, we write $P \underset{L,M}{\overset{\hat{\underline{\alpha}}}{\Longrightarrow}} P'$ whenever there exists some $m \in \mathcal{L}oc$ such that $P \underset{L,M}{\overset{m,\hat{\underline{\alpha}}}{\Longrightarrow}} P'$.

DEFINITION 4.14. A symmetric relation $\mathcal{R} \subseteq \mathcal{P} \times \mathcal{P}$ is an *alternative distributed prioritized weak bisimulation* if for every $\langle P, Q\rangle \in \mathcal{R}$, $\alpha \in A$, $\underline{\alpha} \in \underline{A}$, and $L, M \subseteq \underline{A} \setminus \{\underline{\tau}\}$ the following conditions hold.

(1) $P \underset{M}{\Longrightarrow} P'$ implies $\exists Q'.\, Q \underset{M}{\Longrightarrow} Q'$ and $\langle P', Q'\rangle \in \mathcal{R}$.

(2) $P \stackrel{\hat{\alpha}}{\Longrightarrow} P'$ implies $\exists Q'.\, Q \stackrel{\hat{\alpha}}{\Longrightarrow} Q'$ and $\langle P', Q'\rangle \in \mathcal{R}$.

(3) $P \underset{L,M}{\overset{\hat{\underline{\alpha}}}{\Longrightarrow}} P'$ implies $\exists Q'.\, Q \underset{L,M}{\overset{\hat{\underline{\alpha}}}{\Longrightarrow}} Q'$ and $\langle P', Q'\rangle \in \mathcal{R}$.

We write $P \underline{\approx}_* Q$ if $\langle P, Q\rangle \in \mathcal{R}$ for some alternative distributed prioritized weak bisimulation $\mathcal{R}$.

THEOREM 4.15 (Operational characterization). $\underline{\approx} = \underline{\approx}_*$.

The interested reader can find the proof of this theorem in [55].

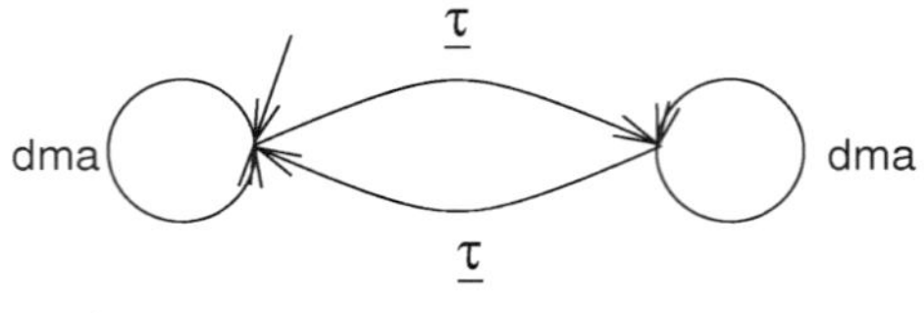

Fig. 5. Semantics of the dma-system Sys.

4.5. *Example*

We now return to the direct-memory-access example system introduced above. The $\mathsf{CCS}^{\mathsf{sl}}$ semantics of Sys, which corresponds to our intuition regarding distributed systems, is given in Figure 5 where we leave out the locations in the transitions' labels.

Recall that the application uses the two memory blocks alternately. Thus, the DMA is expected to be allowed to access the free memory block. Accordingly, the specification of the system can be formalized by $\mathsf{Spec} \stackrel{\text{def}}{=} \mathsf{dma.Spec}$. It is easy to see that the symmetric closure of

$$\{\langle \mathsf{Spec}, \mathsf{Sys}\rangle, \langle \mathsf{Spec}, (\overline{\underline{\mathsf{fetch2}}}.\mathsf{Appl} \mid \mathsf{Block1} \mid \mathsf{Block2}) \setminus \{\underline{\mathsf{fetch1}}, \underline{\mathsf{fetch2}}\}\rangle\}$$

is a distributed prioritized weak bisimulation. Therefore, $\mathsf{Spec} \underline{\approx} \mathsf{Sys}$ as expected, i.e., system Sys meets its specification Spec.

4.6. *Discussion of alternative design decisions*

Up to now we have restricted the number of priority levels in $\mathsf{CCS}^{\mathsf{sl}}$ to two and communication to complementary actions having the same priority. In this section we study the implications of the removal of these restrictions, which leads to a new version of $\mathsf{CCS}^{\mathsf{sl}}$, called $\mathsf{CCS}^{\mathsf{sl}}_{\mathsf{ml}}$ ($\mathsf{CCS}^{\mathsf{sl}}$ with a multi-level priority scheme), that is formally defined in the next section.

Allowing communication between unprioritized actions and complementary prioritized actions raises the question of whether the resulting internal action should be τ or $\underline{\tau}$ (cf. Section 3.7). When dealing with local pre-emption, this decision has no important consequences for sequential communicating processes, i.e., those in *standard concurrent form* [58]; however, it is of obvious importance for processes like $(\underline{a}.\mathbf{0} \mid \overline{a}.\mathbf{0}) + b.\mathbf{0}$ in which one has to decide if the b-transition is enabled. One reasonable view is that a communication should be pre-empted whenever one communication partner is pre-empted, i.e., it cannot engage in a communication. This implies that the *minimal* priority of the complementary actions ought to be assigned to the internal action. To reflect this in the operational semantics, one could replace Rules (Com1), (Com2), and (Com3) for parallel composition by the ones presented in Table 15 plus their symmetric counterparts. The side conditions involve sets $\mathcal{I\!I}(P)$ that include all unprioritized visible actions in which P can initially engage.

Table 15
Modified operational rules

$$\text{Com1}\quad \frac{P \xrightarrow{m,\alpha} P'}{P|Q \xrightarrow{m\cdot L,\alpha} P'|Q}\; \underline{\underline{\mathit{\Pi}}}_{[m]}(P) \cap (\overline{\underline{\mathit{\Pi}}(Q)} \cup \overline{\mathit{\Pi}(Q)}) = \emptyset$$

$$\text{Com3a}\quad \frac{P \xrightarrow{m,a} P' \quad Q \xrightarrow{n,\overline{a}} Q'}{P \mid Q \xrightarrow{\langle m\cdot L, n\cdot R\rangle,\tau} P'|Q'}\; \begin{array}{l} \underline{\underline{\mathit{\Pi}}}_{[m]}(P) \cap (\overline{\underline{\mathit{\Pi}}(Q)} \cup \overline{\mathit{\Pi}(Q)}) = \emptyset \quad \text{and} \\ \underline{\underline{\mathit{\Pi}}}_{[n]}(Q) \cap (\overline{\underline{\mathit{\Pi}}(P)} \cup \overline{\mathit{\Pi}(P)}) = \emptyset \end{array}$$

$$\text{Com3b}\quad \frac{P \xrightarrow{m,\underline{a}} P' \quad Q \xrightarrow{n,\overline{a}} Q'}{P \mid Q \xrightarrow{\langle m\cdot L, n\cdot R\rangle,\tau} P' \mid Q'}\; \underline{\underline{\mathit{\Pi}}}_{[n]}(Q) \cap (\overline{\underline{\mathit{\Pi}}(P)} \cup \overline{\mathit{\Pi}(P)}) = \emptyset$$

It turns out that the largest congruence results concerning distributed prioritized strong bisimulation and distributed prioritized observational congruence can be carried over to the new calculus. However, the new semantics has algebraic shortcomings since parallel composition is *not associative*, as illustrated by the following example. Consider process $(b.\mathbf{0} + \underline{a}.\mathbf{0}) \mid (\overline{a}.\mathbf{0} + \underline{c}.\mathbf{0}) \mid \overline{c}.\mathbf{0}$. In the left-associative case, the initial b-transition is pre-empted, according to Rule (Com1) since $\underline{a}$ may potentially communicate with $\overline{a}$. In the right-associative case, on the other hand, the $\overline{a}$-transition in the second component is pre-empted, and consequently the b-transition is enabled by Rule (Com1). The reason for this problem is that transitions are pre-empted because a process can *potentially* engage in a higher prioritized communication from a comparable location. However, this potential communication cannot take place if the communication partner is itself pre-empted. The same problem also arises when extending $\mathsf{CCS}^{\mathsf{sl}}$ to multiple priority levels, even if communication is only allowed on complementary actions of the same priority, as can be observed by using a slight adaptation of the previous example: $(b:2.\mathbf{0} + a:1.\mathbf{0}) \mid (\overline{a}:1.\mathbf{0} + c:0.\mathbf{0}) \mid \overline{c}:0.\mathbf{0}$.

One can imagine two approaches to fixing the problems with the first (and second) alteration to the theory. One is to change the operational semantics; in particular, the side conditions could be weakened such that an unprioritized transition is only pre-empted when a prioritized action from a comparable location can *actually* engage in a communication. This approach has not yet been investigated in the literature. The second solution follows an idea developed in [24] for a different setting and involves the use of a syntax restriction on processes prohibiting output actions, i.e., actions in $\overline{\Lambda}$, from occurring as initial actions that are in the scope of $+$. Hence, all potential communication partners are also actual ones, and the standard side conditions for parallel composition are sufficient to encode the desired notion of pre-emption. It is important to mention that the proposed syntax restriction still allows one to specify many practically relevant examples within the calculus. Indeed, a similar restriction may be found in the programming language `occam` [45].

4.7. *Extension to multi-level priority schemes*

For $\mathsf{CCS}^{\mathsf{sl}}_{\mathsf{ml}}$ we allow a multi-level priority scheme and communication between complementary actions having potentially different priorities. As seen in the previous section, both of these relaxations yield a semantics for which parallel composition is not associative. However, we have also argued that this problem vanishes if the syntax is restricted such that output actions can never be pre-empted. We adapt the syntax restriction proposed by Camilleri and Winskel [24], which states that initial actions in the scope of a summation operator must be input actions. Therefore, input and output actions are explicitly distinguished in $\mathsf{CCS}^{\mathsf{sl}}_{\mathsf{ml}}$, with the internal action τ being treated as an input action. In the following, we let $a, b, \ldots$ range over the set Λ of input ports and $\overline{a}, \overline{b}, \ldots$ over the set $\overline{\Lambda}$ of output ports. Moreover, we let γ stand for any input action and let α be a representative of $\mathcal{A} =_{\mathrm{df}} \Lambda \cup \overline{\Lambda} \cup \{\tau\}$. Since in the restricted syntax priority values of output actions need never be compared with other priority values, there are no priority values associated with output actions at all. The syntax of $\mathsf{CCS}^{\mathsf{sl}}_{\mathsf{ml}}$ is formally defined by the following BNF for P.

$$\begin{array}{lcl} I & ::= & \mathbf{0} \mid x \mid \gamma : k.I \mid I + I \mid I \oplus I \mid I \mid I \mid I[f] \mid I \setminus L \mid \mu x.I \\ P & ::= & \mathbf{0} \mid x \mid \alpha : k.P \mid I + I \mid P \oplus P \mid P \mid P \mid P[f] \mid P \setminus L \mid \mu x.P \end{array}$$

Here, f is an *injective*, finite relabeling, $L \subseteq \Lambda \cup \overline{\Lambda}$ is a restriction set, and x is a variable taken from a countable domain $\mathcal{V}$. A relabeling satisfies the properties $f(\Lambda) \subseteq \Lambda$, $f(\overline{\Lambda}) \subseteq \overline{\Lambda}$, $f(\tau) = \tau$, and $f(\overline{a}) = \overline{f(a)}$. Thus, in addition to the requirements of a finite relabeling in CCS, relabelings in $\mathsf{CCS}^{\mathsf{sl}}_{\mathsf{ml}}$ may only map input ports to input ports and output ports to output ports. Since actions attached with different priority values do not represent different ports here, relabelings and restriction sets do not deal with priority values. Especially, the priority value of a relabeled transition remains the same, i.e., there is no explicit or implicit mechanism for prioritization or deprioritization (cf. Section 3.5). In the remainder, we let $\mathcal{P}^{\mathsf{sl}}_{\mathsf{ml}}$ denote the set of all $\mathsf{CCS}^{\mathsf{sl}}_{\mathsf{ml}}$ processes.

The semantics of $\mathsf{CCS}^{\mathsf{sl}}_{\mathsf{ml}}$ processes are again labeled transition systems whose transition relations are specified by operational rules. Since transitions labeled by output actions are never pre-empted they do not need to be associated with locations. We first present two auxiliary sets that are useful for presenting the operational rules:

(i) initial output action sets $\overline{\mathbb{I}}(P)$ of a process P, and
(ii) initial input action sets $\mathrm{I}^k_m(P)$ of P with respect to a priority value k and a location m, which are defined to be the least sets satisfying the equations presented in Tables 16 and 17, respectively.

For technical convenience, we remove the complement of output actions in the definition of $\overline{\mathbb{I}}(\cdot)$, and we use the following four abbreviations:

(i) $\mathbb{I}^{<k}_M(P) =_{\mathrm{df}} \bigcup\{\mathrm{I}^l_m(P) \mid m \in M,\ l < k\}$,
(ii) $\mathbb{I}^{<k}_M(P) =_{\mathrm{df}} \mathrm{I}^{<k}_M(P) \setminus \{\tau\}$,

Table 16
Initial output action sets for $\mathsf{CCS}^{\mathsf{sl}}_{\mathsf{ml}}$

$\overline{\mathbb{I}}(\mu x.P)$	$=$	$\overline{\mathbb{I}}(P[\mu x.P/x])$	$\overline{\mathbb{I}}(\overline{a}.P)$	$=$	$\{a\}$
$\overline{\mathbb{I}}(P \mid Q)$	$=$	$\overline{\mathbb{I}}(P) \cup \overline{\mathbb{I}}(Q)$	$\overline{\mathbb{I}}(P \oplus Q)$	$=$	$\overline{\mathbb{I}}(P) \cup \overline{\mathbb{I}}(Q)$
$\overline{\mathbb{I}}(P[f])$	$=$	$\{f(a) \mid a \in \overline{\mathbb{I}}(P)\}$	$\overline{\mathbb{I}}(P \setminus L)$	$=$	$\overline{\mathbb{I}}(P) \setminus (L \cup \overline{L})$

Table 17
Initial input action sets for $\mathsf{CCS}^{\mathsf{sl}}_{\mathsf{ml}}$

$\mathrm{I}^k_m(\mu x.P)$	$= \mathrm{I}^k_m(P[\mu x.P/x])$	$\mathrm{I}^k_\bullet(\gamma:l.P)$	$= \{\gamma \mid k = l\}$
$\mathrm{I}^k_{m\cdot l}(P+Q)$	$= \mathrm{I}^k_m(P)$	$\mathrm{I}^k_{m\cdot L}(P \oplus Q)$	$= \mathrm{I}^k_m(P)$
$\mathrm{I}^k_{n\cdot r}(P+Q)$	$= \mathrm{I}^k_n(Q)$	$\mathrm{I}^k_{n\cdot R}(P \oplus Q)$	$= \mathrm{I}^k_n(Q)$
$\mathrm{I}^k_m(P[f])$	$= \{f(\gamma) \mid \gamma \in \mathrm{I}^k_m(P)\}$	$\mathrm{I}^k_{m\cdot L}(P \mid Q)$	$= \mathrm{I}^k_m(P) \cup \{\tau \mid \mathrm{I}^k_m(P) \cap \overline{\mathbb{II}}(Q) \neq \emptyset\}$
$\mathrm{I}^k_m(P \setminus L)$	$= \mathrm{I}^k_m(P) \setminus (L \cup \overline{L})$	$\mathrm{I}^k_{n\cdot R}(P \mid Q)$	$= \mathrm{I}^k_n(Q) \cup \{\tau \mid \mathrm{I}^k_n(Q) \cap \overline{\mathbb{II}}(P) \neq \emptyset\}$

Table 18
Operational semantics for $\mathsf{CCS}^{\mathsf{sl}}_{\mathsf{ml}}$ with respect to output transitions

$\overline{\mathsf{Act}}$	$\frac{-}{\overline{a}.P \xrightarrow{\overline{a}} P}$	iSum1	$\frac{P \xrightarrow{\overline{a}} P'}{P \oplus Q \xrightarrow{\overline{a}} P'}$	$\overline{\mathsf{Com1}}$	$\frac{P \xrightarrow{\overline{a}} P'}{P \mid Q \xrightarrow{\overline{a}} P' \mid Q}$
$\overline{\mathsf{Rel}}$	$\frac{P \xrightarrow{\overline{a}} P'}{P[f] \xrightarrow{f(\overline{a})} P'[f]}$	$\overline{\mathsf{iSum2}}$	$\frac{Q \xrightarrow{\overline{a}} Q'}{P \oplus Q \xrightarrow{\overline{a}} Q'}$	$\overline{\mathsf{Com2}}$	$\frac{Q \xrightarrow{\overline{a}} Q'}{P \mid Q \xrightarrow{\overline{a}} P \mid Q'}$
$\overline{\mathsf{Rec}}$	$\frac{P[\mu x.P/x] \xrightarrow{\overline{a}} P'}{\mu x.P \xrightarrow{\overline{a}} P'}$	$\overline{\mathsf{Res}}$	$\frac{P \xrightarrow{\overline{a}} P'}{P \setminus L \xrightarrow{\overline{a}} P' \setminus L}\ \overline{a} \notin L \cup \overline{L}$		

(iii) $\mathrm{I}(P) =_{\mathrm{df}} \bigcup\{\mathrm{I}^l_m(P) \mid m \in \mathcal{L}oc,\ l \in \mathcal{N}\}$, and
(iv) $\mathbb{II}(P) =_{\mathrm{df}} \mathrm{I}(P) \setminus \{\tau\}$.

The operational rules for $\mathsf{CCS}^{\mathsf{sl}}_{\mathsf{ml}}$ semantics are formally stated in Table 18 for *output transitions* and in Table 19 for *input transitions*. As expected, the rules for output transitions coincide with the ones for plain CCS [58], whereas the rules for input transitions take local pre-emption into account, thereby using location and priority value information in their side conditions. It is worth taking a closer look at the side conditions of Rules (Sum1) and (Sum2) which differ in principle from the corresponding ones in $\mathsf{CCS}^{\mathsf{sl}}$. They guarantee that an initial $\gamma:l$-transition of a process P is also pre-empted whenever there exists a higher prioritized initial $\gamma:k$-transition of P, i.e., if $k < l$. This new kind of pre-emption reflects that output transitions can communicate with a complementary input transition regardless of its priority value; i.e., if more than one communication partner offering the matching input transition is available from a comparable location, then the one having the highest priority is taken. For this notion of pre-emption to be well-defined, relabelings must be restricted to injective ones, as is pointed out in [24].

The behavioral relations defined for $\mathsf{CCS}^{\mathsf{sl}}$ can be adapted to $\mathsf{CCS}^{\mathsf{sl}}_{\mathsf{ml}}$ in a straightforward fashion, as we demonstrate by the notion of distributed prioritized strong bisimulation.

DEFINITION 4.16. A symmetric relation $\mathcal{R} \subseteq \mathcal{P} \times \mathcal{P}$ is a *distributed prioritized strong bisimulation for* $\mathsf{CCS}^{\mathsf{sl}}_{\mathsf{ml}}$ if for every $\langle P, Q\rangle \in \mathcal{R}$, $\overline{a} \in \overline{\Lambda}$, $\gamma \in \Lambda \cup \{\tau\}$, $k \in \mathcal{N}$, and $m \in \mathcal{L}oc$ the following conditions hold.

Table 19
Operational semantics for $\mathsf{CCS}^{\mathsf{sl}}_{\mathsf{ml}}$ with respect to input transitions

Act	$\dfrac{-}{\gamma:k.P \xrightarrow{\bullet,\gamma:k} P}$	Sum1	$\dfrac{P \xrightarrow{m,\gamma:k} P'}{P+Q \xrightarrow{m\cdot l,\gamma:k} P'}\quad \tau,\gamma \notin \mathrm{I}^{<k}(Q)$
iSum1	$\dfrac{P \xrightarrow{m,\gamma:k} P'}{P \oplus Q \xrightarrow{m\cdot L,\gamma:k} P'}$	Sum2	$\dfrac{Q \xrightarrow{n,\gamma:k} Q'}{P+Q \xrightarrow{n\cdot r,\gamma:k} Q'}\quad \tau,\gamma \notin \mathrm{I}^{<k}(P)$
iSum2	$\dfrac{Q \xrightarrow{n,\gamma:k} Q'}{P \oplus Q \xrightarrow{n\cdot R,\gamma:k} Q'}$	Com1	$\dfrac{P \xrightarrow{m,\gamma:k} P'}{P \mid Q \xrightarrow{m\cdot L,\gamma:k} P' \mid Q}\quad \mathbb{I}^{<k}_{[m]}(P) \cap \overline{\mathbb{I}}(Q) = \emptyset$
Rel	$\dfrac{P \xrightarrow{m,\gamma:k} P'}{P[f] \xrightarrow{m,f(\gamma):k} P'[f]}$	Com2	$\dfrac{Q \xrightarrow{n,\gamma:k} Q'}{P \mid Q \xrightarrow{n\cdot R,\gamma:k} P \mid Q'}\quad \mathbb{I}^{<k}_{[n]}(Q) \cap \overline{\mathbb{I}}(P) = \emptyset$
Rec	$\dfrac{P[\mu x.P/x] \xrightarrow{m,\gamma:k} P'}{\mu x.P \xrightarrow{m,\gamma:k} P'}$	Com3	$\dfrac{P \xrightarrow{m,a:k} P' \quad Q \xrightarrow{\overline{a}} Q'}{P \mid Q \xrightarrow{m\cdot L,\tau:k} P' \mid Q'}\quad \mathbb{I}^{<k}_{[m]}(P) \cap \overline{\mathbb{I}}(Q) = \emptyset$
		Com4	$\dfrac{P \xrightarrow{\overline{a}} P' \quad Q \xrightarrow{n,a:k} Q'}{P \mid Q \xrightarrow{n\cdot R,\tau:k} P' \mid Q'}\quad \mathbb{I}^{<k}_{[n]}(Q) \cap \overline{\mathbb{I}}(P) = \emptyset$
		Res	$\dfrac{P \xrightarrow{m,\gamma:k} P'}{P \setminus L \xrightarrow{m,\gamma:k} P' \setminus L}\quad \gamma \notin L \cup \overline{L}$

(1) $P \xrightarrow{\overline{a}} P'$ implies $\exists Q'.\, Q \xrightarrow{\overline{a}} Q'$ and $\langle P', Q'\rangle \in \mathcal{R}$, and
(2) $P \xrightarrow{m,\gamma:k} P'$ implies $\exists Q', l, n.\, Q \xrightarrow{n,\gamma:l} Q'$, $\mathbb{I}^{<l}_{[n]}(Q) \subseteq \mathbb{I}^{<k}_{[m]}(P)$, and $\langle P', Q'\rangle \in \mathcal{R}$.

We write $P \simeq_{\mathsf{ml}} Q$ if $\langle P, Q\rangle \in \mathcal{R}$ for some distributed prioritized strong bisimulation $\mathcal{R}$ for $\mathsf{CCS}^{\mathsf{sl}}_{\mathsf{ml}}$.

PROPOSITION 4.17. *The relation $\simeq_{\mathsf{ml}}$ is compositional with respect to all operators except summation.*

The proof is by standard techniques [58] and, therefore, is omitted here. The reason for the lack of compositionality with respect to summation is illustrated by the following example: $a:0.\mathbf{0} \simeq_{\mathsf{ml}} a:1.\mathbf{0}$ holds, but $a:0.\mathbf{0} + \tau:0.\mathbf{0} \not\simeq_{\mathsf{ml}} a:1.\mathbf{0} + \tau:0.\mathbf{0}$ since the former process can engage in a transition labeled by action a whereas the latter cannot. This defect can easily be repaired (note the analogy with weak bisimulation [58]) but is not important in the remainder.

4.8. *Camilleri and Winskel's approach*

In this section we briefly review Camilleri and Winskel's approach to priority [24], which we refer to as $\mathsf{CCS}^{\mathsf{cw}}$ (CCS with priority due to Camilleri and Winskel). In contrast to the approaches considered so far, this process algebra with priority does not assign priority values to actions. Instead, there exists a special summation operator $+\!\rangle$, called *prioritized choice*, which favors its left argument over its right one. The syntax of $\mathsf{CCS}^{\mathsf{cw}}$ is given by the following BNF for P.

$$\begin{array}{lll} I & ::= & \mathbf{0} \mid x \mid \gamma.I \mid I+\!\rangle I \mid I+I \mid I \mid I \mid I[f] \mid I\setminus L \mid \mu x.I \\ P & ::= & \mathbf{0} \mid x \mid \alpha.P \mid I+\!\rangle I \mid P+P \mid P \mid P \mid P[f] \mid P\setminus L \mid \mu x.P \end{array}$$

Here, the action γ, the injective, finite relabeling f, and the restriction set L satisfy the constraints mentioned in Section 4.7. Again, closed and guarded terms determine the set $\mathcal{P}^{\mathsf{cw}}$ of $\mathsf{CCS}^{\mathsf{cw}}$ processes. Further, we introduce initial output and input action sets as depicted in Tables 20 and 21, respectively, and write $\mathbb{I}^{\mathsf{cw}}(P)$ for $\mathrm{I}^{\mathsf{cw}}(P)\setminus\{\tau\}$.

The semantics of a $\mathsf{CCS}^{\mathsf{cw}}$ process is given by a labeled transition system whose transition relation possesses transitions of the form $\vdash^{\mathsf{cw}}_{M} P \stackrel{\alpha}{\longrightarrow} P'$, where $M \subseteq \Lambda$. Intuitively, process P can engage in an α-transition to P' whenever the environment does not offer communications on ports in M. Despite notational differences, this is the same underlying principle as for the transition relations defined in the previous sections, which are parameterized by initial action sets. Note that $\alpha \in \overline{\Lambda}$ implies $M = \emptyset$. The $\mathsf{CCS}^{\mathsf{cw}}$ transition relation is formally defined in Table 22, where $f(M)$ stands for $\{f(m) \mid m \in M\}$. Recall that the initial actions of P in $P+\!\rangle Q$ are given preference over the initial actions of Q. As expected, a prioritized τ, i.e., an internal action in which the left argument of $+\!\rangle$ can initially engage, has pre-emptive power over unprioritized actions, i.e., actions in which the right argument of $+\!\rangle$ can initially engage. Thus, the prioritized choice operator $+\!\rangle$ in $\mathsf{CCS}^{\mathsf{cw}}$ is evocative of the summation operator $+$ in $\mathsf{CCS}^{\mathsf{sl}}_{\mathsf{ml}}$. In [24] the operator $+$ stands for nondeterministic choice where priorities arising from the left and the right argument are incomparable. This operator is matched by the distributed summation operator $\oplus$ in $\mathsf{CCS}^{\mathsf{sl}}_{\mathsf{ml}}$. We further investigate the correspondence of these operators in the next section.

Table 20
Initial output action sets for $\mathsf{CCS}^{\mathsf{cw}}$

$\overline{\mathbb{I}}^{\mathsf{cw}}(\overline{a}.P) = \{a\}$	$\overline{\mathbb{I}}^{\mathsf{cw}}(\mu x.P) = \overline{\mathbb{I}}^{\mathsf{cw}}(P[\mu x.P/x])$
$\overline{\mathbb{I}}^{\mathsf{cw}}(P \mid Q) = \overline{\mathbb{I}}^{\mathsf{cw}}(P) \cup \overline{\mathbb{I}}^{\mathsf{cw}}(Q)$	$\overline{\mathbb{I}}^{\mathsf{cw}}(P+Q) = \overline{\mathbb{I}}^{\mathsf{cw}}(P) \cup \overline{\mathbb{I}}^{\mathsf{cw}}(Q)$
$\overline{\mathbb{I}}^{\mathsf{cw}}(P[f]) = \{f(a) \mid a \in \overline{\mathbb{I}}^{\mathsf{cw}}(P)\}$	$\overline{\mathbb{I}}^{\mathsf{cw}}(P\setminus L) = \overline{\mathbb{I}}^{\mathsf{cw}}(P)\setminus(L \cup \overline{L})$

Table 21
Initial input action sets for $\mathsf{CCS}^{\mathsf{cw}}$

$\mathrm{I}^{\mathsf{cw}}(\gamma.P) = \{\gamma\}$	$\mathrm{I}^{\mathsf{cw}}(\mu x.P) = \mathrm{I}^{\mathsf{cw}}(P[\mu x.P/x])$
$\mathrm{I}^{\mathsf{cw}}(P+\!\rangle Q) = \mathrm{I}^{\mathsf{cw}}(P) \cup \mathrm{I}^{\mathsf{cw}}(Q)$	$\mathrm{I}^{\mathsf{cw}}(P+Q) = \mathrm{I}^{\mathsf{cw}}(P) \cup \mathrm{I}^{\mathsf{cw}}(Q)$
$\mathrm{I}^{\mathsf{cw}}(P[f]) = \{f(\gamma) \mid \gamma \in \mathrm{I}^{\mathsf{cw}}(P)\}$	$\mathrm{I}^{\mathsf{cw}}(P\setminus L) = \mathrm{I}^{\mathsf{cw}}(P)\setminus(L \cup \overline{L})$
$\mathrm{I}^{\mathsf{cw}}(P \mid Q) = \mathrm{I}^{\mathsf{cw}}(P) \cup \mathrm{I}^{\mathsf{cw}}(Q) \cup \{\tau \mid \mathrm{I}^{\mathsf{cw}}(P) \cap \overline{\mathbb{I}}^{\mathsf{cw}}(Q) \neq \emptyset\}$	

Table 22
Operational semantics for $\mathsf{CCS}^{\mathsf{cw}}$

Act	$\dfrac{-}{\vdash^{\mathsf{cw}}_{\emptyset} \alpha.P \xrightarrow{\alpha} P}$	Res	$\dfrac{\vdash^{\mathsf{cw}}_{M} P \xrightarrow{\alpha} P'}{\vdash^{\mathsf{cw}}_{M\setminus(L\cup\overline{L})} P\setminus L \xrightarrow{\alpha} P'\setminus L}\ \alpha\notin L\cup\overline{L}$
Sum1	$\dfrac{\vdash^{\mathsf{cw}}_{M} P \xrightarrow{\alpha} P'}{\vdash^{\mathsf{cw}}_{M} P \mathbin{+\!\rangle} Q \xrightarrow{\alpha} P'}$	Sum2	$\dfrac{\vdash^{\mathsf{cw}}_{N} Q \xrightarrow{\alpha} Q'}{\vdash^{\mathsf{cw}}_{N\cup\mathbb{I}^{\mathsf{cw}}(P)} P \mathbin{+\!\rangle} Q \xrightarrow{\alpha} Q'}\ \tau,\alpha\notin \mathbb{I}^{\mathsf{cw}}(P)$
iSum1	$\dfrac{\vdash^{\mathsf{cw}}_{M} P \xrightarrow{\alpha} P'}{\vdash^{\mathsf{cw}}_{M} P+Q \xrightarrow{\alpha} P'}$	Com1	$\dfrac{\vdash^{\mathsf{cw}}_{M} P \xrightarrow{\alpha} P'}{\vdash^{\mathsf{cw}}_{M} P \mid Q \xrightarrow{\alpha} P' \mid Q}\ M\cap\overline{\mathbb{I}}^{\mathsf{cw}}(Q)=\emptyset$
iSum2	$\dfrac{\vdash^{\mathsf{cw}}_{N} Q \xrightarrow{\alpha} Q'}{\vdash^{\mathsf{cw}}_{N} P+Q \xrightarrow{\alpha} Q'}$	Com2	$\dfrac{\vdash^{\mathsf{cw}}_{N} Q \xrightarrow{\alpha} Q'}{\vdash^{\mathsf{cw}}_{N} P \mid Q \xrightarrow{\alpha} P \mid Q'}\ N\cap\overline{\mathbb{I}}^{\mathsf{cw}}(P)=\emptyset$
Rel	$\dfrac{\vdash^{\mathsf{cw}}_{M} P \xrightarrow{\alpha} P'}{\vdash^{\mathsf{cw}}_{f(M)} P[f] \xrightarrow{f(\alpha)} P'[f]}$	Com3	$\dfrac{\vdash^{\mathsf{cw}}_{M} P \xrightarrow{a} P' \quad \vdash^{\mathsf{cw}}_{\emptyset} Q \xrightarrow{\overline{a}} Q'}{\vdash^{\mathsf{cw}}_{M} P \mid Q \xrightarrow{\tau} P' \mid Q'}\ M\cap\overline{\mathbb{I}}^{\mathsf{cw}}(Q)=\emptyset$
Rec	$\dfrac{\vdash^{\mathsf{cw}}_{M} P[\mu x.P/x] \xrightarrow{\alpha} P'}{\vdash^{\mathsf{cw}}_{M} \mu x.P \xrightarrow{\alpha} P'}$	Com4	$\dfrac{\vdash^{\mathsf{cw}}_{\emptyset} P \xrightarrow{\overline{a}} P' \quad \vdash^{\mathsf{cw}}_{N} Q \xrightarrow{a} Q'}{\vdash^{\mathsf{cw}}_{N} P \mid Q \xrightarrow{\tau} P' \mid Q'}\ N\cap\overline{\mathbb{I}}^{\mathsf{cw}}(P)=\emptyset$

Camilleri and Winskel also developed a bisimulation-based semantic theory for $\mathsf{CCS}^{\mathsf{cw}}$. Their notion of strong bisimulation for $\mathsf{CCS}^{\mathsf{cw}}$, as defined below, is a congruence [24].

DEFINITION 4.18. A symmetric relation $\mathcal{R}\subseteq\mathcal{P}\times\mathcal{P}$ is a *distributed prioritized strong bisimulation for* $\mathsf{CCS}^{\mathsf{cw}}$ if for every $\langle P, Q\rangle\in\mathcal{R}$, $\alpha\in\mathcal{A}$, and $M\subseteq\Lambda$ the following condition holds.

$$\vdash^{\mathsf{cw}}_{M} P \xrightarrow{\alpha} P' \quad \text{implies} \quad \exists Q', N.\ \vdash^{\mathsf{cw}}_{N} Q \xrightarrow{\alpha} Q',\ N\subseteq M,\ \text{and}\ \langle P', Q'\rangle\in\mathcal{R}.$$

We write $P\simeq_{\mathsf{cw}} Q$ if $\langle P, Q\rangle\in\mathcal{R}$ for some distributed prioritized strong bisimulation $\mathcal{R}$ for $\mathsf{CCS}^{\mathsf{cw}}$.

4.9. *Relating the different approaches*

We show that the algebras $\mathsf{CCS}^{\mathsf{sl}}_{\mathsf{ml}}$ and $\mathsf{CCS}^{\mathsf{cw}}$ are closely related by providing an embedding of $\mathsf{CCS}^{\mathsf{cw}}$ in $\mathsf{CCS}^{\mathsf{sl}}_{\mathsf{ml}}$. For this purposes we define $\mathcal{N} =_{\mathrm{df}} \{0, 1\}^*$ and the strict order $<$ on priority values to be the lexicographical order on $\mathcal{N}$, where 1 is less than 0.

We now introduce the *translation function* $\xi(\cdot):\mathcal{P}^{\mathsf{cw}}\to\mathcal{P}^{\mathsf{sl}}_{\mathsf{ml}}$ by defining $\xi(P) =_{\mathrm{df}} \xi^{\varepsilon}(P)$, which maps $\mathsf{CCS}^{\mathsf{cw}}$ terms to $\mathsf{CCS}^{\mathsf{sl}}_{\mathsf{ml}}$ terms. The functions $\xi^{k}(P)$, for $k\in\mathcal{N}$, are inductively defined over the structure of $\mathsf{CCS}^{\mathsf{cw}}$ processes as shown in Table 23. We note that the translation function is not *surjective*, e.g., consider $(a:0.\mathbf{0}+b:2.\mathbf{0})+c:1.\mathbf{0}$ on which no $\mathsf{CCS}^{\mathsf{cw}}$ process is mapped. This example also shows that the notion of compositionality

Table 23
Translation function

$\xi^k(\mathbf{0}) =_{\mathrm{df}} \mathbf{0}$	$\xi^k(P + Q) =_{\mathrm{df}} \xi^k(P) \oplus \xi^k(Q)$	$\xi^k(P \setminus L) =_{\mathrm{df}} \xi^k(P) \setminus L$
$\xi^k(x) =_{\mathrm{df}} x$	$\xi^k(P +\rangle Q) =_{\mathrm{df}} \xi^{k0}(P) + \xi^{k1}(Q)$	$\xi^k(P[f]) =_{\mathrm{df}} \xi^k(P)[f]$
$\xi^k(\gamma.P) =_{\mathrm{df}} \gamma : k.\xi^{\varepsilon}(P)$	$\xi^k(P \mid Q) =_{\mathrm{df}} \xi^k(P) \mid \xi^k(Q)$	$\xi^k(\mu x.P) =_{\mathrm{df}} \mu x.\xi^k(P)$
$\xi^k(\overline{a}.P) =_{\mathrm{df}} \overline{a}.\xi^{\varepsilon}(P)$		

in $\mathsf{CCS}^{\mathsf{cw}}$ is more restrictive than the one in $\mathsf{CCS}^{\mathsf{sl}}_{\mathsf{ml}}$, since a comparable summation can only be extended by summands that have a higher or a lower priority than the already considered summands. The following theorem, which is proved in [55], makes the semantic relationship between a $\mathsf{CCS}^{\mathsf{cw}}$ process P and its embedding $\xi(P)$ precise.

THEOREM 4.19. *Let $P, Q \in \mathcal{P}^{\mathsf{cw}}$. Then $P \simeq_{\mathsf{cw}} Q$ if and only if $\xi(P) \simeq_{\mathsf{ml}} \xi(Q)$.*

Consequently, distributed prioritized strong bisimulation for $\mathsf{CCS}^{\mathsf{sl}}_{\mathsf{ml}}$ is also compositional with respect to summation in the sub-calculus of $\mathsf{CCS}^{\mathsf{sl}}_{\mathsf{ml}}$ induced by $\mathsf{CCS}^{\mathsf{cw}}$.

4.10. *Concluding remarks and related work*

A local concept of pre-emption is also considered by Hansson and Orava in [38], where CSP [44] is extended with priority by assigning natural numbers to actions. As for $\mathsf{CCS}^{\mathsf{sl}}$, they use a notion of location, and pre-emption in the calculus is sensitive to locations. Indeed, their work served as an inspiration for $\mathsf{CCS}^{\mathsf{sl}}$. However, the authors only conjecture that their version of strong bisimulation is a congruence, and they provide neither an axiomatization for their behavioral relation nor a theory for observational congruence. One may also criticize their semantics as not truly reflecting distributed computation. In particular, despite having a local pre-emptive semantics they compute a global priority for synchronizations.

After stressing the strong similarity of $\mathsf{CCS}^{\mathsf{sl}}$ to the process algebra $\mathsf{CCS}^{\mathsf{cw}}$ in the previous section, we focus on the algebraic results established in these frameworks. In [24,48] the transition relation is directly annotated with pre-emption potentials. By using this transition relation in the definition of standard strong bisimulation one immediately obtains a congruence. In contrast, [28] starts off by defining *naive* distributed prioritized strong bisimulation using the naive transition relation; the potential of pre-emption is considered subsequently (by introducing the distributed prioritized initial action set condition). Then it is shown that the resulting congruence is the largest congruence in the naive equivalence. Similarly, Jensen [48] defines a naive distributed prioritized weak bisimulation based on the above-mentioned annotated transition relation. His naive weak transition relation corresponds to the distributed prioritized weak transition relation in $\mathsf{CCS}^{\mathsf{sl}}$ if the parameter M is dropped. Because of the difference in the naive transition relations the abstraction result presented here is somewhat stronger than Jensen's, although the observational congruences appear to coincide.

One may wonder about the relationship between $\mathsf{CCS}^{\mathsf{sl}}$ and $\mathsf{CCS}^{\mathsf{sg}}$, i.e., the static priority and global pre-emption language of Section 3. If in $\mathsf{CCS}^{\mathsf{sl}}$ the distributed summation operator is left out and pre-emption is globalized by defining $[m] =_{\mathrm{df}} \mathcal{L}oc$, for all $m \in \mathcal{L}oc$, the operational semantics and the behavioral relations reduce to the corresponding notions presented in Section 3.

Like Camilleri and Winskel, Barrett [8] devises a semantics of `occam`'s priority mechanism that is additionally concerned with fairness aspects. His framework is based on a structural operational semantics augmented with *ready-guard* sets which model possible inputs from the environment. Intuitively, these sets characterize the nature of the contexts in which a transition is enabled. Thus, they correspond to the action sets with which the $\mathsf{CCS}^{\mathsf{sl}}$ and the $\mathsf{CCS}^{\mathsf{cw}}$ transition relations are parameterized. Barrett is not concerned with investigating behavioral relations, but focuses instead on implementing `occam`'s `PRIALT` and `PRIPAR` constructs on the transputer platform.

Other researchers have also extended Hoare's *Communicating Sequential Processes* (CSP) [44] with a concept of static priority. Inspired by the notion of priority in Ada [51], Fidge [33] introduced new versions of the operators for external choice, parallel composition by interleaving, and parallel composition by intersection. Their operational semantics favors execution of the left-hand operands, as was also done by Jensen in [48]. The semantic theory in [33] is based on *failure semantics* which is made sensitive to local pre-emption. For this purpose, traces are augmented with a *preference function* that identifies the priority relation on the initial action sets of a given process. A related approach was presented by Lowe [54]. It differs from [33] in that the underlying algebra is a timed version of CSP [31]. Additionally, Lowe aims at obtaining a fully deterministic language by employing a similar notion of priority as the one proposed by Fidge.

Finally, we remark on the notions of strong and weak bisimulation for $\mathsf{CCS}^{\mathsf{sl}}$. Since the semantic theory reflects local pre-emption, locations occur implicitly in our semantic equivalences. In contrast to work on *location equivalences* [18,25,63], we only *indirectly* consider locations for defining a suitable notion of local pre-emption potential in form of prioritized initial action sets; our objective is not to observe locations but to capture local pre-emption.

5. Dynamic priority and global pre-emption

This section develops a theory in which priorities are dynamic and pre-emption is global. The motivation for this theory originated in a desire to devise a compact model of real-time computation, and we devote significant space to establishing a tight connection between the seemingly different notions of priority and real-time [10]. For this purpose we equip our language with a dynamic priority semantics based on global pre-emption and refer to it as $\mathsf{CCS}^{\mathsf{dg}}$ (CCS with dynamic priority and global pre-emption). The connection with real-time arises when we interpret delays as priorities: the longer the delay preceding an action, the lower its priority. This approach contrasts significantly with more traditional accounts of real-time, where the only notion of pre-emption arises in the context of the *maximal progress assumption* [79], which states that time may only pass if the system under consideration cannot engage in any further internal computation. The main result of this section is the formalization of a one-to-one correspondence between the strong-bisimulation equivalences induced by dynamic priority semantics and real-time semantics.

Unlike the process algebras with priority considered so far, actions in $\mathrm{CCS}^{\mathrm{dg}}$ have priority values that may change as systems evolve. Accordingly, we alter our point of view regarding actions and priorities by separating action names from their priority values, i.e., an action's priority is no longer implicit in its port name. In this vein, we take the set of actions $\mathcal{A}$ to be $\{\alpha, \beta, \ldots\}$. We also allow priority values to come from the full set $\mathbb{N}$ of natural numbers rather than a finite set. Our syntax of processes will then require that each action is equipped with a priority value taken from $\mathbb{N}$.

The structure of this section is as follows. Section 5.1 briefly presents a real-time semantics for our language, whereas the dynamic priority semantics is introduced in Section 5.2. The one-to-one correspondence between dynamic priority semantics and real-time semantics is established in Section 5.3. Finally, Section 5.4 contains our concluding remarks and discusses related work.

5.1. *Real-time semantics*

We first define a real-time semantics, $\mathrm{CCS}^{\mathrm{rt}}$, for the language introduced in Section 2. $\mathrm{CCS}^{\mathrm{rt}}$ explicitly represents timing behavior: the semantics of a process is defined by a labeled transition system that contains explicit *clock transitions* – each representing a delay of one time unit – as well as *action transitions*. With respect to clock transitions, the operational semantics is such that processes willing to communicate with some process running in parallel are able to wait until the communication partner is ready. However, as soon as it is available, the communication must occur, i.e., further idling is prohibited. This assumption is usually referred to as the maximal progress assumption [79] or the *synchrony hypothesis* [13].

Formally, the labeled transition system corresponding to a process P is a four-tuple $\langle \mathcal{P}, \mathcal{A} \cup \{1\}, \longmapsto, P \rangle$, where the alphabet $\mathcal{A} \cup \{1\}$ satisfies $1 \notin \mathcal{A}$. The transition relation for actions coincides with the one for traditional CCS (cf. Table 24). The transition relation $\longmapsto \subseteq \mathcal{P} \times \{1\} \times \mathcal{P}$ for clock transitions is defined in Table 25. We use γ as representative

Table 24
Operational semantics for $\mathrm{CCS}^{\mathrm{rt}}$ (action transitions)

Act	$\dfrac{-}{\alpha:0.P \stackrel{\alpha}{\longmapsto} P}$	Sum1	$\dfrac{P \stackrel{\alpha}{\longmapsto} P'}{P + Q \stackrel{\alpha}{\longmapsto} P'}$	Rec	$\dfrac{P[\mu x.P/x] \stackrel{\alpha}{\longmapsto} P'}{\mu x.P \stackrel{\alpha}{\longmapsto} P'}$
Rel	$\dfrac{P \stackrel{\alpha}{\longmapsto} P'}{P[f] \stackrel{f(\alpha)}{\longmapsto} P'[f]}$	Sum2	$\dfrac{Q \stackrel{\alpha}{\longmapsto} Q'}{P + Q \stackrel{\alpha}{\longmapsto} Q'}$	Res	$\dfrac{P \stackrel{\alpha}{\longmapsto} P'}{P \setminus L \stackrel{\alpha}{\longmapsto} P' \setminus L}\ \alpha \notin L \cup \overline{L}$
Com1	$\dfrac{P \stackrel{\alpha}{\longmapsto} P'}{P\vert Q \stackrel{\alpha}{\longmapsto} P'\vert Q}$	Com2	$\dfrac{Q \stackrel{\alpha}{\longmapsto} Q'}{P\vert Q \stackrel{\alpha}{\longmapsto} P\vert Q'}$	Com3	$\dfrac{P \stackrel{a}{\longmapsto} P' \quad Q \stackrel{\overline{a}}{\longmapsto} Q'}{P\vert Q \stackrel{\tau}{\longmapsto} P'\vert Q'}$

Table 25
Operational semantics for CCS^{rt} (clock transitions)

tNil	$\dfrac{-}{\mathbf{0} \overset{1}{\longmapsto} \mathbf{0}}$	tRec	$\dfrac{P[\mu x.P/x] \overset{1}{\longmapsto} P'}{\mu x.P \overset{1}{\longmapsto} P'}$
tAct1	$\dfrac{-}{\alpha:k.P \overset{1}{\longmapsto} \alpha:(k-1).P}\ k>0$	tAct2	$\dfrac{-}{a:0.P \overset{1}{\longmapsto} a:0.P}$
tSum	$\dfrac{P \overset{1}{\longmapsto} P' \quad Q \overset{1}{\longmapsto} Q'}{P+Q \overset{1}{\longmapsto} P'+Q'}$	tCom	$\dfrac{P \overset{1}{\longmapsto} P' \quad Q \overset{1}{\longmapsto} Q'}{P\vert Q \overset{1}{\longmapsto} P'\vert Q'}\ P\vert Q \overset{\tau}{\not\longmapsto}$
tRel	$\dfrac{P \overset{1}{\longmapsto} P'}{P[f] \overset{1}{\longmapsto} P'[f]}$	tRes	$\dfrac{P \overset{1}{\longmapsto} P'}{P \setminus L \overset{1}{\longmapsto} P' \setminus L}$

of $\mathcal{A} \cup \{1\}$, and write $P \overset{\gamma}{\longmapsto} P'$ instead of $\langle P, \gamma, P' \rangle \in \longmapsto$. If $\gamma \in \mathcal{A}$ we speak of an *action transition*, otherwise of a *clock transition*. Sometimes it is convenient to write $P \overset{\gamma}{\longmapsto}$ for $\exists P' \in \mathcal{P}.\, P \overset{\gamma}{\longmapsto} P'$. In order to ensure maximal progress, our operational semantics is such that $P \overset{1}{\not\longmapsto}$ whenever $P \overset{\tau}{\longmapsto}$, i.e., clock transitions are pre-empted if P can engage in internal computation.

Intuitively, process $\alpha:k.P$, where $k > 0$, may engage in a clock transition and then behave like $\alpha:(k-1).P$. Process $\alpha:0.P$ performs an α-transition to become process P. Moreover, if $\alpha \not\equiv \tau$, it may also idle by executing a clock transition to itself. Time must proceed equally on both sides of summation, i.e., $P + Q$ can engage in a clock transition and, thus, delay the nondeterministic choice if and only if both P and Q can engage in a clock transition. Hence, time is a deterministic concept. Similar to summation, P and Q must synchronize on clock transitions according to Rule (tCom). Its side condition implements maximal progress by ensuring that there is no pending communication between P and Q. Although this condition is negative, our semantics is still well-defined [77]. A semantic theory based on the notion of bisimulation has been developed for CCS^{rt} (cf. [61]). For this section we restrict ourselves to (*strong*) temporal bisimulation, a congruence which is defined as follows.

DEFINITION 5.1 (*Temporal bisimulation*). A symmetric relation $\mathcal{R} \subseteq \mathcal{P} \times \mathcal{P}$ is called a *temporal bisimulation* if for all $\langle P, Q \rangle \in \mathcal{R}$ and $\gamma \in \mathcal{A} \cup \{1\}$ the following holds: $P \overset{\gamma}{\longmapsto} P'$ implies $\exists Q'.\, Q \overset{\gamma}{\longmapsto} Q'$ and $\langle P', Q' \rangle \in \mathcal{R}$. We write $P \sim_{\mathrm{rt}} Q$ if $\langle P, Q \rangle \in \mathcal{R}$ for some temporal bisimulation $\mathcal{R}$.

Observe that CCS^{rt} semantics unfolds every delay value into a sequence of elementary time units. For example, process $a:k.\mathbf{0}$ has $k+2$ states, namely $\mathbf{0}$ and $a:l.\mathbf{0}$, for $0 \leqslant l \leqslant k$ (see also Figure 6 in Section 5.3). Representing $a:k.\mathbf{0}$ by a single transition, which is labeled by $a:k$ and leads to state $\mathbf{0}$, would yield labeled transition systems with fewer states. This idea of compacting the state space of real-time systems can be implemented by viewing k as *priority value* assigned to action a.

5.2. *Dynamic priority semantics*

We introduce $\mathsf{CCS}^{\mathsf{dg}}$, i.e., a dynamic priority semantics for our language presented in Section 2. The notion of pre-emption incorporated in $\mathsf{CCS}^{\mathsf{dg}}$ is similar to $\mathsf{CCS}^{\mathsf{sg}}$, and it naturally encodes the maximal progress assumption employed in $\mathsf{CCS}^{\mathsf{rt}}$ semantics. Formally, the $\mathsf{CCS}^{\mathsf{dg}}$ semantics of a process P is given by a labeled transition system $\langle \mathcal{P}, \mathcal{A} \times \mathbb{N}, \rightarrow, P \rangle$. The presentation of the operational rules for the transition relation $\rightarrow$ requires two auxiliary definitions.

First, we introduce *potential initial action sets* as defined in Table 26, the actions in which a given process can potentially engage. Note that these sets are only supersets of the initial actions of processes because they do not take *pre-emption* into account. However, this is sufficient for our purposes regarding pre-emption, since $\tau \notin \mathrm{I}^{<k}(P)$ if and only if $\nexists l < k.\, P \xrightarrow{\tau:l}$, where $\mathrm{I}^{<k}(P) =_{\mathrm{df}} \mathrm{I}^{k-1}(P)$, for $k > 0$, and $\mathrm{I}^{<0}(P) =_{\mathrm{df}} \emptyset$.

As second auxiliary definition for presenting the transition relation, we introduce a *priority adjustment function* as shown in Table 27. Intuitively, if one parallel component of a process engages in a transition with priority k, then the priority values of all initial actions at every other parallel component have to be decreased by k, i.e., those actions become equally "more urgent." Thus, the semantics of parallel composition employs a kind of *fairness assumption*, and priorities have a *dynamic* character. More precisely, the priority adjustment function applied to a process P and a natural number k, denoted as $[P]^k$, returns a term that is "identical" to P except that the priority values of the initial, top-level actions are decreased by k. Note that a priority value cannot become less than 0 and that "identical" does not mean syntactic equality but syntactic equality *up to unfolding* of recursion.

The operational rules in Table 28 capture the following intuition. Process $a:k.P$ may engage in action a, with priority value $l \geqslant k$, yielding process P. The side condition $l \geqslant k$ means that k does not specify an exact priority but the *maximum* priority of the initial transition of $a:k.P$. Due to the notion of pre-emption incorporated in $\mathsf{CCS}^{\mathsf{dg}}$, $\tau:k.P$ may not perform the initial τ-transition with a lower priority than k. Process $P + Q$ may behave

Table 26
Potential initial action sets for $\mathsf{CCS}^{\mathsf{dg}}$

$\mathrm{I}^k(\alpha:l.P)$	$=$	$\{\alpha \mid l \leqslant k\}$	$\mathrm{I}^k(P \mid Q)$	$=$	$\mathrm{I}^k(P) \cup \mathrm{I}^k(Q) \cup \{\tau \mid \mathrm{I}^k(P) \cap \overline{\mathrm{I}^k(Q)} \neq \emptyset\}$
$\mathrm{I}^k(P+Q)$	$=$	$\mathrm{I}^k(P) \cup \mathrm{I}^k(Q)$	$\mathrm{I}^k(P[f])$	$=$	$\{f(\alpha) \mid \alpha \in \mathrm{I}^k(P)\}$
$\mathrm{I}^k(\mu x.P)$	$=$	$\mathrm{I}^k(P[\mu x.P/x])$	$\mathrm{I}^k(P \setminus L)$	$=$	$\mathrm{I}^k(P) \setminus (L \cup \overline{L})$

Table 27
Priority adjustment function

$[\mathbf{0}]^k$	$=_{\mathrm{df}}$ $\mathbf{0}$,	$[x]^k =_{\mathrm{df}} x$	$[\mu x.P]^k$	$=_{\mathrm{df}}$ $[P[\mu x.P/x]]^k$
$[\alpha:l.P]^k$	$=_{\mathrm{df}}$ $\alpha:(l-k).P$ if $l > k$		$[\alpha:l.P]^k$	$=_{\mathrm{df}}$ $\alpha:0.P$ if $l \leqslant k$
$[P+Q]^k$	$=_{\mathrm{df}}$ $[P]^k + [Q]^k$		$[P \mid Q]^k$	$=_{\mathrm{df}}$ $[P]^k \mid [Q]^k$
$[P[f]]^k$	$=_{\mathrm{df}}$ $[P]^k[f]$		$[P \setminus L]^k$	$=_{\mathrm{df}}$ $[P]^k \setminus L$

Table 28
Operational semantics for $\mathsf{CCS}^{\mathsf{dg}}$

Act1	$\dfrac{-}{a:k.P \xrightarrow{a:l} P}\; l \geqslant k$	Act2	$\dfrac{-}{\tau:k.P \xrightarrow{\tau:k} P}$
Sum1	$\dfrac{P \xrightarrow{\alpha:k} P'}{P+Q \xrightarrow{\alpha:k} P'}\; \tau \notin \mathrm{I}^{<k}(Q)$	Sum2	$\dfrac{Q \xrightarrow{\alpha:k} Q'}{P+Q \xrightarrow{\alpha:k} Q'}\; \tau \notin \mathrm{I}^{<k}(P)$
Com1	$\dfrac{P \xrightarrow{\alpha:k} P'}{P\vert Q \xrightarrow{\alpha:k} P'\vert[Q]^k}\; \tau \notin \mathrm{I}^{<k}(P\vert Q)$	Rel	$\dfrac{P \xrightarrow{\alpha:k} P'}{P[f] \xrightarrow{f(\alpha):k} P'[f]}$
Com2	$\dfrac{Q \xrightarrow{\alpha:k} Q'}{P\vert Q \xrightarrow{\alpha:k} [P]^k\vert Q'}\; \tau \notin \mathrm{I}^{<k}(P\vert Q)$	Res	$\dfrac{P \xrightarrow{\alpha:k} P'}{P\setminus L \xrightarrow{\alpha:k} P'\setminus L}\; \alpha \notin L \cup \overline{L}$
Com3	$\dfrac{P \xrightarrow{a:k} P' \quad Q \xrightarrow{\overline{a}:k} Q'}{P\vert Q \xrightarrow{\tau:k} P'\vert Q'}\; \tau \notin \mathrm{I}^{<k}(P\vert Q)$	Rec	$\dfrac{P[\mu x.P/x] \xrightarrow{\alpha:k} P'}{\mu x.P \xrightarrow{\alpha:k} P'}$

like P if Q does not pre-empt the considered transition by being able to engage in a higher prioritized internal transition. Process $P \mid Q$ denotes the *parallel composition* of P and Q according to an interleaving semantics with synchronized communication on complementary actions of P and Q having the same priority value k, which results in the internal action τ attached with priority value k (cf. Rule (Com3)). The interleaving Rules (Com1) and (Com2) incorporate the dynamic behavior of priority values as explained in the previous paragraph. The side conditions of Rules (Com$i$) implement global pre-emption. The semantics for *relabeling*, *restriction*, and *recursion* is straightforward. As for $\mathsf{CCS}^{\mathsf{rt}}$, we introduce a notion of strong bisimulation, referred to as *prioritized bisimulation*.

DEFINITION 5.2 (*Prioritized bisimulation*). A symmetric relation $\mathcal{R} \subseteq \mathcal{P} \times \mathcal{P}$ is called *prioritized bisimulation* if for every $\langle P, Q\rangle \in \mathcal{R}$, $\alpha \in \mathcal{A}$, and $k \in \mathbb{N}$ the following holds: $P \xrightarrow{\alpha:k} P'$ implies $\exists Q'.\, Q \xrightarrow{\alpha:k} Q'$ and $\langle P', Q'\rangle \in \mathcal{R}$. We write $P \sim_{\mathsf{dg}} Q$ if there exists a prioritized bisimulation $\mathcal{R}$ such that $\langle P, Q\rangle \in \mathcal{R}$.

5.3. *Relating dynamic priority and real-time semantics*

We show that $\mathsf{CCS}^{\mathsf{dg}}$ and $\mathsf{CCS}^{\mathsf{rt}}$ semantics are closely related. The underlying intuition is best illustrated by a simple example dealing with the prefixing operator. Figure 6 depicts the dynamic priority semantics and real-time semantics of process $a:k.\mathbf{0}$. Both transition systems intuitively reflect that process $a:k.\mathbf{0}$ must delay at least k time units before it may engage in the a-transition. According to $\mathsf{CCS}^{\mathsf{rt}}$ semantics, this process consecutively engages in k clock transitions passing the states $a:(k-l).\mathbf{0}$, for $0 \leqslant l \leqslant k$, before it may either continue idling in state $a:0.\mathbf{0}$ or perform the a-transition to inaction process $\mathbf{0}$. Thus,

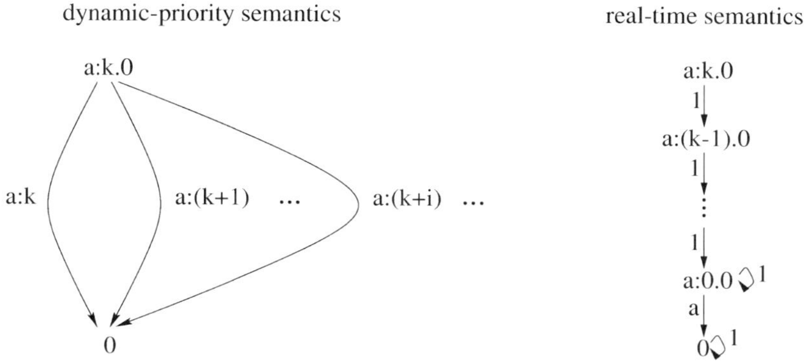

Fig. 6. Relating $\mathsf{CCS}^{\mathsf{dg}}$ semantics and $\mathsf{CCS}^{\mathsf{rt}}$ semantics.

time is explicitly part of states and made visible by clock transitions, each representing a step consuming one time unit.

In contrast, the dynamic priority semantics encodes the delay of at least k time units in the transitions rather than in the states. Hence, it possesses only the two states $a:k.\mathbf{0}$ and $\mathbf{0}$ connected via transitions labeled by $a:l$, for $l \geqslant k$. Although at first sight it seems that the price for saving intermediate states is to be forced to deal with infinite-branching, an upper bound for l can be given. In our example this upper bound is k itself, since a delay by more than k time units only results in idling and does not enable new or disable existing system behavior. Therefore, the dynamic priority transition system of $a:k.\mathbf{0}$ just consists of the two states $a:k.\mathbf{0}$ and $\mathbf{0}$ and a *symbolic* transition labeled by $a:k$, whereas the real-time transition system has $k+2$ states and $k+2$ transitions. The following proposition formally states that $\mathsf{CCS}^{\mathsf{dg}}$ semantics can indeed be understood as an efficient encoding of $\mathsf{CCS}^{\mathsf{rt}}$ semantics. Here, $\overset{1}{\longmapsto}{}^{k}$ denotes k consecutive clock transitions.

PROPOSITION 5.3. *Let* $P, P' \in \mathcal{P}$, $\alpha \in \mathcal{A}$, *and* $k \in \mathbb{N}$. *Then* $P \xrightarrow{\alpha:k} P'$ *if and only if* $\exists P''.\, P \overset{1}{\longmapsto}{}^{k} P'' \overset{\alpha}{\longmapsto} P'$.

Proposition 5.3 is the key to prove the main result of this section.

THEOREM 5.4. *Let* $P, Q \in \mathcal{P}$. *Then* $P \sim_{\mathsf{dg}} Q$ *if and only if* $P \sim_{\mathsf{rt}} Q$.

Consequently, prioritized and temporal bisimulation possess the same properties; in particular, prioritized bisimulation is a congruence for $\mathsf{CCS}^{\mathsf{dg}}$. Proof details can be found in [16].

5.4. *Concluding remarks and related work*

As shown above, real-time semantics can be encoded by dynamic priority semantics. The utility of this encoding stems from the fact that the state space of $\mathsf{CCS}^{\mathsf{dg}}$ models is much smaller and the size of the transition relation is at least not worse, but in practice often better, than the one of corresponding $\mathsf{CCS}^{\mathsf{rt}}$ models. This has been demonstrated by formally modeling and verifying several aspects of the widely-used `SCSI-2` *bus-protocol*, for which the state space of the dynamic priority model is almost an order of magnitude smaller than the one resulting from traditional real-time semantics [16].

Regarding related work, a similar approach to the one presented above has been investigated by Jeffrey [47]. He established a formal relationship between a quantitative real-time process algebra and a process algebra with static priority which is very similar to $\mathsf{CCS}^{\mathsf{sg}}$ introduced in Section 3. Jeffrey also translates real-time to priority based on the idea of time-stamping. In contrast to $\mathsf{CCS}^{\mathsf{rt}}$ semantics, however, a process modeled in Jeffrey's framework may either immediately engage in an action transition or idle forever. This semantics does not allow a process to wait until a communication partner becomes available, but instead forces a "livelock" in such situations. It is only because of this design decision that Jeffrey does not need to choose a dynamic priority framework.

In [22] a variant of CCSR [23], called CCSR92, has been introduced, which allows for the modeling of not only static priority but also dynamic priority. The main focus of CCSR involves the specification and verification of real-time concurrent systems including scheduling behavior. Thus, a notion of dynamic priority, as applied in *priority-inheritance* and *earliest-deadline-first* scheduling algorithms, is important. In [22] dynamic priorities are given as a function of the *history* of the system under consideration. Accordingly, the operational semantics of CCSR92 is re-defined to include the historical context. The authors show that dynamic priorities do, in general, not lead to a compositional semantics and give a sufficient condition that ensures compositionality.

6. Priority in other process-algebraic frameworks

This section completes the discussion of related work by focusing on approaches to priority which either do not fit in our classification scheme presented in Section 1, such as approaches for ACP [5], SCCS [73], and *stochastic* [11,43] or *probabilistic* [49,74,76] process algebras, or are concerned with process-algebraic descriptions of non-process-algebraic languages, such as *Esterel* [12,13] and *Statecharts* [39].

Baeten, Bergstra, and Klop were the first researchers who investigated priority in process algebras [5] by developing a notion of priority in the *Algebra of Communicating Processes* (ACP) [9]. Their work is inspired by the insight that it is essential to incorporate an interrupt mechanism in process-algebraic frameworks, as it enhances their expressive power as specification and verification formalisms for concurrent systems. Therefore, a new unary operator θ together with semantics-defining equations is introduced in [5]. The definition of θ is based on a given partial order $<$ on actions. Intuitively, $\theta(P)$ is the context of P in which action a has precedence over action b, whenever $b < a$, i.e., nondeterministic choices between actions a and b are resolved within $\theta(P)$. Technically, the axiomatic semantics of the new language, notated as a *term rewrite system*, is shown to possess nice algebraic properties such as *confluence* and *termination*. The utility of the theory is demonstrated by simple examples dealing with interrupts, timeouts, and other aspects of system behavior. The approach in [5] differs from most other work presented in this chapter in that the partial order expressing priorities is fixed with respect to the system under consideration, i.e., the same priority relation holds at all states of the system. For example, if $a < b$ at some state of the system, then $a > b$ cannot be valid at another state, i.e., priorities in [5] are not *globally dynamic* in the sense of [73]. It should also be mentioned that the version of ACP used in [5] does not include a designated internal action, cf. action τ in CCS; a fact which simplifies the development of algebraic theories.

Stochastic process algebras [11,43], which enhance the expressiveness of classical process algebras by integrating *performance* descriptions of concurrent systems, also define notions of priority. One example of a popular stochastic process algebra is the *Extended Markovian Process Algebra* (EMPA) [11] whose semantics is given in terms of strong bisimulation and whose static priority approach is adapted from $\mathsf{CCS}^{\mathsf{sg}}$.

Smolka and Steffen [73] have introduced static priority to the *Synchronous Calculus of Communicating Systems* (SCCS) [58]. They extended a probabilistic version of this language, known as PCCS [76], whose semantics is given in terms of *probabilistic bisimulation*. Their work shows that the concept of priority is not only related to real-time, as investigated in Section 5, but also to probability. The idea in [73] is to allow probability guards of value 0 to be associated with alternatives of a probabilistic summation. Such alternatives can be chosen only if the non-zero alternatives are precluded by contextual constraints. Thus, priority may be viewed as an extreme case of probability.

Tofts has investigated another extension of SCCS, the *Weighted Synchronous Calculus of Communicating Systems* (WSCCS) [74]. Its semantics relies upon a notion of *relative frequency* that is suitable for specifying and reasoning about aspects of priority, probability, and time in concurrent systems. In this approach, priority is encoded by means of higher ordinals; a transition has priority over another if their weights are separated at least by a factor of ω. An operator similar to the θ-operator in [5] is defined, which extracts the highest priority transitions enabled at a process state by referring to a global notion of pre-emption. In contrast to [5], Toft's operator allows for different priority structures at different states. His concept of priority yields a simpler operational semantics than the one in [73], but a more complicated definition of bisimulation. For WSCCS, a congruence adapted from strong bisimulation together with an equational characterization, which is sound and complete for finite processes, has been developed.

The concept of pre-emption has also been studied by Berry in *Esterel's zero-delay process calculus* [12], a theoretical version of the Esterel synchronous programming language [13]. The calculus' semantics, which obeys maximal progress [79], interprets processes as deterministic mappings from input sequences to output sequences. Berry emphasizes the importance of pre-emption in control-dominated reactive and real-time programming. He suggests that pre-emption operators be considered as first-class operators that are fully orthogonal with respect to all other primitives, such as concurrency and communication. This is in contrast to the approach chosen in this chapter, in which pre-emption is implicitly encoded as side conditions of operational rules involving nondeterminism. Several examples of useful pre-emption operators are presented and axiomatized in [12], all of which are based on the ideas of *abortion* and *suspension*.

The specification language *Statecharts* [39], for which process-algebraic descriptions of its semantics have been developed [53,56,75], extends communicating finite automata by concepts of *hierarchy*, *concurrency*, and *priority*. In Statecharts static priorities can be expressed via the absence of actions, which are called *events*, by permitting negated actions as guards, which are referred to as *triggers*. As an example, consider the Statecharts-like term $a:b.\,P + \neg b:c.\,Q$ consisting of a nondeterministic choice between a b-transition with guard a to process P, and a c-transition with guard $\neg b$ to Q. Intuitively, the Statechart may only engage in the latter transition if it cannot execute the former one. The reason is that the former transition produces event b which falsifies the guard of the c-transition. Thus, the b-transition is given precedence over the c-transition. Approaches to priority based on negated events (cf. [37]) do not go well with the concept of *hiding*, which is employed in many process algebras and also in an alternative language to Statecharts called *Argos* [57]. Hiding enables one to relabel a visible action into a distinguished invisible action. The difficulty with hiding arises when several events are hidden, i.e., all of them are relabeled to the same event and, thus, have the same implicit priority value attached to them. Hence, hiding may destroy priority schemes. In contrast, in the priority approaches considered in this chapter priorities are assigned to transitions, thereby allowing for a more fine-granular priority mechanism and avoiding the above-mentioned problem.

7. Conclusions and directions for future work

This chapter investigated various aspects of priority in process algebra. The utility of introducing priority to traditional process algebras is to enhance their expressiveness and, thereby, making them more attractive to system designers. In a nutshell, priorities allow one to resolve potential nondeterminism.

Conclusions. We illustrated the most important aspects of priority in a prototypic language that extends Milner's CCS. This language was equipped with several semantics according to whether priorities are static or dynamic and whether the adopted notion of pre-emption is global or local.

In practice it is easy to determine when to use static priority semantics and when to use dynamic priority semantics. For modeling interrupts and prioritized choice constructs a static notion of priority is adequate, whereas for modeling real-time or scheduling behavior, dynamic priorities should be considered. However, static priority approaches also allow for the description of simple scheduling algorithms, as shown in [48] in the presence of a prioritized parallel composition operator. The dynamic priority approach yields a more efficient verification of real-time systems, since the sizes of system models with respect to dynamic priority semantics are often significantly smaller than the ones regarding real-time semantics [16]. If one needs to deal with both interrupt and real-time aspects at the same time, static and dynamic priority approaches should be combined. In this situation each action could be assigned two priority values, the first interpreted as a global priority value for scheduling purposes, and the second interpreted as a local priority value for modeling interrupts, where the first priority value has more weight than the second one.

Suitable guidelines for choosing between a global or a local notion of pre-emption are the following. A semantics obeying global pre-emption is favorable when modeling interrupts and prioritized-choice constructs in concurrent, centralized systems or when specifying real-time and scheduling aspects. Global pre-emption also provides a simple means for enforcing that action sequences are executed atomically. This supports the accurate modeling of certain system behavior and, moreover, helps one to keep system models small [29]. However, when dealing with interrupts or prioritized-choice constructs within distributed systems, the concept of global pre-emption is inadequate. Here, the use of local pre-emption does not only lead to an intuitive but also to an implementable semantics, since it does not require any knowledge about computations that are internal to other, potentially unknown sites (cf. [27]).

In this chapter we equipped the three calculi presented in Sections 3–5 with a bisimulation-based semantics. The re-development of the semantic theory of CCS for the static priority calculi included:

(i) characterizations of the largest congruences contained in the naive adaptations of the standard strong and weak bisimulations,
(ii) encodings of the new behavioral relations as standard strong bisimulations on enriched transition relations, and
(iii) axiomatic characterizations of the prioritized strong bisimulations for the class of finite processes.

For the dynamic priority calculus, strong bisimulation served as a semantic tool for establishing a one-to-one correspondence between dynamic priority and real-time semantics. We want to point out that the semantic theories presented here show that extensions of process algebras by priority do not need to sacrifice the simplicity and the elegance that have made traditional process-algebraic approaches successful.

This chapter also surveyed related approaches to priority which are concerned with process-algebraic calculi other than CCS and its derivatives. We have classified them according to whether priorities are considered to be static or dynamic and whether their concept of pre-emption is global or local. The concept of priority has also been investigated in other concurrency frameworks, most notably in *Petri Nets* [14,72]. In this setting priorities are either expressed explicitly by priority relations over transitions [15] or implicitly via *inhibitor arcs* [46]. Finally, it should be mentioned that priorities can *implicitly* arise when studying causality for *mobile processes* (see, e.g., [32]). In these approaches, priorities eliminate superfluous paths that only present new temporal but not causal system dependencies.

Future work. In addition to the fact that a calculus combining dynamic priority and local pre-emption has not yet been developed, the semantic theories for $\mathsf{CCS}^{\mathsf{sg}}$ and $\mathsf{CCS}^{\mathsf{sl}}$ need to be completed by axiomatizing their *observational* congruences. For *finite* processes, one should be able to establish these axiomatizations using standard techniques [59]. However, for *regular* processes – i.e., the class of finite-state processes that do not contain recursion through static operators – it is not clear how to obtain a completeness result. The point is that existing methods for proving completeness of axiomatizations with respect to observational congruences rely on the possibility to remove or to insert τ-cycles in processes. In the context of pre-emption, however, this would possibly change the pre-emption potential of processes and, thus, is semantically incompatible with the prioritized observational congruences presented here. Recently, a similar problem, which was identified for the temporal process algebra PMC [2] before, has been attacked in [42] and in [20] for a different, more classical process-algebraic setting in which only a *single visible* unprioritized action exists. It remains to be seen whether the techniques used in these papers can be employed for the priority settings investigated in this chapter.

Most process algebras that have been equipped with a notion of priority rely on an interleaving semantics, handshake communication, and a semantic theory based on bisimulation. Therefore, it should be investigated how the approaches and results presented here can be adapted to broadcasting calculi, such as Hoare's CSP [44]. Moreover, since for semantics based on local pre-emption the usual interleaving law is not valid, it is worth pursuing research regarding local pre-emption for non-interleaving semantic frameworks [4,78]. Preliminary considerations have been made in Jensen's thesis [48]. However, the insights obtained by Jensen are restricted to a structural operational semantics for a CCS-based calculus, which is defined using *asynchronous transition systems* [78]. Jensen's results do not involve a behavioral relation such as bisimulation (cf. [63]). Finally, we want to note that – to the best of our knowledge – extensions of higher-order process algebras [60,68] with concepts of priority do not yet exist. Thus, it would be interesting to see if some of the approaches presented here can be carried over.

8. Sources and acknowledgments

Major parts of this chapter have been adapted from several publications by the authors, which include two PhD theses: the results of Section 3 are taken from [26,55,64,65] and the ones of Section 4 from [28,55]; Section 5 heavily borrows from material contained in [16,55].

This work was supported by the National Aeronautics and Space Administration under NASA Contract No. NAS1-97046 while the second author was in residence at the Institute for Computer Applications in Science and Engineering (ICASE), MS 132C, NASA Langley Research Center, Hampton, Virginia 23681-2199, USA. The first author acknowledges support by NSF grants CCR-9257963, CCR-9505662, CCR-9804091, and INT-9603441, AFOSR grant F49620-95-1-0508, and ARO grant P-38682-MA.

Last but not least we would like to thank Girish Bhat, Matthew Hennessy, Michael Mendler, and Bernhard Steffen for many discussions about priority in process algebras, as well as Scott Smolka for carefully proofreading a draft of this chapter.

9. Glossary

The following table lists notation and symbols used in this chapter and is organized according to the chapter's sections.

Notation introduced in Section 2

CCS	Calculus of Communicating Systems
$\mathcal{N}$	finite initial interval of $\mathbb{N}$
λ	port name
Λ_k	ports with priority k (receiving)
λ_k, $\lambda\!:\!k$	representative of Λ_k
$\overline{\Lambda}_k$	complementary ports with priority k (sending)

$\overline{\lambda}_k, \overline{\lambda}:k$	representative of $\overline{\Lambda}_k$
$\tau_k, \tau:k$	invisible, internal action with priority k
$\mathcal{A}_k$	set of actions with priority k
Λ	set of (receiving or input) ports
$\overline{\Lambda}$	set of complementary (i.e., sending or output) ports
$a:k, b:k, \ldots$	representatives of $\Lambda \cup \overline{\Lambda}$
$\mathcal{A}$	set of actions
$\alpha:k, \beta:k, \ldots$	representatives of $\mathcal{A}$
A	set of unprioritized actions (i.e., $\mathcal{A}_0$ in 2-level priority-scheme)
α	representative of A
$\underline{A}$	set of prioritized actions (i.e., $\mathcal{A}_1$ in 2-level priority-scheme)
$\underline{\alpha}$	representative of $\underline{A}$
δ, γ	representatives of $A \cup \underline{A}$
L	subset of $\Lambda \cup \overline{\Lambda}$ (restriction set)
f	finite relabeling
$\mathcal{V}$	set of variables
x	representative of $\mathcal{V}$
$\mathcal{P}$	set of processes
$P, Q, R, \ldots$	representatives of $\mathcal{P}$
$C[X]$	representative of a context
$\mathbf{0}$	nil process
$\alpha:k.$	prefix operator
$+$	nondeterministic choice operator
$\|$	parallel-composition operator
$[f]$	relabeling operator
$\backslash L$	restriction operator
$\mu x.P, x \stackrel{\text{def}}{=} P$	recursion
$\equiv$	syntactic equality
$\mathcal{R}^+$	largest congruence contained in equivalence $\mathcal{R}$

Notation for static priority and global pre-emption (*cf. Section* 3)

$\mathsf{CCS}^{\mathsf{sg}}$	CCS with static priority and global pre-emption
$\underline{\mathcal{I}}(P)$	prioritized initial action set of P
$\underline{\mathcal{II}}(P)$	visible prioritized initial action set of P (i.e., $\underline{\mathcal{I}}(P) \setminus \{\underline{\tau}\}$)
$\stackrel{\gamma}{\longrightarrow}$	(standard) transition relation
$\xrightarrow[L]{\alpha}$	parameterized transition relation
$\stackrel{\gamma}{\Longrightarrow}_{\times}$	naive prioritized weak transition relation
$\stackrel{\underline{\alpha}}{\Longrightarrow}, \stackrel{\alpha}{\underset{L}{\Longrightarrow}}$	prioritized weak transition relation
$\stackrel{\gamma}{\Longrightarrow}_{*}$	alternative prioritized weak transition relation
$\simeq$	prioritized strong bisimulation
$\approx_{\times}$	naive prioritized weak bisimulation

$\underline{\approx}$	prioritized weak bisimulation
$\underline{\approx}^{1}$	prioritized observational congruence
$\underline{\approx}_{*}$	alternative prioritized weak bisimulation
$\underline{\approx}_{\mathrm{pd}}$	extended prioritized weak bisimulation
$\underline{\approx}^{1}_{\mathrm{pd}}$	extended prioritized observational congruence
$\lceil\cdot$	prioritization operator
$\lfloor\cdot$	deprioritization operator
$\vdash$	axiomatic derivation
$\underline{\sqsubseteq}_{\mathrm{i}}$	auxiliary pre-congruence for axiomatization

Notation for static priority and local pre-emption (*cf. Section* 4)

$\mathsf{CCS}^{\mathsf{sl}}$	CCS with static priority and local pre-emption
$\mathcal{A}_{\mathrm{addr}}$	address alphabet $\{L, R, l, r\}$
$Addr$	set of addresses
$\bullet$	address pointing to current process term
$\mathcal{L}oc$	set of transition locations
m, n, o	representatives of $\mathcal{L}oc$
$\bowtie$	location comparability relation
$\oplus$	distributed-summation operator
$\vdash_E$	axiomatic derivation
$\natural$	auxiliary predicate for axiomatization
$\underline{\mathcal{I}}_m(P)$	distributed prioritized initial action set of P with respect to location m
$\underline{\mathcal{I}}_M(P)$	... with respect to locations $m \in M$
$\underline{\mathcal{I}}(P)$	distributed prioritized initial action set of P
$\underline{\mathcal{II}}(P)$	visible distributed prioritized initial action set of P
$\underline{\mathcal{II}}_M(P)$	... with respect to locations $m \in M$ (i.e., $\underline{\mathcal{I}}_M(P) \setminus \{\underline{\tau}\}$)
$\xrightarrow[L]{\alpha}, \xrightarrow[L]{m,\alpha}$	parameterized transition relation
$\stackrel{\alpha}{\Longrightarrow}_{\times}, \stackrel{m,\alpha}{\Longrightarrow}_{\times}$	naive distributed prioritized weak transition relation
$\stackrel{\underline{\alpha}}{\Longrightarrow}, \underset{L,M}{\stackrel{m,\alpha}{\Longrightarrow}}$	distributed prioritized weak transition relation
$\underline{\simeq}$	naive distributed prioritized strong bisimulation
$\underline{\simeq}^{1}$	distributed prioritized strong bisimulation
$\underline{\simeq}_{*}$	alternative distributed prioritized strong bisimulation
$\approx_{\times}$	naive distributed prioritized weak bisimulation
$\underline{\approx}$	distributed prioritized weak bisimulation
$\underline{\approx}^{1}$	distributed prioritized observational congruence
$\underline{\approx}_{*}$	alternative distributed prioritized weak bisimulation

Notation for $\mathsf{CCS}^{\mathsf{sl}}_{\mathsf{ml}}$ (*cf. Section* 4.7)

$\mathsf{CCS}^{\mathsf{sl}}_{\mathsf{ml}}$	$\mathsf{CCS}^{\mathsf{sl}}$ with a multi-level priority-scheme
Λ	set of input ports

$a, b, \ldots$	representatives of Λ
γ	representative of $\Lambda \cup \{\tau\}$
$\overline{\Lambda}$	set of output ports
$\overline{a}, \overline{b}, \ldots$	representatives of $\overline{\Lambda}$
$\mathcal{A}$	set of actions (i.e., $\mathcal{A} =_{\mathrm{df}} \Lambda \cup \overline{\Lambda} \cup \{\tau\}$)
α	representative of $\mathcal{A}$
$\mathcal{P}^{\mathsf{sl}}_{\mathsf{ml}}$	set of $\mathsf{CCS}^{\mathsf{sl}}_{\mathsf{ml}}$ processes
$\xrightarrow{m,\alpha}$	transition relation for $\mathsf{CCS}^{\mathsf{sl}}$
$\underline{\mathcal{I}}(P)$	visible unprioritized initial actions of P
$\overline{\underline{\mathcal{I}}}(P)$	initial output action set for P
$\mathrm{I}^{k}_{m}(P)$	initial input action set for P wrt. priority k and location m
$\mathrm{I}^{<k}_{M}(P)$, $\mathrm{II}^{<k}_{M}(P)$	variants of initial input action sets
$\mathrm{I}(P)$	initial input action set of P
$\mathrm{II}(P)$	initial visible input action set of P (i.e., $\mathrm{I}(P) \setminus \{\tau\}$)
$\simeq_{\mathsf{ml}}$	distributed prioritized strong bisimulation for $\mathsf{CCS}^{\mathsf{sl}}_{\mathsf{ml}}$

Notation for $\mathsf{CCS}^{\mathsf{cw}}$ (*cf. Section* 4.8)

$\mathsf{CCS}^{\mathsf{cw}}$	CCS with priority due to Camilleri and Winskel
$+\rangle$	prioritized-choice operator
$\mathcal{P}^{\mathsf{cw}}$	set of $\mathsf{CCS}^{\mathsf{cw}}$ processes
$\vdash^{\mathsf{cw}}_{M} \xrightarrow{\alpha}$	transition relation for $\mathsf{CCS}^{\mathsf{cw}}$
$\overline{\underline{\mathcal{I}}}^{\mathsf{cw}}(P)$	initial output action set for P
$\mathrm{I}^{\mathsf{cw}}(P)$	initial input action set for P
$\mathrm{II}^{\mathsf{cw}}(P)$	initial visible input action set for P (i.e., $\mathrm{I}^{\mathsf{cw}}(P) \setminus \{\tau\}$)
$\simeq_{\mathsf{cw}}$	distributed prioritized strong bisimulation for $\mathsf{CCS}^{\mathsf{cw}}$
$\xi(\cdot)$	translation function

Notation for dynamic priority and global pre-emption (*cf. Section* 5)

$\mathsf{CCS}^{\mathsf{dg}}$	CCS with dynamic priority and global pre-emption
$\mathcal{A}$	set of actions
$\alpha, \beta, \ldots$	representatives of $\mathcal{A}$
1	clock tick (i.e., one time unit)
γ	representative of $\mathcal{A} \cup \{1\}$
$\xmapsto{\alpha}$	transition relation for action transitions
$\xmapsto{1}$	transition relation for clock transitions
$\xrightarrow{\alpha:k}$	transition relation for dynamic priorities
$\mathrm{I}^{k}(P)$	potential initial action sets for $\mathsf{CCS}^{\mathsf{dg}}$ with respect to P and priority k
$\sim_{\mathsf{rt}}$	temporal bisimulation
$\sim_{\mathsf{dg}}$	prioritized bisimulation
$[P]^{k}$	priority adjustment function

References

[1] L. Aceto, W. Fokkink and C. Verhoef, *Structural operational semantics*, Handbook of Process Algebra, J.A. Bergstra, A. Ponse and S.A. Smolka, eds, Elsevier, Amsterdam (2001), 197–292.

[2] H.R. Andersen and M. Mendler, *Complete axiomatization of observational congruence for PMC*, Technical Report ID-TR:1993-126, Department of Computer Science, Technical University of Denmark, Lyngby, Denmark (1993).

[3] J.C.M. Baeten, ed., *Applications of Process Algebra*, Cambridge Tracts in Theoret. Comput. Sci. 17, Cambridge University Press (1990).

[4] J.C.M. Baeten and T. Basten, *Partial-order process algebra (and its relation to Petri nets)*, Handbook of Process Algebra, J.A. Bergstra, A. Ponse and S.A. Smolka, eds, Elsevier, Amsterdam (2001), 769–872.

[5] J.C.M. Baeten, J.A. Bergstra and J.W. Klop, *Syntax and defining equations for an interrupt mechanism in process algebra*, Fundamenta Informaticae **IX** (1986), 127–168.

[6] J.C.M. Baeten and J.W. Klop, eds, *1st International Conference on Concurrency Theory (CONCUR '90)*, Amsterdam, The Netherlands, Lecture Notes in Comput. Sci. 458, Springer-Verlag (1990).

[7] J.C.M. Baeten and W.P. Weijland, *Process Algebra*, Cambridge Tracts in Theoret. Comput. Sci. 18, Cambridge University Press (1990).

[8] G. Barrett, *The semantics of priority and fairness in occam*, 5th International Conference on Mathematical Foundations of Programming Semantics (MFPS '89), New Orleans, Lousiana, Lecture Notes in Comput. Sci. 442, M. Main, A. Melton, M. Mislove and D. Schmidt, eds, Springer-Verlag (1989), 194–208.

[9] J.A. Bergstra and J.W. Klop, *Algebra of communicating processes with abstraction*, Theoret. Comput. Sci. **37** (1) (1985), 77–121.

[10] J.A. Bergstra, C.A. Middelburg and Y.S. Usenko, *Discrete time process algebra and the semantics of SDL*, This volume.

[11] M. Bernardo and R. Gorrieri, *A tutorial on EMPA: A theory of concurrent processes with nondeterminism, priorities, probabilities and time*, Theoret. Comput. Sci. **202** (1–2) (1998), 1–54.

[12] G. Berry, *Preemption in concurrent systems*, 13the International Conference on Foundations of Software Technology and Theoretical Computer Science (FSTTCS '93), Bombay, India, Lecture Notes in Comput. Sci. 761, R.K. Shyamasundar, ed., Springer-Verlag (1993), 72–93.

[13] G. Berry and G. Gonthier, *The ESTEREL synchronous programming language: Design, semantics, implementation*, Sci. Comput. Programming **19** (2) (1992), 87–152.

[14] E. Best, R. Devillers and M. Koutny, *A unified model for nets and process algebras*, Handbook of Process Algebra, J.A. Bergstra, A. Ponse and S.A. Smolka, eds, Elsevier, Amsterdam (2001), 873–944.

[15] E. Best and M. Koutny, *Petri net semantics of priority systems*, Theoret. Comput. Sci. **96** (1) (1992), 175–215.

[16] G. Bhat, R. Cleaveland and G. Lüttgen, *A practical approach to implementing real-time semantics*, Annals of Software Engineering (Special issue on Real-time Software Engineering) **7** (1999), 127–155.

[17] T. Bolognesi and E. Brinksma, *Introduction to the ISO specification language LOTOS*, Computer Networks and ISDN Systems **14** (1987), 25–59.

[18] G. Boudol, I. Castellani, M. Hennessy and A. Kiehn, *Observing localities*, Theoret. Comput. Sci. **114** (1) (1993), 31–61.

[19] J. Bradfield and C. Stirling, *Modal logics and mu-calculi: An introduction*, Handbook of Process Algebra, J.A. Bergstra, A. Ponse and S.A. Smolka, eds, Elsevier, Amsterdam (2001), 293–330.

[20] M. Bravetti and R. Gorrieri, *A complete axiomatization for observational congruence of prioritized finite-state behaviors*, 27th International Colloquium on Automata, Languages and Programming (ICALP 2000), Geneva, Switzerland, Lecture Notes in Comput. Sci. 1853, U. Montanari, J. Rolim, E. Welzl, eds, Springer-Verlag (2000), 744–755.

[21] P. Brémond-Grégoire, J.-Y. Choi and I. Lee, *A complete axiomatization of finite-state ACSR processes*, Inform. and Comput. **138** (2) (1997), 124–159.

[22] P. Brémond-Grégoire, S. Davidson and I. Lee, *CCSR92: Calculus for communicating shared resources with dynamic priorities*, First North American Process Algebra Workshop (NAPAW '92), Stony Brook, New York, Workshops in Computing, P. Purushothaman and A. Zwarico, eds, Springer-Verlag (1992), 65–85.

[23] P. Brémond-Grégoire, I. Lee and R. Gerber, *A process algebra of communicating shared resources with dense time and priorities*, Theoret. Comput. Sci. **189** (1/2) (1997), 179–219.

[24] J. Camilleri and G. Winskel, *CCS with priority choice*, Inform. and Comput. **116** (1) (1995), 26–37.
[25] I. Castellani, *Process algebras with localities*, Handbook of Process Algebra, J.A. Bergstra, A. Ponse and S.A. Smolka, eds, Elsevier, Amsterdam (2001), 945–1045.
[26] R. Cleaveland and M.C.B. Hennessy, *Priorities in process algebras*, Inform. and Comput. **87** (1/2) (1990), 58–77.
[27] R. Cleaveland, G. Lüttgen and M. Mendler, *An algebraic theory of multiple clocks*, 8th International Conference on Concurrency Theory (CONCUR '97), Warsaw, Poland, Lecture Notes in Comput. Sci. 1243, A. Mazurkiewicz and J. Winkowski, eds, Springer-Verlag (1997), 166–180,
[28] R. Cleaveland, G. Lüttgen and V. Natarajan, *A process algebra with distributed priorities*, Theoret. Comput. Sci. **195** (2) (1998), 227–258.
[29] R. Cleaveland, V. Natarajan, S. Sims and G. Lüttgen, *Modeling and verifying distributed systems using priorities: A case study*, Software – Concepts and Tools **17** (2) (1996), 50–62.
[30] R. Cleaveland and O. Sokolsky, *Equivalence and preorder checking for finite-state systems*, Handbook of Process Algebra, J.A. Bergstra, A. Ponse and S.A. Smolka, eds, Elsevier, Amsterdam (2001), 391–424.
[31] J. Davies and S. Schneider, *A brief history of Timed CSP*, Theoret. Comput. Sci. **138** (2) (1995), 243–271.
[32] P. Degano and C. Priami, *Causality of mobile processes*, International Conference on Automata, Languages and Programming (ICALP '95), Szeged, Hungary, Lecture Notes in Comput. Sci. 944, Z. Fülöp and F. Gécseg, eds, Springer-Verlag (1995), 660–671.
[33] C.J. Fidge, *A formal definition of priority in CSP*, ACM Transactions on Programming Languages and Systems **15** (4) (1993), 681–705.
[34] N. Francez, *Fairness*, Springer-Verlag (1986).
[35] R. Gerber and I. Lee, *CCSR: A calculus for communicating shared resources*, *Applications of Process Algebra*, Cambridge Tracts in Theoretical Comput. Sci. 17, J.C.M. Baeten, ed., Cambridge University Press (1990), 263–277.
[36] R. Gerber and I. Lee, *A resourced-based prioritized bisimulation for real-time systems*, Inform. and Comput. **113** (1) (1994), 102–142.
[37] S.M. German, *Programming in a general model of synchronization*, 3rd International Conference on Concurrency Theory (CONCUR '92), Stony Brook, NY, Lecture Notes in Comput. Sci. 549, R. Cleaveland, ed., Springer-Verlag (1992), 534–549.
[38] H. Hansson and F. Orava, *A process calculus with incomparable priorities*, First North American Process Algebra Workshop (NAPAW '92), Stony Brook, New York, Workshops in Computing, P. Purushothaman and A. Zwarico, eds, Springer-Verlag (1992), 43–64.
[39] D. Harel, *Statecharts: A visual formalism for complex systems*, Sci. Comput. Programming **8** (1987), 231–274.
[40] M.C.B. Hennessy, *An algebraic theory of fair asynchronous communicating processes*, Theoret. Comput. Sci. **49** (1987), 121–143.
[41] M.C.B. Hennessy, *Algebraic Theory of Processes*, MIT Press (1988).
[42] H. Hermanns and M. Lohrey, *Priority and maximal progress are completely axiomatisable*, 9th International Conference on Concurrency Theory (CONCUR '98), Nice, France, Lecture Notes in Comput. Sci. 1466, D. Sangiorgi and R. de Simone, eds, Springer-Verlag (1998), 237–252.
[43] H. Hermanns, M. Rettelbach and T. Weiß, *Formal characterisation of immediate actions in SPA with nondeterministic branching*, Comput. J. **38** (7) (1995), 530–541.
[44] C.A.R. Hoare, *Communicating Sequential Processes*, Prentice Hall (1985).
[45] INMOS Limited, *Occam Programming Manual*, International Series in Computer Science, Prentice Hall (1984).
[46] R. Janicki and M. Koutny, *Semantics of inhibitor nets*, Inform. and Comput. **123** (1) (1995), 1–17.
[47] A. Jeffrey, *Translating timed process algebra into prioritized process algebra*, Symposium on Real-Time and Fault-Tolerant Systems (FTRTFT '92), Lecture Notes in Comput. Sci. 571, J. Vytopil, ed., Springer-Verlag (1992), 493–506.
[48] C.-T. Jensen, *Prioritized and independent actions in distributed computer systems*, Ph.D. Thesis, Aarhus University, Denmark (1994).
[49] B. Jonsson, W. Yi and K.G. Larsen, *Probabilistic extensions of process algebras*, Handbook of Process Algebra, J.A. Bergstra, A. Ponse and S.A. Smolka, eds, Elsevier, Amsterdam (2001), 685–710.

[50] P.C. Kanellakis and S.A. Smolka, *CCS expressions, finite state processes, and three problems of equivalence*, Inform. and Comput. **86** (1) (1990), 43–68.

[51] Kempe Software Capital Enterprises, *Ada95 reference manual: Language and standard libraries* (1995). Available at http://www.adahome.com.

[52] L. Lamport, *What it means for a concurrent program to satisfy a specification: Why no one has specified priority*, 12th Annual ACM Symposium on Principles of Programming Languages (POPL '85), New York, IEEE Computer Society Press (1985), 78–83.

[53] F. Levi, *verification of temporal and real-time properties of Statecharts*, Ph.D. Thesis, University of Pisa-Genova-Udine, Pisa, Italy (1997).

[54] G. Lowe, *Probabilistic and prioritized models of timed CSP*, Theoret. Comput. Sci. **138** (2) (1995), 315–352.

[55] G. Lüttgen, *Pre-emptive modeling of concurrent and distributed systems*, Ph.D. Thesis, University of Passau, Germany, Shaker-Verlag (1998).

[56] G. Lüttgen, M. von der Beeck and R. Cleaveland, *Statecharts via process algebra*, 10th International Conference on Concurrency Theory (CONCUR '99), Eindhoven, The Netherlands, Lecture Notes in Comput. Sci. 1664, J.C.M. Baeten and S. Mauw, eds, Springer-Verlag (1999) 399–414.

[57] F. Maraninchi, *The ARGOS language: Graphical representation of automata and description of reactive systems*, 1991 IEEE Workshop on Visual Languages, Los Alamitos, California, IEEE Computer Society Press (1991).

[58] R. Milner, *Communication and Concurrency*, Prentice Hall (1989).

[59] R. Milner, *A complete axiomatisation for observational congruence of finite-state behaviours*, Inform. and Comput. **81** (2) (1989), 227–247.

[60] R. Milner, J. Parrow and D. Walker, *A calculus of mobile processes, parts I and II*, Inform. and Comput. **100** (1) (1992), 1–77.

[61] F. Moller and C. Tofts, *A temporal calculus of communicating systems*, *Applications of Process Algebra*, Cambridge Tracts in Theoretical Comput. Sci. 17, J.C.M. Baeten, ed., Cambridge University Press (1990), 401–415.

[62] U. Montanari and D. Yankelevich, *Location equivalence in a parametric setting*, Theoret. Comput. Sci. **149** (2) (1995), 299–332.

[63] M. Mukund and M. Nielsen, *CCS, locations and asynchronous transition systems*, 12th International Conference on Foundations of Software Technology and Theoretical Computer Science (FSTTCS '92), New Delhi, India, Lecture Notes in Comput. Sci. 652, R.K. Shyamasundar, ed., Springer-Verlag (1992), 328–341.

[64] V. Natarajan, *Degrees of delay: Semantic theories for priority, efficiency, fairness, and predictability in process algebras*, Ph.D. Thesis, North Carolina State University, Raleigh, North Carolina (1996).

[65] V. Natarajan, L. Christoff, I. Christoff and R. Cleaveland, *Priorities and abstraction in process algebra*, 14th International Conference on Foundations of Software Technology and Theoretical Computer Science (FSTTCS '94), Madras, India, Lecture Notes in Comput. Sci. 880, P.S. Thiagarajan, ed., Springer-Verlag (1994), 217–230.

[66] R. Paige and R.E. Tarjan, *Three partition refinement algorithms*, SIAM J. Comput. **16** (6) (1987), 973–989.

[67] D.M.R. Park, *Concurrency and automata on infinite sequences*, 5th GI Conference on Theoretical Computer Science, Lecture Notes in Comput. Sci. 104, P. Deussen, ed., Springer-Verlag (1981), 167–183.

[68] J. Parrow, *An introduction to the π-calculus*, Handbook of Process Algebra, J.A. Bergstra, A. Ponse and S.A. Smolka, eds, Elsevier, Amsterdam (2001), 479–543.

[69] G.D. Plotkin, *A structural approach to operational semantics*, Technical Report DAIMI-FN-19, Computer Science Department, Aarhus University, Denmark (1981).

[70] K.V.S. Prasad, *Programming with broadcasts*, 4th International Conference on Concurrency Theory (CONCUR '93), Hildesheim, Germany, Lecture Notes in Comput. Sci. 715, E. Best, ed., Springer-Verlag (1993), 173–187.

[71] K.V.S. Prasad, *Broadcasting with priority*, 5th European Symposium on Programming (ESOP '94), Edinburgh, United Kingdom, Lecture Notes in Comput. Sci. 788, D. Sannella, ed., Springer-Verlag (1994), 469–484.

[72] W. Reisig, *Petri Nets: An Introduction*, Springer-Verlag (1985).

[73] S.A. Smolka and B. Steffen, *Priority as extremal probability*, Formal Aspects of Computing **8** (5) (1996), 585–606.

[74] C. Tofts, *Processes with probabilities, priority and time*, Formal Aspects of Computing **6** (5) (1994), 536–564.

[75] A.C. Uselton and S.A. Smolka, *A compositional semantics for Statecharts using labeled transition systems*, 5th International Conference on Concurrency Theory (CONCUR '94), Uppsala, Sweden, Lecture Notes in Comput. Sci. 836, B. Jonsson and J. Parrow, eds, Springer-Verlag (1994), 2–17.

[76] R. van Glabbeek, S.A. Smolka and B. Steffen, *Reactive, generative, and stratified models of probabilistic processes*, Inform. and Comput. **121** (1) (1995), 59–80.

[77] C. Verhoef, *A congruence theorem for structured operational semantics with predicates and negative premises*, Nordic J. Comput. **2** (2) (1995), 274–302.

[78] G. Winskel and M. Nielsen, *Models for concurrency*, Handbook of Logic in Computer Science, Vol. 4, S. Abramsky, D.M. Gabbay and T.S.E. Maibaum, eds, Oxford Science Publications (1995), 1–148.

[79] W. Yi, *CCS + time = an interleaving model for real time systems*, 18th International Colloquium on Automata, Languages and Programming (ICALP '91), Madrid, Spain, Lecture Notes in Comput. Sci. 510, J. Leach Albert, B. Monien and M. Rodríguez-Artalejo, eds, Springer-Verlag (1991), 217–228.

Subject index

action transition, 749
axiomatization of distributed prioritized strong bisimulation, 735
axiomatization of prioritized observational congruence, 728
axiomatization of prioritized strong bisimulation, 720

clock transition, 749

deprioritization operator, 725
distributed summation operator, 735

interrupt, 713
interrupt port, 713

largest congruence of distributed prioritized observational congruence, 739
largest congruence of distributed prioritized strong bisimulation, 734
largest congruence of prioritized observational congruence, 722, 729
location, 730
location comparability relation, 732

maximal progress assumption, 748
may pre-emption, 714
must pre-emption, 714

pre-emption, 713
pre-emption potential, 722
prioritization operator, 725
prioritized action, 717
prioritized choice, 714
prioritized strong bisimulation, 718
priority scheme, 716
priority semantics, 751

real-time, 713, 748
real-time semantics, 749

static priority, 714
synchrony hypothesis, 713

unprioritized action, 717

Part 5
Non-Interleaving Process Algebra

CHAPTER 13

Partial-Order Process Algebra (and its Relation to Petri Nets)

J.C.M. Baeten[1], T. Basten[2,*]

[1]Department of Computing Science, Eindhoven University of Technology, PO Box 513, NL-5600 MB, The Netherlands
E-mail: josb@win.tue.nl

[2]Department of Electrical Engineering, Eindhoven University of Technology, PO Box 513, NL-5600 MB, The Netherlands
E-mail: tbasten@ics.ele.tue.nl

Contents

1. Introduction . . . 771
2. Notation . . . 773
3. Process theory . . . 774
4. Labeled Place/Transition nets . . . 778
5. Process algebra . . . 783
   5.1. The equational theory . . . 783
   5.2. A semantics for the equational theory . . . 793
   5.3. A total-order semantics . . . 801
   5.4. A step semantics . . . 804
   5.5. Concluding remarks . . . 808
6. The causal state operator . . . 809
   6.1. The equational theory . . . 809
   6.2. A semantics . . . 814
   6.3. An algebraic semantics for labeled P/T nets . . . 815
   6.4. Concluding remarks . . . 818
7. Intermezzo on renaming and communication functions . . . 818
8. Modular P/T nets . . . 820
   8.1. Place fusion . . . 820
   8.2. Transition fusion . . . 828
   8.3. Concluding remarks . . . 830
9. Cause addition . . . 831
   9.1. The equational theory . . . 831
   9.2. A semantics . . . 832

*Part of this work was done while the author was employed at the Department of Computing Science, Eindhoven University of Technology, The Netherlands.

HANDBOOK OF PROCESS ALGEBRA
Edited by Jan A. Bergstra, Alban Ponse and Scott A. Smolka

9.3. Buffers 833
9.4. Cause addition and sequential composition 836
9.5. Concluding remarks 842
10. A non-interleaving process algebra 842
10.1. The algebra of non-terminating serializable processes 843
10.2. A brief discussion on causality 852
10.3. Elimination of choice, sequential composition, and communication 853
10.4. Concluding remarks 865
11. Conclusions 866
References 867
Subject index 870

Abstract

To date, many different formalisms exist for describing and analyzing the behavior of concurrent systems. Petri nets and process algebras are two well-known classes of such formalisms. Petri-net theory is well suited for reasoning about concurrent systems in a partial-order framework; it handles causal relationships between actions of concurrent systems in an explicit way. Process algebras, on the other hand, often provide a total-order framework, which means that information about causalities is not always accurate. This chapter illustrates how to develop a partial-order process algebra in the style of ACP. It is shown how to extend such an algebraic theory with a causality mechanism inspired by Petri-net theory. In addition, the chapter clarifies the concepts of interleaving and non-interleaving process algebra; total-order semantics for concurrent systems are often incorrectly referred to as interleaving semantics.

1. Introduction

The behavior of parallel and distributed systems, often called *concurrent systems*, is a popular topic in the literature on (theoretical) computing science. Numerous formal languages for describing and analyzing the behavior of concurrent systems have been developed. The meaning of expressions in such a formal language is often captured in a so-called formal *semantics*. The objective of a formal semantics is to create a precise and unambiguous framework for reasoning about concurrent systems. Along with the development of the large number of formal languages for describing concurrent systems, an almost equally large number of different semantics has been proposed. The existence of so many different semantics for concurrent systems has created a whole new area of research named *comparative concurrency semantics* [43]. The primary purpose of comparative concurrency semantics is the classification of semantics for concurrent systems in a meaningful way.

On a high level of abstraction, the behavior of a concurrent system is often represented by the actions that the system can perform and the ordering of these actions. Such an abstract view of the behavior of a concurrent system is called a *process*. The semantics of a formal language for describing the behavior of concurrent systems defines a process for each expression in the formal language. Two expressions in a formal language describe the same system if and only if they correspond to equivalent processes in the semantics, where the equivalence of processes is determined by a so-called *semantic equivalence*.

The quest of developing formalisms and semantics that are well suited for describing and analyzing the behavior of concurrent systems is characterized by a number of ongoing discussions on classifications of semantic equivalences. One discussion is centered around linear-time semantics versus branching-time semantics. In a linear-time semantics, two processes that agree on the ordering of actions are considered equivalent. However, such processes may differ in their branching structure, where the branching structure of a process is determined by the moments that choices between alternative branches of behavior are made. A branching-time semantics distinguishes processes with the same ordering of actions but different branching structures.

Another discussion focuses on total-order semantics versus partial-order semantics. In a total-order semantics, actions of a process are always totally ordered, whereas in a partial-order semantics, actions may occur simultaneously or causally independent of each other. Partial-order semantics are often referred to as *true-concurrency* semantics, because they are well suited to express concurrency of actions. A typical characteristic of a total-order semantics is that concurrency of actions is equivalent to non-determinism: A process that performs two actions in parallel is equivalent to a process that chooses non-deterministically between the two possible total orderings of the two actions. Thus, a total-order semantics abstracts from the causal dependencies between actions.

Total-order semantics are often confused with *interleaving* semantics. As pointed out in [2], the term "interleaving" originates from one specific class of formal languages for describing concurrent systems, namely process algebras. Process-algebraic theories have in common that processes are represented by terms constructed from action constants and operators such as choice (alternative composition), sequential composition, and parallel composition (merge operator, interleaving operator). A set of axioms or equational laws specifies which processes must be considered equal. Most process-algebraic theories contain some form of *expansion theorem*. An expansion theorem states that parallel composition can be expressed equivalently in terms of choice and sequential composition. A process-algebraic theory with an expansion theorem is called an *interleaving theory*; the semantics of such an algebraic theory is called an *interleaving semantics* or an *interleaving process algebra*. An algebraic theory without an expansion theorem is said to be *non-interleaving*. The semantics of such a theory is a *non-interleaving semantics* or a *non-interleaving process algebra*. As mentioned, most process-algebraic theories are interleaving theories. Very often, such an interleaving theory has a total-order semantics, which causes the confusion between the terms "total-order" and "interleaving". However, in [2], it is shown that it is possible to develop both process-algebraic theories with an interleaving, partial-order semantics and algebraic theories with a non-interleaving, total-order semantics. These examples show that the characterizations interleaving versus non-interleaving and total-order versus partial-order for process algebras are orthogonal. The above discussion also makes clear that, for semantics of non-algebraic languages for describing concurrent systems, the characterization interleaving versus non-interleaving is not meaningful.

This chapter can be situated in the field of comparative concurrency semantics. Its main topic is to illustrate a way to develop a process-algebraic theory with a partial-order semantics. The starting point is an algebraic theory in the style of the Algebra of Communicating Processes (ACP) [13]. The basic semantic equivalence that is used throughout the chapter is bisimilarity [51]. Bisimilarity is often used to provide process-algebraic theories with a semantics that, in the terminology of this chapter, can be characterized as a branching-time, interleaving, total-order semantics. An interesting aspect of bisimilarity is that it can be turned into a semantic equivalence called *step* bisimilarity [49] that provides the basis for a partial-order view on the behavior of concurrent systems. By means of step bisimilarity, it is possible to obtain a process-algebraic theory with a branching-time, interleaving, *partial*-order semantics in a relatively straightforward way. Furthermore, we show how to obtain a *non-interleaving* variant of such a process algebra. This chapter does not discuss variations of process-algebraic theories in the linear-time/branching-time spectrum. The interested reader is referred to [28,29] and [27], Chapter 1 of this Handbook.

Another important theme of this chapter is the study of concepts known from Petri-net theory [53] in the process-algebraic framework developed in this chapter. The Petri-net formalism is a well-known theory for describing and analyzing concurrent systems. Petri nets have been used both as a language for describing concurrent systems and as a semantic framework for providing other languages for describing concurrent systems with a formal semantics. The Petri-net formalism is particularly well suited for creating a partial-order framework for reasoning about the behavior of concurrent systems. The non-interleaving, partial-order process algebra of Section 10 of this chapter is an algebra that incorporates a number of the most important concepts of the Petri-net formalism. In particular, it adopts the standard Petri-net mechanism for handling causalities.

The chapter is written in the style of a tutorial. This means that it is self-contained, focuses on concepts, and contains many (small) examples and detailed explanations. The goal is not to develop a complete framework with a formal description language and semantics that is applicable to the development of large, complex concurrent systems. In addition, we do not claim that the approach of this chapter is the only way to obtain a process-algebraic theory with a partial-order semantics. It is our aim to provide a conceptual understanding of several important concepts that play a role in describing and analyzing the behavior of concurrent systems. A better understanding of these concepts can be useful in the development of formalisms that are sufficiently powerful to support the development of large and complex systems. The chapter combines and extends some of the ideas and results that appeared earlier in [2,4], and [8, Chapter 3].

The remainder of this chapter is organized as follows. Section 2 introduces some notation for bags, which are omnipresent in the remainder of this chapter, and a convenient notation for quantifiers. In Section 3, the basic semantic framework used throughout this chapter is defined. The framework of labeled transition systems is used to formalize the notion of a process and bisimilarity of processes. It is shown how labeled transition systems can be used to obtain both a total-order view of concurrent systems and a partial-order view, where the latter is based on the notion of step bisimilarity. Section 4 introduces a class of Petri nets called labeled P/T nets. The framework of labeled transition systems is used to define both a total-order semantics and a step semantics for labeled P/T nets. Section 5 provides an introduction to standard ACP-style process algebra. As in Section 4, labeled transition systems form the basis for both a total-order and a partial-order framework. In Section 6, the algebraic framework of Section 5 is extended with a class of algebraic operators, called causal state operators. This class of operators is inspired by the Petri-net approach to defining the behavior of concurrent systems. It is shown that the resulting algebraic framework is sufficiently powerful to capture the semantics of labeled P/T nets in algebraic terms. Section 7 contains a brief intermezzo on algebraic renaming and communication functions, which is useful in the remaining sections. Section 8 describes two approaches to modularizing P/T nets. A modular P/T net models a system component that may interact with its environment via a well defined interface. It is explained how modular P/T nets in combination with the algebraic framework of Section 6 can be used to develop a compositional formalism for modeling and analyzing concurrent systems. The resulting formalism is a step towards a framework supporting the development of complex concurrent systems. In Section 9, the algebraic framework of Section 6 is extended with a class of so-called cause-addition operators. Cause-addition operators allow for the explicit specification of causalities in algebraic expressions. This class of operators is inspired by the way causalities are handled in Petri-net theory. The relation between cause addition and sequential composition, which is the most important operator for specifying causal orderings in process algebra, is studied. Section 10 studies a process algebra that incorporates several of the important characteristics of Petri-net theory. It is a non-interleaving, partial-order process algebra that includes the classes of causal state operators and cause-addition operators. In the context of this algebra, the relation between the causality mechanisms of standard ACP-style process algebra and Petri-net theory is investigated. Finally, Section 11 summarizes the most important conclusions of this chapter. It also provides some pointers to related work and it identifies some interesting topics for future study.

2. Notation

Bags. In this chapter, bags are defined as finite multi-sets of elements from some alphabet A. A bag over alphabet A can be considered as a function from A to the natural numbers

$\mathbb{N}$ such that only a finite number of elements from A is assigned a non-zero function value. For some bag X over alphabet A and $a \in A$, $X(a)$ denotes the number of occurrences of a in X, often called the cardinality of a in X. The set of all bags over A is denoted $\mathcal{B}(A)$. Note that any finite set of elements from A also denotes a unique bag over A, namely the function yielding 1 for every element in the set and 0 otherwise. The empty bag, which is the function yielding 0 for any element in A, is denoted $\mathbf{0}$. For the explicit enumeration of a bag, a notation similar to the notation for sets is used, but using square brackets instead of curly brackets and using superscripts to denote the cardinality of the elements. For example, $[a^2 \mid P(a)]$ contains two elements a for every a such that $P(a)$ holds, where P is some predicate on symbols of the alphabet under consideration. To denote individual elements of a bag, the same symbol "$\in$" is used as for sets: for any bag X over alphabet A and element $a \in A$, $a \in X$ if and only if $X(a) > 0$. The sum of two bags X and Y, denoted $X \uplus Y$, is defined as $[a^n \mid a \in A \wedge n = X(a) + Y(a)]$. The difference of X and Y, denoted $X - Y$, is defined as $[a^n \mid a \in A \wedge n = (X(a) - Y(a)) \max 0]$. The binding of sum and difference is left-associative. The restriction of X to some domain $D \subseteq A$, denoted $X \upharpoonright D$, is defined as $[a^{X(a)} \mid a \in D]$. Restriction binds stronger than sum and difference. The notion of subbags is defined as expected: Bag X is a subbag of Y, denoted $X \leqslant Y$, if and only if for all $a \in A$, $X(a) \leqslant Y(a)$.

Quantifiers. For quantifiers, we use the convenient notation of [23] in which the bound variables are represented explicitly. A quantifier is a commutative and associative operator with a unit element. Examples of well-known quantifiers are the logical quantifiers $\forall$ and $\exists$. Given a quantifier $\mathcal{Q}$ and a non-empty list of variables $\bar{x}$, a quantification over $\bar{x}$ is written in the following format: $(\mathcal{Q}\,\bar{x} : D(\bar{x}) : E(\bar{x}))$, where D is a predicate that specifies the *domain* of values over which the variables in $\bar{x}$ range and where $E(\bar{x})$ is the quantified *expression*. Note that also the explicit enumeration of sets and bags could be written in quantifier notation. However, for sets we adhere to the standard notation with curly brackets and for bags we use the notational conventions introduced in the previous paragraph.

Another notational convention concerning quantifiers is the following. Any *binary* commutative and associative operator $\oplus$ with a unit element can be generalized to a quantifier, also denoted $\oplus$, in a straightforward way. Consider, for example, the binary $+$ on natural numbers. The sum of all natural numbers less than 10 can be written as follows: $(+\,x : x \in \mathbb{N} \wedge x < 10 : x)$.

3. Process theory

A natural and straightforward way to describe the behavior of a concurrent system is a *labeled transition system*. A labeled transition system consists of a set of *states* plus a *transition relation* on states. Each transition is labeled with an *action*. An action may be any kind of activity performed by a concurrent system, such as a computation local to some component of the system, a procedure call, the sending of a message, or the receipt of a message. However, for modeling purposes, the details of actions are often not important. The set of states in a labeled transition system is an abstraction of all possible states of a concurrent system. The transition relation describes the change in the state of a concurrent system when some action, represented by the label of the transition, is performed.

The framework of labeled transition systems introduced in this section is used throughout the remainder of this chapter to provide the Petri-net formalism of the next section and the various process-algebraic theories of the other sections with a formal semantics. Labeled transition systems can be used to define both a total-order and a partial-order semantics for concurrent systems. The difference is made by choosing an appropriate definition for the actions of a labeled transition system. In a total-order view, actions are assumed to be atomic entities without internal structure. A system can perform only a single atomic action at a time. Thus, the actions are totally ordered. In a partial-order view, a concurrent system can perform causally independent atomic actions simultaneously. Atomic actions are no longer totally ordered. It is even possible that a system performs the same atomic action several times in parallel, because also different occurrences of the same action may be causally independent. Thus, in a partial-order view on the behavior of concurrent systems, actions are *bags* of atomic actions. In this case, actions are often called *multi-actions* or *steps*. A partial-order semantics of concurrent systems that is based on the notion of steps is often called a *step semantics*.

The basic notion in the framework of labeled transition systems used in this chapter is the so-called *process space*. A process space describes a *set of processes*. A process space is a labeled transition system as described above extended with a termination predicate on states. Each state in a process space can be interpreted as the initial state of a process. A *process* is a labeled transition system extended with a termination predicate and a distinguished initial state. The termination predicate of a process defines in what states the process *can terminate successfully*. If a process is in a state where it cannot perform any actions or terminate successfully, then it is said to be in a *deadlock*. The possibility to distinguish between successful termination and deadlock is often useful.

A predicate on the elements of some set is represented as a subset of the set: It holds for elements in the subset and it does not hold for elements outside the subset.

DEFINITION 3.1 (*Process space*). A *process space* is a quadruple $(\mathcal{P}, \mathcal{A}, \rightarrow, \downarrow)$, where $\mathcal{P}$ is a set of process states, $\mathcal{A}$ is a set of actions, $\_ \stackrel{-}{\longrightarrow} \_ \subseteq \mathcal{P} \times \mathcal{A} \times \mathcal{P}$ is a ternary transition relation, and $\downarrow \_ \subseteq \mathcal{P}$ is a termination predicate.

Let $(\mathcal{P}, \mathcal{A}, \rightarrow, \downarrow)$ be some process space. Each state p in $\mathcal{P}$ uniquely determines a process that consists of all states reachable from p.

DEFINITION 3.2 (*Reachability*). The *reachability relation* $\stackrel{*}{\Longrightarrow} \subseteq \mathcal{P} \times \mathcal{P}$ is defined as the smallest relation satisfying, for any $p, p', p'' \in \mathcal{P}$ and $\alpha \in \mathcal{A}$,

$$p \stackrel{*}{\Longrightarrow} p \quad \text{and}$$
$$(p \stackrel{*}{\Longrightarrow} p' \wedge p' \stackrel{\alpha}{\longrightarrow} p'') \Rightarrow p \stackrel{*}{\Longrightarrow} p''.$$

Process state p' is said to be *reachable* from state p if and only if $p \stackrel{*}{\Longrightarrow} p'$. The set of all states reachable from p is denoted $p*$.

DEFINITION 3.3 (*Process*). Let p be a state in $\mathcal{P}$. The *process* defined by p is the tuple $(p, p*, \mathcal{A}, \rightarrow \cap (p* \times \mathcal{A} \times p*), \downarrow \cap p*)$. Process state p is called the *initial state* of the process.

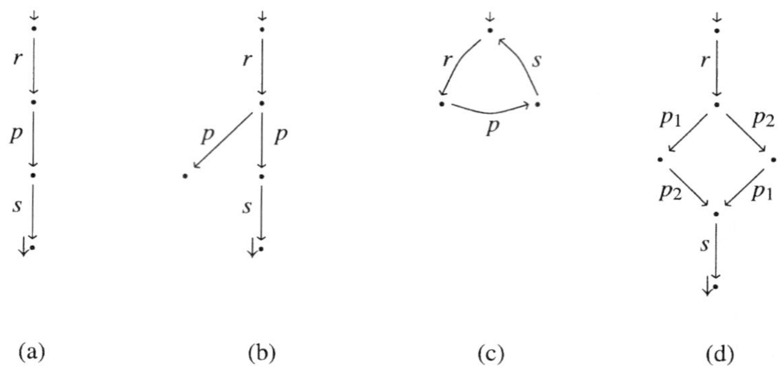

Fig. 1. Some simple examples of processes.

In the remainder, processes are identified with their initial states. Sometimes, it is intuitive to think about elements of $\mathcal{P}$ as states, whereas sometimes it is more natural to see them as processes.

EXAMPLE 3.4. Let $(\mathcal{P}, \mathcal{A}, \rightarrow, \downarrow)$ be some process space, where the set of actions equals a set of atomic actions A that includes the atomic actions p, p_1, p_2, r, and s. Figure 1 shows some examples of processes. Process states are depicted as dots. The initial state of a process is marked with a small incoming arrow. The process in Figure 1(a) is a simple sequential process. It represents, for example, a system that receives a message from its environment, processes this message, sends the processed message to its environment, and terminates successfully. The process in Figure 1(b) is almost the same, except that the processing action may result in a deadlock. The third process is a variant that iterates the behavior of the system in (a) without a possibility of termination. Finally, the process in Figure 1(d) represents a system that performs two processing actions in any order on the incoming message before it returns the processed message to its environment. In a total-order view on concurrent systems, one could also say that the system performs the two processing actions in parallel. There is no way to distinguish concurrency from non-determinism. A total-order semantics abstracts from the causal dependencies between the occurrences of actions.

EXAMPLE 3.5. Consider again a process space $(\mathcal{P}, \mathcal{A}, \rightarrow, \downarrow)$. Let A be the same set of atomic actions as introduced in the previous example. Assume that the set of actions $\mathcal{A}$ is defined as $\mathcal{B}(A)$, the set of bags over A. Figure 2(a) shows the process corresponding to the process in Figure 1(d). It represents a system that receives a message, performs two processing actions in any order, and returns the processed message to the environment. The occurrences of the two processing actions are causally *dependent*. The process of Figure 2(a) does *not* model a system that may perform two causally *independent* processing actions. Such a system is represented by the process depicted in Figure 2(b). It performs the two processing actions in any order *or* simultaneously. Thus, in a step semantics it is possible to distinguish concurrency from non-determinism.

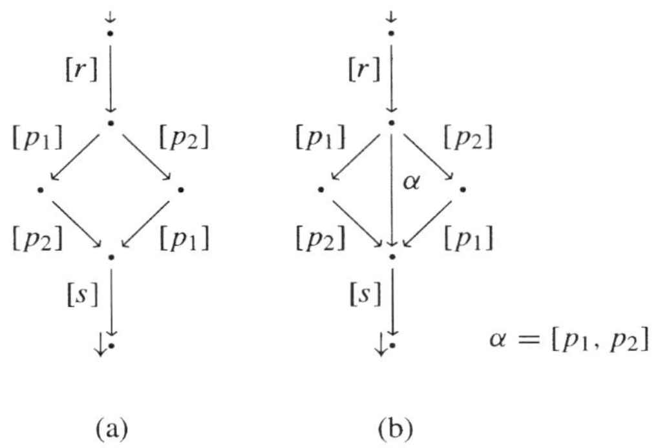

Fig. 2. Concurrency versus non-determinism in a step semantics.

The above example claims that the two processes depicted in Figure 2 are different. This raises the question when precisely processes are equivalent or not. To answer this question, the following definition introduces the so-called bisimilarity relation on processes, which is the basic semantic equivalence used throughout this chapter. It states that two processes are equivalent if and only if they can copy each others behavior and if they have the same termination options. The notion of bisimilarity was originally introduced in [51]. In [28] and [27] (Chapter 1 of this Handbook) bisimilarity is studied in more detail and a comparison with other semantic equivalences is made.

DEFINITION 3.6 (*Bisimilarity*). A binary relation $\mathcal{R} \subseteq \mathcal{P} \times \mathcal{P}$ is called a *bisimulation* if and only if, for any $p, p', q, q' \in \mathcal{P}$ and $\alpha \in \mathcal{A}$,

(i) $p\mathcal{R}q \wedge p \xrightarrow{\alpha} p' \Rightarrow (\exists q' : q' \in \mathcal{P} : q \xrightarrow{\alpha} q' \wedge p'\mathcal{R}q')$,

(ii) $p\mathcal{R}q \wedge q \xrightarrow{\alpha} q' \Rightarrow (\exists p' : p' \in \mathcal{P} : p \xrightarrow{\alpha} p' \wedge p'\mathcal{R}q')$, and

(iii) $p\mathcal{R}q \Rightarrow (\downarrow p \Leftrightarrow \downarrow q)$.

Two processes p and q are called *bisimilar*, denoted $p \sim q$, if and only if there exists a bisimulation $\mathcal{R}$ such that $p\mathcal{R}q$.

PROPERTY 3.7. *Bisimilarity, $\sim$, is an equivalence relation.*

PROOF. Reflexivity of bisimilarity follows from the fact that the identity relation on processes is a bisimulation. Symmetry follows immediately from the definition of bisimilarity. Transitivity follows from the fact that the relation composition of two bisimulations is again a bisimulation. □

An important property of bisimilarity is that it preserves moments of choice in processes as well as deadlocks.

EXAMPLE 3.8. As before, let $(\mathcal{P}, \mathcal{A}, \rightarrow, \downarrow)$ be some process space, with $\mathcal{A}$ defined as the set of atomic actions A of Example 3.4. Consider the three processes depicted in Figure 3. Although all three are different, they appear to behave similarly. In all three cases,

than simple markers. (See the graphical representation of a labeled P/T net in Figure 4.) The state of a net is defined as the distribution of the tokens over the places. In Petri-net theory, the state of a net is often called the *marking* of the net.

DEFINITION 4.4 (*Marked, labeled P/T net*). A *marked*, L-labeled P/T net is a pair (N, m), where $N = (P, T, F, W, \ell)$ is an L-labeled P/T net and where m is a bag over P denoting the marking of the net. The set of all marked, L-labeled P/T nets is denoted $\mathcal{N}$.

The behavior of marked, labeled P/T nets is defined by a so-called *firing rule*, which is simply a transition relation defining the change in the state of a marked net when executing an action. To define a firing rule, it is necessary to formalize when a net is allowed to execute a certain action. The following definitions form the basis for a total-order semantics for labeled P/T nets.

DEFINITION 4.5 (*Transition enabling*). Let (N, m) with $N = (P, T, F, W, \ell)$ be a marked, labeled P/T net in $\mathcal{N}$. A transition $t \in T$ is *enabled*, denoted $(N, m)[t\rangle$, if and only if each of its input places p contains at least as many tokens as the cardinality of p in $\boldsymbol{i}t$. That is, $(N, m)[t\rangle \Leftrightarrow \boldsymbol{i}t \leqslant m$.

EXAMPLE 4.6. In the marked net of Figure 4, transition r is enabled. None of the other transitions is enabled.

When a transition t of a labeled P/T net is enabled, the net can *fire* this transition. Upon firing, t removes $W(p, t)$ tokens from each of its input places p; it adds $W(t, p)$ tokens to each of its output places p. This means that, upon firing t, the marked net (N, m) changes into another marked net $(N, m - \boldsymbol{i}t \uplus \boldsymbol{o}t)$. When firing t, the labeled P/T net executes an *action*, defined by its label. That is, when firing transition t, the P/T net executes the action $\ell(t)$.

DEFINITION 4.7 (*Total-order firing rule*). The *total-order firing rule* $\_[\_\rangle^t\_ \subseteq \mathcal{N} \times L \times \mathcal{N}$ is the smallest relation satisfying for any marked, labeled P/T net (N, m) in $\mathcal{N}$, where $N = (P, T, F, W, \ell)$, and any transition $t \in T$,

$$(N, m)[t\rangle \Rightarrow (N, m)\ [\ell(t)\rangle^t\ (N, m - \boldsymbol{i}t \uplus \boldsymbol{o}t).$$

Tokens that are removed from the marking when firing a transition are often referred to as *consumed* tokens or the *consumption* of a transition; tokens that are added to the marking are referred to as *produced* tokens or the *production* of a transition.

EXAMPLE 4.8. In the P/T net of Figure 4, firing transition r changes the marking from $[1]$ to $[2, 3]$. In the new marking, both transitions p_1 and p_2 are enabled. As a result, *either* p_1 can fire yielding marking $[3, 4]$ *or* p_2 can fire yielding marking $[2, 5]$. After firing one of these two transitions, the other one is the only transition in the net that can still fire. Independent of the order in which p_1 and p_2 are fired the resulting marking is $[4, 5]$. In this marking, transition s is enabled; firing s yields marking $[6]$.

As explained in the introduction to this chapter, a semantics of a formal language for describing concurrent systems defines a process for each expression in the formal language. The definitions given so far provide the following semantics for the language of labeled P/T nets. The usual Petri-net semantics does not distinguish successful termination and deadlock. Recall the notions of a process space and bisimilarity from Definitions 3.1 and 3.6.

DEFINITION 4.9 (*Total-order semantics*). The process space $(\mathcal{N}, L, [\,\rangle^t, \emptyset)$, with the set of marked, labeled P/T nets as processes, the set of labels as actions, the total-order firing rule as the transition relation, and the empty termination predicate, combined with bisimilarity as the semantic equivalence defines a (branching-time) total-order semantics for labeled P/T nets.

EXAMPLE 4.10. In the semantics of Definition 4.9, the process corresponding to the labeled P/T net of Figure 4 is the one depicted in Figure 1(d), except that it does not terminate successfully after the execution of action s. (Note that dots in Figure 1(d) correspond to marked, labeled P/T nets; labeled arrows represent the total-order firing rule $[\,\rangle^t$.) Figure 5 shows two other P/T nets with exactly the same semantics. Intuitively, the P/T net of Figure 4 models a system with true concurrency, whereas the two P/T nets of Figure 5 do not. In the P/T nets of Figure 5, the actions p_1 and p_2 may happen in any order, but not simultaneously. However, as one might expect by now, in the total-order semantics of Definition 4.9, it is not possible to distinguish the three nets.

Note that, if desirable, the process-theoretic framework of the previous section is sufficiently flexible to express that the P/T nets of Figures 4 and 5 terminate successfully after firing transition s. Let $\downarrow_6$ be the termination predicate defined as follows: $\downarrow_6 = \{(N, m) \in \mathcal{N} \mid 6 \in m\}$. In the process space $(\mathcal{N}, L, [\,\rangle^t, \downarrow_6)$, the semantics of all the three P/T nets of Figures 4 and 5 is the process depicted in Figure 1(d).

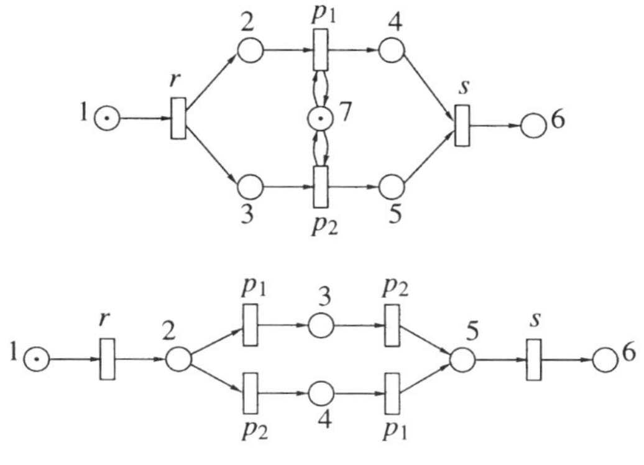

Fig. 5. Two examples of P/T nets without true concurrency.

The total-order semantics of Definition 4.9 can be generalized to a step semantics for labeled P/T nets in a very natural way. It suffices to allow that multiple transitions can fire at the same time.

A marking of a labeled P/T net enables a *set* of transitions if and only if all transitions in the set are simultaneously enabled. The auxiliary functions $\boldsymbol{i}$ and $\boldsymbol{o}$ are lifted to sets of transitions as follows: For any labeled P/T net $N = (P, T, F, W, \ell)$ and set $S \subseteq T$, $\boldsymbol{i}S = (\uplus t : t \in S : \boldsymbol{i}t)$ and $\boldsymbol{o}S = (\uplus t : t \in S : \boldsymbol{o}t)$.

DEFINITION 4.11 (*Set enabling*). Let (N, m) be a marked, labeled P/T net in $\mathcal{N}$, where $N = (P, T, F, W, \ell)$. A set $S \subseteq T$ is *enabled*, denoted $(N, m)[S\rangle$, if and only if $\boldsymbol{i}S \leqslant m$.

Let $N = (P, T, F, W, \ell)$ be an L-labeled P/T net. Any set $S \subseteq T$ that is enabled in some given marking of N determines a possible *step* of N. The transitions in the set S can fire simultaneously; the *bag* of labels of transitions in S forms the step that the net performs when firing S. Note that the set of all subsets of a given set X, also called the powerset of X, is denoted $\mathcal{P}(X)$. The labeling function $\ell : \mathcal{P}(T) \to \mathcal{B}(L)$ on sets of transitions of N is defined as follows: For any set $S \subseteq T$, $\ell(S) = (\uplus t\!: t \in S\!: [\ell(t)])$.

DEFINITION 4.12 (*Step firing rule*). The *step firing rule* $\_\,[\_\rangle^s\,\_ \subseteq \mathcal{N} \times \mathcal{B}(L) \times \mathcal{N}$ is the smallest relation satisfying for any marked, labeled P/T net (N, m) in $\mathcal{N}$, where $N = (P, T, F, W, \ell)$, and any $S \subseteq T$,

$$(N, m)[S\rangle \Rightarrow (N, m)\,[\ell(S)\rangle^s\,(N, m - \boldsymbol{i}S \uplus \boldsymbol{o}S).$$

The step firing rule forms the basis for the following step semantics for labeled P/T nets.

DEFINITION 4.13 (*Step semantics*). The process space $(\mathcal{N}, \mathcal{B}(L), [\,\rangle^s, \emptyset)$, with the bags over L as actions and the step firing rule as the transition relation, combined with bisimilarity as the semantic equivalence defines a step semantics for labeled P/T nets.

Note that the semantic equivalence resulting from Definitions 3.6 (Bisimilarity) and 4.13 (Step semantics) is the step-bisimilarity equivalence introduced in the previous section. Thus, the semantics of Definition 4.13 can be characterized as a branching-time, partial-order semantics.

EXAMPLE 4.14. Recall the two processes depicted in Figure 2. Consider the two variants of these processes that cannot terminate successfully after performing the last action. Assuming the semantics of Definition 4.13 for labeled P/T nets, the process in Figure 2(b) is the semantics of the labeled P/T net of Figure 4. The other process depicts the semantics of the two P/T nets in Figure 5. Thus, in the step semantics of Definition 4.13, it is possible to distinguish concurrency from non-determinism.

This section has given a brief introduction to elementary Petri-net theory. The class of labeled P/T nets has been introduced, as well as two different semantics for this class of nets. Labeled P/T nets handle causal relationships between actions very explicitly via places and tokens and are, therefore, well suited for developing a partial-order framework for reasoning about concurrent systems. The introduction is not exhaustive. In particular, there are several other approaches to defining a partial-order semantics for P/T nets than the one chosen in this section. However, the framework of this section is sufficient for the remainder of this chapter. Good starting points for further reading on Petri nets are [17, 47,52,56–58]. In the remainder of this chapter, we introduce a process-algebraic theory with both a total-order and a step semantics. In addition, we study several concepts from Petri-net theory in the algebraic partial-order framework.

5. Process algebra

The goal of this section is to illustrate how process algebra in the style of the Algebra of Communicating Processes (ACP) can be used to reason about concurrent systems in both a total-order and a partial-order setting. The theory ACP originates from [13]. Good introductions to ACP-style process algebra can be found in [5,6,25]. Other well-known process-algebraic theories are CCS [44,45] and CSP [36]. Specifications and verifications in ACP-style process algebra are based on an equational style of reasoning, whereas CCS and CSP emphasize model-based reasoning. In the latter case, the starting point of the theory is a semantic framework of processes in the style of Section 3. The goal is to find equational laws that are valid in this semantic framework. In the former case, the starting point is an equational theory, which may have several semantic interpretations. For a detailed comparison of ACP, CCS, and CSP, the reader is referred to [6, Chapter 8]. This section presents the theory ACP and a few of its extensions. It illustrates how the communication mechanism of ACP can be used to define a step semantics for concurrent systems. It is also defined when a process-algebraic theory and its semantics are said to be interleaving. Both the total-order and the partial-order framework developed in this section can be characterized as interleaving frameworks.

5.1. *The equational theory*

The Algebra of Communicating Processes. Any ACP-style process-algebraic theory is essentially an *equational theory*. An equational theory consists of a *signature* and a set of *axioms*. The signature defines the *sorts* of the theory, a set of *variables* for each sort, and the *functions* of the theory. Functions and variables can be used to construct *terms*. Terms not containing any variables are called *closed* terms. The axioms of the theory determine which terms are equal. A process-algebraic theory usually has only a single sort; terms of this sort represent processes. A 0-ary function is often called a *constant*; other functions are often called *operators*.

The signature and axioms of the equational theory $\mathrm{ACP}(AC, \gamma)$ are given in Table 1. The theory is parameterized by a set of constants AC, which is a set of actions, and a partial function $\gamma : AC \times AC \to AC$, which is a communication function. The first part of Table 1 lists the sorts in the signature of $\mathrm{ACP}(AC, \gamma)$; the second part defines the constants and the operators in the signature. The third entry of Table 1 gives the variables and lists the

Table 1
The equational theory ACP(AC, γ)

ACP(AC, γ)

$\boldsymbol{P}$;

$AC \subseteq \boldsymbol{P}$; $\delta : \boldsymbol{P}$; $\partial_H : \boldsymbol{P} \to \boldsymbol{P}$; $\_ + \_, \_ \cdot \_, \_ \parallel \_, \_ \mathbin{⌊\!⌊} \_, \_ \mid \_ : \boldsymbol{P} \times \boldsymbol{P} \to \boldsymbol{P}$;

$x, y, z : \boldsymbol{P}$;

$x + y = y + x$	$A1$	$x \parallel y = x \mathbin{⌊\!⌊} y + y \mathbin{⌊\!⌊} x + x \mid y$	$CM1$
$(x + y) + z = x + (y + z)$	$A2$	$e \mathbin{⌊\!⌊} x = e \cdot x$	$CM2$
$x + x = x$	$A3$	$e \cdot x \mathbin{⌊\!⌊} y = e \cdot (x \parallel y)$	$CM3$
$(x + y) \cdot z = x \cdot z + y \cdot z$	$A4$	$(x + y) \mathbin{⌊\!⌊} z = x \mathbin{⌊\!⌊} z + y \mathbin{⌊\!⌊} z$	$CM4$
$(x \cdot y) \cdot z = x \cdot (y \cdot z)$	$A5$	$e \cdot x \mid f = (e \mid f) \cdot x$	$CM5$
$x + \delta = x$	$A6$	$e \mid f \cdot x = (e \mid f) \cdot x$	$CM6$
$\delta \cdot x = \delta$	$A7$	$e \cdot x \mid f \cdot y = (e \mid f) \cdot (x \parallel y)$	$CM7$
		$(x + y) \mid z = x \mid z + y \mid z$	$CM8$
$e \mid f = f \mid e$	$C1$	$x \mid (y + z) = x \mid y + x \mid z$	$CM9$
$(e \mid f) \mid g = e \mid (f \mid g)$	$C2$		
$\delta \mid e = \delta$	$C3$	$e \notin H \Rightarrow \partial_H(e) = e$	$D1$
		$e \in H \Rightarrow \partial_H(e) = \delta$	$D2$
$\gamma(a, b)$ defined $\Rightarrow a \mid b = \gamma(a, b)$	$CF1$	$\partial_H(x + y) = \partial_H(x) + \partial_H(y)$	$D3$
$\gamma(a, b)$ undefined $\Rightarrow a \mid b = \delta$	$CF2$	$\partial_H(x \cdot y) = \partial_H(x) \cdot \partial_H(y)$	$D4$

axioms of the equational theory. Note that new variables may be introduced any time when necessary. An informal explanation of the operators and the axioms is given below.

As mentioned, AC is a set of actions. Terms of sort $\boldsymbol{P}$ represent processes. Each action is a process, namely the process that can only execute the action and then terminates.

The two basic operators of ACP(AC, γ) are $+$ and $\cdot$, denoting alternative composition or choice and sequential composition, respectively. These two operators are elementary in describing the behavior of sequential processes. Sequential composition binds stronger than choice. Choice and sequential composition are axiomatized by Axioms $A1$ through $A5$. Most of these axioms are self-explanatory. Only Axiom $A4$ might need some explanation. It states the *right distributivity* of sequential composition over choice. The converse, left distributivity, is not an axiom of the theory. As a result, processes with different moments of choice are distinguished.

The constant δ stands for *inaction*, often also called *deadlock*. However, the former name is best suited, as follows from Axiom $A6$. It says that a process which can choose between some behavior x and doing nothing is equivalent to the process that has no choice and can only do x. Hence, in the context of a choice, δ is not a true deadlock. Axiom $A7$ shows that δ is a deadlock in the context of a sequential composition.

Axioms $CM1$ through $CM9$ axiomatize the behavior of concurrent processes. Constants e and f range over $AC \cup \{\delta\}$. Thus, an axiom such as $CM2$ containing the constant e is actually an axiom *scheme*; the equational theory ACP(AC, γ) contains one axiom for each possible constant $e \in AC \cup \{\delta\}$.

The parallel-composition operator $\parallel$, often called the *merge* operator, denotes the parallel execution of its operands. It is axiomatized using two auxiliary operators, namely $\mathbin{⌊\!⌊}$, called the *left* merge, and |, called the *communication* merge. The left merge has the same meaning as the merge except that the left process must perform the first action. The communication merge also denotes the parallel execution of its operands, this time with the restriction that they must synchronize on their first action. Sequential composition binds stronger than merge, left merge, and communication merge, whereas choice binds weaker.

The result of the synchronization of two actions is defined by the communication function γ, as expressed by Axioms $CF1$ and $CF2$. Constants a and b range over AC. Operator | and function γ are named *communication* merge and function, respectively, because the standard interpretation in ACP-style process algebra of the synchronization of two actions is (synchronous) communication. However, in the remainder, it is shown that the communication function can also be used to define a step semantics for processes. This means that the synchronization of actions conforms to simultaneous execution.

Axioms $C1$, $C2$, and $C3$, where e, f, and g range over $AC \cup \{\delta\}$, state that the synchronization of actions is commutative and associative and that actions cannot synchronize with the inaction constant δ. To the experienced reader, it might be clear that the combination of Axioms $C1$, $C2$, $C3$, $CF1$, and $CF2$ can only be consistent if the communication function γ is associative and commutative. This requirement becomes explicit in the next subsection, where a model of the equational theory of this subsection is constructed. (See the proof of Theorem 5.13.)

Finally, the equational theory ACP(AC, γ) contains the so-called *encapsulation* operators. For any subset of actions $H \subseteq AC$, the operator ∂_H renames all actions in H in a process to the constant δ, as expressed by Axioms $D1$ through $D4$. Thus, the execution of actions in H is effectively blocked. Constant e ranges over $AC \cup \{\delta\}$. In ACP-style process algebra, an encapsulation operator is usually used to enforce communication between actions. This can be achieved by encapsulating the actions that must participate in a communication when they occur in isolation.

An equational theory such as ACP(AC, γ) of Table 1 provides the basis for reasoning about concurrent processes. The set of axioms of an equational theory defines an equivalence relation on process terms, called *derivability*. For any process terms x and y in some given equational theory X, $X \vdash x = y$ denotes that $x = y$ can be derived from the axioms of X. Derivability in an equational theory is defined as follows. First, the axioms themselves can be derived from the axioms of the theory. Second, since derivability is an equivalence relation, it is reflexive, symmetric, and transitive. Third, if an equation is derivable from the axioms, then also any equation obtained by substituting terms for variables in this equation is derivable. Finally, any equation obtained by replacing a term in an arbitrary context by another derivably equivalent term is also derivable from the theory. The axioms of an equational theory must be chosen in such a way that derivability defines a meaningful equivalence relation on processes. Below, it is explained that in the case of the theory ACP(AC, γ) derivability corresponds to bisimilarity.

EXAMPLE 5.1. Let A be some set of actions that includes the actions p_1, p_2, r, and s; let γ be some arbitrary communication function. It can be shown that ACP(A, γ) $\vdash r \cdot ((p_1 + p_2) \cdot s) = r \cdot (p_1 \cdot s + p_2 \cdot s)$. The first step is to substitute p_1, p_2, and s for the variables x, y, and z in Axiom $A4$, which shows that ACP(A, γ) $\vdash t_1 = t_2$, where t_1 is the term $(p_1 + p_2) \cdot s$ and t_2 denotes the term $p_1 \cdot s + p_2 \cdot s$. The second and final step consists of an application of the context rule explained above: Replacing term t_1 in $r \cdot t_1$ with the equivalent term t_2 yields the desired result.

The intuitive meaning of the operators of ACP(AC, γ), the axioms, and the induced equivalence relation given above can be analyzed further by defining an operational semantics. However, before doing so, a few extensions of ACP(AC, γ) are discussed that are needed in the remainder.

Renaming. The encapsulation operators of ACP(AC, γ) form a subclass of a much larger class of operators, namely the algebraic *renaming operators*. A renaming operator applied to a process term simply renames actions in the term according to a *renaming function*. A renaming function $\boldsymbol{f} : (AC \cup \{\delta\}) \to (AC \cup \{\delta\})$ is any function with the restriction that $\boldsymbol{f}(\delta) = \delta$. This restriction means that the inaction process can be the result of a renaming, but that it cannot be renamed itself. It is included in the domain of renaming functions only for the sake of convenience. The set of all renaming functions is denoted *RF*. The Algebra of Communicating Processes with renaming, abbreviated ACP + RN, is defined in Table 2. It is parameterized with a set of actions *AC* and a communication function γ. The first entry of Table 2 says that the equational theory (ACP + RN)(AC, γ) extends the theory ACP(AC, γ) of Table 1. The other entries of Table 2 have the same meaning as the corresponding entries in Table 1. In Table 2, function $\boldsymbol{f}$ ranges over the set of renaming functions *RF* and constant e ranges over $AC \cup \{\delta\}$.

As mentioned, the encapsulation operators of ACP(AC, γ) belong to the class of renaming operators. Let $H \subseteq AC$ be some set of actions. Let $\boldsymbol{e}(H) : (AC \cup \{\delta\}) \to (AC \cup \{\delta\})$ be a renaming function in *RF* defined as follows: $\boldsymbol{e}(H)(\delta) = \delta$, for any $a \in H$, $\boldsymbol{e}(H)(a) = \delta$, and, for any $a \in AC \setminus H$, $\boldsymbol{e}(H)(a) = a$. It is not difficult to see that the encapsulation operator ∂_H and the renaming operator $\rho_{\boldsymbol{e}(H)}$ are identical.

Iteration. In practice, many concurrent processes exhibit some kind of recursive behavior. A restricted form of recursive behavior is *iterative* behavior. Table 3 defines the Algebra of Communicating Processes with iteration and renaming, abbreviated ACP* + RN.

The *binary Kleene star* * was introduced in ACP-style process algebra in [11]. It is adapted from the original star operator as introduced in [41]. In [11], also the Axioms

Table 2
The equational theory (ACP + RN)(AC, γ)

(ACP + RN)(AC, γ)	
ACP(AC, γ)	
$\rho_f : \boldsymbol{P} \to \boldsymbol{P}$;	
$x, y : \boldsymbol{P}$;	
$\rho_f(e) = \boldsymbol{f}(e)$	*RN*1
$\rho_f(x + y) = \rho_f(x) + \rho_f(y)$	*RN*2
$\rho_f(x \cdot y) = \rho_f(x) \cdot \rho_f(y)$	*RN*3

Table 3
The equational theory $(\mathrm{ACP}^* + \mathrm{RN})(AC, \gamma)$

$(\mathrm{ACP}^* + \mathrm{RN})(AC, \gamma)$	
$(\mathrm{ACP} + \mathrm{RN})(AC, \gamma)$	
$\_^*\_ : \boldsymbol{P} \times \boldsymbol{P} \to \boldsymbol{P}$;	
$x, y, z : \boldsymbol{P}$;	
$x^* y = x \cdot (x^* y) + y$	*BKS*1
$x^* (y \cdot z) = (x^* y) \cdot z$	*BKS*2
$x^* (y \cdot ((x + y)^* z) + z) = (x + y)^* z$	*BKS*3
$\rho_f(x^* y) = \rho_f(x)^* \rho_f(y)$	*BKS*4

*BKS*1 through *BKS*4 appear, although *BKS*4 is formulated only for the specific renaming operators encapsulation and abstraction. Abstraction in process algebra is not covered in this chapter. An extensive study of iteration in ACP-style process algebra can be found in [12], Chapter 5 of this Handbook.

In Table 3, function f ranges over the set of renaming functions RF as introduced above. Axiom *BKS*1 is characteristic for iteration. It states that behavior x is iterated an arbitrary number of times before continuing with behavior y. (Axiom *BKS*3 is a sophisticated axiom which is needed to get a complete axiomatization of bisimilarity in a setting much simpler than the equational theory developed in this subsection; see [11] or [12], Chapter 5 of this Handbook, for more details.) The binary Kleene star has the same binding priority as sequential composition.

Recursive equations. Since often verifications in an equational theory result in an equation that is recursive in some given process, it is useful to have a means to determine solutions for such recursive equations. Table 4 presents the theory $(\mathrm{ACP}^* + \mathrm{RN} + RSP^*)(AC, \gamma)$ which extends $(\mathrm{ACP}^* + \mathrm{RN})(AC, \gamma)$ with a so-called recursion principle. The *Recursive Specification Principle* for the binary Kleene star (RSP^*) is a derivation rule which gives a solution in terms of the binary Kleene star for some restricted set of recursive equations.

Table 4
The equational theory $(\mathrm{ACP}^* + \mathrm{RN} + RSP^*)(AC, \gamma)$

$(\mathrm{ACP}^* + \mathrm{RN} + RSP^*)(AC, \gamma)$	
$(\mathrm{ACP}^* + \mathrm{RN})(AC, \gamma)$	
-	
$x, y, z : \boldsymbol{P}$;	
$\dfrac{x = y \cdot x + z}{x = y^* z}$	RSP^*

Table 5
The equational theory $(\mathrm{ACP}^* + \mathrm{RN} + \mathit{RSP}^* + \mathrm{SC})(AC, \gamma)$

$(\mathrm{ACP}^* + \mathrm{RN} + \mathit{RSP}^* + \mathrm{SC})(AC, \gamma)$	
$(\mathrm{ACP}^* + \mathrm{RN} + \mathit{RSP}^*)(AC, \gamma)$	
-	
$x, y, z : \boldsymbol{P};$	
$x \mid y = y \mid x$	*SC*1
$x \parallel y = y \parallel x$	*SC*2
$x \mid (y \mid z) = (x \mid y) \mid z$	*SC*3
$(x ⌊⌊ y) ⌊⌊ z = x ⌊⌊ (y \parallel z)$	*SC*4
$(x \mid y) ⌊⌊ z = x \mid (y ⌊⌊ z)$	*SC*5
$(x \parallel y) \parallel z = x \parallel (y \parallel z)$	*SC*6

Observe that, compared to theory $(\mathrm{ACP}^* + \mathrm{RN})(AC, \gamma)$, equational theory $(\mathrm{ACP}^* + \mathrm{RN} + \mathit{RSP}^*)(AC, \gamma)$ has no new constants or operators. Another interesting observation is that, in theory $(\mathrm{ACP}^* + \mathrm{RN} + \mathit{RSP}^*)(AC, \gamma)$, the axioms *BKS*2, *BKS*3, and *BKS*4 are derivable from the other axioms of the theory. (It is an interesting exercise for the reader to give the three derivations.)

Standard concurrency. A final extension of the equational theory developed so far in this subsection is the extension with the so-called *axioms of standard concurrency*. Table 5 presents the Algebra of Communicating Processes with iteration, renaming, the Recursive Specification Principle for the binary Kleene star, and standard concurrency, abbreviated $\mathrm{ACP}^* + \mathrm{RN} + \mathit{RSP}^* + \mathrm{SC}$. The equational theory $(\mathrm{ACP}^* + \mathrm{RN} + \mathit{RSP}^* + \mathrm{SC})(AC, \gamma)$ has no new constants or operators. The extension solely consists of six new axioms, formulating some desirable properties of the merge, left merge, and communication merge.

A consequence of the extension of the equational theory with the axioms of standard concurrency is that it allows the straightforward formulation of a general so-called *expansion theorem*. This theorem is very useful for simplifying many calculations. It states how a parallel composition can be expanded into an alternative composition. Recall from Section 2 that the commutative and associative binary process-algebraic operator $+$ with unit element δ generalizes to a quantifier in a standard way. Furthermore, note that the binary operators $\parallel$ and $\mid$ in the equational theory $(\mathrm{ACP}^* + \mathrm{RN} + \mathit{RSP}^* + \mathrm{SC})(AC, \gamma)$ are also commutative and associative but that they do not have unit elements. Nevertheless, general quantifier notation may be used for these two operators. For example, for some positive natural number n, and operands $x_0, \ldots, x_n$, the quantifier notation $(\parallel i : 0 \leqslant i \leqslant n : x_i)$ is used as a shorthand notation for $x_0 \parallel \ldots \parallel x_n$, where redundant brackets are omitted, and the notation $(\parallel i : i = k : x_i)$, for some k with $0 \leqslant k \leqslant n$, denotes x_k.

THEOREM 5.2 (Expansion). *Let $x_0, \ldots, x_n$ be process terms of sort $\boldsymbol{P}$, where n is a positive natural number. Let I be the set $\{0, \ldots, n\}$.*

$$(\mathrm{ACP}^* + \mathrm{RN} + \mathit{RSP}^* + \mathrm{SC})(AC, \gamma) \vdash$$

$$(\|i : i \in I : x_i) = \big(+J : \emptyset \subset J \subset I : (|j : j \in J : x_j) \mathbin{\|\!\_} (\|i : i \in I \setminus J : x_i)\big) \\ + (|i : i \in I : x_i).$$

PROOF. It is straightforward to prove the desired result by induction on n. □

EXAMPLE 5.3. Let A be some set of actions that includes the actions a, b, c, and d. Function $\gamma : A \times A \to A$ is the communication function defining that the communication of actions a and b yields action c. In set notation, $\gamma = \{((a, b), c), ((b, a), c)\}$. In its simplest form, the expansion theorem given above corresponds to Axiom $CM1$ of Table 1. The following derivation in the equational theory $(\mathrm{ACP}^* + \mathrm{RN} + RSP^* + \mathrm{SC})(A, \gamma)$ shows the application of the expansion theorem to a parallel composition of three components:

$$\begin{aligned}
& a \parallel b \parallel d \\
= \quad & \{\text{Theorem 5.2 (Expansion)}\} \\
& a \mathbin{\|\!_} (b \parallel d) + b \mathbin{\|\!_} (a \parallel d) + d \mathbin{\|\!_} (a \parallel b) + (a \mid b) \mathbin{\|\!_} d + (a \mid d) \mathbin{\|\!_} b \\
& + (b \mid d) \mathbin{\|\!_} a + (a \mid b \mid d) \\
= \quad & \{\text{Exercise for the reader}\} \\
& a \cdot (b \cdot d + d \cdot b) + b \cdot (a \cdot d + d \cdot a) + d \cdot (a \cdot b + b \cdot a + c) + c \cdot d.
\end{aligned}$$

Note that the above expansion theorem allows the synchronization of an arbitrary number of processes. In the standard literature on process algebra [5,6], only a simpler version of the expansion theorem is formulated. The simplifying assumption, called the handshaking axiom, is that only pairs of processes can synchronize. In the current setting, it is desirable to allow arbitrary synchronization, because one possible interpretation of synchronization corresponds to concurrency. There is no obvious reason to restrict concurrency to two processes.

Interleaving theory. As explained in the introduction, a process-algebraic theory with an expansion theorem is said to be interleaving. In this paragraph, we formalize the notion of an interleaving theory in terms of more elementary concepts. To the best of our knowledge, such a formalization does not yet exist in the literature on concurrency theory. (The definition of a (non-)interleaving process-algebraic theory in [2] is kept informal.) Therefore, the definition given in this paragraph is phrased in general terms such that it is applicable to any ACP-style equational theory. It can also be easily adapted to other process-algebraic frameworks such as CCS and CSP.

Let $X(AC)$ be some ACP-style equational theory which is parameterized with a set of actions AC. Note that it may have more (unspecified) parameters such as a communication function. In addition, assume that $X(AC)$ has a single process sort $\boldsymbol{P}$ and that its signature contains at least the choice operator $+$, the sequential-composition operator $\cdot$, and the parallel-composition operator $\|$. The following format of process terms forms the basis of the formalization of the notion of an interleaving equational theory. Note that this format is a standard format defined in the process-algebraic literature [6].

DEFINITION 5.4 (*Head normal form*). The set of process terms over the signature of $X(AC)$ in *head normal form*, denoted $\boldsymbol{H}(AC)$, is inductively defined as follows. First, each action constant in AC is an element of $\boldsymbol{H}(AC)$. Second, if the signature of $X(AC)$ contains the inaction constant δ, then also δ is an element of $\boldsymbol{H}(AC)$. Third, for any action a in AC and term x of sort $\boldsymbol{P}$, the term $a \cdot x$ is an element of $\boldsymbol{H}(AC)$. Finally, for any terms y and z in $\boldsymbol{H}(AC)$, the term $y + z$ is an element of $\boldsymbol{H}(AC)$.

An arbitrary term x of sort $\boldsymbol{P}$ is said to *have* a head normal form, denoted $HNF(x)$ if and only if there is a term y in $\boldsymbol{H}(AC)$ such that $X(AC) \vdash x = y$.

EXAMPLE 5.5. Consider the equational theory ACP(AC, γ) of Table 1. As in Example 5.1, let A be some set of actions that includes the actions $p_1, p_2, r,$ and s; let γ be some arbitrary communication function. The closed term $r \cdot ((p_1 + p_2) \cdot s)$ is in head normal form. Term $(p_1 + p_2) \cdot s$ is not in head normal form. However, it has a head normal form, because it is derivably equal to the term $p_1 \cdot s + p_2 \cdot s$, which is in head normal form. The term x of sort $\boldsymbol{P}$, where x is a variable, does not have a head normal form.

An equational theory is said to be interleaving if and only if each parallel composition has a head normal form under the assumption that its operands have head normal forms.

DEFINITION 5.6 (*Interleaving theory*). The equational theory $X(AC)$ is an *interleaving theory* if and only if, for any terms x and y of sort $\boldsymbol{P}$, $HNF(x)$ and $HNF(y)$ implies $HNF(x \parallel y)$.

All the equational theories given so far in this subsection are interleaving theories. The proof of this result uses two auxiliary properties about head normal forms of compositions using the auxiliary operators left merge and communication merge.

THEOREM 5.7. *Let AC be some set of actions; let γ be an arbitrary communication function. The equational theories* ACP(AC, γ), (ACP + RN)(AC, γ), (ACP* + RN)(AC, γ), (ACP* + RN + *RSP**)(AC, γ), *and* (ACP* + RN + *RSP** + SC)(AC, γ) *are all interleaving theories.*

PROOF. The proof is identical for each of the theories. Therefore, let x and y be terms of sort $\boldsymbol{P}$ of any of the above theories such that $HNF(x)$ and $HNF(y)$. It must be proven that $HNF(x \parallel y)$. It follows from Axiom $CM1$ of Table 1 that $x \parallel y = x \mathbin{⌊\!⌊} y + y \mathbin{⌊\!⌊} x + x \mid y$. Definition 5.4 implies that it suffices to prove that $HNF(x \mathbin{⌊\!⌊} y)$, $HNF(y \mathbin{⌊\!⌊} x)$, and $HNF(x \mid y)$, which follows immediately from Properties 5.8 and 5.9. □

PROPERTY 5.8. *Let AC be some set of actions; let γ be an arbitrary communication function. For any terms x and y of sort $\boldsymbol{P}$ of any of the equational theories* ACP(AC, γ), (ACP + RN)(AC, γ), (ACP* + RN)(AC, γ), (ACP* + RN + *RSP**)(AC, γ), *and* (ACP* + RN + *RSP** + SC)(AC, γ), *$HNF(x)$ implies $HNF(x \mathbin{⌊\!⌊} y)$.*

PROOF. Let $\boldsymbol{H}(AC)$ be the set of terms in head normal form of any of the above theories. Let x and y be terms of sort $\boldsymbol{P}$ of the same theory such that $HNF(x)$. It must be proven that

$HNF(x \mathbin{\lfloor\!\lfloor} y)$. According to Definition 5.4, there must be a term in $\boldsymbol{H}(AC)$ that is derivably equal to x. Thus, it suffices to prove that for any $z \in \boldsymbol{H}(AC)$, $HNF(z \mathbin{\lfloor\!\lfloor} y)$. Let z be a term in $\boldsymbol{H}(AC)$. Since the signature of any of the theories in this property contains the inaction constant δ, term z can be in any of four formats. The proof is by induction on the structure of term z. The symbol $\equiv$ denotes syntactical equivalence of terms.

(i) Assume that $z \equiv \delta$. It follows that $\delta \mathbin{\lfloor\!\lfloor} y \overset{CM2,A7}{=} \delta$. Since $\delta \in \boldsymbol{H}(AC)$, it follows that $HNF(\delta \mathbin{\lfloor\!\lfloor} y)$, which completes the proof for the case $z \equiv \delta$.

(ii) Assume that $z \equiv a$, for some action a in AC. It follows that $a \mathbin{\lfloor\!\lfloor} y \overset{CM2}{=} a \cdot y$. Since $a \cdot y \in \boldsymbol{H}(AC)$, it follows that $HNF(a \mathbin{\lfloor\!\lfloor} y)$, which completes the proof also in this case.

(iii) Assume that $z \equiv a \cdot x_1$, for some action a in AC and some term x_1 of sort $\boldsymbol{P}$. It follows that $a \cdot x_1 \mathbin{\lfloor\!\lfloor} y \overset{CM3}{=} a \cdot (x_1 \parallel y)$. Since $a \cdot (x_1 \parallel y) \in \boldsymbol{H}(AC)$, it follows that $HNF(a \cdot x_1 \mathbin{\lfloor\!\lfloor} y)$.

(iv) Finally, assume that $z \equiv z_1 + z_2$, for some terms z_1 and z_2 in $\boldsymbol{H}(AC)$. It follows that $(z_1 + z_2) \mathbin{\lfloor\!\lfloor} y \overset{CM4}{=} z_1 \mathbin{\lfloor\!\lfloor} y + z_2 \mathbin{\lfloor\!\lfloor} y$. By induction, it follows that $HNF(z_1 \mathbin{\lfloor\!\lfloor} y)$ and $HNF(z_2 \mathbin{\lfloor\!\lfloor} y)$, which means that $HNF(z_1 \mathbin{\lfloor\!\lfloor} y + z_2 \mathbin{\lfloor\!\lfloor} y)$. Consequently, $HNF((z_1 + z_2) \mathbin{\lfloor\!\lfloor} y)$, which completes the proof. □

PROPERTY 5.9. *Let AC be some set of actions; let γ be an arbitrary communication function. For any terms x and y of sort $\boldsymbol{P}$ of any of the equational theories* ACP(AC, γ), $(\mathrm{ACP} + \mathrm{RN})(AC, \gamma)$, $(\mathrm{ACP}^* + \mathrm{RN})(AC, \gamma)$, $(\mathrm{ACP}^* + \mathrm{RN} + RSP^*)(AC, \gamma)$, *and* $(\mathrm{ACP}^* + \mathrm{RN} + RSP^* + \mathrm{SC})(AC, \gamma)$, *$HNF(x)$ and $HNF(y)$ implies $HNF(x \mid y)$.*

PROOF. Let $\boldsymbol{H}(AC)$ be the set of terms in head normal form of any of the above theories. Let x and y be terms of sort $\boldsymbol{P}$ of the same theory such that $HNF(x)$ and $HNF(y)$. It must be proven that $HNF(x \mid y)$. According to Definition 5.4, there must be terms in $\boldsymbol{H}(AC)$ that are derivably equal to x and y, respectively. Thus, it suffices to prove that for any terms x_0 and y_0 in $\boldsymbol{H}(AC)$, $HNF(x_0 \mid y_0)$. Let x_0 and y_0 be terms in $\boldsymbol{H}(AC)$. The proof is by induction on the structure of x_0.

(i) Assume that $x_0 \equiv \delta$. It must be shown that $HNF(\delta \mid y_0)$. This result is shown by straightforward induction on the structure of term y_0.

  (a) Assume that $y_0 \equiv \delta$ or that $y_0 \equiv a$, for some action a in AC. It follows that $\delta \mid y_0 \overset{C3}{=} \delta$. Thus, $HNF(\delta \mid y_0)$, which completes the proof in this case.

  (b) Assume that $y_0 \equiv a \cdot y_1$, for some action a in AC and some term y_1 of sort $\boldsymbol{P}$. It follows that $\delta \mid a \cdot y_1 \overset{CM6,C3,A7}{=} \delta$, which completes the proof also in this case.

  (c) Finally, assume that $y_0 \equiv z_1 + z_2$, for some terms z_1 and z_2 in $\boldsymbol{H}(AC)$. It follows that $\delta \mid (z_1 + z_2) \overset{CM9}{=} \delta \mid z_1 + \delta \mid z_2$. By induction, it follows that $HNF(\delta \mid z_1)$ and $HNF(\delta \mid z_2)$, which means that $HNF(\delta \mid (z_1 + z_2))$.

(ii) Assume that $x_0 \equiv a$, for some action a in AC. Again, the desired result that $HNF(a \mid y_0)$ is proven by induction on the structure of y_0.

  (a) Assume that $y_0 \equiv \delta$. It follows that $a \mid \delta \overset{C1,C3}{=} \delta$. Thus, $HNF(a \mid \delta)$.

(b) Assume that $y_0 \equiv b$, for some action b in AC. It follows from Axioms $CF1$ and $CF2$ and the fact that the result of the communication function is an action in AC that $HNF(a \mid b)$.

(c) Assume that $y_0 \equiv b \cdot y_1$, for some action b in AC and some term y_1 of sort $\boldsymbol{P}$. It follows that $a \mid b \cdot y_1 \stackrel{CM6}{=} (a \mid b) \cdot y_1$. It follows from the fact that γ yields an action in AC and Axioms $CF1$, $CF2$, and $A7$ that $HNF((a \mid b) \cdot y_1)$.

(d) Finally, assume that $y_0 \equiv z_1 + z_2$, for some terms z_1 and z_2 in $\boldsymbol{H}(AC)$. It follows that $a \mid (z_1 + z_2) \stackrel{CM9}{=} a \mid z_1 + a \mid z_2$. By induction, it follows that $HNF(a \mid z_1)$ and $HNF(a \mid z_2)$, which means that $HNF(a \mid (z_1 + z_2))$.

(iii) Assume that $x_0 \equiv a \cdot x_1$, for some action a in AC and some term x_1 of sort $\boldsymbol{P}$. As in the previous two cases, the desired result is proven by induction on the structure of y_0. The details are very similar to the proof of the previous case and, therefore, left to the reader.

(iv) Finally, assume that $x_0 \equiv z_1 + z_2$, for some terms z_1 and z_2 in $\boldsymbol{H}(AC)$. It follows that $(z_1 + z_2) \mid y_0 \stackrel{CM8}{=} z_1 \mid y_0 + z_2 \mid y_0$. By induction, it follows that $HNF(z_1 \mid y_0)$ and $HNF(z_2 \mid y_0)$, which means that $HNF((z_1 + z_2) \mid y_0)$. □

EXAMPLE 5.10. Let A be some set of actions that includes the actions a, b, and c; let γ be the communication function $\{((a, b), c), ((b, a), c)\}$. Consider the following derivation (which can be made in any of the equational theories given in this subsection):

$$\begin{array}{ll} & a \parallel b \\ = & \quad \{\text{Axioms } CM1, CM2(2\times)\} \\ & a \cdot b + b \cdot a + a \mid b \\ = & \quad \{\text{Definition } \gamma;\ \text{Axiom } CF1\} \\ & a \cdot b + b \cdot a + c. \end{array}$$

This derivation is characteristic for an interleaving equational theory. It shows how a parallel composition can be rewritten into an alternative composition of all its interleavings, including the interleaving in which the two actions communicate. Clearly, the last term of this derivation is in head normal form. Consequently, term $a \parallel b$ has a head normal form (as required by Theorem 5.7).

Another derivation that is characteristic for an interleaving equational theory is the derivation shown in Example 5.3 that uses the expansion theorem of Theorem 5.2.

It is an interesting observation that all the equational theories given in this subsection are interleaving theories. The proof of this result does not depend on the expansion theorem given in Theorem 5.2. The axiomatization of the merge operator in terms of the auxiliary operators left merge and communication merge is the fundamental reason that a theory is an interleaving theory as defined above. Note that this observation does not contradict our informal definition that an equational theory is an interleaving theory if and only if it has some form of expansion theorem. It is possible to prove an expansion theorem for any of the theories of this subsection, although the convenient formulation of such an expansion theorem is only possible when the theory contains the axioms of standard concurrency given in Table 5.

As already mentioned in the introduction, most process-algebraic theories in the literature are interleaving theories. In particular, the proof of Theorem 5.7 carries over to any of the ACP-style equational theories presented in [5,6,25]. This includes equational theories without communication and theories with recursion or abstraction. To prove that equational theories without communication are interleaving, it is even sufficient to use Property 5.8. Finally, also the equational theories in the CCS framework of [44,45] and the CSP framework of [36] can be characterized as interleaving theories.

Concluding remark. As already mentioned, the motivation to develop the theory $(\mathrm{ACP}^* + \mathrm{RN} + RSP^* + \mathrm{SC})(AC, \gamma)$ of Table 5 is to provide the means to reason about the behavior of concurrent processes. However, so far, the meaning of the algebraic constants, operators, and axioms has only been explained informally. In the next subsection, the semantics of $(\mathrm{ACP}^* + \mathrm{RN} + RSP^* + \mathrm{SC})(AC, \gamma)$ is formalized using the framework of Section 3.

5.2. *A semantics for the equational theory*

A model. As before, assume that AC is a set of action constants and that $\gamma : AC \times AC \rightarrow AC$ is a communication function. The semantics of terms in an equational theory is formalized by defining a so-called *model* of the theory also called an *algebra* for the theory. Since the terms in a single-sorted equational theory such as $(\mathrm{ACP}^* + \mathrm{RN} + RSP^* + \mathrm{SC})(AC, \gamma)$ are supposed to be interpreted as processes, a model of an ACP-style equational theory is also called a *process algebra*.

In general, a model of a single-sorted equational theory X consists of a *domain* of elements plus a number of functions on that domain, called the *signature* of the model. A model $\mathcal{M}$ with domain $\mathcal{D}$ must satisfy the following properties. First, there must exist an interpretation of the functions in the signature of X in terms of the functions in the signature of the model $\mathcal{M}$ that preserves the arity of functions. Second, any equation that is derivable from the axioms of the equational theory must be *valid* in the model, where validity is defined as follows. Let t be a term in the equational theory X; let σ be a mapping from variables in t to elements from domain $\mathcal{D}$, called a variable substitution. The interpretation of t in the model $\mathcal{M}$ under substitution σ, denoted $[\![t]\!]_\sigma$, is obtained by replacing all functions in t by the corresponding functions in $\mathcal{M}$ and by replacing all variables in t by elements of domain $\mathcal{D}$ according to σ. An equation $t_1 = t_2$ in X is *valid* in model $\mathcal{M}$, denoted $\mathcal{M} \models t_1 = t_2$, if and only if, for *all* substitutions σ for the variables in t_1 and t_2, $[\![t_1]\!]_\sigma =_{\mathcal{D}} [\![t_2]\!]_\sigma$, where $=_{\mathcal{D}}$ is the identity on domain $\mathcal{D}$. If $\mathcal{M}$ is a model of an equational theory X, it is also said that X is a *sound axiomatization* of $\mathcal{M}$.

EXAMPLE 5.11. Assume that $\mathcal{M}$ with domain $\mathcal{D}$ is a model of the theory $\mathrm{ACP}(AC, \gamma)$ of Table 1. Recall that the functions in the signature of $\mathrm{ACP}(AC, \gamma)$ consist of the inaction constant δ, the action constants in AC, and the operators listed in Table 1. Assume that, for any function f in the signature of $\mathrm{ACP}(AC, \gamma)$, $\bar{f}$ denotes the corresponding function in the signature of $\mathcal{M}$. Consider the equation $a + \delta = a$, where a is an action in AC. It follows from Axiom $A6$ that this equation is derivable from theory ACP(AC, γ). Since $\mathcal{M}$ is a model of this theory, $a + \delta = a$ must be valid in $\mathcal{M}$, which means that the equality $\bar{a}\bar{+}\bar{\delta} =_{\mathcal{D}} \bar{a}$, where $=_{\mathcal{D}}$ is the identity on domain $\mathcal{D}$, must hold. In fact, Axiom $A6$ itself is derivable from the theory and must, therefore, be valid in $\mathcal{M}$. That is, for any $d \in \mathcal{D}$, the equality $d\bar{+}\bar{\delta} =_{\mathcal{D}} d$ must hold.

The basis for a model of the equational theory $(\mathrm{ACP}^* + \mathrm{RN} + RSP^* + \mathrm{SC})(AC, \gamma)$ is a process space, as defined in Definition 3.1. This means that a set of processes, a set of actions, a transition relation, and a termination predicate need to be defined.

Let $\mathcal{C}(AC)$ be the set of *closed* $(\mathrm{ACP}^* + \mathrm{RN} + RSP^* + \mathrm{SC})(AC, \gamma)$ terms. The set $\mathcal{C}(AC)$ forms the basis for the set of processes in the process space. Since the equational theory $(\mathrm{ACP}^* + \mathrm{RN} + RSP^* + \mathrm{SC})(AC, \gamma)$ has no means to express the process that can perform no actions, but can only terminate successfully, a special process $\surd$, pronounced "tick", is introduced. Thus, the set of processes in the abovementioned process space is the set $\mathcal{C}(AC) \cup \{\surd\}$. The set AC is the set of actions. The termination predicate is the singleton $\{\surd\}$. That is, process $\surd$ is the only process that can terminate successfully. The transition relation $\_ \xrightarrow{\_} \_ \subseteq (\mathcal{C}(AC) \cup \{\surd\}) \times AC \times (\mathcal{C}(AC) \cup \{\surd\})$ can now be defined as the smallest relation satisfying the derivation rules in Table 6. Note that the transition relation has an implicit parameter, namely the communication function γ. It is not difficult to verify that the transition relation conforms to the informal explanation of the operators given in the previous subsection.

The process space $(\mathcal{C}(AC) \cup \{\surd\}, AC, \longrightarrow, \{\surd\})$ can be turned into a model $\mathcal{M}(AC, \gamma)$ of the equational theory $(\mathrm{ACP}^* + \mathrm{RN} + RSP^* + \mathrm{SC})(AC, \gamma)$ as follows.

Recall that bisimilarity, as defined in Definition 3.6, is an equivalence relation on the set of processes $\mathcal{C}(AC) \cup \{\surd\}$. Thus, it is possible to define equivalence classes of processes modulo bisimilarity in the usual way: For any $p \in \mathcal{C}(AC) \cup \{\surd\}$, the equivalence class of p modulo bisimilarity, denoted $[p]_\sim$, is the set $\{q \in \mathcal{C}(AC) \cup \{\surd\} \mid q \sim p\}$. It follows from Definition 3.6 (Bisimilarity) that the special element $[\surd]_\sim$ only contains the process $\surd$. The *domain* of the model under construction is formed by the set of all the equivalence classes of closed terms modulo bisimilarity.[1] The special element $[\surd]_\sim$ is excluded from the domain of model $\mathcal{M}(AC, \gamma)$ for technical reasons: As mentioned, the equational theory $(\mathrm{ACP}^* + \mathrm{RN} + RSP^* + \mathrm{SC})(AC, \gamma)$ has no means to express the process that can only terminate successfully.

It remains to define the constants and operators of $(\mathrm{ACP}^* + \mathrm{RN} + RSP^* + \mathrm{SC})(AC, \gamma)$ on the domain of the model. The interpretation $\bar{c}$ in model $\mathcal{M}(AC, \gamma)$ of some constant c in the signature of $(\mathrm{ACP}^* + \mathrm{RN} + RSP^* + \mathrm{SC})(AC, \gamma)$ is defined as the equivalence class $[c]_\sim$. Note that this definition fulfills the requirement that $\bar{c}$ is a 0-ary function on the domain of $\mathcal{M}(AC, \gamma)$. Also for the operators there is a straightforward way to interpret them in the domain of $\mathcal{M}(AC, \gamma)$, provided that bisimilarity is a *congruence* for all the

[1] In standard process-algebraic terminology, the elements in the domain of a model of some ACP-style equational theory are referred to as processes. However, Definition 3.3 defines a process as some kind of labeled transition system. The domain of model $\mathcal{M}(AC, \gamma)$ consists of equivalence classes of such labeled transition systems. In the literature on concurrency theory, the use of the term "process" for both equivalence classes of labeled transition systems and individual representatives of such equivalence classes is common practice and does not lead to confusion.

operators of $(\mathrm{ACP}^* + \mathrm{RN} + RSP^* + \mathrm{SC})(AC, \gamma)$. That is, the following property must be satisfied. Let $\oplus$ be an arbitrary n-ary operator in the signature of $(\mathrm{ACP}^* + \mathrm{RN} + RSP^* + \mathrm{SC})(AC, \gamma)$, where n is some positive natural number; let $p_1, \ldots, p_n, q_1, \ldots, q_n$ be closed terms in $\mathcal{C}(AC)$ such that $p_1 \sim q_1, \ldots, p_n \sim q_n$. Then, the congruence property requires that $\oplus(p_1, \ldots, p_n) \sim \oplus(q_1, \ldots, q_n)$.

PROPERTY 5.12 (Congruence). *Bisimilarity, $\sim$, is a congruence for the operators of* $(\mathrm{ACP}^* + \mathrm{RN} + RSP^* + \mathrm{SC})(AC, \gamma)$.

PROOF. The property follows from the format of the derivation rules in Table 6. For details, the reader is referred to [5]. Note that the formulation of the derivation rules in Table 6 differs slightly from the formulation of such derivation rules in [5]. However, it is not difficult to verify that the two formulations are equivalent. An extensive treatment of the theory underlying the approach to defining the semantics of (algebraic) languages by means of derivation rules such as those of Table 6 can be found in [1], Chapter 3 of this Handbook. □

Informally, the congruence property says that equivalence classes of processes can be constructed independently of their representatives. Let $p_1, \ldots, p_n$ be closed terms in $\mathcal{C}(AC)$, where n is some positive natural number; for any n-ary operator $\oplus$ in the signature of $(\mathrm{ACP}^* + \mathrm{RN} + RSP^* + \mathrm{SC})(AC, \gamma)$, function $\bar{\oplus}$ is defined on equivalence classes of closed terms as follows: $\bar{\oplus}([p_1]_\sim, \ldots, [p_n]_\sim) = [\oplus(p_1, \ldots, p_n)]_\sim$.

At this point, the construction of model $\mathcal{M}(AC, \gamma)$ of the equational theory $(\mathrm{ACP}^* + \mathrm{RN} + RSP^* + \mathrm{SC})(AC, \gamma)$ is complete. The domain consists of the equivalence classes of closed terms in $\mathcal{C}(AC)$ modulo bisimilarity; the interpretation of any of the constants in the signature of $(\mathrm{ACP}^* + \mathrm{RN} + RSP^* + \mathrm{SC})(AC, \gamma)$ is the corresponding equivalence class; the interpretation of any operator in the signature of the theory is that same operator lifted to equivalence classes of closed terms. Informally, two closed terms in $\mathcal{C}(AC)$ that are derivably equal in the equational theory $(\mathrm{ACP}^* + \mathrm{RN} + RSP^* + \mathrm{SC})(AC, \gamma)$ yield the same equivalence class when they are interpreted in the model $\mathcal{M}(AC, \gamma)$, which in turn implies that the corresponding processes are bisimilar. Thus, the equational theory $(\mathrm{ACP}^* + \mathrm{RN} + RSP^* + \mathrm{SC})(AC, \gamma)$ is a sound axiomatization of bisimilarity.

THEOREM 5.13 (Soundness). *For any closed terms $p, q \in \mathcal{C}(AC)$,*

$$(\mathrm{ACP}^* + \mathrm{RN} + RSP^* + \mathrm{SC})(AC, \gamma) \vdash p = q \;\Rightarrow\; \mathcal{M}(AC, \gamma) \models p = q.$$

PROOF. Since bisimilarity is a congruence for the operators of $(\mathrm{ACP}^* + \mathrm{RN} + RSP^* + \mathrm{SC})(AC, \gamma)$, it suffices to show the validity of each of the axioms. Except for the recursion principle RSP^*, it is not difficult to construct a bisimulation for each axiom. Note that the validity of Axioms $C1$, $C2$, $C3$, $CF1$, and $CF2$ can only be proven if it is assumed that the communication function is commutative and associative.

To prove the validity of RSP^*, assume that p, q, and r are closed terms in $\mathcal{C}(AC)$; let $\mathcal{R}$ be a bisimulation between p and $q \cdot p + r$. It can be shown that the following relation,

Table 6
The transition relation for $(\mathrm{ACP}^* + \mathrm{RN} + RSP^* + \mathrm{SC})(AC, \gamma)$

$a, b : AC;\ f : RF;\ p, p', q, q' : \mathcal{C}(AC);$

$$a \xrightarrow{a} \surd \qquad \frac{p \xrightarrow{a} p'}{p \cdot q \xrightarrow{a} p' \cdot q} \qquad \frac{p \xrightarrow{a} \surd}{p \cdot q \xrightarrow{a} q}$$

$$\frac{p \xrightarrow{a} p'}{p + q \xrightarrow{a} p'} \qquad \frac{q \xrightarrow{a} q'}{p + q \xrightarrow{a} q'} \qquad \frac{p \xrightarrow{a} \surd}{p + q \xrightarrow{a} \surd} \qquad \frac{q \xrightarrow{a} \surd}{p + q \xrightarrow{a} \surd}$$

$$\frac{p \xrightarrow{a} p'}{p \parallel q \xrightarrow{a} p' \parallel q} \qquad \frac{q \xrightarrow{a} q'}{p \parallel q \xrightarrow{a} p \parallel q'} \qquad \frac{p \xrightarrow{a} \surd}{p \parallel q \xrightarrow{a} q} \qquad \frac{q \xrightarrow{a} \surd}{p \parallel q \xrightarrow{a} p}$$

$$\frac{p \xrightarrow{a} p',\ q \xrightarrow{b} q',\ \gamma(a,b) \text{ def.}}{p \parallel q \xrightarrow{\gamma(a,b)} p' \parallel q'} \qquad \frac{p \xrightarrow{a} \surd,\ q \xrightarrow{b} \surd,\ \gamma(a,b) \text{ def.}}{p \parallel q \xrightarrow{\gamma(a,b)} \surd}$$

$$\frac{p \xrightarrow{a} p',\ q \xrightarrow{b} \surd,\ \gamma(a,b) \text{ def.}}{p \parallel q \xrightarrow{\gamma(a,b)} p'} \qquad \frac{p \xrightarrow{a} \surd,\ q \xrightarrow{b} q',\ \gamma(a,b) \text{ def.}}{p \parallel q \xrightarrow{\gamma(a,b)} q'}$$

$$\frac{p \xrightarrow{a} p'}{p \mathbin{⌊\!⌊} q \xrightarrow{a} p' \parallel q} \qquad \frac{p \xrightarrow{a} \surd}{p \mathbin{⌊\!⌊} q \xrightarrow{a} q}$$

$$\frac{p \xrightarrow{a} p',\ q \xrightarrow{b} q',\ \gamma(a,b) \text{ def.}}{p \mid q \xrightarrow{\gamma(a,b)} p' \parallel q'} \qquad \frac{p \xrightarrow{a} \surd,\ q \xrightarrow{b} \surd,\ \gamma(a,b) \text{ def.}}{p \mid q \xrightarrow{\gamma(a,b)} \surd}$$

$$\frac{p \xrightarrow{a} p',\ q \xrightarrow{b} \surd,\ \gamma(a,b) \text{ def.}}{p \mid q \xrightarrow{\gamma(a,b)} p'} \qquad \frac{p \xrightarrow{a} \surd,\ q \xrightarrow{b} q',\ \gamma(a,b) \text{ def.}}{p \mid q \xrightarrow{\gamma(a,b)} q'}$$

$$\frac{p \xrightarrow{a} p',\ f(a) \neq \delta}{\rho_f(p) \xrightarrow{f(a)} \rho_f(p')} \qquad \frac{p \xrightarrow{a} \surd,\ f(a) \neq \delta}{\rho_f(p) \xrightarrow{f(a)} \surd}$$

$$\frac{p \xrightarrow{a} p'}{p * q \xrightarrow{a} p' \cdot (p * q)} \qquad \frac{q \xrightarrow{a} q'}{p * q \xrightarrow{a} q'} \qquad \frac{p \xrightarrow{a} \surd}{p * q \xrightarrow{a} p * q} \qquad \frac{q \xrightarrow{a} \surd}{p * q \xrightarrow{a} \surd}$$

which uses the transitive closure of $\mathcal{R}$, denoted $\mathcal{R}^+$, is a bisimulation between p and $q^* r$, thus proving the validity of RSP^*:

$$\begin{aligned}&\{(p, q^* r)\} \cup \mathcal{R}^+ \cup \{(s, q^* r) \mid s \in \mathcal{C}(AC) \cup \{\surd\} \wedge s\mathcal{R}^+ p\}\\ &\quad \cup \{(s, t \cdot (q^* r)) \mid s \in \mathcal{C}(AC) \cup \{\surd\} \wedge t \in \mathcal{C}(AC) \wedge s\mathcal{R}^+ t \cdot p\}.\end{aligned}$$

In [9], a detailed proof is given (in a setting with silent behavior). □

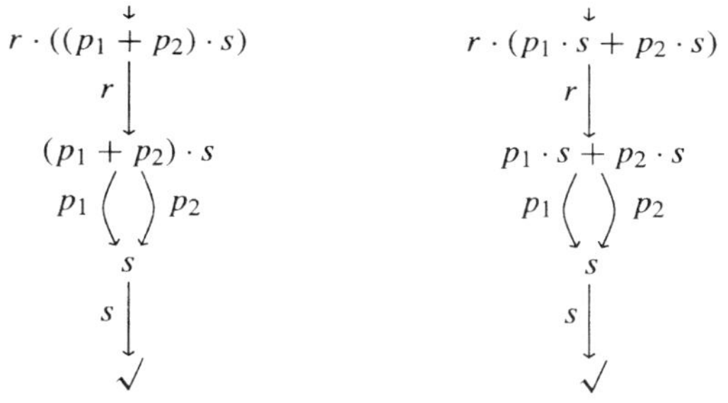

Fig. 6. Visualizing the semantics of closed $(\mathrm{ACP}^* + \mathrm{RN} + RSP^* + \mathrm{SC})(AC, \gamma)$ terms.

EXAMPLE 5.14. Since the basis of the semantics of $(\mathrm{ACP}^* + \mathrm{RN} + RSP^* + \mathrm{SC})(AC, \gamma)$ is a process space, it is possible to visualize the semantics of closed terms in $\mathcal{C}(AC)$. Consider again Example 5.1. Figure 6 depicts the semantics of closed terms $r \cdot ((p_1 + p_2) \cdot s)$ and $r \cdot (p_1 \cdot s + p_2 \cdot s)$. Clearly, these two processes are bisimilar, which conforms to Theorem 5.13 (Soundness) and the conclusion of Example 5.1 that the two closed terms are derivably equal.

Theorem 5.13 is an important result, because it shows that the equational theory $(\mathrm{ACP}^* + \mathrm{RN} + RSP^* + \mathrm{SC})(AC, \gamma)$ provides a meaningful framework for reasoning about the equality of concurrent processes. It is only possible to prove the equality of two processes if they are bisimilar. This raises the question whether it is always possible to prove the equality of two bisimilar processes in the equational theory $(\mathrm{ACP}^* + \mathrm{RN} + RSP^* + \mathrm{SC})(AC, \gamma)$. An equational theory is said to be a *complete* axiomatization of some given model if and only if every equality that is valid in the model is also derivable from the axioms of the theory. To the best of our knowledge, it is still an open problem whether the equational theory $(\mathrm{ACP}^* + \mathrm{RN} + RSP^* + \mathrm{SC})(AC, \gamma)$ is complete for model $\mathcal{M}(AC, \gamma)$. (Note, however, that it follows from the results of [59] that the theory $(\mathrm{ACP}^* + \mathrm{RN} + \mathrm{SC})(AC, \gamma)$, which is obtained by removing the recursion principle RSP^*, is *not* complete for $\mathcal{M}(AC, \gamma)$.)

Interleaving process algebra. In the previous subsection, it has been defined when an equational theory can be characterized as an interleaving theory. As explained in the introduction, a model of an interleaving equational theory is called an interleaving semantics or an interleaving process algebra. However, algebras are well-known mathematical structures (see, for example, [37]) that are not necessarily derived from equational theories as is done in the previous paragraph. Therefore, in this paragraph, we focus on (single-sorted) algebras as mathematical structures in isolation. A single-sorted algebra is simply a structure that has a single domain of elements and a signature of functions of arbitrary arity on that domain. It is not possible to *formally* define when such a single-sorted algebra is a *process* algebra. However, the following informal definition suffices for our purposes. First, the elements of the domain of the algebra should represent processes, where a process is some abstract notion of the behavior of a (concurrent) system. Second, the functions in the signature of the algebra should correspond to meaningful process constants and composition operators on processes. In this paragraph, it is formalized when such a process algebra is an *interleaving* process algebra. In addition, it is shown that the model of the equational theory of Section 5.1 constructed in the previous paragraph is an interleaving process algebra. The formalization of the notion of an interleaving process algebra is based on ACP-style process algebra. However, it can be easily adapted to other process-algebraic frameworks such as CCS and CSP.

Let $\mathcal{A}$ be a process algebra with domain $\mathcal{D}$. The elements of $\mathcal{D}$ are referred to as *processes*. Assume that the signature of $\mathcal{A}$ contains a set of action constants $\mathcal{AC}$. Furthermore, assume that the signature contains at least a (binary) choice operator $\bar{+}$, a sequential composition operator $\bar{\cdot}$, and a parallel-composition operator $\bar{\|}$. In addition, the signature may contain an inaction constant $\bar{\delta}$.

In order to define the notion of an interleaving process algebra, the concept of a process term in head normal form as defined in Definition 5.4 is adapted to the setting of a process algebra. The algebraic framework in this paragraph proves to be simpler than the setting of an equational theory in the previous subsection. The set of processes in a process algebra that have a head normal form can be defined without the use of an auxiliary predicate.

DEFINITION 5.15 (*Head normal form*). The set of processes over the signature of the algebra $\mathcal{A}$ in *head normal form*, denoted $\mathcal{H}$, is inductively defined as follows. First, $\mathcal{AC} \subseteq \mathcal{H}$. Second, if the signature of $\mathcal{A}$ contains an inaction constant $\bar{\delta}$, then also $\bar{\delta}$ is an element of $\mathcal{H}$. Third, for any action $a \in \mathcal{AC}$ and process $d \in \mathcal{D}$, the process $a \mathbin{\bar{\cdot}} d$ is an element of $\mathcal{H}$. Finally, for any processes $u, v \in \mathcal{H}$, process $u \mathbin{\bar{+}} v$ is an element of $\mathcal{H}$.

A process algebra is said to be interleaving if and only if the set of processes in head normal form is closed under parallel composition.

DEFINITION 5.16 (*Interleaving process algebra*). The process algebra $\mathcal{A}$ is an *interleaving process algebra* if and only if, for any processes u and v in $\mathcal{H}$, also $u \mathbin{\bar{\|}} v \in \mathcal{H}$.

Let us recall the construction of the model of the equational theory of Section 5.1 in the previous paragraph. Assume that AC is some set of actions; let γ be a communication function. The model $\mathcal{M}(AC, \gamma)$ of theory $(\text{ACP}^* + \text{RN} + RSP^* + \text{SC})(AC, \gamma)$ is based on the set of closed $(\text{ACP}^* + \text{RN} + RSP^* + \text{SC})(AC, \gamma)$ terms, denoted $\mathcal{C}(AC)$. The domain of $\mathcal{M}(AC, \gamma)$ has been defined as the set of equivalence classes of closed terms in $\mathcal{C}(AC)$ modulo bisimilarity. Furthermore, for any action constant $a \in AC$, the interpretation $\bar{a}$ in the model has been defined as $[a]_\sim$; similarly, the interpretation $\bar{\delta}$ of the inaction constant δ has been defined as $[\delta]_\sim$. Finally, for any n-ary operator $\oplus$ in the signature of $(\text{ACP}^* + \text{RN} + RSP^* + \text{SC})(AC, \gamma)$, where n is some positive natural number, the corresponding function $\bar{\oplus}$ in the signature of $\mathcal{M}(AC, \gamma)$ has been defined as follows: For any closed terms $p_1, \ldots, p_n \in \mathcal{C}(AC)$, $\bar{\oplus}([p_1]_\sim, \ldots, [p_n]_\sim) = [\oplus(p_1, \ldots, p_n)]_\sim$. Clearly, model $\mathcal{M}(AC, \gamma)$ is a process algebra in the general sense defined above.

To prove that the algebra $\mathcal{M}(AC, \gamma)$ is an interleaving process algebra, it is necessary to determine the set of processes in head normal form, as defined by Definition 5.15. The set of action constants $\overline{AC}$ of the algebra $\mathcal{M}(AC, \gamma)$ can be defined in terms of the set of actions AC as follows: $\bar{AC} = \{\bar{a} \mid a \in AC\}$. Furthermore, the signature of the algebra contains, among other functions, the desired choice operator $\bar{+}$ and sequential-composition operator $\bar{\cdot}$. In addition, it contains the inaction constant $\bar{\delta}$. Thus, the set of processes of $\mathcal{M}(AC, \gamma)$ in head normal form, denoted $\mathcal{H}(\bar{AC})$, is defined as in Definition 5.15, where $\bar{AC}$ plays the role of $\mathcal{AC}$. The set of processes in head normal form is parameterized with the set of actions $\bar{AC}$, because $\bar{AC}$ is defined in terms of AC which is a parameter of model $\mathcal{M}(AC, \gamma)$. Thus, varying the value of parameter AC influences the set of processes in head normal form.

Having defined the set $\mathcal{H}(\bar{AC})$, it is possible to prove that $\mathcal{M}(AC, \gamma)$ is an interleaving process algebra. Note that the signature of $\mathcal{M}(AC, \gamma)$ contains a parallel-composition operator, as required by Definition 5.16 (Interleaving process algebra), namely the operator $\bar{\|}$. Essentially, the desired result follows from the fact that $(\mathrm{ACP}^* + \mathrm{RN} + \mathit{RSP}^* + \mathrm{SC})(AC, \gamma)$ is an interleaving equational theory and an auxiliary property stating that each process of $\mathcal{M}(AC, \gamma)$ in head normal form can be expressed in the signature of the equational theory by a closed term that has a head normal form.

THEOREM 5.17. *The process algebra $\mathcal{M}(AC, \gamma)$ is an interleaving process algebra.*

PROOF. Let p and q be closed terms in $\mathcal{C}(AC)$ such that $[p]_\sim, [q]_\sim \in \mathcal{H}(\bar{AC})$. It must be shown that $[p]_\sim \,\bar{\|}\, [q]_\sim \in \mathcal{H}(\bar{AC})$. It follows from Property 5.18 given below that there exist closed terms u and v in $\mathcal{C}(AC)$ such that $\mathit{HNF}(u)$, $\mathit{HNF}(v)$, $[u]_\sim = [p]_\sim$, and $[v]_\sim = [q]_\sim$. Hence, it suffices to prove that $[u]_\sim \,\bar{\|}\, [v]_\sim \in \mathcal{H}_{\mathcal{M}}$. Since $\mathit{HNF}(u)$ and $\mathit{HNF}(v)$, it follows from Theorem 5.7 that $\mathit{HNF}(u \parallel v)$. Definition 5.4 (Head normal form) in the previous subsection implies that there exists a closed term in head normal form $w \in \boldsymbol{H}(AC)$ such that $(\mathrm{ACP}^* + \mathrm{RN} + \mathit{RSP}^* + \mathrm{SC})(AC, \gamma) \vdash u \parallel v = w$. Theorem 5.13 (Soundness) and the definition of validity yield that $[u \parallel v]_\sim = [w]_\sim$. Clearly, it follows from the correspondence between Definitions 5.4 and 5.15 that $[w]_\sim$ is an element of $\mathcal{H}(\bar{AC})$. Since $[w]_\sim = [u \parallel v]_\sim = [u]_\sim \,\bar{\|}\, [v]_\sim$, it is shown that $[u]_\sim \,\bar{\|}\, [v]_\sim$ is an element of $\mathcal{H}(\bar{AC})$, which completes the proof. □

Recall Definition 5.4 from the previous subsection, which formalizes when a closed term defined in the signature of the equational theory $(\mathrm{ACP}^* + \mathrm{RN} + \mathit{RSP}^* + \mathrm{SC})(AC, \gamma)$ has a head normal form. The following property proves the claim made above that any process in the domain of $\mathcal{M}(AC, \gamma)$ that is in head normal form, as defined by Definition 5.15, can be specified in the signature of $(\mathrm{ACP}^* + \mathrm{RN} + \mathit{RSP}^* + \mathrm{SC})(AC, \gamma)$ by means of a closed term that has a head normal form. It also proves the converse, namely that any process in $\mathcal{M}(AC, \gamma)$ that can be specified by means of a closed term that has a head normal form, as defined by Definition 5.4, is indeed a process in head normal form, as defined by Definition 5.15.

PROPERTY 5.18. *For any closed term $p \in \mathcal{C}(AC)$,*

$$[p]_\sim \in \mathcal{H}(\bar{AC}) \Leftrightarrow (\exists q : q \in \mathcal{C}(AC) \wedge \mathit{HNF}(q) : [q]_\sim = [p]_\sim).$$

PROOF. Let p be a closed term in $\mathcal{C}(AC)$.

First, assume that $[p]_\sim \in \mathcal{H}(\bar{AC})$. It must be shown that there exists a closed term $q \in \mathcal{C}(AC)$ such that $HNF(q)$ and $[q]_\sim = [p]_\sim$. The proof is by induction on the structure of processes in $\mathcal{H}(\bar{AC})$, as defined by Definition 5.15.

(i) Assume that $[p]_\sim = \bar{a}$, for some $a \in AC$. It follows from Definition 5.4 (Head normal form) in the previous subsection that $HNF(a)$. The observation that $\bar{a} = [a]_\sim$ completes the proof in this case.

(ii) Assume that $[p]_\sim = \bar{\delta}$. Since $HNF(\delta)$ and $\bar{\delta} = [\delta]_\sim$, term δ satisfies the desired requirements.

(iii) Assume that $[p]_\sim = \bar{a} \mathbin{\bar{\cdot}} [q]_\sim$, for some $a \in AC$ and $q \in \mathcal{C}(AC)$. Since $\bar{a} \mathbin{\bar{\cdot}} [q]_\sim = [a \cdot q]_\sim$ and $HNF(a \cdot q)$, the term $a \cdot q$ satisfies the requirements in this case.

(iv) Assume that $[p]_\sim = [u]_\sim \mathbin{\bar{+}} [v]_\sim$, for some $u, v \in \mathcal{C}(AC)$ such that $[u]_\sim, [v]_\sim \in \mathcal{H}(\bar{AC})$. (Note that terms u and v are not necessarily elements of $\boldsymbol{H}(AC)$, in which case term $u + v$ would have satisfied the requirements.) By induction, it can be derived that there must exist terms r and s in $\mathcal{C}(AC)$ such that $HNF(r)$, $HNF(s)$, $[r]_\sim = [u]_\sim$, and $[s]_\sim = [v]_\sim$. Hence, it follows from Definition 5.4 (Head normal form) that $HNF(r + s)$. Since, in addition, $[u]_\sim \mathbin{\bar{+}} [v]_\sim = [r]_\sim \mathbin{\bar{+}} [s]_\sim = [r + s]_\sim$, term $r + s$ satisfies the requirements, which completes the first part of the proof.

Second, assume that q is a term in $\mathcal{C}(AC)$ such that $HNF(q)$ and $[q]_\sim = [p]_\sim$. It follows from Definition 5.4 (Head normal form) that there is a term r in $\boldsymbol{H}(AC)$ such that $(\mathrm{ACP}^* + \mathrm{RN} + RSP^* + \mathrm{SC})(AC, \gamma) \vdash q = r$. Theorem 5.13 (Soundness) yields that $[q]_\sim = [r]_\sim$. The correspondence between Definitions 5.4 and 5.15 implies that $[r]_\sim$ is an element of $\mathcal{H}(\bar{AC})$. The observation that $[q]_\sim = [p]_\sim = [r]_\sim$ completes the proof. □

EXAMPLE 5.19. Consider again Example 5.10. Let A again be the set of actions that includes the actions a, b, and c; let γ be the communication function $\{((a, b), c), ((b, a), c)\}$. Since the terms $a \parallel b$ and $a \cdot b + b \cdot a + c$ are derivably equal in the equational theory $(\mathrm{ACP}^* + \mathrm{RN} + RSP^* + \mathrm{SC})(A, \gamma)$, it follows from Theorem 5.13 (Soundness) that $[a \parallel b]_\sim = [a \cdot b + b \cdot a + c]_\sim$. That is, the two closed terms specify the same process in the model $\mathcal{M}(A, \gamma)$. (Note that it is also straightforward to obtain this equality by defining a bisimulation between the two terms.) In the algebra $\mathcal{M}(A, \gamma)$, the same process can also be written as $\bar{a} \mathbin{\bar{\parallel}} \bar{b}$ and $\bar{a} \mathbin{\bar{\cdot}} \bar{b} \mathbin{\bar{+}} \bar{b} \mathbin{\bar{\cdot}} \bar{a} \mathbin{\bar{+}} \bar{c}$. Clearly, the latter expression is in head normal form, which means that process $\bar{a} \mathbin{\bar{\parallel}} \bar{b}$ is in head normal form, as required by Theorem 5.17.

An interesting observation is that the proof of Theorem 5.17 does not use any specific characteristics of the equational theory $(\mathrm{ACP}^* + \mathrm{RN} + RSP^* + \mathrm{SC})(AC, \gamma)$ or its model $\mathcal{M}(AC, \gamma)$ other than the fact that the equational theory is an interleaving theory. Thus, the proof carries over to *any* model of *any* interleaving equational theory. This conforms to our informal definition of an interleaving semantics given in Section 1.

Another interesting observation is that Property 5.18 can be strengthened in case the equational theory under consideration is a *complete* axiomatization of an algebra. In that case, any closed term over the signature of the equational theory that specifies a process in head normal form in the algebra must have a head normal form itself.

Concluding remark. Both the equational theory $(\mathrm{ACP}^* + \mathrm{RN} + RSP^* + \mathrm{SC})(AC, \gamma)$ of Section 5.1 and its model $\mathcal{M}(AC, \gamma)$ given in this subsection are parameterized by a set of actions and a communication function. These parameters provide the flexibility to define, for example, a total-order semantics and a step semantics for concurrent systems, as shown in Sections 5.3 and 5.4, respectively. Since Theorem 5.17 proven above is independent of the choice for the parameters AC and γ, both the resulting process algebras are *interleaving* algebras. Thus, the semantics of Section 5.4 is an *interleaving, partial-order* process algebra.

5.3. *A total-order semantics*

In this subsection, the equational theory of Section 5.1 and its model of Section 5.2 are turned into a total-order framework for reasoning about the behavior of concurrent systems. The approach is illustrated by means of a few examples. The essence of a total-order semantics for concurrent systems is that no two actions can occur simultaneously. Thus, in this subsection, it is assumed that actions are atomic.

DEFINITION 5.20 (*Total-order semantics*). Let A be some set of atomic actions. Let $\gamma : A \times A \to A$ be a communication function on atomic actions. The model $\mathcal{M}(A, \gamma)$ defines a (branching-time, interleaving) total-order semantics for $(\mathrm{ACP}^* + \mathrm{RN} + RSP^* + \mathrm{SC})(AC, \gamma)$.

As explained before, a fundamental property of a total-order semantics is that it is impossible to distinguish concurrency and non-determinism. In the setting of the equational theory $(\mathrm{ACP}^* + \mathrm{RN} + RSP^* + \mathrm{SC})(AC, \gamma)$, this means that a (closed) term containing a parallel composition of two component terms must have the same semantics as a term specifying a non-deterministic choice between all possible totally-ordered interleavings of these two component terms.

EXAMPLE 5.21. To illustrate that the semantics of Definition 5.20 is indeed a total-order semantics, assume that A is some set of atomic actions including the actions p_1, p_2, r, and s. Assume that γ is the communication function that is undefined for all the elements in its domain. That is, there is no communication. Consider the following derivation. Since sequential composition is associative (Axiom $A5$), redundant brackets are omitted.

$$
\begin{aligned}
& r \cdot (p_1 \parallel p_2) \cdot s \\
= \quad & \{\text{Axioms } CM1, CM2(2\times)\} \\
& r \cdot (p_1 \cdot p_2 + p_2 \cdot p_1 + p_1 \mid p_2) \cdot s \\
= \quad & \{\text{Definition } \gamma;\ \text{Axioms } CF2, A6\} \\
& r \cdot (p_1 \cdot p_2 + p_2 \cdot p_1) \cdot s.
\end{aligned}
$$

This derivation shows that a term specifying the parallel composition of actions p_1 and p_2 is derivably equal to a term specifying a non-deterministic choice between two totally-ordered alternatives. Theorem 5.13 (Soundness) implies that the two closed terms must correspond to the same process in the semantics of Definition 5.20. Clearly, the semantics of both terms can be visualized by the process depicted in Figure 1(d), where dots correspond to closed terms.

In the previous example, communication plays no role. The following example shows a typical use of the communication function in ACP-style process algebra. Communication is used to enforce synchronization between parallel components.

EXAMPLE 5.22. Let A be some set of atomic actions that includes the actions r_1, r_2, s_2, c_2, and s_3. Let γ be the function $\{((r_2, s_2), c_2), ((s_2, r_2), c_2)\}$. Informally, a process performing an action r_i with $i \in \{1, 2\}$ receives a message of its environment over some port i. If a process performs an action s_i with $i \in \{2, 3\}$, it sends a message to its environment over port i. An action c_2 represents a communication over port 2, which is a synchronization of a send action and a receive action.

A process $B_{1,2}$ that repeatedly receives a message over port 1 and then forwards this message over port 2, thus corresponding to a one-place buffer for messages, can be specified as follows: $B_{1,2} = (r_1 \cdot s_2)^* \delta$. It follows from Axioms $BKS1$ and $A6$ that this specification represents a non-terminating iteration. Similarly, process $B_{2,3} = (r_2 \cdot s_3)^* \delta$ represents a one-place buffer for messages that receives messages over port 2 and sends messages over port 3.

In a number of steps, it can be shown that the composition of the above two one-place buffers yields a two-place buffer that receives messages from its environment over port 1 and sends messages to its environment over port 3. The composition is a parallel composition with the restriction that the two buffers must communicate over port 2. The following auxiliary derivation shows how a non-terminating iteration can be unfolded one cycle:

$$\begin{aligned}
& B_{1,2} \\
= & \qquad \{\text{Substitution}\} \\
& (r_1 \cdot s_2)^* \delta \\
= & \qquad \{\text{Axiom } BKS1\text{; substitution}\} \\
& (r_1 \cdot s_2) \cdot B_{1,2} + \delta \\
= & \qquad \{\text{Axioms } A5, A6\} \\
& r_1 \cdot s_2 \cdot B_{1,2}.
\end{aligned}$$

The next derivation shows how to calculate the first action of the abovementioned parallel composition of the two one-place buffers. To enforce communication over port 2, isolated occurrences of send and receive actions over port 2 must be encapsulated. Therefore, let $H = \{r_2, s_2\}$.

$$\begin{aligned}
& \partial_H(B_{1,2} \parallel B_{2,3}) \\
= & \qquad \{\text{Axiom } CM1\} \\
& \partial_H(B_{1,2} \mathbin{⌊\!⌊} B_{2,3} + B_{2,3} \mathbin{⌊\!⌊} B_{1,2} + B_{1,2} \mid B_{2,3}) \\
= & \qquad \{\text{Derivation above; Axiom } CM3(2\times)\text{; Axiom } CM7\}
\end{aligned}$$

$$
\begin{aligned}
& \partial_H\big(r_1 \cdot (s_2 \cdot B_{1,2} \parallel B_{2,3}) + r_2 \cdot (s_3 \cdot B_{2,3} \parallel B_{1,2}) \\
& \quad + (r_1 \mid r_2) \cdot (s_2 \cdot B_{1,2} \parallel s_3 \cdot B_{2,3})\big) \\
= & \qquad \{\text{Axiom } CF2\text{; Axioms } D3(2\times), D4(3\times), D1(2\times), D2\} \\
& r_1 \cdot \partial_H(s_2 \cdot B_{1,2} \parallel B_{2,3}) + \delta \cdot \partial_H(B_{1,2} \parallel s_3 \cdot B_{2,3}) \\
& \quad + \delta \cdot \partial_H(s_2 \cdot B_{1,2} \parallel s_3 \cdot B_{2,3}) \\
= & \qquad \{\text{Axioms } A6, A7(\text{both } 2\times)\} \\
& r_1 \cdot \partial_H(s_2 \cdot B_{1,2} \parallel B_{2,3}).
\end{aligned}
$$

Similar derivations yield the following results. The receipt of a message over port 1 is followed by a communication action over port 2, after which, in any order, a new message may be received over port 1 and the first message may be send to the environment over port 3.

$$
\begin{aligned}
& \partial_H(s_2 \cdot B_{1,2} \parallel B_{2,3}) \\
& \quad = c_2 \cdot \partial_H(B_{1,2} \parallel s_3 \cdot B_{2,3}) \\
& \quad = c_2 \cdot \big(r_1 \cdot \partial_H(s_2 \cdot B_{1,2} \parallel s_3 \cdot B_{2,3}) + s_3 \cdot \partial_H(B_{1,2} \parallel B_{2,3})\big) \\
& \quad = c_2 \cdot \big(r_1 \cdot s_3 \cdot \partial_H(s_2 \cdot B_{1,2} \parallel B_{2,3}) + s_3 \cdot r_1 \cdot \partial_H(s_2 \cdot B_{1,2} \parallel B_{2,3})\big) \\
& \quad = c_2 \cdot (r_1 \cdot s_3 + s_3 \cdot r_1) \cdot \partial_H(s_2 \cdot B_{1,2} \parallel B_{2,3}).
\end{aligned}
$$

Summarizing the results obtained so far yields the following two equations.

$$
\begin{aligned}
\partial_H(B_{1,2} \parallel B_{2,3}) &= r_1 \cdot \partial_H(s_2 \cdot B_{1,2} \parallel B_{2,3}) \quad \text{and} \\
\partial_H(s_2 \cdot B_{1,2} \parallel B_{2,3}) &= c_2 \cdot (r_1 \cdot s_3 + s_3 \cdot r_1) \cdot \partial_H(s_2 \cdot B_{1,2} \parallel B_{2,3}).
\end{aligned}
$$

The second equation defines a non-terminating iteration, as can be shown by applying Axioms $A6$ and RSP^*:

$$\partial_H(s_2 \cdot B_{1,2} \parallel B_{2,3}) = \big(c_2 \cdot (r_1 \cdot s_3 + s_3 \cdot r_1)\big)^* \delta.$$

Substituting this result in the above equation for $\partial_H(B_{1,2} \parallel B_{2,3})$ yields that

$$\partial_H(B_{1,2} \parallel B_{2,3}) = r_1 \cdot \big(\big(c_2 \cdot (r_1 \cdot s_3 + s_3 \cdot r_1)\big)^* \delta\big).$$

Finally, as in the previous example, it can be shown that the choice in the last result corresponds to a parallel composition:

$$
\begin{aligned}
& r_1 \cdot s_3 + s_3 \cdot r_1 \\
= & \qquad \{\text{Axiom } A6\} \\
& r_1 \cdot s_3 + s_3 \cdot r_1 + \delta \\
= & \qquad \{\text{Axiom } CF2\text{; definition } \gamma\} \\
& r_1 \cdot s_3 + s_3 \cdot r_1 + r_1 \mid s_3 \\
= & \qquad \{\text{Axioms } CM2(2\times), CM1\} \\
& r_1 \parallel s_3.
\end{aligned}
$$

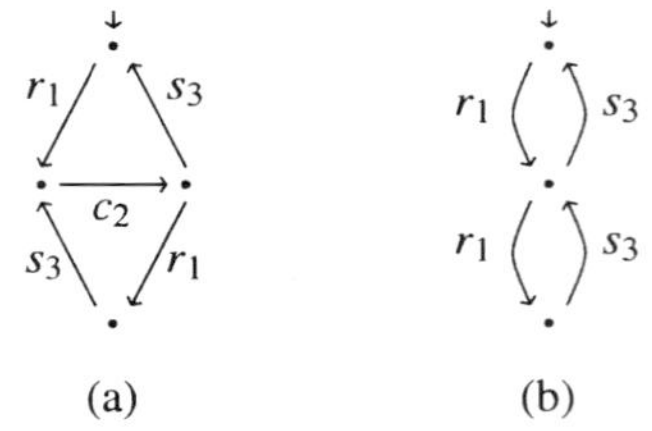

Fig. 7. A two-place buffer in a total-order semantics.

Thus, the final result for the composition of the two one-place buffers is the following:

$$\partial_H(B_{1,2} \parallel B_{2,3}) = r_1 \cdot \big((c_2 \cdot (r_1 \parallel s_3))^* \delta\big).$$

It is not difficult to see that the right-hand side of the above equation specifies the behavior of a two-place buffer. Its semantics is depicted in Figure 7(a), where closed terms are represented by dots. Figure 7(b) shows the process that results from hiding the internal communication action c_2. Clearly, this process exhibits the behavior that is intuitively expected from a two-place buffer. In the current algebraic framework, it is not possible to hide actions. The extension of ACP-style process algebra with an abstraction operator is described in [6,14,25]. It goes beyond the scope of this chapter to treat abstraction in detail.

5.4. *A step semantics*

As in Sections 3 and 4, the basis of a step semantics is to assume that actions are bags of atomic actions. The occurrence of a bag of multiple atomic actions represents the simultaneous execution of its elements. The communication function can be used to specify which actions can be executed in parallel.

DEFINITION 5.23 (*Step semantics*). Let A be some set of atomic actions. Let $\gamma : \mathcal{B}(A) \times \mathcal{B}(A) \to \mathcal{B}(A)$ be some communication function. The model $\mathcal{M}(\mathcal{B}(A), \gamma)$ defines a (branching-time, interleaving) step semantics for $(\mathrm{ACP}^* + \mathrm{RN} + RSP^* + \mathrm{SC})(AC, \gamma)$, *provided* that the communication function γ allows that actions may be executed concurrently.

EXAMPLE 5.24. The most straightforward definition for the communication function γ in the semantics of Definition 5.23 is to define γ as summation of bags: $\gamma = \uplus$. Assume that A includes the atomic actions p_1, p_2, r, and s. Consider the processes specified by $[r] \cdot ([p_1] \cdot [p_2] + [p_2] \cdot [p_1]) \cdot [s]$ and $[r] \cdot ([p_1] \parallel [p_2]) \cdot [s]$. In $(\mathrm{ACP}^* + \mathrm{RN} + RSP^* + \mathrm{SC})(\mathcal{B}(A), \uplus)$, it is not possible to prove that these two terms are equal. The corresponding processes in the semantics of Definition 5.23 can be visualized as in Figure 2. As explained before, these two processes are not bisimilar.

A disadvantage of choosing bag summation for the communication function is that it is no longer possible to do derivations such as the ones in Example 5.22. Choosing bag summation for the communication function means that any action may occur concurrently with any other action. Bag summation cannot be used to enforce communication between actions or to enforce a total ordering on the execution of certain actions that are in conflict with each other. The latter means that actions can neither communicate or happen concurrently. One solution to combine concurrency with communication and conflicts is to carefully define an appropriate communication function.

EXAMPLE 5.25. As in Example 5.22, let A be some set of atomic actions including r_1, r_2, s_2, c_2, and s_3. Recall that the indices of these actions correspond to port numbers.

Informally, the communication function can be specified as follows. The only communication is the simultaneous execution of send action s_2 and receive action r_2, which results in the communication action c_2. Any number of atomic actions that cannot communicate can occur in parallel as long as no two of these actions use the same port. Actions using the same port are in conflict and must, therefore, be totally ordered.

To formalize the definition of the communication function, the auxiliary predicates *conflict* $\subseteq \mathcal{B}(A)$ and *communication* $\subseteq \mathcal{B}(A)$ are defined as follows. For any $\alpha \in \mathcal{B}(A)$,

$$\begin{aligned} \mathit{conflict}(\alpha) &\Leftrightarrow (\exists a : a \in A : [a^2] \leqslant \alpha) \vee [r_2, c_2] \leqslant \alpha \vee [s_2, c_2] \leqslant \alpha \quad \text{and} \\ \mathit{communication}(\alpha) &\Leftrightarrow [s_2, r_2] \leqslant \alpha. \end{aligned}$$

Using these two predicates, the communication function γ is defined as follows. Note that a bag of atomic actions containing no conflicts can contain at most one pair of atomic actions s_2 and r_2. For any $\alpha, \beta \in \mathcal{B}(A)$,

$$\begin{aligned} \neg\mathit{conflict}(\alpha \uplus \beta) \wedge \mathit{communication}(\alpha \uplus \beta) &\Rightarrow \gamma(\alpha, \beta) = (\alpha \uplus \beta) - [s_2, r_2] \uplus [c_2], \\ \neg\mathit{conflict}(\alpha \uplus \beta) \wedge \neg\mathit{communication}(\alpha \uplus \beta) &\Rightarrow \gamma(\alpha, \beta) = \alpha \uplus \beta, \quad \text{and} \\ \mathit{conflict}(\alpha \uplus \beta) &\Rightarrow \gamma(\alpha, \beta) \text{ is undefined.} \end{aligned}$$

It is not difficult to verify that this communication function is commutative and associative.

In the framework of this example, the two one-place buffers of Example 5.22 can be specified as follows: $B_{1,2} = ([r_1] \cdot [s_2]) * \delta$ and $B_{2,3} = ([r_2] \cdot [s_3]) * \delta$. Also in the current setting, it is possible to show that the parallel composition of the two buffers while enforcing communication corresponds to a two-place buffer. To enforce communication, actions that contain isolated send or receive actions over port 2 must be encapsulated. Therefore, let $H \subseteq \mathcal{B}(A)$ be defined as the set $\{\alpha \in \mathcal{B}(A) \mid r_2 \in \alpha \vee s_2 \in \alpha\}$. Calculations very similar to the ones in Example 5.22 lead to the following results:

$$\begin{aligned} \partial_H(B_{1,2} \parallel B_{2,3}) &= [r_1] \cdot \partial_H([s_2] \cdot B_{1,2} \parallel B_{2,3}) \\ \partial_H([s_2] \cdot B_{1,2} \parallel B_{2,3}) &= [c_2] \cdot \partial_H(B_{1,2} \parallel [s_3] \cdot B_{2,3}) \\ &= [c_2] \cdot \big([r_1] \cdot \partial_H([s_2] \cdot B_{1,2} \parallel [s_3] \cdot B_{2,3}) \\ &\qquad + [s_3] \cdot \partial_H(B_{1,2} \parallel B_{2,3}) \\ &\qquad + [r_1, s_3] \cdot \partial_H([s_2] \cdot B_{1,2} \parallel B_{2,3}) \\ &\qquad \big) \\ &= [c_2] \cdot ([r_1] \cdot [s_3] + [s_3] \cdot [r_1] + [r_1, s_3]) \\ &\qquad \cdot \partial_H([s_2] \cdot B_{1,2} \parallel B_{2,3}). \end{aligned}$$

The main difference between the above results and the corresponding results in Example 5.22 is that, in the current setting, the actions r_1 and s_3 can occur concurrently, whereas this is not possible in Example 5.22. Example 5.22 shows how a parallel composition is rewritten into a non-deterministic choice of *totally*-ordered interleavings. In the current example, a parallel composition is rewritten into a non-deterministic choice of *partially*-ordered interleavings, namely the interleavings of all the *steps* that the parallel composition can perform. These observations conform to our remark made earlier that both the total-order framework of the previous subsection and the partial-order framework of this subsection are interleaving frameworks.

Note that one of the above results is a recursive equation. Using *RSP**, the axioms for the three merge operators, and the definition of the communication function, the following result can be derived. Again, the calculations are very similar to the ones in Example 5.22.

$$\partial_H(B_{1,2} \parallel B_{2,3}) = [r_1] \cdot \left(([c_2] \cdot ([r_1] \parallel [s_3]))^* \delta\right).$$

Figure 8(a) visualizes the semantics of the two-place buffer. Figure 8(b) shows the process that results from hiding the internal communication action $[c_2]$.

Observe that the approach of Example 5.25 can be adapted to obtain a total-order setting by specifying that any two atomic actions are in conflict unless they communicate. The result is that any process can only execute steps consisting of a single atomic action.

The attentive reader might have noticed that in the calculations of Example 5.25 no conflicts between atomic actions occur, as specified by the auxiliary predicate *conflict*. Thus, for the above example, the communication function can be simplified. Nevertheless, the definition of conflicts can be meaningful, as the following example shows.

EXAMPLE 5.26. Assume that the set of atomic actions A, the predicates *conflict*, *communication* $\subseteq \mathcal{B}(A)$, the communication function $\gamma : \mathcal{B}(A) \times \mathcal{B}(A) \to \mathcal{B}(A)$, and the set of actions $H \subseteq \mathcal{B}(A)$ are defined as in Example 5.25.

Consider the following two processes: $S = ([r_1] \cdot [s_2] \cdot [r_2])^* \delta$ and $C = (([r_1] \parallel [r_2]) \cdot ([s_2] \mid [s_3]))^* \delta$. Process C, a calculator, models an iterative process of which a cycle starts by receiving two messages over ports 1 and 2. Ports could, for example, be metal wires and messages could be electrical signals. After C has received both signals, it simultaneously

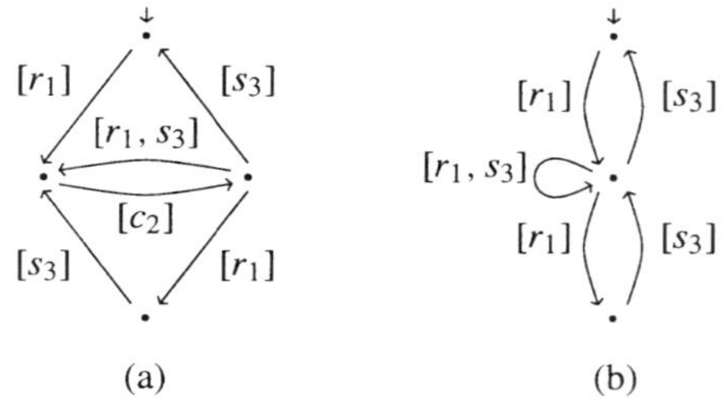

Fig. 8. A two-place buffer in a step semantics.

sends a result signal over port 3 and a confirmation signal over port 2 to indicate that both input signals have been read. Process S is a simple synchronized buffer that forwards a signal from port 1 to port 2; before it reads its next signal over port 1, it waits for a confirmation over port 2.

Intuitively, the parallel composition of processes S and C, enforcing communication over port 2, should yield an iterative process that in each cycle reads two signals over port 1 and sends a result signal over port 3. Recall that the definition of the predicate *conflict* is such that no two atomic actions using the same port can occur simultaneously. The results below confirm that the composed process starts with a single r_1, which is either performed by S or by C. It does not have the option to read its two inputs simultaneously. After the derivation of the first action, the calculations are very similar to the ones in the previous examples.

$$\begin{aligned}
&\partial_H(S \parallel C)\\
&\quad = [r_1] \cdot \partial_H([s_2] \cdot [r_2] \cdot S \parallel C) + [r_1] \cdot \partial_H\big(S \parallel [r_2] \cdot ([s_2] \mid [s_3]) \cdot C\big)\\
&\quad = \big([r_1] \cdot ([r_1] \cdot [c_2] + [c_2] \cdot [r_1] + [r_1, c_2]) + [r_1] \cdot [r_1] \cdot [c_2]\big)\\
&\qquad\quad \cdot \partial_H\big([r_2] \cdot S \parallel ([s_2] \mid [s_3]) \cdot C\big)\\
&\quad = \big([r_1] \cdot ([r_1] \parallel [c_2]) + [r_1] \cdot [r_1] \cdot [c_2]\big) \cdot [c_2, s_3] \cdot \partial_H(S \parallel C)\\
&\quad = ([r_1] \cdot [c_2] \parallel [r_1]) \cdot [c_2, s_3] \cdot \partial_H(S \parallel C).
\end{aligned}$$

Thus, it follows from RSP^* that

$$\partial_H(S \parallel C) = \big(([r_1] \cdot [c_2] \parallel [r_1]) \cdot [c_2, s_3]\big)^* \delta.$$

Ignoring communication actions c_2, the result is indeed a process that, in each cycle, reads two signals over port 1 and sends a result signal over port 3.

The approach to combining communication and concurrency, including conflicts, in the communication function can be generalized to larger examples. However, the resulting communication function may become complex and intuitively difficult to understand. Another solution to combine concurrency with communication in the current framework is to allow arbitrary concurrency between actions, as in Example 5.24, and to use renaming operators to make conflicts and communications explicit. This approach is very similar to the approach to combining concurrency and communication in ACP-style process algebra taken in [4].

EXAMPLE 5.27. Let A be the set of atomic actions defined in Example 5.25. Assume that the communication function γ on bags of atomic actions equals bag summation $\uplus$. Using the auxiliary predicates *conflict* and *communication* of Example 5.25, the renaming functions $\boldsymbol{cf}, \boldsymbol{cm} : (\mathcal{B}(A) \cup \{\delta\}) \to (\mathcal{B}(A) \cup \{\delta\})$ are defined as follows. For any $\alpha \in \mathcal{B}(A)$,

$$\boldsymbol{cf}(\delta) = \delta, \quad \mathit{conflict}(\alpha) \Rightarrow \boldsymbol{cf}(\alpha) = \delta, \quad \text{and} \quad \neg\mathit{conflict}(\alpha) \Rightarrow \boldsymbol{cf}(\alpha) = \alpha$$

and

$$\boldsymbol{cm}(\delta) = \delta, \quad \mathit{communication}(\alpha) \Rightarrow \boldsymbol{cm}(\alpha) = \alpha - [s_2, r_2] \uplus [c_2], \quad \text{and}$$
$$\neg \mathit{communication}(\alpha) \Rightarrow \boldsymbol{cm}(\alpha) = \alpha.$$

Given these definitions, it is not difficult to obtain results similar to Examples 5.25 and 5.26. The basic idea is to start with the parallel composition of component processes, resolve conflicts according to the renaming function $\boldsymbol{cf}$, and enforce communication by renaming actions according to renaming function $\boldsymbol{cm}$ and encapsulating actions with isolated atomic actions that should have participated in a communication. Let $H \subseteq \mathcal{B}(A)$ be the set $\{\alpha \in \mathcal{B}(A) \mid r_2 \in \alpha \vee s_2 \in \alpha\}$, as defined in Example 5.25.

$$\partial_H\big(\rho_{\boldsymbol{cm}}\big(\rho_{\boldsymbol{cf}}(B_{1,2} \parallel B_{2,3})\big)\big) = [r_1] \cdot \big(([c_2] \cdot ([r_1] \parallel [s_3]))^* \delta\big) \quad \text{and}$$
$$\partial_H\big(\rho_{\boldsymbol{cm}}\big(\rho_{\boldsymbol{cf}}(S \parallel C)\big)\big) = \big(([r_1] \cdot [c_2] \parallel [r_1]) \cdot [c_2, s_3]\big)^* \delta,$$

where $B_{1,2}$ and $B_{2,3}$ are the two one-place buffers defined in Example 5.25 and S and C are the synchronized buffer and the calculator of Example 5.26.

Example 5.27 generalizes to larger examples in a rather straightforward way. In general, the definition of a renaming function removing all conflicts from a process term is not very complicated. The most straightforward way to obtain a renaming function handling an arbitrary number of communications is to define such a function as the composition of a number of auxiliary renaming functions that each rename a single pair of communicating atomic actions. Section 7 explains more about renamings and communication functions in a setting with steps.

Finally, observe that, unlike the approach of Example 5.25, the approach of Example 5.27 cannot be adapted to obtain a total-order setting. In the approach of Example 5.25, conflicts and communications are defined in the communication function. Thus, the communication function can be used to enforce a total ordering of (steps of) single atomic actions. In the approach of Example 5.27, the communication function is fixed to bag summation. A total ordering of atomic actions can only be enforced *explicitly* by means of renaming.

5.5. *Concluding remarks*

Summarizing the results of this section, it has been shown how to develop an algebraic framework for equational reasoning both in a total-order setting and in a partial-order setting. In addition, the notion of a (non-)interleaving algebraic theory has been formalized. It has been shown that standard ACP-style process algebra always leads to an interleaving framework. An interesting consequence is that the algebraic framework of Section 5.4 can be characterized as an *interleaving partial-order* theory. This observation conforms to our claim made in the introduction that interleaving versus non-interleaving and total-order versus partial-order are complementary characterizations of process-algebraic frameworks.

The basis of the step semantics for concurrent systems presented in Section 5.4 is the communication function of ACP. It can be used to specify which atomic actions are causally independent and, thus, may occur concurrently. Examples 5.25 and 5.26 show how to combine communication and concurrency in the definition of the communication function. Example 5.27 illustrates another approach that makes conflicts and communications explicit by means of renaming operators. The latter approach generalizes more easily to larger examples than the former. In the next section, a third approach to combining concurrency and communication is studied, which is inspired by the theory of P/T nets developed in Section 4. The basic idea is to use the communication function of ACP solely for the purpose of concurrency, as in Examples 5.24 and 5.27, and to introduce a new operator that can be used – to some extent – to resolve conflicts and to enforce (asynchronous) communication between parallel components. This new operator is inspired by the notion of transition firing in the framework of P/T nets.

6. The causal state operator

In the P/T-net framework of Section 4, the behavior of a P/T net is determined by its marking. The order in which tokens in the marking are consumed and produced determines the order in which actions are executed. Thus, in the P/T-net framework, the mechanism of consuming and producing tokens is the basic mechanism for introducing causal relationships between actions. In this section, the equational theory of the previous section is extended with an algebraic operator, called the *causal state operator*, that is inspired by the idea of consuming and producing tokens. The causal state operator plays an important role in adopting the causality mechanism of Petri nets into process algebra. In ACP-style process algebra, the standard means to enforce causal relationships between actions are sequential composition and communication. The causal state operator is sufficiently powerful to replace to some extent this standard causality mechanism, in particular, the standard communication mechanism. As a result, the communication function can be used to develop a framework with a step semantics, as explained in the previous section, thus combining the strength of communication and concurrency. An interesting application of the equational theory developed in this section is that it can be used to define an algebraic semantics for labeled P/T nets that conforms to the step semantics for labeled P/T nets defined in Section 4.

6.1. *The equational theory*

The causal state operator is a specialization of the so-called *state operator*. The state operator is an operator which has a memory to explicitly describe the state of a process. For a detailed treatment of the state operator, the reader is referred to [3,5,6,25]. A variant of the *causal* state operator first appeared in [4].

Table 7 presents the Algebra of Communicating Processes with iteration and causal state operator, renaming, the Recursive Specification Principle for the binary Kleene star, and standard concurrency, abbreviated $\mathrm{ACP}^*_\lambda + \mathrm{RN} + RSP^* + \mathrm{SC}$. It is parameterized by

Table 7
The equational theory $(\mathrm{ACP}_\lambda^* + \mathrm{RN} + RSP^* + \mathrm{SC})(AC(C, A), \gamma)$

$(\mathrm{ACP}_\lambda^* + \mathrm{RN} + RSP^* + \mathrm{SC})(AC(C, A), \gamma)$	
$(\mathrm{ACP}^* + \mathrm{RN} + RSP^* + \mathrm{SC})(AC(C, A), \gamma)$	
$\lambda_m^I : \boldsymbol{P} \to \boldsymbol{P}$;	
x, y: $\boldsymbol{P}$;	
$\lambda_m^I(\delta) = \delta$	*CSO*1
$\boldsymbol{c}a \lceil I \leqslant m \Rightarrow \lambda_m^I(a) = a$	*CSO*2
$\boldsymbol{c}a \lceil I \not\leqslant m \Rightarrow \lambda_m^I(a) = \delta$	*CSO*3
$\lambda_m^I(a \cdot x) = \lambda_m^I(a) \cdot \lambda^I_{m - \boldsymbol{c}a\lceil I \uplus \boldsymbol{p}a \lceil I}(x)$	*CSO*4
$\lambda_m^I(x + y) = \lambda_m^I(x) + \lambda_m^I(y)$	*CSO*5

a set of actions $AC(C, A)$, which is itself parameterized, and a communication function $\gamma : AC(C, A) \times AC(C, A) \to AC(C, A)$. A detailed explanation is given below.

In order to reason about places, tokens, token consumption, and token production in an equational way, it is necessary to introduce these notions in the equational theory. In a P/T net, the flow relation between places and transitions defines the token consumption and token production of a particular transition. The most straightforward way to introduce the notions of consumption and production in a process-algebraic setting is to include the information in the actions. The two parameters of the set of actions $AC(C, A)$ correspond to a set of so-called *causes* and a set of atomic actions, respectively. Causes are the algebraic equivalent of places and tokens in the P/T-net framework. In the remainder of this section, an action is assumed to be a triple in $\mathcal{B}(C) \times \mathcal{B}(A) \times \mathcal{B}(C)$. That is, $AC(C, A) = \mathcal{B}(C) \times \mathcal{B}(A) \times \mathcal{B}(C)$. The second element of such a triple denotes the actual *step* of the action; the first element represents the *consumption* of causes of the step and the third element its *production*, similar to the notions of token consumption and token production in the P/T-net framework. Three auxiliary functions on actions in $AC(C, A)$, $\boldsymbol{c}, \boldsymbol{p} : AC(C, A) \to \mathcal{B}(C)$ and $\boldsymbol{s} : AC(C, A) \to \mathcal{B}(A)$, are defined. For any $a = (c, \alpha, p)$ in $AC(C, A)$, $\boldsymbol{c}a = c$, $\boldsymbol{p}a = p$, and $\boldsymbol{s}a = \alpha$. Note that the exact structure of actions is not really important as long as these auxiliary functions can be defined.

The causal state operator is, in fact, not a single operator. Theory $(\mathrm{ACP}_\lambda^* + \mathrm{RN} + RSP^* + \mathrm{SC})(AC(C, A), \gamma)$ contains an entire class of causal state operators, namely one operator λ_m^I for each $I \subseteq C$ and $m \in \mathcal{B}(I)$. Set I can be seen as a set specifying *internal* causes. Causes not in I are referred to as *external* causes. Note that m is defined to be a bag of *internal* causes. In the term $\lambda_m^I(x)$, where x is a process term of sort $\boldsymbol{P}$, bag m represents the current state of process x, similar to a marking in P/T-net theory. For this reason, bag m is also referred to as the marking of x. The separation between internal and external causes provides the flexibility to distinguish between communication within a process and communication between the process and its environment. In standard P/T-net theory as introduced in Section 4, places and tokens are not divided into internal and external places and tokens. However, when a notion of modularity is introduced in P/T-net theory, as in

[8, Chapter 3], the separation between internal and external places and tokens becomes meaningful in the P/T-net framework. Section 8 treats this topic in more detail.

The axioms for causal state operators in Table 7 are inspired by the notion of transition firing introduced in Section 4. Constant a ranges over $AC(C, A)$. Axioms $CSO2$, $CSO3$, and $CSO4$ are the most interesting ones. The other two should be clear without further explanation. Axiom $CSO2$ states that an action may occur provided that its consumption of *internal* causes is available in the marking. Axiom $CSO3$ says that if not enough causes are available, an action cannot be executed and results in a deadlock. Axiom $CSO4$ states that the result of executing an action is that the consumption is removed from the marking, whereas the (internal) production is added to the marking. If the marking does not contain enough internal causes to execute the action, then the combination of Axioms $CSO3$ and $A7$ guarantees that the result is a deadlock. The axiomatization assumes that the environment of a process is responsible for consuming and producing external causes.

The intuition behind verifications by means of the causal state operator is the following. Given a number of components, as before, the parallel composition of these components is the starting point. The communication function is defined in such a way that it allows arbitrary concurrency. Thus, the parallel composition defines the largest behavior that is possible when combining the basic components. A causal state operator is used to restrict the composition to those behaviors that are allowed in the initial state of the composition, as defined by the subscript of the state operator. Thus, the causal state operator enforces causal orderings between actions.

As in the previous section, the expansion theorem of Theorem 5.2 can be used to simplify calculations with parallel compositions. It carries over to the theory $(\mathrm{ACP}^*_\lambda + \mathrm{RN} + RSP^* + \mathrm{SC})(AC(C, A), \gamma)$. The proof is identical. The next theorem is a specific instance of the expansion theorem which is often useful in derivations following the above approach.

THEOREM 6.1 (Expansion of iterations). *Let $a_0, \ldots, a_n$ be actions in $AC(C, A)$, where n is a positive natural number. Let I be the set $\{0, \ldots, n\}$.*

$$(\mathrm{ACP}^*_\lambda + \mathrm{RN} + RSP^* + \mathrm{SC})\big(AC(C, A), \gamma\big) \vdash$$
$$(\|i : i \in I : a_i{}^*\delta) = \big(+J : \emptyset \subset J \subseteq I : (|j : j \in J : a_j) \cdot (\|i : i \in I : a_i{}^*\delta)\big).$$

PROOF. It is a straightforward consequence of Theorem 5.2 and the axioms of $(\mathrm{ACP}^*_\lambda + \mathrm{RN} + RSP^* + \mathrm{SC})(AC(C, A), \gamma)$. □

EXAMPLE 6.2. Let C be a set of causes including the causes 1, 2, 3, 4, 5, and 6; let A be a set of atomic actions containing r, p_1, p_2, and s. The following are examples of actions in $AC(C, A)$: $([1], [r], [2, 3])$, $([2], [p_1], [4])$, $([3], [p_2], [5])$, and $([4, 5], [s], [6])$. To better understand this example, it might be helpful to consider the P/T net in Figure 4. To improve readability, in the remainder of this example, square brackets of singleton bags are omitted.

The communication function is used to specify concurrency. In principle, any action may occur in parallel with any other action. To formalize this communication function, the bagsum operator is overloaded to actions in $AC(C, A)$ as follows: For any $a, b \in AC(C, A)$, $a \uplus b = (\boldsymbol{ca} \uplus \boldsymbol{cb}, \boldsymbol{sa} \uplus \boldsymbol{sb}, \boldsymbol{pa} \uplus \boldsymbol{pb})$. Thus, the communication function γ on pairs of actions in $AC(C, A)$ is defined as $\uplus$.

Consider the parallel composition of the non-terminating iterations of the four actions introduced above: $X = (1, r, [2, 3])^*\delta \parallel (2, p_1, 4)^*\delta \parallel (3, p_2, 5)^*\delta \parallel ([4, 5], s, 6)^*\delta$. Parallel composition X specifies the process that may execute any (non-empty) subset of the four actions simultaneously an arbitrary number of times.

As explained, the causal state operator can be used to restrict a general parallel composition such as X to all the behaviors allowed when starting from some given initial state. Assume that all relevant causes are internal. That is, $I = \{1, 2, 3, 4, 5, 6\}$. Furthermore, assume that the initial state of the process contains a single cause 1. In the equational theory $(\mathrm{ACP}^*_\lambda + \mathrm{RN} + \mathit{RSP}^* + \mathrm{SC})(\mathit{AC}(C, A), \uplus)$, the following results can be derived. Theorem 6.1 (Expansion of iterations) is used to simplify the calculations. Note that, according to this theorem, the parallel composition of four iterations can perform *fifteen* different initial actions. However, in the initial state of process X, only one of these fifteen actions can actually occur.

$$
\begin{aligned}
&\lambda^I_1(X)\\
&\quad = (1, r, [2, 3]) \cdot \lambda^I_{[2,3]}(X)\\
&\quad = (1, r, [2, 3]) \cdot \big((2, p_1, 4) \cdot \lambda^I_{[3,4]}(X) + (3, p_2, 5) \cdot \lambda^I_{[2,5]}(X)\\
&\qquad\qquad\qquad\qquad + ([2, 3], [p_1, p_2], [4, 5]) \cdot \lambda^I_{[4,5]}(X)\big)\\
&\quad = (1, r, [2, 3]) \cdot \big((2, p_1, 4) \cdot (3, p_2, 5) + (3, p_2, 5) \cdot (2, p_1, 4)\\
&\qquad\qquad\qquad\qquad + ([2, 3], [p_1, p_2], [4, 5])\big) \cdot \lambda^I_{[4,5]}(X)\\
&\quad = (1, r, [2, 3]) \cdot \big((2, p_1, 4) \parallel (3, p_2, 5)\big) \cdot ([4, 5], s, 6) \cdot \lambda^I_6(X)\\
&\quad = (1, r, [2, 3]) \cdot \big((2, p_1, 4) \parallel (3, p_2, 5)\big) \cdot ([4, 5], s, 6) \cdot \delta.
\end{aligned}
$$

Note that the process $\lambda^I_1(X)$ executes each of the four actions introduced above only once, while the parallel composition X may execute each of the actions an arbitrary number of times. Another interesting observation is that $\lambda^I_1(X)$ ends in a deadlock. It does not terminate successfully. However, the attentive reader might notice the resemblance between this last result and the behavior of the P/T net of Figure 4 in the step semantics of Definition 4.13 (see Example 4.14). As explained in Section 4, the usual semantics of P/T nets does not distinguish deadlock and successful termination. In Section 6.3, an algebraic semantics for labeled P/T nets is defined in which the term $\lambda^I_1(X)$ forms the core of the semantics of the P/T net of Figure 4.

In the final result of the derivation in the above example, consumptions and productions of causes are visible whereas the consumption and production of tokens is not visible in the semantics of labeled P/T nets. However, in the algebraic framework, causes are only needed to enforce a causal ordering between actions. In general, it is not necessary that causes are visible in the final result of derivations as in Example 6.2. Therefore, a new class of algebraic operators, the so-called *cause-abstraction operators*, is introduced. A cause-abstraction operator can be used to hide consumptions and productions of causes. It is a renaming operator, as introduced in Table 2. In order to gain flexibility, a set of (internal) cause identifiers specifies which consumptions and productions must be hidden. For any set of cause identifiers $I \subseteq C$, the corresponding renaming function $\boldsymbol{ca}(I) : (AC(C, A) \cup \{\delta\}) \to (AC(C, A) \cup \{\delta\})$ restricts the bags representing the consumption and the production of an action to those causes not in I. As required, $\boldsymbol{ca}(\delta) = \delta$. Furthermore, for any $(c, \alpha, p) \in AC(C, A)$, $\boldsymbol{ca}(I)(c, \alpha, p) = (c \restriction (C \setminus I), \alpha, p \restriction (C \setminus I))$.

EXAMPLE 6.3. Let C, A, γ, X, and I be defined as in Example 6.2. For the sake of readability, a triple $(\mathbf{0}, \alpha, \mathbf{0}) \in AC(C, A)$ is identified with its second element α. It is not difficult to verify that

$$\rho_{\boldsymbol{ca}(I)}\big(\lambda^{I}_{[1]}(X)\big) = [r] \cdot ([p_1] \parallel [p_2]) \cdot [s] \cdot \delta.$$

It is clear that the right-hand term is an appropriate algebraic expression for the behavior of the labeled P/T net of Figure 4.

As mentioned in the introduction to this section, the causal state operator can be used to replace to some extent the standard communication mechanism in ACP-style process algebra. A good example illustrating the use of communication is the specification of a two-place buffer in terms of two one-place buffers (see Examples 5.22, 5.25, and 5.27). Thus, it is an interesting question how to specify a two-place buffer by means of a causal state operator.

EXAMPLE 6.4. Consider Example 5.27. It is straightforward to adapt the specification of the two-place buffer given in that example to the current setting where actions include input and output causes. Recall that, in Example 5.27, explicit renaming operators are used to resolve conflicts and enforce communications, which has the advantage that the communication function can be used to create a partial-order framework. Hence, assume that parameter γ of the equational theory $(\text{ACP}^*_\lambda + \text{RN} + RSP^* + \text{SC})(AC(C, A), \gamma)$ of this subsection equals bag summation $\uplus$ on actions in $AC(C, A)$. Let A be some set of atomic actions that includes the actions r_1, r_2, s_2, c_2, and s_3; let C be a set of causes including the causes $1, 2, 3, 4$, and 5.

Assume that a bag $\alpha \in \mathcal{B}(A)$ denotes the action $(\mathbf{0}, \alpha, \mathbf{0})$ in $AC(C, A)$. Under this assumption, the two terms $B_{1,2} = ([r_1] \cdot [s_2]) * \delta$ and $B_{2,3} = ([r_2] \cdot [s_3]) * \delta$ specify two one-place buffers. In order to specify a two-place buffer, it is necessary to resolve conflicts and enforce communication. Since conflicts play no role in the behavior of the two-place buffer, it suffices to adapt the renaming function $\boldsymbol{cm}$ of Example 5.27 to the current setting where actions contain information about causes. This renaming function can be lifted to actions in $AC(C, A)$ as follows. For any $(c, \alpha, p) \in AC(C, A)$, $\boldsymbol{cm}(c, \alpha, p) = (c, \boldsymbol{cm}(\alpha), p)$. To enforce communication, actions containing isolated atomic actions that should communicate must be encapsulated. Therefore, let $H \subseteq AC(C, A)$ be the set $\{(c, \alpha, p) \in AC(C, A) \mid r_2 \in \alpha \vee s_2 \in \alpha\}$. The two-place buffer can now be specified as follows:

$$\partial_H\big(\rho_{\boldsymbol{cm}}(B_{1,2} \parallel B_{2,3})\big). \tag{1}$$

in the P/T-net framework. The set of actions is defined as $AC(U, L)$. That is, an action is a triple in $\mathcal{B}(U) \times \mathcal{B}(L) \times \mathcal{B}(U)$.

To obtain an algebraic semantics for labeled P/T nets, each labeled P/T net is translated into a closed $(\mathrm{ACP}^*_\lambda + \mathrm{RN} + \mathit{RSP}^* + \mathrm{SC})(AC(U, L), \gamma)$ term, where γ is some communication function on $AC(U, L)$. The communication function is a parameter of the algebraic semantics in order to gain flexibility. The standard choice for the communication function γ is bag summation $\uplus$ as it is defined on actions in $AC(U, L)$. This means that arbitrary concurrency between actions is allowed.

The algebraic semantics of a labeled P/T net has a second parameter, namely a set of internal causes. As explained before, the distinction between internal and external causes is meaningful when distinguishing between communication within a process and communication between the process and its environment. The standard choice for this second parameter is the entire universe of identifiers U. This means that all causes are considered to be internal.

Assuming the standard choices for the two parameters of the algebraic semantics for P/T nets, it can be shown that the closed term corresponding to a labeled P/T net has the same step semantics as the P/T net (Theorem 6.10 below). Section 8 shows an application of the framework developed in this subsection with different choices for the parameters.

The basis of the algebraic semantics of a labeled P/T net is a closed term that corresponds to the unrestricted behavior of the P/T net, which is the behavior when every transition is always enabled. A causal state operator instantiated with the marking of the P/T net is applied to this term to restrict the behavior to all possible sequences of steps. A cause-abstraction operator, as introduced in Section 6.1, is used to hide consumptions and productions of causes. The unrestricted behavior of a single transition is the nonterminating iteration of its corresponding action in $AC(U, L)$. The unrestricted behavior of an entire P/T net is the parallel composition of all its transitions. Recall that $\boldsymbol{i}$ and $\boldsymbol{o}$ are functions that assign to each node in a labeled P/T net its bag of input and output nodes, respectively. For a transition, the bag of its input places corresponds to its consumption upon firing and the bag of its output places corresponds to its production upon firing. The set of closed $(\mathrm{ACP}^*_\lambda + \mathrm{RN} + \mathit{RSP}^* + \mathrm{SC})(AC(U, L), \gamma)$ terms is abbreviated $\mathcal{C}(U, L)$.

DEFINITION 6.8 (*Algebraic semantics for labeled P/T nets*). Let (N, m), with $N = (P, T, F, W, \ell)$, be a marked, L-labeled P/T net in $\mathcal{N}$, as defined in Definition 4.4. Let γ be some communication function. Assume that $I \subseteq U$ is a subset of identifiers denoting internal causes. The algebraic semantics of (N, m), denoted $[\![N, m]\!]^I_\gamma$, is a closed term in $\mathcal{C}(U, L)$ defined as follows:

$$[\![N, m]\!]^I_\gamma = \rho_{\boldsymbol{ca}(I)}\big(\lambda^I_m(\|\, t : t \in T : (\boldsymbol{i}t, \ell(t), \boldsymbol{o}t)^*\delta)\big).$$

EXAMPLE 6.9. Let $(N, [1])$ be the labeled P/T net of Figure 4. Consider again Examples 6.2 and 6.3. Assume that the set of causes C, the set of atomic actions A, the set of internal causes I, the closed term X, and the communication function γ are defined as in Example 6.2. Assume that the set of labels L equals the set of atomic actions A. The algebraic semantics of the P/T net $(N, [1])$, $[\![N, [1]]\!]^I_\uplus$, is the term $\rho_{\boldsymbol{ca}(I)}(\lambda^I_{[1]}(X))$ of Example 6.3. The semantics of this term in model $\mathcal{M}(C, A, \uplus)$ of the previous subsection, as discussed in Example 6.7, corresponds to the step semantics of the P/T net $(N, [1])$ as defined in Definition 4.13 and explained in Example 4.14.

The correspondence between the step semantics of a labeled P/T net and the step semantics of its algebraic representation, as observed in the above example, is formalized by the following theorem. It uses the standard choices for the two parameters of the algebraic semantics for labeled P/T nets. The theorem states that any step of a labeled P/T net can be simulated by its algebraic semantics and vice versa. Furthermore, since a labeled P/T net cannot terminate successfully, its algebraic semantics cannot terminate successfully either. This means that the algebraic semantics of a labeled P/T net cannot evolve into the special process $\surd$. The symbol $\equiv$ denotes syntactical equivalence of terms.

THEOREM 6.10. *For any labeled P/T nets $(N, m), (N, m') \in \mathcal{N}$, step $\alpha \in \mathcal{B}(L)$, closed term $p \in \mathcal{C}(U, L)$, and action $a \in AC(U, L)$,*

(i) $(N, m)\, [\alpha\rangle^s\, (N, m') \Rightarrow [\![N, m]\!]^U_{\uplus} \xrightarrow{(\mathbf{0},\alpha,\mathbf{0})} [\![N, m']\!]^U_{\uplus}$,

(ii) $[\![N, m]\!]^U_{\uplus} \xrightarrow{a} p \Rightarrow (\exists m', \alpha : m' \in \mathcal{B}(U) \wedge \alpha \in \mathcal{B}(L) :$
$p \equiv [\![N, m']\!]^U_{\uplus} \wedge a = (\mathbf{0}, \alpha, \mathbf{0}) \wedge (N, m)\, [\alpha\rangle^s\, (N, m'))$, *and*

(iii) $[\![N, m]\!]^U_{\uplus} \stackrel{a}{\not\to} \surd$.

PROOF. The proof consists of a straightforward but tedious analysis of the step firing rule of Definition 4.12 and the transition relation of Tables 6 and 8. The proof is almost identical to the proof of Theorem 3.4.8 in [8], where a similar result is proven in a total-order setting. Therefore, for more details, the reader is referred to [8]. □

Assuming that a triple $(\mathbf{0}, \alpha, \mathbf{0}) \in AC(U, L)$ is identified with its second element α, the correspondence between the step semantics of a labeled P/T net and its algebraic semantics can also be formulated as follows. In the process space that is defined as the (component-wise) union of the process spaces $(\mathcal{N}, \mathcal{B}(L), [\ \rangle^s, \emptyset)$ of Definition 4.13 (Step semantics for labeled P/T nets) and $(\mathcal{C}(U, L) \cup \{\surd\}, AC(U, L), \longrightarrow, \{\surd\})$ underlying model $\mathcal{M}(U, L, \uplus)$ of Section 6.2, a labeled P/T net in $\mathcal{N}$ and its algebraic semantics in $\mathcal{C}(U, L)$ are bisimilar.

A consequence of Definition 6.8 is that the theory $(\mathrm{ACP}^*_\lambda + \mathrm{RN} + RSP^* + \mathrm{SC})(AC(U, L), \uplus)$ can be used to reason, in a purely equational way, about the equivalence of labeled P/T nets in a step semantics.

EXAMPLE 6.11. Consider again the marked, labeled P/T nets of Figures 4 and 5. Let $(N, [1])$ be the net of Figure 4 and let $(K, [1, 7])$ and $(M, [1])$ be the two P/T nets of Figure 5. A triple $(\mathbf{0}, \alpha, \mathbf{0}) \in AC(U, L)$ is identified with its second element α. It is derivable from the axioms of $(\mathrm{ACP}^*_\lambda + \mathrm{RN} + RSP^* + \mathrm{SC})(AC(U, L), \uplus)$ that $[\![N, [1]]\!]^U_{\uplus} = [r] \cdot ([p_1] \parallel [p_2]) \cdot [s] \cdot \delta$. and that $[\![K, [1, 7]]\!]^U_{\uplus} = [\![M, [1]]\!]^U_{\uplus} = [r] \cdot ([p_1] \cdot [p_2] + [p_2] \cdot [p_1]) \cdot [s] \cdot \delta$. It is not possible to derive that $[\![N, [1]]\!]^U_{\uplus}$ equals $[\![K, [1, 7]]\!]^U_{\uplus}$ or $[\![M, [1]]\!]^U_{\uplus}$.

Given the definitions so far, it is straightforward to obtain an algebraic semantics for labeled P/T nets that corresponds to the total-order semantics of Definition 4.9. As in

Example 6.7, let $\perp$ be the communication function that is undefined for every pair of actions. Analogously to Theorem 6.10, it is possible to prove that any labeled P/T net $(N, m) \in \mathcal{N}$ and its algebraic semantics $[\![N, m]\!]^U_\perp$ represent corresponding processes in the process spaces $(\mathcal{N}, L, [\,\rangle^t, \emptyset)$ of Definition 4.9 (Total-order semantics for labeled P/T nets) and $(\mathcal{C}(U, L) \cup \{\surd\}, AC(U, L), \longrightarrow, \{\surd\})$ underlying model $\mathcal{M}(U, L, \perp)$ of Section 6.2, respectively.

6.4. *Concluding remarks*

Summarizing this section, it has been shown how to formalize the notion of transition firing known from Petri-net theory in ACP-style process algebra by means of the so-called causal state operator. The causal state operator can be used to some extent as a substitute for the standard communication mechanism of ACP. In this way, the standard communication mechanism can be used to obtain a partial-order framework. An interesting application of the theory of this section is that it can be used to reason in a purely equational (interleaving) way about labeled P/T nets with a step semantics.

An interesting question is to what extent the causal state operator can replace the standard causality mechanism of ACP-style process algebra consisting of sequential composition and communication. In Sections 9 and 10, this topic is investigated in more detail.

7. Intermezzo on renaming and communication functions

In standard process-algebraic theory as introduced in Section 5.1, actions do not have internal structure. Thus, communication functions and renaming functions are defined on actions without making assumptions about the structure that actions might have. However, in Section 5.4 and Section 6, we have seen several examples of actions that are structured and of communication and renaming functions using that structure of actions. This section summarizes several kinds of communication and renaming functions that use the structure of actions and that are of particular interest in the context of this chapter. It is assumed that the set of actions equals $AC(C, A)$ as defined in Section 6.1, with C some set of causes and A some set of atomic actions. Recall from Section 5.1 that a renaming function $f : AC(C, A) \cup \{\delta\} \to AC(C, A) \cup \{\delta\}$ is any function with the restriction that $f(\delta) = \delta$.

Renamings of atomic actions and causes. The first class of renaming functions that is of particular interest are those renaming functions that can be derived from renaming functions on atomic actions. Assume that $f : A \to A$ is a (partial) renaming of atomic actions. It can be turned into a renaming function $f : AC(C, A) \cup \{\delta\} \to AC(C, A) \cup \{\delta\}$ in a standard way. Let the auxiliary function $\bar{f} : A \to A$ be a *total* function defined as follows. For any $a \in A$, $\bar{f}(a)$ equals $f(a)$ if $f(a)$ is defined and a otherwise. Using this auxiliary function, the renaming function f is defined as follows. As required, $f(\delta) = \delta$. Furthermore, for any $(c, \alpha, p) \in AC(C, A)$, $f(c, \alpha, p) = (c, \bar{f}(\alpha), p)$ where $\bar{f}(\alpha) = (\uplus a : a \in \alpha : [(\bar{f}(a))^{\alpha(a)}])$.

Another class of renaming functions are those functions that can be derived from renamings of causes. Let $f : C \to C$ be such a (partial) renaming of causes. The renaming function $\boldsymbol{f}: AC(C, A) \cup \{\delta\} \to AC(C, A) \cup \{\delta\}$ is defined as follows: $\boldsymbol{f}(\delta) = \delta$ and, for any $(c, \alpha, p) \in AC(C, A)$, $\boldsymbol{f}(c, \alpha, p) = (\bar{f}(c), \alpha, \bar{f}(p))$, where the auxiliary function $\bar{f}$ on bags of causes is defined in the same way as the corresponding function on bags of atomic actions above.

Encapsulation operators. As explained in Section 5.1, encapsulation operators are an interesting kind of renaming operators. The basis of an encapsulation operator is a set of actions. Elements in this set are renamed to the inaction constant δ, whereas actions not in this set remain unchanged. It has also been explained how such a set can be transformed into a renaming function, thus showing that encapsulation operators belong to the class of renaming operators.

Encapsulation operators can also be derived from sets of *atomic* actions. Let $H \subseteq A$ be a set of atomic actions. The idea is that the corresponding encapsulation operator blocks actions that contain at least one atomic action of this set. Thus, assume that $\boldsymbol{H} = \{(c, \alpha, p) \in AC(C, A) \mid \alpha \upharpoonright H \neq \mathbf{0}\}$. The encapsulation operator $\partial_{\boldsymbol{H}}$ has the desired effect. The set $\boldsymbol{H}$ can be turned into a renaming function $\boldsymbol{e}(\boldsymbol{H}): AC(C, A) \cup \{\delta\} \to AC(C, A) \cup \{\delta\}$ in the way explained in Section 5.1.

Another set of encapsulation operators can be derived from sets of causes. For any set of causes $H \subseteq C$, the corresponding set $\boldsymbol{H} \subseteq AC(C, A)$ is defined as $\{(c, \alpha, p) \in AC(C, A) \mid (c \uplus p) \upharpoonright H \neq \mathbf{0}\}$. The encapsulation operator $\partial_{\boldsymbol{H}}$ blocks any action that consumes or produces a cause in H. The corresponding renaming function $\boldsymbol{e}(\boldsymbol{H}): AC(C, A) \cup \{\delta\} \to AC(C, A) \cup \{\delta\}$ is defined in the obvious way. Note that it is also straightforward to define encapsulation operators that encapsulate only actions that consume causes in H or only actions that produce causes in H.

Communication functions. Yet another kind of interesting renaming functions are those renaming functions that can be interpreted as explicit communication functions (see Example 5.27). Any associative and commutative (partial) *communication* function on *atomic* actions $cm: A \times A \to A$ can be turned into a renaming function $\boldsymbol{cm}: AC(C, A) \cup \{\delta\} \to AC(C, A) \cup \{\delta\}$ in a standard way. Let, for any $a, b \in A$, the auxiliary function $\bar{cm}(a, b): \mathcal{B}(A) \to \mathcal{B}(A)$ be defined as follows: For any $\alpha \in \mathcal{B}(A)$, $\bar{cm}(a, b)(\alpha) = \alpha - [a^n, b^n] \uplus [cm(a, b)^n]$ if $cm(a, b)$ is defined, $[a, b] \leqslant \alpha$, and $n = \alpha(a) \min \alpha(b)$; $\bar{cm}(a, b)(\alpha) = \alpha$ otherwise. Thus, in case the bag α contains at least one pair of atomic actions a and b, function $\bar{cm}(a, b)$ replaces any occurrence of that pair in α by the atomic action $cm(a, b)$. Another auxiliary function $\bar{cm}: \mathcal{B}(A) \to \mathcal{B}(A)$ is defined as the function composition of all functions $\bar{cm}(a, b)$ with $a, b \in A$. Since cm is associative and commutative, the order of the composition is not important. Function $\bar{cm}$ is a function that renames all pairs of communicating atomic actions. Finally, the desired communication function $\boldsymbol{cm}$ can now be defined as follows: $\boldsymbol{cm}(\delta) = \delta$ and, for any $(c, \alpha, p) \in AC(C, A)$, $\boldsymbol{cm}(c, \alpha, p) = (c, \bar{cm}(\alpha), p)$.

Of course, communication is not necessarily achieved by means of explicit renaming. On the contrary, as explained in Section 5, the standard approach to communication in ACP-style process algebra is to define communications by means of a communication function $\gamma: AC(C, A) \times AC(C, A) \to AC(C, A)$. However, any associative and commutative (partial) communication function on atomic actions $cm: A \times A \to A$ can simply be turned into a communication function γ on actions by means of the renaming function $\mathbf{cm}$ defined above. For any $a, b \in AC(C, A)$, $\gamma(a, b) = \mathbf{cm}(a \uplus b)$.

Cause abstraction. The last class of renaming functions mentioned in this section is the class of renaming functions that forms the basis for the cause-abstraction operators introduced in Section 6.1. Clearly, also these renaming functions use the structure of actions in $AC(C, A)$.

Finally, observe that the definitions of the renaming functions that do not affect causes as well as the definition of the communication functions given in this section can easily be adapted to the simpler framework of Section 5.4, where actions are defined as bags of atomic actions.

8. Modular P/T nets

The framework developed in Section 6 provides the basis for reasoning about the equivalence of labeled P/T nets in an equational way. However, the various examples show that it will be difficult to reason about the behavior of non-trivial P/T-net models of complex concurrent systems. A formalism for modeling and analyzing real-world concurrent systems must support *compositional* reasoning. A formalism is compositional if it allows the system designer to structure a formal model of a system into component models of subsystems in such a way that properties of the system as a whole can be derived from the properties of the components. The goal of this section is to illustrate by means of a few examples in what way compositionality can be achieved in the context developed so far. It is beyond the scope of this chapter to go into much detail. The interested reader is referred to [8, Chapter 3], which describes in detail a formalism supporting the compositional design of concurrent systems. The approach is similar to the one followed in Section 8.1. However, it is important to note that the formalism in [8, Chapter 3] is based on a total-order semantics.

The basis of any compositional formalism based on P/T nets is always some notion of a *modular* P/T net with a mechanism to construct modular P/T nets from other modular P/T nets. The idea is that a modular P/T net models a system component that may interact with its environment via a well defined *interface*. There are several ways to modularize P/T nets based on how the interface of a modular P/T net is defined. In the remainder of this section, two approaches are illustrated. In one approach, the interface of a modular P/T net consists of *places*, whereas in the other approach, it consists of *transitions*. Figure 9 shows two modular-P/T-net models of a one-place buffer that are explained in more detail below.

8.1. *Place fusion*

The most commonly used approach to modularizing Petri nets is the one where interfaces between modular nets are based on places. It appears in the Petri-net frameworks of, for example, [15,35,39], [8, Chapter 3], and [16], Chapter 14 of this Handbook. It is also supported by tools as Design/CPN [40], ExSpect [7], and PEP [18].

Figure 9(a) shows a labeled P/T net of which the set of places is partitioned into two sets by means of a dashed box. Places inside the box are *internal places*, whereas places outside

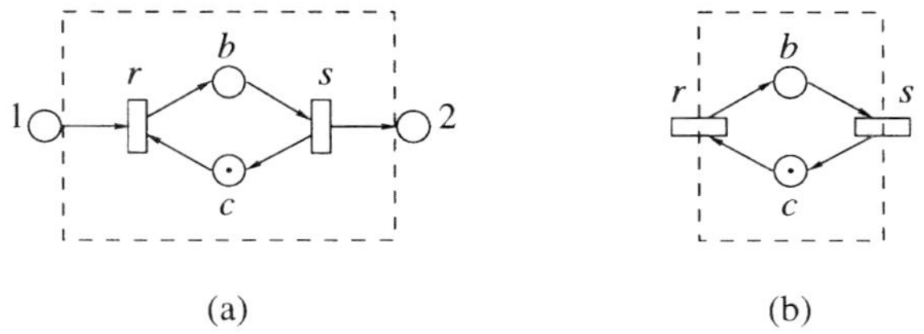

Fig. 9. Modular-P/T-net models of a buffer.

the box are so-called *pins*. Pins are connectors between the system and its environment. Interaction between the system and its environment is established by exchanging tokens via these pins. A P/T net as shown in Figure 9(a) is a very simple example of a modular P/T net. The set of pins defines the interface of the modular net. The mechanism to construct new modular P/T nets from other modular nets is explained in more detail below.

The behavior of a modular P/T net as shown in Figure 9(a) and an ordinary P/T net differs in one important aspect. When considering the behavior of a modular P/T net with an interface consisting of pins, the environment is responsible for consuming tokens from and producing tokens on the pins. If the behavior of the modular P/T net is described without the context of a specific environment, then it is assumed that the environment has always tokens available when they are needed by the modular P/T net; it is also assumed that the environment is always willing to receive any tokens that the modular P/T net might produce. If a modular P/T net is put into a specific context, typically defined by another modular P/T net, the approach followed in this subsection guarantees that the behavior of the modular P/T net is restricted to the behavior allowed by its context.

EXAMPLE 8.1. Consider the modular P/T net of Figure 9(a). As mentioned, it models a one-place buffer. Pins 1 and 2 correspond to ports. The buffer receives messages of its environment over port 1; it sends messages to its environment over port 2. Place b models buffer storage. The number of tokens in place c determines the *capacity* of the buffer. In the example, the capacity is one.

As explained, since no specific environment of the buffer has been given, it is assumed that the environment has always tokens available when they are needed by the buffer. Therefore, in the initial marking depicted in Figure 9(a), transition r is enabled. It has tokens on all its *internal* input places. Transition s is not enabled, because place b is not marked. Firing transition r leads to the marking with a single token in place b. In that marking, only transition s is enabled. Firing transition s, returns the modular P/T net to the state shown in Figure 9(a). The environment has received the token over port 2. Clearly, the behavior of the modular net corresponds to a one-place buffer.

By adding tokens to place c, it is possible to increase the capacity of the buffer. It is not difficult to see that the modular P/T net of Figure 9(a) behaves as an n-place buffer when place c contains n tokens in the initial marking. Note that the modular P/T net is a very abstract model that abstracts from message contents and message ordering. The *number of* tokens in place b corresponds to the *number of* messages in the buffer; the tokens do not correspond to the messages themselves.

It is possible to formalize the behavior of modular P/T nets as illustrated in the above example by means of a (step) firing rule. However, it is also possible to formally define the behavior in the equational theory $(\mathrm{ACP}^*_\lambda + \mathrm{RN} + RSP^* + \mathrm{SC})(AC(U, L), \uplus)$, where U, L, and $AC(U, L)$ are defined as in Section 6.3.

EXAMPLE 8.2. Let I be the set of internal places $\{b, c\}$. Recall the algebraic semantics for labeled P/T nets given in Definition 6.8. The algebraic semantics of the modular P/T net of Figure 9(a) is the following closed term in $\mathcal{C}(U, L)$. Square brackets of singleton bags are omitted.

$$\rho_{\mathbf{ca}(I)}\big(\lambda^I_c\big(([1, c], r, b)^*\delta \parallel (b, s, [c, 2])^*\delta\big)\big).$$

It is straightforward to prove that this term is equivalent to

$$\big((1, r, \mathbf{0}) \cdot (\mathbf{0}, s, 2)\big)^*\delta.$$

Note that only *internal* consumptions and productions are hidden. The reason for not hiding external consumptions and productions will become clear when we consider the construction of modular P/T nets from other modular P/T nets. It provides a way for restricting the behavior of a modular P/T net in a specific context.

The algebraic semantics of the modular P/T net of Figure 9(a) with *two* tokens in place c is the term

$$\rho_{\mathbf{ca}(I)}\big(\lambda^I_{[c^2]}\big(([1, c], r, b)^*\delta \parallel (b, s, [c, 2])^*\delta\big)\big),$$

which can be proven equal to

$$(1, r, \mathbf{0}) \cdot \big(\big((1, r, \mathbf{0}) \parallel (\mathbf{0}, s, 2)\big)^*\delta\big).$$

This last result is the expected term for a two-place buffer.

Finally, consider the variant of the modular P/T net of Figure 9(a) with *three* tokens in c. Its algebraic semantics is the term

$$\rho_{\mathbf{ca}(I)}\big(\lambda^I_{[c^3]}\big(([1, c], r, b)^*\delta \parallel (b, s, [c, 2])^*\delta\big)\big).$$

Let X_i, where $0 \leqslant i \leqslant 3$, denote the closed term $\rho_{\mathbf{ca}(I)}(\lambda^I_{[b^i, c^{3-i}]}(([1, c], r, b) * \delta \parallel (b, s, [c, 2])^*\delta))$. Thus, $X_0 = \rho_{\mathbf{ca}(I)}(\lambda^I_{[c^3]}(([1, c], r, b)^*\delta \parallel (b, s, [c, 2])^*\delta))$. The following results can be obtained:

$$\begin{aligned} X_0 &= (1, r, \mathbf{0}) \cdot X_1, \\ X_1 &= (1, r, \mathbf{0}) \cdot X_2 + (\mathbf{0}, s, 2) \cdot X_0 + (1, [r, s], 2) \cdot X_1, \\ X_2 &= (1, r, \mathbf{0}) \cdot X_3 + (\mathbf{0}, s, 2) \cdot X_1 + (1, [r, s], 2) \cdot X_2, \quad \text{and} \\ X_3 &= (\mathbf{0}, s, 2) \cdot X_2. \end{aligned}$$

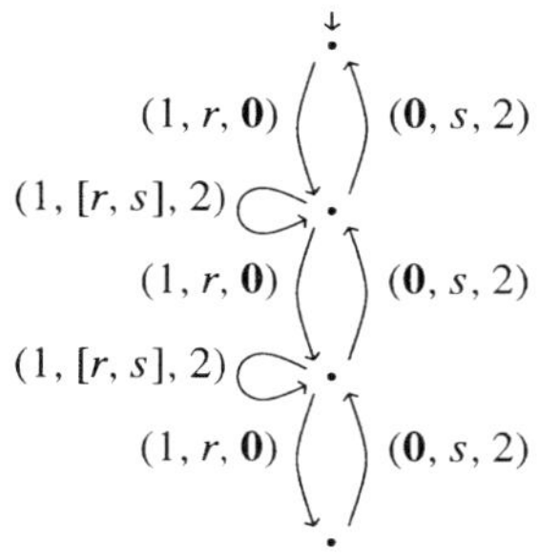

Fig. 10. A three-place buffer in a step semantics.

The semantics of term X_0 is illustrated in Figure 10. Note that the final result for the *two*-place buffer above is a single closed term in $\mathcal{C}(U, L)$ that contains only the sequential-composition, merge, and binary-Kleene-star operators. In particular, it does not contain a causal state operator. There is no such term in $\mathcal{C}(U, L)$ that specifies a three-place buffer. That is, there is no closed term in $\mathcal{C}(U, L)$ that is equivalent to term X_0 and that contains only the sequential-composition, merge, and binary-Kleene-star operators. The reason is that the binary Kleene star is not sufficiently expressive. More information about the expressiveness of the binary Kleene star can be found in [11] and [12], Chapter 5 of this Handbook.

The above example shows how to obtain a modular-P/T-net model of a buffer with arbitrary capacity by varying the marking of the modular net. The examples in the algebraic framework of Section 5 suggest another way to construct buffers with arbitrary capacity, namely by composing a number of one-place buffers. This approach is also possible in the framework of modular P/T nets developed so far.

EXAMPLE 8.3. Figure 11(a) shows two variants of the one-place buffer of Figure 9(a). The left P/T net models a buffer that receives messages of its environment over port 1 and sends messages over port 2. The right P/T net models a buffer communicating via ports 2 and 3. There is one important difference between the models of Figure 11(a) and the model of Figure 9(a). Consider the left P/T net of Figure 11(a). It models a buffer that receives messages via pin im_1 while at the same time acknowledging the receipt to its environment via pin oa_1. The buffer sends messages to the environment via pin om_2; however, it only sends a message when the environment acknowledges via pin ia_2 that it is ready to receive the message. The two acknowledgment pins oa_1 and ia_2 are not present in the modular P/T net of Figure 9(a). They can be used to construct asynchronous communication channels with a finite message capacity between buffers.

To illustrate the composition of modular P/T nets with interfaces consisting of pins, consider the modular P/T net of Figure 11(b). It consists of four components, namely the two one-place buffers of Figure 11(a), which are called *subnets* of the modular P/T net, and two new (internal) places m and a. Place a is marked with a single token. The modular P/T net of Figure 11(b) specifies a specific context for its subnets. It is constructed from

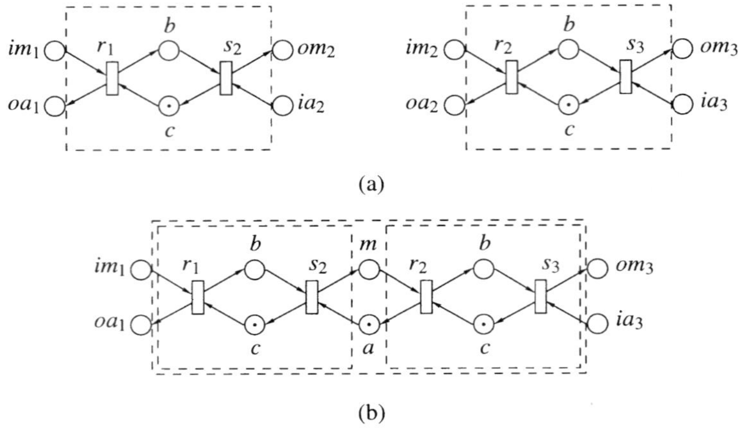

Fig. 11. The modular construction of a three-place buffer.

its four components by means of a so-called *place-fusion* function. A place-fusion function is a (partial) mapping from the pins of the subnets of a modular P/T net onto the places of the modular P/T net. In the example of Figure 11(b), the place-fusion function f maps pins om_2 and im_2 onto place m and pins ia_2 and oa_2 onto place a.

Informally, the modular P/T net of Figure 11(b) models the two one-place buffers of Figure 11(a) that are connected via port 2 which, in this example, is an asynchronous communication channel that can hold a single message. The channel is modeled by places m and a. Place m models the storage for messages and the number of tokens in place a models the storage capacity of the channel. Intuitively, the modular P/T net of Figure 11(b) should behave as a *three*-place buffer. Messages can be stored in the two one-place buffers (the two places marked b) *and* in the channel (place m). The modular P/T net of Figure 11(b) is more concrete than the modular P/T net of Figure 9(a) with three tokens in place c that also models a three-place buffer. Recall that the modular net of Figure 9(a) abstracts from message contents and message ordering. Tokens in the three buffer places of the modular P/T net of Figure 11(b) can be seen as messages, which means that this modular net abstracts from message contents, but not from message ordering.

At this point, the reason for the exchange of acknowledgments between the one-place buffers of Figure 11(a) and the environment should be clear. In the modular P/T net of Figure 11(b) without place a, there would be no upper bound to the number of tokens in place m. Such a modular P/T net conforms to a system where the communication channel between the two one-place buffers has infinite capacity.

In a general framework of modular P/T nets, a modular P/T net may consist of subnets, transitions, places, and arcs between transitions and places. Places may be divided into internal places and pins. Internal places may contain an arbitrary number of tokens. Note that the P/T net of Figure 9(a) as well as the three P/T nets of Figure 11 all belong to this general class of modular P/T nets. In such a general framework of modular P/T nets, it is possible to construct arbitrarily complex P/T-net models using only the simple mechanism of place fusion. Note that pins of the subnets of a modular P/T net may be fused with both its internal places and pins.

The semantics of modular P/T nets as described above can be formalized by defining a process space based on a (step) firing rule. However, it is also possible to define the semantics indirectly via an algebraic semantics.

EXAMPLE 8.4. Recall the algebraic semantics of the modular P/T net of Figure 9(a) given in Example 8.2. The algebraic semantics of the leftmost modular P/T net of Figure 11(a) is the following closed term in $\mathcal{C}(U, L)$. As before, square brackets of singleton bags are omitted. Let I be the set of causes $\{b, c\}$.

$$\rho_{\mathbf{ca}(I)}\big(\lambda_c^I\big(([im_1, c], r_1, [oa_1, b])^* \delta \parallel ([b, ia_2], s_2, [c, om_2])^* \delta\big)\big).$$

It is straightforward to prove that this term is equivalent to

$$\big((im_1, r_1, oa_1) \cdot (ia_2, s_2, om_2)\big)^* \delta.$$

Let $C_{1,2}$ be an abbreviation for this term. The algebraic semantics of the rightmost modular P/T net of Figure 11(a) is the term

$$\rho_{\mathbf{ca}(I)}\big(\lambda_c^I\big(([im_2, c], r_2, [oa_2, b])^* \delta \parallel ([b, ia_3], s_3, [c, om_3])^* \delta\big)\big),$$

which is equivalent to

$$\big((im_2, r_2, oa_2) \cdot (ia_3, s_3, om_3)\big)^* \delta.$$

Let $C_{2,3}$ be an abbreviation for this term. Recall that the exchange of tokens between the modular P/T nets of Figure 11(a) and the environment is represented in the terms $C_{1,2}$ and $C_{2,3}$ explicitly via causes.

It remains to define the algebraic semantics of the modular P/T net of Figure 11(b). Since the goal is to obtain a compositional formalism, it must be based on the terms $C_{1,2}$ and $C_{2,3}$ derived for the two one-place buffers of which it is composed. As explained in Example 8.3, the basis of the construction of this modular P/T net from its components is the place-fusion function f, which maps (some of the) pins of its subnets onto its places. To obtain an algebraic expression for the behavior of subnets in the context of a modular P/T net, it suffices to simply rename consumptions and productions of causes corresponding to pins of subnets according to the place-fusion function. Thus, the place-fusion function in the construction of a modular P/T net corresponds to a renaming operator in its algebraic semantics. Note that the place-fusion function f is simply a (partial) renaming of causes in U. Consequently, the place-fusion function f can be turned into a renaming function $f : AC(U, L) \cup \{\delta\} \to AC(U, L) \cup \{\delta\}$ as explained in Section 7.

Using the renaming function f, the behavior of the modular P/T nets of Figure 11(a) in the context of the modular net of Figure 11(b) is defined by the terms $\rho_f(C_{1,2})$ and $\rho_f(C_{2,3})$, respectively. It follows from the axioms for renaming that

$$\rho_f(C_{1,2}) = \big((im_1, r_1, oa_1) \cdot (a, s_2, m)\big)^* \delta$$

and that

$$\rho_f(C_{2,3}) = \big((m, r_2, a) \cdot (ia_3, s_3, om_3)\big)^* \delta.$$

The basis of the algebraic semantics of the modular P/T net in Figure 11(b) is the parallel composition of the behavior of its subnets in the context of the modular net. A causal state operator instantiated with the marking of the modular P/T net is applied to this term to restrict the behavior to all possible sequences of steps that are allowed in the specific context. The cause-abstraction operator is used to hide internal consumptions and productions of tokens. Thus, the modular P/T net of Figure 11(b) has the following algebraic semantics. Let J be the set of places $\{a, m\}$.

$$\rho_{\mathbf{ca}(J)}\big(\lambda_a^J\big(\rho_f(C_{1,2}) \parallel \rho_f(C_{2,3})\big)\big).$$

Let X_{000} be an abbreviation for this term. The following equations can be derived from the axioms of $(\mathrm{ACP}_\lambda^* + \mathrm{RN} + RSP^* + \mathrm{SC})(AC(U, L), \uplus)$.

$$\begin{aligned}
X_{000} &= (im_1, r_1, oa_1) \cdot X_{100}, \\
X_{100} &= (\mathbf{0}, s_2, \mathbf{0}) \cdot X_{010}, \\
X_{010} &= (im_1, r_1, oa_1) \cdot X_{110} + (\mathbf{0}, r_2, \mathbf{0}) \cdot X_{001} + (im_1, [r_1, r_2], oa_1) \cdot X_{101}, \\
X_{110} &= (\mathbf{0}, r_2, \mathbf{0}) \cdot X_{101}, \\
X_{001} &= (im_1, r_1, oa_1) \cdot X_{101} + (ia_3, s_3, om_3) \cdot X_{000} \\
&\qquad + ([im_1, ia_3], [r_1, s_3], [oa_1, om_3]) \cdot X_{100}, \\
X_{101} &= (\mathbf{0}, s_2, \mathbf{0}) \cdot X_{011} + (ia_3, s_3, om_3) \cdot X_{100} + (ia_3, [s_2, s_3], om_3) \cdot X_{010}, \\
X_{011} &= (im_1, r_1, oa_1) \cdot X_{111} + (ia_3, s_3, om_3) \cdot X_{010} \\
&\qquad + ([im_1, ia_3], [r_1, s_3], [oa_1, om_3]) \cdot X_{110}, \quad \text{and} \\
X_{111} &= (ia_3, s_3, om_3) \cdot X_{110},
\end{aligned}$$

where

$$\begin{aligned}
X_{100} &= \rho_{\mathbf{ca}(J)}\big(\lambda_a^J\big((a, s_2, m) \cdot \rho_f(C_{1,2}) \parallel \rho_f(C_{2,3})\big)\big), \\
X_{010} &= \rho_{\mathbf{ca}(J)}\big(\lambda_m^J\big(\rho_f(C_{1,2}) \parallel \rho_f(C_{2,3})\big)\big), \\
X_{110} &= \rho_{\mathbf{ca}(J)}\big(\lambda_m^J\big((a, s_2, m) \cdot \rho_f(C_{1,2}) \parallel \rho_f(C_{2,3})\big)\big), \\
X_{001} &= \rho_{\mathbf{ca}(J)}\big(\lambda_a^J\big(\rho_f(C_{1,2}) \parallel (ia_3, s_3, om_3) \cdot \rho_f(C_{2,3})\big)\big), \\
X_{101} &= \rho_{\mathbf{ca}(J)}\big(\lambda_a^J\big((a, s_2, m) \cdot \rho_f(C_{1,2}) \parallel (ia_3, s_3, om_3) \cdot \rho_f(C_{2,3})\big)\big), \\
X_{011} &= \rho_{\mathbf{ca}(J)}\big(\lambda_m^J\big(\rho_f(C_{1,2}) \parallel (ia_3, s_3, om_3) \cdot \rho_f(C_{2,3})\big)\big), \quad \text{and} \\
X_{111} &= \rho_{\mathbf{ca}(J)}\big(\lambda_m^J\big((a, s_2, m) \cdot \rho_f(C_{1,2}) \parallel (ia_3, s_3, om_3) \cdot \rho_f(C_{2,3})\big)\big).
\end{aligned}$$

As mentioned in Example 8.3, the modular P/T net of Figure 11(b) should behave as a three-place buffer. This means that the above set of equations for the terms X_{000} through

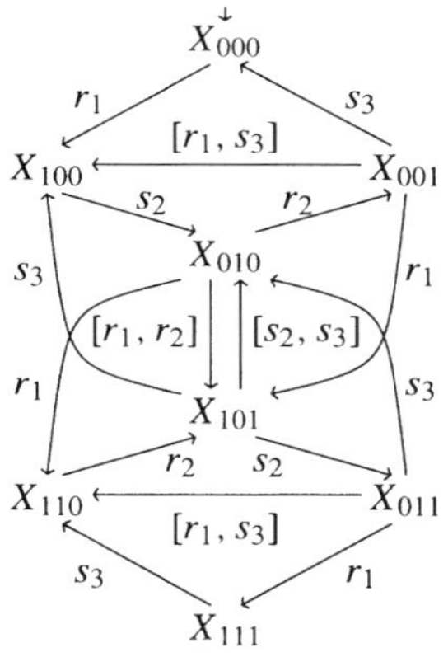

Fig. 12. The semantics of the modular P/T net of Figure 11(b).

X_{111} must also represent the behavior of a three-place buffer. Figure 12 illustrates the semantics of the term X_{000}. For the sake of clarity, consumptions and productions are omitted. Recall that Figure 10 shows a process corresponding to a three-place buffer in a step semantics. Thus, it may be expected that the process in Figure 12 is similar to this process. The first impression might be that this is not the case. However, observe that the modular P/T net of Figure 11(b) contains two transitions that do not have any external effects, namely transitions s_2 and r_2. Hiding these transitions in the process of Figure 12 means that the process states X_{100}, X_{010}, and X_{001} collapse into a single state. The same happens to the process states X_{110}, X_{101}, and X_{011}. A consequence of these abstractions is that the process becomes identical – up to renaming of actions – to the process in Figure 10. Thus, the process of Figure 12 represents the behavior of a three-place buffer indeed. As mentioned earlier, it goes beyond the scope of this chapter to formalize the notion of abstraction.

The examples given so far in this subsection illustrate the essentials of a compositional formalism based on modular P/T nets with interfaces consisting of places in a partial-order framework. The formalism is compositional in the sense that the behavior of a modular P/T net is defined as a closed term in the equational theory $(\mathrm{ACP}^*_\lambda + \mathrm{RN} + RSP^* + \mathrm{SC})(AC(U, L), \uplus)$ that is itself defined in terms of the (parallel) composition of the algebraic semantics of the subnets of the modular P/T net. In [8, Chapter 3], it is explained how such a formalism can be used as the basis for a design method for concurrent systems. The basic idea of this design method is to specify crucial behavioral properties of a concurrent system in process-algebraic terms, model the system by means of a modular P/T net, and verify its behavioral properties via the algebraic semantics of the modular-P/T-net model. The formalism in [8, Chapter 3] includes a formal definition of modular P/T nets. It also formalizes the operational semantics of modular P/T nets in terms of a firing rule. Several algebraic semantics of modular P/T nets are given and it is shown that these semantics correspond to the operational semantics of modular P/T nets as defined by the firing rule (cf. Theorem 6.10). In addition, the notion of abstraction is formalized both in the P/T-net framework and in the algebraic framework. Although the underlying semantics is a total-order semantics, the framework in [8, Chapter 3] can be adapted to the partial-order semantics of this subsection in a straightforward way.

8.2. *Transition fusion*

As explained in the introduction to this section, there are two ways to modularize P/T nets. In the previous subsection, we have concentrated on modular P/T nets with interfaces consisting of places. Another approach is to define interfaces in terms of transitions. This approach is not very common in the Petri-net literature. However, it is closely related to the notion of synchronous communication known from process algebra, as the examples in the remainder of this subsection show.

EXAMPLE 8.5. Figure 13(a) shows two modular P/T nets that are simple variants of the modular P/T net of Figure 9(b). Consider for example the leftmost modular P/T net of Figure 13(a). It will not come as a surprise that it models a one-place buffer that receives messages of its environment over port 1 and sends messages to its environment over port 2. The number of tokens in place b models the number of messages in storage, whereas the number of tokens in place c models the capacity of the buffer. The rightmost modular P/T net of Figure 13(a) models a buffer that communicates with its environment via ports 2 and 3.

In the P/T-net models of Figure 13(a), a dashed box divides the set of transitions of a modular net into *internal transitions* and *transition pins*. The modular P/T nets of Figure 13(a) do not have any internal transitions. Transition pins are connectors between a system and its environment. A modular P/T net with transition pins *in isolation* behaves in exactly the same way as a normal labeled P/T net. However, an environment can interact with a system via *synchronous communication*. The modular P/T net of Figure 13(b) illustrates one possible mechanism to obtain synchronization. Recall that the basic P/T-net framework in this chapter is the class of *labeled* P/T nets. However, for the sake of simplicity, it is assumed that in the modular P/T nets of Figure 13 transition labels are identical to transition identifiers.

The modular net of Figure 13(b) consists of three components, namely the two one-place buffers of Figure 13(a), called its subnets, and one new (internal) transition c_2. Synchro-

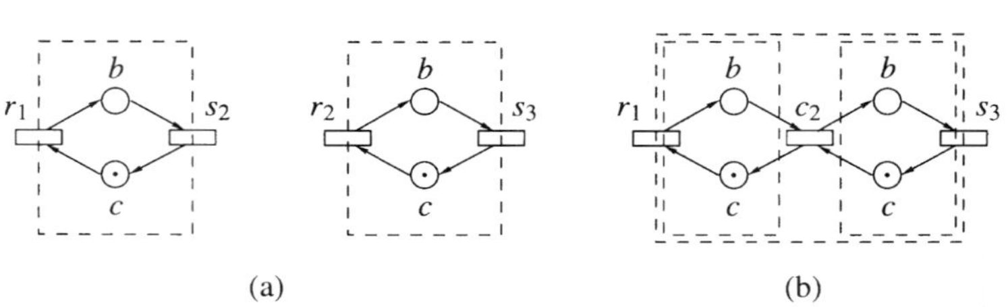

Fig. 13. The modular construction of a two-place buffer.

nization between the two one-place buffers is achieved by means of a *transition-fusion* function. A transition-fusion function is a (partial) mapping from the transition pins of the subnets of a modular P/T net to the transitions of the modular net. In the example, the transition-fusion function cm maps transition pins s_2 and r_2 onto transition c_2. Intuitively, the modular P/T net of Figure 13(b) models a two-place buffer.

The semantics of modular P/T nets with transition pins can be formalized via an algebraic semantics.

EXAMPLE 8.6. As mentioned in Example 8.5, the behavior of the modular P/T nets of Figure 13(a) in isolation is identical to the behavior of normal labeled P/T nets. This means that the algebraic semantics for labeled P/T nets of Definition 6.8 can be used to define the semantics of the modular nets of Figure 13(a). Note that all the places in a modular P/T net with transition pins are internal. Furthermore, the goal is to obtain a step semantics. Thus, we instantiate the parameters I and γ in Definition 6.8 with the universe of identifiers U and bag summation $\uplus$, respectively. The algebraic semantics of the two modular P/T nets of Figure 13(a) are the closed terms

$$\rho_{\mathbf{ca}(U)}\big(\lambda_c^U\,((c, r_1, b)^*\delta \parallel (b, s_2, c)^*\delta)\big)$$

and

$$\rho_{\mathbf{ca}(U)}\big(\lambda_c^U\,((c, r_2, b)^*\delta \parallel (b, s_3, c)^*\delta)\big)$$

in $\mathcal{C}(U, L)$, respectively. As before, square brackets of singleton bags are omitted. Furthermore, a triple $(\mathbf{0}, \alpha, \mathbf{0})$ in $AC(U, L)$ is identified with its second element. It is straightforward to prove that the above two terms are equivalent to

$$(r_1 \cdot s_2)^*\delta$$

and

$$(r_2 \cdot s_3)^*\delta.$$

It remains to define an algebraic semantics for the modular P/T net of Figure 13(b). The basis for the composition of the modular net of Figure 13(b) from its components is the transition-fusion function cm that maps transition pins s_2 and r_2 of the one-place buffers of Figure 13(a) onto transition c_2. Note that this transition-fusion function can be seen as a (partial) communication function on atomic actions in L. Thus, the transition-fusion function can be turned into an algebraic renaming function $\mathbf{cm}: AC(U, L) \cup \{\delta\} \to AC(U, L) \cup \{\delta\}$ following the approach of Section 7. That renaming function can be used to define the algebraic semantics of the modular net of Figure 13(b) in terms of the algebraic expressions derived for the one-place buffers of Figure 13(a). Transition fusion in the framework of modular P/T nets with transition pins corresponds to the synchronous communication mechanism in ACP-style process algebra. Given this observation, the algebraic semantics of the modular P/T net of Figure 13(b) is straightforward. The idea is to take the parallel composition of the algebraic semantics of its subnets, rename actions according to the renaming function constructed from the transition-fusion function, and encapsulate isolated actions corresponding to firings of transition pins. The encapsulation operator that is needed to achieve the latter can be derived from the set of atomic actions $H = \{s_2, r_2\}$. Let $\boldsymbol{H} \subseteq AC(U, L)$ be the set of actions that is constructed from H following the approach of Section 7. The semantics of the modular P/T net of Figure 13(b) is the following closed term in $\mathcal{C}(U, L)$:

$$\partial_{\boldsymbol{H}}\big(\rho_{\boldsymbol{cm}}\big((r_1 \cdot s_2)^* \delta \parallel (r_2 \cdot s_3)^* \delta\big)\big).$$

It is not difficult to prove that this term is equivalent to the term

$$r_1 \cdot \big(\big(c_2 \cdot (r_1 \parallel s_3)\big)^* \delta\big),$$

which is the expected expression for a two-place buffer with an internal communication action c_2.

Note that the definition of the algebraic semantics of the modular P/T net of Figure 13(b) does not use the causal state operator. However, in a general framework of modular P/T nets with transition pins, a modular P/T net may consist of subnets, transitions, and (internal) places. Transitions may be divided into internal transitions and transition pins and places may contain an arbitrary number of tokens. In such a framework, the basis of the algebraic semantics of the modular P/T net is still the parallel composition of the expressions for its subnets, but this composition is restricted to all possible sequences of steps by means of the causal state operator. Finally, note that the modular P/T nets in Example 8.5 are very simple. In a general framework of modular P/T nets with transition pins, the construction of communication functions from the transition-fusion function might not always be straightforward.

8.3. *Concluding remarks*

In this section, we have seen examples of two approaches to modularize P/T nets. One approach is based on fusion of places, whereas the other one is based on fusion of transitions. Both techniques translate to renaming operators when defining the (step) semantics of such modular P/T nets in an algebraic way, where place fusion corresponds to the renaming of causes and where transition fusion corresponds to the notion of synchronous communication. Of course, it is possible to combine the two approaches into one general framework of modular P/T nets with interfaces consisting of (place) pins and transition pins. A formalism of modular P/T nets and its algebraic semantics can be used as the basis for a compositional formalism for modeling and analyzing complex concurrent systems. In [8, Chapter 3], such a compositional formalism based on modular P/T nets with interfaces consisting of places is worked out in detail (in a total-order setting).

9. Cause addition

The previous section illustrates two techniques to modularize P/T nets, namely place fusion and transition fusion. The two techniques are based on the addition of fresh places and transitions, respectively, in order to compose new (modular) P/T nets from other P/T nets. We have seen that the addition of fresh transitions by means of transition fusion corresponds to the notion of synchronous communication in ACP-style process algebra. The new action that results from a synchronous communication in an algebraic expression corresponds to the new transition. It is also possible to define a class of algebraic operators corresponding to the addition of fresh places in P/T nets. Such a class of operators is particularly interesting because places in P/T nets are the standard means to enforce a sequential ordering between actions. Thus, the new operators must be related to the standard sequential-composition operator in ACP-style process algebra. In the remainder of this section, we introduce the class of so-called cause-addition operators and investigate the relation between cause addition and sequential composition. The introduction of cause addition is another step in incorporating the causality mechanism of Petri nets in ACP-style process algebra. Cause addition was first studied in [4].

9.1. *The equational theory*

Table 9 presents the Algebra of Communicating Processes with iteration, cause addition, the causal state operator, renaming, the Recursive Specification Principle for the binary Kleene star, and standard concurrency, abbreviated $\mathrm{ACP}^{*c}_{\lambda} + \mathrm{RN} + RSP^{*} + \mathrm{SC}$. It builds upon the equational theory introduced in Section 6. It is parameterized by a set of actions $AC(C, A)$, with C a set of causes and A a set of atomic actions, and a communication function $\gamma : AC(C, A) \times AC(C, A) \to AC(C, A)$. Recall from Section 6.1 the definitions of $AC(C, A)$ and the three auxiliary functions $\boldsymbol{c}, \boldsymbol{p} : AC(C, A) \to \mathcal{B}(C)$ and $\boldsymbol{s} : AC(C, A) \to \mathcal{B}(A)$: $AC(C, A) = \mathcal{B}(C) \times \mathcal{B}(A) \times \mathcal{B}(C)$ and, for any $a = (c, \alpha, p)$ in $AC(C, A)$, $\boldsymbol{c}a = c$, $\boldsymbol{p}a = p$, and $\boldsymbol{s}a = \alpha$.

Table 9
The equational theory $(\mathrm{ACP}^{*c}_{\lambda} + \mathrm{RN} + RSP^{*} + \mathrm{SC})(AC, \gamma)$

$(\mathrm{ACP}^{*c}_{\lambda} + \mathrm{RN} + RSP^{*} + \mathrm{SC})(AC(C, A), \gamma)$

$(\mathrm{ACP}^{*}_{\lambda} + \mathrm{RN} + RSP^{*} + \mathrm{SC})(AC(C, A), \gamma)$

${}^{c}\_, \_^{c} : \boldsymbol{P} \to \boldsymbol{P};$

$x, y : \boldsymbol{P};$

${}^{c}\delta = \delta$	$IC1$	$\delta^{c} = \delta$	$OC1$
${}^{c}a = (\boldsymbol{c}a \uplus [c], \boldsymbol{s}a, \boldsymbol{p}a)$	$IC2$	$a^{c} = (\boldsymbol{c}a, \boldsymbol{s}a, \boldsymbol{p}a \uplus [c])$	$OC2$
${}^{c}(x + y) = {}^{c}x + {}^{c}y$	$IC3$	$(x + y)^{c} = x^{c} + y^{c}$	$OC3$
${}^{c}(x \cdot y) = {}^{c}x \cdot y$	$IC4$	$(x \cdot y)^{c} = x \cdot y^{c}$	$OC4$

For any cause $c \in C$, the equational theory $(\mathrm{ACP}^{*c}_{\lambda} + \mathrm{RN} + RSP^{*} + \mathrm{SC})(AC(C, A), \gamma)$ contains two new operators when compared to $(\mathrm{ACP}^{*}_{\lambda} + \mathrm{RN} + RSP^{*} + \mathrm{SC})(AC(C, A), \gamma)$, namely input-cause addition ${}^{c}\_$ and output-cause addition $\_^{c}$. The idea is that, for any process term x, ${}^{c}x$ is the process that behaves as x with the one difference that its initial action consumes an extra cause c; x^{c} is the process that behaves as x but produces an extra cause c with its last action. Given this informal interpretation, the axiomatization of cause addition is straightforward. In Table 9, constant a ranges over $AC(C, A)$. Axioms $IC1$ and $OC1$ state that the inaction constant δ does not consume or produce causes. Axiom $IC2$ says that the addition of an input cause to an action means that it is added to the consumption of the action; Axiom $OC2$ states that the addition of an output cause to an action means that it is added to its production. Axioms $IC3$ and $OC3$ say that both cause-addition operators distribute over choice. Finally, Axioms $IC4$ and $OC4$ state that the addition of an input cause to a sequential composition means that the cause is added to the left operand of the sequential composition, whereas the addition of an output cause to a sequential composition means that the cause is added to the right operand.

9.2. *A semantics*

As in Sections 5.2 and 6.2, the basis of a semantics for theory $(\mathrm{ACP}^{*c}_{\lambda} + \mathrm{RN} + RSP^{*} + \mathrm{SC})(AC(C, A), \gamma)$ is a process space. Let $\mathcal{C}(C, A)$ be the set of closed $(\mathrm{ACP}^{*c}_{\lambda} + \mathrm{RN} + RSP^{*} + \mathrm{SC})(AC(C, A), \gamma)$ terms. The set of processes of the process space is defined as the set $\mathcal{C}(C, A) \cup \{\surd\}$. Process $\surd$ is the special process that can perform no actions, but can only terminate successfully. The set $AC(C, A)$ is the set of actions of the process space. The transition relation $\_ \xrightarrow{\_} \_ \subseteq (\mathcal{C}(C, A) \cup \{\surd\}) \times AC(C, A) \times (\mathcal{C}(C, A,) \cup \{\surd\})$ is the smallest relation satisfying the derivation rules in Tables 6, 8, and 10. The termination predicate is the singleton $\{\surd\}$.

The process space $(\mathcal{C}(C, A) \cup \{\surd\}, AC(C, A), \rightarrow, \{\surd\})$ can be turned into a model $\mathcal{M}(C, A, \gamma)$ of the theory $(\mathrm{ACP}^{*c}_{\lambda} + \mathrm{RN} + RSP^{*} + \mathrm{SC})(AC(C, A), \gamma)$ following the approach of Section 5.2. The following congruence property is needed in the construction.

PROPERTY 9.1 (Congruence). *Bisimilarity,* $\sim$, *is a congruence for the operators of* $(\mathrm{ACP}^{*c}_{\lambda} + \mathrm{RN} + RSP^{*} + \mathrm{SC})(AC(C, A), \gamma)$.

PROOF. The desired property follows from the format of the derivation rules in Tables 6, 8, and 10. (See [5] and [1], Chapter 3 of this Handbook, for details.) □

Table 10
The transition relation for cause addition

$a : AC(C, A);\ c : C;\ p, p' : \mathcal{C}(C, A);$

$$\frac{p \xrightarrow{a} p'}{{}^{c}p \xrightarrow{(ca \uplus [c], sa, pa)} p'} \qquad \frac{p \xrightarrow{a} \surd}{{}^{c}p \xrightarrow{(ca \uplus [c], sa, pa)} \surd} \qquad \frac{p \xrightarrow{a} p'}{p^{c} \xrightarrow{a} p'^{c}} \qquad \frac{p \xrightarrow{a} \surd}{p^{c} \xrightarrow{(ca, sa, pa \uplus [c])} \surd}$$

The following theorem states that the equational theory $(\mathrm{ACP}_{\lambda}^{*c} + \mathrm{RN} + RSP^{*} + \mathrm{SC})(AC(C, A), \gamma)$ is a sound axiomatization of the model of closed terms modulo bisimilarity.

THEOREM 9.2 (Soundness). *For any closed terms $p, q \in \mathcal{C}(C, A)$,*

$$(\mathrm{ACP}_{\lambda}^{*c} + \mathrm{RN} + RSP^{*} + \mathrm{SC})\big(AC(C, A), \gamma\big) \vdash p = q \Rightarrow \mathcal{M}(C, A, \gamma) \models p = q.$$

PROOF. Given Theorem 6.6 and Property 9.1, it suffices to show the validity of the Axioms $IC1$ through $IC4$ and $OC1$ through $OC4$. It is straightforward to construct a bisimulation for each case. □

9.3. *Buffers*

The use of cause addition can be best illustrated by means of a few examples. Since the cause-addition operators are inspired by Petri-net theory, (variants of) P/T-net models given earlier are a good source of examples.

EXAMPLE 9.3. Consider the P/T-net model of Figure 14(a). It is a variant of the (modular) P/T net of Figure 9(b) that models a three-place buffer. Definition 6.8 defines the algebraic semantics of such a P/T net as a closed term in the theory $(\mathrm{ACP}_{\lambda}^{*} + \mathrm{RN} + RSP^{*} + \mathrm{SC})(AC(U, L), \gamma)$ where U and L are as defined in Section 4 and where $\gamma : AC(U, L) \times AC(U, L) \to AC(U, L)$ is a parameter that can be used to define a step semantics or a total-order semantics. The standard choice $\uplus$ for this parameter yields a step semantics. In the algebraic semantics of Definition 6.8, the information about the consumption and production of transitions is already included in the actions. However, in the current setting, cause-addition operators can be used to make this information explicit. The step semantics of the P/T net of Figure 14(a) can also be formalized by the following closed $(\mathrm{ACP}_{\lambda}^{*c} + \mathrm{RN} + RSP^{*} + \mathrm{SC})(AC(U, L), \uplus)$ term. As before, a triple $(\mathbf{0}, \alpha, \mathbf{0})$ in $AC(U, L)$ is identified with its second element and square brackets of singleton bags are omitted.

$$\rho_{ca(U)}\big(\lambda_{[1^3]}^{U}({}^{1}r^{2} * \delta \parallel {}^{2}s^{1} * \delta)\big).$$

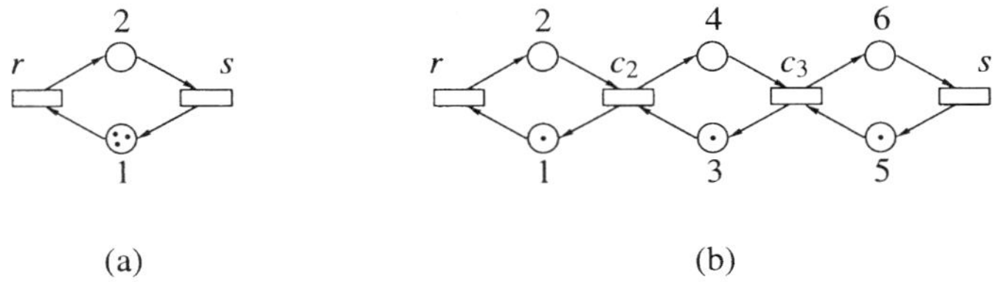

Fig. 14. Two P/T-net models of three-place buffers.

Let X_i, where $0 \leqslant i \leqslant 3$, denote the closed term $\rho_{ca(U)}(\lambda^U_{[1^{3-i},2^i]}({}^1r^{2} * \delta \parallel {}^2s^1 * \delta))$. Thus, X_0 equals the algebraic semantics of the P/T net of Figure 14(a). The following results can be obtained:

$$\begin{aligned}
X_0 &= r \cdot X_1,\\
X_1 &= r \cdot X_2 + s \cdot X_0 + [r, s] \cdot X_1,\\
X_2 &= r \cdot X_3 + s \cdot X_1 + [r, s] \cdot X_2, \quad \text{and}\\
X_3 &= s \cdot X_2.
\end{aligned}$$

Figure 10 visualizes the semantics of X_0 when ignoring consumptions and productions.

EXAMPLE 9.4. Figure 14(b) shows another P/T-net model of a three-place buffer. It can be obtained by combining three one-place buffers by means of transition fusion. As a result, the internal communication actions c_2 and c_3 are explicit in the model. For the P/T net of Figure 14(b), cause-addition operators can be used to define the following alternative algebraic semantics:

$$\rho_{ca(U)}\big(\lambda^U_{[1,3,5]}\big({}^1r^{2} * \delta \parallel {}^2({}^3c_2{}^4)^1 * \delta \parallel {}^5({}^4c_3{}^3)^6 * \delta \parallel {}^6s^5 * \delta\big)\big).$$

Let X_{ijk}, where $0 \leqslant i, j, k \leqslant 1$, denote the closed term

$$\rho_{ca(U)}\big(\lambda^U_{[1^{1-i},2^i,3^{1-j},4^j,5^{1-k},6^k]}\big({}^1r^{2} * \delta \parallel {}^2({}^3c_2{}^4)^1 * \delta \parallel {}^5({}^4c_3{}^3)^6 * \delta \parallel {}^6s^5 * \delta\big)\big).$$

The following results can be obtained from the axioms of $(\mathrm{ACP}^{*c}_{\lambda} + \mathrm{RN} + RSP^* + \mathrm{SC})(AC(U, L), \uplus)$.

$$\begin{aligned}
X_{000} &= r \cdot X_{100},\\
X_{100} &= c_2 \cdot X_{010},\\
X_{010} &= r \cdot X_{110} + c_3 \cdot X_{001} + [r, c_3] \cdot X_{101},\\
X_{110} &= c_3 \cdot X_{101},\\
X_{001} &= r \cdot X_{101} + s \cdot X_{000} + [r, s] \cdot X_{100},\\
X_{101} &= c_2 \cdot X_{011} + s \cdot X_{100} + [c_2, s] \cdot X_{010},\\
X_{011} &= r \cdot X_{111} + s \cdot X_{010} + [r, s] \cdot X_{110}, \quad \text{and}\\
X_{111} &= s \cdot X_{110}.
\end{aligned}$$

Figure 12 visualizes the semantics of term X_{000} when assuming that r_1, s_2, r_2, and s_3 are replaced by r, c_2, c_3, and s, respectively.

As explained in the introduction to this section, cause addition is related to sequential composition.

EXAMPLE 9.5. Consider again the P/T-net model of Figure 14(b). The ordering between transitions r and c_2 is enforced by places 1 and 2. In the algebraic semantics of this P/T net given in the previous example, these two places translate to cause-addition operators. It is possible to replace these cause-addition operators with a sequential composition. Similarly, it is possible to replace the cause-addition operators corresponding to places 5 and 6. The result of these replacements is an alternative closed term in $\mathcal{C}(U, L)$ defining the behavior of a three-place buffer, namely the term

$$\rho_{ca(U)}\big(\lambda_3^U(X \parallel Y)\big),$$

where X and Y are abbreviations for the closed terms

$$\big(r \cdot {}^3c_2{}^4\big) * \delta \quad \text{and} \quad \big({}^4c_3{}^3 \cdot s\big) * \delta.$$

In the P/T-net model of Figure 14(b), the above replacements correspond to *internalizing* the places 1, 2, 5, and 6. The result is a *modular* P/T net with a structure that is very similar to the structure of the modular P/T net of Figure 11(b).

It remains to substantiate our claim that term $\rho_{ca(U)}(\lambda_3^U(X \parallel Y))$ is an alternative expression for the three-place buffer of the previous example. Assuming that X_{000} is an abbreviation for the term $\rho_{ca(U)}(\lambda_3^U(X \parallel Y))$ and assuming that X_{100} through X_{111} are defined as below, exactly the same equations can be derived from the axioms of $(\text{ACP}_\lambda^{*c} + \text{RN} + RSP^* + \text{SC})(AC(U, L), \uplus)$ as the equations for the corresponding terms in Example 9.4. As a consequence, term X_{000} of this example has the same semantics as the term X_{000} of Example 9.4.

$$\begin{aligned}
X_{100} &= \rho_{ca(U)}\big(\lambda_3^U(c_2 \cdot X \parallel Y)\big),\\
X_{010} &= \rho_{ca(U)}\big(\lambda_4^U(X \parallel Y)\big),\\
X_{110} &= \rho_{ca(U)}\big(\lambda_4^U(c_2 \cdot X \parallel Y)\big),\\
X_{001} &= \rho_{ca(U)}\big(\lambda_3^U(X \parallel s \cdot Y)\big),\\
X_{101} &= \rho_{ca(U)}\big(\lambda_3^U(c_2 \cdot X \parallel s \cdot Y)\big),\\
X_{011} &= \rho_{ca(U)}\big(\lambda_4^U(X \parallel s \cdot Y)\big), \quad \text{and}\\
X_{111} &= \rho_{ca(U)}\big(\lambda_4^U(c_2 \cdot X \parallel s \cdot Y)\big).
\end{aligned}$$

Note the structural similarity between the terms X_{000} through X_{111} as defined in this example and those defined in Example 8.4. Another interesting observation is that the term X_{000} of Example 9.4 has three causes in the marking of the state operator and six different cause-addition operators. Term X_{000} of this example has two causes and four cause-addition operators fewer, but it has two sequential-composition operators instead (in the auxiliary terms X and Y). A final observation is that it appears to be impossible to derive the equality of the terms X_{000} of this example and X_{000} of Example 9.4 from the axioms of the equational theory $(\text{ACP}_\lambda^{*c} + \text{RN} + RSP^* + \text{SC})(AC(U, L), \uplus)$. Since these two terms have the same semantics in the model of the previous subsection, this observation means that the equational theory is not complete for this model. It appears that the generalization of RSP^* to the general recursive specification principle RSP, as defined in for example [5, 6,25], is necessary to obtain the desired equality.

Examples 9.4 and 9.5 show that cause-addition operators and sequential-composition operators can replace each other. The relation between cause addition and sequential composition is studied in more detail in the next subsection.

9.4. *Cause addition and sequential composition*

The most straightforward way to specify a causal ordering between two processes in a process-algebraic setting is the sequential-composition operator. In an equational theory with causal state operators and cause addition, causes provide an alternative. In this subsection, it is shown that cause addition and sequential composition are exchangeable notions. The main result is a theorem stating that any sequential-composition operator in a process term can be replaced by a pair of cause-addition operators. This theorem can be seen as a formalization of the statement by Milner in [45] that sequential composition is not necessary as a basic operator in a process-algebraic theory.

The main objective of this subsection is to illustrate the *conceptual* relation between cause addition and sequential composition. Therefore, to facilitate reasoning, we consider an equational theory without iteration. This simplification means that it is only possible to specify processes with bounded behavior. Assume that C is a set of causes and A a set of atomic actions. Let $\gamma : AC(C, A) \times AC(C, A) \to AC(C, A)$ be some communication function. The equational theory considered in this subsection is the theory $(\mathrm{ACP}^c_\lambda + \mathrm{RN})(AC(C, A), \gamma)$. The basis of this equational theory is the theory $(\mathrm{ACP} + \mathrm{RN})(AC(C, A), \gamma)$ of Table 2. This basic theory is extended with the causal state operator of Section 6.1 and cause addition of Section 9.1, yielding the Algebra of Communicating Processes with cause addition, causal state operator, and renaming. Let $\mathcal{C}(C, A)$ be the set of closed $(\mathrm{ACP}^c_\lambda + \mathrm{RN})(AC(C, A), \gamma)$ terms.

It is possible to construct a model of the equational theory $(\mathrm{ACP}^c_\lambda + \mathrm{RN})(AC(C, A), \gamma)$ following the same approach as in earlier sections. However, all the results in this subsection are proven in terms of the equational theory without referring to a particular model. Therefore, the construction of a model is left to the reader.

The main advantage of restricting our attention to processes with bounded behavior is that each process can be expressed in terms of the inaction constant, actions, choice, and action prefix. The latter is a restricted form of sequential composition (also present in, for example, CCS [44,45]) where the left operand is required to be an action. A closed process term that contains only the aforementioned constants and operators is called a *basic* term. Formally, the set of basic terms is defined as follows.

DEFINITION 9.6 (*Basic terms*). The set of *basic* terms over the signature of $(\mathrm{ACP}^c_\lambda + \mathrm{RN})(AC(C, A), \gamma)$, denoted $\mathcal{B}(C, A)$, is inductively defined as follows. The inaction constant is an element of $\mathcal{B}(C, A)$. The set of actions $AC(C, A)$ is contained in $\mathcal{B}(C, A)$. Furthermore, for any $a \in AC(C, A)$ and basic terms $s, t \in \mathcal{B}(C, A)$, also $a \cdot t$ and $s + t$ are elements of $\mathcal{B}(C, A)$.

Note the similarity between basic terms and terms in head normal form, as defined in Definition 5.4. Clearly, any basic term is in head normal form. The converse is not true.

The following theorem states that any closed process term is derivably equal to a basic term.

THEOREM 9.7 (Elimination). *For any closed term* $p \in \mathcal{C}(C, A)$, *there exists a basic term* $t \in \mathcal{B}(C, A)$ *such that* $(\mathrm{ACP}^c_\lambda + \mathrm{RN})(AC(C, A), \gamma) \vdash p = t$.

PROOF. The proof uses the strategy based on term-rewriting techniques explained in [5]. Consider the rewriting system obtained when interpreting Axioms $A3$ through $A7$, $C3$, $CF1$ and $CF2$, and $CM1$ through $CM9$ of Table 1, $RN1$ through $RN3$ of Table 2, $CSO1$ through $CSO5$ of Table 7, and $IC1$ through $IC4$ and $OC1$ through $OC4$ of Table 9 as rewrite rules from left to right. Assuming that this rewriting system is strongly normalizing, it is not difficult to see that each closed term in $\mathcal{C}(C, A)$ has a normal form that is a derivably-equivalent basic term.

To prove that the rewriting system is strongly normalizing, the so-called method of the recursive path ordering can be applied. The basis of the proof is a well-founded ordering on the constants and operators of $(\mathrm{ACP}^c_\lambda + \mathrm{RN})(AC(C, A), \gamma)$. The required ordering is inspired by very similar orderings that are given in [5]. To define the ordering, it is necessary to *rank* the causal state operators, the merge, the left merge, and the communication merge as explained in [5]. For any $I \subseteq C$, $m \in \mathcal{B}(I)$, and $n \in \mathbb{N}$ with $n > 0$, $\lambda^I_{m,n}$ denotes the causal state operator λ^I_m with rank n; for any $n \in \mathbb{N}$ with $n > 1$, $\|_n$, $⌊\!⌊_n$, and $|_n$ denote the merge, left merge, and communication merge with rank n, respectively. The required ordering can now be defined as the transitive closure of the relation $<$ specified as follows: $+ < \cdot$; for any $n \in \mathbb{N}$ with $n > 1$, $\cdot < ⌊\!⌊_n < \|_n < ⌊\!⌊_{n+1}$ and $\cdot < |_n < \|_n < |_{n+1}$; for all $a \in AC(C, A)$, $\delta < a$ and $a < |_2$; for all $f \in RF$, $\cdot < \rho_f$ and, for all $a \in AC(C, A)$, $a < \rho_f$; for all $c \in C$, $\cdot < {}^c\_$, $\cdot < \_^c$, and, for all $a \in AC(C, A)$, $a < {}^c\_$ and $a < \_^c$; for all $I \subseteq C$ and $m \in \mathcal{B}(I)$, $\cdot < \lambda^I_{m,1}$ and, for all $a \in AC(C, A)$, $a < \lambda^I_{m,1}$; finally, for all $I \subseteq C$, $m \in \mathcal{B}(I)$, and $n \in \mathbb{N}$ with $n > 0$, $\lambda^I_{m,n} < \lambda^I_{m,n+1}$. Given this ordering, the details of the proof are straightforward and, hence, omitted. □

Note that Theorem 9.7 implies that any cause-addition operator in a closed term can be eliminated. Thus, cause addition does not add expressive power to an equational theory with choice and sequential composition. However, it is well known that in an interleaving framework choice and sequential composition are sufficient to specify processes with bounded behavior. A more interesting result is the fact that an equational theory with parallel composition, cause addition, causal state operators, and cause abstraction is sufficiently expressive to eliminate sequential composition.

EXAMPLE 9.8. Let r and s be two actions in $AC(C, A)$. Consider the simple sequential process $r \cdot s$. This example shows how to eliminate the occurrence of the sequential-composition operator in this simple process. The idea is to replace the sequential-composition operator with a parallel-composition operator while using two additional causes and a causal state operator to restrict the parallel composition in such a way that it behaves as a sequential composition. Cause abstraction is used to hide the new causes and, thus, to obtain the original sequential process. It is assumed that the communication function equals $\uplus$. Let c be a cause in C that does not occur in the consumption or production of r and s. Consider the term $\rho_{\boldsymbol{ca}(\{c\})}(\lambda_{\mathbf{0}}^{\{c\}}(r^c \parallel {}^c s))$. It specifies that action r produces an extra output cause c, whereas s requires an extra input cause c. As a consequence, the state operator $\lambda_{\mathbf{0}}^{\{c\}}$ enforces a sequential ordering between r and s:

$$
\begin{aligned}
&\rho_{\boldsymbol{ca}(\{c\})}\big(\lambda_{\mathbf{0}}^{\{c\}}(r^c \parallel {}^c s)\big)\\
&\quad = \rho_{\boldsymbol{ca}(\{c\})}\big(\lambda_{\mathbf{0}}^{\{c\}}\big((\boldsymbol{cr}, sr, \boldsymbol{pr} \uplus [c]) \cdot (\boldsymbol{cs} \uplus [c], ss, \boldsymbol{ps})\\
&\qquad\qquad + (\boldsymbol{cs} \uplus [c], ss, \boldsymbol{ps}) \cdot (\boldsymbol{cr}, sr, \boldsymbol{pr} \uplus [c])\\
&\qquad\qquad + (\boldsymbol{cr} \uplus \boldsymbol{cs} \uplus [c], sr \uplus ss, \boldsymbol{pr} \uplus \boldsymbol{ps} \uplus [c])\\
&\qquad\qquad \big)\big)\\
&\quad = r \cdot \rho_{\boldsymbol{ca}(\{c\})}\big(\lambda_{[c]}^{\{c\}}\big((\boldsymbol{cs} \uplus [c], ss, \boldsymbol{ps})\big)\big) + \delta + \delta\\
&\quad = r \cdot s.
\end{aligned}
$$

Note that two assumptions are essential in the above derivation. First, cause c must be a new cause not already occurring in the consumption or production of r and s. If this assumption is not satisfied, the causal state operator $\lambda_{\mathbf{0}}^{\{c\}}$ might disturb the desired behavior. Second, the communication function may not specify any actual communications that affect the new cause c, because such communications can also disturb the desired result. The assumption that the communication function equals $\uplus$, which means that a partial-order framework is obtained, satisfies this restriction. Also any communication function that can be derived from a communication function on atomic actions, as explained in Section 7, is an allowed choice. Yet another possibility is to choose the communication function that is undefined for any pair of actions, which implies a total-order setting.

The remainder of this subsection is devoted to proving that the approach to eliminating a sequential-composition operator illustrated in Example 9.8 can be generalized to arbitrary closed terms in $\mathcal{C}(C, A)$.

The reason for introducing the notion of a basic term and proving the elimination result of Theorem 9.7 is not only to illustrate that cause addition can be eliminated from closed terms. An important consequence of Theorem 9.7 is that properties formulated in terms of closed terms can be rephrased in terms of basic terms without losing generality. Properties on basic terms can often be proved by means of *structural induction*, which is the technique also used in Section 5. Structural induction is also used to prove the main result of this subsection.

In reasoning about cause addition, it is convenient to know the set of causes that occur in a closed term. For this purpose, the *causes function* is introduced. The causes function yields for each closed term in $\mathcal{C}(C, A)$ the set of causes that the corresponding process in the model obtained by following the approach of Section 5.2 consumes or produces during any of its execution paths. The causes function is defined inductively using the structure of basic terms, under the assumption that two terms that are derivably equal have the same set of causes. In combination with the elimination result of Theorem 9.7, this assumption means that it is possible to calculate the causes of arbitrary closed terms in $\mathcal{C}(C, A)$.

DEFINITION 9.9 (*Causes function*). The causes function $causes : \mathcal{C}(C, A) \to \mathcal{P}(C)$ is a function such that, for any closed terms p and q in $\mathcal{C}(C, A)$, $((\mathrm{ACP}^c_\lambda + \mathrm{RN})(AC(C, A), \gamma) \vdash p = q) \Rightarrow causes(p) = causes(q)$. For any $a \in AC(C, A)$ and $p, q \in \mathcal{C}(C, A)$, $causes(\delta) = \emptyset$, $causes(a) = \{c \mid c \in \mathbf{ca} \uplus \mathbf{pa}\}$, $causes(a \cdot p) = causes(a) \cup causes(p)$, $causes(p + q) = causes(p) \cup causes(q)$.

Note that it should be verified that the definition of the causes function is consistent with the axioms of set theory. Inconsistencies arise when the combination of the requirement that derivably equal terms have the same set of causes and the inductive definition allows the derivation of an equality between sets that is not derivable from set theory. It is beyond the scope of this chapter to prove that Definition 9.9 is consistent with set theory.

The definitions given so far are sufficient to formalize the main result of this subsection, namely the elimination of sequential composition by means of cause addition. The proof uses a number of auxiliary properties that are given below. Note that, for the sake of simplicity, the communication function is assumed to be $\uplus$. However, any communication function that does not affect causes is allowed.

THEOREM 9.10 (Elimination of sequential composition). *Let p, q be closed terms in $\mathcal{C}(C, A)$ and c a cause in C; let $I = \{c\}$.*

$$\begin{aligned}
&c \notin causes(p) \cup causes(q) \Rightarrow \\
&\quad (\mathrm{ACP}^c_\lambda + \mathrm{RN})\big(AC(C, A), \uplus\big) \vdash \\
&\qquad p \cdot q = \rho_{\mathbf{ca}(I)}\big(\lambda^I_{\mathbf{0}}(p^c \mathbin{\|\!_} {}^c q)\big) = \rho_{\mathbf{ca}(I)}\big(\lambda^I_{\mathbf{0}}(p^c \parallel {}^c q)\big).
\end{aligned}$$

PROOF. The proof is by induction on the structure of basic terms. Recall that $\equiv$ denotes structural equivalence of terms. It follows from Theorem 9.7 that there exists a basic term $t \in \mathcal{B}(C, A)$ such that $(\mathrm{ACP}^c_\lambda + \mathrm{RN})(AC(C, A), \uplus) \vdash p = t$. Thus, it suffices to show that $(\mathrm{ACP}^c_\lambda + \mathrm{RN})(AC(C, A), \uplus) \vdash t \cdot q = \rho_{\mathbf{ca}(I)}(\lambda^I_{\mathbf{0}}(t^c \mathbin{\|\!\_} {}^c q)) = \rho_{\mathbf{ca}(I)}(\lambda^I_{\mathbf{0}}(t^c \parallel {}^c q))$. It follows from Definition 9.9 (Causes function) that $causes(t) = causes(p)$ and, thus, that $c \notin causes(t)$.

(i) Assume that $t \equiv \delta$. It is straightforward to prove from the axioms of $(\mathrm{ACP}^c_\lambda + \mathrm{RN})$ $(AC(C, A), \uplus)$ that

$$\begin{aligned}
&\rho_{\mathbf{ca}(I)}\big(\lambda^I_{\mathbf{0}}(\delta^c \mathbin{\|\!_} {}^c q)\big) \overset{OC1}{=} \rho_{\mathbf{ca}(I)}\big(\lambda^I_{\mathbf{0}}(\delta \mathbin{\|\!_} {}^c q)\big) \overset{CM2, A7}{=} \rho_{\mathbf{ca}(I)}\big(\lambda^I_{\mathbf{0}}(\delta)\big) \overset{CSO1, RN1}{=} \\
&\delta \overset{A7, t \equiv \delta}{=} t \cdot q,
\end{aligned}$$

which completes the first part of the proof for the case that $t \equiv \delta$. In addition, using the above result, the following derivation can be obtained.

$$\begin{aligned}
&\rho_{\mathbf{ca}(I)}\big(\lambda^I_{\mathbf{0}}(\delta^c \parallel {}^c q)\big) \overset{CM1, OC1}{=} \rho_{\mathbf{ca}(I)}\big(\lambda^I_{\mathbf{0}}(\delta^c \mathbin{\|\!_} {}^c q + {}^c q \mathbin{\|\!_} \delta + \delta \mid {}^c q)\big) \overset{CSO5, RN2}{=} \\
&\rho_{\mathbf{ca}(I)}\big(\lambda^I_{\mathbf{0}}(\delta^c \mathbin{\|\!_} {}^c q)\big) + \rho_{\mathbf{ca}(I)}\big(\lambda^I_{\mathbf{0}}({}^c q \mathbin{\|\!_} \delta)\big) + \rho_{\mathbf{ca}(I)}\big(\lambda^I_{\mathbf{0}}(\delta \mid {}^c q)\big) \overset{9.14, 9.15, RN1}{=} \\
&\rho_{\mathbf{ca}(I)}\big(\lambda^I_{\mathbf{0}}(\delta^c \mathbin{\|\!_} {}^c q)\big) + \delta + \delta \overset{A6}{=} t \cdot q,
\end{aligned}$$

which completes the proof for the case $t \equiv \delta$.

(ii) Assume that $t \equiv a$, for some action $a \in AC(C, A)$. The second step in the following derivation uses the fact that $c \notin causes(t)$; the third step uses that $c \notin causes(q)$.

$$\rho_{ca(I)}\big(\lambda_{\mathbf{0}}^{I}(a^{c} \mathbin{\lfloor\!\lfloor} {}^{c}q)\big) \overset{OC2,CM2}{=} \rho_{ca(I)}\big(\lambda_{\mathbf{0}}^{I}((\mathbf{ca}, \mathbf{sa}, \mathbf{pa} \uplus [c]) \cdot {}^{c}q)\big)$$
$$\overset{CSO4,CSO2,RN3,RN1}{=} a \cdot \rho_{ca(I)}\big(\lambda_{[c]}^{I}({}^{c}q)\big) \overset{9.17,t\equiv a}{=} t \cdot q.$$

Using this result, the following derivation can be obtained.

$$\rho_{ca(I)}\big(\lambda_{\mathbf{0}}^{I}(a^{c} \parallel {}^{c}q)\big) \overset{CM1}{=} \rho_{ca(I)}\big(\lambda_{\mathbf{0}}^{I}(a^{c} \mathbin{\lfloor\!\lfloor} {}^{c}q + {}^{c}q \mathbin{\lfloor\!\lfloor} a^{c} + a^{c} \mid {}^{c}q)\big) \overset{CSO5,RN2}{=}$$
$$\rho_{ca(I)}\big(\lambda_{\mathbf{0}}^{I}(a^{c} \mathbin{\lfloor\!\lfloor} {}^{c}q)\big) + \rho_{ca(I)}\big(\lambda_{\mathbf{0}}^{I}({}^{c}q \mathbin{\lfloor\!\lfloor} a^{c})\big) + \rho_{ca(I)}\big(\lambda_{\mathbf{0}}^{I}(a^{c} \mid {}^{c}q)\big)$$
$$\overset{9.14,9.15,RN1,A6}{=} t \cdot q,$$

which completes the proof for this case.

(iii) Assume that $t \equiv a \cdot s$, for some action $a \in AC(C, A)$ and basic term $s \in \mathcal{B}(C, A)$. The second step in the following derivation uses the fact that $c \notin causes(t)$.

$$\rho_{ca(I)}\big(\lambda_{\mathbf{0}}^{I}((a \cdot s)^{c} \mathbin{\lfloor\!\lfloor} {}^{c}q)\big) \overset{OC4,CM3}{=} \rho_{ca(I)}\big(\lambda_{\mathbf{0}}^{I}(a \cdot (s^{c} \parallel {}^{c}q))\big) \overset{CSO4,CSO2,RN3,RN1}{=}$$
$$a \cdot \rho_{ca(I)}\big(\lambda_{\mathbf{0}}^{I}(s^{c} \parallel {}^{c}q)\big) \overset{\text{Induction}}{=} a \cdot s \cdot q \overset{t\equiv a\cdot s}{=} t \cdot q.$$

Using this result, the following derivation can be made.

$$\rho_{ca(I)}\big(\lambda_{\mathbf{0}}^{I}((a \cdot s)^{c} \parallel {}^{c}q)\big) \overset{CM1,CSO5,RN2}{=}$$
$$\rho_{ca(I)}\big(\lambda_{\mathbf{0}}^{I}((a \cdot s)^{c} \mathbin{\lfloor\!\lfloor} {}^{c}q)\big) + \rho_{ca(I)}\big(\lambda_{\mathbf{0}}^{I}({}^{c}q \mathbin{\lfloor\!\lfloor} (a \cdot s)^{c})\big) + \rho_{ca(I)}\big(\lambda_{\mathbf{0}}^{I}((a \cdot s)^{c} \mid {}^{c}q)\big)$$
$$\overset{9.14,9.15,RN1,A6}{=} t \cdot q.$$

(iv) Assume that $t \equiv u + v$, for basic terms $u, v \in \mathcal{B}(C, A)$.

$$\rho_{ca(I)}\big(\lambda_{\mathbf{0}}^{I}((u + v)^{c} \mathbin{\lfloor\!\lfloor} {}^{c}q)\big) \overset{OC3,CM4}{=} \rho_{ca(I)}\big(\lambda_{\mathbf{0}}^{I}(u^{c} \mathbin{\lfloor\!\lfloor} {}^{c}q + v^{c} \mathbin{\lfloor\!\lfloor} {}^{c}q)\big) \overset{CSO5,RN2}{=}$$
$$\rho_{ca(I)}\big(\lambda_{\mathbf{0}}^{I}(u^{c} \mathbin{\lfloor\!\lfloor} {}^{c}q)\big) + \rho_{ca(I)}\big(\lambda_{\mathbf{0}}^{I}(u^{c} \mathbin{\lfloor\!\lfloor} {}^{c}q)\big) \overset{\text{Induction}(2\times)}{=}$$
$$u \cdot q + v \cdot q \overset{A4,t\equiv u+v}{=} t \cdot q.$$

In addition,

$$\rho_{ca(I)}\big(\lambda_{\mathbf{0}}^{I}((u + v)^{c} \parallel {}^{c}q)\big) \overset{CM1,CSO5,RN2}{=}$$
$$\rho_{ca(I)}\big(\lambda_{\mathbf{0}}^{I}((u + v)^{c} \mathbin{\lfloor\!\lfloor} {}^{c}q)\big) + \rho_{ca(I)}\big(\lambda_{\mathbf{0}}^{I}({}^{c}q \mathbin{\lfloor\!\lfloor} (u + v)^{c})\big) + \rho_{ca(I)}\big(\lambda_{\mathbf{0}}^{I}((u + v)^{c} \mid {}^{c}q)\big)$$
$$\overset{9.14,9.15,RN1,A6}{=} t \cdot q,$$

which completes the proof. □

The above proof uses a number of auxiliary properties that are given below. Example 9.8 can be used to better understand these properties. All the properties are proven by means of structural induction. However, since the proofs are much simpler than the proof of Theorem 9.10, the details are omitted. Note that three of the four properties can be proven for arbitrary communication functions. Only for Property 9.12, it is required that the communication function does not affect causes.

PROPERTY 9.11. *Let p, q be closed terms in $\mathcal{C}(C, A)$, $I \subseteq C$ a set of causes, and c a cause in I.*

$$(\mathrm{ACP}^c_\lambda + \mathrm{RN})\big(AC(C, A), \gamma\big) \vdash \lambda^I_{\mathbf{0}}({}^c p \mathbin{\|\!\lfloor} q) = \delta.$$

PROOF. It follows from Theorem 9.7 that there exists a basic term $t \in \mathcal{B}(C, A)$ such that $(\mathrm{ACP}^c_\lambda + \mathrm{RN})(AC(C, A), \gamma) \vdash p = t$. It suffices to show that $(\mathrm{ACP}^c_\lambda + \mathrm{RN})(AC(C, A), \gamma) \vdash \lambda^I_{\mathbf{0}}({}^c t \mathbin{\|\!\lfloor} q) = \delta$. The proof is by induction on the structure of basic term t. □

PROPERTY 9.12. *Let p, q be closed terms in $\mathcal{C}(C, A)$, $I \subseteq C$ a set of causes, and c a cause in I.*

$$(\mathrm{ACP}^c_\lambda + \mathrm{RN})\big(AC(C, A), \uplus\big) \vdash \lambda^I_{\mathbf{0}}(p \mid {}^c q) = \delta.$$

PROOF. It follows from Theorem 9.7 that there exist basic terms $s, t \in \mathcal{B}(C, A)$ such that $(\mathrm{ACP}^c_\lambda + \mathrm{RN})(AC(C, A), \uplus) \vdash p = s \wedge q = t$. It suffices to show that $(\mathrm{ACP}^c_\lambda + \mathrm{RN})(AC(C, A), \uplus) \vdash \lambda^I_{\mathbf{0}}(s \mid {}^c t) = \delta$. The proof is by induction on the structure of basic term s. The first three cases are proven by induction on the structure of basic term t. The proofs of the second and third case use the fact that the communication function equals $\uplus$. □

PROPERTY 9.13. *Let p be a closed term in $\mathcal{C}(C, A)$, $I \subseteq C$ a set of causes, and m a bag of internal causes in $\mathcal{B}(I)$.*

$$causes(p) \cap I = \emptyset \Rightarrow (\mathrm{ACP}^c_\lambda + \mathrm{RN})\big(AC(C, A), \gamma\big) \vdash \rho_{\mathbf{ca}(I)}\big(\lambda^I_m(p)\big) = p.$$

PROOF. It follows from Theorem 9.7 that there exists a basic term $t \in \mathcal{B}(C, A)$ such that $(\mathrm{ACP}^c_\lambda + \mathrm{RN})(AC(C, A), \gamma) \vdash p = t$. It suffices to show that $(\mathrm{ACP}^c_\lambda + \mathrm{RN})(AC(C, A), \gamma) \vdash \rho_{\mathbf{ca}(I)}(\lambda^I_m(t)) = t$. The proof is by induction on the structure of basic term t. □

PROPERTY 9.14. *Let p be a closed term in $\mathcal{C}(C, A)$, $I \subseteq C$ a set of causes, and c a cause in I.*

$$causes(p) \cap I = \emptyset \Rightarrow (\mathrm{ACP}^c_\lambda + \mathrm{RN})\big(AC(C, A), \gamma\big) \vdash \rho_{\mathbf{ca}(I)}\big(\lambda^I_{[c]}({}^c p)\big) = p.$$

PROOF. It follows from Theorem 9.7 that there exists a basic term $t \in \mathcal{B}(C, A)$ such that $(\mathrm{ACP}^c_\lambda + \mathrm{RN})(AC(C, A), \gamma) \vdash p = t$. It suffices to show that $(\mathrm{ACP}^c_\lambda + \mathrm{RN})(AC(C, A), \gamma) \vdash \rho_{ca(I)}(\lambda^I_{[c]}({}^c t)) = t$. The proof is by induction on the structure of basic term t. The proof of the third case uses Property 9.13. □

9.5. *Concluding remarks*

In this section, it has been shown how cause-addition operators can be used to specify causal orderings between actions in an algebraic expression. The cause-addition operators are inspired by the causality mechanism known from Petri nets. In labeled P/T nets as defined in Section 4, causal relationships are enforced via places. An occurrence of a specific (input or output) cause-addition operator in an algebraic expression corresponds to a place and an accompanying (input or output) arc in a labeled P/T net. In the algebraic framework of Section 6, information about input and output causes of steps is implicit in actions. Cause addition can be used to make causality information explicit. It can be used to adapt the algebraic semantics of labeled P/T nets, as defined in Definition 6.8, accordingly.

Theorem 9.10 shows that the combination of the causal state operator and cause addition can replace sequential composition. It substantiates the statement by Milner in [45] that sequential composition is not necessarily needed as a primitive operator in a process-algebraic theory. Note that Theorem 9.10 is formulated for processes with bounded behavior only. However, we claim that it can be generalized to a setting with iteration and even to a framework allowing general recursion.

10. A non-interleaving process algebra

As mentioned before, Petri-net theory is one of the most well-known examples of a partial-order theory for specifying and analyzing concurrent systems. In Section 5, it has been shown how standard ACP-style process algebra can be turned into a partial-order framework. In Sections 6 and 9, two extensions of the standard process-algebraic framework have been introduced that are inspired by concepts from Petri-net theory, namely the causal state operator and cause addition. Thus, it is possible to give algebraic specifications in the style of Petri-net theory. The basis of such a specification is a parallel composition of component specifications. Causal relationships are specified by means of cause-addition operators. A causal state operator is used to restrict the parallel composition with respect to a specific initial marking of causes and, thus, to enforce an actual ordering of actions.

However, several aspects of the algebraic framework developed so far have no equivalent in Petri-net theory. First, as already mentioned in Section 4, standard Petri-net theory does not distinguish between successful termination and deadlock, whereas such a distinction is made in ACP-style process algebra. Second, the communication-merge operator has no straightforward interpretation in the Petri-net framework of Section 4. Consider, for example, the specification of process C in Example 5.26. It is not possible to specify this process by means of a labeled P/T net as defined in Definition 4.1. Finally, all the equational theories and their models that have been considered so far are interleaving theories and algebras, which means that each parallel composition can be written as a so-called head normal form. The notion of a head normal form is another notion that does not have a straightforward counterpart in Petri-net theory.

In Section 10.1, we develop an algebra of so-called non-terminating serializable processes. The domain of this algebra contains only processes that cannot terminate successfully. Moreover, the signature of this algebra does not contain a communication merge. It turns out that this algebra of non-terminating serializable processes is a non-interleaving process algebra. Thus, the algebra has several important characteristics of a Petri-net framework. Section 10.2 contains a brief discussion on causality in process algebra, in particular, non-interleaving process algebra. In Section 10.3, it is shown that all non-terminating serializable processes with *bounded* behavior can be specified without using choice, sequential composition, or communication. Thus, in the context of non-terminating serializable processes with bounded behavior, the causality mechanism borrowed from Petri-net theory in the form of the causal state operator and cause addition is sufficiently powerful to replace the standard algebraic operators and, thus, the standard causality mechanism in ACP-style process algebra.

10.1. *The algebra of non-terminating serializable processes*

The basis of the algebra of non-terminating serializable processes is the process algebra $\mathcal{M}(C, A, \gamma)$ of Section 9.2, where C is a set of causes, A a set of atomic actions, and γ a communication function on the set of actions $AC(C, A) = \mathcal{B}(C) \times \mathcal{B}(A) \times \mathcal{B}(C)$. In order to obtain a partial-order theory, it is assumed that the communication function equals $\uplus$. The goal is to restrict the domain of $\mathcal{M}(C, A, \uplus)$ in such a way that the resulting set of processes contains only processes that have an intuitive interpretation in terms of labeled P/T nets.

The notion of a non-terminating serializable process is defined in the context of a process space as defined in Definition 3.1. Therefore, let $(\mathcal{P}, \mathcal{A}, \rightarrow, \downarrow)$ be some process space. Let $\stackrel{*}{\Longrightarrow} \subseteq \mathcal{P} \times \mathcal{P}$ be the reachability relation in this process space as defined in Definition 3.2. A process is said to be non-terminating if and only if it cannot terminate successfully. That is, a process is non-terminating if and only if it terminates in a deadlock or it does not terminate because its behavior is unbounded.

DEFINITION 10.1 (*Non-terminating process*). Let p be a process in $\mathcal{P}$. Process p has the option to terminate, denoted $\Downarrow p$, if and only if there is a process p' in $\mathcal{P}$ such that $p \stackrel{*}{\Longrightarrow} p'$ and $\downarrow p'$. Process p is *non-terminating* if and only if $\neg \Downarrow p$.

The notion of *serializability* is only meaningful if processes can perform some kind of *steps*. Therefore, assume that the set of actions in the process space $\mathcal{A}$ equals $AC(C, A)$. The basic idea of serializability is as follows. Assume that a process can perform a step that consists of several atomic actions. The process is serializable only if it can perform all the atomic actions in the step in isolation and in any order. The bags of input and output causes of the step may be distributed over the atomic actions in some arbitrary way.

DEFINITION 10.2 (*Serializable process*). Let p be a process in $\mathcal{P}$. Process p is *serializable* if and only if, for any $p', p'' \in \mathcal{P}$, $a \in AC(C, A)$, and $\alpha_1, \alpha_2 \in \mathcal{B}(A) \setminus \{\mathbf{0}\}$ such that $p \stackrel{*}{\Longrightarrow} p'$ and $\mathbf{s}a = \alpha_1 \uplus \alpha_2$,

$$
\begin{array}{l}
p' \xrightarrow{a} p'' \Rightarrow \\
\quad (\exists\, p''', a_1, a_2 : \\
\qquad p''' \in \mathcal{P} \wedge a_1, a_2 \in AC(C, A) \wedge \mathbf{s}a_1 = \alpha_1 \wedge \mathbf{s}a_2 = \alpha_2 \wedge a = a_1 \uplus a_2 : \\
\qquad p' \xrightarrow{a_1} p''' \xrightarrow{a_2} p'' \\
\quad).
\end{array}
$$

EXAMPLE 10.3. Let p, p_1, p_2, r, and s be atomic actions in A; let $1, 2, 3, 4, 5$, and 6 be causes in C. Figure 15 shows two simple processes (that are variants of processes depicted in Figures 1 and 2). Square brackets of singleton bags are omitted. Clearly, both processes are non-terminating. The process shown in Figure 15(a) corresponds to an unbounded iteration, whereas the process shown in Figure 15(b) terminates in a deadlock. Process (a) is serializable for the simple reason that it cannot perform a step consisting of more than one atomic action. Process (b) is not serializable. After its initial action, it can perform a step $[p_1, p_2]$ or it can perform the sequence of the two steps $[p_1]$ and $[p_2]$. However, it cannot perform the sequence $[p_2]$ followed by $[p_1]$.

In the introduction to this section, it is implicitly claimed that labeled P/T nets as defined in Section 4 correspond to non-terminating serializable processes. Note that Definition 10.2 (Serializability) can be easily adapted to a setting where actions do not contain causes. Consider the step semantics of marked, labeled P/T nets defined in Definition 4.13. Clearly, marked, labeled P/T nets cannot terminate successfully. In addition, it follows from Definition 4.12 (Step firing rule) that all marked, labeled P/T nets define serializable processes. If several transitions in a labeled P/T net are enabled simultaneously, then each of these transitions is also enabled separately and firing one of these transitions does not disable any of the other transitions.

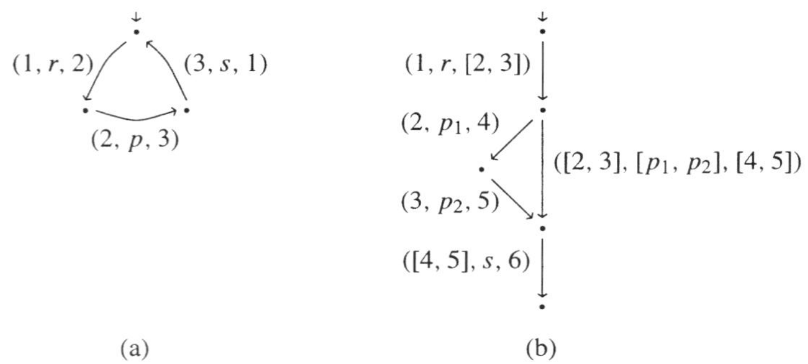

Fig. 15. A non-terminating serializable process (a) and a non-terminating process that is not serializable (b).

EXAMPLE 10.4. Consider the marked, labeled P/T net of Figure 4 and its step semantics discussed in Example 4.14. It is not difficult to see that the P/T net of Figure 4 defines a serializable process. It is interesting to compare this process to the process depicted in Figure 15(b).

Let us return to the process algebra $\mathcal{M}(C, A, \uplus)$. This algebra is a model of the equational theory $(\mathrm{ACP}^{*c}_{\lambda} + \mathrm{RN} + RSP^* + \mathrm{SC})(AC(C, A), \uplus)$ of Section 9.1. Let $\mathcal{C}(C, A)$ be the set of closed $(\mathrm{ACP}^{*c}_{\lambda} + \mathrm{RN} + RSP^* + \mathrm{SC})(AC(C, A), \uplus)$ terms. Let $(\mathcal{C}(C, A) \cup \{\surd\}, AC(C, A), \rightarrow, \{\surd\})$ be the process space defined in Section 9.2, underlying the algebra $\mathcal{M}(C, A, \uplus)$, with reachability relation $\stackrel{*}{\Longrightarrow} \subseteq (\mathcal{C}(C, A) \cup \{\surd\}) \times (\mathcal{C}(C, A) \cup \{\surd\})$. The following result is a direct corollary of Definition 10.1.

COROLLARY 10.5. *For any closed term p in $\mathcal{C}(C, A)$, $\Downarrow p \Leftrightarrow p \stackrel{*}{\Longrightarrow} \surd$.*

As required by Definition 10.2, the set of actions in the process space $(\mathcal{C}(C, A) \cup \{\surd\}, AC(C, A), \rightarrow, \{\surd\})$ corresponds to $AC(C, A)$. The set of all non-terminating serializable processes in this process space is denoted $\mathcal{C}_{\mathrm{ns}}(C, A)$.

EXAMPLE 10.6. Consider again Example 10.3. The process depicted in Figure 15(a) is specified by the term $({}^1r^2 \cdot {}^2p^3 \cdot {}^3s^1) * \delta$, where a triple $(\mathbf{0}, \alpha, \mathbf{0}) \in AC(C, A)$ is identified with its second element and square brackets of singleton bags are omitted. Another term representing the same process is $\lambda_1^{\{1,2,3\}}({}^1r^{2*}\delta \parallel {}^2p^{3*}\delta \parallel {}^3s^{1*}\delta)$. The process of Figure 15(b) is specified by, for example, term $({}^1r^2)^3 \cdot ({}^2{p_1}^4 \cdot {}^3{p_2}^5 + {}^2{p_1}^4 \mid {}^3{p_2}^5) \cdot {}^4({}^5s^6) \cdot \delta$.

Another example of a non-terminating serializable process is the term $({}^1r^2)^3 \cdot ({}^2{p_1}^4 \parallel {}^3{p_2}^5) \cdot {}^4({}^5s^6) \cdot \delta$. Equivalent terms, thus representing the same process, can be found in Example 6.2.

Two other examples of (non-terminating) processes that are not serializable are ${}^2({}^3[p_1, p_2]^4)^5 \cdot \delta$ and $\rho_f({}^2p^3 \cdot \delta)$ where f is the renaming function that renames action $(2, p, 3)$ into action $([2, 3], [p_1, p_2], [4, 5])$. Finally, term C as defined in Example 5.26 corresponds to a (terminating) process that is not serializable (provided the definition of serializability is adapted to the setting without causes).

The next step is to turn the set of non-terminating serializable processes into an algebra. For this purpose, it is necessary to define a signature for the algebra containing a number of meaningful functions on non-terminating serializable processes. The signature of $(\mathrm{ACP}^{*c}_{\lambda} + \mathrm{RN} + RSP^* + \mathrm{SC})(AC(C, A), \uplus)$ might appear to be a suitable starting point. However, not all constants and operators in this signature are useful.

The domain of an algebra must be closed under function application. In other words, a constant in the signature of the algebra of non-terminating serializable processes must, of course, be a non-terminating serializable process; an operator applied to non-terminating serializable processes must yield a non-terminating serializable process again. The consequences of these observations become clear when considering the following. First, the actions in $AC(C, A)$ correspond to processes that can terminate successfully. Thus, these actions are not suitable as constants in our algebra. Second, as illustrated in Example 10.6, processes that can perform isolated actions of which the step contains more than one atomic action are not serializable. Third, the communication merge often yields processes that are not serializable. Finally, some renaming functions, such as the one given in Example 10.6, may yield processes that are not serializable even if they are applied to a serializable process.

The above observations identify a number of constants and operators that cause technical problems. However, in addition, several of the operators in the signature of $(\mathrm{ACP}_\lambda^{*c} + \mathrm{RN} + RSP^* + \mathrm{SC})(AC(C, A), \uplus)$ are not very meaningful in the context of non-terminating processes. The standard ACP-style sequential-composition operator, for example, is not very useful when its left operand is a non-terminating process. The same is true for the binary Kleene star. In addition, also output-cause addition is not very meaningful for non-terminating processes.

Many of the problems can be solved by replacing the sequential-composition operator by a set of action-prefix operators in the style of CCS [44,45] and the binary Kleene star by a class of prefix-iteration operators (see [24] and [12], Chapter 5 of this Handbook). It is important to define a suitable class of actions that can form the basis for these prefix operators. The following definition is inspired by the notion of a transition in a P/T net. A Petri element can be seen as the algebraic equivalent of a transition combined with its bags of input and output places.

DEFINITION 10.7 (*Petri elements*). The set of *Petri elements* over the sets of causes C and atomic actions A, denoted $PE(C, A)$, is inductively defined as follows. For each atomic action a in A, $(\mathbf{0}, [a], \mathbf{0})$ is an element of $PE(C, A)$. For each Petri element e in $PE(C, A)$ and cause c in C, ${}^{c}e$ and e^{c} are elements of $PE(C, A)$.

Note that each Petri element in $PE(C, A)$ is derivably equal to an action in $AC(C, A)$ with a step consisting of only a single atomic action. The converse is also true.

For each Petri element e in $PE(C, A)$, the operator $e \cdot \_$ is an action-prefix operator and $e * \_$ is a prefix-iteration operator. The semantics of action-prefix and prefix-iteration operators follows immediately from the semantics of actions, input-cause addition, output-cause addition, sequential composition, and the binary Kleene star.

Most of the problems identified above are solved if the action constants, the sequential-composition operator, the binary Kleene star, and the output-cause-addition operators are replaced by the above prefix operators. Two problems remain, namely the communication merge and the renaming operators. The solution to the first problem is simply to remove the communication merge from the signature of our algebra. To solve the second problem, recall the subset of renaming functions defined in Section 7. All these renaming functions are derived either from a renaming function or a communication function on atomic actions or from a renaming function on causes. That is, they are all defined in terms of the structure of actions. All these renaming functions using the structure of actions can be used as the basis for a renaming operator in the signature of our algebra. Observe that the renaming function in Example 10.6 is not defined in terms of the structure of actions. Recall from Section 5.1 that RF denotes the set of all renaming functions. Let $SRF \subseteq RF$ be the subset of renaming functions that consists of all the renaming functions introduced in Section 7.

At this point, we have obtained a signature of meaningful operators. It contains the inaction constant δ, the standard ACP operators choice, merge, left merge, and a whole class of renaming operators, including encapsulation operators, explicit communication operators, and cause-abstraction operators, as well as a notion of sequential composition, a notion of iteration, the causal state operator, and input-cause addition. Let Σ denote the signature containing constant δ, the operators $+$, $\|$, and $\lfloor\!\lfloor$, for each $f \in SRF$, the operator ρ_f, for each $e \in PE(C, A)$, the operators $e\,\cdot$ and e^*, for each $I \subseteq C$ and $m \in \mathcal{B}(I)$, the operator λ^I_m, and, for each $c \in C$, the operator ${}^c\_$.

EXAMPLE 10.8. Consider again Example 10.6. The iteration in the term $({}^1r^2 \cdot {}^2p^3 \cdot {}^3s^1)^*\delta$ is not a prefix iteration. However, the equivalent term $\lambda^{\{1,2,3\}}_1({}^1r^2 * \delta \parallel {}^2p^3 * \delta \parallel {}^3s^1 * \delta)$ only contains prefix iterations. The term $({}^1r^2)^3 \cdot ({}^2p_1{}^4 \parallel {}^3p_2{}^5) \cdot {}^4({}^5s^6) \cdot \delta$ contains two Petri elements outside the context of an action prefix and one general sequential composition. Thus, it is not in the restricted signature Σ defined above. It is an interesting exercise to construct an equivalent term that only uses operators of the restricted signature (see also Example 6.2). Finally, it is not difficult to verify that none of the terms in Example 10.6 corresponding to a process that is not serializable is in the restricted signature Σ.

THEOREM 10.9. *The set of non-terminating serializable processes $\mathcal{C}_{\text{ns}}(C, A)$ is closed under the operators in signature Σ.*

PROOF. The proof consists of a tedious, but rather straightforward, analysis of the transition relation defined by the derivation rules in Tables 6, 8, and 10. Let p and q be two closed terms in $\mathcal{C}_{\text{ns}}(C, A)$.

(i) It must be shown that $p + q$ is an element of $\mathcal{C}_{\text{ns}}(C, A)$. First, it must be proven that $\neg \Downarrow p + q$. This follows immediately from the observation that $p + q$ can only perform an action if either p or q can perform that action and the fact that $\neg \Downarrow p$ and $\neg \Downarrow q$. Second, it must be shown that $p+q$ is serializable. The desired result follows from again the observation that $p+q$ can only perform an action if either p or q can perform that action and the fact that both p and q are serializable.

(ii) It must be shown that $p \parallel q$ is an element of $\mathcal{C}_{\text{ns}}(C, A)$. First, it follows immediately from the derivation rules in Table 6 that $\Downarrow p \parallel q$ if and only if both $\Downarrow p$ and $\Downarrow q$. Hence, $\neg \Downarrow p \parallel q$. Second, assume that r is a closed term in $\mathcal{C}(C, A)$ such that $p \parallel q \overset{*}{\Longrightarrow} r$. It follows from the fact that both $\neg \Downarrow p$ and $\neg \Downarrow q$ that r must be of the form $p' \parallel q'$ with $p', q' \in \mathcal{C}(C, A)$ such that $p \overset{*}{\Longrightarrow} p'$ and $q \overset{*}{\Longrightarrow} q'$. Let a be an action in $AC(C, A)$ and let α_1 and α_2 be two non-empty bags in $\mathcal{B}(A) \setminus \{\mathbf{0}\}$ such that $sa = \alpha_1 \uplus \alpha_2$ and r can perform the action a. Three cases can be distinguished. First, there exists a closed term $p'' \in \mathcal{C}(C, A)$ such that $p' \xrightarrow{a} p''$, which means that $r \equiv p' \| q' \xrightarrow{a} p'' \| q'$. Since p is serializable, there exist a closed term $p''' \in \mathcal{C}(C, A)$ and actions $a_1, a_2 \in AC(C, A)$ such that $sa_1 = \alpha_1$, $sa_2 = \alpha_2$, $a = a_1 \uplus a_2$, and $p' \xrightarrow{a_1} p''' \xrightarrow{a_2} p''$. Hence, $p' \parallel q' \xrightarrow{a_1} p''' \parallel q' \xrightarrow{a_2} p'' \parallel q'$, which means that $p \parallel q$ satisfies the serializability requirement in this case. Second, there exists a closed term $q'' \in \mathcal{C}(C, A)$ such that $q' \xrightarrow{a} q''$, which means that $r \equiv p' \parallel q' \xrightarrow{a} p' \parallel q''$. The argument proving that $p \parallel q$ satisfies the serializability requirement in this case is identical to the previous case. Third, there exist closed terms $p'', q'' \in \mathcal{C}(C, A)$ and actions $b, c \in AC(C, A)$ such that $a = \gamma(b, c) = b \uplus c$, $p' \xrightarrow{b} p''$, and $q' \xrightarrow{c} q''$, which means that $r \equiv p' \parallel q' \xrightarrow{a} p'' \parallel q''$. Since $\boldsymbol{s}a = \alpha_1 \uplus \alpha_2$, it follows that there are bags $\alpha_1^b, \alpha_1^c, \alpha_2^b, \alpha_2^c \in \mathcal{B}(A)$ such that $\alpha_1 = \alpha_1^b \uplus \alpha_1^c$, $\alpha_2 = \alpha_2^b \uplus \alpha_2^c$, $\boldsymbol{s}b = \alpha_1^b \uplus \alpha_2^b$, and $\boldsymbol{s}c = \alpha_1^c \uplus \alpha_2^c$. We only consider the case that $\alpha_1^b, \alpha_2^b, \alpha_1^c$, and α_2^c are all non-empty. The serializability of p and q yields that there exist $p''', q''' \in \mathcal{C}(C, A)$ and actions $b_1, b_2, c_1, c_2 \in AC(C, A)$ such that $\boldsymbol{s}b_1 = \alpha_1^b$, $\boldsymbol{s}b_2 = \alpha_2^b$, $b = b_1 \uplus b_2$, $p' \xrightarrow{b_1} p''' \xrightarrow{b_2} p''$, $\boldsymbol{s}c_1 = \alpha_1^c$, $\boldsymbol{s}c_2 = \alpha_2^c$, $c = c_1 \uplus c_2$, and $q' \xrightarrow{c_1} q''' \xrightarrow{c_2} q''$. It easily follows that $r \equiv p' \parallel q' \xrightarrow{b_1 \uplus c_1} p''' \parallel q''' \xrightarrow{b_2 \uplus c_2} p'' \parallel q''$ with $\boldsymbol{s}(b_1 \uplus c_1) = \alpha_1$, $\boldsymbol{s}(b_2 \uplus c_2) = \alpha_2$, and $a = (b_1 \uplus c_1) \uplus (b_2 \uplus c_2)$, which means that also in this case $p \parallel q$ satisfies the serializability requirement.

(iii) It must be shown that $p \mathbin{\lfloor\!\lfloor} q$ is an element of $\mathcal{C}_{\mathrm{ns}}(C, A)$. The proof is almost identical to the proof for the merge operator.

(iv) It must be shown that $\rho_f(p)$ with $f \in SRF$ is an element of $\mathcal{C}_{\mathrm{ns}}(C, A)$. Clearly, $\neg \Downarrow \rho_f(p)$, because $\neg \Downarrow p$. To proof that $\rho_f(p)$ is serializable, assume that r is a closed term in $\mathcal{C}(C, A)$ such that $\rho_f(p) \stackrel{*}{\Longrightarrow} r$. It follows from the fact that $\neg \Downarrow p$ that r must be of the form $\rho_f(p')$ with $p' \in \mathcal{C}(C, A)$ such that $p \stackrel{*}{\Longrightarrow} p'$. Furthermore, assume that fa is an action in $AC(C, A)$ and that α_1 and α_2 are two non-empty bags in $\mathcal{B}(A) \setminus \{\mathbf{0}\}$ such that $\boldsymbol{s}fa = \alpha_1 \uplus \alpha_2$ and r can perform the action fa. For all the renaming functions in SRF, it follows from the fact that p is serializable that there are actions $a, a_1, a_2 \in AC(C, A)$ and closed terms $p'', p''' \in \mathcal{C}(C, A)$ such that

(1) $f(a) = fa$, $p' \xrightarrow{a} p''$, and $r \xrightarrow{fa} \rho_f(p'')$,
(2) $a = a_1 \uplus a_2$ and $p' \xrightarrow{a_1} p''' \xrightarrow{a_2} p''$, and
(3) $\boldsymbol{s}(f(a_1)) = \alpha_1$, $\boldsymbol{s}(f(a_2)) = \alpha_2$, $fa = f(a_1) \uplus f(a_2)$, and $r \xrightarrow{f(a_1)} \rho_f(p''') \xrightarrow{f(a_2)} \rho_f(p'')$.

As a result, $\rho_f(p)$ is serializable.

(v) It must be shown that $e \cdot p$ with $e \in PE(C, A)$ is an element of $\mathcal{C}_{\mathrm{ns}}(C, A)$. It follows from Definition 10.7 (Petri elements) that there exist unique bags of causes $c, d \in \mathcal{B}(C)$ and a unique atomic action $a \in A$ such that $e \cdot p \xrightarrow{(c,[a],d)} p$. Process $e \cdot p$ cannot perform any other action. Thus, it follows easily from the fact that $\neg \Downarrow p$ that also $\neg \Downarrow e \cdot p$. In addition, since the first action of $e \cdot p$ consists of a single atomic action and p is serializable, $e \cdot p$ satisfies the serializability requirement.

(vi) It must be shown that $e * p$ with $e \in PE(C, A)$ is an element of $\mathcal{C}_{\mathrm{ns}}(C, A)$. First, since $\neg \Downarrow p$, also $\neg \Downarrow e * p$. Second, assume that r is a closed term in $\mathcal{C}(C, A)$ such that $e * p \stackrel{*}{\Longrightarrow} r$. It follows from the fact that e is a Petri element and that $\neg \Downarrow p$ that either r must be equal to $e * p$ and all actions executed to reach r from $e * p$ correspond to e or r must be of the form p' with $p' \in \mathcal{C}(C, A)$ such that $p \stackrel{*}{\Longrightarrow} p'$. Thus, serializability of $e * p$ follows easily from the fact that e is a Petri element and that p is serializable.

(vii) It must be shown that $\lambda_m^I(p)$ with $I \subseteq C$ and $m \in \mathcal{B}(I)$ is an element of $\mathcal{C}_{\mathrm{ns}}(C, A)$. Clearly, $\neg \Downarrow \lambda_m^I(p)$, because $\neg \Downarrow p$. The fact that $\lambda_m^I(p)$ is serializable follows from the serializability of p and the observation that, for any two actions $a_1, a_2 \in AC(C, A)$, $\boldsymbol{c}(a_1 \uplus a_2) = \boldsymbol{c}a_1 \uplus \boldsymbol{c}a_2$.

(viii) Finally, it must be shown that ${}^c p$ with $c \in C$ is an element of $\mathcal{C}_{\mathrm{ns}}(C, A)$. Again, $\neg \Downarrow {}^c p$, because $\neg \Downarrow p$. The fact that ${}^c p$ is serializable follows from the serializability of p and the observation that the input-cause addition operator ${}^{c}\_$ only affects the consumption of the first action performed by p. □

The final step in the construction of the algebra of non-terminating serializable processes is to restrict the set of closed terms in $\mathcal{C}_{\mathrm{ns}}(C, A)$ to closed terms over the restricted signature Σ. The set of all non-terminating serializable processes over signature Σ is denoted $\mathcal{C}^{-}_{\mathrm{ns}}(C, A)$. The following result is a direct consequence of Theorem 10.9.

COROLLARY 10.10. *The set $\mathcal{C}^{-}_{\mathrm{ns}}(C, A)$ of all non-terminating serializable processes over signature Σ is closed under the operators in Σ.*

Based on this result, the set of non-terminating serializable processes over signature Σ can be turned into a process algebra following the approach of Section 5.2.

DEFINITION 10.11 (*The algebra of non-terminating serializable processes*). The process algebra $\mathcal{A}_{\mathrm{ns}}(C, A)$ of non-terminating serializable processes is defined as follows. The domain of $\mathcal{A}_{\mathrm{ns}}(C, A)$ consists of equivalence classes of closed terms in $\mathcal{C}^{-}_{\mathrm{ns}}(C, A)$ modulo bisimilarity. The signature of $\mathcal{A}_{\mathrm{ns}}(C, A)$ contains, for each function $\oplus$ in Σ, a function $\bar{\oplus}$ which is function $\oplus$ lifted to equivalence classes of closed terms.

The process algebra $\mathcal{A}_{\mathrm{ns}}(C, A)$ is a subalgebra of the algebra $\mathcal{M}(C, A, \uplus)$ given in Section 9.2. The algebra $\mathcal{A}_{\mathrm{ns}}(C, A)$ incorporates several important features of the Petri-net formalism. An interesting aspect of $\mathcal{A}_{\mathrm{ns}}(C, A)$ is that it is a (branching-time) *non-interleaving, partial-order* process algebra. To prove this claim, it is necessary to slightly adapt the framework developed in Section 5.2. Definition 5.15 (Head normal form) is based on the assumption that the signature of an algebra contains a set of action constants and a general sequential-composition operator, which is not the case for $\mathcal{A}_{\mathrm{ns}}(C, A)$. However, it is straightforward to adapt the definition of a head normal form to the current setting as follows.

DEFINITION 10.12 (*Head normal form*). The set of processes over the signature of the algebra $\mathcal{A}_{\mathrm{ns}}(C, A)$ in *head normal form*, denoted $\mathcal{H}_{\mathrm{ns}}(C, A)$, is inductively defined as follows. First, $\bar{\delta}$ is an element of $\mathcal{H}_{\mathrm{ns}}(C, A)$. Second, for any Petri element e in $PE(C, A)$ and process d in the domain of $\mathcal{A}_{\mathrm{ns}}(C, A)$, the process $e \bar{\cdot} d$ is an element of $\mathcal{H}_{\mathrm{ns}}(C, A)$. Finally, for any processes u and v in $\mathcal{H}_{\mathrm{ns}}(C, A)$, process $u \mathbin{\bar{+}} v$ is an element of $\mathcal{H}_{\mathrm{ns}}(C, A)$.

THEOREM 10.13. *The process algebra $\mathcal{A}_{\mathrm{ns}}(C, A)$ is a non-interleaving, partial-order process algebra.*

PROOF. To prove the theorem, consider the closed term $e \cdot \delta \parallel f \cdot \delta$, where $e \equiv (\mathbf{0}, [a], \mathbf{0})$ and $f \equiv (\mathbf{0}, [b], \mathbf{0})$, for $a, b \in A$, are two Petri elements in $PE(C, A)$. Ignoring consumptions and productions, Figure 16 visualizes this process. Clearly, this process is non-terminating and serializable, which means that it is an element of $\mathcal{C}^{-}_{\mathrm{ns}}(C, A)$. The corresponding process in the algebra $\mathcal{A}_{\mathrm{ns}}(C, A)$ is $e \bar{\cdot} \bar{\delta} \mathbin{\bar{\parallel}} f \bar{\cdot} \bar{\delta}$. Since the process can perform an action $(\mathbf{0}, [a, b], \mathbf{0})$, $\mathcal{A}_{\mathrm{ns}}(C, A)$ is a partial-order algebra. In addition, the process is not

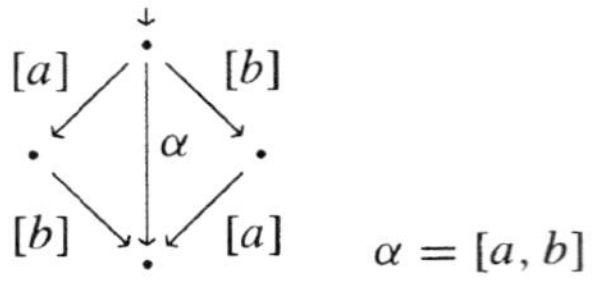

Fig. 16. The visualization of a process in $\mathcal{A}_{ns}(C, A)$ that can perform two atomic actions simultaneously and is not in head normal form.

an element of $\mathcal{H}_{ns}(C, A)$, as defined in Definition 10.12, which means that it is not in head normal form. The process cannot be written in head normal form, because action $(\mathbf{0}, [a, b], \mathbf{0})$ is not a Petri element and because $|$ is not an element of signature Σ. As a consequence, according to Definition 5.16 (Interleaving process algebra), $\mathcal{A}_{ns}(C, A)$ is a non-interleaving process algebra. □

The process algebra $\mathcal{A}_{ns}(C, A)$ is not obtained directly as a model of an equational theory, but indirectly as a subalgebra of a model of an equational theory, namely the theory $(\mathrm{ACP}^{*c}_{\lambda} + \mathrm{RN} + \mathit{RSP}^{*} + \mathrm{SC})(AC(C, A), \uplus)$. An interesting consequence is that this equational theory can be used to reason about the equivalence of processes in the algebra $\mathcal{A}_{ns}(C, A)$. The following theorem states that an equation of two closed terms in $\mathcal{C}^{-}_{ns}(C, A)$ that can be derived from the equational theory $(\mathrm{ACP}^{*c}_{\lambda} + \mathrm{RN} + \mathit{RSP}^{*} + \mathrm{SC})(AC(C, A), \uplus)$ is valid in the algebra $\mathcal{A}_{ns}(C, A)$. That is, the two closed terms specify the same process in the domain of the algebra. (Note that it is straightforward to adapt the notion of validity to the current setting.)

THEOREM 10.14 (Soundness). *For any closed terms* $p, q \in \mathcal{C}^{-}_{ns}(C, A)$,

$$(\mathrm{ACP}^{*c}_{\lambda} + \mathrm{RN} + \mathit{RSP}^{*} + \mathrm{SC})\big(AC(C, A), \uplus\big) \vdash p = q \Rightarrow \mathcal{A}_{ns}(C, A) \models p = q.$$

PROOF. It is a direct consequence of Theorem 9.2 that states that $\mathcal{M}(C, A, \uplus)$ is a model for $(\mathrm{ACP}^{*c}_{\lambda} + \mathrm{RN} + \mathit{RSP}^{*} + \mathrm{SC})(AC(C, A), \uplus)$ and Definition 10.11 that defines $\mathcal{A}_{ns}(C, A)$. □

EXAMPLE 10.15. Consider the closed term $e \cdot \delta \parallel \delta$, where $e \equiv (\mathbf{0}, [a], \mathbf{0})$, for $a \in A$, is a Petri element in $PE(C, A)$. Clearly, this term is an element of $\mathcal{C}^{-}_{ns}(C, A)$. The following derivation shows that it is equivalent to term $e \cdot \delta$ which is also an element of $\mathcal{C}^{-}_{ns}(C, A)$. The interesting aspect of the derivation is that the intermediate term is not an element of $\mathcal{C}^{-}_{ns}(C, A)$.

$$\begin{aligned}
& e \cdot \delta \parallel \delta \\
= \quad & \{CM1, CM3, CM5\} \\
& e \cdot (\delta \cdot \delta + \delta \cdot \delta + \delta \mid \delta) + \delta \cdot e \cdot \delta + (e \mid \delta) \cdot \delta \\
= \quad & \{A6, A7, C1, C3\} \\
& e \cdot \delta.
\end{aligned}$$

It is not difficult to see that the first and last terms of the above derivation correspond to the same process in the algebra $\mathcal{A}_{\mathrm{ns}}(C, A)$.

It is interesting to study a slightly larger example.

EXAMPLE 10.16. Let us return to Example 6.4. This example describes two specifications of a two-place buffer, one specification using the standard communication mechanism of ACP-style process algebra and one specification using the state operator. It is straightforward to adapt the latter specification in such a way that the result is a term in $\mathcal{C}^{-}_{\mathrm{ns}}(C, A)$. Let A be some set of atomic actions that includes the actions r_1, r_2, s_2, c_2, and s_3; let C be a set of causes including the causes $1, 2, 3, 4$, and 5. Let I be the set of causes $\{1, 2, 3, 4, 5\}$.

$$\rho_{\boldsymbol{ca}(I)}\big(\lambda^{I}_{[1,2]}\big({}^{1}({}^{2}r_1{}^{3})^{*}\delta \parallel ({}^{3}c_2{}^{4})^{5\,*}\delta \parallel {}^{4}r_1{}^{1\,*}\delta \parallel {}^{5}s_3{}^{2\,*}\delta\big)\big). \tag{1}$$

The other specification in Example 6.4 is a modular specification. Two terms $B_{1,2}$ and $B_{2,3}$ specify two one-place buffers. Clearly, these terms are not terms over the restricted signature Σ. Therefore, in this example, $B_{1,2}$ and $B_{2,3}$ are defined as follows:

$$\begin{aligned} B_{1,2} &= \rho_{\boldsymbol{ca}(\{1,2\})}\big(\lambda^{\{1,2\}}_{1}({}^{1}r_1{}^{2\,*}\delta \parallel {}^{2}s_2{}^{1\,*}\delta)\big) \quad \text{and} \\ B_{2,3} &= \rho_{\boldsymbol{ca}(\{2,3\})}\big(\lambda^{\{2,3\}}_{2}({}^{2}r_2{}^{3\,*}\delta \parallel {}^{3}s_3{}^{2\,*}\delta)\big). \end{aligned}$$

It is not difficult to verify that the encapsulation operator ∂_H and the renaming operator $\rho_{\boldsymbol{cm}}$ defined in Example 6.4 satisfy the requirements explained in Section 7, which means that they are elements of signature Σ. As a consequence, the two-place buffer can be specified by the following term in $\mathcal{C}^{-}_{\mathrm{ns}}(C, A)$:

$$\partial_H\big(\rho_{\boldsymbol{cm}}(B_{1,2} \parallel B_{2,3})\big). \tag{2}$$

By now, it should no longer be a problem to verify by means of the equational theory $(\mathrm{ACP}^{*c}_{\lambda} + \mathrm{RN} + \mathit{RSP}^{*} + \mathrm{SC})(\mathit{AC}(C, A), \uplus)$ that both specifications (1) and (2) are equivalent to the following term (see also Example 6.4):

$$r_1 \cdot \big((c_2 \cdot (r_1 \parallel s_3))^{*}\delta\big). \tag{3}$$

Clearly, this term is not an element of $\mathcal{C}^{-}_{\mathrm{ns}}(C, A)$ (although it is an element of $\mathcal{C}_{\mathrm{ns}}(C, A)$).

Summarizing, closed terms (1) and (2) have direct interpretations in the algebra $\mathcal{A}_{\mathrm{ns}}(C, A)$. The equivalence of these two terms can be proven via an intermediate term (3), which has no direct interpretation in $\mathcal{A}_{\mathrm{ns}}(C, A)$.

So far, it has been explained how to construct an algebra of non-terminating serializable processes. It has been shown that this algebra is a non-interleaving partial-order process algebra. In that sense, it has the same characteristics as the P/T-net framework of Section 4. It allows algebraic specifications in the style of labeled P/T nets (using parallel compositions, cause addition, and causal state operators as the main operators). In addition, it has been shown that the (interleaving) equational theory of Section 9.1 can be used to reason about processes in the non-interleaving algebra. Example 10.16 illustrates another interesting aspect of the algebra of non-terminating serializable processes. It incorporates both the standard causality mechanism known from process algebra (consisting of sequential composition and communication) and the causality mechanism adopted from Petri nets (consisting of the causal state operator and cause addition). It is interesting to study these two mechanisms in more detail.

10.2. *A brief discussion on causality*

In an interleaving process algebra, each parallel composition can be written equivalently in head normal form. In standard ACP-style process algebra, this means that sequential composition is sufficient to express any causal relationships. That is, communication is not necessary as a primitive causality mechanism. However, in a non-interleaving process algebra, it is no longer possible to express each parallel composition by means of a head normal form. In the algebra $\mathcal{A}_{\text{ns}}(C, A)$ developed in the previous subsection, it is not possible to specify the two-place buffer of Example 10.16 without either communication or a causal state operator. However, the reason is the restricted expressivity of the prefix-iteration operators. In [4], a non-interleaving, partial-order algebra of (possibly terminating) serializable processes is defined in a way very similar to the definition of the non-interleaving algebra of the previous subsection. The algebra in [4] contains the standard algebraic operators, including general sequential composition, plus the causal state operator and cause addition. In addition, the algebra incorporates linear recursion. As a result, in that algebra, it is possible to specify a two-place buffer without communication or a causal state operator. However, it is shown that it is not possible to specify a *three*-place buffer without communication or a causal state operator.

The consequence of this last observation becomes clear when we observe that the technique to construct a non-interleaving process algebra can also be applied to standard ACP-style process algebras. Consider, for example, the standard theory $(\text{ACP}^* + \text{RN} + \mathit{RSP}^* + \text{SC})(\mathcal{B}(A), \uplus)$ as introduced in Section 5, where A is some set of atomic actions. Its model $\mathcal{M}(\mathcal{B}(A), \uplus)$ of Section 5.2 can be reduced to a non-interleaving process algebra following the approach of the previous subsection. The communication merge and the action constants other than the singleton bags are removed from the signature of theory $(\text{ACP}^* + \text{RN} + \mathit{RSP}^* + \text{SC})(\mathcal{B}(A), \uplus)$. This restricted signature and the set of serializable processes over this signature form the basis of a non-interleaving process algebra that is defined in the same way as in Definition 10.11. The interesting aspect is that the signature of the resulting non-interleaving process algebra does not contain causal state operators or cause-addition operators. That is, it only contains the standard process-algebraic causality mechanism consisting of sequential composition and (explicit) communication. Thus, the consequence of the abovementioned result of [4] is that, in a non-interleaving process algebra derived from a standard ACP-style process algebra including linear recursion, it is not possible to specify even a relatively simple process as a three-place buffer without (explicit) communication. That is, in such a non-interleaving process algebra, sequential composition is no longer sufficient to express causal relationships.

The question remains to what extent the causality mechanism adopted from Petri-net theory consisting of cause addition and the causal state operator is sufficiently powerful to replace sequential composition and communication. It is difficult to answer this question for a general algebraic theory including some form of recursion. In [4], it is claimed that the causal state operator and cause addition are sufficiently powerful to replace both communication and sequential composition in a very general setting of (not necessarily serializable) processes with bounded behavior. In the next subsection, we prove a theorem formalizing this claim for non-terminating serializable processes with bounded behavior.

10.3. *Elimination of choice, sequential composition, and communication*

As explained, the aim of this subsection is to prove that, in a process algebra of non-terminating serializable processes with bounded behavior, the ACP-style causality mechanism consisting of communication and sequential composition can be replaced by a causality mechanism consisting of the causal state operator and cause addition. In fact, it is possible to prove even a slightly stronger result: In the proposed setting, the standard ACP operators choice, sequential composition, and (explicit) communication can be eliminated from any process specification. That is, also the choice operator is no longer needed as a primitive operator.

Let C be some set of causes and A some set of atomic actions. Consider the equational theory $(\mathrm{ACP}^c_\lambda + \mathrm{RN} + \mathrm{SC})(AC(C, A), \gamma)$, the Algebra of Communicating Processes with cause addition, causal state operator, renaming, and standard concurrency, where $AC(C, A)$ is the standard set of actions defined before and where γ is some communication function. Theory $(\mathrm{ACP}^c_\lambda + \mathrm{RN} + \mathrm{SC})(AC(C, A), \gamma)$ extends the equational theory of Section 9.4, $(\mathrm{ACP}^c_\lambda + \mathrm{RN})(AC(C, A), \gamma)$, with the standard-concurrency axioms of Table 5.

As before, it is assumed that the communication function equals $\uplus$. Let $\mathcal{C}(C, A)$ be the set of closed $(\mathrm{ACP}^c_\lambda + \mathrm{RN} + \mathrm{SC})(AC(C, A), \uplus)$ terms. Following the approach of Section 5.2, it is possible to construct a model $\mathcal{M}(C, A, \uplus)$ of the equational theory with the underlying process space $(\mathcal{C}(C, A) \cup \{\surd\}, AC(C, A), \rightarrow, \{\surd\})$. Note that this model is an interleaving, partial-order process algebra with a domain that contains only processes with bounded behavior. The set of non-terminating serializable processes in $\mathcal{C}(C, A)$, as defined in Definitions 10.1 (Non-terminating process) and 10.2 (Serializability), is denoted $\mathcal{C}_{\mathrm{bns}}(C, A)$. This set of non-terminating serializable processes can be turned into a non-interleaving, partial-order process algebra $\mathcal{A}_{\mathrm{bns}}(C, A)$ following the approach explained in Section 10.1. Recall that $SRF \subseteq RF$ is the subset of renaming functions that consists of all the renaming functions introduced in Section 7 and that $PE(C, A)$ is the set of Petri elements over C and A as defined in Definition 10.7. The signature of $(\mathrm{ACP}^c_\lambda + \mathrm{RN} + \mathrm{SC})(AC(C, A), \uplus)$ is restricted to the signature Σ_{b} containing inaction constant δ, the operators $+$, $\|$, and $⌊⌊$, for each $f \in SRF$, the operator ρ_f, for each $e \in PE(C, A)$, the action-prefix operator $e\,\cdot$, for each $I \subseteq C$ and $m \in \mathcal{B}(I)$, the operator λ^I_m, and, for each $c \in C$, the operator ${}^c\_$. Note that the signature Σ_{b} differs from the signature Σ introduced in Section 10.1 only with respect to prefix-iteration operators. It follows from (the proof of) Theorem 10.9 that the set of non-terminating serializable processes $\mathcal{C}_{\mathrm{bns}}(C, A)$ (as well as the set of non-terminating serializable processes over the restricted signature Σ_{b}) is closed under the operators in Σ_{b}. Thus, the algebra $\mathcal{A}_{\mathrm{bns}}(C, A)$ can be defined as explained in Definition 10.11 (Process algebra $\mathcal{A}_{\mathrm{ns}}(C, A)$). In addition, it is possible to prove that equational theory $(\mathrm{ACP}_{\lambda}^{c} + \mathrm{RN} + \mathrm{SC})(AC(C, A), \gamma)$ can be used to reason about the equivalence of processes in $\mathcal{A}_{\mathrm{bns}}(C, A)$, similar to Theorem 10.14 for the algebra $\mathcal{A}_{\mathrm{ns}}(C, A)$. Observe that $\mathcal{A}_{\mathrm{bns}}(C, A)$ is a subalgebra of $\mathcal{A}_{\mathrm{ns}}(C, A)$; its signature is contained in the signature of $\mathcal{A}_{\mathrm{ns}}(C, A)$ and it contains all processes in the domain of $\mathcal{A}_{\mathrm{ns}}(C, A)$ with a bounded behavior and no other processes.

EXAMPLE 10.17. Let p_1, p_2, r, and s be atomic actions in A; let $1, 2, 3, 4, 5$, and 6 be causes in C. As before, assume that an action in $AC(C, A)$ with empty consumption and empty production is identified with its second element and that square brackets of singleton bags are omitted. Consider the term $r \cdot (p_1 \parallel p_2) \cdot s \cdot \delta$. It is not difficult to verify that it specifies a non-terminating serializable process with bounded behavior. That is, it is an element of $\mathcal{C}_{\mathrm{bns}}(C, A)$ (although it is not a term over the restricted signature Σ_{b}). Let $I = \{1, 2, 3, 4, 5, 6\}$. It is possible to derive from the axioms of $(\mathrm{ACP}_{\lambda}^{c} + \mathrm{RN} + \mathrm{SC})(AC(C, A), \uplus)$ that the above term is equivalent to term $\rho_{ca(I)}(\lambda_1^I(({}^1r^2)^3 \cdot \delta \parallel {}^2p_1{}^4 \cdot \delta \parallel {}^3p_2{}^5 \cdot \delta \parallel {}^4({}^5s^6) \cdot \delta))$. The latter is a term over the restricted signature Σ_{b}. Furthermore, observe that it does not contain any choice operators or (explicit) communication operators. In addition, it only contains very specific occurrences of action-prefix operators; each occurrence of an action-prefix operator has operand δ.

The main theorem of this subsection (Theorem 10.25 given below) generalizes Example 10.17. That is, it shows that any non-terminating serializable process in $\mathcal{C}_{\mathrm{bns}}(C, A)$ is derivably equal to a term over the restricted signature Σ_{b} that does not contain any occurrences of the choice operator, any explicit communication operator, or any action-prefix operator applied to another term than the inaction constant δ. Note that, for any Petri element $e \in PE(C, A)$, the action prefix $e \cdot \delta$ can be seen as a *constant*, specifying a process that can perform a single action before terminating in a deadlock. In the remainder, we refer to such specific action prefixes as action-prefix constants in order to emphasize that such an action prefix should not be seen as a sequential composition. Theorem 10.25 proves that any term in $\mathcal{C}_{\mathrm{bns}}(C, A)$ is derivably equivalent to a parallel composition of action-prefix constants that is restricted by means of a causal state operator and where any newly-added causes used to enforce causal orderings are hidden by means of a cause-abstraction operator. Interpreting this result in the algebras $\mathcal{A}_{\mathrm{bns}}(C, A)$ and $\mathcal{A}_{\mathrm{ns}}(C, A)$, it means that all processes in the domain of $\mathcal{A}_{\mathrm{bns}}(C, A)$ and, thus, all processes with a bounded behavior in $\mathcal{A}_{\mathrm{ns}}(C, A)$ can be specified without choice, sequential composition, or communication. Theorem 10.25 strengthens our claim that the algebra $\mathcal{A}_{\mathrm{ns}}(C, A)$ of Section 10.1 incorporates some of the most important characteristics of Petri-net theory.

As explained in Section 9.4, an advantage of considering only processes with bounded behavior is that each process can be specified by means of a so-called basic term. Note that the signature of theory $(\mathrm{ACP}_{\lambda}^{c} + \mathrm{RN} + \mathrm{SC})(AC(C, A), \uplus)$ is identical to the signature of theory $(\mathrm{ACP}_{\lambda}^{c} + \mathrm{RN})(AC(C, A), \uplus)$ of Section 9.4. As a result, Definition 9.6 (Basic terms) carries over to the setting of this subsection. The following theorem states the elimination result for theory $(\mathrm{ACP}_{\lambda}^{c} + \mathrm{RN} + \mathrm{SC})(AC(C, A), \uplus)$.

THEOREM 10.18 (Elimination). *For any closed term $p \in \mathcal{C}(C, A)$, there exists a basic term $t \in \mathcal{B}(C, A)$ such that* $(\mathrm{ACP}^c_\lambda + \mathrm{RN} + \mathrm{SC})(AC(C, A), \uplus) \vdash p = t$.

PROOF. The proof is identical to the proof of Theorem 9.7 (Elimination) of Section 9.4. □

The proof of Theorem 10.25 uses induction on the structure of basic terms. Five auxiliary properties are needed in the proof.

The next property is not used in the proof of Theorem 10.25, but in the proof of the first auxiliary result. It is a simple property of bisimilar processes in some arbitrary process space.

PROPERTY 10.19. *Assume that $(\mathcal{P}, \mathcal{A}, \rightarrow, \downarrow)$ is some process space as defined in Definition 3.1. Furthermore, assume that $\stackrel{*}{\Longrightarrow} \subseteq \mathcal{P} \times \mathcal{P}$ is the reachability relation of Definition 3.2. Let p and q be two processes in $\mathcal{P}$; let $\mathcal{R}$ be a bisimulation between p and q as defined in Definition 3.6. For any $p', q' \in \mathcal{P}$,*

(i) $p \stackrel{*}{\Longrightarrow} p' \Rightarrow (\exists q' : q' \in \mathcal{P} : q \stackrel{*}{\Longrightarrow} q' \wedge p'\mathcal{R}q')$ *and*

(ii) $q \stackrel{*}{\Longrightarrow} q' \Rightarrow (\exists p' : p' \in \mathcal{P} : p \stackrel{*}{\Longrightarrow} p' \wedge p'\mathcal{R}q')$.

PROOF. The two properties can be proven by means of straightforward induction on the number of actions needed to reach p' from p and q' from q, respectively. □

Property 10.19 can be used to prove the following result, which is the first auxiliary property needed in the proof of Theorem 10.25. Consider two derivably equivalent closed terms p and q in $\mathcal{C}(C, A)$. Process p is (non-)terminating if and only if process q is (non-)terminating; in addition, p is serializable if and only if q is serializable.

PROPERTY 10.20. *For any closed terms $p, q \in \mathcal{C}(C, A)$ such that* $(\mathrm{ACP}^c_\lambda + \mathrm{RN} + \mathrm{SC})(AC(C, A), \uplus) \vdash p = q$,

(i) $\Downarrow p \Longleftrightarrow \Downarrow q$ *and*

(ii) p *serializable* $\Longleftrightarrow$ q *serializable.*

PROOF. The two properties follow immediately from the fact that $\mathcal{M}(C, A, \uplus)$, with underlying process space $(\mathcal{C}(C, A) \cup \{\surd\}, AC(C, A), \rightarrow, \{\surd\})$, is a model of $(\mathrm{ACP}^c_\lambda + \mathrm{RN} + \mathrm{SC})(AC(C, A), \uplus)$, Property 10.19, and Definitions 3.6 (Bisimilarity), 10.1 (Non-terminating process), and 10.2 (Serializability). □

The second auxiliary property states that any closed term specifying a non-terminating process is derivably equivalent to the sequential composition of a closed term and the inaction constant δ.

PROPERTY 10.21. *For any closed term $p \in \mathcal{C}(C, A)$,*

$$\neg \Downarrow p \Longleftrightarrow \big(\exists q : q \in \mathcal{C}(C, A) : (\mathrm{ACP}^c_\lambda + \mathrm{RN} + \mathrm{SC})\big(AC(C, A), \uplus\big) \vdash p = q \cdot \delta\big).$$

PROOF. First, assume that $\neg \Downarrow p$. It follows from Theorem 10.18 (Elimination) that there exists a basic term $t \in \mathcal{B}(C, A)$ such that $(\mathrm{ACP}^c_\lambda + \mathrm{RN} + \mathrm{SC})(AC(C, A), \uplus) \vdash p = t$. Property 10.20(i) implies that $\neg \Downarrow t$. Thus, it suffices to prove for any basic term $t \in \mathcal{B}(C, A)$ that $\neg \Downarrow t$ implies $(\exists q : q \in \mathcal{C}(C, A) : (\mathrm{ACP}^c_\lambda + \mathrm{RN} + \mathrm{SC})(AC(C, A), \uplus) \vdash t = q \cdot \delta)$. The proof is by induction on the structure of t. The details are straightforward and left to the reader.

Second, assume that $(\exists q : q \in \mathcal{C}(C, A) : (\mathrm{ACP}^c_\lambda + \mathrm{RN} + \mathrm{SC})(AC(C, A), \uplus) \vdash p = q \cdot \delta)$. It must be shown that $\neg \Downarrow p$. The desired result follows immediately from Property 10.20(i) and the observation that $\neg \Downarrow q \cdot \delta$ for any $q \in \mathcal{C}(C, A)$. □

The last three auxiliary properties needed to prove Theorem 10.25 are simple properties concerning the merge, the communication merge, action-prefix constants, and the inaction constant. The proof of Property 10.24 uses Theorem 5.2 (Expansion) which carries over to the current setting.

PROPERTY 10.22. *For any closed term $p \in \mathcal{C}(C, A)$,*

$$(\mathrm{ACP}^c_\lambda + \mathrm{RN} + \mathrm{SC})\big(AC(C, A), \uplus\big) \vdash p \parallel \delta = p \cdot \delta.$$

PROOF. It follows from Theorem 10.18 (Elimination) that it suffices to prove the property for basic terms. The proof by induction on the structure of basic terms is straightforward and left to the reader. □

PROPERTY 10.23. *For any non-empty, finite set of Petri elements $E \subseteq PE(C, A)$,*

$$(\mathrm{ACP}^c_\lambda + \mathrm{RN} + \mathrm{SC})\big(AC(C, A), \uplus\big) \vdash (| \, e : e \in E : e \cdot \delta) = (| \, e : e \in E : e) \cdot \delta.$$

PROOF.

$$\begin{aligned}
& (| \, e : e \in E : e \cdot \delta) \\
= & \qquad \{\text{Definition 10.7 (Petri elements)}, CM7\} \\
& (| \, e : e \in E : e) \cdot (\parallel e : e \in E : \delta) \\
= & \qquad \{\text{Property 10.22}, A7\} \\
& (| \, e : e \in E : e) \cdot \delta.
\end{aligned}$$

□

PROPERTY 10.24. *For any non-empty, finite set of Petri elements $E \subseteq PE(C, A)$,*

$$(\mathrm{ACP}^c_\lambda + \mathrm{RN} + \mathrm{SC})\big(AC(C, A), \uplus\big) \vdash (\parallel e : e \in E : e \cdot \delta) = (\parallel e : e \in E : e) \cdot \delta.$$

PROOF. The proof is by means of natural induction on the size of set E. If E contains only a single element, the property is trivial, because both sides of the equality reduce to the same term. Assume that E contains at least two elements. In the following derivation, applications of axioms of the equational theory $(\mathrm{ACP}^c_\lambda + \mathrm{RN} + \mathrm{SC})(AC(C, A), \uplus)$ are not explicitly mentioned.

$$(\parallel e : e \in E : e \cdot \delta)$$

$$
\begin{aligned}
= \quad & \{\text{Expansion}\} \\
& \big(+D : \emptyset \subset D \subset E : (|d : d \in D : d \cdot \delta) \mathbin{\|\!\!\lfloor} (\|e : e \in E \setminus D : e \cdot \delta)\big) \\
& \quad + (|e : e \in E : e \cdot \delta) \\
= \quad & \{\text{Property 10.23, induction}\} \\
& \big(+D : \emptyset \subset D \subset E : (|d : d \in D : d) \cdot \delta \mathbin{\|\!\!\lfloor} (\|e : e \in E \setminus D : e) \cdot \delta\big) \\
& \quad + (|e : e \in E : e) \cdot \delta \\
= \quad & \{\text{Definition 10.7 (Petri elements), Property 10.22}\} \\
& \big(+D : \emptyset \subset D \subset E : (|d : d \in D : d) \mathbin{\|\!\!\lfloor} (\|e : e \in E \setminus D : e)\big) \cdot \delta \\
& \quad + (|e : e \in E : e) \cdot \delta \\
= \quad & \{\text{Expansion}\} \\
& (\| e : e \in E : e) \cdot \delta.
\end{aligned}
$$

□

At this point, it is possible to prove the main theorem of this subsection. It gives a canonical form for each process term in $\mathcal{C}_{\text{bns}}(C, A)$. This canonical term consists of a parallel composition of action-prefix constants restricted by means of a causal state operator. Causes that do not occur in the original term but are added in the parallel composition to enforce the necessary orderings are hidden by means of a cause-abstraction operator. In the construction of the parallel composition of action-prefix constants, a continuous supply of fresh causes for specifying causal orderings is needed. Therefore, assume that the set of causes C is *infinite*. Note that Definition 9.9 (Causes function) carries over to the current setting.

THEOREM 10.25 (Elimination of choice, sequential composition, and communication). *For any closed term $p \in \mathcal{C}_{\text{bns}}(C, A)$, there are $I \subseteq C, m \in \mathcal{B}(I), n \in \mathbb{N}$, and $e_0, \ldots, e_n \in PE(C, A)$ such that*

$$(\text{ACP}^c_\lambda + \text{RN} + \text{SC})\big(AC(C, A), \uplus\big) \vdash p = \rho_{\boldsymbol{ca}(I)}\big(\lambda^I_m(\| i : 0 \leqslant i \leqslant n : e_i \cdot \delta)\big).$$

PROOF. It follows from Property 10.21 that there exists a closed term $q \in \mathcal{C}(C, A)$ such that $(\text{ACP}^c_\lambda + \text{RN} + \text{SC})(AC(C, A), \uplus) \vdash p = q \cdot \delta$. Property 10.20(ii) implies that $q \cdot \delta$ is serializable. It follows from the derivation rules in Table 6 that also q is serializable. Theorem 10.18 (Elimination) and, again, Property 10.20(ii) yield that there is a basic term $t \in \mathcal{B}(C, A)$ such that $(\text{ACP}^c_\lambda + \text{RN} + \text{SC})(AC(C, A), \uplus) \vdash q = t$ and t is serializable. Thus, to prove the theorem, it suffices to prove the following property: For any *serializable* basic term $t \in \mathcal{B}(C, A)$, there are $I \subseteq C, m \in \mathcal{B}(I), n \in \mathbb{N}$, and $e_0, \ldots, e_n \in PE(C, A)$ such that

$$
\begin{aligned}
& I \cap causes(t) = \emptyset \wedge \\
& \quad (\text{ACP}^c_\lambda + \text{RN} + \text{SC})\big(AC(C, A), \uplus\big) \vdash \\
& \qquad t \cdot \delta = \rho_{\boldsymbol{ca}(I)}\big(\lambda^I_m(\| i : 0 \leqslant i \leqslant n : e_i \cdot \delta)\big).
\end{aligned}
$$

The extra requirement concerning the set of causes I is needed in the proof. Let t be a serializable basic term in $\mathcal{B}(C, A)$. The proof is by induction on the structure of t.

(i) Assume $t \equiv \delta$. Let c be an arbitrary cause in C and e an arbitrary Petri element in $PE(C, A)$. It easily follows that $\{c\} \cap causes(t) = \emptyset$ and that $(\text{ACP}^c_\lambda + \text{RN} + \text{SC})(AC(C, A), \uplus) \vdash \rho_{\boldsymbol{ca}(\{c\})}(\lambda^{\{c\}}_{\mathbf{0}}({}^c e \cdot \delta)) = \delta \cdot \delta = t \cdot \delta$, which completes the proof in this case.

(ii) Assume $t \equiv a$, for some action $a \in AC(C, A)$. It follows from the fact that t is serializable, the derivation rules in Table 6, and Definition 10.2 (Serializability) that $\boldsymbol{sa}$ must be a singleton bag. Consequently, it follows from Definition 10.7 (Petri elements) that there is a Petri element $e \in PE(C, A)$ such that $(\text{ACP}^c_\lambda + \text{RN} + \text{SC})(AC(C, A), \uplus) \vdash a = e$. It is not difficult to prove that $(\text{ACP}^c_\lambda + \text{RN} + \text{SC})(AC(C, A), \uplus) \vdash \rho_{\boldsymbol{ca}(\emptyset)}(\lambda^{\emptyset}_{\mathbf{0}}(e \cdot \delta)) = a \cdot \delta = t \cdot \delta$, which completes the proof also in this case.

(iii) Assume $t \equiv a \cdot s$, for some action $a \in AC(C, A)$ and basic term $s \in \mathcal{B}(C, A)$. Since t is serializable, it follows from Definition 10.2 (Serializability) that $\boldsymbol{sa}$ is a singleton bag and that also s is serializable. Consequently, Definition 10.7 (Petri elements) implies that there is a Petri element $e \in PE(C, A)$ such that $(\text{ACP}^c_\lambda + \text{RN} + \text{SC})(AC(C, A), \uplus) \vdash a = e$. In addition, by induction, it follows that there are $I \subseteq C$, $m \in \mathcal{B}(I)$, $n \in \mathbb{N}$, and $e_0, \ldots, e_n \in PE(C, A)$ such that

$$\begin{aligned} & I \cap causes(s) = \emptyset \wedge \\ & \quad (\text{ACP}^c_\lambda + \text{RN} + \text{SC})\big(AC(C, A), \uplus\big) \vdash \\ & \qquad s \cdot \delta = \rho_{\boldsymbol{ca}(I)}\big(\lambda^I_m(\| i : 0 \leqslant i \leqslant n : e_i \cdot \delta)\big). \end{aligned} \tag{1}$$

Clearly, since C is infinite, it is possible to choose I in such a way that $I \cap causes(a) = \emptyset$ and, hence, $I \cap causes(t) = \emptyset$.

Let $c_0, \ldots, c_n \in C$ be new, distinct causes not in $I \cup causes(t)$. Assume that $N = \{i \mid 0 \leqslant i \leqslant n\}$. Let $I' = I \cup \{c_i \mid i \in N\}$; let $m' = m \uplus [c_i \mid i \in N]$; let $d \equiv (\ldots(e^{c_0})^{\cdots})^{c_n}$. Note that

$$(\text{ACP}^c_\lambda + \text{RN} + \text{SC})\big(AC(C, A), \uplus\big) \vdash d = (\boldsymbol{ca}, \boldsymbol{sa}, \boldsymbol{pa} \uplus [c_i \mid i \in N]). \tag{2}$$

Furthermore, for all $J \subseteq N$ with $J \neq \emptyset$, there is some action $b \in AC(C, A)$ such that

$$\begin{aligned} & (\text{ACP}^c_\lambda + \text{RN} + \text{SC})\big(AC(C, A), \uplus\big) \vdash \\ & \quad (| j : j \in J : e_j) = b \wedge \\ & \qquad (| j : j \in J : {}^{c_j} e_j) = ([c_j \mid j \in J] \uplus \boldsymbol{cb}, \boldsymbol{sb}, \boldsymbol{pb}). \end{aligned} \tag{3}$$

The remainder of this part of the proof is devoted to showing that $t \cdot \delta$ is equivalent to the following term:

$$\rho_{\boldsymbol{ca}(I')}\big(\lambda^{I'}_m\big(d \cdot \delta \parallel (\| i : i \in N : {}^{c_i} e_i \cdot \delta)\big)\big). \tag{4}$$

(Note the correspondence between this term and the construction of Theorem 9.10 (Elimination of sequential composition).) The next derivation shows that the only initial action of term (4) is the action a. Applications of axioms of the equational theory $(\mathrm{ACP}^c_\lambda + \mathrm{RN} + \mathrm{SC})(AC(C, A), \uplus)$ are not explicitly mentioned in the derivation unless no other results are used in a step.

$$
\begin{aligned}
& \rho_{\mathbf{ca}(I')}\big(\lambda_m^{I'}(d\cdot\delta \parallel (\parallel i : i\in N : {}^{c_i}e_i\cdot\delta))\big)\\
= \quad & \{CM1\}\\
& \rho_{\mathbf{ca}(I')}\big(\lambda_m^{I'}(d\cdot\delta \mathbin{⫽} (\parallel i : i\in N : {}^{c_i}e_i\cdot\delta) + (\parallel i : i\in N : {}^{c_i}e_i\cdot\delta) \mathbin{⫽} d\cdot\delta\\
&\qquad + d\cdot\delta \mid (\parallel i : i\in N : {}^{c_i}e_i\cdot\delta))\big)\\
= \quad & \{\text{Expansion}\}\\
& \rho_{\mathbf{ca}(I')}\big(\lambda_m^{I'}(d\cdot\delta \mathbin{⫽} (\parallel i : i\in N : {}^{c_i}e_i\cdot\delta))\big)\\
& + \big(+J : \emptyset\subset J\subset N\\
&\qquad : \rho_{\mathbf{ca}(I')}\big(\lambda_m^{I'}\big(((\mid j : j\in J : {}^{c_j}e_j\cdot\delta) \mathbin{⫽} (\parallel i : i\in N\setminus J : {}^{c_i}e_i\cdot\delta)) \mathbin{⫽} d\cdot\delta\big)\big)\big)\\
& + \rho_{\mathbf{ca}(I')}\big(\lambda_m^{I'}((\mid i : i\in N : {}^{c_i}e_i\cdot\delta) \mathbin{⫽} d\cdot\delta)\big)\\
& + \big(+J : \emptyset\subset J\subset N\\
&\qquad : \rho_{\mathbf{ca}(I')}\big(\lambda_m^{I'}\big(d\cdot\delta \mid ((\mid j : j\in J : {}^{c_j}e_j\cdot\delta) \mathbin{⫽} (\parallel i : i\in N\setminus J : {}^{c_i}e_i\cdot\delta))\big)\big)\big)\\
& + \rho_{\mathbf{ca}(I')}\big(\lambda_m^{I'}(d\cdot\delta \mid (\mid i : i\in N : {}^{c_i}e_i\cdot\delta))\big)\\
= \quad & \{\text{Property 10.23, (2), (3)}\}\\
& \rho_{\mathbf{ca}(I')}\big(\lambda_m^{I'}(d\cdot(\delta \parallel (\parallel i : i\in N : {}^{c_i}e_i\cdot\delta)))\big)\\
& + \big(+J : \emptyset\subset J\subset N\\
&\qquad : \rho_{\mathbf{ca}(I')}\big(\lambda_m^{I'}\big((\mid j : j\in J : {}^{c_j}e_j)\cdot(\delta \parallel (\parallel i : i\in N\setminus J : {}^{c_i}e_i\cdot\delta) \parallel d\cdot\delta)\big)\big)\big)\\
& + \rho_{\mathbf{ca}(I')}\big(\lambda_m^{I'}((\mid i : i\in N : {}^{c_i}e_i)\cdot(\delta \parallel d\cdot\delta))\big)\\
& + \big(+J : \emptyset\subset J\subset N\\
&\qquad : \rho_{\mathbf{ca}(I')}\big(\lambda_m^{I'}\big((d \mid (\mid j : j\in J : {}^{c_j}e_j))\cdot(\delta \parallel (\parallel i : i\in N\setminus J : {}^{c_i}e_i\cdot\delta))\big)\big)\big)\\
& + \rho_{\mathbf{ca}(I')}\big(\lambda_m^{I'}\big((d \mid (\mid i : i\in N : {}^{c_i}e_i))\cdot(\delta\parallel\delta)\big)\big)\\
= \quad & \{(2), (3)\}\\
& a\cdot\rho_{\mathbf{ca}(I')}\big(\lambda_{m'}^{I'}(\delta \parallel (\parallel i : i\in N : {}^{c_i}e_i\cdot\delta))\big)\\
= \quad & \{\text{Properties 10.22 and 10.24}\}\\
& a\cdot\rho_{\mathbf{ca}(I')}\big(\lambda_{m'}^{I'}(\parallel i : i\in N : {}^{c_i}e_i\cdot\delta)\big).
\end{aligned}
$$

The next step is to show the following result:

$$
\begin{aligned}
&(\mathrm{ACP}^c_\lambda + \mathrm{RN} + \mathrm{SC})\big(AC(C, A), \uplus\big) \vdash\\
&\quad \rho_{\mathbf{ca}(I')}\big(\lambda_{m'}^{I'}(\parallel i : i\in N : {}^{c_i}e_i\cdot\delta)\big) = \rho_{\mathbf{ca}(I)}\big(\lambda_m^{I}(\parallel i : i\in N : e_i\cdot\delta)\big). \qquad (5)
\end{aligned}
$$

That is, the extra input causes in the merge quantification in term (4) do not prevent the occurrence of any action of this quantification after the execution of the action corresponding to Petri element d has added the necessary causes to marking m resulting in marking m'. Recall that $N = \{i \mid 0 \leqslant i \leqslant n\}$. It is straightforward to show the desired result if n equals zero, because under this assumption the two quantifications in (5) reduce to single action-prefix constants ${}^{c_0}e_0 \cdot \delta$ and $e_0 \cdot \delta$, respectively. For n greater than zero, the property is proven by means of (strong) natural induction on n. The proof uses the following auxiliary property, which is an immediate corollary of property (3) above. For all $J \subseteq N$ with $J \neq \emptyset$,

$$(\mathrm{ACP}^c_\lambda + \mathrm{RN} + \mathrm{SC})\big(AC(C, A), \uplus\big) \vdash \rho_{\boldsymbol{ca}(I')}\big(\lambda^{I'}_{m'}(\mid j : j \in J : {}^{c_j}e_j)\big) = \rho_{\boldsymbol{ca}(I)}\big(\lambda^{I}_{m}(\mid j : j \in J : e_j)\big). \tag{6}$$

Assume that $n > 0$. To prove property (5) in this case, the following auxiliary notations are needed. Based on Definition 10.7 (Petri element), it is possible to lift the consumption and production functions $\boldsymbol{c}_{\_}$ and $\boldsymbol{p}_{\_}$ from actions to Petri elements. For any $J \subseteq N$, $m(J) = m - (\uplus j : j \in J : \boldsymbol{c}e_j) \uplus (\uplus j : j \in J : \boldsymbol{p}e_j)$ and $m'(J) = m' - (\uplus j : j \in J : \boldsymbol{c}e_j) \uplus (\uplus j : j \in J : \boldsymbol{p}e_j) - [c_j \mid j \in J]$. Note that $m(\emptyset) = m$ and $m'(\emptyset) = m'$. Assume that, for any J with $J \subset N$, $E(J)$ denotes the following equality:

$$\rho_{\boldsymbol{ca}(I')}\big(\lambda^{I'}_{m'(J)}(\| i : i \in N \setminus J : {}^{c_i}e_i \cdot \delta)\big) = \rho_{\boldsymbol{ca}(I)}\big(\lambda^{I}_{m(J)}(\| i : i \in N \setminus J : e_i \cdot \delta)\big).$$

Clearly, property (5) conforms to

$$(\mathrm{ACP}^c_\lambda + \mathrm{RN} + \mathrm{SC})\big(AC(C, A), \uplus\big) \vdash E(\emptyset). \tag{7}$$

The basis of the inductive proof is the following property. For all $i \in N$,

$$(\mathrm{ACP}^c_\lambda + \mathrm{RN} + \mathrm{SC})\big(AC(C, A), \uplus\big) \vdash E(N \setminus \{i\}). \tag{8}$$

Property (8) is easily proven, because in each case the quantifications in the desired equality reduce to single action-prefix constants.

The induction hypothesis is as follows. For all J with $\emptyset \subset J \subset N$,

$$(\mathrm{ACP}^c_\lambda + \mathrm{RN} + \mathrm{SC})\big(AC(C, A), \uplus\big) \vdash E(J). \tag{9}$$

The following derivation shows that $(\mathrm{ACP}^c_\lambda + \mathrm{RN} + \mathrm{SC})(AC(C, A), \uplus) \vdash E(\emptyset)$:

$$\begin{aligned}
& \rho_{\boldsymbol{ca}(I')}\big(\lambda^{I'}_{m'}(\| i : i \in N : {}^{c_i}e_i \cdot \delta)\big) \\
= \quad & \{\text{Property 10.24, expansion, (3)}\} \\
& \big(+J : \emptyset \subset J \subset N : \rho_{\boldsymbol{ca}(I')}\big(\lambda^{I'}_{m'}(\mid j : j \in J : {}^{c_j}e_j)\big) \\
& \qquad \cdot \rho_{\boldsymbol{ca}(I')}\big(\lambda^{I'}_{m'(J)}\big((\| i : i \in N \setminus J : {}^{c_i}e_i) \cdot \delta\big)\big)\big) \\
& + \rho_{\boldsymbol{ca}(I')}\big(\lambda^{I'}_{m'}(\mid i : i \in N : {}^{c_i}e_i)\big) \cdot \delta
\end{aligned}$$

$$
\begin{aligned}
= & \quad \{\text{Property 10.24, (6), (9)}\} \\
& \big(+J : \emptyset \subset J \subset N : \rho_{\boldsymbol{ca}(I)}\big(\lambda^I_m(|j : j \in J : e_j)\big) \\
& \qquad\qquad\qquad \cdot \rho_{\boldsymbol{ca}(I)}\big(\lambda^I_{m(J)}\big((\|i : i \in N \setminus J : e_i) \cdot \delta\big)\big)\big) \\
& + \rho_{\boldsymbol{ca}(I)}\big(\lambda^I_m(|i : i \in N : e_i)\big) \cdot \delta \\
= & \quad \{(3)\text{, expansion, Property 10.24}\} \\
& \rho_{\boldsymbol{ca}(I)}\big(\lambda^I_m(\| \, i : i \in N : e_i \cdot \delta)\big).
\end{aligned}
$$

This derivation proves property (7) and, thus, completes the proof of property (5).

Using property (5), the following derivation can be made:

$$
\begin{aligned}
& a \cdot \rho_{\boldsymbol{ca}(I')}\big(\lambda^{I'}_{m'}(\| \, i : i \in N : {}^{c_i}e_i \cdot \delta)\big) \\
= & \quad \{(5)\} \\
& a \cdot \rho_{\boldsymbol{ca}(I)}\big(\lambda^I_m(\| \, i : i \in N : e_i \cdot \delta)\big) \\
= & \quad \{(1)\} \\
& a \cdot (s \cdot \delta) \\
= & \quad \{t \equiv a \cdot s\} \\
& t \cdot \delta.
\end{aligned}
$$

Hence, combining the results obtained so far shows that

$$
\begin{aligned}
& I' \cap causes(t) = \emptyset \wedge \\
& \quad (\mathrm{ACP}^c_\lambda + \mathrm{RN} + \mathrm{SC})\big(AC(C, A), \uplus\big) \vdash \\
& \qquad t \cdot \delta = \rho_{\boldsymbol{ca}(I')}\big(\lambda^{I'}_m\big(d \cdot \delta \parallel (\| \, i : 0 \leqslant i \leqslant n : {}^{c_i}e_i \cdot \delta)\big)\big),
\end{aligned}
$$

which completes the proof in this case.

(iv) Assume $t \equiv u + v$, for some basic terms $u, v \in \mathcal{B}(C, A)$. Since t is serializable, it follows that u and v are both serializable. By induction, it follows that there are $I_u, I_v \subseteq C$, $m_u \in \mathcal{B}(I_u)$, $m_v \in \mathcal{B}(I_v)$, $n_u, n_v \in \mathbb{N}$, and $e_{u0}, \ldots, e_{un_u}, e_{v0}, \ldots, e_{vn_v} \in PE(C, A)$ such that

$$
\begin{aligned}
& I_u \cap causes(u) = \emptyset \wedge \\
& \quad (\mathrm{ACP}^c_\lambda + \mathrm{RN} + \mathrm{SC})\big(AC(C, A), \uplus\big) \vdash \\
& \qquad u \cdot \delta = \rho_{\boldsymbol{ca}(I_u)}\big(\lambda^{I_u}_{m_u}(\| \, i : 0 \leqslant i \leqslant n_u : e_{ui} \cdot \delta)\big)
\end{aligned}
\tag{1}
$$

and

$$
\begin{aligned}
& I_v \cap causes(v) = \emptyset \wedge \\
& \quad (\mathrm{ACP}^c_\lambda + \mathrm{RN} + \mathrm{SC})\big(AC(C, A), \uplus\big) \vdash \\
& \qquad v \cdot \delta = \rho_{\boldsymbol{ca}(I_v)}\big(\lambda^{I_v}_{m_v}(\| \, j : 0 \leqslant j \leqslant n_v : e_{vj} \cdot \delta)\big).
\end{aligned}
\tag{2}
$$

Since the set of causes C is infinite, it is possible to choose I_u and I_v in such a way that $I_u \cap I_v = \emptyset$, $I_u \cap causes(v) = \emptyset$, and $I_v \cap causes(u) = \emptyset$. As a result, $(I_u \cup I_v) \cap$

$causes(t) = \emptyset$. Let $N_u = \{i \mid 0 \leqslant i \leqslant n_u\}$ and $N_v = \{j \mid 0 \leqslant j \leqslant n_v\}$. Assume that, for any $i \in N_u$ and $j \in N_v$, $c_{ij} \in C$ is a new cause not in $I_u \cup I_v \cup causes(t) \cup \{c_{kl} \mid k \in N_u \wedge l \in N_v \wedge (k \neq i \vee l \neq j)\}$. Let $I' = I_u \cup I_v \cup \{c_{ij} \mid i \in N_u \wedge j \in N_v\}$ and $m' = m_u \uplus m_v \uplus [c_{ij} \mid i \in N_u \wedge j \in N_v]$. Finally, let, for any $i \in N_u$, $d_{ui} \equiv {}^{c_{i0}}(\cdots({}^{c_{in_v}}e_{ui})\ldots)$ and, for any $j \in N_v$, $d_{vj} \equiv {}^{c_{0j}}(\cdots({}^{c_{n_u j}}e_{vj})\ldots)$. The remainder of the proof shows that $t \cdot \delta$ is equivalent to the following term:

$$\rho_{\mathbf{ca}(I')}\big(\lambda^{I'}_{m'}\big((\| i : i \in N_u : d_{ui} \cdot \delta) \parallel (\| j : j \in N_v : d_{vj} \cdot \delta)\big)\big). \tag{3}$$

Informally, the idea of the proof is as follows. New causes are added to each Petri element e_{ui} with $i \in N_u$, namely one cause c_{ij} for all $j \in N_v$, resulting in Petri element d_{ui}; similarly, for all $i \in N_u$ one cause c_{ij} with $j \in N_v$ is added to Petri element e_{vj}. This construction has two important consequences. First, no two Petri elements d_{ui} and d_{uk}, with $i, k \in N_u$ such that $i \neq k$, share any of the newly added causes. The same is true for any two Petri elements d_{vj} and d_{vl}, with $j, l \in N_v$ such that $j \neq l$. As a consequence, the newly added causes do not restrict the occurrence of actions *within* any of the two merge quantifications in term (3). Second, any two Petri elements d_{ui} and d_{vj} with $i \in N_u$ and $j \in N_v$ share exactly one input cause c_{ij}. Thus, assuming the initial marking m', it is not possible that actions of the two quantifications occur simultaneously. In addition, the occurrence of any action in one of the merge quantifications in term (3) effectively disables all actions in the other quantification, thus enforcing a true choice between the two quantifications.

The main line of the proof that term (3) equals $t \cdot \delta$ is as follows. The second step in the derivation uses four properties that are given below. These properties formalize the claims made above.

$$\begin{aligned}
& \rho_{\mathbf{ca}(I')}\big(\lambda^{I'}_{m'}\big((\| i : i \in N_u : d_{ui} \cdot \delta) \parallel (\| j : j \in N_v : d_{vj} \cdot \delta)\big)\big) \\
= \quad & \big\{\text{Axioms of } (\mathrm{ACP}^{\mathrm{c}}_{\lambda} + \mathrm{RN} + \mathrm{SC})\big(AC(C, A), \uplus\big)\big\} \\
& \rho_{\mathbf{ca}(I')}\big(\lambda^{I'}_{m'}\big((\| i : i \in N_u : d_{ui} \cdot \delta) \mathbin{⌊\!⌊} (\| j : j \in N_v : d_{vj} \cdot \delta)\big)\big) + \\
& \rho_{\mathbf{ca}(I')}\big(\lambda^{I'}_{m'}\big((\| j : j \in N_v : d_{vj} \cdot \delta) \mathbin{⌊\!⌊} (\| i : i \in N_u : d_{ui} \cdot \delta)\big)\big) + \\
& \rho_{\mathbf{ca}(I')}\big(\lambda^{I'}_{m'}\big((| i : i \in N_u : d_{ui} \cdot \delta) \mid (| j : j \in N_v : d_{vj} \cdot \delta)\big)\big) \\
= \quad & \{(6), (7), \text{Property 10.23}, (4), (5)\} \\
& \rho_{\mathbf{ca}(I_u)}\big(\lambda^{I_u}_{m_u}(\| i : i \in N_u : e_{ui} \cdot \delta)\big) + \\
& \rho_{\mathbf{ca}(I_v)}\big(\lambda^{I_v}_{m_v}(\| j : j \in N_v : e_{vj} \cdot \delta)\big) + \\
& \delta \\
= \quad & \{(1), (2)\} \\
& u \cdot \delta + v \cdot \delta \\
= \quad & \{t \equiv u + v\} \\
& t \cdot \delta.
\end{aligned}$$

Properties (4) and (5) needed in the above derivation follow immediately from the definitions given earlier. For all $K \subseteq N_u$ with $K \neq \emptyset$, there is some action $a \in AC(C, A)$ such that

$$
\begin{aligned}
&(\mathrm{ACP}^{c}_{\lambda} + \mathrm{RN} + \mathrm{SC})\big(AC(C, A), \uplus\big) \vdash \\
&\quad (| k : k \in K : e_{uk}) = a \wedge \\
&\qquad (| k : k \in K : d_{uk}) = ([c_{kj} \mid k \in K \wedge j \in N_v] \uplus \boldsymbol{ca}, \boldsymbol{sa}, \boldsymbol{pa}).
\end{aligned} \tag{4}
$$

Furthermore, for all $L \subseteq N_v$ with $L \neq \emptyset$, there is some action $b \in AC(C, A)$ such that

$$
\begin{aligned}
&(\mathrm{ACP}^{c}_{\lambda} + \mathrm{RN} + \mathrm{SC})\big(AC(C, A), \uplus\big) \vdash \\
&\quad (| l : l \in L : e_{vl}) = b \wedge \\
&\qquad (| l : l \in L : d_{vl}) = ([c_{il} \mid i \in N_u \wedge l \in L] \uplus \boldsymbol{cb}, \boldsymbol{sb}, \boldsymbol{pb}).
\end{aligned} \tag{5}
$$

The combination of properties (4) and (5) implies that, in the initial marking m', actions from the two merge quantifications in term (3) cannot occur simultaneously, which means that the third alternative in the second expression of the above derivation can be reduced to the inaction constant δ.

It remains to prove the following two properties. They formalize the claims that the newly added causes do not restrict the occurrence of actions within one of the quantifications in term (3) and that the occurrence of an action within one quantification disables any actions from the other quantification.

$$
\begin{aligned}
&(\mathrm{ACP}^{c}_{\lambda} + \mathrm{RN} + \mathrm{SC})\big(AC(C, A), \uplus\big) \vdash \\
&\quad \rho_{\boldsymbol{ca}(I')}\big(\lambda^{I'}_{m'}\big((\| i : i \in N_u : d_{ui} \cdot \delta) \,⌊\!⌊\, (\| j : j \in N_v : d_{vj} \cdot \delta)\big)\big) = \\
&\qquad \rho_{\boldsymbol{ca}(I_u)}\big(\lambda^{I_u}_{m_u}(\| i : i \in N_u : e_{ui} \cdot \delta)\big)
\end{aligned} \tag{6}
$$

and

$$
\begin{aligned}
&(\mathrm{ACP}^{c}_{\lambda} + \mathrm{RN} + \mathrm{SC})\big(AC(C, A), \uplus\big) \vdash \\
&\quad \rho_{\boldsymbol{ca}(I')}\big(\lambda^{I'}_{m'}\big((\| j : j \in N_v : d_{vj} \cdot \delta) \,⌊\!⌊\, (\| i : i \in N_u : d_{ui} \cdot \delta)\big)\big) = \\
&\qquad \rho_{\boldsymbol{ca}(I_v)}\big(\lambda^{I_v}_{m_v}(\| j : j \in N_v : e_{vj} \cdot \delta)\big).
\end{aligned} \tag{7}
$$

For reasons of symmetry, it suffices to prove only property (6). The proof is similar to the proof of property (5) in part (iii) of the proof of this theorem. The following auxiliary notation is needed. For any $K \subseteq N_u$, $m_u(K) = m_u - (\uplus k : k \in K : \boldsymbol{ce}_{uk}) \uplus (\uplus k : k \in K : \boldsymbol{pe}_{uk})$ and $m'(K) = m' - (\uplus k : k \in K : \boldsymbol{cd}_{uk}) \uplus (\uplus k : k \in K : \boldsymbol{pd}_{uk})$. Using this notation, the following auxiliary property can be formulated:

$$
\begin{aligned}
&(\mathrm{ACP}^{c}_{\lambda} + \mathrm{RN} + \mathrm{SC})\big(AC(C, A), \uplus\big) \vdash \\
&\quad \rho_{\boldsymbol{ca}(I')}\big(\lambda^{I'}_{m'(N_u)}(\| j : j \in N_v : d_{vj} \cdot \delta)\big) = \delta.
\end{aligned} \tag{8}
$$

The proof is straightforward and left to the reader.

Recall that $N_u = \{i \mid 0 \leqslant i \leqslant n_u\}$. For $n_u = 0$, property (6) follows rather directly from property (8). Let $n_u > 0$. The basis of the proof is an inductive argument. Assume that, for any $K \subset N_u$, $E(K)$ denotes the following equality.

$$\begin{aligned}&\rho_{\boldsymbol{ca}(I')}\big(\lambda^{I'}_{m'(K)}\big((\| i : i \in N_u \setminus K : d_{ui} \cdot \delta) \mathbin{⌊\!⌊} (\| j : j \in N_v : d_{vj} \cdot \delta)\big)\big) = \\ &\quad \rho_{\boldsymbol{ca}(I_u)}\big(\lambda^{I_u}_{m_u(K)}(\| i : i \in N_u \setminus K : e_{ui} \cdot \delta)\big).\end{aligned}$$

Clearly, property (6) reduces to the following property:

$$(\mathrm{ACP}^c_\lambda + \mathrm{RN} + \mathrm{SC})\big(AC(C, A), \uplus\big) \vdash E(\emptyset). \tag{9}$$

The basis of the inductive proof is the following property, which easily follows from property (8) above. For all $i \in N_u$,

$$(\mathrm{ACP}^c_\lambda + \mathrm{RN} + \mathrm{SC})\big(AC(C, A), \uplus\big) \vdash E(N_u \setminus \{i\}). \tag{10}$$

The induction hypothesis is as follows. For all $K \subset N_u$ with $K \neq \emptyset$,

$$(\mathrm{ACP}^c_\lambda + \mathrm{RN} + \mathrm{SC})\big(AC(C, A), \uplus\big) \vdash E(K). \tag{11}$$

The proof needs one more auxiliary property. For all $K \subseteq N_u$ with $K \neq \emptyset$,

$$\begin{aligned}&(\mathrm{ACP}^c_\lambda + \mathrm{RN} + \mathrm{SC})\big(AC(C, A), \uplus\big) \vdash \\ &\quad \rho_{\boldsymbol{ca}(I')}\big(\lambda^{I'}_{m'}(|k : k \in K : d_{uk})\big) = \rho_{\boldsymbol{ca}(I_u)}\big(\lambda^{I_u}_{m_u}(|k : k \in K : e_{uk})\big).\end{aligned} \tag{12}$$

Consider the following derivation:

$$\begin{aligned}
&\rho_{\boldsymbol{ca}(I')}\big(\lambda^{I'}_{m'}\big((\| i : i \in N_u : d_{ui} \cdot \delta) \mathbin{⌊\!⌊} (\| j : j \in N_v : d_{vj} \cdot \delta)\big)\big) \\
= &\quad \{\text{Property 10.24, expansion}\} \\
&\rho_{\boldsymbol{ca}(I')}\big(\lambda^{I'}_{m'}\big(\\
&\quad \big((+K : \emptyset \subset K \subset N_u : (|k : k \in K : d_{uk}) \mathbin{⌊\!⌊} (\|i : i \in N_u \setminus K : d_{ui})\big) \\
&\quad + (|i : i \in N_u : d_{ui}) \\
&\quad \big) \cdot \delta \mathbin{⌊\!⌊} (\| j : j \in N_v : d_{vj} \cdot \delta)\big)\big) \\
= &\quad \{(4), \text{Properties 10.22 and 10.24}\} \\
&\big(+K : \emptyset \subset K \subset N_u \\
&\qquad : \rho_{\boldsymbol{ca}(I')}\big(\lambda^{I'}_{m'}(|k : k \in K : d_{uk})\big) \\
&\qquad\quad \cdot \rho_{\boldsymbol{ca}(I')}\big(\lambda^{I'}_{m'(K)}\big((\|i : i \in N_u \setminus K : d_{ui} \cdot \delta) \mathbin{⌊\!⌊} (\| j : j \in N_v : d_{vj} \cdot \delta)\big)\big)\big) \\
&\quad + \rho_{\boldsymbol{ca}(I')}\big(\lambda^{I'}_{m'}(|i : i \in N_u : d_{ui})\big) \cdot \rho_{\boldsymbol{ca}(I')}\big(\lambda^{I'}_{m'(N_u)}(\| j : j \in N_v : d_{vj} \cdot \delta)\big) \\
= &\quad \{(12), (11), (8)\} \\
&\big(+K : \emptyset \subset K \subset N_u
\end{aligned}$$

$$: \rho_{\mathbf{ca}(I_u)}\big(\lambda^{I_u}_{m_u}(|k : k \in K : e_{uk})\big) \cdot \rho_{\mathbf{ca}(I_u)}\big(\lambda^{I_u}_{m_u(K)}(\|i : i \in N_u \setminus K : e_{ui} \cdot \delta)\big)\big)$$
$$+ \rho_{\mathbf{ca}(I_u)}\big(\lambda^{I_u}_{m_u}(|i : i \in N_u : e_{ui})\big) \cdot \delta$$
$$= \quad \{\text{Property 10.24, (4), expansion}\}$$
$$\rho_{\mathbf{ca}(I_u)}\big(\lambda^{I_u}_{m_u}(\| i : i \in N_u : e_{ui} \cdot \delta)\big).$$

This derivation proves property (9) and, hence, property (6) for the case that $n_u > 0$. As mentioned, property (7) can be proven by means of a symmetrical argument.

Finally, it suffices to combine the results to prove that

$$I' \cap causes(t) = \emptyset \wedge$$
$$(\mathrm{ACP}^c_\lambda + \mathrm{RN} + \mathrm{SC})\big(AC(C, A), \uplus\big) \vdash$$
$$t \cdot \delta = \rho_{\mathbf{ca}(I')}\big(\lambda^{I'}_{m'}\big((\| i : i \in N_u : d_{ui} \cdot \delta) \parallel (\| j : j \in N_v : d_{vj} \cdot \delta)\big)\big),$$

which completes the proof of the theorem. □

10.4. *Concluding remarks*

In this section, it has been shown how to obtain a (branching-time) non-interleaving, partial-order process algebra that incorporates several important characteristics of the Petri-net formalism of Section 4. The algebra consists of non-terminating serializable processes and contains two causality mechanisms, one based on sequential composition and communication and one based on the causal state operator and cause addition. It has been shown that all non-terminating serializable processes with bounded behavior can be specified using only the latter causality mechanism (Theorem 10.25). That is, each such process has an algebraic specification in the style of Petri-net theory.

It is interesting to look briefly at the generalization of Theorem 10.25 to a setting with a simple form of recursion, namely iteration. It appears that it is possible to generalize Theorem 10.25 to some extent. This generalization is inspired by Definition 6.8 (Algebraic semantics for labeled P/T nets) and Theorem 6.10. Definition 6.8 gives an algebraic expression for any labeled P/T net. Theorem 6.10 proves that the (step) semantics of a labeled P/T net and its algebraic semantics are identical. It is interesting to observe the similarity between the algebraic term in Definition 6.8 and the canonical term in Theorem 10.25. Assuming that cause-addition operators are used to make causes explicit in the expression of Definition 6.8, the only difference between the two terms is that the action-prefix constants in the term in Theorem 10.25 are replaced by prefix-iterations in the term in Definition 6.8. It is our claim that any non-terminating serializable process in set $\mathcal{C}_{\mathrm{ns}}(C, A)$ of Section 10.1 is step bisimilar to a canonical term consisting of a parallel composition of prefix-iterations and restricted by means of a causal state operator where causes added to enforce orderings are hidden. However, it is still an open problem whether any closed term in $\mathcal{C}_{\mathrm{ns}}(C, A)$ is also derivably equivalent to such a term in canonical form, which is a more general result. Recall that Theorem 10.25 is proven by means of induction on the structure of basic terms. This technique does not carry over to a setting with iteration.

A few final remarks are in order. First, it appears to be possible to prove results very similar to Theorem 10.25 for theories concerning processes with bounded behavior that are not necessarily non-terminating and serializable. The interested reader is referred to [4] for more details. Second, it remains an interesting topic for future study to investigate to what extent Theorem 10.25 can be generalized to a setting with linear or even general recursion. Third, an interesting topic for future research is to study axiomatizations of non-interleaving, partial-order process algebras in the style proposed in this section.

11. Conclusions

Concluding remarks. In this chapter, we have studied two major themes. The first theme is the development of a partial-order ACP-style algebraic theory. It has been shown that the standard ACP-style algebraic framework is sufficiently flexible to develop a partial-order theory based on the semantic equivalence of step bisimilarity. The communication mechanism of ACP can be used to specify which actions are causally independent and can, thus, occur concurrently. In addition, the notions of a (non-)interleaving equational theory and a (non-)interleaving process algebra have been formalized. It has been argued that the characterizations interleaving versus non-interleaving and total-order versus partial-order for process-algebraic theories are orthogonal. In particular, it has been shown that the partial-order theory developed in Section 5 is an interleaving theory.

The second major theme is the study of concepts known from Petri-net theory, one of the most well-known approaches to modeling and analyzing concurrent systems by means of partial orders, in a process-algebraic framework. In particular, the causality mechanism of Petri-net theory has been adopted in the partial-order framework of Section 5. The resulting algebraic theory allows for specifications in the style of Petri-net theory. It has been shown that the theory can be used to reason in a purely equational way about the equivalence of labeled P/T nets. In addition, the relation between the causality mechanism adopted from Petri-net theory and the standard algebraic causality mechanism has been investigated. It has been shown that the Petri-net mechanism can replace the standard algebraic mechanism to some extent.

The main contribution of this chapter is that it improves our understanding of some of the most important concepts that play a role in describing and analyzing the behavior of concurrent systems, namely communication and causality.

Related work. The literature on concurrency theory contains several approaches to combining some Petri-net formalism with some process-algebraic theory. One line of research is concerned with the translation of process-algebraic terms into Petri nets. The aim of such an approach is to provide the algebraic theory via a Petri-net semantics with a partial-order semantics. The most well-known example of this approach is the Petri Box Calculus of [15] and [16], Chapter 14 of this Handbook. Other examples of this line of research are [19,21,30,31,46,50,60]. In this chapter, the converse approach is pursued. In particular, in Section 6.3, a translation from labeled P/T nets into algebraic terms is given. Other examples of this approach are [20,22]. The main difference between the approach taken in this chapter and other approaches is that this chapter emphasizes equational reasoning, whereas other approaches often emphasize the semantic framework.

Some other authors have taken an approach to modeling and analyzing concurrent systems based on an explicit representation of concurrent-system behavior in terms of partial orders. Examples of such frameworks can be found in [26,32,38,55]. Other well-known partial-order based theories are trace theory as described in [42] and the theory of event structures presented in [48,61].

Future research. Several topics for future study have already been mentioned earlier in this chapter. At this point, we mention two other general directions for future research. First, it is interesting to study the topics addressed in this chapter in a framework based on some other partial-order semantics. There are partial-order semantics that capture causalities more accurately than the step semantics used in this chapter (see, for example, [54], for more details). Second, it would be interesting to investigate the application of the concepts studied in this chapter in the development and analysis of complex concurrent systems. Such applications require the extension of the framework developed in this chapter with, at least, an abstraction mechanism and a way to reason about data. Abstraction can be included following the approach of [8, Chapter 3]. Extensions of Petri-net theory with data can be found in, for example, [35,39]. An extension of ACP-style process algebra with data is the theory μCRL described in [33] and [34], Chapter 17 of this Handbook.

References

[1] L. Aceto, W.J. Fokkink and C. Verhoef, *Structural operational semantics*, Handbook of Process Algebra, J.A. Bergstra, A. Ponse and S.A. Smolka, eds, Elsevier, Amsterdam (2001), 197–292.

[2] J.C.M. Baeten, *The total order assumption*, First North American Process Algebra Workshop, Proceedings, Workshops in Computing, Stony Brook, New York, USA, August 1992, S. Purushothaman and A. Zwarico, eds, Springer, Berlin (1993), 231–240.

[3] J.C.M. Baeten and J.A. Bergstra, *Global renaming operators in concrete process algebra*, Inform. and Comput. **78** (3) (1988), 205–245.

[4] J.C.M. Baeten and J.A. Bergstra, *Non interleaving process algebra*, CONCUR '93, Proc. 4th International Conference on Concurrency Theory, Proceedings, Hildesheim, Germany, August 1993, Lecture Notes in Comput. Sci. 715, E. Best, ed., Springer, Berlin (1993), 308–323.

[5] J.C.M. Baeten and C. Verhoef, *Concrete process algebra*, Handbook of Logic in Computer Science, Vol. 4, Semantic Modelling, S. Abramsky, D.M. Gabbay and T.S.E. Maibaum, eds, Oxford University Press, Oxford (1995), 149–268.

[6] J.C.M. Baeten and W.P. Weijland, *Process Algebra*, Cambridge Tracts in Theoret. Comput. Sci. 18, Cambridge University Press, Cambridge (1990).

[7] Bakkenist Management Consultants, *ExSpect 6 user manual* (1997). Information available at http://www.exspect.com/.

[8] T. Basten, *In terms of nets: System design with Petri nets and process algebra*, Ph.D. Thesis, Eindhoven University of Technology, Department of Mathematics and Computing Science, Eindhoven, The Netherlands (December 1998).

[9] T. Basten and M. Voorhoeve, *An algebraic semantics for hierarchical P/T nets*, Computing Science Report 95/35, Eindhoven University of Technology, Department of Mathematics and Computing Science, Eindhoven, The Netherlands (December 1995). An extended abstract appeared as [10].

[10] T. Basten and M. Voorhoeve, *An algebraic semantics for hierarchical P/T nets (extended abstract)*, Application and Theory of Petri Nets 1995, Proc. 16th International Conference, Torino, Italy, June 1995, Lecture Notes in Comput. Sci. 935, G. De Michelis and M. Diaz, eds, Springer, Berlin (1995), 45–65.

[11] J.A. Bergstra, I. Bethke and A. Ponse, *Process algebra with iteration and nesting*, Comput. J. **37** (4) (1994), 241–258.

[12] J.A. Bergstra, W.J. Fokkink and A. Ponse, *Process algebra with recursive operations*, Handbook of Process Algebra, J.A. Bergstra, A. Ponse and S.A. Smolka, eds, Elsevier, Amsterdam (2001), 333–389.

[13] J.A. Bergstra and J.W. Klop, *Process algebra for synchronous communication*, Inform. and Control **60** (1–3) (1984), 109–137.

[14] J.A. Bergstra and J.W. Klop, *Algebra of communicating processes with abstraction*, Theoret. Comput. Sci. **37** (1) (1985), 77–121.

[15] E. Best, R. Devillers and J.G. Hall, *The box calculus: A new causal algebra with multi-label communication*, Advances in Petri Nets 1992, Lecture Notes in Comput. Sci. 609, G. Rozenberg, ed., Springer, Berlin (1992), 21–69.

[16] E. Best, R. Devillers and M. Koutny, *A unified model for nets and process algebras*, Handbook of Process Algebra, J.A. Bergstra, A. Ponse and S.A. Smolka, eds, Elsevier, Amsterdam (2001), 873–944.

[17] E. Best and C. Fernández, *Nonsequential processes: A Petri net view*, EATCS Monographs on Theoret. Comput. Sci. 13, Springer, Berlin (1988).

[18] E. Best and B. Grahlmann, *PEP: Documentation and user guide*, Universität Hildesheim, Institut für Informatik (1995). Information available at http://theoretica.informatik.uni-oldenburg.de/~pep.

[19] G. Boudol and I. Castellani, *Flow models of distributed computations: Three equivalent semantics for CCS*, Inform. and Comput. **114** (2) (1994), 247–314.

[20] G. Boudol, G. Roucairol and R. de Simone, *Petri nets and algebraic calculi of processes*, Advances in Petri Nets 1985, Lecture Notes in Comput. Sci. 222 , G. Rozenberg, ed., Springer, Berlin (1985), 41–58.

[21] P. Degano, R. De Nicola and U. Montanari, *A distributed operational semantics for CCS based on condition/event systems*, Acta Informatica **26** (1/2) (1988), 59–91.

[22] C. Dietz and G. Schreiber, *A term representation of P/T systems*, Application and Theory of Petri Nets 1994, Proc. 15th International Conference, Lecture Notes in Comput. Sci. 815, Zaragoza, Spain, June 1994, R. Valette, ed., Springer, Berlin (1994), 239–257.

[23] E.W. Dijkstra and C.S. Scholten, *Predicate Calculus and Program Semantics*, Springer, Berlin (1990).

[24] W.J. Fokkink, *A complete equational axiomatization for prefix iteration*, Inform. Proc. Lett. **52** (6) (1994), 333–337.

[25] W.J. Fokkink, *Introduction to Process Algebra*, Texts in Theoret. Comput. Sci.: An EATCS Series, Springer, Berlin (2000).

[26] J.L. Gischer, *The equational theory of pomsets*, Theoret. Comput. Sci. **61** (2,3) (1988), 199–224.

[27] R.J. van Glabbeek, *The linear time – branching time spectrum I. The semantics of concrete, sequential processes*, Handbook of Process Algebra, J.A. Bergstra, A. Ponse and S.A. Smolka, eds, Elsevier, Amsterdam (2001), 3–99.

[28] R.J. van Glabbeek, *The linear time – branching time spectrum*, CONCUR '90, Proc. Theories of Concurrency: Unification and Extension, Amsterdam, The Netherlands, August 1990, Lecture Notes in Comput. Sci. 458, J.C.M. Baeten and J.W. Klop, eds, Springer, Berlin (1990), 278–297.

[29] R.J. van Glabbeek, *The linear time – branching time spectrum II: The semantics of sequential systems with silent moves (extended abstract)*, CONCUR '93, Proc. 4th International Conference on Concurrency Theory, Hildesheim, Germany, August 1993, Lecture Notes in Comput. Sci. 715, E. Best, ed., Springer, Berlin (1993), 66–81.

[30] R.J. van Glabbeek and F.W. Vaandrager, *Petri net models for algebraic theories of concurrency*, PARLE Parallel Architectures and Languages Europe, Vol. II, Parallel Architectures, Eindhoven, The Netherlands, June 1987, Lecture Notes in Comput. Sci. 259, J.W. de Bakker, A.J. Nijman and P.C. Treleaven, eds, Springer, Berlin (1987), 224–242.

[31] U. Goltz, *CCS and Petri nets*, Semantics of Systems of Concurrent Processes, Proc. LITP Spring School on Theoretical Computer Science, La Roche Posay, France, April 1990, Lecture Notes in Comput. Sci. 469, I. Guessarian, ed., Springer, Berlin (1990), 334–357.

[32] J. Grabowski, *On partial languages*, Fundamenta Informaticae **4** (2) (1981), 427–498.

[33] J.F. Groote and A. Ponse, *The syntax and semantics of μCRL*, Algebra of Communicating Processes 1994, Workshops in Computing, Utrecht, The Netherlands, May 1994, A. Ponse, C. Verhoef and S.F.M. van Vlijmen, eds, Springer, Berlin (1995), 26–62.

[34] J.F. Groote and M.A. Reniers, *Algebraic process verification*, Handbook of Process Algebra, J.A. Bergstra, A. Ponse and S.A. Smolka, eds, Elsevier, Amsterdam (2001), 1151–1208.

[35] K.M. van Hee, *Information Systems Engineering: A Formal Approach*, Cambridge University Press, Cambridge (1994).
[36] C.A.R. Hoare, *Communicating Sequential Processes*, Prentice-Hall International, London (1985).
[37] T.W. Hungerford, *Algebra*, Holt, Rinehart and Winston Inc., New York (1974).
[38] W.P.M. Janssen, *Layered design of parallel systems*, Ph.D. Thesis, University of Twente, Department of Computer Science, Enschede (1994).
[39] K. Jensen, *Coloured Petri Nets. Basic Concepts, Analysis Methods and Practical Use, Vol 1, Basic Concepts*, EATCS Monographs on Theoret. Comput. Sci., Springer, Berlin (1992).
[40] K. Jensen, S. Christensen, P. Huber and M. Holla, *Design/CPN: A reference manual*, Meta Software Corporation (1991). Information available at http://www.daimi.au.dk/designCPN.
[41] S.C. Kleene, *Representation of events in nerve nets and finite automata*, Automata Studies, Annals of Mathematics Studies 34, C.E. Shannon and J. McCarthy, eds, Princeton University Press, Princeton, NJ (1956), 3–41.
[42] A. Mazurkiewicz, *Trace theory*, Petri Nets: Applications and Relationships to Other Models of Concurrency, Advances in Petri Nets 1986, Part II, Proceedings of an Advanced Course, Lecture Notes in Comput. Sci. 255, Bad Honnef, Germany, September 1986, W. Brauer, W. Reisig and G. Rozenberg, eds, Springer, Berlin (1987), 279–324.
[43] A.R. Meyer, *Report on the 5th International Workshop on the Semantics of Programming Languages in Bad Honnef*, Bulletin of the EATCS, No. 27, European Association for Theoret. Comput. Sci. (1985), 83–84.
[44] R. Milner, *A Calculus of Communicating Systems*, Lecture Notes in Comput. Sci. 92, Springer, Berlin (1980).
[45] R. Milner, *Communication and Concurrency*, Prentice-Hall International, London (1989).
[46] U. Montanari and D. Yankelevich, *Combining CCS and Petri nets via structural axioms*, Fundamenta Informaticae **20** (1–3) (1994), 193–229.
[47] T. Murata, *Petri nets: Properties, analysis and applications*, Proc. IEEE **77** (4) (1989), 541–580.
[48] M. Nielsen, G. Plotkin and G. Winskel, *Petri nets, event structures and domains, part I*, Theoret. Comput. Sci. **13** (1) (1981), 85–108.
[49] M. Nielsen and P.S. Thiagarajan, *Degrees of non-determinism and concurrency: A Petri net view*, Foundations of Software Technology and Theoretical Computer Science, Proc. 4th Conference, FST&TCS4, Bangalore, India, December 1984, Lecture Notes in Comput. Sci. 181, M. Joseph and R. Shyamasundar, eds, Springer, Berlin (1984), 89–117.
[50] E.-R. Olderog, *Nets, Terms and Formulas*, Cambridge Tracts in Theoret. Comput. Sci. 23, Cambridge University Press, Cambridge (1991).
[51] D. Park, *Concurrency and automata on infinite sequences*, Theoret. Comput. Sci., Proc. 5th GI-Conference, Karlsruhe, Germany, March 1981, Lecture Notes in Comput. Sci. 104, P. Deussen, ed., Springer, Berlin (1981), 167–183.
[52] J.L. Peterson, *Petri Net Theory and the Modeling of Systems*, Prentice-Hall, Englewood Cliffs, NJ (1981).
[53] C.A. Petri, *Kommunikation mit Automaten*, Ph.D. Thesis, Institut für instrumentelle Mathematik, Bonn, Germany (1962). In German.
[54] L. Pomello, G. Rozenberg and C. Simone, *A survey of equivalence notions for net based systems*, Advances in Petri Nets 1992, Lecture Notes in Comput. Sci. 609, G. Rozenberg, ed., Springer, Berlin (1992), 410–472.
[55] V.R. Pratt, *Modeling concurrency with partial orders*, Internat. J. Parallel Programming **15** (1) (1986), 33–71.
[56] W. Reisig, *Petri Nets: An Introduction*, EATCS Monographs on Theoret. Comput. Sci. 4, Springer, Berlin (1985).
[57] W. Reisig and G. Rozenberg, eds, *Lectures on Petri Nets I: Basic Models*, Advances in Petri Nets, Lecture Notes in Comput. Sci. 1491, Springer, Berlin (1998).
[58] W. Reisig and G. Rozenberg, eds, *Lectures on Petri Nets II: Applications*, Advances in Petri Nets, Lecture Notes in Comput. Sci. 1492, Springer, Berlin (1998).
[59] P. Sewell, *Nonaxiomatisability of equivalences over finite state processes*, Ann. Pure Appl. Logic **90** (1–3) (1997), 163–191.
[60] D. Taubner, *Representing CCS programs by finite predicate/transition nets*, Acta Inform. **27** (6) (1990), 533–565.

[61] G. Winskel, *An introduction to event structures*, Linear Time, Branching Time and Partial Order in Logics and Models for Concurrency, Noordwijkerhout, The Netherlands, May/June 1988, Lecture Notes in Comput. Sci. 354, J.W. de Bakker, W.P. de Roever and G. Rozenberg, eds, Springer, Berlin (1989), 364–397.

Subject index

ACP, 783, 784
ACP + RN, 786
ACP^* + RN, 786
ACP^* + RN + RSP^*, 787
ACP^* + RN + RSP^* + SC, 788, 852
ACP^c_λ + RN, 836
ACP^c_λ + RN + SC, 853
ACP^*_λ + RN + RSP^* + SC, 810
ACP^{*c}_λ + RN + RSP^* + SC, 831, 850
action prefix, 836, 846, 854
action-prefix constants, 854
actions, 774, 775
– atomic, 775
– multi-, 775
algebra, 793, 797
– of non-terminating serializable processes, 843, 845, 849–851, 854
– of non-terminating serializable processes with bounded behavior, 854
algebraic semantics
– for modular P/T nets, 822, 825–827, 829, 830
– for P/T nets, 815–817, 833, 865
atomic actions, 775
axiomatization
– complete, 797
– sound, 793

bags, 773
basic terms, 836–838, 854
binary Kleene star, 786, 846
bisimilarity, 772, 777
– step, 772, 778
bisimulation, 777
branching-time semantics, 771, 778

causal state operator, 809–815, 853
causality, 771, 809, 831, 842, 843, 852, 853, 865, 866
cause abstraction, 812, 820
cause addition, 831–837, 839, 853
– input-, 832
– output-, 832, 846
causes, 810, 812
– external, 810, 811
– internal, 810, 811
causes function, 838, 839

choice, 784, 853
communication, 771, 866
– between modular P/T nets, 823, 824, 828, 829
– between structured actions, 818
– explicit, 807
– in ACP, 785, 802, 809, 813, 814, 829, 852
– in non-interleaving process algebra, 852, 853
– in a partial-order framework, 804, 807, 808
communication merge, 785, 842, 846
comparative concurrency semantics, 771, 772
completeness
– of ACP^* + RN + RSP^* + SC, 797
– of ACP^{*c}_λ + RN + RSP^* + SC, 835
compositionality, 820, 827
concurrency, 771, 776, 778
– in P/T nets, 781
concurrent systems, 771
conflicts, 805–808
consumption, 780, 810
– of causes, 810, 812

deadlock, 775, 842
– in ACP, 784
– of Petri nets, 781
derivability, 785

elimination, 837, 838, 855
– of choice, sequential composition, and communication, 853, 857
– of sequential composition, 839
equational theory, 783
– interleaving, 790
expansion, 772, 788, 792
– of iterations, 811

firing rule, 780
– step, 782
– total-order, 780

handshaking axiom, 789
head normal form, 790, 798–800, 836, 843, 849

interleaving process algebra, 771, 772, 797–799
interleaving semantics, 771, 772, 800
interleaving theory, 772, 789, 790, 792, 800

interleavings, 792
– partially-ordered, 806
– totally-ordered, 806
iteration, 786, 865

labeled P/T nets, 779
labeled transition systems, 774
linear-time semantics, 771

markings, 780, 810
merge, 771, 784
– communication, 785, 842, 846
model, 793
– of $\mathrm{ACP}^* + \mathrm{RN} + RSP^* + \mathrm{SC}$, 794, 795, 798, 801, 804, 852
– of $\mathrm{ACP}_\lambda^c + \mathrm{RN} + \mathrm{SC}$, 853
– of $\mathrm{ACP}_\lambda^* + \mathrm{RN} + RSP^* + \mathrm{SC}$, 814
– of $\mathrm{ACP}_\lambda^{*c} + \mathrm{RN} + RSP^* + \mathrm{SC}$, 832, 850
modular P/T nets, 820, 824, 828, 830
multi-actions, 775
multi-sets, 773

non-determinism, 771, 776, 778
non-interleaving process algebra, 771, 772, 842, 849, 851, 852, 865
non-interleaving semantics, 772
non-interleaving theory, 772

P/T nets, 779, 815, 844
– marked, 780
– modular, 820, 824, 828, 830
parallel composition, 771, 784, 837
partial-order process algebra, 771, 772
partial-order semantics, 771, 772, 775
– for a process-algebraic theory, 772, 866, 867
– for P/T nets, 782
Petri elements, 846
Petri nets, 771, 772, 778, 866
Petri-net semantics, 781
place fusion, 820, 824, 825, 831
prefix iteration, 846
process algebra, 793, 797
– interleaving, 771, 772, 797–799
– interleaving partial-order, 808
– non-interleaving, 771, 772, 842, 849, 851, 852, 865
– of non-terminating serializable processes, 843, 849
– partial-order, 771, 772
– total-order, 772
process space, 775
processes, 771, 775, 794
– non-terminating, 843, 844, 846, 855
– non-terminating serializable, 844, 845, 847, 849, 865
– non-terminating serializable, with bounded behavior, 853, 865
– serializable, 844, 855
– terminating, 855
– with bounded behavior, 836
production, 780, 810
– of causes, 810, 812

quantifiers, 774

Recursive Specification Principle
– for the binary Kleene star (RSP^*), 787
renaming, 786, 846
– of structured actions, 818, 846

semantic equivalence, 771
semantics, 771
– for $\mathrm{ACP}^* + \mathrm{RN} + RSP^* + \mathrm{SC}$, 793
– for $\mathrm{ACP}_\lambda^* + \mathrm{RN} + RSP^* + \mathrm{SC}$, 814
– for $\mathrm{ACP}_\lambda^{*c} + \mathrm{RN} + RSP^* + \mathrm{SC}$, 832
– branching-time, 771, 778
– for modular P/T nets, 822, 825, 829
– for P/T nets, 781, 782, 815, 816
– interleaving, 771, 772, 800
– linear-time, 771
– non-interleaving, 772
– partial-order, 771, 772, 775, 866, 867
– step, 775, 776, 778
– total-order, 771, 772, 775, 776
– true-concurrency, 771
sequential composition, 784, 809, 831, 834–837, 839, 852
– in non-interleaving process algebra, 852, 853
– of non-terminating processes, 846
serializability, 843, 844
soundness
– of $\mathrm{ACP}^* + \mathrm{RN} + RSP^* + \mathrm{SC}$, 795
– of $\mathrm{ACP}_\lambda^* + \mathrm{RN} + RSP^* + \mathrm{SC}$, 815
– of $\mathrm{ACP}_\lambda^{*c} + \mathrm{RN} + RSP^* + \mathrm{SC}$, 833
standard concurrency, 788, 793
state operator, 809
– causal, 809–814
step bisimilarity, 772, 778
step semantics, 775, 776, 778
– for $\mathrm{ACP}^* + \mathrm{RN} + RSP^* + \mathrm{SC}$, 804
– for $\mathrm{ACP}_\lambda^* + \mathrm{RN} + RSP^* + \mathrm{SC}$, 815
– for modular P/T nets, 830
– for P/T nets, 782, 815, 817
steps, 775, 843
– of algebraic terms, 806, 810
– of P/T nets, 782
structural induction, 838

successful termination, 775, 842
– in process algebra, 794
– of Petri nets, 781
synchronization
– in ACP, 785, 789

tick, 794
tokens, 779, 809, 810, 812
total-order process algebra, 772
total-order semantics, 771, 772, 775, 776
– for $\mathrm{ACP}^* + \mathrm{RN} + RSP^* + \mathrm{SC}$, 801
– for $\mathrm{ACP}^*_\lambda + \mathrm{RN} + RSP^* + \mathrm{SC}$, 815
– for P/T nets, 781, 817
transition fusion, 828–830
transition systems, 774
true concurrency, 771

validity, 793

CHAPTER 14

A Unified Model for Nets and Process Algebras

Eike Best[1], Raymond Devillers[2], Maciej Koutny[3]

[1] *Fachbereich Informatik, Carl von Ossietzky Universität, D-26111 Oldenburg, Germany*
E-mail: Eike.Best@informatik.uni-oldenburg.de

[2] *Département d'Informatique, Université Libre de Bruxelles, B-1050 Bruxelles, Belgium*
E-mail: rdevil@ulb.ac.be

[3] *Department of Computing Science, University of Newcastle, Newcastle upon Tyne NE1 7RU, UK*
E-mail: Maciej.Koutny@ncl.ac.uk

Contents

1. Introduction . . . 875
2. Informal introduction to nets, CCS and PBC . . . 877
   2.1. Structure and behaviour . . . 877
   2.2. CCS and PBC . . . 880
3. Elements of an algebra of Petri nets . . . 884
   3.1. Place and transition labellings . . . 884
   3.2. Labelled nets and their semantics . . . 885
   3.3. Equivalence notions . . . 889
   3.4. Plain boxes . . . 892
4. Net refinement . . . 893
   4.1. Operator boxes . . . 894
   4.2. Place and transition names of operator and plain boxes . . . 894
   4.3. Formal definition of net refinement $\Omega(\mathbf{\Sigma})$ . . . 896
5. Recursive equations on boxes . . . 898
   5.1. Solving a recursive equation . . . 899
   5.2. Further recursion examples . . . 903
6. An algebra of nets . . . 905
   6.1. Sos-operator boxes . . . 906
   6.2. Structural operational semantics of composite boxes . . . 908
   6.3. Further restrictions on the application domain . . . 910
7. A generic process algebra and its two consistent semantics . . . 912
   7.1. Syntax . . . 912
   7.2. Denotational semantics . . . 914
   7.3. Structural operational semantics . . . 916
   7.4. Partial order semantics of box expressions . . . 920
   7.5. Consistency results . . . 921

HANDBOOK OF PROCESS ALGEBRA
Edited by Jan A. Bergstra, Alban Ponse and Scott A. Smolka

8. Examples and applications 923
8.1. PBC 923
8.2. The two semantics of PBC 926
8.3. Equivalence properties 928
8.4. Some applications of SOS rules 930
8.5. CCS, TCSP and COSY 933
8.6. Modelling a concurrent programming language 935
9. Conclusions 937
Acknowledgements 939
References 939
Subject index 941

Abstract

This chapter addresses a range of issues that arise when process algebras and Petri nets are combined; in particular, it focusses on compositionality of structure and behaviour, on refinement, and on equivalence notions. A generic algebra of nets and process expressions is defined and equipped with two types of semantics: a Petri net semantics based on step sequences and causal partial orders, and a structural operational semantics based on a system of derivation rules. The main result states that these two semantics are equivalent. A concrete example of this algebraic framework is the Petri Box Calculus (PBC) which is used to convey the basic ideas contained in this chapter.

1. Introduction

Motivation and goals of this chapter. This chapter is about combining and unifying two well studied theories of concurrency: *process algebras* and *Petri nets*. Among the advantages of process algebras, we may identify the following:

- They allow the study of connectives directly related to real programming languages.
- They are compositional by definition, i.e., they enable one to compose larger systems from smaller ones in a structured way.
- They come with a variety of concomitant and derived logics which may facilitate reasoning about important properties of systems.
- They support a variety of algebraic laws which may be used to manipulate systems, either to refine them or to prove them correct with respect to some specification.

The advantages of Petri nets, on the other hand, include the following:

- They sharply distinguish between states and activities (the latter being defined as changes of state), through the distinction made between places (local states) and transitions (local activities).
- Global states and global activities are not basic notions, but are derived from their local counterparts, which, in particular, suits the description of concurrent behaviour by partial orders.
- While being formal, Petri nets also come with a graphical representation which is easy to comprehend and has therefore a wide appeal for practitioners.
- By their representation as bipartite graphs, Petri nets have useful links both to graph theory and to linear algebra, which can be exploited for the verification of systems.

By combining the two models and making this combination as strong and uniform as possible, one may hope to gain added value by retaining their respective advantages, which is, in a nutshell, the goal of this chapter.

The syntax of a process algebra may be viewed as consisting of a set of *process constants* (basic building blocks), a set of *process variables* (useful for expressing hierarchy and recursion) and a set of *process operators* (useful for composing complicated process expressions out of simpler ones). Such a syntax is usually accompanied by a *structural operational semantics* (SOS [45]) specifying, by syntactic induction, the possible executions of a (composite) expression.

More specifically, then, the main goal of this chapter is to impose on the class of (safe) Petri nets an algebraic structure which conforms with the process-algebraic framework just sketched. We will achieve this to the following extent:

- A class of safe labelled Petri nets will be identified (called *plain boxes*) which correspond to (closed) process expressions, and in particular, to process constants (*constant boxes*).
- Process variables correspond one-to-one to net variables.
- Another class of safe labelled Petri nets will be identified (called *operator boxes*) which correspond to process operators in the following way: the arity n of an operator op equals the number of transitions of the corresponding operator box Ω, and the operation of applying op to n arguments corresponds to simultaneously refining the transitions of Ω by n (plain) boxes corresponding to the arguments.

– Under some general circumstances, the two semantics linked in this way (namely, the net-independent SOS rules and the usual net-indigenous semantics) are not just comparable but even fully isomorphic; that is, process expressions and their SOS rules lead to transition systems which are isomorphic to the ones obtained from their Petri net equivalents. We will show that in order to achieve this isomorphism, a property called *factorizability* is instrumental. The class of operator boxes satisfying this property will be called *sos-operator boxes*. All the usual concrete process-algebraic operators are factorizable.

Thus, the process algebra we will ultimately be interested in – the box algebra – is *generic*, in the sense that any operator (not just the usual concrete ones) that can be described by an sos-operator box is guaranteed to have SOS rules which are equivalent with its Petri net semantics.

A historical perspective. This chapter is in the tradition of a long line of research with the theme of giving Petri net semantics to process algebras. This idea was presumably first expressed for path expressions in 1974 [38], and in this context has eventually led to the development of COSY [33]. In our opinion, COSY does qualify as a process algebra in the above sense, though it uses iteration (the unary version of the 'Kleene star') instead of recursion.

In other seminal work, process algebras whose definition has been given independently of any Petri net semantics, have often been provided with translations into nets. In [27], Goltz and Mycroft were the first to translate Milner's *Calculus of Communicating Systems* (CCS [41,42]) into Petri nets. Other translations from process algebras to Petri nets include [6,14,19,25,26,44,48] and many other works. A more complete historical review can be found elsewhere in this Handbook [17]. Also, in [2], a converse approach can be found, namely that of translating elements from Petri nets into a process algebra.

A specific point of departure for the work described in the present chapter has been a concrete process algebra called the *Petri Box Calculus* (PBC [9]). This is a variant of CCS designed with the twin aims of allowing a compositional Petri net semantics [10,16,20–22,30,31,34,35,37], and providing a formal underpinning of a concurrent programming notation [7,12,39].

Structure of this chapter. Before being able to formalize the desired equivalence between process algebras and nets, we need to sort out some issues about static structure and behaviour in both models. This is explained in Section 2, where also some other distinctions between CCS and PBC are listed.

In Section 3, a Petri net framework is presented. This section also contains the definition of plain boxes as the basic elements of a suitable Petri net algebra. A general notion of net refinement, which meets the requirements stated above, is defined in Section 4. This section also contains the definition of operator boxes as the Petri net counterparts of process-algebraic operators. Section 5 describes a fixpoint theory developed in order to solve equations on nets (using net refinement). In Section 6, we identify conditions under which operator application does not lead outside the domain of the algebra, and under which an equivalence between SOS rules and net semantics is possible. This section also contains the definition of sos-operator boxes.

In Section 7, we define the *box algebra*, a generic process-algebraic syntax based on constant boxes, box variables, and sos-operator boxes. We give it both indigenous SOS rules and a box semantics and state the main isomorphism results. In Section 8, finally, we show that this generic algebra has several concrete incarnations, including the PBC, CCS, TCSP [32] and COSY. Section 8 also contains a brief explanation why the box algebra can readily yield a compositional semantics for an imperative concurrent programming language.

In this chapter, all proofs are omitted. The reader is referred to [11] where they can be found. Moreover, the theory presented in [11] is more general, allowing, for instance, infinitary operators, whereas here we restrict our attention to finitary ones.

2. Informal introduction to nets, CCS and PBC

2.1. *Structure and behaviour*

An example. The graph of a Petri net N describes the structure of a dynamic system. The *behaviour* of such a system is defined with respect to a given starting marking (global state) of the graph, called the *initial marking*. In general, a *marking* M of N is a function from the set of places of N (local states) to the set of natural numbers $\mathbf{N} = \{0, 1, 2, \ldots\}$; if $M(s) = n$ then we say that s contains n *tokens*. A transition t (representing some activity) is *enabled by* M if each input place of t contains at least one token. If an enabled transition *occurs*, then this is tantamount to the following change of marking: a token is subtracted from each input place of t; a token is added to each output place of t; and no other places are affected.

We use the notation $M \xrightarrow{t} M'$ (or $(N, M) \xrightarrow{t} (N, M')$, to emphasize N) in order to express that M enables t and M' arises out of M according to the rule just described; M' is then said to be *directly reachable* from M, and the reflexive and transitive closure of this relation gives the overall *reachability* relation. A marking is called *safe* if no place contains more than one token, and a marked net is called *safe* if every marking reachable from the initial marking is safe.

Figure 1 shows an unmarked Petri net, N, and two marked Petri nets, (N, M_0) and (N, M_1). Notice that the underlying net (graph) is the same in all cases, only the markings are different. Moreover, the unmarked net could as well be interpreted as a marked net, namely as one which has the 'empty' marking assigning zero tokens to each place.

Formally, the net N of Figure 1 is a triple (S, T, W) where S is the set of its places, $S = \{s_1, s_2, s_3\}$, T is the set of its transitions, $T = \{a, b\}$, and W assigns a number (in this case, only 0 or 1) to pairs (s, t) and (t, s), where $s \in S$ and $t \in T$, depending on whether an arrow leads from s to t or from t to s, respectively. For instance, $W(s_1, a) = 1$ and $W(b, s_2) = 0$ (i.e., 1 stands for 'arrow' and 0 stands for 'no arrow'). The marking M_0 shown is a function satisfying $M_0(s_1) = 1$, $M_0(s_2) = 0$ and $M_0(s_3) = 0$, while M_1 is a function satisfying $M_1(s_1) = 0$, $M_1(s_2) = 1$ and $M_1(s_3) = 0$. The marking M_0 enables transition a which may occur; if it does, the resulting marking is M_1. Thus, $(N, M_0) \xrightarrow{a} (N, M_1)$.

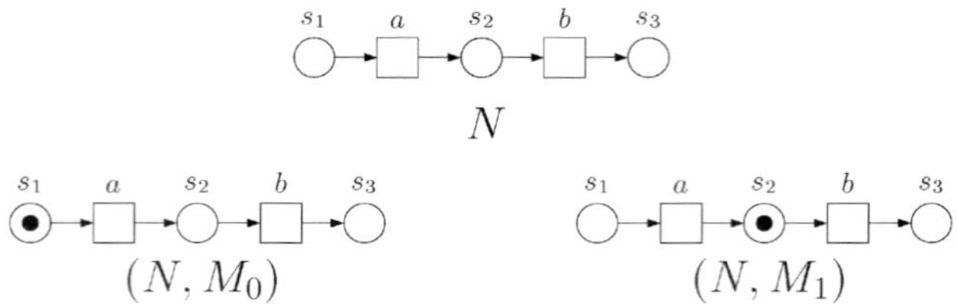

Fig. 1. A Petri net, and the same net with markings M_0 and M_1.

N describes a sequence of two transitions, a and b, and it is therefore related to the CCS expression $a.(b.\mathsf{nil})$. But which of the three systems should correspond to the semantics of $a.(b.\mathsf{nil})$? The answer is clear: (N, M_0). The reason is that only this one has the same *behaviour* as $a.(b.\mathsf{nil})$. The net N by itself has the empty behaviour, and in (N, M_1), b can occur as the first action, but not a as required by $a.(b.\mathsf{nil})$. Consider now what happens when the action a is executed in either model. In CCS, the expression $a.(b.\mathsf{nil})$ is transformed into $b.\mathsf{nil}$. In the net domain, the marked net (N, M_0) is transformed into the net (N, M_1) with the same underlying unmarked net N. But if we were to define a Petri net corresponding to the expression $b.\mathsf{nil}$, then we would not even think of taking the net (N, M_1); rather, a natural choice would be (N, M_1) after deleting s_1 and a.

This example shows that the ways of generating behaviour in net theory and in CCS do not directly correspond to each other. In the latter, the structure of an expression may change, while in the former, only markings change but the structure of a net remains the same through any execution. The difference is a fundamental one, and stems from the underlying ideas of modelling systems. In a net, when a transition has occurred, we may still recover its presence from the static structure of the system, even though it may never again be executed. In a CCS expression, when an action has occurred, we may forget it (unless it is in a loop), precisely because it may never be executed, and hence we may safely change the structure of the expression (expressions with loops may also be changed, for instance by making them longer).

Static and dynamic expressions; structural similarity. PBC respects the division between static and dynamic aspects that can be found in Petri net theory. The basic idea will be to introduce into the syntax of expressions a device that models markings. We distinguish expressions 'without markings' (called static expressions) and expressions 'with markings' (called dynamic expressions): static expressions correspond to unmarked nets, while dynamic expressions correspond to marked nets. Below, E and F will denote static expressions, and G and H will denote dynamic expressions.

For example, $E = a; b$ is a static PBC expression corresponding to the unmarked net N in Figure 1. Making expressions dynamic consists of overbarring or underbarring (parts of) them. For instance, $\overline{E} = \overline{a; b}$ is a dynamic expression which corresponds to the same net as E, but with a marking that lets E be executed from its beginning, i.e., to the 'initial marking' of E. The net corresponding to $\overline{E}$ is (N, M_0). We will also allow dynamic PBC expressions such as $G = \underline{a}; b$. It corresponds to the same underlying expression E, but in a state in which its first action a has just been executed, i.e., it shows the instant at which the final state of a has just been produced. Such a G corresponds to (N, M_1).

This syntactic device of overbarring and underbarring introduces what could, at first glance, be seen as a difficulty. For consider $E = a; b$ in a state in which its second part, b, is just about to be executed, $G' = a; \overline{b}$. Which marked net should this dynamic expression correspond to? There is only one reasonable answer to this question, namely the net (N, M_1) in Figure 1 which, as already observed, corresponds to $G = \underline{a}; b$. Thus, in general, we may have a many-to-one relationship between dynamic expressions and marked nets. Let us use the symbol $\equiv$ to relate dynamic expressions which are not necessarily syntactically equal, but are in the same relationship as G and G'. Then $G \equiv G'$ can be viewed as expressing that 'the state in which the first component of a sequence has terminated is the same as the state in which the second component may begin its execution'. As another example, we observe that $\overline{E} = \overline{a; b}$ also has a dynamic expression which is equivalent but not syntactically equal, namely $H = \overline{a}; b$. And $\overline{E} \equiv H$ can be viewed as saying that 'the initial state of a sequence is the initial state of its first component'.

We do not see this many-to-one relationship as merely incidental, to be overcome, perhaps, by a better syntactic device than that of overbarring and underbarring subexpressions. Instead, we view it as evidence of a fundamental difference between Petri nets and process algebras: expressions of the latter come with in-built structure, while Petri nets do not. For instance, $E = a; b$ is immediately recognized as a sequence of a and b. In the example of Figure 1, one may also recognize the structure to be a sequence. However, this is entirely due to the simplicity of the net. For the graph of an arbitrary net it is far from obvious how it may be seen as constructed from smaller parts, while it is always clear, for any well-formed expression in any process algebra, how it is made up from subexpressions. The fact that an expression of a process algebra always reveals its syntactic structure, has some highly desirable consequences. For example, it is often possible to conduct an argument 'by syntactic induction'; one may build proof systems 'by syntactic definition' around such an algebra; and the availability of operators for the modular construction of systems is usually appreciated by practitioners. On the other hand, one of the disadvantages of such a syntactic view is that some effort has to be invested into the definition of the behaviour of expressions. While this can usually be done inductively, it is still necessary to go through all operators one by one. A non-structured model such a Petri nets has an advantage in this respect; its behaviour (with respect to a marking) is defined by a single rule, namely the transition occurrence rule, which covers all the cases.

In this spectrum, the box algebra covers a middle position, and its expressions come with an in-built structure. On the other hand, the behavioural rules of (dynamic) expressions are, in fact, the Petri net transition rule in disguise. The necessity to consider equivalences, such as the *structural similarity* $\equiv$, can be viewed as being the price that has to be paid for being able to combine advantages of both Petri nets and process algebras.

Notations. We use the standard mathematical notation throughout. Some particular notations are recalled below.

A *partial order* is a pair (X, R) where X is a set and $R \subseteq X \times X$ is a reflexive ($id_X \subseteq R$), transitive ($R \circ R \subseteq R$) and antisymmetric ($R \cap R^{-1} = id_X$) relation.

A *multiset* over a set A is a function $\mathsf{m}: A \to \mathbf{N}$. m is *finite* if the set of all $a \in A$ such that $\mathsf{m}(a) > 0$ is finite. The *cardinality* of a finite m is the number $|\mathsf{m}| = \sum_{a \in A} \mathsf{m}(a)$. The operations of *sum*, (nonnegative) *difference* and *scalar multiplication* for the multisets over A are defined so that, for all $a \in A$ and $n \in \mathbf{N}$,

Sum:	$(\mathsf{m} + \mathsf{m}')(a) = \mathsf{m}(a) + \mathsf{m}'(a)$
Difference:	$(\mathsf{m} - \mathsf{m}')(a) = \max(\mathsf{m}(a) - \mathsf{m}'(a), 0)$
Multiplication:	$(n \cdot \mathsf{m})(a) = n \cdot \mathsf{m}(a)$.

Note that if m and m' are finite then so are $\mathsf{m} + \mathsf{m}$, $\mathsf{m} - \mathsf{m}'$ and $n \cdot \mathsf{m}$. Multiset *inclusion*, $\mathsf{m} \subseteq \mathsf{m}'$, is defined by $\mathsf{m}(a) \leqslant \mathsf{m}'(a)$, for all $a \in A$. In the examples, we will write, e.g., $\{a, a, b\}$ for the multiset defined by $\mathsf{m}(a) = 2$, $\mathsf{m}(b) = 1$, and $\mathsf{m}(c) = 0$ for all $c \in A \backslash \{a, b\}$. The set of all finite multisets over A is denoted by $\mathsf{mult}(A)$. A set of places or transitions will sometimes be identified with the corresponding multiset.

2.2. *CCS and PBC*

In this section we list some further differences between CCS and PBC.

Prefixing and sequential composition. Instead of expressing sequential behaviour by CCS prefixing, PBC uses a true sequential operator, $E; F$, meaning that E is to be executed before F. The sequence can be simulated in CCS by a combination of prefixing, parallel operator and synchronization. One reason for allowing full sequential composition is that in a majority of imperative languages, this is one of the basic operations. Another reason is that in PBC, the semantics of the full sequential composition is no more complicated than that of prefixing.

The CCS's nil process is no longer necessary in PBC. To see why, we compare the CCS way of deriving the complete behaviour of $a.(b.\mathsf{nil})$, and the PBC way of deriving the complete behaviour of $\overline{a; b}$:

$$\text{In CCS:} \quad a.(b.\mathsf{nil}) \xrightarrow{a} b.\mathsf{nil} \xrightarrow{b} \mathsf{nil}.$$
$$\text{In PBC:} \quad \overline{a; b} \equiv \overline{a}; b \xrightarrow{a} \underline{a}; b \equiv a; \overline{b} \xrightarrow{b} a; \underline{b} \equiv \underline{a; b}.$$

In both cases, the action sequence ab is derived, but the PBC way of deriving this behaviour is more symmetric while the CCS way is shorter. The fact that 'execution has ended' is expressed in CCS by the derivation of nil, while in PBC by the derivation of an expression which is completely underbarred, such as $\underline{a; b}$.

In general, if E is a static expression, then we call $\overline{E}$ 'the initial state of E', and $\underline{E}$ 'the final (or terminal) state of E'. However, this terminology is slightly deceptive, for there are expressions E for which no behaviour leads from $\overline{E}$ to $\underline{E}$.

Synchronization. As CCS, the PBC algebra is based upon the idea that in a distributed environment all complex interactions can be decomposed into primitive binary interactions.

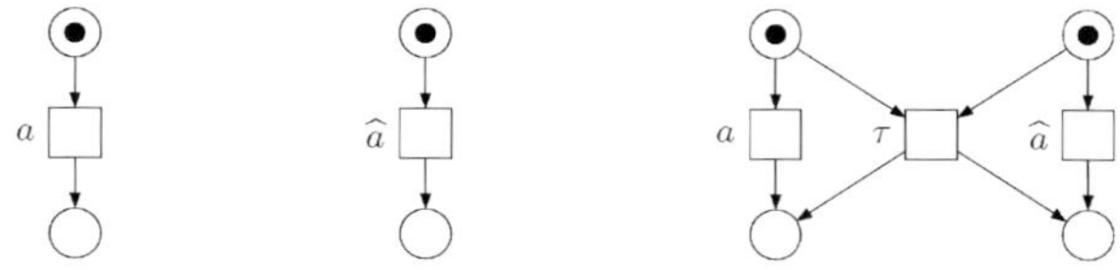

Fig. 2. Activity of a (left), activity of $\hat{a}$ (middle) and activity of $a \,|\, \hat{a}$ (right).

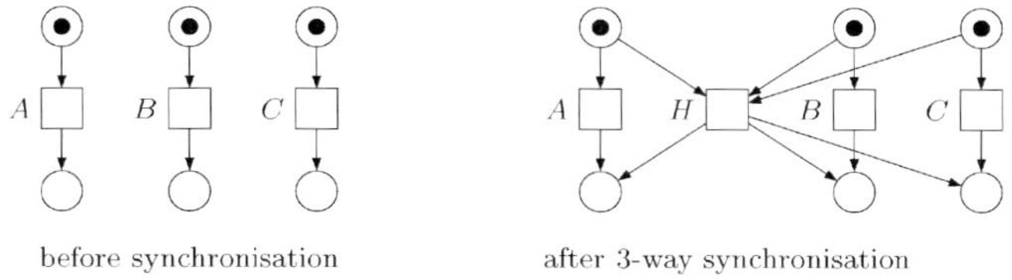

Fig. 3. Handshake between three people.

Suppose that two people, A and B, wish to communicate by exchanging a handshake. In CCS, one could represent the action 'A extends his hand towards B' by a symbol a; the action 'B extends her hand towards A' by another symbol $\hat{a}$ (the 'conjugate'[1] of a); and the fact that A and B are standing face to face by the symbol $|$, yielding the CCS expression $(a.\mathsf{nil}) \,|\, (\hat{a}.\mathsf{nil})$ (abbreviated $(a) \,|\, (\hat{a})$). This expression has the following possible activities:

(i) A may extend his hand and withdraw it, leaving B with the same possibility;
(ii) B may extend her hand and withdraw it, then A does the same; and
(iii) both A and B extend their hands which results in a *handshake*, labelled by τ, the 'silent action'.

Using the existing Petri net semantics of CCS expressions (e.g., in [25,48]), these three possible activities could be represented in Petri net terms as shown on the right hand side of Figure 2.

Now consider how one could describe a handshake between three people A, B and C. Although one might think that 3-way handshakes occur rarely in practice, we will argue in Section 8 that multiway synchronizations can be used well in the modelling of programming languages. In terms of Petri nets, such an activity could be modelled by a three-way transition such as H in Figure 3. A moment's reflection shows that a 3-way synchronization transition like this cannot directly be modelled using the 2-way handshake mechanism of CCS; the reason is that τ cannot be synchronized with any other action (there is nothing like $\hat{\tau}$ in CCS). While one might try to simulate the 3-way-handshake using a series of binary handshakes, PBC proposes instead to extend the CCS framework so that a 3-way (and, in general, n-way) synchronization could be expressed. One possibility, which is akin to the ACP approach [4,3,23], might be to generalize the conjugation operation to a function or a relation that allows the grouping of more than just two actions; however, PBC takes a different approach, which is still based on considering primitive synchronizations to be binary in nature. This works through the usage of structured actions, which we will describe next.

[1] In CCS, conjugation is usually denoted by overbarring; we use hatting instead, because overbarring has already been used to construct dynamic expressions.

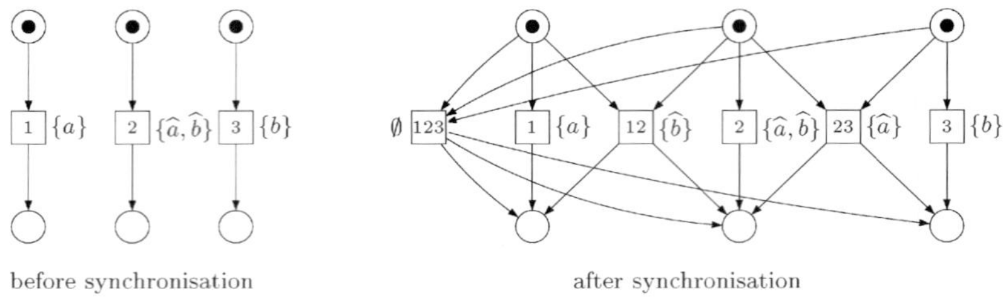

Fig. 4. 3-way handshake in terms of PBC.

To understand the PBC way of describing multiway synchronizations, it is perhaps easiest to interpret a CCS pair of conjugate actions $(a, \hat{a})$ in the original way [41] as denoting two links between two agents that fit together. In this view, a denotes a communication capability in search of a matching $\hat{a}$, and similarly $\hat{a}$ denotes a communication capability in search of a matching a. Once the communication takes place, i.e., a pair of a and $\hat{a}$ is matched, this becomes a private synchronization link (namely, τ) between two agents with no further externally visible communication capability. The PBC model extends this idea in the sense that the result of a synchronization does not have to be completely internal or 'silent', but instead, may contain further communication capabilities that may be linked with other activities. This generalization is achieved by allowing a transition to correspond to a whole (finite) set of communication capabilities, such as for instance, $\emptyset$, $\{a\}$, $\{\hat{a}\}$, $\{a, b\}$, $\{\hat{a}, b\}$, or even $\{a, \hat{a}, b\}$. For instance, an activity $\{\hat{a}\}$ may be synchronized with activity $\{a, b\}$ using the conjugate pair $(a, \hat{a})$. This results in a synchronized activity which is not silent but rather, still has the communication capability $\{b\}$, as the pair $(a, \hat{a})$ is internalized but b remains untouched. For instance, Figure 4 describes a 3-way handshake in terms of the (Petri net view of) PBC. In the figure, a denotes 'A shakes hand with B', $\hat{a}$ denotes 'B shakes hand with A', $\hat{b}$ denotes 'B shakes hand with C' and b denotes 'C shakes hand with B'. Then, by the fact that B performs $\hat{a}$ and $\hat{b}$ simultaneously (expressed by the fact that $\{\hat{a}, \hat{b}\}$ is the label of a transition), the resulting activity 123 (in Figure 4) describes a simultaneous handshake between all three people. Transition 12 describes the handshake only between A and B (with label $\{\hat{b}\}$ and capability to link with C) and transition 23 describes the handshake only between B and C (with label $\{\hat{a}\}$ and capability to link with A). Transition 123 can also be thought of either as a 2-way synchronization between 1 and 23, or as a 2-way synchronisation between 12 and 3.

Let us now consider two actions $\{a, b\}$ and $\{\hat{a}, b\}$. According to the PBC approach, two transitions, say 1 and 2, corresponding, respectively, to these actions can be synchronised using the conjugate pair $(a, \hat{a})$. However, what synchronisation capability should the resulting transition 12 still possess? There are only two meaningful answers: either the set $\{b\}$ or the multiset $\{b, b\}$. The PBC model chooses the second alternative since only then we obtain the (highly desirable and expected) property that the order of a series of consecutive synchronizations is irrelevant.

Therefore, PBC will use a set $\mathsf{A}_{\mathrm{PBC}}$ of primitive actions or *action particles*. There is a *conjugation* function $\hat{}: \mathsf{A}_{\mathrm{PBC}} \to \mathsf{A}_{\mathrm{PBC}}$ with the properties of *involution* ($\hat{} \circ \hat{} = id_{\mathsf{A}_{\mathrm{PBC}}}$) and *discrimination* ($\hat{a} \neq a$). This is the same basic setup as in CCS. However, unlike CCS, the PBC allows as the label of a transition any finite multiset over $\mathsf{A}_{\mathrm{PBC}}$, thus providing a means to specify multiway synchronisation used, in particular, to model simultaneous access to multiple variables in a programming language semantics (see Section 8). That is, the set of *labels* (or, for the more process algebraically inclined, *elementary expressions*, or, to emphasize the fact that more than one action may be combined in a single transition, the set of *multiactions*, or, if no confusion can arise with action particles, *actions*) is defined as $\mathsf{Lab}_{\mathrm{PBC}} = \mathsf{mult}(\mathsf{A}_{\mathrm{PBC}})$.

This specializes to the basic CCS framework as follows: the CCS a and $\hat{a}$ correspond to $\{a\}$ and $\{\hat{a}\}$ in PBC, respectively, and τ to $\emptyset$. No PBC action α, with $|\alpha| > 1$, has a direct CCS equivalent.

Parallel composition. The CCS parallel composition performs the synchronisation for all possible conjugate pairs. For instance, the CCS expression $a.(a.(b.\mathsf{nil})) \mid \hat{a}.(\hat{a}.(\hat{b}.\mathsf{nil}))$ creates five synchronizations (three of them being actually executable); the syntax of CCS does not allow one to express synchronisation using only the pair $(b, \hat{b})$ but not the pair $(a, \hat{a})$. For reasons that will become clear in Section 8, it may be desirable to be able to say precisely which action names are being used for synchronisation.

PBC therefore separates the synchronisation operator from parallel composition, $E \| F$, regarding synchronisation as a unary operator, $E \mathbin{\mathsf{sy}} A$ where $A \subseteq \mathsf{A}_{\mathrm{PBC}}$. The latter makes it very similar to the restriction operator of CCS, but playing an opposite role: while synchronisation adds, restriction removes certain transitions.

Other operators. Most of the remaining PBC operators (except iteration) are akin to their counterparts in CCS.

The choice between two expressions E and F, denoted in PBC by $E \,\square\, F$, specifies behaviour which is, in essence, the union of the behaviours of E and F. As its counterpart in CCS, $+$, the choice allows one to choose nondeterministically between two possible sets of behaviours. The iteration operator obeys the syntax $\langle E * F]$, where E is the looping part (whose execution may be repeated arbitrarily many times), and F is the termination (which may be executed at most once, at termination).

Basic relabelling (simply called 'relabelling' in CCS) is defined with respect to a function f on action particles; $E[f]$ has all the behaviours of E, except that their constituents are relabelled according to f. Restriction, $E \mathbin{\mathsf{rs}} A$, is defined with respect to a set of action particles A, and has all the behaviours of E except those that involve the action particles in $A \cup \widehat{A}$, where $\widehat{A} = \{\hat{a} \mid a{\in}A\}$. Scoping, $[A : E] = (E \mathbin{\mathsf{sy}} A) \mathbin{\mathsf{rs}} A$, is a derived operator; its importance stems from an application to the modelling of blocks in a programming language.

In terms of the Petri net semantics of PBC, it will be seen that the distinction between *control flow* operators and *communication interface* operators manifests itself in a very simple way: the latter (relabelling, synchronisation, restriction and scoping) modify transitions, while the former (sequence, choice, parallel composition and iteration) modify places.

Recursion allows process variables to be defined via equations. E.g., $X \stackrel{\mathrm{df}}{=} \{a\};\ X$ specifies the behaviour of X to be an indefinite succession of the executions of $\{a\}$.

3. Elements of an algebra of Petri nets

We will introduce Petri nets as in [15,43,46], and their semantics as necessary, choosing from concurrency semantics such as: step semantics [24], trace semantics [40], process semantics [8,28], or partial word semantics [29,47,49].

In addition, however, we will also introduce a general compositionality mechanism for combining nets. Combinability of nets will be driven by labellings of both places and transitions. Since such labellings indicate the border between a net and its (potential) surroundings, the resulting combinable objects will be called *boxes*.

More precisely, the compositionality mechanism will be founded upon transition refinement. Each operator on nets will be described by a net Ω whose transitions $t_1, t_2, \ldots, t_n$ can be refined by nets $\Sigma_1, \Sigma_2, \ldots, \Sigma_n$ in the process of forming a new net $\Omega(\Sigma_1, \Sigma_2, \ldots, \Sigma_n)$. This corresponds formally to the application of a process-algebraic operator op to its n arguments.

We shall distinguish plain boxes and operator boxes.

Plain boxes form the Petri net domain upon which various operators are defined. When giving the Petri net semantics of a process algebra (such as PBC or CCS), we will associate a plain box with every (closed) expression. The nets $\Sigma_1, \Sigma_2, \ldots, \Sigma_n$ above will be plain boxes.

Operator boxes are patterns (or functions) defining the ways of constructing new plain boxes out of given ones by refinement. When translating a process algebra into Petri nets, we will aim at associating an operator box with every operator of the process algebra. The net Ω above will be an operator box.

Note that the same type of nets – boxes – serve to describe two seemingly different objects, namely the elements and the operators of the semantical domain. This also corresponds to a process-algebraic idea (and to a general mathematical usage), namely that constants can be viewed as nullary operators.

3.1. *Place and transition labellings*

We introduce the kinds of net labellings we shall need using a simple example. Consider the dynamic expression $\overline{a; (a \| a)}$, a possible Petri net model for which might look like the (safe) net depicted in Figure 5.[2]

Place labellings. In Figure 5, three different kinds of places can be identified. The place s_0 is special in the sense that it contains the token corresponding to the expression in its initial state. And, by symmetry, the places s_3 and s_4, when holding one token each, characterize the terminal state of the expression. The two remaining places, s_1 and s_2, may be considered as internal and contributing to intermediate markings corresponding to intermediate dynamic expressions. The different role of the places will be captured formally by suitable labels, with three possible values $\{\mathsf{e}, \mathsf{i}, \mathsf{x}\}$, corresponding to the three different kinds of places: e for entry places, i for internal places, and x for exit places.

[2] In the examples, we often change notation from $\{a\}$ as an action to a as an action for simplicity, and provided that synchronisation is not involved.

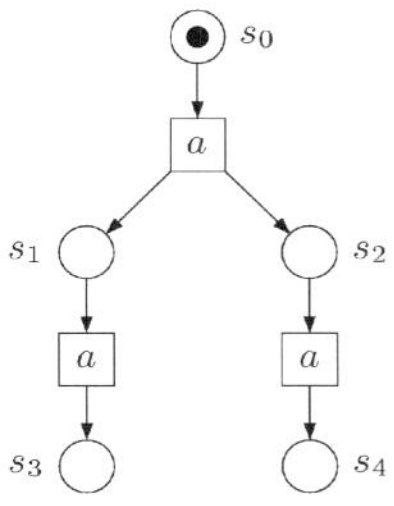

Fig. 5. A marked net corresponding to $\overline{a;\ (a\|a)}$.

Transition labellings. Clearly, the action a in Figure 5 cannot be used to identify the transitions, since there are three of them, all corresponding to the same action a. As a result, to model static or dynamic expressions we need to employ Petri nets with transitions being labelled by action names.

For Petri nets defining net operators, we shall also use transition labellings, but with labels coming from a set of relations, called *(general) relabellings*. This allows combining transitions coming from different nets. And, since action names may be treated as a special kind of ('nullary') relabellings, the latter may be used in full generality.

More formally, we assume a set Lab of *actions* (serving as transition labels) to be given. At this point Lab is an arbitrary set, but when translating PBC we shall consider Lab to be, in particular, the structured set $\mathsf{Lab}_{\mathrm{PBC}}$, and other suitable sets will be used for other concrete process algebras.

A *relabelling* ϱ is a relation

$$\varrho \subseteq \mathsf{mult}(\mathsf{Lab}) \times \mathsf{Lab} \tag{1}$$

such that $(\emptyset, \alpha) \in \varrho$ if and only if $\varrho = \{(\emptyset, \alpha)\}$. The intuition behind a pair (Γ, α) belonging to ϱ is that a (finite) group of transitions whose labels match the argument Γ can be synchronised, yielding a transition with label α.

We distinguish two kinds of relabellings. A *constant* relabelling, $\varrho_\alpha = \{(\emptyset, \alpha)\}$, where α is an action in Lab, can be identified with α itself, so that we may consider the set of actions Lab to be embedded in the set of all relabellings. Constant relabellings will be used for plain boxes. A non-constant relabelling will be called *transformational*; in such a case, the empty set will not be in its domain in order not to 'create an action out of nothing'. Transformational relabellings will be used for operator boxes.

The following specific relabellings are of particular interest. Given a set $A \subseteq \mathsf{Lab}$, the *restriction* $\varrho_A = \{(\{\alpha\}, \alpha) \mid \alpha \in A\}$ only keeps the actions belonging A. And the *identity* relabelling, $\varrho_{id} = \varrho_{\mathsf{Lab}}$ captures the 'keep things as they are' interface (non)change.

3.2. *Labelled nets and their semantics*

A marked net with place and transition labels (*labelled net*, for short) is a tuple

$$\Sigma = (S, T, W, \lambda, M)$$

such that: S and T are disjoint sets of respectively *places* and *transitions*; W is a *weight function* from the set $(S\times T)\cup(T\times S)$ to the set of natural numbers $\mathbf{N}$; λ is a *labelling* function for places and transitions such that $\lambda(s)\in\{\mathsf{e},\mathsf{i},\mathsf{x}\}$, for every place $s\in S$, and $\lambda(t)$ is a relabelling ϱ of the form (1), for every transition $t\in T$; and M is a *marking*, i.e., a multiset over S.

Nets can be represented as directed graphs. Places and transitions are represented, respectively, as circles and as rectangles; the flow relation generated by W is indicated by arcs annotated with the corresponding weights; and markings are shown by placing tokens within circles. As usual, the zero weight arcs will be omitted and the unit weight arcs (or unitary arcs) will be left unannotated. To avoid ambiguity, we will sometimes decorate the various components of Σ with the index Σ; thus, T_Σ denotes the set of transitions of Σ, etc. A net is *finite* if both S and T are finite sets. Figure 6 shows the graph of a labelled net $\Sigma_0=(S_0,T_0,W_0,\lambda_0,M_0)$ defined thus:

$$\begin{aligned}
S_0 &= \{s_0,s_1,s_2,s_3\},\\
T_0 &= \{t_0,t_1,t_2\},\\
W_0 &= \big((TS\cup ST)\times\{1\}\big)\cup\big(\big((S\times T)\setminus ST\cup(T\times S)\setminus TS\big)\times\{0\}\big),\\
\lambda_0 &= \big\{(s_0,\mathsf{e}),(s_1,\mathsf{i}),(s_2,\mathsf{x}),(s_3,\mathsf{e}),(t_0,a),(t_1,b),(t_2,a)\big\},\\
M_0 &= \big\{(s_0,1),(s_1,0),(s_2,0),(s_3,1)\big\},
\end{aligned}$$

where $TS=\{(t_0,s_1),(t_1,s_2),(t_2,s_3)\}$ and $ST=\{(s_0,t_0),(s_1,t_1),(s_3,t_2)\}$.

If the labelling of a place s in a labelled net Σ is e then s is an *entry* place, if i then s is an *internal* place, and if x then s is an *exit* place. By convention, ${}^\circ\Sigma$, Σ° and $\ddot{\Sigma}$ denote respectively the entry, exit and internal places of Σ. For every place (transition) x, we use ${}^\bullet x$ to denote is pre-set, i.e., the set of all transitions (places) y such that there is an arc from y to x, $W(y,x)>0$. The post-set $x^\bullet$ is defined in a similar way. The pre- and post-set notation extends in the usual way to sets R of places and transitions, e.g., ${}^\bullet R=\bigcup\{{}^\bullet r\mid r\in R\}$. In what follows, all nets are assumed to be T-*restricted*, i.e., the pre- and post-sets of each transition are non-empty. For the labelled net of Figure 6 we have ${}^\circ\Sigma_0=\{s_0,s_3\}$, $\Sigma_0^\circ=\{s_2\}$, ${}^\bullet s_0=\emptyset$, $s_0^\bullet=\{t_0\}$ and $\{s_0,s_1\}^\bullet=\{t_0,t_1\}={}^\bullet\{s_1,s_2\}$.

A labelled net Σ is *ex-restricted* if there is at least one entry and at least one exit place, ${}^\circ\Sigma\neq\emptyset\neq\Sigma^\circ$. Σ is *e-directed* (*x-directed*) if the entry (respectively, exit) places are free

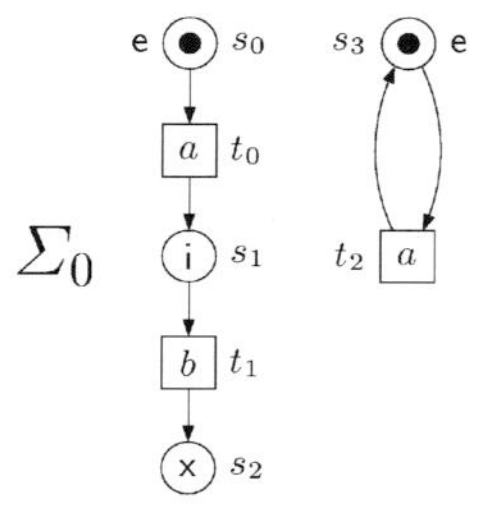

Fig. 6. A labelled net.

from incoming (respectively, outgoing) arcs, i.e., ${}^\bullet({}^\circ\Sigma) = \emptyset$ (respectively, $(\Sigma^\circ)^\bullet = \emptyset$). Σ is *ex-directed* if it is both e-directed and x-directed. Σ is *marked* if $M_\Sigma \neq \emptyset$, and *unmarked* otherwise. The labelled net Σ_0 in Figure 6 is marked, T-restricted, ex-restricted and x-directed, but it is not e-directed.

A labelled net Σ is *simple* if W always returns 0 or 1, and *pure* if for all transitions $t \in T$, ${}^\bullet t \cap t^\bullet = \emptyset$. If Σ is not pure then there are $s \in S$ and $t \in T$ such that $W(s,t) \cdot W(t,s) > 0$; in such a case, the pair $\{s,t\}$ will be called a *side-loop*. The net in Figure 6 is finite and simple, but is not pure as it contains a side-loop $\{s_3, t_2\}$.

We will use three explicit ways of modifying the marking of $\Sigma = (S, T, W, \lambda, M_\Sigma)$. We define $\lfloor \Sigma \rfloor$ as $(S, T, W, \lambda, \emptyset)$; typically, this operation is used when $M_\Sigma \neq \emptyset$, since it erases all tokens. Moreover, we define $\overline{\Sigma}$ and $\underline{\Sigma}$ as, respectively, $(S, T, W, \lambda, {}^\circ\Sigma)$ and $(S, T, W, \lambda, \Sigma^\circ)$. We will call ${}^\circ\Sigma$ the *entry marking*, and Σ° the *exit marking* of Σ. Note also that $\lfloor . \rfloor$, $\overline{(.)}$ and $\underline{(.)}$ are syntactic operations having nothing to do with derivability (reachability) in the sense of the step sequence semantics defined next.

Step sequence semantics. We adopt finite step sequence semantics for a labelled net $\Sigma = (S, T, W, \lambda, M)$, in order to capture the potential concurrency in the behaviour of the system modelled by Σ. A finite multiset of transitions U, called a *step*, is *enabled* by Σ if for every place $s \in S$,

$$M(s) \geqslant \sum_{t \in U} W(s,t) \cdot U(t).$$

We denote this by $\Sigma[U\rangle$, or $M[U\rangle$ if the net is understood from the context. An enabled step U can be *executed*, leading to a follower marking M' defined, for every place $s \in S$, by

$$M'(s) = M(s) - \sum_{t \in U} W(s,t) \cdot U(t) + \sum_{t \in U} W(t,s) \cdot U(t).$$

We will denote this by $M\,[U\rangle\,M'$ or $\Sigma\,[U\rangle\,\Theta$, where Θ is the labelled net (S, T, W, λ, M'). Transition labelling may be extended to steps, through the formula

$$\lambda(U) = \sum_{t \in U} U(t) \cdot \{\lambda(t)\} \in \mathsf{mult}(\mathsf{Lab}).$$

Although we will use the same term 'step' to refer both to a finite set of transitions and to a finite multiset of labels, it will always be clear from the context which one is meant. The notation for label based steps will be $\Sigma\,[\Gamma\rangle_{\mathsf{lab}}\,\Theta$, where $\Gamma = \lambda(U)$.

A *finite step sequence* of Σ is a finite (possibly empty) sequence $\sigma = U_1 \ldots U_k$ of steps for which there are labelled nets $\Sigma_0, \ldots, \Sigma_k$ such that $\Sigma = \Sigma_0$ and for every $1 \leqslant i \leqslant k$, $\Sigma_{i-1}\,[U_i\rangle\,\Sigma_i$. Depending on the need, we shall then use one of the following notations:

$$\Sigma\,[\sigma\rangle\,\Sigma_k \qquad M_\Sigma\,[\sigma\rangle\,M_{\Sigma_k} \qquad \Sigma_k \in [\Sigma\rangle \qquad M_{\Sigma_k} \in [M_\Sigma\rangle.$$

for an operator box Ω (see Section 4.1) with n transitions. The transition refinement part of net refinement serves as a pattern for gluing together an n-tuple of plain boxes $\mathbf{\Sigma}$ – one plain box for every transition in Ω – along their e and x interfaces. The relabelling part of net refinement combines (synchronizes) transitions from $\mathbf{\Sigma}$ and changes the interface of the resulting transition(s) according to the transformations prescribed in Ω's transition labels.

4.1. *Operator boxes*

An *operator box* is a simple finite unmarked box Ω all of whose transition labellings are transformational. We will assume that the set of transitions of Ω is implicitly ordered, $T_\Omega = \{v_1, \ldots, v_n\}$. Then any tuple $\mathbf{\Sigma} = (\Sigma_{v_1}, \ldots, \Sigma_{v_n})$ consisting of plain boxes is called an Ω*-tuple*. For $T \subseteq T_\Omega$, we will denote $\mathbf{\Sigma}_T = \{\Sigma_v \mid v \in T\}$.

Throughout this section, we will refer to an example of net refinement shown in Figure 13, where $\mathbf{\Sigma} = (\Sigma_{v_1}, \Sigma_{v_2}, \Sigma_{v_3})$ and, in Ω:

$$\begin{aligned} \varrho_1 &= \varrho_{id} \backslash \big\{(\{a\}, a)\big\} \cup \big\{(\{a, b\}, g)\big\}, \\ \varrho_2 &= \varrho_{id} \backslash \big\{(\{e\}, e), (\{d\}, d)\big\} \cup \big\{(\{e\}, h)\big\}. \end{aligned}$$

The net $\Omega(\mathbf{\Sigma})$ in this figure is the result of applying combined transition refinement and relabelling, using Ω as an operator box and refining its transitions v_1, v_2 and v_3 by the three plain boxes Σ_{v_1}, Σ_{v_2} and Σ_{v_3}, respectively.

4.2. *Place and transition names of operator and plain boxes*

As far as net refinement as such is concerned, the names (identities) of newly constructed transitions and places are irrelevant, provided that we always choose them fresh. However, in our approach to solving recursive definitions on boxes in Section 5, the *names* of places and transitions are crucial since we use them to define the inclusion order on the domain of labelled nets. A key to the construction of recursive nets is the use of labelled trees as place and transition names. For the example in Figure 13, τ_1 is a tree used to name a place resulting from combining the place e of the operator box Ω, with three places corresponding to the three arcs adjacent to e: the arc incoming from v_1 contributes an exit place x_1 of Σ_{v_1}, the arc outgoing to v_1 contributes an entry place e_1^2 of Σ_{v_1}, and the arc outgoing to v_2 contributes an entry place e_2 of Σ_{v_2}.

We shall assume that there are two disjoint infinite sets of place and transition names, $\mathsf{P_{root}}$ and $\mathsf{T_{root}}$ (such as $e \in S_\Omega$ and $v_3 \in T_\Omega$ in Figure 13). Each name $\eta \in \mathsf{P_{root}} \cup \mathsf{T_{root}}$ (or a pair (t, α), where t is a name in $\mathsf{T_{root}}$ and α is a label in Lab) can be viewed as a special tree with a single root labelled with η (or (t, α)). Moreover, we shall allow more complicated trees as transition and place names, and use a linear notation to express such trees. To this end, the expression $x \lhd \mathcal{S}$, where x is a single root tree labelled with x (x is a name η or a pair (t, α)) and $\mathcal{S}$ is a finite multiset of trees, is a new tree where the trees of the multiset are appended (with their multiplicity) to the root. All our trees will thus be finitely branching.

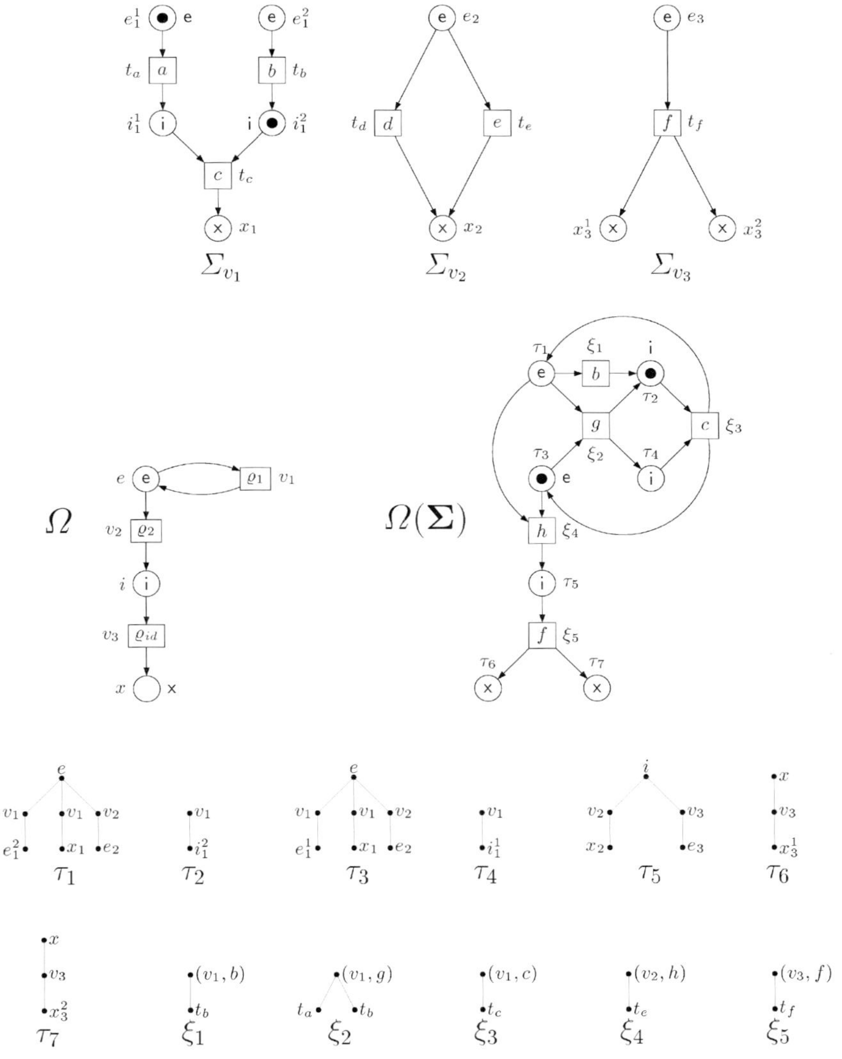

Fig. 13. Running example of net refinement.

If $\mathcal{S} = \{p\}$ is a singleton multiset then $x \lhd \mathcal{S}$ will simply be denoted by $x \lhd p$, and if $\mathcal{S}$ is empty then $x \lhd \mathcal{S} = x$. In Figure 13, $\tau_1 = e \lhd \{v_1 \lhd e_1^2, v_1 \lhd x_1, v_2 \lhd e_2\}$, $\tau_4 = v_1 \lhd i_1^1$ and $\xi_2 = (v_1, g) \lhd \{t_a, t_b\}$.

We shall further assume that in every operator box, all places and transitions are simply names (i.e., single root trees) from respectively P_{root} and T_{root}. For the plain boxes, the trees used as names may be more complex. Each transition tree is a finite tree labelled with elements of T_{root} (at the leaves) and $\mathsf{T}_{\text{root}} \times \mathsf{Lab}$ (elsewhere), and each place tree is a possibly infinite (in depth) tree labelled with names from $\mathsf{P}_{\mathsf{root}}$ and $\mathsf{T}_{\mathsf{root}}$, which has the form:

$$t_1 \lhd t_2 \lhd \cdots \lhd t_n \lhd s \lhd \mathcal{S},$$

where $t_1, \ldots, t_n \in \mathsf{T}_{\mathsf{root}}$ $(n \geqslant 0)$ are transition names and $s \in \mathsf{P}_{\mathsf{root}}$ is a place name (so that no confusion will be possible between transition-trees and place-trees: the latter always have a label from $\mathsf{P}_{\mathsf{root}}$ and the former never). We include all these trees (including the basic ones consisting only of a root as special cases) in our sets of allowed transition and place names, denoted respectively by $\mathsf{P}_{\mathsf{tree}}$ and $\mathsf{T}_{\mathsf{tree}}$. The definition of net refinement will be done in such a way that, provided all the names occurring in Ω are single root trees, and all the names occurring in $\mathbf{\Sigma}$ are in the sets of allowed names, then all the names in $\Omega(\mathbf{\Sigma})$ belong there, too.

4.3. *Formal definition of net refinement $\Omega(\mathbf{\Sigma})$*

Under the assumptions made about the operator box Ω and Ω-tuple of plain boxes $\mathbf{\Sigma}$, the result of a simultaneous substitution of boxes $\mathbf{\Sigma}$ for the transitions in Ω is a labelled net $\Omega(\mathbf{\Sigma})$ whose components are defined thus.

Places. The set of places of $\Omega(\mathbf{\Sigma})$ is defined as the (disjoint) union

$$\begin{aligned} S_{\Omega(\mathbf{\Sigma})} &= \bigcup_{v \in T_\Omega} \mathsf{ST}^v_{\mathsf{new}} \cup \bigcup_{s \in S_\Omega} \mathsf{SP}^s_{\mathsf{new}} \\ &= \bigcup_{v \in T_\Omega} \{v \lhd i \mid i \in \ddot{\Sigma}_v\} \cup \\ &\quad \bigcup_{s \in S_\Omega} \bigl\{s \lhd \bigl(\{v \lhd x_v\}_{v \in {}^\bullet s} + \{w \lhd e_w\}_{w \in s^\bullet}\bigr) \mid x_v \in \Sigma_v^\circ \wedge e_w \in {}^\circ\Sigma_w\bigr\}. \end{aligned}$$

If s is an isolated place, ${}^\bullet s = s^\bullet = \emptyset$, then, by the definition of the tree appending operation, $\mathsf{SP}^s_{\mathsf{new}} = \{s\}$. Notice that the multiset of trees in the definition of a place

$$p = s \lhd \bigl(\{v \lhd x_v\}_{v \in {}^\bullet s} + \{w \lhd e_w\}_{w \in s^\bullet}\bigr) \in \mathsf{SP}^s_{\mathsf{new}} \tag{2}$$

is in fact a set because even in case there is a side-loop between s and $v = w$, then $x_v \neq e_w$ since x_v is an exit place and e_w is an entry place of Σ_v. For the example in Figure 13,

$$\begin{aligned} \mathsf{ST}^{v_1}_{\mathsf{new}} &= \bigl\{v_1 \lhd i \mid i \in \ddot{\Sigma}_{v_1}\bigr\} = \bigl\{v_1 \lhd i \mid i \in \bigl\{i_1^1, i_1^2\bigr\}\bigr\} = \bigl\{v_1 \lhd i_1^1, v_1 \lhd i_1^2\bigr\} \\ &= \{\tau_2, \tau_4\}, \\ \mathsf{SP}^e_{\mathsf{new}} &= \bigl\{e \lhd \bigl(\{v \lhd x_v\}_{v \in \{v_1\}} + \{w \lhd e_w\}_{w \in \{v_1, v_2\}}\bigr) \mid x_v \in \Sigma_v^\circ \wedge e_w \in {}^\circ\Sigma_w\bigr\} \\ &= \bigl\{e \lhd \bigl\{v_1 \lhd e_1^1, v_1 \lhd x_1, v_2 \lhd e_2\bigr\}, e \lhd \bigl\{v_1 \lhd e_1^2, v_1 \lhd x_1, v_2 \lhd e_2\bigr\}\bigr\}. \end{aligned}$$

The following notation is useful in manipulating the tree names upon which a newly constructed place is based. Let y be a transition in Ω. Then for $p = v \lhd i$ in $\mathsf{ST}^{v}_{\mathsf{new}}$, we define $\mathsf{trees}^{y}(p) = \{i\}$ if $v = y$, and $\mathsf{trees}^{y}(p) = \emptyset$ if $v \neq y$; and, for $p \in \mathsf{SP}^{s}_{\mathsf{new}}$ as in (2), we define

$$\mathsf{trees}^{y}(p) = \begin{cases} \{x_y, e_y\} & \text{if } y \in {}^{\bullet}s \cap s^{\bullet}, \\ \{x_y\} & \text{if } y \in {}^{\bullet}s \setminus s^{\bullet}, \\ \{e_y\} & \text{if } y \in s^{\bullet} \setminus {}^{\bullet}s, \\ \emptyset & \text{otherwise.} \end{cases}$$

For the example in Figure 13, $\mathsf{trees}^{v_1}(\tau_1) = \{e_1^2, x_1\}$ and $\mathsf{trees}^{v_1}(\tau_4) = \{i_1^1\}$.

Marking. The marking of a place p in $\Omega(\mathbf{\Sigma})$ is defined in the following way:

$$M_{\Omega(\mathbf{\Sigma})}(p) = \begin{cases} M_{\Sigma_v}(i) & \text{if } p = v \lhd i \in \mathsf{ST}^{v}_{\mathsf{new}}, \\ \sum\limits_{v \in {}^{\bullet}s} M_{\Sigma_v}(x_v) + \sum\limits_{w \in s^{\bullet}} M_{\Sigma_w}(e_w) & \text{if } p \in \mathsf{SP}^{s}_{\mathsf{new}} \text{ is as in (2).} \end{cases} \quad (3)$$

Notice that if $s \in S_\Omega$ is an isolated place then $M_{\Omega(\mathbf{\Sigma})}(s) = 0$, and that $M_{\Omega(\mathbf{\Sigma})}(p)$ is always a natural number since Ω is finite. For the example in Figure 13, $M_{\Omega(\mathbf{\Sigma})}(\tau_2) = M_{\Sigma_{v_1}}(i_1^2) = 1$ and $M_{\Omega(\mathbf{\Sigma})}(\tau_3) = M_{\Sigma_{v_1}}(e_1^1) + M_{\Sigma_{v_1}}(x_1) + M_{\Sigma_{v_2}}(e_2) = 1 + 0 + 0 = 1$.

Transitions. The set of transitions of $\Omega(\mathbf{\Sigma})$ is defined as the union

$$T_{\Omega(\mathbf{\Sigma})} = \bigcup_{v \in T_\Omega} \mathsf{T}^{v}_{\mathsf{new}} = \bigcup_{v \in T_\Omega} \big\{(v, \alpha) \lhd R \mid R \in \mathsf{mult}(T_{\Sigma_v}) \wedge \big(\lambda_{\Sigma_v}(R), \alpha\big) \in \lambda_\Omega(v)\big\}.$$

Notice that the multiset R in $(v, \alpha) \lhd R$ will never be empty since no pair in $\lambda_\Omega(v)$ has the empty multiset as its left argument. In Figure 13,

$$\begin{aligned} \mathsf{T}^{v_1}_{\mathsf{new}} &= \big\{(v_1, \alpha) \lhd R \mid R \in \mathsf{mult}\big(\{t_a, t_b, t_c\}\big) \wedge \big(\lambda_{\Sigma_{v_1}}(R), \alpha\big) \\ &\qquad\qquad \in \varrho_{id} \setminus \big\{(\{a\}, a)\big\} \cup \big\{(\{a, b\}, g)\big\}\big\} \\ &= \big\{(v_1, b) \lhd t_b, (v_1, g) \lhd \{t_a, t_b\}, (v_1, c) \lhd t_c\big\} = \{\xi_1, \xi_2, \xi_3\}. \end{aligned}$$

Similarly as for places, we will denote by $\mathsf{trees}(u)$ the multiset of transitions R upon which a newly constructed transition $u = (v, \alpha) \lhd R \in \mathsf{T}^{v}_{\mathsf{new}}$ is based. In Figure 13, $\mathsf{trees}(\xi_2) = \mathsf{trees}((v_1, g) \lhd \{t_a, t_b\}) = \{t_a, t_b\}$.

Labelling. The label of a node z in $\Omega(\mathbf{\Sigma})$ is defined in the following way:

$$\lambda_{\Omega(\mathbf{\Sigma})}(z) = \begin{cases} \mathsf{i} & \text{if } z \in \mathsf{ST}^{v}_{\mathsf{new}}, \\ \lambda_\Omega(s) & \text{if } z \in \mathsf{SP}^{s}_{\mathsf{new}}, \\ \alpha & \text{if } z = (v, \alpha) \lhd R \in \mathsf{T}^{v}_{\mathsf{new}}. \end{cases}$$

Weights. For a place p and transition u in $\mathsf{T}^{v}_{\mathsf{new}}$, the weight function is given by:

$$W_{\Omega(\mathbf{\Sigma})}(p,u) = \sum_{z \in \mathsf{trees}^{v}(p)} \sum_{t \in \mathsf{trees}(u)} W_{\Sigma_v}(z,t) \cdot \mathsf{trees}(u)(t),$$

$$W_{\Omega(\mathbf{\Sigma})}(u,p) = \sum_{z \in \mathsf{trees}^{v}(p)} \sum_{t \in \mathsf{trees}(u)} W_{\Sigma_v}(t,z) \cdot \mathsf{trees}(u)(t).$$

For the example in Figure 13,

$$\begin{aligned} W_{\Omega(\mathbf{\Sigma})}(\tau_1,\xi_1) &= \sum_{z \in \{e_1^2, x_1\}} \sum_{t \in \{t_b\}} W_{\Sigma_{v_1}}(z,t) = W_{\Sigma_{v_1}}\big(e_1^2, t_b\big) + W_{\Sigma_{v_1}}(x_1,t_b) \\ &= 1+0=1. \end{aligned}$$

PROPOSITION 2. *$\Omega(\mathbf{\Sigma})$ is a plain box which is unmarked if and only if so is each box in $\mathbf{\Sigma}$.*

Notice that the definition of refinement is constructed in such a way that:
- The e/i/x-labelling of $\Omega(\mathbf{\Sigma})$ depends only on the e/i/x-labelling of Ω (and not on that of $\mathbf{\Sigma}$).
- The net structure (connectivity) of $\Omega(\mathbf{\Sigma})$ depends on the e/i/x-labelling of $\mathbf{\Sigma}$ (and, of course, on the structure of all participating nets), but not on the e/i/x-labelling of Ω. More precisely, the e and x labelling of Σ_v's places is used (only) for linking with the pre- and post-places of v in Ω.

At this point, however, it is not guaranteed that if we start from static boxes Ω and $\mathbf{\Sigma}$, $\Omega(\mathbf{\Sigma})$ will be a static box. Indeed, it will turn out that some extra conditions need to be introduced for this to hold.

Net refinement is insensitive to taking isomorphic instances of the nets involved, as well as duplication equivalent instances of the refining boxes. Below we consider two instances of net refinement, $\Omega(\mathbf{\Sigma})$ and $\Omega'(\mathbf{\Sigma}')$.

THEOREM 1. *Let $\sim \in \{\mathsf{iso}, \mathsf{iso}_{\mathsf{ST}}\}$ and ψ be an isomorphism from Ω to Ω' such that $\Sigma_v \sim \Sigma'_{\psi(v)}$, for every $v \in T_\Omega$. Then $\Omega(\mathbf{\Sigma}) \sim \Omega'(\mathbf{\Sigma}')$.*

We do not consider the $\mathsf{iso}_{\mathsf{ST}}$-equivalence for operator boxes, since it does not preserve the number of transitions of operator boxes. Moreover, the behavioural equivalences, $\cong$ and $\approx$, will only be considered for concrete box algebras as they are highly influenced by the particular choice of operator boxes.

5. Recursive equations on boxes

This section presents a method of solving recursive equations in the domain of boxes, following a common practice of using fixpoints and limit constructions. We will apply the standard methods, but also show that if they are applied to Petri nets, a number of specific problems arise which can be solved with help of the tree device for naming new places and transitions in net refinement. To simplify the presentation, we shall consider a *single* equation for a *recursion variable* X, of the form:

$$X \stackrel{\text{df}}{=} \Omega(\mathbf{\Delta}) \tag{4}$$

where Ω is an operator box, and $\mathbf{\Delta}$ is an Ω-tuple of recursion variables and unmarked plain boxes, called *rec-boxes*, Box^{rec}. A rec-box Σ is a *solution* of (4) if $\Sigma = \Omega(\mathbf{\Delta}[\Sigma])$, where $\mathbf{\Delta}[\Sigma]$ is $\mathbf{\Delta}$ with every occurrence of X replaced by Σ.

5.1. *Solving a recursive equation*

We equip the set of rec-boxes with an *ordering relation*, $\sqsubseteq$. For two rec-boxes Σ and Θ, we denote $\Sigma \sqsubseteq \Theta$ if $S_\Sigma \subseteq S_\Theta$, $T_\Sigma \subseteq T_\Theta$, $W_\Sigma = W_\Theta|_{(S_\Sigma \times T_\Sigma) \cup (T_\Sigma \times S_\Sigma)}$ and $\lambda_\Sigma = \lambda_\Theta|_{S_\Sigma \cup T_\Sigma}$. The relation $\sqsubseteq$ is a partial order, and a *chain* is an infinite sequence

$$\chi = \Sigma^0 \quad \Sigma^1 \quad \Sigma^2 \quad \ldots$$

of rec-boxes such that $\Sigma^k \sqsubseteq \Sigma^{k+1}$ for every $k \geqslant 0$. Each such chain has a unique least upper bound or *limit*, denoted by $\bigsqcup \chi$, which is a rec-box given by:

$$\bigsqcup \chi = \left(\bigcup_{k=0}^{\infty} S_{\Sigma^k},\ \bigcup_{k=0}^{\infty} T_{\Sigma^k},\ \bigcup_{k=0}^{\infty} W_{\Sigma^k},\ \bigcup_{k=0}^{\infty} \lambda_{\Sigma^k},\ \emptyset \right).$$

A recursive equation may have different, even non-isomorphic, solutions, each such solution being uniquely identified by the sets of entry and exit places:

THEOREM 2. *Let Σ and Θ be solutions of* (4).
(1) *The sets of transitions of Σ and Θ and their labellings are the same.*
(2) $\Sigma \sqsubseteq \Theta$ *if and only if* ${}^\circ\Sigma \subseteq {}^\circ\Theta$ *and* $\Sigma^\circ \subseteq \Theta^\circ$.
(3) $\Sigma = \Theta$ *if and only if* ${}^\circ\Sigma = {}^\circ\Theta$ *and* $\Sigma^\circ = \Theta^\circ$.

It may even be the case that some solutions are countable while others are uncountable (this is true, for example, for the equation corresponding to $X \stackrel{\text{df}}{=} (a\|b) \,\square\, X$, which is studied in [11]). However, for a recursive system arising in the context of the generic algebra described later, all the solutions can be shown to have *identical* behaviour.

Using fixpoints. The right-hand side of (4) induces a mapping $\mathbf{F}: \mathsf{Box}^{rec} \to \mathsf{Box}^{rec}$ defined by

$$\mathbf{F}(\Sigma) = \Omega\big(\mathbf{\Delta}[\Sigma]\big).$$

Hence solving Equation (4) amounts to finding the fixpoints of $\mathbf{F}$. Following the usual treatment of fixpoint theory, we now proceed by trying to establish the monotonicity and continuity of $\mathbf{F}$. We say that $\mathbf{F}$ is *monotonic* if, for all rec-boxes Σ and Θ, $\Sigma \sqsubseteq \Theta$ implies $\mathbf{F}(\Sigma) \sqsubseteq \mathbf{F}(\Theta)$, and *continuous* if for every chain $\chi = \Sigma^0 \Sigma^1 \ldots$ of rec-boxes,

$$\mathbf{F}\Big(\bigsqcup \chi\Big) = \bigsqcup \mathbf{F}(\Sigma^0)\mathbf{F}(\Sigma^1)\ldots$$

THEOREM 3. $\mathbf{F}$ *is monotonic and even continuous.*

Thus one can attempt to solve (4) by constructing a chain of rec-boxes formed through successive applications of the mapping $\mathbf{F}$ to some initial approximation Υ. Note, though, that we do not necessarily have a complete partial order.

For a rec-box Υ, we denote by $\mathbf{F}^*(\Upsilon)$ the infinite sequence $\mathbf{F}^0(\Upsilon)\mathbf{F}^1(\Upsilon)\mathbf{F}^2(\Upsilon)\ldots$ where $\mathbf{F}^0(\Upsilon) = \Upsilon$ and $\mathbf{F}^n(\Upsilon) = \mathbf{F}(\mathbf{F}^{n-1}(\Upsilon))$, for every $n \geqslant 1$. We then call Υ a *seed* for (4) if $\Upsilon \sqsubseteq \mathbf{F}(\Upsilon)$; in such a case, $\mathbf{F}^*(\Upsilon)$ is a chain of rec-boxes. It turns out that such a chain always yields a solution, and that it suffices to consider ex-boxes as the only seeds.

THEOREM 4.

(1) *If* Υ *is a seed then* $\bigsqcup \mathbf{F}^*(\Upsilon)$ *is a solution of* (4).

(2) *If* Σ *is a solution of* (4) *then* $\Upsilon = \mathsf{ex}({}^\circ\Sigma, \Sigma^\circ)$ *is a seed such that* $\bigsqcup \mathbf{F}^*(\Upsilon) = \Sigma$.

Before looking closer at the seed boxes, we present an example.

Example of the limit construction. Consider the following instance of (4) which intuitively corresponds to the recursive equation $X \stackrel{\mathrm{df}}{=} a; (X \,\Box\, b)$:

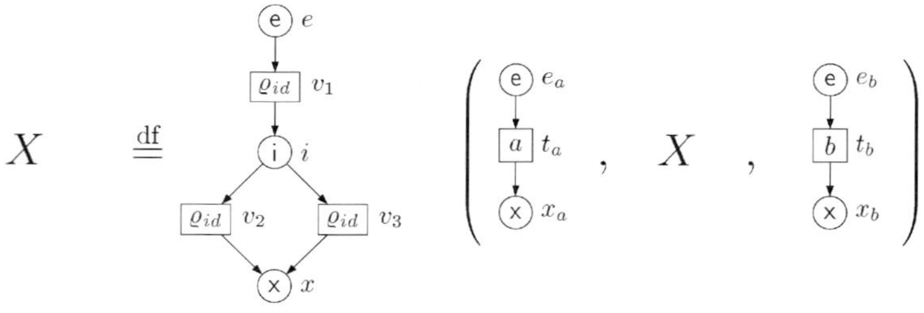

(5)

Let us examine how a solution of the net equation (5) might be constructed using the limit construction. Suppose that there already exists an approximation Υ^0 given as a hypothetical seed. Then inserting that approximation in the place of the X on the right-hand side of (5) yields a new box $\Upsilon^1 = \Omega(\Sigma_a, \Upsilon^0, \Sigma_b)$, where Ω is the operator box in this equation and Σ_a and Σ_b are the two plain boxes, which is a slightly better approximation.

We then construct another approximation $\Upsilon^2 = \Omega(\Sigma_a, \Upsilon^1, \Sigma_b)$ by the same principle, and so on. We already know that if Υ^0 is a seed then the limit of the sequence $\Upsilon^0\Upsilon^1\ldots$ so constructed exists and solves Equation (5). We now claim that the ex-box Υ^0 shown in Figure 14 (with these particular place names, defined on the bottom of the figure) is a good seed for (5). Figure 14 shows also the first four approximations resulting from this seed and, finally, it shows $\Sigma = \bigsqcup \mathbf{F}^*(\Upsilon^0)$, which is the limit of the infinite series of boxes whose first four elements are depicted. The reader may check that this box satisfies (5). To

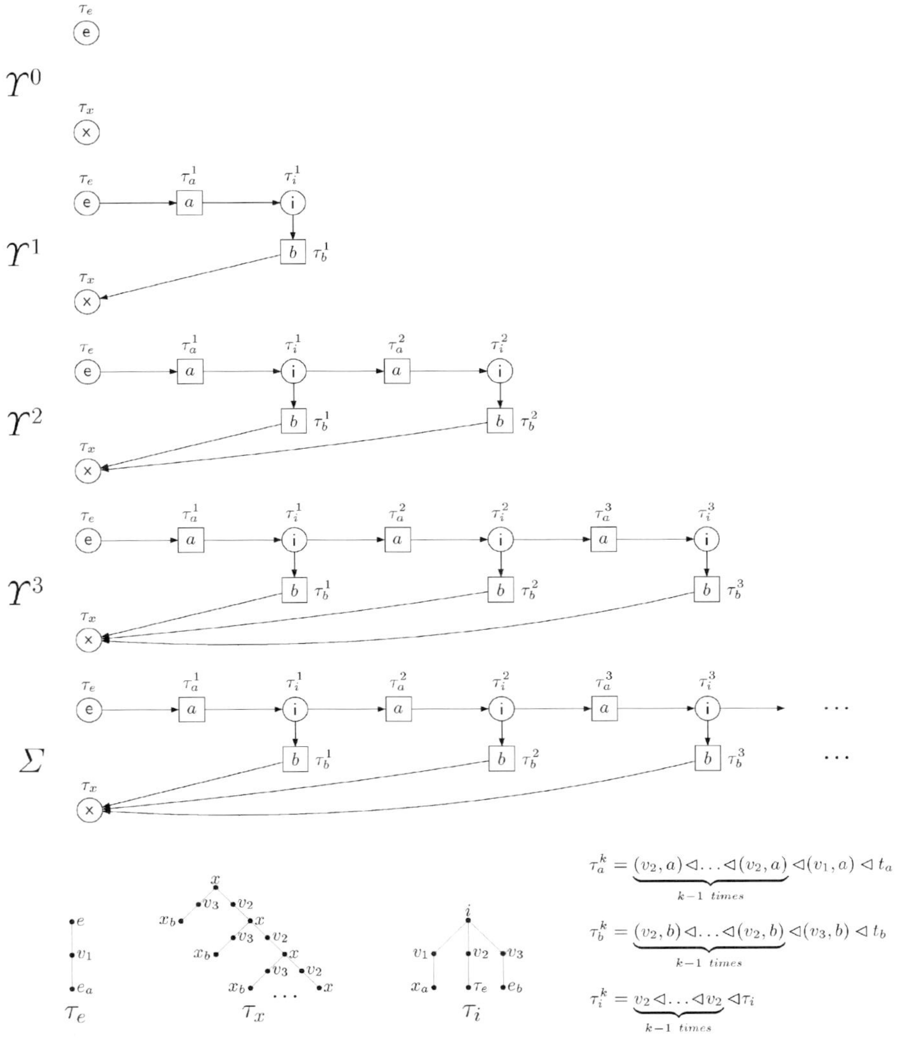

Fig. 14. Four approximations for Equation (5) and the resulting solution.

see this, replace the variable X on the right-hand side of the equation by the box Σ shown in Figure 14, perform net refinement $\Omega(\Sigma_a, \Sigma, \Sigma_b)$, and check that the result is not just isomorphic but even *identical* to Σ, i.e., it has the same places and the same transitions with the same connections and the same labels.

Thus, once an appropriate starting point Υ^0 has been chosen, the approximation of a solution is straightforward (we just keep refining successive approximations into the right-hand side of the equation under consideration and compute the limit) and leads to a solution in all cases.

Deriving seed boxes. To find a seed Υ for (4), we need to ensure that $\Upsilon \sqsubseteq \mathbf{F}(\Upsilon)$ holds for Υ and the next approximation $\mathbf{F}(\Upsilon)$. Moreover, we shall require that ${}^\circ\Upsilon = {}^\circ\mathbf{F}(\Upsilon)$ and $\Upsilon^\circ = \mathbf{F}(\Upsilon)^\circ$, as for such seeds there is a one-to-one correspondence with the solutions of (4) (see Theorem 4(2)). One might ask why the rec-box Υ^0 in Figure 14 was a suitable starting point for solving Equation (5). First, it is a good idea to start always with an ex-box, because these are minimal with respect to the number of transitions and because, as a consequence of Theorem 4(2), all solutions can be so obtained. Moreover, we will only need to ensure that ${}^\circ\Upsilon = {}^\circ\mathbf{F}(\Upsilon)$ and $\Upsilon^\circ = \mathbf{F}(\Upsilon)^\circ$, since then $\Upsilon \sqsubseteq \mathbf{F}(\Upsilon)$ always holds. Let us now fix a hypothetical starting ex-box

$$\Upsilon = \mathsf{ex}(\mathcal{P}, \mathcal{R})$$

and develop conditions for it to be a seed. First, $\mathcal{P}$ and $\mathcal{R}$ should be non-empty and disjoint. Second, with a view to ensure that ${}^\circ\Upsilon = {}^\circ\mathbf{F}(\Upsilon)$ and $\Upsilon^\circ = \mathbf{F}(\Upsilon)^\circ$, some conditions need to be imposed on $\mathcal{P}$ and $\mathcal{R}$, essentially by performing the refinement at the entry and exit places. As both place sets are sets of trees, these conditions can be translated into equalities over sets of place trees.

We define a mapping $\mathbf{f}: 2^{\mathsf{P}_{\mathsf{tree}}} \times 2^{\mathsf{P}_{\mathsf{tree}}} \to 2^{\mathsf{P}_{\mathsf{tree}}} \times 2^{\mathsf{P}_{\mathsf{tree}}}$ associated with $\mathbf{F}$, and through $\mathbf{F}$ with (4), so that for all $P, R \subseteq \mathsf{P}_{\mathsf{tree}}$,

$$\mathbf{f}(P, R) = \Big(\bigcup_{s \in {}^\circ\Omega} \mathbf{f}_s(P, R), \quad \bigcup_{s \in \Omega^\circ} \mathbf{f}_s(P, R) \Big) \tag{6}$$

where each $\mathbf{f}_s(P, R)$ comprises all place trees $s \lhd (\{v \lhd p_v\}_{v \in s^\bullet} + \{w \lhd r_w\}_{w \in {}^\bullet s})$ such that:

$$p_v \in \begin{cases} P & \text{if } \Delta_v = X \\ {}^\circ\Delta_v & \text{if } \Delta_v \neq X \end{cases} \quad \text{and} \quad r_w \in \begin{cases} R & \text{if } \Delta_w = X \\ \Delta_w^\circ & \text{if } \Delta_w \neq X. \end{cases}$$

A pair of sets $(P, R) \in 2^{\mathsf{P}_{\mathsf{tree}}} \times 2^{\mathsf{P}_{\mathsf{tree}}}$ is a *positive fixpoint* of $\mathbf{f}$ if $\mathbf{f}(P, R) = (P, R)$ and $P \neq \emptyset \neq R$. The function $\mathbf{f}$ can be used to calculate the entry and exit places of $\mathbf{F}(\Upsilon)$ since $\mathbf{f}(\mathcal{P}, \mathcal{R}) = ({}^\circ\mathbf{F}(\Upsilon), \mathbf{F}(\Upsilon)^\circ)$. As a result, $\Upsilon = \mathsf{ex}(\mathcal{P}, \mathcal{R})$ is a seed we are looking for if and only if $(\mathcal{P}, \mathcal{R})$ is a positive fixpoint of $\mathbf{f}$. Thus what we aim at amounts to finding positive fixpoints of $\mathbf{f}$. Using the notation introduced in (6), the latter amounts to finding non-empty sets $\mathcal{P}$ and $\mathcal{R}$ solving the following *tree equations*:

$$\begin{aligned} \mathcal{P} &= \bigcup_{s \in {}^\circ\Omega} \mathbf{f}_s(\mathcal{P}, \mathcal{R}), \\ \mathcal{R} &= \bigcup_{s \in \Omega^\circ} \mathbf{f}_s(\mathcal{P}, \mathcal{R}). \end{aligned} \tag{7}$$

The function $\mathbf{f}$ is a monotonic mapping on the domain $(2^{\mathsf{P}_{\mathsf{tree}}} \times 2^{\mathsf{P}_{\mathsf{tree}}}, \subseteq)$ which is a complete lattice with set union as the 'join' and intersection as the 'meet' operation.[3] It therefore follows that Tarski's theorem [18] can be applied.

[3] All operations, including $\subseteq$, are carried out component-wise.

THEOREM 5. *The set of fixpoints of* $\mathbf{f}$ *is a complete lattice whose maximal element,* $\mathsf{fix}_{\mathsf{max}}$, *is positive. Moreover, there is at least one positive fixpoint which is minimal among all the positive fixpoints of* $\mathbf{f}$.

This leads, though not immediately, to the main result of this section.

THEOREM 6. *The recursive equation* (4) *has at least one solution. Moreover, the set of solutions has a unique maximal element and at least one minimal element.*

One might be tempted to always take the minimal fixpoint of $\mathbf{f}$, $\mathsf{fix}_{\mathsf{min}}$, and treat the corresponding solution as *the* solution of (4) since, by Theorem 2(2), such a solution would be minimal. Unfortunately, the minimal fixpoint of $\mathbf{f}$ may not be positive making it unsuitable for our purposes because boxes are ex-restricted. In other words, a minimal solution of (4) that exists by Theorem 6 may not correspond to the minimal fixpoint of $\mathbf{f}$, and it may not be unique, either; for instance, an equation corresponding to $X \stackrel{\text{df}}{=} (a\|b) \,\square\, X$ has an uncountable number of non-isomorphic minimal solutions [11].

There is a simple structural condition guaranteeing that $\mathbf{f}$ is a constant mapping and therefore trivially has a unique fixpoint. The recursive equation (4) is *front-guarded* if no entry place of Ω has a direct access to the variable X; that is, if $\Delta_v \neq X$, for every $v \in {}^{\bullet}({}^{\circ}\Omega) \cup ({}^{\circ}\Omega)^{\bullet}$. Similarly, it is *rear-guarded* if $\Delta_v \neq X$, for every $v \in {}^{\bullet}(\Omega^{\circ}) \cup (\Omega^{\circ})^{\bullet}$. And (4) is *guarded* if it is both front- and rear-guarded.

THEOREM 7. *If* (4) *is guarded then it has a unique solution.*

Solving recursive equations on boxes is insensitive to taking structurally equivalent nets. Below, we can consider two instances of (4), $X \stackrel{\text{df}}{=} \Omega(\mathbf{\Delta})$ and $X \stackrel{\text{df}}{=} \Omega'(\mathbf{\Delta}')$.

THEOREM 8. *Let* $\sim \in \{\mathsf{iso}, \mathsf{iso}_{\mathsf{ST}}\}$ *and* ψ *be an isomorphism from* Ω *to* Ω' *such that* $\Delta_v = X = \Delta'_{\psi(v)}$ *or* $\Delta_v \sim \Delta'_{\psi(v)}$, *for every* $v \in T_{\Omega}$. *Then, for every solution* Σ *of the first equation, there is a solution* Σ' *of the second equation such that* $\Sigma \sim \Sigma'$.

The treatment provided for a single recursive equation is easily lifted to systems of recursive equations on boxes, using standard techniques on Cartesian product spaces (for finite sets of equations) or on function spaces (for infinite sets of equations). The reader is referred to [11] for details.

5.2. *Further recursion examples*

Let us first return to Equation (5). In this case, to find the fixpoints of $\mathbf{f}$ we need to solve the following instance of the tree equations (7):

$$\mathcal{P} = \{e \lhd v_1 \lhd e_a\},$$
$$\mathcal{R} = \big\{x \lhd \{v_2 \lhd t, v_3 \lhd x_b\} \mid t \in \mathcal{R}\big\}.$$

There are exactly two solutions, $\mathsf{fix}_{\mathsf{min}} = (\{\tau_e\}, \emptyset)$ and $\mathsf{fix}_{\mathsf{max}} = (\{\tau_e\}, \{\tau_x\})$, where τ_e and τ_x are as in Figure 14. Since $\mathsf{fix}_{\mathsf{min}}$ is not positive, there is only one solution of Equation (5) which can be obtained from the seed $\Upsilon = \mathsf{ex}(\{\tau_e\}, \{\tau_x\})$, shown in Figure 14.

The first recursion example was front-guarded, but not rear-guarded. The next example is neither front-guarded nor rear-guarded:

$X \overset{\mathrm{df}}{=}$ 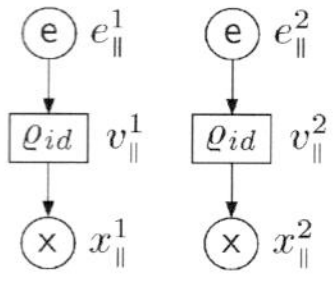

(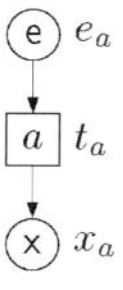

, X) (8)

Intuitively, this corresponds to $X \overset{\mathrm{df}}{=} a \| X$. To construct a seed, we need to solve the following instance of the tree equations for the auxiliary function **f**:

$$\begin{aligned} \mathcal{P} &= \{e^1_\| \lhd v^1_\| \lhd e_a\} \cup \{e^2_\| \lhd v^2_\| \lhd t \mid t \in \mathcal{P}\}, \\ \mathcal{R} &= \{x^1_\| \lhd v^1_\| \lhd x_a\} \cup \{x^2_\| \lhd v^2_\| \lhd t \mid t \in \mathcal{R}\}. \end{aligned} \tag{9}$$

And there is a one-to-one correspondence between the positive solutions of (9) and the solutions of (8). But now we have several positive solutions; for example, the minimal fixpoint containing only finite trees, and the maximal fixpoint which contains infinite trees as well. More precisely, we have $\mathsf{fix}_{\mathsf{min}} = (P, R)$ and $\mathsf{fix}_{\mathsf{max}} = (P \cup \{\tau_e\}, R \cup \{\tau_x\})$, where:

$$\begin{aligned} P &= \{\tau^1_e, \tau^2_e, \tau^3_e, \dots\} \qquad & \tau_e &= e^2_\| \lhd v^2_\| \lhd e^2_\| \lhd v^2_\| \lhd e^2_\| \lhd v^2_\| \cdots, \\ R &= \{\tau^1_x, \tau^2_x, \tau^3_x, \dots\} \qquad & \tau_x &= x^2_\| \lhd v^2_\| \lhd x^2_\| \lhd v^2_\| \lhd x^2_\| \lhd v^2_\| \cdots, \end{aligned}$$

where $\tau^1_e = e^1_\| \lhd v^1_\| \lhd e_a$ and $\tau^1_x = x^1_\| \lhd v^1_\| \lhd x_a$ and $\tau^{k+1}_e = e^2_\| \lhd v^2_\| \lhd \tau^k_e$ and $\tau^{k+1}_x = x^2_\| \lhd v^2_\| \lhd \tau^k_x$, for $k \geqslant 1$. In all, there are four positive fixpoints which determine four starting boxes to begin the approximation of a solution of (8): Σ^{min} using $\mathsf{fix}_{\mathsf{min}}$, $\Sigma^{max:min}$ using $(P \cup \{\tau_e\}, R)$, $\Sigma^{min:max}$ using $(P, R \cup \{\tau_x\})$ and Σ^{max} using $\mathsf{fix}_{\mathsf{max}}$. Depending on the choice, one of the four different solutions of (8) is obtained, shown in Figure 15. It is not hard to check that all four boxes are indeed solutions of (8) in the same strict sense as before. The interleaving behaviour – and also behaviour in terms of finite steps – of the four solutions is the same. However, there are subtle differences between these solutions, which are discussed in [11].

As the next equation – which informally corresponds to $X \overset{\mathrm{df}}{=} X$ – consider the following:

$X \overset{\mathrm{df}}{=}$ 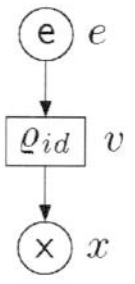

(X) (10)

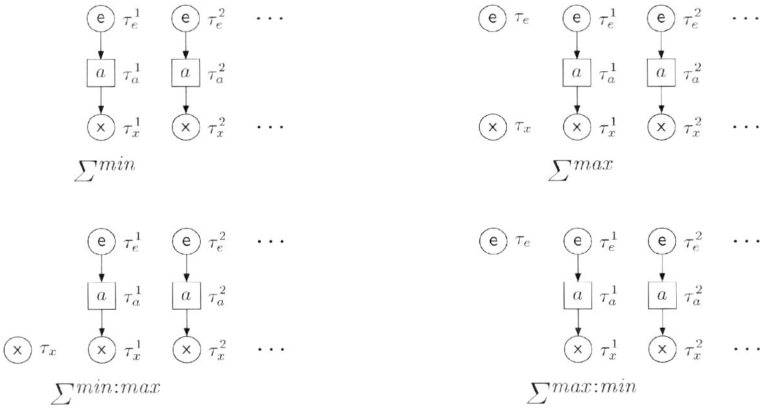

Fig. 15. The solutions of Equation (8).

It leads to the following tree equations for the entry and exit places of the starting box:

$$\mathcal{P} = \{e \lhd v \lhd t \mid t \in \mathcal{P}\},$$
$$\mathcal{R} = \{x \lhd v \lhd t \mid t \in \mathcal{R}\},$$

which means that **f** has a unique positive fixpoint

$$(\mathcal{P}, \mathcal{R}) = \big(\{e \lhd v \lhd e \lhd v \lhd \cdots\}, \{x \lhd v \lhd x \lhd v \lhd \cdots\}\big).$$

In this case, the approximations never change anything, and the box solving the equation is $\mathsf{ex}(\mathcal{P}, \mathcal{R})$. Note that despite the colloquial meaning of $X \stackrel{\text{df}}{=} X$, no box with a transition solves (10). Thus solving recursive equation by boxes with least behaviour is already in-built in our approach.

We feel that this gives an intuitive semantics to the 'recursive call' $X \stackrel{\text{df}}{=} X$ because we do not model procedure entry and procedure exit as separate (silent) actions. Thus, $X \stackrel{\text{df}}{=} X$ essentially means that X is continually 'rewritten' by itself and effects no behaviour at all, be it silent or observable. This is in contrast to the view that $X \stackrel{\text{df}}{=} X$ implies continual entry to a procedure and thus an infinite succession of silent moves; to model such behaviour, we would rather use $X \stackrel{\text{df}}{=} \emptyset; X$ where $\emptyset$ represents a silent action (or even $X \stackrel{\text{df}}{=} \emptyset; X; \emptyset$, to model exit as well as entry, even though none of the transitions coming from the $\emptyset$ following X will be executable).

6. An algebra of nets

In this section, we restrict our attention to a special class of operator boxes, called *sos-operator boxes* (Section 6.1). They are defined in such a way that they allow two results to be true, namely Theorems 9 and 10 in Section 6.2. Even though these two theorems are

The *domain of application* of Ω, denoted by dom_Ω, is then defined as the set comprising every Ω-tuple of static and dynamic boxes $\mathbf{\Sigma}$ whose factorization belongs to fact_Ω, and such that, for every $v \in T_\Omega$:

– If ${}^\bullet v \cap v^\bullet \neq \emptyset$ then Σ_v is ex-exclusive. (Dom1)

– If v is not reversible then Σ_v is x-directed. (Dom2)

The first condition, (Dom1), is meant to prevent the non-safeness of refinements in which a box is refined into a transition which is part of a side-loop (see formula (3) for the marking in net refinement). A typical situation addressed by (Dom1) is the process-algebraic expression $(a\|b)^*$. Translating the unary iteration directly yields a transition v satisfying ${}^\bullet v \cap v^\bullet \neq \emptyset$, and the typical Petri net of $a\|b$ is not ex-exclusive. More careful arguments must be found in order to deal with this situation ([11]; see also Section 9). The second condition, (Dom2), is related to the possible presence of dead transitions in an operator box. Referring again to the sos-operator box in Figure 16, one may observe that if we were allowed to refine the transition v_2 with a non-x-directed box Σ_{v_2}, then some behaviours originating from Σ_{v_2} could be possible from the entry marking of the refined net, while no behaviour beginning at the entry nor exit marking of the operator box can possibly involve v_2.[4]

EXAMPLE (*Continued*). Taking the sos-operator box in Figure 17 and the factorizations in Table 1, we obtain:

$$\mathsf{dom}_{\Omega_0} = \mathsf{Box}^d \times \mathsf{Box}^d \times \mathsf{Box}^s \cup \mathsf{Box}^s \times \mathsf{Box}^s \times \mathsf{Box}^d \cup \mathsf{Box}^s \times \mathsf{Box}^s \times \mathsf{Box}^s.$$

In Figure 17, $\mathbf{\Upsilon} = (\Upsilon_{v_1}, \Upsilon_{v_2}, \Upsilon_{v_3})$ is an Ω_0-tuple with the factorization, $(\{v_1\}, \{v_2\}, \emptyset, \{v_3\})$, which belongs to fact_{Ω_0}. Hence $\mathbf{\Upsilon}$ belongs to the domain of application of the sos-operator box Ω_0. Notice that (Dom1) and (Dom2) are trivially satisfied in this case since Ω_0 is pure and all its transitions are reversible. The result of performing the refinement $\Omega_0(\mathbf{\Upsilon})$ is shown in Figure 17.

6.2. *Structural operational semantics of composite boxes*

A typical structural operational semantics rule in process algebras [1] stipulates that if a process term $expr = op(expr_1, \ldots, expr_n)$ can make a move

$$expr \xrightarrow{act} expr'$$

then *act* can be composed out of some individual moves of the sub-terms $expr_1, \ldots, expr_n$. And conversely, if the sub-terms can individually make moves satisfying some prescribed conditions, then the whole term can make a combined move. We translate this scheme into the present net-theoretic framework.

---

4 Actually, one could have required that no transition be enabled at $\Sigma^\circ_{v_2}$, but x-directedness is a simple structural sufficient condition for this.

Let Ω be an sos-operator box and $\mathbf{\Sigma}$ be an Ω-tuple in its domain, dom_Ω. When dealing with the operational semantics of the compositional box $\Omega(\mathbf{\Sigma})$, we shall use the notation

$$(\Omega : \mathbf{\Sigma}) \xrightarrow{U} (\Omega : \mathbf{\Theta}) \tag{11}$$

to mean that the boxes $\mathbf{\Sigma}$ can individually make moves which, when combined, yield step U and lead to new boxes $\mathbf{\Theta}$. By definition, this will be the case whenever U is a finite set of transitions of $\Omega(\mathbf{\Sigma})$ and, for every transition v in Ω, $U \cap \mathsf{T}^v_{\mathsf{new}} = \{u_1, \ldots, u_k\}$ is a set of transitions such that

$$\Sigma_v \left[\mathsf{trees}(u_1) + \cdots + \mathsf{trees}(u_k)\right\rangle \Theta_v. \tag{12}$$

In fact, the multiset sum in (12) is a disjoint union of k sets.

Instead of expressing behaviours in terms of transitions, it is also possible to express them using actions, through the labelling function

$$\lambda_{\Omega(\mathbf{\Sigma})}(U) = \sum_{u \in U} \{\lambda_{\Omega(\mathbf{\Sigma})}(u)\}$$

which returns multisets rather than sets since different transitions may have the same label.

EXAMPLE (*Continued*). For the boxes in Figure 17 we have

$$(\Omega_0 : \mathbf{\Upsilon}) \xrightarrow{\{w_1, w_3\}} (\Omega_0 : \mathbf{\Theta}), \tag{13}$$

where $\mathbf{\Theta} = (\Theta_{v_1}, \Theta_{v_2}, \Theta_{v_3}) = (\underline{\Upsilon_{v_1}}, \underline{\Upsilon_{v_2}}, \Upsilon_{v_3})$. Indeed, by setting $U = \{w_1, w_3\}$ we obtain:

$$\begin{array}{lll} U \cap \mathsf{T}^{v_1}_{\mathsf{new}} = \{w_1\}, & \mathsf{trees}(w_1) = \{t_{11}, t_{12}\}, & \Upsilon_{v_1} \left[\{t_{11}, t_{12}\}\right\rangle \Theta_{v_1}, \\ U \cap \mathsf{T}^{v_2}_{\mathsf{new}} = \{w_3\}, & \mathsf{trees}(w_3) = \{t_{22}\}, & \Upsilon_{v_2} \left[\{t_{22}\}\right\rangle \Theta_{v_2}, \\ U \cap \mathsf{T}^{v_3}_{\mathsf{new}} = \emptyset, & & \Upsilon_{v_3} \left[\emptyset\right\rangle \Theta_{v_3}. \end{array}$$

Notice that $\{w_1, w_3\}$ is also a valid step for the box $\Omega_0(\mathbf{\Upsilon})$ and $\Omega_0(\mathbf{\Upsilon})\,[\{w_1, w_3\}\rangle\, \Omega_0(\mathbf{\Theta})$.

THEOREM 9. *Let $\mathbf{\Sigma}$ be a tuple in the domain of an sos-operator box Ω. If $\mathbf{\Theta}$ and U are as in* (11)*, then $\mathbf{\Theta} \in \mathsf{dom}_\Omega$ and $\Omega(\mathbf{\Sigma})\,[U\rangle\,\Omega(\mathbf{\Theta})$.*

This theorem states that the transformation captured by (11) is *correct* (or *sound*) with respect to the standard net semantics.

A direct converse of Theorem 9 does not, in general, hold true. For if we consider the tuple of boxes $\mathbf{\Theta}$ as in (13), then $\Omega_0(\mathbf{\Theta})\,[\{w_4\}\rangle$, yet no move at all of the form

$$(\Omega_0 : \mathbf{\Theta}) \xrightarrow{U} (\Omega_0 : \mathbf{\Theta}')$$

for a non-empty U is possible. This is so because, when composing the nets, the tokens contributed by Θ_{v_1} and Θ_{v_2} are inserted into the composed net in such a way that they could all have been contributed by Θ_{v_3} as well. More precisely, we have $\Omega(\boldsymbol{\Theta}) = \Omega(\boldsymbol{\Psi})$, where $\Psi_{v_1} = \lfloor \Upsilon_{v_1} \rfloor$ and $\Psi_{v_2} = \lfloor \Upsilon_{v_2} \rfloor$ and $\Psi_{v_3} = \overline{\Upsilon_{v_3}}$. And the situation is now different since a move

$$(\Omega_0 : \boldsymbol{\Psi}) \xrightarrow{\{w_4\}} (\Omega_0 : \boldsymbol{\Psi}')$$

is possible with $\Psi'_{v_1} = \Psi_{v_1}$, $\Psi'_{v_2} = \Psi_{v_2}$ and $\Psi'_{v_3} = \underline{\Psi_{v_3}}$. Hence the markings in a tuple of boxes $\boldsymbol{\Sigma}$ may need to be re-arranged before attempting to derive a move which is admitted by their composition $\Omega(\boldsymbol{\Sigma})$. Such a re-arrangement is formalized by a similarity relation $\equiv_\Omega$ on Ω-tuples of boxes. This is, of course, very similar to the case discussed earlier, that the token on s_2 in (N, M_1) (see Figure 1) can be interpreted either as an exit token belonging to the a or as an entry token belonging to the b. At this point, we need to define a net-theoretic counterpart of the similarity relation $\equiv$ discussed in Section 2.1.

Let Ω be an sos-operator box and $\boldsymbol{\Sigma}$ and $\boldsymbol{\Theta}$ be Ω-tuples of static and dynamic boxes whose factorizations are respectively μ and κ. Then $\boldsymbol{\Sigma} \equiv_\Omega \boldsymbol{\Theta}$ if μ and κ are factorizations of the same complex marking, $\lfloor \boldsymbol{\Sigma} \rfloor = \lfloor \boldsymbol{\Theta} \rfloor$ and $\Sigma_v = \Theta_v$, for every $v \in \mu_d = \kappa_d$.

$\equiv_\Omega$ is an equivalence relation which is closed in the domain of application of Ω, and relates tuples which yield the same boxes through refinement, $\Omega(\boldsymbol{\Sigma}) = \Omega(\boldsymbol{\Theta})$. A partial converse of the latter property also holds: if $\boldsymbol{\Sigma}$ and $\boldsymbol{\Theta}$ are tuples in the domain of Ω such that $\lfloor \boldsymbol{\Sigma} \rfloor = \lfloor \boldsymbol{\Theta} \rfloor$ and $\Omega(\boldsymbol{\Sigma}) = \Omega(\boldsymbol{\Theta})$, then $\boldsymbol{\Sigma} \equiv_\Omega \boldsymbol{\Theta}$.

With help of the similarity relation $\equiv_\Omega$ it is possible to reverse Theorem 9.

THEOREM 10. *Let $\boldsymbol{\Psi}$ be a tuple in the domain of an sos-operator box Ω. If $\Omega(\boldsymbol{\Psi})\,[U\rangle\, \Sigma$ then there are $\boldsymbol{\Sigma}$ and $\boldsymbol{\Theta}$ in dom_Ω such that $\boldsymbol{\Psi} \equiv_\Omega \boldsymbol{\Sigma}$, $\Omega(\boldsymbol{\Theta}) = \Sigma$ and* (11) *holds.*

This theorem states that the transformation captured by (11) is *complete* with respect to the standard net semantics.

Various important consequences may be derived from Theorems 9 and 10. In particular, they imply that $\Omega(\boldsymbol{\Sigma})$ is a static or dynamic box, and if Σ is a net derivable from $\overline{\Omega(\boldsymbol{\Sigma})}$ or $\underline{\Omega(\boldsymbol{\Sigma})}$, then there is a tuple $\boldsymbol{\Theta}$ in the domain of Ω such that $\lfloor \boldsymbol{\Sigma} \rfloor = \lfloor \boldsymbol{\Theta} \rfloor$ and $\Sigma = \Omega(\boldsymbol{\Theta})$. As a summary, the SOS rule for boxes can be presented in the following way:

$$\frac{(\Omega : \boldsymbol{\Theta}) \xrightarrow{U} (\Omega : \boldsymbol{\Theta}')}{\Omega(\boldsymbol{\Sigma})\,[U\rangle\, \Omega(\boldsymbol{\Sigma}')} \quad \boldsymbol{\Sigma} \equiv_\Omega \boldsymbol{\Theta},\ \boldsymbol{\Sigma}' \equiv_\Omega \boldsymbol{\Theta}'$$

6.3. *Further restrictions on the application domain*

The conditions (Dom1) and (Dom2) imposed on the domain of an sos-operator box are sufficient for the main compositionality results; in particular, for showing the full consistency between the net semantics and operational semantics in the process algebra presented in Section 7.5. So far, however, this is unrelated to any process algebraic considerations, except that the shape of the SOS rule is inspired by process algebra theory. Starting with

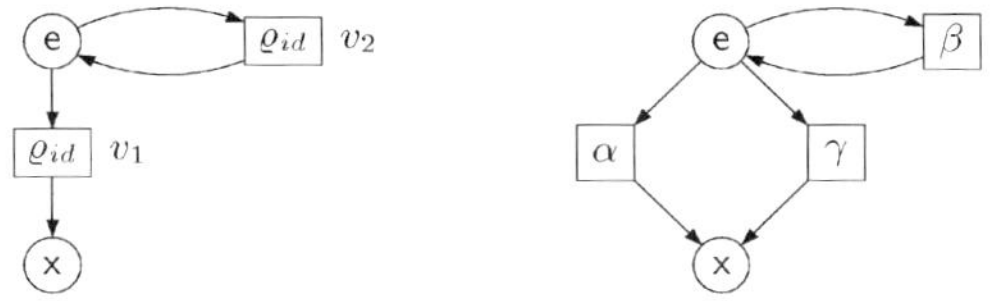

Fig. 18. Operator box Ω^* and a net modelling (tentatively) $\Omega_{\Box}(\mathsf{N}_\alpha, \Omega^*(\mathsf{N}_\beta, \mathsf{N}_\gamma))$.

this section, we shall discuss relationships between operations on nets (realized by operator boxes) and process algebraic operators. In the present section, we address a behavioural consistency issue that pertains to such a discussion.

Consider the operator box Ω^* on the left hand side of Figure 18, whose terminated behaviours are modelled by the expression $v_2^* v_1$ (in other words, any number of repetitions of v_2 followed by a behaviour of v_1). Allowing operator boxes like Ω^* raises the danger of creating composite nets which are not in agreement with an intuitive behavioural meaning of operations specified by operator boxes. For consider an expression such as $\Omega_{\Box}(\mathsf{N}_\alpha, \Omega^*(\mathsf{N}_\beta, \mathsf{N}_\gamma))$ (anticipating, for the moment, the boxes in Figure 20), whose behaviour is, intuitively, 'do either α, or $\beta^*\gamma$'. According to our intuition about the operation specified by Ω^*, one might expect that the net corresponding to this expression could be constructed by first applying Ω^* to N_β and N_γ, and then putting the result in a choice with N_α, yielding the net shown on the right-hand side of Figure 18. This net, however, allows evolutions such as $\{\beta\}\{\beta\}\{\alpha\}$, which do not correspond to what one expects from a choice construct. This phenomenon is neither particular to our approach, nor to Petri nets in general (cf. [4,25]), nor does it invalidate our coherence results (see Section 7.5).

Boxes such as Ω^* are useful in modelling guarded while-loops in programming languages (basically, v_1 corresponds to the negation of the guard(s), and v_2 to the repetitive behaviour). To our experience, in practical programming languages, it is usually ascertained that a loop does not occur initially in an enclosing choice, or in another enclosing loop. We therefore formulate a set of conditions in such a way that these situations, and similar ones, are excluded.

– If $v \neq w$ and ${}^\bullet v \cap {}^\bullet w \neq \emptyset$, then $\Sigma_v, \Sigma_w \in \mathsf{Box}^{edir}$. (Beh1)
– If $v \neq w$ and $v^\bullet \cap w^\bullet \neq \emptyset$, then $\Sigma_v, \Sigma_w \in \mathsf{Box}^{xdir}$. (Beh2)
– If $v^\bullet \cap {}^\bullet w \neq \emptyset$, then $\Sigma_v \in \mathsf{Box}^{xdir}$ or $\Sigma_w \in \mathsf{Box}^{edir}$. (Beh3)
– If ${}^\bullet v \cap \Omega^\circ \neq \emptyset$, then $\Sigma_v \in \mathsf{Box}^{edir}$. (Beh4)
– If $v^\bullet \cap {}^\circ\Omega \neq \emptyset$, then $\Sigma_v \in \mathsf{Box}^{xdir}$. (Beh5)

With (Beh1), it will not be possible to start the execution of Σ_v and later, without finishing it, to enter Σ_w for a conflicting w because we came back to the initial state of Σ_v (notice that, again, the problem really concerns the possibility of coming back to the initial state and e-directedness is a simple structural condition excluding this kind of behaviour). (Beh2) prohibits to finish the execution of Σ_v and afterwards to enter Σ_w from the rear (notice that here the problem concerns the possibility of executing something from an exit marking which, perhaps unexpectedly, makes the situation rather different from the previous one). (Beh3) excludes a situation, combining the previous two problems, where we start the execution of Σ_w and later, without finishing it, we come back to the initial state and (re)enter Σ_v from the rear. (Beh4) excludes the situation whereby we reach the exit marking, start the execution of Σ_v and return to the exit marking without finishing Σ_v, so that a completed history is not composed out of completed histories of the components corresponding to a completed history of Ω. Finally, (Beh5) excludes the situation whereby, from the entry marking of the composed net, we can execute a behaviour originating from Σ_v° which bears no relationship to any possible behaviour of Ω (this intuitively corresponds to the backward reachability in Ω, which is not allowed).

More formally, it can be shown that, for a large and interesting class of operator boxes Ω, if (Beh1)–(Beh5) hold in addition to (Dom1)–(Dom2), then not only SOS compositionality, but also behavioural compositionality is ensured. The latter means that the behaviour of an Ω-composite object is the composition of the behaviours of its components, as determined by the behaviours of Ω. Details can be found in [11].

7. A generic process algebra and its two consistent semantics

This section describes an algebra of process expressions based on the sos-operator boxes, called the *box algebra*. The box algebra is in fact a *meta-model* parameterized by two non-empty, possibly infinite, sets of Petri nets: a set ConstBox of static and dynamic plain boxes which provide a denotational semantics of simple process expressions, including constants, and a disjoint set OpBox of sos-operator boxes which provide interpretation for the connectives. The only assumption about the sos-operator boxes in OpBox and the static boxes in ConstBox is that they have disjoint sets of simple root-only trees as their place and transition names, i.e., for all distinct static and/or operator boxes Σ and Θ in OpBox $\cup$ ConstBox,

$$S_\Sigma \cup T_\Sigma \subseteq \mathsf{P}_{\mathsf{root}} \cup \mathsf{T}_{\mathsf{root}} \quad \text{and} \quad S_\Sigma \cap S_\Theta = T_\Sigma \cap T_\Theta = \emptyset. \tag{14}$$

7.1. *Syntax*

We consider an algebra of process expressions over the signature

$$\mathsf{Const} \cup \left\{\overline{(.)}, \underline{(.)}\right\} \cup \{\mathsf{op}_\Omega \mid \Omega \in \mathsf{OpBox}\},$$

where Const is a fixed non-empty set of *constants* which will be modelled through the boxes in ConstBox, $\overline{(.)}$ and $\underline{(.)}$ are two unary operators, and each op_Ω is a connective of the algebra indexed by an sos-operator box taken from the set OpBox. The set of constants is partitioned into the *static* constants, Const^s, and *dynamic* constants, Const^d; moreover, there are two distinct disjoint subsets of Const^d, denoted by Const^e and Const^x, and respectively called the *entry* and *exit* constants. We also identify three further, not necessarily disjoint subsets of Const, namely: Const^{xcl}, Const^{edir} and Const^{xdir}, called respectively the *ex-exclusive*, *e-directed* and *x-directed* constants. We will also use a fixed set $\mathcal{X}$ of *process variables*. Although we use the symbols $\overline{(.)}$ and $\underline{(.)}$ to denote both operations on boxes and process algebra connectives, it will always be clear from the context what is the intended interpretation.

There are four classes of process expressions corresponding to previously introduced classes of plain boxes: the *entry*, *dynamic*, *exit* and *static* expressions, denoted respectively by Expr^e, Expr^d, Expr^x and Expr^s. Collectively, we will refer to them as the *box expressions*, Expr^{box}. We will also use a counterpart of the notion of the factorization of a tuple of boxes. For an sos-operator box Ω and an Ω-tuple of box expressions $\mathbf{D}$, we define the *factorization* of $\mathbf{D}$ to be the quadruple $\mu = (\mu_e, \mu_d, \mu_x, \mu_s)$ such that $\mu_\delta = \{v \mid D_v \in \mathsf{Expr}^\delta\}$, for $\delta \in \{e, x, s\}$, and $\mu_d = \{v \mid D_v \in \mathsf{Expr}^d \backslash (\mathsf{Expr}^e \cup \mathsf{Expr}^x)\}$. The syntax for the box expressions Expr^{box} is given by:

$$\begin{array}{ll} \mathsf{Expr}^s & E ::= c^s \mid X \mid \mathsf{op}_\Omega(\mathbf{E}), \\ \mathsf{Expr}^e & F ::= c^e \mid \overline{E} \mid \mathsf{op}_\Omega(\mathbf{F}), \\ \mathsf{Expr}^x & G ::= c^x \mid \underline{E} \mid \mathsf{op}_\Omega(\mathbf{G}), \\ \mathsf{Expr}^d & H ::= c^d \mid F \mid G \mid \mathsf{op}_\Omega(\mathbf{H}), \end{array} \tag{15}$$

where $c^\delta \in \mathsf{Const}^\delta$, for $\delta \in \{e, x, s\}$, and $c^d \in \mathsf{Const}^d \backslash (\mathsf{Const}^e \cup \mathsf{Const}^x)$ are constants; $X \in \mathcal{X}$ is a process variable; $\Omega \in \mathsf{OpBox}$ is an sos-operator box; and $\mathbf{E}$, $\mathbf{F}$, $\mathbf{G}$ and $\mathbf{H}$ are Ω-tuples of box expressions. These tuples have to satisfy some conditions determined by the domain of application of the net operator induced by Ω. More precisely, the factorizations of $\mathbf{E}$, $\mathbf{F}$ and $\mathbf{G}$ are respectively factorizations of the complex markings $(\emptyset, \emptyset)$, $({}^\circ\Omega, \emptyset)$ and $(\Omega^\circ, \emptyset)$, and the factorization of $\mathbf{H}$ is a factorization of a reachable complex marking different from $({}^\circ\Omega, \emptyset)$ and $(\Omega^\circ, \emptyset)$. In what follows, expressions in $\mathsf{Const} \cup \overline{\mathsf{Const}^s} \cup \underline{\mathsf{Const}^s}$, will be referred to as *flat*.

The above syntax is liberal in the sense that it only reflects properly the first part of the definition of the domain of an operator box, without reflecting the conditions (Dom1) and (Dom2). We will describe syntactic restrictions implied by (Dom1) and (Dom2) separately.

For every process variable $X \in \mathcal{X}$, there is a unique *defining equation* $X \stackrel{\mathrm{df}}{=} \mathsf{op}_\Omega(\mathbf{L})$ where $\Omega \in \mathsf{OpBox}$ is an sos-operator box and $\mathbf{L}$ is an Ω-tuple of process variables and static constants.

As in the domain of boxes, it is convenient to have a notation for turning an expression into a corresponding static one. We again use $\lfloor . \rfloor$ to denote such an operation. To achieve the desired effect, we assume that for each dynamic constant c there is a corresponding unique static constant denoted by $\lfloor c \rfloor$ and, if C is an expression, then $\lfloor C \rfloor$ is the static expression obtained by removing all occurrences of $\overline{(.)}$ and $\underline{(.)}$, and replacing every occurrence of each dynamic constant c by the corresponding static constant $\lfloor c \rfloor$.

EXAMPLE (*Continued*). We will use the running example based on the boxes depicted in Figure 17 in order to construct a simple algebra of process expressions. More precisely, we will consider the *Do It Yourself* (*DIY*) *algebra* based on the following two sets of boxes:

$$\begin{array}{l} \mathsf{OpBox} = \{\Omega_0\} \quad \text{and} \\ \mathsf{ConstBox} = \{\Phi_1, \Phi_{11}, \Phi_{12}\} \cup \{\Phi_2, \Phi_{21}, \Phi_{22}, \Phi_{23}\} \cup \{\Phi_3\}, \end{array}$$

where $\lfloor \Phi_{ij} \rfloor = \Phi_i = \lfloor \Upsilon_{v_i} \rfloor$, for all i and j, and the following hold:

$$\begin{array}{rll} M_{\Phi_{11}} & = \{q_{11}, q_{14}\} \quad \text{and} \quad M_{\Phi_{12}} = \{q_{13}, q_{12}\} \quad \text{and} \\ M_{\Phi_{2k}} & = \{q_{2k}\} \quad (\text{for } k = 1, 2, 3). \end{array}$$

The constants of the DIY algebra correspond to the boxes in ConstBox:

$$\mathsf{Const}^e = \{c_{21}\}, \quad \mathsf{Const}^s = \{c_1, c_2, c_3\},$$
$$\mathsf{Const}^x = \{c_{23}\}, \quad \mathsf{Const}^d = \{c_{11}, c_{12}, c_{21}, c_{22}, c_{23}\}.$$

Moreover, $\mathsf{Const} = \mathsf{Const}^{edir} = \mathsf{Const}^{xdir}$, $\mathsf{Const}^{xd} = \mathsf{Const} - \{c_1, c_{11}, c_{12}\}$ and $\lfloor c_{ij} \rfloor = c_i$, for every dynamic constant c_{ij}. The syntax of the DIY algebra is obtained by instantiating (15) with concrete constants and operator introduced above, and taking into account the factorizations in fact_{Ω_0} (see Table 1). For example, the syntax for the static and entry DIY expressions is:

$$\begin{array}{ll} \mathsf{Expr}^s & E ::= \mathsf{op}_{\Omega_0}(E, E, E) \mid c_1 \mid c_2 \mid c_3 \mid X, \\ \mathsf{Expr}^e & F ::= \mathsf{op}_{\Omega_0}(F, F, E) \mid c_{21} \mid \overline{E}. \end{array}$$

An example of valid dynamic expression is $\mathsf{op}_{\Omega_0}(\overline{c_1}, c_{22}, c_3)$ which, as we shall later see, corresponds to the net refinement $\Omega_0(\boldsymbol{\Upsilon})$ of Figure 17.

Domain restrictions. We now take into account restrictions implied by (Dom1) and (Dom2). Let Expr^{wf}, Expr^{xcl}, Expr^{edir} and Expr^{xdir} be the largest sets of expressions in Expr^{box} – called respectively the *well formed*, *ex-exclusive*, *e-directed* and *x-directed* box expressions – such that the following hold:

– All ex-exclusive, e-directed and x-directed expressions are well formed. (Expr1)

– If $\mathsf{op}_\Omega(\mathbf{D}) \in \mathsf{Expr}^{wf}$ then, for every $v \in T_\Omega$: (Expr2)
  - $D_v \in \mathsf{Expr}^{wf}$,
  - if ${}^\bullet v \cap v^\bullet \neq \emptyset$ then $D_v \in \mathsf{Expr}^{xcl}$,
  - if v is not reversible then $D_v \in \mathsf{Expr}^{xdir}$.

– If $\mathsf{op}_\Omega(\mathbf{D}) \in \mathsf{Expr}^{xcl}$ then Ω is ex-exclusive and $D_v \in \mathsf{Expr}^{xcl}$, for every $v \in T_\Omega$ satisfying (Expr3)

$${}^\bullet v \cap {}^\circ\Omega \neq \emptyset \neq v^\bullet \cap \Omega^\circ \quad \text{or} \quad v^\bullet \cap {}^\circ\Omega \neq \emptyset \neq {}^\bullet v \cap \Omega^\circ.$$

– If $\mathsf{op}_\Omega(\mathbf{D}) \in \mathsf{Expr}^{edir}$ then Ω is e-directed and, for every $v \in ({}^\circ\Omega)^\bullet$, $D_v \in \mathsf{Expr}^{edir}$. (Expr4)

– If $\mathsf{op}_\Omega(\mathbf{D}) \in \mathsf{Expr}^{xdir}$ then Ω is x-directed and, for every $v \in {}^\bullet(\Omega^\circ)$, $D_v \in \mathsf{Expr}^{xdir}$. (Expr5)

– If $\overline{E} \in \mathsf{Expr}^\delta$, or $\underline{E} \in \mathsf{Expr}^\delta$, or $X \in \mathsf{Expr}^\delta$ and $X \stackrel{\mathrm{df}}{=} E$, then $E \in \mathsf{Expr}^\delta$, for $\delta \in \{wf, xcl, edir, xdir\}$. (Expr6)

– $\mathsf{Const} \subseteq \mathsf{Expr}^{wf}$ and $\mathsf{Const}^\delta = \mathsf{Const} \cap \mathsf{Expr}^\delta$, for $\delta \in \{xcl, edir, xdir\}$. (Expr7)

We will be interested only in well formed expressions, and they will simply be called (box) expressions from now on.

7.2. *Denotational semantics*

Denotational semantics of the box algebra associates a box with every expression. We proceed by syntactic induction.

Constants. Constant expressions are mapped onto constant boxes of corresponding types, i.e., for every constant c and $\delta \in \{e, d, x, s, xd, edir, xdir\}$,

$$c \in \mathsf{Const}^{\delta} \Leftrightarrow \mathsf{box}(c) \in \mathsf{Box}^{\delta} \cap \mathsf{ConstBox}. \tag{16}$$

It assumed that, for every dynamic constant, the underlying box is the same as that for the corresponding static constant, and that it is reachable from the entry or exit marking of the latter; i.e., for every c in Const^d,

$$\lfloor \mathsf{box}(c) \rfloor = \mathsf{box}(\lfloor c \rfloor) \quad \text{and} \quad \mathsf{box}(c) \in \left(\left[\overline{\mathsf{box}(\lfloor c \rfloor)}\right\rangle \cup \left[\underline{\mathsf{box}(\lfloor c \rfloor)}\right\rangle\right).$$

It is also assumed that, for every non-entry and non-exit dynamic box reachable from an initially or terminally marked constant box, there is a corresponding dynamic constant, i.e., for every c in Const^s,

$$\left(\left[\overline{\mathsf{box}(c)}\right\rangle \cup \left[\underline{\mathsf{box}(c)}\right\rangle\right) \setminus \left\{\overline{\mathsf{box}(c)}, \underline{\mathsf{box}(c)}\right\} \subseteq \mathsf{box}\left(\mathsf{Const}^d\right).$$

Variables. With each defining equation $X \stackrel{\mathrm{df}}{=} \mathsf{op}_{\Omega}(\mathbf{L})$, we associate an equation on boxes, $X \stackrel{\mathrm{df}}{=} \Omega(\boldsymbol{\Psi})$, where $\Psi_v = L_v$ if L_v is a process variable (treated here as a box variable), and $\Psi_v = \mathsf{box}(L_v)$ if L_v is a static constant. This creates a system of net equations of the type considered in Section 5, but for a set of recursive variables $\mathcal{X}$ rather than for a single variable X. For such a system, it follows that it has at least one solution in the domain of static boxes. We then fix *any* such solution $\mathsf{sol}: \mathcal{X} \to \mathsf{Box}^s$ and define, for every process variable X, $\mathsf{box}(X) = \mathsf{sol}(X)$. As a result, for every variable X defined by the equation $X \stackrel{\mathrm{df}}{=} \mathsf{op}_{\Omega}(\mathbf{L})$,

$$\mathsf{box}(X) = \Omega\left(\mathsf{box}(\mathbf{L})\right) \in \mathsf{Box}^s.$$

Compound expressions. The definition of the semantical mapping box is completed by considering all the remaining static and dynamic expressions, following the syntax (15). The box mapping is a homomorphism; hence, for every static or dynamic expression $\mathsf{op}_{\Omega}(\mathbf{D})$ and every static expression E,

$$\mathsf{box}\left(\mathsf{op}_{\Omega}(\mathbf{D})\right) = \Omega\left(\mathsf{box}(\mathbf{D})\right) \quad \text{and} \quad \mathsf{box}(\overline{E}) = \overline{\mathsf{box}(E)} \quad \text{and}$$
$$\mathsf{box}(\underline{E}) = \underline{\mathsf{box}(E)}.$$

The semantical mapping always returns a box, and the assumed type consistency between constants and their denotations, (16), carries over to the remaining box expressions.

THEOREM 11. *For every box expression D, $\mathsf{box}(D)$ is a static or dynamic box. Moreover, for every $\delta \in \{e, d, x, s\}$, $D \in \mathsf{Expr}^{\delta}$ if and only if $\mathsf{box}(D) \in \mathsf{Box}^{\delta}$.*

EXAMPLE (*Continued*). In the case of the DIY algebra, for every static constant c_i, $\mathsf{box}(c_i) = \Phi_i$, and for every dynamic constant c_{ij}, $\mathsf{box}(c_{ij}) = \Phi_{ij}$. Thus, for example, the composite box in Figure 17 can be derived in the following way:

$$\begin{aligned}
\mathsf{box}\big(\mathsf{op}_{\Omega_0}(\overline{c_1}, c_{22}, c_3)\big) &= \Omega_0\big(\mathsf{box}(\overline{c_1}), \mathsf{box}(c_{22}), \mathsf{box}(c_3)\big)\\
&= \Omega_0\big(\overline{\mathsf{box}(c_1)}, \Phi_{22}, \Phi_3\big) = \Omega_0\big(\overline{\Phi_1}, \Phi_{22}, \Phi_3\big)\\
&= \Omega_0(\Upsilon_{v_1}, \Upsilon_{v_2}, \Upsilon_{v_3}) = \Omega_0(\boldsymbol{\Upsilon}).
\end{aligned}$$

7.3. *Structural operational semantics*

Similarity relation on expressions. First, we define a structural similarity relation on box expressions, $\equiv$, which provides partial identification of box expressions with the same denotational semantics. It is defined as the least equivalence relation on box expressions such that the following hold.

- For all flat expressions D and H satisfying $\mathsf{box}(D) = \mathsf{box}(H)$, and all defining equations $X \stackrel{\mathrm{df}}{=} \mathsf{op}_\Omega(\mathbf{L})$,

$$\boxed{D \equiv H \qquad X \equiv \mathsf{op}_\Omega(\mathbf{L})} \tag{17}$$

- For every sos-operator box Ω and every factorization μ of $({}^\circ\Omega, \emptyset)$ or $(\Omega^\circ, \emptyset)$, we have respectively:

$$\boxed{\mathsf{op}_\Omega(\mathbf{D}) \equiv \overline{\mathsf{op}_\Omega(\mathbf{H})} \qquad \mathsf{op}_\Omega(\mathbf{D}) \equiv \underline{\mathsf{op}_\Omega(\mathbf{H})}} \tag{18}$$

  where $\mathbf{D}$ and $\mathbf{H}$ are Ω-tuples of expressions such that, for every $v \in T_\Omega$: $D_v = \overline{H_v}$ if $v \in \mu_e$; $D_v = \underline{H_v}$ if $v \in \mu_x$; and $D_v = H_v$ otherwise.
- For every sos-operator box Ω, for every reachable complex marking of Ω, different from $({}^\circ\Omega, \emptyset)$ and $(\Omega^\circ, \emptyset)$, and for every pair of different factorizations μ and κ of that marking,

$$\boxed{\mathsf{op}_\Omega(\mathbf{D}) \equiv \mathsf{op}_\Omega(\mathbf{H})} \tag{19}$$

  where $\mathbf{D}$ and $\mathbf{H}$ are Ω-tuples of expressions for which there is an Ω-tuple of expressions $\mathbf{F}$ such that, for every $v \in T_\Omega$: $D_v = \overline{F_v}$ if $v \in \mu_e$; $D_v = \underline{F_v}$ if $v \in \mu_x$; and $D_v = F_v$ otherwise. Similarly, for every $v \in T_\Omega$: $H_v = \overline{F_v}$ if $v \in \kappa_e$; $H_v = \underline{F_v}$ if $v \in \kappa_x$; and $H_v = F_v$ otherwise. Intuitively, $\mathbf{F}$ corresponds to the common (and thus unchanging) part of $\mathbf{D}$ and $\mathbf{H}$.
- For all static expressions E and F, for every sos-operator box Ω, and for all Ω-tuples of expressions $\mathbf{D}$ and $\mathbf{H}$ with factorizations in fact_Ω:

$$\boxed{\frac{E \equiv F}{\overline{E} \equiv \overline{F}} \qquad \frac{E \equiv F}{\underline{E} \equiv \underline{F}} \qquad \frac{\mathbf{D} \equiv \mathbf{H}}{\mathsf{op}_\Omega(\mathbf{D}) \equiv \mathsf{op}_\Omega(\mathbf{H})}} \tag{20}$$

The structural similarity relation is closed in the domain of box expressions and preserves the denotational semantics. More precisely, whenever a box expression can match one side of a rule, then it is guaranteed that the other side is a box expression too. Thus the rules can be thought of as well formed term rewriting rules. Similar comment applies to the rules of the structural operational semantics.

THEOREM 12. *If D is a box expression and $D \equiv H$ then H is a box expression such that* $\mathsf{box}(D) = \mathsf{box}(H)$.

Thus $\equiv$ is a sound equivalence notion in the domain of box expressions. It is also complete in the sense that $\mathsf{box}(D) = \mathsf{box}(H)$ implies $D \equiv H$ provided that $\lfloor D \rfloor \equiv \lfloor H \rfloor$.

EXAMPLE (*Continued*). The DIY algebra gives rise to five specific rules for the structural equivalence relation (we omit here their symmetric, hence redundant, counterparts). The first two are derived from (17): $\overline{c_2} \equiv c_{21}$ and $\underline{c_2} \equiv c_{23}$. The third and fourth are derived from (18): $\mathsf{op}_{\Omega_0}(\overline{E}, \overline{F}, G) \equiv \overline{\mathsf{op}_{\Omega_0}(E, F, G)}$ and $\mathsf{op}_{\Omega_0}(E, F, \underline{G}) \equiv \underline{\mathsf{op}_{\Omega_0}(E, F, G)}$. Finally, there is a single instance of (19): $\mathsf{op}_{\Omega_0}(\underline{E}, \underline{F}, G) \equiv \mathsf{op}_{\Omega_0}(E, F, \overline{G})$. An application of these rules is illustrated below:

$$\mathsf{op}_{\Omega_0}(\underline{c_1}, c_{23}, c_3) \equiv \mathsf{op}_{\Omega_0}(c_1, c_2, \overline{c_3})$$

$$\overbrace{\mathsf{op}_{\Omega_0}(\underline{c_1}, c_{23}, c_3) \equiv \mathsf{op}_{\Omega_0}(\underline{c_1}, \underline{c_2}, c_3) \quad \mathsf{op}(\underline{c_1}, \underline{c_2}, c_3) \equiv \mathsf{op}_{\Omega_0}(c_1, c_2, \overline{c_3})}$$

$$\overbrace{\underline{c_1} \equiv \underline{c_1} \quad c_{23} \equiv \underline{c_2} \quad c_3 \equiv c_3}$$

Label based operational semantics. The first derivation system has moves of the form $D \xrightarrow{\Gamma} H$, where D and H are expressions and Γ is a finite multiset of labels. Formally, we define a ternary relation $\longrightarrow$ which is the least relation comprising all triples $D \xrightarrow{\Gamma} H$, where D and H are box expressions and $\Gamma \in \mathsf{mult}(\mathsf{Lab} \cup \{\mathsf{skip}, \mathsf{redo}\})$, and the following hold.

– For every static expression E:

$$\boxed{\overline{E} \xrightarrow{\{\mathsf{skip}\}} \underline{E} \quad \underline{E} \xrightarrow{\{\mathsf{redo}\}} \overline{E}} \tag{21}$$

– For all box expressions D, J and H:

$$\boxed{\frac{D \equiv H}{D \xrightarrow{\emptyset} H} \quad \frac{D \xrightarrow{\emptyset} J \xrightarrow{\Gamma} H}{D \xrightarrow{\Gamma} H} \quad \frac{D \xrightarrow{\Gamma} J \xrightarrow{\emptyset} H}{D \xrightarrow{\Gamma} H}} \tag{22}$$

– For every flat expression D and a non-empty step of transitions U enabled at $M_{\mathsf{box}(D)}$, there is an expression H such that $\mathsf{box}(D)[U\rangle\mathsf{box}(H)$, $\lambda_{\mathsf{box}(D)}(U) = \Gamma$ and:

$$\boxed{D \xrightarrow{\Gamma} H}$$

– For every operator box Ω, and all Ω-tuples $\mathbf{D}$ and $\mathbf{H}$ of expressions with factorizations in fact_Ω:

$$\frac{\forall v \in T_\Omega\colon\ D_v \xrightarrow{\Gamma_v^1+\cdots+\Gamma_v^{k_v}} H_v}{\mathsf{op}_\Omega(\mathbf{D}) \xrightarrow{\sum_{v\in T_\Omega}\{\alpha_v^1,\ldots,\alpha_v^{k_v}\}} \mathsf{op}_\Omega(\mathbf{H})} \qquad (\Gamma_v^i,\alpha_v^i)\in\lambda_\Omega(v) \tag{23}$$

The only way to generate a skip or redo is through applying the rules (21), and possibly the rules (22) afterwards; thus, for example, if $D \xrightarrow{\Gamma} H$ and $\mathsf{skip} \in \Gamma$ then $\Gamma = \{\mathsf{skip}\}$. Moreover, $D \xrightarrow{\emptyset} H$ if and only if $D \equiv H$.

A fundamental property of the operational semantics is that it transforms box expressions into box expressions, and the moves generated are all valid steps for the corresponding boxes.

THEOREM 13. *Let D be a box expression. If $D \xrightarrow{\Gamma} H$ then H is a box expression such that* $\mathsf{box}(D)_{\mathsf{sr}}\,[\Gamma\rangle_{\mathsf{lab}}\,\mathsf{box}(H)_{\mathsf{sr}}$. *Conversely, if* $\mathsf{box}(D)_{\mathsf{sr}}\,[\Gamma\rangle_{\mathsf{lab}}\,\Sigma_{\mathsf{sr}}$ *then there is a box expression H such that* $\mathsf{box}(H) = \Sigma$ *and* $D \xrightarrow{\Gamma} H$.

This theorem ascertains the soundness (correctness) and the completeness of the SOS rules for the box algebra with respect to the denotational semantics. In its proof [11], Theorems 9 and 10 are used, along with a string of more technical results about the box-algebraic framework.

EXAMPLE (*Continued*). In the DIY algebra, the operational semantics of flat expressions is given by:

$$\begin{array}{|ccccc|}\hline \overline{c_1} \xrightarrow{\{a\}} c_{12} & \overline{c_2} \xrightarrow{\{c\}} c_{22} & \overline{c_1} \xrightarrow{\{b\}} c_{11} & c_{21} \xrightarrow{\{c\}} c_{22} & \overline{c_1} \xrightarrow{\{a,b\}} \underline{c_1} \\ c_{22} \xrightarrow{\{d\}} \underline{c_2} & c_{12} \xrightarrow{\{b\}} \underline{c_1} & c_{22} \xrightarrow{\{d\}} c_{23} & c_{11} \xrightarrow{\{a\}} \underline{c_1} & \overline{c_3} \xrightarrow{\{e\}} \underline{c_3} \\ \hline \end{array}$$

and the inference rule for the only operator box can be formulated in the following way:

$$\frac{D_1 \xrightarrow{k\cdot\{a,b\}+\Gamma_1} H_1,\ D_2 \xrightarrow{\Gamma_2} H_2,\ D_3 \xrightarrow{\Gamma_3} H_3}{\mathsf{op}_{\Omega_0}(D_1,D_2,D_3) \xrightarrow{k\cdot\{f\}+\Gamma_1+\Gamma_2+\Gamma_3} \mathsf{op}_{\Omega_0}(H_1,H_2,H_3)} \qquad a,b\notin\Gamma_1$$

An application of these rules is shown below:

$$\overline{\mathsf{op}_{\Omega_0}(c_1,c_2,c_3)} \xrightarrow{\{c,f\}} \mathsf{op}_{\Omega_0}(\underline{c_1},c_{22},c_3)$$

$$\overline{\mathsf{op}_{\Omega_0}(c_1,c_2,c_3)} \xrightarrow{\emptyset} \mathsf{op}_{\Omega_0}(\overline{c_1},\overline{c_2},c_3) \qquad \mathsf{op}_{\Omega_0}(\overline{c_1},\overline{c_2},c_3) \xrightarrow{\{c,f\}} \mathsf{op}_{\Omega_0}(\underline{c_1},c_{22},c_3)$$

$$\overline{c_1} \xrightarrow{\{a,b\}} \underline{c_1} \quad \overline{c_2} \xrightarrow{\{c\}} c_{22} \quad c_3 \xrightarrow{\emptyset} c_3$$

The upper step is an application of (22), while the lower step is an application of the preceding inference rule (and hence of (23)).

Transition based operational semantics. Consider the set $\mathsf{T}^{alg}_{\mathsf{tree}}$ of all transition trees in the boxes derived through the box mapping. It is easy to see that each $t \in \mathsf{T}^{alg}_{\mathsf{tree}}$ has a unique label, $\mathsf{lab}(t)$, in all the boxes associated with box expressions in which it occurs.

The operational semantics we will now define has moves of the form $D \xrightarrow{U} H$ such that D and H are box expression and U is a finite subset of $\mathsf{T}^{alg}_{\mathsf{tree}} \cup \{\mathsf{skip}, \mathsf{redo}\}$. The idea here is that U is a valid step for the boxes associated with D and H, after augmenting them with the skip and redo transitions. We will denote, for every such set,

$$\mathsf{lab}(U) = \sum_{t \in U} \{\mathsf{lab}(t)\}$$

assuming that $\mathsf{lab}(\mathsf{skip}) = \mathsf{skip}$ and $\mathsf{lab}(\mathsf{redo}) = \mathsf{redo}$.

In the process of defining a transition based operational semantics, we adopt without any change the structural similarity relation on expressions, $\equiv$. The rules of the operational semantics are also carried forward after changing Γ to U consistently, with the exception of the last rule (23) which is redefined thus:

$$\frac{\forall v \in T_\Omega : D_v \xrightarrow{U_v^1 \uplus \cdots \uplus U_v^{k_v}} H_v}{\mathsf{op}_\Omega(\mathbf{D}) \xrightarrow{U} \mathsf{op}_\Omega(\mathbf{H})} \quad (\mathsf{lab}(U_v^i), \alpha_v^i) \in \lambda_\Omega(v)$$

$$\text{where} \quad U = \bigcup_{v \in T_\Omega} \{(v, \alpha_v^1) \lhd U_v^1, \ldots, (v, \alpha_v^{k_v}) \lhd U_v^{k_v}\}$$

EXAMPLE (*Continued*). In the DIY algebra, the transition based operational semantics of flat expressions is given by the axioms:

$$\begin{array}{lllll}
\overline{c_1} \xrightarrow{\{t_{11}\}} c_{12} & \overline{c_2} \xrightarrow{\{t_{21}\}} c_{22} & \overline{c_1} \xrightarrow{\{t_{12}\}} c_{11} & c_{21} \xrightarrow{\{t_{21}\}} c_{22} & \overline{c_1} \xrightarrow{\{t_{11}, t_{12}\}} \underline{c_1} \\
c_{22} \xrightarrow{\{t_{22}\}} \underline{c_2} & c_{12} \xrightarrow{\{t_{12}\}} \underline{c_1} & c_{22} \xrightarrow{\{t_{22}\}} c_{23} & c_{11} \xrightarrow{\{t_{11}\}} \underline{c_1} & \overline{c_3} \xrightarrow{\{t_{31}\}} \underline{c_3}
\end{array}$$

and the inference rule for the only operator box is:

$$\frac{D \xrightarrow{\{t_1, u_1, \ldots, t_k, u_k, x_1, \ldots, x_l\}} D', \ H \xrightarrow{\{y_1, \ldots, y_m\}} H', \ G \xrightarrow{\{z_1, \ldots, z_n\}} G'}{\mathsf{op}_{\Omega_0}(D, G, H) \xrightarrow{U} \mathsf{op}_{\Omega_0}(D', G', H')}$$

where $k, l, m, n \geqslant 0$; $\{\mathsf{lab}(t_i), \mathsf{lab}(u_i)\} = \{a, b\}$, for every $i \leqslant k$; $\mathsf{lab}(x_i) \notin \{a, b\}$, for every $i \leqslant l$; and the step U is given by:

$$U = \big\{(v_1, f) \lhd \{t_1, u_1\}, \ldots, (v_1, f) \lhd \{t_k, u_k\}\big\} \cup$$

The label based operational semantics of a box expression D is captured by the *transition system* of D, denoted by ts_D and defined in exactly the same way as fts_D, with each transition based move $H \xrightarrow{U} C$ replaced by the corresponding label based move $H \xrightarrow{\Gamma} C$.

We now state two fundamental results which effectively state that the operational and denotational semantics of a box expression capture exactly the same behaviour. Both theorems rely on Theorem 13, but go a step further.

THEOREM 14. *For every box expression D, the relation*

$$\mathsf{iso} = \big\{\big([H]_\equiv, \mathsf{box}(H)\big) \mid [H]_\equiv \text{ is a node of } \mathsf{fts}_D\big\}$$

is an isomorphism between fts_D *and* $\mathsf{fts}_{\mathsf{box}(D)}$ *as well as between* ts_D *and* $\mathsf{ts}_{\mathsf{box}(D)}$.

THEOREM 15. *Let D be a box expression, and π be a partial order.*

(1) *If* $D \xrightarrow{\pi} H$ *then* $\mathsf{box}(D)\,[\pi\rangle_{\mathsf{po}}\,\mathsf{box}(H)$.

(2) *If* $\mathsf{box}(D)\,[\pi\rangle_{\mathsf{po}}\,\Sigma$ *then there is H such that* $\mathsf{box}(H) = \Sigma$ *and* $D \xrightarrow{\pi} H$.

EXAMPLE (*Continued*). In the DIY algebra, Theorem 14 can be illustrated by taking the expression $D = \mathsf{op}_{\Omega_0}(\overline{c_1}, c_{22}, c_3)$ and the corresponding box shown in Figure 17. Figure 19 depicts fts_D and $\mathsf{fts}_{\mathsf{box}(D)}$, where

$$\begin{aligned}
\xi_0 &= \big\{\mathsf{op}_{\Omega_0}(\overline{c_1}, c_{22}, c_3)\big\}, & M_{\Sigma_0} &= \{p_1, p_2, p_6\},\\
\xi_1 &= \big\{\overline{\mathsf{op}_{\Omega_0}(c_1, c_2, c_3)}, \mathsf{op}_{\Omega_0}(\overline{c_1}, \overline{c_2}, c_3), & &\\
&\qquad \mathsf{op}_{\Omega_0}(\overline{c_1}, c_{21}, c_3)\big\}, & M_{\Sigma_1} &= \{p_1, p_2, p_5\},\\
\xi_2 &= \big\{\mathsf{op}_{\Omega_0}(\underline{c_1}, \overline{c_2}, c_3), \mathsf{op}_{\Omega_0}(\underline{c_1}, c_{21}, c_3)\big\}, & M_{\Sigma_2} &= \{p_3, p_4, p_5\},\\
\xi_3 &= \big\{\mathsf{op}_{\Omega_0}(\underline{c_1}, c_{22}, c_3)\big\}, & M_{\Sigma_3} &= \{p_3, p_4, p_6\},\\
\xi_4 &= \big\{\mathsf{op}_{\Omega_0}(\overline{c_1}, c_{23}, c_3), \mathsf{op}_{\Omega_0}(\overline{c_1}, \underline{c_2}, c_3)\big\}, & M_{\Sigma_4} &= \{p_1, p_2, p_7\},
\end{aligned}$$

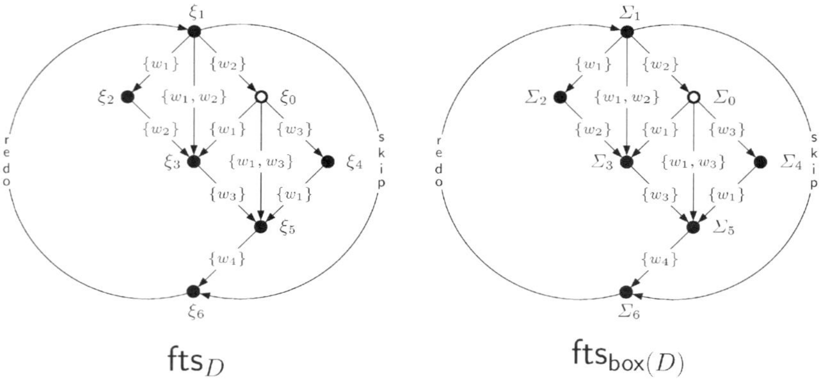

Fig. 19. Isomorphic full transition systems.

$$\xi_5 = \{\mathsf{op}_{\Omega_0}(\underline{c_1}, \underline{c_2}, c_3), \mathsf{op}_{\Omega_0}(\underline{c_1}, c_{23}, c_3), \mathsf{op}_{\Omega_0}(c_1, c_2, \overline{c_3})\}, \qquad M_{\Sigma_5} = \{p_3, p_4, p_7\},$$
$$\xi_6 = \{\underline{\mathsf{op}_{\Omega_0}(c_1, c_2, c_3)}, \mathsf{op}_{\Omega_0}(c_1, c_2, \underline{c_3})\}, \quad M_{\Sigma_6} = \{p_8\}.$$

In Section 3.3, we defined two behavioural equivalences between static and dynamic boxes based on the transition systems they generate. These can be lifted, through the box( ) mapping, to the level of static and dynamic box expressions, and then one can show that they are both preserved by the operators of the box algebra.

8. Examples and applications

In this section, we formalize PBC using the generic framework of the box algebra. We then argue that other process algebra models can be viewed as instances of the general scheme too. Finally, we sketch an application of PBC to a concurrent language with data.

8.1. *PBC*

As a first case study, we discuss a version of PBC which is based on eight operator boxes and two kinds of constants, depicted in Figure 20. The operator boxes are: $\Omega_{[f]}$ (basic relabelling), $\Omega_{\mathsf{sy}\,A}$ (synchronisation), $\Omega_{\mathsf{rs}\,A}$ (restriction), $\Omega_{[A:]}$ (scoping), $\Omega_{;}$ (sequence), $\Omega_{\Box}$ (choice), $\Omega_{\|}$ (parallel composition) and Ω_{*} (iteration), where $f : \mathsf{A}_{\mathrm{PBC}} \to \mathsf{A}_{\mathrm{PBC}}$ is a function and $A \subseteq \mathsf{A}_{\mathrm{PBC}}$ is a set of action particles. (See Section 2.2 for the definition of $\mathsf{A}_{\mathrm{PBC}}$ and other related notions.) All eight operator boxes are sos-operator boxes which are x-directed and have all transitions reversible. Moreover, all except iteration are also pure and e-directed. The specific relabellings used are as follows:

- $\varrho_{\mathsf{sy}\,A}$ is the smallest relabelling which contains ϱ_{id} and such that, if $(\Gamma, \alpha + \{a\}) \in \varrho_{\mathsf{sy}\,A}$ and $(\Delta, \beta + \{\hat{a}\}) \in \varrho_{\mathsf{sy}\,A}$ where $a \in A \cup \widehat{A}$, then $(\Gamma + \Delta, \alpha + \beta) \in \varrho_{\mathsf{sy}\,A}$, as well. This relation captures the multiway synchronisation explained in Section 2.2.
- $\varrho_{[A:]}$ comprises all the pairs $(\Gamma, \alpha) \in \varrho_{\mathsf{sy}\,A}$ such that $a \notin \alpha$, for every $a \in A \cup \widehat{A}$.
- $\varrho_{\mathsf{rs}\,A}$ comprises all the pairs $(\{\alpha\}, \alpha) \in \varrho_{id}$ such that $a \notin \alpha$, for every $a \in A \cup \widehat{A}$.
- $\varrho_{[f]}$ comprises all the pairs $(\{\alpha\}, f(\alpha))$ where f is applied to α element-wise.

The static PBC constants (all ex-directed and ex-exclusive) are: N_α for a multiaction $\alpha \in \mathsf{Lab}_{\mathrm{PBC}}$, and $\mathsf{N}_{[x_{()}]}$ for a data variable x (not a recursion variable). The latter deserve an additional explanation.

Data variables. Let x be a data variable of type D, and $\bullet$ be a special symbol not in D, interpreted as an 'undefined value'. For such a variable, the set of action particles $\mathsf{A}_{\mathrm{PBC}}$ contains the symbols x_{kl} and $\widehat{x_{kl}}$ for all $k, l \in D \cup \{\bullet\}$; moreover, x_{kl} and $\widehat{x_{kl}}$ are conjugate action particles.

The behaviour of x will be represented by a plain box $\mathsf{N}_{[x_{()}]}$. Intuitively, it is composed of an initialization part, a core looping part and a termination part. Such a box can first execute $x_{\bullet u}$, for any value u in D, then a sequence of zero or more executions of x_{uv}, where $u, v \in D$, and possibly terminate by executing $x_{u\bullet}$. Each such execution is carried

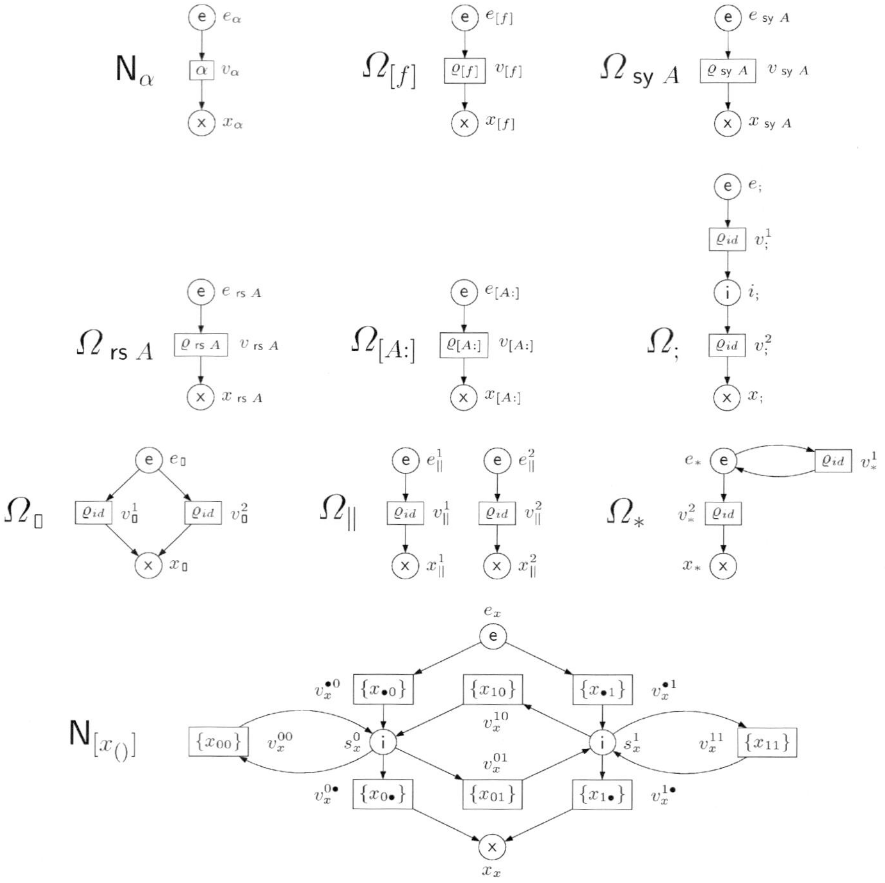

Fig. 20. PBC operator boxes and constants.

out under a restriction that, if x_{ku} and x_{vl} are two consecutively performed actions, then $u = v$.

We use x_{uv} to represent the change of the value of x from u to v, $x_{\bullet u}$ to represent an initialization of x to the value u, and $x_{u\bullet}$ to represent the 'destruction' of x when its current value is u. The net model of a data variable is exemplified in Figure 20 for a binary x, i.e., $D = \{0, 1\}$. In general, we define $\mathsf{N}_{[x_{()}]} = (S_x, T_x, W_x, \lambda_x, \emptyset)$ where:

$$S_x = \{e_x, x_x\} \cup \{s_x^u \mid u \in D\} \quad \text{and}$$
$$T_x = \{v_x^{\bullet u}, v_x^{u\bullet} \mid u \in D\} \cup \{v_x^{uv} \mid u, v \in D\},$$

the weight function W_x returns 1, for all the pairs in the set

$$\{(e_x, v_x^{\bullet u}), (v_x^{\bullet u}, s_x^u), (s_x^u, v_x^{uv}), (v_x^{uv}, s_x^v), (s_x^u, v_x^{u\bullet}), (v_x^{u\bullet}, x_x) \mid u, v \in D\}$$

and 0 otherwise; and the label function is defined, for all the values $u, v \in D$, by:

$$\begin{array}{lll} \lambda_x(e_x) = \mathsf{e}, & \lambda_x(x_x) = \mathsf{x}, & \lambda_x(s_x^u) = \mathsf{i}, \\ \lambda_x(v_x^{\bullet u}) = \{x_{\bullet u}\}, & \lambda_x(v_x^{u \bullet}) = \{x_{u \bullet}\}, & \lambda_x(v_x^{uv}) = \{x_{uv}\}. \end{array}$$

There is also a set of dynamic constant boxes $\mathsf{N}_{[\overline{x_{(u)}}]}$, for every $u \in D$, defined in the same way as $\mathsf{N}_{[x_{()}]}$, but with a single token in the place s_x^u.

PBC syntax I. The following is the syntax (15) for static box expressions generated by the operator boxes and constants in Figure 20.

$$\begin{array}{l} E ::= X \mid \alpha \mid [x_{()}] \mid E \| E \mid E \square E \mid E ; E \mid \langle E * E] \mid \\ \qquad E[f] \mid E \,\mathsf{sy}\, A \mid E \,\mathsf{rs}\, A \mid [A : E]. \end{array} \tag{24}$$

Notice that we have replaced the functional style of denoting box expressions by more common infix and postfix notations. The symbols used should make obvious the association between the boxes in Figure 20 and the PBC term operators.

The *dynamic PBC expressions* are defined below, where E denotes a static PBC expression given by (24):

$$\begin{array}{l} G ::= \overline{E} \mid \underline{E} \mid [\overline{x_{(u)}}] \mid G \| G \mid G \square E \mid E \square G \mid G ; E \mid \\ \qquad E ; G \mid \langle G * E] \mid \langle E * G] \mid G[f] \mid G \,\mathsf{sy}\, A \mid G \,\mathsf{rs}\, a \mid [A : G]. \end{array} \tag{25}$$

The syntaxes (24) and (25) do not yet incorporate the condition (Dom1); the other condition, (Dom2), is trivially satisfied since the operator boxes in Figure 20 have only reversible transitions.

PBC syntax II. The refinement of the syntax (24) – shown below – incorporates (Dom1). It is assumed that the variables $\mathcal{X}^{wf}$ and $\mathcal{X}^{xcl}$ are defined by $\mathcal{X}^{wf} \stackrel{\mathrm{df}}{=} E^{wf}$ and $\mathcal{X}^{xcl} \stackrel{\mathrm{df}}{=} E^{xcl}$, respectively.

$$\begin{array}{l} E^{wf} ::= \mathcal{X}^{wf} \mid E^{wf}[f] \mid E^{wf} \square E^{wf} \mid \langle E^{xcl} * E^{wf}] \mid E^{wf} \| E^{wf} \mid E^{xcl} \\ E^{xcl} ::= \mathcal{X}^{xcl} \mid E^{xcl}[f] \mid E^{xcl} \square E^{xcl} \mid \langle E^{xcl} * E^{xcl}] \mid E^{wf} ; E^{wf} \mid \alpha. \end{array} \tag{26}$$

Expressions defined by E^{wf} and E^{xcl} are the well formed and ex-exclusive PBC expressions, respectively. Moreover, since non-trivial aspects of non-ex-exclusiveness and non-ex-directedness are associated with control flow rather than communication interface operators, we may treat the latter as equivalent; hence we shortened the syntax by arbitrarily selecting only one communication interface operator, viz. basic relabelling. The same comment applies to the constant boxes; since in PBC all these are both ex-exclusive and ex-directed, we included just the basic action. Similar, rather immediate, modifications can be incorporated into the syntax of dynamic PBC expressions.

The refined PBC syntax (26) is an instance of the general box algebra and, as a consequence, *all results formulated in the Sections* 6 *and* 7 *hold for it.*

PBC syntax III. The syntax (26) still admits expressions for which the corresponding boxes may not satisfy the conditions (Beh1–5). We therefore further refine it by introducing a third syntax which identifies a subset of expressions generated by (26) where these conditions have been incorporated. The syntax is based on four syntactic classes E^{β}, and it is assumed that each variable X^{β} is defined by $X^{\beta} \stackrel{\mathrm{df}}{=} E^{\beta}$.

$$
\begin{array}{lllllll}
E^{wf} & ::= \mathcal{X}^{wf} & \mid E^{wf}[f] & \mid E^{wf} \| E^{wf} & \mid \langle E^{edir:xcl} * E^{edir}] & \mid \\
 & \quad E^{xcl} & \mid E^{edir} & & & \\
E^{edir} & ::= \mathcal{X}^{edir} & \mid E^{edir}[f] & \mid E^{edir} \| E^{edir} & \mid E^{edir} \square E^{edir} & \mid E^{edir:xcl} \\
E^{xcl} & ::= \mathcal{X}^{xcl} & \mid E^{xcl}[f] & \mid E^{wf}; E^{wf} & \mid \langle E^{edir:xcl} * E^{edir:xcl}] & \mid E^{edir:xcl} \\
E^{edir:xcl} & ::= \mathcal{X}^{edir:xcl} & \mid E^{edir:xcl}[f] & \mid E^{edir}; E^{wf} & \mid E^{edir:xcl} \square E^{edir:xcl} & \mid \alpha.
\end{array}
$$

The above syntax conforms to the general box algebra scheme (again, we omitted a straightforward adaptation of the syntax for the dynamic PBC expressions) and all composite boxes corresponding to expressions it specifies satisfy the five conditions, (Beh1–5). Referring to the notation used in Section 7.2, the annotations *wf*, *edir* and *xcl* indicate that the respective syntactic entities define expressions which are respectively well formed, e-directed and ex-exclusive. Notice that all expressions are also x-directed since so are the PBC constants and operator boxes in Figure 20.

8.2. *The two semantics of PBC*

The denotational semantics of PBC follows the general pattern, so we omit the details. As to the structural equivalence equations, we provide those specific to the PBC in Table 2.

Table 2
PBC structural equivalence rules

IPAR1	$\overline{E \Vert F} \equiv \overline{E} \Vert \overline{F}$	IPAR2	$\underline{E} \Vert \underline{F} \equiv \underline{E \Vert F}$
IC1L	$\overline{E \square F} \equiv \overline{E} \square F$	IC2L	$\underline{E} \square F \equiv \underline{E \square F}$
IC1R	$\overline{E \square F} \equiv E \square \overline{F}$	IC2R	$E \square \underline{F} \equiv \underline{E \square F}$
IS1	$\overline{E; F} \equiv \overline{E}; F$	IS2	$\underline{E}; F \equiv E; \overline{F}$
IS3	$E; \underline{F} \equiv \underline{E; F}$		
IIT1	$\overline{\langle E * F]} \equiv \langle \overline{E} * F]$	IIT2	$\langle \underline{E} * F] \equiv \langle \overline{E} * F]$
IIT3	$\langle \underline{E} * F] \equiv \langle E * \overline{F}]$	IIT4	$\langle E * \underline{F}] \equiv \underline{\langle E * F]}$
IR1	$\overline{E[f]} \equiv \overline{E}[f]$	IR2	$\underline{E}[f] \equiv \underline{E[f]}$
IRS1	$\overline{E \,\mathsf{rs}\, A} \equiv \overline{E} \,\mathsf{rs}\, A$	IRS2	$\underline{E \,\mathsf{rs}\, A} \equiv \underline{E} \,\mathsf{rs}\, A$
ISY1	$\overline{E \,\mathsf{sy}\, A} \equiv \overline{E} \,\mathsf{sy}\, A$	ISY2	$\underline{E \,\mathsf{sy}\, A} \equiv \underline{E} \,\mathsf{sy}\, A$
ISC1	$\overline{[A : E]} \equiv [A : \overline{E}]$	ISC2	$\underline{[A : E]} \equiv [A : \underline{E}]$

Table 3
PBC operational semantics rules I

AR	$\overline{\alpha} \xrightarrow{\{\alpha\}} \underline{\alpha}$	AR'	$\overline{\alpha} \xrightarrow{\{v_\alpha\}} \underline{\alpha}$
DAT1	$\overline{[x_{()}]} \xrightarrow{\{\{x_{\bullet u}\}\}} [\overline{x_{(u)}}]$	DAT1'	$\overline{[x_{()}]} \xrightarrow{\{v_x^{\bullet u}\}} [\overline{x_{(u)}}]$
DAT2	$[\overline{x_{(u)}}] \xrightarrow{\{\{x_{uv}\}\}} [\overline{x_{(v)}}]$	DAT2'	$[\overline{x_{(u)}}] \xrightarrow{\{v_x^{uv}\}} [\overline{x_{(v)}}]$
DAT3	$[\overline{x_{(u)}}] \xrightarrow{\{\{x_{u\bullet}\}\}} \underline{[x_{()}]}$	DAT3'	$[\overline{x_{(u)}}] \xrightarrow{\{v_x^{u\bullet}\}} \underline{[x_{()}]}$
PAR	$\dfrac{G \xrightarrow{\Gamma} G',\ H \xrightarrow{\Delta} H'}{G \Vert H \xrightarrow{\Gamma+\Delta} G' \Vert H'}$	PAR'	$\dfrac{G \xrightarrow{U} G',\ H \xrightarrow{W} H'}{G \Vert H \xrightarrow{v_\Vert^1 \oplus U \cup v_\Vert^2 \oplus W} G' \Vert H'}$
CL	$\dfrac{G \xrightarrow{\Gamma} H}{G \,\square\, E \xrightarrow{\Gamma} H \,\square\, E}$	CL'	$\dfrac{G \xrightarrow{U} H}{G \,\square\, E \xrightarrow{v_\square^1 \oplus U} H \,\square\, E}$
CR	$\dfrac{G \xrightarrow{\Gamma} H}{E \,\square\, G \xrightarrow{\Gamma} E \,\square\, H}$	CR'	$\dfrac{G \xrightarrow{U} H}{E \,\square\, G \xrightarrow{v_\square^2 \oplus U} E \,\square\, H}$
SL	$\dfrac{G \xrightarrow{\Gamma} H}{G; E \xrightarrow{\Gamma} H; E}$	SL'	$\dfrac{G \xrightarrow{U} H}{G; E \xrightarrow{v_;^1 \oplus U} H; E}$
SR	$\dfrac{G \xrightarrow{\Gamma} H}{E; G \xrightarrow{\Gamma} E; H}$	SR'	$\dfrac{G \xrightarrow{U} H}{E; G \xrightarrow{v_;^2 \oplus U} E; H}$
IT1	$\dfrac{G \xrightarrow{\Gamma} H}{\langle G * E] \xrightarrow{\Gamma} \langle H * E]}$	IT1'	$\dfrac{G \xrightarrow{U} H}{\langle G * E] \xrightarrow{v_*^1 \oplus U} \langle H * E]}$
IT2	$\dfrac{G \xrightarrow{\Gamma} H}{\langle E * G] \xrightarrow{\Gamma} \langle E * H]}$	IT2'	$\dfrac{G \xrightarrow{U} H}{\langle E * G] \xrightarrow{v_*^2 \oplus U} \langle E * H]}$
RR	$\dfrac{G \xrightarrow{\Gamma} H}{G[f] \xrightarrow{f(\Gamma)} H[f]}$	RR'	$\dfrac{G \xrightarrow{U} H}{G[f] \xrightarrow{(v_{[f]}, f) \oplus U} H[f]}$

The rules of the label based and transition based operational semantics, except those which hold for any box algebra, are shown in Tables 3 and 4.

In the Tables 3 and 4, the notations $(v, f) \oplus U$ and $v \oplus U$, where v is a transition name, $f : \mathsf{A}_{\mathrm{PBC}} \to \mathsf{A}_{\mathrm{PBC}}$ is a mapping from labels to labels and $U = \{t_1, \ldots, t_k\}$ is a finite set of transition trees, respectively denote the following sets of transition trees:

$$\big\{\big(v, f\big(\mathsf{lab}(t_1)\big)\big) \lhd t_1, \ldots, \big(v, f\big(\mathsf{lab}(t_k)\big)\big) \lhd t_k\big\} \quad \text{and}$$
$$\big\{\big(v, \mathsf{lab}(t_1)\big) \lhd t_1, \ldots, \big(v, \mathsf{lab}(t_k)\big) \lhd t_k\big\}.$$

Table 4
PBC operational semantics rules II

RS	$\dfrac{G \xrightarrow{\Gamma} H}{G \,\mathsf{rs}\, A \xrightarrow{\Gamma} H \,\mathsf{rs}\, A}$	$\alpha \in \Gamma \Rightarrow (A \cup \widehat{A}) \cap \alpha = \emptyset$
RS'	$\dfrac{G \xrightarrow{U} H}{G \,\mathsf{rs}\, A \xrightarrow{v_{\mathsf{rs}\,A} \oplus U} H \,\mathsf{rs}\, A}$	$\alpha \in \mathsf{lab}(U) \Rightarrow (A \cup \widehat{A}) \cap \alpha = \emptyset$
SY	$\dfrac{G \xrightarrow{\Gamma_1 + \cdots + \Gamma_k} H}{G \,\mathsf{sy}\, A \xrightarrow{\{\alpha_1, \ldots, \alpha_k\}} H \,\mathsf{sy}\, A}$	$(\Gamma_i, \alpha_i) \in \varrho_{\mathsf{sy}\,A}$
SY'	$\dfrac{G \xrightarrow{U_1 \uplus \cdots \uplus U_k} H}{G \,\mathsf{sy}\, A \xrightarrow{\{(v_{\mathsf{sy}\,a}, \alpha_1) \lhd U_1, \ldots, (v_{\mathsf{sy}\,a}, \alpha_k) \lhd U_k\}} H \,\mathsf{sy}\, A}$	$(\mathsf{lab}(U_i), \alpha_i) \in \varrho_{\mathsf{sy}\,A}$
SC	$\dfrac{G \xrightarrow{\Gamma_1 + \cdots + \Gamma_k} H}{[A : G] \xrightarrow{\{\alpha_1, \ldots, \alpha_k\}} [A : H]}$	$(\Gamma_i, \alpha_i) \in \varrho_{[A:]}$
SC'	$\dfrac{G \xrightarrow{U_1 \uplus \cdots \uplus U_k} H}{[A : G] \xrightarrow{\{(v_{[A:]}, \alpha_1) \lhd U_1, \ldots, (v_{[A:]}, \alpha_k) \lhd U_k\}} [A : H]}$	$(\mathsf{lab}(U_i), \alpha_i) \in \varrho_{[A:]}$

Let us verify, for instance, that the rule (23), when instantiated for the sequential composition operator, is equivalent to the two rules SL and SR of Table 3. What we need to do is look closer at the syntax of legal PBC expressions and, moreover, bear in mind that static expressions allow only empty moves. Then (23) can be seen as giving rise to two different rules, where G and H are dynamic, while E and F static expressions:

RS_1	$\dfrac{G \xrightarrow{\Gamma} H,\ E \xrightarrow{\emptyset} F}{G;\, E \xrightarrow{\Gamma} H;\, F}$	RS_2	$\dfrac{E \xrightarrow{\emptyset} F,\ G \xrightarrow{\Gamma} H}{E;\, G \xrightarrow{\Gamma} F;\, H}$

And, if we take $F = E$ (which is always possible since $E \xrightarrow{\emptyset} E$), we obtain SR and SL. Conversely, SR and SL together with the rules (20) and (22) as well as the fact that $E \xrightarrow{\emptyset} F$ implies $E \equiv F$, yield RS_1 and RS_2.

8.3. *Equivalence properties*

As yet we have not discussed any specific algebraic properties of the various box algebra operators with respect to behavioural equivalences. Although we could, in principle, consider these equivalences separately, within the net framework and the process expression framework, in view of the consistency results formulated in Section 7.5, we would do the same thing twice, as every result developed for one would hold (almost) *verbatim* for the other. We therefore present a set of properties – or laws, in CSP terminology – for the static PBC expressions (their extension to the dynamic expressions being immediate). To be more precise, we will say that two static PBC expressions, E and F are *ts-isomorphic*, $E \cong F$, if ts_E and ts_F are isomorphic, and *strongly equivalent*, $E \approx F$, if so are ts_E and ts_F. We now list several laws for PBC.

- Parallel and choice compositions are associative and commutative, while sequential composition is only associative:

$$E\|(F\|F') \cong (E\|F)\|F' \qquad E\|F \cong F\|E$$
$$E \,\square\, (F \,\square\, F') \cong (E \,\square\, F) \,\square\, F' \qquad E \,\square\, F \cong F \,\square\, E$$
$$E;\, (F;\, F') \cong (E;\, F);\, F'.$$

  Choice is also idempotent, in the following sense:

$$E \,\square\, E \approx E \qquad \alpha \,\square\, \alpha \cong \alpha.$$

- Synchronisation and restriction are both commutative, idempotent and insensitive to conjugation; they also distribute over choice, sequence, iteration and, in the case of restriction, parallel composition:

$$E \;\mathsf{sy}\; A \;\mathsf{sy}\; B \cong E \;\mathsf{sy}\; B \;\mathsf{sy}\; A \qquad E \;\mathsf{rs}\; A \;\mathsf{rs}\; B \cong E \;\mathsf{rs}\; B \;\mathsf{rs}\; A$$
$$E \;\mathsf{sy}\; A \;\mathsf{sy}\; A \cong E \;\mathsf{sy}\; A \qquad E \;\mathsf{rs}\; A \;\mathsf{rs}\; A \cong E \;\mathsf{rs}\; A$$
$$E \;\mathsf{sy}\; \widehat{A} \cong E \;\mathsf{sy}\; A \qquad E \;\mathsf{rs}\; \widehat{A} \cong E \;\mathsf{rs}\; A$$
$$(E \,\square\, F) \;\mathsf{sy}\; A \cong (E \;\mathsf{sy}\; A) \,\square\, (F \;\mathsf{sy}\; A) \qquad (E \,\square\, F) \;\mathsf{rs}\; A \cong (E \;\mathsf{rs}\; A) \,\square\, (F \;\mathsf{rs}\; A)$$
$$(E;\, F) \;\mathsf{sy}\; A \cong (E \;\mathsf{sy}\; A);\, (F \;\mathsf{sy}\; A) \qquad (E;\, F) \;\mathsf{rs}\; A \cong (E \;\mathsf{rs}\; A);\, (F \;\mathsf{rs}\; A)$$
$$\langle E * F] \;\mathsf{sy}\; A \cong \big\langle (E \;\mathsf{sy}\; A) * (F \;\mathsf{sy}\; A)\big] \qquad \langle E * F] \;\mathsf{rs}\; A \cong \big\langle (E \;\mathsf{rs}\; A) * (F \;\mathsf{rs}\; A)\big]$$
$$(E\|F) \;\mathsf{rs}\; A \cong (E \;\mathsf{rs}\; A)\|(F \;\mathsf{rs}\; A).$$

  Moreover, synchronisation propagates through parallel composition:

$$(E\|F) \;\mathsf{sy}\; A \cong \big((E \;\mathsf{sy}\; A)\|(F \;\mathsf{sy}\; A)\big) \;\mathsf{sy}\; A.$$

- Scoping is idempotent and insensitive to conjugation; it also distributes over choice, sequence and iteration compositions:

$$\big[A : [A : E]\big] \cong [A : E] \qquad [\widehat{A} : E] \cong [A : E]$$
$$\big[A : (E \,\square\, F)\big] \cong [A : E] \,\square\, [A : F] \qquad \big[A : (E;\, F)\big] \cong [A : E];\, [A : F]$$
$$\big[A : \langle E * F]\big] \cong \big\langle [A : E] * [A : F]\big].$$

- For non-interfering sets of action particles A and B (i.e., $A \cap (B \cup \widehat{B}) = \emptyset$) synchronisation and scoping commute with restriction:

$$(E \;\mathsf{sy}\; A) \;\mathsf{rs}\; B \cong (E \;\mathsf{rs}\; B) \;\mathsf{sy}\; A \qquad [A : E] \;\mathsf{rs}\; B \cong \big[A : (E \;\mathsf{rs}\; B)\big].$$

– Relabelling distributes over the control flow operators:

$$(E;\, F)[f] \cong (E[f]);\, (F[f]) \qquad (E \,\square\, F)[f] \cong (E[f]) \,\square\, (F[f])$$

$$(E \| F)[f] \cong (E[f]) \| (F[f]) \qquad \langle E * F][f] \cong \langle E[f] * F[f]].$$

– For a bijective relabelling f which preserves conjugates (i.e., $f(\hat{a}) = \widehat{f(a)}$) we obtain properties akin to commutativity and distributivity:

$$(E \;\mathsf{sy}\; A)[f] \cong (E[f]) \;\mathsf{sy}\; f(A) \qquad (E[f]) \;\mathsf{sy}\; A \cong (E \;\mathsf{sy}\; f^{-1}(A))[f]$$

$$(E \;\mathsf{rs}\; A)[f] \cong (E[f]) \;\mathsf{rs}\; f(A) \qquad (E[f]) \;\mathsf{rs}\; A \cong (E \;\mathsf{rs}\; f^{-1}(A))[f]$$

$$[A : E][f] \cong [f(A) : (E[f])] \qquad [A : (E[f])] \cong [f^{-1}(A) : E][f].$$

– Successive relabellings can be collapsed, and $[id]$ (where id is the identity relabelling) is a neutral operator:

$$E[f][g] \cong E[g \circ f] \qquad E[id] \cong E.$$

– Iteration can be expressed through recursion (see also Example 3 below):

$$\langle E * F] \approx X \quad \text{where } X \stackrel{\text{df}}{=} (E;\, X) \,\square\, F.$$

It is possible to consider other equivalences in the domain of PBC expressions. For example, two static expressions are *duplication equivalent* if so are the corresponding static boxes. The resulting equivalence was studied in [31] where its sound and complete axiomatization for a variant of the PBC syntax was given.

The applicability of the laws presented in this section can be extended, since they also hold if the behavioural equivalence is expressed using the partial order semantics based on Mazurkiewicz traces. For the majority of PBC laws, this follows from the fact that boxes derived for the constituent expressions are isomorphic.

8.4. *Some applications of SOS rules*

Below we show derivations based on the inference rules and equations for the PBC expressions, indicating which specific rules in Tables 2, 3 and 4 have been applied; other rules introduced in Section 7.3 are applied implicitly. To simplify notation, in the example expressions, a singleton multiset $\{a\} \in \mathsf{Lab}_{\mathrm{PBC}}$ will be abbreviated by a, and similarly $\{b\}$ by b etc. Moreover, we use $\mathsf{sy}\, a$ to denote $\mathsf{sy}\{a\}$.

EXAMPLE 1. We first deduce that

$$\overline{(a;\, b) \,\square\, c} \xrightarrow{\{\{a\}\}\{\{b\}\}} \underline{(a;\, b) \,\square\, c},$$

i.e., that the expression $\overline{(a;\, b) \Box c}$ can do an a-move, and thereafter a terminating b-move:

$$\begin{array}{lllll}
\overline{(a;\, b) \Box c} & \equiv \overline{(a;\, b)} \Box c & \text{IC1L} & \equiv (\overline{a};\, b) \Box c & \text{IS1} \\
& \xrightarrow{\{\{a\}\}} (\underline{a};\, b) \Box c & \text{AR SL CL} & \equiv (a;\, \overline{b}) \Box c & \text{IS2} \\
& \xrightarrow{\{\{b\}\}} (a;\, \underline{b}) \Box c & \text{AR SR CL} & \equiv \underline{(a;\, b)} \Box c & \text{IS3} \\
& \equiv \underline{(a;\, b) \Box c} & \text{IC2L} & &
\end{array}$$

EXAMPLE 2. As an application of the rule for synchronisation, we show that $\overline{E} \xrightarrow{\{\{b,c\}\}} \underline{E}$ where $E = ((\{a,b\} \| \{\hat{a},\hat{a}\}) \| \{a,c\})\,\mathsf{sy}\,a$:

$$\begin{array}{lll}
\overline{E} & \equiv \overline{((\{a,b\} \| \{\hat{a},\hat{a}\}) \| \{a,c\})\,\mathsf{sy}\,a} & \text{ISY1} \\
& \equiv (\overline{(\{a,b\} \| \{\hat{a},\hat{a}\})} \| \overline{\{a,c\}})\,\mathsf{sy}\,a & \text{IPAR1} \\
& \equiv ((\overline{\{a,b\}} \| \overline{\{\hat{a},\hat{a}\}}) \| \overline{\{a,c\}})\,\mathsf{sy}\,a & \text{IPAR1} \\
& \xrightarrow{\{\{b,c\}\}} ((\underline{\{a,b\}} \| \underline{\{\hat{a},\hat{a}\}}) \| \underline{\{a,c\}})\,\mathsf{sy}\,a & \text{AR PAR SY} \\
& \equiv \underline{E} & \text{IPAR2 ISY2}
\end{array}$$

EXAMPLE 3. The next example shows that $\overline{K} \xrightarrow{\{\{a\}\}\{\{b\}\}} \underline{K}$ where K is a process variable with the defining equation $K \stackrel{\text{df}}{=} (a;\, K) \Box b$. To conform with the general scheme, we first replace the defining equation by a set of two equations, $K \stackrel{\text{df}}{=} M \Box b$ and $M \stackrel{\text{df}}{=} a;\, K$. Then the detailed derivation can be carried out as follows:

$$\begin{array}{lllll}
\overline{K} & \equiv \overline{M \Box b} & & \equiv \overline{M} \Box b & \text{IC1L} \\
& \equiv \overline{a;\, K} \Box b & & \equiv (\overline{a};\, K) \Box b & \text{IS1} \\
& \xrightarrow{\{\{a\}\}} (\underline{a};\, K) \Box b & \text{AR SL CL} & \equiv (a;\, \overline{K}) \Box b & \text{IS2} \\
& \equiv (a;\, \overline{(M \Box b)}) \Box b & & \equiv (a;\, (M \Box \overline{b})) \Box b & \text{IC1R} \\
& \xrightarrow{\{\{b\}\}} (a;\, (M \Box \underline{b})) \Box b & \text{AR CR SR CL} & \equiv (a;\, \underline{(M \Box b)}) \Box b & \text{IC2R} \\
& \equiv (a;\, \underline{K}) \Box b & & \equiv \underline{(a;\, K)} \Box b & \text{IS3} \\
& \equiv \underline{M \Box b} & & \equiv \underline{M \Box b} & \text{IC2L} \\
& \equiv \underline{K} & & &
\end{array}$$

Note that $\overline{K}$ can make an infinite sequence of a-moves since $\overline{K} \xrightarrow{\{\{a\}\}} (a;\, \overline{K}) \Box b$.

EXAMPLE 4. The fourth example is similar to that discussed in [4] and involves a non-tail-end recursive expression not directly expressible in CCS. Below we show that the dynamic expression $\overline{L}$, for which the defining equation is $L \stackrel{\text{df}}{=} (L;\, a) \Box a$, can make two successive a-moves, i.e.,

$$\overline{L} \xrightarrow{\{\{a\}\}\{\{a\}\}} \underline{L}.$$

We first transform the equation into the format required by the scheme developed in Section 7: $L \stackrel{\text{df}}{=} N \,\square\, a$ and $N \stackrel{\text{df}}{=} L; a$. The derivation is then obtained in the following way:

$$
\begin{aligned}
\overline{L} &\equiv \overline{N \,\square\, a} \equiv \overline{N} \,\square\, a \equiv \overline{(L; a)} \,\square\, a \equiv (\overline{L}; a) \,\square\, a \\
&\equiv ((\overline{N \,\square\, a}); a) \,\square\, a \equiv ((N \,\square\, \overline{a}); a) \,\square\, a \xrightarrow{\{\{a\}\}} ((N \,\square\, \underline{a}); a) \,\square\, a \\
&\equiv (\underline{(N \,\square\, a)}; a) \,\square\, a \equiv (\underline{L}; a) \,\square\, a \equiv (L; \overline{a}) \,\square\, a \xrightarrow{\{\{a\}\}} (L; \underline{a}) \,\square\, a \equiv \underline{(L; a)} \,\square\, a \\
&\equiv \underline{N} \,\square\, a \equiv \underline{N \,\square\, a} \equiv \underline{L}.
\end{aligned}
$$

By induction, it can be shown that $\overline{L}$ can do n successive a-moves, for every n. But the argument cannot be used to show that it can do an infinite sequence of a-moves; in fact it cannot. This may be contrasted with the previous example.

EXAMPLE 5. We now prove by induction that the dynamic expression $\overline{C}$, where $C \stackrel{\text{df}}{=} a \| C$, can do a Γ_n-move, for every $\Gamma_n = n \cdot \{\{a\}\}$, $n \geqslant 1$. That is, for every $n \geqslant 1$, there is C_n such that

$$\overline{C} \xrightarrow{\Gamma_n} C_n.$$

In the base step, $n = 1$, the above follows from the following derivation:

$$\overline{C} \equiv \overline{a \| C} \equiv \overline{a} \| \overline{C} \xrightarrow{\{\{a\}\}} \underline{a} \| \overline{C} = C_1.$$

In the induction step, by the induction hypothesis, there is C_n such that $\overline{C} \xrightarrow{\Gamma_n} C_n$. Then

$$\overline{C} \equiv \overline{a \| C} \equiv \overline{a} \| \overline{C} \xrightarrow{\{\{a\}\} + \Gamma_n} \underline{a} \| C_n = C_{n+1}$$

which completes the proof.

EXAMPLE 6. We finally turn to the step sequence semantics based on transition trees and the derived partial order semantics. Let E be an expression defined as $((b; a) \| (\hat{a}; c))\, \mathsf{sy}\, a$. Figure 21 shows $\mathsf{box}(\overline{E})$ together with the transition trees which serve as the actual names of transitions. Note that there are four pairs of independent transition trees in the sense of the relation ind^{alg}, namely (τ_1, τ_3), (τ_1, τ_4), (τ_2, τ_3) and (τ_2, τ_4). It is possible to show that $\overline{E}$ can generate the following two step sequences,

$$\overline{E} \xrightarrow{\{\tau_1\}\{\tau_5\}\{\tau_4\}} \underline{E} \quad \text{and} \quad \overline{E} \xrightarrow{\{\tau_1\}\{\tau_2, \tau_3\}\{\tau_4\}} \underline{E}.$$

Below we give a derivation for the first one:

$$
\begin{aligned}
\overline{E} \;&\equiv\; \overline{((b; a) \| (\hat{a}; c))}\, \mathsf{sy}\, a \equiv (\overline{(b; a)} \| \overline{(\hat{a}; c)})\, \mathsf{sy}\, a \equiv ((\overline{b}; a) \| (\overline{\hat{a}}; c))\, \mathsf{sy}\, a \\
&\xrightarrow{\{\tau_1\}} ((\underline{b}; a) \| (\overline{\hat{a}}; c))\, \mathsf{sy}\, a \equiv ((b; \overline{a}) \| (\overline{\hat{a}}; c))\, \mathsf{sy}\, a \xrightarrow{\{\tau_5\}} ((b; \underline{a}) \| (\underline{\hat{a}}; c))\, \mathsf{sy}\, a
\end{aligned}
$$

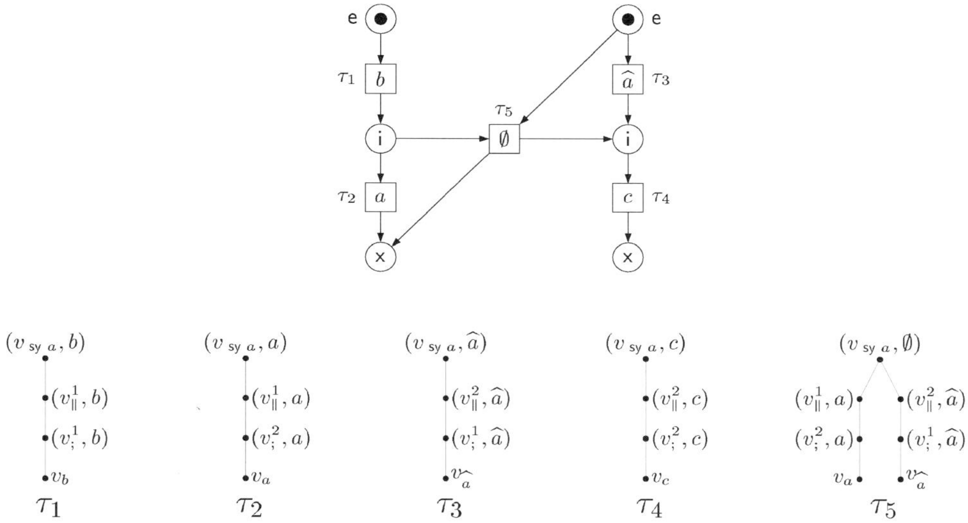

Fig. 21. The box of $\overline{E} = \overline{((b; a)\|(\hat{a}; c))\,\mathsf{sy}\,a}$.

$$\begin{aligned} &\equiv \quad (\underline{(b; a)}\|(\hat{a}; \overline{c}))\,\mathsf{sy}\,a \xrightarrow{\{\tau_4\}} (\underline{(b; a)}\|(\hat{a}; \underline{c}))\,\mathsf{sy}\,a \equiv (\underline{(b; a)}\|\underline{(\hat{a}; c)})\,\mathsf{sy}\,a \\ &\equiv \quad \underline{((b; a)\|(\hat{a}; c))}\,\mathsf{sy}\,a \equiv \underline{E}. \end{aligned}$$

As a result, we obtain that $\overline{E} \xrightarrow{\pi} \underline{E}$ and $\overline{E} \xrightarrow{\pi'} \underline{E}$, where π and π' are the following labelled partial orders:

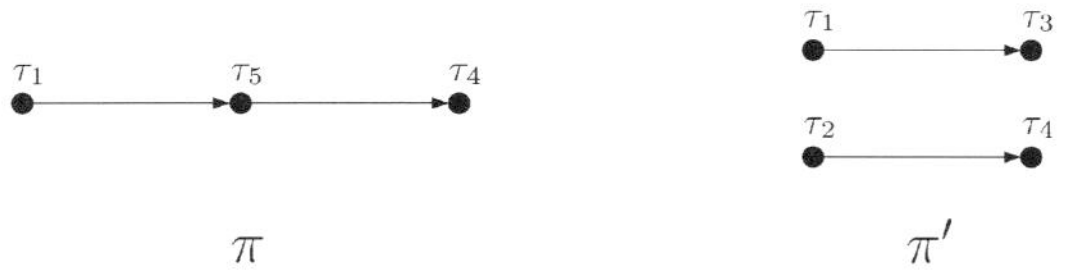

8.5. *CCS, TCSP and COSY*

In this section we outline how CCS, TCSP and COSY could be treated in the general compositionality framework provided by the box algebra. In what follows, by a *simple operator box* we will mean a two-place one-transition operator box similar to $\Omega_{[f]}$ shown in Figure 20. We will denote such an operator box by $\Omega{:}\varrho$, where ϱ is the relabelling of its only transition.

CCS. To model CCS processes we assume that $\mathsf{Lab} = Act \cup \{\tau\}$ is the set of CCS labels and $\hat{\ }{:}Act \to Act$ is a bijection on Act satisfying $\hat{\hat{a}} = a$ and $\hat{a} \neq a$, for all $a \in Act$.

We then define five simple operator boxes, Ω:*restr*(b), Ω:*relab*(h), Ω:*left*, Ω:*right* and Ω:*syn*(CCS), where $b \in Act$ and h is a mapping on labels $h:\mathsf{Lab} \to \mathsf{Lab}$ commuting with $\widehat{(.)}$, defined by the following relabellings:

$$\begin{aligned}
&restr(b) = \{(\{a\}, a) \mid a \in \mathsf{Lab}\backslash\{b\}\},\\
&relab(h) = \{(\{a\}, h(a)) \mid a \in \mathsf{Lab}\},\\
&left = \{(\{a\}, a^L) \mid a \in \mathsf{Lab}\},\\
&right = \{(\{a\}, a^R) \mid a \in \mathsf{Lab}\},\\
&syn(CCS) = \{(\{a^L, \hat{a}^R\}, \tau) \mid a \in Act\} \cup \{(\{a^L\}, a), (\{a^R\}, a) \mid a \in \mathsf{Lab}\}.
\end{aligned}$$

It is assumed that Lab is extended by the labels a^L and a^R, for $a \in \mathsf{Lab}$, but neither a^L nor a^R are allowed in the syntax of CCS expressions; they are auxiliary symbols used to model correctly the semantics of CCS parallel composition. Below, we give a denotational semantics of CCS, through the full translation ϕ_{CCS} for the CCS processes conforming to the standard syntax [42]:

$$\begin{aligned}
&\phi_{CCS}(\mathsf{nil}) = [\{a\} : a] \quad \text{(for any } a),\\
&\phi_{CCS}(a.E) = \mathsf{op}_{\Omega_;}(\mathsf{N}_a, \phi_{CCS}(E)),\\
&\phi_{CCS}(E + F) = \mathsf{op}_{\Omega_\square}(\phi_{CCS}(E), \phi_{CCS}(F)),\\
&\phi_{CCS}(E|F) = \mathsf{op}_{\Omega:syn(CCS)}(\mathsf{op}_{\Omega_\|}(\mathsf{op}_{\Omega:left}(\phi_{CCS}(E)), \mathsf{op}_{\Omega:right}(\phi_{CCS}(F)))),\\
&\phi_{CCS}(E[h]) = \mathsf{op}_{\Omega:relab(h)}(\phi_{CCS}(E)),\\
&\phi_{CCS}(E\backslash a) = \mathsf{op}_{\Omega:restr(a)}(\phi_{CCS}(E)).
\end{aligned}$$

TCSP. The other two models employ synchronisation mechanisms different from those used in CCS and PBC. We will not provide a full translation of TCSP and COSY into a box algebra, leaving this task as an exercise for the reader. Rather, we shall briefly explain how concurrent composition – and thus synchronisation – used in these two models could be handled.

The concurrent composition in TCSP is a binary operator, $E\|_A F$, where $A \subseteq \mathsf{Lab}$ is a set of actions on which $\|_A$ enforces synchronisation. The resulting process can execute an action $a \in A$ if it can be executed simultaneously by the two component processes; this simultaneous execution is denoted by a. The actions outside A can be executed autonomously by the two component processes. Such a synchronisation discipline can be modelled similarly as in CCS. This time, however, we do not assume that the set of labels, Lab, has any special properties. The relevant fragment of the transformation ϕ_{TCSP} from TCSP expressions to box algebra expressions is modelled thus:

$$\phi_{TCSP}(E\|_A F) = \mathsf{op}_{\Omega:TCSP(A)}(\mathsf{op}_{\Omega_\|}(\mathsf{op}_{\Omega:left}(\phi_{TCSP}(E)), \mathsf{op}_{\Omega:right}(\phi_{TCSP}(F)))),$$

where Ω:*TCSP*(A) is a simple operator box with the relabelling defined thus:

$$TCSP(A) = \{(\{a^L, a^R\}, a) \mid a \in A\} \cup \{(\{a^L\}, a), (\{a^R\}, a) \mid a \in (\mathsf{Lab}\backslash A)\}.$$

As before, we assume that the set of labels is temporarily extended by labels a^L and a^R, for each $a \in A$, which are not in the TCSP syntax.

COSY. The concurrent composition in COSY [33] is based on a multiway synchronisation. Consider a path program

$$prog = \mathsf{program} \quad path_1 \quad \ldots \quad path_n \quad \mathsf{endprogram}$$

Here, $prog$ can execute a if it is executed simultaneously in all the paths $path_i$ in which a occurs. To model such a synchronisation mechanism, we first extend $\Omega_\|$ to n-ary ($n \geqslant 1$) parallel composition operator boxes, $\Omega_\|^n$, in the obvious way. Let $A \subseteq \mathsf{Lab}$ be the set of labels occurring in program $prog$ and, for every $a \in A$, let $index(a)$ be the set of all indices i such that a occurs in $path_i$. Define a simple operator $\Omega{:}COSY(prog)$ with the relabelling being given by:

$$COSY(prog) = \bigcup_{a \in A} \Big\{ \Big(\bigcup_{i \in index(a)} \{a^i\}, a \Big) \Big\}$$

and n simple operators, $\Omega{:}index_i$, with the relabellings $index_i = \{(\{a\}, a^i) \mid a \in \mathsf{Lab}\}$, for $i = 1, \ldots, n$. Then the relevant fragment of the transformation ϕ_{COSY} from COSY programs to box algebra expressions is modelled thus:

$$\begin{aligned}\phi_{COSY}(prog) = \mathsf{op}_{\Omega:COSY(prog)}\big(\mathsf{op}_{\Omega_\|^n}\big(&\mathsf{op}_{\Omega:index_1}\big(\phi_{COSY}(path_1)\big), \ldots, \\ &\mathsf{op}_{\Omega:index_n}\big(\phi_{COSY}(path_n)\big)\big)\big).\end{aligned}$$

8.6. *Modelling a concurrent programming language*

In order to illustrate how it is possible to exploit the framework developed so far let us discuss informally how to give, using PBC, a fully compositional semantics to a concurrent programming language using PBC. Consider, for example, the following fragment of a concurrent program:

$\ldots$
begin **var** $x : \{0, 1\}$;
$\quad [\![x{:=}x \oplus 1]\!] \parallel [\![x{:=}y]\!]$
end
$\ldots$

where y is assumed to be declared by **var** $y : \{0, 1\}$ in some outer block, $\oplus$ denotes the addition modulo 2, $\|$ denotes 'shared variable parallelism' (as variable x occurs on both sides of $\|$), and $[\![\ldots]\!]$ delineates an atomic action. Let us discuss how to construct an appropriate, and as small as possible, Petri net modelling this block. Moreover, such a net should be constructed compositionally from nets derived for its three constituents: the declaration **var**x : $\{0, 1\}$, and the two atomic assignments, $[\![x{:=}x \oplus 1]\!]$ and $[\![x{:=}y]\!]$, and should itself be composable, i.e., usable in further compositions with similar nets, e.g., derived for the outer blocks.

As in Section 8.1, we use special action particles, in the form of indexed terms x_{vw} and $\widehat{x_{vw}}$ where $v, w \in \{0, 1\}$. Each such term denotes a change of the value of the program (or data) variable x from v to w, or a test of the value of x, if $v = w$. Then the following will be translations of the two assignments into PBC:

$$
\begin{aligned}
&[\![x{:=}x\oplus 1]\!] \rightsquigarrow \{\widehat{x_{01}}\} \,\square\, \{\widehat{x_{10}}\},\\
&[\![x{:=}y]\!] \rightsquigarrow \{\widehat{x_{00}}, \widehat{y_{00}}\} \,\square\, \{\widehat{x_{10}}, \widehat{y_{00}}\} \,\square\, \{\widehat{x_{01}}, \widehat{y_{11}}\} \,\square\, \{\widehat{x_{11}}, \widehat{y_{11}}\}.
\end{aligned}
\tag{27}
$$

The first expression means that 'either x could be 0, and then it is changed into 1, or x could be 1, and then it is changed into 0', which is a sound semantics of $[\![x{:=}x\oplus 1]\!]$ for a binary variable. Unlike the multiactions in the first line, the two-element multiactions in the second PBC expression are not singletons because $[\![x{:=}y]\!]$ involves both variables, x and y. Each two-element multiaction should be interpreted as denoting two simultaneously executable accesses to these variables, one checking the value of y, the other updating the value of x. For instance, $\{\widehat{x_{10}}, \widehat{y_{00}}\}$ denotes the value of y being checked to be 0 and, simultaneously, the value of x being changed from 1 to 0. The expressions for $[\![x{:=}x \oplus 1]\!]$ and $[\![x{:=}y]\!]$ given in (27) have corresponding Petri nets with respectively two and four transitions labelled by the corresponding multiactions, as shown in Figure 22.

The Petri net corresponding to the declaration of x involves action particles x_{vw}, where x_{vw} is the conjugate of $\widehat{x_{vw}}$, which ensures that the net of a variable can be put in parallel with the net of an atomic action, and then both nets can be synchronised in order to describe a block. The net describing **var**x : $\{0, 1\}$ is shown in Figure 20 and – in abbreviated form – in Figure 23. In the latter, it is arbitrarily assumed that the current value of x is 1, and thus the net contains a token on the corresponding place.

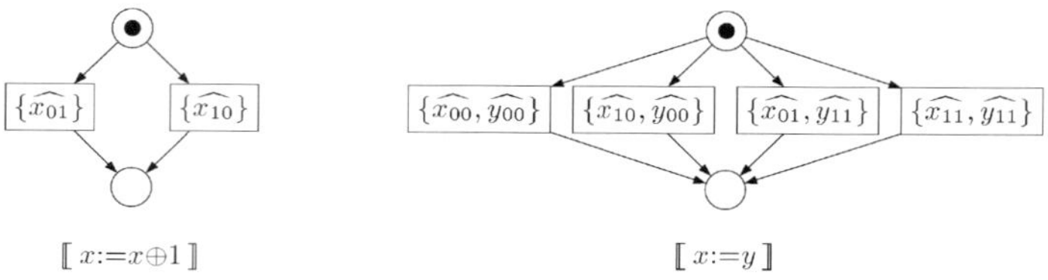

Fig. 22. Petri nets representing $[\![x{:=}x\oplus 1]\!]$ and $[\![x{:=}y]\!]$.

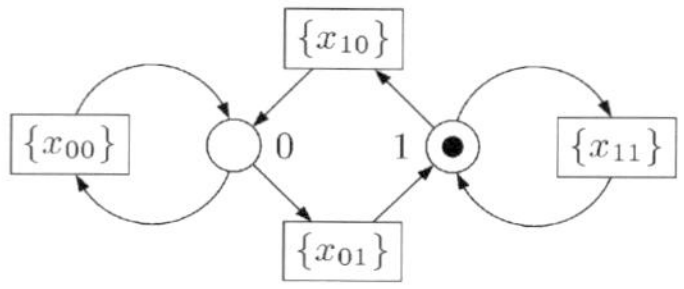

Fig. 23. Part of the Petri net representing a binary variable x, $\mathsf{N}_{[x()]}$.

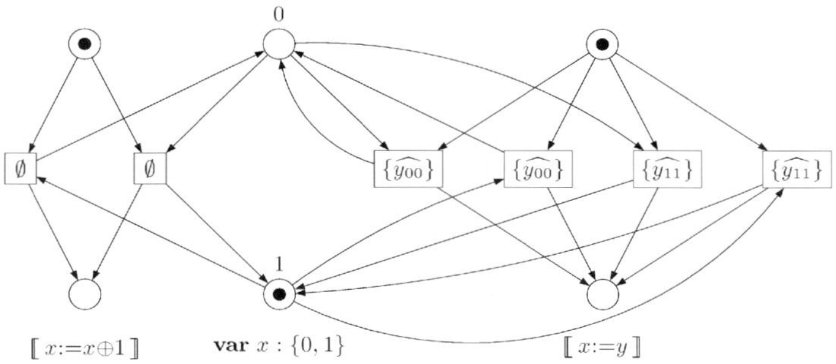

Fig. 24. (Abridged) Petri net representation of a small program fragment.

What was just described is the principal use of conjugation in the PBC semantics of programming languages; we always assume that the command part of a block employs 'hatted' action symbols, while the declaration part uses their 'unhatted' (conjugated) versions in order to describe accesses to variables. A block will be described by putting command part and declaration part in parallel and then synchronizing and restricting (hence scoping) over all variables that are local to it. This leaves only non-local accesses to be still visible in the communication interface, i.e., in the transitions.

In order to describe the block structure of the program fragment (with x being a local and y global variable of the block), the net of **var**x : {0, 1} needs to be synchronised with the nets of $[\![x{:=}x \oplus 1]\!]$ and $[\![x{:=}y]\!]$. For example, the transition labelled $\{x_{10}\}$ coming from the declaration of x is synchronised with the transition labelled $\{\widehat{x_{10}}, \widehat{y_{00}}\}$ coming from the action $[\![x{:=}y]\!]$ using and consuming the conjugate pair $(\widehat{x_{10}}, x_{10})$ but retaining $\widehat{y_{00}}$ (as we have already seen, this mechanism is built into the definition of synchronisation). The resulting transition has the label $\{\widehat{y_{00}}\}$. After that, all the transitions with labels containing $\widehat{x_{uv}}$ and x_{uv} action particles are removed by applying the PBC restriction operator, since the variable x is not known outside its declaring block. The net of the block, obtained through synchronisation and restriction, is shown in Figure 24. This net describes the effect of actions on local variable x directly by its connectivity, and on global variable y indirectly through its labelling. The transitions labelled $\{\widehat{y_{vv}}\}$ can be synchronised again, e.g., with transitions labelled $\{y_{vv}\}$ coming from the declaration of y in an outer block, using and consuming conjugate pairs $(y_{vv}, \widehat{y_{vv}})$ and yielding transitions labelled by ∅, after what is a 3-way synchronisation. During the translation process we have thus seen an effective application of the multiway synchronisation mechanism, in the modelling of an assignment involving two program variables which are declared in different blocks but accessed in the same atomic assignment.

The details and formal treatment are in [11].

9. Conclusions

Brief review of the results. In this chapter, we have knit a tight link between a process-algebraically inspired Structural Operational Semantics framework, and Petri nets and their diverse semantics. In order to obtain our results, both had to be modified: the SOS framework has been brought closer to Petri nets by introducing structural similarity (Sections 2.1 and 6.2), and Petri nets have been made composable by using place and transition labellings (Section 3).

We have shown how composable Petri nets can be understood both as elementary objects (plain boxes, Section 3.4) and as operators (operator boxes, Section 4.1) of a process algebra.

We have defined an operation of net refinement (Section 4.3) which corresponds to the application of a process-algebraic operator to its arguments, and we have demonstrated that recursive net equations can be solved using net refinement (Section 5).

Further, we have shown that in the presence of an additional property (namely factorizability, which distinguishes sos-operator boxes from other operator boxes, Section 6), the domain of composable nets has generic SOS rules (which are, more precisely, in *de Simone format* [1]); see Section 7. The main result states that there is an isomorphism between the SOS-semantics and net-semantics of the resulting algebra (Section 7.5).

Thus, in order to test whether a given process algebra has a (safe) Petri net semantics which is equivalent with SOS rules, one has to check whether or not its constants and operators can be described by plain boxes and sos-operator boxes, respectively; or, alternatively, whether there is a translation into PBC, which is known to have this property (Section 8.1 to 8.4). And the equivalence extends to whatever semantics one wishes to consider, be it step sequences (Section 3.2) or partial orders (Section 7.4), be it either transition-based or label-based (Section 7.5). We have finally demonstrated this for CCS, TCSP, COSY and a concurrent language with data and blocks (Sections 8.5 and 8.6).

Some carefully avoided problems. With the view of developing an equational theory – a modest approximation of which has been given in Section 8.3 –, it is tempting to investigate unit boxes for various common operators.

Regarding the choice operator $\square$, clearly the simplest ex-box $\mathsf{ex} = \mathsf{ex}(\{e\}, \{x\})$ is a unit: $E \mathsf{iso}(\mathsf{ex} \square E) \mathsf{iso}(E \square \mathsf{ex})$.

As for the parallel composition, it is clear that only the empty labelled net o, that is, the net having empty sets as its components, would be a unit for $\|$. (Note that we use o, rather than $\emptyset$, to denote the empty net, since $\emptyset$ is a name for the silent action in PBC.) Thus, we do not have a structural unit for parallel composition, since the empty box has been excluded from consideration by the requirement that boxes are ex-restricted (Section 3.4). The reason for this is that o creates serious problems in other contexts; for instance the sequence $\mathsf{o}; a$ is not safe, for any a, and allows an arbitrarily large step of concurrent a's.

Had we included the empty box o, the treatment of recursion would have become simpler on purely formal grounds, since o would then be the unique least element with respect to $\sqsubseteq$ (Section 5.1), and all approximations could have started with it as a seed. However, consider the recursive equation $X \stackrel{\mathrm{df}}{=} a \square X$. Using the empty seed as starting point, will allow X to make arbitrarily large a-steps. As this is counterintuitive, it makes sense also from a recursion point of view to exclude o.

As for the sequential composition, we also cannot define a structural unit in the framework presented in this paper. However, consider an extension of the notion of a box by allowing a place to be both entry and exit at the same time, i.e., by extending the codomain of the labelling λ on places from $\{\mathsf{e}, \mathsf{i}, \mathsf{x}\}$ (Sections 3.1 and 3.2) to $\{\mathsf{e}, \mathsf{i}, \mathsf{x}, \mathsf{ex}\}$. And then consider the box u consisting only of a place labelled ex and no transitions nor arcs. With a reasonable extension of the definition of net refinement, u will be a structural unit for sequential composition: Eiso(u; E)iso(E; u). Moreover, u has a very simple SOS rule $\underline{\mathsf{u}} \equiv \overline{\mathsf{u}}$ which is fully consistent with it being a unit for sequence.

The extension just mentioned also allows the unary iteration $(E)^*$ to be expressed by an operator box, namely by the box which has a single place labelled ex, a single transition (to be refined by the argument E) and two arcs leading from the place to the transition and back. What is more, $(E)^*$ is a derived operator since $(E)^*$ can be expressed as (E □ u), which is again consistent with the SOS rules for □ and for u. Because the last property is very unusual, and in order not to complicate the definition of net refinement still further, and also since in all applications, the unary iteration plays a minor role (as opposed to iterations with initial commands, or terminal commands, or both, see also Section 6.3), it was decided not to consider this extension for the time being.

In passing, it may be seen that due to the absence of an expansion law, a unit for parallel composition may well differ from a unit for sequential composition. In general, however, we agree with [5] that such units and their interplay deserve further study.

Directions for future work. From the discussion of the previous section, it is apparent that the extension of λ to ex, i.e., allowing places that are both entry and exit, is meaningful and should present only technical but no principal problems. A first step towards this aim is [13] where the definition of transition refinement has been extended to allow such labellings.

Unfortunately, the inclusion of o seems to lead to more fundamental problems. Nevertheless, it may deserve some re-consideration, perhaps by further scrutinizing the definition of net refinement.

There seems to be a further possibility for enlarging the class of allowable operator boxes without losing the essential SOS property. This would allow operators with internal memory, while in the present setup, all boxes are clean and therefore memoryless (Section 3.2).

Finally, an interesting extension is to add real-time. The reader is referred to [36] for a first step in this direction.

Acknowledgements

The work described in this chapter has greatly benefited from cooperation with other people, in particular Javier Esparza, Jon Hall and Richard Hopkins. It has been supported by the Esprit Basic Research Project 3148 DEMON (Design Methods Based on Nets) and Working Group 6067 CALIBAN (Causal Calculi Based on Nets), as well as the Anglo-German Foundation ARC Project 1032 BAT (Box Algebra with Time).

We also thank one reviewer of this paper for a number of comments, and the editors for their support.

References

[1] L. Aceto, W. Fokkink and Ch. Verhoef, *Structural operational semantics*, Handbook of Process Algebra, J.A. Bergstra, A. Ponse and S.A. Smolka, eds, Elsevier, Amsterdam (2001), 197–292.

[2] J.C.M. Baeten and T. Basten, *Partial-order process algebra (and its relation to Petri nets)*, Handbook of Process Algebra, J.A. Bergstra, A. Ponse and S.A. Smolka, eds, Elsevier, Amsterdam (2001), 769–872.

[3] J.C.M. Baeten and C. Verhoef, *Concrete process algebra*, Handbook of Logic in Computer Science S. Abramsky, D.M. Gabbay and T.S.E. Maibaum, eds, Oxford University Press (1995), 149–268.

[4] J.C.M. Baeten and W.P. Weijland, *Process Algebra*, Cambridge Tracts in Theoret. Comput. Sci. 18, Cambridge University Press (1990).

[5] J. Bergstra, W. Fokkink and A. Ponse, *Process algebra with recursive operations*, Handbook of Process Algebra, J.A. Bergstra, A. Ponse and S.A. Smolka, eds, Elsevier, Amsterdam (2001), 333–389.

[6] E. Best, *Semantics of Sequential and Parallel Programs*, Prentice Hall (1996).

[7] E. Best, *Partial order verification with PEP*, Proc. POMIV'96, Partial Order Methods in Verification, G. Holzmann, D. Peled and V. Pratt, eds, American Mathematical Society (1997), 305–328.

[8] E. Best and R. Devillers, *Sequential and concurrent behaviour in Petri net theory*, Theoret. Comput. Sci. **55** (1988), 87–136.

[9] E. Best, R. Devillers and J. Hall, *The Petri box calculus: A new causal algebra with multilabel communication*, Advances in Petri Nets 1992, Lecture Notes in Comput. Sci. 609, G. Rozenberg, ed., Springer-Verlag (1992), 21–69.

[10] E. Best, R. Devillers and M. Koutny, *Petri nets, process algebras and concurrent programming languages*, Advances in Petri Nets, Lectures on Petri Nets II, Applications, Lecture Notes in Comput. Sci. 1492, W. Reisig and G. Rozenberg, eds, Springer-Verlag (1998), 1–84.

[11] E. Best, R. Devillers and M. Koutny, *Petri Net Algebra*, EATCS Monographs on Theoret. Comput. Sci., Springer-Verlag (2000), to be published.

[12] E. Best, W. Frączak, R.P. Hopkins, H. Klaudel and E. Pelz, *M-nets: An algebra of high-level Petri nets, with an application to the semantics of concurrent programming languages*, Acta Inform. **53** (1998), 813–857.

[13] E. Best and A. Lavrov, *Generalised composition operations for high-level Petri nets*, Fundamenta Informaticae **40** (1999), 125–188.

[14] G. Boudol and I. Castellani, *Flow models of distributed computations, three equivalent semantics for CCS*, Inform. and Comput. **114** (2) (1994), 247–314.

[15] G.W. Brams (nom collectif de Ch. André, G. Berthelot, C. Girault, G. Memmi, G. Roucairol, J. Sifakis, R. Valette and G. Vidal-Naquet), *Résaux de Petri, Théorie et Pratique*, Vols. I and II, Editions Masson (1985).

[16] G. Bruns and J. Esparza. *Trapping mutual exclusion in the box calculus*, Theoret. Comput. Sci. **153** (1995), 95–128.

[17] I. Castellani, *Process algebras with localities*, Handbook of Process Algebra, J.A. Bergstra, A. Ponse and S.A. Smolka, eds, Elsevier, Amsterdam (2001), 945–1045.

[18] B.A. Davey and H.A. Priestley, *Introduction to Lattices and Order*, Cambridge Mathematical Textbook (1990).

[19] P. Degano, R. De Nicola and U. Montanari, *A distributed operational semantics for CCS based on C/E systems*, Acta Inform. **26** (1988), 59–91.

[20] R. Devillers, *Construction of S-invariants an S-components for refined Petri boxes*, Proc. ICATPN'93, Lecture Notes in Comput. Sci. 691, M. Ajmone Marsan, ed., Springer-Verlag (1993), 242–261.

[21] R. Devillers, *S-invariant analysis of general recursive Petri boxes*, Acta Inform. **32** (1995), 313–345.

[22] R. Devillers and M. Koutny, *Recursive nets in the box algebra*, Proc. CSD'98, International Conference on Application of Concurrency to System Design, IEEE Press (1998), 239–249.

[23] W.J. Fokkink, *Introduction to Process Algebra*, Texts in Theoret. Comput. Sci., (EATCS Series), Springer-Verlag (2000).

[24] H.J. Genrich, K. Lautenbach and P.S. Thiagarajan, *Elements of general net theory*, Net Theory and Applications, Proc. Advanced Course on General Net Theory of Processes and Systems, Lecture Notes in Comput. Sci. 84, W. Brauer, ed., Springer-Verlag (1980), 21–163.

[25] U. Goltz, *On representing CCS programs by finite Petri nets*, Proc. MFCS'88, Lecture Notes in Comput. Sci. 324, M.P. Chytil, L. Janiga and V. Koubek, eds, Springer-Verlag (1988), 339–350.

[26] U. Goltz and R. Loogen, *A non-interleaving semantic model for nondeterministic concurrent processes*, Fundamentae Informaticae **14** (1991), 39–73.

[27] U. Goltz and A. Mycroft, *On the relationship of CCS and Petri nets*, Proc. ICALP'84, International Conference on Automata, Languages and Programming, Lecture Notes in Comput. Sci. 172, J. Paredaens, ed., Springer-Verlag (1984), 196–208.

[28] U. Goltz and W. Reisig, *The non-sequential behaviour of Petri nets*, Inform. and Control **57** (1983), 125–147.
[29] J. Grabowski, *On partial languages*, Fundamentae Informaticae **4** (1981), 427–498.
[30] J.G. Hall, *An algebra of high-level Petri nets*, Ph.D. Thesis, University of Newcastle upon Tyne (1996).
[31] M. Hesketh and M. Koutny, *An axiomatization of duplication equivalence in the Petri box calculus*, Proc. ICATPN'98, Lecture Notes in Comput. Sci. 1420, J. Desel and M. Silva, eds, Springer-Verlag (1998), 165–184.
[32] C.A.R. Hoare, *Communicating Sequential Processes*, Prentice Hall (1985).
[33] R. Janicki and P.E. Lauer, *Specification and Analysis of Concurrent Systems – the COSY Approach*, EATCS Monographs on Theoret. Comput. Sci., Springer-Verlag (1992).
[34] H. Klaudel and R.-Ch. Riemann, *Consistent equivalence notion for a class of high level Petri nets*, Proc. ISCIS, 11th International Symposium on Computer and Information Science, V. Atalay, U. Halici, K. Inan, N. Yalabik and A. Yazici, eds, Middle East Technical University, Antalya (1996), 7–16.
[35] M. Koutny, *Partial order semantics of box expressions*, Proc. ICATPN'94, Lecture Notes in Comput. Sci. 815, R. Valette, ed., Springer-Verlag (1994), 318–337.
[36] M. Koutny, *A compositional model of time Petri nets*, Proc. ICATPN'2000, Lecture Notes in Comput. Sci. 1825, M. Nielsen and D. Simpson, eds, Springer-Verlag (2000), 303–322.
[37] M. Koutny and E. Best, *Fundamental study: Operational and denotational semantics for the box algebra*, Theoret. Comput. Sci. **211** (1999), 1–83.
[38] P.E. Lauer, *Path expressions as Petri nets, or Petri nets with fewer tears*, Technical Report MRM/70, Computing Laboratory, University of Newcastle upon Tyne (1974).
[39] J. Lilius and E. Pelz, *An M-net semantics for BPN^2 with procedures*, Proc. ISCIS, 11th International Symposium on Computer and Information Science, Antalya, V. Atalay, U. Halici, K. Inan, N. Yalabik and A. Yazici, eds, Middle East Technical University, (1996), 365–374.
[40] A. Mazurkiewicz, *Trace theory*, Advances in Petri Nets 1986, Petri Nets, Applications and Relationships to Other Models of Concurrency, Part II, W. Brauer, W. Reisig and G. Rozenberg, eds, Lecture Notes in Comput. Sci. 255, Springer-Verlag (1987), 279–324.
[41] R. Milner, *A Calculus of Communicating Systems*, Lecture Notes in Comput. Sci. 92, Springer-Verlag (1980).
[42] R. Milner, *Communication and Concurrency*, Prentice-Hall (1989).
[43] T. Murata, *Petri nets: properties, analysis and applications*, Proc. IEEE **77** (1989), 541–580.
[44] E.R. Olderog, *Nets, Terms and Formulas*, Cambridge Tracts in Theoret. Comput. Sci. 23, Cambridge University Press (1991).
[45] G.D. Plotkin, *A structural approach to operational semantics*, Technical Report FN-19, Computer Science Department, University of Aarhus (1981).
[46] W. Reisig, *Petri Nets. An Introduction*, EATCS Monographs on Theoret. Comput. Sci., Springer-Verlag (1985).
[47] P.H. Starke, *Processes in Petri nets*, Elektronische Informationsverarbeitung und Kybernetik **17** (1981), 389–416.
[48] D. Taubner, *Finite representation of CCS and TCSP programs by automata and Petri nets*, Lecture Notes in Comput. Sci. 369, Springer-Verlag (1989).
[49] W. Vogler, *Partial words versus processes: A short comparison*, Advances in Petri Nets 1992, Lecture Notes in Comput. Sci. 609, G. Rozenberg, ed., Springer-Verlag (1992), 292–303.

Subject index

Ω-tuple, 894

ACP, 881
action, 885
– redo, 890
– skip, 890

action particles, 882, 883, 923, 936

Beh1–Beh5, 911
bisimulation
– strong, 891
box, 884, 892

– dynamic, 893
– e-directed, 886
– entry, 893
– ex-directed, 887
– ex-exclusive, 893
– exit, 893
– factorizable, 906
– operator, 884
– plain, 884, 892
– rec-, 899
– seed, 900
– sos operator, 906
– static, 893
– x-directed, 886
box algebra, 912
– denotational semantics, 914
– independence relation ind^{alg}, 921
– semantics
– – label based, 922
– – partial order based, 921
– – transition based, 919
– signature, 912
– structural equivalence, 916
– syntax, 912
– transition trees, 919
box expressions, 913

cardinality of multiset, 880
causality, 889
CCS, 876
chain, 899
clean
– marking, 888
– net, 888
cleanness, 888
communication interface operator, 883
complex marking, 888
constant
– dynamic, 912
– exit, 912
– static, 912
constant boxes, 912
constant relabelling, 885
constants, 912
continuity, 900
control flow operator, 883
COSY, 876, 935
CSP, 934

defining equation, 913
denotational semantics, 914
derivability, 888
DIY algebra, 913
– constants, 914
– denotational semantics, 915
– operational semantics, 918, 919
– similarity relation, 917
– syntax, 914
domain of application of sos-operator box, 908
duplication, 892
dynamic
– box, 893
– constant, 912
– expression, 912

e-directed
– box, 886
– expression, 914
enabled
– step, 887
– transition, 877
entry
– constant, 912
– expression, 912
– marking, 887
– place, 886
entry box, 893
equation
– defining, 913
– front-guarded, 903
– guarded, 903
– rear-guarded, 903
event occurrence, 888
ex-box, 893
ex-directed
– box, 887
ex-exclusive
– box, 893
– expression, 914
– marking, 888
ex-restrictedness, 886
exit
– constant, 912
– expression, 912
– marking, 887
– place, 886
exit box, 893
Expr1–Expr7, 914
expression
– dynamic, 912
– e-directed, 914
– entry, 912
– ex-directed, 914
– ex-exclusive, 914
– exit, 912
– flat, 913
– static, 912, 925

– well formed, 914
– x-directed, 914

factorizable box, 906
factorization
– of a tuple of boxes, 906
– of a tuple of expressions, 913
fixpoint
– positive, 902
full transition system
– of a net, 890
– of an expression, 921

global independence relation, 921
guardedness, 903
– front, 903
– rear, 903

handshake, 881
hatting, 881

identity relabelling, 885
independence relation, 888
– global, 921
initial marking, 877
internal place, 886
isomorphism of nets, 892

labelled net, 885
labelled step, 888
labelling function, 886
limit, 899

marked net, 887
marking, 877, 886
– clean, 888
– complex, 888
– entry, 887
– ex-exclusive, 888
– exit, 887
– initial, 877
– reachable, 877, 888
– safe, 877, 888
Mazurkiewicz traces, 888
monotonicity, 900
multiactions
– $\mathsf{Lab}_{\mathsf{PBC}}$, 883
multiset, 880
– cardinality, 880
– difference, 880
– inclusion, 880
– scalar multiplication, 880
– sum, 880

names of places and transitions, 894
net
– behaviour, 877
– clean, 888
– derivable, 888
– ex-restricted, 886
– finite, 886
– isomorphism, 892
– labelled, 885
– marked, 887
– pure, 887
– safe, 877, 888
– simple, 887
– T-restriction, 886
– transition system ts_{Σ}, 890
– unmarked, 887
net refinement, 893

occurrence
– of transitions, 877
operational semantics for box algebra, 919
operator
– communication interface, 883
– control flow, 883
operator box, 884, 894
– sos, 906
ordering
– on labelled nets, 899
overbarring, 878

partial order, 879
PBC syntax, 925
place, 886
– entry, 886
– exit, 886
– internal, 886
– labelling, 886
plain box, 884
pure net, 887

reachability, 888
– graph, 889
reachable complex marking, 888
reachable marking, 877
rec-box, 899
recursion
– variable, 899
redo, 890
relabelling
– constant, 885
– identity, 885
– relation, 885
– restriction, 885
– transformational, 885

reversible transition, 906

safe
– marking, 877, 888
– net, 877, 888
safeness, 888
seed box, 900
side loop, 887
similarity relation
– for expressions, 916
simple net, 887
skip, 890
solution
– of recursive equation, 899
sos-operator box, 906
– domain of application, 908
sos-operator boxes, 912
static
– box, 893
– constant, 912
– expression, 912, 925
step, 887
– enabled, 887
– executed, 887
– labelled, 888
step sequence
– finite, 887
strong
– bisimulation, 891
– equivalence, 891

T-restricted net, 886
TCSP, 877, 934
token in a marking, 877
transformational relabelling, 885
transition, 886
– enabled, 877
– labelling, 886
– occurrence, 877
– reversible, 906
transition system, 922
– full, 890, 921
– of an expression, 922
– of net, 890
tree equations, 902
ts-isomorphism, 891

underbarring, 878
unmarked net, 887

variable
– recursion, 899

weight function, 886
well formed
– expression, 914

x-directed
– box, 886
– expression, 914

CHAPTER 15

Process Algebras with Localities

Ilaria Castellani

INRIA, 2004 Route des Lucioles, 06902 Sophia-Antipolis, France
E-mail: Ilaria.Castellani@sophia.inria.fr

Contents

1. Introduction . . . 947
2. An abstract view of locations . . . 949
   2.1. Noninterleaving semantics for process algebras . . . 950
   2.2. Motivation for location semantics . . . 960
   2.3. A language for processes with locations . . . 962
   2.4. Static approach . . . 963
   2.5. Dynamic approach . . . 973
   2.6. Equivalence of the two approaches . . . 979
3. Extensions and comparison with other approaches . . . 991
   3.1. The local/global cause semantics . . . 991
   3.2. Comparison with other distributed semantics . . . 995
   3.3. Comparison with causal semantics . . . 1000
   3.4. Extension to the π-calculus . . . 1007
4. A concrete view of locations . . . 1008
   4.1. A concrete location semantics for CCS . . . 1009
   4.2. The location failure approach . . . 1011
   4.3. Combining locations and mobility . . . 1013
5. Conclusions . . . 1034
Acknowledgements . . . 1036
References . . . 1037
Subject index . . . 1043

HANDBOOK OF PROCESS ALGEBRA
Edited by Jan A. Bergstra, Alban Ponse and Scott A. Smolka

Abstract

Process algebras can be enriched with *localities* that explicitly describe the distribution of processes. Localities may represent physical machines, or more generally *distribution units* where processes are grouped according to some criterion like the sharing of resources. In a concurrent process, localities are naturally associated with (groups of) parallel components. These localities then intervene in the semantics of processes and become part, to some extent, of their observable behaviour.

In a first line of research, initiated in the early nineties, localities have been used to give *noninterleaving* semantics for process algebras, and particularly for Milner's calculus CCS. Here localities are used to differentiate parallel components. The resulting semantics, taking into account distribution, is more discriminating than the standard interleaving semantics of the calculus. It is also incomparable with other noninterleaving semantics proposed for CCS, based on the notion of causality.

More recently, localities have appeared in a number of new calculi for describing mobile processes. The idea here is that some "network awareness" is required to model wide-area distributed mobile computation. In these calculi localities are more than simple units of distribution. According to the case, they become units of failure, of communication, of migration or of security.

This chapter reviews in some detail the first body of work, and tries to delineate the main ideas of the more recent studies, which are still, for the most part, at an early stage of development.

1. Introduction

The aim of this chapter is to review some established work on process calculi with explicit localities or locations,[1] as has been developed in the last decade by several authors [28, 29,5,118,38,116,171,117,119,49], as well as to delineate new directions of research in this area that have been spurred by the growing interest in calculi for mobile processes.

The presentation will accordingly be structured in two parts: in the first (Sections 2–3) we shall present the more consolidated work on traditional process calculi, while in the second (Section 4) we will report more informally on recent proposals for combining locations and mobility. The emphasis, as well as most of the technical material, will be on the first part, where locations are used as a means to provide *noninterleaving semantics* for process algebras.

The starting idea of the former work is that *distribution* is an important aspect of parallel systems and should somehow be reflected in their semantics. If systems are modelled in a process calculus equipped with a parallel operator, like Milner's calculus CCS [109,111], this can be achieved by assigning *locations* to parallel components and then observing actions together with the location at which they occur. Using these enriched observations, one obtains so-called *distributed semantics* for the language, which are more discriminating than the standard interleaving semantics. The aim of these distributed semantics is twofold:

(1) give a measure of the *degree of parallelism* present in a system, and
(2) keep track of the *local behaviour* of components within a system.

While (1) can be useful for implementing a system on a given physical architecture, (2) allows for the detection of local properties, like a *local deadlock* or the dependence from a local resource.

The above-mentioned papers are all based on the language CCS, and adopt semantic notions which are variants of the notion of *bisimulation* introduced by Park [126] and Milner [109]. However they differ for the various subsets of CCS they consider. They also vary according to the way locations are assigned to processes: whether statically, before processes are run ([5,118,38,119,116]), or dynamically as the execution proceeds ([28, 29]). We shall refer to the former as "static approaches" and to the latter as "dynamic approaches".

The *static approaches* are further distinguished by the way locations are observed. In [119,49] the identities of locations are significant and the resulting semantics is very *concrete*: processes are equated if they reside on the same set of locations and present the same behaviour at each location. In [5,38], on the other hand, locations are observed up to a concurrency-preserving renaming which is built incrementally along executions. Here the resulting semantics is more *abstract*, identifying processes which are bisimilar in the classical sense and whose computations exhibit the same degree of distribution and the same local behaviours. There is a little subtlety to note here, as processes presenting the same degree of parallelism *in each run* need not consist of the same number of parallel components.

In the *dynamic approach*, which was the first to be developed [28,29], locations are associated with actions rather than with parallel components. They are built incrementally

[1] We shall use the two words interchangeably throughout the chapter.

when actions are performed, and then recorded in the residual processes. Here the location of an action can be viewed as the *history* of that action in the component where it occurs.

Clearly locations do not have the same intuitive meaning in the two approaches. In the static approach they represent *sites*, much as one would expect. In the dynamic approach, on the other hand, the location of an action is a record of its *local causes*, the actions that locally precede it. In spite of this difference in intuition, the dynamic and (abstract) static approach turn out to yield the same semantic notions of *location equivalence* and *location preorder* [5,38]. This means that observing distribution – in conjunction with the usual interleaving behaviour – is essentially the same as observing local causality.

As mentioned previously, these location semantics emerged as particular noninterleaving semantics for process algebras. To provide some historical background, Section 2 will start with a brief survey of such noninterleaving semantics. A comprehensive presentation of the abstract location semantics, both static and dynamic, will then be given in Sections 2.2–2.5. The two approaches will be compared in Section 2.6. Instead, the description of the concrete model of [119,49,46] will be deferred to Section 4.1, as it fits more naturally with the more recent issues examined in that section.

In Section 3 the abstract location semantics is compared with other distributed semantics (like *distributed bisimulation* [37,39,98], the *local mixed-ordering equivalence* of [116], and variants of them [47,48]), as well as with noninterleaving semantics based on causality (like *causal bisimulation* [51,52] and *history preserving bisimulation* [75,78,77]). This comparison will be mainly based on the papers [3,29,100]. In particular, we will recall in some detail the *local/global cause semantics* of [100], which incorporates distributed and causal semantics into a single framework. As it turns out, distributed semantics and causality-based semantics are orthogonal ways of accounting for the parallelism in systems, which only agree on the communication-free fragment of CCS, a class of processes also known as BPP (Basic Parallel Processes).

At the end of Section 3 we present the extension of the location semantics of [29] to the π-calculus, as proposed by Sangiorgi in [145]. An important result of this work is that the location machinery of [29] can be encoded into the standard π-calculus, in such a way that the location semantics reduces to ordinary interleaving semantics.

In Section 4.1, we describe the concrete approach of [119,49,46], where the *identities* of locations are fully observable. Observability of location names is also the standpoint adopted in most of the recent studies on languages for mobility and migration surveyed in Section 4.3.

In Section 4.2, we report on subsequent work dealing with *location failure* and its detection [12,9,140]. This has been developed for extensions of CCS and the π-calculus, with features for creating processes at remote locations, and for testing and killing locations. Contrarily to the previous line of work, here the emphasis is on location failure rather than distribution, and locations are only observable in case of failure. While the previous approaches assumed a global observer for the whole distributed system, here one adopts the perspective of an individual user, wishing to have transparent access to resources irrespective of where they are located. This is the so-called *network transparency* principle. Awareness of locations is only requested in case of failure: a process that has spawned activities at another location wants to be notified in case of failure of this location, in order to stop interacting with it and transfer activities to other locations. This kind of semantics is particularly appropriate for describing mobile applications in distributed systems, where it may be important to distinguish between a process executed locally and a process spawned at a remote location which is liable to failure.

Finally, in Section 4.3, we discuss the use of locations in connection with *mobility* and *migration* in some more recent languages with explicit distribution like the distributed JOIN-calculus [71,69], the distributed π-calculi of [141,92] and [149], the AMBIENT calculus [33], the SEAL calculus [161] and the language Nomadic Pict [150]. These languages address some aspects of large-scale distributed computation in a very explicit way. In particular, most of them abandon the network transparency assumption. For instance, in the AMBIENT calculus the details of routing for migrating processes must be exposed. Similarly, messages are explicitly routed in the distributed π-calculus of [141,92] and in the SEAL calculus. In general, in these models, processes have to know the locations of other processes and resources in order to be able to interact with them.

To conclude this introduction we could say that, while in the first approaches locations were treated simply as *units of distribution*, in these new languages they acquire additional meanings. For instance, in languages dealing with failure detection they become *units of failure*, in that the failure of a site entails the termination of all processes at that site. They constitute *units of communication* in languages with purely local communication, where processes have to convene in a common location in order to interact. In some languages supporting process migration locations are also *units of mobility*, in that the migration of a process from a location provokes the simultaneous migration of all processes placed at the same location. Finally, locations can be used as *units of security*, in the sense that a resource access request emanating from the location of the resource is not submitted to the same security checks as an external request. As their roles progressively accumulate, locations become more and more an integral part of the semantics of the language.

2. An abstract view of locations

This section reviews the approach to abstract locations as described in [28,29] and [5,38]. Generally speaking, this work is concerned with *distributed semantics* for CCS, which account for the spatial distribution of processes among different *sites* or *locations*. Such semantics emerged from a line of research on noninterleaving semantics for process algebras, which started in the early eighties and developed at a steady pace in the subsequent decade. By "noninterleaving semantics" we mean a semantics that does not simulate parallelism by a nondeterministic choice of sequential interleavings. Typically, such reduction of concurrency to interleaving is expressed by Milner's *expansion law* [109], a simple instance of which is:

$$a \mid b = a.b + b.a \tag{1}$$

where "|" and "." stand respectively for parallel and sequential composition, "+" for nondeterministic choice and a, b are any (non-synchronizable) actions. This law is indeed valid for *bisimulation equivalence*, the standard behavioural equality for the calculus CCS. By contrast, semantic models that reject this equation and insist on treating concurrency as primitive are sometimes qualified as "true-concurrency" or "nonsequential" models. Such models mostly evolved from the early ideas of Petri [129]. Paradigmatic models of this kind are *Petri nets* [130,128,138], *event structures* [165,123,168] and *Mazurkiewicz traces* [106, 107]. Similar nonsequential models were studied by Shields [153] and Winkowski [164]. To provide some background for the coming material on location semantics, we shall start with a short review of this area of research, focussing on *operational* semantics for CCS. For a more comprehensive account, covering other process algebras like ACP [14] and the Petri Box Calculus, the reader is referred to the other chapters of Part 5 in this Handbook [17].

2.1. *Noninterleaving semantics for process algebras*

We give here a brief introduction to noninterleaving operational semantics for process algebras. We concentrate on branching-time semantics based on variations of the notion of *bisimulation*, and will mostly restrict our attention to the calculus CCS.

The search for *operational semantics* for process algebras accounting for the concurrent and causal structure of processes, started roughly around the mid-eighties. Previously, there had been a few proposals for interpreting languages like CCS, CSP and TCSP into nonsequential models for concurrent systems such as *Petri nets* [129,130,128] and *event structures* [165,123,168]. Among these early proposals, we will only mention those by Winskel [166], De Cindio et al. [45], and Goltz and Mycroft [83]. Many other interpretations of a similar kind were to be given in subsequent years, on variants of these models or on other semantic domains like *asynchronous transition systems* [15], *causal trees* [51], *trace automata* [13] and *transition systems with independence* [170]. For instance, Petri nets have been further used to give semantics for CCS-like languages by Goltz [83,85, 82], Winskel [167,169], Van Glabbeek and Vaandrager [80], Nielsen [122], Degano, De Nicola and Montanari [61], Degano, Gorrieri and Marchetti [57], Olderog [124,125], Taubner [154,155], Boudol and Castellani [27], Montanari and Gorrieri [86], Montanari and Ferrari [67], and others. Similarly, event structure interpretations have been proposed and studied, among others, by Winskel [167], Goltz and Loogen [105], Degano, De Nicola and Montanari [54], Vaandrager [159], Boudol and Castellani [23,24,27], van Glabbeek and Goltz [78,77], and Langerak [102].

Event-based models like Petri nets and event structures are built from the outset on the primitive notions of *concurrency* and *causality* between events, and therefore naturally provide a noninterleaving semantics for process calculi. The question was then to devise a purely operational semantics for processes, by means of labelled transition systems specified via structural rules (in the style advocated by Plotkin [133], see also [6]), which could express the same information about concurrency and causality.

A natural idea was to generalize the labels of transitions, from simple actions to whole *partial orders* of actions representing nonsequential computations. The representation of concurrent computations as partial orders, where the ordering stands for causality (and the absence of ordering for concurrency), was already standard in nonsequential models like Petri nets, event structures and Mazurkiewicz traces, where these partial orders were called respectively *abstract processes* [84,137], *configurations* and *dependency graphs*.

When dealing with process algebras, a simple way to achieve this generalization of actions is to extract the information about concurrency and causality from the *syntax* of the terms themselves. More precisely, one looks for structural semantic rules by which the constructs of sequential and parallel composition are directly transferred from the terms to the actions labelling their transitions. This idea was successfully applied in [23] to a simple CCS-like language with sequential composition and disjoint parallel composition (not allowing communication): using the notion of *partially ordered multiset*,[2] studied by Grabowski [88], Pratt [134,135] (who coined for it the term "pomset") and Gischer [73], a definition of *pomset transition* was introduced in [23] for this language, with the associated notion of *pomset bisimulation*. This new notion of bisimulation distinguishes the two processes of Equation (1) because the first has a pomset transition

$$a \mid b \xrightarrow{a|b} nil \mid nil$$

which the second cannot simulate. The second process has two possibilities for executing a and b in one step, but in each case the two actions are ordered. One of these pomset transitions is

$$a.b + b.a \xrightarrow{a.b} nil.$$

To generalize the pomset semantics to the full language CCS [24], on the other hand, it appeared necessary to use an indirect construction of a pomset transition starting from a class of permutation-equivalent transition sequences.[3] This construction "a posteriori" of a partial order from an equivalence class of sequences is similar to the synthesis of a partial order from a Mazurkiewicz trace [107]. The main novelty of [24] was the use of enriched transitions for CCS called *proved transitions* (because they were labelled with a representation of their *proof* in the inference system of CCS), from which the concurrency relation between transitions could be derived by syntactic means. For instance, a sequence of proved transitions for the process $a \mid b$ is

$$a \mid b \xrightarrow{|_0 a} nil \mid b \xrightarrow{|_1 b} nil \mid nil.$$

These two transitions are concurrent because they are extracted from two different sides of the parallel operator (i.e., they are inferred using different rules for $|$), and this information is recorded in their labels. The transitions can therefore be commuted and the pomset corresponding to the equivalence class of this sequence is $a \mid b$. On the other hand the process $a.b + b.a$ has the sequence of proved transitions

$$a.b + b.a \xrightarrow{+_0 a} b \xrightarrow{b} nil$$

where the two transitions cannot be commuted, hence the corresponding pomset is $a.b$.

[2] Or *pomset*, a generalization of the notion of partial order, which is necessary to deal with homonymous actions in computations. Formally, pomsets are defined as isomorphism classes of labelled partial orders.

[3] This is because the class of pomsets is not closed with respect to parallel composition when communication is allowed.

Indeed, the permutation semantics of [24] extends the pomset semantics of [23] and can be applied to more interesting examples. For instance the process $(a.\gamma \mid \overline{\gamma}.b)$, where γ and $\overline{\gamma}$ are two synchronizable actions, can do the sequence of proved transitions

$$a.\gamma \mid \overline{\gamma}.b \xrightarrow{|_0 a} \gamma \mid \overline{\gamma}.b \xrightarrow{(\gamma,\overline{\gamma})} nil \mid b \xrightarrow{|_1 b} nil \mid nil.$$

Here the two transitions a and b, although marked as concurrent by their labels, cannot be commuted because they are not adjacent. The pomset generated by this transition sequence is $a.\tau.b$ (where τ represents a generic communication action). Indeed, the communication between γ and $\overline{\gamma}$ creates a *cross-causality* between the actions a and b.

Note that while in the direct pomset semantics of [23] atomic actions are replaced with composite actions representing whole nonsequential computations, in the proved transition semantics actions remain atomic but are decorated with additional information that identifies them uniquely in a term. Pomset transitions are retrieved only in a second step. Indeed, these two proposals by Boudol and Castellani exemplify two different ways of structuring actions: either relax atomicity of actions so as to obtain *compound actions*, or retain portions of their proof in order to pinpoint their identity – we shall speak in this case of *decorated (atomic) actions*.

Another track for defining noninterleaving semantics for process algebras was to structure the *states* – rather than the actions – of the transition system, so as to mark the components responsible for each action. This led first to Degano and Montanari's model of *concurrent histories* [58], concrete computations with structured initial and final states, which could be concatenated while keeping the causality information. From these computations a partial ordering of actions could then be abstracted away. This model was used by Degano, De Nicola and Montanari to define a partial ordering semantics for CCS in [59, 56]. Here the causality relation on actions is recovered by decomposing each CCS process into a set of sequential components called *grapes*, and by causally connecting only those successive actions that originate from the same component. For example the process $a \mid b$ is decomposed into two components $a \mid id$ and $id \mid b$ (here the markers $\mid id$ and $id \mid$ are used to specify the access paths of components inside a term, and therefore to identify them throughout a computation[4]). We have here a computation

$$\{a \mid id, id \mid b\} \xrightarrow{a} \{nil \mid id, id \mid b\} \xrightarrow{b} \{nil \mid id, id \mid nil\}$$

where actions a and b can be seen as emanating from different components and hence are not causally related.[5] On the other hand the process $a.b + b.a$ has only one component, itself. The computation in this case is

$$\{a.b + b.a\} \xrightarrow{a} \{b\} \xrightarrow{b} \{nil\}.$$

[4] Just like the markers $|_0$ and $|_1$ are used to identify occurrences of actions in proved transitions.

[5] This is a slightly simplified rendering of the semantics of [59]. As a matter of fact, also actions were enriched in [59,56] – with a portion of their proof – but this information was somehow redundant, in particular it was not used to derive the partial order computations.

Here actions a and b originate from the same component and thus are causally related. The resulting partial order semantics is the same as that derived from the previously mentioned proved transition systems [24]. In both cases a partial order computation is obtained from a sequential computation together with a relation on transitions (a causality relation here, and a concurrency relation in [24]).

On the other side, the same idea of structuring states led Castellani and Hennessy to the definition of distributed transition systems [37,39], where transitions have distinct *local* and *concurrent* residuals:[6] that is to say, transitions $p \xrightarrow{a} p'$ are replaced by transitions of the form $p \xrightarrow{a} \langle p'_\ell, p'_c \rangle$, where p'_ℓ is what locally follows the action a while p'_c is what is concurrent with it. A notion of *distributed bisimulation*, acting separately on the two residuals, was then proposed to compare these systems. For the two processes of our running example, the distributed transitions would be[7]

$$\begin{aligned} a \mid b &\xrightarrow{a} \langle nil, b \rangle \\ a.b + b.a &\xrightarrow{a} \langle b, nil \rangle \end{aligned}$$

and the distributed bisimulation would tell them apart since the residuals are not pairwise equivalent. It should be pointed out that no partial order computation was synthesized here. Indeed this particular semantics, which will be reviewed in some detail in Section 3.2, departed somehow from the main line of research at the time, which aimed at defining partial ordering transitions. It was the first attempt at a *distributed semantics*, emphasizing local behaviours rather than the global causal behaviour. Unfortunately this distributed semantics was defined only for a subset of CCS without the restriction operator, and its extension to the whole language, investigated by Kiehn in [98], appeared to be difficult. Indeed, the very idea of splitting terms into parts and observing these parts separately did not seem fully compatible with global operators such as restriction.

The partial ordering approach of [59] was pursued in subsequent works by Degano, De Nicola and Montanari [54,61], where the grape semantics was adapted in various ways to match more closely event structure and Petri net interpretations for CCS. The same authors presented a different formulation of a partial ordering semantics for CCS in [60], based on a tree-model called Nondeterministic Measurement Systems (NMS). By applying so-called observation functions to NMS's, different equivalence notions could be obtained, among which a partial ordering equivalence called *NMS-equivalence*, which turned out to be unrelated to pomset bisimulation.[8]

Another rather original tree-model that was put forward in the late eighties, with the express purpose of reconciling the interleaving and partial order approaches, was Darondeau and Degano's model of *causal trees* [51,52]. These are a variant of Milner's synchronization trees, where labels of observable transitions carry an additional element representing the *set of causes* of the corresponding action. More precisely, these labels are of the form

---

[6] This was the original formulation reported in [37]. An alternative formulation in terms of *local* and *global* residuals was considered in [39], as it seemed better suited to deal with the full language CCS.

[7] To be precise the first transition should be $a \mid b \xrightarrow{a} \langle nil, nil \mid b \rangle$, but we are keeping deliberately informal here.

[8] Examples showing the incomparability of the two equivalences are given by Goltz and van Glabbeek in [75] and [79].

(a, K), where K is an encoding – in the form of backward pointers in the tree – of the actions that causally precede a. The bisimulation associated with these new transitions, called *causal bisimulation*, was of course stronger than the usual bisimulation, but also stronger than pomset bisimulation and NMS-equivalence. It may be noted that in causal trees, the causal information is recorded in both the states (nodes) and the actions. In fact, this model illustrates a further way of decorating atomic actions, with parts of their *computational history* rather than parts of their proof.

The initial criterion for these noninterleaving operational semantics for CCS was their agreement with "denotational semantics" given by interpretations into event based semantic domains like event structures and Petri nets. Besides the above-mentioned papers [54, 61], which show the consistency of the grape operational semantics with interpretations of terms into prime event structures and condition/event systems, we can mention also [24, 25], where the pomset semantics is shown to agree with an interpretation of processes into more general event structures, and the work by Olderog [124], where a Petri net semantics for the language CCSP (a mixture of CCS and TCSP) is derived operationally. In [27] the results of [24,25] are extended to Petri nets and asynchronous transition systems. A correspondence between (a variant of) the proved transition semantics and an interpretation of processes as *trace automata* is given by Badouel and Darondeau in [13]. As it were, the search for denotational interpretations of process algebras gave a new impulse to the study of semantic domains, and led to the definition of specific classes of event structures and nets, such as *bundle event structures* [102] and *flow event structures* [25,40],[9] *flow nets* [26], and *Δ-free event structures* [53]. Following the approach of Winskel [167], many of these denotational models had been meanwhile uniformly formalized in a categorical framework, which facilitated their analysis and comparison: several correspondence results between the various models, including Mazurkiewicz traces and transition systems with independence, could be established in this setting [170].[10] We refer to this work and to [147] for a comprehensive survey of such categorical models and their relationships. The unification of models of concurrency into a common algebraic framework had also been a central concern of Ferrari and Montanari's work [65,67].

In the late eighties, a new criterion for assessing equivalences emerged in connection with the issue of *action refinement*. Here, in accordance with the methodology of stepwise refinement, one wishes to allow for a gradual specification of systems by successive refinements of atomic actions into more complicated behaviours. Then, to be applicable along such top-down design of systems, equivalence notions are required to be preserved under action refinement. This means that two equivalent processes should remain equivalent after replacing all occurrences of an action a by a process $r(a)$. At first, this new criterion appeared to be a further argument in favour of "true concurrency" semantics, since all the existing interleaving semantics failed to satisfy it. For instance, as mentioned earlier, the standard bisimulation equivalence $\sim$ satisfies the expansion law

$$a \mid b \sim a.b + b.a.$$

[9] Which are both weaker variants of Wisnkel's *stable event structures*.

[10] The correspondence in [170], as in most other cases, is given by the isomorphism of the computation domains of the models, while in [27] the correspondence between the three considered models was tighter, amounting to the equality of the families of computations.

On the other hand, if action a is refined into the process $a_1.a_2$, the equality no longer holds

$$a_1.a_2 \mid b \not\sim a_1.a_2.b + b.a_1.a_2$$

since after performing a_1 the first process can choose between a_2 and b, while the second cannot. On the basis of this example, the use of partial ordering semantics was advocated by Castellano, de Michelis and Pomello as more appropriate for dealing with action refinement [41]. However, in reaction to this position Goltz and van Glabbeek showed that none of the existing branching-time partial order equivalences was preserved by refinement [75,76]. In particular pomset bisimulation and NMS equivalence[11] were not invariant under refinement. In conclusion it was proved that a finer equivalence called *history preserving bisimulation* – first proposed by Rabinovitch and Trakhtenbrot [136] under the name *BS-equivalence* (Behaviour Structure equivalence) and then reformulated for event structures by Goltz and van Glabbeek – was indeed preserved by refinement. This result was first proven in the setting of prime labelled event structures for a restricted form of refinement [75], and subsequently extended, for more general forms of refinement, to flow and stable event structures [78,77,79], and finally to *configuration structures* [79].

Well before the start of this debate on refinement, Hennessy had proposed in a somewhat confidential publication [89] a semantics for CCS taking into account beginnings and endings of actions, and studied the corresponding bisimulation. This equivalence, later called *split bisimulation*, was reproposed in [90] in a slightly updated form and subsequently studied by Aceto and Hennessy [7,8] as well as other authors. It appeared to be closely related to another equivalence, not conceived for true concurrency but for real-time, introduced by van Glabbeek and Vaandrager in [80] and named *ST-bisimulation*.[12] While ST-bisimulation is preserved by refinement [74], split bisimulation is only preserved by refinement on a subset of CCS [7,81]. Thus ST-bisimulation appeared to be the weakest strengthening of ordinary bisimulation which would be a congruence with respect to refinement. However, since this equivalence is not generally agreed to properly reflect causality (see [79]), it could not be adopted as a representative for partial order semantics.

Meanwhile, refinement had been studied in the setting of Petri nets [162] and causal trees [53],[13] and history preserving bisimulation had been shown to coincide with *concurrent bisimulation* [19] on nets, and with causal bisimulation [3,53] on causal trees. In [3] Aceto also showed the coincidence of causal bisimulation with the *mixed-ordering equivalence* of Degano et al. [55]. These results further consolidated the status of history preserving bisimulation as the best candidate equivalence when both causality and refinement are of concern [79]. It should be pointed out, however, that only the *strong* version of this equivalence was studied, taking into account internal actions as well as visible ones. This is because the weak version of history preserving bisimulation would not be preserved by

[11] Which also coincides with an equivalence proposed by Devillers [63], called weak history preserving bisimulation by Goltz and van Glabbeek [75].

[12] This equivalence is finer than split bisimulation in that it matches action endings with specific beginnings, but it coincides with it on systems without autoconcurrency (i.e., concurrency between actions of the same name).

[13] In fact, refinement was prevalently studied as a substitution operation on the semantic domains used to interpret terms (*semantic refinement*) rather than on the languages themselves (*syntactic refinement*). Among the abovementioned papers, only [7] and [8] give a syntactic treatment of refinement. For other references see [87].

refinement – as is the case, in fact, for most weak equivalences.[14] For more about history preserving bisimulation and the question of action refinement, we refer the reader to [2,79] and to [87].

Most of the noninterleaving semantics reviewed so far are based on the notion of causality, and can therefore be qualified as *causal semantics*. By contrast, *distributed semantics* focus on the local behaviour of parallel components and aim at capturing local dependencies. As mentioned earlier, the distributed bisimulation semantics of Castellani and Hennessy [37,39], initially defined for a subset of CCS without the restriction operator, appeared to be hard to extend to the whole language. The problem can be illustrated by a simple example, where the restriction operator $\backslash\gamma$ has the effect of forcing actions γ and $\overline{\gamma}$ to occur simultaneously (and unobservably). Consider the process:

$$p = (a.b.\gamma \mid \overline{\gamma}.c)\backslash\gamma.$$

Suppose action a is performed. Then, what should be its local and concurrent residuals? Clearly, distributing the restriction operator over the two residuals would unduly limit their behaviour: if we let $p = (a.b.\gamma \mid \overline{\gamma}.c)\backslash\gamma \xrightarrow{a} \langle (b.\gamma)\backslash\gamma,\ (\overline{\gamma}.c)\backslash\gamma\rangle$, then the second residual would be equivalent to *nil*, and the whole process p would be equivalent to $a.b$, which is not what we expect. A variant of this example shows that we could also make unwanted distinctions between processes. Let $q = (a.\gamma.b \mid \overline{\gamma})\backslash\gamma$. Then q should be equivalent to $a.b$, since in both cases actions a and b occur in sequence at the same location. However a transition $q = (a.\gamma.b \mid \overline{\gamma})\backslash\gamma \xrightarrow{a} \langle (\gamma.b)\backslash\gamma,\ \overline{\gamma}\backslash\gamma\rangle$ would yield a local residual equivalent to *nil*, and therefore distinguish q from $a.b$.

The other naïve solution would be to "resolve" the restriction by transforming restricted actions into unobservable τ-actions in the residuals. Again, this would lead to unwanted identifications and distinctions. For instance, we would have in this case a transition $p = (a.b.\gamma \mid \overline{\gamma}.c)\backslash\gamma \xrightarrow{a} \langle b.\tau,\ \tau.c\rangle$ and p would behave like the process $(a.\gamma.b \mid \overline{\gamma}.c)\backslash\gamma$, where action c can occur before action b. On the other hand, the process $(a \mid \overline{\gamma}.b)\backslash\gamma$, which should be equivalent to just a, would be equated with $(a.\gamma \mid \overline{\gamma}.b)\backslash\gamma$.

In fact, it is not clear how to distribute an encapsulation operator such as restriction without losing information about its scope, which is crucial to determine both the local and the overall behaviours. For this reason, subsequent studies on distributed bisimulation tended to keep the residuals in their context (while marking them in some way) rather than observe them in isolation. The alternative formulation in terms of *local* and *global* residuals studied in [39] was a first step in this direction. It did not seem sufficient, however, to overcome the above-mentioned difficulties. Indeed, the later work by Kiehn [98] showed that in the presence of the restriction operator even rather ingenious reformulations of the semantics failed to capture some desired operational distinctions (the later variant considered by De Nicola and Corradini in [48], while getting around Kiehn's counterexample, still suffered from the same drawback).

A few years later, a fresh attack on the problem was resumed with the location semantics of Boudol, Castellani, Hennessy and Kiehn [28,29]. The idea here was to insert locations

[14] In [8], Aceto and Hennessy studied a weak equivalence which is preserved by refinement. However this is essentially a weak ST-bisimulation equivalence.

into processes and observe actions together with the location at which they occur. In the initial work [28,29] locations are created *dynamically* for actions when they are executed, and then recorded in the derivative process. The assignment of locations is such that concurrent actions have disjoint locations, while actions from the same component have related locations (one is a sublocation of the other). For instance the process $a \mid b$ has a sequence of transitions

$$a \mid b \xrightarrow[\ell]{a} \ell :: nil \mid b \xrightarrow[k]{b} \ell :: nil \mid k :: nil$$

where actions a and b occur at different locations. On the other hand the process $a.b + b.a$ has transitions

$$a.b + b.a \xrightarrow[\ell]{a} \ell :: b \xrightarrow[\ell k]{b} l :: k :: nil$$

where the location of b is a *sublocation* of that of a. Note that the idea underlying this semantics is a natural generalization of that of splitting residuals: every component which becomes active is marked with a location which will distinguish it from any other component running in parallel.

Based on this new transition system, notions of *location equivalence* $\approx_\ell$ and *location preorder* $\sqsubseteq\!\!\!\!\sim_\ell$ were introduced. The equivalence $\approx_\ell$ formalizes the idea that two processes are bisimilar, in the classical sense, and moreover are equally distributed and have the same local behaviours in each computation. Formally $\approx_\ell$ is just the ordinary bisimulation associated with the new transition system. It can be easily seen, looking at the transitions above, that

$$a \mid b \not\approx_\ell a.b + b.a.$$

This is expected since the first process is distributed among two different locations while the second is confined within one location. The location preorder $\sqsubseteq\!\!\!\!\sim_\ell$, on the other hand, relates a process with another which is bisimilar but possibly more distributed. Technically, the locations of the second process are required to be *subwords* of the locations of the first. For example

$$a.b + b.a \sqsubseteq\!\!\!\!\sim_\ell a \mid b.$$

Again, this is expected because the first process is less distributed than the second.

In two subsequent papers, Aceto [5] and Castellani [38] proposed an alternative approach where locations are assigned *statically* to parallel components, and observed modulo a concurrency-preserving renaming. This approach, although closer to the intuition, had been initially abandoned in favour of the dynamic one because of some technical difficulties it presented. In the static approach, processes are given an explicit distribution before they are executed, by inserting locations in front of parallel components. For instance, two *distributions* for the processes $a \mid b$ and $a.b + b.a$ would be

$$\ell :: a \mid k :: b \quad \text{and} \quad a.b + b.a$$

where ℓ and k are distinct locations, and processes with no explicit location are supposed to be placed at the location ε of the overall system. The semantics then simply exhibits the locations along transitions. For instance

$$\ell :: a \mid k :: b \xrightarrow[\ell]{a} \ell :: nil \mid k :: b \xrightarrow[k]{b} \ell :: nil \mid k :: nil$$

$$a.b + b.a \xrightarrow[\varepsilon]{a} b \xrightarrow[\varepsilon]{b} nil.$$

Notions of location equivalence and preorder are now redefined in this setting. An immediate observation is that one cannot require here equality of locations. Otherwise one would distinguish processes like $a \mid (b \mid c)$ and $(a \mid b) \mid c$, whose distributions are of the form

$$\ell_1 :: a \mid k_1 :: (\ell_1' :: b \mid k_1' :: c) \quad \text{and} \quad \ell_2 :: (\ell_2' :: a \mid k_2' :: b) \mid k_2 :: c.$$

The idea is then to compare transitions modulo particular *associations* of their locations. For this example the required association would be: $\{(\ell_1, \ell_2\ell_2'), (k_1\,\ell_1', \ell_2\,k_2'), (k_1\,k_1', k_2)\}$.

Interestingly the two semantics, dynamic and static, yield the same notions of location equivalence and preorder, in spite of the fact that locations reflect different intuitions in the two cases. In the static case locations represent *sites*, while in the dynamic case they represent *local causes* for an action. Indeed, in the static case locations may be seen as part of the proof of a transition, in the sense explained earlier, while in the dynamic case they are part of its computational history.

We shall not enter in further details here, since the location semantics will be presented in detail in the next section. Let us just mention that this semantics consistently extends the distributed bisimulation semantics of [39], while being in general incomparable with the causality-based semantics discussed earlier.[15] It therefore offers an interesting complementary point of view about observing parallelism in processes.

Semantics based on static locations were investigated also by other authors. In [118] an equivalence analogous to that of [38] was studied by Mukund and Nielsen for a class of asynchronous transition systems modelling a subset of CCS with guarded sums. It was conjectured to coincide with the dynamic location equivalence of [29], a result that could later be subsumed from those of [38]. A transition system for CCS labelled with static locations, called *spatial transition system*, was considered by Montanari and Yankelevich in [116,171,117]. In this latter work, however, location transitions are only used as a first step to build a second transition system, labelled by *local mixed partial orders* (where the ordering is a mixture of temporal ordering and local causality), which is then used to define the behavioural equivalence. This equivalence, a variant of the mixed-ordering equivalence of [55] called *local mixed-ordering equivalence*, was shown to coincide with location equivalence [171]. The view of location equivalence as capturing local causality was further substantiated by the work of Kiehn [100], who characterized it as a variant of causal bisimulation called *local cause bisimulation*. In the same spirit, in [38] location equivalence was presented as a *local history preserving bisimulation*. The first characterization

[15] Except on the finite fragment of CCS without communication, where the location semantics coincides with the causal semantics of [51] and [75].

was carried out within Kiehn's local/global cause transition system, an enriched transition system which unifies distributed and causal semantics into a single framework, thus shedding light on their relationship. In afterthought, the convergence of these three new formulations of location equivalence was somehow expected since their "global" counterparts (mixed-ordering equivalence, causal bisimulation and history preserving bisimulation) were known to coincide from Aceto's work [3].

A more concrete view of localities was proposed by D. Murphy in [119] (and later extended to unobservable transitions by Corradini and Murphy in [49]), for a subset of CCS with only top-level parallelism. Here again localities are assigned statically to parallel components. The main difference with respect to the other static approaches is that the names of localities are significant here, and processes are considered equivalent only if they reside on the same set of localities and present the same behaviour at each locality. The interest here is centered on implementing a system on a *fixed network of processors*, while in the previous cases the notions of local behaviour and degree of distribution where expressed somewhat abstractly. For instance the two distributed processes

$$(\ell :: a \mid k :: b) \quad \text{and} \quad (\ell :: b \mid k :: a)$$

are distinguished in Murphy's semantics, while they are equated in the other static location semantics.

More recently, Corradini and De Nicola have reformulated in [48] both the dynamic location equivalence and a variant of Kiehn's extended distributed bisimulation [98] within the grape semantics of [56]. Kiehn's extended distributed bisimulation was known to be weaker than location equivalence from [28] and [29]. While exploiting some of Kiehn's ideas, Corradini and De Nicola propose here a slightly stronger equivalence, which however still fails to recover the full power of location equivalence.

We conclude our review of noninterleaving semantics with a remark about structured actions and states. The idea of decorating atomic actions with (portions of) their proof to obtain more expressive semantics for CCS culminated on one side in the above-mentioned *proved transition systems* [24,27], and on the other side in Montanari and Ferrari's *structured transition systems* [65,67,68]. In the first model only actions are structured while in the second both states and transitions are structured.[16] Both these transition systems retain in some sense a maximal information about the structure of the underlying terms, which can be weakened in various ways to obtain specialized semantics like pomset or distributed semantics. For instance the structured transitions of [67] are interpreted through a mapping into an algebra of observations, and the associated notion of bisimulation is *parameterized* with respect to the choice of the algebra. Similarly, proved transitions were exploited by Degano and Priami in [62] to obtain different semantics for CCS. Here, in order to capture the causal bisimulation semantics of [51], the authors use a simpler model based on derivation trees – called *proved trees* – rather than the more general proved transition systems discussed earlier.

[16] In [27] a generalization of the proved transition system called *event transition system* is considered, where also states are structured in order to represent partially executed terms: like the causal trees of Darondeau and Degano and Kiehn's local/global cause transition system, these systems incorporate information about the *computational history* of processes, and this information needs to be recorded in the states in order to be correctly propagated along the transitions.

2.2. *Motivation for location semantics*

In this section we introduce the abstract location semantics proposed in [28,29] and [5, 38], by means of simple examples. These will be described using CCS notation (formally introduced only in the next section), which should be accessible to the unfamiliar reader with the help of comments and pictures.

The distributed structure of CCS processes can be made explicit by assigning different *locations* to their parallel components. This can be done using a *location prefixing* construct $l :: p$, which represents process p residing at a location l. The actions of such a process are observed together with their location. We have for instance

$$(l :: a \mid k :: b) \xrightarrow[l]{a} (l :: nil \mid k :: b) \xrightarrow[k]{b} (l :: nil \mid k :: nil).$$

In CCS, parallelism is not restricted to top-level. It may also appear after the execution of some actions, in which case it is best viewed as a *fork* operation, a future activation of two processes in parallel. Then the locations of actions will not be simple letters $l, k, \ldots$ but rather words $u = l_1 \cdots l_n$, and a "distributed process" will perform transitions of the form $p \xrightarrow[u]{a} p'$. For instance, if the process $(l :: a \mid k :: b)$ is a component of a larger process, we will have

$$\begin{aligned} l' :: c.(l :: a \mid k :: b) \mid k' :: d &\xrightarrow[l']{c} l' :: (l :: a \mid k :: b) \mid k' :: d \\ &\xrightarrow[l'l]{a} l' :: (l :: nil \mid k :: b) \mid k' :: d. \end{aligned}$$

Let us consider a more concrete example, taken from [28]. We may describe in CCS a simple protocol, transferring data one at a time from one port to another, as follows:

$$\begin{aligned} Protocol &\Leftarrow (Sender \mid Receiver)\backslash\alpha, \beta \\ Sender &\Leftarrow in.\overline{\alpha}.\beta.Sender \\ Receiver &\Leftarrow \alpha.out.\overline{\beta}.Receiver, \end{aligned}$$

where $\overline{\alpha}$ represents transmission of a message from the sender to the receiver, and $\overline{\beta}$ is an acknowledgement from the receiver to the sender, signalling that the last message has been processed. In the standard theory of *weak bisimulation equivalence*, usually noted $\approx$, one may prove that this system is equivalent to the following specification

$$PSpec \Leftarrow in.out.PSpec.$$

That is to say, $PSpec \approx Protocol$. The reader familiar with the (weak version of) *causal bisimulation* $\approx_c$ of [51],[17] which is a strengthening of $\approx$, should also be easily convinced that $PSpec \approx_c Protocol$: intuitively, this is because the synchronizations on α, β in *Protocol* create "cross-causalities" between the visible actions *in* and *out*, constraining them to happen alternately in sequence.

[17] This equivalence will be reviewed in Section 3.3.

In a distributed view, on the other hand, it would be reasonable to distinguish these two systems, because *PSpec* is completely sequential and thus performs the actions *in* and *out* at the same location l, what can be represented graphically as follows

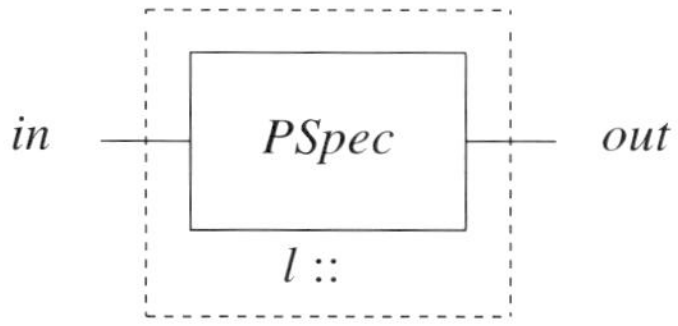

while *Protocol* is a system distributed among two different localities l_1 and l_2, with actions *in* and *out* occurring at l_1 and l_2 respectively. Thus *Protocol* may be represented as

Here the unnamed link represents the communication lines α, β, which are private to the system. We shall see that *PSpec* and *Protocol* will not be equated in the location semantics. On the other hand they will be related by a weaker relation, a *preorder* that orders processes according to their degree of distribution.

Let us consider another example, taken from [29], which describes the solution to a simple mutual exclusion problem. In the system *Mutex*, two processes compete for a device, and a semaphore is used to serialize their accesses to this device

$$\begin{aligned} Mutex &\Leftarrow (Proc \mid Sem \mid Proc)\backslash\{p, v\} \\ Proc &\Leftarrow \overline{p}.enter.exit.\overline{v}.Proc \\ Sem &\Leftarrow p.v.Sem. \end{aligned}$$

As a distributed system, *Mutex* can be pictured as follows, where l_1 and l_2 are again two distinct locations (the location of the semaphore is left implicit, since all its actions are unobservable)

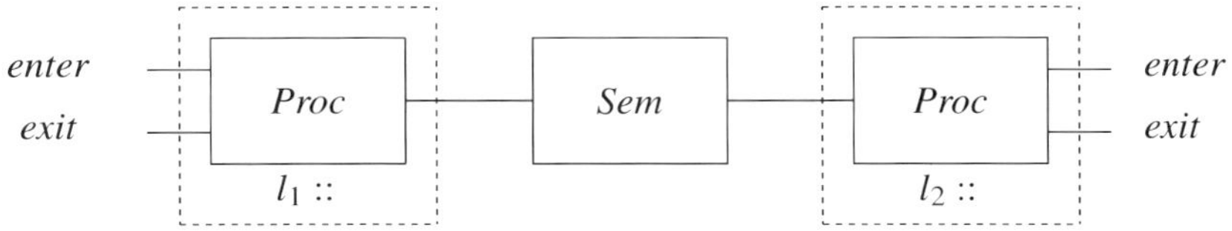

Take now a variant of the system *Mutex*, where one of the processes is faulty and may deadlock after exiting the critical region (the deadlocked behaviour being modelled here as

The set of rules specifying the operational semantics of LCCS is given in Figure 1.[19] The rule for communication (S4) requires some explanation. In the strong location transition system defined here (where no abstraction is made from τ-transitions), we take the location of a communication to be that of the smallest component which includes the two communicating subprocesses: the notation $u \sqcap v$ in rule (S4) stands for the longest common prefix of u and v. For instance we have:

EXAMPLE 2.1.

$$\begin{aligned} l :: \alpha \mid k :: \bar{\alpha}.(l' :: \beta \mid k' :: \bar{\beta}) &\xrightarrow[\varepsilon]{\tau}_s l :: nil \mid k :: (l' :: \beta \mid k' :: \bar{\beta}) \\ &\xrightarrow[k]{\tau}_s l :: nil \mid k :: (l' :: nil \mid k' :: nil). \end{aligned}$$

There are other possibilities for defining locations of τ-actions representing communication. In [116] the location of such an action is the pair of locations of the contributing actions. In the above example, the resulting location would be (l, k) for the first communication, and (kl', kk') for the second. In [28,29], τ-actions have no location at all. In fact, these differences are not important since in all these studies the interest is centered on the *weak location transition system*, where the locations of τ-actions are simply ignored (since the actions themselves are not observable).

The weak location transitions $\overset{a}{\underset{u}{\Longrightarrow}}_s$ and $\overset{\tau}{\Longrightarrow}_s$ are defined by:

$$\begin{aligned} p \overset{\tau}{\Longrightarrow}_s q \iff &\exists u_1, \ldots, u_n,\ p_0, \ldots, p_n,\ n \geqslant 0 \text{ such that} \\ &p = p_0 \xrightarrow[u_1]{\tau}_s p_1 \cdots \xrightarrow[u_n]{\tau}_s p_n = q, \\ p \overset{a}{\underset{u}{\Longrightarrow}}_s q \iff &\exists p_1, p_2 \text{ such that } p \overset{\tau}{\Longrightarrow}_s p_1 \xrightarrow[u]{a}_s p_2 \overset{\tau}{\Longrightarrow}_s q. \end{aligned}$$

The weak location transition system will be the basis for defining notions of *location equivalence* and *location preorder* accounting for the distributed behaviour of CCS processes.

The reader may have noticed, however, that applying the rules of Figure 1 to CCS terms just yields a transition $p \xrightarrow[\varepsilon]{\mu}_s p'$ whenever the standard semantics yields a transition $p \xrightarrow{\mu} p'$. In fact, these rules will not be applied directly to CCS terms. Instead, the idea is to first bring out the parallel structure of CCS terms by assigning locations to their parallel components, thus transforming them into particular LCCS terms called "distributed processes", and then execute these according to the given operational rules.

The set DIS $\subseteq$ LCCS of *distributed processes* is given by the grammar:

$$p ::= nil \mid \mu.p \mid \underbrace{(l :: p \mid k :: q)}_{l \neq k} \mid (p + q) \mid p \backslash \alpha \mid p\langle f \rangle \mid x \mid rec\, x.\ p.$$

Essentially, a distributed process is obtained by inserting a pair of distinct locations in a CCS term wherever there occurs a parallel operator. This is formalized by the following notion of *distribution*.

[19] For convenience we include in the same page the rules for the dynamic transition system, which will only come into play in Section 2.5. These rules (given in Figure 2) can be ignored for the time being.

For each $\mu \in Act_\tau$, $u \in Loc^*$, let $\xrightarrow[u]{\mu}_s$ be the least relation $\xrightarrow[u]{\mu}$ on LCCS processes satisfying the following axiom and rules.

$$
\begin{array}{lll}
(\mathrm{S1}) & \mu.p \xrightarrow[\varepsilon]{\mu} p & \\
(\mathrm{S2}) & p \xrightarrow[u]{\mu} p' & \Rightarrow\ l :: p \xrightarrow[lu]{\mu} l :: p' \\
(\mathrm{S3}) & p \xrightarrow[u]{\mu} p' & \Rightarrow\ p \mid q \xrightarrow[u]{\mu} p' \mid q \\
 & & \quad\ q \mid p \xrightarrow[u]{\mu} q \mid p' \\
(\mathrm{S4}) & p \xrightarrow[u]{\alpha} p',\ q \xrightarrow[v]{\bar{\alpha}} q' & \Rightarrow\ p \mid q \xrightarrow[u \sqcap v]{\tau} p' \mid q' \\
(\mathrm{S5}) & p \xrightarrow[u]{\mu} p' & \Rightarrow\ p + q \xrightarrow[u]{\mu} p' \\
 & & \quad\ q + p \xrightarrow[u]{\mu} p' \\
(\mathrm{S6}) & p \xrightarrow[u]{\mu} p',\ \mu \notin \{\alpha, \bar{\alpha}\} & \Rightarrow\ p\backslash\alpha \xrightarrow[u]{\mu} p'\backslash\alpha \\
(\mathrm{S7}) & p \xrightarrow[u]{\mu} p' & \Rightarrow\ p\langle f\rangle \xrightarrow[u]{f(\mu)} p'\langle f\rangle \\
(\mathrm{S8}) & p[rec\,x.\ p/x] \xrightarrow[u]{\mu} p' & \Rightarrow\ rec\,x.\ p \xrightarrow[u]{\mu} p'
\end{array}
$$

Fig. 1. Static location transitions.

Let $p \xrightarrow[u]{\tau}_d q \Longleftrightarrow_{\mathrm{def}} p \xrightarrow[u]{\tau}_s q$, and for each $a \in Act$, $u \in Loc^*$, let $\xrightarrow[u]{a}_d$ be the least relation $\xrightarrow[u]{a}$ on LCCS processes satisfying rules (S2), (S3), (S5), (S6), (S7), (S8) and the axiom:

$$(\mathrm{D1}) \qquad a.p \xrightarrow[l]{a} l :: p \qquad \text{for any } l \in Loc$$

Fig. 2. Dynamic location transitions.

DEFINITION 2.2. The distribution relation is the least relation $\mathcal{D} \subseteq (\mathrm{CCS} \times \mathrm{DIS})$ satisfying:

- $nil\,\mathcal{D}\,nil$ and $x\mathcal{D}x$.
- $p\mathcal{D}r \ \Rightarrow\ \mu.p\,\mathcal{D}\,\mu.r$
  $p\backslash\alpha\,\mathcal{D}\,r\backslash\alpha$
  $p\langle f\rangle\,\mathcal{D}\,r\langle f\rangle$
  $(rec\,x.\ p)\ \mathcal{D}\ (rec\,x.\ r)$.
- $p\mathcal{D}r \,\&\, q\mathcal{D}s \Rightarrow\ (p \mid q)\mathcal{D}(l :: r \mid k :: s)$, $\forall l, k$ such that $l \neq k$
  $(p+q)\mathcal{D}(r+s)$.

If $p\mathcal{D}r$ we say that r is a *distribution* of p.

Note that the same pair of locations may be used more than once in a distribution. We shall see in fact, at the end of this section, that distributions involving just *two atomic locations* are sufficient for describing the distributed behaviour of CCS processes in the static approach. This will not be the case for the dynamic approach of Section 2.5.

2.4.1. *Static location equivalence.* We shall now introduce an equivalence relation $\approx_\ell^s$ on CCS processes, which is based on a bisimulation-like relation on their distributions. The intuition for two CCS processes p, q to be equivalent is that there exist two distributions of them, say $\bar{p}$ and $\bar{q}$, which perform "the same" location transitions at each step.

However, we shall not go as far as to observe the concrete names of locations: the intention here is that locations should serve to reveal the degree of distribution of processes rather than to specify a placing of processes into a given network. Their precise identities should therefore not be significant. In fact, even if we want to observe distribution, we still aim, to some extent, at an extensional semantics. For instance, we do not want to observe the order in which parallel components have been assembled in a system, nor indeed the number of these components. We are only interested in the number of *active* components in each computation. Then we cannot require the identity of locations in corresponding transitions. For instance, if we want to identify the following processes:

$$a \mid (b \mid c) \quad \text{and} \quad (a \mid b) \mid c, \tag{2}$$

$$a \quad \text{and} \quad a \mid nil, \tag{3}$$

it is clear that, whatever distributions we choose, we must allow corresponding transitions to have different – although somehow related – static locations.

The idea is to compare transitions modulo an *association* between their locations. This was first proposed by Aceto for a subset of CCS called *nets of automata* [5], where parallelism is restricted to top-level, and then generalized by Castellani to full CCS in [38].

In the case of nets of automata the structure of locations is flat and the association is simply a partial bijection between the sets of locations of the two processes. For general CCS processes the association will not be a bijection, nor even a function. For instance, in order to equate the two processes (which should be both equivalent to $a.b.c$):

$$a.(b.c \mid nil) \quad \text{and} \quad a.b.(c \mid nil) \tag{4}$$

we need an association containing the three pairs $(\varepsilon, \varepsilon)$, (l, ε), (l, l'), for some $l, l' \in Loc$.

In fact, the only property we will require of location associations is that they respect independence of locations. To make this precise, let $\ll$ denote the prefix ordering on Loc^*. If $u \ll v$ we say that v is an extension or a *sublocation* of u. If $u \not\ll v$ and $v \not\ll u$, what we indicate by $u \diamond v$, we say that u and v are *independent*. Intuitively, transitions originating from different parallel components will have independent locations. On the other hand, transitions performed in sequence by the same component (like a and b in the two processes of example (4)) will have dependent locations: the location of the second transition will be a sublocation of the location of the first. We shall sometimes by extension talk of independent or dependent transitions.

We introduce now the notion of *consistent location association*, as defined in [38]. This is meant to express the requirement that independent transitions from one process be matched by independent transitions from the other process.

DEFINITION 2.3. A relation $\varphi \subseteq (Loc^* \times Loc^*)$ is a *consistent location association* (*cla*) if:

$$(u, v) \in \varphi \;\&\; (u', v') \in \varphi \;\Rightarrow\; (u \diamond u' \;\Leftrightarrow\; v \diamond v').$$

The following properties of *cla's* are immediate to check:

PROPERTY 2.4 (Properties of cla's).
(1) *If φ is a cla, then φ^{-1} is a cla.*
(2) *If φ is a cla and $\psi \subseteq \varphi$, then ψ is a cla.*
(3) *If φ and ψ are cla's, then $\varphi \circ \psi$ is a cla.*

We will see now that, for a given pair of distributed processes we want to equate, the required *cla* cannot in general be fixed statically, but has to be built incrementally. We will show that this is necessary to ensure such basic identifications as (2) and (3) above.

For consider the two processes, which should certainly be equated if we want parallel composition to absorb the *nil* process, and + to be idempotent, which is the very essence of a bisimulation semantics:

$$(a \mid b) \quad \text{and} \quad (a \mid b) + \big(nil \mid (a \mid b)\big). \tag{5}$$

Any two of these processes will have the following form, where $l \neq k$ and $l_i \neq k_i$ for $i = 1, 2, 3$:

$$(l :: a \mid k :: b) \quad \text{and} \quad (l_1 :: a \mid k_1 :: b) + \big(l_2 :: nil \mid k_2 :: (l_3 :: a \mid k_3 :: b)\big).$$

Suppose we want to equate these two distributions using a fixed location association φ. Clearly we will require φ to contain the pairs (l, l_1), (k, k_1), $(l, k_2 l_3)$, $(k, k_2 k_3)$. But then, if φ is to be consistent, k_2 must be such that $l_1 \ll k_2 l_3$ and $k_1 \ll k_2 k_3$: however this would imply $l_1 = k_2$ and $k_1 = k_2$, contradicting the hypothesis $l_1 \neq k_1$.

Similarly, if we want parallel composition to be associative, we should identify the processes:

$$a \mid (b \mid c) \quad \text{and} \quad a \mid (b \mid c) + (a \mid b) \mid c \tag{6}$$

which have the following generic distributions, where $l \neq k$ and $l_i \neq k_i$ for $i = 1, 2, 3, 4, 5$:

$$\big(l :: a \mid k :: (l_1 :: b \mid k_1 :: c)\big) \quad \text{and}$$

$$\big(l_2 :: a \mid k_2 :: (l_3 :: b \mid k_3 :: c)\big) + \big(l_4 :: (l_5 :: a \mid k_5 :: b) \mid k_4 :: c\big).$$

Here a global location association would have to contain the pairs (l, l_2), (kl_1, k_2l_3), (kk_1, k_2k_3), (l, l_4l_5), (kl_1, l_4k_5), (kk_1, k_4). But this would imply $l_2 = l_4$ and $k_2 = l_4$, whence the contradiction $l_2 = k_2$.

These examples show that, at least if we want to retain a simple notion of location,[20] the location association cannot be fixed once and for all, but has to be built dynamically along computations. In example (5), depending on which summand is chosen in the second process, one will use the association $\varphi_1 = \{(l, l_1), (k, k_1)\}$ or the association $\varphi_2 = \{(l, k_2l_3), (k, k_2k_3)\}$. In example (6), one can choose $\varphi_1 = \{(l, l_2), (kl_1, k_2l_3), (kk_1, k_2k_3)\}$ for one computation, and $\varphi_2 = \{(l, l_4l_5), (kl_1, l_4k_5), (kk_1, k_4)\}$ for the other. Note incidentally that these situations are not due to the phenomenon- known as "symmetric confusion" in Petri nets, and exemplified in CCS by the term $(p \mid q) + r$. Similar examples could be obtained in CCS with guarded sums (the language considered, e.g., in [118]), by just prefixing all summands in the examples above by an action d.

Let us see now how these location associations can be built up dynamically. Let Φ be the set of consistent location associations. We define particular Φ-indexed families of relations S_φ over distributed processes, called *progressive bisimulation families* (although the relations that constitute a family are not themselves bisimulations). The idea is to start with the *empty* association of locations and extend it consistently as the execution proceeds.

DEFINITION 2.5. A *progressive bisimulation family* (*pbf*) is a Φ-indexed family $\mathbf{S} = \{S_\varphi \mid \varphi \in \Phi\}$ of relations over DIS such that, if $p S_\varphi q$ then for all $a \in Act$, $u \in Loc^*$:

(1) $p \overset{a}{\underset{u}{\Longrightarrow}}_s p' \Rightarrow \exists q', v$ such that $q \overset{a}{\underset{v}{\Longrightarrow}}_s q'$, $\varphi \cup \{(u, v)\} \in \Phi$ and $p' \, S_{\varphi \cup \{(u,v)\}} \, q'$.

(2) $q \overset{a}{\underset{v}{\Longrightarrow}}_s q' \Rightarrow \exists p', u$ such that $p \overset{a}{\underset{u}{\Longrightarrow}}_s p'$, $\varphi \cup \{(u, v)\} \in \Phi$ and $p' \, S_{\varphi \cup \{(u,v)\}} \, q'$.

(3) $p \overset{\tau}{\Longrightarrow}_s p' \Rightarrow \exists q'$ such that $q \overset{\tau}{\Longrightarrow}_s q'$ and $p' S_\varphi q'$.

(4) $q \overset{\tau}{\Longrightarrow}_s q' \Rightarrow \exists p'$ such that $p \overset{\tau}{\Longrightarrow}_s p'$ and $p' S_\varphi q'$.

We may now define the *location equivalence* $\approx_\ell^s$ on CCS terms as follows:

DEFINITION 2.6 (*Static location equivalence*). For $p, q \in$ CCS, let $p \approx_\ell^s q$ if and only if for some $\bar{p}, \bar{q} \in$ DIS such that $p \mathcal{D} \bar{p}$ and $q \mathcal{D} \bar{q}$, there exists a progressive bisimulation family $\mathbf{S} = \{S_\varphi \mid \varphi \in \Phi\}$ such that $\bar{p} \, S_\emptyset \, \bar{q}$.

Let us prove that $\approx_\ell^s$ is indeed an equivalence relation. The reader may have noticed that the inverse $\mathcal{D}^{-1}$ of the distribution relation is a function. If we let $\pi =_{\mathrm{def}} \mathcal{D}^{-1}$, then $\pi(p)$ is the pure CCS process underlying the distributed process p. We start by showing that all distributions of the same process are in the relation $S_\emptyset$ for some progressive bisimulation family $\mathbf{S}$.

PROPOSITION 2.7. *Let* $p_1, p_2 \in$ DIS. *Then* $\pi(p_1) = \pi(p_2) \Rightarrow \exists pbf \, \mathbf{S}$ *such that* $p_1 S_\emptyset p_2$.

[20] One could envisage a more structured notion of location, incorporating information about the + operator; this would mark as *conflicting* (another form of dependency) the locations of the first and second summand in the right-hand process of examples (5) and (6). Then, identical locations would be allowed to be matched by conflicting locations and the counterexamples (5) and (6) would not hold anymore. In fact, this would bring us close to the notion of *proved transition* considered in [27], and would certainly complicate the theory as a new relation of conflict would have to be considered on locations.

The proof of this proposition relies on the following definition and lemma.

DEFINITION 2.8. For any $p_1, p_2 \in \mathrm{DIS}$ such that $\pi(p_1) = \pi(p_2)$, let $\varphi(p_1, p_2)$ be the least relation on locations satisfying:

$$\varphi(nil, nil) = \varphi(x, x) = \varphi(\mu.r_1, \mu.r_2) = \varphi\big((r_1 + s_1),\ (r_2 + s_2)\big) = \{(\varepsilon, \varepsilon)\},$$
$$\varphi(r_1\backslash\alpha, r_2\backslash\alpha) = \varphi(r_1\langle f\rangle, r_2\langle f\rangle) = \varphi(r_1, r_2),$$
$$\varphi(r_1[rec\,x.\,r_1/x], r_2[rec\,x.\,r_2/x]) \subseteq \varphi(rec\,x.\,r_1, rec\,x.\,r_2),$$
$$\varphi\big((l_1 :: r_1 \mid k_1 :: s_1),\ (l_2 :: r_2 \mid k_2 :: s_2)\big) = \{(\varepsilon, \varepsilon)\} \cup (l_1, l_2)\cdot\varphi(r_1, r_2) \cup (k_1, k_2)\cdot\varphi(s_1, s_2),$$

where for any relation φ, we let $(l, l')\cdot\varphi =_{\mathrm{def}} \{(lu, l'v) \mid (u, v) \in \varphi\}$.

It may be easily checked that the relation $\varphi(p_1, p_2)$ is a consistent location association. Note that $\varphi(p_1, p_2)$ only relates those locations of p_1 and p_2 which are "statically exhibited", i.e., which do not occur under a dynamic operator. The following lemma establishes the relation between the association $\varphi(p_1, p_2)$ and the transitions of p_1, p_2.

LEMMA 2.9. *Let $p_1, p_2 \in \mathrm{DIS}$ be such that $\pi(p_1) = \pi(p_2)$. Then:*
(1) $p_1 \overset{a}{\underset{u}{\Longrightarrow}}_s p_1' \;\Rightarrow\; \exists\, p_2', v$ *such that* $p_2 \overset{a}{\underset{v}{\Longrightarrow}}_s p_2'$, $\varphi(p_1, p_2) \cup (u, v) \subseteq \varphi(p_1', p_2')$ *and* $\pi(p_1') = \pi(p_2')$
(2) $p_1 \overset{\tau}{\Longrightarrow}_s p_1' \;\Rightarrow\; \exists\, p_2'$ *such that* $p_2 \overset{\tau}{\Longrightarrow}_s p_2'$, $\varphi(p_1, p_2) \subseteq \varphi(p_1', p_2')$ *and* $\pi(p_1') = \pi(p_2')$.

PROOF. By induction on the proofs of transitions of distributed processes. Note that in general $\varphi(p_1, p_2) \subseteq \varphi(p_1', p_2')$ because new parallel structure may appear as the computations proceed. □

We may now prove the above proposition.

PROOF OF PROPOSITION 2.7. Define the family $\mathbf{T} = \{T_\varphi \mid \varphi \in \Phi\}$ by letting:

$$T_\varphi = \big\{(r_1, r_2) \mid \pi(r_1) = \pi(r_2) \text{ and } \varphi \subseteq \varphi(r_1, r_2)\big\}.$$

It is clear that $r_1 T_\emptyset r_2$ for any r_1, r_2 such that $\pi(r_1) = \pi(r_2)$. Let us show that $\mathbf{T}$ is a progressive bisimulation family. Suppose that $r_1 T_\varphi r_2$. If $r_1 \overset{a}{\underset{u}{\Longrightarrow}}_s r_1'$ then by Lemma 2.9 $r_2 \overset{a}{\underset{v}{\Longrightarrow}}_s r_2'$, with $\varphi(r_1, r_2) \cup (u, v) \subseteq \varphi(r_1', r_2')$ and $\pi(r_1') = \pi(r_2')$. We want to show that $\varphi' = \varphi \cup (u, v)$ is a *cla* and that $r_1' T_{\varphi'} r_2'$. But this follows from $\varphi \subseteq \varphi(r_1, r_2)$ and Lemma 2.9, since $\varphi' = \varphi \cup (u, v) \subseteq \varphi(r_1, r_2) \cup (u, v) \subseteq \varphi(r_1', r_2')$. □

Using this proposition, we can now show that:

PROPOSITION 2.10. *The relation $\approx^s_\ell$ is an equivalence on CCS processes.*

PROOF. *Reflexivity*: Consider the family $\mathbf{S} = \{S_\varphi \mid \varphi \in \Phi\}$ defined by:

$$S_\varphi = \begin{cases} \{(q, q) \mid q \in \text{DIS}\} & \text{if } \varphi \subseteq Id, \\ \emptyset & \text{otherwise.} \end{cases}$$

It is clear that $\mathbf{S}$ is a progressive bisimulation family such that $(q, q) \in S_\emptyset$ for any $q \in \text{DIS}$. Hence $p \approx^s_\ell p$ for any $p \in \text{CCS}$.

Symmetry: Let $p \approx^s_\ell q$. This means that for some $\bar{p}, \bar{q} \in \text{DIS}$ such that $\pi(\bar{p}) = p$ and $\pi(\bar{q}) = q$, there exists a progressive bisimulation family $\mathbf{S} = \{S_\varphi \mid \varphi \in \Phi\}$ such that $\bar{p} S_\emptyset \bar{q}$. Define now a family $\mathbf{R} = \{R_\psi \mid \psi \in \Phi\}$ by:

$$R_\psi = \{(r, s) \mid s S_{\psi^{-1}} r\}.$$

Clearly $\bar{q} R_\emptyset \bar{p}$. We show now that $\mathbf{R}$ is a progressive bisimulation family. Suppose $(r, s) \in R_\psi$: this is because $s S_\varphi r$, with $\varphi = \psi^{-1}$. Then $r \overset{a}{\underset{u}{\Longrightarrow}}_s r'$ implies $s \overset{a}{\underset{v}{\Longrightarrow}}_s s'$, with $s' S_{\varphi \cup \{(v,u)\}} r'$. Now $[\varphi \cup \{(v, u)\}]^{-1} = \psi^{-1} \cup \{(u, v)\}$, thus $(r', s') \in R_{\psi^{-1} \cup \{(u,v)\}}$. The case of unobservable transitions is similar. We can then conclude that $q \approx^s_\ell p$.

Transitivity: Let $p \approx^s_\ell r$ and $r \approx^s_\ell q$. This means that for some $\bar{p}, \bar{q}, r_1, r_2 \in \text{DIS}$ such that $\pi(\bar{p}) = p$, $\pi(\bar{q}) = q$, $\pi(r_1) = \pi(r_2) = r$, there exist *pbf's* $\mathbf{R}^1$ and $\mathbf{R}^2$ such that $\bar{p} R^1_\emptyset r_1$, $r_2 R^2_\emptyset \bar{q}$. Moreover, if $\mathbf{T}$ is the *pbf* introduced in the proof of Proposition 2.7, we know (from the same proposition) that $r_1 T_\emptyset r_2$. Hence if we define a family $\mathbf{S} = \{S_\varphi \mid \varphi \in \Phi\}$ as follows:

$$\begin{aligned}(s, s') \in S_\varphi \iff & \exists t_1, t_2 \in \text{DIS} \text{ such that } \pi(t_1) = \pi(t_2); \\ & \exists \textit{pbf's } \mathbf{S}^1, \mathbf{S}^2,\ \exists \textit{cla's } \varphi_1, \psi, \varphi_2, \text{ such that} \\ & \varphi \subseteq \varphi_2 \circ \psi \circ \varphi_1 \text{ and } s S^1_{\varphi_1} t_1 T_\psi\ t_2 S^2_{\varphi_2} s'\end{aligned}$$

it is clear that $\bar{p} S_\emptyset \bar{q}$. Moreover $\mathbf{S}$ is a progressive bisimulation family, since for any u, v, v', w we have: $\varphi \cup \{(u, w)\} \subseteq \varphi_2 \cup \{(v', w)\} \circ \psi \cup \{(v, v')\} \circ \varphi_1 \cup \{(u, v)\}$. □

A pleasant consequence of Proposition 2.7 is that $\approx^s_\ell$ is independent from the particular distributions we choose. If two CCS terms p and q are equivalent, then any two distributions of them are related by $S_\emptyset$, for some progressive bisimulation family $\mathbf{S} = \{S_\varphi \mid \varphi \in \Phi\}$.

COROLLARY 2.11. *For any $p, q \in$ CCS: $p \approx^s_\ell q \iff$ for all $\bar{p}$, $\bar{q} \in$ DIS such that $p \mathcal{D} \bar{p}$ and $q \mathcal{D} \bar{q}$ there exists a progressive bisimulation family $\mathbf{S} = \{S_\varphi \mid \varphi \in \Phi\}$ such that $\bar{p} S_\emptyset \bar{q}$.*

By virtue of this result, we can restrict attention to particular "binary" distributions, systematically associating location 0 to the left operand and location 1 to the right operand of a parallel composition. A distribution of this kind will be called *canonical*. Similarly, elements of $\{0, 1\}^*$ will be called *canonical locations*. These are in fact the locations used in work by Mukund and Nielsen [118] and, with a slightly different notation, by Montanari and Yankelevich [116,171]. In fact, when applied to canonical distributions of CCS terms, the static transition rules give exactly the same transitions as the *spatial transition system* of [116,171] (except for τ-transitions, for which these papers use pairs of locations). Note here that since canonical locations can be extracted directly from the structure of terms (in fact they are just a part of the *proof* of a transition, in the sense of the proved transition systems mentioned in Section 2.1), in the static approach the extension of the language with location prefixing is not really necessary. Using canonical locations only, the whole static location theory could be developed within the standard syntax of CCS. However, it would then not be possible to relax the theory to allow for the placing of more components into the same location, a point which will be discussed at p. 977 (in Section 2.5).

We show now that location equivalence is finer than the ordinary (weak) bisimulation, and that introducing locations only adds discriminations between processes in so far as their distributed aspects are concerned. Let $\mathrm{CCS}_{\mathrm{seq}}$ be the set of sequential processes of CCS, built without the parallel operator. On this language location equivalence $\approx_\ell^s$ reduces to the standard weak bisimulation $\approx$:

PROPOSITION 2.12. *For any processes* $p, q \in \mathrm{CCS}_{\mathrm{seq}}$: $p \approx_\ell^s q \iff p \approx q$.

PROOF. The proof that $p \approx_\ell^s q \;\Rightarrow\; p \approx q$ is straightforward and is left to the reader. We show here that $p \approx_\ell^s q \;\Leftarrow\; p \approx q$. Let R be an ordinary weak bisimulation between $p, q \in \mathrm{CCS}_{\mathrm{seq}}$. Note that the distributions of processes in $\mathrm{CCS}_{\mathrm{seq}}$ coincide with the processes themselves, and therefore all their transitions have the form $p \xrightarrow[\varepsilon]{\mu}_s p'$. Let $S_\emptyset = \{(p, q)\}$. Since the location association $\varphi = \{(\varepsilon, \varepsilon)\}$ is consistent it is clear that the family of relations $\mathbf{S} = \{S_\emptyset\} \cup \{S_\varphi \mid \varphi = \{(\varepsilon, \varepsilon)\}$ and $S_\varphi = R\} \cup \{S_\psi \mid \psi \in \Phi \backslash \{\emptyset, \varphi\},\ S_\psi = \emptyset\}$ is a progressive bisimulation family. □

We give next some simple examples. The reader acquainted with causality-based equivalences such as the causal bisimulation of [52], will notice here the difference between location equivalence and causality-based equivalences (the exact relation between these equivalences will be discussed in Section 3.3).

EXAMPLE 2.13. $a.b + b.a \not\approx_\ell^s (a.\gamma \mid \bar{\gamma}.b)\backslash\gamma + (b.\gamma \mid \bar{\gamma}.a)\backslash\gamma \approx_\ell^s a \mid b$.

By Corollary 2.11, to show $\not\approx_\ell^s$ it is enough to find a pair of distributions for which every association fails to be consistent. Using canonical distributions, it is easy to see that the computation a followed by b yields the association $\varphi = \{(\varepsilon, 0), (\varepsilon, 1)\}$ between the locations of the first two processes, which is *not* consistent. On the other hand, for the second and third process we can build the consistent association $\varphi_1 = \{(0, 0), (1, 1)\}$ for the computation a followed by b, and the consistent association $\varphi_2 = \{(0, 1), (1, 0)\}$ for the computation b followed by a. A related example is:

EXAMPLE 2.14. $[(a.\gamma + b.\bar{\gamma}) \mid (\bar{\gamma}.b + \gamma.a)]\backslash\gamma \approx_\ell^s a \mid b$.

Again, these two processes are intuitively equivalent since they both perform actions a and b in either order at different locations. The construction of the appropriate associations is left to the reader.

EXAMPLE 2.15. $[(\alpha + b) \mid \bar{\alpha}.b]\backslash\alpha \approx^s_\ell b$.

Here the observable behaviour consists for both processes in just one action b occurring at some location. Although the first process may choose its action b from two different components, this cannot be detected by $\approx^s_\ell$. The reader may want to check this fact using canonical distributions.

These examples illustrate the extensional character of $\approx^s_\ell$: from Example 2.14 we see that equivalent processes do not need to be componentwise bisimilar, while Example 2.15 shows that equivalent processes need not even have the same number of parallel components. The following is an example of infinite processes which are not equivalent:

EXAMPLE 2.16. $rec\, x.\, a.x \not\approx^s_\ell (rec\, x.\, a.x \mid rec\, x.\, a.x)$.

Note that in the standard CCS semantics these two processes give rise to isomorphic transition systems. We conclude this section by reconsidering the protocol and mutual exclusion examples discussed in Section 2.2. We let the reader verify the following.

EXAMPLE 2.17. *PSpec* $\not\approx^s_\ell$ *Protocol* and *Mutex* $\not\approx^s_\ell$ *MSpec* $\not\approx_\ell$ *FMutex* $\not\approx^s_\ell$ *Mutex*.

2.4.2. *Static location preorder.* We define now a preorder $\sqsubseteq^s_\ell$ on CCS processes, which formalizes the idea that one process is more sequential or *less distributed* than another. When dealing with noninterleaving semantics, which are more discriminating than the usual bisimulation semantics, it is interesting to have such preorder notions to be able to relate, for instance, a sequential specification with a distributed implementation.

The location preorder $\sqsubseteq^s_\ell$ is obtained by slightly relaxing the notion of consistent association. The intuition for $p \sqsubseteq^s_\ell q$ is that there exist two distributions $\bar{p}$ and $\bar{q}$ of p and q such that whenever $\bar{p}$ can perform two transitions at independent locations, then $\bar{q}$ performs corresponding transitions at locations which are also independent, while the reverse need not be true. This is expressed by the following notion of left-consistency:

DEFINITION 2.18. A relation $\varphi \subseteq (Loc^* \times Loc^*)$ is a left-consistent location association if:

$$(u, v) \in \varphi \;\&\; (u', v') \in \varphi \;\Rightarrow\; (u \diamond u' \;\Rightarrow\; v \diamond v').$$

Now, if Ψ is the set of left-consistent location associations, we may obtain a notion of *progressive pre-bisimulation family* (*ppbf*) on distributed processes of DIS by simply replacing Φ by Ψ in Definition 2.5. Again, this gives rise to a relation on CCS processes:

DEFINITION 2.19 (*Static location preorder*). If $p, q \in$ CCS, let $p \sqsubseteq^s_\ell q$ if and only if for some $\bar{p}, \bar{q} \in$ DIS such that $p\mathcal{D}\bar{p}$ and $q\mathcal{D}\bar{q}$, there exists a progressive pre-bisimulation family $\mathbf{S} = \{S_\psi \mid \psi \in \Psi\}$ such that $\bar{p} S_\emptyset \bar{q}$.

It is easy to see that $p \approx^s_\ell q \;\Rightarrow\; p \sqsubseteq^s_\ell q$. As may be expected the reverse is not true. We have for instance, resuming the examples from the previous section:

EXAMPLE 2.20. $a.b + b.a \sqsubseteq_{\ell}^{s} (a.\gamma \mid \bar{\gamma}.b)\backslash\gamma + (b.\gamma \mid \bar{\gamma}.a)\backslash\gamma$.

EXAMPLE 2.21. $rec\, x.\ a.x \sqsubseteq_{\ell}^{s} (rec\, x.\ a.x \mid rec\, x.\ a.x)$.

EXAMPLE 2.22. $PSpec \sqsubseteq_{\ell}^{s} Protocol$.

The reader may want to check, on the other hand, that the two systems *Mutex* and *FMutex* in the mutual exclusion example of p. 961 are not related by $\sqsubseteq_{\ell}$.

Having introduced both an equivalence $\approx_{\ell}^{s}$ and a preorder $\sqsubseteq_{\ell}^{s}$ based on the same intuition, we may wonder whether $\approx_{\ell}^{s}$ coincides with the equivalence $\simeq_{\ell}^{s} =_{\text{def}} \sqsubseteq_{\ell}^{s} \cap \sqsupseteq_{\ell}^{s}$ induced by the preorder. It is clear that $\approx_{\ell}^{s} \subseteq \simeq_{\ell}^{s}$, since we have both $\approx_{\ell}^{s} \subseteq \sqsubseteq_{\ell}^{s}$ and $\approx_{\ell}^{s} \subseteq \sqsupseteq_{\ell}^{s}$. On the other hand, the kernel of the preorder is weaker than location equivalence, as shown by the following example. Consider the two processes:

$$a.a.a + (a \mid a \mid a) \quad \text{and} \quad a.a.a + a.a \mid a + (a \mid a \mid a).$$

These two processes are not equivalent with respect to $\approx_{\ell}^{s}$, but they are equivalent with respect to $\simeq_{\ell}^{s}$ because $a.a.a \sqsubseteq_{\ell}^{s} a.a \mid a \sqsubseteq_{\ell}^{s} a \mid a \mid a$.

2.5. *Dynamic approach*

We present now the dynamic approach introduced in [29],[21] and in particular the dynamic versions of the location equivalence and preorder, noted $\approx_{\ell}^{d}$ and $\sqsubseteq_{\ell}^{d}$. In the dynamic approach, locations are associated with actions rather than with parallel components. This association is built dynamically, according to the rule:

$$\text{(D1)} \quad a.p \xrightarrow[l]{a}_{d} l :: p \quad \text{for any } l \in Loc.$$

In some sense locations are transmitted from transitions to processes, whereas in the static approach we had the inverse situation. Rule (D1) is the essence of the dynamic location semantics. The remaining rules for observable transitions are just as in the static semantics, see Figure 2. Note that locations increase here at each step, even if the execution goes on within the same parallel component. In fact the location l created by rule (D1) may be seen as an identifier for action a, or more precisely, for that particular occurrence of a. By inspecting the rules, one can easily see that a generic transition has the form

$$p \xrightarrow[ul]{a}_{d} p',$$

where $u \in Loc^*$: here the string u is a record of all action occurrences which causally precede action a (through the prefixing operator), what we shall call the *causal path* (or *access path*) to a. The whole string ul will be called the *dynamic location* of a.

[21] An earlier variant presented in [28] will be discussed in Section 3.2.

Let us see some examples of dynamic transitions. The first and third process of Example 2.13 may both execute an a action followed by a b action. The corresponding sequences of dynamic transitions are (for any choice of locations $l_1, l_2 \in Loc$):

$$a.b + b.a \xrightarrow[l_1]{a}_d l_1 :: b \xrightarrow[l_1 l_2]{b}_d l_1 :: l_2 :: nil, \tag{7}$$

$$a \mid b \xrightarrow[l_1]{a}_d l_1 :: nil \mid b \xrightarrow[l_2]{b}_d l_1 :: nil \mid l_2 :: nil. \tag{8}$$

The observable dynamic transitions $p \xrightarrow[u]{a}_d p'$ are related to the static transitions $p \xrightarrow[u]{a}_s p'$ in a simple way. To see this, let us introduce a few notations. Let $\Delta_k(p)$ be the function that erases the atomic location k in p, wherever it occurs. Formally, $\Delta : (Loc \times \text{LCCS}) \to \text{LCCS}$ is defined by the clauses:

$$\Delta_k(p) = p, \quad \text{if } p \in \text{CCS},$$

$$\Delta_k(l :: p) = \begin{cases} \Delta_k(p) & \text{if } k = l, \\ l :: \Delta_k(p) & \text{otherwise.} \end{cases}$$

together with clauses stating the compatibility of Δ_k with the remaining operators (for instance $\Delta_k(p \mid q) = \Delta_k(p) \mid \Delta_k(q)$, etc.). Also, for any $p \in \text{LCCS}$, let $Loc(p)$ be the set of atomic locations occurring in p, defined in the obvious way. We have then the following correspondence between the two kinds of location transitions:

FACT 2.23. *Let $p \in LCCS$, $l \notin Loc(p)$. Then*:
(i) $p \xrightarrow[ul]{a}_d p' \Rightarrow p \xrightarrow[u]{a}_s \Delta_l(p')$.
(ii) $p \xrightarrow[u]{a}_s p' \Rightarrow \exists\, p''$ *such that* $p \xrightarrow[ul]{a}_d p''$ *and* $\Delta_l(p'') = p'$.

The proof, by induction on the inference of the transition, is left to the reader.

Because of rule (D1), the dynamic location transition system is both infinitely branching and acyclic: it thus gives infinite representations for all regular processes. In fact, while the infinite branching may be overcome easily (through a *canonical* choice of dynamic locations [171,38], as will be explained later), the infinite progression is intrinsic to the dynamic semantics.

For τ-transitions, for which we do not want to introduce additional locations, we simply use the static transition rules. Although this last point differentiates our strong dynamic location transition system from that originally introduced in [29], where no locations were associated with τ-transitions, the resulting *weak (dynamic) location transition system* is the same. The definition of the weak dynamic transitions $\overset{a}{\underset{u}{\Longrightarrow}}_d$ and $\overset{\tau}{\Longrightarrow}_d$ is similar to that of the static transitions $\overset{a}{\underset{u}{\Longrightarrow}}_s$ and $\overset{\tau}{\Longrightarrow}_s$.

For instance, the second process of Example 2.13 has the following sequence of strong dynamic transitions:

$$(a.\gamma \mid \bar{\gamma}.b)\backslash\gamma + (b.\gamma \mid \bar{\gamma}.a)\backslash\gamma \xrightarrow[l_1]{a}_d (l_1 :: \gamma \mid \bar{\gamma}.b)\backslash\gamma \xrightarrow[\varepsilon]{\tau}_d (l_1 :: nil \mid b)\backslash\gamma$$
$$\xrightarrow[l_2]{b}_d (l_1 :: nil \mid l_2 :: nil)\backslash\gamma.$$

From this one can derive (for instance) the following sequence of weak transitions:

$$(a.\gamma \mid \bar{\gamma}.b)\backslash\gamma + (b.\gamma \mid \bar{\gamma}.a)\backslash\gamma \overset{a}{\underset{l_1}{\Longrightarrow}}_d (nil \mid b)\backslash\gamma \overset{b}{\underset{l_2}{\Longrightarrow}}_d (nil \mid nil)\backslash\gamma. \tag{9}$$

We define now the *dynamic location equivalence* $\approx_\ell^d$ and the *dynamic location preorder* $\sqsubset\!\!\!\sim_\ell^d$. Because of the flexibility in the choice of locations offered by rule (D1), these definitions are much simpler than in the static case. In [29] the relations $\approx_\ell^d$ and $\sqsubset\!\!\!\sim_\ell^d$ were obtained as instances of a general notion of *parameterized location bisimulation*. We shall give here directly the instantiated definitions.

2.5.1. *Dynamic location equivalence.* The dynamic location equivalence $\approx_\ell^d$ is just the ordinary bisimulation equivalence associated with the weak dynamic location transition system.

DEFINITION 2.24 (*Dynamic location equivalence*). A relation $R \subseteq \text{LCCS} \times \text{LCCS}$ is called a *dynamic location bisimulation (dlb)* iff for all $(p, q) \in R$ and for all $a \in Act$, $u \in Loc^+$:

(1) $p \overset{a}{\underset{u}{\Longrightarrow}}_d p' \;\Rightarrow\; \exists q'$ such that $q \overset{a}{\underset{u}{\Longrightarrow}}_d q'$ and $(p', q') \in R$.
(2) $q \overset{a}{\underset{u}{\Longrightarrow}}_d q' \;\Rightarrow\; \exists p'$ such that $p \overset{a}{\underset{u}{\Longrightarrow}}_d p'$ and $(p', q') \in R$.
(3) $p \overset{\tau}{\Longrightarrow}_d p' \;\Rightarrow\; \exists q'$ such that $q \overset{\tau}{\Longrightarrow}_d q'$ and $(p', q') \in R$.
(4) $q \overset{\tau}{\Longrightarrow}_d q' \;\Rightarrow\; \exists p'$ such that $p \overset{\tau}{\Longrightarrow}_d p'$ and $(p', q') \in R$.

The largest *dlb* is called *dynamic location equivalence* and denoted $\approx_\ell^d$.

For instance, the three processes of Example 2.13 are related as follows (as it is easy to check by looking at their transition sequences given in (7), (9) and (8) respectively):

EXAMPLE 2.25. $a.b + b.a \not\approx_\ell^d (a.\gamma \mid \bar{\gamma}.b)\backslash\gamma + (b.\gamma \mid \bar{\gamma}.a)\backslash\gamma \approx_\ell^d a \mid b$.

In fact, all the examples given in Section 2.4 for the static location equivalence $\approx_\ell^s$ equally apply to the dynamic location equivalence $\approx_\ell^d$. We refer the reader to [29] for more detailed examples of equivalent and unequivalent processes.

2.5.2. *Dynamic location preorder.* We consider now the location preorder $\sqsubset\!\!\!\sim_\ell^d$. Here, instead of requiring the identity of locations in corresponding transitions, we demand that the locations in the second (more distributed) process be subwords of the locations in the first (more sequential) process. Formally, the *subword* relation $\leqslant_{\text{sub}}$ on Loc^* is defined by:

$$v \leqslant_{\text{sub}} u \iff \exists v_1, \ldots, v_k, \exists w_1, \ldots, w_{k+1} \text{ such that } v = v_1 \cdots v_k \text{ and}$$
$$u = w_1 v_1 \cdots w_k v_k w_{k+1}.$$

DEFINITION 2.26 (*Dynamic location preorder*). A relation $R \subseteq \text{LCCS} \times \text{LCCS}$ is called a *dynamic location pre-bisimulation* (*dlpb*) iff for all $(p, q) \in R$ and for all $a \in Act$, $u \in Loc^+$:

(1) $p \overset{a}{\underset{u}{\Longrightarrow}}_d p' \;\Rightarrow\; \exists v.\; v \leqslant_{\mathrm{sub}} u,\; \exists q'$ such that $q \overset{a}{\underset{v}{\Longrightarrow}}_d q'$ and $(p', q') \in R$.
(2) $q \overset{a}{\underset{v}{\Longrightarrow}}_d q' \;\Rightarrow\; \exists u.\; v \leqslant_{\mathrm{sub}} u,\; \exists p'$ such that $p \overset{a}{\underset{u}{\Longrightarrow}}_d p'$ and $(p', q') \in R$.
(3) $p \overset{\tau}{\Longrightarrow}_d p' \;\Rightarrow\; \exists q'$ such that $q \overset{\tau}{\Longrightarrow}_d q'$ and $(p', q') \in R$.
(4) $q \overset{\tau}{\Longrightarrow}_d q' \;\Rightarrow\; \exists p'$ such that $p \overset{\tau}{\Longrightarrow}_d p'$ and $(p', q') \in R$.

The largest *dlpb* is called *dynamic location preorder* and denoted $\sqsubseteq_\ell^d$.

The intuition is as follows. If p is a sequentialized version of q, then each component of p corresponds to a group of parallel components in q. Thus the local causes of any action of q will correspond to a subset of local causes of the corresponding action of p. This may be easily verified for the following examples:

EXAMPLE 2.27. $a.a.a \sqsubseteq_\ell^d a.a \mid a$ and $a.b + b.a \sqsubseteq_\ell^d a \mid b$.

Let us briefly return to the Protocol example of Section 2.2. This example was already proposed for study under the static semantics (Examples (2.17) and (2.22)), but it is somewhat easier to handle in the dynamic semantics. We want to argue that:

EXAMPLE 2.28. *PSpec* $\not\approx_\ell^d$ *Protocol* but *PSpec* $\sqsubseteq_\ell^d$ *Protocol*.

Recall that the sequential specification *PSpec* was given by $PSpec \Leftarrow in.out.PSpec$, while the system *Protocol* was defined as follows:

$$\begin{aligned} Protocol &\Leftarrow (Sender \mid Receiver)\backslash\alpha, \beta \\ Sender &\Leftarrow in.\overline{\alpha}.\beta.Sender \\ Receiver &\Leftarrow \alpha.out.\overline{\beta}.Receiver. \end{aligned}$$

To see that *PSpec* $\not\approx_\ell^d$ *Protocol*, consider the sequence of transitions of *Protocol*:

$$Protocol \overset{in}{\underset{l}{\Longrightarrow}}_d \overset{out}{\underset{k}{\Longrightarrow}}_d (l :: Sender \mid k :: Receiver)\backslash\alpha, \beta.$$

The specification *PSpec* can only respond by performing the sequence of transitions:

$$PSpec \overset{in}{\underset{l}{\Longrightarrow}}_d \overset{out}{\underset{lk}{\Longrightarrow}}_d l :: k :: PSpec.$$

Clearly, this transition sequence does not match that of *Protocol*. Note however that the locations indexing the transitions of *Protocol* are subwords of those indexing the transitions of *PSpec*. This suggests how to build a dynamic location pre-bisimulation to prove that *PSpec* $\sqsubseteq_\ell^d$ *Protocol*.

All the other examples given in Section 2.4 for the static location preorder $\sqsubseteq_\ell^s$ also hold for $\sqsubseteq_\ell^d$. Indeed, as we shall see in the next section, the dynamic relations $\approx_\ell^d$ and $\sqsubseteq_\ell^d$ coincide with the static relations $\approx_\ell^s$ and $\sqsubseteq_\ell^s$ introduced in the previous section.

Discussion. We now briefly recount how the static and dynamic approaches to locations presented in this section emerged. The static approach may appear more natural than the dynamic one as it matches more closely the intuition behind locations as *units of distribution* for processes. However this approach was initially abandoned in favour of the dynamic one, as it was not clear how to define an equivalence based on static locations. In particular the idea of *dynamic location association* did not immediately come to mind. This idea was first proposed by Aceto in [5] for a class of CCS process with only top level parallelism, called *nets of automata*,[22] and then extended to the whole language CCS by Castellani in [38]. Similarly, the coincidence of the static notions of equivalence and preorder with the dynamic ones (which will be exposed in the next section) was first established for nets of automata in [5] and then for CCS in [38]. Because of the arbitrary nesting of parallelism and prefixing in CCS terms, and of the interplay between sum and parallelism, this extension was not completely straightforward. An intermediate step was taken by Mukund and Nielsen in [118], where a notion of bisimulation equivalence based on static locations was introduced for a class of asynchronous transition systems modelling CCS with guarded sums. This equivalence, similar to that of [38], was already conjectured to coincide with the dynamic location equivalence of [29].

The first definition of the dynamic location semantics appeared in [28]. In that formulation, rule (D1) generated a location *word* u for each action, rather than an atomic location l. As it turned out, the resulting equivalence was slightly too weak,[23] hence this first definition was replaced in [29] by the one presented here.

In [28] a logical characterization is given for the early variant of dynamic location equivalence, which can be easily adapted to the later variant of [29]. An equational characterization for both the location equivalence $\approx_\ell^d$ and the location preorder $\sqsubseteq_\ell^d$ is presented in [29]. This axiomatization makes use of a new prefixing construct $\langle a@ux\rangle.p$, whose semantics is given by:

$$\langle a@ux\rangle.p \xrightarrow[ul]{a}_d p[l/x] \quad \text{for any } l \in Loc.$$

The idea is to reduce terms to normal forms by replacing the ordinary prefixing $a.p$ with the new construct, according to the basic transformation law:

$$a.p = \langle a@x\rangle.x :: p.$$

In fact, the logical characterization is based on a very similar idea. Here modalities of the form $\langle a\rangle_u$ are added to a classical Hennessy–Milner logic, with the following semantics:

$$p \models \langle a\rangle_u \Phi \quad \text{if, for some } p',\ p \overset{a}{\underset{u}{\Longrightarrow}}_d p' \text{ and } p' \models \Phi.$$

Location equivalence also inherits axiomatic characterizations given on various subsets of CCS for other distributed equivalences.[24] For instance an equational characterization was

[22] For this class of processes a consistent location association is just a partial bijection between two sets of locations.

[23] An example of unwanted identification under this equivalence is given in Section 2.6, where this earlier variant is described.

[24] These equivalences will be discussed in Section 2.6.

given for distributed bisimulation in [37,39], for a subset of CCS with communication but without restriction. This made use of auxiliary operators of left-merge and communication merge, taken from [14] and previous work on ACP (interestingly, however, the laws for these operators were somewhat different from those holding in ACP). A proof system for local cause equivalence, which also uses these operators (together with a new operator of cause prefixing, see Section 3.2) and covers all finite CCS processes, is presented by Kiehn in [99]. An axiomatization for a related equivalence may also be found in [66].

We conclude this section with a discussion on the choice of the language LCCS as a basis for the location semantics. The reader may have noticed that, while the relations $\approx^d_\ell$ and $\sqsubseteq^d_\ell$ have been defined on the language LCCS, the static relations $\approx^s_\ell$ and $\sqsubseteq^s_\ell$, on the other hand, have been defined only on CCS. In fact, these relations are defined on CCS by lifting them up from relations on DIS, which is itself a proper sublanguage of LCCS. We shall now briefly comment on this choice, and show that some results, like the coincidence of the static and dynamic notions of equivalence and preorder, would fail to hold on the whole language LCCS.

At first sight, we could envisage to extend the static equivalence $\approx^s_\ell$ to the rest of LCCS by letting, for any terms p and q in LCCS\CCS: $p \approx^s_\ell q$ if and only if $p S_\emptyset q$ for some progressive bisimulation family $\mathbf{S} = \{S_\varphi \mid \varphi \in \Phi\}$ (the notion of progressive bisimulation family being itself generalized). Note that in LCCS the distribution of locations may be quite liberal: we have for instance terms like $(l :: a \mid l :: b)$ and $(l :: a + k :: b)$. With our generalized definition of $\approx^s_\ell$ we would have the following identification:

$$(l :: a \mid l :: b) \approx^s_\ell l :: (a.b + b.a). \tag{10}$$

Now, if this could be argued to be a reasonable equation in a locality based semantics, it would not be compatible with our intended interpretation of $\approx^s_\ell$, which is supposed to preserve the degree of parallelism of processes. In fact, these two processes would *not* be equated in the dynamic semantics:

$$(l :: a \mid l :: b) \not\approx^d_\ell l :: (a.b + b.a) \tag{11}$$

because the first has a computation:

$$(l :: a \mid l :: b) \xrightarrow[ll_1]{a}_d l :: (l_1 :: nil \mid b) \xrightarrow[ll_2]{b}_d l :: (l_1 :: nil \mid l_2 :: nil)$$

which the second is not able to mimic. As it were, the dynamic semantics will *always* distinguish between parallel components, no matter what locations have been inserted in front of them. On the other hand the static semantics can only distinguish between parallel components that have been previously tagged with disjoint locations. This example also suggests that $\approx^s_\ell$ would not be preserved by parallel composition on LCCS: we would have for instance $l :: b \approx^s_\ell k :: b$ but $(l :: a \mid l :: b) \not\approx^s_\ell (l :: a \mid k :: b)$.

Therefore, it seems fair to say that the appropriate language for defining the static relations $\approx^s_\ell$ and $\sqsubseteq^s_\ell$ is DIS, rather than LCCS. On the other hand, a language larger than DIS is needed for the dynamic semantics: for example the term $(a \mid l :: b)$ can arise in the dynamic semantics as a derivative of some CCS term, for instance of the term $(a \mid c.b)$.

This is why LCCS was chosen as the basic language in [29]. It was kept as the reference language in [38], because it is a natural extension of CCS where the dynamic and the static location transitions can be defined in a simple and uniform way.

Another point which is worth discussing is the possibility of relaxing the location semantics to allow for the placing of several parallel components on the same location. Clearly, this would slightly change the focus of the semantics: as it stands, the location semantics prescribes a maximal distribution of processes, identifying their *degree of distribution* with their *degree of parallelism*. In some cases however, it could be interesting to dissociate these two measures, allowing the degree of distribution to be lower than the degree of parallelism. For instance, one may describe in CCS a two-place buffer as a system made of two parallel components. As for the Protocol example discussed earlier, the location semantics would distinguish this system from its sequential specification (see [28] for a detailed description of the buffer example). On the other hand, it may seem excessive to view such an elementary system – representing a simple data structure – as being distributed over different sites. In this and similar cases we may want to allow different parallel components to be sitting together at the same location.

To accommodate this possibility, the static location semantics would require very little modification. It would be enough to omit the requirement that parallel components have different locations in the language DIS. The dynamic semantics, on the other hand, would be slightly more difficult to adapt: since this semantics is sensitive to the degree of parallelism of processes, as shown by example (11), it would be necessary to introduce a kind of *sequential encapsulation* operator $[\![p]\!]$, whose effect would be to conceal all locations in p. The semantics of this operator could be given by:

$$p \xrightarrow[u]{a}_d p' \;\Rightarrow\; [\![p]\!] \xrightarrow[l]{a}_d l :: [\![p']\!] \quad \text{for any } l \in Loc.$$

Then, for instance, the process $[\![a \mid b]\!]$ would have a sequence of transitions:

$$[\![a \mid b]\!] \xrightarrow[l]{a}_d l :: [\![l_1 :: nil \mid b]\!] \xrightarrow[lk]{b}_d l :: k :: [\![l_1 :: nil \mid l_2 :: nil]\!].$$

Introducing this operator, however, would not have exactly the same effect as allowing processes of the form $(l :: p \mid l :: q)$ in the static semantics, since in the latter case the sublocations of l could still be distinguished, while the operator $[\![p]\!]$ conceals locations in all the subterms of p. We leave this point open for speculation.

2.6. *Equivalence of the two approaches*

This section is devoted to showing the equivalence of the static and dynamic approach. More precisely, we shall establish the following theorem.

THEOREM 2.29. *Let $p, q \in CCS$. Then:*

(1) $p \approx_\ell^s q \iff p \approx_\ell^d q$;

(2) $p \sqsubseteq_\ell^s q \iff p \sqsubseteq_\ell^d q$.

To this end, we introduce a new transition system on CCS, called *occurrence transition system*, which in some sense incorporates the information of both location transition systems. This system will serve as an intermediate between the static and the dynamic semantics. The main point will be to prove that starting from a static or a dynamic location computation, one may always reconstruct a corresponding occurrence computation. This means, essentially, that all the information about distribution and local causality is already present in both location transition systems.

The two location transition systems could also be compared directly, without recourse to an auxiliary transition system. However the occurrence transition system is interesting in its own right, since it provides a concrete level of description where the notions of *occurrence* of an action, computational *history* of an occurrence and *computation state* of a process have a precise definition. Moreover, as we shall see, it allows for the definition of a notion of *local history preserving bisimulation*, which turns out to be a third equivalent formulation of location equivalence.

2.6.1. *The occurrence transition system.* We define here a new transition system on CCS, called *occurrence (transition) system*, whose states represent CCS computation states with a "past", and whose labels are occurrences of actions within a computation. This system, which is based on a syntactic notion of *occurrence* of action, is essentially a simplification of the *event (transition) system* introduced in [27] to compare different models of CCS: it is simpler because we do not try to identify uniquely all occurrences of action in a term, as in [27], but only those which can coexist in a computation. Moreover, since we are interested here in weak semantics, we shall not distinguish between different occurrences of τ-actions and we concentrate on *abstract* occurrences, in which τ-actions and communications are absorbed. Formally, the set $\mathcal{O}_\tau$ of *occurrences* is defined as $\mathcal{O}_\tau = \mathcal{O} \cup \{\tau\}$, where the elements of $\mathcal{O}$, the visible occurrences, are given by:

$$e ::= a \mid ae \mid 0e \mid 1e.$$

The meaning of the occurrence constructors is as follows: a denotes an initial occurrence of action a (possibly preceded or followed by some τ actions in the computation), ae denotes the occurrence e after an action a, while $0e$, $1e$ represent the occurrence e at the left, respectively at the right of a parallel operator. Finally the symbol τ is used – with abuse of notation – to represent any occurrence of a τ-action in a computation. We use $e, e', \ldots$ to range over the whole set $\mathcal{O}_\tau$.

We shall see that a visible occurrence $e \in \mathcal{O}$ incorporates both its static and its dynamic location. Note that $\mathcal{O}$ could also be defined as:

$$\mathcal{O} = (Act \cup \{0, 1\})^* Act.$$

Then an occurrence $e \in \mathcal{O}_\tau$ is either τ or a word σa, for some $\sigma \in (Act \cup \{0, 1\})^*$ and $a \in Act$. The *label* of $e \in \mathcal{O}_\tau$ is the action of which e is an occurrence:

DEFINITION 2.30 (*Label*). The function $\lambda : \mathcal{O}_\tau \to Act_\tau$ is defined by:

$$\lambda(\tau) = \tau, \qquad \lambda(\sigma a) = a.$$

This alternative presentation of $\mathcal{O}$ makes it also particularly easy to define the location and the causal path (also called access path in [38]) of a visible occurrence e. The *location* $loc(e)$ of an occurrence $e \in \mathcal{O}$ is the canonical location obtained by projecting e onto $\{0, 1\}$ (here we use $proj_{\Sigma}(w)$ to denote the projection of a word w on an alphabet Σ):

DEFINITION 2.31 (*Location*). The function $loc : \mathcal{O} \rightarrow \{0, 1\}^*$ is defined by:

$$loc(e) = proj_{\{0,1\}}(e).$$

The prefix ordering on $\mathcal{O}$ defines a relation of causality $\preceq$ on visible occurrences:

DEFINITION 2.32 (*Local causality*). The relation $\preceq \subseteq (\mathcal{O} \times \mathcal{O})$ is given by:

$$e \preceq e' \iff e = e' \text{ or } \exists\, e'' \text{ such that } ee'' = e'.$$

We know that $\preceq$ is a partial ordering. We call $\preceq$ *local causality* because it connects occurrences within the same parallel component: $e \preceq e' \;\Rightarrow\; loc(e) \ll loc(e')$. Let $e \prec e'$ stand for ($e \preceq e'$ & $e \neq e'$). For $e \in \mathcal{O}$, we define $\downarrow\! e = \{e' \mid e' \prec e\}$ to be the set of local causes of e. Then the *causal path* of e is the sequence of such causes:

DEFINITION 2.33 (*Causal path*). For $e \in \mathcal{O}$, $path(e) =_{\text{def}} e_1 \cdot \cdots \cdot e_n$, where $\{e_1, \ldots, e_n\} = \downarrow\! e$ and $e_i \prec e_{i+1}$, $1 \leqslant i < n$.

For instance, if $e = 0a10b11c$, then $\downarrow\! e = \{0a, 0a10b\}$ and $path(e) = (0a) \cdot (0a10b)$. We call $e \in \mathcal{O}$ an *initial occurrence* if $\downarrow\! e = \emptyset$ (equivalently $path(e) = \varepsilon$). An initial occurrence has always the form $e = loc(e) \cdot \lambda(e)$. More generally, if $\eta' \ll \eta$ and η/η' is the residual of η after η', defined by $\eta/\eta' = \eta''$ if $\eta = \eta'\eta''$, we have the following characterization for visible occurrences:

FACT 2.34. *An occurrence $e \in \mathcal{O}$ is completely determined by its label, location and causal path. Namely, if $\lambda(e) = a$, $loc(e) = \eta$ and $path(e) = e_1 \cdot \cdots \cdot e_n$, $n \geqslant 1$, the occurrence e is given by $e = (loc(e_1) \cdot \lambda(e_1)) \cdot (loc(e_2)/loc(e_1) \cdot \lambda(e_2)) \cdot \cdots \cdot (\eta/loc(e_n) \cdot a)$. If $path(e) = \varepsilon$ then $e = \eta \cdot a$.*

Note that $loc(e)$ may be seen as the *static location* of an occurrence e, while $path(e) \cdot \lambda(e)$ represents its *dynamic location*. We define now the relation of *concurrency* on visible occurrences:

DEFINITION 2.35 (*Concurrency*). The relation $\smile \subseteq (\mathcal{O} \times \mathcal{O})$ is defined by:

$$e \smile e' \iff \begin{cases} \textit{either} & e = \sigma 0 e_0 \;\&\; e' = \sigma 1 e'_0, \\ \textit{or} & e = \sigma 1 e_0 \;\&\; e' = \sigma 0 e'_0, \end{cases}$$

where $\sigma \in (Act \cup \{0, 1\})^*$, $e_0, e'_0 \in \mathcal{O}$.

Clearly, the relation $\smile$ is symmetric and irreflexive, and $e \smile e' \iff loc(e) \diamond loc(e')$.

The occurrences e will be the labels of the occurrence transition system. Let us now shift attention to the states of this transition system. As we said, these states are meant to represent processes with a *past*. The past records the observable guards which have been passed along a computation. Formally, the set $\mathcal{S}$ of *computation states* is given by:

$$\xi \ ::= \ nil \mid \mu.p \mid p+q \mid x \mid rec\,x.\,p \mid \widehat{a}.\xi \mid (\xi \mid \xi') \mid \xi \backslash \alpha \mid \xi\langle f\rangle$$

where $p, q \in$ CCS. The construct $\widehat{a}.\xi$ is used to represent the state ξ with "past" a, that is, after a guard a has been passed. The idea is that any transition labelled by a visible occurrence will introduce a "hat" in the resulting state. The basic operational rule is:

$$\text{(O1)} \quad a.p \xrightarrow{a} \widehat{a}.p.$$

On the other hand an invisible occurrence τ does not leave any trace in the past. This is expressed by the rule:

$$\text{(O1}'\text{)} \quad \tau.p \xrightarrow{\tau} p.$$

The hats recorded in states are used to build up "deep" occurrences along computations, according to the rule:

$$\text{(O2)} \quad \xi \xrightarrow{e} \xi',\ e \neq \tau \ \Rightarrow \ \widehat{a}.\xi \xrightarrow{ae} \widehat{a}.\xi'.$$

Occurrences of the form ie, for $i = 0, 1$, originate from parallel terms $\xi \mid \xi'$:

$$\text{(O3)} \quad \xi \xrightarrow{e} \xi',\ e \neq \tau \ \Rightarrow \ \xi \mid \xi'' \xrightarrow{0e} \xi' \mid \xi'',\ \xi'' \mid \xi \xrightarrow{1e} \xi'' \mid \xi'.$$

For defining the whole occurrence system we need a few more notations. First, we extend action relabellings f to occurrences by letting: $f(ae) = f(a)f(e)$ and $f(ie) = if(e)$ for $i = 0, 1$. Moreover, we shall use an auxiliary function for defining the communication rule. In the occurrence system communication arises from concurrent occurrences with complementary labels. However the resulting τ-occurrence should not contribute to the past, since this only keeps track of observable actions. Thus we need to take back the hats introduced by the synchronizing occurrences. To this end we introduce a function $\delta_e(\xi)$, which erases the hat corresponding to occurrence e in ξ (somewhat similar to the function $\Delta_k(p)$ used in Section 2.4). The partial function $\delta : (\mathcal{O} \times \mathcal{S}) \rightarrow \mathcal{S}$ is given by:

$$\begin{aligned}
\delta_a(\widehat{a}.p) &= p, \\
\delta_{ae}(\widehat{a}.\xi) &= \widehat{a}.\delta_e(\xi), \\
\delta_{0e}(\xi \mid \xi') &= \delta_e(\xi) \mid \xi', \\
\delta_{1e}(\xi \mid \xi') &= \xi \mid \delta_e(\xi'), \\
\delta_e(\xi \backslash \alpha) &= \delta_e(\xi) \backslash \alpha, \\
\delta_{f(e)}(\xi\langle f\rangle) &= \delta_e(\xi)\langle f\rangle.
\end{aligned}$$

We have now all the elements to define the occurrence system for CCS. The rules specifying this system are listed in Figure 3. Note that the condition $\lambda(e) \neq \{\alpha, \bar{\alpha}\}$ in rule (O6) could be strengthened to $proj_{\{\alpha,\bar{\alpha}\}}(e) = \varepsilon$, to prevent transitions like $\widehat{a}\,.\,b.p \xrightarrow{ab} \widehat{a}\,.\,\widehat{b}.p$. However this would make no difference for states ξ obtained via an occurrence computation from a CCS term (more will be said on this point below). The *weak occurrence system* is now given by:

$$\xi \overset{\tau}{\Longrightarrow} \xi' \iff \exists\, \xi_0, \ldots, \xi_n,\ n \geqslant 0 \text{ such that } \xi = \xi_0 \xrightarrow{\tau} \xi_1 \cdots \xrightarrow{\tau} \xi_n = \xi'$$
$$\xi \overset{e}{\Longrightarrow} \xi' \iff \exists\, \xi_1, \xi_2 \text{ such that } \xi \overset{\tau}{\Longrightarrow} \xi_1 \xrightarrow{e} \xi_2 \overset{\tau}{\Longrightarrow} \xi'.$$

Let us examine some properties of this weak occurrence system. It is clear that any term ξ gives an intensional representation of a CCS computation state. In fact from each state ξ one may extract the set of visible occurrences that have led to it. Obviously, this set should be empty for a CCS term. Formally, the set of *past occurrences* of a term ξ is defined by:

$$\begin{aligned} occ(p) &= \emptyset, \quad \text{if } p \in \text{CCS}, \\ occ(\widehat{a}.\xi) &= \{a\} \cup a \cdot occ(\xi), \\ occ(\xi \mid \xi') &= 0 \cdot occ(\xi) \cup 1 \cdot occ(\xi'), \\ occ(\xi \backslash \alpha) &= \{e \in occ(\xi) \mid \lambda(e) \neq \alpha, \bar{\alpha}\}, \\ occ(\xi \langle f \rangle) &= f\big(occ(\xi)\big). \end{aligned}$$

For states ξ reachable from a CCS term, the clause for restriction reduces to $occ(\xi \backslash \alpha) = occ(\xi)$. In fact, one may easily verify that such states, what we shall call CCS *computation states*, are exactly those ξ whose subterms $\xi' \backslash \alpha$ satisfy $\alpha, \bar{\alpha} \notin \lambda(occ(\xi'))$. Moreover for a CCS computation state ξ each relabelling f is injective on $occ(\xi)$ (this justifies, for instance, the last clause in the definition of the function δ above).

REMARK 2.36. If $\delta_e(\xi)$ is defined, then $occ(\delta_e(\xi)) = occ(\xi) - \{e\}$.

The next statements explain how visible occurrences are generated along computations, and how they are related.

LEMMA 2.37. *Let ξ be a* CCS *computation state. Then*:

$$(1) \qquad \xi \xrightarrow{e} \xi',\ e \neq \tau \;\Rightarrow\; \begin{cases} \text{(i) } occ(\xi') = occ(\xi) \cup \{e\}, \\ \text{(ii) } \forall e' \in occ(\xi)\text{: } e' \prec e \text{ or } e' \smile e), \\ \text{(iii) } {\downarrow} e \subseteq occ(\xi). \end{cases}$$

$$(2) \qquad \xi \xrightarrow{\tau} \xi' \Rightarrow occ(\xi') = occ(\xi).$$

COROLLARY 2.38. *Let $p \in$ CCS. If $p \overset{e_1}{\Longrightarrow} \xi_1 \cdots \overset{e_n}{\Longrightarrow} \xi_n$, where $\forall i$: $e_i \neq \tau$, then*:

(1) $\forall i$: $occ(\xi_i) = \{e_1, \ldots, e_i\}$.

(2) $i < j \Rightarrow e_i \prec e_j$ or $e_i \smile e_j$.

The following proposition shows how local causality may be recovered from static locations along a computation:

PROPOSITION 2.39. *Let $p \in CCS$. If $p \overset{e_1}{\Longrightarrow} \xi_1 \cdots \overset{e_n}{\Longrightarrow} \xi_n$, where $\forall i$: $e_i \neq \tau$, then*:

$$e_i \prec e_j \iff i < j \text{ and } loc(e_i) \ll loc(e_j).$$

PROOF. ($\Leftarrow$) Since $i < j$, by Corollary 2.38 either $e_i \prec e_j$ or $e_i \smile e_j$. But it cannot be $e_i \smile e_j$, since this would imply $loc(e_i) \diamond loc(e_j)$. Thus $e_i \prec e_j$.

($\Rightarrow$) If $e_i \prec e_j$, then it cannot be $j < i$, because of Corollary 2.38 again. Moreover, since e_i is a prefix of e_j, also $loc(e_i)$ is a prefix of $loc(e_j)$. □

We proceed now to define a notion of bisimulation on the weak occurrence system. Once again we use a notion of consistency and progressive bisimulation family.

DEFINITION 2.40. A *consistent occurrence aliasing* is a partial injective function $g : \mathcal{O} \to \mathcal{O}$ which satisfies, for any e, e' on which it is defined:

(i) $\lambda(e) = \lambda(g(e))$.

(ii) $e' \prec e \iff g(e') \prec g(e)$.

Let $\mathcal{G}$ be the set of consistent occurrence aliasings on $\mathcal{O}$.

DEFINITION 2.41. A *progressive o-bisimulation family* is a $\mathcal{G}$-indexed family of relations over $\mathcal{S}$, $\mathbf{R} = \{R_g \mid g \in \mathcal{G}\}$, such that if $\xi_0 R_g \xi_0'$ then for all $e \in \mathcal{O}$:

(1) $\xi_0 \overset{e}{\Longrightarrow} \xi_1 \;\Rightarrow\; \exists e', \xi_1'$ such that $\xi_0' \overset{e'}{\Longrightarrow} \xi_1'$, $g \cup \{(e, e')\} \in \mathcal{G}$ and $\xi_1 R_{g\cup\{(e,e')\}} \xi_1'$.

(2) $\xi_0' \overset{e'}{\Longrightarrow} \xi_1' \;\Rightarrow\; \exists e, \xi_1$ such that $\xi_0 \overset{e}{\Longrightarrow} \xi_1$, $g \cup \{(e, e')\} \in \mathcal{G}$ and $\xi_1 R_{g\cup\{(e,e')\}} \xi_1'$.

(3) $\xi_0 \overset{\tau}{\Longrightarrow} \xi_1 \;\Rightarrow\; \exists \xi_1' \in \mathcal{S}$ such that $\xi_0' \overset{\tau}{\Longrightarrow} \xi_1'$ and $\xi_1 R_g \xi_1'$.

(4) $\xi_0' \overset{\tau}{\Longrightarrow} \xi_1' \;\Rightarrow\; \exists \xi_1 \in \mathcal{S}$ such that $\xi_0 \overset{\tau}{\Longrightarrow} \xi_1$ and $\xi_1 R_g \xi_1'$.

These relations induce an equivalence $\approx^{occ}$ on CCS processes as follows:

DEFINITION 2.42 (*Equivalence on the occurrence system*). For any $p, q \in$ CCS, let $p \approx^{occ} q$ iff $p R_\emptyset q$ for some progressive o-bisimulation family $\mathbf{R} = \{R_g \mid g \in \mathcal{G}\}$.

The reader familiar with the notion of *history preserving bisimulation* (cf., e.g., [75]) may have noticed the similarity with our definition of $\approx^{occ}$. In fact history preserving bisimulation is itself a "progressive" notion, and it is clear that a consistent occurrence aliasing g is nothing else than an isomorphism between two partially ordered sets of occurrences. In Section 3.3 we give a formal definition of *local history preserving bisimulation* on the occurrence system (so-called because the ordering is that of local causality), and show that it is a direct reformulation of the equivalence $\approx^{occ}$.

The preorder $\sqsubseteq\!\!\!\!\raisebox{-1ex}{$\sim^{occ} is obtained using the same definition, after weakening the notion of *consistency* as follows:

DEFINITION 2.43. A *right-consistent occurrence aliasing* is a partial injective function $g : \mathcal{O} \to \mathcal{O}$ which satisfies, for any e, e' on which it is defined:

(i) $\lambda(e) = \lambda(g(e))$.

(ii) $g(e') \prec g(e) \Rightarrow e' \prec e$.

The main result, presented in the next sections, is that the equivalence $\approx^{occ}$ coincides with both $\approx^s_\ell$ and $\approx^d_\ell$, and similarly that $\sqsubseteq\!\!\!\!\sim^{occ}$ coincides with both $\sqsubseteq\!\!\!\!\sim^s_\ell$ and $\sqsubseteq\!\!\!\!\sim^d_\ell$. The proofs rely on the above properties of the occurrence system, and on proving conversion lemmas between the different kinds of transitions. Most of these proofs will be omitted or only outlined. The reader is referred to [38] for a full exposition.

2.6.2. *Occurrence semantics = static location semantics.* We establish here the relationship between $\approx^{occ}$ and $\approx^s_\ell$. We saw in Section 2.4 that $\approx^s_\ell$ can be defined in terms of canonical distributions. We recall that these are distributions always associating location 0 to the left operand and 1 to the right operand of a parallel composition. Let CDIS denote the set of these canonical distributions, and $\eta, \zeta \in \{0, 1\}^*$ range over canonical locations. With each state ξ, we associate a distributed process $dis(\xi) \in$ CDIS as follows:

$$
\begin{aligned}
&dis(\xi) = \xi, \quad \text{if } \xi = nil \text{ or } \xi = x,\\
&dis(a.p) = a.\, dis(p),\\
&dis(p + q) = dis(p) + dis(q),\\
&dis(rec\,x.\ p) = rec\,x.\, dis(p),\\
&dis(\widehat{a}.\xi) = dis(\xi),\\
&dis(\xi \mid \xi') = 0 :: dis(\xi) \mid 1 :: dis(\xi'),\\
&dis(\xi\backslash\alpha) = dis(\xi)\backslash\alpha,\\
&dis(\xi\langle f\rangle) = dis(\xi)\langle f\rangle.
\end{aligned}
$$

Thus $dis(\xi)$ is the canonical distribution of the CCS term underlying ξ. We can now give the conversion lemma between occurrence transitions and static location transitions:

LEMMA 2.44 (Conversion: static $\leftrightarrow$ occurrence). *Let $p, p' \in$ CDIS and $\xi, \xi' \in \mathcal{S}$. Then:*

(i) $\xi \overset{\tau}{\Longrightarrow} \xi' \Rightarrow dis(\xi) \overset{\tau}{\Longrightarrow}_s dis(\xi')$.

(ii) $p \overset{\tau}{\Longrightarrow}_s p' \Rightarrow \forall\xi$ *such that* $dis(\xi) = p\ \exists\xi'$ *such that* $\xi \overset{\tau}{\Longrightarrow} \xi'$ *and* $dis(\xi') = p'$.

(iii) $\xi \overset{e}{\Longrightarrow} \xi' \Rightarrow dis(\xi) \underset{\eta}{\overset{a}{\Longrightarrow}}_s dis(\xi')$, *where* $a = \lambda(e)$ *and* $\eta = loc(e)$.

(iv) $p \underset{\eta}{\overset{a}{\Longrightarrow}}_s p' \Rightarrow \forall\xi$ *such that* $dis(\xi) = p\ \exists e, \xi'$ *such that* $\lambda(e) = a,\ loc(e) = \eta$, $dis(\xi') = p'$ *and* $\xi \overset{e}{\Longrightarrow} \xi'$.

Using Lemma 2.44 and Proposition 2.39 we may now prove our equivalence result.

THEOREM 2.45. *For any $p, q \in$ CCS: $p \approx^s_\ell q \iff p \approx^{occ} q$.*

PROOF OUTLINE. ($\Rightarrow$) Suppose $p \approx_\ell^s q$, and let $\mathbf{S} = \{S_\varphi \mid \varphi \in \Phi\}$ be a progressive bisimulation family such that $dis(p)\, S_\emptyset\, dis(q)$. Define a $\mathcal{G}$-indexed family of relations $\mathbf{R} = \{R_g \mid g \in \mathcal{G}\}$ as follows:

$$(\xi, \xi') \in R_g \iff \text{there exist occurrence transition sequences}$$

$$p \stackrel{\tau}{\Longrightarrow} \xi_0 \stackrel{e_1}{\Longrightarrow} \cdots \stackrel{e_n}{\Longrightarrow} \xi_n = \xi$$

$$q \stackrel{\tau}{\Longrightarrow} \xi'_0 \stackrel{e'_1}{\Longrightarrow} \cdots \stackrel{e'_n}{\Longrightarrow} \xi'_n = \xi'$$

such that

(1) $g = \{(e_1, e'_1), \ldots, (e_n, e'_n)\}$

(2) $dis(\xi)\, S_\varphi\, dis(\xi')$, where $\varphi = \{(loc(e), loc(e')) \mid (e, e') \in g\}$.

Clearly $(p, q) \in R_\emptyset$. It may be shown that $\mathbf{R}$ is a progressive o-bisimulation family.

($\Leftarrow$) Suppose $p \approx^{occ} q$, and let $\mathbf{R} = \{R_g \mid g \in \mathcal{G}\}$ be a progressive o-bisimulation family such that $p R_\emptyset q$. Consider the family $\mathbf{S} = \{S_\varphi \mid \varphi \in \Phi\}$ of relations over CDIS given by:

$$(r, s) \in S_\varphi \iff \text{there exist occurrence transition sequences}$$

$$p \stackrel{\tau}{\Longrightarrow} \xi_0 \stackrel{e_1}{\Longrightarrow} \cdots \stackrel{e_n}{\Longrightarrow} \xi_n = \xi$$

$$q \stackrel{\tau}{\Longrightarrow} \xi'_0 \stackrel{e'_1}{\Longrightarrow} \cdots \stackrel{e'_n}{\Longrightarrow} \xi'_n = \xi'$$

such that

(1) $dis(\xi) = r, \quad dis(\xi') = s$,

(2) $\varphi = \{(loc(e_i), loc(e'_i)) \mid 1 \leqslant i \leqslant n\}$,

(3) $\xi R_g \xi' \quad$ for $g = \{(e_i, e'_i) \mid 1 \leqslant i \leqslant n\}$.

It may be shown that $\mathbf{S}$ is a progressive bisimulation family. □

Using a variation of this proof, a similar result can be established for the preorders:

THEOREM 2.46. *For any $p, q \in$ CCS: $p \sqsubseteq_\ell^s q \iff p \sqsubseteq^{occ} q$.*

This concludes the comparison between the static location semantics and the occurrence transition semantics.

2.6.3. *Occurrence semantics = dynamic location semantics.* We turn now to the relation between $\approx^{occ}$ and $\approx_\ell^d$. To establish the coincidence of the two equivalences, we will use a property of $\approx_\ell^d$ which was first pointed out by Kiehn in [100], namely that $\approx_\ell^d$ only depends on computations where *distinct* atomic locations are chosen at each step. We start by recalling some definitions and results from [29]:

DEFINITION 2.47. A *location renaming* is a mapping $\rho : Loc \to Loc^*$. For any $p \in$ LCCS, let $p[\rho]$ denote the process obtained by replacing all occurrences of l in p with $\rho(l)$, for any $l \in Loc$.

We use the notation $\rho\{u/l\}$ for the renaming which maps l to u and acts like ρ on $Loc \setminus \{l\}$. Also, we shall abbreviate $p[id\{u/l\}]$ to $p\{u/l\}$. In what follows, we shall mainly consider alphabetic renamings $\rho : Loc \to Loc$. Note that any partial function $f : Loc \to Loc$ may be seen as a location renaming $\rho : Loc \to Loc$, by letting:

$$\rho(l) = \begin{cases} l & \text{if } f(l) \text{ is not defined,} \\ f(l) & \text{otherwise.} \end{cases}$$

For instance the empty function $\emptyset$ corresponds to the identity renaming *id*. In the following we shall freely use the renaming notation $p[f]$ whenever f is a partial function $f : Loc \to Loc$.

The following lemma relates the transitions of $p[\rho]$ with those of p.

LEMMA 2.48. *Let $p \in LCCS$. Then for any alphabetic location renaming $\rho : Loc \to Loc$:*

(1) (a) $p \xrightarrow[u]{\tau}_d p' \Rightarrow \exists v$ *such that* $\rho(u) \ll v$ *and* $p[\rho] \xrightarrow[v]{\tau}_d p'[\rho]$.

(b) $p \overset{\tau}{\Longrightarrow}_d p' \Rightarrow p[\rho] \overset{\tau}{\Longrightarrow}_d p'[\rho]$.

(2) (a) $p[\rho] \xrightarrow[v]{\tau}_d p' \Rightarrow \exists u, p''$ *such that* $\rho(u) \ll v$, $p''[\rho] = p'$ *and* $p \xrightarrow[v]{\tau}_d p''$.

(b) $p[\rho] \overset{\tau}{\Longrightarrow}_d p' \Rightarrow \exists p''$ *such that* $p''[\rho] = p'$ *and* $p \overset{\tau}{\Longrightarrow}_d p''$.

(3) (a) $p \xrightarrow[ul]{a}_d p', l \notin Loc(p) \Rightarrow \forall k \in Loc, p[\rho] \xrightarrow[vk]{a}_d p'[\rho\{k/l\}]$, *where* $v = \rho(u)$.

(b) *Same as* (a)*, with weak transitions.*

(4) (a) $p[\rho] \xrightarrow[vl]{a}_d p' \Rightarrow \exists u$ *such that* $\rho(u) = v$ *and* $\forall k \notin Loc(p) \exists p''$ *such that* $p''[\rho\{l/k\}] = p'$ *and* $p \xrightarrow[uk]{a}_d p''$.

(b) *Same as* (a)*, with weak transitions.*

We recall now Kiehn's definition for $\approx_\ell^d$, and show that it is equivalent to the original one.

NOTATION 2.49. Let LCCS^ν be the set of LCCS processes whose atomic locations are all distinct.

DEFINITION 2.50 (*ν-dynamic location equivalence*). A relation $R \subseteq (\mathrm{LCCS}^\nu \times \mathrm{LCCS}^\nu)$ is a *ν-dynamic location bisimulation (ν-dlb)* iff for all $(p, q) \in R$ and for all $a \in Act, u \in Loc^*$:

(1) $p \overset{a}{\underset{ul}{\Longrightarrow}}_d p', l \notin Loc(p) \cup Loc(q) \Rightarrow \exists q'$ such that $q \overset{a}{\underset{ul}{\Longrightarrow}}_d q'$ and $(p', q') \in R$.

(2) $q \overset{a}{\underset{ul}{\Longrightarrow}}_d q', l \notin Loc(p) \cup Loc(q) \Rightarrow \exists p'$ such that $p \overset{a}{\underset{ul}{\Longrightarrow}}_d p'$ and $(p', q') \in R$.

(3) $p \overset{\tau}{\Longrightarrow}_d p' \Rightarrow \exists q'$ such that $q \overset{\tau}{\Longrightarrow}_d q'$ and $(p', q') \in R$.

(4) $q \overset{\tau}{\Longrightarrow}_d q' \Rightarrow \exists p'$ such that $p \overset{\tau}{\Longrightarrow}_d p'$ and $(p', q') \in R$.

The largest ν-dlb is called ν-dynamic location equivalence and denoted $\approx_\ell^\nu$.

FACT 2.51. *For any processes* $p, q \in CCS$: $p \approx_\ell^d q \Longleftrightarrow p \approx_\ell^\nu q$.

PROOF OUTLINE. The $\Rightarrow$ direction is trivial. For the $\Leftarrow$ direction, Lemma 2.48 is used in a straightforward way. □

We proceed now to show that $\approx^{occ} = \approx_\ell^\nu$. To do this, we need to establish a conversion between occurrence transitions and dynamic location transitions. We start by converting terms ξ into LCCS terms which represent the same state of computation. The idea is to replace every "hat" in ξ by a canonical atomic location representing uniquely the corresponding occurrence. The simplest way to do this is to take the occurrences themselves as *canonical (dynamic) locations*. We shall then assume, from now onwards, that $\mathcal{O} \subseteq Loc$. We also introduce, for any $\gamma \in Act \cup \{0, 1\}$, a renaming ρ_γ which prefixes by γ all the occurrences appearing as locations in p, namely $\rho_\gamma(e) = \gamma e$. Then the canonical LCCS process $proc(\xi)$ corresponding to a computation state $\xi \in \mathcal{S}$ is defined by:

$$proc(\xi) = \xi, \quad \text{if } \xi \in \text{CCS},$$

$$proc(\widehat{a}.\xi) = a :: proc(\xi)[\rho_a],$$

$$proc(\xi \mid \xi') = proc(\xi)[\rho_0] \mid proc(\xi')[\rho_1],$$

$$proc(\xi \backslash \alpha) = proc(\xi)\backslash\alpha,$$

$$proc(\xi\langle f\rangle) = proc(\xi)[f]\langle f\rangle.$$

Note that in the last clause the first f is a location renaming, while the second is an action relabelling. We have for instance: $proc(\widehat{a}.\widehat{b}.nil \mid c.nil) = 0a :: 0ab :: nil \mid c.nil$. Similarly, if f is the relabelling given by $f(a) = a_1$, $f(b) = b_1$, $f(c) = c_1$, then $proc((\widehat{a}.\widehat{b}.nil \mid c.nil)\langle f\rangle) = (0a :: 0ab :: nil \mid c.nil)[f]\langle f\rangle = 0a_1 :: 0a_1b_1 :: nil \mid c_1.nil$. It can be easily checked that $Loc(proc(\xi)) = occ(\xi)$ and $proc(\xi) \in \text{LCCS}^\nu$.

We give next the conversion lemma between occurrence transitions and dynamic location transitions, followed by the coincidence result:

LEMMA 2.52 (Conversion: dynamic $\leftrightarrow$ occurrence). *If* $\xi, \xi' \in \mathcal{S}$, $f : occ(\xi) \to Loc$, *then*:

(i) $\xi \stackrel{\tau}{\Longrightarrow} \xi' \Rightarrow proc(\xi)[f] \stackrel{\tau}{\Longrightarrow}_d proc(\xi')[f]$.

(ii) $proc(\xi)[f] \stackrel{\tau}{\Longrightarrow}_d p' \Rightarrow \exists \xi'$ *such that* $proc(\xi')[f] = p'$ *and* $\xi \stackrel{\tau}{\Longrightarrow} \xi'$.

(iii) $\xi \stackrel{e}{\Longrightarrow} \xi' \Rightarrow \forall l \in Loc : proc(\xi)[f] \stackrel{a}{\underset{ul}{\Longrightarrow}}_d proc(\xi')[f\{l/e\}]$, *where* $a = \lambda(e)$ *and* $u = f(path(e))$.

(iv) $proc(\xi)[f] \stackrel{a}{\underset{ul}{\Longrightarrow}}_d p' \Rightarrow \exists e, \xi'$ *such that* $\lambda(e) = a$, $f(path(e)) = u$, $\xi \stackrel{e}{\Longrightarrow} \xi'$ *and* $proc(\xi')[f\{l/e\}] = p'$.

THEOREM 2.53. *For any* $p, q \in \text{CCS}$: $p \approx^{occ} q \Longleftrightarrow p \approx_\ell^d q$.

PROOF OUTLINE. ($\Rightarrow$) Let $p \approx^{occ} q$. Then there exists a $\mathcal{G}$-indexed family of relations $\mathbf{R} = \{R_g \mid g \in \mathcal{G}\}$ such that $p R_\emptyset q$. Define a relation S on processes by:

$(r, s) \in S \Longleftrightarrow$ there exist occurrence transition sequences

$$p \stackrel{\tau}{\Longrightarrow} \xi_0 \stackrel{e_1}{\Longrightarrow} \cdots \stackrel{e_n}{\Longrightarrow} \xi_n = \xi$$

$$q \stackrel{\tau}{\Longrightarrow} \xi'_0 \stackrel{e'_1}{\Longrightarrow} \cdots \stackrel{e'_n}{\Longrightarrow} \xi'_n = \xi'$$

and there exist two functions:

$$f_1 : \{e_1, \ldots, e_n\} \to Loc, \qquad f_2 : \{e'_1, \ldots, e'_n\} \to Loc$$

such that

(1) $\forall i \in \{1, \ldots, n\} : f_1(e_i) = f_2(e'_i)$,

(2) $proc(\xi)[f_1] = r$, $proc(\xi')[f_2] = s$,

(3) $\xi R_g \xi'$ for $g = \{(e_1, e'_1), \ldots, (e_n, e'_n)\}$.

It may be shown that S is a dynamic location bisimulation.

($\Leftarrow$) to prove this direction, we use the alternative definition $\approx_\ell^\nu$ of dynamic location equivalence. Suppose $p \approx_\ell^\nu q$. Define a $\mathcal{G}$-indexed family of relations $\mathbf{R} = \{R_g \mid g \in \mathcal{G}\}$ as follows:

$(\xi, \xi') \in R_g \Longleftrightarrow$ there exist occurrence transition sequences

$$p \stackrel{\tau}{\Longrightarrow} \xi_0 \stackrel{e_1}{\Longrightarrow} \cdots \stackrel{e_n}{\Longrightarrow} \xi_n = \xi$$

$$q \stackrel{\tau}{\Longrightarrow} \xi'_0 \stackrel{e'_1}{\Longrightarrow} \cdots \stackrel{e'_n}{\Longrightarrow} \xi'_n = \xi'$$

such that

(1) $g = \{(e_1, e'_1), \ldots, (e_n, e'_n)\}$,

(2) $\exists$ injection $f : \{e'_1, \ldots e'_n\} \to Loc$ such that:

$$proc(\xi)[f \circ g] \approx_\ell^\nu proc(\xi')[f].$$

Note that $proc(\xi)[f \circ g], proc(\xi')[f] \in \mathrm{LCCS}^\nu$, given that f and g are injections. Also, it is easy to see that $(p, q) \in R_\emptyset$, since $proc(p)[\emptyset] = p$ and $proc(q)[\emptyset] = q$ (recall that the partial function $\emptyset$ corresponds to the identity renaming id). It may be shown that $\mathbf{R} = \{R_g \mid g \in \mathcal{G}\}$ is a progressive o-bisimulation family. $\square$

We state now the analogous result for the preorders. To prove

$$p \sqsubseteq_\ell^d q \Rightarrow p \sqsubseteq^{occ} q$$

For each $e \in \mathcal{O}_\tau$ let $\stackrel{e}{\longrightarrow} \subseteq (\mathcal{S} \times \mathcal{S})$ be the least binary relation satisfying the following axioms and rules.

(O1)	$a.p \stackrel{a}{\longrightarrow} \widehat{a}.p$	
(O1′)	$\tau.p \stackrel{\tau}{\longrightarrow} p$	
(O2)	$\xi \stackrel{e}{\longrightarrow} \xi', \quad e \neq \tau$	$\Rightarrow \widehat{a}.\xi \stackrel{a\,e}{\longrightarrow} \widehat{a}.\xi'$
(O3)	$\xi \stackrel{e}{\longrightarrow} \xi', \quad e \neq \tau$	$\Rightarrow \xi \mid \xi'' \stackrel{0\,e}{\longrightarrow} \xi' \mid \xi''$
		$\xi'' \mid \xi \stackrel{1\,e}{\longrightarrow} \xi'' \mid \xi'$
(O2′)	$\xi \stackrel{\tau}{\longrightarrow} \xi'$	$\Rightarrow \widehat{a}.\xi \stackrel{\tau}{\longrightarrow} \widehat{a}.\xi'$
(O3′)	$\xi \stackrel{\tau}{\longrightarrow} \xi'$	$\Rightarrow \xi \mid \xi'' \stackrel{\tau}{\longrightarrow} \xi' \mid \xi''$
		$\xi'' \mid \xi \stackrel{\tau}{\longrightarrow} \xi'' \mid \xi'$
(O4)	$\left.\begin{array}{l} \xi_0 \stackrel{e_0}{\longrightarrow} \xi_0', \quad \xi_1 \stackrel{e_1}{\longrightarrow} \xi_1' \\ \text{and} \quad \lambda(e_0) = \overline{\lambda(e_1)} \end{array}\right\}$	$\Rightarrow \xi_0 \mid \xi_1 \stackrel{\tau}{\longrightarrow} \delta_{e_0}(\xi_0') \mid \delta_{e_1}(\xi_1')$
(O5)	$p \stackrel{e}{\longrightarrow} \xi$	$\Rightarrow p + q \stackrel{e}{\longrightarrow} \xi$
		$q + p \stackrel{e}{\longrightarrow} \xi$
(O6)	$\xi \stackrel{e}{\longrightarrow} \xi', \quad \lambda(e) \notin \{\alpha, \bar{\alpha}\}$	$\Rightarrow \xi \backslash \alpha \stackrel{e}{\longrightarrow} \xi' \backslash \alpha$
(O7)	$\xi \stackrel{e}{\longrightarrow} \xi'$	$\Rightarrow \xi \langle f \rangle \stackrel{f(e)}{\longrightarrow} \xi' \langle f \rangle$
(O8)	$p\,[rec\,x.\ p/x] \stackrel{e}{\longrightarrow} \xi$	$\Rightarrow rec\,x.\ p \stackrel{e}{\longrightarrow} \xi$

Fig. 3. Occurrence transition system.

one uses a preorder $\sqsubseteq^{v}_{\ell}$ (the obvious variant of $\approx^{v}_{\ell}$) in place of $\sqsubseteq^{d}_{\ell}$. The proof is an easy adaptation of that for the equivalences.

THEOREM 2.54. *For any $p, q \in$ CCS: $p \sqsubseteq^{d}_{\ell} q \Longleftrightarrow p \sqsubseteq^{occ} q$.*

This concludes our comparison between the dynamic location semantics and the occurrence transition semantics.

Putting together Theorems 2.45 and 2.53, we obtain Theorem 2.29 (1), stating the coincidence of $\approx^{s}_{\ell}$ and $\approx^{d}_{\ell}$. Similarly the coincidence of $\sqsubseteq^{s}_{\ell}$ and $\sqsubseteq^{d}_{\ell}$, stated in Theorem 2.29(2), follows from Theorems 2.46 and 2.54.

In the light of these results, the location equivalence and preorder will be simply denoted $\approx_\ell$ and $\sqsubseteq_\ell$ in the following, unless an explicit reference to their specific definition is required.

3. Extensions and comparison with other approaches

We compare here the location semantics presented in the previous Section with other non-interleaving semantics that have been proposed for CCS in the literature, all based on variations of the notion of bisimulation. An extension of the location semantics to the π-calculus is also presented in Section 3.4. This section is mainly addressed to readers with a keen interest in understanding the various noninterleaving semantics for CCS. The reader preferring to keep to the main exposition of location-based semantics may safely skip to Section 4.3, which surveys more recent approaches to process algebras with localities.

To carry out our comparison we will rely on the *local/global cause transition system* of Kiehn [100], which records both local and global causes of actions along computations. This is a convenient formalism for our purpose, as it brings together the ideas of distributed and causal semantics. Intuitively, distributed semantics deal with *local causality*, which is induced by the structure of terms, while causal semantics are concerned with *global causality*, which is determined by the flow of control of computations. We shall see in fact that both the location semantics of [29] and the causal semantics of [51] can be characterized as particular instantiations of the local/global cause-semantics.

As announced already in Section 2.2, distributed and causal semantics are in general incomparable. However they coincide over the communication and restriction-free fragment of CCS, a class of processes often referred to as BPP (Basic Parallel Processes). This convergence result, first proved for finite processes in [100,3], has been recently generalized to recursive BPP processes by Kiehn [97]. For easy reference we recall here the syntax of BPP:

$$p ::= nil \mid \mu.p \mid (p \mid q) \mid (p+q) \mid x \mid rec\,x.\,p.$$

In BPP, communication is not allowed in parallel composition. The same language, where communication is allowed, is referred to as CPP (Communicating Parallel Processes). We will use $\mathrm{BPP_f}$ (respectively $\mathrm{CPP_f}$) to denote the finite fragment of BPP (respectively CPP).

We start by introducing the local/global cause transition system, which will serve as our reference model in this section.

3.1. *The local/global cause semantics*

We present here the *local/global cause* (*lgc*) *transition system*, which was proposed by Kiehn in [100] (and further studied in [97]) as a unifying framework for studying distributed and causal semantics. This transition system also provides a new, stronger semantics arising from the joint observation of local and global causality.

The *lgc*-semantics is defined on an extended syntax, which we call here LGCCS. This includes, besides the operators of CCS, a construct of *cause prefixing* $\langle \Lambda, \Gamma \rangle :: p$, where

Λ and Γ represent the sets of *local causes*, respectively *global causes*, of the process p. Intuitively, these causes have been generated along the computation leading to state p. Local causes correspond to actions witch precede p via the standard prefixing operator, while global causes may also be *inherited* from another component by means of a communication. Thus $\Lambda \subseteq \Gamma$. A CCS process p is identified with the extended process $\langle \emptyset, \emptyset \rangle :: p$. Formally, Λ and Γ are sets of pairs of the form (a, i), where a is a visible action in Act and $i \in \mathbb{N}$. For simplicity, a pair (a, i) is rendered as a_i (this will imply $a_i \neq a_j$ for all $i \neq j$). We let $\mathcal{K} = \{a_i \mid a \in Act,\ i \in \mathbb{N}\}$ be the set of all causes. By construction $\mathcal{K} \cap Act = \emptyset$. We extend the complementation function to causes by setting $\overline{(a, i)} = (\bar{a}, i)$. We also use $\mathcal{K}_{\mathrm{L}}(p)$, respectively $\mathcal{K}_{\mathrm{G}}(p)$, to denote the set of causes occurring in some Λ, respectively Γ, inside p. Finally, we let $\mathcal{K}(p) = \mathcal{K}_{\mathrm{L}}(p) \cup \mathcal{K}_{\mathrm{G}}(p)$ denote the set of all causes occurring in p.

Note the formal similarity between cause prefixing $\langle \Lambda, \Gamma \rangle :: p$ and location prefixing $l :: p$. In fact the semantics will operate much in the same way as the location semantics, exhibiting the causes of actions in transitions. These transitions have the general form $p \xrightarrow{a_i, \langle \Lambda, \Gamma \rangle} p'$, representing an occurrence of action a with set of local causes Λ and set of global causes Γ. The most notable difference with the (dynamic) location semantics is in the rule for communication, which propagates the global causes of each communicating process to the other partner. Again, causes are only assigned to visible actions. Thus τ-transitions have the form $p \xrightarrow{\tau} p'$ and are inferred using the standard rules of CCS.

Formally the semantics is given by the rules in Figure 4. We write $p \xrightarrow{obs} p'$ when the actual label of the transition is not important. In the communication rule (LG4) the notation $\{\Gamma_q / a_i\}$ is used to indicate the *cause replacement* of a_i by Γ_q in p': this is the process obtained by removing a_i from all local cause sets occurring in p' and replacing a_i by the elements of Γ_q in all the global cause sets in p'.

Based on these transitions, one has the following notion of *lgc*-bisimulation:

Definition 3.1 (*Local/global cause bisimulation*). A relation $R \subseteq \mathrm{LGCCS} \times \mathrm{LGCCS}$ is a local/global cause bisimulation (*lgc*-bisimulation) iff for all $(p, q) \in R$ and for each label $obs \in \{\tau\} \cup (\mathcal{K} \times 2^{\mathcal{K}} \times 2^{\mathcal{K}})$:

(1) $p \xrightarrow{obs} p' \Rightarrow \exists q'$ such that $q \xrightarrow{obs} q'$ and $(p', q') \in R$.

(2) $q \xrightarrow{obs} q' \Rightarrow \exists\, p'$ such that $p \xrightarrow{obs} p'$ and $(p', q') \in R$.

The largest *lgc*-bisimulation is called local/global cause (bisimulation) equivalence and denoted $\sim_{\ell gc}$.

Local/global cause equivalence requires the equality of both local and global causes in matching transitions. If this condition is relaxed by allowing $obs = a_i, \langle \Lambda, \Gamma \rangle$ to be matched by another label $obs' = a_i, \langle \Lambda, \Gamma' \rangle$, one obtains a weaker equivalence notion called *local cause equivalence* and denoted by $\sim_{\ell c}$. Similarly, if $obs = a_i, \langle \Lambda, \Gamma \rangle$ is allowed to be matched by $obs' = a_i, \langle \Lambda', \Gamma \rangle$, one obtains a weaker equivalence called *global cause equivalence* and written $\sim_{gc}$. In [100] a parametric definition is given in place of Definition 3.1, where labels *obs* are filtered through an observation function. Then the three equivalences $\sim_{\ell gc}$, $\sim_{\ell c}$ and $\sim_{gc}$, as well as the ordinary bisimulation equivalence $\sim$, can be obtained as special instances of this parametric definition. In fact, the easiest way to

$$\text{(LG1)} \qquad a.p \xrightarrow{a_i,\langle\emptyset,\emptyset\rangle} \langle\{a_i\},\{a_i\}\rangle :: p \qquad\qquad a_i \notin \mathcal{K}(p)$$

$$\text{(LG2)} \qquad p \xrightarrow{a_i,\langle\Lambda',\Gamma'\rangle} p' \Rightarrow \langle\Lambda,\Gamma\rangle :: p \xrightarrow{a_i,\langle\Lambda\cup\Lambda',\Gamma\cup\Gamma'\rangle} \langle\Lambda,\Gamma\rangle :: p'$$

$$\text{(LG3)} \qquad p \xrightarrow{a_i,\langle\Lambda,\Gamma\rangle} p', \quad a_i \notin \mathcal{K}(q) \Rightarrow \begin{array}{l} p \mid q \xrightarrow{a_i,\langle\Lambda,\Gamma\rangle} p' \mid q \\ q \mid p \xrightarrow{a_i,\langle\Lambda,\Gamma\rangle} q \mid p' \end{array}$$

$$\text{(LG4)} \qquad \left.\begin{array}{l} p \xrightarrow{a_i,\langle\Lambda_p,\Gamma_p\rangle} p', q \xrightarrow{\bar{a}_j,\langle\Lambda_q,\Gamma_q\rangle} q' \\ a_i \notin \mathcal{K}(q),\ \bar{a}_j \notin \mathcal{K}(p), \\ \mathcal{K}_{\mathrm{L}}(p) \cap \mathcal{K}_{\mathrm{L}}(q) = \emptyset \end{array}\right\} \Rightarrow p \mid q \xrightarrow{\tau} p'\{\Gamma_q/a_i\} \mid q'\{\Gamma_p/\bar{a}_j\}$$

$$\text{(LG5)} \qquad p \xrightarrow{obs} p' \qquad \Rightarrow \begin{array}{l} p+q \xrightarrow{obs} p' \\ q+p \xrightarrow{obs} p' \end{array}$$

$$\text{(LG6)} \qquad \left.\begin{array}{l} p \xrightarrow{a_i,\langle\Lambda_q,\Gamma_q\rangle} p' \\ a \notin \{\alpha,\bar{\alpha}\},\ \Gamma_q \cap \{\alpha,\bar{\alpha}\} = \emptyset \end{array}\right\} \Rightarrow p\backslash\alpha \xrightarrow{a_i,\langle\Lambda_q,\Gamma_q\rangle} p'\backslash\alpha$$

$$\text{(LG7)} \qquad \left.\begin{array}{l} p \xrightarrow{a_i,\langle\Lambda_q,\Gamma_q\rangle} p' \\ f(a)_j \notin \mathcal{K}(p\langle f\rangle) \end{array}\right\} \Rightarrow p\langle f\rangle \xrightarrow{f(a)_j,\langle f(\Lambda_q), f(\Gamma_q)\rangle} p'\langle f\rangle$$

$$\text{(LG8)} \qquad p[rec\,x.\ p/x] \xrightarrow{obs} p' \qquad \Rightarrow rec\,x.\ p \xrightarrow{obs} p'$$

Fig. 4. Local/global cause transitions.

Transitions $p \xrightarrow{\tau} p'$ are inferred from (LG4) for communication, from the standard rules of CCS for the remaining operators, and from the following rule for cause prefixing:

$$\text{(LG9)} \qquad p \xrightarrow{\tau} p' \Rightarrow \langle\Lambda,\Gamma\rangle :: p \xrightarrow{\tau} \langle\Lambda,\Gamma\rangle :: p'$$

Fig. 5. Rules for τ-transitions.

look upon lc-equivalence and gc-equivalence is to regard them as the bisimulations associated with the *reduced* lgc-transition system where cause prefixing $\langle \Lambda, \Gamma \rangle :: p$ is replaced by $\Lambda :: p$ and $\Gamma :: p$ respectively (and the operational rules are simplified accordingly). We shall call these reduced transition systems the lc-transition system and gc-transition system.

The weak lgc-transitions are derived from the strong transitions in the standard way, and induce notions of weak lgc-equivalence, weak lc-equivalence, and weak gc-equivalence, denoted respectively $\approx_{\ell gc}$, $\approx_{\ell c}$ and $\approx_{gc}$.

We recall from [100] the main results concerning these equivalences. The first is a simple consequence of the definitions.

PROPOSITION 3.2. *Let $p, q \in$ LGCCS. If $p \approx_{\ell gc} q$ then $p \approx_{\ell c} q$ and $p \approx_{gc} q$.*

The reverse of this proposition is not true. The following example, taken from [100], shows that $\approx_{\ell gc}$ is strictly finer than the intersection of $\approx_{\ell c}$ and $\approx_{gc}$.

EXAMPLE 3.3. For the following processes p and q, we have $p \approx_{\ell c} q$ and $p \approx_{gc} q$ but not $p \approx_{\ell gc} q$:

$$p = a \mid b.c + (a.\gamma \mid \bar{\gamma}.b.\delta \mid \bar{\delta}.c)\backslash\gamma, \delta, \qquad q = p + (a.\gamma \mid \bar{\gamma}.b.c)\backslash\gamma, \delta.$$

In [100] it is also shown that local cause equivalence and global cause equivalence are incomparable, even on the simple sublanguage without restriction:

PROPOSITION 3.4. *Let $p, q \in$ LGCCS. Then $p \approx_{\ell c} q \not\Rightarrow p \approx_{gc} q$, and $p \approx_{gc} q \not\Rightarrow p \approx_{\ell c} q$.*

The proof is given by the example:

EXAMPLE 3.5.

$$(a.\gamma \mid \bar{\gamma}.b) + a.b \;\; \begin{matrix} \not\approx_{\ell c} \\ \approx_{gc} \end{matrix} \;\; (a.\gamma \mid \bar{\gamma}.b),$$

$$(a.\gamma \mid \bar{\gamma}.b) + (b.\gamma \mid \bar{\gamma}.a) + a \mid b \;\; \begin{matrix} \approx_{\ell c} \\ \not\approx_{gc} \end{matrix} \;\; (a.\gamma \mid \bar{\gamma}.b) + (b.\gamma \mid \bar{\gamma}.a).$$

On the other hand, the two equivalences coincide when no communication is allowed. This is expected since the only rule that differentiates the lc-transition system and the gc-transition system is the rule for communication.

PROPOSITION 3.6. *Let $p, q \in$ BPP. Then $p \approx_{\ell c} q \Longleftrightarrow p \approx_{gc} q$.*

In fact this coincidence already holds for the strong equivalences $\sim_{\ell c}$ and $\sim_{gc}$. The main results of [100] are the characterizations of location equivalence $\approx_{\ell}$ and causal bisimulation $\approx_c$ [51] for CCS processes, respectively as the local cause equivalence $\approx_{\ell c}$ and the global cause equivalence $\approx_{gc}$. We simply state here these results, referring to Kiehn's paper for the proofs.

PROPOSITION 3.7 (Characterization of location equivalence).
Let $p, q \in$ CCS. *Then* $p \approx_{\ell} q \Longleftrightarrow p \approx_{\ell c} q$.

PROPOSITION 3.8 (Characterization of causal bisimulation).
Let $p, q \in$ CCS. *Then* $p \approx_{c} q \Longleftrightarrow p \approx_{gc} q$.

A couple of comments may be helpful here. As regards Proposition 3.7, we noted already the similarity between the *lc*-semantics and the location semantics. This becomes even more evident if we consider the *lc*-transition system to be given by the rules of Figure 4 where $\langle \Lambda, \Gamma \rangle :: p$ is replaced by just $\Lambda :: p$. There remain some differences though: in the location transition system locations are not required to be new at each step;[25] moreover, local causes (locations) are recorded together with their occurrence ordering, while in the *lc*-transition system they are just collected into a set. In fact for general LGCCS processes location equivalence is finer than *lc*-equivalence. For instance, using $l :: p$ as an abbreviation for $\langle \{l\}, \emptyset \rangle :: p$, the two processes $l :: k :: a$ and $k :: l :: a$ are local cause equivalent but not location equivalent. Indeed, the fact that $\approx_{\ell} = \approx_{\ell c}$ for CCS processes does not mean that the occurrence ordering of causes is immaterial here, but rather that for cause-free processes this ordering is already accounted for by the bisimulation relations.

The relationships between the *gc*-transition system and the causal transition system of [51,52] will be discussed in Section 3.3, where causal bisimulation is reviewed.

3.2. *Comparison with other distributed semantics*

We examine here how location equivalence relates to an earlier version introduced in [28], called *loose location equivalence*. We also compare it with *distributed bisimulation* ([39, 37,98,48]), the first distribution based equivalence proposed in the literature, and with the *local mixed-ordering equivalence* of [171,117].

We shall arrive at the following picture for distribution based equivalences: for finite restriction-free processes of CPP_{f} they all coincide; on the whole language CCS, distributed bisimulation equivalence and loose location equivalence are incomparable, and both are weaker than location equivalence and local mixed-ordering equivalence, which coincide.

3.2.1. *Loose location equivalence.* We start by comparing the dynamic location equivalence $\approx_{\ell}^{d}$ with its earlier formulation given in [28]. While the definition of the two equivalences is formally the same, the underlying location transition systems are slightly different. Transitions in [28] are more general in that the location allocated at each step is a word $u \in Loc^*$ instead of an atomic location $l \in Loc$. To avoid confusion the transitions of [28]

[25] We have seen already in Section 2.6 that the location equivalence and preorder for CCS processes do not change if new locations are created at each step. Indeed, this was established by Kiehn in [100] precisely for the purpose of comparing location equivalence with *lc*-equivalence.

will be called *loose location transitions*, and denoted $\overset{a}{\underset{u}{\rightsquigarrow}}$. For the loose transition system the rule (D1) of Figure 2 is replaced by

$$\text{(LD1)} \quad a.p \overset{a}{\underset{u}{\rightsquigarrow}} u :: p \quad u \in Loc^*.$$

The rules concerning the other process constructors, the τ-transitions and weak transitions are the same as for the dynamic location transition system of Section 2.5. We denote the weak loose transitions by $p \overset{a}{\underset{u}{\rightsquigarrow\!\!\!\Rightarrow}} p'$. The location equivalence based on these transitions is called *loose location equivalence* and denoted $\simeq_{\ell\ell}$.

Although the difference between $(D1)$ and $(LD1)$ may seem quite inoffensive (and indeed it was thought to be so when the formulation in [28] was chosen), it turns out that the loose transition system gives more freedom for relating process behaviours. While for the location equivalence $\approx_\ell$, based on atomic allocation, we implicitly require the equality of the last allocated locations, this is not true for loose location equivalence. The latter can introduce more than one atomic location in one step and thus is able to fill up "missing locations". The following example (due to Rob Van Glabbeek) shows that loose location equivalence $\simeq_{\ell\ell}$ can equate processes which are distinguished by location equivalence $\approx_\ell$. Let p and q represent respectively the processes $(l :: \alpha \mid \bar{\alpha}.b)\backslash\alpha$ and $(l :: (\alpha + b) \mid \bar{\alpha}.b)\backslash\alpha$. Then the move of q

$$\big(l :: (\alpha + b) \mid \bar{\alpha}.b\big)\backslash\alpha \xrightarrow[lk]{b} (l :: k :: nil \mid \bar{\alpha}.b)\backslash\alpha$$

which is also a loose move, can only be matched by a loose move of p, introducing the location lk in one step:

$$(l :: \alpha \mid \bar{\alpha}.b)\backslash\alpha \overset{b}{\underset{lk}{\rightsquigarrow\!\!\!\Rightarrow}} (l :: nil \mid lk :: nil)\backslash\alpha.$$

Indeed we have $p \approx_{\ell\ell} q$ but $p \not\approx_\ell q$. This also implies that the CCS terms $(a.\alpha \mid \bar{\alpha}.b)\backslash\alpha$ and $(a.(\alpha + b) \mid \bar{\alpha}.b)\backslash\alpha$ are loosely location equivalent but not location equivalent. Now, intuitively it is not desirable to equate these processes, since in the first process action b is not locally dependent upon action a.

From this example we see that the two location equivalences are different for CCS terms. However the property of filling up "missing locations" only comes into play when processes containing the restriction operator are considered. It can be shown, in fact, that for finite restriction-free processes the two equivalences coincide [29]:

PROPOSITION 3.9. *Let* $p, q \in \mathrm{CPP_f}$. *Then* $p \simeq_{\ell\ell} q \iff p \approx_\ell q$.

On the full language, on the other hand, location equivalence is finer than loose location equivalence.

PROPOSITION 3.10. *Let* $p, q \in \mathrm{CCS}$. *Then* $p \approx_\ell q \Rightarrow p \simeq_{\ell\ell} q$.

In conclusion, $\approx_\ell$ is stronger than $\simeq_{\ell\ell}$ and captures more precisely the intuition about local dependencies. This explains why in [29] the first formulation of location equivalence $\simeq_{\ell\ell}$ was abandoned in favour of $\approx_\ell$. The interested reader is referred to this paper for more details on these results.

3.2.2. *Distributed bisimulation.* We now turn to the relationship between $\approx_\ell$ and distributed bisimulation. Distributed bisimulation was the first attempt to define a semantics for CCS taking the distributed nature of processes into account. It was introduced by Castellani and Hennessy in [37,39] and further studied by Kiehn in [98,97] and by Corradini and De Nicola in [48].

The basic idea is similar to that of location equivalence. The capabilities of observers are increased so that they can observe actions together with the location where they are performed. However, locations are not explicitly introduced here; instead, there are multiple observers which can move from one location to another as the computation proceeds. When observing an action, the current observer moves to its location and appoints a new observer for the remainder of the process. Thus the number of observers increases by one each time a visible action is performed. In this way locations are observed much in the same way as in the dynamic approach, but without assigning names to them.

We use here the original definition of distributed bisimulation, as given in [37] for the subset $\mathrm{CPP_f}$ of CCS. This is based on a *distributed transition system* where transitions give rise to a pair of residuals, a *local residual* and a *concurrent residual* (or *remote residual*). Intuitively, the first represents what locally follows the action, while the second is the part of the process which is concurrent with the action.[26] We shall give here a slightly different presentation of the operational semantics of [37], which is taken from [98].

We will distinguish the local residual by including it in brackets $[\ ]$, as for instance in $(a.nil \mid [b.nil]) \mid c.\bar{b}.nil$. The syntax for "processes with a local component" is given by the following grammar, where p stands for any $\mathrm{CPP_f}$-term,

$$P ::= [p] \mid (P \mid p) \mid (p \mid P).$$

Let $\mathcal{P}_{[\ ]}$ denote this language. We use $\mathcal{C}[p], \mathcal{D}[q], \ldots$ to range over $\mathcal{P}_{[\ ]}$. In $\mathcal{C}[p]$, p is the *local process* and $\mathcal{C}$ is its *context*. For example in $(a.nil \mid [b.nil]) \mid c.\bar{b}.nil$ the local process is $p = b.nil$ and the context is $\mathcal{C} = (a.nil \mid .) \mid c.\bar{b}.nil$. We also use the notation $\mathcal{C}(p)$ to represent the CCS process obtained by simply embedding p in $\mathcal{C}$ without the $[\ ]$-brackets. Processes in $\mathcal{P}_{[\ ]}$ are used to exhibit the location associated with the last visible action. Visible transitions have the form $p \xrightarrow{a}_d \mathcal{C}[p']$. The observer who sees action a moves to its location, and from then onwards only observes the local component p'. The newly appointed observer takes care of the remote component $\mathcal{C}(nil)$.

The rules for (strong) distributed transitions are given in Figure 6. These rules apply only to restriction-free terms.

Weak transitions are derived by the rule

$$\text{(WD)} \quad p \stackrel{\varepsilon}{\Longrightarrow} p_1 \xrightarrow{a}_d \mathcal{C}[p'] \stackrel{\varepsilon}{\Longrightarrow} \mathcal{D}[p''] \quad \text{implies} \quad p \stackrel{a}{\Longrightarrow}_d \mathcal{D}[p''],$$

[26] An alternative formulation in terms of *local* and *global* residuals was proposed in [39].

For each $a \in Act$ let $\xrightarrow{a}_d \subseteq (\mathrm{CPP_f} \times \mathcal{P}_{[\ ]})$ be the least binary relation satisfying the following axiom and rules.

(D1) $\quad a.p \xrightarrow{a}_d [p]$

(D2) $\quad p \xrightarrow{a}_d \mathcal{C}[p'] \;$ implies $\; p+q \xrightarrow{a}_d \mathcal{C}[p']$
$\qquad\qquad\qquad\qquad\qquad\qquad q+p \xrightarrow{a}_d \mathcal{C}[p']$

(D3) $\quad p \xrightarrow{a}_d \mathcal{C}[p'] \;$ implies $\; p \mid q \xrightarrow{a}_d \mathcal{C}[p'] \mid q$
$\qquad\qquad\qquad\qquad\qquad\qquad q \mid p \xrightarrow{a}_d q \mid \mathcal{C}[p']$

Fig. 6. Distributed transitions.

where transitions $\mathcal{C}[p'] \overset{\varepsilon}{\Longrightarrow} \mathcal{D}[p'']$ are defined in the usual way from the transitions $\xrightarrow{\tau}$, which in turn are based on the relations $\xrightarrow{\mu}$ given by the standard operational rules of CCS together with the rule:

(D4) $\quad p \xrightarrow{\mu} p' \quad$ implies $\quad [p] \xrightarrow{\mu} [p']$.

For example we can derive

$$(a \mid \bar{a}.b) \mid c.\bar{b} \xrightarrow{\bar{a}}_d (a \mid [b]) \mid c.\bar{b}$$
$$(a \mid \bar{a}.b) \mid c.\bar{b} \overset{c}{\Longrightarrow}_d (nil \mid nil) \mid [nil].$$

The main difference between the location and the distributed semantics does not lie in the underlying transition systems but in the way the equivalences work; in the first case observing a location results in assigning a name to it, while in the latter it results in splitting the process into a local part and a remote part. Note however that in the distributed case we have at each step only two "locations", while in the location semantics the number of locations appearing in a computation state is unbounded.

DEFINITION 3.11 (*Distributed bisimulation equivalence*). A symmetric relation $R \subseteq \mathrm{CPP_f} \times \mathrm{CPP_f}$ is called a *distributed bisimulation* iff $R \subseteq D(R)$ where $(p,q) \in D(R)$ iff

(1) $p \overset{\varepsilon}{\Longrightarrow} p'$ implies $q \overset{\varepsilon}{\Longrightarrow} q'$ for some $q' \in \mathrm{CPP_f}$ such that $(p', q') \in R$.

(2) $p \overset{a}{\Longrightarrow}_d \mathcal{C}[p']$ implies $q \overset{a}{\Longrightarrow}_d \mathcal{D}[q']$ for some $\mathcal{D}[q'] \in \mathcal{P}_{[\ ]}$ such that $(p', q') \in R$ and $(\mathcal{C}(nil), \mathcal{D}(nil)) \in R$.

Two distributed processes p and q are said to be *distributed bisimulation equivalent*, $p \approx_d q$, if there is a distributed bisimulation R such that $(p,q) \in R$.

As an example of non-equivalent processes consider $a \mid b$ and $a.b + b.a$. If $a.b + b.a \overset{a}{\Longrightarrow}_d [b]$ then $a \mid b$ can match this move only with $a \mid b \overset{a}{\Longrightarrow}_d [nil] \mid b$. So we have to compare the local subprocesses, b and nil, and the remaining ones, i.e., $\mathcal{C}(nil) = nil$ and

$\mathcal{D}(nil) = nil \mid b$. Obviously in both cases the equivalence does not hold. On the other hand we have

$$p = (a \mid \bar{a}.b) \mid c.\bar{b} + c.\bar{b} \mid b \approx_d (a \mid \bar{a}.b) \mid c.\bar{b} = q.$$

For example the transition $p \stackrel{c}{\Longrightarrow}_d [\bar{b}] \mid b$ can be matched $q \stackrel{c}{\Longrightarrow}_d nil \mid b \mid [\bar{b}]$; it is obvious that the processes at the current locations are equivalent and so are the remaining processes $nil \mid b$ and $nil \mid b \mid nil$.

We next state the coincidence of $\approx_\ell$ and $\approx_d$ on $\mathrm{CPP_f}$. The main difference between these two equivalences resides in the fact that distributed bisimulation acts separately on the two residuals of a transition. The crux of the proof is a decomposition property for location equivalence, which allows located terms themselves to be split. This property says that if $u :: p \mid q \approx_\ell u :: r \mid s$ then $p \approx_\ell r$ and $q \approx_\ell s$. A proof of this property, as well as of the following coincidence result, may be found in [28].

PROPOSITION 3.12. *Let* $p, q \in \mathrm{CPP_f}$. *Then* $p \approx_d q \Longleftrightarrow p \approx_\ell q$.

A consequence of this proposition is that results available for distributed bisimulation can be reused for location equivalence. For instance, from [37,39] one gains a complete axiomatization for location equivalence on $\mathrm{CPP_f}$, which is significantly different from that given in [29].

In [98] the definition of distributed bisimulation was generalized by Kiehn to all finite CCS processes. We will not give the definition of this *generalized distributed bisimulation* here, but simply mention that this equivalence is weaker than $\approx_\ell$. This is shown by a rather involved example in [98], which we therefore do not report here.

Intuitively the difference between the two equivalences is that distributed bisimulation may "forget" locations which have been observed in previous (not immediately preceding) states. Hence, a transition of some component of one process can be matched by a transition of a component of the other process which originally was "at a different location". This shows that location equivalence is a better choice than generalized distributed bisimulation when looking for an extension of distributed bisimulation to general CCS processes.

A further proposal for generalizing distributed bisimulation to full CCS was presented by Corradini and De Nicola in [48]. Here distributed bisimulation is reformulated within the grape semantics of Degano, De Nicola and Montanari, allowing for a more symmetric treatment of the local and global residuals with respect to [98]. The resulting equivalence is finer than Kiehn's one, but still fails to recover the full power of location equivalence. An example is provided to show that the grape distributed bisimulation is coarser than location equivalence. Indeed, as observed by Kiehn in [97], this example illustrates quite clearly that using a pair of locations (which is essentially what distributed bisimulation does) is not enough to capture location equivalence. It is also argued by Kiehn that by suitably generalizing this example, one could show that *no finite number of locations* would be enough to capture location equivalence.

Since Kiehn's example from [98] may also be used to show that generalized distributed bisimulation is weaker than loose location equivalence, which in turn is weaker than location equivalence, the relationships among these three equivalences can be summarized as follows: for finite and restriction-free processes of CPP_f they all coincide; on the whole language CCS, distributed bisimulation equivalence and loose location equivalence are incomparable, while location equivalence is finer than both of them.

3.2.3. *Local mixed ordering equivalence.* A notion of bisimulation based on canonical static locations,[27] for a class of asynchronous transition systems modelling a subset of CCS with guarded sums, was proposed by Mukund and Nielsen in [118]. The resulting equivalence notion is essentially the same as the static location equivalence reviewed in Section 2.4. Indeed, this notion was conjectured in [118] to coincide with the dynamic location equivalence $\approx_\ell^d$, a result which was to be subsumed by those of [38].

Another transition system labelled with canonical static locations, called *spatial transition system*, was defined by Montanari and Yankelevich for the whole language CCS in [171,117]. Here location transitions are used as a first step to build a second transition system, labelled by *mixed partial orders* (where the ordering is a combination of temporal ordering and local causality), which is then used to define the behavioural equivalence. This equivalence, a variant of the mixed-ordering equivalence of [55] called *local mixed-ordering equivalence*, was shown to coincide with location equivalence, applying techniques similar to those used in Section 2.6. In [171,117], the authors also consider the equivalence obtained by ignoring the temporal ordering and retaining only the local causality ordering: this equivalence, which they call *abstract* location equivalence and provide with some intuitive justification, does not seem to coincide with any other known notion. The *local mixed-ordering equivalence* is obtained in [117] as an instance of a parametric approach; it thus inherits some general results of this approach, such as a complete axiomatization [66].

We shall not elaborate further on the notions of [118] and [117] here, and refer the reader to these papers for further details. Let us only note that in both cases, since canonical locations can be extracted from the syntax of terms, there is no need to extend the language with a location prefixing construct. A disadvantage of such implicit locations, on the other hand, could be the difficulty of adapting the semantics to allow the placing of several parallel components in the same location (see discussion at p. 31 (in Section 2.5)).

Another reformulation of the location semantics of [29] was given by Corradini and De Nicola [48], within the grape semantics of [56]. The resulting equivalence is called *maximal distribution equivalence* and shown to coincide with location equivalence. In fact, their formulation is conceptually very close to the original one, once it is observed that sequential components themselves (*grapes* in that approach) can be used to represent localities.

3.3. *Comparison with causal semantics*

We compare here the location semantics, which is designed to reflect distribution in space, with a number of semantics aimed at reflecting *global causality*. This notion of causality covers both the local causality determined by the structure of terms, and the *inherited*

[27] That is, words over $\{0, 1\}$, see the definition at p. 25 (Section 2.4.1).

causality induced by the flow of control among different components, which may vary according to the computations. Semantics which are based on such notion of global causality will be collectively referred to as *causal semantics*.

We shall see that, in general, equivalence notions based on causal semantics, such as pomset bisimulation and causal bisimulation, are incomparable with equivalences based on distribution. This is already apparent from the characterizations given by Kiehn [100], reviewed in the previous section. On the other hand, on the sublanguage BPP all these equivalences reduce to the same one, apart from pomset bisimulation which is less discriminating than the others.

3.3.1. *Pomset bisimulation.* We relate here the location equivalence $\approx_\ell$ and preorder $\sqsubseteq\!\!\!\sim_\ell$ with similar notions associated with *pomset semantics*. This semantics, where transitions are labelled with partially ordered multisets (pomsets) of actions, was studied in [23,24] for a subset of CCS without communication and restriction, essentially the class BPP (in fact, a variation of BPP with full sequential composition in place of action prefixing), and extended to the whole language in [25]. The partial ordering semantics proposed by Degano, De Nicola and Montanari in [59,60,54] were also designed to generate pomset transitions for CCS.

On the language BPP, the pomset semantics can be defined directly by means of structural rules. We give the rules for finite processes in Figure 7.

Note how rules (P2) and (P5) are used to build pomsets u, v of actions on top of transitions. As an operation on pomsets, prefixing $a.u$ is interpreted as generating causality between a and the actions of u, and parallel composition $u \mid v$ as generating concurrency between the actions of u and the actions of v.

The *pomset bisimulation* on BPP, noted $\sim_{pom}$, is just the ordinary bisimulation equivalence associated with this transition system [23]. Weak transitions can then be defined by

(P1)	$a.p \xrightarrow{a} p$	
(P2)	$p \xrightarrow{u} p'$	$\Rightarrow a.p \xrightarrow{a.u} p'$
(P3)	$p \xrightarrow{u} p'$	$\Rightarrow p+q \xrightarrow{u} p'$
		$q+p \xrightarrow{u} p'$
(P4)	$p \xrightarrow{u} p'$	$\Rightarrow p \mid q \xrightarrow{u} p' \mid q$
		$q \mid p \xrightarrow{u} q \mid p'$
(P5)	$p \xrightarrow{u} p', q \xrightarrow{v} q' \Rightarrow p \mid q \xrightarrow{u \mid v} p' \mid q'$	

Fig. 7. Pomset transitions.

restricting to visible actions the pomsets labelling the strong transitions. The corresponding weak pomset bisimulation is written $\approx_{pom}$.

We have seen in the previous section that on BPP location equivalence coincides with distributed bisimulation, which in turn was known from [37] to be finer than pomset bisimulation on this language. The following is an example of processes which are distinguished by location (and distributed) bisimulation but equated by pomset bisimulation.

EXAMPLE 3.13.

$$p = a.(b+c) + (a \mid b) \;\underset{\approx_{pom}}{\not\approx_{\ell}}\; a.(b+c) + (a \mid b) + (a.b) = q.$$

These two processes are pomset bisimilar because the behaviour of the additional summand of q can be simulated by p: the transition $q \xrightarrow{a} b$ can be matched by the summand $(a \mid b)$ of p, while the transition $q \xrightarrow{a.b} nil$ is matched by the first summand of p.

Using the result of [37], the situation can be summarized as follows.

PROPOSITION 3.14. *Let $p, q \in BPP$. Then $p \approx_{\ell} q \Rightarrow p \approx_{pom} q$ but $p \approx_{pom} q \not\Rightarrow p \approx_{\ell} q$.*

Intuitively, location and distributed bisimulation are stronger than pomset bisimulation on BPP because they also capture the property that matching computations should always be *extensions* of smaller matching computations.

As soon as communication and restriction are introduced in the language, location equivalence becomes incomparable with pomset bisimulation. We have for instance:

EXAMPLE 3.15.

$$p = (a.\alpha + b.\beta \mid \bar{\alpha}.b + \bar{\beta}.a) \;\underset{\not\approx_{pom}}{\approx_{\ell}}\; (a.\alpha + b.\beta \mid \bar{\alpha}.b + \bar{\beta}.a) + (a \mid b) = q.$$

Similar examples were given in Section 3.2 to differentiate local cause equivalence from global cause equivalence (Example 3.5). These examples also hold for pomset bisimulation, replacing $\approx_{gc}$ with $\approx_{pom}$. For general CCS processes we thus have the following.

PROPOSITION 3.16. *Let $p, q \in$ CCS. Then $p \approx_{\ell} q \not\Rightarrow p \approx_{pom} q$ and $p \approx_{pom} q \not\Rightarrow p \approx_{\ell} q$.*

Let us now turn to the preorder relations. To our knowledge the first definition of a preorder expressing the idea that one process is more sequential than another was proposed by Aceto in [1] for a subset of CCS. This preorder is based on the pomset transition semantics: one process is more sequential than another, in notation $p \sqsubseteq_{pom} q$, when the pomsets labelling the transitions of p are more sequential than those labelling the transitions of q (this "more sequential than" ordering on pomsets had already been introduced by Grabowski [88] and Gischer [73]). Thus the intuition underlying Aceto's preorder is different from that of the location preorder $\sqsubseteq_{\ell}$, in the same way as pomset bisimulation is different from location equivalence.

Indeed, for the preorders we have a similar situation as for the equivalences: the causality-based preorder $\sqsubseteq_{pom}$ and the distribution-based preorder $\sqsubseteq_{\ell}$ are already differentiated on BPP. The following example, suggested by Aceto and already reported in [29], shows that $\sqsubseteq_{pom} \not\subseteq \sqsubseteq_{\ell}$. Let:

$$\begin{aligned} p &= a.b.c + c.a.b + (a \mid b \mid c), \\ q &= p + a.b \mid c. \end{aligned}$$

Then $p \sqsubseteq_{pom} q$ but $p \not\sqsubseteq_{\ell} q$ (while we have both $q \sqsubseteq_{pom} p$ and $q \sqsubseteq_{\ell} p$). To see why $p \not\sqsubseteq_{\ell} q$ consider the move $q \xrightarrow[l]{a} l :: b \mid c$, due to the summand $a.b \mid c$ of q. Now p has two ways to match this transition, namely $p \xrightarrow[l]{a} l :: b.c$ and $p \xrightarrow[l]{a} l :: nil \mid b \mid c$, but neither of them is appropriate. Another example showing that $p \sqsubseteq_{pom} q \not\Rightarrow p \sqsubseteq_{\ell} q$ is Example 3.13.

On BPP, it would be plausible that $p \sqsubseteq_{\ell} q$ implies $p \sqsubseteq_{pom} q$ (although this has not been shown, to our knowledge). On full CCS, the two preorders $\sqsubseteq_{pom}$ and $\sqsubseteq_{\ell}$ are incomparable. Example 3.15 can be reused here. For the two processes p and q of this example we have $q \sqsubseteq_{\ell} p$ but $q \not\sqsubseteq_{pom} p$. In conclusion we have:

PROPOSITION 3.17. *Let $p, q \in CCS$. Then $p \sqsubseteq_{\ell} q \not\Rightarrow p \sqsubseteq_{pom} q$ and $p \sqsubseteq_{pom} q \not\Rightarrow p \sqsubseteq_{\ell} q$.*

3.3.2. *Causal bisimulation.* The causal bisimulation semantics was introduced by Darondeau and Degano in [51], and further investigated in relation to action refinement in [53]. The causal transition system is again defined on an extended language. The new construct here has the form $K \Rightarrow p$, representing process p with set of (global) causes K. The set K is the exact analogue of the global cause set Γ in the lgc-transition system, which in fact was specifically designed to encompass both the causal and the location transition systems.

As stated already in Section 3.2, causal bisimulation coincides with the global cause equivalence of Kiehn. Although the language is essentially the same as that considered by Kiehn, the transition systems present a number of differences.

The first difference is that causes are introduced here in the form of *backward pointers* in a computation. A term $K \Rightarrow p$ represents a CCS process in some computation state, where it has "accumulated" the causes K. A process in its initial state has the form $\emptyset \Rightarrow p$. In $K \Rightarrow p$ the set K consists of numbers $n, m, \ldots$ representing the positions of the causes of p (as backward displacements along the computation). This backward referencing requires an updating of pointers at each step, which is not needed in the gc-transition system.

Assuming, for the sake of comparison, that backward pointers are replaced by uniquely named causes, as in Kiehn's system, there remain a couple of differences between the two semantics. In the causal transition system the new cause introduced for an action does not appear in the label of the transition, but only in the resulting state. Instead of rules (LG1), (LG2), we have the following rule

$$\text{(C1)} \quad K \Rightarrow a.p \xrightarrow{a,K} K \cup \{a_i\} \Rightarrow p, \quad a_i \notin K.$$

Finally, the rule for communication resorts here to an auxiliary transition system, whose transitions have the form

$$p \xrightarrow{a,K,K'} p'.$$

Intuitively, the set K' represents new causes which may be inherited from another component via a communication. These new causes are "guessed" by a kind of early instantiation scheme through the rule:

$$\text{(CC1)} \quad K \Rightarrow a.p \xrightarrow{a,K,K'} K \cup K' \Rightarrow p.$$

Note that no new cause is created here to represent the action a in the rest of the computation. This is because the transitions $p \xrightarrow{a,K,K'} p'$ are only introduced in order to be used in the communication rule:

$$\text{(C2)} \quad p \xrightarrow{a,K,K'} p', q \xrightarrow{a,K',K} q' \quad \text{imply} \quad p \mid q \xrightarrow{\tau} p' \mid q'.$$

The remaining rules are essentially the same as in the global cause transition system.

Because of these few differences in formulation, the coincidence of $\sim_{gc}$ and $\approx_c$ is not completely straightforward, although intuitively expected. This coincidence allows us to merge results obtained for the two equivalences. For instance, the examples given in Section 3.2 to differentiate local and global cause equivalence may be applied to location and causal bisimulation. A similar example is the following. Let $r = (a.\alpha + b.\beta \mid \bar{\alpha}.b + \bar{\beta}.a)$. Then

$$\begin{array}{ccccc} r + (a \mid b) & \approx_\ell & r & \not\approx_\ell & r + a.b. \\ & \not\approx_c & & \approx_c & \end{array}$$

Note also that

$$(a \mid b) \approx_\ell (r \backslash \alpha, \beta) \approx_c a.b + b.a.$$

These absorption phenomena, respectively of $(a \mid b)$ in r with respect to $\approx_\ell$, and of $a.b$ in r with respect to $\approx_c$, clearly show the difference between the two equivalences: the former equates processes with the same parallel structure, while the latter equates processes with the same causal structure. Another interesting example, involving the restriction operator, is the following

$$p = (a.\alpha.c \mid b.\bar{\alpha}.d) \backslash \alpha \; \underset{\not\approx_\ell}{\overset{\approx_c}{}} \; (a.\alpha.d \mid b.\bar{\alpha}.c) \backslash \alpha = q.$$

Here $p \approx_c q$ since in both processes actions c and d causally depend on a and b. On the other hand $p \not\approx_\ell q$ since in p action d is not spatially dependent upon action a. One can see that in a distributed view, the "cross-causalities" induced by communication are observed as purely temporal dependencies, while in the causal approach they are assimilated to local causalities.

Let us now revisit the mutual exclusion and protocol examples from Section 2. It is easy to see that $\mathit{Mutex} \approx_c \mathit{MSpec} \approx_c \mathit{FMutex}$, because the synchronizations on p, v in *Mutex* and *FMutex* create cross-causalities between the critical regions of the two competing processes. The same holds for the protocol example, for which we have $\mathit{PSpec} \approx_c \mathit{Protocol}$.

Causal bisimulation has been shown to coincide with *history preserving bisimulation* and with an instance of "NMS equivalence" in [78]. The coincidence of causal and distributed bisimulations in absence of communication (on $\mathrm{BPP_f}$) was also shown in [4]. Finally, causal bisimulation has been studied in relation to action refinement in [52,53]. It was shown to be indeed robust with respect to action refinement.

The model obtained by unfolding the causal transition system, called *causal trees*, has been used to define a direct "denotational" interpretation of CCS terms, which agrees with a standard event structure semantics; it has also been taken as the basis for a complete axiomatization for causal bisimulation [51]. Indeed, one of the initial motivations of Darondeau and Degano's work was the search for a *tree-model* on which standard results from the usual theory of bisimulation could be carried over.

3.3.3. *History-preserving bisimulation.* We noted already that the original definitions of history-preserving bisimulation were given on event structures or similar semantic domains. More precisely, history-preserving bisimulation was originally defined in [136] and [75] for prime event structures, and extended in [78] and [3] to flow and stable event structures respectively. More recently, this equivalence has also been studied on *configurations structures* [79]. Two variants of the notion of history preserving bisimulation have also been considered: *hereditary* history preserving bisimulation [16] and *strong* history preserving bisimulation [95]. Both are finer than the original notion.

Using the occurrence transition system of Section 2.6, we may define a notion of *local history preserving bisimulation* for CCS processes. Essentially, a history-preserving bisimulation is a bisimulation which preserves, at each state of computation, the partially ordered set of events that led to that state. Moreover, it also accounts for the way this partial ordering is *extended* along a computation.

The following definition, taken from [38], differs from that of [75] and [3] in two respects: it is "syntactic", in that it is defined directly on (an enrichment of) the CCS transition system, and it is based on the local rather than the global causality ordering.

The occurrence system provides a notion of *state* (or configuration) for CCS terms. For $p \in \mathrm{CCS}$, define:

$$\mathit{States}(p) = \left\{\xi \mid \exists\, e_i, \xi_i \text{ such that } p \stackrel{\tau}{\Longrightarrow} \xi_0 \stackrel{e_1}{\Longrightarrow} \cdots \stackrel{e_n}{\Longrightarrow} \xi_n = \xi\right\}.$$

Recall that each state ξ has an associated set of events $occ(\xi)$, ordered by the local causality relation $\preceq$. Unlike the global causality ordering in flow and stable event structures, which is relative to a configuration, the local causality ordering $\preceq$, which is essentially a static notion, is the same for all computations.

DEFINITION 3.18 (*Local history preserving bisimulation*). Let $p, q \in \mathrm{CCS}$. A relation

$$R \subseteq States(p) \times States(q) \times \wp\big(occ\big(States(p)\big) \times occ\big(States(q)\big)\big)$$

is a *local history preserving bisimulation* (*lhp-bisimulation*) between p and q if $(p, q, \emptyset) \in R$ and whenever $(\xi_0, \xi_0', g) \in R$ then:

(1) g is an isomorphism between $(occ(\xi_0), \preceq)$ and $(occ(\xi_0'), \preceq)$.

(2) (a) $\xi_0 \stackrel{e}{\Longrightarrow} \xi_1 \;\Rightarrow\; \exists e', \xi_1'$ such that $\xi_0' \stackrel{e'}{\Longrightarrow} \xi_1'$ and $(\xi_1, \xi_1', g \cup (e, e')) \in R$.

(b) $\xi_0' \stackrel{e'}{\Longrightarrow} \xi_1' \;\Rightarrow\; \exists e, \xi_1$ such that $\xi_0 \stackrel{e}{\Longrightarrow} \xi_1$ and $(\xi_1, \xi_1', g \cup (e, e')) \in R$.

(3) (a) $\xi_0 \stackrel{\tau}{\Longrightarrow} \xi_1 \;\Rightarrow\; \exists \xi_1'$ such that $\xi_0' \stackrel{\tau}{\Longrightarrow} \xi_1'$ and $(\xi_1, \xi_1', g) \in R$.

(b) $\xi_0' \stackrel{\tau}{\Longrightarrow} \xi_1' \;\Rightarrow\; \exists \xi_1$ such that $\xi_0 \stackrel{\tau}{\Longrightarrow} \xi_1$ and $(\xi_1, \xi_1', g) \in R$.

We say that p and q are local history preserving equivalent, $p \approx^{lhp} q$, if there exists a local history preserving bisimulation between them.

We have then the following characterization:

FACT 3.19. *For any processes $p, q \in \mathrm{CCS}$: $p \approx^{lhp} q \iff p \approx^{occ} q$.*

PROOF. It should be clear that if $\mathbf{R} = \{R_g \mid g \in \mathcal{G}\}$ is a progressive bisimulation family such that $p R_\emptyset q$, then the relation:

$$S = \big\{(\xi, \xi', g) \mid \xi \in States(p),\ \xi' \in States(q),\ g \in \mathcal{G} \text{ and } \xi R_g \xi'\big\}$$

is a lhp-bisimulation between p and q. In fact, if $\xi R_g \xi'$ for $\xi \in States(p)$ and $\xi' \in States(q)$, then g is an occurrence aliasing such that $occ(\xi) \subseteq dom(g)$ and $occ(\xi') \subseteq range(g)$, that is an isomorphism between $occ(\xi)$ and $occ(\xi')$.

Similarly, if S is a lhp-bisimulation between p and q then the family $\mathbf{R} = \{R_g \mid g \in \mathcal{G}\}$ defined by:

$$R_g = \big\{(\xi, \xi') \mid (\xi, \xi', g) \in S\big\}$$

is a progressive bisimulation family such that $p R_\emptyset q$. □

In the light of the results of Section 2.6, we have then also:

COROLLARY 3.20. *For any processes $p, q \in \mathrm{CCS}$: $p \approx_\ell^d q \iff p \approx^{lhp} q \iff p \approx_\ell^s q$.*

A similar notion of *local history preserving preorder*, $\sqsubseteq\!\!\!\!\sim^{lhp}$, can be obtained by requiring g, in Definition 3.18, to be a bijection between $(occ(\xi_0), \preceq)$ and $(occ(\xi_0'), \preceq)$ whose inverse is a homomorphism. This preorder may be shown to coincide with $\sqsubseteq\!\!\!\!\sim_\ell$.

Note that the local history-preserving bisimulation just defined is *weak*, in the sense that it ignores τ-actions. This distinguishes it from the other notions of history-preserving bisimulation, which are *strong*. Indeed, as noted already in Section 2.1, the notion of history-preserving bisimulation was designed to be robust under refinement, and most weak equivalences are not. Nevertheless, it could be worth investigating a similar syntactic definition for the usual notion of (global) history preserving bisimulation, by taking a slightly more concrete notion of occurrence where communications are pairs of visible occurrences, and adopting the corresponding global causality ordering (as defined in [27]).

3.4. *Extension to the π-calculus*

In this section we give an informal account of Sangiorgi's extension of the location equivalence $\approx_\ell$ to the π-calculus, as presented in [145]. An important result of this research is that the location machinery of [29] can be *implemented* in the π-calculus by means of a fully abstract encoding, in such a way that the study of location equivalence $\approx_\ell$ reduces to that of ordinary bisimulation equivalence.

The π-calculus is an extension of CCS proposed by Milner, Parrow and Walker in [114] to deal with mobile processes, building on earlier work by Engberg and Nielsen [64]. In the π-calculus, processes interact by exchanging *channel names* (also called simply *names*) over channels. In particular, a restricted name can be sent outside its original scope, which is then enlarged to include the receiver process, a phenomenon-known as *scope extrusion*. The combination of channel transmission and scope extrusion is the most powerful feature of the π-calculus with respect to its predecessor CCS. Using these mechanisms, it is possible to model the creation of *unique* names and simulate the transmission of resources and processes. For a formal definition of the π-calculus, the reader is referred to Parrow's chapter [127] in this Handbook. The syntax of (a core version of) the language may also be found at p. 1017.

The language used by Sangiorgi is in fact the *polyadic* variant of the π-calculus, proposed by Milner in [112], which allows names to be transmitted in tuples rather than one at a time. This variant is particularly useful for writing programming examples as well as encoding other languages.

The version of location equivalence used by Sangiorgi is a generalization of the *dynamic location equivalence* $\approx_\ell^d$ examined in Section 2.5 (hereafter referred to simply as $\approx_\ell$). To accommodate the definition of location equivalence, the syntax and operational semantics of the π-calculus are enriched with location names, exactly in the same way as for the language LCCS examined in Section 2. This adaptation can be done quite smoothly, as name-passing does not introduce additional complications in this respect.

We shall now give an outline of Sangiorgi's encoding, which maps *located* π-processes into more complex π-processes that incorporate the location information. This encoding exploits in a crucial way the π-calculus ability to generate and transmit unique names. We shall use $\mathcal{E}[\![P]\!]$ to denote the encoding of the located process P. As can be expected, locations will be represented by in $\mathcal{E}[\![P]\!]$. Let a, b, x, y, z range over names, and $\langle\tilde{a}\rangle$ represent a tuple of names $a_1, \ldots, a_k$. The idea is the following.

Suppose that $P \xrightarrow[ul]{\overline{a}\langle\tilde{b}\rangle}_d P'$. Then $\mathcal{E}[\![P]\!]$ will send an additional name x which will serve as a "handle" for the observer, enabling him to access both the location ul and the causal path u of action $\overline{a}$. Assume that P is located at y. Then we will have approximately:

$$\mathcal{E}[\![P]\!] \xrightarrow{\overline{a}\langle\tilde{b}x\rangle} (\nu z)(Loc\langle xzy\rangle \mid \mathcal{E}[\![P']\!]) \tag{12}$$

where the process $Loc\langle xzy\rangle$ implements the location mechanism and can be interrogated by the observer using name x. Here, y is the name representing the causal path u of action $\overline{a}$, while z represents the location ul of the derivative P' in $\mathcal{E}[\![P']\!]$. The process $Loc\langle xzy\rangle$ is defined by:

$$Loc\langle xzy\rangle = \overline{x}\langle z\rangle \mid !z.\overline{y}.$$

In fact, to avoid interference with the environment, the name x needs to be unique in (12).

We give now the formal encoding for input and output processes, where the parameter y represents the location u of the encoded process (which is initially the empty string ε):

$$\mathcal{E}[\![\overline{a}\langle\tilde{b}\rangle.P]\!]\langle y\rangle = (\nu x)\overline{a}\langle\tilde{b}x\rangle.(\nu z)(Loc\langle xzy\rangle \mid \mathcal{E}[\![P']\!]\langle z\rangle), \tag{13}$$

$$\mathcal{E}[\![a(\tilde{b}).P]\!]\langle y\rangle = a(\tilde{b}x).(\nu z)(Loc\langle xzy\rangle \mid \mathcal{E}[\![P]\!]'\langle z\rangle). \tag{14}$$

These two clauses give the essence of the translation; further details are omitted. The main result is that this encoding is *fully abstract*, as expressed by the following theorem by Sangiorgi [145]. Here $\approx^c$ and $\approx^c_\ell$ are the congruences associated respectively with the early bisimulation equivalence of the π-calculus and with the location equivalence $\approx_\ell$.

THEOREM 3.21. *For any located π-calculus processes P, Q, $P \approx^c_\ell Q \Longleftrightarrow \mathcal{E}[\![P]\!] \approx^c \mathcal{E}[\![Q]\!]$.*

Note that this is essentially an expressiveness result about the π-calculus. In this sense it falls in the same class as the results by Milner and Sangiorgi [113,143] on the encoding of the λ-calculus and higher-order calculi, and that by Walker [163] on the encoding of object-oriented languages.

4. A concrete view of locations

In this section we consider a *concrete* approach to the use of localities in process algebras, where *location identities* are significant and cannot be abstracted away as in the abstract approaches examined in Section 2. In the abstract approaches locations were used essentially to *differentiate* parallel components. Their exact names were immaterial, as long as they bore the same dependence or independence relation with each other. In the concrete approach we shall describe now, on the other hand, location names are important; indeed the topmost locations of a process may be viewed here as *physical addresses* in a network.

4.1. *A concrete location semantics for CCS*

We start with the basic and most concrete approach to processes with explicit locations, as was originally proposed by Murphy in [119] and further pursued by Corradini and Murphy in [49] and [46]. Although this concrete approach might be seen as the first step towards a theory of processes with locations, we postponed its presentation because it is easier to formalize once the general set-up of location equivalence has been understood. Indeed, this approach followed, chronologically, the more abstract approaches exposed in the previous sections. It also responds to a rather different concern: that of placing parallel processes onto a network of processors with a fixed geometry. In this context, the identities of locations are significant, as they may refer to physical processors. Accordingly, the semantics is very concrete: processes are equated if they reside on the same set of locations and present the same behaviour at each location.

The language considered in [119] and [49,46] is a simple subset of CCS, essentially a finite combination of regular processes by means of the static operators of CCS, parallel composition, restriction and renaming.[28] This is the same class of processes considered by Aceto in [5] under the name *nets of automata*, and we shall retain this terminology here. In this language all the occurrences of the parallel operator appear at top level, and locations are assigned to parallel components in the same way as in the static approach examined in Section 2.4, using the location prefix construct $\ell :: p$. The only difference is that here the two operands of a parallel composition are allowed to be placed at the same location. This is required for solving the placing problem, in case the number of processes is larger than that of processors. It is then immediately clear that locations will not serve as a measure of parallelism in this model. Also, since there is no nesting of parallelism in nets of automata, the structure of locations will be *flat*. Formally, we have the following two-level syntax.

The set of *sequential processes* is given by the grammar:

$$p ::= nil \mid \mu.p \mid (p + p) \mid x \mid rec\, x.\ p.$$

Then the set of (*distributed*) *nets of automata* is given by:

$$n ::= l :: p \mid (n \mid n) \mid n \backslash \alpha \mid n\langle f \rangle.$$

In concentrating on this simple language, the authors were motivated by implementation concerns: they wanted to rule out distributed choices like $(\ell :: p + k :: q)$, and also terms of the form $(\ell :: p \parallel k :: q) + r$, since in both cases a synchronization between possibly distant sites is required.

The original work [119] presents a strong transition semantics for this language, with the associated strong bisimulation, while [49] and [46] consider the corresponding weak semantics, abstracting from τ-transitions. We shall mainly consider the latter here.

[28] The language studied in [119] and [49] is in fact a slight variant of CCS, with an asymmetric notion of communication where reception is directed (tagged with the location from which the value is expected to come) while emission is not. This choice was motivated by a similar asymmetry present in several synchronization protocols. Here, for the sake of simplicity and ease of comparison with other approaches, we take a more standard CCS-like language, as was done already by Corradini in [46].

Formally, the semantics of nets of automata is given by the same rules used for static location transitions in Figure 1: the transitions are simply indexed by the location at which they occur. Then the *weak located equivalence* of Corradini and Murphy [49,46], which we denote here by $\approx_{CM}$, is defined as the ordinary weak bisimulation equivalence associated with these transitions. We have for instance:

$$(\ell :: a \parallel \ell :: b) \approx_{CM} \ell :: (a.b + b.a), \tag{15}$$

$$(\ell :: a \parallel k :: b) \not\approx_{CM} (\ell :: b \parallel k :: a), \tag{16}$$

$$\big(\ell :: (\alpha + b) \mid k :: \bar{\alpha}.b\big)\backslash\alpha \not\approx_{CM} \ell :: b. \tag{17}$$

Note that the nets of examples (16) and (17) are equated by location equivalence $\approx_\ell$ (we refer here to the static location equivalence $\approx_\ell^s$ defined in Section 2). As for the nets of example (15), these are *not* admissible distributed processes in the static approach of Section 2.4, and therefore $\approx_\ell$ is not defined on them.[29]

In fact, it can be easily shown that the equivalence $\approx_{CM}$ is *stronger* than location equivalence $\approx_\ell$ on the intersection of the two languages, that is, on the language of nets of automata where parallel components have different locations). A proof of this fact may be found in [49]. In the light of this result it is fair to say that the model of Murphy and Corradini reflects a *concrete view of localities*.

It may be remarked that this concrete model could very nearly be simulated in ordinary CCS, by using the relabelling operator to distinguish actions of different parallel components. This is only approximately true, because one should then adopt a slightly more flexible communication discipline to allow relabelled complementary actions like α_i and $\bar{\alpha}_j$ to synchronize. For instance one could use a *communication function* as in [14] or a *synchronization algebra* as in [166]. Modulo this little twist in defining communication, one would indeed be able to define a simple encoding $\mathcal{E}[\![n]\!]$ of distributed nets of automata into CCS, such that

$$n_1 \approx_{CM} n_2 \Longleftrightarrow \mathcal{E}[\![n_1]\!] \approx \mathcal{E}[\![n_2]\!],$$

where $\approx$ is the ordinary weak bisimulation of CCS. Note on the other hand that this would not hold for location equivalence $\approx_\ell$, even in the simple setting of nets of automata, as shown by the example:

$$\big[(a.\gamma + b.\bar{\gamma}) \mid (\bar{\gamma}.b + \gamma.a)\big]\backslash\gamma \approx_\ell a \mid b.$$

It should be clear that any relabelling here would distinguish the two actions a (as well as the two actions b) in the first process, and therefore the two relabelled processes would be not be identified by $\approx$.

Concerning located semantics, two additional results were presented in [119] and [49]: a calculus of *efficient placings* of processes into networks of processors, and an axiomatization, along standard lines, of the weak located equivalence $\approx_{CM}$.

[29] Note on the other hand that the dynamic location equivalence $\approx_\ell^d$ could be applied to these terms and would distinguish them.

An intermediary location semantics, which could be called "super-static", was also considered by Corradini in [46]: the idea was to use a static assignment of locations, as in the calculus here, but compare transitions modulo a *static* injective renaming of their locations. This semantics would equate the two systems of example (16) above, while still distinguishing the systems of example (17). Indeed, it not difficult to show that the resulting equivalence is strictly included between the located equivalence $\approx_{CM}$ and the location equivalence $\approx_{\ell}$ of Section 2.

To conclude this section, let us mention another study on distributed CCS conducted by Khrisnan in [96], which is also based on a concrete view of locations. The author considers an extension of CCS with locations and explicit message passing primitives. The semantics he proposes for the language is rather different from those presented here, and we shall not attempt a detailed comparison; it is also more directly driven by practical concerns, like the limitations on the degree of parallelism imposed by a physical architecture. Applications of the calculus to the description of configuration and routing in distributed systems are also discussed.

As pointed out already, the names of localities are important in the concrete approach, since localities are meant to represent physical sites in a distributed architecture. This leads quite naturally to the treatment of locations in more recent and powerful languages for distributed *mobile* computation. These languages will be the subject of Section 4.3.

4.2. *The location failure approach*

In this section we briefly review the location failure approach, as has been developed by Amadio and Prasad in [12] and further pursued by Amadio in [9], and by Riely and Hennessy in [140]. Location failure has been considered also in other calculi, for instance in the distributed JOIN calculus [71]. In these latter calculi, however, locations have more complex meanings, hence their treatment is deferred to Section 4.3.

The standpoint here is rather different from the one taken so far, where a global observer for a whole distributed system was implicitly assumed, and the issue was to identify the degree of distribution of the system. In the "location failure" approach, one starts from the hypothesis that locations – nodes in a distributed system – are liable to failure. The perspective adopted here is that of an individual programmer or process, wishing to have transparent access to resources irrespective of where they are located, at least as long as these locations are running. On the other hand, some awareness of locations is requested in case of failure: a process that has spawned activities at another location wants to be notified in case of failure of this location, in order to stop interacting with it and transfer activities to other locations.

The language considered in [140] is based on pure CCS with no value-passing, while those of [12,9] build on more powerful formalisms, respectively the programming language FACILE of [157] and the asynchronous π-calculus [94,21], which allow for the transmission of processes, respectively names, in communications.

We will concentrate here on [140] and [9], which have been developed independently but appear to share several features. Common to both studies is the idea to enrich the basic language (CCS and the asynchronous π-calculus[30], respectively) with new operators for *spawning* a process at a remote location, *killing* a location and *testing* the status of a location. Locations $\ell \in Loc$ have two possible states, dead and alive. In the case of [9] locations are also transmissible values – in fact they are just a particular category of π-calculus names.

Formally, *basic processes* are built with the standard operators on the underlying calculus together with the following new operators (taking a blending of the constructs of [9] and [140]):

- $\mathsf{spawn}\,(\ell\,,p)$, which creates a new process p at location ℓ;
- $\mathsf{kill}\;\ell$, which kills location ℓ, entailing the termination of all processes running at ℓ;
- $\mathsf{if}\;\ell\;\mathsf{then}\;p\;\mathsf{else}\;q$, which evolves to either p or q, depending on whether ℓ is alive or dead.

Then *distributed processes* are built on top of basic processes using the static operators of the language as well as an operator for *locating* processes, $\ell :: p$, which represents process p running at location ℓ (this is analogous to the construct encountered in the previous sections, so we use the same notation). We therefore have a two-level syntax similar to that of [119,49], although with a richer basic language.

The operational semantics, which we shall not report here, is then defined directly on distributed processes: as can be expected it makes use of the additional information about the state of locations (for instance the process $\ell :: p$ can only execute if ℓ is alive), but it is otherwise very close to the standard semantics of the underlying languages. In particular, unlike all the semantics considered so far in this paper, it does not record the location at which actions are performed. In fact, the common intention of the works [9] and [140] is that distribution should be completely *transparent* in the absence of failure.

We should point out, however, that the names of locations are significant in this model, just like in the concrete approach of [119,49]: for instance the two systems $(\ell :: a \parallel k :: b)$ and $(\ell :: b \parallel k :: a)$, which would be equated by the location equivalence $\approx_\ell$ of Section 2, can be distinguished here by an observer which has the capability of killing ℓ or k.

The semantic notion which is adopted in [9] and [140] is a form of contextual equivalence called *barbed equivalence*, essentially a bisimulation based on silent transitions and basic observations called "barbs", which is further closed with respect to contexts. For more details on this kind of equivalence see Parrow's chapter [127] in this Handbook. In [140] the barbed equivalence is given a more tractable formulation as a bisimulation relation called *located-failure equivalence*, noted $\approx_{LF}$. An example of distributed processes that are equated by $\approx_{LF}$ is:

$$\begin{aligned} P_1 &= \big(\ell :: \bar{\alpha} \parallel k :: (\alpha.b)\big)\backslash\alpha, \\ Q_1 &= \ell :: \big(\mathsf{spawn}\,(k\,,b)\big)\backslash\alpha. \end{aligned}$$

Intuitively, these are equivalent because even an observer capable of killing ℓ is not able to tell them apart. In [140], located-failure equivalence is also given an alternative symbolic characterization (along the lines of [91]), which further improves its tractability.

---

[30] The reader unfamiliar with the π-calculus is referred to Parrow's chapter [127] in this Handbook. However, knowledge of the π-calculus is not strictly necessary for the discussion in this section.

Given the similarity of the underlying languages, the equivalence $\approx_{LF}$ can be easily compared with the concrete located equivalence $\approx_{CM}$ of [49] and the location equivalence $\approx_{\ell}$ of [29]. It turns out that $\approx_{LF}$ is neither stronger nor weaker than either of them. Here is an example, taken from [140], of processes which are both $\approx_{CM}$-equivalent and $\approx_{\ell}$-equivalent but are distinguished by $\approx_{LF}$:

$$P_2 = \big(\ell :: \alpha \parallel k :: (\bar{\alpha} + \tau.b)\big)\backslash\alpha,$$
$$Q_2 = \big(\ell :: (\alpha + \tau) \parallel k :: (\bar{\alpha}.b)\big)\backslash\alpha.$$

These are not $\approx_{LF}$-equivalent because if ℓ is killed by some context, then P_2 is capable of doing a b-transition which Q_2 cannot match. In [140] one may find a slightly larger example, due to F. Corradini, showing that two $\approx_{LF}$-equivalent processes need not be $\approx_{CM}$-equivalent nor $\approx_{\ell}$-equivalent. Therefore $\approx_{LF}$ reflects indeed a different perception of localities.

The basic language considered by Amadio in [9] is a "disciplined" asynchronous π-calculus where for each channel name there is at most one receiver, which can be viewed as a *server* for that channel. Calculi of this kind are often said to have the *unique receiver property*. To emphasize this property, Amadio calls his language the π_1-calculus, and its located version the $\pi_{1\ell}$-calculus. The $\pi_{1\ell}$-calculus will be further discussed in Section 4.3.

In conclusion of this section, we can remark that in the location-failure approach, locations are *units of failure* as well as units of distribution; and while distribution in itself is not observable, failure may be observed (as the absence of certain communication capabilities) and thus reveal some parts of the distribution.

4.3. *Combining locations and mobility*

We give now an overview of some recent work combining locations and mobility. Most of this work is still at an early stage and therefore our account will be essentially descriptive, trying to isolate those features that are more relevant to our subject of interest. We will only consider languages which are presented as process algebras or strongly based on them, namely distributed versions of the π-calculus [114] and the JOIN calculus [70], the Nomadic π-calculus of [151], and new calculi expressly designed with distribution and security concerns in mind, such as the AMBIENT calculus of Cardelli and Gordon [33] and the SEAL calculus of Castagna and Vitek [161]. We shall also briefly touch upon a distributed calculus based on the coordination language Linda, the calculus LLINDA of De Nicola, Ferrari and Pugliese [120].

It can be noted [148] that while most early work on process calculi was centered around modelling protocols (or other concurrent systems) and reasoning about them, these new process calculi have been more specifically designed to be the basis of programming languages. Indeed, a few of them have already given rise to prototypical programming languages such as the Join Calculus Language [72], NOMADIC PICT [152] and KLAIM [121].

With the advent of the *network* as a computing support, *mobility* is bound to play an important role in programming. However theoretical models for mobile computation, such as the π-calculus [114] (see [127]), were already proposed and thoroughly studied before the wide generalization of network programming that we witness today. This partly explains why, although the very idea of mobility seems to presuppose the existence of different sites among which to move, the notion of *distribution* among sites was not originally present in calculi for mobile processes. Only very recently have "distributed versions" of these calculi been defined and started to be investigated. In fact, there are two aspects to the notion of mobility as it is currently used:

(1) The *transmission* of processes or links to processes as *arguments* of communication along channels. This form of mobility, which could perhaps be better qualified as "process-passing" and "name-passing" (or "link mobility"), is exemplified by the π-calculus and its higher-order variants, CHOCS [156] and HOPI [143,144], as well as by the more practical languages derived from them such as FACILE [157] and PICT [132].

(2) The actual *movement* of processes from one place to another, what is commonly called *migration*. This is the kind of mobility emphasized in formalisms such as the AMBIENT calculus and the SEAL calculus, and also accounted for in the distributed JOIN calculus [71,69] and in distributed versions of the π-calculus such as [9,149,92] and NOMADIC PICT [151,152].

Strictly speaking, only the second form of mobility requires an explicit distribution of processes among localities, as a kind of "reference frame" for the movement to take place. For the first form, it may indeed be sufficient to assume an implicit distribution given by the parallel structure of processes, since mobility is here only a facet of communication.

To fix ideas, we will say that a model or language is "distribution-oriented" or simply "distributed" if it explicitly integrates some notion of *site*, *locality* or *domain*. In this sense, current developments of the π-calculus such as Amadio's $\pi_{1\ell}$-calculus [9], the distributed π-calculi *dpi* of Sewell [149] and Dπ of Hennessy and Riely [141,92], the distributed JOIN calculus [71,69] and NOMADIC PICT [152] are distributed, and so are newly proposed calculi such as the AMBIENT calculus [33] and the SEAL calculus [161]. These are essentially all the models we shall examine in this section. Now, what exactly is a locality in these models? According to the model localities may have different attributes, but some common features can be identified:

- A locality is the host of a number of processes and resources. Sometimes, the locality itself is assimilated to a particular process (the "locality process" of the $\pi_{1\ell}$-calculus) which plays the role of a controller for the processes in its scope.
- Localities can be observed or controlled. The need for controlling localities is often motivated by failure detection or security concerns. The observation of localities is required to implement migration.
- A locality can be named, and as such designated as the target of a communication (if communication across localities is allowed) or the destination of a migration.

Moreover, localities have a bearing on the handling of communication. In a distributed calculus there are two forms of communication: *local communication*, between processes at the same locality, and *distant communication*, between processes at different localities.

Distant communication can be transparent – in the sense that distant destinations are reachable as if they were local (distributed JOIN calculus, Amadio's calculus, Sewell's *dpi*-calculus), restricted to the immediate neighbourhood (SEAL calculus), or even completely forbidden (AMBIENT calculus, Hennessy and Riely's Dπ-calculus). In the last case only local communication is possible, and processes have to convene in a common location in order to communicate. Typically, a process wishing to send a message will have to move to the location of the destination process. Clearly this choice only makes sense in calculi with a migration primitive, where distant communication can be simulated.

Depending on the form of distant communication allowed, one can distinguish between *global communication*, *proximity communication* and *purely local communication* models. In most of these models both local and distant communication are *asynchronous* (in the sense that the sending of a message is non-blocking,[31] see also [127]). In general, distributed calculi have tended to adopt asynchronous rather than synchronous communication, as the former seems more primitive, easier to implement, and closer to the asynchronous message delivery mechanisms of actual networks. In the SEAL calculus, on the other hand, both local and distant communication are synchronous. This choice may be due to the fact that the primary concern of this calculus being security, a more rigid communication discipline is preferred (the implementation issue is less crucial here since only short-range communication is allowed). Similarly, communication is synchronous in the $D\pi$-calculus of Hennessy and Riely, where only local communication is allowed.

The structure of locations can be *flat* or *hierarchical*, in which case they can be organized in two levels or arbitrarily nested in a tree. In a hierarchy of locations, the first level may be seen as representing physical machines[32] while lower levels correspond to software agents. Locations are flat in Amadio's $\pi_{1\ell}$-calculus, in Hennessy and Riely's $D\pi$-calculus, and in NOMADIC PICT. They have a two-level structure in Sewell's *dpi*-calculus and a tree structure in the distributed JOIN calculus, the AMBIENT calculus and the SEAL calculus. Hierarchical locations have a relevance in the presence of failures, since the failure of a location entails the failure of all its sublocations. In a hierarchy of locations, moreover, locations may be themselves migrating entities: in this case a migrating location will bring together its sublocations. Structured locations also allow for a finer control of mobility and a more flexible handling of security issues when modelling networks partitioned into administrative or "trust" domains.

Locations are the reference frame for migration. Migration is intended as the movement of processes or some other entity having executable content (such as locations themselves), as opposed to simple data transfer. One can distinguish between *subjective* and *objective* migration. In the first case the migration instruction comes from the migrating entity itself (possibly from a process enclosed in it, if this entity is for instance a location), while in the second it is outside its control.

A further classification can be made into *strong* and *weak* migration. In the first case a process can be interrupted during its execution and moved together with its current environment to another location where execution is pursued. In this case, where "active" code can be moved, one often speaks of *migrating agent*. In the case of weak mobility, instead, only "passive" code can be moved. Finally, strong migration can be *blocking* for

[31] This is not to be confused with the use of "asynchronous" and "synchronous" in reference to Milner's calculi CCS and SCCS (Synchronous CCS) [110]. In SCCS synchrony means that parallel components proceed together in lock-step.

[32] Although, strictly speaking, this would imply a different treatment of topmost locations, to forbid them to move or be dynamically generated. This issue is explicitly addressed only in NOMADIC PICT, where the flat space of locations is fixed from the start, and the *dpi*-calculus, where new locations may only be created as sublocations of existing ones.

the migrating entity, in the sense that this has to suspend execution until the migration is completed, or *asynchronous*, as is the case for the migration of localities, which can still receive messages while being subject to a migration instruction.

In the next sections, we shall examine in some detail a selection of these models for distributed mobile computation.

4.3.1. *Distributed π-calculi.* The π-calculus does not have an explicit notion of location, although its constructs are powerful enough, as shown in Section 2.6, to simulate the location machinery of [29]. Indeed the π-calculus, while being one of the most influential foundational calculi for mobility, does not provide for direct process mobility. The movement of processes is represented indirectly via the movement of channels that refer to them: roughly speaking, two processes move close to each other when they acquire a common channel. It could then be natural to view the notion of *name scoping* as an abstraction of that of location. However this analogy cannot be brought very far, since processes may share several names with other processes. For instance one cannot define two processes to be co-located whenever they share a private name, since this notion of "co-localization" would not be transitive. A more convincing interpretation of name scoping would then be as "trust domains" (that is, groups of processes that trust each other), since these are allowed to overlap and membership to a common trust domain is not expected to be transitive.

To directly model distributed aspects such as migration of agents and failure of machines, one must add to the π-calculus primitives for grouping processes into units of migration or units of failure. An early proposal was presented by Amadio and Prasad in [12], to model the distributed module of the FACILE programming language [157]. More recently, enrichments of the π-calculus with explicit locations have been studied by Amadio [9], Hennessy and Riely [141,92,93,142], Sewell [149], and Sewell et al. [151,152]. The emphasis in this work is in developing type systems for mobile distributed computation, building on existing type systems for the π-calculus. These type systems may serve a variety of purposes. For instance, Amadio's type system is used to ensure the uniqueness of receptors on communication channels. The type systems of Hennessy and Riely's [141,92] aim at regulating the use of resources according to permissions expressed by types; these type systems are further refined in [93,142], to deal with the case where some components of the network, representing malicious agents, may not be assumed to be well-typed. The goal of Sewell's type system [149] is to distinguish local and remote channels in order to allow efficient implementation of communication. The language Nomadic Pict [151,152] stands a little aside from this track: it is conceived as a distributed layer over the language Pict, offering high-level global communication while providing at the same time its implementation in terms of low-level located communication.

Most of these distributed π-calculi use a flat space of locations, and a primitive for the migration of processes. In the Dπ-calculus, this primitive is coupled with a purely local communication mechanism, while in the other calculi communication is global. Other models like the distributed JOIN calculus, the SEAL calculus and the AMBIENT calculus, although strongly inspired by the π-calculus, have rather distinct characteristics. Here the space of locations is tree–structured, and the hierarchy is relevant for migration as the units of migration are locations. In these calculi the emphasis is put respectively on implementation, security and administration[33] issues. Regulation on the use of channels and migration is imposed mostly by syntactic means. Type systems have not been a main concern in these models, although they are starting to be actively investigated for the AMBIENT calculus.

We proceed now to a more detailed description of the various models. Most of them are based on the mini π-calculus studied by Milner in [113], or on its asynchronous version [21,94]. For convenience we recall here the syntax of these calculi, referring to Parrow's chapter [127] in this Handbook for more details. The syntax of the mini π-calculus is given by:

$$P, Q ::= 0 \mid x(u).P \mid \overline{x}\langle u\rangle.P \mid P \mid Q \mid (\nu x)P \mid !P.$$

The asynchronous π-calculus is obtained by replacing the output prefix construct $\overline{x}\langle u\rangle.P$ by simple messages $\overline{x}\langle u\rangle$, thus preventing output actions from blocking a process. Syntactic variants may be used: for instance replication may be replaced by general recursion or by *replicated inputs*, that is processes of the form $!x(u).P$. The *polyadic* version may be preferred, where input and output prefixes are generalized to allow transmission of tuples of names, noted $x(\overrightarrow{y})$ and $\overline{x}\langle\overrightarrow{y}\rangle$.

The semantics of these calculi is given by first defining a structural congruence $\equiv$, which is used to predispose terms for communication, and then a set of *reduction rules* describing process execution. This kind of semantics, which factors out structural rearrangements of terms, was first proposed by Milner in [113], taking inspiration from the *Chemical Abstract Machine* (*CHAM*) style of semantics advocated by Berry and Boudol [18].

The $\pi_{1\ell}$-calculus. The first extension of the π-calculus with locations was proposed by Amadio and Prasad in [12], to model an abstraction of the failure semantics of the language FACILE. The language considered in that paper, called the π_ℓ-calculus, is based on a variant of the synchronous polyadic π-calculus with guarded sums; moreover, a *flat* space of locations is introduced, and failure primitives similar to those described in Section 4.2 are provided. We shall not describe the calculus in detail, but only mention its main features, some of which reappear in later calculi.

In the π_ℓ-calculus *locations* are added as a new category of *names*, which can be dynamically created and transmitted in communication; all channels and processes are *located*. The reduction semantics of the π_ℓ-calculus is defined relative to an environment $\mathcal{L}$, describing the assignment of locations to channels. Communication is *global*: it may occur between any two complementary channels irrespective of their location. In general, locations intervene only in the failure semantics of the language. Although all locations are "failure-independent" from each other (since they are unstructured), channels are dependent on the location where they are placed: the failure of a location prevents any further communication on its channels.

The abstract semantics of the calculus is given by defining a *barbed equivalence* (see Parrow's chapter [127] for a definition of this notion) on configurations $(\mathcal{L}, p)$. Although the observation with failures is rather different from the usual observation of the π-calculus, an encoding of the π_ℓ-calculus into the π-calculus is given, which is proved adequate with respect to the barbed equivalences of the two languages.

[33] In the sense of modelling navigation through administrative domains.

The $\pi_{1\ell}$-calculus studied subsequently by Amadio in [9] (this is the version of the calculus we shall refer to in the following) is based on the *asynchronous* rather than the synchronous π-calculus. More precisely, the $\pi_{1\ell}$-calculus builds on the π_1-calculus, a typed asynchronous π-calculus satisfying the *unique receiver* property. This property, which requires each channel to have at most one receiver, was already mentioned in Section 4.2. As we shall see, it is a common requirement in mobile distributed calculi, as it simplifies the implementation of distant communication. It also brings the π-calculus closer to object-oriented programming, where interaction arises when an object calls a method of another uniquely determined object. In the π_1-calculus, the unique receiver property is enforced by means of a simple type system. A barbed congruence is also defined on the π_1-calculus, which is characterized as a variant of the asynchronous bisimulation of [11]. The located $\pi_{1\ell}$-calculus adds to π_1 the model of locations and failures of the π_ℓ-calculus, together with a primitive $\mathsf{spawn}(\ell, p)$ for process migration. The failure primitives of the $\pi_{1\ell}$-calculus have already been discussed in Section 4.2. We shall only add here a few remarks about semantic and typing issues.

The $\mathsf{spawn}(\ell, p)$ construct provides a form of *objective* migration: the process containing the instruction causes process p to move to location ℓ. A distinctive feature of the $\pi_{1\ell}$-calculus is that locations have an associated controller process, called *locality-process*, which records the status of its controlled location and handles messages addressed to the location, in particular all spawn, test and halt requests.[34] In the $\pi_{1\ell}$-calculus communication is global, just like in the π_ℓ-calculus. However the asynchronous setting allows for a more elegant communication rule. In fact, the movement of a message towards its destination is described in two steps: first, the message is exported in the ether (that is, out of its current location[35]), and then it enters the location of the destination process (which is uniquely determined, because of the unique receiver property). Although divided in two steps, the routing of messages is completely transparent: indeed, local communication is implemented in exactly the same way. Note that the movement of messages towards their destination provides another implicit form of migration, besides that given by the $\mathsf{spawn}(\ell, p)$ primitive.

To conclude, let us mention a few results established for the $\pi_{1\ell}$-calculus. The type system proposed in [9] ensures both the unique receiver property and the invariability of the location of channel receivers. Similar properties, as we shall see, will be imposed by syntactic restrictions in the distributed JOIN calculus. Indeed the $\pi_{1\ell}$-calculus shares several ideas with the distributed JOIN calculus, like the localization of receptors and the transparency of distant communication and migration in the absence of failures. Another notable result of [9], which extends the analogous result for the π_ℓ-calculus recalled earlier, is that the $\pi_{1\ell}$-calculus can be encoded into the π_1-calculus, which in turn is sufficiently expressive to encode the full asynchronous π-calculus. Although applied to different models of locations, these results are also similar in spirit to the encoding of the location semantics of [29] into the π-calculus, given by Sangiorgi in [145] (see Section 3.4).

[34] In Section 4.2 we omitted the controller process, for the sake of simplicity. In fact, this controller may be seen as a way of implementing side-conditions on the status of locations in the semantic rules.

[35] What requires this location to be alive, a condition ensured by the controller process. In general, liveness is required for the *source* location of a communication or a migration, but not for the *target* location.

The Dπ*-calculus of Hennessy–Riely.* The Dπ-calculus of Hennessy and Riely is a typed synchronous π-calculus where types are used for resource access control. The calculus offers constructs for process migration and local communication. In its simplest form [92], a *flat* space of locations is introduced. In the richer variant presented in an earlier paper [141], which we may call the Dπ^+-calculus, structured locations and a primitive for moving locations were also considered; in addition the issue of site failure was addressed there. Since the treatment of structured locations and failures in [141] is essentially the same as in the distributed JOIN-calculus described in Section 4.3.2, we shall concentrate here on the simpler version of [92], which is also the one retained by the authors in subsequent papers [93, 142]. At the end of this section we will briefly discuss an asynchronous "receptive" version of the Dπ-calculus, recently studied by Amadio, Boudol and Lhoussaine [10].

The basic entities of the Dπ-calculus are threads (or processes), systems and resources. *Threads* p, q are processes of the polyadic π-calculus, with an additional operator $\mathtt{go\ to}\ \ell.p$ to allow movement of code p towards location ℓ. This operator is similar to the construct $\mathsf{spawn}\,(\ell\,,\,p)$ considered by the authors in [140] and by Amadio in [12,9] (see Section 4.2). *Locations* ℓ, k have a flat structure. They are just a particular category of π-calculus names, and can be created via the operator $(\nu\ell)p$ and transmitted in communication.

Systems P, Q are collections of *located threads* of the form $\ell[p]$, also called *agents*. More precisely, systems are obtained by combining located threads via the static operators of parallel composition and name creation $(\mathtt{new}\ u : T)\ P$, an operator similar to restriction which allows sharing of information among different locations.

The *resources* of the calculus are channels, supporting synchronous communication between agents. Channels are also *located* and communication is restricted to being *purely local*: it is only allowed between agents placed at the same location, using channels that have been allocated at that location. Thus processes wishing to interact first have to move to a common location. This means that in the Dπ-calculus, as in other languages adopting the "go and communicate" paradigm, locations are used as *units of communication*.

The reduction semantics is given directly on systems P. It is defined modulo a structural congruence $\equiv$ that allows co-located threads to be split and reassembled, that is $\ell[p \mid q] \equiv \ell[p] \mid \ell[q]$, and empty agents to be garbage collected, $\ell[0] \equiv 0$. The rules for code movement and communication[36] are the following:

$$\begin{aligned}
\ell[\mathtt{goto}\ k.p] &\rightarrow k[p],\\
\ell[\overline{a}\langle u\rangle.p] \mid \ell[a(x).q] &\rightarrow \ell[p] \mid \ell[q\{^u/_x\}].
\end{aligned}$$

Then, assuming a is a channel declared in location k, remote communication can be "implemented" using migration and local communication as follows:

$$\ell[\mathtt{goto}\ k.\overline{a}\langle u\rangle.p \mid r] \mid k[a(x).q] \rightarrow^* \ell[r] \mid k[p \mid q\{^u/_x\}].$$

Note that in Dπ, the units of migration are simple threads of the form $\ell[\mathtt{goto}\ k.p]$. If such a thread runs in parallel with other threads located at ℓ, these will stay put while

[36] For simplicity we consider here the monadic version of the calculus.

p moves to k. Moreover process p will only be activated when reaching its destination. Hence migration is *weak* in Dπ, that is, only passive code can be moved. This makes a difference with other calculi such as NOMADIC PICT, where migrating agents may contain several threads in parallel, which continue to run until the migration takes effect.

Local communication is the main parting point of the Dπ-calculus with respect to Amadio's $\pi_{1\ell}$-calculus (examined in the previous section), at least as far as the untyped languages are concerned. The *type systems* of the two languages, on the other hand, are rather different. While Amadio's type system is used to ensure uniqueness of receptor channels, types in the Dπ-calculus are designed to control the use of resources. The idea is that agents should not be allowed to use a non-local resource before being granted the appropriate *capability*. The type system of Dπ is based on a new notion of *location type*, which describes the set of resources available to an agent at a particular location. Resources, that is channels, are themselves equipped with capabilities described by types. The types of channels describe the values they can transmit. They may be further refined to prescribe the way in which channels should be used, thus extending the input/output types studied in previous work by Pierce and Sangiorgi [131].

More formally, a location type has the form $\mathtt{loc}\{\overrightarrow{a}:\overrightarrow{A}\}$, where $\overrightarrow{a}$ is a tuple of distinct channel names and $\overrightarrow{A}$ is the associated tuple of capabilities. For instance the location $\ell:\mathtt{loc}\{a:A, b:B\}$ offers two channels a and b with capabilities A and B respectively. A simple *channel type* has the form $\mathtt{chan}\langle T\rangle$, where T is the type of the transmitted value. This value may be a location name, a *local* channel name, or a *remote* channel name marked with its location of creation, written $\ell[a]$. Compound names $\ell[a]$ may only be used as arguments in communication (not as transmission channels), and their types are of the form $\mathtt{loc}[A]$.

The syntax is then enriched with type information for received values and created names. The structural congruence $\equiv$ allows manipulations on these terms, like for instance $\ell[(\mathtt{new}\, a:A)\; p] \equiv (\mathtt{new}\, \ell[a]:\mathtt{loc}[A])\ell[p]$. A new semantic rule for communication of remote channels is added:

$$\ell[\overline{a}\langle k[b]\rangle.p] \mid \ell[a(z[x]:\mathtt{loc}[A]).q] \to \ell[p] \mid \ell[q\{^{k,b}/_{z,x}\}].$$

For process $q\{^{k,b}/_{z,x}\}$ to be well-typed, it is necessary that the types of names k and b match those of z and x. Thus the typing ensures that names are received with the same capabilities they are sent with. In fact, thanks to subtyping, names can be more generally transmitted with a *subset* of the sender's capabilities. This means that a process sending a name can control the use that will be made of it, which is indeed the property the calculus was trying to ensure.

A remote procedure call may be described in the calculus as follows:

$$\ell[(\nu r)\ \mathtt{goto}\ k.\overline{a}\langle \ell[r]\rangle \mid r(X).p] \mid k[a(z[x]:\mathtt{loc}[A]).\ \mathtt{goto}\ z.\overline{x}\langle V\rangle].$$

We shall not elaborate further on the type system of [92] here. Let us just recall that this type system, which is rather sophisticated, satisfies the standard soundness properties of subject reduction (type preservation) and type safety. In later work [93,142] the authors have adapted this type system to guarantee that even a partially typed system (such as can be an open network with untrusted components) will continue to satisfy the same property of safe access to resources.

To conclude, we will mention a result established in [10] by Amadio, Boudol and Lhoussaine, for an *asynchronous* version of the Dπ-calculus. Combining ideas from previous studies by Amadio [9], Boudol [22] and Sangiorgi [146], the authors propose a simple type system for this calculus, ensuring localization of receptors and *receptiveness* of channel names (a kind of input-enabledness). They show that well-typed systems enjoy a local deadlock-freedom property called "message deliverability". This property, which states that every migrating message will find a receiver at its target locality, is particularly desirable in an asynchronous setting, where there is no control on message consumption.

The dpi *calculus of Sewell.* The *dpi*-calculus studied by Sewell in [149] is an asynchronous π-calculus extended with location and migration primitives similar to those of the Dπ-calculus. Its hierarchical model of locations, on the other hand, as well as its global communication discipline, are inspired from those of the distributed JOIN calculus. A few syntactic variations with respect to the Dπ-calculus and the distributed JOIN calculus are worth mentioning:

(1) the *dpi*-calculus adopts a two-level rather than a tree-structured hierarchy of locations;
(2) locations are associated with each elementary subterm of a system, for instance every prefix is located;
(3) when used for a location u, the construct for name creation ($\mathtt{new}\ u : @_\ell\ T$) only allows the creation of u as a *sublocation* of an existing location ℓ.

This means that new locations cannot be created at the topmost level, which is interpreted as a set of physical addresses.

We shall not dwell longer on the description of the language here, but rather sketch the main contribution of the paper [149], which is the study of a type system distinguishing *local* and *global* channels with the aim of localizing communication whenever possible. The idea is that restricting the use of some channels to be local may be useful both for efficient implementation (these channels do not need to be globally unique nor registered with a central name server) and for robustness against external attacks.

The proposed type system generalises that of Pierce and Sangiorgi [131], where channel types are annotated with input and output *capabilities*. Here the input/output types of channels are refined to indicate whether their use is allowed to be global or restricted to be local. Formally, a channel type has the form $\updownarrow_{(i,o)} T$, where the input and output capabilities i and o are taken from the set $\{\mathtt{G}, \mathtt{L}, -\}$. In such a channel's type the input (respectively output) capability can be *global* (G), in which case it can be exercised at any location, *local* (L), in which case it may only be exercised at the channel's location, or absent (-), in which case the channel cannot be used at all for input (respectively output). For instance, if x is a channel located at ℓ of type $\updownarrow_{(\mathtt{L},\mathtt{G})} T$, then x can be used for output anywhere, but for input only at location ℓ. Such a channel may be used, for instance, to send requests to a server located at ℓ. Note that in calculi enjoying the *unique receiver property*, such as the $\pi_{1\ell}$-calculus and the JOIN calculus (see next section), all channels would have such a type.

The type system ensures that channel names are used with the appropriate capabilities and that local capabilities are not sent outside the location they refer to. Subtyping allows for the transmission of subcapabilities. A type preservation result (subject reduction) is proven, and a method for inferring the *most local capabilities* of a channel is proposed. In the light of these results, it can be said that the *dpi*-calculus allows for compile-time optimization of communication, while retaining the expressiveness of global channel communication. The same idea of offering both local and global communication while clearly distinguishing between them, may be found also in the language NOMADIC PICT [152].

4.3.2. *The distributed Join calculus.* The JOIN calculus was introduced by Fournet and Gonthier in [70], followed shortly later by its distributed version [71]. As it was presented by its authors in [70], the JOIN calculus is "an offspring of the π-calculus, in the asynchronous branch of the family".[37] This calculus was specifically designed to facilitate distributed implementations of channel mechanisms. Compared to the asynchronous π-calculus, it offers a more explicit notion of places of interaction. The distributed JOIN calculus [71,69] goes farther still, adding to the original calculus notions of *named location* and *distributed failure*.

We start with a description of the basic JOIN calculus, and then move to its distributed version. We will mainly follow the presentation of [104], which gives a good overview of the two calculi and the initial results that have been established for them.

The main feature of the JOIN calculus is that every input channel is persistent and enclosed in a *definition*, a construct that combines scope restriction and replicated reception.

In the π-calculus, there is no constraint on the use of receptor channels, and it is possible to write:

$$x(y).P \mid x(z).Q \mid \overline{x}a.$$

In practice, if the two receptors $x(y).P$ and $x(z).Q$ are far from each other, one faces a *distributed consensus problem*, since the two processes have to agree about who takes the value. The JOIN calculus tries to avoid this problem by forcing all the receptors on a given channel to be grouped together in a *definition* of the form:

$$\texttt{def}(x\langle y\rangle \rhd P) \wedge (x\langle z\rangle \rhd Q)\ \texttt{in}\ x\langle a\rangle. \tag{18}$$

In a definition $\texttt{def}\ D\,\texttt{in}\ R$ the notation $x\langle y\rangle$ is overloaded to mean reception of y on x in the D-part (declaration[38] part) and emission of y on x in the R-part (the running part, or body of the definition). The def construct delimits the scope of its defined names, that is the receptor channels declared in the D-part. Moreover, the receptors $x\langle y\rangle$ are *permanently defined* and correspond to π-calculus replicated inputs $!x(y).P$. Thus the analogue of definition (18) in the π-calculus is:

$$(\nu x)(!x(y).P \mid !x(z).Q \mid \overline{x}a).$$

An important point to note is that in the JOIN calculus every receptor channel is *statically defined*. Since their names are *bound* by their definition, receptors cannot be renamed:

[37] This branch was to grow quite loaded in subsequent years.

[38] The D-part of a definition construct was also called *definition* in [70,71]. For clarity, we prefer to use "declaration" here.

two different receptors will never be equated.[39] Moreover, receptors are replicated in the definition where they occur, and thus are always available. This gives a precise meaning to the statement that receptors are *localized* in the JOIN calculus (even in its basic version).

As in the asynchronous π-calculus [21,94], communication is *asynchronous* in the JOIN calculus: the syntax only allows simple messages $\overline{x}\langle\overrightarrow{y}\rangle$, without continuation. To provide some synchronization mechanism, the calculus allows *join-patterns* in definitions. These are groups of guards expecting to receive a corresponding group of messages atomically. A *join pattern* J is either the reception of a tuple of names, $x\langle\overrightarrow{v}\rangle$, or a composition $J' \mid J''$ of such receptions. In a join pattern, the names of the receptors are supposed to be all distinct. A *declaration* D can then be an elementary declaration $J \triangleright P$, or a conjunction $D' \wedge D''$ of such declarations. It can also be empty. In each conjunct $J \triangleright P$, the names of the join pattern J are binding for the continuation process P. Join patterns in different conjuncts need not have disjoint receptor names. For instance, in example (18) above, the same receptor x was used in two different conjuncts.

Besides messages and the construct $\mathtt{def}\ D\ \mathtt{in}\ R$, the calculus has constructs for termination and parallel composition. An example of a JOIN process is the following:

$$\mathtt{def}(x\langle y\rangle \mid t\langle u\rangle \triangleright P) \wedge (x\langle z\rangle \mid t\langle v\rangle \triangleright Q)\ \mathtt{in}\ x\langle a\rangle \mid t\langle c\rangle \mid x\langle b\rangle.$$

According to our informal description of the $\mathtt{def}$ construct, this process may reduce to:

$$\mathtt{def}(x\langle y\rangle \mid t\langle u\rangle \triangleright P) \wedge (x\langle z\rangle \mid t\langle v\rangle \triangleright Q)\ \mathtt{in}\ P\{^{a,c}/_{y,u}\} \mid x\langle b\rangle.$$

The notation $P\{^{a,c}/_{y,u}\}$ indicates the simultaneous substitution of a for y and c for u. In this reduction the two messages $x\langle a\rangle$ and $t\langle c\rangle$ have been consumed by the first join-pattern, triggering the process P. Alternatively, the two messages $x\langle a\rangle$ and $t\langle c\rangle$ (or also $x\langle b\rangle$ and $t\langle c\rangle$) could have been used to trigger process Q. Thus join-patterns may be a source of nondeterminism in the language. Note on the other hand that if another message on t had been available, both P and Q could have been triggered in two successive reduction steps.

The general form of the reduction above is given by the following rule:

$$\mathtt{def}\ D \wedge J \triangleright P\ \mathtt{in}\ J\sigma \mid Q \rightarrow \mathtt{def}\ D \wedge J \triangleright P\ \mathtt{in}\ P\sigma \mid Q \tag{19}$$

where the substitution σ is an instantiation of the received values of the pattern J. The semantics of the JOIN calculus is completely determined by the reduction rule (19), together with standard contextual rules and the use of a structural congruence $\equiv$ which allows, for instance, all occurrences of $\mathtt{def}$ to be pushed in front of a term.[40]

The distributed JOIN calculus (DJOIN for short) extends the basic calculus with explicit *locations* and primitives for *migration*. Locations $a, b, \ldots$ are introduced in the syntax by

---

[39] Indeed the JOIN calculus satisfies a property similar to the *output only capability* introduced explicitly in later calculi [108]. This property says that a transmitted channel name can never be used by the recipient as a receptor channel.

[40] In fact, the semantics of the JOIN calculus was originally defined using a *Chemical Abstract Machine* (CHAM) in the style proposed by Berry and Boudol [18]. The CHAM for the JOIN calculus is said to be *reflexive* in that the definition mechanism of the calculus allows new sets of rules to be added to the reduction rules of the machine.

adding a new clause for declarations, the *located declaration* $a[D : P]$, which represents a declaration of input channels *located* at a. The process P is used to "initialize" location a when it becomes active. Just like channels, locations are defined together with their code: in $a[D : P]$, the name a is a declared name, whose scope ranges over the whole def expression which contains it. Unlike a receptor, however, a location can only be declared *once* in a definition. For instance, the syntax does not allow the expression def $a[D : P] \wedge a[D' : Q] \rhd R$ in S.

Receptors are required to be declared in the join-patterns of a *unique location*. For instance, the expression def $a[x\langle y\rangle \rhd P : Q] \wedge b[x\langle z\rangle \rhd Q : R]$ in T is forbidden. On the other hand def $a[x\langle y\rangle \rhd P \wedge x\langle z\rangle \rhd Q : R]$ in S is a correct definition.

Like in the distributed π-calculi examined previously (see Section 4.3.1), in the DJOIN calculus locations are first class objects and can be transmitted in communication. It is worth stressing that locations are *globally declared* in the DJOIN calculus: just like a receptor channel, a location a declared in some definition cannot be declared in any other definition. It can only be used by a process in another definition if it has become known to it (by effect of a communication). Distant communication is transparent in the calculus: a message located at b will automatically move to the unique location a of its receptor.

As the syntax allows nested located declarations, locations are *hierarchically structured* in a tree. At the top of the tree, there is an unlocated root. Since location names are required to be *globally unique*, every proper node in the tree corresponds to a different location a.[41] The location path φa from the root to a node a represents the position of a in the location tree. If location b has position $\varphi a \psi b$, then b is a sublocation of a. The path φa is not necessary to identify a location a (the name a is sufficient, by the uniqueness assumption); it is useful however to express consistency conditions in the semantic rules, like the fact that a location does not migrate to a sublocation.

Migration is expressed by a new process construct go $\langle a, k\rangle$. A process executing the instruction go $\langle a, k\rangle$ causes its *current location* , say b, to become an immediate sublocation of location a; a null message $k\langle\ \rangle$ is emitted on channel k upon termination of the migration. This message notifies that the migration is completed: it can be used, for instance, to render a migration blocking. If the position of a in the location tree is given by φa, the new position of location b will be φab. For the migration to take effect, it is necessary that b does not occur in φa, that is b is not a superlocation of a. This is because in the DJOIN calculus, like in other calculi with structured locations, a migrating location b moves together with all its sublocations (the subtree rooted at b). This means that the location tree may be substantially reconfigurated at each migration. Note that migration is *subjective* (since the go instruction applies to its enclosing location), and that the units of migration are the locations themselves.[42]

Let us consider an example of reduction in the DJOIN calculus. Without entering into technical details, and borrowing ideas from [141] and the *open* JOIN calculus [20], we may assume a runtime environment $\mathcal{L}$ which records the position of locations and declarations at any given time. Then the reduction relation would be defined on configurations $(\mathcal{L}, P)$.

[41] This contrasts with other models with structured locations, where a general location is represented by the whole path from the root to a node and an atomic location may appear in different paths.

[42] In fact, the movement of messages towards their receptor's location could also be seen as a simple form of *process* migration. This point will be discussed further in Section 4.3.5.

For instance, supposing the environment $\mathcal{L}$ defines the position of b to be ψb, and the position of a to be a string φa not containing b, we would have the following reduction:

$$(\mathcal{L}, \texttt{def}\ b[D : P \mid \texttt{go}\ \langle a, k\rangle] \wedge D'\ \texttt{in}\ R) \rightarrow$$
$$(\mathcal{L}', \texttt{def}\ b[D : P \mid k\langle\ \rangle] \wedge D'\ \texttt{in}\ R) \tag{20}$$

where $\mathcal{L}'$ is obtained from $\mathcal{L}$ by replacing all positions of the form $\psi b \psi'$ by $\varphi a b \psi'$. In other words, the effect of this reduction is to move the tree rooted at b to become an immediate subtree of the tree rooted at a.

To summarize, the DJOIN calculus offers *global access* to locations as well as to channels. Once a process knows the name of a channel (respectively a location), it can make direct use of it independently of its location (respectively its position in the location tree). The distributed infrastructure required to implement such global mechanisms is simplified by the assumption of global uniqueness of locations and localization of receptors.

In other calculi, such as Amadio's $\pi_{1\ell}$-calculus considered in the previous section, the localization of receptors was ensured by means of a static type system. The DJOIN calculus makes the choice of handling locations directly, without recourse to a type system.

In calculi with global communication and migration one expects a *network transparency* result in the absence of node failures. As for the location failure calculi examined in Section 4.2, this result also holds for the DJOIN calculus, on the condition that systems are free of *circular migration* (such migration would appear, for instance, in the process of example (20) if location b were replaced by a). More precisely, under this condition two processes of the DJOIN calculus are equivalent if and only if they are equivalent in the basic JOIN calculus once stripped of all location information.

Clearly, one cannot hope for transparency results in the presence of location failures. Like the calculi for location failure discussed in Section 4.2, the DJOIN calculus has been equipped with a way of marking locations (as dead or alive), and constructs for *halting* a location and *testing* its status. As in Amadio's $\pi_{1\ell}$-calculus, it is still possible to send a message or migrate to a dead location. On the other hand emission or migration from a dead location is ruled out. Rules for communication and migration are then conditioned on the *liveness* of the source location. As in other models where locations are structured, a location fails with all its sublocations. Thus the dependence between locations is meaningful for failure as well as for migration.

While the location failure calculi in [12] and [140] are synchronous, Amadio's $\pi_{1\ell}$-calculus and the DJOIN calculus are asynchronous. As observed in [104], in an asynchronous world the assumption of a global knowledge of the liveness of every location may be unrealistic. For a detailed discussion about the treatment of failures in the DJOIN calculus, and more generally about results concerning the calculus, the reader is invited to consult Fournet's thesis [69].

Compared to other calculi described in this section, the JOIN and DJOIN calculi have a relatively developed theory. Standard equivalence relations (barbed equivalences) have been defined, although reasoning techniques still have to be worked out. Encodings of the JOIN calculus into and from the asynchronous π-calculus have been given in [70,71,69]. The paper [20] presents a characterization of the barbed congruence of the JOIN calculus as an asynchronous bisimulation, inspired from [11]. A language based on the JOIN calculus has been implemented, the Join Calculus Language presented in [72].

4.3.3. *The Ambient calculus.* The ambient calculus of Cardelli and Gordon [33] is a process calculus which emphasizes process mobility rather than communication. In fact, it appears that the mobility primitives are sufficient to give the calculus its full expressiveness, and the communication primitives are mostly added for convenience.

An *ambient* is an area delimited by a boundary, where multi-threaded computation takes place. Each ambient has a *name*, a collection of local *processes* (or threads) and a collection of *subambients*. Thus ambients can be arbitrarily nested in a tree structure. Processes may cause the enclosing ambient to move inside and outside other ambients. The form of migration provided by the calculus is therefore *subjective*.[43] Processes may also "open" an ambient at the same level, that is dissolve its boundary causing its contents to spread into the parent ambient. Finally, processes within the same ambient may communicate with each other asynchronously, by releasing and consuming messages in the local ether. The arguments of communication may be names, *capabilities* (that is, permissions to operate on ambients with a given name), or sequences of such arguments. We shall not consider further the communication primitives here.

Ambients move as a whole, bringing together all their threads and subambients. They are therefore *units of migration*. The tree structure of ambients allows for a close control of mobility: an ambient can only enter a sibling ambient and exit a parent ambient. Similarly, a process can only open an ambient which is enclosed in the same parent ambient. This form of mobility, which could be called *proximity mobility*, is well-suited to model navigation through a hierarchy of *administrative domains*, which was the original inspiration for the calculus. The ability to move and open ambients is regulated by capabilities, possibly acquired through communication. These capabilities model the notion of *authorization*, necessary to regulate access to resources and, before that, the crossing of administrative domains.

Formally, an ambient is written $n[P]$, where n is the name of the ambient and P its contents, described by the following syntax, where M represents one of the capabilities $\texttt{in}\, n$, $\texttt{out}\, n$, $\texttt{open}\, n$ (and the communication primitives are omitted):

$$P, Q ::= 0 \mid M.P \mid P \mid Q \mid (\nu n)P \mid !P \mid n[P].$$

In $n[P]$, it is understood that P is actively running, even when the surrounding ambient is moving. This fact is expressed by a semantic rule saying that any reduction of P can also take place within n:

$$P \to Q \Rightarrow n[P] \to n[Q].$$

The *mobility constructs* are $\texttt{in}\, n.P$ (to enter an ambient named n), $\texttt{out}\, n.P$ (to exit an ambient named n), and $\texttt{open}\, n.P$ (to open an ambient named n). Their semantics is given

[43] Indeed the terminology "subjective/objective" for migration was introduced by Cardelli and Gordon.

by the following reduction rules:

$$\begin{aligned} n[\mathtt{in}\, m.P \mid Q] \mid m[R] &\to m[n[P \mid Q] \mid R], \\ m[n[\mathtt{out}\, m.P \mid Q] \mid R] &\to n[P \mid Q] \mid m[R], \\ \mathtt{open}\, m.P \mid m[Q] &\to P \mid Q. \end{aligned}$$

Clearly, the in and out operations are duals of each other. The open operation, on the other hand, has a more definitive and disruptive effect. However, its use is regulated by a capability which must have been given out by the ambient to which it is applied.[44]

Ambient *names* can be created, and used to name multiple ambients. Ambients with the same name may coexist both at the same level and at different levels in the hierarchy. In case a mobility instruction matches several ambients, one of them is chosen nondeterministically. Ambients with the same name have separate identities, and an empty ambient is still observable through its name. This is expressed by the following inequalities (where $\equiv$ stands for semantic equality):

$$\begin{aligned} n[P] \mid n[Q] &\not\equiv n[P \mid Q], \\ n[\mathbf{0}] &\not\equiv \mathbf{0}. \end{aligned}$$

An interesting equality which holds for ambients is the so-called *perfect firewall equation*:

$$(\nu n)n[P] \equiv \mathbf{0} \quad \text{if } n \notin fn(P).$$

This law says that a process enclosed in an ambient whose name is not known by the environment (and cannot be used by the process itself) is equivalent to the terminated process.

In many ways, an ambient resembles a *named location*. Like a location, it supports internal communication and can receive ambients and messages (represented themselves by ambients) from other places. However, ambients have a more general meaning. For instance, they can move across places while carrying their running contents with them. In this respect they behave rather like *mobile agents*. Moreover, as we have seen, the handling of names is rather liberal: the same name can be borne by several ambients within the same scope. In this way, ambients may be used to model *service providers*, of which there may be more than one, as long as the same service is granted.

As mentioned earlier, one of the initial motivations for the AMBIENT calculus was to model firewall crossing in large networks. In fact, the calculus is rather ambitious, and intends to model a variety of distributed programming concepts such as channels, network nodes, packets, services and software agents. Some examples may be found in [33]. In the same work, it is also shown how the calculus can be used to encode languages such as the asynchronous π-calculus and some λ-calculi. Current work on the AMBIENT calculus is focussed on type systems for controlling the type of values transmitted within ambients [32] and controlling mobility. In [103], a modified calculus called Safe Ambients is

[44] Note that the open operation provides a form of *objective* migration, of the contents of an ambient into its father ambient.

proposed, with a type system for controlling mobility and preventing so-called "dangerous interferences". Modal logics accounting for the spatial structure of ambients via the use of "spatial" modalities are also under investigation [34].

4.3.4. *The Seal calculus.* Like the AMBIENT calculus, the SEAL calculus of Castagna and Vitek [161] is designed to model distributed computing over large scale open networks. As some of the previous formalisms examined in this section, the SEAL calculus takes the stand of exposing the network topology and handing over control of localities to the system programmer. However, compared to other calculi, the SEAL calculus focusses more particularly on the *security issues* raised by the new paradigm of network programming: it also aims to provide programmers with protection mechanisms for their resources and against malevolent hosts. Among the calculi we have seen, only the Dπ-calculus of Hennessy and Riely and the AMBIENT calculus had a strong concern in security.

The SEAL calculus may be concisely described as a polyadic *synchronous* π-calculus with hierarchical mobile locations and resource access control. The main entities of the calculus are *processes*, *locations* and *resources*.

A *process* P is a term of the mini synchronous π-calculus, where channels are tagged with the relative location to which they belong. This will be explained more precisely below, when discussing the use of channels for supporting communication and mobility. Moreover a process can also be a named location containing a process, that is a *seal* $n[P]$.

Locations are called *seals* in the calculus. A seal is written $n[P]$, where n is a name (not necessarily unique) and P is the process running inside the seal. Like channel names, seal names can be transmitted in communication and restricted upon. Since a seal encapsulates a process, and a seal is itself a process, it is clear that seals are hierarchically structured. Processes which are not contained in a seal are assumed to be placed at the root's location, thus all processes are – explicitly or implicitly – located at some seal. In $n[P]$, the seals contained in P are called the *children* of n, while n is called their *parent*. Seals are also the migration units of the calculus. The migration of a seal is ordered by the parent seal, and a seal can only be moved one level up or one level down the hierarchy. This is achieved, as we will see, by transmitting the name of the seal to be moved.

As in the Dπ-calculus, the only *resources* of the SEAL calculus are channels. Channels are used, as usual, for the communication of names. The SEAL calculus provides both *local communication*, between processes running in the same seal, and *linear proximity communication*, between processes placed in the parent-child relationship. No direct communication is possible, for instance, between sibling seals. For distant communication, the routing of messages in the hierarchy of seals must be explicitly programmed.

Channels are located in seals (inherited from the enclosing process),[45] and processes have access to local channels as well as to their parent's and children's channels. To enable processes to refer to proximity channels, channel names are annotated with tags referring to their relative place in the hierarchy. Formally, *tags* are defined by $\eta ::= \star \mid \uparrow \mid n$, denoting respectively the current seal, the parent seal, and a child seal bearing name n. Then input and output actions for transmitting channel names are given by:

$$a ::= \overline{x}^{\eta}(\overrightarrow{y}) \mid x^{\eta}(\overrightarrow{y}).$$

[45] Thus channels may be located in different seals.

Two co-located processes communicate via a local channel. We have for instance:

$$\overline{x}^{\star}(\overrightarrow{z}).P \mid x^{\star}(\overrightarrow{y}).Q \to P \mid Q\{\overrightarrow{z}/\overrightarrow{y}\}.$$

For immediate upward or downward communication, a channel of either of the two partners may be chosen. Therefore the communication will always involve a channel tagged $\star$ and a complementary channel tagged with $\uparrow$ or some name n. For instance we have:

$$n[\overline{x}^{\uparrow}(\overrightarrow{z}).P] \mid x^{\star}(\overrightarrow{y}).Q \to n[P] \mid Q\{\overrightarrow{z}/\overrightarrow{y}\}. \tag{21}$$

In fact, this is only approximately true. Because of the security orientation of the calculus, a further ingredient will be required for non-local communication to take place. The idea is that allowing non-local processes (like $\overline{x}^{\uparrow}(\overrightarrow{z}).P$ in this example) to use local channels (like channel x of process $x^{\star}(\overrightarrow{y}).Q$) constitutes a threat to security. Indeed the first process may have migrated to location n from some untrusted location. Then it should not be allowed to access the local channel x, unless it is given explicit permission to do so.

To control inter-seal communication, the calculus proposes a protection mechanism called *portal*. The idea is the following: for seal A to be able to use a channel x of seal B, seal B must first *open* a portal for A at x. A portal is a *linear access permission* for a channel: it can only be used once. Formally, a new construct $\mathtt{open}_\eta\ x.P$ (respectively $\mathtt{open}_\eta\ \overline{x}.P$) is introduced: it denotes a process which opens a portal for seal η allowing it to perform an input (respectively output) on local channel x, and then continues like P. The correct reduction replacing (21) is then:

$$n[\overline{x}^{\uparrow}(\overrightarrow{z}).P] \mid x^{\star}(\overrightarrow{y}).Q \mid \mathtt{open}_n\ \overline{x}.0 \to n[P] \mid Q\{\overrightarrow{z}/\overrightarrow{y}\} \mid 0. \tag{22}$$

The last important point to examine is *mobility of seals*. As mentioned earlier, seals may be transmitted over channels, and that is the way mobility is modelled in the calculus. The prefixing actions for exchanging seal names will be written $\overline{x}^{\eta}\{y\}$ and $x^{\eta}\{\overrightarrow{y}\}$, to distinguish them from the actions used for transmission of channel names. Note that an output action can only send a single seal name, for a reason that will soon be made clear. The complete syntax for prefixing actions is then:

$$a ::= \overline{x}^{\eta}(\overrightarrow{y}) \mid x^{\eta}(\overrightarrow{y}) \mid \overline{x}^{\eta}\{y\} \mid x^{\eta}\{\overrightarrow{y}\}.$$

A seal can only be moved by a process contained in the parent seal. A seal *move action*, represented by an output $\overline{x}^{\eta}\{y\}$, requires that a seal with name y be present in the current seal, otherwise it blocks. If several seals match the name, then one of them is chosen nondeterministically. Note the difference with the ambient calculus: migration is *objective* here (a process causes a seal at the same level to move), while it was subjective in the ambient calculus (a process was causing the whole enclosing ambient to move).

The synchronization of a move action $\overline{x}^{\eta}\{y\}$ with a receive action $x^{\eta}\{\overrightarrow{z}\}$ where $\overrightarrow{z} = z_1, \ldots, z_k$ has the effect of creating k copies of seal y under the names $z_1, \ldots, z_k$. The seal y will disappear from the location of the sender and the copies $z_1, \ldots, z_k$ will appear at the location of the receptor. Thus seals may be both *renamed* and *duplicated* during a migration. This may be useful respectively for protection and fault tolerance.

Regarded as a tree rewriting operation on the hierarchy of seals, a move $\overline{x}^{\eta}\{y\}$ has the effect of disconnecting a subtree rooted at seal y and grafting it either onto the parent of y, or onto one of y's children, or back onto y itself. Each of these moves can be accompanied by renaming and introduce copies. For instance, suppose $P = \overline{x}^{\uparrow}\{y\}.P'$ and $R = x^{\star}\{z\}.R'$. Then we have the following reduction:

$$R \mid n[P \mid m[Q] \mid y[S]] \rightarrow R' \mid z[S] \mid n[P' \mid m[Q]] \tag{23}$$

where P causes the seal named y to move into its parent seal (the root seal in this case) renaming it with z.[46] This is an example of upward migration. Note that the location of the transmission channel is not important: the same reduction would have taken place if $P = \overline{x}^{\star}\{y\}.P'$ and $R = x^{n}\{z\}.R'$. The direction of the migration is always from the sender to the receptor seal. The following is an example of downward migration, assuming $P = \overline{x}^{m}\{y\}.P'$ and $Q = x^{\star}\{z\}.Q'$:

$$P \mid m[Q] \mid y[S] \rightarrow P' \mid m[Q' \mid z[S]]. \tag{24}$$

The two reductions (23) and (24) show that the transmission of a seal name is really a `spawn` operation in disguise. Its semantics is rather different from that of channel transmission. The protection mechanism by means of portals, on the other hand, is the same as for channel names, thus the formulation of migration as a form of communication is convenient.

Note that as regards protection, the direction (upward or downward) in the hierarchy of seals makes no difference: portals are used in the same way to protect a seal from its parent seal or from a subseal. For migration instead this direction is important, since a (process in the) parent can instruct a child to move but not vice versa.

The SEAL calculus was designed to provide protection against attacks from visitor hosts as well as attacks from the environment. A detailed discussion about the protection mechanisms of the calculus against various kinds of typical attacks (like disclosure of information, denial of service and Trojan horses) may be found in [160]. Some programming examples are presented in [161]. An implementation of the calculus is also under development.

We conclude with a brief comparison with the AMBIENT calculus, which was one of the sources of inspiration for the SEAL calculus. We have already remarked that seal mobility is objective while ambient mobility is subjective. Moreover, mobility of ambients is controlled by capabilities which can be transmitted and used more than once. By contrast, the SEAL calculus gives full control of mobility to the environment; moreover, this control is fine-grained since permissions must be renewed for each non-local access. Furthermore, the SEAL calculus does not allow boundaries to be dissolved: the `open` operation of the AMBIENT calculus is considered dangerous in that it allows a lower-level agent, which should be subject to some control, to step at the local level. Finally, while the AMBIENT calculus focusses on migration rather than communication, the SEAL calculus gives priority

[46] Again this is not completely exact, and a portal is needed to authorize the use of non-local channels, just as for channel name transmission.

to communication, and indeed implements migration via communication. Like the AMBIENT calculus, the SEAL calculus also satisfies the *perfect firewall equation* $(\nu x)x[P] \equiv 0$, which in this case holds without requirements on P. This equation states that an arbitrary process can be completely isolated from the rest of the system, i.e., prevented from any communication with it. It is a useful security property as it guarantees that a seal trapped in a firewall cannot disclose any information.

4.3.5. *Nomadic Pict.* The language NOMADIC PICT [152] of Sewell, Wojciechowski and Pierce is an extension of PICT, a strongly-typed high-level concurrent language based on the asynchronous π-calculus [114,94,21], which was developed by Pierce and Turner [158, 132]. PICT supports a wide range of high-level constructs including data structures, higher-order functional programming, concurrent control structures and objects. The vocation of NOMADIC PICT is to "complete" PICT to make it suitable for wide-area distributed programming. In particular, NOMADIC PICT accommodates notions of location, agent and migration. The language is layered in two *levels of abstraction*:

- A *low level* that makes distribution and network communication clear; at this level one has *location-dependent* constructs, like agent migration and asynchronous communication between located agents.
- A *high level* where the distributed infrastructure can be ignored; this level provides *location-independent* constructs to allow mobile agents to communicate without having to explicitly track each other's movements.

In this brief description we shall not make a sharp distinction between the language NOMADIC PICT itself [152] and the underlying calculus, the Nomadic π-calculus studied in [151]. We shall essentially present the calculus, but using mostly the notation of [152].

The main entities of the language are sites, agents and channels. *Sites* may be thought of as physical machines: they are unstructured and each site has a unique name s. *Agents* are units of executing computation that can migrate between sites. At any moment, an agent is located at a particular site s. Every agent has a name a, which is unique in the whole network. *Channels* are the support for communication: they are used for asynchronous communication both within an agent and between agents (possibly located at different sites). Channel names c can be created dynamically, as in the π-calculus. New names of agents can also be created dynamically, while names of sites are fixed once and for all. All names can be transmitted in communication.

Low-level language. The syntax is rigidly structured into the three categories of sites, agents and processes. *Sites* are assumed to be given: they are the set of machines, or rather an instantiation of the runtime system on these machines. Sites contain a number of *agents* running in parallel: the initial assignment of agents to sites is also supposed to be known. Finally, *processes* always appear as the body of some agent. They are built with the usual constructs of the asynchronous π-calculus (with typed channels), together with two new primitives for agent creation and migration:

$\texttt{agent}\ a = P\ \texttt{in}\ Q$	agent creation
$\texttt{migrate to}\ s.P$	agent migration.

The effect of the first instruction in the body of some agent b is to create a new agent $a[P]$ at the same site, with name a and body P. After the creation, both P and Q start executing in parallel with the rest of agent b. The new name a is unique and is binding in both P and Q.

The execution of the second instruction in an agent a makes the whole agent move to site s. After the migration, P starts executing at site s in parallel with the rest of agent a.

More formally, assuming an environment $\mathcal{L}$ recording the sites of agents, with $\mathcal{L}(a) = s_a$, the semantics of these constructs is given by:

$$\big(\mathcal{L}, a[(\texttt{agent}\ b = P\ \texttt{in}\ Q) \mid R]\big) \rightarrow \big(\mathcal{L}, (\nu b@s_a)(b[P] \mid a[Q \mid R])\big),$$
$$\big(\mathcal{L}, a[(\texttt{migrate to}\ s.P) \mid Q]\big) \rightarrow \big(\mathcal{L}\{a \mapsto s\}, a[P \mid Q]\big).$$

The notation $(\nu b@s_a)$ represents creation of a new name b at site s_a, while $\mathcal{L}\{a \mapsto s\}$ is environment update. The semantics of the construct $(\nu a@s)$ is given by the rule:

$$\frac{(\mathcal{L}\{a \mapsto s\}, P) \rightarrow (\mathcal{L}'\{a \mapsto s'\}, P')}{(\mathcal{L}, (\nu a@s)P) \rightarrow (\mathcal{L}', (\nu a@s')P')}.$$

Note that migration is *subjective*, as in the AMBIENT and DJOIN calculi. It is is also *asynchronous*, since processes inside the moving agent a – other than the prefixed process P – continue to run until the migration takes place. Note also that the `migrate` instruction is the only means provided by the syntax to explicitly specify the site of an agent.

Direct communication on channels is only supported between processes running in the same agent. For interaction between agents, the low-level language provides two distinct constructs for *local* and *distant* communication:

$$\left.\begin{array}{ll} \langle a\rangle c!v & \text{send to local agent} \\ \langle a@s\rangle c!v & \text{send to distant agent} \end{array}\right\} \quad \text{send to located agent.}$$

The construct $\langle a\rangle c!v$ in an agent b attempts to deliver the message $c!v$ ("value" v on channel c) to agent a on the same site as b. The construct $\langle a@s\rangle c!v$, instead, tries to deliver the message $c!v$ to agent a on site s. Both operations fail silently if agent a is not found on the expected site. The implicit assumption is that at the low level agents must know each other's location in order to communicate successfully. In fact, these two constructs can be derived from another instruction available in the low-level language:

$$\texttt{if local}\ \langle a\rangle c!v\ \texttt{then}\ P\ \texttt{else}\ Q \quad \text{test-and-send to local agent.}$$

This instruction in the body of an agent b has two possible effects: if agent a is present on the same site, the message $c!v$ is delivered to a and process P is spawned in parallel with the rest of b ; otherwise the message is discarded and Q is spawned in parallel with the rest of b. This construct, which reintroduces some implicit synchrony in the language, is mainly used for implementing the global communication mechanism of the high-level language. Formally, its semantics is given by the rule:

$$(\mathcal{L}, a[\texttt{if local } \langle b\rangle c!v \texttt{ then } P \texttt{ else } Q]) \to \begin{cases} (\mathcal{L}, b[c!v] \mid a[P]) & \text{if } \mathcal{L}(a) = \mathcal{L}(b), \\ (\mathcal{L}, a[Q]) & \text{otherwise,} \end{cases}$$

where again a structural congruence is used for splitting and grouping parts of an agent: $b[R] \mid b[S] \equiv b[R \mid S]$. The test-and-send is the only *primitive* communication construct of the low-level language, as the previous two constructs can be encoded as follows:

$$\langle a\rangle c!v = \texttt{if local } \langle a\rangle c!v \texttt{ then } nil \texttt{ else } nil$$

$$\begin{aligned}\langle a@s\rangle c!v =\ & \texttt{if local } \langle a\rangle c!v \texttt{ then } nil \texttt{ else} \\ & \texttt{agent } b = \texttt{migrate to } s.\langle a\rangle c!v \texttt{ in } nil.\end{aligned}$$

High-level language. The high-level language is obtained by adding a single location-independent construct to the low-level language, namely one for *global communication*:

$\langle a@?\rangle c!v$ send to agent anywhere.

The effect of such a global output is to deliver the message $c!v$ to agent a, no matter what its current location and what further migrations it can undergo.

The idea is that at the high level an agent must be able to access all other agents uniformly, without having to locate them explicitly. It is important to note, on the other hand, that the global communication construct is *not primitive* in the language, and has to be implemented using the low-level communication primitives.

The low-level primitives remain available in the high-level language, for interacting with agents whose locations are predictable. This possibility of choice is interesting since global communication is potentially expensive to implement.

As can be seen, the model of NOMADIC PICT is very simple. The units of migration are agents, while sites are simply units of distribution.

The calculus LLinda. Linda [35] is a coordination language where processes share a global environment called a *tuple space*, and interact by fetching and releasing tuples asynchronously in this space. Distributed versions of Linda have been proposed, where multiple tuple spaces are introduced and remote operations on them are provided. A variant considered in [36] also allows nested tuple spaces. The calculus LLinda of De Nicola, Ferrari and Pugliese [120] combines a core distributed Linda with a set of operators borrowed from Milner's calculus CCS. Both tuple spaces and operations over tuples are located. A primitive for spawning a process at a remote tuple space is added (generalizing an existing operation on tuples). The located tuple spaces of LLinda have a flat structure, and can be accessed from other tuple spaces without restrictions. Some typical examples (like a remote procedure call and a remote server) are provided to illustrate the use of the language for remote programming. An evolution of LLinda, the calculus KLAIM studied subsequently by the same authors [121], is also equipped with a type system for checking access rights of mobile agents. A prototype language based on KLAIM has been implemented.

Discussion. To summarize, the models reviewed in this section can be broadly divided in two classes as regards their treatment of locations:

- Models based on a *flat space of locations*, where the migrating entities are processes or agents. These are essentially the distributed π-calculi, excluding the *dpi*-calculus. Migration is mostly weak in these calculi; it is only strong (and asynchronous) in NOMADIC PICT. Research has focussed on type systems guaranteeing a correct or safe use of names, be they locations or channels. In some cases, as in the $D\pi$-calculus and in its asynchronous version, locations are also *units of communication*. In the $\pi_{1\ell}$-calculus, the only calculus in this category to account for failure, locations are also *units of failure*.
- Models based on *hierarchical locations*, where the migrating entities are locations themselves. The representation of migration is more sophisticated here, and better suited for handling failure and security issues. In all these models locations are *units of migration* and migration is strong and asynchronous. In the DJOIN calculus,[47] locations are also *units of failure*. In the AMBIENT calculus and the SEAL calculus the issue of failure is neglected (or judged inappropriate for large networks), and priority is given to security questions. In the SEAL calculus locations are clearly treated as *units of security*. In the AMBIENT calculus locations are *units of communication* and it is more debatable whether they can be seen as units of security. In some of these models, a form of process migration is also provided. This is the case for instance in the $D\pi^+$-calculus. In the AMBIENT calculus, process migration is introduced by the `open` construct, which allows processes at one location to diffuse into their father's location. In the DJOIN calculus and in the *dpi*-calculus, message movement towards a distant destination may be seen as an elementary form of process migration.

All these languages provide an explicit migration primitive (which in the SEAL calculus is rendered as communication of location names). In all models locations have names which can be transmitted in communication. In all cases except for NOMADIC PICT, where they are fixed, locations may also be dynamically created; the *dpi*-calculus forbids creation of new locations at the top level, since this is meant to represent a fixed architecture of physical machines.

The main sources for this survey on distributed calculi for mobility were the original papers cited in the various sections. The references [30] and [148] were also of help.

5. Conclusions

In this chapter we have reviewed two main approaches to process algebras with locations, which we called *abstract* and *concrete*. These approaches respond to different concerns: in the first case locations are introduced to distinguish parallel components, as the basis for particular noninterleaving semantics. Locations are treated here as "units of sequential computation", although parallel subcomputations may be spawned at sublocations as the execution proceeds. In the second approach locations are used to support a new style

[47] As in other calculi inspired from it, like the *dpi*-calculus and the $D\pi^+$-calculus.

of *network-aware programming*. In general, locations contain multi-threaded computation here, and their space may vary as a consequence of location creation and migration.

In the first approach transitions are indexed by the location where they occur, while in the second approach there is a variety of possible observations of locations, ranging from indexed transitions (as in Murphy and Corradini's model) to more implicit observations. Locations may even become completely unobservable under certain conditions (for instance in the DJOIN and $\pi_{1\ell}$ calculi in the absence of failures).

Since locations are directly reflected in the semantics in the abstract approach and not in the concrete approach (if we except Murphy and Corradini's work), the reader might be puzzled by our terminology "abstract vs concrete": this refers to the fact that in the "abstract" case locations are observed modulo a relabelling of their names, while in the "concrete" case the names of locations, even though their observation is more restricted, have an absolute meaning and cannot be abstracted upon.

Although locations do not have the same role in the two approaches, the *models* used to represent them, be they flat or structured, are not substantially different. Within the abstract approaches, techniques have been developed for reasoning about traditional (CCS-like) process calculi enriched with locations, and a body of results have been obtained. As regards the integration of locations and mobility, on the other hand, interesting calculi have been proposed, with various foci, but the theory has not yet stabilized. The question is then: to which extent can techniques and results developed for a simple process calculus (CCS) in the first approach be extended to the more sophisticated calculi for mobile computation addressed by the concrete approach?

In exposing the abstract approach, we have concentrated on two different presentations of location semantics, based respectively on static and dynamic notions of location. The relation between the two formulations has been examined in full detail. The connection with other work on noninterleaving semantics for CCS has also been discussed in some depth. Equational and logical characterizations for the location equivalence and preorder were also mentioned. Very little has been said, on the other hand, about questions of decidability and verification for location equivalence and other distributed equivalences. We shall briefly touch upon such issues here. As regards *decidability*, a first decision procedure for distributed bisimulation, based on a tableau method, was presented by Christensen in [42], for a recursive fragment of CCS with parallelism known as BPP (Basic Parallel Processes, see Section 3.2). The correctness of this procedure was based on a cancellation property of the form $p \mid q \sim_d p \mid r \Rightarrow q \sim_d r$, which does not hold for other noninterleaving equivalences such as location equivalence or causal bisimulation. Building on work by Christensen et al. [43,44], an alternative method for the decision of such equivalences over BPP was proposed by Kiehn and Hennessy in [101]. An extension of this decision procedure to the weak versions of the equivalences over representative subclasses of BPP is also offered in the same work.

Another concern was to find *compact representations* for the location transition system and efficient verification procedures for location equivalence. In its original dynamic formulation of [29], the location semantics associates an infinitely branching transition system with any non-trivial CCS process. As soon as recursion is involved, this transition system also becomes infinite state. We thus lose the correspondence between regular behaviours and finite transition system representations that we have in the standard CCS semantics.

Now, while infinite-branching can be easily removed by choosing canonical locations at each step (as in the numbered transition system of Yankelevich [171] or the occurrence transition system considered in Section 2.6), the infinite progression is really intrinsic to the dynamic location semantics (as to any other *history dependent* semantics).

In its dynamic formulation, location equivalence involves the creation of new locations at each step. It thus requires the use of specific algorithms. In particular these algorithms work only on pairs of processes – following the so-called *on the fly* verification technique – and do not allow for the definition of minimal representatives for equivalence classes of processes. To overcome this problem, Montanari and Pistore have proposed in [115] a model called *location automata*: these are enriched transition systems where the information on locations appears on the labels, and is used in such a way that location equivalence on terms can be reduced to the standard bisimulation on the corresponding location automata. Moreover, location automata provide *finite representations* for all finitary CCS processes (those where no parallelism appears under recursion). This has the pleasant consequence that standard algorithms can be applied for the verification of location equivalence, with a worst case complexity which is comparable to that of verifying ordinary bisimulation, and *minimal representatives* for processes can be defined.

Recently, Bravetti and Gorrieri have proposed a technique that improves on Montanari and Pistore's model in that it offers a *compositional* interpretation of processes into a variation of location automata [31]. Although this technique has been mainly applied to ST-bisimulation, it may reveal useful also for location bisimulation.

As regards the static definition of location equivalence, a simple algorithm was proposed by Aceto in [5] for checking this equivalence on a subset of CCS (nets of automata). This line of investigation was not further pursued, however, and no experimentation was conducted on the static formulation of the equivalence for full CCS.

Concerning the concrete approach to localities, and particularly the new formalisms for combining locations and mobility, our account in Section 4.3 was admittedly very focussed and far from complete. For the sake of uniformity, we have concentrated on a family of calculi derived from the π-calculus. However, mobility of agents and processes is a vast and rapidly evolving field, and several other languages for mobile agents have been proposed in the literature in the last decade. To mention but a few, proposals like Telescript, Obliq, Oz, Kali Scheme, Odyssey, Voyager and Agent Tcl all bear some relation with the calculi examined here.

Aspects of distribution have also been studied in other settings; most of the languages cited in the previous paragraph are based on the object-oriented paradigm; for a treatment of locations in the context of concurrent constraint programming see for instance [139].

Acknowledgements

I would like to thank the anonymous referee for helpful comments on a previous version of this chapter. Many thanks also to Luca Aceto, Roberto Amadio, Gérard Boudol, Rob van Glabbeek, Matthew Hennessy, Astrid Kiehn, Alban Ponse and Davide Sangiorgi for giving me useful feedbacks on various parts of this material. This work was partly supported by the European Working Group CONFER2 and by the French RNRT Project MARVEL.

References

[1] L. Aceto, *On Relating Concurrency and Nondeterminism*, Proceedings MFPS-91, Lecture Notes in Comput. Sci. 598 (1991).

[2] L. Aceto, *Action-refinement in Process Algebras*, Distinguished Dissertations in Computer Science, Cambridge University Press (1992).

[3] L. Aceto, *History preserving, causal and mixed-ordering equivalence over stable event structures (Note)*, Fundamenta Informaticae **17** (4) (1992), 319–331.

[4] L. Aceto, *Relating distributed, temporal and causal observations of simple processes*, Fundamenta Informaticae **17** (4) (1992), 369–397.

[5] L. Aceto, *A static view of localities*, Formal Aspects of Computing **6** (1994), 201–222.

[6] L. Aceto, W.J. Fokkink and C. Verhoef, *Structural operational semantics*, Handbook of Process Algebra, J.A. Bergstra, A. Ponse and S.A. Smolka, eds, Elsevier, Amsterdam (2001), 197–292.

[7] L. Aceto and M. Hennessy, *Towards action-refinement in process algebras*, Inform. and Comput. **103** (2) (1993), 204–269.

[8] L. Aceto and M. Hennessy, *Adding action refinement to a finite process algebra*, Inform. and Comput. **115** (2) (1994), 179–247.

[9] R. Amadio, *An asynchronous model of locality, failure, and process mobility*, Proceedings COORDINATION-97, Lecture Notes in Comput. Sci. 1282 (1997).

[10] R. Amadio, G. Boudol and C. Lhoussaine, *The receptive distributed Pi-calculus*, Proceedings FST-TCS-99, Lecture Notes in Comput. Sci. 1738 (1999).

[11] R. Amadio, I. Castellani and D. Sangiorgi, *On bisimulations for the asynchronous π-calculus*, Theoret. Comput. Sci. **195** (1998), 291–324.

[12] R. Amadio and S. Prasad, *Localities and failures*, Proceedings FST-TCS-94, Lecture Notes in Comput. Sci. 880 (1994), 205–216.

[13] E. Badouel and P. Darondeau, *Structural operational specifications and trace automata*, Proceedings CONCUR-92, Lecture Notes in Comput. Sci. 630 (1992).

[14] J.C.M. Baeten and W.P. Weijland, *Process Algebra*, Cambridge Tracts in Theoret. Comput. Sci. 18, Cambridge University Press (1990).

[15] M. Bednarczyk, *Categories of asynchronous systems*, Ph.D. Thesis, University of Sussex (1988).

[16] M. Bednarczyk, *Hereditary history-preserving bisimulation, or what is the power of the future perfect in program logics*, Research Report, Institute of Computer Science, Polish Academy of Sciences, Gdansk (1991).

[17] J. Bergstra, A. Ponse and S. Smolka, eds, *Handbook of Process Algebra*, Elsevier, Amsterdam (2001).

[18] G. Berry and G. Boudol, *The chemical abstract machine*, Theoret. Comput. Sci. **96** (1992), 217–248.

[19] E. Best, R. Devillers, A. Kiehn and L. Pomello, *Concurrent bisimulations in Petri nets*, Acta Inform. **28** (3) (1991) 231–264.

[20] M. Boreale, C. Fournet and C. Lanve, *Bisimulations in the Join-calculus*, Proceedings PROCOMET-98, 1998.

[21] G. Boudol, *Asynchrony and the π-calculus*, Research Report 1702, INRIA, Sophia-Antipolis (1992).

[22] G. Boudol, *The pi-calculus in direct style*, Proceedings POPL-97, 1997.

[23] G. Boudol and I. Castellani, *On the semantics of concurrency: Partial orders and transition systems*, Proceedings TAPSOFT-87, Lecture Notes in Comput. Sci. 249 (1987), 123–137.

[24] G. Boudol and I. Castellani, *Concurrency and atomicity*, Theoret. Comput. Sci. **59** (1988), 25–84.

[25] G. Boudol and I. Castellani, *Permutation of transitions: An event structure semantics for CCS and SCCS*, Proceedings REX School/Workshop on Linear Time, Branching Time and Partial Order in Logics and Models for Concurrency, Noordwijkerhout, Lecture Notes in Comput. Sci. 354 (1988).

[26] G. Boudol and I. Castellani, *Flow models of distributed computations: Event structures and nets*, Report 1482, INRIA, 1991. Previous version by G. Boudol in: Proceedings LITP Spring School, La Roche-Posay, Lecture Notes in Comput. Sci. 469 (1990).

[27] G. Boudol and I. Castellani, *Flow models of distributed computations: Three equivalent semantics for CCS*, Inform. and Comput. **114** (2) (1994), 247–314.

[28] G. Boudol, I. Castellani, M. Hennessy and A. Kiehn, *Observing localities*, Theoret. Comput. Sci. 114 (1993), 31–61.

[29] G. Boudol, I. Castellani, M. Hennessy and A. Kiehn, *A theory of processes with localities*, Formal Aspects of Computing **6** (1994), 165–200.

[30] G. Boudol, F. Germain and M. Lacoste, *Projet MARVEL: Document d'analyse des langages et modèles de la mobilité*, Document of the French RNRT Project MARVEL (1999).

[31] M. Bravetti and R. Gorrieri, *Deciding and axiomatizing ST-bisimulation for a process algebra with recursion and action refinement*, Proceedings EXPRESS-99, ENTCS 27 (1999).

[32] L. Cardelli, G. Ghelli and A. D. Gordon, *Mobility types for mobile ambients*, Proceedings ICALP-98, Lecture Notes in Comput. Sci. 1644 (1998), 230–239.

[33] L. Cardelli and A. D. Gordon, *Mobile ambients*, Proceedings FoSSaCS-98, Lecture Notes in Comput. Sci. 1378 (1998).

[34] L. Cardelli and A.D. Gordon, *Anytime, anywhere: Modal logics for mobile ambients*, Proceedings POPL-2000, ACM Press (2000).

[35] N. Carriero and D. Gelernter, *Linda in context*, JACM **32** (4) (1989), 444–458.

[36] N. Carriero, D. Gelernter and L. Zuck, *Bauhaus Linda*, Proceedings Object-Based Models and Languages for Concurrent Systems, Lecture Notes in Comput. Sci. 924 (1995).

[37] I. Castellani, *Bisimulations for concurrency*, Ph.D. Thesis, University of Edinburgh (1988).

[38] I. Castellani, *Observing distribution in processes: Static and dynamic localities*, Internat. J. Found. Comput. Sci. **4** (6) (1995), 353–393.

[39] I. Castellani and M. Hennessy, *Distributed bisimulations*, JACM **36** (4) (1989), 887–911.

[40] I. Castellani and G.Q. Zhang, *Parallel product of event structures*, Theoret. Comput. Sci. **179** (1–2) (1997), 203–215.

[41] L. Castellano, G. De Michelis and L. Pomello, *Concurrency vs interleaving: An instructive example*, Bull. EATCS **31** (1987), 12–15.

[42] S. Christensen, *Distributed bisimilarity is decidable for a class of infinite state-space systems*, Proceedings CONCUR-92, Lecture Notes in Comput. Sci. 630 (1992).

[43] S. Christensen, *Decidability and decomposition in process algebras*, Ph.D. Thesis, University of Edinburgh (1993).

[44] S. Christensen, Y. Hirshfeld and F. Moller, *Bisimulation equivalence is decidable for all parallel processes*, Proceedings LICS-93, IEEE Computer Society Press (1993), 143–157.

[45] F. De Cindio, G. De Michelis, L. Pomello and C. Simone, *A Petri net model for CSP*, Proceedings CIL-81 (1981).

[46] F. Corradini, *Space, time and nondeterminism in process algebras*, Ph.D. Thesis, Università di Roma La Sapienza (1996).

[47] F. Corradini and R. De Nicola, *On the relationships between four partial ordering semantics*, Fundamenta Informaticae **34** (4) (1996), 349–384.

[48] F. Corradini and R. De Nicola, *Locality based semantics for process algebras*, Acta Inform. **34** (1997), 291–324.

[49] F. Corradini and D. Murphy, *Located processes and located processors*, Report SI/RR-94/02, Università di Roma La Sapienza (1994).

[50] F. Corradini and R. De Nicola, *Distribution and locality of concurrent systems*, Proceedings ICALP-94, Lecture Notes in Comput. Sci. 820 (1994).

[51] Ph. Darondeau and P. Degano, *Causal trees*, Proceedings ICALP-89, Lecture Notes in Comput. Sci. 372 (1989), 234–248.

[52] Ph. Darondeau and P. Degano, *Causal trees: Interleaving + causality*, Proceedings LITP Spring School, La Roche-Posay, Lecture Notes in Comput. Sci. 469 (1990).

[53] Ph. Darondeau and P. Degano, *Refinement of actions in event structures and causal trees*, Theoret. Comput. Sci. **118** (1) (1993), 21–48.

[54] P. Degano, R. De Nicola and U. Montanari, *On the consistency of truly concurrent operational and denotational semantics*, Proceedings LICS-88, IEEE Computer Society Press (1988).

[55] P. Degano, R. De Nicola and U. Montanari, *Partial orderings descriptions and observations of nondeterministic concurrent processes*, Proceedings REX School/Workshop on Linear Time, Branching Time and Partial Order in Logics and Models for Concurrency, Noordwijkerhout, Lecture Notes in Comput. Sci. 354 (1989).

[56] P. Degano, R. De Nicola and U. Montanari, *A partial ordering semantics for CCS*, Theoret. Comput. Sci. **75** (3) (1990), 223–262.

[57] P. Degano, R. Gorrieri and S. Marchetti, *An exercise in concurrency: A CSP process as a C/E system*, Advances in Petri Nets 1988, Lecture Notes in Comput. Sci. 154 (1988).

[58] P. Degano and U. Montanari, *Concurrent histories: A basis for observing distributed systems*, J. Comput. System Sci. **34** (2/3) (1987), 422–461.

[59] P. Degano, R. De Nicola and U. Montanari, *Partial ordering derivations for CCS*, Proceedings FCT-85, Lecture Notes in Comput. Sci. 199 (1985).

[60] P. Degano, R. De Nicola and U. Montanari, *Observational equivalences for concurrency models*, Formal Description of Programming Languages III, North Holland (1987), 105–132.

[61] P. Degano, R. De Nicola and U. Montanari, *A distributed operational semantics for CCS based on condition/event systems*, Acta Inform. **26** (1/2) (1988), 59–91.

[62] P. Degano and C. Priami, *Proved trees*, Proceedings ICALP-92, Lecture Notes in Comput. Sci. 623 (1992).

[63] R. Devillers, *On the definition of a bisimulation notion based on partial words*, Petri Net Newsletter **29** (1988), 16–19.

[64] U. Engberg and M. Nielsen, *A calculus of communicating systems with label-passing*, Report DAIMI PB-208, Computer Science Department, University of Aarhus, Denmark (1986).

[65] G. Ferrari, *Unifying models of concurrency*, Ph.D. Thesis, University of Pisa (1990).

[66] G. Ferrari, R. Gorrieri and U. Montanari, *An extended expansion theorem*, Proceedings TAPSOFT-91, Lecture Notes in Comput. Sci. 494 (1991).

[67] G. Ferrari and U. Montanari, *Towards the unification of models of concurrency*, Proceedings CAAP-90, Lecture Notes in Comput. Sci. 431 (1990).

[68] G. Ferrari and U. Montanari, *Parametrized structured operational semantics*, Fundamenta Informaticae **34** (1998), 1–31.

[69] C. Fournet, *The Join-calculus: A calculus for distributed mobile programming*, Ph.D. Thesis, École Polytechnique (1998).

[70] C. Fournet and G. Gonthier, *The reflexive chemical abstract machine and the Join-calculus*, Proceedings POPL-96 (1996), 372–385.

[71] C. Fournet, G. Gonthier, J.-J. Lévy, L. Maranget and D. Rémy, *A calculus of mobile agents*, Proceedings CONCUR-96, Lecture Notes in Comput. Sci. 1119 (1996).

[72] C. Fournet and L. Maranget, *The Join-calculus language, Release 1.04* (documentation and user's manual) (1999).

[73] J.L. Gischer, *Partial orders and the axiomatic theory of shuffle*, Ph.D. Thesis, Stanford University (1984).

[74] R.J. van Glabbeek, *The refinement theorem for ST-bisimulation semantics*, Proceedings IFIP TC2 Working Conference on Programming Concepts and Methods, Israel, April 1990, M. Broy and C.B. Jones, eds, North-Holland (1990), 27–52.

[75] R.J. van Glabbeek and U. Goltz, *Equivalence notions for concurrent systems and refinement of actions*, Proceedings MFCS-89, Lecture Notes in Comput. Sci. 379 (1989).

[76] R.J. van Glabbeek and U. Goltz, *Partial order semantics for refinement of actions – neither necessary nor always sufficient but appropriate when used with care*, Bull. EATCS **38** (1989).

[77] R.J. van Glabbeek and U. Goltz, *Equivalences and refinement*, Proceedings LITP Spring School, La Roche-Posay, Lecture Notes in Comput. Sci. 469 (1990).

[78] R.J. van Glabbeek and U. Goltz, *Refinement of actions in causality based models*, Proceedings REX Workshop on Stepwise Refinement of Distributed Systems: Models, Formalism, Correctness, Lecture Notes in Comput. Sci. 430 (1990).

[79] R.J. van Glabbeek and U. Goltz, *Refinement of actions and equivalence notions for concurrent systems*, Hildesheimer Informatik-Berichte 6/98, Institut für Informatik, Universität Hildesheim (1998). To appear in: Acta Inform.

[80] R.J. van Glabbeek and F.W. Vaandrager, *Petri net models for algebraic theories of concurrency*, Proceedings PARLE-87, Lecture Notes in Comput. Sci. 259 (1987).

[81] R.J. van Glabbeek and F.W. Vaandrager, *The difference between splitting in n and $n + 1$*, Inform. and Comput. **136** (2) (1997), 109–142.

[82] U. Goltz, *On representing CCS programs by finite Petri nets*, Proc. MFCS-88, Lecture Notes in Comput. Sci. 324 Springer (1988), 339–350.

[83] U. Goltz and A. Mycroft, *On the relationship of CCS and Petri nets*, Proceedings ICALP-84, Lecture Notes in Comput. Sci. 172 (1984).
[84] U. Goltz and W. Reisig, *The nonsequential behaviour of Petri nets*, Inform. and Control **57** (2–3) (1983), 125–147.
[85] U. Goltz and W. Reisig, *CSP programs as nets with individual tokens*, Advances in Petri nets 1984, Lecture Notes in Comput. Sci. 188 (1985).
[86] R. Gorrieri and U. Montanari, *SCONE: A simple calculus of nets*, Proceedings CONCUR-90, Lecture Notes in Comput. Sci. 458 (1990).
[87] R. Gorrieri and A. Rensink, *Action refinement*, Handbook of Process Algebra, J.A. Bergstra, A. Ponse and S.A. Smolka, eds, Elsevier, Amsterdam (2001), 1047–1147.
[88] J. Grabowski, *On partial languages*, Fundamenta Informaticae **4** (1) (1981), 427–498.
[89] M. Hennessy, *On the relationship between time and interleaving*, Unpublished draft, Sophia-Antipolis (1980).
[90] M. Hennessy, *Axiomatising finite concurrent processes*, SIAM J. Computing **17** (5) (1988).
[91] M. Hennessy and H. Lin, *Symbolic bisimulations*, Theoret. Comput. Sci. **138** (1995).
[92] M. Hennessy and J. Riely, *Resource access control in systems of mobile agents*, Proceedings HLCL-98, ENTCS 16 (1998).
[93] M. Hennessy and J. Riely, *Type-safe execution of mobile agents in anonymous networks*, Proceedings of the 1999 Workshop on Secure Internet Programming, Lecture Notes in Comput. Sci. 1603 (1999).
[94] K. Honda and M. Tokoro, *On asynchronous communication semantics*, Proceedings Workshop on Object-Based Concurrent Computing, Lecture Notes in Comput. Sci. 612 (1992).
[95] A. Joyal, M. Nielsen and G. Winskel, *Bisimulation and open maps*, Proceedings LICS-93, IEEE Computer Society Press (1993), 418–427.
[96] P. Khrisnan, *Distributed CCS*, Proceedings CONCUR-91, Lecture Notes Comput. Sci. 527 (1991).
[97] A. Kiehn, *Concurrency in process algebras*, Habilitation Thesis, Technische Universität München (1999).
[98] A. Kiehn, *Distributed bisimulations for finite CCS*, Report 7/89, University of Sussex (1989).
[99] A. Kiehn, *Proof systems for cause based equivalences*, Proceedings MFCS-93, Lecture Notes Comput. Sci. 711 (1993).
[100] A. Kiehn, *Comparing locality and causality based equivalences*, Acta Inform. **31** (1994), 697–718.
[101] A. Kiehn and M. Hennessy, *On the decidability of noninterleaving process equivalences*, Fundamenta Informaticae **30** (1997), 23–43.
[102] R. Langerak, *Bundle event structures: A noninterleaving semantics for LOTOS*, Technical Report, University of Twente (1991).
[103] F. Levi and D. Sangiorgi, *Controlling interference in ambients*, Proceedings POPL-2000, ACM Press (2000).
[104] J.-J. Lévy, *Some results in the Join-calculus*, Proceedings TACS-97, Lecture Notes in Comput. Sci. 1281 (1997).
[105] R. Loogen and U. Goltz, *Modelling nondeterministic concurrent processes with event structures*, Fundamenta Informaticae **XIV** (1) (1991), 39–74.
[106] A. Mazurkiewicz, *Concurrent program schemes and their interpretation*, Report DAIMI PB-78, Aarhus University (1977).
[107] A. Mazurkiewicz, *Trace theory*, Advances in Petri Nets 1986, Lecture Notes in Comput. Sci. 255 (1987).
[108] M. Merro and D. Sangiorgi, *On asynchrony in name-passing calculi*, Proc. ICALP-98, Lecture Notes in Comput Sci. 1443, Springer-Verlag (1998).
[109] R. Milner, *A Calculus of Communicating Systems*, Lecture Notes in Comput. Sci. 92, Springer-Verlag (1980).
[110] R. Milner, *Calculi for synchrony and asynchrony*, Theoret. Comput. Sci. **25** (1983), 269–310.
[111] R. Milner, *Communication and Concurrency*, Prentice-Hall (1989).
[112] R. Milner, *The polyadic π-calculus: A tutorial*, Technical Report ECS-LFCS-91-180, LFCS, University of Edinburgh (1991).
[113] R. Milner, *Functions as processes*, Math. Structures Comput. Sci. **2** (2) (1992), 119–141.
[114] R. Milner, J. Parrow and D. Walker, *A calculus of mobile processes, Parts 1–2*, Inform. and Comput. **100** (1) (1992), 1–77.

[115] U. Montanari and M. Pistore, *Efficient minimization up to location equivalence*, Proceedings ESOP-96, Lecture Notes in Comput. Sci. 1058 (1996).
[116] U. Montanari and D. Yankelevich, *A parametric approach to localities*, Proceedings ICALP-92, Lecture Notes in Comput. Sci. 623 (1992).
[117] U. Montanari and D. Yankelevich, *Location equivalence in a parametric setting*, Theoret. Comput. Sci. **149** (1995), 299–332.
[118] M. Mukund and M. Nielsen, *CCS, Locations and asynchronous transition systems*, Proceedings FST-TCS-92, Lecture Notes in Comput. Sci. 652 (1992).
[119] D. Murphy, *Observing located concurrency*, Proceedings MFCS-93, Lecture Notes in Comput. Sci. 711 (1993).
[120] R. De Nicola, G. Ferrari and R. Pugliese, *Locality based Linda: Programming with explicit localities*, Proceedings TAPSOFT-FASE-97, Lecture Notes in Comput. Sci. 1214 (1997).
[121] R. De Nicola, G. Ferrari and R. Pugliese, *Klaim: A kernel language for agents interaction and mobility*, IEEE Trans. on Software Engrg. **24** (5) (1998), 315–330.
[122] M. Nielsen, *CCS and its relationship to net theory*, Advances in Petri Nets 1986, Lecture Notes in Comput. Sci. 255 (1987).
[123] M. Nielsen, G. Plotkin and G. Winskel, *Petri nets, event structures and domains, Part I*, Theoret. Comput. Sci. **13** (1) (1981), 85–108.
[124] E.-R. Olderog, *Operational Petri net semantics for CCSP*, Advances in Petri Nets 1987, Lecture Notes Comput. Sci. 266 (1987).
[125] E.-R. Olderog, *Nets, Terms and Formulas: Three Views of Concurrent Processes and Their Relationship*, Cambridge Tracts in Theoret. Comput. Sci. 23, Cambridge University Press (1991).
[126] D. Park, *Concurrency and automata on infinite sequences*, Proceedings 5th GI-Conference on Theoretical Computer Science, Lecture Notes in Comput. Sci. 104 (1981).
[127] J. Parrow, *An introduction to the π-calculus*, Handbook of Process Algebra, J.A. Bergstra, A. Ponse and S.A. Smolka, eds, Elsevier, Amsterdam (2001), 479–543.
[128] J. Peterson, *Petri Net Theory and the Modeling of Systems*, Prentice-Hall (1981).
[129] C.A. Petri, *Kommunication mit Automaten*, Schriften des Institutes für Instrumentelle Mathematik, Bonn (1962).
[130] C.A. Petri, *Nonsequential processes*, Research Report 77-05, GMD, Sankt Augustin (1977).
[131] B. Pierce and D. Sangiorgi, *Typing and subtyping for mobile processes*, Logic in Computer Science, 1993. Full version in: Mathem. Structures Comput. Sci. **6** (5) (1996).
[132] B.C. Pierce and D.N. Turner, *Pict: A programming language based on the pi-calculus*, Technical Report CSCI 476, Indiana University (1997). To appear in: Proof, Language and Interaction: Essays in Honour of Robin Milner, G. Plotkin, C. Stirling and M. Tofte, eds, MIT Press.
[133] G.D. Plotkin, *A structural approach to operational semantics*, Report DAIMI FN-19, Computer Science Department, Aarhus University (1981).
[134] V.R. Pratt, *The pomset model of parallel processes: Unifying the temporal and the spatial*, Proceedings of Seminar on Concurrency, Lecture Notes in Comput. Sci. 197 (1985).
[135] V.R. Pratt, *Modelling concurrency with partial orders*, Internat. J. Parallel Programming **15** (1) (1986).
[136] A. Rabinovich and B.A. Trakhtenbrot, *Behaviour structures and nets*, Fundamenta Informaticae **XI** (4) (1988), 357–404.
[137] W. Reisig, *On the semantics of Petri nets*, Formal Models in Programming, North-Holland (1985), 347–372.
[138] W. Reisig, *Petri Nets*, EATCS Monographs on Theoretical Computer Science (1985).
[139] J.-H. Réty, *Distributed concurrent constraint programming*, Fundamenta Informaticae **34** (1998), 323–346.
[140] J. Riely and M. Hennessy, *Distributed processes and location failures*, Proceedings ICALP-97, Lecture Notes in Comput. Sci. 1256 (1997).
[141] J. Riely and M. Hennessy, *A typed language for distributed mobile processes*, Proceedings POPL-98 (1998).
[142] J. Riely and M. Hennessy, *Trust and partial typing in open systems of mobile agents*, Proceedings POPL-99 (1999).

[143] D. Sangiorgi, *Expressing mobility in process algebras: First-order and higher order paradigms*, Ph.D. Thesis, University of Edinburgh (1992).

[144] D. Sangiorgi, *From π-calculus to higher-order π-calculus – and back*, Proc. TAPSOFT-93, Lecture Notes Comput. Sci. 668 (1993).

[145] D. Sangiorgi, *Locality and interleaving semantics in calculi for mobile processes*, Theoret. Comput. Sci. **155** (1996), 39–83.

[146] D. Sangiorgi, *The name discipline of uniform receptiveness*, Theoret. Comput. Sci. **221** (1999), 457–493.

[147] V. Sassone, M. Nielsen and G. Winskel, *Models for concurrency*, Theoret. Comput. Sci. **170** (1–2) (1996), 297–348.

[148] P. Sewell, *A Brief Introduction to Applied π*, Notes from lectures at the MATHFIT Instructional Meeting on Recent Advances in Semantics and Types for Concurrency, Imperial College (1998).

[149] P. Sewell, *Global/local subtyping and capability inference for a distributed π-calculus*, Proceedings ICALP-98, Lecture Notes in Comput. Sci. 1443 (1998), 695–706.

[150] P. Sewell, P.T. Wojciechowski and B.C. Pierce, *Location independence for mobile agents*, Proceedings of the 1998 Workshop on Internet Programming Languages (1998).

[151] P. Sewell, P.T. Wojciechowski and B.C. Pierce, *Location-independent communication for mobile agents: A two-level architecture*, Technical Report 462, Computer Laboratory, University of Cambridge (1998).

[152] P. Sewell, P.T. Wojciechowski and B.C. Pierce, *Nomadic Pict: Language and infrastructure design for mobile agents*, Proceedings of ASA/MA-99 (First International Symposium on Agent Systems and Applications/Third International Symposium on Mobile Agents) (1999).

[153] M.W. Shields, *Concurrent machines*, The Comput. J. **28** (1985), 449–465.

[154] D.A. Taubner, *Finite Representation of CCS and TCSP Programs by Automata and Petri Nets*, Lecture Notes in Comput. Sci. 369 (1989).

[155] D.A. Taubner, *Representing CCS programs by finite predicate/transition nets*, Acta Inform. **27** (1990).

[156] B. Thomsen, *Plain CHOCS, a second generation calculus for higher-order processes*, Acta Inform. **30** (1993), 1–59.

[157] B. Thomsen, L. Leth, S. Prasad, T.M. Kuo, A. Kramer, F. Knabe and A. Giacalone, *Facile Antigua release programming guide*, Technical Report ECRC-93-20, ECRC, Munich (1993).

[158] D.N. Turner, *The polymorphic pi-calculus: Theory and implementation*, Ph.D. Thesis, University of Edinburgh (1996).

[159] F.W. Vaandrager, *A simple definition of parallel composition for prime event structures*, Technical Report CS-R8903, CWI, Amsterdam (1989).

[160] J. Vitek and G. Castagna, *Mobile computations and hostile hosts*, Proceedings of the 10th JFLA (Journées Francophones des Langages Applicatifs) (1999).

[161] J. Vitek and G. Castagna, *Seal: A framework for secure mobile computations*, Workshop on Internet Programming Languages, D. Tsichritzis, ed. (1999).

[162] W. Vogler, *Behaviour preserving refinements of Petri nets*, Proceedings Workshop on Graph-Theoretic Concepts in Computer Science, Lecture Notes in Comput. Sci. 246 (1987).

[163] D. Walker, *Objects in the π-calculus*, Inform. and Comput. **116** (1995), 253–271.

[164] J. Winkowski, *Behaviours of concurrent systems*, Theoret. Comput. Sci. **12** (1980), 39–60.

[165] G. Winskel, *Events in computation*, Ph.D. Thesis, University of Edinburgh (1980).

[166] G. Winskel, *Event structure semantics for CCS and related languages*, Proceedings ICALP-82, Lecture Notes in Comput. Sci. 140 (1982).

[167] G. Winskel, *Categories of models for concurrency*, Proceedings Seminar on Concurrency, Lecture Notes in Comput. Sci. 197 (1984).

[168] G. Winskel, *Event structures*, Advances in Petri Nets 1986, Lecture Notes in Comput. Sci. 255 (1987).

[169] G. Winskel, *Petri nets, algebras, morphisms and compositionality*, Inform. and Control **72** (1987), 197–238.

[170] G. Winskel and M. Nielsen, *Models for concurrency*, Handbook of Logic in Computer Science, Vol. 4, Oxford (1995), 1–148.

[171] D. Yankelevich, *Parametric views of process description languages*, Ph.D. Thesis, University of Pisa (1993).

Subject index

ν-dynamic location equivalence, 987
π-calculus, 948, 1007, 1013, 1017
$\pi_{1\ell}$-calculus, 1013, 1014, 1017

abstract, 947, 1034
access permission, 1029
access to resources, 1011
action refinement, 954, 956
actions, 959
agent creation, 1031
agent migration, 1031
agents, 1019, 1031
ambient calculus, 1026
AMBIENT calculus, 949, 1013
asynchronous, 1015, 1016, 1021–1023
asynchronous π-calculus, 1011, 1017, 1018, 1021, 1025
asynchronous bisimulation, 1018, 1026
asynchronous transition systems, 950, 977, 1000
atomic actions, 954
atomic locations, 962

barbed equivalence, 1012, 1017
bisimulation, 947
bisimulation equivalence, 949, 954, 1008
blocking, 1015
BPP (Basic Parallel Processes), 948, 991, 1001, 1002
BS-equivalence, 955

calculus LLinda, 1033
calculus LLINDA, 1013
canonical, 970, 974
canonical (dynamic) locations, 988
canonical distributions, 971, 985
canonical locations, 970, 971
canonical static locations, 1000
capabilities, 1020, 1021, 1026
causal bisimulation, 948, 954, 959, 960, 971, 995, 1003
causal path, 973, 981, 1008
causal semantics, 956, 962, 1001
causal transition system, 1003
causal trees, 950, 953, 1005
causality, 946, 950
CCS, 946, 980, 996, 1009–1011
channel names, 1007
channel type, 1020
channels, 1019, 1028
Chemical Abstract Machine, 1017, 1023
circular migration, 1025

communication, 1023
compound actions, 952
concrete, 947, 1010, 1011, 1034
concrete location, 1009
concurrency, 950, 981
concurrent bisimulation, 955
condition/event systems, 954
configurations structures, 955, 1005
consistent location association, 967, 969
consistent occurrence aliasing, 984
CPP (Communicating Parallel Processes), 991, 996, 997, 999
cross-causalities, 952, 960, 1005

Dπ, 1014
Dπ-calculus, 1019
deadlock-freedom, 1021
decidability, 1035
decorated (atomic) actions, 952
decorating, 959
decorating atomic actions, 954
DIS, 964, 978
distant communication, 1014, 1032
distributed π-calculi, 949
distributed bisimulation, 948, 953, 997, 998, 1002
distributed failure, 1022
distributed processes, 964, 1010, 1012
distributed semantics, 947, 949, 953, 956, 995
distributed transition system, 997
distributed JOIN calculus, 949, 1011, 1014, 1023
distribution, 947, 964, 970, 1014
distribution units, 946
distributions, 957, 967
dpi-calculus, 1021
dynamic location, 973, 986
dynamic location equivalence, 975
dynamic location preorder, 975

equational characterization, 977
equivalence, 992
equivalence on the occurrence system, 984
event (transition) system, 980
event structures, 950
expansion law, 949, 954

failures, 1015, 1018, 1035
flat, 1009, 1015
flat space of locations, 1019, 1034
flow, 1005

general locations, 963
generalized distributed bisimulation, 999

global, 1017
global causality, 991, 1000
global cause equivalence, 992, 1003
global causes, 992, 1003
global channels, 1021
global communication, 1015, 1016, 1021, 1022, 1033
grape semantics, 953, 999, 1000
grapes, 952

hierarchical, 1015, 1021
hierarchical locations, 1034
hierarchically structured, 1024
history-preserving bisimulation, 948, 955, 959, 984, 1005, 1007

independent, 966
interleaving, 963
interleaving semantics, 946
invisible actions, 962

JOIN calculus, 1013, 1022

killing, 1012
KLAIM, 1013

label, 980
LCCS, 962, 963, 978
left-consistent location association, 972
local, 1021, 1022, 1032
local causality, 981, 984, 991
local cause bisimulation, 958
local cause equivalence, 978, 992
local causes, 948, 958, 992
local channel, 1020, 1029
local communication, 1014, 1020, 1028
local deadlock, 947, 962
local history preserving bisimulation, 958, 980, 984, 1006
local mixed-ordering equivalence, 948, 958, 1000
local/global cause, 992
local/global cause bisimulation, 992
local/global cause semantics, 948, 991
local/global cause transition system, 991
localities, 946, 947, 1010
localization of receptors, 1018, 1021, 1025
located equivalence, 1010, 1013
located-failure equivalence, 1012
location, 981, 997
location automata, 1036
location equivalence, 948, 957, 991, 995, 1007, 1008, 1010, 1012
location failures, 948, 1011, 1025
location prefixing, 960, 962, 963
location preorder, 948, 957
location renaming, 987
location semantics, 956, 960, 973, 985, 986
location type, 1020
locations, 947, 949, 974, 1011, 1023
logical characterization, 977
loose location equivalence, 995

maximal distribution equivalence, 1000
Mazurkiewicz traces, 950
migrating agent, 1015
migration, 949, 1014, 1015, 1024
migration units, 1028
mixed-ordering equivalence, 955, 959
mobile agents, 1027
mobility, 949, 1013, 1036
modalities, 977
Mutex, 961, 972, 973, 1005

name scoping, 1016
name-passing, 1014
names, 962, 1007
nets of automata, 966, 977, 1009, 1010, 1036
network, 1013
network awareness, 946
network transparency, 948, 1025
network-aware, 1035
NMS equivalence, 953, 955, 1005
Nomadic Pict, 949, 1031
NOMADIC PICT, 1013
noninterleaving semantics, 946, 947

objective, 1015, 1029
objective migration, 1018, 1027
occurrence, 980
occurrence equivalence, 984
occurrence semantics, 985, 986
occurrence transition system, 980
operational semantics, 950

partial ordering semantics, 952, 955, 1001
partial orders, 950
partially ordered multiset, 951
past occurrences, 983
perfect firewall equation, 1027, 1031
permutation semantics, 952
permutation-equivalent, 951
Petri nets, 950
polyadic, 1007, 1017, 1028
pomset, 951
pomset bisimulation, 951, 1001
pomset semantics, 952
preorder, 991
prime event structures, 955, 1005

process algebras, 946
process migration, 1024, 1034
progressive bisimulation family, 968
progressive o-bisimulation family, 984
progressive pre-bisimulation family, 972
Protocol, 960, 961, 972, 973, 976, 1005
proved transition systems, 959
proved transitions, 951, 968
proved trees, 959
proximity mobility, 1026

remote channels, 1020
remote procedure call, 1020, 1033
resource access control, 1019, 1028
resources, 947, 1019, 1028
right-consistent occurrence aliasing, 985
routing, 949, 1018, 1028

SEAL calculus, 949, 1013, 1028
security, 1028
sites, 948, 949, 958, 1014, 1031
"spatial" modalities, 1028
spatial transition system, 958, 971, 1000
spawn, 1030
SPAWN, 1018
spawning, 1012
split bisimulation, 955
ST-bisimulation, 955, 1036
stable event structures, 955, 1005
static location equivalence, 966, 968
static location preorder, 972
static locations, 981, 984, 985
strong, 1015
strong location transition, 964
structured locations, 1019
structured transition systems, 959
subjective, 1015, 1024, 1026, 1032
sublocation, 957, 966
subwords, 957
synchronous π-calculus, 1028
synchronous communication, 1015

testing, 1012, 1025
trace automata, 950, 954
transition systems with independence, 950
transparency, 1018
transparent, 1012, 1014
true-concurrency, 950
two atomic locations, 966
type systems, 1016, 1020

unique receiver, 1018
unique receiver property, 1013, 1021
units of communication, 946, 949, 1019, 1034
units of distribution, 949, 977, 1013, 1033
units of failure, 946, 949, 1013, 1034
units of migration, 946, 1024, 1026, 1034
units of mobility, 949
units of security, 946, 949, 1034

verification, 1035
visible actions, 962

weak, 1015, 1020
weak (dynamic) location transition system, 974
weak bisimulation, 971, 1010
weak bisimulation equivalence, 960
weak location transition system, 964
weak occurrence system, 983

CHAPTER 16

Action Refinement

Roberto Gorrieri[1], Arend Rensink[2]

[1] *Dipartimento di Scienze dell'Informazione, Università di Bologna, Mura Anteo Zamboni 7, I-40127 Bologna, Italy*
E-mail: gorrieri@cs.unibo.it

[2] *Department of Computer Science, University of Twente, Postbus 217, NL-7500 AE Enschede, The Netherlands*
E-mail: rensink@cs.utwente.nl

Contents

1. Introduction . . . 1049
   1.1. What is action refinement about? . . . 1049
   1.2. Refinement operator versus hierarchy of descriptions . . . 1050
   1.3. Atomic versus non-atomic action refinement . . . 1051
   1.4. Syntactic versus semantic action refinement . . . 1052
   1.5. Interleaving versus true concurrency . . . 1053
   1.6. Strict versus relaxed forms of refinement . . . 1056
   1.7. Vertical implementation . . . 1057
   1.8. Overview of the chapter . . . 1058
   1.9. Whither action refinement? . . . 1059
2. Sequential systems . . . 1060
   2.1. The sequential language . . . 1060
   2.2. Operational semantics . . . 1063
   2.3. Denotational semantics . . . 1064
   2.4. Behavioural semantics and congruences . . . 1067
   2.5. Application: A very simple data base . . . 1069
3. Atomic refinement . . . 1070
   3.1. Parallel composition and atomizer . . . 1071
   3.2. Denotational semantics . . . 1074
   3.3. Congruences and axiomatizations . . . 1075
   3.4. Application: Critical sections . . . 1076
4. Non-atomic refinement: An event-based model . . . 1077
   4.1. Event annotations and operational semantics . . . 1078
   4.2. Operational event-based semantics . . . 1080
   4.3. Stable event structures . . . 1083
   4.4. Denotational event-based semantics . . . 1085
   4.5. Compatibility of the semantics . . . 1088

HANDBOOK OF PROCESS ALGEBRA
Edited by Jan A. Bergstra, Alban Ponse and Scott A. Smolka

4.6. Application: A simple data base . . . 1089
5. Non-atomic refinement: Other observational congruences . . . 1090
5.1. Pomsets . . . 1091
5.2. Causal links . . . 1094
5.3. Splitting actions . . . 1095
5.4. ST-semantics . . . 1099
5.5. Application: A simple data base . . . 1105
5.6. ω-completeness . . . 1105
6. Semantic versus syntactic substitution . . . 1106
6.1. Finite sequential systems . . . 1107
6.2. Recursive sequential systems . . . 1110
6.3. Atomic refinement . . . 1111
6.4. Synchronization . . . 1112
6.5. Application: A simple data base . . . 1116
7. Dependency-based action refinement . . . 1117
7.1. Dependencies in linear time . . . 1118
7.2. Application: Critical sections . . . 1121
7.3. Dependencies in branching time . . . 1121
7.4. Application: A simple data base . . . 1126
7.5. The dual view: Localities . . . 1127
8. Vertical implementation . . . 1128
8.1. Refinement functions . . . 1130
8.2. Vertical delay bisimulation . . . 1131
8.3. Requirements for vertical implementation . . . 1134
8.4. Further developments . . . 1138
8.5. Application: A simple data base . . . 1140
Acknowledgements . . . 1141
References . . . 1141
Subject index . . . 1147

Abstract

In this chapter, we give a comprehensive overview of the research results in the field of action refinement during the past 12 years. The different approaches that have been followed are outlined in detail and contrasted to each other in a uniform framework. We use two running examples to discuss their effects, benefits and disadvantages. The chapter contains results only; appropriate references are given to the original papers containing the proofs.

1. Introduction

1.1. *What is action refinement about?*

A widely accepted approach to specify the behaviour of concurrent systems relies on state/transition abstract machines, such as labelled transition systems: an activity, supposed to be atomic at a certain abstraction level, can be represented by a transition, the label of which is the name of the activity itself. Once these *atomic actions* are defined, one technique to control the complexity of a concurrent system specification is by means of (horizontal) modularity: a complex system can be described as composed of smaller subsystems. Indeed, this is the main achievement of process algebras: the specification is given as a term whose subterms denotes subcomponents; the specification, as well as its analysis, can be done component-wise, focussing on few details at a time.

However, from a software engineering viewpoint, the resulting theory may in many cases still be unsuitable, as the abstraction level is fixed once and for all by the given set of atomic actions. In the development of software components, it may be required to compare systems that belong to conceptually different abstraction levels (where the change of the level is usually accompanied by a change in the sets of actions they perform) in order to verify if they realize essentially the same functionality. Once the sets of actions at the different abstraction levels are defined, a technique (orthogonal to the previous one) for controlling the complexity of concurrent system specifications is by means of *vertical* modularity: a complex system can be first described succinctly as a simple, abstract specification and then refined stepwise to the actual, complex implementation; the specification, as well as the analysis on it, can be done level by level, focussing each time on the relevant details introduced by passing from the previous level to the current one. This well-known approach is sometimes referred to as *hierarchical specification methodology*. It has been successfully developed for sequential systems, yielding, for instance, a technique known as *top-down systems design*, where a high-level "instruction" is macro-expanded to a lower level "module" until the implementation level is reached (see, e.g., [141]).

In the context of process algebra, this refinement strategy amounts to introducing a mechanism for transforming high-level primitives/actions into lower level processes (i.e., processes built with lower level actions).

EXAMPLE 1.1. As a running (toy) example in this chapter, we consider a data base which can be *queried* using an operation *qry* and *updated* using an operation *upd*. Both are atomic, i.e., once invoked their effect is as if they finish immediately, and no concurrently invoked action can interfere with them. The latter operation is then transformed into a transaction consisting of two phases, *req* in which the update is requested and *cnf* in which it is confirmed. The question addressed in this chapter is what the behaviour of the data base on the resulting lower level of abstraction should be.

There are several ways to go about studying this issue, depending on some choices that can be taken. These are discussed in the following subsections.

1.2. *Refinement operator versus hierarchy of descriptions*

In traditional programming languages, there is an operator that supports a hierarchical specification methodology: the declaration (and call) of a procedure, given by (some syntactical variant of) "`let` $a = u$ `in` t". This specifies that the abstract name a is declared to equal its body u in the scope t. So, whenever a is encountered during the execution of t, u is executed in its place. Similarly, one way to support vertical modularity in process algebra is by introducing an explicit operator, called action refinement and written $t[a \to u]$, which plays a role similar to that of procedure call: it is nothing but a declaration, introducing the name a for its body u in the scope t. The discussion about the possible meanings of the refinement operator is postponed to Sections 1.3 and 1.4; here we simply recall that the main problem faced by the advocates of this approach, the so-called *congruence problem*, is to find an observational equivalence which respects the refinement operator. A non-exhaustive list of papers following this approach in process algebra is [8,9,13,19, 31,37,38,48,50,69,89,110,121,126] and in semantic models [22,42,52,62,63,93,130–132].

Most of this chapter is devoted to a study of the operator for action refinement within process algebra. However, also another approach to support vertical modularity is discussed in this chapter: a hierarchy of descriptions, equipped with a suitable implementation relation establishing an ordering among them. Typically, a concurrent system, described at several levels of detail, can be seen as a collection of different albeit related systems. Each of these systems may be described in a particular (process algebraic) language. Therefore, in order to relate the various systems, it is necessary that we are able to correctly relate the different languages. The implementation of a language into another language may be often seen as the definition of the primitives of the former as derived operators of the latter. Of course, if we assume that all the systems are described in the same language (as we do in this chapter, as described in Section 1.7), the task is easier. Anyway, some work is needed: a suitable partition of the action set by abstraction levels, a *refinement function* associating lower level processes to high level actions, and an *implementation relation* that states when a low level process implements a high level process, according to the refinement function. This "vertical" implementation relation is not to be confused with existing "horizontal" implementation relations, such as trace or testing pre-orders, which rather reflect the idea that a given system implements another on the *same* abstraction level, by being closer to an actual implementation, for instance more deterministic.

Although action refinement as an operator or through a hierarchy of descriptions are solutions to the same problem, a comparison is not easy. On the one hand, a single language with a mechanism for hierarchy among its operators is a quite appealing approach, as it permits to define the horizontal and the vertical modularities in a uniform way. Indeed, this is in the line of the development of sequential languages: the definition of control abstraction mechanisms, such as procedures and functions, or of data abstraction mechanisms, such as abstract data types, should be considered a standard way of "internalizing", in the usual horizontal modularity, concepts that are typical of the vertical one. On the other hand, this approach has also some disadvantages: no clear separation of the abstraction levels in the specification (free combination of horizontal and vertical operations) and no clear distinction of what actions are of what level (confusion may be risky, as we will discuss in Section 1.4 and, much more extensively, in Section 6). Another major difference between the two approaches is the following. According to the former approach, given a specification S and a refinement function, there is *only one* possible implementation I; hence, there is no need to develop a notion of *correctness* of the implementation: the implementation I is what one obtains by applying the operator of refinement to S. On the contrary, the latter approach may admit several different implementations for a given specification, namely all those that are correct according to the vertical implementation relation. See Section 1.7 for a further discussion. For the time being, we concentrate on the interpretation of action refinement as an operator.

1.3. *Atomic versus non-atomic action refinement*

The basic approaches to action refinement can be divided into two main groups. On the one hand, there is *atomic refinement* [13,19,48,72,75,97], where one takes the point of view that actions are atomic and their refinements should in some sense preserve this atomicity. On the other hand, there is a more liberal notion of refinement – called *non-atomic* – according to which atomicity is always relative to the current level of abstraction, and may in a sense be destroyed by refinement.

To better explain this issue, let us consider one simple example: $a \mid\mid\mid b$, representing the parallel composition of a and b, and the refinement of a by $a_1; a_2$. Figure 1 shows the labelled transition systems for $a \mid\mid\mid b$ and $(a \mid\mid\mid b)[a \rightarrow a_1; a_2]$ when refinement is atomic and non-atomic, respectively. In this figure, black dots represent abstract observable states (i.e., states that are relevant for observation and where every atomic action is completed) and white dots denote concrete invisible states (i.e., states that represent intermediate execution steps at a lower level of abstraction and that cannot be seen by an external observer). It is easy to see the difference: if the refinement is atomic, there is no observable state in between the execution of a_1 and a_2 (all-or-nothing); moreover, action b cannot be executed in between their execution (non-interruptible).

The atomic approach seems well-suited in some cases. For instance, when implementing one language into another one, one needs to implement the primitives of the former as compound programs of the latter. In this case, keeping atomicity in the implementation

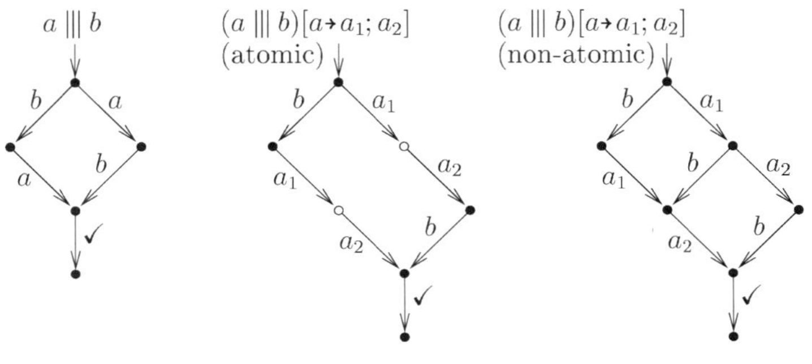

Fig. 1. Atomic and non-atomic refinement of $a \mid\mid\mid b$.

may be a vital feature for correctness. One example of this is [77], where Milner's CCS is mapped to a finer grained calculus; each transition of CCS is implemented as a suitable transaction (sequence of transitions executed atomically) of the latter calculus. Another example supporting atomic refinement is when mutual exclusion on a shared resource is necessary to prevent faulty behaviour. For instance, when refining an abstract write operation a on a shared variable x as a complex process u, we would like to allow further, possibly parallel, reading or writing operations on x only after the completion of u. Further arguments in favour of atomic refinement may be found in [19].

On the other hand, also the non atomic approach has its own adherents; see [9,31,50,63, 89,93,110,121,133]. Actually, this approach is on the whole more popular than the former. For instance, in the example of Figure 1, if b is an action completely independent of a it seems unreasonable to impose the restriction that b stays idle while executing the sequence $a_1 a_2$. In our opinion, the choice between atomic and non-atomic refinement should be driven by the application at hand.

1.4. *Syntactic versus semantic action refinement*

There are essentially two interpretations of action refinement, which we call *syntactic* and *semantic*. In the syntactic approach, the treatment of action refinement closely resembles the copy rule for procedure call (i.e., inlining the body for the calling action) in sequential programming. For instance, [110] exploits a *static* copy rule (syntactic replacement before execution); instead [8] follows the *dynamic* copy rule: as soon as a is to occur while executing t, the first action of u is performed reaching, say, a state u', after which that occurrence of a in t is syntactically replaced by u'. In either case, the semantics of $t[a \to u]$ is, *by definition*, the semantics of the term $t\{u/a\}$. Among other things, this implies that the process algebra is to be equipped with an operation of sequential composition (rather than the more standard action prefix) as studied, for instance, in the context of ACP [11], since otherwise it would not be closed under the necessary syntactic substitution.

In the *semantic* interpretation, a substitution operation is defined in the semantic domain used to interpret terms. Then the semantics of $t[a \to u]$ can be defined using this operator. For example, when using event structures as semantic domains, an event structure $\mathcal{E} = [\![u]\!]$, representing the semantics of u, would be substituted for every a-labelled event d in the event structure $[\![t]\!]$. The refinement operation preserves the semantic embedding of events: e.g., if d is in conflict with an event e, then all the events of $\mathcal{E}$ will be in conflict with e, and similarly for the order relation. Investigations of such refinement operators can be found in, e.g., [17,31,32,41,42,50,93,60,63,64,68,121,131,133] over the semantic domains of Prime, Free, Flow and Stable Event Structures, Configuration Structures, Families of Posets, Synchronization Trees, Causal Trees, ST Trees and Petri Nets. The advocates of the semantic substitution approach (to which the authors of this chapter belong) claim that the starting point is introducing a notion of semantic refinement as pure substitution in a semantic model, which, usually, is not very difficult. In contrast, the hard part is finding the operational definition for the syntactic operator of refinement that correctly implements semantic substitution, i.e., a concurrent counterpart of the copy rule in the sequential case.

These two approaches are inherently different; simple examples showing this are given in Section 6. Essentially, syntactically substituting u for a in t produces a confusion of the abstraction levels that is not possible with semantic substitution; such a confusion may originate new communications between processes that were not possible at the higher level, and, conversely, may destroy at the lower level some communications established at the higher level. Conceptually, this is an undesirable situation which, in general, prevents the definition of an algebraic theory for action refinement. Technically, syntactic refinement corresponds to a *homomorphism* between algebras whose signature is given by the language (since such a homomorphism is essentially generated by a mapping from actions to subterms, which is required to distribute over all the operators); on the other hand, semantic refinement defines an operation on the semantic model which is *compositional*. As the two approaches do not coincide, we cannot expect to be able, in general, to define compositional homomorphisms. In Section 6, we compare the two approaches with the aim to identify under which restrictions they yield the same result. That is, we report about conditions under which the following diagram commutes:

$$\begin{array}{ccc} t[a \to u] & \xrightarrow[\text{syntactic ref.}]{} & t\{u/a\} \quad \textit{syntax} \\ \text{semantic ref.}\Big\downarrow & & \Big\downarrow \\ [\![t]\!][a \to [\![u]\!]] & \longrightarrow & [\![t\{u/a\}]\!] \quad \textit{semantics} \end{array} \tag{1}$$

Not only does the result give a clearer understanding of the theory of action refinement, but also it is interesting for applications of action refinement to know when semantic refinement can be implemented by the conceptually simpler syntactic substitution.

It should be mentioned that a dual approach to the one described above is followed in [94]: they take syntactic refinement as a starting point and adapt the language (using an operator for *self-synchronization*) so that semantic refinement coincides with it.

Concerning atomic refinement, one can observe that this is naturally definable as a form of semantic refinement on trees labelled on sequences of actions, but can be also represented via syntactic substitution, provided that the process replacing the action is atomized (see Sections 3 and 6 for more details).

1.5. *Interleaving versus true concurrency*

As mentioned above, when action refinement is an operator of the language, a natural key problem is that of finding a notion of equivalence that is a congruence for such operator. Formally, given a candidate equivalence notion $\simeq$, we want to find the coarsest relation $\equiv$, contained in $\simeq$, that is a congruence for all the operators of the language; to be precise:

- whenever $u_1 \equiv u_2$, then $t[a \to u_1] \equiv t[a \to u_2]$;
- whenever $t_1 \equiv t_2$, then $t_1[a \to u] \equiv t_2[a \to u]$.

The first half of the congruence problem turns out to be easy: the main requirement is that one makes a clear distinction between deadlock (where a system can do nothing at all) and termination (where a system can do nothing except terminate, i.e., relinquish control); see also [61]. This distinction is easily made if one models termination as a special action (in this paper denoted $\checkmark$). The reason why a congruence has to make this distinction may be understood by considering an example. Let $t = a; b$, $u_1 = c$ – denoting the execution of c

leading to successful termination – and $u_2 = c; \mathbf{0}$ – denoting the execution of c leading to deadlock. u_1 and u_2 are equivalent when ignoring termination; however, $t[a \to u_1]$ can perform b after c, while $t[a \to u_2]$ cannot.

A solution to the second half of the congruence problem, on the other hand, may either be easy (in cases where one can stay within interleaving semantics) or rather difficult (in cases where one is forced to move to truly concurrent semantics), depending on the assumptions and the algebraic properties one wants to impose on the operator.

- One can define semantic action refinement as an operator on transition systems (e.g., see Section 3 below) that is well-defined up to strong bisimilarity. (For the case where internal moves are abstracted, the situation is slightly less straightforward: the construction is not well-defined up to the standard (rooted) weak bisimilarity, instead one has to resort to (rooted) *delay* or *branching* bisimilarity. A detailed discussion can be found in Section 2.4 below.) At any rate, it should be clear that there is no intrinsic reason why an interleaving relation cannot be a congruence for action refinement.
- However, the operator referred to above fails to satisfy a very intuitive and important property: namely, it does not distribute over parallel composition. If one wants this property to hold, the easiest way is to adopt *atomic* refinement instead, as argued in [48] (see above and Section 3); both strong and weak bisimilarity are congruences for this operator. The price is that one has either to distinguish concrete and abstract states or to use action sequences rather than single actions as transition labels.
- Atomic refinement is not always appropriate. If one requires a non-atomic refinement operator that distributes over parallel composition, this can still be defined on standard transition systems, provided that refinement is disallowed for all actions that decide choices, as well as all actions that occur concurrently with themselves; see [38]. Once more, strong bisimilarity is a congruence for the resulting operator. As a consequence of the limitations on refinable actions, this operator no longer distributes over *choice*. We do not discuss this operator any further.
- If one wants action refinement to distribute over parallel composition *and* to be non-atomic *and* to be applicable to all actions, irregardless of their position in a term, *then* it becomes necessary to use a model that is more expressive than standard transition systems. Among the earliest observations of this fact are [115,32]. Since it has received widespread attention in the literature, we will discuss this issue in more detail.

Let us consider $a \mathbin{|||} b$ and $a; b + b; a$, which are equivalent in interleaving semantics: the former represents the concurrent execution of the actions a and b, the latter their execution in either order. When refining a to $a_1; a_2$ (and distributing the refinement over parallel and sequential composition), the resulting processes become equivalent to, respectively, $(a_1; a_2) \mathbin{|||} b$ and $a_1; a_2; b + b; a_1; a_2$; these are no longer equivalent in interleaving semantics, as only the former can execute the sequence $a_1\, b\, a_2$. It follows that the required congruence $\equiv$ cannot equate $a \mathbin{|||} b$ and $a; b + b; a$.

A solution to this problem, which has received a large amount of attention in the literature on action refinement, is to move to so-called *truly concurrent* semantics, i.e., semantic models that contain more information about the concurrency of the system's activities than the standard interleaving semantics. For instance, *event-based models* (inspired by Winskel's *event structures*, see [139,140]) have been investigated for this purpose in [42, 50,63,64,121]. In Section 4 we present an example of an event-based truly concurrent (operational and denotational) semantics for a language with parallel composition, synchronization and refinement.

Isomorphism of event-based models gives rise to a congruence; however, it has been argued that this relation is now too strong (rather than too weak as in the case of the interleaving relations), in that it makes more distinctions than strictly necessary. This can be repaired by interpreting the event-based model up to a weaker relation than isomorphism, such as for instance *history-preserving bisimilarity* (see [60,116,46]); or, alternatively, by considering less distinguishing models such as *causal trees* (see [39,42]). It turns out that the *minimal* amount of information one must add (giving rise to the *coarsest* congruences contained in existing interleaving relations) is to distinguish the (related) beginnings and endings of all actions. This is called the *ST*-principle, after the name chosen by Van Glabbeek and Vaandrager in [65], where it appeared for the first time. In Section 5 we give an overview of truly concurrent observational criteria and the congruence properties to which they give rise; another, very systematic and detailed summary can be found in [64].

As pointed out by Meyer [105,106] and Vogler [134], the issue of finding congruences with respect to action refinement is also relevant to another, quite different area, namely that of completeness in the presence of process variables. This, too, is briefly discussed in Section 5.

It should be noted that the issue of finding the fully abstract model for action refinement can also be avoided altogether, by interpreting terms as *functions* on denotations rather than basic denotations. Namely, assume *Act* is the set of actions, *Lang* the language under consideration and $\mathcal{M}^{flat}$ is the space of denotational models for the flat (i.e., refinement-free) language fragment $Lang^{flat}$, with denotational constructions $\overline{op}$ for all operators *op* of $Lang^{flat}$. Based on these, one can define a new denotational model $\mathcal{M} = (Act \to \mathcal{M}^{flat}) \to \mathcal{M}^{flat}$: i.e., objects of $\mathcal{M}$ are *functions* that yield a "flat" model when provided with an arbitrary mapping $f : Act \to \mathcal{M}^{flat}$ that "pre-evaluates" (in fact, refines) all action occurrences.

Denotational constructions $\overline{op}$ on $\mathcal{M}$ are obtained by pointwise extension from $\mathcal{M}^{flat}$:

$$\overline{op}(M_1, \ldots, M_n) = \lambda f.\overline{op}\big(M_1(f), \ldots, M_n(f)\big)$$

for all $M_i \in \mathcal{M}$ $(i = 1, \ldots, n)$, except if *op* is actually a constant action a (interpreted as a nullary operator), in which case

$$\overline{op} = \lambda f.f(a).$$

$\mathcal{M}$ allows a straightforward definition of refinement, as an operator $\_[a \to \_] : (\mathcal{M} \times \mathcal{M}) \to \mathcal{M}$:

$$M_1[a \to M_2] = \lambda f.M_1\big(f \pm (a \mapsto M_2(f))\big)$$

where $f \pm (a \mapsto M)$ with $M \in \mathcal{M}^{flat}$ denotes the function $Act \to \mathcal{M}^{flat}$ mapping a to M and all $b \neq a$ to $f(b)$. This immediately gives rise to a corresponding semantic function $[\![\_]\!] : Lang \to \mathcal{M}$.

This approach is followed in [83]. An advantage is that, since $\mathcal{M}^{flat}$ is itself not required to be compositional for refinement, it can still consist of basic interleaving models, such as labelled transition systems or (in [83]) traces. On the other hand, this interpretation of terms as functions is intensional: it does not give rise to a concrete representation of behaviour. This is in sharp contrast to the spirit of the rest of the paper. We will not attempt a further comparison.

1.6. *Strict versus relaxed forms of refinement*

Non-atomic refinement is more flexible than atomic refinement, because it allows the concurrent execution of a process u refining an action a with the actions in t independent of a. For instance, in $(a \,|||\, b)[a \to a_1; a_2]$, b can be executed in between a_1 and a_2 only if refinement is non-atomic (see Figure 1).

Nonetheless, there are other desirable forms of flexibility that non-atomic refinement, as defined above, is unable to offer. For instance, consider $t = (a; b)[a \to a_1; a_2]$ where actually only a_1 is a necessary precondition (i.e., a cause) for the occurrence of b. According to the definition of non-atomic refinement, t is equivalent to $a_1; a_2; b$, which fails to show that b may be executed independently of a_2. Unfortunately, in the semantic domains usually considered, it happens that the causal context of actions is tightly preserved, hence enforcing causality in the refined system also when not strictly necessary. We could say that traditional refinement is too *strict*: it forces all abstract causalities to be inherited in the implementation.

A possible solution, proposed by Janssen, Poel and Zwiers [91] and Wehrheim [136, 126] is to introduce a certain degree of *relaxation* of the causal ordering during refinement. Technically, this is achieved by means of a *dependency relation* over the universe of actions, combined with a weak form of sequential composition that may allow the execution of actions of the second component if they are independent of those occurring in the first component. In the example above, one may declare that only a_1 and b are dependent, hence in $a_1 \cdot a_2 \cdot b$ (where $\cdot$ is weak sequential composition), b can be performed before a_2.

A possible relaxation of another kind concerns choice rather than sequential composition. Traditional action refinement requires that alternative actions, once refined, give rise to strictly alternative processes. A different view is taken in [51], where refinement is defined over event structures and the conflict relation is not respected tightly by refinement, e.g., conflicting events may be refined to processes that are not completely mutually exclusive. The intuitive motivation for this is that, in a competition, many actions are anyway performed by both processes before the decision of which will be served is taken.

Yet another form of relaxation concerning choice requires that not all options specified by a refinement function must indeed be offered by the refined system. For instance, if we refine a to $a'; b + a'; c$, the abstract system $a; d$ is implemented by the concrete system $(a'; b + a'; c); d$. An interesting alternative is to take the decision about which option to implement during the refinement step, hence allowing $a'; b; d$ or $a'; c; d$ as an implementation, or to turn the nondeterministic choice into a deterministic one, hence allowing $a'; (b + c); d$ as an implementation. This kind of design step is in line with the sort of transformation allowed by standard implementation relations (such as trace or failure inclusion). We do not know of any paper dealing with such relaxed forms of action refinement (see [124] for a discussion on the problems raised by such an approach in presence of communication).

1.7. *Vertical implementation*

Another way to obtain a more relaxed notion of action refinement, already briefly mentioned in Section 1.2, is by abandoning the notion of an operator and regarding action refinement as an implementation relation instead. There is a long tradition in defining process refinement theories based on the idea that a process I is an implementation of another process S if I is more directly executable, in particular more deterministic according to the chosen semantics. Examples can be found in, for example, [27,43,108]; see also [12] for a collection of papers in this line. As these theories do not take changes in the level of abstraction on which S and I are described into account, we call such implementation relations *horizontal*. On the other hand, as also pointed above, almost no theory has been developed to compare systems that realize essentially the same functionality but belong to conceptually different abstraction levels. For this purpose, we have introduced the concept of *vertical* implementation. Some sensible criteria that any vertical implementation relation should satisfy are listed below:

(1) It is parametric with respect to a *refinement* function r that maps abstract actions of the specification to concrete processes and thus fixes the implementation of the basic building blocks of the abstract system.
(2) It is flexible enough
  (i) to offer several possible implementations for any given specification, and
  (ii) not to require that the ordering of abstract actions is tightly preserved at the level of their implementing processes, i.e., refinement need not to be strict (as discussed in Section 1.6, above).
(3) It is a generalization of existing *horizontal* implementation relations; i.e., if the refinement function is the identity, then the vertical implementation should collapse to the horizontal implementation. So, the theory of horizontal and vertical implementations can be integrated uniformly.

As we have seen in the previous sections, the classic work on action refinement in process algebra, where it is interpreted as a substitution-like operation on the syntactic or semantic domain, satisfies few of these requirements. In particular, the consequence of the classic approach is that there is *only one* possible implementation for a given specification; in other words, the action refinement function is used as a *prescriptive tool* to specify the only way abstract actions are to be implemented. However, there is no deep motivation for this functional point of view (only one implementation) in favour of a relational one (more than one implementation). For instance, when considering $(a \mid\mid\mid b)[a \to a_1; a_2]$, there is no real reason for not accepting also $a_1; a_2; b + b; a_1; a_2$ as a possible, more sequential implementation; similarly, for $(a; b + b; a)[a \to a_1; a_2]$ the more parallel implementation $a_1; a_2 \mid\mid\mid b$ could be admitted.

The concept of action refinement through vertical implementation has a striking consequence. The congruence problem (discussed at length in Section 1.5 above) simply disappears: since a specification may admit non-equivalent implementations, *a fortiori* two equivalent specifications need not to have equivalent implementations. Hence, there is no longer a need to move to truly concurrent semantics. This has the advantage of allowing to reuse most existing techniques developed for interleaving semantics. In particular, it is a natural requirement that vertical implementation may collapse to some horizontal relation, by hiding all the actions that were refined, reminiscent of the interface refinement principle discussed in [26]. This makes it possible to mix vertical refinement with established methods for horizontal implementation.

Some of the basic ideas behind this approach were proposed first (in a restrictive setting) in [72] and later (independently) in [118,119]. See Section 8, based on [123], for more details.

1.8. *Overview of the chapter*

We can classify the choices we have made in the material of this chapter according to the discussion above.

Operator versus hierarchy. In all the next sections, except the last one (Section 8), we consider some process algebra enriched with an operator for action refinement.

Atomic versus non-atomic. In all the sections that deal with the operator of action refinement but one, we stick to non-atomic refinement. Atomic refinement is dealt with in Section 3, where we consider a process algebra with an operator for parallel composition but without communication, and also (briefly) in Section 7 in the context of action dependencies.

Syntactic versus semantic. Throughout the chapter we use semantic refinement, as the operation of action refinement is always defined on a semantic domain (trees or event structures), with the exception of Section 6, where we report sufficient conditions to guarantee that the two approaches are the same.

Interleaving versus true concurrency. We discuss the simplest cases (sequential systems in Section 2 and atomic refinement in Section 3) using interleaving semantics; then, when parallel composition comes into play and refinement is non-atomic, we move to truly concurrent semantics (Sections 4 and 5).

Strict versus relaxed. All the sections on the operation deal with strict refinement, except Section 7, where we discuss other approaches using a dependency relation to ensure some form of relaxation of the causality relation.

Vertical implementation is covered in Section 8.

The chapter is organized as follows: Section 2 studies non-atomic refinement for the class of sequential systems. We introduce many concepts that will be used throughout the paper, such as well-formedness conditions on admissible terms, allowable refinements and some standard interleaving behavioural equivalences. Section 3 deals with a larger language with parallel composition (but without communication) under the assumption that refinement is atomic. The operational semantics is not standard, making use of concrete invisible states; the denotational semantics uses trees labelled on sequences. Section 4 deals with non-atomic refinement for a full-fledged process algebra. The denotational semantics is given in terms of stable event structures and the operational semantics is very concrete: process terms and transitions are tagged by event annotations. Then, observational semantics and congruence issues for the full language are discussed in Section 5. In Section 6 we present some conditions ensuring that syntactic and semantic action refinement coincide. Relaxed forms of refinement are recalled in Section 7, while Section 8 reports on the issue of vertical implementation relations.

As much as possible, we have presented all the results of this chapter in a single format, by using a common underlying process algebraic language in which the different approaches have been formulated. Predictably, however, various theories come with their own specific concepts, operators and limitations, making a fully integrated presentation impossible. To enable the reader to find his way among the different language fragments and well-formedness conditions used in the various sections, Table 16 gives an overview.

Note that in this chapter, we have limited ourselves to a process algebraic understanding of action refinement. Thus, we have not included a comparison with methods addressing the same concerns in other fields of theoretical computer science, such as logic-based, state-based or stream-based refinement; e.g., [10,28,29,54,96,98].

1.9. *Whither action refinement?*

After enjoying a broad interest in the years 1989–1995 and thereabouts (as evidenced by the sheer number of papers quoted above, as well as the Ph.D. Theses [2,33,36,53,56,71, 90,92,118,128,137]), the subject of action refinement has left the central stage of research in process algebra. Yet we feel that the basic principle underlying action refinement, concerning changes in the level of abstraction and the grain of atomicity at which a system is described, is not less relevant now than it was a decade ago. Indeed, the same principle is very fundamental in, for instance, object-based systems, where the method interface of an object on the one hand and its implementation on the other are prime candidates for a description in terms of action refinement.

If we analyze the kind of work that has received the most attention during the aforementioned period, it can be seen that the main issue that has been studied, almost to the exclusion of everything else, is the congruence question for the refinement operator. This has greatly enhanced the insight in the relative advantages of various truly concurrent semantics, some of which are now also being used in other contexts (e.g., ST-semantics for stochastic process algebras, see [23,25]); and similar for variations on weak (interleaving) bisimulation (see, for instance, [68]). Nevertheless, with the benefit of hindsight, the concentration on this one subject can also be seen to have had some disadvantages: on the one hand, the refinement operator itself has not found much practical use, whereas on the other, this singular focus has prevented the development of alternative approaches, such as the use of action refinement as a correctness criterion rather than an operator. This is what we are currently trying to remedy with the concept of vertical implementation.

Having realized this change in perspective, it becomes clear that there is an enormous amount of work yet to be done. In particular, no attention at all has yet been paid to the extremely important question of how to integrate data refinement and action refinement. Similarly, the only aspect of atomicity addressed so far is the (non-)interference of atomic actions; at least as important is their all-or-nothing nature, in particular the possibility to *abort* an atomic action that (on a concrete level) has already started. Remarkably, in the context of process algebra the latter issue has not been addressed by *any* theory of action refinement we know. Furthermore, although we have begun to explore some of the possible variations on vertical bisimulation, similar studies should be initiated for vertical testing, since the lack of atomicity of test primitives is causing hitherto unexplored problems. Finally, the most challenging task is to apply the emerging theory to realistic applications, be it in the form of case studies of any size or by integrating it with the design phase of some software engineering lifecycle.

2. Sequential systems

We start our technical presentation by addressing the simplest case: a process algebra without parallelism and communication. In this case, we can easily accommodate the new construct of action refinement within interleaving semantics.

2.1. *The sequential language*

We assume the existence of a universe *Act* of visible actions, ranged over by $a, b, \ldots$, and an invisible action $\tau \notin Act$; we write $Act_\tau = Act \cup \{\tau\}$, ranged over by α. Furthermore, we assume a set *Var* of process variables, ranged over by x, y, z. The language of *sequential processes* (or *agents*), denoted $Lang^{seq}$ and ranged over by t, u, v, is the set of *well-formed* terms generated by the following grammar:

$$\mathsf{T} ::= \mathbf{0} \mid \mathbf{1} \mid \alpha \mid \mathsf{V} + \mathsf{V} \mid \mathsf{T}; \mathsf{V} \mid \mathsf{T}/A \mid \mathsf{T}[a \to \mathsf{V}] \mid x \mid \mu x.\mathsf{V},$$

where $a \in Act$, $\alpha \in Act_\tau$, $A \subseteq Act$ and $x \in Var$ are arbitrary. In this grammar, T is an arbitrary term whereas V stands for a so-called *virgin operand*, on which we impose the condition that it may not contain an *auxiliary operator* – where the constant $\mathbf{1}$ is the only auxiliary operator of $Lang^{seq}$. These matters are discussed below in more detail.

We first go into the intuitive meaning of the operators, at the same time informally introducing the concepts of well-formedness and guardedness.

- $\mathbf{0}$ is a deadlocked process that cannot proceed.
- $\mathbf{1}$ is a terminated process, that is, a process that immediately terminates with a transition labelled $\checkmark \notin Act_\tau$, expressed by $\mathbf{1} \xrightarrow{\checkmark} \mathbf{0}$.
  $\mathbf{1}$ is an *auxiliary operator*, in the sense that we expect the "user" of the language to write down terms not containing $\mathbf{1}$. Rather, it will be needed to express the semantics of other operators (notably, sequential composition). The fact that an operator is auxiliary influences the well-formedness of terms; see below.
- α can execute action α, and then terminates, i.e., $\alpha \xrightarrow{\alpha} \mathbf{1}$ ($\xrightarrow{\checkmark} \mathbf{0}$).
- $t + u$ indicates a CCS-like choice between the behaviours described by the sub-terms t and u. The choice is decided by the first action that occurs from either sub-term, after which the other sub-term is discarded.

We call the operands t and u *virgin*, because (due to the semantics of choice) it is clear that they are untouched, in the sense of not having participated in any transition – otherwise the choice would have been resolved. An important general well-formedness condition on terms is that *virgin operands may not contain auxiliary operators*. Specifically, in this case, we require that t and u contain no occurrence of **1**. (The technical consequence of this condition is that neither operand can be terminated, and hence no $\checkmark$-transition can resolve a choice. This circumvents some intricate technical problems, for instance in the definition of a denotational event-based model; see, e.g., [118, Example 3.7]. A different solution with essentially the same consequence was chosen in [7,49], where a choice may terminate only if *both* operands can do so.)

- $t; u$ is the sequential composition of t and u, i.e., t proceeds until it terminates properly, after which u takes over; if t does not terminate properly, u is not enabled. This semantics is in the line of ACP [11], and differs from the one in, e.g., [109], where u starts after t is deadlocked. We have chosen the former as it is the only one distinguishing correctly between deadlock and termination.
  Again, u is a *virgin operand*: once it participates in a transition, t is discarded from the term.
  The two main motivations for using sequential composition instead of CCS action prefixing are as follows. On the one hand, the operational (sometimes also the denotational) semantics for action refinement can be naturally defined only by means of a sequential composition operator, as we will see later on; on the other hand, syntactic substitution, which is the way action refinement is implemented in many papers, is naturally defined by means of such an operator.
- t/A behaves as t, except that the actions in A are *hidden*, i.e., turned into the internal action τ that cannot be observed by any external observer.
- $t[a \to u]$ is a process t where a is refined to u. The operand u is *virgin*. Since, by well-formedness, virgin operands may not contain auxiliary operators, it is certain that the refinement of an action is an agent that is not terminated; this prevents the so-called *forgetful refinement* (the implementation of an action by **1**), which is not only technically difficult, as discussed, e.g., in [31,118], but also counter-intuitive.
  Apart from forbidding forgetful refinements, we are still rather generous in allowing some types of refining agents that are rather questionable, such as deadlocked or never-ending ones. Intuitively, it seems natural to require that an action, which by itself cannot deadlock, is implemented by a process that cannot deadlock either, since otherwise refinement would introduce deadlocks. However, imposing such a semantic restriction is an unnecessary burden; on the one hand, there is no technical problem in managing deadlocked refining agents; on the other hand, it is not easy to characterize syntactically a large class of deadlock-free processes (while, semantically, the problem is even undecidable).
  Similar arguments hold for infinite refinement: intuitively, an action is certainly accomplished in a finite amount of time; therefore its refinement should eventually terminate, as required, e.g., in [48,90]. A typical term satisfying this requirement under a suitable fairness assumption is $\mu x.a; x + b$. Again, however, allowing arbitrary recursion in refining agents does not complicate matters, whereas restricting it to some special cases would.

Table 1
Free variables and syntactic substitution (where *op* is an arbitrary operator)

t	$fv(t)$	$t\{u/x\}$
$op(t_1,\dots,t_n)$	$\bigcup_{1\leqslant i\leqslant n} fv(t_i)$	$op(t_1\{u/x\},\dots,t_n\{v/x\})$
y	$\{y\}$	$\begin{cases} u & \text{if } y = x \\ y & \text{otherwise} \end{cases}$
$\mu y.t_1$	$fv(t_1)\setminus\{y\}$	$\mu z.(t_1\{z/y\}\{u/x\})$ where $z \notin \{x\}\cup fv(t,u)$

- $x \in Var$ is a process variable, presumably bound by some encompassing recursive operator (see next item), or to be replaced by substitution. The variables of t that are not bound are called *free*: we write $fv(t)$ for the free variables of a term t and $fv(t,u)$ for $fv(t) \cup fv(u)$. A term t is called *closed* if $fv(t) = \emptyset$. The free variables can be *instantiated* by substitution: $t\{u/x\}$ denotes the substitution within the term t of every free occurrence of x by the term u. The free variables and their instantiation are formally defined in Table 1.
- $\mu x.t$ with $x \in Var$ is a recursive term. It can be understood through its unfolding, $t\{\mu x.t/x\}$. The variable x is considered to be *bound* in $\mu x.t$, meaning that it cannot be affected by substitution. Therefore, the identity of bound variables is considered irrelevant; in fact, we apply the standard technique of identifying all terms up to renaming of the bound variables, meaning that if y is a fresh variable not occurring free in t, then $\mu x.t$ and $\mu y.t\{y/x\}$ are identified in all contexts. As a further well-formedness condition, we require that all recursion variables are *guarded*, in the sense defined below.
  In $\mu x.t$, the recursion body t is considered to be a *virgin operand*, and hence by well-formedness may not contain the auxiliary operator $\mathbf{1}$.

We will use $Lang^{seq,fin}$ to denote the recursion-free fragment of $Lang^{seq}$.

Guardedness. The notion of guardedness is used to simplify the proof of correspondence of various kinds of operational and denotational semantics developed in this chapter for different fragments of *Lang*. As usual, the idea is that the semantic model of a recursive term corresponds to a fixpoint of the denotational function generated by its body, and this fixpoint is unique if the variable is guarded in the term; see for instance [108]. Therefore, if one defines the denotational semantics as the fixpoint, and on the other hand, proves that the operational semantics also gives rise to a fixpoint, then the two must coincide.

DEFINITION 2.1. We first define what it means for a term to be a *guard*.
- $\mathbf{1}$ and x are *not* guards;
- $\mathbf{0}$, α and $t+u$ are guards;
- $t[a \to u]$ is a guard if t is a guard;
- For all other operators *op*, $op(t_1,\dots,t_n)$ is a guard if one of the t_i is a guard.

Next, we define what it means for a variable to be guarded in a term.
- x is guarded in y ($\in Proc$) if $x \neq y$;
- x is guarded in $t;u$ if x is guarded in t and either t is a guard or x is guarded in u;
- x is guarded in $\mu y.t$ if either $x = y$ or x is guarded in t;
- For all other operators *op*, x is guarded in $op(t_1,\dots,t_n)$ if x is guarded in all t_i.

(Note that the last clause "for all operators *op*" includes the case where *op* is a constant, i.e., $\mathbf{0}$, $\mathbf{1}$ or α.) A typical example of guarded recursion is $\mu x.(\mathbf{1}; a); x$, and of non-guarded recursion, $\mu x.x \mathbin{|||} a$.

Well-formedness. To summarize the well-formedness conditions on $Lang^{seq}$:

- A virgin operand may contain no auxiliary operators. (The virgin operands are the ones denoted V in the grammar of $Lang^{seq}$; the only auxiliary operator in $Lang^{seq}$ is $\mathbf{1}$.)
- Recursion is allowed on guarded variables only.

2.2. *Operational semantics*

A labelled transition system (lts, for short) is a tuple $\langle Lab, N, \rightarrow\rangle$, where Lab is a set of labels (ranged over by λ), N a set of nodes (ranged over by n) and $\rightarrow \subseteq N \times Lab \times N$ a (labelled) transition relation. $(n, \lambda, n') \in \rightarrow$ is more often denoted $n \xrightarrow{\lambda} n'$. Sometimes we use transition systems with initial states, $\langle Lab, N, \rightarrow, \iota\rangle$, where $\iota \in N$. Unless explicitly stated otherwise, Lab will equal $Act_{\tau\checkmark} = Act_\tau \cup \{\checkmark\}$ in this chapter; we usually omit it. N will often correspond to the terms of a language – for instance, $Lang^{seq}$. Furthermore, the transition systems we use in this chapter all satisfy the following special properties regarding $\checkmark$-labelled transitions:

Termination is deterministic. If $n \xrightarrow{\checkmark} n'$ and $n \xrightarrow{\lambda} n''$, then $\lambda = \checkmark$ and $n'' = n'$.

Termination is final. If $n \xrightarrow{\checkmark} n'$, then there is no $\lambda \in Lab$ such that $n' \xrightarrow{\lambda}$.

Finally, some more terminology regarding transition systems:

- $n \in N$ is called *terminated* if $n \xrightarrow{\checkmark}$;
- $n \in N$ is called *deadlocked* if there is no $\lambda \in Lab$ such that $n \xrightarrow{\lambda}$;
- $n \in N$ is called *potentially deadlocking* if either n is deadlocked or there is a $n \xrightarrow{\lambda} n'$ with $\lambda \neq \checkmark$ such that n' is potentially deadlocking.

One of the main uses of lts's in this chapter is to provide operational semantics. The definition of the transition relation is according to Plotkin's SOS approach [114], meaning that it is the least relation generated by a set of axioms and rules concerning the transition predicate. For $Lang^{seq}$, the operational rules are given in Table 2. Most operational rules are standard; the only unusual ones are those for refinement. If the action to be refined is executed by t, then the first action of the refinement u is executed instead, followed by the whole residual u'; only when u' is terminated, the computation will proceed with t' subject to the refinement.

Note that, due to the well-formedness of terms (especially the condition that $\mathbf{1}$ does not occur in the operands of choice), termination is indeed deterministic and final.

The essential consequence of guardedness can be expressed in terms of its operational semantics if we apply the operational rules also on open terms, as follows:

PROPOSITION 2.2. *Assume x to be guarded in t.*

(1) $t\{u/x\} \xrightarrow{\lambda} t'$ *if and only if* $t \xrightarrow{\lambda} t''$ *for some* t'' *with* $t' = t''\{u/x\}$.
(2) $\mu x.t \xrightarrow{\alpha} t'$ *if and only if* $t \xrightarrow{\alpha} t''$ *for some* t'' *with* $t' = t''\{\mu x.t/x\}$.

Table 2
Transition rules for $Lang^{seq}$

$$\frac{}{\mathbf{1} \xrightarrow{\checkmark} \mathbf{0}} \qquad \frac{}{\alpha \xrightarrow{\alpha} \mathbf{1}} \qquad \frac{t \xrightarrow{\alpha} t'}{t+u \xrightarrow{\alpha} t'} \qquad \frac{u \xrightarrow{\alpha} u'}{t+u \xrightarrow{\alpha} u'}$$

$$\frac{t \xrightarrow{\alpha} t'}{t;u \xrightarrow{\alpha} t';u} \qquad \frac{t \xrightarrow{\checkmark} t' \quad u \xrightarrow{\alpha} u'}{t;u \xrightarrow{\alpha} u'} \qquad \frac{t \xrightarrow{\lambda} t' \quad \lambda \notin A}{t/A \xrightarrow{\lambda} t'/A} \qquad \frac{t \xrightarrow{a} t' \quad a \in A}{t/A \xrightarrow{\tau} t'/A}$$

$$\frac{t \xrightarrow{a} t' \quad u \xrightarrow{\alpha} u'}{t[a \to u] \xrightarrow{\alpha} u';(t'[a \to u])} \qquad \frac{t \xrightarrow{\lambda} t' \quad \lambda \neq a}{t[a \to u] \xrightarrow{\lambda} t'[a \to u]} \qquad \frac{t\{\mu x.t/x\} \xrightarrow{\alpha} t'}{\mu x.t \xrightarrow{\alpha} t'}$$

2.3. *Denotational semantics*

We now present a denotational semantics for $Lang^{seq}$, using edge-labelled trees as a model. An edge-labelled tree $\mathcal{T}$ is an lts with initial state $\langle N, \to, \iota\rangle$, which is connected, acyclic, has no $\to$-predecessor for ι and precisely one $\to$ -predecessor for all other nodes. The trees that are used to refine actions are always non-terminated. The class of all trees is denoted $\mathbf{T}$.

A tree can be easily obtained from an lts with initial state through the *unfolding* operation. The nodes of the unfolding correspond to *paths* through $\mathcal{T}$, starting from the initial state and including all intermediate nodes and labels. Formally, given the lts $\mathcal{T} = \langle N, \to, \iota\rangle$, its unfolding, $Unf\,\mathcal{T}$ is the tree $\langle N', \to', \iota\rangle$, where:

- $N' = \{n_1\lambda_1 n_2\lambda_2 \ldots \lambda_{k-1} n_k \in (N\,Lab)^* N \mid n_1 = \iota,\ \forall 1 \leqslant i \leqslant k\colon n_i \xrightarrow{\lambda_i} n_{i+1}\}$.
- $\to' = \{(\vec{n}\, n_1, \lambda, \vec{n}\, n_1\, \lambda\, n'_1) \mid n_1 \xrightarrow{\lambda} n'_1\}$.

We are now ready to define the operations on trees.

Deadlock. $\mathcal{T}_\perp = \langle\{\mathbf{0}\}, \emptyset, \mathbf{0}\rangle$.

Termination. $\mathcal{T}_\checkmark = \langle\{\mathbf{1}, \mathbf{0}\}, \{(\mathbf{1}, \checkmark, \mathbf{0})\}, \mathbf{1}\rangle$.

Single action. $\mathcal{T}_\alpha = \langle\{\alpha, \mathbf{1}, \mathbf{0}\}, \{(\alpha, \alpha, \mathbf{1}), (\mathbf{1}, \checkmark, \mathbf{0})\}, \alpha\rangle$.

Sequential composition. $\mathcal{T}_1; \mathcal{T}_2 = Unf\langle N, \to, \iota\rangle$, where

- $N = N_1 \cup \{(n, n') \mid n \xrightarrow{\checkmark}_1, n' \in N_2\}$;
- $\to = \{(n, \alpha, n') \mid n \xrightarrow{\alpha}_1 n' \not\xrightarrow{\checkmark}_1\}$
  $\cup \{(n, \alpha, (n', \iota_2)) \mid n \xrightarrow{\alpha}_1 n' \xrightarrow{\checkmark}_1\}$
  $\cup \{((n, n'), \lambda, (n, n'')) \mid n \xrightarrow{\checkmark}_1, n' \xrightarrow{\lambda}_2 n''\}$;
- $\iota = \iota_1$ if $\iota_1 \not\xrightarrow{\checkmark}_1$, otherwise $\iota = (\iota_1, \iota_2)$ (i.e., when $\mathcal{T}_1$ is isomorphic to $\mathcal{T}_\checkmark$).

Choice. If $\iota_i \not\xrightarrow{\checkmark}_i$ for $i = 1, 2$ and $N_1 \cap N_2 = \emptyset$, then $\mathcal{T}_1 + \mathcal{T}_2 = Unf\langle N, \to, \iota\rangle$, where

- $N = N_1 \cup N_2 \cup \{(\iota_1, \iota_2)\}$;
- $\to = \to_1 \cup \to_2 \cup \{((\iota_1, \iota_2), \alpha, n) \mid \iota_1 \xrightarrow{\alpha}_1 n\} \cup \{((\iota_1, \iota_2), \alpha, n) \mid \iota_2 \xrightarrow{\alpha}_2 n\}$;

- $\iota = (\iota_1, \iota_2)$.

Hiding. $\mathcal{T}/A = \langle N, \to', \iota \rangle$, where

- $\to' = \{(n, \lambda, n') \mid n \xrightarrow{\lambda} n', \lambda \notin A\} \cup \{(n, \tau, n') \mid n \xrightarrow{\alpha} n', \alpha \in A\}$.

Refinement. If $\iota_2 \not\xrightarrow{\checkmark}_2$, then $\mathcal{T}_1[a \to \mathcal{T}_2] = \mathit{Unf}\langle N, \to, \iota \rangle$, where

- $N = \{(n_1, n_2) \mid n_1 \in N_1, n_2 \in N_{R(n_1)}\}$;
- $\to = \{((n_1, n_2), \mu, (n'_1, n'_2)) \mid n_1 \xrightarrow{\lambda}_1 n'_1, n_2 \xrightarrow{\checkmark}_{R(n_1)}, \iota_{R(n'_1)} \xrightarrow{\mu}_{R(n'_1)} n'_2\}$
  $\cup \{((n_1, n_2), \alpha, (n_1, n'_2)) \mid n_2 \xrightarrow{\alpha}_{R(n_1)} n'_2\}$;
- $\iota = (\iota_1, \iota_{R(\iota_1)})$

and $R : N_1 \to \mathbf{T}$ is defined as follows:

$$R : n \mapsto \begin{cases} \mathcal{T}_{\checkmark} & \text{if } n = \iota_1, \\ \mathcal{T}_\lambda & \text{if } n' \xrightarrow{\lambda} n \text{ for } \lambda \neq a, \\ \mathcal{T}_2 & \text{if } n' \xrightarrow{a} n. \end{cases}$$

Some comments on the operations above are mandatory. In the operation of sequential composition, the second argument is "reproduced" in as many copies as the number of the nodes that are terminated. The choice operation is nothing but a coalesced sum of trees. Note that these operations are both defined with the help of the *unfold* operation on trees; this simplifies the presentation. The refinement operation is rather unusual, but it holds also when the two trees are infinite as well as when the second tree is potentially deadlocking. A function R is defined, associating to each node n of $\mathcal{T}_1$ a tree depending on the label λ of its incoming transition (which is uniquely determined since $\mathcal{T}_1$ is a tree): $R(n)$ equals $\mathcal{T}_\lambda$ if λ is anything but the action to be refined, $\mathcal{T}_2$ if λ is the action to be refined, and $\mathcal{T}_{\checkmark}$ if there is no incoming transition (n is the initial state). Then each edge $n \xrightarrow{\lambda}_1 n'$ of $\mathcal{T}_1$ is replaced by the tree $R(n')$.

EXAMPLE 2.3. Consider the following trees (where the node identities are made explicit):

$$\mathcal{T}_1 = A \xrightarrow{a} B \xrightarrow{\checkmark} C \qquad \mathcal{T}_{\checkmark} = 0 \xrightarrow{\checkmark} 1 \qquad \mathcal{T}_b = 2 \xrightarrow{b} 3 \xrightarrow{\checkmark} 4$$
$$\phantom{\mathcal{T}_1 = A} \searrow^{b} D \xrightarrow{a} E$$

$$\mathcal{T}_2 = 5 \xrightarrow{b} 6 \xrightarrow{\checkmark} 7$$
$$\phantom{\mathcal{T}_2 = 5} \searrow^{c} 8 \xrightarrow{d} 9$$

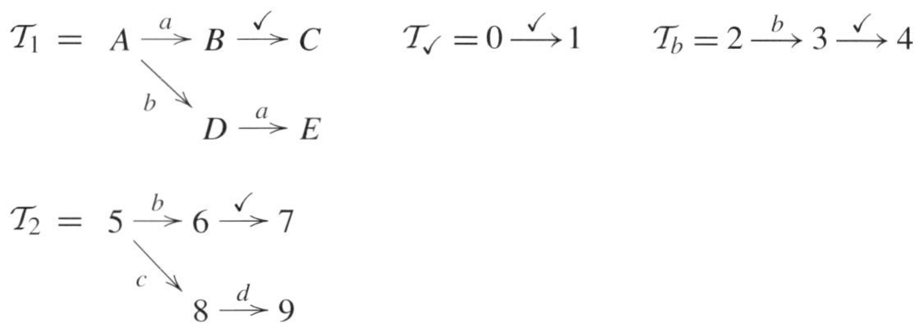

Note that $\mathcal{T}_1$ is a model of $a + b; a; \mathbf{0}$ and $\mathcal{T}_2$ is a model of $b + c; d; \mathbf{0}$. The refinement $\mathcal{T}_1[a \to \mathcal{T}_2]$ is then given by the tree

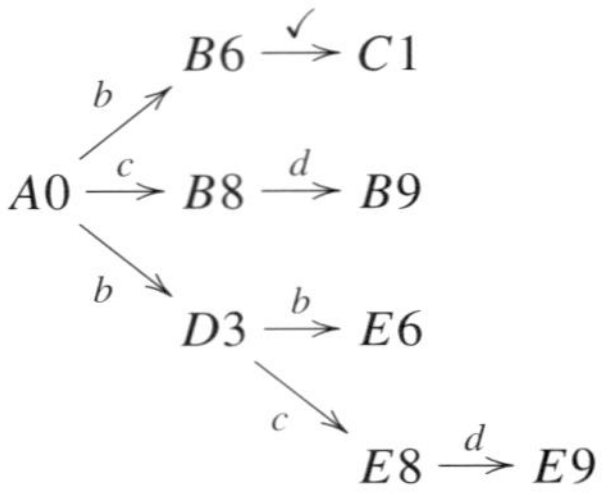

The first thing to make sure of is that the above constructions indeed yield trees, satisfying all the conditions listed above. Since we have defined them through unfolding, for all the operators the proof is straightforward.

PROPOSITION 2.4. *Each of the operators above applied to trees, when defined, again yields a tree.*

In order to deal with recursion, we use an approximation ordering over the class $\mathbf{T}$, which is essentially the one used in [138]. We assume that $\mathcal{T}_i = \langle N_i, \to_i, \iota_i \rangle$ for $i = 1, 2$.

$$\mathcal{T}_1 \sqsubseteq \mathcal{T}_2 :\Longleftrightarrow N_1 \subseteq N_2, \quad \to_1 = \to_2 \cap (N_1 \times Lab \times N_1), \quad \iota_1 = \iota_2$$

(where the symbol ":$\Longleftrightarrow$" stands for "is defined to hold if and only if"). Then the following property states that this gives rise to an appropriate semantic domain, namely a (bottom-less) complete partial order, containing a least upper bound for all $\sqsubseteq$-chains. The proof is omitted as it is a variation of the one in [138, Theorem 3.4].

PROPOSITION 2.5. $\langle \mathbf{T}, \sqsubseteq \rangle$ *is a complete partial order with minimal elements* $\langle \{n\}, \emptyset, n \rangle$ *for all* n *and least upper bounds* $\bigsqcup_i \mathcal{T}_i = \langle \bigcup_i N_i, \bigcup_i \to_i, \iota \rangle$ *for all* $\sqsubseteq$*-directed sets* $\{\mathcal{T}_i\}_i \subseteq \mathbf{T}$ *(with* $\iota_i = \iota$ *for all* i*).*

The absence of a bottom element is not deleterious, since the minimal elements do just as well as the starting point of approximation; in particular, note that $\mathcal{T}_\perp$ is a minimal element of $\mathbf{T}$. The next observation is that all the operations defined above are $\sqsubseteq$-continuous functions on $\mathbf{T}$. Indeed, as termination and deadlock are dealt with properly, also the operations of sequential composition (notably on its first argument) and refinement are continuous. Hence, recursion can be computed as the limit of the chain of its approximations. We define approximations of recursive terms, $\mu^i x.t$ for all $i \in \mathbb{N}$, in the standard way (see [138]):

$$\begin{aligned} \mu^0 x.t &= \mathbf{0}, \\ \mu^{i+i} x.t &= t\{\mu^i x.t/x\}. \end{aligned}$$

The denotational tree semantics of $Lang^{seq}$ is then given in Table 3. In order to express in

Table 3
Denotational tree semantics; $\overline{op}$ is the semantic counterpart of op

$$\begin{aligned}
[\![\mathbf{0}]\!] &= \mathcal{T}_{\perp} \\
[\![\mathbf{1}]\!] &= \mathcal{T}_{\surd} \\
[\![\alpha]\!] &= \mathcal{T}_{\alpha} \\
[\![op(t_1, \ldots, t_n)]\!] &= \overline{op}([\![t_1]\!], \ldots, [\![t_n]\!]) \quad (op \text{ any non-nullary operator}) \\
[\![\mu x.t]\!] &= \textstyle\bigcup_{i \in \mathbb{N}} [\![\mu^i x.t]\!]
\end{aligned}$$

what sense the operational and denotational semantics coincide, we first have to go into the issue of semantic relations among transition systems.

2.4. *Behavioural semantics and congruences*

In an interleaving operational semantics such as the above, a widely accepted equivalence relation on processes is *strong bisimilarity*; see [108]. Here we recall its definition.

DEFINITION 2.6. Let $\mathcal{T}$ be a transition system.

- A *bisimulation* over $\mathcal{T}$ is a symmetric relation $\rho \subseteq N \times N$ such that for all $n_1 \rho n_2$ and $n_1 \xrightarrow{\lambda} n_1'$, there exists n_2' such that $n_2 \xrightarrow{\lambda} n_2'$ and $n_1' \rho n_2'$;
- *Strong bisimilarity* over $\mathcal{T}$, denoted $\sim$, is the largest bisimulation over $\mathcal{T}$.

Moreover, we can also compare states from different transition systems by taking the (disjoint) union of the transition systems and then applying the above definition. A standard proof (cf. [11,108]) shows that bisimulation is a congruence over $Lang^{seq}$. Alternatively, one can use the "SOS format" theory (e.g., [18,44,81]; see [6] for an overview) for this purpose (the rules in Table 2 are in fact in the De Simone format of [44]).

PROPOSITION 2.7. *$\sim$ is a congruence over $Lang^{seq}$.*

Since our operational rules are in the De Simone format, [4] guarantees there is a complete axiomatization for $Lang^{seq,fin}$, notably also for the operator of action refinement. It turns out that the set of axioms is nothing but the usual set for ACP (see [11]), together with some axioms stating that refinement distributes over the other operators (see Section 6).

Furthermore, an inductive argument very similar to the one used to prove Proposition 2.7 establishes the compatibility of the operational and denotational semantics.

PROPOSITION 2.8. *$t \sim [\![t]\!]$ for all $t \in Lang^{seq}$.*

PROOF (*Sketch*). By induction on the structure of t. In fact, for those tree constructions that make use of the unfolding operator (choice, sequential composition and refinement) the induction step can already be proved on the transition system obtained *before* unfolding. Since (it is easily shown that) $Unf\,\mathcal{T} \sim \mathcal{T}$ for every lts $\mathcal{T}$, this suffices.

As an example, we will sketch the case for refinement. Let $\mathcal{T}_i = [\![t_i]\!]$ for $i = 1, 2$, and assume $t_i \sim \mathcal{T}_i$. Now $t_1[a \to t_2]$ is proved by the fact (of which we omit the proof) that the following relation is a bisimulation:

$$\rho = \big\{(t_1[a \to t_2], (\iota_1, \iota_{\mathcal{T}_\checkmark}))\big\} \cup \big\{(u_2; u_1[a \to t_2], (n_1, n_2)) \mid u_1 \sim n_1, u_2 \sim n_2\big\}.$$

For the case of recursion, the proof relies on well-formedness, in particular guardedness of recursion: using Proposition 2.2 and also Proposition 2.7, it can be proved that $t\{u/x\} \sim u$ implies $u \sim \mu x.t$ for every guarded recursive term $\mu x.t$. Since by construction, $[\![\mu x.t]\!]$ is a fixpoint of the semantic function that maps all $[\![u]\!]$ to $[\![t\{u/x\}]\!]$, we are done. □

τ-abstracting equivalences. The action τ represents an internal activity that should be considered invisible to some extent. In the literature, the most widely adopted τ-insensitive equivalence is (rooted) *weak bisimilarity*; see, e.g., [108]. To recall the definition, let $\Rightarrow = \xrightarrow{\tau}{}^*$.

DEFINITION 2.9. Let $\mathcal{T}$ be a labelled transition system.

- A *weak bisimulation* over $\mathcal{T}$ is a symmetric relation $\rho \subseteq N \times N$ such that for all $n_1 \rho n_2$ and $n_1 \xrightarrow{\lambda} n_1'$, one of the following holds:
  - $\lambda = \tau$ and $n_1' \rho n_2$;
  - there exists n_2' such that $n_2 \Rightarrow \xrightarrow{\lambda} \Rightarrow n_2'$ and $n_1' \rho n_2'$.
- A *root* of a binary relation $\rho \subseteq N \times N$ is a relation $\hat{\rho} \subseteq \rho$ such that for all $n_1 \hat{\rho} n_2$
  - if $n_1 \xrightarrow{\tau} n_1'$, then there exists n_2' such that $n_2 \Rightarrow \xrightarrow{\tau} \Rightarrow n_2'$ and $n_1' \rho n_2'$;
  - if $n_2 \xrightarrow{\tau} n_2'$, then there exists n_1' such that $n_1 \Rightarrow \xrightarrow{\tau} \Rightarrow n_1'$ and $n_1' \rho n_2'$.
- *Weak bisimilarity* over $\mathcal{T}$, denoted $\approx_{\mathsf{w}}$, is the largest weak bisimulation over $\mathcal{T}$, and *rooted weak bisimilarity*, denoted $\simeq_{\mathsf{w}}$, is the largest root of $\approx_{\mathsf{w}}$.

Unfortunately, (weak and) rooted bisimilarity are not congruences for action refinement, as illustrated by the following example, originally due to [67] (subsumed by [68]), which essentially shows that the third τ-law of [87] does not hold in presence of refinement.

EXAMPLE 2.10. Consider $t = a; (b + \tau)$ and $t' = a; (b + \tau) + a$. It is not difficult to observe that $t \simeq_{\mathsf{w}} t'$ (in fact, this is an instance of the third τ-law of [108]). Unfortunately, $t[a \to a_1; a_2] \not\simeq_{\mathsf{w}} t'[a \to a_1; a_2]$ because $t'[a \to a_1; a_2] \xrightarrow{a_1} a_2; (\mathbf{1}[a \to a_1; a_2])$, hence reaching a state where b is no longer possible, while no bisimilar state can be reached from $t[a \to a_1; a_2]$ with a transition labelled a_1.

This example shows that $\simeq_{\mathsf{w}}$ is not preserved by refinement, because the branching structure of processes is not preserved enough by rooted weak bisimilarity. For this reason, Van Glabbeek and Weijland proposed in [67] a finer equivalence, called *branching* bisimilarity, which is a congruence for action refinement. However, the natural question is to single out the *coarsest* congruence for action refinement inside (rooted) weak bisimilarity. The answer is not branching bisimilarity, but rather (rooted) *delay* bisimilarity, based on an equivalence in [107] and provided with this name in [68]. We recall the definitions and the precise results.

DEFINITION 2.11. Let $\mathcal{T}$ be a labelled transition system.

- A *delay* [*branching*] *bisimulation* over $\mathcal{T}$ is a symmetric relation $\rho \subseteq N \times N$ such that for all $n_1 \rho n_2$ and $n_1 \xrightarrow{\lambda} n'_1$, one of the following conditions holds:
  - $\lambda = \tau$ and $n'_1 \rho n_2$;
  - there exist n'_2, n''_2 such that $n_2 \Rightarrow n''_2 \xrightarrow{\lambda} n'_2$ such that $n'_1 \rho n'_2$ [and $n_1 \rho n''_2$].
- A *delay* [*branching*] *root* of a binary relation $\rho \subseteq N \times N$ is a relation $\hat{\rho} \subseteq \rho$ such that for all $n_1 \hat{\rho} n_2$
  - if $n_1 \xrightarrow{\tau} n'_1$, then there exist n'_2, n''_2 such that $n_2 \Rightarrow n''_2 \xrightarrow{\tau} n'_2$ and $n'_1 \rho n'_2$ [and $n_1 \rho n''_2$];
  - if $n_2 \xrightarrow{\tau} n'_2$, then there exist n'_1, n''_1 such that $n_1 \Rightarrow n''_1 \xrightarrow{\tau} n'_1$ and $n'_1 \rho n'_2$ [and $n''_1 \rho n_2$].
- *Delay* [*branching*] *bisimilarity* over $\mathcal{T}$, denoted $\approx_{\mathsf{d}}$ [$\approx_{\mathsf{b}}$], is the largest delay [branching] bisimulation over $\mathcal{T}$, and *rooted delay [branching] bisimilarity*, denoted $\simeq_{\mathsf{d}}$ [$\simeq_{\mathsf{b}}$], is the largest delay [branching] root of $\approx_{\mathsf{d}}$ [$\approx_{\mathsf{b}}$].

PROPOSITION 2.12.

(1) $\simeq_{\mathsf{d}}$ *is the coarsest congruence over* $Lang^{seq}$ *contained in* $\simeq_{\mathsf{w}}$;

(2) $\simeq_{\mathsf{b}}$ *is a congruence over* $Lang^{seq}$ *contained in* $\simeq_{\mathsf{d}}$.

The proof of congruence of Clause (1) is originally due to [52], while the proof that it is the *coarsest* congruence is due to [34]. Walker in [135] proved that the specific axioms of delay bisimilarity are the first and the second τ-laws of rooted weak bisimilarity. Finally, the proof of Clause 2 can be found in [41,68], whereas the transitivity of $\simeq_{\mathsf{b}}$ is separately addressed in [15].

2.5. *Application: A very simple data base*

We now give a very simple example of the use of action refinement. In this and similar examples further on, for the sake of readability we use an alternative representation of recursion: instead of writing recursion operators within terms, we write process invocations, where the processes are defined elsewhere as part of the terms' global context. Thus, instead of $\mu x.t$ we may invoke a process X, provided $X := t\{X/x\}$ is a process definition. See also Milner [108]. Moreover, every semantics we discuss in this chapter equates the terms $\mathbf{1}; t$ and t for arbitrary terms t; for that reason, it is always safe to treat the two terms as equal, and in examples we will usually do so.

Consider a distributed data base that can be queried and updated. We first deal with the case where the state of the data base is completely abstracted away from. The behaviour of the system is given by $Data^1_S$, defined by

$$Data^1_S := (qry + upd); Data^1_S.$$

The operational semantics of $Data^1_S$ is depicted in Figure 2. Now suppose that updating is refined so that it consists of two separate stages, in which the update is *requested* and *confirmed*, respectively. In our setting, this can be expressed by refining action *upd* to *req*; *cnf*,

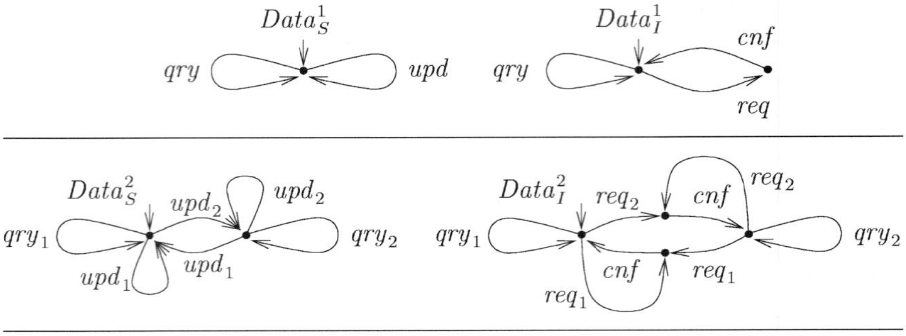

Fig. 2. Specification and refinement of a data base with 1 state (above) and 2 states (below).

thus obtaining the process $Data_I^1 := Data_S^1[upd \to req; cnf]$, also depicted in Figure 2. Note that in $Data_I^1$, the query operation cannot be performed in between the two stages of the update operation.

If we make this slightly more realistic by taking into consideration that the state of the data base can take different values, say from 1 up to n (with initial state 1), we arrive at the following specification:

$$Data_S^n := State_1,$$

$$State_i := qry_i; State_i + \sum_{k=1}^{n} upd_k; State_k \quad (\text{for } i = 1, \ldots, n).$$

Hence $Data_S^n$ specifies that after an update action, where a value is written, any number of consecutive queries can be performed, each of which reads the value just written. The behaviour of $Data_S^n$ for $n = 2$ is again depicted in Figure 2. Refining the actions upd_i to $req_i; cnf$ results in

$$Data_I^2 := Data_S^2[upd_1 \to req_1; cnf][upd_2 \to req_2; cnf].$$

Like in the case of $Data_I^1$, between request and confirmation querying the data base is disabled. If it is desired that querying be enabled at that point (for instance because the confirmation action does not change the data base state, so it is safe to read it), this requires a more flexible notion of action refinement; see Sections 7 and 8.

3. Atomic refinement

The basic idea underlying atomic action refinement is the following. An abstract specification describes a system in terms of executions of basic actions, that – by their nature – are intrinsically atomic; hence, when a specification is made more detailed via action refinement, the atomicity of an abstract action should be preserved by the concrete process implementing that abstract action. The atomic execution of a process means that it enjoys the following two properties:

All-or-nothing: the concrete process is either executed completely, or not at all; this implies that the process is not observable during its execution, but only before and after.

Non-interruptible: no other process can interrupt its execution; this implies that the atomic process is never interleaved with others.

Observe that action refinement in Section 2 is not atomic: the problem described in Example 2.10 simply disappears if we assume that $a_1; a_2$ is executed atomically, as the intermediate state where only a_1 has been performed is not observable. (Hence, as we will see, rooted weak bisimilarity is a congruence for atomic refinement.) There are many real problems where the assumption of atomicity is vital, e.g., when mutual exclusion on a shared resource is necessary to prevent faulty behaviour. For more discussion see also Section 1.3.

In order to understand atomic action refinement, it is useful to enhance the language with a mechanism for making the execution of processes atomic. For this purpose, we add an atomizer construct. The operational model has to be extended accordingly: it is necessary to divide the set of states into the *abstract* (or observable) ones, in which no atomic process is running, and the *concrete* (or unobservable) ones, which correspond to the intermediate states of some atomic process. With these extensions, we can easily accommodate atomic action refinement within interleaving semantics. Papers following this approach include [13,19,72,75,97]. Moreover, [48] takes an intermediate position where action refinement is non-interruptible but not all-or-nothing; that is, the concrete states are observable. Other papers dealing with some (weaker) form of atomicity in process algebra are [16,111].

In the above discussion, we have implicitly assumed that it is possible to put processes in parallel; otherwise, the notion of interruption would not be meaningful. To make the assumption valid, we enlarge the process algebra of the previous section with an operator for parallel composition; however, for the sake of simplicity, we ignore communication for now. The problems due to communication are treated extensively in Section 4, and are not essentially different for atomic and non-atomic refinement.

3.1. *Parallel composition and atomizer*

The language we consider in this section, denoted $Lang^{atom}$, extends $Lang^{seq}$ of Section 2.1 with several new operators (underlined):

$$\mathsf{T} ::= \mathbf{0} \mid \mathbf{1} \mid \alpha \mid \mathsf{V} + \mathsf{V} \mid \mathsf{T}; \mathsf{V} \mid \underline{\mathsf{T} \mathbin{|||} \mathsf{T} \mid \langle \mathsf{V} \rangle \mid *\mathsf{T}} \mid \mathsf{T}/A \mid \mathsf{T}[a \to \mathsf{V}] \mid x \mid \mu x.\mathsf{V}.$$

See also Table 16 for a complete overview of the different languages used in this chapter. The definitions of free variables and syntactic substitution (Table 1) and of guardedness (Definition 2.1) extend directly to the new operators. Their intuitive meaning and associated well-formedness conditions are as follows:

- $t \mathbin{|||} u$ is the parallel composition of the behaviours described by t and u, where all the actions can be done by either sub-term in isolation (no communication), except for the termination action $\checkmark$ on which t and u have to synchronize. We impose as a well-formedness condition that either t or u (or both) are *abstract* (where the notion of an abstract state is defined below).
- $\langle t \rangle$ behaves like t, except that the intermediate states of the execution are not observable. In particular, if t has no path ending in a $\checkmark$-transition, $\langle t \rangle$ does not offer any observable behaviour. The operand of $\langle t \rangle$ is *virgin*, and hence may contain no auxiliary operator (either $\mathbf{1}$ or $*$).
- $*t$ is an auxiliary operator expressing the residual of some atomized process; its meaning differs from that of t only in that the former is considered to be a *concrete* term (unless it is terminated), while the latter is an *abstract* term (provided it contains no more occurrences of $*$). Being an auxiliary operator, due to well-formedness $*$ may not occur in any virgin operand.

The operational semantics for $\langle t \rangle$ can be given at two different description levels. For illustration, in Figure 3 we depict two alternative semantics for $\langle a_1; a_2 \rangle \mathbin{|||} b$. One possible semantics describes the intermediate states of the execution; this requires to distinguish observable states and unobservable ones (depicted as white circles) in the labelled transition system, as only the former states should be considered when defining behavioural semantics. This line has been followed in, e.g., [75] and also in [76,72] where an axiomatization of the low level operator used for the atomizer (called *strong prefixing*) is given. The alternative semantics is more abstract, as the intermediate states are not described in the operational model, at the price of labelling transitions with action sequences. This line has been proposed in, e.g., [72]. Here we follow the former approach in the operational semantics (although the atomizer construct presented here is more general than in the papers cited above) and the second approach in the denotational semantics, where we use trees labelled on sequences.

Abstract terms. In the well-formedness condition imposed above on parallel composition as well as in the operational rules for parallel composition introduced below, we use the concept of an *abstract* term. The set of abstract terms is defined inductively by the following rules:

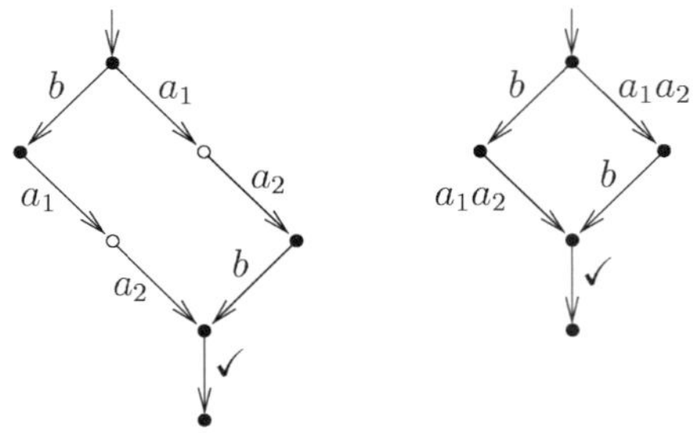

Fig. 3. Two representations for $\langle a_1; a_2 \rangle \mathbin{|||} b$.

- $\mathbf{0}$, $\mathbf{1}$, any variable $x \in \mathit{Var}$ and any action $\alpha \in \mathit{Act}_{\checkmark}$ are abstract;
- if $t \xrightarrow{\checkmark}$, then $*t$ is abstract;
- if t and u are abstract, then $t + u$, $t; u$, t/A, $t \,|||\, u$, $\langle t \rangle$, $t[a \to u]$ and $\mu x.t$ are abstract.

All the terms in Lang^{atom} that are not abstract are called *concrete*. As we will see, if $*t$ is abstract, which is the case if and only if $t \xrightarrow{\checkmark}$, then both $*t$ and t are bisimilar to $\mathbf{1}$. It follows that the abstract states are roughly (namely, up to bisimilarity) given by the $*$-less fragment of Lang^{atom}. On the other hand, concrete states are terms with some non-terminated subterm in the scope of a $*$-operator.

Well-formedness. To summarize the well-formedness conditions on Lang^{atom}:

- No virgin operand (V in the grammar) contains an auxiliary operator ($\mathbf{1}$ or $*$);
- At least one of the operands of parallel composition must be abstract;
- Recursion is allowed on guarded variables only.

Operational semantics. The labelled transition system is $\langle \mathit{Lang}^{atom}, \to \rangle$, where $\to$ is the transition relation defined by the rules in Table 4 (only the rules for the new operators and the two rules for refinement are reported). Let us comment on the rules. The rules for parallel composition give priority to the concrete component, if any: if only one of the two components is abstract, this has to remain idle. Observe that there is no way to reach a state where both components are concrete, starting from an initial abstract state (in fact, such a state would not be well-formed). Moreover, if both components are abstract, the rules allow both to proceed. The rule for the atomizer is simple: $\langle t \rangle$ does what t does, but the reached state is concrete (if not properly terminated). A concrete state $*t$ does what t does; the only difference is that the reached state is still concrete (if not properly terminated). The rules for sequential composition are responsible for the $*$ clean-up, when reaching an abstract state. The rule for refinement shows that the residual u' is not only to be non-interruptible (as in the corresponding rule of the previous section), but that the execution is to be all-or-nothing (by making the intermediate states concrete).

EXAMPLE 3.1. Consider $t = \langle a \rangle; b$ and $u = \langle a; b \rangle$. It is easy to see that $t \xrightarrow{a} (*\mathbf{1}); b \xrightarrow{b} \mathbf{1}$ where all the states are abstract, while $u \xrightarrow{a} *(\mathbf{1}; b) \xrightarrow{b} *\mathbf{1}$, where the intermediate state

Table 4
Transition rules

$$\frac{t \xrightarrow{\alpha} t' \quad u \text{ abstract}}{t \,|||\, u \xrightarrow{\alpha} t' \,|||\, u} \qquad \frac{u \xrightarrow{\alpha} u' \quad t \text{ abstract}}{t \,|||\, u \xrightarrow{\alpha} t \,|||\, u'} \qquad \frac{t \xrightarrow{\checkmark} t' \quad u \xrightarrow{\checkmark} u'}{t \,|||\, u \xrightarrow{\checkmark} t' \,|||\, u'}$$

$$\frac{t \xrightarrow{\alpha} t'}{\langle t \rangle \xrightarrow{\alpha} *t'} \qquad \frac{t \xrightarrow{\lambda} t'}{*t \xrightarrow{\lambda} *t'}$$

$$\frac{t \xrightarrow{\lambda} t' \quad \lambda \neq a}{t[a \to u] \xrightarrow{\lambda} t'[a \to u]} \qquad \frac{t \xrightarrow{a} t' \quad u \xrightarrow{\alpha} u'}{t[a \to u] \xrightarrow{\alpha} (*u'); t'[a \to u]}$$

$*(\mathbf{1}; b)$ is concrete. So, $t \mid\mid\mid c$ will allow the execution of c in between a and b, while $u \mid\mid\mid c$ will forbid this behaviour.

3.2. *Denotational semantics*

We present a denotational semantics for the $*$-less fragment of $Lang^{atom}$, using trees (see Section 2.3 for the precise definition of a tree) labelled on $Act_\tau^+ \cup \{\checkmark\}$ as a model. We use w to range over Act_τ^+.

The definitions for the basic operations are the same as in Section 2.3, except that sequential composition and choice have to be adapted to sequences. Here we report only the definitions for the two new operations (parallel composition and the atomizer) and for refinement. (Note that $*$ is not modelled denotationally.)

Parallel. If $N_1 \cap N_2 = \emptyset$, then $\mathcal{T}_1 \mid\mid\mid \mathcal{T}_2 = Unf\langle N, \rightarrow, \iota\rangle$ where:

- $N = N_1 \times N_2$;
- $\rightarrow\, = \{((n_1, n_2), w, (n_1', n_2)) \mid n_1 \xrightarrow{w}_1 n_1'\} \cup \{((n_1, n_2), w, (n_1, n_2')) \mid n_2 \xrightarrow{w}_2 n_2'\}$;
- $\iota = (\iota_1, \iota_2)$.

Atomizer. $\langle \mathcal{T} \rangle = Unf\langle N, \rightarrow', \iota\rangle$ where:

- $\rightarrow' = \{(\iota, w_1 \cdots w_k, n) \mid k \geqslant 1, \iota \xrightarrow{w_1} \cdots \xrightarrow{w_k} n \xrightarrow{\checkmark}\} \cup \{(n, \checkmark, n') \mid n \xrightarrow{\checkmark} n'\}$.

Action Refinement. If $\iota_2 \not\xrightarrow{\checkmark}_2$, then $\mathcal{T}_1[a \rightarrow \mathcal{T}_2] = Unf\langle N_1, \rightarrow, \iota_1\rangle$ where:

- $\rightarrow\, = \{(n, w', n') \mid n \xrightarrow{w}_1 n', w' \in R(w)\} \cup \{(n, \checkmark, n') \mid n \xrightarrow{\checkmark} n'\}$

  in which $R(w)$ is the set of expansions of w, defined as follows:

$$
\begin{aligned}
R : \varepsilon &\mapsto \{\varepsilon\} \\
\alpha w &\mapsto \{\alpha\} \cdot R(w) \quad \text{if } a \neq \alpha \\
a w &\mapsto \{w' \mid \iota_2 \xrightarrow{w'\checkmark}_2\} \cdot R(w)
\end{aligned}
$$

  and $\cdot$ is the concatenation operation on sets of strings.

The operation of parallel composition is nothing but the unfolding of the lts obtained by making the (asynchronous) product of the two trees. The operation of atomising a tree yields a tree that is of depth two (counting $\checkmark$). This indeed ensures that the resulting behaviour is atomic, as it is performed in one single step (followed by $\checkmark$). Moreover, all the paths in $\mathcal{T}$ that do not reach a terminated node are simply omitted; if all paths are so, then $\langle \mathcal{T} \rangle$ is isomorphic to $\mathcal{T}_\perp$ (the tree modelling the deadlock constant $\mathbf{0}$). Finally, the refinement operation is a sort of relabelling: an arc labelled w is replaced by possibly many arcs, connecting the same two end nodes, each one with a label obtained by macro-expansion of w where every occurrence of a in w is replaced by one of the possible terminated sequences of $\mathcal{T}$ (i.e., the steps of $\langle \mathcal{T} \rangle$); subsequently, the resulting transition system is unfolded. The denotational semantics for the $*$-less fragment of $Lang^{atom}$ is obtained as a direct extension of Table 3 to the new operators. An example is given in Figure 4.

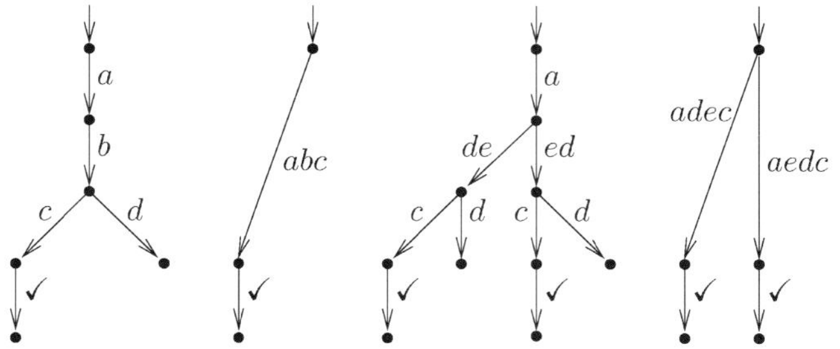

Fig. 4. Atomic trees denoting $t = a;\, b;\, (c + d;\, \mathbf{0})$, $\langle t \rangle$, $t[b \rightarrow d \,|||\, e]$ and $\langle t \rangle [b \rightarrow d \,|||\, e]$.

3.3. *Congruences and axiomatizations*

As done in the previous section, strong bisimulation equivalence is used to compare the two semantics. However, the operational semantics is quite different, as states may be also concrete. So, we first need to define which long steps are to be used in the definition of bisimulation for the operational semantics.

To this aim, from the transition relation defined operationally over $Lang^{atom}$ (Tables 2 and 4), we derive a "long step" transition relation over the *abstract* terms in $Lang^{atom}$, denoted $\rightarrow$ as well but labelled by $Act_\tau^+ \cup \{\checkmark\}$. Long step transitions relate a pair of abstract states as follows: $t \xrightarrow{w} t'$ iff $t \xrightarrow{\alpha_1} t_1 \xrightarrow{\alpha_2} t_2 \cdots \xrightarrow{\alpha_k} t'$, where all t_i are concrete ($i = 1, \ldots, k-1$) and $w = \alpha_1 \cdots \alpha_k$. Bisimulation equivalence is thus defined by setting $Lab = Act_\tau^+ \cup \{\checkmark\}$ in Definition 2.6; it relates only those abstract states that are able to match their long steps. Under this interpretation, the following holds:

PROPOSITION 3.2.
(1) $t \sim [\![t]\!]$ *for all* $*$*-less* $t \in Lang^{atom}$;
(2) $\sim$ *is a congruence over the* $*$*-less fragment of* $Lang^{atom}$.

In order to prove the first clause, for each long step of the operational semantics one finds a corresponding single step in the denotational semantics, and vice versa; the correspondence is set by an inductive argument on the term structure (taking into account the guardedness of recursive terms), much like for Proposition 2.8.

As strong bisimilarity is a congruence, it is natural to look for axioms for the atomizer and refinement operators. Some typical axioms for the atomizer are reported in Table 5; the axiomatic treatment of refinement is deferred to Section 6. Another axiom system for a language with atomizer is presented in [97]; since the assumptions there are quite different, it is no surprise that also the axioms are quite different.

Passing to τ-insensitive equivalence, we need to abstract the observable behaviour of a one-step sequence w, by removing all the occurrences of τ in it. Let $w \setminus \tau$ denote the resulting sequence; e.g., $a\tau b \setminus \tau = ab \setminus \tau = ab\tau \setminus \tau = ab$ and $\tau\tau\tau \setminus \tau = \varepsilon$. Then we define the relation $\Rightarrow$ as follows:

$$t \Rightarrow u :\Longleftrightarrow t \xrightarrow{w_1} t_1 \cdots \xrightarrow{w_n} u$$

Table 5
Some typical axioms for $\langle \_ \rangle$, sound modulo $\sim$ (above the line) and $\simeq_{\mathrm{w}}$ (all)

$\langle t; \mathbf{0} \rangle$	$= \mathbf{0}$
$\langle \alpha \rangle$	$= \alpha$
$\langle \langle t_1 \rangle; \langle t_2 \rangle \rangle$	$= \langle t_1; t_2 \rangle$
$\langle t_1; (t_2 + t_3) \rangle$	$= \langle t_1; t_2 \rangle + \langle t_1; t_3 \rangle$
$\langle \tau; t \rangle = \langle t \rangle$	

where $w_i \setminus \tau = \varepsilon$ for $1 \leqslant i \leqslant n$. Rooted weak bisimilarity (see Definition 2.9), in our setting of transition labelled on sequences, is defined modulo the removal of τs; that is, u has to simulate a transition $t \xrightarrow{w} t'$ in one of the following ways:

- either $w \setminus \tau = \varepsilon$ and (t', u) is in the relation;
- or $u \Rightarrow \xrightarrow{w'} \Rightarrow u'$ with $w' \setminus \tau = w \setminus \tau$ and (t', u') is in the relation.

We observe that $\simeq_{\mathrm{w}}$ is a congruence for atomic refinement. In fact, the counterexample reported in Example 2.10 simply disappears if refinement is atomic.

EXAMPLE 3.3. Consider again $t = a; (b + \tau)$ and $t' = a; (b + \tau) + a$ such that $t \simeq_{\mathrm{w}} t'$. According to the rules for atomic refinement and rooted weak bisimilarity, $t[a \to a_1; a_2] \simeq_{\mathrm{w}} \langle a_1; a_2 \rangle; (b + \tau) \simeq_{\mathrm{w}} \langle a_1; a_2 \rangle; (b + \tau) + \langle a_1; a_2 \rangle \simeq_{\mathrm{w}} t'[a \to a_1; a_2]$.

Summarizing, we have the following result:

PROPOSITION 3.4. *$\simeq_{\mathrm{w}}$ is a congruence over the $*$-less fragment of $Lang^{atom}$.*

Table 5 contains a typical axiom for $Lang^{atom}$ up to $\simeq_{\mathrm{w}}$.

3.4. *Application: Critical sections*

Consider two processes repeatedly entering a critical section. Abstractly, the activities in the critical section are modelled as a single action, cs_i for $i = 1, 2$, and the system is specified by $Sys_S = Proc_1 \,|||\, Proc_2$ where the processes $Proc_i$ are given by

$$Proc_i := cs_i; Proc_i.$$

Now suppose cs_1 is refined into $a_1; b_1$ and cs_2 into $b_2 + c_2$; that is, the abstract system is refined into $Sys_I = Sys_S[cs_1 \to a_1; b_1][cs_2 \to b_2 + c_2]$. Due to the nature of atomic refinement, the refined critical sections do not overlap: in particular, b_2 and c_2 cannot occur in between a_1 and b_1. The behaviour of the abstract and refined systems are given in Figure 5.

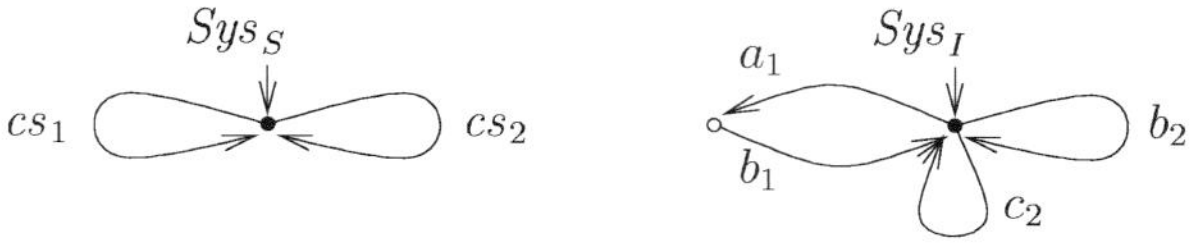

Fig. 5. Critical sections and their refinement: $Sys_I = Sys_S[cs_1 \to a_1; b_1][cs_2 \to b_2 + c_2]$.

Unfortunately, the atomicity of the refinement can also affect actions for which it was not intended. This can be seen by considering extensions of the above processes that also have a non-critical section, ncs_i:

$$Proc'_i := ncs_i; cs_i; Proc_i.$$

If we refine the cs_i-actions in $Proc'_1 \mathbin{|||} Proc'_2$ in the same way as above, then it turns out that not only the critical section (cs_2) but also the non-critical section (ncs_2) of $Proc'_2$ is prevented from proceeding during the critical section of $Proc'_1$ (between a_1 and b_1). This may not be the intended behaviour. However, in order to avoid it, more information regarding the dependencies between actions is needed than the current framework provides: in particular, it must be made clear that, in contrast to cs_1 and cs_2, ncs_1 and cs_2 are independent and may safely overlap. In Section 7.2 we show how one can use action dependencies to solve this type of problem.

4. Non-atomic refinement: An event-based model

In this and the next section, we consider refinement in a fully general language, comparable with CCS, CSP or ACP. Compared to Section 3, we extend parallel composition with synchronization, resulting in terms $t \|_A u$ where $A \subseteq Act$ is the *synchronization set*; $t \mathbin{|||} u$ is a special case where $A = \emptyset$. On the other hand, we drop the atomizer construct. The resulting language, *Lang*, is thus generated by the following grammar (where the new operator with respect to $Lang^{seq}$ of Section 2 is underlined):

$$\mathsf{T} ::= \mathbf{0} \mid \mathbf{1} \mid \alpha \mid \mathsf{V} + \mathsf{V} \mid \mathsf{T}; \mathsf{V} \mid \underline{\mathsf{T} \|_A \mathsf{T}} \mid \mathsf{T}/A \mid \mathsf{T}[a \to \mathsf{V}] \mid x \mid \mu x.\mathsf{V}.$$

See also Table 16 for a complete overview of the different languages used in this chapter. As before, well-formedness implies that the auxiliary operator **1** does not occur in virgin operands (V in the grammar) and that recursion is guarded. A more extensive discussion of *Lang* follows below. In this section, we develop an event-based operational and denotational semantics for *Lang*; in the next section, we review the relevant congruence questions studied over the last decade. Note that, except for the refinement operator, *Lang* essentially corresponds to a fragment of ACP [11]; in particular, the parallel composition (which is based on TCSP, see [113]) has a standard operational semantics given in Table 6.

It is clear from the discussion in the introduction that the standard lts semantics is not discriminating enough to provide a compositional model for non-atomic refinement in the

Table 6
Transition rules for synchronization

$$\frac{t \xrightarrow{\alpha} t' \quad \alpha \notin A}{t \|_A u \xrightarrow{\lambda} t' \|_A u} \qquad \frac{u \xrightarrow{\alpha} u' \quad \alpha \notin A}{t \|_A u \xrightarrow{\lambda} t \|_A u'} \qquad \frac{t \xrightarrow{\lambda} t' \quad u \xrightarrow{\lambda} u' \quad \lambda \in A_{\checkmark}}{t \|_A u \xrightarrow{\lambda} t' \|_A u'}$$

general setting of *Lang*. To amend this, we add information in the form of *events*. Events serve to give a closer identification of which occurrence of an action is being executed; as a consequence, the same occurrence can be recognized among different transitions or different runs.

For the purpose of formalization, we assume the existence of a global universe of events, *Evt*, with $* \notin Evt$ ($*$ being a special event used to model synchronization and refinement of event structures), which is closed under pairing, also with $*$; i.e., $(Evt_* \times Evt_*) \setminus (*, *) \subseteq Evt$ where $Evt_* = Evt \cup \{*\}$. The set of *singular* (i.e., *non-paired*) events is denoted $Evt^\bullet = Evt \setminus (Evt_* \times Evt_*)$; this set is assumed to be countably infinite. Evt_* is ranged over by d, e; subsets of Evt_* are denoted E, F, G.

In the operational semantics, events are used to enrich the transition labels: these are now taken from $Act_{\tau\checkmark}^{Evt} = (Evt \times Act_\tau) \cup \{\checkmark\}$ rather than just $Act_{\tau\checkmark}$; i.e., the non-termination transitions have associated event identities. This allows us to distinguish the independent execution of actions from their interleaving. Denotationally, we capture the behaviour of processes through *event structures*.

4.1. *Event annotations and operational semantics*

An immediate technical question is how event identities are generated and kept distinct for different occurrences within a given model, given the fact that the terms of *Lang* do not contain any information to this purpose. There are several possible answers:

- By interpreting models up to isomorphism, and implicitly or explicitly selecting isomorphic representatives to guarantee distinctness of events; see, e.g., Winskel [140].
- By generating event identities automatically from the term structure, as in [47,21]; see also [78,23] for an application of the same principle in a timed semantics for process algebras.
- By adding information about event identities explicitly to the terms, through their *annotation* at "compile time", i.e., before evaluating them; see Langerak [100].

The first method is less suitable for operational semantics, since there not all event identities are known beforehand; hence we do not follow this approach. For event-based semantics, the third method is mathematically less sophisticated but gives rise to an (in our opinion) simpler presentation. For that reason, instead of studying *Lang* directly, we consider the annotated language *Lang*(*Evt*) generated by the following grammar (new or adapted operators are underlined):

$$t ::= \mathbf{0} \mid \mathbf{1} \mid \underline{{}_e\alpha} \mid \mathsf{V} + \mathsf{V} \mid \mathsf{T}; \mathsf{V} \mid \mathsf{T} \|_A \mathsf{T} \mid \mathsf{T}/A \mid \underline{\mathsf{T}[a \to \mathsf{V}, \vec{e} \to \vec{\mathsf{T}}]} \mid \underline{{}_e x} \mid \underline{{}_e[\mathsf{T}]} \mid \mu x.\mathsf{V}$$

where $e \in Evt^\bullet$. As an additional well-formedness condition, we require that the event annotations are compatible. To make this precise, we introduce a set $E_t \subseteq Evt^\bullet$ of events syntactically occurring within t, which is only defined if the annotations are compatible.

- $\mathbf{0}$ and $\mathbf{1}$ are the deadlock, respectively termination constants as before. We define $E_\mathbf{0} = E_\mathbf{1} = \emptyset$.
- ${}_e\alpha$ denotes the action $\alpha \in Act_\tau$ annotated with the event $e \in Evt^\bullet$ used to model its occurrence. We define $E_{{}_e\alpha} = \{e\}$.
- $t + u$ denotes choice; we require $E_t \cap E_u = \emptyset$ and define $E_{t+u} = E_t \cup E_u$.
- $t; u$ denotes sequential composition; we require $E_t \cap E_u = \emptyset$ and define $E_{t;u} = E_t \cup E_u$.
- $t \|_A u$ denotes the parallel composition of t and u with synchronization over A, meaning that actions in A (and also the termination action $\checkmark$) may only be executed by t and u simultaneously, and all others independently by either operand. We use $t \,|||\, u = t \|_\emptyset u$ as a special case. As for the previous binary operators, we require $E_t \cap E_u = \emptyset$ and define $E_{t \|_A u} = E_t \cup E_u$.
- t/A denotes hiding; we define $E_{t/A} = E_t$.
- $t[a \to u, \vec{e} \to \vec{v}]$ denotes the refinement of the actions a occurring in t by u; the latter may not contain the termination constant $\mathbf{1}$. Moreover, $\vec{e} \to \vec{v}$ denotes a vector of *pending refinements* $e_1 \to v_1, \ldots, e_n \to v_n$ (with $n = |\vec{e}| = |\vec{v}|$), where the e_i identify distinct occurrences of a in t (i.e., $e_i = e_j$ implies $i = j$) that are currently being executed, or actually refined; the v_i represent the corresponding non-terminated states reached during the execution of the refinement body u. We require $E_t \cap E_u = \emptyset$ and $E_{v_i} \subseteq E_u$ for all $1 \leqslant i \leqslant n$, and define $E_{t[a \to u, \vec{e} \to \vec{v}]} = E_t \cup E_u$.
  In the sequel, we write $\{\vec{e}\} = \{e_1, \ldots, e_n\}$ for the set of events in $\vec{e}$.
- ${}_e x$ denotes the process variable $x \in Var$, annotated with an event $e \in Evt^\bullet$. This annotation is used in order to prevent the re-occurrence of event identities in recursive unfoldings. We define $E_{{}_e x} = \{e\}$.
- ${}_e[t]$, with $e \in Evt^\bullet$, models the unfolding of the process invocation ${}_e x$ within the body of a recursive definition $\mu x.t$; its effect is the "relocation" of the events in t to the fresh range $\{e\} \times Evt$. We require t to be well-formed (in particular, E_t should be defined) and define $E_{{}_e[t]} = \{e\}$. The notion of syntactic substitution is adapted accordingly; see below.
- $\mu x.t$ denotes the recursive binding of the (guarded) process variable x in t. We let $E_{\mu x.t} = E_t$.

The definition of the free variables and substitution is extended to *Lang*(*Evt*) (see Table 1); the only interesting new case is ${}_e x$, for which $fv({}_e x) = \{x\}$ and ${}_e x\{t/x\} = {}_e[t]$. The latter preserves the event annotation of process variables, and shows the reason why we need terms of the form ${}_e[t]$ at all. For instance, the first three approximations of the annotated term $\mu x.{}_0 a;\, {}_1 x$ (according to the definition in Section 3) are given by $\mathbf{0}$, ${}_0 a;\, {}_1[\mathbf{0}]$ and ${}_0 a;\, {}_1[{}_0 a;\, {}_1[\mathbf{0}]]$.

Well-formedness. To summarize the well-formedness conditions on *Lang*(*Evt*):

- No virgin operand (V in the grammar) contains an occurrence of an auxiliary operator ($\mathbf{1}$);
- Recursion is allowed on guarded variables only;
- The set E_t of event annotations is defined.

We then have the following property (which is straightforward to prove):

PROPOSITION 4.1. *For all well-formed* $t, u \in Lang(Evt)$, $t\{u/x\}$ *is well-formed and* $E_{t\{u/x\}} = E_t$.

Annotating and stripping. Annotated terms are no more than an auxiliary device to give an event-based semantics for *Lang*. To make the connection between *Lang* and *Lang*(*Evt*) more explicit, consider a partial function *strip* : *Lang*(*Evt*)*Lang* that is undefined on terms ${}_e[t]$ and on $t[a \to u, \vec{e} \to \vec{v}]$ for nonempty $\vec{e}$, and otherwise removes all event annotations. It should be clear that *strip* is surjective, meaning that there is an annotated term for any $t \in Lang$; however, it is far from injective, since there are many ways to annotate t in a well-formed manner. For instance, one particularly simple method is to assume $\mathbb{N} \subseteq Evt^\bullet$ and consecutively number the subterms to be annotated, from left to right; e.g., $(a; b\|_a c; a)[a \to \mu x.d; x]$ is annotated according to this method as $({}_0a; {}_1b\|_{a2}c; {}_3a)[a \to \mu x.{}_4d; {}_5x]$. When defining the semantics, we will make sure that the chosen annotation of a term makes no difference; that is, if $strip(t) = strip(u)$ for $t, u \in Lang(Evt)$ then t and u will be equivalent.

4.2. *Operational event-based semantics*

A major advantage of annotated terms is that their operational semantics is a straightforward extension of the standard operational semantics (given in Tables 2 and 4), except for the refinement operator, which now has to be treated in full generality. The presentation below is based on [121].

Event transition systems. As stated above, in order to model the behaviour of *Lang*(*Evt*) we take transition labels from $Act^{Evt}_{\tau\surd}$ $(= (Evt \times Act_\tau) \cup \{\surd\})$ rather than $Act_{\tau\surd}$. The idea is that the event identifier uniquely identifies the action occurrence; in other words, on the level of events the transition system is deterministic. Furthermore, if two events can be executed in either order (in a given state), we call them *independent* (in that state); the state reached is independent of the ordering, and moreover, any set of pairwise independent events will remain independent if one of them is executed. Formally:

DEFINITION 4.2. An event transition system is an $Act^{Evt}_{\tau\surd}$-labelled transition system such that for all $n \in N$

- If $n \xrightarrow{e,\alpha_1} n'_1$ and $n \xrightarrow{e,\alpha_2} n'_2$, then $\alpha_1 = \alpha_2$ and $n'_1 = n'_2$.
- e_1 and e_2 are called n*-independent* if $n \xrightarrow{e_1,\alpha_1} \xrightarrow{e_2,\alpha_2} n'$ and $n \xrightarrow{e_2,\alpha_2} \xrightarrow{e_1,\alpha_1} n''$; if e_1 and e_2 are n-independent, then $n' = n''$.
- If e_1, e_2, e_3 are pairwise n-independent and $n \xrightarrow{e_1,\alpha_1} n'$, then e_2 and e_3 are n'-independent.

It follows that if a set of events F is pairwise n-independent for some node n, then starting in n, the events in F can be executed in arbitrary order, and the state reached is independent of the order of execution. We use $n \xrightarrow{F}$ to denote that F is pairwise n-independent. For instance, we will see that ${}_0a \,\|\, {}_1b \xrightarrow{\{(0,*),(*,1)\}}$, signalling the independence of a and b, but $t = {}_0a;\ {}_1b + {}_2b;\ {}_3a \overset{\{0,2\}}{\not\longrightarrow}$ although $t \xrightarrow{0,a}$ and $t \xrightarrow{2,b}$, since here the actions a and b are not independent.

To compare the behaviour of event transition systems, we define an extension of strong bisimulation which allows events to be converted bijectively between the behaviours under comparison, in order to abstract away sufficiently from the precise occurrence identifiers.

DEFINITION 4.3. Let $\langle Act_{\tau\surd}^{Evt}, N, \rightarrow\rangle$ be an event transition system.

- For any bijective $\phi : Evt \rightarrow Evt$, a ϕ-simulation is a relation $\rho \subseteq N \times N$ such that for all $n_1 \rho n_2$:
  - if $n_1 \xrightarrow{e,\alpha} n_1'$ then $n_2 \xrightarrow{\phi(e),\alpha} n_2'$ with $n_1' \rho n_2'$.
  - if $n_1 \xrightarrow{\surd} n_1'$ then $n_2 \xrightarrow{\surd} n_2'$ with $n_1' \rho n_2'$.
- n_1 and n_2 are called *event bisimilar*, denoted $n_1 \sim_{\mathsf{ev}} n_2$, if there is a ϕ-simulation ρ (for some ϕ) such that ρ^{-1} is a ϕ^{-1}-simulation.

Note that, in contrast to more sophisticated relations such as history-preserving bisimilarity (cf. [60]), in the above definition there is a single mapping ϕ establishing a static, one-to-one relation between the events of the transition systems under comparison. This makes for a very discriminating equivalence: for instance, for event transition systems generated by stable event structures, event bisimilarity coincides with isomorphism of the event structures' configurations (see the discussion after Proposition 4.15). Indeed, we will have ${}_1a + {}_2a \not\sim_{\mathsf{ev}} {}_3a$, even though both ${}_1a \sim_{\mathsf{ev}} {}_3a$ and ${}_2a \sim_{\mathsf{ev}} {}_3a$.

The event-based operational semantics of *Lang*(*Evt*) is given in Table 7. For the fragment without refinement (the part of the table above the line), the semantics is unoriginal; see [99] for a very similar set of rules.

Extended refinement functions. The definition of the operational semantics of refinement requires some care. During the execution of a refined action, the *remainder* of the refinement must be "stored" somewhere in the term. For that purpose, we have extended refinements from terms $t[a \rightarrow u]$ to terms $t[a \rightarrow u, e_1 \rightarrow v_1, \ldots, e_k \rightarrow v_k]$ for arbitrary finite k; for all $1 \leqslant i \leqslant k$, e_i is an event of t under refinement, and v_i is the current remainder. The events in $\{\vec{e}\}$ are called *busy*. The remainder is kept only as long as it is not terminated; furthermore, during the same period, the event e_i is not actually executed, i.e., it is kept in the term t. The latter is necessary to make sure that events that causally follow e_i are not executed prematurely, before the refinement of e_i is terminated.

Note that the e_i may very well resolve choices within t. Since we do not let t execute the e_i as long as the corresponding v_i is not terminated, there is the danger that an event that actually conflicts with one of the busy events, say e_i, is executed as well, despite the fact that the choice has actually been resolved in favour of e_i. In order to prevent this from happening, only events of t that are independent of all the e_i may occur.

Table 7
Operational event-based semantics; $\lambda \in Act^{Evt}_{\tau\checkmark}$ arbitrary

$$\frac{}{\mathbf{1} \xrightarrow{\checkmark} \mathbf{0}} \qquad \frac{}{{}_e\alpha \xrightarrow{e,\alpha} \mathbf{1}} \qquad \frac{t \xrightarrow{e,\alpha} t'}{t+u \xrightarrow{e,\alpha} t'} \qquad \frac{u \xrightarrow{e,\alpha} u'}{t+u \xrightarrow{e,\alpha} u'} \qquad \frac{t \xrightarrow{e,\alpha} t'}{t;u \xrightarrow{e,\alpha} t';u} \qquad \frac{t \xrightarrow{\checkmark} t' \quad u \xrightarrow{\lambda} u'}{t;u \xrightarrow{\lambda} u'}$$

$$\frac{t \xrightarrow{e,\alpha} t' \quad \alpha \notin A}{t\|_A u \xrightarrow{(e,*),\alpha} t'\|_A u} \qquad \frac{u \xrightarrow{e,\alpha} u' \quad \alpha \notin A}{t\|_A u \xrightarrow{(*,e),\alpha} t\|_A u'} \qquad \frac{t \xrightarrow{d,\alpha} t' \quad u \xrightarrow{e,\alpha} u' \quad \alpha \in A}{t\|_A u \xrightarrow{(d,e),\alpha} t'\|_A u'}$$

$$\frac{t \xrightarrow{\checkmark} t' \quad u \xrightarrow{\checkmark} u'}{t\|_A u \xrightarrow{\checkmark} t'\|_A u'} \qquad \frac{t \xrightarrow{\lambda} t' \quad \lambda \notin Evt \times A}{t/A \xrightarrow{\lambda} t'/A} \qquad \frac{t \xrightarrow{e,a} t' \quad a \in A}{t/A \xrightarrow{e,\tau} t'/A}$$

$$\frac{t \xrightarrow{d,\alpha} t'}{e[t] \xrightarrow{(e,d),\alpha} e[t']} \qquad \frac{t \xrightarrow{\checkmark} t'}{e[t] \xrightarrow{\checkmark} e[t']} \qquad \frac{t\{\mu x.t/x\} \xrightarrow{\lambda} t'}{\mu x.t \xrightarrow{\lambda} t'}$$

$$\frac{t \xrightarrow{d,\alpha} t' \xrightarrow{\{\vec{e}\}} \quad \alpha \neq a}{t[a \to u, \vec{e} \to \vec{v}] \xrightarrow{(d,*),\alpha} t'[a \to u, \vec{e} \to \vec{v}]} \qquad \frac{t \xrightarrow{\checkmark} t'}{t[a \to u] \xrightarrow{\checkmark} t'[a \to u]}$$

$$\frac{t \xrightarrow{d,a} t' \xrightarrow{\{\vec{e}\}} \quad u \xrightarrow{e',\alpha} u' \not\xrightarrow{\checkmark}}{t[a \to u, \vec{e} \to \vec{v}] \xrightarrow{(d,e'),\alpha} t[a \to u, \vec{e} \to \vec{v}, d \to u']} \qquad \frac{t \xrightarrow{d,a} t' \xrightarrow{\{\vec{e}\}} \quad u \xrightarrow{e',\alpha} u' \xrightarrow{\checkmark}}{t[a \to u, \vec{e} \to \vec{v}] \xrightarrow{(d,e'),\alpha} t'[a \to u, \vec{e} \to \vec{v}]}$$

$$\frac{v_i \xrightarrow{e',\alpha} v' \not\xrightarrow{\checkmark}}{t[a \to u, \vec{e} \to \vec{v}] \xrightarrow{(e_i,e'),\alpha} t[a \to u, (\vec{e} \to \vec{v}) \pm (e_i \to v')]} \qquad \frac{t \xrightarrow{e_i,a} t' \quad v_i \xrightarrow{e',\alpha} v' \xrightarrow{\checkmark}}{t[a \to u, \vec{e} \to \vec{v}] \xrightarrow{(e_i,e'),\alpha} t'[a \to u, (\vec{e} \to \vec{v}) \setminus e_i]}$$

Some notation: for all $1 \leqslant i \leqslant k$, $(\vec{e} \to \vec{v}) \pm (e_i \to v')$ denotes the replacement of the residual $e_i \to v_i$ by $e_i \to v'$, and $(\vec{e} \to \vec{v}) \setminus e_i$ denotes the removal of the residual $e_i \to v_i$ from the vector.

The rules in Table 7 (below the line) can be understood as follows:

- The first rule concerns the execution of a non-refined action; the premise $t' \xrightarrow{\{\vec{e}\}}$ in this rule and others makes sure that the event d is not in conflict with any of events e_i currently under refinement. The event identity is changed from d to $(d, *)$, for uniformity with events that are properly refined.
- The second rule concerns termination; note that there can be no remaining refinements.
- The third and fourth rules concern the start of a new refinement instance; the latter deals with the case where the refinement immediately terminates again (i.e., it consisted of a single action only).
- The last two rules deal with the continuation of a busy event, which either remains busy (if the remainder is still not terminated) or disappears from the scene (if the remainder is terminated).

EXAMPLE 4.4. Consider the derivable transitions of the term $t = ({}_0b;\ {}_1a\|_a({}_2c+{}_3a))[a \to {}_4d;\ {}_5d]$.

- The parallel composition gives rise to

$$\begin{array}{ccccccc} {}_0b;\,{}_1a\|_a({}_2c+{}_3a) & \xrightarrow{(0,*),b} & \mathbf{1};\,{}_1a\|_a({}_2c+{}_3a) & \xrightarrow{(1,3),a} & \mathbf{1}\|_a\mathbf{1} & \xrightarrow{\checkmark} & \mathbf{0}\|_a\mathbf{0} \\ {\scriptstyle (*,2),c}\downarrow & & {\scriptstyle (*,2),c}\downarrow & & & & \\ {}_0b;\,{}_1a\|_a\mathbf{1} & \xrightarrow{(0,*),b} & \mathbf{1};\,{}_1a\|_a\mathbf{1} & & & & \end{array}$$

- Taking the refinement into account, we get

$$\begin{array}{ccccc} ({}_0b;\,{}_1a\|_a({}_2c+{}_3a))[a\to {}_4d;\,{}_5d] & \xrightarrow{((0,*),*),b} & (\mathbf{1};\,{}_1a\|_a({}_2c+{}_3a))[a\to {}_4d;\,{}_5d] & \xrightarrow{((1,3),4),d} & \cdots \\ {\scriptstyle ((*,2),*),c}\downarrow & & {\scriptstyle ((*,2),*),c}\downarrow & & \\ ({}_0b;\,{}_1a\|_a\mathbf{1})[a\to {}_4d;\,{}_5d] & \xrightarrow{((0,*),*),b} & (\mathbf{1};\,{}_1a\|_a\mathbf{1})[a\to {}_4d;\,{}_5d] & & \end{array}$$

where the upper right hand transition leads to

$$\begin{aligned} &(\mathbf{1};\,{}_1a\|_a({}_2c+{}_3a))[a\to {}_4d;\,{}_5d] \\ &\quad\xrightarrow{((1,3),4),d} (\mathbf{1};\,{}_1a\|_a({}_2c+{}_3a))[a\to {}_4d;\,{}_5d,\,(1,3)\to \mathbf{1};\,{}_5d] \\ &\quad\xrightarrow{((1,3),5),d} (\mathbf{1}\|_a\mathbf{1})[a\to {}_4d;\,{}_5d] \xrightarrow{\checkmark} (\mathbf{0}\|_a\mathbf{0})[a\to {}_4d;\,{}_5d]. \end{aligned}$$

In particular, note that $(\mathbf{1};\,{}_1a\|_a({}_2c+{}_3a))[a\to {}_4d;\,{}_5d,\,(1,3)\to \mathbf{1};\,{}_5d] \stackrel{((*,2),*),c}{\not\longrightarrow}$.

The first property to be proved is that the operational semantics indeed gives rise to an event transition system as defined in Definition 4.2. Furthermore, it is important to show that the associated equivalence (in this case, event bisimilarity) is a congruence. The proof is a straightforward variation on the standard case (modified by the fact that event identities may be converted).

PROPOSITION 4.5. $\sim_{\mathsf{ev}}$ *is a congruence over Lang(Evt).*

Moreover, it can be shown that all annotations of a given term in *Lang* are event bisimilar; i.e., the following holds (proved by induction on term structure):

PROPOSITION 4.6. *For all* $t, u \in$ *Lang(Evt), if strip*$(t) =$ *strip*(u) *then* $t \sim_{\mathsf{ev}} u$.

4.3. *Stable event structures*

The denotational semantics for *Lang(Evt)* we present here is based on the *stable event structures* of Winskel [140], extended with a termination predicate as in [64].

DEFINITION 4.7. A stable event structure is a tuple $\mathcal{E} = \langle E, \#, \vdash, \mathit{Ter}, \ell\rangle$, where

- $E \subseteq Evt_*$ is a set of *events*;
- $\# \subseteq E \times E$ is an irreflexive and symmetric *conflict relation*. The reflexive closure of # is denoted $\#^=$. The set of finite, consistent (i.e., conflict-free) subsets of E will be denoted $Con = \{F \subseteq_{fin} E \mid \nexists d, e \in F\colon d \# e\}$.
- $\vdash \subseteq Con \times E$ is an *enabling relation*, which satisfies

  Saturation: If $F \vdash e$ and $F \subseteq G \in Con$, then $G \vdash e$;

  Stability: If $F \vdash e$, $G \vdash e$ and $F \cup G \cup \{e\} \in Con$, then $F \cap G \vdash e$.

  The set of *initial events* will be denoted $Ini = \{e \in E \mid \emptyset \vdash e\}$.
- $Ter \subseteq Con$ is a *termination predicate* on sets of events, which satisfies

  Completeness: If $F \in Ter$ and $F \subseteq G \in Con$, then $F = G$.

- $\ell : E \to Act_\tau$ is a *labelling function*.

The class of stable event structures is denoted **ES**. In the sequel, we drop the qualifier "stable" and just talk about event structures. We sometimes write $E_\mathcal{E}$, $\#_\mathcal{E}$ etc. for the components of an event structure $\mathcal{E}$ and E_i, $\#_i$ etc. for the components of an event structure $\mathcal{E}_i$. As a further notational convention, we write $e_1, \ldots, e_n \vdash e$ for $\{e_1, \ldots, e_n\} \vdash e$ and $\vdash e$ for $\emptyset \vdash e$. Two event structures $\mathcal{E}_1, \mathcal{E}_2$ are said to be isomorphic, denoted $\mathcal{E}_1 \cong \mathcal{E}_2$, if there is a bijection $\phi : E_1 \to E_2$ (called an isomorphism) such that $d \#_1 e$ iff $\phi(d) \#_2 \phi(e)$, $F \vdash_1 e$ iff $\phi(F) \vdash_2 \phi(e)$, $F \in Ter_1$ iff $\phi(F) \in Ter_2$ and $\ell_1 = \ell_2 \circ \phi$.

The intuition behind stable event structures is that an action a may occur if an event $e \in E$ with $\ell(e) = a$ is enabled, meaning that $F \vdash e$ for the set F of events that have occurred previously. If two events are enabled simultaneously, i.e., $F \vdash d$ and $F \vdash e$, then either $d \# e$, meaning that after d or e has occurred, the other is ruled out; or d and e are independent. A more formal definition of the meaning is given through the *configurations* of an event structure; see further below.

With respect to the standard definition in [140], the only modification is the extension with the termination predicate. For this, we have used a solution due to [64]: termination is modelled by a predicate on sets of events that may hold only for the *complete* consistent sets (i.e., those that are maximal with respect to $\subseteq$). Note that this completeness condition implies an analogy to the conditions for the enabling relation:

Saturation: If $F \in Ter$ and $F \subseteq G \in Con$, then $G \in Ter$;

Stability: If $F, G \in Ter$ and $F \cup G \in Con$, then $F \cap G \in Ter$.

In fact, $F \in Ter$ can also be interpreted as $F \vdash e_\checkmark$ for some special termination event $e_\checkmark$ signalling termination – where, however, we do not really model $e_\checkmark$ explicitly. As a consequence of the completeness of *Ter*, the following holds:

LEMMA 4.8. *If* $\mathcal{E} \in \mathbf{ES}$ *such that* $\emptyset \in Ter$, *then* $\mathcal{E} = \langle \emptyset, \emptyset, \emptyset, \{\emptyset\}, \emptyset \rangle$.

In order to deal with recursion, just as for the tree semantics in the previous sections we use an approximation ordering over the class of models, in this case **ES**. It is essentially the one used in [140].

$$\begin{aligned}\mathcal{E}_1 \sqsubseteq \mathcal{E}_2 \;:\Longleftrightarrow\; & E_1 \subseteq E_2,\\ & \#_1 = \#_2 \cap (E_1 \times E_1),\\ & \vdash_1 = \vdash_2 \cap (Con_1 \times E_1),\\ & Ter_1 = Ter_2 \cap Con_1,\\ & \ell_1 = \ell_2 \upharpoonright E_1.\end{aligned}$$

Moreover, for an arbitrary set $\{\mathcal{E}_i\}_{i\in I} \subseteq \mathbf{ES}$, the component-wise union of the $\mathcal{E}_i$ is written $\bigsqcup_{i\in I} \mathcal{E}_i$. We then recall the following property (see [140, Theorem 4.4]):

PROPOSITION 4.9. *$\langle \mathbf{ES}, \sqsubseteq\rangle$ is a complete partial order, with bottom element $\langle\emptyset,\emptyset,\emptyset,\emptyset,\emptyset\rangle$ and least upper bounds $\bigsqcup_i \mathcal{E}_i = \langle \bigcup_i E_i, \bigcup_i \#_i, \bigcup_i \vdash_i, \bigcup_i Ter_i, \bigcup_i \ell_i\rangle$ for all $\sqsubseteq$-directed sets $\{\mathcal{E}_i\}_i \subseteq \mathbf{ES}$.*

4.4. *Denotational event-based semantics*

We now define a number of partial operations on **ES**, corresponding to the operators of *Lang*(*Evt*). By relying on the annotation of terms, we will make sure that the operations are only applied where they are defined. For arbitrary $F \subseteq Evt_* \times Evt_*$, we use the following notation:

$$\begin{aligned}\pi_i(F) &= \{e_i \mid (e_1,e_2) \in F\} \quad (i=1,2),\\ F(d) &= \{e \mid (d,e) \in F\}.\end{aligned}$$

Thus π_i projects onto the ith component, and $F(d)$ is a function-like use of F (albeit set-valued, since F is actually a relation). The definition of the constructions is inspired by [30,64,140], except for our treatment of termination. Note especially the construction for refinement, $\mathcal{E}[\mathcal{R}]$: this relies on a function $\mathcal{R}$ mapping all the events of $\mathcal{E}$ (and not just the a-labelled ones as in the operational semantics) to (non-terminated) event structures.

DEFINITION 4.10. We use the following constructions on event structures:

Deadlock. $\mathcal{E}_\perp = \langle\emptyset,\emptyset,\emptyset,\emptyset,\emptyset\rangle$.

Termination. $\mathcal{E}_\surd = \langle\emptyset,\emptyset,\emptyset,\{\emptyset\},\emptyset\rangle$.

Single action. $\mathcal{E}_{e,\alpha} = \langle\{e\},\emptyset,\{(\emptyset,e),(\{e\},e)\},\{\{e\}\},\{(e,\alpha)\}\rangle$ for arbitrary $e \in Evt_*$.

Sequential composition. If $E_1 \cap E_2 = \emptyset$, then $\mathcal{E}_1;\mathcal{E}_2 = \langle E,\#,\vdash,Ter,\ell\rangle$ such that

- $E = E_1 \cup E_2$;
- $\# = \#_1 \cup \#_2$;
- $\vdash = \{(F,e) \mid F\cap E_1 \vdash_1 e,\, F\cap E_2 \in Con_2\} \cup \{(F,e) \mid F \cap E_1 \in Ter_1,\, F\cap E_2 \vdash_2 e\}$;
- $Ter = \{F \subseteq E \mid F\cap E_1 \in Ter_1,\, F\cap E_2 \in Ter_2\}$;

- $\ell = \ell_1 \cup \ell_2$.

Choice. If $E_1 \cap E_2 = \emptyset$ and $\emptyset \notin Ter_1 \cup Ter_2$, then $\mathcal{E}_1 + \mathcal{E}_2 = \langle E, \#, \vdash, Ter, \ell \rangle$ such that

- $E = E_1 \cup E_2$;
- $\# = \#_1 \cup \#_2 \cup (E_1 \times E_2) \cup (E_2 \times E_1)$;
- $\vdash = \vdash_1 \cup \vdash_2 \cup (Con_1 \times Ini_2) \cup (Con_2 \times Ini_1)$;
- $Ter = Ter_1 \cup Ter_2$;
- $\ell = \ell_1 \cup \ell_2$.

Parallel composition. $\mathcal{E}_1 \|_A \mathcal{E}_2 = \langle E, \#, \vdash, Ter, \ell \rangle$ such that

- $E = (\bigcup_{a \in A} \ell_1^{-1}(a) \times \ell_2^{-1}(a)) \cup (\ell_1^{-1}(Act_\tau \setminus A) \times \{*\}) \cup (\{*\} \times \ell_2^{-1}(Act_\tau \setminus A))$;
- $\#^{=} = \{((d_1, d_2), (e_1, e_2)) \in E \times E \mid d_1 \#_1^{=} e_1 \neq * \vee d_2 \#_2^{=} e_2 \neq *\}$;
- $\vdash = \{(F, (e_1, e_2)) \in Con \times E \mid \forall i \in \{1, 2\}\colon e_i = * \vee \pi_i(F) \setminus * \vdash_i e_i\}$;
- $Ter = \{F \in Con \mid \forall i \in \{1, 2\}\colon \pi_i(F) \setminus * \in Ter_i\}$;
- $\ell = \{((e_1, e_2), \alpha) \in E \times Act_\tau \mid \alpha = \ell_1(e_1) \vee \alpha = \ell_2(e_2)\}$.

Hiding. $\mathcal{E}_1 / A = \langle E_1, \#_1, \vdash_1, Ter_1, \ell \rangle$, where $\ell(e) = \tau$ if $e \in \ell_1^{-1}(A)$ and $\ell(e) = \ell_1(e)$ otherwise.

Refinement. If $\mathcal{R}\colon E_1 \to (\mathbf{ES} \setminus \mathcal{E}_{\surd})$, then $\mathcal{E}_1[\mathcal{R}] = \langle E, \#, \vdash, Ter, \ell \rangle$ such that

- $E = \bigcup_{e \in E_1} (\{e\} \times E_{\mathcal{R}(e)})$;
- $\# = \{((d_1, d_2), (e_1, e_2)) \in E \times E \mid d_1 \#_1 e_1 \vee (d_1 = e_1 \wedge d_2 \#_{\mathcal{R}(d_1)} e_2)\}$;
- $\vdash = \{(F, (e_1, e_2)) \in Con \times E \mid ready(F) \vdash_1 e_1, F(e_1) \vdash_{\mathcal{R}(e_1)} e_2\}$;
- $Ter = \{F \in Con \mid ready(F) \in Ter_1\}$;
- $\ell = \{((e_1, e_2), \alpha) \in E \times Act_\tau \mid \alpha = \ell_{\mathcal{R}(e_1)}(e_2)\}$,

where for all $F \in Con$, $ready(F) = \{d \in \pi_1(F) \mid F(d) \in Ter_{\mathcal{R}(d)}\}$ is the set of events from $\mathcal{E}_1$ whose refinement has reached a terminated configuration.

Relocation. $e \times \mathcal{E}_1 = \langle E, \#, \vdash, Ter, \ell \rangle$ such that

- $E = \{e\} \times E_1$;
- $\# = \{((e, d_1), (e, e_1)) \mid d_1 \#_1 e_1\}$;
- $\vdash = \{(\{e\} \times F, (e, d)) \mid F \vdash_1 d\}$;
- $Ter = \{\{e\} \times F \mid F \in Ter_1\}$;
- $\ell = \ell_1 \circ \pi_2$.

EXAMPLE 4.11. Consider again the term $t = ({}_0b; {}_1a \|_a ({}_2c + {}_3a))[a \to {}_4d; {}_5d]$ of Example 4.4.

- The subterm ${}_0b; {}_1a$ is modelled by the structure $\mathcal{E}_1 = \mathcal{E}_{0,b}; \mathcal{E}_{1,a}$ with $E_1 = \{0, 1\}$, no conflicts, enablings $\vdash_1 0$ and $0 \vdash_1 1$, termination predicate $\{\{0, 1\}\}$ and labelling $0 \to b, 1 \to a$. Analogous for ${}_4d; {}_5d$.
- The subterm ${}_2c + {}_3a$ is modelled by the structure $\mathcal{E}_2 = \mathcal{E}_{2,c} + \mathcal{E}_{3,a}$ with $E_2 = \{2, 3\}$, conflict $2 \#_2 3$, enablings $\vdash_2 2$ and $\vdash_2 3$, termination predicate $\{\{2\}, \{3\}\}$ and labelling $2 \to c, 3 \to a$.

- The parallel composition yields $\mathcal{E}_3 = \mathcal{E}_1 \|_a \mathcal{E}_2$ with $E_3 = \{(0,*), (*,2), (1,3)\}$, conflict $(*,2) \#_3 (1,3)$, enablings $\vdash_3 (0,*)$, $(0,*) \vdash_3 (1,3)$ and $\vdash_3 (*,2)$, termination predicate $\{\{(0,*),(1,3)\}\}$ and labelling $(0,*) \to b$, $(*,2) \to c$, $(a,3) \to a$.
- The refinement in t gives rise to $\mathcal{R} : E_3 \to \mathbf{ES}$ with $\mathcal{R}(0,*) = \mathcal{E}_{(0,*),b}$, $\mathcal{R}(*,2) = \mathcal{E}_{(*,2),c}$ and $\mathcal{E}_{(1,3)} = \mathcal{E}_{4,d}; \mathcal{E}_{5,d}$. t is then modelled by $\mathcal{E} = \mathcal{E}_3[\mathcal{R}]$ with $E = \{e_1, e_2, e_3, e_4\}$ where

$$e_1 = \big((0,*),*\big), \qquad e_2 = \big((*,2),*\big), \qquad e_3 = \big((1,3),4\big), \qquad e_4 = \big((1,3),5\big)$$

  and with conflicts $e_2 \# e_3$ and $e_2 \# e_4$, enablings $\vdash e_1$, $e_1 \vdash e_3$, $e_3 \vdash e_4$ and $\vdash e_2$, termination predicate $\{\{e_1, e_3, e_4\}\}$ and labelling $e_1 \to b$, $e_2 \to c$, $e_3 \to d$, $e_4 \to d$.

The first thing to make sure of is that the above constructions indeed yield event structures, and moreover, that an isomorphic argument gives rise to an isomorphic result.

PROPOSITION 4.12. *Each of the operators in Definition* 4.10 *maps into* **ES** *and is well-defined up to isomorphism.*

For refinement, this was proved in [64]; for the other operators except sequential composition, see [30] (except for the termination predicate, whose construction, however, is very similar to that of enabling). For sequential composition, finally, the proof is straightforward.

The denotational event structure semantics of *Lang*(*Evt*) is given in Table 8. It should be noted that the disjointness requirements in the denotational constructions for choice and sequential composition are guaranteed to be satisfied due to the well-formedness of annotated terms. The most interesting definition is that for refinement: in principle, the semantic refinement function maps the events of the term being refined to the event structures obtained as the denotational semantics of the syntactic refinement function; however, the events that are in conflict with some busy event of the refinement are mapped to the

Table 8
Denotational event structure semantics; $\overline{op}$ is the semantic counterpart of *op*

$$
\begin{aligned}
[\![\mathbf{0}]\!] &= \mathcal{E}_\perp \\
[\![\mathbf{1}]\!] &= \mathcal{E}_\surd \\
[\![{}_e\alpha]\!] &= \mathcal{E}_{e,\alpha} \\
[\![t[a \to u, \vec{e} \to \vec{v}]]\!] &= [\![t]\!][\mathcal{R}], \text{ where } \mathcal{R}(d) = \begin{cases} \mathcal{E}_{*,b} & \text{if } \ell_{[\![t]\!]}(d) = b \neq a \text{ and} \\ & \nexists e_i \in \{\vec{e}\}\colon d \,\#_{[\![t]\!]}\, e_i, \\ [\![u]\!] & \text{if } \ell_{[\![t]\!]}(d) = a \text{ and} \\ & \nexists e_i \in \{\vec{e}\}\colon d \,\#^{=}_{[\![t]\!]}\, e_i, \\ [\![v_i]\!] & \text{if } d = e_i, \\ \mathcal{E}_\perp & \text{otherwise} \end{cases} \\
[\![{}_e[t]]\!] &= e \times [\![t]\!] \\
[\![\mu x.t]\!] &= \textstyle\bigcup_{i \in \mathbb{N}} [\![\mu^i x.t]\!] \\
[\![op(t_1, \ldots, t_n)]\!] &= \overline{op}([\![t_1]\!], \ldots, [\![t_n]\!]) \quad \text{for all other operators } op
\end{aligned}
$$

deadlocked structure instead, to model the fact that, even if such events have not yet been removed from t (as we have seen in Example 4.4, choices in t are not resolved syntactically until the refinement has terminated), they nevertheless play no further role in the overall behaviour.

The following is the denotational counterpart to Proposition 4.6, namely that two annotations of the same basic term give rise to isomorphic event structures:

PROPOSITION 4.13. *For all $t, u \in Lang(Evt)$, if $strip(t) = strip(u)$ then $[\![t]\!] \cong [\![u]\!]$.*

4.5. *Compatibility of the semantics*

In order to compare the event-based operational and denotational semantics, we must first decide on a relation up to which we wish them to be compatible. Isomorphism of the denotational semantics is certainly much stronger than (event) bisimilarity of the operational semantics; in fact, this was also already true for the sequential language in Section 2. We therefore choose event bisimilarity as the compatibility criterion. (Similar compatibility results of event-based operational and denotational semantics, albeit not including refinement, were given in [45,102].

This means that we have to generate event transition systems from event structures. The most natural way to do this is to regard the event structure's *configurations* as states and define transitions from each configuration to all its direct $\subset$-successors. Formally:

DEFINITION 4.14. Let $\mathcal{E} \in \mathbf{ES}$.

- An *event trace* of $\mathcal{E}$ is a sequence $e_1 \cdots e_n \in E^*$ such that $\neg(e_i \#^= e_j)$ for all $1 \leqslant i < j \leqslant n$ (it is duplication- and conflict-free) and $e_1, \ldots, e_i \vdash e_{i+1}$ for all $1 \leqslant i < n$ (it is secured).
- A *configuration* of $\mathcal{E}$ is a set $F \subseteq E$ such that $F = \{e_1, \ldots, e_n\}$ for some event trace $e_1 \cdots e_n$.
- An event transition between configurations, $F \xrightarrow{e,\alpha} G$, holds iff $e \notin F$, $G = F \cup \{e\}$ and $\alpha = \ell(e)$. Furthermore, there is a transition $F \xrightarrow{\checkmark} \bullet$ for each $F \in Ter$, where $\bullet$ is a special state introduced only for this purpose.

For arbitrary $\mathcal{E}$, we denote $\mathcal{T}(\mathcal{E}) = \langle \mathcal{C}(\mathcal{E}) \cup \{\bullet\}, \rightarrow, \emptyset \rangle$ (where $\mathcal{C}(\mathcal{E})$ is the set of configurations of $\mathcal{E}$ and $\rightarrow$ the transition relation defined above). Note that $\mathcal{T}(\mathcal{E})$ is indeed an event transition system in the sense of Definition 4.2. For instance, the event structure $\mathcal{E}$ developed in Example 4.11 gives rise to the following transition system $\mathcal{T}(\mathcal{E})$:

$$
\begin{array}{ccccccccc}
\emptyset & \xrightarrow{e_1,b} & \{e_1\} & \xrightarrow{e_3,d} & \{e_1,e_3\} & \xrightarrow{e_4,d} & \{e_1,e_3,e_4\} & \xrightarrow{\checkmark} & \bullet \\
{\scriptstyle e_2,c}\downarrow & & \downarrow{\scriptstyle e_2,c} & & & & & & \\
\{e_2\} & \xrightarrow{e_1,b} & \{e_1,e_2\} & & & & & &
\end{array}
$$

It is easy to see that this is event bisimilar to the transition system generated by the operational semantics (Example 4.4). Since information is lost in the generation of configurations from event structures, one can expect event bisimilarity to be strictly weaker than event structure isomorphism. The following proposition states that it is indeed weaker.

PROPOSITION 4.15. *If* $\mathcal{E}_1 \cong \mathcal{E}_2$ *then* $\mathcal{T}(\mathcal{E}_1) \sim_{\mathsf{ev}} \mathcal{T}(\mathcal{E}_2)$.

To see that the reverse implication does not hold, consider the event structure semantics of the terms ${}_1a;\, {}_2b;\, {}_3c$ and ${}_1a;\, ({}_2b \,|||\, {}_3c)\,\|_{b,c}\, {}_4b;\, {}_5c$. In the first of these, the third, c-labelled event (with identity 3) is enabled by the b-labelled event 2 only; i.e., $2 \vdash 3$. In the second, on the other hand, the c-labelled event (here $(3,5)$) is enabled by the combination of a- and b-events; i.e., $(1,*), (2,4) \vdash (3,5)$. (In fact, event bisimilarity of event structures coincides with isomorphism of the *sets of configurations* of those event structures.)

We can now state the compatibility result of the event-based operational and denotational semantics of *Lang*.

PROPOSITION 4.16. *For any* $t \in \mathit{Lang}(\mathit{Evt})$, $t \sim_{\mathsf{ev}} \mathcal{T}([\![t]\!])$.

4.6. *Application: A simple data base*

To show the potential of non-atomic refinement of synchronizing actions, we consider a variation on the data base example of Section 2.5. Namely, we assume that, in addition to updating and querying the data base, one may also make a backup copy of it. We abstract away from the state of the data base; extending it to the n-state version (see Section 2.5) makes no essential difference to the current example (except for making the transition system more complicated). The behaviour of the date base is specified by Data_S, defined by

$$\begin{aligned} \mathit{Data}_S &:= \mathit{State} \,\|_{\mathit{upd}}\, \mathit{Backup}, \\ \mathit{State} &:= (\mathit{qry} + \mathit{upd}); \mathit{State}, \\ \mathit{Backup} &:= (\mathit{copy} + \mathit{upd}); \mathit{Backup}. \end{aligned}$$

Now suppose that on a more concrete level, both the update and the copy operations are split into two phases: $\mathit{upd} \to \mathit{req}; \mathit{cnf}$ (as in Section 2.5) and $\mathit{copy} \to \mathit{back}; \mathit{copy}$ (where *back* is a request to make a backup copy and *copy* is re-used on the concrete level to stand for the actual operation of copying). Thus, the refined behaviour is specified by

$$\mathit{Data}_I := \mathit{Data}_S[\mathit{upd} \to \mathit{req}; \mathit{cnf}][\mathit{copy} \to \mathit{back}; \mathit{copy}].$$

The abstract and refined behaviour are shown in Figure 6. Note that *qry* and *copy* in Data_S are independent (which is evident from the fact that these actions give rise to a diamond in the event transition system). As a consequence of this independence, in Data_I *qry* may occur while the (refined) backup operation is in progress, i.e., between *back* and the concrete *copy*, whereas *req* is disabled at that time; on the other hand, *qry* is *not* enabled while the (refined) update operation is in progress, i.e., between *req* and *cnf*. In fact, Data_I is

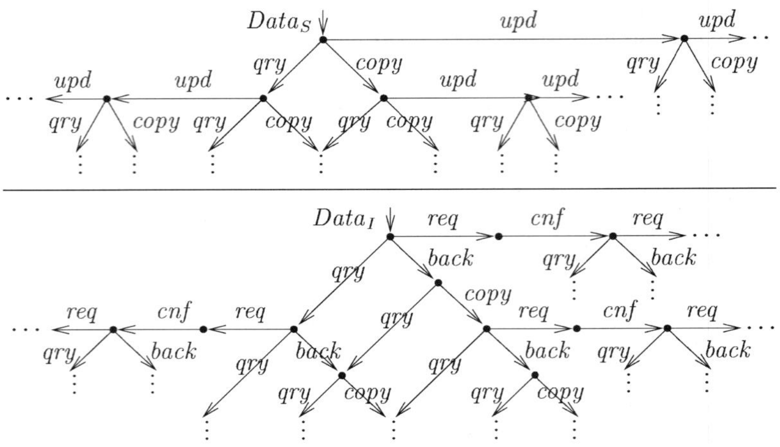

Fig. 6. An initial fragment of the event-based operational semantics of $Data_S$ and $Data_I$ (event identities omitted).

(both interleaving and event) bisimilar to the process where the refinements are interpreted syntactically, given by $State^{flat} \|_{req,cnf} Backup^{flat}$ where

$$\begin{aligned} State^{flat} &:= (qry + req;\, cnf);\, State^{flat}, \\ Backup^{flat} &:= (back;\, copy + req;\, cnf);\, Backup^{flat}. \end{aligned}$$

The relation between semantic and syntactic refinement is discussed extensively in Section 6. A final observation is that $Data_S$ is interleaving bisimilar (but not event bisimilar) to a process where all actions are specified sequentially, given by $Data_S^{seq}$ with the following definition:

$$Data_S^{seq} := (qry + copy + upd);\, Data_S^{seq}.$$

If we refine $Data_S^{seq}$ in the same way as $Data_S$, the resulting behaviour is not interleaving bisimilar to $Data_I$: in the refinement of $Data_S^{seq}$, *qry* cannot occur between *back* and *copy*. Once again, this is evidence of the fact that interleaving bisimilarity is not a congruence for the refinement operator.

5. Non-atomic refinement: Other observational congruences

In the previous section, we have chosen a particular modelling principle (namely, action occurrences are events) and shown that this can be used to give semantics to a process algebra with action refinement; moreover, the corresponding notion of (event) bisimulation is a congruence. However, one may wonder if this congruence is the most suitable one. For instance, event bisimilarity distinguishes between the terms "$(a + b); c$" and "$a; c + b; c$"; for most purposes this distinction is not relevant.

A large part of the literature on action refinement is devoted to the quest for alternative congruences, in particular ones that are weaker (less distinguishing) than event bisimilarity. As a special instance, one can search for the *coarsest* congruence for action refinement contained within a given interleaving relation. There are two ways to go about this quest: to define equivalences over event structures that are weaker than isomorphism and investigate the induced relation on terms, or to develop alternative models altogether. An impressive overview based on the first of these methods has been drawn up by Van Glabbeek and Goltz in [64].

In this section, we review three possible bases for alternative models: *pomsets* (Section 5.1), *causal links* (Section 5.2) and *splitting* (Section 5.3). Pomsets give rise to a congruence on the level of traces but not when combined with bisimilarity. In contrast, causal links can be combined with bisimilarity into a congruence. Finally, with some care, splitting can be used to obtain coarsest congruences. In fact, in Section 5.4 we present an operational semantics directly based on the idea of splitting.

The material in this section is based on the language *Lang* introduced in Section 4; the semantics in Section 5.4 uses an extension of it, denoted $Lang^{ST}$.

5.1. *Pomsets*

The use of pomsets for the modelling of distributed system behaviour has been proposed first by Grabowski [79] (who, among other things, introduces the special class of so-called *N-free* or *series-parallel* pomsets, also investigated by Aceto in [1], as an auxiliary device to study Petri nets) and strongly advocated by Pratt [115] (who introduced the term "pomset"); other references are Jónsson [95] (who investigates pomsets from a more mathematical perspective), Gischer [55] (who works out Pratt's framework in more detail) and Rensink [122]. Furthermore, also *Mazurkiewicz traces* (see [103,104]) are tightly connected to a certain class of pomsets.

Mathematically, a pomset is an isomorphism class of labelled partial orders. For the current purpose, the elements of the partial orders are events and their labels are action names. The basic definitions are listed below.

- A labelled partial order (lpo) is a tuple $p = \langle E, <, \ell\rangle$, where $E \subseteq Evt$ is a set of events causally ordered by the irreflexive and transitive relation $< \subseteq E \times E$, and $\ell : E \to Act_{\tau\surd}$ is a labelling function.
- If $p_i = \langle E_i, <_i, \ell_i\rangle$ is an lpo for $i = 1, 2$, then p_1 and p_2 are called isomorphic if there is a bijection $f : E_1 \to E_2$ such that $d <_1 e$ iff $f(d) <_2 f(e)$ for all $d, e \in E_1$ and $\ell_1 = \ell_2 \circ f$.
- The isomorphism class of a poset $\langle E, <, \ell\rangle$ is called a *pomset*. The class of all pomsets is denoted *Pom*.

By taking isomorphism classes of posets, the precise identities of events are abstracted away from *before* they can be used to pinpoint the exact moment at which choices are resolved. Consequently, a pomset-based model is more abstract than an event-based model and thus induces a weaker equivalence over *Lang*.

Pomset transitions from event transitions. In fact, from the event transition systems used for the operational semantics in the previous section (Definition 4.2), one can generate pomset-labelled transition systems, in the following fashion:

- Each sequence of event-action-pairs, say

$$(e_1, \alpha_1)(e_2, \alpha_2) \cdots (e_n, \alpha_n) \in (Evt \times Act_\tau)^*,$$

can be regarded as an lpo that happens to be totally ordered: namely $p = \langle E, <, \ell \rangle$ where $E = \{e_1, \ldots, e_n\}$, $e_i < e_j$ iff $i < j$ and $\ell(e_i) = \alpha_i$ for all $1 \leq i, j \leq n$.
- For each pair of states s, s' between which there is a path not containing a $\checkmark$-labelled transition, one can generate an lpo by considering *all* paths from s to s' and intersecting their ordering. That is, let $p_1, \ldots, p_n$ with $p_i = \langle E_i, <_i, \ell_i \rangle$ be the set of totally ordered lpo's generated from transition paths between s and s' in the fashion described above; then $E_i = E_j$ and $\ell_i = \ell_j$ for all $1 \leq i, j \leq n$ due to the properties of the transition system. This gives rise to another lpo $q = \langle E, <, \ell \rangle$ with $E = E_1$, $< = \bigcap_{1 \leq i \leq n} <_i$ and $\ell = \ell_1$; we let $s \xrightarrow{q} s'$.
- $\checkmark$-labelled transitions are left unchanged.
- Each lpo thus obtained can be turned into a pomset by taking its isomorphism class.

The idea is that in the event-based view, each sequence of transitions corresponds to the execution of a causally ordered set of labelled events; hence an lpo. Due to the construction of the transition system, there is an exact correspondence between the paths between the states and the linearizations of this lpo. The original lpo can therefore be reconstructed through the intersection of all these linearizations.

Pomset-labelled transition systems can be compared in a number of ways; for instance, one may look only at the pomset traces, or one may apply the natural notion of bisimilarity.

Pomset traces. The pomset traces of a term of *Lang* are given by the outgoing pomset transitions of the initial state. There is a distinguished class of *terminated* pomset traces, which are the ones leading to a state with an outgoing $\checkmark$-transition. We call two terms *pomset trace equivalent* if they give rise to equal sets of terminated and non-terminated pomsets.

As an alternative to this roundabout construction, one can directly define an algebra of (sets of) pomsets; in fact, this is the subject of most of the pomset papers cited above. For the synchronization-free fragment of *Lang*^{fin} (i.e., containing $\_\|_A\_$ only for $A = \emptyset$, but including refinement), such an algebra is for instance defined in [110]. The extension to recursion is straightforward; the extension to synchronization, although technically awkward, is also unproblematic – see for instance [116]. (The awkwardness is due to that fact that the synchronization of two pomsets is in general not a single-valued operation: for instance, the synchronization of

$$\boxed{\begin{array}{c} a \\ a \to b \end{array}} \quad \text{and} \quad \boxed{a \to a}$$

on a yields

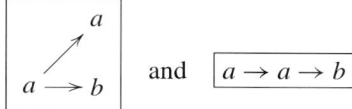

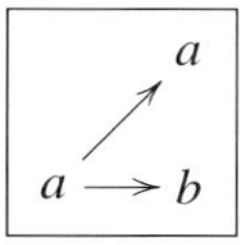

and $\boxed{a \to a \to b}$

as two possible outcomes.)

We do not go into detail here concerning the definition of the operators, except to remark that the refinement operator on sets of pomsets crucially depends on the ability to distinguish the terminated pomsets from the non-terminated ones (see also Section 1.5). For instance, if one ignores this distinction then $a; \mathbf{0}$ and a give rise to the same sets of pomsets, and yet $(b; c)[b \to a; \mathbf{0}]$ and $(b; c)[b \to a]$ should not be pomset trace equivalent. Since in the restricted setting of [110], the terminated pomsets coincide with the maximal ones, the distinction is easy there.

It follows from the existence of an algebraic operator on the semantic model that pomset trace equivalence is a congruence for refinement. This fact was first stated in [32]. As a consequence of the construction of pomsets from event transition systems and the existence of a full *Lang*-algebra over sets of pomsets, this can be extended to the following result.

THEOREM 5.1. *Pomset trace equivalence is a congruence over Lang that is coarser than* $\sim_{\mathsf{ev}}$.

Pomset bisimilarity. The above construction of pomsets from event transition systems actually gives rise to a much richer model than just pomset traces: namely, a transition system labelled on $Pom \cup \{\checkmark\}$. Having a transition system, we can once more apply the principles of bisimulation. We call two terms *pomset bisimilar* if their associated pomset-labelled transition systems are strongly bisimilar (see also Boudol and Castellani [20]). Surprisingly (as first observed by Van Glabbeek and Goltz in [60]), pomset bisimilarity is *not* a congruence for refinement. A simple counter-example is given by the pomset bisimilar terms

$$t = a; (b + c) + a \mathbin{|||} b \quad \text{and} \quad u = t + a; b,$$

whose refinements $t[a \to a_1; a_2]$ and $u[a \to a_1; a_2]$ are *not* pomset bisimilar: after the execution of a_1 there is a behaviour that only $u[a \to a_1; a_2]$ may show, namely a state where a_2 and b are executable only sequentially and c is not executable at all. For more details see [64]; an extensive discussion can also be found in [133].

As an alternative to the indirect construction of pomset transition systems from event transition systems, [20] presents an operational semantics directly on pomset transition systems, albeit only for the synchronization- and refinement-free fragment of $Lang^{fin}$. The extension to synchronization and recursion is once more straightforward; however, from the fact that pomset bisimilarity is not a congruence for refinement, it follows that an operational semantics for refinement will be difficult to find.

5.2. *Causal links*

A different modelling principle is obtained if one takes not the events themselves but the *causality* between them as the essential notion. Operationally, this can be done by adding a set of *causal links* to each transition that point back to those previous transitions on which the current one depends. This idea was first worked out in the *causal trees* of Darondeau and Degano [39]. They use a particular simple way to implement the causal links: namely, the links are given as positive natural numbers corresponding to the number of transitions one must count back from the current one. Note that for this to work, it is essential that each state have a unique predecessor, i.e., that the model be a tree. In our setting, there is an additional class of $\checkmark$-labelled transitions. Thus, a causal tree is a tree-shaped transition system labelled on $(\mathbf{2}^{\mathbb{N}^{>0}} \times Act_\tau) \cup \{\checkmark\}$ (where $\mathbb{N}^{>0}$ denotes the set of positive natural numbers).

Causal transitions from event transitions. The event transition systems of the previous section (Definition 4.2) give rise to causal trees in the following way. Let $\mathcal{T}$ be an arbitrary event transition system.

- The states of the causal tree derived from $\mathcal{T}$ are vectors of events $\vec{e} = e_1 \cdots e_k$ (with $e_i \in Evt$ for $1 \leqslant i < k$ and $e_k \in Evt \cup \{\checkmark\}$) corresponding to traces of t. To be precise, $\vec{e}$ is a state iff $\iota \xrightarrow{e_1,\alpha_1} \cdots \xrightarrow{e_{k-1},\alpha_{n-1}} n'$ for some n' and either $n' \xrightarrow{e_k,\alpha_k} n''$ (if $e_k \in Evt$) or $n' \xrightarrow{\checkmark} n''$ (if $e_k = \checkmark$) for some n''. Note that if n'' exists, then it is uniquely determined by $\vec{e}$; we denote $n'' = \iota_{\vec{e}}$.
- The outgoing non-termination transitions of $\vec{e}$ are defined by induction on $|\vec{e}|$. Each event transition $\iota_{\vec{e}} \xrightarrow{e',\alpha} n'$ gives rise to a causal transition $\vec{e} \xrightarrow{K,\alpha} \vec{e}\, e'$, where the set K of causal links is determined in one of the following two ways.
  - If $\iota_{e_1 \cdots e_{k-1}} \xrightarrow{e',\alpha}$ (i.e., e' was already enabled in the direct e_k-predecessor of $\iota_{\vec{e}}$) then (due to the inductive definition) we already had derived the causal transition

    $$e_1 \cdots e_{k-1} \xrightarrow{K',\alpha} e_1 \cdots e_{k-1} e'$$

    for some set $K' \subseteq \mathbb{N}^{>0}$. We let $K = K' + 1$ $(= \{i + 1 \mid i \in K'\})$.
  - If $\iota_{e_1 \cdots e_{k-1}} \stackrel{e',\alpha}{\not\longrightarrow}$, then we had already derived the causal transition

    $$e_1 \cdots e_{k-1} \xrightarrow{K',\alpha} \vec{e}$$

    for some $K' \subseteq \mathbb{N}^{>0}$. We let $K = (K' + 1) \cup \{1\}$.
  - Termination transitions are given by $\vec{e} \xrightarrow{\checkmark} \vec{e}\checkmark$ whenever $\iota_{\vec{e}} \xrightarrow{\checkmark}$.

Having obtained causal trees from terms of *Lang*, one can again study induced relations over *Lang*, generated by different interpretations of causal trees. For instance, we call two terms *causally bisimilar* if their respective causal trees are bisimilar. It was first observed in [42] that causal bisimilarity is a congruence for action refinement. This can again be generalized to the following result:

PROPOSITION 5.2. *Causal bisimilarity is a congruence over Lang, which is weaker than* $\sim_{\mathsf{ev}}$.

A similar property holds for causal traces (the linear-time counterpart of causal trees), which in fact constitute an alternative representation for pomset traces, studied in [120].

Instead of recovering the causal links from a more concrete semantic model, as we have done here, it is possible to define a *Lang*-algebra of causal trees (see [40,42]) or to use causal trees as an operational model. With respect to the latter, see [50] for a structural operational semantics for (essentially) $Lang^{fin}$, which can easily be extended to deal with recursion. Since the semantics is given in SOS style, that paper also reports a set of axioms for the calculus.

As a final remark, we recall that there are several alternative (indeed historically preceding) characterizations of causal bisimilarity. Rabinovich and Trakhtenbrot first introduced the relation in [116] as *behaviour structure bisimilarity*. In [60], the same relation is characterized as *history preserving bisimilarity* over event structures. Finally, the *mixed ordering bisimilarity* of [46] also gives rise to the same equivalence. See [3,64] for further details concerning these correspondences.

5.3. *Splitting actions*

Having found (in causal bisimilarity) a bisimulation-based equivalence that is a congruence with respect to action refinement, a natural question is if this is the *coarsest* of its kind, i.e., the coarsest congruence for action refinement contained in (standard) interleaving bisimilarity. An analogous question can be posed for pomset trace or causal trace equivalence. It turns out that the answer to either question is "no": in order to obtain coarsest congruences, one needs to add information to the interleaving model in a way that is different from either pomsets or causal links.

EXAMPLE 5.3. There exist processes with distinct causal structures that yet cannot be distinguished up to bisimilarity by any refinement context. For instance, consider the bisimilar (but not causally bisimilar) terms $a \mathbin{|||} b; \mathbf{0}$ (where a and b are causally independent) and $(a; b \mathbin{|||} b) \|_b b$ (where there is alternative way to perform b, causally following a). It is not difficult to convince oneself that there is no way to make them non-bisimilar (in the interleaving sense) by refining their actions.

So, what is the coarsest congruence within strong bisimilarity? Intuitively, by refining an action we are implicitly assuming that such an action is no longer non-interruptible, as the execution of its refinement can be interleaved with the execution of other concurrent processes. In the example above, after refining a into $a_1; a_2$, the existence of the sequence $a_1\, b\, a_2$ shows the interruptability of the refined a. Hence, a natural question is whether the actions could not *a priori* be described as split into phases: for instance, a^+ for the beginning and a^- for the ending of an arbitrary action a. If one interprets the resulting transition systems up to strong bisimilarity, $a \mathbin{|||} b; \mathbf{0}$ and $a; b; \mathbf{0} + b; a; \mathbf{0}$ (for example) are distinguished whereas $a \mathbin{|||} b; \mathbf{0}$ and $(a; b \mathbin{|||} b) \|_b b$ are not (see Figure 7).

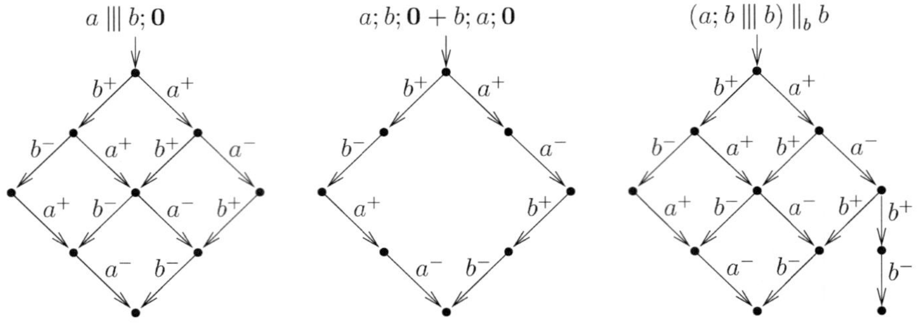

Fig. 7. Three *split*$_2$ transition systems.

This is the basic idea behind the so-called *split* semantics, probably introduced for the first time in [84] (although the principle is already mentioned by Hoare in [88]). Actions can be split in any number of phases, say n, giving rise to a family of *split*$_n$-equivalences for every interleaving equivalence. In fact, splitting into n phases can be seen as a restricted form of action refinement where the refinement body can only consist of sequences of length n.

Split bisimilarity. For (essentially) the synchronization-free fragment of *Lang* (or alternatively a language with CCS-like communication but without restriction), Aceto and Hennessy [8] established that strong *split*$_2$-bisimilarity is the coarsest congruence for action refinement contained in strong bisimilarity. Actually, in this language, *split*$_n$-bisimilarity coincides with *split*$_{n+1}$-bisimilarity for any $n \geqslant 2$.

However, as soon as synchronization is included, *split*$_2$-bisimilarity is no longer a congruence: splitting actions in more and more phases yields more and more discriminating equivalences. Van Glabbeek and Vaandrager [66] present an example, parametric in n, to show that the *split*-bisimilarities form a strict chain. The example is based on confusion on identity of auto-concurrent actions (i.e., multiple occurrences of the same action, concurrently executed) that *split*$_n$ semantics may bring to light when there are more than $n-1$ such actions. A simplified example that clarifies the difference between *split*$_2$ and *split*$_3$ is the so-called *owl example*, originally proposed by Van Glabbeek in 1991.

EXAMPLE 5.4. Consider $t = (t_1 \,\|_S\, t_2)[\phi]$, where $S = \{c', c'', d, d', e, e'\}$ and ϕ relabels c' and c'' into c, d' into d and e' into e (leaving all the other labels fixed).

$$\begin{aligned} t_1 &= a; \big((c; (c'; e' \,|||\, d' + e) + c''; d' \,|||\, e') \,|||\, d; \mathbf{0}\big), \\ t_2 &= b; \big((c; (c''; d' \,|||\, e' + d) + c'; e' \,|||\, d') \,|||\, e; \mathbf{0}\big). \end{aligned}$$

Symmetrically, we define $u = (u_1 \,\|_S\, u_2)[\phi]$, where

$$\begin{aligned} u_1 &= a; \big((c; (c'; d' \,|||\, e' + d) + c''; e' \,|||\, d') \,|||\, e; \mathbf{0}\big), \\ u_2 &= b; \big((c; (c''; e' \,|||\, d' + e) + c'; d' \,|||\, e') \,|||\, d; \mathbf{0}\big). \end{aligned}$$

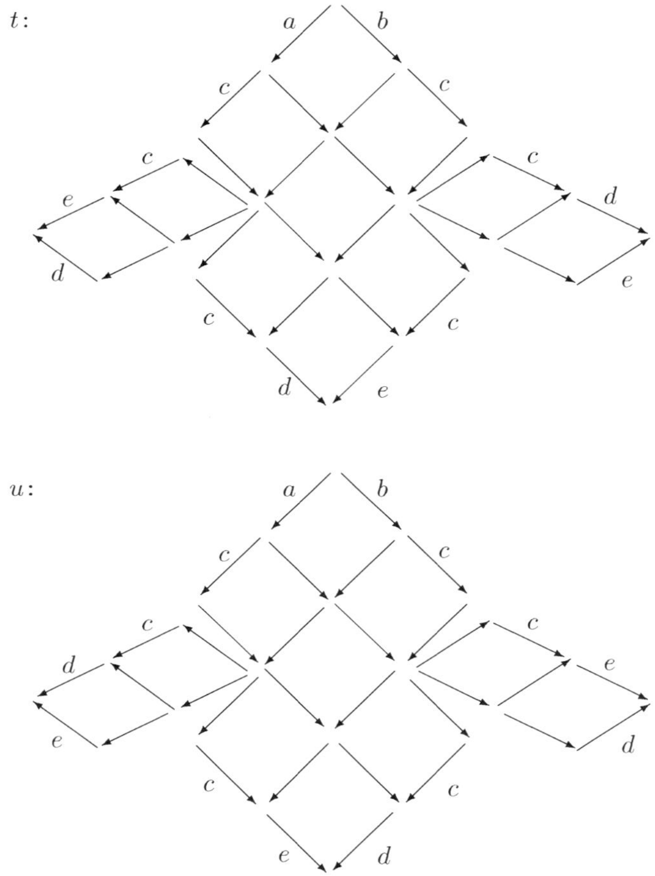

Fig. 8. The "owl" example.

The two interleaving transition systems are shown in Figure 8. The labels of the inner transitions are omitted for improving readability: they can be assigned by looking at the label of the parallel outer transitions. It is not difficult to see that the two transition systems are interleaving bisimilar. For instance, the sequence $b\,a\,c$ leads to bisimilar states in both systems: the "wing" of t is simulated by the "body" of u, and vice versa.

The $split_2$ transition systems for t and u can be guessed by considering that all the "diamonds" in the graph are divided into four sub-diamonds. The two resulting graphs are bisimilar (so t and u are $split_2$ bisimilar) by following an argument similar to the above. The crucial point is reached after the sequence $a_1\,a_2\,c_1\,b_1\,b_2\,c_1$. In that state (which is the same in both graphs), the c_2 executed by t is matched by the 'symmetric' c_2 in the graph for u. This means that whenever t completes the c causally dependent of, let say, a then u completes the c caused by b.

The same game cannot be played when splitting actions into three phases; indeed, it is always possible to recognize *observationally* in which of the two directions we are moving, i.e., which of the two c's we are going to complete. This informal consideration is illustrated by the sequence

$$ac_1\, bc_2\, c_1\, c_3 e$$

which can be executed by t but not by u (unsplit actions stand for the consecutive execution of their three phases).

Two natural questions then arise:

- What is the limit of this chain of $split_n$-bisimulation equivalences?
- How does the chain limit relate to the coarsest congruence problem?

It turns out (see Gorrieri and Laneve [73,74]) that the limit of strong $split_n$-bisimilarity is an equivalence known in the literature as strong ST-bisimilarity, originally due to Van Glabbeek and Vaandrager [65]. The ST-principle improves over *split* by including a mechanism for reconstructing the individuality of each action occurrence. Typically, this is done by associating to each action one unique identifier that is used when it is necessary to know the right beginning for the currently executed ending. By the limit theorem of [74], it follows that strong ST-bisimilarity is the coarsest congruence for splitting of actions; and in fact, the result extends to general refinement (see [57,133] for a more general congruence result, [30] for a coarsest congruence result for any form of refinement in the model of stable event structures and the survey paper [64] for another proof of coarsest congruence on several event-based models). Hence, strong ST-bisimilarity is the required coarsest congruence contained in strong bisimilarity.

It has also been proved in [9] that rooted weak ST-bisimilarity (called refine equivalence in that paper) is the coarsest congruence for (syntactic) action refinement contained in rooted weak bisimilarity.

Split traces. If one starts the investigation by considering trace semantics [88] instead of bisimulation, then the answers to the two questions above are slightly different. In the full language *Lang*, $split_n$-trace equivalence again gives rise to a strict chain of equivalences. However, as shown in [101], the limit of the chain in this case is not ST-trace equivalence.

EXAMPLE 5.5. Consider $t = a; b \mid\mid\mid b; c; \mathbf{0}$ and $u = (a; b; c \mid\mid\mid b; c) \|_c c$. The only pomset trace of t is

$$\boxed{\begin{array}{l} a \to b \\ b \to c \end{array}}$$

whereas u additionally has

$$\boxed{\begin{array}{l} a \to b \to c \\ b \end{array}}$$

However, t and u cannot be distinguished on the basis of their $split_n$-traces. To see this, first note that the only possible difference can be that u has a trace that t cannot match; in particular, a trace that is a linearization of

$$\boxed{\begin{array}{l} a \to b \to c \\ b \end{array}}$$

Thus, such a distinguishing trace must reflect the fact that c is caused by the upper b (the one following a) rather than the lower b. However, no matter in how many phases we split b, this distinction cannot be made: in particular, any $split_n$-trace $b_1 \cdots b_i \, a_1 \cdots a_n \, b_1 \cdots b_n \, c_1$ of u can be matched by t because the $b_{i+1} \cdots b_n$-subsequence might belong to the first b-occurrence as well as the second.

On the other hand, t and u are not ST-trace equivalent. For instance, although both processes can perform the $split_2$-trace $b_1 \, a_1 \, a_2 \, b_1 \, b_2 \, c_1$, in t the b_2 must be the end of the first b_1 whereas in u it may also be the end of the second b_1.

The limit of $split_n$-trace semantics has been characterized in [134] and called *swap*-equivalence. It turns out that *swap*-equivalence is not a congruence for general action refinement. For instance, although (as we argued) t and u in Example 5.5 are *swap*-equivalent, refining b into $v = b_1; b_2 + b_3$ shows up their difference, since $u[b \to v]$ has trace $b_1 \, a \, b_3 \, c$ which $t[b \to v]$ cannot match. Vogler has proved in [131,132] that the coarsest congruence for action refinement contained in trace equivalence is ST-trace equivalence.

Further split semantics. In the linear time – branching time spectrum [59], in between bisimilarity and trace equivalence there are many other equivalences, among which failure [27] (or testing [43]) equivalence is the most relevant. The issues above have been investigated for this case as well: Vogler in [131,132] proves that ST-failure semantics is a congruence for action refinement, and in [134] conjectures that it is also the coarsest congruence.

Most of the work mentioned above has been developed on semantic models rather than process algebra. In addition to the work already cited above, papers dealing with operational *split*- and ST-semantics include [31,37]. Characterizations of ST-failure (or testing) semantics in process algebra are reported in [5] on a simple process algebra, and in [85] on CCS.

5.4. *ST-semantics*

We go somewhat deeper into the ST-semantics of terms. Below, we present a conversion from event transition systems into ST-transition systems as well as a (compatible) operational ST-semantics for *Lang* – including, of course, the refinement operator.

ST-transition systems. It is clear from the above discussion that the main idea in ST-semantics is to distinguish the start and end of actions, in such a way that there is an unambiguous association between the two. There are several different ways to encode this association.

- The most straightforward method is to associate with the start of the action a unique fresh identifier, much like the event annotation, and annotate the end of the action with the same identifier. The disadvantage of this method is that the choice of identifier is free, and therefore it cannot be guaranteed that the same identifier is used in bisimilar states; one therefore needs a more elaborate, indexed version of bisimulation like we already had for event bisimilarity (Definition 4.3); see, for instance, [57,9,23].
- The above method can be improved by somehow fixing the choice of the fresh identifier so that it depends uniquely on the set of identifiers currently in use for "outstanding" action occurrences, i.e., actions that have started but not finished. This can be done for instance by imposing a total ordering on the set of all identifiers. One can then safely assume that the choice of identifier is always the same in bisimilar states. For an example, see [24].
- Alternatively, one can assume a tree-like structure and let the end of the action point back to the start through a natural number indicating the number of transitions one has to count back to find the associated start. This idea was used, e.g., in [73,74,31]. The disadvantage (which in fact also holds for causal transition systems as defined in Section 5.2) is that the resulting transition systems are necessarily infinite state for all recursive behaviours.
- Here we use yet another principle, developed more recently in [24], which is in some sense a mixture of the above: the end actions indeed carry counters, but instead of transitions it is the number of outstanding occurrences (of the same action) that is counted. As a consequence, ordinary bisimulation can be used to compare *ST*-transition systems. Moreover, the operational semantics not only generates finite state transition systems for the refinement-free fragment of *Lang* whenever the standard interleaving semantics does (which is also the case for the semantics of [23]), but also provides finite state models for action refinement.

As labels in the *ST*-semantics, for all $a \in Act$ we use a^+ to denote the start of a fresh occurrence of a, and a_i^- with $i \in \mathbb{N}^{>0}$ to denote the end of the ith occurrence of a that was still outstanding (counting backwards, that is, the most recent occurrence is numbered 1). We denote $Act^{ST} = \{a^+ \mid a \in Act\} \cup \{a_i^- \mid a \in Act, i \in \mathbb{N}^{>0}\}$ and $Act^{ST}_{\tau\checkmark} = Act^{ST} \cup \{\tau, \checkmark\}$.

DEFINITION 5.6. An *ST*-transition system is an $Act^{ST}_{\tau\checkmark}$-labelled transition system, such that for all states $n \in N$ and action $a \in Act$ there is a number $O(n, a)$ of outstanding occurrences where

- $O(\iota, a) = 0$ for all $a \in Act$;
- If $n \xrightarrow{a^+} n'$ then $O(n', a) = O(n, a) + 1$ and $O(n', b) = O(n, b)$ for all $a \neq b$;
- If $n \xrightarrow{a_i^-} n'$ then $i \leqslant O(n, a)$, $O(n', a) = O(n, a) - 1$ and $O(n', b) = O(n, b)$ for all $a \neq b$;
- If $n \xrightarrow{\tau} n'$ then $O(n', a) = O(n, a)$ for all $a \in Act$;
- $n \xrightarrow{a_i^-}$ for all $a \in Act$ and $1 \leqslant i \leqslant O(n, a)$.

(Note that this uniquely determines $O(n,a)$ for all reachable states.) As stated above, to compare ST-transition systems we will use ordinary bisimulation.

ST-transitions from event transitions. The event transition systems used in the previous section (Definition 4.2) give rise to ST-transition systems in the following way. Let $\mathcal{T}$ be an arbitrary event transition system.

- The states of the ST-transition system derived from $\mathcal{T}$ are tuples $(n,\vec{e})$ for some node $n \in N$ and finite vector of events $\vec{e} = e_1 \cdots e_k$ such that $n \xrightarrow{e_1,\alpha_1} \cdots \xrightarrow{e_k,\alpha_k} n'$ for some n' and $n \xrightarrow{e_i,\alpha_i}$ for all $1 \leqslant i \leqslant k$. (In other words, the events in $\vec{e}$ are all enabled in n and can also be executed serially; this means that they are pairwise independent.) Due to the properties of event transition systems, the node n' is uniquely determined by n and $\vec{e}$; we denote $n' = n_{\vec{e}}$. It follows that also $n' = n_{\vec{d}}$ for an arbitrary permutation $\vec{d}$ of $\vec{e}$.
- The outgoing transitions of $(n,\vec{e})$ are the following (where for all $1 \leqslant i \leqslant k$, α_i is the action associated with e_i, and $\vec{e} \setminus i$ denotes the removal of the ith element of $\vec{e}$):
  - If $n_{\vec{e}} \xrightarrow{e',a}$ such that e' is independent, in n, of all e_i, then $(n,\ \vec{e}) \xrightarrow{a^+} (n,\ \vec{e}\,e')$;
  - If $n_{\vec{e}} \xrightarrow{e',\tau}$ such that e' is independent, in n, of all e_i, then $(n,\ \vec{e}) \xrightarrow{\tau} (n,\ \vec{e}\,e')$;
  - For all $1 \leqslant i \leqslant k$, if $\alpha_i \in Act$ then let $a = \alpha_i$ and $m = |\{i \leqslant j \leqslant k \mid \alpha_j = a\}|$; then $(n,\ \vec{e}) \xrightarrow{a_m^-} (n_{e_i},\ \vec{e} \setminus i)$.
  - For all $1 \leqslant i \leqslant k$, if $\alpha_i = \tau$ then $(n,\ \vec{e}) \xrightarrow{\tau} (n_{e_i},\ \vec{e} \setminus i)$.
  - If $\vec{e} = \varepsilon$ and $n \xrightarrow{\checkmark} n'$, then $(n,\ \vec{e}) \xrightarrow{\checkmark} (n',\ \varepsilon)$.

The number of outstanding a-occurrences of an ST-transition system constructed in this way equals the number of a's in the $\vec{e}$: that is, $O((n,\vec{e}),a) = |\{1 \leqslant j \leqslant k \mid \alpha_j = a\}|$ for all states $(n,\vec{e})$. It is not difficult to check that this construction indeed gives rise to an ST-transition system in the sense of Definition 5.6.

Operational ST-semantics. In addition to the indirect construction of ST-transition systems through the event-based operational semantics, presented above, we will also give a direct ST-operational semantics for *Lang*. To be precise, we use a language $Lang^{ST}$ with the following grammar (changes with respect to *Lang* are underlined; see also Table 16 for a complete overview of the different languages used in this chapter).

$$\mathsf{T} ::= \mathbf{0} \mid \mathbf{1} \mid \alpha \mid \underline{\alpha^-} \mid \mathsf{V}+\mathsf{V} \mid \mathsf{T};\mathsf{V} \mid \underline{\mathsf{T}\|_{A,M}\mathsf{T}} \mid \mathsf{T}/A \mid \underline{\mathsf{T}[a \to \mathsf{V}, \vec{\mathsf{T}}]_M} \mid x \mid \mu x.\mathsf{V}.$$

We briefly discuss the new operators.

- α^- is an auxiliary operator denoting the intermediate state reached by α after it has started; depending on whether α is visible or not, it performs the corresponding end action or another τ-action. (That is to say, τ's are split as well, mainly to preserve some intuitive axioms up to bisimilarity.) As usual with auxiliary operators, α^- is not allowed in virgin operands (V in the grammar).
- $t\|_{A,M}\,u$ denotes parallel composition extended with a mapping $M : Act \to \{0,1\}^*$ to be discussed below.
- $t[a \to u, \vec{v}]_M$ denotes refinement, extended with a vector $\vec{v}$ consisting of *pending* (or *busy*) refinements, entirely analogous to the busy refinements used in the previous sec-

tion, except that here we do not need the identities of the events being refined. (In fact, those identities are encoded in the position of the pending refinement within the vector $\vec{v}$.) We use $\vec{v} \setminus i$ to denote the removal of the ith element of the vector, and $\vec{v} \pm (i, v')$ to denote the replacement of the ith element by v'. Again, M is a mapping, this time from Act to $\{0, \ldots, |\vec{v}|\}^*$, whose purpose is explained below.

In the last two operators, M may be omitted if it maps all $a \in Act$ to the empty string; thus, *Lang* is embedded in $Lang^{ST}$. The well-formedness conditions on $Lang^{ST}$ thus come down to the usual:

- No virgin operand (V in the grammar) contains an auxiliary operator (**1** or α^-);
- Recursion is allowed on guarded variables only.

The challenge in the definition of the operational semantics of $Lang^{ST}$ is to adjust the counters in the a_i^--labels of the operands' transitions in such a way that they always point to the correct start action. For the flat, sequential fragment of *Lang* this presents no particular problem, since there the order in which new actions are started and old ones are finished is directly derived from the operands. In the case of synchronization and refinement, however, we are faced with the problem that there is more than one operand that is *active*, i.e., may have outstanding action occurrences: in the case of parallel composition, both operands are active, and in the case of refinement, both the term being refined and all pending refinements are active. This means that the numbering of the outstanding action occurrences in the combined behaviour changes with respect to the numbering in the individual (active) operands. For instance, in a term $a \,|||\, a$, the left hand side may do a^+ and be ready to do a_1^-, after which the right hand side may also do a^+ and be ready to do a_1^-; however, in the combined behaviour the a_1^--actions should be distinguished since they refer to different start actions. We use a reverse numbering of outstanding actions in which the last action started is numbered 1, etc. This means that in the above example, the left hand side's a_1^--transition should be renumbered, in the combined behaviour, to a_2^-.

To do this kind of renumbering systematically, we append a mapping M to the synchronization and refinement operators, which to every action a associates a string w of *active operand positions*: the ith element of w equals p precisely if the ith outstanding a-occurrence in the combined behaviour originated from the active operand in position p. For instance, in the behaviour $a \,|||\, a$ considered above, after an a^+-transition of the left hand side (operand 0) the associated string is "0" whereupon after the a^+-transition of the right hand side (operand 1), the associated string becomes "1 0" (note that the "1" is inserted *in front* of the string; this is in order to implement the reverse numbering of outstanding actions, mentioned above).

As is apparent from this example, we use natural numbers to identify active operand positions, starting with 0. Thus, the active operand positions of $t \,\|_A\, u$ are $\{0, 1\}$, and those of $t[a \to u, \vec{v}]$ are $\{0, \ldots, |\vec{v}|\}$. The association string for each action is given by a string w of active operand positions. We use $w[i]$ to denote the ith element of w, $w \setminus i$ to denote the removal of the ith element from w (where $1 \leqslant i \leqslant |w|$ in both cases) and $p \cdot w$ to denote the insertion of the operand position $p \in \mathbb{N}$ in front of w. Furthermore, we use mappings $M : Act \to \mathbb{N}^*$ (with only finitely many non-empty images) to associate such strings to actions, and use the following operations to adjust such mappings:

$$[a,p]\cdot M : b \mapsto \begin{cases} p\cdot M(a) & \text{if } b=a,\\ M(b) & \text{otherwise,}\end{cases}$$

$$M\setminus(a,i) : b \mapsto \begin{cases} M(a)\setminus i & \text{if } b=a,\\ M(b) & \text{otherwise.}\end{cases}$$

The purpose of M is to determine the renumbering of action occurrences. We use $\widehat{M}(p,a_i^-)$ to denote the number that the ith outstanding a-action of the pth operand gets according to M. To determine this number, one must count the p's in $M(a)$: the index of the ith occurrence of p in $M(a)$ is the number we are looking for. Formally: $\widehat{M}(p,a_i^-)=m$ such that $M(a)[m]=p$ and $|\{j\leqslant m \mid M(a)[j]=p\}|=i$. For instance, if $M(a)=2\,1\,0\,2$, then $\widehat{M}(2,a_1^-)=1$ and $\widehat{M}(2,a_2^-)=4$.

The operational rules for termination, choice, hiding and recursion are as before (see Table 2) and omitted here. The operational rules for simple actions and synchronization are presented in Table 9. The meaning of the operational rules for "$t\,\|_{A,M}\,u$" is the following. When t performs a^+ ($a\notin A$), $M(a)$ is extended to reflect the renumbering of this new outstanding occurrence; on the other hand, when t (operand 0) performs a_i^-, the occurrence i is renumbered according to $M(a)$ and the occurrence is subsequently removed from $M(a)$. This removal automatically adjusts the renumbering scheme so that the remaining outstanding occurrences are always numbered consecutively, as required in Definition 5.6.

The number of rules for synchronization has grown from 3 to 7; this is because we have to distinguish start actions, end actions and τ-actions. For the case of refinement we even have 10 rules, given in Table 10. There are four groups of rules concerning terms of the form $t[a\to u,\vec{v}]_M$ (compare also the event-based rules in Table 7):

Table 9
Operational ST-semantics for simple actions and synchronization

$$\frac{}{a\xrightarrow{a^+}a^-}\qquad\frac{}{a^-\xrightarrow{a_1^-}\mathbf{1}}\qquad\frac{}{\tau\xrightarrow{\tau}\tau^-}\qquad\frac{}{\tau^-\xrightarrow{\tau}\mathbf{1}}$$

$$\frac{t\xrightarrow{a^+}t'\quad a\notin A}{t\,\|_{A,M}\,u\xrightarrow{a^+}t'\,\|_{A,[a,0]\cdot M}\,u}\qquad\frac{u\xrightarrow{a^+}u'\quad a\notin A}{t\,\|_{A,M}\,u\xrightarrow{a^+}t\,\|_{A,[a,1]\cdot M}\,u'}$$

$$\frac{t\xrightarrow{a_i^-}t'\quad a\notin A\quad m=\widehat{M}(0,a_i^-)}{t\,\|_{A,M}\,u\xrightarrow{a_m^-}t'\,\|_{A,M\setminus(a,m)}\,u}\qquad\frac{u\xrightarrow{a_i^-}u'\quad a\notin A\quad m=\widehat{M}(1,a_i^-)}{t\,\|_{A,M}\,u\xrightarrow{a_m^-}t\,\|_{A,M\setminus(a,m)}\,u'}$$

$$\frac{t\xrightarrow{\tau}t'}{t\,\|_{A,M}\,u\xrightarrow{\tau}t'\,\|_{A,M}\,u}\qquad\frac{u\xrightarrow{\tau}u'}{t\,\|_{A,M}\,u\xrightarrow{\tau}t\,\|_{A,M}\,u'}$$

$$\frac{t\xrightarrow{\lambda}t'\quad u\xrightarrow{\lambda}u'\quad \mathit{type}(\lambda)\in A\cup\{\checkmark\}}{t\,\|_{A,M}\,u\xrightarrow{\lambda}t'\,\|_{A,M}\,u'}$$

Table 10
Operational ST-semantics for the refinement operator

$$\frac{t \xrightarrow{b^+} t' \quad b \neq a}{t[a \to u,\, \vec{v}]_M \xrightarrow{b^+} t'[a \to u,\, \vec{v}]_{[b,0]\cdot M}} \qquad \frac{t \xrightarrow{b_i^-} t' \quad b \neq a \quad m = \widehat{M}(0, a_i^-)}{t[a \to u,\, \vec{v}]_M \xrightarrow{b_m^-} t'[a \to u,\, \vec{v}]_{M\setminus(b,m)}}$$

$$\frac{t \xrightarrow{\lambda} t' \quad \lambda \in \{\tau, \checkmark\}}{t[a \to u,\, \vec{v}]_M \xrightarrow{\lambda} t'[a \to u,\, \vec{v}]_M}$$

$$\frac{t \xrightarrow{a^+} t' \quad u \xrightarrow{b^+} u'}{t[a \to u,\, \vec{v}]_M \xrightarrow{b^+} t'[a \to u,\, \vec{v}\,u']_{[b,|\vec{v}|+1]\cdot M}} \qquad \frac{t \xrightarrow{a^+} t' \quad u \xrightarrow{\tau} u'}{t[a \to u,\, \vec{v}]_M \xrightarrow{\tau} t'[a \to u,\, \vec{v}\,u']_M}$$

$$\frac{v_p \xrightarrow{b^+} v'}{t[a \to u,\, \vec{v}]_M \xrightarrow{b^+} t[a \to u,\, \vec{v} \pm (p, v')]_{[b,p]\cdot M}}$$

$$\frac{v_p \xrightarrow{b_i^-} v' \not\xrightarrow{\checkmark} \quad m = \widehat{M}(p, b_i^-)}{t[a \to u,\, \vec{v}]_M \xrightarrow{b_m^-} t[a \to u,\, \vec{v} \pm (p, v')]_{M\setminus(b,m)}} \qquad \frac{v_p \xrightarrow{\tau} v' \not\xrightarrow{\checkmark}}{t[a \to u,\, \vec{v}]_M \xrightarrow{\tau} t[a \to u,\, \vec{v} \pm (p, v')]_M}$$

$$\frac{t \xrightarrow{a_p^-} t' \quad v_p \xrightarrow{b_i^-} v'_p \xrightarrow{\checkmark} \quad m = \widehat{M}(p, b_i^-)}{t[a \to u,\, \vec{v}]_M \xrightarrow{b_m^-} t'[a \to u,\, \vec{v} \setminus p]_{M\setminus(b,m)}} \qquad \frac{t \xrightarrow{a_p^-} t' \quad v_p \xrightarrow{\tau} v'_p \xrightarrow{\checkmark}}{t[a \to u,\, \vec{v}]_M \xrightarrow{\tau} t'[a \to u,\, \vec{v} \setminus p]_M}$$

- The first group deals with non-a-transitions of t; these are analogous to the non-synchronizing transitions of parallel composition.
- The second group deals with a^+-transitions of t. This starts a new refinement; the initial action of u is either a start action or an internal action. The list of pending refinements is extended, creating a new active operand position.
- The third group deals with transitions of a pending refinement v_p in $\vec{v}$ (where the index p corresponds to the active operand positions) that do not lead to a terminated state. These are also comparable to the non-synchronizing transitions of parallel composition.
- The fourth group deals with transitions of a pending refinement v_p that do lead to a terminated state. Such a transition can only be labelled by an end action or an internal action. The corresponding end action of t can now also occur, and the remainder of the pending refinement can be discarded.

A careful investigation of the operational semantics reveals that it yields an ST-transition system in the sense of Definition 5.6. Moreover, (standard) bisimilarity, applied to the ST-semantics, is a congruence over $Lang^{ST}$. For an independent proof of this property as well as a complete axiomatization of ST-bisimilarity over a large class of recursive processes see [24]. Furthermore, the semantics satisfies the finiteness properties claimed above. Let $[\![t]\!]^{op}_{ST}$ denote the operational ST-semantics of t.

THEOREM 5.7 (see [24]). *Let $t, u \in Lang$ be arbitrary.*

(i) *If t is a refinement-free term with finite state interleaving semantics, then $[\![t]\!]^{op}_{ST}$ is finite state.*

(ii) *If* $[\![t]\!]^{op}_{ST}$ *and* $[\![u]\!]^{op}_{ST}$ *are finite state, then* $[\![t[a \to u]]\!]^{op}_{ST}$ *is finite state.*

As a final result of this section, we have that the indirect and direct ST-semantics of *Lang* coincide. To be precise: if we denote the event-based operational semantics of a term $t \in \mathit{Lang}(\mathit{Evt})$ by $[\![t]\!]^{op}_{ev}$ and the conversion of an event transition system $\mathcal{T}$ into an ST-transition system by $ST(\mathcal{T})$, the following can be proved by induction on the structure of t:

THEOREM 5.8. *For arbitrary* (*well-formed*) $t \in \mathit{Lang}(\mathit{Evt})$,

$$ST([\![t]\!]^{op}_{ev}) \sim [\![\mathit{strip}(t)]\!]^{op}_{ST}.$$

PROOF (*Sketch*). A state $(n, \vec{e})$ from $ST([\![t]\!]^{op}_{ev})$ corresponds to a term of $\mathit{Lang}(\mathit{Evt})$ (n is a derivative of t) plus a sequence $\vec{e}$ of initial events of n that are considered to have started but not yet finished. The events in $\vec{e}$ are thus independent in the sense of Definition 4.2. $(n, \vec{e})$ can therefore

- start a new action, say α, precisely if $n \xrightarrow{d,\alpha} n'$ for some d and n' such that $n' \xrightarrow{\{\vec{e}\}}$;
- finish any of the actions corresponding to events in $\{\vec{e}\}$; the index is derived from the ordering in $\vec{e}$;
- terminate if n is terminated and $\vec{e}$ is empty.

The correspondence with derivatives of $\mathit{strip}(t)$ in the ST-semantics is intricate especially for synchronization and refinement terms: the mapping M used in the ST-semantics is obtained by unraveling the vector $\vec{e}$ in order to determine the source of each event (and vice versa to reconstruct the events from the mapping M). □

5.5. *Application: A simple data base*

To show the advantages of the above ST-semantics, both in providing a congruence for refinement and in yielding finite state systems, we return to the example of Section 4.6. Figure 9 shows the ST-semantics of the abstract system $\mathit{Data}_S := \mathit{State} \parallel_{upd} \mathit{Backup}$ and the refinement $\mathit{Data}_I := \mathit{Data}_S[\mathit{upd} \to \mathit{req}; \mathit{cnf}][\mathit{copy} \to \mathit{back}; \mathit{copy}]$, where

$$\begin{aligned} \mathit{State} &:= (\mathit{qry} + \mathit{upd}); \mathit{State}, \\ \mathit{Backup} &:= (\mathit{copy} + \mathit{upd}); \mathit{Backup}. \end{aligned}$$

The independence of the *qry*- and *copy*-actions, noted before in Section 4.6, is apparent from the ST-transition system by the fact that *qry* may be started during *copy*, i.e., between copy^+ and copy^-_1 (and vice versa); on the other hand, *qry* may not be started during *upd*. As a consequence, in Data_I *qry* may interrupt the refinement of *copy* at any time, but may not interrupt the refinement of *upd*.

5.6. ω-*completeness*

As pointed out by Meyer in [105,106], the issue of coarsest congruences is relevant to another, quite different subject as well, namely that of ω-*completeness* of equational theories.

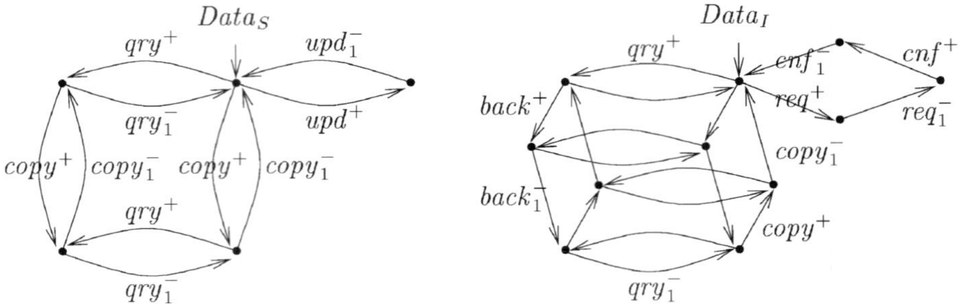

Fig. 9. Operational *ST*-semantics of $Data_S$ and $Data_I$.

To explain this, we first recall the syntactic interpretation of action refinement discussed in Section 1.4 above, and note that the syntactic substitution of an action by a term is quite similar to the syntactic substitution of a process variable by a term. Also recall that (in the standard interpretation) two open terms (i.e., with free process variables) are considered equivalent iff their instantiations are equivalent; in other words, if the equivalence cannot be invalidated by some substitution of terms for process variables. But this in turn corresponds precisely (given the similarity between action substitution and variable substitution just pointed out) to the second congruence property listed at the beginning of this sub-section.

An equational theory is called ω-complete if any valid equation between open terms is derivable. We have seen just now that an equation between open terms is valid iff all instantiations are valid; this in turn is the case (interpreting variables as actions) iff the terms are equivalent under all action refinements, i.e., iff they are equivalent in the coarsest congruence. Summarizing, we have:

In a setting where syntactic and semantic action refinement coincide, a complete axiomatization of the coarsest refinement congruence contained in $\simeq$ gives rise to an ω-complete axiomatization of $\simeq$, when interpreting all actions as variables.

A similar idea is pursued in [134]: if we consider expressions built with variables (which stand for languages) and the operations of concatenation, choice (union) and parallel composition (shuffle), then substitution of languages for variables can be seen as action refinement; validity of equations of such expressions can be checked using *ST* trace semantics. Analogously, if variables take values on words, validity of equations can be checked using *swap*-equivalence.

6. Semantic versus syntactic substitution

We have stated in the introduction that there are two fundamentally different ways to regard action refinement: as an operator in the signature, or as a homomorphism (in fact, an automorphism) of the algebra establishing a transformation of terms. Since the semantic effect of the action refinement operator $t[a \to u]$ is understood as the substitution, in some sense, of the action a by the behaviour of u in the *semantic* model of t, whereas a homomorphism effectively replaces a by u *syntactically* in t, the two interpretations are actually quite close; yet they do not always coincide. In the previous sections, we have considered the semantic interpretation of action refinement; below, we study the syntactic interpretation and investigate the difference between the two. The material of this section is mainly based on [69].

Note that if the two interpretations coincide, we automatically have an axiomatization for the action refinement operator: namely, it distributes over all other operators. In fact, this distribution property is a strong point in favour of the syntactic interpretation. On the other hand, the precise notion of syntactic substitution has to be considered carefully, as it may depend on the operators available in the language and on the kind of refinement we are interested in (atomic or non-atomic).

In the above discussion, we first relied on the algebraic notion of an automorphism and then stated that this corresponds to the syntactic substitution of actions by terms. The connection between the concepts of automorphism and syntactic substitution is established by regarding the term algebra generated by the signature as the algebra over which the automorphism is defined. However, there are several respects in which this algebraic concept does not quite cover the notion of syntactic substitution we need.

- It does not tell us how to deal with recursion, which is not an operator in the usual algebraic sense but rather a fixpoint constructor, i.e., a higher-order operator. As it turns out, the straightforward syntactic substitution within the body of the recursion (which does not affect the process variables bound by recursion) gives rise to a satisfactory solution.
- It does not take operators into account that *bind* actions, such as hiding and (to some degree) synchronization. Strictly speaking, an automorphism f on the term algebra would transform each term t/A into $op(f(t))$, where op is the f-image of the unary operator $\_/A$; the natural choice of op is $\_/A'$ for some A'. As we will see below, in some cases this is not an appropriate way to model action refinement.

Hence, although the intuition of an automorphism is initially helpful to understand action refinement algebraically, it is actually only an approximation. In the remainder of this section we will no longer appeal to this intuition.

6.1. *Finite sequential systems*

Let us first consider again the simple sequential language of Section 2, restricted, moreover, to closed, finite terms (i.e., containing no process variables or recursion operators). We call a term *flat* if it contains no occurrence of the refinement operator. In defining the syntactic substitution of a term for an action in $Lang^{seq,fin}$, we already encounter the problem with respect to the hiding operator, mentioned above.

EXAMPLE 6.1. Suppose we want to substitute the action a by the term $b; c$ within t.

(1) If $t = a/\{a\}$ (specifying an invisible action), then a naive substitution gives rise to $(b; c)/\{a\}$. Hence, after refinement we suddenly obtain visible behaviour where before there was none.

(2) A slightly more sophisticated refinement of $t = a/\{a\}$ would yield $(b; c)/\{b, c\}$ (i.e., the set of hidden actions is adapted as well). This is still unexpected, since the refined system has two invisible steps instead of just one (and thus the refinement is not well-defined up to strong bisimilarity). Worse yet, using this principle $(a; b)/\{a\}$ would give rise to $(b; c; b)/\{b, c\}$, where the abstract b-action disappears even though it is not refined itself.
(3) If $t = a/\{b\}$ (specifying the execution of a), then a naive substitution gives rise to $(b; c)/\{b\}$. This starts with an invisible action rather than the expected execution of b.

Technically, these problems all have to do with the fact that, in our semantics, hiding acts like a *binder* of the hidden actions, just like restriction does in CCS (see [9]). Moreover, precisely the same problem occurs for action substitution within non-flat terms, since the refinement operator is a binder, too: in the term $t[a \to u]$, all occurrences of a in t are bound by the refinement. By analogy with the syntactic substitution of free variables within the body of recursion, defined in Table 1, here we have to make sure that there is no confusion of "free" and "bound" actions during substitution. For that purpose, we first have to have a clear notion of the "free actions" of a term. In the terminology of Milner [108], these are given by the *sort* (called the *alphabet* in [11,88]). The term sort is defined in Table 11. The sort is somewhat comparable to the *type* of a term in functional languages. It is an upper bound to the actions that the process will do during its lifetime. This is expressed by the following key proposition (in the type analogy corresponding to *subject reduction*):

PROPOSITION 6.2. *If* $t \xrightarrow{\alpha} t'$ *then* $\mathcal{S}(t') \cup \{\alpha \mid \alpha \neq \tau\} \subseteq \mathcal{S}(t)$.

To guarantee the necessary absence of confusion, we take a rather drastic solution: the syntactic substitution of a term for an action will be undefined whenever a "bound action",

Table 11
Sort and action substitution

t	$\mathcal{S}(t)$	$t\{u/a\}$
α	$\begin{cases} \emptyset & \text{if } \alpha = \tau \\ \{\alpha\} & \text{otherwise} \end{cases}$	$\begin{cases} u & \text{if } \alpha = a \\ \alpha & \text{otherwise} \end{cases}$
t_1/A	$\mathcal{S}(t_1) \setminus A$	$t_1\{u/a\}/A$ if $a \notin A$ and $\mathcal{S}(u) \cap A = \emptyset$
$t_1[b \to t_2]$	$\begin{cases} \mathcal{S}(t_1) & \text{if } b \notin \mathcal{S}(t_1) \\ (\mathcal{S}(t_1) \setminus b) \cup \mathcal{S}(t_2) & \text{otherwise} \end{cases}$	$t_1\{u/a\}[b \to t_2\{u/a\}]$ if $a \neq b$ and $b \notin \mathcal{S}(u)$
$op(t_1, \ldots, t_n)$	$\mathcal{S}(t_1) \cup \cdots \cup \mathcal{S}(t_n)$	$op(t_1\{u/a\}, \ldots, t_n\{u/a\})$ (other operators)
x	$\mathcal{S}_x$	x
$\mu x.t_1$	$\mathcal{S}_x$	$\mu x.(t_1\{u/a\})$

i.e., a hidden action or an action being refined, either equals the action being replaced by substitution or is contained in the sort of the term substituted for it. The definition of action substitution is given in Table 11; see also [9].

The partiality of action substitution is subsequently smoothed out of existence as follows. Let $\equiv$ be the smallest congruence over $Lang^{seq}$ such that

$$\begin{aligned} t/(A \uplus \{a\}) &\equiv t\{b/a\}/(A \uplus \{b\}) \quad \big(b \notin \mathcal{S}(t)\big) \\ t[a \to u] &\equiv t\{b/a\}[b \to u] \quad \big(b \notin \mathcal{S}(t)\big) \end{aligned}$$

for an arbitrary b such that the substitutions on the right hand side are defined. It is not difficult to prove that, given a large enough universe of actions Act, for all $t\{u/a\}$ there is a $t' \equiv t$ such that $t'\{u/a\}$ is defined; moreover, as a consequence of the fact that $\equiv$ is a congruence, if $t \equiv t'$ then $t\{u/a\} \equiv t'\{u/a\}$ whenever both are defined. In other words, action substitution is total and well-defined modulo $\equiv$. Since $\equiv$ is finer than $\sim$ and in fact finer than any semantic equivalence we consider in this chapter, it follows that we may always pick an appropriate $\equiv$-representative such that action substitution is defined. In the sequel, we implicitly assume that such a representative has been chosen, and proceed as if action substitution is total. (A similar assumption concerning free variables is made in [14] and elsewhere in the literature on functional programming.)[1]

We now set out to show that (with the definition of action substitution developed above), in $Lang^{seq,fin}$ there is no difference between the semantic interpretation of the action refinement operator (see Table 2) and its interpretation as action substitution (see Table 11).

To formulate the result we are after, we define a *flattening* function over sequential terms, which removes all refinement operators from a given process expression by performing the corresponding syntactic substitution.

DEFINITION 6.3. The *flattening* of a term $t \in Lang^{seq,fin}$, denoted $\mathit{flat}(t)$, is defined inductively on the structure of t as follows:

$$\begin{aligned} \mathit{flat}\big(op(t_1, \ldots, t_n)\big) &:= op\big(\mathit{flat}(t_1), \ldots, \mathit{flat}(t_n)\big), \\ \mathit{flat}(t[a \to u]) &:= \mathit{flat}(t)\{\mathit{flat}(u)/a\}. \end{aligned}$$

The coincidence of syntactic and semantic substitution is then concisely stated by the equation $t = \mathit{flat}(t)$. The first result of this kind is the following:

THEOREM 6.4. *$t \sim \mathit{flat}(t)$ for any $t \in Lang^{seq,fin}$.*

PROOF (*Sketch*). The idea is to prove that refinement satisfies the defining equations of substitution (Table 11) except for that for refinement (and recursion); that is, to show that the following equations hold up to $\sim$:

$$\alpha[a \to u] = \begin{cases} u & \text{if } \alpha = a, \\ \alpha & \text{otherwise,} \end{cases}$$

[1] An alternative solution would have been to build the necessary conversion of bound actions into the definition of action substitution, as we have done for variable substitution in Table 1. However, we find the present definition somewhat clearer.

$$(t/A)[a \to u] = t[a \to u]/A \quad \text{if } a \notin A \text{ and } \mathcal{S}(u) \cap A = \emptyset,$$
$$op(t_1, t_2)[a \to u] = op(t_1[a \to u], t_2[a \to u]) \quad \text{for other operators } op.$$

(These equations in fact hold up to $\sim_{\mathsf{ev}}$.) Using these equations, it is straightforward to show that for all *flat* terms $t \in Lang^{seq,fin}$ the following holds up to $\sim$ (respectively, $\sim_{\mathsf{ev}}$):

$$t[a \to u] = t\{u/a\}$$

(by induction over the structure of t). The actual theorem then follows by the definition of *flat* and the fact that $\sim$ is a congruence (using a further induction over the structure of t). □

6.2. *Recursive sequential systems*

In extending Theorem 6.4 the full sequential language $Lang^{seq}$, we encounter several problems. The major one of these is that it is in general not possible to flatten terms of $Lang^{seq}$ in which refinement and recursion are arbitrarily mixed.

EXAMPLE 6.5. Consider the term $t = \mu x.(a; x[a \to a; b])$. This specifies a process having as partial runs all prefixes of the infinite string $a\, b^0\, a\, b^1\, a\, b^2\, a\, b^3 \cdots$. This set of partial runs is not context-free; it follows that the behaviour of t cannot be specified in the flat fragment of $Lang^{seq}$, not even modulo trace equivalence, let alone modulo any stronger semantics. In other words, t admits no flattening within $Lang^{seq}$.

A similar situation occurs with $t = \mu x.a[a \to a; (b + x)]$, where recursion and refinement are mixed in a different way but to much the same effect.

In order to avoid problems of this kind, for the purpose of this section we restrict ourselves to instances of the refinement operator in which both operands of refinement are closed terms.[2]

Another, more technical issue is that the notions of sort and syntactic action substitution have yet to be defined for process variables and recursive terms.

- To define the sort of open terms, we impose the requirement that all process variables are implicitly sorted; i.e., there is a pre-defined sort $\mathcal{S}_x \subseteq Act$ for all $x \in Var$. Furthermore, $\mathcal{S}(\mu x.t)$ is set to $\mathcal{S}_x$. We then impose the following *well-sortedness* condition on recursive terms (in addition to the existing well-formedness conditions): $\mu x.t$ is well-sorted only if $\mathcal{S}(t) \subseteq \mathcal{S}_x$.
- With respect to action substitution, the natural solution is to define $(\mu x.t)\{u/a\}$ to equal $\mu x.t\{u/a\}$ and $x\{u/a\}$ to equal x. At first sight, this invalidates the correspondence between action refinement and action substitution, since it gives rise to $\mu x.(a; x\{b/a\}) = \mu x.a; x$ which clearly does not have the same behaviour as $\mu x.(a; x[a \to b])$. However, we are saved by the new requirement that the operands of refinement must be closed terms; this precisely rules out terms of the form $x[a \to b]$.

[2] This is essentially the same restriction we used in [69], except that there we worked with process environments rather than recursion operators. In that paper, we justified the restriction by another well-formedness condition, namely that the sorts of t and u in $t[a \to u]$ have to be disjoint.

The resulting definitions of the sort and action substitution for process variables and recursive terms were already given in Table 11.

Well-formedness. To summarize the well-formedness conditions in this section:

- No virgin operand contains an occurrence of an auxiliary operator;
- The operands of refinement must be closed terms;
- Recursion is allowed on guarded variables only;
- Recursion must be well-sorted.

The "correctness criterion" for the sort of recursive terms is that Proposition 6.2 now also holds for all closed terms of $Lang^{seq}$. As for action substitution, to prove the corresponding extension of Theorem 6.4 we have to adapt the notion of flattening to process variables and recursive terms as well:

$$\begin{aligned} \mathit{flat}(x) &= x, \\ \mathit{flat}(\mu x.t) &= \mu x.\mathit{flat}(t). \end{aligned}$$

We now come to the correspondence result for the full sequential language.

THEOREM 6.6. *$t \sim \mathit{flat}(t)$ for any $t \in Lang^{fin}$.*

PROOF (*Sketch*). We have seen that Theorem 6.4 essentially consisted of proving $t[a \to u] \sim t\{u/a\}$ by induction on the structure of t. Since t is bound to be closed, for the purpose of the current theorem we only have to extend the existing proof with the case for $t = \mu x.t_1$; i.e., we show

$$(\mu x.t_1)[a \to u] \sim \mu x.t_1[a \to u].$$

The proof consists of showing that

$$\rho = \big\{(u_1\{\mu x.t_1/x\}[a \to u], u_1\{\mu x.t_1[a \to u]/x\)\} fv(u_1) \subseteq \{x\}\big\}$$

is a bisimulation. The proof obligation follows as a special case (taking $u_1 = x$). □

6.3. *Atomic refinement*

We now consider the language $Lang^{atom}$ with atomic refinement, discussed in Section 3. This entails taking two new operators into account: parallel composition (without synchronization) and the atomizer. The definition of the sort and action substitution for these new operators follows from the "other operators"-clause of Table 11.

The definition of flattening in $Lang^{atom}$ is a variation on the one we had before: the difference is that the effect of *atomically* refining an action a by u is not the straight execution of u but rather the *atomic* execution of u, i.e., the execution of $\langle u \rangle$. This gives rise to a function flat^{atom}, whose definition for the flat fragment of Lang^{atom} is analogous to the definition of *flat* (Definition 6.3) but for the refinement operator is given by

$$\mathit{flat}^{atom}(t[a \to u]) := \mathit{flat}^{atom}(t)\{\langle \mathit{flat}^{atom}(u)\rangle / a\}.$$

This is actually the key point distinguishing atomic and non-atomic refinement. We obtain the analogous result to Theorem 6.6; the proof is also analogous, and hence omitted.

THEOREM 6.7. *$t \sim \mathit{flat}^{atom}(t)$ for any $t \in \mathit{Lang}^{atom}$.*

6.4. *Synchronization*

In the full language with non-atomic refinement, we encounter the operator for parallel composition with synchronization, indexed by the set of synchronizing actions. When we do action substitution in such a term, this explicit occurrence of action names gives rise to the same type of confusion as for hiding (see Example 6.1).

EXAMPLE 6.8. Consider the replacement of a by $b;\, c$ in the term $t \in \mathit{Lang}$.

(1) If $t = a \|_a a$, specifying a single execution of a, then a naive substitution gives rise to $t\{b;\, c/a\} = b;\, c \|_a b;\, c$, which can do two times $b;\, c$ in parallel and hence is not an intuitively correct refinement of t.

(2) A slightly more sophisticated substitution in t above yields $b;\, c \|_{b,c} b;\, c$, which is more reasonable; however, according to the same principle, $c;\, a \|_a a$, which can perform c, gives rise to $c;\, b;\, c \|_{b,c} b;\, c$, which is deadlocked.

(3) If $t = a;\, c \|_{b,c} c$, then the substitution gives rise to $t\{b;\, c/a\} = b;\, c;\, c \|_{b,c} c$; this is not a reasonable refinement since the former can do $a;\, c$ whereas the latter is deadlocked.

However, in contrast to hiding, in the case of synchronization the actions are not bound; rather, our operational semantics (Section 4) specifies that actions remain visible after synchronization, and are therefore still available for further synchronization or, indeed, refinement. For this reason, although action substitution for parallel composition needs a side condition that is comparable to the one for hiding in Table 11, we cannot argue (as we did before) that substitution is still totally defined modulo $\equiv$. Sort and action substitution for synchronization are defined in Table 12. Note that when replacing a synchronizing action, the synchronization set is adapted accordingly; this corresponds to the "more sophisticated" solution pointed out in Example 6.8.2.

However, our problems are not yet over. Consider $t = t_1 \|_A t_2$. As long as we do not substitute actions in A, and moreover take care that A and $\mathcal{S}(u)$ are disjoint, the desired distributivity property $t[a \to u] = t_1[a \to u] \|_A t_2[a \to u]$ holds up to $\sim_{\mathsf{ev}}$; hence everything goes smoothly (see the proof sketch of Theorem 6.4). However, when a occurs in A, then $t\{u/a\}$ and $t[a \to u]$ may be different.

EXAMPLE 6.9. We show some t and u such that $t[a \to u] \not\sim t\{u/a\}$.

Table 12
Sort and action substitution for parallel composition with synchronization

t	$\mathcal{S}(t)$	$t\{u/a\}$	
$t_1 \parallel_A t_2$	$\mathcal{S}(t_1) \cup \mathcal{S}(t_2)$	$\begin{cases} t_1\{u/a\} \parallel_{(A\setminus\{a\})\cup\mathcal{S}(u)} t_2\{u/a\} \\ t_1\{u/a\} \parallel_A t_2\{u/a\} \end{cases}$	if $a \in A$ and $\mathcal{S}(u) \cap \mathcal{S}(t) \subseteq A$ if $a \notin A$ and $\mathcal{S}(u) \cap A = \emptyset$

- Consider $t = a \parallel_a a$ and $u = b + b;\, c$. t always terminates after having performed a, and so $t[a \to u]$ terminates after either b or $b\,c$, but $t\{u/a\}$ does not always terminate:

$$t\{u/a\} = (b + b;\, c) \parallel_{b,c} (b + b;\, c) \xrightarrow{b} \mathbf{1} \parallel_{b,c} c,$$

  where the right hand term is deadlocked.
- Consider $t = (a \parallel\!\parallel a;\, b) \parallel_a (a \parallel\!\parallel a;\, b)$ and $u = c;\, d$. In $t[a \to u]$, there is a c-transition after which in every state reachable by $c\,d$, either no b or two consecutive b's are enabled:

$$\begin{aligned} t[a \to u] &= ((_0a \parallel\!\parallel {_1a}) \parallel_a (_2a \parallel\!\parallel {_3a};\, {_4b}))[a \to {_5c};\, {_6d}] \\ &\xrightarrow{(e,5),c} ((_0a \parallel\!\parallel {_1a}) \parallel_a (_2a \parallel\!\parallel {_3a};\, {_4b}))[a \to {_5c};\, {_6d},\, e \to {_6d}], \end{aligned}$$

  where $e = ((0, *), (2, *))$. On the other hand, for all $t\{u/a\} \xrightarrow{c} t'$ there is a $t' \xrightarrow{c}\xrightarrow{d}\xrightarrow{b} t''$ such that $t'' \stackrel{b}{\not\longrightarrow}$: in particular, after $t\{u/a\} \xrightarrow{a} (d \parallel\!\parallel c;\, d;\, b) \parallel_{c,d} (d \parallel\!\parallel c;\, d;\, b)$ $(= t')$, which intuitively corresponds to the transition above, we can continue with

$$\begin{aligned} t' &\xrightarrow{c} (d \parallel\!\parallel d;\, b) \parallel_{c,d} (d \parallel\!\parallel d;\, b) \xrightarrow{d} (\mathbf{1} \parallel\!\parallel d;\, b) \parallel_{c,d} (d \parallel\!\parallel b) \\ &\xrightarrow{b} (\mathbf{1} \parallel\!\parallel d;\, b) \parallel_{c,d} (d \parallel\!\parallel \mathbf{1}) \quad (= t'') \end{aligned}$$

  such that $t'' \stackrel{b}{\not\longrightarrow}$.

In [69] we discussed necessary and sufficient conditions under which refinement and substitution of a synchronizing action give rise to event bisimilar models. (To be precise, in [69] we considered isomorphism of configurations; as we stated before, however, this actually coincides with event bisimilarity.) The crucial part is to find conditions for the following distributivity property (which is analogous to the rule for action substitution in Table 12):

$$(t_1 \parallel_{A \uplus \{a\}} t_2)[a \to u] \sim_{\mathsf{ev}} t_1[a \to u] \parallel_{A \uplus \mathcal{S}(u)} t_2[a \to u].$$

This is indeed precisely the property that fails to be satisfied in both instances of Example 6.9. The solution given in [69] consists of a number of fairly involved semantic constraints, which are in general undecidable; rather than repeating the constraints here, we give a decidable "approximation" or "estimate", which is a simplification of the one in [69].

Table 13
Auto-concurrent actions and determinism

t	$\mathcal{C}(t)$		$\mathcal{D}(t)$
0	$\emptyset$		**true**
1	$\emptyset$		**true**
α	$\{\alpha \mid \alpha \neq \tau\}$		$\alpha \neq \tau$
$t_1 + t_2$	$\mathcal{C}(t_1) \cup \mathcal{C}(t_2)$		$\mathcal{D}(t_1) \wedge \mathcal{D}(t_2) \wedge \nexists a\colon (t_1 \xrightarrow{a}) \wedge (t_2 \xrightarrow{a})$
$t_1; t_2$	$\mathcal{C}(t_1) \cup \mathcal{C}(t_2)$		$\mathcal{D}(t_1) \wedge \mathcal{D}(t_2)$
$t_1 \parallel_A t_2$	$\mathcal{C}(t_1) \cap \mathcal{C}(t_2)$ $\cup (\mathcal{C}(t_1) \cup \mathcal{C}(t_2)) \setminus A$ $\cup (\mathcal{S}(t_1) \cap \mathcal{S}(t_2)) \setminus A$		$\mathcal{D}(t_1) \wedge \mathcal{D}(t_2) \wedge (\mathcal{S}(t_1) \cap \mathcal{S}(t_2) \subseteq A)$
t_1 / A	$\mathcal{C}(t_1) \setminus A$		$\mathcal{D}(t_1) \wedge (\mathcal{S}(t_1) \cap A = \emptyset)$
$t_1[a \to t_2]$	$(\mathcal{C}(t_1) \setminus a) \cup \mathcal{S}(t_2)$	if $a \in \mathcal{C}(t_1)$	$\mathcal{D}(t_1) \wedge (a \notin \mathcal{S}(t_1) \vee \mathcal{D}(t_2))$
	$\mathcal{C}(t_1) \cup \mathcal{C}(t_2)$	if $a \in \mathcal{S}(t_1) \setminus \mathcal{C}(t_1)$	
	$\mathcal{C}(t_1)$	otherwise	
x	$\mathcal{C}_x$		$\mathcal{D}_x$
$\mu x.t_2$	$\mathcal{C}_x$		$\mathcal{D}_x$

The decidable approximations, which are given in Table 13 in the form of functions over *Lang*, provide sufficient conditions for the actual semantic constraints; the idea is the same as for the sort of a term (see Tables 11 and 12), which provides an approximation of the actions that may be performed by the term (see Proposition 6.2). The intention of the functions defined in Table 13 is as follows:

- $\mathcal{C}(t) \subseteq Act$ approximates the set of auto-concurrent actions of t, i.e., the ones that are executed in more than one parallel component without being synchronized. This is an over-estimate, due to the fact that one or more occurrences of an action may fail to be executed (as for a in $a \parallel_b b; a$) or are serialized (as for a in $a; b \parallel_b b; a$).
- $\mathcal{D}(t) \in \mathbb{B}$ is a Boolean indicating whether t is deterministic and contains no internal actions. This is an under-estimate: terms for which $\mathcal{D}(t)$ does not hold may yet be deterministic – for example, $a \parallel_b b; a$.

For recursive terms, the definition of these functions relies on pre-defined sets $\mathcal{C}_x$ and predicates $\mathcal{D}_x$ for all $x \in Var$, just as the definition of the sort in Table 11. We extend the well-sortedness condition to imply that these sets and predicates are in correspondence with the recursion body:

Extended well-sortedness. t is well-sorted if for all subterms $\mu x.u$, in addition to $\mathcal{S}(u) \subseteq \mathcal{S}_x$ we also have $\mathcal{C}(u) \subseteq \mathcal{C}_x$ and $\mathcal{D}_x \Rightarrow \mathcal{D}(u)$.

In the remainder of this section, we only consider terms that are well-sorted in this extended sense. The following proposition expresses the semantic properties that are guaranteed by the functions in Table 13.

PROPOSITION 6.10. *Let $t \in Lang$ be arbitrary.*

- *$a \in \mathcal{C}(t)$ if there exists a t' reachable from t such that $t' \xrightarrow{e_1,a} \xrightarrow{e_2,a}$ and $t' \xrightarrow{e_2,a} \xrightarrow{e_1,a}$ (in the event-based semantics of Section 4).*

- *$\mathcal{D}(t)$ implies that for all t' reachable from t, $t' \stackrel{\tau}{\not\longrightarrow}$ and if $t' \stackrel{a}{\longrightarrow} t_1''$ and $t' \stackrel{a}{\longrightarrow} t_2''$ then $t_1'' = t_2''$.*

For the proof see [69]. An important further property of these functions is that they are insensitive to the flattening of terms, in the following sense:

PROPOSITION 6.11. *If $t \in Lang$ such that $flat(t)$ is defined, then $\mathcal{S}(t) = \mathcal{S}(flat(t))$, $\mathcal{C}(t) = \mathcal{C}(flat(t))$ and $\mathcal{D}(t) \iff \mathcal{D}(flat(t))$.*

For the proof see [69]. We now define the concept of *u/a-compatibility* over the flat fragment of *Lang*, which establishes a sufficient condition guaranteeing that the refinement of a by u and the substitution of a by u coincide.

(1) t/A is u/a-compatible if $a \notin A$, $\mathcal{S}(u) \cap A = \emptyset$ and t is u/a-compatible;
(2) $t = t_1 \parallel_A t_2$ is u/a-compatible if t_1 and t_2 are u/a-compatible and one of the following holds:
   (a) $a \notin A$ and $\mathcal{S}(u) \cap A = \emptyset$;
   (b) $a \in A \setminus \mathcal{S}(t)$ and $\mathcal{S}(u) \cap \mathcal{S}(t) = \emptyset$;
   (c) $a \in A \setminus \mathcal{C}(t)$, $\mathcal{S}(u) \cap \mathcal{S}(t) = \emptyset$ and $\mathcal{D}(u)$;
   (d) $a \in A$ and $u = \sum_{i=1}^{n} a_i$ such that $\mathcal{S}(u) \cap \mathcal{S}(t) = \emptyset$;
(3) $\mu x.t$ is u/a-compatible if t is u/a-compatible;
(4) For all other operators *op*, $op(t_1, \ldots, t_n)$ is u/a-compatible if all t_i are u/a-compatible.

The interesting part is the requirement for parallel composition. This consists of the case where the substituted action is not in the synchronization set (Clause 2(a)), or, if it is in the synchronization set, the case where it is never performed (Clause 2(b)), where it is performed but not auto-concurrent, and the refinement is deterministic (Clause 2(c)) and where it is auto-concurrent and the refinement is a choice of atomic actions (Clause 2(d)). As a consequence of the main theorem of [69], we have the following property:

PROPOSITION 6.12. *If $t \in Lang$ is flat and u/a-compatible, then $t\{u/a\}$ is defined and $t[a \to u] \sim_{\mathsf{ev}} t\{u/a\}$.*

Building on substitution compatibility, we now define the property of *reducibility* of terms, which establishes a sufficient condition guaranteeing that every action refinement operator in a term can be converted into an action substitution (i.e., the term can be flattened) without changing its semantics.

- $\mu x.t$ is reducible if t is reducible;
- $t[a \to u]$ is reducible if t and u are reducible and *flat*(t) is u/a-compatible.
- For all other operators *op*, $op(t_1, \ldots, t_n)$ is reducible if all t_i are reducible.

Here, the interesting case is that of refinement. The following theorem extends Theorem 6.6 to *Lang*: a term and its flattening coincide *if* the term is reducible. The proof is by induction on the term structure, using Proposition 6.12.

THEOREM 6.13. *If $t \in Lang$ is reducible, then $flat(t)$ is defined and $t \sim_{\mathsf{ev}} flat(t)$.*

By "tuning" the functions $\mathcal{S}$, $\mathcal{C}$ and $\mathcal{D}$, it is, of course, possible to come to an improved characterization of reducibility, i.e., one which provides a better estimate of the underlying semantic property; however, as mentioned above, decidable necessary and sufficient criteria for the coincidence of syntactic and semantic refinement do not exist.

6.5. *Application: A simple data base*

We apply Theorem 6.13 to the example of Sections 4.6 and 5.5. This consists of a process $Data_S := State \parallel_{upd} Backup$, where

$$\begin{aligned} State &:= (qry + upd); State, \\ Backup &:= (copy + upd); Backup. \end{aligned}$$

We then refine *upd* and *copy*, resulting in $Data_I := Data_S[upd \to req; cnf][copy \to back; copy]$. We show $Data_I$ to be reducible, in the sense defined above. For this purpose, we must check that the terms $Data_S[upd \to req; cnf]$ and *back*; *copy* are reducible and that $flat(Data_S[upd \to req; cnf])$ is $(back; copy)/copy$-compatible.

- To see that $Data_S[upd \to req; cnf]$ is reducible, we must check that $Data_S$ and *req*; *cnf* are reducible (which is trivially true, since both are already flat) and that $flat(Data_S)$ $(= Data_S)$ is $(req; cnf)/upd$-compatible. The latter comes down to checking that $State \parallel_{upd} Backup$ (the body of $Data_S$) is $(req; cnf)/upd$-compatible.
  For this purpose, we must check that *State* and *Backup* are $(req; cnf)/upd$-compatible (which is trivially true, since neither contains parallel composition) and that *upd* and *req*; *cnf* are compatible with the synchronization set. This is indeed the case, since $\mathcal{C}(State \parallel_{upd} Backup) = \emptyset$ and hence $upd \in \{upd\} \setminus \mathcal{C}(State \parallel_{upd} Backup)$, and furthermore, $\mathcal{D}(req; cnf)$; hence compatibility clause 2(c) holds.
- *back*; *copy* is trivially reducible, since it is a flat term.
- $flat(Data_S[upd \to req; cnf]) = Data'$ where

$$\begin{aligned} Data' &:= State' \parallel_{req,cnf} Backup', \\ State' &:= (qry + req; cnf); State', \\ Backup' &:= (back; copy + req; cnf); Backup'. \end{aligned}$$

  To see that this is $(back; copy)/copy$-compatible, the main point is checking that $State' \parallel_{req,cnf} Backup'$ is so. This is indeed the case, since $copy \notin \{upd\}$; hence compatibility clause 2(a) holds.

According to Theorem 6.13, since $Data_I$ is reducible it can be flattened, i.e., it is equivalent (event bisimilar, and hence also *ST*-bisimilar and interleaving bisimilar) to the process obtained by carrying out the refinements syntactically, given by $Data_I^{flat} := State^{flat} \parallel_{req,cnf} Backup^{flat}$ where

$$\begin{aligned} State^{flat} &:= (qry + req; cnf); State^{flat}, \\ Backup^{flat} &:= (back; copy + req; cnf); Backup^{flat}. \end{aligned}$$

This can be verified by comparing the behaviour of $Data_I^{flat}$ with that of $Data_I$ as depicted in Figure 6 (taking the event-based semantics modulo event bisimilarity) and Figure 9 (taking the ST-semantics modulo standard bisimilarity).

7. Dependency-based action refinement

Although intuitively, action refinement is closely connected to the concept of top-down design, in practice the steps one would like to make in top-down design do not always correspond completely to the constructions allowed by action refinement. An example is the sequential ordering of tasks at different levels of abstraction. On an abstract level of design, one may specify that two activities, which on that level are regarded as atomic, are causally ordered – for instance because one of them produces an output that is consumed by the other. When one refines this specification in a top-down fashion, taking into account, among other things, that the previously atomic actions actually consist of several sub-activities, then the causal dependency does not necessarily hold between *all* respective sub-activities; rather, it may be the case that the output is produced somewhere *during* the refinement of the first action (not necessarily at the very end), and consumed *during* the refinement of the second (not necessarily at the very beginning).

Motivating example. As a clarifying example (originally inspired by [99]), consider a buffer of capacity 1, which allows the alternating *put* and *get* of a data value. In fact, consider the case where only a single *put* and *get* occur. Abstractly, it is clear that a causal dependency exists between *put* and *get*, induced by the flow of data. Now consider a design step driven by the fact that the data values are too large for a single buffer cell, and consequently it is decided to use two buffers in parallel: data values are split in two and a piece is put in either (implementation-level) buffer. This comes down to refining *put* by $put_1 \,|||\, put_2$ and *get* by $get_1 \,|||\, get_2$.

If we treat this design step through the application of the refinement operator to the original specification, *put*; *get*, we arrive at the behaviour

$$(put;\, get)[put \rightarrow put_1 \,|||\, put_2][get \rightarrow get_1 \,|||\, get_2] \sim_{\mathsf{ev}} (put_1 \,|||\, put_2);\, (get_1 \,|||\, get_2).$$

This, however, is *not* necessarily the intended implementation-level behaviour: there is an ordering between put_1 and get_2 and between put_2 and get_1, whereas the data flow only imposes a causal ordering between put_i and get_i for $i = 1, 2$.

In this section, we review the solutions that have been proposed for this problem, first by Janssen, Poel and Zwiers [91] in a linear time setting and later in [89,90,136,126] for branching time. These solutions are based on an idea taken from Mazurkiewicz traces [103, 104], namely to recognize explicit *dependencies* between actions. That is, rather than working with actions that are (from the point of view of the formalism) completely uninterpreted, as before, now we assume some further knowledge about them, in the form of a (reflexive and symmetric) *dependency relation* over the action universe *Act*. Action refinement takes dependencies into account by imposing causal ordering only between dependent actions, regardless of the ordering specified on the abstract level.

For instance, in the example above, put_i and get_i are dependent (for $i = 1, 2$) but put_1 and get_2 as well as put_2 and get_1 are independent. Thus, the behaviour of the refinement, given by the term $(put_1 \mid\mid\mid get_1); (put_2 \mid\mid\mid get_2)$, would be equivalent to $put_1; get_1 \mid\mid\mid put_2; get_2$, which, in view of the assumptions, is a reasonable design. On the other hand, if the more complete ordering above was actually intended (for whatever reason), this can still be obtained by explicitly specifying that all get_i-actions are dependent on all put_j-actions.

7.1. *Dependencies in linear time*

The idea of action dependencies was transferred to a process-algebraic setting by Janssen, Poel and Zwiers [91], using an approach again advocated in [142]. We can reconstruct the basic elements of the setup of [91] as follows.

- They assume a global dependency relation $D \subseteq Act \times Act$, which is reflexive and symmetric but not necessarily transitive. The underlying intuition is that $a\ D\ b$ if a and b *share resources*, and so the order in which they are executed may influence the outcome of the computation. We also write $a\ D\ A$ if $\exists b \in A$: $a\ D\ b$.
  The complement of D is given by $I \subseteq Act \times Act$. It follows that $a\ I\ b$ only if a and b can be executed independently (or in either order). We also write $a\ I\ A$ if $\neg(a\ D\ A)$, i.e., if $\forall b \in A$: $a\ I\ b$.
- They define the *weak sequential composition* of two processes t and u, denoted $t \cdot u$, in such a way that only the dependent actions from first and second component are ordered. As a special case, if $a\ I\ b$ for all $a \in \mathcal{S}(t)$ and $b \in \mathcal{S}(u)$, then the semantics of $t \cdot u$ and $t \mid\mid\mid u$ coincide; or more generally, if the sorts of t_1 and u_2, respectively u_1 and t_2 are independent then the following *communication closed layers* law holds:

  $$(t_1 \mid\mid\mid u_2) \cdot (t_2 \mid\mid\mid u_2) = (t_1 \cdot t_2) \mid\mid\mid (u_1 \cdot u_2).$$

  In fact, in [91,142] strong and weak sequential composition are both present within the same formalism; since the novelty is in the latter, we ignore the former in this section.
- They require that *refinement preserves independencies*, in the sense that if a is refined into u and $a\ I\ b$ for some b, then also $b\ I\ \mathcal{S}(u)$. (This requirement is not explicitly stated in [91], but it is necessary for the theory to be sound.)
- They do not consider invisible actions or hiding. The technical reason for this is that the information added by the dependency relation would be lost upon hiding. Although one can envisage an approach in which there is a family of invisible actions in which the dependency relation is somehow retained, this has not been worked out. In the remainder of this section, we therefore ignore hiding and invisible actions.

This results in a semantics where refinement distributes over weak sequential composition:

$$(t_1 \cdot t_2)[a \to u] = t_1[a \to u] \cdot t_2[a \to u].$$

Taking the example at the beginning of this section and interpreting the abstract sequential composition *put*; *get* as the weak $put \cdot get$ instead, and assuming $put_i\ I\ get_j$ for $i \neq j$, it follows that refinement does not impose more ordering than necessary:

$$(put \cdot get)[put \rightarrow put_1 \mathbin{|||} put_2][get \rightarrow get_1 \mathbin{|||} get_2]$$
$$= (put_1 \mathbin{|||} put_2) \cdot (get_1 \mathbin{|||} get_2) = put_1 \cdot get_1 \mathbin{|||} put_2 \cdot get_2.$$

Janssen et al. [91] use a compositional, *linear-time* semantics. The linear time nature has to do with the fact that they consider sets of traces as their semantic model, in which the moment of choice is not recorded. The traces are partially ordered (hence pomsets, see Section 5.1) in a special manner, closely related to Mazurkiewicz traces: two occurrences are ordered if and only if the associated actions are dependent. This means that even if parallel execution is specified, dependent actions in different operands are interleaved; thus, the traces of $a \mathbin{|||} b$ with $a\ D\ b$ are given by $\boxed{a \rightarrow b}$ and $\boxed{b \rightarrow a}$ and not by $\boxed{\substack{a \\ b}}$. As usual in a trace-based semantics, the choice operator is modelled by taking the union of trace sets.

Atomic refinement. The above informal discussion of the refinement operator does not precisely correspond to the solution of [91]. Namely, within the setting of action dependencies, one has once more the choice between *atomic* and *non-atomic* refinement, discussed at length earlier in this chapter (Sections 1.3 and 3). It is the former that has been worked out in [91]; in particular, they also include the atomizer $\langle t \rangle$ we discussed in Section 3. This gives rise to a language $Lang_D^{lin}$ with weak sequential composition and atomizer, generated by the following grammar:[3]

$$\mathsf{T} ::= \mathbf{0} \mid \mathbf{1} \mid a \mid \mathsf{T} + \mathsf{T} \mid \underline{\mathsf{T} \cdot \mathsf{T}} \mid \mathsf{T} \mathbin{|||} \mathsf{T} \mid \langle \mathsf{T} \rangle \mid \mathsf{T}[a \rightarrow \mathsf{T}] \mid x \mid \mu x.\mathsf{T}.$$

See also Table 16 for a complete overview of the different languages used in this chapter. In the linear-time setting of [91], there is no need to forbid termination in choice or refinement terms; correspondingly, we have not distinguished "virgin operands" in the above grammar. Instead, as stated above, the theory is sound only if we impose as a well-formedness condition that *refinement has to be D-compatible*:

DEFINITION 7.1. $t[a \rightarrow u]$ is said to be *D-compatible* if $a\ I\ b$ implies $\mathcal{S}(u)\ I\ b$ for all $b \in Act$.

In this setting the atomic execution of a term t comes down to the following. In a term of the form $\langle t \rangle \mathbin{|||} u$, those actions a of u that are *dependent* on some action (in the sort) of t are scheduled either weakly before or weakly after t, where the adjective "weakly" implies that a can still overlap with any initial or final fragment of t of which it is completely independent; but not in the middle (that is, after one action b of t with $a\ D\ b$ but before another action c of t with $a\ D\ c$). On the other hand, if a is *independent* of t, i.e., $a\ I\ \mathcal{S}(t)$, then it may be executed at any time during the execution of $\langle t \rangle$.[4]

[3] In fact, only a special case of recursion is considered in [91], namely the Kleene star for weak sequential composition. We conjecture, however, that the theory carries over to full recursion without a problem.

[4] Thus, in terms of the discussion in Section 3, here the execution of an atomic process is non-interruptible for dependent actions but not for independent ones, and in contrast to the operator worked out in Section 3, here the execution of an atomic process is not all-or-nothing.

EXAMPLE 7.2. If $b\ D\ a_i$ for $i = 1, 2$, then $a_1\, b\, a_2$ is not an allowed execution sequence of either $\langle a_1 \cdot a_2 \rangle \mathbin{|||} b$ or $\langle a_1 \mathbin{|||} a_2 \rangle \mathbin{|||} b$; if furthermore $b\ I\ a_3$ then $a_1\, a_2\, b\, a_3$ is an allowed trace of $\langle a_1 \cdot a_2 \cdot a_3 \rangle \mathbin{|||} b$. Finally, if $c\ I\ a_i$ for $i = 1, 2, 3$ then $\langle a_1 \cdot a_2 \cdot a_3 \rangle \mathbin{|||} c$ puts no constraints whatsoever on the moment of execution of c with respect to the a_i.

The main motivation given in [91] for choosing atomic over non-atomic action refinement is to make refinement distribute over parallel composition, i.e., to satisfy

$$(t_1 \mathbin{|||} t_2)[a \to u] = t_1[a \to u] \mathbin{|||} t_2[a \to u].$$

The following example shows why this is incompatible with non-atomic refinement. (Recall also Section 1.5, where we made exactly the same point in connection with standard interleaving semantics.)

EXAMPLE 7.3. Consider $Act = \{a, a_1, a_2, b\}$ with $a\ D\ b$ and $a_i\ D\ b$ for $i = 1, 2$. As we saw above, the semantics of $a \mathbin{|||} b$ is given by $\{\boxed{a \to b}, \boxed{b \to a}\}$; therefore $(a \mathbin{|||} b)[a \to a_1 \cdot a_2]$ is modelled by $\{\boxed{a_1 \to a_2 \to b}, \boxed{b \to a_1 \to a_2}\}$. In other words, the semantics satisfies

$$(a \mathbin{|||} b)[a \to a_1 \cdot a_2] = a_1 \cdot a_2 \cdot b + b \cdot a_1 \cdot a_2.$$

On the other hand, if we interpret action refinement non-atomically (meaning that $a[a \to t] = t$), then the above distributivity property implies

$$(a \mathbin{|||} b)[a \to a_1 \cdot a_2] = a[a \to a_1 \cdot a_2] \mathbin{|||} b[a \to a_1 \cdot a_2] = (a_1 \cdot a_2) \mathbin{|||} b.$$

This contradicts the previous equality, since $(a_1 \cdot a_2) \mathbin{|||} b \neq a_1 \cdot a_2 \cdot b + b \cdot a_1 \cdot a_2$. With atomic refinement (which satisfies $a[a \to t] = \langle t \rangle$ instead), we obtain $(a \mathbin{|||} b)[a \to a_1 \cdot a_2] = \langle a_1 \cdot a_2 \rangle \mathbin{|||} b$, which is fine, since indeed $\langle a_1 \cdot a_2 \rangle \mathbin{|||} b = a_1 \cdot a_2 \cdot b + b \cdot a_1 \cdot a_2$.

The details of the semantics of [91] are outside the scope of this paper; suffice it to say that it gives rise to a pre-order $\sqsubseteq^D_{\mathrm{tr}} \subseteq Lang^{lin}_D \times Lang^{lin}_D$ with the following property:

PROPOSITION 7.4. *$\sqsubseteq^D_{\mathrm{tr}}$ is a pre-congruence over $Lang^{lin}_D$.*

Furthermore, just as in the case of $Lang^{atom}$, action refinement distributes over the other operators (see [91, Theorem 3.2]) and thus corresponds to atomized action substitution; that is, the following equality holds up to $\simeq^D_{\mathrm{tr}} = \sqsubseteq^D_{\mathrm{tr}} \cap \sqsupseteq^D_{\mathrm{tr}}$ (compare with [91, Corollary 3.3]):

$$t[a \to u] = t\{\langle u \rangle / a\}.$$

This implies that the following also holds (compare with Theorem 6.7):

THEOREM 7.5. *$t \simeq^D_{\mathrm{tr}} flat^{atom}(t)$ for any $t \in Lang^{lin}_D$.*

7.2. *Application: Critical sections*

As an example of the use of action dependencies, consider again the system of Section 3.4: $Sys_S := Proc_1 \,|||\, Proc_2$ where for $i = 1, 2$

$$Proc_i := ncs_i\,;\, cs_i\,;\, Proc_i\,.$$

The non-critical sections, ncs_i, do not depend on any action of the other process; i.e., we have $ncs_i \; I \; ncs_j$ and $ncs_i \; I \; cs_j$ for $i \neq j$. As a consequence, in the semantics of Sys_S the occurrences of ncs_i are unordered with respect to cs_j (for $i \neq j$); for instance, a valid trace is

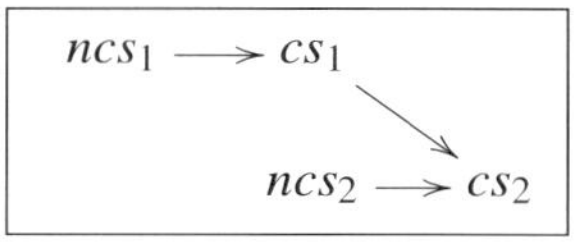

Now consider the same refinement as in Section 3.4:

$$Sys_I := Sys_S[cs_1 \rightarrow a_1\,;\, b_1][cs_2 \rightarrow b_2 + c_2].$$

In order for the refinements to be D-consistent, we must have $ncs_i \; I \; a_j$ whenever $i \neq j$; moreover, we assume all actions of the critical sections are mutually dependent. This time, the implementation does not suffer from the problem discussed in Section 3.4, where we found that the non-critical sections may not overlap with the critical sections. For instance, a valid trace of Sys_I is

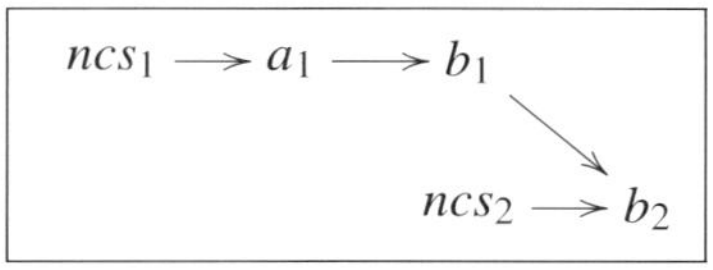

where ncs_2 is unordered with respect to a_1 and b_1 and hence may be scheduled in between them.

7.3. *Dependencies in branching time*

As we have seen above (Example 7.3), the main motivation in [91] for considering atomic refinement was to obtain distributivity over parallel composition. If we abandon this property, there is no intrinsic difficulty in modelling either non-atomic refinement or parallel composition with synchronization in a setting with action dependencies. For this purpose, we define the language $Lang_D^{bra}$, with the following grammar:

$$\mathsf{T} ::= \underline{\mathbf{0}_A} \mid a \mid \mathsf{T} + \mathsf{T} \mid \underline{\mathsf{T} \cdot \mathsf{T}} \mid \mathsf{T} \parallel_A \mathsf{T} \mid \underline{\mathsf{T}[\vec{a} \rightarrow \vec{\mathsf{T}}]} \mid x \mid \mu x.\mathsf{T}.$$

(Note that internal actions and hiding are again omitted, for reasons discussed above. See also Table 16 for a complete overview of the different languages used in this chapter.) The new constant $\mathbf{0}_A$ denotes *partial termination* and generalizes both $\mathbf{0}$ and $\mathbf{1}$; it is discussed below in some detail. In contrast to $\mathbf{1}$, however, $\mathbf{0}_A$ is not considered an auxiliary operator; in fact, $Lang_D^{bra}$ has no auxiliary operators and hence the concept of "virgin operand" plays no role in the well-formedness of terms of $Lang_D^{bra}$. Instead, we impose several well-formedness conditions to do with the dependency relation; see below. Another new aspect of $Lang_D^{bra}$ is that its refinement operator maps a *vector* of actions to refining agents, instead of just one: the motivation for this is also given below.

Operational semantics. Table 14 contains a branching time operational semantics for $Lang_D^{bra}$, insofar it deviates from the semantics given before (Tables 2 and 6). The semantics for the flat fragment of $Lang_D^{bra}$ was developed in [125] and extended with refinement in [136]; the solution presented here is based on an improved version in [126]. We briefly discuss the salient points.

Partial termination. Compared with the linear time case, there are some interesting complications having to do with the interplay between weak sequential composition and choice.

EXAMPLE 7.6. Consider actions $a, b, c \in Act$ such that $a\ D\ c$ and $b\ I\ c$, and consider the process $(a + b) \cdot c$. The linear time behaviour of this process consists of a trace where a and c are executed in sequence, and a trace where b and c are executed independently. The

Table 14
Transition rules for $Lang_D^{bra}$

$$\frac{b\ I\ a}{a \xrightarrow{\surd_b} a} \qquad \frac{a\ I\ A}{\mathbf{0}_A \xrightarrow{\surd_a} \mathbf{0}_A}$$

$$\frac{t \xrightarrow{\surd_a} t' \quad u \xrightarrow{\surd_a} u'}{t+u \xrightarrow{\surd_a} t'+u'} \qquad \frac{t \xrightarrow{\surd_a} t' \quad u \not\xrightarrow{\surd_a}}{t+u \xrightarrow{\surd_a} t'} \qquad \frac{t \not\xrightarrow{\surd_a} \quad u \xrightarrow{\surd_a} u'}{t+u \xrightarrow{\surd_a} u'}$$

$$\frac{t \xrightarrow{\surd_a} t' \quad u \xrightarrow{\surd_a} u'}{t \cdot u \xrightarrow{\surd_a} t' \cdot u'} \qquad \frac{t \xrightarrow{\surd_a} t' \quad u \xrightarrow{\surd_a} u'}{t \parallel_A u \xrightarrow{\surd_a} t' \parallel_A u'} \qquad \frac{t \xrightarrow{\surd_a} t'}{t[\vec{a} \to \vec{u}] \xrightarrow{\surd_a} t'[\vec{a} \to \vec{u}]}$$

$$\frac{a\ I\ \mathcal{S}_x}{\mu x.t \xrightarrow{\surd_a} \mu x.t} \qquad \frac{t\{\mu x.t/x\} \xrightarrow{\surd_a} t' \quad a\ D\ \mathcal{S}_x}{\mu x.t \xrightarrow{\surd_a} t'}$$

$$\frac{t \xrightarrow{a} t'}{t \cdot u \xrightarrow{a} t' \cdot u} \qquad \frac{t \xrightarrow{\surd_a} t' \quad u \xrightarrow{a} u'}{t \cdot u \xrightarrow{a} t' \cdot u'}$$

$$\frac{t \xrightarrow{b} t' \quad b \notin \{\vec{a}\}}{t[\vec{a} \to \vec{u}] \xrightarrow{b} t'[\vec{a} \to \vec{u}]} \qquad \frac{t \xrightarrow{a_i} t' \quad u_i \xrightarrow{b} u'}{t[\vec{a} \to \vec{u}] \xrightarrow{b} u' \cdot t'[\vec{a} \to \vec{u}]}$$

corresponding branching time behaviour implies that c can be executed initially, but *this resolves the choice* between a and b.

In capturing these effects operationally, one has to take into account that

- the second operand of weak sequential composition can be active even if the first has not yet terminated;
- actions performed by the second operand of weak sequential composition may resolve choices in the first.

For this purpose, we have introduced the concept of *partial termination*: instead of a deadlock constant $\mathbf{0}$ and a termination constant $\mathbf{1}$, we now have a family of constants $\mathbf{0}_A$, where A stands for the *deadlock alphabet*. $\mathbf{0}_A$ is deadlocked for all actions a that are dependent on some $b \in A$, and terminated for all other actions. The notion of "termination for a given action" is captured by a family of new transition labels $\surd_a$ for $a \in A$: $t \xrightarrow{\surd_a} t'$ expresses that t is terminated for a, where invoking the termination possibly resolves some choices by which the term becomes t'.

For instance, we have that $a \xrightarrow{\surd_b} a$ if $b \; I \; a$ and $\mathbf{0}_A \xrightarrow{\surd_a} \mathbf{0}_A$ if $a \; I \; A$. The latter implies that the deadlock constant $\mathbf{0}$ of *Lang* corresponds to $\mathbf{0}_{Act}$ (no termination at all) whereas the termination constant $\mathbf{1}$ corresponds to $\mathbf{0}_\emptyset$ (termination for all actions). We continue using $\mathbf{0}$ and $\mathbf{1}$ as abbreviations for these special cases.

From the rules in Table 14, it can be seen that partial termination resolves choice if precisely one of the operands terminates and the other cannot. In Example 7.6, for instance, $b \xrightarrow{\surd_c} b$ but $a \stackrel{\surd_c}{\not\longrightarrow}$, and hence $a + b \xrightarrow{\surd_c} b$; we obtain the derivation

$$\cfrac{\cfrac{\cfrac{}{a \stackrel{\surd_c}{\not\longrightarrow}} \qquad \cfrac{b \; I \; c}{b \xrightarrow{\surd_c} b}}{a + b \xrightarrow{\surd_c} b} \qquad \cfrac{}{c \xrightarrow{c} \mathbf{1}}}{(a + b) \cdot c \xrightarrow{c} b \cdot \mathbf{1}}$$

which is indeed the expected behaviour. (To avoid generating infinite state spaces for virtually all recursive terms, one then also has to add a mechanism to get rid of superfluous $\mathbf{1}$'s, such as in the target term of the above transition; this can be done for instance by normalizing terms with respect to some structural congruence that includes the axiom $t \cdot \mathbf{1} \equiv t$.)

Recursion. The termination behaviour of recursion has a problem connected to the negative premise in two of the rules for choice, if one takes the natural extension of the standard operational rule for recursion to partial termination:

$$\frac{t\{\mu x.t/x\} \xrightarrow{\surd_a} t'}{\mu x.t \xrightarrow{\surd_a} t'}.$$

EXAMPLE 7.7. Let $a, b, c \in Act$ with $a \; I \; c$ and $b \; I \; c$, and consider the process $t = \mu x.(a \cdot$

$x + b)$ with $\mathcal{S}_x = \{a, b\}$. If we assume $t \not\xrightarrow{\surd_c}$, we obtain a contradiction:

$$\dfrac{\dfrac{\dfrac{t \not\xrightarrow{\surd_c}}{a \cdot t \not\xrightarrow{\surd_c}} \qquad \dfrac{b \mathrel{I} c}{b \not\xrightarrow{\surd_c} b}}{a \cdot t + b \xrightarrow{\surd_c} b}}{t \xrightarrow{\surd_c} b}\ .$$

On the other hand, no transition of the form $t \xrightarrow{\surd_c} t'$ can be derived, since this would require an "infinite unfolding" of the recursion.

Problems due to negative premises have been studied in depth in [80,58]. A crucial point in our solution is to take the sort of process variables (see Table 11) into account within the operational rules:

- A recursive term always terminates, without unfolding, for any action that is independent of the entire sort of the recursion variable. For instance, for the term $t = \mu x.(a \cdot x + b)$ in Example 7.7 we obtain $t \xrightarrow{\surd_c} t$.
- The previous rule (in Example 7.7) for the partial termination of recursion is restricted to $\surd_a$-transitions with $a \mathrel{D} \mathcal{S}_x$, because otherwise we could unfold recursive terms arbitrarily often in the derivation of a $\surd_a$-transition; for instance, for the same term t above

$$\dfrac{\dfrac{\dfrac{t \xrightarrow{\surd_c} t}{a \cdot t \xrightarrow{\surd_c} a \cdot t}}{a \cdot t + b \xrightarrow{\surd_c} a \cdot t + b}}{t \xrightarrow{\surd_c} a \cdot t + b}\ .$$

A related problem is that, due to the changed nature of termination, the standard notion of guardedness (Definition 2.1) is not appropriate any more, but needs to be strengthened.

DEFINITION 7.8. We first define what it means for a term to be a *D-guard* for an action $a \in Act$.

- $\mathbf{0}_A$ is a D-guard for a if $a \mathrel{D} b$ for some $b \in A$;
- x is a D-guard for a if $a \mathrel{D} b$ for some $b \in \mathcal{S}_x$;
- b is a D-guard for a if $a \mathrel{D} b$;
- $t + u$ is a D-guard for a if both t and u are D-guards for a;
- $t[b \to u]$ is a D-guard for a if t is a D-guard for a;
- For all other operators op, $op(t_1, \ldots, t_n)$ is a D-guard for a if one of the t_i is a D-guard for a.

Next, we define what it means for a variable to be D-guarded in a term.

- x is D-guarded in y ($\in Proc$) if $x \neq y$;
- x is D-guarded in $t; u$ if x is D-guarded in t and either t is a D-guard for all $a \in \mathcal{S}_x$ or x is D-guarded in u;
- x is D-guarded in $\mu y.t$ if either $x = y$ or x is D-guarded in t;

- For all other operators op, x is D-guarded in $op(t_1, \ldots, t_n)$ if x is D-guarded in all t_i.

For instance, the term $\mu x.(a \cdot x + b)$ in Example 7.7 is D-guarded if $a\ D\ b$ but not if $a\ I\ b$ (because then $a \xrightarrow{\surd_b} a$ but $b\ D\ b \in \mathcal{S}_x$). We require D-guardedness as a necessary condition for well-formedness.

Refinement. The operational rules for refinement in Table 14 are essentially the same as for the sequential language $Lang^{seq}$ (see Table 2), except that we now use weak rather than strong sequential composition. However, for these rule to make sense, the refinement should be compatible with the dependency relation, in a stronger sense than in the linear time case (Definition 7.1).

DEFINITION 7.9. $t[a \to u]$ is said to be *strictly D-compatible* if it is D-compatible and, moreover, for all $b \in Act$, $a\ D\ b$ implies

(i) $u \stackrel{\surd_b}{\not\longrightarrow}$ and
(ii) $c\ D\ b$ for all $u \xrightarrow{c}$.

The additional condition expresses that dependencies should be preserved as well as independencies: if a and b are dependent then u must be deadlocked for b (ensuring that if b has been specified sequentially after a then b cannot occur before u has started) and b must be deadlocked for all initial actions of u (implying the dual property, namely that if a is to occur after b then no initial action of u can occur before b).

The interesting thing about the current setting is that the operational rules remain correct even in the context of parallel composition with synchronization: interleaving bisimulation is a congruence for action refinement. (Of course, what we have here is not really the standard interleaving setting, since the action dependencies provide additional information just like the event annotations in the event-based semantics of Section 4; however, since here the information is provided globally rather than locally in each individual transitions, it is less visible.)

Unfortunately, strict D-compatibility is more restrictive than one would like. For instance, the term $t[get \to u]$ where $t = (put \cdot get)[put \to put_1 \mathbin{|||} put_2]$ and $u = get_1 \mathbin{|||} get_2$ is not well-formed: according to D-compatibility, due to the fact that $get_2\ D\ put_2$ we must have $get\ D\ put_2$; however, since $u \xrightarrow{get_1}$ but nevertheless $get_1\ I\ put_2$, condition (ii) of strong D-compatibility is violated. A similar violation occurs if we first refine *get* and then *put*. In fact, this particular design step can only be understood if we assume both refinements to take place *at the same time*. Looking back, this is in fact what we silently assumed in the discussion.

Thus, rather than considering single-action refinements $\_[a \to u]$, we should allow simultaneous refinements $\_[a_1 \to u_1, \ldots, a_n \to u_n]$ (with $a_i \neq a_j$ for $i \neq j$), also written in vector notation as $\_[\vec{a} \to \vec{u}]$. Strict D-compatibility is extended accordingly.

DEFINITION 7.10. $t[\vec{a} \to \vec{u}]$ is said to be strictly D-compatible if for all $b \in Act$

(1) $a_i\ I\ b$ implies $\mathcal{S}(u_i)\ I\ b$;
(2) $a_i\ D\ b$ implies (i) $u_i \stackrel{\surd_b}{\not\longrightarrow}$ and (ii) $c\ D\ b$ for all $u_i \xrightarrow{c}$.

This indeed holds for $t = get; put$ and the refinement $put \to put_1 \,|||\, put_2, get \to get_1 \,|||\, get_2$ under consideration; for instance, $get\ D\ put$ and indeed for all $r(put) \xrightarrow{c}$ (meaning $c = put_i$ for $i \in \{1, 2\}$) we have $r(get) \not\xrightarrow{\surd_c}$.

Well-formedness. To summarize the well-formedness conditions on $Lang_D^{bra}$:
- Recursion is required to be D-guarded;
- Refinement is required to be strictly D-compatible.

7.1. *Behavioural semantics.* Strong bisimilarity (Definition 2.6) is once more a congruence; for the proof see [126].

PROPOSITION 7.11. *$\sim$ is a congruence over $Lang_D^{bra}$.*

As an example, consider again the buffer discussed at the start of this section, and let $put_i\ I\ get_j$ for $i \neq j$. Using the operational rules, we can derive for instance

$$\begin{aligned}
&(put \cdot get)[put \to put_1 \,|||\, put_2, get \to get_1 \,|||\, get_2] \\
&\xrightarrow{put_1} (\mathbf{1} \,|||\, put_2) \cdot (\mathbf{1}; get)[put \to put_1 \,|||\, put_2, get \to get_1 \,|||\, get_2] \\
&\xrightarrow{get_1} (\mathbf{1} \,|||\, put_2) \cdot (\mathbf{1} \,|||\, get_2) \cdot (\mathbf{1}; \mathbf{1})[put \to put_1 \,|||\, put_2, get \to get_1 \,|||\, get_2].
\end{aligned}$$

This shows that a low-level get_i-action can occur directly after the corresponding put_i-action, without having to wait for other put_j-actions to be executed as well; thus, the desired flexibility is achieved.

Another issue reported in [126] is the development of a denotational partial-order model for $Lang_D^{bra}$, for the purpose of showing that the above interleaving semantics is compatible with existing interpretations of action refinement. The denotational model is event-based, based on the families of posets model of [117], extended to take partial termination into account. The details are beyond the scope of this section; suffice it to say that one can define an isomorphism over the denotational model that is a congruence for $Lang_D^{bra}$ and strictly stronger than $\sim$, and moreover, there is a straightforward transformation from the denotational model into labelled transition systems (including $\surd_a$-transitions).

The fact that in the presence of action dependencies, the interleaving paradigm is strong enough to give rise to a compositional model for action refinement was discussed separately in [70].

7.4. *Application: A simple data base*

Once more consider the data base example of Section 2.5 (which does not include the *copy*-operation of Section 4.6). The 2-state version of this data base was specified by $Data_S^2 := State_1$ where for $i = 1, 2$

$$State_i := qry_i; State_i + upd_1; State_1 + upd_2; State_2.$$

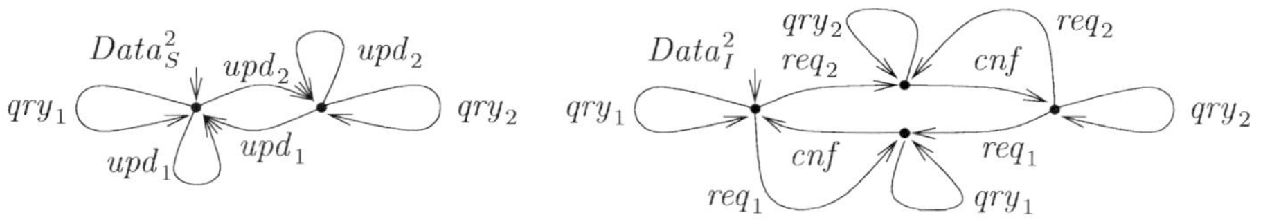

Fig. 10. Refinement of $Data_S^2$ using action dependencies: $qry_i\ D\ req_j$ but $qry_i\ I\ cnf$.

We then refined upd_i into $req_i; cnf$ for $i = 1, 2$, where req_i is a request to change the state to i and cnf is the confirmation that this has been done. In the standard setting of Section 2.5 this automatically implies that qry is disabled in between req_i and cnf (see Figure 2). If one wants to allow qry also at that point (because the change to the state has already been made, so it is safe to read it), this can be achieved using action dependencies by specifying $qry_i\ I\ cnf$ for $i = 1, 2$ (and all other actions mutually dependent), and setting $r = (upd_1 \to req_1 \cdot cnf, upd_2 \to req_2 \cdot cnf)$ as before. Note that r is strongly D-consistent: we have $qry_i\ D\ upd_j$ for all i, j, and indeed

(i) $r(upd_j) \stackrel{\surd_{qry_i}}{\not\longrightarrow}$ and

(ii) $qry_i\ D\ a_j$ for all $r(upd_j) \xrightarrow{a_j}$ (namely, $a_j = req_j$).

The behaviour of the system $Data_S^2$ and its refinement $Data_I^2 := Data_S^2[r]$ is depicted in Figure 10. It should be noted that the previous solution, which did not allow the overlap, can still be derived if we set $qry_i\ D\ cnf$ instead.

7.5. *The dual view: Localities*

Another way to integrate the concept of action dependencies into process algebra is by taking the dual view of *localities*. In the context of action refinement, this was worked out by Huhn in [89,90]. We briefly recapitulate the main points of this approach.

The basic assumption is that a system is described by the parallel composition of a finite, indexed set of sequential sub-systems, such that all actions performed by any given sub-system are mutually dependent, and this is the only possible source of dependencies. Actions are refined *locally* at each sequential sub-system, possibly differently at each location; refinement has to preserve and reflect the locality of actions.

More precisely, one considers a finite set of locations Loc and a mapping $\ell : Act \to \mathbf{2}^{Loc}$ defining at what localities an action may occur; then $a\ D\ b$ for $a, b \in Act$ if and only if $\ell(a) \cap \ell(b) \neq \emptyset$. Let us denote $Act_l = \{a \in Act \mid l \in \ell(a)\}$ for the alphabet of location l. Overall system behaviour is specified by terms of the form $\Pi_{l \in Loc} t_l$, denoting the parallel composition of sequential processes t_l, where $\mathcal{S}(t_l) \subseteq Act_l$ for all $l \in Loc$, with implicit synchronization over all common actions. The semantics of such a parallel composition is given by an ordinary labelled transition system.

Action refinement is interpreted in this context by a *family* of images for a given abstract action, namely one for each location that partakes in the action. Thus, in general, an action is mapped simultaneously to a different concrete process at each location. Operationally, therefore, the effect of a global action refinement is captured by the local action refinements, which in turn are refinements of a sequential system and hence analogous to the treatment in Section 2. In particular, at each location $l \in \ell(a)$, the essential rule for the behaviour of a refined system is given by

$$\frac{t_l \xrightarrow{a} t_l' \quad u_l \xrightarrow{b} u_l'}{t_l[a \to u_l] \xrightarrow{b} u_l'; t_l'[a \to u_l]},$$

where the subscript $_l$ indicates that we are dealing with the process at location l. There are some restrictions in [90] on the allowable (families of) refinement functions, which are closely related to the preservation of (in)dependencies imposed in the previous subsection.

- Only the locations $l \in \ell(a)$ may refine a, and $\mathcal{S}(u_l) \subseteq Act_l$ for all $l \in Loc$. This implies that refinement is D-compatible (Definition 7.1).
- None of the local refinements may map onto a terminated process. This goes somewhat in the direction of strict D-compatibility: If $a\ D\ b$ then a and b share a location, say l; since $r_l(a) \stackrel{\checkmark}{\not\longrightarrow}$ for all such shared l, it follows that b cannot "overtake" $r(a)$.

Refinement on the system level is interpreted by distribution to the local level: that is, one may specify a global refinement $\_[a \to \vec{u}]$ where $\vec{u}$ is a vector of local images u_l; $(\Pi_l t_l)[a \to \vec{u}]$ is then defined to correspond to $\Pi_l(t_l[a \to u_l])$.

If we use $Lang^{loc}$ to denote the resulting locality-based language, then the following result is an immediate consequence of the format of the operational rules:

PROPOSITION 7.12. *$\sim$ is a congruence over $Lang^{loc}$.*

Furthermore, the fact that action refinement is essentially performed on local, sequential systems implies that action refinement and action substitution coincide (see Section 6.2):

THEOREM 7.13. *$t \sim flat(t)$ for any $t \in Lang^{loc}$.*

The motivating example of this section can be dealt with by considering $Loc = \{1, 2\}$, i.e., a system with two localities, and assuming $\ell(put) = \ell(get) = \{1, 2\}$. The specification is then given by $put; get \,|||\, put; get$ (one sequential term for each location), and the local refinements by $put \to put_l$ and $get \to get_l$ for $l = 1, 2$. Refinement then yields

$$(put; get)[put \to put_1][get \to get_1] \,|||\, (put; get)[put \to put_2][get \to get_2]$$

which is equivalent to the desired solution, $put_1; get_1 \,|||\, put_2; get_2$.

8. Vertical implementation

Finally, we present a different, more radical solution to the lack of flexibility of the action refinement operator that we noted in the previous section. The presentation is based on

[119,123]. It consists of re-interpreting action refinement so that it does not take the form of an *operator* (effectively a function $\_[a \to u]: Lang \to Lang$ for all u) but rather of a *relation* $\leqslant^{a\to u} \subseteq Lang^{flat} \times Lang^{flat}$, where $Lang^{flat}$ is the flat fragment of $Lang$ (see Section 4), i.e., without the refinement operator. (See also Table 16 for a complete overview of the different languages used in this chapter.) That is, if a system is abstractly specified by a term t_1, and a design step is taken that maps an abstract action a to a term u, the result is not uniquely determined but instead can be any term t_2 satisfying the requirement

$$t_1 \leqslant^{a\to u} t_2.$$

$\leqslant^{a\to u}$ is called a *vertical implementation relation*, in analogy with the usual implementation relations (i.e., pre-orders) such as trace or failure inclusion (see, e.g., [27,43]). This notion is interesting especially if we consider *simultaneous* refinements (see also Section 7.3); thus, actually we will consider indexed relations of the form $\leqslant^{\vec{a}\to\vec{u}} \subseteq Lang^{flat} \times Lang^{flat}$ where $\vec{a} \to \vec{u}$ is a finite list of refinement mappings. In this section, we will often denote such vectors by r, and occasionally interpret r as a function $r: Act \to Lang^{flat}$ that maps all a_i onto u_i and is the identity everywhere else.

For instance, consider again the small buffer example used as motivation in the previous section. We started with a specification *put*; *get* and a (simultaneous) refinement given by $r = put \to put_1 \,|||\, put_2, get \to get_1 \,|||\, get_2$; by fixing a dependency relation, we arrived at a uniquely determined implementation. Using the concept of vertical implementation, on the other hand, there could be a number of correct implementations; for instance

$$\begin{aligned}
put;\, get &\leqslant^r (put_1 \,|||\, put_2);\, (get_1 \,|||\, get_2),\\
put;\, get &\leqslant^r (put_1;\, get_1 \,|||\, put_2);\, get_2,\\
put;\, get &\leqslant^r put_1;\, get_1 \,|||\, put_2;\, get_2.
\end{aligned}$$

A major advantage of this approach is that we can stay in the realm of interleaving semantics, just as in the case of dependency-based refinement. This is due to the fact that the congruence question simply disappears once we have abandoned the idea of refinement as an operator. For instance, the following design steps all give rise to perfectly valid vertical implementations (compare Section 1.3):

$$\begin{aligned}
a \,|||\, b &\leqslant^{a\to a_1;a_2} a_1;\, a_2 \,|||\, b,\\
a \,|||\, b &\leqslant^{a\to a_1;a_2} a_1;\, a_2;\, b + b;\, a_1;\, a_2,\\
a;\, b + b;\, a &\leqslant^{a\to a_1;a_2} a_1;\, a_2 \,|||\, b,\\
a;\, b + b;\, a &\leqslant^{a\to a_1;a_2} a_1;\, a_2;\, b + b;\, a_1;\, a_2.
\end{aligned}$$

As a consequence, it is possible to formulate a single, unified framework in which the standard concept of implementation (which, in contrast, one might call *horizontal*) and vertical implementation are integrated. In fact, given that there are many variations on the standard concept of implementation (based, e.g., on traces, testing or bisimulation), one can imagine analogous variations on vertical implementation.

Below, we propose a specific vertical implementation relation, based on delay bisimulation (see Section 2.4). Another vertical implementation relation, based on weak bisimulation, was worked out in [123,124]; a testing-based relation was proposed in [118]. We then formulate some properties that one would naturally expect a vertical implementation relation to satisfy, inspired by properties of horizontal implementation (e.g., transitivity, monotonicity); we assert that vertical delay bisimulation indeed satisfies these properties.

8.1. *Refinement functions*

Vertical implementation relations are a new area of study, and the results that have been achieved hold only for a restricted class of refinement functions. For the purpose of this section, we therefore impose some strong restrictions on the allowable refinements. See [124] for a more extensive discussion.

First of all, we only consider refinement images in a restricted sub-language $Lang^{ref}$, generated by the following grammar:

$$\mathsf{T} ::= a \mid \mathsf{T};\mathsf{T} \mid \mathsf{T}+\mathsf{T}.$$

(Note that this limits actions to $a \in Act$, disallowing invisible actions.) Furthermore, we impose as a well-formedness condition on $Lang^{ref}$ that terms must be *distinct*, in the following inductively defined sense:

- Every action $a \in Act$ is said to be distinct;
- $t_1; t_2$ and $t_1 + t_2$ are said to be distinct if t_1 and t_2 are distinct and $\mathcal{S}(t_1) \cap \mathcal{S}(t_2) = \emptyset$.

Furthermore, given a set of actions $A \subseteq Act$, a simultaneous refinement $r : Act \to Lang^{ref}$ (using only well-formed terms as images) is called *distinct on* A if $\mathcal{S}(r(a)) \cap \mathcal{S}(r(b)) = \emptyset$ for $a \in A$ and $b \in Act \setminus \{a\}$. This has the consequence that, given a b-transition performed by the r-image of some action $a \in A$, both the action a and the position in $r(a)$ where b occurs are uniquely determined. In the remainder of this section, we will implicitly only consider vertical implementation relations $\leqslant^r$ between pairs t_1 and t_2 such that r is distinct on $\mathcal{S}(t_1)$.

Note that the image of $put \to put_1 \,|||\, put_2$ is not in $Lang^{ref}$ (and neither is the image of get); hence we cannot treat the motivating example in its current form. Below, we consider the somewhat simpler case of $r = (put \to put_1; put_2$ and $get \to get_1; get_2)$; we still expect the following valid vertical implementations.

$$put;\, get \leqslant^r put_1; get_1; put_2; get_2,$$
$$put;\, get \leqslant^r put_1; (get_1 \,|||\, put_2); get_2.$$

We furthermore use the following refinement function-related concepts:

- For every refinement function r, the function $\mathcal{S}_r : Act \to \mathbf{2}^{Act}$ maps all actions to the alphabet of their r-images; that is, for all $a \in A$

$$\mathcal{S}_r : a \mapsto \mathcal{S}\big(r(a)\big).$$

This is also extended to sets of actions $A \subseteq Act$:

$$\mathcal{S}_r : A \mapsto \bigcup_{a \in A} \mathcal{S}_r(a).$$

- For every refinement function r and set of actions A, r/A denotes the function that is the identity on A and coincides with r elsewhere; that is,

$$r/A : a \mapsto \begin{cases} a & \text{if } a \in A, \\ r(a) & \text{otherwise.} \end{cases}$$

- For all refinement functions r_1 and r_2, $r_1 + r_2$ denotes the function that maps all $a \in Act$ to the choice between the r_i-images:

$$(r_1 + r_2) : a \mapsto r_1(a) + r_2(a).$$

8.2. *Vertical delay bisimulation*

We develop a vertical implementation relation based on a variant of weak bisimulation called *delay bisimulation*, which we already introduced in Definition 2.11.[5] We actually drop the qualifier "delay" in the remainder of this section, when this does not give rise to confusion.

An important extension with respect to the standard notion of bisimulation is that we have to take into account that in any given state of the implementation, there may be associated refined actions whose execution has not yet terminated. These will be collected in a (multi)set of *residual* (or *pending*) *refinements* and will be used to parameterize the bisimulation. Thus, bisimulations are not binary but ternary relations.[6] Furthermore, in contrast to standard bisimulation, the simulation directions from specification to implementation and vice versa are no longer symmetric. To simulate the abstract transitions of the specification by the implementation (with the set of pending refinements as an intermediary), we define the concept of *down-simulation*; the other direction is called an *up-simulation*. Finally, we also require a *residual simulation* which captures how the pending refinements are processed by the implementation.

Residual sets. A residual set for r will be a multiset of non-terminated proper derivatives of r-images, formally represented by a function $R : Lang^{seq} \to \mathbb{N}$. We will write $t \in R$ if $R(t) > 0$. To be precise, the collection of residual sets for r is defined as follows:

$$RS(r) = \left\{ R : Lang^{seq} \to \mathbb{N} \mid \forall u' \in R \colon \exists a \in Act, \sigma \in Act^+ \colon r(a) \stackrel{\sigma}{\Longrightarrow} u' \stackrel{\surd}{\not\to} \right\}.$$

[5] Previously (in [123]), we took weak bisimulation as a basis for vertical extension; in doing so, we encountered some problems that we conjectured to be related the fact that weak bisimulation is not a congruence for the refinement operator in the sequential language $Lang^{seq}$ (see Section 2). Since delay bisimulation is the coarsest congruence for refinement contained in weak bisimulation (Proposition 2.12), its vertical extension is smoother than that of weak bisimulation.

[6] Other ternary bisimulation-based relations are, for instance, *history-preserving* bisimulation [62], and *symbolic* bisimulation [86].

We use the following constructions over residual sets:

$$\begin{aligned}
\emptyset : u &\mapsto 0, \\
[t] : u &\mapsto \begin{cases} 1 & \text{if } u = t \text{ and } t \stackrel{\checkmark}{\not\longrightarrow}, \\ 0 & \text{otherwise}, \end{cases} \\
R_1 \oplus R_2 : u &\mapsto R_1(u) + R_2(u), \\
R_1 \ominus R_2 : u &\mapsto \max\{R_1(u) - R_2(u), 0\}.
\end{aligned}$$

The behaviour of a residual set corresponds to the synchronization-free parallel composition of its elements. Formally:

$$R \xrightarrow{\alpha} R' :\iff \exists t \in R \colon \exists (t \xrightarrow{\alpha} t') \colon R' = (R \ominus [t]) \oplus [t'].$$

Note that terminated terms do not contribute to the residual set, and $\alpha \neq \tau$. The reason why we can ignore terminated terms is that (due to well-formedness) it is certain that such terms no longer display any operational behaviour (in terms of Section 2, termination is final).

Notation for ternary relations. In the following, we will often work with ternary relations of the form $\rho \subseteq N \times N \times RS(r)$. We use the notation $n_1 \rho^R n_2$ to abbreviate $(n_1, n_2, R) \in \rho$; in other words, ρ^R is interpreted as the binary relation $\{(n_1, n_2) \mid (n_1, n_2, R) \in \rho\}$.

DEFINITION 8.1. Let $\mathcal{T}$ be a labelled transition system. A *strict down-simulation up to r* over $\mathcal{T}$ is a ternary relation $\rho \subseteq N \times N \times RS(r)$ such that for all $n_1 \rho^{\emptyset} n_2$, if $n_1 \xrightarrow{\alpha} n_1'$ then one of the following holds:

(1) $\alpha = \tau$ and $n_1' \rho^{\emptyset} n_2$;
(2) $\alpha \in \{\tau, \checkmark\}$ and $\exists (n_2 \Rightarrow\xrightarrow{\alpha} n_2')$ such that $n_1' \rho^{\emptyset} n_2'$;
(3) $\alpha \in Act$ and $\forall (r(\alpha) \xrightarrow{c} u') \colon \exists (n_2 \Rightarrow\xrightarrow{c} n_2')$ such that $n_1' \rho^{[u']} n_2'$.

Note that down-simulation only imposes conditions on triples (n_1, n_2, R) where $R = \emptyset$, i.e., there are no pending refinements. For such triples, apart for the simulation of internal and termination transitions (which are standard for delay bisimulation), it is required that any action a of the specification (n_1) is refined into a term $r(a)$ of which all initial actions are simulated by the implementation (n_2), such that the successors of n_1 and n_2 are again related, with the successor of $r(a)$ pending. The reason for the additive "strict" will become clear later this section.

DEFINITION 8.2. Let $\mathcal{T}$ be a labelled transition system. An *up-simulation up to r* over $\mathcal{T}$ is a ternary relation $\rho \subseteq N \times N \times RS(r)$ such that for all $n_1 \rho^R n_2$, if $n_2 \xrightarrow{\gamma} n_2'$ then one of the following holds:

(1) $\gamma = \tau$ and $n_1 \rho^R n_2'$;
(2) $\gamma \in \{\tau, \checkmark\}$ and $\exists (n_1 \Rightarrow\xrightarrow{\gamma} n_1')$ such that $n_1' \rho^R n_2'$;
(3) $\exists a \in Act \colon \exists (n_1 \Rightarrow\xrightarrow{a} n_1')$ and $\exists (r(a) \xrightarrow{\gamma} u')$ such that $n_1' \rho^{R \oplus [u']} n_2'$;
(4) $\exists (n_1 \Rightarrow n_1')$ and $\exists (R \xrightarrow{\gamma} R')$ such that $n_1' \rho^{R'} n_2'$.

Thus, if the implementation's move γ is not an internal or termination transition, either it corresponds to the initial concrete action of a refined abstract action a (in which case, since r is distinct, a is uniquely determined), or it is an action of a residual refinement (which again is uniquely determined), in which case the specification is allowed to move silently. Note that these two cases (Clauses (3) and (4) in the definition) are mutually exclusive, again due to the distinctness of r.

DEFINITION 8.3. Let $\mathcal{T}$ be a labelled transition system. A *strict residual simulation* over $\mathcal{T}$ is a ternary relation $\rho \subseteq N \times N \times RS(r)$ such that for all $R \neq \emptyset$, if $n_1 \rho^R n_2$ then $\forall (R \xrightarrow{\gamma} R')$: $\exists (n_2 \Rightarrow \xrightarrow{\gamma} n_2')$ such that $n_1 \rho^{R'} n_2'$.

This specifies that any move of a pending refinement has to be (delay-)simulated by the implementation, while the specification remains as it is. This implies that pending refinements can be "worked off" in any possible order, or indeed in parallel, by the implementation. This property can be construed as an operational formulation of *non-interruptability*: that which is started can always be finished (see also Section 3). The combination of down-simulation, up-simulation and residual simulation gives rise to vertical bisimulation.

DEFINITION 8.4. Let $\mathcal{T}$ be a labelled transition system.

- A *strict vertical bisimulation up to r* over $\mathcal{T}$ is a ternary relation $\rho \subseteq N \times N \times RS(r)$ that is both a strict down-simulation up to r, an up-simulation up to r and a strict residual simulation.
- *Strict vertical bisimilarity up to r* over $\mathcal{T}$, denoted $\precsim_{\forall}^{r}$, is the largest strict vertical bisimulation up to r, and *rooted strict vertical bisimilarity up to r*, denoted $\preceq_{\forall}^{r}$, is the largest delay root of $\precsim_{\forall}^{r,\emptyset}$.

The subscript "$\forall$" is related to the adjective "strict" and the $\forall$-quantifiers in the definitions of strict down- and residual simulation; see also below. The following is an example of an actual strict vertical bisimulation.

EXAMPLE 8.5. Let $r : a \mapsto a_1; a_2$. The following figure shows strict vertical bisimulations proving $a; b \preceq_{\forall}^{r} a_1; a_2; b$ and $a; b \preceq_{\forall}^{r} a_1; (a_2 \parallel\!| b)$, where the dotted lines connect related states and their labelling is the residual set indexing the relation:

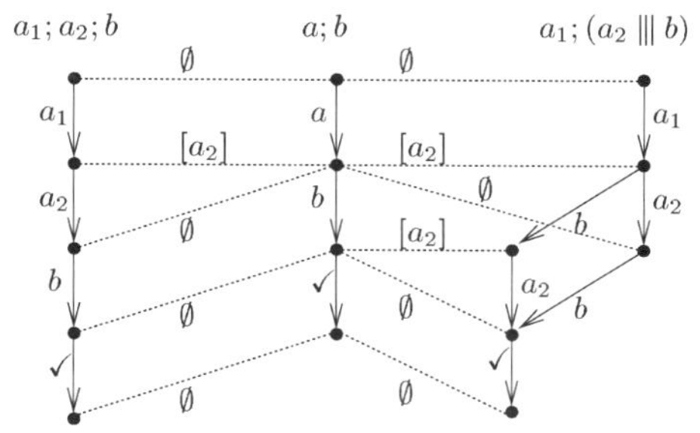

However, this does not yet quite allow the design step we used in our (modified) motivating example: that is, $put_1; (get_1 \mid\mid\mid put_2); get_2$ is not a vertical implementation of $put; get$. The reason is that the sequence

$$put_1; (get_1 \mid\mid\mid put_2); get_2 \xrightarrow{put_1} (get_1 \mid\mid\mid put_2); get_2 \xrightarrow{get_1} (\mathbf{1} \mid\mid\mid put_2); get_2$$

of the implementation cannot be strictly bisimulated: the only possibility would be

$$put; get \xrightarrow{put} get \xrightarrow{get} \mathbf{1}$$

with residual sets $R_0 = \emptyset$ (initially), $R_1 = [put_2]$ (after the first step) and $R_2 = [put_2] \oplus [get_2]$ (after the second step). Since $R_2 \xrightarrow{get_2}$, strict residual simulation requires that the implementation can do get_2 already, which is not the case.

To improve on this, we formulate a second, weaker notion of down-simulation and residual simulation, which state that not *all* transitions of a refinement image, respectively *all* transitions of the residual have to be simulated, but rather *at least one*.

DEFINITION 8.6. Let $\mathcal{T}$ be a labelled transition system and assume $\rho \subseteq N \times N \times RS(r)$.

- ρ is a *lax down-simulation up to r* if for all $n_1 \rho^{\emptyset} n_2$ and $n_1 \xrightarrow{\alpha} n_1'$, one of the following holds:
  (1) $\alpha = \tau$ and $n_1' \rho^{\emptyset} n_2$;
  (2) $\alpha \in \{\tau, \checkmark\}$ and $\exists (n_2 \Rightarrow\xrightarrow{\alpha} n_2')$ such that $n_1' \rho^{\emptyset} n_2'$;
  (3) $\alpha \in Act$ and $\exists (r(\alpha) \xrightarrow{c} u')$: $\exists (n_2 \Rightarrow\xrightarrow{c} n_2')$ such that $n_1' \rho^{[u']} n_2'$.
- ρ is a *lax residual simulation* if for all $R \neq \emptyset$, if $n_1 \rho^R n_2$ then $\exists (R \xrightarrow{\gamma} R')$: $\exists (n_2 \Rightarrow\xrightarrow{\gamma} n_2')$ such that $n_1 \rho^{R'} n_2'$.
- ρ is a *lax vertical bisimulation up to r* if it is both a lax down-simulation up to r, an up-simulation up to r and a lax residual simulation.
- *Lax vertical bisimilarity up to r*, denoted $\precsim_{\exists}^{r}$, is the largest lax vertical bisimulation up to r, and *rooted lax vertical bisimilarity up to r*, denoted $\preceq_{\exists}^{r}$, is the largest delay root of $\precsim_{\exists}^{r,\emptyset}$.

Note that the $\forall$-quantifiers in Definitions 8.1 and 8.3 have been turned into $\exists$; this is precisely what makes $\preceq_{\exists}^{r}$ weaker than $\preceq_{\forall}^{r}$. Now we indeed have $put; get \preceq_{\exists}^{r} put_1; (get_1 \mid\mid\mid put_2); get_2$ (where $r = (put \to put_1; put_2, get \to get_1; get_2)$ as before); witness Figure 11. Due to the nature of our refinement functions, which are distinct and therefore cannot deadlock, the lax relation is indeed (strictly) weaker than the strong one.

PROPOSITION 8.7. *For all distinct r,* $\preceq_{\forall}^{r} \subseteq \preceq_{\exists}^{r}$.

8.3. *Requirements for vertical implementation*

Apart from our one motivating example, one would like to have objective criteria to decide between the strict and lax versions of vertical bisimulation; and indeed to know whether these relations can be made part of a larger framework for design, as discussed at the start

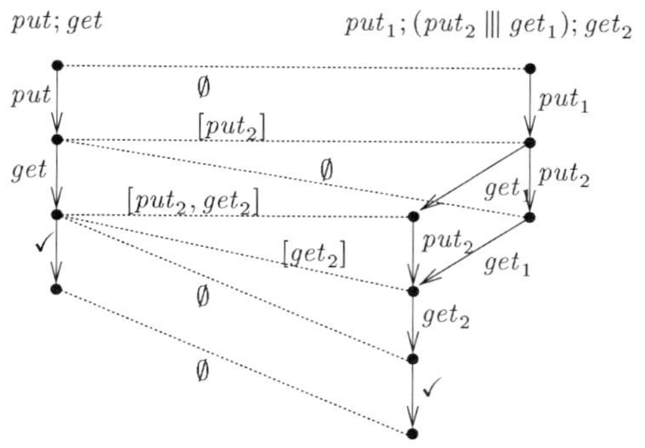

Fig. 11. A lax vertical bisimulation.

Table 15
Proof rules for vertical implementation

$\dfrac{}{t \sqsubseteq^{id} t}\,\mathrm{R}_1$	$\dfrac{t \sqsubseteq^{id} u}{t \sqsubseteq u}\,\mathrm{R}_2$	$\dfrac{t \sqsubseteq t' \quad t' \sqsubseteq^{r} u' \quad u' \sqsubseteq u}{t \sqsubseteq^{r} u}\,\mathrm{R}_3$
$\dfrac{}{\mathbf{0} \sqsubseteq^{r} \mathbf{0}}\,\mathrm{R}_4$ $\quad$ $\dfrac{}{\mathbf{1} \sqsubseteq^{r} \mathbf{1}}\,\mathrm{R}_5$	$\dfrac{}{\alpha \sqsubseteq^{r} r(\alpha)}\,\mathrm{R}_6$	$\dfrac{t_1 \sqsubseteq^{r} u_1 \quad t_2 \sqsubseteq^{r} u_2}{t_1 + t_2 \sqsubseteq^{r} u_1 + u_2}\,\mathrm{R}_7$
$\dfrac{t_1 \sqsubseteq^{r} u_1 \quad t_2 \sqsubseteq^{r} u_2}{t_1 ; t_2 \sqsubseteq^{r} u_1 ; u_2}\,\mathrm{R}_8$	$\dfrac{t \sqsubseteq^{r} u}{t/A \sqsubseteq^{r/A} u/\mathcal{S}_r(A)}\,\mathrm{R}_9$	$\dfrac{t_1 \sqsubseteq^{r} u_1 \quad t_2 \sqsubseteq^{r} u_2}{t_1 \parallel_A t_2 \sqsubseteq^{r} u_1 \parallel_{\mathcal{S}_r(A)} u_2}\,\mathrm{R}_{10}$
$\dfrac{r(a) = u_1 ; u_2 \quad t \sqsubseteq^{r} v}{a ; t \sqsubseteq^{r} u_1 ; (u_2 \parallel\!\mid v)}\,\mathrm{R}_{11}$	$\dfrac{r(a) = u_1 ; u_2 \quad t \sqsubseteq^{r} v_1 ; v_2}{a ; t \sqsubseteq^{r} u_1 ; (u_2 \parallel\!\mid v_1) ; v_2}\,\mathrm{R}_{12}$	$\dfrac{t \sqsubseteq^{r_1} u}{t \sqsubseteq^{r_1 + r_2} u}\,\mathrm{R}_{13}$

of this section. To answer these questions, we formulate a set of general proof rules for vertical implementation, and investigate to what degree either version of vertical bisimulation satisfies them.

The rules are collected in Table 15. Note that they feature a generic relation symbol $\sqsubseteq^{r}$ standing for any vertical implementation relation, and a generic symbol $\sqsubseteq$ standing for a horizontal implementation relation. In this context, $\sqsubseteq$ is called the *basis* of $\sqsubseteq^{r}$, and $\sqsubseteq^{r}$ a *vertical extension* of $\sqsubseteq$. We briefly discuss the rules.

- The first group of rules (R_1–R_3) expresses our basic assumption of working modulo the horizontal implementation relation. R_1 states that every term implements itself as long as no refinement takes place; R_2 says that $\sqsubseteq^{id}$ (where *id* is the identity refinement, mapping each action to itself) is compatible with the horizontal implementation relation; while R_2 explains the interplay between horizontal and vertical implementation. Note that, as a consequence, we also have the derived rule

$$\frac{t \sqsubseteq u}{t \sqsubseteq^{id} u}$$

that, in conjunction with R_2, ensures that $\sqsubseteq$ and $\sqsubseteq^{id}$ are indeed the same relation. Note also that R_1 and R_2 imply that $\sqsubseteq$ is reflexive, whereas R_1–R_3 together imply that $\sqsubseteq$ is transitive; hence $\sqsubseteq$ is a pre-order, which indeed is the standard requirement for horizontal implementation relations.

- The second group of rules (R_4–R_{10}) essentially express congruence of vertical implementation with respect to the operators of $Lang^{flat}$. For instance, if the refinement functions in these rules are set to *id*, then the properties expressed by the rules collapse to the standard pre-congruence properties of $\sqsubseteq$ for the operators of $Lang^{flat}$. (In other words, the horizontal implementation relation $\sqsubseteq$ needs to be a pre-congruence, at least.)
  Rules R_4 and R_5 simply express that deadlock and termination are independent of the abstraction level. R_6 is the core of the relationship between the refinement function r and the vertical implementation relation: it expresses the basic expectation that $r(a)$ should be an implementation for a. R_7–R_{10} are quite obvious, as they inductively go into the structure of the components. Note that in R_{10}, the synchronization set A of the specification is refined in the implementation.
  Rule R_9 is similar, with the proviso that the refinement images of the actions that are hidden is set to the identity. An interesting special case of R_9 is given by the following derived rule:

$$\frac{t \sqsubseteq^r u \quad A = \{a \mid r(a) \neq a\}}{t/A \sqsubseteq^{id} u/\mathcal{S}_r(A)}.$$

  Hence, by hiding all the actions that are properly refined, the vertical implementation is turned back into a horizontal implementation relation.
- The last group of rules (R_{11}–R_{13}) describes design steps that are different from the ones using the congruence-like rules of the previous group. In fact, these rules capture the gain over the traditional refinement operator: R_{11} and R_{12} allow certain parts of refinement images to overlap in the implementation, even if the specification imposes an ordering between the corresponding abstract actions, whereas R_{13} allows refinements to be implemented partially.

It seems natural to require that a vertical implementation relation satisfies at least rules R_1–R_{10}, and as many of the other rules as possible. In [124], we proved that this is the case for rooted vertical weak bisimulation; the proof carries over to rooted strict vertical bisimulation ($\preceq^r_\forall$).

THEOREM 8.8. *$\preceq^r_\forall$ satisfies the rules in Table* 15 *except* R_{12} *and* R_{13}.

Unfortunately, lax vertical bisimilarity (Definition 8.6) does not satisfy the parallel composition rule.

EXAMPLE 8.9. Consider $r : a \mapsto a_1 + a_2$; let $t = a$ and $u_i = a_i$ for $i = 1, 2$. We have $t \preceq^r_\exists u_1$ and $t \preceq^r_\exists u_2$; however, $a_1 \parallel_{a_1,a_2} a_2 \simeq_\mathsf{d} \mathbf{0}$ and hence $t \parallel_a t \simeq_\mathsf{d} a \not\preceq^r_\exists \mathbf{0} \simeq_\mathsf{d} u_1 \parallel_{\mathcal{S}_r(a)} u_2$.

The reason lies in the existential nature of the lax down-simulations (and also of the lax residual simulations, although the above example does not show that): the implementations of the parallel components both have to have *some* matching transition, but they need not be the same. However, this is the only proof rule that is problematic, and then only for synchronization actions.

THEOREM 8.10. *$\preceq_{\exists}^{r}$ satisfies the rules in Table 15, where R_{10} is replaced by the following special case*:

$$\frac{t_1 \sqsubseteq^r u_1 \quad t_2 \sqsubseteq^r u_2 \quad r/A = id}{t_1 \parallel_A t_2 \sqsubseteq^r u_1 \parallel_A u_2}.$$

The situation is not yet very satisfactory: for lax vertical implementation, the failure of R_{10} is a grave disadvantage, since synchronization is the basic interaction mechanism between system components: if one cannot synchronize on an action after its refinement, the formalism is not very useful. On the other hand, strong vertical implementation does not allow some design steps that seem intuitively reasonable and attractive – such as the one of our motivating example.

The situation is alleviated, however, by the fact that strong and lax refinement can be *combined* so as to merge the advantages of both to a large degree. Namely, it is possible to implement one operand of a synchronization using the lax vertical bisimulation, and the other using the strong version; the resulting implementation will still be a vertical implementation, in the lax version. Formally:

THEOREM 8.11. *Let $A \subseteq Act$ and assume $r/A = id$. If $t_1 \preceq_{\forall}^{r} u_1$ and $t_2 \preceq_{\exists}^{r} u_2$, then $t_1 \parallel_A t_2 \preceq_{\exists}^{r} u_1 \parallel_{\mathcal{S}_r(A)} u_2$.*

EXAMPLE 8.12. To see the possible interplay between strict and lax vertical implementation, consider a booking agent, whose function on an abstract level consists of a continuous series of *book* actions, and two customers that concurrently invoke this action. Algebraic specifications are given by

$$\begin{aligned} Agent_S &:= book; Agent_S, \\ Users_S &:= book \parallel\!\mid book. \end{aligned}$$

Now consider a refinement $r : book \mapsto req; (yes + no)$, specifying that a booking consists of a request, which may be granted (*yes*) or refused (*no*). A possible implementation that is lax for the agent and strict for the users is given by

$$\begin{aligned} Agent_I &:= req; (yes \parallel\!\mid req); Empty, \\ Empty &:= (no \parallel\!\mid req); Empty, \\ Users_I &:= req; (yes + no) \parallel\!\mid req; (yes + no). \end{aligned}$$

The agent will receive the request of a second customer in parallel with the reply to the first, but replies *yes* to the first one only. The customers take either yes or no for an answer. We have $Agent_S \preceq_{\exists}^{r} Agent_I$ and $Users_S \preceq_{\forall}^{r} Users_I$; thus, according to Theorem 8.11, $Agent_S \parallel_{book} Users_S \preceq_{\exists}^{r} Agent_I \parallel_{req,yes,no} Users_I$.

Informally, the reason why this works is that every (down- or residual) simulated transition that exists on the lax side exists on the strict side as well; hence they can be combined into a simulated transition of the synchronized behaviour. We arrive at the following schema regarding the synchronization of vertical implementations:

- Strict with strict yields strict;
- Strict with lax yields lax;
- Lax with lax does not yield a valid implementation.

8.4. *Further developments*

We briefly review two issues regarding vertical implementation, worked out in [124] for the strict case, that have been omitted from the presentation above.

Open terms and recursion. Table 15 contains no proof rule for recursion. Thus, for instance, we cannot prove the relation $Agent_S \preceq^r_\exists Agent_I$ in Example 8.12 algebraically. Intuitively, a proof rule would take the form

$$\frac{t \sqsubseteq^r u}{\mu x.t \sqsubseteq^r \mu x.u}.$$

However, here t and u are (in general) open terms; we have not discussed how to interpret vertical bisimulation over such. In fact, the most natural definition, which says that $t \sqsubseteq^r u$ iff all closed instances of t and u are related, is not satisfactory: for then the relation $x \sqsubseteq^r x$ would be inconsistent except for $r = id$. In fact, in $t \sqsubseteq^r u$, the occurrences of x in t (on the abstract side) should be interpreted differently from the occurrences of x in u (the concrete side). The solution developed for this in [124] (which we will not go into here) consists of adding an environment to the proof rules which records precisely these differences in interpretation. The above proof rule for recursion, extended appropriately with such environments, is sound for strict and lax vertical bisimilarity, provided we restrict ourselves to strongly guarded terms (meaning that any occurrence of the recursion variable in the body of a recursive term is guarded by a visible action).

Action substitution. When the rules in Table 15 are extended with a rule for recursion as discussed above, it can be seen that (simultaneous) action substitution in the line of Section 6 will automatically give rise to a correct implementation (with respect to any vertical implementation relation that satisfies all these rules). More precisely: let $t\{\vec{u}/\vec{a}\}$ denote the simultaneous substitution of all a_i-occurrences in t by u_i, defined through a straightforward generalization of Table 11. Then for any vertical implementation relation $\leqslant^{\vec{a}\to\vec{u}}$ and any term $t \in Lang^{flat}$, we have

$$t \leqslant^{\vec{a}\to\vec{u}} t\{\vec{u}/\vec{a}\}.$$

This provides a link to the interpretation of action refinement as an operator, since in Section 6 we have seen that this operator can in many cases also be mimicked by action substitution.

Table 16
Overview of language fragments and well-formedness conditions

Language		*Lang*	*seq*	*atom*	*Evt*	*ST*	D, lin	D, bra	*loc*	*flat*	*ref*
Section		4	2	3	4	5.4	7.1	7.3	7.5	8	8
$\mathbf{0}$		✓	✓	✓	✓	✓	✓		✓	✓	
$\mathbf{0}_A$								✓			
$\mathbf{1}$	1)	✓	✓	✓	✓	✓	✓		✓	✓	
α		✓	✓	✓		✓				✓	
${}_e\alpha$					✓						
α^-	1)					✓					
a							✓	✓	✓		✓
$\mathsf{V} + \mathsf{V}$		✓	✓	✓	✓ [2]	✓	✓ [3]	✓ [3]	✓	✓	✓ [4]
$\mathsf{T}; \mathsf{V}$		✓	✓	✓	✓ [2]	✓			✓	✓	✓ [4]
$\mathsf{T} \cdot \mathsf{T}$							✓	✓			
$\mathsf{T} \parallel_A \mathsf{T}$		✓			✓ [2]			✓		✓	
$\mathsf{T} \mid\mid\mid \mathsf{T}$				✓ [5]			✓				
$\mathsf{T} \parallel_{A,M} \mathsf{T}$						✓					
$\Pi\vec{\mathsf{T}}$									✓		
$\langle \mathsf{V} \rangle$				✓			✓ [3]				
$*\mathsf{T}$	1)			✓							
T/A		✓	✓	✓	✓	✓				✓	
$\mathsf{T}[a \to \mathsf{V}]$		✓ [6]	✓ [6]	✓ [6]			✓ [3][7]				
$\mathsf{T}[a \to \mathsf{V}, \vec{e} \to \vec{\mathsf{T}}]$					✓ [2]						
$\mathsf{T}[a \to \mathsf{V}, \vec{\mathsf{T}}]_M$						✓					
$\mathsf{T}[\vec{a} \to \vec{\mathsf{V}}]$								✓ [3][8]			
$\mathsf{T}[a \to \vec{\mathsf{V}}]$									✓		
x		✓	✓	✓		✓	✓	✓	✓	✓	
${}_e x$					✓						
${}_e[\mathsf{T}]$					✓						
$\mu x.\mathsf{V}$	9)	✓ [10]	✓ [10]	✓	✓	✓	✓ [3]	✓ [3] [11]	✓	✓	

Well-formedness conditions (conditions in the left hand column apply to all languages):

1) auxiliary operator; may not occur in virgin operands (V in the grammar);

2) event annotations are compatible; 3) operands are *not* virgin (i.e., may contain auxiliary operators);

4) terms are distinct (Section 8.1); 5) either operand is abstract (Section 3.1);

6) in Section 6, both operands are closed terms; 7) refinement is D-compatible (Definition 7.1);

8) refinement is strictly D-compatible (Definition 7.10); 9) x is guarded in V (Definition 2.1); in Section 6, $\mu x.\mathsf{V}$ is well-sorted;

10) not allowed in the fragments $Lang^{fin}$ and $Lang^{seq,fin}$; 11) x is *D-guarded* in V (Definition 7.8).

8.5. *Application: A simple data base*

Let us consider once more the data base example of Section 2.5 (not including the *copy* operation considered in Section 4.6). We will show that, under *lax* vertical refinement, the n-state version of the data base, $Data_S^n$, is an implementation of the state-less (or 1-state) version, $Data_S^1$. The specifications are

$$\begin{aligned} Data_S^1 &:= (qry + upd); Data_S^1, \\ Data_S^n &:= State_1, \\ State_i &:= qry_i; State_i + \sum_{k=1}^{n} upd_k; State_k \quad (\text{for } i = 1, \dots, n). \end{aligned}$$

Let r map qry to $\sum_{i=1}^{n} qry_i$ and upd to $\sum_{i=1}^{n} upd_i$; then it follows that $Data_S^1 \preceq_\exists^r Data_S^n$. In fact, this implementation relation can be proved algebraically, using the rules of Table 15 extended to cope with recursion as in [124] (briefly discussed above). The main point is that (in lax refinement) $qry \preceq_\exists^r qry_i$ for arbitrary i, i.e., it is not necessary to implement all qry-actions whenever qry is specified on the abstract level.

If we also consider a user of the data base who actually queries it, this can be modelled by a process $User_S$ synchronizing with $Data_S^1$ over qry. Since $Data_S^n$ is a lax vertical implementation of $Data_S^1$, in order to refine the interaction with the user, we have to implement $User_S$ in the *strong* sense, meaning that whenever $User_S$ specifies a query, this must be implemented as the choice between all possible qry_i-actions. In fact, this is a reasonable requirement: since the user does not know the state of the data base at the moment he issues the query, he must accept all possible answers.

In a sense, the above example is not a case of proper action refinement, since all r-images terminate after a single action. Alternatively, also the *refined* n-state database can be derived as an implementation of $Data_S^1$. For instance, for the case where $n = 2$, both the implementation $Data_I^{2,seq}$ obtained in Section 2.5 (see Figure 2) and $Data_I^{2,D}$ obtained in Section 7.4 (see Figure 10) are valid lax implementations of $Data_S^1$ under the refinement r

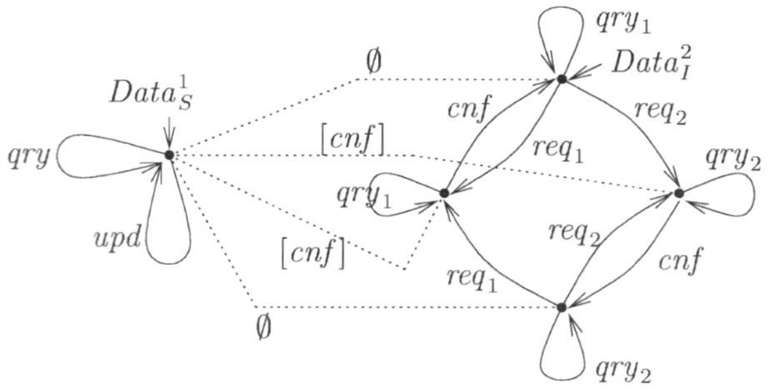

Fig. 12. Lax vertical implementation of a state-less as a 2-state data base, while refining *upd*.

defined by

$$upd \mapsto (req_1 + req_2); cnf,$$
$$qry \mapsto qry_1 + qry_2.$$

For instance, Figure 12 shows the lax vertical bisimulation proving $Data_S^1 \preceq_\exists^r Data_I^{2,D}$.

Acknowledgements

We wish to thank Mario Bravetti, Pierpaolo Degano, Thomas Firley, Walter Vogler and Heike Wehrheim for their comments on preliminary versions of this chapter.

References

[1] L. Aceto, *Full abstraction for series-parallel pomsets*, TAPSOFT '91, Vol. 1, Lecture Notes in Comput. Sci. 493, S. Abramsky and T.S.E. Maibaum, eds, Springer-Verlag (1991), 1–25.
[2] L. Aceto, *Action refinement in process algebras*, Ph.D. Thesis, University of Sussex, Cambridge University Press (1992).
[3] L. Aceto, *History-preserving, causal and mixed-ordering equivalence over stable event structures*, Fundamenta Informaticae **17** (4) (1992), 319–331.
[4] L. Aceto, B. Bloom and F.W. Vaandrager, *Turning SOS rules into equations*, Inform. and Comput. **111** (1) (1994), 1–52.
[5] L. Aceto and U. Engberg, *Failures semantics for a simple process language with refinement*, Foundations of Software Technology and Theoretical Computer Science, S. Biswas and K.V. Nori, eds, Lecture Notes in Comput. Sci. 590, Springer-Verlag (1991), 89–108.
[6] L. Aceto, W. Fokkink and C. Verhoef, *Structural operational semantics*, Handbook of Process Algebra, J.A. Bergstra, A. Ponse and S.A. Smolka, eds, Elsevier, Amsterdam (2001), 197–292.
[7] L. Aceto and M.C.B. Hennessy, *Termination, deadlock, and divergence*, J. ACM **39** (1) (1992), 147–187.
[8] L. Aceto and M.C.B. Hennessy, *Towards action-refinement in process algebras*, Inform. and Comput. **103** (1993), 204–269.
[9] L. Aceto and M.C.B. Hennessy, *Adding action refinement to a finite process algebra*, Inform. and Comput. **115** (1994), 179–247.
[10] R.J.R. Back, *A method for refining atomicity in parallel algorithms*, PARLE '89, Vol. I: Parallel Architectures, Lecture Notes in Comput. Sci. 366, E. Odijk, M. Rem and J.-C. Syre, eds, Springer-Verlag (1989), 199–216.
[11] J.C.M. Baeten and W.P. Weijland, *Process Algebra*, Cambridge University Press (1990).
[12] J.W. de Bakker, W.-P. de Roever and G. Rozenberg, eds, *Linear Time, Branching Time and Partial Order in Logics and Models for Concurrency*, Lecture Notes in Comput. Sci. 354, Springer-Verlag (1989).
[13] J.W. de Bakker and E.P. de Vink, *Bisimulation semantics for concurrency with atomicity and action refinement*, Fundamenta Informaticae **20** (1994), 3–34.
[14] H.P. Barendregt, *The Lambda Calculus*, 2nd edn., Studies in Logic and the Foundations of Math. 103, North-Holland, Amsterdam (1984).
[15] T. Basten, *Branching bisimilarity is an equivalence indeed!*, Inform. Process. Lett. **58** (1998), 141–147.
[16] J.A. Bergstra and J.W. Klop, *Process algebra for synchronous communication*, Inform. and Control **60** (1984), 109–137.
[17] E. Best, R. Devillers and J. Esparza, *General refinement and recursion operators for the Petri box calculus*, STACS-93, Lecture Notes in Comput. Sci. 665, P. Enjalbert, A. Finkel and K.W. Wagner, eds, Springer-Verlag (1993), 130–140.

[18] B. Bloom, S. Istrail and A.R. Meyer, *Bisimulation can't be traced*, J. ACM **42** (1) (1995), 232–268.

[19] G. Boudol, *Atomic actions*, Bull. Eur. Ass. Theoret. Comput. Sci. **38** (1989), 136–144.

[20] G. Boudol and I. Castellani, *On the semantics of concurrency: Partial orders and transition systems*, TAPSOFT '87, Vol. 1, Lecture Notes in Comput. Sci. 249, H. Ehrig, R. Kowalski, G. Levi and U. Montanari, eds, Springer-Verlag (1987), 123–137.

[21] G. Boudol and I. Castellani, *Flow models of distributed computations: Three equivalent semantics for CCS*, Inform. and Comput. **114** (2) (1994), 247–314.

[22] W. Brauer, R. Gold and W. Vogler, *A survey of behaviour and equivalence preserving refinements of Petri nets*, Advances in Petri Nets 1990, Lecture Notes in Comput. Sci. 483, G. Rozenberg, ed., Springer-Verlag (1990), 1–46.

[23] M. Bravetti, M. Bernardo and R. Gorrieri, *Towards performance evaluation with general distributions in process algebras*, Concur '98: Concurrency Theory, Lecture Notes in Comput. Sci. 1466, D. Sangiorgi and R. De Simone, eds, Springer-Verlag (1998), 405–422.

[24] M. Bravetti and R. Gorrieri, *Deciding and axiomatizing ST bisimulation for a process algebra with recursion and action refinement*, Technical Report UBLCS-99-1, Department of Computer Science, University of Bologna (February 1999).

[25] M. Bravetti and R. Gorrieri, *Interactive generalized semi-Markov processes*, Proc. of the 7th Int. Workshop on Process Algebras and Performance Modeling, J. Hillston and M. Silva, eds, Prensas Universitarias de Zaragoza (1999), 83–98. Url: http://www.cs.unibo.it/~bravetti/papers/papm99.ps.

[26] E. Brinksma, B. Jonsson and F. Orava, *Refining interfaces of communicating systems*, TAPSOFT '91, Vol. 2, Lecture Notes in Comput. Sci. 494, S. Abramsky and T.S.E. Maibaum, eds, Springer-Verlag (1991), 297–312.

[27] S.D. Brookes, C.A.R. Hoare and A.W. Roscoe, *A theory of communicating sequential processes*, J. ACM **31** (3) (1984), 560–599.

[28] M. Broy, *(Inter-)action refinement: The easy way*, Program Design Calculi, NATO ASI Series F: Computer and System Sci. 118, M. Broy, ed. (1993), 121–158.

[29] M. Broy, *Compositional refinement of interactive systems*, J. ACM **44** (6) (1997), 850–891.

[30] N. Busi, *Raffinamento di azioni in linguaggi concorrenti*, Master's Thesis, Università degli Studi di Bologna (1992). In Italian.

[31] N. Busi, R.J. van Glabbeek and R. Gorrieri, *Axiomatising ST bisimulation equivalence*, Programming Concepts, Methods and Calculi, IFIP Trans. A-56, E.-R. Olderog, ed., IFIP (1994), 169–188.

[32] L. Castellano, G. De Michelis and L. Pomello, *Concurrency vs. interleaving: An instructive example*, Bull. Eur. Ass. Theoret. Comput. Sci. **31** (1987), 12–15. Note.

[33] F. Cherief, *Contribution à la sémantique du parallélisme: Bisimulations pour le raffinement et le vrai parallélisme*, Ph.D. Thesis, University of Grenoble (1992). In French.

[34] F. Cherief and P. Schnoebelen, *τ-bisimulation and full abstraction for refinement of actions*, Inform. Process. Lett. **40** (1991), 219–222.

[35] W.R. Cleaveland, ed., *Concur '92*, Lecture Notes in Comput. Sci. 630, Springer-Verlag (1992).

[36] R. Costantini, *Abstraktion in ereignisbasierten Modellen verteilter Systeme*, Ph.D. Thesis, University of Hildesheim (1995). In German.

[37] J.P. Courtiat and D.E. Saïdouni, *Relating maximality-based semantics to action refinement in process algebras*, Formal Description Techniques VII, IFIP WG 6.1, D. Hogrefe and S. Leue, eds, Chapman-Hall (1994), 293–308.

[38] I. Czaja, R.J. van Glabbeek and U. Goltz, *Interleaving semantics and action refinement with atomic choice*, *Advances in Petri Nets 1992*, Lecture Notes in Comput. Sci. 609, G. Rozenberg, ed., Springer-Verlag (1992), 89–109.

[39] P. Darondeau and P. Degano, *Causal trees*, Automata, Languages and Programming, Lecture Notes in Comput. Sci. 372, G. Ausiello, M. Dezani-Ciancaglini and S. Ronchi Della Rocca, eds, Springer-Verlag (1989), 234–248.

[40] P. Darondeau and P. Degano, *Causal trees = interleaving + causality*, Semantics of Systems of Concurrent Processes, Lecture Notes in Comput. Sci. 469, I. Guessarian, ed., Springer-Verlag (1990), 239–255.

[41] P. Darondeau and P. Degano, *About semantic action refinement*, Fundamenta Informaticae **XIV** (1991), 221–234.

[42] P. Darondeau and P. Degano, *Refinement of actions in event structures and causal trees*, Theoret. Comput. Sci. **118** (1993), 21–48.
[43] R. De Nicola and M.C.B. Hennessy, *Testing equivalences for processes*, Theoret. Comput. Sci. **34** (1984), 83–133.
[44] R. De Simone, *Higher-level synchronizing devices in Meije-SCCS*, Theoret. Comput. Sci. **37** (1985), 245–267.
[45] P. Degano, R. De Nicola and U. Montanari, *On the consistency of 'truly concurrent' operational and denotational semantics*, Third Annual Symposium on Logic in Computer Science, IEEE, Computer Society Press (1988), 133–141.
[46] P. Degano, R. De Nicola and U. Montanari, *Partial orderings descriptions and observations of nondeterministic concurrent processes*, Linear Time, Branching Time and Partial Order in Logics and Models for Concurrency, Lecture Notes in Comput. Sci. 354, J.W. de Bakker, W.-P. de Roever and G. Rozenberg, eds, Springer-Verlag (1989), 438–466.
[47] P. Degano, R. De Nicola and U. Montanari, *A partial ordering semantics for CCS*, Theoret. Comput. Sci. **75** (1991), 223–262.
[48] P. Degano and R. Gorrieri, *Atomic refinement for process description languages*, Mathematical Foundations of Computer Science 1991, Lecture Notes in Comput. Sci. 520, A. Tarlecki, ed., Springer-Verlag (1991), 121–130. Extended abstract; full version: Technical Report 17-91, Hewlett-Packard Pisa Science Center.
[49] P. Degano and R. Gorrieri, *An operational definition of action refinement*, Technical Report TR-28/92, Università di Pisa (1992).
[50] P. Degano and R. Gorrieri, *A causal operational semantics of action refinement*, Inform. and Comput. **122** (1995), 97–119.
[51] P. Degano, R. Gorrieri and G. Rosolini, *A categorical view of process refinement*, Semantics: Foundations and Applications, Lecture Notes in Comput. Sci. 666, J.W. de Bakker, W.-P. de Roever and G. Rozenberg, eds, Springer-Verlag (1992), 138–153.
[52] R. Devillers, *Maximality preserving bisimulation*, Theoret. Comput. Sci. **102** (1992), 165–183.
[53] J.V. Echagüe Zappettini, *Sémantique des systèmes réactifs: Raffinement, bisimulations et sémantique opérationnelle structurée dans les systeèmes de transitions asynchrones*, Ph.D. Thesis, Institut National Polytechnique de Grenoble (1993). In French.
[54] R. Gerth, R. Kuiper and J. Segers, *Interface refinement in reactive systems*, Concur '92, Lecture Notes in Comput. Sci. 630, W.R. Cleaveland, ed., Springer-Verlag (1992), 77–93.
[55] J.L. Gischer, *The equational theory of pomsets*, Theoret. Comput. Sci. **61** (1988), 199–224.
[56] R.J. van Glabbeek, *Comparative concurrency semantics and refinement of actions*, Ph.D. Thesis, Free University of Amsterdam (1990).
[57] R.J. van Glabbeek, *The refinement theorem for ST-bisimulation semantics*, Programming Concepts and Methods, IFIP, North-Holland Publishing Company (1990).
[58] R.J. van Glabbeek, *The meaning of negative premises in transition system specifications II*, Automata, Languages and Programming, Lecture Notes in Comput. Sci. 1099, F. Meyer auf der Heide and B. Monien, eds, Springer-Verlag (1995), 502–513. Full report version: STAN-CS-TN-95-16, Department of Computer Science, Stanford University.
[59] R.J. van Glabbeek, *The linear time – branching time spectrum*. Chapter 1 of this volume. To be checked.
[60] R.J. van Glabbeek and U. Goltz, *Equivalence notions for concurrent systems and refinement of actions*, Mathematical Foundations of Computer Science 1989, Lecture Notes in Comput. Sci. 379, A. Kreczmar and G. Mirkowska, eds, Springer-Verlag (1989), 237–248. Revised and extended in [64].
[61] R.J. van Glabbeek and U. Goltz, *A deadlock-sensitive congruence for action refinement*, SFB-Bericht 342/23/90 A, Technische Universität München, Institut für Informatik (November 1990).
[62] R.J. van Glabbeek and U. Goltz, *Equivalences and refinement*, Semantics of Systems of Concurrent Processes, Lecture Notes in Comput. Sci. 469, I. Guessarian, ed., Springer-Verlag (1990), 309–333. Revised and extended in [64].
[63] R.J. van Glabbeek and U. Goltz, *Refinement of actions in causality based models*, Stepwise Refinement of Distributed Systems – Models, Formalisms, Correctness, Lecture Notes in Comput. Sci. 430, J.W. de Bakker, W.-P. de Roever and G. Rozenberg, eds, Springer-Verlag (1990), 267–300. Revised and extended in [64].

[64] R.J. van Glabbeek and U. Goltz, *Refinement of actions and equivalence notions for concurrent systems*, Hildesheimer Informatik-Bericht 6/98, University of Hildesheim (1998). To appear in: Acta Informatica.

[65] R.J. van Glabbeek and F.W. Vaandrager, *Petri Net models for algebraic theories of concurrency*, PARLE – Parallel Architectures and Languages Europe, Vol. II: Parallel Languages, Lecture Notes in Comput. Sci. 259, J.W. de Bakker, A.J. Nijman and P.C. Treleaven, eds, Springer-Verlag (1987), 224–242.

[66] R.J. van Glabbeek and F.W. Vaandrager, *The difference between splitting in n and $n+1$*, Inform. and Comput. **136** (2) (1997), 109–142.

[67] R.J. van Glabbeek and W.P. Weijland, *Refinement in branching time semantics*, Proc. AMAST Conference, Iowa State University (1989), 197–201.

[68] R.J. van Glabbeek and W.P. Weijland, *Branching time and abstraction in bisimulation semantics*, J. ACM **43** (3) (1996), 555–600.

[69] U. Goltz, R. Gorrieri and A. Rensink, *Comparing syntactic and semantic action refinement*, Inform. and Comput. **125** (2) (1996), 118–143.

[70] U. Goltz and H. Wehrheim, *Modelling causality via action dependencies in branching time semantics*, Inform. Process. Lett. **59** (1996), 179–184.

[71] R. Gorrieri, *Refinement, atomicity and transactions for process description languages*, Ph.D. Thesis, Report no. TD-2/91, Università di Pisa (1991).

[72] R. Gorrieri, *A hierarchy of system descriptions via atomic linear refinement*, Fundamenta Informaticae **16** (1992), 289–336.

[73] R. Gorrieri and C. Laneve, *The limit of $split_n$-bisimulations for CCS agents*, Mathematical Foundations of Computer Science 1991, Lecture Notes in Comput. Sci. 520, A. Tarlecki, ed., Springer-Verlag (1991), 170–180.

[74] R. Gorrieri and C. Laneve, *Split and ST bisimulation semantics*, Inform. and Comput. **116** (1) (1995), 272–288.

[75] R. Gorrieri, S. Marchetti and U. Montanari, *A^2CSS: Atomic actions for CCS*, Theoret. Comput. Sci. **72** (1990), 203–223.

[76] R. Gorrieri and U. Montanari, *Towards hierarchical specification of systems: A proof system for strong prefixing*, Internat. J. Found. Comput. Sci. **11** (3) (1990), 277–293.

[77] R. Gorrieri and U. Montanari, *On the implementation of concurrent calculi in net calculi: Two case studies*, Theoret. Comput. Sci. **141** (1995), 195–252.

[78] R. Gorrieri, M. Roccetti and E. Stancampiano, *A theory of processes with durational actions*, Theoret. Comput. Sci. **140** (1995), 73–94.

[79] J. Grabowski, *On partial languages*, Fundamenta Informaticae **IV** (2) (1981), 427–498.

[80] J.F. Groote, *Transition system specifications with negative premises*, Theoret. Comput. Sci. **118** (1993), 263–299.

[81] J.F. Groote and F.W. Vaandrager, *Structured operational semantics and bisimulation as a congruence*, Inform. and Comput. **100** (1992), 202–260.

[82] I. Guessarian, ed., *Semantics of Systems of Concurrent Processes*, Lecture Notes in Comput. Sci. 469, Springer-Verlag (1990).

[83] J.I. den Hartog, E.P. de Vink and J.W. de Bakker, *Full abstractness of an interleaving semantics for action refinement*, Rapport IR-443, Faculteit der Wiskunde en Informatica, Vrije Universiteit Amsterdam (February 1998).

[84] M.C.B. Hennessy, *Axiomatising finite concurrent processes*, SIAM J. Comput. **15** (5) (1988), 997–1017.

[85] M.C.B. Hennessy, *Concurrent testing of processes*, Acta Inform. **32** (6) (1995), 509–543.

[86] M.C.B. Hennessy and H. Lin, *Symbolic bisimulations*, Theoret. Comput. Sci. **138** (2) (1995), 353–389.

[87] M.C.B. Hennessy and R. Milner, *Algebraic laws for nondeterminism and concurrency*, J. ACM **23** (1) (1985), 137–161.

[88] C.A.R. Hoare, *Communicating Sequential Processes*, Prentice-Hall (1985).

[89] M. Huhn, *Action refinement and property inheritance in systems of sequential agents*, Concur '96: Concurrency Theory, Lecture Notes in Comput. Sci. 1119, U. Montanari and V. Sassone, eds, Springer-Verlag (1996), 639–654.

[90] M. Huhn, *On the hierarchical design of distributed systems*, Ph.D. Thesis, University of Hildesheim (1997).

[91] W. Janssen, M. Poel and J. Zwiers, *Action systems and action refinement in the development of parallel systems*, Concur '91, Lecture Notes in Comput. Sci. 527, J.C.M. Baeten and J.F. Groote, eds, Springer-Verlag (1991), 298–316.

[92] L. Jategaonkar, *Observing "true" concurrency*, Ph.D. Thesis, Massachusetts Institute of Technology (1993). Available as report MIT/LCS/TR-618.

[93] L. Jategaonkar and A.R. Meyer, *Testing equivalences for Petri nets with action refinement*, Concur '92, Lecture Notes in Comput. Sci. 630, W.R. Cleaveland, ed., Springer-Verlag (1992), 17–31.

[94] L. Jategaonkar and A.R. Meyer, *Self-synchronization of concurrent processes*, Proc. 8th Annual IEEE Symposium on Logic in Computer Science, IEEE, Computer Society Press (1993), 409–417.

[95] B. Jónsson, *Arithmetic of ordered sets*, Ordered Sets, I. Rival, ed., Reidel Pub. Co. (1982), 3–41.

[96] S. Katz, *Refinement with global equivalence proofs in temporal logic*, DIMACS Series in Discrete Math. and Theoret. Comput. Sci. **29** (1997), 59–78.

[97] P.M.W. Knijnenburg and J.N. Kok, *The semantics of the combination of atomized statements and parallel choice*, Formal Aspects of Computing **9** (5/6) (1997), 518–536.

[98] L. Lamport, *On interprocess communication, Part I: Basic formalisms*, Distributed Computing **1** (1986), 77–85.

[99] R. Langerak, *Transformations and semantics for LOTOS*, Ph.D. Thesis, University of Twente (November 1992).

[100] R. Langerak, *Bundle event structures: A non-interleaving semantics for LOTOS*, Formal Description Techniques – V, IFIP WG 6.1, M. Díaz and R. Groz, eds, North-Holland Publishing Company (1993), 203–218.

[101] K.S. Larsen, *A fully abstract model for a process algebra with refinements*, Master's Thesis, Aarhus University (1988).

[102] R. Loogen and U. Goltz, *Modelling nondeterministic concurrent processes with event structures*, Fundamenta Informaticae **XIV** (1991), 39–73.

[103] A. Mazurkiewicz, *Concurrent program schemes and their interpretations*, DAIMI Report PB-78, Aarhus University (1977).

[104] A. Mazurkiewicz, *Basic notions of trace theory*, Linear Time, Branching Time and Partial Order in Logics and Models for Concurrency, Lecture Notes in Comput. Sci. 354, J.W. de Bakker, W.-P. de Roever and G. Rozenberg, eds, Springer-Verlag (1989), 285–363.

[105] A. Meyer, *Observing truly concurrent processes*, Theoretical Aspects of Computer Software, Lecture Notes in Comput. Sci. 789, M. Hagiya and J.C. Mitchell, eds, Springer-Verlag (April 1994), 886.

[106] A. Meyer, *Concurrent process equivalences: Some decision problems*, STACS-95, Lecture Notes in Comput. Sci. 900, E.W. Mayr and C. Puech, eds, Springer-Verlag (1995), 349.

[107] R. Milner, *A modal characterisation of observable machine-behaviour*, CAAP '81, Lecture Notes in Comput. Sci. 112, E. Astesiano and C. Böhm, eds, Springer-Verlag (1981).

[108] R. Milner, *Communication and Concurrency*, Prentice-Hall (1989).

[109] F. Moller, *Axioms for concurrency*, Ph.D. Thesis, University of Edinburgh (1989). Available as report CST-59-89 and ECS-LFCS-89-84.

[110] M. Nielsen, U. Engberg and K.S. Larsen, *Fully abstract models for a process language with refinement*, Linear Time, Branching Time and Partial Order in Logics and Models for Concurrency, Lecture Notes in Comput. Sci. 354, J.W. de Bakker, W.-P. de Roever and G. Rozenberg, eds, Springer-Verlag (1989), 523–549.

[111] A. Obaid and L. Logrippo, *An atomic calculus of communicating systems*, Protocol Specification, Testing, and Verification, VII, IFIP WG 6.1, H. Rudin and C.H. West, eds, North-Holland Publishing Company (1987).

[112] E.-R. Olderog, ed., *Programming Concepts, Methods and Calculi*, IFIP Trans. A-56, IFIP (1994).

[113] E.-R. Olderog and C.A.R. Hoare, *Specification-oriented semantics for communicating processes*, Acta Inform. **23** (1986), 9–66.

[114] G.D. Plotkin, *A structural approach to operational semantics*, Technical Report DAIMI FN-19, Aarhus University (1981).

[115] V.R. Pratt, *Modeling concurrency with partial orders*, Internat. J. Parallel Programming **15** (1) (1986), 33–71.

[116] A. Rabinovich and B.A. Trakhtenbrot, *Behaviour structure and nets*, Fundamenta Informaticae **XI** (4) (1988), 357–404.

[117] A. Rensink, *Posets for configurations!*, Concur '92, Lecture Notes in Comput. Sci. 630, W.R. Cleaveland, ed., Springer-Verlag (1992), 269–285.

[118] A. Rensink, *Models and methods for action refinement*, Ph.D. Thesis, University of Twente, Enschede, Netherlands (August 1993).

[119] A. Rensink, *Methodological aspects of action refinement*, Programming Concepts, Methods and Calculi, IFIP Trans. A-56, E.-R. Olderog, ed., IFIP (1994), 227–246.

[120] A. Rensink, *Causal traces*, Hildesheimer Informatik-Bericht 39/95, Institut für Informatik, University of Hildesheim (December 1995).

[121] A. Rensink, *An event-based SOS for a language with refinement*, Structures in Concurrency Theory, Workshops in Computing, J. Desel, ed., Springer-Verlag (1995), 294–309.

[122] A. Rensink, *Algebra and theory of order-deterministic pomsets*, Notre Dame Journal of Formal Logic **37** (2) (1996), 283–320.

[123] A. Rensink and R. Gorrieri, *Action refinement as an implementation relation*, TAPSOFT '97: Theory and Practice of Software Development, Lecture Notes in Comput. Sci. 1214, M. Bidoit and M. Dauchet, eds, Springer-Verlag (1997), 772–786. Revised in [124].

[124] A. Rensink and R. Gorrieri, *Vertical bisimulation*, Hildesheimer Informatik-Bericht 9/98, University of Hildesheim (June 1998). To appear in: Inform. and Comput.

[125] A. Rensink and H. Wehrheim, *Weak sequential composition in process algebras*, Concur '94: Concurrency Theory, Lecture Notes in Comput. Sci. 836, B. Jonsson and J. Parrow, eds, Springer-Verlag (1994), 226–241.

[126] A. Rensink and H. Wehrheim, *Dependency-based action refinement*, Mathematical Foundations of Computer Science 1997, Lecture Notes in Comput. Sci. 1295, I. Prívara and P. Ruzicka, eds, Springer-Verlag (1997), 468–477. Extended report version: CTIT Technical Report 99-02, University of Twente. To appear in: Acta Inform.

[127] G. Rozenberg, ed., *Advances in Petri Nets 1992*, Lecture Notes in Comput. Sci. 609, Springer-Verlag (1992).

[128] D. E. Saïdouni, *Sémantique de maximalité: application au raffinement d'actions dans LOTOS*, Ph.D. Thesis, LAAS-CNRS, Toulouse (1996). In French.

[129] A. Tarlecki, ed., *Mathematical Foundations of Computer Science 1991*, Lecture Notes in Comput. Sci. 520, Springer-Verlag (1991).

[130] R. Valette, *Analysis of Petri Nets by stepwise refinement*, J. Comput. System. Sci. **18** (1979), 35–46.

[131] W. Vogler, *Failures semantics based on interval semiwords is a congruence for refinement*, Distribut. Comput. **4** (1991), 139–162.

[132] W. Vogler, *Modular Construction and Partial Order Semantics of Petri Nets*, Lecture Notes in Comput. Sci. 625, Springer-Verlag (1992).

[133] W. Vogler, *Bisimulation and action refinement*, Theoret. Comput. Sci. **114** (1993), 173–200.

[134] W. Vogler, *The limit of* $Split_n$*-language equivalence*, Inform. and Comput. **127** (1) (1996), 41–61.

[135] D. J. Walker, *Bisimulation and divergence*, Inform. and Comput. **85** (1990), 202–241.

[136] H. Wehrheim, *Parametric action refinement*, Programming Concepts, Methods and Calculi, IFIP Trans. A–56, E.-R. Olderog, ed., IFIP (1994), 247–266. Full report version: Hildesheimer Informatik-Berichte 18/93, Institut für Informatik, University of Hildesheim (November 1993).

[137] H. Wehrheim, *Specifying reactive systems with action dependencies*, Ph.D. Thesis, University of Hildesheim (1996).

[138] G. Winskel, *Synchronization trees*, Theoret. Comput. Sci. **34** (1984), 33–82.

[139] G. Winskel, *Event structures*, Petri Nets: Applications and Relationships to Other Models of Concurrency, Lecture Notes in Comput. Sci. 255, W. Brauer, W. Reisig and G. Rozenberg, eds, Springer-Verlag (1987), 325–392.

[140] G. Winskel, *An introduction to event structures*, Linear Time, Branching Time and Partial Order in Logics and Models for Concurrency, Lecture Notes in Comput. Sci. 354, J.W. de Bakker, W.-P. de Roever and G. Rozenberg, eds, Springer-Verlag (1989), 364–397.

[141] N. Wirth, *Program development by stepwise refinement*, Comm. ACM **14** (4) (1971), 221–227.

[142] J. Zwiers and W. Janssen, *Partial order based design of concurrent systems*, A Decade of Concurrency, Lecture Notes in Comput. Sci. 803, J.W. de Bakker, W.-P. de Roever and G. Rozenberg, eds, Springer-Verlag (1994), 622–684.

Subject index

Part 6
Tools and Applications

CHAPTER 17

Algebraic Process Verification

J.F. Groote[1], M.A. Reniers[2]

[1] *CWI, P.O. Box 94079, NL-1090 GB Amsterdam, The Netherlands, Faculty of Mathematics and Computing Science, Eindhoven University of Technology, P.O. Box 513, NL-5600 MB Eindhoven, The Netherlands*
E-mail: JanFriso.Groote@cwi.nl

[2] *Faculty of Mathematics and Computing Science, Eindhoven University of Technology, P.O. Box 513, NL-5600 MB Eindhoven, The Netherlands CWI, P.O. Box 94079, NL-1090 GB Amsterdam, The Netherlands*
E-mail: M.A.Reniers@tue.nl

Contents

HANDBOOK OF PROCESS ALGEBRA
Edited by Jan A. Bergstra, Alban Ponse and Scott A. Smolka

Abstract

This chapter addresses the question how to verify distributed and communicating systems in an effective way from an explicit process algebraic standpoint. This means that all calculations are based on the axioms and principles of the process algebras. The first step towards such verifications is to extend process algebra (ACP) with equational data types which adds required expressive power to describe distributed systems. Subsequently, linear process operators, invariants, the cones and foci method, the composition of many similar parallel processes, and the use of confluence are explained, as means to verify increasingly complex systems. As illustration, verifications of the serial line interface protocol (SLIP) and the IEEE 1394 tree identify protocol are included.

1. Introduction

The end of the seventies, beginning of the eighties showed the rise of process algebras such as CCS (Calculus of Communicating Systems) [38], CSP (Communicating Sequential Processes) [27], and ACP (Algebra of Communicating Processes) [7,8]. The basic motivation for the introduction of process algebras was the need to describe and study programs that are dynamically interacting with their environment [3,37]. Before this time the mathematical view on programs was that of deterministic input/output transformers: a program starts with some input, runs for a while, and if it terminates, yields the output. Such programs can be characterized by partial functions from the input to the output. This view is quite suitable for simple 'batch processing', but it is clearly inadequate for commonly used programs such as operating systems, control systems or even text editors. These programs are constantly obtaining information from the environment that is subsequently processed and communicated. The development of distributed computing, due to the widespread availability of computer networks and computing equipment, makes that proper means to study interacting systems are needed.

Process algebras allow for a rather high level view on interacting systems. They assume that we do not know the true nature of such systems. They just regard all such systems as *processes*, objects in some mathematical domain. A process is best viewed as some object describing all the potential behaviour a program or system can execute. We only assume that certain (uninterpreted) actions a, b, $c, \ldots$ are processes, and that we can combine processes using a few operators, such as the sequential composition operator or the parallel composition operator. A number of axioms restrict these operators, just to guarantee that they satisfy the basic intuitions about them. This basically constitutes a process algebra: a domain of processes, and a set of operators satisfying certain axioms. Unfortunately, axioms are not always sufficient, and more general 'principles' are employed. All these principles, however, adhere to the abstract view on processes.

There are many approaches in the literature that, contrary to the process algebra approach, study processes as concrete objects, such as failure traces [43], traces decorated with actions that cannot be executed at certain moments, Mazurkiewicz traces [36], which contain an explicit indication of parallelism, event structures [52], Petri nets [31,45], objects in metric spaces [11], etc. etc. A partial overview of process models is given in [49,50]. A very useful perspective, which we employ for illustrations, is the view of a process as an automaton of which the transitions are labelled with actions. Each traversal through the automaton is a run of the process. This view allows one to compactly depict the operational behaviour of a process.

In this chapter, we want to increase our understanding of processes by manipulating them, proving their properties, or proving that certain processes have the same behaviour. We stress again that we do this strictly from the abstract process algebraic perspective. This means that all our calculations in this chapter are based on the axioms and principles.

There are two major difficulties one runs into if one tries to do process algebraic verifications in this way, applied to more than just trivial examples, namely restricted expressivity and lack of effective proof methodologies.

The basic reason for the expressivity problem is that basic process algebras cannot explicitly deal with data. Often this problem is circumvented by annotating data in the subscripts of process variables. A consequence of this is that the number of process variables becomes large or infinite, which is less elegant. Furthermore, it is impossible to communicate data taken from infinite data domains. This generally is dealt with by considering only finite, but sufficiently large data domains. A true problem, however, is that for larger verifications the majority of the calculations tend to shift to the data. The role of data as second class citizens hinders its effective manipulation. This has a direct repercussion on the size and difficulty of the systems that can be handled.

The other problem is that the axioms and principles are very elementary. This means that although there are very many ways to prove some property of processes, finding a particular proof turns out to be an immense task. What is called for are *proof methodologies*, i.e., recipes and guidelines that lead in reasonable time to relatively short proofs.

We have addressed the first problem by extending one of the basic process algebras, with data. The result is μCRL (micro Common Representation Language). Basically, it is a minimal extension to ACP with equational abstract data types. Special care has been taken to keep the language sufficiently small, to allow study of the language itself, and sufficiently rich to precisely and effectively describe all sorts of protocols, distributed algorithms and, in general all communicating systems.

In μCRL process variables and actions can be parameterized with data. Data can influence the course of a process through a conditional (if-then-else)construct, and alternative quantification is added to express choices over infinite data sorts. Recently, the language has been extended with features to express time [23], but time is not addressed in this chapter. μCRL has been the basis for the development of a proof theory [20], and a toolset [10] allowing to simulate μCRL specifications and to perform all forms of finite state analysis on them. Using the toolset, it is even possible to do various forms of symbolic process manipulations, on the basis of the axioms, and currently this is an area under heavy investigation.

A lot of effort went into the specification and (manual) verification of various interactive systems [2,4,16,18,26,32–34]. When doing so, we developed a particular methodology of verification, culminating in the cones and foci technique [24], which enabled an increase in the order of magnitude of systems that could be analyzed. As we strictly clung to the basic axioms and principles of process algebra, it was relatively easy (but still time consuming) to check our proofs using proof checkers such as Coq [12], PVS [47] and Isabelle [41,42] (for an overview see [19]).

The first observation we ran into was that proving systems described in the full μCRL process syntax is inconvenient, despite the fact that the language is concise. Therefore, a normal form that is both sufficiently powerful to represent all systems denotable in μCRL and that is very straightforward was required. We took Linear Process Operators or Linear Process Equations as the normal form. This format resembles I/O automata [35], extended finite state machines [30] or Unity processes [9]. It is explained in Section 3.1.

An obvious verification problem that we often encounter is to prove an implementation adhering to its specification. We found this to be hard for basically the same reasons in all instances we studied. By equivalence we generally understand rooted branching bisimilarity. In this case the verification task is roughly that visible actions in the implementation should be matched by visible actions in corresponding states in the specification and vice versa. The difficulty is that often an action in the specification can only be matched in the

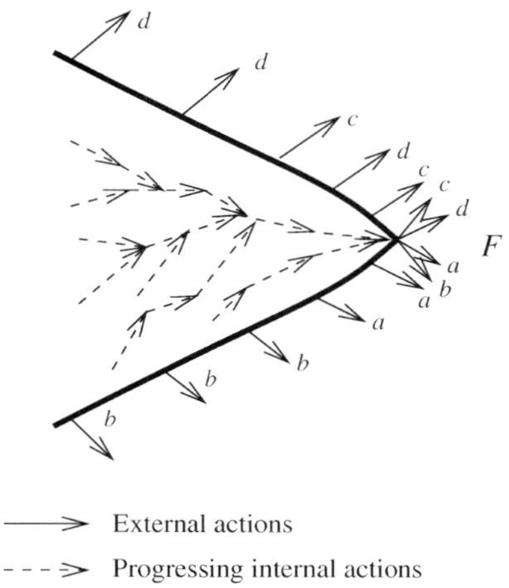

Fig. 1. A cone and a focus point.

implementation by first doing a large number of internal steps. The implementation contains many cone-like structures as sketched in Figure 1, where internal actions in the cone precede the external actions at the edge of the cone. The cones and foci method employs these structures, as summarized in the generalized equality theorem (see Section 3.3), reducing the proof of equality to a proof of properties of data parameters that are relatively easy to handle.

In this method the notion of invariant has been introduced. Despite the fact that invariants are the most important technical means to carry out sequential program verification, they were virtually absent in process algebra. Although it can be shown that formally invariants are not needed, in the sense that each process algebraic proof using invariants can be transformed into one without explicit use of this notion, we believe that invariants are important. The reason for this is that it allows to split a proof in on the one hand finding appropriate invariants, and on the other hand proving an equivalence or property.

Another difficulty is that one often needs to prove properties of distributed systems that consist of an arbitrary number, say n, similar processes. It turns out that calculating the parallel composition of these n processes with induction on n is cumbersome (see, e.g., [33]). The reason for this is that it requires to describe the behaviour of only a subset of the n processes. However, if one looks at the problem from a different angle, the parallel composition becomes a simple mechanical procedure. This is exactly the topic of Section 5.

Section 7 deals with confluence which is one of the most obvious structures occurring in distributed systems. In Milner's seminal book on process algebra [38], a full chapter was devoted to the subject. In Section 7 it is shown how confluence can be used to simplify the behaviour of a system considerably, after which it is much easier to understand and analyze it.

We feel slightly unsatisfied that only the topics mentioned above are addressed in this chapter. There is much more known and to be known about process verification. We could not include the large number of potentially effective techniques about which ideas exist, but which have not yet developed sufficiently in the process algebra context to be included. One may think about classifying distributed systems in several categories with similar structures, the use of symmetry in distributed systems, and the use of history and especially prophecy variables. Even if we would have attempted to include such ideas, we would soon be incomplete, as we feel that algebraic process verification is only at the brink of its development.

The next section starts with a thorough explanation of the language μCRL. We subsequently address the topics mentioned above. Interleaved with these we prove two distributed systems correct, as an illustration of the method.

The material presented in this chapter is based on a number of publications. Section 2 is based on [21]. The cones and foci method and the general equality theorem presented in Section 3 are taken from [24]. The verification of the SLIP protocol in Section 4 is slightly adapted from [19]. The linearization of a number of similar processes presented in Section 5 is taken from [22]. The example of the Tree Identify Protocol of IEEE 1394 (Section 6) is taken from [48]. Section 7 on confluence is based on [25].

2. Process algebra with data: μCRL

In this section we describe μCRL, which is a process algebra with data. The process algebra μCRL is used in different contexts and for different purposes. On the one hand it is used as a formal specification language with a strict syntax and (static) semantics. As such it can be used as input for a formal analysis toolset. On the other hand it is a mathematical notation, with the flexibility to omit obvious definitions, to only sketch less relevant parts, introduce convenient ad hoc notation, etc. In this section we stick quite closely to μCRL as a formal language. In the subsequent sections we are much less strict, and take a more mathematical approach.

2.1. *Describing data types in μCRL*

In μCRL there is a simple, yet powerful mechanism for specifying data. We use equationally specified abstract data types with an explicit distinction between constructor functions and 'normal' functions. The advantage of having such a simple language is that it can easily be explained and formally defined. Moreover, all properties of a data type must be defined explicitly, and henceforth it is clear which assumptions can be used when proving properties of data or processes. A disadvantage is of course that even the simplest data types must be specified each time, and that there are no high level constructs that allow compact specification of complex data types. Still, thus far these shortcomings have not outweighed the advantage of the simplicity of the language.

Each data type is declared using the keyword **sort**. Therefore, a data type is also called a data sort. Each declared sort represents a non-empty set of data elements. Declaring the sort of the Booleans is simply done by:

sort **Bool**

Table 1
Basic axioms for **Bool**

Bool1	$\neg(\mathsf{t} = \mathsf{f})$
Bool2	$\neg(b = \mathsf{t}) \rightarrow b = \mathsf{f}$

Because Booleans are used in the if-then-else construct in the process language, the sort **Bool** must be declared in every μCRL specification.

Elements of a data type are declared by using the keywords **func** and **map**. Using **func** one can declare all elements in a data type defining the structure of the data type. For example, by

$$\begin{array}{ll}\textbf{sort} & \textbf{Bool}\\ \textbf{func} & \mathsf{t}, \mathsf{f} :\rightarrow \textbf{Bool}\end{array}$$

one declares that t (true) and f (false) are the only elements of sort **Bool**. We say that t and f are the constructors of sort **Bool**. Not only the sort **Bool**, but also its elements t and f must be declared in every specification; they must be distinct, and the only elements of **Bool**. This is expressed in axioms Bool1 and Bool2 in Table 1. In axiom Bool2 and elsewhere we use a variable b that can only be instantiated with data terms of sort **Bool**. If in a specification t and f can be proven equal, for instance if the specification contains an equation $\mathsf{t} = \mathsf{f}$, we say that the specification is inconsistent and it looses any meaning. We often write ϕ and $\neg\phi$ instead of $\phi = \mathsf{t}$ and $\phi = \mathsf{f}$, respectively.

It is now easy to declare the natural numbers using the constructors 0 and successor S.

$$\begin{array}{ll}\textbf{sort} & Nat\\ \textbf{func} & 0 :\rightarrow Nat\\ & S : Nat \rightarrow Nat\end{array}$$

This says that each natural number can be written as 0 or the result of a number of applications of the successor function to 0.

If a sort D is declared without any constructor with target sort D, then it is assumed that D may be arbitrarily large. In particular D may contain elements that cannot be denoted by terms. This can be extremely useful, for instance when defining a data transfer protocol, that can transfer data elements from an arbitrary domain D. In such a case it suffices to declare in μCRL:

$$\textbf{sort} \quad D$$

The keyword **map** is used to declare additional functions for a domain of which the structure is already given. For instance declaring a function $\wedge$ on the Booleans, or, declaring the $+$ on natural numbers, can be done by adding the following lines to a specification in which *Nat* and **Bool** have already been declared:

$$\begin{array}{ll}\textbf{map} & \wedge : \textbf{Bool} \times \textbf{Bool} \to \textbf{Bool} \\ & + : Nat \times Nat \to Nat\end{array}$$

By adding plain equations between terms assumptions about the functions can be added. For the two functions declared above, we could add the equations:

$$\begin{array}{ll}\textbf{var} & x : \textbf{Bool} \\ & n, n' : Nat \\ \textbf{rew} & x \wedge \mathsf{t} = x \\ & x \wedge \mathsf{f} = \mathsf{f} \\ & n + 0 = n \\ & n + S(n') = S(n + n')\end{array}$$

Note that before each group of equations starting with the keyword **rew** we must declare the variables that are used.

The machine readable syntax of μCRL only allows prefix notation for functions, but we use infix or even postfix notation, if we believe that this increases readability. Moreover, we use common mathematical symbols such as $\wedge$ and $+$ in data terms, which are also not allowed by the syntax of μCRL, for the same reason.

Functions may be overloaded, as long as every term has a unique sort. This means that the name of the function together with the sort of its arguments must be unique. For example, it is possible to declare $max : Nat \times Nat \to Nat$ and $max : \textbf{Bool} \times \textbf{Bool} \to \textbf{Bool}$, but it is not allowed to declare a function $f : \textbf{Bool} \to \textbf{Bool}$ and $f : \textbf{Bool} \to Nat$. Actually, the overloading rule holds in general in μCRL. The restrictions on declarations are such that every term is either an action, a process name or a data term, and if it is a data term, it has a unique sort.

Although we have that every term of a data sort equals a term that is only built from the constructor functions this does not mean that we always know which constructor term this will be. For example, if we introduce an additional function 2 for the sort *Nat* by means of the declaration

$$\textbf{map} \quad 2 : \to Nat$$

this does not give us which constructor term equals the constant 2. This information can be added explicitly by adding an equation

$$\textbf{rew} \quad 2 = S\big(S(0)\big)$$

When we declare a sort D, it must be nonempty. Therefore, the following declaration is invalid.

$$\begin{array}{ll} \textbf{sort} & D \\ \textbf{func} & f : D \to D \end{array}$$

It declares that D is a domain in which all the terms have the form $f(f(f(\ldots)))$, i.e., an infinite number of applications of f. Such terms do not exist, and therefore D must be empty. This problem can also occur with more than one sort. For example, sorts D and E with constructors from D to E and E to D. Fortunately, it is easy to detect such problems and therefore it is a static semantic constraint that such empty sorts are not allowed (see [21]).

In proving the equality of data terms we can use the axioms, induction on the constructor functions of the data types and all deduction rules of equational logic. An abstract data type can be used to prove elementary properties. We explain here how we can prove data terms equal with induction, and we also show how we can prove data terms to be nonequal.

An easy and very convenient axiom is Bool2. It says that if a Boolean term b is not equal to t, it must be equal to f or in other words that there are at most two Boolean values. Applying this axiom boils down to a case distinction, proving a statement for the values t and f, and concluding that the property must then universally hold. We refer to this style of proof by the phrase 'induction on Booleans'.

A typical example is the proof of $b \wedge b = b$. Using induction on **Bool**, it suffices to prove that this equality holds for $b = \mathsf{t}$ and $b = \mathsf{f}$. In other words, we must show that $\mathsf{t} \wedge \mathsf{t} = \mathsf{t}$ and $\mathsf{f} \wedge \mathsf{f} = \mathsf{f}$. These are trivial instances of the defining axioms for $\wedge$ listed above.

Note that the sort **Bool** is the only sort for which we explicitly state that the constructors t and f are different. For other sorts, like *Nat*, there are no such axioms.

The division between constructors and mappings gives us general induction principles. If a sort D is declared with a number of constructors, then we may assume that every term of sort D can be written as the application of a constructor to a number of arguments chosen from the respective argument sorts. So, if we want to prove a property $p(d)$ for all d of sort D, we only need to provide proofs for $p(c_n(d_1, \ldots, d_n))$ for each n-ary constructor $c_n : S_1 \times \cdots \times S_n \to D$ and each d_i a term of sort S_i. If any of the arguments of c_n, say argument d_j, is of sort D then, as d_j is smaller than d, we may use that $p(d_j)$. If we apply this line of argumentation, we say we apply induction on D.

Suppose we have declared the natural numbers with constructors zero and successor, as done above. We can for instance derive that $0 + n = n$ for all n. We apply induction on *Nat*. First, we must show that $0 + 0 = 0$, considering the case where $n = 0$. This is a trivial instance of the first axiom on addition. Secondly, we must show $0 + S(n') = S(n')$, assuming that n has the form $S(n')$. In this case we may assume that the property to be proven holds already for n', i.e., $0 + n' = n'$. Then we obtain:

$$0 + S(n') = S(0 + n') = S(n').$$

As an example, we define a sort *Queue* on an arbitrary non-empty domain D, with an empty queue $[\,]$, and *in* to insert an element of D into the queue. The arbitrary non-empty domain is obtained by the specification of sort D without constructors.

$$\begin{array}{ll}
\textbf{sort} & D, Queue \\
\textbf{func} & [\,] : \to Queue \\
 & in : D \times Queue \to Queue
\end{array}$$

We extend this with auxiliary definitions *toe* to get the first element from a queue, *untoe* to remove the first element from a queue, *isempty* to check whether a queue is empty and $++$ to concatenate two queues.

$$\begin{array}{ll}
\textbf{map} & toe : Queue \to D \\
 & untoe : Queue \to Queue \\
 & isempty : Queue \to \mathbf{Bool} \\
\textbf{var} & d, d' : D \\
 & q, q' : Queue \\
\textbf{rew} & toe\big(in(d, [\,])\big) = d \\
 & toe\big(in\big(d, in(d', q)\big)\big) = toe\big(in(d', q)\big) \\
 & untoe\big(in(d, [\,])\big) = [\,] \\
 & untoe\big(in\big(d, in(d', q)\big)\big) = in\big(d, untoe\big(in(d', q)\big)\big) \\
 & isempty([\,]) = \mathsf{t} \\
 & isempty\big(in(d, q)\big) = \mathsf{f} \\
 & [\,] {+}{+} q = q \\
 & in(d, q) {+}{+} q' = in(d, q {+}{+} q')
\end{array}$$

A queue q_1 from which the last element has been removed can be given by $untoe(q_1)$ and a queue q_2 into which the last element of q_1 has been inserted is given by $in(toe(q_1), q_2)$. Now we prove

$$\neg isempty(q_1) \to untoe(q_1) {+}{+} in\big(toe(q_1), q_2\big) = q_1 {+}{+} q_2.$$

Suppose that $\neg isempty(q_1)$. We prove the proposition by induction on the structure of queue q_1.

Base. Suppose that $q_1 = [\,]$. Then $isempty([\,]) = \mathsf{t}$, which contradicts the assumption that $\neg isempty(q_1)$.

Induction step. Suppose that $q_1 = in(d, q_1')$ for some $d{:}D$ and $q_1'{:}Queue$. By induction we have $\neg isempty(q_1') \to untoe(q_1') {+}{+} in(toe(q_1'), q_2) = q_1' {+}{+} q_2$. Then we can distinguish the following two cases for q_1':

- $q_1' = [\,]$. In this case we have

$$\begin{aligned}
& untoe(q_1) {+}{+} in\big(toe(q_1), q_2\big) \\
& \quad = untoe\big(in(d, [\,])\big) {+}{+} in\big(toe\big(in(d, [\,])\big), q_2\big)
\end{aligned}$$

$$= [\,]{+}{+}in(d, q_2)$$
$$= in(d, q_2)$$
$$= in(d, [\,]{+}{+}q_2)$$
$$= in(d, [\,]){+}{+}q_2$$
$$= q_1{+}{+}q_2.$$

- $q_1' = in(d', q_1'')$. In this case we have

$$\begin{aligned} &untoe(q_1){+}{+}in\big(toe(q_1), q_2\big)\\ &\quad = untoe\big(in\big(d, in(d', q_1'')\big)\big){+}{+}in\big(toe\big(in\big(d, in(d', q_1'')\big)\big), q_2\big)\\ &\quad = in\big(d, untoe\big(in(d', q_1'')\big)\big){+}{+}in\big(toe\big(in(d', q_1'')\big), q_2\big)\\ &\quad = in\big(d, untoe(q_1'){+}{+}in\big(toe(q_1'), q_2\big)\big)\\ &\quad = in(d, q_1'{+}{+}q_2)\\ &\quad = in(d, q_1'){+}{+}q_2\\ &\quad = q_1{+}{+}q_2. \end{aligned}$$

Note that we used that $untoe(q_1'){+}{+}in(toe(q_1'), q_2) = q_1'{+}{+}q_2$. This is allowed as we can derive that $isempty(q_1') = isempty(in(d', q_1'')) = \mathsf{f}$.

Using the previous proposition we can easily prove that

$$\neg isempty(q) \to untoe(q){+}{+}in\big(toe(q), [\,]\big) = q$$

for all q:*Queue*. For if we take $q_1 = q$ and $q_2 = [\,]$ we obtain:

$$\neg isempty(q) \to untoe(q){+}{+}in\big(toe(q), [\,]\big) = q{+}{+}[\,].$$

Assuming that we can prove $q{+}{+}[\,] = q$, it is not hard to see that we thus have obtained $\neg isempty(q) \to untoe(q){+}{+}in(toe(q), [\,]) = q$. The property $q{+}{+}[\,] = q$ for all q:*Queue* can be proven with induction on the structure of q:

Base. Suppose that $q = [\,]$. Clearly $q{+}{+}[\,] = [\,]{+}{+}[\,] = [\,] = q$.

Induction step. Suppose that $q = in(d, q')$. By induction we have $q'{+}{+}[\,] = q'$. Then $q{+}{+}[\,] = in(d, q'){+}{+}[\,] = in(d, q'{+}{+}[\,]) = in(d, q') = q$.

In μCRL it is possible to establish when two data terms are not equal. This is for instance required in order to establish that two processes cannot communicate. There is a characteristic way of proving that terms are not equal, namely by assuming that they are equal, and showing that this implies $\mathsf{t} = \mathsf{f}$, contradicting axiom Bool1.

We give an example showing that the natural numbers zero (0) and one ($S(0)$) are not equal. We assume that the natural numbers with a 0 and successor function S are appropriately declared. In order to show zero and one different, we need a function that relates *Nat* to **Bool**. Note that if there is no such function, there are models of the data type *Nat* where zero and one are equal. For our function we choose 'less than or equal to', notation $\leqslant$, on the natural numbers, defined as follows:

$$\begin{array}{ll}\textbf{map} & \leqslant : Nat \times Nat \to \textbf{Bool} \\ \textbf{var} & n, m : Nat \\ \textbf{rew} & 0 \leqslant n = \mathsf{t} \\ & S(n) \leqslant 0 = \mathsf{f} \\ & S(n) \leqslant S(m) = n \leqslant m \end{array}$$

Now assume $0 = S(0)$. Clearly, $0 \leqslant 0 = \mathsf{t}$. On the other hand, using the assumption, we also find $0 \leqslant 0 = S(0) \leqslant 0 = \mathsf{f}$. So, we can prove $\mathsf{t} = \mathsf{f}$. Hence, we may conclude $0 \neq S(0)$.

This finishes the most important aspects of the data types. There are several standard libraries available [51,40] of which some also contain numerous provable identities. The general theory about abstract data types is huge, see for instance [14].

2.2. *Describing processes in μCRL*

2.2.1. *Actions.* Actions are abstract representations of events in the real world that is being described. For instance sending the number 3 can be described by $send(3)$ and boiling food can be described by *boil*(*food*) where 3 and *food* are terms declared by a data type specification. An action consists of an action name possibly followed by one or more data terms within brackets. Actions are declared using the keyword **act** followed by an action name and the sorts of the data with which it is parameterized. Below, we declare the action name *timeout* without parameters, an action a that is parameterized with Booleans, and an action b that is parameterized with pairs of natural numbers and data elements. The set of all action names that are declared in a μCRL specification is denoted by Act.

$$\begin{array}{ll}\textbf{act} & timeout \\ & a : \textbf{Bool} \\ & b : Nat \times D \end{array}$$

In accordance with mainstream process algebras, actions in μCRL are considered to be atomic. If an event has a certain positive duration, such as boiling food, then it is most appropriate to consider the action as the beginning of the event. If the duration of the event is important, separate actions for the beginning and termination of the event can be used.

In the tables with axioms we use the letters a and a' for action names, and in order to be concise, we give each action a single argument, although in μCRL these actions may have zero or more than one argument. The letter c is used for actions with an argument, and for the constants δ and τ, which are explained in Section 2.2.3 and Section 2.2.8 respectively.

2.2.2. *Alternative and sequential composition.* The two elementary operators for the construction of processes are the *sequential composition* operator, written as $p \cdot q$ and the *alternative composition* operator, written as $p + q$. The process $p \cdot q$ first performs the

Table 2
Basic axioms for μCRL

A1	$x + y = y + x$
A2	$x + (y + z) = (x + y) + z$
A3	$x + x = x$
A4	$(x + y) \cdot z = x \cdot z + y \cdot z$
A5	$(x \cdot y) \cdot z = x \cdot (y \cdot z)$

actions of p, until p terminates, and then continues with the actions in q. It is common to omit the sequential composition operator in process expressions. The process $p + q$ behaves like p or q, depending on which of the two performs the first action. Using the actions declared above, we can describe that $a(3, d)$ must be performed, except if a time out occurs, in which case $a(\mathrm{t})$ must happen.

$$a(3, d) + \mathit{timeout} \cdot a(\mathrm{t})$$

Observe that the sequential composition operator binds stronger than the alternative composition operator.

In Table 2 axioms A1–A5 are listed describing the elementary properties of the sequential and alternative composition operators. For instance, the axioms A1, A2 and A3 express that $+$ is commutative, associative and idempotent. In these and other axioms we use variables x, y and z that can be instantiated by process terms.

For processes we use the shorthand $x \subseteq y$ for $x + y = y$ and we write $x \supseteq y$ for $y \subseteq x$. This notation is called *summand inclusion*. It is possible to divide the proof of an equality into proving two inclusions, as the following lemma shows.

LEMMA 2.1. *For arbitrary μCRL-terms x and y we have: if $x \subseteq y$ and $y \subseteq x$, then $x = y$.*

PROOF. Suppose $x \subseteq y$ and $y \subseteq x$. By definition we thus have
(1) $x + y = y$, and
(2) $y + x = x$.
Thus we obtain: $x \overset{(2)}{=} y + x \overset{\mathrm{A1}}{=} x + y \overset{(1)}{=} y$. □

2.2.3. *Deadlock.* The language μCRL contains a constant δ, expressing that no action can be performed, for instance in case a number of computers are waiting for each other, and henceforth not performing anything. This constant is called *deadlock*. A typical property for δ is $p + \delta = p$; the choice in $p + q$ is determined by the first action performed by either p or q, and therefore one can never choose for δ. In other words, as long as there are alternatives deadlock is avoided. In Table 3 the axioms A6 and A7 characterize the main properties of δ.

2.2.4. *Process declarations.* Process expressions are expressions built upon actions indicating the order in which the actions may happen. In other words, a process expression represents the potential behaviour of a certain system.

Table 3
Axioms for deadlock

A6	$x + \delta = x$
A7	$\delta \cdot x = \delta$

In a μCRL specification process expressions appear at two places. First, there can be a single occurrence of an initial declaration, of the form

init p

where p is a process expression indicating the initial behaviour of the system that is being described. The **init** section may be omitted, in which case the initial behaviour of the system is left unspecified.

The other place where process expressions may occur are in the right hand side of process declarations, which have the form:

proc $X(x_1 : s_1, \ldots, x_n : s_n) = p$

Here X is the process name, the x_i are variables, not clashing with the name of a constant function or a parameterless process or action name, and the s_i are sort names. In this rule, process $X(x_1, \ldots, x_n)$ is declared to have the same behaviour as the process expression p.

The equation in a process declaration must be considered as an equation in the ordinary mathematical sense. This means that in a declaration such as the one above an occurrence of $X(u_1, \ldots, u_n)$ may be replaced by $p(u_1/x_1, \ldots, u_n/x_n)$, or vice versa, $p(u_1/x_1, \ldots, u_n/x_n)$ may be replaced by $X(u_1, \ldots, u_n)$.

An example of a process declaration is the following clock process which repeatedly performs the action *tick* and displays the current time. In this example and also in later examples we assume the existence of a sort *Nat* with additional operators which represents the natural numbers. We simply write 1 instead of $S(0)$, 2 instead of $S(S(0))$, etc. Furthermore, we assume that the standard functions on naturals are defined properly. Examples of such functions are $+$, $\leqslant$, $<$, $>$, etc.

act *tick*
display : *Nat*
proc $Clock(t{:}Nat) = tick \cdot Clock(t+1) + display(t) \cdot Clock(t)$
init $Clock(0)$

2.2.5. *Conditionals.* The process expression $p \lhd b \rhd q$ where p and q are process expressions, and b is a data term of sort **Bool**, behaves like p if b is equal to t (true) and behaves like q if b is equal to f (false). This operator is called the *conditional operator*, and operates precisely as an *then_if_else* construct. Through the conditional operator data influences process behaviour. For instance a counter, that counts the number of a actions that

Table 4
Axioms for conditionals

C1	$x \lhd \mathsf{t} \rhd y = x$
C2	$x \lhd \mathsf{f} \rhd y = y$

Table 5
Axioms for alternative quantification

SUM1	$\sum_{d:D} x = x$
SUM3	$\sum X = \sum X + Xd$
SUM4	$\sum_{d:D}(Xd + Yd) = \sum X + \sum Y$
SUM5	$(\sum X) \cdot x = \sum_{d:D}(Xd \cdot x)$
SUM11	$(\forall_{d:D}\ (Xd = Yd)) \to \sum X = \sum Y$

occur, issuing a b action and resetting the internal counter after 10 a's, can be described by:

$$\textbf{proc}\quad \mathit{Counter}(n{:}\mathit{Nat}) = a \cdot \mathit{Counter}(n+1) \lhd n < 10 \rhd b \cdot \mathit{Counter}(0)$$

The conditional operator is characterized by the axioms C1 and C2 in Table 4. All the properties of conditionals that we need are provable from these axioms and Bool1, Bool2. The conditional operator binds stronger than the alternative composition operator and weaker than the sequential composition operator.

LEMMA 2.2. *The following identities hold for arbitrary μCRL-terms x, y, z and for arbitrary Boolean terms b, b_1, b_2.*

(1) $x \lhd b \rhd y = x \lhd b \rhd \delta + y \lhd \neg b \rhd \delta$;
(2) $x \lhd b_1 \vee b_2 \rhd \delta = x \lhd b_1 \rhd \delta + x \lhd b_2 \rhd \delta$;
(3) $(b = \mathsf{t} \to x = y) \to x \lhd b \rhd z = y \lhd b \rhd z$.

2.2.6. *Alternative quantification.* The *sum operator* or *alternative quantification* $\sum_{d:D} P(d)$ behaves like $P(d_1) + P(d_2) + \cdots$, i.e. as the possibly infinite choice between $P(d_i)$ for any data term d_i taken from D. This is generally used to describe a process that is reading some input. E.g. in the following example we describe a single-place buffer, repeatedly reading a natural number n using action name r, and then delivering that value via action name s.

$$\textbf{proc}\quad \mathit{Buffer} = \sum_{n:\mathit{Nat}} r(n) \cdot s(n) \cdot \mathit{Buffer}$$

Note that alternative quantification binds stronger than the alternative composition operator and weaker than the conditional operator.

In Table 5 the axioms for the sum operator are listed. The sum operator $\sum_{d:D} p$ is a difficult operator, because it acts as a binder just like the lambda in the lambda calculus (see, e.g., [1]). This introduces a range of intricacies with substitutions. In order to avoid having to deal with these explicitly, we allow the use of explicit lambda operators and variables representing functions from data to process expressions.

In the tables the variables x, y and z may be instantiated with process expressions and the capital variables X and Y can be instantiated with functions from some data sort to process expressions. The sum operator $\sum$ expects a function from a data sort to a process expression, whereas $\sum_{d:D}$ expects a process expression. Moreover, we take $\sum_{d:D} p$ and $\sum \lambda d{:}D.p$ to be equivalent.

When we substitute a process expression p for a variable x or a function $\lambda d{:}D.p$ for a variable X in the scope of a number of sum operators, no variable in p may become bound by any of these sum operators. So, we may not substitute the action $a(d)$ for x in the left hand side of SUM1 in Table 4, because this would cause d to become bound by the sum operator. So, SUM1 is a concise way of saying that if d does not appear in p, then we may omit the sum operator in $\sum_{d:D} p$.

As another example, consider axiom SUM4. It says that we may distribute the sum operator over a plus, even if the sum binds a variable. This can be seen by substituting for X and Y $\lambda d{:}D.a(d)$ and $\lambda d{:}D.b(d)$, where no variable becomes bound. After β-reduction, the left hand side of SUM4 becomes $\sum_{d:D}(a(d) + b(d))$ and the right hand side is $\sum_{d:D} a(d) + \sum_{d:D} b(d)$. In conformity with the λ-calculus, we allow α-conversion in the sum operator, and do not state this explicitly. Hence, we consider the expressions $\sum_{d:D} p(d)$ and $\sum_{e:D} p(e)$ as equal.

The axiom SUM3 allows to split single summand instances from a given sum. For instance the process expressions $\sum_{n:Nat} a(n)$ and $\sum_{n:Nat} a(n) + a(2)$ are obviously the same, as they allow an $a(n)$ action for every natural number n. Using SUM3 we can prove them equal. Instantiate X with $\lambda n.a(n)$ and d with 2. We obtain:

$$\sum \lambda n.a(n) = \sum \lambda n.a(n) + \big(\lambda n.a(n)\big)2.$$

By β-reduction this reduces to $\sum_{n:Nat} a(n) = \sum_{n:Nat} a(n) + a(2)$.

We show how we can eliminate a finite sum operator in favour of a finite number of alternative composition operators. Such results always depend on the fact that a data type is defined using constructors. Therefore, we need induction in the proof, which makes it appear quite involved. This apparent complexity is increased by the use of axioms SUM3 and SUM11. Consider the equality

$$\sum_{n:Nat} r(n) \lhd n \leqslant 2 \rhd \delta = r(0) + r(1) + r(2), \tag{1}$$

assuming that the natural numbers together with the $\leqslant$ relation have been appropriately defined. The result follows in a straightforward way by the following lemma that we prove first.

LEMMA 2.3. *For all $m{:}Nat$ we find (S is the successor function)*:

$$\sum_{n:Nat} Xn = X0 + \sum_{m:Nat} XS(m).$$

PROOF. Using Lemma 2.1 we can split the proof into two summand inclusions.

($\subseteq$) We first prove the following statement with induction on n:

$$Xn \subseteq X0 + \sum_{m:Nat} XS(m).$$

- ($n = 0$) Trivial using A3.
- ($n = S(n')$)

$$\begin{aligned} & X0 + \sum_{m:Nat} XS(m) \\ \overset{\text{SUM3}}{=} \ & X0 + \sum_{m:Nat} XS(m) + XS(n') \\ \supseteq \ & Xn. \end{aligned}$$

So the statement has been proven without assumptions on n (i.e. for all n). Hence, application of SUM11, SUM4 and SUM1 yields:

$$\sum_{n:Nat} Xn \subseteq X0 + \sum_{m:Nat} XS(m),$$

as was to be shown.

($\supseteq$) Using SUM3 it immediately follows that for all m

$$\sum_{n:Nat} Xn \supseteq X0 + XS(m).$$

So, SUM11, SUM4 and SUM1 yield:

$$\sum_{n:Nat} Xn \supseteq X0 + \sum_{m:Nat} XS(m).$$

□

Equation (1) can now easily be proven by:

$$\begin{aligned} & \sum_{n:Nat} r(n) \lhd n \leqslant 2 \rhd \delta \\ \overset{\text{Lemma 2.3}}{=} \ & r(0) \lhd 0 \leqslant 2 \rhd \delta + \sum_{n':Nat} r(n'+1) \lhd n'+1 \leqslant 2 \rhd \delta \\ \overset{\text{Lemma 2.3}}{=} \ & r(0) + r(1) \lhd 1 \leqslant 2 \rhd \delta + \sum_{n'':Nat} r(n''+2) \lhd n''+2 \leqslant 2 \rhd \delta \\ \overset{\text{Lemma 2.3}}{=} \ & r(0) + r(1) + r(2) \lhd 2 \leqslant 2 \rhd \delta + \sum_{n''':Nat} r(3+n''') \lhd n'''+3 \leqslant 2 \rhd \delta \\ = \ & r(0) + r(1) + r(2). \end{aligned}$$

All the identities on data that we have used in the proof above can be proved from the axioms on natural numbers in Section 2.1.

An important law is *sum elimination*. It states that the sum over a data type from which only one element can be selected can be removed. This lemma occurred for the first time in [17]. Note that we assume that we have a function *eq* available, reflecting equality between terms.

LEMMA 2.4 (Sum elimination). *Let D be a sort and $eq: D \times D \to$ **Bool** a function such that for all $d, e{:}D$ it holds that $eq(d,e) = \mathsf{t}$ iff $d = e$. Then*

$$\sum_{d:D} Xd \lhd eq(d,e) \rhd \delta = Xe.$$

PROOF. According to Lemma 2.1 it suffices to prove summand inclusion in both directions.

($\subseteq$) Using Lemma 2.2.2 above we find:

$$Xe = Xe \lhd eq(d,e) \rhd \delta + Xe \lhd \neg eq(d,e) \rhd \delta.$$

Using SUM11 and SUM4 we find:

$$\sum_{d:D} Xe = \sum_{d:D} Xd \lhd eq(d,e) \rhd \delta + \sum_{d:D} Xe \lhd \neg eq(d,e) \rhd \delta.$$

Using SUM1 and the summand inclusion notation we obtain:

$$\sum_{d:D} Xd \lhd eq(d,e) \rhd \delta \subseteq Xe.$$

($\supseteq$) By applying SUM3, and the assumption that $eq(e,e) = \mathsf{t}$, we find:

$$\sum_{d:D} Xd \lhd eq(d,e) \rhd \delta \supseteq Xe \lhd eq(e,e) \rhd \delta = Xe. \qquad \square$$

LEMMA 2.5. *If there is some $e{:}D$ such that $b(e)$ holds, then*

$$x = \sum_{d:D} x \lhd b(d) \rhd \delta.$$

2.2.7. *Encapsulation.* Sometimes, we want to express that certain actions cannot happen, and must be blocked, i.e., renamed to δ. Generally, this is only done when we want to force this action into a communication. The *encapsulation* operator ∂_H ($H \subseteq \mathsf{Act}$) is specially designed for this task. In $\partial_H(p)$ it prevents all actions of which the action name is mentioned in H from happening. Typically,

$$\partial_{\{b\}}\big(a \cdot b(3) \cdot c\big) = a \cdot \delta,$$

where a, b and c are action names. The properties of ∂_H are axiomatized in Table 6.

Table 6
Axioms for encapsulation.

DD	$\partial_H(\delta) = \delta$	
D1	$\partial_H(a(d)) = a(d)$	if $a \notin H$
D2	$\partial_H(a(d)) = \delta$	if $a \in H$
D3	$\partial_H(x + y) = \partial_H(x) + \partial_H(y)$	
D4	$\partial_H(x \cdot y) = \partial_H(x) \cdot \partial_H(y)$	
SUM8	$\partial_H(\sum X) = \sum_{d:D} \partial_H(Xd)$	

Table 7
Axioms for internal actions and abstraction

B1	$c \cdot \tau = c$	
B2	$x \cdot (\tau \cdot (y + z) + y) = x \cdot (y + z)$	
TID	$\tau_I(\delta) = \delta$	
TIT	$\tau_I(\tau) = \tau$	
TI1	$\tau_I(a(d)) = a(d)$	if $a \notin I$
TI2	$\tau_I(a(d)) = \tau$	if $a \in I$
TI3	$\tau_I(x + y) = \tau_I(x) + \tau_I(y)$	
TI4	$\tau_I(x \cdot y) = \tau_I(x) \cdot \tau_I(y)$	
SUM9	$\tau_I(\sum X) = \sum_{d:D} \tau_I(Xd)$	
DT	$\partial_H(\tau) = \tau$	

2.2.8. *Internal actions and abstraction.* Abstraction is an important means to analyze communicating systems. It means that certain actions are made invisible, so that the relationship between the remaining actions becomes clearer. A specification can be proven equal to an implementation, consisting of a number of parallel processes, after abstracting from all communications between these components.

The *internal action* is denoted by τ. It represents an action that can take place in a system, but that cannot be observed directly. The internal action is meant for analysis purposes, and is hardly ever used in specifications, as it is very uncommon to specify that something unobservable must happen.

Typical identities characterising τ are $a \cdot \tau \cdot p = a \cdot p$, with a an action and p a process expression. It says that it is impossible to tell by observation whether or not internal actions happen after the a. Sometimes, the presence of internal actions can be observed, due to the context in which they appear. For example, $a + \tau \cdot b \neq a + b$, as the left hand side can silently execute the τ, after which it only offers a b action, whereas the right hand side can always do an a. The difference between the two processes can be observed by insisting in both cases that the a happens. This is always successful in $a + b$, but may lead to a deadlock in $a + \tau \cdot b$.

The natural axiom for internal actions is B1 in Table 7. Using the *parallel composition* operator (Section 2.2.9) and encapsulation, B1 can be used to prove all closed instantiations of B2 [50], and therefore B2 is also a natural law characterising internal actions. The semantics that is designed around these axioms is rooted branching bisimulation. The axioms in all other tables hold in strong bisimulation semantics, which does not abstract from internal actions. The first semantics abstracting from internal actions is weak bisimulation [38]. Weak bisimulation relates strictly more processes than rooted branching bisimulation, which in turn relates more processes than strong bisimulation. It is a good habit to prove results in the strongest possible semantics, as these results automatically carry over to all weaker variants. We do not consider these semantics explicitly in this section. The reader is referred to for instance [8,49,39].

In order to abstract from actions, the *abstraction* operator τ_I $(I \subseteq \mathsf{Act})$ is introduced, where I is a set of action names. The process $\tau_I(p)$ behaves as the process p, except that all actions with action names in I are renamed to τ. This is clearly characterized by the axioms in Table 7.

2.2.9. *Parallel processes.* The parallel composition operator can be used to put processes in parallel. The behaviour of $p \parallel q$ is the arbitrary interleaving of actions of the processes p and q, assuming for the moment that there is no communication between p and q. For example the process $a \parallel b$ behaves like $a \cdot b + b \cdot a$.

The parallel composition operator allows us to describe intricate processes. For instance a bag reading natural numbers using action name r and delivering them via action name s can be described by:

$$\begin{array}{ll} \textbf{act} & r, s : Nat \\ \textbf{proc} & Bag = \sum_{n:Nat} r(n) \cdot (s(n) \parallel Bag) \end{array}$$

Note that the elementary property of bags, namely that at most as many numbers can be delivered as have been received in the past, is satisfied by this description.

It is possible to let processes p and q in $p \parallel q$ communicate. This can be done by declaring in a communication section that certain action names can synchronize. This is done as follows:

$$\textbf{comm} \quad a \mid b = c$$

This means that if actions $a(d_1, \ldots, d_n)$ and $b(d_1, \ldots, d_n)$ can happen in parallel, they may synchronize and this synchronization is denoted by $c(d_1, \ldots, d_n)$. If two actions synchronize, their arguments must be exactly the same. In a communication declaration it is required that action names a, b and c are declared with exactly the same data sorts. It is not necessary that these sorts are unique. It is for example perfectly right to declare the three actions both with a sort *Nat* and with a pair of sorts $D \times \mathbf{Bool}$.

If a communication is declared as above, synchronization is another possibility for parallel processes. For example the process $a \parallel b$ is now equivalent to $a \cdot b + b \cdot a + c$. Often, this is not quite what is desired, as the intention generally is that a and b do not happen on their own. For this, the encapsulation operator can be used. The process $\partial_{\{a,b\}}(a \parallel b)$ is equal to c.

Axioms that describe the parallel composition operator are in Table 8. In this table the communications between action names from the communication section are represented by the *communication function* γ. In order to formulate the axioms two auxiliary parallel

Table 8
Axioms for parallelism in μCRL

CM1	$x \parallel y = x \mathbin{\|\!\!\lfloor} y + y \mathbin{\|\!\!\lfloor} x + x \mid y$
CM2	$c \mathbin{\|\!\!\lfloor} x = c \cdot x$
CM3	$c \cdot x \mathbin{\|\!\!\lfloor} y = c \cdot (x \parallel y)$
CM4	$(x + y) \mathbin{\|\!\!\lfloor} z = x \mathbin{\|\!\!\lfloor} z + y \mathbin{\|\!\!\lfloor} z$
SUM6	$(\sum X) \mathbin{\|\!\!\lfloor} x = \sum_{d:D}(Xd \mathbin{\|\!\!\lfloor} x)$
CF	$a(d) \mid a'(e) = \begin{cases} \gamma(a, a')(d) \lhd eq(d, e) \rhd \delta & \text{if } \gamma(a, a') \text{ defined} \\ \delta & \text{otherwise} \end{cases}$
CD1	$\delta \mid c = \delta$
CD2	$c \mid \delta = \delta$
CT1	$\tau \mid c = \delta$
CT2	$c \mid \tau = \delta$
CM5	$c \cdot x \mid c' = (c \mid c') \cdot x$
CM6	$c \mid c' \cdot x = (c \mid c') \cdot x$
CM7	$c \cdot x \mid c' \cdot y = (c \mid c') \cdot (x \parallel y)$
CM8	$(x + y) \mid z = x \mid z + y \mid z$
CM9	$x \mid (y + z) = x \mid y + x \mid z$
SUM7	$(\sum X) \mid x = \sum_{d:D}(Xd \mid x)$
SUM7′	$x \mid (\sum X) = \sum_{d:D}(x \mid Xd)$

composition operators have been defined. The left merge $\mathbin{\|\!\!\lfloor}$ is a binary operator that behaves exactly as the parallel composition operator, except that its first action must come from the left hand side. The communication merge $\mid$ is also a binary operator behaving as the parallel composition operator, except that the first action must be a synchronization between its left and right operand. The core law for the parallel composition operator is CM1 in Table 8. It says that in $x \parallel y$ either x performs the first step, represented by the summand $x \mathbin{\|\!\!\lfloor} y$, or y can do the first step, represented by $y \mathbin{\|\!\!\lfloor} x$, or the first step of $x \parallel y$ is a communication between x and y, represented by $x \mid y$. All other axioms in Table 8 are designed to eliminate the parallel composition operators in favour of the alternative composition and the sequential composition operator. The operators for parallel composition ($\parallel$, $\mathbin{\|\!\!\lfloor}$, and $\mid$) bind stronger than the conditional operator and weaker than the sequential composition operator.

Data transfer between parallel components occurs very often. As an example we describe a simplified instance of data transfer. One process sends a natural number n via action name s, and another process reads it, via action name r and then announces it via action name a. Using an encapsulation and an abstraction operator we force the processes to communicate, and make the communication internal. Of course we expect the process p to be equal to $\tau \cdot a(n)$.

var $n{:}\,Nat$

act $r, s, c, a : Nat$

comm $r \mid s = c$

proc $p = \tau_{\{c\}}\left(\partial_{\{r,s\}}\left(s(n) \parallel \sum_{m:Nat} r(m) \cdot a(m)\right)\right)$

Table 9
Axioms for renaming in μCRL

RD	$\rho_R(\delta) = \delta$
RT	$\rho_R(\tau) = \tau$
R1	$\rho_R(a(d)) = R(a)(d)$
R3	$\rho_R(x + y) = \rho_R(x) + \rho_R(y)$
R4	$\rho_R(x \cdot y) = \rho_R(x) \cdot \rho_R(y)$
SUM10	$\rho_R(\sum X) = \sum_{d:D} \rho_R(Xd)$

Assuming that *eq* is an equality function on natural numbers, we have

$$\begin{aligned} p &= \tau_{\{c\}}\Big(\partial_{\{r,s\}}\Big(s(n) \parallel \sum_{m:Nat} r(m) \cdot a(m)\Big)\Big) \\ &= \tau_{\{c\}}\Big(\partial_{\{r,s\}}\Big(s(n) \cdot \sum_{m:Nat} r(m) \cdot a(m) + \sum_{m:Nat} r(m) \cdot \big(s(n) \parallel a(m)\big) \\ &\qquad + \sum_{m:Nat} c(m) \cdot a(m) \lhd eq(n,m) \rhd \delta\Big)\Big) \\ &= \tau_{\{c\}}\Big(\sum_{m:Nat} c(m) \cdot a(m) \lhd eq(n,m) \rhd \delta\Big) \\ &= \sum_{m:Nat} \tau \cdot a(m) \lhd eq(n,m) \rhd \delta \\ &= \tau \cdot a(n). \end{aligned}$$

2.2.10. *Renaming.* In some cases it is efficient to reuse a given specification with different action names. This allows, for instance, the definition of generic components that can be used in different configurations. We introduce a *renaming operator* ρ_R. The subscript R is a sequence of renamings of the form $a \to b$, meaning that action name a must be replaced by b. This sequence of renamings is not allowed to contain distinct entries that replace the same action name. For example the subscript $a \to b, a \to c$ is not allowed. So, clearly, $\rho_R(p)$ is the process p with its action names replaced in accordance with R. An equational characterization of the renaming operator may be found in Table 9.

3. A strategy for verification

In process algebra it is common to verify the correctness of a description (the implementation) by proving it equivalent in some sense, e.g., with respect to rooted branching bisimulation, to a more abstract specification. When data is introduced into the descriptions, proving equivalence is more complex, since data can considerably alter the flow of control in the process. The cones and foci technique of [24] addresses this problem. A requirement of the cones and foci proof method is that the processes are defined by a linear equation

(Definition 3.1). The linearization of process terms is a common transformation in process algebra. Informally, all operators other than $\cdot$, $+$ and the conditional are eliminated. Therefore, we first present the linear process operator.

3.1. *Linear process operators*

We start out with the definition of 'linear process operator'. The advantage of the linear format is that it is simple. It only uses a few simple process operators in a restricted way. In particular, it does not contain the parallel composition operator. In general a linear process operator can easily be obtained from a μCRL description, including those containing parallel composition operators, without undue expansion (see also Section 5). Other formats, such as transition systems or automata, generally suffer from exponential blow up when the parallel composition operator is eliminated. This renders them unusable for the analysis of most protocols. We use *linear process operators* and *linear process equations*.

DEFINITION 3.1. A *linear process operator* (*LPO*) over data sort D is an expression of the form

$$\lambda p.\lambda d{:}D.\sum_{i\in I}\sum_{e_i:E_i} c_i\big(f_i(d,e_i)\big)\cdot p\big(g_i(d,e_i)\big) \lhd b_i(d,e_i) \rhd \delta$$

for some finite index set I, action names $c_i \in \mathsf{Act} \cup \{\tau\}$, data sorts E_i, D_i, and functions $f_i : D \times E_i \to D_i$, $g_i : D \times E_i \to D$, and $b_i : D \times E_i \to \mathbf{Bool}$. (We assume that τ has no parameter.)

Here D represents the state space, c_i are the action names, f_i represents the action parameters, g_i is the state transformation and b_i represents the condition determining whether an action is enabled. Note that the bound variable p ranges over processes parameterized with a datum of sort D. We use a meta-sum notation $\sum_{i\in I} p_i$ for $p_1 + p_2 + \cdots + p_n$ assuming $I = \{1, \ldots, n\}$; the p_i's are called *summands* of $\sum_{i\in I} p_i$. For $I = \emptyset$ we define $\sum_{i\in I} p_i = \delta$. We generally use letters Φ, Ψ, and Ξ to refer to LPOs.

According to the definition in [5], an LPO may have summands that allow termination. We have omitted these here, because they hardly occur in actual specifications and obscure the presentation of the theory. Moreover, it is not hard to add them if necessary.

LPOs have been defined as having a single data parameter. Most LPOs that we consider have several parameters, but these may be reduced to one parameter by means of Cartesian products and projection functions. Often, parameter lists get rather long. Therefore, we use the following notation for updating elements in the list. Let $\vec{d}$ abbreviate the vector $d_1, \ldots, d_n$. A summand of the form $\sum_{e_i:E_i} c_i(f_i(\vec{d}, e_i)) \cdot p(d_i'/d_i) \lhd b_i(\vec{d}, e_i) \rhd \delta$ in the definition of a process $p(\vec{d})$ abbreviates $\sum_{e_i:E_i} c_i(f_i(\vec{d}, e_i)) \cdot p(d_1, \ldots, d_{i-1}, d_i', d_{i+1}, \ldots d_n) \lhd b_i(\vec{d}, e_i) \rhd \delta$. Here, the parameter d_i is updated to d_i' in the recursive call. This notation is extended in the natural way to multiple updates. If no parameter is updated, we write the summand as $\sum_{e_i:E_i} c_i(f_i(\vec{d}, e_i)) \cdot p \lhd b_i(\vec{d}, e_i) \rhd \delta$.

Given a process operator Ψ, the associated linear process equation (LPE) can be written as $X(d) = \Psi X d$. Conversely, given a linear process equation $X(d) = p$, the associated LPO can be written as $\lambda X.\lambda d{:}D.p$. As a consequence we can choose whether to use linear process operators or equations at each point. Notions defined for LPOs carry over to LPEs in a straightforward manner and vice versa.

As an example consider the unreliable data channel that occurs in the alternating bit protocol [8], usually specified by:

$$\textbf{proc} \quad K = \sum_{d:D} \sum_{b:Bit} r(\langle d, b\rangle) \cdot \big(j \cdot s(\langle d, b\rangle) + j' \cdot s_3(ce)\big) \cdot K$$

The channel K reads frames consisting of a datum from some data type D and an alternating bit. It either delivers the frame correctly, or loses or garbles it. In the last case a checksum error ce is sent. The non-deterministic choice between the two options is modeled by the actions j and j'. If j is chosen the frame is delivered correctly and if j' happens it is garbled or lost.

The process K can be transformed into linear format by introducing a special variable i_k indicating the state of the process K. Just before the r action this state is 1. Directly after it, the state is 2. The state directly after action j is 3, and the state directly after j' is 4. We have indicated these states in the equation by means of encircled numbers:

$$\textbf{proc} \quad K = {}_{①}\sum_{d:D} \sum_{b:Bit} r(\langle d, b\rangle) \cdot {}_{②}(j \cdot {}_{③}s(\langle d, b\rangle) + j' \cdot {}_{④}s_3(ce)) \cdot {}_{①}K$$

With some experience it is quite easy to see that the channel K has the following linear description:

$$\begin{aligned}
\textbf{proc} \quad & K(d{:}D, b{:}Bit, i_k{:}Nat) \\
& = \sum_{d':D} \sum_{b':Bit} r(\langle d', b'\rangle) \cdot K\big(d'/d, b'/b, 2/i_k\big) \lhd eq(i_k, 1) \rhd \delta \\
& \quad + j \cdot K(3/i_k) \lhd eq(i_k, 2) \rhd \delta \\
& \quad + j' \cdot K(4/i_k) \lhd eq(i_k, 2) \rhd \delta \\
& \quad + s(\langle d, b\rangle) \cdot K(1/i_k) \lhd eq(i_k, 3) \rhd \delta \\
& \quad + s(ce) \cdot K(1/i_k) \lhd eq(i_k, 4) \rhd \delta.
\end{aligned}$$

Note that we have deviated from the pure LPO format: in the last four summands there is no summation over the data types D and Bit, in the second and third summand j and j' do not carry a parameter and in the first summand there are actually two sum operators. This is easily remedied by introducing dummy summands and dummy arguments, and pairing of variables. Note that linear process equations are not very readable, and therefore, they are less suited for specification purposes.

3.2. *Proof principles and elementary lemmata*

In order to verify recursive processes, we need auxiliary rules. The axioms presented in the previous section are not sufficiently strong to prove typical recursive properties. We introduce here the principles L-RDP (*Linear Recursive Definition Principle*) and CL-RSP (*Convergent Linear Recursive Specification Principle*). All the methods that we present in the sequel are derived from these rules.[1]

Processes can be defined as fixed points for convergent LPOs and as solutions for LPEs. In this chapter we use the term *solution* for both.

DEFINITION 3.2. A *solution* of an LPO Φ is a process p, parameterized with a datum of sort D, such that for all $d{:}D$ we have $p(d) = \Phi pd$.

DEFINITION 3.3. An LPO Φ written as in Definition 3.1 is called *convergent* iff there is a well-founded ordering $<$ on D such that for all $i \in I$ with $c_i = \tau$ and for all $e_i{:}E_i$, $d{:}D$ we have that $b_i(d, e_i)$ implies $g_i(d, e_i) < d$.

For each LPO Φ, we assume an axiom which postulates that Φ has a canonical solution. Then, we postulate that every *convergent* LPO has at most one solution. In this way, convergent LPOs define processes. The two principles reflect that we only consider process algebras where every LPO has at least one solution and converging LPOs have precisely one solution.

DEFINITION 3.4. The *Linear Recursive Definition Principle* (*L-RDP*) says that every linear process operator Ψ has at least one solution, i.e., there exists a p such that for all $d{:}D$ we have $p(d) = \Psi pd$.

The *Convergent Linear Recursive Specification Principle* (*CL-RSP*) [5] says that every convergent linear process operator has at most one solution, i.e. for all p and q if $p = \Psi p$ and $q = \Psi q$, then for all $d{:}D$ we have $p(d) = q(d)$.

The following theorem, proven in [5], says that if an LPO is convergent in the part of its state space that satisfies an invariant I, then it has at most one solution in that part of the state space. It has been shown to be equivalent to CL-RSP in [5].

DEFINITION 3.5. An *invariant* of an LPO Φ written as in Definition 3.1 is a function $I : D \to \mathbf{Bool}$ such that for all $i \in I$, $e_i{:}E_i$, and $d{:}D$ we have:

$$b_i(d, e_i) \wedge I(d) \to I\big(g_i(d, e_i)\big).$$

THEOREM 3.6 (Concrete Invariant Corollary). *Let Φ be an LPO. If, for some invariant I of Φ, the LPO $\lambda p.\lambda d.\Phi pd \lhd I(d) \rhd \delta$ is convergent and for some processes q, q', parameterized by a datum of sort D, we have $I(d) \to q(d) = \Phi qd$ and $I(d) \to q'(d) = \Phi q'd$, then $I(d) \to q(d) = q'(d)$.*

[1] Elsewhere we also use Koomen's Fair Abstraction Rule, but as we avoid processes with internal loops, we do not need KFAR here.

To develop the theory it is convenient to work with a particular form of LPOs, which we call *(action) clustered*.[2] Clustered LPOs contain, for each action a, at most one summand starting with an a. Thus clustered LPOs can be defined by summation over a finite index set I of actions.

DEFINITION 3.7. Let $Act \subseteq \mathsf{Act} \cup \{\tau\}$ be a finite set of action names. A *clustered linear process operator* (*C-LPO*) over Act is an expression of the form

$$\Phi = \lambda p.\lambda d{:}D. \sum_{a \in Act} \sum_{e_a : E_a} a\big(f_a(d, e_a)\big) \cdot p\big(g_a(d, e_a)\big) \lhd b_a(d, e_a) \rhd \delta.$$

The first part of the following theorem states that it is no restriction to assume that LPOs are clustered. The second part is a prelude on the general equality theorem, as it requires that, for each action, the sorts in the sum operators preceding this action are the same in specification and implementation. A proof is given in [24].

THEOREM 3.8.

(1) *Every convergent LPO Φ can be rewritten to a C-LPO Φ' with the same solution, provided all occurrences of the same action have parameters of the same type.*

(2) *Consider convergent C-LPOs Φ, Ψ such that action a occurs both in Φ and in Ψ (with parameters of the same data type). There exist convergent C-LPOs Φ', Ψ' having the same solutions as Φ, Ψ, respectively, such that a occurs in Φ' and Ψ' in summands with summation over the same sort E_a.*

The two summands $s(\langle d, b\rangle) \cdot K(1/i_k) \lhd eq(i_k, 3) \rhd \delta$ and $s(ce) \cdot K(1/i_k) \lhd eq(i_k, 4) \rhd \delta$ of the channel K can be grouped together as

$$s\big(if(eq(i_k, 3), \langle d, b\rangle, ce)\big) \cdot K(1/i_k) \lhd eq(i_k, 3) \vee eq(i_k, 4) \rhd \delta.$$

Here we assume that ce is of the same sort as the pair $\langle d, b\rangle$.

3.3. *The general equality theorem*

In this section, we are concerned with proving equality of solutions of C-LPOs Φ and Ψ. The C-LPO Φ defines an implementation and the C-LPO Ψ defines the specification of a system. We use the *cones and foci* proof method of [24].

We assume that τ-steps do not occur in the specification Ψ. We want to show that after abstraction of internal actions in a set *Int* the solution of Φ is equal to the solution of Ψ. We assume that Φ cannot perform an infinite sequence of internal actions. It turns out to be convenient to consider Φ where the actions in *Int* are already renamed to τ. Hence, we speak about a C-LPO Ξ which is Φ where actions in *Int* have been abstracted from (so

[2] At some places clustered LPOs have been called deterministic. However, this is a bad name, as the process underlying a 'deterministic' LPO is not at all a deterministic process, i.e., a process that can for each action a always do at most one a transition.

$\tau_{Int}(\Phi) = \Xi$). Note that Ξ is convergent, and hence defines a process. We fix the C-LPOs Ξ and Ψ as follows (where the action names are taken from a set Act):

$$\Xi = \lambda p.\lambda d{:}D_\Xi.\sum_{a\in Act}\sum_{e_a:E_a} a\big(f_a(d,e_a)\big)\cdot p\big(g_a(d,e_a)\big) \lhd b_a(d,e_a) \rhd \delta,$$

$$\Psi = \lambda q.\lambda d{:}D_\Psi.\sum_{a\in Act\setminus\{\tau\}}\sum_{e_a:E_a} a\big(f'_a(d,e_a)\big)\cdot q\big(g'_a(d,e_a)\big) \lhd b'_a(d,e_a) \rhd \delta.$$

The issue that we consider is how to prove the solutions of Ξ and Ψ equal.

The main idea of the cones and foci proof method is that there are usually many internal events in the implementation, but they are only significant in that they must progress somehow towards producing a visible event which can be matched with a visible event in the specification. A state of the implementation where no internal actions are enabled is called a *focus point*, and there may be several such points in the implementation. Focus points are characterized by a Boolean condition on the data of the process called the *focus condition*. The focus condition $FC_\Xi(d)$ is the negation of the condition which allows τ actions to occur. The focus condition $FC_\Xi(d)$ is true if d is a focus point and false otherwise.

DEFINITION 3.9. The *focus condition* $FC_\Xi(d)$ of Ξ is the formula $\neg\exists_{e_\tau:E_\tau}(b_\tau(d,e_\tau))$.

The *cone* belonging to a focus point is the part of the state space from which the focus point can be reached by internal actions; imagine the transition system forming a cone or funnel pointing towards the focus. Figure 1 in Section 1 visualizes the core observation underlying this method.

The final element in the proof method is a *state mapping* $h: D_\Xi \to D_\Psi$ between the data states of the implementation and the data states of the specification. It explains how the data parameter that encodes states of the specification is constructed out of the data parameter that encodes states of the implementation. This mapping is surjective, but almost certainly not injective, since the data of the specification is likely to be simpler than that of the implementation. So in this respect we have a refinement, but in terms of actions we have an equivalence.

In order to prove implementation and specification rooted branching bisimilar, the state mapping should satisfy certain properties, which we call *matching criteria* because they serve to match states and transitions of implementation and specification. They are inspired by numerous case studies in protocol verification, and reduce complex calculations to a few straightforward checks. If these six criteria are satisfied then the specification and the implementation can be said to be rooted branching bisimilar under the General Equality Theorem (Theorem 3.11). The general forms of the matching criteria are given in Definition 3.10. Given the particular actions, conditions and mapping for a system, the matching criteria can be mechanically derived. Of course, the choice of mapping requires some thought, as does the subsequent proof of the criteria.

Now we formulate the criteria. We discuss each criterion directly after the definition.

DEFINITION 3.10. Let $h: D_\Xi \to D_\Psi$ be a state mapping. The following criteria referring to Ξ, Ψ and h are called the *matching criteria*. We refer to their conjunction by $C_{\Xi,\Psi,h}(d)$.

For all $e_\tau{:}E_\tau$, $a \in Act \setminus \{\tau\}$, and $e_a{:}E_a$:

$$\begin{array}{ll}
(1) & \Xi \text{ is convergent.} \\
(2) & b_\tau(d, e_\tau) \rightarrow h(d) = h\big(g_\tau(d, e_\tau)\big). \\
(3) & b_a(d, e_a) \rightarrow b'_a\big(h(d), e_a\big). \\
(4) & FC_\Xi(d) \wedge b'_a\big(h(d), e_a\big) \rightarrow b_a(d, e_a). \\
(5) & b_a(d, e_a) \rightarrow f_a(d, e_a) = f'_a\big(h(d), e_a\big). \\
(6) & b_a(d, e_a) \rightarrow h\big(g_a(d, e_a)\big) = g'_a\big(h(d), e_a\big).
\end{array}$$

Criterion (1) says that Ξ must be convergent. In effect this does not say anything else than that in a cone every internal action τ constitutes progress towards a focus point. Criterion (2) says that if in a state d in the implementation an internal step can be done (i.e., $b_\tau(d, e_\tau)$ is valid) then this internal step is not observable. This is described by saying that the state before the τ-step and the state after the τ-step both relate to the same state in the specification. Criterion (3) says that when the implementation can perform an external step, then the corresponding state in the specification must also be able to perform this step. Note that in general, the converse need not hold. If the specification can perform an a-action in a certain state e, then it is only necessary that in every state d of the implementation such that $h(d) = e$ an a-step can be done *after some internal actions*. This is guaranteed by criterion (4). It says that in a focus point of the implementation, an action a in the implementation can be performed if it is enabled in the specification. Criteria (5) and (6) express that corresponding external actions carry the same data parameter (modulo h) and lead to corresponding states.

Assume that r and q are solutions of Ξ and Ψ, respectively. Using the matching criteria, we would like to prove that, for all $d{:}D$, $C_{\Xi,\Psi,h}(d)$ implies $r(d) = q(h(d))$.

In fact we prove a more complicated result. This has two reasons. The first one is that the statement above is not generally true. Consider the case where d is a non-focus point of Ξ. In this case, $r(d)$ can perform a τ-step. Since q cannot perform τ-steps, $r(d)$ cannot be equal to $q(h(d))$. Therefore, in the setting of rooted branching bisimulation we can for non-focus points d only prove $\tau \cdot r(d) = \tau \cdot q(h(d))$.

The second reason why we need a more complicated result is of a very general nature. A specification and an implementation are in general only equivalent for the reachable states in the implementation. A common tool to exclude non-reachable states is an invariant. Therefore we have added an invariant to the theorem below. For a proof of this theorem we refer to [24].

THEOREM 3.11 (General Equality Theorem). *Let Ξ be a C-LPO and let Ψ be a C-LPO that does not contain τ-steps. Let h be a state mapping. Assume that r and q are solutions of Ξ and Ψ, respectively. If I is an invariant of Ξ and $I(d) \rightarrow C_{\Xi,\Psi,h}(d)$ for all $d{:}D_\Xi$, then*

$$\forall_{d:D_\Xi}\Big(I(d) \rightarrow r(d) \triangleleft FC_\Xi(d) \triangleright \tau \cdot r(d) = q\big(h(d)\big) \triangleleft FC_\Xi(d) \triangleright \tau \cdot q\big(h(d)\big)\Big).$$

3.4. *Pre-abstraction and idle loops*

The proof strategy presented in the previous section can only be applied to systems for which the implementation is convergent. This is an all too serious restriction. In this section we present a generalization of the proof strategy which is also capable to deal with idle loops.

The most important concept in this generalization is the *pre-abstraction function*. A pre-abstraction function divides the occurrences of the internal actions into *progressing* and *non-progressing* internal actions. The progressing internal actions are the ones for which the pre-abstraction function gives true and the non-progressing actions are the ones for which the pre-abstraction function gives false.

DEFINITION 3.12. Let Φ be a C-LPO and let *Int* be a finite set of action names. A *pre-abstraction function* ξ is a mapping that yields for every action $a \in Int$ an expression of sort $D \times E_a \to \mathbf{Bool}$. The partial pre-abstraction function ξ is extended to a total function on Act by assuming $\xi(\tau)(d, e_\tau) = \mathsf{t}$ and $\xi(a)(d, e_a) = \mathsf{f}$ for all $a \in \mathsf{Act} \setminus Int$.

The pre-abstraction function ξ defines from which internal actions we abstract. If the pre-abstraction function of an action yields true (progressing internal action), the action is replaced by τ, while if the pre-abstraction function yields false the action remains unchanged.

In a nutshell, the adaptation of the proof strategy is that instead of renaming all internal actions into τ-actions we only rename the progressing internal actions. As a consequence the notions of convergence and focus point need to be adapted. Instead of considering τ-actions, all internal actions involved in the pre-abstraction must be considered.

DEFINITION 3.13. Let Φ be a C-LPO with internal actions *Int* and let ξ be a pre-abstraction function. The C-LPO Φ is called *convergent with respect to* ξ iff there exists a well-founded ordering $<$ on D such that for all $a \in Int \cup \{\tau\}$, $d{:}D$ and all $e_a{:}E_a$ such that $\xi(a)(d, e_a)$ we have that $b_a(d, e_a)$ implies $g_a(d, e_a) < d$.

DEFINITION 3.14. Let ξ be a pre-abstraction function. The *focus condition of* Φ *relative to* ξ is defined by

$$FC_{\Phi,Int,\xi}(d) = \forall_{a \in Int \cup \{\tau\}} \forall_{e_a:E_a} \big(\xi(a)(d, e_a) \to \neg b_a(d, e_a)\big).$$

DEFINITION 3.15. Let Φ be a C-LPO over $Ext \cup Int \cup \{\tau\}$ (pairwise disjoint) and let Ψ be a C-LPO over *Ext*. Let $h : D_\Phi \to D_\Psi$ and let ξ be a pre-abstraction function. The following six criteria are called the *matching criteria for idle loops* and their conjunction is denoted by $CI_{\Phi,\Psi,\xi,h}(d)$. For all $i \in Int \cup \{\tau\}$, $e_i{:}E_i$, $a \in Ext$, and $e_a{:}E_a$:

(1) Φ is convergent with respect to ξ.

(2) $b_i(d, e_i) \to h(d) = h\big(g_i(d, e_i)\big)$.

(3) $b_a(d, e_a) \to b'_a\big(h(d), e_a\big)$.

(4) $FC_{\Xi,Int,\xi}(d) \wedge b'_a(h(d), e_a) \to b_a(d, e_a)$.
(5) $b_a(d, e_a) \to f_a(d, e_a) = f'_a(h(d), e_a)$.
(6) $b_a(d, e_a) \to h(g_a(d, e_a)) = g'_a(h(d), e_a)$.

THEOREM 3.16 (Equality theorem for idle loops). *Let Φ be a C-LPO over $Ext \cup Int \cup \{\tau\}$ (pairwise disjoint) and Ψ a C-LPO over Ext. Let $h: D_\Phi \to D_\Psi$ and let ξ be a pre-abstraction function. Assume that r and q are solutions of Φ and Ψ respectively. If I is an invariant of Φ and $I(d) \to CI_{\Phi,\Psi,\xi,h}(d)$ for all $d{:}D_\Phi$, then*

$$\forall_{d:D_\Phi}\big(I(d) \to \tau \cdot \tau_{Int}(r(d)) = \tau \cdot q(h(d))\big).$$

A proof of this theorem can be found in [24].

4. Verification of the Serial Line Interface Protocol

In this section we give a completely worked out example of a simple protocol to illustrate the use of μCRL and the general equality theorem. The Serial Line Interface Protocol (SLIP) is one of the protocols that is very commonly used to connect individual computers via a modem and a phone line. It allows only one single stream of bidirectional information.

Basically, the SLIP protocol works by sending blocks of data. Each block is a sequence of bytes that ends with the special `end` byte. Confusion can occur when the `end` byte is also part of the ordinary data sequence. In this case, the `end` byte is 'escaped', by placing an `esc` byte in front of the `end` byte. Similarly, to distinguish an ordinary `esc` byte from the escape character `esc`, each `esc` in the data stream is replaced by two `esc` characters.

In our modeling of the protocol, we ignore the process of dividing the data into blocks, but only look at the insertion and removal of `esc` characters into the data stream. For simplicity we assume that all occurrences of `end` and `esc` bytes have to be 'escaped'. We model the system by three components: a sender (S), inserting escape characters, a channel (C), modeling the medium along which data is transferred, and a receiver (R), removing the escape characters (see Figure 2). We let the channel be a buffer of capacity one in this example.

We use four data types *Nat*, **Bool**, *Byte*, and *Queue* to describe the SLIP protocol and its external behaviour. The sort *Nat* contains the natural numbers. With 0 and S we denote the zero element and the successor function on *Nat* as before. Numerals (e.g., 3) are used as abbreviations. The function $eq: Nat \times Nat \to \mathbf{Bool}$ is true when its arguments represent the same number. The sort **Bool** contains exactly two constants t (true) and f (false) and we assume that all required Boolean connectives are defined in a proper way.

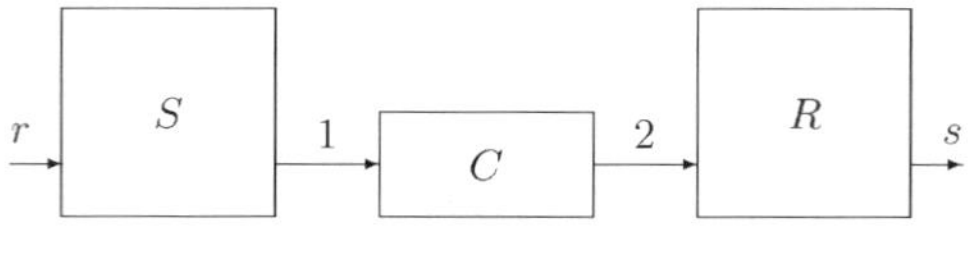

Fig. 2. Architecture of the SLIP protocol.

The sort *Byte* contains the data elements to be transferred via the SLIP protocol. As the definition of a byte as a sequence of 8 bits is very detailed and actually irrelevant we only assume about *Byte* that it contains at least two not necessarily different constants $\mathtt{esc}$ and $\mathtt{end}$, and a function $eq: Byte \times Byte \to \mathbf{Bool}$ that represents equality.

$$\begin{array}{ll}\textbf{sort} & Byte\\ \textbf{map} & \mathtt{esc}: \to Byte\\ & \mathtt{end}: \to Byte\\ & eq: Byte \times Byte \to \mathbf{Bool}\end{array}$$

Furthermore, to describe the external behaviour of the system, we use the sort *Queue* as described in Section 2.1, but this time we take the elements of the queue from the sort *Byte*. Additionally, we use the auxiliary function *len* which yields the number of elements in a queue.

$$\begin{array}{ll}\textbf{sort} & Queue\\ \textbf{map} & len: Queue \to Nat\\ \textbf{var} & b: Byte\ q: Queue\\ \textbf{rew} & len([\,]) = 0\\ & len\big(in(b,q)\big) = S\big(len(q)\big)\end{array}$$

The processes are defined by guarded process declarations for the channel C, the sender S and the receiver R (cf. Figure 2). The equation for the channel C expresses that first a byte b is read using a read action r_1 via port 1, and subsequently this value is sent via port 2 using action s_2. After this the channel is back in its initial state, ready to receive another byte. The encircled numbers can be ignored for the moment. They serve to explicitly indicate the state of these processes and are used later.

$$\textbf{proc}\quad C =_{⓪} \sum_{b:Byte} r_1(b) \cdot_{①} s_2(b) \cdot_{⓪} C$$

Using the r action the sender reads a byte from a protocol user, who wants to use the service of the SLIP protocol to deliver this byte elsewhere. It is obvious that if b equals $\mathtt{esc}$ or $\mathtt{end}$, first an additional $\mathtt{esc}$ is sent to the channel (via action s_1) before b itself is sent. Otherwise, b is sent without prefix.

$$\textbf{proc}\quad S =_{⓪} \sum_{b:Byte} r(b) \cdot_{①} \big(s_1(\mathtt{esc}) \cdot_{②} s_1(b) \cdot_{⓪} S \lhd eq(b,\mathtt{end}) \vee eq(b,\mathtt{esc}) \rhd s_1(b) \cdot_{⓪} S\big)$$

The receiver is equally straightforward. After receiving a byte b from the channel (via r_2) it checks whether it is an $\mathtt{esc}$. If so, it removes it and delivers the trailing $\mathtt{end}$ or $\mathtt{esc}$. Otherwise, it just delivers b. Both the sender and the receiver repeat themselves indefinitely, too.

$$\textbf{proc} \quad R = {}_{⓪}\sum_{b:Byte} r_2(b) \cdot {}_{①}\Big(\Big(\sum_{b':Byte} r_2(b') \cdot {}_{②}s(b') \cdot {}_{⓪}R\Big) \lhd eq(b, \texttt{esc}) \rhd s(b) \cdot {}_{⓪}R\Big)$$

The SLIP protocol is defined by putting the sender, channel and receiver in parallel. We let the actions r_1 and s_1 communicate and the resulting action is called c_1. Similarly, r_2 and s_2 communicate into c_2.

$$\textbf{comm} \quad s_1 \mid r_1 = c_1$$
$$s_2 \mid r_2 = c_2$$

The encapsulation operator $\partial_{\{r_1,s_1,r_2,s_2\}}$ forbids the actions r_1, s_1, r_2 and s_2 to occur on their own by renaming these actions to δ. In this way the actions are forced to communicate. The abstraction operation $\tau_{\{c_1,c_2\}}$ abstracts from these communications by renaming them to the internal action τ. For the SLIP protocol the external actions are r and s.

$$\textbf{proc} \quad SLIP = \tau_{\{c_1,c_2\}}\big(\partial_{\{r_1,s_1,r_2,s_2\}}(S \parallel C \parallel R)\big)$$

We want to obtain a better understanding of the protocol, because although rather simple, it is not straightforward to understand its external behaviour completely. Data that is read at r is of course delivered in sequence at s without loss or duplication of data. So, the protocol behaves like a kind of queue. The reader should now, before reading further, take a few minutes to determine the size of this queue. Simultaneously, one byte can be stored in the receiver, one in the channel and one in the sender. If an esc or end is in transfer, it matters in which of the processes it is stored. If the esc or end byte is stored in the sender, no leading esc is produced yet. Hence three bytes can be stored in this case. If the esc or end byte is stored in the channel, it must have been sent there by the sender. Obviously the sender in this case has first sent a leading esc byte. This byte is either stored in the receiver or removed by the receiver. In both cases the receiver contains no byte that is visible in the environment. Hence in this case at most two bytes can be stored. Finally, if the end or esc byte is stored in the receiver, the leading esc byte produced by the sender has been removed already by the receiver. Hence three bytes can be stored in this case (assuming that no other esc or end byte is in transit). So, the conclusion is that the queue behaves as a queue of size three, except when an esc or end occurs at the second position in the queue (the channel), in which case the size is at most two. For this purpose the auxiliary predicate *full* is defined.

$$\textbf{map} \quad full : Queue \to \textbf{Bool}$$
$$\textbf{var} \quad q : Queue$$
$$\textbf{rew} \quad full(q) = eq\big(len(q), 3\big) \vee \big(eq(len(q), 2) \wedge \big(eq\big(toe\big(untoe(q)\big), \texttt{esc}\big) \vee eq\big(toe\big(untoe(q)\big), \texttt{end}\big)\big)\big)$$

Using this predicate we obtain the description of the external behaviour of the SLIP protocol below: If the queue is not full, an additional byte b can be read. If the queue is not empty an element can be delivered.

$$\begin{aligned}\textbf{proc}\quad & \mathit{Spec}(q{:}\mathit{Queue})\\ & = \sum_{b:\mathit{Byte}} r(b)\cdot \mathit{Spec}\big(\mathit{in}(b,q)\big) \lhd \neg \mathit{full}(q) \rhd \delta\\ & \quad + \mathit{sl}\big(\mathit{toe}(q)\big)\cdot \mathit{Spec}\big(\mathit{untoe}(q)\big) \lhd \neg \mathit{isempty}(q) \rhd \delta\end{aligned}$$

The theorem that we are interested in proving is:

THEOREM 4.1. *We have* $\mathit{SLIP} = \mathit{Spec}([\,])$.

PROOF. This follows directly from Lemma 4.2 and Lemma 4.4 that are given below. □

We describe the linear equation for *SLIP*. We have numbered the different summands for easy reference. Note that the specification is already linear.

$$\begin{aligned}&\textbf{proc}\quad \mathit{LinImpl}(b_s{:}\mathit{Byte}, s_s{:}\mathit{Nat}, b_c{:}\mathit{Byte}, s_c{:}\mathit{Nat}, b_r{:}\mathit{Byte}, s_r{:}\mathit{Nat})\\ &\text{(a)}\quad = \sum_{b:\mathit{Byte}} r(b)\cdot \mathit{LinImpl}(b,1,b_c,s_c,b_r,s_r) \lhd \mathit{eq}(s_s,0) \rhd \delta\\ &\text{(b)}\quad\quad + \tau\cdot \mathit{LinImpl}(b_s,2,\mathtt{esc},1,b_r,s_r)\\ &\quad\quad\quad \lhd \mathit{eq}(s_c,0)\wedge \mathit{eq}(s_s,1)\wedge (\mathit{eq}(b_s,\mathtt{end})\vee \mathit{eq}(b_s,\mathtt{esc})) \rhd \delta\\ &\text{(c)}\quad\quad + \tau\cdot \mathit{LinImpl}(b_s,0,b_s,1,b_r,s_r)\\ &\quad\quad\quad \lhd \mathit{eq}(s_c,0)\wedge \big(\mathit{eq}(s_s,2)\vee \big(\mathit{eq}(s_s,1)\wedge \neg\big(\mathit{eq}(b_s,\mathtt{end})\\ &\quad\quad\quad\quad \vee\, \mathit{eq}(b_s,\mathtt{esc})\big)\big)\big) \rhd \delta\\ &\text{(d)}\quad\quad + \tau\cdot \mathit{LinImpl}(b_s,s_s,b_c,0,b_c,1)\\ &\quad\quad\quad \lhd \mathit{eq}(s_r,0)\wedge \mathit{eq}(s_c,1) \rhd \delta\\ &\text{(e)}\quad\quad + \tau\cdot \mathit{LinImpl}(b_s,s_s,b_c,0,b_c,2)\\ &\quad\quad\quad \lhd \mathit{eq}(s_r,1)\wedge \mathit{eq}(b_r,\mathtt{esc})\wedge \mathit{eq}(s_c,1) \rhd \delta\\ &\text{(f)}\quad\quad + s(b_r)\cdot \mathit{LinImpl}(b_s,s_s,b_c,s_c,b_r,0)\\ &\quad\quad\quad \lhd \mathit{eq}(s_r,2)\vee \big(\mathit{eq}(s_r,1)\wedge \neg \mathit{eq}(b_r,\mathtt{esc})\big) \rhd \delta\end{aligned}$$

We obtained this form by identifying three explicit states in the sender and receiver, and two in the channel. These have been indicated by encircled numbers in the defining equations of these processes. The states of these processes are indicated by the variables s_s, s_r and s_c respectively. Each of the three processes also stores a byte in certain states. The bytes for each process are indicated by b_s, b_r and b_c. The τ in summand (b) comes from abstracting from $c_1(\mathtt{esc})$, in summand (c) it comes from $c_1(b_s)$, in (d) from $c_2(b_c)$ and in (e) from $c_2(b_c)$.

The following lemma says that *LinImpl*, the linear equation for *SLIP*, indeed equals the description of *SLIP*.

LEMMA 4.2. *For any* b_1, b_2, b_3: *Byte it holds that*

$$LinImpl(b_1, 0, b_2, 0, b_3, 0) = SLIP.$$

We list below a number of invariants of *LinImpl* that are sufficient to prove the results in the sequel. The proof of the invariants is straightforward, except that we need invariant 2 to prove invariant 3.

LEMMA 4.3. *The following expressions are invariants for LinImpl*:

$$(1) \quad s_s \leqslant 2 \wedge s_c \leqslant 1 \wedge s_r \leqslant 2;$$

$$(2) \quad eq(s_s, 2) \to \big(eq(b_s, \mathsf{esc}) \vee eq(b_s, \mathsf{end})\big);$$

$$\begin{aligned}(3) \quad & \neg eq(s_s, 2) \to \big(\big(eq(s_c, 0) \wedge \neg\big(eq(s_r, 1) \wedge eq(b_r, \mathsf{esc})\big)\big) \vee \\ & \qquad \big(eq(s_c, 1) \wedge \big(\big(eq(s_r, 1) \wedge eq(b_r, \mathsf{esc})\big) \leftrightarrow \\ & \qquad\qquad \big(eq(b_c, \mathsf{esc}) \vee eq(b_c, \mathsf{end})\big)\big)\big)\big) \wedge \\ & eq(s_s, 2) \to \big(\big(eq(s_c, 1) \wedge eq(b_c, \mathsf{esc}) \wedge \neg\big(eq(s_r, 1) \wedge eq(b_r, \mathsf{esc})\big)\big) \vee \\ & \qquad \big(eq(s_c, 0) \wedge eq(s_r, 1) \wedge eq(b_r, \mathsf{esc})\big)\big).\end{aligned}$$

The next step is to relate the implementation and the specification using a state mapping $h: Nat \times Byte \times Nat \times Byte \times Nat \times Byte \to Queue$. For this, we use the auxiliary function *cadd* (conditional add). The expression $cadd(c, b, q)$ yields a queue with byte b added to q if Boolean c equals true. If c is false, it yields q itself.

$$\begin{array}{ll}\mathbf{map} & cadd: \mathbf{Bool} \times Byte \times Queue \to Queue \\ \mathbf{var} & b: Byte\ q: Queue \\ \mathbf{rew} & cadd(\mathsf{f}, b, q) = q \\ & cadd(\mathsf{t}, b, q) = in(b, q)\end{array}$$

The state mapping is in this case:

$$\begin{aligned}& h(b_s, s_s, b_c, s_c, b_r, s_r) \\ & \quad = cadd\big(\neg eq(s_s, 0), b_s, \\ & \qquad cadd\big(eq(s_c, 1) \wedge \big(\neg eq(b_c, \mathsf{esc}) \vee \big(eq(s_r, 1) \wedge eq(b_r, \mathsf{esc})\big)\big), b_c, \\ & \qquad cadd\big(eq(s_r, 2) \vee \big(eq(s_r, 1) \wedge \neg eq(b_r, \mathsf{esc})\big), b_r, [\,]\big)\big)\big).\end{aligned}$$

So, the state mapping constructs a queue out of the state of the implementation, containing at most b_s, b_c and b_r in that order. The byte b_s from the sender is in the queue if the sender is not about to read a new byte ($\neg eq(s_s, 0)$). The byte b_c from the channel is in the queue if the channel is actually transferring data ($eq(s_c, 1)$) and if it does not contain an escape character indicating that the next byte must be taken literally. Similarly, the byte b_r from the receiver must be in the queue if it is not empty and b_r is not an escape character.

The focus condition of the SLIP implementation can easily be extracted and is (slightly simplified using the invariant):

$$\begin{aligned}&\big(eq(s_c,0) \rightarrow eq(s_s,0)\big) \wedge \\ &\big(eq(s_c,1) \rightarrow \big(\neg eq(s_r,0) \wedge \big(eq(s_r,1) \rightarrow \neg eq(b_r,\mathtt{esc})\big)\big)\big).\end{aligned}$$

LEMMA 4.4. *For all* b_1, b_2, b_3:*Byte*

$$Spec([\,]) = LinImpl(b_1, 0, b_2, 0, b_3, 0).$$

PROOF. We apply Theorem 3.11 by taking *LinImpl* for p, *Spec* for q and the state mapping and invariant provided above. We simplify the conclusion by observing that the invariant and the focus condition are true for $s_s = 0$, $s_c = 0$ and $s_r = 0$. By moreover using that $h(b_1, 0, b_2, 0, b_3, 0) = [\,]$, the lemma is a direct consequence of the generalized equation theorem. We are only left with checking the matching criteria:

(1) The measure $13 - s_s - 3s_c - 4s_r$ decreases with each τ step.

(2) (b) Distinction on s_r; use invariant. (c) Distinguish different values of s_s; use invariant. (d) Trivial. (e) Trivial.

(3) (a) Let $\vec{s}$ denote the tuple $(b_s, s_s, b_c, s_c, b_r, s_r)$. We must show that $s_s = 0$ implies $\neg full(h(\vec{s}))$. From $s_s = 0$ and the definitions of h and *full* observe that $full(h(\vec{s}))$ can only be the case if $s_c = 1 \wedge (b_c \neq \mathtt{esc} \vee (s_r = 1 \wedge b_r = \mathtt{esc})) \wedge (s_r = 2 \vee (s_r = 1 \wedge b_r \neq \mathtt{esc}))$ and $b_c = \mathtt{esc} \vee b_c = \mathtt{end}$. In all other cases we easily find that $\neg full(h(\vec{s}))$. If $b_c = \mathtt{esc}$ we find $b_r = \mathtt{esc}$ and $s_r = 2$. We also have $s_c = 1$. Using the invariant we obtain $b_c \neq \mathtt{esc} \wedge b_c \neq \mathtt{end}$. This leads to a contradiction and therefore $\neg full(h(\vec{s}))$. If $b_c = \mathtt{end}$ then using the invariant we find that $s_r = 1$ and $b_r = \mathtt{esc}$. This contradicts the above assumption. Therefore also in this case $\neg full(h(\vec{s}))$. (f) Trivial.

(4) (a) Let $\vec{s}$ denote the tuple $(b_s, s_s, b_c, s_c, b_r, s_r)$. We must show that if the focus condition and $\neg full(h(\vec{s}))$ hold, then $eq(s_s, 0)$. The proof proceeds by deriving a contradiction under the assumption $\neg eq(s_s, 0)$. If $eq(s_s, 1)$ it follows from the invariant and the focus condition that $len(h(\vec{s})) = 3$, contradicting that $\neg full(h(\vec{s}))$. If $eq(s_s, 2)$, then $len(h(\vec{s})) = 2$, $toe(untoe(h(\vec{s}))) = b_s$ and $eq(b_s, \mathtt{esc}) \vee eq(b_s, \mathtt{end})$ in a similar way. Also in this case this contradicts $\neg full(h(\vec{s}))$.

(f) Let $\vec{s}$ denote the tuple $(b_s, s_s, b_c, s_c, b_r, s_r)$. We must show that the invariant, the focus condition and the statement $\neg isempty(h(\vec{s}))$ imply $eq(s_r, 2) \vee (eq(s_r, 1) \wedge \neg eq(b_r, \mathtt{esc}))$. Assume FC and, towards using contraposition, $\neg eq(s_r, 2) \wedge (\neg eq(s_r, 1) \vee eq(b_r, \mathtt{esc}))$. Using the invariant we deduce $eq(s_r, 0) \vee (eq(b_r, \mathtt{esc}) \wedge eq(s_r, 1)$. By the second conjunct of the FC (contraposition), we obtain $\neg eq(s_c, 1)$, so by the invariant, $eq(s_c, 0)$, and by the first conjunct of FC, $eq(s_s, 0)$ holds. By the definition of the state mapping h, we easily see that $h(\vec{s}) = [\,]$.

(5) (a) Trivial. (f) Use $toe(cadd(c_1, b_1, cadd(c_2, b_2, in(b_3, [\,])))) = b_3$.

(6) (a) Trivial using definitions. (f) Idem. □

5. Calculating with $n+1$ similar processes

5.1. *Introduction*

Distributed algorithms are generally configured as an arbitrarily large but finite set of processors that run similar programs. Using μCRL this can be neatly described. Assume that the individual processes are given by $P(k)$, where k:*Nat* is the index of the process. The following equation puts $n+1$ of these processes in parallel:

$$Sys(n\colon Nat) = P(0) \lhd eq(n,0) \rhd \bigl(Sys(n-1) \parallel P(n)\bigr). \tag{2}$$

Clearly, the process $Sys(n)$ stands for $P(0) \parallel P(1) \parallel \ldots \parallel P(n)$.

We find descriptions along this line in verifications of the bakery protocol [17], Milner's scheduler [33], a leader election protocol [16], grid protocols [6], and a summing protocol [18].

The description in Equation (2) gives rise to two issues. The first one is whether the equation *defines* in a proper way that *Sys* is the parallel composition of the processes $P(k)$. It is clear that the parallel composition of processes $P(k)$ is a solution for *Sys*. In this section we show that, assuming the principle CL-RSP (see Definition 3.4), this is the only solution for *Sys* in (2). So, an equation in the form of (2) is indeed a proper definition.

The second issue is to extract a linear process equation from the description in (2). We show in general terms how given a linear format for the processes $P(k)$, a process in linear format equivalent to $Sys(n)$ can be given.

5.2. *Linearization of two different parallel processes*

To provide the reader of this section with a basic understanding of the issues involved in the linearization of the parallel composition of processes, we provide a simple example first. Two LPEs will be composed and a linear process equation for their parallel composition will be given.

We provide here the linearization of the parallel composition of two linear processes $P(d)$ and $P'(d')$. The parameters d and d' of certain arbitrary sorts D and D' denote the parameters of the LPEs. We assume that the processes are defined by linear equations of the following form:

$$P(d{:}D) = \sum_{i\in I}\sum_{e_i:E_i} a_i\bigl(f_i(d,e_i)\bigr)\cdot P\bigl(g_i(d,e_i)\bigr) \lhd c_i(d,e_i) \rhd \delta,$$

and

$$P'(d'{:}D') = \sum_{i'\in I'}\sum_{e'_{i'}:E'_{i'}} a'_{i'}\bigl(f'_{i'}(d',e'_{i'})\bigr)\cdot P'\bigl(g'_{i'}(d',e'_{i'})\bigr) \lhd c'_{i'}(d',e'_{i'}) \rhd \delta.$$

We also assume that these equations are convergent, since by CL-RSP this guarantees that they define unique processes.

Now consider the LPE $Q(d, d')$ defined by the following equation:

$$
\begin{aligned}
&Q(d{:}D, d'{:}D') \\
&\quad = \sum_{i \in I} \sum_{e_i : E_i} a_i\big(f_i(d, e_i)\big) \cdot Q\big(g_i(d, e_i), d'\big) \lhd c_i(d, e_i) \rhd \delta \\
&\qquad + \sum_{i' \in I'} \sum_{e'_{i'} : E'_{i'}} a'_{i'}\big(f'_{i'}(d', e'_{i'})\big) \cdot Q\big(d, g'_{i'}(d', e'_{i'})\big) \lhd c'_{i'}(d', e'_{i'}) \rhd \delta \\
&\qquad + \sum_{i \in I} \sum_{i' \in I'} \sum_{e_i : E_i} \sum_{e'_{i'} : E'_{i'}} \gamma\big(a_i\big(f_i(d, e_i)\big), a'_{i'}\big(f'_{i'}(d', e'_{i'})\big)\big) \cdot \\
&\qquad\quad Q\big(g_i(d, e_i), g'_{i'}(d', e'_{i'})\big) \lhd c_i(d, e_i) \wedge c'_{i'}(d', e'_{i'}) \rhd \delta.
\end{aligned}
$$

The first summand describes the cases that an action from $P(d)$ can be executed. The second summand describes the cases that an action from $P'(d')$ can be executed. The third summand describes the cases that the two processes try to communicate. In each summand the change of the state is only due to the original processes involved in the action that is executed.

Although the process Q is strictly speaking not an LPE, it is obvious that an LPE Q' can be given that is equivalent. This is left as an exercise to the reader.

THEOREM 5.1. *Let P, P', and Q be the LPEs given above. For all $d{:}D$ and $d'{:}D'$ we have*

$$P(d) \parallel P'(d') = Q(d, d').$$

5.3. *Linearization of parallel processes*

We shall now describe the linearization of $n+1$ parallel linear processes $P(k, d)$. The natural number k ($0 \leqslant k \leqslant n$) is the index of the process and the parameter d of some arbitrary sort D denotes the other parameters. We assume that each process $P(k, d)$ is defined using a linear equation of the form:

$$
\begin{aligned}
&P(k{:}Nat, d{:}D) \\
&\quad = \sum_{i \in I} \sum_{e_i : E_i} a_i\big(f_i(k, d, e_i)\big) \cdot P\big(k, g_i(k, d, e_i)\big) \lhd c_i(k, d, e_i) \rhd \delta. \qquad (3)
\end{aligned}
$$

We also assume that this equation is convergent, as this guarantees that this equation defines a unique process.

In order to define the parallel composition we need to determine the 'state space' of the LPE Q. The state space of Q is built from the state spaces of the LPEs $P(k, d)$ that are composed in parallel. As we do not know in advance the number of processes composed (i.e., n) we define a new sort *DTable*, which is a table indexed by natural numbers containing elements of the sort D. In this table the kth entry represents the state space of process

$P(k,d)$. In order to do so, we also need an auxiliary function *if* : **Bool** $\times D \times D \to D$ for *if* − *then* − *else*.

$$\begin{array}{ll}
\textbf{map} & \mathit{if}: \textbf{Bool} \times D \times D \to D \\
\textbf{var} & d, d' : D \\
\textbf{rew} & \mathit{if}(\mathsf{t}, d, d') = d \\
 & \mathit{if}(\mathsf{f}, d, d') = d'
\end{array}$$

In the sequel we assume an equality function $eq: D \times D \to \textbf{Bool}$.

The constant *emT* of sort *DTable* is the empty table. The function *upd* places a new entry with an index in the table and the function *get* gets a specific entry from the table using its index. We characterize these operators by one single equation. Note that we do not specify what happens if an element is read from the empty table. We refer to the characterising axiom for tables as the *table axiom*. Besides this axiom, we use the fact that tables are generated from the empty table by the update function. This allows us the use of induction on these two operations.

$$\begin{array}{ll}
\textbf{sort} & \mathit{DTable} \\
\textbf{func} & \mathit{emT}: \to \mathit{DTable} \\
 & \mathit{upd}: \mathit{Nat} \times D \times \mathit{DTable} \to \mathit{DTable} \\
\textbf{map} & \mathit{get}: \mathit{Nat} \times \mathit{DTable} \to D \\
\textbf{var} & n, m: \mathit{Nat} \\
 & d: D \\
 & dt: \mathit{DTable} \\
\textbf{rew} & \mathit{get}\big(n, \mathit{upd}(m, d, dt)\big) = \mathit{if}\big(eq(n, m), d, \mathit{get}(n, dt)\big)
\end{array}$$

In the remainder we write $dt[i]$ instead of $get(i, dt)$.

We can use the following process definition to put n of the processes $P(k,d)$ in parallel.

$$\begin{aligned}
&\mathit{Sys}(n: \mathit{Nat}, dt: \mathit{DTable}) \\
&\quad = P(0, dt[0]) \lhd eq(n, 0) \rhd \big(P(n, dt[n]) \parallel \mathit{Sys}(n-1, dt)\big).
\end{aligned} \tag{4}$$

We assume that there is a commutative and associative communication function γ that explains how two actions can synchronize. In case two actions do not synchronize it yields δ. In this section we assume the so-called handshaking axiom, that says that no more than two actions can synchronize. In other words, for all actions a_1, a_2 and a_3, $\gamma(a_1, \gamma(a_2, a_3)) = \delta$ (cf. [8]).

In this section we work towards a linear description of $\mathit{Sys}(n, dt)$ (Lemma 5.3). As a bonus we get that $\mathit{Sys}(n, dt)$ has at most one solution (Corollary 5.5). We also provide an alternative transformed linear description which we believe to be more convenient in concrete instances (Theorem 5.6).

The following lemma will be useful in the calculations in the sequel.

LEMMA 5.2. *For* c_1, c_2:**Bool**, d, d_1, d_2, d_3: *D*, m, m_1, m_2, n: *Nat, and dt*: *DTable*:

(1) $\neg(c_1 \wedge c_2) \rightarrow \mathit{if}(c_1, d_1, \mathit{if}(c_2, d_2, d_3)) = \mathit{if}(c_2, d_2, \mathit{if}(c_1, d_1, d_3))$;

(2) $m > n \rightarrow \mathit{Sys}(n, dt) = \mathit{Sys}(n, \mathit{upd}(m, d, dt))$;

(3) $m_1 \neq m_2 \rightarrow \mathit{upd}(m_1, d_1, \mathit{upd}(m_2, d_2, dt))[n] = \mathit{upd}(m_2, d_2, \mathit{upd}(m_1, d_1, dt))[n]$.

PROOF. The first two facts are straightforwardly proven by induction on c_1 and n, respectively. The last fact follows directly by the table axiom and the first item of this lemma. □

Below we present the main lemma of this section. It gives an expansion of *Sys*. As has been stated above, we find the form of this expansion not very convenient as it has the condition $k_1 > k_2$ in its second group of summands. The more convenient form in Theorem 5.6 restricts the number of alternatives by requiring $i_2 \leqslant i_1$. However, contrary to the linearization in Theorem 5.6 we can prove this lemma by induction on n.

LEMMA 5.3. *The process Sys as defined in Equations* (3) *and* (4) *is a solution for Q in Equation* (5) *below.*

$$
\begin{aligned}
&Q(n{:}\,\mathit{Nat}, dt{:}\,\mathit{DTable})\\
&\quad= \sum_{i \in I} \sum_{k:\mathit{Nat}} \sum_{e_i:E_i} a_i\big(f_i(k, dt[k], e_i)\big) \cdot Q\big(n, \mathit{upd}\big(k, g_i(k, dt[k], e_i), dt\big)\big)\\
&\qquad \lhd c_i(k, dt[k], e_i) \wedge k \leqslant n \rhd \delta\\
&\qquad + \sum_{i_1 \in I} \sum_{i_2 \in I} \sum_{k_1:\mathit{Nat}} \sum_{k_2:\mathit{Nat}} \sum_{e_{i_1}:E_{i_1}} \sum_{e_{i_2}:E_{i_2}} \gamma(a_{i_1}, a_{i_2})\big(f_{i_1}(k_1, dt[k_1], e_{i_1})\big) \cdot \qquad (5)\\
&\qquad Q\big(n, \mathit{upd}\big(k_1, g_{i_1}(k_1, dt[k_1], e_{i_1}), \mathit{upd}\big(k_2, g_{i_2}(k_2, dt[k_2], e_{i_2}), dt\big)\big)\big)\\
&\qquad \lhd c_{i_1}(k_1, dt[k_1], e_{i_1}) \wedge c_{i_2}(k_2, dt[k_2], e_{i_2}) \wedge\\
&\qquad\quad \mathit{eq}\big(f_{i_1}(k_1, dt[k_1], e_{i_1}), f_{i_2}(k_2, dt[k_2], e_{i_2})\big) \wedge k_1 > k_2 \wedge k_1 \leqslant n \rhd \delta.
\end{aligned}
$$

LEMMA 5.4. *Equation* (5) *is convergent.*

PROOF. As (3) is convergent, there is a well-founded relation $<$ on $\mathit{Nat} \times D$, such that if $c_i(k, d, e_i)$ and $a_i = \tau$, then $(k, g_i(k, d, e_i)) < (k, d)$.

Using $<$ we can define a well-founded relation $\prec$ on $\mathit{Nat} \times \mathit{DTable}$ as follows:

$$
\begin{aligned}
(n_1, dt_1) \prec (n_2, dt_2) \text{ iff } & \mathit{eq}(n_1, n_2)\\
& \wedge \text{ for all } 0 \leqslant k \leqslant n_1 : (k, dt_1[k]) \leqslant (k, dt_2[k])\\
& \wedge \text{ for some } 0 \leqslant k \leqslant n_1 : (k, dt_1[k]) < (k, dt_2[k])
\end{aligned}
$$

where $(k_1, d_1) \leqslant (k_2, d_2)$ iff $(k_1, d_1) < (k_2, d_2)$, or $\mathit{eq}(k_1, k_2)$ and $\mathit{eq}(d_1, d_2)$.

It is easy to see that $\prec$ is a convergence witness for Equation (5). □

COROLLARY 5.5. *Equation* (4) *has at most one solution for the variable Sys.*

PROOF. Lemma 5.3 says that any solution for *Sys* in (4) is a solution for Q in (5). By Lemma 5.4 and CL-RSP there is at most one solution for Q in (5). Henceforth, Equation (4) has at most one solution, too. □

We consider the following theorem the main result of this section, as it provides a linear equivalent of Equation (4) that is easy to use and to obtain. We assume that there is a total reflexive ordering $\leqslant$ on I. As I is an index set, this is a very reasonable assumption.

THEOREM 5.6. *The process Sys as defined in Equations* (3) *and* (4) *is the* (*unique*) *solution for Q in the* (*convergent*) *equation below*; *so* $Sys(n, dt) = Q(n, dt)$ *for all* $n{:}\,Nat$ *and* $dt{:}\,DTable$.

$$
\begin{aligned}
&Q(n{:}\,Nat, dt{:}\,DTable)\\
&\quad = \sum_{i\in I}\sum_{k:Nat}\sum_{e_i:E_i} a_i\big(f_i(k, dt[k], e_i)\big)\,Q\big(n, upd\big(k, g_i(k, dt[k], e_i), dt\big)\big)\\
&\qquad \lhd c_i(k, dt[k], e_i) \wedge k \leqslant n \rhd \delta\\
&\qquad + \sum_{i_1\in I}\ \sum_{i_2\in I\wedge i_2\leqslant i_1}\ \sum_{k_1:Nat}\ \sum_{k_2:Nat}\ \sum_{e_{i_1}:E_{i_1}}\ \sum_{e_{i_2}:E_{i_2}} \gamma(a_{i_1}, a_{i_2})\big(f_{i_1}(k_1, dt[k_1], e_{i_1})\big)\cdot\\
&\qquad Q\big(n, upd(k_1, g_{i_1}(k_1, dt[k_1], e_{i_1}), upd\big(k_2, g_{i_2}(k_2, dt[k_2], e_{i_2}), dt\big)\big)\big)\\
&\qquad \lhd c_{i_1}(k_1, dt[k_1], e_{i_1}) \wedge c_{i_2}(k_2, dt[k_2], e_{i_2}) \wedge\\
&\qquad\quad eq\big(f_{i_1}(k_1, dt[k_1], e_{i_1}), f_{i_2}(k_2, dt[k_2], e_{i_2})\big) \wedge\\
&\qquad\quad \neg eq(k_1, k_2) \wedge k_1 \leqslant n \wedge k_2 \leqslant n \rhd \delta.
\end{aligned}
$$

6. The Tree Identify Protocol of IEEE 1394 in μCRL

We apply the cones and foci technique (Section 3.3) and the linearization of a number of similar parallel processes (Section 5) to a fragment of the software for a high performance serial multimedia bus, the IEEE standard 1394 [29], also known as "Firewire".

Briefly, IEEE 1394 connects together a collection of systems and devices in order to transport all forms of digitalized video and audio quickly, reliably, and inexpensively. Its architecture is scalable, and it is "hot-pluggable", so a designer or user can add or remove systems and peripherals easily at any time. The only requirement is that the form of the network should be a tree (other configurations lead to errors).

The protocol is subdivided into layers, in the manner of OSI, and further into phases, corresponding to particular tasks, e.g., data transmission or bus master identification. Various parts of the standard have been verified using various formalisms and proof techniques. For example, the operation of sending packets of information across the network is described using μCRL in [34] and shown to be faulty using E-LOTOS in [46]. The former is essentially a description only, with five correctness properties stated informally, but not formalized or proved. The exercise of [46] is based on the μCRL description, adding another layer of the protocol and carrying out the verification as suggested, using the tool CADP [15].

In this section we concentrate on the tree identify phase of the physical layer which occurs after a bus reset in the system, e.g., when a node is added to or removed from the network. The purpose of the tree identify protocol is to assign a (new) root, or leader, to the network. Essentially, the protocol consists of a set of negotiations between nodes to establish the direction of the parent-child relationship. Potentially, a node can be a parent to many nodes, but a child of at most one node. A node with no parent (after the negotiations are complete) is the leader. The tree identify protocol must ensure that a leader is chosen, and that it is the only leader chosen.

The specification of the external behaviour of the protocol merely announces that a single leader has been chosen. In the implementation, nodes are specified individually and negotiate with their neighbours to determine the parent-child relationship. Communication is by handshaking. These descriptions may be found in Section 6.1. They were derived with reference to the transition diagram in Section 4.4.2.2 of the standard [29].

We prove, using the cones and foci technique, that the implementation has the same behaviour with respect to rooted branching bisimulation as the specification, thereby showing that the implementation behaves as required, i.e., a single leader is chosen. The proofs may be found in Section 6.2.

Several papers deal with the formal description and analysis of parts of the P1394 protocol. See, e.g., [13,34,46,48].

6.1. *Description of the Tree Identify Protocol*

The μCRL data definitions used here (e.g., *Nat*, *NatSet*, *NatSetList*) are assumed and not presented; they are straightforward and examples of these or similar types may be found in [34].

The most abstract specification of the tree identify protocol is the one which merely reports that a leader has been found. The network is viewed as a whole, and no communications between nodes are specified. We define

$$Spec = leader \cdot \delta.$$

In the description of the implementation each node in the network is represented by a separate process. Individual nodes are specified below as processes *Node*. Nodes are described by three parameters:

- a natural number i: the identification number of the node. This is used to parameterize communications between nodes, and is not changed during any run of the protocol;
- a set of natural numbers p: the set of node identifiers of potential parents of the node. The initial value is the set of all neighbours, decreasing to either a singleton (containing the parent node) or the empty set (indicating that the node is the elected leader);
- a natural number s: the current state of the node. We use two state values: 0 corresponds to "still working" and 1 to "finished". The initial value is 0.

The identification number of nodes has been introduced to aid specification and does not appear in [29]. In reality a device has a number of ports and knows whether or not a port is connected to another node; there is no need for node identifiers.

A node can send and receive messages: an action $s(i, j)$ is the sending of a request by node i to node j to become its parent, and an action $r(i, j)$ is the receiving of a parent request from node i by node j. When the nodes of the network are composed in parallel, these two actions synchronize with each other to produce a c action. An action $c(i, j)$ is the establishment of a child-parent relation between node i and node j, where i is the child and j is the parent.

We define the actions and the communications by the following μCRL specification:

$$\begin{array}{ll} \textbf{act} & s, r, c: Nat \times Nat \\ & leader \\ \textbf{comm} & s \mid r = c \end{array}$$

If a node is still active and its set of potential parents is empty, it declares itself leader by the execution of the *leader* action. By definition, nodes in state 1 are equivalent to deadlock. An individual node with identification number i is defined by means of the process $Node(i, p, s)$. The processes $Node(i, p, s)$ are defined by the following LPE.

DEFINITION 6.1 (*Implementation of a node*).

$$\begin{aligned} & Node(i{:}Nat,\ p{:}NatSet,\ s{:}Nat) \\ & \quad = leader \cdot Node(i, p, 1) \lhd s = 0 \land isempty(p) \rhd \delta \\ & \qquad + \sum_{j:Nat} r(j, i) \cdot Node(i, p \setminus \{j\}, s) \lhd s = 0 \land j \in p \rhd \delta \\ & \qquad + \sum_{j:Nat} s(i, j) \cdot Node(i, p, 1) \lhd s = 0 \land p = \{j\} \rhd \delta. \end{aligned}$$

The implementation then consists of the parallel composition of $n + 1$ nodes where the loose send and receive actions are encapsulated: $H = \{s, r\}$. This implementation is described by the process $Imp(n, P_0)$, with P_0 describing the configuration of the network:

$$Imp(n{:}Nat,\ P_0{:}NatSetList) = \partial_H\big(Nodes(n, P_0)\big),$$

where

$$\begin{aligned} Nodes(n, P_0) \ = \ & Node(0, P_0[0], 0) \lhd n = 0 \rhd \\ & \big(Node(n, P_0[n], 0) \parallel Nodes(n-1, P_0)\big). \end{aligned}$$

P_0 is a list of sets of connections for all nodes indexed by node number; it gives the initial values for the sets of potential parents. Initially all nodes are in state 0.

6.2. *Correctness of the implementation*

As mentioned earlier, the protocol operates correctly only on tree networks, i.e., under the assumption of a good network topology. Networks with loops will cause a timeout in the real protocol, and unconnected nodes will simply be regarded as another network. The property of *GoodTopology* is formalized below.

DEFINITION 6.2. Given n:*Nat*, the maximal node identifier in the network, and a list P_0:*NatSetList* giving a set of neighbours for all nodes in the network, the conjunction of the following properties is called *GoodTopology*(n, P_0):
- P_0 is symmetric: $\forall_{i,j}\ (i \in P_0[j] \leftrightarrow j \in P_0[i])$.
- P_0 is a tree, i.e., it is a connected graph with no loops.

As a preliminary step to applying the cones and foci proof method, the process *Spec* defined in Section 6.1 must be translated into linear form. In order to do so, a data parameter must be added on which to base a state mapping from the data of process *Imp*. We define

$$\textit{L-Spec}(b{:}\mathbf{Bool}) = \textit{leader} \cdot \textit{L-Spec}(\mathsf{f}) \lhd b \rhd \delta.$$

Then, the process *L-Spec*(t) and the original specification *Spec* are equivalent.

THEOREM 6.3. *Let Spec and L-Spec be as above. Then L-Spec*(t) = *Spec.*

PROOF. The following computation clearly establishes the equivalence: $\textit{L-Spec}(\mathsf{t}) = \textit{leader} \cdot \textit{L-Spec}(\mathsf{f}) \lhd \mathsf{t} \rhd \delta = \textit{leader} \cdot \textit{L-Spec}(\mathsf{f}) = \textit{leader} \cdot (\textit{leader} \cdot \textit{L-Spec}(\mathsf{f}) \lhd \mathsf{f} \rhd \delta) = \textit{leader} \cdot \delta = \textit{Spec}$. □

The linearization of *Imp* is given by the following LPE for *L-Imp*.

$$\begin{aligned}
&\textit{L-Imp}(n{:}\textit{Nat}, P{:}\textit{NatSetList}, S{:}\textit{NatList}) \\
&\quad = \sum_{i:\textit{Nat}} \textit{leader} \cdot \textit{L-Imp}(1/S[i]) \lhd S[i] = 0 \wedge \textit{isempty}(P[i]) \wedge i \leqslant n \rhd \delta \\
&\qquad + \sum_{i,j:\textit{Nat}} c(j,i) \cdot \textit{L-Imp}\big((P[i] \setminus \{j\})/P[i], 1/S[j]\big) \\
&\qquad \lhd S[j] = 0 \wedge P[j] = \{i\} \wedge S[i] = 0 \wedge j \in P[i] \wedge i \neq j \wedge i, j \leqslant n \rhd \delta.
\end{aligned}$$

This linearization can be derived straightforwardly from the definition of individual nodes using the linearization technique of Section 5.

THEOREM 6.4. *Let S_0:NatList be the list of initial state values for the nodes, so for all i:Nat we have $S_0[i] = 0$. Then Imp(n, P_0) = L-Imp(n, P_0, S_0).*

The proof of correctness also requires an invariant on the data states of the implementation. The invariant $I(n, P, S)$ is the conjunction of the invariants listed below. These invariants hold in every state that can be reached from the initial state (n, P_0, S_0). The variables i and j are universally quantified over $\{0, \ldots, n\}$. The notation *singleton*(X) represents the predicate $|X| = 1$, i.e., it expresses that the set X contains precisely one element.

$$\mathrm{I}_1\colon\quad S[i] = 0 \vee S[i] = 1.$$

I_2: $j \in P_0[i] \leftrightarrow j \in P[i] \vee i \in P[j]$.

I_3: $j \in P_0[i] \wedge j \notin P[i] \rightarrow S[j] = 1$.

I_4: $S[i] = 1 \rightarrow singleton(P[i]) \vee isempty(P[i])$.

I_5: $j \in P[i] \wedge S[i] = 0 \rightarrow S[j] = 0 \wedge i \in P[j]$.

The proofs of these invariants are straightforward, and omitted here. Invariant I_1 expresses that each of the components is in state 0 or in state 1. Invariant I_2 expresses that if the nodes i and j are connected initially, then at all times one is a potential parent of the other. This means that no connections are removed. Invariant I_3 expresses that the potential parent relationship between a node j and a node i is only destroyed if i becomes the parent node of j. Invariant I_4 states that if a node is done there is at most one potential parent left. Invariant I_5 expresses that if a node is still busy and has another node as its potential parent then also this potential parent is still busy and considers the other node as a potential parent.

The linearization of *L-Imp* is not sufficient to allow us to apply Theorem 3.11. A prerequisite for applying the cones and foci technique is that the indices of the alternative quantifications preceding a visible action must be the same in the specification and the implementation; clearly this is not the case. The summation over the node identifiers preceding the *leader* action in *L-Imp* correctly reflects that any node can be the root, i.e., there are multiple foci. However, it is not important *which* node is the root, only that one is chosen, and the Boolean condition guarding the *leader* action in *L-Imp* ensures that this is the case. We adapt the specification in such a way that the *leader* action there is also preceded by an alternative quantification over the node identifiers. Clearly the linear specification obtained in this way is equal to the old linear specification.

$$L\text{-}Spec(b{:}\mathbf{Bool}) = \sum_{i:Nat} leader \cdot L\text{-}Spec(\mathsf{f}) \lhd b \wedge i \leqslant n \rhd \delta.$$

The theorem to be demonstrated can now be stated as:

THEOREM 6.5. *Under the assumption of GoodTopology*(n, P_0) *and* $I(n, P_0, S_0)$,

$$\tau \cdot L\text{-}Spec(\mathsf{t}) = \tau \cdot \tau_{\{c\}}\big(L\text{-}Imp(n, P_0, S_0)\big).$$

In the special case where $n = 0$ (*there is only one node in the network*) *we have*

$$L\text{-}Spec(\mathsf{t}) = \tau_{\{c\}}\big(L\text{-}Imp(n, P_0, S_0)\big).$$

This is a direct instantiation of Theorem 3.11 with the initial state, because in the initial state the focus condition (defined below) is true if and only if $n = 0$. In order to prove Theorem 6.5 the matching criteria must be satisfied. To show that the matching criteria hold we first define the focus condition and the state mapping for $\tau_{\{c\}}(L\text{-}Imp)$. The focus condition *FC* is the condition under which no more τ steps can be made, i.e., it is the negation of the condition for making a τ step:

$$FC(n, P, S) = \forall_{i,j \leqslant n}\big(S[i] = 1 \vee P[i] \neq \{j\} \vee S[j] = 1 \vee i \notin P[j] \vee i = j\big).$$

The state mapping h is a function mapping data states of the implementation into data states of the simple specification. In this case h is defined so that it is t before the visible *leader* action occurs and f afterwards:

$$h(n, P, S) = \neg\big(\forall_{i \leqslant n}(S[i] = 1)\big).$$

If a node can do the leader action then all other nodes are in state 1. So, if a node declares itself the leader, then it is the first to do so, and because after this all nodes are in state 1, there will be no subsequent leader action.

LEMMA 6.6 (Uniqueness of root).

$$\forall_{i \leqslant n}\big(isempty(P[i])\big) \rightarrow \forall_{j \leqslant n}(j \neq i \rightarrow S[j] = 1).$$

PROOF. We assume nodes $i, j \leqslant n$ such that $i \neq j \wedge isempty(P[i]) \wedge S[j] = 0$, and derive a contradiction. By *GoodTopology* there is a path of distinct nodes $i = i_0 \ldots i_m = j$, such that $\forall_{k<m}\ (i_{k+1} \in P_0[i_k])$. By I_2 and $isempty(P[i_0])$ we see that $i_0 \in P[i_1]$. Then by I_3 $S[i_1] = 1$, and by I_4 $singleton(P[i_1])$. In a similar way we derive for all $0 < k \leqslant m$ that $P[i_k] = \{i_{k-1}\}$ and $S[i_k] = 1$. So in particular $S[j] = 1$. □

The matching criteria. Given the particulars of *L-Imp*, *L-Spec*, *FC* and h, the matching criteria may be mechanically derived from the general forms of Definition 3.10. The instantiated matching criteria are stated below, together with the proofs that they hold.

(1) The implementation is convergent. Each τ step decreases the number of nodes i for which $S[i] = 0$ by one.

(2) In any data state $d = (n, P, S)$ of the implementation, the execution of an internal step leads to a state with the same h-image.
Suppose an internal action is possible, i.e., there are nodes $i, j \leqslant n$ such that

$$S[i] = 0 \wedge P[i] = \{j\} \wedge S[j] = 0 \wedge i \in P[j] \wedge\ i \neq j.$$

We see that $S[i] = 0$ and $S[j] = 0$ and hence $h(d) = \mathsf{t}$. We have to show that if we reach a state $d' = (n, P', S')$ by the communication between nodes i and j, then $h(d') = \mathsf{t}$. We easily find that $S'[k] = S[k]$ for $0 \leqslant k \leqslant n$ and $k \neq j$ and $S'[j] = 1$. Hence $S'[i] = 0$. Therefore $h(d') = \mathsf{t}$.

(3) If the implementation can do the *leader* action, then so can the specification:

$$\big(\exists_{i \leqslant n}\big(S[i] = 0 \wedge isempty(P[i])\big)\big) \rightarrow \exists_{i \leqslant n} b.$$

Trivial.

(4) If the specification can do the *leader* action and the implementation cannot do an internal action, then the implementation must be able to do the *leader* action:

$$(\exists_{i \leqslant n} b) \wedge FC \rightarrow \big(\exists_{i \leqslant n}\big(S[i] = 0 \wedge isempty(P[i])\big)\big).$$

The specification can do the *leader* action if it is in a state where its only parameter b equals t. This means that for the corresponding state d in the implementation we have $S[i] \neq 1$ for some $i \leqslant n$. Using invariant I_1 we thus obtain $S[i] = 0$. Now we only have to show that $isempty(P[i])$.
So suppose that $\neg isempty(P[i])$. Let $i_1 \in P[i]$. From invariant I_5 it follows that $S[i_1] = 0 \wedge i \in P[i_1]$. From $FC \wedge S[i] = 0 \wedge S[i_1] = 0 \wedge i_1 \in P[i]$ it follows that $P[i_1] \neq \{i\}$. Thus there exists $i_2 \in P[i_1]$, $i_2 \neq i$ such that $S[i_2] = 0 \wedge i_1 \in P[i_2]$. In this way an infinite sequence $i = i_0, i_1, i_2, i_3, \ldots$ can be constructed such that for all $k{:}Nat$ we have $S[k] = 0 \wedge i_k \in P[i_{k+1}] \wedge i_k \neq i_{k+2}$. By I_2 we see that this infinite path is also a path in P_0. This contradicts *GoodTopology*.

(5) The implementation and the specification perform external actions with the same parameter. Trivial; the action *leader* involves no data.

(6) After the implementation and the specification perform the *leader* action, the mapping h still holds: if the implementation can reach data state d' by the execution of the *leader* action, then $h(d') = \mathsf{f}$.
Assume that the implementation can perform the *leader* action: i.e., $S[i] = 0 \wedge isempty(P[i])$ for some $i \leqslant n$. Then also the specification can do the *leader* action by item (3) Hence $b = \mathsf{t}$. After the execution of the leader action the state of the specification is given by $b = \mathsf{f}$. Then by Lemma 6.6 we see that all nodes other than i are in state 1. We also see that by the execution of the *leader* action the state of node i becomes 1. So after the action all nodes are in state 1, so then the value of h will be f.

By Theorem 3.11 it follows that for all n, P, S

$$\begin{aligned} I(n,P,S) \rightarrow\ & L\text{-}Imp(n,P,S) \lhd FC(n,P,S) \rhd \tau \cdot L\text{-}Imp(n,P,S) \\ & = L\text{-}Spec\big(h(n,P,S)\big) \lhd FC(n,P,S) \rhd \tau \cdot L\text{-}Spec\big(h(n,P,S)\big). \end{aligned}$$

Instantiation of this equation gives

$$\begin{aligned} I(n,P_0,S_0) \rightarrow\ & L\text{-}Imp(n,P_0,S_0) \lhd FC(n,P_0,S_0) \rhd \tau \cdot L\text{-}Imp(n,P_0,S_0) \\ & = L\text{-}Spec\big(h(n,P_0,S_0)\big) \lhd FC(n,P_0,S_0) \rhd \\ & \quad \tau \cdot L\text{-}Spec\big(h(n,P_0,S_0)\big), \end{aligned}$$

which reduces to

$$\begin{aligned} & L\text{-}Imp(n,P_0,S_0) \lhd n = 0 \rhd \tau \cdot L\text{-}Imp(n,P_0,S_0) \\ & \quad = L\text{-}Spec(\mathsf{t}) \lhd n = 0 \rhd \tau \cdot L\text{-}Spec(\mathsf{t}). \end{aligned}$$

7. Confluence for process verification

7.1. *Introduction*

In his seminal book [38] Milner devotes a chapter to the notions strong and observation confluence in process theory. Many other authors have confirmed the importance of confluence. For example, in [28,44] the notion is used for on-the-fly reduction of finite state spaces and in [38] it has been used for the verification of protocols.

We felt that a more general treatment of the notion of confluence is in order. The first reason for this is that the treatment of confluence has always been somewhat ad hoc in the setting of process theory. This strongly contrasts with for instance term rewriting, where confluence is one of the major topics. In particular, we want to clarify the relation with *τ-inertness*, which says that if $s \xrightarrow{\tau} s'$, then s and s' are branching bisimilar.

The second reason is that we want to develop systematic ways to prove distributed systems correct in a precise and formal fashion. In this way we want to provide techniques to construct fault free distributed systems. For this purpose the language μCRL is used. Experience with several protocols gave rise to the development of new and the adaptation of existing techniques to make systematic verification possible [4,5]. Employing confluence also belongs to these techniques. It appears to enable easier verification of distributed systems, which in essence boils down to the application of τ-inertness.

In Section 7.2, we address the relationship between confluence and τ-inertness on transition systems. We introduce strong and weak confluence. We establish that strong confluence implies τ-inertness and we establish that, for convergent transition systems, weak confluence implies τ-inertness.

To be able to deal with systems with idle loops, for instance communication protocols over unreliable channels, we distinguish between progressive and non progressive internal actions. This leads to a notion of weak progressive confluence that only considers the progressing internal actions. We find that weak progressive confluence is enough to guarantee τ-inertness for transition systems that are convergent with respect to progressing internal steps.

In Section 7.3, we direct our attention to establishing confluence. It does hardly make sense to establish confluence directly on transition systems, because these are generally far too large to be represented. Therefore, we try to establish confluence on processes described by LPEs [5] because these can represent large transition systems in a compact symbolic way. In Section 7.4, we show how we can use τ-inertness and confluence to reduce state spaces both on transition systems and on linear processes. Finally, we provide an example illustrating that the application of confluence often reduces the size of state spaces considerably and simplifies the structure of distributed systems, while preserving branching bisimulation. This is in general a very profitable preprocessing step before analysis, testing or simulation of a distributed system.

7.2. *Confluence and τ-inertness*

Throughout this section we fix the set of actions A, which contains an internal action τ.

DEFINITION 7.1. A *transition system* is a pair $(S, \longrightarrow)$ with S a set and $\longrightarrow \subseteq S \times \mathsf{A} \times S$. The set of triples $\longrightarrow$ induces a binary relation $\xrightarrow{a} \subseteq S \times S$ for each $a \in \mathsf{A}$ as follows: for all $s, t \in S$ we have $(s, t) \in \xrightarrow{a}$ iff $(s, a, t) \in \longrightarrow$. We write $s \xrightarrow{a} t$ instead of $(s, a, t) \in \longrightarrow$ and $(s, t) \in \xrightarrow{a}$. The relation $\xrightarrow{a^*} \subseteq S \times S$ denotes the reflexive, transitive closure of the relation $\xrightarrow{a}$.

A transition system $(S, \longrightarrow)$ is called *convergent* iff there is no infinite sequence of the form $s_1 \xrightarrow{\tau} s_2 \xrightarrow{\tau} s_3 \xrightarrow{\tau} \cdots$.

A relation $R \subseteq S \times S'$ is called a *branching bisimulation* on $(S, \longrightarrow\!\!\triangleright)$ and $(S', \longrightarrow\!\!\blacktriangleright)$ iff for all $s \in S$ and $s' \in S'$ such that $s R s'$ we have
(1) $s \stackrel{a}{\longrightarrow\!\!\triangleright} t \to (a = \tau \wedge t R s') \vee (\exists_{u, u'}(s' \stackrel{\tau^*}{\longrightarrow\!\!\blacktriangleright} u \stackrel{a}{\longrightarrow\!\!\blacktriangleright} u' \wedge s R u \wedge t R u'))$, and
(2) $s' \stackrel{a}{\longrightarrow\!\!\blacktriangleright} t' \to (a = \tau \wedge s R t') \vee (\exists_{u, u'}(s \stackrel{\tau^*}{\longrightarrow\!\!\triangleright} u \stackrel{a}{\longrightarrow\!\!\triangleright} u' \wedge u R s' \wedge u' R t'))$.

We say that R is a branching bisimulation on $(S, \longrightarrow\!\!\triangleright)$ iff R is a branching bisimulation on $(S, \longrightarrow\!\!\triangleright)$ and $(S, \longrightarrow\!\!\triangleright)$. The union of all branching bisimulations is denoted as $\underline{\leftrightarrow}_b$.

Next, we present three different notions of confluence on transition systems, namely *strong confluence*, *weak confluence* and *weak progressive confluence*. We investigate whether or not these different notions of confluence are strong enough to serve as a condition for

$$s \stackrel{\tau}{\longrightarrow\!\!\triangleright} t \to s \underline{\leftrightarrow}_b t \tag{6}$$

to hold. Transition systems that satisfy Equation (6) for all $s, t \in S$ are called τ-*inert with respect to* $\underline{\leftrightarrow}_b$.

DEFINITION 7.2. A transition system $(S, \longrightarrow\!\!\triangleright)$ is called *strongly confluent* iff for all pairs $s \stackrel{a}{\longrightarrow\!\!\triangleright} t$ and $s \stackrel{\tau}{\longrightarrow\!\!\triangleright} s'$ of different steps there exists a state t' such that $t \stackrel{\tau}{\longrightarrow\!\!\triangleright} t'$ and $s' \stackrel{a}{\longrightarrow\!\!\triangleright} t'$. In a diagram:

$$\begin{array}{ccc} s & \stackrel{a}{\longrightarrow\!\!\triangleright} & t \\ \tau \downarrow & & \downarrow \tau \\ s' & \stackrel{a}{-\!-\!\triangleright} & t' \end{array}$$

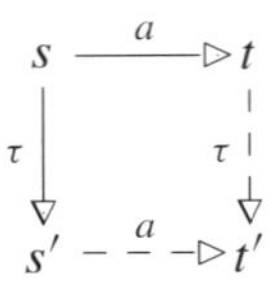

Omitting the word 'different' in Definition 7.2 would give a stronger notion. This can be seen as follows: the transition system represented by $s \stackrel{\tau}{\longrightarrow\!\!\triangleright} t$ is strongly confluent, but would not be strongly confluent when the word 'different' was omitted.

THEOREM 7.3. *Strongly confluent transition systems are τ-inert with respect to $\underline{\leftrightarrow}_b$.*

The converse of Theorem 7.3 is obviously not valid. A transition system that is τ-inert with respect to $\underline{\leftrightarrow}_b$, is not necessarily strongly confluent. As a counter-example one can take the transition system

$$t \stackrel{\tau}{\triangleleft\!\!\longleftarrow} s \stackrel{\tau}{\longrightarrow\!\!\triangleright} u. \tag{7}$$

This counter-example means that strong confluence is actually a stronger notion than we need since we are primarily interested in τ-inertness (with respect to $\underline{\leftrightarrow}_b$). Hence we introduce a weaker notion of confluence, which differs from strong confluence in that it allows zero or more τ-steps in the paths from t to t' and from s' to t'.

DEFINITION 7.4. A transition system $(S, \longrightarrow)$ is called *weakly confluent* iff for each pair $s \xrightarrow{a} t$ and $s \xrightarrow{\tau} s'$ of different steps one of the following holds:

- there exist states t', t_1', t_2' such that $t \xrightarrow{\tau^*} t'$ and $s' \xrightarrow{\tau^*} t_1' \xrightarrow{a} t_2' \xrightarrow{\tau^*} t'$. In a diagram:

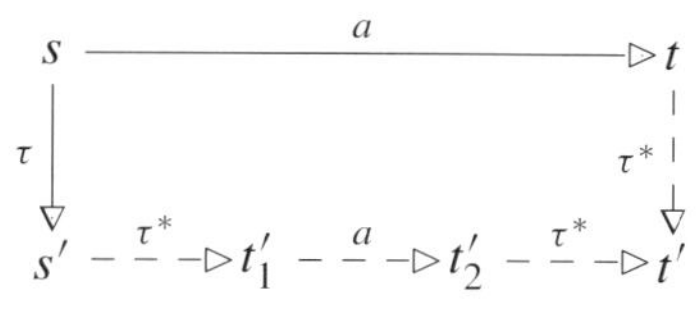

- $a = \tau$ and there exists a state t' such that $t \xrightarrow{\tau^*} t'$ and $s' \xrightarrow{\tau^*} t'$. In a diagram:

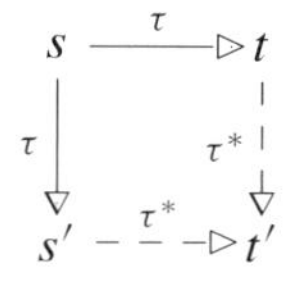

Weak confluence is too weak to serve as a condition for Equation (6) to hold, i.e., weakly confluent transition systems are not necessarily τ-inert with respect to $\underline{\leftrightarrow}_b$. However, if we also assume that the transition system is convergent, then weak confluence implies τ-inertness.

THEOREM 7.5. *Let $(S, \longrightarrow)$ be convergent and weakly confluent, then $(S, \longrightarrow)$ is τ-inert with respect to $\underline{\leftrightarrow}_b$.*

Theorem 7.5 relies on convergence of the transition system in question. However, many realistic examples of protocol specifications correspond to transition systems that are not convergent. As soon as a protocol internally consists in some kind of correction mechanism (e.g., retransmissions in a data link protocol) the specification of that protocol will contain an idle loop.

Since we strongly believe in the importance of applicability to realistic examples, we considered the requirement that the transition system is convergent a serious drawback. Therefore, we distinguish, as in Section 3.4, between progressive internal actions, denoted by $\tau_>$ and non-progressive internal actions, denoted by $\tau_<$. This enables us to formulate a notion of confluence, which is sufficiently strong for our purposes and only relies on convergence of the progressive τ-steps.

CONVENTION 7.6. We use the following notations:

- $s \xrightarrow{\tau_>} t$ for a progressive τ-step from s to t,
- $s \xrightarrow{\tau_<} t$ for a non-progressive τ-step from s to t,
- $s \xrightarrow{\tau} t$ for $s \xrightarrow{\tau_>} t$ or $s \xrightarrow{\tau_<} t$.

From now on we try to prove $\tau_>$*-inertness with respect to* $\leftrightarrow_b$, instead of τ-inertness with respect to $\leftrightarrow_b$. In a formula:

$$s \xrightarrow{\tau_>} t \to s \leftrightarrow_b t \tag{8}$$

It should be noted that the definition of branching bisimulation is not affected by the distinction of progressive and non-progressive τ-steps. We first provide the definition of weak $>$-confluence.

DEFINITION 7.7. A transition system $(S, \longrightarrow)$ is called *weakly* $>$*-confluent* (pronounce: weakly progressive confluent) iff for each pair $s \xrightarrow{\tau_>} s'$ and $s \xrightarrow{a} t$ of different steps one of the following holds:

- there exist states t', s_1', s_2' such that $t \xrightarrow{\tau_>^*} t'$ and $s \xrightarrow{\tau_>^*} s_1' \xrightarrow{a} s_2' \xrightarrow{\tau_>^*} t'$, or
- $a = \tau$ and there exists a state t such that $t \xrightarrow{\tau_>^*} t'$ and $s \xrightarrow{\tau_>^*} t$.

THEOREM 7.8. *Let* $(S, \longrightarrow)$ *be* $>$*-convergent and weakly* $>$*-confluent, then* $(S, \longrightarrow)$ *is* $\tau_>$*-inert with respect to* $\leftrightarrow_b$.

7.3. *Confluence of linear process equations*

We want to use the notion confluence to verify the correctness of processes. In order to do so, we must be able to determine whether a transition system is confluent. This is in general not possible, because the transition systems belonging to distributed systems are often too large to be handled as plain objects. In order to manipulate with large state spaces, processes described by C-LPEs can be used as in these the state space is compactly encoded using data parameters.

In this section we describe how a C-LPE can be shown to be confluent. In the next section we show how confluence is used to reduce the size of state spaces.

Recall the definition of a clustered linear process equation of Definition 3.7. A clustered linear process equation is an expression of the form

$$p(d) = \sum_{a \in Act} \sum_{e_a : E_a} a\big(f_a(d, e_a)\big) \cdot p\big(g_a(d, e_a)\big) \lhd b_a(d, e_a) \rhd \delta. \tag{9}$$

We assume that the internal action τ ($\tau_>$ and $\tau_<$ if progressing and non-progressing τ's are distinguished) has no data parameter.

It is straightforward to see how a linear process equation determines a transition system. The process $p(d)$ can perform an action $a(f_a(d, e_a))$ for every $a \in Act$ and every data element e_a of sort E_a, provided the condition $b_a(d, e_a)$ holds. The process then continues as $p(g_a(d, e_a))$. Hence, the notions defined in the previous sections carry over directly. Thus, if $b_a(d, e_a)$ holds then

$$p(d) \xrightarrow{a(f_a(d, e_a))} p\big(g_a(d, e_a)\big).$$

As we distinguish between progressing and non-progressing τ's, we use the notion convergence with respect to the progressing τ's (i.e., $\tau_>$).

DEFINITION 7.9. A clustered linear process equation as given in Equation (9) is called $>$-convergent iff there is a well-founded ordering $<$ on D such that for all $a \in Act$ with $a = \tau_>$ and for all $d:D$, $e_a:D_a$ we have that $b_a(d, e_a)$ implies $g_a(d, e_a) < d$.

Note that this definition of progressive convergence is in line with the normal notion of convergence (Definition 3.3) in the sense that only the progressive τ-steps are considered.

We provide sufficient criteria for p to be strongly confluent. Let p be the clustered linear process equation given in Equation (9).

THEOREM 7.10. *The process p as defined in Equation (9) is strongly confluent if for all $d:D$, $a \in Act$, $e_a:E_a$, $e_\tau:E_\tau$ such that*
  (i) *if $a = \tau$ then $g_a(d, e_a) \neq g_\tau(d, e_\tau)$,*
 (ii) $b_a(d, e_a)$, *and*
(iii) $b_\tau(d, e_\tau)$
the following property holds: there exist $e'_a:E_a$ and $e'_\tau:E_\tau$ such that

$$f_a(d, e_a) = f_a\big(g_\tau(d, e_\tau), e'_a\big)$$
$$\wedge\, b_a\big(g_\tau(d, e_\tau), e'_a\big)$$
$$\wedge\, b_\tau\big(g_a(d, e_a), e'_\tau\big)$$
$$\wedge\, g_a\big(g_\tau(d, e_\tau), e'_a\big) = g_\tau\big(g_a(d, e_a), e'_\tau\big).$$

The criteria can best be understood via the following diagram.

$$\begin{array}{ccc}
p(d) & \xrightarrow{\ \tau\ } & p(g_\tau(d, e_\tau)) \\
{\scriptstyle a(f_a(d,e_a))}\Big\downarrow & {\scriptstyle a(f_a(g_\tau(d,e_\tau),e'_a))}\Big\downarrow & \\
p(g_a(d, e_a)) & \dashrightarrow^{\tau} & p(g_\tau(g_a(d, e_a), e'_\tau)) = p(g_a(g_\tau(d, e_\tau), e'_a))
\end{array}$$

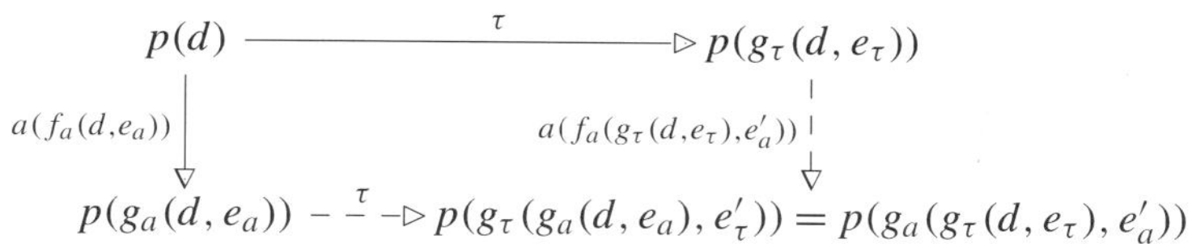

Note that in this diagram $p(g_a(d, e_a))$ and $p(g_\tau(d, e_\tau))$ are supposed to be different if $a = \tau$ (see condition (i) in the above definition).

Now that we have derived a rather simple condition for strong confluence we turn our attention to weak progressive confluence. This is more involved, because we must now speak about sequences of transitions. In order to keep notation compact, we introduce some notation. Let $\sigma, \sigma', \ldots$ range over lists of pairs $\langle a, e_a \rangle$ with $a \in Act$ and $e_a:E_a$. We define $\mathcal{G}_d(\sigma)$ with $d \in D$ by induction over the length of σ:

$$\mathcal{G}_d(\lambda) = d,$$
$$\mathcal{G}_d\big(\sigma\langle a, e_a\rangle\big) = g_a\big(\mathcal{G}_d(\sigma), e_a\big).$$

Each σ determines an execution fragment:

$$\underbrace{p(d) \longrightarrow\!\!\triangleright\, p(\mathcal{G}_d(\sigma))}_{\text{determined by } \sigma} \xrightarrow{a(f_a(\mathcal{G}_d(\sigma), e_a))}\!\!\triangleright\; p(\mathcal{G}_d(\sigma\langle a, e_a\rangle))$$

is the execution fragment determined by $\sigma\langle a, e_a\rangle$. This execution fragment is allowed for $p(d)$ iff the conjunction $\mathcal{B}_d(\sigma)$ of all conditions associated to the actions in σ evaluates to *true*. The Boolean $\mathcal{B}_d(\sigma)$ is also defined by induction to the length of σ:

$$\begin{aligned} \mathcal{B}_d(\lambda) &= \mathit{true}, \\ \mathcal{B}_d(\sigma\langle a, e_a\rangle) &= \mathcal{B}_d(\sigma) \wedge b_a(\mathcal{G}_d(\sigma), e_a). \end{aligned}$$

We write $\pi_1(\sigma)$ for the sequence of actions that is obtained from σ by applying the first projection to all its elements. For example, $\pi_1(\langle a, e_a\rangle\langle b, e_b\rangle) = ab$.

In the following theorem we provide sufficient criteria for a C-LPE p to be weakly $>$-confluent. Due to its generality the theorem looks rather complex. However, in those applications that we considered, the lists that are existentially quantified were mainly empty, which trivializes the main parts of the theorem.

THEOREM 7.11. *The process p as defined in Equation* (9) *is weakly $>$-confluent if p is $>$-convergent and for all $d{:}D$, $a \in Act$, $e_a{:}E_a$, $e_{\tau_>}{:}E_{\tau_>}$ such that*

(i) *if $a = \tau_>$ then $g_a(d, e_a) \neq g_{\tau_>}(d, e_\tau)$,*

(ii) *$b_a(d, e_a)$, and*

(iii) *$b_{\tau_>}(d, e_{\tau_>})$*

the following property holds: there exist $\sigma_1, \sigma_2, \sigma_3$ and $e'_a{:}E'_a$ such that

$$\begin{aligned} &\pi_1(\sigma_i) = \tau_>^* \quad \text{for all } i = 1, 2, 3 \\ &\wedge f_a(d, e_a) = f_a(\mathcal{G}_{g_{\tau_>}(d, e_{\tau_>})}(\sigma_1), e'_a) \\ &\wedge \mathcal{B}_{g_a(d, e_a)}(\sigma_3) \\ &\wedge \mathcal{B}_{g_{\tau_>}(d, e_{\tau_>})}(\sigma) \\ &\wedge \mathcal{G}_{g_a(d, e_a)}(\sigma_3) = \mathcal{G}_{g_{\tau_>}(d, e_{\tau_>})}(\sigma), \end{aligned}$$

where $\sigma = \sigma_1\langle a, e'_a\rangle\sigma_2$, or $a = \tau$ and $\sigma = \sigma_1\sigma_2$.

7.4. *State space reduction*

In this section, we employ the results about confluence and τ-inertness that we have obtained thus far to achieve state space reductions and to simplify the behaviour of processes. First, we present the results on transition systems in general, and then on linear process equations. This is done as for transition systems the results are easy to understand. However, as argued in the previous section, the results can be applied more conveniently in the setting of linear process equations.

DEFINITION 7.12. Let $T_1 = (S, \longrightarrow\!\!\!\!\rhd)$ and $T_2 = (S, \longrightarrow\!\!\!\!\blacktriangleright)$ be transition systems. We call T_2 a *Tau-Prioretized-reduction* (*TP-reduction*) of T_1 iff

(i) $\longrightarrow\!\!\!\!\blacktriangleright \subseteq \longrightarrow\!\!\!\!\rhd$,

(ii) for all $s, s' \in S$ if $s \xrightarrow{a}\!\!\!\!\rhd s'$ then $s \xrightarrow{a}\!\!\!\!\blacktriangleright s'$ or $s \xrightarrow{\tau_>}\!\!\!\!\blacktriangleright s''$ for some s''.

Clearly, T_2 can be obtained from T_1 by iteratively removing transitions from states as long as these keep at least one outgoing progressive τ-step. It does not need any comment that this may considerably reduce the state space of T_1, especially because large parts may become unreachable.

The following theorem says that if T_1 is $\tau_>$-inert with respect to $\underline{\leftrightarrow}_b$, then a TP-reduction maintains branching bisimulation. As confluence implies τ-inertness, this theorem explains how confluence can be used to reduce the size of transition systems.

THEOREM 7.13. *Let $T_1 = (S, \longrightarrow\!\!\!\!\rhd)$ and $T_2 = (S, \longrightarrow\!\!\!\!\blacktriangleright)$ be transition systems. If T_1 is $\tau_>$-inert with respect to $\underline{\leftrightarrow}_b$ and T_2 is a $>$-convergent TP-reduction of T_1 then for each state $s \in S$: $s \underline{\leftrightarrow}_b s$.*

As has been shown in the previous section, weak $>$-confluence can relatively easy be determined on C-LPEs. We provide a way to reduce the complexity of a C-LPE. Below we reformulate the notion of a TP-reduction on C-LPEs. We assume that p is a C-LPE according to Equation (9) and that the data sort E_τ is ordered by some total ordering $\prec$.

DEFINITION 7.14. The *TP-reduction* of p is the linear process

$$p_r(d) = \sum_{a \in Act} \sum_{e_a : E_a} a\big(f_a(d, e_a)\big) \cdot p_r\big(g_a(d, e_a)\big) \lhd b_a(d, e_a) \wedge c_a(d, e_a) \rhd \delta,$$

where

$$c_a(d, e_a) = \begin{cases} \neg \exists_{e_{\tau_>} : E_{\tau_>}} b(d, e_{\tau_>}) & \text{if } a \neq \tau_>, \\ \neg \exists_{e_{\tau_>} : E_{\tau_>}} (e_a \prec e_{\tau_>} \wedge b(d, e_{\tau_>})) & \text{if } a = \tau_>. \end{cases}$$

Note that for the sake of conciseness, we use an existential quantification in the condition $c_a(d, e_a)$, which does not adhere to the formal definition of μCRL.

THEOREM 7.15. *If the linear process p is $>$-convergent and weakly $>$-confluent, then for all $d{:}D$*

$$p(d) \underline{\leftrightarrow}_b p_r(d).$$

7.5. *An example: Concatenation of two queues*

We illustrate how we apply the theory by means of an example, where the structure of the processes is considerably simplified by a confluence argument. Consider the following linear process $Q(q_1, q_2)$ describing the concatenation of two queues q_1 and q_2. The architecture of this process is given in Figure 3.

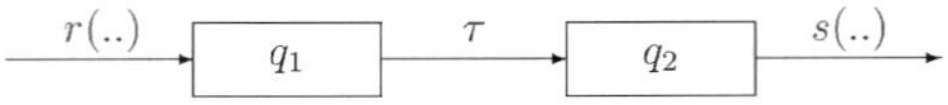

Fig. 3. Architecture of the concatenation of the queues.

$$Q(q_1, q_2) = \sum_{d:D} r(d) \cdot Q(in(d, q_1), q_2) \triangleleft true \triangleright \delta$$
$$+ \tau \cdot Q(untoe(q_1), in(toe(q_1), q_2)) \triangleleft \neg isempty(q_1) \triangleright \delta$$
$$+ s(toe(q_2)) \cdot Q(q_1, untoe(q_2)) \triangleleft \neg isempty(q_2) \triangleright \delta.$$

As we can see, the process $Q(q_1, q_2)$ can always read a datum and insert it in q_1. If q_2 is not empty then the 'toe' of q_2 can be sent. The internal action τ removes the first element of q_1 and inserts it in q_2.

Using Theorem 7.10 we can straightforwardly prove that $Q(q_1, q_2)$ is strongly confluent. Let us consider the strong confluence in more detail with respect to the read action $r(d)$.

$$\begin{array}{ccc} Q(q_1, q_2) & \stackrel{\tau}{\longrightarrow} & Q(untoe(q_1), in(toe(q_1), q_2)) \\ {\scriptstyle r(d)} \downarrow & & \\ Q(in(d, q_1), q_2) & & \end{array}$$

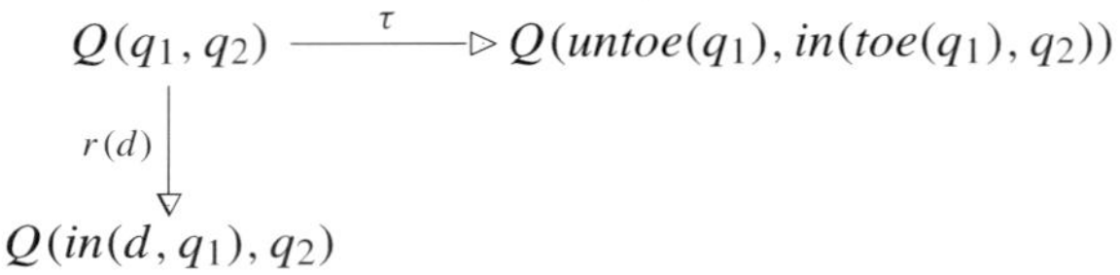

This situation can only occur if both the read action $r(d)$ and the τ action are enabled: t and $\neg isempty(q_1)$. To establish strong confluence in this specific case we must find:

$$\begin{array}{c} Q(untoe(q_1), in(toe(q_1), q_2)) \\ {\scriptstyle r(d')} \downarrow \\ Q(in(d', untoe(q_1)), in(toe(q_1), q_2)) \\ Q(in(d, q_1), q_2) \stackrel{\tau}{\dashrightarrow} Q(untoe(in(d, q_1)), in(toe(in(d, q_1)), q_2)) \end{array}$$

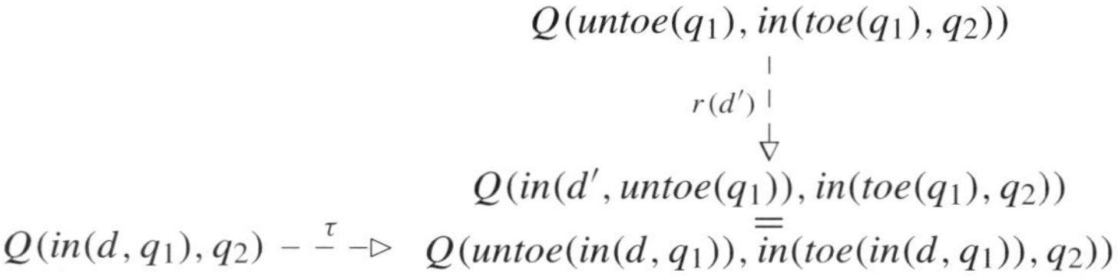

for some $d':D$ such that the data parameters of the two read actions are equal ($d = d'$). Furthermore, we need that the states resulting after the execution of these actions are equal. The state after the $r(d')$ action is given by $(in(d', untoe(q_1)), in(toe(q_1), q_2))$ and after the τ action by $(untoe(in(d, q_1)), in(toe(in(d, q_1)), q_2))$ respectively. With the observation that the equality of states is defined to be pairwise equality we obtain the following condition: for all queues q_1, q_2 and $d:D$

$$\neg isempty(q_1) \rightarrow \exists_{d':D} (d = d'$$
$$\wedge \neg isempty(in(d, q_1))$$
$$\wedge in(d', untoe(q_1)) = untoe(in(d, q_1))$$
$$\wedge in(toe(q_1), q_2) = in(toe(in(d, q_1)), q_2)$$
$$).$$

Similarly, we can formulate the following conditions for the action s. For all queues q_1, q_2

$$\begin{aligned}
&\neg isempty(q_2) \wedge \neg isempty(q_1) \\
&\quad \rightarrow (\ \ toe(q_2) = toe\big(in(toe(q_1), q_2)\big) \\
&\qquad \wedge \neg isempty\big(in\big(toe(q_1), q_2\big)\big) \\
&\qquad \wedge \neg isempty(q_1) \\
&\qquad \wedge untoe(q_1) = untoe(q_1) \\
&\qquad \wedge untoe\big(in\big(toe(q_1), q_2\big)\big) = in\big(toe(q_1), untoe(q_2)\big) \\
&\qquad).
\end{aligned}$$

With the appropriate axioms for queues, the validity of these facts is easily verified.

For the $a = \tau$ we find that the precondition $a = \tau \rightarrow g_a(d, e_a) \neq g_\tau(d, e_\tau)$ is instantiated to $\tau = \tau \rightarrow (untoe(q_1) \neq untoe(q_1) \vee in(toe(q_1), q_2) \neq in(toe(q_1), q_2))$, which is a contradiction.

Now, by Theorem 7.15, the following TP-reduced version (see Definition 7.14) of $Q(q_1, q_2)$ is branching bisimilar to $Q(q_1, q_2)$.

$$\begin{aligned}
&Q_r(q_1, q_2) \\
&\quad = \sum_{d:D} r(d) \cdot Q_r\big(in(d, q_1), q_2\big) \triangleleft isempty(q_1) \triangleright \delta \\
&\qquad + \tau \cdot Q_r\big(untoe(q_1), in\big(toe(q_1), q_2\big)\big) \triangleleft \neg isempty(q_1) \triangleright \delta \\
&\qquad + s\big(toe(q_2)\big) \cdot Q_r\big(q_1, untoe(q_2)\big) \triangleleft \neg isempty(q_2) \wedge isempty(q_1) \triangleright \delta.
\end{aligned}$$

Note that after the TP-reduction q_1 never contains more than one element!

Acknowledgements

We would like to thank Harm van Beek, Michiel van Osch, Piet Rodenburg, and Mark van der Zwaag for their valuable comments on several versions of this chapter.

References

[1] H.P. Barendregt, *The Lambda Calculus*, North-Holland (1981).

[2] M.A. Bezem, R.N. Bol and J.F. Groote, *Formalizing process algebraic verifications in the calculus of constructions*, Formal Aspects of Comput. **9** (1997), 1–48.

[3] H. Bekič, *Towards a mathematical theory of processes*, Technical Report TR25.125, IBM Laboratory, Vienna (1971).

[4] M.A. Bezem and J.F. Groote, *A correctness proof of a one-bit sliding window protocol in μCRL*, The Comput. J. **37** (4) (1994), 289–307.

[5] M.A. Bezem and J.F. Groote, *Invariants in process algebra with data*, CONCUR'94: Concurrency Theory, Uppsala, Sweden, Lecture Notes in Comput. Sci. 836, B. Jonsson and J. Parrow, eds, Springer-Verlag (1994), 401–416.

[6] J.A. Bergstra, J.A. Hillebrand and A. Ponse, *Grid protocols based on synchronous communication*, Sci. Comput. Programming **29** (1997), 199–233.

[7] J.A. Bergstra and J.W. Klop, *Process algebra for synchronous communication*, Inform. and Control **60** (1–3) (1984), 109–137.

[8] J.C.M. Baeten and W.P. Weijland, *Process Algebra*, Cambridge Tracts in Theoret. Comput. Sci. 18, Cambridge University Press (1990).

[9] K.M. Chandy and J. Misra, *Parallel Program Design: A Foundation*, Addison-Wesley (1989).

[10] CWI, *μCRL: A language and toolset to study communicating processes with data* (2000). http://www.cwi.nl/~mcrl/.

[11] J.W. de Bakker and E. de Vink, *Control Flow Semantics*, MIT Press (1996).

[12] G. Dowek, A. Felty, H. Herbelin, G. Huet, C. Murthy, C. Parent, C. Paulin-Mohring and B. Werner, *The Coq proof assistant user's guide, Version 5.9*, Technical Report, INRIA-Rocquencourt and CNRS-ENS Lyon (1993).

[13] M.C.A. Devillers, W.O.D. Griffioen, J.M.T. Romijn and F.W. Vaandrager, *Verification of a leader election protocol – formal methods applied to IEEE 1394*, Technical Report, Computing Science Institute, University of Nijmegen (1997).

[14] H. Ehrig and B. Mahr, *Fundamentals of Algebraic Specification: Equations and Initial Semantics*, EATCS Monographs on Theoret. Comput. Sci. 6, Springer-Verlag (1985).

[15] J.-C. Fernandez, H. Garavel, A. Kerbrat, R. Mateescu, L. Mounier and M. Sighireanu, *CADP (CAESAR?ALDEBARAN Development Package): A protocol validation and verification toolbox*, Proc. CAV'96, Lecture Notes in Comput. Sci. 1102, R. Alur and T.A. Henzinger, eds, Springer-Verlag (1996), 437–440.

[16] L.-Å. Fredlund, J.F. Groote and H. Korver, *Formal verification of a leader election protocol in process algebra*, Theoret. Comput. Sci. **177** (1997), 459–486.

[17] J.F. Groote and H. Korver, *Correctness proof of the bakery protocol in μCRL*, Algebra of Communicating Processes, Workshops in Computing, A. Ponse, C. Verhoef and S.F.M. van Vlijmen, eds, Springer-Verlag (1994), 63–86.

[18] J.F. Groote, F. Monin and J. Springintveld, *A computer checked algebraic verification of a distributed summing protocol*, Technical Report 97/14, Department of Mathematics and Computing Science, Eindhoven University of Technology (1997).

[19] J.F. Groote, F. Monin and J. van de Pol, *Checking verifications of protocols and distributed systems by computer*, Proc. CONCUR'98, Sophia Antipolis, Lecture Notes in Comput. Sci. 1466, D. Sangiorgi and R. de Simone, eds, Springer-Verlag (1998), 629–655.

[20] J.F. Groote and A. Ponse, *Proof theory for μCRL: A language for processes with data*, Semantics of Specification Languages, Proceedings of the International Workshop on Semantics of Specification Languages, Utrecht, The Netherlands, 25–27 October 1993, Workshops in Computing, D.J. Andrews, J.F. Groote and C.A. Middelburg, eds, Springer-Verlag (1994), 232–251.

[21] J.F. Groote and A. Ponse, *The syntax and semantics of μCRL*, Algebra of Communicating Processes, Utrecht 1994, Workshops in Computing, A. Ponse, C. Verhoef and S.F.M. van Vlijmen, eds, Springer-Verlag (1995), 26–62.

[22] J.F. Groote, *A note on* n *similar parallel processes*, ERCIM Workshop on Formal Methods for Industrial Critical Systems, S. Gnesi and D. Latella, eds, Cesena, Italy (1997), 65–75.

[23] J.F. Groote, *The syntax and semantics of timed μCRL*, Technical Report SEN-R9709, CWI, Amsterdam (1997).

[24] J.F. Groote and J. Springintveld, *Focus points and convergent process operators. A proof strategy for protocol verification*, Technical Report 142, University Utrecht, Department of Philosophy (1995).

[25] J.F. Groote and M.P.A. Sellink, *Confluence for process verification*, Theoret. Comput. Sci. **170** (1–2) (1996), 47–81.

[26] J.F. Groote and J.C. van de Pol, *A bounded retransmission protocol for large data packets. A case study in computer checked verification*, Proc. AMAST'96, Munich, Germany, Lecture Notes in Comput. Sci. 1101, M. Wirsing and M. Nivat, eds, (1996), 536–550.

[27] C.A.R. Hoare, *Communicating Sequential Processes*, International Series in Comput. Sci., Prentice-Hall International (1985).

[28] G.J. Holzmann and D. Peled, *An improvement in formal verification*, Proc. FORTE '94, Berne, Switzerland (1994).

[29] IEEE Computer Society, *IEEE Standard for a high performance serial bus* (1996).

[30] ITU-T, *Recommendation Z.100: Specification and Description Language (SDL)*, ITU-T, Geneva (June 1994).

[31] K. Jensen, *Coloured Petri Nets. Basic Concepts, Analysis Methods and Practical Use*, EATCS Monographs on Theoret. Comput. Sci., Springer-Verlag, Berlin (1992).

[32] J.J.T. Kleijn, M.A. Reniers and J.E. Rooda, *A process algebra based verification of a production system*, Proc. 2nd IEEE Conference on Formal Engineering Methods, Brisbane, Australia, December 1998, J. Staples, M.G. Hinchley and S. Liu, eds, IEEE Computer Society Press (1998), 90–99.

[33] H. Korver and J. Springintveld, *A computer-checked verification of Milner's scheduler*, Proc. Internat. Symposium on Theoretical Aspects of Computer Software (TACS'94), Sendai, Japan, Lecture Notes in Comput. Sci. 789, M. Hagiya and J.C. Mitchell, eds, Springer-Verlag (1994), 161–178.

[34] S.P. Luttik, *Description and formal specification of the link layer of P1394*, Technical Report SEN-R9706, CWI, Amsterdam (1997).

[35] N.A. Lynch, *Distributed Algorithms*, Morgan Kaufmann (1996).

[36] A. Mazurkiewicz, *Basic notions of trace theory*, Linear Time, Branching Time and Partial Orders in Logics and Models for Concurrency, Lecture Notes in Comput. Sci. 354, J.W. de Bakker, W.-P. de Roever and G. Rozenberg, eds, Springer-Verlag (1988), 285–363.

[37] R. Milner, *A mathematical model of computing agents*, Proceedings Logic Colloquium 1973, H.E. Rose and J.C. Shepherdson, eds, North-Holland (1973), 158–173.

[38] R. Milner, *A Calculus of Communicating Systems*, Lecture Notes in Comput. Sci. 92, Springer-Verlag (1980).

[39] R. Milner, *Communication and Concurrency*, International Series in Computer Science, Prentice-Hall International (1989).

[40] S. Mauw and G.J. Veltink, *A process specification formalism*, Fundamenta Informaticae **XIII** (1990), 85–139.

[41] L.C. Paulson, *Isabelle: the next 700 theorem provers*, Logic and Computer Science, P. Odifreddi, ed., Academic Press (1990), 361–386.

[42] L.C. Paulson, *Isabelle: A Generic Theorem Prover*, Lecture Notes in Comput. Sci. 828, Springer-Verlag (1994).

[43] I.C.C. Phillips, *Refusal testing*, Theoret. Comput. Sci. **50** (1987), 241–284.

[44] H. Qin, *Efficient verification of determinate processes*, CONCUR'91, Proc. 2nd International Conference on Concurrency Theory, Amsterdam, The Netherlands, Lecture Notes in Comput. Sci. 527, J.C.M. Baeten and J.F. Groote, eds, Springer-Verlag (1991), 471–494.

[45] W. Reisig, *Petri Nets: An Introduction*, EATCS Monographs in Theoret. Comput. Sci. 4, Springer-Verlag, Berlin (1985).

[46] M. Sighireanu and R. Mateescu, *Validation of the link layer protocol of the IEEE-1394 serial bus (FireWire): An experiment with E-LOTOS*, Technical Report 3172, INRIA (1997).

[47] N. Shankar, S. Owre and J.M. Rushby, *The PVS proof checker: A reference manual,* Technical Report, Computer Science Laboratory, SRI International, Menlo Park, CA (1993).

[48] C. Shankland and M.B. van der Zwaag, *The tree identify protocol of IEEE 1394 in μCRL*, Formal Aspects of Comput. **10** (1998), 509–531.

[49] R.J. van Glabbeek, *The linear time – branching time spectrum*, CONCUR'90 – Theories of Concurrency: Unification and Extension, Amsterdam, Lecture Notes in Comput. Sci. 458, J.C.M. Baeten and J.W. Klop, eds, Springer-Verlag (1990), 278–297.

[50] R.J. van Glabbeek, *The linear time – branching time spectrum II. The semantics of sequential processes with silent moves*, CONCUR'93, Proc. International Conference on Concurrency Theory, Hildesheim, Germany, Lecture Notes in Comput. Sci. 715, E. Best, ed., Springer-Verlag (1993), 66–81.

[51] J.J. van Wamel, *Verification techniques for elementary data types and retransmission protocols*, Ph.D. Thesis, University of Amsterdam (1995).

[52] G. Winskel, *Event structures*, Petri Nets: Applications and Relationships to Other Models of Concurrency, Lecture Notes in Comput. Sci. 255, Brauer, Reissig and G. Rozenberg, eds, Springer-Verlag (1987), 325–392.

Subject index

CHAPTER 18

Discrete Time Process Algebra and the Semantics of SDL

J.A. Bergstra[1,3], C.A. Middelburg[2,3,*], Y.S. Usenko[4,*]

[1] *Programming Research Group, University of Amsterdam, P.O. Box 41882, 1009 DB Amsterdam, The Netherlands*

[2] *Computing Science Department, Eindhoven University of Technology, P.O. Box 513, 5600 MB Eindhoven, The Netherlands*

[3] *Department of Philosophy, Utrecht University, P.O. Box 80126, 3508 TC Utrecht, The Netherlands*

[4] *Cluster of Software Engineering, CWI P.O. Box 94079, 1090 GB Amsterdam, The Netherlands*

E-mails: janb@wins.uva.nl, keesm@win.tue.nl, ysu@cwi.nl

Contents

*The work presented in this chapter has been partly carried out while the second and third author were at UNU/IIST (United Nations University, International Institute for Software Technology) in Macau.

HANDBOOK OF PROCESS ALGEBRA
Edited by Jan A. Bergstra, Alban Ponse and Scott A. Smolka

Abstract

We present an extension of discrete time process algebra with relative timing where recursion, propositional signals and conditions, a counting process creation operator, and the state operator are combined. Except the counting process creation operator, which subsumes the original process creation operator, these features have been developed earlier as largely separate extensions of time free process algebra. The change to the discrete time case and the combination of the features turn out to be far from trivial. We also propose a semantics for a simplified version of SDL, using this extension of discrete time process algebra to describe the meaning of the language constructs. This version covers all behavioural aspects of SDL, except for communication via delaying channels – which can easily be modelled. The semantics presented here facilitates the generation of finitely branching transition systems for SDL specifications and thus it enables validation.

1. Introduction

In this chapter, we present an extension of discrete time process algebra with relative timing where recursion, propositional signals and conditions, a counting process creation operator, and the state operator are combined. We also propose a semantics for a simplified version of SDL, called φ^-SDL, using this extension of discrete time process algebra with relative timing to describe the meaning of the language constructs. The choice of a process algebra in the style of ACP [8,9] as the basis of the presented semantics is obvious. This algebraic approach to concurrency represents a large body of relevant theory. In particular, many features of φ^-SDL are related to topics that have been studied extensively in the framework of ACP. Besides, the axiom system and operational semantics of an ACP-style process algebra facilitate advances in the areas of validation and verification.

We take the remainder of this introductory section to introduce φ^-SDL, to motivate the choices made in the selection of this dramatically simplified version of SDL, and to describe its close connection with full SDL. We also explain the need for a semantics that deals properly with the time related aspects of SDL in case one intends to validate SDL specifications, or to justify design steps made using SDL by formal verification.

1.1. *Background*

At present, SDL [10,18] is widely used in telecommunications for describing structure and behaviour of generally complex systems at different levels of abstraction. It originated from an informal graphical description technique already commonly used in the telecommunications field at the time of the first computer controlled telephone switches. Our starting-point is the version of SDL defined in [33], the ITU-T Recommendation Z.100 published in 1994. There, a subset of SDL, called Basic SDL, is identified and used to describe the meaning of the language constructs of SDL that are not in Basic SDL. This subset is still fairly complicated.

φ^-SDL is a simplified version of Basic SDL.[1] The following simplifications have been made:

- blocks and channels are removed;
- all variables are revealed and they can be viewed freely;
- timer setting is regarded as just a special use of signals;
- timer setting is based on discrete time.

Besides, φ^-SDL does not deal with the specification of abstract data types. An algebraic specification of all data types used in a φ^-SDL specification is assumed as well as an initial algebra semantics for it. The pre-defined data types **Boolean** and **Natural**, with the obvious interpretation, should be included.

We decided to focus in φ^-SDL on the behavioural aspects of SDL. We did so for the following two reasons. Firstly, the structural aspects of SDL are mostly of a static nature and therefore not very relevant from a semantic point of view. Secondly, the part of SDL that deals with the specification of abstract data types is well understood – besides, it can

[1] This subset is called φ^-SDL, where φ stands for flat, as it does not cover the structural aspects of SDL, and $^-$ indicates that delaying channels are left out.

easily be isolated and treated as a parameter.[2] For practical reasons, we also chose not to include initially procedures, syntypes with a range condition and process types with a bound on the number of instances that may exist simultaneously. Similarly, the **any** expression is omitted as well. Services are not supported by φ^{-}SDL for the following reasons: the semantics of services is hard to understand, ETSI forbids for this reason their use in European telecommunication standards (see [31]), and the SDL community discusses its usefulness.

In [14], we introduced a simplified version of SDL, called φSDL, which covers all behavioural aspects of SDL, including communication via delaying channels. φ^{-}SDL is φSDL without communication via delaying channels. The process algebra semantics of φSDL proposed in [14] made clear that φSDL specifications can always be transformed to semantically equivalent ones in φ^{-}SDL. Apart from the data type definitions, SDL specifications can be transformed to φSDL specifications, and hence to φ^{-}SDL specifications, provided that no use is made of facilities that are not included initially. The transformation from SDL to φSDL has, apart from some minor adaptations, already been given. The first part of the transformation is the mapping for the shorthand notations of SDL which is given informally in the ITU-T Recommendation Z.100 [33] and defined in a fully precise manner in its Annex F.2 [34]. The second and final part is essentially the mapping *extract-dict* defined in its Annex F.3 [35].

The semantics of φ^{-}SDL agrees with the semantics of SDL as far as reasonably possible. This means in the first place that obvious errors in [35] have not been taken over. For example, the intended effect of SDL's create and output actions may sometimes be reached with interruption according to [35] – allowing amongst other things that a process ceases to exist while a signal is sent to it instantaneously. Secondly, the way of dealing with time is considered to be unnecessarily complex and inadequate in SDL and has been adapted as explained below.

In SDL, real numbers are used for times and durations. So when a timer is set, its expiration time is given by a real number. However, the time considered is the system time which proceeds actually in a discrete manner: the system receives ticks from the environment which increase the system time with a certain amount (how much real time they represent is left open). Therefore, the timer is considered to expire when the system receives the first tick that indicates that its expiration time has passed. So nothing is lost by adopting in φ^{-}SDL a discrete time approach, using natural numbers for times and durations, where the time unit can be viewed as the time between two ticks but does not really rely upon the environment. This much simpler approach also allows us to remove the original inadequacy to relate the time used with timer setting to the time involved in waiting for signals by processes.

We generally had to make our own choices with respect to the time related aspects of SDL, because they are virtually left out completely in the ITU-T Recommendation Z.100. Our choices were based on communications with various practitioners from the telecommunications field using SDL, in particular the communications with Leonard Pruitt [26].

[2] The following is also worth noticing: (1) ETSI discourages the use of abstract data types other than the predefined ones in European telecommunication standards (see [31]); (2) ASN.1 [32] is widely used for data type specification in the telecommunications field, and there is an ITU-T Recommendation, Z.105, for combining SDL and ASN.1 (see [36]).

They provided convincing practical justification for the premise of our current choices: communication with the environment takes a good deal of time, whereas internal processing takes a negligible deal of time. Ease of adaptation to other viewpoints on time in SDL is guaranteed relatively well by using a discrete time process algebra, PA^{-}_{drt} (see [5]) without immediate deadlock, as the basis of the presented semantics.

In the telecommunications field, SDL is increasingly used for describing generally complex telecommunications systems, including switching systems, services and protocols, at different levels of abstraction – from initial specification till implementation. Initial specification of systems is done with the intention to analyse the behavioural properties of these systems and thus to validate the specification. There is also a growing need to verify whether the properties represented by one specification are preserved in another, more concrete, specification and thus to justify design steps. However, neither SDL nor the tools and techniques that are used in conjunction with SDL provide appropriate support for validation of SDL specifications and verification of design steps made using SDL. The main reason is that the semantics of SDL according to the ITU-T Recommendation Z.100 is at some points inadequate for advanced validation and formal verification. In particular, the semantics of time related features, such as timers and delaying channels, is insufficiently precise. Moreover, the semantics is at some other points unnecessarily complex. Consequently, rules of logical reasoning, indispensable for formal verification, have not yet been developed and most existing analysis tools, e.g., GEODE [2] and SDT [37], offer at best a limited kind of model checking for validation.

Prerequisites for advanced validation and formal verification is a dramatically simplified version of SDL and an adequate semantics for it. Only after that possibilities for advanced analysis can be elaborated and proof rules for formal verification devised. The language φ^{-}SDL and the presented semantics for it are primarily intended to come up to these prerequisites.

1.2. *Organization of this chapter*

The structure of this chapter is as follows. In Section 2, we present the extension of discrete time process algebra with relative timing that is used for the process algebra semantics of φ^{-}SDL proposed in Section 4. An overview of φ^{-}SDL is given in Section 3. Following the overview, in Section 4, we present the proposed semantics of φ^{-}SDL in two steps. First, a semantics of φ^{-}SDL process definitions, which are the main elements of φ^{-}SDL specifications, is given. This semantics abstracts from dynamic aspects of process behaviour such as process creation and process execution in a state. A semantics of φ^{-}SDL system definitions, i.e. complete φ^{-}SDL specifications, is then given in terms of the semantics of φ^{-}SDL process definitions using the counting process creation operator and the state operator. In Section 5, we give an overview of related work and we explain how the semantics presented in Section 4 can be used to transform φ^{-}SDL specifications to transition systems that can be used for advanced validation. There are appendices about notational conventions used (Appendix A) and details concerning the contexts used to model scope in the presented semantics (Appendix B). Small examples of specification in φ^{-}SDL and the meaning of the process definitions being found in these examples are also presented (in Section 3 and Section 4, respectively).

2. Process algebra

2.1. *Introduction*

In this section, we present an extension of discrete time process algebra with relative timing where recursion, propositional signals and conditions, a counting process creation operator, and the state operator are combined. Its signature, axioms and a structural operational semantics are given and it is shown that strong bisimulation equivalence is a congruence for all operations. Except the counting process creation operator, which subsumes the original process creation operator, these features have been developed earlier as largely separate extensions of time free process algebra. However, both the change to the discrete time case and the combination of the features turn out to be far from trivial. Besides, some of the features are slightly adapted versions of the original ones in order to meet the needs of the semantics of φ^{-}SDL.

In Section 2.2, we present discrete relative time process algebra without immediate deadlock and delayable actions ($\mathrm{PA}^{-}_{\mathrm{drt}}$-ID). In Section 2.3, we add propositional signals and conditions to $\mathrm{PA}^{-}_{\mathrm{drt}}$-ID. In the discrete relative time case, this addition requires some axioms to be refined. We introduce a new guarded command operator that yields a deadlock in the current time slice if the condition does not hold at the start, i.e. waiting is no option if the condition does not hold. In Section 2.4, we add recursion to the extension presented in Section 2.3. The main definitions related to recursion, such as the definitions of recursive specification, solution and guardedness, are given here for the case with relative timing in discrete time as well as propositional signals and conditions.

In Section 2.5 and 2.6, we describe the counting process creation operator and the state operator, respectively, for the discrete relative time case in the presence of propositional signals and conditions. The counting process creation operator is a straighforward extension of the original process creation operator that allows to assign a unique "process identification value" to each process created. The state operator presented here allows to deal with conditions whose truth depends on the state and with state changes due to progress of time to the next time slice.

The main reference to discrete time process algebra in the style of ACP is [5]. The features with which it is combined here are discussed as separate extensions of time free process algebra in [6] (propositional signals and conditions, state operator), [8] (recursion) and [11] (process creation). Our discussion of axioms is concentrated on the crucial axioms for the discrete time case and each of these features, and on the alterations and additions needed if all this is combined. For a systematic introduction to process algebra in the style of ACP, the reader is referred to [8] and [9].

2.2. *Discrete relative time process algebra*

In this subsection, we present discrete relative time process algebra without immediate deadlock and delayable actions. The term discrete time is used here to indicate that time is divided into time slices and timing of actions is done with respect to the time slices in which they are performed – within a time slice there is only the order in which actions are performed. Additionally, performance of actions and passage to the next time slice are separated here. This corresponds to the two-phase functioning scheme for modeling timed processes [25]. Note that it means that processes are supposed to be capable of performing certain actions, like in time free process algebra, as well as passing to the next time slice. A coherent collection of versions of ACP with timing where performance of actions and passage of time are separated, is presented in [7].

First we treat the basic discrete relative time process algebra $\mathrm{BPA}^{-}_{\mathrm{drt}}$-ID. Then we treat $\mathrm{PA}^{-}_{\mathrm{drt}}$-ID, the extension of $\mathrm{BPA}^{-}_{\mathrm{drt}}$-ID with parallel composition in which no communication between processes is involved. $\mathrm{ACP}^{-}_{\mathrm{drt}}$-ID, the extension of $\mathrm{BPA}^{-}_{\mathrm{drt}}$-ID with parallel composition in which synchronous communication between processes is involved, will not be treated. $\mathrm{BPA}^{-}_{\mathrm{drt}}$-ID, $\mathrm{PA}^{-}_{\mathrm{drt}}$-ID and $\mathrm{ACP}^{-}_{\mathrm{drt}}$-ID are presented in detail in [27]. We also present the extension of $\mathrm{PA}^{-}_{\mathrm{drt}}$-ID with encapsulation, described before for the discrete relative time case without immediate deadlock in [4].

2.2.1. *Basic process algebra.* In $\mathrm{BPA}^{-}_{\mathrm{drt}}$-ID, we have the sort P of processes, the constants $\underline{\underline{a}}$ (one for each action a) and $\underline{\underline{\delta}}$, the unary operator σ_{rel} (time unit delay), and the binary operators $\cdot$ (sequential composition) and $+$ (alternative composition). The constants $\underline{\underline{a}}$ stand for a in the current time slice. Similarly, the constant $\underline{\underline{\delta}}$ stands for a deadlock in the current time slice. The process $\sigma_{\mathsf{rel}}(x)$ is the process x delayed till the next time slice. The process $x \cdot y$ is the process x followed after successful termination by the process y. The process $x + y$ is the process that proceeds with either the process x or the process y, but not both. We also have the auxiliary unary operator ν_{rel} (now) in $\mathrm{BPA}^{-}_{\mathrm{drt}}$-ID. This operator makes an elegant axiomatization of $\mathrm{PA}^{-}_{\mathrm{drt}}$-ID possible. The process $\nu_{\mathsf{rel}}(x)$ is the part of x that is not delayed till the next time slice.

It is assumed that a fixed but arbitrary set A of *actions* has been given.

SIGNATURE OF $\mathrm{BPA}^{-}_{\mathrm{drt}}$-ID. The signature of $\mathrm{BPA}^{-}_{\mathrm{drt}}$-ID consists of the *undelayable action* constants $\underline{\underline{a}} : \to \mathsf{P}$ (for each $a \in A$), the *undelayable deadlock* constant $\underline{\underline{\delta}} : \to \mathsf{P}$, the *alternative composition* operator $+ : \mathsf{P} \times \mathsf{P} \to \mathsf{P}$, the *sequential composition* operator $\cdot : \mathsf{P} \times \mathsf{P} \to \mathsf{P}$, the *time unit delay* operator $\sigma_{\mathsf{rel}} : \mathsf{P} \to \mathsf{P}$, and the *now* operator $\nu_{\mathsf{rel}} : \mathsf{P} \to \mathsf{P}$.

We assume that an infinite set of variables (of sort P) has been given. Given the signature of $\mathrm{BPA}^{-}_{\mathrm{drt}}$-ID, terms of $\mathrm{BPA}^{-}_{\mathrm{drt}}$-ID, often referred to as process expressions, are constructed in the usual way. The need to use parentheses is reduced by ranking the precedence of the operators. Throughout this chapter we adhere to the following precedence rules: (i) all unary operators have the same precedence, (ii) unary operators have a higher precedence than binary operators, (iii) the operator $\cdot$ has the highest precedence amongst the binary operators, (iv) the operator $+$ has the lowest precedence amongst the binary operators, and (v) all other binary operators have the same precedence. We will also use the following abbreviation. Let $(p_i)_{i \in \mathcal{I}}$ be an indexed set of terms of $\mathrm{BPA}^{-}_{\mathrm{drt}}$-ID where $\mathcal{I} = \{i_1, \ldots, i_n\}$. Then we write $\sum_{i \in \mathcal{I}} p_i$ for $p_{i_1} + \ldots + p_{i_n}$. We further use the convention that $\sum_{i \in \mathcal{I}} p_i$ stands for $\underline{\underline{\delta}}$ if $\mathcal{I} = \emptyset$.

We denote variables by $x, x', y, y', \ldots$. We use the convention that $a, a', b, b', \ldots$ denote elements of $A \cup \{\delta\}$ in the context of an equation, but elements of A in the context of an operational semantics rule. Furthermore, H denotes a subset of A. We write A_δ for $A \cup \{\delta\}$.

Table 1
Axioms of $\mathrm{BPA}^{-}_{\mathrm{drt}}$-ID ($a \in A_\delta$)

$x + y = y + x$	A1	$\sigma_{\mathsf{rel}}(x) + \sigma_{\mathsf{rel}}(y) = \sigma_{\mathsf{rel}}(x + y)$	DRT1
$(x + y) + z = x + (y + z)$	A2	$\sigma_{\mathsf{rel}}(x) \cdot y = \sigma_{\mathsf{rel}}(x \cdot y)$	DRT2
$x + x = x$	A3	$\underline{\underline{\delta}} \cdot x = \underline{\underline{\delta}}$	DRT3
$(x + y) \cdot z = x \cdot z + y \cdot z$	A4	$x + \underline{\underline{\delta}} = x$	DRT4A
$(x \cdot y) \cdot z = x \cdot (y \cdot z)$	A5	$\nu_{\mathsf{rel}}(\underline{\underline{a}}) = \underline{\underline{a}}$	DCS1
		$\nu_{\mathsf{rel}}(x + y) = \nu_{\mathsf{rel}}(x) + \nu_{\mathsf{rel}}(y)$	DCS2
		$\nu_{\mathsf{rel}}(x \cdot y) = \nu_{\mathsf{rel}}(x) \cdot y$	DCS3
		$\nu_{\mathsf{rel}}(\sigma_{\mathsf{rel}}(x)) = \underline{\underline{\delta}}$	DCS4

AXIOMS OF $\mathrm{BPA}^{-}_{\mathrm{drt}}$-ID. The axiom system of $\mathrm{BPA}^{-}_{\mathrm{drt}}$-ID consists of the equations A1–A5, DRT1–DRT4A and DCS1–DCS4 given in Table 1.

Axioms DRT1 and DRT2 represent the interaction of time unit delay with alternative composition and sequential composition, respectively. Axiom DRT1, called the time factorization axiom, expresses that passage to the next time slice by itself can not determine a choice. Axiom DRT2 expresses that timing is relative to the performance of the previous action.

In [27], a structural operational semantics of $\mathrm{BPA}^{-}_{\mathrm{drt}}$-ID is presented and proofs are given of the soundness and completeness of the axiom system of $\mathrm{BPA}^{-}_{\mathrm{drt}}$-ID for the set of closed $\mathrm{BPA}^{-}_{\mathrm{drt}}$-ID terms modulo (strong) *bisimulation* equivalence. This notion is precisely defined in [8]. Roughly, bisimilarity of two processes means that if one process is capable of doing a certain step, i.e. performing some action or passing to the next time slice, and next going on as a certain follow-up process then the other process is capable of doing the same step and next going on as a process bisimilar to the follow-up process.

2.2.2. *Parallel composition.* In $\mathrm{PA}^{-}_{\mathrm{drt}}$-ID, we have, in addition to sequential and alternative composition, parallel composition of processes. In $\mathrm{PA}^{-}_{\mathrm{drt}}$-ID, unlike in $\mathrm{ACP}^{-}_{\mathrm{drt}}$-ID, parallel composition does not involve communication between processes. The parallel composition operator $\|$ of $\mathrm{PA}^{-}_{\mathrm{drt}}$-ID is called free merge to indicate that no communication is involved. The process $x \parallel y$ is the process that proceeds simultaneously with the processes x and y. In order to get a finite axiomatization, we also have the auxiliary operator $\lfloor\!\lfloor$ (left merge) in $\mathrm{PA}^{-}_{\mathrm{drt}}$-ID. The processes $x \lfloor\!\lfloor y$ and $x \parallel y$ are the same except that $x \lfloor\!\lfloor y$ must start with a step of x.

SIGNATURE OF $\mathrm{PA}^{-}_{\mathrm{drt}}$-ID. The signature of $\mathrm{PA}^{-}_{\mathrm{drt}}$-ID is the signature of $\mathrm{BPA}^{-}_{\mathrm{drt}}$-ID extended with the *free merge* operator $\| : \mathsf{P} \times \mathsf{P} \to \mathsf{P}$ and the *left merge* operator $\lfloor\!\lfloor : \mathsf{P} \times \mathsf{P} \to \mathsf{P}$.

We will use the following abbreviation. Let $(p_i)_{i \in \mathcal{I}}$ be an indexed set of terms of $\mathrm{PA}^{-}_{\mathrm{drt}}$-ID where $\mathcal{I} = \{i_1, \ldots, i_n\}$. Then, we write $\|_{i \in \mathcal{I}}\, p_i$ for $p_{i_1} \parallel \ldots \parallel p_{i_n}$.

Table 2
Additional axioms for $\mathrm{PA}^{-}_{\mathrm{drt}}$-ID ($a \in A_\delta$)

$x \parallel y = x \mathbin{\underline{\|}} y + y \mathbin{\underline{\|}} x$	DRTM1
$\underline{\underline{a}} \mathbin{\underline{\|}} x = \underline{\underline{a}} \cdot x$	DRTM2
$\underline{\underline{a}} \cdot x \mathbin{\underline{\|}} y = \underline{\underline{a}} \cdot (x \parallel y)$	DRTM3
$(x + y) \mathbin{\underline{\|}} z = x \mathbin{\underline{\|}} z + y \mathbin{\underline{\|}} z$	DRTM4
$\sigma_{\mathsf{rel}}(x) \mathbin{\underline{\|}} \nu_{\mathsf{rel}}(y) = \underline{\underline{\delta}}$	DRTM5
$\sigma_{\mathsf{rel}}(x) \mathbin{\underline{\|}} (\nu_{\mathsf{rel}}(y) + \sigma_{\mathsf{rel}}(z)) = \sigma_{\mathsf{rel}}(x \mathbin{\underline{\|}} z)$	DRTM6

Table 3
Axioms for encapsulation ($a \in A_\delta$)

$\partial_H(\underline{\underline{a}}) = \underline{\underline{a}}$ if $a \notin H$	DRTD1
$\partial_H(\underline{\underline{a}}) = \underline{\underline{\delta}}$ if $a \in H$	DRTD2
$\partial_H(x \cdot y) = \partial_H(x) \cdot \partial_H(y)$	DRTD3
$\partial_H(x + y) = \partial_H(x) + \partial_H(y)$	DRTD4
$\partial_H(\sigma_{\mathsf{rel}}(x)) = \sigma_{\mathsf{rel}}(\partial_H(x))$	DRTD5

AXIOMS OF $\mathrm{PA}^{-}_{\mathrm{drt}}$-ID. The axiom system of $\mathrm{PA}^{-}_{\mathrm{drt}}$-ID consists of the axioms of $\mathrm{BPA}^{-}_{\mathrm{drt}}$-ID and the equations DRTM1–DRTM6 given in Table 2.

Axioms DRTM5 and DRTM6 represent the interaction between time unit delay and left merge. These axioms express that passage to the next time slice of parallel processes must synchronize.

In [27], a structural operational semantics of $\mathrm{PA}^{-}_{\mathrm{drt}}$-ID is presented and proofs are given of the soundness and completeness of the axiom system of $\mathrm{PA}^{-}_{\mathrm{drt}}$-ID for the set of closed $\mathrm{PA}^{-}_{\mathrm{drt}}$-ID terms modulo (strong) bisimulation equivalence.

2.2.3. *Encapsulation.* We extend the signature of $\mathrm{PA}^{-}_{\mathrm{drt}}$-ID with the encapsulation operator $\partial_H : \mathsf{P} \to \mathsf{P}$. This operator turns all undelayable actions $\underline{\underline{a}}$, where a in $H \subseteq A$, into undelayable deadlock. The encapsulation operator is defined by the equations DRTD1–DRTD5 given in Table 3.

An operational semantics of encapsulation is presented in [4].

2.3. *Propositional signals and conditions*

In [6], process algebra with propositional signals and conditions is introduced for the time free case. In this subsection, we adapt it for discrete relative time. The result is referred to by $\mathrm{PA}^{\mathrm{psc}}_{\mathrm{drt}}$. In later sections, we will call propositional signals "propositions" in order to avoid ambiguity with signals of φ^{-}SDL.

In process algebra with propositional signals and conditions, propositions are used both as signals that are emitted by processes and as conditions that are imposed on processes to proceed. Condition testing is looked upon as signal inspection. The intuition is that the signal emitted by a process, as well as each of its logical consequences, holds at the start of the process. The signal emitted by a process is called its root signal.

Like in the time free case we have $\perp$ (non-existence) as additional constant and $\curlywedge$ (root signal emission) and $:\rightarrow$ (guarded command) as additional operators. Like the constant $\underline{\underline{\delta}}$, the constant $\perp$ stands for a process that is incapable of doing any step and incapable of terminating successfully. In addition, going on as $\perp$ after performing an action is impossible. The process $\phi \curlywedge x$ is the process x where the proposition ϕ holds at its start. Broadly speaking, the process $\phi :\rightarrow x$ is the process that may proceed as the process $\nu_{\mathsf{rel}}(x)$ if the proposition ϕ holds at its start, but may also proceed as the process $\sigma_{\mathsf{rel}}(\phi :\rightarrow y)$ in case $x = \nu_{\mathsf{rel}}(x) + \sigma_{\mathsf{rel}}(y)$. In other words, with the guarded command operator $:\rightarrow$, it is possible to wait till a proposition holds. This agrees with the original intention of the operator to make actions conditional.

We also have a non-waiting version of the guarded command operator, namely the operator $\underset{\cdot}{:}\rightarrow$ (strict guarded command). The process $\phi \underset{\cdot}{:}\rightarrow x$ is the process that proceeds as the process x if the proposition ϕ holds at its start, and otherwise yields a deadlock in the current time slice. Both guarded commands agree with the one in the time free case for processes that are not capable of passing to the next time slice.

Lifting propositional signals and conditions to the discrete time case requires the axioms for left merge to be adapted. If these axioms are not adapted, the equation $\sigma_{\mathsf{rel}}(x \mathbin{⌊\!⌊} y) = \underline{\underline{\delta}}$ becomes derivable – unless axiom NE1 (Table 5) is not adopted. Without adaptations, the root signal of $x \mathbin{⌊\!⌊} y$ would be the root signal of the process x. With the chosen adaptations, the root signal of $x \mathbin{⌊\!⌊} y$ is the conjunction of the root signals of the processes x and y. Thus, different from the time free case, the process $x \mathbin{⌊\!⌊} y$ is neither capable of performing an action nor capable of passing to the next time slice if the root signal of y is equal to f. This difference is not relevant to the free merge operator because the root signal of $x \parallel y$ is the conjunction of the root signals of x and y anyhow.

It is assumed that a fixed but arbitrary set B_{at} of *atomic propositions* has been given. From now on we have, in addition to the sort P of processes, the sort B of propositions over B_{at}; with constants t, f (true, false) and operators $\neg$, $\vee$, $\wedge$, $\rightarrow$, $\leftrightarrow$ (negation, disjunction, conjunction, implication, bi-implication). In case B_{at} is empty, B represents the boolean algebra over the set $\mathbb{B} = \{\mathsf{t}, \mathsf{f}\}$.

We denote propositions by $\phi, \psi, \ldots$. In derivations we may always use logical equivalences of (classical) propositional logic. So we are actually using equivalence classes of formulas, with respect to logical equivalence, instead of the formulas themselves.

A *valuation* v of atomic formulas is a function $v : B_{\mathrm{at}} \rightarrow \mathbb{B}$. Any valuation v can be extended to B in the usual homomorphic way, i.e.:

$$v(\kappa) = \kappa \text{ for the constants } \kappa \in \{\mathsf{t}, \mathsf{f}\},$$
$$v(\neg\phi) = \neg v(\phi),$$
$$v(\phi \mathbin{o} \psi) = v(\phi) \mathbin{o} v(\psi) \text{ for the binary operators } o \in \{\vee, \wedge, \rightarrow, \leftrightarrow\}.$$

We will use the same name for a valuation v and its extension to B. If a proposition ϕ is satisfied by a valuation v ($v(\phi) = \mathsf{t}$), we write $v \models \phi$ to indicate this.

SIGNATURE OF $\mathrm{PA}^{\mathrm{psc}}_{\mathrm{drt}}$. The signature of $\mathrm{PA}^{\mathrm{psc}}_{\mathrm{drt}}$ is the signature of the $\mathrm{PA}^{-}_{\mathrm{drt}}$-ID extended with the encapsulation operator, the *non-existence* constant $\bot : \to \mathsf{P}$, the *strict guarded command* operator $:\!\!\to\ : \mathsf{B} \times \mathsf{P} \to \mathsf{P}$, the *root signal emission* operator $\,^{\wedge}\!\!\blacktriangle : \mathsf{B} \times \mathsf{P} \to \mathsf{P}$ and the *weak guarded command* operator $:\!\!\to: \mathsf{B} \times \mathsf{P} \to \mathsf{P}$.

AXIOMS OF $\mathrm{PA}^{\mathrm{psc}}_{\mathrm{drt}}$. The axiom system of $\mathrm{PA}^{\mathrm{psc}}_{\mathrm{drt}}$ consists of the axioms of $\mathrm{BPA}^{-}_{\mathrm{drt}}$-ID and the equations DRTM1, DRTM4 from Table 2, the equations DRTM2′, DRTM3′, DRTM5′, DRTM6′ from Table 4, the equations NE1–NE3 from Table 5, the equations SGC1–SGC6, MSGC from Table 6, the equations SRSE1–SRSE7, MSRSE from Table 7, the equations DCS5, DCS6, DRTM7 from Table 8, the equations DGC1–DGC8 from Table 9, the equations DRTD1–DRTD5 from Table 3, and the equations PD1–PD3 from Table 10.

The axioms A3, DRT3 and DRT4A of $\mathrm{BPA}^{-}_{\mathrm{drt}}$-ID (Table 1) are derivable from the axioms SGC1, SGC2, SGC4 and SGC6 (Table 6).

Table 4
Adapted axioms for left merge ($a \in A_\delta$)

$\underline{\underline{a}} \mathbin{\|\!\lfloor} x = \underline{\underline{a}} \cdot x + \partial_A(\nu_{\mathsf{rel}}(x))$	DRTM2′
$\underline{\underline{a}} \cdot x \mathbin{\|\!\lfloor} y = \underline{\underline{a}} \cdot (x \parallel y) + \partial_A(\nu_{\mathsf{rel}}(y))$	DRTM3′
$\sigma_{\mathsf{rel}}(x) \mathbin{\|\!\lfloor} \nu_{\mathsf{rel}}(y) = \partial_A(\nu_{\mathsf{rel}}(y))$	DRTM5′
$\sigma_{\mathsf{rel}}(x) \mathbin{\|\!\lfloor} (\nu_{\mathsf{rel}}(y) + \sigma_{\mathsf{rel}}(z)) = \sigma_{\mathsf{rel}}(x \mathbin{\|\!\lfloor} z) + \partial_A(\nu_{\mathsf{rel}}(y))$	DRTM6′

Table 5
Axioms for non-existence ($a \in A_\delta$)

$x + \bot = \bot$	NE1
$\bot \cdot x = \bot$	NE2
$\underline{\underline{a}} \cdot \bot = \underline{\underline{\delta}}$	NE3

Table 6
Axioms for strict guarded command

$\mathsf{t} :\!\!\to x = x$	SGC1
$\mathsf{f} :\!\!\to x = \underline{\underline{\delta}}$	SGC2
$\phi :\!\!\to (x + y) = (\phi :\!\!\to x) + (\phi :\!\!\to y)$	SGC3
$(\phi \vee \psi) :\!\!\to x = (\phi :\!\!\to x) + (\psi :\!\!\to x)$	SGC4
$\phi :\!\!\to (\psi :\!\!\to x) = (\phi \wedge \psi) :\!\!\to x$	SGC5
$\phi :\!\!\to (x \cdot y) = (\phi :\!\!\to x) \cdot y$	SGC6
$(\phi :\!\!\to x) \mathbin{\|\!\lfloor} y = \phi :\!\!\to (x \mathbin{\|\!\lfloor} y)$	MSGC

Table 7
Axioms for root signal emission

$(\phi \mathbin{{}^{\wedge}\!\!\blacktriangle} x) \cdot y = \phi \mathbin{{}^{\wedge}\!\!\blacktriangle} (x \cdot y)$	SRSE1
$(\phi \mathbin{{}^{\wedge}\!\!\blacktriangle} x) + y = \phi \mathbin{{}^{\wedge}\!\!\blacktriangle} (x + y)$	SRSE2
$\phi \mathbin{{}^{\wedge}\!\!\blacktriangle} (\psi \mathbin{{}^{\wedge}\!\!\blacktriangle} x) = (\phi \wedge \psi) \mathbin{{}^{\wedge}\!\!\blacktriangle} x$	SRSE3
$\mathsf{t} \mathbin{{}^{\wedge}\!\!\blacktriangle} x = x$	SRSE4
$\mathsf{f} \mathbin{{}^{\wedge}\!\!\blacktriangle} x = \bot$	SRSE5
$\phi \mathrel{\vdots\!\rightarrow} (\psi \mathbin{{}^{\wedge}\!\!\blacktriangle} x) = (\phi \rightarrow \psi) \mathbin{{}^{\wedge}\!\!\blacktriangle} (\phi \mathrel{\vdots\!\rightarrow} x)$	SRSE6
$\phi \mathbin{{}^{\wedge}\!\!\blacktriangle} (\phi \mathrel{\vdots\!\rightarrow} x) = \phi \mathbin{{}^{\wedge}\!\!\blacktriangle} x$	SRSE7
$(\phi \mathbin{{}^{\wedge}\!\!\blacktriangle} x) \mathbin{\|\!\lfloor} y = \phi \mathbin{{}^{\wedge}\!\!\blacktriangle} (x \mathbin{\|\!\lfloor} y)$	MSRSE

Table 8
Additional axioms for time unit delay and now operator

$\nu_{\mathsf{rel}}(\phi \mathrel{\vdots\!\rightarrow} x) = \phi \mathrel{\vdots\!\rightarrow} \nu_{\mathsf{rel}}(x)$	DCS5
$\nu_{\mathsf{rel}}(\phi \mathbin{{}^{\wedge}\!\!\blacktriangle} x) = \phi \mathbin{{}^{\wedge}\!\!\blacktriangle} \nu_{\mathsf{rel}}(x)$	DCS6
$\sigma_{\mathsf{rel}}(x) \mathbin{\|\!\lfloor} ((\phi \mathrel{\vdots\!\rightarrow} y) + z) = (\phi \mathrel{\vdots\!\rightarrow} (\sigma_{\mathsf{rel}}(x) \mathbin{\|\!\lfloor} (y + z))) + (\neg\phi \mathrel{\vdots\!\rightarrow} (\sigma_{\mathsf{rel}}(x) \mathbin{\|\!\lfloor} z))$	DRTM7

Table 9
Axioms for weak guarded command ($a \in A_\delta$)

$\phi \mathrel{:\rightarrow} \underline{\underline{\delta}} = \underline{\underline{\delta}}$	DGC1
$\phi \mathrel{:\rightarrow} \underline{\underline{a}} = \phi \mathrel{\vdots\!\rightarrow} \underline{\underline{a}}$	DGC2
$\phi \mathrel{:\rightarrow} (\underline{\underline{a}} \cdot x) = (\phi \mathrel{:\rightarrow} \underline{\underline{a}}) \cdot x$	DGC3
$\phi \mathrel{:\rightarrow} (\psi \mathrel{\vdots\!\rightarrow} x) = \psi \mathrel{\vdots\!\rightarrow} (\phi \mathrel{:\rightarrow} x)$	DGC4
$\phi \mathrel{:\rightarrow} (x + y) = (\phi \mathrel{:\rightarrow} x) + (\phi \mathrel{:\rightarrow} y)$	DGC5
$\phi \mathrel{:\rightarrow} \sigma_{\mathsf{rel}}(x) = \sigma_{\mathsf{rel}}(\phi \mathrel{:\rightarrow} x)$	DGC6
$\phi \mathrel{:\rightarrow} \nu_{\mathsf{rel}}(x) = \nu_{\mathsf{rel}}(\phi \mathrel{:\rightarrow} x)$	DGC7
$\phi \mathrel{:\rightarrow} (\psi \mathbin{{}^{\wedge}\!\!\blacktriangle} x) = \psi \mathbin{{}^{\wedge}\!\!\blacktriangle} (\phi \mathrel{:\rightarrow} x)$	DGC8

Table 10
Additional axioms for encapsulation

$\partial_H(\phi \mathrel{\vdots\!\rightarrow} x) = \phi \mathrel{\vdots\!\rightarrow} \partial_H(x)$	PD1
$\partial_H(\phi \mathrel{:\rightarrow} x) = \phi \mathrel{:\rightarrow} \partial_H(x)$	PD2
$\partial_H(\phi \mathbin{{}^{\wedge}\!\!\blacktriangle} x) = \phi \mathbin{{}^{\wedge}\!\!\blacktriangle} \partial_H(x)$	PD3

The axioms DRTM2′, DRTM3′, DRTM5′ and DRTM6′ (Table 4) are the axioms DRTM2, DRTM3, DRTM5 and DRTM6 of $\text{PA}^{-}_{\text{drt}}$-ID (Table 2) where a summand is added to the right hand side of the axioms, viz. $\partial_A(\nu_{\mathsf{rel}}(x))$ in case of DRTM2 and $\partial_A(\nu_{\mathsf{rel}}(y))$ in case of the other axioms. Thus is expressed that the root signal of the left merge of two processes is always the conjunction of the root signals of both processes.

The axioms NE1–NE3, SGC1–SGC6, SRSE1–SRSE7, MSGC and MSRSE (Tables 5–7) are straightforward reformulations of corresponding axioms for the time free case, i.e. axioms of PA_{ps}, given in [6]. The constants a and the constant δ have been replaced by the constants $\underline{\underline{a}}$ and the constant $\underline{\underline{\delta}}$, respectively; and the operator $:\rightarrow$ has been replaced by the operator $\mathrel{\vdots\!\rightarrow}$. The axioms NE1 and NE2 (Table 5) are derivable from the axiom A1 of $\text{BPA}^{-}_{\text{drt}}$-ID and the axioms SRSE1, SRSE2 and SRSE5 (Table 7). Axiom NE3 expresses that going on as $\bot$ after performing an action is impossible. We do not have $\sigma_{\mathsf{rel}}(\bot) = \underline{\underline{\delta}}$, an equation expressing that going on as $\bot$ after passing to the next time slice is impossible. The reason is that $\sigma_{\mathsf{rel}}(x) = \underline{\underline{\delta}}$ would become derivable. In view of this mismatch between $\underline{\underline{a}} \cdot \bot$ and $\sigma_{\mathsf{rel}}(\bot)$, in retrospect, axiom NE3 may be considered to be a wrong choice in [6]. Axiom SRSE5 expresses that a process where falsity holds at its start is non-existent. The crucial axioms are SRSE6 and SRSE7 which represent the interaction between the root signal emission operator and the strict guarded command. Axiom SRSE6 expresses that if a proposition holds at the start of a process and that process is guarded by another proposition then at the start of the whole the former proposition holds or the latter proposition does not hold. Axiom SRSE7 expresses that it is superfluous to guard a process by a proposition if the proposition holds at the start of the whole.

The additional axioms DCS5, DCS6 and DRTM7 (Table 8) are needed because propositional signals and conditions are lifted to the discrete time case. The strict guarded command is non-waiting, i.e. we do not have $\phi \mathrel{\vdots\!\rightarrow} \sigma_{\mathsf{rel}}(x) = \sigma_{\mathsf{rel}}(\phi \mathrel{\vdots\!\rightarrow} x)$. Root signal emission is non-persistent, i.e. we do not have $\phi \mathbin{{}^{\wedge\!\blacktriangle}} \sigma_{\mathsf{rel}}(x) = \sigma_{\mathsf{rel}}(\phi \mathbin{{}^{\wedge\!\blacktriangle}} x)$. Axiom DRTM7 is necessary for the elimination of parallel composition. This axiom expresses that if processes are capable of passing to the next time slice conditionally then their parallel composition can do so if all conditions concerned hold. Note that from DRTM7 we can derive $\sigma_{\mathsf{rel}}(x) \mathbin{\lfloor\!\lfloor} (\phi \mathrel{\vdots\!\rightarrow} y) = \phi \mathrel{\vdots\!\rightarrow} (\sigma_{\mathsf{rel}}(x) \mathbin{\lfloor\!\lfloor} y)$.

The axioms DGC1–DGC8 (Table 9) define the weak guarded command with which, unlike with the strict guarded command, it is possible to wait till a proposition holds. From these axioms we can derive

$$x = \nu_{\mathsf{rel}}(x) \;\Rightarrow\; \phi :\rightarrow x = (\phi \mathrel{\vdots\!\rightarrow} x) + \partial_A(\nu_{\mathsf{rel}}(x))$$
$$x = \nu_{\mathsf{rel}}(x) + \sigma_{\mathsf{rel}}(y) \;\Rightarrow\; \phi :\rightarrow x = (\phi \mathrel{\vdots\!\rightarrow} \nu_{\mathsf{rel}}(x)) + \sigma_{\mathsf{rel}}(\phi :\rightarrow y) + \partial_A(\nu_{\mathsf{rel}}(x))$$

which gives a full picture of the differences between the two guarded commands.

SEMANTICS OF $\text{PA}^{\text{psc}}_{\text{drt}}$. We shall give a structural operational semantics for $\text{PA}^{\text{psc}}_{\text{drt}}$ using rules in the style of Plotkin to define the following unary and binary relations on the closed terms of $\text{PA}^{\text{psc}}_{\text{drt}}$:

a unary relation	$\_ \xrightarrow{v,a} \surd$	for each valuation v and $a \in A$,
a binary relation	$\_ \xrightarrow{v,a} \_$	for each valuation v and $a \in A$,
a binary relation	$\_ \xrightarrow{v,\sigma} \_$	for each valuation v,
a unary relation	$v \in [\mathsf{s}_\rho(\_)]$	for each valuation v.

These relations can be explained as follows:

$t \xrightarrow{v,a} \surd$: under valuation v, t is capable of first performing a in the current time slice and then terminating successfully;
$t \xrightarrow{v,a} t'$: under valuation v, t is capable of first performing a in the current time slice and then proceeding as t';
$t \xrightarrow{v,\sigma} t'$: under valuation v, t is capable of first passing to the next time slice and then proceeding as t';
$v \in [\mathsf{s}_\rho(t)]$: v makes the root signal of t true.

The rules have the form $\frac{p_1, \ldots, p_m}{c_1, \ldots, c_n}\ s$, where s is optional. They are to be read as "if p_1 and ... and p_m then c_1 and ... and c_n, provided s". As usual, $p_1, \ldots, p_m$ and $c_1, \ldots, c_n$ are called the premises and the conclusions, respectively. The conclusions are positive formulas of the form $t \xrightarrow{v,a} \surd$, $t \xrightarrow{v,a} t'$, $t \xrightarrow{v,\sigma} t'$ or $v \in [\mathsf{s}_\rho(t)]$, where t and t' are open terms of $\mathrm{PA}^{\mathrm{psc}}_{\mathrm{drt}}$. The premises are positive formulas of the above forms or negative formulas of the form $t \overset{v,\sigma}{\not\twoheadrightarrow}$. A negative formula $t \overset{v,\sigma}{\not\twoheadrightarrow}$ means that for all closed terms t' of $\mathrm{PA}^{\mathrm{psc}}_{\mathrm{drt}}$ not $t \xrightarrow{v,\sigma} t'$. The rules are actually rule schemas. The optional s is a side-condition restricting the valuations over which v ranges, the actions over which a ranges, the propositions over which ϕ ranges, and the sets of actions over which H ranges. If $m = 0$ and there is no side-condition, the horizontal bar is left out.

The signature of $\mathrm{PA}^{\mathrm{psc}}_{\mathrm{drt}}$ together with the rules that will be given constitute a term deduction system in *panth* format as defined in [29]. It is known from [29] that if a term deduction system in panth format is *stratifiable*, (strong) bisimulation equivalence is a congruence for the operators in the signature concerned. For a comprehensive introduction to rule formats guaranteeing that bisimulation equivalence is a congruence, the reader is referred to [1].

Let T be a term deduction system and $\mathrm{PF}(T)$ be the set of positive formulas occurring in the rules of T. Then a mapping $S : \mathrm{PF}(T) \to \alpha$ for an ordinal α is called a *stratification* for T if for all rules $\frac{P}{C}$ of T, formulas c in C, and closed substitutions σ the following conditions hold:

for all positive formulas p in P, $S(\sigma(p)) \leqslant S(\sigma(c))$;
for all negative formulas $t \neg R$ in P, $S(\sigma(tRt')) < S(\sigma(c))$ for all closed terms t';
for all negative formulas $\neg Pt$ in P, $S(\sigma(Pt)) < S(\sigma(c))$.

Recall that the rules that will be given are actually rule schemas. Within the framework of term deduction systems, the instances of the rule schemas that satisfy the stated side-conditions should be taken as the rules under consideration. For the rest, we continue to use the word rule in the broader sense.

A structural operational semantics of $\mathrm{PA}^{\mathrm{psc}}_{\mathrm{drt}}$ is described by the rules given in Tables 11, 12, 13 and 14. Note that we write $t \overset{v,\sigma}{\not\twoheadrightarrow}$ instead of $t \neg \xrightarrow{v,\sigma}$.

All rules are in panth format. In order to prove the fact that strong bisimulation is a congruence, we only have to find a stratification. We define a stratification S as follows:

$$S(t \xrightarrow{v,\sigma} t') = n_+(t) \text{ and } S(t \xrightarrow{v,a} \surd) = S(t \xrightarrow{v,a} t') = S(v \in [\mathsf{s}_\rho(t)]) = 0,$$

Table 11
Rules for basic operators of $\mathrm{PA}^{\mathrm{psc}}_{\mathrm{drt}}$ $(a \in A)$

$$\underline{\underline{a}} \xrightarrow{v,a} \surd \qquad \sigma_{\mathsf{rel}}(x) \xrightarrow{v,\sigma} x$$

$$\frac{x \xrightarrow{v,a} \surd}{\phi \mathrel{\vdots\!\!\to} x \xrightarrow{v,a} \surd}\ v \models \phi \qquad \frac{x \xrightarrow{v,a} x'}{\phi \mathrel{\vdots\!\!\to} x \xrightarrow{v,a} x'}\ v \models \phi \qquad \frac{x \xrightarrow{v,\sigma} x'}{\phi \mathrel{\vdots\!\!\to} x \xrightarrow{v,\sigma} x'}\ v \models \phi$$

$$\frac{x \xrightarrow{v,a} \surd}{\phi \mathrel{:\!\to} x \xrightarrow{v,a} \surd}\ v \models \phi \qquad \frac{x \xrightarrow{v,a} x'}{\phi \mathrel{:\!\to} x \xrightarrow{v,a} x'}\ v \models \phi \qquad \frac{x \xrightarrow{v,\sigma} x'}{\phi \mathrel{:\!\to} x \xrightarrow{v,\sigma} \phi \mathrel{:\!\to} x'}$$

$$\frac{x \xrightarrow{v,a} \surd}{\phi \mathbin{/\!\!\blacktriangle} x \xrightarrow{v,a} \surd}\ v \models \phi \qquad \frac{x \xrightarrow{v,a} x'}{\phi \mathbin{/\!\!\blacktriangle} x \xrightarrow{v,a} x'}\ v \models \phi \qquad \frac{x \xrightarrow{v,\sigma} x'}{\phi \mathbin{/\!\!\blacktriangle} x \xrightarrow{v,\sigma} x'}\ v \models \phi$$

$$\frac{x \xrightarrow{v,a} \surd,\ w \in [\mathsf{s}_\rho(y)]}{x \cdot y \xrightarrow{v,a} y} \qquad \frac{x \xrightarrow{v,a} x'}{x \cdot y \xrightarrow{v,a} x' \cdot y} \qquad \frac{x \xrightarrow{v,\sigma} x'}{x \cdot y \xrightarrow{v,\sigma} x' \cdot y}$$

$$\frac{x \xrightarrow{v,a} \surd,\ v \in [\mathsf{s}_\rho(y)]}{x + y \xrightarrow{v,a} \surd,\ y + x \xrightarrow{v,a} \surd} \qquad \frac{x \xrightarrow{v,a} x',\ v \in [\mathsf{s}_\rho(y)]}{x + y \xrightarrow{v,a} x',\ y + x \xrightarrow{v,a} x'}$$

$$\frac{x \xrightarrow{v,\sigma} x',\ y \xrightarrow{v,\sigma} y'}{x + y \xrightarrow{v,\sigma} x' + y'} \qquad \frac{x \xrightarrow{v,\sigma} x',\ y \stackrel{v,\sigma}{\not\longrightarrow},\ v \in [\mathsf{s}_\rho(y)]}{x + y \xrightarrow{v,\sigma} x',\ y + x \xrightarrow{v,\sigma} x'}$$

Table 12
Rules for parallel composition $(a \in A)$

$$\frac{x \xrightarrow{v,a} \surd,\ v \in [\mathsf{s}_\rho(y)]}{x \parallel y \xrightarrow{v,a} y,\ y \parallel x \xrightarrow{v,a} y,\ x \mathbin{\lfloor\!\lfloor} y \xrightarrow{v,a} y}$$

$$\frac{x \xrightarrow{v,a} x',\ v \in [\mathsf{s}_\rho(y)],\ w \in [\mathsf{s}_\rho(x')],\ w \in [\mathsf{s}_\rho(y)]}{x \parallel y \xrightarrow{v,a} x' \parallel y,\ y \parallel x \xrightarrow{v,a} y \parallel x',\ x \mathbin{\lfloor\!\lfloor} y \xrightarrow{v,a} x' \parallel y}$$

$$\frac{x \xrightarrow{v,\sigma} x',\ y \xrightarrow{v,\sigma} y'}{x \parallel y \xrightarrow{v,\sigma} x' \parallel y',\ x \mathbin{\lfloor\!\lfloor} y \xrightarrow{v,\sigma} x' \mathbin{\lfloor\!\lfloor} y'}$$

Table 13
Rules for $v \in [\mathsf{s}_\rho(\_)]$

$$v \in [\mathsf{s}_\rho(\underline{\underline{a}})] \qquad v \in [\mathsf{s}_\rho(\underline{\underline{\delta}})] \qquad v \in [\mathsf{s}_\rho(\sigma_{\mathsf{rel}}(x))]$$

$$\frac{v \in [\mathsf{s}_\rho(x)]}{v \in [\mathsf{s}_\rho(\phi \mathrel{\vdots\!\!\to} x)]} \qquad \frac{}{v \in [\mathsf{s}_\rho(\phi \mathrel{\vdots\!\!\to} x)]}\ v \not\models \phi$$

$$\frac{v \in [\mathsf{s}_\rho(x)]}{v \in [\mathsf{s}_\rho(\phi \mathrel{:\!\to} x)]} \qquad \frac{v \in [\mathsf{s}_\rho(x)]}{v \in [\mathsf{s}_\rho(\phi \mathbin{/\!\!\blacktriangle} x)]}\ v \models \phi$$

$$\frac{v \in [\mathsf{s}_\rho(x)]}{v \in [\mathsf{s}_\rho(x \cdot y)]} \qquad \frac{v \in [\mathsf{s}_\rho(x)],\ v \in [\mathsf{s}_\rho(y)]}{v \in [\mathsf{s}_\rho(x + y)],\ v \in [\mathsf{s}_\rho(x \parallel y)],\ v \in [\mathsf{s}_\rho(x \mathbin{\lfloor\!\lfloor} y)]}$$

Table 14
Rules for now operator and encapsulation ($a \in A$)

$$\frac{x \xrightarrow{v,a} \surd}{\nu_{\mathsf{rel}}(x) \xrightarrow{v,a} \surd} \qquad \frac{x \xrightarrow{v,a} x'}{\nu_{\mathsf{rel}}(x) \xrightarrow{v,a} x'} \qquad \frac{v \in [\mathsf{s}_\rho(x)]}{v \in [\mathsf{s}_\rho(\nu_{\mathsf{rel}}(x))]}$$

$$\frac{x \xrightarrow{v,a} \surd}{\partial_H(x) \xrightarrow{v,a} \surd}\ a \notin H \qquad \frac{x \xrightarrow{v,a} x'}{\partial_H(x) \xrightarrow{v,a} \partial_H(x')}\ a \notin H$$

$$\frac{x \xrightarrow{v,\sigma} x'}{\partial_H(x) \xrightarrow{v,\sigma} \partial_H(x')} \qquad \frac{v \in [\mathsf{s}_\rho(x)]}{v \in [\mathsf{s}_\rho(\partial_H(x))]}$$

where $n_+(t)$ stands for the number of occurrences of $+$ in t. So $S(F)$ is the number of occurrences of $+$ in the terms t occurring as the left-hand side of the formulas in F that have the form $t \xrightarrow{v,\sigma} t'$. It is straightforward to prove that the mapping S is a stratification. We have to check all rules. This is trivial except for the only rule with a negative formula in its premises, viz. the last rule of Table 11. The number of occurrences of $+$ in the conclusion of that rule is strictly greater than the number of occurrences of $+$ in the negative formula in the premises.

Note that the two rules for alternative composition concerning passage to the next time slice (Table 11) have complementary conditions. Together they enforce that the choice between two processes that both can pass to the next time slice is postponed till after the passage to the next time slice. This corresponds to the property reflected by the axiom DRT1 of $\mathrm{BPA}^{-}_{\mathrm{drt}}$-ID (Table 1).

In order to rule out processes that are capable of performing an action and then going on as $\bot$, there are premises in the first rule for sequential composition (Table 11) and the second rule for parallel composition (Table 12) concerning the existence of valuations that makes the root signal of certain processes true. This corresponds to the property reflected by the axiom NE3 of $\mathrm{PA}^{\mathrm{psc}}_{\mathrm{drt}}$ (Table 5).

The rule for time unit delay (Table 11), shows that $\sigma_{\mathsf{rel}}(t)$ is capable of passing to the next time slice under all valuations instead of only the valuations under which t is capable of doing things. This excludes persistency of root signal emission and waiting of the strict guarded command because all rules for these operators restrict the valuations under which the resulting processes are capable of doing things.

2.4. *Recursion*

In this subsection, we add recursion to $\mathrm{PA}^{\mathrm{psc}}_{\mathrm{drt}}$. Recursive specification, solution, guardedness, etc. are defined in a similar way as for BPA in [8].

Let V be a set of variables (of sort P). A *recursive specification* $E = E(V)$ in $\mathrm{PA}^{\mathrm{psc}}_{\mathrm{drt}}$ is a set of equations $E = \{X = s_X \mid X \in V\}$ where each s_X is a $\mathrm{PA}^{\mathrm{psc}}_{\mathrm{drt}}$ term that only contains variables from V. We shall use $X, X', Y, Y', \ldots$ for variables bound in a recursive specification. A *solution* of a recursive specification $E(V)$ is a set of processes $\{\langle X|E\rangle \mid$

$X \in V\}$ in some model of $\mathrm{PA}^{\mathrm{psc}}_{\mathrm{drt}}$ such that the equations of $E(V)$ hold if, for all $X \in V$, X stands for $\langle X|E\rangle$. Mostly, we are interested in one particular variable $X \in V$.

We can now introduce the equational theory of $\mathrm{PA}^{\mathrm{psc}}_{\mathrm{drt}}\mathrm{rec}$.

SIGNATURE OF $\mathrm{PA}^{\mathrm{psc}}_{\mathrm{drt}}\mathrm{rec}$. The signature of $\mathrm{PA}^{\mathrm{psc}}_{\mathrm{drt}}\mathrm{rec}$ consists of the signature of $\mathrm{PA}^{\mathrm{psc}}_{\mathrm{drt}}$ extended with a constant $\langle X|E\rangle : \to \mathsf{P}$ for each $X \in V$ and each recursive specification $E(V)$.

Let t be an open term in $\mathrm{PA}^{\mathrm{psc}}_{\mathrm{drt}}$ and $E = E(V)$ be a recursive specification. Then we write $\langle t|E\rangle$ for t with, for all $X \in V$, all occurrences of X in t replaced by $\langle X|E\rangle$.

AXIOMS OF $\mathrm{PA}^{\mathrm{psc}}_{\mathrm{drt}}\mathrm{rec}$. Axioms of $\mathrm{PA}^{\mathrm{psc}}_{\mathrm{drt}}\mathrm{rec}$ The axiom system of $\mathrm{PA}^{\mathrm{psc}}_{\mathrm{drt}}\mathrm{rec}$ consists of the axioms of $\mathrm{PA}^{\mathrm{psc}}_{\mathrm{drt}}$ and an equation $\langle X|E\rangle = \langle s_X|E\rangle$ for each $X \in V$ and each recursive specification $E(V)$.

Let t be a term of $\mathrm{PA}^{\mathrm{psc}}_{\mathrm{drt}}$ containing a variable X. We call an occurrence of X in t *guarded* if t has a subterm of the form $\underline{\underline{a}} \cdot t'$ or $\sigma_{\mathsf{rel}}(t')$ with t' a $\mathrm{PA}^{\mathrm{psc}}_{\mathrm{drt}}$ term containing this occurrence of X. We call a recursive specification *guarded* if all occurrences of all its variables in the right-hand sides of all its equations are guarded or it can be rewritten to such a recursive specification using the axioms of $\mathrm{PA}^{\mathrm{psc}}_{\mathrm{drt}}$ and its equations. An interesting form of guarded recursive specification is linear recursive specification. We call a recursive specification $E(V)$ *linear* if each equation in E has the form

$$X = \sum_{i<n} \phi_i :\to \underline{\underline{a_i}} \cdot X_i + \sum_{i<m} \psi_i :\to \underline{\underline{b_i}} + \sum_{i<k} \chi_i :\to \sigma_{\mathsf{rel}}(X'_i) + \xi \mathbin{\hat{}\!\!\blacktriangle} \underline{\underline{\delta}}$$

for certain actions a_i and b_i, propositions ϕ_i, ψ_i, χ_i and ξ, and variables $X, X_i, X'_i \in V$. Note that, without loss of generality we can assume that for all i and j such that $i \neq j$: $\underline{\underline{a_i}} \cdot X_i \not\equiv \underline{\underline{a_j}} \cdot X_j$, $\underline{\underline{b_i}} \not\equiv \underline{\underline{b_j}}$, $X'_i \not\equiv X'_j$ and $\chi_i \not\equiv \chi_j$. We can also assume that ϕ_i, ψ_i and χ_i are not f.

PRINCIPLES OF $\mathrm{PA}^{\mathrm{psc}}_{\mathrm{drt}}\mathrm{rec}$. The *(restricted) recursive definition principle* ($\mathrm{RDP}^{(-)}$) is the assumption that every (guarded) recursive specification has a solution. The *recursive specification principle* (RSP) is the assumption that every guarded recursive specification has at most one solution.

Note that the axioms $\langle X|E\rangle = \langle s_X|E\rangle$ for a fixed E express that the constants $\langle X|E\rangle$ make up a solution of E, i.e. RDP holds for any model of $\mathrm{PA}^{\mathrm{psc}}_{\mathrm{drt}}\mathrm{rec}$. The conditional equations $E \Rightarrow X = \langle X|E\rangle$ express that this solution is the only one. So RSP can be described by means of conditional equations – as already mentioned in [28].

SEMANTICS OF $\mathrm{PA}^{\mathrm{psc}}_{\mathrm{drt}}\mathrm{rec}$. A structural operational semantics of $\mathrm{PA}^{\mathrm{psc}}_{\mathrm{drt}}\mathrm{rec}$ is described by the rules for $\mathrm{PA}^{\mathrm{psc}}_{\mathrm{drt}}$ and the rules given in Table 15.

Table 15
Rules for recursion ($a \in A$)

$$\frac{\langle s_X | E\rangle \xrightarrow{v,a} \surd}{\langle X | E\rangle \xrightarrow{v,a} \surd} \quad \frac{\langle s_X | E\rangle \xrightarrow{v,a} y}{\langle X | E\rangle \xrightarrow{v,a} y}$$

$$\frac{\langle s_X | E\rangle \xrightarrow{v,\sigma} y}{\langle X | E\rangle \xrightarrow{v,\sigma} y} \quad \frac{v \in [\mathsf{s}_\rho(\langle s_X | E\rangle)]}{v \in [\mathsf{s}_\rho(\langle X | E\rangle)]}$$

The rules added for recursion are also in panth format. We define a stratification S as follows:

$$S(t \xrightarrow{v,\sigma} t') = \omega \cdot n_{\mathrm{sol}}(t) + n_+(t) \text{ and } S(t \xrightarrow{v,a} \surd) = S(t \xrightarrow{v,a} t') = S(v \in [\mathsf{s}_\rho(t)]) = 0,$$

where $n_{\mathrm{sol}}(t)$ stands for the number of unguarded occurrences of constants $\langle X | E\rangle$ in t and $n_+(t)$ stands for the number of occurrences of $+$ in t. The addition of the summand $\omega \cdot n_{\mathrm{sol}}(t)$ for formulas $t \xrightarrow{v,\sigma} t'$ solves the problem that the conclusion of the rule for recursion concerning passage to the next time slice does not contain occurrences of $+$.

Let $E = \{X = s_X \mid X \in V\}$ be a recursive specification. Then roughly, the rules for recursion come down to looking upon $\langle X | E\rangle$ as the process s_X with, for all $X' \in V$, all occurrences of X' in s_X replaced by $\langle X' | E\rangle$.

2.5. *Counting process creation*

In this subsection, we introduce the counting process creation operator E^n_Φ that is used for the semantics of φ^-SDL. This operator subsumes the original process creation operator introduced in [11]. The latter process creation operator was used in [14] for the semantics of φSDL. But the approach used there does not guarantee that a unique process identification is assigned to each created process.

It is assumed that a fixed but arbitrary set D of *data* has been given together with a function $\Phi : \mathbb{N} \times D \to \mathsf{P}$, and that there exist actions $cr(d)$ and $\overline{cr}(n, d)$ for all $d \in D$ and $n \in \mathbb{N}$. The process creation operator E^n_Φ allows, given the function Φ, the use of actions $cr(d)$ to create processes $\Phi(n, d)$.

The counting process creation operator $\mathrm{E}^n_\Phi : \mathsf{P} \to \mathsf{P}$ is defined by the equations PRCR1–PRCR8 given in Table 16. The crucial axiom is PRCR4. It expresses that counting process creation applied to a process, with the counter set to n, leaves the action $\underline{\underline{\overline{cr}}}(n, d)$ as a trace and starts a process $\Phi(n, d)$ in parallel with the remaining process when it comes across an undelayable process creation action $\underline{\underline{cr}}(d)$. Besides, it increases the counter by one. The counting process creation operator E^n_Φ is an extension of the process creation operator E_ϕ from [11]. We can write $\mathrm{E}^n_\Phi = \mathrm{E}_\phi$ if $\Phi(n, d) = \phi(d)$ for all $n \in \mathbb{N}$ and $d \in D$.

A structural operational semantics for the counting process creation operator is described by the rules given in Table 17. The stratification introduced for $\mathrm{PA}^{\mathrm{psc}}_{\mathrm{drt}}$rec still works if we add the rules for the counting process creation operator.

Table 16
Axioms for counting process creation ($a \in A_\delta$)

$\mathrm{E}^n_\Phi(\underline{\underline{a}}) = \underline{\underline{a}}$ if $a \neq cr(d)$ for $d \in D$	PRCR1
$\mathrm{E}^n_\Phi(\underline{\underline{cr}}(d)) = \overline{\underline{\underline{cr}}}(n,d) \cdot \mathrm{E}^{n+1}_\Phi(\Phi(n,d))$	PRCR2
$\mathrm{E}^n_\Phi(\underline{\underline{a}} \cdot x) = \underline{\underline{a}} \cdot \mathrm{E}^n_\Phi(x)$ if $a \neq cr(d)$ for $d \in D$	PRCR3
$\mathrm{E}^n_\Phi(\underline{\underline{cr}}(d) \cdot x) = \overline{\underline{\underline{cr}}}(n,d) \cdot \mathrm{E}^{n+1}_\Phi(\Phi(n,d) \parallel x)$	PRCR4
$\mathrm{E}^n_\Phi(x + y) = \mathrm{E}^n_\Phi(x) + \mathrm{E}^n_\Phi(y)$	PRCR5
$\mathrm{E}^n_\Phi(\sigma_{\mathsf{rel}}(x)) = \sigma_{\mathsf{rel}}(\mathrm{E}^n_\Phi(x))$	PRCR6
$\mathrm{E}^n_\Phi(\phi :\rightarrow x) = \phi :\rightarrow \mathrm{E}^n_\Phi(x)$	PRCR7
$\mathrm{E}^n_\Phi(\phi \mathbin{{}^\wedge\!\blacktriangle} x) = \phi \mathbin{{}^\wedge\!\blacktriangle} \mathrm{E}^n_\Phi(x)$	PRCR8

Table 17
Rules for counting process creation ($a \in A$)

$$\frac{x \xrightarrow{v,a} \surd}{\mathrm{E}^n_\Phi(x) \xrightarrow{v,a} \surd}\; a \neq cr(d) \qquad \frac{x \xrightarrow{v,cr(d)} \surd,\; w \in [\mathsf{s}_\rho(\Phi(n,d))]}{\mathrm{E}^n_\Phi(x) \xrightarrow{v,\overline{cr}(n,d)} \mathrm{E}^{n+1}_\Phi(\Phi(n,d))}$$

$$\frac{x \xrightarrow{v,a} x'}{\mathrm{E}^n_\Phi(x) \xrightarrow{v,a} \mathrm{E}^n_\Phi(x')}\; a \neq cr(d) \qquad \frac{x \xrightarrow{v,cr(d)} x',\; w \in [\mathsf{s}_\rho(\Phi(n,d))],\; w \in [\mathsf{s}_\rho(x')]}{\mathrm{E}^n_\Phi(x) \xrightarrow{v,\overline{cr}(n,d)} \mathrm{E}^{n+1}_\Phi(\Phi(n,d) \parallel x')}$$

$$\frac{x \xrightarrow{v,\sigma} x'}{\mathrm{E}^n_\Phi(x) \xrightarrow{v,\sigma} \mathrm{E}^n_\Phi(x')} \qquad \frac{v \in [\mathsf{s}_\rho(x)]}{v \in [\mathsf{s}_\rho(\mathrm{E}^n_\Phi(x))]}$$

2.6. *State operator*

In this subsection, we introduce a state operator for $\mathrm{PA}^{\mathrm{psc}}_{\mathrm{drt}}$. It generalizes and extends the state operator for $\mathrm{ACP_{ps}}$ proposed in [6]: the truth value of propositional signals and conditions may depend upon the state and passage to the next time slice may have an effect on the state. The possibility of non-deterministic behaviour as the result of applying the operator is included as well, like for the extended state operator Λ (see, e.g., [8]).

It is assumed that a fixed but arbitrary set S of states has been given, together with functions:

$$\begin{aligned}
&act &&: A \times S \rightarrow \mathcal{P}_{\mathrm{fin}}(A)\\
&\mathit{eff} &&: A \times S \times A \rightarrow S\\
&\mathit{eff}_\sigma &&: S \rightarrow \mathcal{P}_{\mathrm{fin}}(S) \setminus \{\emptyset\}\\
&sig &&: S \rightarrow \mathsf{B}\\
&val &&: B_{\mathrm{at}} \times S \rightarrow \mathbb{B}
\end{aligned}$$

The state operator λ_s ($s \in S$) allows, given these functions, processes to interact with a state. The process $\lambda_s(x)$ is the process x executed in a state s. The function *act* gives, for each action a and state s, the set of actions that may be performed if a is executed in state s. The function *eff* gives, for each action a, state s and action a', the state that results when a is executed in state s and a' is the action that is actually performed as the result of the

Table 18
Axioms for state operator ($a \in A$)

$\lambda_s(\underline{\underline{a}}) = sig(s) \mathbin{\wedge\!\!\blacktriangle} \sum_{a' \in act(a,s)} \underline{\underline{a'}}$	SO1
$\lambda_s(\underline{\underline{a}} \cdot x) = sig(s) \mathbin{\wedge\!\!\blacktriangle} \sum_{a' \in act(a,s)} (\underline{\underline{a'}} \cdot \lambda_{eff(a,s,a')}(x))$	SO2
$\lambda_s(x + y) = \lambda_s(x) + \lambda_s(y)$	SO3
$\lambda_s(\sigma_{\mathsf{rel}}(x)) = sig(s) \mathbin{\wedge\!\!\blacktriangle} \sigma_{\mathsf{rel}}(\sum_{s' \in eff_\sigma(s)} \lambda_{s'}(x))$	SO4
$\lambda_s(\phi :\rightarrow x) = sig(s) \mathbin{\wedge\!\!\blacktriangle} (val(\phi, s) :\rightarrow \lambda_s(x))$	SO5
$\lambda_s(\phi \mathbin{\wedge\!\!\blacktriangle} x) = val(\phi, s) \mathbin{\wedge\!\!\blacktriangle} \lambda_s(x)$	SO6

execution. The function eff_σ gives, for each state s, the set of states that may result when time passes to the next time slice in state s. The function sig gives, for each state s, the propositional signal that holds at the start of any process executed in state s. The function val gives, for each state s, the valuation $val(\_, s)$ of the atomic propositions in state s. The valuation $val(\_, s)$ can be extended to all propositions in the usual homomorphic way as any other valuation. We will use the notation $val(\_, s)$ to refer to the extension as well.

The state operator $\lambda_s : \mathsf{P} \to \mathsf{P}$ is defined by the equations SO1–SO6 given in Table 18. The axioms SO1–SO3 are straightforward reformulations – in the same way as for $\mathrm{PA}^{\mathrm{psc}}_{\mathrm{drt}}$ – of corresponding axioms given in [6] for the time free case. The additional axiom SO4 expresses how passage to the next time slice has influence on the execution of a process in a state. The axioms SO5 and SO6 are also reformulations of corresponding axioms given in [6]. In these axioms the proposition ϕ has been replaced by $val(\phi, s)$. Thus the case is covered where the truth value of propositional signals and conditions may depend upon the state. Note that from SO5 we can derive $\lambda_s(\underline{\underline{\delta}}) = \underline{\underline{\delta}}$.

A structural operational semantics for the state operator is described by the rules given in Table 19. Note that the rules added for the state operator have a common side-condition given at the bottom of the table. The stratification introduced for $\mathrm{PA}^{\mathrm{psc}}_{\mathrm{drt}}\mathrm{rec}$ still works if we add the rules for the state operator.

Table 19
Rules for state operator ($a \in A$)

$$\frac{x \xrightarrow{v,a} \surd}{\lambda_s(x) \xrightarrow{v',a'} \surd}\; a' \in act(a, s)$$

$$\frac{x \xrightarrow{v,a} x',\; w \in [\mathsf{s}_\rho(x')]}{\lambda_s(x) \xrightarrow{v',a'} \lambda_{eff(a,s,a')}(x')}\; a' \in act(a, s) \wedge w \models sig(eff(a, s, a'))$$

$$\frac{x \xrightarrow{v,\sigma} x'}{\lambda_s(x) \xrightarrow{v',\sigma} \lambda_{s'}(x')}\; s' \in eff_\sigma(s) \qquad \frac{v \in [\mathsf{s}_\rho(x)]}{v' \in [\mathsf{s}_\rho(\lambda_s(x))]}$$

for all v, v' and s such that $v' \models sig(s) \wedge \forall\phi \bullet v \models \phi \leftrightarrow v' \models val(\phi, s)$

3. Overview of φ^-SDL

3.1. *Introduction*

In this section, we give an overview of φ^-SDL, i.e. φSDL without delaying channels. φSDL is a small subset of SDL, introduced in [14], which covers all behavioural aspects of SDL, including communication via delaying channels, timing and process creation. Leaving out delaying channels simplifies the presentation. Besides work on the process algebra semantics of φSDL made clear that φSDL specifications can always be transformed to semantically equivalent ones in φ^-SDL. At the end of Section 4 is shown how to model a delaying channel by means of a φ^-SDL process.

In Sections 3.2, 3.3 and 3.4, the syntax of φ^-SDL is described by means of production rules in the form of an extended BNF grammar (the extensions are explained in Appendix A). The meaning of the language constructs of the various forms distinguished by these rules is explained informally. Some peculiar details of the semantics, inherited from full SDL, are left out to improve the comprehensibility of the overview. These details will, however, be taken into account in Section 4, where the process algebra semantics of φ^-SDL is presented. In Section 3.5, some remarks are made about the context-sensitive conditions for syntactic correctness of φ^-SDL specifications. The syntactic differences with full SDL are summarized in this section as well. In Section 3.6 some examples of φ^-SDL specifications are given.

In line with full SDL, we can define a graphical representation for φ^-SDL specifications. We pay no attention to this practically important point because it is not relevant to the subject of this chapter.

3.2. *System definition*

First of all, the φ^-SDL view of a system is explained in broad outline.

Basically, a system consists of *processes* which communicate with each other and the environment by sending and receiving *signals* via *signal routes*. A process proceeds in parallel with the other processes in the system and communicates with these processes in an asynchronous manner. This means that a process sending a signal does not wait until the receiving process consumes it, but it proceeds immediately. A process may also use local *variables* for storage of values. A variable is associated with a value that may change by assigning a new value to it. A variable can only be assigned new values by the process to which it is local, but it may be viewed by other processes. Processes can be distinguished by unique addresses, called *pid values* (process identification values), which they get with their creation.

A signal can be sent from the environment to a process, from a process to the environment or from a process to a process. A signal may carry values to be passed from the sender to the receiver; on consumption of the signal, these values are assigned to local variables of the receiver. A signal route is a unidirectional communication path for sending signals from the environment to a process, from one process to another process or from a process to the environment. If a signal is sent to a process via a signal route it will be instantaneously delivered to that process.

```
<decision> ::=
  decision {<expr> | any} ;
    ( [ <ground expr> ] ) : <transition>
    {( [ <ground expr> ] ) : <transition>}+
  enddecision
```

A state definition **state** st**; save** $s_1,\ldots,s_m$**;**$alt_1 \ldots alt_n$ defines a state st. The signals of the types $s_1, \ldots, s_m$ are saved for the state. The input guard of each of the transition alternatives $alt_1, \ldots, alt_n$ gives a type of signals that may be consumed in the state; the corresponding transition is the one that is initiated on consumption of a signal of that type. The alternatives with **input none;** instead of an input guard are the spontaneous transitions that may be made from the state. No signals are saved for the state if **save** $s_1,\ldots,s_m$**;** is absent.

An input guard **input** s**(**$v_1,\ldots,v_n$**);** may consume a signal of type s and, on consumption, it assigns the carried values to the variables $v_1, \ldots, v_n$. If the signals of type s carry no value, **(**$v_1,\ldots,v_n$**)** is left out.

A transition $a_1 \ldots a_n$ **nextstate** st**;** performs the actions $a_1, \ldots, a_n$ in sequential order and ends with entering the state st. Replacing **nextstate** st by the keyword **stop** yields a transition ending with process termination. Replacing it by the decision dec leads instead to transfer of control to one of two or more transition branches.

An output action **output** s**(**$e_1,\ldots,e_n$**) to** e **via** $r_1,\ldots,r_m$**;** sends a signal of type s carrying the current values of the expressions $e_1, \ldots, e_n$ to the process with the current (pid) value of the expression e as its address, via one of the signal routes $r_1, \ldots, r_m$. If the signals of type s carry no value, **(**$e_1,\ldots,e_n$**)** is left out. If **to** e is absent, the signal is sent via one of the signal routes $r_1, \ldots, r_m$ to an arbitrary process of its receiver type. The output action is called an output action with explicit addressing if **to** e is present. Otherwise, it is called an output action with implicit addressing.

A set action **set(**e**,**s**(**$e_1,\ldots,e_n$**));** sets a timer that expires, unless it is set again or reset, at the current (time) value of the expression e with sending a signal of type s that carries the current values of the expressions $e_1, \ldots, e_n$.

A reset action **reset(**s**(**$e_1,\ldots,e_n$**));** de-activates the timer identified with the signal type s and the current values of the expressions $e_1, \ldots, e_n$.

An assignment task action **task** v**:=**e**;** assigns the current value of the expression e to the local variable v.

A create action **create** X**(**$e_1,\ldots,e_n$**);** creates a process of type X and passes the current values of the expressions $e_1, \ldots, e_n$ to the newly created process. If no values are passed on creation of processes of type X, **(**$e_1,\ldots,e_n$**)** is left out.

A decision **decision** e**;(**e_1**):**$tr_1 \ldots$**(**e_n**):**tr_n **enddecision** transfers control to the transition branch tr_i ($1 \leqslant i \leqslant n$) for which the value of the expression e_i equals the current value of the expression e. Non-existence of such a branch results in an error. A non-deterministic choice can be obtained by replacing the expression e by the keyword **any** and removing all the expressions e_i.

3.4. *Values*

The value of expressions in φ^-SDL may vary according to the last values assigned to variables, including local variables of other processes. It may also depend on timers being active, the system time, etc.

Syntax:

<expr> ::=
<operator nm> [**(** <expr> {**,** <expr>}* **)**]
| **if** <boolean expr> **then** <expr> **else** <expr> **fi**
| <variable nm>
| **view (** <variable nm> **,** <pid expr> **)**
| **active (** <signal nm> [**(** <expr> {**,** <expr>}* **)**] **)**
| **now** | **self** | **parent** | **offspring** | **sender**

An operator application $op(e_1,\ldots,e_n)$ evaluates to the value yielded by applying the operation op to the current values of the expressions $e_1,\ldots,e_n$.

A conditional expression **if** e_1 **then** e_2 **else** e_3 **fi** evaluates to the current value of the expression e_2 if the current (Boolean) value of the expression e_1 is true, and the current value of the expression e_3 otherwise.

A variable access v evaluates to the current value of the local variable v of the process evaluating the expression.

A view expression **view**(v,e) evaluates to the current value of the local variable v of the process with the current (pid) value of the expression e as its address.

An active expression **active**$(s(e_1,\ldots,e_n))$ evaluates to the Boolean value true if the timer identified with the signal type s and the current values of the expressions $e_1,\ldots,e_n$ is currently active, and false otherwise.

The expression **now** evaluates to the current system time.

The expressions **self**, **parent**, **offspring** and **sender** evaluate to the pid values of the process evaluating the expression, the process by which it was created, the last process created by it, and the sender of the last signal consumed by it. Natural numbers are used as pid values. The pid value 0 is a special pid value that never refers to any existing process – in full SDL this pid value is denoted by **null** – and the pid value 1 is reserved for the environment. The expressions **parent**, **offspring** and **sender** evaluate to 0 in case there exists no parent, offspring and sender, respectively.

3.5. *Miscellaneous issues*

3.5.1. *Context-sensitive syntactic rules.* We remain loose about the context-sensitive conditions for syntactic correctness of φ^-SDL specifications. For the most part, they are as usual: only defined names may be used, use of names must agree with their definitions, types of expressions must be in accordance with the type information in the definitions, etc. There is one condition that needs particular attention: signals of the same type may not be used for both signal sending and timer setting/resetting. All φ^{-}SDL specifications that are obtained by semantics preserving transformations of syntactically correct specifications in full SDL will be syntactically correct φ^{-}SDL specifications as well.

3.5.2. *Syntactic differences with SDL.* Syntactically, φ^{-}SDL is not exactly a subset of SDL. The syntactic differences are as follows:
- variable definitions occur at the system level instead of inside process definitions;
- signal route definitions and process definitions occur at the system level instead of inside block definitions;
- formal parameters in process definitions are variable names instead of pairs of variable names and sort names;
- signal names are used as timer names.

These differences are all due to the simplifications mentioned in Section 1.

3.6. *Examples*

We give three small examples to illustrate how systems are specified in φ^{-}SDL. The examples concern simple repeaters and routers.

3.6.1. *Repeater.* The first example concerns a simple repeater, i.e. a system that simply passes on what it receives. The system, called `Repeater`, consists of only one process, viz. `rep`, which communicates signals `s` with the environment via the signal routes `fromenv` and `toenv`. The process has only one state.

```
system Repeater
  signal s;

  signalroute fromenv from env to rep with s;
  signalroute toenv from rep to env with s;

  process rep(1);
    start;
      nextstate pass;
    state pass;
      input s;
        output s via toenv;
        nextstate pass;
  endprocess;
endsystem;
```

3.6.2. *Address driven router.* The second example concerns address driven routing of data. The system, called `AddrRouter`, consists of three processes, one instance of `rtr` and two instances of `rep`. The latter two processes are created by the former process. The process `rtr` consumes signals `s(a)`, delivered via signal route `fromenv`, and passes them to one of the instances of `rep` (via signal route `rs`) depending on the value `a`. The instances of `rep` then pass the signals received from `rtr` to the environment via the signal route `toenv`.

```
system AddrRouter
  signal s(Bool);

  signalroute fromenv from env to rtr with s;
  signalroute rs from rtr to rep with s;
  signalroute toenv from rep to env with s;

  dcl a Bool; dcl rep1 Nat; dcl rep2 Nat;

  process rtr(1);
    start;
      create rep; task rep1 := offspring;
      create rep; task rep2 := offspring;
      nextstate route;
    state route;
      input s(a);
        decision a;
          (False):
            output s(a) to rep1 via rs;
            nextstate route;
          (True):
            output s(a) to rep2 via rs;
            nextstate route;
        enddecision;
  endprocess;

  process rep(0);
    start;
        nextstate pass;
    state pass;
      input s(a);
        output s(a) via toenv;
        nextstate pass;
  endprocess;
endsystem;
```

3.6.3. *Load driven router.* The last example concerns load driven routing of data. The system, called `LoadRouter`, consists of three processes, one instance of `rtr` and two instances of `trep`. The latter two processes are created by the former process. The process `rtr` consumes signals `s`, delivered via signal route `fromenv`, and passes them to one of the instances of `trep` (via signal route `rs`) depending on their load. The instances of `trep` then pass the signals received from `rtr` to the environment via the signal routes `toenv`. Either delivers the data consumed after a fixed time `delay`. One repeater delivers twice as fast as the other one. Only two load factors are considered for each of the two repeaters: one indicating its idleness and one indicating the opposite.

```
system LoadRouter
  signal s; signal t;

  signalroute fromenv from env to rtr with s;
  signalroute rs from rtr to trep with s;
  signalroute toenv from trep to env with s;
```

```
    dcl idle Bool; dcl delay Nat;
    dcl rep1 Nat; dcl rep2 Nat;

    process rtr(1);
      start;
        create trep(10); task rep1 := offspring;
        create trep(20); task rep2 := offspring;
        nextstate route;
      state route;
        input s;
          decision view(idle,rep1) <-> view(idle,rep2);
            (True):
              output s via rs; nextstate route;
            (False):
              decision view(idle,rep1)
                (True):
                  output s to rep1 via rs;
                  nextstate route;
                (False):
                  output s to rep2 via rs;
                  nextstate route;
              enddecision
          enddecision
    endprocess;

    process trep(0) fpar delay;
      start;
        task idle := True;
        nextstate get;
      state get;
        input s;
          task idle := False;
          set(now + delay, t);
          nextstate put;
      state put;
        save s;
        input t;
          output s via toenv;
          task idle := True;
          nextstate get;
     endprocess;
  endsystem;
```

Most features of φ^{-}SDL are used in this example. Of the main features, only spontaneous transitions, i.e. transition alternatives with **input none;** instead of an input guard, are missing. In Section 4.6.3, use is made of this feature to define a process modeling a delaying channel.

4. Semantics of φ^-SDL

4.1. *Introduction*

In this section, we propose a process algebra semantics for φ^-SDL. This semantics is presented in two steps.

In Section 4.3, a semantics for φ^-SDL process definitions is given. This semantics abstracts from dynamic aspects of process behaviour such as process creation and process execution in a state. It describes the meaning of φ^-SDL process definitions by means of finite guarded recursive specifications in $\mathrm{BPA}^{\mathrm{psc}}_{\mathrm{drt}}$. The counting process creation operator and the state operator are not needed for this semantics. Preceeding, in Section 4.2, the actions and atomic conditions used are introduced. These actions and conditions are parametrized by expressions with values that depend on the state in which the action or condition concerned is executed.

In Section 4.6, a semantics of φ^-SDL system definitions is given in terms of the semantics for φ^-SDL process definitions using the counting process creation operator and the state operator. Preceeding, in Section 4.5, all the details of the instance of the state operator needed for this semantics are given. Included are the definitions of the state space and the functions that describe how the actions and conditions used for the semantics of φ^-SDL process definitions interact with the state when this instance of the state operator is applied. The interaction with the environment is another aspect covered by the semantics of φ^-SDL system definitions. For this purpose an environment process is introduced as well.

The semantics of φ^-SDL is described by means of a set of equations recursively defining interpretation functions for all syntactic categories. The special notation used is explained in Appendix A. We will be lazy about specifying the range of each interpretation function, since this is usually clear from the context. Many of the interpretations are functions from natural numbers to process expressions or equations. These process expressions and equations will simply be written in their display form. If an optional clause represents a sequence, its absence is always taken to stand for an empty sequence. Otherwise, it is treated as a separate case. The semantics is defined using contextual information κ extracted from the φ^-SDL specification concerned. This is further described in Appendix B.

The special notation used for parametrized actions, conditions and propositions is explained in Appendix A, and so is the uncustomary notation as regards sets, functions and sequences. The words action and process are used in connection with both ACP-style process algebras and versions of SDL, but with slightly different meanings. In case it is not clear from the context, these words will be preceded by ACP if they should be taken in the ACP sense, and by SDL otherwise.

Data types. We mentioned before that φ^-SDL does not deal with the specification of abstract data types. We assume a fixed algebraic specification covering all data types used and an initial algebra semantics, denoted by $\mathcal{A}$, for it. The data types **Boolean** and **Natural**, with the obvious interpretation, must be included. We will write $Sort_{\mathcal{A}}$ and $Op_{\mathcal{A}}$ for the set of all sort names and the set of all operation names, respectively, in the signature of $\mathcal{A}$. We additionally assume a constant name, i.e. a nullary operation name, in $Op_{\mathcal{A}}$, referred to as $\underline{n}$, for each $n \in \mathbb{N}$. We will write U for $\bigcup_{T \in Sort_{\mathcal{A}}} T^{\mathcal{A}}$, where $T^{\mathcal{A}}$ is the interpretation of the sort name T in $\mathcal{A}$. We have that $\mathbb{B}, \mathbb{N} \subset U$ because of our earlier requirement that **Boolean**, **Natural** $\in Sort_{\mathcal{A}}$. We assume that nil $\notin U$. In the sequel, we will use for each $op \in Op_{\mathcal{A}}$ an extension to $U \cup \{\text{nil}\}$, also denoted by op, such that $op(t_1, \ldots, t_n) = \text{nil}$ if not all of the t_is are of the appropriate sort. Thus, we can change over from the many-sorted case to the one-sorted case for the description of the meaning of the language constructs of φ^-SDL. We can do so without loss of generality, because it can (and should) be statically checked that only terms of appropriate sorts occur.

4.2. *Actions and conditions*

In the semantics of φ^-SDL process definitions, which will be presented in Section 4.3, ACP actions and conditions are used. Here, we introduce the actions and atomic conditions concerned.

The actions and atomic conditions, used to describe the meaning of φ^-SDL process definitions, are parametrized by various domains. These domains depend upon the specific variable names, signal names and process names introduced in the system definition concerned. These sets of names largely make up the context ascribed to the system definition by means of the function $\{\![\_]\!\}$ defined in Appendix B. For convenience, we define these sets for arbitrary contexts κ (the notation concerning contexts introduced in Appendix B is used):

$$\begin{aligned} V_\kappa &= vars(\kappa) \cup \{\textbf{parent}, \textbf{offspring}, \textbf{sender}\} \\ S_\kappa &= sigs(\kappa) \\ P_\kappa &= procs(\kappa) \end{aligned}$$

Most arguments of the parametrized actions and conditions introduced here are expressions that originate in φ^-SDL expressions or objects that are somehow composed of such expressions and variable, signal and process names. The reason of this is that the value of the original φ^-SDL expressions may vary according to the last values assigned to the variables referred to, the status of the timers referred to, etc. In other words, these expressions stand for values that are not known until the action or condition concerned is executed in the appropriate state.

The syntax of the expressions concerned, called value expressions, is as follows:

<value expr> ::=
<operator nm> [(<value expr> {, <value expr>}*)]
| *cond* (<boolean value expr> , <value expr> , <value expr>)
| *value* (<variable nm> , <pid value expr>)
| *active* (<signal nm> [(<value expr> {, <value expr>}*)] , <pid value expr>)
| *now*

where the terminal productions of <operator nm>, <variable nm> and <signal nm> are assumed to yield the sets $Op_{\mathcal{A}}$, V_κ and S_κ, respectively. $ValE_\kappa$ denotes the set of all syntactically correct value expressions. The forms of value expressions distinguished above correspond to operator applications, conditional expressions, view expressions, active expressions, and the expression **now**, respectively, in φ^{-}SDL.

We define now the set $SigP_\kappa$ of signal patterns, the set $SigE_\kappa$ of signal expressions, the set $SaveSet_\kappa$ of save sets and the set $PrCrD_\kappa$ of process creation data. Further on, we will look at a signal as an object that consists of the name of the signal and the sequence of values that it carries. Signal patterns and signal expressions are like signals, but variables and value expressions, respectively, are used instead of values. A save set is just a finite set of signal names. A process creation datum consists of the name of the process to be created, its formal parameters, expressions denoting its actual parameters and the pid value of its creator.

$$\begin{array}{ll}
SigP_\kappa & = \{(s, \langle v_1, \ldots, v_n\rangle) \mid (s, \langle T_1, \ldots, T_n\rangle) \in sigds(\kappa), v_1, \ldots, v_n \in V_\kappa\} \\
SigE_\kappa & = \{(s, \langle e_1, \ldots, e_n\rangle) \mid (s, \langle T_1, \ldots, T_n\rangle) \in sigds(\kappa), e_1, \ldots, e_n \in ValE_\kappa\} \\
SaveSet_\kappa & = \mathcal{P}_{\mathrm{fin}}(S_\kappa) \\
PrCrD_\kappa & = \{(X, \langle v_1, \ldots, v_n\rangle, \langle e_1, \ldots, e_n\rangle, e) \mid \\
& \qquad X \in P_\kappa, v_1, \ldots, v_n \in V_\kappa, e_1, \ldots, e_n, e \in ValE_\kappa\}
\end{array}$$

We write $pnm(d)$, where $d = (X, vs, es, e) \in PrCrD_\kappa$, for X. Each process creation datum contains the formal parameters for the process type concerned. The alternative would be to make the association between process types and their formal parameters itself a parameter of the state operator, which is very unattractive.

The following actions are used:

$$\begin{array}{ll}
input & : SigP_\kappa \times SaveSet_\kappa \times ValE_\kappa \\
outpute & : SigE_\kappa \times ValE_\kappa \times ValE_\kappa \\
outputi & : SigE_\kappa \times ValE_\kappa \times (P_\kappa \cup \{\mathbf{env}\}) \\
set & : ValE_\kappa \times SigE_\kappa \times ValE_\kappa \\
reset & : SigE_\kappa \times ValE_\kappa \\
ass & : V_\kappa \times ValE_\kappa \times ValE_\kappa \\
cr & : PrCrD_\kappa \\
stop & : ValE_\kappa \\
inispont & : ValE_\kappa \\
tt & :
\end{array}$$

Cr_κ denotes the set of all *cr* actions, and Act_κ^{-} denotes the set of all *input*, *outpute*, *outputi*, *set*, *reset*, *ass*, *stop* and *inispont* actions. The *tt* action is a special action with no observable effect whatsoever. The other actions correspond to the input guards, the SDL actions, the terminator **stop** and the void guard **input none**. The last argument of each action is the pid value of the process from which the action originates, except for the *outpute* and *outputi* actions where the pid value concerned is available as the second argument. The *outpute* and *outputi* actions correspond to **output** actions with explicit addressing and implicit addressing, respectively, in φ^{-}SDL. The last argument of these actions refers to the receiver. With an *outpute* action, the receiver is fully determined by the pid value given as the last argument. With an *outputi* action, the receiver is not fully determined; it is an arbitrary process of the given type.

The conditions used are built from the following atomic conditions:

$$\begin{array}{ll} \textit{waiting} & : \textit{SaveSet}_{\kappa} \times \textit{ValE}_{\kappa} \\ \textit{type} & : \textit{ValE}_{\kappa} \times (P_{\kappa} \cup \{\textbf{env}\}) \\ \textit{hasinst} & : P_{\kappa} \cup \{\textbf{env}\} \\ \textit{eq} & : \textit{ValE}_{\kappa} \times \textit{ValE}_{\kappa} \end{array}$$

$AtCond_{\kappa}$ denotes the set of all *waiting*, *type*, *hasinst* and *eq* conditions. A condition $waiting(ss, e)$ is used to test whether the process with the pid value denoted by e has to wait for a signal if the signals with names in ss are withhold from being consumed. A condition $type(e, X)$ is used to test whether X is the type of the process with the pid value denoted by e. A condition $hasinst(X)$ is used to test whether there exists a process of type X. A condition $eq(e_1, e_2)$ is used to test whether the expressions e_1 and e_2 denote the same value. These conditions are used to give meaning to the state definitions, output actions and decisions of φ^-SDL.

4.3. *Semantics of process definitions*

We describe now the meaning of φ^-SDL process definitions and their constituents. The meaning of the process definitions occurring in the examples from Section 3.6 is presented in Section 4.4.

4.3.1. *Process definitions.* The meaning of each process definition occurring in a system definition consists of the process name introduced and a family of processes, one for each possible pid value, which are described by means of finite guarded recursive specifications in $\mathrm{BPA}^{\mathrm{psc}}_{\mathrm{drt}}$. The meaning of a process definition is expressed in terms of the meaning of its start transition and its state definitions.

$$\begin{aligned} &[\![\, \textbf{process}\; X(k);\, \textbf{fpar}\; v_1, \ldots, v_m;\, \textbf{start};\, tr\; d_1 \ldots d_n\; \textbf{endprocess};]\!]^{\kappa} := \\ &\quad (X, \{i \mapsto \langle X | \{X = [\![tr]\!]_i^{\kappa'}, [\![d_1]\!]_i^{\kappa'}, \ldots, [\![d_n]\!]_i^{\kappa'}\}\rangle \mid i \in \mathbb{N}\}) \\ &\quad \text{where } \kappa' = \mathit{updscopeunit}(\kappa, X) \end{aligned}$$

The recursive specification of the process of type X with pid value i describes how it behaves at its start (the equation $X = [\![tr]\!]_i^{\kappa'}$) and how it behaves from each of the n states in which it may come while it proceeds (the equations $[\![d_1]\!]_i^{\kappa'}, \ldots, [\![d_n]\!]_i^{\kappa'}$).

In the remainder of this section, we will be loose in the explanation of the meaning of the constituents of process definitions about the fact that there is always a family of meanings, one for each possible pid value.

4.3.2. *States and transitions.* The meaning of a state definition, occurring in the scope of a process definition, is a process described by an equation of the form $Z = s_Z$ where Z is a variable corresponding to the state and s_Z is a $\mathrm{BPA}^{\mathrm{psc}}_{\mathrm{drt}}$ term that only contains variables corresponding to states introduced in the process definition concerned. The equation describes how a process of the type being defined behaves from the state. The meaning of the state definition is expressed in terms of the meaning of its transition alternatives, which are process expressions describing the behaviour from the state being defined for the individual signal types of which instances may be consumed and, in addition, possibly for some spontaneous transitions. The meaning of each transition alternative is in turn expressed in terms of the meaning of its input guard, if the alternative is not a spontaneous transition, and its transition.

$$[\![\,\textbf{state}\; st;\,\textbf{save}\; s_1,\ldots,s_m;alt_1\;\ldots\;alt_n\,]\!]_i^{\kappa} \;:=$$
$$X_{st} = \underline{\underline{tt}} \cdot ([\![alt_1]\!]_i^{\kappa'} + \ldots + [\![alt_n]\!]_i^{\kappa'} + waiting(\{s_1,\ldots,s_m\},\underline{i}) :\rightarrow \sigma_{\mathsf{rel}}(X_{st}))$$
$$\text{where } X = scopeunit(\kappa),\quad \kappa' = updsaveset(\kappa,\{s_1,\ldots,s_m\})$$

$$[\![\,\textbf{input}\; s(v_1,\ldots,v_n);tr]\!]_i^{\kappa} \;:=\; \underline{\underline{input}}((s,\langle v_1,\ldots,v_n\rangle),ss,\underline{i})\cdot[\![tr]\!]_i^{\kappa}$$
$$\text{where } ss = saveset(\kappa)$$

$$[\![\,\textbf{input none};\,tr]\!]_i^{\kappa} \;:=\; \underline{\underline{inispont}}(\underline{i})\cdot[\![tr]\!]_i^{\kappa}$$

The equation that corresponds to a state definition describes that a process of type X behaves from the state st as one of the given transition alternatives, and that this behaviour is possibly delayed till the first future time slice in which there is a signal to consume if there are no more signals to consume in the current time slice. Entering a state is supposed to take place by way of some internal action – thus it is precluded that a process is in more than one state. We use process names with state name subscripts, such as X_{st} above, as variables. Notice that, in the absence of spontaneous transitions, a delay becomes inescapable if there are no more signals to consume in the current time slice. The process expression that corresponds to a guarded transition alternative expresses that the transition tr is initiated on consumption of a signal of type s. In case of an unguarded transition alternative, the process expression expresses that the transition tr is initiated spontaneously, i.e. without a preceding signal consumption – with **sender** set to the value of **self** (see Section 4.5).

The meaning of a transition is described by a process expression – a $\mathrm{BPA}_{\mathrm{drt}}^{\mathrm{psc}}$ term to be precise. It is expressed in terms of the meaning of its actions and its transition terminator.

$$[\![a_1\ldots a_n\;\textbf{nextstate}\;st;]\!]_i^{\kappa} \;:=\; [\![a_1]\!]_i^{\kappa}\cdot\ldots\cdot[\![a_n]\!]_i^{\kappa}\cdot X_{st}$$
$$\text{where } X = scopeunit(\kappa)$$

$$[\![a_1\ldots a_n\;\textbf{stop};]\!]_i^{\kappa} \;:=\; [\![a_1]\!]_i^{\kappa}\cdot\ldots\cdot[\![a_n]\!]_i^{\kappa}\cdot\underline{\underline{stop}}(\underline{i})$$

$$[\![a_1\ldots a_n\;dec;]\!]_i^{\kappa} \;:=\; [\![a_1]\!]_i^{\kappa}\cdot\ldots\cdot[\![a_n]\!]_i^{\kappa}\cdot[\![dec]\!]_i^{\kappa}$$

The process expression that corresponds to a transition terminated by **nextstate** st expresses that the transition performs the actions $a_1,\ldots,a_n$ in sequential order and ends with entering state st – i.e. goes on behaving as defined for state st. In case of termination by **stop**, the process expression expresses that it ends with ceasing to exist; and in case of termination by a decision dec, that it goes on behaving as described by decision dec.

The meaning of a decision is described by a process expression as well. It is expressed in terms of the meaning of its expressions and transitions.

$$[\![\textbf{decision } e;(e_1)\text{:}tr_1 \ldots (e_n)\text{:}tr_n \textbf{ enddecision}]\!]_i^\kappa := eq([\![e]\!]_i, [\![e_1]\!]_i) :\rightarrow [\![tr_1]\!]_i^\kappa + \ldots + eq([\![e]\!]_i, [\![e_n]\!]_i) :\rightarrow [\![tr_n]\!]_i^\kappa$$

$$[\![\textbf{decision any; ()}\text{:}tr_1 \ldots \textbf{()}\text{:}tr_n \textbf{ enddecision}]\!]_i^\kappa := [\![tr_1]\!]_i^\kappa + \ldots + [\![tr_n]\!]_i^\kappa$$

The process expression that corresponds to a decision with a question expression e expresses that the decision transfers control to the transition tr_i for which the value of e equals the value of e_i. In case of a decision with **any** instead, the process expression expresses that the decision transfers non-deterministically control to one of the transitions $tr_1, \ldots, tr_n$.

4.3.3. *Actions.* The meaning of an SDL action is described by a process expression, of course. It is expressed in terms of the meaning of the expressions occurring in it. It also depends on the occurring names (variable names, signal names, signal route names and process names – dependent on the kind of action).

$$[\![\textbf{output } s(e_1, \ldots, e_n) \textbf{ to } e \textbf{ via } r_1, \ldots, r_m;]\!]_i^\kappa := type([\![e]\!]_i, X_1) \vee \ldots \vee type([\![e]\!]_i, X_m) :\rightarrow \underline{\underline{outpute}}((s, \langle [\![e_1]\!]_i, \ldots, [\![e_n]\!]_i \rangle), \underline{i}, [\![e]\!]_i) + \neg(type([\![e]\!]_i, X_1) \vee \ldots \vee type([\![e]\!]_i, X_m)) :\rightarrow \underline{\underline{tt}}$$

where for $1 \leqslant j \leqslant m$: $X_j = rcv(\kappa, r_j)$

$$[\![\textbf{output } s(e_1, \ldots, e_n) \textbf{ via } r_1, \ldots, r_m;]\!]_i^\kappa := \underline{\underline{outputi}}((s, \langle [\![e_1]\!]_i, \ldots, [\![e_n]\!]_i \rangle), \underline{i}, X_1)) + \ldots + \underline{\underline{outputi}}((s, \langle [\![e_1]\!]_i, \ldots, [\![e_n]\!]_i \rangle), \underline{i}, X_m)) + \neg(hasinst(X_1) \wedge \ldots \wedge hasinst(X_m)) :\rightarrow \underline{\underline{tt}}$$

where for $1 \leqslant j \leqslant m$: $X_j = rcv(\kappa, r_j)$

$$[\![\textbf{set } (e, s(e_1, \ldots, e_n));]\!]_i^\kappa := \underline{\underline{set}}([\![e]\!]_i, (s, \langle [\![e_1]\!]_i, \ldots, [\![e_n]\!]_i \rangle), \underline{i})$$

$$[\![\textbf{reset } (s(e_1, \ldots, e_n));]\!]_i^\kappa := \underline{\underline{reset}}((s, \langle [\![e_1]\!]_i, \ldots, [\![e_n]\!]_i \rangle), \underline{i})$$

$$[\![\textbf{task } v := e;]\!]_i^\kappa := \underline{\underline{ass}}(v, [\![e]\!]_i, \underline{i})$$

$$[\![\textbf{create } X(e_1, \ldots, e_n);]\!]_i^\kappa := \underline{\underline{cr}}((X, fpars(\kappa, X), \langle [\![e_1]\!]_i, \ldots, [\![e_n]\!]_i \rangle, \underline{i}))$$

All cases except the ones for output actions are straightforward. The cases of output actions need further explanation. In the case of an output action with explicit addressing, the process with pid value e must be of the receiver type associated with one of the signal routes $r_1, \ldots, r_m$. Therefore, the condition $type([\![e]\!]_i, X_1) \vee \ldots \vee type([\![e]\!]_i, X_m)$ is used. If the process with pid value e is not of the receiver type associated with any of the signal routes $r_1, \ldots, r_m$, or a process with that pid value does not exist, the signal is simply discarded and no error occurs. This is expressed by the summand $\neg(type([\![e]\!]_i, X_1) \vee \ldots \vee type([\![e]\!]_i, X_m)) :\rightarrow \underline{\underline{tt}}$. In the case of an output action with implicit addressing, first an arbitrary choice from the signal routes $r_1, \ldots, r_m$ is made and thereafter an arbitrary choice from the existing processes of the receiver type for the chosen signal route is made. Therefore, there is a summand for the receiver type of each signal route. If a process of the receiver type for the chosen signal route does not exist, the signal is simply discarded and no error occurs. This is expressed by the summand $\neg(hasinst(X_1) \wedge \ldots \wedge hasinst(X_m)) :\!\rightarrow \underline{\underline{tt}}$. Note that the signal may already be discarded if there is one signal route for which there exists no process of its receiver type.

4.3.4. *Values.* The meaning of a φ^-SDL expression is given by a translation to a value expression of the same kind. There is a close correspondence between the φ^-SDL expressions and their translations. Essential of the translation is that $\underline{i}$ is added where the local states of different processes need to be distinguished. Consequently, a variable access v is just treated as a view expression **view**$(v,$ **self**$)$. For convenience, the expressions **parent**, **offspring** and **sender** are also regarded as variable accesses.

$$[\![op(e_1,\ldots,e_n)]\!]_i := op([\![e_1]\!]_i,\ldots,[\![e_n]\!]_i)$$

$$[\![\textbf{if}\, e_1 \,\textbf{then}\, e_2 \,\textbf{else}\, e_3 \,\textbf{fi}]\!]_i := cond([\![e_1]\!]_i, [\![e_2]\!]_i, [\![e_3]\!]_i)$$

$$[\![v]\!]_i := value(v, \underline{i})$$

$$[\![\textbf{view}\,(v,e)]\!]_i := value(v, [\![e]\!]_i)$$

$$[\![\textbf{active}\,(s(e_1,\ldots,e_n))]\!]_i := active((s, \langle [\![e_1]\!]_i,\ldots,[\![e_n]\!]_i\rangle), \underline{i})$$

$$[\![\textbf{now}]\!]_i := now$$

$$[\![\textbf{self}]\!]_i := \underline{i}$$

$$[\![\textbf{parent}]\!]_i := value(\textbf{parent}, \underline{i})$$

$$[\![\textbf{offspring}]\!]_i := value(\textbf{offspring}, \underline{i})$$

$$[\![\textbf{sender}]\!]_i := value(\textbf{sender}, \underline{i})$$

4.4. *Examples*

We present the meaning of the process definitions occurring in the examples from Section 3.6. To be more precise, we give for each process definition a constant of the form $\langle X|E\rangle$ that stands for the process of the type concerned with pid value i.

It is clear that there are many similarities with the original process definitions in φ^-SDL. There is an equation for each state, the right hand side of each equation has a summand for each transition alternative of the corresponding state, etc. In addition, there is always a summand in which the time unit delay operator appears; this summand allows a delay to a future time slice to occur if there is no input to be read from the input queue of the process concerned. The main difference between the φ^-SDL process definitions and the description of their meaning in $\mathrm{BPA}^{\mathrm{psc}}_{\mathrm{drt}}\mathrm{rec}$ is that the latter can be subjected to equational reasoning using the axioms of $\mathrm{BPA}^{\mathrm{psc}}_{\mathrm{drt}}\mathrm{rec}$.

4.4.1. *Repeater.*

The `rep` process with pid value i is

$$
\begin{array}{l}
\langle \mathtt{rep} | \\
\quad \{\mathtt{rep} = \mathtt{rep}_{\mathtt{pass}}, \\
\quad \mathtt{rep}_{\mathtt{pass}} = \underline{\underline{tt}} \cdot (\underline{\underline{input}}((\mathtt{s}, \langle\rangle), \emptyset, \underline{i}) \cdot \\
\qquad\qquad (\underline{\underline{outputi}}((\mathtt{s}, \langle\rangle), \underline{i}, \mathbf{env}) + \neg hasinst(\mathbf{env}) :\rightarrow \underline{\underline{tt}}) \cdot \mathtt{rep}_{\mathtt{pass}} + \\
\qquad\qquad waiting(\emptyset, \underline{i}) :\rightarrow \sigma_{\mathsf{rel}}(\mathtt{rep}_{\mathtt{pass}})) \\
\quad \} \\
\rangle
\end{array}
$$

4.4.2. *Address driven router.*

The `rtr` process with pid value i is

$$
\begin{array}{l}
\langle \mathtt{rtr} | \\
\quad \{\mathtt{rtr} = \underline{\underline{cr}}((\mathtt{rep}, \langle\rangle, \langle\rangle, \underline{i})) \cdot \underline{\underline{ass}}(\mathtt{rep1}, value(\mathbf{offspring}, \underline{i}), \underline{i}) \cdot \\
\qquad\qquad \underline{\underline{cr}}((\mathtt{rep}, \langle\rangle, \langle\rangle, \underline{i})) \cdot \underline{\underline{ass}}(\mathtt{rep2}, value(\mathbf{offspring}, \underline{i}), \underline{i}) \cdot \mathtt{rtr}_{\mathtt{route}}, \\
\quad \mathtt{rtr}_{\mathtt{route}} = \underline{\underline{tt}} \cdot (\underline{\underline{input}}((\mathtt{s}, \langle \mathtt{a} \rangle), \emptyset, \underline{i}) \cdot \\
\qquad\qquad (eq(value(\mathtt{a}, \underline{i}), \mathtt{False}) :\rightarrow (type(value(\mathtt{rep1}, \underline{i}), \mathtt{rep}) :\rightarrow \\
\qquad\qquad \underline{\underline{outpute}}((\mathtt{s}, \langle value(\mathtt{a}, \underline{i}) \rangle), \underline{i}, value(\mathtt{rep1}, \underline{i})) + \\
\qquad\qquad \neg type(value(\mathtt{rep1}, \underline{i}), \mathtt{rep}) :\rightarrow \underline{\underline{tt}}) \cdot \mathtt{rtr}_{\mathtt{route}} + \\
\qquad\qquad eq(value(\mathtt{a}, \underline{i}), \mathtt{True}) :\rightarrow (type(value(\mathtt{rep2}, \underline{i}), \mathtt{rep}) :\rightarrow \\
\qquad\qquad \underline{\underline{outpute}}((\mathtt{s}, \langle value(\mathtt{a}, \underline{i}) \rangle), \underline{i}, value(\mathtt{rep2}, \underline{i})) + \\
\qquad\qquad \neg type(value(\mathtt{rep2}, \underline{i}), \mathtt{rep}) :\rightarrow \underline{\underline{tt}}) \cdot \mathtt{rtr}_{\mathtt{route}}) + \\
\qquad\qquad waiting(\emptyset, \underline{i}) :\rightarrow \sigma_{\mathsf{rel}}(\mathtt{rtr}_{\mathtt{route}})) \\
\quad \} \\
\rangle
\end{array}
$$

The `rep` process with pid value i is

$$
\begin{array}{l}
\langle \mathtt{rep} | \\
\quad \{\mathtt{rep} = \mathtt{rep}_{\mathtt{pass}}, \\
\quad \mathtt{rep}_{\mathtt{pass}} = \underline{\underline{tt}} \cdot (\underline{\underline{input}}((\mathtt{s}, \langle \mathtt{a} \rangle), \emptyset, \underline{i}) \cdot \\
\qquad\qquad (\underline{\underline{outputi}}((\mathtt{s}, \langle \mathtt{a} \rangle), \underline{i}, \mathbf{env}) + \neg hasinst(\mathbf{env}) :\rightarrow \underline{\underline{tt}}) \cdot \mathtt{rep}_{\mathtt{pass}} + \\
\qquad\qquad waiting(\emptyset, \underline{i}) :\rightarrow \sigma_{\mathsf{rel}}(\mathtt{rep}_{\mathtt{pass}})) \\
\quad \} \\
\rangle
\end{array}
$$

4.4.3. *Load driven router.*

The `rtr` process with pid value i is

$$
\begin{array}{l}
\langle \mathtt{rtr} | \\
\quad \{\mathtt{rtr} = \underline{\underline{cr}}((\mathtt{trep}, \langle \mathtt{delay} \rangle, \langle \mathtt{10} \rangle, \underline{i})) \cdot \underline{\underline{ass}}(\mathtt{rep1}, value(\mathbf{offspring}, \underline{i}), \underline{i}) \cdot \\
\qquad\qquad \underline{\underline{cr}}((\mathtt{trep}, \langle \mathtt{delay} \rangle, \langle \mathtt{20} \rangle, \underline{i})) \cdot \underline{\underline{ass}}(\mathtt{rep2}, value(\mathbf{offspring}, \underline{i}), \underline{i}) \cdot \mathtt{rtr}_{\mathtt{route}}, \\
\quad \mathtt{rtr}_{\mathtt{route}} = \underline{\underline{tt}} \cdot (\underline{\underline{input}}((\mathtt{s}, \langle\rangle), \emptyset, \underline{i}) \cdot
\end{array}
$$

$$
\begin{array}{l}
(eq(value(\mathtt{idle}, value(\mathtt{rep1}, \underline{i})) \leftrightarrow value(\mathtt{idle}, value(\mathtt{rep2}, \underline{i})), \mathtt{True}) :\rightarrow \\
\quad (\underline{\underline{outputi}}((\mathtt{s}, \langle\rangle), \underline{i}, \mathtt{trep}) + \neg hasinst(\mathtt{trep}) :\rightarrow \underline{\underline{tt}}) \cdot \mathtt{rtr}_{\mathtt{route}} + \\
eq(value(\mathtt{idle}, value(\mathtt{rep1}, \underline{i})) \leftrightarrow value(\mathtt{idle}, value(\mathtt{rep2}, \underline{i})), \mathtt{False}) :\rightarrow \\
(eq(value(\mathtt{idle}, value(\mathtt{rep1}, \underline{i})), \mathtt{True}) :\rightarrow \\
\quad (type(value(\mathtt{rep1}, \underline{i}), \mathtt{rep}) :\rightarrow \underline{\underline{outpute}}((\mathtt{s}, \langle\rangle), \underline{i}, value(\mathtt{rep1}, \underline{i})) + \\
\quad \neg type(value(\mathtt{rep1}, \underline{i}), \mathtt{rep}) :\rightarrow \underline{\underline{tt}}) \cdot \mathtt{rtr}_{\mathtt{route}} + \\
\quad eq(value(\mathtt{idle}, value(\mathtt{rep1}, \underline{i})), \mathtt{False}) :\rightarrow \\
\quad (type(value(\mathtt{rep2}, \underline{i}), \mathtt{rep}) :\rightarrow \underline{\underline{outpute}}((\mathtt{s}, \langle\rangle), \underline{i}, value(\mathtt{rep2}, \underline{i})) + \\
\quad \neg type(value(\mathtt{rep2}, \underline{i}), \mathtt{rep}) :\rightarrow \underline{\underline{tt}}) \cdot \mathtt{rtr}_{\mathtt{route}})) + \\
waiting(\emptyset, \underline{i}) :\rightarrow \sigma_{\mathsf{rel}}(\mathtt{rtr}_{\mathtt{route}}))
\end{array}
$$

}

⟩

The `trep` process with pid value i is

$$
\begin{array}{l}
\langle \mathtt{trep} | \\
\{\mathtt{trep} = \underline{\underline{ass}}(\mathtt{idle}, \mathtt{True}) \cdot \mathtt{trep}_{\mathtt{get}}, \\
\;\mathtt{trep}_{\mathtt{get}} = \underline{\underline{tt}} \cdot (\underline{\underline{input}}((\mathtt{s}, \langle\rangle), \emptyset, \underline{i}) \cdot \\
\qquad \underline{\underline{ass}}(\mathtt{idle}, \mathtt{False}) \cdot \underline{\underline{set}}(now + \mathtt{delay}, \mathtt{t}) \cdot \mathtt{trep}_{\mathtt{put}} + \\
\qquad waiting(\emptyset, \underline{i}) :\rightarrow \sigma_{\mathsf{rel}}(\mathtt{trep}_{\mathtt{get}})), \\
\;\mathtt{trep}_{\mathtt{put}} = \underline{\underline{tt}} \cdot (\underline{\underline{input}}((\mathtt{t}, \langle\rangle), \{\mathtt{s}\}, \underline{i}) \cdot \\
\qquad (\underline{\underline{outputi}}((\mathtt{s}, \langle\rangle), \underline{i}, \mathbf{env}) + \neg hasinst(\mathbf{env}) :\rightarrow \underline{\underline{tt}}) \cdot \\
\qquad \underline{\underline{ass}}(\mathtt{idle}, \mathtt{True}) \cdot \mathtt{trep}_{\mathtt{get}} + \\
\qquad waiting(\{\mathtt{s}\}, \underline{i}) :\rightarrow \sigma_{\mathsf{rel}}(\mathtt{trep}_{\mathtt{put}}))
\end{array}
$$

}

⟩

4.5. *Interaction with states*

In the semantic of φ^{-}SDL process definitions, ACP actions are used to give meaning to input guards, SDL actions, the terminator **stop** and the void guard **input none**. Thus, the facilities for storage, communication, timing and process creation offered by these language constructs are not fully covered; the ACP actions are meant to interact with a system state. In the semantics of φ^{-}SDL system definitions, which will be presented in Section 4.6, the state operator mentioned in Section 2 is used to describe this. First, we will describe the state space, the actions that may appear as the result of executing the above-mentioned actions in a state, and the result of executing processes, built up from these actions, in a state from the state space.

4.5.1. *State space, actions and propositional signals.* The state space, used to describe the meaning of system definitions, depends upon the specific variable names, signal names and process names introduced in the system definition concerned. That is, the sets V_κ, S_κ and P_κ are used here as well.

First, we define the set Sig_κ of signals and the set $ExtSig_\kappa$ of extended signals. A signal consist of the name of the signal and the sequence of values that it carries. An extended signal contains, in addition to a signal, the pid values of its sender and receiver.

$$\begin{aligned} Sig_\kappa &= \{(s, \langle u_1, \ldots, u_n \rangle) \mid (s, \langle T_1, \ldots, T_n \rangle) \in sigds(\kappa), u_1, \ldots, u_n \in U\} \\ ExtSig_\kappa &= Sig_\kappa \times \mathbb{N}_1 \times \mathbb{N}_1 \end{aligned}$$

We write $snm(sig)$ and $vals(sig)$, where $sig = (s, vs) \in Sig_\kappa$, for s and vs, respectively. We write $sig(xsig)$, $snd(xsig)$ and $rcv(xsig)$, where $xsig = (sig, i, i') \in ExtSig_\kappa$, for sig, i and i', respectively. Note that 0 is excluded as pid value of the sender or receiver of a signal because it is a special pid value that never refers to any existing process.

The local state of a process includes a storage which associates local variables with the values assigned to them, an input queue where delivered signals are kept until they are consumed, and a component keeping track of the expiration times of active timers. We define the set Stg_κ of storages, the set $InpQ_\kappa$ of input queues and the set $Timers_\kappa$ of timers as follows:

$$\begin{aligned} Stg_\kappa &= \bigcup_{V \subseteq V_\kappa} (V \to U) \\ InpQ_\kappa &= ExtSig_\kappa{}^* \\ Timers_\kappa &= \bigcup_{T \in \mathcal{P}_{\mathrm{fin}}(Sig_\kappa)} (T \to (\mathbb{N} \cup \{\mathsf{nil}\})) \end{aligned}$$

We will follow the convention that the domain of a function from Stg_κ does not contain variables with which no value is associated because a value has never been assigned to them. We will also follow the convention that the domain of a function from $Timers_\kappa$ contains precisely the active timers. While an expired timer is still active, its former expiration time will be replaced by nil. The basic operations on Stg_κ and $Timers_\kappa$ are general operations on functions: function application, overriding ($\oplus$) and domain subtraction ($\lhd$). Overriding and domain subtraction are defined in Appendix A. In so far as the facilities for communication are concerned, the basic operations on $InpQ_\kappa$ are the functions

$$\begin{aligned} getnxt &: InpQ_\kappa \times SaveSet_\kappa \to ExtSig_\kappa \cup \{\mathsf{nil}\}, \\ rmvfirst &: InpQ_\kappa \times Sig_\kappa \to InpQ_\kappa, \\ merge &: \mathcal{P}_{\mathrm{fin}}(ExtSig_\kappa) \to \mathcal{P}_{\mathrm{fin}}(InpQ_\kappa) \end{aligned}$$

defined below. The value of $getnxt(\sigma, ss)$ is the first (extended) signal in σ with a name different from the ones in ss. The value of $rmvfirst(\sigma, sig)$ is the input queue σ from which the first occurrence of the signal sig has been removed. Both functions are used to describe the consumption of signals by SDL processes. The function $getnxt$ is recursively defined by

$$\begin{aligned} & getnxt(\langle\rangle, ss) = \mathsf{nil} \\ & getnxt((sig, i, i') \mathbin{\&} \sigma, ss) = (sig, i, i') && \textbf{if } snm(sig) \notin ss \\ & getnxt((sig, i, i') \mathbin{\&} \sigma, ss) = getnxt(\sigma, ss) && \textbf{if } snm(sig) \in ss \end{aligned}$$

and the function $rmvfirst$ is recursively defined by

$$\begin{aligned} & rmvfirst(\langle\rangle, sig) = \langle\rangle \\ & rmvfirst((sig, i, i') \mathbin{\&} \sigma, sig) = \sigma \\ & rmvfirst((sig, i, i') \mathbin{\&} \sigma, sig') = (sig, i, i') \mathbin{\&} rmvfirst(\sigma, sig') \quad \textbf{if } sig \neq sig' \end{aligned}$$

For each process, signals noticing timer expiration have to be merged when time progresses to the next time slice. The function $merge$ is used to describe this precisely. For a given set of extended signals it gives the set of all possible sequences of them. It is inductively defined by

$$\langle\rangle \in merge(\emptyset)$$
$$(sig, i, i') \notin \Sigma \wedge \sigma \in merge(\Sigma) \Rightarrow (sig, i, i') \,\&\, \sigma \in merge(\{(sig, i, i')\} \cup \Sigma)$$

We define now the set $\mathcal{L}_\kappa$ of local states. The local state of a process contains, in addition to the above-mentioned components, the name of the process. Thus, the type of the process concerned will not get lost. This is important, because a signal may be sent to an arbitrary process of a process type.

$$\mathcal{L}_\kappa = Stg_\kappa \times InpQ_\kappa \times Timers_\kappa \times P_\kappa$$

We write $stg(L)$, $inpq(L)$, $timers(L)$ and $ptype(L)$, where $L = (\rho, \sigma, \theta, X) \in \mathcal{L}_\kappa$, for ρ, σ, θ and X, respectively.

The global state of a system contains, besides a local state for each existing process, a component keeping track of the system time. To keep track of the system time, natural numbers suffice. We define the state space $\mathcal{G}_\kappa$ of global states as follows:

$$\mathcal{G}_\kappa = \mathbb{N} \times \bigcup_{I \in \mathcal{P}_{\mathrm{fin}}(\mathbb{N}_2)} (I \to \mathcal{L}_\kappa)$$

We write $now(G)$ and $lsts(G)$, where $G = (n, \Sigma) \in \mathcal{G}_\kappa$, for n and Σ, respectively. We write $exists(i, G)$, where $i \in \mathbb{N}$ and $G \in \mathcal{G}_\kappa$, for $i \in dom(lsts(G))$. Note that 1 is excluded from being used as pid value of an existing process of the system because it is a special pid value that is reserved for the environment.

Every state from $\mathcal{G}_\kappa$ produces a proposition which is considered to hold in the state concerned. In this way, the state of a process is made partly visible. These propositions are built from the following atomic propositions:

$$value : V_\kappa \times U \times \mathbb{N}_2$$
$$active : Sig_\kappa \times \mathbb{N}_2$$

$AtProp_\kappa$ denotes the set of all *value* and *active* propositions. We write $Prop_\kappa$ for the set of all propositions that can be built from $AtProp_\kappa$. An atomic proposition of the form $value(v, u, i)$ is intended to indicate that u is the value of the local variable v of the process with pid value i. An atomic proposition of the form $active(sig, i)$ is intended to indicate that the timer of the process with pid value i identified with signal *sig* is active. By using only atomic propositions of these forms, the state of a process can not be made fully visible via the proposition produced. The proposition produced by each state, given by the function *sig* defined in Section 4.5.4, makes only visible the value of all local variables and the set of active timers for all existing processes.

Below, we introduce the additional actions that are used for the semantics of φ^-SDL system definitions. Like most of the actions used for the semantics of φ^-SDL process definitions, these actions are parametrized. The following additional actions are used:

$$\overline{cr} : \mathbb{N} \times PrCrD_\kappa$$
$$input' : ExtSig_\kappa$$
$$output' : ExtSig_\kappa$$
$$set' : \mathbb{N} \times Sig_\kappa \times \mathbb{N}_2$$
$$reset' : Sig_\kappa \times \mathbb{N}_2$$

$\overline{Cr}_\kappa$ denotes the set of all $\overline{cr}$ actions; and Act'_κ denotes the set of all *input′*, *output′*, *set′* and *reset′* actions. The $\overline{cr}$ actions appears as the result of applying the process creation operator E^n_Φ to *cr* actions. The *input′*, *output′*, *set′* and *reset′* actions appear as the result of applying the state operator λ_G to *input*, *outpute*/*outputi*, *set* and *reset* actions, respectively (see Section 4.5.4).

4.5.2. *State transformers and observers.* In the process algebra semantics of φ^-SDL process definitions, presented in Section 4.3, ACP actions are used to describe the meaning of input guards, SDL actions, the terminator **stop** and the void guard **input none**. These ACP actions are meant to interact with a state from $\mathcal{G}_\kappa$. Later on, we will define the result of executing a process, built up from these actions, in a state from $\mathcal{G}_\kappa$. That is, we will define the relevant state operator. This will partly boil down to describing how the actions, and the progress of time (modeled by the time unit delay operator σ_{rel}), transform states. For the sake of comprehensibility, we will first define matching state transforming operations.

In addition, we will define some state observing operations. Two of the state observing operations are used directly to define the action and effect function of the state operator and three others are used to define the valuation function of the state operator. The remaining two are used to define the signal function of the state operator as well as an auxiliary evaluation function needed for the value expressions that occur as arguments, and as components of arguments, of the above-mentioned actions and conditions (see Section 4.2).

State transformers. In general, the state transformers change one or two components of the local state of one process. The notable exception is *csmsig*, which is defined first. It may change all components except, of course, the process type. This is a consequence of the fact that the facilities for storage, communication and timing are rather intertwined on the consumption of signals in φ^-SDL. For each state transformer it holds that everything remains unchanged if an attempt is made to transform the local state of a non-existing process. This will not be explicitly mentioned in the explanations given below.

The function $csmsig : ExtSig_\kappa \times V_\kappa{}^* \times \mathcal{G}_\kappa \to \mathcal{G}_\kappa$ is used to describe how the ACP actions corresponding to the input guards of φ^-SDL transform states.

$$\begin{aligned}
&csmsig((sig, i, i'), \langle v_1, \ldots, v_n\rangle, G) = \\
&\quad (now(G), lsts(G) \oplus \{i' \mapsto (\rho, \sigma, \theta, X)\}) \quad \textbf{if } exists(i', G) \\
&\quad G \quad \textbf{otherwise}
\end{aligned}$$

$$\begin{aligned}
\text{where } \rho &= stg(lsts(G)_{i'}) \oplus \{v_1 \mapsto vals(sig)_1, \ldots, v_n \mapsto vals(sig)_n, \textsf{sender} \mapsto i\}, \\
\sigma &= rmvfirst(inpq(lsts(G)_{i'}), sig), \\
\theta &= \{sig\} \lhd timers(lsts(G)_{i'}), \\
X &= ptype(lsts(G)_{i'})
\end{aligned}$$

$csmsig((sig, i, i'), \langle v_1, \ldots, v_n\rangle, G)$ deals with the consumption of signal *sig* by the process with pid value i'. It transforms the local state of the process with pid value i', the process by which the signal is consumed, as follows:

- the values carried by *sig* are assigned to the local variables $v_1, \ldots, v_n$ of the consuming process and the sender's pid value (i) is assigned to **sender**;
- the first occurrence of *sig* in the input queue of the consuming process is removed;
- if *sig* is a timer signal, it is removed from the active timers.

Everything else is left unchanged.

The function *sndsig* : $ExtSig_\kappa \times \mathcal{G}_\kappa \to \mathcal{G}_\kappa$ is used to describe how the ACP actions corresponding to the output actions of φ^-SDL transform states.

$$
\begin{aligned}
&\mathit{sndsig}((\mathit{sig}, i, i'), G) = \\
&\quad \begin{cases} (\mathit{now}(G), \mathit{lsts}(G) \oplus \{i' \mapsto (\rho, \sigma, \theta, X)\}) & \textbf{if } \mathit{exists}(i', G) \\ G & \textbf{otherwise} \end{cases}
\end{aligned}
$$

$$
\begin{aligned}
\text{where } \rho &= \mathit{stg}(\mathit{lsts}(G)_{i'}), \\
\sigma &= \mathit{inpq}(\mathit{lsts}(G)_{i'}) \frown \langle(\mathit{sig}, i, i')\rangle, \\
\theta &= \mathit{timers}(\mathit{lsts}(G)_{i'}), \\
X &= \mathit{ptype}(\mathit{lsts}(G)_{i'})
\end{aligned}
$$

$\mathit{sndsig}((\mathit{sig}, i, i'), G)$ deals with passing signal *sig* from the process with pid value i to the process with pid value i'. It transforms the local state of the process with pid value i', the process to which the signal is passed, as follows:

- *sig* is put into the input queue of the process to which the signal is passed (unless $i' = 1$, indicating that the signal is passed to the environment).

Everything else is left unchanged.

The function *settimer* : $\mathbb{N} \times Sig_\kappa \times \mathbb{N}_2 \times \mathcal{G}_\kappa \to \mathcal{G}_\kappa$ is used to describe how the ACP actions corresponding to the set actions of φ^-SDL transform states.

$$
\begin{aligned}
&\mathit{settimer}(t, \mathit{sig}, i, G) = \\
&\quad \begin{cases} (\mathit{now}(G), \mathit{lsts}(G) \oplus \{i \mapsto (\rho, \sigma, \theta, X)\}) & \textbf{if } \mathit{exists}(i, G) \\ G & \textbf{otherwise} \end{cases}
\end{aligned}
$$

$$
\begin{aligned}
\text{where } \rho &= \mathit{stg}(\mathit{lsts}(G)_i), \\
\sigma &= \begin{cases} \mathit{rmvfirst}(\mathit{inpq}(\mathit{lsts}(G)_i), \mathit{sig}) & \textbf{if } t \geqslant \mathit{now}(G) \\ \mathit{rmvfirst}(\mathit{inpq}(\mathit{lsts}(G)_i), \mathit{sig}) \frown \langle(\mathit{sig}, i, i)\rangle & \textbf{otherwise}, \end{cases} \\
\theta &= \begin{cases} \mathit{timers}(\mathit{lsts}(G)_i) \oplus \{\mathit{sig} \mapsto t\} & \textbf{if } t \geqslant \mathit{now}(G) \\ \mathit{timers}(\mathit{lsts}(G)_i) \oplus \{\mathit{sig} \mapsto \mathsf{nil}\} & \textbf{otherwise}, \end{cases} \\
X &= \mathit{ptype}(\mathit{lsts}(G)_i)
\end{aligned}
$$

$\mathit{settimer}(t, \mathit{sig}, i, G)$ deals with setting a timer, identified with signal *sig*, to time t. If t has not yet passed, it transforms the local state of the process with pid value i, the process to be notified of the timer's expiration, as follows:

- the occurrence of *sig* in the input queue originating from an earlier setting, if any, is removed;
- *sig* is included among the active timers with expiration time t; thus overriding an earlier setting, if any.

Otherwise, it transforms the local state of the process with pid value i as follows:

- *sig* is put into the input queue after removal of its occurrence originating from an earlier setting, if any;
- *sig* is included among the active timers without expiration time.

Everything else is left unchanged.

The function *resettimer* : $Sig_\kappa \times \mathbb{N}_2 \times \mathcal{G}_\kappa \to \mathcal{G}_\kappa$ is used to describe how the ACP actions corresponding to the reset actions of φ^-SDL transform states.

$$
\begin{aligned}
&\mathit{resettimer}(\mathit{sig}, i, G) = \\
&\quad \begin{cases} (\mathit{now}(G), \mathit{lsts}(G) \oplus \{i \mapsto (\rho, \sigma, \theta, X)\}) & \textbf{if } \mathit{exists}(i, G) \\ G & \textbf{otherwise} \end{cases}
\end{aligned}
$$

$$\text{where } \begin{aligned} \rho &= \mathit{stg}(\mathit{lsts}(G)_i), \\ \sigma &= \mathit{rmvfirst}(\mathit{inpq}(\mathit{lsts}(G)_i), \mathit{sig}), \\ \theta &= \{\mathit{sig}\} \lhd \mathit{timers}(\mathit{lsts}(G)_i), \\ X &= \mathit{ptype}(\mathit{lsts}(G)_i) \end{aligned}$$

$\mathit{resettimer}(\mathit{sig}, i, G)$ deals with resetting a timer, identified with signal sig. It transforms the local state of the process with pid value i, the process that would otherwise have been notified of the timer's expiration, as follows:

- the occurrence of sig in the input queue originating from an earlier setting, if any, is removed;
- if sig is an active timer, it is removed from the active timers.

Everything else is left unchanged.
Notice that $\mathit{settimer}(t, \mathit{sig}, i, G)$ and $\mathit{settimer}(t, \mathit{sig}, i, \mathit{resettimer}(\mathit{sig}, i, G))$ have the same effect. In other words, *settimer* resets implicitly. In this way, at most one signal from the same timer will ever occur in an input queue. Furthermore, the context-sensitive conditions for syntactic correctness of φ^-SDL specifications imply that timer signals and other signals are kept apart: not a single signal can originate from both timer setting and customary signal sending. Thus, resetting, either explicitly or implicitly, will solely remove signals from input queues that originate from timer setting.

The function $\mathit{assignvar} : V_\kappa \times U \times \mathbb{N}_2 \times \mathcal{G}_\kappa \to \mathcal{G}_\kappa$ is used to describe how the ACP actions corresponding to the assignment task actions of φ^-SDL transform states.

$$\begin{aligned} &\mathit{assignvar}(v, u, i, G) = \\ &\quad \begin{cases} (\mathit{now}(G), \mathit{lsts}(G) \oplus \{i \mapsto (\rho, \sigma, \theta, X)\}) & \textbf{if } \mathit{exists}(i, G) \\ G & \textbf{otherwise} \end{cases} \end{aligned}$$

$$\text{where } \begin{aligned} \rho &= \mathit{stg}(\mathit{lsts}(G)_i) \oplus \{v \mapsto u\}, \\ \sigma &= \mathit{inpq}(\mathit{lsts}(G)_i), \\ \theta &= \mathit{timers}(\mathit{lsts}(G)_i), \\ X &= \mathit{ptype}(\mathit{lsts}(G)_i) \end{aligned}$$

$\mathit{assignvar}(v, u, i, G)$ deals with assigning value u to variable v. It transforms the local state of the process with pid value i, the process to which the variable is local, as follows:

- u is assigned to the local variable v, i.e. v is included among the variables in the storage with value u; thus overriding an earlier assignment, if any.

Everything else is left unchanged.

The function $\mathit{createproc} : \mathbb{N}_2 \times \mathit{PrCrD}'_\kappa \times \mathcal{G}_\kappa \to \mathcal{G}_\kappa$ is used to describe how the ACP actions corresponding to the create actions of φ^-SDL transform states. The elements of PrCrD'_κ are like process creation data, i.e. elements of PrCrD_κ, but values are used instead of value expressions: $\mathit{PrCrD}'_\kappa = \{(X, \langle v_1, \ldots, v_n\rangle, \langle u_1, \ldots, u_n\rangle, i) \mid X \in P_\kappa,\ v_1, \ldots, v_n \in V_\kappa,\ u_1, \ldots, u_n \in U,\ i \in \mathbb{N}_2\}$.

$$\begin{aligned} &\mathit{createproc}(i', (X', \langle v_1, \ldots, v_n\rangle, \langle u_1, \ldots, u_n\rangle, i), G) = \\ &\quad \begin{cases} (\mathit{now}(G), \mathit{lsts}(G) \oplus \{i \mapsto (\rho, \sigma, \theta, X), i' \mapsto (\rho', \sigma', \theta', X')\}) & \textbf{if } \mathit{exists}(i, G) \\ G & \textbf{otherwise} \end{cases} \end{aligned}$$

$$
\begin{aligned}
\text{where } \rho &= stg(lsts(G)_i) \oplus \{\textbf{offspring} \mapsto i'\},\\
\sigma &= inpq(lsts(G)_i),\\
\theta &= timers(lsts(G)_i),\\
X &= ptype(lsts(G)_i),\\
\rho' &= \{v_1 \mapsto u_1, \ldots, v_n \mapsto u_n, \textbf{parent} \mapsto i, \textbf{offspring} \mapsto 0, \textbf{sender} \mapsto 0\},\\
\sigma' &= \langle\rangle,\\
\theta' &= \{\}
\end{aligned}
$$

$createproc(i', (X', \langle v_1, \ldots, v_n\rangle, \langle u_1, \ldots, u_n\rangle, i), G)$ deals with creating a process of type X'. It transforms the local state of the process with pid value i, the parent of the created process, as follows:

- the pid value of the created process (i') is assigned to **offspring**.

Besides, it creates a new local state for the created process which is initiated as follows:

- the values $u_1, \ldots, u_n$ are assigned to the local variables $v_1, \ldots, v_n$ of the created process and the parent's pid value (i) is assigned to **parent**;
- X' is made the process type.

Everything else is left unchanged.

The function $stopproc : \mathbb{N}_2 \times \mathcal{G}_\kappa \to \mathcal{G}_\kappa$ is used to describe how the ACP actions corresponding to **stop** in φ^-SDL transform states.

$$stopproc(i, G) = (now(G), \{i\} \lhd lsts(G))$$

$stopproc(i, G)$ deals with terminating the process with pid value i. It disposes of the local state of the process with pid value i. Everything else is left unchanged.

The function $inispont : \mathbb{N}_2 \times \mathcal{G}_\kappa \to \mathcal{G}_\kappa$ is used to describe how the ACP actions used to initiate spontaneous transitions transform states.

$$
\begin{aligned}
&inispont(i, G) =\\
&\quad (now(G), lsts(G) \oplus \{i \mapsto (\rho, \sigma, \theta, X)\}) \quad \textbf{if } exists(i, G)\\
&\quad G \quad \textbf{otherwise}
\end{aligned}
$$

$$
\begin{aligned}
\text{where } \rho &= stg(lsts(G)_i) \oplus \{\textbf{sender} \mapsto i\},\\
\sigma &= inpq(lsts(G)_i),\\
\theta &= timers(lsts(G)_i),\\
X &= ptype(lsts(G)_i)
\end{aligned}
$$

$inispont(i, G)$ deals with initiating spontaneous transitions. It transforms the local state of the process with pid value i, the process for which a spontaneous transition is initiated, by assigning i to **sender**. Everything else is left unchanged.

The function $unitdelay : \mathcal{G}_\kappa \to \mathcal{P}_{\text{fin}}(\mathcal{G}_\kappa)$ is used to describe how progress of time transforms states. In general, these transformations are non-deterministic – how signals from expiring timers enter input queues is not uniquely determined. Therefore, this function yields for each state a set of possible states.

$$
\begin{aligned}
&G' \in unitdelay(G) \leftrightarrow\\
&\quad now(G') = now(G) + 1 \wedge\\
&\quad \forall i \in dom(lsts(G)) \bullet\\
&\qquad stg(lsts(G')_i) = stg(lsts(G)_i) \wedge\\
&\qquad (\exists \sigma \in InpQ \bullet
\end{aligned}
$$

$$\begin{aligned}
&\quad inpq(lsts(G')_i) = inpq(lsts(G)_i) \frown \sigma \wedge \\
&\quad \sigma \in merge(\{(sig, i, i) \mid timers(lsts(G)_i)(sig) \leqslant now(G)\}) \wedge \\
&timers(lsts(G')_i) = \\
&\quad timers(lsts(G)_i) \oplus \{sig \mapsto \mathsf{nil} \mid timers(lsts(G)_i)(sig) \leqslant now(G)\} \wedge \\
&ptype(lsts(G')_i) = ptype(lsts(G)_i)
\end{aligned}$$

unitdelay(G) transforms the global state as follows:
- the system time is incremented by one unit;
- for the local state of each process:
  - its storage is left unchanged;
  - the signals that correspond to expiring timers are put into the input queue in a non-deterministic way;
  - for each of the expiring timers, the expiration time is removed;
  - its process type is left unchanged.

State observers. In general, the state observers examine one component of the local state of some process. The only exceptions are *has_instance* and *instances*, which may even examine the process type component of all processes. If an attempt is made to observe the local state of a non-existing process, each non-boolean-valued state observer yields nil and each boolean-valued state observer yields f. This will not be explicitly mentioned in the explanations given below.

The function $contents : V_\kappa \times \mathbb{N}_2 \times \mathcal{G}_\kappa \to U \cup \{\mathsf{nil}\}$ is used to describe the value of expressions of the form $value(v, e)$, which correspond to the variable accesses and view expressions of φ^-SDL.

$$\begin{aligned}
contents(v, i, G) &= \begin{cases} \rho(v) & \textbf{if } exists(i, G) \wedge v \in dom(\rho) \\ \mathsf{nil} & \textbf{otherwise} \end{cases} \\
\text{where } \rho &= stg(lsts(G)_i)
\end{aligned}$$

$contents(v, i, G)$ yields the current value of the variable v that is local to the process with pid value i.

The function $is\_active : Sig_\kappa \times \mathbb{N}_2 \times \mathcal{G}_\kappa \to \mathbb{B}$ is used to describe the value of expressions of the form $active(sig, e)$, which correspond to the active expressions of φ^-SDL.

$$\begin{aligned}
is_active(sig, i, G) &= \begin{cases} \mathsf{t} & \textbf{if } exists(i, G) \wedge sig \in dom(\theta) \\ \mathsf{f} & \textbf{otherwise} \end{cases} \\
\text{where } \theta &= timers(lsts(G)_i)
\end{aligned}$$

$is\_active(sig, i, G)$ yields true iff *sig* is an active timer signal of the process with pid value i.

The function $is\_waiting : SaveSet_\kappa \times \mathbb{N}_2 \times \mathcal{G}_\kappa \to \mathbb{B}$ is used to describe the value of conditions of the form $waiting(\{s_1, \ldots, s_n\}, e)$, which are used to give meaning to the state definitions of φ^-SDL.

$$\begin{aligned}
is_waiting(ss, i, G) &= \begin{cases} \mathsf{t} & \textbf{if } exists(i, G) \wedge getnxt(\sigma, ss) = \mathsf{nil} \\ \mathsf{f} & \textbf{otherwise} \end{cases} \\
\text{where } \sigma &= inpq(lsts(G)_i)
\end{aligned}$$

$is\_waiting(ss, i, G)$ yields true iff there is no signal in the input queue of the process with pid value i that has a name different from the ones in ss.

The function $type : \mathbb{N}_1 \times \mathcal{G}_\kappa \to P_\kappa \cup \{\textsf{env}\} \cup \{\textsf{nil}\}$ is used to describe the value of conditions of the form $type(e, X)$, which are used to give meaning to the output actions with explicit addressing of φ^-SDL.

$$\begin{array}{rcll} type(i, G) & = & X & \textbf{if } exists(i, G) \\ & & \textsf{env} & \textbf{if } i = 1 \\ & & \textsf{nil} & \textbf{otherwise} \\ \text{where } X & = & ptype(lsts(G)_i) & \end{array}$$

$type(i, G)$ yields the type of the process with pid value i. Different from the other state observers, it yields a result if $i = 1$ as well, viz. **env**.

The function $has\_instance : (P_\kappa \cup \{\textsf{env}\}) \times \mathcal{G}_\kappa \to \mathbb{B}$ is used to describe the value of conditions of the form $hasinst(X)$, which are used to give meaning to the output actions with implicit addressing of φ^-SDL.

$$\begin{array}{rcll} has\_instance(X, G) & = & \textsf{t} & \textbf{if } \exists i \in \mathbb{N}_1 \bullet (i = 1 \vee exists(i, G)) \wedge type(i, G) = X \\ & & \textsf{f} & \textbf{otherwise} \end{array}$$

$has\_instance(X, G)$ yields true iff there exists a process of type X.

The function $nxtsig : SaveSet_\kappa \times \mathbb{N}_2 \times \mathcal{G}_\kappa \to ExtSig_\kappa \cup \{\textsf{nil}\}$ is used to describe the result of executing *input* actions, which are used to give meaning to the input guards of φ^-SDL.

$$\begin{array}{rcll} nxtsig(ss, i, G) & = & getnxt(\sigma, ss) & \textbf{if } exists(i, G) \\ & & \textsf{nil} & \textbf{otherwise} \\ \text{where } \sigma & = & inpq(lsts(G)_i) & \end{array}$$

$nxtsig(ss, i, G)$ yields the first signal in the input queue of the process with pid value i that has a name different from the ones in ss.

The function $instances : (P_\kappa \cup \{\textsf{env}\}) \times \mathcal{G}_\kappa \to \mathcal{P}_{\text{fin}}(\mathbb{N}_2) \cup \{\{1\}\}$ is used to describe the result of executing *outputi* actions, which correspond to the output actions with implicit addressing of φ^-SDL.

$$\begin{array}{rcll} instances(X, G) & = & \{i \in dom(lsts(G)) \mid type(i, G) = X\} & \textbf{if } X \neq \textsf{env} \\ & & \{1\} & \textbf{otherwise} \end{array}$$

$instances(X, G)$ yields the set of pid values of all existing processes of type X if $X \neq$ **env** and the singleton set $\{1\}$ otherwise.

4.5.3. *Evaluation of value expressions.* The function $eval_G$ is used to evaluate value expressions in a state G. The state observers *contents* and *is_active* defined in Section 4.5.2 are used to define the evaluation function.

$$\begin{array}{ll} eval_G(op(e_1, \ldots, e_n)) = & \\ \quad op(eval_G(e_1), \ldots, eval_G(e_n)) & \textbf{if } eval_G(e_1) \neq \textsf{nil} \wedge \cdots \wedge eval_G(e_n) \neq \textsf{nil} \\ \quad \textsf{nil} & \textbf{otherwise} \end{array}$$

$$eval_G(cond(e_1, e_2, e_3)) = \begin{array}{ll} eval_G(e_2) & \textbf{if } eval_G(e_1) = \mathsf{t} \\ eval_G(e_3) & \textbf{if } eval_G(e_1) = \mathsf{f} \\ \mathsf{nil} & \textbf{otherwise} \end{array}$$

$$eval_G(value(v, e)) = \begin{array}{ll} contents(v, eval_G(e), G) & \textbf{if } eval_G(e) \in \mathbb{N}_2 \\ \mathsf{nil} & \textbf{otherwise} \end{array}$$

$$eval_G(active(s(e_1, \ldots, e_n), e)) =$$
$$\begin{array}{ll} is\_active(sig, eval_G(e), G) & \textbf{if } eval_G(e_1) \neq \mathsf{nil} \wedge \cdots \wedge eval_G(e_n) \neq \mathsf{nil} \wedge \\ & \quad eval_G(e) \in \mathbb{N}_2 \\ \mathsf{nil} & \textbf{otherwise} \end{array}$$
$$\text{where } sig = (s, \langle eval_G(e_1), \ldots, eval_G(e_n) \rangle)$$

$$eval_G(now) = now(G)$$

In all of these cases, if the value of at least one of the subexpressions occurring in an expression is undefined in the state concerned, the expression will be undefined, i.e. yield nil.

We extend $eval_G$ to sequences of value expressions and signal expressions as follows:

$$eval_G(\langle e_1, \ldots, e_n \rangle) =$$
$$\begin{array}{ll} \langle eval_G(e_1), \ldots, eval_G(e_n) \rangle & \textbf{if } eval_G(e_1) \neq \mathsf{nil} \wedge \cdots \wedge eval_G(e_n) \neq \mathsf{nil} \\ \mathsf{nil} & \textbf{otherwise} \end{array}$$

$$eval_G((s, es)) = \begin{array}{ll} (s, eval_G(es)) & \textbf{if } eval_G(es) \neq \mathsf{nil} \\ \mathsf{nil} & \textbf{otherwise} \end{array}$$

4.5.4. *Definition of the state operator.* In this subsection, we define the functions *act*, *eff*, eff_σ, *sig* and *val* in completion of the definition of the state operator used to describe the meaning of φ^-SDL system definitions. Recall that for this state operator $S = \mathcal{G}_\kappa$.

Action and effect functions. In the current application of $\mathrm{PA}^{\mathrm{psc}}_{\mathrm{drt}}$, $A = Cr_\kappa \cup Act_\kappa \cup Act'_\kappa \cup \{\mathit{tt}\}$, where $Act_\kappa = Act^-_\kappa \cup \overline{Cr}_\kappa$. The actions in Act_κ are actions that may change the state in which they are executed. The actions in Act'_κ are actions that are performed as the result of the execution of actions in Act_κ in a state. The actions in Cr_κ and $\overline{Cr}_\kappa$ are the process creation actions and the actions that are left as a trace of the process creations that occur, respectively – note that we use the data set $D = PrCrD_\kappa$ for process creation. The action and effect functions are trivial for actions that are not in Act_κ.

In order to keep the definitions comprehensible, we will use the following abbreviations. Let *e* be a value expression, let *es* be a sequence of value expressions, and let *se* be a signal expression. Then we write e' for $eval_G(e)$, es' for $eval_G(es)$ and se' for $eval_G(se)$.

$$act(input((s, \langle v_1, \ldots, v_n \rangle), ss, e), G) =$$
$$\{input'((sig, i, j)) \mid snm(sig) = s, nxtsig(ss, e', G) = (sig, i, j), e' \in \mathbb{N}_2\}$$

$$\mathit{eff}(input((s, \langle v_1, \ldots, v_n \rangle), ss, e), G, a) =$$
$$\begin{array}{ll} csmsig(nxtsig(ss, e', G), \langle v_1, \ldots, v_n \rangle, G) & \textbf{if } a \in act(input((s, \langle v_1, \ldots, v_n \rangle), ss, e), G) \\ G & \textbf{otherwise} \end{array}$$

$$
\begin{array}{l}
act(outpute(se, e_1, e_2), G) \;=\\
\quad \{output'((se', e_1', e_2')) \mid se' \neq \mathsf{nil}, e_1' \in \mathbb{N}_2, e_2' \in \mathbb{N}_1\}\\
eff(outpute(se, e_1, e_2), G, a) \;=\\
\quad \begin{array}{ll} sndsig((se', e_1', e_2'), G) & \textbf{if } a \in act(outpute(se, e_1, e_2), G)\\ G & \textbf{otherwise} \end{array}
\end{array}
$$

$$
\begin{array}{l}
act(outputi(se, e, X), G) \;=\\
\quad \{output'((se', e', i)) \mid se' \neq \mathsf{nil}, e' \in \mathbb{N}_2, i \in instances(X, G)\}\\
eff(outputi(se, e, X), G, a) \;=\\
\quad \begin{array}{ll} sndsig((se', e', i), G) & \textbf{if } a \in act(outputi(se, e, X), G) \wedge a = output'((se', e', i))\\ G & \textbf{otherwise} \end{array}
\end{array}
$$

$$
\begin{array}{lcl}
act(set(e_1, se, e_2), G) & = & \{set'(e_1', se', e_2') \mid e_1' \in \mathbb{N}, se' \neq \mathsf{nil}, e_2' \in \mathbb{N}_2\}\\
eff(set(e_1, se, e_2), G, a) & = & \begin{array}{ll} settimer(e_1', se', e_2', G) & \textbf{if } a \in act(set(e_1, se, e_2), G)\\ G & \textbf{otherwise} \end{array}
\end{array}
$$

$$
\begin{array}{lcl}
act(reset(se, e), G) & = & \{reset'(se', e') \mid se' \neq \mathsf{nil}, e' \in \mathbb{N}_2\}\\
eff(reset(se, e), G, a) & = & \begin{array}{ll} resettimer(se', e', G) & \textbf{if } a \in act(reset(se, e), G)\\ G & \textbf{otherwise} \end{array}
\end{array}
$$

$$
\begin{array}{lcl}
act(ass(v, e_1, e_2), G) & = & \{tt \mid e_1' \neq \mathsf{nil}, e_2' \in \mathbb{N}_2\}\\
eff(ass(v, e_1, e_2), G, a) & = & \begin{array}{ll} assignvar(v, e_1', e_2', G) & \textbf{if } a \in act(ass(v, e_1, e_2), G)\\ G & \textbf{otherwise} \end{array}
\end{array}
$$

$$
\begin{array}{l}
act(\overline{cr}(i, (X, vs, es, e)), G) \;=\; \{tt \mid i \in \mathbb{N}_2, es' \neq \mathsf{nil}, e' \in \mathbb{N}_2\}\\
eff(\overline{cr}(i, (X, vs, es, e)), G, a) \;=\\
\quad \begin{array}{ll} createproc(i, (X, vs, es', e'), G) & \textbf{if } a \in act(\overline{cr}(i, (X, vs, es, e)), G)\\ G & \textbf{otherwise} \end{array}
\end{array}
$$

$$
\begin{array}{lcl}
act(stop(e), G) & = & \{tt \mid e' \in \mathbb{N}_2\}\\
eff(stop(e), G, a) & = & \begin{array}{ll} stopproc(e', G) & \textbf{if } a \in act(stop(e), G)\\ G & \textbf{otherwise} \end{array}
\end{array}
$$

$$
\begin{array}{lcl}
act(inispont(e), G) & = & \{tt \mid e' \in \mathbb{N}_2\}\\
eff(inispont(e), G, a) & = & \begin{array}{ll} inispont(e', G) & \textbf{if } a \in act(inispont(e), G)\\ G & \textbf{otherwise} \end{array}
\end{array}
$$

For all actions $a \in Cr_\kappa \cup Act_\kappa' \cup \{tt\}$:

$$
\begin{array}{lcl}
act(a, G) & = & a\\
eff(a, G, a') & = & G
\end{array}
$$

The effect of applying the state operator to a process of the form $\sigma_{\mathsf{rel}}(x)$ is described by mean of the function eff_σ.

$$eff_\sigma(G) = unitdelay(G)$$

Signal function. First, we define the function $atoms : \mathcal{G}_\kappa \to \mathcal{P}_{\text{fin}}(AtProp_\kappa)$ giving for each state the set of atomic propositions that hold in that state. It is inductively defined by

$$contents(v, i, G) = u \;\Rightarrow\; value(v, u, i) \in atoms(G)$$
$$is\_active(sig, i, G) = \mathsf{t} \;\Rightarrow\; active(sig, i) \in atoms(G)$$

We now define the function $sig : \mathcal{G}_\kappa \to Prop_\kappa$, giving the propositions produced by the states. as follows:

$$sig(G) \;=\; \bigwedge\nolimits_{\phi \in atoms(G)} \phi$$

So $sig(G)$ is the conjunction of all atomic propositions that hold in state G.

Valuation function. In the current application of $\mathrm{PA}^{\mathrm{psc}}_{\mathrm{drt}}$, $B_{\mathrm{at}} = AtCond_\kappa \cup AtProp_\kappa$.
The function *val* is defined in the following way:

$$val(eq(e_1, e_2), G) = \begin{array}{l} \mathsf{t} \ \textbf{if}\ e'_1 = e'_2 \wedge e'_1 \neq \mathsf{nil} \wedge e'_2 \neq \mathsf{nil} \\ \mathsf{f} \ \textbf{otherwise} \end{array}$$

$$val(waiting(ss, e), G) = \begin{array}{l} \mathsf{t} \ \textbf{if}\ is\_waiting(ss, e', G) = \mathsf{t} \wedge e' \in \mathbb{N}_2 \\ \mathsf{f} \ \textbf{otherwise} \end{array}$$

$$val(type(e, X), G) = \begin{array}{l} \mathsf{t} \ \textbf{if}\ type(e', G) = X \wedge e' \in \mathbb{N}_1 \\ \mathsf{f} \ \textbf{otherwise} \end{array}$$

$$val(hasinst(X), G) = has\_instance(X, G)$$

$$val(value(v, u, i), G) = \begin{array}{l} \mathsf{t} \ \textbf{if}\ value(v, u, i) \in atoms(G) \\ \mathsf{f} \ \textbf{otherwise} \end{array}$$

$$val(active(sig, i), G) = \begin{array}{l} \mathsf{t} \ \textbf{if}\ active(sig, i) \in atoms(G) \\ \mathsf{f} \ \textbf{otherwise} \end{array}$$

Note that the sets $AtCond_\kappa$ and $AtProp_\kappa$ are disjoint. The elements of $AtCond_\kappa$ are used as conditions with a truth value that depends upon the state. The elements of $AtProp_\kappa$ are not used as conditions and their valuation in a state is not needed for the semantics of φ^-SDL.

4.6. *Semantics of system definitions*

In this subsection, we present a semantics for φ^-SDL system definitions. It relies heavily upon the specifics of the state operator defined in Section 4.5.

According to the semantics presented here, the meaning of a φ^-SDL system definition is a process described by a process expression – a term of $\mathrm{PA}^{\mathrm{psc}}_{\mathrm{drt}}$ extended with the counting process creation operator and the state operator to be precise. It is given in terms of the semantics of φ^-SDL process definitions presented in Section 4.3 and the contextual information extracted by means of the function $\{\![\_]\!\}$ defined in Appendix B.

4.6.1. *System definition.* The semantics of a φ^-SDL system definition depends on a parameter *Env* representing the environment of the system. This environment *Env* has to be described by a $\mathrm{PA}^{\mathrm{psc}}_{\mathrm{drt}}$ term.

$$[\![\,\textbf{system}\ S;D_1\ldots D_n\ \textbf{endsystem;}]\!](\mathit{Env}) := \lambda_{G_0}(\mathrm{E}_{\Phi}^{n_0+2}(P)\parallel \mathit{Env})$$

$$\begin{aligned}
\text{where } P &= \|_{i\in\{1,\ldots,n_0\}}\ \Phi(i+1, (pt(i+1), \langle\rangle, \langle\rangle, 0)),\\
\Phi &= \{(i,d)\mapsto \Psi(i) \mid \exists D\in\{D_1,\ldots,D_n\}\bullet [\![D]\!]^{\kappa} = (pnm(d), \Psi)\},\\
G_0 &= (0, \{i+1\mapsto L_0(i+1)\mid i\in\{1,\ldots,n_0\}\}),\\
L_0(i) &= (\{\textbf{parent}\mapsto 0, \textbf{offspring}\mapsto 0, \textbf{sender}\mapsto 0\}, \langle\rangle, \{\}, pt(i)),\\
n_0 &= \textstyle\sum_{X\in procs(\kappa)} init(\kappa, X),\\
\kappa &= \{\![\textbf{system}\ S;D_1\ldots D_n\ \textbf{endsystem;}]\!\}
\end{aligned}$$

$$\begin{aligned}
\text{and }\ & pt:\{1+1,\ldots,n_0+1\}\to procs(\kappa) \text{ is such that}\\
& \forall X\in procs(\kappa)\bullet \operatorname{card}(pt^{-1}(X)) = init(\kappa, X).
\end{aligned}$$

The function *pt* is used to assign pid values for the processes created during system start-up. It maps a set of pid values to the types of the processes with these pid values.

The process expression that corresponds to a system definition expresses that, for each of the process types defined, the given initial number of processes are created and these processes are executed in parallel, starting in the state G_0, while they receive signals via signal routes from the environment *Env*. G_0 is the state in which the system time is zero and there is a local state for each of the processes that is created during system start-up. Recall that the pid value 1 is reserved for the environment. The mapping Φ from pid values and process creation data to process expressions is derived from the meaning of the process definitions occurring in the system definition. This mapping is used by the counting process creation operator, which is needed for process creation after system start-up.

4.6.2. *Environments.* The semantics of φ^-SDL system definitions describes the meaning of a system definition for an arbitrary process *Env* that represents the behaviour of the environment. Here we explain how the environment's behaviour is described by a $\mathrm{PA}^{\mathrm{psc}}_{\mathrm{drt}}$ process.

Some general assumptions have to be made about the behaviour of the environment of any system described using φ^-SDL. Further assumptions may be made about the behaviour of the environment of a specific system described using φ^-SDL, including ones that facilitate analysis of the system concerned. The general assumptions made about environments are:

- the environment can only send signals that are defined in the system definition concerned;
- the environment can only send signals to processes to which it is connected by signal routes;
- the environment can send only a finite number of signals during a time slice.

Besides, the viewpoint is taken that the processes that constitute a system are not observable to its environment. This leads to the use of output actions with implicit addressing in representing the environment's behaviour.

The set EnvSig_{κ} of possible environment signals is determined by the specific types of signals and signal routes introduced in the system definition concerned. It can be obtained from the environment signal description yielded by applying the function *envsigd* (defined in Appendix B) to the context ascribed to the system definition. For an arbitrary context κ, the set of environment signals is obtained as follows:

$$EnvSig_\kappa = \bigcup_{((s,\langle T_1,\ldots,T_n\rangle),X)\in envsigd(\kappa)} \{((s,\langle t_1,\ldots,t_n\rangle),1,X) \mid t_1 \in T_1^{\mathcal{A}},\ldots,t_n \in T_n^{\mathcal{A}}\}$$

It is clear that in general the set $EnvSig_\kappa$ is infinite because signals can carry values from infinite domains. Besides, the environment can send an arbitrary signal from $EnvSig_\kappa$. To represent this we need the alternative composition of an infinite number of alternatives. This can be done using the operator $\sum_{d:D}$ of μCRL [22]. However, the combination of this operator with the extension of discrete time process algebra we are using has not been investigated thoroughly. Besides, the potential unbounded non-determinism introduced by this operator does not allow to use conventional validation techniques.

A process that satisfies the three above-mentioned assumptions can be defined in the following way:

$$\begin{aligned} Env_\kappa &= \sum_{n:\mathbb{N}} Env_n \\ Env_0 &= \sigma_{\mathsf{rel}}(Env_\kappa) \\ Env_{n+1} &= Env_n + \sum_{osig:EnvSig_\kappa} \underline{\underline{outputi}}(osig) \cdot Env_n \end{aligned}$$

In order to use an environment process as a parameter of the presented semantics of φ^-SDL, it has to be given as a process in $\mathrm{PA}^{\mathrm{psc}}_{\mathrm{drt}}$. Below we define such an environment process. We call it a standard environment process for the semantics of φ^-SDL. It is determined by two restrictions:

- the set of signals which the environment can send to the system is restricted to a finite subset $ES \subseteq EnvSig_\kappa$;
- the maximal number of signals which can be sent in one time slice is bounded by a natural number N_s.

A standard environment is defined as follows:

$$\begin{aligned} Env^{\mathrm{st}}_\kappa &= \sum_{n\in\{0,\ldots,N_s\}} Env'_n \\ Env'_0 &= \sigma_{\mathsf{rel}}(Env^{\mathrm{st}}_\kappa) \\ Env'_{n+1} &= Env'_n + \sum_{osig\in ES} \underline{\underline{outputi}}(osig) \cdot Env'_n \end{aligned}$$

One may also define another environment process for a specific system. In any case the process representing the system's environment must satisfy the above-mentioned assumptions.

4.6.3. *Delaying channels.* The process algebra semantics of φ^-SDL makes clear how to model a delaying channel by means of a φ^-SDL process. Below a φ^-SDL process definition of such a process, called `ch`, is given. It is assumed that the process can only receive signals of type `es`.

The process consumes signals `es(rcv,v1,...,vn)` and passes them on after an arbitrary delay, possibly zero, with `rcv` replaced by `snd`. Each signal consumed carries the pid value of the ultimate receiver, and this pid value is replaced by the one of the original sender when the signal is passed on. This is needed because the original sender and ultimate receiver have now an intermediate receiver and intermediate sender, respectively. The `decision` construct is used to find out whether the original sender used implicit or explicit addressing. Note that we write `Null` for `0`, i.e. the special pid value that never refers to any existing process.

```
process ch(1);
  start;
    nextstate receive;
  state receive;
    input es(rcv,v1,...,vn);
      task snd := sender;
      nextstate deliver;
  state deliver;
    save es;
    input none;
      decision rcv = Null;
        (True):
          output es(snd,v1,...,vn) via sr_out;
          nextstate receive;
        (False):
          output es(snd,v1,...,vn) to rcv via sr_out;
          nextstate receive;
      enddecision;
endprocess;
```

In state `deliver`, there will never be signals to consume because all signals are withhold from being consumed by means of `save es`. This means that the behaviour from this state may be delayed till any future time slice. The total lack of signals to consume does not preclude the process to proceed, because the only transition alternative is a spontaneous transition, i.e. it is not initiated by the consumption of a signal.

5. Closing remarks

In this section, we sum up what has been achieved. We also describe in broad outline what is anticipated to be achieved more, thus hinting at topics for future research. But, to begin with, we present an overview of related work.

5.1. *Related work*

In [17], a foundation for the semantics of SDL, based on streams and stream processing functions, has been proposed. This proposal indicates that the SDL view of a system gives an interesting type of asynchronous dataflow networks, but the treatment of time in the proposal is however too sketchy to be used as a starting point for a semantics of the time related features of SDL. Besides, process creation is not covered. In [15], we present a process algebra model of asynchronous dataflow networks as a semantic foundation for SDL. The model is close to the concepts around which SDL has been set up. However, we are not able to cover process creation too.

An operational semantics for a subset of SDL, which covers timing, has been given in [21]. Many relevant details are worked out, but it is not quite clear whether time is treated satisfactory. This is largely due to the intricacy of the operational semantics. At the outset, we have also tried shortly to give an operational semantics for φ^{-}SDL, but we found that it is very difficult. Our experience is that the main motivations for the rules describing an operational semantics are unavoidably intuitive. This may already lead to an undesirable semantics in relatively simple cases. For example, working on $\mathrm{PA}^{\mathrm{psc}}_{\mathrm{drt}}$, we have seriously considered to have the premise $w \in [\mathsf{s}_\rho(x)]$ in the rule for time unit delay (see Table 11). However, this plausible option would render all delayed processes bisimilar to deadlock in the current time slice, i.e. $\sigma_{\mathsf{rel}}(x) = \underline{\underline{\delta}}$ would hold.

It is likely that, if we had taken parallel composition with communication, we would have been able to use special processes, put in parallel with the other ones, instead of the counting process creation operator and the state operator. The approach to use special processes is followed in [35]. There, it is largely responsible for the exceptional intricacy of the resulting semantics, which, by the way, has kept various obvious errors unnoticed for a long time. Amongst other things for this reason, we have chosen to use the counting process creation operator and the state operator instead. Besides, the approach to use special processes brings along a lot of internal communication that is irrelevant from a semantic point of view. Of course, in case an ACP-style process algebra is used as the basis of the presented semantics, there is the option to use an abstraction operator to abstract from the internal communication (for abstraction in process algebra, see, e.g., [9]). However, it seems far from easy to elaborate the addition of abstraction to $\mathrm{PA}^{\mathrm{psc}}_{\mathrm{drt}}$ or its adaptation to parallel composition with communication.

For a subset of SDL, called μSDL, both an operational semantics and a process algebra semantics has been given in [20]. The operational semantics of μSDL is related to the one presented in [21], but time is not treated. The basis of the corresponding process algebra semantics is a time free process algebra, essentially μCRL [22] extended with the state operator. Like in [35], special processes are used for channels and input queues although there is no need for that with the state operator at one's disposal. Interesting is that first the intended meaning of the language constructs is laid down in an operational semantics so that it can be reasonably checked later whether the process algebra semantics reflects the intentions. However, the SDL facilities for storage, timing and process creation are not available at all in μSDL; and the facilities for communication are only partially available.

In [19], BSDL is proposed as a basis for the semantics of SDL. BSDL is developed from scratch, using Object-Z, but it does not seem to fit in very well with SDL. In [24], it is proposed to use Duration Calculus [30] to describe the meaning of the language constructs of SDL. Thus an interesting semantic link is made between SDL and Duration Calculus, but it seems a little bit odd to view the main semantic objects used, viz. traces, as phases of system behaviour, called state variables in Duration Calculus, of which the duration is the principal attribute.

5.2. *Conclusions and future work*

We have presented an extension of discrete time process algebra with relative timing and we have proposed a semantics for the core of SDL, using this extension to describe the meaning of the language constructs. The operational semantics and axiom system of this ACP-style process algebra facilitates advances in the areas of validation of SDL specifications and verification of design steps made using SDL, respectively. At present, we focus on validation. We do so because practically useful advanced tools and techniques are within reach now while there is a tremendous need for them.

For validation purposes, we have to transform φ^{-}SDL specifications to transition systems in accordance with the process algebra semantics. Generating transition systems from finite linear recursive specifications in $\mathrm{BPA}^{\mathrm{psc}}_{\mathrm{drt}}$ (see Section 2.4) is straightforward. In Section 4, the meaning of a φ^{-}SDL process definition is described by a finite guarded recursive specification in $\mathrm{BPA}^{\mathrm{psc}}_{\mathrm{drt}}$ that can definitely be brought into linear form. The meaning of a φ^{-}SDL system definition is given in terms of the meaning of the occurring process definitions using the parallel composition operator, the counting process creation operator and the state operator. An obvious direction is to devise, for each of these operators on processes, a corresponding syntactic operator on finite linear recursive specifications that, under certain conditions, yields a linear recursive specification of the process that results from applying the operator on processes to the process(es) defined by the recursive specification(s) to be operated on. Of course, we look for non-restrictive conditions, but finiteness of the state space, a finite upper bound on the total number of process creations and a finite upper bound on the number of signals per time slice originating from the environment are inescapable. It goes without saying that we have to show the correctness of these syntactic operators. For that purpose, we have available the axioms presented in Section 2 and RSP (see Section 2.4).

In [12], timed frames, which are closely related to the kind of transition systems presented by the operational semantics of, for example, $\mathrm{BPA}^{-}_{\mathrm{drt}}$-ID and $\mathrm{PA}^{-}_{\mathrm{drt}}$-ID, are studied in a general algebraic setting and results concerning the connection between timed frames and discrete time processes with relative timing are given. In [16], a model for $\mathrm{BPA}^{-}_{\mathrm{drt}}$-IDlin ($\mathrm{BPA}^{-}_{\mathrm{drt}}$-ID with finite linear recursive specifications) is presented that gives an interpretation of its constants and operators on timed frames; and it is shown that the bisimulation model induced by the original structured operational semantics is isomorphic to the timed frame model. It is plausible that these results can be extended to $\mathrm{BPA}^{\mathrm{psc}}_{\mathrm{drt}}$lin – timed frames support propositional signals. This would mean that we can transform φ^{-}SDL specifications to timed frames. In that case, we can basically check whether a system described in φ^{-}SDL satisfies a property expressed in TFL [13], an expressive first-order logic proposed for timed frames. We are considering to look for a fragment of TFL that is suitable to serve as a logic for φ^{-}SDL and to adapt an existing model checker to φ^{-}SDL and this logic – and thus to automate the checking. In particular, the model checker MEC [3] seems suited for this purpose – at least for small-scale systems. A fragment of Duration Calculus may be considered as well, since in [23] validity for Duration Calculus formulas in timed frames is defined.

The extension of discrete time process algebra with relative timing, used to describe the meaning of the language constructs of φ^{-}SDL, is fairly large and rather intricate. Theoretically interesting general properties, such as elimination, conservativity, completeness, etc. have yet to be established. We think that we are near the limit of what can be made provably free from defect. Still, owing to the nontrivial state space taken for the state operator, the presented semantics for φ^-SDL uses an excursion outside the realm of process algebra that is not negligible. We wish to have abstraction added to the process algebra used in order to provide a more abstract semantics for φ^-SDL, but right now we consider the consequences of this addition too difficult to grasp. All this suggests the option to develop a special process algebra that is closer to the concepts around which SDL has been set up. Of course, there is also the alternative to simplify SDL by removing SDL features that introduce semantic complexities but do not serve any practical purpose. The presented semantics of φ^-SDL may assist in identifying such cases.

A. Notational conventions

Meta-language for syntax. The syntax of φ^-SDL is described by means of production rules in the form of an *extended* BNF grammar. The curly brackets "{" and "}" are used for grouping. The asterisk "*" and the plus sign "+" are used for zero or more repetitions and one or more repetitions, respectively, of curly bracketed groups. The square brackets "[" and "]" are also used for grouping, but indicate that the group is optional. An underlined part included in a nonterminal symbol does not belong to the context-free syntax; it describes a context-sensitive condition.

Meta-language for semantics. The semantics of φ^-SDL is described by means of a set of equations recursively defining interpretation functions for all syntactic categories. For each syntactic category, the corresponding interpretation function gives a meaning to each language construct c belonging to the category. We use the notation $[\![c]\!]$ or $[\![c]\!]^\kappa$ for applications of all interpretation functions. The exact interpretation function is always clear from the context. If contextual information κ is needed for the interpretation, it is provided by an additional argument and the notation $[\![c]\!]^\kappa$ is used.

Special action, condition and proposition notation. We write $a : D_1 \times \cdots \times D_n$ to indicate that a is an action parametrized by $D_1 \times \cdots \times D_n$. This means that there is an action, referred to as $a(d_1, \ldots, d_n)$, for each $(d_1, \ldots, d_n) \in D_1 \times \cdots \times D_n$. For atomic conditions and propositions, we use analogous notations.

Special set, function and sequence notation. We write $\mathcal{P}(A)$ for the set of all subsets of A, and we write $\mathcal{P}_{\text{fin}}(A)$ for the set of all finite subsets of A. We use abbreviations $\mathbb{N}_1$ and $\mathbb{N}_2$ for the sets $\mathbb{N} \setminus \{0\}$ and $\mathbb{N} \setminus \{0, 1\}$, respectively.

We write $f : A \to B$ to indicate that f is a total function from A to B, that is $f \subseteq A \times B \wedge \forall x \in A \bullet \exists_1 y \in B \bullet (x, y) \in f$. We write $dom(f)$, where $f : A \to B$, for A. We also write $A \to B$ for the set of all functions from A to B. For an (ordered) pair (x, y), where x and y are intended for argument and value of some function, we use the notation $x \mapsto y$ to emphasize this intention. The binary operators $\lhd$ (domain subtraction) and $\oplus$ (overriding) on functions are defined by

$$A \lhd f \;=\; \{x \mapsto y \mid x \in dom(f) \wedge x \notin A \wedge f(x) = y\}$$
$$f \oplus g \;=\; (dom(g) \lhd f) \cup g$$

For a function $f : A \to B$ presenting a family B indexed by A, we use the notation f_i (for $i \in A$) instead of $f(i)$.

Functions are also used to present sequences; as usual we write $\langle x_1, \ldots, x_n \rangle$ for the sequence presented by the function $\{1 \mapsto x_1, \ldots, n \mapsto x_n\}$. The unary operators *hd* and *tl* stand for selection of head and tail, respectively, of sequences. The binary operator $\frown$ stands for concatenation of sequences. We write $x \,\&\, t$ for $\langle x \rangle \frown t$.

B. Contextual information

The meaning of a language construct of φ^-SDL generally depends on the definitions in the scope in which it occurs. Contexts are primarily intended for modeling the scope. The context that is ascribed to a complete φ^-SDL specification is also used to define the state space used to describe its meaning. The context of a language construct contains all names introduced by the definitions of variables, signal types, signal routes and process types occurring in the specification on hand and additionally:

- if the language construct occurs in the scope of a process definition, the name introduced by that process definition, called the *scope unit*;
- if the language construct occurs in the scope of a state definition, the set of names occurring in the **save** part of that state definition, called the *save set*.

The names introduced by the definitions are in addition connected with their static attributes. For example, a name of a variable is connected with the name of the sort of the values that may be assigned to it; and a name of a process type is connected with the names of the variables that are its formal parameters and the number of processes of this type that have to be created during the start-up of the system.

$$Context =$$
$$\mathcal{P}_{\mathrm{fin}}(VarD) \times \mathcal{P}_{\mathrm{fin}}(SigD) \times \mathcal{P}_{\mathrm{fin}}(RouteD) \times \mathcal{P}_{\mathrm{fin}}(ProcD) \times (ProcId \cup \{\mathsf{nil}\}) \times \mathcal{P}_{\mathrm{fin}}(SigId)$$

where

$$\begin{aligned} VarD &= VarId \times SortId \\ SigD &= SigId \times SortId^* \\ RouteD &= RouteId \times (ProcId \cup \{\textbf{env}\}) \times (ProcId \cup \{\textbf{env}\}) \times \mathcal{P}_{\mathrm{fin}}(SigId) \\ ProcD &= ProcId \times VarId^* \times \mathbb{N} \end{aligned}$$

For language constructs that do not occur in a process definition, the absence of a scope unit will be represented by nil and, for language constructs that do not occur in a state definition, the absence of a save set will be represented by $\emptyset$. We write $vards(\kappa)$, $sigds(\kappa)$, $routeds(\kappa)$, $procds(\kappa)$, $scopeunit(\kappa)$ and $saveset(\kappa)$, where $\kappa = (V, S, R, P, X, ss) \in Context$, for V, S, R, P, X and ss, respectively. We write $vars(\kappa)$ for $\{v \mid \exists T \bullet (v, T) \in vards(\kappa)\}$. The abbreviations $sigs(\kappa)$ and $procs(\kappa)$ are used analogously.

We make use of the following functions on *Context*:

$$
\begin{array}{ll}
rcv & : Context \times RouteId \rightarrow ProcId \cup \{\mathbf{env}\}, \\
fpars & : Context \times ProcId \rightarrow VarId^*, \\
init & : Context \times ProcId \rightarrow \mathbb{N}, \\
updscopeunit & : Context \times ProcId \rightarrow Context, \\
updsaveset & : Context \times \mathcal{P}_{\text{fin}}(SigId) \rightarrow Context, \\
envsigd & : Context \rightarrow \mathcal{P}_{\text{fin}}(SigD \times ProcId)
\end{array}
$$

The first three functions are partial functions, but they will only be applied in cases where the result is defined. The function *rcv* is used to extract the receiver type of a given signal route from the context. This function is inductively defined by

$$(r, X_1, X_2, ss) \in routeds(\kappa) \;\Rightarrow\; rcv(\kappa, r) = X_2$$

The functions *fpars* and *init* are used to extract the formal parameters and the initial number of processes, respectively, of a given process type from the context. These functions are inductively defined by

$$
\begin{array}{l}
(X, vs, k) \in procds(\kappa) \;\Rightarrow\; fpars(\kappa, X) = vs, \\
(X, vs, k) \in procds(\kappa) \;\Rightarrow\; init(\kappa, X) = k
\end{array}
$$

The functions *updscopeunit* and *updsaveset* are used to update the scope unit and the save set, respectively, of the context. These functions are inductively defined by

$$
\begin{array}{l}
\kappa = (V, S, R, P, X, ss) \;\Rightarrow\; updscopeunit(\kappa, X') = (V, S, R, P, X', ss), \\
\kappa = (V, S, R, P, X, ss) \;\Rightarrow\; updsaveset(\kappa, ss') = (V, S, R, P, X, ss')
\end{array}
$$

The function *envsigd* is used to determine the possible environment signals, i.e. signals that the system can receive via signal routes from the environment. It is inductively defined by

$$
\begin{array}{l}
s \in ss \wedge (s, \langle T_1, \ldots, T_n \rangle) \in sigds(\kappa) \wedge (r, \mathbf{env}, X_2, ss) \in routeds(\kappa) \;\Rightarrow \\
\quad ((s, \langle T_1, \ldots, T_n \rangle), X_2) \in envsigd(\kappa)
\end{array}
$$

The context ascribed to a complete φ^-SDL specification is a minimal context in the sense that the contextual information available in it is common to all contexts on which language constructs occurring in it depend. The additional information that may be available applies to the scope unit for language constructs occurring in a process definition and the save set for language constructs occurring in a state definition. The context ascribed to a complete specification is obtained by taking the union of the corresponding components of the partial contexts contributed by all definitions occurring in it, except for the scope unit and the save set which are permanently the same – nil and $\emptyset$, respectively.

$$
\begin{array}{l}
\{\![\mathbf{system}\ S;D_1 \ldots D_n\ \mathbf{endsystem;}]\!\} := \\
\quad (vards(\{\![D_1]\!\}) \cup \ldots \cup vards(\{\![D_n]\!\}), \\
\quad sigds(\{\![D_1]\!\}) \cup \ldots \cup sigds(\{\![D_n]\!\}), \\
\quad routeds(\{\![D_1]\!\}) \cup \ldots \cup routeds(\{\![D_n]\!\}), \\
\quad procds(\{\![D_1]\!\}) \cup \ldots \cup procds(\{\![D_n]\!\}), \\
\quad \mathsf{nil}, \emptyset)
\end{array}
$$

$$\{\![\mathbf{dcl}\ v\ T;]\!\} := (\{(v, T)\}, \emptyset, \emptyset, \emptyset, \mathsf{nil}, \emptyset)$$

$\{\![\textbf{signal}\, s(T_1, \ldots, T_n);]\!\} := (\emptyset, \{(s, \langle T_1, \ldots, T_m \rangle)\}, \emptyset, \emptyset, \mathsf{nil}, \emptyset)$

$\{\![\textbf{signalroute}\, r\, \textbf{from}\, X_1\, \textbf{to}\, X_2\, \textbf{with}\, s_1, \ldots, s_n;]\!\} :=$
$(\emptyset, \emptyset, \{(r, X_1, X_2, \{s_1, \ldots, s_n\})\}, \emptyset, \mathsf{nil}, \emptyset)$

$\{\![\textbf{process}\, X(k); \textbf{fpar}\, v_1, \ldots, v_m; \textbf{start};\, tr\, d_1 \ldots d_n\, \textbf{endprocess};]\!\} :=$
$(\emptyset, \emptyset, \emptyset, \{(X, \langle v_1, \ldots, v_m \rangle, k)\}, \mathsf{nil}, \emptyset)$

Acknowledgements

For one month all three authors were at UNU/IIST (United Nations University, International Institute for Software Technology) in Macau. During that period, some of the more important corrections and technical changes of the material presented here were made. We thank Dines Bjørner, the former director of UNU/IIST, for bringing us together in Macau. The third author thanks Radu Şoricuţ and Bogdan Warinschi for their helpful comments and discussions.

References

[1] L. Aceto, W.J. Fokkink and C. Verhoef, *Structural operational semantics*, Handbook of Process Algebra, J.A. Bergstra, A. Ponse and S.A. Smolka, eds, Elsevier, Amsterdam (2001), 197–292.

[2] B. Algayres, Y. Lejeune, F. Hugonnet and F. Hantz, *The AVALON project: A validation enviroment for SDL/MSC descriptions*, SDL '93: Using Objects, O. Færgemand and A. Sarma, eds, North-Holland (1993), 221–235.

[3] A. Arnold, *MEC, a system for constructing and analysing transition systems*, Automatic Verification Methods for Finite State Systems, Lecture Notes in Comput. Sci. 407, J. Sifakis, ed., Springer-Verlag (1990), 117–132.

[4] J.C.M. Baeten and J.A. Bergstra, *Discrete time process algebra* (*extended abstract*), CONCUR'92, Lecture Notes in Comput. Sci. 630, W.R. Cleaveland, ed., Springer-Verlag (1992), 401–420. Full version: Report P9208b, Programming Research Group, University of Amsterdam.

[5] J.C.M. Baeten and J.A. Bergstra, *Discrete time process algebra*, Formal Aspects of Computing **8** (1996), 188–208.

[6] J.C.M. Baeten and J.A. Bergstra, *Process algebra with propositional signals*, Theoret. Comput. Sci. **177** (1997), 381–405.

[7] J.C.M. Baeten and C.A. Middelburg, *Process algebra with timing: Real time and discrete time*, Handbook of Process Algebra, J.A. Bergstra, A. Ponse and S.A. Smolka, eds, Elsevier, Amsterdam (2001), 627–684.

[8] J.C.M. Baeten and C. Verhoef, *Concrete process algebra*, Handbook of Logic in Computer Science, Vol. IV, S. Abramsky, D. Gabbay and T.S.E. Maibaum, eds, Oxford University Press (1995), 149–268.

[9] J.C.M. Baeten and W.P. Weijland, *Process Algebra*, Cambridge Tracts Theoret. Comput. Sci. 18, Cambridge University Press (1990).

[10] F. Belina, D. Hogrefe and A. Sarma, *SDL with Applications from Protocol Specification*, Prentice-Hall (1991).

[11] J.A. Bergstra, *A process creation mechanism in process algebra*, Applications of Process Algebra, J.C.M. Baeten, ed., Cambridge Tracts Theoret. Comput. Sci. 17, Cambridge University Press (1990), 81–88.

[12] J.A. Bergstra, W.J. Fokkink and C.A. Middelburg, *Algebra of timed frames*, Internat. J. Comput. Math. **61** (1996), 227–255.

[13] J.A. Bergstra, W.J. Fokkink and C.A. Middelburg, *A logic for signal inserted timed frames*, Logic Group Preprint Series 155, Utrecht University, Department of Philosophy (1996).

[14] J.A. Bergstra and C.A. Middelburg, *Process algebra semantics of φSDL*, Logic Group Preprint Series 129, Utrecht University, Department of Philosophy (March 1995).
[15] J.A. Bergstra, C.A. Middelburg and R. Şoricuţ, *Discrete time network algebra for a semantic foundation of SDL*, Research Report 98, United Nations University, International Institute for Software Technology (1997).
[16] J.A. Bergstra, C.A. Middelburg and B. Warinschi, *Timed frame models for discrete time process algebras*, Research Report 100, United Nations University, International Institute for Software Technology (1997).
[17] M. Broy, *Towards a formal foundation of the specification and description language SDL*, Formal Aspects of Computing **3** (1991), 21–57.
[18] J. Ellsberger, D. Hogrefe and A. Sarma, *SDL, Formal Object-Oriented Language for Communicating Systems*, Prentice-Hall (1997).
[19] J. Fischer, S. Lau and A. Prinz, *A short note about BSDL*, SDL Newsletter **18** (1995), 15–21.
[20] A. Gammelgaard and J.E. Kristensen, *A correctness proof of a translation from SDL to CRL*, SDL '93: Using Objects, O. Færgemand and A. Sarma, eds, Elsevier (North-Holland) (1993), 205–219. Full version: Report TFL RR 1992-4, Tele Danmark Research.
[21] J.C. Godskesen, *An operational semantics model for Basic SDL (extended abstract)*, SDL '91: Evolving Methods, O. Færgemand and R. Reed, eds, Elsevier (North-Holland), (1991), 15–22. Full version: Report TFL RR 1991-2, Tele Danmark Research.
[22] J.F. Groote and A. Ponse, *The syntax and semantics of μCRL*, Algebra of Communicating Processes 1994, Workshop in Computing Series, A. Ponse, C. Verhoef and S.F.M. van Vlijmen, eds, Springer-Verlag (1995), 26–62.
[23] C.A. Middelburg, *Truth of duration calculus formulae in timed frames*, Fund. Inform. **36** (1998), 235–263.
[24] S. Mørk, J.C. Godskesen, M.R. Hansen and R. Sharp, *A timed semantics for SDL*, FORTE IX: Theory, Application and Tools, R. Gotzhein and J. Bredereke, eds, Chapman & Hall (1997), 295–309.
[25] X. Nicollin and J. Sifakis, *The algebra of timed processes ATP: Theory and application*, Inform. Comput. **114** (1994), 131–178.
[26] L. Pruitt, 1994. Personal Communications.
[27] M.A. Reniers and J.J. Vereijken, *Completeness in discrete-time process algebra*, Computing Science Report 96-15, Eindhoven University of Technology, Department of Mathematics and Computing Science (1996).
[28] F.W. Vaandrager, *Algebraic Techniques for Concurrency and their Applications*, Ph.D. thesis, Centre for Mathematics and Computer Science, Amsterdam (1989).
[29] C. Verhoef, *A congruence theorem for structured operational semantics with predicates and negative premises*, Nordic J. Comput. **2** (1995), 274–302.
[30] Zhou Chaochen, C.A.R. Hoare and A.P. Ravn, *A calculus of durations*, Inform. Process. Lett. **40** (1991), 269–276.
[31] *Methods for Testing and Specification (MTS); use of SDL in European telecommunications standards rules for testability and facilitating validation*, ETSI Document ETS 300 414 (1995).
[32] *Specification of Abstract Syntax Notation One (ASN.1)*, ITU-T Recommendation X.208 (1988).
[33] *Specification and Description Language (SDL)*, ITU-T Recommendation Z.100 (1994).
[34] *Specification and Description Language (SDL) – SDL formal definition: Static semantics*, ITU-T Recommendation Z.100 F2 (1994). Annex F.2 to Recommendation Z.100.
[35] *Specification and Description Language (SDL) – SDL formal definition: Dynamic semantics*, ITU-T Recommendation Z.100 F3 (1994). Annex F.3 to Recommendation Z.100.
[36] *SDL combined with ASN.1 (SDL/ASN.1)* ITU-T Recommendation Z.105 (1995).
[37] *SDT user's guide*, TeleLOGIC AB, Sweden (1992).

Subject index

CHAPTER 19

A Process Algebra for Interworkings

S. Mauw, M.A. Reniers

Faculty of Mathematics and Computing Science, Eindhoven University of Technology, P.O. Box 513, NL-5600 MB Eindhoven, The Netherlands

CWI, P.O. Box 94079, NL-1090 GB Amsterdam, The Netherlands

Emails: sjouke@win.tue.nl, M.A.Reniers@tue.nl

Contents

HANDBOOK OF PROCESS ALGEBRA
Edited by Jan A. Bergstra, Alban Ponse and Scott A. Smolka

Abstract

The Interworking language (IW) is a graphical formalism for displaying the communication behaviour of system components. In this chapter, we develop a formal semantics for the Interworking language. This semantics must support the analysis of (collections of) Interworking diagrams and allow to express the relation between diagrams. We will explain how techniques from process algebra can be successfully applied to this problem. Thereto, we introduce process operators for expressing the relationship between Interworking diagrams. We define a number of process algebras with increasing complexity. For each of these we prove completeness with respect to an operational semantics.

1. Introduction

1.1. *History and motivation*

The Interworking language (IW) is a graphical formalism for displaying the communication behaviour of system components. It was developed in order to support the informal diagrams used at Philips Kommunikations Industrie (Nürnberg) which were used for requirements specification and design. Before discussing the rationale behind the IW language, we first show a simple Interworking diagram[1] in Figure 1. The name of the Interworking is displayed in the upper left corner of the diagram. This Interworking describes the interaction behaviour of three entities, which are called s, *medium* and r. Each entity is represented by a vertical line, which, when read from top to bottom, describes the successive interactions in which this entity takes part. A message exchange is represented by an arrow. The diagram shows that the three entities exchange four messages. First, s sends a *req* message to *medium*. Next, the same message is being sent from *medium* to r. Then, r sends a message *reply* back to *medium*, which sends the same message to s.

This example shows the basic use of Interworkings. It describes one scenario of interaction between communicating entities. In general, when using IW for requirements specification, a collection of Interworkings is needed, containing a description of the most interesting scenarios. Often there is one main scenario, complemented with a number of scenarios describing exceptional behaviour. Using Interworkings in this way, the scenarios express alternative behaviours.

There are, however, more reasons for having to deal with large collections of Interworkings for the description of a distributed system. First, the specified scenario can be too long to physically or logically fit in one diagram. Such a large scenario is then decomposed into a number of sub scenarios which are "sequentially" linked to each other.

A second reason is that the horizontal size of the system, or more precisely the number of distinct entities, may be too large to fit in a single diagram. This gives rise to a collection of sub scenarios which denote the behaviour of different parts of the system. Each part then describes the behaviour of just a number of (logically related) entities. Of course, there must be a means to express that entities from distinct parts exchange messages with each other. The scenarios of these parts are linked to each other in a parallel way.

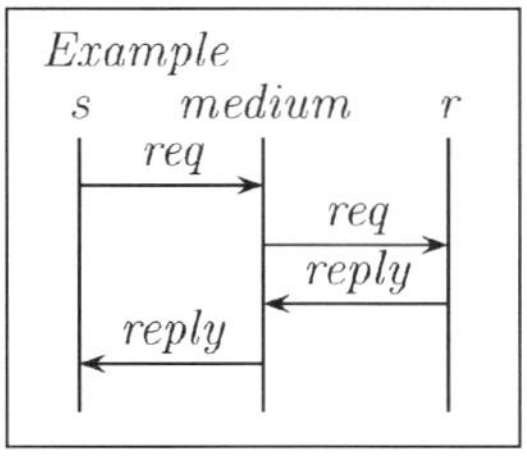

Fig. 1. An example Interworking diagram.

[1] The Interworking diagrams in this chapter are drawn with the *MSC Macro package* which can be obtained at http://www.win.tue.nl/~sjouke/mscpackage.html.

Due to the above mentioned reasons, in practice a system description using Interworkings often consists of a large collection of diagrams. Practical experience showed that it was very hard to maintain such large collections by hand. First of all, manually drawing and updating diagrams is an expensive activity. Secondly, the relation between the diagrams in a collection is only implicit. Some diagrams describe alternatives, some describe successive behaviour, and some describe parallel behaviour. The third problem, assuming the relation between the diagrams to be known, is that if one diagram changes also several related diagrams must be updated. A consistent update of a large collection of Interworkings could not be achieved manually. A final problem was that there existed different interpretations of the meaning of even simple Interworkings.

These observations lead to the conclusion that when using Interworkings in the traditional and informal way it was not possible to take full advantage of the language. Therefore, the Interworking language needs a complete and explicit definition.

Not only the development of an explicit language is motivated in this way, but also the need for a formal semantics of Interworkings. This semantics must support the analysis of (collections of) Interworking diagrams and allow to express the relation between diagrams. Moreover, since tool support is needed, the semantics must allow for easy derivation of (prototype) tools.

In this chapter, we will explain how techniques from process algebra can be successfully applied to this problem. Thereto, we introduce process operators for expressing the relationship between Interworking diagrams. As explained above, there are three possible relations between Interworkings: alternative composition, sequential composition, and parallel composition. The most interesting is the *interworking sequencing* operator for composing Interworkings sequentially. Later in this chapter it will be explained why the standard process algebra operator for sequencing is not appropriate for Interworkings. The operator for parallel composition of Interworkings, is derived from the standard interleaving operator with synchronization. For the alternative composition operator there are different choices. For a discussion on this choice we refer to Section 2.3.

1.2. *Interworkings and similar languages*

The Interworking language is not a unique and isolated language. It is very natural and intuitive to express the behaviour of a distributed system in such a graphical way. In fact, informal IW-like drawings are encountered very often in system design.

Therefore, the Interworking language is a member of a large class of similar graphical notations, most of which are only informally defined, such as Signal Sequence Charts, Use Cases, Information Flow Diagrams, Message Flow and Arrow Diagrams. In object oriented design, a similar notation, called Sequence Diagrams, is used. They play an important role in the description of Use Cases in UML [31]. Interworkings are also related to Message Sequence Charts (MSC), see [12], which are standardized by the International Telecommunication Union (ITU). The main difference is that Interworkings describe synchronous communication, whereas Message Sequence Charts describe asynchronous communication. The semantics of MSC as described in [23,28] is also very similar to the semantics of IW.

Traditionally, the main application area for IW and similar languages is the field of telecommunication systems. This is mainly due to the distributed nature of these systems. However, more and more applications outside the telecommunication world can be found, e.g., the description of work flows in business organizations [1].

The main reason why IW-like diagrams are so popular is the fact that they can be understood easily. This is due to their intuitive and graphical appearance. The diagrams can be used in different stages of the design of a software system. The main application is during requirements engineering, where they are used to capture initial requirements about the interactions in a system. Furthermore, they play a role in documentation, simulation and testing.

The results of this chapter cannot completely be transferred to similar languages. This is mainly because IW describes synchronous communication, whereas most similar languages consider asynchronous communication. Nevertheless, the approach taken in this chapter is generic. It is at the basis of the semantics definition of Message Sequence Charts, as standardized by the ITU in Annex B to Recommendation Z.120 [13].

1.3. *Purpose and structure of this chapter*

This chapter serves several purposes. First, it shows the process algebraic approach in defining the semantics of a scenario language. This typically entails the use of a number of operators which describe the ways in which scenarios or fragments of scenarios are combined. The meaning of such a diagram is then described by a process algebraic expression, which can be analyzed using standard techniques.

Secondly, this chapter shows the development of non-standard operators in process algebra, needed for some domain specific application. These newly introduced operators will probably have little application outside the realm of scenarios. On the other hand, the interworking sequencing operator already received attention in a more general context, and was named *weak sequential composition* (see [29]).

Thirdly, we show in detail which (proof) obligations occur when introducing new operators. We both give an operational and an algebraic definition, and prove their correspondence.

This chapter is subdivided as follows.

First, we will introduce the Interworking language and the operators for combining Interworkings (Section 2). Next, we formally define the operators involved. We will not simply give one process algebra containing all operators, but we will formalize the operators in a modular way. This yields a collection of process algebras, for which we obtain some additional proof obligations, such as conservativity. The first process algebra (defined in Section 3) only contains the operator for sequential composition. This operator suffices to give a formal semantics of Interworking diagrams. In Section 4 we define the theory of the basic process algebra operators ($+$ and $\cdot$) which we enrich with partial deadlocks. Next, in Section 5, these process algebras are combined. The following two sections deal with the introduction of the interworking merge operator. In Section 6 we first define a parameterized version of this operator, the E-interworking merge. The general interworking merge operator is defined in Section 7, which yields the final process algebra for Interworkings.

Every operator is both defined algebraically and by means of an operational semantics. The relation between these descriptions is given in several soundness and completeness theorems.

The treatment of Interworkings in the current chapter is mainly on a theoretic level. We will not introduce graphical and linear syntax of the language, and we will not present a mapping from Interworking diagrams to process algebra expressions (for a thorough treatment see [24]). Our main goal is to define the theory needed to formally understand Interworkings. Neither will we explain methodological aspects of the use of Interworkings or supporting tools. For a description of a prototype tool set based on these semantical definitions, we refer to [26].

2. Interworkings

An Interworking specification consists of a collection of Interworking diagrams. The relation between these diagrams is defined by means of operators. An Interworking diagram specifies (part of) a single scenario and the operators can be used to compose simple scenarios into more complex scenarios. We consider operators for sequential composition, alternative composition and parallel composition of Interworkings.

In this section we will only give an informal explanation of syntax and semantics of Interworkings. Simple examples show the relevant properties, which are formalized in the sections to come.

We will not give a formal definition of the graphical syntax of Interworkings, since for our purposes an informal and intuitive mapping from Interworkings to the semantical domain suffices. There exists a textual representation of Interworkings too, but we will not discuss this. Consult [24] for more information on this topic.

2.1. *Interworking diagrams*

An example of an Interworking diagram is shown in Figure 2. Such a diagram consists of a number of vertical lines and horizontal arrows, surrounded by a frame. The name of the Interworking diagram (*Co-operation*) is in the upper left corner of the frame. The vertical lines denote the entities of which (part of) the behaviour is being described. Above the lines are the names of these entities. Here we have four entities, called a, b, c, and d.

The arrows denote the exchange of messages between the entities. Interworkings describe synchronous communication, which means that an arrow represents one single event. The order in which the communications take place is also expressed in the diagram. On every entity axis, time runs from top to bottom and the events connected to an entity axis are causally ordered in this way. However, there is no global time axis and the only way to synchronize the behaviour of the entities is by means of a message exchange. So, message k causally precedes message m. And because m precedes o, we have that k also precedes o. Messages k and l are not causally related; they may occur in any order. In our semantical treatment we assume an interleaved model of operation, which means that k and l cannot occur simultaneously.

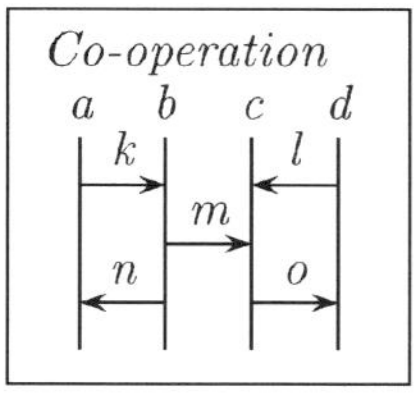

Fig. 2. An example Interworking diagram.

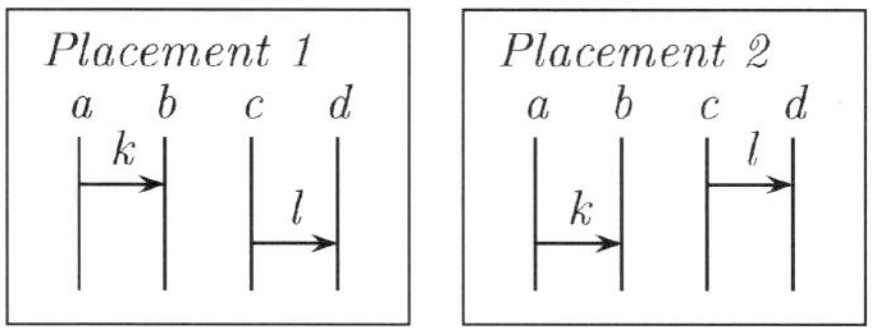

Fig. 3. Two semantically equivalent Interworkings.

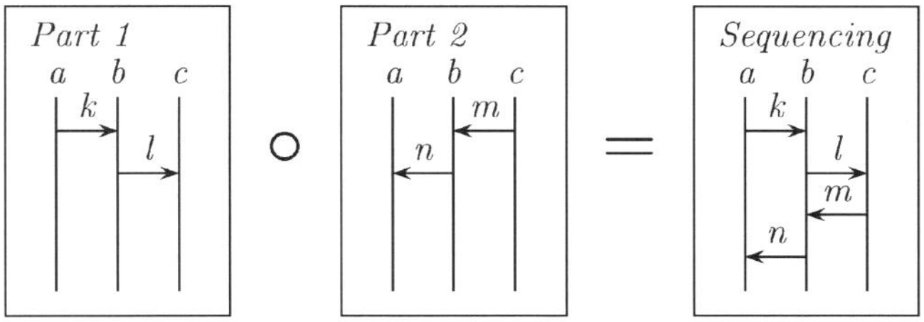

Fig. 4. Sequential composition of two Interworkings.

The fact that the time lines of all entities are independent, implies that the vertical placement of two messages which are not causally related has no semantical meaning. Therefore, the Interworkings from Figure 3 have identical semantics.

A special case in our semantics is the *empty Interworking*. This is an Interworking which describes no behaviour at all and contains no entities. In the next sections the empty Interworking is denoted by ε.

2.2. *Sequencing*

Sequential composition is the easiest way to compose two Interworkings. Intuitively, sequential composition can be considered as the concatenation of two Interworkings, thereby connecting the corresponding entity axes. Figure 4 shows the sequential composition of two Interworkings. The circle denotes the sequencing operator.

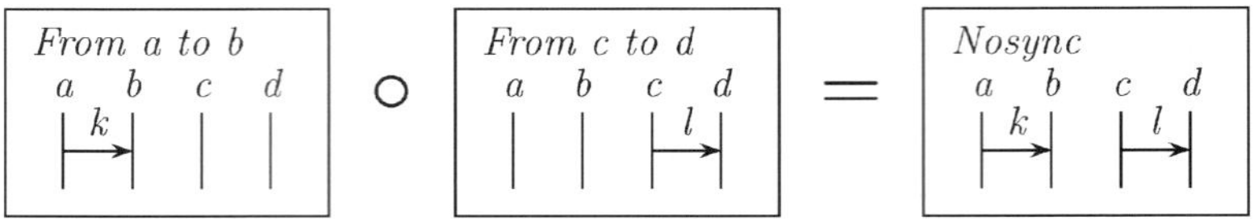

Fig. 5. No synchronization through sequential composition.

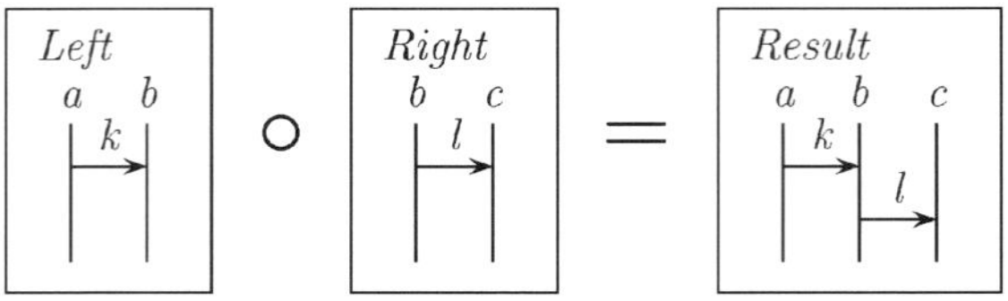

Fig. 6. Sequential composition with different entity sets.

One must take into account that there is no (implicit) synchronization between the entities at the point where the two Interworkings are concatenated. For this reason, the operator for sequential composition of Interworkings is called the *weak* sequential composition operator (or *interworking sequencing*). Although we will also introduce an operator for strong sequential composition of Interworkings in our semantical treatment, this operator is not part of the Interworkings language. Figure 5 shows that the weak sequential composition of two unrelated messages gives an Interworking where these two messages still are unordered.

In the previous examples, the two composed Interworkings contained the same set of entities. This is not a requirement for sequential composition. The Interworking resulting from a sequential composition simply contains all entities from its constituents, as shown in Figure 6.

Given the above interpretation of Interworking diagrams and sequential composition, the following observation is apparent. Every Interworking diagram is equivalent to the sequential composition of all its events. Look, e.g., at Interworking *Co-operation* (Figure 2) which is the sequential composition of five simple Interworking diagrams, each containing one arrow. The order in which these Interworkings are composed should of course correspond to the causal ordering of the original Interworking. So, if K, L, M, N, and O are Interworking diagrams containing the messages k, l, m, n, and o, respectively, then $L \circ K \circ M \circ N \circ O$ would be an example of such an expression. An alternative for this expression is $K \circ L \circ M \circ O \circ N$.

2.3. *Alternatives*

In theoretical approaches to MSC-related languages different operators for alternative composition are used. These are the delayed choice operator ($\mp$, see [3]) and the non-deterministic choice operator ($+$, see [6]). In the standardized semantics of MSC [28] the

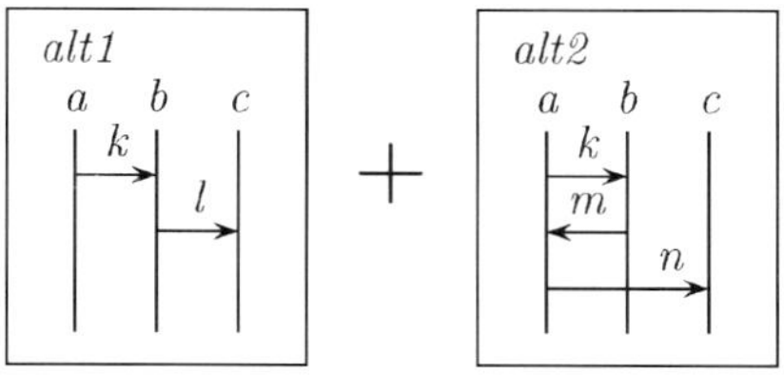

Fig. 7. Alternative composition of two Interworking diagrams.

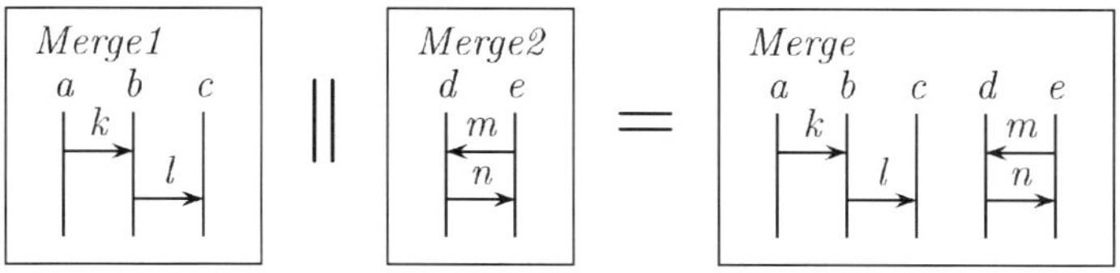

Fig. 8. Merge of Interworking diagrams without shared entities.

delayed choice operator is used. The essential difference between these two operators is that non-deterministic choice determines the moment of choice between the alternatives at the place where it occurs, whereas the delayed choice postpones the moment of choice to the place where the alternatives start to differ. The latter leads to a trace semantics (if non-deterministic choice is not present as well). As a consequence also all other operators in which a choice is manifest (such as parallel composition) must be changed to adopt the delayed interpretation of choices [28]. In our opinion the use of the delayed choice is only interesting if non-deterministic choice is present too. If the delayed choice is the only alternative composition operator of interest, then a better solution is to adopt a trace theoretical approach towards the semantics. In the process algebra approach of this handbook it seems more appropriate to study the non-deterministic choice operator.

Hence, the operator which expresses the fact that two Interworkings describe alternative scenarios is denoted by $+$. Figure 7 contains an example of the choice between two alternative Interworking diagrams. This expression describes the non-deterministic choice between the two given scenarios. Both scenarios start with message k, but the first continues with message l and the second with messages m and n.

Notice that the class of Interworking diagrams is not closed under application of the $+$-operator. The behaviour defined in Figure 7 cannot be expressed without application of the $+$.

2.4. *Merge*

Whereas the sequencing operator is used for vertical composition of Interworkings, the merge operator is used for horizontal composition.

In the case that the two operands have no entities in common, the merge of two Interworking diagrams is simply their juxtaposition, as illustrated in Figure 8.

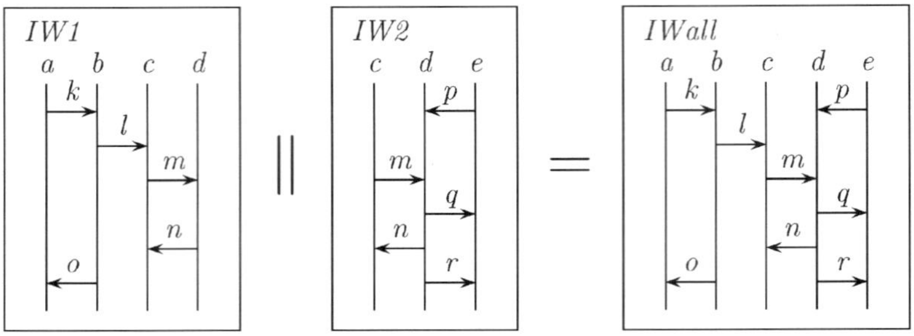

Fig. 9. Merge of Interworking diagrams with two shared entities.

In the case that the two operands do share some entities, the situation is a bit more complicated. Suppose, for example, that the two Interworking diagrams have two entities in common, as in Figure 9. Then the messages exchanged between the shared entities must be identical for both operands. The resulting Interworking contains only one occurrence of every shared entity. Also the messages exchanged between the shared entities, which must occur in the same order in both operands, appear only once in the resulting Interworking. In Figure 9 the two operands share the entities c and d with shared messages m and n.

In the case that the two operands do not describe identical behaviour with respect to the shared entities, as in Figure 10, a *deadlock* occurs. The resulting Interworking contains the parallel behaviour of the operands, up to the point where the behaviours on the shared entities start to diverge. At this point the deadlock occurs, denoted by two horizontal bars. Such a deadlock only covers entities which are blocked. This means that we do not have the *global deadlock* as used elsewhere, but a partial deadlock. We refer to this partial deadlock as deadlock. An entity shows no behaviour after it has entered a deadlock situation. All behaviour which is causally dependent on a communication which causes the deadlock, is also blocked. In Figure 10 this means that, since messages x and n do not match, a deadlock occurs on entities c and d. Moreover, since message r is causally dependent upon message n, the deadlock extends to entity e. In the following sections, such a deadlock will be denoted by δ_E, where E is the set of deadlocked entities. If a deadlock occurs as a consequence of merging two Interworkings, we say that the two operands are *merge-inconsistent*.

This explanation of the merge operator generalizes easily to the case where the operands have more than two entities in common. However, the case where they share only one entity yields a different situation. It is clear that this shared entity should occur only once in the resulting Interworking, but what happens with the events that this entity takes part in? This situation occurs in Figure 11. There is no reason to introduce a causal ordering between the messages l and m, and therefore the result cannot be a single Interworking diagram. The result of the merge in Figure 11 contains two alternative Interworking diagrams, which together describe all possible orderings of l and m.

Care has to be taken to correctly handle entities which are included in an Interworking diagram but which do not take part in any communication, so-called *empty entities*. In the case that such an entity occurs in the set of shared entities, it cannot be discarded. Figure 12

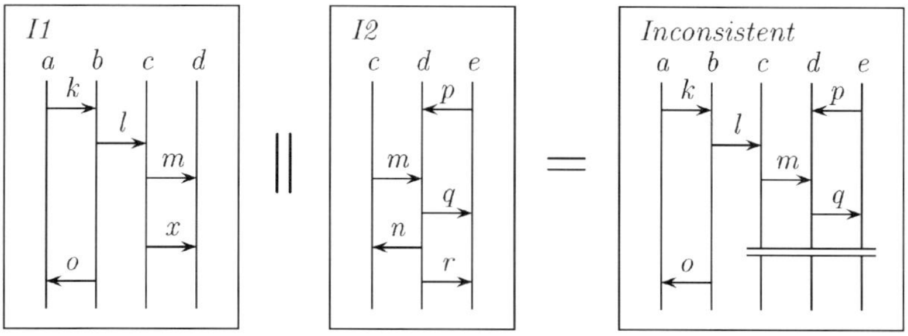

Fig. 10. Merge of two inconsistent Interworking diagrams.

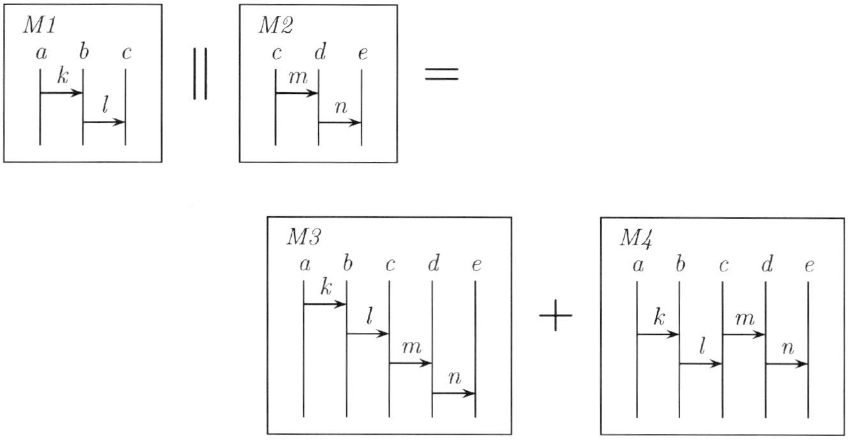

Fig. 11. Consistent merge.

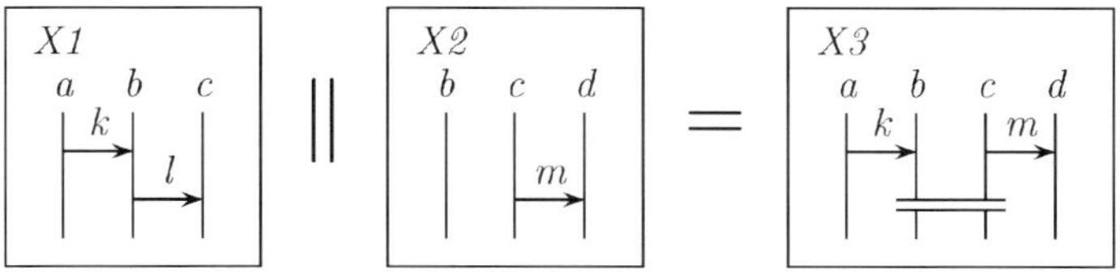

Fig. 12. Inconsistent Interworking diagrams with empty entity.

shows an example. Entity b occurs in both operands, but in the second operand there is no behaviour associated to b. Because in the first operand a message l is sent to b, a deadlock occurs.

The situation would be quite different if we would omit entity b from the second operand. Then the two operands would be merge-consistent. This is shown in Figure 13.

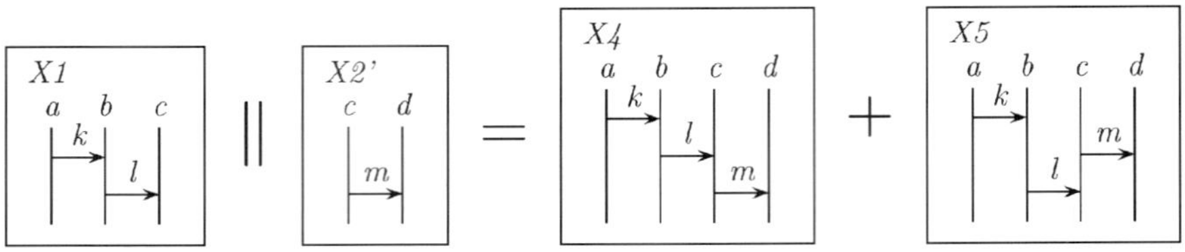

Fig. 13. Empty entity removed.

3. Semantics of Interworkings

In this section we will present a simple process algebra that can be used for reasoning about the equality of Interworking diagrams. Based on the textual syntax of Interworking diagrams a process term is generated as follows. With every message in the Interworking diagram an atomic action is associated. A deadlock that covers the entities from a set E is denoted by δ_E. The atomic actions are combined by means of interworking sequencing. The process algebra is called *IWD*($\circ$).

We assume the existence of sets *EID* and *MID* of names of entities and messages, respectively. Actually, these can be considered as parameters of the process algebra. A message is characterized by the name of the sender, the name of the receiver, and the message name. These messages form the set of atomic actions.

DEFINITION 1 (*Atomic actions*). The set A of *atomic actions* is given by

$$A = \{c(i, j, m) \mid i, j \in EID,\ m \in MID\}.$$

DEFINITION 2 (*Signature of IWD*($\circ$)). The *signature* Σ_{IWD} of the process algebra *IWD*($\circ$) consists of the atomic actions $a \in A$, the deadlock constants δ_E ($E \subseteq EID$), the empty process ε, and the binary operation interworking sequencing $\circ_{\mathrm{iw}}$.

The set of all (open) terms over the signature Σ_{IWD} is denoted as $\mathcal{O}(\Sigma_{IWD})$. The set of all closed terms over the signature Σ_{IWD} is denoted as $\mathcal{C}(\Sigma_{IWD})$. We will use similar notations for other signatures.

We provide the process algebra with an operational semantics by associating a term deduction system to it. We will first summarize the terminology related to term deduction systems. For a formal definition of term deduction systems and related notions we refer to [5]. A term deduction system is a structure (Σ, D) where Σ is a signature and D a set of deduction rules. The set of deduction rules is parameterized by a set of relation symbols and a set of predicate symbols. If P is such a predicate symbol, R such a relation symbol, and $s, t \in \mathcal{O}(\Sigma)$, then the expressions Ps and sRt are called formulas. A deduction rule is of the form $\frac{H}{C}$ where H is a set of formulas, called *hypotheses*, and C is a formula, called the *conclusion*.

In the term deduction systems used in this chapter we use relations $\_ \xrightarrow{-} \_ \subseteq \mathcal{O}(\Sigma) \times A \times \mathcal{O}(\Sigma)$ and the predicate $\_\downarrow \subseteq \mathcal{O}(\Sigma)$. The formula $x \xrightarrow{a} x'$ expresses that the process x can perform an action a and thereby evolves into the process x'. The formula $x \downarrow$ expresses that process x has an option to terminate immediately and successfully.

In the remainder of this chapter we use the following shorthands: $x \xrightarrow{a}$ represents the predicate that $x \xrightarrow{a} x'$ for some x', $x \not\xrightarrow{a} x'$ represents the proposition that $x \xrightarrow{a} x'$ is not derivable from the deduction system, $x \not\xrightarrow{a}$ represents $\neg(x \xrightarrow{a})$, and $x \not\rightarrow$ represents $x \not\xrightarrow{a}$ for all $a \in A$. Similarly we use $x \not\downarrow$ to represent $\neg(x \downarrow)$.

A proof of a formula ϕ is a well-founded upwardly branching tree of which the nodes are labeled by formulas such that the root is labeled by the formula ϕ and if χ is the label of a node and $\{\chi_i \mid i \in I\}$ is the set of labels belonging to the nodes directly above it, then

$$\frac{\{\chi_i \mid i \in I\}}{\chi}$$

is an instantiation of a deduction rule.

The term deduction system for the process algebra *IWD*($\circ$) consists of the signature Σ_{IWD} and the deduction rules given in Table 1.

Before we can give the operational description of the interworking sequencing operator we first define the *active entities* associated with a process term representing an Interworking diagram. The active entities of an Interworking diagram are those entities which are involved in a communication or in a deadlock.

DEFINITION 3 (*Active entities*). For $i, j \in EID$, $m \in MID$, $E \subseteq EID$, and $x, y \in \mathcal{C}(\Sigma_{IWD})$ we define the mapping $AE : \mathcal{C}(\Sigma_{IWD}) \to \mathbb{P}(EID)$ inductively as follows:

$$\begin{aligned} AE\big(c(i, j, m)\big) &= \{i, j\}, \\ AE(\varepsilon) &= \emptyset, \\ AE(\delta_E) &= E, \\ AE(x \circ_{\mathrm{iw}} y) &= AE(x) \cup AE(y). \end{aligned}$$

The operational semantics of the process algebra *IWD*($\circ$) is given by the deduction rules in Table 1 and the equations defining the active entities in Definition 3. These equations can easily be written as deduction rules. The empty process does not execute any actions, but it terminates successfully and immediately. The fact that it does not execute any action is visible by the impossibility of deriving that it can execute an action. The process a can execute the action a and in doing so evolves into the empty process ε. The process δ_E cannot execute any actions nor can it terminate successfully. The interworking sequencing of two processes terminates if and only if both processes can terminate. The process $x \circ_{\mathrm{iw}} y$ executes an action a if x can execute action a or if y can execute a and this action is not related to an active entity of x ($AE(a) \cap AE(x) = \emptyset$). This expresses the intuition that the first operand may always perform its actions, while the second operand may only perform actions which are not blocked because they are causally dependent on actions from the first operand.

The Interworking from Figure 14 can semantically be represented by the process term

$$c(p, q, m) \circ_{\mathrm{iw}} \big(c(r, s, o) \circ_{\mathrm{iw}} c(q, r, n)\big).$$

Table 1
Deduction rules for interworking sequencing ($a \in A$)

$$\frac{}{\varepsilon \downarrow} \qquad \frac{}{a \xrightarrow{a} \varepsilon} \qquad \frac{x \downarrow \quad y \downarrow}{x \circ_{\text{iw}} y \downarrow}$$

$$\frac{x \xrightarrow{a} x'}{x \circ_{\text{iw}} y \xrightarrow{a} x' \circ_{\text{iw}} y} \qquad \frac{AE(a) \cap AE(x) = \emptyset \quad y \xrightarrow{a} y'}{x \circ_{\text{iw}} y \xrightarrow{a} x \circ_{\text{iw}} y'}$$

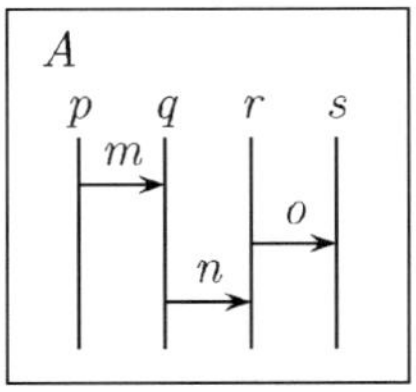

Fig. 14. Example of an Interworking diagram.

Then the following is a derivation of the fact that first the communication of message o can take place:

$$\frac{AE(c(r,s,o)) \cap AE(c(p,q,m)) = \emptyset \qquad \dfrac{c(r,s,o) \xrightarrow{c(r,s,o)} \varepsilon}{c(r,s,o) \circ_{\text{iw}} c(q,r,n) \xrightarrow{c(r,s,o)} \varepsilon \circ_{\text{iw}} c(q,r,n)}}{c(p,q,m) \circ_{\text{iw}} (c(r,s,o) \circ_{\text{iw}} c(q,r,n)) \xrightarrow{c(r,s,o)} c(p,q,m) \circ_{\text{iw}} (\varepsilon \circ_{\text{iw}} c(q,r,n))}.$$

Two processes x and y are considered equivalent if they can mimic each others behaviour in terms of the predicates and relations that are used in the term deduction system. In this case these are the execution of actions, the termination of a process, and the active entities of a process. As a consequence of introducing partial deadlock constants, we must be able to distinguish deadlocks over different sets of entities. This is the reason that we require that two processes are equivalent only if they have the same active entities. This type of equivalence is usually called *strong bisimilarity*, but we call it IWD-bisimilarity.

DEFINITION 4 (*IWD-bisimilarity*). Let Σ be a signature. A symmetric relation $R \subseteq \mathcal{C}(\Sigma) \times \mathcal{C}(\Sigma)$ is called an *IWD-bisimulation* iff for all p, q such that pRq we have

(1) $AE(p) = AE(q)$;

(2) if $p \downarrow$, then $q \downarrow$;

(3) if $p \xrightarrow{a} p'$ for some $a \in A$ and p', then there exists a q' such that $q \xrightarrow{a} q'$ and $p'Rq'$.

Two processes x and y are called *IWD-bisimilar*, notation $x \underline{\leftrightarrow}_{\text{iwd}} y$, iff there exists an IWD-bisimulation R such that xRy. The notation $R : x \underline{\leftrightarrow}_{\text{iwd}} y$ expresses that R is an IWD-bisimulation that relates x and y.

THEOREM 1 (Equivalence). *IWD-bisimilarity is an equivalence relation.*

PROOF. We must prove that IWD-bisimilarity is reflexive, symmetric, and transitive.

(1) $\underline{\leftrightarrow}_{\mathrm{iwd}}$ is reflexive. Let $R = \{(p, p) \mid p \in \mathcal{C}(\Sigma_{IWD})\}$. Clearly, R is an IWD-bisimulation.

(2) $\underline{\leftrightarrow}_{\mathrm{iwd}}$ is symmetric. Suppose that $p \underline{\leftrightarrow}_{\mathrm{iwd}} q$. This means that there exists an IWD-bisimulation R such that pRq. Since any IWD-bisimulation is symmetrical we also have qRp. Hence $q \underline{\leftrightarrow}_{\mathrm{iwd}} p$.

(3) $\underline{\leftrightarrow}_{\mathrm{iwd}}$ is transitive. Suppose $p \underline{\leftrightarrow}_{\mathrm{iwd}} q$ and $q \underline{\leftrightarrow}_{\mathrm{iwd}} r$. Thus there exist IWD-bisimulations R_1 and R_2 such that pR_1q and qR_2r. Let $R = (R_1 \circ R_2)^S$. For a relation ρ on X, the notation ρ^S denotes the symmetric closure of ρ. It is not hard to show that R is an IWD-bisimulation and pRr. Hence $p \underline{\leftrightarrow}_{\mathrm{iwd}} r$. □

THEOREM 2 (Congruence). *IWD-bisimilarity is a congruence for interworking sequencing.*

PROOF. The term deduction system for *IWD*($\circ$) is in path format. From [4], we then have that IWD-bisimilarity is a congruence for interworking sequencing. The path format is a syntactical restriction on the form of the deduction rules and can be easily checked. □

In Table 2 we present the axioms of the process algebra *IWD*($\circ$).

The first three axioms express straightforward properties. The axioms $\circ_{\mathrm{iw}}1$ and $\circ_{\mathrm{iw}}3$ describe the propagation of partial deadlocks through the Interworking diagram. The first of these is illustrated in Figure 15 for $E = \{p, q\}$ and $a = c(q, r, m)$.

For deriving equalities between process terms we can use all instantiations of the axioms and the usual laws of equational logic. These are reflexivity, symmetry, transitivity, and Leibniz's rule.

As a simple example, we present the derivation that the empty process is a right unit for interworking sequencing. The fact that it is a left unit is put forward as an axiom.

LEMMA 1 (Properties). *For $x \in \mathcal{O}(\Sigma_{IWD})$ we have $x \circ_{\mathrm{iw}} \varepsilon = x$.*

PROOF. As $AE(\varepsilon) = \emptyset$, we have $AE(x) \cap AE(\varepsilon) = \emptyset$. Then, using the axioms Comm. $\circ_{\mathrm{iw}}$ and Idem. $\circ_{\mathrm{iw}}$, we have $x \circ_{\mathrm{iw}} \varepsilon = \varepsilon \circ_{\mathrm{iw}} x = x$. □

Thus far we have presented an operational semantics and a process algebra on the signature Σ_{IWD}. Ideally, there is a strong connection between these. In this case we will first show that every pair of derivably equal closed *IWD*($\circ$)-terms is IWD-bisimilar. This relation between an equational theory and its model is usually referred to as *soundness* of the equational theory with respect to the operational semantics. It can also be stated from the point of view of the operational semantics: the set of closed *IWD*($\circ$)-terms modulo IWD-bisimilarity is a *model* of the equational theory. Later we will also present a relation in the other direction: every pair of IWD-bisimilar closed *IWD*($\circ$)-terms is also derivably equal. This result is referred to as *completeness*.

Table 2
Axioms of $IWD(\circ)$ ($a \in A$, $E, F \in EID$)

Idem. $\circ_{iw}$	$\varepsilon \circ_{iw} x = x$	
Comm. $\circ_{iw}$	$x \circ_{iw} y = y \circ_{iw} x$	if $AE(x) \cap AE(y) = \emptyset$
Ass. $\circ_{iw}$	$(x \circ_{iw} y) \circ_{iw} z = x \circ_{iw} (y \circ_{iw} z)$	
$\circ_{iw}1$	$\delta_E \circ_{iw} a = \delta_{E \cup AE(a)}$	if $AE(a) \cap E \neq \emptyset$
$\circ_{iw}3$	$\delta_E \circ_{iw} \delta_F = \delta_{E \cup F}$	

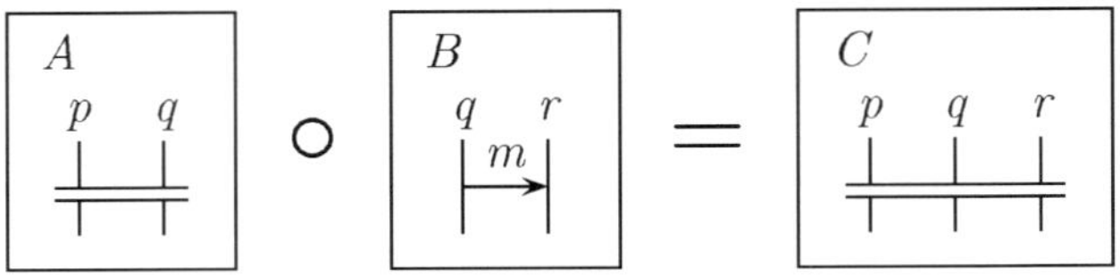

Fig. 15. Propagation of partial deadlocks.

THEOREM 3 (Soundness). *$IWD(\circ)$ is a sound axiomatization of IWD-bisimilarity on closed $IWD(\circ)$-terms.*

PROOF. Due to the congruence of IWD-bisimilarity with respect to all operators from the signature of $IWD(\circ)$, it suffices to prove soundness of all closed instantiations of the axioms in isolation. We give an IWD-bisimulation for each of the axioms. These are the following

- for axiom Idem. $\circ_{iw}$: $R = \{(\varepsilon \circ_{iw} p, p) \mid p \in \mathcal{C}(\Sigma_{IWD})\}^S$;
- for axiom Comm. $\circ_{iw}$: $R = \{(p \circ_{iw} q, q \circ_{iw} p) \mid p, q \in \mathcal{C}(\Sigma_{IWD}), AE(p) \cap AE(q) = \emptyset\}^S$;
- for axiom Ass. $\circ_{iw}$: $R = \{(p \circ_{iw} (q \circ_{iw} r), (p \circ_{iw} q) \circ_{iw} r) \mid p, q, r \in \mathcal{C}(\Sigma_{IWD})\}^S$;
- for axiom $\circ_{iw}1$: $R = \{(\delta_E \circ_{iw} a, \delta_{E \cup AE(a)}) \mid AE(a) \cap E \neq \emptyset\}^S$;
- For axiom $\circ_{iw}3$: $R = \{(\delta_E \circ_{iw} \delta_F, \delta_{E \cup F}) \mid E, F \subseteq EID\}^S$. □

The proof of completeness consists of a number of steps. First we define the notion of basic term and prove that every closed term is derivably equal to some basic term. The introduction of basic terms makes it easier to perform inductive reasoning on the structure of a closed term.

DEFINITION 5 (*Basic terms*). The set of *basic terms* is the smallest set such that

(1) ε is a basic term;
(2) for $E \subseteq EID$, δ_E is a basic term;
(3) for $a \in A$ and x a basic term, $a \circ_{iw} x$ is a basic term.

The set of all basic terms over the signature Σ_{IWD} is denoted $\mathcal{B}(\Sigma_{IWD})$.

THEOREM 4 (Elimination). *For every closed term there is a basic term which is derivably equal.*

PROOF. By induction on the structure of closed term x.

(1) $x \equiv \varepsilon$. This is a basic term.

(2) $x \equiv \delta_E$ for some $E \subseteq EID$. This is a basic term.

(3) $x \equiv a$ for some $a \in A$. Then, using Lemma 1, $a = a \circ_{\text{iw}} \varepsilon$ which is a basic term.

(4) $x \equiv x_1 \circ_{\text{iw}} x_2$ for some $x_1, x_2 \in \mathcal{C}(\Sigma_{IWD})$. By induction we have the existence of basic terms b_1 and b_2 such that $x_1 = b_1$ and $x_2 = b_2$. By induction on the structure of basic term b_1.

(a) $b_1 = \varepsilon$. Then $x = x \circ_{\text{iw}} y = b_1 \circ_{\text{iw}} b_2 = \varepsilon \circ_{\text{iw}} b_2 = b_2$ which is a basic term.

(b) $b_1 = \delta_E$ for some $E \subseteq EID$. By induction on the structure of basic term b_2.

(i) $b_2 = \varepsilon$. Then $x = x_1 \circ_{\text{iw}} x_2 = \delta_E \circ_{\text{iw}} \varepsilon = \delta_E$, which is a basic term.

(ii) $b_2 = \delta_F$ for some $F \subseteq EID$. Then $x = x_1 \circ_{\text{iw}} x_2 = b_1 \circ_{\text{iw}} b_2 = \delta_E \circ_{\text{iw}} \delta_F = \delta_{E \cup F}$, which is a basic term.

(iii) $b_2 = a_2 \circ_{\text{iw}} b_2'$ for some $a_2 \in A$ and $b_2' \in \mathcal{B}(\Sigma_{IWD})$. By induction we have the existence of a basic term c such that $\delta_E \circ_{\text{iw}} b_2' = c$. Also by induction we have the existence of a basic term c' such that $\delta_{E \cup AE(a_2)} \circ_{\text{iw}} b_2' = c'$. If $AE(a_2) \cap E \neq \emptyset$, then $x = x_1 \circ_{\text{iw}} x_2 = b_1 \circ_{\text{iw}} b_2 = \delta_E \circ_{\text{iw}} (a_2 \circ_{\text{iw}} b_2') = (\delta_E \circ_{\text{iw}} a_2) \circ_{\text{iw}} b_2' = \delta_{E \cup AE(a_2)} \circ_{\text{iw}} b_2' = c'$, which is a basic term. If $AE(a) \cap E = \emptyset$, then $x = x_1 \circ_{\text{iw}} x_2 = b_1 \circ_{\text{iw}} b_2 = \delta_E \circ_{\text{iw}} (a_2 \circ_{\text{iw}} b_2') = (\delta_E \circ_{\text{iw}} a_2) \circ_{\text{iw}} b_2' = (a_2 \circ_{\text{iw}} \delta_E) \circ_{\text{iw}} b_2' = a_2 \circ_{\text{iw}} (\delta_E \circ_{\text{iw}} b_2') = a_2 \circ_{\text{iw}} c$ which is a basic term.

(c) $b_1 = a_1 \circ_{\text{iw}} b_1'$ for some $a_1 \in A$ and $b_1' \in \mathcal{B}(\Sigma_{IWD})$. By induction we have the existence of a basic term c such that $b_1' \circ_{\text{iw}} b_2 = c$. Then $x = x_1 \circ_{\text{iw}} x_2 = b_1 \circ_{\text{iw}} b_2 = (a_1 \circ_{\text{iw}} b_1') \circ_{\text{iw}} b_2 = a_1 \circ_{\text{iw}} (b_1' \circ_{\text{iw}} b_2) = a_1 \circ_{\text{iw}} c$, which is a basic term. □

The next step towards the proof of completeness is the following lemma. It provides a link between axiomatic reasoning and reasoning in the (operational) model. The proof of this lemma requires the notion of *norm* of a closed term. It counts the number of actions and sequencing operators occurring in the term.

DEFINITION 6 (*Norm*). For $E \subseteq EID$, $a \in A$, and $x, y \in \mathcal{C}(\Sigma_{IWD})$ we define the mapping $|\_| : \mathcal{C}(\Sigma_{IWD}) \to \mathbb{N}$ inductively as follows:

$$\begin{aligned} |\varepsilon| &= 0, \\ |\delta_E| &= 0, \\ |a| &= 1, \\ |x \circ_{\text{iw}} y| &= |x| + |y| + 1. \end{aligned}$$

LEMMA 2. *For all $x, x' \in \mathcal{C}(\Sigma_{IWD})$ and $a \in A$ we have*

(1) *if $x \xrightarrow{a} x'$, then $|x'| < |x|$;*

(2) *if $x \downarrow$, then $x = \varepsilon$;*

(3) *if $x \xrightarrow{a} x'$, then $x = a \circ_{\text{iw}} x'$;*

(4) *if $x \not\downarrow$, $x \not\rightarrow$, then $x = \delta_{AE(x)}$.*

PROOF.

(1) By induction on the structure of closed term x. Suppose $x \xrightarrow{a} x'$.

(a) $x \equiv \varepsilon$. This case cannot occur.

(b) $x \equiv \delta_E$ for some $E \subseteq EID$. This case cannot occur.

(c) $x \equiv b$ for some $b \in A$. Then necessarily $b \equiv a$ and $x' \equiv \varepsilon$. Observe that

$$|x'| = |\varepsilon| = 0 < 1 = |b| = |x|.$$

(d) $x \equiv x_1 \circ_{\mathrm{iw}} x_2$ for some $x_1, x_2 \in \mathcal{C}(\Sigma_{IWD})$. We can distinguish two cases for $x_1 \circ_{\mathrm{iw}} x_2 \xrightarrow{a} x'$.

(i) $x_1 \xrightarrow{a} x_1'$ for some $x_1' \in \mathcal{C}(\Sigma_{IWD})$ such that $x' \equiv x_1' \circ_{\mathrm{iw}} x_2$. By induction we have that $|x_1'| < |x_1|$. Thus we obtain

$$|x'| = |x_1' \circ_{\mathrm{iw}} x_2| = |x_1'| + |x_2| + 1 < |x_1| + |x_2| + 1 = |x_1 \circ_{\mathrm{iw}} x_2| = |x|.$$

(ii) $AE(a) \cap AE(x_1) = \emptyset$ and $x_2 \xrightarrow{a} x_2'$ for some $x_2' \in \mathcal{C}(\Sigma_{IWD})$ such that $x' \equiv x_1 \circ_{\mathrm{iw}} x_2'$. By induction we have that $|x_2'| < |x_2|$. Thus we obtain

$$|x'| = |x_1 \circ_{\mathrm{iw}} x_2'| = |x_1| + |x_2'| + 1 < |x_1| + |x_2| + 1 = |x_1 \circ_{\mathrm{iw}} x_2| = |x|.$$

(2) By induction on the structure of closed term x. Suppose $x\downarrow$.

(a) $x \equiv \varepsilon$. Trivial.

(b) $x \equiv \delta_E$ for some $E \subseteq EID$. This case cannot occur.

(c) $x \equiv a$ for some $a \in A$. This case cannot occur.

(d) $x \equiv x_1 \circ_{\mathrm{iw}} x_2$ for some $x_1, x_2 \in \mathcal{C}(\Sigma_{IWD})$. As $x\downarrow$, we have $x_1\downarrow$ and $x_2\downarrow$. By induction we then have $x_1 = \varepsilon$ and $x_2 = \varepsilon$. Then $x = x_1 \circ_{\mathrm{iw}} x_2 = \varepsilon \circ_{\mathrm{iw}} \varepsilon = \varepsilon$.

(3) By induction on the structure of closed term x. Suppose $x \xrightarrow{a} x'$.

(a) $x \equiv \varepsilon$. This case cannot occur.

(b) $x \equiv \delta_E$ for some $E \subseteq EID$. This case cannot occur.

(c) $x \equiv b$ for some $b \in A$. Then necessarily $b \equiv a$ and $x' \equiv \varepsilon$. Then,

$$x = b = a = a \circ_{\mathrm{iw}} \varepsilon = a \circ_{\mathrm{iw}} x'.$$

(d) $x \equiv x_1 \circ_{\mathrm{iw}} x_2$ for some $x_1, x_2 \in \mathcal{C}(\Sigma_{IWD})$. For $x_1 \circ_{\mathrm{iw}} x_2 \xrightarrow{a} x'$ two cases can be distinguished:

(i) $x_1 \xrightarrow{a} x_1'$ for some $x_1' \in \mathcal{C}(\Sigma_{IWD})$ such that $x' \equiv x_1' \circ_{\mathrm{iw}} x_2$. By induction we then have $x_1 = a \circ_{\mathrm{iw}} x_1'$. Then,

$$x = x_1 \circ_{\mathrm{iw}} x_2 = \left(a \circ_{\mathrm{iw}} x_1'\right) \circ_{\mathrm{iw}} x_2 = a \circ_{\mathrm{iw}} \left(x_1' \circ_{\mathrm{iw}} x_2\right) = a \circ_{\mathrm{iw}} x'.$$

(ii) $x_2 \xrightarrow{a} x_2'$ and $AE(a) \cap AE(x_1) = \emptyset$ for some $x_2' \in \mathcal{C}(\Sigma_{IWD})$ such that $x' \equiv x_1 \circ_{\mathrm{iw}} x_2'$. By induction we have $x_2 = a \circ_{\mathrm{iw}} x_2'$. Then,

$$\begin{aligned} x &= x_1 \circ_{\mathrm{iw}} x_2 = x_1 \circ_{\mathrm{iw}} \left(a \circ_{\mathrm{iw}} x_2'\right) = (x_1 \circ_{\mathrm{iw}} a) \circ_{\mathrm{iw}} x_2' \\ &= (a \circ_{\mathrm{iw}} x_1) \circ_{\mathrm{iw}} x_2' = a \circ_{\mathrm{iw}} \left(x_1 \circ_{\mathrm{iw}} x_2'\right) = a \circ_{\mathrm{iw}} x'. \end{aligned}$$

(4) By induction on $|x|$ and case analysis on the structure of x. Suppose $x \not\downarrow$ and $x \not\rightarrow$.

(a) $x \equiv \varepsilon$. This case cannot occur.

(b) $x \equiv \delta_E$ for some $E \subseteq EID$. Trivial as $AE(x) = AE(\delta_E) = E$ and $x = \delta_E = \delta_{AE(x)}$.

(c) $x \equiv b$ for some $b \in A$. This case cannot occur as $b \xrightarrow{b}$ contradicts the assumption that $x \not\rightarrow$.

(d) $x \equiv x_1 \circ_{\mathrm{iw}} x_2$ for some $x_1, x_2 \in \mathcal{C}(\Sigma_{IWD})$. If $x_1 \downarrow$ then we find $x_1 = \varepsilon$. As $x \not\downarrow$, we also find $x_2 \not\downarrow$. As $x_1 \circ_{\mathrm{iw}} x_2 \not\rightarrow$, we find $x_1 \not\rightarrow$, and $x_2 \stackrel{a}{\not\rightarrow} \vee AE(a) \cap AE(x_1) \neq \emptyset$ for all $a \in A$. As $x_1 = \varepsilon$, we find $AE(a) \cap AE(x_1) = AE(a) \cap AE(\varepsilon) = \varepsilon$. Therefore, we must have $x_2 \stackrel{a}{\not\rightarrow}$. By induction (note that $|x_2| < |x|$) we thus have $x_2 = \delta_{AE(x_2)}$. Then

$$x = x_1 \circ_{\mathrm{iw}} x_2 = \varepsilon \circ_{\mathrm{iw}} \delta_{AE(x_2)} = \delta_{AE(x_2)} = \delta_{AE(x_1) \cup AE(x_2)} = \delta_{AE(x)}.$$

If $x_1 \not\downarrow$, then we have by induction $x_1 = \delta_{AE(x_1)}$ as we also have $x_1 \stackrel{a}{\not\rightarrow}$. First, suppose that $x_2 \downarrow$. Then $x_2 = \varepsilon$ and we obtain $x = x_1 \circ_{\mathrm{iw}} x_2 = \delta_{AE(x_1)} \circ_{\mathrm{iw}} \varepsilon = \delta_{AE(x_1)} = \delta_{AE(x_1) \cup AE(x_2)} = \delta_{AE(x)}$. Second, suppose $x_2 \not\downarrow$. Again we can distinguish two cases:

(i) $x_2 \stackrel{a}{\not\rightarrow}$ for all $a \in A$. As $|x_2| < |x|$, we can apply the induction hypothesis and obtain $x_2 = \delta_{AE(x_2)}$. Thus,

$$x = x_1 \circ_{\mathrm{iw}} x_2 = \delta_{AE(x_1)} \circ_{\mathrm{iw}} \delta_{AE(x_2)} = \delta_{AE(x_1) \cup AE(x_2)} = \delta_{AE(x)}.$$

(ii) $x_2 \xrightarrow{a} x_2'$ for some $a \in A$. Then we have $x_2 = a \circ_{\mathrm{iw}} x_2'$. As $x_1 \circ_{\mathrm{iw}} x_2 \stackrel{a}{\not\rightarrow}$, we must have $AE(a) \cap AE(x_1) \neq \emptyset$. Then,

$$\begin{aligned} x &= x_1 \circ_{\mathrm{iw}} x_2 = \delta_{AE(x_1)} \circ_{\mathrm{iw}} \left(a \circ_{\mathrm{iw}} x_2'\right) = (\delta_{AE(x_1)} \circ_{\mathrm{iw}} a) \circ_{\mathrm{iw}} x_2' \\ &= \delta_{AE(x_1) \cup AE(a)} \circ_{\mathrm{iw}} x_2'. \end{aligned}$$

Note that $|x_2'| < |x_2|$. Observe that

$$\begin{aligned} \left|\delta_{AE(x_1) \cup AE(a)} \circ_{\mathrm{iw}} x_2'\right| &= \left|x_2'\right| + 1 < |x_2| + 1 \leqslant |x_1| + |x_2| + 1 \\ &= |x_1 \circ_{\mathrm{iw}} x_2|. \end{aligned}$$

Hence we can apply the induction hypothesis to obtain

$$\begin{aligned} \delta_{AE(x_1) \cup AE(a)} \circ_{\mathrm{iw}} x_2' &= \delta_{AE(x_1) \cup AE(a) \cup AE(x_2')} = \delta_{AE(x_1) \cup AE(x_2)} \\ &= \delta_{AE(x)}. \end{aligned}$$ □

THEOREM 5 (Completeness). *IWD(∘) is a complete axiomatization of IWD-bisimilarity on closed IWD(∘)-terms.*

PROOF. Suppose that $x \leftrightarroweq_{\mathrm{iwd}} y$. Then we must prove that $x = y$. By the elimination theorem we have the existence of a basic term x' such that $x = x'$. As the axioms are sound,

we also have $x \underline{\leftrightarrow}_{iwd} x'$. Hence it suffices to prove $x' = y$. We do this by induction on the structure of basic term x'.
(1) $x' = \varepsilon$. Then $x' \downarrow$. Since $x' \underline{\leftrightarrow}_{iwd} y$, we also have $y \downarrow$. By Lemma 2.2 we then have $y = \varepsilon$. Hence $x' = \varepsilon = y$.
(2) $x' = \delta_E$ for some $E \subseteq EID$. Then $x' \not\downarrow$ and $x' \xrightarrow{a} \!\!\!/\,$ for all $a \in A$. As $x' \underline{\leftrightarrow}_{iwd} y$, also $y \not\downarrow$ and $y \xrightarrow{a} \!\!\!/\,$. We also have $AE(y) = AE(x') = AE(\delta_E) = E$. By Lemma 2.2 we have $y = \delta_E$. So $x' = \delta_E = y$.
(3) $x' = a \circ_{iw} x''$ for some $a \in A$ and $x'' \in \mathcal{B}(\Sigma_{IWD})$. Then $x' \xrightarrow{a} \varepsilon \circ_{iw} x''$. Since $x' \underline{\leftrightarrow}_{iwd} y$ we also have $y \xrightarrow{a} y'$ for some y' such that $\varepsilon \circ_{iw} x'' \underline{\leftrightarrow}_{iwd} y'$. Then, using transitivity of IWD-bisimilarity and the soundness of Idem. $\circ_{iw}$, also $x'' \underline{\leftrightarrow}_{iwd} y'$. By induction we then have $x'' = y'$. By Lemma 2.2 we have $y = a \circ_{iw} y'$. Then $x' = a \circ_{iw} x'' = a \circ_{iw} y' = y$. □

4. Sequential and alternative composition

In the previous section we have defined a sound and complete axiomatization of Interworking diagrams. For this purpose we needed to introduce the interworking sequencing operator only. If we want to extend this theory with other operators, we first have to introduce the Basic Process Algebra operators $+$ and $\cdot$. This section is devoted to the development of the process algebra $BPA(\delta_E)$ without interworking sequencing. In the next section, the interworking sequencing is added to this algebra.

The $+$ is called *alternative composition* and $\cdot$ is called *sequential composition*. The process $x + y$ can execute either process x or process y, but not both. The process $x \cdot y$ starts executing process x, and upon termination thereof starts the execution of process y. Operationally these operators are described by the deduction rules given in Table 3. In this table we assume that $a \in A$ and $E \subseteq EID$. The theory presented in this section is very similar to standard Basic Process Algebra with deadlock and empty process $BPA_{\delta\varepsilon}$ (see, e.g., [5]).

DEFINITION 7 (*Active entities*). For $i, j \in EID$, $m \in MID$, $E \subseteq EID$, $x, y \in \mathcal{C}(\Sigma_{BPA(\delta_E)})$, and $\odot \in \{+, \cdot\}$, we define the mapping $AE : \mathcal{C}(\Sigma_{BPA(\delta_E)}) \to \mathbb{P}(EID)$ inductively as follows:

$$AE(c(i, j, m)) = \{i, j\},$$
$$AE(\varepsilon) = \emptyset,$$
$$AE(\delta_E) = E,$$
$$AE(x \odot y) = AE(x) \cup AE(y).$$

The alternative composition of two terms can terminate if either one of these terms can terminate. It can perform every action that its operands can perform, but by doing so a choice is made. A sequential composition can terminate if both operands can terminate. It can perform all actions from its first operand and if the first operand can terminate, it can perform the actions from the second operand.

Again, we first need to prove that IWD-bisimilarity is a congruence for all operators in the process algebra.

Table 3
Deduction rules for alternative and sequential composition ($a \in A$)

$$\frac{x\downarrow}{x+y\downarrow} \qquad \frac{y\downarrow}{x+y\downarrow} \qquad \frac{x\downarrow \quad y\downarrow}{x\cdot y\downarrow}$$

$$\frac{x \xrightarrow{a} x'}{x+y \xrightarrow{a} x'} \qquad \frac{y \xrightarrow{a} y'}{x+y \xrightarrow{a} y'} \qquad \frac{x \xrightarrow{a} x'}{x\cdot y \xrightarrow{a} x'\cdot y} \qquad \frac{x\downarrow \quad y \xrightarrow{a} y'}{x\cdot y \xrightarrow{a} y'}$$

Table 4
Axioms of alternative and sequential composition ($E \subseteq EID$)

A1	$x + y = y + x$	
A2	$(x + y) + z = x + (y + z)$	
A3	$x + x = x$	
A4	$(x + y) \cdot z = x \cdot z + y \cdot z$	
A5	$(x \cdot y) \cdot z = x \cdot (y \cdot z)$	
A6	$x + \delta_E = x$	if $E \subseteq AE(x)$
A7	$\delta_E \cdot x = \delta_{E \cup AE(x)}$	
A8	$\varepsilon \cdot x = x$	
A9	$x \cdot \varepsilon = x$	

THEOREM 6 (Congruence). *IWD-bisimilarity is a congruence for alternative composition and sequential composition.*

PROOF. The term deduction system for $BPA(\delta_E)$ is in path format. From [4], we then have that IWD-bisimilarity is a congruence for all operators. □

These operators are axiomatized by the axioms from Table 4. In these axioms the variables x, y and z denote arbitrary process terms. In order to reduce the number of parentheses in processes we have the following priorities on operators: unary operators bind stronger that binary operators; $\cdot$ binds stronger than all other binary operators and $+$ binds weaker than all other binary operators.

Axioms A1–A5 express straightforward properties, such as commutativity, associativity, and idempotency of alternative composition, distributivity of alternative composition over sequential composition, and associativity of sequential composition. Axioms A6 and A7 characterize the deadlock constant. A6 states that if an entity has the choice between performing an action and deadlocking, it will never deadlock. Axiom A7 expresses that after a deadlock no more actions can occur. The scope of the deadlock is thereby extended to include all entities on which blocked actions occur. Axioms A8 and A9 express the standard behaviour of the empty process.

The proof of soundness is straightforward.

THEOREM 7 (Soundness). *$BPA(\delta_E)$ is a sound axiomatization of IWD-bisimilarity on closed $BPA(\delta_E)$-terms.*

PROOF. In this and other soundness proofs we use I to denote the diagonal relation. If "$s = t$ if b" represents either one of A1–A4, A6–A8, then the relation $R = \{(s,t), (t,s) \mid b\} \cup I$ is an IWD-bisimulation for that axiom. For the axioms A5 and A9 the IWD-bisimulations are given by $R = \{((p \cdot y) \cdot z, p \cdot (y \cdot z)) \mid p \in \mathcal{C}(\Sigma_{BPA(\delta_E)})\}^S \cup I$ and $R = \{(p \cdot \varepsilon, p) \mid p \in \mathcal{C}(\Sigma_{BPA(\delta_E)})\}^S$, respectively. □

The proof of completeness consists of a number of steps. First we define basic terms and prove the elimination property. Next, we formulate a lemma which relates semantical properties to equational properties, and, finally, we prove completeness.

DEFINITION 8 (*Basic terms*). The set of basic terms is the smallest set that satisfies:
(1) ε is a basic term;
(2) for $E \subseteq EID$, δ_E is a basic term;
(3) for $a \in A$ and x a basic term, $a \cdot x$ is a basic term;
(4) if x and y are basic terms, then $x + y$ is a basic term.
The set of all basic terms of the process algebra $BPA(\delta_E)$ is denoted by $\mathcal{B}(\Sigma_{BPA(\delta_E)})$.

The following lemma expresses that we can always combine multiple deadlock alternatives into one deadlock alternative.

LEMMA 3. *For $E, F \subseteq EID$ we have $\delta_E + \delta_F = \delta_{E \cup F}$.*

PROOF. Consider the following derivation:

$$\begin{aligned}\delta_E + \delta_F &= (\delta_E + \delta_F) + \delta_{E\cup F} = \delta_{E\cup F} + (\delta_E + \delta_F) = (\delta_{E\cup F} + \delta_E) + \delta_F \\ &= \delta_{E\cup F} + \delta_F = \delta_{E\cup F}.\end{aligned}$$ □

As alternative composition is idempotent, commutative and associative, and $\delta_\emptyset$ is a zero element for it, we can define a generalized alternative composition operator. For finite index set I, the notation $\sum_{i \in I} x_i$ represents the alternative composition of the process terms x_i. If $I = \emptyset$, then $\sum_{i \in I} x_i = \delta_\emptyset$. If $I = \{i_1, \ldots, i_n\}$ for $n \geqslant 1$, then

$$\sum_{i \in I} x_i = x_{i_1} + x_{i_2} + \cdots + x_{i_n}.$$

Then we can easily establish that every basic term is of the form

$$\sum_{i \in I} a_i \cdot x_i + \sum_{j \in J} \delta_{E_j} + \sum_{k \in K} \varepsilon$$

for some finite index sets I, J, K, $a_i \in A$, $E_j \subseteq EID$ and basic terms x_i of a similar form. For convenience in proofs to follow we combine the deadlock alternatives into one alternative by using Lemma 3:

$$\sum_{i \in I} a_i \cdot x_i + \delta_E + \sum_{k \in K} \varepsilon,$$

where $E = \bigcup_{j \in J} E_j$. The summand $\sum_{k \in K} \varepsilon$ is only used to indicate presence ($K \neq \emptyset$) or absence ($K = \emptyset$) of a termination option.

THEOREM 8 (Elimination). *For every closed term there is a derivably equal basic term.*

PROOF. We prove this theorem by induction on the structure of closed term x.

(1) $x \equiv \varepsilon$. Trivial as ε is a basic term.

(2) $x \equiv \delta_E$ for some $E \subseteq EID$. Trivial as δ_E is a basic term.

(3) $x \equiv a$ for some $a \in A$. Then $x = a = a \cdot \varepsilon$, which is a basic term.

(4) $x \equiv x_1 \cdot x_2$ for some $x_1, x_2 \in \mathcal{C}(\Sigma_{BPA(\delta_E)})$. By induction we have the existence of basic terms b_1 and b_2 such that $x_1 = b_1$ and $x_2 = b_2$. By induction on the structure of basic term b_1 we will prove that there exists a basic term c such that $b_1 \cdot b_2 = c$.
   (a) $b_1 \equiv \varepsilon$. Then $b_1 \cdot b_2 = \varepsilon \cdot b_2 = b_2$.
   (b) $b_1 \equiv \delta_E$ for some $E \subseteq EID$. Then $b_1 \cdot b_2 = \delta_E \cdot b_2 = \delta_{E \cup AE(b_2)}$.
   (c) $b_1 \equiv a \cdot b_1'$ for some $a \in A$ and $b_1' \in \mathcal{B}(\Sigma_{BPA(\delta_E)})$. By induction we have the existence of basic term c' such that $b_1' \cdot b_2 = c_1'$. Then $b_1 \cdot b_2 = (a \cdot b_1') \cdot b_2 = a \cdot (b_1' \cdot b_2) = a \cdot c_1'$.
   (d) $b_1 \equiv b_1' + b_1''$ for some $b_1', b_1'' \in \mathcal{B}(\Sigma_{BPA(\delta_E)})$. By induction we have the existence of basic terms c' and c'' such that $b_1' \cdot b_2 = c'$ and $b_1'' \cdot b_2 = c_2$. Then $b_1 \cdot b_2 = (b_1' + b_1'') \cdot b_2 = b_1' \cdot b_2 + b_1'' \cdot b_2 = c' + c''$.

   Observe that in each case we have the existence of basic term c such that $b_1 \cdot b_2 = c$. Hence $x = x_1 \cdot x_2 = b_1 \cdot b_2 = c$, which is a basic term.

(5) $x \equiv x_1 + x_2$ for some $x_1, x_2 \in \mathcal{C}(\Sigma_{BPA(\delta_E)})$. By induction we have the existence of basic terms b_1 and b_2 such that $x_1 = b_1$ and $x_2 = b_2$. Then $x = x_1 + x_2 = b_1 + b_2$, which is a basic term. □

LEMMA 4. *For all $x, x' \in \mathcal{C}(\Sigma_{BPA(\delta_E)})$ and $a \in A$ we have*

(1) *if $x \downarrow$, then $x = \varepsilon + x$;*

(2) *if $x \xrightarrow{a} x'$, then $x = a \cdot x' + x$.*

PROOF.

(1) We will prove this by induction on the structure of x. Suppose $x \downarrow$.
   (a) $x \equiv \varepsilon$. Then trivially $x = \varepsilon = \varepsilon + \varepsilon = \varepsilon + x$.
   (b) $x \equiv \delta_E$ for some $E \subseteq EID$. Then we have a contradiction as $\delta_E \not\downarrow$.
   (c) $x \equiv a$ for some $a \in A$. Then we also have a contradiction as $a \not\downarrow$.
   (d) $x \equiv x_1 + x_2$ for some $x_1, x_2 \in \mathcal{C}(\Sigma_{BPA(\delta_E)})$. Then we have $x_1 \downarrow$ or $x_2 \downarrow$. By induction we then have $x_1 = \varepsilon + x_1$ or $x_2 = \varepsilon + x_2$. In both cases we find $x = x_1 + x_2 = \varepsilon + x_1 + x_2 = \varepsilon + x$.
   (e) $x \equiv x_1 \cdot x_2$ for some $x_1, x_2 \in \mathcal{C}(\Sigma_{BPA(\delta_E)})$. Then we have $x_1 \downarrow$ and $x_2 \downarrow$. By induction we then have $x_1 = \varepsilon + x_1$ and $x_2 = \varepsilon + x_2$. Therefore, $x = x_1 \cdot x_2 = (\varepsilon + x_1) \cdot x_2 = \varepsilon \cdot x_2 + x_1 \cdot x_2 = x_2 + x_1 \cdot x_2 = \varepsilon + x_2 + x_1 \cdot x_2 = \varepsilon + \varepsilon \cdot x_2 + x_1 \cdot x_2 = \varepsilon + (\varepsilon + x_1) \cdot x_2 = \varepsilon + x_1 \cdot x_2 = \varepsilon + x$.

(2) We will prove this by induction on the structure of x. Suppose $x \xrightarrow{a} x'$.
   (a) $x \equiv \varepsilon$. Then we have a contradiction as $\varepsilon \not\xrightarrow{a}$.
   (b) $x \equiv \delta_E$ for some $E \subseteq EID$. Then we also have a contradiction as $\delta_E \not\xrightarrow{a}$.

(c) $x \equiv b$ for some $b \in A$. Then necessarily $b \equiv a$ and $x' \equiv \varepsilon$. Hence $x = b = b + b = a + b = a \cdot \varepsilon + b = a \cdot x' + x$.

(d) $x \equiv x_1 + x_2$ for some $x_1, x_2 \in \mathcal{C}(\Sigma_{BPA(\delta_E)})$. Then we have $x_1 \xrightarrow{a} x'$ or $x_2 \xrightarrow{a} x'$. By induction we then have $x_1 = a \cdot x' + x_1$ or $x_2 = a \cdot x' + x_2$. Then $x = x_1 + x_2 = a \cdot x' + x_1 + x_2 = a \cdot x' + x$.

(e) If $x \equiv x_1 \cdot x_2$ for some $x_1, x_2 \in \mathcal{C}(\Sigma_{BPA(\delta_E)})$. We can distinguish two cases.
First, $x_1 \xrightarrow{a} x_1'$ for some $x_1' \in \mathcal{C}(\Sigma_{BPA(\delta_E)})$ such that $x' \equiv x_1' \cdot x_2$. By induction we then have $x_1 = a \cdot x_1' + x_1$. Therefore, $x = x_1 \cdot x_2 = (a \cdot x_1' + x_1) \cdot x_2 = (a \cdot x_1') \cdot x_2 + x_1 \cdot x_2 = a \cdot (x_1' \cdot x_2) + x = a \cdot x' + x$.
Second, $x_1 \downarrow$ and $x_2 \xrightarrow{a} x'$. By induction we have $x_2 = a \cdot x' + x_2$. From the first part of this lemma we have $x_1 = \varepsilon + x_1$. Therefore, $x = x_1 \cdot x_2 = (\varepsilon + x_1) \cdot x_2 = \varepsilon \cdot x_2 + x_1 \cdot x_2 = x_2 + x_1 \cdot x_2 = a \cdot x' + x_2 + x_1 \cdot x_2 = a \cdot x' + \varepsilon \cdot x_2 + x_1 \cdot x_2 = a \cdot x' + (\varepsilon + x_1) \cdot x_2 = a \cdot x' + x_1 \cdot x_2 = a \cdot x' + x$. □

THEOREM 9 (Completeness). *$BPA(\delta_E)$ is a complete axiomatization of IWD-bisimilarity on closed $BPA(\delta_E)$-terms.*

PROOF. Suppose that $x \underline{\leftrightarrow}_{iwd} y$. By the elimination theorem and the soundness of the axioms we can assume, without loss of generality, that x is a basic term. By congruence and the soundness of axiom A3 it suffices to prove that if $x + y \underline{\leftrightarrow}_{iwd} y$ then $x + y = y$. This can be seen as follows. From $x \underline{\leftrightarrow}_{iwd} y$ we obtain $x + y \underline{\leftrightarrow}_{iwd} y + y$ using congruence of $\underline{\leftrightarrow}_{iwd}$ with respect to $+$, and reflexivity of $\underline{\leftrightarrow}_{iwd}$. Using the soundness of axiom A3 we have $y + y \underline{\leftrightarrow}_{iwd} y$. By transitivity of $\underline{\leftrightarrow}_{iwd}$ we obtain $x + y \underline{\leftrightarrow}_{iwd} y$. Then $x + y = y$. Similarly we can obtain $y + x = x$. Therefore, $x = y + x = x + y = y$.

We prove this by induction on the structure of basic term x.

(1) $x \equiv \varepsilon$. Then $x \downarrow$. So $x + y \downarrow$. Therefore, $y \downarrow$. Then, by Lemma 4.1, we have $y = \varepsilon + y$. So we find $x + y = \varepsilon + y = y$.

(2) $x \equiv \delta_E$. Then $AE(x + y) = AE(\delta_E + y) = E \cup AE(y)$. As $x + y \underline{\leftrightarrow}_{iwd} y$, we must also have $AE(y) = AE(x + y) = E \cup AE(y)$. Thus we obtain $E \subseteq AE(y)$. Then $x + y = \delta_E + y = y + \delta_E = y$.

(3) $x \equiv a \cdot x'$. Then $x \xrightarrow{a} \varepsilon \cdot x'$. So $x + y \xrightarrow{a} \varepsilon \cdot x'$. Therefore, $y \xrightarrow{a} y'$ for some y' such that $\varepsilon \cdot x' \underline{\leftrightarrow}_{iwd} y'$. By the soundness of axiom A8 we find $x' \underline{\leftrightarrow}_{iwd} y'$. By induction we then have $x' = y'$. By Lemma 4.4 we have $y = a \cdot y' + y$. Then $x + y = a \cdot x' + y = a \cdot y' + y = y$.

(4) $x \equiv x_1 + x_2$. Using $x_1 + x_2 + y \underline{\leftrightarrow}_{iwd} y$ implies $x_1 + y \underline{\leftrightarrow}_{iwd} y$ and $x_2 + y \underline{\leftrightarrow}_{iwd} y$. By induction we then have $x_1 + y = y$ and $x_2 + y = y$. Then $x + y = (x_1 + x_2) + y = x_1 + (x_2 + y) = x_1 + y = y$. □

5. The interworking sequencing

In Section 3 we have introduced the interworking sequencing and in Section 4 we have defined alternative and sequential composition operators. When combining these operators into one single theory, we need to express the relation between the interworking sequencing

Table 5
Deduction rules for auxiliary operators for interworking sequencing ($a \in A$)

$$\frac{x\downarrow \quad y\downarrow}{x\ \mathbf{L}{\circ_{\mathrm{iw}}}\ y\downarrow} \qquad \frac{x\downarrow \quad y\downarrow}{x\ \mathbf{R}{\circ_{\mathrm{iw}}}\ y\downarrow}$$

$$\frac{x \xrightarrow{a} x'}{x\ \mathbf{L}{\circ_{\mathrm{iw}}}\ y \xrightarrow{a} x' \circ_{\mathrm{iw}} y} \qquad \frac{AE(a) \cap AE(x) = \emptyset \quad y \xrightarrow{a} y'}{x\ \mathbf{R}{\circ_{\mathrm{iw}}}\ y \xrightarrow{a} x \circ_{\mathrm{iw}} y'}$$

on the one hand and alternative and sequential composition on the other hand. By introducing the auxiliary operators $\mathbf{L}{\circ_{\mathrm{iw}}}$ and $\mathbf{R}{\circ_{\mathrm{iw}}}$, we can express the interworking sequencing in terms of the alternative and sequential composition operators. The process algebra obtained in this way is called $IWD(\circ, \cdot, +)$. It is a conservative extension of both the process algebra $IWD(\circ)$ from Section 3 and the process algebra $BPA(\delta_E)$ from Section 4. Furthermore, all axioms formulated for interworking sequencing in the theory $IWD(\circ)$ can now be derived for closed terms.

The intuition of the auxiliary operators is as follows. The process $x\ \mathbf{L}{\circ_{\mathrm{iw}}}\ y$ behaves like the process $x \circ_{\mathrm{iw}} y$ with the restriction that the first action to be executed must originate from process x. The process $x\ \mathbf{R}{\circ_{\mathrm{iw}}} y$ also behaves like the process $x \circ_{\mathrm{iw}} y$ but this time with the restriction that the first action to be executed must be from process y. In this case, the first action from y can only be executed if it is not blocked by any of the actions from x.

These definitions resemble the use of the left-merge operator in *PA* to define the merge operator. That we need two auxiliary operators instead of one is caused by the fact that interworking sequencing is not commutative.

DEFINITION 9 (*Active entities*). For $i, j \in EID$, $m \in MID$, $E \subseteq EID$, $x, y \in \mathcal{C}(\Sigma_{IWD(\circ,\cdot,+)})$, and $\odot \in \{\circ_{\mathrm{iw}}, \mathbf{L}{\circ_{\mathrm{iw}}}, \mathbf{R}{\circ_{\mathrm{iw}}}, +, \cdot\}$, we define the mapping $AE : \mathcal{C}(\Sigma_{IWD(\circ,\cdot,+)}) \to \mathbb{P}(EID)$ inductively as follows:

$$\begin{aligned} AE\big(c(i, j, m)\big) &= \{i, j\}, \\ AE(\varepsilon) &= \emptyset, \\ AE(\delta_E) &= E, \\ AE(x \odot y) &= AE(x) \cup AE(y). \end{aligned}$$

The operational semantics of the interworking sequencing is given already in Table 1. The operational semantics of the auxiliary operators is given in Table 5. The rules follow from the intuitive explanation above. The termination behaviour of the interworking sequencing is incorporated in both auxiliary operators. This is not necessary but facilitates the axiomatization of these operators and the proof of the auxiliary proposition in the proof of Proposition 1.

THEOREM 10 (Congruence). *IWD-bisimilarity is a congruence for* $\mathbf{L}{\circ_{\mathrm{iw}}}$ *and* $\mathbf{R}{\circ_{\mathrm{iw}}}$.

Table 6
Axioms for interworking sequencing and auxiliary operators ($a \in A$, $E \subseteq EID$)

S	$x \circ_{\mathrm{iw}} y = x \mathbf{L}\!\circ_{\mathrm{iw}} y + x \mathbf{R}\!\circ_{\mathrm{iw}} y$	
L1	$\varepsilon \mathbf{L}\!\circ_{\mathrm{iw}} \varepsilon = \varepsilon$	
L2	$\varepsilon \mathbf{L}\!\circ_{\mathrm{iw}} \delta_E = \delta_E$	
L3	$\varepsilon \mathbf{L}\!\circ_{\mathrm{iw}} a \cdot x = \delta_{AE(a \cdot x)}$	
L4	$\varepsilon \mathbf{L}\!\circ_{\mathrm{iw}} (x + y) = \varepsilon \mathbf{L}\!\circ_{\mathrm{iw}} x + \varepsilon \mathbf{L}\!\circ_{\mathrm{iw}} y$	
L5	$\delta_E \mathbf{L}\!\circ_{\mathrm{iw}} x = \delta_{E \cup AE(x)}$	
L6	$a \cdot x \mathbf{L}\!\circ_{\mathrm{iw}} y = a \cdot (x \circ_{\mathrm{iw}} y)$	
L7	$(x + y) \mathbf{L}\!\circ_{\mathrm{iw}} z = x \mathbf{L}\!\circ_{\mathrm{iw}} z + y \mathbf{L}\!\circ_{\mathrm{iw}} z$	
R1–4	$x \mathbf{R}\!\circ_{\mathrm{iw}} \varepsilon = \varepsilon \mathbf{L}\!\circ_{\mathrm{iw}} x$	
R5	$x \mathbf{R}\!\circ_{\mathrm{iw}} \delta_E = \delta_{E \cup AE(x)}$	
R6a	$x \mathbf{R}\!\circ_{\mathrm{iw}} a \cdot y = a \cdot (x \circ_{\mathrm{iw}} y)$	if $AE(a) \cap AE(x) = \emptyset$
R6b	$x \mathbf{R}\!\circ_{\mathrm{iw}} a \cdot y = \delta_{AE(x) \cup AE(a \cdot y)}$	if $AE(a) \cap AE(x) \neq \emptyset$
R7	$x \mathbf{R}\!\circ_{\mathrm{iw}} (y + z) = x \mathbf{R}\!\circ_{\mathrm{iw}} y + x \mathbf{R}\!\circ_{\mathrm{iw}} z$	

PROOF. The term deduction system is in path format and hence IWD-bisimilarity is a congruence for all operators. □

The axioms defining the interworking sequencing in terms of alternative and sequential composition are given in Table 6. The first axiom, S, states that the first action from $x \circ_{\mathrm{iw}} y$ can either come from x (via the term $x \mathbf{L}\!\circ_{\mathrm{iw}} y$) or from y (via the term $x \mathbf{R}\!\circ_{\mathrm{iw}} y$). Axioms L1–L7 define the operator $\mathbf{L}\!\circ_{\mathrm{iw}}$ using the structure of basic terms. As stated before, $x \mathbf{L}\!\circ_{\mathrm{iw}} y$ behaves like the process $x \circ_{\mathrm{iw}} y$ with the restriction that the first action to be executed comes from x. This is expressed clearly in axiom L6. The relation between $\mathbf{L}\!\circ_{\mathrm{iw}}$ and $\circ_{\mathrm{iw}}$ also explains the distributive law L7 and the absorption law L5. Axioms L1–L4 define the termination behaviour of $\mathbf{L}\!\circ_{\mathrm{iw}}$: $x \mathbf{L}\!\circ_{\mathrm{iw}} y$ can only terminate if both operands can terminate. A deadlock occurs if the left operand is ε and the right operand cannot terminate (axioms L2, L3).

The definition of $\mathbf{R}\!\circ_{\mathrm{iw}}$ is similar. The intuition behind the operator $\mathbf{R}\!\circ_{\mathrm{iw}}$ is that the right operand may only execute actions which are not blocked by the left operand. Therefore, we make a distinction using the condition $AE(a) \cap AE(x) = \emptyset$ (see axioms R6a and R6b).

THEOREM 11 (Soundness). *The axioms given in Table* 6 *are sound with respect to IWD-bisimilarity on closed IWD*$(\circ, \cdot, +)$*-terms.*

PROOF. If "$s = t$ if b" represents either one of S, L1–L5, L7, R1–4, R5, R6b, or R7, then the relation $R = \{(s, t), (t, s) \mid b\} \cup I$ is an IWD-bisimulation for that axiom.

For the axioms L6 and R6a the IWD-bisimulations are given by

$$R = \big\{\big(a \cdot x \mathbf{L}\!\circ_{\mathrm{iw}} y, a \cdot (x \circ_{\mathrm{iw}} y)\big), \big(\varepsilon \cdot x \circ_{\mathrm{iw}} y, \varepsilon \cdot (x \circ_{\mathrm{iw}} y)\big), (\varepsilon \cdot x \circ_{\mathrm{iw}} q, x \circ_{\mathrm{iw}} q) \\ \mid q \in \mathcal{C}(\Sigma_{IWD(\circ, \cdot, +)})\big\}^S \cup I$$

and

$$R = \{(x \; \mathbf{R}\circ_{\mathrm{iw}} a \cdot y, a \cdot (x \circ_{\mathrm{iw}} y)), (x \circ_{\mathrm{iw}} \varepsilon \cdot y, \varepsilon \cdot (x \circ_{\mathrm{iw}} y)), (p \circ_{\mathrm{iw}} \varepsilon \cdot q, p \circ_{\mathrm{iw}} q) \\ \mid p, q \in \mathcal{C}(\Sigma_{IWD(\circ,\cdot,+)}), AE(a) \cap AE(x) = \emptyset\}^S \cup I$$

respectively. □

We will consider basic terms as in Definition 8 of Section 4. To prove the elimination property we will need the following lemma.

LEMMA 5. *For arbitrary basic terms b_1 and b_2 we have the existence of basic terms c_1, c_2 and c_3 such that $b_1 \; \mathbf{L}\circ_{\mathrm{iw}} b_2 = c_1$, $b_1 \; \mathbf{R}\circ_{\mathrm{iw}} b_2 = c_2$ and $b_1 \circ_{\mathrm{iw}} b_2 = c_3$.*

PROOF. These statements are proven simultaneously with induction on the structure of basic terms b_1 and b_2. The details of the proofs are omitted. □

THEOREM 12 (Elimination). *For every closed IWD$(\circ, \cdot, +)$-term x there is a derivably equal basic term s.*

PROOF. This theorem is proven by induction on the structure of closed *IWD*$(\circ, \cdot, +)$-term x. The only interesting cases are the following: $x \equiv x' \; \mathbf{L}\circ_{\mathrm{iw}} x''$, $x \equiv x' \; \mathbf{R}\circ_{\mathrm{iw}} x''$, and $x \equiv x' \circ_{\mathrm{iw}} x''$ for closed *IWD*$(\circ, \cdot, +)$-terms x' and x''. In all cases we find the existence of basic terms b_1 and b_2 such that $x_1 = b_1$ and $x_2 = b_2$. Using the previous lemma we find the desired result. □

Next, we prove that the process algebra *IWD*$(\circ, \cdot, +)$ is a conservative extension of the process algebra *BPA*(δ_E). This means that every equality between closed terms from the signature of *BPA*(δ_E) is also derivable from the process algebra *IWD*$(\circ, \cdot, +)$, and also that in the process algebra *IWD*$(\circ, \cdot, +)$ only those equalities are derivable. The proof of this theorem uses the approach of [34].

THEOREM 13 (Conservativity). *The process algebra IWD$(\circ, \cdot, +)$ is a conservative extension of the process algebra BPA(δ_E).*

PROOF. The conservativity follows from the following observations:
(1) IWD-bisimilarity is definable in terms of predicate and relation symbols only,
(2) *BPA*(δ_E) is a complete axiomatization of IWD-bisimilarity on closed *BPA*(δ_E)-terms,
(3) *IWD*$(\circ, \cdot, +)$ is a sound axiomatization of IWD-bisimilarity on closed *IWD*$(\circ, \cdot, +)$-terms (see Theorem 11),
(4) The term deduction system for *BPA*(δ_E) is pure, well-founded and in path format, and
(5) The term deduction system for *IWD*$(\circ, \cdot, +)$ is in path format. □

THEOREM 14 (Completeness). *The process algebra IWD$(\circ, \cdot, +)$ is a complete axiomatization of IWD-bisimilarity on closed IWD$(\circ, \cdot, +)$-terms.*

PROOF. By the General Completeness Theorem of [33], the completeness of the process algebra $IWD(\circ, \cdot, +)$ follows immediately from the properties mentioned in the proof of Theorem 13 and the fact that $IWD(\circ, \cdot, +)$ has the elimination property for $BPA(\delta_E)$ (see Theorem 8). □

In Section 3 we have given a direct axiomatization of interworking sequencing, while in this section we have expressed interworking sequencing in terms of alternative and sequential composition. We will prove that the axioms used in the direct axiomatization are still valid in the current setting for closed terms.

As a consequence of the fact that IWD-bisimilarity is a congruence for all operators in the signature and the fact that for every closed term there exists a derivably equal basic term, we can prove equalities for closed terms with induction.

PROPOSITION 1 (Commutativity of $\circ_{\mathrm{iw}}$). *For closed $IWD(\circ, \cdot, +)$-terms x and y such that $AE(x) \cap AE(y) = \emptyset$ we have*

$$x \circ_{\mathrm{iw}} y = y \circ_{\mathrm{iw}} x.$$

PROOF. Suppose that $AE(x) \cap AE(y) = \emptyset$. We prove the statements $x\ \mathbf{L}\!\circ_{\mathrm{iw}} y = y\ \mathbf{R}\!\circ_{\mathrm{iw}} x$ and $x \circ_{\mathrm{iw}} y = y \circ_{\mathrm{iw}} x$ simultaneously with induction on the structure of basic terms x and y. First we present the proof of $x\ \mathbf{L}\!\circ_{\mathrm{iw}} y = y\ \mathbf{R}\!\circ_{\mathrm{iw}} x$.

(1) $x \equiv \varepsilon$. Trivial by axiom R1–4.

(2) $x \equiv \delta_E$ for some $E \subseteq EID$. Then $x\ \mathbf{L}\!\circ_{\mathrm{iw}} y = \delta_E\ \mathbf{L}\!\circ_{\mathrm{iw}} y = \delta_{E \cup AE(y)} = y\ \mathbf{R}\!\circ_{\mathrm{iw}} \delta_E = y\ \mathbf{R}\!\circ_{\mathrm{iw}} x$.

(3) $x \equiv a \cdot x'$ for some $a \in A$ and $x' \in \mathcal{B}(\Sigma_{IWD(\circ,\cdot,+)})$. As $AE(a \cdot x') \cap AE(y) = \emptyset$ implies $AE(x') \cap AE(y) = \emptyset$, we have by induction that $x' \circ_{\mathrm{iw}} y = y \circ_{\mathrm{iw}} x'$. Then

$$\begin{aligned} x\ \mathbf{L}\!\circ_{\mathrm{iw}} y &= a \cdot x'\ \mathbf{L}\!\circ_{\mathrm{iw}} y = a \cdot (x' \circ_{\mathrm{iw}} y) = a \cdot (y \circ_{\mathrm{iw}} x') \\ &= y\ \mathbf{R}\!\circ_{\mathrm{iw}} a \cdot x' = y\ \mathbf{R}\!\circ_{\mathrm{iw}} x. \end{aligned}$$

Note that we have also used that $AE(a \cdot x') \cap AE(y) = \emptyset$ implies $AE(a) \cap AE(y) = \emptyset$.

(4) $x \equiv x' + x''$ for some $x', x'' \in \mathcal{B}(\Sigma_{IWD(\circ,\cdot,+)})$. As $AE(x' + x'') \cap AE(y) = \emptyset$ implies $AE(x') \cap AE(y) = \emptyset$ and $AE(x'') \cap AE(y) = \emptyset$, we have by induction $x'\ \mathbf{L}\!\circ_{\mathrm{iw}} y = y\ \mathbf{R}\!\circ_{\mathrm{iw}} x'$ and $x''\ \mathbf{L}\!\circ_{\mathrm{iw}} y = y\ \mathbf{R}\!\circ_{\mathrm{iw}} x''$. Then

$$\begin{aligned} x\ \mathbf{L}\!\circ_{\mathrm{iw}} y &= (x' + x'')\ \mathbf{L}\!\circ_{\mathrm{iw}} y = x'\ \mathbf{L}\!\circ_{\mathrm{iw}} y + x''\ \mathbf{L}\!\circ_{\mathrm{iw}} y = y\ \mathbf{R}\!\circ_{\mathrm{iw}} x' + y\ \mathbf{R}\!\circ_{\mathrm{iw}} x'' \\ &= y\ \mathbf{R}\!\circ_{\mathrm{iw}} (x' + x'') = y\ \mathbf{R}\!\circ_{\mathrm{iw}} x. \end{aligned}$$

Then we have $x \circ_{\mathrm{iw}} y = x\ \mathbf{L}\!\circ_{\mathrm{iw}} y + x\ \mathbf{R}\!\circ_{\mathrm{iw}} y = y\ \mathbf{R}\!\circ_{\mathrm{iw}} x + y\ \mathbf{L}\!\circ_{\mathrm{iw}} x = y \circ_{\mathrm{iw}} x$. □

PROPOSITION 2 (Unit element). *For closed $IWD(\circ, \cdot, +)$-terms x we have*

$$\varepsilon \circ_{\mathrm{iw}} x = x \circ_{\mathrm{iw}} \varepsilon = x.$$

PROOF. First, we prove $\varepsilon \circ_{\mathrm{iw}} x = x$ with induction on the structure of basic term x.

(1) $x \equiv \varepsilon$. Then $\varepsilon \circ_{\mathrm{iw}} x = \varepsilon \circ_{\mathrm{iw}} \varepsilon = \varepsilon \mathbin{\mathbf{L}\!\circ_{\mathrm{iw}}} \varepsilon + \varepsilon \mathbin{\mathbf{R}\!\circ_{\mathrm{iw}}} \varepsilon = \varepsilon + \varepsilon = \varepsilon = x$.

(2) $x \equiv \delta_E$ for some $E \subseteq EID$. Then $\varepsilon \circ_{\mathrm{iw}} x = \varepsilon \circ_{\mathrm{iw}} \delta_E = \varepsilon \mathbin{\mathbf{L}\!\circ_{\mathrm{iw}}} \delta_E + \varepsilon \mathbin{\mathbf{R}\!\circ_{\mathrm{iw}}} \delta_E = \delta_E + \delta_{E \cup AE(\varepsilon)} = \delta_E = x$.

(3) $x \equiv a \cdot x'$ for some $a \in A$ and $x' \in \mathcal{B}(\Sigma_{IWD(\circ,\cdot,+)})$. By induction we have $\varepsilon \circ_{\mathrm{iw}} x' = x'$. Then $\varepsilon \circ_{\mathrm{iw}} x = \varepsilon \circ_{\mathrm{iw}} a \cdot x' = \varepsilon \mathbin{\mathbf{L}\!\circ_{\mathrm{iw}}} a \cdot x' + \varepsilon \mathbin{\mathbf{R}\!\circ_{\mathrm{iw}}} a \cdot x' = \delta_{AE(a \cdot x')} + a \cdot (\varepsilon \circ_{\mathrm{iw}} x') = \delta_{AE(a \cdot x')} + a \cdot x' = a \cdot x'$.

(4) $x \equiv x_1 + x_2$ for some $x_1, x_2 \in \mathcal{B}(\Sigma_{IWD(\circ,\cdot,+)})$. By induction we have $\varepsilon \circ_{\mathrm{iw}} x_1 = x_1$ and $\varepsilon \circ_{\mathrm{iw}} x_2 = x_2$. Then $\varepsilon \circ_{\mathrm{iw}} x = \varepsilon \circ_{\mathrm{iw}} (x_1 + x_2) = \varepsilon \mathbin{\mathbf{L}\!\circ_{\mathrm{iw}}} (x_1 + x_2) + \varepsilon \mathbin{\mathbf{R}\!\circ_{\mathrm{iw}}} (x_1 + x_2) = \varepsilon \mathbin{\mathbf{L}\!\circ_{\mathrm{iw}}} x_1 + \varepsilon \mathbin{\mathbf{L}\!\circ_{\mathrm{iw}}} x_2 + \varepsilon \mathbin{\mathbf{R}\!\circ_{\mathrm{iw}}} x_1 + \varepsilon \mathbin{\mathbf{R}\!\circ_{\mathrm{iw}}} x_2 = \varepsilon \circ_{\mathrm{iw}} x_1 + \varepsilon \circ_{\mathrm{iw}} x_2 = x_1 + x_2 = x$.

Then, using the commutativity of $\circ_{\mathrm{iw}}$ and the fact that $AE(x) \cap AE(\varepsilon) = \emptyset$ we easily find $x \circ_{\mathrm{iw}} \varepsilon = \varepsilon \circ_{\mathrm{iw}} x = x$. □

PROPOSITION 3 (Associativity of $\circ_{\mathrm{iw}}$). *For closed IWD($\circ, \cdot, +$)-terms x, y, and z, we have*

$$(x \circ_{\mathrm{iw}} y) \circ_{\mathrm{iw}} z = x \circ_{\mathrm{iw}} (y \circ_{\mathrm{iw}} z).$$

PROOF. Without loss of generality we can assume that x, y, and z are basic terms. To prove this theorem the following propositions are proven simultaneously with induction on the general form of the basic terms x, y, and z.

$$\begin{aligned}
(x \mathbin{\mathbf{L}\!\circ_{\mathrm{iw}}} y) \mathbin{\mathbf{L}\!\circ_{\mathrm{iw}}} z &= x \mathbin{\mathbf{L}\!\circ_{\mathrm{iw}}} (y \circ_{\mathrm{iw}} z), && (1)\\
(x \mathbin{\mathbf{R}\!\circ_{\mathrm{iw}}} y) \mathbin{\mathbf{L}\!\circ_{\mathrm{iw}}} z &= x \mathbin{\mathbf{R}\!\circ_{\mathrm{iw}}} (y \mathbin{\mathbf{L}\!\circ_{\mathrm{iw}}} z), && (2)\\
(x \circ_{\mathrm{iw}} y) \mathbin{\mathbf{R}\!\circ_{\mathrm{iw}}} z &= x \mathbin{\mathbf{R}\!\circ_{\mathrm{iw}}} (y \mathbin{\mathbf{R}\!\circ_{\mathrm{iw}}} z), && (3)\\
(x \circ_{\mathrm{iw}} y) \circ_{\mathrm{iw}} z &= x \circ_{\mathrm{iw}} (y \circ_{\mathrm{iw}} z). && (4)
\end{aligned}$$

This way of proving associativity of interworking sequencing is similar to the way in which associativity of parallel composition is proven in *ACP*. Similar equations in the setting of *ACP* are usually called the Axioms of Standard Concurrency [6].

Let

$$\begin{aligned}
x &= \sum_{i \in I} a_i \cdot x_i + \delta_E + \sum_{k \in K} \varepsilon,\\
y &= \sum_{l \in L} b_l \cdot y_l + \delta_F + \sum_{n \in N} \varepsilon,\\
z &= \sum_{o \in O} c_o \cdot z_o + \delta_G + \sum_{q \in Q} \varepsilon,
\end{aligned}$$

for some finite index sets I, K, L, N, O, Q, $a_i, b_l, c_o \in A$, $E, F, G \subseteq EID$ and basic terms x_i, y_l, z_o.

The following identities are used in the proofs of these four equations. Their proofs are omitted.

$$(\varepsilon \mathbin{\mathbf{L}\!\circ_{\mathrm{iw}}} y) \mathbin{\mathbf{L}\!\circ_{\mathrm{iw}}} z = \varepsilon \mathbin{\mathbf{L}\!\circ_{\mathrm{iw}}} (y \circ_{\mathrm{iw}} z), \qquad (a)$$

$$(x \,\mathbf{R}\!\circ_{\mathrm{iw}} \varepsilon) \,\mathbf{L}\!\circ_{\mathrm{iw}} z = x \,\mathbf{R}\!\circ_{\mathrm{iw}} (\varepsilon \,\mathbf{L}\!\circ_{\mathrm{iw}} z), \tag{b}$$

$$(x \circ_{\mathrm{iw}} y) \,\mathbf{R}\!\circ_{\mathrm{iw}} \varepsilon = x \,\mathbf{R}\!\circ_{\mathrm{iw}} (y \,\mathbf{R}\!\circ_{\mathrm{iw}} \varepsilon). \tag{c}$$

PROOF OF (1).

$$\begin{aligned}
& (x \,\mathbf{L}\!\circ_{\mathrm{iw}} y) \,\mathbf{L}\!\circ_{\mathrm{iw}} z \\
= {} & \{\text{ass. } x,\ \text{distribution laws}\} \\
& \sum_{i \in I} (a_i \cdot x_i \,\mathbf{L}\!\circ_{\mathrm{iw}} y) \,\mathbf{L}\!\circ_{\mathrm{iw}} z + (\delta_E \,\mathbf{L}\!\circ_{\mathrm{iw}} y) \,\mathbf{L}\!\circ_{\mathrm{iw}} z + \sum_{k \in K} (\varepsilon \,\mathbf{L}\!\circ_{\mathrm{iw}} y) \,\mathbf{L}\!\circ_{\mathrm{iw}} z \\
= {} & \{\text{L6, L5}\} \\
& \sum_{i \in I} a_i \cdot \big((x_i \circ_{\mathrm{iw}} y) \circ_{\mathrm{iw}} z\big) + \delta_{E \cup AE(y) \cup AE(z)} + \sum_{k \in K} (\varepsilon \,\mathbf{L}\!\circ_{\mathrm{iw}} y) \,\mathbf{L}\!\circ_{\mathrm{iw}} z \\
= {} & \{\text{Induction hypothesis (4)},\ AE(y) \cup AE(z) = AE(y \circ_{\mathrm{iw}} z),\ \text{(a)}\} \\
& \sum_{i \in I} a_i \cdot \big(x_i \circ_{\mathrm{iw}} (y \circ_{\mathrm{iw}} z)\big) + \delta_{E \cup AE(y \circ_{\mathrm{iw}} z)} + \sum_{k \in K} \varepsilon \,\mathbf{L}\!\circ_{\mathrm{iw}} (y \circ_{\mathrm{iw}} z) \\
= {} & \{\text{L6, L5}\} \\
& \sum_{i \in I} a_i \cdot x_i \circ_{\mathrm{iw}} (y \circ_{\mathrm{iw}} z) + \delta_E \,\mathbf{L}\!\circ_{\mathrm{iw}} (y \circ_{\mathrm{iw}} z) + \sum_{k \in K} \varepsilon \,\mathbf{L}\!\circ_{\mathrm{iw}} (y \circ_{\mathrm{iw}} z) \\
= {} & \{\text{distribution laws, ass. } x\} \\
& x \,\mathbf{L}\!\circ_{\mathrm{iw}} (y \circ_{\mathrm{iw}} z).
\end{aligned}$$

PROOF OF (2). Let $L' = \{l \in L \mid AE(b_l) \cap AE(x) = \emptyset\}$ and $L'' = L \setminus L'$.

$$\begin{aligned}
& (x \,\mathbf{R}\!\circ_{\mathrm{iw}} y) \,\mathbf{L}\!\circ_{\mathrm{iw}} z \\
= {} & \{\text{ass. } y,\ \text{distribution laws}\} \\
& \sum_{l \in L} (x \,\mathbf{R}\!\circ_{\mathrm{iw}} b_l \cdot y_l) \,\mathbf{L}\!\circ_{\mathrm{iw}} z + (x \,\mathbf{R}\!\circ_{\mathrm{iw}} \delta_F) \,\mathbf{L}\!\circ_{\mathrm{iw}} z + \sum_{n \in N} (x \,\mathbf{R}\!\circ_{\mathrm{iw}} \varepsilon) \,\mathbf{L}\!\circ_{\mathrm{iw}} z \\
= {} & \{\text{R6a, R6b, L6, R5, L5}\} \\
& \sum_{l \in L'} b_l \cdot \big((x \circ_{\mathrm{iw}} y_l) \circ_{\mathrm{iw}} z\big) + \sum_{l \in L''} \delta_{AE(x) \cup AE(b_l \cdot y_l) \cup AE(z)} + \delta_{F \cup AE(x) \cup AE(z)} \\
& \quad + \sum_{n \in N} (x \,\mathbf{R}\!\circ_{\mathrm{iw}} \varepsilon) \,\mathbf{L}\!\circ_{\mathrm{iw}} z \\
= {} & \{\text{Induction hypothesis (4), (b)}\} \\
& \sum_{l \in L'} b_l \cdot \big(x \circ_{\mathrm{iw}} (y_l \circ_{\mathrm{iw}} z)\big) + \sum_{l \in L''} \delta_{AE(x) \cup AE(b_l \cdot y_l) \cup AE(z)} + \delta_{F \cup AE(x) \cup AE(z)} \\
& \quad + \sum_{n \in N} x \,\mathbf{R}\!\circ_{\mathrm{iw}} (\varepsilon \,\mathbf{L}\!\circ_{\mathrm{iw}} z) \\
= {} & \{\text{R6a, R6b, L6, R5, L5}\} \\
& \sum_{l \in L} x \,\mathbf{R}\!\circ_{\mathrm{iw}} (b_l \cdot y_l \,\mathbf{L}\!\circ_{\mathrm{iw}} z) + x \,\mathbf{R}\!\circ_{\mathrm{iw}} (\delta_F \,\mathbf{L}\!\circ_{\mathrm{iw}} z) + \sum_{n \in N} x \,\mathbf{R}\!\circ_{\mathrm{iw}} (\varepsilon \,\mathbf{L}\!\circ_{\mathrm{iw}} z)
\end{aligned}$$

$$= \{\text{distribution laws, ass. } y\}$$
$$x \ \mathbf{R}\circ_{\mathrm{iw}}(y \ \mathbf{R}\circ_{\mathrm{iw}} z).$$

PROOF OF (3). Let $O' = \{o \in O \mid AE(c_o) \cap AE(x \circ_{\mathrm{iw}} y) = \emptyset\}$ and $O'' = O \setminus O'$.

$$(x \circ_{\mathrm{iw}} y) \ \mathbf{R}\circ_{\mathrm{iw}} z$$
$$= \{\text{ass. } z, \text{ distribution laws}\}$$
$$\sum_{o \in O} (x \circ_{\mathrm{iw}} y) \ \mathbf{R}\circ_{\mathrm{iw}} c_o \cdot z_o + (x \circ_{\mathrm{iw}} y) \ \mathbf{R}\circ_{\mathrm{iw}} \delta_G + \sum_{q \in Q} (x \circ_{\mathrm{iw}} y) \ \mathbf{R}\circ_{\mathrm{iw}} \varepsilon$$
$$= \{\text{R6a, R6b, R5}\}$$
$$\sum_{o \in O'} c_o \cdot \big((x \circ_{\mathrm{iw}} y) \circ_{\mathrm{iw}} z_o\big) + \sum_{o \in O''} \delta_{AE(x \circ_{\mathrm{iw}} y) \cup AE(c_o \cdot z_o)} + \delta_{G \cup AE(x \circ_{\mathrm{iw}} y)}$$
$$+ \sum_{q \in Q} (x \circ_{\mathrm{iw}} y) \ \mathbf{R}\circ_{\mathrm{iw}} \varepsilon$$
$$= \big\{\text{Induction hypothesis (4), } AE(x \circ_{\mathrm{iw}} y) = AE(x) \cup AE(y), \text{ (c)}\big\}$$
$$\sum_{o \in O'} c_o \cdot \big(x \circ_{\mathrm{iw}} (y \circ_{\mathrm{iw}} z_o)\big) + \sum_{o \in O''} \delta_{AE(x) \cup AE(y) \cup AE(c_o \cdot z_o)}$$
$$+ \delta_{G \cup AE(x) \cup AE(y)} + \sum_{q \in Q} x \ \mathbf{R}\circ_{\mathrm{iw}}(y \ \mathbf{R}\circ_{\mathrm{iw}} \varepsilon)$$
$$= \{\text{R6a, R6b, R5}\}$$
$$\sum_{o \in O} x \ \mathbf{R}\circ_{\mathrm{iw}}(y \ \mathbf{R}\circ_{\mathrm{iw}} c_o \cdot z_o) + x \ \mathbf{R}\circ_{\mathrm{iw}}(y \ \mathbf{R}\circ_{\mathrm{iw}} \delta_G) + \sum_{q \in Q} x \ \mathbf{R}\circ_{\mathrm{iw}}(y \ \mathbf{R}\circ_{\mathrm{iw}} \varepsilon)$$
$$= \{\text{distribution laws, ass. } z\}$$
$$x \ \mathbf{R}\circ_{\mathrm{iw}}(y \ \mathbf{R}\circ_{\mathrm{iw}} z).$$

PROOF OF (4).

$$(x \circ_{\mathrm{iw}} y) \circ_{\mathrm{iw}} z$$
$$= \{\text{S}\}$$
$$(x \circ_{\mathrm{iw}} y) \ \mathbf{L}\circ_{\mathrm{iw}} z + (x \circ_{\mathrm{iw}} y) \ \mathbf{R}\circ_{\mathrm{iw}} z$$
$$= \{\text{S}\}$$
$$(x \ \mathbf{L}\circ_{\mathrm{iw}} y + x \ \mathbf{R}\circ_{\mathrm{iw}} y) \ \mathbf{L}\circ_{\mathrm{iw}} z + (x \circ_{\mathrm{iw}} y) \ \mathbf{R}\circ_{\mathrm{iw}} z$$
$$= \{\text{L7}\}$$
$$(x \ \mathbf{L}\circ_{\mathrm{iw}} y) \ \mathbf{L}\circ_{\mathrm{iw}} z + (x \ \mathbf{R}\circ_{\mathrm{iw}} y) \ \mathbf{L}\circ_{\mathrm{iw}} z + (x \circ_{\mathrm{iw}} y) \ \mathbf{R}\circ_{\mathrm{iw}} z$$
$$= \{\text{Induction hypotheses (1), (2), (3)}\}$$
$$x \ \mathbf{L}\circ_{\mathrm{iw}}(y \circ_{\mathrm{iw}} z) + x \ \mathbf{R}\circ_{\mathrm{iw}}(y \ \mathbf{L}\circ_{\mathrm{iw}} z) + x \ \mathbf{R}\circ_{\mathrm{iw}}(y \ \mathbf{R}\circ_{\mathrm{iw}} z)$$
$$= \{\text{R7}\}$$
$$x \ \mathbf{L}\circ_{\mathrm{iw}}(y \circ_{\mathrm{iw}} z) + x \ \mathbf{R}\circ_{\mathrm{iw}}(y \ \mathbf{L}\circ_{\mathrm{iw}} z + y \ \mathbf{R}\circ_{\mathrm{iw}} z)$$

$$= \{S\}$$
$$x \mathbf{L}\circ_{iw}(y \circ_{iw} z) + x \mathbf{R}\circ_{iw}(y \circ_{iw} z)$$
$$= \{S\}$$
$$x \circ_{iw} (y \circ_{iw} z). \qquad \square$$

Finally, we give two more identities. They correspond to the axioms $\circ_{iw}1$ and $\circ_{iw}3$ from Table 2.

PROPOSITION 4. *For $E, F \subseteq EID$ and $a \in A$ such that $AE(a) \cap E \neq \emptyset$ we have*

$$\delta_E \circ_{iw} a = \delta_{E \cup AE(a)},$$
$$\delta_E \circ_{iw} \delta_F = \delta_{E \cup F}.$$

PROOF. For the first identity consider

$$\begin{aligned} \delta_E \circ_{iw} a &= \delta_E \mathbf{L}\circ_{iw} a + \delta_E \mathbf{R}\circ_{iw} a = \delta_{E \cup AE(a)} + \delta_E \mathbf{R}\circ_{iw} a \cdot \varepsilon \\ &= \delta_{E \cup AE(a)} + \delta_{E \cup AE(a)} = \delta_{E \cup AE(a)}, \end{aligned}$$

and for the second consider

$$\delta_E \circ_{iw} \delta_F = \delta_E \mathbf{L}\circ_{iw} \delta_F + \delta_E \mathbf{R}\circ_{iw} \delta_F = \delta_{E \cup F} + \delta_{E \cup F} = \delta_{E \cup F}. \qquad \square$$

Observe that we have now shown that all identities on closed *IWD*($\circ$)-terms that are derivably equal in the process algebra *IWD*($\circ$), are also derivably equal in the process algebra *IWD*($\circ, \cdot, +$).

6. The E-interworking merge

Now that we have defined the process algebras *BPA*(δ_E) and *IWD*($\circ, \cdot, +$) which include operators for alternative and sequential composition, we aim at extending them with the merge operator. For technical reasons, we do this in two steps: First we will define the E-interworking merge in this section and in the next section we will extend the obtained process algebra to its final shape.

The E-interworking merge of x and y, denoted by $x \parallel_{iw}^E y$, is the parallel execution of the processes x and y with the restriction that the processes must synchronize on all atomic actions which are defined on entities from the set E. This set E is static, which means that it remains unchanged during calculations on a term which contains the E-interworking merge operator. The resulting process algebra is called *IWD*($\circ, +, \cdot, \parallel^E$).

The deduction rules defining the operational semantics of the E-interworking merge are given in Table 7. The E-interworking merge of two processes can terminate if and only if both operands can terminate. The second and third rule in Table 7 say that if an operand can perform an action, the merge can perform the same action, provided that the action is not supposed to synchronize (i.e., the sender and receiver are not both in E). The fourth rule expresses the behaviour of a merge in case a synchronized action is possible.

Table 7
Deduction rules for E-interworking merge ($a \in A$, $E \subseteq EID$)

$$\frac{x\downarrow \quad y\downarrow}{x \parallel^E_{\mathrm{iw}} y\downarrow} \qquad \frac{x \xrightarrow{a} x' \quad AE(a) \not\subseteq E}{x \parallel^E_{\mathrm{iw}} y \xrightarrow{a} x' \parallel^E_{\mathrm{iw}} y} \qquad \frac{y \xrightarrow{a} y' \quad AE(a) \not\subseteq E}{x \parallel^E_{\mathrm{iw}} y \xrightarrow{a} x \parallel^E_{\mathrm{iw}} y'}$$

$$\frac{x \xrightarrow{a} x' \quad y \xrightarrow{a} y' \quad AE(a) \subseteq E}{x \parallel^E_{\mathrm{iw}} y \xrightarrow{a} x' \parallel^E_{\mathrm{iw}} y'}$$

Table 8
Deduction rules for auxiliary operators of
E-interworking merge ($a \in A$, $E \subseteq EID$)

$$\frac{x\downarrow \quad y\downarrow}{x \lfloor\!\lfloor^E_{\mathrm{iw}} y\downarrow} \qquad \frac{x \xrightarrow{a} x' \quad AE(a) \not\subseteq E}{x \lfloor\!\lfloor^E_{\mathrm{iw}} y \xrightarrow{a} x' \parallel^E_{\mathrm{iw}} y}$$

$$\frac{x \xrightarrow{a} x' \quad y \xrightarrow{a} y' \quad AE(a) \subseteq E}{x \mid^E_{\mathrm{iw}} y \xrightarrow{a} x' \parallel^E_{\mathrm{iw}} y'}$$

DEFINITION 10 (*Active entities*). For $i, j \in EID$, $m \in MID$, $E \subseteq EID$, $x, y \in \mathcal{C}(\Sigma_{IWD(\circ,+,\cdot,\parallel^E)})$, and $\odot \in \{\circ_{\mathrm{iw}}, \mathbf{L}\circ_{\mathrm{iw}}, \mathbf{R}\circ_{\mathrm{iw}}, +, \cdot, \parallel^E_{\mathrm{iw}}, \lfloor\!\lfloor^E_{\mathrm{iw}}, \mid^E_{\mathrm{iw}} \mid E \subseteq EID\}$, we define the mapping $AE : \mathcal{C}(\Sigma_{IWD(\circ,+,\cdot,\parallel^E)}) \to \mathbb{P}(EID)$ inductively as follows:

$$\begin{aligned} AE\big(c(i, j, m)\big) &= \{i, j\}, \\ AE(\varepsilon) &= \emptyset, \\ AE(\delta_E) &= E, \\ AE(x \odot y) &= AE(x) \cup AE(y). \end{aligned}$$

For the axiomatization of the E-interworking merge we need two auxiliary operators, similar to the axiomatization of the communication merge of *ACP*. These additional operators are $\lfloor\!\lfloor^E_{\mathrm{iw}}$ (E-interworking left-merge) and $\mid^E_{\mathrm{iw}}$ (E-interworking synchronization merge). The process $x \lfloor\!\lfloor^E_{\mathrm{iw}} y$ behaves like the process $x \parallel^E_{\mathrm{iw}} y$ with the restriction that the first action must come from process x and that action cannot synchronize with an action from y. The process $x \mid^E_{\mathrm{iw}} y$ behaves as the process $x \parallel^E_{\mathrm{iw}} y$ with the restriction that the first action must be a synchronization. This is formalized by the deduction rules in Table 8. The term deduction system $T(IWD(\circ, +, \cdot, \parallel^E))$ consists of the deduction rules of Tables 1, 3, 5, 7 and 8.

Table 9 presents the axioms defining the E-interworking merge and its auxiliary operators. Axiom M states that either one of the two operands executes a non-synchronized action ($x \lfloor\!\lfloor^E_{\mathrm{iw}} y + y \lfloor\!\lfloor^E_{\mathrm{iw}} x$), or that a synchronized action takes place ($x \mid^E_{\mathrm{iw}} y$). The definition of the $\lfloor\!\lfloor^E_{\mathrm{iw}}$ operator (LM1–LM7) is very similar to the definition of the $\mathbf{L}\circ_{\mathrm{iw}}$ operator

Table 9
Axioms of E-interworking merge ($a, b \in A$, $E, F \in EID$)

M	$x \Vert^{E}_{\mathrm{iw}} y = x \mathbin{\Vert\!\!\!\lfloor}^{E}_{\mathrm{iw}} y + y \mathbin{\Vert\!\!\!\lfloor}^{E}_{\mathrm{iw}} x + x \mid^{E}_{\mathrm{iw}} y$	
LM1	$\varepsilon \mathbin{\Vert\!\!\!\lfloor}^{E}_{\mathrm{iw}} \varepsilon = \varepsilon$	
LM2	$\varepsilon \mathbin{\Vert\!\!\!\lfloor}^{E}_{\mathrm{iw}} \delta_F = \delta_F$	
LM3	$\varepsilon \mathbin{\Vert\!\!\!\lfloor}^{E}_{\mathrm{iw}} a \cdot x = \delta_{AE(a \cdot x)}$	
LM4	$\varepsilon \mathbin{\Vert\!\!\!\lfloor}^{E}_{\mathrm{iw}} (x + y) = \varepsilon \mathbin{\Vert\!\!\!\lfloor}^{E}_{\mathrm{iw}} x + \varepsilon \mathbin{\Vert\!\!\!\lfloor}^{E}_{\mathrm{iw}} y$	
LM5	$\delta_F \mathbin{\Vert\!\!\!\lfloor}^{E}_{\mathrm{iw}} x = \delta_{F \cup AE(x)}$	
LM6a	$a \cdot x \mathbin{\Vert\!\!\!\lfloor}^{E}_{\mathrm{iw}} y = a \cdot (x \Vert^{E}_{\mathrm{iw}} y)$	if $AE(a) \not\subseteq E$
LM6b	$a \cdot x \mathbin{\Vert\!\!\!\lfloor}^{E}_{\mathrm{iw}} y = \delta_{AE(a \cdot x) \cup AE(y)}$	if $AE(a) \subseteq E$
LM7	$(x + y) \mathbin{\Vert\!\!\!\lfloor}^{E}_{\mathrm{iw}} z = x \mathbin{\Vert\!\!\!\lfloor}^{E}_{\mathrm{iw}} z + y \mathbin{\Vert\!\!\!\lfloor}^{E}_{\mathrm{iw}} z$	
CM1	$\varepsilon \mid^{E}_{\mathrm{iw}} x = \delta_{AE(x)}$	
CM2	$x \mid^{E}_{\mathrm{iw}} \varepsilon = \delta_{AE(x)}$	
CM3	$\delta_F \mid^{E}_{\mathrm{iw}} x = \delta_{F \cup AE(x)}$	
CM4	$x \mid^{E}_{\mathrm{iw}} \delta_F = \delta_{F \cup AE(x)}$	
CM5a	$a \cdot x \mid^{E}_{\mathrm{iw}} b \cdot y = a \cdot (x \Vert^{E}_{\mathrm{iw}} y)$	if $a \equiv b \wedge AE(a) \subseteq E$
CM5b	$a \cdot x \mid^{E}_{\mathrm{iw}} b \cdot y = \delta_{AE(a \cdot x) \cup AE(b \cdot y)}$	if $a \not\equiv b \vee AE(a) \not\subseteq E$
CM6	$(x + y) \mid^{E}_{\mathrm{iw}} z = x \mid^{E}_{\mathrm{iw}} z + y \mid^{E}_{\mathrm{iw}} z$	
CM7	$x \mid^{E}_{\mathrm{iw}} (y + z) = x \mid^{E}_{\mathrm{iw}} y + x \mid^{E}_{\mathrm{iw}} z$	

in Section 5 (Table 6, axioms L1–L7). The only difference is that axiom L6 is unconditional, whereas axiom LM6b has to take care of eliminating actions which are supposed to synchronize. Axioms CM1–CM7 define the $\mid^{E}_{\mathrm{iw}}$ operator. This operator enables all actions that can be performed by both operands and which must synchronize. In all other cases it yields a deadlock, where the scope of the deadlock can be derived from the operands.

It turns out that IWD-bisimilarity is a congruence for the operators in the signature of *IWD*$(\circ, +, \cdot, \Vert^{E})$. Furthermore, *IWD*$(\circ, +, \cdot, \Vert^{E})$ is a sound and complete axiomatization of IWD-bisimilarity on closed *IWD*$(\circ, +, \cdot, \Vert^{E})$-terms. The proofs are based on the meta-theory presented in [5,34].

THEOREM 15 (Congruence). *IWD-bisimilarity is a congruence for E-interworking merge and the auxiliary operators.*

PROOF. The term deduction system is in path format and hence IWD-bisimilarity is a congruence for all operators. □

THEOREM 16 (Soundness). *The process algebra IWD*$(\circ, +, \cdot, \Vert^{E})$ *is a sound axiomatization of IWD-bisimilarity on closed IWD*$(\circ, +, \cdot, \Vert^{E})$*-terms.*

PROOF. For the axioms LM1–LM5, LM6b, CM1–CM4, and CM5b the IWD-bisimulations that witness the soundness are trivial. If "$s = t$ if b" represents such an axiom, then the IWD-bisimulation is $R = \{(s,t), (t,s) \mid b\}$.

For the axioms M, LM7, CM6, and CM7 the IWD-bisimulation is given by $R = \{(s,t), (t,s) \mid b\} \cup I$ if the axiom is given as "$s = t$ if b".

For the axioms LM6a and CM5a the IWD-bisimulations are

$$R = \big\{\big(a \cdot x \mathbin{\|\!\lfloor}^E_{\mathrm{iw}} y, a \cdot \big(x \parallel^E_{\mathrm{iw}} y\big)\big), \big(\varepsilon \cdot x \parallel^E_{\mathrm{iw}} y, \varepsilon \cdot \big(x \parallel^E_{\mathrm{iw}} y\big)\big), \big(\varepsilon \cdot x \parallel^E_{\mathrm{iw}} q, x \parallel^E_{\mathrm{iw}} q\big) \\ \mid AE(a) \not\subseteq E, q \in \mathcal{C}(\Sigma_{IWD(\circ,+,\cdot,\|^E)})\big\}^S \cup I$$

and

$$R = \big\{\big(a \cdot x \mid^E_{\mathrm{iw}} b \cdot y, a \cdot \big(x \parallel^E_{\mathrm{iw}} y\big)\big), \big(\varepsilon \cdot x \parallel^E_{\mathrm{iw}} \varepsilon \cdot y, \varepsilon \cdot \big(x \parallel^E_{\mathrm{iw}} y\big)\big), \\ \big(p \parallel^E_{\mathrm{iw}} \varepsilon \cdot y, p \parallel^E_{\mathrm{iw}} y\big), \big(\varepsilon \cdot x \parallel^E_{\mathrm{iw}} q, x \parallel^E_{\mathrm{iw}} q\big) \\ \mid 2a \equiv b, AE(a) \subseteq E, p, q \in \mathcal{C}(\Sigma_{IWD(\circ,+,\cdot,\|^E)})\big\}^S \cup I$$

respectively. □

LEMMA 6. *For arbitrary basic terms b_1 and b_2 we have the existence of basic terms c_1, c_2, and c_3 such that $b_1 \mathbin{\|\!\lfloor}^E_{\mathrm{iw}} b_2 = c_1$, $b_1 \mid^E_{\mathrm{iw}} b_2 = c_2$ and $b_1 \parallel^E_{\mathrm{iw}} b_2 = c_3$.*

PROOF. These statements are proven simultaneously with induction on the total number of symbols of the basic terms b_1 and b_2.

(1) By case distinction on the structure of basic term b_1.

(a) $b_1 \equiv \varepsilon$. By case distinction on the structure of basic term b_2.

(i) $b_2 \equiv \varepsilon$. Then $b_1 \mathbin{\|\!\lfloor}^E_{\mathrm{iw}} b_2 = \varepsilon \mathbin{\|\!\lfloor}^E_{\mathrm{iw}} \varepsilon = \varepsilon$.

(ii) $b_2 \equiv \delta_{F_2}$ for some $F_2 \subseteq EID$. Then $b_1 \mathbin{\|\!\lfloor}^E_{\mathrm{iw}} b_2 = \varepsilon \mathbin{\|\!\lfloor}^E_{\mathrm{iw}} \delta_{F_2} = \delta_{F_2}$.

(iii) $b_2 \equiv a_2 \cdot b_2'$ for some $a_2 \in A$ and $b_2' \in \mathcal{B}(\Sigma_{IWD(\circ,+,\cdot,\|^E)})$. Then

$$b_1 \mathbin{\|\!\lfloor}^E_{\mathrm{iw}} b_2 = \varepsilon \mathbin{\|\!\lfloor}^E_{\mathrm{iw}} a_2 \cdot b_2' = \delta_{AE(a_2 \cdot b_2')}.$$

(iv) $b_2 \equiv b_2' + b_2''$ for some $b_2', b_2'' \in \mathcal{B}(\Sigma_{IWD(\circ,+,\cdot,\|^E)})$. By induction we have the existence of basic terms c_1' and c_1'' such that $\varepsilon \mathbin{\|\!\lfloor}^E_{\mathrm{iw}} b_2' = c_1'$ and $\varepsilon \mathbin{\|\!\lfloor}^E_{\mathrm{iw}} b_2'' = c_1''$. Then

$$b_1 \mathbin{\|\!\lfloor}^E_{\mathrm{iw}} b_2 = \varepsilon \mathbin{\|\!\lfloor}^E_{\mathrm{iw}} \big(b_2' + b_2''\big) = \varepsilon \mathbin{\|\!\lfloor}^E_{\mathrm{iw}} b_2' + \varepsilon \mathbin{\|\!\lfloor}^E_{\mathrm{iw}} b_2'' = c_1' + c_1''.$$

(b) $b_1 \equiv \delta_{F_1}$ for some $F_1 \subseteq EID$. Then $b_1 \mathbin{\|\!\lfloor}^E_{\mathrm{iw}} b_2 = \delta_{F_1} \mathbin{\|\!\lfloor}^E_{\mathrm{iw}} b_2 = \delta_{F_1 \cup AE(b_2)}$.

(c) $b_1 \equiv a_1 \cdot b_1'$ for some $a_1 \in A$ and $b_1' \in \mathcal{B}(\Sigma_{IWD(\circ,+,\cdot,\|^E)})$. By induction we have the existence of basic term c_1' such that $b_1' \parallel^E_{\mathrm{iw}} b_2 = c_1'$. Then, if $AE(a) \not\subseteq E$,

$$b_1 \mathbin{\|\!\lfloor}^E_{\mathrm{iw}} b_2 = a_1 \cdot b_1' \mathbin{\|\!\lfloor}^E_{\mathrm{iw}} b_2 = a_1 \cdot \big(b_1' \parallel^E_{\mathrm{iw}} b_2\big) = a_1 \cdot c_1'.$$

If $AE(a_1) \subseteq E$, then $b_1 \mathbin{\|\!\lfloor}^E_{\mathrm{iw}} b_2 = a_1 \cdot b_1' \mathbin{\|\!\lfloor}^E_{\mathrm{iw}} b_2 = \delta_{AE(a_1 \cdot b_1') \cup AE(b_2)}$.

(d) $b_1 \equiv b_1' + b_1''$ for some $b_1', b_1'' \in \mathcal{B}(\Sigma_{IWD(\circ,+,\cdot,\|^E)})$. By induction we have the existence of basic terms c_1' and c_1'' such that $b_1' \mathbin{⌊\!⌊}^E_{\mathrm{iw}} b_2 = c_1'$ and $b_1'' \mathbin{⌊\!⌊}^E_{\mathrm{iw}} b_2 = c_1''$. Then

$$b_1 \mathbin{⌊\!⌊}^E_{\mathrm{iw}} b_2 = (b_1' + b_1'') \mathbin{⌊\!⌊}^E_{\mathrm{iw}} b_2 = b_1' \mathbin{⌊\!⌊}^E_{\mathrm{iw}} b_2 + b_1'' \mathbin{⌊\!⌊}^E_{\mathrm{iw}} b_2 = c_1' + c_1''.$$

(2) By case distinction on the structure of basic term b_1.

(a) $b_1 \equiv \varepsilon$. Then $b_1 \mid^E_{\mathrm{iw}} b_2 = \varepsilon \mid^E_{\mathrm{iw}} b_2 = \delta_{AE(b_2)}$.

(b) $b_1 \equiv \delta_{F_1}$ for some $F_1 \subseteq EID$. Then $b_1 \mid^E_{\mathrm{iw}} b_2 = \delta_{F_1} \mid^E_{\mathrm{iw}} b_2 = \delta_{F_1 \cup AE(b_2)}$.

(c) $b_1 \equiv a_1 \cdot b_1'$ for some $a_1 \in A$ and $b_1' \in \mathcal{B}(\Sigma_{IWD(\circ,+,\cdot,\|^E)})$. By case distinction on the structure of basic term b_2.

(i) $b_2 \equiv \varepsilon$. Then $b_1 \mid^E_{\mathrm{iw}} b_2 = b_1 \mid^E_{\mathrm{iw}} \varepsilon = \delta_{AE(b_1)}$.

(ii) $b_2 \equiv \delta_{F_2}$ for some $F_2 \subseteq EID$. Then $b_1 \mid^E_{\mathrm{iw}} b_2 = b_1 \mid^E_{\mathrm{iw}} \delta_{F_2} = \delta_{AE(b_1) \cup F_2}$.

(iii) $b_2 \equiv a_2 \cdot b_2'$ for some $a_2 \in A$ and $b_2' \in \mathcal{B}(\Sigma_{IWD(\circ,+,\cdot,\|^E)})$. By induction we have the existence of basic term c_3 such that $b_1' \|^E_{\mathrm{iw}} b_2' = c_3$. Then, if $a_1 \equiv a_2 \wedge AE(a_1) \subseteq E$,

$$b_1 \mid^E_{\mathrm{iw}} b_2 = a_1 \cdot b_1' \mid^E_{\mathrm{iw}} a_2 \cdot b_2' = a_1 \cdot (b_1' \|^E_{\mathrm{iw}} b_2') = a_1 \cdot c_3.$$

If $a_1 \not\equiv a_2 \vee AE(a_1) \not\subseteq E$, then

$$b_1 \mid^E_{\mathrm{iw}} b_2 = a_1 \cdot b_1' \mid^E_{\mathrm{iw}} a_2 \cdot b_2' = \delta_{AE(a_1 \cdot b_1') \cup AE(a_1 \cdot b_2')}.$$

(iv) $b_2 \equiv b_2' + b_2''$ for some $b_2', b_2'' \in \mathcal{B}(\Sigma_{IWD(\circ,+,\cdot,\|^E)})$. By induction we have the existence of basic terms c_2' and c_2'' such that $b_1 \mid^E_{\mathrm{iw}} b_2' = c_2'$ and $b_1 \mid^E_{\mathrm{iw}} b_2'' = c_2''$. Then

$$b_1 \mid^E_{\mathrm{iw}} b_2 = b_1 \mid^E_{\mathrm{iw}} (b_2' + b_2'') = b_1 \mid^E_{\mathrm{iw}} b_2' + b_1 \mid^E_{\mathrm{iw}} b_2'' = c_2' + c_2''.$$

(d) $b_1 \equiv b_1' + b_1''$ for some $b_1', b_1'' \in \mathcal{B}(\Sigma_{IWD(\circ,+,\cdot,\|^E)})$. By induction we have the existence of basic terms c_2' and c_2'' such that $b_1' \mid^E_{\mathrm{iw}} b_2 = c_2'$ and $b_1'' \mid^E_{\mathrm{iw}} b_2 = c_2''$. Then

$$b_1 \mid^E_{\mathrm{iw}} b_2 = (b_1' + b_1'') \mid^E_{\mathrm{iw}} b_2 = b_1' \mid^E_{\mathrm{iw}} b_2 + b_1'' \mid^E_{\mathrm{iw}} b_2 = c_2' + c_2''.$$

(3) By the previous two items we have the existence of basic terms c_1', c_1'', and c_2 such that $b_1 \mathbin{⌊\!⌊}^E_{\mathrm{iw}} b_2 = c_1'$, $b_2 \mathbin{⌊\!⌊}^E_{\mathrm{iw}} b_1 = c_1''$, and $b_1 \mid^E_{\mathrm{iw}} b_2 = c_2$. Then,

$$b_1 \|^E_{\mathrm{iw}} b_2 = b_1 \mathbin{⌊\!⌊}^E_{\mathrm{iw}} b_2 + b_2 \mathbin{⌊\!⌊}^E_{\mathrm{iw}} b_1 + b_1 \mid^E_{\mathrm{iw}} b_2 = c_1' + c_1'' + c_2. \qquad \square$$

THEOREM 17 (Elimination). *For every closed IWD*($\circ, +, \cdot, \|^E$)*-term x there is a derivably equal basic term s.*

PROOF. This theorem is proven by induction on the structure of closed $IWD(\circ, +, \cdot, \|^E)$-term x. All cases except for $x \equiv x' \mathbin{\|\!\!\_}^E_{\mathrm{iw}} x''$, $x \equiv x' \,|^E_{\mathrm{iw}}\, x''$, and $x \equiv x' \,\|^E_{\mathrm{iw}}\, x''$ have already been treated in the proof of Theorem 12. In the remaining three cases we find the existence of basic terms b_1 and b_2 such that $x_1 = b_1$ and $x_2 = b_2$. Using the previous lemma we find the desired result. □

THEOREM 18 (Conservativity). *The process algebra $IWD(\circ, +, \cdot, \|^E)$ is a conservative extension of the process algebra $IWD(\circ, \cdot, +)$.*

PROOF. The proof of this theorem uses the approach of [34]. The conservativity follows from the following observations:

(1) IWD-bisimilarity is definable in terms of predicate and relation symbols only,
(2) $IWD(\circ, \cdot, +)$ is a complete axiomatization of IWD-bisimilarity on closed $IWD(\circ, \cdot, +)$-terms,
(3) $IWD(\circ, +, \cdot, \|^E)$ is a sound axiomatization of IWD-bisimilarity on closed $IWD(\circ, +, \cdot, \|^E)$-terms (see Theorem 16),
(4) the term deduction system for $IWD(\circ, \cdot, +)$ is pure, well-founded and in path format, and
(5) the term deduction system for $IWD(\circ, +, \cdot, \|^E)$ is in path format. □

THEOREM 19 (Completeness). *The process algebra $IWD(\circ, +, \cdot, \|^E)$ is a complete axiomatization of IWD-bisimilarity on closed $IWD(\circ, +, \cdot, \|^E)$-terms.*

PROOF. By the General Completeness Theorem of [33], the completeness of the process algebra $IWD(\circ, +, \cdot, \|^E)$ follows immediately from the properties mentioned in the proof of Theorem 18 and the fact that $IWD(\circ, +, \cdot, \|^E)$ has the elimination property for $BPA(\delta_E)$(see Theorem 8) and hence also for $IWD(\circ, \cdot, +)$. □

When defining an operator for parallel composition, several properties are desirable, such as commutativity, the existence of a unit element, and associativity (under some condition). The proof of associativity in the process algebra is quite complicated.

PROPOSITION 5 (Commutativity $\|^E_{\mathrm{iw}}$). *For closed $IWD(\circ, +, \cdot, \|^E)$-terms x, y, and a set of entities E we have*

$$x \,|^E_{\mathrm{iw}}\, y = y \,|^E_{\mathrm{iw}}\, x,$$
$$x \,\|^E_{\mathrm{iw}}\, y = y \,\|^E_{\mathrm{iw}}\, x.$$

PROOF. The propositions are proven simultaneously with induction on the general structure of basic terms x and y. Let

$$x = \sum_{i \in I} a_i \cdot x_i + \delta_E + \sum_{k \in K} \varepsilon,$$
$$y = \sum_{l \in L} b_l \cdot y_l + \delta_F + \sum_{n \in N} \varepsilon,$$

for some finite index sets I, K, L, N, $a_i, b_l \in A$, $E, F \subseteq EID$ and basic terms x_i, y_l. Then,

$$\begin{aligned}
x \mid_{\mathrm{iw}}^{E} y &= \sum_{i\in I}\sum_{l\in L} a_i \cdot x_i \mid_{\mathrm{iw}}^{E} b_l \cdot y_l + x \mid_{\mathrm{iw}}^{E} \delta_F + \sum_{n\in N} x \mid_{\mathrm{iw}}^{E} \varepsilon + \delta_E \mid_{\mathrm{iw}}^{E} y + \sum_{k\in K} \varepsilon \mid_{\mathrm{iw}}^{E} y \\
&= \sum_{i\in I} \sum_{l\in L, a_i \equiv b_l, AE(a_i)\subseteq E} a_i \cdot \left(x_i \parallel_{\mathrm{iw}}^{E} y_l\right) \\
&\quad + \sum_{i\in I} \sum_{l\in L, a_i \not\equiv b_l \vee AE(a_i)\not\subseteq E} \delta_{AE(a_i\cdot x_i)\cup AE(b_l\cdot y_l)} \\
&\quad + \delta_{AE(x)\cup F} + \sum_{n\in N} \delta_{AE(x)} + \delta_{E\cup AE(y)} + \sum_{k\in K} \delta_{AE(y)} \\
&= \sum_{l\in L} \sum_{i\in I, b_l \equiv a_i, AE(b_l)\subseteq E} b_l \cdot \left(y_l \parallel_{\mathrm{iw}}^{E} x_i\right) \\
&\quad + \sum_{l\in L} \sum_{i\in I, b_l \not\equiv a_i \vee AE(b_l)\not\subseteq E} \delta_{AE(b_l\cdot y_l)\cup AE(a_i\cdot x_i)} \\
&\quad + \delta_F \mid_{\mathrm{iw}}^{E} x + \sum_{n\in N} \varepsilon \mid_{\mathrm{iw}}^{E} x + y \mid_{\mathrm{iw}}^{E} \delta_E + \sum_{k\in K} y \mid_{\mathrm{iw}}^{E} \varepsilon \\
&= \sum_{l\in L}\sum_{i\in I} b_l \cdot y_l \mid_{\mathrm{iw}}^{E} a_i \cdot x_i + \delta_F \mid_{\mathrm{iw}}^{E} x + \sum_{n\in N} \varepsilon \mid_{\mathrm{iw}}^{E} x + y \mid_{\mathrm{iw}}^{E} \delta_E + \sum_{k\in K} y \mid_{\mathrm{iw}}^{E} \varepsilon \\
&= y \mid_{\mathrm{iw}}^{E} x
\end{aligned}$$

and

$$x \parallel_{\mathrm{iw}}^{E} y = x \mathbin{\|\!\lfloor}_{\mathrm{iw}}^{E} y + y \mathbin{\|\!\lfloor}_{\mathrm{iw}}^{E} x + x \mid_{\mathrm{iw}}^{E} y = y \mathbin{\|\!\lfloor}_{\mathrm{iw}}^{E} x + x \mathbin{\|\!\lfloor}_{\mathrm{iw}}^{E} y + y \mid_{\mathrm{iw}}^{E} x = y \parallel_{\mathrm{iw}}^{E} x. \qquad \square$$

PROPOSITION 6. *For closed IWD*($\circ, +, \cdot, \parallel^{E}$)*-terms x we have*

$$\begin{aligned}
x \parallel_{\mathrm{iw}}^{\emptyset} \varepsilon &= x, \\
\varepsilon \parallel_{\mathrm{iw}}^{\emptyset} x &= x.
\end{aligned}$$

PROOF. The first proposition is proven with induction on the general structure of basic term x. Let

$$x = \sum_{i\in I} a_i \cdot x_i + \delta_E + \sum_{k\in K} \varepsilon,$$

for some finite index sets I, K, $a_i \in A$, $E \subseteq EID$ and basic terms x_i. The induction hypothesis is $x_i \parallel_{\mathrm{iw}}^{\emptyset} \varepsilon = x_i$ for all $i \in I$. Then,

$$\begin{aligned}
x \mathbin{\|\!\lfloor}_{\mathrm{iw}}^{\emptyset} \varepsilon &= \sum_{i\in I} a_i \cdot x_i \mathbin{\|\!\lfloor}_{\mathrm{iw}}^{\emptyset} \varepsilon + \delta_E \mathbin{\|\!\lfloor}_{\mathrm{iw}}^{\emptyset} \varepsilon + \sum_{k\in K} \varepsilon \mathbin{\|\!\lfloor}_{\mathrm{iw}}^{\emptyset} \varepsilon \\
&= \sum_{i\in I} a_i \cdot \left(x_i \parallel_{\mathrm{iw}}^{\emptyset} \varepsilon\right) + \delta_E + \sum_{k\in K} \varepsilon
\end{aligned}$$

$$
\begin{aligned}
&= \sum_{i\in I} a_i \cdot x_i + \delta_E + \sum_{k\in K} \varepsilon \\
&= x
\end{aligned}
$$

and

$$
\begin{aligned}
\varepsilon \mathbin{\lfloor\!\lfloor}^{\emptyset}_{\mathrm{iw}} x &= \varepsilon \mathbin{\lfloor\!\lfloor}^{\emptyset}_{\mathrm{iw}} \left(\sum_{i\in I} a_i \cdot x_i + \delta_E + \sum_{k\in K} \varepsilon\right) \\
&= \sum_{i\in I} \varepsilon \mathbin{\lfloor\!\lfloor}^{\emptyset}_{\mathrm{iw}} a_i \cdot x_i + \varepsilon \mathbin{\lfloor\!\lfloor}^{\emptyset}_{\mathrm{iw}} \delta_E + \sum_{k\in K} \varepsilon \mathbin{\lfloor\!\lfloor}^{\emptyset}_{\mathrm{iw}} \varepsilon \\
&= \sum_{i\in I} \delta_{AE(a_i \cdot x_i)} + \delta_E + \sum_{k\in K} \varepsilon.
\end{aligned}
$$

Using these two subcomputations we obtain:

$$
\begin{aligned}
x \parallel^{\emptyset}_{\mathrm{iw}} \varepsilon &= x \mathbin{\lfloor\!\lfloor}^{\emptyset}_{\mathrm{iw}} \varepsilon + \varepsilon \mathbin{\lfloor\!\lfloor}^{\emptyset}_{\mathrm{iw}} x + x \mid^{\emptyset}_{\mathrm{iw}} \varepsilon \\
&= x + \sum_{i\in I} \delta_{AE(a_i \cdot x_i)} + \delta_E + \sum_{k\in K} \varepsilon + \delta_{AE(x)} \\
&= x.
\end{aligned}
$$

The other part of the proposition is obtained using the commutativity of $\parallel^{\emptyset}_{\mathrm{iw}}$. □

The following proposition serves our needs in proving interworking merge associative in the next section.

PROPOSITION 7 (Associativity of $\parallel^{E}_{\mathrm{iw}}$). *For closed IWD*$(\circ, +, \cdot, \parallel^{E})$*-terms* x, y, *and* z, *and sets of entities* E_1, E_2, *and* E_3 *such that* $AE(x) \subseteq E_1$, $AE(y) \subseteq E_2$, *and* $AE(z) \subseteq E_3$, *we have*

$$
\left(x \parallel^{E_1\cap E_2}_{\mathrm{iw}} y\right) \parallel^{(E_1\cup E_2)\cap E_3}_{\mathrm{iw}} z = x \parallel^{E_1\cap(E_2\cup E_3)}_{\mathrm{iw}} \left(y \parallel^{E_2\cap E_3}_{\mathrm{iw}} z\right).
$$

PROOF. Without loss of generality we can assume that x, y, and z are basic terms. We use the following shorthands: $S = E_1 \cap E_2 \cap E_3$, $E_{12} = (E_1 \cap E_2) \setminus S$, $E_{23} = (E_2 \cap E_3) \setminus S$, and $E_{13} = (E_1 \cap E_3) \setminus S$. In Figure 16 these sets are indicated in a Venn diagram.

We prove the following propositions simultaneously with induction on the structure of the basic terms x, y, and z.

$$
\left(x \mathbin{\lfloor\!\lfloor}^{E_{12}\cup S}_{\mathrm{iw}} y\right) \mathbin{\lfloor\!\lfloor}^{E_{13}\cup E_{23}\cup S}_{\mathrm{iw}} z = x \mathbin{\lfloor\!\lfloor}^{E_{12}\cup E_{13}\cup S}_{\mathrm{iw}} \left(y \parallel^{E_{23}\cup S}_{\mathrm{iw}} z\right), \tag{5}
$$

$$
\left(x \mid^{E_{12}\cup S}_{\mathrm{iw}} y\right) \mid^{E_{13}\cup E_{23}\cup S}_{\mathrm{iw}} z = x \mid^{E_{12}\cup E_{13}\cup S}_{\mathrm{iw}} \left(y \mid^{E_{23}\cup S}_{\mathrm{iw}} z\right), \tag{6}
$$

$$
\left(x \mid^{E_{12}\cup S}_{\mathrm{iw}} y\right) \mathbin{\lfloor\!\lfloor}^{E_{13}\cup E_{23}\cup S}_{\mathrm{iw}} z = x \mid^{E_{12}\cup E_{13}\cup S}_{\mathrm{iw}} \left(y \mathbin{\lfloor\!\lfloor}^{E_{23}\cup S}_{\mathrm{iw}} z\right), \tag{7}
$$

$$
\left(x \parallel^{E_{12}\cup S}_{\mathrm{iw}} y\right) \parallel^{E_{13}\cup E_{23}\cup S}_{\mathrm{iw}} z = x \parallel^{E_{12}\cup E_{13}\cup S}_{\mathrm{iw}} \left(y \parallel^{E_{23}\cup S}_{\mathrm{iw}} z\right). \tag{8}
$$

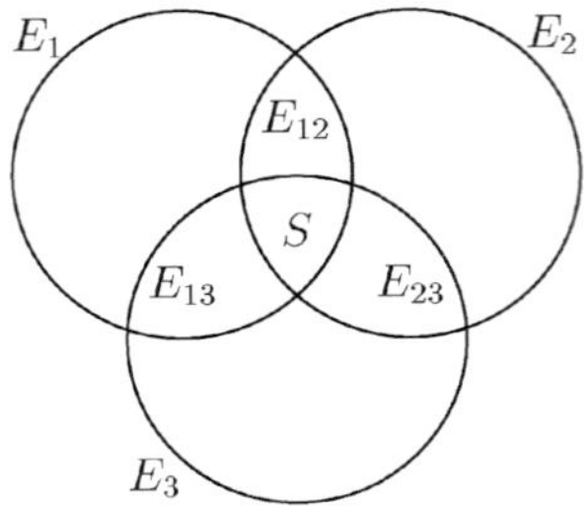

Fig. 16. Explanation of shorthands.

Let

$$x = \sum_{i\in I} a_i \cdot x_i + \delta_E + \sum_{k\in K} \varepsilon,$$

$$y = \sum_{l\in L} b_l \cdot y_l + \delta_F + \sum_{n\in N} \varepsilon,$$

$$z = \sum_{o\in O} c_o \cdot z_o + \delta_G + \sum_{q\in Q} \varepsilon,$$

for some finite index sets I, K, L, N, O, Q, $a_i, b_l, c_o \in A$, $E, F, G \subseteq EID$ and basic terms x_i, y_l, z_o.

We only give the proofs for (6) and (8). The proof for (6) uses induction on the general form of basic terms x, y, and z.

From $AE(x) \subseteq E_1, AE(y) \subseteq E_2$, and $AE(z) \subseteq E_3$ we obtain $AE(x_i) \subseteq E_1, AE(y_l) \subseteq E_2$, and $AE(z_o) \subseteq E_3$ for all $i \in I$, $l \in L$, and $o \in O$. This means that we are allowed to use

$$\big(x_i \parallel_{\mathrm{iw}}^{E_{12}\cup S} y_l\big) \parallel_{\mathrm{iw}}^{E_{13}\cup E_{23}\cup S} z_o = x_i \parallel_{\mathrm{iw}}^{E_{12}\cup E_{13}\cup S} \big(y_l \parallel_{\mathrm{iw}}^{E_{23}\cup S} z_o\big).$$

First, we give a number of subcomputations. These are used in proving Equation (6).

Subcomputation 1:

$$\begin{aligned}
&\big(a_i \cdot x_i \mid_{\mathrm{iw}}^{E_{12}\cup S} b_l \cdot y_l\big) \mid_{\mathrm{iw}}^{E_{13}\cup E_{23}\cup S} c_o \cdot z_o \\
&= \begin{cases} a_i \cdot (x_i \parallel_{\mathrm{iw}}^{E_{12}\cup S} y_l) \mid_{\mathrm{iw}}^{E_{13}\cup E_{23}\cup S} c_o \cdot z_o & \text{if } a_i \equiv b_l \wedge AE(a_i) \subseteq E_{12} \cup S, \\ \delta_{AE(a_i\cdot x_i)\cup AE(b_l\cdot y_l)} \mid_{\mathrm{iw}}^{E_{13}\cup E_{23}\cup S} c_o \cdot z_o & \text{otherwise} \end{cases} \\
&= \begin{cases} a_i \cdot ((x_i \parallel_{\mathrm{iw}}^{E_{12}\cup S} y_l) \parallel_{\mathrm{iw}}^{E_{13}\cup E_{23}\cup S} z_o) & \text{if } a_i \equiv b_l \wedge AE(a_i) \subseteq E_{12} \cup S \\ & \wedge\, a_i \equiv c_o \wedge AE(a_i) \subseteq E_{13} \cup E_{23} \cup S, \\ \delta_{AE(a_i)\cup AE(x_i)\cup AE(y_l)\cup AE(c_o\cdot z_o)} & \text{if } a_i \equiv b_l \wedge AE(a_i) \subseteq E_{12} \cup S \\ & \wedge\, (a_i \not\equiv c_o \vee AE(a_i) \not\subseteq E_{13} \cup E_{23} \cup S), \\ \delta_{AE(a_i\cdot x_i)\cup AE(b_l\cdot y_l)\cup AE(c_o\cdot z_o)} & \text{otherwise} \end{cases}
\end{aligned}$$

$$= \begin{cases} a_i \cdot ((x_i \parallel_{\text{iw}}^{E_{12} \cup S} y_l) \parallel_{\text{iw}}^{E_{13} \cup E_{23} \cup S} z_o) & \text{if } a_i \equiv b_l \wedge AE(a_i) \subseteq E_{12} \cup S \\ & \wedge\, a_i \equiv c_o \wedge AE(a_i) \subseteq E_{13} \cup E_{23} \cup S, \\ \delta_{AE(a_i \cdot x_i) \cup AE(b_l \cdot y_l) \cup AE(c_o \cdot z_o)} & \text{otherwise} \end{cases}$$

$$= \begin{cases} a_i \cdot ((x_i \parallel_{\text{iw}}^{E_{12} \cup S} y_l) \parallel_{\text{iw}}^{E_{13} \cup E_{23} \cup S} z_o) & \text{if } a_i \equiv b_l \equiv c_o \wedge AE(a_i) \subseteq S, \\ \delta_{AE(a_i \cdot x_i) \cup AE(b_l \cdot y_l) \cup AE(c_o \cdot z_o)} & \text{otherwise} \end{cases}$$

$$= \begin{cases} a_i \cdot (x_i \parallel_{\text{iw}}^{E_{12} \cup E_{13} \cup S} (y_l \parallel_{\text{iw}}^{E_{23} \cup S} z_o)) & \text{if } a_i \equiv b_l \equiv c_o \wedge AE(a_i) \subseteq S, \\ \delta_{AE(a_i \cdot x_i) \cup AE(b_l \cdot y_l) \cup AE(c_o \cdot z_o)} & \text{otherwise} \end{cases}$$

$$= \begin{cases} a_i \cdot (x_i \parallel_{\text{iw}}^{E_{12} \cup E_{13} \cup S} (y_l \parallel_{\text{iw}}^{E_{23} \cup S} z_o)) & \text{if } a_i \equiv b_l \wedge AE(a_i) \subseteq E_{12} \cup E_{13} \cup S \\ & \wedge\, b_l \equiv c_o \wedge AE(b_l) \subseteq E_{23} \cup S, \\ \delta_{AE(a_i \cdot x_i) \cup AE(b_l) \cup AE(y_l) \cup AE(z_o)} & \text{if } (a_i \not\equiv b_l \vee AE(a_i) \not\subseteq E_{12} \cup E_{13} \cup S) \\ & \wedge\, b_l \equiv c_o \wedge AE(b_l) \subseteq E_{23} \cup S, \\ \delta_{AE(a_i \cdot x_i) \cup AE(b_l \cdot y_l) \cup AE(c_o \cdot z_o)} & \text{otherwise} \end{cases}$$

$$= \begin{cases} a_i \cdot x_i \mid_{\text{iw}}^{E_{12} \cup E_{13} \cup S} b_l \cdot (y_l \parallel_{\text{iw}}^{E_{23} \cup S} z_o) & \text{if } b_l \equiv c_o \wedge AE(b_l) \subseteq E_{23} \cup S, \\ a_i \cdot x_i \mid_{\text{iw}}^{E_{12} \cup E_{13} \cup S} \delta_{AE(b_l \cdot y_l) \cup AE(c_o \cdot z_o)} & \text{otherwise} \end{cases}$$

$$= a_i \cdot x_i \mid_{\text{iw}}^{E_{12} \cup E_{13} \cup S} \left(b_l \cdot y_l \mid_{\text{iw}}^{E_{23} \cup S} c_o \cdot z_o\right).$$

Subcomputation 2:

$$\begin{aligned} \left(\delta_E \mid_{\text{iw}}^{E_{12} \cup S} y\right) \mid_{\text{iw}}^{E_{13} \cup E_{23} \cup S} z &= \delta_{E \cup AE(y)} \mid_{\text{iw}}^{E_{13} \cup E_{23} \cup S} z \\ &= \delta_{E \cup AE(y) \cup AE(z)} \\ &= \delta_E \mid_{\text{iw}}^{E_{12} \cup E_{13} \cup S} \left(y \mid_{\text{iw}}^{E_{23} \cup S} z\right). \end{aligned}$$

Similarly we obtain

$$\left(x \mid_{\text{iw}}^{E_{12} \cup S} \delta_F\right) \mid_{\text{iw}}^{E_{13} \cup E_{23} \cup S} z = x \mid_{\text{iw}}^{E_{12} \cup E_{13} \cup S} \left(\delta_F \mid_{\text{iw}}^{E_{23} \cup S} z\right)$$

and

$$\left(x \mid_{\text{iw}}^{E_{12} \cup S} y\right) \mid_{\text{iw}}^{E_{13} \cup E_{23} \cup S} \delta_G = x \mid_{\text{iw}}^{E_{12} \cup E_{13} \cup S} \left(y \mid_{\text{iw}}^{E_{23} \cup S} \delta_G\right).$$

Subcomputation 3:

$$\begin{aligned} \left(\varepsilon \mid_{\text{iw}}^{E_{12} \cup S} y\right) \mid_{\text{iw}}^{E_{13} \cup E_{23} \cup S} z &= \delta_{AE(y)} \mid_{\text{iw}}^{E_{13} \cup E_{23} \cup S} z \\ &= \delta_{AE(y) \cup AE(z)} \\ &= \varepsilon \mid_{\text{iw}}^{E_{12} \cup E_{13} \cup S} \left(y \mid_{\text{iw}}^{E_{23} \cup S} z\right). \end{aligned}$$

Similarly we obtain

$$\left(x \mid_{\text{iw}}^{E_{12} \cup S} \varepsilon\right) \mid_{\text{iw}}^{E_{13} \cup E_{23} \cup S} z = x \mid_{\text{iw}}^{E_{12} \cup E_{13} \cup S} \left(\varepsilon \mid_{\text{iw}}^{E_{23} \cup S} z\right)$$

and

$$\left(x \mid_{\mathrm{iw}}^{E_{12} \cup S} y\right) \mid_{\mathrm{iw}}^{E_{13} \cup E_{23} \cup S} \varepsilon = x \mid_{\mathrm{iw}}^{E_{12} \cup E_{13} \cup S} \left(y \mid_{\mathrm{iw}}^{E_{23} \cup S} \varepsilon\right).$$

Then, using these subcomputations, we obtain

$$\begin{aligned}
&\left(x \mid_{\mathrm{iw}}^{E_{12} \cup S} y\right) \mid_{\mathrm{iw}}^{E_{13} \cup E_{23} \cup S} z \\
&= \sum_{i \in I} \sum_{l \in L} \sum_{o \in O} \left(a_i \cdot x_i \mid_{\mathrm{iw}}^{E_{12} \cup S} b_l \cdot y_l\right) \mid_{\mathrm{iw}}^{E_{13} \cup E_{23} \cup S} c_o \cdot z_o \\
&\quad + \left(\delta_E \mid_{\mathrm{iw}}^{E_{12} \cup S} y\right) \mid_{\mathrm{iw}}^{E_{13} \cup E_{23} \cup S} z + \left(x \mid_{\mathrm{iw}}^{E_{12} \cup S} \delta_F\right) \mid_{\mathrm{iw}}^{E_{13} \cup E_{23} \cup S} z \\
&\quad + \left(x \mid_{\mathrm{iw}}^{E_{12} \cup S} y\right) \mid_{\mathrm{iw}}^{E_{13} \cup E_{23} \cup S} \delta_G + \sum_{k \in K} \left(\varepsilon \mid_{\mathrm{iw}}^{E_{12} \cup S} y\right) \mid_{\mathrm{iw}}^{E_{13} \cup E_{23} \cup S} z \\
&\quad + \sum_{n \in N} \left(x \mid_{\mathrm{iw}}^{E_{12} \cup S} \varepsilon\right) \mid_{\mathrm{iw}}^{E_{13} \cup E_{23} \cup S} z + \sum_{q \in Q} \left(x \mid_{\mathrm{iw}}^{E_{12} \cup S} y\right) \mid_{\mathrm{iw}}^{E_{13} \cup E_{23} \cup S} \varepsilon \\
&= \sum_{i \in I} \sum_{l \in L} \sum_{o \in O} a_i \cdot x_i \mid_{\mathrm{iw}}^{E_{12} \cup E_{13} \cup S} \left(b_l \cdot y_l \mid_{\mathrm{iw}}^{E_{23} \cup S} c_o \cdot z_o\right) \\
&\quad + \delta_E \mid_{\mathrm{iw}}^{E_{12} \cup E_{13} \cup S} \left(y \mid_{\mathrm{iw}}^{E_{23} \cup S} z\right) + x \mid_{\mathrm{iw}}^{E_{12} \cup E_{13} \cup S} \left(\delta_F \mid_{\mathrm{iw}}^{E_{23} \cup S} z\right) \\
&\quad + x \mid_{\mathrm{iw}}^{E_{12} \cup E_{13} \cup S} \left(y \mid_{\mathrm{iw}}^{E_{23} \cup S} \delta_G\right) + \sum_{k \in K} \varepsilon \mid_{\mathrm{iw}}^{E_{12} \cup E_{13} \cup S} \left(y \mid_{\mathrm{iw}}^{E_{23} \cup S} z\right) \\
&\quad + \sum_{n \in N} x \mid_{\mathrm{iw}}^{E_{12} \cup E_{13} \cup S} \left(\varepsilon \mid_{\mathrm{iw}}^{E_{23} \cup S} z\right) + \sum_{q \in Q} x \mid_{\mathrm{iw}}^{E_{12} \cup E_{13} \cup S} \left(y \mid_{\mathrm{iw}}^{E_{23} \cup S} \varepsilon\right) \\
&= x \mid_{\mathrm{iw}}^{E_{12} \cup E_{13} \cup S} \left(y \mid_{\mathrm{iw}}^{E_{23} \cup S} z\right).
\end{aligned}$$

Finally, Equation (8) is proven as follows:

$$\begin{aligned}
&\left(x \parallel_{\mathrm{iw}}^{E_{12} \cup S} y\right) \parallel_{\mathrm{iw}}^{E_{13} \cup E_{23} \cup S} z \\
&= \left(x \parallel_{\mathrm{iw}}^{E_{12} \cup S} y\right) \mathbin{⌊\!⌊}_{\mathrm{iw}}^{E_{13} \cup E_{23} \cup S} z + z \mathbin{⌊\!⌊}_{\mathrm{iw}}^{E_{13} \cup E_{23} \cup S} \left(x \parallel_{\mathrm{iw}}^{E_{12} \cup S} y\right) + \left(x \parallel_{\mathrm{iw}}^{E_{12} \cup S} y\right) \mid_{\mathrm{iw}}^{E_{13} \cup E_{23} \cup S} z \\
&= \left(x \mathbin{⌊\!⌊}_{\mathrm{iw}}^{E_{12} \cup S} y + y \mathbin{⌊\!⌊}_{\mathrm{iw}}^{E_{12} \cup S} x + x \mid_{\mathrm{iw}}^{E_{12} \cup S} y\right) \mathbin{⌊\!⌊}_{\mathrm{iw}}^{E_{13} \cup E_{23} \cup S} z + z \mathbin{⌊\!⌊}_{\mathrm{iw}}^{E_{13} \cup E_{23} \cup S} \left(x \parallel_{\mathrm{iw}}^{E_{12} \cup S} y\right) \\
&\quad + \left(x \mathbin{⌊\!⌊}_{\mathrm{iw}}^{E_{12} \cup S} y + y \mathbin{⌊\!⌊}_{\mathrm{iw}}^{E_{12} \cup S} x + x \mid_{\mathrm{iw}}^{E_{12} \cup S} y\right) \mid_{\mathrm{iw}}^{E_{13} \cup E_{23} \cup S} z \\
&= \left(x \mathbin{⌊\!⌊}_{\mathrm{iw}}^{E_{12} \cup S} y\right) \mathbin{⌊\!⌊}_{\mathrm{iw}}^{E_{13} \cup E_{23} \cup S} z + \left(y \mathbin{⌊\!⌊}_{\mathrm{iw}}^{E_{12} \cup S} x\right) \mathbin{⌊\!⌊}_{\mathrm{iw}}^{E_{13} \cup E_{23} \cup S} z + \left(x \mid_{\mathrm{iw}}^{E_{12} \cup S} y\right) \mathbin{⌊\!⌊}_{\mathrm{iw}}^{E_{13} \cup E_{23} \cup S} z \\
&\quad + z \mathbin{⌊\!⌊}_{\mathrm{iw}}^{E_{13} \cup E_{23} \cup S} \left(x \parallel_{\mathrm{iw}}^{E_{12} \cup S} y\right) + \left(x \mathbin{⌊\!⌊}_{\mathrm{iw}}^{E_{12} \cup S} y\right) \mid_{\mathrm{iw}}^{E_{13} \cup E_{23} \cup S} z + \left(y \mathbin{⌊\!⌊}_{\mathrm{iw}}^{E_{12} \cup S} x\right) \mid_{\mathrm{iw}}^{E_{13} \cup E_{23} \cup S} z \\
&\quad + \left(x \mid_{\mathrm{iw}}^{E_{12} \cup S} y\right) \mid_{\mathrm{iw}}^{E_{13} \cup E_{23} \cup S} z \\
&= x \mathbin{⌊\!⌊}_{\mathrm{iw}}^{E_{12} \cup E_{13} \cup S} \left(y \parallel_{\mathrm{iw}}^{E_{23} \cup S} z\right) + y \mathbin{⌊\!⌊}_{\mathrm{iw}}^{E_{12} \cup E_{23} \cup S} \left(x \parallel_{\mathrm{iw}}^{E_{13} \cup S} z\right) + x \mid_{\mathrm{iw}}^{E_{12} \cup E_{13} \cup S} \left(y \mathbin{⌊\!⌊}_{\mathrm{iw}}^{E_{23} \cup S} z\right) \\
&\quad + z \mathbin{⌊\!⌊}_{\mathrm{iw}}^{E_{13} \cup E_{23} \cup S} \left(y \parallel_{\mathrm{iw}}^{E_{12} \cup S} x\right) + z \mid_{\mathrm{iw}}^{E_{13} \cup E_{23} \cup S} \left(x \mathbin{⌊\!⌊}_{\mathrm{iw}}^{E_{12} \cup S} y\right) + z \mid_{\mathrm{iw}}^{E_{13} \cup E_{23} \cup S} \left(y \mathbin{⌊\!⌊}_{\mathrm{iw}}^{E_{12} \cup S} x\right) \\
&\quad + x \mid_{\mathrm{iw}}^{E_{12} \cup E_{13} \cup S} \left(y \mid_{\mathrm{iw}}^{E_{23} \cup S} z\right)
\end{aligned}$$

$$
\begin{aligned}
&= x \mathbin{\lfloor\!\lfloor}^{E_{12}\cup E_{13}\cup S}_{\mathrm{iw}} \big(y \parallel^{E_{23}\cup S}_{\mathrm{iw}} z\big) + y \mathbin{\lfloor\!\lfloor}^{E_{12}\cup E_{23}\cup S}_{\mathrm{iw}} \big(z \parallel^{E_{13}\cup S}_{\mathrm{iw}} x\big) + x \mid^{E_{12}\cup E_{13}\cup S}_{\mathrm{iw}} \big(y \mathbin{\lfloor\!\lfloor}^{E_{23}\cup S}_{\mathrm{iw}} z\big) \\
&\quad + \big(z \mathbin{\lfloor\!\lfloor}^{E_{23}\cup S}_{\mathrm{iw}} y\big) \mathbin{\lfloor\!\lfloor}^{E_{12}\cup E_{13}\cup S}_{\mathrm{iw}} x + \big(z \mid^{E_{13}\cup S}_{\mathrm{iw}} x\big) \mathbin{\lfloor\!\lfloor}^{E_{12}\cup E_{23}\cup S}_{\mathrm{iw}} y + \big(z \mid^{E_{23}\cup S}_{\mathrm{iw}} y\big) \mathbin{\lfloor\!\lfloor}^{E_{12}\cup E_{13}\cup S}_{\mathrm{iw}} x \\
&\quad + x \mid^{E_{12}\cup E_{13}\cup S}_{\mathrm{iw}} \big(y \mid^{E_{23}\cup S}_{\mathrm{iw}} z\big) \\
&= x \mathbin{\lfloor\!\lfloor}^{E_{12}\cup E_{13}\cup S}_{\mathrm{iw}} \big(y \parallel^{E_{23}\cup S}_{\mathrm{iw}} z\big) + \big(y \mathbin{\lfloor\!\lfloor}^{E_{23}\cup S}_{\mathrm{iw}} z\big) \mathbin{\lfloor\!\lfloor}^{E_{12}\cup E_{13}\cup S}_{\mathrm{iw}} x + x \mid^{E_{12}\cup E_{13}\cup S}_{\mathrm{iw}} \big(y \mathbin{\lfloor\!\lfloor}^{E_{23}\cup S}_{\mathrm{iw}} z\big) \\
&\quad + \big(z \mathbin{\lfloor\!\lfloor}^{E_{23}\cup S}_{\mathrm{iw}} y\big) \mathbin{\lfloor\!\lfloor}^{E_{12}\cup E_{13}\cup S}_{\mathrm{iw}} x + \big(x \mid^{E_{13}\cup S}_{\mathrm{iw}} z\big) \mathbin{\lfloor\!\lfloor}^{E_{12}\cup E_{23}\cup S}_{\mathrm{iw}} y + \big(y \mid^{E_{23}\cup S}_{\mathrm{iw}} z\big) \mathbin{\lfloor\!\lfloor}^{E_{12}\cup E_{13}\cup S}_{\mathrm{iw}} x \\
&\quad + x \mid^{E_{12}\cup E_{13}\cup S}_{\mathrm{iw}} \big(y \mid^{E_{23}\cup S}_{\mathrm{iw}} z\big) \\
&= x \mathbin{\lfloor\!\lfloor}^{E_{12}\cup E_{13}\cup S}_{\mathrm{iw}} \big(y \parallel^{E_{23}\cup S}_{\mathrm{iw}} z\big) + \big(y \mathbin{\lfloor\!\lfloor}^{E_{23}\cup S}_{\mathrm{iw}} z + z \mathbin{\lfloor\!\lfloor}^{E_{23}\cup S}_{\mathrm{iw}} y + y \mid^{E_{23}\cup S}_{\mathrm{iw}} z\big) \mathbin{\lfloor\!\lfloor}^{E_{12}\cup E_{13}\cup S}_{\mathrm{iw}} x \\
&\quad + x \mid^{E_{12}\cup E_{13}\cup S}_{\mathrm{iw}} \big(y \mathbin{\lfloor\!\lfloor}^{E_{23}\cup S}_{\mathrm{iw}} z\big) + x \mid^{E_{12}\cup E_{13}\cup S}_{\mathrm{iw}} \big(z \mathbin{\lfloor\!\lfloor}^{E_{23}\cup S}_{\mathrm{iw}} y\big) + x \mid^{E_{12}\cup E_{13}\cup S}_{\mathrm{iw}} \big(y \mid^{E_{23}\cup S}_{\mathrm{iw}} z\big) \\
&= x \mathbin{\lfloor\!\lfloor}^{E_{12}\cup E_{13}\cup S}_{\mathrm{iw}} \big(y \parallel^{E_{23}\cup S}_{\mathrm{iw}} z\big) + \big(y \parallel^{E_{23}\cup S}_{\mathrm{iw}} z\big) \mathbin{\lfloor\!\lfloor}^{E_{12}\cup E_{13}\cup S}_{\mathrm{iw}} x \\
&\quad + x \mid^{E_{12}\cup E_{13}\cup S}_{\mathrm{iw}} \big(y \mathbin{\lfloor\!\lfloor}^{E_{23}\cup S}_{\mathrm{iw}} z + z \mathbin{\lfloor\!\lfloor}^{E_{23}\cup S}_{\mathrm{iw}} y + y \mid^{E_{23}\cup S}_{\mathrm{iw}} z\big) \\
&= x \mathbin{\lfloor\!\lfloor}^{E_{12}\cup E_{13}\cup S}_{\mathrm{iw}} \big(y \parallel^{E_{23}\cup S}_{\mathrm{iw}} z\big) + \big(y \parallel^{E_{23}\cup S}_{\mathrm{iw}} z\big) \mathbin{\lfloor\!\lfloor}^{E_{12}\cup E_{13}\cup S}_{\mathrm{iw}} x + x \mid^{E_{12}\cup E_{13}\cup S}_{\mathrm{iw}} \big(y \parallel^{E_{23}\cup S}_{\mathrm{iw}} z\big) \\
&= x \parallel^{E_{12}\cup E_{13}\cup S}_{\mathrm{iw}} \big(y \parallel^{E_{23}\cup S}_{\mathrm{iw}} z\big). \qquad \square
\end{aligned}
$$

By taking $E_1 = E_2 = E_3$ we obtain associativity of $\parallel^{E}_{\mathrm{iw}}$.

The final property which we prove is the correspondence between the $\circ_{\mathrm{iw}}$ and $\parallel^{E}_{\mathrm{iw}}$ operators. This formalizes the resemblance between the axiomatic definitions of these operators.

PROPOSITION 8. *For closed IWD*$(\circ, +, \cdot, \parallel^{E})$*-terms x and y such that $AE(x) \cap AE(y) = \emptyset$ we have*

$$x \parallel^{\emptyset}_{\mathrm{iw}} y = x \circ_{\mathrm{iw}} y.$$

PROOF. Let

$$
\begin{aligned}
x &= \sum_{i\in I} a_i \cdot x_i + \delta_E + \sum_{k\in K} \varepsilon, \\
y &= \sum_{l\in L} b_l \cdot y_l + \delta_F + \sum_{n\in N} \varepsilon
\end{aligned}
$$

for some finite index sets I, K, L, N, $a_i, b_l \in A$, $E, F \subseteq EID$ and basic terms x_i, y_l. The induction hypotheses are $x_i \parallel^{\emptyset}_{\mathrm{iw}} y = x_i \circ_{\mathrm{iw}} y$ for all $i \in I$ and $x \parallel^{\emptyset}_{\mathrm{iw}} y_l = x \circ_{\mathrm{iw}} y_l$ for all $l \in L$. Then

$$
\begin{aligned}
x \mathbin{\lfloor\!\lfloor}^{\emptyset}_{\mathrm{iw}} y &= \sum_{i\in I} a_i \cdot x_i \mathbin{\lfloor\!\lfloor}^{\emptyset}_{\mathrm{iw}} y + \delta_E \mathbin{\lfloor\!\lfloor}^{\emptyset}_{\mathrm{iw}} y + \sum_{k\in K} \varepsilon \mathbin{\lfloor\!\lfloor}^{\emptyset}_{\mathrm{iw}} y \\
&= \sum_{i\in I} a_i \cdot \big(x_i \parallel^{\emptyset}_{\mathrm{iw}} y\big) + \delta_{E\cup AE(y)} + \sum_{k\in K} \varepsilon \mathbin{\lfloor\!\lfloor}^{\emptyset}_{\mathrm{iw}} y
\end{aligned}
$$

$$
\begin{aligned}
&= \sum_{i\in I} a_i \cdot (x_i \circ_{\mathrm{iw}} y) + \delta_{E\cup AE(y)} + \sum_{k\in K}\sum_{n\in N} \varepsilon + \delta_{AE(y)} \\
&= \sum_{i\in I} a_i \cdot (x_i \circ_{\mathrm{iw}} y) + \delta_E \; \mathbf{L}\!\circ_{\mathrm{iw}} y + \sum_{k\in K} \varepsilon \; \mathbf{L}\!\circ_{\mathrm{iw}} y \\
&= \sum_{i\in I} a_i \cdot x_i \; \mathbf{L}\!\circ_{\mathrm{iw}} y + \delta_E \; \mathbf{L}\!\circ_{\mathrm{iw}} y + \sum_{k\in K} \varepsilon \; \mathbf{L}\!\circ_{\mathrm{iw}} y \\
&= x \; \mathbf{L}\!\circ_{\mathrm{iw}} y.
\end{aligned}
$$

In the following computations we use that $AE(b_l) \subseteq AE(y)$ and $AE(x) \cap AE(y) = \emptyset$ imply $AE(b_l) \cap AE(x) = \emptyset$:

$$
\begin{aligned}
y \mathbin{\|\!\_}^{\emptyset}_{\mathrm{iw}} x &= \sum_{l\in L} b_l \cdot y_l \mathbin{\|\!\_}^{\emptyset}_{\mathrm{iw}} x + \delta_F \mathbin{\|\!\_}^{\emptyset}_{\mathrm{iw}} x + \sum_{n\in N} \varepsilon \mathbin{\|\!\_}^{\emptyset}_{\mathrm{iw}} x \\
&= \sum_{l\in L} b_l \cdot \left(y_l \parallel^{\emptyset}_{\mathrm{iw}} x\right) + \delta_{F\cup AE(x)} + \sum_{n\in N}\sum_{k\in K} \varepsilon + \delta_{AE(x)} \\
&= \sum_{l\in L} b_l \cdot \left(x \parallel^{\emptyset}_{\mathrm{iw}} y_l\right) + \delta_{F\cup AE(x)} + \sum_{n\in N}\sum_{k\in K} \varepsilon + \delta_{AE(x)} \\
&= \sum_{l\in L} b_l \cdot (x \circ_{\mathrm{iw}} y_l) + x \; \mathbf{R}\!\circ_{\mathrm{iw}} \delta_F + \sum_{n\in N} x \; \mathbf{R}\!\circ_{\mathrm{iw}} \varepsilon \\
&= \sum_{l\in L} x \; \mathbf{R}\!\circ_{\mathrm{iw}} b_l \cdot y_l + x \; \mathbf{R}\!\circ_{\mathrm{iw}} \delta_F + \sum_{n\in N} x \; \mathbf{R}\!\circ_{\mathrm{iw}} \varepsilon \\
&= x \; \mathbf{R}\!\circ_{\mathrm{iw}} y, \\
x \mid^{\emptyset}_{\mathrm{iw}} y &= \sum_{i\in I}\sum_{l\in L} a_i \cdot x_i \mid^{\emptyset}_{\mathrm{iw}} b_l \cdot y_l + \delta_E \mid^{\emptyset}_{\mathrm{iw}} y + x \mid^{\emptyset}_{\mathrm{iw}} \delta_F + \sum_{k\in K} \varepsilon \mid^{\emptyset}_{\mathrm{iw}} y + \sum_{n\in N} x \mid^{\emptyset}_{\mathrm{iw}} \varepsilon \\
&= \sum_{i\in I}\sum_{l\in L} \delta_{AE(a_i\cdot x_i)\cup AE(b_l\cdot y_l)} + \delta_{E\cup AE(y)} + \delta_{F\cup AE(x)} + \delta_{AE(y)} + \delta_{AE(x)} \\
&= \delta_{AE(x)\cup AE(y)},
\end{aligned}
$$

and therefore

$$
\begin{aligned}
x \parallel^{\emptyset}_{\mathrm{iw}} y &= x \mathbin{\|\!\_}^{\emptyset}_{\mathrm{iw}} y + y \mathbin{\|\!\_}^{\emptyset}_{\mathrm{iw}} x + x \mid^{\emptyset}_{\mathrm{iw}} y \\
&= x \; \mathbf{L}\!\circ_{\mathrm{iw}} y + x \; \mathbf{R}\!\circ_{\mathrm{iw}} x + \delta_{AE(x)\cup AE(y)} \\
&= x \circ_{\mathrm{iw}} y + \delta_{AE(x)\cup AE(y)} \\
&= x \circ_{\mathrm{iw}} y.
\end{aligned}
$$

□

7. Process algebra for Interworkings

In the previous section we introduced the E-interworking merge. This operator was parameterized with the set of entities on which the processes should synchronize. In order for the interworking merge to be generally applicable, the set E must be determined from the actual operands of the interworking merge. Therefore, we have to generalize the E-interworking merge to the interworking merge operator.

There is a technical complication which makes this generalization non-trivial: we have to explicitly attribute every process term with the set of entities that it contains. The reason for this is revealed by the examples in Figures 12 and 13 (see Section 2.4).

Using the definitions from the previous sections, interworking $X2$ from Figure 12 has the following semantical representation: $c(c, d, m)$. There is no explicit mention of the empty entity b. Indeed, this interpretation is exactly the same as the interpretation of interworking $X2'$ from Figure 13.

In a context with only $+$ and $\circ_{\mathrm{iw}}$ operators, this identification would be completely harmless, however Figures 12 and 13 show that there is a merge context which makes a distinction between $X2$ and $X2'$.

The reason for this anomaly is that we did not take empty entities into consideration. Therefore, in order to properly define the interworking merge, we have to extend our semantical representation with information about the entities contained.

There are several ways to achieve this. A first option would be to attribute the empty process ε with a set of entities. Empty entity b would then be represented by $\varepsilon_{\{b\}}$. A second option would be to label a complete process term with the set of entities which it ranges over. The semantical representation of $X2$ would then become $\langle c(c, d, m), \{b, c, d\}\rangle$, whereas $X2'$ would be represented by $\langle c(c, d, m), \{c, d\}\rangle$.

For technical reasons we choose to elaborate on the second option. An Interworking with a dynamical behaviour denoted by x over the entities from E is denoted by $\langle x, E\rangle$. Such a tuple $\langle x, E\rangle$ will be called an entity-labeled process.

DEFINITION 11 (*Signature*). The signature of the process algebra $IWE(\circ, +, \|)$ consists of the operators $\langle \_, \_\rangle$, $+$, $\circ_{\mathrm{iw}}$, and $\|_{\mathrm{iw}}$.

For $\langle x, E\rangle$ to be a well-formed expression we do not require that the active entities from x are all contained in E. All entities in E which are not active entities in x are empty entities. The active entities of $\langle x, E\rangle$ can be determined from x solely. The complete set of entities of $\langle x, E\rangle$, denoted by $Ent(\langle x, E\rangle)$, contains the active entities from x and the entities from E.

DEFINITION 12 (*Active entities, entities*). For closed $IWD(\circ, +, \cdot, \|^{E})$-term x, $E \subseteq EID$, and closed $IWE(\circ, +, \|)$-terms s and t we define the mappings $AE\colon \mathcal{C}(\Sigma_{IWE(\circ,+,\|)}) \to \mathbb{P}(EID)$ and $Ent\colon \mathcal{C}(\Sigma_{IWE(\circ,+,\|)}) \to \mathbb{P}(EID)$ inductively as follows:

$$
\begin{aligned}
AE\big(\langle x, E\rangle\big) &= AE(x),\\
AE(s + t) &= AE(s) \cup AE(t),\\
AE(s \circ_{\mathrm{iw}} t) &= AE(s) \cup AE(t),\\
AE(s \,\|_{\mathrm{iw}}\, t) &= AE(s) \cup AE(t),\\
Ent\big(\langle x, E\rangle\big) &= E \cup AE(x),\\
Ent(s + t) &= Ent(s) \cup Ent(t),
\end{aligned}
$$

Table 10
Operational semantics of entity-labeled processes ($a \in A$, $E \subseteq EID$, x, x' $IWD(\circ, +, \cdot, \|^E)$-terms, s, s', t, t' entity-labeled processes)

$$\frac{x\downarrow}{\langle x, E\rangle\downarrow} \qquad \frac{x \xrightarrow{a} x'}{\langle x, E\rangle \xrightarrow{a} \langle x', E \cup AE(x)\rangle}$$

$$\frac{s\downarrow}{s+t\downarrow} \qquad \frac{t\downarrow}{s+t\downarrow} \qquad \frac{s \xrightarrow{a} s'}{s+t \xrightarrow{a} s' \circ_{\mathrm{iw}} \langle \varepsilon, Ent(t)\rangle} \qquad \frac{t \xrightarrow{a} t'}{s+t \xrightarrow{a} t' \circ_{\mathrm{iw}} \langle \varepsilon, Ent(s)\rangle}$$

$$\frac{s\downarrow \quad t\downarrow}{s \circ_{\mathrm{iw}} t\downarrow} \qquad \frac{s \xrightarrow{a} s'}{s \circ_{\mathrm{iw}} t \xrightarrow{a} s' \circ_{\mathrm{iw}} t} \qquad \frac{AE(a) \cap AE(s) = \emptyset \quad t \xrightarrow{a} t'}{s \circ_{\mathrm{iw}} t \xrightarrow{a} s \circ_{\mathrm{iw}} t'}$$

$$\frac{s \xrightarrow{a} s' \quad AE(a) \not\subseteq Ent(s) \cap Ent(t)}{s \,\|_{\mathrm{iw}}\, t \xrightarrow{a} s' \,\|_{\mathrm{iw}}\, t} \qquad \frac{t \xrightarrow{a} t' \quad AE(a) \not\subseteq Ent(s) \cap Ent(t)}{s \,\|_{\mathrm{iw}}\, t \xrightarrow{a} s \,\|_{\mathrm{iw}}\, t'}$$

$$\frac{s\downarrow \quad t\downarrow}{s \,\|_{\mathrm{iw}}\, t\downarrow} \qquad \frac{s \xrightarrow{a} s' \quad t \xrightarrow{a} t' \quad AE(a) \subseteq Ent(s) \cap Ent(t)}{s \,\|_{\mathrm{iw}}\, t \xrightarrow{a} s' \,\|_{\mathrm{iw}}\, t'}$$

$$\begin{aligned} Ent(s \circ_{\mathrm{iw}} t) &= Ent(s) \cup Ent(t), \\ Ent(s \,\|_{\mathrm{iw}}\, t) &= Ent(s) \cup Ent(t). \end{aligned}$$

On entity-labeled processes we define the operators interworking sequencing and interworking merge. The set of all entity-labeled processes is called $IWE(\circ, +, \|)$. The definition of the interworking sequencing on entity-labeled processes is straightforward.

Before we give axioms for the process algebra $IWE(\circ, +, \|)$, we define a operational semantics. The operational semantics of entity-labeled processes, as expressed in Table 10, is similar to the operational semantics of non-labeled processes.

The first two rules relate the domains of non-labeled processes and entity labeled processes. In the second rule we have to take care that we do not loose information about the involved entities after executing an action. It may happen that some active entity from x which does not occur in E is not active anymore in x' since the last action from that entity has been executed. Therefore, we have to extend the entity label of x' with the active entities of x. The rules for the interworking merge correspond to the rules for the E-interworking merge but the condition $AE(a) \subseteq E$ is replaced by $AE(a) \subseteq Ent(s) \cap Ent(t)$. The set $Ent(s) \cap Ent(t)$ contains the shared entities from s and t, so this is the set of entities which should synchronize.

For the "correctness" of the deduction rules for interworking merge it is necessary that the set of entities of a process does not change by executing actions (Lemma 8). This is guaranteed by the deduction rules. We first prove that the set of active entities does not expand due to the execution of actions.

LEMMA 7. *For all $a \in A$ and closed $IWD(\circ, +, \cdot, \|^E)$-terms x and x' we have: if $x \xrightarrow{a} x'$, then $AE(x) \supseteq AE(x')$.*

PROOF. This lemma is proven with induction on the structure of closed $IWD(\circ, +, \cdot, \|^E)$-term x. Suppose that $x \xrightarrow{a} x'$.

(1) $x \equiv \varepsilon$. This case cannot occur as $\varepsilon \not\xrightarrow{a}$.

(2) $x \equiv \delta_E$ for some $E \subseteq EID$. This case cannot occur as $\delta_E \not\xrightarrow{a}$.

(3) $x \equiv b$ for some $b \in A$. Then it must be the case that $b \equiv a$ and $x' \equiv \varepsilon$. Clearly $AE(x) = AE(b) \supseteq \emptyset = AE(\varepsilon) = AE(x')$.

(4) $x \equiv x_1 + x_2$ for some closed $IWD(\circ, +, \cdot, \|^E)$-terms x_1 and x_2. Then it must be the case that either $x_1 \xrightarrow{a} x'$ or $x_2 \xrightarrow{a} x'$. By induction we thus have either $AE(x_1) \supseteq AE(x')$ or $AE(x_2) \supseteq AE(x')$. In either case we have $AE(x) = AE(x_1 + x_2) = AE(x_1) \cup AE(x_2) \supseteq AE(x')$.

(5) $x \equiv x_1 \cdot x_2$ for some closed $IWD(\circ, +, \cdot, \|^E)$-terms x_1 and x_2. Then we can distinguish two cases. First, $x_1 \xrightarrow{a} x_1'$ for some closed $IWD(\circ, +, \cdot, \|^E)$-term x_1' such that $x' \equiv x_1' \cdot x_2$. By induction we have $AE(x_1) \supseteq AE(x_1')$. Then $AE(x) = AE(x_1 \cdot x_2) = AE(x_1) \cup AE(x_2) \supseteq AE(x_1') \cup AE(x_2) = AE(x_1' \cdot x_2) = AE(x')$. Second, $x_1 \downarrow$ and $x_2 \xrightarrow{a} x'$. By induction we have $AE(x_2) \supseteq AE(x')$. Then $AE(x) = AE(x_1 \cdot x_2) = AE(x_1) \cup AE(x_2) \supseteq AE(x_1) \cup AE(x') \supseteq AE(x')$.

(6) $x \equiv x_1 \circ_{\mathrm{iw}} x_2$ for some closed $IWD(\circ, +, \cdot, \|^E)$-terms x_1 and x_2. Then we can distinguish two cases. First, $x_1 \xrightarrow{a} x_1'$ for some closed $IWD(\circ, +, \cdot, \|^E)$-term x_1' such that $x' \equiv x_1' \circ_{\mathrm{iw}} x_2$. By induction we have $AE(x_1) \supseteq AE(x_1')$. Then $AE(x) = AE(x_1 \circ_{\mathrm{iw}} x_2) = AE(x_1) \cup AE(x_2) \supseteq AE(x_1') \cup AE(x_2) = AE(x_1' \circ_{\mathrm{iw}} x_2) = AE(x')$. Second, $AE(a) \cap AE(x_1) = \emptyset$ and $x_2 \xrightarrow{a} x_2'$ for some closed $IWD(\circ, +, \cdot, \|^E)$-term x_2' such that $x' \equiv x_1 \circ_{\mathrm{iw}} x_2'$. By induction we have $AE(x_2) \supseteq AE(x_2')$. Then $AE(x) = AE(x_1 \circ_{\mathrm{iw}} x_2) = AE(x_1) \cup AE(x_2) \supseteq AE(x_1) \cup AE(x_2') = AE(x_1 \circ_{\mathrm{iw}} x_2') = AE(x')$.

(7) $x \equiv x_1 \|_{\mathrm{iw}}^E x_2$ for some $E \subseteq EID$ and closed $IWD(\circ, +, \cdot, \|^E)$-terms x_1 and x_2. Then we can distinguish three cases. First, $AE(a) \not\subseteq E$ and $x_1 \xrightarrow{a} x_1'$ for some closed $IWD(\circ, +, \cdot, \|^E)$-term x_1' such that $x' \equiv x_1' \|_{\mathrm{iw}}^E x_2$. By induction we have $AE(x_1) \supseteq AE(x_1')$. Then $AE(x) = AE(x_1 \|_{\mathrm{iw}}^E x_2) = AE(x_1) \cup AE(x_2) \supseteq AE(x_1') \cup AE(x_2) = AE(x_1' \|_{\mathrm{iw}}^E x_2) = AE(x')$. Second, $AE(a) \not\subseteq E$ and $x_2 \xrightarrow{a} x_2'$ for some closed $IWD(\circ, +, \cdot, \|^E)$-term x_2' such that $x' \equiv x_1 \|_{\mathrm{iw}}^E x_2'$. By induction we have $AE(x_2) \supseteq AE(x_2')$. Then $AE(x) = AE(x_1 \|_{\mathrm{iw}}^E x_2) = AE(x_1) \cup AE(x_2) \supseteq AE(x_1) \cup AE(x_2') = AE(x_1 \|_{\mathrm{iw}}^E x_2') = AE(x')$. Third, $AE(a) \subseteq E$, $x_1 \xrightarrow{a} x_1'$, and $x_2 \xrightarrow{a} x_2'$ for some closed $IWD(\circ, +, \cdot, \|^E)$-terms x_1' and x_2' such that $x' \equiv x_1' \|_{\mathrm{iw}}^E x_2'$. By induction we have $AE(x_1) \supseteq AE(x_1')$ and $AE(x_2) \supseteq AE(x_2')$. Then $AE(x) = AE(x_1 \|_{\mathrm{iw}}^E x_2) = AE(x_1) \cup AE(x_2) \supseteq AE(x_1') \cup AE(x_2') = AE(x_1' \|_{\mathrm{iw}}^E x_2') = AE(x')$. □

LEMMA 8. *For all closed $IWE(\circ, +, \|)$-terms s and t and all $a \in A$ we have: if $s \xrightarrow{a} s'$, then $Ent(s) = Ent(s')$.*

PROOF. This lemma is proven with induction on the structure of closed $IWE(\circ, +, \|)$-term s.

(1) $s \equiv \langle x, E\rangle$ for some closed $IWD(\circ, +, \cdot, \|^E)$-term x and $E \subseteq EID$. Then $s \xrightarrow{a} s'$ must be due to $x \xrightarrow{a} x'$ for some x' such that $s' \equiv \langle x', E \cup AE(x)\rangle$. Clearly we have $Ent(s) = Ent(\langle x, E\rangle) = E \cup AE(x)$ and $Ent(s') = Ent(\langle x', E \cup AE(x)\rangle) = E \cup AE(x) \cup AE(x')$. Using Lemma 7 we obtain $Ent(s) = Ent(s')$.
(2) $s \equiv s_1 + s_2$ for some closed $IWE(\circ, +, \|)$-terms s_1 and s_2. We can distinguish two cases. First, $s_1 \xrightarrow{a} s_1'$ for some closed $IWE(\circ, +, \|)$-term s_1' such that $s' \equiv s_1' \circ_{\mathrm{iw}} \langle \varepsilon, Ent(s_2)\rangle$. By induction we have $Ent(s_1) = Ent(s_1')$. Therefore, $Ent(s) = Ent(s_1 + s_2) = Ent(s_1) \cup Ent(s_2) = Ent(s_1') \cup Ent(s_2) = Ent(s_1') \cup Ent(\langle \varepsilon, Ent(s_2)\rangle) = Ent(s_1' \circ_{\mathrm{iw}} \langle \varepsilon, Ent(s_2)\rangle) = Ent(s')$. Second, $s_2 \xrightarrow{a} s_2'$ for some closed $IWE(\circ, +, \|)$-term s_2' such that $s' \equiv s_2' \circ_{\mathrm{iw}} \langle \varepsilon, Ent(s_1)\rangle$. This case is symmetrical to the first case.
(3) $s \equiv s_1 \circ_{\mathrm{iw}} s_2$ for some closed $IWE(\circ, +, \|)$-terms s_1 and s_2. We can distinguish two cases. First, $s_1 \xrightarrow{a} s_1'$ for some closed $IWE(\circ, +, \|)$-term s_1' such that $s' \equiv s_1' \circ_{\mathrm{iw}} s_2$. By induction we have $Ent(s_1) = Ent(s_1')$. Therefore, $Ent(s) = Ent(s_1 \circ_{\mathrm{iw}} s_2) = Ent(s_1) \cup Ent(s_2) = Ent(s_1') \cup Ent(s_2) = Ent(s_1' \circ_{\mathrm{iw}} s_2) = Ent(s')$. Second, $AE(a) \cap AE(s_2) = \emptyset$ and $s_2 \xrightarrow{a} s_2'$ for some closed $IWE(\circ, +, \|)$-term s_2' such that $s' \equiv s_1 \circ_{\mathrm{iw}} s_2'$. By induction we have $Ent(s_2) = Ent(s_2')$. Therefore, $Ent(s) = Ent(s_1 \circ_{\mathrm{iw}} s_2) = Ent(s_1) \cup Ent(s_2) = Ent(s_1) \cup Ent(s_2') = Ent(s_1 \circ_{\mathrm{iw}} s_2') = Ent(s')$.
(4) $s \equiv s_1 \|_{\mathrm{iw}} s_2$ for some closed $IWE(\circ, +, \|)$-terms s_1 and s_2. We can distinguish three cases. First, $AE(a) \not\subseteq Ent(s_1) \cap Ent(s_2)$ and $s_1 \xrightarrow{a} s_1'$ for some closed $IWE(\circ, +, \|)$-term s_1' such that $s' \equiv s_1' \|_{\mathrm{iw}} s_2$. By induction we have $Ent(s_1) = Ent(s_1')$. Therefore, $Ent(s) = Ent(s_1 \|_{\mathrm{iw}} s_2) = Ent(s_1) \cup Ent(s_2) = Ent(s_1') \cup Ent(s_2) = Ent(s_1' \|_{\mathrm{iw}} s_2) = Ent(s')$. Second, $AE(a) \not\subseteq Ent(s_1) \cap Ent(s_2)$ and $s_2 \xrightarrow{a} s_2'$ for some closed $IWE(\circ, +, \|)$-term s_2' such that $s' \equiv s_1 \|_{\mathrm{iw}} s_2'$. By induction we have $Ent(s_2) = Ent(s_2')$. Therefore, $Ent(s) = Ent(s_1 \|_{\mathrm{iw}} s_2) = Ent(s_1) \cup Ent(s_2) = Ent(s_1) \cup Ent(s_2') = Ent(s_1 \|_{\mathrm{iw}} s_2') = Ent(s')$. Third, $AE(a) \subseteq Ent(s_1) \cap Ent(s_2)$, $s_1 \xrightarrow{a} s_1'$, and $s_2 \xrightarrow{a} s_2'$ for some closed $IWE(\circ, +, \|)$-terms s_1' and s_2' such that $s' \equiv s_1' \|_{\mathrm{iw}} s_2'$. By induction we have $Ent(s_1) = Ent(s_1')$ and $Ent(s_2') = Ent(s_2)$. Therefore, $Ent(s) = Ent(s_1 \|_{\mathrm{iw}} s_2) = Ent(s_1) \cup Ent(s_2) = Ent(s_1') \cup Ent(s_2') = Ent(s_1' \|_{\mathrm{iw}} s_2') = Ent(s')$. □

Next, we adapt the definition of IWD-bisimilarity to take into account the set of entities of a process.

DEFINITION 13 (*Entity bisimilarity*). A symmetric relation R on closed $IWE(\circ, +, \|)$-terms is an *entity bisimulation*, if and only if, for every pair $(s, t) \in R$ and $a \in A$, the following conditions hold:
(1) $AE(s) = AE(t)$,
(2) if $s\downarrow$, then $t\downarrow$,
(3) if $s \xrightarrow{a} s'$, then there is a closed $IWE(\circ, +, \|)$-term t' such that $t \xrightarrow{a} t'$ and $(s', t') \in R$,
(4) $Ent(s) = Ent(t)$.
The closed $IWE(\circ, +, \|)$-terms s and t are *entity bisimilar*, notation $s \leftrightarrow t$, if and only if there exists an entity bisimulation R relating them.

THEOREM 20 (Equivalence). *Entity bisimilarity is an equivalence relation.*

PROOF. The proof is similar to the proof that IWD-bisimilarity is an equivalence (Theorem 1) and therefore omitted. □

THEOREM 21 (Congruence). *Entity bisimilarity is a congruence for the function symbols in the signature of* $IWE(\circ, +, \|)$ *which are defined on* $IWE(\circ, +, \|)$*-terms.*

PROOF. Suppose $R : x \underline{\leftrightarrow}_{\mathrm{iwd}}\, y$ and $E_x = E_y$. Now we must prove that there exists an entity bisimulation R' such that $R' : \langle x, E_x\rangle \underline{\leftrightarrow} \langle y, E_y\rangle$. Let $R' = \{(\langle p, E\rangle, \langle q, F\rangle) \mid pRq, E = F\}$. Let p and q be closed $IWD(\circ, +, \cdot, \|^E)$-terms such that pRq and $E, F \subseteq EID$ such that $E = F$. Since $p \underline{\leftrightarrow}_{\mathrm{iwd}}\, q$ we have $AE(p) = AE(q)$.

(1) $AE(\langle p, E\rangle) = AE(p) = AE(q) = AE(\langle q, F\rangle)$.

(2) Suppose that $\langle p, E\rangle \xrightarrow{a} s$ for some closed $IWE(\circ, +, \|)$-term s. This must be due to $p \xrightarrow{a} p'$ for some closed $IWD(\circ, +, \cdot, \|^E)$-term p' such that $s \equiv \langle p', E \cup AE(p)\rangle$. As $p \underline{\leftrightarrow}_{\mathrm{iwd}}\, q$, we have the existence of closed $IWD(\circ, +, \cdot, \|^E)$-term q' such that $q \xrightarrow{a} q'$ and $p'Rq'$. Then we also obtain $\langle q, F\rangle \xrightarrow{a} \langle q', F \cup AE(q)\rangle$. Clearly $\langle p', E \cup AE(p)\rangle R' \langle q', F \cup AE(q)\rangle$.

(3) Suppose that $\langle p, E\rangle \downarrow$. This must be due to $p \downarrow$. As $p \underline{\leftrightarrow}_{\mathrm{iwd}}\, q$, we have $q \downarrow$. Therefore, $\langle q, F\rangle \downarrow$.

(4) $Ent(\langle p, E\rangle) = E \cup AE(p) = F \cup AE(q) = AE(\langle q, F\rangle)$.

Suppose $R_1 : s_1 \underline{\leftrightarrow} t_1$ and $R_2 : s_2 \underline{\leftrightarrow} t_2$. Let $R = \{(s_1 + t_1, s_2 + t_2), (p_1 \circ_{\mathrm{iw}} \langle \varepsilon, E\rangle, q_1 \circ_{\mathrm{iw}} \langle \varepsilon, E\rangle), (p_2 \circ_{\mathrm{iw}} \langle \varepsilon, E\rangle, q_2 \circ_{\mathrm{iw}} \langle \varepsilon, E\rangle) \mid p_1 R_1 q_1, p_2 R_2 q_2, E \subseteq EID\}$. Obviously, this relation is an entity bisimulation.

Suppose $R_1 : s_1 \underline{\leftrightarrow} t_1$ and $R_2 : s_2 \underline{\leftrightarrow} t_2$. Let $R = \{(p_1 \circ_{\mathrm{iw}} p_2, q_1 \circ_{\mathrm{iw}} q_2) \mid p_1 R_1 q_1, p_2 R_2 q_2\}$. Obviously this relation R is an entity bisimulation. The proof is similar to the proof that IWD-bisimilarity is a congruence for interworking sequencing (see Theorem 2).

Suppose $R_1 : s_1 \underline{\leftrightarrow} t_1$ and $R_2 : s_2 \underline{\leftrightarrow} t_2$. Let $R = \{(p_1 \|_{\mathrm{iw}} p_2, q_1 \|_{\mathrm{iw}} q_2) \mid p_1 R_1 q_1, p_2 R_2 q_2\}$. Obviously this relation R is an entity bisimulation. □

As was done in [25], the interworking merge is expressed in terms of the E-interworking merge operator and the common entities of the operands. The axioms for entity-labeled processes are given in Table 11 for $E, F \subseteq EID$. The extension of $IWD(\circ, \cdot, +)$ with entity-labeled processes is denoted by $IWE(\circ, +, \|)$.

Table 11
Axioms of entity-labeled processes ($E, F \subseteq EID$)

*IWE*1	$\langle x, E\rangle = \langle x, E \cup AE(x)\rangle$	
*IWE*2	$\langle x, E\rangle + \langle y, F\rangle = \langle x + y, E \cup F\rangle$	
*IWE*3	$\langle x, E\rangle \circ_{\mathrm{iw}} \langle y, F\rangle = \langle x \circ_{\mathrm{iw}} y, E \cup F\rangle$	
*IWE*4	$\langle x, E\rangle \|_{\mathrm{iw}} \langle y, F\rangle = \langle x \|_{\mathrm{iw}}^{E\cap F} y, E \cup F\rangle$	if $AE(x) \subseteq E$ and $AE(y) \subseteq F$

Axiom *IWE*1 describes the convention discussed before that the entity-part of an $IWE(\circ, +, \|)$-term contains at least the empty entities of the Interworking. Axioms *IWE*2–*IWE*4 describe how the other operators on $IWE(\circ, +, \|)$-terms can be defined in terms of their counterparts on $IWD(\circ, +, \cdot, \|^E)$-terms. It is also possible to define entity bisimulation in terms of IWD-bisimilarity of the process-parts and set equality of the entity-parts. Also for our final process algebra, $IWE(\circ, +, \|)$, we prove soundness and completeness.

THEOREM 22 (Soundness). *The process algebra* $IWE(\circ, +, \|)$ *is a sound axiomatization of IWD-bisimilarity on closed* $IWD(\circ, +, \cdot, \|^E)$*-terms. The process algebra* $IWE(\circ, +, \|)$ *is a sound axiomatization of entity bisimulation on closed* $IWE(\circ, +, \|)$*-terms.*

PROOF. For the first proposition observe that we did not add any axioms relating closed $IWD(\circ, +, \cdot, \|^E)$-terms. We will prove the second proposition. Since entity bisimulation is a congruence for the closed $IWE(\circ, +, \|)$-terms (Theorem 21) we only have to show that the axioms from Table 11 are sound. Thereto, we provide an entity bisimulation relation for each axiom. For *IWE*1, the relation $R = \{(\langle x, E\rangle, \langle x, E \cup AE(x)\rangle)\}^S \cup I$ is an entity bisimulation. For the axiom *IWE*2 the relation $R = \{(\langle p, E\rangle + \langle q, F\rangle, \langle p+q, E \cup F\rangle), (\langle p, E\rangle \circ_{\mathrm{iw}} \langle \varepsilon, F\rangle, \langle p, E \cup F\rangle) \mid p, q \in \mathcal{C}(\Sigma_{IWD(\circ,+,\cdot,\|^E)}), E, F \subseteq EID\}^S$ is an entity bisimulation. For axiom *IWE*3 the relation $R = \{(\langle p, E'\rangle \circ_{\mathrm{iw}} \langle y, F\rangle, \langle p \circ_{\mathrm{iw}} y, E' \cup F \cup AE(y)\rangle), (\langle x, E\rangle \circ_{\mathrm{iw}} \langle q, F'\rangle, \langle x \circ_{\mathrm{iw}} q, E' \cup F' \cup AE(x)\rangle) \mid p, q \in \mathcal{C}(\Sigma_{IWD(\circ,+,\cdot,\|^E)}), E', F' \subseteq EID\}^S$ is an entity bisimulation. For *IWE*4, the relation $R = \{(\langle p, E\rangle \|_{\mathrm{iw}} \langle q, F\rangle, \langle p \|_{\mathrm{iw}}^{E \cap F} q, E \cup F\rangle \mid p, q \in \mathcal{C}(\Sigma_{IWD(\circ,+,\cdot,\|^E)}), E, F \subseteq EID, AE(p) \subseteq E, AE(q) \subseteq F\}^S$ is an entity bisimulation. □

DEFINITION 14 (*Basic terms*). The set of basic terms is the smallest set that satisfies: if x is a closed $IWD(\circ, +, \cdot, \|^E)$-term and $E \subseteq EID$ such that $AE(x) \subseteq E$, then $\langle x, E\rangle$ is a basic $IWE(\circ, +, \|)$-term. The set of all basic terms over the signature of $IWE(\circ, +, \|)$ is denoted by $\mathcal{B}(\Sigma_{IWE(\circ,+,\|)})$.

THEOREM 23 (Elimination). *For every closed* $IWE(\circ, +, \|)$*-term* s *there exists a basic* $IWE(\circ, +, \|)$*-term* t *such that* $IWE(\circ, +, \|) \vdash s = t$.

PROOF. This theorem is proven with induction on the structure of a closed $IWE(\circ, +, \|)$-term. First, consider the case $s \equiv \langle x, E\rangle$ for some closed $IWD(\circ, +, \cdot, \|^E)$-term x and $E \subseteq EID$. Then $s = \langle x, E\rangle = \langle x, E \cup AE(x)\rangle$. Clearly $AE(x) \subseteq E \cup AE(x)$ and hence $\langle x, E \cup AE(x)\rangle$ is a basic $IWE(\circ, +, \|)$-term. Then, consider the case $s \equiv s_1 + s_2$ for some closed $IWE(\circ, +, \|)$-terms s_1 and s_2. By induction we have the existence of basic terms $\langle x_1, E_1\rangle$ and $\langle x_2, E_2\rangle$ for some closed $IWD(\circ, +, \cdot, \|^E)$-terms x_1 and x_2 and $E_1, E_2 \subseteq EID$ such that $AE(x_1) \subseteq E_1$ and $AE(x_2) \subseteq E_2$. Then, $s = s_1 + s_2 = \langle x_1, E_1\rangle + \langle x_2, E_2\rangle = \langle x_1 + x_2, E_1 \cup E_2\rangle$. Clearly $AE(x_1 + x_2) \subseteq E_1 \cup E_2$. Next, consider the case $s \equiv s_1 \circ_{\mathrm{iw}} s_2$ for some closed $IWE(\circ, +, \|)$-terms s_1 and s_2. By induction we have the existence of basic terms $\langle x_1, E_1\rangle$ and $\langle x_2, E_2\rangle$ for some closed $IWD(\circ, +, \cdot, \|^E)$-terms x_1 and x_2 and $E_1, E_2 \subseteq EID$ such that $AE(x_1) \subseteq E_1$ and $AE(x_2) \subseteq E_2$. Then $s = s_1 \circ_{\mathrm{iw}} s_2 = \langle x_1, E_1\rangle \circ_{\mathrm{iw}} \langle x_2, E_2\rangle = \langle x_1 \circ_{\mathrm{iw}} x_2, E_1 \cup E_2\rangle$. Clearly $AE(x_1 \circ_{\mathrm{iw}} x_2) = AE(x_1) \cup AE(x_2) \subseteq E_1 \cup E_2$. Finally, consider the case $s \equiv s_1 \|_{\mathrm{iw}} s_2$ for some s_1, s_2 closed $IWE(\circ, +, \|)$-terms. By induction we have the existence of basic terms $\langle x_1, E_1\rangle$ and $\langle x_2, E_2\rangle$ for some closed *IWD*$(\circ, +, \cdot, \|^E)$-terms x_1 and x_2 and $E_1, E_2 \subseteq EID$ such that $AE(x_1) \subseteq E_1$ and $AE(x_2) \subseteq E_2$. Then $s = s_1 \|_{\mathrm{iw}} s_2 = \langle x_1, E_1\rangle \|_{\mathrm{iw}} \langle x_2, E_2\rangle = \langle x_1 \|_{\mathrm{iw}}^{E_1 \cap E_2} x_2, E_1 \cup E_2\rangle$. Clearly $AE(x_1 \|_{\mathrm{iw}} x_2) = AE(x_1) \cup AE(x_2) \subseteq E_1 \cup E_2$. □

LEMMA 9. *For basic IWE*$(\circ, +, \|)$*-terms* $\langle x, E\rangle$ *and* $\langle y, F\rangle$ *we have*

$$\langle x, E\rangle \underline{\leftrightarrow} \langle y, F\rangle \quad \textit{iff} \quad x \underline{\leftrightarrow}_{\mathrm{iwd}}\, y \textit{ and } E = F.$$

PROOF. First, suppose that $R : \langle x, E\rangle \underline{\leftrightarrow} \langle y, F\rangle$. Let $R' = \{(p, q) \mid \langle p, E'\rangle R \langle q, F'\rangle, E' = F'\}$. As $\langle x, E\rangle R \langle y, F\rangle$, $AE(x) \subseteq E$, and $AE(y) \subseteq F$, we have $E = E \cup AE(x) = Ent(x) = Ent(y) = F \cup AE(y) = F$. We will prove that R' is an IWD-bisimulation.

(1) $AE(p) = AE(\langle p, E'\rangle) = AE(\langle q, F'\rangle) = AE(q)$.

(2) $p \downarrow$ iff $\langle p, E'\rangle \downarrow$ iff $\langle q, F'\rangle \downarrow$ iff $q \downarrow$.

(3) Suppose that $p \xrightarrow{a} p'$ for some closed *IWD*$(\circ, +, \cdot, \|^E)$-term p'. Then $\langle p, E'\rangle \xrightarrow{a} \langle p', E' \cup AE(p)\rangle$. So we have $\langle q, F'\rangle \xrightarrow{a} \langle q', F' \cup AE(q')\rangle$ for some closed *IWD*$(\circ, +, \cdot, \|^E)$-term q' such that $\langle p', E' \cup AE(p)\rangle R \langle q', F' \cup AE(q')\rangle$. From this we obtain that $E' \cup AE(p') = F' \cup AE(q')$. Thus $p' R' q'$.

The proof in the other direction is trivial. □

THEOREM 24 (Completeness). *The process algebra IWE*$(\circ, +, \|)$ *is a complete axiomatization of entity bisimulation on closed IWE*$(\circ, +, \|)$*-terms.*

PROOF. By the elimination theorem (Theorem 23) we only have to prove this theorem for basic *IWE*$(\circ, +, \|)$-terms. Let $\langle x, E_1\rangle$ and $\langle y, E_2\rangle$ be basic *IWE*$(\circ, +, \|)$-terms such that $\langle x, E_1\rangle \underline{\leftrightarrow} \langle y, E_2\rangle$. By Lemma 9 we have $x \underline{\leftrightarrow}_{\mathrm{iwd}}\, y$ and $E_1 = E_2$. Since *IWD*$(\circ, +, \cdot, \|^E)$ is a complete axiomatization of *IWD*$(\circ)$-bisimilarity on closed *IWD*$(\circ, +, \cdot, \|^E)$-terms, we have $x = y$, and hence $\langle x, E_1\rangle = \langle y, E_2\rangle$. □

THEOREM 25 (Conservativity). *The process algebra IWE*$(\circ, +, \|)$ *is a conservative extension of the process algebra IWD*$(\circ, +, \cdot, \|^E)$.

PROOF. With respect to *IWD*$(\circ, +, \cdot, \|^E)$-terms, the process algebra *IWE*$(\circ, +, \|)$ and the process algebra *IWD*$(\circ, +, \cdot, \|^E)$ have exactly the same axioms. Then clearly the same equalities can be derived between closed *IWD*$(\circ, +, \cdot, \|^E)$-terms. □

We end our treatment of the semantics of Interworkings with some properties of Interworkings. The interworking sequencing is commutative under the assumption that the active entities of the operands are disjoint. Furthermore, it is associative. The interworking merge is both commutative and associative.

PROPOSITION 9 (Unit elements). *For closed IWE*$(\circ, +, \|)$ *terms s,*

$$s \circ_{\mathrm{iw}} \langle \varepsilon, \emptyset\rangle = s, \tag{9}$$

$$\langle \varepsilon, \emptyset\rangle \circ_{\mathrm{iw}} s = s, \tag{10}$$

$$s \parallel_{\mathrm{iw}} \langle \varepsilon, \emptyset \rangle = s, \tag{11}$$

$$\langle \varepsilon, \emptyset \rangle \parallel_{\mathrm{iw}} s = s. \tag{12}$$

PROOF. By the elimination theorem it is allowed to restrict the proof of the statements to basic terms. Let $s = \langle x, E \rangle$ for some closed $IWD(\circ, +, \cdot, \parallel^E)$-term x and $E \subseteq EID$ such that $AE(x) \subseteq E$. Then

$$\begin{aligned}
s \circ_{\mathrm{iw}} \langle \varepsilon, \emptyset \rangle &= \langle x, E \rangle \circ_{\mathrm{iw}} \langle \varepsilon, \emptyset \rangle = \langle x \circ_{\mathrm{iw}} \varepsilon, E \cup \emptyset \rangle = \langle x, E \rangle = s, \\
\langle \varepsilon, \emptyset \rangle \circ_{\mathrm{iw}} s &= \langle \varepsilon, \emptyset \rangle \circ_{\mathrm{iw}} \langle x, E \rangle = \langle \varepsilon \circ_{\mathrm{iw}} x, \emptyset \cup E \rangle = \langle x, E \rangle = s, \\
s \parallel_{\mathrm{iw}} \langle \varepsilon, \emptyset \rangle &= \langle x, E \rangle \parallel_{\mathrm{iw}} \langle \varepsilon, \emptyset \rangle = \big\langle x \parallel_{\mathrm{iw}}^{E \cap \emptyset} \varepsilon, E \cup \emptyset \big\rangle = \big\langle x \parallel_{\mathrm{iw}}^{\emptyset} \varepsilon, E \big\rangle = \langle x, E \rangle = s, \\
\langle \varepsilon, \emptyset \rangle \parallel_{\mathrm{iw}} s &= \langle \varepsilon, \emptyset \rangle \parallel_{\mathrm{iw}} \langle x, E \rangle = \big\langle \varepsilon \parallel_{\mathrm{iw}}^{\emptyset \cap E} x, \emptyset \cup E \big\rangle = \big\langle \varepsilon \parallel_{\mathrm{iw}}^{\emptyset} x, E \big\rangle = \langle x, E \rangle = s.
\end{aligned}$$

□

PROPOSITION 10 (Commutativity and associativity of $\circ_{\mathrm{iw}}$ and $\parallel_{\mathrm{iw}}$). *For closed $IWE(\circ, +, \parallel)$-terms s, t, u we have*

$$s \circ_{\mathrm{iw}} t = t \circ_{\mathrm{iw}} s, \quad \textit{if } AE(s) \cap AE(t) = \emptyset \tag{13}$$

$$(s \circ_{\mathrm{iw}} t) \circ_{\mathrm{iw}} u = s \circ_{\mathrm{iw}} (t \circ_{\mathrm{iw}} u), \tag{14}$$

$$s \parallel_{\mathrm{iw}} t = t \parallel_{\mathrm{iw}} s, \tag{15}$$

$$(s \parallel_{\mathrm{iw}} t) \parallel_{\mathrm{iw}} u = s \parallel_{\mathrm{iw}} (t \parallel_{\mathrm{iw}} u). \tag{16}$$

PROOF. By the elimination theorem it is allowed to restrict the proof of the statements to basic terms. Let $s = \langle x_1, E_1 \rangle$, $t = \langle x_2, E_2 \rangle$, and $u = \langle x_3, E_3 \rangle$ for some $E_1, E_2, E_3 \subseteq EID$ and closed $IWD(\circ, +, \cdot, \parallel^E)$-terms x_1, x_2, and x_3 such that $AE(x_1) \subseteq E_1$, $AE(x_2) \subseteq E_2$, and $AE(x_3) \subseteq E_3$. Then

$$\begin{aligned}
s \circ_{\mathrm{iw}} t &= \langle x_1, E_1 \rangle \circ_{\mathrm{iw}} \langle x_2, E_2 \rangle = \langle x_1 \circ_{\mathrm{iw}} x_2, E_1 \cup E_2 \rangle \\
&= \langle x_2 \circ_{\mathrm{iw}} x_1, E_2 \cup E_1 \rangle = \langle x_2, E_2 \rangle \circ_{\mathrm{iw}} \langle x_1, E_1 \rangle \\
&= t \circ_{\mathrm{iw}} s, \\
(s \circ_{\mathrm{iw}} t) \circ_{\mathrm{iw}} u &= \big(\langle x_1, E_1 \rangle \circ_{\mathrm{iw}} \langle x_2, E_2 \rangle\big) \circ_{\mathrm{iw}} \langle x_3, E_3 \rangle \\
&= \langle x_1 \circ_{\mathrm{iw}} x_2, E_1 \cup E_2 \rangle \circ_{\mathrm{iw}} \langle x_3, E_3 \rangle \\
&= \big\langle (x_1 \circ_{\mathrm{iw}} x_2) \circ_{\mathrm{iw}} x_3, (E_1 \cup E_2) \cup E_3 \big\rangle \\
&= \big\langle x_1 \circ_{\mathrm{iw}} (x_2 \circ_{\mathrm{iw}} x_3), E_1 \cup (E_2 \cup E_3) \big\rangle \\
&= \langle x_1, E_1 \rangle \circ_{\mathrm{iw}} \langle x_2 \circ_{\mathrm{iw}} x_3, E_2 \cup E_3 \rangle \\
&= \langle x_1, E_1 \rangle \circ_{\mathrm{iw}} \big(\langle x_2, E_2 \rangle \circ_{\mathrm{iw}} \langle x_3, E_3 \rangle\big) \\
&= s \circ_{\mathrm{iw}} (t \circ_{\mathrm{iw}} u), \\
s \parallel_{\mathrm{iw}} t &= \langle x_1, E_1 \rangle \parallel_{\mathrm{iw}} \langle x_2, E_2 \rangle = \big\langle x_1 \parallel_{\mathrm{iw}}^{E_1 \cap E_2} x_2,\ E_1 \cup E_2 \big\rangle \\
&= \big\langle x_2 \parallel_{\mathrm{iw}}^{E_2 \cap E_1} x_1,\ E_2 \cup E_1 \big\rangle = \langle x_2, E_2 \rangle \parallel_{\mathrm{iw}} \langle x_1, E_1 \rangle \\
&= t \parallel_{\mathrm{iw}} s,
\end{aligned}$$

$$
\begin{aligned}
(s \parallel_{\mathrm{iw}} t) \parallel_{\mathrm{iw}} u &= \bigl(\langle x_1, E_1\rangle \parallel_{\mathrm{iw}} \langle x_2, E_2\rangle\bigr) \parallel_{\mathrm{iw}} \langle x_3, E_3\rangle \\
&= \bigl\langle x_1 \parallel_{\mathrm{iw}}^{E_1\cap E_2} x_2, E_1 \cup E_2\bigr\rangle \parallel_{\mathrm{iw}} \langle x_3, E_3\rangle \\
&= \bigl\langle \bigl(x_1 \parallel_{\mathrm{iw}}^{E_1\cap E_2} x_2\bigr) \parallel_{\mathrm{iw}}^{(E_1\cup E_2)\cap E_3} x_3, (E_1 \cup E_2) \cup E_3\bigr\rangle \\
&= \bigl\langle x_1 \parallel_{\mathrm{iw}}^{E_1\cap (E_2\cup E_3)} \bigl(x_2 \parallel_{\mathrm{iw}}^{E_2\cap E_3} x_3\bigr), (E_1 \cup E_2) \cup E_3\bigr\rangle \\
&= \langle x_1, E_1\rangle \parallel_{\mathrm{iw}} \bigl\langle x_2 \parallel_{\mathrm{iw}}^{E_2\cap E_3} x_3, E_2 \cup E_3\bigr\rangle \\
&= \langle x_1, E_1\rangle \parallel_{\mathrm{iw}} \bigl(\langle x_2, E_2\rangle \parallel_{\mathrm{iw}} \langle x_3, E_3\rangle\bigr) \\
&= s \parallel_{\mathrm{iw}} (t \parallel_{\mathrm{iw}} u). \qquad \square
\end{aligned}
$$

PROPOSITION 11. *For closed IWE*$(\circ, +, \parallel)$*-terms* s *and* t *such that* $Ent(s) \cap Ent(t) = \emptyset$ *we have*

$$s \parallel_{\mathrm{iw}} t = s \circ_{\mathrm{iw}} t.$$

PROOF. By the elimination theorem it is allowed to restrict the proof of the statements to basic terms. Let $s = \langle x, E\rangle$ and $t = \langle y, F\rangle$ for some $E, F \subseteq EID$ and closed $IWD(\circ, +, \cdot, \parallel^E)$-terms x and y such that $AE(x) \subseteq E$ and $AE(y) \subseteq F$.

$$
\begin{aligned}
s \parallel_{\mathrm{iw}} t &= \langle x, E\rangle \parallel_{\mathrm{iw}} \langle y, F\rangle = \bigl\langle x \parallel_{\mathrm{iw}}^{\emptyset} y, E \cup F\bigr\rangle \\
&= \langle x \circ_{\mathrm{iw}} y, E \cup F\rangle = \langle x, E\rangle \circ_{\mathrm{iw}} \langle y, F\rangle. \qquad \square
\end{aligned}
$$

8. Conclusions

The starting point of the application described in this chapter was the informal drawing technique, called Interworkings. After analyzing the informal meaning of the language and the way in which users applied this language, our aim was to formalize the Interworking language.

The assets of having a formal semantics are well-known. We mention the following. Formalization yields a thorough understanding of the language and the aspects of the application domain which can be modeled; it allows for an unambiguous interpretation of expressions in the language; it enables formal analysis; and it can be used to derive, or even automatically generate supporting tools.

These points directly addressed the problems that users were confronted with when applying the language. The language organically grew from a collection of examples and it was not clear which constructs were exactly part of the language. For some diagrams even specialists disagreed on the exact interpretation. It was not clear under which precise conditions two Interworkings could be merged. And, finally, in order to efficiently work with collections of Interworkings tool support was required.

The research carried out helped to solve these issues to a large extent. The kernel of the work was the description of the formal semantics of the language by means of process algebra. This is the part of the research covered in this chapter.

Our choice was to use process algebra for the formal definition of Interworkings. This worked out quite successfully. The process algebraic approach even proved suitable to define the semantics of a similar, but much larger language (MSC'96). Although it showed very beneficial, we do not advocate that the process algebraic approach is the best or even the only suitable approach towards the formalization of sequence chart languages. Other techniques, such as Petri nets and partial orders, have also been successfully applied, and when considering only the core of these sequence chart languages, the several approaches do not differ too much with respect to expressivity and simplicity. Only when extending the sequence chart language with specific features, such as recursion and interrupts, some approaches offer a more natural way of modeling.

The work presented here only describes the part of the project which has to do with the theoretical foundations of the project. The main point here was to identify the basic Interworking constructs and operators, and to give their operational and algebraic semantics. The extension with a theory of refinement or the derivation of computer tools is not in the focus of this handbook.

Although already an overwhelming variety of operators has been described in process algebra literature, we have introduced yet more operators. This is typical for the process algebraic approach. For a specific application domain a specific algebra is needed. In the case of sequence charts, the standard operators for sequential and parallel composition do not properly describe the user's intuition. Because the synchronization implied by strong sequential composition is in contradiction with intuition, we developed the interworking sequencing. Because the standard parallel composition operator could not deal with overlapping areas of an Interworking, we had to investigate a variation: interworking merge. Even though these are newly invented operators, their definitions resemble the definition of well-studied operators.

This approach of defining new operators and variations on existing operators has been illustrated in this chapter. We have treated all proof obligations, such as soundness and completeness in full detail. We have especially taken care of setting up our theory in a modular way. This means that we have first defined the kernel of the theory (i.e., the semantics of single Interworking diagrams) and subsequently extended this with other operators.

The kernel of our theory just consists of the interworking sequencing operator. This single operator already allows for the definition of the semantics of Interworking diagrams. After that, we defined the basic process algebra consisting of the standard operators for alternative and sequential composition, extended with a special constant for expressing partial deadlocks. The alternative composition operator is used to express alternative scenarios. This process algebra is independent of the previous one, and the next module simply consisted of the combination of these two theories. The interworking sequencing can now be expressed in terms of the other operators. The axioms defining the interworking sequencing in the first process algebra are now derivable properties. Finally, we extended this algebra with the interworking merge operator. This required two separate steps. First we introduced the E-interworking merge, which is parameterized by the set of entities which should synchronize. And next, we extended the semantical interpretation of Interworkings in order to be able to define the unparameterized interworking merge. This modular approach is illustrated in Figure 17.

In our opinion, such a modular approach brings several assets. A mathematical theory, just like a piece of software, requires maintenance. Parts of the theory may become obsolete due to new insights or new extensions may be required due to additional user re-

Besides the literature on the semantics of MSC based on process algebra, we also mention some other approaches. In [11], an MSC is transformed into a Petri net. In [14], a semantics of Message Flow Graphs is presented that translates an MSC into a Büchi automaton. In [2], Alur, Holzmann and Peled, present a partial order semantics for Basic Message Sequence Charts.

In the ITU Recommendation Z.120, the only assumption about communication between entities is that it is asynchronous and that sending of a message occurs before its receipt. In [8], communication is discussed based on FIFO buffers. A variety of communication models is obtained by considering different ways of connecting entities through buffers. A hierarchy of these communication models is presented based on the possibility of implementing MSCs in the communication models. One of the communication models is identified with the synchronous communication in Interworkings.

References

[1] W.M.P. van der Aalst, *Interorganizational workflow: An approach based on Message Sequence Charts and Petri nets*, Systems Anal. – Modelling – Simulation **34** (3) (1999), 335–367.

[2] R. Alur, G.J. Holzmann and D. Peled, *An analyzer for Message Sequence Charts*, Software – Concepts and Tools **17** (2) (1996), 70–77.

[3] J.C.M. Baeten and S. Mauw, *Delayed choice: an operator for joining Message Sequence Charts*, Formal Description Techniques VII, Proc. 7th IFIP WG 6.1 International Conference on Formal Description Techniques, Berne, D. Hogrefe and S. Leue, eds, Chapman & Hall (1995), 340–354.

[4] J.C.M. Baeten and C. Verhoef, *A congruence theorem for structured operational semantics with predicates*, CONCUR'93, International Conference on Concurrency Theory, Lecture Notes in Comput. Sci. 715, E. Best, ed., Springer-Verlag (1993), 477–492.

[5] J.C.M. Baeten and C. Verhoef, *Concrete process algebra*, Semantic Modelling, Handbook of Logic in Computer Science 4, S. Abramsky, D.M. Gabbay and T.S.E. Maibaum, eds, Oxford University Press (1995), 149–268.

[6] J.C.M. Baeten and W.P. Weijland, *Process Algebra*, Cambridge Tracts Theoret. Comput. Sci. 18, Cambridge University Press (1990).

[7] J. van den Brink and W.O.D. Griffioen, *Formal semantics of Interworkings with discrete absolute time*, Algebra of Communicating Processes, Utrecht 1994 Workshops in Computing A. Ponse, C. Verhoef and S.F.M. van Vlijmen, eds, Springer-Verlag (1995), 106–123.

[8] A. Engels, S. Mauw and M.A. Reniers, *A hierarchy of communication models for Message Sequence Charts*, Formal Description Techniques and Protocol Specification, Testing and Verification Proc. FORTE X and PSTV XVII '97, Osaka, Japan, T. Mizuno, N. Shiratori, T. Higashino and A. Togashi, eds, Chapman & Hall (1997), 75–90.

[9] L.M.G. Feijs, *Synchonous sequence charts in action*, Information and Software Technology **39** (1997), 583–606.

[10] L.M.G. Feijs, *Generating FSMs from Interworkings*, Distrib. Comput. **12** (1) (1999), 31–40.

[11] J. Grabowski, P. Graubmann and E. Rudolph, *Towards a Petri net based semantics definition for Message Sequence Charts*, SDL'93 – Using Objects, Proc. 6th SDL Forum, Darmstadt, O. Færgemand and A. Sarma, eds, North-Holland, Amsterdam (1993), 179–190.

[12] ITU-T, *Recommendation Z.120: Message Sequence Chart (MSC)*, ITU-T, Geneva (1993).

[13] ITU-T, *Recommendation Z.120 Annex B: Algebraic semantics of Message Sequence Charts*, ITU-T, Geneva (1995).

[14] P.B. Ladkin and S. Leue, *Interpreting message flow graphs*, Formal Aspects of Computing **7** (5) (1995), 473–509.

[15] J. de Man, *Towards a formal semantics of Message Sequence Charts*, SDL'93 – Using Objects, Proc. 6th SDL Forum, Darmstadt, O. Færgemand and A. Sarma, eds, North-Holland, Amsterdam (1993) 157–165.

[16] S. Mauw and E.A. van der Meulen, *Generating tools for Message Sequence Charts*, SDL'95 – with MSC in CASE, Proc. 7th SDL Forum, Oslo, R. Bræk and A. Sarma, eds, North-Holland, Amsterdam (1995), 51–62.

[17] S. Mauw and M.A. Reniers, *An algebraic semantics of Basic Message Sequence Charts*, Comput. J. **37** (4) (1994), 269–277.

[18] S. Mauw and M.A. Reniers, *An algebraic semantics of Message Sequence Charts*, Technical Report CSN 94/23, Eindhoven University of Technology, Department of Computing Science, Eindhoven (1994).

[19] S. Mauw and M.A. Reniers, *Empty Interworkings and refinement – semantics of Interworkings revised*, Technical Report CSR 95-12, Eindhoven University of Technology, Department of Computing Science (1995).

[20] S. Mauw and M.A. Reniers, *Empty Interworkings and refinement – semantics of Interworkings revised*, ACP'95, Proc. Second Workshop on Algebra of Communicating Processes Computing Science Reports CSR 95/14, 3 Eindhoven University of Technology, Department of Computing Science (1995), 67–385.

[21] S. Mauw and M.A. Reniers, *Refinement in Interworkings*, CONCUR'96, Pisa, Italy, Lecture Notes in Comput. Sci. 1119, U. Montanari and V. Sassone, eds, Springer-Verlag (1996), 671–686.

[22] S. Mauw and M.A. Reniers, *High-level Message Sequence Charts*, SDL'97: Time for Testing – SDL, MSC and Trends, Proc. 8th SDL Forum, Evry, France, A. Cavalli and A. Sarma, eds, North-Holland, Amsterdam (1997), 291–306.

[23] S. Mauw and M.A. Reniers, *Operational semantics for MSC96*, Computer Networks amd ISDN Systems **31** (17) (1999), 1785–1799. Special issue on Advanced topics on SDL and MSC, A. Cavalli, ed.

[24] S. Mauw, M. van Wijk, and T. Winter, *Syntax and semantics of synchronous Interworkings*, Technical Report RWB-508-re-92436, Information and Software Technology, Philips Research (1992).

[25] S. Mauw, M. van Wijk and T. Winter, *A formal semantics of synchronous Interworkings*, SDL'93 – Using Objects, Proc. 6th SDL Forum, Darmstadt, O. Færgemand and A. Sarma, eds, North-Holland, Amsterdam (1993), 167–178.

[26] S. Mauw and T. Winter, *A prototype toolset for Interworkings*, Philips Telecommunication Review **51** (3) (1993), 41–45.

[27] M.A. Reniers, *An algebraic semantics of Message Sequence Charts*, M.S. thesis, Department of Mathematics and Computing Science, Eindhoven University of Technology (1994).

[28] M.A. Reniers, *Message Sequence Chart: Syntax and semantics*, Ph.D. thesis, Eindhoven University of Technology (1999).

[29] A. Rensink and H. Wehrheim, *Weak sequential composition in process algebras*, CONCUR'94: Concurrency Theory, Lecture Notes in Comput. Sci. 836, Uppsala, B. Jonsson and J. Parrow, eds, Springer-Verlag (1994), 226–241.

[30] E. Rudolph, P. Graubmann and J. Grabowski, *Message Sequence Chart: Composition techniques versus OO-techniques – 'tema con variazioni'*, SDL'95 – with MSC in CASE, Proc. 7th SDL Forum, Oslo, R. Bræk and A. Sarma, eds, North-Holland, Amsterdam (1995), 77–88.

[31] J. Rumbaugh, I. Jacobson and G. Booch, *The Unified Modeling Language Reference Manual*, Addison-Wesley (1999).

[32] J.J. Vereijken, *Discrete-time process algebra*, Ph.D. thesis, Eindhoven University of Technology (1997).

[33] C. Verhoef, *A general conservative extension theorem in process algebra*, Programming Concepts, Methods and Calculi (PROCOMET'94), IFIP Transactions A: Computer Science and Technology 56, E.-R. Olderog, ed., Elsevier Science B.V. (1994), 149–168.

[34] C. Verhoef, *A congruence theorem for structured operational semantics with predicates and negative premises*, Nordic J. Comput. **2** (2) (1995), 274–302.

[35] J.L.M. Vrancken, *The algebra of communicating processes with empty process*, Theoret. Comput. Sci. **177** (2) (1997), 287–328.

Subject index

Author Index

Roman numbers refer to pages on which the author (or his/her work) is mentioned. Italic numbers refer to reference pages. Numbers between brackets are the reference numbers. No distinction is made between first and co-author(s).